Standard Handbook
for Civil Engineers

Other McGraw-Hill Handbooks of Interest

Standard Handbook for Civil Engineers

FREDERICK S. MERRITT Editor
Consulting Engineer, Syosset, N.Y.

Second Edition

McGraw-Hill Book Company

New York   St. Louis   San Francisco   Auckland   Bogotá   Düsseldorf
Johannesburg   London   Madrid   Mexico   Montreal
New Delhi   Panama   Paris   São Paulo   Singapore
Sydney   Tokyo   Toronto

Library of Congress Cataloging in Publication Data
Main entry under title:

Standard handbook for civil engineers.

Includes bibliographical references and index.
1. Civil engineering—Handbooks, manuals, etc.
I. Merritt, Frederick S.
TA151.S8   1976   624   75-25850
ISBN 0-07-041510-2

    67890   KPKP   8543210

The editors for this book were Harold B. Crawford and Virginia Fechtmann,
the designer was Naomi Auerbach, and the production supervisor
was Teresa F. Leaden. It was set in Caledonia
by Typographic Sales, Inc.

Printed and bound by The Kingsport Press.

Contents

SECTION 3

SPECIFICATIONS, by Joseph Goldbloom and John J. White **3-1**

SECTION 4

CONSTRUCTION MANAGEMENT, by J. B. Bonny **4-1**

SECTION 5

CONSTRUCTION MATERIALS, by Russel C. Jones . . . .

PORTLAND CEMENT

AGGREGATES

CONCRETE

OTHER CEMENTING MATERIALS

ATOMIC BASIS OF BEHAVIOR

STRENGTHENING MECHANISMS

STRUCTURAL STEELS

SECTION 6
STRUCTURAL THEORY, by Frederick S. Merritt   6-1

x   **Contents**

SECTION 8
CONCRETE DESIGN AND CONSTRUCTION, by Lev Zetlin and
Donald Griff   8-1

SECTION 9
STRUCTURAL-STEEL DESIGN AND CONSTRUCTION, by Robert O. Disque
and Frank W. Stockwell, Jr. **9-1**

SECTION 10
COLD-FORMED STEEL DESIGN AND CONSTRUCTION, by Paul S. Buker and Don S. Wolford

SECTION 11
WOOD DESIGN AND CONSTRUCTION, by Maurice J. Rhude 11-1

SECTION 12
SURVEYING, by Russell C. Brinker 12-1

SECTION 13

EARTHWORK, by Charles H. Sain

SECTION 14

MUNICIPIAL AND REGIONAL PLANNING, by Gustav J. Requardt, Kenneth A. McCord, and Frederick R. Knoop, Jr.

SECTION 15
BUILDING ENGINEERING, by Frederick S. Merritt 15-1

SECTION 16
HIGHWAY ENGINEERING, by Richard Duttenhoeffer, Bruce E. Podwal, and Viktoras A. Kirkyla 16-1

SECTION 17

BRIDGE ENGINEERING, by John J. Kozak and Joachim F. Leppmann **17-1**

GENERAL DESIGN CONSIDERATIONS

STEEL BRIDGES

CONCRETE BRIDGES

SECTION 18

AIRPORT ENGINEERING, by Herbert H. Howell **18-1**

SECTION 19

RAIL-TRANSPORTATION ENGINEERING, by G. M. Magee 19-1

SECTION 20

TUNNEL ENGINEERING, by John O. Bickel 20-1

SECTION 21

WATER ENGINEERING, by Samuel B. Nelson 21-1

FLUID MECHANICS

PIPE FLOW

PIPE STRESSES

CULVERTS

OPEN-CHANNEL FLOW

SECTION 22
ENVIRONMENTAL ENGINEERING, by William T. Ingram 22-1

SECTION 23

Index follows Section 23

Contributors

John O. Bickel, Associate Consultant to Parsons, Brinckerhoff, Quade & Douglas, Consulting Engineers, New York, N.Y. *(Tunnel Engineering)*

J. B. Bonny, Late President and Chairman, Morrison-Knudsen Company, Inc., Boise, Idaho *(Construction Management)*

Russell C. Brinker, Visiting Professor of Civil Engineering, New Mexico State University *(Surveying)*

Paul S. Buker, Development Coordinator, Building Systems, Armco Steel Corporation, Middletown, Ohio *(Cold-formed Steel Design and Construction)*

David Carsen, Director, Omnidata Services, Inc., New York, N.Y. *(Computers in Civil Engineering)*

Robert O. Disque, Chief Engineer, American Institute of Steel Construction, New York, N.Y. *(Structural-Steel Design and Construction)*

Richard Duttenhoeffer, Partner, Parsons, Brinckerhoff, Quade & Douglas, Consulting Engineers, New York, N.Y. *(Highway Engineering)*

Joseph Goldbloom, Chief Specifications Engineer, and Associate Consultant, Parsons, Brinckerhoff, Quade & Douglas, Consulting Engineers, New York, N.Y. *(Specifications)*

Donald Griff, Vice President, Lev Zetlin Associates, New York, N.Y. *(Concrete Design and Construction)*

Herbert H. Howell, Airport Consultant, St. Louis, Mo. *(Airport Engineering)*

William T. Ingram, Consulting Engineer, Whitestone, N.Y. *(Environmental Engineering)*

Russell C. Jones, Chairman, Department of Civil Engineering, The Ohio State University, Columbus, Ohio *(Construction Materials)*

Viktoras A. Kirklya, Department Head, Highway—Civil Department, Parsons, Brinckerhoff, Quade & Douglas, Consulting Engineers, New York, N.Y. *(Highway Engineering)*

Frederick R. Knoop, Jr., Partner, Whitman, Requardt and Associates, Engineers and Consultants, Baltimore, Md. *(Municipal and Regional Planning)*

John J. Kozak, Chief, Division of Structures, California Department of Transportation, Sacramento, Calif. *(Bridge Engineering)*

Joachim F. Leppmann, Consulting Engineer, Berkeley, Calif. *(Bridge Engineering)*

Kenneth A. McCord, Partner, Whitman, Requardt and Associates, Engineers and Consultants, Baltimore, Md. *(Municipal and Regional Planning)*

G. M. Magee, Railroad Engineering Consultant, Retired Assistant Vice President, Research Department, Association of American Railroads, Chicago, Ill. *(Rail-Transportation Engineering)*

Frederick S. Merritt, Consulting Engineer, Syosset, N.Y. *(Building Engineering, Geotechnical Engineering, Structural Engineering)*

Frank Muller, Director, Contracts and Construction Management Services, Parsons, Brinckerhoff, Quade & Douglas, Inc., Consulting Engineers, New York, N.Y. *(Design Management)*

Samuel B. Nelson, Former Director of Public Works, State of California; Retired General Manager and Chief Engineer, Department of Water and Power, City of Los Angeles; Former General Manager, Southern California Rapid Transit District; Vice President, Daniel, Mann, Johnson & Mendenhall; Vice Chairman, Board of Directors, Metropolitan Water District of Southern California; and Member, California Water Commission *(Water Engineering)*

Bruce E. Podwal, Associate, Parsons, Brinckerhoff, Quade & Douglas, Consulting Engineers, New York, N.Y. *(Highway Engineering)*

Alonzo DeF. Quinn, Consulting Engineer, Centerport, N.Y. *(Harbor Engineering)*

Gustav J. Requardt, Partner, Whitman, Requardt and Associates, Engineers and Consultants, Baltimore, Md. *(Municipal and Regional Planning)*

Maurice J. Rhude, President, Sentinel Structures, Inc., Peshtigo, Wis. *(Wood Design and Construction)*

Charles H. Sain, Consulting Engineer, Birmingham, Ala. *(Earthwork)*

Frank W. Stockwell, Jr., Assistant Chief Engineer, American Institute of Steel Construction, New York, N.Y. *(Structural-Steel Design and Construction)*

Charles P. C. Tung, Director, Omnidata Services, Inc., New York, N.Y. *(Computers in Civil Engineering)*

John J. White, Former Chief Specifications Engineer, and Associate Consultant, Parsons, Brinckerhoff, Quade & Douglas, Consulting Engineers, New York, N.Y. *(Specifications)*

Don S. Wolford, Principal Research Associate, Armco Steel Corporation, Middletown, Ohio *(Cold-formed Steel Design and Construction)*

Lev Zetlin, Consulting Engineer; President, Lev Zetlin Associates, Inc., New York, N.Y.; University Professor of Civil Engineering and Architecture, University of Virginia *(Concrete Design and Construction)*

Preface

Civil engineering is that field of engineering concerned with planning, design, and construction for environmental control, development of natural resources, buildings, transportation facilities, and other structures required for the health, welfare, safety, employment, and pleasure of mankind.

The second edition of the "Standard Handbook for Civil Engineers" was made necessary by broad, major advances throughout civil engineering. Because of these advances, several sections of the book had to be completely rewritten, and others required important additions, considerable changes, and some deletions.

In preparing this edition, we maintained the objectives of the first edition: We set out to provide in a single volume information that would be of greatest usefulness to everyone engaged in civil engineering, especially to those who have to make decisions affecting planning, design, and construction. We wanted the handbook to meet the needs of consulting engineers, public works engineers, architects, contractors, educators, material and equipment suppliers, inspectors, construction labor, students, and many others. As before, we were faced with an extremely difficult problem in selecting subject matter, because there was such a wealth of material available that each section could readily be expanded into a thick handbook.

We decided to adopt the same solution that gained ready acceptance for the first edition:
• The handbook is comprehensive, but each topic is treated as briefly as clarity permits.
• Information incorporated is of a nature that should be valuable in making decisions—characteristics of construction materials and equipment, essentials of stress analysis and structural theory, basic principles of civil engineering and their application, recommended construction practices and why they are used, and cost estimating.
• Frequent reference is made to other sources where additional detailed information can be obtained.

• Each section is written for the nonspecialist in the field, on the assumption that the specialist prefers to seek answers in a more detailed text dealing exclusively with that field.   Emphasis is placed on fundamentals rather than on tables of design data.

• Tables of design data, building codes, standard specifications, and similar material that may be obtained easily from trade associations, technical societies, and government agencies are referred to but not reprinted in this book.

• The practical approach is stressed throughout.   Methods are presented that are as simple and short as possible.

Also, as in the first edition, more than half the book is devoted to the specialty fields of civil engineering, such as building, bridge, highway, and environmental engineering.   The rest of the book deals with engineering common to those fields, such as design and construction management, computer applications, specifications, geotechnical engineering, and structural engineering.   In addition, to help you locate needed information speedily, a detailed table of contents has been provided at the front of the book and an extensive index at the back.

The sections of the handbook dealing with design and construction with concrete, structural steel, cold-formed steel, and wood have been rewritten for the second edition because of major changes in standard design specifications.

Since publication of the first edition, use of high-speed computers and other electronic devices has spread throughout all areas of civil engineering. This development is reflected in nearly all sections of the second edition. In the first edition, the section *Computers in Civil Engineering* had anticipated this expansion of computer applications, but there were some unforeseen developments in recent years, especially in use of terminals, networks, and programmable calculators.   Hence, the section had to be updated to cover these applications.   Similarly, *Specifications* has been revised to cover use of master specifications and computer-controlled automatic typewriters for speedier and accurate production of project specifications.   Also, the finite-element method, made feasible by high-speed computers, has been added to *Structural Theory*, offering structural engineers still another analytic tool.   (The section, however, retains the descriptions of methods suitable for manual calculation that were given in the first edition.)   Furthermore, discussions of electronic surveying devices and equipment have been added to *Surveying*.

Greater emphasis on environmental control and prevention of pollution in recent years has brought about changes in civil engineering practice and has had an impact on many sections of this handbook.   One important effect has been changing the title of the *Sanitary Engineering* section to *Environmental Engineering*, with a corresponding rewriting of the text.

Other sections also have been revised to deal with new developments and advances in the fields covered.   *Earthwork* has been updated to describe new equipment or models and new methods.   In *Building Engineering*, fire protection and acoustics have been expanded.   New authors have brought a fresh viewpoint to *Highway Engineering*. Revisions of *Rail-Transportation Engineering* extend the treatment of rapid transit and discuss new types of vehicles for high-speed movement of passengers.   *Bridge Engineering, Airport Engineering, Tunnel Engineering,* and *Water Engineering* have been revised to reflect current practice.

In preparing this reference work, the contributors drew heavily on numerous sources of information.   Many of these are credited or given as references throughout the book, but space limitations preclude mentioning them all. The editor and the contributors wish to acknowledge their indebtedness to these sources and to express their gratitude.

The editor is especially grateful to the contributors, not only because he appreciates the great value of their contributions but also because he is keenly aware of their considerable sacrifices in taking time to prepare their sections. Also, the editor wishes to acknowledge and express gratitude for the help he received from a Board of Consultants composed of Walter S. Douglas, Rolf Eliasson, Prof. T. Y. Lin, Samuel B. Nelson, Alonzo DeF. Quinn, and Edward E. White in preparation of the first edition; their influence, in consequence, extended to the development of the second edition.

We all hope that you will find this edition even more useful than the previous one.

Frederick S. Merritt, Editor

Computers in Civil Engineering

CHARLES P. C. TUNG and DAVID CARSEN

Directors, Omnidata Services, Inc., New York, N.Y.

Electronic computers with high speeds, large memory capacities, and sophisticated monitor systems may be used in every phase of civil engineering from original concept to final construction. But engineers must select the most economical method of use. To do this, they need an understanding of the basic features and principles of computer operation.

1-1. Advantages of Electronic Computers. Computers will never quite live up to expectations of users unless they realize that what is involved is not just high-speed computing equipment but a whole new concept of computing for engineering. An electronic computer, in the hands of competent personnel, can become the production equivalent of a staff of designers or, with the addition of plotting equipment, a squad of draftsmen; or it can aid a planner by providing economic justification for large projects. This does not mean that the machine is endowed with human qualities but simply that it has the ability to compute, with lightning speed, solutions established by programmed systems.

In any project, the work may be divided into two components: original thinking and detail thinking. Most, if not all, of the detail thinking can be done by machine in the major branches of civil engineering. In addition, large areas of original thinking have been invaded by computers because of their ability to test and reject solutions based on established criteria of design.

An examination of the benefits that can be derived from using computers shows that a civil engineer must learn to work with these machines to remain in a competitive position, either as an individual or as a business enterprise. Correct usage of computers saves time, manpower, and money. But these obvious savings, important as they are, do not give a complete concept of the benefits that can be obtained from computers.

For example, a program can be written for design of composite wide-flange bridge stringers varying in span from 30 to 100 ft, in increments of 1 ft, and in spacing from 5 to 10 ft, in increments of 3 in. These results, printed by a computer in tabular form, eliminate the need for ever again designing a composite wide-flange bridge stringer. This type of design aid can be prepared for such diverse problems as concrete and steel columns under direct stress and bending, welded girders for bridges and buildings, retaining walls and bridge abutments, and prestressed-concrete beams for bridges and buildings,· to mention only a few.

In addition, whole new areas of accurate solutions, previously considered too difficult to be obtained by normal manual methods, have become rather simple procedures on the computer. Any engineer who has tried to solve complex space frames with 20 or more unknowns or variables knows how formidable a task this can be. On the computer, the answers are achieved in a matter of minutes.

Alternative designs to determine maximum economy of construction have always been limited because of high design costs. With the computer, the entire gamut of possible alternative designs can be studied in depth for a small part of the costs normally associated with extensive redesigns.

But even these examples do not begin to delineate the increase in design ability computers give engineers. With computers, engineers can do in hours what used to take weeks manually. They can arrive at an earthwork balance for a 50-mile highway in 5 hr, design a 40-story building in steel and concrete (for an economic comparison) in a morning, or complete final designs and estimates for the most economical arrangement of substructure and superstructure for a $10-million bridge in three 8-hr working days. Thus, using computers, engineers are in a position to provide services to clients far beyond the limits of the largest firms in existence that do not use computers. By the same token, firms that cannot provide these services are at a great disadvantage.

1-2. Digital and Analog Computers. There are two basic types of computers: *digital* and *analog* computers. A digital computer performs its calculations by numerical operations, such as addition, subtraction, multiplication, and division, similar to the conventional desk calculator. The operations, however, are performed in millionths of a second in a specified sequential order. This order is established by a formal set of instructions called a *program*. An analog computer solves problems by analogy.

The digital computer has five basic components:

1. Input mechanism, which introduces the problem to the computer in the form of data and instructions.

2. Memory or data-storage unit, which stores data or instructions. This unit is similar in concept to a simple tape recorder. Information can be permanently retained or erased and subsequent information substituted.

3. Arithmetic-logic register, which performs the internal calculations by directing information to the computer section from memory, processing it, and directing it back to memory. It is in the arithmetic-logic register that the logic of the computer, or "understanding," resides in the form of the ability to accept or reject solutions.

4. Control unit, which synchronizes all operations of the computer.

5. Output mechanism, which presents the computer results.

Since the function of a digital computer is to make only arithmetic calculations, it is apparent that this type of computer is best suited to problems that can be expressed in mathematical terms. A survey of all the branches of civil engineering shows that most analysis and design problems have solutions that can be expressed in mathematical forms. Hence, for the normal applications in general civil engineering practice, digital computers are the logical choice. Since accounting, construction scheduling, and traffic analysis are based on the simplest mathematical concepts, these specialized areas can also be considered part of the digital-computer domain.

Analog computers perform in an entirely different manner. They establish a quantity by measuring it in the same way that a slide rule measures a number or a speedometer measures a distance. They often solve a problem by obtaining the answer from an electrical model that uses electronic circuitry to simulate the mathematical forms of the problem. In this type of computer, voltages represent the engineering quantities in the problem, such as forces, strains, or distances.

All the basic components of an analog computer act in unison to perform the basic mathematical operations. Its important elements are operational amplifiers, input and feedback impedances, coefficient potentiometers, and for more complex problems, peripheral devices, such as function generators, multipliers, diodes, and relays. Visualizing analog computers as basically integrating mechanisms gives a clearer picture of most practical applications. Solutions of simultaneous differential equations; plotting capabilities, especially as related to highway cross sections; and mathematical models dealing with variables of wide range are general problems particularly adaptable to analog computers.

1-3. Central Processing and Auxiliary Equipment. The size and extent of the computer system selected should reflect the user's ability to derive maximum benefits from it. Every engineering organization can get exactly what it needs because the module concept of computers provides wide flexibility in capacity and performance.

The range of possible users of computing equipment runs from small contractors and consulting firms to large state and Federal bureaus. Realistically, it may be difficult for a small firm to support an in-house computer installation. An alternative is to rent time on another's computer or to share a computer with other firms.

Perhaps a 20-designer office or a contractor with 10 projects simultaneously under construction is the smallest organization that can afford its own computer. For such firms, a small-capacity digital computer will service all its needs. Equipment might include a central processing unit (4^k memory capacity) renting at $700 to $1,000 per month and a key punch at $70 per month. (These and following costs are for 1975 and are approximate.) As the consulting firm grows in size to 75 or more designers or the contractor takes on more projects, additional equipment will be needed to take care of the larger volumes of data processing required. In these instances, an 8^k memory capacity with storage disk, card read-in, and printer can be considered since the monthly rent will increase to about $1,500. These added pieces of equipment, however, raise the scope and ability of the central processing unit to a very high degree.

The 4^k unit is a small-sized computer. The working area, or *core*, of the machine can store only 4,000 words. This machine will have an output speed about that of a very fast typewriter, or less than 150 words per minute. An increase in the core capacity by attachment of a disk drive adds storage of about one-half million words per disk. If the disk drive can handle more than one disk, two disks will increase the storage capacity of the original machine to nearly 1 million words. An interchange of disks can increase this capacity even more. This increase in internal speed and capacity should then be matched by a comparable printout speed. The use of a high-speed printer will do this by increasing the output speed from 150 words a minute to 600 lines a minute and up. Therefore, while the smallest unit may rent for less than $1,000 per month, a substantial unit with tremendous increase in processing ability will cost little more, a bit over $1,500 per month.

This procedure of adding modular units and minor peripheral equipment, such as additional key punches or a key-punch verifier, can continue with larger units renting for $5,000 to $10,000 per month. But a consultant or a contractor doing multi-million-dollar projects will, under normal circumstances, find data-processing needs satisfied by machines renting from $1,500 to $2,500 per month. A large state highway department should, with correct usage and satisfactory programs, get maximum economic benefits from computers in the $5,000 to $10,000 per month range. A summary of basic equipment with approximate costs is listed in Table 1-1.

Several small, specialized machines with somewhat limited capacities and capabilities are also available. Some of these sell for as low as $20,000; others can be rented for $500 to $600 per month. Although these machines may be economical for specialized firms, such as those dealing exclusively with surveys and property work, they are not suitable for the broader problems encountered in the general civil engineering field (see also Arts. 1–5 and 1–11).

Table 1-1. Basic Computer Equipment

Type of equipment	Monthly rental
Small central processing unit (4^k to 8^k)	$ 700–$1,500
Medium central processing unit (8^k to 32^k)	$2,500–$6,000
Disk drive	*
Card-read punch	*
High-speed line printer (300–600 lines per minute)	*
Key punch	$70
Key-punch verifier	$90

*Included in cost noted for central processing unit.

1-4. Applicability of Computer Operation. Despite comparatively low computer cost, economic benefits derivable from computers may be elusive. The main reason is that probably the single largest expense that a user incurs is that of programming. Program libraries available for use in the civil engineering field are extremely limited. A user must therefore analyze the type. of work planned for the computer before making a decision to install an in-house machine. Certain fields with large amounts of repetitive-type calculations can use simple programs, which are available to a limited degree or can be written for a relatively small cost. As computer applications become more sophisticated and the parameters of the problems increase to include judgment factors, programming costs rise swiftly.

Table 1-2 assigns a probable value to the percent of design calculations that could be done on a computer in the major fields of civil engineering. Allied fields such as construction scheduling and accounting are also included. The table shows that although all the branches of civil engineering can benefit from computer usage, the major applications are for highways, bridges, buildings, construction scheduling, and accounting. Since all these fields require highly repetitive calculations, real savings can be attained with computers.

Table 1-2. Potential Computer Applications in Civil Engineering Design and Construction

Field	% of design calculations applicable to computer usage
Highways	95
Bridges:	
Simple spans (all types)	90
Continuous and cantilever spans:	
Girders	85
Trusses	75
Prestressed concrete (all types)	65
Rigid frames	70
Arches (fixed or hinged)	65
Movable	50
Suspension	50
Buildings:	
Structural	80
Heating, ventilating, air-conditioning	40
Surveying	90
Hydraulics	70
Sewerage and water supply	50
Soil mechanics and foundations	40
Stress analysis	90
Cost accounting and payroll	95
Critical-path method (construction)	90

The decision to use a computer must be based on three fundamental facts: percentage of work that can be done on the computer; availability of programs for doing the work or cost of getting programs; and cost of computer usage.

1-5. Selecting a Computer. (See also Arts. 1-3 and 1-4.) Once the decision is made to install an in-house computer, several other important decisions have to be made. Type of machine, speed, memory capacity, peripheral equipment, available programs, and maintenance service —all enter into the final choice.

Selection of a digital or analog computer will be determined entirely by the function to be performed. Consulting engineers, contractors, and most public agencies involved in civil and military works will find almost all their normal requirements satisfied by the digital computer. Users involved exclusively in plotting or setting up models for specific areas, such as economic benefits, traffic studies, graphic displays, or large real estate subdivision layouts, will normally make better use of an analog computer. If volume and type of work warrant the expense, then a combined system with both a digital and an analog computer for computing and plotting would be in order.

With the large selection of machines available, it is impossible to state that any individual piece of equipment is preferable to another. Considering small units ranging in rental cost from $500 to $1,500 per month, the choice will probably be determined by features other than just actual machine operation.   Reliability of maintenance service, availability of usable programs, and software support, such as monitor system and delivery schedule, are very important considerations. Small machines serve a sensitive economic market; they must start paying off immediately and maintain a good record of use with a minimum amount of down time.

Choosing one of the large machines ranging in cost from $3,000 to $6,000 per month is more difficult.   Almost all computer manufacturers have equipment of excellent caliber in this price range.   The user should not get too involved in making detailed studies of the use of magnetic tape vs. storage disks or whether drum storage is better than disk storage.   These are operational details that one soon learns to live with.   Availability of programs and ease of creating new programs, however, should play an important part in the decision.   It is important to see whether there is a large users group with common exchange of programs.   Since speed and memory capacity of most machines are a direct function of cost, there will probably be no startling difference between different makes in any one price range; but there will be big increases in capacity for each $1,000 invested in increasing this specific function (see Art. 1-3).

A good rule-of-thumb measure for determining computer-equipment cost would be $12,000 to $17,000 per year per 50 people; so that a 100-person office could safely assume a yearly computer rental from $25,000 to $35,000.   (Costs are for 1975.)

Equipment in the price bracket above $6,000 per month can be considered economically feasible only by public agencies or the very largest private companies.   Since this investment is such a big one at the start, this type of user, of necessity, must have a very sophisticated approach to programs and cost of new programs.   It is logical, then, for users in this group to concern themselves with machine capabilities, paying particular attention to such items as input and output speed, internal speed, programming logic, memory capacity, magnetic tape, storage disks, and basic control mechanisms.   Here, the variation is considerable between different manufacturers, and detailed studies to compare operating characteristics are very much in order.

1-6. Buy, Lease, or Rent Computer Time.   Of the three alternatives open to any civil-engineer user, the least desirable, except under unusual circumstances, is outright purchase of a computer.   Apart from the large capital outlay, purchasing involves the commitment to operate and maintain equipment that becomes outmoded in a short time.   The computer field develops so rapidly that today's marvel becomes a clumsy handicap tomorrow.   Part of the loss reflected in this rapid obsolescence is the simple cost for space.   Newer computers may occupy much less space than older equipment.   Owning a large piece of equipment, therefore, forces a firm to allocate space for the machine permanently.   Also, expensive air-conditioning may be required, which may not be needed by newer equipment.   In addition, purchase of a computer forces the user to expend time and expense on programming that becomes obsolete almost as quickly as the machine.

The decision whether to lease equipment for an in-house installation or rent time at a service bureau depends on the economic benefits involved.   The following analysis of these alternatives may help in making this decision.   Assume a machine is leased.   A minimum practical size for civil engineering problems normally is one with an 8^k memory capacity and one storage disk, renting annually for about $15,000 (Art. 1-3).   Because available program libraries (Art. 1-9) are either too limited or too outmoded, additional salary cost for programming must be added.   At a yearly base wage of $12,000, plus 50% for overhead, for one programmer, annual cost of the computer installation rises to about $35,000.   A key-punch operator, plus office supplies and minor overhead items, brings this minimum cost to $50,000.   If, on the average, only 20% of the entire work on projects handled by the firm can be adapted to a computer, then annual fees of at least $250,000 would be required to offset the cost of the installation.   Since there must be some factor of safety, it would be more realistic to assume that a $300,000 yearly billing is necessary before an in-house installation can be profitable.   It must be recognized, however, that any evaluation must include the fact that an extensive working library must be readily available to make the required saving.   Since, in many cases, this library will be either nonexistent or relatively limited, a time factor of 1 to 2 years for preparing programs must be included in any realistic appraisal.   In such instances, it might be more beneficial to add a programmer to the staff and develop a library before leasing computer equipment.

Renting time on a computer may be the cheapest procedure for some firms, but it presents problems.   Unless the firm has a program library, it may not be able to use the services of a computer service bureau.   Relatively few service bureaus are dedicated to civil engineering applications.   As new equipment comes on the market, the capabilities of these bureaus diminish momentarily because of the time required for updating available programs.   For the smaller firm, however, this method of using computers should still prove to be the most reasonable. (See also Arts. 1-10 and 1-11.)

Joint ownership or cooperative efforts on the part of several users can combine some of the best features of leasing computers or renting time. Several such ventures have been made with success. But there have been failures, attributable mainly to cost of programs.

1-7. Personnel Organization for Computer Operation. For the small firm that rents time at an available computer installation, the only personnel required is a programmer knowledgeable in civil engineering theory and practice. Ideally, this position should be filled by an engineer trained for computer use and with several years' experience in the field for which the computer will be used. Emphasis should be placed, however, on programming ability and experience above everything else. It is impossible to write a complete set of specifications to guide in selection of this type of personnel, but one qualification is a must: an understanding of the type of work to be programmed, coupled with a thorough grasp of the mathematical concepts involved.

The function of the programmer is to create a program library and train the staff in program applications. The importance of this position must not be underestimated if real computer benefits are to be realized. Since useful programs are not always available (Art. 1-9), the major portion of the programmer's efforts must be devoted to writing required programs. Apart from the cost involved, this will, in the end, provide the best results since all such programs will be designed for specific needs and maximum benefit. It must be understood, however, that considerable time, perhaps years, is required to create a useful program library.

For the large firm considering an in-house installation, additional personnel is required. One important addition is a systems engineer, whose function is to assist the programmer, run the computer, guide the engineers, and schedule the work for the computer. The systems engineer must know thoroughly every feature of available programs and should be capable of fitting engineering needs to the program capabilities. This type of personnel should also be a trained engineer (preferably civil) and have abilities and background like those of programmers, but his or her forte should be organization and scheduling. In addition, the in-house installation requires the services of a key-punch operator.

With key-punch personnel to handle the increased volume of data, the basic staff of one programmer and one systems engineer can service the needs of private firms or public agencies up to 200-designer size. With the increase of just one additional programmer and one additional systems engineer, large users can increase production considerably.

It is axiomatic in computer usage that large programming staffs do not produce large libraries, only expensive ones. For maximum economic benefits, large staffs should never be used to replace knowledgeable personnel.

Normally, small design firms, with 20 to 50 designers, gain the most from the "open-shop" practice of letting all the engineers prepare their own data input and run the computer themselves. This allows the entire technical staff to become expert and could be a large morale factor. For large contractors and design firms, as well as public organizations, this type of organization may be unwieldy. The "closed-shop" practice may work better. Here, a permanent staff, generally consisting of a programmer and a systems engineer, serves the design staff. However, with this type of organization, there is always the danger that designers, feeling too distant from the computer, lose interest and turn their problems over to the computer staff. This should be avoided. It should be mandatory for designers to solve their own problems by preparing all required data.

No matter which type of organization is used for computer operation, a programmer should be assigned to supervise and coordinate the entire system. Although the programmer's function is primarily one of devising required programs, the programmer is also ideally suited to act as the guide and mentor of the design staff.

1-8. Principles of Programming. Many problems in civil engineering can be adapted to computer applications. Determination of problems suitable for computer solution and the steps required for translation into working programs require careful analysis. The major phases of development in selection and preparation of a program are as follows:

Determination of the suitability of a problem for computer application.
Definition of the problem in mathematical terms.
Selection of a method of solution for computer.
Analysis of the problem for logical flow of programming sequences.
Coding the problem to translate it into a program.
Testing and checkout.

Although many programmers will find this step-by-step procedure useful, the more experienced and sophisticated practitioner may rearrange and combine several steps without materially affecting the end result.

To illustrate the principles involved in each phase, a sample program is given in Table 1-3 on pp. 1-8 and 1-9. It deals with an elementary problem in highway engineering: computation of theoretical grade elevations from a given profile line.

Suitability of a problem for computer application may be determined with the aid of these criteria:

Will there be enough repetition of the problem to justify writing the program? Is the problem so complex that manual computation is impractical? The expense of creating a computer program must be weighed against the benefits that can be derived from it.

Is the program within the capacities of computer and programmers? If complete solution of a problem is beyond the capacity of the available computer or too big a task for the available programming personnel, perhaps the problem can be broken up and solutions obtained by computer for the parts. Solutions to component parts of the problem should be given first priority.

Correct evaluation of problem suitability requires good engineering judgment and sound programming knowledge. The failure of many programming systems can be traced directly to initial improper evaluation of this step.

The problem treated in the sample program, computation of theoretical grade elevations from a given profile line, is a very common one in highway engineering. The scope of the program is limited for quick solution on a small-sized computer. The program can be expanded to include many other features, such as computation of elevations for highway-curve transitions and location of even-foot contour points along a highway with multiple lanes. As in any complex engineering project, a sophisticated and all-inclusive programming system must always start with a simple, straightforward program like the sample in Table 1-3.

Definition of the problem in mathematical terms is essential for writing a program. In general, the following are necessary: statement of given data, answers required, and desired accuracy of the computed results.

For the sample program, the problem is defined by:

Given profile data: This information consists of the station of the PVI (*PVIS*), PVI elevation (*PVIE*), and PVI vertical curve length (*PVIL*) (see Fig. 1-1) at each PVI point (intersection of tangents to a vertical curve) on the profile line. All units are in feet and decimals of a foot. A minimum of 2 and a maximum of 50 PVI points may be given at one time. Station *SB* for a beginning station, station *SE* for an ending station, and *SI*, the distance between successive stations, all in feet, must also be given. (Stations are assumed equally spaced.)

Required answers: These are the theoretical grade elevations *TGE* for all stations between the beginning and ending stations.

Accuracy of data: Computer input and output must observe the usual rules for good highway engineering practice.

Selection of a method of solution should consider programming costs as well as cost of computer time. Often, a problem can be solved by several methods. A programmer who knows the capability of his computer and the theoretical background of a problem is likely to choose an optimum approach. But with the high speed of computers, the difference in cost of machine time for solution by two methods may be negligible compared with the difference in programming costs. Emphasis should be placed on saving time in programming.

After a method of solution for computer has been selected, all mathematical formulas must be completely defined for the problem.

Formulas for detailed computations of the sample problem follow on pp. 1-10 and 1-11.

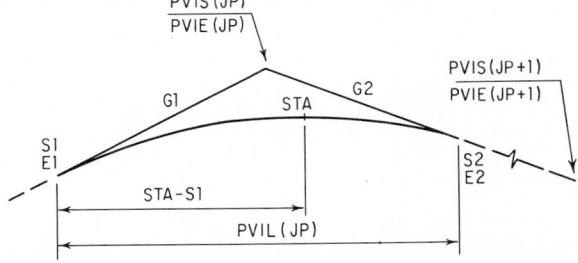

Fig. 1-1. Case 1 (*LP* = 1) of sample problem: determination of elevations along a vertical-curve segment of a profile.

Table 1-3. Computer Program in FORTRAN for Computation of Vertical Profile

```
C*        SAMPLE PROGRAM FOR COMPUTATION OF TGE (THEORETICAL
C         GRADE ELEVATION) FROM A GIVEN PROFILE LINE OF KNOWN
C         PVIS (PVI STATIONS), PVIE (PVI ELEVATIONS) AND PVIL (PVI
C         VERTICAL CURVE LENGTHS) MAXIMUM NUMBER OF CONSECUTIVE
C         PVI IS SET AT 50
          DIMENSION PVIS(50),PVIE(50),PVIL(50)
C         READ INPUT DATA
C         NPVI (NUMBER OF PVI) MUST BE BETWEEN 2 AND 50
C         SB,SE AND SI ARE BEGINNING STATION,ENDING STATION AND
C         INCREMENT STATION, RESPECTIVELY
       1  READ 2,NPVI,SB,SE,SI
       2  FORMAT (I4,3F8.2)
       3  FORMAT (10F8.2)
          IF (NPVI − 2) 4,10,6
C         PRINT ERROR MESSAGE AND STOP
       4  PRINT 5
       5  FORMAT (32HINVALID DATA, PUSH START TO READ)
          PAUSE
          GO TO 1
       6  IF (NPVI − 50) 10,10,4
      10  READ 3, (PVIS(I),I = 1,NPVI)
          READ 3,(PVIE(I),I = 1,NPVI)
          READ 3,(PVIL(I),I = 1,NPVI)
C         PUNCH GIVEN PROFILE DATA
C         LOGICAL SEQUENCE OF THE GIVEN PROFILE DATA IS ALSO
C         CHECKED
          PUNCH 11
      11  FORMAT (/ /6H NPVI,14X,4HPVIS,8X,4HPVIE,8X,4HPVIL/)
          DO 14 I = 1,NPVI
          IF (I = 1) 13,13,12
      12  IF (PVIS(I) − .5*PVIL(I) − PVIS(I − 1) − .5*PVIL(I − 1)) 4,13,13
      13  IST = PVIS(I)/100.
          FST = IST
          RST = PVIS(I) − FST*100.
      14  PUNCH 15,I,IST,RST,PVIE(I),PVIL(I)
      15  FORMAT (I6,I9,2H +,F7.3,F12.2,F12.3)
          PUNCH 16
      16  FORMAT (/ /6X,7HSTATION,6X,10HPERC GRADE,2X,9HPROF
         1 ELEV/)
C         SET STA = SB
C         JP IS THE INDEX FOR NTH PVI WHICH IS USED IN COMPUTATION S
C         LA IS A SWITCH DEVICE USED TO SET UP THE FIRST PROFILE
C         SEGMENT FOR CASE 2 ONLY
          STA = SB
          JP = 0
          LA = 1
      20  JP = JP +1
          IF (JP + 1 − NPVI) 23,23,4
      21  S1 = S2
          E1 = E2
C         TEST FOR SEGMENT CASE
C         PRESENT CASE 1 WILL GO TO NEXT CASE 2, PRESENT CASE 2 WILL
C         GO TO NEXT CASE 1
          GO TO (22,20),LP
C         SET UP PROFILE SEGMENT FOR CASE 2 (LP = 2)
      22  G1 = G2
          PCG1 = PCG2
          S2 = PVIS(JP + 1) − .5*PVIL(JP + 1)
          E2 = PVIE(JP + 1) − .5*G1*PVIL(JP + 1)
```

```
        LP = 2
        GO TO 40
C       SET UP PROFILE SEGMENT FOR CASE 1(LP = 1)
   23   S2 = PVIS(JP) + .5*PVIL(JP)
        G2 = (PVIE(JP + 1) - PVIE(JP))/(PVIS(JP + 1) - PVIS(JP))
        IF(G2) 24,25,25
   24   PCG2 = G2*100. - .00005
        GO TO 26
   25   PCG2 = G2*100. + .00005
   26   E2 = PVIE(JP) + .5*G2*PVIL(JP)
        GO TO (33,30),LA
   30   IF(PVIL(JP)) 4,31,32
   31   CP = 0.
        GO TO 33
   32   CP = (G2 - G1)/(2.*PVIL(JP))
   33   LP = 1
        GO TO (34,40),LA
   34   LA = 2
        GO TO 21
C       TEST STA FOR LIMIT OF SEGMENT
   40   IF (STA - S1) 4,42,41
   41   IF (STA - S2) 42,42,21
C       COMPUTE TGE ACCORDING TO CASE 1 OR 2
C       PCG (PERCENTAGE GRADE) OF THE STATION COMPUTED IS ALSO
C       DETERMINED
C       ALL DATA OUTPUT ITEMS ARE PROPERLY ROUNDED
   42   GO TO (43,47),LP
   43   IF (STA - PVIS(JP)) 44,45,45
   44   PCG = PCG1
        GO TO 46
   45   PCG = PCG2
   46   TGE = E1 + (G1 + CP*(STA - S1))*(STA - S1)
        GO TO 50
   47   PCG = PCG1
        TGE = E1 + G1*(STA - S1)
   50   IF (TGE) 51,52,52
   51   TGE = TGE - .00005
        GO TO 53
   52   TGE = TGE + .00005
   53   IST = STA/100.
        FST = IST
        RST = STA - FST*100.
        IF (RST) 54,55,55
   54   RST = RST - .0005
        GO TO 56
   55   RST = RST + .0005
C       PUNCH STA, PERCENT GRADE AND TGE
   56   PUNCH 57,IST,RST,PCG,TGE
   57   FORMAT (I8,2H +,F7.3,F12.4,F11.4)
C       INCREMENT STA BY SI
        STA = STA + SI
C       TEST FOR STA LESS, EQUAL OR GREATER THAN SE
        IF (STA - SE) 40,40,60
C       PRINT END MESSAGE AND STOP
   60   PRINT 61
   61   FORMAT (38HEND OF PROFILE RUN, PUSH START TO READ)
        PAUSE
        GO TO 1
        END
```

---

*C indicates explanatory text.

Profile Segment, Case 1 (Fig. 1-1). LP = 1   (*LP* = 1 designates the curve portion of the profile.) The station of the end station, ft, is computed from

$$S2 = PVIS(JP) + 0.5PVIL(JP) \tag{1-1}$$

where *JP* = index number of the PVI under consideration
  PVIS = station of PVI, ft
  PVIL = vertical-curve length, ft
The grade of the second profile tangent (Fig. 1-1) is

$$G2 = \frac{PVIE(JP + 1) - PVIE(JP)}{PVIS(JP + 1) - PVIS(JP)} \tag{1-2}$$

where *JP* + 1 = index number of the next *PVI*
  PVIE = elevation of PVI, ft
Elevation, ft, at *S*2 is

$$E2 = PVIE(JP) + 0.5G2[PVIL(JP)] \tag{1-3}$$

Since the vertical curve between *S*1 and *S*2 is a parabola, its equation is

$$y = a + bx + cx^2 \tag{1-4}$$

where *x* = *STA* − *S*1
  y = *TGE*, ft
  STA = station of any point on parabola, ft
  *S*1 = station of beginning station, ft
  a, b, c = coefficients defining the parabola
Differentiating Eq. (1-4) yields

$$\frac{dy}{dx} = b + 2cx \tag{1-5}$$

From known boundary conditions: When *x* = 0, *y* = *E*1, the elevation at *S*1; therefore, *a* = *E*1. When *x* = 0, *dy*/*dx* = *G*1, the grade of the first profile tangent; therefore, *b* = *G*1. When *x* = *PVIL*(*JP*), *dy*/*dx* = *G*2; hence,

$$G2 = b + 2cx = G1 + 2c[PVIL(JP)]$$

and *c* = 0.5(*G*2 − *G*1)/*PVIL*(*JP*).
  The theoretical grade elevation, ft, at any point *STA* then may be computed from the parabolic equation [Eq. (1-4)] in the form

$$TGE = E1 + G1(STA - S1) + CP(STA - S1)^2 \tag{1-6a}$$

or
$$TGE = E1 + [G1 + CP(STA - S1)](STA - S1) \tag{1-6b}$$

where $CP = \dfrac{0.5(G2 - G1)}{PVIL(JP)}$

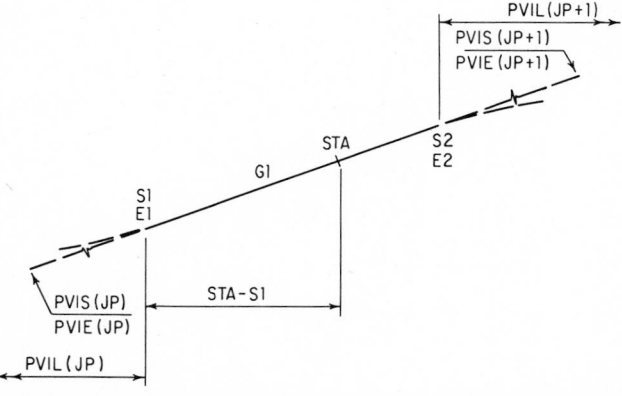

Fig. 1-2. Case 2 (*LP* = 2) of sample problem: determination of elevations along a constant grade (tangent).

Profile Segment, Case 2 (Fig. 1-2).   LP = 2   (*LP* = 2 designates the tangent portion of the profile.)   The station of the end station is

$$S2 = PVIS(JP + 1) - 0.5PVIL(JP + 1) \tag{1-7}$$

The elevation, ft, at $S2$ is

$$E2 = PVIE(JP + 1) - 0.5G1[PVIL(JP + 1)] \tag{1-8}$$

The theoretical grade elevation, ft, at any point *STA* on the tangent is

$$TGE = E1 + G1(STA - S1) \tag{1-9}$$

Analysis of the problem for logical flow of programming sequences follows the selection of a method of solution.   The objective is to identify component operations and their interrelationships.   The extent of analysis for a problem varies considerably, depending on both the programmer and the problem. , The analysis may consist of a list of steps in mathematical form for

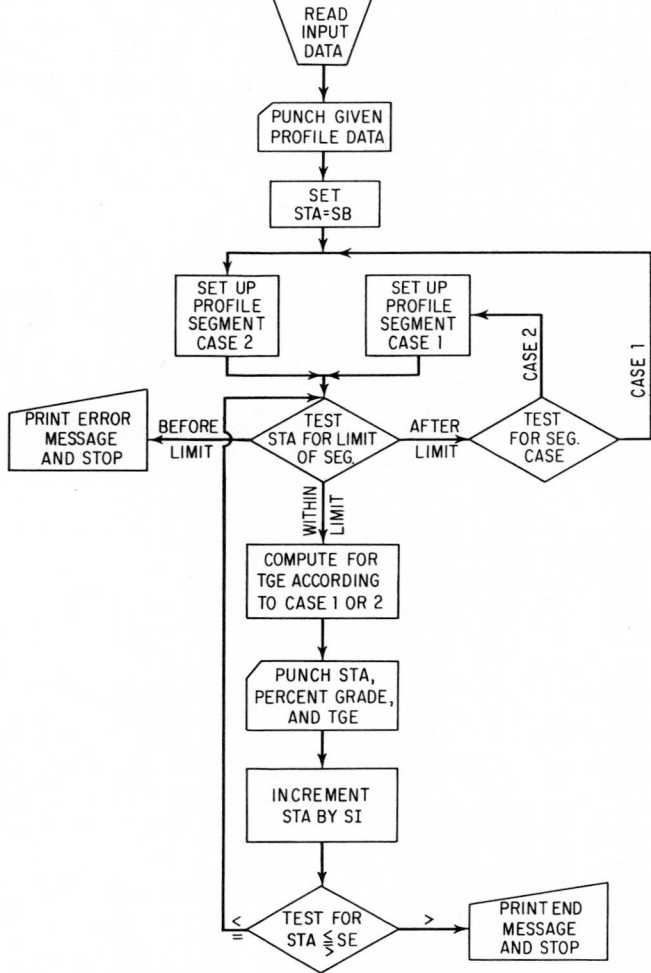

Fig. 1-3. Flowchart for developing program for sample problem of determining theoretical grade elevations *TGE* along a profile.

simple problems or, for more complex problems, a picture of the process (or flowchart), which establishes the parts of the problem and their sequence. Flowcharts can be general statements describing different operations or detailed breakdowns of every step. Full documentation of a problem always is desirable for future reference and an invaluable help to other users of the program.

Figure 1-3 shows a flowchart for the sample program (Table 1-3). The method of general statements describing different operations is employed. There is ample program narrative material within the program to serve as cross reference to the flowchart. This technique serves to clarify the documentation further. In Fig. 1-3, SB represents the station, ft, at the beginning of the profile, SE the station, ft, at the end, and SI the distance, ft, between stations. Initial steps cover input of data and start of the computation with SB. Case 2, a tangent, is assumed initially. The step, Test STA for Limit of Seg, follows to determine from the given data whether the beginning station is actually on a tangent. If the station lies before the start of the profile segment, an error has occurred. The machine reports this and stops. If the station lies after the end of the profile segment, the computer determines whether the segment is a curve (case 1) or a tangent (case 2). And if the station lies between the beginning and end of the profile segment, the computer computes the desired data for the station. Then, the computer repeats the cycle for the next station. When the station equals or exceeds SE, the computation ends.

Coding the problem to translate it into a program involves writing a detailed set of instructions for a specific computer. In general, instructions may be written in machine, symbolic, and FORTRAN languages. Some computers have languages similar to FORTRAN, but they are too specialized in nature to warrant discussion here.

Machine language employs number codes. A code usually is applicable only to a specific computer model. An example of a typical addition instruction is 2100500-09400. This instruction tells the machine to call for the number stored at 09400 and add it to the number stored at 00500, while 21 indicates that the numbers should be added.

Symbolic language, a more convenient form of coding, is subsequently transformed into machine language for the computer to perform. An example of a simple instruction to add labor cost to an existing total cost in a symbolic programming system is A COST, LABOR.

FORTRAN (abbreviation of FORmula TRANslation) is an excellent language for an engineer to communicate with a computer. FORTRAN is not the natural language of a computer, nor is it the natural language of the engineer. It can be regarded as a compromise between the two. To satisfy the computer, FORTRAN uses symbols that the computer can understand and requires that the rules for their use be closely followed. To satisfy the engineer, it eliminates as many of the detailed computer control operations as possible from the job of writing programs and uses a problem-statement format close to that of mathematical equations.

The engineer codes a problem in the FORTRAN language into statements. The statements are translated into the natural language of the computer (machine language) by the computer itself, with the aid of a program called FORTRAN processor. The resulting machine-language program then is used to obtain the solution of the problem.

The major advantage of a FORTRAN program is that it is not limited to a specific computer. A FORTRAN program can be readily converted from one computer to another. Generally, a small modification to conform to a specific FORTRAN version or adjustment to a specific monitor system may be all that is required. Thus, when a change is made from one machine to another, the programmer has very little work to do for conversion. The sample program (Table 1-3) was written in FORTRAN for the IBM 1620 computer and 1620 Monitor System.

Testing and checkout (debugging) is the last step in preparing a program. Any errors discovered while executing the program must be corrected. In debugging, the program is tried out on a set of test data, usually simple, so that hand computations can be made for comparison and verification. If the program requires several decisions, the test data should cover each possibility for a complete checkout of the program. Continued satisfactory use of the program over a considerable length of time (field test) serves as the final checking procedure of the program.

Table 1-4 gives typical input and output for the sample program. (A portion of the output that is repetitive is omitted to save space.)

1-9. Programs: Type, Need, and Availability. Success in use of computers depends more on availability of programs than on any other feature. Because of the very large range of problems in civil engineering and the many specialties that can be computer-oriented, a large investment in time, money, and human labor is required to develop a working library.

Although each branch of civil engineering has its own theoretical and practical requirements,

Table 1-4. Typical Input and Output for Program in Table 1-3

DATA INPUT OF THE SAMPLE TEST PROBLEM

```
 11   550000 3410000     10000
549500   676500 1163000 1486500 1627000 1924500 2221000 2503500 2708500
326650 3416500
   37908     35601    32201    24009    27593    42993    46271    37811    39511
   22411     26911
    000     30000    50000   100000    40000   120000   100000    60000   100000
  160000       000
```

DATA OUTPUT OF THE SAMPLE TEST PROBLEM

NPVI	PVIS	PVIE	PVIL
1	54 + 95.000	379.08	0.000
2	54 + 65.000	356.01	300.000
3	116 + 30.000	322.01	500.000
4	148 + 65.000	240.09	1000.000
5	164 + 70.000	275.93	400.000
6	192 + 45.000	429.93	1200.000
7	222 + 10.000	462.71	1000.000
8	250 + 35.000	378.11	600.000
9	270 + 85.000	395.11	1000.000
10	326 + 65.000	224.11	1600.000
11	341 + 65.000	269.11	0.000

STATION	PERC* GRADE	PROF ELEV
55 + 0.000	− 1.8165	378.9892
56 + 0.000	− 1.8165	377.1726
57 + 0.000	− 1.8165	375.3561
58 + 0.000	− 1.8165	373.5396
59 + 0.000	− 1.8165	371.7230
60 + 0.000	− 1.8165	369.9065
61 + 0.000	− 1.8165	368.0900
62 + 0.000	− 1.8165	366.2734
63 + 0.000	− 1.8165	364.4569
64 + 0.000	− 1.8165	362.6404
65 + 0.000	− 1.8165	360.8238
66 + 0.000	− 1.8165	359.0073
67 + 0.000	− 1.8165	357.3253
68 + 0.000	− .6989	356.0118
69 + 0.000	− .6989	355.0707
70 + 0.000	− .6989	354.3677
71 + 0.000	− .6989	353.6688
72 + 0.000	− .6989	352.9699

INTERMEDIATE LINES OF OUTPUT NOT SHOWN

296 + 0.000	− 3.0645	318.0374
297 + 0.000	− 3.0645	314.9729
298 + 0.000	− 3.0645	311.9084
299 + 0.000	− 3.0645	308.8439
300 + 0.000	− 3.0645	305.7794
301 + 0.000	− 3.0645	302.7149
302 + 0.000	− 3.0645	299.6503
303 + 0.000	− 3.0645	296.5858
304 + 0.000	− 3.0645	293.5213
305 + 0.000	− 3.0645	290.4568
306 + 0.000	− 3.0645	287.3923
307 + 0.000	− 3.0645	284.3278
308 + 0.000	− 3.0645	281.2632
309 + 0.000	− 3.0645	278.1987
310 + 0.000	− 3.0645	275.1342
311 + 0.000	− 3.0645	272.0697
312 + 0.000	− 3.0645	269.0052
313 + 0.000	− 3.0645	265.9407
314 + 0.000	− 3.0645	262.8761

Table 1-4. Typical Input and Output for Program in Table 1-3 *(Continued)*

STATION	PERC* GRADE	PROF ELEV
315 + 0.000	− 3.0645	259.8116
316 + 0.000	− 3.0645	256.7471
317 + 0.000	− 3.0645	253.6826
318 + 0.000	− 3.0645	250.6181
319 + 0.000	− 3.0645	247.5768
320 + 0.000	− 3.0645	244.8344
321 + 0.000	− 3.0645	242.4711
322 + 0.000	− 3.0645	240.4868
323 + 0.000	− 3.0645	238.8816
324 + 0.000	− 3.0645	237.6554
325 + 0.000	− 3.0645	236.8082
326 + 0.000	− 3.0645	236.3401
327 + 0.000	3.0000	236.2510
328 + 0.000	3.0000	236.5409
329 + 0.000	3.0000	237.2098
330 + 0.000	3.0000	238.2578
331 + 0.000	3.0000	239.6848
332 + 0.000	3.0000	241.4909
333 + 0.000	3.0000	243.6760
334 + 0.000	3.0000	246.2401
335 + 0.000	3.0000	249.1600
336 + 0.000	3.0000	252.1600
337 + 0.000	3.0000	255.1600
338 + 0.000	3.0000	258.1600
339 + 0.000	3.0000	261.1600
340 + 0.000	3.0000	264.1600
341 + 0.000	3.0000	267.1600

*PERC = PERCENT

computer programs written for unrelated areas of work often are interrelated, for example, continuity analysis of structures by converging approximations and pipe-network analysis by converging approximations. The procedures for the structural program are nearly identical to those for the hydraulic program. Consequently, it is possible to adapt programs written for one field of engineering for use in another field. It is vital, therefore, that a thorough study be made of available programs before an attempt is made to write a new one. Chances are very good that needed programs are available, though they may require remodeling.

An examination of existing programs shows that an excellent working program library is available for such fields as surveying; land subdivisions; highways; structures, especially buildings and bridges; foundation design; stress analysis; construction scheduling; and cost accounting. Areas of weakness, however, exist for water supply, sanitary engineering, hydrology, and the more sophisticated phases of soil mechanics. But these gaps are being closed gradually.

One of the more severe brakes on continuing development of *software*, such as programs, is the rapid improvement of computer *hardware*, or equipment, that has taken place. Because each new computer requires at least a review and often rewriting of existing programs, considerable loss of time and money occurs when an engineering organization upgrades or downgrades its computer capacity. To circumvent this handicap, new concepts in programming are being evolved. Small, isolated, problem-oriented programs are being discarded in favor of vast systems capable of handling entire branches of civil engineering. New and simplified languages are being used to delay or avoid obsolescence.

Two different approaches to these systems have given excellent results. One approach, known as ICES, developed by the Massachusetts Institute of Technology, applies to the general civil engineering field. It includes subsystems for geometry, highway design, structural design, bridge design, soil mechanics, traffic analysis, and critical-path scheduling. This system will become an invaluable addition to all future civil engineering program libraries.

The other approach, taken by a computer center, Omnidata Services, Inc., New York, develops one system for each branch of civil engineering. The objective is to enable each system to solve every conceivable problem likely to be met in its respective branch. When this approach reaches its final development, it will be possible to have a library of 12 or 13 systems to cover the entire field of civil engineering.

Many programs are available for older-type computers. These programs require considerable modification for use with new computers. Nevertheless, they can serve as the basis for a useful but limited library. Sources for such programs are varied, and caution must be used in making selections. Programs written in FORTRAN should be chosen for easier updating. A strange program should never be accepted, however, without a thorough check of basic theory and accurate comparisons with known solutions.

The following list of agencies dedicated to program development may prove useful:

HEEP . . . . .Highway Engineers Exchange Programs is an organization of state highway departments and private consultants dedicated to development of programs for highways and bridges.

Federal Highway
Administration . . . . .This agency, headquartered in Washington, D.C., maintains a list of programs available to the public that is useful for traffic analysis and highway work.

CEPA . . . . .Civil Engineers Programmed Applications is a user group dedicated to developing programs for the IBM 1130 computer.

COMMON . . . . .A small-systems group sponsored by IBM for program development for the 1130 and 360 series of computers.

State highway departments . .Many state highway departments have written extensive programs in FORTRAN. In most instances, requests for such programs have been honored, particularly from engineers involved in road building.

Engineering universities . . . . .Colleges sponsor many programs and may make them available on request.

Service bureaus and private
"software" specialists . . . . . . .This last group can be considered the consultant's consultant. Specialists of proven caliber may seem very expensive, but a careful analysis of costs vs. benefits may favor their engagement. Many large computer users turn to "software" specialists to solve in-house problems.

1-10. Terminals and Networks. Article 1-6 pointed out that an alternative to an in-house computer is to rent computer time. A convenient way of doing this is by **teleprocessing,** that is, transmitting information to a computer over telephone wires. This procedure requires that a small terminal connected to a remote computer be available in the user's office. With many such terminals so connected, the computer is the heart and brains of a computation network.

The relatively low hardware cost (about $150 per month) for the average small terminal gives the user easy and ordinarily immediate access to a large computer. The low cost of the terminal, however, may be accompanied by relatively high processing costs if transmitting speeds are slow. The low-cost terminals operating at speeds as low as 10 to 15 characters per second that have been used in the past, both for transmitting input data and printing final output, require long periods of *connect time.* For such terminals, *on-line time* costs often negate the low hardware expense. In addition, slow-speed transmission increases the possibility of line errors induced in the telephone wires. As a result, teleprocessing at low speed cannot successfully meet the needs of the small firms for which in-house computer facilities are not economical.

Ideally, an in-house terminal installation should supply the user with the equivalent of a real computer of the size and speed capable of doing all the user's computations at a hardware and processing cost not exceeding the ranges listed in Table 1-5. The various elements necessary

Table 1-5. Approximate Limiting Cost Ranges for Computer Usage

Size of firm*	Yearly cost†
10–20	$6,000–$8,000
20–50	$8,000–$12,000
50–80	$12,000–$18,000
80–100	$18,000–$24,000
For each additional 25 employees over 100	Add $4,000

*Includes only the technical employees on the staff.
†As of 1975.

for achieving these cost ranges include a low-cost, high-speed terminal, a substantial program library (either developed in-house or available on the network computer), and a skilled individual who is familiar with the use of large computer systems and knows what practical programs are available.

Consequently, high-speed terminals are preferable. Such terminals are available, ranging from $150 to $300 per month in rental cost and capable of transmitting input data at speeds ranging from 60 to 120 characters per second. For engineering computations, these translate to speeds of 50 to 100 lines per minute. Roughly, such equipment is the equivalent of a small IBM 1130.

An additional feature that is vital to successful operation of these terminals is the ability to input a complete set of data, print it to permit a check of the data (before submittal to the central processing unit), and then transmit it from magnetic tape on the terminal to the network computer. Terminals operating with magnetic tape cassettes fulfill these requirements and provide flexibility for the user.

The following is a general procedure for teleprocessing with low connect time and at low cost. An engineer lists data input on standard input forms. A terminal operator inserts the data on cassette magnetic tape and at the same time prepares a printout for checking. After the data review and final corrections have been made, the data input may be transmitted to the computer. (Transmission may be accomplished at night, however, because many computer networks have greatly reduced rates for such hours.) As soon as input transmittal is finished, the operator installs an output tape at the terminal, with instructions to the central processing unit to return the output directly to this tape or, if the output is voluminous, have it delivered by messenger.

Over the years, many large networks have increased the capacities of their civil engineering application programs. These programs, supplemented by the individual user's own programs, offer a substantial base for terminal-oriented computer use.

1-11. Electronic Programmable Calculators.   Paralleling the growth in power, speed, and memory capacity of large-capacity electronic computers, there have been a rapid development and growth in the use of electronic calculators, or slow-speed minicomputers. With continuing improvements, these calculators are overtaking the capabilities of the lower range of electronic computers.

During the early stages of conversion from mechanical drives to electronic operation, the memory capacity of electronic calculators was too small to be considered a practical alternative to electronic computers. But with peripheral hardware such as disk drives, magnetic tape cassette memory, plotters, and line printers readily available, these augmented calculators can match the ability of small computers (8 to 32^k range). Partly overcoming the handicaps of low speed and capacity by permitting random access to memory, the new hardware provides users with the power of a small- to medium-sized computer at a much lower cost. Calculators more recently developed use storage disks with storage capacities of 1,000,000 to 2,500,000 bytes. These disks can increase storage capacity to as high as 600,000 items of data of 12 digits each.

In addition to this enlargement of data storage capability, addition of suitable interface equipment enables an electronic programmable calculator to be used as an in-house terminal as part of a computation network. Such peripheral equipment allows users to obtain all the advantages of a small- to medium-sized computer plus a relatively fast terminal at a relatively low cost.

Electronic programmable calculators, within prescribed areas of work, can also perform as ably as the larger electronic computers. Despite their limitations of relatively slow speed and present inability to accept programs written in such languages as FORTRAN or COBOL, these calculators have a high cost-benefit ratio and user applicability. Since the rental costs of electronic calculators may be as little as one-third to one-half those of electronic computers of comparable size and ability, these limitations may be acceptable for many engineering organizations.

The small but powerful desktop electronic calculators are suitable for mathematical and statistical computations for all fields of engineering. These devices offer ease of operation, relatively small space requirements, and simplicity of programming. The only feature that tends to limit their applicability is the need for man-machine communication via a very simple but limited language, such as BASIC. This simplicity of language, which is advantageous for simple applications, is a severe handicap for computations involving sophisticated concepts. However, there are such wide fields in all the engineering disciplines requiring relatively limited mathematical procedures that use of electronic calculators is eminently practical.

Surveying and geometric calculations, for example, are well suited to calculator solution. The mathematical functions involved, such as trigonometric functions, are simple, and the uncomplicated procedures required can be expressed in BASIC as easily as in FORTRAN. Similarly, highway layout, land subdivision, field stakeouts, and innumerable similar simple applications definitely warrant use of an electronic programmable calculator. Payrolls, cost accounting, and progress reports, too, can be economically processed. Structural analysis and design of simple structures with few members, which require a somewhat higher level of language competence, also can be programmed in BASIC, but not as simply as surveying computations.

However, when analysis or design of complex structures is desired, coupled with varying design code requirements, a simple language such as BASIC becomes inadequate. This is particularly true where the logic requires a selection process that involves an intricate series of commands, each of which is the result of considerable calculation and accommodation to code requirements.

Nevertheless, an electronic programmable calculator is an invaluable piece of hardware for an average engineering firm because much of its work is relatively straightforward and routine.

1-12. Bibliography for Computer Applications.

1. Arden, B. W., and K. N. Astill, "Numerical Algorithms: Origins and Applications," Addison-Wesley Publishing Company, Inc., Reading, Mass.

2. Carnahan, B., H. A. Luther, and J. O. Wilkes, "Applied Numerical Methods," John Wiley & Sons, Inc., New York.

3. Chapin, N., "Computers: A Systems Approach," Van Nostrand Reinhold Company, New York.

4. Desai, C. S., and J. F. Abel, "Introduction to the Finite Element Method," Van Nostrand Reinhold Company, New York.

5. Dunn, W. L., "Introduction to Digital Computer Problems Using FORTRAN IV," McGraw-Hill Book Company, New York.

6. Fenves, S. J., "Computer Methods in Civil Engineering," Prentice-Hall, Inc., Englewood Cliffs, N.J.

7. Hausner, A., "Analog and Analog Hybrid Computer Programming," Prentice-Hall, Inc., Englewood Cliffs, N.J.

8. Jenkins, W., "Matrix and Digital Computer Methods in Structural Analysis," McGraw-Hill Book Company, New York.

9. Kazmier, L. J., and A. S. Philippakis, "Fundamentals of EDP and FORTRAN," McGraw-Hill Book Company, New York.

10. Kelly, L. G., "Handbook of Numerical Methods and Applications," Addison-Wesley Publishing Company, Inc., Reading, Mass.

11. Kuo, S. S., "Computer Applications of Numerical Methods," Addison-Wesley Publishing Company, Inc., Reading, Mass.

12. Merritt, F. S., "Modern Mathematical Methods in Engineering" and "Applied Mathematics in Engineering Practice," McGraw-Hill Book Company, New York.

13. Pipes, L. A., and S. A. Hovanessian, "Matrix Computer Methods in Engineering," John Wiley & Sons, Inc., New York.

14. Rubenstein, M. F., "Matrix Computer Analysis of Structures," Prentice-Hall, Inc., Englewood Cliffs, N.J.

15. Schick, W., and C. J. Merz, Jr., "FORTRAN for Engineers," McGraw-Hill Book Company, New York.

16. Spindel, P. D., "Computer Applications in Civil Engineering," Van Nostrand Reinhold Company, New York.

17. "STRESS: A Reference Manual," The M.I.T. Press, Cambridge, Mass.

Design Management

FRANK MULLER

**Director, Contracts and Construction Management Services,
Parsons, Brinckerhoff, Quade & Douglas, Inc.
New York, N.Y.**

Design management is concerned with an engineer's sphere of activity. It is important, therefore, to consider the variety and types of design activities to which professionals devote their efforts.

Basically, the engineer's role is to harness scientific principles and other knowledge to benefit humanity. In fulfillment of this role, design management is concerned with proper utilization of human labor, energy, and technical skills to serve present and future needs of the world's economy.

2-1. Where Engineers Are Employed. Principal fields of employment for engineers include:

Academic. For many engineers, the teaching profession is both the first and final career. Many others, however, devote to teaching a few years of their careers or sometimes part of their time, for example, teaching evening courses.

Many educators also serve as advisers to industry and consulting firms. Thus, they move into the designer's sphere of activity. Furthermore, many university departments are retained by government and industry for research projects. As a consequence, the departments, in essence, act as private firms performing professional services. The university administrators have to work within budgets and have contracts to negotiate, reimbursable expenses to determine, and schedules to meet. They also have to contend with other administrative matters that are part of design management.

Industry. Industrial firms that handle any substantial volume of business have engineers on their staff. The role of such engineers, however, varies. A firm with productive capacity and thus plant facilities must have a plant engineer and staff to insure proper maintenance and operation of the plant. In many industries, the plant engineers also serve their employers in the design field. For instance, if new equipment is to be installed in an existing plant, not only must space be provided, but engineering questions must be answered. Typically: Are the foundations adequate to carry the added loads? Are new utility services required? Is the present power supply adequate? Further, a new building may have to be constructed to house the equipment. Thus, a plant engineer's normal activities and responsibilities often lead to the design field.

Because of their size, growth, and specialized needs, many industries have their own engineering and design departments. Such a department fulfills the same professional function as a private engineering firm but with one basic difference. The industry engineer serves one client, whereas the design firm serves many. Concerned with many of the same administrative matters as a design firm, an engineering department can be organized like a design firm. The engineering department will be organized to operate efficiently in meeting the specialized needs of only its industrial employer.

Government.   Like engineers in industry, government engineers serve only one client—their employer. The Federal government is the largest single employer of architects and engineers. In addition, most states, counties, cities, towns, and public bodies have engineers and architects on their staffs or in their employ. These professionals perform a variety of functions encompassing both design and administrative activity.

The agencies or authorities maintain engineering and architectural departments that perform basic design work and thus act as in-house professional service firms. Such organizations do not need to retain outside private consultants, except for specialized tasks. In addition, these agencies, whether they have in-house design capability or not, employ professionals who work on a variety of different administrative levels. These levels include administration and supervision of projects as well as review of basic design and construction activities. Administration of the engineering projects requires the services of professionals on all levels, starting with junior staff members and extending up to top-level administrators and officials charged with responsibility for implementation of the public projects.

In public service, the engineer may be the designer or the client.

Engineer-Contractor.   The term as used herein refers to the construction firm that identifies itself as both an engineer designer and contractor. Although there are many who use the title engineer-contractor and perform only the actual construction, we are concerned here with the firm that truly undertakes "turnkey" projects—both design and construction under a single contract.

Process and utility industries are the commonest users of the turnkey contract. These industries are primarily interested in the final product, such as number of barrels of oil refined or number of kilowatt-hours produced. The engineering staff of the company building a plant establishes design criteria that the engineer-contractor has to meet. Because of the specialized nature of these industries, the engineer-contractor employs designers with knowledge of particular processes to develop the most economical and efficient design. All engineer-contractors bid on their own designs. Other turnkey operations include those that combine land acquisition, design, and construction for commercial and industrial buildings, and can even include financing.

The design is accomplished by the same organization, or division within the organization, that constructs the building or facility. Depending on a variety of factors, there are advantages and disadvantages of this combined service as compared with the division of responsibility between a design firm and a construction company.

Consulting Engineer.   A consulting engineer has been defined as a "professional experienced in the application of scientific principles to engineering problems." As professionals, consulting engineers owe a duty to the public as well as to their clients. In addition to rendering a professional service, the consulting engineer also operates a business. Consulting engineering is practiced by sole practitioners, partnerships, and corporations, many with large staffs of professionals, draftsmen, and other supporting personnel. Regardless of the form of the engineer's organization, the final product a client receives retains the same professional characteristics and meets the same professional standards. Consulting engineers usually have several clients, and they must select methods of operation to suit their own and their clients' needs best.

Others.   There are numerous specialty firms that practice in private industry but limit their activity to specific or specialized services. These groups can often be categorized as consulting engineers, engineer-contractors, or industrial firms, depending on the extent and scope of the service rendered. One example is the construction management firm. This type of firm may be an independent professional organization or a division of a consulting engineering or architectural firm that renders expanded engineering-construction supervision and management functions. Construction management firms can also be set up by contractors who, in addition to their normal construction activity, adopt a total management capability in which they represent the owner in planning, purchasing, and implementing the entire project.

2-2. Forms of Consulting Engineering Organization.   Consulting engineers may practice as individuals, partnerships, or corporations.

Individual Proprietorship.   This form of organization is the simplest, has the fewest legal complications, and enables the proprietor to exercise direct control over the operation.   As a one-person operation, however, this type of practice has distinct limitations since its activity essentially can be restricted to the efforts of the individual.

Though conducting a business as a sole proprietor, a consulting engineer may have several employees.   Thus, as an employer, the consulting engineer is operating a business and has to handle the problems associated with a business enterprise.   Also, since consulting engineers represent the legal entities conducting their businesses, they are responsible for all obligations of a business and all contracts are entered into in their names.   Consulting engineers are personally responsible for all debts and can be liable for these to the extent of all their assets, business or personal.   All profits, however, are earned by the consulting engineers, and they are not required to distribute earnings, as in a partnership, or be concerned with the declaration of dividends, as with a corporation.

Partnership.   The commonest form for a consulting engineering organization is a *partnership*, that is, an association of two or more professionals who combine forces and talents to serve their clients on a more comprehensive scale and, by offering more services, to serve a wider clientele.   Typically, each partner is responsible for a specific area.   The management of the business, depending on its complexity, is assigned to one partner, the *management partner*.

A partnership retains the identity of the individual professional, and basically its legal structure is similar to that of the individual proprietorship.   Instead of one individual assuming all contractual obligations, liabilities, and debts, and earning all profits, these are shared by the partners.   The partners, however, may not necessarily share equally in the business.   Interest can be worked out among the partners as desired.   For instance, one partner may own more than 50% and thus have a position comparable to that of the majority stockholder of a corporation.

Although widely used, partnerships, from the business point of view, have several disadvantages.   These have caused many firms to incorporate in states where practice by corporation is not restricted.   One disadvantage of partnerships is that each partner is liable to the extent of his total personal assets for the wrongful act of any partner in the ordinary course of business. Another disadvantage is that a partnership terminates on death or retirement of one partner unless other provisions are made in the partnership agreement.   Furthermore, a partnership does not have the flexibility of a corporation for comprehensive employee-benefit programs and provision for key employee participation.

Although a partnership as an entity does not pay taxes, the partners pay taxes on the profits as individuals.   This is not necessarily a disadvantage, but it can be a prime consideration in the choice of an operating organization.

Corporations.   Most states permit the formation of professional engineering corporations. But usually a corporation can be formed for the purpose of practicing engineering only under certain conditions.   These require that ownership and management of the company be totally vested in professionals or at least that majority interests be held by professionals.   In many states, legislation permitting the formation of such corporations has been passed to give professionals, not only in engineering but also in other professions, the benefits and protection of conducting business as a corporation.   While permitting such corporate practice, the legislation includes requirements so structured that the public is protected from unqualified persons conducting a professional practice under a corporate guise.

With such protective requirements, professional identity can be maintained in corporate practice.   Therefore, if conditions warrant and state law permits, engineering organizations should consider the corporate form of practice.   The advantages that are attained, however, are mainly business ones.   The management structure of the organization is clarified.   Responsibility is defined.   The area of employee fringe benefits becomes more diversified.   Opportunities exist for profit sharing, for realistic retirement plans, and for employees to buy into the firm.   Also, personal liability of the principals is limited to the assets of the corporation, although the principals continue to be responsible for their own professional acts and cannot use the corporate structure as a shield from liability for professional errors and omissions.

Each form of practice has to be evaluated on its own merits.   A corporate structure for an individual practitioner with a small practice may not be warranted, but one with a large volume of business that can be assigned to subordinates may find a corporation advantageous.   For some firms, the tax advantages of a corporation may be more beneficial than operating as a partnership. (For Federal income tax purposes, a small business corporation, meeting certain requirements, can elect to be taxed as a partnership, a practice advantageous for a small corporation.)

2-3. Clients for Engineering Services.   Each client and each project has particular needs.

Types of clients include:

Federal Government.   As the largest single employer of engineers and largest contractor for services and products, the Federal government is a potential client for most design firms.   To qualify for consideration by any of the government branches, a firm must file periodically with agencies from which work may be obtained a questionnaire detailing the firm's organization, key personnel (education and experience), special areas of competence, and experience (including completed projects).   Preparation of such data is time-consuming.   Many agencies, however, have standardized their requirements so that the same form can be used for many filings.

Within the Federal government the standard questionnaire for architects-engineers, SF 251, is utilized by a majority of the agencies retaining professional services.   This form presents in summary fashion basic data describing the experience and qualifications of individual professionals and identifies their firm's qualifications and past experience by projects and areas of expertise.   In addition, many agencies have established computerized data banks utilizing the information contained in standard qualification forms to simplify both their records and the search for qualified professional firms to serve specific project needs.   When an agency needs outside design services, it searches its qualification files and selects a group of firms with the particular capabilities necessary for the project.   Interviews follow, leading to selection of a firm and contract negotiation.

Other-than-Federal Public Work.   Public work other than that performed for the Federal government is in the province of states, counties, cities, and municipalities.   The contracting party varies, depending on the nature of the work and its scope.   Usually, engineering work is under the jurisdiction of an agency's engineering department.   Sometimes, however, states or cities establish authorities to administer, construct, operate, and maintain projects.   Many states, for example, have separate authorities for construction and operation of limited-access toll roads, for ports, for bridges and tunnels, and for public buildings such as schools and colleges.   These authorities, as well as the public bodies, have different methods of operation.   Some perform all or nearly all design in-house; they engage outside consultants infrequently.   Others retain consulting engineers for most design.

Considerable areas of engineering activity lie within public authorities or regional public agencies, such as transportation, sewer, or water authorities, established within regions for the implementation of specific tasks.   Such agencies either retain consultants to perform the necessary engineering for implementation of their public projects or establish in-house capability to perform the same functions.

Industry and Commerce.   With the growth of industrial and commercial corporations and their capital expenditures, the potential market for design firms in industrial and office construction has grown.   Also, industry can be a client for housing developments and related facilities, such as streets, water supply, and sewerage systems.

Professionals.   Many consulting engineers have as clients only other construction-industry professionals.   Most often, these engineers specialize in one facet of design, such as structural or mechanical engineering.   These firms render most of their services under contract to architects or engineers who are retained by clients for complete design of a facility but within their own firms have the capability of only their own specialties.

In addition, professionals serve each other within their own fields of competence.   Engineers may retain other engineers as consultants to supplement their own capabilities, either to take advantage of specialized knowledge or experience or for independent checks on their firm's analysis and calculations.

Other Clients.   Sometimes, an owner may engage an engineer for projects that may require a few hours' attention or for the design of an entire facility.   Professionals such as lawyers consult engineers as much as engineers seek professional counsel from lawyers.   Also, engineers are often called on to give testimony on technical matters as expert witnesses.

2-4. Scope of Engineering Services.   The range of activity of engineers in design covers a broad spectrum from brief advice to inspection of construction and includes preparation of plans and specifications.   Many firms, although qualified to render a variety of services, may limit the scope of services offered as well as specialize in a particular field.   For example, some engineers offer only structural-design services or foundation consulting.

Following is a brief summary of services rendered by engineering firms:

Advice and Consultation.   This phase may comprise no more than an expression of the consultant's opinion based on experience and technical knowledge.   Normally, detailed engineering design is not an element in this phase; but the engineer may advise a client on the merits of undertaking a new project and on its related technical consideration.   Or this phase may just be the rendering of an opinion on the advisability of undertaking further studies to determine the need for repairs or rehabilitation of an existing structure.

Technical Investigation and Analysis.   After consultation, the engineer may undertake detailed studies, such as physical exploration, including soil borings, topographic surveys, and hydrographic studies.   Possible methods of construction may be considered.   Preparation of a feasibility report may follow.   This report usually considers economic as well as engineering aspects.   Both aspects have to be explored to enable an owner to decide whether or not to undertake a project.

Planning.   If, on the basis of a feasibility report or other information, the owner decides to proceed with the construction project, the planning phase is started.   Planning must be considered separately from design.   If, for instance, a plant or complex of structures is being developed, planning includes rough preliminary sketches and a master plan of the proposed project.   With a master plan, the owner can develop a project in stages and schedule construction according to his available funds.

Design.   This phase can be subdivided into schematic, preliminary, and final designs.   There can be a review with the owner at the end of each stage, or the review can be continuous to enable the owner to visualize the implementation of requirements and allow additions and changes to be made as the need arises.   The completed design documents consist of detailed plans and specifications and contracts for construction (Arts. 3-2 and 3-5).   The designer's role, however, does not end at completion of final design.   Normally, the designer acts as the owner's representative in taking construction bids, awarding contracts, and administering construction contracts.

Construction Consultation and Inspection.   During the construction phase, the engineer's role covers field consultation and on-site inspection.   Field consultation, normally part of the designer's obligation under the design contract, includes periodic visits to the site, issuance of clarifying drawings (if required), and check of equipment catalogs and contractor's shop drawings for compliance with the contract documents.   On-site representation consists of a resident engineer and a staff, the size (or existence) of which depends on the nature, magnitude, and complexity of the project.   The resident engineer's prime function is to make sure that the contractor complies with the design.   By accomplishing this, the resident engineer is able to maintain quality control.   Also, as the engineering representative on the site, the resident engineer is able to assist the contractors and subcontractors when questions arise requiring interpretation of the plans.   Resident engineers are the links between the field and design offices.

Construction Management.   Due to the growth, complexity, and inflationary costs' spiral of construction, construction management services have evolved both as traditional consulting or contracting services and as management of construction projects.   A construction manager, often retained at about the same time as the project designer, may commence tasks at the beginning of design.   The services of a construction manager may include basic program review and analysis, design review and evaluation, scheduling (CPM and PERT), cost estimating, value engineering, bid analysis, contractor selection, detailed construction inspection, coordination of trades and separate construction contractors, cost control, and program management.   Acting as an owner's agent, the construction manager can perform all or some of these tasks to assure the owner of project and budgetary controls.

Other Services.   Among the other services rendered by engineering firms are preparation of technical reports; investigation surveys, such as land and property surveys to establish title to property; evaluation and rate studies; appraisal of property and building values; expert testimony in court; and services to industry, financial institutions, and public bodies in the economic field.

2-5. Selection of Consultants.   (See also Art. 2-3.)   A consultant prefers not to submit bids for services.   This is primarily a matter of ethics and has become an essential element in his profession.   The logic for this is self-evident.   Since consultants render professional services, it is impossible to set a comparative basis for evaluating competitive bids.   Furthermore, if consultants were selected on a price basis, the owner, in retaining the lowest bidder regardless of professional qualifications, would risk purchasing an incomplete or incompetent service.   Since the fee paid a consultant is a small percentage of the total cost of a project, an owner is well advised to pay properly for such services and obtain the best professional services available.   Fees are based on accepted standards and yardsticks in the industry ("Manual and Report on Engineering Practice," No. 45, American Society of Civil Engineers, and "Compilation of Fees," American Consulting Engineers Council).

Fee negotiation and competitive pricing have been studied by various government agencies and challenged in the courts as a result of antitrust administrative rulings issued by the Justice Department.   One consequence has been that the American Society of Civil Engineers removed from its Code of Ethics a provision making bidding for the supply of professional services unethical.   Another consequence has been that the following sequence of steps in selection of a professional consultant by an owner has become common practice:

1. Review of the qualifications of several firms and evaluation with respect to requirements for the project.   An owner may have knowledge from past experiences of such firms; but if not, the owner may contact professional organizations, such as the American Consulting Engineers Council or American Society of Civil Engineers, for a recommended list of firms.   Owners without past experience in selecting consultants would be well advised to confer with associates in their own industries for a list of recommended firms.

2. Selection of up to six (normally three) firms with the experience and knowledge for undertaking the assignment.

3. Request of the selected firms for an indication of interest and detailed data pertaining to their qualifications and ability to undertake the project.   With this submission, the firms are also asked to submit information concerning size of staff, availability of personnel to be assigned to the particular project, and their experience in similar lines of work.   The firms are also interviewed.

4. Selection of the firm most qualified to undertake the project.   In addition, the owner should list one or two additional firms, in order of their desirability, in case a contract cannot be negotiated with the first choice.

5. Notification of the firm chosen of its selection, negotiation of a fee, and execution of an agreement for professional services to be rendered.   If a mutually agreeable fee cannot be arrived at, negotiation with this firm terminates and negotiations then begin with the No. 2 selection.   (For ethical reasons, to avoid conflict of interest, a consultant will not negotiate with a prospective owner if negotiations are still pending with another firm.   As a consequence, the negotiations with the first firm must be terminated.)

In determining the firm most qualified to undertake a project, an owner should consider technical qualifications, ability to absorb the additional workload in relation to the firm's capability and existing workload, experience, reputation, financial standing, and past accomplishments in related fields.

Since the cost of any service is important to an owner, an equitable fee for the services to be rendered has to be established.   A caveat for owners is: "You receive only the professional services you pay for." If the fee is cut, services rendered are reduced.   In the development of a project, it is important for an owner to receive complete and competent professional advice.   If this is done, owners can be assured that their projects will be designed economically and efficiently and the fee paid for proper professional services will be a wise investment.

2-6. Fees for Services.   (See also Art. 2-5.)   There are three basic methods for determining fees: lump sum, percentage of construction value, and cost plus.

Lump-sum Fee.   A fixed fee is arrived at by estimating the man-hours and expenses anticipated for rendering the service.   Although easily underestimated in importance, since fees often are based on percentage of construction value, estimating the anticipated cost of a consultant's service is essential.   When the scope of a project is specifically outlined, the consultant can evaluate anticipated costs for services by analyzing the demands of the project and drawing on experience and on knowledge of the firm's capabilities.   The consultant can translate the project into man-hours required and compute the cost.   To the cost of labor must be added overhead, any expenses beyond those normally included in the overhead factor, any unusual elements that might add to costs, and anticipated profit.   When such an estimate has been made, it can be converted into a percentage of the anticipated construction cost of the project.   Although the fixed fee may be established by using accepted industry percentages as a yardstick, the contract is negotiated for a lump sum regardless of the eventual construction cost of the project.   Only if there is a change in the scope of services initially agreed on will there be a possible change in the fee.

A variation of this form of payment is the lump-sum fee plus expenses.   Such a payment method is used if there are extraordinary expenses, for instance, a more-than-normal amount of travel to a distant site, or if subsoil investigation and surveys are included in the consultant's scope of work.

Percentage of Construction Value.   Professional societies and public contracting agencies, such as the Federal government, set predetermined acceptable percentages of construction value for consultants' fees.   These percentages are intended to be used as a guide by the contracting parties in determining a fee.   The fee schedule lists recommended ranges for the percentage fee depending on the nature, size, and complexity of the work to be undertaken.   Each fee is a matter for negotiation between the contracting parties since each project has its particular needs and requirements.   Thus it must be cautioned that the recommended-fee schedules should be used only as guides, as they are intended, not as fixed inflexible rates.

If a percentage fee is negotiated between the parties, it is of great importance to define what amount will be used for the construction value.   Will it be the estimated value or the actual construction value based on the contractor's low bid?   If the fee is to be based on the estimated value, will the preliminary or detailed estimate govern?   If the fee is to be based on the low bid, the design contract must state that the contractor's bid be bona fide since contractors sometimes make mistakes and submit improper bids.   Furthermore, the design contract should provide for a payment method if, for some reason, construction does not proceed and no bids are available to establish a construction value for fee-payment purposes.

Cost-Plus Fee.   The cost-plus type of contract is normally used when the scope of work cannot be readily defined.   Then, the owner agrees to reimburse the consultant for costs plus a fee.   The reimbursable costs consist of technical payroll and actual expenditures, such as travel, subsistence while away from home, long-distance telephone calls, and other costs incurred directly for the project.   Normally, the fee is determined by a factor applied against payroll cost.   The factor compensates the consultant for management, overhead, indirect costs, and fee.   Principals, partners, or officers, if engaged in actual production work (technical, as differentiated from administrative), are reimbursed for their services in the same manner as employees on the payroll.

In a variation of this payment method, a time factor (hourly or daily) is used with wage rates to reimburse a consultant for costs, overhead, and fee.   For example, owner and consultant may agree on a rate of pay for a category of employee and multiply this rate by an overhead-and-fee factor.   If a designer's average rate were set at \$5 per hour and the overhead factor at 150%, the payment provision in the contract would state that reimbursement to the consultant for the designer's time would be at \$12.50 per hour (\$5 + 1.5 × \$5).   Rates also would be set for other categories of personnel to be employed on the project.

Additional cost-plus arrangements most commonly used by Federal and other public agencies establish both a basis for identifying all allowable costs and a fixed fee at the time of contract negotiation.   Although calculated as a percentage (frequently 10%) of estimated costs, this fee remains fixed (a lump sum) for the contract unless there is a change in the scope of work.   The fixed fee covers profit and nonallowable costs.   Allowable costs are reimbursed as incurred for the prosecution of the work.   Such costs include direct labor, direct project costs, and overhead and indirect costs attributed to the labor base.   Federal Procurement Regulations spell out in great detail categories of costs, both allowable and nonallowable.   All such costs are subject to audit and verification by government audit agencies.   Contractors or consultants who contract with the Federal government conduct yearly audits in which they verify and agree on the cost basis to be utilized.

Some owners engage consultants on a retainer basis.   However, this reimbursement method is not a substitute for payment of fees as previously described.   An owner who has a continuing need for engineering advice and consultation may retain a professional engineer for a period of time, normally on an annual basis.   The owner is free to call on the consultant for professional assistance on a continuing basis, such as attending periodic planning and development meetings.   If, however, the service required becomes more than consultation and design of a project is called for, the retainer would not be sufficient compensation.   A separate fee would be negotiated.

2-7. Design Costs.   (See also Art. 2-6.)   The major expense incurred in design is the cost of technical personnel or productive payroll (designers and draftsmen).   Also, a firm has an administrative or indirect payroll.   Included in this category are administrative assistants, business managers, controllers, bookkeepers, accountants, office managers, and secretaries.   Other charges may be placed in two categories: direct and indirect expenses.

Direct expenses are those costs incurred due to work on a particular project and therefore are treated as bookkeeping charges to that project.   These include long-distance travel and local transportation, hotels, meals and other living expenses while away from home, reproduction and printing costs, computer rental, stationery and drafting supplies, and long-distance telephone and telegraph costs.   If other engineers are retained by the designer for consultation and to undertake part of the design, the cost of their services also would be a direct expense.

Indirect expenses essentially constitute a firm's overhead.   Although they may not all be necessarily fixed costs, indirect expenses are not directly attributable to a particular project.

Typical indirect costs incurred by a design firm are office rent; transportation, hotel, meals, living expenses, stationery, drafting supplies, and telephone and telegraph costs—all not chargeable to a project; insurance and depreciation; contributions; entertainment; new business and sales expenses; legal and accounting costs; microfilm expense; and miscellaneous taxes.

Certain indirect costs are fixed, and in essence they are independent of the firm's volume of business. An example is the rent, or if the firm owns the office space occupied, the cost of maintaining such space, such as real estate taxes, cleaning, utilities, and guards. During the term of a lease or during the period of ownership, cost of space is the same regardless of the volume of business or number of employees on the payroll at a particular time. In contrast, a variable expense depends on the nature and amount of business. An example is transportation. If, for instance, a firm owns cars, the expense of maintaining and operating them can be controlled by disposing of some or all of them or by adding more cars, depending on the volume of business and the particular need.

Consultants must be concerned with the efficient use of their technical staff in order to maintain peak productivity and with control of nontechnical costs in order to keep them consistent with the volume of business.

2-8. Project Methods and Standards.   For efficient operation, a firm should establish standard methods and systems. This does not mean that once a procedure is established it is inviolate; it is subject to improvement and refinement. But within reason, the standard procedures should be adhered to on all projects. Without standardization, the result would be more than wasting of time; the firm would be unable to operate efficiently within available budgets.

A code number should be assigned to identify each project. A commonly used system identifies the project by a series of numbers, including the year (calendar or fiscal) in which a project is started. This number should be used on all work, whether a final drawing, rough calculation,

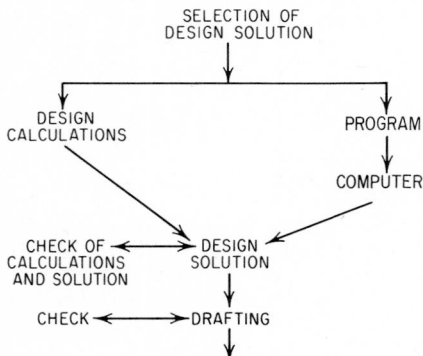

Fig. 2-1. Typical design procedure.

or correspondence. All costs and charges pertaining to the project should also be identified by this number.

A standard procedure for the performance of all work should be established. This includes a procedure for checking calculations and a system for preparation and approval of drawings, from draftsmen's work to the final authorized signature. Regardless of what internal procedure is established, the ultimate objective is the same: to operate economically and efficiently. After a design problem has been evaluated and analyzed and a method of solution established, a typical design procedure would be as indicated in Fig. 2-1.

Since many specifications are similar to each other in outline and technical provisions, standardization of specifications can be most useful. This does not necessarily mean that the firm should prepare "canned" specifications for use interchangeably on all projects. Each project has different requirements; but the various sections of the specifications should be prepared in a consistent manner on all projects. For instance, in a concrete specification, a typical section might contain the following major paragraphs: scope of work, related work (cross reference to other specification sections), general, material (cement, sand, aggregates, etc.), reinforcing steel, formwork, concrete strength and mixing, and concrete placement. Each paragraph has to be tailored to fit the requirements of a project—pier, bridge, or building. Many of the provisions, however, may be essentially the same in many instances, for example, the provisions for the quality of material in one geographic area.

For simplification, the firm may adopt standard specifications prepared by technical societies for an item, such as structural concrete. These specifications require the designer to insert requirements for a specific project but eliminate the necessity of writing anew for each project sections that are substantially the same for all projects.

2-9. Scheduling Design. Without proper scheduling, a firm may find that its operation is as inefficient as if no standard procedures were used. For design, the firm is essentially scheduling manpower needs. This task becomes more important with the number of projects to be handled at the same time. A properly run firm should be able to schedule its work so as not to take on more than it can adequately handle with a stable size of staff.

For scheduling total workload, individual project scheduling is essential. The simplest and commonest device for this. purpose is the *bar chart*, a graphic representation of manpower (represented by bars) plotted against time. By studying such a chart, one can quickly determine job start and completion dates and when and for what manpower needs will be greatest.

Scheduling devices such as the critical-path method (CPM) and program evaluation and review technique (PERT) have a definite place in programming design-manpower requirements. Although the design project for which a complete CPM or PERT study would be employed is unusual, modification or limited use of these programming devices is warranted in many cases. A complete computer CPM program, including scheduling costs as well as time and evaluating the economics of "crash" programs, would be used only on the most complex projects. Because more thorough planning is required, use of the basic CPM and PERT activity diagram (Art. 4-28) can often result in a better schedule project than if a bar chart were used. With the use of a bar chart, the start or completion of activities represented by a bar can be extended a week or more without affecting the basic schedule. A CPM or PERT diagram does not permit this since the diagraming of the activities interrelates them all and the change in time in one activity can affect all.

2-10. Production Control. Once a project is undertaken, the work involved has to be completed regardless of time or cost. Still the firm must operate within a budget since design can be performed efficiently. A designer does not deal with a tangible product for which the firm can establish a cost per unit and operate on a production-line basis. Nor should the firm go to the extreme of establishing a control in such a manner that the cost becomes more important than the product.

Cost control in its simplest form is a matter of bookkeeping. The firm should keep records of all costs relating to each project. At the end of a project, therefore, the firm should know the costs and income received and whether the work was done at a profit or loss. When a firm undertakes a new project similar in nature and size to one completed previously, a record is available to guide the new activities. Such cost accounting can be refined to varying degrees.

Also, it is well to know one's financial standing as to work on hand before completion of the work, since it may be years before some projects are completed. During the course of a project, the firm should project the costs and income based on percent completion at a particular time to determine whether they are in line. Such projections should be made periodically to gain a picture of the financial condition of the firm's operation at a particular time.

Cost accounting serves an additional purpose: It establishes controls during the programming of the work. These controls enable the firm to determine where productivity and efficiency need improvement before the end of the project when it is too late.

A professional firm, like any business, is concerned with making a profit. Maintaining a proper profit margin is essential to survival and growth. Such a profit margin varies with the size of a firm and number of principals; whether principals are on a salary, as in a corporation, or not, as in a partnership. Cost control is an important tool in assisting managers to insure the required profit margin to keep the firm operating efficiently.

(T. G. Hicks, "Successful Engineering Management," McGraw-Hill Book Company, New York.)

2-11. Internal Organization of a Design Firm. Basically, an engineering firm consists of technical departments and administrative and support staff. Figures 2-2 to 2-4 are illustrations of typical consulting-firm organizations.

Technical Departments. Depending on the size of the firm, the technical department can be divided into divisions, such as structural, civil, mechanical, and electrical engineering and architectural divisions. These can be subdivided and overlapped under the direction of a job

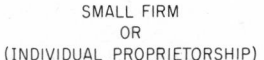

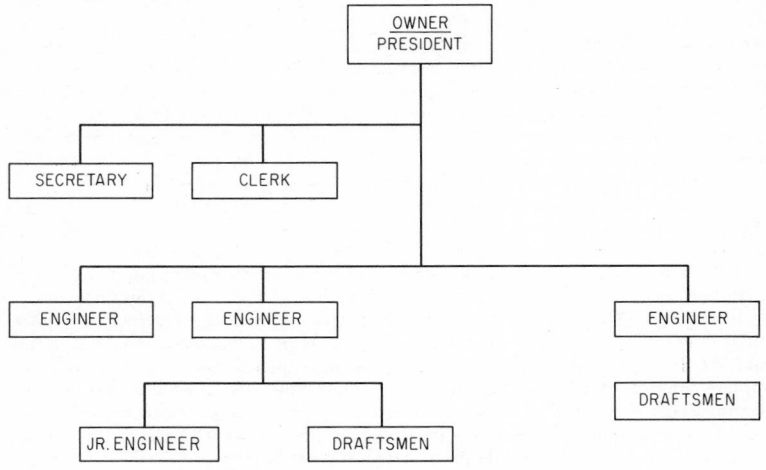

Fig. 2-2. Typical organization of a sole proprietorship.

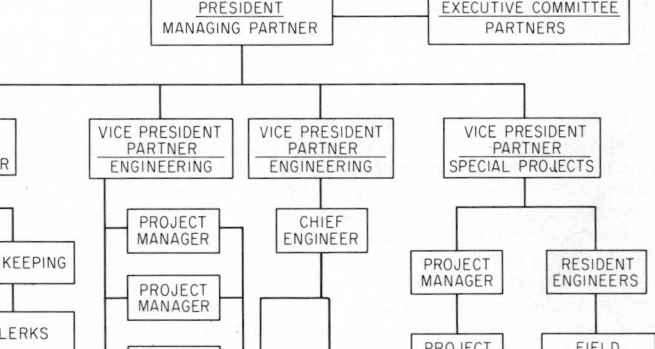

Fig. 2-3. Typical organization chart for a consulting firm.

captain, project manager, or project partner for particular projects.   (In very small firms, many functions are performed by one individual, including the proprietor.)

There are numerous ways of organizing a technical department (see, for example, Figs. 2-2 to 2-4).   The most important consideration in any organization is communication.   Whenever

CONSULTING FIRM
(PARTNERSHIP OR CORPORATION)

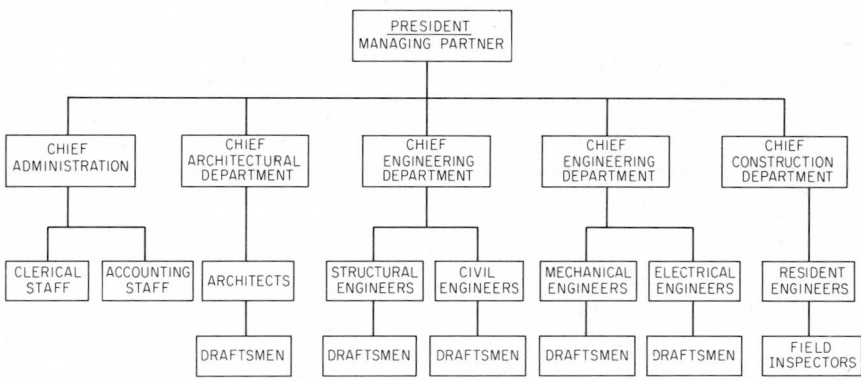

Fig. 2-4. Alternative form of organization for a consulting firm.

a firm is formed or expanded or new departments are established, communication should be considered of prime importance.   The flow of information between line levels should be well defined.   Furthermore, there should always be one individual who acts as project manager or captain in a position to coordinate all activities whether they are only those of departments within the organization or those of outside contractors or consultants involved in the project.

Many firms also have a separate construction or construction management department.   It consists of the project and construction managers, resident engineers and inspectors required on a project site, and project engineers rendering field consultation services and coordinating the efforts of field personnel.   Instead of establishing a separate department for this function, some firms have the project engineers for design in the various design divisions continue in the same capacity through the construction phase; they draw on a nucleus of field personnel for back-up, as necessary, for on-site inspection.

The primary supporting functions are new business development, personnel management, accounting, and office management.

New Business Development.   Professionals do not sell services directly but instead must apprise the market of their availability.   The firm has to prepare qualification data (Arts. 2-3 and 2-5), which can range from completion of standard prequalification forms to preparation of elaborate brochures, supplemented with extensive project descriptions and photographs.   Although a new client may make the initial contact and retain a design firm without prior communication, a design organization cannot rely on this manner of receiving new business.   As a consequence, client contact is an essential part of an organization's operation.

Client contact can be limited to impersonal contact by mail or can range to active sales efforts, where an employee or principal (or even a staff if the size of the firm warrants it) makes personal calls on potential clients.   The name of a firm has to be kept active.   Sales efforts, however, should not be a substitute for the quality of service.

In the face of intense competition and the need for growth and diversity of a firm, the search for new markets and development of new business are vital functions.

Personnel Management.   Employers have certain legal obligations.   They must pay payroll taxes, such as Social Security and state unemployment and disability, and they must withhold taxes from employees' earnings.   These requirements result in administrative burdens involving the filing of forms and reports.   Also, there are insurance obligations and statutory requirements, such as workmen's compensation.

Employers may wish to give employees the opportunity to subscribe to other forms of insurance as a group, and they may pay all or part of the costs for other benefits.   In the competitive market for skilled personnel, such fringe benefits must often be added to the basic wage.   Determination and administration of such benefits are another time-consuming function that the firm has to assume.   Fringe benefits include group life insurance, accident and major medical insurance, medical and hospitalization protective insurance (Blue Cross–Blue Shield), salary-continuance program in case of inability to work for extended periods of time due to accident or illness, and pension plan. Either employees or employer may pay the full cost of these, or they may share the cost.

Employers must have firm wage and salary policies. Besides paying the so-called going wage, they must establish policies for salary reviews and increases, salary ranges for various types of positions, bonuses, and whether to initiate a profit-sharing plan. Primarily, however, employers must give employees opportunity for advancement. Also, they must give recognition for efforts on behalf of the firm. If employers can instill the pride of accomplishment and profession, they will have efficient and happy work forces.

Accounting. (See also Art. 2-10.) The importance of cost accounting and cost control should not be minimized. To operate efficiently, a firm must be able at all times to evaluate and analyze its financial situation. For this, the firm has to maintain proper accounts. The compiling and recording of all transactions relating to the financial aspects of a business are the basic responsibility of accounting. The recording of financial transactions has to be orderly for proper interpretation. This is necessary to make possible preparation of financial statements and to provide information on the economic health of the business.

The method or extent of bookkeeping varies with each firm's size and needs. Normally, the double-entry system (classification of accounts into assets, liabilities, and net worth) is used. Each firm maintains journals and ledgers. The journal is a daily record of all transactions, debits, and credits. The ledgers carry the journal entries in specific accounts. Again, the number and extent of ledgers required vary with the firm.

A consulting firm has to decide how it is going to maintain its books for tax purposes, whether it be on a cash or accrual basis. On a cash basis, income is recorded when cash is received and expenditures are recorded when they are made. On an accrual basis, income is reported when earned and expenditures (or debits) when incurred, regardless of the time the cash transaction takes place. When tax considerations are significant in the business operation, the choice of accounting system is of primary importance; as is evident, a firm's cash and accrued statement at the same time could be quite different.

Although it is poor business practice to take a particular action solely because of the tax consequences, tax considerations are important in a consulting firm's business practice. The initial decision of which form of organization to operate under should take into account the different tax consequences on individuals, partnerships, and corporations. Depending on income, a corporation may pay a large Federal income tax; in addition, its dividends are taxed. A partnership pays no tax on its income, but the partners, who receive no salaries, are taxed as individuals on their share of the firm's earnings. State and local taxes should also be considered when establishing and operating a design practice.

Payroll is a consulting firm's largest expenditure. Payroll costs should be identified as direct (technical) and indirect (administrative) (Art. 2-7). Records of direct costs, preferably by department, should be maintained for each project. Also, identifiable direct expenses, such as travel, subsistence, and other allowances, long-distance telephone and telegraph, and reproduction costs, should be accounted for and identified as a job expense. Major indirect or overhead expenses should also be identifiable to enable management to analyze indirect costs and their relation to fees earned during a specified period.

In addition to internal accounting, it is customary and advisable to have an audited financial report prepared by a certified public accounting firm at the end of each fiscal year.

From its inception on, an engineering firm is concerned with finances. For one thing, a consultant will not be reimbursed for a firm's services the day after they were rendered. Terms of payment depend on contract conditions. Payments may be on a monthly basis, or the first payment may not be due until 25% (or another percentage) of the work has been completed. Also, the final payment may not be received for a long time after all expenditures have been made. This sums up to one basic need: capital.

Consulting engineers must have capital to start and operate their organizations. The source of capital may be a loan or earnings. But regardless of the source, there must be proper financing to meet financial obligations that cannot be deferred until accounts are paid.

Office Management. The administrative staff's primary function is internal operation of the firm. Personnel on the administrative staff may include an office manager, secretaries, typists, receptionist, file clerks, and office employees. The number of employees and degrees of responsibility vary with the size of firm. However small the firm may be, though, the basic administrative duties have to be fulfilled. Letters have to be typed; so do reports. Files have to be maintained, telephones answered and messages taken, and plans reproduced. Although all the elements that constitute office management are secondary to design, the primary function of the firm, they should not be neglected. A poorly typed letter can make a poor first impression on its recipient, who may be a potential client. A first impression of a firm can also be made by the manner in which the telephone is answered. So although the administrative duties are routine in most offices, they should be handled as competently as the technical work. The administrative positions should be filled by competent, properly trained personnel.

In an engineering firm, there is a substantial amount of reproduction of plans and specifications and duplication of reports. The mechanics of providing the necessary reproductions is best handled by a separate department within the firm. Whether the work is done on office-owned equipment or is sent out to a printing company is a matter of economics determined by the volume of the firm. In addition, office services must encompass selection of the most economical and efficient office systems for the firm. Basic equipment, such as automatic typewriters, telephone aids, and filing systems, is continually being improved and becoming more sophisticated. The office manager should be familiar with these systems and their availability and should be able to judge their applicability to the firm's needs.

(T. G. Hicks, "Successful Engineering Management," McGraw-Hill Book Company, New York.)

2-12. Professional Societies. The role of professional societies, such as the American Society of Civil Engineers and various associations of consulting engineers, initially was determined by their existence as organizations of individuals rather than of firms. These societies at first were concerned mainly with technical matters and very little with business affairs. Although the medical, legal, and accounting professions each have one major society that speaks for the profession, this is not the case for civil engineers. They are generally represented by the American Society of Civil Engineers, American Consulting Engineers Council, or National Society of Professional Engineers. These societies, however, collaborate with each other on matters of common interest.

In a complex and progressive economic society, few firms other than industry giants have the resources to stay abreast of all the latest developments, keep informed of all current legislation, state and Federal, and be aware of all administrative rules, regulations, and factors influencing their day-to-day activities. An association can fill these needs. By serving these needs, professional associations are playing a more important role than previously.

In prior years, also, a design firm was "on its own." It had little knowledge, if any, of the activities of its competitors, or even of its closest associates. Today a firm still is on its own in the competitive marketplace, but it can pool its resources in associations that represent the profession and industry. United action and sharing of information advance the interests of individual firms.

Activities of professional groups now include:

Legislation. Maintaining a file and index on current legislation; representing and filing position papers with Congress and state legislatures on pending bills in which association members have a vital interest.

Government Relations. Liaison with various administrative agencies, Federal, state, and municipal. This area could include assistance to member firms interested in capitalizing on opportunities abroad.

Liaison with Industry. Maintaining contact with other organizations and establishing joint committees to study and evaluate areas of common interest.

Publications. Initiating and distributing to members documents reporting current activities and areas of importance and concern.

Insurance. Establishing group insurance policies (life, accident, health, etc.) to give smaller members advantages of larger group plans; advising member firms in fields of common concern, such as professional-liability insurance, an area of increasing concern due to numerous third-party suits against consulting engineers.

Engineering Practice. Acting as a pool and distribution center for information on the latest technical developments and areas of interest to the profession; sponsoring continuing education programs.

Section **3**

Specifications

JOSEPH GOLDBLOOM
Chief Specifications Engineer and Associate Consultant
Parsons, Brinckerhoff, Quade & Douglas, New York, N.Y.

and
JOHN J. WHITE
Former Chief Specifications Engineer, and Associate Consultant
Parsons, Brinckerhoff, Quade & Douglas, New York, N.Y.

Specifications have been essential to construction since ancient times. Before man could write, the specifications were oral. When drawings were first introduced, the specifications appeared on the drawings. As man started to use more and more sophisticated tools and materials and society became more complicated, it became necessary to furnish written specifications in greater detail to cover the many facets of construction work.

One of the first recorded written specifications is found in the Bible. In the Book of Genesis, Chapter 6, verses 14-16, we read the specifications the Lord gave to Noah for the construction of the ark: "Make thee an ark of gopherwood, rooms shalt thou make in the ark, and shalt pitch it within and without with pitch. And this is the fashion which thou shalt make it of: The length of the ark shall be three hundred cubits, the breadth of it fifty cubits, and the height of it thirty cubits. A window shalt thou make to the ark, and in a cubit shalt thou finish it above; and the door of the ark shalt thou set in the side thereof; with lower, second and third stories shalt thou make it."

In the intervening passage of time, man has prepared many plans and specifications but none save those prepared for the ark served their intended purpose better; the ark survived the terrible destruction of the entire world by flood, as recorded in the Scriptures.

3-1. Composition of Specifications. Specifications comprise two major parts: (1) the requirements under which a bidder prepares and submits a bid or proposal for performing the work, as well as the contractual requirements under which the successful bidder (the contractor) should perform the work; and (2) the detailed requirements for the work to be accomplished. This combination of requirements, together with the contract drawings and bidding documents, comprises the contract documents. When faced with the task of preparing specifications for an engineering project, the engineer must consider many factors. Among the most important of these are:

Nature of the owner's business—private industry or public body.

Magnitude of the project.

Estimated duration of construction period.

Does the owner require the engineer to adhere to a standard form of specifications, or will the engineer have a free hand in preparing the type of specifications?

Does the owner have an attorney who will review the legal aspects of the specifications?

Does the owner have an engineering staff, such as that for a state department of transportation, which will review the specifications?

Also, the engineer must realize that courts of law recognize the status of contractual relations between owner and contractor as that between free and independent individuals, not as that between a principal and agent. The specifications must support this relationship by refraining from prescribing construction methods and exercising control over the contractor's workers.

After the basic conditions have been established, engineers may find that they are obligated to prepare complete contract documents for projects. The principal parts of these documents usually consist of the following:

Advertisement for bids (notice to contractors, or invitation to bid).

Information to bidders.

Proposal form.

Contract-agreement form.

Bond forms.

General provisions, or general conditions.

Special provisions, or special conditions.

Technical specifications.

For general guidance, forms for all but the last two are available from sponsoring agencies, such as the American Consulting Engineers Council, American Institute of Architects, American Society of Civil Engineers, National Society of Professional Engineers, Associated General Contractors of America, and General Services Administration. An example of a specification prepared for a public agency having standard documents is given in Art. 3-20. (For a discussion of general provisions, see Art. 3-8.)

3-2. Contract Documents and Contracting Procedures. The implementation of contracts between owners and contractors for construction work requires that certain legal formalities be observed by the parties. Such steps are evidenced by executed written documents, which, together with the plans and specifications, constitute the contract documents. The nature and content of the contract documents vary with the owner agency that sponsors the improvement and the procedure employed for the receipt of bids.

It is standard practice for government and other public agencies at all levels to provide for public letting of contracts for public works. In such cases, sealed bids are invited by advertising in various news media for stated periods. After bids are opened, publicly read aloud, tabulated, and evaluated, the low bidder is determined.

It is customary to issue the plans and specifications to prospective bidders who apply and pay stated charges. In most cases, proposals must be accompanied by a proposal guaranty in the form of a certified check or surety bond, or both, to insure that the successful bidder will enter into a contract. If an award is made, the guaranty is returned. If the low bidder fails to execute the contract, the amount of the certified check will be forfeited as liquidated damages and obligations of the surety under the bond will be enforced as compensation to the owner for the cost of awarding the contract to the next lowest bidder or for the added cost of readvertising. As a general rule, proposals are acceptable from competent bidders (evidenced by statements of experience and financial responsibility submitted to the owner.)

Under the foregoing procedure, the contract documents generally comprise the advertisement (instruction to bidders may be included or may be separately provided); proposal, properly executed; contractor's progress schedule; resolution of award of contract; executed form of contract; contract bond; plans and specifications; supplementary agreements; change orders; letters or other information, including addenda (Art. 3-5); and all provisions required by law to be inserted in the contract, whether actually inserted or not. All the documents constitute one legal instrument.

When required by law, a noncollusion affidavit must accompany the submission of the proposal. This affidavit certifies that the bid has been submitted without collusion or fraud and that no member of the government agency or officer or employee of the owner is directly or indirectly interested in the bid.

3-3. Types of Contracts. Construction contracts for public works are almost always let on a competitive-bid basis. Usually, such contracts are of either of two types—unit price or lump sum—depending on the method of paying the contractor. Contracts for construction for private owners may be either competitive-bid or negotiated, but in either case, they generally are of the same two types (see also Art. 4-12).

Unit-price Contract.   When it is not possible to delineate on the drawings the exact limits for the various items of work included in the contract, the work is broken down for payment purposes into major elements with respect to the kind of work and trades involved.   Each element designated as a payment item, with its number of units estimated by the engineer and called estimated quantity, is listed in the proposal, and the bidders are required to write in a bid for each unit.   An example is the number of cubic yards of concrete to be bid at a unit price per cubic yard.

The total bid is obtained by summation of the amounts, in dollars, for all items scheduled in the proposal, arrived at by multiplying the estimated number of units for each item by the corresponding unit-price bid.   The total bid becomes the basis for comparison of all bids received to establish the low bid upon which the award of contract to the successful bidder will be made.   Payment to the contractor will be made on the basis of the measured actual quantity of each item incorporated in the work at the contract unit price (see also Art. 4-14).

Lump-sum Contract.   When it is possible to delineate accurately on the drawings the limits of work comprised in the contract, whereby the bidder may make a precise quantity survey as the basis for a bid, a lump-sum contract is employed.   For such a contract, it is imperative that the drawings and specifications be comprehensive and show in complete detail all features and requirements of the work.   Compensation to the contractor is made on the basis of a lump-sum, or fixed-price, bid to cover all work and services required by the drawings and specifications (see also Art. 4-13).

Contract with Lump Sum and Unit Prices.   It is not unusual to combine unit-price and lump-sum bids in the same contract; for example, an entire structure completely detailed on the drawings will be listed in the proposal as a lump-sum item, whereas unit prices may be required for features of variable quantities, such as excavation or lengths of bearing piles.

Negotiated Contract.   On occasion, public works contracts and, more often, private works contracts are negotiated.   These contracts may be prepared on the basis of one or more of several different methods of payment.   Some of the more widely used are:

Lump sum or unit price or combination.
Cost reimbursable with a ceiling price and fixed fee.
Cost reimbursable plus a fixed fee.
Cost reimbursable plus a percentage of cost.
Construction-management contract.

In addition, incentives may be added.

For a negotiated contract, the owner chooses a contractor recognized for dependability, experience, and skill, and in direct negotiation establishes the terms of the agreement between them and the amount of the fee to be paid.   Factors contributing to the selection of a contractor are ordinarily determined by the prequalification or qualification procedures of public agencies from questionnaires and investigation.   Such questionnaires are readily adaptable for use on contracts to be negotiated by private owners.

A *negotiated lump-sum or unit-price agreement* is negotiated around the engineer's estimate.   A fixed percentage for overhead and profit is determined and agreed to, and the labor and material prices of the contractor and those of the engineer's estimate are adjusted by mutual agreement.

A *cost-reimbursable agreement with a ceiling price* is one where the contractor receives reimbursement for all costs as prescribed in the agreement up to a maximum cost.   The contractor receives a fixed fee, which will not vary with the cost of the work and otherwise is negotiated similarly to the cost-plus-fixed-fee type of agreement.

The determination of the fee to be paid the contractor under a *cost-plus-fixed-fee agreement*, which will be fair and reasonable to both parties to the contract, requires definitive plans, an estimate of the cost of construction, and a knowledge of the magnitude and complexities of the work, the estimated time of completion, and the amount of work to be done under subcontracts.   The terms of the contract must therefore set forth the methods for control and approval of expenditures and for determination of the actual cost.

Under a *cost-plus-percentage-of-cost contract*, the contractor's profit is based on a fixed percentage of the actual cost of the work.   This form is less desirable than the fixed fee since the contractor's compensation increases with increase in construction cost.   This creates a situation where there would be no incentive for the contractor to effect any economies during construction.

A *construction management agreement* requires the contractor to divide the work into segments, usually by trade. The contractor takes bids for the work from a group of subcontractors and awards the work to them. The prime contractor usually performs a certain prescribed segment of the work and coordinates the work of others. The owner reimburses the prime contractor for all the work of the subcontractors and for the contractor's work plus a small profit and pays a negotiated fee for management of the subcontracts.

Incentive-type contracts vary. The basic premise is that the owner will pay bonuses for economic construction and for earlier completion and that the contractor may have to suffer for inefficiency and late completion.

3-4. Specialty Contracts. Special situations sometimes dictate a departure from the ordinary contract-letting procedure (Art. 3-2). Examples are contracts for the procurement and installation of highly specialized equipment and machinery, such as toll-collection facilities and communication systems.

In such cases, instead of advertising publicly for bids, the owner invites proposals from a selected group of contractors especially qualified and generally recognized as specialists for the manufacture and installation of such facilities. When competition is possible, it is so arranged. The contract documents prepared by the owner's engineer in such instances are as described in Art. 3-2, with certain exceptions. Since advertisement is not resorted to, this and related items of the documents are not included, but the contracting procedure is substantially that followed for contracts publicly bid.

3-5. Contract Revisions. For various reasons, revisions of the contract documents become necessary between issuance of the invitation or advertisement for proposals and the termination of the contract. Such revisions may be classified as addenda, stipulations, change orders, and supplementary agreements.

Addenda are revisions of the contract documents made during the bidding period. They mainly are concerned with changes in the contract drawings and specifications due to errors or omissions, with the necessity for clarification of parts of these documents, as revealed by questions raised by prospective bidders, or with changes required by the owner. An addendum is also issued to notify bidders when a bid-opening date has been postponed.

Addenda should be delivered sufficiently in advance of the bid-opening date to permit all persons to whom contract drawings and specifications have been issued to make the necessary adjustments in their proposals. Bidders must acknowledge receipt of all addenda; otherwise, their bids will not be accepted.

Stipulation is a written instrument in which the bidder awarded a contract agrees, at the time of execution of the contract, to a modification of the contract terms proposed by the owner.

Change order is a written order to the contractor, approved by the owner and signed by the engineer, for a change in the work from that originally shown by the drawings and specifications. Usually, under a change order, the work is considered as being within the general scope of the contract. The owner, represented by the engineer, may issue the order to the contractor unilaterally, with payment provided for by negotiated price, unit prices, or force account.

A change order may apply to changes affecting lump-sum work and to increases and decreases in quantities of work to be performed under the various items in the contract. The changes in quantity will be evaluated at the contract unit prices and the contract total amount adjusted accordingly. But if the total cost change amounts to more than a specified percentage, say, 25%, of the total contract price, a supplementary agreement acceptable to both parties to the contract should be executed before they proceed with the affected work.

Supplementary agreement is a written agreement used for modifying work considered outside the general scope and terms of the contract or for changes in work within the scope of the contract but exceeding a stipulated percentage of the original amount of the contract. The agreement must be signed by both parties to the contract.

3-6. Standard Specifications. Government agencies and many other public bodies sponsoring public works publish "standard specifications" establishing a uniformity of administrative procedure and of quality of constructed facilities, as evidenced by specific requirements of materials and workmanship. A sponsor's standard specifications usually contain information for prospective bidders, general requirements governing contractual procedures and performance of work by a contractor, and technical specifications covering construction of the particular work which lies within their jurisdiction. Highways, bridges, buildings, and water and sanitary works are examples of the types of improvements for which agencies may have standard specifications. Standard specifications, published periodically, may be updated in the interim by issuance of amendments, revisions, or supplements.

In order that the specifications for a particular contract may be completely adaptable to the work of that contract, the standard specifications almost always require modifications and additions. The assembled modifications and additions are known as supplementary specifications, special provisions, or special conditions. In conjunction with the standard specifications, they comprise the specifications for the work (see also Art. 3-20).

3-7. Master Specifications. Whereas published standard specifications are commonplace with government and other agencies (Art. 3-6), master specifications are useful tools for design organizations that serve private clients. A master specification covers a particular item of construction, such as excavation and embankment, concrete structures, or structural steel. It contains requirements for all possible conditions and construction that can be anticipated for that particular item. Master specifications are prepared in-house. (Engineers who work primarily for agencies that impose their own standards as the basic text for project specifications will find only limited uses for master specifications.)

In applying a master specification to a specific project, the specifications writer deletes those requirements that do not apply to the project. Thus, use of a master specification not only effects a reduction in the time required to produce a contract specification, but also serves as a checklist for the writer and minimizes errors and omissions. Another important advantage of using a master specification is that the edited text can be used directly for review without waiting for typing to be completed. In editing a master specification, however, failure to delete nonapplicable provisions results in both encumbering and increasing the length of the project specifications. In addition, nonapplicable provisions are confusing to contractors and others using the final documents.

To remain effective, a master specification should be periodically updated to incorporate current practices or new developments. Out-of-date information can never be considered acceptable in project specifications.

3-8. General Provisions of Specifications. The general provisions set forth the rights and responsibilities of the parties to the construction contract (owner and contractor) and the surety, the requirements governing their business and legal relationships, and the authority of the engineer. These articles are often miscalled "the legals" or the "boilerplate."

When a contracting agency has published standard specifications, the specifications for a project comprise these standards and the modifications and additions necessary for the particular requirements of the project, generally called the *special provisions.*

There are primarily three principal subdivisions of the specifications: general provisions, materials specifications, and construction requirements. The materials specifications and construction requirements, taken together, fall into the general category of "technical specifications," which also includes provisions for measurement and payment for the work. The sequence and arrangement of these features may vary at the option of the owner. But in all cases, general provisions come first.

On privately owned work, where generally there are no owner-published standard specifications, the specifications are especially tailored to fit the requirements of the project. A substantial part of standard general provisions will be found to be pertinent to such contracts. Requirements peculiar to the nature of the work are added, as necessary. Parts of the general provisions that pertain to legal requirements inherent in a public agency's corporate existence naturally will not be included in a contract for privately owned construction.

The general provisions, usually designated as Division I of the specifications, may be set forth in detail under the following major sections:

Definitions and abbreviations, which include definition of terms and abbreviations used in the specifications.

Bidding requirements, which include preparation and submission of bids and other pertinent information for bidders (Art. 3-14).

Contract and subcontract procedure, which includes award and execution of the contract, requirements for contract bond, submission of progress schedule, recourse for failure to execute the contract, and provisions for subletting and assigning contracts.

Scope of the work, which includes a statement describing the work to be performed; requirements for maintenance and protection of highway and railroad traffic, where involved; cleaning up before final acceptance of the project; and availability of space for contractor's plant, equipment, and storage at the construction site. Also, a limit is set on the permissible deviation of actual quantities from estimated quantities of the proposal without change in contract price.

Control of the work, which includes the authority of the engineer, plans and specifications and working drawings, construction stakes, lines, and grades; inspection procedures; relations with other contractors at or adjacent to the site; provision of a field office and other facilities for the engineer needed in administration of the contract and control of the work; materials inspection, sampling, and testing; handling of unauthorized or defective work; contractor's claims for additional compensation or extension of time; and acceptance of work upon completion of project.

Legal and Public Relations.   This section of the general provisions deals with legal aspects that determine the relations between the contractor and the owner agency and between the contractor and the general public. It sets up the requirements to be observed and protective measures to be taken by the contractor in order that the liabilities for actions arising out of the prosecution of the work are properly oriented and provided for.   Topics included are the disclaimer of any personal liability upon the contracting officer of the agency, the engineer, and their respective authorized representatives in carrying out the provisions of the contract or in exercising any power or authority granted them by virtue of their position; in such matters they act as agents and representatives of the owner agency, such as Federal government, state department, municipality, or authority.

Other features of legal and public relations that control contractors' procedures are as follows: damage claims; laws, ordinances, and regulations; responsibility for work; explosives; sanitary provisions; public safety and convenience; accident prevention; property damage; public utilities.

Damage Claims.   Indemnification and save-harmless provisions are invoked to protect owners and their agents.   The protection extends to suits and costs of every kind and description and all damages to which they may be subjected by reason of injury to person or property of others resulting from the performance of the contract work or through negligence of the contractor, use of improper or defective machinery, implements, or appliances, or any act or omission on the part of the contractor or contractor's agents, employees, or servants. These provisions are made to apply to subcontractors, material suppliers, and laborers performing work on the project.

These requirements may be implemented by requiring the contractor to provide insurance of specified character and in specified amounts as will provide adequate protection for the contractor, the owners, their successors, officers, agents, and servants and for others lawfully on the site of the work against all claims, liabilities, damages, and accidents.   However, neither approval nor failure to disapprove insurance furnished by the contractor is to release him of full responsibility for all liability inherent in the indemnification and save-harmless provisions.   Generally included in the insurance to be carried by the contractor, each, when applicable, in required minimum amounts of coverage established on the basis of loss in any one occurrence, are:

Workmen's Compensation Insurance, statutory, as applicable.   It should be extended where warranted to include obligations under the Longshoremen's and Harborworkers' Compensation Act and Admiralty law.

Contractor's Comprehensive General Liability, including Contractual Liability, with Bodily Injury Liability and Property Damage Liability.   It should be augmented, by the prime contractor when there are subcontractors concerned, by Contractor's Protective Liability Insurance on the prime contractor's behalf and Comprehensive General Liability on behalf of each subcontractor.   Policies should provide coverages for explosion, collapse, and other underground hazards (XCU coverage) when such hazards are incident to the work. To cover a lapse of time between the contractors' completion of the work and the owner's acceptance, the policies should bear endorsement for completed operations coverages. Also Contractual Liability Insurance policies should bear endorsements noting acceptance by the underwriters of the indemnification and save-harmless clauses.

Comprehensive Automobile Liability providing coverage of all owned or rented vehicles and automotive construction equipment and with coverages of Bodily Injury Liability and Property Damage Liability.

Builder's Risk providing coverage of loss due to damage to a structure from fire, wind, etc.

Owner's Protective Public Liability and Property Damage Insurance, a separate original Public Liability, and Property Damage Insurance (Owner's Protective) should be provided by the contractor, designating the owner, successors, officers, agents, and servants as the named insured with respect to all operations performed by the contractor.

Protection and Indemnity Insurance, or comparable coverage, should be carried by contractors, where applicable, with respect to all watercraft used by or operated for them, chartered or otherwise, covering bodily injury liability and property-damage liability. (See also Arts. 4-29 to 4-33.)

Insurance is a specialized field. Hence, the specifying of insurance coverage should be left to those experienced in that field.

Laws, Ordinances, and Regulations. The pertinent Federal and state laws, rules, and regulations, and local ordinances that affect those engaged or employed on the project, the materials or equipment used, or the conduct of the work are cited. All necessary permits and licenses for the conduct of the work are specified to be procured by the contractors at their expense.

Responsibility for Work. Contractors are required to assume full responsibility for materials and equipment employed in the construction of the project. They are prohibited from making claim against the owner for damages to such materials or equipment from any cause whatsoever. Until final acceptance, the contractor is responsible for damage to or destruction of the project or any part thereof due to any cause, except for damage caused by owner-operated equipment engaged in deicing or snow-removal operations, as on a highway or bridge. The contractor is required to make good all work damaged or destroyed, except that caused by others, before final acceptance of the project and to include all costs thereof in the prices bid for the various scheduled items in the proposal.

Explosives. The use, handling, and storage of explosives are required to conform to regulations of government agencies controlling these features of the work. Proper means are required to be used to avoid blasting damage to public and private property.

Sanitary Provisions. The contractor is required to provide and maintain suitable sanitary facilities for personnel in accordance with the requirements of Federal, state, and local agencies having jurisdiction.

Public Safety and Convenience. This article provides that the contractor conduct the work so as to inconvenience as little as possible the public and residents adjacent to the project and provide protection for persons and property. The contractor must install temporary crossings to give access to private property. Also measures must be taken to prevent deposits of earth or other materials on roads and thoroughfares on which hauling equipment is operated and to remove promptly such deposits, if they occur, and thoroughly clean the surfaces. The contractor shall employ construction methods and means to keep flying dust to a minimum.

Accident prevention. This article provides for observance of safety provisions outlined in the rules and regulations of public agencies functioning in this field. It is the contractor's responsibility to provide safe working conditions on the project. The contractor is held fully responsible for the safe prosecution of the work at all times.

Property damage. This article defines the obligations of the contractor when entering upon or using private property in carrying out the work and in connection with any damage to that property.

Public Utilities. Through this article the contractor's attention is directed to the possibility of encountering public and private utility installations that either are obstructions to the prosecution of the work and need to be moved out of the way or, if not, must be properly protected during construction. It sets up the procedures to be followed and establishes costs to be absorbed by the contractor as well as the utility companies and the public agency in accordance with agency policy and laws dealing with such situations.

Abatement of Soil Erosion, Water Pollution, and Air Pollution. Through this article, the contractors are reminded of their responsibility for minimizing erosion of soils and preventing silting and muddying of streams, irrigation systems, impoundments, and adjacent lands. Pollutants such as fuels, lubricants, and other harmful materials are not to be discharged into or near streams, impoundments, or channels. No burning of any material is permitted.

Prosecution and Progress. This section of the general provisions deals with such pertinent considerations as commencement and prosecution of the work, time of completion of the contract, suspension of the work, unavoidable delays, annulment and default of contract, liquidated damages, and extension of time.

Commencement and Prosecution of the Work. This article establishes the date on which work is to start and from which contract time is to run; that construction must proceed in a manner and sequence insuring completion established by the contractor's progress schedule previously approved by the engineer; whatever limitations of operations there may be at the site of work, including traffic, work by others, and schedule of stage completion; and that the ability, adequacy, and character of workers, construction methods, and equipment be competent for full prosecution of the work to completion in the time and manner specified.

Time of Completion.   It is advantageous to specify time of completion in calendar days from date of commencement of work rather than working days, for the actual designation of a working day is often a cause of contention.   Herein may be specified stage completion when it is to the owner's advantage to have occupancy of a part of the work prior to completion of the entire contract or where a priority of construction of a particular feature of the work is essential to subsequent procedures.

Suspension of the Work.   This article covers the usual conditions under which the owner may suspend work, in whole or in part, for such period of time as may be deemed necessary, without breach of contract, and the period of time that suspension may be effected without allowance of compensation.   These conditions may include weather or other conditions that are unfavorable for prosecution of the work and the contractor's failure to perform in accordance with provisions of the contract or to correct conditions unsafe for workers or the general public.

Unavoidable Delays.   For delays for any reason beyond the control of the contractor, other than those caused by suspension of the work, the contractor may be granted an extension in the contract time.   This citation, however, gives the contractor no right or claim to additional compensation unless the contract specifically provides for such compensation.

Annulment and Default of Contract.   Provision is made for terminating the contract as follows:

For annulment: A public-office holder acting in the public interest or a national or state agency ordering a work stoppage may result in the owner's annulment of a contract.   With a contractor not in default, settlement is usually made for work completed and proper costs of work in progress and for moving from the site, with no allowances for anticipated profit. Also, the owner may annul a contract when a contractor is found to have compensated others for soliciting a public contract, thus violating the warranty of noncollusion with others.

For default: When a project or any part of it has been abandoned, is unnecessarily delayed, or cannot be completed by the contractor within the time specified, or on which the contractor willfully violates terms of the contract or carries out the contract in bad faith, the owner usually has just cause to declare the contractor to be in default on the contract and to notify the contractor to discontinue work on the project.   When a contractor is in default, the owner may make use of contractor-furnished material and equipment in completing the project through the contractor's surety or by other means considered necessary for completion of the contract in an acceptable manner.   All costs, over and above contract costs, for completing the project are recoverable from the contractor or his surety.

Liquidated Damages.   Provision is made for the contractor to pay the owner a sum of money for each day of delay in completing the contract beyond the date due.   This agreement on damages prior to breach of contract avoids litigation and dispute over almost undeterminable actual damage, while providing an incentive to the contractor to complete work on time. When the specified sum of money is unsupportable as representative of the actual damage suffered by the owner in added costs, it becomes, in fact, a penalty for delayed completion and unenforceable in the courts.

Extension of Time.   This article establishes certain of the conditions which will be considered just cause for an extension of the time stipulated in the contract for completion of the project.   These may include change orders adding to the work of the contract, suspension of work, or delay of work for other than normal weather conditions.

Measurement and Payment.   This section of the general provisions provides for measurement of quantities of the completed work; scope of payment; change of plans and consequent methods of payment; payment; procedures for partial and final payments; termination of contractor's responsibility; and guaranty against defective work.

Measurement of Quantities.   This article stipulates that all completed work of the contract will be measured for payment by the engineer according to United States standard measures.

Scope of Payment.   This article establishes that payment for a measured quantity at the unit-price bid will constitute full compensation for performing and completing the work and for furnishing all labor, materials, tools, equipment, and all else necessary and incidental thereto.

Change of Plans.   Provision is made for payments pertinent to changes in the work; i.e., the measured quantities of work completed or of materials furnished which are greater than or less than the corresponding estimated quantities scheduled in the proposal and the quantitative limits of such changes permitted by change orders; the context of the change order, inclusive of kind and character of work, materials to be furnished, and changes in contract time of completion; supplementary agreement for changes in contract prices of scheduled items and the performance of work not identified with any scheduled item in the proposal.

Payment.   This article establishes the procedure by which payment will be made for the actual quantity of authorized work completed and accepted under each item scheduled in the proposal either at the unit price bid or at the unit price stipulated in the supplementary agreement.

The procedure usually provides for partial payments on account to be made periodically. These are based on approximate quantities of work completed during the preceding period, as measured by the engineer and attested to by certificates for payment.   The owner may retain a percentage of the amount of each certificate, pending completion of the contract. Upon completion and acceptance of the contract, a final certificate of cost prepared by the engineer and approved by the owner determines the total amount of money due the contractor and from which previous payments on account will be deducted.   Final payment is made upon satisfactory representation by the contractor that there are no outstanding claims against him filed with the owner, that the contractor has satisfied or satisfactorily arranged for payment of all due obligations incurred personally and by subcontractors in carrying out the project, and that whatever guaranty bond may be required has been posted.

Termination of Contractor Responsibility.   This article establishes that upon completion and acceptance of all work included in the contract and payment of final certificate, the project is considered complete and the contractor is released from further obligation and requirements.

Guaranty against Defective Work.   A guaranty period is established for all or portions of the work, together with an amount of guaranty, usually calculated as a percentage of the contract cost.   A guaranty bond shall be furnished by the contractor and conditioned to replace all work and all materials that were not performed or furnished according to the terms and performance requirements of the contract and to make good defects that become apparent before the end of the guaranty period.

3-9. Technical Specifications.   These specifications may take several forms.   One or more of these forms may be selected to serve best the purpose for which the specifications are prepared.   Types of technical specifications in common use are:

Materials and workmanship specifications.

Material procurement specifications.

Performance specifications (procurement).

Materials and Workmanship Specifications.   This type of specification is almost universally used on construction contracts.   It is comprehensive in its coverage of the principal factors entering into the prosecution and completion of the work covered by the contract.   These factors include the general and special conditions affecting the performance of the work, material requirements, construction details, measurement of quantities under the scheduled items of work, and method of payment for these items.

Material Procurement Specifications.   These specifications are used on projects of considerable magnitude requiring many separate general construction contracts, usually in simultaneous operation and under which the types of construction are similar. For example, material procurement specifications may be desirable for a highway of considerable length involving the construction of grade-crossing structures of structural steel or precast and prestressed concrete items.   In such cases, it has often been found advantageous to separate contracts for the structural steel or prestressed concrete from the general contracts for the overall project.   This procedure insures uniformity and availability of the materials.   It facilitates construction by scheduling deliveries to coincide with the general contractors' needs for these items at any particular location throughout the entire project.   A similar procedure may also be used for the procurement of other construction materials in quantity.

The specifications for contracts of this nature contain, besides fabrication processes, all the elements of materials and workmanship specifications, except for the field construction details.   If erection of the items is to be included in the procurement specifications, the procedure is the same as for materials and workmanship specifications.

Performance Specifications. These specifications are used to a great extent in procurement contracts for machinery and plant operating equipment, as distinct from material procurement contracts. Contracts for machinery and equipment may be let separately by the owner prior to a construction contract under which installation will be made. The purpose is to insure delivery to the job in time for installation within the scheduled construction sequence. Advance letting of procurement contracts is usually necessary because of the great amount of time consumed in the manufacture of such items. In general, performance specifications, in addition to defining the materials entering into the manufacture of equipment, with all the pertinent physical and chemical properties, prescribe those characteristics that evidence equipment capability under actual operating conditions. Thus, the specifications must be complete in defining quality, function, and other requirements that must be met. Since a performance specification requires samples, tests, affidavits, and other supporting evidence of compliance, it tends to increase contractor's costs for furnishing the items and engineer's costs for checking submitted data. It also adds to the designer's responsibility for an unsatisfactory or inadequate product.

Requirements for tests and certification of the results are set up in the specifications in accordance with test procedures established by the appropriate industry associations.

When not critical from the standpoint of manufacture and delivery schedules, machinery and equipment may be covered by the construction specifications. For typical technical specifications, see Art. 3-21.

3-10. Materials Specifications. (See also Arts. 3-9 and 3-12.) Under this division of the specifications are prescribed the various materials of construction to be used in the work and their properties. The principal properties to be considered in the preparation of specifications of materials for construction are:

1. Physical properties, such as strength, durability, hardness, and elastic properties.
2. Chemical composition.
3. Electrical, thermal, and acoustical properties.
4. Appearance, including color, texture, pattern, and finishes.

Materials specifications should also include procedures and requirements to be met in inspections, tests, and analyses made by the manufacturer during manufacture and processing of the material and later by the owner. Note should be made as to whether a material is to be inspected at the shop or mill during manufacture and the number of test specimens, identified with the material proposed to be furnished, that will be furnished to the owner for test.

In addition, the specifications should cover the protection necessary in the interval between manufacture and processing of the materials and their incorporation in the work. Some materials are subject to deterioration or damage, under certain conditions of exposure, during stages of transportation, handling, and storage.

3-11. Reference Standards. Standards published as reference specifications for construction materials and processes by professional engineering societies, government agencies, and industry associations are widely followed for construction work. The recommendations of these organizations form the basis of current construction practice, particularly with regard to quality of materials and, in some cases, fabrication practices and construction methods.

3-12. Arrangement and Composition of Technical Specifications. (See also Art. 3-21.) The general provisions, as Division 1 of the specifications (Art. 3-8), are followed by the various divisions of the technical specifications in numerical order and in a sequence generally based on a logical order of construction stages for progressing the work. For example, on a highway project, successive divisions may be:

Division 2, Earthwork: (*a*) Clearing and grubbing; (*b*) excavation and embankment; (*c*) trench, culvert, and structure excavation; (*d*) borrow.

Division 3, Bases and Subbases: (*a*) Subbase course; (*b*) bituminous stabilized course; (*c*) soil cement course; (*d*) hydrated-lime stabilized subgrade.

Division 4, Bituminous Pavements: (*a*) Sheet-asphalt pavement; (*b*) asphaltic-concrete pavement; (*c*) road-mix pavement; (*d*) penetration macadam; (*e*) bituminous surface treatment; (*f*) stabilized-gravel surface course.

Division 5, Rigid Pavement: (*a*) Portland-cement-concrete pavement; (*b*) portland-cement-concrete base for pavement; (*c*) continuously reinforced portland-cement-concrete pavement.

Division 6, Structures: (*a*) Bearing piles; (*b*) structural concrete; (*c*) bar reinforcement; (*d*) structural steel; (*e*) bridge railing; (*f*) structural timber; (*g*) painting.

Division 7, Miscellaneous Construction: (*a*) Curbs and sidewalks; (*b*) culverts and storm drains; (c) guard rails; (d) fencing; (e) riprap; (f) topsoiling and seeding; (g) roadway lighting.

Division 8, Materials: Specifications for materials and material test requirements.

With respect to the last division, its location in the specifications is a matter of choice. With some agencies, this division precedes the work-item specifications; with others, it may follow those specifications; and often the material requirements are embodied in the text of the specifications for each of the work items covered. This last arrangement might be the case with private-owner project specifications that do not include standard specifications prepared and issued by the owner.

The detailed specifications for each section, for example, bearing piles, under Division 6, are generally arranged under the following subheadings:

1. Description of Work.
2. Materials.
3. Construction Requirements.
4. Measurement for Payment.
5. Basis of Payment.

The last two items may be combined under a single heading: Measurement and Payment.

Description of Work. Under this heading, a concise statement is made of the nature and extent of the work included in the section and its pertinent features, including the requirement that performance conform to the plans and specifications.

Materials. This article presents the requirements for the various materials involved in the performance of the work of the section. If a separate division on materials has been included as a part of the technical specifications, simple references to it for properties are made (see also Art. 3-10). If such a division is not included, reference to standard specifications of the professional engineering societies, government agencies, and industry associations will be appropriate. When manufactured products are not listed in available reference standards, it is customary to name several of proved quality and performance. Usually, three are specified by name and manufacture, any one of which will be considered acceptable for use on the work.

"Or Equal." When a given construction material or piece of equipment does not lend itself readily to standard-specification designation or easily describable specifications, public bodies require the names of at least two or three suppliers or the name of one supplier with the added phrases "or equal," "or approved equal," "or equal as approved by the engineer." The theory behind this requirement is that it promotes fair competition and that it complies with the law. In many instances, the procedure originates in the office of an attorney general or other public official and is based upon a ruling that competition is a requirement of most public-works laws. In private-ownership practice, the main reason for use of this procedure is to obtain the best product for a client at the most economical price.

The engineer is presumed to have sufficient knowledge of the properties of materials and characteristics of equipment to decide what is best suited for the job. The owner engages the engineer, not the contractor, for decisions of this nature. Seldom do contractors offer a substitute material or piece of equipment with any thought other than to save cost to themselves.

The least of the bad effects of the "or equal" clause is that the engineer is prevented from detailing foundations and spaces assigned for equipment until the merits of the substitute offered by the contractor are evaluated and a selection has been made. Investigating the merits of an offered substitution after the award of contract places an unwarranted burden on the engineer and specifications writers at the busiest phase of their work and is not conducive to the proper determination of the merits of the substitute. It is likely to result in a hasty, perhaps unfortunate, approval.

Commonly overlooked is the additional cost to the engineer in investigating and evaluating the merit of substitutions offered by the contractor. Some specifications stipulate in such cases that the contractor shall reimburse the engineer for such costs. Except where the owner waives the requirements of the specifications to take advantage of the low cost of a substitute and thereby relieves the engineer of his or her responsibility, the specifications should require the contractor to assume full responsibility for compliance with all applicable provisions of the specifications upon approval of a substitution; such approval should be given in writing.

Specifications writers have coped with the substitution problem by requiring the bidder to offer substitutions along with the proposal. Under this scheme, the specifications prescribe the exact material or equipment required. Bidders must describe the substitutions in detail in accompanying specifications, drawings, and cuts and by other means. They must also stipulate the amount by which they will reduce or increase their base bids upon acceptance of the substitutions. Aside from curbing substitutions intended solely for the contractor's benefit, this plan permits the engineer to evaluate the substitutions concurrently with all other features of the proposal and to do so free from the pressure that prevails after the award of contract.

The ideal solution would be not to call for the "or equal." This omission, if used properly, can produce for the owner, private or public, a facility that is most economical and free from excessive maintenance costs. This can be accomplished by specifying more than two suppliers of each item and accepting substitutions only before the award of the contract.

Construction Requirements. The primary purpose of this article in the detailed specifications for each work item is to prescribe the operations involved in its construction without relieving the contractor of responsibility for the satisfactory accomplishment of the end result. Among the principal features to be stressed are quality of workmanship and finish, with consideration given to practical limitations in tolerances, clearances, and other limiting factors. Necessary precautions should be given for the protection of the work or adjacent property. Methods of inspection and tests applicable to the work, with particulars as to off-site inspection at mill or shop, as well as inspection at the site, should be specified.

Specifications for workmanship should indicate the results to be attained insofar as practicable. Thereby, the contractor obtains latitude in selection of construction procedures. In some instances, however, it may be necessary to designate methods to insure satisfactory completion of the work, for example, compaction of earth embankments or shop and field welding procedures on steel structures. It may also be necessary to specify precautions and restrictions for purposes of protection and coordination of the work as a whole or when a definite sequence in construction operations is made necessary by design conditions or to meet conditions contemplated by the owner.

Measurement and Payment. This heading combines measurement for payment and basis of payment. Every contract, regardless of type, must include provisions for payment. For a unit-price contract, the quantity of work completed under each bid item listed in the proposal must be measured by applying an appropriate unit of measurement. Some items, such as assembled units, are measured by the number required; others are measured by linear foot, square yard, cubic yard, pound, or gallon, as applicable.

The quantity to be considered for payment should be clearly defined so as to cover all deductions to be made for deficiencies and unauthorized work performed beyond the limits delineated on the plans or ordered by the engineer. Partial and final payments for the actual quantity of work completed and accepted can then be computed. To determine the payment due, each such quantity is multiplied by the corresponding unit price bid by the contractor and the products are totaled.

It is essential for payment purposes that the specifications define precisely each bid item per unit of measurement (cubic yard, linear foot, etc.). The specifications should clearly and fully state all the work and incidentals that should be included by the bidder in the item for which the unit price is to be submitted. When there are operations closely associated with a particular item of work for which separate payment is provided, the specifications should make this clear to avoid controversy or double payment for the work.

It is not uncommon in a unit-price contract to include items for which lump-sum prices are required. These are subject to all the conditions governing unit-price items, except measurement for payment and the right of the owner to vary the quantity of work without change order. The cost of all work and materials necessary to complete the construction of the lump-sum bid items, as delineated on the drawings and required by the specifications, must be included in the lump-sum bid. Work associated with construction of lump-sum-bid items but not made a part thereof must be indicated as being included for payment under other bid items.

To facilitate partial payments for work performed on lump-sum items, as well as for contract lump-sum bids, the contractor should be required to submit a breakdown for the component parts of the work. The breakdown should include quantities for the different types of work or trades involved and unit prices applicable to each. When extended and summarized, the prices should equal the lump-sum bid for the completed item or contract. Such a schedule must be approved by the engineer before it becomes effective.

3-13. Advertisement or Invitation for Bids. (See also Art. 3-2.) It is standard practice for government and public agencies to provide for the public letting of contracts for public works.   Sealed bids are invited by advertising in newspapers and engineering publications for legally required periods.   The advertisement should contain the following information: issuing office, date of issue, date for receipt of bids and time of opening of bids, brief description of work (identification of project), location of project, quantities of major items of work, office where plans and specifications can be obtained and charges for them, proposal security, and rights reserved to the owner.   For private work, an invitation for bids is issued by the owner to a selected group of contractors.   It conveys much of the information that would be included in an advertisement as it may apply to the particular project.

3-14. Bidding Requirements.   (See also Art. 3-2.)   Bidding requirements for public-works contracts are usually defined in the general provisions of the standard specifications for the particular agency.   The object of these requirements is to advise prospective bidders of the routine to be followed in the submission of a bid and their eligibility to do so.   The principal points covered are:

Prequalification or Qualification.   For a bid to be acceptable, the bidder must have been either prequalified with the contracting agency for capability and financial standing, by submission of documents furnishing required information (updated to reflect the situation at bid time), or otherwise qualified along the same lines by furnishing evidence thereof with the bid.   Some states require that contractors be licensed.   In that case, a record of the contractor's license is filed automatically with the contracting agency.

Preparation and Delivery of Proposal.   (See also Art. 4-23.)   Instructions for preparing a proposal on forms furnished by the contracting agency are given to avoid irregularities, which could nullify the bid.   Proposals must be signed and signatures legally acknowledged before being placed in envelopes furnished for the purpose and then sealed.   Receipt of all addenda issued during the bidding period must be acknowledged on the proposal form, where provision is usually made for this purpose.   Information requested of the bidder on the exterior of the envelope must be entered in the spaces provided.   A bid may be delivered by mail or messenger but must be received before the time set for opening; otherwise, it may not be accepted.

Proposal Guaranty.   (See also Arts. 4-23 and 4-34.)   Public agencies always require a guaranty that the bidder will execute the contract agreement if awarded the contract.   The guaranty may be in the form of a surety bond or certified check for a stated percentage of the bid.   Usually, this is 10%, with maximum limit of a fixed amount; but this could vary to serve the interest of the particular agency.   Sometimes both a surety bond and certified check are required.   The amount of the surety bond may vary from 50% of the bid price down to 5% at the discretion of the contracting agency.

Proposal guaranties must accompany the proposal.   Bid securities are returned to all but the lowest three bidders within a short time after bids have been opened.   Those of the lowest three bidders are returned after a contract has been executed.

Noncollusion Affidavit.   A noncollusion affidavit is generally required by public agencies by law.

3-15. Evaluation and Comparison of Bids.   Following the opening of bids, a public announcement is made of the prices bid for the various items listed in the proposal.   These data then are tabulated, the totals for each item verified, and their summation, establishing the total amounts of bids, checked for each bid submitted.   Comparison of the total amounts of the bids establishes the lowest bid and those that follow in the order of increasing amounts.

3-16. Award and Execution of Contract.   Having verified all specified submissions, such as licensing, prequalification statements, and noncollusion affidavits, and having established the low bidder, the owner officially notifies the successful bidder of the award of the contract; the bidder is expected to execute the contract agreement within a specified time.   It is a requisite for this final step in the contracting procedure that the successful bidder furnish a performance and payment bond acceptable to the contracting agency.   The amount of the bond equals the total amount of the bid.   The bond is a guaranty to the owner that all the work required to be performed will be faithfully carried out according to the terms of the contract; also it guarantees that the contractor will pay all lawful claims for payment to subcontractors, material suppliers, and labor for all work done and materials supplied in the performance of the work under the contract.

The bond must also provide that the owner be saved harmless, defended and indemnified against and from all suits and costs of any kind and damages to which he may be put by reason of injury to the person or property of others resulting from performance of the work or through negligence of the contractor.   In addition, the owner must be shielded in the same way from all suits and actions which may be brought or instituted by subcontractors, material suppliers, or laborers who have performed work or furnished material on the project and on account of any claims, or amount recovered, by infringement of patents or copyrights. The requirement of the contractor to indemnify and save harmless the owner may be implemented by insurance, by retaining a percentage of the contract amount until final acceptance of the work, and by the contract bonds (see also Art. 4-34).

For private owners, the procedures for submitting, receiving, and opening bids are more informal since they are not subject to the laws governing such procedures for public-works contracts. The manner in which these steps are handled is entirely at the discretion of the owner or engineer. Bid securities are not required. Advertisement for bids is not usually employed. Instead, a Notice to Contractors is issued to a selected group of contractors, known to the owner to be qualified. This notice is accompanied by Instructions to Bidders and a proposal form when competitive bids are required. The Instructions to Bidders generally include the information necessary for preparing and delivering the proposal. Noncollusion affidavits are not required. Tabulation and evaluation of bids and award and execution of contract usually follow the procedure for public-works contracts, modified to suit the owner's particular needs.

3-17. Specifications Writing: Style and Form. Preparation of the specifications for a construction contract starts with an overall analysis of requirements based on a survey of the proposed work, conditions under which it must be accomplished, materials, details of construction, and owner's administrative procedures. The analysis provides the various items for appropriate distribution among the contract documents. Also, a close study of the contract drawings will reveal that which is insufficiently shown and needs to be supplemented in the specifications. A descriptive outline of such a distribution or a proposed table of contents with subheadings facilitates and expedites the work of the specifications writer in assembling the documents.

A basic format for specifications may be oriented for a particular project and its sponsor. There should be a title page identifying the documents and a table of contents listing the various sections of general and special provisions by number, title, and page. Cross references in a section should be made by title only.

Specifications usually are written in the traditional style of composition, grammatically correct. They should go into as much detail as necessary, qualitatively and quantitatively, to convey that which is required and therefore agreed to. Chances for misunderstandings and disputes, which frequently result in expensive litigation, should be kept to a minimum. Ambiguity and verbosity should be avoided. A good specification is clear, concise, and easily understood. It gives little cause for doubt of the intentions of the parties concerned and leaves nothing to be taken for granted. The courts have traditionally interpreted ambiguous requirements against the party who prepared them.

Inasmuch as the specifications, in conjunction with the drawings, are the means employed to guide the contractor in producing the desired end product, it is essential that they be correlated so that conflicts and misunderstandings of the requirements may be avoided. Instructions more readily described in words belong in the technical specifications, whereas information that can be more effectively portrayed graphically should appear on the drawings. Information on the drawings should not be duplicated in the specifications, or vice versa, because there may be a discrepancy between the information provided in the two documents that may cause trouble.

Since specifications complement the drawings, the special provisions and Standard Specifications together should leave no doubt as to the quality and quantity of the required work. The function of the drawings is to show location, dimensions, scope, configuration, and detail of the required work, while the function of the specifications is to define the minimum requirements of quality of material and workmanship, prescribe tests by which these shall be established, and describe methods of measurement and payment.

When preparing contract documents for a project for which there is an owner's standard specifications, for example, for a project of a public agency, the specifications writer will be obliged to incorporate these specifications by reference and to identify and establish this standard in the special provisions. It is not unusual to cite sections of the standard specifications by reference at the beginning of each applicable section of the special provisions, with a paragraph similar to the following:

All work shall be in accordance with Standard Specifications (list section number and name), as amended herein.

However, in the text of a section of the special provisions, references may be made to one or more of the provisions of the standard specifications or to standard specifications other than the owner's, in whole or in part.

Special provisions therefore modify, restrict, or add to the standard specifications, where necessary, and admit such options and alternatives as may be permitted. Portions of the standard specifications should not be repeated in the special provisions, and repeated references in special provisions to a standard specifications section should be avoided. Redundancy leads to error.

Of major importance in coordination and interpretation of contract documents is the establishment of an order of precedence. It is usual to provide that the contract drawings govern over the standard specifications and that the special provisions govern over the standard specifications and the contract drawings. Thus, in the preparation of special provisions care must be exercised to avoid conflict with the other contract documents and to insure a definite and clear description of the required work. Care must also be taken to avoid duplication of information in the special provisions or in both the drawings and special provisions to preclude conflict and errors, especially in the event of changes. It is well not to specify both the method to be used and the desired results thereof, for a conflict may relieve the contractor of responsibility.

Each technical section usually begins with a brief description of the work included in it. Work contingent upon, but not included in, the work specified under a particular section may be referenced as "Related work specified under other sections" rather than "Work not included." Each section should be complete, with description of materials, workmanship, and requirements for testing clearly defined. All separate bid items must be mentioned, with methods of measurement and payment specified for each item.

The contract documents should be fair to owner, bidders, contractor, and others concerned. Any aspect of the work not clearly defined in the specifications or on the drawings will result in time and effort wasted during bidding or during construction, higher contract prices to include "contingencies," and in all probability arguments over extras, with ensuing delays.

Following are some applicable general considerations in writing specifications: Be specific, not indefinite. Be brief; avoid unnecessary words or phrases. Give all facts necessary; avoid repetition. Specify in the positive form. Use correct grammar. Direct rather than suggest. Use short rather than long sentences. Do not specify both methods and results. Do not specify requirements in conflict. Do not justify a requirement. Avoid sentences that require other than the simplest punctuation. Also, avoid words that are likely to be unfamiliar to users of the specifications, especially if the words have more than one meaning.

Be particularly careful when requiring approval by the engineer. Specific approval by the engineer of the contractor's equipment, methods, temporary construction, or safety standards, in certain situations, can relieve the contractor of responsibility under the terms of the contract.

When preparing the Construction Details of a specification, arrange the material in the sequence in which the work will be done. For example, specify the curing of concrete after specifying formwork, concrete mixing, and concrete placing. When inserting a reference to a national standard, such as a standard ASTM specification, read the standard first to assure yourself that it contains nothing that conflicts with job requirements.

The measurement and payment portion of a specification is most important to both the contractor and the owner. Every item of work to be done by the contractor must be accounted for, whether it be measured and paid for separately or included for payment in another item.

Refer only to the principals to the contract: the owner, as represented by the engineer, or the contractor. Do not refer to other contractors, subcontractors, bidders, etc.

Refer to "these" specifications rather than "this" specification; use the plural.

Workmanship should be in accordance with and materials should conform to a reference specification.

Use the phrase "at no additional cost to the owner" only when there is a definite possibility of the contractor's not understanding that he is to bear a certain expense. Liberal use of the phrase might imply that other work specified is not at his expense.

Use the word "shall" for requirements placed on the contractor and the word "will" for expressions of intent on the part of the owner.

Do not confuse the meaning of words; proper word usage is of utmost importance.

Do not use indefinite words when more exact words may be substituted.

Repeated use of stock phrases and stereotyped expressions should be avoided. Specifications should not be encumbered with legal phrases that obscure their meaning or subordinate their function to that of a legal document.

Streamlined Specifications.   As an alternative to the traditional style, specifications may be written in a streamlined form, which is a simplification of style by shortening of sentence structure wherever practicable.   Properly employed, streamlining may be a major improvement.   In general, streamlining consists in omitting from the specifications, without a change in meaning, those words having no legal significance.   Only necessary provisions are retained.   A good long-form specification can be streamlined without the slightest adulteration, and yet streamlining can reduce its bulk by one-third or more.

The technique of streamlining specifications may be adopted as a simplification of style, productive of a distinctive form of writing specifications, whereas the general format remains the same.   However, it should be noted that this style is more readily adaptable to building-construction contracts wherein each section of the technical specifications relates directly to a particular construction trade.

Some of the aspects and considerations in streamlining specifications advocated by Ben John Small ("The Case for Streamlined Specifications," *The Construction Specifier,* July, 1949) are:

The term "streamlining" should not be interpreted to mean that it refers to a specification lacking thoroughness or that streamlining is synonymous with specifications devoid of the three C's (Clarity-Conciseness-Comprehensiveness).   Any specification long or short must be equipped with the requisite C's if it is to associate properly with its other relatives, which constitute the family of Contract Documents, such as the Agreement, General Conditions, the Drawings, etc.

Streamlining offers no cure for ineptitude in writing specifications, such as conflicting repetitions, giving contradictory instructions, etc.   What it does, affirmatively, is to translate the writer's knowledge of construction and materials into simple, readable expressions subject to less misinterpretation.   The most important part of streamlining is a statement that not only explains the use of the streamlined specification format but states only once in the entire specifications the requisite mandatory provisions that are usually repeated ad nauseam in traditional specifications.   By requisite mandatory provisions we mean expressions such as "The Contractor shall . . . ," "The Contractor must . . . ," "The Contractor may . . . ." These expressions tell the contractor to do something in different ways, which in a dispute could bring as many interpretations.   The explanatory statement of streamlined specifications should be included as an article in the General Conditions, such as:

ART. 64—SPECIFICATIONS EXPLANATION

(a) The Specifications are of the abbreviated, simplified or streamlined type and include incomplete sentences.   Omissions of words or phrases, such as "The Contractor shall," "in conformity therewith," "shall be," "as noted on the Drawings," "according to the plans," "a," "an," "the," and "all" are intentional.   Omitted words or phrases shall be supplied by inference in the same manner as they are when a "note" occurs on the Drawings.

(b) The Contractor shall provide all items, articles, materials, operations, or methods listed, mentioned, or scheduled either on the Drawings or specified herein, or both, including all labor, materials, equipment, and incidentals necessary and required for their completion.

(c) Whenever the words "approved," "satisfactory," "directed," "submitted," "inspected," or similar words or phrases are used, it shall be assumed that the words "Engineer or his representative" follow the verb as the object of the clause, such as "approved by the Engineer or his representative."

(d) All references to standard specifications or manufacturer's installation directions shall mean the latest edition thereof unless specifically noted otherwise.

3-18. Automatic Typewriters and Computers.   Specification automation systems in use range from automatic typewriters to more sophisticated systems utilizing large computers. All systems involve the processes of storage and retrieval in which the specifications writer has a stored bank of information that is retrievable on call.   This information is stored in a manner that enables it to be easily modified and reproduced accurately and efficiently.

Hooking an automatic typewriter up with a small computer permits storage of larger volumes of information and quick retrieval.   In using an automatic typewriter, the typist produces normal, finished pages of material, and, meanwhile, the machine lodges the same material in its memory in the form of cards or tape (magnetic or paper).   Magnetic tape or cards can be reused indefinitely.   The memory can be played back easily on the same machine with fresh paper, and what was stored in the memory will spill out at twice the speed of manual typing exactly as it was originally lodged.

A first step in establishing a system is preparation of master specifications for storage (Art. 3-7). The master specifications copy is used by the specifications writer as the basis for preparation of the project specifications. The writer edits the master and deletes inapplicable sections. The typist, in turn, working from the edited copy, has the machine type the master specifications automatically to the point where the specifications writer has made changes. The typist then stops the machine and types the changes manually. After this is done, the machine is restarted to continue automatic typing of the specifications. The only proofreading required is that of the changed portion.

One of the primary tasks of the specifications writer in the use of storage and retrieval systems is to constantly upgrade and update the master specifications. Through the use of automated specifications, it is possible to improve the quality of specifications, especially their clarity, and avoid omissions.

3-19. Construction Specifications Institute. The Construction Specifications Institute (CSI) is an organization devoted to improvement of construction specifications. Its members consist principally of architects and representatives of material manufacturers, with a small percentage of engineers.

The Construction Specifications Institute has promulgated the CSI Format, a widely used system for arranging all the specifications for building construction into 16 divisions. Each division, in turn, is broken down into sections, with each section covering a particular construction activity. Specification information distributed by manufacturers usually is also arranged in the CSI Format.

In addition to the Format, the CSI has compiled a *Manual of Practice* containing much useful information for specifications writers in general. Among these documents are titles such as "An Introduction to Specifications and Related Contract Documents," "The Use of Reference Specifications," "Proprietary Specifications," "Performance Specifications," "Specification Language," and "Specification Writing Procedures."

The efforts of the Construction Specifications Institute have been directed principally toward establishment of specification standards for building construction. One example was the restructuring by the American Concrete Institute, in collaboration with CSI, of ACI Standard 301-72, "Specifications for Structural Concrete for Buildings," in conformity with the CSI Format.

3-20. Example of Special Provisions. In the following, an example is given of a special provision. Used to amend a standard specification for riprap, it was part of the project specifications for construction of low-level trestle approaches to a lift span for a highway crossing over water in Virginia. The standard specification was incorporated in the then-current edition of "Road and Bridge Specifications" of the Virginia Department of Highways as Sec. 418, Riprap. This standard specification began as follows.

SECTION 418—RIPRAP

Sec. 418.01. Description—This work shall consist of the construction of the specified type of riprap in accordance with the plans and these specifications.

Sec. 418.02. Materials—Unless otherwise modified herein, all materials shall conform to the applicable requirements of Division II of these specifications. (Division II deals with materials requirements.) Grading A, B, or C sand may be used in mortared or grouted riprap.

Sec. 418.03. Dry Riprap, Class I, for Slopes—Unless otherwise specified, all stones used in this class of riprap shall weigh between 50 and 150 lb each and at least 60% of them shall weigh more than 100 lb each.

The stones shall be placed upon a slope not steeper than the natural angle of repose of the fill material. The stones shall be laid with joints as close as practicable. The courses shall be laid from the bottom of the bank upward, the larger stones being placed in the lower courses. Open joints shall be filled with spalls.

The section then continued with similar detailed requirements for the following:

> Sec. 418.04. Dry Riprap, Class 2, for Slopes
> Sec. 418.05. Mortared Riprap for Slopes
> Sec. 418.06. Grouted Riprap for Slopes
> Sec. 418.07. Stone Riprap for Foundation Protection
> Sec. 418.08. Dumped Riprap
> Sec. 418.09. Erosion Control Stone
> Sec. 418.10. Concrete Riprap in Bags
> Sec. 418.11. Concrete Slab Riprap

In addition to the presentation of material requirements in subsections labeled (*a*) and (*b*),

Sec. 418.08 also dealt with excavation and fine grading, bedding, and construction methods in subsequent sections. For example:

(*d*) *Bedding:* Riprap bedding of the thickness indicated on the plans shall be placed on the embankment to form a backing for the riprap. Riprap bedding shall be spread uniformly on the prepared base, in a satisfactory manner, to the lines indicated on the plans or as directed. Placing of material by methods which will tend to segregate particles sizes within the bedding will not be permitted. Any damage to the surface of the bedding base during placing of the bedding shall be repaired before proceeding with the work. Compaction of the bedding material will not be required but it shall be finished to present a reasonably even surface free from mounds or depressions.

The standard specification concluded with provisions for method of measurement and payment:

Sec. 418.12. Method of Measurement—Stone riprap for slope walls will be measured in square yards.

Stone riprap for foundation protection will be measured in units of volume or weight as specified.

Bedding for dumped riprap will be measured in cubic yards in place to the lines shown in the plans or as specified.

Dumped riprap will be measured in tons as determined on approved scales.

Erosion control stone will be measured in cubic yards.

When the ton unit is specified, the quantity shall be determined on scales equipped with a dial and an automatic printer, all of which have been approved and sealed in accordance with Sec. 109. (Section 109, Measurement and Payment, sets forth the standards for scales when they are used in determining weight for the purpose of payment.) When the material is transported by rail, the weight shall be evidenced by railroads bills of lading.

Concrete slab riprap will be measured in units of square yards of riprap actually placed.

Sec. 418.13. Basis of Payment—Riprap will be paid for at the contract unit price for the type specified, which payment shall be full compensation for furnishing and laying the riprap, including reinforcing steel, mortar, or grout when specified, excavation as necessary, and all labor, tools, equipment and incidentals necessary to complete the work.

Payment will be made under:

Pay item	*Pay unit*
Dry riprap, class	Square yard
Mortared riprap	Square yard
Grouted riprap	Square yard
Stone riprap	Cubic yard or ton
Dumped riprap, type	Ton
Bedding for dumped riprap	Cubic yard
Concrete riprap in bags	Cubic yard
Concrete slab riprap	Square yard
Erosion control treatment, type	Cubic yard

The preceding standard specification, Section 418, Riprap, was modified by the Special Provisions of the trestle project specifications, to meet the particular job requirements, as follows:

SECTION 418—RIPRAP

Sec. 418.01 Description. This section is amended as follows:

This work shall consist of the construction of rock dikes and rock slope protection, on a bedding course, at both abutments.

Existing riprap removed under Section 303, Excavation and Embankment, that complies with the requirements of this Section, may be re-used in the work.

Sec. 418.02. Materials. This Section is amended as follows:

Bedding material shall consist of crushed gravel or quarry tailings meeting the requirements of Section 205.02 (which deals with material and grading requirements) except that material shall also be free from adherent coatings and injurious amounts of thin or laminated pieces; shall consist of particles of which a minimum of 20% by weight of those retained on a No. 4 sieve shall have at least one fractured face by artificial crushing; and shall be graded as follows:

Sieve size	% Passing by weight
8 in.	100
3 in.	65–95
1½ in.	50–80
¾ in.	35–65
No. 4	15–40
No. 16	0–20
No. 50	0–5

Rock Dike material shall be reasonably well graded, conforming to the requirements of Section 205.02 for riprap stone, having a minimum density of 162 lb per cu ft and weighing between 250 and 1,500 lb each, except that 10% by weight may consist of pieces weighing 10 to 250 lb each. Neither the width nor thickness of any piece shall be less than ⅓ of its length.

Sec. 418.03 Dry Riprap, Class 1, for Slopes. This Section is amended as follows:

Bedding for dry riprap shall be placed and spread as specified in Section 418.08 (*d*), Bedding.

Sec. 418.08. Dumped Riprap. This Section is amended by adding the following subsection:

(*f*) *Rock Dike:* Bedding for rock dikes shall conform to the requirements for riprap bedding and shall be placed within the limits shown on the plans.

Rock dikes shall be constructed in a manner to produce a reasonably well graded mass of rock with a minimum of large voids, to the limits shown on the plans.

Sec. 418.12. Method of Measurement. This Section is amended as follows:

Measurement for bedding shall include bedding for Dry Riprap, Class 1 and for Rock Dikes.

Dry Riprap will be measured in square yards for the area actually covered in accordance with the plans, specifications and orders of the engineer.

Rock Dikes will be measured in tons (2,000 lb) for the quantity furnished and placed in accordance with the plans, specifications and orders of the engineer.

Existing riprap removed under Section 303 and incorporated in the work under this Section, will be measured and paid for under this Section.

Sec. 418.13. Basis of Payment.

Payment will be made under:

Pay item	Pay unit
Bedding	Cubic yard
Dry riprap, class 1	Square yard
Rock dike	Ton

3-21. Example of a Technical Specification. The following example illustrates a technical specification that was part of the project specifications prepared for the construction of a wharf and approach trestles in the Caribbean area.

SECTION T3—STEEL PIPE PILES

1. Description The work specified in this Section includes the furnishing and driving of closed-end steel pipe piles, including protective coating, test piles, load tests and concrete fill, all as shown on the plans and as specified herein.

2. Materials

a. Pipe for piles shall be new, seamless, steel pipe conforming to the requirements of ASTM Designation A252-69, Grade 2. Pipe shall be eighteen (18) inches outside diameter with a wall thickness of one-half (½) inch, ordered in double random lengths. Ends of pipe sections shall be perpendicular to the longitudinal axis and shall be beveled as shown on the plans, where required for welded splices. Mill certificates for chemical composition and two certified copies of the records of the physical tests performed on the newly manufactured pipe in accordance with the above ASTM requirements shall be furnished before any driving is started.

b. Steel Points for pile tips shall be of cast steel conforming to the requirements of ASTM Designation A27-71, Grade 65-35. They shall be a standard 60° point with inside flange and two interior cross ribs. Each point shall be marked with the manufacturer's name or identification number. The Contractor shall submit to the Engineer for approval, details of the point he proposes to use.

c. Splice Rings as shown on the plans, shall be of structural steel conforming to the requirements of ASTM Designation A36-69.

d. Concrete for piles shall be 3,500 psi conforming to the requirements of Section T5, Concrete.

e. Reinforcement for cages in the top of piles shall conform to the requirements of Section T5, Concrete.

f. Welding Electrodes shall conform to the requirements of the American Welding Society "Specifications for Mild-Steel Covered-Arc Welding Electrodes."

g. Protective Coatings shall consist of the following:

(1) Inorganic zinc-rich paint (1 coat), self-curing, with zinc pigment packaged separately, to be mixed at time of application. Zinc dust content to be 75% by weight of total non-volatile content. Acceptable products are Mobilzinc No. 7 by Mobil Chemical Co., No. 92 Tneme-Zinc by Tnemec Co., or Zinc-Rich 220 by USS Chemicals, Div. of U.S. Steel Corp.

(2) Coal-tar epoxy coating (2 coats), to be a two-component amine or polyamide-epoxy coal-tar product, black in color. Acceptable products are Amercoat No. 78 Ameron Corrosion Control Div., Tar-Coat No. 78-J-2 Val-Chem by Mobil Chemical Co., or Tarset No. C-200 by USS Chemicals.

(3) Both the zinc-rich paint and coal-tar epoxy shall conform to the applicable requirements of Federal Spec. MIL-P-23236.

3. Construction Details

a. Protective Coatings. Zinc-rich paint and coal-tar epoxy shall be applied to exterior surfaces of pipe piles, including splice areas, within the respective limits shown on the plans. The Contractor shall apply the protective coatings to a sufficient length of pile sections to insure that the pile when driven to its required resistance, will be protected within the required limits.

Prior to the application of the zinc-rich paint and coal-tar epoxy, bare surfaces shall be blast cleaned to white metal in accordance with the Steel Structures Painting Council Specification No. SP-5.

The zinc-rich paint shall be applied in the shop to a dry-film thickness of 2 mils. The coal-tar epoxy may be applied in the shop or in the field and shall have a total dry film thickness of 16 mils. Coated pile sections shall not be stored in direct sunlight longer than one month without a tarpaulin covering.

Care shall be taken while handling coated pile sections during loading, transporting, unloading and placing, so that the protective coating is not penetrated or removed. Coated pile sections shall be inspected before placing in the leads and any damaged surfaces shall be repaired and recoated to the satisfaction of the Engineer.

The Contractor's attention is directed to the "Hazardous Warning Label" on the coal-tar epoxy products and the manufacturer's literature regarding the use of protective clothing, gloves, creams and goggles during mixing, application and cleanup.

The cured coal-tar epoxy coating will be tested by the Engineer to determine resistance to film removal by a mechanical force, as follows:

(1) Lay a sharp wood chisel almost flat on the coating surface in line with the pipe length.

(2) Drive the chisel using a hammer, through the coating and along the substrate.

(3) If the coating film is acceptably bonded to the surface, considerable force will be required to lift a layer of the film.

(4) Portions of the coating should remain in the valleys of the blast pattern adhering to surface for an acceptable test.

(5) The tested area shall be repaired as per these specifications by the Contractor.

(6) The number of tests will be limited to two (2) acceptable tests for each shipment or for each day's field application of coating.

b. Preparation for Driving

(1) Piles shall not be driven in any area until all necessary excavation or grading has been completed.

(2) *Pile Points:* The tip of every pile shall be closed with an approved pile point, welded in place to produce a watertight joint.

(3) *Splices:* The number of splices shall be kept to the practical minimum. The number and location of splices will be subject to the approval of the Engineer. Splices shall be made with full strength butt welds utilizing an internal steel back-up splice ring as shown on the plans. Should the Contractor desire to use an alternate splice design, he shall submit full details of his proposed splice to the Engineer for approval. All splices shall be watertight.

(4) *Welding:* Welding shall conform to the applicable requirements of the current edition of the American Welding Society "Specifications for Welded Highway and Railway Bridges." Welders shall be qualified for the work, as prescribed in the AWS Specifications.

c. Equipment for Driving: All equipment shall be subject to the approval of the Engineer. Piles shall be driven with a single-acting hammer which shall develop a manufacturer's rated energy per blow at full stroke of not less than 30,000 foot-pounds. The striking weight shall be not less than 10,000 pounds.

Sufficient boiler or compressor capacity must be provided at all times to maintain the rated speed of the hammer during the full time of driving a pile. The valve mechanism and other parts of the hammer shall be maintained in first-class condition so that the length of stroke for which the hammer is designed, will be obtained.

Piles shall be driven with leads constructed in such a manner as to afford freedom of movement of the hammer.   Leads shall be held in position by guys or stiff braces to give the required support to the pile during driving.   Inclined leads shall be used for driving battered piles.   Leads shall be of sufficient length, as the use of a follower will not be permitted.

Water jets shall not be used for pile penetration unless authorized by the Engineer.   When water jets are authorized, the Contractor shall submit to the Engineer for approval full details of his proposed jetting operation.   In no event shall a pile be jetted within ten feet of its anticipated final tip elevation.

d. Accuracy of Driving: Completed piles at the cut-off elevation shall not vary from the plan locations by more than three (3) inches.   Piles shall be driven with a variation of not more than one-eighth (⅛) inch per foot from the vertical or from the batter shown on the plans or as directed by the Engineer.

Piles shall not be subjected to force in order to place them in correct alignment or horizontal position.   Piles exceeding the allowable tolerances will be considered unacceptable unless the Contractor submits a satisfactory working plan showing the corrective work he proposes.   Such work shall not proceed until the working plan has been approved by the Engineer.

e. Defective Piles: Piles damaged by reason of internal defects or by improper handling or driving will be rejected.   Corrective measures shall be submitted by the Contractor to the Engineer for approval.   Approved corrective measures undertaken by the Contractor shall be at no additional cost to the owner.

f. Limitations of Driving: The Contractor's attention is directed to the existence of cement-waste fill material in the proposed work area, as indicated in the boring logs.   All piles shall penetrate this layer.   The Contractor shall take the necessary measures to accomplish this penetration subject to the approval of the Engineer.

g. Lengths of Piles: The lengths of piles indicated in the Proposal are for estimating purposes only.   The actual lengths of piles necessary will be determined in the field by driving the pile sections to the required resistance established by the test piles and pile load tests.

h. Pile Cut-offs: Pile cut-offs may be used in other piles.   However, useable cut-offs must be at least ten (10) feet in length and only one cut-off length will be permitted in any one pile.

i. Driving: Driving of a pile shall be continuous as far as practicable.   When driving is resumed after an interruption, the blow count shall not be taken into consideration until the temporary set of the pile resulting from the interruption has been broken.

Piles shall not be driven within 60 feet of concrete that is less than 7 days old.

Piles shall be driven for the last six inches to the resistance determined from the test piles and pile load tests and as established by the Engineer.

All piles forced up by any cause shall be driven down again as directed by the Engineer and any such costs shall be included in the unit price bid for the piles.

j. Inspection: The Contractor shall have available at all times a suitable drop-light for the inspection of each pile throughout its entire length.

k. Concrete: No concrete shall be placed in a pile until it has been inspected and accepted by the Engineer.   Accumulations of water in the pile shall be removed before concrete placement.   Concrete, 3,500 psi shall be mixed and conveyed as specified in Section T5, Concrete.   Concrete shall be placed continuously in each pile to the extent that there will be no cold joints.   The slump shall not exceed 3 inches.   Special care shall be exercised in filling the piles to prevent honeycomb and air pockets from forming in the concrete.   Internal vibration and other means shall be used to the maximum depth practicable, to consolidate the concrete.

Should the Contractor be unable to remove water from within the pile to enable the concrete to be placed in "the dry," he shall submit details of his proposed tremie operation for filling the pile.

l. Cutting off: The tops of piles shall be cut off at the elevations shown on the plans.

m. Reinforcement: The tops of piles shall be reinforced as shown on the plans.   The reinforcing steel shall be secured in such a manner as to insure its proper location in the finished piles.

n. Test Piles: Test piles shall be driven at the locations shown on the plans or directed by the Engineer, for determining approximate pile lengths.   In addition, test piles will be load tested to verify the bearing value of the driven pile.

The initial test pile shall be driven to the design bearing value of 150 tons, using the following formula:

$$P = \frac{2WH}{S + 0.1}$$

where P = safe bearing capacity, pounds
$\quad\quad W$ = weight, pounds, of striking parts of hammer
$\quad\quad H$ = height of fall, feet
$\quad\quad S$ = average penetration, in. per blow for the last 10 to 20 blows

o. Pile Load Tests: Load tests shall be performed in accordance with the requirements of ASTM Designation D1143-69, "Load-Settlement Relationship for Individual Vertical Piles Under Static Axial Load," as modified herein:

(1) Pretest Information specified in Section 2, will not be required.

(2) Under Section 5, Procedure:

(*a*) A time period of at least 7 days shall elapse between driving and loading the test pile.

(*b*) The test pile shall be filled with concrete at least 3 days before loading.

(*c*) No further loading beyond 200% of the design working load of 150 tons will be required.

(*d*) Intermediate loads shall not be removed.

(*e*) The full test load shall remain in place a minimum of 24 hours, as determined by the Engineer.

(*f*) A final rebound reading shall be recorded 24 hours after the entire test load has been removed.

(*g*) The increase in loading shall be applied at a uniform rate with no sudden load impact. Reducing the load shall be handled in the same manner.

The Contractor shall submit to the Engineer full details of his proposed method of performing the load tests, including arrangement of equipment.

The safe bearing capacity of the test pile will be considered as one-half that test load which produces a permanent settlement of the top of the pile of not more than one-quarter (¼) inch.

4. Method of Measurement

a. The quantity of 18-in. steel pipe piles to be paid for will be the number of linear feet of piles, including test piles in the completed structure, installed in accordance with the plans and specifications, measured from the point of the pile to cut-off.

b. The quantity of Pile Load Tests to be paid for will be the number of completed tests performed in accordance with the plans and specifications.

5. Basis of Payment

a. The unit price bid per linear foot of 18-in. steel pipe piles shall include the cost of furnishing all labor, materials and equipment necessary to complete the work, including protective coatings, pile points, splices, concrete, reinforcement, jetting when authorized, corrective measures, unused pile cut-offs and test piles.

b. The unit price bid per each Pile Load Test shall include the cost of furnishing all labor, materials and equipment necessary to complete the work including the removal of all temporary materials and equipment.

3-22 Qualifications for Specifications Engineers. A review of the character and function of specifications bears witness to the knowledge specifications writers must have of the proposed work and the conditions under which it must be accomplished, the materials and methods of construction that may be used, and the owner's prescribed procedures for administering the contract. In addition to technical skill, a major requisite of a specifications writer is ability to convey full understanding of the contract to others: engineers, constructors, workers, lawyers, financiers, and the general public. Writing ability is an important element because specifications are of little value unless they can be clearly understood.

Specifications writers for civil construction should be graduate civil engineers with some design and broad field experience. Mechanical and electrical engineers and architects should prepare the technical input to the specifications for their respective fields.

A specifications engineer should have a minimum of 10 years' exposure to construction practices, preferably as a representative of the owner. At least 3 to 5 years should have been spent as a resident engineer, interpreting, enforcing, and defending the project specifications. The specifications engineer will thus have acquired an appreciation of the part that specifications play in the development and successful completion of projects.

Basically, contractors want to know what they are required to do under the terms of a contract and how they are to be paid for it. The more clearly and simply that this information can be presented in the contract documents, the less the likelihood of problems, delays, and claims developing on the job.

The Construction Committee of the U.S. Committee on Large Dams stated in Paper 8781, published by the American Society of Civil Engineers:

The proper framing of a set of construction specifications is not easy. Engineering specialists called specifications writers are normally employed for that purpose, and their work requires good judgment, a broad knowledge of the technical aspects of the job, and appreciation of the construction problems entailed, plus the ability to express clearly and concisely all of the terms, conditions, and provisions necessary to present an accurate picture to the constructor. It is a very large order.

Section **4**

Construction Management

J. B. BONNY

Late President, and Chairman, Morrison-Knudsen Company, Inc., Boise, Idaho

In essence, construction is a combination of organizations, engineering science, studied guesses, and calculated risks. From their very nature, construction operations must be performed at the site of the project. Construction is a dynamic, restless, compelling business.

Two basic factors, however, help to stabilize the construction business. In prosperous times, there is immediate and widespread increase in demand for contractors' services from both government and private industry; during periods of recession, Federal and state governments tend to accelerate public-works programs to offset economic downswings. Another inherent element of stability is the industry's mobility, making it less subject to regional economic slumps.

Construction is essentially a service industry. The construction of a project involves thousands of details and complex, interwoven relationships among owners, architects, engineers, general contractors, specialty contractors, manufacturers, material dealers, equipment distributors, governmental bodies and agencies, labor, and others.

4-1. Role of Contractors. A contractor should have two prime objectives: First, the contractor must complete a service for the owner that is satisfactory and on time; second, the contractor must make a profit. The contractor assumes responsibility for delivery of the properly completed facility at a specified time and cost, and in so doing, accepts legal, financial, and managerial obligations.

Technological advances are resulting in more complex facilities. Hence, there is increasing necessity for skillful coordination of all construction operations to attain maximum efficiency, speed, and economy. Thus, the professional function of managing and coordinating construction operations and performing the work with his own experienced organization makes the contractor a key figure in the economy.

4-2. Forms of Business Organization. Contractors, subcontractors, and speculative builders in the United States constitute about one-half million private business enterprises, ranging in size from very small to very large. Their operations may be generally classified as (1) residential and light commercial, (2) industrial and processing plants, and (3) heavy construction, including dams, highways, railroads, airports, bridges, and river and harbor improvements.

These contracting entities employ the usual business forms. Perhaps the greater number are sole proprietorships, where one person owns or controls the enterprise. Many others are partnerships, where two or more individuals form a voluntary association to carry on a business for profit.   The corporate form has a particular appeal to both large- and small-scale enterprises operating in the construction field.   To the large enterprise, corporate structure offers an easier way to finance itself by dividing ownership into many small units that can be sold to a wide economic range of purchasers, including those with only small savings to invest.   In addition to assisting financing operations, the corporate device brings a limited liability to the persons interested in the enterprise and a perpetual succession not affected by the death of any particular owner or by the transfer of any owner's interest.   Because of these features, the corporate vehicle is also used by numerous small contractors.

4-3. Functional Organization Factors.   The type of organization employed to carry out construction is influenced by considerations peculiar to that industry.   Many of these are unlike those affecting manufacturing, merchandising, or distribution of goods.   This is due largely to the degree of mobility required, type of risk inherent in the particular type of construction, and geographical area to be served.

Each product that a construction team makes, it makes only once; the next time its work will be done at a new location, to a new pattern, and under new, although often similar, specifications.   Further, from the very inception of each construction project, the contractor is wholly devoted to working himself out of business at that particular location.   His purpose is to complete the undertaking as quickly and economically as possible and then move out.   In contrast, most other businesses strive for expansion and growth from roots put down at specific and presumably permanent plant sites or business locations.

Although it is true that construction operations can be broken down into repetitive-type functions, such as driving a nail by applying impact to the head of the nail or moving a yard of earth, so many variables are involved from job to job that direct application of industrial engineering principles, usually accepted and employed in industry, is often impractical.

The problems of construction differ from those of industrial-type businesses.   The solutions can best be developed within the construction industry itself, recognizing the unique character of the construction business, which calls for extreme flexibility in its operations.   Based on foundations resting within the industry itself, the construction industry has erected organizational structures under which most successful contractors find it necessary to operate.   They tend to take executives away from the conference table and put them in close touch with the field.   This avoids the type of organizational bureaucracy that hinders rapid communication between office and field and delays vital decisions by management.

More than in any other business, success or failure in construction is determined by the quality of leadership.   Construction management is fundamentally the management of people, the ability to gather people into a compact group who respect their boss and cooperate with each other.   To achieve respect and loyalty of people, it is necessary for the management to be preeminently fair in business dealings and in relations with employees.   Any deviation from this has an adverse effect, and the organization quickly falls apart.   Employees must be given the opportunity to take responsibility in keeping with their capacity and given full credit for good performance; also, they must be given help when needed.

Although they are at the bottom of organization charts, those who actually do the work with their own hands—who use the tools and operate the machines—are in the front line of the productive effort.   These forces usually are organized by crafts or specialty work classifications.

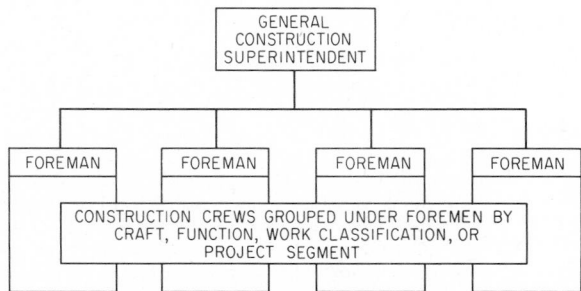

Fig. 4-1. Basic work-performing unit; also, organization for small construction company.

Each unit is directed by a foreman who reports to a general construction superintendent (Fig. 4-1).

The construction superintendent is in charge of all actual construction, including direction of the production forces, recommendation of construction methods, and selection of personnel, equipment, and materials needed to accomplish the work. Superintendents supervise and coordinate the work of the various craft superintendents and foremen. They report to management, or in cases where the magnitude or complexity of the project warrants, to a project manager, who, in turn, reports to management. To enable the general construction superintendent and project manager to achieve efficient on-the-job production of completed physical facilities, they must be backed up by others not in the direct line of production.

4-4. Organization of Construction Firms. Figure 4-1 is representative of the operation of a small contracting business where the sole proprietor or owner serves as general construction superintendent. Such owners operate their businesses with limited office help for payroll preparation. They may do their own estimating and make commitments for major purchases, but often they use outside accounting and legal services.

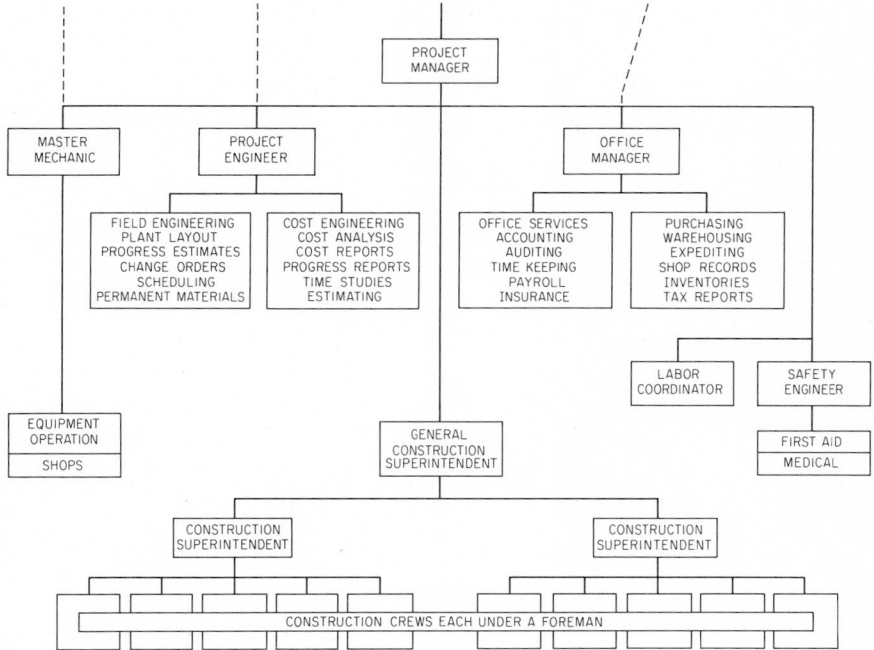

Fig. 4-2. Project organization, with that for the lowest level as shown in Fig. 4-1.

As business expands and the owner undertakes larger and more complex jobs, more crafts, functions, or work classifications are involved than can be properly supervised by one person. Accordingly, additional foremen and crews may be grouped under as many craft superintendents as required. They report to the general construction superintendent, who, in turn, reports to the project manager, who still may be the owner (Fig. 4-2).

Along with this expansion of field forces, the owner of a one-person business next finds that the volume and complexities of the growing business require specialized support personnel. They have to perform such services as:

1. Purchasing, receiving, and warehousing of permanent materials to be incorporated into the completed project, as well as purchasing, receiving, and warehousing of goods and supplies consumed or required by the contractor in doing the work.

2. Timekeeping and payroll, with all the ramifications arising out of Federal income tax and Social Security legislation, and detail involved in contracts with organized labor.

3. Accounting and auditing, financing, and tax reporting.
4. Engineering estimating, cost control, plant layout, etc.
5. Accident prevention, labor relations, etc.

To correlate the operation of supporting staff required for general administration of the business and servicing its field forces, the head of the organization will need overall management with freedom from the direct demands of on-the-job supervision of construction operations. This problem may be solved by employing a general construction superintendent or project manager or by entering into a partnership with an outside person capable of filling that position, with the owner taking the overall management position.

Although the superintendent reports to the project manager, the latter and staff must be devoted to servicing the superintendent's every need to achieve the common goal of profitable completion of the project. This calls for good communication and close cooperation between them.

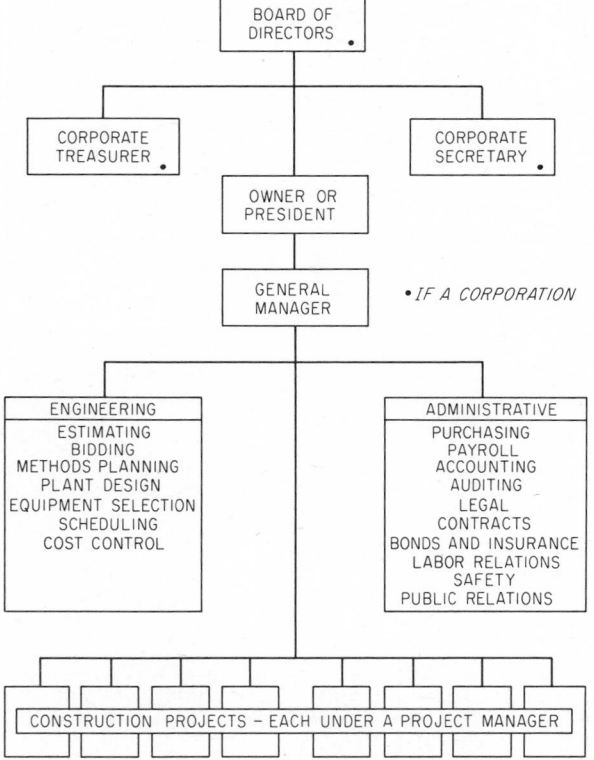

Fig. 4-3. Headquarters-type organization, with that for the lowest level as shown in Fig. 4-2.

Further growth may find the company operating construction jobs simultaneously at a number of locations. Arrangements for the operation of this type of business take the form of an expanded headquarters, or centralized, organization to administer and control the jobs and service the general construction superintendent or project manager at each location (Fig. 4-3). This concept contemplates, in general, delegation to the field of those duties and responsibilities that cannot best be executed by the headquarters function.

Accordingly, the various jobs will usually have a project manager in charge (Fig. 4-2). On small jobs, or in those cases where the general construction superintendent is in direct charge, the project manager will be accompanied by service personnel to perform the functions that must be conducted in the field, such as timekeeping, warehousing, and engineering layout. In the performance of their work, these service personnel take general instructions from their department heads at the company headquarters and immediate direction from the general superintendent or project manager. In this type of organization, all operations are directly coordinated from a central headquarters.

Some large construction firms, whose operations are regional, nationwide, or worldwide in scope, delegate considerable authority to operate the business to districts or divisions formed on a geographical or functional basis (Fig. 4-4). District managers, themselves frequently corporate officers, are responsible to the general management of the home office for their actions. But they are free to conduct the business within their jurisdiction with less detailed supervision although within definite confines of well-established company policies. The headquarters office maintains overall administrative control and close communication but constructs projects by and through its district organizations (Fig. 4-5).

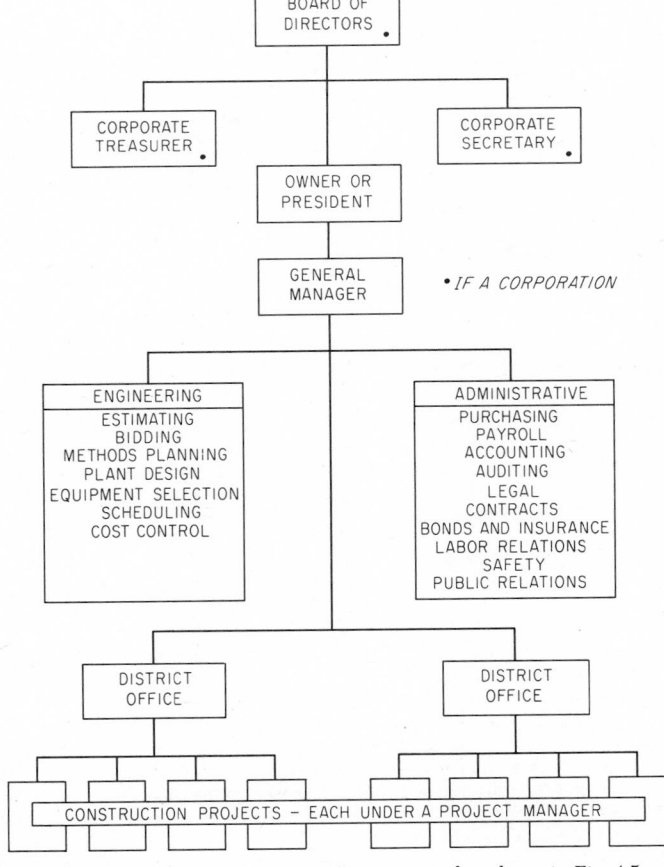

Fig. 4-4. District-type organization, with district offices organized as shown in Fig. 4-5 and projects as indicated in Fig. 4-2.

The home office of a decentralized, or district-type, business is organized much the same as a headquarters-type business but on a less extensive scale because of the delegation of responsibilities to district offices.

As construction businesses grow, each business develops organizational arrangements best suited to its particular type of work and field of operations. The tendency for large firms is toward greater decentralization in administration and operation.

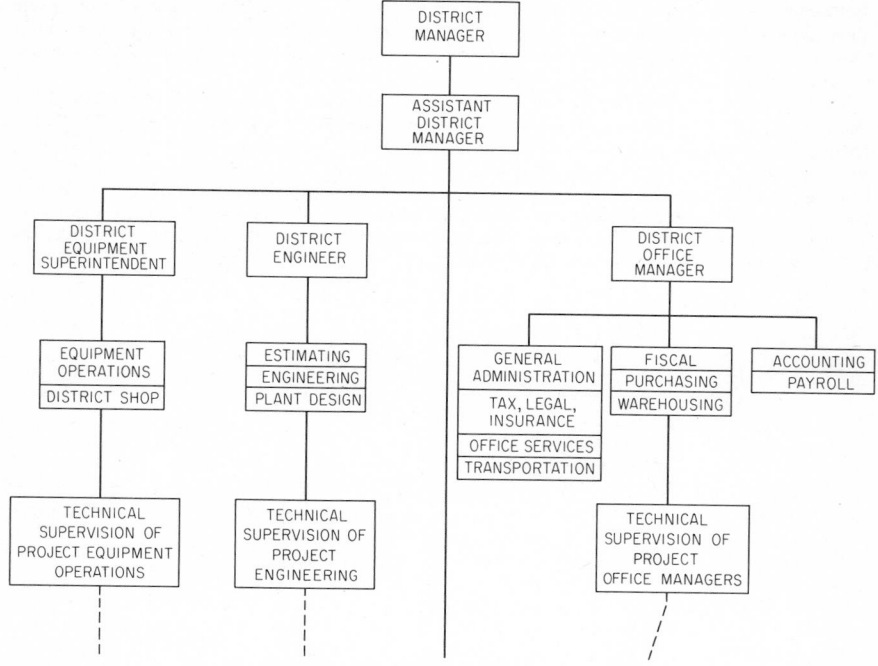

Fig. 4-5. District organization for construction company.

Although the district-type organization in some cases may be slightly more costly from the standpoint of supervisory, administrative, and overhead expense, it has offsetting advantages over headquarters-type organizations arising out of closer proximity with the work and familiarity with local conditions. These factors make it possible for the contractor to maintain closer supervisory control of the work and render better service, and they contribute to successful competition for new work.

Administratively, the success or failure of the district operation can be charged to a considerable degree to the district manager. With delegation and independence goes the natural tendency of district managers to develop along independent lines, creating thereby, in varying degree, problems of direction and control by the home office.

District-type operation requires more numerous competent and loyal employees. Individual ability must be allowed to develop and must be rewarded. There is a compelling urge toward visible and substantial achievement among construction workers that perhaps is stronger than the motivation in many other lines of endeavor. This drive toward accomplishment engenders competence in itself, for it brings out the best in people while weeding out those who would compromise with the daily challenges of the construction business.

By careful analysis of the advantages and disadvantages of headquarters and district organization, based upon conditions prevailing in each contractor's organization, the best form for business operations can be determined.

Survival in the highly competitive construction market is evidence that a contractor's organizational makeup has merit, whatever form it may take. Organization charts (Figs. 4-1 to 4-5) are indicative only of the highlights of organization and function employed within the broad spectrum of the industry.

(J. B. Bonny, "Handbook of Construction Management and Organization," Van Nostrand Reinhold Company, New York; G. E. Deatherage, "Constuction Company Organization and Management," McGraw-Hill Book Company, New York.)

4-5. Joint Ventures. Since risk is an important factor in construction, it is only prudent to spread it as widely as possible. One safeguard is a joint venture with other contractors whenever the financial hazard of any particular project makes such action expedient. In brief, a joint venture is a short-term partnership arrangement wherein each of two or more participating construction companies is committed to a predetermined percentage of a contract and each shares proportionately in the final profit or loss.   One of the participating companies acts as the manager or sponsor of the project.

The joint-venture method of doing work has lent stability to the industry without curtailing the ability of individual business firms to grow and without restraining competition.   This method uses the same operating principles adopted by insurance companies, and it contributes to a consistency of profit.   Through the joint-venture method, a large loss on a major contract that could be disastrous to an individual company becomes absorbable when shared on a per-centage basis and participation in a greater number of profitable contracts becomes possible.

Experience has demonstrated that a board of directors or operating committee can establish policy but cannot build a project.   Recognizing this truth, joint-venture partners agree that one of the partners will run the operation as sponsor.   This determination may be made on the basis of the partner having the greatest competence in the type of work contemplated by the proposed joint venture.   Often, the sponsor is the firm that organized the group to bid the work, is the most enthusiastic about undertaking it, is the least busy when the job comes up, or has at the time the largest amount of available equipment or competent supervisors.   The sponsor usually draws no extra fee but normally has the largest share of investment in the venture. Sometimes the sponsor has a controlling interest.

In recent years, because of the offering of extremely large projects for firm-price proposals, sometimes in the hundreds of millions of dollars in size and frequently in foreign countries, it has appeared imprudent for any one company sponsoring the joint-venture proposal to take for itself a major portion of the risk.   Under these circumstances, the practice has arisen where the proposed partners select a sponsor who, by reason of having only a relatively minor portion as participation, is allowed a small management fee, chargeable as part of cost, after which the sponsor shares proportionately with the other partners.

4-6. Project Division into Subcontracts.   A subcontract is an agreement between the general contractor and a subcontractor under which the subcontractor agrees to carry out certain por-tions of a project.   Normally, the portions subcontracted are of a specialty nature, such as elec-trical installations, plumbing, heating, ventilating, structural-steel erection, drilling, and grouting.   However, subcontracts may cover a portion of the prime contractor's principal undertaking, such as a section of excavation work on a highway.   The extent to which subcon-tractors are used varies with the business practice of individual contractors as well as with the kind of work involved.

4-7. Limitations on Subcontracting.   Construction contracts often limit the proportion of the contract that a prime contractor may subcontract.   For example, contracts covering Federal-aid highway construction usually require the prime contractor to perform not less than 50% of the contract value with his own organization.   The percentage may be reduced if found to be in the public interest.   Also, some states and municipalities have imposed by statute similar limitations on general contractors on state and local public works.   Such restrictions on sub-contracting are designed to prevent prime contractors from functioning as brokers instead of bona fide builders and to eliminate difficulties associated with subcontracting an excessive por-tion of a prime contract.

Actually, the practice of subcontracting has much to recommend it, especially in building construction, because it is frequently not practical for the prime contractor to maintain within his organization the supervision, equipment, and tools for all crafts and trades.   For some of these, the prime contractor may have only occasional need—often for relatively minor portions of a project.   By subcontracting specialty work, this burden can be eliminated while at the same time each subcontractor is afforded the opportunity to organize, equip, and provide continuity of employment for personnel required in his specialized line of work.

4-8. Approval of Owner for Subcontracts.   Although prime contracts often provide for approval of subcontractors as to fitness and responsibility, the making of a subcontract establishes only indirect relationships between owner and subcontractor.   The basis upon which subcontract agreements are drawn on fixed-price work is of no concern to the owner because the prime con-tractor, by terms of the agreement with the owner, assumes complete responsibility.   Under cost-plus prime contracts, however, subcontracts are items of reimbursable cost; as such, their terms, particularly the monetary considerations involved, are properly subject to approval of the owner.

4-9. Subcontract Agreements.   To achieve a fair distribution of risks and provide protective techniques for the benefit of both parties, it is necessary for subcontracts to be carefully drawn. The prime contractor wishes to be assured that the subcontractor will perform in a timely and efficient manner. On the other hand, the subcontractor wishes to be assured of being promptly and fairly compensated and that no onerous burdens of performance or of administration will be imposed.

Basic problems arise where parties fail to come to agreement with respect to at least the essentials of the transaction, including the scope of work to be performed, price to be paid, and performance. The subcontract must include the regulatory requirements of the prime contract and, in addition, must include appropriate arrangements for price, delivery, and specifications. It is insufficient to assume that writing a subcontract a purchase order binds that subcontractor to the terms of the prime contractor's agreement. Not only should subcontracts be explicit in respect to observance of the prime contract, but subcontractors should be fully informed by being furnished with the prime-contract plans, specifications, and other construction documents necessary for a complete understanding of the obligations to which they are bound.

4-10. Retained Earnings. Prime contracts require, as a rule, that a percentage—usually 10%—of the contractor's earnings be retained by the owner until final completion of the job and acceptance by the owner. Unless otherwise arranged, the provisions of the prime contract regarding payment and retainage pass into the subcontract by way of the usual stipulation that makes the subcontract subject to all the requirements of the prime contract.

For subcontractors whose work, such as site clearing, access-road building, or excavation, is performed in the early construction stages of a project, the standard retainage provisions may result in their having to wait a long time after completion of their work to collect the retained percentage. So the retainage on the general run of subcontracts, particularly those for work in the early phases of a project, often is reduced to a nominal amount after completion of the subcontractor's work. Justification for waiting until final completion of the job and acceptance by the owner may exist, however, under subcontracts for installed equipment carrying performance guarantees or for other items with vital characteristics.

To avoid the possibility of disputes over retained earnings, it is well for the subcontract to be specific in the matter of payment and release of retained earnings.

4-11. Industry Subcontracting Practice. Although subcontract agreements customarily define the sequence in which the work is to be done and fix time limits on the performance of the work, prime contractors are reluctant to delegate by means of subcontracts portions of a project where failure to perform might have serious consequences insofar as completion of the project as a whole is concerned, for example, the construction of a tunnel for diversion of water in dam construction.

In the heavy-construction industry, the greater the risk of loss from failure to perform, the less work is subcontracted. Such damages as may be recovered under subcontract agreements for lack of performance are usually small recompense for the overall losses arising out of the detrimental effect on related operations and upon execution of the construction project as a whole.

This situation has given rise to a trade practice in the heavy-construction industry whereby the prime contractor often has built up a following of subcontractors known for their ability to complete commitments properly and on time and generally to cooperate with and fit into the contractor's job operating team. Instead of openfield solicitation of subcontractor bids, subcontracts are often negotiated or competition is limited to a few such firms, with the result that the same subcontractors may follow the prime contractor from job to job.

Mutual confidence plays a large part in successful subcontracting. However, commitments should be fully documented and formalized by written subcontract agreements with performance-bond coverage comparable to the performance-bond requirements of the prime contract. Insurance should be provided of the kind and in the amount appropriate to the risk involved. Failure to have done so, on occasion, has given rise to much of the difficulty encountered in subcontracting work and has reflected unfavorably on a system that fundamentally has considerable merit.

4-12. Nature of a Contract. A contract is an agreement that creates an obligation. Its essentials are competent parties, subject matter, a legal consideration, mutuality of agreement, and mutuality of obligation [17 C.J.S. Contracts, Sec. 1 (1)]. A construction contract is an agreement to construct a definite project in accordance with plans and specifications for an agreed sum and to complete it, ready for use and occupancy, within a certain time.

Although contracts may be expressed or implied, oral or written, agreements between owners and contractors are almost universally reduced to writing. Their forms may vary from the simple acceptance of an offer to the usual fully documented contracts in which the complete plans, specifications, and other instruments used in bidding, including the contractor's proposal, are made a part of the contract by reference.

Recognizing that there are advantages to standardization and simplification of construction contracts, the Joint Conference on Standard Construction Contracts prepared standard documents for construction contracts intended to be fair to both parties. The American Institute of Architects also has developed standard contract documents. And the Contract Committees of the American Society of Municipal Engineers and the Associated General Contractors of America have proposed and approved a Standard Code for Municipal Construction.

Contractors generally secure business by submission of proposals in response to invitations to bid or by negotiations initiated by either party without formal invitation or competitive bidding. Agencies and instrumentalities of the Federal government and most state and municipal governments, however, are generally required by law to let contracts only on the basis of competitive bidding. However, certain Federal agencies, for reasons of security or in emergency, may restrict bidders to a selected list and, in these cases, may not open bids in public.

Normally, competitive bidding leads to fixed-price contracts. These may set either a lump-sum price for the job as a whole or unit prices to be paid for the number of prescribed units of work actually performed. Although negotiated contracts may be on a lump-sum or unit-price basis, they often take other forms embodying devices for making possible start of construction in the absence of complete plans and specifications, for early-completion bonus, or for profit-sharing arrangements as incentives to the contractor, etc. (see also Art. 3-3).

4-13. Lump-sum Contracts. Where the type of construction is such that division of the classifications of work makes breakdown into measurable units impractical, a lump-sum contract is usually employed. Most building construction is accomplished by this method. The contractor agrees to construct the project for a fixed price.

Successful use of this type of contract requires completely detailed plans and specifications describing the work to be done. If plans and specifications are indefinite, the contractor is forced to increase the bid to cover the worst conditions anticipated or to gamble on the uncertainties. Changes and extra-work orders issued after the contract is signed make the work more involved and costly to the owner than would otherwise have been necessary. If these difficulties can be avoided, the owner enjoys the advantage of knowing in advance what the exact cost of the work will be upon completion. Also, the owner can be assured that profit considerations will motivate the contractor to complete the work in the least practicable time (see also Art. 3-3).

4-14. Unit-price Contracts. When the volume of work cannot be exactly determined in advance, a unit-price contract has many advantages and is generally used. Unit-price contracts are particularly well adapted to heavy construction work, such as highways, bridges, dams, and river and harbor improvements, where large quantities of relatively few types of construction are involved. This type of contract sets a price for each unit of work.

Unit-price contracts offer all the advantages of competitive bidding yet allow reasonable variation in the quantities of the various items of work without formal change orders. Plans and specifications must be complete in all respects to enable the contractor to assess the magnitude and complexities of the project. The quantities of work on which competitive bids are received are estimated quantities, determined by the owner's engineer to indicate the size of the undertaking and for comparison of bids. Payment to the contractor is made for the number of units of each item of work actually performed in the field.

In order that the contractor may have protection against a wide variation between the quantity of work bid and the quantity actually performed without benefit of price adjustment, the contract customarily provides that the unit prices will apply within a range, such as 25% below and 25% above the specified quantities.

The owner, either directly or through an engineer, must support the necessary field force for determination of quantities since these become the basis of payment to the contractor. In the unit-price contract the ultimate cost to the owner is not known until completion of the project when the units of each item of work have been measured.

It is not uncommon to employ both lump-sum and unit-price features in a single contract, for instance, on a hydroelectric project where the powerhouse structure is built on a lump-sum basis while foundations, dam structure, and tunnel work are done on a unit-price basis (see also Art. 3-3).

4-15. Negotiated Contracts. For various reasons, owners may prefer to enter directly into negotiations with one or more selected contractors rather than invite competitive bids. After a study of the qualifications, experience, plant and equipment, financial resources, and possible schemes for accomplishing the proposed work, a contract is entered into without open-field competition or the formal receipt and opening of bids. The benefits of negotiated contracts are normally limited to private construction. Public bodies, except during war emergency, are required by law to contract for their public-works projects only after receipt of and on the basis of competitive bids.

4-16. Contract Claims. In many lump-sum and unit-price contracts, difficulties arise over claims that develop because the contract drawings and specifications are incomplete or unclear and changes or clarifications become necessary. In other words, if the buyer fails to say exactly what he wants, the seller has difficulty in naming a price (see also Art. 3-5.)

4-17. Cost-plus Contracts. Although lump-sum or unit-price contracts can be negotiated as readily as any other type, often some form of cost-plus arrangement is used. This is usually done to cope with special problems, such as those encountered where owners desire to start construction without waiting for plans and specifications to be developed to the point where competitive bids can be received, or where the project is of an unusual nature involving new techniques or experimental technology, is in a remote or relatively inaccessible geographic location, or has other features making the risks involved difficult to appraise. Under this arrangement, the owner guarantees to reimburse the contractor for all costs and pay a fee for services. The owner thus assumes the risk for which, in the case of a fixed-price contract, the contractor would necessarily make some charge.

Under the original form of cost-plus contract, the contractor's fee was computed on the basis of percentage of the cost of construction. This accomplished the purpose of permitting start of construction before completion of plans and satisfied requirements of owners acquainted with the competence and integrity of the contractors whom they selected. However, in cost-plus contracts between public bodies and contractors where less mutual confidence could exist, abuses might occur, resulting in higher costs.

To remedy that situation, *cost-plus-fixed-fee contracts* were developed. The contractor is reimbursed for the cost plus a fixed amount, as a fee for accomplishment of the work. After the scope of the work has been clearly defined and both parties have agreed on the estimated cost, the amount of the contractor's fee is determined in relation to character and volume of work involved and the duration of the project. Thereafter the fee remains fixed, regardless of any fluctuation in actual cost of the project. There is no incentive for the contractor to inflate the cost under this type of contract since the contractor's fee is unaffected thereby. But maximum motivation toward efficiency and speedy completion inherent in fixed-price contracts may be lacking.

If sufficiently detailed plans and specifications are available to permit establishment of a reasonable maximum price, that feature is often incorporated into cost-plus-fixed-fee contracts. The contractor is reimbursed for the actual cost of the work plus a fixed fee if the limitation established in the contract is not exceeded. In case the cost exceeds the maximum, the contractor suffers the loss. If the cost underruns the agreed maximum, the contractor may or may not share in the savings, depending upon whether that feature was made a part of the agreement.

Where revenue-producing projects such as electric power plants, commercial rental structures, or manufacturing facilities are involved, an additional feature may be incorporated in cost-plus-fixed-fee contracts: In addition to a basic fee, the contractor is paid a share of the revenue or earnings accruing to the owner by reason of completion in less time than prescribed in the contract. (This feature sometimes is extended to call for a penalty, deducted from the basic fee, for failure to complete within the specified time.)

A profit-sharing clause is sometimes written into the cost-plus-fixed-fee contracts as an incentive for the contractor to keep cost to the minimum, allowing the contractor a share of the savings if the actual cost, upon completion, underruns the estimated cost. This provision may also be accompanied by a penalty to be assessed against the contractor's fee in case the actual cost exceeds the agreed estimated cost.

A fundamental requirement for all cost-plus-fixed-fee contract agreements is a definition of cost. A clear distinction should be made between reimbursable costs and costs that make up the contractor's general expense, payable out of the contractor's fee. Some contracts, which would otherwise run smoothly, become difficult because of failure to define cost clearly. Usually, only the cost directly and solely assignable to the project is reimbursed to the contractor. Therefore, the contractor's central office overhead and general expense, salaries of principals and headquarters staff, and interest on capital attributable to the project frequently come out of the fee, although a fixed allowance in cost for contractor's home-office expense may be allowed.

Cost-plus-fixed-fee contracts do not guarantee a profit to the contractor. They may also result, particularly in government cost-plus-fixed-fee contracts, in unusually high on-job overhead occasioned by frequent government requirements for onerous and cumbersome procedures in accountability and accounting.

4-18. General Requirements of Estimating. The two most important requisites for success in the construction business are efficient management of work in progress and correct estimating. Costs cannot be forecast exactly. But the contractor who can approach most nearly to an accurate forecast of cost will bid intelligently a high percentage of the time and will be most successful over a period of years.

Construction estimates are prepared to determine the probable cost of constructing a project. Such estimates are almost universally prepared by contractors prior to submitting bids or entering into contracts for important projects. To be of value, an estimate must be based on a detailed mental picture of the entire operation; that is, it is necessary to plan the job and picture just how it is going to be done. Accordingly, it is wise to have the general construction superintendent or project manager who will be in charge of the job take part in the preparation of the job estimate.

Each job has its own individuality and is to be looked upon as a new problem. Nothing of value is obtained if an estimate is based only on what other contractors have been bidding for similar work. Accordingly, it is not the duty of the estimator to try to guess what others may bid. The estimate should be based on a known definite way of doing the work or on firm figures from reliable subcontractors. Also, it is not the function of the estimator to make the final decision as to the bid. Therefore, no matter how experienced the estimator may be in estimating or performing work and no matter how accurate the estimate may be, the estimate is of relatively little value unless it is put together in such a way that the person whose decision is final can follow its detail quickly and easily and be able to judge where, how, and to what extent the items of work have been covered.

An estimate that is poorly assembled has little value except that its grand total represents the opinions of the person or persons who prepared the estimate. However, if others can follow the detail rapidly and in a short time develop a worthwhile opinion of their own as to the accuracy of the total, the value of the estimate is greatly multiplied.

It is desirable to standardize the arrangement and mechanical makeup of estimates to as great an extent as possible without encroaching on the individuality of the estimator. The mechanical makeup of an estimate must be simple because conditions usually require that it be prepared in a short time—sometimes 2 or 3 days when the estimator would like to have a month. These conditions do not change; it will always be necessary to make estimates quickly.

4-19. Relation of Estimating to Cost Accounting. Estimating and cost accounting should be very closely tied together. The estimate should be prepared in such a way that if the bid is successful, the estimate can be used as the framework for the cost accounts.

Estimating should be based on cost records to whatever extent may be reasonable in the particular case. But, prominently in the picture, there should also be a continuous study of new equipment, methods, and cost-cutting possibilities. The data most valuable, when used with due consideration of surrounding conditions and possible improvement, are cost records of the detail of operations rather than of operations as a whole. Cost records and estimated costs for the labor portion of an operation should be expressed in man-hours as well as in dollars. A clear and complete narrative description of all the circumstances affecting the work should be made a part of the cost records prepared for use in future bidding; otherwise, the usefulness of the data is greatly reduced.

The need for good production and cost records is emphasized by an increasing reluctance of some engineers and owners to make decisions and adjustments on the job. The resultant tendency is to throw the settlement of ordinary items of business into arbitration or into court, where basic information is a fundamental requirement.

Normally, cost records in full detail are not available with sufficient promptness to be of substantial value on the job on which the costs are incurred. It is very desirable, however, that a current check on operating costs be maintained. This may be done by less formal procedures and still be adequate to provide timely information on undesirable deviations in progress and cost.

4-20. Forms for Estimating. Preparation of estimates is facilitated by standardization of forms for recording construction methods, equipment, and procedures that the estimator proposes as best adapted to the various items of work, to record calculations of the estimated cost of performing the work, and to summarize the estimated cost of the project. It is unnecessary and impractical to provide detailed printed forms for all types of work. A few simple forms are all that are needed.

4-21. Job Progress Schedule. The first thing to be done when beginning the preparation of an estimate is to make a time schedule of the proposed operation and set up a tentative plan for and methods of doing the work. It is necessary to study the plans and specifications in detail before visiting the site of the project. This study should proceed far enough to establish a tentative progress schedule for the more important or governing items of work.

The progress schedule should show all items affecting the progress of the work and consider the length of construction season at the particular site. Where applicable, it should consider the most advantageous date or the required date for river diversion, when deliveries of new or specialized construction plant or equipment can be obtained, possible delivery dates on critical items of contractor-furnished permanent materials, delivery dates of major items of permanent equipment to be furnished by the owner, and other controlling factors. Using the preceding dates, production rates for the controlling items of work must be determined and the type, number, and size of the various units of construction plant and equipment needed to complete the work, as required by this schedule, must be tentatively decided upon. Progress schedules can be prepared in several forms. Figure 4-6 shows a form that can be adapted to fit most conditions (see also Arts. 4-25 to 4-28).

Based upon the progress schedule, a brief narrative description of the job should be written, which calls attention to indefinite, hazardous, and uncertain features as well as items likely to increase and decrease in quantity. Also, the description should include a statement of the total man-hours of labor and the total machine-hours for important equipment estimated to be required in doing the work. In addition, it should include peak labor requirements and controlling delivery requirements for important items of material and equipment. Finally, the description should contain a statement of cash requirements derived from scheduled income and expenditures.

4-22. Prebid Site Investigations and Observations. No contractor should bid a job without first thoroughly examining the site. This should be done early enough for the owner to have sufficient time to issue addenda to the plans and specifications, if required, to clarify questionable items.

Before visiting the site, the contractor should prepare a checklist of items to be investigated, including, where applicable: transportation facilities, electric power supply, water supply, source of construction materials, type of material to be encountered in required excavation or borrow pits, possible property damage from blasting and other operations of the contractor, interference from traffic, available labor supply (number and length of shifts per week being worked in the vicinity), areas available for construction of special plant, location of waste-disposal areas and access thereto, and weather records if not otherwise available.

It sometimes proves helpful to take pictures of critical areas of the site at the time the investigation is made. Frequently, questionable items that were not covered on the original visit can be cleared up by referring to the photographs. They are sometimes of great value to the engineers doing the takeoff work and can prove to be of assistance in explaining the job to others reviewing the estimate who have not visited the site.

4-23. Nature and Significance of a Proposal. Contractors obtain most of their business from offers submitted in response to invitations to bid issued by owners, both public and private (Art. 3-13). Inasmuch as award is usually made to the "lowest bidder" or "lowest responsible bidder," the contractor is constantly faced with the likelihood of failing to secure the business if a bid is too high. On the other hand, the contractor risks financial loss in executing the work if a bid is low enough so that the contract is awarded. It can be seen, therefore, that the submission of a proposal is a commitment of far-reaching significance. The contractor is responsible for the consequences of such mistakes as may be made as well as those risks inherent in construction over which the contractor may have no control.

A proposal is an offer made by the contractor to the owner to perform the work required by the contract documents for a stated sum of money. Further, the proposal is a promise by the contractor that upon acceptance of the proposal by the owner, the contractor will enter into a contract and perform the work for the stated remuneration. Note that the proposal and its acceptance, together with the monetary consideration, constitute the essential elements of a contract between competent parties. Ordinarily, a proposal is effective until it is rejected by the owner. Most owners, however, provide in their invitations for bid that award of contract will be made within a stipulated period of time, such as 30 days after the opening date.

By furnishing the form of proposal to be used by contractors in submitting bids and stipu-

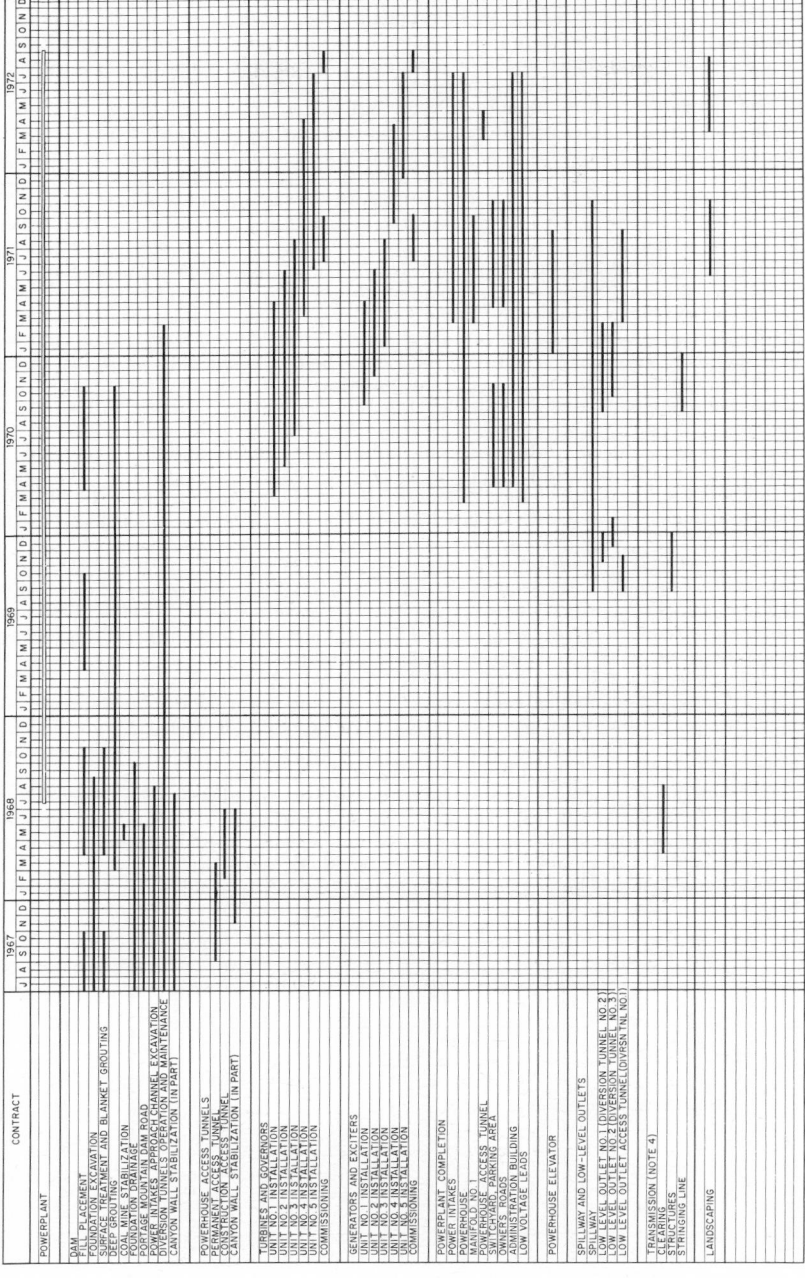

Fig. 4-6. Bar-chart progress schedule.

lating how it must be completed, the owner intends to put all bids on the same basis, thereby permitting equitable comparison and selection for award of contract. Although the time allotted for preparation of the estimate and submission of bid is seldom regarded as sufficient by the contractor, it is nonetheless incumbent upon the contractor to prepare the proposal in strict conformity with instructions in the invitation to bidders and other documents. Failure to do so may result in disqualification of the bid on the grounds of irregularity, with a resulting loss of time and money expended in the preparation of the bid.

4-24. Bid Alternatives. In addition to the basic bid, the owner may call for prices on alternative materials, equipment, or work items. These prices may be either added to or deducted from the base bid. This device is generally employed as a means of insuring that an award can be made within the amount of the owner's available funds. It serves, also, as an aid to selection by the owner after having the benefit of firm prices on the various alternatives. Accordingly, figures quoted by the contractor on alternatives should be complete within themselves, including overhead and profit.

4-25. Scheduling to Save Money. Time is less tangible an ingredient of construction than labor or material but nonetheless real and important. Money and time are related in many ways.

To the owner of revenue-producing facilities, such as electric generating installations, processing plants, rental buildings, petroleum pipelines, or wharves, docks, and other improvements for shipping, reduction in time required for completion of construction results in less interest on investment during the period of construction. Also, increased income accrues to the extent that completion time is shortened, thereby permitting earnings to begin at an earlier date.

To the contractor, reduction in time for completing the job means, likewise, a reduction in interest charges on cash invested during construction. Also, the shorter the time to complete the job, the less will be the supervisory, administrative, and overhead expense. In addition, benefits accrue from shortened time because it permits earlier release of equipment for use on other work.

Construction scheduling consists essentially in arranging the several operations involved in the construction of a project in the sequence required to accomplish completion in the minimum period of time consistent with economy. To insure completion within the contract time limit and to attempt to reduce the time required for doing the job, it is necessary to program each unit of the project within itself and properly relate each unit to all the others (see also Art. 4-21).

4-26. Scheduling with a Rectangular Bar Chart. Progress schedules show starting and completion dates for the various elements of a project. For unit-price work, the bid-item breakdown is normally used. On lump-sum contracts, subdivision according to that used in estimating the work is common. Schedules may be prepared in either tabular or graphical form, although the graphical form is generally used because of ease in visualization.

The most widely used graphical representation of the work schedule is the rectangular bar chart (Fig. 4-6). It shows starting and completion dates for each item of work. It indicates the items on which work must proceed concurrently, the items that overlap others and by how much, and the items that must be completed before work on others can begin.

Progress schedules should be prepared at the outset of the job as an aid in coordinating work

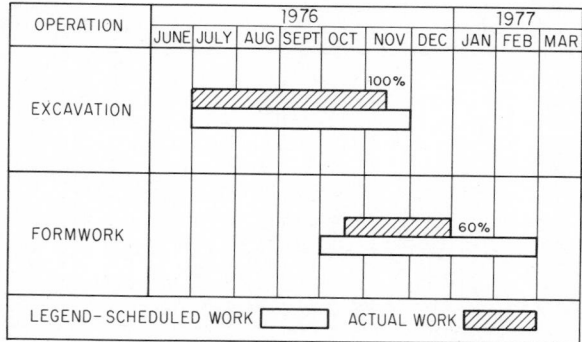

Fig. 4-7. Rectangular-bar progress schedule.

by all departments of the contractor's organization (Art. 4-21). For instance, the progress schedule is a convenient way to advise the purchasing agent of necessary material delivery dates. Construction contracts often require the contractor to submit a progress schedule to the owner for approval within a specified time after award of the contract and before construction is started. The importance of this requirement often is emphasized in the contract by provisions to the effect that failure to submit a satisfactory schedule shall be just cause for annulment of the award and forfeiture of the proposal guarantee.

For comparing performance of work with that scheduled, a bar is often placed above the schedule bar showing actual start and completion dates. The chart in Fig. 4-7 indicates that excavation started on the date programmed and was completed ahead of time whereas formwork began late. At the close of December, formwork was 60% complete. This method has the advantage of simplicity, but it fails to disclose the rate of progress required by the schedule or whether actual performance is ahead of or behind schedule.

4-27. Triangular Bar Chart. (See also Art. 4-26.) The concept of rate of progress is introduced in Fig. 4-8, which deals with the same items charted in Fig. 4-7. In Fig. 4-8, horizontal distances represent time allotted for doing the work and vertical distances represent percentage of completion. Therefore, the sloping lines indicate the rate of progress.

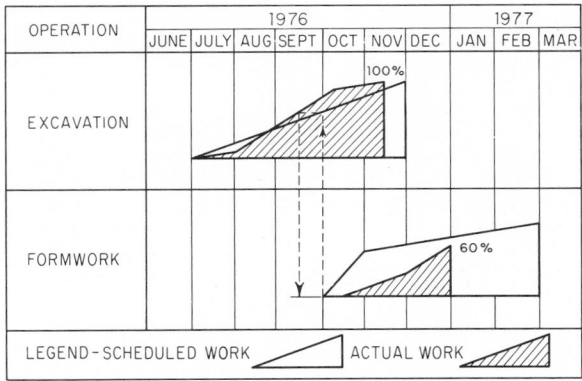

Fig. 4-8. Triangular-bar progress schedule.

For example, excavation was scheduled to proceed from start to finish at a uniform rate (straight sloping line). Work started on time, progressed slowly at first, and tapered off at the end (crosshatched area). Greater production scheduled midway in the operation was sufficient, however, to bring the item to completion 15 days early. The date on which formwork could have begun was advanced by reason of the accelerated rate of excavation from Oct. 1 to Sept. 15 (dashed lines).

Instead of being stepped up to take advantage of the time gained on excavation, formwork was late in getting started and progressed slowly until Dec. 1. Then, it was speeded up, but the 60% completion reached at the close of December falls short of scheduled requirements. (In practice, the time gained on excavation would doubtless have been captured and put to beneficial use by arranging start of formwork on Sept. 15, ½ month ahead of schedule.)

The effect of time gained or lost on any one item reflects into many other work items. As a result, frequent revision is necessary if progress schedules are to be kept currently accurate in all respects. Formalized revision of the overall progress schedule, however, is often rendered unnecessary because contractor dependency upon it is gradually supplanted by such intimate acquaintance with the operations that controlling and critical factors become common knowledge and all concerned know what must be done and when.

Critical items often are subjected to detailed analysis and scheduling. This may take the form of three-dimensional schematics, expanded views, stage-construction drawings, concrete-pouring diagrams, and similar devices as aids to visualization. After that, further scheduling, such as concrete-pouring programs, earthwork-quantity movement schedules, or programming of piping runs may be devised and utilized as required.

4-28. Critical-Path Method of Scheduling (CPM). The critical-path method has been developed as a tool of management useful in specialized situations. It is required by several Federal and state agencies on some contracts. CPM is based upon planning and job analysis going far beyond that necessary for bidding a job. In addition to the step-by-step breakdown of the job into its component operations and the plotting of sequential relationships, the planners must know how long each operation will take, the lead time required for procurement of materials and equipment, how long it will take to prepare shop drawings and obtain their approval, and how long it will take for fabrication and delivery after approval of shop drawings. They must know about special tests required and the time needed to make them.

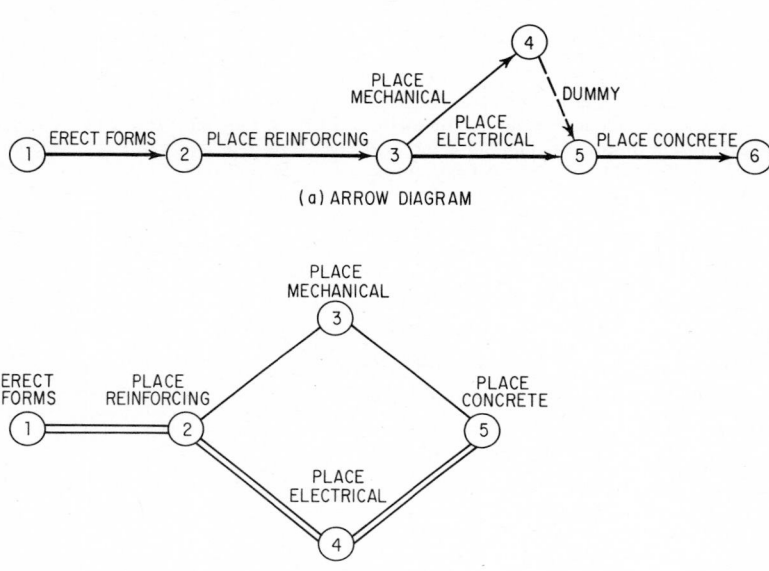

Fig. 4-9. Simple network for CPM plotted with activities represented by (a) arrows and (b) nodes.

After the project has been broken down into all its activities, they are listed or plotted in such a way that all sequential relationships are shown. Activities may be represented by arrows (Fig. 4-9a) or by circles, or nodes, connected by sequence lines (Fig. 4-9b). Analysis may be by manual methods or electronic computer for establishing a realistic time schedule, pinpointing the operations whose completion times are responsible for establishing the overall project duration, settling change orders by determining the operations affected and the effect on project duration, establishing the proper sequence of work operations, and determining the status of work in progress in relation to the number of days behind or ahead of schedule.

An *arrow diagram* (Fig. 4-9a) is drawn by setting the tail of an arrow representing an activity, such as placing concrete, at the tip of an arrow representing the immediately preceding activity, such as placing electrical conduit and outlets. The nodes (tips and tails) are assigned unique numbers to identify the activities (1-2, 2-3, and so forth). Each node represents the completion of the preceding activities and the start of the following activities. Sometimes, a dummy arrow is needed to complete the network.

A *precedence diagram* (Fig. 4-9b) is drawn by setting the node representing an activity to the right of the node representing an immediately preceding activity. Each node is assigned a number greater than that of any preceding activity. The nodes are connected by lines to indicate the sequence of the work. Precedence diagrams are simpler to draw and analyze than arrow diagrams.

In either type of diagram, the critical path is the sequence of operations requiring the most time to complete. The critical path determines the duration of the project. To shorten the project, it is necessary to decrease the time required for one or more activities on the critical path (critical activities). These activities have zero *total float*. Total float is the difference between time required and time available to execute an activity. It is equivalent to the difference between earliest and latest start (or finish) times for an activity. Table 4-1 shows the calculation of float for the simple network in Fig. 4-9. Float is determined in two steps: a forward and a backward pass over the network.

Table 4-1. Float Calculations for Critical-Path Method

Activity number		Dura-tion, days	Early start date	Early finish date	Late start date	Late finish date	Total float, days
Arrow diagram	Precedence diagram						
1–2	1	2	0	2	0	2	0
2–3	2	1	2	3	2	3	0
3–4	3	1	3	4	4	5	1
3–5	4	2	3	5	3	5	0
5–6	5	1	5	6	5	6	0
4–5	. . .	0	4	4	5	5	1

The forward pass starts with the early start (or scheduled) date for the first activity, Erect Forms. In this case, the date is 0. Addition of the duration of this activity, 2 days, to the early start time yields the early finish date, 2, which is also the early start date for the next activity, Place Reinforcing. Its early finish date is obtained by adding its duration, 1 day, to the early start date. The forward pass continues with computation of early start and finish times for all subsequent activities. Where one activity follows several others, its early start date is the largest of the early finish dates of those activities.

The backward pass determines late start and finish dates. It begins with the late finish date of the final activity, Place Concrete, which is set equal to the early finish date, 6, of that activity. Subtraction of the duration, 1 day, from the late finish date yields the late start date, 5, which is also the late finish date of preceding activities, Place Mechanical and Place Electrical and their late start dates are found by subtracting their durations from the late finish dates. Where one activity precedes several others, its late finish date is the smallest of the late start dates of those activities. The backward pass continues until late start and finish dates are computed for all activities. Then, the float can be found for each activity as the difference between early and late start times. Critical activities (those with zero total float) are connected by heavy arrows in Fig. 4-9a and by double lines in Fig. 4-9b to indicate the critical path.

Eighty percent of the effort in CPM is spent in activity analysis and network preparation, which requires, essentially, construction judgment and know-how, practical experience, and common sense. The method, accordingly, rests on the same foundation as conventional methods of planning and scheduling. Constantly increasing complexity of projects necessitates careful and complete planning. There is danger, however, of overdoing the detail to the point where essential facts become obscured and results are placed in doubt. On occasion, a project may be divided for study of construction sequence into so many operations that it becomes necessary to turn to an electronic computer for relief from the burden of detail. Extensive breakdown is helpful in some cases, but this should be done with extreme caution because correct answers for each event, based on experienced judgment, must be available before the problem can be fed into the computer. What comes out of it is no better than the construction judgment of those who programmed the problem.

(J. B. Bonny, "Handbook of Construction Management and Organization," Van Nostrand Reinhold Company, New York; J. J. Moder and C. R. Phillips, "Project Management with CPM and PERT," Van Nostrand Reinhold Company, New York; J. J. O'Brien, "CPM in Construction Management," McGraw-Hill Book Company, New York.)

4-29. Liability Insurance. Law, contracts, and common sense require that responsible contractors be adequately protected by liability insurance in all phases of their operations.

Required by Law. Most states require users of highways to furnish evidence of automobile bodily-injury and property-damage liability insurance in basic minimum limits. This is particularly true of businesses that have trucks or other heavy equipment on the highways or public roads. Special permits to move heavy equipment on the highways generally require somewhat higher limits of protection.

A contractor who operates in foreign nations generally finds that the liability-insurance requirements are even more stringent than those in the United States and that automobile liability insurance must be procured from an insurance company headquartered in the nation in which he operates.

Required by Contract. Almost without exception, construction contracts require the contractor to carry comprehensive liability insurance. The purpose is to protect the contractor, owner, and owner's engineers against all liability for bodily injuries or third-party property damage arising out of or in connection with the performance of the contract. Occasionally, the contract requires a separate Owner's Protective Liability Insurance policy. Also, when a contractor operates alongside or across the property of a railroad company, a Railroad Protective Liability Insurance policy is generally required.

Required by Common Sense. Regardless of the coverages required by law or by contract, the prudent contractor should carry liability-insurance protection in substantial amounts. The very nature of the construction industry subjects a contractor to the possibility of substantial risk of liability to third parties. In certain situations, particularly where the contractor is using explosives, risk may approach absolute liability.

Serious accidents may involve a number of injuries or substantial property damage. There are numerous cases of judgments against contractors in excess of $1 million. Even the small contractor engaged in a small construction job, conceivably, can have an occurrence that will generate claims in excess of $1 million. Good judgment would dictate that a $1 million limit currently is the minimum limit for liability-insurance protection, with large and hazardous jobs requiring substantially higher protection—at times exceeding $10 million.

4-30. Property Insurance. In addition to liability insurance, contractors must protect themselves against damage or loss to their own property and to the projects on which they are working.

Contractor's Equipment, Plant, Temporary Buildings, Materials and Supplies Insurance. Almost all assets of the typical construction contractor consist of contractor's equipment, construction plant, temporary buildings, materials, and supplies. Common sense dictates that contractors keep their property insured. Ordinarily, the contractor's heavy equipment and vehicles are purchased on conditional sales contracts or are leased under agreements that require the contractor to maintain insurance against physical damage to the equipment and vehicles, with losses payable to the contractor and the secured owners as their respective interests may appear at time of loss.

The contractor can maintain separate coverage for heavy equipment, automobile fire, theft, and collision coverage on his highway trucks and automobiles and fire and extended coverage insurance on plant and temporary buildings. However, "piecemeal" coverages do not provide sound all-risk protection on all property. Furthermore, the premiums in the aggregate often add up to more than the cost of a single all-risk blanket coverage on all property. Obviously, too, the risks to which a contractor's property is subject stem from different and more varied sources than the risks of a merchant or manufacturer. For example, a contractor engaged in the construction of a dam may have little risk from fire or the usual extended perils, but risk from flood may be great. Yet, flood is generally a standard excepted peril in most property coverages.

Contractors' property insurance should be in an amount sufficient to cover the total values of property subject to any conceivable risk at one location. A contractor who has a normal recurrence of property losses may reduce the cost of insurance by arranging a deductible in an amount that approaches a normal loss recurrence. Ordinarily, the deductibles are based on the value of equipment at risk. A deductible of $1,000 on equipment valued in excess of $5,000 may be adequate to protect the ordinary contractor against calamitous loss and still be sufficient to provide coverage at the most reasonable premium cost. On equipment valued in excess of $10,000, a deductible of $2,500 is reasonable. Generally, small tools, materials, and supplies can be covered in the same policy at a more reasonable premium than would be charged for a separate policy covering the contractor's inventory of these items.

Builders' All-Risk Insurance. Invariably, the construction contract places full responsibility (and liability) on the contractor for protection of the project work and for repair or replacement of damage until the completed project work has been accepted by the owner. Occasionally, the owner carries "course of construction" insurance in which the contractor is an additional insured. In these situations, the contractor should make sure to be relieved of the responsibility for repair or replacement of damaged work. A contractor who assumes such responsibility, as is usual, should carry Builders' All-Risk Insurance.

Perhaps the most serious risk of damage to the work arises from the contractor's operations, such as failure of hoisting machinery or negligent operation of heavy equipment. The contractor's liability insurance would not protect him in such a situation because risks arising from the contractor's negligence or failure of machinery used by him are excluded under the standard "care, custody, and control" exclusion in the liability-insurance policy. Likewise, fire and extended coverage insurance, being restricted to the specific perils named, would not insure the contractor against loss resulting from operation of equipment, blasting, or other causes of risk usual to the contractor's operations.

Builders' All-Risk Insurance generally protects against any natural occurrence, act of God, or damage caused by human error. The possible loss can be substantial in amount, and the policy limit should be adequate to cover the largest conceivable loss. Inasmuch as the contractor's main concern is protection against catastrophe loss, the contractor should require a high limit but can afford a substantial deductible which will permit the purchase of this important coverage at the most reasonable cost.

4-31. Workmen's Compensation and Employee Benefits Insurance. In all states, Canada, and most foreign nations, Workmen's Compensation Insurance is required by law. The construction industry is regarded as "extra-hazardous" in the terminology employed in Workmen's Compensation Laws. Premiums are based on the classifications of work in which each craft of construction workmen is engaged. The premium rates vary greatly from as high as $75 per $100 of payroll (a rate applied to miners engaged in pressure-tunnel work in New York City) to as low as 5 cents per $100 for office employees engaged solely in clerical work. Obviously, the cost of Workmen's Compensation Insurance is an important factor in preparation of a bid.

Employer's Liability Insurance is automatically included in most Workmen's Compensation Insurance policies. Although Workmen's Compensation is, without exception, the sole remedy of an injured workman or of the family of one who dies as the result of an industrial injury, there may be occasions where, because of liability assumed by contract or otherwise, a contractor may be required to defend an action at law or pay a judgment based on injuries to an employee or a subcontractor's employee.

In seven states of the United States, commonly called the *monopolistic-fund states,* and in all provinces of Canada, Workmen's Compensation Insurance is required to be carried with the state or provincial fund. In these states and provinces, Employer's Liability Insurance is generally neither required by law nor furnished by the funds. The prudent contractor will carry a special Employer's Liability Insurance policy with a private carrier when operating in these states and provinces.

Also, the contractor who is engaged in work bordering on a waterway or navigable stream should carry insurance for protection against liability under the Longshoremen's and Harbor Workers' Compensation Act and the Jones Act. These coverages can generally be provided by endorsement to the standard Workmen's Compensation policy at little or no additional premium.

Other coverages the contractor may wish to consider but which are generally elective are group hospital, surgical and medical plans, and group term life and accidental death and dismemberment coverages. Often, these coverages will be provided by jointly administered employer-union benefit plans created by collective bargaining in the construction industry. The union plans, of course, are limited solely to the contractor's employees covered by a collective-bargaining agreement. It is up to the contractor to decide whether to provide similar coverage for salaried, managerial, engineering, and clerical personnel.

4-32. Miscellaneous Insurance Coverages. The miscellaneous insurance needs of contractors vary with the type and scope of their operations. Among those considered essential, however, are consequential loss insurance, fidelity and forgery insurance, and money and securities insurance.

Consequential Loss Insurance. The contractor soon discovers that physical-damage protection on the construction work in progress or on contractor's equipment will pay only a portion of his out-of-pocket financial loss. On permanent project work, builders' all-risk coverage will reimburse the actual cost of restoring the work. This recovery, of course, is limited to the original value of the work, and the deductible, which is generally substantial, will be applied. No allowance is made for extra overhead incurred for the time required to repair or replace the damaged work, overtime expense, etc.; it is almost always excluded under the terms of the builders' all-risk coverage. A contractor may procure a form of "business interruption" insurance that will pay the contractor any extra expense for extended overhead, etc., arising out of a builders'-risk type of loss.

The contractor who loses the use of equipment through physical damage must provide substitute equipment for the time damaged equipment is being repaired.   Often, the contractor can obtain insurance with contractor's equipment coverage that will cover rental expense of replacement equipment.

Fidelity and Forgery Insurance.   A contractor who has delegated authority with respect to the firm's business and financial affairs to one or more employees should carry fidelity insurance in a limit adequate to cover such sums as the employees may deal with.   Likewise, the prudent contractor will carry depositor's forgery insurance to protect against financial loss caused by forgery of checks against banking accounts.

Money and Securities Insurance.   Ordinarily, the contractor keeps only small sums of cash in his office or at job offices.   Sometimes and in some states contractors meet their payroll in cash.   In such situations, it is advisable to carry money and securities coverage, which will protect the contractor against loss by outside theft, including burglary and robbery.   This coverage should carry a limit equal to the largest sum of cash on hand at any one location.

4-33. "Coverage Boosters" and "Cost Savers."   A prudent selection of insurance plans, coupled with an active safety program, will materially lessen the contractor's overall insurance costs.

Blanket Coverages and Package Plans.   One of the basic concepts of insurance is "risk spreading."   The more a risk can be spread, geographically or otherwise, the more economical will be the premium.   Therefore, a contractor who insures all operations under a single policy against a common risk, whether it be liability, physical damage, or fidelity, etc., will enjoy the broadest protection at the lowest cost.   On builders'-risk insurance, for instance, some of the contractor's operations may be quite hazardous; others may be virtually risk-free.   In such a situation, the contractor is able to maintain builders'-risk coverage at a reasonable rate on a hazardous project by charging all operations at the same premium rate, simply because his low-risk work contributes to the overall cost.   The same analogy may be made with respect to other coverages.

"Quantity discounts" come into play on package plans where a number of different coverages are combined in one policy.   Ordinarily, many "fringe coverages" can be included at no additional premium.   A package plan with which most individuals are familiar is the "Homeowner's policy."   Although no similar "contractor's package policy" is a "shelf item" in the insurance industry, nevertheless the contractor's agent or broker should be able to arrange some packaging of coverages.

Contractor's Safety Program.   Contractors should always be aware of one of the best cost savers available to them, namely, a good safety program.   The largest insurance expenditure, by far, is the Workmen's Compensation premium.   Almost every underwriter of Workmen's Compensation Insurance offers substantial discounts, dividends, or retrospective premium-return plans that are based on favorable low-accident experience.   A contractor often can carry on a safety program at a cost much lower than the dividends earned from Workmen's Compensation premium returns.   For the small contractor, almost every Workmen's Compensation carrier provides regular safety inspection and safety educational materials and services.

On large projects with substantial payrolls, contractors can generally avail themselves of a retrospective premium-return plan, which, essentially, is a "cost-plus" insurance program.   With a retrospective plan, the contractor pays the cost of injuries plus a modest amount to cover the insurance carrier's administrative expense and premium against a catastrophe or multiple-injury accident.

4-34. Bonds.   Bonds are not insurance.   A surety bond is equivalent to a cosigned promissory note.   The principal on a surety bond, as on a promissory note, is primarily liable to the obligee.   The surety, as is a cosigner, is liable only in the event that the principal fails to discharge the obligation undertaken.

The obligation undertaken in a contractor's surety bond runs in favor of the owner.   And the owner, alone, is protected.   The contractor, as principal, has no protection under a bond.   On the contrary, the contractor is ultimately liable and fully obligated, not only to the owner, but also to the surety company that issued the bond.

Contractors should read in full the applications they sign for bids, performance, or payment bonds.   They will discover that they have pledged, transferred, and conveyed their entire assets and all contract revenues to the surety as security against the surety having to pay any amount or discharge any obligation under the bond.   The smaller contractor pledges not only business but home and personal assets as well.   If the contractor is an incorporated firm and its assets and income are insufficient to afford adequate security, the surety company will insist that the individual stockholders of the contracting company pledge sufficient personal assets to indemnify adequately the surety against loss.

The contractor pays a premium for a bond similar to interest on a promissory note. The premium charged depends on the type of construction to be performed, the time that the bond will be in effect, and the amount or contract price of the project to be built.

Almost all public construction and most larger private projects will require bid, performance, and payment bonds (Arts. 3-4 and 3-6). Prudent contractors, intending to submit a bid, will inquire of their surety companies whether they will write bid bonds for them. Generally, surety companies will not write a bid bond on a project without being satisfied as to the contractor's financial capacity. Once so satisfied, the surety, by issuing its bid bond, indicates its intention to write the performance and payment bonds if the contractor's bid is successful and a contract is awarded.

Bid bonds are generally based on the amount of the bid. For the most part, they run from 5 to 20% of the amount of the accompanying bid. This amount represents the damages or costs that the owner will incur if the bidder fails to enter into a contract and the work has to be readvertised for bids, or the difference in cost between the low bid submitted by a defaulting bidder and the next responsible bid, where the work must be awarded to the next lowest bidder.

Performance and payment bonds are usually in the full contract amount, or at least 50% of the contract amount. If, during the course of a project, the contractor defaults or becomes insolvent and is financially unable to carry on the work, the owner will require the surety to complete the work and to pay for labor, materials, and supplies. In such event, the surety, in discharging its obligations under the bond, has first claim as a secured creditor against the contractor's assets. Ultimately, the surety company's loss is the cost of completing the work less the recovery it can make from the contractor's assets.

4-35. Bibliography

1. Bonny, J. B.: "Handbook of Construction Management and Organization," Van Nostrand Reinhold Company, New York.

2. Clough, R. H.: "Construction Contracting," John Wiley & Sons, Inc., New York.

3. Deatherage, G. E.: "Construction Company Organization and Management," McGraw-Hill Book Company, New York.

4. Dunham, C. W., and R. D. Young: "Contracts, Specifications and Law for Engineers," McGraw-Hill Book Company, New York.

5. Havers, J., and F. W. Stubbs: "Handbook of Heavy Construction," McGraw-Hill Book Company, New York.

6. Kostro, George: "The Triangular Bar Chart," *Civil Engineering*, vol. 34, no. 11, p. 67, November, 1964.

7. Merritt, F. S.: "Building Construction Handbook," McGraw-Hill Book Company, New York.

8. Moder, J. J., and C. R. Phillips: "Project Management with CPM and PERT," Van Nostrand Reinhold Company, New York.

9. O'Brien, J. J.: "CPM in Construction Management," McGraw-Hill Book Company, New York.

10. Paul, Jack: "United States Government Contracts and Subcontracts," American Law Institute, Philadelphia, Pa.

11. Peurifoy, R. L.: "Construction Planning, Equipment and Methods," McGraw-Hill Book Company, New York.

12. Pulver, H. E.: "Construction Estimates and Costs," McGraw-Hill Book Company, New York.

Section **5**

Construction Materials

RUSSEL C. JONES*

Chairman, Department of Civil Engineering,
The Ohio State University

Part 1. CEMENTITIOUS MATERIALS

Cementitious materials include the variety of inorganic nonmetallic products that may be mixed with water or another liquid to form a paste. The paste, which is temporarily plastic and may be molded, may or may not have aggregate added to it. Later, it hardens or sets to a rigid mass.

The simple cementing materials, such as limes and plasters, are produced by driving off a liquid or gas from some natural mineral. Their cementing properties arise from the reabsorption of the liquid or gas that has been expelled and the formation of the same chemical compounds of which the original raw material was composed.

The more complex hydraulic cements derive their cementing properties from formation of new chemical compounds during the manufacturing process. The term *hydraulic* applied to cements means capable of developing strength and hardening in the presence of water.

Portland-cement concrete is the most important construction material employing a cement. Understanding of the factors affecting the constituents of concrete, the portland cement and aggregates, is essential to a fundamental understanding of the production and behavior of concrete.

PORTLAND CEMENT

5-1. Manufacture. Portland cements are made by blending a mixture of calcareous (lime-containing) materials and argillaceous (clayey) materials. The raw materials are carefully proportioned to provide the desired amounts of lime, silica, aluminum oxide, and iron oxide. After grinding to facilitate burning, the raw materials are fed into a long rotary kiln, which is maintained at a temperature around 2700° F. The raw materials, burned together, react chemically to form hard, walnut-sized pellets of a new material, clinker. This reaction is shown in Eq. (5-1), where the four major cement compounds are shown as the reaction products.

*With minor revisions for the second edition by Frederick S. Merritt, Consulting Engineer.

The clinker, after discharge from the kiln and cooling, is ground to a fine powder (not less than 1,600 sq cm per gram specific surface). During this grinding process, a retarder (usually a few percent of gypsum) is added to control the rate of setting when the cement is eventually hydrated. The resulting fine powder is portland cement.

Since portland cement is derived from unrefined raw materials, additional compounds are usually present in addition to the major, essential compounds shown in Eq. (5-1).

$$\begin{array}{l} (CaO + CO_2) \quad + \quad (SiO_2 + Al_2O_3 + Fe_2O_3 + H_2O) \; + \; heat \\ \text{(lime + carbon dioxide)} \; + \quad \text{(silica + alumina + ferric oxide + water)} \; + \; \text{heat} \\ \quad \text{(limestone)} \qquad\qquad\qquad \text{(clay)} \\ \rightarrow \; (3CaO \cdot SiO_2 \; + \; 2CaO \cdot SiO_2 \; + \; 3CaO \cdot Al_2O_3 \; + 4CaO \cdot Al_2O_3 \cdot Fe_2O_3) \\ \quad \text{(tricalcium silicate} \; + \; \text{dicalcium silicate} \; + \; \text{tricalcium aluminate} \; + \; \text{tetracalcium aluminoferrite)} \\ \qquad\qquad\qquad\qquad\qquad\qquad \text{(cement)} \end{array} \qquad (5\text{-}1)$$

Since cement is a mixture of many compounds, representation by a chemical formula is impractical. Four compounds, however, make up more than 90% of cement, by weight: tricalcium silicate (C_3S), dicalcium silicate (C_2S), tricalcium aluminate (C_3A), and tetracalcium aluminoferrite (C_4AF). Each of these four compounds is identifiable in the highly magnified microstructure of portland-cement clinker, and each has characteristic properties that it contributes to the final mixture.

5-2. Hydration of Cement.   When water is added to portland cement, the basic compounds present (Art. 5-1) are transformed to new compounds by chemical reactions [Eq. (5-2)].

Tricalcium silicate + water → tobermorite gel + calcium hydroxide
Dicalcium silicate + water → tobermorite gel + calcium hydroxide
Tetracalcium aluminoferrite + water + calcium hydroxide
$$\rightarrow \text{calcium aluminoferrite hydrate} \qquad (5\text{-}2)$$
Tricalcium aluminate + water + calcium hydroxide
$$\rightarrow \text{tetracalcium aluminate hydrate}$$
Tricalcium aluminate + water + gypsum →calcium monosulfoaluminate

Two calcium silicates, which constitute about 75% of portland cement by weight, react with the water to produce two new compounds: tobermorite gel and calcium hydroxide. In fully hydrated portland-cement paste, the calcium hydroxide accounts for 25% of the weight and the tobermorite gel makes up about 50%. The third and fourth reactions in Eqs. (5-2) show how the other two major compounds in portland cement combine with water to form reaction products. The final reaction involves gypsum, the compound added to portland cement during grinding of the clinker to control set.

Each product of the hydration reaction plays a role in the mechanical behavior of the hardened paste. The most important of these, by far, is the compound called *tobermorite gel*, which is the main cementing component of cement paste. This gel has a composition and crystal structure similar to those of a naturally occurring mineral, called tobermorite, named for the area where it was discovered, Tobermory in Scotland. The gel is an extremely finely divided substance with a coherent structure.

The average diameter of a grain of portland cement as ground from the clinker is about 10 microns. The particles of the hydration product, tobermorite gel, are on the order of a thousandth of that size. Particles of such small size can be observed only by using the magnification available in an electron microscope. The enormous surface area of the gel (about 3 million sq cm per gram) results in attractive forces between particles since atoms on each surface are attempting to complete their unsaturated bonds by adsorption. These forces cause particles of tobermorite gel to adhere to each other and to particles of aggregate introduced into the cement paste. Thus, tobermorite gel forms the heart of hardened cement paste and concrete in that it cements everything together.

Each of the four major compounds in portland cement makes a contribution to the behavior of the cement as it proceeds from the plastic to the hardened state after hydration. Knowledge of the behavior of each of the major compounds upon hydration permits the amounts of each to be adjusted during manufacture to produce desired properties in the cement.

Tricalcium silicate (C_3S) is primarily responsible for the high early strength of hydrated portland cement. It undergoes initial and final set within a few hours. The reaction of C_3S with water gives off a large quantity of heat (heat of hydration). The rate of hardening of cement paste is directly related to the heat of hydration; the faster the set, the greater the exotherm. Hydrated C_3S compound attains most of its strength in 7 days.

Dicalcium silicate (C_2S) is found in three different forms, designated alpha, beta, and gamma. Since the alpha phase is unstable at room temperature and the gamma phase shows no hardening when hydrated, only the beta phase is important in portland cement.

Beta C_2S takes several days to set. It is primarily responsible for the later-developing strength of portland-cement paste. Since the hydration reaction proceeds slowly, the heat of hydration is low. The beta C_2S compound in portland cement generally produces little strength until after 28 days, but the final strength of this compound is equivalent to that of the C_3S.

Tricalcium aluminate (C_3A) exhibits an instantaneous or flash set when hydrated. It is primarily responsible for the initial set of portland cement and gives off large amounts of heat upon hydration. The gypsum added to the portland cement during grinding in the manufacturing process combines with the C_3A to control the time to set. The C_3A compound shows little strength increase after 1 day. Although hydrated C_3A alone develops a very low strength, its presence in hydrated portland cement produces more desirable effects. An increased amount

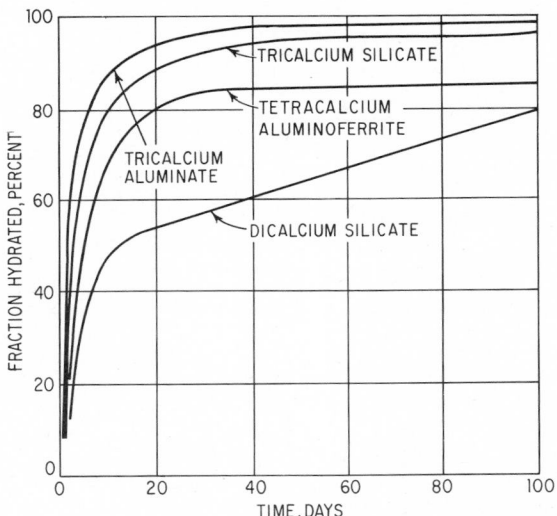

Fig. 5-1. Curves show speed of hydration of the major compounds in portland cement. (*Based on data reported by S. Brunauer and L. E. Copeland, Scientific American, April, 1964.*)

of C_3A in portland cement results in faster sets and also decreases the resistance of the final product to sulfate attack.

Tetracalcium aluminoferrite (C_4AF) is similar to C_3A in that it hydrates rapidly and develops only low strength. Unlike C_3A, however, it does not exhibit a flash set.

The speed of hydration of the four major compounds in portland cement is shown in Fig. 5-1. Since amount of heat of hydration is directly related to hydration speed, this diagram also gives an indication of the exotherm from each hydrated compound.

In addition to composition, speed of hydration is affected by fineness of grinding, amount of water added, and temperatures of the constituents at the time of mixing. To achieve faster hydration, cements are ground finer. Increased initial temperature and the presence of a sufficient amount of water also speed the reaction rate.

5-3. Types of Portland Cement. Portland cements are normally made in five types, the properties of these types being standardized on the basis of the ASTM Standard Specification for Portland Cement (C150). Distinction between the types is made on the basis of both chemical and physical requirements. Some requirements, extracted from ASTM C150, are shown in Table 5-1. Most cements will exceed the strength requirements of the specification by a comfortable margin.

Type I, general-purpose cement, is the one commonly used for structural purposes when the special properties specified for the other four types of cement are not required.

Table 5-1. Chemical and Physical Requirements for Portland Cement

Type	I	II	III	IV	V
Name	General-purpose	Modified General-purpose	High-Early-strength	Low-heat	Sulfate-resist-ing
C_3S,* max, %				35	
C_2S,* min, %				40	
C_3A,* max, %		8	15	7	5
Compressive strength, psi; mortar cubes of 1 part cement and 2.75 parts graded standard sand, by weight after:					
3 days					
Standard	1,200	1,000	3,000		
Air-entraining	900	750	2,500		
7 days					
Standard	2,100	1,800		800	1,500
Air-entraining	1,500	1,400			
28 days					
Standard				2,000	3,000

NOTE: See ASTM C150 for more detailed requirements.
*See Art. 5-2.

Type II, modified general-purpose cement, is used where a moderate exposure to sulfate attack is anticipated or where a moderate heat of hydration is required. These characteristics are attained by placing limitations on the C_3A and C_3S content of the cement (Art. 5-1). Type II cement gains strength a little more slowly than Type I but ultimately will reach equal strength.

Type III, high-early-strength cement, is designed for use when early strength is needed because of a particular construction situation. Concrete made with Type III cement develops in 3 days the same strength that it takes 28 days to develop in concretes made with Type I or Type II cement. This high early strength is achieved by increasing the C_3S and C_3A content of the cement and by finer grinding. No minimum is placed upon the fineness by specification, but a practical limit occurs when the particles are so small that minute amounts of moisture will prehydrate the cement during handling and storage. Since it has high heat evolution, Type III cement should not be used in large masses. With 15% C_3A it has poor sulfate resistance. The C_3A content may be limited to 8% to obtain moderate sulfate resistance or to 5% when high sulfate resistance is required.

Type IV, low-heat-of-hydration cement, has been developed for mass-concrete applications. If Type I cement is used in large masses that cannot lose heat by radiation, it will liberate enough heat during hydration to raise the temperature of the concrete as much as 50 or 60°F. This results in a relatively large increase in dimensions while the concrete is still plastic, and later differential cooling after hardening will cause shrinkage cracks to develop. Low heat of hydration in Type IV cement is achieved by limiting the compounds that make the greatest contribution to heat of hydration, C_3A and C_3S. Since these compounds also produce the early strength of cement paste, their limitation results in a paste that gains strength relatively slowly. The heat of hydration of Type IV cement usually is about 80% of that of Type II, 65% of that of Type I, and 55% of that of Type III during the first week of hydration. The percentages are slightly higher after about 1 year.

Type V, sulfate-resisting cement, is specified where there is extensive exposure to sulfates. Typical applications include hydraulic structures exposed to water with high alkali content and structures subjected to seawater exposure. The sulfate resistance of Type V cement is achieved by reducing the C_3A content to a minimum, since that compound is most susceptible to sulfate attack.

Typical proportions of the major compounds in the five types of portland cement are shown in Table 5-2. These average percentages were obtained by x-ray diffraction and analysis of several cements. The relative strengths of concretes made with each of the five types of cement are compared in Table 5-3 at three different ages; at each age, the strength values have been normalized for comparison with Type I concrete.

(Types IV and V are specialty cements not normally carried in dealer's stocks. They are

Table 5-2. Typical Proportions of Major Compounds in Portland Cement, Percent

Compound	Type of cement				
	I General-purpose	II Modified general-purpose	III High-early-strength	IV Low heat	V Sulfate-resisting
C_3S	53	47	58	26	38
C_2S	24	32	16	54	43
C_3A	8	3	8	2	4
C_4AF	8	12	8	12	8
Total	93	94	90	94	93

Table 5-3. Relative Strengths of Concrete as a Function of Cement Type

Types of portland cement	Compressive strength, % of Type I portland-cement concrete		
	3 days	28 days	3 months
I. General-purpose	100	100	100
II. Modified	80	85	100
III. High-early-strength	190	130	115
IV. Low-heat	50	65	90
V. Sulfate-resisting	65	65	85

usually obtainable for use on a large job if advance arrangements are made with a cement manufacturer.)

Air-entraining portland cements are available for the manufacture of concrete for exposure to severe frost action. These cements are available in Types I, II, and III but not in Types IV and V. When an air-entraining agent has been added to the cement by the manufacturer, it is designated Type IA, IIA, or IIIA.

AGGREGATES

5-4. Desirable Characteristics. Aggregates comprise about 75%, by volume, of a typical concrete mix. The term *aggregate* includes the natural sands and gravels and crushed stone used for making mortars and concretes, and also is applied to special materials used in producing light- and heavyweight concretes.

Cleanliness, soundness, strength, and particle shape are important in any aggregate. Aggregates are considered clean if they are free of excess clay, silt, mica, organic matter, chemical salts, and coated grains. An aggregate is physically sound if it retains dimensional stability under temperature or moisture change and resists weathering without decomposition. To be considered adequate in strength, an aggregate should be able to develop the full strength of the cementing matrix. When wear resistance is important, the aggregate should be hard and tough. Flat or elongated particles have a detrimental effect on the workability of concrete, resulting in the necessity of more highly sanded mixes and the consequent use of more cement and water.

Several processes have been developed for improving the quality of aggregates that do not meet desired specifications. Washing may be used to remove particle coatings or to change aggregate gradation. Heavy-media separation, using a variable-specific-gravity liquid, such as a suspension of water and finely ground magnetite and ferrosilicon, can be used to improve coarse aggregates. Deleterious lightweight material is removed by flotation, and heavyweight particles settle out. Hydraulic jigging, where lighter particles are carried upward by pulsations caused by air or rubber diaphragms, also provides a means for separation of lighter particles. Soft, friable particles can be separated from hard, elastic particles by a process called *elastic fractionation*. Aggregates are dropped onto an inclined hardened-steel surface, and their quality is measured by the distance they bounce.

5-5. Gradation.   The grading and maximum size of aggregate are important because of their effect on relative proportions, workability, economy, porosity, and shrinkage.   The particle-size distribution is determined by separation with a series of standard screens.   The standard sieves used are Nos. 4, 8, 16, 30, 50, and 100 for fine aggregate and 6, 3, 1½, ¾, and ⅜ in., and No. 4 for coarse aggregate.

Fineness modulus is an index used to describe the fineness or coarseness of aggregate.   The *F.M.* of a sand is computed by adding the cumulative percentages retained on the six standard sieves and dividing the sum by 100.   For example, the following computation shows a typical sand analysis:

Screen No.	Individual percentages retained	Cumulative percentages retained
4	1	1
8	18	19
16	20	39
30	19	58
50	18	76
100	16	92
Pan	8	
	100	285

$$F.M. = 285/100 = 2.85.$$

The fineness modulus is not an indication of grading since an infinite number of gradings will give the same value for fineness modulus.   It does, however, give a measure of the coarseness or fineness of the material.   Values of *F.M.* from 2.50 to 3.00 are normal.

Changes in sand grading over an extreme range have little effect on the compressive strengths of mortars and concretes when water-cement ratio and slump are held constant.   Such changes in sand grading, however, do cause the cement content to vary inversely with the fineness modulus of the sand.   Although this cement-content change is small, the grading of sand has a large influence on the workability and finishing quality of concrete.

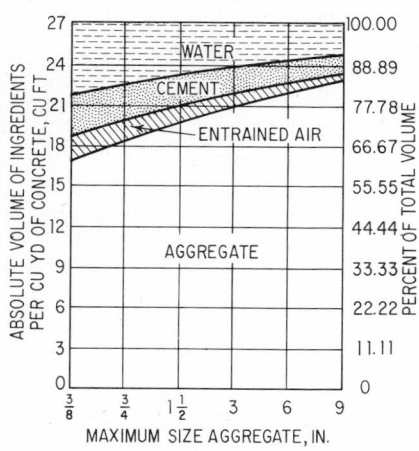

Fig. 5-2. Variations in amounts of water, cement, and entrained air in concrete mixes with maximum sizes of aggregates. The chart is based on natural aggregates of average grading in mixes with a water-cement ratio of 0.54 by weight, 3-in. slump, and recommended air contents. (*From "Concrete Manual," U.S. Bureau of Reclamation.*)

Coarse aggregate is usually graded up to the largest size practical for a job, with a normal upper limit of 6 in.   As shown in Fig. 5-2, the larger the maximum size of coarse aggregate, the less will be the water and cement required to produce concrete of a given quality.

A grading chart is useful for depicting the size distribution of aggregate particles in both the fine and coarse ranges.   Figure 5-3 illustrates grading curves for sand, gravel, and combined aggregate, showing recommended limits and typical size distributions.

5-6. Lightweight Aggregates.   Lightweight aggregates are produced by expanding clay, shale, slate, perlite, obsidian, and vermiculite with heat; by expanding blast-furnace slag through special cooling processes; from natural deposits of pumice, scoria, volcanic cinders, tuff, and diatomite; and from industrial cinders.   The strength of concrete made with lightweight aggregates is roughly proportional to its weight, which may vary from 35 to 115 lb per cu ft.   Lightweight concrete has better fire resistance and heat- and sound-insulation properties than ordinary concrete, and it offers savings in structural supports and decreased foundations due to decreased dead loads.   Structural concrete with lightweight aggregates costs 30 to 50% more, however, than that made with ordinary aggregates and has greater porosity and more drying shrinkage.

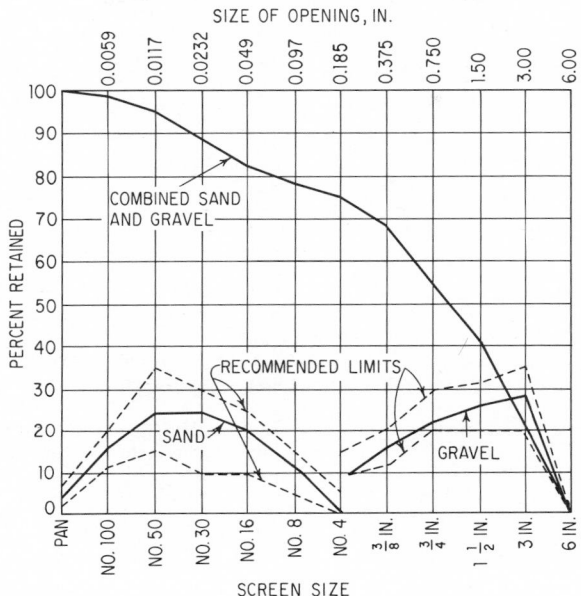

Fig. 5-3. Recommended and typical size distributions of natural aggregates for concrete mixes. Note that if No. 16 is 20% or less, No. 8 may be increased to 20%. (*From "Concrete Manual," U.S. Bureau of Reclamation.*)

Resistance to weathering is about the same for both types of concrete.   Lightweight concrete can also be made with foaming agents, such as aluminum powder, which generate a gas while the concrete is still plastic and may be expanded.

5-7. Heavy Aggregate.   In the construction of atomic reactors, large amounts of heavyweight concrete are used for shielding and structural purposes.   Heavy aggregates are used in shielding concrete because gamma-ray absorption is proportional to density.   Heavy concrete may vary between the 150 lb per cu ft weight of conventional sand-and-gravel concrete and the theoretical maximum of 384 lb per cu ft where steel shot is used as fine aggregate and steel punchings as coarse aggregate.   In addition to manufactured aggregates from iron products, various quarry products and ores, such as barite, limonite, and magnetite, have been used as heavy aggregates.

Table 5-4 shows the specific gravity of several heavy aggregates and the unit weights of concrete made with these aggregates.   Since the introduction of high-density aggregates causes difficulty in mixing and placing operations due to segregation, grouting techniques are usually used in place of conventional methods.

CONCRETE

5-8. Constituents.   Concrete is a mixture of portland cement, fine aggregate, coarse aggregate, air, and water.   It is a temporarily plastic material, which can be cast or molded, but is later converted to a solid mass by chemical reaction.   The user of concrete desires adequate strength, placeability, and durability at minimum cost.   The concrete designer may vary the proportions of the five constituents of concrete over wide limits in attaining these aims.   The principal variables are the water-cement ratio, cement-aggregate ratio, size of coarse aggregate, ratio of fine aggregate to coarse aggregate, type of cement, and use of admixtures.

Table 5-4. Heavy Aggregates for High-density Concrete

Aggregate	Specific gravity	Unit weight of concrete, lb per cu ft	
		Conventional placement	Grouting
Sand and stone		150	
Magnetite	4.30–4.34	220	346
Barite	4.20–4.31		232
Limonite	3.75–3.80		263
Ferrophosphorus	6.28–6.30	300	
Steel shot or punchings	7.50–7.78		384

Established basic relationships and laboratory tests provide guidelines for approaching optimum combinations. ACI 211.1, "Recommended Practice for Selecting Proportions for Normal-Weight Concrete," and ACI 211.2, "Recommended Practice for Selecting Proportions for Structural Lightweight Concrete," American Concrete Institute, P.O. Box 4754, Redford Station, Detroit, Mich. 48219, provide data for mix design under a wide variety of specified conditions.

5-9. Admixtures. An admixture is any material other than aggregates, portland cement, or water that is added as an ingredient of concrete immediately before or during mixing. Many concrete admixtures are available to modify, improve, or give special properties to concrete mixtures. Admixtures should be used only when they offer a needed improvement not economically attainable by adjusting the basic mixture. Since improvement of one characteristic often results in an adverse effect on other characteristics, admixtures must be used with care.

Air-entraining agents increase the resistance of concrete to frost action by introducing numerous tiny air bubbles into the hardened cement paste. These bubbles act as stress relievers for stresses induced by freezing and thawing. Air-entraining agents are usually composed of natural or synthetic soaps. In addition to increasing durability of the hardened cement, they also decrease the amount of water required and increase the workability of the mix. One unfavorable feature is the decrease in strength in the hardened concrete, which varies with the

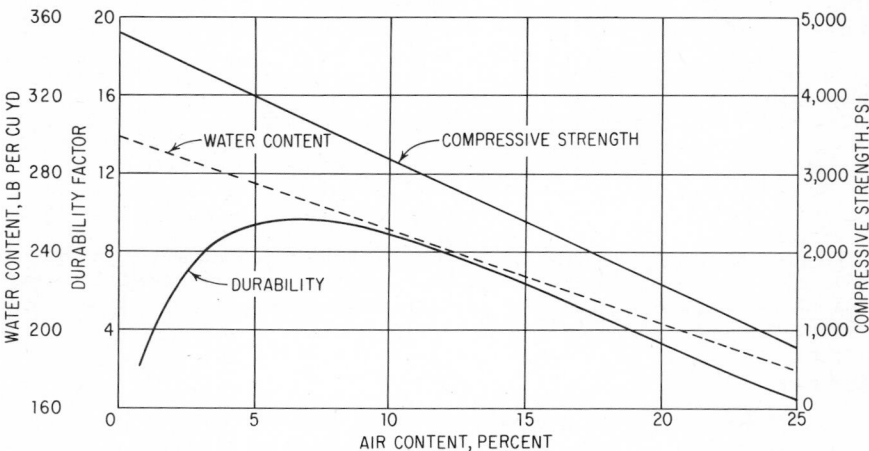

Fig. 5-4. Concrete strength, durability, and required water content vary with air content. Water-cement ratio, slump, and sand percentage held constant. Platte River aggregate, 1½ in. maximum size. (*From "Concrete Manual," U.S. Bureau of Reclamation.*)

percentage of air entrained and may be about 20% for usual mixes. The main characteristics of air entrainment in concrete are shown in Fig. 5-4. Air contents are usually controlled to between 2 and 6%.

Calcium chloride is used to accelerate the set and development of strength of concrete. It offers advantages in cold-weather concreting by speeding the set at low temperature and reducing the time that protection is necessary. When used in usual amounts (less than 2% by weight of cement) it does not, however, act as an antifreeze agent by lowering the freezing point. When 2% calcium chloride is used under normal conditions, it will reduce the initial set time from 3 hrs to 1 hr and the final set time from 6 to 2 hrs, and at 70° F it will double the 1-day strength. Use of calcium chloride as an admixture improves workability, reduces bleeding, and results in a more durable concrete surface. Problems in its use may arise from impairment of volume stability (drying shrinkage may be increased as much as 50%) and from an increase in the rate of heat liberation.

Pozzolans can be used in combination with or for partial replacement of portland cement. Several advantages can be gained through their use: better workability with less total water, cost reduction through cement savings, reduction in heat of hydration, increased resistance to sulfates, and prevention of calcium hydroxide leaching. Pozzolanic materials are siliceous substances (e.g., fly ash or pumice) that will react with lime in the presence of water. They are often used in mass-concrete applications, where the saving on cement cost and the reduction in heat liberation are particularly significant. Disadvantages of pozzolans may include slow development of final strength, increased drying shrinkage, and impaired durability.

Water-reducing and set-controlling admixtures are available as patented preparations under a variety of trade names. Dispersing agents cover each cement particle in the mix with negative or positive charges, causing the cement particles to repel each other and make it easier for the water to reach each particle. Set retardation is attributed to the formation of an alumina-silica-gel precipitate, which is deposited on the cement grains and slows the hydration process. These agents generally improve workability and may also entrain air. Setting times for cements can be increased by several hours by these agents, and the amount of mixing water needed may be reduced by 5% or more. Compressive strength may be improved by these admixtures because of the water reduction.

(Report by ACI Committee 212, "Guide for Use of Admixtures in Concrete," American Concrete Institute.)

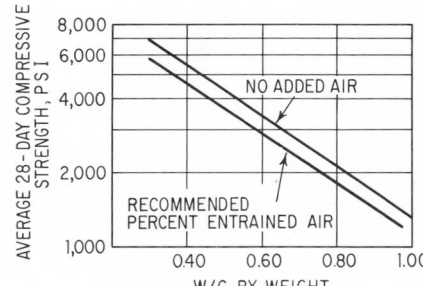

Fig. 5-5. Concrete strength decreases with increase in water-cement ratio for concrete with and without entrained air. (*From "Concrete Manual," U.S. Bureau of Reclamation.*)

5-10. Factors Affecting Strength. The term hydraulic, used in conjunction with portland cement, means that it is capable of developing strength in the presence of water. In contrast, nonhydraulic cementing materials, such as gypsum, which develops strength by losing water, have no strength in the presence of water.

Water-cement ratio is the prime factor affecting the strength of concrete. Figure 5-5 shows how W/C, expressed as a ratio by weight, affects the compressive strength for both air-entrained and non-air-entrained concrete. Strength decreases with an increase in W/C in both cases.

Cement content itself affects the strength of concrete, with strength decreasing as cement content is decreased. In air-entrained concrete, this strength decrease can be partly overcome by taking advantage of the increased workability due to air entrainment, which permits a reduction in the amount of water. Strength vs. cement-content curves for two air-entrained concretes and non-air-entrained concretes are shown in Fig. 5-6. Because of the water-reduction

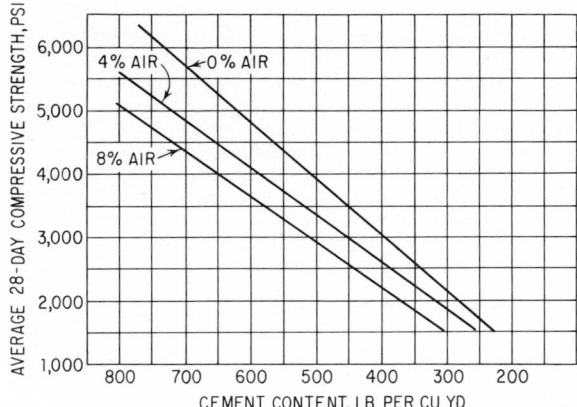

Fig. 5-6. Concrete strength increases with cement content but decreases with additions of air. Chart was drawn for concretes with ¾-in. maximum size aggregates, 43% sand, and a 3-in. maximum slump. (*From "Concrete Manual," U.S. Bureau of Reclamation.*)

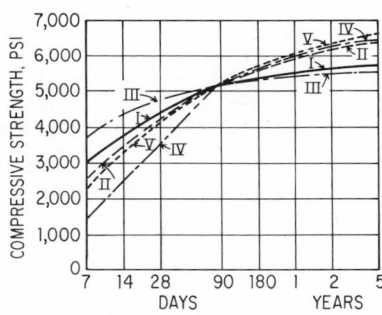

Fig. 5-7. Rates of strength development vary for concretes made with different types of cement. Tests were made on 6 × 12-in. cylinders, fog-cured at 70° F. The cylinders were made from comparable concretes containing 1½-in. maximum size aggregates and six bags of cement per cu yd. (*From "Concrete Manual," U.S. Bureau of Reclamation.*)

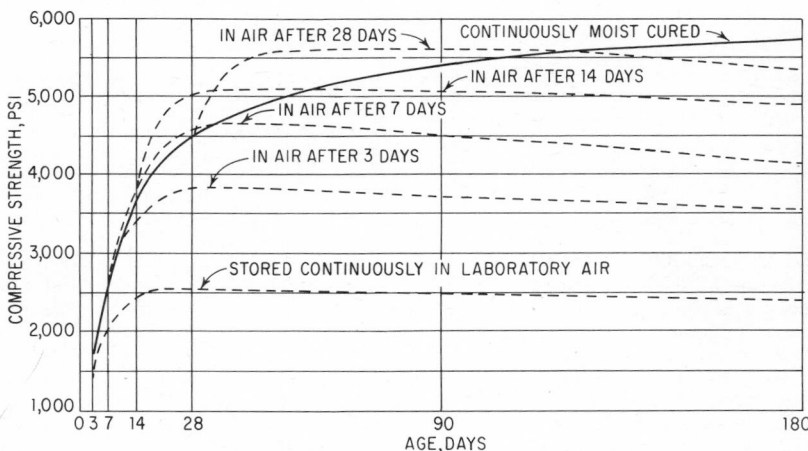

Fig. 5-8. Concrete compressive strength varies with moist-curing conditions. Mixes tested had a water-cement ratio of 0.50, slump of 3.5 in., cement content of 556 lb per cu yd, sand content of 36%, and air content of 4%. (*From "Concrete Manual," U.S. Bureau of Reclamation.*)

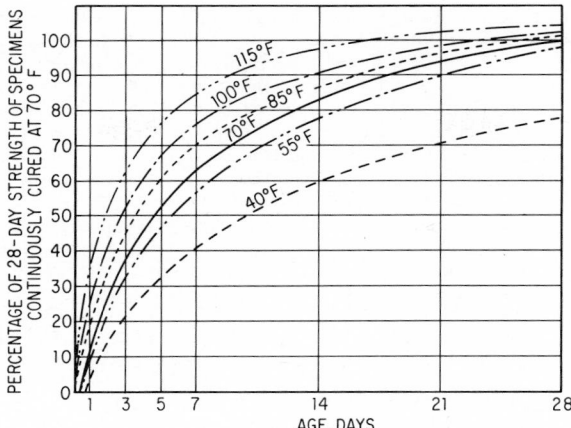

Fig. 5-9. Compressive strength of concrete at a given age increases with curing temperature between 40 and 115° F. Specimens were cast, sealed, and maintained at indicated temperatures. The concrete had a water-cement ratio of 0.50 and contained 606 lb of Type II cement per cu yd and 40% sand, no added air. (*From "Concrete Manual," U.S. Bureau of Reclamation.*)

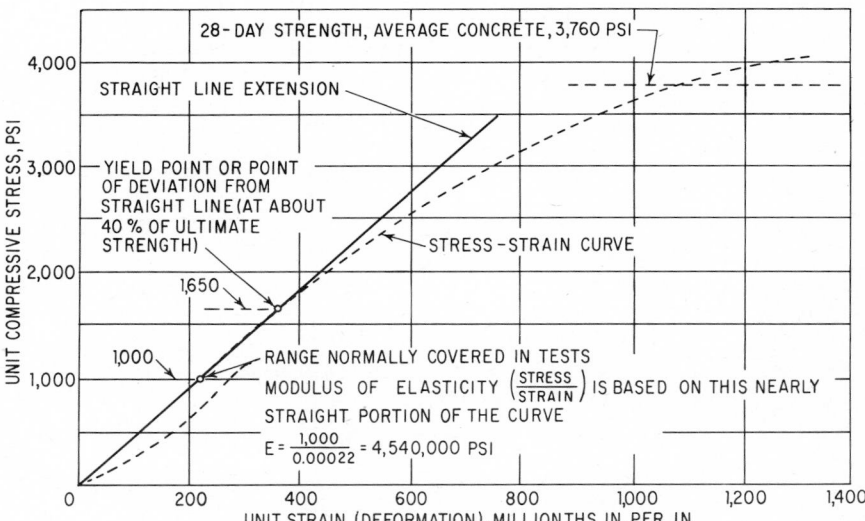

Fig. 5-10. Typical stress-strain diagram for cured concrete that has been moderately preloaded. (*From "Concrete Manual," U.S. Bureau of Reclamation.*)

possibility, the strengths of air-entrained concrete do not fall as far below those for non-air-entrained concrete as those previously indicated in Fig. 5-5.

Type of cement affects the rate at which strength develops and the final strength. Figure 5-7 shows how concretes made with each of the five types of portland cement compare when made and cured under similar conditions.

Curing conditions are vital in the development of concrete strength. Since cement-hydration reactions proceed only in the presence of an adequate amount of water, moisture must be maintained in the concrete during the curing period. Figure 5-8 shows how the strength of concrete is detrimentally affected by early transfer from a moist atmosphere to a dry one. Curing temperature also affects concrete strength, as shown in Fig. 5-9. Longer periods of moist curing are required at lower temperatures to develop a given strength.   Although continued curing at elevated temperatures results in faster strength development up to 28 days, at later ages the trend is reversed; concrete cured at lower temperatures develops higher strengths.

5-11. Properties.   Concrete is not a linearly elastic material; the stress-strain relation for continuously increasing loading plots as a curved line.   For concrete that has hardened thoroughly and has been moderately preloaded, however, the stress-strain curve is practically a straight line within the range of usual working stresses.   As shown in Fig. 5-10, a modulus of elasticity can be determined from this portion of the curve.   The elastic modulus for ordinary concretes at 28 days ranges from 2,000,000 to 6,000,000 psi.

In addition to the elastic deformation that results immediately upon application of a load to concrete, deformation continues to increase with time under a sustained load.   This plastic flow, or creep, continues for an indefinite time.   It proceeds at a continuously diminishing rate and approaches a limiting value which may be one to three times the initial elastic deformation. Although increasing creep-deformation measurements have been recorded for periods in excess of 10 years, more than half of the ultimate creep usually takes place within the first 3 months after loading.   Typical creep curves are shown in Fig. 5-11, where the effects of water-cement

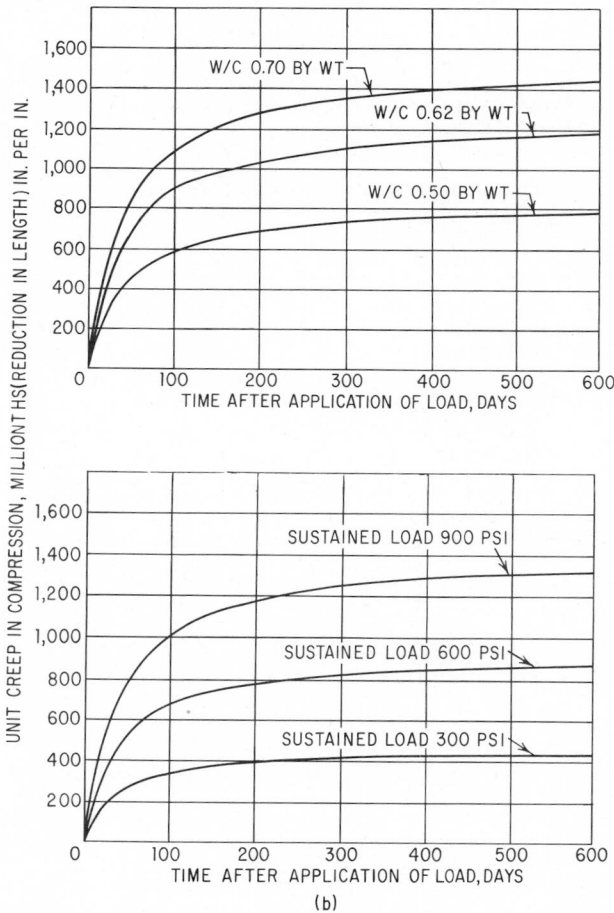

Fig. 5-11. Creep of concrete increases with increase in water-cement ratio or sustained load. (*a*) Effect of water-cement ratio on creep (applied load constant). (*b*) Effect of intensity of applied load on creep (concretes identical). (*From "Concrete Manual," U.S. Bureau of Reclamation.*)

ratio and load intensity are illustrated. Upon unloading, an immediate elastic recovery takes place, followed by a plastic recovery of lesser amount than the creep on first loading.

Volume changes play an important part in the durability of concrete. Excessive or differential volume changes can cause cracking as a result of shrinkage and insufficient tensile strength, or spalling at joints due to expansion. Swelling and shrinking of concrete occur with changes in moisture within the cement paste.

Hardened cement paste contains minute pores of molecular dimensions between particles of tobermorite gel and larger pores between aggregations of gel particles. The volume of pore space in a cement paste depends upon the initial amount of water mixed with the cement; any excess water gives rise to additional pores, which weaken the structure of the cement paste. Movements of moisture into and out of this pore system cause volume changes. The drying shrinkage of concrete is about ½ in. per 100 ft.

The thermal coefficient of expansion of concrete varies mainly with the type and amount of coarse aggregate used. The cement paste has a minor effect. An average value used for estimating is 5.5×10^{-6} in. per in. per °F.

OTHER CEMENTING MATERIALS

5-12. Hydraulic Cements. Although portland cements are the commonest of the modern hydraulic cements, several other kinds are in common use.

Aluminous cements are prepared by fusing a mixture of limestone and bauxite in an electric furnace or blast furnace, then cooling and grinding the resulting clinker. The bauxite is very high in alumina compared with the clays used in the manufacture of portland cements. High-alumina cements have a high early strength, with most of the ultimate strength attained within 24 hr. The total amount of heat of hydration is about the same as that of portland cement, but because of the speed of the chemical reaction it is developed rapidly. Concrete made with aluminous cements normally shows good resistance to attack by sea water and sulfate-bearing waters, and great resistance to high temperatures.

Natural cements are produced by calcining (making powdery by action of heat) a naturally occurring clayey limestone having approximately the proper proportions for portland cement. Calcination is carried out at a temperature no higher than that necessary to drive off the carbonic acid gas. This temperature, usually about 2370° F, is not high enough to cause any fusion. Since natural cements are derived from naturally occurring materials and no adjustment of the composition is made, the composition and properties vary widely. Tensile and compressive strengths of mortars and concretes made with natural cements are usually only one-third to one-half those obtained with standard portland cement.

White portland cement produces mortars of brilliant white color for use in architectural applications. To obtain this white color in the cement, it is necessary to use raw materials with a low iron oxide content, to use fuel free of pyrite, and to burn at a temperature above that for normal portland cement. The physical properties generally conform to the requirements of a Type I portland cement.

Hydraulic lime is made by calcining a limestone containing silica and alumina to a temperature short of incipient fusion in much the same way as quicklime is produced. In slaking (hydration), just sufficient water is provided to hydrate the free lime (CaO) formed but to leave unhydrated sufficient calcium silicates to give the dry powder its hydraulic cement properties. Because of the low silicate and high lime contents, hydraulic limes are relatively weak. They are principally used in masonry mortars.

5-13. Limes. Lime is a simple cementing material produced by driving off water from natural materials. Its cementing properties are caused by the reabsorption of the expelled water and the formation of the same chemical compounds of which the original raw material was composed.

Quicklime is the product of calcination (making powdery by heating) of limestone containing large proportions of calcium carbonate ($CaCO_3$) and some magnesium carbonate ($MgCO_3$). The calcination evaporates the water in the stone, heats the limestone to a high enough temperature for chemical dissociation, and drives off carbon dioxide as a gas, leaving the oxides of calcium and magnesium. The resulting calcium oxide (CaO), called *quicklime*, has a great affinity for water.

Quicklime intended for use in construction must first be combined with the proper amount of water to form a lime paste, a process called *slaking*. When quicklime is mixed with from two to three times its weight of water, the calcium oxide combines with the water to form calcium hydroxide, and sufficient heat is evolved to bring the entire mass to a boil.   The resulting product is a suspension of finely divided calcium hydroxide (and magnesium oxide) which, upon cooling, stiffens to a putty.   This putty, after a period of seasoning, is used principally in masonry mortar, to which it imparts workability.   It may also be used as an admixture in concrete to improve workability.

Hydrated limes are prepared from quicklimes by the addition of a limited amount of water during the manufacturing process.   Hydrated lime was developed so that greater control could be exercised over the slaking operation by having it carried out during manufacture rather than on the construction job.   After the hydration process ceases to evolve heat, a fine, dry powder is left as the resulting product.

Hydrated lime can be used in the field in the same manner as quicklime, as a putty or paste, but it does not require a long seasoning period.   It can also be mixed with sand while dry, before water is added.   Hydrated lime can be handled more easily than quicklime because it is not so sensitive to moisture.   The plasticity of mortars made with hydrated limes, although better than that obtained with most cements, is not nearly so high as that of mortars made with an equivalent amount of slaked quicklime putty.

5-14. Gypsum Cements.   Mineral gypsum, when pure, consists of crystalline calcium sulfate dihydrate ($CaSO_4 \cdot 2H_2O$).   When it is heated to temperatures above 212°F but not exceeding 374°F, three-fourths of the water of crystallization is driven off.   The resulting product, called *plaster of paris*, is a fine, white powder.   When recombined with water, it sets rapidly and attains strength on drying by reforming the original calcium sulfate dihydrate.   Plaster of paris is used as a molding or gaging plaster or is combined with fiber or sand to form a "cement" plaster.   Gypsum plasters have a strong set and gain their full strength when dry.

5-15. Oxychloride Cements.   Magnesium oxychloride cements are formed by a reaction between lightly calcined magnesium oxide (MgO) and a strong aqueous solution of magnesium chloride ($MgCl_2$).   The resulting product is a dense, hard cementing material with a crystalline structure.   This oxychloride cement, or Sorel cement, develops better bonding with aggregate than portland cement.   It is often mixed with colored aggregate in making flooring compositions or used to bond wood shavings or sawdust in making partition block or tile.   It is moderately resistant to water but should not be used under continuously wet conditions.   A similar oxychloride cement is made by mixing zinc oxide and zinc chloride.

5-16. Cementitious Materials References

1. American Society for Testing and Materials; "Annual Standards," Philadelphia.
2. Bureau of Reclamation: "Concrete Manual," Government Printing Office, Washington, D.C.
3. Gypsum Association: "Manual of Gypsum Lathing and Plastering," Evanston, Ill.
4. Lea, F. M., and C. H. Desch: "The Chemistry of Cement and Concrete," Edward Arnold (Publishers) Ltd., London.
5. Portland Cement Association: "Concrete for Hydraulic Structures," "Concrete for Massive Structures."   "Design and Control of Concrete Mixtures," Skokie, Ill.

Part 2. METALLIC MATERIALS

Regularity of atomic-level structure has made possible better understanding of the microscopic and atomic-level foundations of the mechanical properties of metals than of other kinds of materials.   Attempts to explain macroscopic behavior on the basis of micromechanisms are relatively successful for metallic materials.

ATOMIC BASIS OF BEHAVIOR

5-17. Elastic Deformation.   Metals consist of atoms bonded together in large, regular aggregations.   Energy is required to separate the atoms because the collective potential energy of the aggregation is lower than the sum of the potential energies of the separate atoms if they were at great distances from each other.

Metallic bonds between atoms are due to the sharing of electrons in unsaturated covalent bonds.   In regular covalent bonds, when atoms with incomplete outer electron shells come close together, the unpaired electrons of each atom tend to pair up with similarly unpaired electrons in the other atoms.   If the number of electrons in the outer shell is much smaller than the number of states available in the shell, however, each atom forms electron pairs with its neighbors in a statistical sense only.   This kind of bond, found to occur only between metal atoms, is called an *unsaturated covalent bond,* or simply a *metallic bond.*

Because of this unsaturated type of bonding, each metal atom tends to surround itself with as many similar atoms as it can.   This leads to close-packed arrays of atoms in metallic structures. The three usual types of packing in metals, given names descriptive of the short-range order of the atomic structure, are *face-centered cubic* (fcc), *body-centered cubic* (bcc), and *hexagonal close-packed* (hcp).

The elastic behavior of metallic materials under limited loadings can be explained in terms of interatomic bonding.   Consider two atoms in their ground states, infinitely far apart. Potential energy is inversely proportional to some power of the distance of separation; so the potential energy of interaction is zero at this initial spacing.   Since the atoms have electrical charges, they either attract or repel each other as they approach.   The potential energy due to attraction is defined as negative since the atoms do the work of attraction.   Repulsive energy is considered positive because external work must be done to bring two atoms together.   The total potential energy, shown in Fig. 5-12a, is the sum of these two terms.

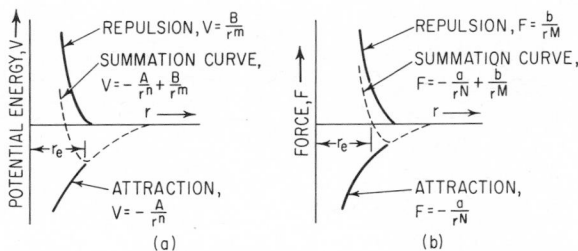

Fig. 5-12. Variation of (a) energy and (b) forces of attraction and repulsion with distance between atoms.

Since force is the derivative of potential energy, the forces of atomic interaction can be derived. The force vs. interatomic spacing relationships are shown in Fig. 5-12b.   Attractive forces are essentially due to the attraction of negative and positive charges on the two atoms.   At a separation of a few atomic diameters, the repulsive forces between the like charges of the atomic nuclei start to assert themselves.   At the equilibrium separation r_e, the forces of attraction just equal the forces of repulsion, and the potential energy is at a minimum.   If the atoms try to move closer than r_e, the repulsive force increases much more rapidly than the attractive force as the electron clouds begin to overlap.   If the atoms are pulled apart to a separation slightly greater than r_e, when released, they tend to go back to the equilibrium spacing, at which the potential energy is a minimum.

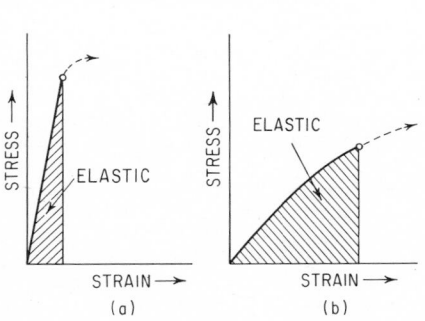

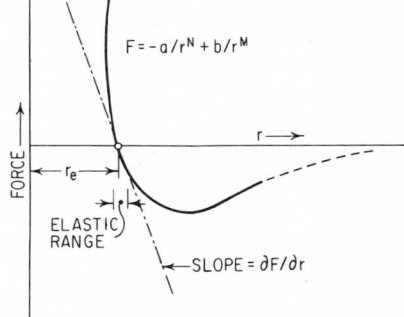

Fig. 5-13. Stress-strain diagram for metals may be (a) linear or (b) nonlinear elastic. Metals recover shape when returned to original stress state when stressed within the elastic range.

Fig. 5-14. Variation of force with distance between atoms is nearly linear in the region of equilibrium separation, providing the basis for linear elastic behavior of metals.

The deformation of materials under applied load is elastic if the change in shape is entirely recovered when the material is returned to its original stress state.   Elastic load-deformation relationships may or may not be linear, as shown in Fig. 5-13, but many metals behave linearly.

The shape of the stress-strain curve for small elastic deformations is based on the shape of the summation curve for interatomic force vs. interatomic spacing, which is redrawn in Fig. 5-14. If the summation curve and its tangent are practically coincident over the range of elastic strains, stress may be considered proportional to strain in the elastic range.   This is usually the case for metals and other crystalline materials.   The macroscopic modulus of elasticity thus has its basis in the limited stretching of the atomic bonds when the force vs. interatomic spacing curve is essentially linear near the equilibrium atomic spacing.   Strongly bonded materials exhibit higher elastic moduli than do weakly bonded materials.

5-18. Yield and Plastic Deformation.   Ductile crystalline materials often fail by the slip of adjacent planes of atoms over each other.   This mode of failure occurs when the resolved shear stress on some slip plane reaches a critical value before any possible brittle-fracture mode has been activated.   If the shear stress to move one plane of atoms past another plane could be computed from atomic-bonding considerations, the strength of a material under a given external loading system could be predicted.

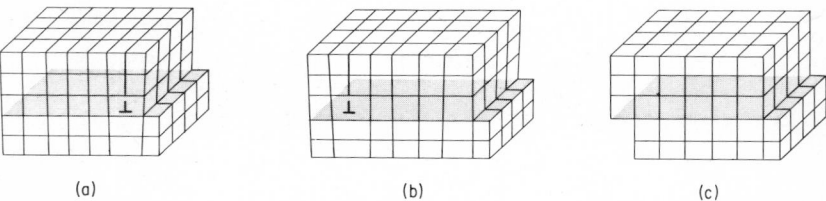

(a)   (b)   (c)

Fig. 5-15. Slip resulting from the movement of a pure edge dislocation through a crystalline lattice. (a) Dislocation (inverted T) just introduced. (b) Dislocation moving across lattice by stepwise breaking and reforming of bonds. (c) Permanent deformation left as dislocation sweeps entirely across slip plane.

Slip on atomic planes actually proceeds in a stepwise manner, not by the gross slipping of whole atomic planes over each other.   This stepwise slip is described in terms of *dislocations*, which are imperfections in the crystalline lattice at the atomic scale.   A *pure edge dislocation* is the discontinuity at the end of an extra half plane of atoms inserted in the crystal lattice, as represented by an inverted T in Fig. 5-15a.   Such discontinuities arise as accidents of growth during crystal formation and as a result of multiplication mechanisms during deformation. Under applied loading, an edge dislocation moves across the slip plane in a stepwise manner, breaking and reforming bonds as it moves.   As shown in Fig. 5-15b and c, this movement results in plastic deformation equivalent to the sliding of one whole plane of atoms across another by one atomic dimension.   This dislocation mechanism is the one by which yield begins in metals and by which plastic deformation continues.

A second type of pure dislocation, known as a *screw dislocation*, is associated with shear deformations in crystalline structures.   In general, dislocations in real crystalline lattices are mixed dislocations with both edge and screw components.

The elastic portion of a stress-strain curve, based on bond stretching at the atomic scale (Art. 5-17), ends with the onset of plastic deformation at the yield point.   Yielding is associated with the irreversible movement of dislocations with which plastic straining begins.   Beyond the yield point the material no longer returns to exactly its initial state with load removal; some plastic deformation remains.

A dislocation is surrounded by an elastic stress field that results in forces between dislocations and in interactions with other irregularities in the crystalline structure.   The general effect of the interaction of dislocations with each other and with other obstacles after yielding is a work hardening of the material, that is, an increase in the stress required to continue plastic deformation.   This arises from the increased difficulty of moving dislocations, with their surrounding stress fields, through the stress fields of other irregularities in the crystalline lattice.

Metals can be strengthened if ways can be found to keep dislocations from beginning to move or if obstacles to the movement can slow or stop them once the dislocations have begun to move. In addition to the strain hardening that results from interactions of moving dislocations, other means may be used to strengthen metals at the atomic level.

STRENGTHENING MECHANISMS

5-19. Cold Working. Plastic deformation in metals is characterized by a phenomenon known as *strain hardening* (Art. 5-18). When metals are deformed beyond the elastic limit, a permanent change in shape occurs. If a metal is loaded beyond its yield point, unloaded, then loaded again, the elastic limit is raised. This phenomenon, represented in Fig. 5-16, indicates that a metal can be strengthened by deformation previous to its loading in an engineering application.

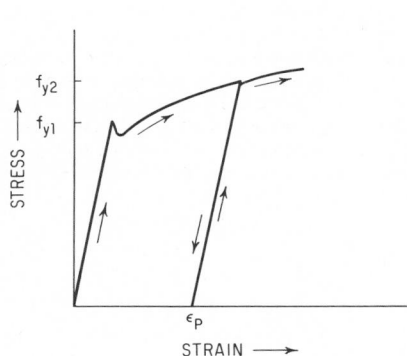

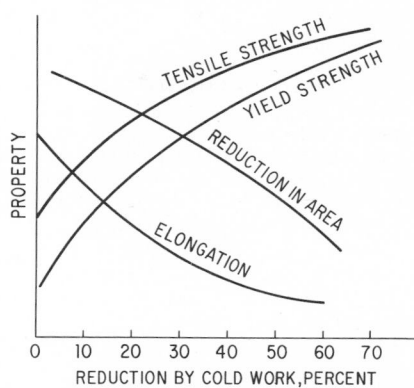

Fig. 5-16. Stress-strain curve for metal stressed beyond the elastic limit, unloaded, and reloaded. Yield stress on second loading is higher than on first.

Fig. 5-17. Variation of tensile properties with amount of cold work. (*From G. E. Dieter, "Mechanical Metallurgy," McGraw-Hill Book Company, New York.*)

Strain hardening is caused by dislocations interacting with each other and with barriers to their motion through the crystal lattice. Because of the operation of dislocation-multiplication mechanisms, the number of dislocations present increases from 10^6 to 10^8 per sq cm to about 10^{12} per sq cm during straining. This increase in the number of dislocations means that more obstacles are present to impede additional plastic deformation.

Dislocations piling up at obstacles on the slip plane cause strain hardening due to a back stress opposing the applied stress. The obstacles at which dislocations may be blocked during plastic deformation include foreign atoms in the lattice, precipitate particles, intersection of slip planes where dislocations combine to block each other, and grain boundaries.

Plastic deformation that is carried out in a temperature range and over a time interval such that the strain hardening is not relieved is called *cold work*. Cold working is employed to harden and strengthen metals and alloys that do not respond to heat treatment. Figure 5-17 shows the variation of tensile properties of a metal with amount of cold work. Note that although the strength increases considerably, the ductility, as measured by elongation, decreases greatly.

5-20. Solid-solution Hardening. Strengthening produced by dispersed, atomic-size lattice defects in a metal is referred to as solid-solution hardening. Substitutional and interstitial impurity atoms are the commonest varieties of such defects. Whenever a dislocation (Art. 5-18) encounters an irregularity within a crystal lattice, hardening occurs.

Solute atoms introduced into solid solution in a pure metal produce an alloy stronger than the original metal. If the solute and solvent atoms are roughly similar, the solute atoms will occupy lattice points in the crystal lattice of the solvent atoms. This forms a substitutional solid solution. If the solute atoms are considerably smaller than the solvent atoms, they occupy interstitial positions in the solvent lattice. Such elements as carbon, nitrogen, oxygen, hydrogen, and boron commonly form such interstitial solid solutions.

Solute atoms are generally not randomly distributed throughout the solvent lattice. They tend to group preferentially at dislocations, stacking faults, and grain boundaries. These solute atoms form "atmospheres" around dislocations by interacting with the stress field of the dislocations. The atoms above a positive edge dislocation (Fig. 5-15a) are compressed (i.e., in the region where the extra half plane has been inserted), and the atoms below the slip plane are in a tensile region. The strain energy of distortion of the crystal is reduced by large atoms collecting in the expanded region and small atoms collecting in the compressed region. Interstitial atoms also can collect in the expanded region below the slip plane of a positive edge dislocation to reduce further the strain energy of the system. Once this atmosphere has formed around a dislocation, the local energy is lower than it would be if the dislocation were forced out into a crystal region away from the solute atoms. Thus, a higher stress will be required to make the dislocation move than would be required if there were no interaction between the dislocation and the solute atoms.

If the applied stress becomes high enough, the dislocation is torn away from its atmosphere, and it is then free to move at a lower stress. The high upper yield point in carbon steel, followed by flow at a lower yield stress, is associated with this type of dislocation interaction with solute atoms.

Other types of interactions between solute atoms and dislocations are more complex and less fully understood. Instead of a dislocation being attracted to solute atoms and becoming difficult to pull away, a dislocation moved through a lattice may be repelled by solute atoms, and so extra work must be done to pass them. Because of the valency effects in solid solution, electrical interactions must also be considered. It appears, though, that the strengthening due to electrical interactions is only one-sixth to one-third as much as that due to elastic interaction between solute atoms and dislocations.

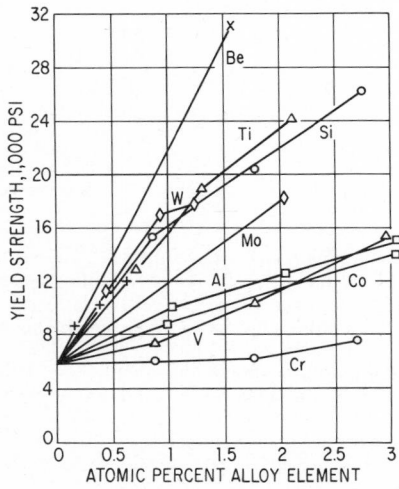

Fig. 5-18. Yield strength of iron increases because of solid-solution alloy additions of chromium (Cr), vanadium (V), cobalt (Co), aluminum (Al), molybdenum (Mo), silicon (Si), tungsten (W), titanium (Ti), and beryllium (Be). (*From C. E. Lacy and M. Gensamer, Transactions of the American Society for Metals, vol. 32, p. 88, 1944.*)

Studies of the effect of solid-solution alloying additions on strength have been made for most of the metal alloys of engineering interest. The increase of strength of iron due to the addition of alloying elements, which form solid solutions, is shown in Fig. 5-18 for several different alloying elements.

5-21. Precipitation Hardening. Dispersion hardening is the strengthening produced by a finely dispersed insoluble second phase in a matrix of metal atoms. These second-phase particles act as obstacles to the movement of dislocations (Art. 5-18). Thus, higher stresses are required to cause plastic deformation when dislocations must overcome these obstacles to move across slip planes.

Since a dislocation has a strain energy of lattice distortion associated with it per unit length, any increase in length of the dislocation line due to bending of it requires expenditure of extra energy. Figure 5-19a and b shows how a dislocation line is bent as it approaches two precipitate particles. In Fig. 5-19c, the line is bent to a critical radius as it is partly restrained by the precipitates. Finally, the line passes through the obstacle area (Fig. 5-19d and e).

Note that small loops of dislocation are left around each particle as the dislocation line breaks through. This makes it even more difficult for additional dislocations to overcome the same obstacles as deformation continues. The stress needed to force a dislocation line between the obstacles is inversely proportional to the spacing between them. Thus, the finer and more closely spaced the dispersed particles, the greater the stress required to cause deformation.

One method of producing this type of strengthening, precipitation hardening, or age hardening is by a heat-treatment process. In an alloy such as copper-aluminum, a greater amount of the alloying element can be put into solid solution at an elevated temperature than at room temperature. If the temperature is reduced, a supersaturation of alloying atoms results. If the solid solution is cooled slowly, the excess solute atoms will leave the solution by migrating to areas of disorder, such as grain boundaries, and forming large precipitates. Because of slow cooling, enough diffusion takes place that large precipitates that are not spaced closely enough to be effective in strengthening are formed. If rapid cooling follows the solutionizing treatment, however, the excess alloying atoms are retained in solid solution. In such a rapid quench, there is no time for diffusion to the grain boundaries to occur. Once the supersaturated solid solution exists at room temperature, it may be aged at room temperature or some slightly elevated temperature to allow precipitates to form on a very fine scale throughout the host metal. These fine precipitate particles are effective in blocking dislocation movement and thus strengthen and harden the metal. Figure 5-20 shows how the properties of an aluminum alloy change during a precipitation heat treatment.

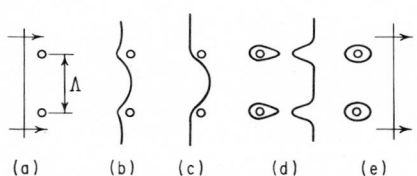

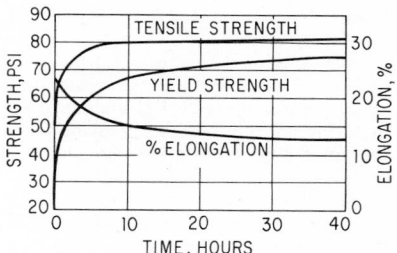

Fig. 5-19. Passage of a dislocation through obstacles in a dispersion-hardened matrix. (a) As dislocation line approaches particles, it bends more and more (b) and (c). After it breaks through (d) and (e), it leaves small loops of dislocations around the particles. (*After Orowan.*)

Fig. 5-20. Changes in mechanical properties during precipitation heat treatment of 7076 aluminum alloy at 250° F. (*After Dix.*)

A continuation of the process of local segregation of alloying atoms over a long time leads to overaging, or softening. The continued growth of precipitates, in which small, closely spaced areas combine through diffusion to produce large precipitates, leaves a structure with less resistance to dislocation movement.

5-22. Grain Size. Although single crystals of metals are specially grown for research investigations, commercial grades of metals are polycrystalline materials. Each grain in a polycrystalline metal is a small volume of atoms stacked in such a way that the atomic planes are essentially parallel. Each grain has an orientation quite different from that of neighboring grains. The interfaces between individual grains, called *grain boundaries*, are areas of great atomic misfit. Because of changes in orientation and the disruption of regular atomic structure at grain boundaries, dislocations are greatly inhibited in their motion at these areas. The more numerous the grain boundaries, the higher will be the strength of the metal.

Decreasing the average size of the grains in a polycrystalline metal increases the strength by increasing the number of grain-boundary obstacles to dislocation movement. Yield strength can be related to the reciprocal square root of grain size:

$$f_y = f_i + kl^{-1/2} \tag{5-3}$$

where f_y = yield stress
l = grain size
f_i and k = constants

This type of relationship, with the constants evaluated in each case, is valid for most metals. Grain size can be controlled by the heating and rolling operations in the production of structural metals.

STRUCTURAL STEELS

5-23. Classifications. High-strength steels are used in many civil engineering projects. New steels are generally introduced under trademarks by their producers, but a brief check into their composition, heat treatment, and properties will normally allow them to be related to other existing materials. Following are some working classifications that allow comparison of new products with those that have been standardized.

General classifications allow the currently available structural steels to be grouped into four major categories, some of which have further subcategories. The steels that rely on carbon as the main alloying element are called *structural carbon steels*. The older grades in this category were the workhorse steels of the construction industry for many years, and the newer, improved carbon steels still account for the bulk of structural tonnage.

Two subcategories can be grouped together in the general classification *low-alloy carbon steels*. To develop higher strengths than ordinary carbon steels, the low-alloy steels contain moderate proportions of one or more alloying elements in addition to carbon. The *columbium-vanadium-bearing* steels are higher-yield-strength metals produced by addition of small amounts of these two elements to low-carbon steels.

There are two kinds of *heat-treated steels* on the market for construction applications. *Heat-treated carbon* steels are available in either normalized or quenched-and-tempered condition, both relying essentially on carbon alone for strengthening. *Heat-treated constructional alloy* steels are quenched and tempered steels containing moderate amounts of alloying elements in addition to carbon.

Another general category, *maraging steels*, consists of high-nickel alloys containing little carbon. These alloys are heat-treated to age the iron-nickel martensite. Maraging steels are unique in that they are the first constructional-grade steels that are essentially carbon-free. They rely entirely on other alloying elements to develop their high strength. This class of steels probably represents the opening of a door to the development of a whole field of carbon-free alloys.

ASTM specification designations are usually used to classify the structural steels that have been in use long enough to be codified (Table 5-5). These specifications cover production variables, such as process, chemical content, and heat treatment, as well as performance minima in tensile and hardness properties.

Table 5-5. Classification of Structural Steels by ASTM Number and Yield Strength

General classification	ASTM No.	Yield strength,* ksi
Structural carbon steels	A36	36
High-strength low-alloy steels	A440	
	A441	42–50
	A242	
	A588	
Columbium-vanadium-bearing steels	A572	42–65
Heat-treated carbon steels	None yet	50–80
Heat-treated constructional alloy steels	A514	
	A517	90–100
	A633	
Maraging steels	None yet	200–300

*Yield strength is generally designated as a function of the thickness of the section. These values of yield strength are for minimum section sizes.

Yield-strength grouping can be developed as another classification scheme (Table 5-5). Five general groupings can be made, based mainly upon order of appearance:

36 ksi. This is the basic structural carbon steel for use in riveted, bolted, or welded construction of bridges and buildings and for general structural purposes.

42 to 65 ksi. The high-strength low-alloy steels were developed in the 1930s to push minimum yield values above those available with structural carbon steels.

50 to 80 ksi. Steels developed in the late 1950s and early 1960s for filling the gap between the 60- and 100-ksi yield strengths.

90 to 100 ksi. Heat-treated alloy steels introduced in the early 1950s for constructional use.

200 to 300 ksi. The currently fertile area of development where the steel industry is competing with other kinds of materials in the race to produce higher strength-to-weight ratio materials.

See also Art. 5-25.

Chemical-content comparison of carbon and other alloying elements can be used to distinguish one structural steel from another. Most structural steels, except for the maraging steels, contain carbon in amounts between 0.10 and 0.28%. The older steels have few alloying elements and are usually classified as carbon steels. Steels containing moderate amounts of alloying elements, with less than about 2% of any one constituent element, are called *low-alloy steels*. Steels containing larger percentages of alloying elements, such as the 18% nickel maraging steels, are designated as *high-alloy steels*. Specified chemical compositions of the codified structural steels are listed in ASTM specifications. Typical chemical compositions of other structural steels are available from steel producers.

A basic numbering system sometimes is used to describe the carbon and alloy content of steels. In the American Iron and Steel Institute numbering system for low-alloy steels, the first two numbers indicate the alloy content and the last two numbers indicate the nominal carbon content in units of 0.01%. A brief description of some AISI steels is given in Table 5-6. Complete listings of AISI steels, with composition limits and hardenability bands, can be found

Table 5-6. AISI Numbering System for Some Low-alloy Steels

Alloy steel	AISI No.	Average alloy content
Plain carbon steels	10xx	None
Manganese steels	13xx	1.75% Mn
Nickel-chromium steels	31xx	1.25% Ni, 0.65% Cr
Molybdenum steels	40xx	0.25% Mo
Chromium-molybdenum steels	41xx	0.9% Cr, 0.2% Mo
Nickel-chromium-molybdenum steels	43xx	1.8% Ni, 0.8% Cr, 0.25% Mo
	86xx	0.55% Ni, 0.5% Cr, 0.2% Mo
	87xx	0.55% Ni, 0.5% Cr, 0.25% Mo
Nickel-molybdenum steels	46xx	1.8% Ni, 0.25% Mo
Chromium steels	51xx	0.9% Cr
Chromium-vanadium steels	61xx	0.9% Cr, 0.15% V
Silicon-manganese steels	92xx	2.0% Si, 0.85% Mn
Boron-intensified steels	xxBxx	0.0005% B min
Electric-furnace steels	Exxxx	
Hardenable steels	xxxxH	Subject to controlled hardenability-band limits

in vol. 1 of "Metals Handbook" (American Society for Metals). A typical AISI designation, 4130, has the following meaning:

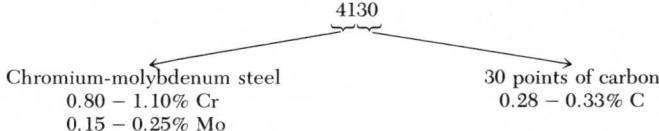

Chromium-molybdenum steel	30 points of carbon
0.80 − 1.10% Cr	0.28 − 0.33% C
0.15 − 0.25% Mo	

Also specified are: 0.40 to 0.60% Mn, 0.040% P maximum, 0.040% S maximum, 0.20 to 0.35% Si.

Heat treatment can be used as another means of classification. The older structural carbon steels and high-strength low-alloy steels are not specially heat-treated, but their properties are controlled by the hot-rolling process. The heat-treated, constructional alloy and carbon steels rely on a quenching and tempering process for development of their high-strength properties. The ASTM A514 steels are heat-treated by quenching in water or oil from not less than 1650°F and then tempering at not less than 1100°F. The heat-treated carbon steels are subjected to a similar quenching and tempering sequence: austenizing, water quenching, and then tempering at temperatures between 1000 and 1300°F.

The typical heat treatment of the maraging steels involves annealing at 1500°F for 1 hr, air-cooling to room temperature, and then aging at 900°F for 3 hr. The aging treatment in the maraging steels may be varied to obtain different strength levels. Figure 5-21 shows how heat treatment and other classifications interrelate.

5-24. Microstructure. Mechanical properties observed and measured at the macroscopic scale are based upon the constituent microstructure of the steel. Although there are variations

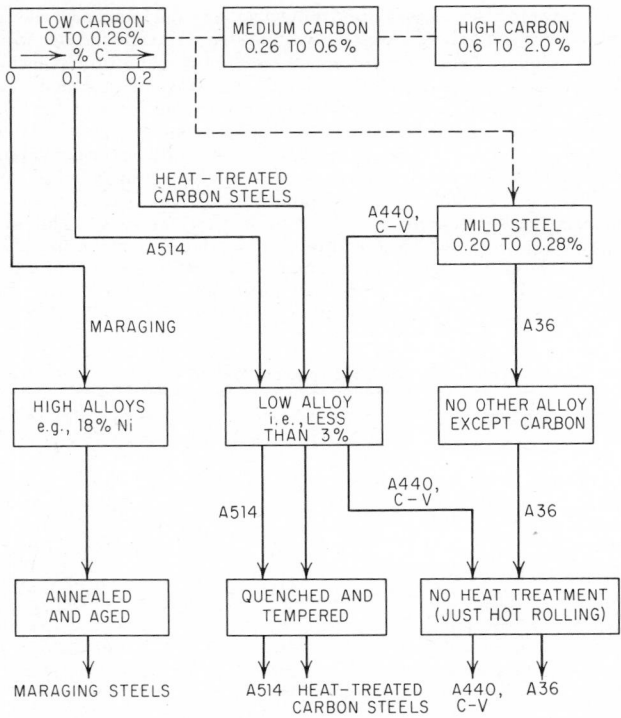

Fig. 5-21. General classifications of structural steel with respect to alloy content and heat treatment. (*After Krokosky.*)

in the details of microstructure of a particular type of steel as the chemical composition and heat treatment vary within allowable limits, general characteristics of microstructure can be described for each of the broad classifications of structural steels (Art. 5-23).

If steel is cooled very slowly from its high temperature or molten condition to room temperature, it takes a characteristic form depending on the percentage of carbon present in the iron matrix. The forms present at any temperature and composition are readily displayed on the iron-carbon diagram shown in Fig. 5-22. This is an equilibrium diagram; it represents the situation for a given temperature and composition only if sufficient time has elapsed for the material to reach thermodynamic equilibrium. In many structural steels, nonequilibrium structures are purposely produced to obtain desired mechanical properties.

The structure of iron is different in each of its phases, just as ice, water, and steam have different structures in their respective stable temperature ranges. Ferrite, or alpha iron, is the body-centered-cubic structure iron found at room temperature. Ferrite has a low solubility of carbon since the carbon atom is too small for a substitutional solid solution and too large for an extensive interstitial solid solution (see Art. 5-20). Austenite, or gamma iron, is the face-centered-cubic modification of iron that is stable between 1670 and 2550° F. (These temperatures are for no carbon. See Fig. 5-22 for the entire range of stability of the gamma phase.) The face-centered-cubic structure has larger interstices than the ferrite, and hence can have more carbon in the structure. The maximum solubility is 2% carbon by weight. Delta iron is the body-centered-cubic form of iron that is stable above 2550° F. The relative solubilities of carbon in the iron matrix play an important part in the nonequilibrium structures that result from certain heat treatments of steel.

The combination of iron and carbon represented by the vertical line at 6.67% carbon content in Fig. 5-22 is called *cementite* (or Fe_3C, iron carbide). Carbon in excess of the solubility limit in iron forms this second phase, in which the crystal lattice contains iron and carbon atoms

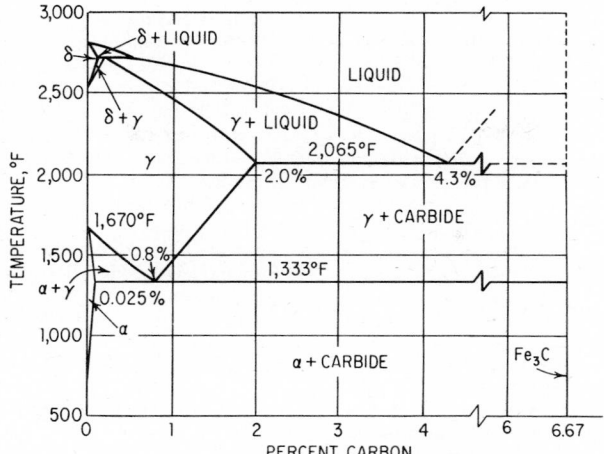

Fig. 5-22. Iron-carbon equilibrium diagram.

in a 3:1 ratio. The iron-carbon eutectoid reaction, occurring as a dip in Fig. 5-22 at 0.8% carbon, involves the simultaneous formation of ferrite and carbide from austenite of eutectoid composition. Since the ferrite and Fe_3C form simultaneously, they are intimately mixed. The mixture, called *pearlite*, is a lamellar structure composed of alternate layers of ferrite and carbide.

The nonequilibrium structures produced by heat treatment can be represented on a time-temperature-transformation (TTT) plot. A typical TTT curve for a 1080 steel is shown in Fig. 5-23. When the temperature is decreased below the point where the gamma phase (austenite) is stable, there is a driving force for transformation to the body-centered-cubic alpha phase (ferrite). This transformation takes some time, as shown on the TTT curve, and the time and temperature path followed determines the kind of structure formed.

If the temperature is maintained just below the transformation temperature, a coarse pearlite is formed because of high diffusion rates, which allow the excess carbon atoms to combine into large areas of Fe_3C. At somewhat lower temperatures, where diffusion rates are not so high, a fine pearlite is formed. If the unstable austenite is cooled quickly enough to prevent diffusion, the carbon present remains in solution instead of segregating out as a carbide. The resulting body-centered structure is tetragonal rather than cubic because of the strain in the lattice due to the excess carbon atoms. Since no diffusion occurs in the formation of this structure, which is called *martensite* (M in Fig. 5-23), there is essentially no time lag for this reaction.

The start of the martensitic transformation is labeled M_s, and the finish M_f. Martensite is metastable, and its existence does not alter the validity of the iron-carbon equilibrium diagram. With sufficient time at temperatures below the eutectoid temperature, the supersaturated

Fig. 5-23. Isothermal-transformation curve for a plain carbon (0.80%) steel.

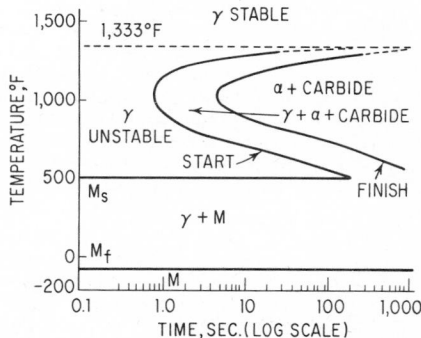

solution of carbon in iron transforms to an alpha-plus-carbide mixture called tempered martensite. The resulting microstructure is not lamellar like that of pearlite.

The rapid quenching of austenite to miss the "nose" on the TTT curve to form martensite is an important step in the heat treatment of steels. The ensuing tempering at somewhat elevated temperatures produces steels of good toughness and high strength for construction applications.

TTT curves are also called *isothermal transformation* (IT) curves because of the way they are produced. Such curves are obtained by heating small samples into the austenite temperature range long enough for complete transformation, then quenching to various lower temperatures and holding. Samples are then quenched to room temperature at various times, and the stages of transformation noted. Although the IT diagram is produced by observation of isothermal transformations, it is often used as an indication of results to be expected from nonisothermal transformations. The "Atlas of Isothermal Transformation Diagrams" (U.S. Steel Corp.) is a useful compilation of IT diagrams for a wide variety of steels.

Structural carbon steels contain about 0.2% carbon, an amount greater than that which can be dissolved in body-centered-cubic ferrite at room temperature. Little heat treatment is used with these steels, with control over the microstructure achieved by chemical composition and hot-rolling practice. Structural shapes are usually subjected to a low-temperature hot-rolling process, which results in a small, uniform grain size. Upon cooling, the final product is a fine ferrite plus pearlite (a lamellar aggregate of ferrite and iron carbide) structure.

High-strength low-alloy steels derive their strength increase from a finer microstructure and from solid-solution strengthening (Art. 5-20). Alloying elements delay the transformation of the austenite to pearlite and contribute elements that go into solution in the ferrite. This solid-solutioning strengthens the ferrite.

Heat-treated carbon steels are subjected to a water quench from the austenite phase. The resulting low-temperature transformation products (martensite) are high in strength but very brittle. Tempering at about 1200°F leads to improved toughness and ductility, with little loss in yield strength. This tempering results in the formation of a uniform structure consisting of a dense dispersion of carbides in a ferrite matrix.

Heat-treated constructional alloys are usually tempered martensitic structures. The M_s (martensitic transformation temperature) is about 700°F for these steels. The presence of alloying elements pushes back the nose of the isothermal transformation curve, thus allowing for more complete hardening. These steels are tempered at about 1200°F, at which temperature the carbide-forming elements present (C, V, Mo) assist in the formation of various stable alloy carbides. The alloy carbides form a fine dispersion, strengthening the steel by dispersion hardening (Art. 5-21).

Maraging steels are still being studied to determine the strengthening mechanism. Present evidence indicates that the strengthening is due to formation of a finely dispersed nickel-based precipitate. During the aging process in 18% nickel maraging steels, extremely fine particles form on dislocation sites. These precipitates are responsible for the extremely high strength of the maraging steels. The difference in mechanical behavior between these nickel-based precipitates and the carbide precipitates found in heat-treated carbon steels seems to account for the superior toughness of the maraging steels.

5-25. Mechanical Properties. Tensile strength of structural steels generally lies between about 60 and 80 ksi for the carbon and low-alloy grades and between 105 and 135 ksi for the quenched-and-tempered alloy steels (A514). Yield strengths are listed in Table 5-5. Elongation in 2 in., a measure of ductility, generally exceeds 20%, except for A514 steels. Modulus of elasticity usually is close to 29,000 ksi.

The high-strength low-alloy steels have become almost as important in construction as the carbon steels. The A242 series, in addition to having a yield strength considerably higher than the structural carbon steels, also have four to six times the corrosion resistance of A36 carbon steel without copper. A441 is a manganese-vanadium steel with 0.20% minimum copper content and is intended primarily for welded construction. It has about twice the corrosion resistance of carbon steels. A588 steels have similar properties, but different chemistry makes possible a 50-ksi yield strength in thicknesses up to 4 in., whereas the yield strength of A441 steels decreases from 50 to 46 ski for thicknesses greater than ¾ in. and to 42 ksi for thicknesses over 1½ in.

The foremost property of the A514 steels is their high yield strength, which is almost three times that of A36. The heat-treated constructional alloy steels also exhibit good toughness over a wide range of temperatures and excellent resistance to atmospheric corrosion.

ASTM has also prepared a general specification, A709, for structural steel for bridges, encompassing previously generally used grades; a similar specification may be developed for buildings.

Carbon-free iron-nickel martensite, the base material for maraging, is relatively soft and ductile compared with carbon-containing martensite. But iron-nickel martensite becomes hard, strong, and tough when aged. Thus maraging steels can be fabricated while they are in a comparatively ductile martensitic condition, and later strengthened by a simple aging treatment.

Typical stress-strain curves for several types of steels are shown in Fig. 5-24.

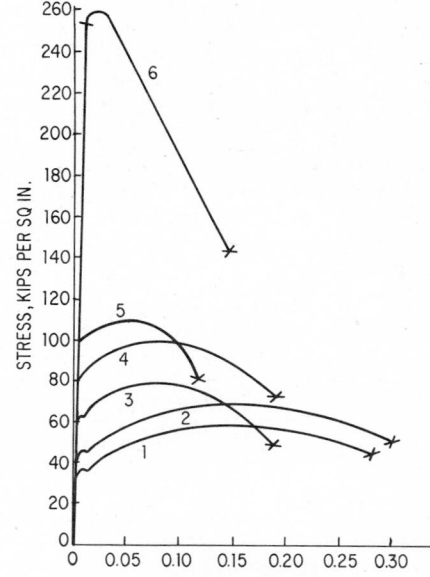

Fig. 5-24. Typical stress-strain curves for structural steels: (1) ASTM A36. (2) ASTM A242, A440, and A441. (3) Grade 60 columbium-vanadium bearing steel. (4) Heat-treated carbon steel. (5) ASTM A514. (6) 18% nickel (250) maraging steel.

5-26. Welding. Fusion welding is a process for joining metals either by melting them together or by fusing them while a filler metal is deposited in the joint between them. During welding, the part of the base metal near the joint and all the filler metal are molten. Because of the good thermal conductivity of metal, a temperature gradient is developed, varying from the melting point at the fusion zone to the ambient temperature at some distance from the weld zone.

The heat required can be produced by burning together such gases as oxygen and acetylene in a welding torch but is more usually supplied by an electric arc. The arc may be struck either between the work and a consumable electrode, which also serves as the filler material, or between the work and a nonconsumable electrode, with external filler metal added.

A protective environment is usually provided to insure weld soundness. This inert atmosphere may be formed by the decomposition of coatings on the welding electrodes or may be provided by separate means. Several welding processes are in common use today. Shielded metal-arc welding may employ coated electrodes or have bare electrodes passing through a separately maintained flux pool (submerged arc welding). Consumable metal-arc inert-gas welding is done under the protection of an inert shielding gas coming from a nozzle. Tungsten-arc inert-gas welding also employs inert shielding gas but uses a virtually nonconsumed tungsten electrode. On joints where filler metals are required with a tungsten arc, a filler rod is fed into the weld zone and melted with the base metal, as in the oxyacetylene process. These processes can be used manually or in semiautomatic or automatic equipment where the electrode may be fed continuously.

Some structural steels are intended specifically for welded construction. They may be used instead of relying on special welding techniques for steels customarily joined by riveting and bolting. The higher-strength structural steels, including heat-treated types, are also readily weldable if the correct different procedures for welding them are rigorously followed. These procedures may include use of a welding process with rapid cooling rates so that the properties of the heat-affected zone approach those of the steel in the quenched condition, minimum preheat and interpass temperatures, and postweld heat treatment. Procedures appropriate for the welding of higher-strength structural steels should be based on recommendations of the steel producer.

In the welding of steels, precautions are required to minimize pickup of hydrogen by the weld metal and heat-affected zone. Hydrogen tends to embrittle the steel and to cause cracking underneath the deposited weld bead. In addition to providing a shielding atmosphere, it may be necessary to bake the electrodes to insure that moisture content is low at time of use.

The weld zone, and a region for some distance on either side of it, may have poorer mechanical properties than the unheated portions of the base metal. The fusion zone itself usually is thickened somewhat during welding; so in a tensile test the failure is most likely to occur in the heat-affected zone adjacent to the fusion zone. The metallurgical response of a steel to the thermal cycle of welding can be predicted from a cooling transformation diagram for the steel. If proper welding procedures are designed and used so that the metallurgical response of the heat-affected zone is nearly the same as that of the base metal, joint efficiencies near 100% can be expected.

ALUMINUM ALLOYS

5-27. Classifications. Wrought-aluminum alloys are designated by a four-digit index. The first digit identifies the alloy type, according to the following code:

```
Pure aluminum, 99.00% min and greater . . . . . 1xxx
Copper . . . . . . . . . . . . . . . . . . . . . . . . . . . . . . . . . . 2xxx
Manganese . . . . . . . . . . . . . . . . . . . . . . . . . . . . . . .3xxx
Silicon . . . . . . . . . . . . . . . . . . . . . . . . . . . . . . . . . . .4xxx
Magnesium and silicon . . . . . . . . . . . . . . . . . . . .6xxx
Zinc . . . . . . . . . . . . . . . . . . . . . . . . . . . . . . . . . . . . . .7xxx
Other elements . . . . . . . . . . . . . . . . . . . . . . . . . . .8xxx
```

The second digit signifies specific alloy modifications, and the last two digits identify the specific aluminum alloy or indicate the aluminum purity.

These wrought-aluminum alloys may be heat-treatable or non-heat-treatable. Alloys are heat-treatable if the dissolved alloying elements are less soluble in the solid state at ordinary temperatures than at elevated temperatures. This makes age hardening possible. Cold working or other forms of strain hardening may also be employed to strengthen aluminum alloys. The temper of an alloy is indicated by adding a symbol to the alloy designation, as follows:

```
-F . . . . . As fabricated, no control of temper
-O  . . . . Annealed (recrystallized)
-H  . . . . Cold-worked
-T . . . . . Heat-treated
```

The letters H and T are usually followed by additional numbers indicating more details of the treatment. H1 designates an alloy that has been strain-hardened only, while H2 designates one that has been strain-hardened and then partially annealed. A second number following the H indicates increasing amounts of strain hardening on a scale from 2 to 9. H3 indicates an alloy that has been strain-hardened and stabilized by suitable annealing. The various tempers produced by heat treatment are indicated by T followed by a number, as follows:

```
-T3 . . . . Solution heat treatment followed by strain hardening; different amounts of strain hard-
            ening are indicated by a second digit.
-T4 . . . . Solution heat treatment followed by natural aging at room temperature.
-T5 . . . . Artificial aging after an elevated-temperature fabrication process.
-T6 . . . . Solution heat treatment followed by artificial aging.
-T7 . . . . Solution heat treatment followed by stabilization with an overaging heat treatment.
-T8 . . . . Solution heat treatment, strain hardening, and then artificial aging.
-T9 . . . . Solution heat treatment, artificial aging, and then strain hardening.
```

As an example of the application of this system, consider alloy 7075. Its nominal composition is 5.6% zinc, 1.6% copper, 2.5% magnesium, 0.3% chromium, and the remainder aluminum and impurity traces. If it is designated 7075-O, it is in a soft condition produced by annealing at 775°F for a few hours. If it is designated in a hard temper, 7075-T6, it has been solution heat-treated at 870°F and aged to precipitation-harden it at 250°F for about 25 hr.

5-28. Mechanical Properties. Aluminum alloys find uses in structural applications because the strength-to-weight ratio is often more favorable than that of other materials. Aluminum structures also need a minimum of maintenance since aluminum stabilizes in most atmospheres.

Wrought-aluminum alloys for structural applications are usually precipitation-hardened to strengthen them (Art. 5-21). Typical properties of some aluminum alloys frequently used in structural applications are shown in Table 5-7. The range of properties from the soft to the hardest available condition is shown.

Table 5-7. Properties of Selected Structural Aluminum Alloys

Alloy desig- nation	Principal alloying elements	Hardening process	Range of properties (soft to hard conditions)		
			Tensile strength, ksi	Yield strength, ksi	Elonga- tion in 2 in., %
2014	4.4% Cu, 0.8% Si, 0.8% Mn, 0.4% Mg	Precipitation	27–70	14–60	18–13
2024	4.5% Cu, 1.5% Mg, 0.6% Mn	Precipitation	27–72	11–57	20–13
5456	5.0% Mg, 0.7% Mn, 0.15% Cu, 0.15% Cr	Cold working	45–51	23–37	24–16
6061	1.0% Mg, 0.6% Si, 0.25% Cu, 0.25% Cr	Precipitation	18–45	8–40	25–12
7075	5.5% Zn, 2.5% Mg, 1.5% Cu, 0.3% Cr	Precipitation	33–83	15–73	17–11
Clad 7075	Layer of pure aluminum bonded to surface of alloy to increase cor- rosion resistance	Precipitation	32–76	14–67	17–11

5-29. Welding of Aluminum. All wrought-aluminum alloys are weldable but with different degrees of care required. The entire class of non-heat-treatable wrought alloys can be welded with little difficulty.

The tensile strength of sound, as-deposited weld metal is about 20 to 40 ksi, the as-cast an- nealed strength of the aluminum filler materials. For annealed, non-heat-treatable alloys, joints can always be made to fail in the base metal as long as the thicker weld bead is left in place. For hard-rolled tempers, the base metal in the heat-affected zone is softened by the welding heat; so joint efficiency is less than 100%. With heat-treatable alloys in the 6000 series, 100% efficiency can be obtained if the welded structure can be solution- and precipitation-heat- treated after welding. Nearly 100% efficiency can also be obtained without the solution heat treatment if a high-speed welding technique (such as inert-gas shielded metal arc) is used to limit heat flow into the base metal, and a precipitation heat treatment is used after welding. In the 2000 and 7000 series, such practices produce less improvement. Weld strengths in general range from about 60 to 100% of the strength of the alloy being welded.

5-30. Metals References

1. Aluminum Company of America: "Alcoa Structural Handbook" and "Welding Alcoa Aluminum," Pittsburgh, Pa.

2. The Aluminum Association: "Aluminum Standards and Data." New York.

3. American Society for Metals: "Metals Handbook," vol. 1, "Properties and Selection of Materials"; vol. 2, "Heat Treating, Cleaning and Finishing," Metals Park, Ohio.

4. American Welding Society: "Welding Handbook," Miami, Fla.

5. Dieter, George E.: "Mechanical Metallurgy," McGraw-Hill Book Company, New York.

6. Hanson, A., and J. G. Parr: "The Engineer's Guide to Steel," Addison-Wesley Publishing Company, Inc., Reading, Mass.

7. Merritt, F. S.: "Structural Steel Designers' Handbook," McGraw-Hill Book Company, New York.

8. Van Lancker, M.: "Metallurgy of Aluminum Alloys," John Wiley & Sons, Inc., New York.

Part 3. ORGANIC MATERIALS

Organic materials are perhaps the oldest construction materials. Through many generations of use, people have found ways of getting around some of the limitations of naturally occurring organic construction materials. Plywood, for instance, has overcome the problem of the highly directional properties of wood. In addition to improving natural materials, technologists have developed many synthetic polymers (plastics), which are important in current construction.

WOOD

5-31. Basis of Properties.   Wood is a natural polymer composed of cells in the shape of long thin tubes with tapered ends. The cell wall consists of crystalline cellulose aligned parallel to the axis of the cell. Typical natural cellulose has several thousand $C_6H_{10}O_5$ molecular units in each chain. The cellulose crystals are bonded together by a complex amorphous lignin composed of carbohydrate compounds. Wood substance is 50 to 60% cellulose and 20 to 35% lignin, the remainder being other carbohydrates and mineral matter.

The tree trunk grows by a process of developing concentric layers of cells outside the already established wood and under the bark. The annual growth cycle, caused by seasonal variations in temperature and moisture, produces the familiar rings and grain of wood. Cells formed in the spring have thin walls, and the wood has an open texture. The cells formed during summer growth have thicker walls, resulting in a closer texture and stronger wood substance.

Most of the cells in wood are oriented vertically, but some are radially oriented to serve as reinforcement against spreading of the vertical fibers under the natural compressive loading of the tree trunk. Because of its directed cell structure, wood has greater strength and stiffness in the longitudinal direction than in other directions.

As the tree grows, the cells in the central part of the trunk cease functioning as living cells and become part of the dead heartwood. As the tree grows in size, the heartwood increases in thickness until it represents a greater portion of the trunk than the surrounding living sapwood. The density of the wood substance is about the same for all species: 1.56. The bulk density of the gross wood is much lower, however, because of voids (cavity cells) and accidental cracks in the cell structure. For common woods, the density varies from the 0.12 of balsa to the 0.74 of oak. The strength and hardness of these woods vary similarly.

The cell wall has a high affinity for moisture because cellulose contains numerous hydroxyl groups, which are strongly hydrophilic. When exposed to moisture, often in the form of air with a high relative humidity, the cell walls in the wood absorb large amounts of water and swell. This process causes the intermolecular forces between the cellulose macromolecules to be neutralized by the absorbed water, thus reducing the strength and rigidity of the wood.

The moisture present in green wood consists of water absorbed in the cell walls and water contained in the cell cavities. As the wood dries, water is first removed from the cell cavities. At a point called the *fiber-saturation point*, the cavities are empty, while the cell walls are still fully saturated with water. On further drying in normal air, this moisture decrease continues until an equilibrium moisture content is reached. At an atmosphere of 60% relative humidity in 70°F air, the moisture content of wood stabilizes at about 11%. Although kiln drying can lower the moisture content of the wood 2 to 6% more, the decrease is not permanent and the moisture content will go back to about 11% when returned to normal air.

Dimensional changes due to swelling and shrinking resulting from atmospheric moisture changes occur only at moisture contents below the fiber-saturation point. Additional moisture fills cell cavities but causes no appreciable dimensional changes. When dimensional changes occur, they take place in radial and tangential directions, transverse to the long axis of the wood, because the cell walls swell or shrink in the direction perpendicular to the long dimension of the fibers. Wood is seasoned before it is put into service, so that it comes to equilibrium under atmospheric conditions.

5-32. Mechanical Properties of Wood.   Wood has three mutually perpendicular axes of symmetry: longitudinal, or parallel to the grain, tangential, and radial. Strength and elastic properties differ in these three directions because of the directional cell structure of the wood. Values of modulus of elasticity in the two directions perpendicular to the grain are only one-twentieth to one-twelfth the value parallel to the grain. Table 5-8 compares the elastic and shear moduli of some typical woods in the longitudinal, tangential, and radial directions. These perpendicular moduli are important in the design of composite materials containing wood.

Table 5-8. Moduli of Various Woods*

Species	Longitudinal modulus E_L, 10^3 psi	Young's modulus ratios		Modulus of rigidity ratios		
		E_T/E_L †	E_R/E_L †	G_{LR}/E_L †	G_{LT}/E_L †	G_{RT}/E_L †
Ash	2,180	0.064	0.109	0.057	0.041	0.017
Balsa	550	0.015	0.046	0.054	0.037	0.005
Birch, yellow	2,075	0.050	0.078	0.074	0.067	0.017
Douglas fir	2,280	0.050	0.068	0.064	0.078	0.007
Poplar, yellow	1,407	0.043	0.092	0.075	0.069	0.011
Walnut	1,630	0.056	0.106	0.085	0.062	0.021

*These data are for specific values of specific gravity and moisture content for each wood species. From U.S. Forest Products Laboratory, "Wood Handbook."

†E_T = modulus of elasticity, psi, in tangential direction, E_R = modulus in radial direction, G_{LR} = shear modulus in a plane normal to the tangential direction, G_{LT} = shear modulus in a plane normal to the radial direction, and G_{RT} = shear modulus in a plane normal to the longitudinal direction.

The principal mechanical properties of some woods commonly used in structural applications are shown in Table 5-9. Note that increasing moisture content reduces all the strength and stiffness properties. To obtain allowable stresses from values of ultimate strength, such as those listed in Table 5-9, the following must be considered: (1) Scatter in strengths of individual pieces may be as much as 25% above and below the average. (2) Tabulated strengths are based on tests conducted over very short periods of time. Over typical service periods, wood under sustained load may fail at about nine-sixteenths of the load recorded in a standard test. (3) The modulus of rupture of a standard 2-in.-deep flexural-test specimen is greater than that of a deep

Table 5-9. Average Strength Properties of Woods Commonly Used in Structural Applications*

Results of tests on small, clear specimens in the green and air-dry condition

Common name of species	Moisture content, %		Modulus of elasticity in bending, kips per sq in.		Proportional limit in compression parallel to grain, psi		Compressive strength parallel to grain, psi		Proportional limit in compression perpendicular to grain, psi		Shearing strength parallel to grain, psi	
	Green	Air-dry	Green	Air-dry	Green	Air-dry	Green	Air-dry	Green	Air-dry	Green	Air-dry
Softwoods:												
Cedar, western red	37	12	920	1,120	2,470	4,360	2.750	5,020	340	610	710	860
Cedar, Atlantic white	55	12	750	930	1,660	2,740	2,390	4,700	300	500	690	800
Cypress, southern	91	12	1,180	1,440	3,100	4,740	3,580	6,360	500	900	810	1,000
Douglas fir (coast type)	38	12	1,570	1,950	3,130	5,850	3,860	7,430	440	870	930	1,160
Fir, white	115	12	1,030	1,380	2,390	3,590	2,710	5,350	370	600	750	930
Hemlock, eastern	111	12	1,070	1,200	2,600	4,020	3,080	5,410	440	800	850	1,060
Hemlock, western	74	12	1,220	1,490	2,480	5,340	2,990	6,210	390	680	810	1,170
Larch, western	58	12	1,530	1,960	3,010	5,620	3,990	8,110	420	980	900	1,410
Pine, southern yellow:												
longleaf	63	12	1,600	1,990	3,430	6,150	4,300	8,440	590	1,190	1,040	1,500
Pine, western white	54	12	1,170	1,510	2,430	4,480	2,650	5,620	290	540	640	850
Redwood, virgin	112	12	1,180	1,340	3,700	4,560	4,200	6,150	520	860	800	940
Spruce, Sitka	42	12	1,230	1,570	2,240	4,780	2,670	5,610	340	710	760	1,150
Hardwoods:												
Ash, white	42	12	1,460	1,770	3,190	5,790	3,990	7,410	810	1,410	1,380	1,950
Birch, yellow	67	12	1,500	2,010	2,620	6,130	3,380	8,170	530	1,190	1,110	1,880
Maple, sugar	58	12	1,550	1,830	2,850	5,390	4,020	7,830	800	1,810	1,460	2,330
Oak, red (northern)	80	12	1,350	1,820	2,360	4,580	3,440	6,760	760	1,250	1,210	1,780
Oak, white	68	12	1,250	1,780	3,090	4,760	3,560	7,440	830	1,320	1,250	2,000
Poplar, yellow	83	12	1,220	1,580	2,070	3,730	2,660	5,540	300	560	790	1,190

*From U.S. Forest Products Laboratory, "Wood Handbook."

beam. (4) A safety factor (perhaps $^3/_5$) should be applied to average ultimate strengths to provide allowable stresses. (5) Blemishes (such as knots and checks) further reduce the allowable stresses in actual wood members. (See Sec. 11 for engineering design in timber.)

5-33. Commercial Grades of Wood. Lumber is graded to enable a user to buy the quality that best suits a particular purpose. The grade of a piece of lumber is based on the number, character, and location of strength-reducing features and on factors affecting durability and utility. The best grades are virtually free of blemishes, but the other grades, which comprise the great bulk of lumber, contain numerous knots and other features that affect quality to varying degrees. Various associations of lumber manufacturers assume jurisdiction over the grading of certain species. Two principal sets of grading rules are employed for hardwood and softwood.

Hardwood is graded according to rules adopted by the National Hardwood Lumber Association. Since most hardwood boards are cut into smaller pieces to make a fabricated product, the grading rules are based on the proportion of a given piece that can be cut into smaller pieces. Usable material must have one clear face, and the reverse face must be sound.

Softwood is classified and graded under rules adopted by a number of regional lumber manufacturers' associations. American lumber standards for softwood lumber were formulated as a result of conferences organized by the U.S. Department of Commerce to improve and simplify the grading rules. These standards, issued in pamphlet form by the Department of Commerce, have resulted in more uniform practices throughout the country. Softwood lumber is classified according to use, size, and process of manufacture.

Use classifications include: (1) yard lumber, intended for general building purposes; (2) structural lumber, which is limited to the larger sizes and intended for use where minimum working stresses are required; and (3) factory and shop lumber, intended to be cut up for use in further manufacture.

Lumber classified according to manufacture includes: (1) rough lumber, which is in the undressed condition after sawing; (2) surfaced lumber, which is surface-finished by running through a planer; and (3) worked lumber, which has been matched or molded.

All softwood lumber is graded into two general categories, select and common, on the basis of appearance and characteristics. Structural lumber is graded according to strength for each species.

5-34. Improvement of Properties. Because of its high anisotropy and hygroscopic properties, wood has limitations in use as a structural material. Various techniques are employed to improve the strength or dimensional stability of wood in service atmospheres. Preservatives may be applied to combat decay and attack by animal organisms. Thin sheets of wood may be bonded together to build up a modified wood structure. The sheets can be effectively impregnated to fill the cell cavities. As a further modification, the thin-sheet structure may be compressed during the period of bond curing to increase the density and strength. Such treatments improve the chemical resistance, decay resistance, and dimensional stability of the wood.

Plywood is made by bonding together a number of thin sheets or veneers of wood. The grain in adjacent plies is oriented at right angles, and an odd number of plies is used. The main purpose of plywood is to overcome the directional properties of wood, thereby obtaining a material more uniform in all directions. Plywood shows greater resistance to checking and splitting than lumber and has better dimensional stability because of reduced shrinkage and swelling.

Plywood is classed as interior or exterior, depending on the type of adhesive used to bond the plies together. Interior grade usually is bonded with water-soluble glues and thus has limited resistance to moisture. Exterior grade is completely waterproof in that it can withstand prolonged immersion in water without disintegration.

PLASTICS

5-35. Formation and Structure of Polymers. The synonymous terms *plastics* and *synthetic resins* denote synthetic organic high polymers. *Polymers* are compounds in which the basic molecular-level subunits are long-chain molecules. The word *plastic* has been adopted as a general name for this group of materials because all are capable of being molded at some stage in their manufacture.

The mechanisms by which polymerization (the formation of long-chain molecules) takes place can be grouped into two general categories: addition and condensation. In *addition polymerization*, succeeding mers (the smallest repetitive unit in a polymer) are added to the molecules to increase the average molecular size and weight. This reaction generally occurs by the breaking of double bonds between atoms in monomers and the forming of two single bonds in their place. In contrast, *condensation polymerization* produces a by-product as well as the growing polymer molecules. The nonpolymerizable molecule that results as a by-product in condensation reactions is usually water or some other simple molecule. Condensation polymers may or may not be linear in the character of their structure.

In addition polymerization, the simultaneous polymerization of two or more chemically different monomers can be employed to form a polymer containing both monomers in one chain. Such *copolymers* frequently have more desirable physical and mechanical properties than either of the polymers that have been combined. The range of properties available through copolymerization means that the engineer can have plastics tailor-made to specific requirements.

Polymers may be formed in either an amorphous or crystalline state, depending on the relative arrangements of the long-chain molecules. An amorphous (without form) state is characterized by a completely random arrangement of molecules. A crystalline state in a polymer consists of crystalline regions, called *crystallites*, embedded in an amorphous matrix.

The ordered crystalline regions do not extend throughout the structure, and thus crystalline polymers differ from typical crystalline solids, such as metals. Several typical features of polymers favor noncrystallinity: (1) molecular chains are very long and branched; (2) side groups of atoms are randomly arranged along the chains; (3) copolymer chains contain two or more kinds of mers; and (4) low-molecular-weight additives, called *plasticizers*, tend to separate the main chains from each other. Crystallization in polymers is thus usually imperfect at best and may be completely lacking.

Crystallization causes a denser packing of polymer molecules, thus increasing the inter-molecular forces. The resulting polymers have greater strength and stiffness and a higher softening point than amorphous polymers of the same chemical structure and molecular weight.

The mechanical behavior of a plastic is greatly affected by the internal structure of the polymer. Monomers with two reaction sites, called *bifunctional*, tend to produce linear polymers. In such linear structures, only van der Waals bonds hold adjacent chains together, and so slip can occur between molecules. Other molecules with three or four reaction sites (tri- or tetra-functional) tend to form network or framework polymers. Slip does not occur as readily in such a three-dimensionally bonded structure as in linear polymers.

Cross-linking, a common variation in the growth of polymers, ties the chains of molecules together at intervals by primary bonds. For effective cross-linking, there must be normally unsaturated carbon atoms present within the polymer chain since cross-linking takes place through such connecting points. Cross-linking greatly restricts the movement between adjacent polymer chains and thus alters the mechanical properties of the material. A cross-linked polymer has higher tensile strength, more recoverable deformation (elasticity), and less elongation at failure. The vulcanization of natural rubber with sulfur is the classic example of the kind of transformation that cross-linking can effect—from tire treads to battery cases.

Three-dimensional structures can also be formed from chain polymers by *branching*, where main chains are bifurcated into two chains. The extent of branching can be controlled in the production process. If branching is extensive enough, it restricts the movement between adjacent chains by causing intertangling.

5-36. Deformation of Polymers. The elastic moduli of plastics generally range from 10^4 to 10^6 psi, considerably lower than for metals. The greater strains observed when plastics are loaded result from the fact that there is bond straightening in polymers as well as bond lengthening. Network polymer structures are more rigid than linear structures and thus show higher moduli.

Permanent deformation in plastics occurs as slip between adjacent molecular chains. In polymers where the chains are bonded only by van der Waals forces, deformation occurs by slippage at the bonds between molecules rather than by the breaking of the molecular bonds in a chain. Such slip between polymer chains cannot occur in framework polymers. These three-dimensionally primary-bonded polymers are thus generally stronger but do not show much plastic deformation prior to failure.

Application of stress to a plastic favors crystallization since the molecular chains are pulled into closer alignment and proximity. Thus, the properties of polymers may change as high stresses are applied. This phenomenon of orientation is employed to produce plastics with different properties in one direction than in others. Drawing, which orients the molecular chains in the direction of drawing, produces strength in the longitudinal direction that is several times that of undrawn material.

Plastics are divided into two large categories on the basis of their thermal behavior: thermoplastic and thermosetting materials.

Thermoplastics become extremely plastic, that is, easily deformable, at elevated temperatures. They become hard again on cooling. They can be so softened by heating and hardened by cooling any number of times. Thermoplastic resins deform easily under applied pressure, particularly at elevated temperatures, and so are used to make molded products.

Thermosetting materials are either originally soft or soften at once upon heating, but upon further heating they harden permanently. The final, continuous framework structure of thermosetting resins may develop from the condensation polymerization mechanism or may harden by the formation of primary bonds between molecular chains as thermal energy is applied. The completion of polymerization, which is accelerated at higher temperatures, provides a permanent set to the thermosetting resins. In general, thermosetting plastics are stronger than thermoplastic resins, particularly at elevated temperatures.

Amorphous polymers have a characteristic temperature at which the properties make a drastic change, called the **glass transition temperature.** The transition from glassy behavior to rubbery behavior occurs over a 10 to 20°C temperature range. On the high-temperature side of this transition, the molecular segments are free to move past each other; and on the low-temperature side, they are rigidly confined.

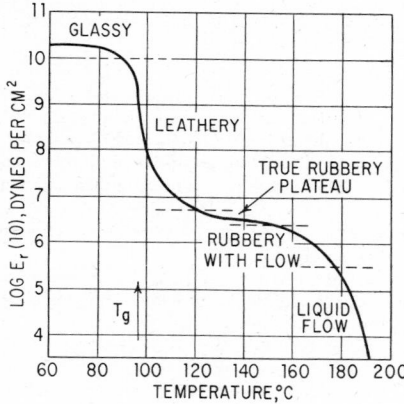

Fig. 5-25. Transition-temperature phenomena in polystyrene. (*After Tobolsky.*)

Several properties of the polymer change at T_g, the glass transition temperature. Figure 5-25 shows how the relaxed modulus (with the viscoelastic component of strain relaxed out) of polystyrene varies with temperature in the region of the transition temperature. The coefficient of expansion, heat capacity, diffusion coefficient, and strength also are similarly affected. The value of T_g for a given polymer can be controlled by several techniques. Slower cooling from the melt or the addition of plasticizers (low-molecular-weight polymers) can lower T_g. Rapid cooling from the melt and the addition of fillers tend to raise T_g for a given material. Copolymerization also can be used to obtain desired values of T_g. Copolymers have T_g values intermediate between those of the pure homopolymers.

Polymers are viscoelastic in that they are subject to time-dependent phenomena. Plastic materials, subjected to a steady load, creep to greater strains than under shorttime loading. If the material is instead stretched to a given elongation, the stress necessary to maintain the elongation will diminish with time. Both creep and stress relaxation are accelerated at higher temperatures, where the molecular chains have more thermal energy to assist in reorientation or slipping. Since the properties are time-dependent, the rate of loading of a polymer can affect the observed behavior. Increased loading rates produce steeper stress-strain curves, indicating that the material is stiffer when the time for molecular readjustments is decreased.

Plasticizers and fillers may be added to polymers to change their basic properties. Plasticizers are low-molecular-weight (short-chain) substances added to reduce the average molecular weight of a polymer and thus make it more flexible. Fillers may be added, particularly to the softer plastics, to stiffen them, increase their strength and impact properties, or improve their resistance to heat.   Wood flour, mica, asbestos fibers, and chopped fibers or fabric may be used as filler material for polymers.

5-37. Thermosetting Plastics. (See also Art. 5-36.) These materials, once soft, harden permanently upon further heating. In the following, the principal varieties of thermosets are described briefly and their main applications noted. (For detailed data on the properties of these plastics, see the annual Encyclopedia Issue of *Modern Plastics*.)

Phenol formaldehydes provide the greatest variety of thermosetting molded plastic articles. They are used for chemical decorative, electrical, mechanical, and thermal applications of all kinds. Hard and rigid, they change slightly, if at all, on aging indoors but on outdoor exposure lose their bright surface gloss. However, the outdoor-exposure characteristics of the more durable formulations are otherwise generally good. Phenol formaldehydes have good electrical properties, do not burn readily, and do not support combustion. They are strong, light in weight, and generally pleasant to the eye and touch. Light colors normally are not obtainable because of the dark-brown basic color of the resin. They have low water absorption and good resistance to attack by most commonly found chemicals.

Furan resins are similar to phenolics in many respects. Tough and durable, they have many industrial uses, such as for large aggregate-filled molds for shaping light metals.

Cast phenolics were once used in large quantities for brilliantly colored parts, but today they are used principally in industrial applications, including molds.

Epoxy and polyester casting resins are used for a variety of purposes. For example, electronic parts with delicate components are sometimes cast completely in these materials to give them complete and continuous support and resistance to thermal and mechanical shock. Some varieties must be cured at elevated temperatures; others can be formulated to be cured at room temperatures. One of the outstanding attributes of the epoxies is their excellent adhesion to a variety of materials, including such metals as copper, brass, steel, and aluminum.

Polyester molding materials, when compounded with fibers (particularly glass fibers) or with various mineral fillers (including clay), can be formulated into putties or premixes that are easily compression- or transfer-molded into parts having high impact resistance.

Melamine formaldehyde materials are unaffected by common organic solvents, greases, and oils, as well as most weak acids and alkalies. Their water absorption is low. They are insensitive to heat and are highly flame-resistant, depending on the filler. Electrical properties are particularly good, especially resistance to arcing. Unfilled materials are highly translucent and have unlimited color possibilities. Principal fillers are alpha cellulose for general-purpose compounding; minerals to improve electrical properties, particularly at elevated temperatures; chopped fabric to afford high shock resistance and flexural strength; and cellulose, used mainly for electrical purposes.

Alkyds are customarily combined with mineral or glass fillers, the latter for high impact strength. Extreme rapidity and completeness of cure permit rapid production of large numbers of parts from relatively few molds. Because electrical properties, especially resistance to arcing, are good, many of the applications for alkyd molding materials are in electrical applications.

Urea formaldehydes, like the melamines, offer unlimited translucent to opaque color possibilities, light fastness, good mechanical and electrical properties, and resistance to organic solvents as well as to mild acids and alkalies. Although there is no swelling or change in appearance, the water absorption of urea formaldehyde is relatively high, and therefore it is not recommended for applications involving long exposure to water. Occasional exposure to water has no deleterious effect. Strength properties are good.

Silicones, unlike other plastics, are based on silicon rather than carbon. As a consequence, their inertness and durability under a wide variety of conditions are outstanding. As compared with the phenolics, their mechanical properties are poor, and consequently glass fibers are added. Molding is more difficult than with other thermosetting materials. Unlike most other resins, they may be used in continuous operations at 400°F; they have very low water absorption; their dielectric properties are excellent over an extremely wide variety of chemical attack; and under outdoor conditions their durability is particularly outstanding. In liquid solutions, silicones are used to impart moisture resistance to masonry walls and to fabrics. They also form the basis for a variety of paints and other coatings capable of maintaining flexibility and inertness to attack at high temperatures in the presence of ultraviolet sunlight and ozone. Silicone rubbers maintain their flexibility at much lower temperatures than other rubbers.

5-38. Thermoplastic Resins. (See also Art. 5-36.) Materials in this category can be repeatedly softened by heating and hardened by cooling. The main varieties of thermoplastics are described briefly in the following. (For detailed data on the properties of these plastics, see the annual Encyclopedia Issue of *Modern Plastics*.)

Acrylics in the form of large transparent sheets are used in aircraft enclosures and building construction. Although not so hard as glass, acrylics have perfect clarity and transparency. They are the most resistant of the transparent plastics to sunlight and outdoor weathering, and they have an optimum combination of flexibility and rigidity, with resistance to shattering. A wide variety of transparent, translucent, and opaque colors can be produced. Sheets of acrylic are readily formed to complex shapes. They are used for such applications as transparent windows, outdoor and indoor signs, parts of lighting equipment, decorative and functional automotive parts, reflectors, household-appliance parts, and similar applications. Acrylics can be used as large sheets, molded from molding powders, or cast from the liquid monomer.

Polyethylene, in its unmodified form, is a flexible, waxy, translucent plastic maintaining flexibility at very low temperatures, in contrast with many other thermoplastic materials. The heat-distortion point of the older, low-density polyethylenes is low; these plastics are not recommended for uses at temperatures above 150° F. Newer, high-density materials have higher heat-distortion temperatures; some may be heated to temperatures above 212° F. The heat-distortion point may rise well above 250° F for plastics irradiated with high-energy beams. Unlike most plastics, polyethylene is partly crystalline. It is highly inert to solvents and corrosive chemicals of all kinds at ordinary temperatures. Usually, low moisture permeability and absorption are combined with excellent electrical properties. Its density is lower than that of any other commercially available nonporous plastic. When compounded with black pigment, its weathering properties are good. Polyethylene is widely used as a primary insulating material on wire and cable and has been used as a replacement for the lead jacket on communication cables and other cables. It is widely used also as thin flexible film for packaging, particularly of food, and as corrosion-proof lining for tanks and other chemical equipment.

Polypropylene, a polyolefin, is similar in many ways to its counterpart, polyethylene, but is generally harder, stronger, and more temperature-resistant. It has a great many uses, among them for complete water cisterns for water closets in plumbing systems abroad.

Polytetrafluoroethylene, with the very active element fluorine in its structure, is a highly crystalline linear-type polymer, unique among organic compounds in its chemical inertness and resistance to change at high and low temperatures. It has an extremely low dielectric-loss factor. In addition its other electrical properties are excellent. Its outstanding property is extreme resistance to attack by corrosive agents and solvents of all kinds. At temperatures well above 500° F, polytetrafluoroethylene can be held for long periods with practically no change in properties except loss in tensile strength. Service temperatures are generally maintained below 480° F. This material is not embrittled at low temperatures, and its films remain flexible at temperatures below − 100° F. It is difficult to mold because it has no true softening temperature. Therefore a modified form in which one chlorine is substituted for fluorine is employed: **polymonochlorotrifluoroethylene.** Like the silicones, these fluorocarbons are difficult to wet; consequently they have high moisture repellence and are often used as parting agents or where sticky materials, such as candy, must be handled.

Teflon is polytetrafluoroethylene that is used in bridges as beam seats or bearings and in buildings calling for resistance to extreme conditions, or for applications requiring low friction. In steam lines, for example, supporting pads of Teflon permit the line to slide easily over the pad as expansion and contraction with changes in temperature cause the line to lengthen and shorten. The temperatures involved have little or no effect. Mechanical properties are only moderately high, and reinforcement may be necessary to prevent creep and squeeze-out under heavy loads.

Polyurethane is used as thermal insulation in the form of foam, either prefoamed or foamed in place. The latter is particularly useful in irregular spaces. When blown with fluorocarbons, the foam has exceptional thermal conductivity and therefore is widely used in thin-walled refrigerators. Other uses include field-applied or baked-on clear or colored coatings and finishes for floors, walls, furniture, and casework generally. The rubbery form is employed for sprayed or troweled-on roofing and for gaskets and caulking compounds.

Polyvinyl fluoride has much of the superior inertness to chemical and weathering attack typical of the fluorocarbons. Among other uses, it is used as thin-film overlays for building boards to be exposed outdoors.

Polyvinyl formal resins are used principally as a base for tough, water-resistant insulating enamel for electric wire. **Polyvinyl butyral** is the tough interlayer in safety glass. In its cross-linked and plasticized form, polyvinyl butyral is used extensively in coating fabrics for raincoats, upholstery, and other heavy-duty moisture-resistant applications.

Vinyl chloride polymers and copolymers vary from hard and rigid to highly flexible. Polyvinyl chloride is naturally hard and rigid but can be plasticized to any required degree of flexibility, as in raincoats and shower curtains. Copolymers, including vinyl chloride plus vinyl acetate, are naturally flexible without plasticizers. Nonrigid vinyl plastics are widely used as insulation and jacketing for electric wire and cable because of their electrical properties and their resistance to oil and water. Thin films are used for rainwear and similar applications, whereas heavy-gage films and sheets are used widely for upholstery. Vinyl chlorides are used for floor coverings in the form of tile and sheet because of their abrasion resistance and relatively low water absorption. The rigid materials are used for tubing, pipe, and many other applications which require resistance to corrosion and action of many chemicals, especially acids and alkalies; they are attacked by a variety of organic solvents, however. Like all thermoplastics, vinyl chlorides soften at elevated temperatures; their maximum recommended temperature is about 140°F, although at low loads they may be used at temperatures as high as 180°F.

Vinylidene chloride is highly resistant to most inorganic chemicals and to organic solvents generally. It is impervious to water on prolonged immersion, and its films are highly resistant to moisture-vapor transmission. It can be sterilized, if not under load, in boiling water, and its mechanical properties are good. Vinylidene chloride is not recommended for uses involving high-speed impact, shock resistance, or flexibility at subfreezing temperatures. It should not be used in applications requiring continuous exposure to temperatures in excess of 170°F.

Polystyrene formulations constitute a large and important segment of the entire field of thermoplastic materials. Numerous modified polystyrenes provide a relatively wide range of properties. Polystyrene is one of the lightest of the presently available commercial plastics. It is relatively inexpensive and easily molded and has good dimensional stability and good stability at low temperatures. It is brilliantly clear when transparent but can be produced in an infinite range of colors. Water absorption is negligible even after long immersion. Electrical characteristics are excellent. It is resistant to most corrosive chemicals, such as acids, and a variety of organic solvents, although it is attacked by others. Polystyrenes, as a class, are considerably more brittle and less extendable than many other thermoplastic materials; but these properties are markedly improved by copolymerization. Under some circumstances, they tend to develop fine cracks, known as *craze marks*, on exposure, particularly outdoors. This is true of many other thermoplastics, especially when highly stressed.

Nylon, in molded form, is used in increasing quantities for impact and high resistance to abrasion. It is employed in small gears, cams, and other machine parts because even when unlubricated, nylon is highly resistant to wear. Its chemical resistance, except to phenols and mineral acids, is excellent. Extruded nylon is coated onto electric wire, cable, and rope for abrasion resistance. Applications like hammerheads indicate its impact resistance.

5-39. Cellulose Derivatives. Cellulose is a naturally occurring high polymer found in all woody plant tissue and in such materials as cotton. It can be modified by chemical processes into a variety of thermoplastic materials, which, in turn, may be still further modified with plasticizers, fillers, and other additives to provide a wide variety of properties. The oldest of all plastics is cellulose nitrate.

Cellulose acetate is the basis of safety film, developed to overcome the highly flammable nature of cellulose nitrate. Starting as film, sheet, or molding powder, it is made into a variety of items, such as transparent packages and a large variety of general-purpose items. Depending on the plasticizer content, it may be hard and rigid or soft and flexible. Moisture absorption of this and all other cellulosics is relatively high, and they are therefore not recommended for long-continued outdoor exposure. But cellulose acetate film, reinforced with metal mesh, is widely used for temporary enclosures of buildings during construction.

Cellulose acetate butyrate, a butyrate copolymer, is inherently softer and more flexible than cellulose acetate and requires less plasticizer to achieve a given degree of softness and flexibility. It is made in the form of clear transparent sheet and film, or in the form of molding powders, which can be molded by standard injection-molding procedures into a wide variety of products. Like the other cellulosics, this material is inherently tough and has good impact resistance. It has infinite colorability, like the other cellulosics. Cellulose acetate butyrate tubing is used for such applications as irrigation and gas lines.

Ethyl cellulose is similar to cellulose acetate and acetate butyrate in its general properties. Two varieties, general-purpose and high-impact, are common; high-impact ethyl cellulose is made for better-than-average toughness at normal and low temperatures.

Cellulose nitrate, one of the toughest of the plastics, is widely used for tool handles and similar applications requiring high-impact strength. Its high flammability requires great caution, particularly when the plastic is in the form of film. Most commercial photographic film is made of cellulose nitrates rather than safety film. Cellulose nitrate is the basis of most of the widely used commercial lacquers for furniture and similar items.

ASPHALT

5-40. Chemistry of Asphalt. Asphalt materials have been known and used in road and building construction since ancient times. Early asphalt was of natural origin, found in pools and asphalt lakes, but current supplies come mainly from the residues of refined petroleum.

Asphalt, which is the black or dark brown petroleum derivative, is distinct from tar, the residue from destructive distillation of coal. Asphalt consists of hydrocarbons and their derivatives and is completely soluble in carbon disulfide (CS_2). It is the residue of petroleums after the evaporation, by natural or artificial means, of their most volatile components.

Asphalt is colloidal in nature. The components of the highest molecular weights constitute the disperse phase (micelles), and the constituents of lower molecular weights comprise the continuous (intermicellar) phase. Asphaltenes are the solid, hard, dark-brown or black particles precipitated when asphalt is dissolved in a large quantity of solvent. Maltenes constitute the fraction of asphalt that remains dissolved as the asphaltenes precipitate from the solvent solution. In undiluted asphalt, the maltenes form a viscous, dark-brown oil.

The percentages of asphaltenes and maltenes present in an asphalt can be determined in a given solvent and must be defined in terms of that solvent to make sense. For example, Table 5-10 shows the fractional components of asphalt after a hundred-fold dilution with *n*-pentane.

Table 5-10. Components of a Diluted Asphalt

Fraction	Molecular weight	C/H	Amount, %
Asphaltenes	10,000	9–10	10–20
Maltenes:			
a. Resins	800	8	10
b. Oils	600	6	70–80

In the detailed structure of the colloid, the resins immediately surround the asphaltenes, and the oils surround that composite. Since it is difficult to determine the proportions of various hydrocarbons present in asphalt, the ratio between the number of carbon and hydrogen atoms (C/H ratio) is used to characterize the chemical composition of fractions of the asphalt. The ratio gives an indication of the degree of saturation of the hydrocarbon mixture and can be correlated with the properties of different asphalts.

Depending on the degree of aromaticity of the maltenes and the nature of the concentration of the asphaltenes, two types of structures might form: (1) sol-type asphalts, in which the micelles in the asphalt move freely with respect to each other; and (2) gel-type asphalts, in which the micelles, by mutual attraction, form a structure throughout the bituminous mass. Sol-type asphalts have high ductility, high susceptibility to temperature change, no measurable elasticity, and a high rate of age hardening. Gel-type asphalts have low ductility, low susceptibility to temperature change, no measurable elasticity, and a low rate of age hardening. Intermediate asphalts, with a structure between sols and gels, are called "meds."

5-41. Bituminous Pavements. Asphalt refined to meet specifications for paving purposes is called *asphalt cement.* At normal temperatures, it is a semisolid, with the degree of solidity measured by a penetration test. It is heated until liquefied before being blended with aggregate in paving mixtures.

If asphalt is so soft that a penetration test is not an appropriate means for measuring consistency, it is called *liquid asphalt.* Several types of liquid asphalt are produced:

1. Rapid curing (RC) asphalt, which is liquefied with naphtha or gasoline, both of which are highly volatile and evaporate quickly to leave asphalt cement.

2. Medium curing (MC) asphalt, which is asphalt cement liquefied with a kerosene diluent.

3. Slow curing (SC) asphalt, which is blended with a low-volatile oil.

4. Emulsified asphalt, which is produced by mixing together water with an emulsifying agent and asphalt cement. This heterogeneous system of spherical globules in the water medium hardens as the water evaporates.

Asphalt pavements are composed of asphalt, aggregate, and voids (2 to 7% air). A typical makeup of an asphalt-aggregate composite is given in Table 5-11.

Table 5-11. Typical Asphalt-Aggregate Composite

	Weight, %	Volume, %
Asphalt	6	14.4
Coarse aggregate	53	43.7
Fine aggregate	35	33.4
Mineral dust	6	4.9
Air	. . .	3.6

An asphalt pavement carries applied load by particle friction and interlock. Its strength is a function of the surface texture (particularly of the fine aggregate) and density (compactness) of the aggregate. A rough surface texture is desirable.

Dense mixtures are obtained by using well-graded aggregates, where the fine aggregate fills the voids in the coarser aggregate structure. Coarse aggregate is that retained on a No. 8 sieve, fine aggregate passes a No. 8 sieve, and mineral dust passes a No. 200 sieve. The asphalt cement binds the aggregate particles together and waterproofs the pavement. The air voids allow for expansion of the asphalt cement or compaction of the composite by providing space for the asphalt cement to move into instead of pushing the aggregate farther apart.

5-42. Asphalt Building Products. Because of its water-resistant qualities and durability, asphalt is used for many building applications. For dampproofing (mopped-on coating only) and waterproofing (built-up coating of one or more plies), three types of asphalt are used: Type A, an easy-flowing, soft, adhesive material for use under ground or in other moderate-temperature applications; Type B, a less susceptible asphalt for use above ground where temperatures do not exceed 125°F; and Type C, for use above ground where exposed on vertical surfaces to direct sunlight or in other areas where temperatures exceed 125°F.

Asphalt and asphalt products are also used extensively in roofing applications. Asphalt is used as a binder between layers in built-up roofing and as the impregnating agent in roofing felts, roll roofing, and shingles.

5-43. Organic Materials References

1. The Asphalt Institute: "The Asphalt Handbook," College Park, Maryland.
2. Baer, Eric, ed.: "Engineering Design for Plastics," Van Nostrand Reinhold Company, New York.
3. Forest Products Laboratory, Forest Service, U.S. Department of Agriculture: "Wood Handbook," Superintendent of Documents, Washington, D.C.
4. "Modern Plastics Encyclopedia," Plastics Catalog Corp., New York.
5. National Lumber Manufacturers Association: "National Design Specification for Stress-grade Lumber and Its Fastenings," Washington, D.C.
6. Nielsen, Lawrence E.: "Mechanical Properties of Polymers," Van Nostrand Reinhold Company, New York.

Part 4. COMPOSITE MATERIALS

Well-known products such as plywood, reinforced concrete, and pneumatic tires are evidence that the concept of composite materials has been applied for many years. But new families of composites with expanding ranges and a variety of properties are continually being created. Composite materials for structural applications are particularly important where higher strength-to-weight and stiffness-to-weight ratios are desired than can be had with basic materials.

TYPES OF COMPOSITES

5-44. Materials Combinations and Forms. Composites can be classified in seven basic material combinations and three primary forms. The materials categories are permutations of combinations of the three basic kinds of materials: metal-metal, metal-inorganic, metal-organic, inorganic-inorganic, inorganic-organic, organic-organic, metal-inorganic-organic. Here inorganic applies to nonmetallic, inorganic materials, such as ceramics, glasses, and minerals. No limitation on the number of phases embodied in a composite is intended by these designations. Thus metal-organic includes composites with two metallic phases and one organic phase or four-phase composites having two metallic and two organic components.

The three primary forms of composite structures are shown in Fig. 5-26. *Matrix systems* are characterized by a discontinuous phase, such as particles, flakes, or fibers, or combinations of these, in a continuous phase or matrix (see also Arts. 5-45 and 5-48). *Laminates* are characterized by two or more layers bonded together. As a rule, strengthening is less an objective than other functional requirements in the design of laminated composites (see also Art. 5-50). *Sandwich structures* are characterized by a single, low-density core, such as honeycomb or foamed material, between two faces of comparatively higher density. A sandwich may have several cores or an open face (see also Arts. 5-46 and 5-49). One primary form of composite may contain another. The faces of a sandwich, for example, might consist of a laminate or matrix system.

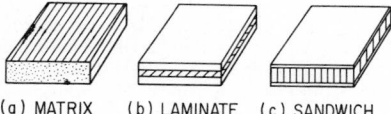

(a) MATRIX   (b) LAMINATE   (c) SANDWICH

Fig. 5-26. Primary forms of composite materials.

(L. J. Broutman and R. H. Krock, "Modern Composite Materials," Addison-Wesley Publishing Company, Reading, Mass.)

5-45. Matrix Systems. (See also Art. 5-44.) Most important among the matrix systems are steel-reinforced concrete and those containing fibers or fiberlike material, such as whiskers, that enhance strength. Here advantage is taken of the high strengths available in some materials, especially when produced in the form of fine filaments a few microns in diameter.

Dominant among the fiber-based structural composites are those based on continuous filaments, of which glass-reinforced plastics are typical. Whisker composites are another group based on the extremely high strength available from materials in fine fibrous form. Alumina whiskers can now be made with strengths consistently ranging from 1 to 3×10^6 psi. Silver has been strengthened from its normal level of 25,000 to 230,000 psi, with a 24% (by volume) addition of these whiskers. Similarly, a 50% gain has been obtained with a 12% addition to an 80-20 nickel-chromium alloy (see also Arts. 5-47 and 5-48).

5-46. Sandwich Systems. (See also Art. 5-44.) A primary objective of most sandwich composites is superior structural performance. To this end, the core separates and stabilizes the faces against buckling under edgewise compression, torsion, or bending. Other considerations, such as heat resistance and electrical requirements, dictate the choice of materials. Cores are usually lightweight materials. Typical forms of core material are honeycomb structures (metal, glass-reinforced plastic, or resin-impregnated paper) and foams (generally plastic, but they may be ceramic). Synthetic organic adhesives are employed to assemble sandwich components, except when thermal considerations preclude them (e.g., epoxies, phenolics, polyesters).

5-47. Continuous Filament Composites. (See also Art. 5-44.) Glass filaments have dominated this field ever since they were first introduced in quantity about 40 years ago. Most popular among the commercially available continuous filaments is E-glass, produced in a diameter of about 10 microns. Typical properties of E-glass include a strength, when made, of 500,000 psi, an elastic modulus of 10.5×10^6 psi, and a density of 0.092 lb per cu in.

The fibers are converted into yarns, rovings, and woven fabrics in a variety of configurations. Matrix materials employed with glass fibers generally have been synthetic resins, largely the polyester, phenolic, and epoxy families.

The attributes of glass-fiber-reinforced plastic make it an important structural material. Its mechanical properties are competitive with metals, considering density. It exhibits great freedom from corrosion, although it is not wholly immune to deterioration. The dielectric properties are very good. It may be fabricated in complex shapes, in limited quantities, with comparatively inexpensive tooling. The properties of a typical glass-fabric epoxy-resin composite are shown in Table 5-12.

Table 5-12. Mechanical Properties of a Glass-reinforced Plastic*

Property	Tested dry	Tested wet
Tensile strength, psi	54,000	51,000
Tensile modulus, psi	3.1×10^6	
Compressive strength, psi, edgewise	57,000	56,000
Compressive modulus, psi	3.1×10^6	
Flexural strength, psi	71,000	68,000
Shear strength, psi	17,000	

Reinforcement: glass fabric. Thickness: ⅛ in. Resin content: 32% by weight. Specific gravity: 1.75.
*P. M. Goodwin, "Composite Materials," *Machine Design*, July 18, 1963, p. 190.

Woven glass-fiber materials fill a definite need, but the weaving process adds expense and tends to degrade the filaments. Systems of handling the filaments without weaving thus offer distinct advantages. Filament winding processes involve just such a system. In this method, bundles of filaments in the form of roving or bands of roving are laid in place by winding on a mandrel under controlled conditions in a predetermined pattern. Filament winding is basically applicable to bodies of revolution. A unique advantage of filament winding is the ability to adjust directional strength to that actually needed.

A variety of filaments can be used to obtain various composite properties and efficiencies: E-glass, Al_2O_3 glass, silica, beryllium, boron, and steel. Filament geometry presents still another degree of freedom. One example is hollow filament, which offers more stiffness than solid filament for the same weight. Also, matrix-filament ratios can be adjusted. And filament-alignment possibilities are infinite.

To compare structural materials, parameters reflecting the interdependence of strength and weight or stiffness and weight are used. Specific strength σ/ρ, where σ is the yield strength for metals and the composite tensile strength for plastics and ρ is the density (in pounds per cubic inch), is a measure of the structural efficiency of materials in tension. Another parameter, $\sigma^{1/2}/\rho$, measures material efficiency for a rectangular beam of fixed length and width under a given load, where strength controls. Where deflection controls, $E^{1/3}/\rho$, where E is the modulus of elasticity (in pounds per square inch), applies to such a beam. It also applies to long, thin-walled cylinders of fixed length and diameter under external pressure. In Table 5-13, the efficiencies of various composites are compared with these parameters.

Table 5-13. Efficiencies of Composites Relative to Steel*

	Density, lb/cu in.	Strength, 10^3 psi	Modulus of elasticity, psi	Relative efficiencies σ/ρ	$\sigma^{1/2}/\rho$	$E^{1/3}/\rho$
Steel (4340)	0.28	200	30,000,000	1.00	1.00	1.00
Titanium (7A1-4Mo)	0.16	170	17,000,000	1.49	1.61	1.45
Aluminum (7075)	0.10	70	10,400,000	0.97	1.64	1.94
Glass mat, polyester	0.060	20	2,000,000	0.47	1.47	1.89
Glass fabric, epoxy	0.063	54	3,100,000	1.20	2.30	2.09
Glass filament wound, epoxy	0.075	68	3,500,000	1.27	2.18	1.82

NOTES: 1. Strength means yield stress in metals and composite tensile strength for plastics.
2. The filament-wound reinforced plastic had a matrix-to-filament volume ratio of 1/2 and 50/50 rectangular filament alignment.
*P. M. Goodwin, "Composite Materials," *Machine Design*, July 18, 1963, p. 190.

MECHANICS OF COMPOSITE BEHAVIOR

5-48. Matrix Systems. (See also Arts. 5-44, 5-45, and 5-47.) *Fibrous reinforcement* can range from short chopped fibers added randomly to a matrix to continuous parallel filaments. The binding matrix provides transverse stiffness and strength and transfers stress between filaments. A simple analysis of a fibrous reinforced matrix system might follow the method of transformed section used in the design of reinforced concrete (Art. 8-21). Such an analysis would be limited, however, because the bond strength of the matrix-reinforcement interface is not taken into consideration. The details of the interaction between fibers and matrix are complex, involving strain concentrations and local cracking, but must be taken into consideration in adequate analysis and design procedures.

Particulate fillers in composite systems involve several important design considerations. The size, shape, and size distribution of the aggregate determine the mechanical interaction between aggregate particles and with the matrix. The aggregate-to-matrix volume ratio and packing of the aggregate determine the consistency and properties of the mix. The surface area, wetting, and distribution of the aggregate affect the adhesion between the two phases. Volume changes on curing or loading may cause residual or transient internal stresses, particularly at the matrix-particle interface. These interface stresses must be compared with the bond strength between filler and matrix.

In asphalt composites, the mechanical interaction of the aggregate plays an important role in the behavior of the composite. Such tools as photoelastic analysis of an inclusion in a matrix have helped to give analytical approaches to such composites as concrete, where the matrix forms a solid continuum.

5-49. Sandwich Structures. In sandwich composites, the total thickness of the skins is usually less than one-fifth the thickness of the core. The properties of the structure (strength, stiffness, etc.) can be determined in terms of the properties of the face and the core, by simple mathematical analysis. Several different types of failure modes are possible in sandwich structures, depending on the type of loading and type of core: simple buckling, where the core and skins fail as a unit; shear crimping, where the composite fails as a unit; face dimpling (in honeycomb core structures), where the skin buckles between points of attachment to the core; and face wrinkling, where the face separates from or bulges into a solid core.

5-50. Laminates. Orthotropic materials are laminates whose principal axes are mutually perpendicular, with strengths different in these three directions. Because of a carry-over of terminology from wood technology, the three principal axes are called *longitudinal, transverse,* and *radial.* If the direction of loading is not parallel to one of the principal directions, a new system of descriptive coordinates can be defined in the loading directions. The two systems can be related by Mohr-circle relationships.

A single orthotropic ply loaded in directions not parallel to the natural axes can be analyzed, and after a fairly complicated development, elastic properties in the loading directions can be described. More complicated is the analysis of a plate composed of N layers, with the natural axes parallel in various layers. In the most general case, consider a plate of N layers, each orthotropic, natural axes of various layers not parallel, layers having various thicknesses, loading not parallel to any natural axes (but in their plane), and the elastic properties different in each layer, as well as different in orthotropic directions within each layer. A simplified analysis of this problem, neglecting stresses and strains in the thickness direction, is available from computer solutions of the many complicated expressions.

5-51. Composite Materials References

1. Corten, H. T.: "Reinforced Plastics," in Eric Baer, ed., "Engineering Design for Plastics," Van Nostrand Reinhold Company, New York.
2. Faupel, Joseph H.: "Engineering Design," John Wiley & Sons, Inc., New York.

Part 5. ENVIRONMENTAL INFLUENCES

Materials are usually subjected to atmospheres other than ideal inert conditions. They may encounter low or elevated temperatures, corrosion or oxidation, or irradiation by nuclear particles. Exposure to such environmental influences can affect the mechanical properties of the materials to such an extent that they do not meet service requirements.

THERMAL EFFECTS

5-52. Elevated or Low Temperatures. Variations in temperature are often divided into two classifications: *elevated temperatures* (above room temperature) and *low temperatures* (below room temperature). This can be misleading, since critical temperatures for the material itself may be high or low compared with room temperature. The lower limit of interest for all materials is absolute zero. The upper limit is the melting point for ceramics and metals, or melting or disintegration points for polymers and woods. Other critical temperatures include those for recrystallization in metals, softening and flow in thermoplastics, ductile-brittle transitions, and fictive temperature in glass. These temperatures mark the dividing lines between ranges in which materials behave in certain characteristic ways.

The immediate effect of thermal changes on materials is reflected in their mechanical properties, such as yield strength, viscous flow, and ultimate strength. For most materials there is a general downward trend of both yield and ultimate strength with increasing temperature. Sometimes, however, behavior irregularities in such are caused by structural changes (e.g., polymorphic transformations). Low-temperature behavior is usually defined on the basis of transition from ductile to brittle behavior. This phenomenon is particularly important in body-centered-cubic metals which show well-defined transition temperatures.

Table 5-14. Service Temperatures of Some Refractory Materials*

Material	Melting point, $°F$	Max service temp, $°F$	% absolute melting point
Refractories:			
Zirconia (ZrO_2)	4710	4530	96
Alumina ceramics	3686	3000	83
Tungsten	6170	2550	45
Molybdenum	4760	2650	59
Metals and alloys:			
Titanium	3100	1200	46
Steel (austenitic)	2800	1600	63
Aluminum	1220	550	60
Lead	620		
Plastics:			
Nylon	455	290	83
Acrylics	300	170	83

*After C. W. Richards, "Engineering Materials Science," Wadsworth Publishing Co., San Francisco, 1961.

Porous materials exhibit a special low-temperature effect: freezing and thawing. Concrete, for example, almost always contains water in its pores. Below $32°F$, the water is transformed to ice, which has a larger volume. The resulting swelling causes cracking. Thus, repeated thawing and freezing have a weakening effect on concrete. Brick is another, similar example.

5-53. Refractory Materials. Materials whose melting points are very high relative to room temperature are called *refractories*. They may be either metallic or nonmetallic (ceramic) but are usually the latter. Generally, refractories are defined as those materials having melting points above $3000°F$. Their absolute maximum service temperatures may be as high as 90% of their absolute melting temperatures. Table 5-14 lists maximum service temperatures of several materials compared with their melting points.

CORROSION AND OXIDATION

5-54. Mechanisms of Corrosion. The usually accepted definition for corrosion is limited to metals and involves some kind of chemical reaction: Corrosion is destruction of a metal by chemical or electrochemical reaction with its environment. There are other forms of degradation of materials similar to this: Solvents attack organic materials, sodium hydroxide dissolves glass, plastics may swell or crack, wood may split or decay, and portland cement may leach away. So the definition could be broadened: corrosion is deterioration and loss of material due to chemical attack.

The simplest corrosion is by means of chemical solution, where an engineering material is dissolved by a strong solvent (e.g., when a rubber hose through which gasoline flows is in contact with hydrocarbon solvents).

Wet corrosion occurs by mechanisms essentially electrochemical in nature. This process requires that the liquid in contact with the metallic material be an electrolyte. Also, there must exist a difference of potential either between two dissimilar metals or between different areas on the surface of a metal. Many variables modify the course and extent of the electrochemical reactions, but it is usually possible to explain the various forms of corrosion by referring to basic electrochemical mechanisms.

Corrosion of metals is one phenomenon now understood in some detail. Corrosion as a chemical reaction is a characteristic of metals associated with the freedom of their valence electrons. It is this very freedom that produces the metallic bond that makes metals useful by allowing electric conduction. Being loosely bound to their atoms, the electrons in metals are easily removed in chemical reactions. In the presence of nonmetals, such as oxygen, sulfur, or chlorine, with their incomplete valence shells, there is a tendency for metals to form a compound, thus corroding the metal.

Galvanic corrosion occurs when two dissimilar metals are in electrical contact with each other and are exposed to an electrolyte. A less noble metal will dissolve and form the anode, whereas the more noble metal will act as the cathode. The corrosion current flows at the expense of the anode metal, which is corroded, whereas the cathode metal is protected from the attack. A galvanic series lists metals in their order of corroding tendencies in a given environment and enables the probable corroding element to be identified. In seawater, for example, magnesium and zinc corrode more than steels, and lead, copper, and nickel corrode less than steels. Thus, in a galvanic cell of steel and nickel in seawater, the steel would be anodic (corroded) and the nickel cathodic (protected).

Corrosion by a gas involves the reaction between a metal and the molecules of a gas. The gas molecules are absorbed on the surface of the metal and react with the surface atoms to form corrosion products, such as oxides or salts. The corrosion products always form a layer or film on the surface of the metal. If the volume of the corrosion product is greater than that of the metal consumed in the reaction, the layer must be compressed to fit the surface. The result is a nonporous, protective shield over the metal surface. If the volume of corrosion product is less than that of the metal consumed, the layer must expand to cover the surface. Here, the result is a porous covering, which offers little or no protection against further corrosion.

Several types of corrosion are accelerated by the presence of some mechanical action. For example, if a local disorder is produced on a surface, the local energy is increased and the distorted material tends to become more anodic. The result is a localized decrease in resistance to corrosion. Examples of this stress corrosion include localized attack of cold-worked areas, such as sharp bends and punched holes; slip bands, which act as paths for internal corrosion across crystals; and stress-corrosion cracking, in which a metal under constant stress fails in tension after a time.

Pits and other surface irregularities produced by corrosion have the same effect on fatigue as other stress raisers, thus leading to corrosion fatigue. The constant reversal of strain has the effect of breaking any passivating film that may form on the surface. Thus, the corrosion fatigue strength of stainless steel may be as low as that of plain-carbon steels. With the formation of fatigue cracks at corrosion pits, the stress concentration at the crack tip further increases corrosion rate. Corrosion products fill the crack, exerting wedging action.

Other forms of corrosion include fretting corrosion due to mechanical wear in a corrosive atmosphere, cavitation damage serving to accelerate corrosion due to surface roughening, underground corrosion resulting from soil acidity, microbiological corrosion due to the metabolic activity of various microorganisms, and selective corrosion leading to the deterioration of alloys.

Concrete deterioration is generally attributable in part to chemical reactions between alkalies in the cement and mineral constituents of the concrete aggregates. Deterioration of concrete also results from contact with various chemical agents, which attack it in one of three forms: (1) corrosion resulting from the formation of soluble products that are removed by leaching; (2) chemical reactions producing products that disrupt the concrete because their volume is greater than that of the cement paste from which they were formed; and (3) surface deterioration by the crystallizing of salts in the pores of the concrete under alternate wetting and drying.

5-55. Corrosion Control and Prevention.   Proper selection of materials and sound engineering design are the best means of controlling and preventing corrosion. For example, avoid use of dissimilar metals in contact where galvanic corrosion may result. Also, alloying can be used to improve chemical resistance.

Modifying the environment may also control corrosion. Such techniques as dehumidification and purification of the atmosphere or the addition of alkalies to neutralize the acidic character of a corrosive environment are typical of this approach. Inhibitors that effectively decrease the corrosion rate when added in small amounts to a corrosive environment may be used to prevent or control the anodic and cathodic reactions in electrochemical cells.

In corrosion, galvanic cells are formed in which certain areas become anodes and others cathodes. Electric current flows through the electrolyte, and metal at the anode is dissolved or corroded. Cathodic protection reverses these currents and thereby makes cathodic all the metal to be protected. The procedure is to insert a new anode in the system, whose potential overcomes the potential of the original anode plus the resistance of the electrical elements. In this way, corrosion is concentrated in the new anode, which can be periodically replaced.

Application of protective coatings also furthers corrosion prevention and control. Three types of coating are often employed: mechanical protection, separating the electrode from electrolyte (paints, grease, fired enamels); galvanic protection by being anodic to the base metal (zinc coating on galvanized iron); and passivators, which shift the base metal toward the cathodic end of the electromotive series.

Several types of preservatives are employed to combat deterioration in woods: oily preservatives, such as coal-tar creosote; water-soluble salts, such as zinc chloride, sodium fluoride, copper salts, and mercuric salts; and solvent-soluble organic materials, such as pentachlorphenol. These preservatives may be applied by brushing, dipping, or pressure injection. Pressure treatments, by far the most effective, may be classified as either full or empty-cell. In the full-cell treatment, a partial vacuum is first drawn to remove the air from the wood cells; then the preservative is pumped in under pressure. In the empty-cell treatment, air pressure in the cells restricts the pressure-applied preservative to the cell walls.

IRRADIATION

5-56. Types of Irradiation. Radiation affects materials in a variety of ways because of the diversity of types of radiation and the differences in materials. Radiation effects may be either detrimental or beneficial to the desirable properties of a material. Some effects of radiation have been known for some time (e.g., discoloration of glass after long exposure to light; exposure of photographic film to electromagnetic radiation), while others have been more recently observed with the advent of nuclear reactors and accelerators (very high-energy radiation).

Radiation may be divided into two general groups:

1. Electromagnetic radiation, which is considered to be wavelike in nature (e.g., radio, heat, light, x-ray, gamma rays). These waves can also be considered as energy packets, called *photons.*

2. Radiation that is particulate in nature [e.g., accelerated protons (H^+), neutrons, electrons (beta rays), and helium nuclei (alpha rays)]. These rays, although particulate, have many of the characteristics of waves.

The passage of a neutron through a solid is much like the passage of a molecule through a gas. As a neutral particle, it is neither attracted nor repelled by other particles, and so it can interact with the solid only by direct collisions with atomic nuclei. The range of neutron energies goes from less than 1 ev (electron volt) to 10^6 ev (1 mev).

The quantity of radiation is measured in terms of neutron flux density, the number of neutrons traversing one unit of area per unit of time. For slow neutrons, this quantity ranges from 10^{11} to 4×10^{14} neutrons per cm per sec. Total irradiation is measured in terms of integrated neutron flux density. If n is the number of neutrons per unit volume and v is their average velocity, then nv is the neutron flux density. If t is the total time of exposure, then nvt is the integrated neutron flux density. Exposures of the order of 10^{18} to 10^{20} neutrons per sq cm are commonly used in experimental work.

5-57. Effects of Radiation. The principal effect of radiation on materials arises from the extra energy it supplies, which assists in the breaking of existing bonds and the rearrangement of the atoms into new structures. In metals, heavy particles with sufficient radiant energy, such as fission fragments and fast neutrons, may displace atoms from the lattice, resulting in vacancies, interstitial atoms, and dislocations. These imperfections affect the physical and mechanical properties of metals. The general effect is similar to that brought about by precipitation hardening or by cold work.

The hardening effects, like strain hardening, can be removed by annealing, which allows vacancies and interstitials to become mobile enough to recombine. In some metals, if the metal is held at high enough temperature while being irradiated (common in reactors), little hardening will actually occur. A disturbing development is that radiation embrittlement of steels cannot be depended upon to anneal out at ordinary reactor operating temperatures. Consequently, other materials (aluminum, titanium, and zirconium) are used for structural components in reactors.

In polymers, radiation damage seems to be a function of the actual radiation energy absorbed by the material-regardless of the nature of the radiation. The energy imparted causes excitation and ionization of the molecules, which produce free radicals and ions. These molecular fragments may recombine with each other or with displaced electrons and oxygen from the air, causing either an increase or a decrease in the molecular weight of the polymer. Thus, some polymers show an increased hardness, a higher softening point, and brittleness when irradiated, whereas others become soft. Most polymers lose strength through radiation damage.

5-58. Environmental Influences References

1. Clauss, F. J.: "Engineer's Guide to High-Temperature Materials," Addison-Wesley Publishing Company, Inc., Reading, Mass.

2. Fontana, M. G., and N. D. Greene: "Corrosion Engineering," McGraw-Hill Book Company, New York.

3. Hanson, A., and J. G. Parr: "The Engineer's Guide to Steel," Addison-Wesley Publishing Company, Inc., Reading, Mass.

4. Kircher, J. F., and R. E. Bowman: "Effects of Radiation on Materials and Components," Van Nostrand Reinhold Company, New York.

5. Uhlig, Herbert H.: "Corrosion and Corrosion Control," John Wiley & Sons, Inc., New York.

Structural Theory

FREDERICK S. MERRITT
Consulting Engineer, Syosset, N.Y.

Structural theory describes the behavior of structures under various types of loads. It predicts the strength and deformations of structures. Design formulas and methods based on structural theory, when verified by laboratory and field tests and observations of structures under service conditions, insure that a structure subjected to specified loads will not suffer structural damage. Such damage exists when any portion of a structure is unable to function satisfactorily. It may be indicated by excessive elastic deformation, inelastic deformation or yielding, fracture, or collapse.

To serve design and analysis needs, structural theory relates properties and arrangements of materials and the behavior of structures made of them. But if structural theory were to take into account every variable involved, it would become too complicated for practical use in most cases. So general practice is to make simplifying assumptions that yield consistent and sufficiently accurate results. Experience, experiments, and basic understanding often are required to determine whether a given theory or method is applicable to a particular structure.

EQUILIBRIUM

6-1. Types of Load. Loads are the external forces acting on a structure. Stresses are the internal forces that resist the loads.

Tensile forces tend to stretch a component, **compressive forces** tend to shorten it, and **shearing forces** tend to slide parts of it past each other.

Loads also may be classified as static or dynamic. **Static loads** are forces that are applied slowly and then remain nearly constant. One example is the weight, or dead load, of a floor system. **Dynamic loads** vary with time. They include repeated loads, such as alternating forces from oscillating machinery; moving loads, such as trucks or trains on bridges; impact loads, such as that from a falling weight striking a floor or the shock wave from an explosion impinging on a wall; and seismic loads or other forces created in a structure by rapid movements of its supports.

Loads may be considered distributed or concentrated. **Uniformly distributed loads** are forces that are, or for practical purposes may be considered, constant over a surface of the supporting member.   Dead weight of a rolled-steel beam is a good example.   **Concentrated loads** are forces that have such a small contact area as to be negligible compared with the entire surface area of the supporting member.   For example, a beam supported on a girder, for all practical purposes, may be considered a concentrated load on the girder.

In addition loads may be axial, eccentric, or torsional.   An **axial load** is a force whose resultant passes through the centroid of a section under consideration and is perpendicular to the plane of the section.   An **eccentric load** is a force perpendicular to the plane of the section under consideration but not passing through the centroid of the section, thus bending the supporting member.   **Torsional loads** are forces that are offset from the shear center of the section under consideration and are inclined to or in the plane of the section, thus twisting the supporting member.

Also, loads are classified in accordance with the nature of the source.   For example: **Dead loads** include materials, equipment, constructions, or other elements of weight supported in, on, or by a structural element, including its own weight, that are intended to remain permanently in place.   **Live loads** include all occupants, materials, equipment, constructions, or other elements of weight supported in, on, or by a structural element that will or are likely to be moved or relocated during the expected life of the structure. **Impact loads** are a fraction of the live loads used to account for additional stresses and deflections resulting from movement of the live loads.   **Wind loads** are maximum forces that may be applied to a structural element by wind in a mean recurrence interval, or a set of forces that will produce equivalent stresses.   Mean recurrence intervals generally used are 25 years for structures with no occupants or offering negligible risk to life, 50 years for ordinary permanent structures, and 100 years for permanent structures with a high degree of sensitivity to wind and an unusually high degree of hazard to life and property in case of failure.   **Snow loads** are maximum forces that may be applied by snow accumulation in a mean recurrence interval.   **Seismic loads** are forces that produce maximum stresses or deformations in a structural element during an earthquake, or equivalent forces.

Probable maximum loads should be used in design.   For buildings, minimum design load should be that specified for expected conditions in the local building code or, in the absence of an applicable local code, in the American National Standard "Building Code Requirements for Minimum Design Loads in Buildings and Other Structures," A58.1, American National Standards Institute, New York.   For highways and highway bridges, minimum design loads should be those given in "Standard Specifications for Highway Bridges," American Association of State Highway and Transportation Officials, Washington, D.C.   For railways and railroad bridges, minimum design loads should be those given in "Manual for Railway Engineering," American Railway Engineering Association, Chicago.

6-2.   Static Equilibrium.   If a structure and its components are so supported that after a small deformation occurs no further motion is possible, they are said to be in equilibrium. Under such circumstances, external forces are in balance and internal forces, or stresses, exactly counteract the loads.

Since there is no translatory motion, the vector sum of the external forces must be zero. Since there is no rotation, the sum of the moments of the external forces about any point must be zero.   For the same reason, if we consider any portion of the structure and the loads on it, the sum of the external and internal forces on the boundaries of that section must be zero.   Also, the sum of the moments of these forces must be zero.

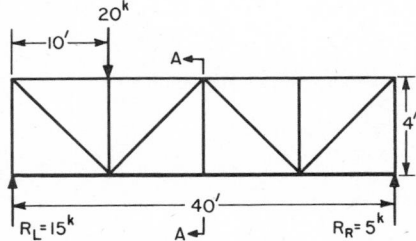

Fig. 6-1. Truss in equilibrium under load. Upward-acting forces, or reactions, R_L and R_R, equal the 20-kip downward-acting force.

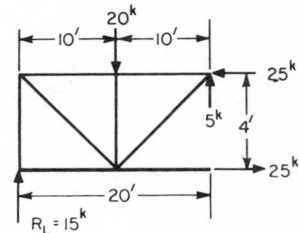

Fig. 6-2. Section of truss kept in equilibrium by stresses in its components.

In Fig. 6-1, for example, the sum of the forces R_L and R_R needed to support the truss is equal to the 20-kip load on the truss (1 **kip** = 1 kilopound = 1,000 lb = 0.5 ton). Also, the sum of the moments of the external forces is zero about any point; about the right end, for instance, it is $40 \times 15 - 30 \times 20 = 600 - 600$.

Figure 6-2 shows the portion of the truss to the left of section AA. The internal forces at the cut members balance the external load and hold this piece of the truss in equilibrium.

When the forces act in several directions, it generally is convenient to resolve them into components parallel to a set of perpendicular axes that will simplify computations. For example, for forces in a single plane, the most useful technique is to resolve them into horizontal and vertical components. Then, for a structure in equilibrium, if H represents the horizontal components, V the vertical components, and M the moments of the components about any point in the plane,

$$\Sigma H = 0 \qquad \Sigma V = 0 \qquad \text{and} \qquad \Sigma M = 0 \qquad (6\text{-}1)$$

These three equations may be used to determine three unknowns in any nonconcurrent coplanar force system, such as the truss in Figs. 6-1 and 6-2. They may determine the magnitude of three forces for which the direction and point of application already are known, or the magnitude, direction, and point of application of a single force. Suppose, for the truss in Fig. 6-1, the reactions at the supports are to be computed. Take the sum of the moments about the right support and equate them to zero to determine the left reaction: $40R_L - 30 \times 20 = 0$, from which $R_L = {}^{600}/_{40} = 15$ kips. To find the right reaction, take moments about the left support and equate the sum to zero: $10 \times 20 - 40R_R = 0$, from which $R_R = 5$ kips. As an alternative, equate the sum of the vertical forces to zero to obtain R_R after finding R_L: $20 - 15 - R_R = 0$, from which $R_R = 5$ kips.

STRESS AND STRAIN

6-3. Unit Stress. It is customary to give the strength of a material in terms of unit stress, or internal force per unit of area. Also, the point at which yielding starts generally is expressed as a *unit stress*. Then, in some design methods, a safety factor is applied to either of these stresses to determine a unit stress that should not be exceeded when the member carries design loads. That unit stress is known as the *allowable stress*, or *working stress*.

In working-stress design, to determine whether a structural member has adequate load-carrying capacity, the designer generally has to compute the maximum unit stress produced by design loads in the member for each type of internal force—tensile, compressive, or shearing—and compare it with the corresponding allowable unit stress.

When the loading is such that the unit stress is constant over a section under consideration, the stress may be computed by dividing the force by the area of the section. But, in general, the unit stress varies from point to point. In those cases, the unit stress at any point in the section is the limiting value of the ratio of the internal force on any small area to that area, as the area is taken smaller and smaller.

6-4. Unit Strain. Sometimes in the design of a structure, the designer may be more concerned with limiting deformation or strain than with strength. Deformation in any direction is the total change in the dimension of a member in that direction. *Unit strain* in any direction is the deformation per unit of length in that direction.

When the loading is such that the unit strain is constant over the length of a member, it may be computed by dividing the deformation by the original length of the member. In general, however, unit strain varies from point to point in a member. Like a varying unit stress, it represents the limiting value of a ratio (Art. 6-3).

6-5. Stress-Strain Relations. When a material is subjected to external forces, it will develop one or more of the following types of strain: linear elastic, nonlinear elastic, visco-elastic, plastic, and anelastic. Many structural materials exhibit linear elastic strains under design loads. For these materials, unit strain is proportional to unit stress until a certain stress, the proportional limit, is exceeded (point A in Fig. 6-3a, b, and c). This relationship is known as **Hooke's law.**

For axial tensile or compressive loading, this relationship may be written

$$f = E\epsilon \qquad \text{or} \qquad \epsilon = \frac{f}{E} \qquad (6\text{-}2)$$

where f = unit stress
    ϵ = unit strain
    E = Young's modulus of elasticity
Within the elastic limit, there is no permanent residual deformation when the load is removed. Structural steels have this property.

In nonlinear elastic behavior, stress is not proportional to strain but there is no permanent residual deformation when the load is removed. The relation between stress and strain may take the form

$$\epsilon = \left(\frac{f}{K}\right)^{n} \tag{6-3}$$

where K = a pseudoelastic modulus determined by test
    n = constant determined by test
Viscoelastic behavior resembles linear elasticity. The major difference is that in linear elastic behavior, the strain stops increasing if the load does; but in viscoelastic behavior, the strain continues to increase though the load becomes constant, and a residual strain remains when the load is removed. This is characteristic of many plastics.

Anelastic deformation is time-dependent and completely recoverable. Strain at any time is proportional to change in stress. Behavior at any given instant depends on all prior stress changes. The combined effect of several stress changes is the sum of the effects of the several stress changes taken individually.

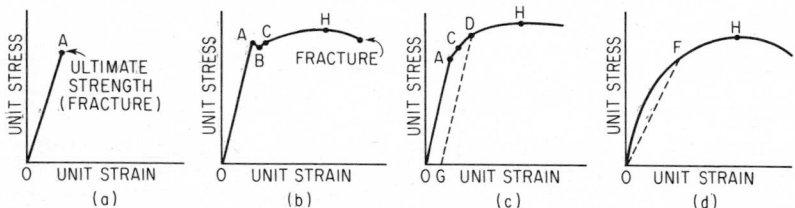

Fig. 6-3. Relationship of unit stress and unit strain for various materials. (*a*) Brittle. (*b*) Linear elastic with a distinct proportional limit. (*c*) Linear elastic with an indistinct proportional limit. (*d*) Nonlinear.

Plastic strain is not proportional to stress, and a permanent deformation remains on removal of the load. In contrast with anelastic behavior, plastic deformation depends primarily on the stress and is largely independent of prior stress changes.

When materials are tested in axial tension and corresponding stresses and strains are plotted, stress-strain curves similar to those in Fig. 6-3 result. Figure 6-3*a* is typical of a brittle material, which deforms in accordance with Hooke's law up to fracture. The other curves in Fig. 6-3 are characteristic of ductile materials; because strains increase rapidly near fracture with little increase in stress, they warn of imminent failure, whereas brittle materials fail suddenly.

Figure 6-3*b* is typical of materials with a marked proportional limit *A*. When this is exceeded, there is a sudden drop in stress, then gradual stress increase with large increases in strain to a maximum before fracture. Figure 6-3*c* is characteristic of materials that are linearly elastic over a substantial range but have no definite proportional limit. And Fig. 6-3*d* is a representative curve for materials that do not behave linearly at all.

Modulus of elasticity *E* is given by the slope of the straight-line portion of the curves in Fig. 6-3*a*, *b*, and *c*. It is a measure of the inherent rigidity or stiffness of a material. For a given geometric configuration, a material with a larger *E* deforms less under the same stress.

At the termination of the linear portion of the stress-strain curve, some materials, such as low-carbon steel, develop an upper and lower **yield point** (*A* and *B* in Fig. 6-3*b*). These points mark a range in which there appears to be an increase in strain with no increase or a small decrease in stress. This behavior may be a consequence of inertia effects in the testing machine and the deformation characteristics of the test specimen. Because of the location of the yield points, the yield stress sometimes is used erroneously as a synonym for proportional limit and elastic limit.

The **proportional limit** is the maximum unit stress for which Hooke's law is valid. The **elastic limit** is the largest unit stress that can be developed without a permanent set remaining after removal of the load (C in Fig. 6-3). Since the elastic limit is always difficult to determine and many materials do not have a well-defined proportional limit, or even have one at all, the offset yield strength is used as a measure of the beginning of plastic deformation.

The **offset yield strength** is defined as the stress corresponding to a permanent deformation, usually 0.01% (0.0001 in. per in.) or 0.20% (0.002 in. per in.). In Fig. 6-3c, the yield strength is the stress at D, the intersection of the stress-strain curve and a line GD parallel to the straight-line portion and starting at the given unit strain. This stress sometimes is called the **proof stress.**

For materials with a stress-strain curve similar to that in Fig. 6-3d, with no linear portion, a **secant modulus,** represented by the slope of a line, such as OF, from the origin to a specified point on the curve, may be used as a measure of stiffness. An alternative measure is the **tangent modulus,** the slope of the stress-strain curve at a specified point.

Ultimate tensile strength is the maximum axial load observed in a tension test divided by the original cross-sectional area. Characterized by the beginning of necking down, a decrease in cross-sectional area of the specimen, or local instability, this stress is indicated by H in Fig. 6-3.

Ductility is the ability of a material to undergo large deformations without fracture. It is measured by elongation and reduction of area in a tension test and expressed as a percentage. Ductility depends on temperature and internal stresses, as well as on the characteristics of the material; a material that may behave ductilely under one set of conditions may have a brittle failure at lower temperatures or under tensile stresses in two or three perpendicular directions.

Modulus of rigidity, or shearing modulus of elasticity, is defined by

$$G = \frac{v}{\gamma} \tag{6-4}$$

where G = modulus of rigidity
v = unit shearing stress
γ = unit shearing strain
It is related to the modulus of elasticity in tension and compression E by the equation

$$G = \frac{E}{2(1 + \mu)} \tag{6-5}$$

where μ is a constant known as Poisson's ratio (Art. 6-7).

Toughness is the ability of a material to absorb large amounts of energy. Related to the area under the stress-strain curve, it depends on both strength and ductility. Because of the difficulty of determining toughness analytically, often toughness is measured by the energy required to fracture a specimen, usually notched and sometimes at low temperatures, in impact tests. Charpy and Izod, both applying a dynamic load by pendulum, are the tests most commonly used.

Hardness is a measure of the resistance a material offers to scratching and indentation. A relative numerical value usually is determined for this property in such tests as Brinell, Rockwell, and Vickers. The numbers depend on the size of an indentation made under a standard load. Scratch resistance is measured on the Mohs scale by comparison with the scratch resistance of 10 minerals arranged in order of increasing hardness from talc to diamond.

Creep is a gradual flow or change in dimension under sustained constant load. **Relaxation** is a decrease in load or stress under a sustained constant deformation.

If stresses and strains are plotted in an axial tension test as a specimen enters the inelastic range and then is unloaded, the curve during unloading, if the material was elastic, will descend parallel to the straight portion of the curve (for example, DG in Fig. 6-3c). Completely unloaded, the specimen will have a permanent set (OG). This also will occur in compression tests.

If the specimen now is reloaded, strains will be proportional to stresses (the curve will practically follow DG) until the curve rejoins the original curve at D. Under increasing load, the reloading curve will coincide with that for a single loading. Thus, loading the specimen into the inelastic range, but not to ultimate strength, increases the apparent elastic range. The phenomenon, called **strain hardening,** or work hardening, appears to increase the yield strength.

But if the reloading is in compression, the compressive yield strength would be decreased. This is called the **Bauschinger effect.** This, however, is present only for relatively small strains. For large inelastic tensile strains initially, reloading in compression increases yield strength to some extent. But if this reloading is continued to a higher stress than that reached in the initial loading in tension, the yield strength will not show any increase in subsequent tension loading.

(F. R. Shanley, "Strength of Materials," McGraw-Hill Book Company, New York; J. H. Faupel, "Engineering Design," John Wiley & Sons, Inc., New York; J. E. Dorn, "Mechanical Behavior of Materials at Elevated Temperatures," McGraw-Hill Book Company, New York; M. J. Sinnott, "The Solid State for Engineers," John Wiley & Sons, Inc., New York.)

6-6. Constant Unit Stress.   The simplest cases of stress and strain are those in which the unit stress and strain are constant.   Stresses due to an axial tension or compression load, a centrally applied shear, or a bearing load are examples.   These conditions are illustrated in Figs. 6-4 to 6-7.

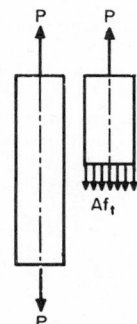

Fig. 6-4. Tension member.

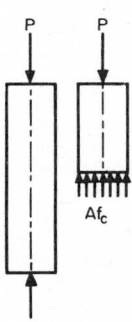

Fig. 6-5. Compression member.

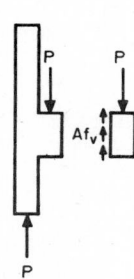

Fig. 6-6. Bracket in shear.

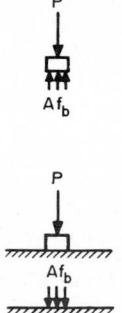

Fig. 6-7. Bearing load.

For constant unit stress, the equation of equilibrium may be written

$$P = Af \tag{6-6}$$

where P = load, lb

A = cross-sectional area (normal to load) for tensile or compressive forces, or area on which sliding may occur for shearing forces, or contact area for bearing loads, sq in.

f = a tensile, compressive, shearing, or bearing unit stress, psi

For torsional stresses, see Art. 6-75.

Unit strain for the axial tensile and compressive loads is given by

$$\epsilon = \frac{e}{L} \tag{6-7}$$

where ϵ = unit strain, in. per in.

e = total lengthening or shortening of the member, in.

L = original length of the member, in.

Application of Hooke's law and Eq. (6-6) to Eq. (6-7) yields a convenient formula for the deformation:

$$e = \frac{PL}{AE} \tag{6-8}$$

where P = load on member, lb
A = its cross-sectional area, sq in.
E = modulus of elasticity of the material, psi
[Since long compression members tend to buckle, Eqs. (6-6) to (6-8) are applicable only to short members. See Arts. 6-71 to 6-73.]

Although tension and compression strains represent a simple stretching or shortening of a member, shearing strain is a distortion due to a small rotation. The load on the small rectangular portion of the member in Fig. 6-6 tends to distort it into a parallelogram. The unit shearing strain is the change in the right angle, measured in radians. (See also Art. 6-5.)

6-7. Poisson's Ratio. When a material is subjected to axial tensile or compressive loads, it deforms not only in the direction of the loads but also normal to them. Under tension, the cross section of a member decreases, and under compression, it increases. The ratio of the unit lateral strain to the unit longitudinal strain is called *Poisson's ratio.*

Within the elastic range, Poisson's ratio is a constant for a material. For materials such as concrete, glass, and ceramics, it may be taken as 0.25; for structural steel, 0.3. It gradually increases beyond the proportional limit and tends to approach a value of 0.5. (F. R. Shanley, "Strength of Materials," McGraw-Hill Book Company, New York.)

Assume, for example, that a steel hanger with an area of 2 sq in. carries a 40-kip (40,000-lb) load. The unit stress is 40,000/2, or 20,000 psi. The unit tensile strain, with modulus of elasticity of steel E = 30,000,000, is 20,000/30,000,000, or 0.00067 in. per in. With Poisson's ratio as 0.3, the unit lateral strain is -0.3×0.00067, or a shortening of 0.00020 in. per in.

6-8. Thermal Stresses. When the temperature of a body changes, its dimensions also change. Forces are required to prevent such dimensional changes, and stresses are set up in the body by these forces.

If α is the coefficient of expansion of the material and T the change in temperature, the unit strain in a bar restrained by external forces from expanding or contracting is

$$\epsilon = \alpha T \tag{6-9}$$

According to Hooke's law, the stress f in the bar is

$$f = E\alpha T \tag{6-10}$$

where E = modulus of elasticity

When a circular ring, or hoop, is heated, then slipped over a cylinder of slightly larger diameter d than d_1, the original hoop diameter, the hoop will develop a tensile stress on cooling. If the diameter is very large compared with the hoop thickness, so that radial stresses can be neglected, the unit tensile stresses may be assumed constant. The unit strain will be

$$\epsilon = \frac{\pi d - \pi d_1}{\pi d_1} = \frac{d - d_1}{d_1}$$

and the hoop stress will be

$$f = \frac{(d - d_1)E}{d_1} \tag{6-11}$$

6-9. Axial Stresses in Composite Members. In a homogeneous material, the centroid of a cross section lies at the intersection of two perpendicular axes so located that the moments of the areas on opposite sides of an axis about that axis are zero. To find the centroid of a cross section containing two or more materials, the moments of the products of the area A of each material and its modulus of elasticity E should be used, in the elastic range.

Consider now a prism composed of two materials, with modulus of elasticity E_1 and E_2, extending the length of the prism. If the prism is subjected to a load acting along the centroidal axis, then the unit strain ϵ in each material will be the same. From the equation of equilibrium and Eq. (6-8), noting that the length L is the same for both materials,

$$\epsilon = \frac{P}{A_1 E_1 + A_2 E_2} = \frac{P}{\Sigma AE} \tag{6-12}$$

where A_1 and A_2 are the cross-sectional areas of each material and P the axial load. The unit stresses in each material are the products of the unit strain and its modulus of elasticity:

$$f_1 = \frac{PE_1}{\Sigma AE} \qquad f_2 = \frac{PE_2}{\Sigma AE} \tag{6-13}$$

6-10. Stresses in Pipes and Pressure Vessels. In a cylindrical pipe under internal radial pressure, the circumferential unit stresses may be assumed constant over the thickness of the pipe if the diameter is relatively large compared with the thickness (at least 15 times as large). Then, the circumferential unit stress, in pounds per square inch, is given by

$$f = \frac{pR}{t} \tag{6-14}$$

where p = internal pressure, psi
  R = average radius of the pipe, in. (see also Art. 21-14)
  In a closed cylinder, the pressure against the ends will be resisted by longitudinal stresses in the cylinder. If the cylinder is thin, these stresses, psi, are given by

$$f_z = \frac{pR}{2t} \tag{6-15}$$

Equation (6-15) also holds for the stress in a thin spherical tank under internal pressure p with R the average radius.
  In a thick-walled cylinder, the effect of radial stresses f_r becomes important. Both radial and circumferential stresses may be computed from Lamé's formulas:

$$f_r = p \frac{r_i^2}{r_o^2 - r_i^2} \left(1 - \frac{r_o^2}{r^2} \right) \tag{6-16}$$

$$f = p \frac{r_i^2}{r_o^2 - r_i^2} \left(1 + \frac{r_o^2}{r^2} \right) \tag{6-17}$$

where r_i = internal radius of cylinder, in.
  r_o = outside radius of cylinder, in.
  r = radius to point where stress is to be determined
The equations show that if the pressure p acts outward, the circumferential stress f will be tensile (positive) and the radial stress compressive (negative). The greatest stresses occur at the inner surface of the cylinder ($r = r_i$):

$$\text{Max } f_r = -p \tag{6-18}$$

$$\text{Max } f = \frac{k^2 + 1}{k^2 - 1} p \tag{6-19}$$

where $k = r_o/r_i$. Maximum shear stress is given by

$$\text{Max } f_v = \frac{k^2}{k^2 - 1} p \tag{6-20}$$

For a closed cylinder with thick walls, the longitudinal stress is approximately

$$f_z = \frac{p}{r_i(k^2 - 1)} \tag{6-21}$$

But because of end restraints, this stress will not be correct near the ends.
  (S. Timoshenko and J. N. Goodier, "Theory of Elasticity," McGraw-Hill Book Company, New York; F. R. Shanley, "Strength of Materials," McGraw-Hill Book Company, New York.)
  6-11. Strain Energy. Stressing a bar stores energy in it. For an axial load P and a deformation e, the energy stored is

$$U = \frac{1}{2} Pe \tag{6-22}$$

assuming the load is applied gradually and the bar is not stressed beyond the proportional limit. The equation represents the area under the load-deformation curve up to the load P. Application of Eqs. (6-2) and (6-6) to Eq. (6-22) yields another useful equation for energy, in inch-pounds:

$$U = \frac{f^2}{2E} AL \tag{6-23}$$

where f = unit stress, psi
    E = modulus of elasticity of the material, psi
    A = cross-sectional area, sq in.
    L = length of the bar, in.

Since AL is the volume of the bar, the term $f^2/2E$ gives the energy stored per unit of volume. It represents the area under the stress-strain curve up to the stress f.

Modulus of resilience is the energy stored per unit of volume in a bar stressed by a gradually applied axial load up to the proportional limit. This modulus is a measure of the capacity of the material to absorb energy without danger of being permanently deformed. It is important in designing members to resist energy loads.

Equation (6-22) is a general equation that holds true when the **principle of superposition** applies (the total deformation produced at a point by a system of forces is equal to the sum of the deformations produced by each force). In the general sense, P in Eq. (6-22) represents any group of statically interdependent forces that can be completely defined by one symbol, and e is the corresponding deformation.

The strain-energy equation can be written as a function of either the load or the deformation. For axial tension or compression, strain energy, in inch-pounds, is given by

$$U = \frac{P^2L}{2AE} \qquad U = \frac{AEe^2}{2L} \qquad \text{(6-22a)}$$

where P = axial load, lb
    e = total elongation or shortening, in.
    L = length of member, in.
    A = cross-sectional area, sq in.
    E = modulus of elasticity, psi
For pure shear:

$$U = \frac{V^2L}{2AG} \qquad U = \frac{AGe^2}{2L} \qquad \text{(6-22b)}$$

where V = shearing load, lb
    e = shearing deformation, in.
    L = length over which deformation takes place, in.
    A = shearing area, sq. in.
    G = shearing modulus, psi
For torsion:

$$U = \frac{T^2L}{2JG} \qquad U = \frac{JG\phi^2}{2L} \qquad \text{(6-22c)}$$

where T = torque, in.-lb
    ϕ = angle of twist, radians
    L = length of shaft, in.
    J = polar moment of inertia of the cross section, in.^{4}
    G = shearing modulus, psi
For pure bending (constant moment):

$$U = \frac{M^2L}{2EI} \qquad U = \frac{EI\theta^2}{2L} \qquad \text{(6-22d)}$$

where M = bending moment, in.-lb
    θ = angle of rotation of one end of the beam with respect to the other, radians
    L = length of beam, in.
    I = moment of inertia of the cross section, in.^{4}
    E = modulus of elasticity, psi (see also Art. 6-55)
For beams carrying transverse loads, the total strain energy is the sum of the energy for bending and that for shear.

STRESSES AT A POINT

Tensile and compressive stresses sometimes are referred to as *normal stresses* because they act normal to the cross section. Under this concept, tensile stresses are considered positive normal stresses and compressive stresses negative.

6-12. Stress Notation.   Consider a small cube extracted from a stressed member and placed with three edges along a set of x, y, z coordinate axes.   The notations used for the components of stress acting on the sides of this element and the directions assumed as positive are shown in Fig. 6-8.

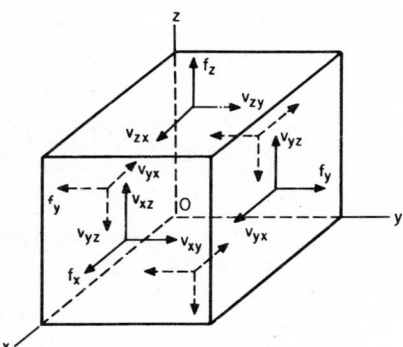

Fig. 6-8. Stresses at a point in a rectangular coordinate system.

For example, for the sides of the element perpendicular to the z axis, the normal component of stress is denoted by f_z.   The shearing stress v is resolved into two components and requires two subscript letters for a complete description.   The first letter indicates the direction of the normal to the plane under consideration.   The second letter gives the direction of the component of stress.   Thus, for the sides perpendicular to the z axis, the shear component in the x direction is labeled v_{zx} and that in the y direction v_{zy}.

6-13. Stress Components.   If, for the small cube in Fig. 6-8, moments of the forces acting on it are taken about the x axis, and assuming the lengths of the edges as dx, dy, and dz, the equation of equilibrium requires that

$$(v_{zy} dx\, dy)\, dz = (v_{yz} dx\, dz)\, dy$$

(Forces are taken equal to the product of the area of the face and the stress at the center.)   Two similar equations can be written for moments taken about the y axis and the z axis.   These equations show that

$$v_{xy} = v_{yx} \qquad v_{zx} = v_{xz} \qquad v_{zy} = v_{yz} \tag{6-24}$$

Thus, components of shearing stress on two perpendicular planes and acting normal to the intersection of the planes are equal.   Consequently, to describe the stresses acting on the coordinate planes through a point, only six quantities need be known: the three normal stresses f_x, f_y, f_z and three shearing components $v_{xy} = v_{yx}$, $v_{zx} = v_{xz}$, and $v_{zy} = v_{yz}$.

If only normal stresses are acting, the unit strains in the x, y, and z directions are

$$\epsilon_x = \frac{1}{E} \left[f_x - \mu(f_y + f_z) \right]$$

$$\epsilon_y = \frac{1}{E} \left[f_y - \mu(f_x + f_z) \right] \tag{6-25}$$

$$\epsilon_z = \frac{1}{E} \left[f_z - \mu(f_x + f_y) \right]$$

where μ = Poisson's ratio.   If only shearing stresses are acting, the distortion of the angle between edges parallel to any two coordinate axes depends only on shearing-stress components parallel to those axes.   Thus, the unit shearing strains are (see Art. 6-5)

$$\gamma_{xy} = \frac{1}{G}\, v_{xy} \qquad \gamma_{yz} = \frac{1}{G}\, v_{yz} \qquad \gamma_{zx} = \frac{1}{G}\, v_{zx} \tag{6-26}$$

6-14. Two-dimensional Stress.   When the six components of stress necessary to describe the stresses at a point are known (Art. 6-13), the stresses on any inclined plane through the same point can be determined.   For two-dimensional stress, only three stress components need be known.

Assume, for example, that at a point O in a stressed plate, the components f_x, f_y, and v_{xy} are known (Fig. 6-9). To find the stresses on any other plane through the z axis, take a plane parallel to it close to O, so that this plane and the coordinate planes form a tiny triangular prism. Then, if

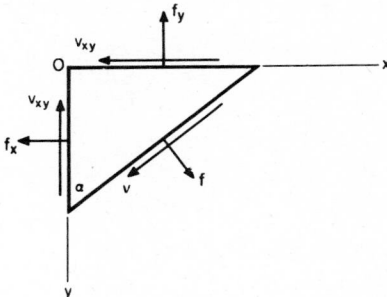

Fig. 6-9. Stresses at a point on a plane inclined to the axes.

α is the angle the normal to the plane makes with the x axis, the normal and shearing stresses on the inclined plane, to maintain equilibrium, are

$$f = f_x \cos^2 \alpha + f_y \sin^2 \alpha + 2v_{xy} \sin \alpha \cos \alpha \tag{6-27}$$

$$v = v_{xy}(\cos^2 \alpha - \sin^2 \alpha) + (f_y - f_x) \sin \alpha \cos \alpha \tag{6-28}$$

(See also Art. 6-17.)

6-15. Principal Stresses. If a plane at a point O in a stressed plate is rotated, it will reach a position for which the normal stress on it is a maximum or a minimum. The directions of maximum and minimum normal stress are perpendicular to each other. And on the planes in those directions, there are no shearing stresses.

The directions in which the normal stresses become maximum or minimum are called *principal directions*, and the corresponding normal stresses are called *principal stresses*. To find the principal directions, set the value of v given by Eq. (6-28) equal to zero. Then, the normals to the principal planes make an angle with the x axis given by

$$\tan 2\alpha = \frac{2v_{xy}}{f_x - f_y} \tag{6-29}$$

If the x and y axes are taken in the principal directions, $v_{xy} = 0$. In that case, Eqs. (6-27) and (6-28) simplify to

$$f = f_x \cos^2 \alpha + f_y \sin^2 \alpha \tag{6-30}$$

$$v = \frac{1}{2} (f_y - f_x) \sin 2\alpha \tag{6-31}$$

where f_x and f_y are the principal stresses at the point, and f and v are, respectively, the normal and shearing stress on a plane whose normal makes an angle α with the x axis.

If only shearing stresses act on any two perpendicular planes, the state of stress at the point is said to be one of pure shear or simple shear. Under such conditions, the principal directions bisect the angles between the planes on which these shearing stresses act. The principal stresses are equal in magnitude to the pure shears.

6-16. Maximum Shearing Stress at a Point. The maximum unit shearing stress occurs on each of two planes that bisect the angles between the planes on which the principal stresses at a point act. The maximum shear equals half the algebraic difference of the principal stresses:

$$\text{Max } v = \frac{f_1 - f_2}{2} \tag{6-32}$$

where f_1 is the maximum principal stress and f_2 the minimum.

6-17. Mohr's Circle. As explained in Art. 6-14, if the stresses on any plane through a point in a stressed plate are known, the stresses on any other plane through the point can be computed. This relationship between the stresses may be represented conveniently on Mohr's circle (Fig. 6-10). In this diagram, normal stress f and shear stress v are taken as rectangular coordinates.

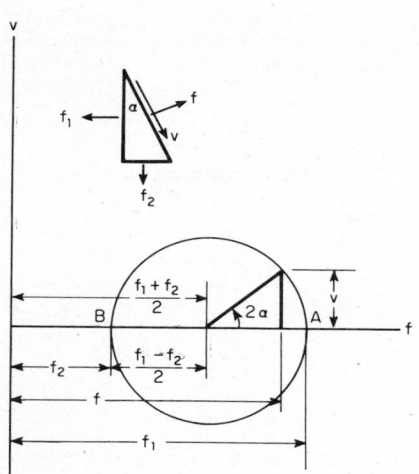

Fig. 6-10. Mohr's circle for stresses at a point— constructed from known principal stresses f_1 and f_2 in a plane.

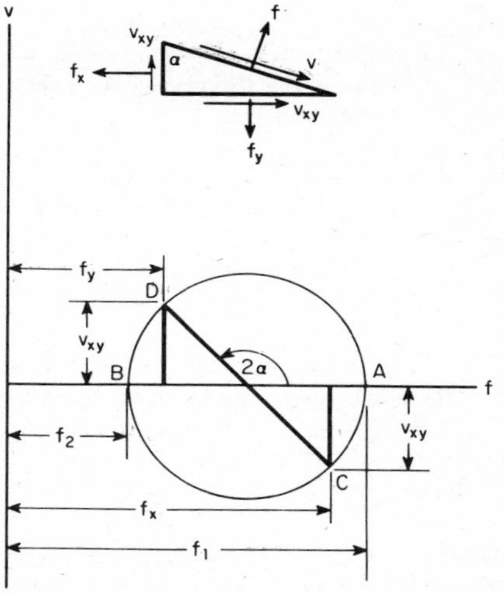

Fig. 6-11. Stress circle constructed from two known normal positive stresses f_x and f_y and a known shear v_{xy}.

Then, for each plane through the point there will correspond a point on the circle the coordinates of which are the values of f and v for the plane.

Given the principal stresses f_1 and f_2 (Art. 6-15), to find the stresses on a plane making an angle α with the plane on which f_1 acts: Mark off the principal stresses on the f axis (points A and B in Fig. 6-10). Measure tensile stresses to the right of the v axis and compressive stresses to the left. Construct a circle passing through A and B and having its center on the f axis. This is the Mohr's circle for the given stresses at the point under consideration. Draw a radius making an angle 2α with the f axis, as indicated in Fig. 6-10. The coordinates of the intersection with the circle represent the normal and shearing stresses acting on the plane f and v.

Given the stresses on any two perpendicular planes f_x, f_y, and v_{xy}, but not the principal stresses f_1 and f_2, to draw the Mohr's circle: Plot the two points representing the known stresses with respect to the f and v axes (points C and D in Fig. 6-11). The line joining these points is a diameter of the circle. So bisect CD to find the center of the circle and draw the circle. Its intersections with the f axis determine f_1 and f_2.

(S. Timoshenko and J. N. Goodier, "Theory of Elasticity," McGraw-Hill Book Company, New York; I. S. Sokolnikoff, "Mathematical Theory of Elasticity," McGraw-Hill Book Company, New York.)

STRAIGHT BEAMS

6-18. Beam-and-Girder Framing.   Bridge decks and floors and roofs of buildings frequently are supported on a rectangular grid of flexural members. Different names often are given to the components of the grid, depending on the type of structure and the part of the structure supported on the grid. In general, though, the members spanning between main supports are called **girders** and those they support are called **beams** (Fig. 6-12). Hence, this type of framing is known as beam-and-girder framing.

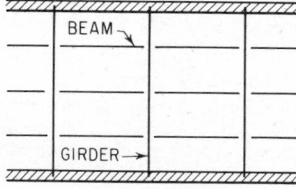

Fig. 6-12. Beam-and-girder framing.

In bridges, the smaller structural members parallel to the direction in which traffic moves may be called **stringers** and the transverse members **floor beams.** In building roofs, the grid components may be referred to as **purlins** and **rafters;** and in floors, they may be called **joists** and **girders.**

Beam-and-girder framing usually is used for relatively short spans and where shallow members are desired to provide ample headroom underneath.

6-19. Types of Beams. There are many ways in which beams may be supported. Some of the commoner methods are shown in Figs. 6-13 to 6-19. The beam in Fig. 6-13 is called a simply supported beam, or **simple beam.** It has supports near its ends that restrain it only against vertical movement. The ends of the beam are free to rotate. When the loads have a horizontal component, or when change in length of the beam due to temperature may be important, the supports may also have to prevent horizontal motion. In that case, horizontal restraint at one support generally is sufficient. The distance between the supports is called the **span.** The load carried by each support is called a **reaction.**

The beam in Fig. 6-14 is a **cantilever.** It has a support only at one end. The support provides restraint against rotation and horizontal and vertical movement. Such a support is called a **fixed end.** Placing a support under the free end of the cantilever produces the beam in Fig. 6-15. Fixing the free end yields a **fixed-end beam** (Fig. 6-16); no rotation or vertical movement can occur at either end. In actual practice, however, a fully fixed end can seldom be obtained. Most support conditions are intermediate between those for a simple beam and those for a fixed-end beam.

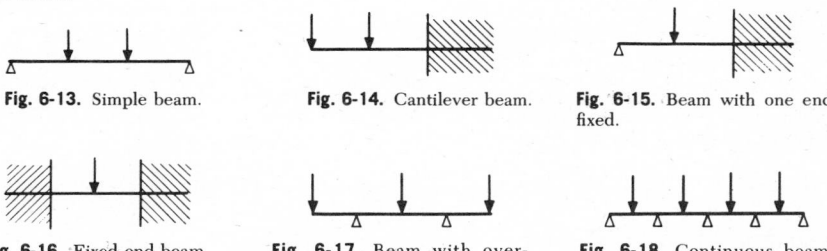

Fig. 6-13. Simple beam.   **Fig. 6-14.** Cantilever beam.   **Fig. 6-15.** Beam with one end fixed.

Fig. 6-16. Fixed-end beam.   **Fig. 6-17.** Beam with over-hangs.   **Fig. 6-18.** Continuous beam.

Fig. 6-19. Hung-span (suspended-span) construction.

Figure 6-17 shows a beam that overhangs both its simple supports. The overhangs have a free end, like a cantilever, but the supports permit rotation.

Two types of beams that extend over several supports are illustrated in Figs. 6-18 and 6-19. Figure 6-18 shows a **continuous beam.** The one in Fig. 6-19 has one or two hinges in certain spans; it is called **hung-span,** or suspended-span, construction. In effect, it is a combination of simple beams and beams with overhangs.

Reactions for the beams in Figs. 6-13, 6-14, and 6-17 and the type of beam in Fig. 6-19 with hinges suitably located may be found from the equations of equilibrium. They are classified as **statically determinate beams** for that reason.

The equations of equilibrium, however, are not sufficient to determine the reactions of the beams in Figs. 6-15, 6-16, and 6-18. For those beams, there are more unknowns than equations. Additional equations must be obtained based on a knowledge of the deformations; for example, that a fixed end permits no rotation. Such beams are classified as **statically indeterminate.** Methods for finding the stresses in that type of beam are given in Arts. 6-52 to 6-70.

6-20. Reactions. As pointed out in Art. 6-19, the loads imposed by a simple beam on its supports can be found by application of the equations of equilibrium [Eq. (6-1)]. Consider, for example, the 60-ft-long beam with overhangs in Fig. 6-20. This beam carries a uniform load of 200 lb per lin ft over its entire length and several concentrated loads. The span is 36 ft.

To find reaction R_1, take moments about R_2 and equate the sum of the moments to zero (assume clockwise rotation to be positive, counterclockwise, negative):

$$-2{,}000 \times 48 + 36R_1 - 4{,}000 \times 30 - 6{,}000 \times 18 + 3{,}000 \times 12 - 200 \times 60 \times 18 = 0$$

$$R_1 = 14{,}000 \text{ lb}$$

In this calculation, the moment of the uniform load was found by taking the moment of its resultant, 200×60, which acts at the center of the beam.

To find R_2, proceed in a similar manner by taking moments about R_1 and equating the sum to zero, or equate the sum of the vertical forces to zero. Generally it is preferable to use the moment equation and apply the other equation as a check.

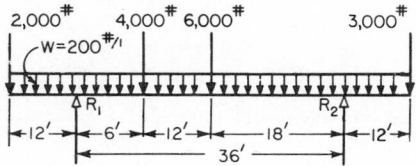

Fig. 6-20. Beam with overhangs loaded with both uniform and concentrated loads.

As an alternative procedure, find the reactions due to uniform and concentrated loads separately and sum the results. Make use of the fact that the reactions due to symmetrical loading are equal, to simplify the calculation. To find R_2 by this procedure, take half the total uniform load

$$0.5 \times 200 \times 60 = 6,000 \text{ lb}$$

and add it to the reaction due to the concentrated loads, found by taking moments about R_1, dividing by the span, and summing:

$$-2,000 \times \frac{12}{36} + 4,000 \times \frac{6}{36} + 6,000 \times \frac{18}{36} + 3,000 \times \frac{48}{36} = 7,000 \text{ lb}$$

$$R_2 = 6,000 + 7,000 = 13,000 \text{ lb}$$

Check to see that the sum of the reactions equals the total applied load:

$$14,000 + 13,000 = 2,000 + 4,000 + 6,000 + 3,000 + 200 \times 60$$
$$27,000 = 27,000$$

Reactions for simple beams with various loads are given in Figs. 6-33 to 6-38.

To find the reactions of a continuous beam, first determine the end moments and shears (Arts. 6-60 to 6-70); then if the continuous beam is considered as a series of simple beams with these applied as external loads, the beam will be statically determinate and the reactions can be determined from the equations of equilibrium. (For an alternative method see Art. 6-59.)

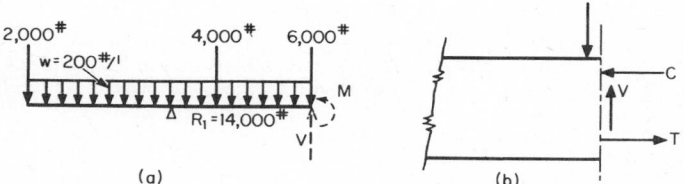

(a)   (b)

Fig. 6-21. Sections of beam kept in equilibrium by internal stresses.

6-21. Internal Forces. At every section of a beam in equilibrium internal forces act to prevent motion. For example, assume the beam in Fig. 6-20 cut vertically just to the right of its center. Adding the external forces, including the reaction, to the left of this cut (see Fig. 6-21a) yields an unbalanced downward load of 4,000 lb. Evidently, at the cut section, an upward-acting internal force of 4,000 lb must be present to maintain equilibrium. Also, taking moments of the external forces about the section yields an unbalanced moment of 54,000 ft-lb. To maintain equilibrium, there must be an internal moment of 54,000 ft-lb resisting it.

This internal, or resisting, moment is produced by a couple consisting of a force C acting on the top part of the beam and an equal but opposite force T acting on the bottom part (Fig. 6-21b). For this type of beam and loading, the top force is the resultant of compressive stresses acting over the upper portion of the beam, and the bottom force is the resultant of tensile stresses acting over the bottom part. The surface at which the stresses change from compression to tension—where the stress is zero—is called the **neutral surface**.

6-22. Shear Diagrams. As explained in Art. 6-21, at a vertical section through a beam in equilibrium external forces on one side of the section are balanced by internal forces. The unbalanced external vertical force at the section is called the shear. It equals the algebraic sum of the forces that lie on either side of the section. For forces on the left of the section, those acting upward are considered positive and those acting downward negative. For forces on the right of the section, signs are reversed.

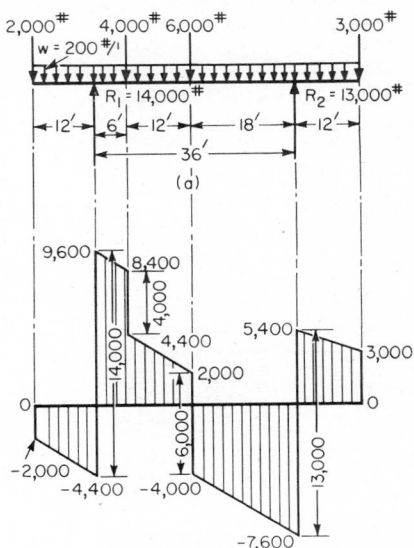

Fig. 6-22. Shear diagram for beam of Fig. 6-20.

A shear diagram represents graphically the shear at every point along the length of a beam. The shear diagram for the beam in Fig. 6-20 is shown in Fig. 6-22b. The beam is drawn to scale and the loads and reactions are located at the points at which they act. Then, a convenient zero axis is drawn horizontally from which to plot the shears to scale. Start at the left end of the beam, and directly under the 2,000-lb load there, scale off $-2,000$ from the zero axis. Next, determine the shear just to the left of the next concentrated load, the left support: $-2,000 - 200 \times 12 = -4,400$ lb. Plot this downward under R_1. Note that in passing from just to the left of the support to just to the right, the shear changes by the magnitude of the reaction, from $-4,400$ to $-4,400 + 14,000$, or $9,600$ lb. So plot this value also under R_1. Under the 4,000-lb load, plot the shear just to the left of it, $9,600 - 200 \times 6$, or $8,400$ lb, and the shear just to the right, $8,400 - 4,000$, or $4,400$ lb. Proceed in this manner to the right end, where the shear is 3,000 lb, equal to the load on the free end.

To complete the diagram, the points must be connected. Straight lines can be used because shear varies uniformly for a uniform load (see Fig. 6-24b).

Shear diagrams for various loading conditions on simple beams are shown in Figs. 6-33 to 6-38.

6-23. Bending-moment Diagrams. About a vertical section through a beam in equilibrium, there is an unbalanced moment due to external forces. It is called *bending moment*. For forces on the left of the section, clockwise moments are considered positive and counterclockwise moments negative. For forces on the right of the section, the signs are reversed. Thus, when the bending moment is positive, the bottom of a simple beam is in tension and the top is in compression.

A bending-moment diagram represents graphically the bending moment at every point along the length of the beam. Figure 6-23c is the bending-moment diagram for the beam with concentrated loads in Fig. 6-23a. The beam is drawn to scale, and the loads and reactions are located at the points at which they act. Then, a horizontal line is drawn to represent the zero axis from which to plot the bending moments to scale. Note that the bending moment at both supports for this simple beam is zero. Between the supports and the first load the bending moment is proportional to the distance from the support since the bending moment in that region equals the reaction times the distance from the support. Hence, the bending-moment diagram for this portion of the beam is a sloping straight line.

To find the bending moment under the 6,000-lb load, consider only the forces to the left of it, in this case only the reaction R_1. Its moment about the 6,000-lb load is $7,000 \times 10$, or 70,000 ft-lb. The bending-moment diagram, then, between the left support and the first concentrated load is a straight line rising from zero at the left end of the beam to 70,000, plotted, to a convenient scale, under the 6,000-lb load.

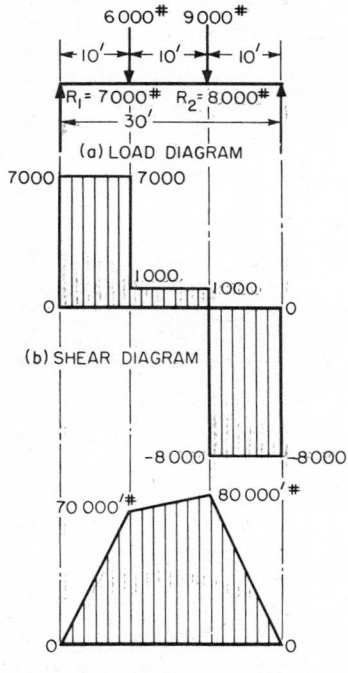

(a) LOAD DIAGRAM

(b) SHEAR DIAGRAM

(c) BENDING MOMENT DIAGRAM

Fig. 6-23. Shear and moment diagram for beam with concentrated loads.

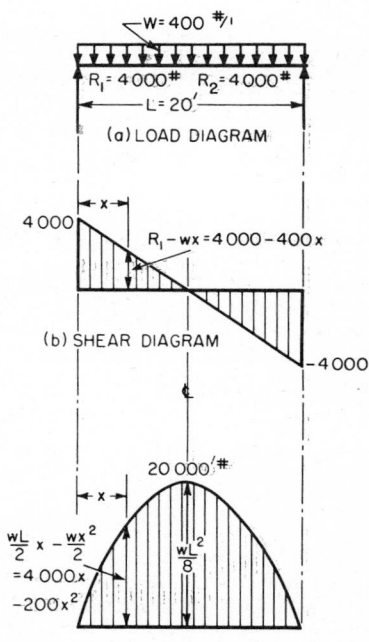

(a) LOAD DIAGRAM

(b) SHEAR DIAGRAM

(c) BENDING MOMENT DIAGRAM

Fig. 6-24. Shear and moment diagram for uniformly loaded beam.

To find the bending moment under the 9,000-lb load, add algebraically the moments of the forces to its left: $7,000 \times 20 - 6,000 \times 10 = 80,000$ ft-lb. (This result could have been obtained more easily by considering only the portion of the beam on the right, where the only force present is R_2, and reversing the sign convention: $8,000 \times 10 = 80,000$ ft-lb.) Since there are no other loads between the 6,000- and 9,000-lb loads, the bending-moment diagram between them is a straight line.

If the bending moment and shear are known at any section, the bending moment at any other section can be computed if there are no unknown forces between the sections. The rule is:

The bending moment at any section of a beam equals the bending moment at any section to the left, plus the shear at that section times the distance between sections, minus the moments of intervening loads. If the section with known moment and shear is on the right, the sign convention must be reversed.

For example, the bending moment under the 9,000-lb load in Fig. 6-23a also could have been determined from the moment under the 6,000-lb load and the shear just to the right of that load.

As indicated in the shear diagram (Fig. 6-23b), that shear is 1,000 lb. Thus, the moment is given by $70,000 + 1,000 \times 10 = 80,000$ ft-lb.

Bending-moment diagrams for simple beams with various loadings are shown in Figs. 6-33 to 6-38. To obtain bending-moment diagrams for loading conditions that can be represented as a sum of the loadings shown, sum the bending moments at corresponding locations on the beam as given on the diagrams for the component loads.

For a simple beam carrying a uniform load, the bending-moment diagram is a parabola (Fig. 6-24c). The maximum moment occurs at the center and equals $wL^2/8$ or $WL/8$, where w is the load per linear foot and $W = wL$ is the total load on the beam.

The bending moment at any section of a simply supported, uniformly loaded beam equals one-half the product of the load per linear foot and the distances to the section from both supports:

$$M = \frac{w}{2} x(L - x) \tag{6-33}$$

6-24. Shear-Moment Relationship. The slope of the bending-moment curve at any point on a beam equals the shear at that point. If V is the shear, M the moment, and x the distance along the beam,

$$V = \frac{dM}{dx} \tag{6-34}$$

Since maximum bending moment occurs when the slope changes sign, or passes through zero, maximum moment (positive or negative) occurs at the point of zero shear.

Integration of Eq. (6-34) yields

$$M_1 - M_2 = \int_{x_2}^{x_1} V \, dx \tag{6-35}$$

Thus, the change in bending moment between any two sections of a beam equals the area of the shear diagram between ordinates at the two sections.

6-25. Moving Loads and Influence Lines. Influence lines are a useful device for solving problems involving moving loads. An influence line indicates the effect at a given section of a unit load placed at any point on the structure.

For example, to plot the influence line for bending moment at a point on a beam, compute the moments produced at that point as a unit load moves along the beam and plot these moments under the corresponding positions of the unit load. Actually, the unit load need

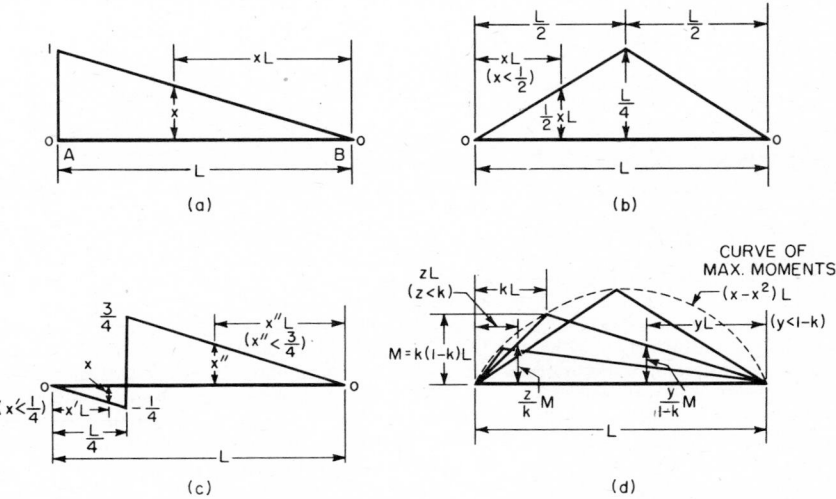

Fig. 6-25. Influence lines for (a) reaction at A; (b) midspan bending moment; (c) quarter-point shear; and (d) bending moments at several points in a beam.

not be placed at every point along the beam. The equation of the influence line can be determined in many cases by placing the load at an arbitrary point and computing the bending moment in general terms. (See also Art. 6-58.)

To draw the influence line for reaction at A for a simple beam AB (Fig. 6-25a), place a unit load at an arbitrary distance xL from B. The reaction at A due to this load is $1 \ xL/L = x$. Then, $R_A = x$ is the equation of the influence line. It represents a straight line sloping downward from unity at A, when the unit load is at that end of the beam, to zero at B, when the load is at the opposite end (Fig. 6-25a).

Figure 6-25b shows the influence line for bending moment at the center of a beam. It resembles in appearance the bending-moment diagram for a load at the center of the beam, but its significance is entirely different. Each ordinate gives the moment at midspan for a load at the location of the ordinate. The diagram indicates that if a unit load is placed at a distance xL from one end, it produces a bending moment of $xL/2$ at the center of the span.

Figure 6-25c shows the influence line for shear at the quarter point of a beam. When the load is to the right of the quarter point, the shear is positive and equal to the left reaction. When the load is to the left, the shear is negative and equals the right reaction. Thus, to produce maximum shear at the quarter point, loads should be placed only to the right of the quarter point, with the largest load at the quarter point, if possible. For a uniform load, maximum shear results when the load extends from the right end of the beam to the quarter point.

Suppose, for example, that a 60-ft crane girder is to carry wheel loads of 20 and 10 kips, 5 ft apart. For maximum shear at the quarter point, place the 20-kip wheel there and the 10-kip wheel 5 ft to the right. The corresponding ordinates of the influence line (Fig. 6-25c) are ¾ and $^{40}/_{45} \times$ ¾. Hence, the maximum shear is $20 \times$ ¾ $+ 10 \times {}^{40}/_{45} \times$ ¾ $= 21.7$ kips.

Figure 6-25d shows influence lines for bending moment at several points on a beam. The apexes of the triangular diagrams fall on a parabola, as indicated by the dash line. From the diagram, it can be concluded that the maximum moment produced at any section by a single concentrated load moving along a beam occurs when the load is at that section. And the magnitude of the maximum moment increases when the section is moved toward midspan, in accordance with the equation for the parabola given in Fig. 6-25d.

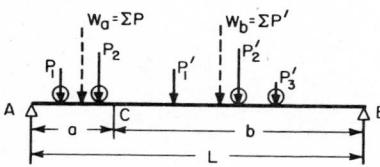

Fig. 6-26. Moving loads on simple beam AB placed for maximum moment at C.

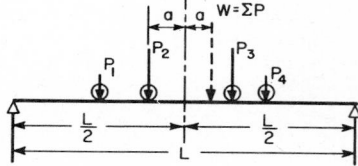

Fig. 6-27. Moving loads placed for maximum moment in a simple beam.

6-26. Maximum Bending Moment. When a span is to carry several moving concentrated loads, an influence line is useful in determining the position of the loads for which bending moment is a maximum at a given section (see Art. 6-25). For a simple beam, maximum bending moment will occur at a section C as loads move across the beam when one of the loads is at C. The load to place at C is the one for which the expression $W_a/a - W_b/b$ (Fig. 6-26) changes sign as that load passes from one side of C to the other. (W_a is the sum of the loads on one side of C and W_b the sum of the loads on the other side of C.)

When several concentrated loads move along a simple beam, the maximum moment they produce in the beam may be near but not necessarily at midspan. To find the maximum moment, first determine the position of the loads for maximum moment at midspan. Then, shift the loads until the load P_2 (Fig. 6-27) that was at the center of the beam is as far from midspan as the resultant of all the loads on the span is on the other side of midspan. Maximum moment will occur under P_2. When other loads move on or off the span during the shift of P_2 away from midspan, it may be necessary to investigate the moment under one of the other loads when it and the new resultant are equidistant from midspan.

6-27. Bending Stresses in a Beam. The commonly used flexure formula for computing bending stresses in a beam is based on the following assumptions:

1. The unit stress parallel to the bending axis at any point of a beam is proportional to the unit strain in the same direction at the point. Hence, the formula holds only within the proportional limit.

2. The modulus of elasticity in tension is the same as that in compression.

3. The total and unit axial strain at any point are both proportional to the distance of that point from the neutral surface. (Cross sections that are plane before bending remain plane after bending. This requires that all fibers have the same length before bending, thus that the beam be straight.)

4. The loads act in a plane containing the centroidal axis of the beam and are perpendicular to that axis. Furthermore, the neutral surface is perpendicular to the plane of the loads. Thus, the plane of the loads must contain an axis of symmetry of each cross section of the beam. (The flexure formula does not apply to a beam with cross sections loaded unsymmetrically.)

5. The beam is proportioned to preclude prior failure or serious deformation by torsion, local buckling, shear, or any cause other than bending.

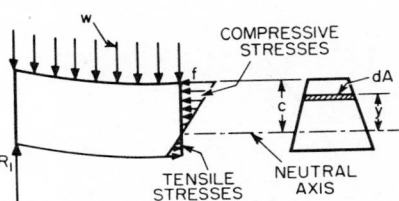

Fig. 6-28. Unit stresses due to bending on a beam section.

Equating the bending moment to the resisting moment due to the internal stresses at any section of a beam yields

$$M = \frac{fI}{c} \tag{6-36}$$

where M = bending moment at the section, in.-lb

f = normal unit stress at a distance c, in., from the neutral axis (Fig. 6-28) psi

I = moment of inertia of cross section with respect to neutral axis, in.^{4}

Generally, c is taken as the distance to the outermost fiber to determine maximum f.

6-28. Moment of Inertia. The neutral axis in a symmetrical beam coincides with the centroidal axis; that is, at any section the neutral axis is so located that

$$\int y \, dA = 0 \tag{6-37}$$

where dA is a differential area parallel to the axis (Fig. 6-28), y is its distance from the axis, and the summation is taken over the entire cross section.

Moment of inertia with respect to the neutral axis is given by

$$I = \int y^2 \, dA \tag{6-38}$$

Values of I for several common cross sections are given in Fig. 6-29. Values for standard structural-steel sections are listed in the American Institute of Steel Construction "Steel Construction Manual." When the moments of inertia of other types of sections are needed, they can be computed directly by application of Eq. (6-38) or by breaking the section up into components for which the moment of inertia is known.

With the following formula, the moment of inertia of a section can be determined from that of its components:

$$I' = I + Ad^2 \tag{6-39}$$

where I = moment of inertia of component about its centroidal axis, in.^{4}

I' = moment of inertia of component about parallel axis, in.^{4}

A = cross-sectional area of component, sq in.

d = distance between centroidal and parallel axes, in.

The formula enables you to compute the moment of inertia of a component about the centroidal axis of a section from the moment of inertia about the component's centroidal axis, usually obtainable from Fig. 6-29 or the AISC manual. By summing up the transferred moments of inertia for all the components, you obtain the moment of inertia of the section.

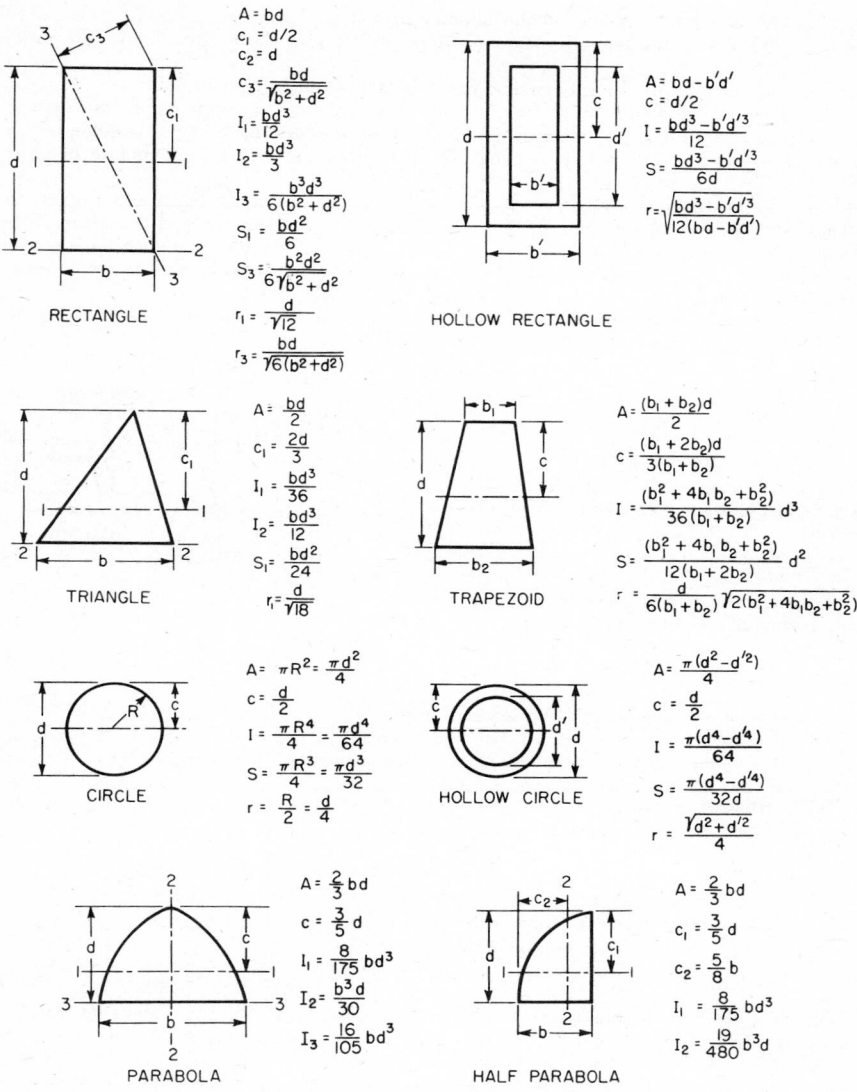

Fig. 6-29. Geometric properties of sections.

When the moments of inertia of an area with respect to any two perpendicular axes are known, the moment of inertia with respect to any other axis passing through the point of intersection of the two axes may be obtained through the use of Mohr's circle as for stresses (Fig. 6-11). In this analog, I_x corresponds with f_x, I_y with f_y, and the **product of inertia** I_{xy} with v_{xy} (Art. 6-17)

$$I_{xy} = \int xy\, dA \tag{6-40}$$

The two perpendicular axes through a point about which the moments of inertia are a maximum or a minimum are called the principal axes. The product of inertia is zero for the principal axes.

6-29. Section Modulus. The ratio $S = I/c$, relating bending moment and maximum bending stresses within the elastic range in a beam [Eq. (6-36)], is called the *section modulus*. I is the moment of inertia of the cross section about the neutral axis and c the distance from the neutral axis to the outermost fiber. Values of S for common types of sections are given in Fig. 6-29. Values for standard structural-steel sections are listed in the American Institute of Steel Construction "Steel Construction Manual."

6-30. Shearing Stresses in a Beam. Vertical shear at any section in a beam is resisted by nonuniformly distributed, vertical unit stresses (Fig. 6-30). At every point in the section, there also is a horizontal unit stress, which is equal in magnitude to the vertical unit shearing stress there [see Eq. (6-24)].

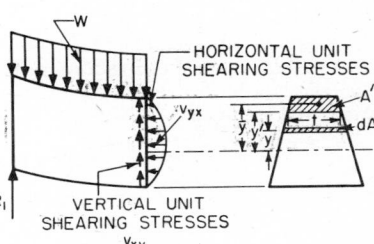

Fig. 6-30. Unit shearing stresses on a beam section.

At any distance y' from the neutral axis, both the horizontal and vertical shearing unit stresses are equal to

$$v = \frac{V}{It} A'\overline{y} \tag{6-41}$$

where V = vertical shear at the cross section, lb
  t = thickness of beam at distance y' from neutral axis, in.
  I = moment of inertia of section about neutral axis, in.^{4}
  A' = area between the outermost surface and the surface for which the shearing stress is being computed, sq in.
  $\overline{y}$ = distance of center of gravity of this area from the neutral axis, in.

For a rectangular beam, with width $t = b$ and depth d, the maximum shearing stress occurs at middepth. Its magnitude is

$$v = \frac{V}{(bd^3/12)b} \frac{bd}{2} \frac{d}{4} = \frac{3}{2} \frac{V}{bd}$$

That is, the maximum shear stress is 50% greater than the average shear stress on the section. Similarly, for a circular beam, the maximum is one-third greater than the average. For an I or wide-flange beam, however, the maximum shear stress in the web is not appreciably greater than average for the web section alone, assuming that the flanges take no shear.

6-31. Combined Shear and Bending Stress. For deep beams on short spans and beams with low tensile strength, it sometimes is necessary to determine the maximum normal stress f' due to a combination of shear stress v and bending stress f. This maximum or principal stress (Art. 6-15) occurs on a plane inclined to that of v and of f. From Mohr's circle (Fig. 6-11) with $f = f_x$, $f_y = 0$, and $v = v_{xy}$,

$$f' = \frac{f}{2} + \sqrt{v^2 + \left(\frac{f}{2}\right)^2} \tag{6-42}$$

6-32. Beam Stresses in the Plastic Range. (See also Arts. 6-91 and 6-92.) When bending stresses in a beam exceed the proportional limit and stress no longer is proportional to strain, the distribution of bending stresses over a cross section ceases to be linear. As the bending moment increases, the outer fibers deform with little change in stress, while the fibers not stressed beyond the proportional limit continue to take more stress.

If the actual stress distribution over the cross section is plotted, then, for equilibrium, the area under the curve for the tensile stresses must equal the area under the curve for the compressive stresses. Also, the moments of the areas under these two curves about the neutral axis must equal the bending moment.

Modulus of rupture is the stress computed from the flexure formula [Eq. (6-36)] corresponding to the maximum bending moment a beam sustains at failure. Usually, it is considerably higher than the actual maximum unit stress in the beam but it sometimes is used to compare the strength of beams with the same cross section and material.

6-33. Beam Deflections. The **elastic curve** is the position taken by the longitudinal centroidal axis of a beam when it deflects under load. The radius of curvature at any point of this curve is

$$R = \frac{EI}{M} \tag{6-43}$$

where M = bending moment at the point
  E = modulus of elasticity
  I = moment of inertia of the cross section about the neutral axis

Since the slope of the elastic curve is very small, $1/R$ is approximately d^2y/dx^2, where y is the deflection of the beam at a distance x from the origin of coordinates. Hence, Eq. (6-43) may be rewritten

$$M = EI\frac{d^2y}{dx^2} \tag{6-44}$$

To obtain the slope and deflection of a beam, this equation may be integrated, with M expressed as a function of x. Constants introduced during the integration must be evaluated in terms of known points and slopes of the elastic curve.

After integration, Eq. (6-44) yields

$$\theta_B - \theta_A = \int_A^B \frac{M}{EI}\,dx \tag{6-45}$$

in which θ_A and θ_B are the slopes of the elastic curve at any two points A and B. If the slope is zero at one of the points, the integral in Eq. (6-45) gives the slope of the elastic curve at the other. The integral represents the area of the bending-moment diagram between A and B with each ordinate divided by EI.

The **tangential deviation** t of a point on the elastic curve is the distance of this point, measured in a direction perpendicular to the original position of the beam, from a tangent drawn at some other point on the curve.

$$t_B - t_A = \int_A^B \frac{Mx}{EI}\,dx \tag{6-46}$$

Equation (6-46) indicates that the tangential deviation of any point with respect to a second point on the elastic curve equals the moment about the first point of the area of the M/EI diagram between the two points. The moment-area method for determining beam deflections is a technique employing Eqs. (6-45) and (6-46).

Moment-area Method. Suppose, for example, the deflection at midspan is to be computed for a beam of uniform cross section with a concentrated load at the center (Fig. 6-31). Since the deflection at midspan for this loading is the maximum for the span, the slope of the elastic curve at midspan is zero; that is, the tangent is parallel to the undeflected position of the beam. Hence, the deviation of either support from the midspan tangent equals the deflection at the center of the beam. Then, by the moment-area theorem [Eq. (6-46)], the deflection y_c is given by the moment about either support of the area of the M/EI diagram included between an ordinate at the center of the beam and that support:

$$y_c = \left(\frac{1}{2}\frac{PL}{4EI}\frac{L}{2}\right)\frac{L}{3} = \frac{PL^3}{48EI}$$

Suppose, now, the deflection y at any point D at a distance xL from the left support (Fig. 6-31) is to be determined. Note that from similar triangles, $xL/L = DE/t_{AB}$, where DE is the distance from the undeflected position of D to the tangent to the elastic curve at support A, and t_{AB} is the tangential deviation of B from that tangent. But DE also equals $y + t_{AD}$, where t_{AD} is the tangential deviation of D from the tangent at A. Hence,

$$y + t_{AD} = xt_{AB}$$

This equation is perfectly general for the deflection of any point of a simple beam, no matter how loaded. It may be rewritten to give the deflection directly:

$$y = xt_{AB} - t_{AD} \qquad (6\text{-}47)$$

But t_{AB} is the moment of the area of the M/EI diagram for the whole beam about support B. And t_{AD} is the moment about D of the area of the M/EI diagram included between ordinates at A and D. So at any point x of the beam in Fig. 6-31, the deflection is

$$y = x \left[\frac{1}{2} \frac{PL}{4EI} \frac{L}{2} \left(\frac{L}{3} + \frac{2L}{3} \right) \right] - \frac{1}{2} \frac{PLx}{2EI} (xL) \frac{xL}{3} = \frac{PL^3}{48EI} x(3 - 4x^2)$$

It also is noteworthy that, since the tangential deviations are very small distances, the slope of the elastic curve at A is given by

$$\theta_A = \frac{t_{AB}}{L} \qquad (6\text{-}48)$$

This holds, in general, for all simple beams regardless of the type of loading.

Conjugate-beam Method. The procedure followed in applying Eq. (6-47) to the deflection of the loaded beam in Fig. 6-31 is equivalent to finding the bending moment at D with the M/EI diagram serving as the load diagram. The technique of applying the M/EI diagram as a load and determining the deflection as a bending moment is known as the conjugate-beam method.

The conjugate beam must have the same length as the given beam; it must be in equilibrium with the M/EI load and the reactions produced by the load; and the bending moment at any section must be equal to the deflection of the given beam at the corresponding section. The last requirement is equivalent to specifying that the shear at any section of the conjugate beam with the M/EI load be equal to the slope of the elastic curve at the corresponding section of the given beam. Figure 6-32 shows the conjugates for various types of beams.

Deflection Computations. Deflections for several types of loading on simple beams are given in Figs. 6-33 and 6-35 to 6-38 and for cantilevers and beams with overhangs in Figs. 6-39 to 6-44.

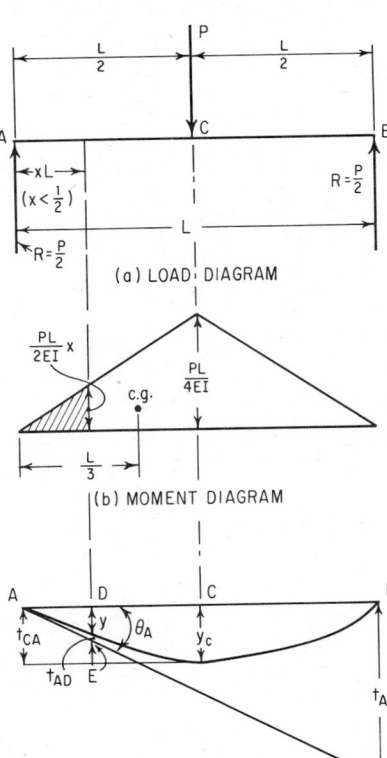

Fig. 6-31. Elastic curve for a simple beam and tangential deviations at ends.

When a beam carries several different types of loading, the most convenient method of computing its deflection usually is to find the deflections separately for the uniform and concentrated loads and add them up.

For several concentrated loads, the easiest method of obtaining the deflection at a point on a beam is to apply the reciprocal theorem (Art. 6-58). According to this theorem, if a concentrated load is applied to a beam at a point A, the deflection the load produces at point B equals the deflection at A for the same load applied at B ($d_{AB} = d_{BA}$). So place the loads one at a time

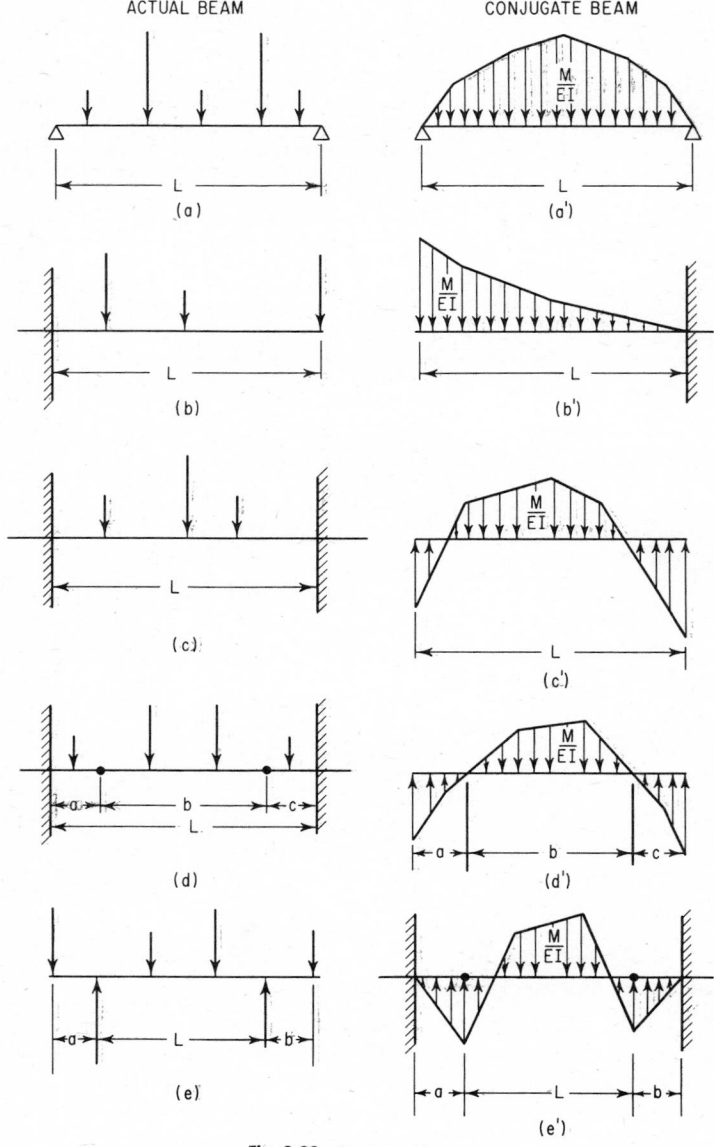

Fig. 6-32. Conjugate beams.

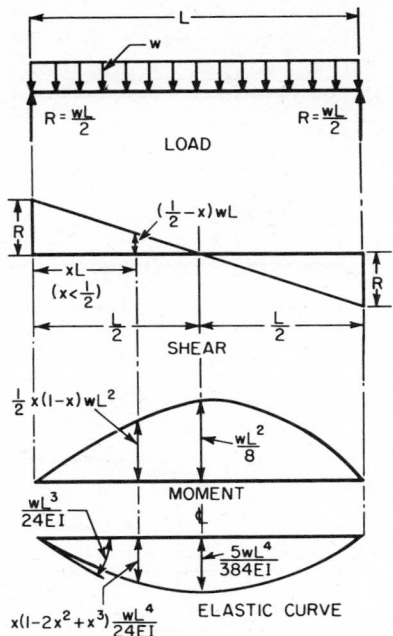

Fig. 6-33. Shears, moments, deflections for full uniform load on a simply supported prismatic beam.

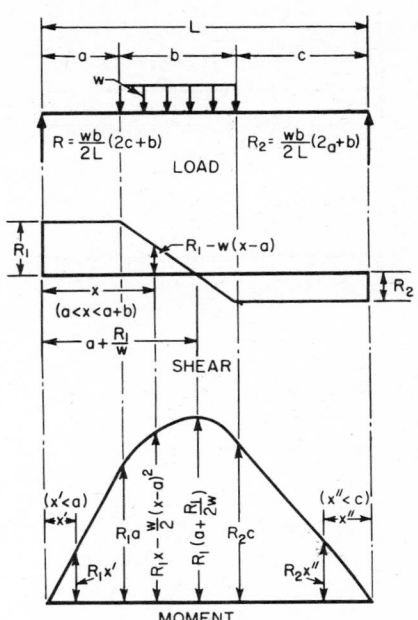

Fig. 6-34. Shears and moments for uniformly distributed load over part of a simply supported beam.

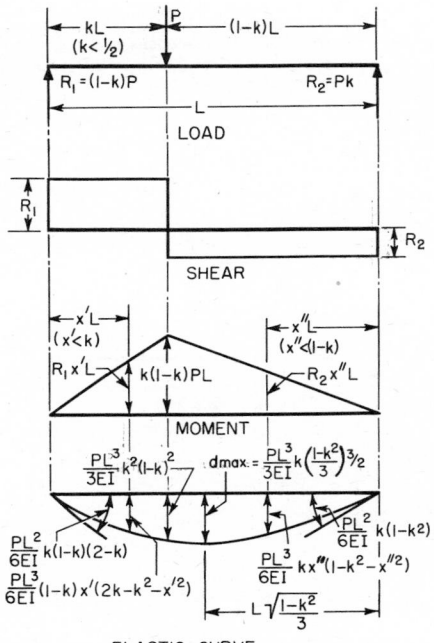

Fig. 6-35. Shears, moments, deflections for a concentrated load at any point of a simply supported prismatic beam.

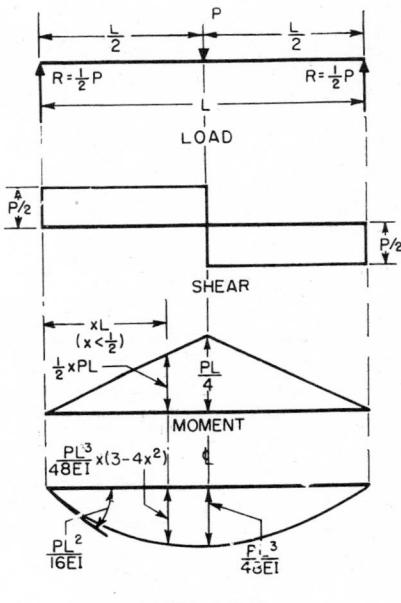

Fig. 6-36. Shears, moments, deflections for a concentrated load at midspan of a simply supported prismatic beam.

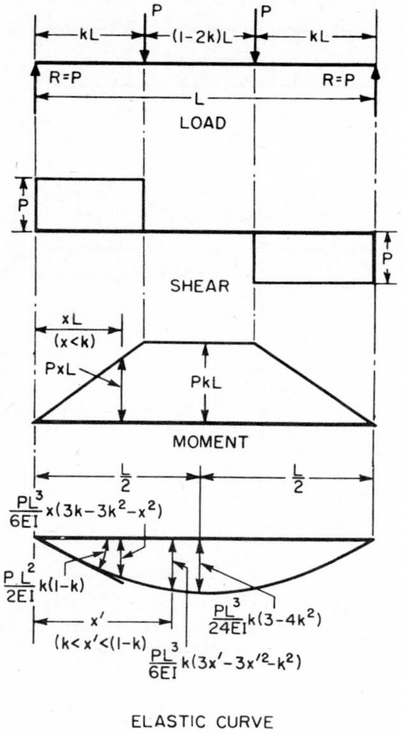

Fig. 6-37

LOAD: kL — $(1-2k)L$ — kL; P, P; $R=P$, $R=P$; L

SHEAR: P; P

MOMENT: xL $(x<k)$; PxL; PkL; $\dfrac{L}{2}$, $\dfrac{L}{2}$

ELASTIC CURVE:
$$\frac{PL^3}{6EI}x(3k-3k^2-x^2)$$
$$\frac{PL^2}{2EI}k(1-k)$$
x' $(k<x'<(1-k))$
$$\frac{PL^3}{6EI}k(3x'-3x'^2-k^2)$$
$$\frac{PL^3}{24EI}k(3-4k^2)$$

Fig. 6-37. Shears, moments, deflections for two equal concentrated loads on a simply supported prismatic beam.

Fig. 6-38

$W=nP$

LOAD: $a=\dfrac{1}{n+1}$; $\dfrac{L}{2}$, $\dfrac{L}{2}$; aL P aL P aL P aL P aL; $R=\dfrac{1}{2}nP$, $R=\dfrac{1}{2}nP$; L

SHEAR: R; P, P, P, P, P; R

MOMENT: $\dfrac{L}{2}$; maL; $\dfrac{L}{2}$
$$\frac{PL}{2}\frac{m(n-m+1)}{n+1}$$
$$\frac{PL}{8}(n+1)\quad\text{(FOR n AN ODD NUMBER)}$$
$$\frac{PL}{8}\cdot\frac{n(n+2)}{n+1}\quad\text{(FOR n AN EVEN NUMBER)}$$

ELASTIC CURVE:
$$\frac{PL^2}{24EI}\frac{n(n+2)}{n+1}$$
$$\frac{PL^3}{384EI}\frac{n(n+2)(5n^2+10n+6)}{(n+1)^3}\quad\text{(FOR n AN EVEN NUMBER)}$$
$$\frac{PL^3}{384EI}\frac{5n^2+10n+1}{n+1}\quad\text{(FOR n AN ODD NUMBER)}$$

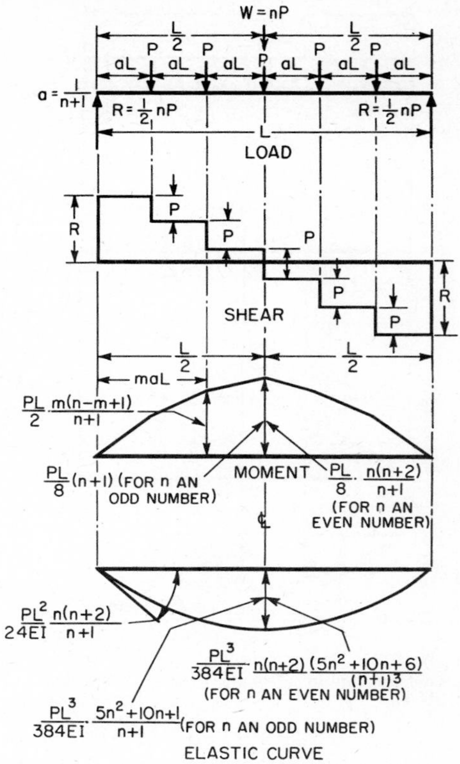

Fig. 6-38. Shears, moments, deflections for several equal loads equally spaced on a simply supported prismatic beam.

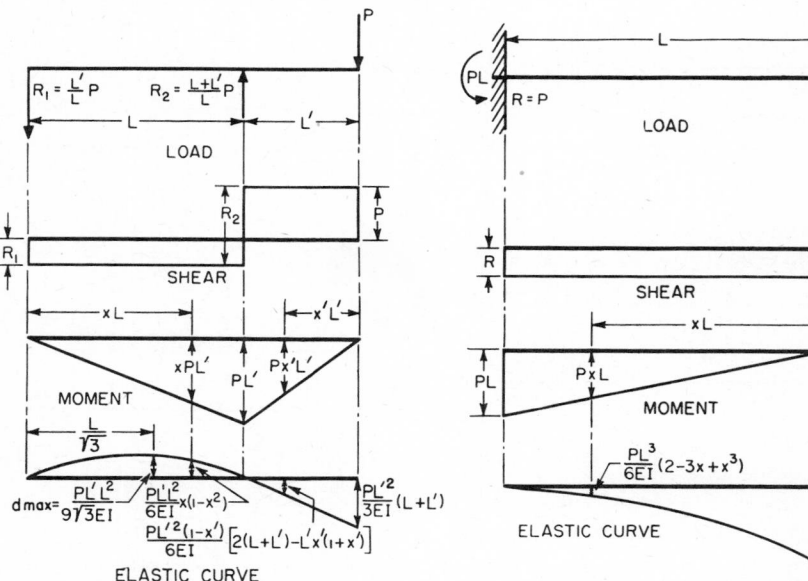

Fig. 6-39

LOAD: $R_1=\dfrac{L'}{L}P$, $R_2=\dfrac{L+L'}{L}P$; P; L, L'

SHEAR: R_1; R_2; P

MOMENT: xL; $x'L'$; xPL'; PL'; $Px'L'$; $\dfrac{L}{\sqrt{3}}$

ELASTIC CURVE:
$$d_{max}=\frac{PL'L^2}{9\sqrt{3}EI}$$
$$\frac{PL'L^2}{6EI}x(1-x^2)$$
$$\frac{PL'^2(1-x')}{6EI}\left[2(L+L')-L'x'(1+x')\right]$$
$$\frac{PL'^2}{3EI}(L+L')$$

Fig. 6-39. Shears, moments, deflections for a concentrated load on a beam overhang.

Fig. 6-40

LOAD: PL; $R=P$; L; P

SHEAR: R; P; xL

MOMENT: PL; PxL
$$\frac{PL^3}{6EI}(2-3x+x^3)$$

ELASTIC CURVE: $\dfrac{PL^3}{3EI}$

Fig. 6-40. Shears, moments, deflections for a concentrated load on the end of a prismatic cantilever.

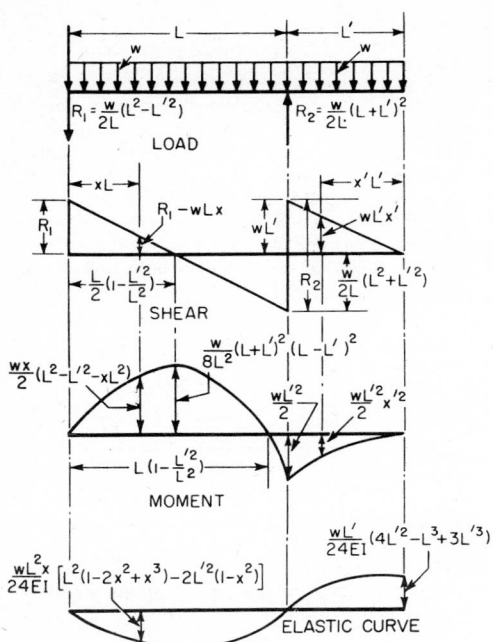

Fig. 6-41. Shears, moments, deflections for a uniform load over the full length of a beam with overhang.

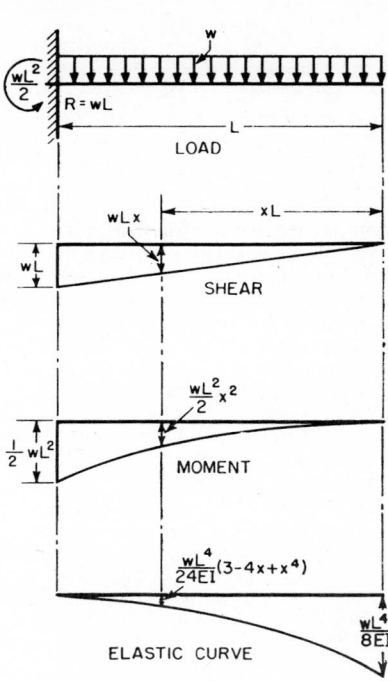

Fig. 6-42. Shears, moments, deflections for a uniform load over the full length of a cantilever.

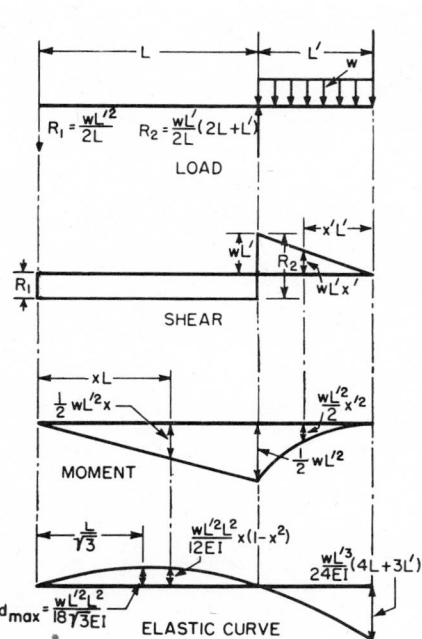

Fig. 6-43. Shears, moments, deflections for a uniform load on a beam overhang.

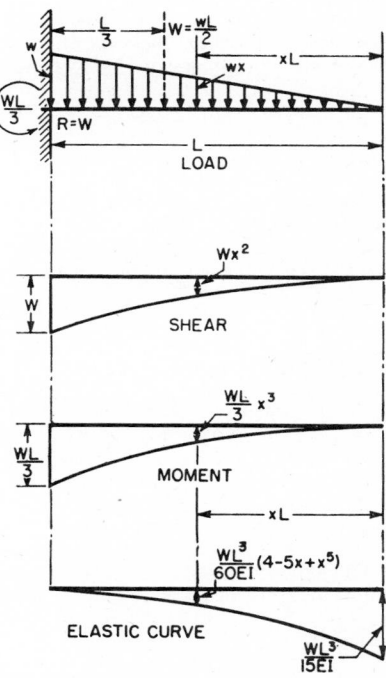

Fig. 6-44. Shears, moments, deflections for triangular loading on a prismatic cantilever.

at the point for which the deflection is to be found, and from the equation of the elastic curve determine the deflections at the actual locations of the loads. Then, sum these deflections.

Suppose, for example, the midspan deflection is to be computed. Assume each load in turn applied at the center of the beam and compute the deflection at the point where it originally was applied from the equation of the elastic curve given in Fig. 6-36. The sum of these deflections is the total midspan deflection.

Another method for computing deflections is presented in Art. 6-55. This method also may be used to determine the deflection of a beam due to shear.

6-34. Combined Axial and Bending Loads. For short beams, subjected to both transverse and axial loads, the stresses are given by the principle of superposition if the deflection due to bending may be neglected without serious error. That is, the total stress is given with sufficient accuracy at any section by the sum of the axial stress and the bending stresses. The maximum stress, psi, equals

$$f = \frac{P}{A} + \frac{Mc}{I} \qquad (6\text{-}49)$$

where P = axial load, lb
A = cross-sectional area, sq in.
M = maximum bending moment, in.-lb
c = distance from neutral axis to outermost fiber at the section where maximum moment occurs, in.
I = moment of inertia about neutral axis at that section, in.[4]

When the deflection due to bending is large and the axial load produces bending stresses that cannot be neglected, the maximum stress is given by

$$f = \frac{P}{A} + (M + Pd)\frac{c}{I} \qquad (6\text{-}50)$$

where d is the deflection of the beam. For axial compression, the moment Pd should be given the same sign as M, and for tension, the opposite sign, but the minimum value of $M + Pd$ is zero. The deflection d for axial compression and bending can be obtained by applying Eq. (6-44). (S. Timoshenko and J. M. Gere, "Theory of Elastic Stability," McGraw-Hill Book Company, New York; Friedrich Bleich, "Buckling Strength of Metal Structures," McGraw-Hill Book Company, New York.) But it may be closely approximated by

$$d = \frac{d_o}{1 - (P/P_c)} \qquad (6\text{-}51)$$

where d_o = deflection for the transverse loading alone, in.
P_c = critical buckling load, $\pi^2 EI/L^2$ (see Art. 6-71), lb

6-35. Eccentric Loading. If an eccentric longitudinal load is applied to a bar in the plane of symmetry, it produces a bending moment Pe, where e is the distance, in., of the load P from the centroidal axis. The total unit stress is the sum of the stress due to this moment and the stress due to P applied as an axial load:

$$f = \frac{P}{A} \pm \frac{Pec}{I} = \frac{P}{A}\left(1 \pm \frac{ec}{r^2}\right) \qquad (6\text{-}52)$$

where A = cross-sectional area, sq in.
c = distance from neutral axis to outermost fiber, in.
I = moment of inertia of cross section about neutral axis, in.[4]
r = **radius of gyration** = $\sqrt{I/A}$, in.

Figure 6-29 gives values of the radius of gyration for several cross sections.

If there is to be no tension on the cross section under a compressive load, e should not exceed r^2/c. For a rectangular section with width b and depth d, the eccentricity, therefore, should be less than $b/6$ and $d/6$; i.e., the load should not be applied outside the middle third. For a circular cross section with diameter D, the eccentricity should not exceed $D/8$.

When the eccentric longitudinal load produces a deflection too large to be neglected in computing the bending stress, account must be taken of the additional bending moment Pd, where d is the deflection, in. This deflection may be computed by using Eq. (6-44) or closely approximated by

$$d = \frac{4eP/P_c}{\pi(1 - P/P_c)} \qquad (6\text{-}53)$$

P_c is the critical buckling load $\pi^2 EI/L^2$ (see Art. 6-71), lb.

If the load P does not lie in a plane containing an axis of symmetry, it produces bending about the two principal axes through the centroid of the section. The stresses, psi, are given by

$$f = \frac{P}{A} + \frac{Pe_x c_x}{I_y} + \frac{Pe_y c_y}{I_x}$$ (6-54)

where A = cross-sectional area, sq in.

e_x = eccentricity with respect to principal axis YY, in.

e_y = eccentricity with respect to principal axis XX, in.

c_x = distance from YY to outermost fiber, in.

c_y = distance from XX to outermost fiber, in.

I_x = moment of inertia about XX, in.[4]

I_y = moment of inertia about YY, in.[4]

The principal axes are the two perpendicular axes through the centroid for which the moments of inertia are a maximum or a minimum and for which the products of inertia are zero.

6-36. Unsymmetrical Bending. When a beam is subjected to loads that do not lie in a plane containing a principal axis of each cross section, unsymmetrical bending occurs. Assuming that the bending axis of the beam lies in the plane of the loads, to preclude torsion (see Art. 6-37), and that the loads are perpendicular to the bending axis, to preclude axial components, the stress, psi, at any point in a cross section is

$$f = \frac{M_x y}{I_x} + \frac{M_y x}{I_y}$$ (6-55)

where M_x = bending moment about principal axis XX, in.-lb

M_y = bending moment about principal axis YY, in.-lb

x = distance from point where stress is to be computed to YY axis, in.

y = distance from point to XX axis, in.

I_x = moment of inertia of cross section about XX, in.[4]

I_y = moment of inertia about YY, in.[4]

If the plane of the loads makes an angle θ with a principal plane, the neutral surface will form an angle α with the other principal plane such that

$$\tan \alpha = \frac{I_x}{I_y} \tan \theta$$ (6-56)

6-37. Beams with Unsymmetrical Sections. The derivation of the flexure formula $f = Mc/I$ (Art. 6-27) assumes that a beam bends, without twisting, in the plane of the loads and that the neutral surface is perpendicular to the plane of the loads. These assumptions are correct for beams with cross sections symmetrical about two axes when the plane of the loads contains one of these axes. They are not necessarily true for beams that are not doubly symmetrical. The reason is that in beams that are doubly symmetrical the bending axis coincides with the centroidal axis, whereas in unsymmetrical sections the two axes may be separate. In the latter case, if the plane of the loads contains the centroidal axis but not the bending axis, the beam will be subjected to both bending and torsion.

The **bending axis** is the longitudinal line in a beam through which transverse loads must pass to preclude the beam's twisting as it bends. The point in each section through which the bending axis passes is called the **shear center**, or center of twist. The shear center also is the center of rotation of the section in pure torsion (Art. 6-74). Its location depends on the dimensions of the section.

If a beam has an axis of symmetry, the shear center lies on it. In doubly symmetrical beams, the shear center lies at the intersection of the two axes of symmetry and hence coincides with the centroid.

For any section composed of two narrow rectangles, such as a T beam or an angle, the shear center may be taken as the intersection of the longitudinal center lines of the rectangles.

For a channel section with one axis of symmetry, the shear center is outside the section at a distance from the centroid equal to $e(1 + h^2 A/4I)$, where e is the distance from the centroid to the center of the web, h is the depth of the channel, A the cross-sectional area, and I the moment of inertia about the axis of symmetry. (The web lies between the centroid and the shear center.)

Locations of shear centers for several other sections are given in Friedrich Bleich, "Buckling Strength of Metal Structures," chap. 3, McGraw-Hill Book Company, New York, 1952.

GRAPHIC-STATICS FUNDAMENTALS

6-38. Representation of a Force.   Since a force is completely determined when it is known in magnitude, direction, and point of application, any force may be represented by the length, direction, and position of a straight line.   The length of line to a given scale represents the magnitude of the force.   The position of the line parallels the line of action of the force.   And an arrowhead on the line indicates the direction in which the force acts.

Graphically represented, a force may be designated by a letter, sometimes followed by a subscript, such as P_1 and P_2 in Fig. 6-45.   Or each extremity of the line may be indicated by a letter and the force referred to by means of these letters (Fig. 6-45a).   The order of the letters indicates the direction of the force; in Fig. 6-45a, referring to P_1 as OA indicates it acts from O toward A.

Forces are concurrent when their lines of action meet.   If they lie in the same plane, they are coplanar.

6-39. Parallelogram of Forces.   The **resultant** of several forces is a single force that would produce the same effect on a rigid body.   The resultant of two concurrent forces is determined by the parallelogram law:

If a parallelogram is constructed with two forces as sides, the diagonal represents the resultant of the forces (Fig. 6-45a).

The resultant is said to be equal to the sum of the forces, sum here meaning vectorial sum, or addition by the parallelogram law.   Subtraction is carried out in the same manner as addition, but the direction of the force to be subtracted is reversed.

If the direction of the resultant is reversed, it becomes the **equilibrant,** a single force that will hold the two given forces in equilibrium.

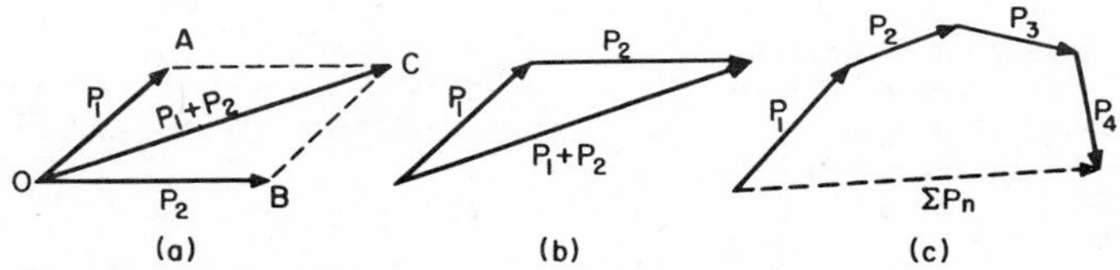

Fig. 6-45. Addition of forces by (a) parallelogram law; (b) triangle construction; and (c) polygon construction.

6-40. Resolution of Forces.   Any force may be resolved into two components acting in any given directions.   To resolve a force into two components, draw a parallelogram with the force as a diagonal and sides parallel to the given directions.   The sides then represent the components.

The procedure is: (1) Draw the given force.   (2) From both ends of the force draw lines parallel to the directions in which the components act.   (3) Draw the components along the parallels through the origin of the given force to the intersections with the parallels through the other end.   Thus, in Fig. 6-45a, P_1 and P_2 are the components in directions OA and OB of the force represented by OC.

6-41. Force Polygons.   Examination of Fig. 6-45a indicates that a step can be saved in adding forces P_1 and P_2.   The same resultant could be obtained by drawing only the upper half of the parallelogram.   Hence, to add two forces, draw the first force; then draw the second force at the end of the first one.   The resultant is the force drawn from the origin of the first force to the end of the second force, as shown in Fig. 6-45b.

This diagram is called a force triangle.   Again, the equilibrant is the resultant with direction reversed.   If it is drawn instead of the resultant, the arrows representing the direction of the forces will all point in the same direction around the triangle.   From the force triangle, an important conclusion can be drawn:

If three forces meeting at a point are in equilibrium, they will form a closed force triangle.

To add several forces P_1, P_2, P_3, . . . , P_n, draw P_2 from the end of P_1, P_3 from the end of P_2, etc.   The force required to complete the force polygon is the resultant (Fig. 6-45c).

If a group of concurrent forces is in equilibrium, they will form a closed force polygon.

6-42. Equilibrium Polygon.   When forces are coplanar but not concurrent, the force polygon will yield the magnitude and direction of the resultant but not its point of application.   To complete the solution, the easiest method generally is to employ an auxiliary force polygon, called an equilibrium, or funicular (string), polygon.   Sides of this polygon represent the lines of action of certain components of the given forces; more specifically, they take the configuration of a weightless string holding the forces in equilibrium.

In Fig. 6-46a, the forces P_1, P_2, P_3, and P_4 acting upon the given body are not in equilibrium. The magnitude and direction of their resultant R are obtained from the force polygon $abcde$. The line of action of this resultant may be obtained as follows:

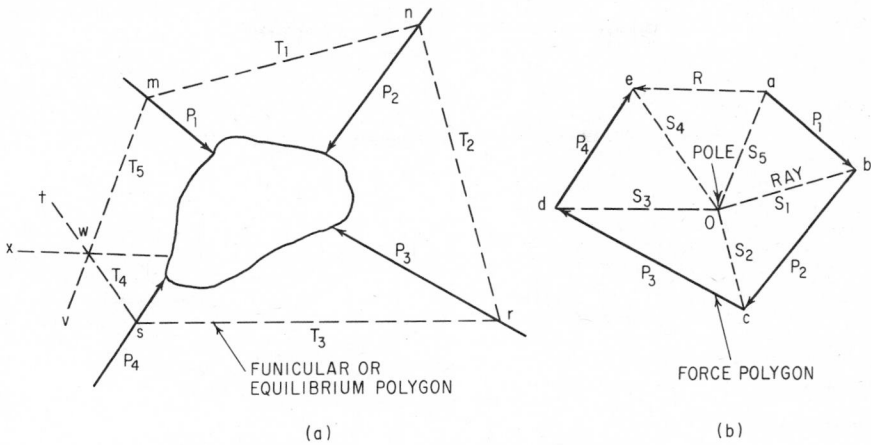

Fig. 6-46. Force and equilibrium polygons for a system of forces.

From any point O in the force polygon, draw a line to each of the vertices of the polygon. Since the lines Oa and Ob form a closed triangle with the force P_1, they represent two forces S_5 and S_1 that hold P_1 in equilibrium—two forces that may replace P_1 in a force diagram. So, as in Fig. 6-46a, at any point m on the line of action of P_1, draw lines mn and mv parallel to S_1 and S_5, respectively, to represent the lines of action of these forces. Similarly, S_1 and S_2 represent two forces that may replace P_2. The line of action of S_1 already is indicated by the line mn, and it intersects P_2 at n. So through n draw a line parallel to S_2, intersecting P_3 at r. Through r, draw rs parallel to S_3; and through s, draw st parallel to S_4. Lines mv and st, parallel to S_5 and S_4, respectively, represent the lines of action of S_5 and S_4. But these two forces form a closed force triangle with the resultant ae (Fig. 6-46b), and therefore, the three forces must be concurrent. Hence, the line of action of the resultant must pass through the intersection w of the lines mv and st. The resultant of the four given forces is thus fully determined. A force of equal magnitude but acting in the opposite direction, from e to a, will hold P_1, P_2, P_3, and P_4 in equilibrium.

The polygon $mnrsw$ is called an *equilibrium polygon*. Point O is called the *pole*, and $S_1 \ldots S_5$ are called the *rays of the force polygon*.

6-43. Beam and Truss Reactions by Graphics. Reactions of simple beams and trusses can be found with the aid of an equilibrium polygon (Art. 6-42). First, a force polygon is constructed to obtain the magnitude and direction of the resultant of the loads, which is equal in magnitude but opposite in direction to the sum of the reactions. Second, rays are drawn to the vertices of the polygon from a conveniently located pole. These rays are used to construct all but one side of an equilibrium polygon. The closing side is the common line of action of two equal but opposite forces that act with a pair of rays already drawn to hold the reactions in equilibrium. Therefore, draw a line through the pole parallel to the closing side. The intersection with the resultant in the force polygon separates it into two forces, which are equal to the reactions sought.

For example, suppose the reactions were to be obtained graphically for the beam (or truss) in Fig. 6-47a. As a first step, construct force polygon $ABCD$ (Fig. 6-47b) with the loads P_1, P_2, and P_3. Since the loads are parallel, the force polygon is a straight line. Select a pole O, and draw rays from it to the extremities of the forces. Note that the sum of the reactions equals AD, the length of the line polygon.

Next, construct the equilibrium polygon as follows: Start with a convenient point on the line of action of R_1, the left reaction, and draw a line oa parallel to ray OA in Fig. 6-47b. Locate its intersection with the line of action of P_1 (Fig. 6-47d). Through this intersection, draw ob, a line parallel to OB. At the intersection with P_2, draw oc parallel to OC, and finally, through the intersection with P_3, a line od parallel to OD, which terminates on the line of action of R_2.

Now, draw the closing line *oe* of the equilibrium polygon between the terminal and the starting point. The last step is to draw through the pole (Fig. 6-47*b*) *OE*, a line parallel to *oe*, the closing line of the equilibrium polygon, cutting the force polygon resultant at *E*. Then, $DE = R_2$, and $EA = R_1$. (See Art. 6-20 for an analytical method of determining reactions.)

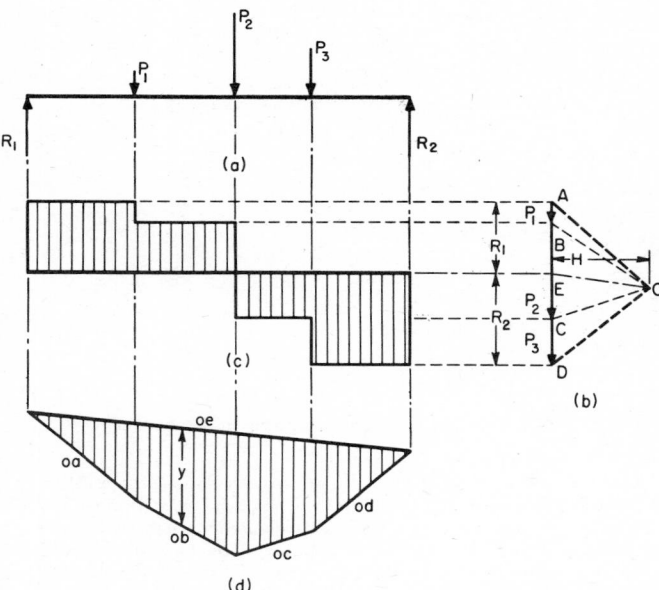

Fig. 6-47. Shear and moment diagrams obtained by graphic statics. (*a*) Loaded beam; (*b*) force polygon; (*c*) shear diagram; and (*d*) equilibrium polygon and bending-moment diagram.

6-44. Shear and Moment Diagrams by Graphics. The shear at any section of a beam equals the algebraic sum of the loads and reactions on the left of the section, upward-acting forces being considered positive, downward forces negative. If the forces are arranged in the proper order, the shear diagram may be obtained directly from the force polygon after the reactions have been determined.

For example, the shear diagram for the beam in Fig. 6-47*a* can be easily obtained from the force polygon *ABCDE* (in this case, a line, because the loads are parallel) in Fig. 6-47*b*. The zero axis is a line (Fig. 6-47*c*) parallel to the beam through *E*. As indicated in Fig. 6-47*c*, the ordinates of the shear diagram are marked off, starting with R_1 along the line of action of the left reaction, by drawing lines parallel to the zero axis through the extremities of the forces in the force polygon. (See Art. 6-22 for an analytical method of determining shears.)

The moment of a force about a point can be obtained from the equilibrium and force polygons. In the equilibrium diagram, draw a line parallel to the force through the given point. Measure the intercept of this line between the two adjoining funicular-polygon sides (extended if necessary) that originate at the given force. The moment is the product of this intercept and the distance of the force-polygon pole from the force. The intercept should be measured to the same linear scale as the beam and load positions, and the pole distance to the same scale as the forces in the force polygon.

As a consequence of this relationship between the sides of the funicular polygon, each ordinate (parallel to the forces) multiplied by the pole distance equals the bending moment at the corresponding section of the beam or truss. In Fig. 6-47*d*, for example, the equilibrium polygon, to scale, is the bending-moment diagram for the beam in Fig. 6-47*a*. At any section, the bending moment equals the ordinate *y* multiplied by the pole distance *H*. (See Art. 6-23 for an analytical method.)

STRESSES IN TRUSSES

A truss is a coplanar system of structural members joined together at their ends to form a stable framework.   Usually, analysis of a truss is based on the assumption that the joints are hinged. Neglecting small changes in the lengths of the members due to loads, the relative positions of the joints cannot change.   Stresses due to joint rigidity or deformations of the members are called **secondary stresses.**

Three bars pinned together to form a triangle represent the simplest type of truss.   Some of the commoner types of trusses are shown in Fig. 6-48.

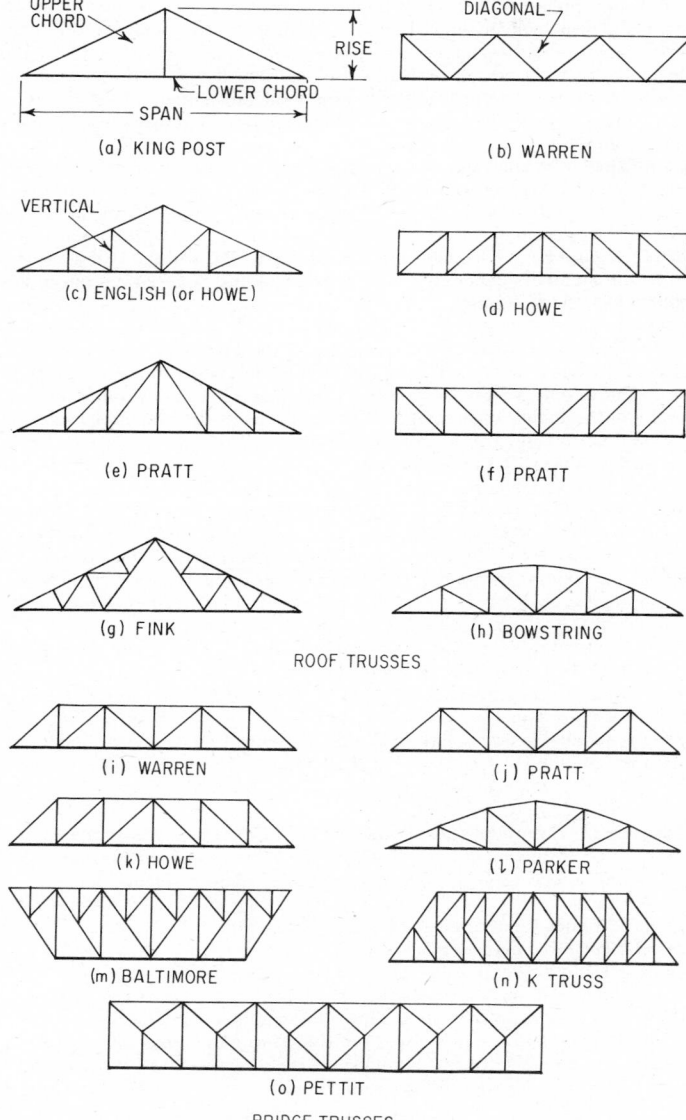

ROOF TRUSSES

BRIDGE TRUSSES

Fig. 6-48. Common types of trusses.

The top members are called the **upper chord,** the bottom members the **lower chord,** and the verticals and diagonals **web members.**

Trusses act like long, deep girders with cutout webs. Roof trusses have to carry not only their own weight and the weight of roof framing but also wind loads, snow loads, suspended ceilings and equipment, and a live load to take care of construction, maintenance, and repair loading. Bridge trusses have to support their own weight and that of deck framing and deck, live loads imposed by traffic (automobiles, trucks, railroad trains, pedestrians, etc.) and impact caused by live load, plus wind on structural members and vehicles. **Deck trusses** carry the live load on the upper chord, and **through trusses** on the lower chord.

Loads generally are applied at the intersection of members, or panel points, so that the members will be subjected principally to direct stresses. To simplify stress analysis, the weight of the truss members is apportioned to upper- and lower-chord panel points. The members are assumed to be pinned at their ends, even though this may actually not be the case. If, however, the joints are of such nature as to restrict relative rotation substantially, then the "secondary" stresses set up as a result should be computed and superimposed on the stresses obtained with the assumption of pin ends.

6-45. Bow's Notation. In analyzing trusses, especially in graphical analysis, Bow's notation is useful in identifying truss members, loads, and stresses. Capital letters are placed in the spaces between truss members and between forces. Each member and load is then designated by the letters on opposite sides of it. For example, in Fig. 6-49a, the upper-chord members are *AF, BH, CJ,* and *DL.* The loads are *AB, BC,* and *CD,* and the reactions are *EA* and *DE.* Stresses in the members generally are designated by the same letters but in lowercase.

6-46. Graphical Analysis of Trusses. Stresses in trusses may be determined by the graphical methods explained in Arts. 6-38 to 6-43. The general method is based on the assumption that at every joint of a truss the lines of action of loads and of stresses in members meet at a point. Then, since the loads and stresses are in equilibrium, the vectors, or arrows, representing them form a closed polygon. But in only a few cases are the stresses at a joint known initially; hence, the polygon for each joint usually is incomplete. The requirement that the forces form a polygon, however, permits determination of up to two unknowns. Thus, the procedure for analyzing trusses graphically is to start with those joints with only two unknown stresses (the lines of action are known because they lie along the longitudinal axes of the truss members). Solving those joints yields the magnitude of stresses that now can be used to solve other joints. By proceeding in this manner from joint to joint, all the stresses in a truss generally can be determined.

Figure 6-49b, c, d, and e show how the polygon of force may be applied at each joint of a truss to determine the two unknown stresses. The solution presumes that the reactions are known. They may be computed analytically or graphically, with load and funicular polygons, as explained in Arts. 6-20 and 6-43.

For convenience, use Bow's notation (Art. 6-45) to designate loads and truss members, as shown in Fig. 6-49a. Start with joint 1, where the reaction is known and there are only two unknown stresses, *af* in upper chord *AF* and *fe* in lower chord *FE.* Represent the reaction *ea* by an upward vertical arrow equal to 12 kips to a convenient scale. Through *a* draw a line parallel to *AF* and through *e* parallel to *FE,* to form the force triangle for joint 1 (Fig. 6-49b). The intersection is *f,* determining the stresses *af* and *fe* to the same scale as *ea.* Note that *fe* acts away from the joint and thus represents a tensile stress in *FE; af* acts toward the joint and represents a compressive stress in *AF.*

Move next to joint 2. There, the stress in the vertical *FG* is zero, because there are no forces with a component in the direction of *FG.* Now, with the stresses *af* and *fg* known, solution of joint 3 by completing a force polygon becomes possible. There remain only two unknown stresses, *bh* in *BH* and *hg* in *HG,* the load *ab* = 8 kips being given. Figure 6-49c shows the solution for joint 3. Joints 4 and 5 are solved similarly in Fig. 6-49d and e.

Examination of these force polygons indicates that each stress occurs in two force polygons. Hence, the graphical solution can be shortened by combining the polygons. The combination of the various polygons for all the joints into one stress diagram is known as a Maxwell diagram.

Maxwell Diagram. The procedure for a Maxwell diagram consists of first constructing the force polygon for the loads and reactions and then completing the force polygons for each joint on the same diagram. To make it easy to determine whether the stresses are compression or tension, plot loads and reactions in the force polygon in the order in which they are passed in going clockwise around the truss. Similarly, in drawing the force polygon for each joint, plot the forces in a clockwise direction around the joint. If these rules are followed, the order of the letters indicates the direction of the forces. Going around joint 1 clockwise, for example, we find, in Fig. 6-49*b*, the reaction *EA* as an upward-acting vertical force *ea* (*e* to *a*), the top-chord stress *af* acting toward the joint (*a* to *f*) and the bottom-chord stress *fe* acting away from the joint (*f* to *e*).   Hence, *af* is compressive, *fe* tensile.

To construct a Maxwell diagram for the truss in Fig. 6-49*a*, lay off the loads and reactions in clockwise order (Fig. 6-49*f*).   The force polygon *abcde* is a straight line because all the forces are vertical.   To solve joint 1, draw a line through *a* in Fig. 6-49*f* parallel to *AF* and a second line through *e* parallel to *FE*.   Their intersection is *f* .   At joint 2, since the stress in the vertical *FG* is zero, *g* coincides with *f*.

To solve joint 3, start with the known stress *fa* and proceed clockwise around the joint.   To complete the force polygon, draw a line through *b* parallel to *BH* and a line through *g* parallel to *GH*.   Their intersection locates *h*.   Similarly, the force polygon for joint 4 is completed by drawing a line through *c* parallel to *CJ* and a line through *h* parallel to *JH*.

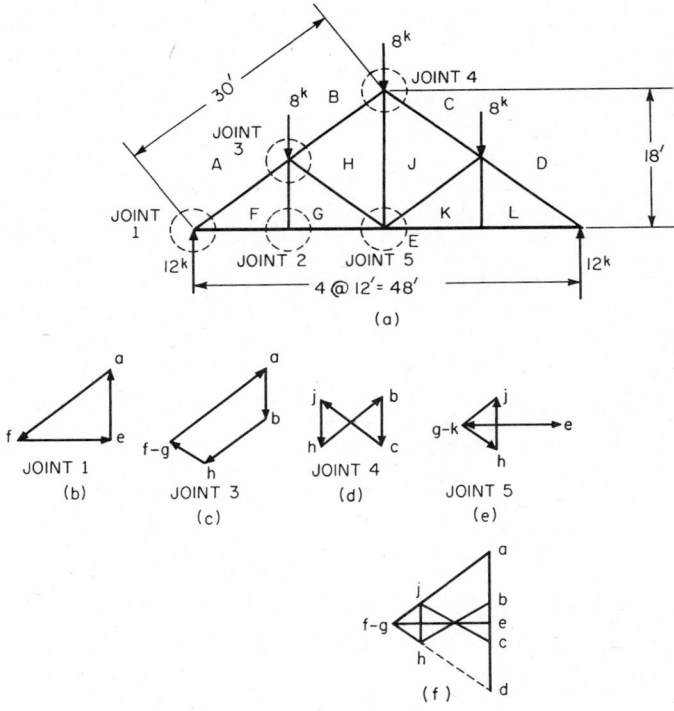

Fig. 6-49. Graphical determination of stresses at each joint of the truss in (*a*) may be expedited by constructing the single Maxwell diagram in (*f*).

Wind loads on a roof truss with a sloping top chord are assumed to act normal to the roof. In that case, the load polygon will be an inclined line or a true polygon.   The reactions are computed generally on the assumption either that both are parallel to the resultant of the wind loads or that one end of the truss is free to move horizontally and therefore will not resist the horizontal components of the loads.   The stress diagram is plotted in the same manner as for vertical loads after the reactions have been found.

Some trusses are complex and require special methods of analysis.   For methods of solving these, see C. H. Norris and J. B. Wilbur, "Elementary Structural Analysis," McGraw-Hill Book Company, New York; or T. Au, "Elementary Structural Mechanics," Prentice-Hall, Inc., Englewood Cliffs, N. J.   These references also present a graphical method of obtaining the deflections of a truss.   An analytical method is given in Art. 6-56.

6-47. Method of Sections for Truss Stresses.   A convenient method of computing the stresses in truss members is to isolate a portion of the truss by a section so chosen as to cut only as many members with unknown stresses as can be evaluated by the laws of equilibrium applied to that portion of the truss. The stresses in the members cut by the section are treated as external forces and must hold the loads on that portion of the truss in equilibrium. Compressive forces act toward each joint or panel point, and tensile forces away from the joint.

Joint Isolation.   A choice of section that often is convenient is one that isolates a joint with only two unknown stresses. Since the stresses and load at the joint must be in equilibrium, the sum of the horizontal components of the forces must be zero, and so must be the sum of the vertical components. Since the lines of action of all the forces are known (the stresses act along the longitudinal axes of the truss members), we can therefore compute two unknown magnitudes of stresses at each joint by this method.

To apply it to joint 1 of the truss in Fig. 6-49a, first equate the sum of the vertical components to zero. This equation shows that the vertical component of the top chord must be equal and opposite to the reaction, 12 kips (see Fig. 6-49b). The stress in the top chord at this joint, then, must be a compression equal to $12 \times {}^{30}/_{18} = 20$ kips. Next, equate the sum of the horizontal components to zero. This equation indicates that the stress in the bottom chord at the joint must be equal and opposite to the horizontal component of the top chord. Hence, the stress in the bottom chord must be a tension equal to $20 \times {}^{24}/_{30} = 16$ kips.

Taking a section around joint 2 in Fig. 6-49a reveals that the stress in the vertical is zero, since there are no loads at the joint and the bottom chord is perpendicular to the vertical. Also, the stress must be the same in both bottom-chord members at the joint, since the sum of the horizontal components must be zero.

After joints 1 and 2 have been solved, a section around joint 3 cuts only two unknown stresses, S_{BH} in top chord BH (see Bow's notation, Art. 6-45) and S_{HG} in diagonal HG. Application of the laws of equilibrium to this joint yields the following two equations, one for the vertical components and the second for the horizontal components:

$$\Sigma V = 0.6 S_{FA} - 8 - 0.6 S_{BH} + 0.6 S_{HG} = 0$$
$$\Sigma H = 0.8 S_{FA} - 0.8 S_{BH} - 0.8 S_{HG} = 0$$

Both unknown stresses are assumed to be compressive, i.e., acting toward the joint. The stress in the vertical does not appear in these equations because it already was determined to be zero. The stress in FA, S_{FA}, was found from analysis of joint 1 to be 20 kips. Simultaneous solution of the two equations yields $S_{HG} = 6.7$ kips and $S_{BH} = 13.3$ kips. (If these stresses had come out with a negative sign, it would have indicated that the original assumption of their directions was incorrect; they would, in that case, be tensile forces instead of compressive forces.)

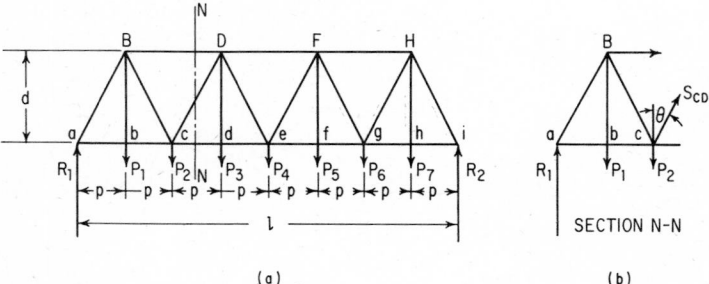

(a)   (b)

Fig. 6-50. Vertical section through truss enables determination of stress in diagonal.

Parallel-chord Trusses.   A convenient section for determining the stresses in diagonals of parallel-chord trusses is a vertical one, such as N-N in Fig. 6-50a. The sum of the forces acting on that portion of the truss to the left of N-N equals the vertical component of the stress in diagonal cD (see Fig. 6-50b). Thus, if θ is the acute angle between cD and the vertical,

$$R_1 - P_1 - P_2 + S \cos \theta = 0$$

But $R_1 - P_1 - P_2$ is the algebraic sum of all the external vertical forces on the left of the section and is the vertical shear in the section. It may be designated as V. Therefore,

$$V + S \cos \theta = 0 \quad \text{or} \quad S = -V \sec \theta$$

From this it follows that, for trusses with horizontal chords and single-web systems, the stress in any web member, other than subverticals, equals the vertical shear in the member multiplied by the secant of the angle that the member makes with the vertical.

Fig. 6-51. Stress in truss diagonal determined by taking a vertical section and computing moments about the intersection of top and bottom chords.

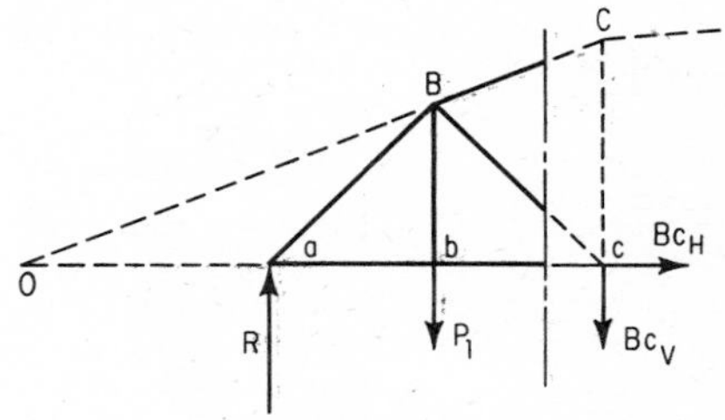

Nonparallel Chords. A vertical section also can be used to determine the stress in diagonals when the chords are not parallel, but the previously described procedure must be modified. Suppose, for example, that the stress in the diagonal Bc of the Parker truss in Fig. 6-51 is to be found. Take a vertical section to the left of joint c. This section cuts BC, the top chord, and Bc, both of which have vertical components, as well as the horizontal bottom chord bc. Now, extend BC and bc until they intersect, at O. If O is used as the center for taking moments of all the forces, the moments of the stresses in BC and bc will be zero, since the lines of action pass through O. Since Bc remains the only stress with a moment about O, Bc can be computed from the fact that the sum of the moments about O must equal zero, for equilibrium.

Generally, the calculation can be simplified by determining first the vertical component of the diagonal and from it the stress. So resolve Bc into its horizontal and vertical components Bc_H and Bc_V, at c, so that the line of action of the horizontal component passes through O. Taking moments about O yields

$$(Bc_v \times Oc) - (R \times Oa) + (P_1 \times Ob) = 0$$

from which Bc_V may be determined. The actual stress in Bc is Bc_V multiplied by the secant of the angle that Bc makes with the vertical.

The stress in verticals, such as Cc, can be found in a similar manner. But take the section on a slope so as not to cut the diagonal but only the vertical and the chords. The moment equation about the intersection of the chords yields the stress in the vertical directly, since it has no horizontal component.

Subdivided Panels. In a truss with parallel chords and subdivided panels, such as the one in Fig. 6-52a, the subdiagonals may be either tension or compression. In Fig. 6-52a, the subdiagonal Bc is in compression and $d'E$ is in tension. The vertical component of the stress in any

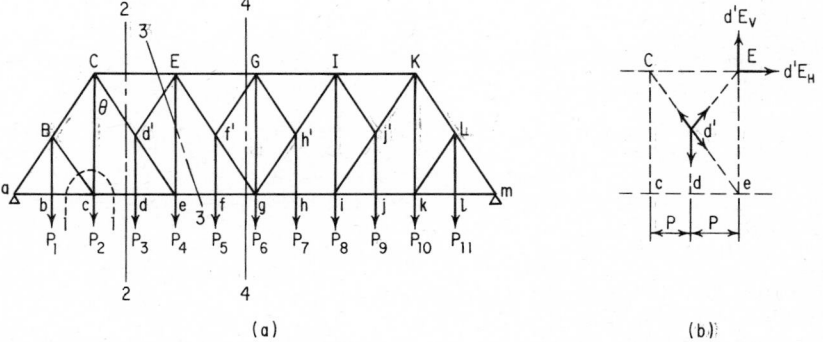

Fig. 6-52. Sections taken through truss with subdivided panels for finding stresses in web members.

subdiagonal equals half the stress in the vertical at the intersection of the subdiagonal and main diagonal.

This can be proved as follows: Take a circular section around d' in Fig. 6-52a. Both the upper and lower portions of the main diagonal Ce are in tension. The stress in the vertical $d'd$ also is tension and equal to the load applied at d. The direction of the stresses Cd', $d'e$, and $d'd$ with reference to joint d' is shown in Fig. 6-52b. Resolve the unknown stress $d'E$ into components at any point along its line of action, such as E. Then, if moments are taken about C, the moments of Cd', $d'e$, and $d'E_H$ are zero, since the lines of action of these stresses pass through C. Equating the sum of the moments about C of the stresses at d' to zero yields

$$(d'd \times p) - (d'E_V \times 2p) = 0 \qquad \text{or} \qquad d'E_V = \frac{1}{2}d'd$$

Generally, in determining the stresses in the long verticals and main diagonals, you must include the stresses in the subdiagonals. For examples, take a section 1-1 (Fig. 6-52a) and equate to zero the sum of the vertical components of the forces. Since the stress in Bc is compressive and its vertical component equals $\frac{1}{2}P_1$, the equation for the stress in Cc is $Cc - \frac{1}{2}P_1 - P_2 = 0$. If you take a vertical section 2-2, the equation for the stress in Cd' is the same as for a diagonal of a truss with a single system of web members, i.e., $V_2 \sec \theta$. If you take a sloping section 3-3, the equation for the stress in the vertical Ee is $Ee + \frac{1}{2}P_3 + V_3 = 0$. And if you take a vertical section 4-4, the equation for the stress in $f'g$ is $-f'g \cos \theta + \frac{1}{2}P_5 + V_4 = 0$.

For a truss with inclined chords and subdivided panels, the vertical component of the stress in a subdiagonal is not equal to half the stress in the vertical at the intersection with the main diagonal as for parallel-chord trusses. For example, the stress in $d'E$ for a truss with nonparallel chords is $d'd \times l/h$, where l is the length of $d'E$ and h the length of Ee. The stress in $d'e$ can be found from its vertical component, which can be determined by taking a vertical section just to the left of E, resolving the stress in $d'E$ into horizontal and vertical components at E and the stress in $d'e$ into components at e, then equating to zero the moments about the intersection of the top and bottom chords.

6-48. Moving Loads on Trusses and Girders. To minimize bending stresses in truss members, framing is arranged to transmit loads to panel points. Usually, in bridges, loads are transmitted from a slab to stringers parallel to the trusses, and the stringers carry the load to transverse floor beams, which bring it to truss panel points. Similar framing generally is used for bridge girders.

In many respects, analysis of trusses and girders is similar to that for beams—determination of maximum end reaction for moving loads, for example, and use of influence lines. For girders, maximum bending moments and shears at various sections must be determined for moving loads, as for beams; and as indicated in Art. 6-47, stresses in truss members may be determined by taking moments about convenient points or from the shear in a panel. But girders and trusses differ from beams in that analysis must take into account the effect at critical sections of loads between panel points, since such loads are distributed to the nearest panel points; hence, in some cases, influence lines differ from those for beams.

Stresses in Verticals. The maximum total stress in a load-bearing stiffener of a girder or in a truss vertical, such as Bb in Fig. 6-53a, equals the maximum reaction of the floor beam at the panel point. The influence line for the reaction at b is shown in Fig. 6-53b. It indicates that for maximum reaction a uniform load of w lb per lin ft should extend a distance of $2p$, from a to c, where p is the length of a panel. In that case, the stress in Bb equals wp.

Maximum floor beam reaction for concentrated moving loads occurs when the total load between a and c, W_1 (Fig. 6-53c), equals twice the load between a and b. Then, the maximum live-load stress in Bb is

$$r_b = \frac{W_1 g - 2Pg'}{p} = \frac{W_1(g - g')}{p}$$

where g is the distance of W_1 from c, and g' is the distance of P from b.

Stresses in Diagonals. For a truss with parallel chords and single-web system, stress in a diagonal, such as Bc in Fig. 6-53a, equals the shear in the panel multiplied by the secant of the angle θ the diagonal makes with the vertical. The influence diagram for stress in Bc, then, is the shear influence diagram for the panel multiplied by $\sec \theta$, as indicated in Fig. 6-53d. For maximum tension in Bc, loads should be placed only in the portion of the span for which the influence diagram is positive (crosshatched in Fig. 6-53d). For maximum compression, the loads should be placed where the diagram is negative (minimum shear).

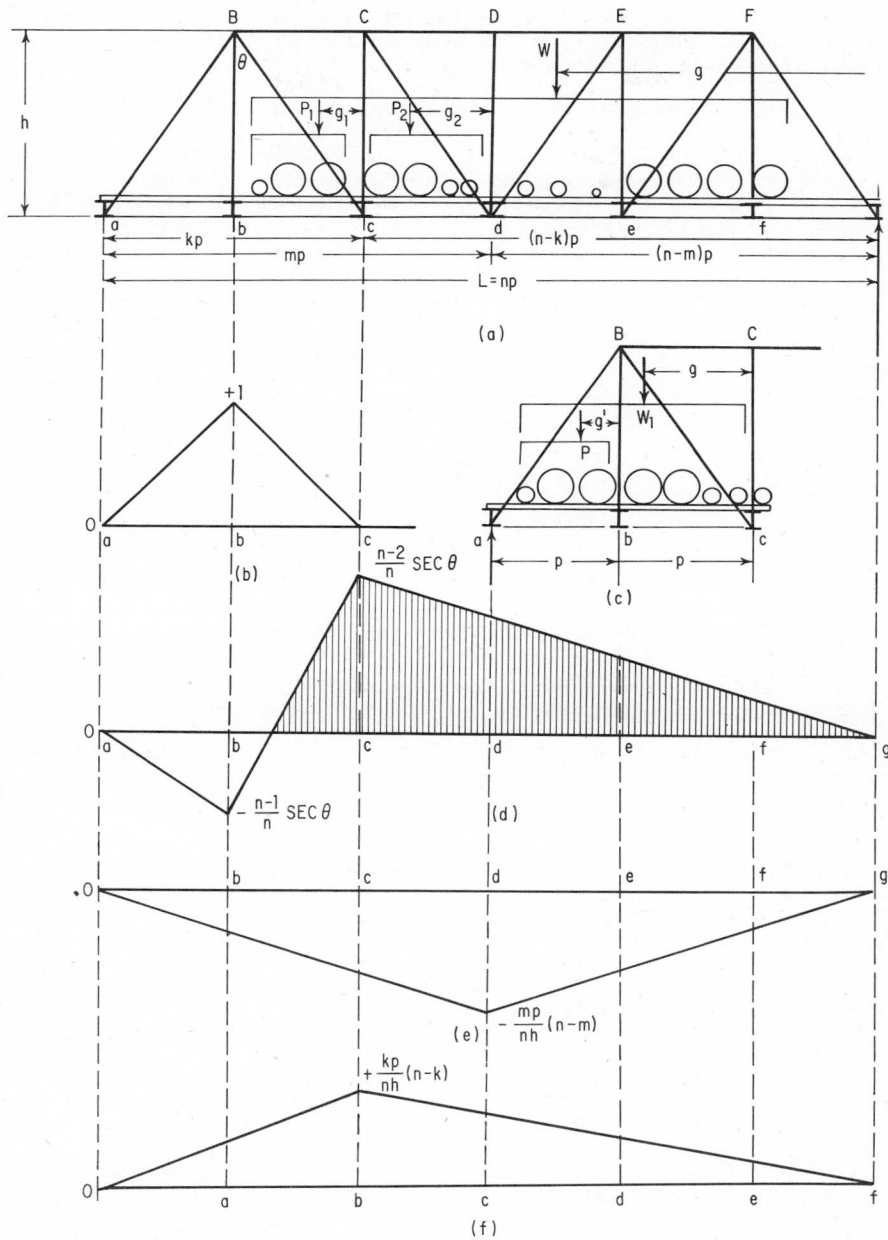

Fig. 6-53. Stresses in a truss due to moving loads determined with influence lines.

A uniform load, however, cannot be placed over the full positive or negative portions of the span, to get a true maximum or minimum. Any load in the panel is transmitted to the panel points at both ends of the panel and decreases the shear. True maximum shear occurs for Bc when the uniform load extends into the panel a distance x from c equal to $(n - k)p/(n - 1)$, where n is the number of panels in the truss and k the number of panels from the left end of the truss to c.

For maximum stress in Bc due to moving concentrated loads, the loads must be placed to produce maximum shear in the panel, and this may require several trials with different wheels placed at c (or, for minimum shear, at b). When the wheel producing maximum shear is at c, the loading will satisfy the following criterion: When the wheel is just to the right of c, W/n is greater than P_1, where W is the total load on the span and P_1 the load in the panel (Fig. 6-53a); when the wheel is just to the left of c, W/n is less than P_1.

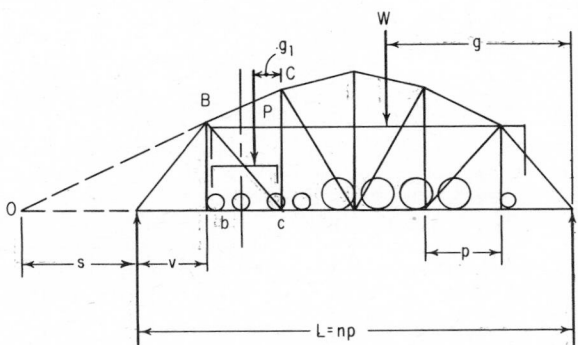

Fig. 6-54. Moving loads on a truss with inclined chord.

For a truss with inclined chords and single-web system, stress in a diagonal, such as Bc in Fig. 6-54, is determined by taking moments about the intersection O of chord BC and the bottom chord. Maximum stress in Bc will occur with a heavy wheel at c, when the following criterion is satisfied: When the wheel is just to the right of c, W/n is greater than $P(1 + v/s)$; when the wheel is just to the left of c, W/n is less than $P(1 + v/s)$, where W is the total load on the span, n the total number of panels in the truss, P the load in the panel through which the section is passed, v the distance from the left support to the left end of the panel through which the section cutting the web member is passed, and s the distance from O to the left support.

Stresses in Chords. Stresses in truss chords, in general, can be determined from the bending moment at a panel point. So the influence diagram for chord stress has the same shape as that for bending moment at an appropriate panel point. For example, Fig. 6-53e shows the influence line for stress in upper chord CD (minus signifies compression). The ordinates are proportional to the bending moment at d, since the stress in CD can be computed by considering the portion of the truss just to the left of d and taking moments about d. Figure 6-53f similarly shows the influence line for stress in bottom chord cd.

For maximum stress in a truss chord under uniform load, the load should extend the full length of the truss.

For maximum chord stress due to moving concentrated loads, the loads must be placed to produce maximum bending moment at the appropriate panel point, and this may require several trials with different wheels placed at the panel point. Usually, maximum moment will be produced with the heaviest grouping of wheels about the panel point.

In all trusses with verticals, the loading producing maximum chord stress will satisfy the following criterion: When the critical wheel is just to the right of the panel point, Wm/n is greater than P, where mp is the distance of the panel point from the left end of the truss and P is the sum of the loads to the left of the panel point; when the wheel is just to the left of the panel point, Wm/n is less than P.

In a truss without verticals, the maximum stress in the loaded chord is determined by a different criterion. For example, the moment center for the lower chord bc (Fig. 6-55) is panel point C, at a distance c from b. When the critical load is at b or c, the following criterion will be satis-

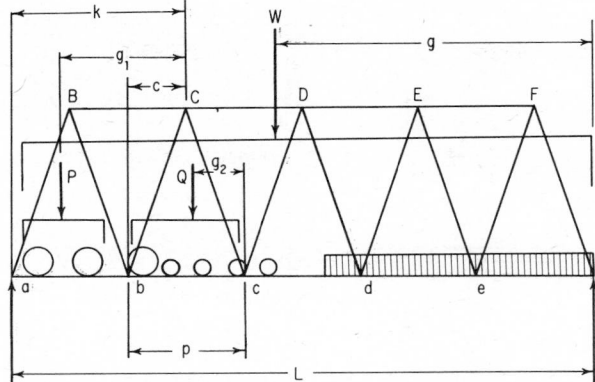

Fig. 6-55. Moving loads on a truss without verticals.

fied: When the wheel is just to the right of b or c, Wk/L is greater than $P + Qc/p$; when the wheel is just to the left of b or c, Wk/L is less than $P + Qc/p$, where W is the total load on the span, Q the load in panel bc, P the load to the left of bc, and k the distance of the center of moments C from the left support. The moment at C is $Wgk/L - Pg_1 - Qcg_2/p$, where g is the distance of the center of gravity of the loads W from the right support, g_1 the distance of the center of gravity of the loads P from C, and g_2 the distance of the center of gravity of the loads Q from c, the right end of the panel.

(C. H. Norris and J. B. Wilbur, "Elementary Structural Analysis," McGraw-Hill Book Company, New York; Tung Au, "Elementary Structural Mechanics," Prentice-Hall, Inc., Englewood Cliffs, N. J.)

6-49. Counters. For long-span bridges, it often is economical to design the diagonals of trusses for tension only. But in the panels near the center of a truss, maximum shear due to live loads plus impact may exceed and be opposite in sign to the dead-load shear, thus inducing compression in the diagonal. If the tension diagonal is flexible, it will buckle. Hence, it becomes necessary to place in such panels another diagonal crossing the main diagonal (Fig. 6-56). Such diagonals are called *counters*.

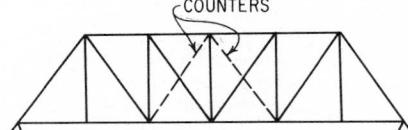

Fig. 6-56. Truss with counters.

Designed only for tension, a counter is assumed to carry no stress under dead load, because it would buckle slightly. It comes into action only when the main diagonal is subjected to compression. Hence, the two diagonals never act together.

Although the maximum stresses in the main members of a truss are the same whether or not counters are used, the minimum stresses in the verticals are affected by the presence of counters. In most trusses, however, the minimum stresses in the verticals where counters are used are of the same sign as the maximum stresses and hence have no significance.

6-50. Stresses in Trusses Due to Lateral Forces. To resist lateral forces on bridge trusses, trussed systems are placed in the planes of the top and bottom chords, and the ends, or **portals,** also are braced as low down as possible without impinging on headroom needed for traffic (Fig. 6-57). In analyzing the lateral trusses, wind loads may be assumed as all applied on the windward chord or as applied equally on the two chords. In the former case, the stresses in the lateral struts are one-half panel load greater than if the latter assumption were made; but this is of no practical consequence.

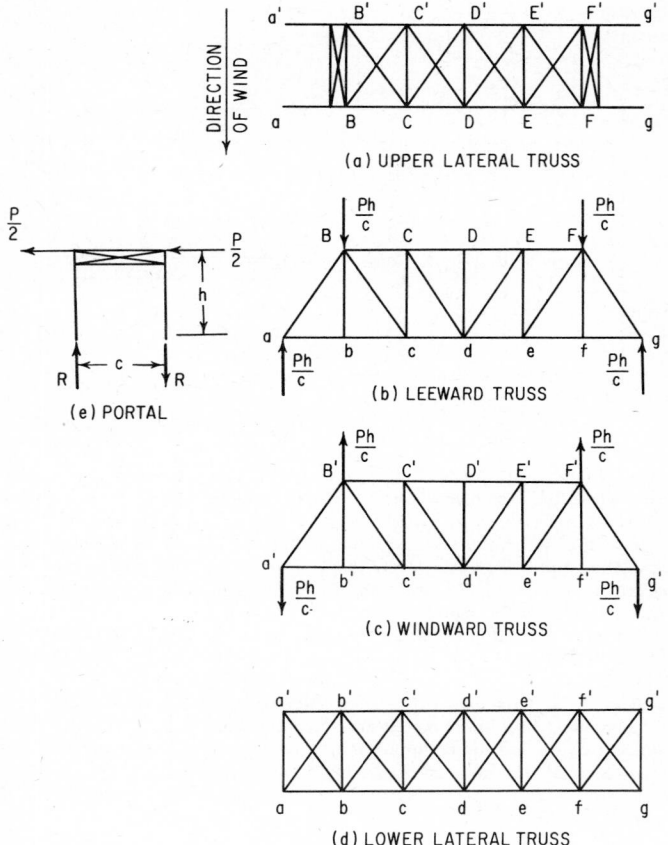

Fig. 6-57. Lateral trusses for bracing top and bottom chords of bridge trusses.

Where the diagonals are considered as tension members only, counter stresses need not be computed, since reversal of wind direction gives greater stresses in the members concerned than any partial loading from the opposite direction. When a rigid system of diagonals is used, the two diagonals of a panel may be assumed to be equally stressed. Stresses in the chords of the lateral truss should be combined with those in the chords of the main trusses due to dead and live loads.

In computing stresses in the lateral system for the loaded chords of the main trusses, the wind on the live load should be added to the wind on the trusses. Hence, the wind on the live load should be positioned for maximum stress on the lateral truss. Methods described in Art. 6-47 can be used to compute the stresses on the assumption that each diagonal takes half the shear in each panel.

When the main trusses have inclined chords, the lateral systems between the sloping chords lie in several planes, and the exact determination of all the wind stresses is rather difficult. The stresses in the lateral members, however, may be determined without significant error by considering the lateral truss flattened into one plane. Panel lengths will vary, but the panel loads will be equal and may be determined from the horizontal panel length.

Since some of the lateral forces are applied considerably above the horizontal plane of the end supports of the bridge, these forces tend to overturn the structure (Fig. 6-57e). The lateral forces of the upper lateral system (Fig. 6-57a) are carried to the portal struts, and the horizontal loads at these points produce an overturning moment about the horizontal plane of the supports. In Fig. 6-57e, P represents the horizontal load brought to each portal strut by the upper lateral bracing, h the depth of the truss, and c the distance between trusses. The overturning moment produced at each end of the structure is Ph, which is balanced by a reaction couple Rc. The value of the reaction R is then Ph/c. An equivalent effect is achieved on the main trusses if loads equal to Ph/c are applied at B and F and at B' and F', as shown in Fig. 6-57b and c. These loads produce stresses in the end posts and in the lower-chord members, but the web members are not stressed.

The lateral force on the live load also causes an overturning moment. It may be treated in a similar manner. But there is a difference as far as the web members of the main truss are concerned. Since the lateral force on the live load produces an effect corresponding to the position of the live load on the bridge, equivalent panel loads, rather than equivalent reactions, must be computed. If the distance from the resultant of the wind force to the plane of the loaded chord is h', the equivalent vertical panel load is Ph'/c, where P is the horizontal panel load due to the lateral force.

6-51. Complex Trusses. The method of sections may not provide a direct solution for some trusses with inclined chords and multiple-web systems. But if the truss is stable and statically determinate, a solution can be obtained by applying the equations of equilibrium to a section taken around each joint. The stresses in the truss members are obtained by solution of the simultaneous equations.

Since two equations of equilibrium can be written for the forces acting at a joint (Art. 6-47), the total number of equations available for a truss is $2n$, where n is the number of joints. If r is the number of horizontal and vertical components of the reactions, and s the number of stresses, $r + s$ is the number of unknowns.

If $r + s = 2n$, the unknowns can be obtained from solution of the simultaneous equations. If $r + s$ is less than $2n$, the structure is unstable (but the structure may be unstable even if $r + s$ exceeds $2n$). If $r + s$ is greater than $2n$, there are too many unknowns; the structure is statically indeterminate.

GENERAL TOOLS FOR STRUCTURAL ANALYSIS

For some types of structures, the equilibrium equations are not sufficient to determine the reactions or the internal stresses. These structures are called *statically indeterminate*.

For the analysis of such structures, additional equations must be written based on a knowledge of the elastic deformations. Hence, methods of analysis that enable deformations to be evaluated in terms of unknown forces or stresses are important for the solution of problems involving statically indeterminate structures. Some of these methods, like the method of virtual work, also are useful in solving complicated problems involving statically determinate systems.

6-52. Virtual Work. A virtual displacement is an imaginary, small displacement of a particle consistent with the constraints upon it. Thus, at one support of a simply supported beam, the virtual displacement could be an infinitesimal rotation $d\theta$ of that end, but not a vertical movement. However, if the support is replaced by a force, then a vertical virtual displacement may be applied to the beam at that end.

Virtual work is the product of the distance a particle moves during a virtual displacement and the component in the direction of the displacement of a force acting on the particle. If the displacement and the force are in opposite directions, the virtual work is negative. When the displacement is normal to the force, no work is done.

Suppose a rigid body is acted upon by a system of forces with a resultant R. Given a virtual displacement ds at an angle α with R, the body will have virtual work done on it equal to $R \cos \alpha \, ds$. (No work is done by the internal forces. They act in pairs of equal magnitude but opposite direction, and the virtual work done by one force of a pair is equal and opposite in sign to the work done by the other force.) If the body is in equilibrium under the action of the forces, then $R = 0$, and the virtual work also is zero.

Thus, the principle of virtual work may be stated:

If a rigid body in equilibrium is given a virtual displacement, the sum of the virtual work of the forces acting on it must be zero.

As an example of how the principle may be used, let us apply it to the determination of the reaction R of the simple beam in Fig. 6-58a. First, replace the support by an unknown force R. Next, move that end of the beam upward a small amount dy as in Fig. 6-58b. The displacement under the load P will be $x \, dy/L$, upward. Then, by the principle of virtual work, $R \, dy - Px \, dy/L = 0$, from which $R = Px/L$.

The principle also may be used to find the reaction R of the more complex beam in Fig. 6-58c. The first step again is to replace the support by an unknown force R.   Next, apply a virtual downward displacement dy at hinge A (Fig. 6-58d). The displacement under the load P will be $x\,dy/c$, and at the reaction R will be $a\,dy/(a + b)$. According to the principle of virtual work, $-Ra\,dy/(a + b) + Px\,dy/c = 0$; thus, $R = Px(a + b)/ac$. In this type of problem, the method has the advantage that only one reaction need be considered at a time and internal forces are not involved.

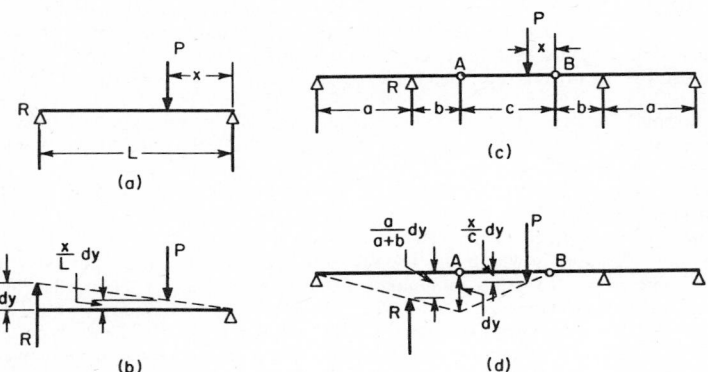

Fig. 6-58. Virtual work applied to determination of a simple-beam reaction (a) and (b) and the reaction of a beam with suspended span (c) and (d).

Strain Energy.   When an elastic body is deformed, the virtual work done by the internal forces equals the corresponding increment of the strain energy dU, in accordance with the principle of virtual work.

Assume a constrained elastic body acted upon by forces $P_1, P_2, \ldots$, for which the corresponding deformations are $e_1, e_2, \ldots$.   Then, $\Sigma P_n\,de_n = dU$.   The increment of the strain energy due to the increments of the deformations is given by

$$dU = \frac{\partial U}{\partial e_1}\,de_1 + \frac{\partial U}{\partial e_2}\,de_2 + \cdots$$

In solving a specific problem, a virtual displacement that is most convenient in simplifying the solution should be chosen.   Suppose, for example, a virtual displacement is selected that affects only the deformation e_n corresponding to the load P_n, other deformations being unchanged. Then, the principle of virtual work requires that

$$P_n\,de_n = \frac{\partial U}{\partial e_n}\,de_n$$

This is equivalent to

$$\frac{\partial U}{\partial e_n} = P_n \qquad\qquad (6\text{-}57)$$

which states that the partial derivative of the strain energy with respect to a specific deformation gives the corresponding force.

Suppose, for example, the stress in the vertical bar in Fig. 6-59 is to be determined.   All bars are made of the same material and have the same cross section A.   If the vertical bar stretches an amount e under the load P, the inclined bars will each stretch an amount $e \cos \alpha$.   The strain energy in the system is [from Eqs. (6-22a)]

$$U = \frac{AE}{2L}\,(e^2 + 2e^2 \cos^3 \alpha)$$

and the partial derivative of this with respect to e must be equal to P; that is,

$$P = \frac{AE}{2L}\,(2e + 4e \cos^3 \alpha) = \frac{AEe}{L}\,(1 + 2 \cos^3 \alpha)$$

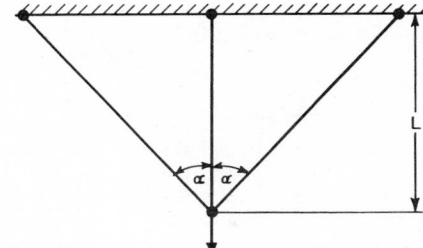

Fig. 6-59. Indeterminate truss.

Noting that the force in the vertical bar equals AEe/L, we find from the above equation that the required stress equals $P/(1 + 2\cos^3 \alpha)$.

6-53. Castigliano's Theorem.   If strain energy is expressed as a function of statically independent forces, the partial derivative of the strain energy with respect to a force gives the deformation corresponding to that force:

$$\frac{\partial U}{\partial P_n} = e_n \qquad (6\text{-}58)$$

(See also Art. 6-52.)

This is known as Castigliano's first theorem.   (His second theorem is the principle of least work.)

6-54. Method of Least Work.   Castigliano's second theorem, also known as the method of least work, states:

The strain energy in a statically indeterminate structure is the minimum consistent with equilibrium.

As an example of the use of the method of least work, an alternative solution will be given for the stress in the vertical bar in Fig. 6-59 (see Art. 6-52).   Calling this stress X, we note that the stress in each of the inclined bars must be $(P - X)/2\cos\alpha$.   Using Eq. (6-22a), we can express the strain energy in the system in terms of X:

$$U = \frac{X^2 L}{2AE} + \frac{(P - X)^2 L}{4AE \cos^3 \alpha}$$

Hence, the internal work in the system will be a minimum when

$$\frac{\partial U}{\partial X} = \frac{XL}{AE} - \frac{(P - X)L}{2AE \cos^3 \alpha} = 0$$

Solving for X gives the stress in the vertical bar as $P/(1 + 2\cos^3 \alpha)$, as in Art. 6-52.

6-55. Dummy Unit-Load Method.   The strain energy for pure bending is $U = M^2 L/2EI$ [see Eq. (6-22d)].   To find the strain energy due to bending stress in a beam, we can apply this equation to a differential length dx of the beam and integrate over the entire span.   Thus,

$$U = \int_0^L \frac{M^2 \, dx}{2EI} \qquad (6\text{-}59)$$

If we let M represent the bending moment due to a generalized force P, the partial derivative of the strain energy with respect to P is the deformation d corresponding to P.   Differentiating Eq. (6-59) gives

$$d = \int_0^L \frac{M}{EI} \frac{\partial M}{\partial P} \, dx \qquad (6\text{-}60)$$

The partial derivative in this equation is the rate of change of bending moment with the load P. It equals the bending moment m produced by a unit generalized load applied at the point where the deformation is to be measured and in the direction of the deformation.   Hence, Eq. (6-60) can also be written as

$$d = \int_0^L \frac{Mm}{EI} \, dx \qquad (6\text{-}61)$$

To find the vertical deflection of a beam, we apply a dummy unit load vertically at the point where the deflection is to be measured and substitute the bending moments due to this load and the actual loading in Eq. (6-61). Similarly, to compute a rotation, we apply a dummy unit moment.

As a simple example, let us apply the dummy unit-load method to the determination of the deflection at the center of a simply supported, uniformly loaded beam of constant moment of inertia (Fig. 6-60a). As indicated in Fig. 6-60b, the bending moment at a distance x from one end is $(wL/2)x - (w/2)x^2$. If we apply a dummy unit load vertically at the center of the beam (Fig. 6-60c), where the vertical deflection is to be determined, the moment at x is $x/2$, as indicated in Fig. 6-60d. Substituting in Eq. (6-61) gives

$$d = 2 \int_0^{L/2} \left(\frac{wL}{2} x - \frac{w}{2} x^2 \right) \frac{x}{2} \frac{dx}{EI} = \frac{5wL^4}{384EI}$$

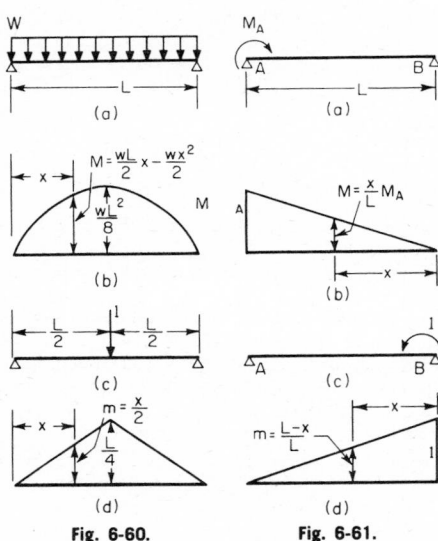

Fig. 6-60. Dummy unit-load method applied to a uniformly loaded beam (a) to find the mid-span deflection, (b) moment diagram for the uniform load, (c) unit load at midspan, (d) moment diagram for unit load.

Fig. 6-61. End rotation at B in beam AB (a) due to end moment at A determined by dummy unit-load method, (b) moment diagram for end moment, (c) unit moment applied at beam end, (d) moment diagram for unit moment.

Fig. 6-60.   **Fig. 6-61.**

As another example, let us apply the method to finding the end rotation at one end of a simply supported, prismatic beam produced by a moment applied at the other end. In other words, the problem is to find the end rotation at B, θ_B, in Fig. 6-61a, due to M_A. As indicated in Fig. 6-61b, the bending moment at a distance x from B due to M_A is $M_A x/L$. If we apply a dummy unit moment at B (Fig. 6-61c), it will produce a moment at x of $(L - x)/L$ (Fig. 6-61d). Substituting in Eq. (6-61) gives

$$\theta_B = \int_0^L M_A \frac{x}{L} \frac{L - x}{L} \frac{dx}{EI} = \frac{M_A L}{6EI}$$

To determine the deflection of a beam due to shear, Castigliano's theorem can be applied to the strain energy in shear:

$$U = \int \int \frac{v^2}{2G} \, dA \, dx$$

where v is the shearing unit stress, G the modulus of rigidity, and A the cross-sectional area.

6-56. Truss Deflections by Dummy Unit-Load Method. Article 6-55 shows how the dummy unit-load method applies to determination of beam deflections. The method also may be adapted to computation of truss deformations.

The strain energy in a truss is given by

$$U = \sum \frac{S^2 L}{2AE} \qquad (6-62)$$

which represents the sum of the strain energy for all the members of the truss. S is the stress in each member due to the loads, L the length of each, A the cross-sectional area, and E the modulus of elasticity. Application of Castigliano's first theorem (Art. 6-53) and differentiation inside the summation sign yield the deformation:

$$d = \sum \frac{SL}{AE} \frac{\partial S}{\partial P} \tag{6-63}$$

where, as in Art. 6-55, P represents a generalized load. The partial derivative in this equation is the rate of change of axial stress with P. It equals the axial stress u produced in each member of the truss by a unit load applied at the point where the deformation is to be measured and in the direction of the deformation. Consequently, Eq. (6-63) also can be written

$$d = \sum \frac{SuL}{AE} \tag{6-64}$$

To find the vertical deflection at any point of a truss, apply a dummy unit vertical load at the panel point where the deflection is to be measured. Substitute in Eq. (6-64) the stresses in each member of the truss due to this load and the actual loading. Similarly, to find the rotation of any joint, apply a dummy unit moment at the joint, compute the stresses in each member of the truss, and substitute in Eq. (6-64). When it is necessary to determine the relative movement of two panel points in the direction of a member connecting them, apply dummy unit loads in opposite directions at those points.

It should be noted that members that are not stressed by the actual loads or the dummy loads do not enter into the calculation of a deformation.

As an example of the application of Eq. (6-64), let us compute the midspan deflection of the truss in Fig. 6-62a. The stresses in kips due to the 20-kip load at every lower-chord panel point are shown in Fig. 6-62a, and the ratios of length of members in inches to their cross-sectional areas in square inches are given in Table 6-1. We apply a dummy unit vertical load at L_2, where the deflection is required. Stresses u due to this load are shown in Fig. 6-62b and the table.

Table 6-1 also contains the computations for the deflection. Members not stressed by the 20-kip loads or the dummy unit loads are not included. Taking advantage of the symmetry of the truss, the values are tabulated for only half the truss and the sum is doubled. Also, to reduce the amount of calculation, the modulus of elasticity E, which is equal to 30,000,000, is not included until the very last step, since it is the same for all members.

6-57. Statically Indeterminate Trusses. A truss is statically indeterminate when the number of unknown quantities exceeds the number of independent equations of static equilibrium that

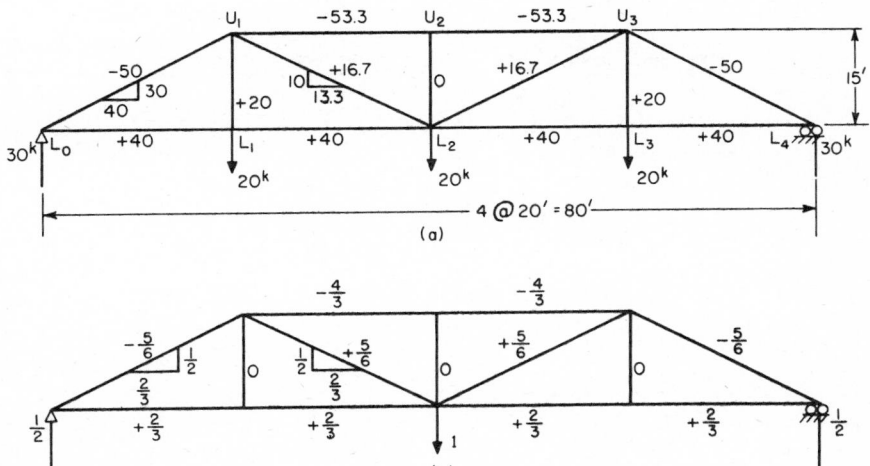

Fig. 6-62. Dummy unit-load method applied to a loaded truss (a) to find midspan deflection, (b) unit load applied at midspan.

Table 6-1. Midspan Deflection of Truss of Fig. 6-62

Member	L/A	S	u	SuL/A
L_0L_2	160	$+40$	$+2/3$	4,267
L_0U_1	75	-50	$-5/6$	3,125
U_1U_2	60	-53.3	$-4/3$	4,267
U_1L_2	150	$+16.7$	$+5/6$	2,083
				13,742

$$d = \sum \frac{SuL}{AE} = \frac{2 \times 13,742,000}{30,000,000} = 0.916 \text{ in.}$$

may be written for the structure or portions of the structure. As noted in Art. 6-51, if n is the number of joints in a truss, r the number of horizontal and vertical components of the reactions, and s the number of stresses, the truss is statically indeterminate if $r + s$ is greater than $2n$.

The truss in Fig. 6-63a is statically indeterminate. It has 4 joints, 3 reaction components, and 6 members, so that $(r + s = 3 + 6) > (2n = 2 \times 4)$. To determine the stresses in this truss, it is necessary to add to the equations of equilibrium an equation based on a knowledge of the elastic deformations of the truss. For this purpose, remove member ac to make the truss statically determinate and instead apply unknown stresses S_1 at both a and c, as indicated in Fig. 6-63b. Then, determine S_1 by equating the sum of the horizontal deflection at c of the statically determinate truss and the elongation of ac to zero.

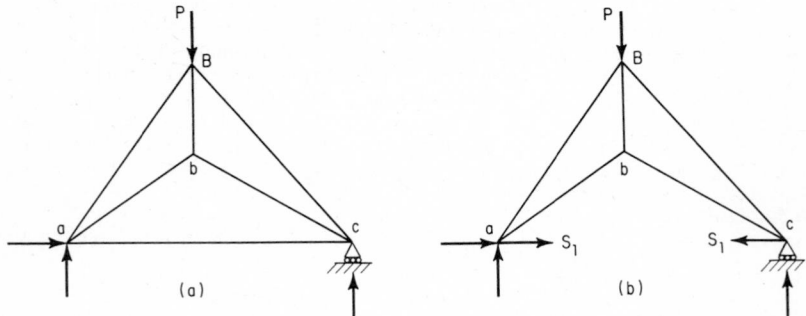

Fig. 6-63. Statically indeterminate truss (a) made determinate (b).

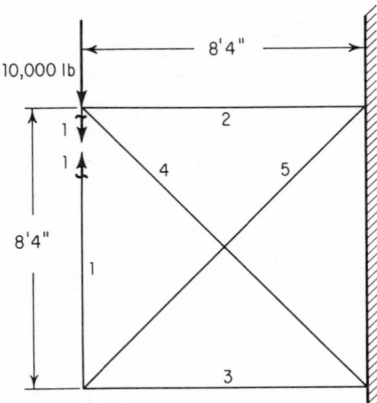

Fig. 6-64. Truss with two diagonals made determinate by cutting a member.

Let the stress in any member due to the load P and S_1 be $S' + uS_1$, where S' is the stress due to P, u the stress in any member due to unit force applied at a and c, and uS_1 the stress in any member due to S_1. Applying Eq. (6-64), we find that the horizontal displacement of c (assuming a fixed) due to both P and S_1 is

$$\delta = \sum_2^n \frac{(S' + uS_1)uL}{AE} = \sum_2^n \frac{S'uL}{AE} + \sum_2^n \frac{S_1 u^2 L}{AE}$$

where L is the length of the member, A the cross-sectional area, and E the modulus of elasticity. From Eq. (6-8), the elongation of ac is $S_1 L_1/A_1 E = S_1 u^2 L_1/A_1 E$, since $u = 1$ in ac. Then, since the sum of δ and the elongation of ac is zero,

$$\sum_2^n \frac{S'uL}{AE} + \sum_2^n \frac{S_1 u^2 L}{AE} + \frac{S_1 u^2 L_1}{A_1 E} = \sum_2^n \frac{S'uL}{AE} + S_1 \sum_1^n \frac{u^2 L}{AE} = 0$$

Solving for S_1 and assuming E constant, we obtain

$$S_1 = -\frac{\displaystyle\sum_2^n (S'uL/A)}{\displaystyle\sum_1^n (u^2 L/A)} \tag{6-65}$$

With S_1 known, the stress in each member of the truss can be determined by adding uS_1 for the member to S'.

As an example of the use of Eq. (6-65), let us determine the stresses in the truss of Fig. 6-64. The truss has 4 reaction components, 5 members (considering each of the main diagonals as one member), and 4 joints (not counting the intersection of the diagonals). Hence, $(r + s = 4 + 5) > (2n = 2 \times 4)$. The truss is statically indeterminate, with one unknown more than equations for static equilibrium.

Assume member 1 as the redundant member and apply a pair of unit forces near the upper end of the member in such a direction as to cause tension in the member. Equation (6-65) is solved in Table 6-2 with E a constant. From Table 6-2, $S_1 = -717,000/168.4 = -4,300$ lb.

Table 6-2. Computation of Stresses in Truss of Fig. 6-64

Member	A, sq in.	L, in.	S', lb	$S'L/A$	u, lb	$S'uL/A$	u^2L/A	S_1u	$S = S' + S_1u$, lb
1	4	100	0	....	$+1.00$	....	$+25.0$	$-4,300$	$-4,300$
2	4	100	$+10,000$	$+250,000$	$+1.00$	$+250,000$	$+25.0$	$-4,300$	$+5,700$
3	4	100	0	....	$+1.00$	....	$+25.0$	$-4,300$	$-4,300$
4	6	141	$-14,100$	$-331,000$	-1.41	$+467,000$	$+46.7$	$+6,000$	$-8,100$
5	6	141	0	....	-1.41	....	$+46.7$	$+6,000$	$+6,000$
					Σ	$+717,000$	$+168.4$		

The true stress in each member of the complete frame is given in the last column of the table.
(C. H. Norris and J. B. Wilbur, "Elementary Structural Analysis," McGraw-Hill Book Company, New York.)

6-58. Reciprocal Theorem and Influence Lines.   Consider a structure loaded by a group of independent forces A, and suppose that a second group of forces B is added. The work done by the forces A acting over the displacements due to B will be W_{AB}.

Now, suppose the forces B had been on the structure first, and then load A had been applied. The work done by the forces B acting over the displacements due to A will be W_{BA}.

The reciprocal theorem states that $W_{AB} = W_{BA}$.

Some very useful conclusions can be drawn from this equation. For example, there is the reciprocal deflection relationship:

The deflection at a point A due to a load at B equals the deflection at B due to the same load applied at A.   Also, the rotation at A due to a load (or moment) at B equals the rotation at B due to the same load (or moment) applied at A.

Another consequence is that deflection curves also may be influence lines, to some scale, for reactions, shears, moments, or deflections (**Mueller-Breslau principle**). For example, suppose the influence line for a reaction is to be found; that is, we wish to plot the reaction R as a unit load moves over the structure, which may be statically indeterminate. For loading condition A, we analyze the structure with a unit load on it at a distance x from some reference point. For loading condition B, we apply a dummy unit vertical load upward at the place where the reaction is to be determined, deflecting the structure off the support. At a distance x from the reference point, the displacement is d_{xR}, and over the support the displacement is d_{RR}. Hence, $W_{AB} = -1\, d_{xR} + R d_{RR}$. On the other hand, W_{BA} is zero, since loading condition A provides no displacement for the dummy unit load at the support in condition B. Consequently, from the reciprocal theorem, $W_{AB} = W_{BA} = 0$; and hence

$$R = \frac{d_{xR}}{d_{RR}}$$

Since d_{RR}, the deflection at the support due to a unit load applied there, is a constant, R is proportional to d_{xR}. So the influence line for a reaction can be obtained from the deflection curve resulting from a displacement of the support (Fig. 6-65a). The magnitude of the reaction is obtained by dividing each ordinate of the deflection curve by d_{RR}.

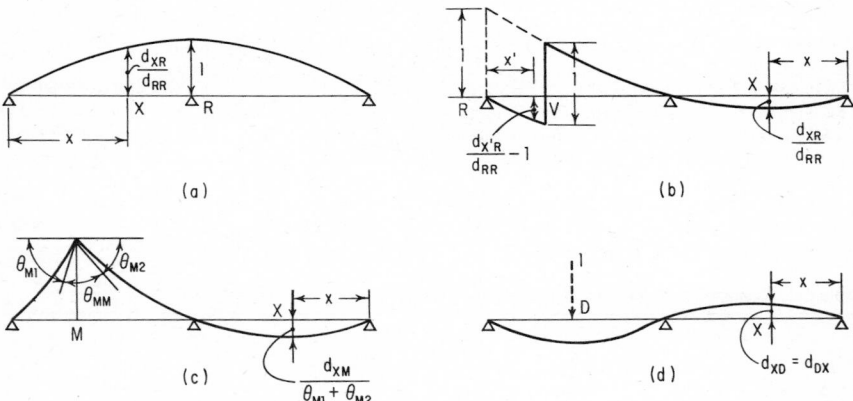

Fig. 6-65. Influence lines for a continuous beam obtained from deflection curves. (a) Reaction at R. (b) Shear at V. (c) Moment at M. (d) Deflection at D.

Similarly, the influence line for shear can be obtained from the deflection curve produced by cutting the structure and shifting the cut ends vertically at the point for which the influence line is desired (Fig. 6-65b).

The influence line for bending moment can be obtained from the deflection curve produced by cutting the structure and rotating the cut ends at the point for which the influence line is desired (Fig. 6-65c).

And finally it may be noted that the deflection curve for a load of unity is also the influence line for deflection at that point (Fig. 6-65d).

CONTINUOUS BEAMS AND FRAMES

Continuous beams and frames are statically indeterminate. Bending moments in them are functions of the geometry, moments of inertia, and modulus of elasticity of individual members as well as of loads and spans. While these moments can be determined by the methods described in Arts. 6-52 to 6-58, methods specially developed for beams and frames are available that often make analysis simpler. The following articles describe some of these methods.

6-59. General Method of Analysis. Continuous beams and frames consist of members that can be treated as simple beams, the ends of which are prevented by moments from rotating freely. Member LR in the continuous beam in Fig. 6-66a, for example, can be isolated, as shown in Fig.

6-66*b*, and the elastic restraints at the ends replaced by couples M_L and M_R. In this way, *LR* is converted into a simply supported beam acted upon by end moments and transverse loads.

The bending-moment diagram for *LR* is shown at the left in Fig. 6-66*c*. Treating *LR* as a simple beam, we can break this diagram down into three simple components, as shown at the right of the equals sign in Fig. 6-66*c*. Thus, the bending moment at any section equals the simple-beam moment due to the transverse loads, plus the simple-beam moment due to the end moment at *L*, plus the simple-beam moment due to the end moment at *R*.

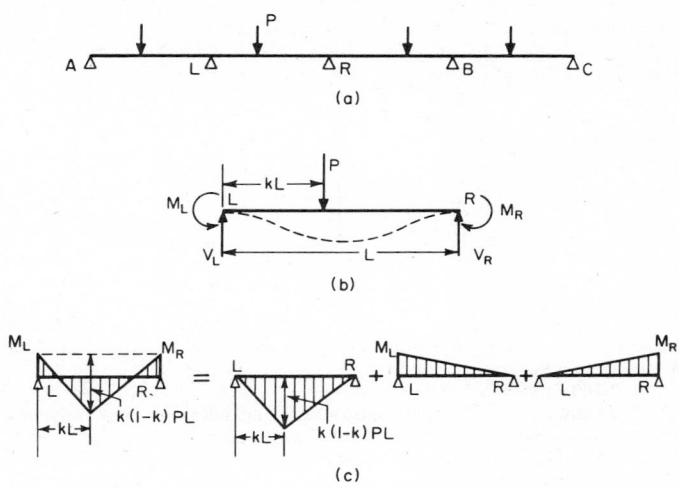

Fig. 6-66. Any span of a continuous beam (*a*) can be treated as a simple beam, as shown in (*b*) and (*c*). In (*c*) the moment diagram is decomposed into basic components.

Once M_L and M_R have been determined, the shears may be computed by taking moments about any section. Similarly, if the reactions or shears are known, the bending moments may be calculated.

A general method for determining the elastic forces and moments exerted by redundants, supports, and moments is as follows: Remove as many redundant supports or members as necessary to make the structure statically determinate. Compute, for the actual loads, the

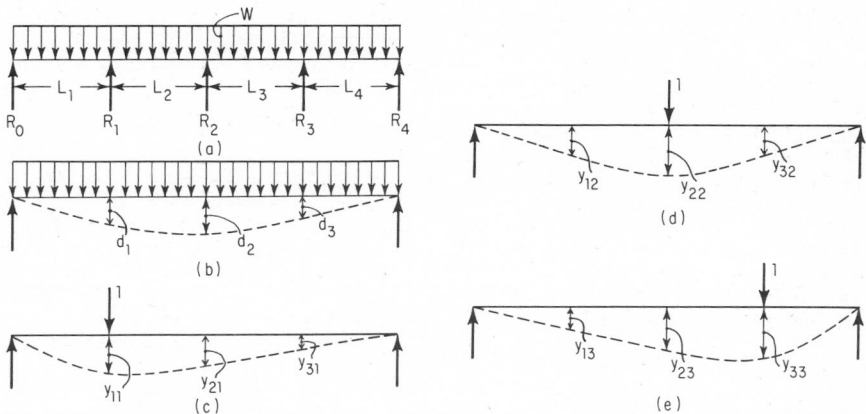

Fig. 6-67. Reactions of continuous beam (*a*), found by making the beam statically determinate. (*b*) Deflections computed with interior supports removed. (*c*), and (*d*), and (*e*) Deflections calculated for unit load over each removed support to obtain equations for each redundant.

deflections or rotations of the statically determinate structure in the direction of the forces and couples exerted by the removed supports and members. Then, in terms of these forces and couples, compute the corresponding deflections or rotations the forces and couples produce in the statically determinate structure. Finally, for each redundant support or member, write equations that give the known rotations or deflections of the original structure in terms of the deformations of the statically determinate structure.

For example, one method of finding the reactions of the continuous beam in Fig. 6-67a is to remove the interior supports temporarily. The beam then will have deflections d_1, d_2, and d_3 at those supports (Fig. 6-67b). Next, place a unit load at the location of each support in succession (Fig. 6-67c, d, and e). Let y_{mn} denote the simple-beam deflection, where m represents the support where the deflection is measured and n the support at which the unit load is applied. Since the continuous beam has no deflections at its supports, we can now write one equation for each support equating the downward deflection of the loaded simple beam to its upward deflection due to the unknown reactions:

$$d_1 = y_{11}R_1 + y_{12}R_2 + y_{13}R_3$$
$$d_2 = y_{21}R_1 + y_{22}R_2 + y_{23}R_3 \qquad (6\text{-}66)$$
$$d_3 = y_{31}R_1 + y_{32}R_2 + y_{33}R_3$$

Solution of these equations yields R_1, R_2, and R_3, and R_0 and R_4 can be obtained from these reactions by applying the equations of equilibrium.

For continuous beams and frames with a large number of redundants, this method becomes unwieldy because of the number of simultaneous equations. Special methods, like moment distribution, are preferable in such cases.

See also Arts. 6-100 to 6-103.

6-60. Sign Convention. For moment distribution, the following sign convention is most convenient: A moment acting at an end of a member or at a joint is positive if it tends to rotate the end or joint, clockwise; negative, if it tends to rotate the joint counterclockwise. Hence, in Fig. 6-66b, M_R is positive and M_L is negative.

Similarly, the angular rotation at the end of a member is positive if in a clockwise direction, negative if counterclockwise. Thus, a positive end moment produces a positive end rotation in a simple beam. (See also Art. 6-59.)

For ease in visualizing the shape of the elastic curve under the action of loads and end moments, plot bending-moment diagrams on the tension side of each member. Hence, if an end moment is represented by a curved arrow, the arrow will point in the direction in which the moment is to be plotted.

6-61. Fixed-end Moments. A beam so restrained at its ends that no rotation is produced there by the loads is called a fixed-end beam, and the end moments are called fixed-end moments. Actually, it would be very difficult to construct a beam with ends that are truly fixed. The concept of fixed ends, however, is useful in determining the moments in continuous beams and frames.

Fixed-end moments may be expressed as the product of a coefficient and WL, where W is the total load on the span L. The coefficient is independent of the properties of other members of the structure. Thus, any member of a continuous beam or frame can be isolated from the rest of the structure and its fixed-end moments computed. Then, the actual moments in the beam can be found by applying a correction to each fixed-end moment.

Fixed-end moments may be determined conveniently by the moment-area method or the conjugate-beam method (Art. 6-33).

Fixed-end moments for several common types of loading on beams of constant moment of inertia (prismatic beams) are given in Fig. 6-68. Also, the curves in Fig. 6-70 enable fixed-end moments to be computed easily for any type of loading on a prismatic beam. Before the curves can be entered, however, certain characteristics of the loading must be calculated. These include $\bar{x}L$, the location of the center of gravity of the loading with respect to one of the loads; $G^2 = \Sigma b_n^2 P_n / W$, where $b_n L$ is the distance from each load P_n to the center of gravity of the loading (taken positive to the right); and $S^3 = \Sigma b_n^3 P_n / W$ (see case 8, Fig. 6-69). These values are given in Fig. 6-69 for some common types of loading.

The curves in Fig. 6-70 are entered at the bottom with the location a of the center of gravity of the loading with respect to the left end of the span. At the intersection with the proper G curve, proceed horizontally to the left to the intersection with the proper S line, then vertically to the horizontal scale indicating the coefficient m by which to multiply WL to obtain the fixed-end moment. The curves solve the equations:

$$m_L = \frac{M_L{}^F}{WL} = G^2[1 - 3(1 - a)] + a(1 - a)^2 + S^3 \qquad (6\text{-}67a)$$

$$m_R = \frac{M_R{}^F}{WL} = G^2(1 - 3a) + a^2(1 - a) - S^3 \qquad (6\text{-}67b)$$

where $M_L{}^F$ is the fixed-end moment at the left support and $M_R{}^F$ at the right support.

As an example of the use of the curves, find the fixed-end moments in a prismatic beam of 20-ft span carrying a triangular loading of 100 kips, similar to the loading shown in case 4, Fig. 6-69, distributed over the entire span, with the maximum intensity at the right support.

Case 4 gives the characteristics of the loading: $y = 1$; the center of gravity is $L/3$ from the right support; so $a = 0.67$, $G^2 = {}^1/_{18} = 0.056$, and $S^3 = -{}^1/_{135} = -0.007$. To find $M_R{}^F$, we enter Fig. 6-70 at the bottom with $a = 0.67$ on the upper scale and proceed vertically to the estimated location of the intersection of the coordinate with the $G^2 = 0.06$ curve. Then we move horizontally to the intersection with the line for $S^3 = -0.007$, as indicated by the dash line in Fig. 6-70. Referring to the scale at the top of the diagram, we find the coefficient m_R to be 0.10. Similarly, with $a = 0.67$ on the lowest scale, we find the coefficient m_L to be 0.07. Hence, the fixed-end moment at the right support is $0.10 \times 100 \times 20 = 200$ ft-kips, and at the left support $-0.07 \times 100 \times 20 = -140$ ft-kips.

6-62. Fixed-end Stiffness. To correct a fixed-end moment to obtain the end moment for the actual conditions of end restraint in a continuous structure, the end of the member must be permitted to rotate. The amount it will rotate depends on its stiffness, or resistance to rotation.

The fixed-end stiffness of a beam is defined as the moment required to produce a rotation of unity at the end where it is applied, while the other end is fixed against rotation. It is represented by $K_R{}^F$ in Fig. 6-71.

For prismatic beams, the fixed-end stiffnesses for both ends equal $4EI/L$, where E is the modulus of elasticity, I the moment of inertia of the cross section about the centroidal axis, and L the span (generally taken center to center of supports). When deformations need not be calculated, only the relative values of K^F for each member need be known; hence, only the ratio of

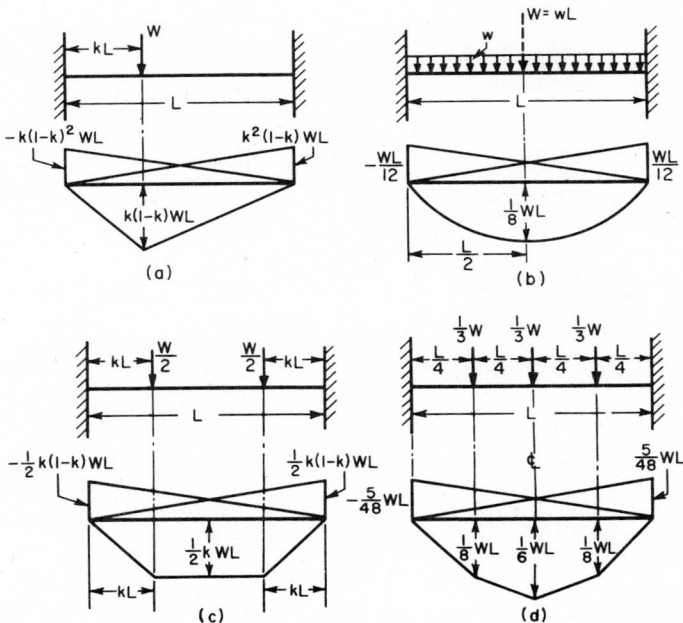

Fig. 6-68. Fixed-end moments for a prismatic beam: (a) for a concentrated load; (b) for a uniform load; (c) for two equal concentrated loads; (d) for three equal concentrated loads.

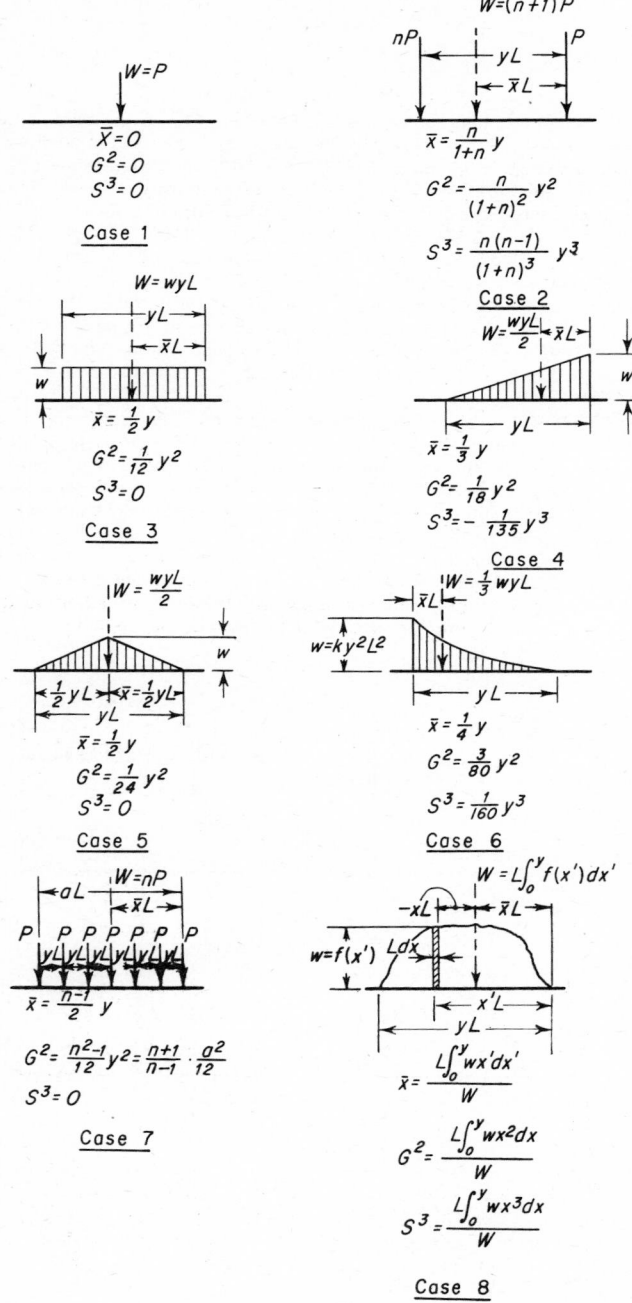

Fig. 6-69. Characteristics of loadings.

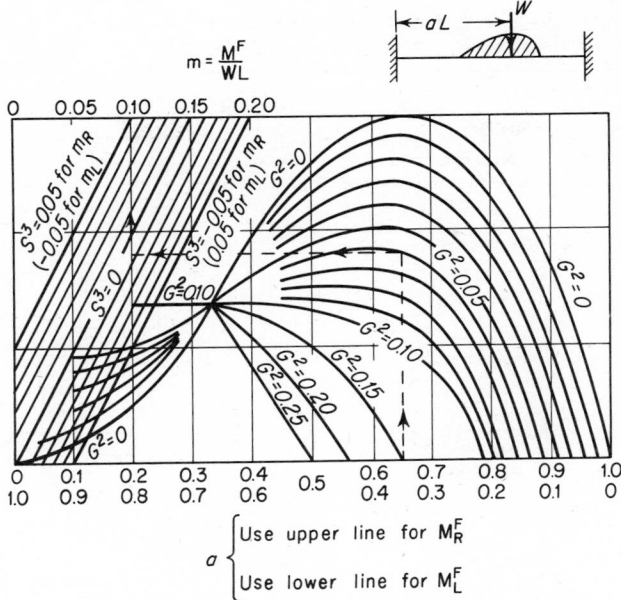

$$m = \frac{M^F}{WL}$$

Fig. 6-70. Chart for fixed-end moments due to any type of loading.

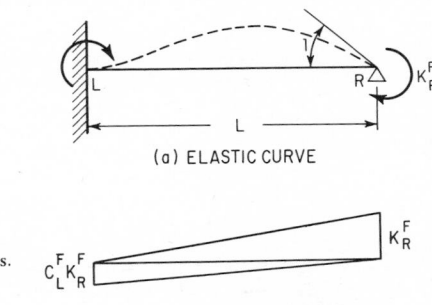

(a) ELASTIC CURVE

Fig. 6-71. Fixed-end stiffness.

(b) MOMENT DIAGRAM

I to L has to be computed. (For simply supported, prismatic beams, the actual stiffness is $3EI/L$, or three-fourths the fixed-end stiffness.)

For beams of variable moment of inertia, the fixed-end stiffness may be calculated by methods presented in Art. 6-69 or obtained from tables, such as those in the "Handbook of Frame Constants," published by the Portland Cement Association, Skokie, Ill.; and J. M. Gere, "Moment Distribution Factors for Beams of Tapered I-section," American Institute of Steel Construction, New York.

6-63. Fixed-end Carry-over Factor. When a moment is applied at one end of a continuous beam, a resisting moment is induced at the far end if that end is restrained by other beams or columns against rotation (Fig. 6-71). The ratio of the resisting moment at a fixed end to the applied moment is called the fixed-end carry-over factor C^F.

For prismatic beams, the fixed-end carry-over factor toward either end is 0.5. It should be noted that the applied moment and the resisting moment have the same sign (Fig. 6-71a); that is, if the applied moment acts in a clockwise direction, the carryover moment also acts clockwise.

For beams of variable moment of inertia, the fixed-end carry-over factor may be calculated by methods presented in Art. 6-69 or obtained from tables, such as those referred to in Art. 6-62.

6-64. Moment Distribution by Converging Approximations. The frame in Fig. 6-72 consists of four prismatic members rigidly connected together at O and fixed at ends A, B, C, and D. If an external moment U is applied at O, the sum of the end moments in each member at O must be equal to U. Furthermore, all members must rotate at O through the same angle θ, since they are assumed to be rigidly connected there. Hence, by the definition of fixed-end stiffness (Art. 6-62), the proportion of U induced in or "distributed" to the end of each member at O equals the ratio of the stiffness of that member to the sum of the stiffnesses of all the members at O. This ratio is called the distribution factor at O for the member.

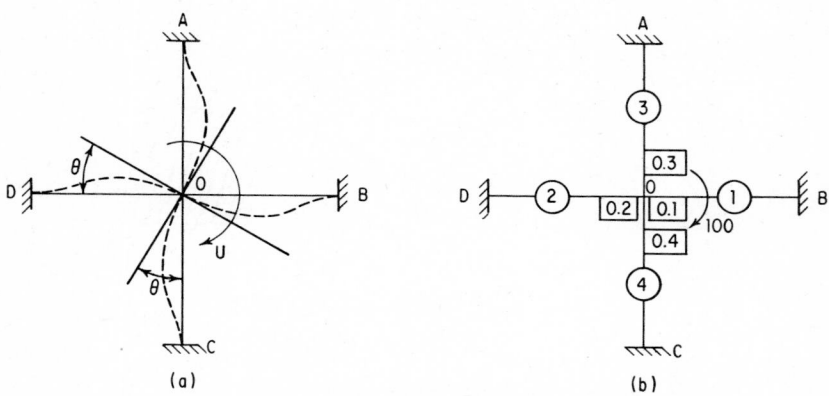

Fig. 6-72. Joint in simple frame rotated by moment. (*a*) Elastic curve. (*b*) Stiffness and distribution factors.

Suppose a moment of 100 ft-kips is applied at O, as indicated in Fig. 6-72*b*. The relative stiffness (or I/L) is assumed as shown in the circle on each member. The distribution factors for the moment at O are computed from the stiffnesses and shown in the boxes. For example, the distribution factor for OA equals its stiffness divided by the sum of the stiffnesses of all the members at the joint: $3/(3 + 1 + 4 + 2) = 0.3$. Hence, the moment induced in OA at O is $0.3 \times 100 = 30$ ft-kips. Similarly, OB gets 10 ft-kips, OC 40 ft-kips, and OD 20 ft-kips.

Because the far ends of these members are fixed, one-half of these moments are carried over to them (Art. 6-63). Thus $M_{AO} = 0.5 \times 30 = 15$; $M_{BO} = 0.5 \times 10 = 5$; $M_{CO} = 0.5 \times 40 = 20$; and $M_{DO} = 0.5 \times 20 = 10$.

Most structures consist of frames similar to the one in Fig. 6-72, or even simpler, joined together. Though the ends of the members may not be fixed, the technique employed for the frame in Fig. 6-72 can be applied to find end moments in such continuous structures.

Span with Simple Support. Before the general method is presented, one shortcut is worth noting. Advantage can be taken when a member has a hinged end to reduce the work in distributing moments. This is done by using the true stiffness of the member instead of the fixed-end stiffness. (For a prismatic beam, the stiffness of a member with a hinged end is three-fourths the fixed-end stiffness; for a beam with variable moment of inertia, it is equal to the fixed-end stiffness times $1 - C_L{}^F C_R{}^F$, where $C_L{}^F$ and $C_R{}^F$ are the fixed-end carry-over factors to each end of the beam.) Naturally, the carry-over factor toward the hinge is zero.

Moment Release and Distribution. When beam ends are neither fixed nor pinned, but restrained by elastic members, moments can be distributed by a series of converging approximations. At first, all joints are locked against rotation. As a result, the loads will create fixed-end moments (Art. 6-61) at the ends of every loaded member. At each joint, the unbalanced moment, a moment equal to the algebraic sum of the fixed-end moments at the joint, is required to hold it fixed. But if the joint actually is not fixed, the unbalanced moment does not exist. It must be removed by applying an equal but opposite moment. One joint at a time is unlocked by applying a moment equal but opposite in sign to the unbalanced moment. The unlocking moment must be distributed to the members at the joint in proportion to their fixed-end stiffnesses. As a result, the far end of each member should receive a "carry-over" moment equal to the distributed moment times a carry-over factor (Art. 6-63).

After all joints have been released at least once, it generally will be necessary to repeat the process—sometimes several times—before the corrections to the fixed-end moments become negligible. To reduce the number of cycles, start the unlocking of joints with those having the greatest unbalanced moments.

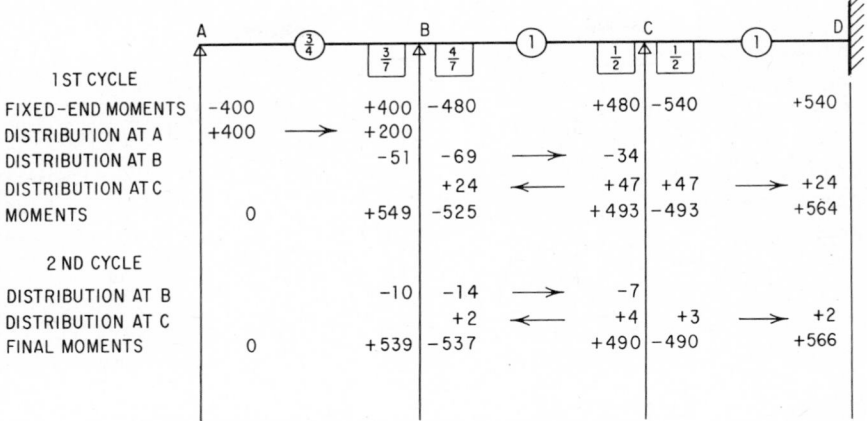

Fig. 6-73. Moment distribution by converging approximations.

Example. Suppose the end moments are to be found for the continuous beam *ABCD* in Fig. 6-73, given the fixed-end moments on the first line of the figure. The I/L values for all spans are given equal; therefore, the relative fixed-end stiffness for all members is unity. But since A is a hinged end, the computation can be shortened by using the actual relative stiffness, which is ¾. Relative stiffnesses for all members are shown in the circle on each member. The distribution factors are shown in the boxes at each joint.

Begin the computation with removal of the unbalance in fixed-end moments (first line in Fig. 6-73). The greatest unbalanced moment, by inspection, occurs at hinged end A and is −400; so unlock this joint first. Since there are no other members at the joint, distribute the full unlocking moment of +400 to AB at A and carry over one-half to B. The unbalance at B now is +400 − 480 plus the carry-over of +200 from A, or a total of +120. Hence, a moment of −120 must be applied and distributed to the members at B by multiplying by the distribution factors in the corresponding boxes.

The net moment at B could be found now by adding the fixed-end and distributed moments at the joint. But it generally is more convenient to delay the summation until the last cycle of distribution has been completed.

After B is unlocked, the moment distributed to BA need not be carried over to A, because the carry-over factor toward the hinged end is zero. But half the moment distributed to BC is carried over to C.

Similarly, unlock joint C and carry over half the distributed moments to B and D, respectively. Joint D should not be unlocked, since it actually is a fixed end. Thus, the first cycle of moment distribution has been completed.

Carry out the second cycle in the same manner. Release joint B, and carry over to C half the distributed moment in BC. Finally, unlock C to complete the cycle. Add the fixed-end and distributed moments to obtain the final moments.

6-65. Continuous Frames. In continuous frames, maximum end moments and maximum interior moments are produced by different combinations of loadings. For maximum end moment in a beam, live load should be placed on that beam and on the beam adjoining the end for which the moment is to be computed. Spans adjoining these two should be assumed to be carrying only dead load.

For maximum midspan moments, the beam under consideration should be fully loaded, but adjoining spans should be assumed to be carrying only dead load.

The work involved in distributing moments due to dead and live loads in continuous frames in buildings can be greatly simplified by isolating each floor. The tops of the upper columns and the bottoms of the lower columns can be assumed fixed. Furthermore, the computations can be

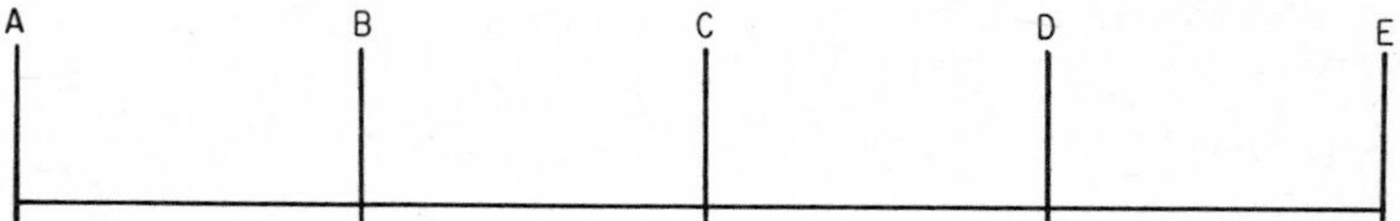

	A		B			C			D		E
1. STIFFNESS		1			1			1			1	
2. DISTRIBUTION FACTOR	0.33		0.25	0.25		0.25	0.25		0.25	0.25		0.33
3. F.E.M. DEAD LOAD	—		+91	-37		+37	-70		+70	-59		—
4. F.E.M. TOTAL LOAD	-172	+99	+172	-78	+73	+78	-147	+85	+147	-126	+63	+126
5. CARRY OVER	-17	+11	+29	-1	+1	-2	-11	+7	+14	-21	+13	+7
6. ADDITION	-189	+18	+201	-79	-1	+76	-158	+9	+161	-147	+5	+133
7. DISTRIBUTION	+63		-30	-30		+21	+21		-4	-4		-44
8. MAX. MOMENTS	-126	+128	+171	-109	+73	+97	-137	+101	+157	-151	+81	+89

Fig. 6-74. Moment distribution in a continuous frame by converging approximations.

condensed considerably by following the procedure recommended in "Continuity in Concrete Building Frames," published by the Portland Cement Association, Skokie, Ill., and illustrated in Fig. 6-74.

Figure 6-74 presents the complete calculation for maximum end and midspan moments in four floor beams AB, BC, CD, and DE. Columns are assumed to be fixed at the story above and below. None of the beam or column sections is known to begin with; so as a start, all members will be assumed to have a fixed-end stiffness of unity, as indicated on the first line of the calculation.

Column Moments. The second line gives the distribution factors (Art. 6-64) for each end of the beams; column moments will not be computed until moment distribution to the beams has been completed. Then, the sum of the column moments at each joint may be easily computed, since they are the moments needed to make the sum of the end moments at the joint equal to zero. The sum of the column moments at each joint can then be distributed to each column there in proportion to its stiffness. In this example, each column will get one-half the sum of the column moments.

Fixed-end moments at each beam end for dead load are shown on the third line, just above the horizontal line, and fixed-end moments for live plus dead loads on the fourth line. Corresponding midspan moments for the fixed-end condition also are shown on the fourth line, and like the end moments will be corrected to yield actual midspan moments.

Maximum End Moments. For maximum end moment at A, beam AB must be fully loaded, but BC should carry dead load only. Holding A fixed, we first unlock joint B, which has a total-load fixed-end moment of $+172$ in BA and a dead-load fixed-end moment of -37 in BC. The releasing moment required, therefore, is $-(172 - 37)$, or -135. When B is released, a moment of -135×0.25 is distributed to BA. One half of this is carried over to A, or $-135 \times 0.25 \times 0.5 = -17$. This value is entered as the carry-over at A on the fifth line in Fig. 6-74. Joint B then is relocked.

At A, for which we are computing the maximum moment, we have a total-load fixed-end moment of -172 and a carry-over of -17, making the total -189, shown on the sixth line. To release A, a moment of $+189$ must be applied to the joint. Of this, 189×0.33, or 63, is distributed to AB, as indicated on the seventh line. Finally, the maximum moment at A is found by adding lines 6 and 7: $-189 + 63 = -126$.

For maximum moment at B, both AB and BC must be fully loaded, but CD should carry only dead load. We begin the determination of the maximum moment at B by first releasing joints A and C, for which the corresponding carry-over moments at BA and BC are $+29$ and $-(+78 - 70) \times 0.25 \times 0.5 = -1$, shown on the fifth line in Fig. 6-74. These bring the total fixed-end moments in BA and BC to $+201$ and -79, respectively. The releasing moment required is $-(201 - 79) = -122$. Multiplying this by the distribution factors for BA and BC when joint B is released, we find the distributed moments, -30, entered on line 7. The maximum end moments finally are obtained by adding lines 6 and 7: $+171$ at BA and -109 at BC. Maximum moments at C, D, and E are computed and entered in Fig. 6-74 in a similar manner. This procedure is equivalent to two cycles of moment distribution.

Maximum Midspan Moments. The computation of maximum midspan moments in Fig. 6-74 is based on the assumption that in each beam the midspan moment is the sum of the simple-beam midspan moment and one-half the algebraic difference of the final end moments (the span carries full load but adjacent spans only dead load). Instead of starting with the simple-beam moment, however, we begin, for convenience, with the midspan moment for the fixed-end condition and then apply two corrections. In each span, these corrections equal the carry-over moments entered on line 5 for the two ends of the beam multiplied by a factor.

For beams with variable moment of inertia, the factor is $\pm\frac{1}{2}(1/C^F + D - 1)$, where C^F is the fixed-end carry-over factor toward the end for which the correction factor is being computed and D the distribution factor for that end. The plus sign is used for correcting the carry-over at the right end of a beam, and the minus sign for the carry-over at the left end. For prismatic beams, the correction factor becomes $\pm\frac{1}{2}(1 + D)$.

For example, to find the corrections to the midspan moment in AB, we first multiply the carry-over at A on line 5, -17, by $-\frac{1}{2}(1 + 0.33)$. The correction, $+11$, also is entered on the fifth line. Then, we multiply the carry-over at B, $+29$, by $+\frac{1}{2}(1 + 0.25)$ and enter the correction, $+18$, on line 6. The final midspan moment is the sum of lines 4, 5, and 6: $+99 + 11 + 18 = +128$. Other midspan moments in Fig. 6-74 are obtained in a similar manner.

Approximate methods for determining wind and seismic stresses in tall buildings are given in Arts. 15-8 and 15-9.

6-66. Moment-Influence Factors. For certain types of structures, particularly those for which different types of loading conditions must be investigated, it may be more convenient to find maximum end moments from a table of moment-influence factors. This table is made up by listing for the end of each member in a structure the moment induced in that end when a moment (for convenience, $+1,000$) is applied to each joint successively. Once this table has been pre-

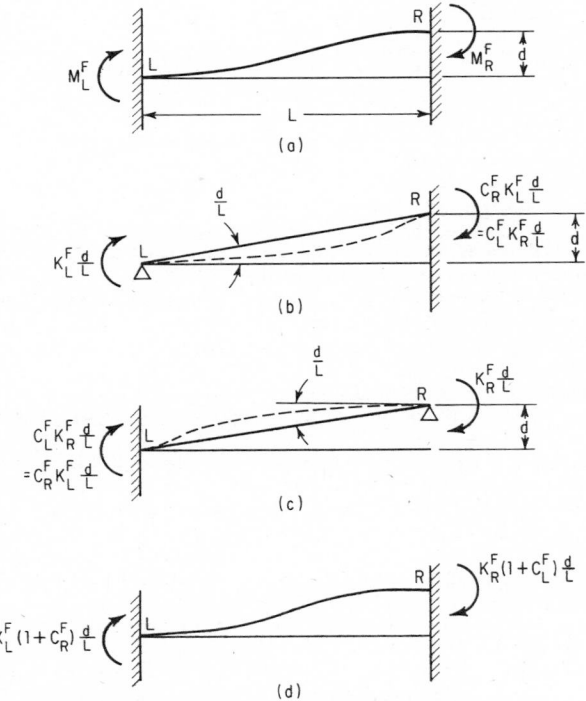

Fig. 6-75. Moments due to deflection of a fixed-end beam.

pared, no additional moment distribution is necessary for computing the end moments due to any loading condition.

For a specific loading pattern, the moment at any beam end M_{AB} may be obtained from the moment-influence table by multiplying the entries under AB for the various joints by the actual unbalanced moments at those joints divided by 1,000 and summing. (See also Art. 6-68 and Tables 6-3 and 6-4.)

6-67. Deflection of Supports. For some structures, it is convenient to know the effect of a movement of a support normal to the original position. But the moment-distribution method is based on the assumption that such movement of a support does not occur. The method, however, can be modified to evaluate end moments resulting from a support movement.

The procedure is to distribute moments as usual, assuming no deflection at the supports. This implies that additional external forces are exerted at the supports to prevent movement. These forces can be computed. Then, equal and opposite forces are applied to the structure to produce the final configuration, and the moments that they induce are distributed as usual. These moments added to those obtained with undeflected supports yield the final moments.

To apply this procedure, it is first necessary to know the fixed-end moments for a beam with supports at different levels. In Fig. 6-75a, the right end of a beam with span L is at a height d above the left end. To find the fixed-end moments, we first deflect the beam with both ends hinged, then fix the right end, leaving the left end hinged, as in Fig. 6-75b. Noting that a line connecting the two supports makes an angle approximately equal to d/L (its tangent) with the original position of the beam, we apply a moment at the hinged end to produce an end rotation there equal to d/L. By the definition of stiffness (Art. 6-62), this moment equals $K_L^F d/L$. The carry-over to the right end is C_R^F times this.

By the law of reciprocal deflections, the fixed-end moment at the right end of a beam due to a rotation of the other end equals the fixed-end moment at the left end of the beam due to the same rotation at the right end. Therefore, the carry-over moment for the right end also equals $C_L^F K_R^F d/L$ (see Fig. 6-75c). By adding the end moments for the loading conditions in Fig. 6-75b and c, we obtain the end moments in Fig. 6-75d, which is equivalent to the deflected beam in Fig. 6-75a:

$$M_L^F = K_L^F (1 + C_R^F) \frac{d}{L} \tag{6-68}$$

$$M_R^F = K_R^F (1 + C_L^F) \frac{d}{L} \tag{6-69}$$

In a similar manner, the fixed-end moment can be found for a beam with one end hinged and the supports at different levels:

$$M^F = K \frac{d}{L} \tag{6-70}$$

where K is the actual stiffness for the end of the beam that is fixed; for beams of variable moment of inertia K equals the fixed-end stiffness times $(1 - C_L^F C_R^F)$.

6-68. Procedure for Sidesway. The problem of computing sidesway moments in rigid frames is conveniently solved by the following method:

1. Apply forces to the structure to prevent sidesway while the fixed-end moments due to loads are distributed.

2. Compute the moments due to these forces.

3. Combine the moments obtained in steps 1 and 2 to eliminate the effect of the forces that prevented sidesway.

Example—Horizontal Axial Load. Suppose the rigid frame in Fig. 6-76 is subjected to a 2,000-lb horizontal load acting to the right at the level of beam BC. The first step is to compute the moment-influence factors by applying moments of $+1,000$ at joints B and C (Art. 6-66), assuming sidesway is prevented.

Since there are no intermediate loads on the beam and columns, the only fixed-end moments that need be considered are those in the columns due to lateral deflection of the frame caused by the horizontal load.

This deflection, however, is not known initially. So we assume an arbitrary deflection, which produces a fixed-end moment of $-1,000M$ at the top of column CD. M is an unknown constant to be determined from the fact that the sum of the shears in the deflected columns must equal the 2,000-lb load. The same deflection also produces a moment of $-1,000M$ at the bottom of CD [see Eqs. (6-68) and (6-69)].

Table 6-3. Moment-influence Factors for Fig. 6-76

Member	+1,000 at B	+1,000 at C
AB	351	−105
BA	702	−210
BC	298	210
CB	70	579
CD	−70	421
DC	−35	210

From the geometry of the structure, we furthermore note that the deflection of B relative to A equals the deflection of C relative to D. Then, according to Eqs. (6-68) and (6-69), the fixed-end moments of the columns of this frame are proportional to the stiffnesses of the columns and hence are equal in AB to $-1,000M \times {}^6/_2 = -3,000M$. The column fixed-end moments are entered in the first line of Table 6-4, the moment-collection table for Fig. 6-76.

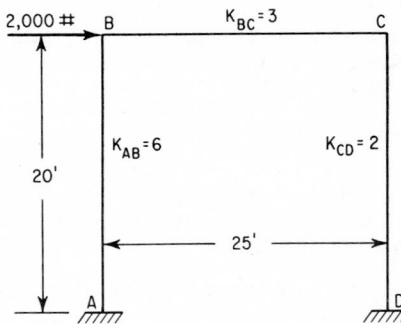

Fig. 6-76. Laterally loaded rigid frame.

Table 6-4. Moment-collection Table for Fig. 6-76

Remarks	AB	BA	BC	CB	CD	DC
1. Sidesway *FEM*	−3,000M	−3,000M	....	....	−1,000M	−1,000M
2. Distribution for B.	+1,053M	+2,106M	+894M	+210M	−210M	−105M
3. Distribution for C.	−105M	−210M	+210M	+579M	+421M	+210M
4. Final sidesway M.	−2,052M	−1,104M	+1,104M	+789M	−789M	−895M
5. For 2,000-lb horizontal	−17,000	−9,100	+9,100	+6,500	−6,500	−7,400
6. 4,000-lb vertical *FEM*	....	....	−12,800	+3,200		
7. Distribution for B.	+4,490	+8,980	+3,820	+897	−897	−448
8. Distribution for C.	+336	+672	−672	−1,853	−1,347	−673
9. Moments with no sidesway	+4,826	+9,652	−9,652	+2,244	−2,244	−1,121
10. Sidesway M	−4,710	−2,540	+2,540	+1,810	−1,810	−2,060
11. For 4,000-lb vertical	+116	+7,112	−7,112	+4,054	−4,054	−3,181

In the deflected position of the frame, joints B and C are unlocked in succession. First, we apply a releasing moment of $+3,000M$ at B. We distribute it by multiplying by 3 the entries in the column marked "$+1,000$ at B" in the table of moment-influence factors for Fig. 6-76. Similarly, a releasing moment of $+1,000M$ is applied at C and distributed with the aid of the moment-influence factors. The distributed moments are entered in the second and third lines of the moment-collection table. The final moments are the sum of the fixed-end moments and the distributed moments and are given in the fourth line, in terms of M.

Isolating each column and taking moments about one end, we find that the overturning moment due to the shear equals the sum of the end moments. We have one such equation for each column. Adding these equations, noting that the sum of the shears equals 2,000 lb, we obtain

$$-M(2,052 + 1,104 + 789 + 895) = -2,000 \times 20$$

from which we find $M = 8.26$. This value is substituted in the sidesway totals (line 4) in the moment-collection table to yield the end moments for the 2,000-lb horizontal load (line 5).

Example—Vertical Load on Beam. Suppose a vertical load of 4,000 lb is applied to BC of the rigid frame in Fig. 6-76, 5 ft from B. The same moment-influence factors and moment-collection table can again be used to determine the end moments with a minimum of labor:

The fixed-end moment at B, with sidesway prevented, is $-12,800$, and at C $+3,200$ (Fig. 6-68a). With the joints still locked, the frame is permitted to move laterally an arbitrary amount, so that in addition to the fixed-end moments due to the 4,000-lb load, column fixed-end moments of $-3,000M$ at A and B and $-1,000M$ at C and D are induced. The moment-collection table already indicates in line 4 the effect of relieving these column moments by unlocking joints B and C. We now have to superimpose the effect of releasing joints B and C to relieve the fixed-end moments for the vertical load. This we can do with the aid of the moment-influence factors. The distribution is shown in lines 7 and 8 of Table 6-4, the moment-collection table. The sums of the fixed-end moments and distributed moments for the 4,000-lb load are shown in line 9.

The unknown M can be evaluated from the fact that the sum of the horizontal forces acting on the columns must be zero. This is equivalent to requiring that the sum of the column end moments equal zero:

$$-M(2,052 + 1,104 + 789 + 895) + 4,826 + 9,652 - 2,244 - 1,121 = 0$$

from which $M = 2.30$. This value is substituted in line 4 of Table 6-4 to yield the sidesway moments for the 4,000-lb load (line 10). Addition of these moments to the totals for no sidesway (line 9) gives the final moments (line 11).

This procedure permits analysis of one-story bents with straight beams by solution of one equation with one unknown, regardless of the number of bays. If the frame is multistory, the procedure can be applied to each story. Since an arbitrary horizontal deflection is introduced at each floor or roof level, there are as many unknowns and equations as there are stories. (For approximate methods for determining wind and seismic stresses in tall buildings, see Arts. 15-8 and 15-9.)

The procedure is more difficult to apply to bents with curved or polygonal members between the columns. The effect of the change in the horizontal projection of the curved or polygonal portion of the bent must be included in the calculations. In many cases, it may be easier to analyze the bent as a curved beam (arch).

6-69. Single-cycle Moment Distribution. In the method of moment distribution by converging approximations (Art. 6-64), all joints but the one being unlocked are considered fixed. In distributing moments, the stiffnesses and carry-over factors used are based on this assumption (Arts. 6-62 and 6-63). If, however, actual stiffnesses and carry-over factors are employed, moments can be distributed throughout continuous frames in a single cycle.

Formulas for actual stiffnesses and carry-over factors can be written in several simple forms. The equations given in this article were chosen to permit use of existing tables for beams of variable moment of inertia that are based on fixed-end stiffnesses and fixed-end carry-over factors.

Considerable simplification of the formulas results if they are based on the simple-beam stiffness of members of continuous frames. This value can always be obtained from tables of fixed-end properties by multiplying the fixed-end stiffness by $(1 - C_L{}^F C_R{}^F)$, where $C_L{}^F$ is the fixed-end carry-over factor to the left and $C_R{}^F$ is the fixed-end carry-over factor to the right.

To derive the basic constants needed, apply a unit moment to one end of a member, isolated from the structure and simply supported (Fig. 6-77a). The end rotation at the support where the moment is applied is α, and at the far end, the rotation is β. By the dummy-load method (Art. 6-55), if x is measured from the β end,

$$\alpha = \int_0^L \frac{x^2}{EI_x} \, dx \tag{6-71}$$

$$\beta = \int_0^L \frac{x(L - x)}{EI_x} \tag{6-72}$$

in which I_x is the moment of inertia at a section a distance of x from the β end, and E is the modulus of elasticity.

Simple-beam stiffness K of the member is the moment required to produce a rotation of unity at the end where it is applied (Fig. 6-77b). Hence, at each end of a member, $K = 1/\alpha$.

For prismatic beams, K has the same value for both ends and equals $3EI/L$. For haunched beams, K for each end can be obtained from tables for fixed-end stiffnesses (Art. 6-62) or by numerical integration of Eq. (6-71).

While the value of α, and consequently of K, is different at opposite ends of an unsymmetrical beam, the value of β is the same for both ends, in accordance with the law of reciprocal deflections. Now, apply a moment J at one end of the simply supported beam to produce a rotation of unity at the other end (Fig. 6-77c). This moment will be equal to $1/\beta$ and will have the same value regardless at which end it is applied. K/J equals the **fixed-end carry-over factor** for the beam.

J equals $6EI/L$ for prismatic beams. For haunched beams, it can be computed by numerical integration of Eq. (6-72).

Actual stiffness S of the end of an unloaded span is the moment producing a rotation of unity at the end where it is applied when the other end of the beam is restrained against rotation by the other members of the structure (Fig. 6-77d).

The bending-moment diagram for a moment S_L applied at the left end of a member of a continuous frame is shown in Fig. 6-77e. As indicated, the moment carried over to the far end is $C_R S_L$, where C_R is the carry-over factor to the right. At L, the rotation produced by S_L alone

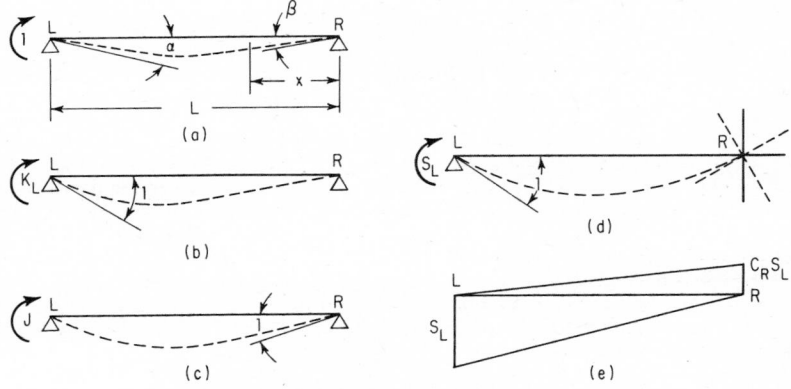

Fig. 6-77. End rotations of simple beams for determining stiffness.

is S_L/K_L, and by $C_R S_L$ alone, $-C_R S_L/J$. By definition of stiffness, the sum of these angles must equal unity:

$$\frac{S_L}{K_L} - \frac{C_R S_L}{J} = 1$$

Solving for S_L and noting that $K_L/J = C_L{}^F$, the fixed-end carry-over factor to the left, we find the formula for the stiffness of the left end of a member:

$$S_L = \frac{K_L}{1 - C_L{}^F C_R} \tag{6-73}$$

Similarly, the stiffness of the right end of a member is

$$S_R = \frac{K_R}{1 - C_R{}^F C_L} \tag{6-74}$$

For prismatic beams, the stiffness formulas reduce to

$$S_L = \frac{K}{1 - C_R/2} \quad \text{and} \quad S_R = \frac{K}{1 - C_L/2} \tag{6-75}$$

where $K = 3EI/L$.

Estimate of Stiffness. When the far end of a beam is hinged, the carry-over factor is zero; the stiffness equals K. When the far end of a prismatic beam is fully fixed against rotation, the carry-over factor equals ½. Hence, the fixed-end stiffness equals $4K/3$. This indicates that the effect of partial restraint on prismatic beams is to vary the stiffness between K for no restraint and $1.33K$ for full restraint. Because of this small variation, in many cases, an estimate of the actual stiffness of a beam may be sufficiently accurate.

Restraint R at the end of an unloaded beam in a continuous frame is the moment applied at that end to produce a unit rotation in all the members of the joint. Since the sum of the moments at the joint must be zero, R must equal the sum of the stiffnesses of the adjacent ends of the members connected to the given beam at that joint.

Furthermore, the moment induced in or "distributed" to any of these other members bears the same ratio to the applied moment as the stiffness of the member does to the restraint. Consequently, *end moments are distributed at a joint in proportion to the stiffnesses of the members.*

Actual carry-over factors can be computed by modifying the fixed-end carry-over factors. In Fig. 6-77d and e, by definition of restraint, the rotation at joint R is $-C_R S_L/R_R$, which must equal the rotation of the beam at R due to the moments at L and R. The rotation of the beam due to $C_R S_L$ alone equals $C_R S_L/K_R$, and the rotation due to S_L alone, $-S_L/J$. Hence

$$-\frac{C_R S_L}{R_R} = \frac{C_R S_L}{K_R} - \frac{S_L}{J}$$

Solving for C_R and noting that $K_R/J = C_R{}^F$, the fixed-end carry-over factor to the right, we find the actual carry-over factor to the right:

$$C_R = \frac{C_R{}^F}{1 + K_R/R_R} \tag{6-76}$$

Similarly, the actual carry-over factor to the left is

$$C_L = \frac{C_L{}^F}{1 + K_L/R_L} \tag{6-77}$$

In analyzing a continuous beam, we generally know the carry-over factors toward the ends of the first and last spans. Starting with these values, we can calculate the rest of the carry-over factors and the stiffnesses of the members. But in many frames, there are no end conditions known in advance. To analyze these structures, we must assume several carry-over factors.

This will not complicate the analysis, because in many cases it will be found unnecessary to correct the values of C based on assumed carry-over factors of preceding spans. The reason is that C is not very sensitive to the restraint at far ends of adjacent members. When carry-over factors are estimated, the greatest accuracy will be attained if the choice of assumed values is restricted to members subject to the greatest restraint.

Estimated Carry-over Factors. A very good approximation to the carry-over factor for prismatic beams may be obtained from the following formula, which is based on the assumption that far ends of adjacent members are subject to equal restraint:

$$C = \frac{\Sigma K - K}{2(\Sigma K - \delta K)} \tag{6-78}$$

where ΣK is the sum of the K values of all the members of the joint toward which the carry-over factor is acting; K is the simple-beam stiffness of the member for which the carry-over factor is being computed; and δ is a factor that varies from zero for no restraint to ¼ for full restraint at the far ends of the connecting members. Since δ varies within such narrow limits, it affects C very little.

To illustrate the estimating and calculation of carry-over factors, the carry-over factors and stiffnesses in the clockwise direction will be computed for the frame in Fig. 6-78a. Relative I/L, or K values, are shown in the circles.

A start will be made by estimating C_{AB}. Arbitrarily taking $\delta = ⅛$, we apply Eq. (6-78) with $K = 3$ and $\Sigma K = 3 + 2 = 5$ and find $C_{AB} = 0.216$, as noted in Fig. 6-78a. The stiffness S_{AB} then equals $3/(1 - 0.216/2) = 3.37$. Noting that the restraint $R_{AD} = S_{AB}$, we can now use the exact formula, Eq. (6-77), to obtain the carry-over factor from D to A: $C_{DA} = 0.5/(1 + 6/3.37) = 0.180$. Continuing around the frame in this manner, we return to C_{AB} and recalculate it with Eq. (6-77), obtaining 0.221. This differs only slightly from the estimated value. The change in C_{DA} due to the new value of C_{AB} is negligible.

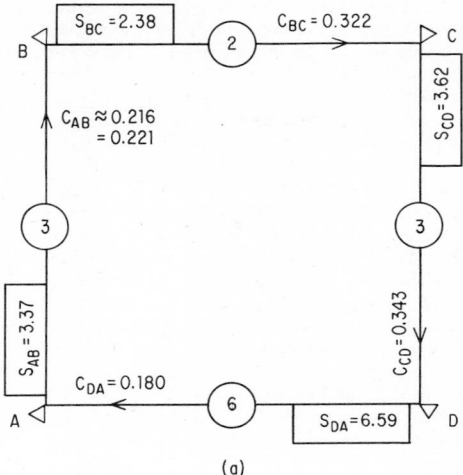

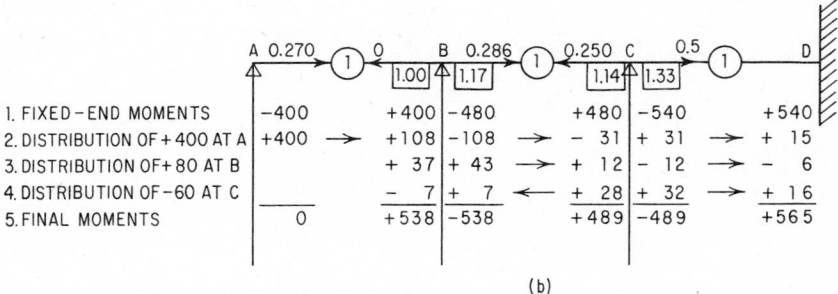

Fig. 6-78. Moments in a quadrangular frame (*a*) and a continuous beam (*b*).

If a bending moment of 1,000 ft-lb were introduced at *A* in *AB*, it would induce a moment of $1{,}000C_{AB} = 221$ ft-lb at *B*; $221 \times 0.322 = 71$ ft-lb at *C*; $71 \times 0.343 = 24$ ft-lb at *D*; $24 \times 0.180 = 4$ ft-lb at *A*, etc.

Example—Continuous Beam. To demonstrate how the moments in a continuous beam would be computed, the end moments will be determined for the beam in Fig. 6-78*b*, which is identical with the one in Fig. 6-73, for which moments were obtained by converging approximations. Relative I/L, or K values, are shown in the circles on each span.

Since *A* is a hinged end, $C_{BA} = 0$. $S_{BA} = 1/(1 - 0) = 1$. Since there is only one member joined to *BC* at *B*, $R_{BC} = S_{BA}$, and $C_{CB} = 0.5/(1 + 1/1) = 0.250$. With this value, we compute $S_{CB} = 1.14$. To obtain the carry-over factors in the opposite direction, we start with $C_{CD} = 0.5$, since we know *D* is a fixed end. This enables us to compute $S_{CD} = 1.33$ and the remainder of the beam constants.

The fixed-end moments are given on the first line of calculations in Fig. 6-78*b*. We start the distribution by unlocking *A* by applying a releasing moment of $+400$. Since *A* is a hinge, the full 400 is given to *A* and $400 \times 0.270 = 108$ is carried over to *B*. If several members had been connected to *AB* at *B*, this moment with sign changed would be distributed to them in proportion to their stiffnesses. But since only one member connects to *B*, the moment at *BC* is -108. Next, $-108 \times 0.286 = -31$ is carried over to *C*. Finally, a moment of $+15$ is carried to *D*.

Then, joint *B* is unlocked. The unbalanced moment of $400 - 480 = -80$ is counteracted with a moment of $+80$. This is distributed to *BA* and *BC* in proportion to their stiffnesses (shown in the boxes at the joint). *BC*, for example, gets $80 \times 1.17/(1.17 + 1) = 43$. The carry-over to *C* is $43 \times 0.286 = 12$, and to *D*, -6.

Similarly, the unbalanced moment at C is counteracted and distributed to CB and CD in porportion to their stiffnesses, shown in the boxes at C, then carried over to B and D. The final moments are the sum of the fixed-end and distributed moments.

Shortcuts. On occasion, advantage can be taken of certain properties of loads and structures to save work in distribution by using carry-over factors as the ratio of end moments in loaded members. For example, suppose it is obvious, from symmetry of loading and structure, that there will be no end rotation at an interior support. The part of the structure on one side of the support can be isolated and the moments distributed only in this part, with the carry-over factor toward the support taken as C^F.

Again, suppose it is evident that the final end moments at opposite ends of a span must be equal in magnitude and sign. Isolate the structure on each side of this beam and distribute moments only in each part, with the carry-over factor for this span taken as 1.

6-70. Method for Checking Moment Distribution. End moments computed for a continuous structure must satisfy both the laws of equilibrium and the requirements of continuity. At each joint, therefore, the sum of the moments must be zero (or equal to an external moment applied there). In addition, the end of every member connected there must rotate through the same angle. It is a simple matter to determine the sum of the moments; but further calculation is needed to prove that the moments yield the same rotation for the end of each member at a joint. The following method not only will indicate that the requirements of continuity are satisfied but also will tend to correct automatically any mistakes that may have been made in computing the end moments.

Consider a joint O made of several members OA, OB, OC, etc. The members are assumed to be loaded and the calculation of end moments due to the loads to have started with fixed-end moments. For any one of the members, say OA, the end rotation at O for the fixed-end condition was zero; that is,

$$\frac{M_{OA}{}^F}{K_{OA}} - \frac{M_{AO}{}^F}{J_{OA}} - \phi_{OA} = 0 \tag{6-79}$$

where $M_{OA}{}^F$ = fixed-end moment at O
$M_{AO}{}^F$ = fixed-end moment at A
K_{OA} = simple-beam stiffness at O (see Art. 6-69)
J_{OA} = moment required at A to produce a unit rotation at O when the span is considered simply supported
ϕ_{OA} = simple-beam end rotation at O due to loads

For the final end moments, the rotation at O is

$$\theta = \frac{M_{AO}}{K_{OA}} - \frac{M_{AO}}{J_{OA}} - \phi_{OA} \tag{6-80}$$

Subtracting Eq. (6-79) from Eq. (6-80) and multiplying by K_{OA} yields

$$K_{OA}\theta = M_{OA} - M_{OA}{}^F - C_{AO}{}^F M'_{OA} \tag{6-81}$$

In Eq. (6-81) the carry-over factor toward O, $C_{AO}{}^F$, has been substituted for K_{OA}/J_{OA}, and M'_{OA} for $M_{AO} - M_{AO}{}^F$. An analogous expression can be written for each of the other members at O. Summing these equations, we obtain

$$\theta \Sigma K_O = \Sigma M_O - \Sigma M_O{}^F - \Sigma C_O{}^F M'_O \tag{6-82}$$

With this value of θ, we can solve for each of the final end moments at O and thus determine the equations that will check the joint for continuity. For example, using Eq. (6-82) with Eq. (6-81) yields

$$M_{OA} = M_{OA}{}^F + C_{AO}{}^F M'_{OA} - m_{OA} \tag{6-83}$$

$$m_{OA} = \frac{K_{OA}}{\Sigma K_O}(-\Sigma M_O + \Sigma M_O{}^F + \Sigma C_O{}^F M'_O) \tag{6-84}$$

Similar equations can be written for the other members at O by substituting the proper letter for A in the subscripts.

If the calculations based on these equations are carried out in table form, the equations prove to be surprisingly simple (see Tables 6-5 and 6-6).

For prismatic beams, the terms $C^F M'$ become half the change in the fixed-end moment at the far end of each member at a joint.

Examples. Suppose we want to check the end moments in the beam in Fig. 6-78b. Each joint and the ends of the members connected there are listed in Table 6-5, and a column is provided for the summation of the various terms for each joint. K values are given on line 1, the end moments to be checked on line 2, and the fixed-end moments on line 3. On line 4 is entered one-half the difference obtained when the fixed-end moment is subtracted from the final moment at the far end of each member. $-m$ is placed on line 5 and the corrected end moment on line 6. The $-m$ values are obtained from the summation columns by adding line 2 to the negative of the sum of lines 3 and 4 and distributing the result to the members of the joint in proportion to the K values. The corrected moment M is the sum of lines 3, 4, and 5.

Table 6-5. Continuity Check for Fig. 6-78b

	A		B			C			D	
	AB	Σ	BA	BC	Σ	CB	CD	Σ	DC	Σ
1. K	1	1	1	1	2	1	1	2	1	∞
2. M	0	0	+538	−538	0	+489	−489	0	+565	
3. M^F	−400	−400	+400	−480	−80	+480	−540	−60	+540	
4. $C^F M'$			+200	+5	+205	−29	+13	−16	+26	
5. $-m$			−62	−63	−125	+38	+38	+76	0	
6. Check	0	0	+538	−538	0	+489	−489	0	+566	

CHECK WHEN MOMENTS ARE INCORRECT

	AB	Σ	BA	BC	Σ	CB	CD	Σ	DC	Σ
2. Wrong M	0	0	+560	−560	0	+450	−450	0	+530	
3. M^F	−400	−400	+400	−480	−80	+480	−540	−60	+540	
4. $C^F M'$			+200	−15	+185	−40	−5	−45	+45	
5. $-m$			−53	−52	−105	+53	+52	+105	0	
6. New M	0	0	+547	−547	0	+493	−493	0	+585	

SECOND CYCLE

	AB	Σ	BA	BC	Σ	CB	CD	Σ	DC	Σ
7. Trial M	0	0	+547	−547	0	+493	−493	0	+585	
8. M^F	−400	−400	+400	−480	−80	+480	−540	−60	+540	
9. $C^F M'$			+200	+3	+203	−34	+12	−22	+24	
10. $+m$			−62	−61	−123	+41	+41	+82	0	
11. New M	0	0	+538	−538	0	+487	−487	0	+564	

Assume that a mistake was made in computing the end moments for Fig. 6-78b giving the results shown in line 2 of the second part of Table 6-5 for the fixed-end moments on line 3. The correct moments can be obtained as follows:

At joint B, the sum of the incorrect moments is zero, as shown in the summation column on line 2. The sum of the fixed-end moments at B is -80, as indicated on line 3. For BA, $\frac{1}{2}M'$ is obtained from lines 2 and 3 of the column for AB: $\frac{1}{2} \times (0 + 400) = +200$, which is entered on line 4. The line 4 entry for BC is obtained from CB: $\frac{1}{2} \times (450 - 480) = -15$. The sum of the line 4 values at B is therefore $200 - 15 = +185$. Entered in the summation column, this is then added to the summation value on line 3, the sign is changed, and the number on line 2 (in this case zero) is added to the sum, giving -105. This is inserted in the summation column on line 5. The values for BA and BC on line 5 are obtained by multiplying -105 by the ratio of the K value of each member to the sum of the K values at the joint; that is, for BA, $-m = -105 \times \frac{1}{2} = -53$. The corrected moment for BA, the sum of lines 3, 4, and 5, is $+400 + 200 - 53 = +547$. Since this differs from the value on line 2, it indicates that one or more of the moments on that line were incorrect. The other corrected moments are found in the same way and are shown on line 6.

A comparison with Fig. 6-78b shows that the new moments, though incorrect, are closer to the right answer than those with which we started. Even closer results can be obtained by repeating the calculations.   Convergence, however, can be obtained much more quickly by starting first with fixed ends and joints that appear to have been most in error.   Then, use the corrected values obtained for these in correcting adjacent joints.

For example, for the second cycle shown in the table, calculations were started with joint D, which is a fixed end.   Using the values obtained at the end of the first cycle to compute M', we find the corrected value for DC to be $+564$, which is very close to the exact final moment.   Then, we move to joint C.   For M, we use the value obtained at the end of the first cycle.   But M' for CD is based on the end moment just computed: $+564 - 540 = +24$, and half of this is placed on line 9 under CD.   Continuing in this manner, we obtain moments that check closely the final moments in the first part of the table.

The procedure is useful also for estimating the effect of changing the stiffness of one or more members.

Check for Sidesway.   The checking equations can be generalized to include the effect of the movement d of a support in a direction normal to the initial position of a span of length L:

$$K_{OA}\theta = M_{OA} - M_{OA}{}^F - C_{AO}{}^F M'_{OA} + K_{OA}\frac{d}{L} \tag{6-85}$$

$$M_{OA} = M_{OA}{}^F + C_{AO}{}^F M'_{OA} - K_{OA}\frac{d}{L} - m_{OA} \tag{6-86a}$$

$$m_{OA} = \frac{K_{OA}}{\Sigma K_O}\left(-\Sigma M_O + \Sigma M_O{}^F + \Sigma C_O{}^F M'_O - \Sigma K_O\frac{d}{L}\right) \tag{6-86b}$$

For each span with a support movement, the term Kd/L can be obtained from the fixed-end

Table 6-6. Continuity Check for Fig. 6-76

	A		B			C			D	
	AB	Σ	BA	BC	Σ	CB	CD	Σ	DC	Σ
1. K	7	∞	6	3	9	3	2	5	2	∞
2. M	$+116$		$+7,110$	$-7,110$	0	$+4,050$	$-4,050$	0	$-3,180$	
3. M^F	0		0	$-12,800$	$-12,800$	$+3,200$	0	$+3,200$	0	
4. $C^F M'$	$+3,555$		$+60$	$+430$	$+490$	$+2,840$	$-1,590$	$+1,250$	$-2,030$	
5. $-Kd/L$	$-3,450$		$-3,450$	0	$-3,450$	0	$-1,150$	$-1,150$	$-1,150$	
6. $-m$	0		$+10,510$	$+5,250$	$+15,760$	$-1,980$	$-1,320$	$-3,300$	0	
7. M	$+105$		$+7,120$	$-7,120$	0	$+4,060$	$-4,060$	0	$-3,180$	

moment due to this deflection alone; for OA, for example, Kd/L can be obtained by multiplying $M_{OA}{}^F$ by $(1 - C_{AO}{}^F C_{OA}{}^F)/(1 + C_{OA}{}^F)$.   [See Eqs. (6-68) and (6-73).]   For prismatic beams, this factor reduces to ½.   Equation (6-85) is called a **slope-deflection equation.**

In Table 6-6, the solution for the bent in Fig. 6-76 is checked for the condition in which a 4,000-lb vertical load is placed 5 ft from B on span BC.   The computations are similar to those in the preceding table, except that the terms $-Kd/L$ are included for the columns to account for sidesway.   These values are obtained from the sidesway fixed-end moments in the moment-collection table for Fig. 6-76.   For BA, for example, $Kd/L = ½ \times 3,000M$, with $M = 2.30$, as found in the solution.   The check indicates that the original solution was sufficiently accurate for a slide-rule computation.   If line 7 had contained a different set of moments, the shears would have had to be investigated again.   A second cycle could be carried out by distributing the unbalance to the columns to obtain new Kd/L values.

BUCKLING OF COLUMNS

Columns are compression members whose cross-sectional dimensions are small compared with their length in the direction of the compressive force.   Failure of such members occurs because of instability when a certain load (called the critical or **Euler load**) is equaled or exceeded.   The member may bend, or buckle, suddenly and collapse.

Hence, the strength of a column is determined not by the unit stress in Eq. (6-6) $(P = Af)$, but by the maximum load it can carry without becoming unstable.   The condition of instability is characterized by disproportionately large increases in lateral deformation with slight increase in load. It may occur in slender columns before the unit stress reaches the elastic limit.

6-71. Equilibrium of Columns. Figure 6-79 represents an axially loaded column with ends unrestrained against rotation. If the member is initially perfectly straight, it will remain straight as long as the load P is less than the critical load P_c. If a small transverse force is applied, it will deflect, but it will return to the straight position when this force is removed. Thus, when P is less than P_c, internal and external forces are in stable equilibrium.

If $P = P_c$ and a small transverse force is applied, the column again will deflect, but this time, when the force is removed, the column will remain in the bent position (dashed line in Fig. 6-79).

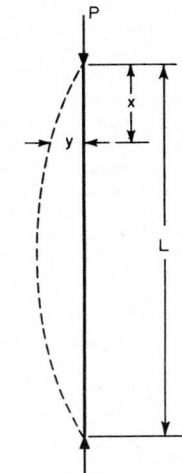

Fig. 6-79. Buckling of a column.

The equation of this elastic curve can be obtained from Eq. (6-44):

$$EI \frac{d^2y}{dx^2} = -P_c y \tag{6-87}$$

in which E = modulus of elasticity, psi

I = least moment of inertia of the cross section, in.[4]

y = deflection of the bent member from the straight position at a distance x from one end, in.

This assumes that the stresses are within the elastic limit.

Solution of Eq. (6-87) gives the smallest value of the Euler load as

$$P_c = \frac{\pi^2 EI}{L^2} \tag{6-88}$$

Equation (6-88) indicates that there is a definite magnitude of an axial load that will hold a column in equilibrium in the bent position when the stresses are below the elastic limit. Repeated application and removal of small transverse forces or small increases in axial load above this critical load will cause the member to fail by buckling. Internal and external forces are in a state of unstable equilibrium.

It is noteworthy that the Euler load, which determines the load-carrying capacity of a column, depends on the stiffness of the member, as expressed by the modulus of elasticity, rather than on the strength of the material of which it is made.

By dividing both sides of Eq. (6-88) by the cross-sectional area A, sq in., and substituting r^2 for I/A (r is the radius of gyration of the section), we can write the solution of Eq. (6-87) in terms of the average unit stress on the cross section:

$$\frac{P_c}{A} = \frac{\pi^2 E}{(L/r)^2} \tag{6-89}$$

This holds only for the elastic range of buckling; that is, for values of the **slenderness ratio** L/r above a certain limiting value that depends on the properties of the material.

Effects of End Conditions. Equation (6-89) was derived on the assumption that the ends of the column are free to rotate. It can be generalized, however, to take into account the effect of end conditions:

$$\frac{P_c}{A} = \frac{\pi^2 E}{(kL/r)^2} \tag{6-90a}$$

where k is a factor that depends on the end conditions. For a pin-ended column, $k = 1$; for a column with both ends fixed, $k = \frac{1}{2}$; for a column with one end fixed and one end pinned, k is about 0.7; and for a column with one end fixed and one end free from all restraint, $k = 2$. When a column has different restraints or different radii of gyration about its principal axes, the largest value of kL/r for a principal axis should be used in Eq. (6-90a).

Inelastic Buckling. Equations (6-88) to (6-90), having been derived from Eq. (6-87), the differential equation for the elastic curve, are based on the assumption that the critical average stress is below the elastic limit when the state of unstable equilibrium is reached. In members with slenderness ratio L/r below a certain limiting value, however, the elastic limit is exceeded before the column buckles. As the axial load approaches the critical load, the modulus of elasticity varies with the stress. Hence, Eqs. (6-88) to (6-90), based on the assumption that E is a constant, do not hold for these short columns.

After extensive testing and analysis, prevalent engineering opinion favors the Engesser equation for metals in the inelastic range:

$$\frac{P_t}{A} = \frac{\pi E_t}{(kL/r)^2} \tag{6-90b}$$

This differs from Eq. (6-90a) only in that the tangent modulus E_t (the actual slope of the stress-strain curve for the stress P_t/A) replaces E, the modulus of elasticity in the elastic range. P_t is the smallest axial load for which two equilibrium positions are possible, the straight position and a deflected position.

Eccentric Loading. Under eccentric loading, the maximum unit stress in short compression members is given by Eqs. (6-52) and (6-54), with the eccentricity e increased by the deflection given by Eq. (6-53). For columns, the stress within the elastic range is given by the **secant formula:**

$$f = \frac{P}{A} \left(1 + \frac{ec}{r^2} \sec \frac{kL}{2r} \sqrt{\frac{P}{AE}} \right) \tag{6-91}$$

When the slenderness ratio L/r is small, the formula approximates Eq. (6-52).

(S. Timoshenko and J. M. Gere, "Theory of Elastic Stability," McGraw-Hill Book Company, New York; B. G. Johnston, "Column Research Council Guide to Design Criteria for Metal Compression Members," John Wiley & Sons, Inc., New York; F. Bleich, "Buckling Strength of Metal Structures," McGraw-Hill Book Company, New York.)

6-72. Column Curves. The result of plotting the critical stress in columns for various values of slenderness ratios (Art. 6-71) is called a column curve. For axially loaded, initially straight columns, it consists of two parts: the Euler critical values [Eq. (6-89)] and the Engesser, or tangent-modulus, critical values [Eq. (6-91)], with $k = 1$.

The second part of the curve is greatly affected by the shape of the stress-strain curve for the material of which the column is made, as indicated in Fig. 6-80. The stress-strain curve for a material, such as an aluminum alloy or high-strength steel, which does not have a sharply defined yield point, is shown in Fig. 6-80a. The corresponding column curve is plotted in Fig. 6-80b. In contrast, Fig. 6-80c presents the stress-strain curve for structural steel, with a sharply defined yield point, and Fig. 6-80d the related column curve. This curve becomes horizontal as the critical stress approaches the yield strength of the material and the tangent modulus becomes zero, whereas the column curve in Fig. 6-80b continues to rise with decreasing values of the slenderness ratio.

Examination of Fig. 6-80d also indicates that slender columns, which fall in the elastic range, where the column curve has a large slope, are very sensitive to variations in the factor k, which represents the effect of end conditions. On the other hand, in the inelastic range, where the column curve is relatively flat, the critical stress is relatively insensitive to changes in k. Hence, the effect of end conditions is of much greater significance for long columns than for short columns.

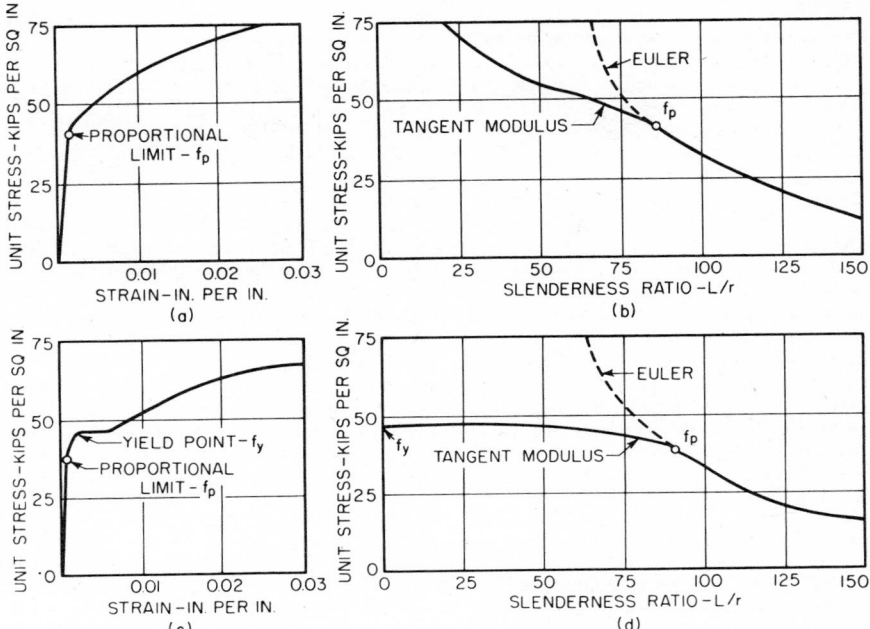

Fig. 6-80. Column curves. (*a*) Stress-strain curve for a material without a sharply defined yield point; (*b*) column curve for the material in (*a*); (*c*) stress-strain curve for a material with a sharply defined yield point; (*d*) column curve for the material in (*c*).

6-73. Behavior of Actual Columns.

For many reasons, columns in structures behave differently from the ideal column assumed in deriving Eqs. (6-88) to (6-91). A major consideration is the effect of accidental imperfections, such as nonhomogeneity of materials, initial crookedness, and unintentional eccentricities of the axial load. These effects can be taken into account by a proper choice of safety factor.

There are, however, other significant conditions that must be considered in any design procedure: continuity in framed structures and eccentricity of the load. Continuity affects column action in two ways. The restraint and sidesway at column ends determine the value of k, and bending moments are transmitted to the columns by adjoining structural members.

Because of the deviation of the behavior of actual columns from the ideal, columns generally are designed by empirical formulas. Separate equations usually are given for short columns, intermediate columns, and long columns, and still other equations for combinations of axial load and bending moment.

Furthermore, a column may fail not by buckling of the member as a whole but, as an alternative, by buckling of one of its components. Hence, when members like I beams, channels, and angles are used as columns, or when sections are built up of plates, the possibility that the critical load on a component (leg, half flange, web, lattice bar) will be less than the critical load on the column as a whole should be investigated.

Similarly, the possibility of buckling of the compression flange or the web of a beam should be investigated.

Local buckling, however, does not always result in a reduction in the load-carrying capacity of a column. Sometimes, it results in a redistribution of the stresses, which enables the member to carry additional load.

For more details on column action, see S. Timoshenko and J. M. Gere, "Theory of Elastic Stability," McGraw-Hill Book Company, New York; B. G. Johnston, "Column Research Council Guide to Design Criteria for Metal Compression Members," John Wiley & Sons, Inc., New York; F. Bleich, "Buckling Strength of Metal Structures," McGraw-Hill Book Company, New York.

TORSION

Forces that cause a member to twist about a longitudinal axis are called torsional loads. Simple torsion is produced only by a couple, or moment, in a plane perpendicular to the axis.

If a couple lies in a nonperpendicular plane, it can be resolved into a torsional moment, in a plane perpendicular to the axis, and bending moments, in planes through the axis.

6-74. Shear Center. The point in each normal section of a member through which the axis passes and about which the section twists is called the shear center. (The location of the shear center in some commonly used shapes is given in Art. 6-37.) If the loads on a beam, for example, do not pass through the shear center, they cause the beam to twist.

6-75. Stresses Due to Torsion. Simple torsion is resisted by internal shearing stresses. These can be resolved into radial and tangential shearing stresses, which being normal to each other also are equal (see Art. 6-13). Furthermore, on planes that bisect the angles between the planes on which the shearing stresses act, there also occur compressive and tensile stresses. The magnitude of these normal stresses is equal to that of the shear. Therefore, when torsional loading is combined with other types of loading, the maximum stresses occur on inclined planes and can be computed by the methods of Arts. 6-14 and 6-17.

Circular Sections. If a circular shaft (hollow or solid) is twisted, a section that is plane before twisting remains plane after twisting. Within the proportional limit, the shearing stress at any point in a transverse section varies with the distance from the center of the section. The maximum shear, psi, occurs at the circumference and is given by

$$v = \frac{Tr}{J} \tag{6-92}$$

where T = torsional moment, in.-lb
r = radius of section, in.
J = polar moment of inertia, in.[4]
Polar moment of inertia of a cross section is defined by

$$J = \int \rho^2 dA \tag{6-93}$$

where ρ = radius from shear center to any point in section
dA = differential area at the point
In general, J equals the sum of the moments of inertia about any two perpendicular axes through the shear center. For a solid circular section, $J = \pi r^4/2$. For a hollow circular section with diameters D and d, $J = \pi(D^4 - d^4)/32$.

Within the proportional limit, the angular twist between two points L in. apart along the axis of a circular bar is, in radians (1 radian = 57.3°):

$$\theta = \frac{TL}{GJ} \tag{6-94}$$

where G is the shearing modulus of elasticity (see Art. 6-5).

Noncircular Sections. If a shaft is not circular, a plane transverse section before twisting does not remain plane after twisting. The resulting warping increases the shearing stresses in some parts of the section and decreases them in others, compared with the shearing stresses that would occur if the section remained plane. Consequently, shearing stresses in a noncircular section are not proportional to distances from the shear center. In elliptical and rectangular sections, for example, maximum shear occurs on the circumference at a point nearest the shear center.

For a solid rectangular section, this maximum may be expressed in the following form:

$$v = \frac{T}{kb^2d} \tag{6-95}$$

where b = short side of the rectangle, in.
d = long side, in.
k = a constant depending on the ratio of these sides:

d/b = 1.0	1.5	2.0	2.5	3	4	5	10	∞
k = 0.208	0.231	0.246	0.258	0.267	0.282	0.291	0.312	0.333

(S. Timoshenko and J. N. Goodier, "Theory of Elasticity," McGraw-Hill Book Company, New York.)

Hollow Tubes. If a thin-shell hollow tube is twisted, the shearing force per unit of length on a cross section (**shear flow**) is given approximately by

$$H = \frac{T}{2A} \tag{6-96}$$

where A is the area enclosed by the mean perimeter of the tube, sq in. And the unit shearing stress is given approximately by

$$v = \frac{H}{t} = \frac{T}{2At} \tag{6-97}$$

where t is the thickness of the tube, in. For a rectangular tube with sides of unequal thickness, the total shear flow can be computed from Eq. (6-96) and the shearing stress along each side from Eq. (6-97), except at the corners, where there may be appreciable stress concentration.

Channels and I Beams. For a narrow rectangular section, the maximum shear is very nearly equal to

$$v = \frac{T}{\frac{1}{3}b^2 d} \tag{6-98}$$

This formula also can be used to find the maximum shearing stress due to torsion in members, such as I beams and channels, made up of thin rectangular components. Let $J = \frac{1}{3}\Sigma b^3 d$, where b is the thickness of each rectangular component and d the corresponding length. Then, the maximum shear is given approximately by

$$v = \frac{Tb'}{J} \tag{6-99}$$

where b' is the thickness of the web or the flange of the member. Maximum shear will occur at the center of one of the long sides of the rectangular part that has the greatest thickness. (F. B. Seely and J. O. Smith, "Advanced Mechanics of Materials," John Wiley & Sons, Inc., New York; I. S. Sokolnikoff, "Mathematical Theory of Elasticity," McGraw-Hill Book Company, New York.)

STRESSES IN ARCHES

An arch is a curved beam, the radius of curvature of which is very large relative to the depth of section. It differs from a straight beam in that: (1) loads induce both bending and direct compressive stress in an arch; (2) arch reactions have horizontal components even though all loads are vertical; and (3) deflections have horizontal as well as vertical components.

The necessity of resisting the horizontal components of the reactions is an important consideration in arch design. Sometimes these forces are taken by tie rods between the supports, sometimes by heavy abutments or buttresses.

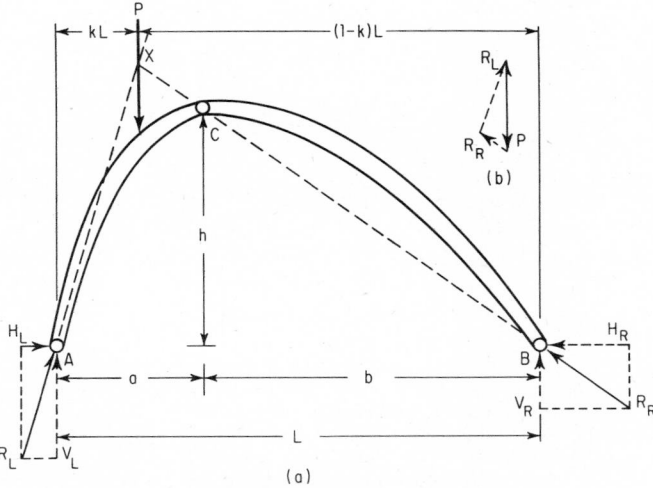

Fig. 6-81. Three-hinged arch.

Arches may be built with fixed ends, as can straight beams, or with hinges at the supports. They may also be built with an internal hinge, usually located at the uppermost point, or crown.

6-76. Three-hinged Arches. An arch with an internal hinge and hinges at both supports (Fig. 6-81) is statically determinate. There are four unknowns—two horizontal and two vertical components of the reactions—but four equations based on the laws of equilibrium are available: (1) The sum of the horizontal forces must be zero. (In Fig. 6-81, $H_L = H_R = H$.) (2) The sum of the moments about the left support must be zero. ($V_R = Pk$.) (3) The sum of the moments about the right support must be zero. [$V_L = P(1 - k)$.] (4) The bending moment at the crown hinge must be zero (not to be confused with the sum of the moments about the crown, which also must be equal to zero but which would not lead to an independent equation for the solution of the reactions). Hence, for the right half of the arch in Fig. 6-81, $Hh - V_R b = 0$, and $H = V_R b/h$. The influence line for H is a straight line, varying from zero for loads over the supports to a maximum of Pab/Lh for a load at C.

Reactions and stresses in three-hinged arches can be determined graphically by using the principles presented in Arts. 6-38 to 6-44 and 6-46, and by taking advantage of the fact that the bending moment at the crown hinge is zero. For example, in Fig. 6-81a, the load P is applied to segment AC of the arch. Then, since the bending moment at C must be zero, the line of action of the reaction R_R at B must pass through the crown hinge. It intersects the line of action of P at X. The line of action of the reaction R_L at A also must pass through X, since P and the two reactions are in equilibrium. By constructing a force triangle with the load P and the lines of action of the reactions thus determined, you can obtain the magnitude of the reactions (Fig. 6-81b). After the reactions have been found, the stresses can be computed from the laws of statics or, in the case of a trussed arch, determined graphically.

6-77. Two-hinged Arches. When an arch has hinges at the supports only (Fig. 6-82a), it is statically indeterminate; there is one more unknown reaction component than can be determined

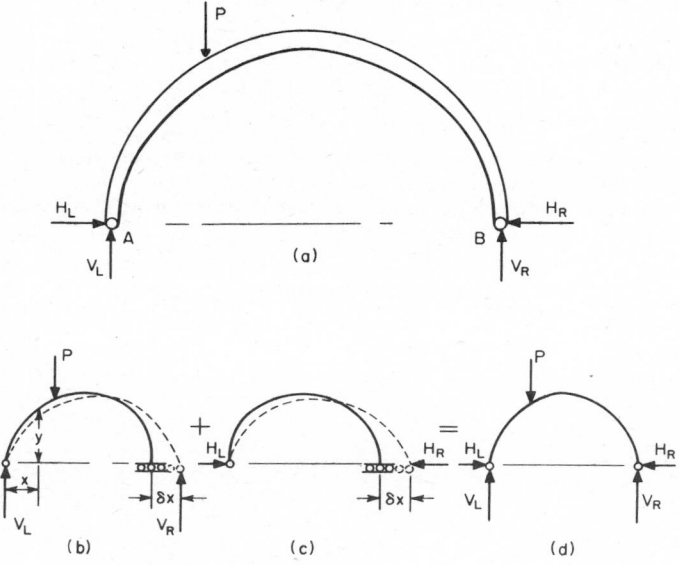

Fig. 6-82. Two-hinged arch.

by the three equations of equilibrium. Another equation can be written from knowledge of the elastic behavior of the arch. One procedure is to assume that one of the supports is on rollers. The arch then is statically determinate, and the reactions and horizontal movement of the support can be computed for this condition (Fig. 6-82b). Next, the horizontal force required to return the movable support to its original position can be calculated (Fig. 6-82c). Finally, the reactions for the two-hinged arch (Fig. 6-82d) are obtained by superimposing the first set of reactions on the second.

For example, if δx is the horizontal movement of the support due to the loads on the arch, and if $\delta x'$ is the horizontal movement of the support due to a unit horizontal force applied to the support, then

$$\delta x + H \; \delta x' = 0 \qquad (6\text{-}100a)$$

$$H = -\frac{\delta x}{\delta x'} \qquad (6\text{-}100b)$$

where H is the unknown horizontal reaction. (When a tie rod is used to take the thrust, the right-hand side of the first equation is not zero, but the elongation of the rod HL/A_sE_s, where L is the length of the rod, A_s its cross-sectional area, and E_s its modulus of elasticity. To account for the effect of an increase in temperature t, add to the left-hand side $EctL$, where E is the modulus of elasticity of the arch, c the coefficient of expansion.)

The dummy-unit-load method can be used to compute δx and $\delta x'$ (Art. 6-55):

$$\delta x = \int_A^B \frac{My \; ds}{EI} - \int_A^B \frac{N \; dx}{AE} \qquad (6\text{-}101)$$

where M = bending moment at any section due to loads
$\quad y$ = ordinate of section measured from immovable end of arch
$\quad I$ = moment of inertia of arch cross section
$\quad A$ = cross-sectional area of arch
$\quad ds$ = differential length along arch axis
$\quad dx$ = differential length along the horizontal
$\quad N$ = normal thrust on the cross section due to loads

$$\delta x' = -\int_A^B \frac{y^2 ds}{EI} - \int_A^B \frac{\cos \alpha \; dx}{AE} \qquad (6\text{-}102)$$

where α = the angle the tangent to axis at the section makes with horizontal.

Equations (6-101) and (6-102) do not include the effects of shear deformation and curvature, which usually are negligible. Unless the thrust is very large, the second term on the right-hand side of Eq. (6-101) also can be dropped.

In most cases, integration is impracticable. The integrals generally must be evaluated by approximate methods. The arch axis is divided into a convenient number of elements of length Δs, and the functions under the integral sign are evaluated for each element. The sum of these terms is approximately equal to the integral. Thus, for the usual two-hinged arch

$$H = \frac{\displaystyle\sum_A^B (My \; \Delta s/EI)}{\displaystyle\sum_A^B (y^2 \Delta s/EI) + \sum_A^B (\cos \alpha \; \Delta x/AE)} \qquad (6\text{-}103)$$

6-78. Fixed Arches. An arch is considered fixed when translation and rotation are prevented at the supports (Fig. 6-83a). Such an arch is statically indeterminate; there are six reaction components and only three equations are available from conditions of equilibrium. Three more equations must be obtained from a knowledge of the elastic behavior of the arch.

One way to determine the reactions is to consider the arch cut at the crown, forming two cantilevers. First, the horizontal and vertical deflections and rotation produced at the end of each half arch by the loads are computed (Fig. 6-83b). Next, the deflection components and rotation at those ends are found for unit vertical force, unit horizontal force, and unit moment applied separately at the ends. These deformations, multiplied, respectively, by V, the unknown vertical shear; H, the unknown horizontal thrust at the crown; and M, the unknown moment there, yield the deformations caused by the unknown forces at the crown. Adding these deformations algebraically to the corresponding ones produced by the loads gives the net movement of the free end of each half arch. Since these ends must deflect and rotate the same amount to maintain continuity, three equations can be written for determination of V, H, and M. The various deflections can be computed by the dummy-unit-load method [Eq. (6-61)], as demonstrated in Art. 6-77 for two-hinged arches.

Elastic Center. The solution of the equations, however, can be simplified considerably if the center of coordinates is shifted to the elastic center of the arch and the coordinate axes are properly oriented. If the unknown forces and moments V, H, and M are determined at the

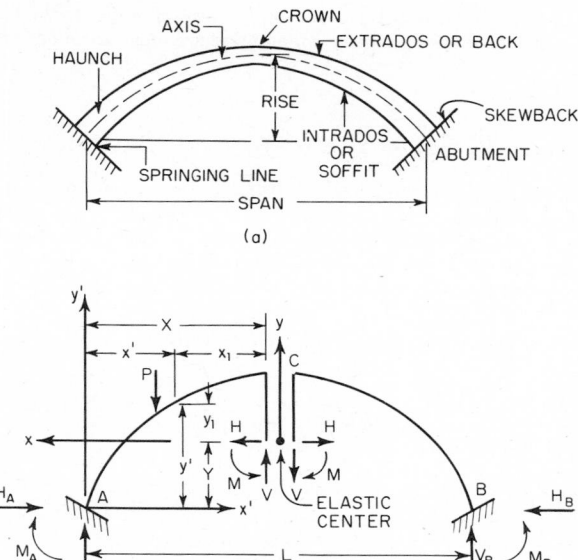

Fig. 6-83. Fixed-end arch.

elastic center, each equation will contain only one unknown. When the unknowns at the elastic center are determined, the shears, thrusts, and moments at any point on the arch can be found by application of the laws of equilibrium.

Determination of the location of the elastic center of an arch is equivalent to finding the center of gravity of an area. Instead of an increment of area dA, however, an increment of length ds multiplied by a width $1/EI$ must be used. (E is the modulus of elasticity, I the moment of inertia.) Since, in general, integrals are difficult or impracticable to evaluate, the arch axis usually is divided into a convenient number of elements of length Δs and numerical integration is used, as described in Art. 6-77. Then, if the origin of coordinates is temporarily chosen at A, the left support of the arch in Fig. 6-83b, and if x' is the horizontal distance to a point on the arch and y' the vertical distance, the coordinates of the elastic center are

$$X = \frac{\sum\limits_{A}^{B} (x' \, \Delta s/EI)}{\sum\limits_{A}^{B} (\Delta s/EI)} \qquad Y = \frac{\sum\limits_{A}^{B} (y' \, \Delta s/EI)}{\sum\limits_{A}^{B} (\Delta s/EI)} \qquad (6\text{-}104)$$

If the arch is symmetrical about the crown, the elastic center lies on a normal to the tangent at the crown. In that case, there is a savings in calculations by taking the origin of the temporary coordinate system at the crown and measuring coordinates parallel to the tangent and the normal. To determine Y, the distance of the elastic center from the crown, Eq. (6-104) can be used with the summations limited to the half arch between crown and either support. For a symmetrical arch also, the final coordinate system should be chosen parallel to the tangent and normal to the crown.

After the elastic center has been located, the origin of a new coordinate system should be taken at the center. For convenience, the new coordinates x_1, y_1 may be taken parallel to those in the temporary system. Then, for an unsymmetrical arch, the final coordinate axes should be chosen so that the x axis makes an angle α, measured clockwise, with the x_1 axis such that

$$\tan 2\alpha = \frac{2\sum_{A}^{B}(x_1 y_1 \,\Delta s/EI)}{\sum_{A}^{B}(x_1{}^2 \,\Delta s/EI) - \sum_{A}^{B}(y_1{}^2 \,\Delta s/EI)} \tag{6-105}$$

The unknown forces H and V at the elastic center should be taken parallel, respectively, to the final x and y axes.

Forces at Elastic Center. For a coordinate system with origin at the elastic center and axes oriented to satisfy Eq. (6-105),

$$H = \frac{\sum_{A}^{B}(M'y \,\Delta s/EI)}{\sum_{A}^{B}(y^2/\,\Delta s/EI)}$$

$$V = \frac{\sum_{A}^{B}(M'x \,\Delta s/EI)}{\sum_{A}^{B}(x^2\,\Delta s/EI)} \tag{6-106}$$

$$M = \frac{\sum_{A}^{B}(M' \,\Delta s/EI)}{\sum_{A}^{B}(\Delta s/EI)}$$

where M' is the average bending moment on each element due to loads. To account for the effect of an increase in temperature t, add $EctL$ to the numerator of H, where c is the coefficient of expansion and L the distance between abutments.

(S. Timoshenko and D. H. Young, "Theory of Structures," McGraw-Hill Book Company, New York; S. F. Borg and J. J. Gennaro, "Modern Structural Analysis," Van Nostrand Reinhold Company, New York; G. Winter and A. H. Nilson, "Design of Concrete Structures," McGraw-Hill Book Company, New York; V. Leontovich, "Frames and Arches," McGraw-Hill Book Company, New York.)

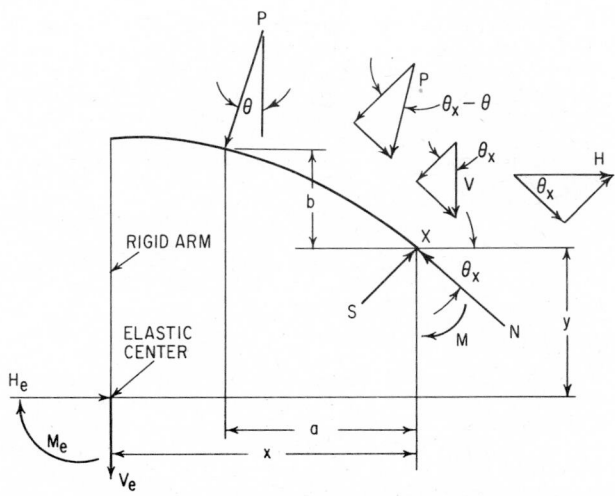

Fig. 6-84. Portion of an arch, including the elastic center.

6-79. Stresses in Arch Ribs. When the reactions have been found for an arch (Arts. 6-76 to 6-78), the principal forces acting on any cross section can be found by applying the equations of equilibrium. For example, consider the portion of a fixed arch in Fig. 6-84, where the forces at the elastic center H_e, V_e, and M_e are known and the forces acting at point X are to be found. The load P, H_e, and V_e may be resolved into components parallel to the axial thrust N and the shear S at X, as indicated in Fig. 6-84. Then, by equating the sum of the forces in each direction to zero, we get

$$N = V_e \sin \theta_x + H_e \cos \theta_x + P \sin (\theta_x - \theta)$$
$$S = V_e \cos \theta_x - H_e \sin \theta_x + P \cos (\theta_x - \theta) \tag{6-107}$$

And by taking moments about X and equating to zero, we obtain

$$M = V_e x + H_e y - M_e + Pa \cos \theta + Pb \sin \theta \tag{6-108}$$

The shearing unit stress on the arch cross section at X can be determined from S with the aid of Eq. (6-41). The normal unit stresses can be calculated from N and M with the aid of Eq. (6-49).

In designing an arch, it may be necessary to compute certain secondary stresses, in addition to those caused by live, dead, wind, and snow loads. Among the secondary stresses to be considered are those due to temperature changes, rib shortening due to thrust or shrinkage, deformation of tie rods, and unequal settlement of footings. The procedure is the same as for loads on the arch, with the deformations producing the secondary stresses substituted for or treated the same as the deformations due to loads.

CATENARIES

6-80. Stresses in Cables. When a cable is suspended from two points of support, the end reactions have a horizontal as well as a vertical component. If the cable carries only vertical loads, the horizontal component of the tension at any point in the cable is equal to the horizontal component of the end reaction. The vertical components of the reactions can be computed by taking moments about the supports in the same way as for beams, provided the cable supports are at the same level. When the supports are not at the same level, another equation, derived from a knowledge of the shape of the cable, is required.

Generally, it is safe to assume that the thickness of the cable is negligible compared with the span. Hence, bending stresses may be neglected, and the bending moment at any point is equal to zero. As a consequence, the shape assumed by the cable is similar to that of the bending-moment diagram for a simply supported beam carrying the same loads. Stresses in the cable are directed along the axis.

At the lowest point of a cable, the vertical shear either is zero or changes sign on either side of the low point.

The horizontal component of the reaction can be computed from the fact that the bending moment is zero at a point on the cable at which the sag is known (usually the low point).

For example, suppose a cable spanning 30 ft between supports at the same level is permitted to sag 2 ft in supporting a 12-kip load at a third point. Taking moments about the supports, we find the vertical components of the reactions equal to 8 and 4 kips, respectively. Setting the bending moment at the low point equal to zero, we can write the equation $8 \times 10 - 2H = 0$; from which the horizontal component H of the reaction is found to be 40 kips. The maximum tension in the cable then is

$$T = \sqrt{8^2 + 40^2} = 40.8 \text{ kips}$$

(F. S. Merritt, "Structural Steel Designers' Handbook," McGraw-Hill Book Company, New York.)

THIN-SHELL STRUCTURES

A structural shell is a curved surface structure. Usually, it is capable of transmitting loads in more than two directions to supports. It is highly efficient structurally when it is so shaped, proportioned, and supported that it transmits the loads without bending or twisting.

A shell is defined by its middle surface, halfway between its extrados, or outer surface, and intrados, or inner surface. Thus, depending on the geometry of the middle surface, it might be a type of dome, barrel arch, cone, or hyperbolic paraboloid. Its thickness is the distance, normal to the middle surface, between extrados and intrados.

6-81. Thin-shell Analysis. A thin shell is a shell with a thickness relatively small compared with its other dimensions. But it should not be so thin that deformations would be large compared with the thickness.

The shell should also satisfy the following conditions: Shearing stresses normal to the middle surface are negligible. Points on a normal to the middle surface before it is deformed lie on a straight line after deformation. And this line is normal to the deformed middle surface.

Calculation of the stresses in a thin shell generally is carried out in two major steps, both usually involving the solution of differential equations. In the first, bending and torsion are neglected (membrane theory, Art. 6-82). In the second step, corrections are made to the previous solution by superimposing the bending and shear stresses that are necessary to satisfy boundary conditions (bending theory, Art. 6-87).

6-82. Membrane Theory for Thin Shells. Thin shells usually are designed so that normal shears, bending moments, and torsion are very small, except in relatively small portions. In the membrane theory, these stresses are ignored.

Despite the neglected stresses, the remaining stresses are in equilibrium, except possibly at boundaries, supports, and discontinuities. At any interior point, the number of equilibrium conditions equals the number of unknowns. Thus, in the membrane theory, a thin shell is statically determinate.

The membrane theory does not hold for concentrated loads normal to the middle surface, except possibly at a peak or valley. The theory does not apply where boundary conditions are incompatible with equilibrium. And it is inexact where there is geometric incompatibility at the boundaries. The last is a common condition, but the error is very small if the shell is not very flat. Usually, disturbances of membrane equilibrium due to incompatibility with deformations at boundaries, supports, or discontinuities are appreciable only in a narrow region about each source of disturbance. Much larger disturbances result from incompatibility with equilibrium conditions.

To secure the high structural efficiency of a thin shell, select a shape, proportions, and supports for the specific design conditions that come as close as possible to satisfying the membrane theory. Keep the thickness constant; if it must change, use a gradual taper. Avoid concentrated and abruptly changing loads. Change curvature gradually. Keep discontinuities to a minimum. Provide reactions that are tangent to the middle surface. At boundaries, insure, to the extent possible, compatibility of shell deformations with deformations of adjoining members, or at least keep restraints to a minimum. Make certain that reactions along boundaries are equal in magnitude and direction to the shell forces there.

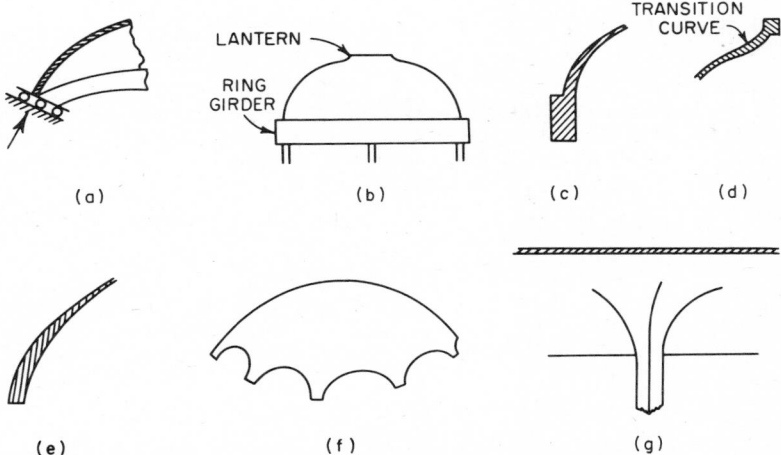

Fig. 6-85. Special provisions made at supports and boundaries of thin shells to meet requirements of the membrane theory include (*a*) a device to insure a reaction tangent to the middle surface; (*b*) stiffened edges such as the ring girder at the base of a dome; (*c*) gradually increased shell thickness at a stiffening member; (*d*) a transition curve at changes in section; (*e*) a stiffened edge obtained by thickening the shell; (*f*) scalloped edges; and (*g*) a flared support.

Means usually adopted to satisfy these requirements at boundaries and supports are illustrated in Fig. 6-85. In Fig. 6-85*a*, the slope of the support and provision for movement normal to the middle surface insure a reaction tangent to the middle surface. In Fig. 6-85*b*, a stiff rib, or ring girder, resists unbalanced shears and transmits normal forces to columns below. The enlarged view of the ring girder in Fig. 6-85*c* shows gradual thickening of the shell to reduce the abruptness of the change in section. The stiffening ring at the lantern in Fig. 6-85*d*, extending around the opening at the crown, projects above the middle surface, for compatibility of strains, and connects through a transition curve with the shell; often, the rim need merely be thickened when the edge is upturned, and the ring can be omitted. In Fig. 6-85*e*, the boundary of the shell is a thickened edge. In Fig. 6-85*f*, a scalloped shell provides gradual tapering for transmitting the loads to the supports, at the same time providing access to the shell enclosure. And in Fig. 6-85*g*, a column is flared widely at the top to support a thin shell at an interior point.

Even when the conditions for geometric compatibility are not satisfactory, the membrane theory is a useful approximation. Furthermore, it yields a particular solution to the differential equations of the bending theory.

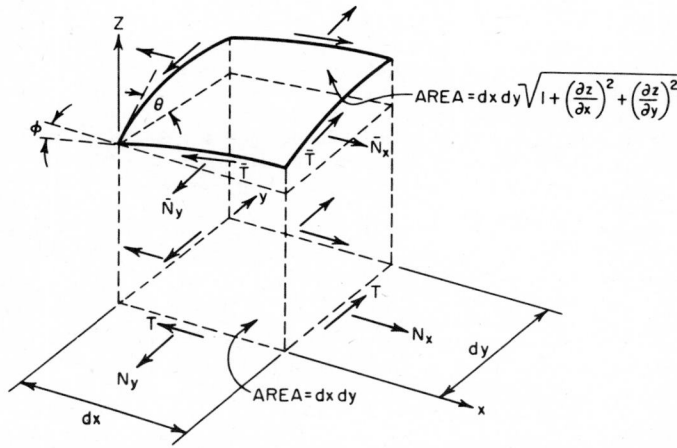

Fig. 6-86. Shears and normal forces acting on an element of the middle surface of a thin shell in the membrane theory.

6-83. Membrane Forces in Shells of General Shape. By applying the equilibrium conditions to a small element of the middle surface, one obtains the membrane forces. The sides of the element are acted upon by normal forces and shears, as indicated in Fig. 6-86. For simplification, the equilibrium equations are usually expressed in terms of the projection of these forces on the *xy* plane. Thus:

$$\bar{N}_x = N_x \frac{\cos \phi}{\cos \theta} \qquad \bar{N}_y = N_y \frac{\cos \theta}{\cos \phi} \qquad \bar{T} = T \tag{6-109}$$

The forces act per unit of length of the edges of the element and over the full thickness of the shell, and hence are called unit forces. The subscripts indicate the coordinate axis they parallel.

Assume that the loading on the element per unit area in the *xy* plane is given by its components X, Y, and Z. Then, for equilibrium in the directions of the three coordinate axes:

$$\frac{\partial N_x}{\partial x} + \frac{\partial T}{\partial y} + X = 0 \qquad \frac{\partial N_y}{\partial y} + \frac{\partial T}{\partial x} + Y = 0$$

$$\frac{\partial}{\partial x}\left(N_x \frac{\partial z}{\partial x} + T \frac{\partial z}{\partial y}\right) + \frac{\partial}{\partial y}\left(N_y \frac{\partial z}{\partial y} + T \frac{\partial z}{\partial x}\right) + Z = 0 \tag{6-110}$$

where $\tan \phi = \partial z / \partial x$ and $\tan \theta = \partial z / \partial y$. These equations can be reduced to a single equation by introduction of a stress function $F(x,y)$ such that

$$N_x = \frac{\partial^2 F}{\partial y^2} - \int X \, dx \qquad N_y = \frac{\partial^2 F}{\partial x^2} - \int Y \, dy \qquad T = -\frac{\partial^2 F}{\partial x \, \partial y} \tag{6-111}$$

where the lower limits of the integrals are constants and the upper limits are x and y, respectively. The equilibrium equation for the general case of a thin shell then is

$$\frac{\partial^2 F}{\partial x^2}\frac{\partial^2 z}{\partial y^2} - 2\frac{\partial^2 F}{\partial x \, \partial y}\frac{\partial^2 z}{\partial x \, \partial y} + \frac{\partial^2 F}{\partial y^2}\frac{\partial^2 z}{\partial x^2} = P$$

$$P = -Z + X\frac{\partial z}{\partial x} + Y\frac{\partial z}{\partial y} + \frac{\partial^2 z}{\partial x^2}\int X \, dx + \frac{\partial^2 z}{\partial y^2}\int Y \, dy \tag{6-112}$$

Hence, the membrane forces are determined when this equation is solved for F. (S. Timoshenko and S. Woinowsky-Krieger, "Theory of Plates and Shells," McGraw-Hill Book Company, New York.)

6-84. Membrane Forces in a Hyperbolic-Paraboloid Shell. Equations (6-112) of Art. 6-83 can be used to determine the membrane forces in a thin shell with a hyperbolic-paraboloid middle surface. Assume, for example, that the equation of the middle surface is

$$z = cxy$$

where c is a constant, and the loading on the shell, shown in Fig. 6-87, is vertical, say a snow load, $-Z_s$. Then, $P = -Z = Z_s$. (Note that when x is a constant or y is a constant, the intercept with the surface is a straight line, though the surface is curved.)

Differentiation of the surface equation gives

$$\frac{\partial^2 z}{\partial x^2} = \frac{\partial^2 z}{\partial y^2} = 0 \qquad \frac{\partial^2 z}{\partial x \, \partial y} = c$$

Substitution of these derivatives in Eqs. (6-111) and (6-112) yields

$$-2\frac{\partial^2 F}{\partial x \, \partial y}c = 2Tc = -Z = Z_s$$

and this gives

$$T = \frac{Z_s}{2c} \tag{6-113}$$

Furthermore, N_x can be any function of y, and N_y any function of x.

Suppose the shell to be stiffened along its edges by a diaphragm with negligible resistance normal to its plane. Then, $N_x = \bar{N}_x = 0$ and $N_y = \bar{N}_y = 0$, since the edges are free of normal forces.

Now assume the load to be the weight of the shell, a constant $-Z_w$ per unit area of the surface. The area of the horizontal projection of an element of the surface equals $1/(1 + c^2x^2 + c^2y^2)^{1/2}$ times the area of the element. Then, the horizontal projections of the unit forces are

$$T = \frac{Z_w}{2c}(1 + c^2x^2 + c^2y^2)^{1/2}$$

$$N_x = -\frac{Z_w y}{2}\log_e[cx + (1 + c^2x^2 + c^2y^2)^{1/2}] + f_1(y) \tag{6-114}$$

$$N_y = -\frac{Z_w x}{2}\log_e[cy + (1 + c^2x^2 + c^2y^2)^{1/2}] + f_2(x)$$

where $f_1(y)$ is an arbitrary function of y and $f_2(x)$ an arbitrary function of x; logarithms are taken to the base e. The actual unit forces $\bar{N}_x$ and $\bar{N}_y$ can be calculated by applying Eqs. (6-109) to Eqs. (6-114), noting that $\tan\phi = -cy$ and $\tan\theta = -cx$.

(S. Timoshenko and S. Woinowsky-Krieger, "Theory of Plates and Shells," McGraw-Hill Book Company, New York.)

6-85. Elliptical-Paraboloid Membrane Forces. Equations (6-112) of Art. 6-83 also can be used to obtain membrane forces in an elliptical paraboloid (parabolas in vertical planes, ellipses in horizontal),

$$z = \frac{h_x x^2}{a^2} + \frac{h_y y^2}{b^2}$$

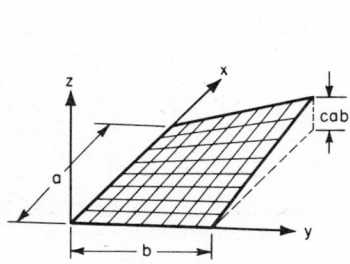

Fig. 6-87. Hyperbolic-paraboloid thin shell.

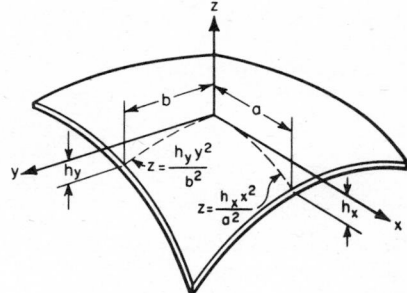

Fig. 6-88. Elliptical-paraboloid thin shell.

rectangular in plan (Fig. 6-88). The solution usually is given as a Fourier series, summed for $n = 1, 3, 5, \ldots, \infty$.

Suppose the shell to be subjected to a uniform load p over its horizontal projection. Assume that forces normal to the boundaries vanish. Then, the unit forces, acting per unit of length over the thickness of the shell, are

$$\bar{N}_x = -\frac{pb^2}{2h_x}\left(\frac{a^4 + 4h_x^2x^2}{b^4 + 4h_y^2y^2}\right)^{1/2}\left[1 + \frac{4}{\pi}\sum_n(-1)^{(n+1)/2}\frac{\cosh(n\pi x/c)}{n\cosh(n\pi a/c)}\cos\frac{n\pi y}{2b}\right]$$

$$\bar{N}_y = \frac{2pa^4}{\pi h_x b^2}\left(\frac{b^4 + 4h_y^2y^2}{a^4 + 4h_x^2x^2}\right)^{1/2}\sum_n(-1)^{(n+1)/2}\frac{\cosh(n\pi x/c)}{n\cosh(n\pi a/c)}\cos\frac{n\pi y}{2b} \qquad (6\text{-}115)$$

$$T = \frac{2pa^2}{\pi h_x}\sum_n(-1)^{(n+1)/2}\frac{\sinh(n\pi x/c)}{n\cosh(n\pi a/c)}\sin\frac{n\pi y}{2b}$$

where $c = 2a(h_y/h_x)^{1/2}$

The series for T does not converge at the corners. Actually, bending moments and torsion will exist in those regions, counteracting the huge membrane shears.

6-86. Membrane Forces in Shells of Revolution. In a shell of revolution, the middle surface results from the rotation of a plane curve about an axis, called the "shell axis." A plane through the axis cuts the surface in a meridian. A plane normal to the axis cuts the surface in a circle, called a parallel.

The membrane forces in a shell of revolution usually are taken parallel to the x, y, z axes of a cartesian coordinate system at each point of the middle surface. The x axis is tangent to the

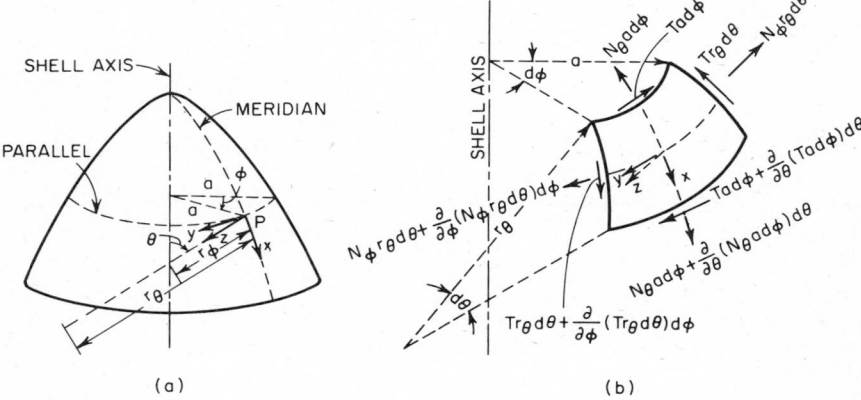

     (a)   (b)

Fig. 6-89. Shell of revolution. (*a*) Coordinate system used in analysis; (*b*) unit forces acting on an element.

meridian, the y axis is tangent to the parallel, and the z axis lies in the direction of the normal to the surface (Fig. 6-89).

A given point P on the surface is determined by the angle θ between the shell axis and the normal through P, and by the angle ϕ between the radius through P of the parallel on which it lies and a fixed, reference direction. Let r_θ be the radius of curvature of the meridian, and let r_ϕ, the length of the shell normal between P and the shell axis, be the radius of curvature of the normal section at P. If a is the radius of the parallel through P, then $a = r_\phi \sin \theta$.

Consider the element of the middle surface of the shell in Fig. 6-89b. Its edges are subjected to unit normal and shear forces, acting per unit of length of edge over the thickness of the shell. Assume that the loading on the element per unit of area is given by its X, Y, Z components. Then, the equations of equilibrium are

$$\frac{\partial}{\partial \theta} (N_\theta r_\phi \sin \theta) + \frac{\partial T}{\partial \phi} r_\theta - N_\phi r_\theta \cos \theta + X r_\theta r_\phi \sin \theta = 0$$

$$\frac{\partial N_\phi}{\partial \phi} r_\theta + \frac{\partial}{\partial \theta} (T r_\phi \sin \theta) + T r_\theta \cos \theta + Y r_\theta r_\phi \sin \theta = 0 \qquad (6\text{-}116)$$

$$N_\theta r_\phi + N_\phi r_\theta + Z r_\theta r_\phi = 0$$

Symmetrical Loads. When the loads also are symmetrical about the shell axis, Eqs. (6-116) take a simpler form. If a is the radius of a parallel and R the resultant of the total vertical load on the shell above that parallel, then the first equation can be written in the alternative form:

$$N_\theta = - \frac{R}{2\pi a} \sin \theta = - \frac{R}{2\pi r_\phi} \sin^2 \theta \qquad (6\text{-}117a)$$

Substitution of this in the third equation yields

$$N_\phi = \frac{R}{2\pi r_\theta} \sin^2 \theta - Z r_\phi \qquad (6\text{-}117b)$$

Because of rotational symmetry

$$T = 0 \qquad (6\text{-}117c)$$

For a **spherical shell**, $r_\theta = r_\phi = r$. If the load p is uniform over the horizontal projection of the shell, then Eqs. (6-117) indicate that the unit shear $T = 0$; with $R = \pi a^2 p$, the unit meridional thrust $N_\theta = - pr/2$; and the unit hoop force $N_\phi = - (pr/2) \cos 2\theta$. Thus, there is a constant meridional compression throughout the shell. The hoop forces are compressive in the upper half of the shell and tensile in the lower half, vanishing at $\theta = 45°$.

Again for a spherical dome, if the load w is uniform over the area of the shell, then Eqs. (6-117) give $T = 0$; with $R = 2\pi r^2 (1 - \cos \theta) w$, the unit meridional thrust $N_\theta = - wr/(1 + \cos \theta)$; and the unit hoop force $N_\phi = wr[1/(1 + \cos \theta) - \cos \theta]$. In this case, the compression along the meridian increases with θ. The hoop forces are compressive in the upper part of the shell, become zero at $\theta = 51°50'$, and turn to tension for larger values of θ. Usually, at the lower dome boundary, these tensile hoop forces are resisted by a ring girder, and since shell and girder under the membrane theory will have different strains, bending stresses will be imposed on the shell.

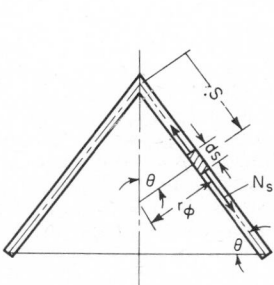

Fig. 6-90. Conical thin shell.

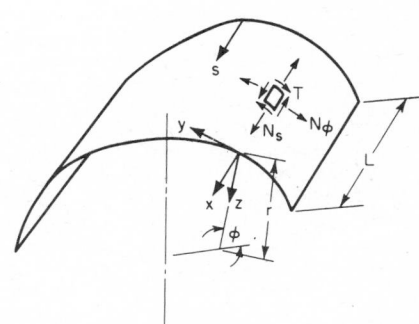

Fig. 6-91. Cylindrical thin shell.

When there is an opening around the crown of the dome, as in Fig. 6-85b, the upper edge may be thickened or reinforced with a ring girder to resist the hoop forces. If $2\theta_0$ is the angle of the opening and P the vertical load per unit of length of the compression ring, then

$$N_\theta = -wr \, \frac{\cos\theta_0 - \cos\theta}{\sin^2\theta} - P \, \frac{\sin\theta_0}{\sin^2\theta}$$

$$N_\phi = wr \left(\frac{\cos\theta_0 - \cos\theta}{\sin^2\theta} - \cos\theta \right) + P \, \frac{\sin\theta_0}{\sin^2\theta}$$

For a **conical shell,** parallel circles are located by the coordinate s, measured from the vertex along a generator, instead of the angle θ, and the unit meridional force is denoted by N_s (Fig. 6-90). The equilibrium equations are

$$\frac{\partial}{\partial s}(N_s s) + \frac{\partial T}{\partial\phi}\frac{1}{\cos\theta} - N_\phi + Xs = 0$$

$$\frac{\partial N_\phi}{\partial\phi}\frac{1}{\cos\theta} + \frac{\partial}{\partial s}(Ts) + T + Ys = 0 \qquad (6\text{-}118)$$

$$N_\phi + Zs\cot\theta = 0$$

For a vertical load p uniform over the horizontal projection of a conical shell, Eqs. (6-118) give

$$T = 0 \qquad N_s = -\frac{ps}{2}\cot\theta \qquad \text{and} \qquad N_\phi = -ps\,\frac{\cos^2\theta}{\sin\theta}$$

Suppose that the cone has an opening at the top at $s = s_0$, and the shell is subjected to a vertical uniform load w over its area. Then,

$$T = 0 \qquad N_s = -\frac{s^2 - s_0^2}{2s}\frac{w}{\sin\theta} \qquad \text{and} \qquad N_\phi = -ws\,\frac{\cos^2\theta}{\sin\theta}$$

For a horizontal load q uniform over the vertical projection of the cone with lantern,

$$T = -q\,\frac{s^3 - s_0^3}{3s^2}\sin\phi$$

$$N_s = -\frac{qs}{2}\left[\cos\theta - \frac{1}{3\cos\theta} - \frac{s_0^2}{s^2}\left(\cos\theta - \frac{1}{\cos\theta}\right) - \frac{s_0^3}{s^3}\frac{2}{3\cos\theta} \right]\cos\phi$$

and $\qquad N_\phi = -qs\cos\theta\cos\phi$

For a **cylindrical shell** with horizontal axis, a point on the surface is located by its distance s along a generator from one end and by the angle ϕ which the radius of curvature at the point makes with the horizontal (Fig. 6-91). The equilibrium equations are

$$\frac{\partial N_s}{\partial s}r + \frac{\partial T}{\partial\phi} + Xr = 0 \qquad \frac{\partial N_\phi}{\partial\phi} + \frac{\partial T}{\partial s}r + Yr = 0 \qquad N_\phi + Zr = 0 \qquad (6\text{-}119)$$

The third equation indicates that $N_\phi = 0$ when $Z = 0$. Hence, for uniform vertical loads over the horizontal projection of the shell area, vertical reactions vanish along the longitudinal edges of a cylindrical shell if the tangent to the directrix at the edges is vertical. Shells with a semicircular, semielliptical, or cycloidal directrix have this property.

Parabolic Shells. Equations (6-119) can be used to demonstrate an unusual property of parabolic vaults. The equation of the parabola can be rewritten from the common form $u = v^2/2r_0$, where r_0 is the radius of curvature at the vertex ($\phi = \pi/2$), into the more convenient form $r = r_0/\sin^3\phi$. Now, for a vertical load p uniform over the horizontal projection, $X = 0$, while $Y = -p\sin\phi\cos\phi$ and $Z = p\sin^2\phi$. Then, solution of Eqs. (6-119) yields $T = 0$ and $N_s = 0$, so loads are transmitted to the supports only in the ϕ direction. $N_\phi = -pr_0/\sin\phi$. Thus, under this type of loading, a parabolic vault behaves like a parabolic arch. The same results are obtained with a catenary-shaped vault under vertical loading uniform over the shell area.

Semicircular Shells. For a semicircular barrel arch, r is a constant. If the vertical load w is uniform over the shell area, $X = 0$, $Y = w\cos\phi$, and $Z = w\sin\phi$. Then, Eqs. (6-119) give $T = -w(L - 2s)\cos\phi$, $N_s = -w(s/r)(L - s)\sin\phi$, and $N_\phi = -wr\sin\phi$. Note that no supports are needed along the longitudinal edges, since N_ϕ is zero there, but provision must be made to resist the shear. Such provision, usually in the form of an edge member, generally creates bending stresses due to incompatibility of deformations at the shell boundary. The shears transmit the loads to the edges $s = 0$ and $s = L$, where supports must be provided.

Other Shapes. For membrane stresses in a variety of shells, see W. Flügge, "Stresses in Shells," Springer-Verlag, New York; S. Timoshenko and S. Woinowsky-Krieger, "Theory of Plates and Shells," McGraw-Hill Book Company, New York; D. P. Billington, "Thin Shell Concrete Structures," McGraw-Hill Book Company, New York.

6-87. Bending Theory for Thin Shells. When equilibrium conditions are not satisfied or incompatible deformations exist at boundaries, bending and torsion stresses arise in the shell. Sometimes, the design of the shell and its supports can be modified to reduce or eliminate these stresses (Art. 6-82). When the design cannot eliminate them, provision must be made for the shell to resist them.

But even for the simplest types of shells and loading, the stresses are difficult to compute. In bending theory, a thin shell is statically indeterminate; deformation conditions must supplement equilibrium conditions in setting up differential equations for determining the unknown forces and moments. Solution of the resulting equations may be tedious and time-consuming, if indeed solution is possible.

In practice, therefore, shell design relies heavily on the designer's experience and judgment. The designer should consider the type of shell, material of which it is made, and support and boundary conditions, and then decide whether to apply a bending theory in full, use an approximate bending theory, or make a rough estimate of the effects of bending and torsion. (Note that where the effects of a disturbance are large, these change the normal forces and shears computed by the membrane theory.) For domes, for example, the usual procedure is to use as a support a deep, thick girder or a heavily reinforced or prestressed tension ring, and the shell is gradually thickened in the vicinity of this support (Fig. 6-85c).

Circular barrel arches, with ratio of radius to distance between supporting arch ribs less than 0.25, may be designed as beams with curved cross section. Secondary stresses, however, must be taken into account. These include stresses due to volume change of rib and shell, rib shortening, unequal settlement of footings, and temperature differentials between surfaces.

Bending theory for cylinders and domes is given in W. Flügge, "Stresses in Shells," Springer-Verlag, New York; S. Timoshenko and S. Woinowsky-Krieger, "Theory of Plates and Shells," McGraw-Hill Book Company, New York; "Design of Cylindrical Concrete Shell Roofs," Manual of Practice No. 31, American Society of Civil Engineers.

6-88. Stresses in Thin Shells. The results of the membrane and bending theories are expressed in terms of unit forces and unit moments, acting per unit of length over the thickness of the shell. To compute the unit stresses from these forces and moments, usual practice is to assume normal forces and shears to be uniformly distributed over the shell thickness and bending stresses to be linearly distributed.

Then, normal stresses can be computed from equations of the form

$$f_x = \frac{N_x}{t} + \frac{M_x}{t^3/12} z$$

where z = distance from middle surface
t = shell thickness
M_x = unit bending moment about an axis parallel to direction of unit normal force N_x

Similarly, shearing stresses produced by central shears and twisting moments may be calculated from equations of the form

$$v_{xy} = \frac{T}{t} \pm \frac{D}{t^3/12} z$$

where D = twisting moment

Normal shearing stresses may be computed on the assumption of a parabolic stress distribution over the shell thickness:

$$v_{xz} = \frac{V}{t^3/6} \left(\frac{t^2}{4} - z^2 \right)$$

where V = unit shear force normal to middle surface

For axes rotated with respect to those used in the thin-shell analysis, use Eqs. (6-27) and (6-28) to transform stresses or unit forces and moments from the given to the new axes.

FOLDED PLATES

A folded-plate structure consists of a series of thin planar elements, or flat plates, connected to one another along their edges. Usually used on long spans, especially for roofs, folded plates derive their economy from the girder action of the plates and the mutual support they give one another.

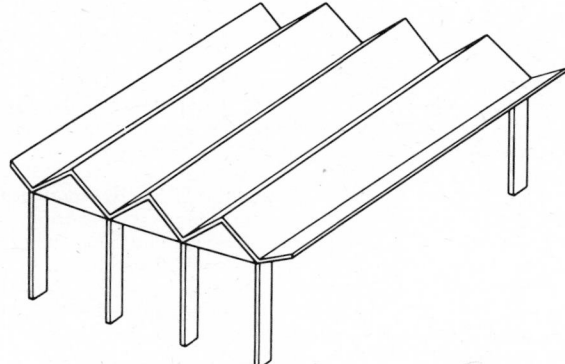

Fig. 6-92. Folded-plate structure.

Longitudinally, the plates may be continuous over their supports. Transversely, there may be several plates in each bay (Fig. 6-92). At the edges, or folds, they may be capable of transmitting both moment and shear or only shear.

6-89. Folded-plate Theory. A folded-plate structure has a two-way action in transmitting loads to its supports. Transversely, the elements act as slabs spanning between plates on either side. The plates then act as girders in carrying the load from the slabs longitudinally to supports, which must be capable of resisting both horizontal and vertical forces.

If the plates are hinged along their edges, the design of the structure is relatively simple. Some simplification also is possible if the plates, though having integral edges, are steeply sloped or if the span is sufficiently long with respect to other dimensions that beam theory applies. But there are no criteria for determining when such simplification is possible with acceptable accuracy. In general, a reasonably accurate analysis of folded-plate stresses is advisable.

Several good methods are available (D. Yitzhaki, "The Design of Prismatic and Cylindrical Shell Roofs," North Holland Publishing Company, Amsterdam, available in the United States from W. S. Heinman Books, 400 East 72d Street, New York, N.Y.; "Phase I Report on Folded-plate Construction," Proceedings Paper 3741, *Journal of the Structural Division, American Society of Civil Engineers,* December, 1963; and A. L. Parme and J. A. Sbarounis, "Direct Solution of Folded Plate Concrete Roofs," Portland Cement Association, 33 West Grand Avenue, Skokie, Ill.). They all take into account the effects of plate deflections on the slabs and usually make the following assumptions:

The material is elastic, isotropic, and homogeneous. The longitudinal distribution of all loads on all plates is the same. The plates carry loads transversely only by bending normal to their planes and longitudinally only by bending within their planes. Longitudinal stresses vary linearly over the depth of each plate. Supporting members, such as diaphragms, frames, and beams, are infinitely stiff in their own planes and completely flexible normal to their own planes. Plates have no torsional stiffness normal to their own-planes. Displacements due to forces other than bending moments are negligible.

Regardless of the method selected, the computations are rather involved; so it is wise to carry out the work in a well-organized table. The Yitzhaki method (Art. 6-90) offers some advantages over others in that the calculations can be tabulated, it is relatively simple, it requires the solution of no more simultaneous equations than one for each edge for simply supported plates, it is flexible, and it can easily be generalized to cover a variety of conditions.

6-90. Yitzhaki Method for Folded Plates. Based on the assumptions and general procedure given in Art. 6-89, the Yitzhaki method deals in two ways with the slab and plate systems that comprise a folded-plate structure. In the first, a unit width of slab is considered continuous over supports immovable in the direction of the load (Fig. 6-93b). The strip usually is taken where the longitudinal plate stresses are a maximum. Secondly, the slab reactions are taken as loads on the plates, which now are assumed to be hinged along the edges (Fig. 6-93c). Thus, the slab reactions cause angle changes in the plates at each fold. Continuity is restored by applying an unknown moment to the plates at each edge. The moments can be determined from the fact that at each edge the sum of the angle changes due to the loads and to the unknown moments must equal zero.

The angle changes due to the unknown moments have two components. One is the angle change at each slab end, now hinged to an adjoining slab, in the transverse strip of unit width. The second is the angle change due to deflection of the plates. The method assumes that the angle change at each fold varies in the same way longitudinally as the angle changes along the other folds.

The method will be demonstrated for the symmetrical folded-plate structure in Fig. 6-93a. To simplify the explanation, the tabulated calculations will be broken up into several steps, though, in practice, the table is suitable for continuous calculation from start to finish. Dimensions and geometric data are given in Table 6-7A.

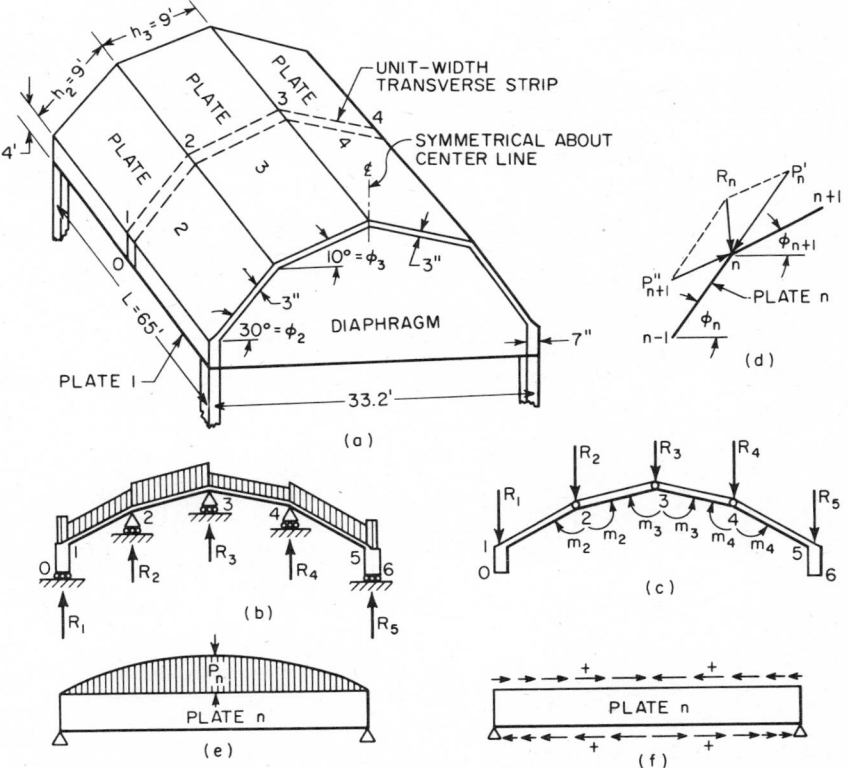

Fig. 6-93. Folded plate is analyzed by first considering a transverse strip (a) as a continuous slab on supports that do not settle (b). Then, the slabs are assumed hinged (c) and acted upon by the reactions computed in the first step and unknown moments to correct for this assumption. In the longitudinal direction, the plates act as deep girders (e) with shears along the edges, positive directions shown (f). Slab reactions are resolved into plate forces, parallel to the planes of the plates (d).

Table 6-7A. Geometric Data for Symmetric Folded Plate

Plate	h, in.	t, in.	a, in.	ϕ, deg	$\cos \phi$	$\tan \phi$	A, sq in.	W, lb per ft
1 and 6	48	7	0	90	0	. . . . .	326	350
2 and 5	108	3	93.5	30	0.866	0.577	324	337.5
3 and 4	108	3	106.4	10	0.985	0.1763	324	337.5

In the table, h is the depth in inches, t the thickness in inches, a the horizontal projection of the depth in inches, and A the cross-sectional area of a plate in square inches. ϕ is the angle each plate makes with the horizontal, in degrees. W is the vertical load in pounds per linear foot of plate and thus also the total load in pounds on a 12-in.-wide slab.

Step 1. Compute the loads on a 12-in.-wide transverse strip at midspan. For the edge beam, $W = 350$ lb per lin ft. For a load of 37.5 lb per sq ft on the plates, each slab in the strip carries 337.5 lb.

Step 2. Consider the strip as a continuous slab supported at the folds (Fig. 6-93b), and compute the bending moments by moment distribution. To take advantage of the symmetrically loaded, symmetrical structure, edge 3 is taken as fixed, and end moments are distributed only in the left half of the strip. Also, since edge 1 is free to rotate, the relative stiffness of plate 2 is taken as three-fourths its actual relative stiffness to eliminate the need for distributing moments to edge 1. The computations are shown in Table 6-7B.

Table 6-7B. Slab on Rigid Supports

Property	Plate 1		Plate 2		Plate 3	
	Edge 0	Edge 1	Edge 1	Edge 2	Edge 2	Edge 3
Stiffness Ratio				¾ × 1	1	Fixed
				0.428	0.572	0
M^F, in.-lb	0	0	0	3,945	−2,996	2,996
				−406	−543	−272
M, in.-lb	0	0	0	3,539	−3,539	2,744

Table 6-7C. Slab Reactions and Plate Loads

Property	Plate 1		Plate 2		Plate 3	
	Edge 0	Edge 1	Edge 1	Edge 2	Edge 2	Edge 3
Shears	0	350	168.8	168.8	168.8	337.5
$\pm \Sigma M / a_n$	0	0	− 37.8	37.8	7.7	− 15.4
Reaction	0		481		383.1	322.1
k					0.401	0.3526
R/k					957	913
From R_{n-1}		0		0		− 972
From R_n		481		1,105		927
Plate load		481		1,105		− 45
M, ft-lb		254,000		584,000		− 23,800

Step 3. From the end moments M found in step 2, compute slab reactions and plate loads. Reactions (positive upward) at the nth edge are

$$R_n = V_n + V_{n+1} + \frac{M_{n-1} + M_n}{a_n} - \frac{M_n + M_{n+1}}{a_{n+1}} \qquad (6\text{-}120)$$

where V_n, V_{n+1} = shears at both sides of edge n
M_n = moment at edge n, in.-lb

M_{n-1} = moment at edge $(n-1)$, in.-lb
M_{n+1} = moment at edge $(n+1)$, in.-lb

The calculations are shown in Table 6-7C.

For the vertical beam, the shear is zero at edge 0, and $W = 350$ at edge 1. For the plates, the shear is $W/2 = 168.8$ at edges 1 and 2, and $W = 337.5$ at edge 3 to include the load from plate 4. These values are shown on the first line of the table. The entries on the second line are obtained by dividing the sum of the end moments in each plate by a, the horizontal projection of the plate. Entries on the right for each edge are written with sign changed. The reactions are obtained by adding the values for each edge on the first and second lines of the table.

Let $k = \tan \phi_n - \tan \phi_{n+1}$, where ϕ is positive as shown in Fig. 6-93a. Then, the load (positive downward) on the nth plate is

$$P_n = \frac{R_n}{k_n \cos \phi_n} - \frac{R_{n-1}}{k_{n-1} \cos \phi_n} \tag{6-121}$$

[Figure 6-93d shows the resolution of forces at edge n; edge $(n-1)$ is similar.] Equation (6-121) does not apply for the case of a vertical reaction on a vertical plate, for R/k is the horizontal component of the reaction.

In the table, the two components of Eq. (6-121) are shown on separate lines after the values of R/k, the second component being entered first, with sign changed. The components are summed to obtain the plate loads in pounds per linear foot.

Step 4. Calculate the midspan (maximum) bending moment in each plate. In this example, each plate is a simple beam and $M = PL^2/8$ ft-lb, where L is the span in feet. The moments for each plate are entered in Table 6-7C after the plate loads. These moments are assumed to vary parabolically along the edges.

Step 5. Determine the free-edge longitudinal stresses at midspan. In each plate, these can be computed from

$$f_{n-1} = \frac{72M}{Ah} \qquad f_n = -\frac{72M}{Ah} \tag{6-122}$$

where f is the stress in psi, M the moment in ft-lb from step 4, and tension is taken as positive, compression as negative.

Step 6. Apply a shear to adjoining edges to equalize the stresses there. Compute the adjusted stresses by converging approximations, similar to moment distribution. To do this, distribute the unbalanced stress at each edge in proportion to the reciprocals of the areas of the plates, and use a carry-over factor of $-\frac{1}{2}$ to distribute the stress to a far edge. Edge 0, being a free edge, requires no distribution of the stress there. Edge 3, because of symmetry, may be treated the same, and distribution need be carried out only in the left half of the structure. The calculations are shown in Table 6-7D.

Table 6-7D. Plate Stresses for Slab Loads

Property	Plate 1		Plate 2		Plate 3	
	Edge 0	Edge 1	Edge 1	Edge 2	Edge 2	Edge 3
1/A ratio		0.491	0.509	0.5	0.5	
Free-edge f	1,133	−1,133	1,200	−1,200	−49	49
Distribution	−573	1,145	−1,188	594		
			−139	278	−279	139
	34	−68	71	−36		
			−9	18	−18	9
	2	−5	4	−2		
				1	−1	
Adjusted f	596	−61	−61	−347	−347	197

Step 7. Compute the midspan edge deflections. In general, the vertical component δ, in inches, can be computed from

$$\frac{E}{L^2} \delta_n = \frac{15}{k_n} \left(\frac{f_{n-1} - f_n}{a_n} - \frac{f_n - f_{n+1}}{a_{n+1}} \right) \tag{6-123a}$$

where E = modulus of elasticity, psi

$k = \tan \phi_n - \tan \phi_{n-1}$, as in step 3

The factor E/L^2 is retained for convenience; it is eliminated by dividing the simultaneous angle equations by it. For a vertical plate, the vertical deflection is given by

$$\frac{E}{L^2} \delta_n = \frac{15(f_{n-1} - f_n)}{h_n} \tag{6-123b}$$

The calculations are shown in Table 6-7E.

Table 6-7E. Edge Rotations for Plate Loads

Property	Plate 1		Plate 2		Plate 3	
	Edge 0	Edge 1	Edge 1	Edge 2	Edge 2	Edge 3
$\pm \Delta f/(a$ or $h)$		13.70	3.06	5.11	−5.11	−5.11
$E\delta_n/L^2$			205.6		305.6	−435
$\pm E\Delta\delta/L^2 a$				1.07	6.95	−13.90
$E\theta_P/L^2$					8.02	−13.90

On the first line of the table are entered the results of dividing the difference of the stresses at the edges of each plate by the depth for the vertical edge beam and by the horizontal projection of the depth for the other plates. The results for the top edge of a plate are entered with sign changed. The second line, giving $E\delta_n/L^2$, is obtained from the first line by multiplying by 15 for the edge beam or by adding the entries for each plate, and then multiplying by $15/k_n$.

Step 8. Compute the midspan angle change θ_P at each edge. This can be determined from

$$\frac{E}{L^2} \theta_P = -\frac{\delta_{n-1} - \delta_n}{a_n} + \frac{\delta_n - \delta_{n+1}}{a_{n+1}} \tag{6-124}$$

by continuing the calculations of step 7 in the table. The third line of Table 6-7E is obtained by subtracting the successive entries on the second line and then dividing by the horizontal projection of the plate. The left-hand entry at each edge is written with sign changed. The fourth line, giving $E\theta_P/L^2$, is the sum of the third-line entries for each edge.

Step 9. To correct the edge rotations with a symmetrical loading, apply an unknown moment of $+1,000m_n \sin (\pi x/L)$, in.-lb (positive when clockwise) to plate n at edge n. Apply $-1,000m_n \sin (\pi x/L)$ to its counterpart, plate n' at edge n'. Also, apply $-1,000m_n \sin (\pi x/L)$ to plate $(n + 1)$ at edge n and $+1,000m_n \sin (\pi x/L)$ to its counterpart; x is the distance along an edge from the end of a plate. (The sine function is assumed to make the loading vary longitudinally in approximately the same manner as the deflections.) At midspan, the absolute value of these moments is $1,000m_n$.

The 12-in.-wide transverse strip at midspan, hinged at the supports, will then be subjected at the supports to moments of $1,000m_n$. Compute the rotations thus caused in the slabs from

$$\frac{E}{L^2} \theta''_{n-1} = \frac{166.7 h_n m_n}{L^2 t_n^3}$$

$$\frac{E}{L^2} \theta''_n = \frac{333.3 m_n}{L^2} \left(\frac{h_n}{t_n^3} + \frac{h_{n+1}}{t^3_{n+1}} \right) \tag{6-125}$$

$$\frac{E}{L^2} \theta''_{n+1} = \frac{166.7 h_{n+1} m_n}{L^2 t^3_{n+1}}$$

The results of this calculation are shown in Table 6-7F.

Because edge 1, in effect, is hinged in the actual structure, the correction moments need be applied only at edges 2 and 3. And the angle changes in the slabs due to these moments need be computed only at those edges. Results are given in terms of m_2, the moment at edge 2, and m_3, the moment at edge 3.

Table 6-7F. Edge Rotations, Reactions, and Plate Loads for Unknown Moments

	Plate 1		Plate 2		Plate 3	
	Edge 0	Edge 1	Edge 1	Edge 2	Edge 2	Edge 3
$E\theta''/L^2$ for m_2				$0.631m_2$		$0.316m_2$
$E\theta''/L^2$ for m_3				$0.158m_3$		$0.631m_3$
R for m_2			$-10.70m_2$	$20.08m_2$		$-18.76m_2$
R for m_3				$-9.38m_3$		$18.76m_3$
k				0.401		0.3526
R/k for m_2				$50.3m_2$		$-53.2m_2$
R/k for m_3				$-23.4m_3$		$53.2m_3$
P for m_2	$-10.70m_2$		$58.1m_2$		$-105.2m_2$	
P for m_3	0		$-27.1m_3$		$77.8m_3$	
M for m_2	$-4{,}580m_2$		$24{,}840m_2$		$-45{,}100m_2$	
M for m_3	0		$-11{,}620m_3$		$33{,}300m_3$	

Step 10. Compute the slab reactions and plate loads due to the unknown moments. The reactions are

$$R_{n-1} = -\frac{1{,}000m_n}{a_n} \qquad R_n = 1{,}000m_n\left(\frac{1}{a_n} + \frac{1}{a_{n+1}}\right) \qquad R_{n+1} = -\frac{1{,}000m_n}{a_{n+1}} \qquad (6\text{-}126)$$

The plate loads are

$$P_n = \frac{1}{\cos\phi_n}\left(\frac{R_n}{k_n} - \frac{R_{n-1}}{k_{n-1}}\right) \qquad (6\text{-}127)$$

The calculations are shown as a continuation of Table 6-7F.

Step 11. Assume that the loading on each plate is $P_n \sin(\pi x/L)$ (Fig. 6-93e), and calculate the midspan (maximum) bending moment. For a simple beam,

$$M = PL^2/\pi^2 \qquad \text{ft-lb}$$

The moments also are shown in Table 6-7F in terms of m_2 and m_3.

Step 12. Using Eqs. (6-122), compute the free-edge longitudinal stresses at midspan. Then, as in step 6, apply a shear at each edge to equalize the stresses. Determine the adjusted stresses by converging approximations. The results of this calculation are shown in Table 6-7G.

Table 6-7G. Plate Stresses for Unknown Moments

Stress	Plate 1		Plate 2		Plate 3	
	Edge 0	Edge 1	Edge 1	Edge 2	Edge 2	Edge 3
Free-edge f, for m_2	$-20.5m_2$	$20.5m_2$	$51.1m_2$	$-51.1m_2$	$-92.6m_2$	$92.6m_2$
Adjusted f, for m_2	$-31.2m_2$	$41.9m_2$	$41.9m_2$	$-66.3m_2$	$-66.3m_2$	$79.5m_2$
Free-edge f, for m_3			$-23.8m_3$	$23.8m_3$	$68.5m_3$	$-68.5m_3$
Adjusted f, for m_3	$9.2m_3$	$-18.4m_3$	$-18.4m_3$	$41.4m_3$	$41.4m_3$	$-54.9m_3$

Step 13. Compute the vertical component of the edge deflections at midspan from

$$\frac{E}{L^2}\delta_n = \frac{144}{\pi^2 k_n}\left(\frac{f_{n-1} - f_n}{a_n} - \frac{f_n - f_{n+1}}{a_{n+1}}\right) \qquad (6\text{-}128a)$$

or for a vertical plate from

$$\frac{E}{L^2}\delta_n = \frac{144(f_{n-1} - f_n)}{\pi^2 h_n} \qquad (6\text{-}128b)$$

The results of this calculation are shown in Table 6-7H. The procedure is the same as step 7.

Table 6-7H. Edge Rotations for Unknown Moments

Rotation	Plate 1		Plate 2		Plate 3	
	Edge 0	Edge 1	Edge 1	Edge 2	Edge 2	Edge 3
$\pm \Delta f/(a \text{ or } h)$ for m_2		$-1.522m_2$	$1.155m_2$	$1.368m_2$	$-1.368m_2$	$-1.368m_2$
$E\delta/L^2$, for m_2		$-22.2m_2$		$92.0m_2$		$-113.2m_2$
$\mp E \Delta\delta/L^2 a$, for m_2				$1.222m_2$	$1.927m_2$	$-3.854m_2$
$E\theta'/L^2$, for m_2				$3.149m_2$		$-3.854m_2$
$\pm \Delta f/(a \text{ or } h)$ for m_3		$0.574m_3$	$-0.640m_3$	$-0.904m_3$	$0.904m_3$	$0.904m_3$
$E\delta/L^2$, for m_3		$8.4m_3$		$-56.2m_3$		$74.8m_3$
$\mp E \Delta\delta/L^2 a$, for m_3				$-0.691m_3$	$-1.230m_3$	$2.460m_3$
$E\theta'/L^2$, for m_3				$-1.921m_3$		$2.460m_3$

Step 14. Using Eq. (6-124), determine the midspan angle change θ' at each edge. The computations are given in the continuation of Table 6-7H and follow the procedure of step 8.

Step 15. At each edge, set up an equation by putting the sum of the angle changes equal to zero. Thus, after dividing through by E/L^2: $\theta_p + \theta'' + \Sigma\theta' = 0$. Solve these simultaneous equations for the unknown moments.

The sum of the angles for edge 2 is

$$8.02 + 0.631m_2 + 0.158m_3 + 3.149m_2 - 1.921m_3 = 0$$

The sum of the angles for edge 3 is

$$-13.90 + 0.316m_2 + 0.631m_3 - 3.854m_2 + 2.460m_3 = 0$$

The solution is $m_2 = -0.040$ and $m_3 = 4.45$.

Step 16. Determine the actual reactions, loads, stresses, and deflections by substituting for m_2 and m_3 the values just found. For example, the final longitudinal stress at edge 0 is 596 $-31.2(-0.040) + 9.2(4.45) = 638$ psi. Similarly, the final longitudinal stress at edge 1 is -145 psi; at edge 2, -160 psi; and at edge 3, -50 psi.

Step 17. Compute the shear stresses. The shear stress at edge n in pounds (Fig. 6-93f) is

$$T_n = T_{n-1} - \frac{f_{n-1} + f_n}{2} A_n \tag{6-129}$$

In the example, $T_0 = 0$; so the shears at the edges can be obtained successively, since the stresses f are known. The calculations are shown in Table 6-7I.

Table 6-7I. Shearing Stresses

Stress	Plate 1		Plate 2		Plate 3	
	Edge 0	Edge 1	Edge 1	Edge 2	Edge 2	Edge 3
$-(f_{n-1} + f_n) A_n/2$		$-83,500$		$49,500$		$34,000$
Midspan edge shear, lb		$-83,500$		$-34,000$		0
Edge shear, psi, $x = 0$	0	-61.2	-143	-58	-58	0
$0.75PL/A$	69.9		148.3		46.0	
0.25 edge shear	0	-15.3	-35.8	-14.5	-14.5	0
Middepth shear, psi	54.6		98.0		31.5	

For a uniformly loaded folded plate, the shearing stress S, psi, at any point on an edge n is approximately

$$S = \frac{2T_{max}}{3Lt}\left(\frac{1}{2} - \frac{x}{L}\right) \tag{6-130}$$

with a maximum at plate ends of

$$S_{max} = \frac{T_{max}}{3Lt} \tag{6-131}$$

The edge shears computed with Eq. (6-131) are given in Table 6-7I.

The shear stress, psi, at middepth (not always a maximum) is

$$v_n = \left(\frac{3P_nL}{2A_n} + \frac{S_{n-1}+S_n}{2}\right)\left(\frac{1}{2} - \frac{x}{L}\right) \tag{6-132}$$

and has its largest value at $x = 0$:

$$v_{max} = \frac{0.75P_nL}{A_n} + \frac{S_{n-1}+S_n}{4} \tag{6-133}$$

The middepth shears also are given in Table 6-7I.

For more details, see D. Yitzhaki and Max Reiss, "Analysis of Folded Plates," *Proceedings Paper 3303, Journal of the Structural Division, American Society of Civil Engineers,* October, 1962.

ULTIMATE STRENGTH OF DUCTILE MEMBERS AND FRAMES

When an elastic material, such as structural steel, is loaded with a gradually increasing load, stresses are proportional to strains up to the yield point. If the material, like steel, also is ductile, then it continues to carry load beyond the yield point, though strains increase rapidly with little increase in load (Fig. 6-94a).

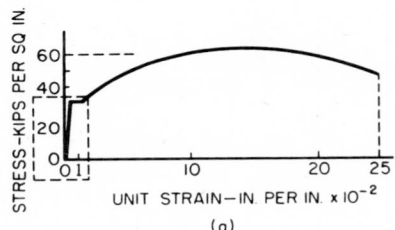
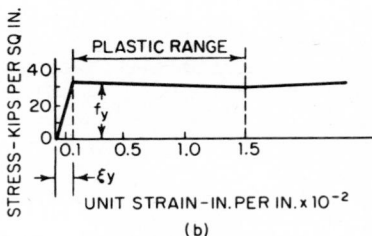

Fig. 6-94. Stress-strain relationship for a ductile material generally is similar to the curve shown in (a). To simplify plastic analysis, the portion of (a) enclosed by the dashed lines is approximated by the curve shown in (b), which extends to the range where strain hardening begins.

Similarly, a beam made of an elastic material continues to carry more load after the stresses in the outer fibers reach the yield point. However, the stresses will no longer vary with distance from the neutral axis; so the flexural formula [Eq. (6-36)] no longer holds. However, if simplifying assumptions are made, approximating the stress-strain relationship beyond the elastic limit, the load-carrying capacity of the beam can be computed with satisfactory accuracy.

6-91. Theory of Plastic Behavior. For a ductile material, the idealized stress-strain relationship in Fig. 6-94b may be assumed. Stress is proportional to strain until the yield-point stress f_y is reached, after which strain increases at a constant stress.

For a beam of this material, the following assumptions will also be made: plane sections remain plane, strains thus being proportional to distance from the neutral axis; properties of the material in tension are the same as those in compression; its fibers behave the same in flexure as in tension; and deformations remain small.

Strain distribution across the cross section of a rectangular beam, based on these assumptions, is shown in Fig. 6-95a. At the yield point, the unit strain is ϵ_y and the curvature ϕ_y, as indicated in (1). In (2), the strain has increased several times, but the section still remains plane. Finally, at failure, (3), the strains are very large and nearly constant across upper and lower halves of the section.

Corresponding stress distributions are shown in Fig. 6-95b. At the yield point (1), stresses vary linearly and the maximum is f_y. With increase in load, more and more fibers reach the yield point, and the stress distribution becomes nearly constant, as indicated in (2). Finally, at failure (3), the stresses are constant across the top and bottom parts of the section and equal to the yield-point stress.

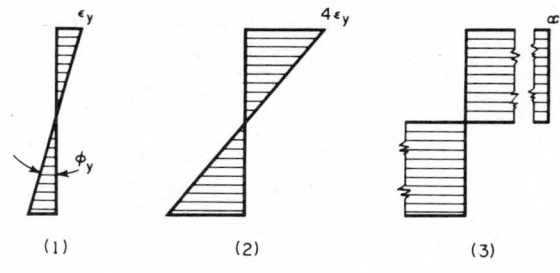

(a) STRAIN DISTRIBUTION

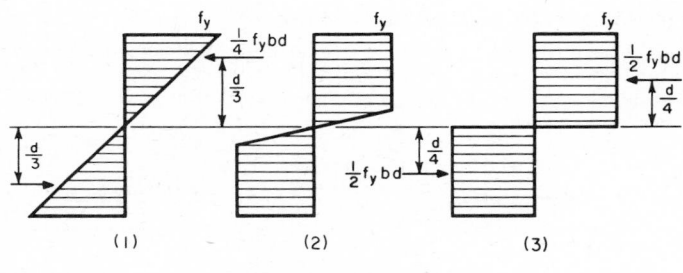

(b) STRESS DISTRIBUTION

Fig. 6-95. Strain distribution is shown in (a) and stress distribution in (b) for a cross section of a beam as it is loaded beyond the yield point, assuming the idealized stress-strain relationship in Fig. 6-94b. Stage (1) shows the conditions at the elastic limit of the outer fibers; (2) after yielding starts; and (3) at ultimate load.

The resisting moment at failure for a rectangular beam can be computed from the stress diagram for stage 3. If b is the width of the member and d its depth, then the ultimate moment for a rectangular beam is

$$M_P = \frac{bd^2}{4} f_y \qquad (6\text{-}134)$$

Since the resisting moment at stage 1 is $M_y = f_y bd^2/6$, the beam carries 50% more moment before failure than when the yield-point stress is first reached in the outer fibers ($M_P/M_y = 1.5$).

A circular section has an M_P/M_y ratio of about 1.7, while a diamond section has a ratio of 2. The average wide-flange rolled-steel beam has a ratio of about 1.14.

The relationship between moment and curvature in a beam can be assumed to be similar to the stress-strain relationship in Fig. 6-94b. Curvature ϕ varies linearly with moment until $M_y = M_P$ is reached, after which ϕ increases indefinitely at constant moment. That is, a **plastic hinge** forms.

This ability of a ductile beam to form plastic hinges enables a fixed-end or continuous beam to carry more load after M_P occurs at a section, because a redistribution of moments takes place. Consider, for example, a uniformly loaded fixed-end beam. In the elastic range, the end moments are $M_L = M_R = WL/12$, while the midspan moment M_C is $WL/24$. The load when the yield point is reached in the outer fibers is $W_y = 12M_y/L$. Under this load, the moment capacity of the ends of the beam is nearly exhausted; plastic hinges form there when the moment equals M_P. As load is increased, the ends then rotate under constant moment and the beam deflects like a simply supported beam. The moment at midspan increases until the moment capacity at that section is exhausted and a plastic hinge forms. The load causing that condition is the ultimate load W_u since, with three hinges in the span, a link mechanism is formed and the member continues to deform at constant load. At the time the third hinge is formed, the moments at ends and center are all equal to M_P. Therefore, for equilibrium, $2M_P = W_u L/8$, from which $W_u = 16M_P/L$. Since, for the idealized moment-curvature relationship, M_P was assumed equal to M_y, the carrying capacity due to redistribution of moments is 33% greater.

6-92. Upper and Lower Bounds for Ultimate Loads. Methods for computing the ultimate strength of continuous beams and frames may be based on two theorems that fix upper and lower limits for load-carrying capacity.

Upper-bound Theorem. A load computed on the basis of an assumed link mechanism will always be greater than or at best equal to the ultimate load.

Lower-bound Theorem. The load corresponding to an equilibrium condition with arbitrarily assumed values for the redundants is smaller than or at best equal to the ultimate loading—provided that everywhere moments do not exceed M_P.

Equilibrium Method. The equilibrium method, based on the lower-bound theorem, usually is easier for simple cases. The steps involved are: (1) Select redundants that if removed would leave the structure determinate. (2) Draw the moment diagram for the determinate structure. (3) Sketch the moment diagram for an arbitrary value of each redundant. (4) Combine the moment diagrams, forming enough peaks so that the structure will act as a link mechanism if plastic hinges are formed at those points. (5) Compute the value of the redundants from the equations of equilibrium, assuming that at the peaks $M = M_P$. (6) See that there are sufficient plastic hinges to form a mechanism and that M is everywhere less than or equal to M_P.

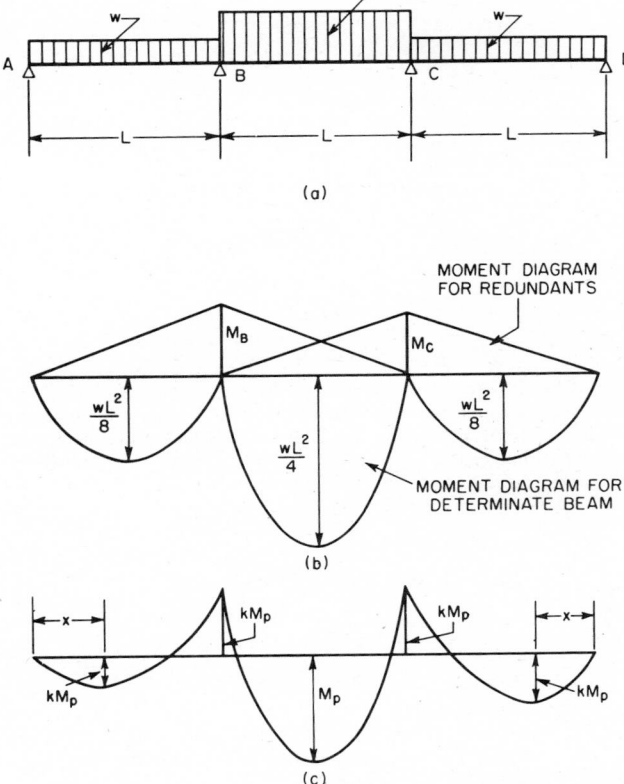

Fig. 6-96. Continuous beam shown in (a) carries twice as much uniform load in the center span as in the side span. In (b) are shown the moment diagrams for this loading condition with redundants removed and for the redundants. The two moment diagrams are combined in (c), producing peaks at which plastic hinges are assumed to form.

Consider, for example, the continuous beam $ABCD$ in Fig. 6-96a with three equal spans, uniformly loaded, the center span carrying double the load of the end spans. Assume that the ratio of the plastic moment for the end spans is k times that for the center span ($k < 1$). For what value of k will the ultimate strength be the same for all spans?

Figure 6-96b shows the moment diagram for the beam made determinate by ignoring the moments at B and C, and the moment diagram for end moments M_B and M_C applied to the determinate beam. Figure 6-96c gives the combined moment diagram. If plastic hinges form at all the peaks, a link mechanism will exist.

Since at B and C the joints can develop only the strength of the weakest beam, a plastic hinge will form when $M_B = M_C = kM_P$. A plastic hinge also will form at the center of span BC when the midspan moment is M_P. For equilibrium to be maintained, this will occur when

$$M_P = wL^2/4 - \tfrac{1}{2}M_B - \tfrac{1}{2}M_C = wL^2/4 - kM_P$$

from which

$$M_P = \frac{wL^2}{4(1 + k)}$$

Maximum moment will occur in the interior of spans AB and CD when

$$x = \frac{L}{2} - \frac{M}{wL} \quad \text{or if } M = kM_P, \text{ when } x = \frac{L}{2} - \frac{kM_P}{wL}$$

A plastic hinge will form at this point when the moment equals kM_P. For equilibrium, therefore,

$$kM_P = \frac{w}{2}\, x(L - x) - \frac{x}{L} kM_P = \frac{w}{2}\left(\frac{L}{2} - \frac{kM_P}{wL}\right)\left(\frac{L}{2} + \frac{kM_P}{wL}\right) - \left(\frac{1}{2} - \frac{kM_P}{wL^2}\right)kM_P$$

This leads to the quadratic equation:

$$\frac{k^2 M_P^2}{wL^2} - 3kM_P + \frac{wL^2}{4} = 0$$

When the value of M_P previously computed is substituted in this equation, it becomes

$$7k^2 + 4k = 4 \quad \text{or} \quad k(k + {}^4/_7) = {}^4/_7$$

from which $k = 0.523$. And the ultimate load is

$$wL = \frac{4M_P(1 + k)}{L} = 6.1\,\frac{M_P}{L}$$

Mechanism Method. The mechanism method, based on the upper-bound theorem, requires the following steps: (1) Determine the location of possible hinges (points of maximum moment). (2) Pick combinations of hinges to form possible link mechanisms. Certain types are classed as elementary, such as the mechanism formed when hinges are created at the ends and center of a fixed-end or continuous beam, or when hinges form at top and bottom of a column subjected to lateral load, or when hinges are formed at a joint in the members framing into it, permitting the joint to rotate freely. If there are n possible plastic hinges and x redundant forces or moments, then there are $n - x$ independent equilibrium equations and $n - x$ independent mechanisms. Both elementary mechanisms and possible combinations of them must be investigated. (3) Apply a virtual displacement to each possible mechanism in turn and compute the internal and external work (Art. 6-52). (4) From the equality of internal and external work, compute the critical load for each mechanism. The mechanism with the lowest critical load is the most probable, and its load is the ultimate, or limit load. (5) Make an equilibrium check to ascertain that moments everywhere are less than or equal to M_P.

As an example of an application of the mechanism method, let us find the ultimate load for the rigid frame of constant section throughout in Fig. 6-97a. Assume that the vertical load at midspan is equal to 1.5 times the lateral load.

Maximum moments can occur at five points—A, B, C, D, and E—and plastic hinges may form there. Since this structure has three redundants, the number of elementary mechanisms is $5 - 3 = 2$. These are shown in Fig. 6-97b and c. A combination of these mechanisms is shown in Fig. 6-97d.

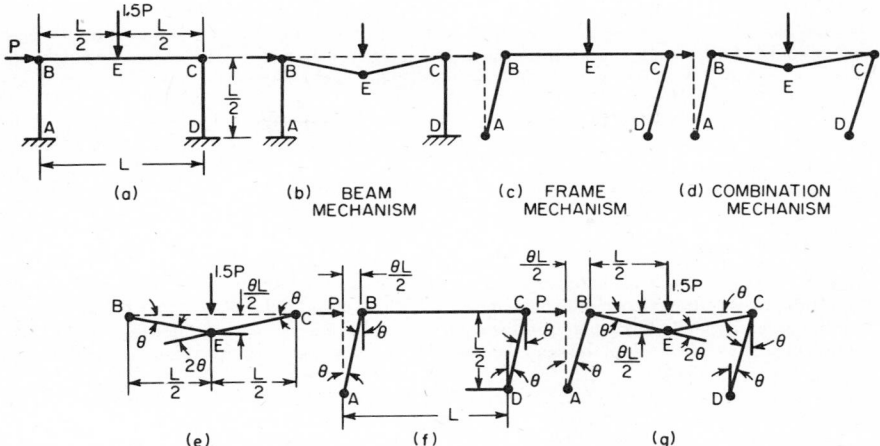

Fig. 6-97. Ultimate-load possibilities for a rigid frame of constant section with fixed bases.

Let us first investigate the beam alone, with plastic hinges at B, E, and C. As indicated in Fig. 6-97e, apply a virtual rotation θ to BE at B. The center deflection, then, is $\theta L/2$. Since C is at a distance of $L/2$ from E, the rotation at C must be θ and at E, 2θ. The external work, therefore, is $1.5P$ times the midspan deflection, $\theta L/2$, and equals $\frac{3}{4}\theta PL$. The internal work is the sum of the work at each hinge, or

$$M_P\theta + 2M_P\theta + M_P\theta = 4M_P\theta$$

Equating internal and external work:

$$\tfrac{3}{4}\theta PL = 4M_P\theta$$

from which $P = 5.3M_P/L$.

Next, let us investigate the frame mechanism, with plastic hinges at A, B, C, and D. As shown in Fig. 6-97f, apply a virtual rotation θ to AB. The point of application of the lateral load P will then move a distance $\theta L/2$, and the external work will be $\theta PL/2$. The internal work, being the sum of the work at the hinges, will be $4M_P\theta$. Equating internal and external work,

$$\theta PL/2 = 4M_P\theta$$

from which $P = 8M_P/L$.

Finally, let us compute the critical load for the combination beam-frame mechanism in Fig. 6-97g. Again, apply a virtual rotation θ to AB, moving B horizontally and E vertically a distance of $\theta L/2$. The external work, therefore, equals

$$\frac{\theta PL}{2} + \frac{3}{4}\theta PL = \frac{5\theta PL}{4}$$

The internal work is the sum of the work at hinges A, E, C, and D and is equal to

$$M_P\theta + 2M_P\theta + 2M_P\theta + M_P\theta = 6M_P\theta$$

Equating internal and external work,

$$5\theta PL/4 = 6M_P\theta$$

from which $P = 4.8M_P/L$.

The combination mechanism has the lowest critical load. Now an equilibrium check must be made to insure that the moments everywhere in the frame are less than M_P. If they are, then the combination mechanism is the correct solution.

Consider first the length EC of the beam as a free body. Taking moments about E, noting that the moments at E and C equal M_P, the shear at C is computed to be $4M_P/L$. Therefore, the shear at B is $1.5 \times 4.8M_P/L - 4M_P/L = 3.2M_P/L$.   By taking moments about B, with BE considered as a free body, the moment at B is found to be $0.6M_P$.   Similar treatment of the columns as free bodies indicates that nowhere is the moment greater than M_P.   Therefore, the combination mechanism is the correct solution, and the ultimate load for the frame is $4.8M_P/L$ laterally and $7.2M_P/L$ vertically at midspan.

(R. O. Disque, "Applied Plastic Design in Steel," Van Nostrand Reinhold Company, New York; "Plastic Design in Steel—A Guide and Commentary," M & R No. 41, American Society of Civil Engineers, New York; M. R. Horne, "Plastic Theory of Structures," The MIT Press, Cambridge, Mass.)

STRUCTURAL DYNAMICS

Article 6-1 noted that loads can be classified as static or dynamic and that the distinguishing characteristic was the rate of application of load.   If a load is applied slowly, it may be considered static.   Since dynamic loads may produce stresses and deformations considerably larger than those caused by static loads of the same magnitude, it is important to know reasonably accurately what is meant by slowly.

A useful definition can be given in terms of the natural period of vibration of the structure or member to which the load is applied.   If the time in which a load rises from zero to its maximum value is more than double the natural period, the load may be treated as static.   Loads applied more rapidly may be dynamic.   Structural analysis and design for such loads are considerably different from and more complex than those for static loads.

In general, exact dynamic analysis is possible only for relatively simple structures, and only when both the variation of load and resistance with time are a convenient mathematical function. Therefore, in practice, adoption of approximate methods that permit rapid analysis and design is advisable.   And usually, because of uncertainties in loads and structural resistance, computations need not be carried out with more than a few significant figures, to be consistent with known conditions.

6-93. Properties of Materials under Dynamic Loading.   In general, mechanical properties of structural materials improve with increasing rate of load application.   For low-carbon steel, for example, yield strength, ultimate strength, and ductility rise with increasing rate of strain. Modulus of elasticity in the elastic range, however, is unchanged.   For concrete, the dynamic ultimate strength in compression may be much greater than the static strength.

Since the improvement depends on the material and the rate of strain, values to use in dynamic analysis and design should be determined by tests approximating the loading conditions anticipated.

Under many repetitions of loading, though, a member or connection between members may fail because of "fatigue" at a stress smaller than the yield point of the material.   In general, there is little apparent deformation at the start of a fatigue failure.   A crack forms at a point of high stress concentration.   As the stress is repeated, the crack slowly spreads, until the member ruptures without measurable yielding.   Though the material may be ductile, the fracture looks brittle.

Endurance Limit.   Some materials (generally those with a well-defined yield point) have what is known as an **endurance limit**.   This is the maximum unit stress that can be repeated, through a definite range, an indefinite number of times without causing structural damage. Generally, when no range is specified, the endurance limit is intended for a cycle in which the stress is varied between tension and compression stresses of equal value.   For a different range, if f is this endurance limit, f_y the yield point, and r the ratio of the minimum stress to the maximum, then the relationship between endurance stresses is given approximately by

$$f_{max} = \frac{2f}{(1 - r) + (f/f_y)(1 + r)} \qquad (6\text{-}135)$$

A range of stress may be resolved into two components—a steady, or mean, stress and an alternating stress.   The endurance limit sometimes is defined as the maximum value of the alternating stress that can be superimposed on the steady stress an indefinitely large number of times without causing fracture.   If f is the endurance limit for completely reversed stresses, s the steady unit stress, and f_u the ultimate tensile stress, then the alternating stress may be obtained from a relationship of the type

$$f_a = f\left(1 - \frac{s^n}{f_u}\right) \qquad (6\text{-}136)$$

where n lies between 1 and 2, depending on the mechanical properties of the material.

Improvement of Fatigue Strength. Design of members to resist repeated loading cannot be executed with the certainty with which members can be designed to resist static loading. Stress concentrations may be present for a wide variety of reasons, and it is not practicable to calculate their intensities. But sometimes it is possible to improve the fatigue strength of a material or to reduce the magnitude of a stress concentration below the minimum value that will cause fatigue failure.

In general, avoid design details that cause severe stress concentrations or poor stress distribution. Provide gradual changes in section. Eliminate sharp corners and notches. Do not use details that create high localized constraint. Locate unavoidable stress raisers at points where fatigue conditions are the least severe. Place connections at points where stress is low and fatigue conditions are not severe. Provide structures with multiple load paths or redundant members, so that a fatigue crack in any one of the several primary members is not likely to cause collapse of the entire structure.

Fatigue strength of a material may be improved by cold-working the material in the region of stress concentration, by thermal processes, or by prestressing it in such a way as to introduce favorable internal stresses. Where fatigue stresses are unusually severe, special materials may have to be selected with high energy absorption and notch toughness.

(J. H. Faupel, "Engineering Design," John Wiley & Sons, Inc., New York; C. H. Norris et al., "Structural Design for Dynamic Loads," McGraw-Hill Book Company, New York; W. H. Munse, "Fatigue of Welded Steel Structures," Welding Research Council, 345 East 47th Street, New York, N.Y. 10017; Almen and Black, "Residual Stresses and Fatigue in Metal," McGraw-Hill Book Company, New York.)

6-94. Natural Period of Vibration. A preliminary step in dynamic analysis and design is determination of this period. It can be computed in many ways, including by application of the laws of conservation of energy and momentum or Newton's second law of motion, $F = M(dv/dt)$, where F is force, M mass, v velocity, and t time. But in general, an exact solution is possible only for simple structures. Therefore, it is general practice to seek an approximate —but not necessarily inexact—solution by analyzing an idealized representation of the actual member or structure. Setting up this model and interpreting the solution requires judgment of a high order.

Natural period of vibration is the time required for a structure to go through one cycle of free vibration, that is, vibration after the disturbance causing the motion has ceased.

To compute the natural period, the actual structure may be conveniently represented by a system of masses and massless springs, with additional resistances provided to account for energy losses due to friction, hysteresis, and other forms of damping. In simple cases, the masses may be set equal to the actual masses; otherwise, equivalent masses may have to be computed (Art. 6-99). The spring constants are the ratios of forces to deflections.

For example, a single mass on a spring (Fig. 6-98b) may represent a simply supported beam with mass that may be considered negligible compared with the load W at midspan (Fig. 6-98a). The spring constant k should be set equal to the load that produces a unit deflection at midspan; thus, $k = 48EI/L^3$, where E is the modulus of elasticity, psi; I the moment of inertia, in.[4]; and L the span, in., of the beam. The idealized mass equals W/g, where g is the acceleration due to gravity, 386 in. per sec².

Also, a single mass on a spring (Fig. 6-98d) may represent the rigid frame in Fig. 6-98c. In that case, $k = 2 \times 12EI/h^3$, where I is the moment of inertia, in.[4], of each column and h the column height, in. The idealized mass equals the sum of the masses on the girder and the girder mass. (Weight of columns and walls is assumed negligible.)

Degree of a System. The spring and mass in Fig. 6-98b and d form a one-degree system. The degree of freedom of a system is determined by the least number of coordinates needed to define the positions of its components. In Fig. 6-98, only the coordinate y is needed to locate the mass and determine the state of the spring. In a two-degree system, such as one comprising two masses connected to each other and to the ground by springs and capable of movement in only one direction, two coordinates are required to locate the masses.

One-degree System. If the mass with weight W, lb, in Fig. 6-98 is isolated, as shown in Fig. 6-98e, it will be in dynamic equilibrium under the action of the spring force $-ky$ and the inertia force $(d^2y/dt^2)(W/g)$. Hence, the equation of motion is

$$\frac{W}{g}\frac{d^2y}{dt^2} + ky = 0 \tag{6-137}$$

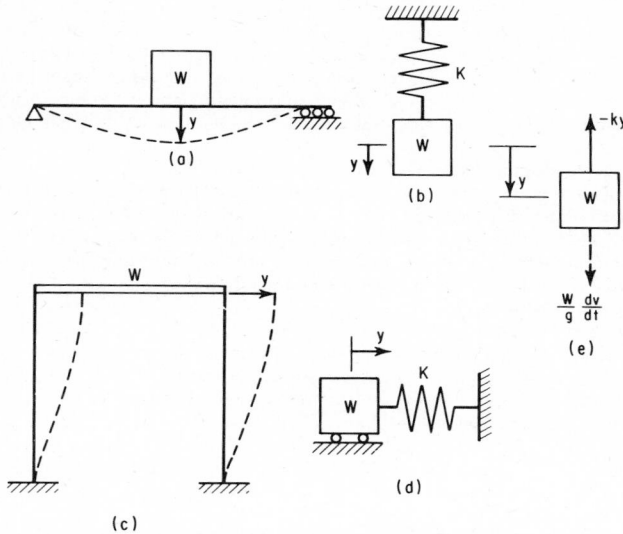

Fig. 6-98. Mass on weightless spring (b) or (d) may represent the motion of a beam (a) or a rigid frame (b) in free vibration.

This may be written in the more convenient form

$$\frac{d^2y}{dt^2} + \frac{kg}{W}\, y = \frac{d^2y}{dt^2} + \omega^2 y = 0 \tag{6-138}$$

The solution is

$$y = A \sin \omega t + B \cos \omega t \tag{6-139}$$

where A and B are constants to be determined from initial conditions of the system, and

$$\omega = \sqrt{\frac{kg}{W}} \tag{6-140}$$

is the **natural circular frequency,** radians per second.

The motion defined by Eq. (6-139) is harmonic. Its natural period in seconds is

$$T = \frac{2\pi}{\omega} = 2\pi \sqrt{\frac{W}{gk}} \tag{6-141}$$

Its **natural frequency** in cycles per second is

$$f = \frac{1}{T} = \frac{1}{2\pi} \sqrt{\frac{kg}{W}} \tag{6-142}$$

If, at time $t = 0$, the mass has an initial displacement y_0 and velocity v_0, substitution in Eq. (6-130) yields $A = v_0/\omega$ and $B = y_0$. Hence, at any time t, the mass is completely located by

$$y = \frac{v_0}{\omega} \sin \omega t + y_0 \cos \omega t \tag{6-143}$$

The stress in the spring can be computed from the displacement y, because the spring force equals $-ky$.

Multi-degree Systems.   In multiple-degree systems, an independent differential equation of motion can be written for each degree of freedom. Thus, in an N-degree system with N masses, weighing W_1, W_2, . . . , W_N lb, and N^2 springs with constants k_{rj} ($r = 1, 2, \ldots, N$; $j = 1, 2, \ldots, N$), there are N equations of the form

$$\frac{W_r}{g}\frac{d^2y_r}{dt^2} + \sum_{j=1}^{N} k_{rj}y_j = 0 \qquad r = 1, 2, \ldots, N \tag{6-144}$$

Simultaneous solution of these equations reveals that the motion of each mass can be resolved into N harmonic components. They are called the fundamental, second, third, etc., harmonics. Each set of harmonics for all the masses is called a **normal mode** of vibration.

There are as many normal modes in a system as degrees of freedom. Under certain circumstances, the system could vibrate freely in any one of these modes. During any such vibration, the ratio of displacement of any two of the masses remains constant. Hence, the solutions of Eqs. (6-144) take the form

$$y_r = \sum_{n=1}^{N} a_{rn} \sin \omega_n(t + \tau_n) \tag{6-145}$$

where a_{rn} and τ_n are constants to be determined from the initial conditions of the system and ω_n is the natural circular frequency for each normal mode.

Natural Periods. To determine ω_n, set $y_1 = A_1 \sin \omega t$; $y_2 = A_2 \sin \omega t$ .... Then, substitute these and their second derivatives in Eqs. (6-144). After dividing each equation by $\sin \omega t$, the following N equations result:

$$\left(k_{11} - \frac{W_1}{g}\omega^2\right)A_1 + k_{12}A_2 + \cdots + k_{1N}A_N = 0$$

$$k_{21}A_1 + \left(k_{22} - \frac{W_2}{g}\omega^2\right)A_2 + \cdots + k_{2N}A_N = 0 \tag{6-146}$$

$$\cdots\cdots\cdots\cdots\cdots\cdots\cdots\cdots\cdots\cdots\cdots\cdots$$

$$k_{N1}A_1 + k_{N2}A_2 + \cdots + \left(k_{NN} - \frac{W_N}{g}\omega^2\right)A_N = 0$$

If there are to be nontrivial solutions for the amplitudes $A_1, A_2, \ldots, A_N$, the determinant of their coefficients must be zero. Thus,

$$\begin{vmatrix} k_{11} - \dfrac{W_1}{g}\omega^2 & k_{12} & \cdots & k_{1N} \\[2ex] k_{21} & k_{22} - \dfrac{W_2}{g}\omega^2 & \cdots & k_{2N} \\[2ex] \cdots\cdots\cdots\cdots\cdots\cdots\cdots\cdots\cdots\cdots \\[2ex] k_{N1} & k_{N2} & \cdots & k_{NN} - \dfrac{W_N}{g}\omega^2 \end{vmatrix} = 0 \tag{6-147}$$

Solution of this equation for ω yields one real root for each normal mode. And the natural period for each normal mode can be obtained from Eq. (6-141).

Modal Amplitudes. If ω for a normal mode now is substituted in Eqs. (6-146), the amplitudes $A_1, A_2, \ldots, A_N$ for that mode can be computed in terms of an arbitrary value, usually unity, assigned to one of them. The resulting set of modal amplitudes defines the **characteristic shape** for that mode.

The normal modes are mutually orthogonal; that is,

$$\sum_{r=1}^{N} W_r A_{rn} A_{rm} = 0 \tag{6-148}$$

where W_r is the rth mass out of a total of N, A represents the characteristic amplitude of a normal mode, and n and m identify any two normal modes. Also, for a total of S springs

$$\sum_{s=1}^{S} k_s y_{sn} y_{sm} = 0 \tag{6-149}$$

where k_s is the constant for the sth spring and y represents the spring distortion.

Stodola-Vinanello Method. When there are many degrees of freedom, the preceding procedure for free vibration becomes very lengthy. In such cases, it may be preferable to solve Eqs.

(6-146) by numerical, trial-and-error procedures, such as the Stodola-Vianello Method. In that method, the solution converges first on the highest or lowest mode. Then, the other modes are determined by the same procedure after elimination of one of the equations by use of Eq. (6-148). The procedure requires assumption of a characteristic shape, a set of amplitudes A_{r1}. These are substituted in one of Eqs. (6-146) to obtain a first approximation of ω^2. With this value and with $A_{N1} = 1$, the remaining $(N - 1)$ equations are solved to obtain a new set of A_{r1}. Then, the procedure is repeated until assumed and final characteristic amplitudes agree.

Rayleigh Method. Because even the Stodola-Vianello method is lengthy for many degrees of freedom, the Rayleigh approximate method may be used to compute the fundamental mode. The frequency obtained by this method, however, may be a little on the high side.

The Rayleigh method also starts with an assumed set of characteristic amplitudes A_{r1} and depends for its success on the small error in natural frequency produced by a relatively larger error in the shape assumption. Next, relative inertia forces acting at each mass are computed: $F_r = W_r A_{r1}/A_{N1}$, where A_{N1} is the assumed displacement at one of the masses. These forces are applied to the system as a static load and displacements B_{r1} due to them calculated. Then, the natural frequency can be obtained from

$$\omega^2 = \frac{g \sum_{r=1}^{N} F_r B_{r1}}{\sum_{r=1}^{N} W_r B_{r1}^2} \tag{6-150}$$

where g is the acceleration due to gravity, 386 in. per sec². For greater accuracy, the computation can be repeated with B_{r1} as the assumed characteristic amplitudes.

When the Rayleigh method is applied to beams, the characteristic shape assumed initially may be chosen conveniently as the deflection curve for static loading.

The Rayleigh method may be extended to determination of higher modes by the Schmidt orthogonalization procedure, which adjusts assumed deflection curves to satisfy Eq. (6-148). The procedure is to assume a shape, remove components associated with lower modes, then use the Rayleigh method for the residual deflection curve. The computation will converge on the next higher mode. The method is shorter than the Stodola-Vianello procedure when only a few modes are needed.

For example, suppose the characteristic amplitudes A_{r1} for the fundamental mode have been obtained, and the natural frequency for the second mode is to be computed. Assume a value for the relative deflection of the rth mass A_{r2}. Then, the shape with the fundamental mode removed will be defined by the displacements

$$a_{r2} = A_{r2} - c_1 A_{r1} \tag{6-151}$$

where c_1 is the participation factor for the first mode.

$$c_1 = \frac{\sum_{r=1}^{N} W_r A_{r2} A_{r1}}{\sum_{r=1}^{N} W_r A_{r1}^2} \tag{6-152}$$

Substitute a_{r2} for B_{r1} in Eq. (6-150) to find the second-mode frequency and, from deflections produced by $F_r = W_r a_{r2}$, an improved shape. (For more rapid convergence, A_{r2} should be selected to make c_1 small.) The procedure should be repeated, starting with the new shape.

For the third mode, assume deflections A_{r3} and remove the first two modes:

$$a_{r3} = A_{r3} - c_1 A_{r1} - c_2 A_{r2} \tag{6-153}$$

The participation factors are determined from

$$c_1 = \frac{\sum_{r=1}^{N} W_r A_{r3} A_{r1}}{\sum_{r=1}^{N} W_r A_{r1}^2} \qquad c_2 = \frac{\sum_{r=1}^{N} W_r A_{r3} A_{r2}}{\sum_{r=1}^{N} W_r A_{r2}^2} \tag{6-154}$$

Use a_{r3} to find an improved shape and the third-mode frequency.

Distributed Mass. For some structures with mass distributed throughout, it sometimes is easier to solve the dynamic equations based on distributed mass than the equations based on equivalent lumped masses. A distributed mass has an infinite number of degrees of freedom and normal modes. Every particle in it can be considered a lumped mass on springs connected to other particles. Usually, however, only the fundamental mode is significant, though sometimes the second and third modes must be taken into account.

For example, suppose a beam weighs w lb per lin ft and has a modulus of elasticity E, psi, and moment of inertia I, in.^{4}. Let y be the deflection at a distance x from one end. Then, the equation of motion is

$$EI \frac{\partial^4 y}{\partial x^4} + \frac{w}{g} \frac{\partial^2 y}{\partial t^2} = 0 \qquad (6\text{-}155)$$

(This equation ignores the effects of shear and rotational inertia.) The deflection y_n for each mode, to satisfy the equation, must be the product of a harmonic function of time $f_n(t)$ and of the characteristic shape $Y_n(x)$, a function of x with undetermined amplitude. The solution is

$$f_n(t) = c_1 \sin \omega_n t + c_2 \cos \omega_n t \qquad (6\text{-}156a)$$

where ω_n is the natural circular frequency and n indicates the mode, and

$$Y_n(x) = A_n \sin \beta_n x + B_n \cos \beta_n x + C_n \sinh \beta_n x + D_n \cosh \beta_n x \qquad (6\text{-}156b)$$

where

$$\beta_n = \sqrt[4]{\frac{w \omega_n{}^2}{EIg}} \qquad (6\text{-}156c)$$

Simple Beam. For a simple beam, the boundary (support) conditions for all values of time t are $y = 0$ and bending moment $M = EI\, \partial^2 y / \partial x^2 = 0$. Hence, at $x = 0$ and $x = L$, the span length, $Y_n(x) = 0$ and $d^2 Y_n / dx^2 = 0$. These conditions require that

$$B_n = C_n = D_n = 0$$

and $\beta_n = n\pi/L$, to satisfy Eq. (6-156b). Hence, according to Eq. (6-156c), the natural circular frequency for a simply supported beam is

$$\omega_n = \frac{n^2 \pi^2}{L^2} \sqrt{\frac{EIg}{w}} \qquad (6\text{-}157)$$

The characteristic shape is defined by

$$Y_n(x) = \sin \frac{n\pi x}{L} \qquad (6\text{-}158)$$

The constants c_1 and c_2 in Eq. (6-156a) are determined by the initial conditions of the disturbance. Thus, the total deflection, by superposition of modes, is

$$y = \sum_{n=1}^{\infty} A_n(t) \sin \frac{n\pi x}{L} \qquad (6\text{-}159)$$

where $A_n(t)$ is determined by the load (see Art. 6-97).

Equations (6-156) apply to spans with any type of end restraints. Figure 6-99 shows the characteristic shape and gives constants for determination of natural circular frequency ω and natural period T for the first four modes of cantilever, simply supported, fixed-end, and fixed-hinged beams. To obtain ω, select the appropriate constant from Fig. 6-99 and multiply it by $\sqrt{EI/wL^4}$. To get T, divide the appropriate constant by $\sqrt{EI/wL^4}$.

To determine the characteristic shapes and natural periods for beams with variable cross section and mass, use the Rayleigh method. Convert the beam into a lumped-mass system by dividing the span into elements and assuming the mass of each element to be concentrated at its center. Also, compute all quantities, such as deflection and bending moment, at the center of each element. Start with an assumed characteristic shape and apply Eq. (6-150).

Methods are available for dynamic analysis of continuous beams. (G. S. Rogers, "An Introduction to the Dynamics of Framed Structures," John Wiley & Sons, Inc., New York; D. G. Fertis and E. C. Zobel, "Transverse Vibration Theory," The Ronald Press Company, New York.) But even for beams with constant cross section, these procedures are very lengthy. Generally, approximate solutions are preferable.

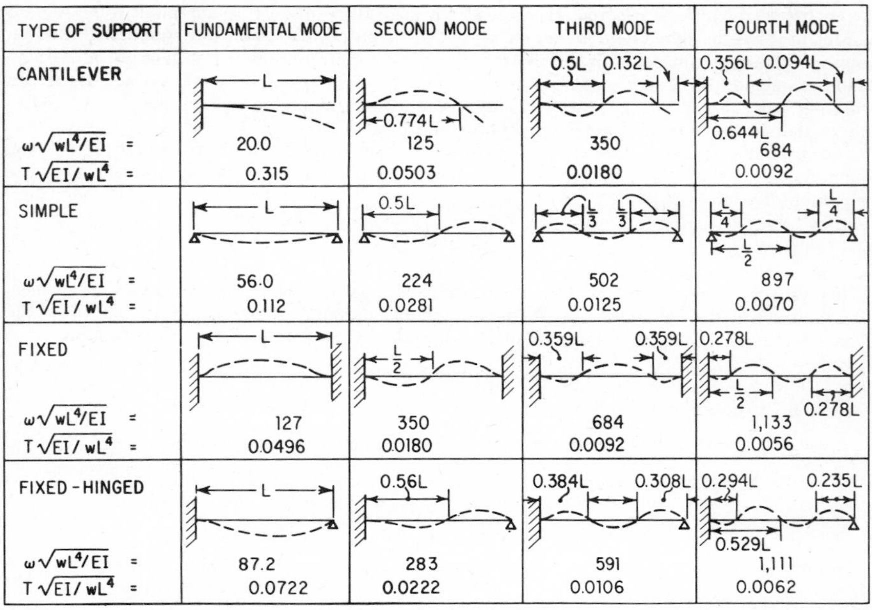

TYPE OF SUPPORT	FUNDAMENTAL MODE	SECOND MODE	THIRD MODE	FOURTH MODE
CANTILEVER			0.5L 0.132L	0.356L 0.094L
		←0.774L→		0.644L
$\omega\sqrt{wL^4/EI}$ =	20.0	125	350	684
$T\sqrt{EI/wL^4}$ =	0.315	0.0503	0.0180	0.0092
SIMPLE	L	0.5L	$\frac{L}{3}$ $\frac{L}{3}$	$\frac{L}{4}$ $\frac{L}{4}$
				$\frac{L}{2}$
$\omega\sqrt{wL^4/EI}$ =	56.0	224	502	897
$T\sqrt{EI/wL^4}$ =	0.112	0.0281	0.0125	0.0070
FIXED	L	$\frac{L}{2}$	0.359L 0.359L	0.278L
				$\frac{L}{2}$ 0.278L
$\omega\sqrt{wL^4/EI}$ =	127	350	684	1,133
$T\sqrt{EI/wL^4}$ =	0.0496	0.0180	0.0092	0.0056
FIXED–HINGED	L	0.56L	0.384L 0.308L	0.294L 0.235L
				0.529L
$\omega\sqrt{wL^4/EI}$ =	87.2	283	591	1,111
$T\sqrt{EI/wL^4}$ =	0.0722	0.0222	0.0106	0.0062

Fig. 6-99. Coefficients for computing natural circular frequencies and natural periods of vibration of prismatic beams.

(J. M. Biggs, "Introduction to Structural Dynamics," McGraw-Hill Book Company, New York; N. M. Newmark and E. Rosenblueth, "Fundamentals of Earthquake Engineering," Prentice-Hall, Inc., Englewood Cliffs, N. J.)

6-95. Lagrange's Equation. This is a basic tool for analyzing dynamic systems. It applies to both lumped-mass and distributed-mass systems. But it is a device for writing the equations of motion and not a method of solution. The usual form of Lagrange's equation is

$$\frac{d}{dt}\left(\frac{\partial K}{\partial q_i'}\right) - \frac{\partial K}{\partial q_i} + \frac{\partial U}{\partial q_i} - \frac{\partial W_c}{\partial q_i} = \frac{\partial W_e}{\partial q_i} \tag{6-160}$$

where q_i = a generalized coordinate (displacement in x, y, or z directions, angular rotation about x, y, or z axes)

q_i' = derivative of q_i with respect to time

t = time

K = kinetic energy

U = strain energy

W_c = work done by damping forces

W_e = work done by external forces

Equation (6-160) provides one equation of motion for each coordinate.

For example, for a one-degree system, there is one coordinate, $q = y$. The kinetic energy $K = \frac{1}{2}M(dy/dt)^2$, where M = mass. Strain energy $U = \frac{1}{2}ky^2$, where k is the spring constant (Art. 6-94). If the damping force is assumed proportional to velocity, the work done $W_c = (-c\,dy/dt)y$, where c is the damping coefficient. If a force varying with time $F(t)$ is applied to the system, $W_e = F(t)y$. Noting that $\partial K/\partial y = 0$, since kinetic energy is a function of velocity, not displacement, substitution in Eq. (6-160) yields the equation of motion

$$\frac{d}{dt}\frac{dy}{dt}M + ky + c\frac{dy}{dt} = F(t)$$

6-96. Impact and Sudden Loads. Under impact, there is an abrupt exchange or absorption of energy and drastic change in velocity. Stresses caused in the colliding members may be several times larger than stresses produced by the same weights applied statically.

An approximation of impact stresses in the elastic range can be made by neglecting the inertia of the body struck and the effect of wave propagation and assuming that the kinetic energy is converted completely into strain energy in that body.   Consider a prismatic bar subjected to an axial impact load in tension.   The energy absorbed per unit of volume when the bar is stressed to the proportional limit is called the **modulus of resilience.**   It is given by $f_y^2/2E$, where f_y is the yield stress and E the modulus of elasticity, both in psi.   Below the proportional limit, the stress, psi, due to an axial load U, in.-lb, is

$$f = \sqrt{\frac{2UE}{AL}} \tag{6-161}$$

where A is the cross-sectional area, sq in., and L the length of bar, in.

This equation indicates that energy absorption of a member may be improved by increasing its length or area.   Sharp changes in cross section should be avoided, however, because of associated high stress concentrations.   Also, uneven distribution of stress in a member due to changes in section should be avoided.   For example, if part of a member is given twice the diameter of another part, the stress in the larger portion is one-fourth that in the smaller.   Since the energy absorbed is proportional to the square of the stress, the energy taken per unit of volume by the larger portion is therefore only one-sixteenth that absorbed by the smaller.   So despite the increase in volume due to doubling of the diameter, the larger portion absorbs much less energy than the smaller.   Thus, energy absorption would be larger with a uniform stress distribution throughout the length of the member.

If a static axial load W would produce a tensile stress f' in the bar and an elongation e', in., then the axial stress produced when W falls a distance h, in., is

$$f = f' + f' \sqrt{1 + \frac{2h}{e'}} \tag{6-162}$$

if f is within the proportional limit.   The elongation due to this impact load is

$$e = e' + e' \sqrt{1 + \frac{2h}{e'}} \tag{6-163}$$

These equations indicate that the stress and deformation due to an energy load may be considerably larger than those produced by the same weight applied gradually.

The same equations hold for a beam with constant cross section struck by a weight at midspan, except that f and f' represent stresses at midspan and e and e', midspan deflections.

According to Eqs. (6-162) and (6-163), a sudden load ($h = 0$) causes twice the stress and twice the deflection as the same load applied gradually.

For very long members, the effect of wave propagation should be taken into account.   Impact is not transmitted instantly to all parts of the struck body.   At first, remote parts remain undisturbed, while particles struck accelerate rapidly to the velocity of the colliding body.   The deformations produced move through the struck body in the form of elastic waves.   The waves travel with a constant velocity, ft per sec,

$$c = 68.1 \sqrt{\frac{E}{\rho}} \tag{6-164}$$

where E = modulus of elasticity, psi
ρ = density of the struck body, lb per cu ft

Impact Waves.   If an impact imparts a velocity v, ft per sec, to the particles at one end of a prismatic bar, the stress, psi, at that end is

$$f = 0.0147v \sqrt{E\rho} \tag{6-165}$$

if f is in the elastic range.   In a compression wave, the velocity of the particles is in the direction of the wave.   In a tension wave, the velocity of the particles is in the opposite direction to the wave.

In the plastic range, Eqs. (6-164) and (6-165) hold, but with E as the tangent modulus of elasticity.   Hence, c is not a constant and the shape of the stress wave changes as it moves.   The elastic portion of the stress wave moves faster than the wave in the plastic range.   Where they overlap, the stress and irrecoverable strain are constant.

(The impact theory is based on an assumption difficult to realize in practice—that contact takes place simultaneously over the entire end of the bar.)

At a free end of a bar, a compressive stress wave is reflected as an equal tension wave, and a tension wave as an equal compression wave. The velocity of the particles at the free end equals $2v$.

At a fixed end of a bar, a stress wave is reflected unchanged. The velocity of the particles at the fixed end is zero, but the stress is doubled, because of the superposition of the two equal stresses on reflection.

For a bar with a fixed end struck at the other end by a moving mass weighing W_m lb, the initial compressive stress, psi, is, from Eq. (6-165),

$$f_o = 0.0147 v_o \sqrt{E\rho} \qquad (6\text{-}166)$$

where v_o is the initial velocity of the particles, ft per sec, at the impacted end of the bar and E and ρ the modulus of elasticity, psi, and density, lb per cu ft, of the bar. As the velocity of W_m decreases, so does the pressure on the bar. Hence, decreasing compressive stresses follow the wave front. At any time $t < 2L/c$, where L is the length of the bar, in., the stress at the struck end is

$$f = f_o e^{-2\alpha t/\tau} \qquad (6\text{-}167)$$

where $e = 2.71828$; α is the ratio of W_b, the weight of the bar, to W_m; and $\tau = 2L/c$.

When $t = \tau$, the wave front with stress f_o arrives back at the struck end, assumed still to be in contact with the mass. Since the velocity of the mass cannot change suddenly, the wave will be reflected as from a fixed end. During the second interval, $\tau < t < 2\tau$, the compressive stress is the sum of two waves moving away from the struck end and one moving toward this end.

Maximum stress from impact occurs at the fixed end. For α greater than 0.2, this stress is

$$f = 2f_o(1 + e^{-2\alpha}) \qquad (6\text{-}168a)$$

For smaller values of α, it is given approximately by

$$f = f_o \left(1 + \sqrt{\frac{1}{\alpha}}\right) \qquad (6\text{-}168b)$$

Duration of impact, time it takes for the stress at the struck end to drop to zero, is approximately

$$T = \frac{\pi L}{c\sqrt{\alpha}} \qquad (6\text{-}169)$$

for small values of α.

When W_m is the weight of a falling body, velocity at impact is $\sqrt{2gh}$, when it falls a distance h, in. Substitution in Eq. (6-166) yields $f_o = \sqrt{2EhW_b/AL}$, since $W_b = \rho AL$ is the weight of the bar. Putting $W_b = \alpha W_m$; $W_m/A = f'$, the stress produced by W_m when applied gradually, and $E = f'L/e'$, where e' is the elongation for the static load, gives $f_o = f' \sqrt{2h\alpha/e'}$. Then, for values of α smaller than 0.2, the maximum stress, from Eq. (6-168b), is

$$f = f' \left(\sqrt{\frac{2h\alpha}{e'}} + \sqrt{\frac{2h}{e'}}\right) \qquad (6\text{-}170)$$

For larger values of α, the stress wave due to gravity acting on W_m during impact should be added to Eq. (6-168a). Thus, for α larger than 0.2,

$$f = 2f'(1 - e^{-2\alpha}) + 2f' \sqrt{\frac{2h\alpha}{e'}} (1 + e^{-2\alpha}) \qquad (6\text{-}171)$$

Equations (6-170) and (6-171) correspond to Eq. (6-162), which was developed without taking wave effects into account. For a sudden load, $h = 0$, Eq. (6-171) gives for the maximum stress $2f'(1 - e^{-2\alpha})$, not quite double the static stress, the result indicated by Eq. (6-162). (See also Art. 6-97.)

(S. Timoshenko and J. N. Goodier, "Theory of Elasticity," McGraw-Hill Book Company, New York; S. Timoshenko and D. H. Young, "Engineering Mechanics," McGraw-Hill Book Company, New York; D. D. Barkan, "Dynamics of Bases and Foundations," McGraw-Hill Book Company, New York; J. H. Faupel, "Engineering Design," John Wiley & Sons, Inc., New York.)

6-97. Dynamic Analysis of Simple Structures.   (See also Arts. 6-93 to 6-96.) As noted in Art. 6-94, an approximate solution based on an idealized representation of an actual member or structure is advisable for dynamic analysis and design.   Generally, the actual structure may be conveniently represented by a system of masses and massless springs, with additional resistances to account for damping. In simple cases, the masses may be set equal to the actual masses; otherwise, equivalent masses may be substituted for the actual masses (Art. 6-99).   The spring constants are the ratios of forces to deflections (see Art. 6-94).

Usually, for structural purposes, the data sought are the maximum stresses in the springs and their maximum displacements and the time of occurrence of the maximums. This time generally is computed in terms of the natural period of vibration of the member or structure or in terms of the duration of the load. Maximum displacement may be calculated in terms of the deflection that would result if the load were applied gradually.

The term D by which the static deflection e', spring forces, and stresses are multiplied to obtain the dynamic effects is called the **dynamic load factor.** Thus, the dynamic displacement is

$$y = De' \tag{6-172}$$

and the maximum displacement y_m is determined by the maximum dynamic load factor D_m, which occurs at time t_m.

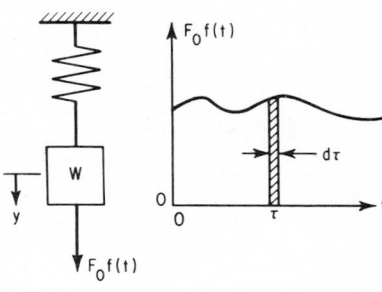

Fig. 6-100. One-degree system acted on by varying force.

(a)   (b)

One-degree System. Consider the one-degree-of-freedom system in Fig. 6-100a.   It may represent a weightless beam with a mass weighing W lb applied at midspan and subjected to a varying force $F_o f(t)$, or a rigid frame with a mass weighing W lb at girder level and subjected to this force.   The force is represented by an arbitrarily chosen constant force F_o times $f(t)$, a function of time.

If the system is not damped, the equation of motion in the elastic range is

$$\frac{W}{g} \frac{d^2 y}{dt^2} + ky = F_o f(t) \tag{6-173}$$

where k is the spring constant and g the acceleration due to gravity, 386 in. per sec². The solution consists of two parts. The first, called the complementary solution, is obtained by setting $f(t) = 0$. This solution is given by Eq. (6-139). To it must be added the second part, the particular solution, which satisfies Eq. (6-173).

The general solution of Eq. (6-173), arrived at by treating an element of the force-time curve (Fig. 6-100b) as an impulse, is

$$y = y_o \cos \omega t + \frac{v_o}{\omega} \sin \omega t + e' \omega \int_0^t f(\tau) \sin \omega (t - \tau) \, d\tau \tag{6-174}$$

where y = displacement of mass from equilibrium position, in.
 y_o = initial displacement of mass ($t = 0$), in.
 $\omega = \sqrt{kg/W}$ = natural circular frequency of free vibration
 k = spring constant = force producing unit deflection, lb per in.
 v_o = initial velocity of mass, in. per sec
 $e' = F_o/k$ = displacement under static load, in.
A closed solution is possible if the integral can be evaluated.

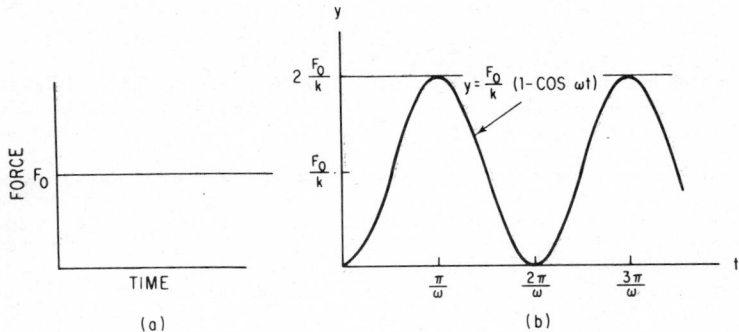

Fig. 6-101. Harmonic vibrations (*b*) result when constant force (*a*) is applied to an undamped one-degree system such as the one in Fig. 6-100*a*.

Assume, for example, the mass is subjected to a suddenly applied force F_o that remains constant (Fig. 6-101*a*). If y_o and v_o are initially zero, the displacement y of the mass at any time t can be obtained from the integral in Eq. (6-174) by setting $f(t) = 1$:

$$y = e'\omega \int_0^t \sin \omega(t - \tau)\, d\tau = e'(1 - \cos \omega t) \tag{6-175}$$

The dynamic load factor $D = 1 - \cos \omega t$.   It has a maximum value $D_m = 2$ when $t = \pi/\omega$. Figure 6-101*b* shows the variation of displacement with time.   For values of D_m for some other types of forces, see Table 6-8, p. 6-112.

Multidegree Systems.   A multidegree lumped-mass system may be analyzed by the modal method after the natural frequencies of the normal modes have been determined (Art. 6-94). This method is restricted to linearly elastic systems in which the forces applied to the masses have the same variation with time.   For other cases, numerical analysis must be used.

In the modal method, each normal mode is treated as an independent one-degree system. For each degree of the system, there is one normal mode.   A natural frequency and a characteristic shape are associated with each mode.   In each mode, the ratio of the displacements of any two masses is constant with time.   These ratios define the characteristic shape.   The modal equation of motion for each mode is

$$\frac{d^2 A_n}{dt^2} + \omega_n{}^2 A_n = \frac{gf(t) \sum\limits_{r=1}^{j} F_r \phi_{rn}}{\sum\limits_{r=1}^{j} W_r \phi_{rn}{}^2} \tag{6-176}$$

where A_n = displacement in the nth mode of an arbitrarily selected mass
   ω_n = natural frequency of the nth mode
   $F_r f(t)$ = varying force applied to the rth mass
   W_r = weight of the rth mass
   j = number of masses in the system
   ϕ_{rn} = ratio of the displacement in the nth mode of the rth mass to A_n
   g = acceleration due to gravity
We define the modal static deflection as

$$A_n' = \frac{g \sum\limits_{r=1}^{j} F_r \phi_{rn}}{\omega_n{}^2 \sum\limits_{r=1}^{j} W_r \phi_{rn}{}^2} \tag{6-177}$$

Then, the response for each mode is given by

$$A_n = D_n A_n' \tag{6-178}$$

where D_n is the dynamic load factor.   Since D_n depends only on ω_n and $f(t)$, the variation of force with time, solutions for D_n obtained for one-degree systems also apply to multidegree systems. Thus, the dynamic load factor for each mode may be obtained from Table 6-8. The total deflection at any point is the sum of the displacements for each mode, $\Sigma A_n \phi_{rn}$, at that point.

Beams. The response of beams to dynamic forces can be determined in a similar way. The modal static deflection is defined by

$$A_n' = \frac{\displaystyle\int_0^L p(x)\phi_n(x)\, dx}{\omega_n^2 \dfrac{w}{g}\displaystyle\int_0^L \phi_n^2(x)\, dx} \tag{6-179}$$

where $p(x)$ = load distribution on the span $[p(x)f(t)$ is the varying force]
$\quad\phi_n(x)$ = characteristic shape of the nth mode (see Art. 6-94)
$\quad L$ = span length
$\quad w$ = uniformly distributed weight on the span
The response of the beam then is given by Eq. (6-178) and the dynamic deflection is the sum of the modal components, $\Sigma A_n \phi_n(x)$.

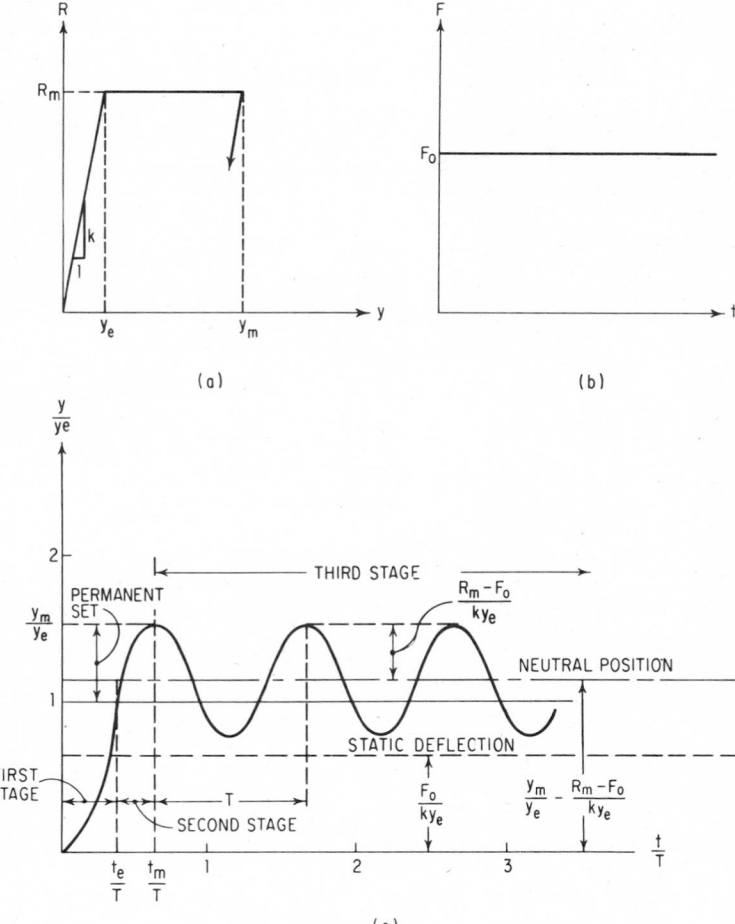

(a)   (b)

(c)

Fig. 6-102. Response in the plastic range of a one-degree system with resistance characteristics in (*a*) to a constant force (*b*) is shown in (*c*).

Nonlinear Responses. When the structure does not react linearly to loads, the equations of motion can be solved by numerical analysis if resistance is a unique function of displacement. Sometimes, the behavior of the structure can be represented by an idealized resistance-displacement diagram that makes possible a solution in closed form. Figure 6-102a shows such a diagram.

Elastic-Plastic Responses. Resistance is assumed linear ($R = ky$) until a maximum R_m is reached. After that, R remains equal to R_m for increases in y substantially larger than the displacement y_e at the elastic limit. Thus, some portions of the structure deform into the plastic range. Figure 6-102a, therefore, may be used for ductile structures only rarely subjected to severe dynamic loads. When this diagram can be used for designing such structures, more economical designs can be produced than for structures limited to the elastic range, because of the high energy-absorption capacity of structures in the plastic range.

For a one-degree system, Eq. (6-173) can be used as the equation of motion for the initial sloping part of the diagram (elastic range). For the second stage, $y_e < y < y_m$, where y_m is the maximum displacement, the equation is

$$\frac{W}{g}\frac{d^2y}{dt^2} + R_m = F_o f(t) \qquad (6\text{-}180)$$

For the unloading stage, $y < y_m$, the equation is

$$\frac{W}{g}\frac{d^2y}{dt^2} + R_m - k(y_m - y) = F_o f(t) \qquad (6\text{-}181)$$

Suppose, for example, the one-degree undamped system in Fig. 6-100a behaves in accordance with the bilinear resistance function of Fig. 6-102a and is subjected to a suddenly applied constant load (Fig. 6-102b). With zero initial displacement and velocity, the response in the first stage ($y < y_e$), according to Eq. (6-175), is

$$y = e'(1 - \cos \omega t_1) \qquad (6\text{-}182)$$

$$\frac{dy}{dt} = e'\omega \sin \omega t_1$$

Equation (6-175) also indicates that y_e will be reached at a time t_e such that $\cos \omega t_e = 1 - y_e/e'$.

For convenience, let $t_2 = t - t_e$ be the time in the second stage; thus, $t_2 = 0$ at the start of that stage. Since the condition of the system at that time is the same as the condition at the end of the first stage, the initial displacement is y_e and the initial velocity $e'\omega \sin \omega t_e$. The equation of motion is

$$\frac{W}{g}\frac{d^2y}{dt^2} + R_m = F_o \qquad (6\text{-}183)$$

The solution, taking into account initial conditions after integrating, for $y_e < y < y_m$ is

$$y = \frac{g}{2W}(F_o - R_m)t_2^2 + e'\omega t_2 \sin \omega t_e + y_e \qquad (6\text{-}184)$$

Maximum displacement occurs at the time

$$t_m = \frac{W\omega e'}{g(R_m - F_o)} \sin \omega t_e \qquad (6\text{-}185)$$

and can be obtained by substituting t_m in Eq. (6-184).

The third stage, unloading after y_m has been reached, can be determined from Eq. (6-181) and conditions at the end of the second stage. The response, however, is more easily found by noting that the third stage consists of an elastic, harmonic residual vibration. In this stage, the amplitude of vibration is $(R_m - F_o)/k$, since this is the distance between the neutral position and maximum displacement, and in the neutral position the spring force equals F_o. Hence, the response can be obtained directly from Eq. (6-175) by substituting $y_m - (R_m - F_o)/k$ for e', because the neutral position, $y = y_m - (R_m - F_o)/k$, occurs when $\omega t_3 = \pi/2$. The solution is

$$y = y_m - \frac{R_m - F_o}{k} + \frac{R_m - F_o}{k} \cos \omega t_3 \qquad (6\text{-}186)$$

where $t_3 = t - t_e - t_m$.

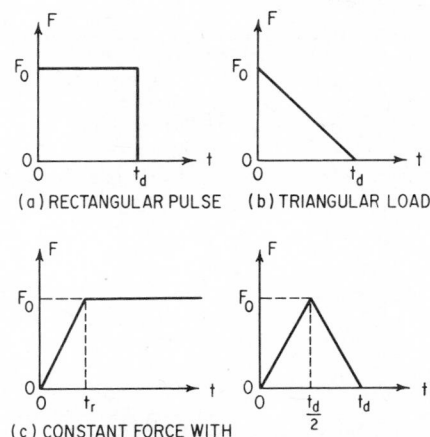

Fig. 6-103. Idealized force-time relationships.

(a) RECTANGULAR PULSE   (b) TRIANGULAR LOAD

(c) CONSTANT FORCE WITH SHORT RISE TIME   (d) TRIANGULAR PULSE

Response in the three stages is shown in Fig. 6-102c. In that diagram, however, to represent a typical case, the coordinates have been made nondimensional by expressing y in terms of y_e and the time in terms of T, the natural period of vibration.

Figure 6-103 shows some idealized force-time functions. When actual conditions approximate these, maximum displacement and time of its occurrence may be obtained from Table 6-8, which was calculated on the assumption of a bilinear resistance-displacement relationship (Fig. 6-102a). For each type of load, as load duration or time of rise varies, Table 6-8 lists the dynamic load factors in the elastic range and the ratio $R_m/F_o = y_e/e'$ in the plastic range. Values in the plastic range are given for several values of y_m/y_e.

(J. M. Biggs, "Introduction to Structural Dynamics," McGraw-Hill Book Company, New York; G. L. Rogers, "Dynamics of Framed Structures," John Wiley & Sons, Inc., New York; D. G. Fertis and E. C. Zobel, "Transverse Vibration Theory," The Ronald Press Company, New York; N. M. Newmark and E. Rosenbleuth, "Fundamentals of Earthquake Engineering," Prentice-Hall, Inc., Englewood Cliffs, N.J.)

6-98. Resonance and Damping. Damping in structures, due to friction and other causes, resists motion imposed by dynamic loads. Generally, the effect is to decrease the amplitude and lengthen the period of vibrations. If damping is large enough, vibration may be eliminated.

When maximum stress and displacement are the prime concern, damping may not be of great significance for short-time loads. These maximums usually occur under such loads at the first peak of response, and damping, unless unusually large, has little effect in a short period of time. But under conditions close to resonance, damping has considerable effect.

Resonance is the condition of a vibrating system under a varying load such that the amplitude of successive vibrations increases. Unless limited by damping or changes in the condition of the system, amplitudes may become very large.

Two forms of damping generally are assumed in structural analysis, viscous and constant (Coulomb). For viscous damping, the damping force is taken proportional to the velocity but opposite in direction. For Coulomb damping, the damping force is assumed constant and opposed in direction to the velocity.

Viscous Damping. For a one-degree system (Arts. 6-94 to 6-97), the equation of motion for a mass weighing W lb and subjected to a force F varying with time but opposed by viscous damping is

$$\frac{W}{g}\frac{d^2y}{dt^2} + ky = F - c\frac{dy}{dt} \tag{6-187}$$

where y = displacement of the mass from equilibrium position, in.

k = spring constant, lb per in.

t = time, sec

c = coefficient of viscous damping

g = acceleration due to gravity = 386 in. per sec^2

Table 6-8. Maximum Response and Occurrence Time for Short-duration Loads*

RECTANGULAR PULSES (FIG. 6-103*a*)

t_d/T	Elastic response		$y_m/y_e = 1.5$		$y_m/y_e = 2.0$		$y_m/y_e = 5.0$		$y_m/y_e = 10.0$	
	D_m	t_m/T	R_m/F_o	t_m/T	R_m/F_o	t_m/T	R_m/F_o	t_m/T	R_m/F_o	t_m/T
0.1	0.6	0.3	0.5	0.3	0.4	0.4	0.2	0.5	0.2	0.7
0.2	1.2	0.4	0.8	0.4	0.6	0.4	0.4	0.5	0.3	0.8
0.3	1.6	0.4	1.2	0.4	1.0	0.5	0.5	0.6	0.4	0.8
0.4	1.9	0.5	1.4	0.5	1.1	0.5	0.7	0.7	0.5	0.9
0.5	2.0	0.5	1.5	0.6	1.2	0.6	0.8	0.8	0.6	0.9
1.0	2.0	0.5	1.5	0.6	1.3	0.8	1.1	1.1	0.9	1.3
10.0	2.0	0.5	1.5	0.6	1.3	0.8	1.1	1.7	1.1	3.0

TRIANGULAR LOADS (FIG. 6-103*b*)

t_d/T	Elastic response		$y_m/y_e = 1.5$		$y_m/y_e = 2.0$		$y_m/y_e = 5.0$		$y_m/y_e = 10.0$	
	D_m	t_m/T	R_m/F_o	t_m/T	R_m/F_o	t_m/T	R_m/F_o	t_m/T	R_m/F_o	t_m/T
0.1	0.3	0.3	0.2	0.3	0.2	0.4	0.1	0.5		
0.2	0.6	0.3	0.4	0.4	0.3	0.4	0.2	0.6	0.1	0.8
0.3	0.8	0.4	0.6	0.4	0.5	0.5	0.3	0.7	0.2	0.9
0.4	1.0	0.4	0.7	0.4	0.6	0.5	0.4	0.7	0.3	0.9
0.5	1.2	0.4	0.9	0.4	0.7	0.6	0.4	0.7	0.3	0.9
1.0	1.5	0.5	1.1	0.5	1.0	0.6	0.7	0.8	0.5	1.0
5.0	2.0	0.5	1.4	0.6	1.3	0.6	1.0	1.4	0.9	2.0
10.0	2.0	0.5	1.4	0.7	1.3	0.7	1.1	1.5	1.0	2.5

CONSTANT FORCE WITH SHORT RISE TIMES (FIG. 6-103*c*)

t_r/T	Elastic response		$y_m/y_e = 1.5$		$y_m/y_e = 2.0$		$y_m/y_e = 5.0$		$y_m/y_e = 10.0$	
	D_m	t_m/T	R_m/F_o	t_m/T	R_m/F_o	t_m/T	R_m/F_o	t_m/T	R_m/F_o	t_m/T
0.1	2.0	0.5	1.5	0.8	1.3	1.0	1.1	1.7	1.1	3.5
0.5	1.7	0.8	1.3	1.0	1.2	1.1	1.1	2.6	1.0	6.0
1.0	1.0	1.0	1.0	1.0	1.0	1.0	1.0	1.0	1.0	1.0
1.5	1.2	1.4	1.0	2.2	1.0	3.0	1.0	3.0		
2.0	1.0	2.0	1.0	2.0	1.0	2.0	1.0	2.0		

TRIANGULAR PULSES (FIG. 6-103*d*)

t_d/T	Elastic response		$y_m/y_e = 1.5$		$y_m/y_e = 2.0$		$y_m/y_e = 5.0$		$y_m/y_e = 10.0$	
	D_m	t_m/T	R_m/F_o	t_m/T	R_m/F_o	t_m/F	R_m/F_o	t_m/F	R_m/F_o	t_m/F
0.1	0.3	0.3	0.2	0.3	0.2	0.4				
0.2	0.6	0.4	0.5	0.4	0.4	0.4	0.2	0.6	0.2	0.6
0.3	1.0	0.4	0.7	0.4	0.5	0.4	0.3	0.6	0.2	0.9
0.4	1.2	0.5	0.8	0.5	0.7	0.5	0.4	0.7	0.3	0.9
0.5	1.3	0.5	0.9	0.6	0.8	0.6	0.5	0.7	0.3	1.0
1.0	1.6	0.7	1.1	0.7	1.0	0.8	0.7	1.0	0.5	1.3
5.0	1.1	2.7	1.0	3.0	0.9	3.1	0.8	3.5	0.8	3.9
10.0	1.0	5.0	1.0	5.5	0.9	6.0	0.9	6.3	0.8	7.0

*SOURCE: U.S. Army Corps of Engineers, "Design of Structures to Resist the Effects of Atomic Weapons," Manual EM 1110-345-415.

Let us set $\beta = cg/2W$ and consider those cases in which $\beta < \omega$, the natural circular frequency [Eq. (6-140)], to eliminate unusually high damping (overdamping). Then, for initial displacement y_o and velocity v_o, the solution of Eq. (6-187) with $F = 0$ is

$$y = e^{-\beta t}\left(\frac{v_o + \beta y_o}{\omega_d}\sin\omega_d t + y_o\cos\omega_d t\right) \tag{6-188}$$

where $\omega_d = \sqrt{\omega^2 - \beta^2}$ and $e = 2.71828$. Equation (6-188) represents a decaying harmonic motion with β controlling the rate of decay and ω_d the natural frequency of the damped system.
When $\beta = \omega$

$$y = e^{-\omega t}[v_o t + (1 + \omega t)y_o] \tag{6-189}$$

which indicates that the motion is not vibratory. Damping producing this condition is called critical, and the critical coefficient is

$$c_d = \frac{2W\beta}{g} = \frac{2W\omega}{g} = \sqrt{\frac{kW}{g}} \tag{6-190}$$

Damping sometimes is expressed as a percent of critical (β as a percent of ω).

For small amounts of viscous damping, the damped natural frequency is approximately equal to the undamped natural frequency minus $\frac{1}{2}\beta^2/\omega$. For example, for 10% critical damping ($\beta = 0.1\omega$), $\omega_d = \omega[1 - \frac{1}{2}(0.1)^2] = 0.995\omega$. Hence, the decrease in natural frequency due to damping generally can be ignored.

Damping sometimes is measured by **logarithmic decrement,** the logarithm of the ratio of two consecutive peak amplitudes during free vibration.

$$\text{Logarithmic decrement} = 2\pi\beta/\omega \tag{6-191}$$

For example, for 10% critical damping, the logarithmic decrement equals 0.2π. Hence, the ratio of a peak to the following peak amplitude is $e^{0.2\pi} = 1.87$.

The complete solution of Eq. (6-187) with initial displacement y_o and velocity v_o is

$$y = e^{-\beta t}\left(\frac{v_o + \beta y_o}{\omega_d}\sin\omega_d t + y_o\cos\omega_d t\right) + e'\frac{\omega^2}{\omega_d}\int_0^t f(\tau)e^{-\beta(t-\tau)}\sin\omega_d(t - \tau)\,d\tau \tag{6-192}$$

where e' is the deflection that the applied force would produce under static loading. Equation (6-192) is identical to Eq. (6-174) when $\beta = 0$.

Unbalanced rotating parts of machines produce pulsating forces that may be represented by functions of the form $F_o\sin\alpha t$. If such a force is applied to an undamped one-degree system, Eq. (6-174) indicates that if the system starts at rest the response will be

$$y = \frac{F_o g}{W}\left(\frac{1/\omega^2}{1 - \alpha^2/\omega^2}\right)\left(\sin\alpha t - \frac{\alpha}{\omega}\sin\omega t\right) \tag{6-193}$$

And since the static deflection would be $F_o/k = F_o g/W\omega^2$, the dynamic load factor is

$$D = \frac{1}{1 - \alpha^2/\omega^2}\left(\sin\alpha t - \frac{\alpha}{\omega}\sin\omega t\right) \tag{6-194}$$

If α is small relative to ω, maximum D is nearly unity; thus, the system is practically statically loaded. If α is very large compared with ω, D is very small; thus, the mass cannot follow the rapid fluctuations in load and remains practically stationary. Therefore, when α differs appreciably from ω, the effects of unbalanced rotating parts are not too serious. But if $\alpha = \omega$, resonance occurs; D increases with time. Hence, to prevent structural damage, measures must be taken to correct the unbalanced parts to change α, or to change the natural frequency of the vibrating mass, or damping must be provided.

The response as given by Eq. (6-193) consists of two parts, the free vibration and the forced part. When damping is present, the free vibration is of the form of Eq. (6-188) and is rapidly damped out. Hence, the free part is called the **transient response,** and the forced part, the **steady-state response.** The maximum value of the dynamic load factor for the steady-state response D_m is called the **dynamic magnification factor.** It is given by

$$D_m = \frac{1}{\sqrt{(1 - \alpha^2/\omega^2)^2 + (2\beta\alpha/\omega^2)^2}} \tag{6-195}$$

With damping, then, the peak values of D_m occur when $\alpha = \omega\sqrt{1 - \beta^2/\omega^2}$ and are approximately equal to $\omega/2\beta$. For example, for 10% critical damping,

$$D_m = \omega/0.2\omega = 5$$

So even small amounts of damping significantly limit the response at resonance.

Coulomb Damping. For a one-degree system with Coulomb damping, the equation of motion for free vibration is

$$\frac{W}{g}\frac{d^2y}{dt^2} + ky = \pm F_f \qquad (6\text{-}196)$$

where F_f is the constant friction force and the positive sign applies when the velocity is negative. If initial displacement is y_o and initial velocity is zero, the response in the first half cycle, with negative velocity, is

$$y = \left(y_o - \frac{F_f}{k}\right)\cos \omega t + \frac{F_f}{k} \qquad (6\text{-}197)$$

equivalent to a system with a suddenly applied constant force. For the second half cycle, with positive velocity, the response is

$$y = \left(-y_o + 3\frac{F_f}{k}\right)\cos \omega\left(t - \frac{\pi}{\omega}\right) - \frac{F_f}{k}$$

If the solution is continued with the sign of F_f changing in each half cycle, the results will indicate that the amplitude of positive peaks is given by $y_o - 4nF_f/k$, where n is the number of complete cycles, and the response will be completely damped out when $t = ky_oT/4F_f$, where T is the natural period of vibration, or $2\pi/\omega$.

Analysis of the steady-state response with Coulomb damping is complicated by the possibility of frequent cessation of motion.

(J. P. Den Hartog, "Mechanical Vibrations," McGraw-Hill Book Company; L. S. Jacobsen and R. S. Ayre, "Engineering Vibrations," McGraw-Hill Book Company; D. D. Barkan, "Dynamics of Bases and Foundations," McGraw-Hill Book Company; W. C. Hurty and M. F. Rubinstein, "Dynamics of Structures," Prentice-Hall, Englewood Cliffs, N. J.)

6-99. Approximate Design for Dynamic Loading. (See also Arts. 6-93 to 6-98.) Complex analysis and design methods seldom are justified for structures subjected to dynamic loading because of lack of sufficient information on loading, damping, resistance to deformation, and other factors. In general, it is advisable to represent the actual structure and loading by idealized systems that permit a solution in closed form.

Whenever possible, represent the actual structure by a one-degree system consisting of an equivalent mass with massless spring. For structures with distributed mass, simplify the analysis in the elastic range by computing the response only for one or a few of the normal modes. In the plastic range, treat each stage—elastic, elastic-plastic, and plastic—as completely independent; for example, a fixed-end beam may be treated, when in the elastic-plastic stage, as a simply supported beam.

Choose the parameters of the equivalent system to make the deflection at a critical point, such as the location of the concentrated mass, the same as it would be in the actual structure. Stresses in the actual structure should be computed from the deflections in the equivalent system.

Compute an assumed shape factor ϕ for the system from the shape taken by the actual structure under static application of the loads. For example, for a simple beam in the elastic range with concentrated load at midspan, ϕ may be chosen, for $x < L/2$, as $(Cx/L^3)(3L^2 - 4x^2)$, the shape under static loading, and C may be set equal to 1 to make ϕ equal to 1 when $x = L/2$. For plastic conditions (hinge at midspan), ϕ may be taken as Cx/L, and C set equal to 2, to make $\phi = 1$ when $x = L/2$.

For a structure with concentrated forces, let W_r be the weight of the rth mass, ϕ_r the value of ϕ at the location of that mass, and F_r the dynamic force acting on W_r. Then, the equivalent weight of the idealized system is

$$W_e = \sum_{r=1}^{j} W_r \phi_r^2 \qquad (6\text{-}199)$$

where j is the number of masses. The equivalent force is

$$F_e = \sum_{r=1}^{j} F_r \phi_r \qquad (6\text{-}200)$$

For a structure with continuous mass, the equivalent weight is

$$W_e = \int w\phi^2 dx \tag{6-201}$$

where w is the weight in lb per lin ft. The equivalent force is

$$F_e = \int q\phi dx \tag{6-202}$$

for a distributed load q, lb per lin ft.

The resistance of a member or structure is the internal force tending to restore it to its unloaded static position. For most structures, a bilinear resistance function, with slope k up to the elastic limit and zero slope in the plastic range (Fig. 6-102a), may be assumed. For a given distribution of dynamic load, maximum resistance of the idealized system may be taken as the total load with that distribution that the structure can support statically. Similarly, stiffness is numerically equal to the total load with the given distribution that would cause a unit deflection at the point where the deflections in the actual structure and idealized system are equal. Hence, the equivalent resistance and stiffness are in the same ratio to the actual as the equivalent forces to the actual forces.

Let k be the actual spring constant, g the acceleration due to gravity, 386 in. per sec², and

$$W' = \frac{W_e}{F_e} \Sigma F \tag{6-203}$$

where ΣF represents the actual total load. Then, the equation of motion of an equivalent one-degree system is

$$\frac{d^2y}{dt^2} + \omega^2 y = g \frac{\Sigma F}{W'} \tag{6-204}$$

and the natural circular frequency is

$$\omega = \sqrt{\frac{kg}{W'}} \tag{6-205}$$

The natural period of vibration equals $2\pi/\omega$. Equations (6-204) and (6-205) have the same form as Eqs. (6-138), (6-140), and (6-173). Consequently, the response can be computed as indicated in Arts. 6-94 to 6-97.

Whenever possible, select a load-time function for ΣF to permit use of a known solution, such as those in Table 6-8, Art. 6-97.

For preliminary design of a one-degree system loaded into the plastic range by a suddenly applied force that remains substantially constant up to the time of maximum response, the following approximation may be used for that response:

$$y_m = \frac{y_e}{2(1 - F_o/R_m)} \tag{6-206}$$

where y_e is the displacement at the elastic limit, F_o the average value of the force, and R_m the maximum resistance of the system. This equation indicates that for purely elastic response, R_m must be twice F_o; whereas, if y_m is permitted to be large, R_m may be made nearly equal to F_o, with greater economy of material.

For preliminary design of a one-degree system subjected to a sudden load with duration t_d less than 20% of the natural period of the system, the following approximation can be used for the maximum response:

$$y_m = \frac{1}{2} y_e \left[\left(\frac{F_o}{R_m} \omega t_d \right)^2 + 1 \right] \tag{6-207}$$

where F_o is the maximum value of the load and ω the natural frequency. This equation also indicates that the larger y_m is permitted to be, the smaller R_m need be.

For a beam, the spring force of the equivalent system is not the actual force, or reaction, at the supports. The real reactions should be determined from the dynamic equilibrium of the complete beam. This calculation should include the inertia force, with distribution identical with the assumed deflected shape of the beam. For example, for a simply supported beam with uniform load, the dynamic reaction in the elastic range is $0.39R + 0.11F$, where R is the resistance, which varies with time, and $F = qL$ is the load. For a concentrated load F at midspan, the dynamic reaction is $0.78R - 0.28F$. And for concentrated loads $F/2$ at each third point, it is $0.62R - 0.12F$. (Note that the sum of the coefficients equals 0.50, since the dynamic-reaction equations must hold for static loading, when $R = F$.) These expressions also can be used for fixed-end beams without significant error. If high accuracy is not required, they also can be used for the plastic range.

Structures usually are designed to resist the dynamic forces of earthquakes by use of equivalent static loads. Many building codes specify such a method. See, for example, Art. 15-4.

(J. M. Biggs, "Structural Dynamics," McGraw-Hill Book Company, New York; G. L. Rogers, "Dynamics of Framed Structures," John Wiley & Sons, Inc., New York; U.S. Army Corps of Engineers, "Design of Structures to Resist the Effects of Atomic Weapons," Manual EM 1110-345-415.)

FINITE-ELEMENT METHODS

From the basic principles given in preceding articles, systematic methods have been developed for determining the behavior of a structure from a knowledge of the behavior under load of its components. In these methods, called finite-element methods, a structural system is considered an assembly of a finite number of finite-size components, or elements. These are assumed to be connected to each other only at discrete points, called nodes. From the characteristics of the elements, such as their stiffness or flexibility, the characteristics of the whole system can be derived. With these known, internal stresses and strains throughout can be computed.

Choice of elements to be used depends on the type of structure. For example, for a truss with joints considered hinged, a natural choice of element would be a bar, subjected only to axial forces. For a rigid frame, the elements might be beams subjected to bending and axial forces, or to bending, axial forces, and torsion. For a thin plate or shell, elements might be triangles or rectangles, connected at vertices. For three-dimensional structures, elements might be beams, bars, tetrahedrons, cubes, or rings.

For many structures, because of the number of finite elements and nodes, analysis by a finite-element method requires mathematical treatment of large amounts of data and solution of numerous simultaneous equations. For this purpose, the use of high-speed computers is advisable.

The mathematics of such analyses is usually simpler and more compact when the data are handled in matrix form. Matrix notation is especially convenient in indicating the solution of simultaneous linear equations. For example, suppose a set of equations is represented in matrix notation by $AX = B$. Multiplication of both sides of the equation by inverse A^{-1} yields $A^{-1}AX = A^{-1}B$. Since $A^{-1}A = I$, the identity matrix, and $IX = X$, the solution of the equations is given by $X = A^{-1}B$. (F. S. Merritt, "Modern Mathematical Methods for Engineers," McGraw-Hill Book Company, New York.)

The methods used for analyzing structures generally may be classified as force (flexibility) or displacement (stiffness) methods.

In analysis of statically indeterminate structures by force methods, forces are chosen as redundants, or unknowns. The choice is made in such a way that equilibrium is satisfied. These forces are then determined from the solution of equations that insure compatibility of all displacements of elements at each node. After the redundants have been computed, stresses and strains throughout the structure can be found from equilibrium equations and stress-strain relations.

In displacement methods, displacements are chosen as unknowns. The choice is made in such a way that geometric compatibility is satisfied. These displacements are then determined from the solution of equations that insure that forces acting at each node are in equilibrium. After the unknowns have been computed, stresses and strains throughout the structure can be found from equilibrium equations and stress-strain relations.

In choosing a method, the following should be kept in mind: In force methods, the number of unknowns equals the degree of indeterminacy. In displacement methods, the number of unknowns equals the degrees of freedom of displacement at nodes. The fewer the unknowns, the fewer the calculations required.

6-100. Force-Displacement Relations. Assume a right-handed cartesian coordinate system, with axes x, y, z. Assume also at each node of a structure to be analyzed a system of base unit vectors, e_1 in the direction of the x axis, e_2 in the direction of the y axis, and e_3 in the direction of the z axis. Forces and moments acting at a node are resolved into components in the directions of the base vectors. Then, the forces and moments at the node may be represented by the vector $P_i e_i$, where P_i is the magnitude of the force or moment acting in the direction of e_i. This vector, in turn, may be conveniently represented by a column matrix P. Similarly, the displacements

—translations and rotation—of the node may be represented by the vector $\Delta_i e_i$, where Δ_i is the magnitude of the displacement acting in the direction of e_i. This vector, in turn, may be represented by a column matrix Δ.

To conserve space, column vectors, when given in terms of their components, will be represented here by their transposes (rows interchanged with columns), represented by a superscript T. Thus, the vector $P_i e_i$ will be represented by the row matrix $P^T = [P_1 P_2 \ldots P_n]$. Also, vectors and matrices will be indicated by boldface symbols.

For compactness, and because, in structural analysis, similar operations are performed on all nodal forces, all the loads, including moments, acting on all the nodes may be combined into a single column matrix P. Similarly, all the nodal displacements may be represented by a single column matrix Δ.

If the independent loads and displacements are listed in the appropriate order in each matrix, a matrix equation relating them can be written:

$$\Delta = FP \qquad (6\text{-}208)$$

For an elastic structure, F is the **flexibility matrix**. It is square (has the same number of rows as columns) and symmetric ($F = F^T$). A typical element F_{ij} of F gives the deflection of a node in the direction of displacement Δ_i when a unit force acts at the same or another node in the direction of force P_j. The jth column of F, therefore, contains all the nodal displacements when one force P_j is set equal to unity and all other independent forces are zero.

Multiplication of both sides of Eq. (6-208) by F^{-1} yields $F^{-1}\Delta = F^{-1}FP = IP = P$, or

$$P = K\Delta \qquad (6\text{-}209)$$

where K = inverse of flexibility matrix = F^{-1}

For an elastic structure, K is the **stiffness matrix**. It too is square and symmetric. A typical element K_{ij} of K gives the force at a node, in the direction of load P_i, that results when the same or another node is given a unit displacement in the direction of displacement Δ_j. The jth column of K, therefore, contains the nodal forces that produce a unit displacement of the node at which Δ_j occurs and in the direction of Δ_j, but no other nodal displacements throughout the structure.

For some structures, the flexibility or stiffness matrix can be constructed from the definition. In other cases, the matrix may have to be synthesized from the flexibility or stiffness matrices of finite elements comprising the structures. The procedures are explained in the following articles.

6-101. Matrix Force (Flexibility) Method. In this article, the structure is assumed to consist of n finite elements connected only at the nodes. The flexibility matrices of the elements are assumed known. Right-handed cartesian coordinate axes, x, y, z, are chosen for the structure, with base unit vectors e_i parallel to these axes at each node, as in Art. 6-100. The flexibility matrix for the whole structure can be constructed from the element flexibility matrices, as will be demonstrated.

For the ith element, nodal forces and displacements are related by

$$\delta_i = f_i S_i \qquad i = 1, 2, \ldots n \qquad (6\text{-}210)$$

where δ_i = matrix of displacements of the nodes of the ith element in directions of base vectors

f_i = flexibility matrix of the ith element

S_i = matrix of forces, moments, torques acting at nodes of the ith element in directions of base vectors

For compactness, this relationship between nodal displacements and forces for each element can be combined into a single matrix equation applicable to all the elements:

$$\delta = fS \qquad (6\text{-}211)$$

where δ = matrix of all nodal displacements for all elements

S = matrix of all forces acting at the nodes of all elements

$$f = \begin{bmatrix} f_1 & 0 & \ldots 0 \\ 0 & f_2 & \ldots 0 \\ \multicolumn{3}{c}{\cdots\cdots\cdots} \\ 0 & 0 & \ldots f_n \end{bmatrix} \qquad (6\text{-}212)$$

Element forces S can be determined from the loads P at the nodes to meet equilibrium requirements at the nodes:

$$S = a_0 P \tag{6-213}$$

where a_0 is a matrix of influence coefficients. The jth column of a_0 contains the element forces when one load $P_j = 1$, and all other loads are set equal to zero.

From Eqs. (6-211) and (6-213),

$$\delta = f a_0 P \tag{6-214}$$

From energy relations, it can be shown that

$$\Delta = a_0{}^T \delta \tag{6-215}$$

where Δ = matrix of displacements at the nodes of the structure

$a_0{}^T$ = transpose of a_0 = matrix a_0 with rows and columns interchanged

Finally, from Eqs. (6-214) and (6-215) comes the relationship that permits computation of the flexibility matrix of the whole structure from the flexibility matrices of its elements:

$$\Delta = a_0{}^T f a_0 P \tag{6-216}$$

Since also, by Eq. (6-208), $\Delta = FP$,

$$F = a_0{}^T f a_0 \tag{6-217}$$

Equations (6-208) and (6-210) to (6-217) can be used to compute all forces and displacements in a statically determinate structure.

If the structure is statically indeterminate, however, a_0 cannot be developed from the equilibrium equations alone. Forces and displacements, though, can be determined from the following equations:

First, the structure is reduced, made statically determinate by replacing some constraints with redundant forces, represented by a matrix X. All element forces S are related to loads P at the nodes of the reduced structure by

$$S = a_0 P + a_1 X \tag{6-218}$$

where a_0 = matrix of influence coefficients for loads acting on reduced structure

a_1 = matrix of influence coefficients for redundants acting on reduced structure

The ith column of a_1 contains the element forces resulting from a unit load applied in the direction of X_i at the node where X_i acts. Substitution of Eq. (6-218) in Eq. (6-211) yields

$$\delta = f S = f(a_0 P + a_1 X) \tag{6-219}$$

Generally, the displacements associated with redundants are zero. Hence, by Eq. (6-215),

$$a_1{}^T \delta = 0 \tag{6-220}$$

Solving for the redundants from Eqs. (6-219) and (6-220) yields

$$X = -D^{-1} D_0 P \tag{6-221}$$

where $D_0 = a_1{}^T f a_0$

D^{-1} = inverse of matrix D

$D = a_1{}^T f a_1$

Substitution of the known values of the redundants in Eq. (6-218) permits computation of the element forces from

$$S = (a_0 - a_1 D^{-1} D_0) P \tag{6-222}$$

Methods of developing flexibility matrices of finite elements are given in Art. 6-103.

(H. I. Laursen, "Matrix Analysis of Structures," McGraw-Hill Book Company, New York; O. C. Zienkiewicz, "The Finite Element Method in Engineering Science," McGraw-Hill Book Company, New York; C. S. Desai and J. F. Abel, "Introduction to the Finite Element Method," Van Nostrand Reinhold Company, New York; M. F. Rubinstein, "Structural Systems—Statics, Dynamics and Stability," Prentice-Hall, Englewood Cliffs, N.J.)

6-102. Matrix Displacement (Stiffness) Method. In this article, the structure is assumed to consist of n finite elements connected only at the nodes. The stiffness matrices of the elements are assumed known. Right-handed cartesian coordinate axes, x, y, z, are chosen for the structure with base unit vectors e_i parallel to these axes at each node, as in Art. 6-100. Forces and displacements are resolved into components along e_i. The stiffness matrix for the whole structure can be constructed from the element stiffness matrices, as will be demonstrated.

For the ith element, nodal forces and displacements are related by

$$S_i = k_i \delta_i \qquad i = 1, 2, \ldots n \tag{6-223}$$

where S_i = matrix of forces, including moments and torques, acting at the nodes of the ith element

    k_i = stiffness matrix of the ith element

    δ_i = matrix of displacements of the nodes of the ith element

For compactness, this relationship between nodal displacements and forces for each element can be combined into a single matrix equation applicable to all the elements:

$$S = k\delta \tag{6-224}$$

where S = matrix of all forces acting at the nodes of all elements

    δ = matrix of all nodal displacements for all elements

$$k = \begin{bmatrix} k_1 & 0 & \ldots & 0 \\ 0 & k_2 & \ldots & 0 \\ \multicolumn{4}{c}{\dotfill} \\ 0 & 0 & \ldots & k_n \end{bmatrix} \tag{6-225}$$

Element nodal displacements δ can be determined from the displacements Δ of the nodes of the structure to insure geometric compatibility:

$$\delta = b_0 \Delta \tag{6-226}$$

where b_0 is a matrix of influence coefficients. The jth column of b_0 contains the element nodal displacements when the node where Δ_j occurs is given a unit displacement in the direction of Δ_j, and no other nodes are displaced.

From Eqs. (6-224) and (6-226),

$$S = kb_0 \Delta \tag{6-227}$$

From energy relationships, it can be shown that

$$P = b_0{}^T S \tag{6-228}$$

where P = matrix of all the loads acting at the nodes of the structure

    $b_0{}^T$ = transpose of b_0 = matrix b_0 with rows and columns interchanged

Finally, from Eqs. (6-227) and (6-228) comes the relationship that permits computation of the stiffness matrix of the whole structure from the stiffness matrices of its elements:

$$P = b_0{}^T k b_0 \Delta \tag{6-229}$$

Since also, by Eq. (6-209), $P = K\Delta$,

$$K = b_0{}^T k b_0 \tag{6-230}$$

With the stiffness matrix known, the nodal displacements can be computed from the nodal loads. Multiplication of both sides of Eq. (6-209) by K^{-1} gives

$$\Delta = K^{-1} P \tag{6-231}$$

This permits member forces to be computed from Eq. (6-227).

For structures with a large number of nodal displacements, inversion of the stiffness matrix K directly may be impracticable. The size of K, however, often can be reduced by taking advantage initially of conditions at nodes where loads do not act in the directions of permissible displacements.

Let X be the matrix of the unknown displacements at those nodes. Then, as in Eq. (6-226),

$$\delta = b_0 \Delta + b_1 X \tag{6-232}$$

where b_0 = matrix of influence coefficients for displacements of nodes where loads act in directions of displacements

    b_1 = matrix of influence coefficients for displacements of nodes where loads do not act in directions of displacements

The jth column of b_1 contains the element nodal displacements when the node where X_j occurs is given a unit displacement in the direction of X_j, and no other nodes are displaced. Substitution of Eq. (6-232) in Eq. (6-224) yields

$$S = k(b_0 \Delta + b_1 X) \tag{6-233}$$

Since $P_j = 0$ for those displacements not associated with loads, Eq. (6-228) indicates that

$$\mathbf{b_1}^T \mathbf{S} = \mathbf{0} \qquad (6\text{-}234)$$

Solving for the unknown displacements from Eq. (6-233) and (6-234) gives

$$\mathbf{X} = -\mathbf{B}^{-1} \mathbf{B_0} \mathbf{\Delta} \qquad (6\text{-}235)$$

where $\mathbf{B_0} = \mathbf{b_1}^T \mathbf{k} \mathbf{b_0}$
$\quad \mathbf{B}^{-1} = $ inverse of matrix $\mathbf{B}$
$\quad \mathbf{B} = \mathbf{b_1}^T \mathbf{k} \mathbf{b_1}$

Substitution for $\mathbf{X}$ in Eq. (6-232) gives the element nodal displacements:

$$\mathbf{\delta} = (\mathbf{b_0} - \mathbf{b_1}\mathbf{B}^{-1}\mathbf{B_0})\mathbf{\Delta} = \mathbf{b_2}\mathbf{\Delta} \qquad (6\text{-}236)$$

where $\mathbf{b_2} = \mathbf{b_0} - \mathbf{b_1}\mathbf{B}^{-1}\mathbf{B_0}$

From Eqs. (6-233), (6-235), and (6-236), the element member forces are given by

$$\mathbf{S} = \mathbf{k}\mathbf{b_2}\mathbf{\Delta} \qquad (6\text{-}237)$$

And from Eq. (6-228),

$$\mathbf{P} = \mathbf{b_0}^T \mathbf{k} \mathbf{b_2} \mathbf{\Delta} \qquad (6\text{-}238)$$

Since also, by Eq. (6-209), $\mathbf{P} = \mathbf{K}\mathbf{\Delta}$, the stiffness matrix for the whole structure is given by

$$\mathbf{K} = \mathbf{b_0}^T \mathbf{k} \mathbf{b_2} \qquad (6\text{-}239)$$

This matrix may be considerably smaller than that given by Eq. (6-230). As before, nodal displacements can be computed from Eq. (6-231) after $\mathbf{K}$ has been inverted.

Methods of developing stiffness matrices of finite elements are described in Art. 6-103. (See bibliography at the end of Art. 6-101.)

6-103. Element Flexibility and Stiffness Matrices. As indicated in Arts. 6-101 and 6-102, the relationship between independent forces and displacements at nodes of finite elements in a structure is determined by flexibility matrices $\mathbf{f_i}$ [Eq. (6-210)] or stiffness matrices $\mathbf{k_i}$ [Eq. (6-223)] of the elements. In some cases, the components of these matrices can be developed from the defining equations.

The jth column of a flexibility matrix of a finite element, for which we will use the symbol $\mathbf{f}$ without subscript in this article, contains all the nodal displacements of the element when one force S_j is set equal to unity and all other independent forces are set equal to zero.

The jth column of a stiffness matrix of a finite element, for which we will use the symbol $\mathbf{k}$ without subscript in this article, consists of the forces acting at the nodes of the element to produce a unit displacement of the node at which δ_j occurs and in the direction of δ_j but no other nodal displacements of the element.

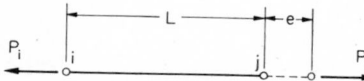

Fig. 6-104. Elastic bar in tension.

Bars with Axial Stress Only. As an example of the use of the definitions of flexibility and stiffness, consider the simple case of an elastic bar under tension applied by axial forces P_i and P_j at nodes i and j, respectively (Fig. 6-104). The bar might be the finite element of a truss, such as a diagonal or a hanger. Connections to other members are made at nodes i and j, which can transmit only forces in the directions i to j or j to i.

For equilibrium, $P_i = P_j = P$. Displacement of node j relative to node i is e. From Eq. (6-8), $e = PL/AE$, where L is the initial length of the bar, A the bar cross-sectional area, and E the modulus of elasticity. Setting $P = 1$ yields the flexibility of the bar,

$$f = \frac{L}{AE} \qquad (6\text{-}240)$$

Setting $e = 1$ gives the stiffness of the bar,

$$k = \frac{AE}{L} \qquad (6\text{-}241)$$

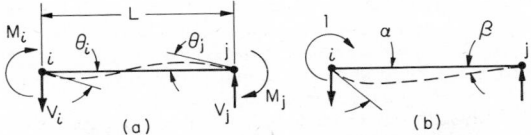

Fig. 6-105. Beam subjected to end moments and shears.

Beams with Bending Only. As another example of the use of the definition to determine element flexibility and stiffness matrices, consider the simple case of an elastic beam in bending applied by moments M_i and M_j at nodes i and j, respectively (Fig. 6-105). The beam might be a finite element of a rigid frame. Connections to other members are made at nodes i and j, which can transmit moments and forces normal to the beam.

Nodal displacements of the element can be sufficiently described by rotations θ_i and θ_j relative to the straight line between nodes i and j. For equilibrium, forces $V_j = -V_i$ normal to the beam are required at nodes j and i, respectively, and $V_j = (M_i + M_j)/L$, where L is the span of the beam. Thus, M_i and M_j are the only independent forces acting. Hence, Eqs. (6-210) and (6-223) can be written for this element as

$$\boldsymbol{\theta} = \begin{bmatrix} \theta_i \\ \theta_j \end{bmatrix} = f \begin{bmatrix} M_i \\ M_j \end{bmatrix} = \mathbf{fM} \tag{6-242}$$

$$\mathbf{M} = \begin{bmatrix} M_i \\ M_j \end{bmatrix} = k \begin{bmatrix} \theta_i \\ \theta_j \end{bmatrix} = \mathbf{k\boldsymbol{\theta}} \tag{6-243}$$

The flexibility matrix $\mathbf{f}$ then will be a 2×2 matrix. The first column can be obtained by setting $M_i = 1$ and $M_j = 0$ (Fig. 6-105b). The resulting angular rotations are given by Eqs. (6-71) and (6-72). For a beam with constant moment of inertia I and modulus of elasticity E, the rotations are $\alpha = L/3EI$ and $\beta = -L/6EI$. Similarly, the second column can be developed by setting $M_i = 0$ and $M_j = 1$.

The flexibility matrix for a beam in bending then is

$$\mathbf{f} = \begin{bmatrix} \dfrac{L}{3EI} & -\dfrac{L}{6EI} \\ -\dfrac{L}{6EI} & \dfrac{L}{3EI} \end{bmatrix} = \frac{L}{6EI} \begin{bmatrix} 2 & -1 \\ -1 & 2 \end{bmatrix} \tag{6-244}$$

The stiffness matrix, obtained in a similar manner or by inversion of $\mathbf{f}$, is

$$k = \begin{bmatrix} \dfrac{4EI}{L} & \dfrac{2EI}{L} \\ \dfrac{2EI}{L} & \dfrac{4EI}{L} \end{bmatrix} = \frac{2EI}{L} \begin{bmatrix} 2 & 1 \\ 1 & 2 \end{bmatrix} \tag{6-245}$$

Beams Subjected to Bending and Axial Forces. For a beam subjected to nodal moments M_i and M_j and axial forces P, flexibility and stiffness are represented by 3×3 matrices. The load-displacement relations for a beam of span L, constant moment of inertia I, modulus of elasticity E, and cross-sectional area A are given by

$$\begin{bmatrix} \theta_i \\ \theta_j \\ e \end{bmatrix} = f \begin{bmatrix} M_i \\ M_j \\ P \end{bmatrix} \qquad \begin{bmatrix} M_i \\ M_j \\ P \end{bmatrix} = k \begin{bmatrix} \theta_i \\ \theta_j \\ e \end{bmatrix} \tag{6-246}$$

where e = axial displacement. In this case, the flexibility matrix is

$$f = \frac{L}{6EI} \begin{bmatrix} 2 & -1 & 0 \\ -1 & 2 & 0 \\ 0 & 0 & \eta \end{bmatrix} \tag{6-247}$$

where $\eta = 6I/A$, and the stiffness matrix, with $\psi = A/I$, is

$$k = \frac{EI}{L} \begin{bmatrix} 4 & 2 & 0 \\ 2 & 4 & 0 \\ 0 & 0 & \psi \end{bmatrix} \tag{6-248}$$

Loads between Nodes. When loads act along a beam, they should be replaced by equivalent loads at the nodes—simple-beam reactions and fixed-end moments, both with signs reversed. The equivalent forces should be used in constructing a_0 for Eqs. (6-213) and (6-218) for determining redundants; but the fixed-end moments should not be included in a_0 in Eq. (6-222) for computing element forces. When the element forces are obtained from Eq. (6-227) or (6-237), the final element forces then are determined by adding the fixed-end moments and simple-beam reactions to the solution of Eq. (6-227) or (6-237).

General Methods for Computing Flexibility. Each component f_{ij} of a flexibility matrix equals a displacement at a node due to a unit force at the same or another node. When the matrix for a finite element cannot readily be constructed from this definition, other means must be used.

For some elements, the components of the flexibility matrix may be obtained by integrating over the whole volume of the element

$$f_{ij} = \int_V \boldsymbol{\sigma}^T \boldsymbol{\varepsilon} \, dV \tag{6-249}$$

where $\boldsymbol{\sigma}^T$ = transpose of the matrix of the stresses produced in the element by a unit force at the node where displacement f_{ij} occurs and in the direction of that displacement

$\boldsymbol{\varepsilon}$ = matrix of the unit strains produced in the element by a unit force at the same or another node

dV = differential volume

For use in the integral, note that for a three-dimensional, homogeneous, elastic material with modulus of elasticity E, shearing modulus G, and Poisson's ratio μ, stresses σ and unit strains ϵ in the direction of x, y, z Cartesian coordinate axes are related by

$$\boldsymbol{\varepsilon} = \boldsymbol{\phi}\boldsymbol{\sigma} \tag{6-250}$$

$$\boldsymbol{\varepsilon}^T = [\epsilon_x \epsilon_y \epsilon_z \gamma_{xy} \gamma_{xz} \gamma_{yz}] \tag{6-251a}$$

$$\boldsymbol{\sigma}^T = [\sigma_x \sigma_y \sigma_z \tau_{xy} \tau_{xz} \tau_{yz}] \tag{6-251b}$$

$$\boldsymbol{\phi} = \frac{1}{E} \begin{bmatrix} 1 & -\mu & -\mu & 0 & 0 & 0 \\ -\mu & 1 & -\mu & 0 & 0 & 0 \\ -\mu & -\mu & 1 & 0 & 0 & 0 \\ 0 & 0 & 0 & \dfrac{E}{G} & 0 & 0 \\ 0 & 0 & 0 & 0 & \dfrac{E}{G} & 0 \\ 0 & 0 & 0 & 0 & 0 & \dfrac{E}{G} \end{bmatrix} \tag{6-251c}$$

See Arts. 6-12 to 6-14.

For some elements, the components of the flexibility matrix may be obtained by differentiating twice the strain energy U in the element due to forces at the nodes:

$$f_{ij} = \frac{\partial^2 U}{\partial S_i \partial S_j} \tag{6-252}$$

where S_i = force acting at the node where displacement f_{ij} occurs and in the direction of that displacement

S_j = force acting at the same or another node

Flexibility of Thin Triangular Element. For an elastic, triangular element with thickness t very small compared with other dimensions, the terms in Eq. (6-250) become

$$\boldsymbol{\varepsilon}^T = [\epsilon_x \epsilon_y \gamma_{xy}] \tag{6-253a}$$

$$\boldsymbol{\sigma}^T = [\sigma_x \sigma_y \tau_{xy}] \tag{6-253b}$$

$$\boldsymbol{\phi} = \frac{1}{E} \begin{bmatrix} 1 & -\mu & 0 \\ -\mu & 1 & 0 \\ 0 & 0 & \dfrac{E}{G} \end{bmatrix} \tag{6-253c}$$

See Arts. 6-12 to 6-14.

Assume that forces S_i act only at the vertices (nodes) of the triangle and in the directions of the x and y axes. Of the six possible forces that can act on the element, only the three forces S_1, S_2, S_3, shown in Fig. 6-106, are independent. The other three can be computed from these from equilibrium conditions. The independent forces can be formed into a column matrix $\mathbf{S}$ for loads on the element. This matrix is related to the nodal displacements $\boldsymbol{\delta}$ of the element by Eq. (6-211), $\boldsymbol{\delta} = \mathbf{fS}$, where $\mathbf{f}$ is the flexibility matrix of the element. In this case, $\boldsymbol{\delta}$ also is a column vector with three components. Thus, $\mathbf{f}$ is a 3×3 matrix.

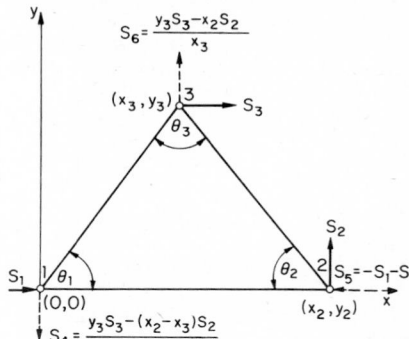

Fig. 6-106. Forces acting at nodes of thin triangular element.

A key assumption will now be made: *Stresses in the element are constant.*

This places constraints on the displacements and forces of the structure composed of such elements. As a result, equilibrium or geometric compatibility may be violated, and the idealized structure may then be stiffer or more flexible than the actual structure. The approximation may be improved by such refinements as selecting an element with more nodes (triangle with nodes at vertices and midpoints of sides) or a smaller element. (The former choice usually requires fewer calculations for about the same degree of accuracy.)

For constant σ_x, σ_y, and τ_{xy}, the stresses and nodal forces are related in terms of the nodal coordinates by

$$\boldsymbol{\sigma} = \boldsymbol{\alpha}\mathbf{S} \tag{6-254}$$

$$\boldsymbol{\alpha} = \frac{2}{t} \begin{bmatrix} -\dfrac{1}{y_3} & 0 & \dfrac{x_3 - x_2}{x_2 y_3} \\[2ex] 0 & -\dfrac{1}{x_3} & \dfrac{y_3}{x_2 x_3} \\[2ex] 0 & 0 & \dfrac{1}{x_2} \end{bmatrix} \tag{6-255}$$

From energy relations, it can be shown that for integration over the volume of the element

$$\int_V \boldsymbol{\varepsilon}^T \boldsymbol{\sigma} \, dV = \boldsymbol{\delta}^T \mathbf{S} \tag{6-256}$$

Substituting Eq. (6-254) in Eq. (6-256) and transposing matrices results in the relation between strains and nodal displacements:

$$\boldsymbol{\delta} = \int_V \boldsymbol{\alpha}^T \boldsymbol{\varepsilon} \, dV \tag{6-257}$$

By Eqs. (6-250) and (6-254), $\boldsymbol{\varepsilon} = \boldsymbol{\phi}\boldsymbol{\sigma} = \boldsymbol{\phi}\boldsymbol{\alpha}\mathbf{S}$. Hence, Eq. (6-257) can be written

$$\boldsymbol{\delta} = \left[\int_V \boldsymbol{\alpha}^T \boldsymbol{\phi}\boldsymbol{\alpha} \, dV \right] \mathbf{S} \tag{6-258}$$

But also, $\boldsymbol{\delta} = \mathbf{fS}$. Hence, the flexibility matrix is given by

$$\mathbf{f} = \int_V \boldsymbol{\alpha}^T \boldsymbol{\phi}\boldsymbol{\alpha} \, dV = \int_V \boldsymbol{\alpha}^T \boldsymbol{\phi}\boldsymbol{\alpha} t \, dA = \frac{x_2 y_3}{2} t\boldsymbol{\alpha}^T \boldsymbol{\phi}\boldsymbol{\alpha} \tag{6-259}$$

Substitution for $\boldsymbol{\alpha}$ from Eq. (6-255) and for $\boldsymbol{\phi}$ from Eq. (6-253c) gives, with $E/G = 2(1 + \mu)$,

$$\mathbf{f} = \frac{2}{Et} \begin{bmatrix} \dfrac{x_2}{y_3} & -\dfrac{\mu x_2}{x_3} & \dfrac{x_2 - x_3}{y_3} + \dfrac{\mu y_3}{x_3} \\[2ex] -\dfrac{\mu x_2}{x_3} & \dfrac{x_2 y_3}{x_3^2} & \dfrac{\mu(x_3 - x_2)}{x_3} - \dfrac{y_3^2}{x_3^2} \\[2ex] \dfrac{x_2 - x_3}{y_3} + \dfrac{\mu y_3}{x_3} & \dfrac{\mu(x_3 - x_2)}{x_3} - \dfrac{y_3^2}{x_3^2} & \dfrac{(x_3 - x_2)^2}{x_2 y_3} + \dfrac{2y_3(\mu x_2 + x_3)}{x_2 x_3} + \dfrac{y_3^3}{x_2 x_3^2} \end{bmatrix} \tag{6-260}$$

Flexibility Matrix for Rotated Axes. If the orientation of the coordinate axes for Eq. (6-260) differs from that of the axes chosen for the whole structure, the flexibility matrix and nodal forces and displacements must be transformed accordingly. Assume that the triangular element is rotated counterclockwise through an angle θ relative to the coordinate axes of the structure. Then, new nodal forces $\mathbf{S}$ in the directions of those axes and the original nodal forces

$$\mathbf{S}_o = [S_1 S_4 S_2 S_5 S_3 S_6]^T$$

are related by

$$\mathbf{S}_o = \mathbf{RS} \tag{6-261}$$

where $\mathbf{R}$ is a diagonal matrix with elements

$$\mathbf{R}_{ij} = \begin{bmatrix} \cos \theta & \sin \theta \\ -\sin \theta & \cos \theta \end{bmatrix} \text{ for } i = j \qquad \mathbf{R}_{ij} = \mathbf{0} \text{ for } i \neq j \tag{6-262}$$

From Eq. (6-261), the relation between original nodal forces and new nodal forces can be expressed as

$$[S_1 S_2 S_3]_o^T = \mathbf{r}[S_1 S_2 S_3]^T \tag{6-263}$$

The transformed flexibility matrix $\mathbf{f}$ then is given in terms of the flexibility matrix $\mathbf{f}_o$ of Eq. (6-260) by

$$\mathbf{f} = \mathbf{r}^T \mathbf{f}_o \mathbf{r} \tag{6-264}$$

General Methods for Computing Stiffness. Each component k_{ij} of a stiffness matrix equals a force at a node when a unit displacement occurs only at the same or another node. When the matrix for a finite element cannot readily be developed from this definition, other means must be used.

For some elements, the components of the stiffness matrix can be obtained by integrating over the whole volume of the element

$$k_{ij} = \int_V \boldsymbol{\varepsilon}^T \boldsymbol{\sigma} \, dV \tag{6-265}$$

where $\boldsymbol{\varepsilon}^T$ = transpose of the matrix of unit strains produced in the element when the node at which force k_{ij} acts is given a unit displacement in the direction of the force while no other displacements occur

$\boldsymbol{\sigma}$ = matrix of the stresses in the element when a unit displacement occurs at the same or another node while no other displacements occur

dV = differential volume

(See Arts. 6-12 to 6-14.)

For some elements, the components of the stiffness matrix can be obtained by differentiating the strain energy U in the element because of nodal displacements:

$$k_{ij} = \frac{\partial^2 U}{\partial \delta_i \partial \delta_j} \tag{6-266}$$

where δ_i = displacement of the node where k_{ij} acts, in the direction of k_{ij}

δ_j = displacement of the same or another node

Stiffness of Thin Triangular Element. For an elastic triangular element with modulus of elasticity E, shearing modulus G, Poisson's ratio μ, and thickness t very small relative to other dimensions, stresses σ are related to strains ϵ by

$$\boldsymbol{\sigma} = \boldsymbol{\kappa \varepsilon} \tag{6-267}$$

$$\boldsymbol{\sigma}^T = [\sigma_x \sigma_y \tau_{xy}] \tag{6-268a}$$

$$\boldsymbol{\varepsilon}^T = [\epsilon_x \epsilon_y \gamma_{xy}] \tag{6-268b}$$

$$\kappa = \frac{E}{1 - \mu^2} \begin{bmatrix} 1 & \mu & 0 \\ \mu & 1 & 0 \\ 0 & 0 & \frac{1 - \mu}{2} \end{bmatrix} \tag{6-268c}$$

See Arts. 6-12 to 6-14.

Assume that forces S_i act only at the vertices (nodes) of the triangle and in the directions of the x, y axes. A displacement δ_i in the direction of each axis can occur at each node, as shown in Fig. 6-107. Hence, six nodal displacements can be chosen. They can be formed into a displacement matrix $\boldsymbol{\delta}$. This matrix is related to the matrix S of nodal forces by Eq. (6-223), $S = k\boldsymbol{\delta}$, where k is the stiffness matrix of the element.

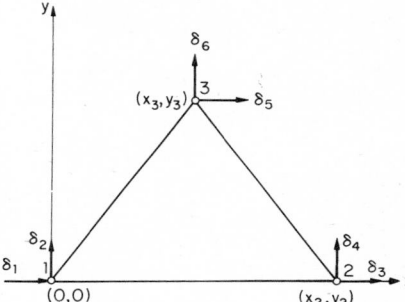

Fig. 6-107. Displacements at nodes of thin triangular elements.

A key assumption will now be made: *Unit strains in the element are constant.* (This places constraints on the displacements and forces, as discussed in connection with the development of the flexibility matrix of a triangular element.) Because

$$\boldsymbol{\varepsilon}^T = \begin{bmatrix} \dfrac{\partial \delta_x}{\partial x} & \dfrac{\partial \delta_y}{\partial y} & \dfrac{\partial \delta_x}{\partial y} + \dfrac{\partial \delta_y}{\partial x} \end{bmatrix} \tag{6-269}$$

where $\boldsymbol{\varepsilon}^T$ = transpose of the matrix of unit strains produced in the element when the node at which force k_{ij} acts is given a unit displacement in the direction of the force while no other displacements occur

δ_x = component of displacement within the element in the direction of x axis

δ_y = component of displacement within the element in the direction of y axis

this assumption is equivalent to requiring that displacements vary linearly within the element. In matrix form, this relation can be written:

$$\begin{bmatrix} \delta_x \\ \delta_y \end{bmatrix} = \mathbf{Xc} = \begin{bmatrix} 1 & x & y & 0 & 0 & 0 \\ 0 & 0 & 0 & 1 & x & y \end{bmatrix} [c_1 c_2 c_3 c_4 c_5 c_6]^T \tag{6-270}$$

where c_i = constant. The strains then, by Eq. (6-269), are given by

$$\boldsymbol{\varepsilon} = \mathbf{Cc} = \begin{bmatrix} 0 & 1 & 0 & 0 & 0 & 0 \\ 0 & 0 & 0 & 0 & 0 & 1 \\ 0 & 0 & 1 & 0 & 1 & 0 \end{bmatrix} [c_1 c_2 c_3 c_4 c_5 c_6]^T \tag{6-271}$$

Nodal displacements can be obtained from Eq. (6-270) on substitution of the nodal coordinates (Fig. 6-107):

$$\boldsymbol{\delta} = \begin{bmatrix} 1 & 0 & 0 & 0 & 0 & 0 \\ 0 & 0 & 0 & 1 & 0 & 0 \\ 1 & x_2 & 0 & 0 & 0 & 0 \\ 0 & 0 & 0 & 1 & x_2 & 0 \\ 1 & x_3 & y_3 & 0 & 0 & 0 \\ 0 & 0 & 0 & 1 & x_3 & y_3 \end{bmatrix} \begin{bmatrix} c_1 \\ c_2 \\ c_3 \\ c_4 \\ c_5 \\ c_6 \end{bmatrix} \tag{6-272}$$

Solving for the constants yields

$$
\mathbf{c} = \frac{1}{x_2 y_3}
\begin{bmatrix}
x_2 x_3 & 0 & 0 & 0 & 0 & 0 \\
-y_3 & 0 & 0 & y_3 & 0 & 0 \\
x_3 - x_2 & 0 & -x_3 & 0 & x_2 & 0 \\
0 & x_2 y_3 & 0 & 0 & 0 & 0 \\
0 & -y_3 & 0 & y_3 & 0 & 0 \\
0 & x_3 - x_2 & 0 & -x_3 & 0 & x_2
\end{bmatrix}
\begin{bmatrix}
\delta_1 \\ \delta_2 \\ \delta_3 \\ \delta_4 \\ \delta_5 \\ \delta_6
\end{bmatrix}
\tag{6-273}
$$

Substitution of Eq. (6-273) in Eq. (6-271) develops the relation between unit strains and nodal displacements:

$$
\boldsymbol{\varepsilon} = \boldsymbol{\beta}\boldsymbol{\delta}
\tag{6-274}
$$

$$
\boldsymbol{\beta} = \frac{1}{x_2 y_3}
\begin{bmatrix}
-y_3 & 0 & 0 & y_3 & 0 & 0 \\
0 & x_3 - x_2 & 0 & -x_3 & 0 & x_2 \\
x_3 - x_2 & -y_3 & -x_3 & y_3 & x_2 & 0
\end{bmatrix}
\tag{6-275}
$$

From energy relations, it can be shown that for integration over the whole volume of the element

$$
\int_V \boldsymbol{\sigma}^{\mathrm{T}} \boldsymbol{\varepsilon}\, dV = \mathbf{S}^{\mathrm{T}}\boldsymbol{\delta}
\tag{6-276}
$$

Substituting Eq. (6-274) in (6-276) and transposing matrices results in the relation between stresses and nodal forces:

$$
\mathbf{S} = \int_V \boldsymbol{\beta}^{\mathrm{T}}\boldsymbol{\sigma}\, dV
\tag{6-277}
$$

By Eqs. (6-267) and (6-274), $\boldsymbol{\sigma} = \boldsymbol{\kappa}\boldsymbol{\varepsilon} = \boldsymbol{\kappa}\boldsymbol{\beta}\boldsymbol{\delta}$. Hence, Eq. (6-277) can be written as

$$
\mathbf{S} = \left[\int_V \boldsymbol{\beta}^{\mathrm{T}}\boldsymbol{\kappa}\boldsymbol{\beta}\, dV \right] \boldsymbol{\delta}
\tag{6-278}
$$

But also, $\mathbf{S} = \mathbf{k}\boldsymbol{\delta}$. Consequently, according to Eq. (6-278), the stiffness matrix for the triangular element is given by

$$
\mathbf{k} = \int_V \boldsymbol{\beta}^{\mathrm{T}}\boldsymbol{\kappa}\boldsymbol{\beta}\, dV = \int_V \boldsymbol{\beta}^{\mathrm{T}}\boldsymbol{\kappa}\boldsymbol{\beta} t\, dA = \frac{x_2 y_3}{2}\, t\boldsymbol{\beta}^{\mathrm{T}}\boldsymbol{\kappa}\boldsymbol{\beta}
\tag{6-279}
$$

Substitution for $\boldsymbol{\beta}$ from Eq. (6-275) and for $\boldsymbol{\kappa}$ from Eq. (6-268c) gives

$$
\mathbf{k} = \frac{Et}{2(1-\mu^2)x_2 y_3}
\begin{bmatrix}
y_3^2 + x_{23}^2\lambda & x_{23}y_3(\mu+\lambda) & -y_3^2 + x_3 x_{23}\lambda \\
x_{23}y_3(\mu+\lambda) & x_{23}^2 + y_3^2\lambda & -y_3(x_{23}\mu - x_3\lambda) \\
-y_3^2 + x_3 x_{23}\lambda & -y_3(x_{23}\mu - x_3\lambda) & y_3^2 + x_3^2\lambda \\
y_3(x_3\mu - x_{23}\lambda) & x_3 x_{23} - y_3^2\lambda & -x_3 y_3(\mu+\lambda) \\
-x_2 x_{23}\lambda & -x_2 y_3\lambda & -x_2 x_3\lambda \\
-x_2 y_3\mu & -x_2 x_{23} & x_2 y_3\mu
\end{bmatrix}
$$

$$
\begin{bmatrix}
y_3(x_3\mu - x_{23}\lambda) & -x_2 x_{23}\lambda & -x_2 y_3\mu \\
x_3 x_{23} - y_3^2\lambda & -x_2 y_3\lambda & -x_2 x_{23} \\
-x_3 y_3(\mu+\lambda) & -x_2 x_3\lambda & x_2 y_3\mu \\
x_3^2 + y_3^2\lambda & x_2 y_3\lambda & -x_2 x_3 \\
x_2 y_3\lambda & x_2^2\lambda & 0 \\
-x_2 x_3 & 0 & x_2^2
\end{bmatrix}
\tag{6-280}
$$

where $\lambda = (1 - \mu)/2$

$x_{23} = x_2 - x_3$

Stiffness Matrix for Rotated Axes. If the orientation of the coordinate axes for Eq. (6-280) differs from that of the axes chosen for the whole structure, the stiffness matrix and nodal displacements must be transformed accordingly. Assume that the triangular element is rotated counterclockwise through an angle θ relative to the coordinate axes of the structure. Then, new nodal displacements $\boldsymbol{\delta}$ in the direction of those axes are related to the original nodal displacements $\boldsymbol{\delta}_o$ used in determining Eq. (6-280) by

$$
\boldsymbol{\delta}_o = \mathbf{R}\boldsymbol{\delta}
\tag{6-281}
$$

where $\mathbf{R}$ is the diagonal matrix with elements given by Eq. (6-262). The transformed stiffness matrix $\mathbf{k}$ then is given in terms of the flexibility matrix $\mathbf{k_0}$ of Eq. (6-280) by

$$\mathbf{k} = \mathbf{R}^\mathrm{T}\mathbf{k}_o\mathbf{R} \qquad\qquad (6\text{-}282)$$

(See bibliography at the end of Art. 6-101.)

Section **7**

Geotechnical Engineering

FREDERICK S. MERRITT
Consulting Engineer, Syosset, N.Y.

Geotechnical engineering deals with analysis of soil behavior and design and construction of substructures, those parts of structures that transmit the loads of the structures into the earth. Geotechnical engineering also involves necessary treatment of underlying material to insure adequate load-carrying capacity without undesirable deformations. In addition, geotechnical engineering deals with measures for execution of construction below or at grade without damage to adjacent property or injury to persons engaged in the work or present in the vicinity.

Geotechnical engineers, thus, must have a thorough knowledge of soils and their behavior—geology and soil mechanics—of structural theory, and of materials suitable for substructure construction. Engineering geology aids in classifying soils and understanding soil characteristics.

Soil mechanics applies the laws of physics to engineering problems dealing with soils. Soils are sediments and other unconsolidated accumulations of solid particles produced by disintegration of rocks, and mixtures of such particles with organic substances (Art. 7-2).

Because of the complexity of substructure design and construction, consultants specializing in this field often are engaged to assist architects, engineers, and contractors. This is especially desirable since inadequate foundations may lead to extensive damage to or destruction of the superstructures. Also, since knowledge of foundation practices in a geographic area is very important, because theory may not be adequate for a specific problem, engagement of experts familiar with practice in this area is advisable.

7-1. Lessons from Failures. Foundation inadequacies or failures usually are costly. They often are discovered only after the structure has been completed and in use, perhaps for years. Usually, the trouble is due to improper engineering analysis and neglect of conditions that could have been foreseen. Whatever the cause, the remedy is expensive. On the other hand, an overly conservative foundation also can be costly. So geotechnical engineering requires a balancing of conservative design and economics, based on an accurate determination and analysis of site conditions and superstructure loads.

When foundations give trouble, the cause usually is one or more of the following: inadequate site investigation; improper interpretation of exploration findings; faulty foundation design; poor workmanship during construction; insufficient provision for the effects of exceptional thermal and biological conditions, rainfall, and floods. Analyses of reported foundation failures indicate that, in addition to the precautionary steps necessary for good, safe super-structure design and construction, certain steps are required to insure good, safe, economical substructure design and construction. These include:

Make a thorough site investigation. This applies to small as well as large structures. Experience shows that there is more trouble with one-story buildings, such as houses, shopping centers, and small factories, than with major structures, because many engineers do not consider a small building to be heavy enough to settle.

Inform those making the investigation and tests of the nature of the structure and founda-tion requirements. Require close collaboration between the contractors, designers, and those making the tests and investigation. Inform the geotechnical engineers of unexpected changes in the soil encountered during construction, especially in load-bearing capacity, and of increases in the load on the soil due to superstructure changes. See what can be learned from the engineers responsible for design of foundations of neighboring structures. On completion of the structure, compare actual settlements with those computed.

See that soil samples are properly extracted and handled. Take special care with those for cohesionless or stratified soils.

Determine whether groundwater is present. If it is, determine the elevation of the groundwater table and estimate future variations in level. Take into account possible changes in surface-water runoff due to removal of vegetation. Consider the possibility of subsidence from lowering of the groundwater table.

Investigate the effects of uplift, seepage, scour, and water flow.

Determine the effects on the soil of load from the superstructure, including probable future changes in the nature or distribution of the load. For example, a change in a building from a static to a vibratory load may cause a foundation to fail. A change in load distribution may produce instability or overload a foundation. Consider also possible changes in the load-carrying capacity of the subsoil due to vibration, scour, groundwater, or erection of struc-tures nearby.

Design foundations to avoid differential settlements, since such settlements may seriously affect the structure, or take them into account in design. For example, use articulation instead of continuity, or separate parts with unequal settlement, or use a rigid foundation.

Select a foundation appropriate both for the superstructure to be supported and for the soil it rests on. For example, abutments on bedrock or on battered and vertical piles are suitable for arches. But for a concrete girder continuous over several spans, supporting some reactions on spread footings and others on piles could cause serious trouble, due to differential settlements.

Avoid use of different types of foundation for a structure where the effects of differential settlement may be serious.

Design against tipping over. Failures have occurred because of overloading, unbalanced loading, or variations in depth and type of subsoil. Failures also have taken place because of release of pressure along one side of a structure with no corresponding release on the other. Such a condition might occur along construction for a subway or a deep intercepting sewer.

Interpret the findings of a site investigation and tests prudently. Misinterpretation can lead to a poor design. Be especially careful in use of interpolation and extrapolation. For example, below-grade rock may not slope evenly from one part of a site to another; it may have sharp peaks and abrupt depressions, undetected by insufficient borings. Bearing strata may vary in thickness and deform in different amounts under the same load at dif-ferent parts of the site. Consider the effects of such strata on a structure for which the same size footing is always used for the same load.

Investigate the possibility of stress superposition on the soil under a foundation causing overstressing. Such a condition may arise when footings are placed so close that high pressures from one overlap those from others. Or it may arise when an adjoining fill is placed; for example, behind a bridge abutment.

When a structure is to be supported on inclined soil strata, consider the possibility of soil movement under the load of the structure.

Design below-grade walls to resist initial and future earth pressures. See that construction is carried out in accordance with design assumptions.

Either avoid foundation construction that may seriously affect adjacent structures or provide adequate support for them, both during and after construction. Prevent loss of lateral support for an existing foundation or flow of subsoil from under it. Do not set a footing below a nearby one without first adequately underpinning the existing footing. Construct the deepest foundations first.

Keep construction under close supervision to prevent shoddy work, unsuitable methods, and use of inadequate equipment. See that excavations are adequately braced. Bracing should resist lateral pressure of retained earth, nonuniform pressure distributions from different soil layers, vibrations from vehicles and machines, superimposed loads on the retained soil, and other loads that may induce oblique stresses.

Check cofferdams for adequate bracing. Be sure that they are watertight and extend to a sufficient depth.

Be careful in removing water from an excavation that the surrounding soil is not disturbed and that adjacent structures are not endangered.

Avoid pile driving where the vibration may cause the soil to consolidate or liquefy and lose bearing capacity. Do not use piles where they are exposed to deterioration, for example, steel piles where they may corrode or wood piles where they may rot or be attacked by borers. Consider the possibility that future lowering of the water table will expose untreated wood piles to decay.

Consider the effect of drag or negative friction on piles in place where there are layers of compressible material or where fill is placed on compressible material.

Provide a reasonable pile spacing. Too close a spacing is at best uneconomical and at worst unsafe. See that piles are placed properly and not bent or damaged during driving.

(C. Szechy, "Foundation Failures," Concrete Publications, Ltd., 14 Dartmouth St., London, S.W.1; T. McKaig, "Building Failures," McGraw-Hill Book Company, New York; J. Feld, "Lessons from Failures of Concrete Structures," American Concrete Institute, Detroit, Mich., and The Iowa State University Press, Ames, Iowa.)

7-2. Soil Characteristics. Soils may consist of rock, rock particles, minerals derived from rock, organic matter, or a mixture of two or more of these materials. Any of these materials or mixtures may extend from the surface down to bedrock. Or they may lie in layers, with several layers of different types between the surface and bedrock.

Bedrock is sound, hard rock lying in the position where it was formed and not underlain by any material other than rock. Basically, rock may be igneous, sedimentary, or metamorphic. Igneous rocks were formed by solidification of molten material. Typical igneous rocks include granite, basalt, diorite, rhyolite, and andesite. Sedimentary rocks were created from deposits, chiefly by water, but sometimes by air and ice. Common sedimentary rocks include conglomerate, sandstone, shale, and limestone. Metamorphic rocks were produced by heat and pressure acting on existing rocks to create materials with new characteristics. Gneiss, quartzite, slate, and marble are typical metamorphic rocks.

Bedrock usually is capable of withstanding high pressures. Thus, it is highly desirable for foundations when it can be reached economically. (Some building codes permit a bearing pressure on it of 60 tons per sq ft. In Chicago, 100 tons per sq ft is allowed on the Niagara limestone underlying the city.) But care must be taken in site exploration that boulders or weathered or disintegrated rock is not mistaken for bedrock. Also, it should be borne in mind that bedrock is not always a satisfactory foundation material. It may be cracked or lie in inclined strata. Movements along the joints may occur under load or when the joints get wet. Limestone may have sharp peaks and valleys or contain hollow pockets, caves, and clay layers. Some rocks may be physically or chemically unstable, especially when exposed to air or water; some shales, slates, and limestones fall in this category. So in making a site investigation, learn as much as possible about the nature of bedrock, in addition to its distances below the surface throughout the site.

Weathered rock is some stage in the deterioration of bedrock into overburden, or soil. Usually, it is found at the interface of bedrock and overburden. It may have seams filled with rock fragments or claylike materials. Layers of soil or disintegrated rock may separate pieces of weathered rock from the main strata. Sometimes, water, ice, or wind carries away much of the fine material. Weathered rock should not be trusted to carry heavy loads. It should not be used for sides of steep, exposed cuts, because of the danger of slides.

Boulders are rock fragments over 10 in. in maximum dimension. *Cobbles* are 2 to 10 in. in size; *pebbles,* 4 mm to 2 in. Boulders encountered in weak soil should not be trusted to carry heavy loads, and in any soil, they may tend to tip when loaded. They interfere with driving of piles or caissons. Where it is uneconomical to remove boulders or set the foundations above them, it may be necessary to relocate the foundations or even to select another site for the project.

Gravel denotes unconsolidated rock fragments from 2 mm to 6 in. in size. If composed of hard, sound rock, gravel makes a good foundation material. (Shaly gravel is an exception.) But the gravel should not be underlain by a weak stratum or subject to scour. Some building codes permit bearing pressures up to 10 tons per sq ft on gravel strata, depending on compactness.

Sand consists of rock particles from 0.05 to 2 mm in size. (The lower limit may be different in some soil-classification systems.) A stratum with particles in the range from 0.05 to 0.25 mm may be classified as fine sand; from 0.2 to 0.6 mm, as medium sand; and from 0.25 to 2 mm, as coarse sand. Usually, compact sands make a good foundation material, especially if confined. Some building codes allow up to 6 tons per sq ft bearing pressure on compact sands. Coarse to medium sands may be allowed higher pressures than fine sands: 4 tons per sq ft compared with 2.

Water may convert very fine sands to quicksand, which may flow out from under even a very lightly loaded foundation. Such materials may be used to support a foundation only when they are confined by the surrounding soil.

Where loosely consolidated sand deposits, or even gravel deposits, are encountered, allowable loads should be reduced to prevent excessive settlement.

Silt and clay are fine-grained soils; individual particles cannot be readily distinguished with the unaided eye. In some classification systems, they are distinguished by particle size. In other systems, plasticity is the determining characteristic, that is, the ability to deform rapidly without cracking, crumbling, or volume change and with relatively small rebound when the deforming force is removed.

Silt, then, in one classification system, consists of rock particles from 0.005 to 0.05 mm in size. In another system, it is a fine-grained inorganic soil that cannot be made plastic by adjustment of water content and that exhibits little or no strength when air-dried. Organic silt is silt mixed with finely divided organic matter. Malodorous, it frequently underlies lakes, river deltas, and backwaters.

Silt is an unreliable foundation material when wet, though some building codes permit up to 2 tons per sq ft bearing pressure on dry, confined silts in a consolidated state. Low frictional resistance permits silt to flow, however, when it is wet and unconfined. Because of its low-permeability and slow-drainage characteristics, silt is one of the most difficult materials to excavate.

Clay, in one classification system, consists of inorganic particles less than 0.005 mm in size. In another system, it is a fine-grained inorganic soil that can be made plastic by adjustment of water content and that exhibits considerable strength when air-dried. It loses its plasticity when dried and its strength when wetted. Also, it shrinks and cracks when drying and expands when moisture is restored. A varved clay contains thin silt strata. Formed by periodic variations in sedimentation, it often is alternately light and dark in color.

Clays may be classified as soft, medium, and stiff, depending on moisture content and prior consolidation. They make satisfactory foundation materials under the right circumstances. Some building codes, for example, allow up to 5 tons per sq ft on hard, dry, consolidated clay (compressed for a long time under glaciers or other overburden), but only 1.5 tons per sq ft for stiff clay and 1 ton per sq ft for medium clay. But such pressures should not be used in design if increases in moisture content may occur.

Because of its cohesiveness, clay can stand on steep slopes temporarily. It also can transmit moderate lateral pressures around a small excavation. Furthermore, because clay is impermeable, it can keep groundwater out of an excavation. Deep excavations, however, may blow up if the clay bottom is subjected to hydrostatic pressure from underlying pervious layers.

Hardpan consists of cohering material containing rock fragments. Hardpans of glacial origin are composed of particles, ranging in size from colloidal clay to boulders, that were at least partly cemented together by high pressures. Other common hardpans may be mixtures of sand, gravel, and clay; cemented sand and gravel; or cemented silt and sand. Dry cemented hardpans will not disintegrate when submerged in water, but those with clay as the binder will. The suitability of hardpan as a foundation material depends on its consolidation and the characteristics of underlying strata. Some building codes allow up to 12 tons per sq ft on hardpan overlying rock.

Till is a glacial deposit of mixtures and pockets of clay, silt, sand, gravel, and boulders. Some glacial tills, highly compressed, or indurated, are very hard in their natural state and almost impervious. Usually well graded, they are excellent for construction of earth dams and embankments.   Allowable bearing pressures may range up to 10 tons per sq. ft for buildings.   Loose tills, however, vary in character and may cause uneven settlements.

Loam, or topsoil, is a mixture of humus (organic matter) and sand, silt, or clay.   While good for agricultural purposes, it is not desirable for foundations.

Adobe is a heavy-textured alluvial clay found in arid regions of the southwestern United States.   Though stiff and cohesive when dense and moist, it is not a desirable foundation material because it loses these properties when wet.

Loess is a uniform, cohesive, wind-blown deposit of fine-grained soil.   Often light brown in color, it consists of particles ranging in size from 0.01 to 0.05 mm.   It becomes impervious and difficult to compact when worked.   Bond between particles is due to a calcareous binder. Hence, when saturated, a loess deposit may settle, depending on its initial density and solubility of binder.   Loess can stand unsupported on nearly vertical cuts, but inclined cuts erode badly.   Allowable bearing pressures for building foundations may range up to 2 tons per sq ft.

Soils that generally should not be used for foundations include **gumbo,** a claylike material, very sticky when wet; **mud,** or **muck,** a sticky mixture of earth and water, very weak or fluid; **peat,** partly decayed organic material, as found in swamps, with very poor bearing capacity; and **bentonite,** decomposed volcanic ash, which swells when wetted.

Soils whose strength depends on a soluble binder may be used as a foundation material only where water can be excluded.   For example, *caliche,* a conglomerate of clay, sand, silt, and gravel, cemented together by calcium carbonate, may lose the binder by leaching. Hence, footings may settle and steep cuts break down.

Soils often occur as mixtures.   Their classification depends on the basic soils that determine their behavior, but not necessarily on the relative amounts of these soils present. Generally, a mixture is denoted by the name of the soil with primary effect on behavior and an adjective based on the soil with secondary effect.   In one classification system, for example, mixtures with more than 20% clay, by dry weight, are classified as clay, because

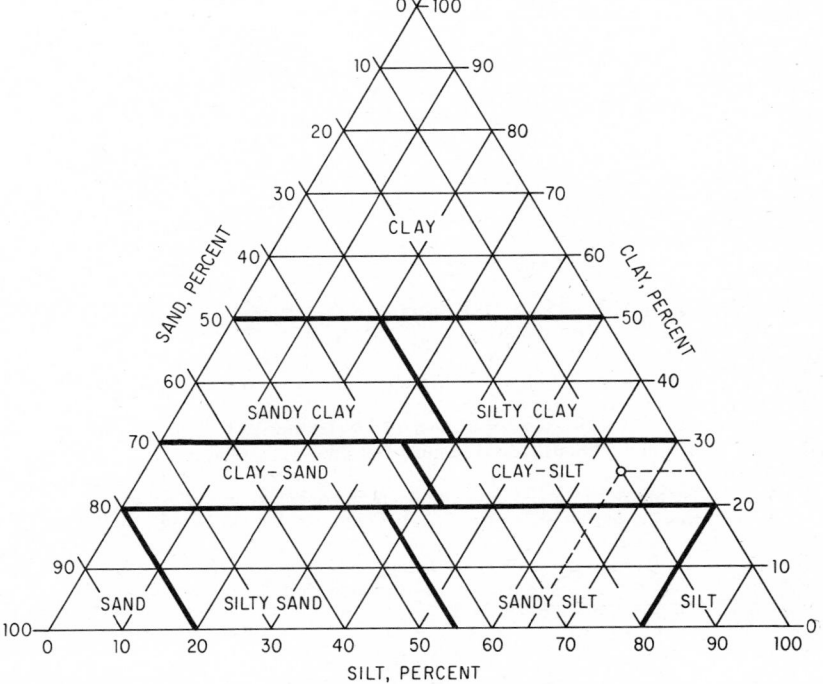

Fig. 7-1. Classification chart for mixed soils. *(Lower Mississippi Valley Division, U.S. Corps of Engineers.)*

such amounts of clay dominate the behavior of the mixture. If the secondary fraction is sand, the mixture is classified as sandy clay; if the secondary fraction is silt, the mixture is silty clay. When more than 30% clay is present, no adjective is used.

Figure 7-1 shows diagrammatically a classification system developed by the Lower Mississippi Valley Division, U.S. Corps of Engineers. It assumes the following size limits: sand, 2.0 to 0.05 mm; silt, 0.05 to 0.005 mm; and clay, less than 0.005 mm. Percentages are based on dry weight. In this system, a mixture with 50% or more clay is classified as clay; with 80% or more silt, as silt; and with 80% or more sand, as sand. Clay is the primary material in mixtures with 30% or more clay; for example, a mixture with 40% clay and 40% sand is a sandy clay. A mixture with 25% clay and 65% silt is a clay-silt (intersection of dash lines in Fig. 7-1). (See also Art. 18-33.)

For an alternative, more precise method of classifying soils, see American Society for Testing and Materials, "Standard Method for Classification of Soils for Engineering Purposes," D2487.

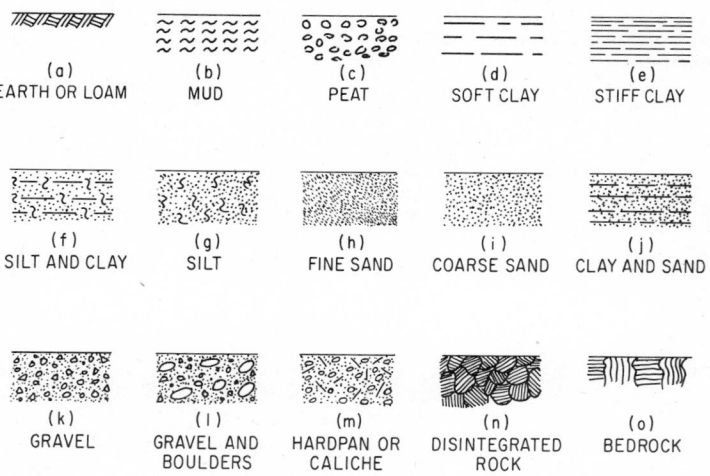

Fig. 7-2. Symbols for soils.

Figure 7-2 presents some suggested symbols for representing soils on drawings.

(R. F. Legget, "Geology and Engineering," McGraw-Hill Book Company, New York; D. P. Krynine and W. R. Judd, "Principles of Engineering Geology and Geotechnics," McGraw-Hill Book Company, New York; T. W. Lambe and R. V. Whitman, "Soil Mechanics," John Wiley & Sons, Inc., New York; H. F. Winterkorn and H. Fang, "Foundation Engineering Handbook," Van Nostrand Reinhold Company, New York.)

7-3. Index Properties of Soils. (See also Art. 7-2.) Easily observed physical properties of soils frequently suffice as indexes of behavior. Such index properties include textural appearance, density, moisture content, consistency, permeability, compressibility, and shearing strength. [See also American Society for Testing and Materials "Recommended Practice for Description of Soils (Visual-Manual Procedure)," D2488.]

Soil texture, or surface appearance, depends on particle size, shape, and gradation. (Color rarely is significant in foundation engineering, except as an indication of the presence of certain minerals.) As noted in Art. 7-2, boulders, gravel, sand, silt, and clay may be distinguished by particle size, and mixed soils may be classified in terms of these basic classes, for example, as sandy clays or clay-sand.

Particles vary not only in size but also in surface area and shape. Some significant properties, such as adsorption, are a function of surface area per unit of mass, or specific surface. Several different characteristics of sands and clays thus may be explained by the fact that the specific surface of sand is about 0.001 that of clay.

Particle shape influences maximum density, compressibility, shearing strength, and other properties. Gravel, sand, and silt particles generally have three principal dimensions of the same order of magnitude; such particles are called *bulky*. Bulky particles may be angular or nearly round.   Coarse-grained soils with angular particles have greater strength and bearing capacity than those with round particles. They generally also have greater bearing capacity than clays, because platelike particles predominate in most clays.   Accumulations of such particles usually are more compressible than soils composed of bulky particles.

Soil density is a measure of the concentration or packing of particles in a soil mass.   It also is an index of compressibility.   Less dense, or loosely packed, soils are much more compressible than those with higher density.   Soil density may be expressed numerically as void ratio and porosity.   Void ratio is given by

$$e = \frac{V_v}{V_s} \tag{7-1}$$

where V_v = volume of voids (all space not occupied by solids) in a soil mass
$\quad V_s$ = volume of solids in the mass

$$V_v + V_s = V_t \tag{7-2}$$

where V_t = total volume of the mass.
$\quad$ Porosity, percent, is given by

$$n = \frac{V_v}{V_t} 100 \tag{7-3}$$

(Note that a soil with high porosity may not necessarily be highly pervious. For example, clay, which has very high porosity, has very low permeability.   Passage of water through a soil depends on size, shape, and continuity of voids.)

Relative density D_d, percent, is a measure of the compactness of a soil with a void ratio e, when the maximum void ratio is e_{max} and the minimum e_{min}.

$$D_d = \frac{e_{max} - e}{e_{max} - e_{min}} 100 \tag{7-4}$$

Percentage compaction usually is used to measure soil density in a fill.   Generally, the maximum Proctor weight (dry, lb per cu ft), determined by a standard laboratory test, is set up as the job standard.   Or American Association of State Highway and Transportation Officials Method of Test T180, "Moisture-Density Relations of Soils Using a 10-lb Rammer and an 18-in. Drop" (ASTM D1557), or AASHTO test method T99 (ASTM D698), with a 5½-lb hammer, may be used.   Percentage compaction is 100 times the ratio of the unit dry weight, lb per cu ft, of a section of fill to the job standard.   Normally, from 90 to 100% compaction is specified.

Moisture content or water content is an important influence on soil behavior.   Water content, dry-weight basis, percent, is given by

$$w = \frac{W_w}{W_s} 100 \tag{7-5}$$

where W_w = weight of water in a soil mass, lb
$\quad W_s$ = weight of solids in the mass, lb

$$W_w + W_s = W_t \tag{7-6}$$

where W_t = wet weight of the mass, lb.
$\quad$ Degree of saturation, percent, is given by

$$S = \frac{V_w}{V_v} 100 \tag{7-7}$$

where V_w = volume of water in the mass
$\quad V_v$ = total void volume
$\quad$ Saturation, density, and moisture content are related:

$$Se = wG \tag{7-8}$$

where G = specific gravity of the solids in the mass.
$\quad$ Thus, in a fully saturated soil, void ratio is proportional to water content.

Consistency describes the condition of fine-grained soils—soft, firm, or hard.   Shearing strength and bearing capacity vary significantly with consistency.

Consistency generally is measured by the Atterberg system. This recognizes four states: liquid, plastic, semisolid, and solid. A soil, such as clay, will pass through these states as moisture content changes. When a soil that is fluid loses water, there is a stage where it ceases to behave as a liquid. The moisture content at this stage is called the *liquid limit*. With further removal of moisture, the plastic state terminates. The moisture content at that stage is the *plastic limit*. Finally, the *shrinkage limit* is reached when volume change with decrease in water content ceases.

Plasticity index is the difference between liquid and plastic limits. *Shrinkage index* is the difference between plastic and shrinkage limits. These indexes may be used for soil classification. (A. Casagrande, "Classification and Identification of Soils," *Proceedings, American Society of Civil Engineers,* June 1947.)

Use of the Atterberg system is limited principally to fills and soils that are to be compacted because the laboratory tests for determining the limits are conducted on remolded specimens. Remolding may significantly change properties of natural soil. Nevertheless, application of the Atterberg system yields some generally useful information on fine-grained soils: Liquid limit of a soil increases with increase in clay or organic content. Increase in finely divided material raises the liquid limit of inorganic soils. The lower the plasticity index, the greater the permeability and compressibility at a given liquid limit.

Permeability is the ability of a soil to conduct or discharge water under a hydraulic gradient. This property depends on soil density, degree of saturation, and particle size. Coarse-grained soils are highly pervious and have high permeability coefficients; fine-grained soils are much less pervious and have low coefficients (Table 7-1). The coefficient of permeability k, cm per sec, is defined by Darcy's law:

$$Q = kiA \qquad (7-9)$$

where Q = rate of flow of water through a soil mass, cu cm per sec
   i = hydraulic gradient, or total head lost per unit of flow distance, cm per cm
   A = total cross-sectional area of soil through which flow takes place, sq cm

Darcy's law usually is used to estimate the flow through saturated soil of gravitational water (water free to move under gravity forces, as contrasted with water moved by capillary action or held by adsorption). Water also may flow through soils by osmosis, where solutions with different concentrations occur. For example, osmotic flow may take place when clay that is under compression is unloaded and draws in water in contact with it. Osmotic flow also may account for some part of frost heaving that occurs when ground freezes; additional water, if available, is drawn into the freezing zone.

Table 7-1. Permeability and Drainage Characteristics of Soils

Soil type	Approximate coefficient of permeability k, cm per sec	Drainage characteristic
Clean gravel	5–10	Good
Clean coarse sand	0.4–3	Good
Clean medium sand	0.05–0.15	Good
Clean fine sand	0.004–0.02	Good
Silty sand and gravel	10^{-5}–0.01	Poor to good
Silty sand	10^{-5}–10^{-4}	Poor
Sandy clay	10^{-6}–10^{-5}	Poor
Silty clay	10^{-6}	Poor
Clay	10^{-7}	Poor
Colloidal clay	10^{-9}	Poor

Osmosis sometimes is used to advantage in draining silt. Two electrodes are installed in saturated ground and a direct current is applied. The groundwater normally moves to the cathode, where it can be pumped out through a wellpoint. This process, called **electro-osmosis,** is based on an ion-exchange reaction. Positive hydrogen ions, produced by electrolysis of the water, replace positive ions in fine-grained soils as those ions move to the cathode. Movement of the ions induces flow of the water in the same direction. Electroosmosis is effective with silts, which usually are difficult to drain by open pumping or with wellpoints.

Soil compressibility is important in geotechnical engineering mainly as an indication of the prospects of settlement. Such deformation ordinarily occurs because of change of position of particles in a soil mass. Compression packs them closer. (In a saturated soil, however, compaction cannot occur without discharge of water.) Soil compressibility is indicated by the slope of a compression diagram, in which void ratio is plotted against pressure. Compressibility depends on type of soil, density, history of previous loading, handling of sample, and magnitude of the stress increment relative to the existing loading at any point.

Some soil formations take a long time to consolidate, or reach equilibrium, under load. The primary cause of this delay probably is hydrodynamic lag, movement of water out of the loaded soil.

Shearing strength is the shear stress in a soil mass at failure or when continuous displacement occurs at relatively constant stress. Shearing strength often is an important factor in determining ultimate bearing capacity of soil, stability of embankments, and pressure against retaining walls. It varies with type of soil, depth, and structural disturbance. It also varies with seasonal changes in groundwater level, capillary saturation, moisture content, and seepage. Usually, shearing strength is determined in laboratory tests, with specimens under constant normal load.

Tests show that for cohesionless soils—sand, silt, gravel—shearing strength varies almost directly with the normal pressure. The relationship is given approximately by

$$f_v = f_n \tan \phi \tag{7-10}$$

where f_v = shear strength, tons per sq ft
  f_n = average normal pressure, tons per sq ft
  ϕ = angle of internal friction (Table 7-2)

Dense cohesionless soils can sustain a shearing stress larger than the ultimate strength; that is, the stress-strain curve reaches a maximum and then follows a downward path until the ultimate strength is reached. Nevertheless, ultimate strength is the same for a soil, whether it is loose or dense. The reason is that the shearing forces bring the soil when loose and the soil when dense to about the same density. As shear increases, the void ratio of the initially loose soil decreases and that of the initially dense soil increases. Ultimately, dense and loose soils reach about the same void ratio, the *critical void ratio*. (Very little volume change takes place in a shear test on a specimen initially at the critical void ratio.)

Pore pressures affect shearing strength. If drainage of a cohesionless soil is prevented, shearing strength of the soil when loose and saturated will be less than when it is partly saturated or dry; but shearing strength of the soil when dense and saturated will be greater than when it is unsaturated. If drainage is unobstructed, shearing strength of a loose soil when saturated or partly saturated will be greater than that of the undrained soil.

Shearing strength of cohesionless soils is nearly independent of the rate at which shear force is applied.

Cohesionless materials develop the major part of their resistance to shear by friction between solids. Also, interlocking of particles contributes to the resistance. But strength due to interlocking is lost when sliding starts. In approximate calculations, therefore, the effect of interlocking may be ignored and shearing strength calculated from Eq. (7-10). Table 7-2 lists approximate values of ϕ for this purpose.

Table 7-2. Approximate Friction Angles for Cohesionless Soils

Soil type	ϕ, deg	$\tan \phi$
Silt, or uniform fine to medium sand	26–30	0.5–0.6
Well-graded sand	30–34	0.6–0 7
Sand and gravel	32–36	0.6–0.7

Shearing strength of clays derives principally from cohesion and adhesion. Cohesion is the uniting of two clay masses; adhesion, the uniting of clay with other materials.

Shearing strength of saturated fine clays (particles 0.002 mm in size or smaller) at constant water content depends mainly on viscous friction and initially also on interlocking or interference of particles on a microscopic scale. Thus, shearing strength is nearly independent of normal pressure but depends on rate of shearing and varies directly with the contact area. Shearing strength also depends on consistency. A clay at or near the liquid limit has little or no strength, whereas the same clay with less moisture may have high resistance to shear. Figure 7-3 shows how shearing strength may vary with consistency when drainage is prevented.

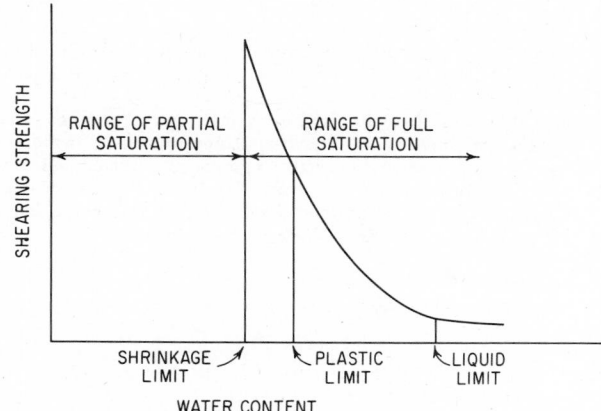

Fig. 7-3. Shearing strength of saturated clays decreases with increase in water content.

Consolidation of saturated fine clay by loading it changes the consistency when drainage can occur. Strength is thus improved, the amount depending on the compressibility of the clay. The greater the compressibility, the greater the increase in strength. In normally consolidated clays, undrained shear resistance increases approximately linearly with depth.

Shearing strength of saturated clays varies inversely with particle size and probably with specific surface, at a given water content. Particle shape also plays a role in shear resistance, as does orientation. Clays with platy particles have lower strength than those with needle-shaped particles. Random orientation of nonround particles increases shear resistance compared with orientation parallel to the shear plane.

Some clays that are firm or hard when undisturbed become soft when disturbed or remolded without change in water content. This characteristic is indicated by degree of sensitivity:

$$S_t = \frac{\text{shear strength, undisturbed sample}}{\text{shear strength, remolded sample}} \qquad (7\text{-}11)$$

For most clays, S_t ranges between 2 and 4. For sensitive clays it falls between 4 and 8, and for extra-sensitive clays, it exceeds 8. A similar loss of strength is exhibited by some clays with high water content when they are agitated. These thixotropic materials generally regain their strength after agitation ceases. Sensitivity is important where foundations may be subject to vibrations or where construction operations, such as pile driving, may disturb natural clay formations.

Shearing strength of mixed soils derives from cohesion and solid friction. Small quantities of silt and sand in a fine clay produce friction; small quantities of clay in silt or sand introduce cohesion. Thus, for such soils as silty clay, sand-clay, and silty sand, shearing strength may be estimated from Coulomb's law:

$$f_v = c + f_n \tan \phi \qquad (7\text{-}12)$$

where c = unit cohesion
$\quad f_n$ = normal pressure on sliding surface
$\quad \phi$ = angle of internal friction

In practice, Coulomb's law is difficult to apply. For one thing, c is a variable affected by many factors. For another, ϕ is affected by interlocking as well as solid friction, and by changes in clay consistency due to consolidation, changes in water content, rate of shear, structural disturbance, and many other factors. Hence, shearing strength of mixed soils often is determined by tests without attempting to separate contributions of cohesion and friction to strength.

(R. F. Scott and J. J. Schoustra, "Soil: Mechanics and Engineering," McGraw-Hill Book Company, New York; G. P. Tschebotarioff, "Soil Mechanics, Foundations, and Earth Structures," McGraw-Hill Book Company, New York.)

7-4. Site Investigations. (See also Art. 7-1) Field investigations to determine surface and subsurface conditions at a site should be made as soon as possible. Often, money can be saved if explorations are made before a site is purchased. They may reveal foundation conditions undesirable for the type of structure to be erected, in which case another site may be selected.

Numerous techniques may be used for site investigations. They vary in cost from relatively low-cost visual inspection to costly subsurface explorations and laboratory tests. How many of these techniques should be used and to what extent they should be employed depends on the type of structure, its cost, relative cost of an overly conservative design, findings of preliminary explorations, and probability and nature of damage that might occur from lack of more accurate information.

Visual inspection is an essential preliminary step. It should provide data on surface soils, surface water, slopes, accessibility for equipment for subsurface exploration, availability of water for drilling equipment, existing structures on the site, former structures, and adjacent construction. It should also determine whether underground utilities may cross the site.

Research should yield much useful information at low cost. It includes questioning local people about the history of the site, floods, and groundwater conditions. Geological maps and reports of the U.S. Geological Survey and pedological maps and reports of the U.S. Department of Agriculture should be examined. So should records of borings made nearby, if they are obtainable. Architects, engineers, and contractors for adjacent construction should be questioned on their experiences. Information should be obtained on type and depth of foundations for those structures and on behavior, including settlement records. And especially in cities, maps showing underground structures and utilities should be studied. If present, such construction should be avoided in foundation work or protected. Furthermore, research should unearth information on existing or former structures on the site, including type, location, and depth of foundations and their behavior.

Geological surveys, when made by a specialist, can provide much information on soil and rock conditions. The surveys identify distinctive landforms, such as ancient shorelines, lake beds, glacial deposits, terraces, and weathered remnants of rock formations. From these observations, the specialist often can deduce the nature of the materials in various parts of the site. Geological analysis is especially useful as a preliminary to subsurface explorations, but usually cost and time restrict its application to large projects, such as dams, levees, highways, and airports.

Aerial surveys extend geological analysis beyond the limitations of ground reconnaissance. They are economical for studies of a large area. They identify the landforms, drainage patterns, shape of erosion channels, type of vegetation, and land use, from which soil and rock characteristics of the area can be deduced. The technique is not applicable to heavily wooded or built-up areas.

Geophysical methods provide some subsurface information and are useful in planning a complete subsurface exploration. Assisted by a few borings for calibration and checking, they can map a large area faster and at less cost than methods involving borings alone. One geophysical method relates electrical resistivity to significant soil characteristics. Others relate shock waves to these characteristics. These seismic methods are based on the fact that the velocity of shock waves varies with the material through which they travel. For locating rock about 100 ft or more below grade, seismic methods are faster and less costly than borings. For locating boundaries of different strata, seismic methods require more borings for calibration than does the electrical-resistivity method.

Probing, or sounding, derives subsurface information by driving a rod or pipe into the ground and measuring penetration resistance. It is a low-cost method but is likely to supply inadequate and misleading information, especially on depth and nature of bedrock.

Jetting, using water to aid penetration, enables a pipe or hollow rod to penetrate hard or gravelly formations, where probing would be unsuccessful. Changes in color of discharge water indicate changes in soil types. Jetting has the same disadvantages as probing.

Augers provide subsurface data by bringing up material for examination. But it is difficult to establish the depth at which the material was obtained. Also, the auger disturbs the soil so much that little or no information can be obtained on the character of the soil in its natural state.

Test pits permit visual examination of soil in place and provide information on difficulty of excavation. They also make it possible to obtain an undisturbed sample of the soil manually. Cost of digging a test pit, however, increases with depth, rapidly if done manually. The work can be speeded and costs cut with a bulldozer for pits or trenches 8 to 10 ft deep, or with a backhoe for pits up to about 15 ft deep. Nevertheless, costs frequently limit test pits to relatively shallow depths.

Wash boring is a drilling process in which a hole is formed in the ground for soil sampling or rock drilling. The hole may be sustained with a steel casing or with drilling mud, a slurry of water and clay. Information on subsoil characteristics may be obtained from the resistance of the casing to driving and from samples of the soil brought up in wash water. This method of soil sampling yields unsatisfactory results, because the jetting disturbs the soil and the wash water leaves coarse particles behind.

Rotary drilling is another method of forming a hole in the ground for soil sampling or rock drilling. Powered equipment rotates a bit capable of reducing to chips the most compact soil formations or rock. Water or drilling mud fed to the bit brings the cuttings to the surface. A rotary drilling rig usually is larger and more powerful than a wash-boring rig, and often is better for deep drilling. But the "wet" samples obtained by the rotary drill, like those from wash borings, provide little useful information.

Table 7-3. Correlation of Standard-boring-spoon* Penetration with Soil Consistency and Strength

Soil consistency	Number of blows per ft on spoon	Unconfined compressive strength, tons per sq ft
Sand:		
Loose	15	
Medium compact	16–30	
Compact	30–50	
Very compact	Over 50	
Clay:		
Very soft	3 or less	0.3 or less
Soft	4–12	0.3–1.0
Stiff	12–35	1.0–4
Hard	Over 35	4 or more

*2-in. spoon, 140-lb hammer, 30-in. fall.

"**Dry**" **sampling** yields more satisfactory results. The object generally is to obtain a complete sample of the natural soil. When necessary, an undisturbed sample can be secured, but at greater cost. Any of several different types of sample spoons, or samplers, may be used to obtain a dry sample from a borehole. Thin-walled types cause less soil disturbance, but thick-walled types may be preferable in stiff, nonplastic soils. Some samplers have sectional liners for collecting samples to permit delivery to a laboratory without manual handling of the soil.

Most frequently, the borehole for dry sampling is formed with a wash-boring rig. A steel casing is driven, a section at a time. Usually, the number of blows per foot is recorded as an indication of the resistance of the soil. The borehole is cleaned before a sample is obtained. The sampler is driven into the bottom of the hole with a free-falling weight. Use of a standard 140-lb weight falling 30 in. on a 2-in.-OD sampler and recording the number of blows per foot of penetration give a better measure of soil-penetration resistance than the casing driving because of the friction on the casing, which increases irregularly with length. Actually, blows per 6 in. for an 18-in. run are preferable, to offset the disturbance of washing in the first 6 in. of the run.

Sampler penetration data often are used for soil classification. Table 7-3 shows a system that has been used for a 140-lb weight falling 30 in. on a standard 2-in.-OD spoon.

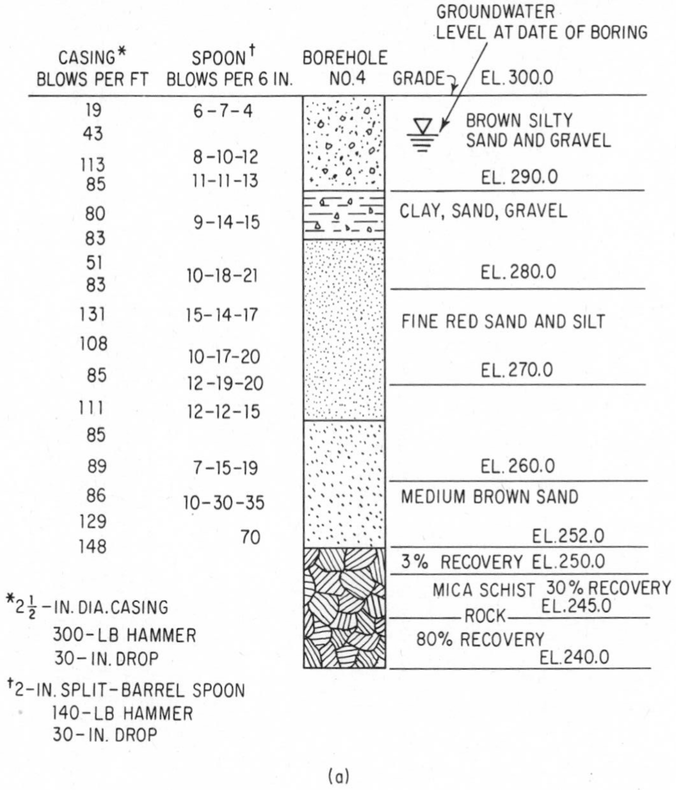

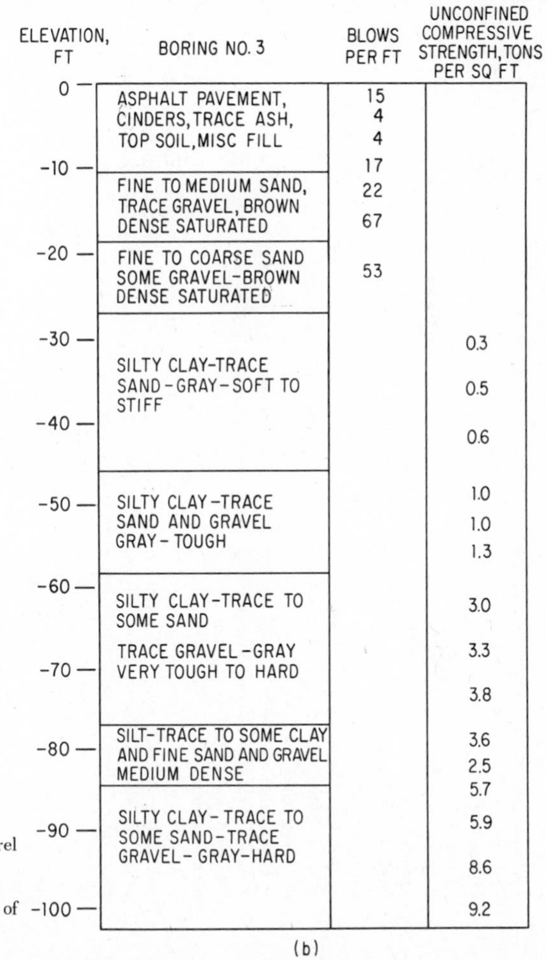

(b)

ELEVATION, FT	BORING NO. 3	BLOWS PER FT	UNCONFINED COMPRESSIVE STRENGTH, TONS PER SQ FT
0	ASPHALT PAVEMENT, CINDERS, TRACE ASH, TOP SOIL, MISC FILL	15 4 4	
-10	FINE TO MEDIUM SAND, TRACE GRAVEL, BROWN DENSE SATURATED	17 22 67	
-20	FINE TO COARSE SAND SOME GRAVEL–BROWN DENSE SATURATED	53	
-30	SILTY CLAY–TRACE SAND–GRAY–SOFT TO STIFF		0.3 0.5
-40			0.6
-50	SILTY CLAY–TRACE SAND AND GRAVEL GRAY–TOUGH		1.0 1.0 1.3
-60	SILTY CLAY–TRACE TO SOME SAND		3.0 3.3
-70	TRACE GRAVEL–GRAY VERY TOUGH TO HARD		3.8
-80	SILT–TRACE TO SOME CLAY AND FINE SAND AND GRAVEL MEDIUM DENSE		3.6 2.5 5.7
-90	SILTY CLAY – TRACE TO SOME SAND–TRACE GRAVEL–GRAY–HARD		5.9 8.6 9.2
-100			

(a)

GROUNDWATER LEVEL AT DATE OF BORING   EL. 300.0

CASING* BLOWS PER FT	SPOON† BLOWS PER 6 IN.	BOREHOLE NO. 4	
19	6-7-4		GRADE — BROWN SILTY SAND AND GRAVEL
43	8-10-12		EL. 290.0
113	11-11-13		CLAY, SAND, GRAVEL
85	9-14-15		EL. 280.0
80	10-18-21		FINE RED SAND AND SILT
83	15-14-17		EL. 270.0
51	10-17-20		EL. 260.0
83	12-19-20		MEDIUM BROWN SAND
131	12-12-15		EL. 252.0
108	7-15-19		EL. 250.0
85	10-30-35		3% RECOVERY EL. 250.0
111	70		MICA SCHIST 30% RECOVERY EL. 245.0
85			ROCK
89			80% RECOVERY EL. 240.0
86			
129			
148			

*2½–IN. DIA. CASING
300–LB HAMMER
30–IN. DROP

†2–IN. SPLIT–BARREL SPOON
140–LB HAMMER
30–IN. DROP

Fig. 7-4a. Typical boring logs. Data include blows per foot on the casing and blows per 6 in. on a split-barrel sampling spoon.

Fig. 7-4b. Data include blows per foot on a spoon in granular material and unconfined compressive strength of cohesive material.

Figure 7-4 indicates the kind of information provided by a boring log. Figure 7-4a shows a log that reports blows per foot on the casing and blows per 6 in. on the spoon for 18-in. runs. The record also shows weight and fall of hammers used in driving casing and spoon, size of casing, and size and type of spoon. If laboratory tests are made, the data should be reported on the log.

In clayey soils, for example, in Chicago, Detroit, and Cleveland, Shelby tubes, thin-wall spoons, are forced into the clay to obtain undisturbed samples. These are subjected to unconfined compression tests in a soils laboratory. The results are of greater value to soils engineers than are spoon blows. (Shearing strength equals half the unconfined compressive strength.) The technique is useful mainly in cohesive soils. Figure 7-4b shows a log that reports blows per foot on the spoon in sandy soils and unconfined compressive strength in clayey soils.

Percentage recovery of rock noted in Fig. 7-4a is the ratio of length of core obtained to distance drilled. Generally, it is safe to assume that the higher the recovery, the better the condition of the rock. But bear in mind that recovery also depends on the care in taking rock samples.

To obtain rock cores, rotary drilling with shot or diamond bits is used. Either type is satisfactory in hard rock. Diamond bits, however, produce better small-diameter cores. In soft rock, shot drilling is not satisfactory, because the shot, used as an abrasive to cut the rock, often breaks it up. The diamond bit is annular, with small commercial diamonds, or bortz, on the cutting edge. This type of bit usually is used in diameters up to 3 in.

Groundwater conditions should be reported when a site investigation is made. One method is to set up a permanent observation well and take weekly or monthly readings. Another way is to take readings in boreholes for at least 24 hr after completion of the holes. When there is little or no change in water level in a hole in granular material in successive readings, the level probably is that of the groundwater table on that day. It may fluctuate considerably with the seasons, however. Since foundation design and cost may be affected by the amount of fluctuation, efforts should be made to obtain data for estimating it. (Water levels in holes in impervious soils, such as clays, are of little value; the water usually comes from that used in drilling.)

The number, location, and depth of test pits and borings that should be made depend on the type and size of structure, loading, and soils encountered. At least one boring should extend into bedrock. Preferably, it should be the first one drilled, to obtain information on all the strata early in the investigation. Drilling should be carried at least 5 ft into rock; 10 ft would be better, to insure that a boulder or weathered rock is not mistaken for bedrock. Where cuts are to be made, sufficient holes should be drilled to ascertain that rock does not extend above the bottom of the cut. If rock is found, additional borings are necessary to determine the rock profile.

In general, borings need not be more than 100 ft deep or deeper than the smallest horizontal dimension of the structure to be supported, except where poor material overlies bedrock. When soils in the first 50 ft or so below grade have high bearing capacity, extending more than one borehole to greater depth usually is not necessary. When weak soils are encountered, the borings should extend deep enough for determination of the thickness of the weak layer and to insure that there are no weak layers under the good ones.

Local building codes may specify the minimum number of borings or test pits in terms of building area, for example, one for every 2,500 sq ft. They also may indicate the depth to which holes should be carried.

More than the minimum number of borings should be made when there is evidence of fill or previous construction on the site, when settlement must be held within close limits, or when subsurface conditions vary considerably. It usually is more economical to do a complete boring investigation than to design for the worst conditions.

In-place soil tests are advisable when they will yield important data. Load tests are frequently made; penetration, compressibility, and shear tests are occasionally carried out in the field. They may be executed in test pits or at the bottom of casings driven into the subsoil to be investigated. After the casing has been cleaned out, the exposed bottom may be tested with statically loaded cone-shaped plungers or tight-fitting bearing plates. Or the vane shear test may be made. In this test, a rod with two to four vertical steel plates, or vanes, at the tip is inserted into the ground and rotated. The torques necessary to start and maintain rotation are measured. These values can be correlated with shear resistance and friction and have been used to determine soil-bearing capacities and pile-friction resistance, for estimating pile lengths. Results generally have been reliable in soft and fine-grained soils, but they may be erratic when pebbles or stones are present.

Load tests usually are made on a small area of soil in a pit at the level of the foundations to be constructed. Load is applied in increments to a bearing plate by jacks or weights. Settlements are determined after each increment, generally for 24 hr or until settlement stops. The data are used to plot a load-settlement curve. Local building codes may describe a procedure for making such tests, or ASTM D1194 may be used as a guide.

Load-test results should be interpreted with extreme caution and generally with the aid of borings that provide information on deep subsoil layers. Caution is necessary because of the short duration of the test and the difficulty of correlating the behavior of a small loaded area of soil with the large area under the actual structure. A load test can indicate the bearing capacity of a compact granular soil. But it may provide no information on an underlying weak stratum that may not be able to resist the stresses of the actual structure. Furthermore, the usual short-duration load test cannot be extrapolated to apply to the same pressure acting on a plastic soil over a large area for a long time.

To be meaningful, the tests should simulate actual foundation conditions. They should be carried out at or near places where important loads will be applied and at the elevation of the foundations. If the test plate is applied at the surface of the ground or of a relatively large excavated area, the test may indicate low shearing resistance. The ground offers little lateral support to the plate, since it is not embedded. In contrast, excessive embedment may indicate higher shearing resistance than will be provided to the actual foundations.

Building codes generally require load tests when engineers want to use a design load higher than presumptive bearing values would justify. The codes specify that settlement observations should be made for the design load and an overload, either 50 or 100% of the design load. Usually, the design load is accepted if, under it, settlement does not exceed a specified amount, usually ¾ in. But also, under the overload, settlement must be nearly proportional to settlement under the design load. Such field tests may yield erroneous results when stresses from adjacent foundations may be superimposed on those from the proposed foundations or when there is a weak substratum. Stresses imposed by a relatively small test plate through the bearing stratum on the weak subsoil may be very small, whereas the stresses imposed by the actual foundations may cause excessive settlement.

(T. W. Lambe and R. V. Whitman, "Soil Mechanics," John Wiley and Sons, Inc., New York; A. G. Leonards, "Foundation Engineering," McGraw-Hill Book Company, New York; R. F. Scott and J. J. Schoustra, "Soil: Mechanics and Engineering," McGraw-Hill Book Company, New York.)

7-5. Significance of Laboratory Soil-Test Results. Laboratory tests are made to identify soils, determine their properties, and predict their behavior under a proposed structure. Tests are most useful and economical when at least a preliminary design of the substructure is available as a guide in selecting locations for sampling and for estimating loads to be supported. Numerous tests are available, but only a few are generally necessary. Foundations requiring more than the ordinary tests are those with very heavy or dynamic loads and those on weak or uncertain soils.

Mechanical analyses determine particle-size distribution in a soil sample. Distribution of coarser particles is determined by sieving. Distribution of material finer than the openings in a 200- or 270-mesh sieve is found by sedimentation, usually by hydrometer test. The findings generally are most useful when presented graphically, as a grain-size accumulation curve or a gradation curve. Particle size, mm, is plotted to a logarithmic scale, and percent of particles smaller (finer) than the size shown is plotted arithmetically. The shape of the curve usually indicates whether gradation is good or poor.

A material with good gradation is relatively stable, resistant to erosion or scour, can readily be compacted to a very dense condition, and can develop high shear resistance and bearing capacity. Poor gradation characterizes a material with insufficient coarse particles, insufficient fine particles, or skip grading. Such materials have less supporting power. One measure of gradation, obtainable from a gradation curve, is the Hazen uniformity coefficient

$$C_u = \frac{D_{60}}{D_{10}} \tag{7-13}$$

where D_{10} = particle diameter at the 10% finer point
D_{60} = particle diameter at the 60% finer point

Typical coefficients are 1.2 to 2 for fine or medium, uniform sand and uniform, inorganic silt. For well-graded materials, the coefficient ranges from 5 to 10 for silty sand, 4 to 6 for fine to coarse sand, 15 to 300 for silty sand and gravel, and 25 to 1,000 for a mixture of clay, silt, sand, and gravel.

D_{10} is known as *effective size*. It has been used to estimate the permeability coefficient of artificially graded filter sands.

As indicated in Art. 7-2, soils may be distinguished by particle size. And for granular soils, permeability, frost action, compaction, and shear resistance may be estimated from a mechanical analysis.

Density determinations measure the relative volumes of voids and solids in a soil. Relative density [Eq. (7-4)] indicates the compressibility of the soil; loosely packed soils are more compressible than the more compact.

Compaction tests, such as the standard Proctor or the modified AASHTO [test method AASHTO T180 (ASTM D1557) or T99 (ASTM D698)], determine the maximum unit weight or minimum void ratio that can be obtained for a soil, generally one to be used for a fill. For a given soil, density depends on compaction effort and moisture content. For a given moisture content, density increases with compaction effort. For a given compaction effort, density reaches a maximum at the *optimum water content;* the greater the compaction effort, the smaller the optimum moisture content. When the maximum density is known, a job standard can be established to insure that soil compacted in the field will have a desired minimum strength, permeability, or compressibility. Often, densities of at least 95% of maximum are specified for compacted fills.

In-place density tests are made to correlate field-compaction results with the specified density. One method is with a penetrometer, by checking readings obtained in the field against those made in the laboratory. Another method is to measure the weight and volume of a sample.

Moisture-content determinations provide useful data for estimating soil compaction and compressibility. Knowledge of the degree of saturation is important, because no volume change can occur in a saturated soil mass without intake or discharge of water. Hence, pervious soils, such as coarse sand, though saturated, compress almost as rapidly as load can be applied, if the load is applied gradually. In saturated fine-grained soils through which water moves slowly, compressibility varies with time. Partly saturated soils generally compress nearly as rapidly as loading is applied. Water content also influences the shearing strength of clay; cohesion of a saturated clay decreases with increase in water content. Furthermore, water content is important to know for retaining-wall backfills; a saturated backfill imposes higher lateral pressures than one that is drained.

Atterberg-limit tests determine for fine-grained soils the water contents at the boundaries between liquid, plastic, semisolid, and solid states. Thus, if the water content of a clayey soil falls between the plastic and liquid limits, the soil is in a plastic state. The plasticity index, difference between liquid and plastic limits, and liquid limit can be used for soil particles. At a given liquid limit, the lower the plasticity index, the more likely it is that organic material is present and the greater the permeability and compressibility. Also, shearing strength of a saturated clay depends on consistency: a clay that has little or no strength at the liquid limit may have substantial strength at lower moisture content.

Permeability tests may be conducted in laboratory or field to determine the permeability coefficient k. The tests are made with a permeameter, through which water flows under conditions for which Darcy's law applies; k can be computed from Eq. (7-9). Permeability is not a constant for a soil. The coefficient changes with change in density, particle size, gradation, or degree of saturation. Nevertheless, it sometimes is used for estimating subsurface water flow, such as artesian flow, flow under sheet-piling, and seepage through an earth dam. (See also Table 7-1, Art. 7-3.)

Confined compression tests are conducted in consolidometers, ring-type compression devices, which restrict lateral deformation. The object is to obtain information pertinent to the behavior of foundations where volume change of soil can occur under compression only through change in thickness. Load is applied in increments and maintained constant until settlement stops. Test results are converted to show the relationship between pressure, tons per sq ft, and void ratio. Often, the results are plotted with pressure to a logarithmic scale and void ratio to an arithmetic scale, because the relationship frequently then plots nearly linear.

Compressibility of a soil is indicated by the slope of the curve. This varies with density, history of previous loading, handling of the sample, and magnitude of the stress increment relative to the existing loading at any point. Furthermore, neither void ratio nor compressibility is a single-valued function of pressure. When test load is applied to a relatively loose soil, the pressure-void-ratio plot, or virgin compression curve, has a relatively steep slope. The expansion, or rebound, curve for removal of the load is much flatter (Fig. 7-5). On application of load again, the plot, like the rebound curve, is much flatter than the virgin compression curve. At the pressure where initial loading was stopped, the void ratio is less than that attained before. If loading is further increased, the plot becomes nearly an extension of the virgin compression curve. Thus, if tests are made on undisturbed samples, because of the removal of the pressure of overburden, the pressure-void-ratio plot will resemble the reload curve of Fig. 7-5.

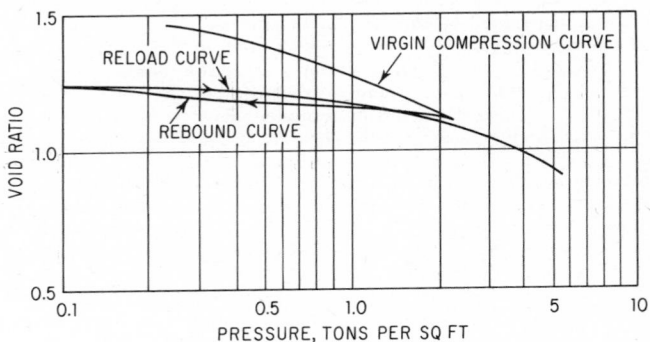

Fig. 7-5. Compression diagram for a remolded soil specimen.

Since compressibility is different for each branch of a compression curve, interpretation of test results must be based on conditions under which compression occurs. The coefficient of compressibility, the slope of the virgin compression curve, indicates the compressibility of soil not affected by past cycles of loading and unloading. Compression index, applicable to the straight-line portion of that curve, is the ratio of the change in void ratio to the corresponding change in logarithm of pressure. Field compression diagrams, curves indicating the amount of compression to be expected in a natural soil formation, may be constructed with the aid of such data from laboratory compressibility tests.

Unconfined compression tests are used to estimate the shearing strength of cohesive soils (Art. 7-4). When a cylindrical specimen is subjected to axial loading without lateral support, the shear stress (unit cohesion) when failure occurs equals approximately half the compressive stress, the unconfined compressive strength of the soil.

Consolidation tests are made on saturated silts and clays to determine the rate of volume change under constant load. (Partly saturated soils and fully saturated coarse-grained soils, which lose water when compressed, deform almost as rapidly as loads are applied during construction.) Slow settlement under load on fine-grained soils is mainly due to hydrodynamic lag, delay in movement of water out of the soil mass. Hence, consolidation tests are carried out much like confined compression tests, with porous enclosure to permit water to ooze from the specimen. But a record is kept of volume or thickness change with time.

Average percent consolidation U at any time equals the percent change in thickness or settlement at the time. From theoretical considerations, U is a function of a time factor T. (For $0 < U < 0.6$, approximately $T = \pi U^2/4$; for $U = 0.7$, $T = 0.40$; for $U = 0.9$, $T = 0.85$.) Then, the time t months (or years) to reach a given settlement or U may be computed from

$$t = \frac{TL^2}{c_v} \qquad (7\text{-}14)$$

where L = length, ft, of longest drainage path to an outlet; for example, the ground surface or a drained pervious stratum

c_v = coefficient of consolidation, sq ft per month (or year)

The coefficient of consolidation may be computed from

$$c_v = \frac{k(1 + e)}{62.4a_v}$$ (7-15)

where k = coefficient of permeability
e = void ratio
a_v = coefficient of compressibility

But usually c_v is estimated from laboratory-obtained time-consolidation curves and Eq. (7-14). When this is done, the coefficient of permeability may be obtained from Eq. (7-15).

Direct shear tests are made in the laboratory to obtain data for determining the bearing capacity of soils and the stability of embankments. (For in-place testing, see description of vane shear test in Art. 7-4.) In direct shear tests, shear forces are applied along a plane through a sample, usually at a constant rate of strain. Also, except for soft clays, a constant normal pressure is maintained on the sample. The shearing strength of granular materials varies with the normal pressure. The shearing strength of saturated clays depends on the rate of strain. When stress and strain observed in a test are plotted, the result generally is a curve from the origin to failure, where displacement occurs without increase in stress. In a consolidated or preconsolidated test on saturated clays, drainage is permitted when the normal load is applied and application of the shearing forces is delayed until full consolidation occurs.

Triaxial compression tests are another means of determining shearing strength of a soil. Usually, liquid pressure is applied along the sides of a sample and an axial load is applied by a piston. Triaxial tests are superior to direct shear tests because of better control over intake and discharge of water from the specimen. This control permits observation of the effects of pore pressures on shearing strength.

California bearing-ratio tests are used to evaluate subgrades for pavements. The tests may be carried out in field or laboratory. They determine the resistance to penetration of a subgrade soil relative to that of a standard crushed-rock base. CBR, percent, equals $0.1f$, where f is the loading intensity, psi, on a 3-sq-in. penetrometer at 0.1-in. penetration. (Loading intensity for 0.1-in. penetration of the standard base is taken as 1,000 psi.) Before the test, the soil sample must be compacted approximately to the density and moisture content that will be specified for the subgrade.

Plate bearing tests are field tests also for evaluating subgrades for pavements. Made by loading a 30-in.-diameter bearing plate and recording the settlements, these tests determine the subgrade modulus k, psi per in., which is indicative of the supporting capacity of the subgrade. The modulus is the slope of the load-settlement curve in a range of settlements that might occur under the loaded pavement. Some government agencies and highway departments specify 0.05 in., others 0.1 in. for the settlement for determination of k. Hence, k equals loading intensity, psi, on the plate that causes a settlement of 0.05 in. (or 0.10 in.) divided by 0.05 (or 0.10). As in CBR tests, the subgrade to be tested should be at the specified percent compaction. Variations in subgrade conditions have much greater effect on the modulus computed than choice of settlement for computing k.

(R. F. Scott and J. J. Schoustra, "Soil: Mechanics and Engineering," McGraw-Hill Book Company, New York; Standard Specifications for Highway Materials and Methods of Sampling and Testing, Part II, American Association of State Highway and Transportation Officials, Washington, D.C.; "Soils Manual for the Design of Asphalt Pavement Structures," MS-10, The Asphalt Institute, College Park, Md.; G. B. Tschebotarioff, "Soil Mechanics, Foundations, and Earth Structures," McGraw-Hill Book Company, New York; T. W. Lambe and R. V. Whitman, "Soil Mechanics," John Wiley & Sons, Inc., New York.)

7-6. Foundation Loads and Pressures. Foundations should be designed to support the weight of structure, live load, and wind, and also the following loads when they are present: earth pressure, weight of fill on elements of the structure, traction forces, centrifugal forces, snow and ice, seismic forces, hydrostatic and hydrodynamic loads, and surcharge, or load transmitted to supporting soil other than through structural components. Also, the design should provide for the moments produced by the loads.

For a building, the maximum live load need not be the sum of fully loaded floor and roof areas. Building codes generally permit a reduction, sometimes up to 60%, depending on the area to be supported. Usually, the foundation live load can be taken as the design live load on the column or wall to be supported, plus any reduced live load applied directly to the footings. Special analysis is necessary when the live load is vibratory or heavy impact is involved.

For a bridge, distinction should be made between normal and maximum live load. Normal live load is the live load that is likely to be transmitted to the foundations throughout the greater portion of the useful life of the structure. Maximum live load is the greatest live load that may be anticipated.

In design for soil bearing capacity, usual practice is to design for the combination of loads that produces maximum effect. The safety factor generally is reduced for combinations that include transient loads, such as wind and earthquake. Often, the safety factor is 3 for dead, live, and snow loads, and 2 for combinations including transient loads.

Design for Settlement. Usual practice is to ignore transient loads. (On some poorly drained, saturated clays, a load on a structure for weeks or months may be considered transient, because of the long time required for consolidation.) To keep differential settlements small, footings are proportioned to apply equal pressures to the soil, usually under dead plus normal live loads. The assumption that equal intensities will produce equal settlements, however, is erroneous. Its accuracy, varying with type of soil, depends on the footings being nearly equal in size and having about the same shape. Usually, for the same intensity of loading, large footings settle more than small ones; square footings, more than rectangular ones. A surface through the settled footings of a building often is nearly dish-shaped. Overlapping pressures from footings cause the building interior to settle more than the exterior. Corners settle less than the center of the walls.

Each footing also should be large enough to keep total settlement within tolerable limits for the structure. When differential or large settlements are inevitable because of soil conditions, the structure should be kept free of adjacent structures and should be designed with slip and hinged joints to permit relative movements.

Pressure Distribution. Pressures used in computing bearing capacity or settlement are those in excess of the pressure due to the weight of the soil above. Thus, consideration also should be given to pressures superimposed by adjacent foundations.

Usual practice in foundation design, except for pile foundations, is to assume that the bearing pressure at the bottom of a foundation or on a parallel plane below is constant for concentric loads or varies linearly for eccentric loads. While this assumption may not be true, more accurate but more complicated theories rarely are justified by knowledge of future loadings and of present and future soil characteristics. The assumption of constant or linear pressure distribution has been the basis of design for many foundations which have behaved satisfactorily for decades. Especially for rigid foundations and for foundations on soils with an allowable bearing pressure of 3 tons per sq ft or more, the error in using it is unimportant. For flexible foundations on weaker soils, the assumption of constant or

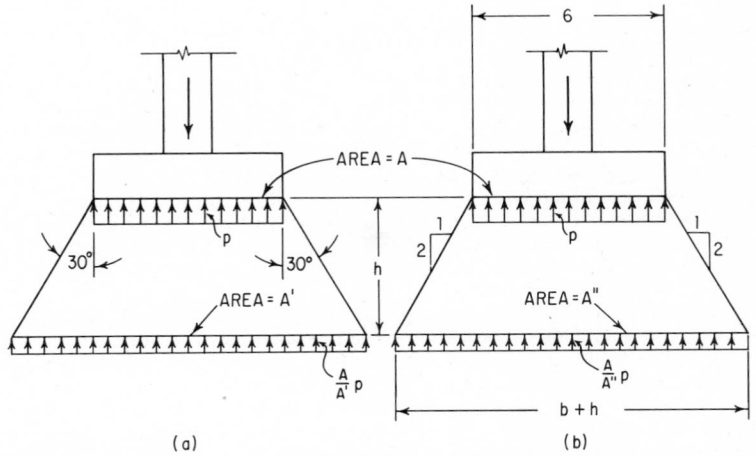

Fig. 7-6. Assumed pressure distributions under a footing. (*a*) 30° spread. (*b*) One-on-two spread.

linear variation of pressure may be unduly conservative. For example, some engineers relate the pressures at the bottom of a footing to its deflections, an assumption that generally results in calculation of smaller bending moments than for constant or linear pressure distributions.

Another common assumption in foundation design is that the pressure spreads out with depth from the bottom of the foundation at an angle of 30° with the vertical or at a slope of 1 on 2 (Fig. 7-6). Thus, for a total load on a foundation of P, the pressure at the base of the footing would be assumed to be $p = P/A$, where A is the area of the footing. As shown in Fig. 7-6b, the pressure at depth h may be taken as P/A'', or, for a square footing, as $P/(b + h)^2$. When a weak stratum underlies the stratum on which a footing is founded, this method may be used to estimate the pressure on the lower layer. Also, the method can be adapted to determination of the pressure resulting from superimposition of load from adjacent foundations. For example, for two footings, below the plane where the 30° or 1-on-2 surfaces for the footings overlap, pressures may be computed as though the two footings were one.

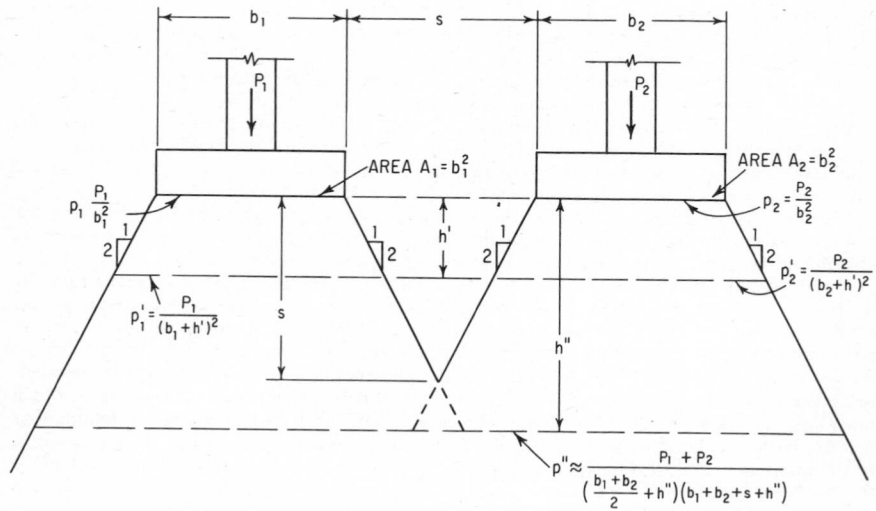

Fig. 7-7. Assumed pressure distribution under adjacent square footings.

Figure 7-7 shows two square footings spaced a distance s apart. If pressure is assumed to spread from each footing on a slope of 1 on 2, the pressure zones will overlap at a distance s below the footings. On a plane above the overlap, the pressure under each footing generally is assumed to be the load on the footing divided by the area of the plane included in the pressure zone for that footing. On a plane below the overlap, the pressure generally is assumed to be the sum of the loads on the footings divided by the area of the plane included in the combined pressure zones. For large overlaps, however, it is more conservative to compute the pressure under each footing as if it were isolated and then add the pressures. Results of these calculations should be considered as approximate average pressures; actual distribution generally will not be uniform.

Allowable bearing pressures on soils and rock may be determined with the aid of laboratory and field tests. For buildings, presumptive bearing capacities in local building codes may be used as a guide in the absence of tests. Table 7-4 is an example, taken from the Code Manual supplementing the New York State Building Construction Code. The allowable pressures in the table apply to footing bases at the ground surface or where there is no permanent lateral support for the soil. For footings, below the surface, higher pressures may be used on sand, gravel, and medium stiff to stiff clay.

Settlement Estimates. Settlements may be estimated from tests. In field tests made with loads distributed over relatively small areas, however, the effects of weak soil layers or saturated clays at some distance below grade may not be disclosed. Preferably, design should be based on pressure-settlement curves determined from confined compression tests and consolidation tests on soil samples. If the data are given in terms of void ratio, they may

Table 7-4. Allowable Bearing Values on Soils, Tons per Sq Ft*

Massive crystalline bedrock: granite, gneiss, traprock—in sound condition	100
Foliated rock: schist and slate—in sound condition	40
Sedimentary rock: hard shales, siltstones, sandstones—in sound condition	15
Exceptionally compacted gravels or sands	10
Compact gravel or sand-gravel mixtures	6
Loose gravel; compact coarse sand	4
Loose coarse sand or sand-gravel mixtures; compact fine sand, or wet, confined coarse sand	3
Loose fine sand or wet, confined fine sand	2
Stiff clay	4
Medium stiff clay	2
Soft clay	1

*Code Manual, New York State Building Construction Code. See also Table 7-3.

be converted to settlement with

$$\Delta h = \frac{(e_i - e_p)12h}{1 + e_i} \tag{7-16}$$

where Δh = total expected settlement under load, in.

e_i = initial voids ratio

e_p = voids ratio under pressure p

h = thickness of soil layer, ft

Suppose settlement data are plotted for a soil layer 10 ft thick. Then, settlement on a soil layer considerably deeper may be estimated by dividing the layer into 10-ft-thick sections, calculating the pressure in the middle of each section, determining from the pressure-settlement curve the settlement for each section, and adding the settlements. In computing pressures, they may be assumed to spread from a footing on a 30° or 1-on-2 slope.

As an alternative, the pressure at the center of the thick layer may be computed. Then, the settlement for a 10-ft layer with this pressure may be obtained from the pressure-settlement curves. Finally, the total settlement may be obtained by multiplying this settlement by the ratio of the layer thickness, ft, to 10 ft. The two methods do not give exactly the same results, but they should be close enough for practical purposes. Either technique also may be adapted to computation of settlements where different soil layers underlie a footing, if pressure-settlement curves are available for each type of soil.

(S. Johnson and T. Kavanagh, "The Design of Foundations for Buildings," McGraw-Hill Book Company, New York; J. E. Bowles, "Foundation Analysis and Design," McGraw-Hill Book Company, New York; N. M. Newmark, "Simplified Computation of Vertical Pressures in Elastic Foundations" and "Influence Charts for Computation of Stresses in Elastic Foundations," *University of Illinois, Engineering Experiment Station Bulletins*, vol. 33, no. 4, Sept. 24, 1935, and Series 338, Nov. 10, 1942; H. F. Winterkorn and H. Fang, "Foundation Engineering Handbook," Van Nostrand Reinhold Company, New York.)

7-7. Spread Footings and Mats. (See also Art. 7-6.) The purpose of spread footings and mats is to distribute loads over a large enough area so that the soil can support the loads safely and without excessive settlement. Usually, such foundations are made of reinforced concrete, though sometimes plain concrete is used for spread footings. Concrete for footings should be placed on undisturbed soil. If possible, the last few inches of soil should be excavated just before concrete placement starts.

Independent or isolated spread footings usually are placed under concentrated loads, such as columns or piers (Fig. 7-8). Such spread footings generally are rectangular or square. To prevent overturning and unequal stress distribution in the soil, the centroid of a footing is placed as nearly as possible under the resultant of the loads on it. The area of the footing must be large enough to insure that the bearing capacity of the soil is not exceeded and that maximum settlement is within acceptable limits. Also, the footing must be sized so that differential settlements will not be excessive. For this purpose, usual practice is to proportion the footings for a structure so that the unit pressure under each is the same for working loads, usually dead load plus normal live load.

Combined footings are spread footings that support two or more concentrated loads (Fig. 7-9). Conditions where such footings may be used instead of an isolated footing under each load include: Two piers or columns so close that isolated footings would overlap or nearly overlap. Property line so close to one column that a spread footing kept within the line would be eccentrically loaded. (In that case, the eccentricity can be reduced or eliminated with a combined footing under the outer column and an interior one.) Presence of uplift, overturning, or opposing horizontal forces makes tying of two footings together desirable. Combined footings, however, should be avoided on soils sensitive to variations in load.

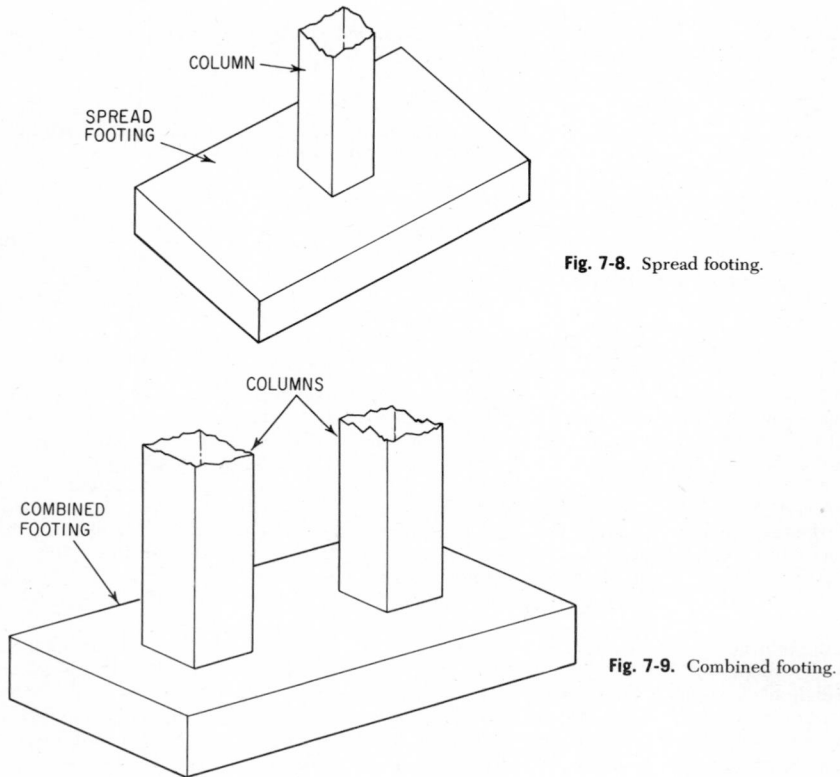

Fig. 7-8. Spread footing.

Fig. 7-9. Combined footing.

A combined footing may be a rectangular or trapezoidal flat slab or an inverted T beam, symmetrical about a line between the two loads. The centroid of the footing should be set as nearly as possible under the resultant of the loads. The footing should be shaped and sized to keep the soil pressure within desirable limits, assuming linear variation in pressure under eccentric loading. The design also should maintain bearing pressure under the entire footing area under all loading conditions.

Cantilever footings sometimes are used to support two columns (Fig. 7-10). Consisting of a pad under each column with a strap or lever between them, they sometimes are more economical than combined footings because less concrete and reinforcing steel are required. Usually, the pressure under the pads is assumed to be uniform, and they are sized, shaped, and placed so that their centroid lies under the resultant of the loads. The strap is designed as a rigid beam, ignoring soil pressure under it.

Continuous footings (Fig. 7-11) are used under walls or to support a row of columns or piers. Often, the concentrated loads are supported on a wall, which distributes them over the footing. In the design of a wall footing, generally a typical 1-ft-long section of footing is selected and designed to transmit the soil pressure transversely to the wall. The resulting design is used for the rest of the wall unless significantly different conditions are encountered, such as changes in loading or dimensions of wall. (Basement walls for buildings should be designed to resist active lateral earth pressures.)

If the concentrated loads are supported directly on the footing, assumption of uniform soil presure may lead to an unduly conservative design of the footing. Yet, more complex assumptions, such as one relating soil pressures to footing deflections, may not be warranted because the accuracy of the deflection calculations for the concrete substructure is questionable. One design method sometimes used allots each column load to a footing area symmetrically located about the column. The area is sized to develop the allowable bearing pressure of the soil. Portions of the footing between such column areas may be considered unloaded. Thus, bending moments in the continuous footing are much smaller than for the assumption of uniform pressure.

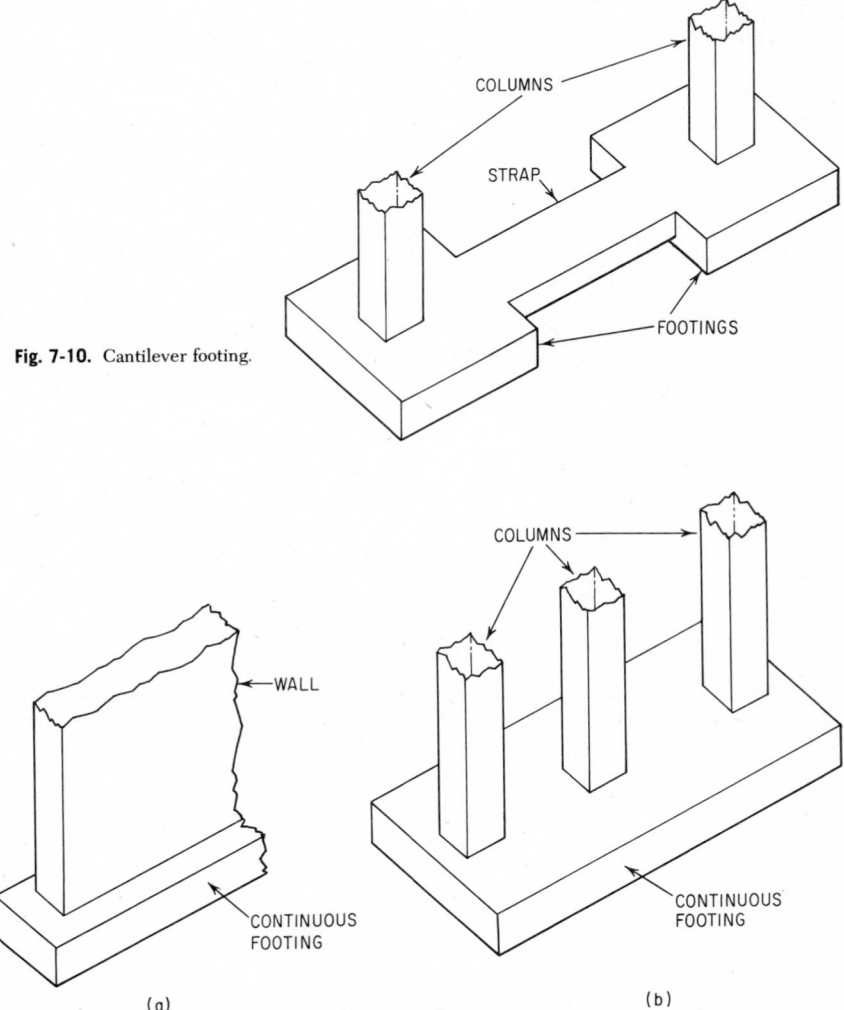

Fig. 7-10. Cantilever footing.

Fig. 7-11. Continuous footings: (*a*) for a wall; (*b*) for several columns.

Mat, or raft, foundations are continuous under an entire structure. They may be used to insure watertightness. Or they may be chosen because the bearing capacity of the soil is low and isolated spread footings would be so large that it becomes economical to join them. Many types of mat foundations have been used: a very thick, reinforced-concrete flat plate; a thick, inverted flat slab; inverted beam-and-girder construction; inverted ribbed slab; mat with walls; and flat plate with depressed footings under the columns. As for design of continuous linear footings, the assumption of uniform pressure under a mat may lead to an unduly conservative design. A more economical design results if column loads are allotted to areas symmetrical about the columns. Each area is sized to develop the allowable soil pressure under the column load. (These loaded areas sometimes take the form of continuous column strips.) Portions of the footing between such areas are assumed unloaded. Resulting bending moments are smaller than for the assumption of uniform loading on the mat. Also, the unloaded portions may be made thinner than the rest of the mat and may contain less reinforcing steel.

Floating foundations are mats built on poor soils and designed for low soil pressure. Such foundations should be used only with great caution, since the ground may be settling under its own weight, and fills around the structure may cause even more settlement.

Generally, mats should be made watertight because they may be subject to uplift from water pressure. But even if there is no uplift, a waterproof layer is advisable to prevent loss of moisture from the soil through the concrete. Movement of water out of the soil may increase the rate of settlement.

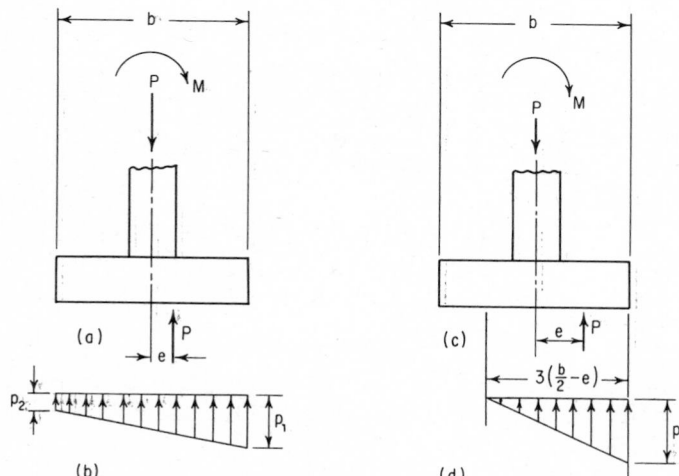

Fig. 7-12. Footings subjected to overturning.

Footings subjected to overturning forces usually are designed on the assumption of linear variation of soil pressure. For a continuous footing, for example, with a vertical axial load P and overturning moment M normal to the longitudinal axis of the footing (Fig. 7-12a), the pressure diagram generally is assumed as shown in Fig. 7-12b for full bearing. This occurs when the resultant of the forces lies within the kern of the footing; in this case, within the middle third. The maximum and minimum pressures then are, for a 1-ft length of footing,

$$p_1 = \frac{P}{b}\left(1 + \frac{6e}{b}\right) \tag{7-17a}$$

$$p_2 = \frac{P}{b}\left(1 - \frac{6e}{b}\right) \tag{7-17b}$$

where b = width of footing

e = eccentricity = M/P

When the resultant lies outside the middle third of the footing, that is, when $e > b/6$, p_2 would be negative according to Eq. (7-17b). Since the soil cannot resist much tension, it is generally assumed that there is bearing over only a portion of the footing. In that case, the pressure diagram becomes triangular, as shown in Fig. 7-12d. The pressures extend over a distance of $3(b/2 - e)$. The maximum pressure is

$$p = \frac{2P}{3(b/2 - e)} \tag{7-18}$$

Treatment of a square or rectangular spread footing with overturning moment M about one principal axis is similar. For full bearing on the footing, the maximum and minimum pressures are

$$p_1 = \frac{P}{ab}\left(1 + \frac{6e}{b}\right) \tag{7-19a}$$

$$p_2 = \frac{P}{ab}\left(1 - \frac{6e}{b}\right) \tag{7-19b}$$

where b = length of footing in the plane of overturning
a = length of the other side of the footing
When $e > b/6$, the resultant lies outside the middle third, and full bearing does not exist. In that case, the pressure diagram is a triangular prism, and the maximum pressure is

$$p = \frac{2P}{3a(b/2 - e)} \tag{7-20}$$

For square or rectangular footings subjected to overturning about two principal axes and for unsymmetrical footings, the procedure is generally similar. Eccentricities of the loading, e_1 and e_2, are determined about the two principal axes. Then, for bearing over the entire footing, the maximum soil pressure is computed from

$$p = \frac{P}{A}\left(1 + \frac{e_1 c_1}{r_1^2} + \frac{e_2 c_2}{r_2^2}\right) \tag{7-21}$$

where A = area of footing
c_1 = distance from one principal axis to outermost point on footing
r_1 = radius of gyration of footing area about that axis
c_2 = distance from second principal axis to the outermost point
r_2 = radius of gyration of footing area about that axis
When pressure exists over only a portion of the footing, the maximum pressure may be approximated by trial.

A safety factor of at least 1.5 should be provided against overturning about the edge of the footing. Thus, for 1.5 or more times M, the resultant force should lie within the footing, and the allowable pressure on the soil should not be exceeded.

Sliding caused by horizontal forces also should be investigated. If frictional resistance will not be adequate, the footing should be keyed, tied to footings subjected to opposing forces, or otherwise anchored to prevent movement.

(L. Zeevaert, "Foundation Engineering for Difficult Subsoil Conditions," Van Nostrand Reinhold Company, New York; J. E. Bowles, "Foundation Analysis and Design," McGraw-Hill Book Company, New York.)

7-8. Pile Foundations. Piles are slender underground columns, generally placed in groups. They may support their loads through bearing at the tip, friction along their sides, adhesion to the soil, or a combination of these means. Thus, the behavior of a pile foundation depends on the strength of the piles and the bearing and shear capacities of the soil.

Placement Tolerances. Pile behavior depends on how closely field conditions accord with design assumptions. For example, design may assume the load to be shared equally by each pile in a group. To insure this behavior, the tops of the piles usually are embedded in a rigid reinforced-concrete footing, or cap. Still, if the assumption of equal loading is to be valid, the piles must be installed close to the positions called for in the design, with their centroid under the resultant of the loads.

A deviation of 3 in. from the designated position may be permitted without reduction in load capacity of the pile if no pile in the group is carrying more than 110% of its normal permitted value. If any pile is overloaded because of poor placement of piles, additional balancing piles should be added to the group, or the footing should be combined with an adjacent footing to redistribute the load.

Deviation of piles from the vertical can be measured with a plumb bob inside a pile casing. Determination of the plumbness of solid piles is difficult, because there may be a curve or bend in them not evident from the ground surface. A tolerance of 2% of the pile length is permissible. If excessive deviation occurs, the resulting horizontal components of the loads on the out-of-plumb piles should be counteracted, with ties to adjacent column caps if necessary. Additional piles may have to be added to the group to compensate for any loss in load-carrying capacity. Driving of piles out of plumb, however, does not affect pile capacity as much as driving piles out of position.

Advantages. Pile foundations are relatively costly. They usually are selected only when soil conditions are such as to make spread or mat foundations impractical or uneconomical. Piles can be used when other types of foundations are impractical because piles can penetrate weak soils and transmit loads to an underlying stratum with sufficient bearing capacity. Also, piles can distribute loads over a sufficiently large vertical area of relatively weak soil to enable it to support the loads safely. Furthermore, they can be sloped, or battered, to resist horizontal forces.

Types of Support. Piles often are classified as end-bearing piles or friction piles, depending on the predominant means of transmitting loads to the soil. Usually, however, driven piles transmit loads through both end bearing and friction. As they are driven, the piles displace soils downward at the tip and laterally at the sides.

Piles should be driven into a good bearing stratum. Relatively weak soils should not be relied on to be sufficiently compacted below the tip to provide appreciable bearing capacity.

Taper of a pile provides some bearing, but only in good material. Also, significant resistance to penetration may be provided along the pile sides by silt, clay, sand, or gravel. But these materials, in turn, also must resist sliding along a surface surrounding the pile a short distance away. Thus, soil shearing capacity is important in determining the capacity of a friction pile. While the area for frictional resistance or adhesion is the embedded surface area of a pile, the dependable area for shear resistance is the minimum area circumscribing the pile. For example, for an H pile, the effective shear surface is the bounding rectangular prism. Soil gets wedged between the flanges during pile driving and after that moves with the pile.

Uplift. Piles often are used to resist uplift. Resistance to uplift is developed through friction along the sides of the piles. The capacity of such piles also depends on their strength in tension and the shearing resistance of the soil. When enlarged-tip piles are used, the weight of a cone of earth above the enlargement adds to the uplift resistance. The safe friction values to use for a specific project should be determined by uplift tests. In the absence of test data, the Standard Specifications for Highway Bridges, of the American Association of State Highway and Transportation Officials, permit an intermittent uplift of 40% of the presumptive downward capacity of a pile. But the tension pile must be adequately anchored at the top, and friction forces along the sides should not exceed the weight of soil, buoyancy considered, surrounding the pile.

Minimum Length. A friction pile should extend at least 10 ft below ground or any level to which the ground may be excavated in the future. When the soil has a presumptive bearing capacity of less than 2 tons per sq ft, however, the minimum length should be greater.

Lateral Bracing. Since compression piles are slender, they require support against buckling below grade. This support usually is provided by the soil. Even a moderately dense mud or a soft silt or clay serves this function. But if piles pass through water or fluid soil, they should be designed as long columns. Sometimes, it may be necessary to provide lateral stays and bracing between pile caps for such piles.

Above grade, piles also must be braced against lateral displacement. Reinforced-concrete caps over a pile group rigidly connected to at least two other pile groups in radial directions not less than 60° apart should be used for this purpose. For a footing with only one or two piles or for a footing requiring extra lateral stability because of loading or weak soil, the pile cap should be braced with struts to adjacent caps. (Struts in the weak direction only may be adequate for two-pile footings.) Minimum dimension of the struts should be one-twentieth the clear distance between caps, and depth should be at least 8 in. As an alternative, the bracing may be the floor of a building, a continuous reinforced-concrete mat at least 6 in. thick. The mat may be supported by and anchored to the caps, or the piles may be embedded at least 3 in. in it. The mat should not depend on the soil for direct support of its own weight and loads.

Bracing between caps also may be necessary for piles driven in soils that consolidate considerably during driving or for piles installed through great depths of unconsolidated silts and mucks. When such soils consolidate in the future, their weight may hang on the piles and cause excessive settlement and lateral movement.

Effects on Soil. Pile driving alters soil properties. Coarse-grained soils, such as sand and gravel, become more compact, even if saturated. (Sometimes, piles are driven just to compact such soils.) Silt and clay, when not saturated, also become more compact. When saturated, however, they move during pile driving in the direction of least resistance, usually upward, and undergo little compaction. (Heaving of the ground surface about a pile driven through such soils is evidence of this underground action.) Disturbance of these soils often results in a reduction in soil strength.

Shock waves and vibrations emanate from a pile being driven. Their effect on nearby structures depends on the type of soil and type of pile. A solid pile of wood, concrete, or closed-end pipe displaces more soil and causes far more disturbance than a low-displacement pile, such as a steel H beam or open-end pipe. Usually, low-displacement piles should be used close to existing structures, and the piles should be driven with care. Jacking, though more costly than hammering, is a more desirable method of placing piles where vibration will be harmful.

Vibration, whether from pile driving or from loads on a completed structure, may cause friction piles already driven to settle.

When a high-displacement pile is driven in saturated clay, movement of the soil may heave and tilt previously driven piles in the group. Piles not out of alignment should be redriven to insure adequate capacity. Additional piles should be driven to replace those out of alignment or damaged.

Pile Spacing. This may be adjusted to minimize the influence of a pile on adjacent ones and on the soil surrounding them. Generally, piles may be spaced, c to c, as close as twice the average diameter, or 1.75 times the diagonal of the piles, but end-bearing piles not less than 24 in. c to c, and friction piles not less than 30 in. When more than four friction piles are driven in a group, the minimum spacing in soils with less than 6-ton-per-sq-ft bearing capacity should be increased at least 10% for each interior pile. But the spacing need not be more than 42 in., unless larger spacing is desirable for other reasons. For example, for cast-in-place concrete piles for which a steel casing is driven, a sufficiently large spacing is desirable to avoid damaging adjacent empty casings or uncured concrete in nearby casings. A minimum spacing of 42 in. is desirable for such piles.

As the number of friction piles in a group is increased, the average capacity per pile decreases if normal (close) spacing is maintained. An increase in spacing will reduce or eliminate the decrease in capacity. Below-normal spacing will reduce the average pile capacity even more. The reduction occurs because of the superposition in the soil of shear and compressive stresses produced by frictional and end-bearing forces of the piles. Thus, the maximum compression under a group of piles is greater than that under an isolated single pile. And the shears induced in the soil by one pile of a group add to those created by adjacent piles. While estimates can be made of the reduction in capacity due to below-normal spacing or grouping of piles, driving the piles until the tips reach a bearing stratum not underlain by weak soil is a better procedure, where feasible. The piles may be longer, but they can carry more load, and the foundation will be more dependable than one with friction piles.

Caps. Pile footings should be sized to accommodate the selected pile spacing. The distance from the side of any pile to the nearest edge of a concrete footing should be at least 9 in. Pile tops should project at least 6 in., and preferably 12 in., into the concrete after all damaged pile material has been removed. Reinforcement generally is set 3 in. above the top of the piles. Thickness at the footing edge should be at least 12 in. above reinforcing, or 14 in. for unreinforced concrete.

The pile caps should be designed as structural members that distribute the superstructure load to the piles, which act as reacting concentrated loads. (The soil directly below the footings should not be assumed to aid the piles in supporting the load.) Generally, when the footings are made thick enough to resist shear and bending, they also are so rigid that when the centroid of the pile group lies under the resultant of the load on a footing, the piles may be considered equally loaded.

Lateral Loads. Capped piles can resist some horizontal force without lateral bracing or batter piles. The amount can be estimated from the combined stresses in the piles due to

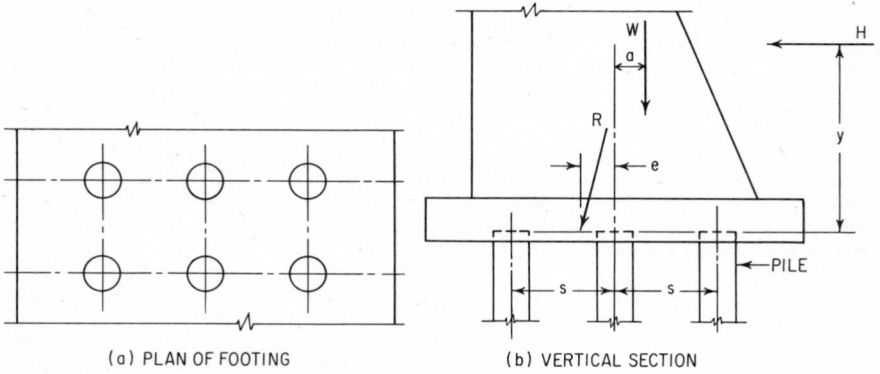

(a) PLAN OF FOOTING   (b) VERTICAL SECTION

Fig. 7-13. Pile foundation subjected to eccentric loading.

axial load and bending or the passive resistance of the earth to movement of the upper portion of the piles. The piles may be assumed to act as cantilevers, fixed and laterally supported about 5 ft below the ground surface in firm soils and 10 ft below in soft soils.

In the design of footings with vertical piles subjected to eccentric vertical loads or moderate horizontal force, such as the footing in Fig. 7-13, the load on a pile may be computed from

$$P = \frac{W}{n} \pm \frac{Mx}{\Sigma x^2} \qquad (7\text{-}22)$$

where W = total vertical load on a length of footing supported on a typical pile group
n = number of piles in the group
M = overturning moment about a plane through the top of the piles $= Hy - Wa$
H = horizontal load on the length of footing
a = distance from centroidal axis of piles to W
y = distance from top of piles to H
x = distance from centroidal axis of piles to each pile center

The plus sign is used when overturning induces compression in a pile. Equation (7-22) assumes all piles are identical, driven to the same extent and in the same soil. It ignores the moment of inertia of each pile about its own axis.

For footings subjected to large horizontal forces, such as the footing in Fig. 7-14, batter piles may be used. Since a limiting slope for such piles is about 1 on 2, for practical reasons, they can carry nearly the same vertical load as the vertical piles while resisting horizontal forces. To simplify calculations in design, batter piles and vertical piles are assumed to carry the same vertical loads. Equation (7-22) can be used to compute these loads.

For footings subjected to overturning, design usually involves trial and error. Because of the uncertainties in footing, pile, and soil behavior and interaction, tedious computations are not warranted and may not yield results any more accurate than simple calculations based on reasonable assumptions.

Pile Capacity. The load on any pile should not exceed the capacity of the pile as a structural member or its capacity to transmit its load to the soil. Nor should the load exceed the capacity of the soil to support it. In determining these capacities, consideration should be given to the difference between the supporting capacity of a single pile and a group of piles. The capacity of underlying strata to support the load of a pile group should be ascertained. The effect of driving additional piles should be evaluated. And account should be taken of the possibility of scour and uplift and their effects.

Piles should be designed for the computed load as columns. See Art. 7-9 for allowable stresses and maximum design loads.

To insure that friction piles have adequate capacity for transmitting load to the soil, the load may be determined by one or more of the following methods: Driving and loading test piles; pile-driving experience in the vicinity; tests of the soil strata through which the piles are to be driven. If possible, the findings from site investigations and experience should be com-

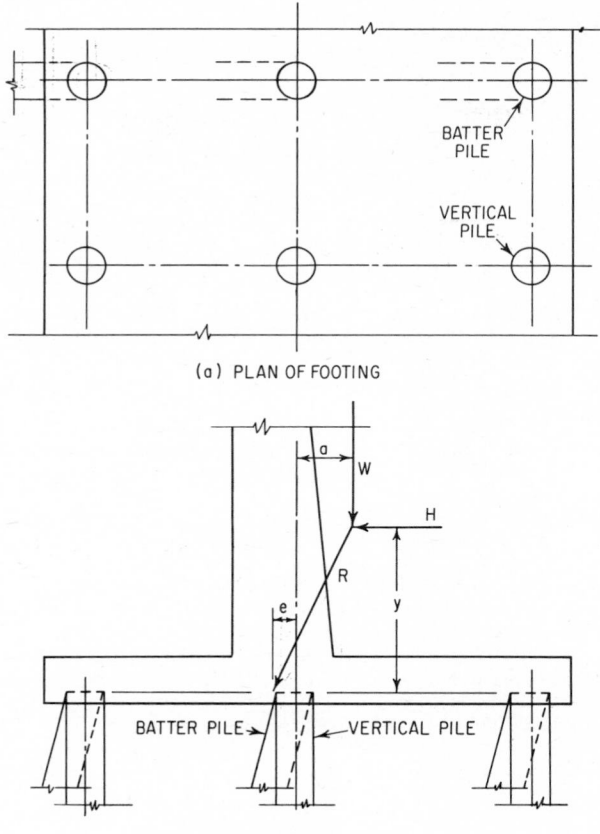

(a) PLAN OF FOOTING

(b) VERTICAL SECTION

Fig. 7-14. Pile foundation with batter piles.

pared with records obtained elsewhere with similar materials through which piles of known capacity were driven.

Capacity of soil to support load from a pile should be determined preferably by test loading and subgrade investigation. If possible and economically feasible, piles should be driven into a bearing stratum. Sufficient borings should be made to determine the thickness and quality of the stratum and whether it is underlain by firm soil or rock. If there is soft material a short distance below the top of the bearing stratum, the piles should be driven to a deeper bearing stratum or the load per pile reduced because of group action. For friction piles, the soil capacity should be estimated from soil tests and verified by load tests. Reductions should be made in load per pile for group action.

When reductions are necessary, the total load on the footing can be supported by adding more piles of the same length or by increasing pile length. Often, fewer piles of greater length and spacing will be more economical. Piles under the same footing should have the same length, but pile groups for separated footings may differ in length.

The AASHTO Specifications for Highway Bridges suggest the Converse-Labarre formula for computing the ratio of the allowable load per pile in a group to the allowable load on a single pile:

$$E = 1 - \phi \frac{(n - 1)m + (m - 1)n}{90mn} \qquad (7\text{-}23)$$

where $\phi = \arctan (d/s)$, deg
$\quad d =$ average pile diameter, in.
$\quad s =$ center-to-center spacing of piles, in.
$\quad n =$ number of piles in each row
$\quad m =$ number of rows in each group

Other formulas that have been proposed require reductions that may differ considerably from those given by Eq. (7-23). None appears applicable to all types of soils and piles, and none accounts for effects of pile length. For large projects, load tests on pile groups are advisable; for small projects, for which tests may be too costly, longer piles may be desirable.

Load Tests. In a load test, a static load is placed on a platform supported on a pile or group of piles. Means are provided for measuring the test load and the pile settlement under each increment of load. As an alternative, hydraulic jacks with suitable yokes and pressure gages may be used.

The AASHTO Specifications for Highway Bridges require that at least one pile be tested for each group of 100 piles. For buildings, the foundation area should be divided into sections in which soil conditions, as indicated by borings, are nearly uniform and with a maximum area of 15,000 sq ft. At least two piles in different parts of each section, preferably near a test boring, should be load-tested.

Codes generally require that in a load test settlements under the design load should be determined. Then, settlements under a 50 or 100% overload should be measured, as well as the rebound on removal of the load. It is desirable, however, to develop a load-settlement diagram with more than two points.

According to the AASHTO specifications, the safe allowable load is 50% of that load which, after a continuous application of 48 hr, produces a permanent settlement not greater than ¼ in. measured at the top of the pile. This maximum settlement should not be increased by continuous application of the test load for 60 hr or longer.

For buildings, the commonest rule is that the safe load is half that load causing a total settlement not exceeding 1 in. and a net settlement of 0.01 in. per ton of test load in 24 hr, after deducting rebound. When the load-settlement diagram shows a definite yield point, the safe load may be taken as 50% of the test load at yield point. For a specific project, the local building code should be consulted.

While load tests are helpful, the results should be analyzed carefully. Soil conditions under the rest of the substructure may not be the same as those under the test piles. Unless a pile group is tested, tests on single piles do not indicate the reduction necessary for a pile group. Short-time tests do not show the effects of long-time consolidation of the soil.

Pile Formulas. A pile-driving formula is a widely used means of evaluating soil capacity. Formulas, however, have many limitations and should be used only in conjunction with adequate boring information and pile load tests. The usual formula gives the ultimate or the safe load of a single pile as determined from energy applied to a pile in driving it and the consequent penetration into the ground. Some commonly used formulas are listed in Table 7-5.

Because of factors omitted from such formulas and uncertainties in values to be used for terms included, ultimate loads obtained by formula may not agree closely with those obtained in static-load tests to failure. The same formula at times may give ultraconservative results, and at other times, unsafe loads. Furthermore, variations in subgrade conditions may give very misleading results. For example, a pile driven into a dense layer of sand overlying soft clay will drive very hard. Pile-driving formulas will indicate excellent load-carrying capacity. But because of the underlying soft layer, the pile actually will have low capacity. Hence, complete reliance cannot be placed on pile-driving formulas alone.

(R. D. Chellis, "Pile Foundations," McGraw-Hill Book Company, New York; L. Zeevaert, "Foundation Engineering for Difficult Subsoil Conditions," Van Nostrand Reinhold Company, New York.)

7-9. Pile Characteristics. In selection of piles for a foundation, initial cost is only one consideration. The potential high cost of repairing damage due to excessive settlement, or of adding piles later, or of replacing piles that have rotted or corroded should also be taken into account.

The cheapest pile does not necessarily yield the lowest-cost pile foundation. The number and length required, for example, may make use of such a pile impractical. Selection of the lowest-cost suitable pile requires consideration of numerous conditions: destructive influences, permissible settlement, loads, length of pile required, bearing value per pile, availability of

Table 7-5. Some Pile-driving Formulas

Engineering News formula for gravity hammers:

$$R = \frac{2WH}{S + 1} \tag{7-24a}$$

Engineering News formula for single-acting steam hammers:

$$R = \frac{2WH}{S + 0.1} \tag{7-24b}$$

Engineering News formula for double-acting steam hammers:

$$R = \frac{2H(W + Ap)}{S + 0.1} \tag{7-24c}$$

Modified *Engineering News* formula:

$$R = \frac{2WH}{S + 0.1(P/W)} \tag{7-24d}$$

where R = safe load on a single pile, lb
  W = weight, lb, of striking parts of hammer
  H = height of fall less twice the height of bounce, ft
  S = average penetration, in. per blow, for the last 5 to 10 blows for gravity hammers and the last 10 to 20 blows for steam hammers
  A = area of piston, sq in.
  p = steam pressure at hammer, psi
  P = weight of pile as driven, lb
Hiley formula (takes into account weight of pile):

$$R_u = \frac{e_f W_r h}{S + 0.5C} \frac{W_r + e^2 W_p}{W_r + W_p} \tag{7-24e}$$

where R_u = ultimate load on pile, lb
  e_f = hammer efficiency: 100% for diesel hammers and gravity hammers released by trigger; 80 to 90% for single-acting hammers; 65 to 85% for double-acting hammers; 75 to 85% for differential-acting hammers (see manufacturers' recommendations; also, allow for wear, weather, etc.)
  W_r = weight, lb, of striking parts of hammer
  W_p = weight of pile as driven, lb
  h = fall of hammer, in.
  S = average penetration, in. per blow, for the last 5 blows for gravity hammers and the last 20 blows for other types
  e = coefficient of restitution of cap blocks or cushions or butt of pile
  C = sum of temporary compression allowances, in., for pile, pile head and cap, and ground

New York City Building Code, 1968, Minimum Driving Resistance for Friction Piles, Except Timber Piles*

Pile capacity, tons	20			30			40			50				60			
Rated hammer energy, ft-kips	15	19	24	15	19	24	15	19	24	15	19	24	32	15	19	24	32
Resistance, blows per ft	19	15	11	30	23	18	44	32	24	72	49	35	24	96	63	44	30

*Ratio of pile weight to hammer weight should not exceed 3.5. For timber piles, the *Engineering News* formulas may be used, with the following limitations:

Pile capacity, tons	Up to 20	20–25	25–30	More than 30 Single-acting hammer	More than 30 Double-acting hammer
Rated hammer energy, ft-kips	7.5–12	9–12	14–16	12–16	15–20

pile types, cost of pile in place, ease and speed of installing, variations in pile length and cost of bringing pile tops to desired elevation, cost of concrete pile caps, level of water table, and problems arising from construction in water.

End-bearing piles may be structurally designed for a desired bearing capacity. Allowable stresses are lower than for compression members of the same material in the superstructure, to allow for subgrade conditions. Table 7-6 lists allowable stresses often used, but local building codes or highway or railway specifications should be consulted for a specific structure. These documents sometimes also limit the maximum design load on a type of pile, as indicated in Table 7-6, unless tests prove that higher loads are safe. Tests are desirable because the capacity of a pile is not determined by its strength alone, but also by its ability to distribute its load to the soil and the capacity of the soil to support the load (Art. 7-8).

Table 7-6. Maximum Design Loads on Single Piles

Type of pile	Allowable stresses when bearing on rock, psi	Maximum design load, tons		
		For piles on rock	For piles on soft rock, hardpan, or gravel over rock	Friction piles
Timber piles		25†	25†	30
Closed-end pipe, cast-in-place concrete,				
and compacted concrete piles		120	80	60
Concrete	$0.33f'_c$*			
Reinforcing steel	$0.40f_y$*			
Steel shell over ¼ in. thick	12,000			
Open-end concrete-filled pipe:				
18-in.-diameter or larger		250	80	60
Under 18 in. diameter		200	80	80
Concrete	$0.33f'_c$*			
Steel shell	12,000			
Steel bearing piles	12,000	150	80	60
Drilled-in caissons		No limit		
Concrete	$0.25f'_c$*			
Steel shell	12,000			
Steel core	18,000			

*f'_c = 28-day compressive strength of concrete test cylinder, psi
 f_y = yield strength of steel reinforcing; limited to 40,000 psi
†Some building codes limit timber piles to as low as 12 tons; others place no limitation on design load when bearing on rock or other hard stratum other than allowable stresses on pile cross section at upper surface of soil supporting the pile.

Piles usually are made of commonly used structural materials: wood, steel, concrete, or combinations of these materials. Many types have been or are patented. Under certain circumstances, these are more economical or can function where others cannot.

Wood piles generally are economical when friction piles are required. But other types of piles may be less costly for heavy loads. Generally, timber piles are available in lengths up to 75 ft in the eastern and central parts of the United States and exceeding 110 ft in the northwest. Nearly every type of wood can be used. If kept permanently below water, the material will not decay. Hence, when wood piles are used, they should be cut off below the water table, allowance being made for future lowering of the groundwater. But in sea or brackish water, the wood may be attacked by marine borers. Parts above water may be damaged by fungi or destroyed by fire.

Treated wood piles have greater resistance to decay and attack by borers and insects than untreated wood piles. But the treatment makes the wood more brittle and may not provide immunity under severe conditions. Sometimes, additional protection is given wood piles above ground by painting them or encasing them in metal or concrete. The usual preservative treatment is pressure creosoting. Final retention should be at least 12 lb of creosote per cu ft of wood. (Bark should be removed before treatment.) The cut tops of installed piles should be treated with three coats of hot creosote. When treated piles extend above ground level, the preservative may have to be replenished after a time.

Wood piles are relatively flexible and have high resistance to shock. Hence, they often are used for small piers, ferry slips, fenders, and dolphins.

Hard driving may fracture wood piles and broom the butts. Hence, wood piles should not be used where it is necessary to drive a pile into or through rock, rock fill, a field of boulders, or other obstructions.

A steel shoe will protect the pile tip when it is driven into gravel or small boulders. When a shoe is used, the minimum dimension is taken as that at the upper end of the shoe. Generally, the minimum tip diameter for a wood pile is 6 in.

The pile should taper uniformly from butt to tip. Also, a straight line from center of butt to center of tip should nowhere lie outside the pile. Lack of straightness can cause the pile to buckle or crumple during driving.

Cast-in-place concrete piles can carry heavier loads in end bearing than timber and therefore may be less costly in foundations of heavily loaded structures. Some types of cast-in-place piles also have the advantage that inspection of the full length in the ground is possible before concrete is placed. Simple, inexpensive adjustment of length is another advantage, particularly over precast-concrete piles.

Cast-in-place piles have the disadvantage of displacing ground when driven. This can adversely affect adjacent structures and piles already positioned. Hence, the order of driving cast-in-place piles and the spacing should be planned to avoid damaging uncured concrete in nearby piles. The Uniform Building Code of the International Conference of Building Officials recommends that no pile be driven within 4.5 average pile diameters of a pile filled with concrete less than 24 hr old.

Cast-in-place concrete piles usually are constructed by placing concrete in permanent steel pipes or mandrel-driven steel shells. The pipe or shell prevents earth from mixing with the fresh concrete, serves as a form for the concrete, and distributes the pile load to the soil. The concrete and the pipe or shells, which may range from $3/16$ to $1/2$ in. in thickness, form a composite member. When the casing is more than $1/4$ in. thick, the steel may be considered to carry some of the load (see Table 7-6), with $1/16$ in. deducted from the thickness when corrosion may be expected. Reinforcing is not placed in the concrete in this type of pile, because no allowance may be made for the load carried by reinforcing.

The shells, either tapered or cylindrical corrugated tubes, are driven with a removable mandrel inside that stiffens them. The mandrels may be solid, tapered, or step tapered. Or they may be cylindrical and may collapse pneumatically or mechanically to permit withdrawal from corrugated cylindrical shells. If the piles are more than 60 ft long, pipe lengths are fastened to the bottom of the casing.

Tapered shells are primarily used for friction piles. Cylindrical shells and pipe may be used for friction or end-bearing piles.

Pipe piles may be used closed or open end. The open-end type usually is driven to bearing on rock and is preferred where displacement of the ground will affect adjacent structures or piles. Soil inside the pipe is removed by blowing with water and compressed-air jets or with augers. If the pipes are large enough, a worker may be lowered to inspect them after they have been cleaned of all foreign matter. Otherwise, a lamp may be lowered to permit inspection from the top. If water is present, a 3- to 4-ft-thick tremie seal of grout should be placed in the bottom. The water should then be pumped from the pile so that the concrete core can be placed in the "dry."

Uncased piles may be simple cylinders, or cylindrical shafts with a pedestal or base at the bottom. Ordinarily, the simple cylinder pile is made by first driving into the ground a heavy steel pipe with a removable core or detachable shoe to keep soil from entering. When the pipe has been positioned in the ground, the core is removed. Next, the pipe is filled with concrete, then withdrawn. To insure continuity of shaft, the pipe may be withdrawn while pressure is exerted on the concrete by the weight of the pile-driver hammer and the core, by compressed air, or by the reaction to the pipe withdrawal force.

Cast-in-place piles without an outside shell are not widely used because experience with them has been bad. The reason for this is the difficulty of placing the concrete without intrusion of soil or necking of the piles. They should be used only with great care in concreting to insure that the shell-less pile is uniform and continuous in section. Hence, other types of piles generally can be placed faster and are more reliable.

Drilled piles are uncased concrete piles made by drilling holes and filling them with concrete. The holes may be belled out at the bottom if necessary for load distribution on a bearing stratum. These piles are suitable in cohesive soils where soil will not fall into or water enter the holes. Advantages include economy of material, no soil displacement, no heave, rapid installation, less noise and vibration than pile driving, and visual inspection of the complete pile before concreting. Temporary casings may be used to keep water or unstable ground out of the hole, but great care must be exercised in removing the casings so as not to damage the newly placed concrete.

The holes may be made with augers or bucket-type drills. The bells may be excavated mechanically with a belling bucket.

Displacement piles are a type of concrete pile suitable for areas where steel is expensive or in short supply. This type of pile should be used only where it can bear on good granular material, because the compaction under the pile tips is dissipated in clay and silts. Before the casing, which usually is 20 in. in diameter, is driven, a plug of 3 to 5 ft of no-slump concrete mix is rammed into the bottom (Fig. 7-15a). The concrete plug is hammered into the ground by a heavy weight dropped on the concrete. Friction between the dry concrete and the casing pulls the casing down, too, displacing the soil (Fig. 7-15b). At the desired depth, the pipe is held in place while the plug is hammered out of the casing (Fig. 7-15c). As the concrete spreads down and outward, additional concrete is added to form a pedestal. Then, the casing is gradually withdrawn while successive charges of concrete are rammed down to form the pile shaft (Fig. 7-15d). Installation of a permanent shell before concreting the shaft will prevent necking of the shaft and poor concrete caused by intrusion of soil.

Precast-concrete piles generally cost more to fabricate than cast-in-place piles. The precast piles have to be handled, transported, and sometimes stored. They have to be reinforced or prestressed to prevent damage when they are handled, transported, or driven. Required lengths must be determined fairly accurately before the piles are cast; cutting off piles that are too long is wasteful and costly, while splicing piles that are too short is time-consuming and costly. So time and money usually must be spent in driving test piles to predetermine pile lengths. Another disadvantage of precast piles is the possibility of damage when they are driven through boulders. The lower portion may crack or the tip spall where the damage cannot be detected.

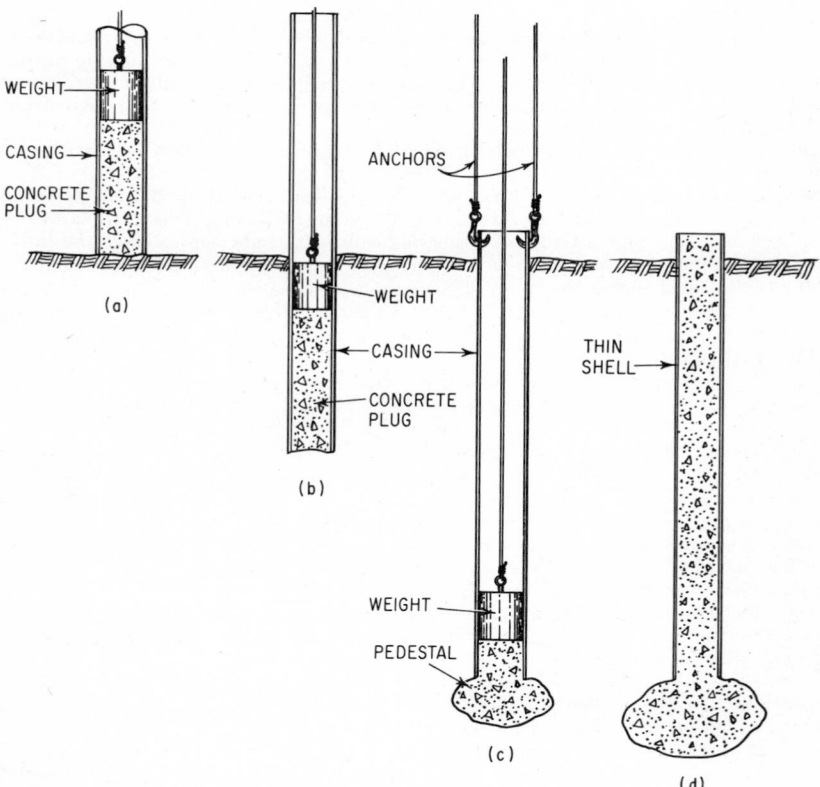

Fig. 7-15. Steps in producing a concrete displacement pile.

But precast piles are advantageous for docks, piers, and bulkheads, mainly because of the difficulty of casting concrete in open water. They also may be economical for bridge approaches, trestles, and viaducts where the piles extend above the ground or water to serve also as columns. Usually made of high-strength concrete, precast piles are not likely to be damaged by the driving of adjacent piles. Reinforced or prestressed, they are suitable for use as tension piles, as well as end-bearing and friction piles. Immediately after completion of driving, precast piles are available for supporting the full design load and hence can be used to support erection loads. Prestressing is especially desirable for exposed piles, to keep cracks from opening and exposing embedded steel to corrosive influences.

Because precast piles are heavy, a heavy hammer with short strokes generally is desirable for driving them. Use of a water jet is advantageous for driving into or through dense granular materials. For hard driving, the piles may be equipped with steel driving shoes to prevent spalling at the tip.

Composite timber-concrete piles are used to obtain the lower costs of wood piles without the necessity of deep excavation to cut them off below the water table. The wood portion, forming the lower part of the composite pile, is driven until the top is just above grade. A concrete-pile follower then is used to drive the wood pile below the water table. Almost any type of concrete pile can serve as the follower. Successful performance of the composite pile, however, depends on the joint. It must be able to keep out mud and water and to resist tension and bending. Many types of patented joints are available.

Steel bearing piles are widely used because they can carry heavy loads and serve as friction and end-bearing piles. Standard H-pile sections, proportioned to prevent local crippling and buckling under design loads, should be used. Flanges and web of a standard shape have the same thickness, and flange width and depth of section are nearly equal. Allowable unit stresses and maximum design load for bearing on rock without load tests are listed in Table 7-6. Table 7-7 gives the capacity at 12,000 psi for standard H piles.

Table 7-7. Capacities of Standard Steel Bearing Piles*

Nominal depth, in.	Weight, lb per ft	Area, sq in.	Capacity, tons at 12,000 psi
8	36	10.60	63.60
10	42	12.35	74.10
10	57	16.76	100.56
12	53	15.58	93.48
12	74	21.76	130.56
14	73	21.46	128.76
14	89	26.19	157.14
14	102	30.01	180.06
14	117	34.44	206.64

*For maximum allowable capacity without load tests, see Table 7-6.

One important advantage of steel bearing piles is their low displacement of ground during driving. This permits them to be driven near structures, where high-displacement piles are not advisable because of shock waves and vibrations. Steel piles can be used in long lengths, and adjustments in length are relatively easy. They can be driven through hard soils. But they may be deflected or damaged by boulders unless flanges and web at the tip are protected by a shoe or welded plates.

Steel piles may be economical for docks, piers, bulkheads, bridges and bridge approaches, trestles, and viaducts where the piles extend above the water to serve also as columns. Because they can be withdrawn for reuse, steel piles are used temporarily during construction to support falsework or to retain earth cuts.

Steel-plate caps have been used at the tops of steel piles to distribute the reaction of the piles over the concrete of a footing. Tests show, however, that in a massive footing, where thickness of concrete above the pile tops is 2 or 3 ft, such bearing devices are not necessary. Hence, the trend in practice is away from the use of plate caps.

Steel piles exposed to the air, alternate wetting and drying, acids, or electrolysis will corrode. Where application of a coating is feasible, the steel should be protected with paint or bitumastic. Encasement in dense, reinforced concrete is beneficial, especially at the ground line or in the tidal range. Cathodic protection may be satisfactory for piles in the ground or below low tide but may not work for piles in the tidal range or splash zone.

Sand drains, or piles, are used for compaction, stabilization, and draining of compressible material. Replacing piles, the drains are economical for supporting loads on soil with much smaller bearing capacity than sand. They actually are cylindrical piles of sand in the ground, often 18 to 20 in. in diameter and 6 to 10 ft apart. They act as wicks for water squeezed from the soil, thus accelerating settlement of the poor material by allowing pore water to drain. Holes for the sand may be drilled with an auger or driven with a pile, which is retracted. After the holes are filled with sand, a drainage blanket and a surcharge are placed over the area to expedite consolidation of the soil. If, however, the compressible layer is permeable or thin, for example, a clay layer less than about 30 ft thick, preloading with a surcharge may sufficiently accelerate consolidation without sand drains.

Drilled-in caissons are composite steel-concrete piles socketed into rock. Ranging in diameter from 24 to 42 in., they have been used to support loads as large as 2,800 tons (Fig. 7-16a). These piles can be extended to great depths, for example, 200 ft. They can reach these depths at less cost and faster than pneumatic caissons, which are limited to a maximum air pressure of 50 psi, or depths of 115 ft. Drilled-in caissons can be passed through boulders or obstacles that would stop driven piles. And they can be installed close to existing structures.

A drilled-in caisson consists of a ½-in.-thick steel pipe, a structural steel core, and concrete (Fig. 7-16b to d). The pipe casing, with reinforced cutting edge, is driven and churndrilled, if necessary, open-ended to rock. Meanwhile, material inside the pipe is cleaned out. Boulders encountered may be drilled out. When rock is reached, a socket is drilled into the rock. The socket should be long enough to seal the casing in the rock and to transfer the design load from the core to the rock through bond. Next, the steel core is inserted in the casing and grouted into the rock socket. After that, the casing is filled with concrete. When the load does not warrant a full-length steel core, a stub core can be used at the bottom to distribute the load from the casing and fix the bottom of the caisson in the rock.

Sheetpiles are vertical interlocking sections that are driven into the ground at the bottom to form a wall or enclosure above excavation level. To resist lateral pressures, they may brace each other or may be supported by a bracing system.

Sheetpiles may be made of wood, steel, or concrete. Applications include cofferdams, retaining walls, bulkheads, shoring, trench sheathing, cutoff walls under dams, bridge and building foundations, sea walls, and dock and wharf walls.

Wood sheetpiles may consist of a single, double, or triple thickness of planks. A single or double thickness may be adequate for supporting earth cuts, but a triple thickness resists driving better. Some types of wood sheetpiles come with provision for connection along vertical edges.

Steel sheetpiles may be ordered to manufacturers' standard specifications or to American Society for Testing and Materials Standard Specification for Steel Sheet Piling A328. Sections available include straight-web, arched-web, Z, and trapezoid (Larssen). They come with handling or extracting holes along the web center line at one end or both ends. Vertical edges are shaped to interlock, permitting relative vertical movement so they can be driven in succession but without separating horizontally.

Corrugated-steel sheetpiles are lightweight and low-cost. They can be driven by hand or light hammers. Applications include sheathing for trenches and shallow foundation excavations.

Precast-concrete planks, reinforced or prestressed, and with provision for connection along the vertical edges, also serve as sheetpiles. When a wall or enclosure must be watertight, the joints are grouted after the piles have been driven. Expansion and contraction joints with flexible filler may be provided at intervals of 25 to 50 ft. Or a special unit, solid below ground and split above, may be used, with filler in the split.

(R. D. Chellis, "Pile Foundations," McGraw-Hill Book Company, New York; ACI Committee 543, "Recommendations for Design, Manufacture, and Installation of Concrete Piles," *American Concrete Institute Journal*, August 1973, p. 509.)

7-10. Pile-Installation Equipment. Piles are installed in the ground by displacing soil, pushing it down or to the side, and by excavating the soil, forcing it upward. Principal displacement, or pile-driving, methods are hammering, jacking, and vibrating a pile or pile component into place. Excavating methods include preexcavating, jetting, excavation with a bucket, and hand digging, to form a hole in which a pile can be made or placed.

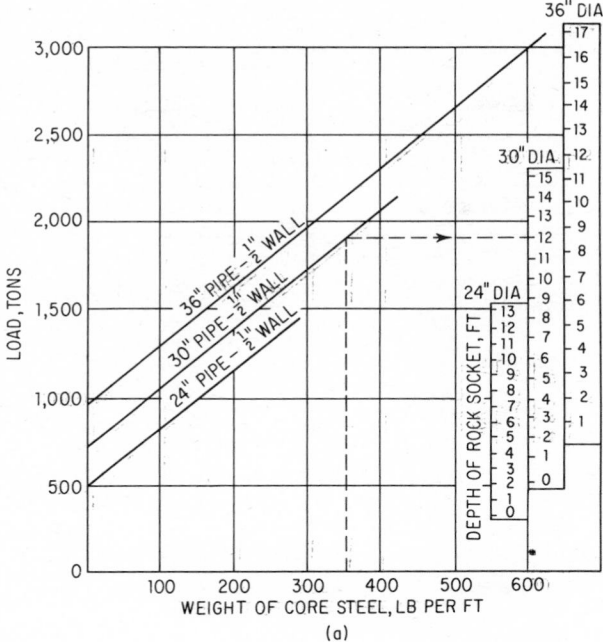

WEIGHT OF CORE STEEL, LB PER FT

(a)

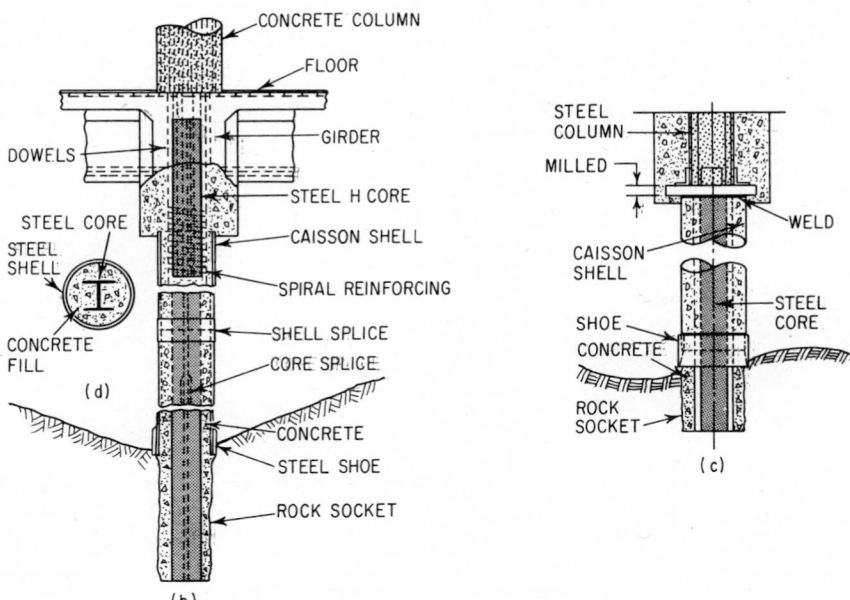

(b)

(c)

(d)

Fig. 7-16. Drilled-in caissons. (a) Curves show variation of capacity with shell diameter and weight of steel core. (b) Typical section of caisson for a concrete superstructure. (c) Typical section of caisson for a steel superstructure. (d) Typical caisson cross section.

Table 7-8. Rated Values of Pile-driving Hammers

Hammer type	Rated energy, ft-lb	Blows per minute	Ram weight, lb
1. Drop Hammers			
Vulcan: Hammers weighing from 500 to 3,000 lb; drop 12 to 48 in.			
2. Single-acting Hammers			
McKiernan Terry No.			
OS60	180,000	55	60,000
OS40	120,000	55	40,000
OS30	90,000	60	30,000
OS20	60,000	60	20,000
OS14	37,500	60	14,000
OS10	32,500	55	10,000
OS8	26,000	55	8,000
OS5	16,250	60	5,000
OS3	9,000	65	3,000
Vulcan No.			
O60	180,000	62	60,000
O40	120,000	60	40,000
O30	90,000	55	30,000
O20	60,000	60	20,000
O14	42,000	60	14,000
O10	32,500	50	10,000
O8	26,000	50	8,000
O	24,375	50	7,500
O6	19,500	60	6,500
1	15,000	60	5,000
2	7,260	70	3,000
3. Double- and Differential-acting Hammers			
Super-Vulcan No.			
400C	113.486	100	40,000
200C	50,200	98	20,000
140C	36,000	103	14,000
80C* and 80M †	24,450	111	8,000
65C	19,200	117	6,500
50C* and 50M †	15,100	120	5,000
30C* and 30M †	7,260	133	3,000
DGH900	4,000	238	900
18C	3,600	150	1,800
DGH100B	386	303	100
McKiernan-Terry No.			
C8	26,000	77–85	8,000
C826	24,000	85–95	8,000
11B3	19,150	95	5,000
C5	16,000	100–110	5,000
10B3	13,100	105	3,000
C3	9,000	130–140	3,000
9B3	8,750	145	1,600
7	4,150	225	800
6.5	3,200	280	600
6	2,500	275	400
5	1,000	300	200
3	385	400	68

Table 7-8. Rated Values of Pile-driving Hammers *(Continued)*

Hammer type	Rated energy, ft-lb	Blows per minute	Ram weight, lb
	4. DIESEL HAMMERS		
Delmag No.			
D44 (single-acting)	87,000	37–56	9,500
D30 (single-acting)	54,140	39–60	6,600
D22 (single-acting)	39,700	42–60	4,850
D12 (single-acting)	22,500	42–60	2,750
D5 (single-acting)	9,100	42–60	1,100
McKiernan-Terry No.			
DE40 (single-acting)	43,000	48–52	4,000
DA35 (single-acting)	35,500	48	2,800
DA35 (double-acting)	21,000	82	2,800
DE30 (single-acting)	30,100	48–52	2,800
DE20 (single-acting)	18,800	48–52	2,000
DE10 (single-acting)	9,900	48–52	1,100
Link-Belt Speeder No.			
520 (partial double-acting)	30,000 max	80–84	5,070
440 (double-acting)	18,200	86–90	4,000
312 (partial double-acting)	18,000 max	100–105	3,855
180 (partial double-acting)	8,100	90–95	1,725
105 (partial double-acting)	7,500	90–98	1,445
	5. VIBRATORY HAMMERS		
McKiernan-Terry No.			
V14 (hydraulic)	150 hp	1,500–1,850 vpm	
L. B. Foster No.			
2-60 (tandem)	240 hp	700–1,020 vpm	
2-75E (electric)	150 hp	700–1,020 vpm	
2-60 (electric)	120 hp	700–1,020 vpm	
2-50 (electric)	100 hp	700–1,020 vpm	
2-40H (hydraulic)	114 hp	0–1,800 vpm	
2-40E (electric)	80 hp	955–1,100 vpm	
2-35 (electric)	70 hp	890–1,120 vpm	
2-20E (electric)	40 hp	955–1,100 vpm	
2-17 (electric)	34 hp	1,090-1,290 vpm	
2-3 (electric)	6 hp	1,800 vpm	

*Open type
†Closed type
Courtesy L. B. Foster Company, Pittsburgh, Pa.

Hammers may be used for pile driving in nearly every type of soil. Exceptions include sites with many large boulders, and fills with stones and broken concrete slabs. But hammering should not be done near existing structures because of possible damage from vibration. Hammering also may damage or move nearby piles (Art. 7-8).

Various types of hammers are available (Table 7-8). For some applications, all types may be suitable, but none is best for all conditions. In general, a heavy hammer should be used for heavy piles; a light hammer, for light piles. The blow should not be so powerful that it overstresses the piles. A high-energy blow with low velocity at impact usually gives best results in dense strata, such as stiff clay, shale, hardpan, and compact gravel. In soils that readjust rapidly after being disturbed, rapid blows give best results. (See also Table 7-5.)

Drop hammers consist of a weight, guides, supporting framework, and means for raising and dropping the weight. The weight is raised along the fixed guides, or leads, by a cable that goes over a sheave atop the framework to a drum or geared shaft. At a desired height, the weight is dropped.

Drop hammers are slower than other types of hammers. Between blows, the pile is at rest and may become hard to restart. In soft soils, where high drops can be used, pile driving with a drop hammer is fast. But in hard soils, where the drop has to be reduced to avoid over-stressing the pile, driving is slow.

Pile hammers use steam or compressed air for operation. Moving along fixed leads, or leads swinging or hanging from a crane, the weight is attached to a piston and cylinder device through which pressure is applied.

In a single-acting hammer, steam or air raises the weight. But when released, it falls free. The weight usually is heavy, the drop small, and energy delivered high. This type of hammer delivers more blows per minute than a drop hammer.

In double-acting hammers, steam or air imparts additional energy to the weight when it falls. These hammers use a lighter weight but give more blows per minute than single-acting hammers, thus reducing friction, inertia, and point resistance more in some soils. The weight may slide along fixed or hanging leads, or the hammer, suspended from a crane, may rest on the pile butt. For proper efficiency, however, the pile-driving equipment should have leads.

Differential-acting hammers are modified double-acting hammers. The weight is lifted by the difference in pressure on two pistons. Number of blows per minute is about the same as for double-acting hammers. Weight and free fall are comparable with those of single-acting hammers. Pile driving with a differential-acting hammer generally is much faster than with the same-size single-acting hammer and requires less steam or air.

Diesel hammers use rapid combustion of a fuel directly to drive a pile down and the weight, or ram, up. A unit contains a fuel-injection system, ram, anvil, and vertical cylinder. Leads may be fixed or hanging. To start the hammer, a cable lifts the ram and a trip drops it. As it falls, it actuates a fuel pump to inject fuel between the ram and the anvil. The impact ignites the fuel, and the explosion drives the pile down, the ram up.

Energy determination is difficult with diesel hammers. The operator can vary the fuel supply. Energy is lost because of friction and compression of air and gases in the combustion chamber. Length of stroke varies with pile resistance.

The variation in drop of the ram may be an advantage in hard driving, a disadvantage in soft soil. Where resistance is high, the stroke is large and high energy per blow is delivered. In soft soil, where pile resistance is low, the impact may not be sufficient to ignite the fuel. The hammer's reaction to pile resistance also is advantageous as a safety measure: A broken pile, or pile falling out of the leads, offering little resistance to the dropping ram, may cause the hammer to stop.

Diesel hammers are longer and operate more slowly than correspondingly rated double-acting hammers, but weigh less and are more mobile. With no boilers, compressors, or hoses, diesel hammers require less preparatory and closing-down work.

Underwater Driving. When piles must be driven below water, a hammer may be kept above by attaching a follower, or pile extension, to the pile top. Driving can be done underwater with an enclosed steam hammer, with a steam-exhaust hose extending above the water surface. Air pressure at about 0.5 psi per ft of water depth in the lower part of the hammer housing will keep water out of the casing.

Pile-driving plants may be specialized rigs or specially equipped cranes. Cranes need a boom for handling pile leads and the hammer. Also, they need a drum or winch head with a line to the head of the boom for positioning the piles. The specialized rigs consist of a framework and platform for supporting the equipment needed in pile driving, for example, for a steam hammer, an engine, boiler, winches, and hammer.

Jacking piles into the ground is more costly than hammering. But jacking may be necessary to avoid damage from hammering vibrations or because of insufficient headroom. Often, the method is used for underpinning. Piles are forced down by hydraulic jacks reacting against heavy weights. An existing structure may be used for the reaction if it is strong and heavy enough.

Vibratory pile driving is another method of installing or extracting piles. High- and low-frequency vibrators have been developed for the purpose (Table 7-8). The method has worked well in sand and soft soils, but it cannot punch through glacial till or thin rock layers, where hammers and steel piles are suitable. In the vibration method, a vibrator applies to a pile longitudinal vibrations, which weaken friction and adhesion along the pile sides. Some soils become viscous, or even fluid, and dead weight moves the pile down. Where the method works, it is fast and produces little noise. But equipment charges and possible savings in labor should be checked.

Preexcavating often is required to prevent heave caused by pile driving. In some soils, preexcavation is faster and the substructure can be completed sooner. A hole about 2 in. larger in diameter than the pile is drilled with a soil auger or a churn drill, and the pile is installed in the hole.

An **auger** has a rotating powered bit with a continuous helical cutter. A **churn drill** has a steel bit that is mechanically lifted and dropped to disintegrate soil or rock.

In wet rotary preexcavation, a special hollow bit is used. Water or slurry, fed down through the drill stem, emerges from ports in the bit and mixes with the cuttings. Carrying the cuttings up, the fluid clears the hole. The hydraulic pressure keeps the hole open and fills the voids in the side walls with slurry.

Jetting sinks piles by washing out some soil and lubricating the soil. The method is used to speed pile driving. Because jetting disturbs the soil, the method should be limited to end-bearing piles, and the piles should be driven the last 5 ft. Water is pumped under high pressure through pipes in or alongside the piles and discharged near the tip. The water then flows upward, bringing soil with it. Compressed-air jets sometimes are used to assist this upward flow.

Granular soils displaced by jetting usually slide back against the pile when the water flow ceases, but clays and silts may stay disturbed and lubricated for a long time.

Water pipes used in jetting usually are 2 to 4 in. in diameter. Nozzles range from ½ to 1½ in. in diameter. Pressures range from 100 to over 300 psi, and the flow per pile may exceed 1,000 gpm. Two- to four-stage bronze-fitted centrifugal pumps generally are used for jetting. They are driven by gasoline or diesel engines.

(R. D. Chellis, "Pile Foundations," McGraw-Hill Book Company, New York; R. L. Peurifoy, "Construction Planning, Equipment, and Methods," McGraw-Hill Book Company, New York.)

7-11. Caissons. These are load-bearing enclosures sunk into the ground, usually to protect excavation for a foundation, aid construction of the substructure, and serve as part of the permanent structure. Sometimes, a caisson is used to enclose a subsurface space to be used for such purposes as a pump well, machinery pit, or access to a deeper shaft or tunnel. Several caissons may be aligned to form a bridge pier, bulkhead, sea wall, foundation wall for a building, or impervious core wall for an earth dam.

In foundations, caissons are used to facilitate construction of shafts or piers extending from near the surface of land or water to a bearing stratum. This type of construction can carry heavy loads to great depths. Built of common structural materials, they may have any shape in cross section. They range in size from about that of a pile to over 100 ft in length and width. Some small ones are considered caisson piles (Art. 7-9). For some of the smaller caissons used for foundation shafts, the casing generally is not assigned any load-carrying capacity, or it may be withdrawn as concrete fills the hole.

Caissons often are installed by sinking them under their own weight or with a surcharge. The operation is assisted by jacking, jetting, excavating, and undercutting. Care must be taken during this operation to maintain alignment. The caissons may be built up as they sink, to permit construction to be carried out at the surface, or they may be completely prefabricated. Types of caissons used for foundation work are as follows:

Chicago caissons have been used for a long time for constructing foundation shafts through a thick layer of clay to hardpan or rock. The method is useful where the soil is sufficiently stiff to permit excavation for short distances without caving. A circular pit is dug about 5 ft deep and lined with wood staves. This vertical lagging is braced with two rings made with steel channels. Then, 5 ft of soil is removed, and the operation is repeated. If the ground is poor, shorter lengths are dug until the bearing stratum is reached. If necessary, the caissons can be belled at the bottom to carry large loads. Finally, the hole is filled with concrete. Minimum economical diameter for hand digging is 4 ft.

Sheeted piers or caissons are similarly constructed, but the vertical lagging of wood or steel is driven down during or before excavation. This system usually is used for shallow depths in wet ground.

In dry ground, horizontal wood sheeting may be used. This is economical and necessary where there is inadequate vertical clearance. Louvered construction should be used to provide drainage and to permit packing behind the wood sheeting where soil will not maintain a vertical face long enough to permit insertion of the next sheet. This type of construction requires overexcavating so that the wood sheets can be placed. Openings must be wide enough between sheets to allow backfilling and tamping, to correct the excavation irregularities and equalize pressure on all sides. Small blocks may be inserted between successive sheets to leave packing gaps. If the excavation is large, soldier beams, vertical cantilevers, can be driven to break up the long sheeting spans.

Drilled caissons are used primarily in cohesive soils. With the use of bentonite, sand layers also can be penetrated. These caissons are limited in use where boulders or wet granular materials are encountered.

Drilled caissons may be constructed much like drilled piles (Arts. 7-9 and 7-10). Diameters may range from 18 in. to over 10 ft.

Foundation and cutoff walls may be constructed of such caissons by overlapping them. One way is to place a first group of caissons in line slightly less than two diameters apart. Then, drilled caissons are installed between the first ones, before the concrete gains full strength. This method should be used with care in wet ground, because it is difficult to effect a seal between the caissons. If the caissons go out of plumb, large voids will be present in the wall.

Benoto caissons are slower to place and more expensive than drilled caissons, except in wet granular material and where soil conditions are too tough for augers or rotating-bucket diggers. Benoto caissons up to 39 in. in diameter may be sunk through water-bearing sands, hardpan, and boulders to depths of 150 ft. Excavation is done with a hammer grab, a single-line orange-peel bucket, inside a temporary, cylindrical steel casing. The hammer grab is dropped to cut into or break up the soil. After impact, the blades close around the soil. Then, the bucket is lifted out and discharged. Boulders are broken up with heavy, percussion-type drills. Rock is drilled out by churn drills.

The casing is bolted together in 20-ft-deep sections, starting with a cutting edge. A hydraulic attachment on the machine oscillates the casing continuously to ease sinking and withdrawal, while jacks force the casing into the ground. As concrete is placed, the jacks withdraw the casing in a way that allows excellent concreting of the caisson.

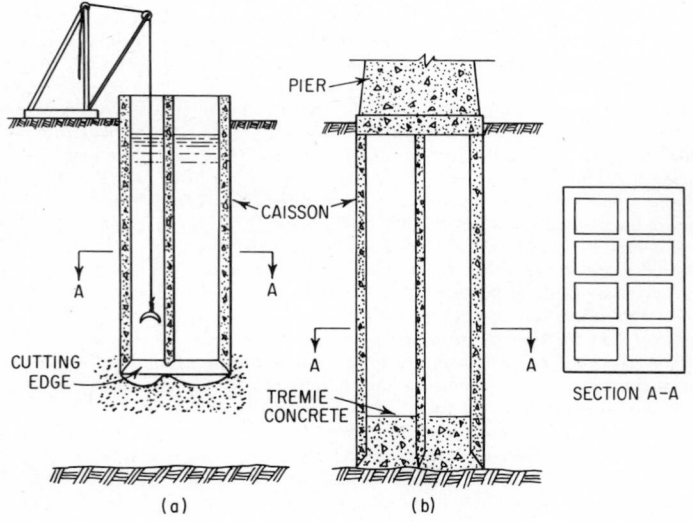

Fig. 7-17. Construction with an open concrete caisson.

Open caissons are enclosures without top and bottom during the lowering process. When used for pump wells and shafts, they often are cylindrical; for bridge piers, they usually are rectangular and compartmented. The compartments serve as dredging wells, pipe passages, and access shafts (Fig. 7-17a). Dredging wells usually have 12- to 16-ft clear openings to facilitate excavation with clamshell or orange-peel buckets.

An open caisson may be a braced steel shell that is filled with concrete, except for the wells, as it is sunk into place. Or a caisson may be constructed entirely of concrete.

Friction along the caisson sides may range from 300 to over 1,000 psf. So despite steel cutting edges at the wall bottoms, the caisson may not sink. Water and compressed-air jets may be used to lubricate the soil to decrease the friction.   For the purpose, vertical jetting pipes should be embedded in the outer walls.

If the caisson does not sink under its own weight with the aid of jets when soil within has been removed down to the cutting edge, the caisson must be weighted.   One way is to build it higher, to its final height, if necessary.   Otherwise, a platform may have to be built on top and weights piled on it.   This can be expensive.

Care must be taken to undercut the edges evenly, or else the caisson will tip.   Obstructions and variations in the soil also can cause uneven sinking.

When the caisson reaches the bearing strata, the bottom is sealed with concrete (Fig. 7-17b). The plug may be placed by tremie or made by injecting grout into the voids of coarse aggregate.

When a caisson must be placed through water, marine work sometimes may be converted to a land job by construction of a sand island.   Fill is placed until it projects above the water surface. Then, the caisson is constructed and sunk as usual on land.

Pneumatic caissons contain at the base a working chamber with compressed air at a pressure equal to the hydrostatic pressure of the water in the soil.   Without the balancing pressure, the water would force soil from below up into a caisson.   A working chamber clear of water also permits hand work to remove obstructions that buckets, air lifts, jets, and divers cannot.   Thus, the downward course of the caisson can be better controlled.   But sinking may be slower and more expensive, and compressed-air work requires precautions against safety and health hazards.

Access to the working chamber for workers, materials, and equipment is through air locks, usually placed at the top of the caisson (Fig. 7-18).   Steel access cylinders 3 ft in diameter connect the air locks with the working chamber in large caissons.

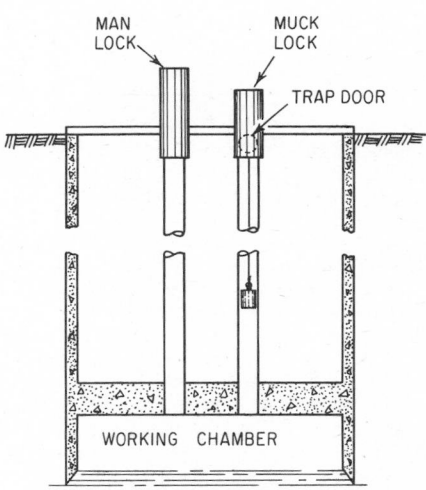

Fig. 7-18. Pneumatic caisson. Pressure in working chamber is above atmosphere.

Entrance to the working chamber requires only a short stay for a worker in an air lock.   But the return stop may be lengthy, depending on the pressure in the chamber, to avoid the bends, or caisson disease.   This is caused by air bubbles in muscles, joints, and the blood.   Slow decompression gives the body time to eliminate the excess air.   In addition to slow decompression, it is necessary to limit the hours worked at various pressures (Table 7-9) and the maximum pressure to 50 psi above atmospheric.   (There are, however, several modifications of these limitations.)   The restriction on pressure limits the maximum depth at which compressed-air work can be done to 115 ft.   A medical, or recompression, lock is also required on the site for treatment of workers attacked by the bends.

Floating caissons are used when it is desirable to fabricate caissons on land, tow them into position, and sink them through water.   They are constructed much like open or pneumatic caissons, but with a "false" bottom, "false" top, or buoyant cells.   When floated into position, a caisson must be kept in alignment as it is lowered.   A number of means may be used for the purpose, including anchors, templates supported on temporary piles, anchored barges, and cofferdams. Sinking generally is accomplished by adding concrete to the walls. When the cutting edges reach the bottom, the temporary bulkheads at the base, or false bottoms, are removed, since buoyancy no longer is necessary. With false tops, buoyancy is controlled with compressed air, which can be released when the caisson sits on the bottom. With buoyant cells, buoyancy is gradually lost as the cells are filled with concrete.

Closed-box caissons are similar to floating caissons, except the top and bottom are permanent. Constructed on land, of steel or reinforced concrete, they are towed into position. Sometimes, the site can be dredged in advance to expose soil that can safely support the caisson and loads that will be imposed on it. Where loads are heavy, however, this may not be practicable. Then, the box caisson may have to be supported on piles; but allowance can be made for its buoyancy. This type of caisson has been used for breakwaters, sea walls, and bridge-pier foundations.

Table 7-9. Allowable Working Hours per 24-hr Period in Compressed Air

Pressure above atmospheric, psi	Maximum allowable total working hours	Allowable working time in first period, hr	Minimum interval of rest in open air, hr	Allowable working time in second period, hr
15–26	6	3	1	3
26–33	4	2	2	2
33–38	3	1.5	3	1.5
38–43	2	1	4	1
43–48	1.5	0.75	5	0.75
48–50	1	0.5	6	0.5

Potomac caissons have been used in wide tidal rivers with deep water underlain by deep, soft deposits of sand and silt. Large timber mats are placed on the river bottom, to serve as a template for piles and to retain tremie concrete. Long, steel pipe or H piles are driven in clusters, vertical and battered, as required. Prefabricated steel or concrete caissons are set on the mat over the pile clusters, to serve as permanent forms for concrete shafts to be supported on the piles. Then, concrete is tremied into the caissons. Since the caissons are used only as forms, construction need not be so heavy as for conventional construction, where they must withstand launching and sinking stresses, and cutting edges are not required.

(G. A. Leonards, "Foundation Engineering," McGraw-Hill Book Company, New York; H. F. Winterkorn and H. Fang, "Foundation Engineering Handbook," Van Nostrand Reinhold Company, New York.)

7-12. Cofferdams. These are temporary walls or enclosures for protecting an excavation. Generally, one of the most important functions is to permit work to be carried out on a nearly dry site.

Cofferdams should be planned so that they can be easily dismantled for reuse. Since they are temporary, safety factors can be small, 1.25 to 1.5, when all probable loads are accounted for in the design. But design stresses should be kept low when stresses, unit pressure, and bracing reactions are uncertain. Design should allow for construction loads and the possibility of damage from construction equipment. For cofferdams in water, the design should provide for dynamic effect of flowing water and impact of waves. The height of the cofferdam should be adequate to keep out floods that occur frequently.

Earth dikes, when fill is available, are likely to be the cheapest type of cofferdam for keeping water out of an excavation. If impervious material is not easily obtained, however, a steel sheetpile cutoff wall may have to be driven along the dike, to permit pumps to handle the leakage. With an impervious core in the dike, wellpoints, deep-well pumps, or sumps and ditches may be able to keep the excavation unwatered.

Timber cribs also are relatively cheap cofferdams. Built on shore, they can be floated to the site and sunk by filling with rock. The water side may be faced with wood boards for watertightness (Fig. 7-19). For greater watertightness, two lines of cribs may be used to support two lines of wood sheeting between which clay is tamped to form a "puddle" wall. Design of timber cribs should provide ample safety against overturning and sliding.

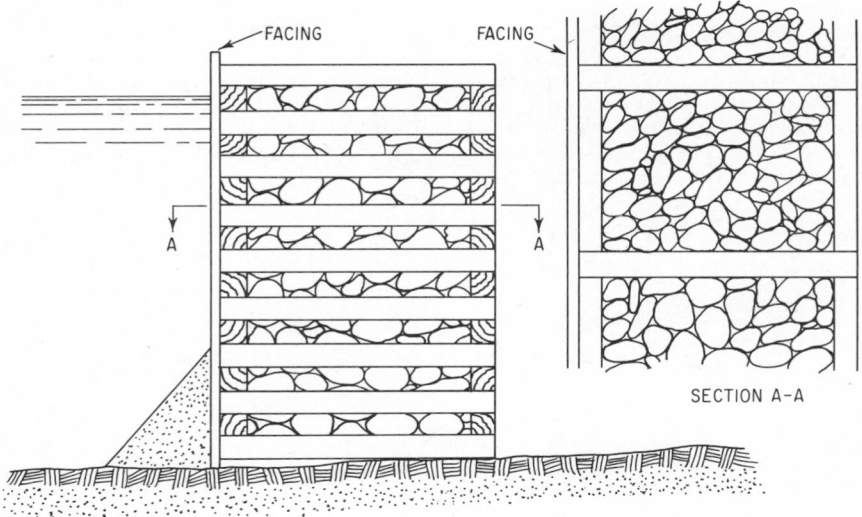

ELEVATION

Fig. 7-19. Timber crib with stone filling.

Double-wall cofferdams may be erected in water to enclose large areas. They consist of two lines of sheetpiles tied to each other. The space between is filled with sand (Fig. 7-20). For sheetpiles driven to irregular rock, or gravel, or onto boulders, the bottom of the space between walls may be filled with a thick layer of tremie concrete to seal gaps below the tips of the sheeting. Double-wall cofferdams are likely to be more watertight than single-wall and can be used to greater depths.

A berm may be placed against the outside face of a cofferdam for stability. If so, it should be protected against erosion. For this purpose, riprap, woven mattresses, streamline fins or jetties, or groins may be used. If the cofferdam rests on rock, a berm need be placed on the inside only if required to resist sliding, overturning, or shearing. On sand, an ample berm must be provided so that water has a long path to travel to enter the cofferdam. (The amount of percolation is proportional to the length of path and the head.) Otherwise, the inside face of the cofferdam may settle, and the cofferdam may overturn as water percolates under the cofferdam and causes a quick, or boiling, excavation bottom. An alternative to a wide berm is wider spacing of the

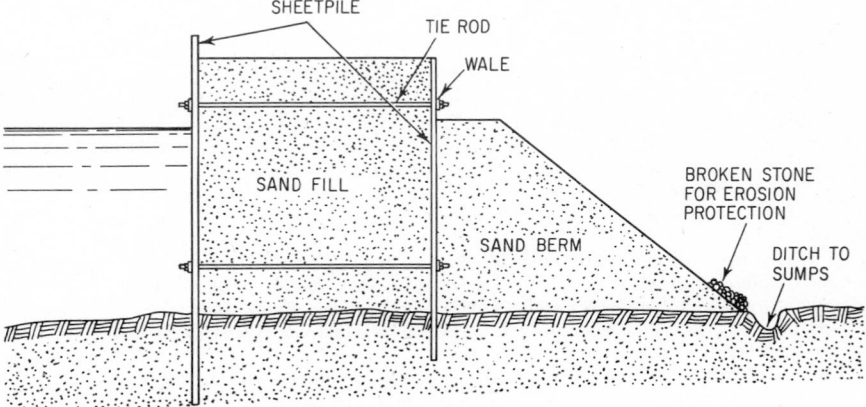

Fig. 7-20. Double-wall cofferdam.

cofferdam walls. This is more expensive but has the added advantage that the top of the fill can be used by construction equipment and for construction plant.

Cellular cofferdams, used in construction of dams, locks, wharves, and bridge piers, are suitable for enclosing large areas in deep water. These enclosures are composed of relatively wide units. Average width of a cellular cofferdam on rock should be 0.7 to 0.85 times the head of water against the outside. As with double-wall cofferdams, however, when constructed on sand, a cellular cofferdam should have an ample berm on the inside to prevent the excavation bottom from becoming quick (Fig. 7-21*d*).

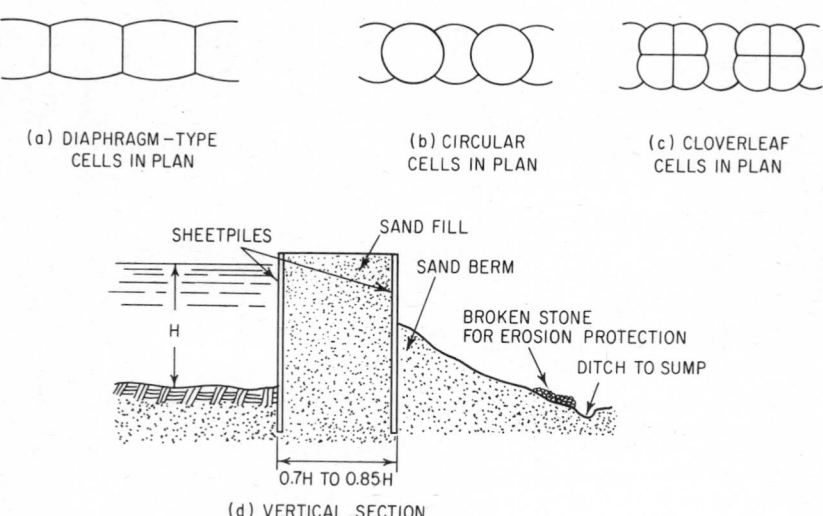

(a) DIAPHRAGM-TYPE
CELLS IN PLAN

(b) CIRCULAR
CELLS IN PLAN

(c) CLOVERLEAF
CELLS IN PLAN

(d) VERTICAL SECTION

Fig. 7-21. Cellular sheetpile cofferdams.

Cells are formed with interlocking steel sheetpiles. One type of cell consists of circular arcs connected by straight diaphragms (Fig. 7-21*a*). Another type comprises circular cells connected by circular arcs (Fig. 7-21*b*). Still another type is the cloverleaf, composed of large circular cells subdivided by straight diaphragms (Fig. 7-21*c*). The cells are filled with sand. The internal shearing resistance of the sand contributes substantially to the strength of the cofferdam. For this reason, it is unwise to fill a cofferdam with clay or silt. Weepholes on the inside sheetpiles drain the fill, thus relieving the hydrostatic pressure on those sheets and increasing the shear strength of the fill.

In circular cells, lateral pressure of the fill causes only ring tension in the sheetpiles. Maximum stress in the pile interlocks usually is limited to 8,000 lb per lin in. This, in turn, limits the maximum diameter of the circular cells. Because of numerous uncertainties, this maximum generally is set at 60 ft. When larger-size cells are needed, the cloverleaf type may be used.

Circular cells are preferred to the diaphragm type because each circular cell is a self-supporting unit. It may be filled completely to the top before construction of the next cell starts. (Unbalanced fills in a cell may distort straight diaphragms.) When a circular cell has been filled, the top may be used as a platform for construction of the next cell. Also, circular cells require less steel per linear foot of cofferdam. The diaphragm type, however, may be made as wide as desired.

In driving the sheetpiles, care must be taken to avoid breaking the interlocks. Accurately set and plumbed against a structurally sound template, the sheetpiles should be driven in short increments, so that when uneven bedrock or boulders are encountered, driving can be stopped before the cells or interlocks are damaged. Also, starting and driving lightly all the piles in a cell until it is completed can reduce jamming troubles with the last piles to be installed for the cell.

Single-wall cofferdams form an enclosure with only one line of sheeting. If there is no water problem, they may be built with soldier piles (vertical cantilevers) and horizontal wood lagging (Fig. 7-22). If there is a water problem, the cofferdam may be constructed of sheetpiles. While

Fig. 7-22. Soldier beams and wood lagging retain the sides of an excavation.

they require less wall material than double-wall or cellular cofferdams, single-wall cofferdams generally require bracing on the inside. Also, unless the bottom is driven into a thick, impervious layer, they may leak excessively at the bottom. There may also be leakage at interlocks. Furthermore, there is danger of flooding and collapse due to hydrostatic forces when these cofferdams are unwatered.

For marine applications, therefore, it is advantageous to excavate, drive piles, and place a seal of tremie concrete without unwatering single-wall sheetpile cofferdams. Often, it is advisable to predredge the area before the cofferdam is constructed, to facilitate placing of bracing and to remove obstructions to pile driving. Also, if blasting is necessary, it would severely stress the sheeting and bracing if done after they were installed.

For buildings, single-wall cofferdams must be carefully installed. Small movements and consequent loss of ground usually must be prevented to avoid damaging neighboring structures, streets, and utilities. Therefore, the cofferdams must be amply braced. Sheeting close to an existing structure should not be a substitute for underpinning.

Cantilevered sheetpiles may be used for shallow single-wall cofferdams in water or on land where small lateral movement will not be troublesome. Embedment of the piles in the bottom must be deep enough to insure stability. Design usually is based on the assumptions that lateral passive resistance varies linearly with depth and the point of inflection is about two-thirds the embedded length below the surface.

Cofferdams may be braced in many ways. Figure 7-23 shows some commonly used methods. Circular cofferdams may be braced with horizontal rings (Fig. 7-23a). For small rectangular cofferdams, horizontal braces, or wales, along side and end walls may be connected to serve only as struts. For larger cofferdams, diagonal bracing (Fig. 7-23b) or cross-lot bracing (Fig. 7-23d and e) is necessary. When space is available at the top of an excavation, pile tops can be anchored with concrete dead men (Fig. 7-23c). Where rock is close, the wall can be tied back with tensioned wires or bars anchored in grouted sockets in the rock (Fig. 7-24).

Horizontal cross braces should be spaced to minimize interference with excavation, form construction, concreting, and pile driving. Spacing of 12 to 18 ft is common. Piles and wales selected should be strong enough as beams to permit such spacing. In marine applications, divers often have to install the wales and braces underwater. To reduce the amount of such work, tiers of bracing may be prefabricated and lowered into the cofferdam from falsework or from the top set of wales and braces, which is installed above the water surface. In some cases, it may be advantageous to prefabricate and erect the whole cage of bracing before the sheetpiles are driven. Then, the cage, supported on piles, can serve also as a template for driving the sheetpiles.

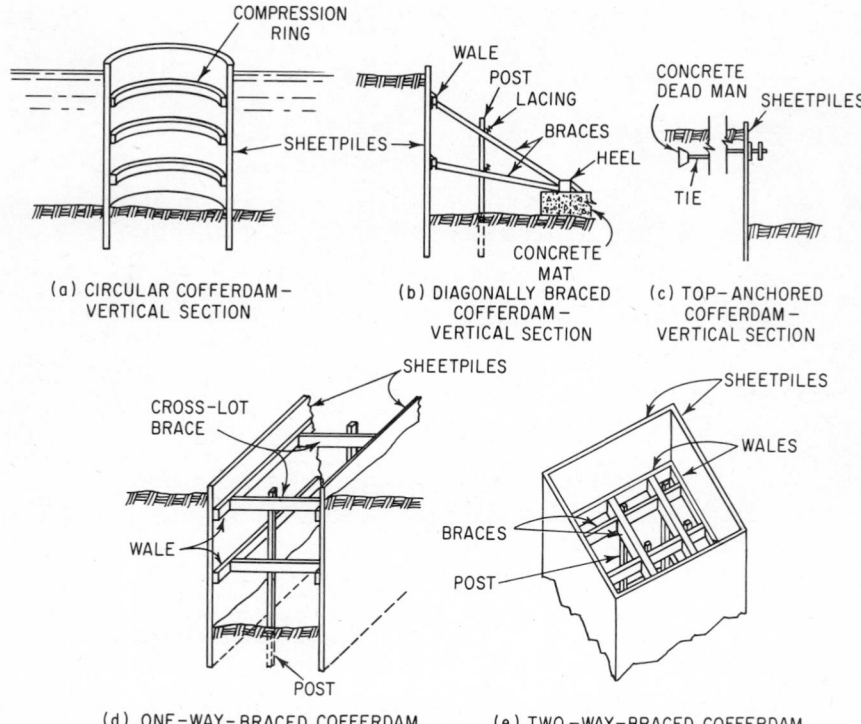

(a) CIRCULAR COFFERDAM—VERTICAL SECTION

(b) DIAGONALLY BRACED COFFERDAM—VERTICAL SECTION

(c) TOP—ANCHORED COFFERDAM—VERTICAL SECTION

(d) ONE—WAY—BRACED COFFERDAM

(e) TWO—WAY—BRACED COFFERDAM

Fig. 7-23. Types of cofferdam bracing.

All wales and braces should be forced into bearing with the sheeting by wedges and jacks.

When pumping cannot control leakage into a cofferdam, excavation may have to be carried out in compressed air. This requires a sealed working chamber, access shafts, and air locks, as for pneumatic caissons (Art. 7-11). Other techniques, such as use of a tremie concrete seal or chemical solidification or freezing of the soil, if practicable, however, will be more economical.

Braced sheetpiles may be designed as continuous beams subjected to uniform loading for earth and to loading varying linearly with depth for water (Art. 7-13). (Actually, earth pressure depends on the flexibility of the sheeting and relative stiffness of supports. An exact calculation of such pressures and the stresses induced generally is impractical.) Wales may be designed for uniform loading. Allowable unit stresses in the wales, struts, and ties may be taken at half the elastic limit for the materials, because the construction is temporary and the members are exposed to view. Distress in a member can easily be detected, and remedial steps taken quickly.

Soldier beams and horizontal wood sheeting are a variation of single-wall cofferdams often used where impermeability is not required. The soldier beams, or piles, are driven vertically into the ground to below the bottom of the proposed excavation. Spacing usually ranges from 5 to 10 ft (Table 7-10). (The wood lagging can be used in the thicknesses shown in Table 7-10 because of arching of the earth between successive soldier beams.)

As excavation proceeds, the wood boards are placed horizontally between the soldiers (Fig. 7-22). Louvers, or packing spaces, 1 to 2 in. high are left between the boards so that earth can be tamped behind them to hold them in place. Hay may also be stuffed behind the boards to keep the ground from running through the gaps. The louvers permit important drainage of water, to relieve hydrostatic pressure on the sheeting and thus allow use of a lighter bracing system. The soldiers may be braced directly with horizontal or inclined struts; or wales and braces may be used.

Table 7-10. Usual Maximum Spans of Horizontal Sheeting with Soldier Piles, Ft

Nominal thickness of sheeting, in.	In well-drained soils	In cohesive soils with low shear resistance
2	5	4.5
3	8.5	6
4	10	8

Advantages of soldier-beam construction include fewer piles; the sheeting does not have to extend below the excavation bottom, as do sheetpiles; and the soldiers can be driven more easily in hard ground than can sheetpiles. Varying the spacing of the soldiers permits underground utilities to be avoided. Use of heavy sections for the piles allows wide spacing of wales and braces. But the soldiers and lagging, as well as sheetpiles, are no substitute for underpinning; it is necessary to support and underpin even light adjoining structures.

Liner-plate cofferdams may be used for excavating circular shafts. The plates are placed in horizontal rings as excavation proceeds. Stamped from steel plate, usually about 16 in. high and 3 ft long, light enough to be carried by one man, liner plates have inward-turned flanges along all edges. Top and bottom flanges provide a seat for successive rings. End flanges permit easy bolting of adjoining plates in a ring. The plates also are corrugated for added stiffness. Large-diameter cofferdams may be constructed by bracing the liner plates with steel beam rings.

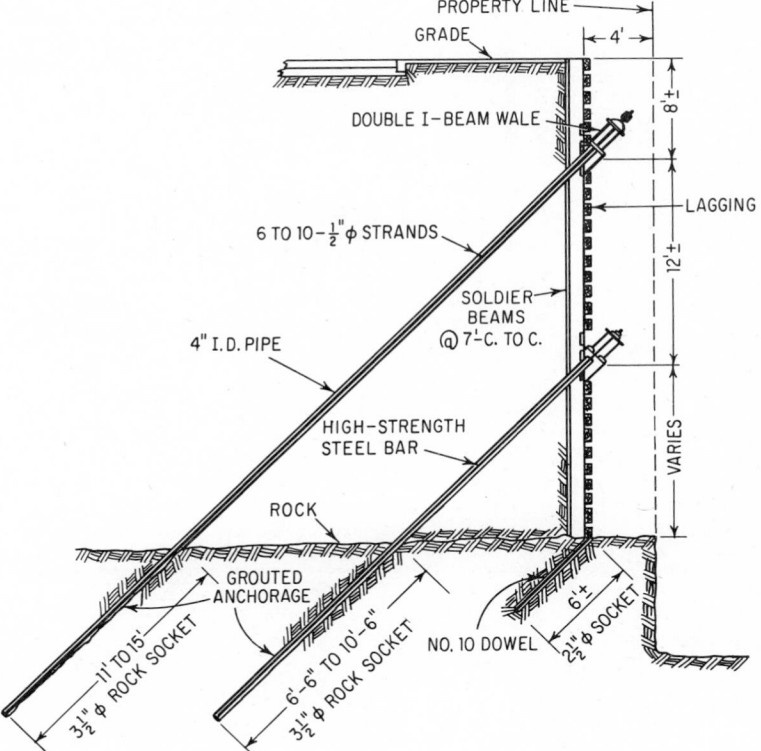

Fig. 7-24. Vertical section showing prestressed tie-backs for soldier beams.

Vertical-lagging cofferdams, with horizontal-ring bracing, also may be used for excavating circular shafts.   The method is similar to that used for Chicago caissons (Art. 7-11).   It is similarly restricted to soils that can stand without support in depths of 3 to 5 ft for a short time.

Slurry trenches may be used for constructing concrete walls.   The method permits building a wall in a trench without the earth sides collapsing.   While excavation proceeds for a 24- to 36-in.-wide trench, the hole is filled with a bentonite slurry with a specific gravity of 1.05 to 1.10 (Fig. 7-25a).   The fluid pressure against the sides and caking of bentonite on the sides prevent the earth walls of the trench from collapsing.   Excavation is carried out a section at a time.   A section may be 20 ft long and as much as 100 ft deep.   When the bottom of the wall is reached in a section, reinforcing is placed in that section.   (Tests have shown that the bond

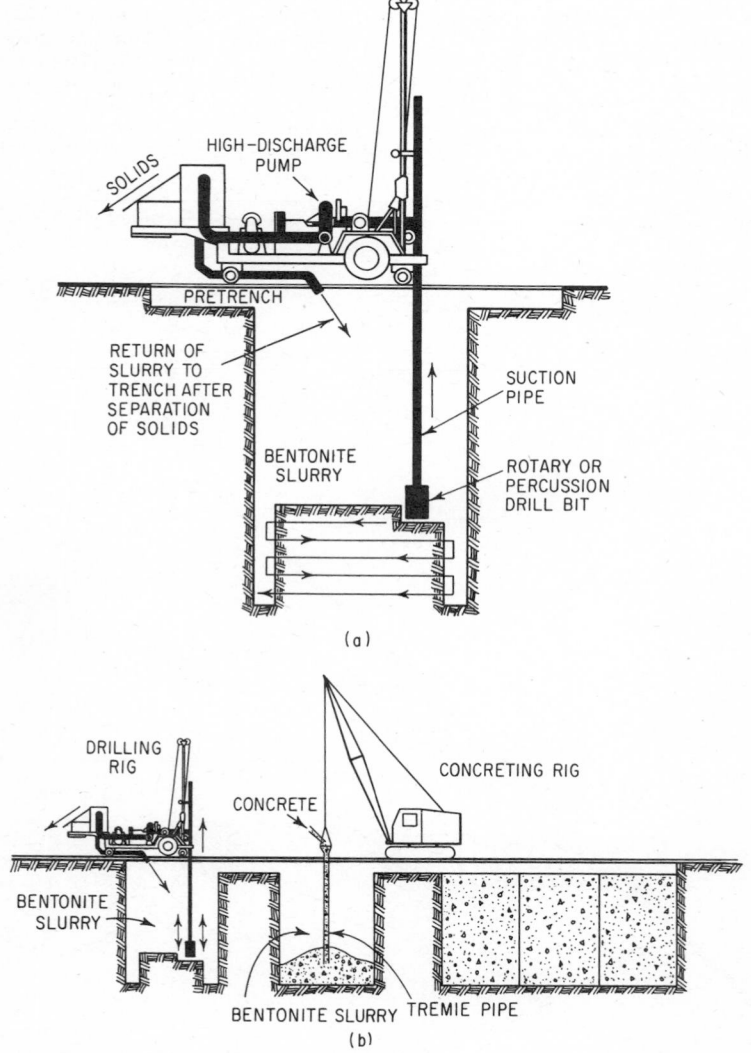

Fig. 7-25. Slurry-trench method for constructing a continuous concrete wall. (*a*) Excavating one section. (*b*) Concreting one section while another is excavated.

of the reinforcing to concrete is not materially reduced by the bentonite.) Then, concrete is tremied into the trench, replacing the slurry, which may flow into the next section to be excavated or be pumped into tanks for reuse in the next section (Fig. 7-25b). The method has been used to construct cutoffs for dams, cofferdams, foundations, walls of buildings and shafts.

Chemical grouting is used to solidify soils. Although it is effective in water-bearing soils, the method is not suitable in silts and clays, which are impervious to the chemicals. One method, the Joosten process, calls for successive injection of sodium silicate and calcium chloride, which react to form a hard, insoluble, calcium silicate gel. In sand, the resulting product resembles sandstone. Another method is based on oxidation of lignin with a chromium salt; a premixed single solution is injected, and time for gel formation may be controlled. This solution offers the advantage of low viscosity, which enables the liquid to penetrate less permeable soils. It is, however, expensive and has low strength.

Freezing is another means of solidifying water-bearing soils where obstructions or depth preclude pile driving. It can be used for deep shaft excavations and requires little material for temporary construction; the refrigeration plant has high salvage value. But freezing the soil may take a very long time. Also, holes have to be drilled below the bottom of the proposed excavation for insertion of refrigeration pipes.

(L. White and E. A. Prentis, "Cofferdams," Columbia University Press, New York; G. A. Leonards, "Foundation Engineering," McGraw-Hill Book Company, New York; H. F. Winterkorn and H. Fang, "Foundation Engineering Handbook," Van Nostrand Reinhold Company, New York.)

7-13. Lateral Earth Pressure. Water exerts against a vertical surface a horizontal pressure equal to the vertical pressure. At any level, the vertical pressure equals the weight of a 1-sq-ft column of water above that level. Hence, the horizontal pressure p, psf, at any level is

$$p = wh \qquad (7\text{-}25a)$$

where w = unit weight of water, lb per cu ft
h = depth of water, ft

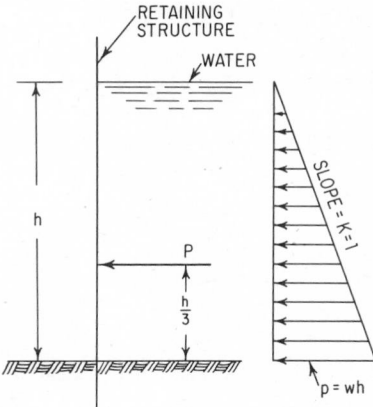

Fig. 7-26. Pressure diagram for water.

The pressure diagram is triangular (Fig. 7-26). Equation (7-25a) also can be written

$$p = Kwh \qquad (7\text{-}25b)$$

where K = pressure coefficient = 1.00

The resultant, or total, pressure, lb per lin ft, represented by the area of the hydrostatic-pressure diagram, is

$$P = K\frac{wh^2}{2} \qquad (7\text{-}26)$$

It acts at a distance $h/3$ above the base of the triangle.

Soil also exerts lateral pressure. But the amount of this pressure depends on the type of soil, its compaction or consistency, and its degree of saturation, and on the resistance of the structure 'to the pressure. Also, the magnitude of passive pressure differs from that of active pressure.

Active pressure tends to move a structure in the direction in which the pressure acts. Passive pressure opposes motion of a structure.

Free-standing walls retaining cuts in sand tend to rotate slightly around the base. Behind such a wall, a wedge of sand *ABC* (Fig. 7-27a) tends to shear along plane *AC*. C. A. Coulomb determined that the ratio of sliding resistance to sliding force is a minimum when *AC* makes an angle of $45° + \phi/2$ with the horizontal, where ϕ is the angle of internal friction of the soil, deg.

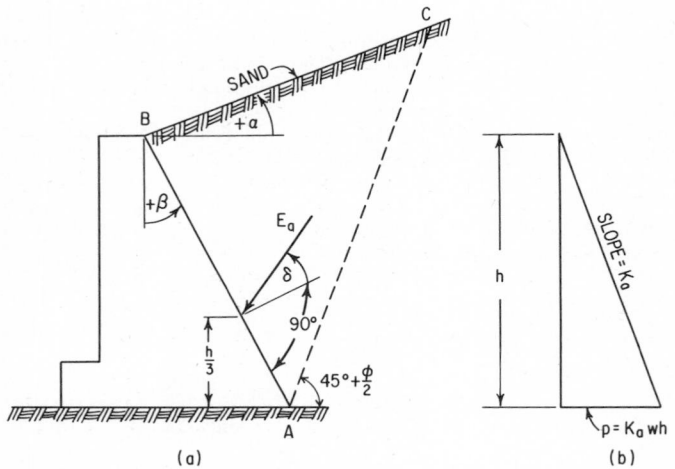

Fig. 7-27. Free-standing wall with sand backfill (*a*) is subjected to triangular pressure distribution (*b*).

For triangular pressure distribution (Fig. 7-27b), the active lateral pressure of a cohesionless soil at a depth *h*, ft, is

$$p = K_a w h \tag{7-27}$$

where K_a = coefficient of active earth pressure
  w = unit weight of soil, lb per cu ft
The total active pressure, lb per lin ft, is

$$E_a = K_a \frac{wh^2}{2} \tag{7-28}$$

Because of frictional resistance to sliding at the face of the wall, E_a is inclined at an angle δ with the normal to the wall, where δ is the angle of wall friction, deg (Fig. 7-27a). If the face of the wall is vertical, the horizontal active pressure equals $E_a \cos \delta$. If the face makes an angle β with the vertical (Fig. 7-27a), the horizontal active pressure equals $E_a \cos(\delta + \beta)$. The resultant acts at a distance of $h/3$ above the base of the wall.

If the ground slopes upward from the top of the wall at an angle α, deg, with the horizontal, then for cohesionless soils

$$K_a = \frac{\cos^2(\phi - \beta)}{\cos^2 \beta \cos(\delta + \beta)\left[1 + \sqrt{\dfrac{\sin(\phi + \delta)\sin(\phi - \alpha)}{\cos(\delta + \beta)\cos(\alpha - \beta)}}\right]^2} \tag{7-29}$$

The effect of wall friction on K_a is small and usually is neglected. For $\delta = 0$,

$$K_a = \frac{\cos^2(\phi - \beta)}{\cos^3 \beta \left[1 + \sqrt{\dfrac{\sin \phi \sin(\phi - \alpha)}{\cos \beta \cos(\alpha - \beta)}}\right]^2} \tag{7-30}$$

Table 7-11. Active-Lateral-Pressure Coefficients K_a

$\phi =$		10°	15°	20°	25°	30°	35°	40°
	$\alpha = 0$	0.70	0.59	0.49	0.41	0.33	0.27	0.22
	$\alpha = 10°$	0.97	0.70	0.57	0.47	0.37	0.30	0.24
$\beta = 0$	$\alpha = 20°$	. . .	. . .	0.88	0.57	0.44	0.34	0.27
	$\alpha = 30°$	. . .	. . .	. . .	. . .	0.75	0.43	0.32
	$\alpha = \phi$	0.97	0.93	0.88	0.82	0.75	0.67	0.59
	$\alpha = 0$	0.76	0.65	0.55	0.48	0.41	0.43	0.29
	$\alpha = 10°$	1.05	0.78	0.64	0.55	0.47	0.38	0.32
$\beta = 10°$	$\alpha = 20°$	. . .	. . .	1.02	0.69	0.55	0.45	0.36
	$\alpha = 30°$	. . .	. . .	. . .	. . .	0.92	0.56	0.43
	$\alpha = \phi$	1.05	1.04	1.02	0.98	0.92	0.86	0.79
	$\alpha = 0$	0.83	0.74	0.65	0.57	0.50	0.43	0.38
	$\alpha = 10°$	1.17	0.90	0.77	0.66	0.57	0.49	0.43
$\beta = 20°$	$\alpha = 20°$	. . .	. . .	1.21	0.83	0.69	0.57	0.49
	$\alpha = 30°$	. . .	. . .	. . .	. . .	1.17	0.73	0.59
	$\alpha = \phi$	1.17	1.20	1.21	1.20	1.17	1.12	1.06
	$\alpha = 0$	0.94	0.86	0.78	0.70	0.62	0.56	0.49
	$\alpha = 10°$	1.37	1.06	0.94	0.83	0.74	0.65	0.56
$\beta = 30°$	$\alpha = 20°$	. . .	. . .	1.51	1.06	0.89	0.77	0.66
	$\alpha = 30°$	. . .	. . .	. . .	. . .	1.55	0.99	0.79
	$\alpha = \phi$	1.37	1.45	1.51	1.54	1.55	1.54	1.51

Table 7-12. Angles of Internal Friction and Unit Weights of Soils

Type of soil	Density or consistency	Angle of internal friction ϕ, deg	Unit weight w, lb per cu ft
Coarse sand or sand and gravel	Compact	40	140
	Loose	35	90
Medium sand	Compact	40	130
	Loose	30	90
Fine silty sand or sandy silt	Compact	30	130
	Loose	25	85
Uniform silt	Compact	30	135
	Loose	25	85
Clay-silt	Soft to medium	20	90–120
Silty clay	Soft to medium	15	90–120
Clay	Soft to medium	0–10	90–120

Table 7-11 lists values of K_a determined from Eq. (7-30). Approximate values of ϕ and unit weights for various soils are given in Table 7-12.

For level ground at the top of the wall ($\alpha = 0$):

$$K_a = \frac{\cos^2 (\phi - \beta)}{\cos^3 \beta \left(1 + \dfrac{\sin \phi}{\cos \beta}\right)^2}$$

If, in addition, the back face of the wall is vertical ($\beta = 0$), Rankine's equation is obtained:

$$K_a = \frac{1 - \sin \phi}{1 + \sin \phi} \tag{7-32a}$$

Coulomb derived the trigonometric equivalent:

$$K_a = \tan^2 \left(45° - \frac{\phi}{2}\right) \tag{7-32b}$$

When information on the value of the angle of wall friction is not available, δ may be taken equal to $\phi/2$ for determining the horizontal component of E_a.

Note: Even light compaction may permanently increase the earth pressure into the passive range. This may be compensated for in wall design by use of a safety factor of at least 2.5.

Unyielding walls retaining cuts in sand, such as the abutment walls of a rigid-frame concrete bridge or foundations walls braced by floors, do not allow shearing resistance to develop in the sand along planes that can be determined analytically. For such walls, triangular pressure diagrams may be assumed, and K_a may be taken equal to 0.5. But only sand or gravel should be permitted for the backfill, and compaction should be light within 5 to 10 ft of the walls. See Note for free-standing walls.

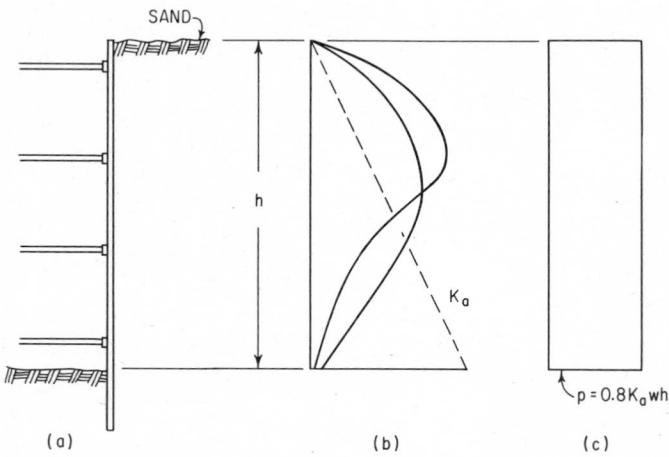

Fig. 7-28. Braced wall retaining sand (*a*) may have to resist pressures of the type shown in (*b*). Uniform pressure distribution (*c*) may be assumed for design.

Braced walls retaining cuts in sand (Fig. 7-28*a*) are subjected to earth pressure gradually and develop resistance in increments as excavation proceeds and braces are installed. Such walls tend to rotate about a point in the upper portion. Hence, the active pressures do not vary linearly with depth. Field measurements have yielded a variety of curves for the pressure diagram, of which two types are shown in Fig. 7-28*b*. Consequently, some authorities have recommended a trapezoidal pressure diagram, with a maximum ordinate

$$p = 0.8K_a wh \qquad (7\text{-}33)$$

K_a may be obtained from Table 7-11. The total pressure exceeds that for a triangular distribution.

Figure 7-29 shows earth-pressure diagrams developed for a sandy soil and a clayey soil. In both cases, the braced wall is subjected to a 3-ft-deep surcharge, and height of wall is 34 ft. For the sandy soil (Fig. 7-29*a*), Fig. 7-29*b* shows the pressure diagram assumed. The maximum pressure can be obtained from Eq. (7-33), with $h = 34 + 3 = 37$ ft and K_a assumed as 0.30 and w as 110 lb per cu ft:

$$p_1 = 0.8 \times 0.3 \times 110 \times 37 = 975 \text{ lb per sq ft}$$

The total pressure is estimated as

$$P = 0.8 \times 975 \times 37 = 28{,}900 \text{ lb per lin ft}$$

The equivalent maximum pressure for a trapezoidal diagram for the 34-ft height of the wall then is

$$p = \frac{28{,}900}{0.8 \times 34} = 1{,}060 \text{ lb per sq ft}$$

Assumption of a uniform distribution (Fig. 7-28c), however, simplifies the calculations and has little or no effect on the design of the sheeting and braces, which should be substantial to withstand construction abuses. Furthermore, trapezoidal loading terminating at the level of the excavation may not apply if piles are driven inside the completed excavation. The shocks may temporarily decrease the passive resistance of the sand in which the wall is embedded and lower the inflection point. This would increase the span between the inflection point and the lowest brace and increase the pressure on that brace. Hence, uniform pressure distribution may be more applicable than trapezoidal for such conditions.

See Note for free-standing walls.

Flexible bulkheads retaining sand cuts are subjected to active pressures that depend on the fixity of the anchorage. If the anchor moves sufficiently or the tie from the anchor to the upper portion of the bulkhead stretches enough, the bulkhead may rotate slightly about a point near the bottom. In that case, the sliding-wedge theory may apply. The pressure distribution may be taken as triangular, and Eqs. (7-27) to (7-32) may be used. But if the anchor does not yield, then pressure distributions much like those in Fig. 7-28b for a braced cut may occur. Either a trapezoidal or uniform pressure distribution may be assumed, with maximum pressure given by Eq. (7-33). Stresses in the tie should be kept low because it may have to resist unanticipated pressures, especially those resulting from a redistribution of forces from soil arching. The safety factor for design of ties and anchorages should be at least twice that used in conventional design.

Free-standing walls retaining plastic-clay cuts (Fig. 7-30a) may have to resist two types of active lateral pressure, both with triangular distribution. If the shearing resistance is due to cohesion only, a clay bank may be expected to stand with a vertical face without support for a height, ft, of

$$h' = \frac{2c}{w} \tag{7-34}$$

where $2c$ = unconfined compressive strength of the clay, psf
w = unit weight of clay, lb per cu ft

So if there is a slight rotation of the wall about its base, the upper portion of the clay cut will stand vertically without support for a depth h'. Below that, the pressure will increase linearly with depth as if the clay were a heavy liquid (Fig. 7-30b):

$$p = wh - 2c$$

The total pressure, lb per lin ft, then is

$$E_a = \frac{w}{2} \left(h - \frac{2c}{w} \right)^2 \tag{7-35}$$

It acts at a distance $(h - 2c/w)/3$ above the base of the wall. These equations assume wall friction is zero, the back face of the wall is vertical, and the ground is level.

This condition is likely to be temporary. In time, the clay will consolidate. The pressure distribution may become approximately triangular (Fig. 7-30d) from the top of the wall to the base. The pressures then may be calculated from Eqs. (7-27) to (7-32) with an apparent angle of internal friction for the soil (for example, see the values of ϕ in Table 7-12). The wall should be designed for the pressures producing the highest stresses and overturning moments.

Note: The finer the backfill material, the more likely it is that pressures greater than active will develop, because of plastic deformations, water-level fluctuation, temperature changes, and other effects. As a result, it would be advisable to use in design at least the coefficient for earth pressure at rest:

$$K_o = 1 - \sin \phi \tag{7-36}$$

The safety factor should be at least 2.5.

Clay should not be used behind retaining walls, where other economical alternatives are open. The swelling type especially should be avoided because it can cause high pressures and progressive shifting or rotation of the wall.

For a mixture of cohesive and cohesionless soils, the pressure distribution may temporarily be as shown in Fig. 7-30c. The height, ft, of the unsupported vertical face of the clay is

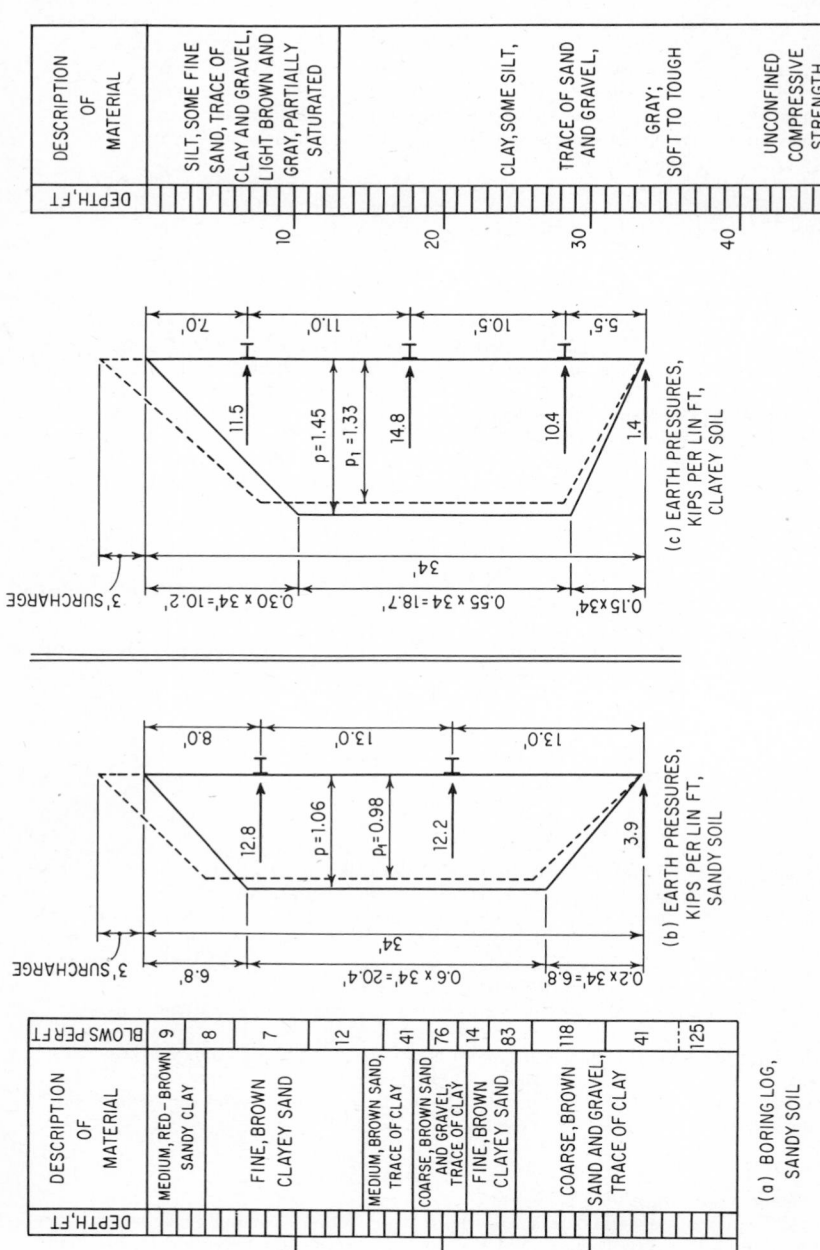

Fig. 7-29. Assumed trapezoidal pressure diagrams for a braced wall.

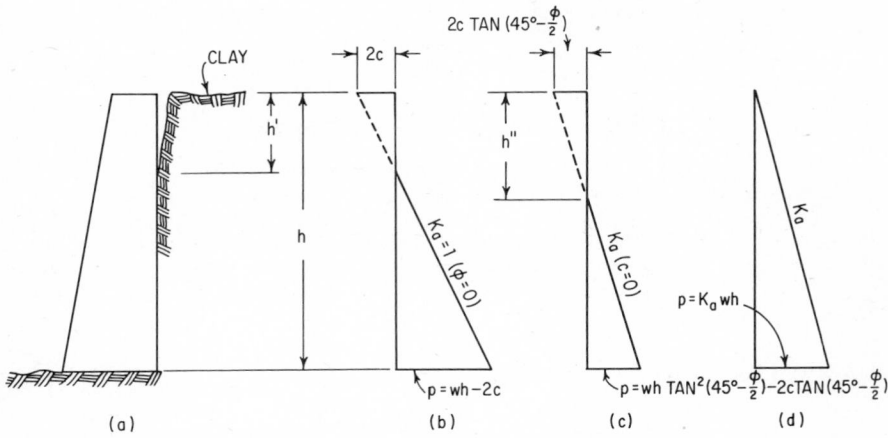

Fig. 7-30. Free-standing wall retaining clay (a) may have to resist the pressure distributions shown in (b) and (d). For mixed soils, the distribution may approximate that in (c).

$$h'' = \frac{2c}{w \tan (45° - \phi/2)} \tag{7-37}$$

The pressure at the base is

$$p = wh \tan^2 \left(45° - \frac{\phi}{2}\right) - 2c \tan \left(45° - \frac{\phi}{2}\right) \tag{7-38a}$$

The total pressure, lb per lin ft, is

$$E_a = \frac{w}{2} \left[h \tan \left(45° - \frac{\phi}{2}\right) - \frac{2c}{w} \right]^2 \tag{7-38b}$$

It acts at a distance $(h - h'')/3$ above the base of the wall.

Braced walls retaining clay cuts (Fig. 7-31a) also may have to resist two types of active lateral pressure. As for sand, the pressure distribution may temporarily be approximated by a trapezoidal diagram (Fig. 7-31b). On the basis of field observations, R. B. Peck has

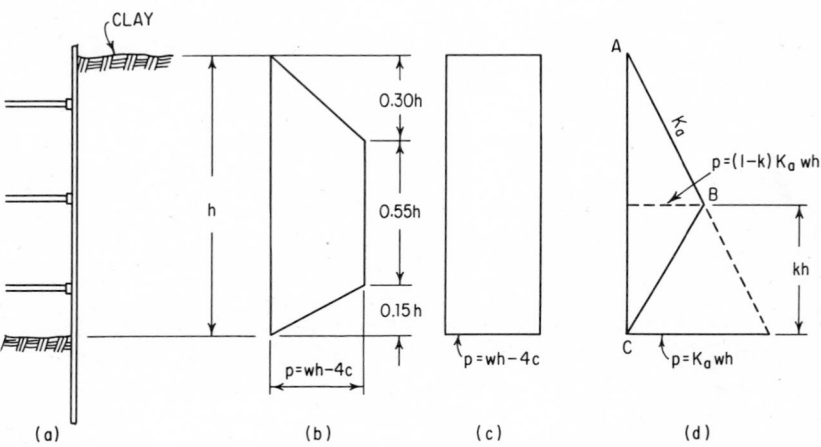

Fig. 7-31. Braced wall retaining clay (a) may have to resist pressures approximated by the pressure distributions in (b) and (d). Uniform pressure distribution (c) may be assumed in design.

recommended a maximum pressure of

$$p = wh - 4c$$

and a total pressure, lb per lin ft, of

$$E_a = \frac{1.55h}{2}(wh - 4c) \qquad (7\text{-}39)$$

[R. B. Peck, "Earth Pressure Measurements in Open Cuts, Chicago (Ill.) Subway," *Transactions, American Society of Civil Engineers,* 1943, pp. 1008–1036.]

Figure 7-29c shows a trapezoidal earth-pressure diagram determined for the clayey-soil condition of Fig. 7-29d. The weight of the soil is taken as 120 lb per cu ft; c is assumed as zero and the active-lateral-pressure coefficient as 0.3. Height of the wall is 34 ft, surcharge 3 ft. Then, the maximum pressure, obtained from Eq. (7-27) since the soil is clayey, not pure clay, is

$$p_1 = 0.3 \times 120 \times 37 = 1,330 \text{ lb per sq ft}$$

From Eq. (7-39) with the above assumptions, the total pressure is

$$P = \frac{1.55}{2} \times 37 \times 1,330 = 38,100 \text{ lb per lin ft}$$

The equivalent maximum pressure for a trapezoidal diagram for the 34-ft height of wall is

$$p = \frac{38,100}{34} \times \frac{2}{1.55} = 1,450 \text{ lb per sq ft}$$

To simplify calculations, a uniform pressure distribution may be used instead (Fig. 7-31c).

If after a time the clay should attain a consolidated equilibrium state, the pressure distribution may be better represented by a triangular diagram ABC (Fig. 7-31d), as suggested by G. P. Tschebotarioff. The peak pressure may be assumed at a distance of $kh = 0.4h$ above the excavation level for a stiff clay; that is, $k = 0.4$. For a medium clay, k may be taken as 0.25, and for a soft clay, as zero. For computing the pressures, K_a may be estimated from Table 7-11 with an apparent angle of friction obtained from laboratory tests or approximated from Table 7-12. The wall should be designed for the pressures producing the highest stresses and overturning moments.

See also Note for free-standing walls.

Flexible bulkheads retaining clay cuts and anchored near the top similarly should be checked for two types of pressures. When the anchor is likely to yield slightly or the tie to stretch, the pressure distribution in Fig. 7-31d with $k = 0$ may be applicable. For an unyielding anchor, any of the pressure distributions in Fig. 7-31 may be assumed, as for a braced wall. The safety factor for design of ties and anchorages should be at least twice that used in conventional design. See also Note for free-standing walls.

Backfill placed against a retaining wall should preferably be gravel to facilitate drainage. Also, weepholes should be provided through the wall near the bottom and a drain installed along the footing, to conduct water from the back of the wall and prevent buildup of hydrostatic pressures.

Saturated or submerged soil imposes substantially greater pressure on a retaining wall than dry or moist soil. The active lateral pressure for a soil-fluid backfill is the sum of the hydrostatic pressure and the lateral soil pressure based on the buoyed unit weight of the soil. This weight roughly may be 60% of the dry weight.

Surcharge, or loading imposed on a backfill, increases the active lateral pressure on a wall and raises the line of action of the total, or resultant, pressure. A surcharge w_s, psf, uniformly distributed over the entire ground surface may be taken as equivalent to a layer of soil of the same unit weight w as the backfill and with a thickness of w_s/w. The active lateral pressure, psf, due to the surcharge, from the backfill surface down, then will be $K_a w_s$. This should be added to the lateral pressures that would exist without the surcharge. K_a may be obtained from Table 7-11.

Passive pressures of cohesionless soils, resisting movement of a wall or anchor, develop because of internal friction in the soils. Because of friction between soil and wall, the failure surface is curved, not plane as assumed in the Coulomb sliding-wedge theory. Use of the

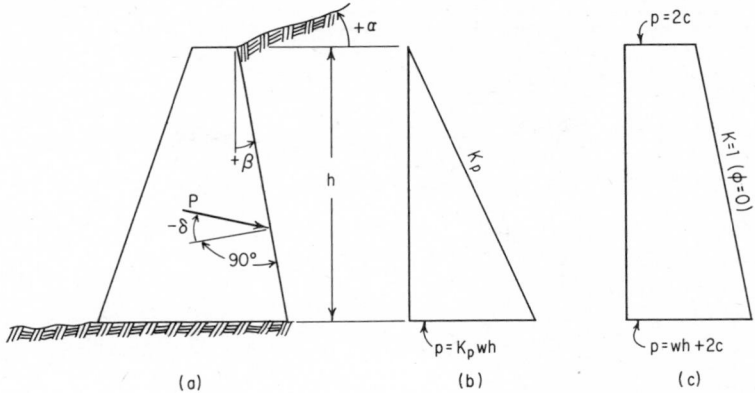

Fig. 7-32. Passive pressures on a wall (a) may vary as shown in (b) for sand or in (c) for clay.

Coulomb theory yields unsafe values of passive pressure when the effects of wall friction are included.

Total passive pressure, lb per lin ft, on a wall or anchor extending to the ground surface (Fig. 7-32a) may be expressed for sand in the form

$$P = K_p \frac{wh^2}{2} \tag{7-40}$$

where K_p = coefficient of passive lateral pressure
      w = unit weight of soil, lb per cu ft
      h = height of wall or anchor to ground surface, ft

The pressure distribution usually assumed for sand is shown in Fig. 7-32b. Table 7-13 lists values of K_p for a vertical wall face ($\beta = 0$) and horizontal ground surface ($\alpha = 0$), for curved surfaces of failure. (Many tables and diagrams for determining passive pressures are given in A. Caquot and J. Kérisel, "Tables for Calculation of Passive Pressure, Active Pressure, and Bearing Capacity of Foundations," Gauthier-Villars, Paris.)

Since a wall usually transmits a downward shearing force to the soil, the angle of wall friction δ correspondingly is negative (Fig. 7-32a). For embedded portions of structures, such as anchored sheetpile bulkheads, δ and the angle of internal friction ϕ of the soil reach their peak values simultaneously in dense sand. For those conditions, if specific information is not available, δ may be assumed as $-\frac{2}{3}\phi$ (for $\phi > 30°$). For such structures as a heavy anchor block subjected to a horizontal pull or thrust, δ may be taken as $-\phi/2$ for dense sand. For those cases, the wall friction develops as the sand is pushed upward by the anchor and is unlikely to reach its maximum value before the internal resistance of the sand is exceeded.

When wall friction is zero ($\delta = 0$), the failure surface is a plane inclined at an angle of $45° - \phi/2$ with the horizontal. The sliding-wedge theory then yields

$$K_p = \frac{\cos^2(\phi + \beta)}{\cos^3 \beta \left[1 - \sqrt{\dfrac{\sin\phi \, \sin(\phi + \alpha)}{\cos\beta \, \cos(\alpha - \beta)}}\right]^2} \tag{7-41}$$

When the ground is horizontal ($\alpha = 0$):

$$K_p = \frac{\cos^2(\phi + \beta)}{\cos^3 \beta \, (1 - \sin\phi/\cos\beta)^2} \tag{7-42}$$

If, in addition, the back face of the wall is vertical ($\beta = 0$):

$$K_p = \frac{1 + \sin\phi}{1 - \sin\phi} = \tan^2\left(45° + \frac{\phi}{2}\right) = \frac{1}{K_a} \tag{7-43}$$

The first line of Table 7-13 lists values obtained from Eq. (7-43).

Table 7-13. Passive Lateral-Pressure Coefficients K_p*

$\phi =$	10°	15°	20°	25°	30°	35°	40°
$\delta = 0$	1.42	1.70	2.04	2.56	3.00	3.70	4.60
$\delta = -\phi/2$	1.56	1.98	2.59	3.46	4.78	6.88	10.38
$\delta = -\phi$	1.65	2.19	3.01	4.29	6.42	10.20	17.50

*For vertical wall face ($\beta = 0$) and horizontal ground surface ($\alpha = 0$).

Continuous anchors below the ground surface, when subjected to horizontal pull or thrust, develop in sand ($\phi = 33°$) passive pressures, lb per lin ft, of about

$$P = 1.5wh^2 \qquad (7\text{-}44)$$

where h = distance from bottom of anchor to the surface, ft.

This holds for ratios of h to height d, ft, of anchor of 1.5 to 5.5, and assumes a horizontal ground surface and vertical anchor face. For a square anchor within the same range of h/d, approximately

$$P = \left(2.50 + \frac{h}{8d}\right)^2 d\,\frac{wh^2}{2} \qquad (7\text{-}45)$$

where P = the passive lateral pressure, lb
d = length and height of the anchor, ft

Passive pressures of cohesive soils, resisting movement of a wall or anchor extending to the ground surface, depend on the unit weight of the soil w and its unconfined compressive strength $2c$, psf. At a distance h, ft, below the surface, the passive lateral pressure, psf, is

$$p = wh + 2c \qquad (7\text{-}46)$$

The total pressure, lb per lin ft, is

$$P = \frac{wh^2}{2} + 2ch \qquad (7\text{-}47)$$

and acts at a distance, ft, above the bottom of the wall or anchor of

$$\frac{h(wh + 6c)}{3(wh + 4c)}$$

The pressure distribution for plastic clay is shown in Fig. 7-32c.

Continuous anchors below the ground surface, when subjected to horizontal pull or thrust, develop in plastic clay passive pressures, lb per lin ft, of about

$$P = cd\left[8.7 - \frac{11,600}{(h/d + 11)^3}\right] \qquad (7\text{-}48)$$

where h = distance from bottom of anchor to surface, ft
d = height of anchor, ft

Equation (7-48) is based on tests made with horizontal ground surface and vertical anchor face.

Safety factors should be applied to the passive pressures computed from Eqs. (7-40) to (7-48) for design use. Experience indicates that a safety factor of 2 is satisfactory for clean sands and gravels. For clay, a safety factor of 3 may be desirable because of uncertainties as to effective shearing strength.

(G. P. Tschebotarioff, "Soil Mechanics, Foundations, and Earth Structures," McGraw-Hill Book Company, New York; K. Terzaghi and R. B. Peck, "Soil Mechanics in Engineering Practice," John Wiley & Sons, Inc., New York; Leo Casagrande, "Comments on Conventional Design of Retaining Structures," *Journal of the Soil Mechanics and Foundations Division,* American Society of Civil Engineers, 1973, pp. 181–198; H. F. Winterkorn and H. Fang, "Foundation Engineering Handbook," Van Nostrand Reinhold Company, New York.)

7-14. Vertical Earth Pressure on Conduit. The vertical load on an underground conduit depends principally on the weight of the prism of soil directly above it. But the load also is affected by vertical shearing forces along the sides of this prism. Caused by differential settlement of the prism and adjoining soil, the shearing forces may be directed up or down. Hence, the load on the conduit may be less or greater than the weight of the soil prism directly above it.

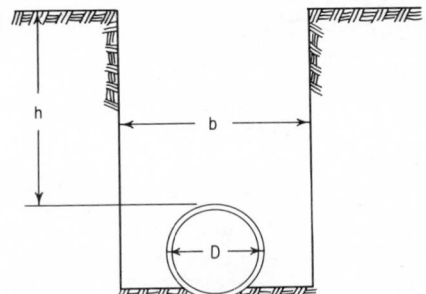

Fig. 7-33. Ditch conduit.

Conduits are classified as ditch or projecting, depending on installation conditions that affect the shears. A ditch conduit is a pipe set in a relatively narrow trench dug in undisturbed soil (Fig. 7-33). Backfill then is placed in the trench up to the original ground surface. A projecting conduit is a pipe over which an embankment is placed.

A projecting conduit may be positive or negative, depending on the extent of the embankment vertically. A positive projecting conduit is installed in a shallow bed with the pipe top above the surface of the ground. Then, the embankment is placed over the pipe (Fig. 7-34a). A negative projecting conduit is set in a narrow, shallow trench with the pipe top below the original ground surface (Fig. 7-34b). Then, the ditch is backfilled, after which the embankment is placed. The load on the conduit is less when the backfill is not compacted.

Load on underground pipe also may be reduced by the imperfect-ditch method of construction. This starts out as for a positive projecting conduit, with the pipe at the original ground surface. The embankment is placed and compacted for a few feet above the pipe. But then, a trench as wide as the conduit is dug down to it through the compacted soil. The trench is backfilled with a loose, compressible soil (Fig. 7-34c). After that, the embankment is completed.

The load, lb per lin ft, on a rigid ditch conduit may be computed from

$$W = C_D w h b \tag{7-49}$$

and on a flexible ditch conduit from

$$W = C_D w h D \tag{7-50}$$

where C_D = load coefficient for ditch conduit
w = unit weight of fill, lb per cu ft
h = height of fill above top of conduit, ft
b = width of ditch at top of conduit, ft
D = outside diameter of the conduit, ft

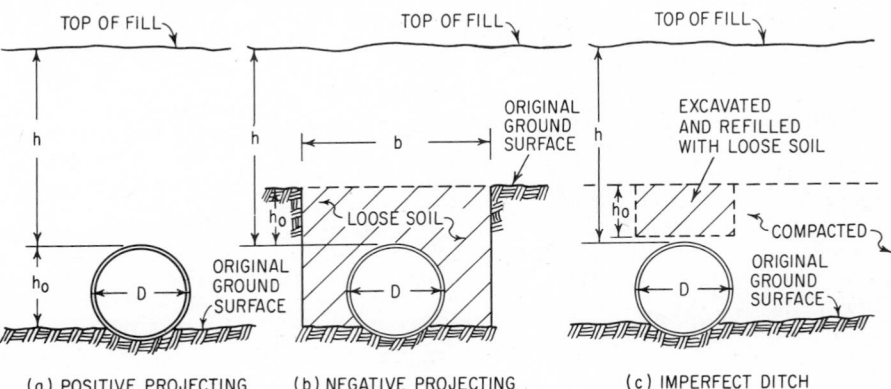

(a) POSITIVE PROJECTING CONDUIT

(b) NEGATIVE PROJECTING CONDUIT

(c) IMPERFECT DITCH CONDUIT

Fig. 7-34. Projecting conduit.

From the equilibrium of vertical forces, including shears, acting on the backfill above the conduit, C_D may be determined:

$$C_D = \frac{1 - e^{-kh/b}}{k} \frac{b}{h} \tag{7-51}$$

where $e = 2.718$
$k = 2K_a \tan \theta$
K_a = coefficient of active earth pressure [Eq. (7-32) and Table 7-11]
θ = angle of friction between fill and adjacent soil ($\theta \leq \phi$, the angle of internal friction of the fill)
Table 7-14 gives values of C_D for $k = 0.33$ for cohesionless soils, $k = 0.30$ for saturated topsoil, and $k = 0.26$ and 0.22 for clay (usual maximum and saturated).

Table 7-14. Load Coefficients C_D for Ditch Conduit

h/b	Cohesionless soils	Saturated topsoil	Clay	
			$k = 0.26$	$k = 0.22$
1	0.85	0.86	0.88	0.89
2	0.75	0.75	0.78	0.80
3	0.63	0.67	0.69	0.73
4	0.55	0.58	0.62	0.67
5	0.50	0.52	0.56	0.60
6	0.44	0.47	0.51	0.55
7	0.39	0.42	0.46	0.51
8	0.35	0.38	0.42	0.47
9	0.32	0.34	0.39	0.43
10	0.30	0.32	0.36	0.40
11	0.27	0.29	0.33	0.37
12	0.25	0.27	0.31	0.35
Over 12	$3.0b/h$	$3.3b/h$	$3.9b/h$	$4.5b/h$

Vertical load, lb per lin ft, on conduit installed by tunneling may be estimated from

$$W = C_D b(wh - 2c) \tag{7-52}$$

where c = cohesion of the soil, or half the unconfined compressive strength of the soil, psf
The load coefficient C_D may be computed from Eq. (7-51) or obtained from Table 7-14 with b = maximum width of tunnel excavation, ft, and h = distance from tunnel top to ground surface, ft.

For a ditch conduit, shearing forces extend from the pipe top to the ground surface. For a projecting conduit, however, if the embankment is sufficiently high, the shear may become zero at a horizontal plane below grade, the plane of equal settlement. Load on a projecting conduit is affected by the location of this plane.

Vertical load, lb per lin ft, on a positive projecting conduit may be computed from

$$W = C_P whD \tag{7-53}$$

where C_P = load coefficient for positive projecting conduit. Formulas have been derived for C_P and the depth of the plane of equal settlement. These formulas, however, are too lengthy for practical application, and the computation does not appear to be justified by the uncertainties in actual relative settlement of the soil above the conduit. Tests may be made in the field to determine C_P. If so, the possibility of an increase in earth pressure with time should be considered. For a rough estimate, C_P may be assumed as 1 for flexible conduit and 1.5 for rigid conduit.

The vertical load, lb per lin ft, on negative projecting conduit may be computed from

$$W = C_N whb \tag{7-54}$$

where C_N = load coefficient for negative projecting conduit
h = height of fill above top of conduit, ft
b = horizontal width of trench at top of conduit, ft

The load on an imperfect ditch conduit may be obtained from

$$W = C_N whD \tag{7-55}$$

where D = outside diameter of conduit, ft

Formulas have been derived for C_N, but they are complex, and insufficient values are available for the parameters involved. As a rough guide, C_N may be taken as 0.9 when depth of cover exceeds conduit diameter. (See also Art. 10-33.)

Superimposed surface loads increase the load on an underground conduit. The magnitude of the increase depends on the depth of the pipe below grade and the type of soil. For moving loads, an impact factor of about 2 should be applied. A superimposed uniform load w', psf, of large extent may be treated for projecting conduit as an equivalent layer of embankment with a thickness, ft, of w'/w. For ditch conduit, the load due to the soil should be increased by $bw'e^{-kh/b}$, where $k = 2K_a \tan \theta$, as in Eq. (7-51). The increase caused by concentrated loads can be estimated by assuming the loads to spread out linearly with depth, at an angle of about 30° with the vertical (Fig. 7-6).

(G. A. Leonards, "Foundation Engineering," McGraw-Hill Book Company, New York; M. G. Spangler, "Soil Engineering," International Textbook Company, Scranton, Pa.; "Handbook of Steel Drainage and Highway Construction Products," American Iron and Steel Institute, New York.)

7-15. Soil Improvement. Soil for foundations can be altered to conform to desired characteristics. Whether this should be done depends on the relative cost of alternatives.

Investigations of soil and groundwater conditions on a site should indicate whether soil improvement, or stabilization, is needed. Tests may be necessary to determine which of several applicable techniques may be feasible and economical. Table 7-15 lists some conditions for which soil improvement should be considered and the methods that may be used.

As indicated in the table, soil stabilization may increase strength, increase or decrease permeability, reduce compressibility, improve stability, or decrease heave due to frost or swelling. Main techniques used are mechanical and chemical stabilization.

Mechanical stabilization rearranges, adds, or removes soil particles. The objective usually is to modify density, water content, or gradation. Particles may be rearranged by blending the layers of a stratified soil, remolding an undisturbed soil, or densifying a soil. Often, the desired improvement can be obtained by drainage alone. But compaction plus water control is even more effective.

Compaction is the process of increasing soil density by mechanical action, which remolds or structurally changes a soil. The denser a soil, the lower its compressibility is. In the field, heavy equipment compacts a thin layer of soil by passing over it several times. Laboratory tests determine the optimum water content, the amount of moisture required for maximum soil density, and nearly this quantity is maintained during compaction. (See Arts. 7-5 and 13-19.)

Compaction effort is the number of passes made with a specific machine of given weight and at a given speed. For a given compaction effort, density varies with moisture content. For a given moisture content, increasing the compaction effort increases the density and reduces the permeability.

As an alternative to heavy equipment (Art. 13-4), other means may be used to compact granular soils, for example, vibration or shock (blasting with explosives). By the Vibroflotation process, granular soils have been compacted to depths of more than 50 ft under favorable conditions. In this process, a vibrator is jetted into the ground to the desired depth, then set into action. The vibrator is withdrawn in 1-ft increments as the sand compacts laterally. When a crater forms at the surface, soil is added and is, in turn, compacted. When one place has been compacted, the process is repeated, usually about 10 ft away, until the whole area has been stabilized.

Consolidation, applied to saturated cohesive soils, gives the same end result as compaction, but with static loading and removal of pore water. When a soil consolidates, it comes into equilibrium with the load on it. For a saturated clay, this usually is a very slow process (Art. 7-5). A foundation may be preconsolidated nevertheless if a static load, usually a pile of earth, is applied to it and left in place for many months. If provision is made for removal of pore water, consolidation can be speeded. With a surcharge exceeding the load the foundation is expected to support in service, settlement of the soil under the reduced loading, after removal of the surcharge, can be made very small.

Mixing an existing soil with select material or removing selected sizes of particles from an existing soil can change its properties considerably. Adding clay to a cohesionless soil in a nonfrost region, for example, may make the soil suitable as a base course for a road (if drainage is not too greatly impaired). Adding clay to a pervious soil may reduce its permeability sufficiently to permit its use as a reservoir bottom. Washing particles finer than 0.02 mm from gravel makes the soil less susceptible to frost heave (desirable upper limit for this soil fraction: 3%).

Table 7-15. Where Soil Improvement May Be Economical

Soil deficiency	Probable type of failure	Probable cause	Possible remedies
Slope instability	Slides on slope	Pore-water pressure	Drain; flatten slope; freeze
		Loose granular soil	Compact
		Weak soil	Mix or replace with select material
	Mud flow	Excessive water	Exclude water
	Slides—movement at toe	Toe instability	Place toe fill, and drain
Low bearing capacity	Excessive settlement	Saturated clay	Consolidate with surcharge, and drain
		Loose granular soil	Compact; drain; increase footing depth; mix with chemicals
		Weak soil	Superimpose thick fill; mix or replace with select material; inject or mix with chemicals; freeze (if saturated); fuse with heat (if unsaturated)
Heave	Excessive rise	Frost	For buildings: place foundations below frost line; insulate refrigeration-room floors; refrigerate to keep ground frozen. For roads: Remove fines from gravel; replace with nonsusceptible soil
		Expansion of clay	Exclude water; replace with granular soil
Excessive permeability	Seepage	Pervious soil or fissured rock	Mix or replace soil with select material; inject or mix soil with chemicals; construct cutoff wall with grout; enclose with sheetpiles and drain
"Quick" bottom	Loss of strength	Flow under cofferdam	Add berm against cofferdam inner face; increase width of cofferdam between lines of sheeting; drain with wellpoints outside the cofferdam

Drainage is effective in soil stabilization because strength of a soil generally decreases with an increase in amount and pressure of pore water. Drainage may be accomplished by gravity, pumping, compression with an external load on the soil, electroosmosis, heating, or freezing.

Pumping often is used for draining excavations (Art. 7-16). For permanent stabilization of slopes, however, advantage must be taken of gravity flow. Usually, intercepting drains, laid approximately along contours, suffice. Vertical wells may be used to relieve artesian pressures. Where mud flows may occur, water must be excluded from the area. Surface and subsurface flow must be intercepted and conducted away at the top of the area. Also, cover, such as heavy mulching and planting, should be placed over the entire surface to prevent water from percolating downward into the soil.

Electrical drainage adapts the principle that water flows to the cathode when a direct current passes through saturated soil (Art. 7-3). The water may be pumped out at the cathode. Electroosmosis is relatively expensive and therefore usually is limited to special conditions, such as drainage of silts, which ordinarily are hard to drain by other methods.

Vertical sand drains, or piles (Art. 7-9), may be used to compact loose, saturated cohesionless soils or to consolidate saturated cohesive soils. They provide an escape channel for water squeezed out of the soil by an external load. A surcharge of pervious material placed over the ground surface also serves as part of the drainage system, as well as part of the fill, or external load. Usually, the surcharge is placed before the sand piles are formed, to support equipment, such as pile drivers, over the soft soil. Fill should be placed in thin layers to avoid formation of mud flows, which might shear the sand drains and cause mud waves. Analyses should be made of embankment stability at various stages of construction.

Thermal stabilization, still in the experimental stage, generally is costly and is restricted to conditions for which other methods are not suitable. Heat has been used to strengthen nonsaturated loess and to decrease the compressibility of cohesive soils. One technique is to burn liquid or gas fuel in a borehole. Another is to inject into the soil through pipes spaced about 10 ft apart a mixture of liquid fuel and air under pressure. Then, the mixture is burned for about 10 days, producing a solidified soil.

Freezing a wet soil converts it into a rigid material with considerable strength, but it must be kept frozen. The method is excellent for a limited excavation area, for example, freezing the ground to sink a shaft. For the purpose, a network of pipes must be placed in the ground and a liquid, usually brine, at low temperature circulated through the pipes. Care must be taken that the freezing does not spread beyond the area to be stabilized and cause heaving damage.

Chemical stabilization, including use of portland cement and bitumens, meets many needs. In surface treatments, it supplements mechanical stabilization to make the effects more lasting. In subsurface treatments, chemicals may be used to improve bearing capacity or decrease permeability.

Soil-cement, a mixture of portland cement and soil, is suitable for subgrades, base courses, and pavements of roads not carrying heavy traffic ("Essentials of Soil-Cement Construction," Portland Cement Association). Bitumen-soil mixtures are extensively used in road and airfield construction and sometimes as a seal for earth dikes (Guide Specifications for Highway Construction, American Association of State Highway and Transportation Officials, National Press Building, Washington, D.C. 20004). Hydrated, or slaked, lime may be used alone as a soil stabilizer, or with fly ash, portland cement, or bitumen ("Lime Stabilization of Roads," National Lime Association, 925 15th St., N.W., Washington, D.C. 20006). Calcium or sodium chloride is used as a dust palliative and an additive in construction of granular base and wearing courses for roads ("Calcium Chloride for Stabilization of Bases and Wearing Courses," Calcium Chloride Institute, Ring Building, Washington, D.C. 20036).

Grouting, with portland cement or other chemicals, often is used to fill rock fissures, decrease soil permeability, form underground cutoff walls to eliminate seepage, and stabilize soils at considerable depth. The chemicals may be used to fill the voids in the soil, to cement the particles, or to form a rocklike material. The process, however, generally is suitable only in pervious soils. Also, rapid setting of chemicals may interfere with thorough injection of the underground region. Chemicals used include sodium silicate and salts or acids; chrome-lignin; and low-viscosity organics ("Chemical Grouting," *Journal of the Soil Mechanics and Foundations Division, American Society of Civil Engineers,* vol. 83, November 1957).

(R. A. Barron, "Consolidation of Fine-grained Soils by Drain Wells," *Transactions, American Society of Civil Engineers,* vol. 113, p. 718, 1948; L. Casagrande, "Electro-osmotic Stabilization of Soils," *Journal, Boston Society of Civil Engineers,* vol. 39, p. 51, 1952; K. Terzaghi and R. B. Peck, "Soil Mechanics in Engineering Practice," John Wiley & Sons, Inc., New York; G. P. Tschebotarioff, "Soil Mechanics, Foundations, and Earth Structures," McGraw-Hill Book Company, New York; H. F. Winterkorn and H. Fang, "Foundation Engineering Handbook," Van Nostrand Reinhold Company, New York.)

7-16. Dewatering Methods for Excavations. The main purpose of dewatering is to enable construction to be carried out under relatively dry conditions. But proper drainage also stabilizes excavated slopes, reduces lateral loads on sheeting and bracing, reduces required air pressure in tunneling, makes excavated material lighter and easier to handle, prevents loss of soil below slopes or from the bottom of the excavation, and prevents a "quick" or "boiling" bottom. In addition, permanent lowering of the groundwater table or relief of artesian pressure may allow a less expensive design for the structure, especially when the soil consolidates or becomes compact. If lowering of the water level or pressure relief is temporary, however, the improvement of the soil should not be considered in design. Increases in strength and bearing capacity may be lost when the soil again becomes saturated.

To keep an excavation reasonably dry, the groundwater table should be kept at least 2 ft, and preferably 5 ft, below the bottom in most soils.

Site investigations should yield information useful for deciding on the most suitable and economical dewatering method.   Important is a knowledge of the types of soil in and below the site, probable groundwater levels during construction, permeability of the soils, and quantities of water to be handled.   A pumping test may be desirable for estimating capacity of pumps needed and drainage characteristics of the ground.

Many methods have been used for dewatering excavations.   Those used most often are listed in Table 7-16 with conditions for which they generally are most suitable.   (See also Art. 7-15.)

In many small excavations, or where there are dense or cemented soils, water may be collected in ditches or sumps at the bottom and pumped out.   This is the most economical method of dewatering, and the sumps do not interfere with future construction as does a comprehensive wellpoint system.   But the seepage may slough the slopes, unless they are stabilized with gravel, and may hold up excavation while the soil drains.   Also, springs may develop in fine sand or silt and cause underground erosion and subsidence of the ground surface.

For sheetpile-enclosed excavations in pervious soils, it is advisable to intercept water before it enters the enclosure.   Otherwise, the water will put high pressures on the sheeting.   Seepage also can cause the excavation bottom to become quick, overloading the bottom bracing, or create piping, undermining the sheeting.   Furthermore, pumping from the inside of the cofferdam is likely to leave the soil to be excavated wet and tough to handle.

Wellpoints often are used for lowering the water table in pervious soils.   They are not suitable, however, in soils that are so fine that they will flow with the water or in soils with low permeability.   Also, other methods may be more economical for deep excavations, very heavy flows, or considerable lowering of the water table (Table 7-16).

Table 7-16. Methods for Dewatering Excavations

Saturated-soil conditions	Dewatering method probably suitable
Surface water	Ditches; dikes; sheetpiles and pumps or underwater excavation and concrete tremie seal
Gravel	Underwater excavation; grout curtain; gravity drainage with large sumps with gravel filters
Sand (except very fine sand)	Gravity drainage
Waterbearing strata near surface; water table does not have to be lowered more than 15 ft	Wellpoints with vacuum and centrifugal pumps
Waterbearing strata near surface; water table to be lowered more than 15 ft, low pumping rate	Wellpoints with jet-eductor pumps
Excavations 30 ft or more below water table; artesian pressure; high pumping rate; large lowering of water table—all where adequate depth of pervious soil is available for submergence of well screen and pump	Deep wells, plus, if necessary, wellpoints
Sand underlain by rock near excavation bottom	Wellpoints to rock, plus ditches, drains, automatic "mops"
Sand underlain by clay	Wellpoints in holes 3 or 4 ft into the clay, backfilled with sand
Silt; very fine sand (permeability coefficient between 0.001 and 0.00001 cm per sec)	For lifts up to 15 ft, wellpoints with vacuum; for greater lifts, wells with vacuum; sumps
Silt or silty sand underlain by pervious soil	At top of excavation, and extending to the pervious soil, vertical sand drains plus wellpoints or wells
Clay-silts, silts	Electro-osmosis
Clay underlain by pervious soil	At top of excavation, wellpoints or deep wells extending to pervious soil
Dense or cemented soils; small excavations	Ditches and sumps

Wellpoints are metal well screens about 2 to 3 in. in diameter and up to about 4 ft long.   A pipe connects each wellpoint to a header, from which water is pumped to discharge (Fig. 7-35). Each pump usually is a combined vacuum and centrifugal pump.   Spacing of wellpoints generally ranges from 3 to 12 ft c to c.

A wellpoint may be jetted into position or set in a hole made with a hole puncher or heavy steel casing.   Accordingly, wellpoints may be self-jetting or plain-tip.   To insure good drainage in fine and dirty sands or in layers of silt or clay, the wellpoint and riser should be surrounded by filter sand to just below the water table. The space above the filter should be sealed with silt or clay to keep air from getting into the wellpoint through the filter.

Wellpoints generally are relied on to lower the water table 15 to 20 ft. Deep excavations may be dewatered with multistage wellpoints, with one row of wellpoints for every 15 ft of depth. Or when the flow is less than about 15 gpm per wellpoint, a single-stage system of wellpoints may be installed above the water table and operated with jet-eductor pumps atop the wellpoints. These pumps can lower the water table up to about 100 ft, but they have an efficiency of only about 30%.

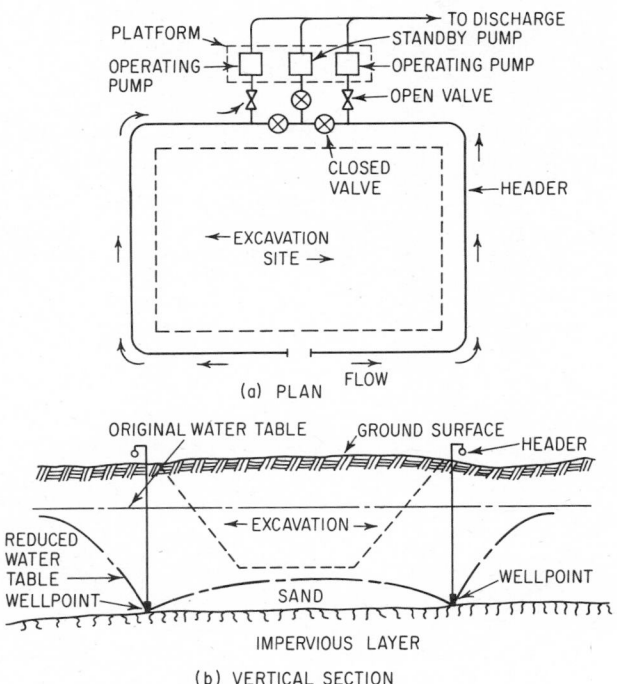

Fig. 7-35. Wellpoint system for dewatering an excavation.

Deep wells may be used in pervious soils for deep excavations, large lowering of the water table, and heavy water flows. They may be placed along the top of an excavation to drain it, to intercept seepage before pressure makes slopes unstable, and to relieve artesian pressure before it heaves the excavation bottom.

Usual spacing of wells ranges from 20 to 250 ft. Diameter generally ranges from 6 to 20 in. Well screens may be 20 to 75 ft long, and they are surrounded with a sand-gravel filter. Generally, pumping is done with a submersible or vertical turbine pump installed near the bottom of each well.

Figure 7-36 shows a deep-well installation used for a 300-ft-wide by 600-ft-long excavation for a building of the Smithsonian Institution, Washington, D.C. Two deep-well pumps lowered the general water level in the excavation 20 ft. The well installation proceeded as follows: (1) Excavation to water level (elevation 0.0). (2) Driving of sheetpiles around the well area (Fig. 7-36a). (3) Excavation underwater inside the sheetpile enclosure to elevation −37.0 ft (Fig. 7-36b). Bracing installed as digging progressed. (4) Installation of a wire-mesh-wrapped timber frame, extending from elevation 0.0 to −37.0 (Fig. 7-36c). Weights added to sink the frame. (5) Backfilling of space between sheetpiles and mesh with ³/₁₆- to ⅜-in. gravel. (6) Removal of sheetpiles. (7) Installation of pump and start of pumping.

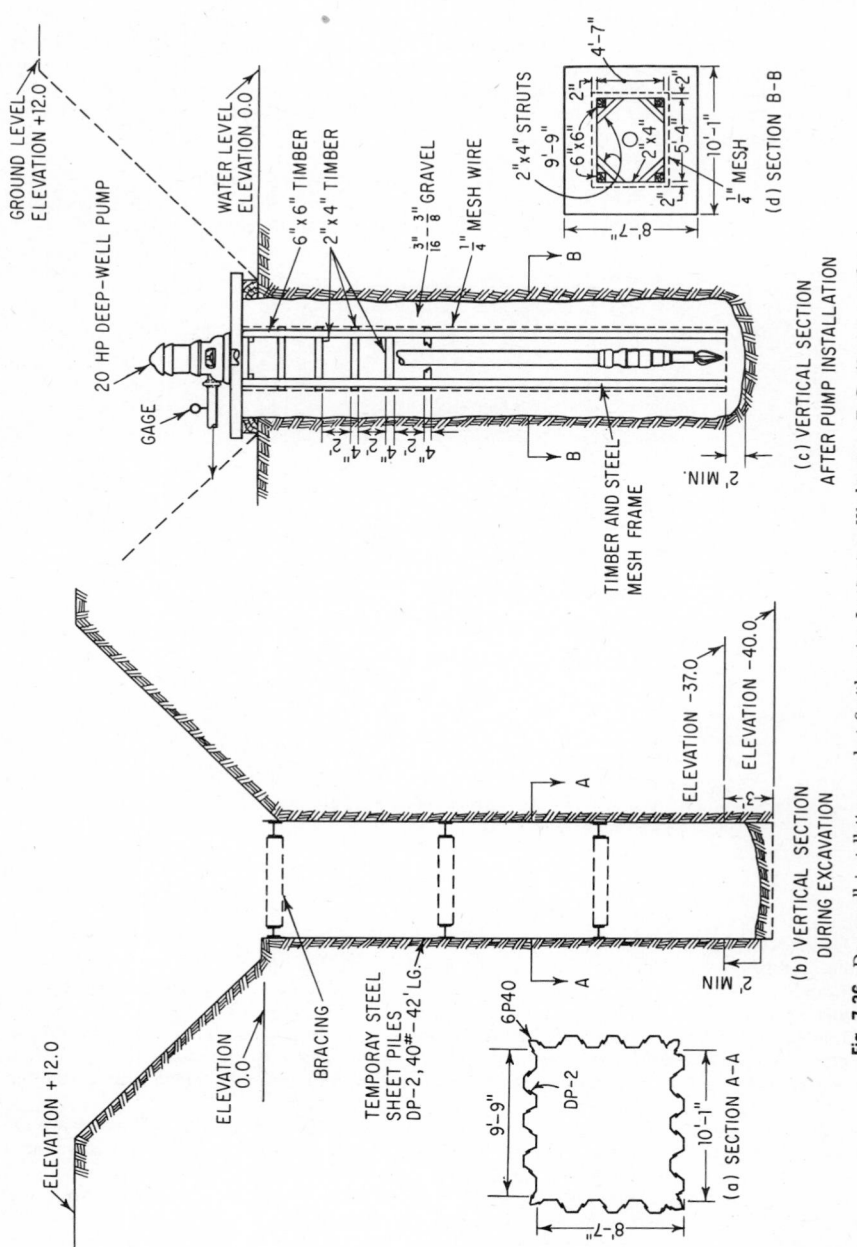

Fig. 7-36. Deep-well installation used at Smithsonian Institution, Washington, D.C. (*Spencer, White & Prentis, Inc.*)

Vacuum well or wellpoint systems may be used to drain silts with low permeability (coefficient between 0.001 and 0.00001 cm per sec). In these systems, wells or wellpoints are closely spaced, and a vacuum is held with vacuum pumps in the well screens and sand filters. At the top, the filter, well, and risers should be sealed to a depth of 5 ft with bentonite or an impervious soil to prevent loss of the vacuum. Water drawn to the well screens is pumped out with submersible or centrifugal pumps.

Where a pervious soil underlies silts or silty sands, vertical sand drains and deep wells can team up to dewater an excavation. Installed at the top, and extending to the pervious soil, the sand piles intercept seepage and allow it to drain down to the pervious soil. Pumping from the deep wells relieves the pressure in that deep soil layer.

For some silts and clay-silts, electrical drainage with wells or wellpoints may work, whereas gravity methods may not (Art. 7-15). In saturated clays, thermal or chemical stabilization may be necessary.

Small amounts of surface water may be removed from excavations with "mops." Surrounded with gravel to prevent clogging, these drains are connected to a header with suction hose or pipe. For automatic operation, each mop should be opened and closed by a float and float valve.

When structures on silt or soft material are located near an excavation to be dewatered, care should be taken that lowering of the water table does not cause them to settle. It may be necessary to underpin the structures or to pump discharge water into recharge wells near the structures to maintain the water table around them.

(G. A. Leonards, "Foundation Engineering," McGraw-Hill Book Company, New York; L. Zeevaert, "Foundation Engineering for Difficult Subsoil Conditions," Van Nostrand Reinhold Company, New York; H. F. Winterkorn and H. Fang, "Foundation Engineering Handbook," Van Nostrand Reinhold Company, New York.)

7-17. Underpinning. The general methods and main materials used to give additional support at or below grade to structures are called underpinning. Usually, the added support is applied at or near the footings.

Underpinning may be remedial or precautionary. Remedial underpinning adds foundation capacity to an inadequately supported structure. Precautionary underpinning is provided to obtain adequate foundation capacity to sustain higher loads or to accommodate changes in ground conditions. Usually, this type of underpinning is required for the foundations of a structure when deeper foundations are to be constructed nearby for an addition or another structure. Loss of ground, even though very small, into an adjoining excavation may cause excessive settlement of existing foundations.

Presumably, an excavation influences an existing substructure when a plane through the outermost foundations, on a 1-on-1 slope for sand or a 1-on-2 slope for unconsolidated silt or soft clay, penetrates the excavation. For a cohesionless soil, underpinning exterior walls within a 1-on-1 slope usually suffices; interior columns are not likely to be affected if farther from the edge of the excavation than half the depth of the cut.

The commonly accepted procedures of structural and foundation design should be used for underpinning. Data for computing dead loads may be obtained from plans of the structure or a field survey. Since underpinning is applied to existing structures, some of which may be old, engineers in charge of underpinning design and construction should be familiar with older types of construction as well as the most modern.

Before underpinning starts, the engineers should investigate and record existing defects in the structure. Preferably, the engineers should be accompanied in this investigation by a representative of the owner. The structure should be thoroughly inspected, from top to bottom, inside (if possible) and out. The report should include names of inspectors, dates of inspection, and description and location of defects. Photographs are useful in verifying written descriptions of damaged areas. The engineers should mark existing cracks in such a way that future observations would indicate whether they are continuing to open or spread.

Underpinning generally is accompanied by some settlement. If design and field work are good, the settlement may be limited to about ¼ to ⅜ in. But as long as settlement is uniform in a structure, damage is unlikely. Differential settlement should be avoided. To check on settlement, elevations of critical points, especially columns and walls, should be measured frequently during underpinning. Since movement may also occur laterally, the plumbness of walls and columns also should be checked.

One of the first steps in underpinning usually is digging under a foundation. This decreases its load-carrying capacity temporarily. Hence, preliminary support may be necessary until underpinning is installed. This support may be provided by shores, needles, grillages, and piles. Sometimes, it is desirable to leave them in place as permanent supports.

Generally, it is advisable to keep preliminary supports at a minimum, for economy and to avoid interference with other operations. For the purpose, advantage may be taken of arching action and of the ability of a structure to withstand moderate overloads. Also, columns centrally supported on large spread footings need not be shored when digging is along an edge and involves only a small percent of the total footing area. A large part of the column load is supported by the soil directly under the column.

When necessary, weak portions of a structure, especially masonry, should be repaired or strengthened before underpinning starts.

Shores, installed vertically or on a slight incline, often are used to support walls or piers while underpinning pits are dug (Fig. 7-37a). Good bearing should be provided at top and bottom of the shores. One way of providing bearing at the top is to cut a niche and mortar a steel bearing plate against the upper face. An alternative to the plate is a Z shape, made by removing diagonally opposite half flanges from an H beam. When the top of the shore is cut to fit between flange and web of the Z, movement of the shore is restrained. For a weak masonry wall, the load may have to be distributed over a larger area. One way of doing this is to insert a few lintel angles about 12 in. apart vertically and bolt them to a vertical, heavy timber or steel distributing beam. The horizontal leg of an angle on the beam then transmits the load to a shore.

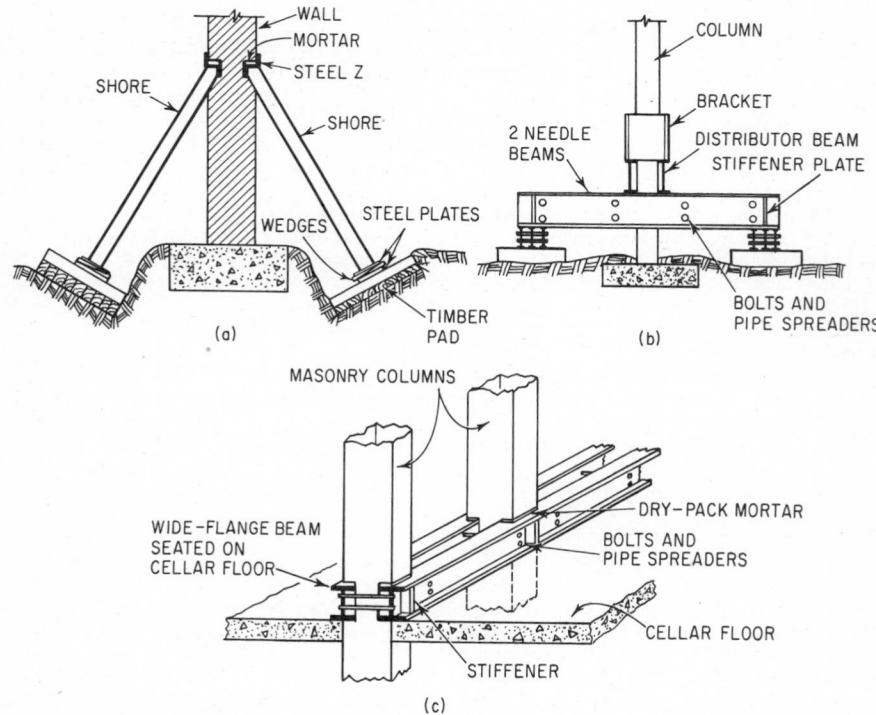

Fig. 7-37. Temporary supports used in underpinning: (a) shores; (b) needle beams; (c) grillage.

Inclined shores on only one side of a wall require support at the base for horizontal as well as vertical forces. One way is to brace the shores against an opposite wall at the floor. Preferably, the base of each shore should sit on a footing perpendicular to the axis of the shore. Sized to provide sufficient bearing on the soil, the footing may be made of heavy timbers, steel beams, or reinforced concrete, depending on the load on the shore.

Loads may be transferred to a shore by wedges or jacks. Oak wedges are suitable for light loads; forged steel wedges and bearing plates are desirable for heavy loads. Jacks, however, offer greater flexibility in length adjustments and allow corrections during underpinning for settlement of shore footings.

Needles are the beams installed horizontally to transfer the load of a wall or column to either or both sides of its foundation, to permit digging of underpinning pits (Fig. 7-37b). These beams are more expensive than shores, which transmit the load directly into the ground. Needles usually are steel wide-flange beams, sometimes plate girders, used in pairs, with bolts and pipe spreaders between the beams. This arrangement provides resistance to lateral buckling and torsion. The needles may be prestressed with jacks to eliminate settlement when the load is applied.

The load from steel columns may be transmitted through brackets to the needles. For masonry walls, the needles may be inserted through niches. The load should be transferred from the masonry to the needles through thin wood fillers. These crush when the needles deflect and maintain nearly uniform bearing.

Wedges may be placed under the ends of the needles to shift the load from the member to be supported to those beams. The beam ends may be carried on timber pads, which distribute the load over the soil.

Grillages, which have considerably more bearing area on the ground than needles, often are used as an alternative to needles and shores for closely spaced columns. A grillage may be installed horizontally on soil at foundation level to support and tie together two or more column footings, or it may rest on a cellar floor (Fig. 7-37c). These preliminary supports may consist of two or more steel beams, tied together with bolts and pipe spreaders, or of a steel-concrete composite. Also, grillages sometimes are used to strengthen or repair existing footings by reinforcing them and increasing their bearing area. The grillages may take the form of dowels or of encircling concrete or steel-concrete beams. They should be adequately cross-braced against buckling and torsion. Holes should be made in steel beams to be embedded in concrete, to improve bond.

Pit Underpinning. After preliminary supports have been installed and weak construction strengthened or repaired, underpinning may start. The commonest method of underpinning a foundation is to construct concrete piers down to deeper levels with adequate bearing capacity and to transfer the load to the piers by wedging up with dry packing. To build the piers, pits must be dug under the foundation. Because of the danger of loss of ground and consequent settlement where soils are saturated, the method usually is restricted to dry subsoil.

When piers have to be placed close together, a continuous wall may be constructed instead. But the underpinning wall should be built in short sections, usually about 5 ft long, to avoid undermining the existing foundation. Alternate sections are built first, and then the gaps are filled in.

Underpinning pits rarely are larger than about 5 ft square in cross section. Minimum size for adequate working room is 3 × 4 ft. Access to the pit is provided by an approach pit started alongside the foundation and extending down about 6 ft. The pits must be carefully sheeted and adequately braced to prevent loss of ground, which can cause settlement of the structure.

In soils other than soft clay, 2-in.-thick wood planks installed horizontally may be used to sheet pits up to 5 ft square, regardless of depth. Sides of the pits should be trimmed back no more than absolutely necessary. The boards, usually 2 × 8's, are installed one at a time with 2-in. spaces between them vertically. Soil is repacked through these louvers to fill the voids behind the boards. In running sands, hay may be stuffed behind the boards to block the flow. Corners of the sheeting often are nailed with vertical 2 × 4 in. wood cleats.

In soft clay, sheeting must be tight and braced against earth pressure. Chicago caissons, or a similar type, may be used (Art. 7-11).

In water-bearing soil with a depth not exceeding about 5 ft, vertical sheeting can sometimes be driven to cut off the water. For the purpose, light steel or tongue-and-groove wood sheeting may be used. The sheeting should be driven below the bottom of the pit a sufficient distance to prevent boiling of the bottom due to hydrostatic pressure. With water cut off, the pit can be pumped dry and excavation continued.

After a pit has been dug to the desired level, it is filled with concrete to within 3 in. of the foundation to be supported. The gap is dry-packed, usually by ramming stiff mortar in with a 2 × 4 pounded by an 8-lb hammer. The completed piers should be laterally braced if soil is excavated on one side to a depth of more than about 6 ft. An example of pit underpinning is the work done in the restoration of the White House, in which a cellar and subcellar were created (Fig. 7-38).

Pile Underpinning. If water-bearing soil more than about 5 ft deep underlies a foundation, the structure may have to be underpinned with piles. Driven piles generally are preferred to jacked piles because of lower cost. The feasibility of driven piles, however, depends on availability of at least 12 ft of headroom and space alongside the foundations. Thus, driven piles often can be used to underpin interior building columns when headroom is available. But they are hard to install for exterior walls unless there is ample space alongside the walls. For very lightly loaded structures, brackets may be attached to underpinning piles to support the structure. But such construction puts bending into the piles, reducing their load-carrying capacity.

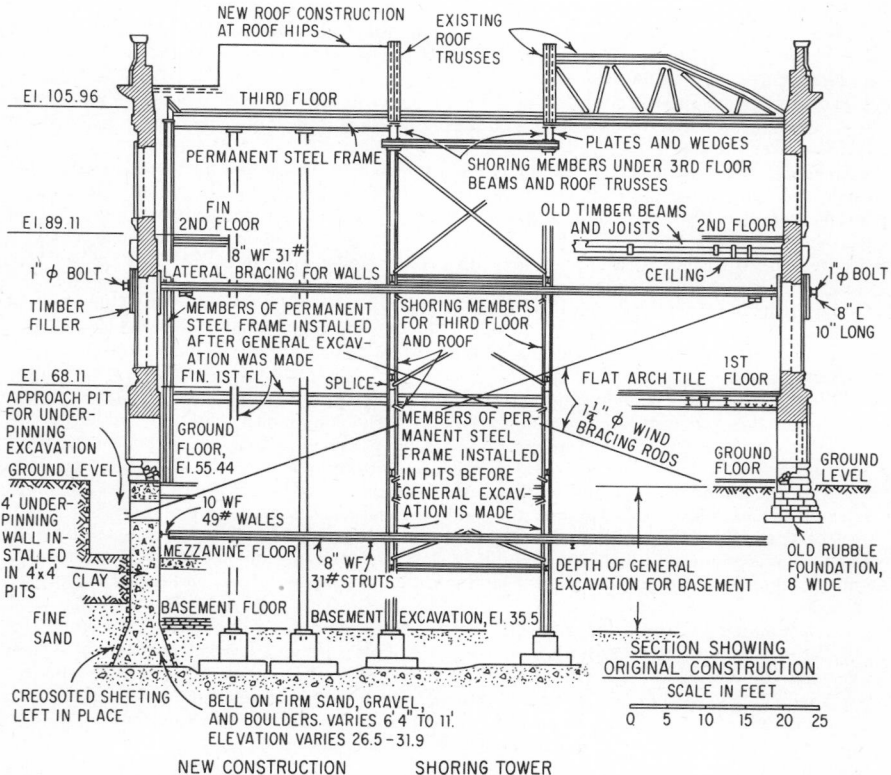

Fig. 7-38. Vertical section through White House during restoration. Pit underpinning was used for walls. *(Spencer, White & Prentis, Inc.)*

Driven piles usually are 12- to 14-in.-diameter steel pipe, ⅜ in. thick. They are driven open-ended, to reduce vibration, and in lengths determined by available headroom. Joints may be made with cast-steel sleeves. After soil has been removed from the pipe interior, it is filled with concrete.

Jacked piles require less headroom and may be placed under a footing. Also made of steel pipe installed open-ended, these piles are forced down by hydraulic jacks reacting against the footing. The operation requires an approach pit under the footing to obtain about 6 ft of headroom.

Pretest piles, originally patented by Spencer, White & Prentis, New York City, are used to prevent the rebound of piles when jacking stops and subsequent settlement when the load of the structure is transferred to the piles. A pipe pile is jacked down, in 4-ft lengths, to the desired depth. The hydraulic jack reacts against a steel plate mortared to the underside of the footing to be supported. After the pile has been driven to the required depth and cleaned out, it is filled with concrete and capped with a steel bearing plate. Then two hydraulic jacks atop the pile overload it 50%. As the load is applied, a bulb of pressure builds up in the soil at the pile bottom. This pressure stops downward movement of the pile. While the jacks maintain the load, a short length of beam is wedged between the pile top and the steel plate under the footing. Then, the jacks are unloaded and removed. The load, thus, is transferred without further settlement. Later, the space under the footing is concreted. Figure 7-39 shows how pretest piles were used for underpinning existing structures during construction of a subway in New York City.

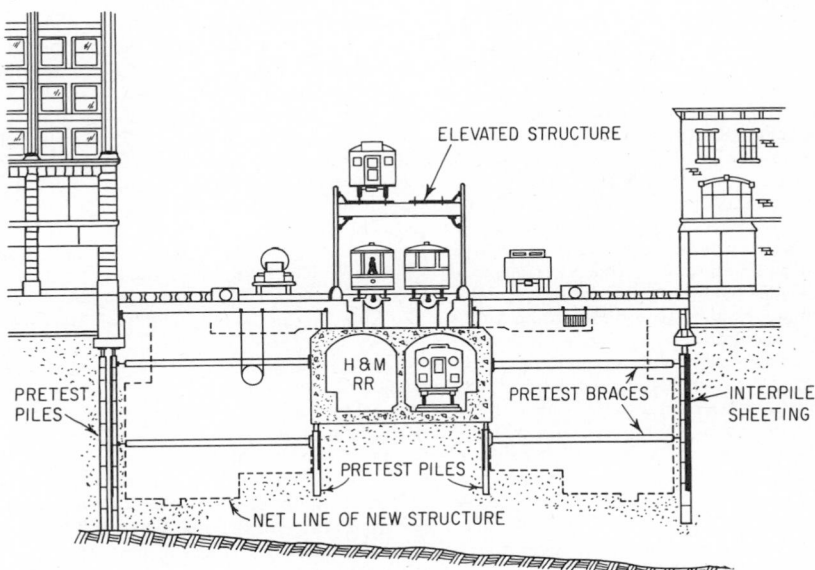

Fig. 7-39. Pretest piles supported existing buildings, old elevated railway, and Hudson and Manhattan Railroad tunnel during construction of subway under Avenue of the Americas, New York City. (*Spencer, White & Prentis, Inc.*)

Miscellaneous Methods. Spread footings may be pretested in much the same way as piles. The weight of the structure is used to jack down the footings, which then are wedged in place, and the gap is concreted. The method may be resorted to for unconsolidated soils where a high water table makes digging under a footing unsafe or where a firm stratum is deep down.

A form of underpinning may be used for slabs on ground. When concrete slabs settle, they may be restored to the proper elevation by mud jacking. In this method (which will not prevent future settlement) a fluid grout is pumped under the slab through holes in it, raising it. Pressure is maintained until the grout sets. The method also may be used to fill voids under a slab.

In loose sandy soils, Vibroflotation (Art. 7-15) may be used for underpinning. One difficulty with this method is that the structure has to be shored before underpinning starts, and the structure and shoring must be isolated from the vibrating equipment and soil being compacted.

Chemical or thermal stabilization (Arts. 7-12 and 7-15) sometimes may be used to assist underpinning.

(E. A. Prentis and L. White, "Underpinning," Columbia University Press, New York; G. A. Leonards, "Foundation Engineering," McGraw-Hill Book Company, New York; H. F. Winterkorn and H. Fang, "Foundation Engineering Handbook," Van Nostrand Reinhold Company, New York.)

7-18. Slope Stability Analysis. Under the constant force of gravity, soil on slopes tends to move downward. This movement is resisted by passive pressure of the soil below, friction, and cohesion, as well as by surface elements, such as vegetation and walls. When the forces tending to cause movement exceed those resisting it, a slide occurs. Stability analysis is used to predict such slides before they occur, to estimate the safety factor against slides, and to explain why slides have occurred.

Approximate analyses often are possible. Their accuracy depends on how closely the assumptions on which they are based reflect actual conditions. The difficulty in making precise analyses arises from the fact that they must evaluate the effects of the worst conditions that may develop. This requires a knowledge of soil conditions to considerable depth and of significant changes that may occur. These include possible changes in weight of soil, frictional resistance to sliding, cohesion, passive pressure, and pore water pressures.

Rotational slides occur when soil masses shift along failure surfaces approximating circular cylinders. The Swedish circle method of analysis assumes a circular arc for the line of failure projected on a vertical section. The analysis determines the "worst" circle, the one for which the safety factor is smallest. This is done usually by trial and error. An initial circle is assumed, extending from some point atop the slope to the toe or beyond. The slope is divided into vertical sections extending down to the failure surface. For each section, the sliding and resisting forces along the failure surface are computed. Each type of force is added. The safety factor is determined as the ratio of the sum of the sliding forces to the sum of the resisting forces.

If water may be present, its effect should be taken into account. It may reduce weight of soil in a section, decrease friction, destroy cohesion, apply uplift. Length of sections may be chosen for convenience in dealing with changes in types of soil.

A similar procedure is followed to find the safety factor for additional trial circles. For slopes as flat as 1:3, the worst circle generally extends below the base of the slope and beyond the toe. For slopes between 1:2.5 and 1:1.5, the worst circle usually swings upward from the toe and with larger radius. For much steeper slopes, the Culmann method of stability analysis is simpler. It employs a plane as the failure surface.

Translatory slides occur when soil masses move along natural planes of weakness, often nearly horizontal. A soil mass, in this case, may be assumed to comprise a large mass subjected to active pressure from a wedge below the slope top. Movement is resisted by passive pressure from a wedge at the toe and shear along the base of the mass.

Flow slides occur when a soil mass becomes liquid. Cohesionless soils on a slope will become unstable when fully saturated, relatively loose, and subjected to shock, vibration, or shearing strain. Clay frequently becomes a *mud flow* because of intake of water, which softens the clay or reduces cohesion. But a mud flow may occur also because of shearing strain or other structural disturbance, without change in water content. Stability analysis of mud flows is very difficult because of the necessity of estimating the effects of future changes. Investigations of the possibility of mud flows on natural clay slopes should include reconnaissance of the surrounding area, including studies of aerial photographs, to see if such slides have occurred before.

(B. K. Hough, "Basic Soils Engineering," The Ronald Press Company, New York; K. Terzaghi and R. B. Peck, "Soil Mechanics in Engineering Practice," John Wiley & Sons, Inc., New York; H. F. Winterkorn and H. Fang, "Foundation Engineering Handbook," Van Nostrand Reinhold Company, New York.)

Concrete Design and Construction

LEV ZETLIN

Consulting Engineer; President, Lev Zetlin Associates, Inc., New York, N.Y.;
University Professor of Civil Engineering and Architecture,
University of Virginia

and

DONALD GRIFF

Consulting Engineer; Vice President, Lev Zetlin Associates, Inc., New York, N.Y.

Concrete made with portland cement is widely used as a construction material because of its many favorable characteristics. One of the most important is a high strength-cost ratio in many applications. Another is that concrete, while plastic, may be cast in forms easily at ordinary temperatures to produce almost any desired shape. The exposed face may be developed into a smooth or rough, hard surface, capable of withstanding the wear of truck or airplane traffic, or it may be treated to create desired architectural effects. In addition, concrete has high resistance to fire and penetration of water.

But concrete also has disadvantages. An important one is that quality control sometimes is not so good as for other construction materials, because concrete often is manufactured in the field under conditions where responsibility for its production cannot be pinpointed. Another is that concrete is a relatively brittle material. Its tensile strength is small compared with its compressive strength. This disadvantage, however, can be offset by reinforcing or prestressing concrete with steel. The combination of the two materials, reinforced concrete, possesses many of the best properties of each. It finds use in a wide variety of constructions, including building frames, floors, roofs, and walls; bridges; pavements; piles; dams; and tanks.

8-1. Important Properties of Concrete. The characteristics of portland-cement concrete can be varied to a considerable extent through control over its ingredients. Thus, for a specific structure, it is economical to use a concrete that has exactly the characteristics needed, though weak in others. For example, concrete for a building frame should have high compressive strength, whereas concrete for a dam should be durable and watertight, and strength can be relatively small.

Workability is an important property for many applications of concrete. Difficult to evaluate, workability, in essence, is the ease with which the ingredients can be mixed and the resulting mix handled, transported, and placed with little loss in homogeneity. One characteristic of workability that engineers frequently try to measure is consistency, or fluidity. For this purpose, they often make a slump test.

In the slump test, a specimen of the mix is placed in a mold shaped as the frustum of a cone, 12 in. high, with 8-in.-diameter base and 4-in.-diameter top (ASTM Specification C143). When the mold is removed, the change in height of the specimen is measured. When the test is made in accordance with the ASTM specification, the change in height may be taken as the slump. (As measured by this test, slump decreases as temperature increases; so the temperature of the mix at time of test should be specified, to avoid erroneous conclusions.)

Tapping the slumped specimen gently on one side with a tamping rod after completing the test may give additional information on the cohesiveness, workability, and placeability of the mix ("Concrete Manual," Bureau of Reclamation, Government Printing Office, Washington, D.C. 20402). A well-proportioned, workable mix will settle slowly, retaining its original identity. A poor mix will crumble, segregate, and fall apart.

Slump of a given mix may be increased by adding water, or increasing the percentage of fines (cement or aggregate), or entraining air, or incorporating an admixture that reduces water requirements. But these changes affect other properties of the concrete, sometimes adversely. In general, the slump specified should yield the desired consistency with the least amount of water and cement.

Durability is another important property of concrete. Concrete should be capable of withstanding the weathering, chemical action, and wear to which it will be subjected in service. Much of the weather damage sustained by concrete is attributable to cycles of freezing and thawing. Resistance of concrete to such damage can be improved by increasing the watertightness; or by entrainment of 2 to 6% air, by use of an air-entraining agent; or by applying a protective coating to the surface.

Chemical agents, such as inorganic acids, acetic and carbonic acids, and sulfates of calcium, sodium, magnesium, potassium, aluminum, and iron, will disintegrate or damage concrete. When contact between these agents and concrete may occur, the concrete should be protected with a resistant coating; for resistance to sulfates, Type V portland cement may be used (Art. 5-3). Resistance to wear usually is achieved by use of a high-strength, dense concrete, made with hard aggregates.

Watertightness is an important property of concrete that can often be improved by reducing the amount of water in the mix. Excess water leaves voids and cavities after evaporation, and if they are interconnected, water can penetrate or pass through the concrete. Entrained air (minute bubbles) usually increases watertightness, as also does prolonged thorough curing.

Volume change is another characteristic of concrete that should be taken into account. Expansion due to chemical reactions between the ingredients of concrete may cause buckling, and drying shrinkage may cause cracking.

Expansion due to alkali-aggregate reaction can be avoided by selection of nonreactive aggregates. If reactive aggregates must be used, expansion may be reduced or eliminated by addition of pozzolanic material, such as fly ash, to the mix. Expansion due to heat of hydration of cement can be reduced by keeping cement content as low as possible, using Type IV cement (Art. 5-3), and chilling the aggregates, water, and concrete in the forms. Expansion due to increases in air temperature may be decreased by producing concrete with a lower coefficient of expansion, usually by use of coarse aggregates with a lower coefficient of expansion.

Drying shrinkage can be reduced principally by cutting down on water in the mix. But less cement also will reduce shrinkage, as will adequate moist curing. Addition of pozzolans, however, unless enabling a reduction in water, will increase drying shrinkage.

Autogenous volume change, a result of chemical reaction and aging within the concrete, and usually shrinkage rather than expansion, is relatively independent of water content. This type of shrinkage may be decreased by using less cement, and sometimes also by using a different cement.

Whether volume change will damage the concrete often depends on the restraint present. For example, a highway slab that cannot slide on the subgrade while shrinking may crack; a building floor that cannot contract because it is anchored to relatively stiff girders also may crack. Hence, consideration should always be given to eliminating restraints or resisting the stresses they may cause.

Strength is a property of concrete that nearly always is of concern. Usually, it is determined by the ultimate strength of a specimen in compression, but sometimes flexural or tensile capacity is the criterion. Since concrete usually gains strength over a long period of time, the compressive strength at 28 days is commonly used as a measure of this property. In the United States, it is general practice to determine the compressive strength of concrete by testing specimens in the form of standard cylinders made in accordance with ASTM Specification C192 or C31. C192 is intended for research testing or for selecting a mix (laboratory specimens). C31 applies to work in progress (field specimens). The tests should be made as recommended in ASTM C39. Sometimes, however, it is necessary to determine the strength of concrete by taking drilled cores; in that case, ASTM C42 should be adopted. (See also American Concrete Institute Standard 214, Recommended Practice for Evaluation of Compression Test Results of Field Concrete.)

The 28-day compressive strength of concrete can be estimated from the 7-day strength by a formula proposed by W. A. Slater (*Proceedings of the American Concrete Institute*, 1926):

$$S_{28} = S_7 + 30 \sqrt{S_7} \tag{8-1}$$

where S_{28} = 28-day compressive strength, psi
S_7 = 7-day strength, psi

For S_{28}, Psi	S_7 Should Be at Least, Psi
4,000	2,500
3,500	2,120
3,000	1,750
2,500	1,390
2,000	1,040

Concrete strength is influenced chiefly by the water-cement ratio; the higher this ratio, the lower the strength. In fact, the relationship is approximately linear when expressed in terms of the variable C/W, the ratio of cement to water by weight: $S_{28} = 2,700 C/W - 760$ (for a workable mix).

Strength may be increased by decreasing water-cement ratio, using higher-strength aggregates, grading the aggregates to produce a smaller percentage of voids in the concrete, moist-curing the concrete after it has set, adding a pozzolan, such as fly ash, vibrating the concrete in the forms, and sucking out excess water with a vacuum from concrete in the forms. The short-time strength may be increased by use of Type III (high-early-strength) portland cement (Art. 5-3) and of accelerating admixtures, such as calcium chloride, and by increasing curing temperatures; but long-time strengths may not be affected. Strength-increasing admixtures generally accomplish their objective by reducing water requirements for the desired workability. (See also Art. 5-10.)

The stress-strain diagram for concrete of a specified compressive strength is a curved line (Fig. 8-1). Maximum stress is reached at a strain of 0.002 in. per in., after which the curve descends.

The modulus of elasticity generally used in design for concrete is a secant modulus. In ACI 318-71, Building Code Requirements for Reinforced Concrete, it is determined by

$$E_c = w^{1.5} 33 \sqrt{f'_c} \tag{8-2}$$

where w = weight of concrete, lb per cu ft
f'_c = specified compressive strength at 28 days, psi
For ordinary concrete, with w = 145 lb per cu ft,

$$E_c = 57,000 \sqrt{f'_c} \tag{8-3}$$

The modulus increases with age, as does the strength. (See also Art. 5-11.)

Tensile strength of concrete is much lower than compressive strength and regardless of the type of test usually has poor correlation with f'_c. As determined in flexural tests, the tensile strength (modulus of rupture—not the true strength) is about $7\sqrt{f'_c}$ for the higher-strength concretes and $10\sqrt{f'_c}$ for the lower-strength concretes.

Creep is strain that occurs under a constant long-time load. The concrete continues to deform, but at a rate that diminishes with time. It is approximately proportional to the stress at working loads and increases with increasing water-cement ratio. It decreases with increase in relative humidity. In design of reinforced-concrete beams for allowable stress, the effects of

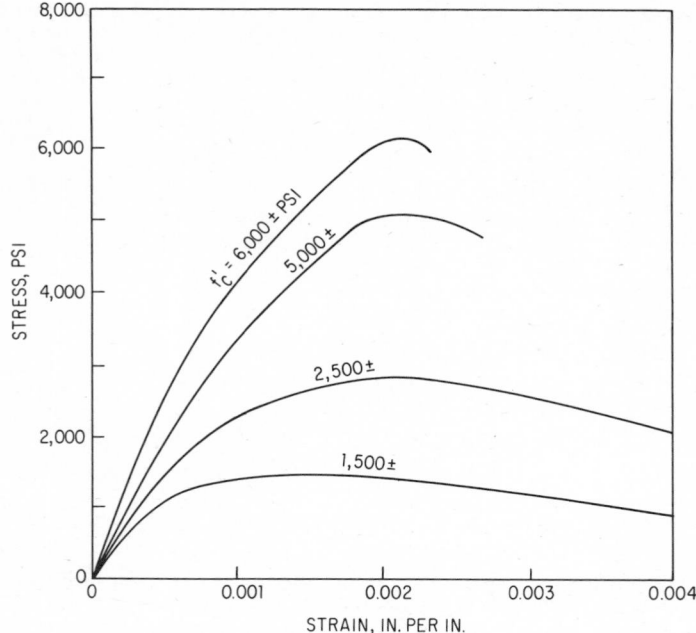

Fig. 8-1. Stress-strain curves for concrete.

creep are taken into account by reducing the modulus of elasticity of the concrete, usually by 50%.   In design of prestressed-concrete beams, creep may be taken as 100% of the elastic strain for concrete in a very humid atmosphere to 300% for concrete in a very dry atmosphere.   Part of the creep is recoverable on removal of the load.   (See also Art. 5-11.)

Weight per cubic foot of ordinary sand-and-gravel concrete usually is about 145 lb.   It may be slightly lower if the maximum size of coarse aggregate is less than 1½ in.   It can be increased by using denser aggregate.   And it can be decreased by using lightweight aggregate, increasing the air content, or incorporating a foaming, or expanding, admixture.   (See also Arts. 5-1 to 5-3, 5-8, 5-9, and 5-12.)

(G. E. Troxell, H. E. Davis, and J. W. Kelly, "Composition and Properties of Concrete," McGraw-Hill Book Company, New York; D. F. Orchard, "Concrete Technology," John Wiley & Sons, Inc., New York; J. J. Waddell, "Concrete Construction Handbook," McGraw-Hill Book Company, New York.)

8-2. Lightweight Concrete.   Concrete lighter in weight than ordinary sand-and-gravel concrete is used principally to reduce dead load, or for thermal insulation, or nailability, or fill. Disadvantages of structural concretes include higher cost, need for more care in placing, greater porosity, and more drying shrinkage.   For a given percentage of cement, usually the lighter the concrete, the lower the strength.

Lightweight concrete generally is made by using lightweight aggregates or by use of gas-forming or foaming agents, such as aluminum powder, which are added to the mix.   The light-weight aggregates are produced by expanding clay, shale, slate, diatomaceous shale, perlite, obsidian, and vermiculite with heat; and by special cooling of blast-furnace slag.   They also are obtained from natural deposits of pumice, scoria, volcanic cinders, tuff, and diatomite, and from industrial cinders (Art. 5-6).   Usual ranges of weights obtained with some lightweight aggregates are given in Table 8-1.

Production of lightweight-aggregate concretes is more difficult than that of ordinary concrete because aggregates vary in absorption of water, specific gravity, moisture content, and amount and grading of undersize.   Frequent unit-weight and slump tests are necessary so that cement and water content of the mix can be adjusted, if uniform results are to be obtained.   Also, the concretes usually tend to be harsh and difficult to place and finish because of the porosity and angularity of the aggregates. Sometimes, the aggregates may float to the surface. Workability can be improved by increasing the percentage of fine aggregates or by using an air-entraining admixture to incorporate from 4 to 6% air. (See also ACI 613A, Recommended Practice for Selecting Proportions for Structural Lightweight Concrete, American Concrete Institute.)

Table 8-1. Approximate Weights of Lightweight Concretes

Aggregate	Concrete weight, lb per cu ft
Cinders:	
Without sand	85
With sand	110–115
Shale or clay	90–110
Pumice	90–100
Scoria	90–110
Perlite	50–80
Vermiculite	35–75

To improve uniformity of moisture content of aggregates and reduce segregation during stock-piling and transportation, lightweight aggregate should be wetted 24 hr before use. Dry aggregate should not be put into the mixer, because the aggregate will continue to absorb moisture after it leaves the mixer and thus cause the concrete to segregate and stiffen before placement is completed. Continuous water curing is especially important with lightweight concrete.

Others types of lightweight concretes may be made with organic aggregates, or by omission of fines, or gap grading, or replacing all or part of the aggregates with air or gas. Nailing concrete usually is made with sawdust, though expanded slag, pumice, perlite, and volcanic scoria also are suitable. A good nailing concrete can be made with equal parts by volume of portland cement, sand, and pine sawdust, and sufficient water to produce a slump of 1 to 2 in. The sawdust should be fine enough to pass through a ¼-in. screen and coarse enough to be retained on a No. 16 screen. (Bark in the sawdust may retard setting and weaken the concrete.) The behavior of this type of concrete depends on the type of tree from which the sawdust came. Hickory, oak, or birch may not give good results ("Concrete Manual," U.S. Bureau of Reclamation, Government Printing Office, Washington, D.C.). Some insulating lightweight concretes are made with wood chips as aggregate.

For no-fines concrete, 20 to 30% entrained air replaces the sand. Pea gravel serves as the coarse aggregate. This type of concrete is used where low dead weight and insulation are desired and strength is not of importance. No-fines concrete may weigh from 105 to 118 lb per cu ft and have a compressive strength from 200 to 1,000 psi.

A porous concrete may be made by gap grading or single-size aggregate grading. It is used where drainage is desired or for light weight and low conductivity. For example, drain tile may be made with a No. 4 to ⅜- to ½-in. aggregate and a low water-cement ratio; just enough cement is used to bind the aggregates into a mass resembling popcorn.

Gas and foam concretes usually are made with admixtures. Foaming agents include sodium lauryl sulfate, alkyl aryl sulfonate, certain soaps, and resins. In another process, the foam is produced by the type of foaming agents used to extinguish fires, such as hydrolyzed waste protein. Foam concretes range in weight from 20 to 110 lb per cu ft.

Aluminum powder, when used as an admixture, expands concrete by producing hydrogen bubbles. Generally, about ¼ lb of the powder per bag of cement is added to the mix, sometimes with an alkali, such as sodium hydroxide or trisodium phosphate, to speed the reaction.

The heavier cellular concretes have sufficient strength for structural purposes, such as floor slabs and roofs. The lighter ones are weak but provide good thermal and acoustic insulation, or are useful as fill; for example, they are used over structural floor slabs to embed electrical conduit.

("Guide for Cast-in-place Low-density Concrete," "Guide for Structural Lightweight-Aggregate Concrete," and "Recommended Practice for Selecting Proportions for Structural Lightweight Concrete," American Concrete Institute.)

8-3. Heavyweight Concrete. Concrete weighing up to about 385 lb per cu ft can be produced by using heavier-than-ordinary aggregate. Theoretically, the upper limit can be achieved with steel shot as fine aggregate and steel punchings as coarse aggregate. (See also Art. 5-7.) The heavy concretes are used principally in radiation shields and counterweights.

Concrete made with barite develops an optimum density of 232 lb per cu ft and compressive strength of 6,000 psi; with limonite and magnetite, densities from 210 to 224 lb per cu ft and strengths of 3,200 to 5,700 psi; with steel punchings and sheared bars as coarse aggregate and steel shot for fine aggregate, densities from 250 to 288 lb per cu ft and strengths of about 5,600 psi. Gradings and mix proportions are similar to those used for conventional concrete. These concretes usually do not have good resistance to weathering or abrasion.

8-4. Proportioning and Mixing Concrete. Components of a mix should be selected to produce a concrete with the desired characteristics for the service conditions and with adequate workability at the lowest cost. For economy, the amount of cement should be kept to a minimum. Generally, this objective is facilitated by selection of the largest-size coarse aggregate consistent with job requirements and good gradation, to keep the volume of voids small. The smaller this volume, the less cement paste needed to fill the voids.

The water-cement ratio, for economy, should be as large as feasible to yield a concrete with the desired compressive strength, durability, and watertightness and without excessive shrinkage. Water added to a stiff mix improves workability, but an excess of water has deleterious effects (Art. 8-1).

A concrete mix is specified by indicating the weight, in pounds, of water, sand, coarse aggregate, and admixture to be used per 94-lb bag of cement. In addition, type of cement, fineness modulus of the aggregates, and maximum sizes of aggregates should be specified. Sometimes, a mix is indicated briefly by the ratio, by weight, of cement to sand to coarse aggregate; for example, 1:2:4; plus the minimum cement content per cubic yard of concrete.

Because of the large number of variables involved, it usually is advisable to proportion concrete mixes by making and testing trial batches. A start is made with the selection of the water-cement ratio. Then, several trial batches are made with varying ratios of aggregates to obtain the desired workability with the least cement. The aggregates used in the trial batches should have the same moisture content as the aggregates to be used on the job. The amount of mixing water to be used must include water that will be absorbed by dry aggregates or must be reduced by the free water in wet aggregates. The batches should be mixed by machine, if possible, to obtain results close to those that would be obtained in the field. Observations should be made of the slump of the mix and appearance of the concrete. Also, tests should be made to evaluate compressive strength and other desired characteristics. After a mix has been selected, some changes may have to be made after some field experience with it.

Table 8-2 estimates the 28-day compressive strength that may be attained with various water-cement ratios, with and without air entrainment. It should be noted that air entrainment permits a reduction in water, so a lower water-cement ratio for a given workability is feasible with air entrainment.

Table 8-3 lists recommended maximum sizes of aggregate for various types of construction.

These tables may be used with Table 8-4 for proportioning concrete mixes for small jobs where time or other conditions do not permit proportioning by the trial-batch method. Start with mix B in Table 8-4 corresponding to the selected maximum size of aggregate. Add just enough water for the desired workability. If the mix is undersanded, change to mix A; if over-

Table 8-2. Estimated Compressive Strength of Concrete for Various Water-Cement Ratios*

Water-cement ratio by weight	28-day compressive strength	
	Air-entrained concrete	Non-air-entrained concrete
0.40	4,300	5,400
0.45	3,900	4,900
0.50	3,500	4,300
0.55	3,100	3,800
0.60	2,700	3,400
0.65	2,400	3,000
0.70	2,200	2,700

*"Concrete Manual," U.S. Bureau of Reclamation.

Table 8-3. Recommended Maximum Sizes of Aggregate*

Minimum dimension of section, in.	Maximum size, in., of aggregate for		
	Reinforced-concrete beams, columns, walls	Heavily reinforced slabs	Lightly reinforced or unreinforced slabs
5 or less	. . .	¾–1½	¾–1½
6–11	¾–1½	1½	1½–3
12–29	1½–3	3	3–6
30 or more	1½–3	3	6

*"Concrete Manual," U.S. Bureau of Reclamation.

sanded change to mix C. Weights are given for dry sand. For damp sand, increase the weight of sand 10 lb, and for very wet sand, 20 lb, per bag of cement.

Admixtures may be used to control specific characteristics of concrete. Major types of admixtures include set accelerators, water reducers, air entrainers, and waterproofing compounds. In general, admixtures are helpful in improving concrete quality, and their use should be recommended. But some admixtures, if not administered properly, could have undesirable side effects. Hence, every engineer should be familiar with admixtures and their chemical components, as well as their advantages and limitations. Moreover, admixtures should be used in accordance with manufacturers' recommendations and, if possible, under the supervision of a manufacturer's representative. Many admixtures are covered by specifications of the American Society for Testing and Materials.

Set accelerators are used in cold weather, when it takes too long for concrete to set naturally. Calcium chloride is one chemical that has been used for this purpose. If not used in the right quantities, however, it could have harmful effects on reinforcement and concrete. Engineers, therefore, must know how to use it properly.

Water reducers lubricate the mix. If only the amount of water that is needed to achieve specified strength were used, concrete without a water-reducing admixture usually would be very stiff and unworkable. Most of the water in a normal concrete mix is needed for workability of the mix. If, however, the workability of the mix is improved by a chemical lubricant, less water is needed. With the same cement content but less water, the concrete attains

Table 8-4. Typical Concrete Mixes*

Maximum size of aggregate, in.	Mix designation	Bags of cement per cu yd of concrete	Aggregate, lb per bag of cement		
			Sand		Gravel or crushed stone
			Air-entrained concrete	Concrete without air	
½	A	7.0	235	245	170
	B	6.9	225	235	190
	C	6.8	225	235	205
¾	A	6.6	225	235	225
	B	6.4	225	235	245
	C	6.3	215	225	265
1	A	6.4	225	235	245
	B	6.2	215	225	275
	C	6.1	205	215	290
1½	A	6.0	225	235	290
	B	5.8	215	225	320
	C	5.7	205	215	345
2	A	5.7	225	235	330
	B	5.6	215	225	360
	C	5.4	205	215	380

*"Concrete Manual," U.S. Bureau of Reclamation.

greater strength.   Reduction of the quantity of water, however, permits a proportionate decrease in cement and thus reduces shrinkage of the hardened concrete.   An additional advantage of a water-reducing admixture is easier placement of concrete.   This, in turn, helps the workers and reduces the possibility of honeycombed concrete.   Some water-reducing admixtures act also as retarders of concrete set.   This is helpful in hot weather and in integrating consecutive pours of concrete.

Air-entraining agents entrain minute bubbles of air in concrete.   This increases resistance of concrete to freezing and thawing.   Therefore, air-entraining agents are extensively used in exposed concrete.

Set-retarding, water-reducing, and air-entraining admixtures may be combined into one compound.   This, however, should be done by an admixture manufacturer, not by a concrete contractor.

Waterproofing chemicals may be added to a concrete mix.   But often they are applied as surface treatments.   Silicones, for example, are used on hardened concrete as a water repellent. If applied properly and uniformly over a concrete surface, they can be effective against rainwater, preventing it from penetrating the surface.   (Some silicone coatings discolor with age. Most lose their effectiveness after a number of years; when that happens, the surface should be covered with a new coat of silicone for continued protection.)   Epoxies also may be used as water repellents.   They are much more durable, but they also may be much more costly.   Epoxies have many other uses in concrete, such as protection of wearing surfaces, patching compounds for cavities and cracks, and glue for connecting pieces of hardened concrete.

Mixing.   Components for concrete generally are stored in batching plants before being fed to a mixer.   These plants consist of weighing and control equipment and hoppers, or bins, for storing cement and aggregates.   Proportions are controlled by manually operated or automatic scales.   Mixing water is measured out from measuring tanks or with the aid of water meters.

Machine mixing is used wherever possible to achieve uniform consistency of each batch. Good results are obtained with the revolving-drum-type mixer, commonly used in the United States, and countercurrent mixers, with mixing blades rotating in the direction opposite to that of the drum.

Mixing time, measured from the time the ingredients, including water, are in the drum, should be at least 1.5 min for a 1-cu-yd mixer, plus 0.5 min for each cubic yard of capacity over 1 cu yd.   But overmixing may remove entrained air and increase fines, thus requiring more water to maintain workability.   So it is advisable also to set a maximum on mixing time.   As a guide, use three times the minimum mixing time.

Ready-mixed concrete is batched in central plants and delivered to various job-sites in trucks, usually in mixers mounted on the trucks.   The concrete may be mixed en route or after arrival at the site.   Though concrete may be kept plastic and workable for as long as 1½ hr by slow revolving of the mixer, better control of mixing time can be maintained if water is added and mixing started after arrival of the truck at the job, where the operation can be inspected.

(Guide for Use of Admixtures in Concrete; ACI 211.1, Recommended Practice for Selecting Proportions for Normal Weight Concrete; ACI 211.2, Recommended Practice for Selecting Proportions for Structural Lightweight Concrete; and ACI 304, Recommended Practice for Measuring, Mixing, Transporting, and Placing Concrete, American Concrete Institute, P.O. Box 4754, Redford Station, Detroit, Mich. 48219; G. E. Troxell, H. E. Davis, and J. W. Kelly, "Composition and Properties of Concrete," McGraw-Hill Book Company, New York; D. F. Orchard, "Concrete Technology," John Wiley & Sons, Inc., New York; J. J. Waddell, "Concrete Construction Handbook," McGraw-Hill Book Company, New York.)

8-5. Concrete Placement.   When concrete is discharged from the mixer, precautions should be taken to prevent segregation because of uncontrolled chuting as it drops into buckets, hoppers, carts, or forms.   Such segregation is more likely to occur with nontilting mixers with discharge chutes that allow the concrete to pass in relatively small streams than with tilting mixers.   To prevent segregation, a baffle or, better still, a section of downpipe should be inserted at the end of the chutes so that the concrete will fall vertically into the center of the receptacle.

Steel buckets, when selected for the job conditions and properly operated, handle and place concrete very well.   But they should not be used if they have to be hauled so far that there will be noticeable separation, bleeding, or loss of slump exceeding 1 in.   The discharge should be controllable in amount and direction.

Rail cars and trucks sometimes are used to transport concrete after it is mixed.   But there is a risk of stratification, with a layer of water on top, coarse aggregate on the bottom.   Most effective prevention is use of dry mixes and air entrainment.   If stratification occurs, the concrete should be remixed either as it passes through the discharge gates or by passing small quantities of compressed air through the concrete en route.

Chutes frequently are used for concrete placement. But the operation must be carefully controlled to avoid segregation and objectionable loss of slump. The slope must be constant under varying loads and sufficiently steep to handle the stiffest concrete to be placed. Long chutes should be shielded from sun and wind to prevent evaporation of mixing water. Control at the discharge end is of utmost importance to prevent segregation; discharge should be vertical, preferably through a short length of downpipe.

Tremies, or elephant trunks, deposit concrete under water. These are tubes about 1 ft or more in diameter at the top, flaring slightly at the bottom. They should be long enough to reach the bottom. When concrete is being placed, the tremie is always kept full of concrete, with the lower end immersed in the concrete just deposited. The tremie is raised as the level of concrete rises. Concrete should never be deposited through water unless confined.

Belt conveyors for placing concrete also present segregation and loss-of-slump problems. These may be reduced by adoption of the same precautions as for transportation by trucks and placement with chutes.

Sprayed concrete (shotcrete or gunite) is applied directly onto a form by an air jet. A "gun," or mechanical feeder, mixer, and compressor comprise the principal equipment for this method of placement. Compressed air and the dry mix are fed to the gun, which jets them out through a nozzle equipped with a perforated manifold. Water flowing through the perforations is mixed with the dry mix before it is ejected. Because sprayed concrete can be placed with a low water-cement ratio, it usually has high compressive strength. The method is especially useful for building up shapes without a form on one side.

Pumping is a suitable method for placing concrete, but it seldom offers advantages over other methods. Curves, lifts, and harsh concrete reduce substantially maximum pumping distance. For best performance, an agitator should be installed in the pump feed hopper to prevent segregation.

Barrows are used for transporting concrete very short distances, usually from a hopper to the forms. In the ordinary wheelbarrow, a worker can move 1½ to 2 cu ft of concrete 25 ft in 3 min.

Concrete carts serve the same purpose as wheelbarrows but put less load on the transporter. Heavier and wider, the carts can handle 4.5 cu ft. Motorized carts with ½ cu yd capacity also are available.

Regardless of the method of transportation or equipment used, the concrete should be deposited as nearly as possible in its final position. Concrete should not be allowed to flow into position, because then less durable mortar concentrates in ends and corners where durability is most important.

Vibration of concrete in the forms is desirable because it eliminates voids. The consolidation also insures close contact of the concrete with the forms, reinforcement, and other embedded items. It usually is accomplished with electric or pneumatic vibrators.

For consolidation of structural concrete and tunnel-invert concrete, immersion vibrators are recommended. Oscillation should be at least 7,000 vibrations per minute when the vibrator head is immersed in the concrete. Precast concrete of relatively small dimensions and concrete in tunnel arch and side walls may be vibrated with vibrators rigidly attached to the forms and operating at 8,000 vibrations per min or more. Concrete in canal and lateral linings should be vibrated at more than 4,000 vibrations per min, with the immersion type, though external vibration may be used for linings less than 3 in. thick. For mass concrete, with 3- and 6-in. coarse aggregate, vibrating heads should be at least 4 in. in diameter and operate at frequencies of at least 6,000 vibrations per min when immersed. Each cubic yard should be vibrated for at least 1 min. A good small vibrator can handle from 5 to 10 cu yd per hr and a large, two-person, heavy-duty type, about 50 cu yd per hr, in uncramped areas.

("Concrete Manual," U.S. Bureau of Reclamation, Government Printing Office, Washington, D.C.; ACI 311, Recommended Practice for Concrete Inspection; ACI 304, Recommended Practice for Measuring, Mixing, Transporting, and Placing Concrete; and ACI 506, Recommended Practice for Shotcreting; also, Placing Concrete by Pumping Methods; and Preplaced Aggregate Concrete for Structural and Mass Concrete, American Concrete Institute, Detroit, Mich.)

8-6. Construction Joint. A construction joint is formed when unhardened concrete is placed against concrete that has become so rigid that the new concrete cannot be incorporated in the old by vibration. Generally, steps must be taken to insure bond between the two.

The first step is to clean the exposed surface. Next, green cutting generally is advisable, especially if the exposed concrete is not of the highest quality. This requires use of a jet of air and water, at about 100 psi, to remove laitance and inferior surface concrete. Then, for a final cleanup, the surface should be sandblasted or brushed vigorously with fine-wire brooms before the new concrete is placed. Sandblasting without initial cleanup can produce excellent joints on horizontal surfaces of mass concrete that was placed with a slump of 2 in. or less, but the surface must be protected from excessive traffic. After sandblasting, the surface should be thoroughly washed and allowed to dry.

Also before new concrete is deposited, the surface should be coated with ½ in. of mortar of the same proportions as that in the concrete. The mortar should be scrubbed into the surface with wire brooms or sprayed with an air gun. The first layer of new concrete should be placed before the ½-in. layer of mortar has dried.

("Concrete Manual," U. S. Bureau of Reclamation, Government Printing Office, Washington, D. C.; ACI 304, Recommended Practice for Measuring, Mixing, Transporting, and Placing Concrete, and "ACI Manual of Concrete Inspection," American Concrete Institute, Detroit, Mich.)

8-7. Finishing of Unformed Concrete Surfaces.   After concrete has been consolidated, screeding, floating, and the first troweling should be performed with as little working and manipulation of the surface as possible. Excessive manipulation draws inferior fines and water to the top and can cause checking, crazing, and dusting.

To avoid bringing fines and water to the top in the rest of the finishing operations, each step should be delayed as long as possible. If water accumulates, it should be removed by blotting with mats or draining, or it should be pulled off with a loop of hose, and the next finishing operation should be delayed until the water sheen disappears. Do not work neat cement into wet areas to dry them.

Screeds are guides for a straightedge to bring a concrete surface to a desired elevation or for a template to produce a desired curved shape. The screeds must be sufficiently rigid to resist distortion as the concrete is spread. They may be made of lumber or steel pipe.

For floors, screeding is followed by hand floating with wood floats or power floating. Permitting a stiffer mix with a higher percentage of large-size aggregate, power-driven floats with revolving disks and vibrators produce a sounder, more durable surface than wood floats. Floating may begin as soon as the concrete surface has hardened sufficiently to bear a person's weight without leaving an indentation. The operation continues until hollows and humps are removed or, if the surface is to be troweled, until a small amount of mortar is brought to the top.

If a finer finish is desired, the surface may be steel-troweled, by hand or by powered equipment. This is done as soon as the floated surface has hardened enough so that excess fine material will not be drawn to the top. Heavy pressure during troweling will produce a dense, smooth, watertight surface. Do not permit sprinkling of cement or cement and sand on the surface to absorb excess water or facilitate troweling. If an extra hard finish is desired, the floor should be troweled again when it has nearly hardened.

Concrete surfaces dust to some extent and may benefit from treatment with certain chemicals. They penetrate the pores to form crystalline or gummy deposits. Thus, they make the surface less pervious and reduce dusting by acting as plastic binders or by making the surface harder. Poor-quality concrete floors may be improved more by such treatments than high-quality concrete, but the improvement is likely to be temporary and the treatment will have to be repeated periodically.

("Concrete Manual," U. S. Bureau of Reclamation, Government Printing Office, Washington, D. C.)

8-8. Forms for Concrete.   Formwork retains concrete until it has set and produces the desired shapes and sometimes, also, desired surface finishes. Forms must be supported on falsework of adequate strength and sufficient rigidity to keep deflections within acceptable limits. The forms, too, must be strong and rigid, to meet dimensional tolerances. But they must be tight in addition; otherwise mortar will leak out during vibration and cause unsightly sand streaks and rock pockets. Yet, they must be low-cost and often easily demountable to permit reuse. These requirements are met by steel, reinforced plastic, concrete, and plain or coated lumber and plywood.

Unsightly bulges and offsets at horizontal joints should be avoided. This can be done by resetting forms with only 1 in. of form lining overlapping the existing concrete below the line made by a grade strip. Also, the forms should be tied and bolted close to the joint to keep the lining snug against existing concrete (Fig. 8-2). If a groove along a joint will not be aesthetically

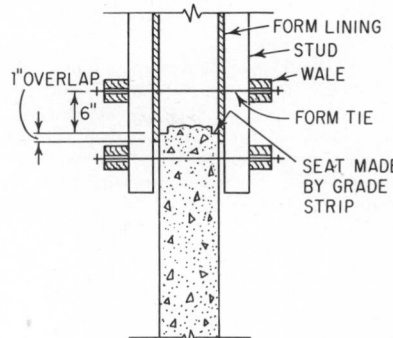

Fig. 8-2. Form at horizontal joint in concrete wall.

objectionable, forming of a groove along the joint will obscure the unsightliness often associated with construction joints.

Where form ties have to pass through the concrete, they should be as small in cross section as possible. (The holes they form sometimes have to be plugged to stop leaks.) Ends of form ties should be removed without spalling adjacent concrete.

Plastic coatings, proper oiling, or effective wetting can protect forms from deterioration, weather, and shrinkage before concreting. Form surfaces should be clean. They should be treated with a suitable form oil or other coating that will prevent the concrete from sticking to them. A straight, refined, pale, paraffin-base mineral oil usually is acceptable for wood forms; synthetic castor oil and some marine-engine oils are examples of compounded oils that give good results on steel forms. The oil or coating should be brushed or sprayed evenly over the forms. It

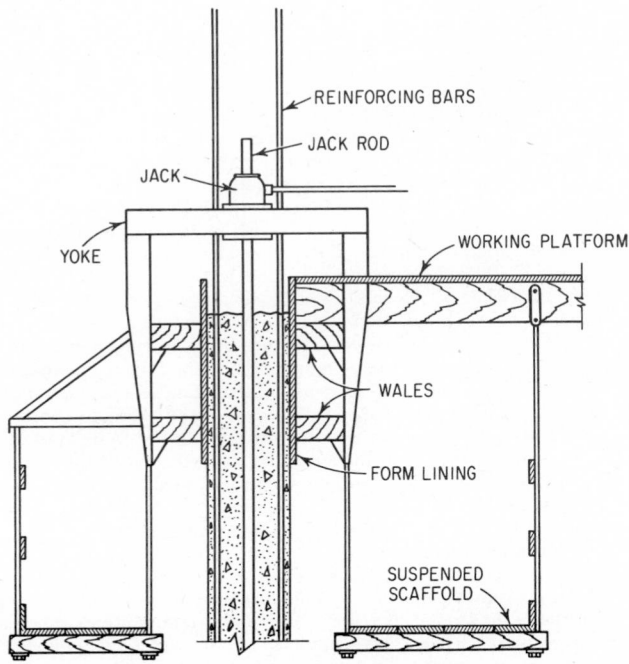

Fig. 8-3. Slip form for concrete wall.

should not be permitted to get on construction joint surfaces or reinforcing bars, because it will interfere with bond.

Forms should provide ready access for placement and vibration of concrete and for inspection.

Generally, forms are stationary. But, for some applications, such as highway pavements, precast-concrete slabs, silos, and service cores of buildings, use of continuous moving forms—sliding forms or slip forms—is advantageous.

A slip form for vertical structures consists principally of a form lining or sheathing about 4 ft high, wales or ribs, yokes, working platforms, suspended scaffolds, jacks, climbing rods, and control equipment (Fig. 8-3). Spacing of the sheathing is slightly larger at the top to permit easy upward movement. The wales hold the sheathing in alignment, support the working platforms and scaffolds, and transmit lifting forces from yokes to sheathing. Each yoke has a horizontal cross member perpendicular to the wall and connected to a jack. From each end of the member, vertical legs extend downward on opposite sides of and outside the wall. The lower end of each leg is attached to a bottom wale. The jack pulls the slip form upward by climbing a vertical steel rod, usually about 1 in. in diameter, embedded in the concrete. The suspended scaffolds provide access for finishers to the wall. Slip-form climbing rates range upward from about 2 to about 12 in. per hr.

Stationary forms should be removed only after the concrete has attained sufficient strength so that there will be no noticeable deformation or damage to the concrete. If supports are removed before beams or floors are capable of carrying superimposed loads, they should be reshored until they have gained sufficient strength.

Early removal of forms generally is desirable to permit quick reuse, start curing as soon as possible, and allow repairs and surface treatment while the concrete is still green and conditions are favorable for good bond. In cold weather, however, forms should not be removed while the concrete is still warm. Rapid cooling of the surface will cause checking and surface cracks. For this reason also, curing water applied to newly stripped surfaces should not be much cooler than the concrete.

(R. L. Peurifoy, "Formwork for Concrete Structures," McGraw-Hill Book Company, New York; "Concrete Manual," U.S. Bureau of Reclamation, Government Printing Office, Washington, D.C.; ACI 347, Recommended Practice for Concrete Formwork, "ACI Manual of Concrete Inspection" and "Formwork for Concrete," American Concrete Institute, Detroit, Mich.)

8-9. Curing Concrete. While more than enough mixing water for hydration is incorporated in normal concrete mixes, drying of the concrete after initial set may delay or prevent complete hydration. Curing includes all operations after concrete has set that improve hydration. Properly done for a sufficiently long period, curing produces stronger, more watertight concrete.

Methods may be classified as maintenance of a moist environment by addition of water, sealing in the water in the concrete, and those hastening hydration.

Maintenance of a moist environment by addition of water is the commonest field procedure. Generally, exposed concrete surfaces are kept continuously moist by spraying or ponding or by a covering of earth, sand, or burlap kept moist. Concrete made with ordinary and sulfate-resistant cements (Types I, II, and V) should be cured this way for at least 14 days; that made with low-heat cement (Type IV) for at least 21 days. Concrete made with high-early-strength cement should be kept moist until sufficient strength has been attained, as indicated by test cylinders.

Precast concrete and concrete placed in cold weather often are steam-cured in enclosures. While this is a form of moist curing, hydration is speeded by the higher-than-normal temperature, and the concrete attains a high early strength. Temperatures maintained usually range between 100 and 165°F. Higher temperatures produce greater strengths shortly after steam curing commences, but there are severe losses in strength after 2 days. A delay of 1 to 6 hr before steam curing will produce concrete with higher 24-hr strength than if the curing starts immediately after the concrete is cast. This "preset" period allows early cement reactions to occur and development of sufficient hardness to withstand the more rapid temperature curing to follow. Length of the preset period depends on the type of aggregate and temperature; the period should be longer for ordinary aggregate than for lightweight and for higher temperatures. Duration of steam curing depends on the concrete mix, temperature, and desired results.

Autoclaving, or high-pressure steam curing, maintains concrete in a saturated atmosphere at temperatures above the boiling point of water. Generally, temperatures range from 325 to 375°F at pressures from 80 to 170 psig. Main application is for concrete masonry. Advantages claimed are high early strength, reduced volume change in drying, better chemical resistance, and lower susceptibility to efflorescence. As for steam curing, a preset period of 1 to 6 hr is desirable. This is followed by single- or two-stage curing. Single-stage curing consists of a pressure buildup of at least 3 hr, 8 hr at maximum steam pressure, and rapid pressure release (20 to 30 min). The rapid release vaporizes moisture from the block. In two-stage curing, the concrete products are placed in kilns for the duration of the preset period. Saturated steam then is introduced into the kiln. After the concrete has developed sufficient strength to permit handling, the products are removed from the kiln, set in a compact arrangement, and placed in the autoclave.

Curing concrete by sealing the water in can be accomplished by either covering the concrete or coating it with a waterproof membrane. When coverings, such as heavy building paper or plastic sheets, are used, care must be taken that the sheets are sealed airtight and that corners and edges are adequately protected against loss of moisture. Coverings can be placed as soon as the concrete has been finished.

Coating concrete with a sealing compound generally is done by spraying to insure a continuous membrane. Brushing may damage the concrete surface. Sealing compound may be applied after the surface has stiffened so that it will no longer respond to float finishing. But in hot climates, it may be desirable, before spraying, to moist-cure for 1 day surfaces exposed to the sun. Surfaces from which forms have been removed should be saturated with water before spraying with compound. But the compound should not be applied to either formed or unformed surfaces until the moisture film on them has disappeared; spraying should be started as soon as the surfaces assume a dull appearance. The coating should be protected against damage. Continuity must be maintained for at least 28 days.

White or gray pigmented compound often is used for sealing because it facilitates inspection and reflects heat from the sun. Temperatures with white pigments may be decreased as much as 40°F, reducing cracking caused by thermal changes.

Surfaces of ceilings and walls inside buildings require no curing other than that provided by forms left in place at least 4 days. But wood forms are not acceptable for moist-curing outdoor concrete. Water should be applied at the top, for example, by a soil-soaker hose, and allowed to drip down between the forms and the concrete.

("Concrete Manual," U.S. Bureau of Reclamation, Government Printing Office, Washington, D.C.; ACI 517, Recommended Practice for Atmospheric Pressure Steam Curing of Concrete; "Low-pressure Steam Curing," Report of ACI Committee 517; and "High-pressure Steam Curing: Modern Practice, and Properties of Autoclaved Products," Report of ACI Committee 516, American Concrete Institute, Detroit, Mich.)

8-10. Cold-weather Concreting. Methods used during cold weather should prevent damage to concrete from freezing and thawing at an early age, maintain proper curing conditions, and limit temperature changes. They should also allow the concrete to develop early strength, thus permitting removal of forms and start of curing. The lower the temperature, the slower concrete gains strength and the longer the curing period.

In the fall, after the first frost, air-entrained concrete should be protected from freezing for at least 48 hr after casting. After the mean daily temperature falls below 40°F for more than 1 day, the temperature of the air-entrained concrete placed in the forms should exceed 40°F for mass concrete (maximum size aggregate 3 in. or more) and 50°F for thin sections. These temperatures should be maintained for at least 3 days, and protection against freezing provided for 3 more days. In early spring, after the mean daily temperature exceeds 40°F for more than 3 successive days, placement at and maintenance of the minimum temperatures may be discontinued. But concrete should be protected against freezing for at least 48 hr after casting.

For concrete without entrained air, the duration of protection should be twice as long for maximum durability. Nevertheless, durability will not be so good as that for air-entrained concrete, nor will additional protection bring it up to that level.

Use of additional cement, high-early-strength cement, or 1% calcium chloride by weight of cement will increase strength during the protective period or reduce the period of protection required to attain a specific strength. Calcium chloride in permissible amounts will not lower the freezing point of concrete significantly, nor will other chemicals in the mix. So use of such admixtures does not justify a reduction in the amount of protective cover, heat, or other winter protection. Also, calcium chloride should not be used in concrete that will be subjected to sulfates or that contains embedded aluminum or galvanized metal.

While freezing is a danger to concrete, so is overheating the concrete to prevent it. By accelerating chemical action, overheating can cause excessive loss of slump, raise the water requirement for a given slump, and increase thermal shrinkage. Rarely will mass concrete leaving the mixer have to be at more than 55°F and thin-section concrete at more than 75°F.

To obtain the minimum temperatures for concrete mixes in cold weather, heat the water and, if necessary, the aggregates. Mixing water should be heated to at least 140°F, under such control and in sufficient quantity to avoid fluctuations in temperature from batch to batch. To avoid flash set of the cement and loss of entrained air due to the heated water, aggregates and water should be placed in the mixer before the cement and air-entraining agent so that the colder aggregates will reduce the water temperature to below 80°F.

When heating of aggregates is necessary, it is best done with steam or hot water in pipes. Use of steam jets is objectionable because of resulting variations in moisture content of the aggregates. For small jobs, aggregates may be heated over culvert pipe in which fires are maintained. But care must be taken not to overheat.

Before concrete is placed in the forms, the interior should be cleared of ice, snow, and frost. This may be done with steam under canvas or plastic covers.

Concrete should not be placed on frozen earth. It would lower the concrete temperature below the minimum and may cause settlement on thawing. The subgrade may be protected from freezing by a covering of straw and tarpaulins or other insulating blankets. If it does freeze, the subgrade must be thawed deep enough so that it will not freeze back up to the concrete during the required protection period.

The usual method of protecting concrete after it has been cast is to enclose the structure with tarpaulins or plastic and heat the interior. Since corners and edges are especially vulnerable to low temperatures, the enclosure should enclose corners and edges, not rest on them. The enclosure must be not only strong but also windproof. If wind can penetrate it, required concrete temperatures may not be maintained despite high fuel consumption. Heat may be supplied by live or piped steam, salamanders, stoves, or warm air blown in through ducts from heaters outside the enclosure. But strict fire-prevention measures should be enforced. When dry heat is used, the concrete should be kept moist to prevent it from drying.

Concrete also may be protected with insulation. For example, pavements may be covered with layers of straw, shavings, or dry earth. For structures, forms may be insulated.

When protection is discontinued or when forms are removed, precautions should be taken that the drop in temperature of the concrete will be gradual. Otherwise, the concrete may crack and deteriorate. Surface temperatures of thin sections may be permitted to drop 40°F per day without damage; but for mass concrete, the drop should not exceed 20°F per day, because of the likelihood of large temperature differences between the interior and the surface.

("Concrete Manual," U. S. Bureau of Reclamation, Government Printing Office, Washington, D. C.; ACI 306, Recommended Practice for Cold-weather Concreting, American Concrete Institute, Detroit, Mich.)

8-11. Hot-weather Concreting. The higher the temperature, the more rapid will be the hydration of cement, the faster the evaporation of mixing water, the lower the concrete strength, and the larger the volume change. Unless precautions are taken, setting and rate of hardening will accelerate, shortening the available time for placing and finishing the concrete. Quick stiffening encourages undesirable additions of mixing water, or retempering, and may also result in inadequate consolidation and cold joints. The tendency to crack is increased because of rapid evaporation of water, increased drying shrinkage, or rapid cooling of the concrete from its high initial temperature. If an air-entrained concrete is specified, control of the air content is more difficult. And curing becomes more critical.

To avoid impairing the quality of concrete, its temperature during placement should be as much below 90°F as economically feasible. Use cold water or a mixture of ice and water in the mix to attain this objective. Also, cool the coarse aggregate with cold-air blasts or with refrigerated water by sprinkling or inundation. Shade materials and facilities not otherwise protected from the heat. Insulate the mixing drums or cool them with water sprays or wet burlap coverings. Also, insulate water-supply lines and tanks, or at least paint them white. Do not use cement with a temperature exceeding 170°F. And if necessary, work only at night.

Set-retarding admixtures counteract the accelerating effect of high temperature and lessen the need for increase in mixing water. Their use should be considered when the weather is so hot that the temperature of concrete being placed is consistently above 75°F.

Before concrete is cast in hot weather, forms, reinforcing, and subgrade should be sprinkled with cool water. Concrete should be speedily placed and finished, to minimize slump loss.

Continuous water curing gives best results in hot weather. Curing should be started as soon as the concrete has hardened sufficiently to withstand surface damage. Water should be applied to formed surfaces while forms are still in place. Surfaces without forms should be kept moist by wet curing for at least 24 hr. Moist coverings are effective in eliminating evaporation loss from concrete, by protecting it from sun and wind. If moist curing is discontinued after the first day, the surface should be protected with a curing compound (Art. 8-9).

(ACI 305, Recommended Practice for Hot-weather Concreting, American Concrete Institute, Detroit, Mich.)

8-12. Contraction and Expansion Joints. Contraction joints are used mainly to control locations of cracks caused by shrinkage of concrete after it has hardened. If the concrete, while shrinking, is restrained from moving, by friction or attachment to more rigid construction, cracks are likely to occur at points of weakness. Contraction joints, in effect, are deliberately made weakness planes. They are formed in the expectation that if a crack occurs it will be along the neat geometric pattern of a joint, and thus irregular, unsightly cracking will be prevented. Such joints are used principally in floors, roofs, pavements, and walls.

A contraction joint is an indentation in the concrete. Width may be ¼ or ⅜ in. and depth one-sixth to one-fourth the thickness of the slab. The indentation may be made with a saw cut while the concrete still is green but before appreciable shrinkage stress develops. Or the joint may be formed by insertion of a strip of joint material before the concrete sets or by grooving the surface during finishing. Spacing of joints depends on the mix, strength and thickness of the concrete, and the restraint to shrinkage. The indentation in highway and airport pavements usually is filled with a sealing compound.

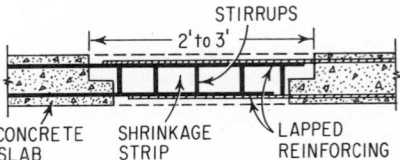

Fig. 8-4. Vertical section through shrinkage strip in concrete slab.

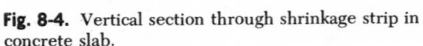

CONCRETE SLAB   SHRINKAGE STRIP   LAPPED REINFORCING

Control joints or shrinkage strips sometimes are used instead of contraction joints in floors. Or alternate bays may be cast in checkerboard pattern. The aim is to leave gaps when a floor is cast and to concrete them later, after much of the shrinkage in the concrete already placed has occurred. The gaps for shrinkage strips are about 2 or 3 ft wide extending across the full width or length of a slab (Fig. 8-4). Slab reinforcement should be lapped to permit free movement of concrete on each side of the strip. Stirrups sometimes are placed around the lapped bars to confine the concrete.

Expansion joints are used to help prevent cracking due to thermal dimension changes in concrete. They usually are placed where there are abrupt changes in thickness, offsets, or changes in types of construction, for example, between a bridge pavement and a highway pavement. Expansion joints provide a complete separation between two parts of a slab. The opening must be large enough to prevent buckling or other undesirable deformation due to expansion of the concrete.

To prevent the joint from being jammed with dirt and becoming ineffective, the opening is sealed with a compressible material. For watertightness, a flexible water stop should be placed across the joint. And if load transfer is desired, dowels should be embedded between the parts separated by the joint. The sliding ends of the dowels should be enclosed in a close-fitting metal cap or thimble, to provide space for movement of the dowel during expansion of the concrete. This space should be at least ¼ in. longer than the width of the joint.

(Guide to Joint Sealants for Concrete Structures, and ACI 325, Recommended Practice for Design of Concrete Pavements, American Concrete Institute, Detroit, Mich.)

8-13. Types of Reinforcing Steel. Because of the low tensile strength of concrete, steel is embedded in it to resist tensile stresses. Steel, however, also is used to take compression, in beams and columns, to permit use of smaller members. It serves other purposes, too. It controls strains due to temperature and shrinkage and distributes load to the concrete and other reinforcing steel. It can be used to prestress the concrete. And it ties other reinforcing together for easy placement or to resist lateral stresses.

Most reinforcing is in the form of bars or wires. Their surfaces may be smooth or deformed. The latter type is generally used because it produces better bond with the concrete because of the raised pattern on the steel.

Bars range in diameter from ¼ to 2¼ in. Sizes are designated by numbers, which are approximately eight times the nominal diameters. (See the latest edition of ASTM Specifications for Steel Bars for Concrete Reinforcement. These also list the minimum yield points and tensile strengths for each type of steel.)   Use of bars with yield points over 60,000 psi for flexural reinforcement is limited because special measures are required to control cracking and deflection.

Wires usually are used for reinforcing concrete pipe and, in the form of welded-wire fabric, for slab reinforcement.   The latter consists of a rectangular grid of uniformly spaced wires, welded at all intersections, and meeting the minimum requirements of ASTM A185 and A497.   Fabric offers the advantages of easy, fast placement of both longitudinal and transverse reinforcement and excellent crack control, because of high mechanical bond with the concrete.   (Deformed wires are designated by D followed by a number equal to the nominal area, sq in., times 100.)   Bars and rods also may be prefabricated into grids, by clipping or welding (ASTM A184).

Sometimes, metal lath is used for reinforcing concrete, for example, in thin shells.   It may serve as both form and reinforcing when concrete is applied by spray (gunite or shotcrete).

8-14. Bending and Placing Reinforcing Steel.   Bars are shipped by a mill to a fabricator in uniform long lengths and in bundles of 5 or more tons.   The fabricator transports them to the job straight and cut to length or cut and bent.

Bends may be required for beam and girder reinforcing, longitudinal reinforcing of columns where they change size, stirrups, column ties and spirals, and slab reinforcing.   Dimensions of standard hooks and typical bends and tolerances for cutting and bending are given in ACI 315, "Manual of Standard Practice for Detailing Reinforced Concrete Structures," American Concrete Institute, Detroit, Mich.

Some preassembling of reinforcing steel is done in the fabricating shop or on the job.   Beam, girder, and column steel often is wired into frames before placement in the forms.   Slab reinforcing may be clipped or welded into grids, or mats, if not supplied as welded-wire fabric.

Some rust is permissible on reinforcing if it is not loose and there is no appreciable loss of cross-sectional area.   In fact, rust, by creating a rough surface, will improve bond between the steel and concrete.   But the bars should be free of loose rust, scale, grease, oil, or other coatings that would impair bond.

Bars should not be bent or straightened in any way that will damage them.   If heat is necessary for bending, the temperature should not be higher than that indicated by a cherry-red color, and the steel should be allowed to cool slowly, not quenched.

Reinforcing should be supported and tied in the locations and positions called for in the plans.   The steel should be inspected before concrete is placed.   Neither the reinforcing nor other parts to be embedded should be moved out of position before or during the casting of the concrete.

Bars and wire fabric should not be kinked or have unspecified curvatures when positioned.   Kinked and curved bars, including those misshaped by workers walking on them, may cause the hardened concrete to crack when the bars are tensioned by service loads.

Usually, reinforcing is set on wire bar supports, preferably galvanized for exposed surfaces.   Lower-layer bars in slabs usually are supported on bolsters consisting of a horizontal wire welded to two legs about 5 in. apart.   The upper layer generally is supported on bolsters with runner wires on the bottom so that they can rest on bars already in place.   Or individual or continuous high chairs can be used to hold up a support bar, often a No. 5, at appropriate intervals, usually 5 ft.   An individual high chair is a bar seat that looks roughly like an inverted U braced transversely by another inverted U in a perpendicular plane.   A continuous high chair consists of a horizontal wire welded to two inverted-U legs 8 or 12 in. apart.   Beam and joist chairs have notches to receive the reinforcing.   These chairs usually are placed at 5-ft intervals.

While it is essential that reinforcement be placed exactly where called for in the plans, some tolerances are necessary.   Reinforcement in beams and slabs should be within $\pm\frac{1}{4}$ in. of the specified distance from the tension or compression face.   Lengthwise, a cutting tolerance of ±1 in. and a placement tolerance of ±2 in. are normally acceptable.   If length of embedment is critical, the designer should specify bars 3 in. longer than the computed minimum to allow for accumulation of tolerances.   Spacing of reinforcing in wide slabs and tall walls may be permitted to vary $\pm\frac{1}{2}$ in., or slightly more if necessary to clear obstructions, so long as the required number of bars are present.

Lateral spacing of bars in beams and columns, spacing between multiple reinforcement layers, and concrete cover over stirrups, ties, and spirals in beams and columns should never be less than that specified but may exceed it by $\frac{1}{4}$ in.   A variation in setting of an individual stirrup or column hoop of 1 in. may be acceptable, but the error should not be permitted to accumulate.   (CRSI Recommended Practice for Placing Reinforcing Bars, and "Manual of Standard Practice," Concrete Reinforcing Steel Institute, 180 North La Salle St., Chicago, Ill. 60601.)

8-15. Spacing, Splicing, and Cover for Reinforcing.   In buildings, the minimum clear distance between parallel bars should be 1 in. for bars up to No. 8 and the nominal bar diameter for larger bars. For columns, however, the clear distance between longitudinal bars should be at least 1.5 in. for bars up to No. 8 and 1.5 times the nominal bar diameter for larger bars. And the clear distance between multiple layers of reinforcement in building beams and girders should be at least 1 in.; upper-layer bars should be directly above corresponding bars below. These minimum-distance requirements also apply to the clear distance between a contact splice and adjacent splices or bars.

A common requirement for minimum clear distance between parallel bars in highway bridges is 1.5 times the diameter of the bars, and spacing center to center should be at least 1.5 times the maximum size of coarse aggregate.

Many codes and specifications relate the minimum bar spacing to maximum size of coarse aggregate. This is done with the intention of providing enough space for all the concrete mix to pass between the reinforcing. But if there is space to place concrete between layers of steel and between the layers and the forms, and the concrete is effectively vibrated, experience has shown that bar spacing or form clearance does not have to exceed the maximum size of coarse aggregate to insure good filling and consolidation. That portion of the mix which is molded by vibration around bars, and between bars and forms, is not inferior to that which would have filled those parts had a larger bar spacing been used. The remainder of the mix in the interior, if consolidated layer after layer, is superior because of its reduced mortar and water content ("Concrete Manual," U.S. Bureau of Reclamation, Government Printing Office, Washington, D.C.).

Bundled Bars. Groups of parallel reinforcing bars bundled in contact to act as a unit may be used only when they are enclosed by ties or stirrups. Four bars are the maximum permitted in a bundle, and all must be deformed bars. If full-length bars cannot be used between supports, then there should be a stagger of at least 40 bar diameters between any discontinuities. Also, the length of lap should be increased 20% for a three-bar bundle and 33% for a four-bar bundle. In determining minimum clear distance between a bundle and parallel reinforcing, the bundle should be treated as a single bar of equivalent area.

Maximum Spacing. In walls and slabs in buildings, except for concrete-joist construction, maximum spacing, center to center, of principal reinforcement should be 18 in. or three times the wall or slab thickness, whichever is smaller.

Tension Development Lengths. For bars in tension, the basic development length l_d, in., for No. 11 and smaller bars is defined as

$$l_d = \frac{0.04 A_b f_y}{\sqrt{f'_c}} \qquad (8\text{-}4a)$$

where A_b = area of bar, sq in.

f_y = yield strength of bar steel, psi

f'_c = 28-day compressive strength of concrete, psi

But l_d should not be less than $0.0004 d_b f_y$, where d_b = bar diameter, in., or less than 12 in. except in computation of lap splices or web anchorage.

For No. 14 bars,

$$l_d = 0.085 \frac{f_y}{\sqrt{f'_c}} \qquad (8\text{-}4b)$$

For No. 18 bars,

$$l_d = 0.11 \frac{f_y}{\sqrt{f'_c}} \qquad (8\text{-}4c)$$

and for deformed wire,

$$l_d = 0.03 d_b \frac{f_y}{\sqrt{f'_c}} \qquad (8\text{-}4d)$$

For top reinforcement, with horizontal bars so placed that more than 12 in. of concrete is cast below the bars, the basic development length should be increased by 40%.

Development lengths for tension bars are given in Table 8-5a.

Added development length is also required where f_y exceeds 60 ksi or when lightweight concrete is used.

Where reinforcing bars being developed are spaced laterally at least 6 in. on centers and at least 3 in. from the side face of a member, the required development length l_d may be reduced by 20%.

Where bars are enclosed within a spiral formed by a bar at least ¼ in. in diameter and with not more than a 4-in. pitch, the required development length l_d may be reduced by 25%.

Compression Development Lengths. For bars in compression, the basic development length l_d is defined as

$$l_d = \frac{0.02 f_y d_b}{\sqrt{f_c'}} \tag{8-5}$$

but l_d should not be less than 8 in. or $0.0003 f_y d_b$. See Table 8-5b.

For f_y greater than 60 ksi or concrete strengths less than 3,000 psi, the required development length should be increased.

Table 8-5. Minimum Development Lengths l_d, In.*

a. TENSION DEVELOPMENT IN NORMAL-WEIGHT CONCRETE†

Bar size no.	$f_c' = 3,000$ psi		$f_c' = 3,750$ psi		$f_c' = 4,000$ psi		$f_c' = 5,000$ psi		$f_c' = 6,000$ psi	
	Top‡ bars	Other bars	Top‡ bars	Other bars	Top‡ bars	Other bars	Top‡ bars	Other bars	Top‡ bars	Other bars
3	13	12	13	12	13	12	13	12	13	12
4	17	12	17	12	17	12	17	12	17	12
5	21	15	21	15	21	15	21	15	21	15
6	27	19	25	18	25	18	25	18	25	18
7	37	26	33	24	32	23	29	21	29	21
8	48	35	43	31	42	30	38	27	34	25
9	61	44	55	39	53	38	48	34	43	31
10	78	56	70	50	67	48	60	43	55	39
11	96	68	86	61	83	59	74	53	68	48
14	130	93	117	83	113	81	101	72	92	66
18	169	120	151	108	146	104	131	93	119	85

b. COMPRESSION DEVELOPMENT IN NORMAL-WEIGHT CONCRETE§

Bar size no.	f_c' (Normal-weight concrete)			
	3,000 psi	3,750 psi	4,000 psi	Over 4,444 psi¶
3	8	8	8	8
4	11	10	10	9
5	14	12	12	11
6	17	15	14	14
7	19	17	17	16
8	22	20	19	18
9	25	22	22	20
10	28	25	24	23
11	31	28	27	25
14	37	33	32	31
18	50	44	43	41

*Courtesy Concrete Reinforcing Steel Institute.
†1. For bars enclosed in standard column spirals, use $0.75 l_d$.
 2. For bars, such as usual temperature bars, spaced 6 in. or more, use $0.8\ l_d$.
 3. Longer embedments for lightweight concrete are generally required, depending on the tensile splitting strength f_{ct}.
 4. Standard 90 or 180° end hooks may be used to replace part of the required embedment.
‡Horizontal bars with more than 12 in. of concrete below.
§For embedments enclosed by spirals use 0.75 length given in a, above but not less than 8 in.
¶For $f_c' > 4,444$ psi, minimum embedment = $18 d_b$.

Bar Lap Splices. Because of the difficulty of transporting very long bars, reinforcement cannot always be continuous. When splices are necessary, it is advisable that they should be made where the tensile stress is less than half the permissible stress.

Bars up to No. 11 in size may be spliced by overlapping them and wiring them together.

Bars spliced by noncontact lap splices in flexural members should not be spaced transversely farther apart than one-fifth the required lap length or 6 in.

Welded Splices. These other positive connections should be used for bars larger than No. 11 and are an acceptable alternative for smaller bars. Welding should conform to AWS D12.1, Reinforcing Steel Welding Code, American Welding Society, 2501 N.W. 7th St., Miami, Fla. 33125. Bars should be butted and welded so that the splice develops in tension at least 125% of their specified yield strength. Other positive connections should be equivalent in strength.

Tension Lap Splices. The length of lap for bars in tension should conform to the following, with l_d taken as the tensile development length for the full yield strength f_y of the reinforcing steel [Eq. (8-4)]:

Class A splices (lap of l_d) are permitted where the maximum computed stress in the bar is always less than $0.5f_y$ and no more than three-fourths of the bars are lap spliced within a required lap length.

Class B splices (lap of $1.3l_d$) are required where either

1. The maximum computed stress in the bar is always less than $0.5f_y$ and more than three-fourths of the bars are lap spliced within a required lap length, or

2. Splices are used in regions of maximum moment and no more than one-half of the bars are spliced within a required lap length.

Class C splices (lap of $1.7l_d$) are required in regions of maximum moment where more than one-half of the bars are lap spliced within a lap length.

Class D splices (lap of $2l_d$) are required for tension tie members.

Where feasible, splices in tension tie members and splices in regions of high stress should be staggered and made with full welded connections.

Compression Lap Splices. For a bar in compression, the minimum length of a lap splice should be the largest of the development length l_d given by Eq. (8-5), or 12 in., or $0.0005f_y d_b$, for steel yield strength f_y of 60 ksi or less, where d_b is the bar diameter.

For tied compression members where the ties have an area, sq in., of at least $0.0015hs$ in the vicinity of the lap, the lap length may be reduced to 83% of the preceding requirements but not to less than 12 in. (h is the overall thickness of the member, in., and s is the tie spacing, in.).

For spirally reinforced compression members, the lap length may be reduced to 75% of the basic required lap but not to less than 12 in.

In columns where reinforcing bars are offset and one bar of a splice has to be bent to lap and contact the other one, the slope of the bent bar should not exceed 1 in 6. Portions of the bent bar above and below the offset should be parallel to the column axis. The design should account for a horizontal thrust at the bend taken equal to at least 1.5 times the horizontal component of the nominal stress in the inclined part of the bar. This thrust should be resisted by steel ties, or spirals, or members framing into the column. This resistance should be provided within a distance of 6 in. of the point of the bend.

Where column faces are offset 3 in. or more, vertical bars must be lapped by separate dowels.

In columns, a minimum tensile strength at each face equal to one-fourth the area of vertical reinforcement multiplied by f_y should be provided at horizontal cross sections where splices are located.

Welded-Wire Fabric. Wire reinforcing normally is spliced by lapping. When the tensile stress at the splice is less than half the permissible stress, the overlap measured between outermost cross wires should be at least 2 in. Where stresses are higher, the overlap should equal the spacing of the cross wires plus 2 in.

Slab Reinforcement. Structural floor and roof slabs with principal reinforcement in only one direction should be reinforced for shrinkage and temperature stresses in a perpendicular direction. The crossbars may be spaced at a maximum of 18 in. or five times the slab thickness. The ratio of reinforcement area of these bars to gross concrete area should be at least 0.0025 for plain bars, 0.0020 for deformed bars with less than 60,000 psi yield strength, and 0.0018 for deformed bars with 60,000 psi yield strength and welded-wire fabric with welded intersections in the direction of stress not more than 12 in. apart.

Concrete Cover. To protect reinforcement against fire and corrosion, thickness of concrete cover over the outermost steel should be at least that given in Table 8-6.

Table 8-6. Cast-in-Place Concrete Cover for Steel Reinforcement (Non-prestressed)

1. Concrete deposited against the ground, 3 in.
2. Concrete exposed to seawater, 4 in.; except precast-concrete piles, 3 in.
3. Concrete exposed to the weather or in contact with the ground after form removal, 2 in. for bars larger than No. 5 and 1½ in. for No. 5 or smaller.
4. Unexposed concrete slabs, walls, or joists, ¾ in. for No. 11 and smaller, 1½ in. for No. 14 and No. 18 bars. Beams, girders, and columns, 1½ in. Shells and folded-plate members, ¾ in. for bars larger than No. 5, and ½ inch for No. 5 and smaller.

(ACI 318, Building Code Requirements for Reinforced Concrete, American Concrete Institute, Detroit, Mich.; Standard Specifications for Highway Bridges, American Association of State Highway and Transportation Officials, National Press Building, Washington, D.C.)

8-16. Tendons.   High-strength steel is required for prestressing concrete to make the stress loss due to creep and shrinkage of concrete and to other factors a small percentage of the applied stress. This type of loss does not increase as fast as increase in stress in the prestressing steel, or tendons.

Tendons should have specific characteristics in addition to high strength to meet the requirements of prestressed concrete. They should elongate uniformly up to initial tension for accuracy in applying the prestressing force. After the yield strength has been reached, the steel should continue to stretch as stress increases, before failure occurs. American Society for Testing and Materials specifications for prestressing wire and strands, A421 and A416, set the yield strength at 80 to 85% of the tensile strength. Furthermore, the tendons should exhibit little or no creep, or relaxation, at the high stresses used.

ASTM A421 covers two types of uncoated, stress-relieved, high-carbon-steel wire commonly used for linear prestressed-concrete construction. Type BA wire is used for applications in which cold-end deformation is used for end anchorages, such as buttonheads. Type WA wire is intended for end anchorages by wedges and where no cold-end deformation of the wire is involved. The wire is required to be stress-relieved by a continuous-strand heat treatment after it has been cold-drawn to size. Type BA usually is furnished 0.196 and 0.250 in. in diameter, with an ultimate strength of 240,000 psi and yield strength (at 1% extension) of 192,000 psi. Type WA is available in two sizes and also 0.192 and 0.276 in. in diameter, with ultimate strengths ranging from 250,000 for the smaller diameters to 235,000 psi for the largest. Yield strengths range from 200,000 for the smallest to 188,000 psi for the largest (Table 8-7).

For pretensioning, where the steel is tensioned before the concrete is cast, wires usually are used individually, as is common for reinforced concrete. For posttensioning, where the tendons are tensioned and anchored to the concrete after it has attained sufficient strength, the wires generally are placed parallel to each other in groups, or cables, sheathed or ducted to prevent bond with the concrete.

A seven-wire strand consists of a straight center wire and six wires of slightly smaller diameter winding helically around and gripping it. High friction between the center and outer wires is important where stress is transferred between the strand and concrete through bond. ASTM A416 covers strand with ultimate strengths of 250,000 and 270,000 psi (Table 8-7).

Galvanized strands sometimes are used for posttensioning, particularly when the tendons may not be embedded in grout. Sizes normally available range from a 0.600-in.-diameter seven-wire strand, with 46,000-lb breaking strength, to 1¹¹/₁₆-in.-diameter strand, with 352,000-lb breaking strength. The cold-drawn wire comprising the strand is stress-relieved when galvanized, and stresses due to stranding are offset by prestretching the strand to about 70% of its ultimate strength.

Hot-rolled alloy-steel bars used for prestressing concrete generally are not so strong as wire or strands. The bars usually are stress-relieved, then cold-stretched to at least 90% of ultimate strength to raise the yield point. The cold stretching also serves as proof stressing, eliminating bars with defects.

(H. K. Preston and N. J. Sollenberger, "Modern Prestressed Concrete," McGraw-Hill Book Company, New York; J. R. Libby, "Modern Prestressed Concrete," Van Nostrand Reinhold Company, New York.)

Table 8-7. Properties of Tendons

Diameter, in.	Area, sq in.	Weight per 1,000 ft, lb	Ultimate strength
	UNCOATED TYPE WA WIRE		
0.276	0.05983	203.2	235,000 psi
0.250	0.04909	166.7	240,000 psi
0.196	0.03017	102.5	250,000 psi
0.192	0.02895	98.3	250,000 psi
	UNCOATED TYPE BA WIRE		
0.250	0.04909	166.7	240,000 psi
0.196	0.03017	102.5	240,000 psi
	UNCOATED SEVEN-WIRE STRANDS 250 GRADE		
$\frac{1}{4}$	0.04	122	9,000 lb
$\frac{5}{16}$	0.058	197	14,500 lb
$\frac{3}{8}$	0.080	272	20,000 lb
$\frac{7}{16}$	0.108	367	27,000 lb
$\frac{1}{2}$	0.144	490	36,000 lb
	270 GRADE		
$\frac{3}{8}$	0.085	290	23,000 lb
$\frac{7}{16}$	0.115	390	31,000 lb
$\frac{1}{2}$	0.153	520	41,300 lb

8-17. Fabrication of Prestressed-Concrete Members.   Prestressed concrete may be produced much like high-strength reinforced concrete, either cast in place or precast.   Prestressing offers several advantages for precast members, which have to be transported from casting bed to final position and handled several times.   Prestressed members are lighter than reinforced members of the same capacity, both because higher-strength concrete generally is used and because the full cross section is effective.   In addition, the prestressing normally counteracts handling stresses.   And if a prestressed member survives the full prestress and handling, the probability of its failing under service loads usually is very small.

Two general methods of prestressing are in common use, pretensioning and posttensioning, and both may be used for the same member.   Pretensioning, where the tendons are tensioned before embedment in the concrete and stress transfer from steel to concrete usually is by bond, is especially useful for mass production of precast elements.   Often, elements may be fabricated in long lines, by stretching the tendons (Art. 8-16) between abutments at the ends of the lines. By use of tiedowns and struts, the tendons may be draped in a vertical plane to develop upward and downward components on release.   After the tendons have been jacked to their full stress, they are anchored to the abutments.   The casting bed over which the tendons are stretched usually is made of a smooth-surface concrete slab with easily stripped side forms of steel.   (Forms for pretensioned members must permit them to move on release of the tendons.)   Separators are placed in the forms to divide the long line into members of required length and provide space for cutting the tendons.   After the concrete has been cast and has attained its specified strength, generally after a preset period and steam curing, side forms are removed.   Then, the tendons are detached from the anchorages at the ends of the line and relieved of their stress. Restrained from shortening by bond with the concrete, the tendons compress it.   At this time, it is safe to cut the tendons between the members and remove the members from the forms.

Posttensioning frequently is used for cast-in-place members and long-span flexural members. Cables or bars (Art. 8-16) are placed in the forms in flexible ducts to prevent bond with the concrete.   They may be draped in a vertical plane to develop upward and downward forces when tensioned. After the concrete has been placed and has attained sufficient strength, the tendons are tensioned by jacking against the member, then are anchored to it. Grout may be pumped into the duct to establish bond with the concrete and protect the tendons against corrosion. Applied at pressures of 75 to 100 psi, a typical grout consists of 1 part portland cement, 0.75 parts sand (capable of passing through a No. 30 sieve), and 0.75 parts water, by volume.

Concrete with higher strengths than ordinarily used for reinforced concrete offers economic advantages for prestressed concrete. In reinforced concrete, much of the concrete in a slab or beam is assumed to be ineffective because it is in tension and likely to crack under service loads. In prestressed concrete, the full section is effective because it is always under either compression or very low tension. Furthermore, high-strength concrete develops higher bond stresses with the tendons, greater bearing strength to withstand the pressure of anchorages, and a higher modulus of elasticity. The last indicates reductions in initial strain and camber when prestress is applied initially and in creep strain. The reduction in creep strain reduces the loss of prestress with time. Generally, concrete with a 28-day strength of 5,000 psi or more is advantageous for prestressed concrete.

Concrete cover over prestressing steel, ducts, and nonprestressed steel should be at least 3 in. for concrete surfaces in contact with the ground; 1½ in. for prestressing steel and main reinforcing bars, and 1 in. for stirrups and ties in beams and girders; 1 in. in slabs and joists exposed to the weather; and ¾ in. for unexposed slabs and joists. In extremely corrosive atmospheres or other severe exposures, the amount of protective cover should be increased.

Minimum clear spacing between pretensioning steel at the ends of a member should be four times the diameter of individual wires and three times the diameter of strands. Some codes also require that the spacing be at least 1⅓ times the maximum size of aggregate. (See also Art. 8-15.) Away from the ends of a member, prestressing steel or ducts may be bundled. Concentrations of steel or ducts, however, should be reinforced to control cracking.

Prestressing force may be determined by measuring tendon elongation, or by checking jack pressure on a recently calibrated gage, or by the use of a recently calibrated dynamometer. If several wires or strands are stretched simultaneously, the method used should be such as to induce approximately equal stress in each.

Splices should not be used in parallel-wire cables, especially if a splice has to be made by welding, which would weaken the wire. Failure is likely to occur during tensioning.

Strands may be spliced, if necessary, when the coupling will develop the full strength of the tendon, not cause it to fail under fatigue loading, and does not replace sufficient concrete to weaken the member. Individual wires in a strand may be butt-welded during fabrication, but there should not be more than one such joint in any 150-ft length of strand. Large strands are never spliced because they can be supplied sufficiently long without a splice, though individual wires may be welded.

High-strength bars are generally spliced mechanically. The couplers should be capable of developing the full strength of the bars without decreasing resistance to fatigue and without replacing an excessive amount of concrete.

Anchor fittings are different for pretensioned and posttensioned members. For pretensioned members, the fittings hold the tendons temporarily against anchors outside the members, and therefore can be reused. In posttensioning, the fittings usually anchor the tendons permanently to the members.

In pretensioning, the tendons may be tensioned one at a time to permit the use of relatively light jacks, in groups, or all simultaneously. A typical stressing arrangement consists of a stationary anchor post, against which jacks act, and a moving crosshead, which is pushed by the jacks and to which the tendons are attached. Usually, the tendons are anchored to a thick steel plate that serves as a combination anchor plate and template. It has holes through which the tendons pass to place them in the desired pattern. Various patented grips are available for anchoring the tendons to the plate. Generally, they are a wedge or chuck type capable of developing the full strength of the tendons.

A variety of patented fittings also are available for anchoring in posttensioned members. Such fittings should be capable of developing the full strength of the tendons under static and fatigue loadings. The fittings also should spread the prestressing force over the concrete or transmit it to a bearing plate. Sufficient space must be provided for the fittings in the anchor zone.

Generally, all the wires of a parallel-wire cable are anchored with a single fitting (Figs. 8-5 and 8-6). The type shown in Fig. 8-6 requires that the wires be cut to exact length and a button-head be cold-formed on the ends for anchoring.

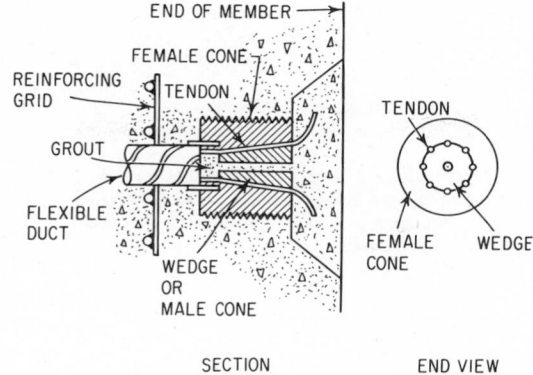

Fig. 8-5. Conical-wedge anchorage for prestressing wires.

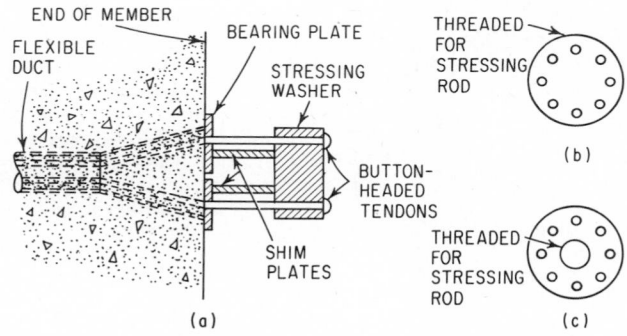

Fig. 8-6. End anchorage for button-headed wires. Stressing head for tensioning wires may be threaded externally (*b*) or internally (*c*) for attachment to jack.

The wedge type in Fig. 8-5 requires a double-acting jack. One piston, with the wires wedged to it, stresses them, and a second piston forces the male cone into the female cone to grip the tendons. Normally, a hole is provided in the male cone for grouting the wires. After final stress is applied, the anchorage may be embedded in concrete to prevent corrosion and improve appearance.

With the buttonhead type, a stressing rod may be screwed over threads on the circumference of a thick, steel stressing washer (Fig. 8-6*b*) or into a center hole in the washer (Fig. 8-6*c*). The

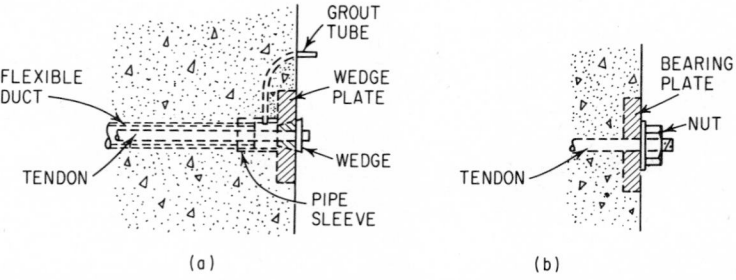

Fig. 8-7. End anchorages for bars. (*a*) Conical wedge. (*b*) Nut and washer at threaded end, acting against bearing plate.

rod then is bolted to a jack.   When the tendons have been stressed, the washer is held in position by steel shims inserted between it and a bearing plate embedded in the member.   The jack pressure then can be released and the jack and stressing rod removed.   Finally, the anchorage is embedded in concrete.

Posttensioning bars may be anchored individually with steel wedges (Fig. 8-7a) or by tightening a nut against a bearing plate (Fig. 8-7b).   The former has the advantage that the bars do not have to be threaded.

Posttensioning strands normally are shop-fabricated in complete assemblies, cut to length, anchor fittings attached, and sheathed in flexible duct.   Swaged to the strands, the anchor fittings have a threaded stud projecting from the end.   The threaded stud is used for jacking the stress into the strand and for anchoring by tightening a nut against a bearing plate in the member (Fig. 8-8).

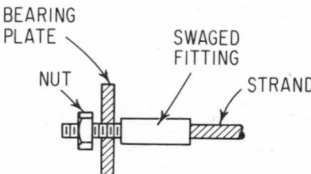

BEARING
PLATE

SWAGED
FITTING

NUT

STRAND

Fig. 8-8. Swaged fitting for anchoring strands.   Prestress is maintained by tightening the nut against the bearing plate.

To avoid overstressing and failure in the anchorage zone, the anchorage assembly must be placed with care.   Bearing plates should be placed perpendicular to the tendons to prevent eccentric loading.   Jacks should be centered for the same reason and so as not to scrape the tendons against the plates.   The entire area of the plates should bear against the concrete.

Prestress normally is applied with hydraulic jacks.   The amount of prestressing force is determined by measuring tendon elongation and comparing with an average load-elongation curve for the steel used.   In addition, the force thus determined should be checked against the jack pressure registered on a recently calibrated gage or by use of a recently calibrated dynamometer. Discrepancies of less than 5% may be ignored.

When prestressed-concrete beams do not have a solid rectangular cross section in the anchorage zone, an enlarged end section, called an end block, may be necessary to transmit the prestress from the tendons to the full concrete cross section a short distance from the anchor zone. End blocks also are desirable for transmitting vertical and lateral forces to supports and to provide adequate space for the anchor fittings.

The transition from end block to main cross section should be gradual (Fig. 8-9).   Length of end block, from beginning of anchorage area to the start of the main cross section, should be at least 24 in.   The length normally ranges from three-fourths the depth of the member for deep beams to the full depth for shallow beams.   The end block should be reinforced vertically and horizontally to resist tensile bursting and spalling forces induced by the concentrated loads of the tendons.   In particular, a grid of reinforcing should be placed directly behind the anchorages to resist spalling.

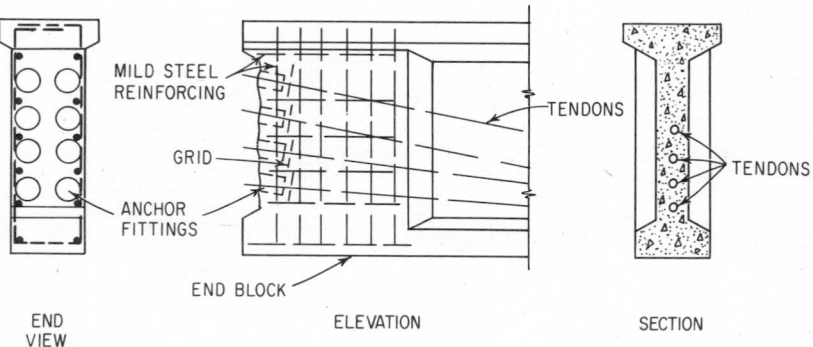

MILD STEEL
REINFORCING

GRID

ANCHOR
FITTINGS

TENDONS

TENDONS

END BLOCK

END
VIEW

ELEVATION

SECTION

Fig. 8-9. Transition from end block of a prestressed-concrete beam to main cross section.

Ends of pretensioned beams should be reinforced with vertical stirrups over a distance equal to one-fourth the beam depth. The stirrups should be capable of resisting in tension a force equal to at least 4% of the prestressing force.

Control of camber is important for prestressed members. Camber tends to increase with time because of creep. If a prestressed beam or slab has an upward camber under prestress and long-time loading, the camber will tend to increase upward. Excessive camber should be avoided, and for deck-type structures, such as highway bridges and building floors and roofs, the camber of all beams and girders of the same span should be the same.

Computation of camber with great accuracy is difficult, mainly because of the difficulty of ascertaining with accuracy the modulus of elasticity of the concrete, which varies with time. Other difficult-to-evaluate factors also influence camber: departure of the actual prestressing force from that calculated, effects of long-time loading, influence of length of time between prestressing and application of full service loads, methods of supporting members after removal from the forms, and influence of composite construction.

When camber is excessive, it may be necessary to use concrete with higher strength and modulus of elasticity (change from lightweight to ordinary concrete); or increase the moment of inertia of the section; or use partial prestressing, that is, decrease the prestressing force and add reinforcing steel to resist the tensile stresses; or use a larger prestressing force with less eccentricity.

To insure uniformity of camber, a combination of pretensioning and posttensioning is desirable for precast members. Sufficient prestress may be applied initially to permit removal of the member from the forms and transportation to a storage yard. After the member has increased in strength but before erection, additional prestress is applied by posttensioning to bring the camber to the desired value. During storage, the member should be supported in the same manner as it will be in the structure.

For handling prestressed precast members, inserts usually are embedded in the concrete. Stresses should be computed for use of these inserts as pickup points, and the member should not be handled by pickup at other points. Nor should it be supported upside down, sideways, or at points that would create greater stresses than it will sustain under service loads.

(H. K. Preston and N. J. Sollenberger, "Modern Prestressed Concrete," McGraw-Hill Book Company, New York; J. R. Libby, "Modern Prestressed Concrete," Van Nostrand Reinhold Company, New York.)

8-18. Precast Concrete. When concrete products are made in other than their final position, they are considered precast. They may be unreinforced, reinforced, or prestressed. They include in their number a wide range of products: block, brick, pipe, plank, slabs, conduit, joists, beams and girders, trusses and truss components, curbs, lintels, sills, piles, pile caps, and walls.

Precasting often is chosen because it permits efficient mass production of concrete units. With precasting, it usually is easier to maintain quality control and produce higher-strength concrete than with field concreting. Formwork is simpler, and a good deal of falsework can be eliminated. Also, since precasting normally is done at ground level, workers can move about more freely. But sometimes these advantages are more than offset by the cost of handling, transporting, and erecting the precast units. Also, joints may be troublesome and costly.

Design of precast products follows the same rules, in general, as for cast-in-place units. However, ACI 318, Building Code Requirements for Reinforced Concrete (American Concrete Institute, Detroit, Mich.), permits the concrete cover over reinforcing steel to be as low as ⅜ in. for slabs, walls, or joists not exposed to weather. (See also ACI 512, Recommended Practice for Manufactured Reinforced Concrete Floor and Roof Units.) Also, ACI Standard 525, Minimum Requirements for Thin-Section Precast Concrete Construction, permits the cover for units not exposed to weather to be only ⅜ in. Furthermore, for surfaces exposed to the weather, or in contact with the ground or water, cover for main reinforcement in beams, girders, and columns need only be ½ in.; and reinforcement in slabs and secondary reinforcement in beams, girders, and columns need have only ⅜ in. of cover. Hence, this standard permits units to be as thin as 1 in., reinforced with welded-wire fabric. But dense, watertight concrete must be used. For standard concrete, minimum cover should be that given in Table 8-6.

Concrete for precast elements not exposed to weather or in contact with the ground should have a minimum 28-day strength of 4,000 psi. Exposed concrete should have a 5,000-psi strength. Aggregate is restricted to a maximum of ¾ in. or two-thirds the minimum clear distance between parallel reinforcing bars. In thin elements, spacing of wires in welded-wire fabric may not exceed 2 in.

Precast units must be designed for handling and erection stresses, which may be more severe than those they will be subjected to in service. Normally, inserts are embedded in the concrete for picking up the units. They should be picked up by these inserts, and when set down, they should be supported right side up, in such a manner as not to induce stresses higher than the units would have to resist in service. (See also Art. 8-17.)

For precast beams, girders, joists, columns, slabs, and walls, joints usually are made with cast-in-place concrete. Often, in addition, steel reinforcing projecting from the units to be joined is welded together. ("Suggested Design of Joints and Connections in Precast Structural Concrete," Report by ACI-ASCE Committee 512, American Concrete Institute, Detroit, Mich.)

8-19. Lift-slab Construction. A type of precasting used in building construction involves the casting of floor and roof slabs at or near ground level and lifting them to their final position; hence the name lift-slab construction. It offers many of the advantages of precasting (Art. 8-18) and eliminates many of the storing, handling, and transporting disadvantages. It normally requires fewer joints than other types of precast building systems.

Typically, columns are erected first, but not necessarily for the full height of the building. Near the base of the columns, floor slabs are cast in succession, one atop another, with a parting compound between them to prevent bond. The roof slab is cast last, on top. Usually, the construction is flat plate, and the slabs have uniform thickness; but waffle slabs or other types also can be used. Openings are left around the columns, and a steel collar is slid down each column for embedment in every slab. The collar is used for lifting the slab, connecting it to the column, and reinforcing the slab against shear.

To raise the slabs, jacks are set atop the columns. They turn threaded rods that pass through the collars and do the lifting. As each slab reaches its final position, it is wedged in place and the collars are welded to the columns.

8-20. Ultimate-Strength Theory for Reinforced Concrete Beams. For consistent, safe, economical design of beams, their actual load-carrying capacity should be known. The safe load then can be determined by dividing this capacity by a safety factor. Or the design load can be multiplied by the safety factor to indicate what the capacity of the beams should be. It should be noted, however, that under design loads, stresses and deflections may be computed with good approximation on the assumptions of a linear stress-strain diagram and a cracked cross section.

ACI 318, Building Code Requirements for Reinforced Concrete (American Concrete Institute, Detroit, Mich.), provides for design by ultimate-strength theory. Bending moments in members are determined as if the structure were elastic. Ultimate-strength theory is used to design critical sections, those with the largest bending moments. The ultimate strength of each section is computed, and the section is designed for this capacity with an appropriate safety factor.

Stress Redistribution. The ACI Code recognizes that, below ultimate load, a redistribution of stress occurs in continuous beams, frames, and arches, which allows the structure to carry loads higher than those indicated by elastic analysis. The code permits an increase or decrease of up to 10% in the negative moments calculated by elastic theory at the supports of continuous flexural members. But these modified moments must also be used for determining the mo-

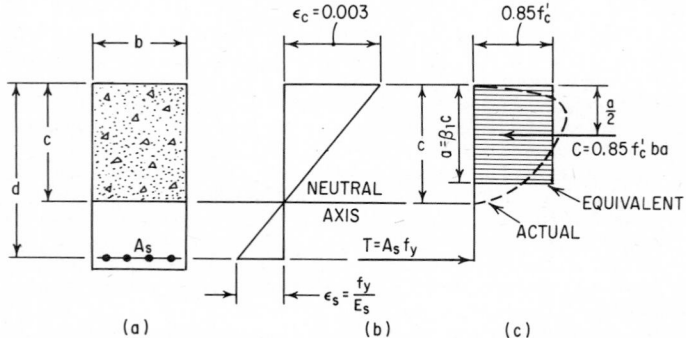

Fig. 8-10. Stresses and strains on a reinforced concrete section (*a*) at ultimate load. (*b*) Strain diagram. (*c*) Actual and assumed equivalent-stress diagrams.

ments at other sections for the same loading conditions. [The modifications, however, are permissible only for relatively small steel ratios at each support. The steel ratios ρ or $\rho - \rho'$ (see Arts. 8-23, 8-24, and 8-28 to 8-31) should be less than half ρ_b, the steel ratio for balanced conditions (concrete strength equal to steel strength) at ultimate load.] For example, suppose elastic analysis of a continuous beam indicates a maximum negative moment at a support of $wL^2/12$ and a maximum positive moment at midspan of $wL^2/8 - wL^2/12$, or $wL^2/24$. Then, the code permits the negative moment to be decreased to $0.9wL^2/12$, if the positive moment is increased to $wL^2/8 - 0.9wL^2/12$, or $1.2wL^2/24$.

Design Assumptions. Ultimate strength of any section of a reinforced concrete beam may be computed assuming the following:

Strain in the concrete is directly proportional to the distance from the neutral axis (Fig. 8-10b).

Except in anchorage zones, strain in reinforcing steel equals strain in adjoining concrete.

At ultimate strength, maximum strain at the extreme compression surface equals 0.003 in. per in.

When the reinforcing steel is not stressed to its yield strength f_y, the steel stress is 29,000,000 psi times the steel strain, in. per in. After the yield strength has been reached, the stress remains constant at f_y, though the strain increases.

Tensile strength of the concrete is negligible.

At ultimate strength, concrete stress is not proportional to strain. The actual stress distribution may be represented by an equivalent rectangle that yields ultimate strengths in agreement with numerous, comprehensive tests (Fig. 8-10c).

The ACI Code recommends that the compressive stress for the equivalent rectangle be taken as $0.85f'_c$, where f'_c is the 28-day compressive strength of the concrete. The stress is assumed constant from the surface of maximum compressive strain over a depth $a = \beta_1 c$, where c is the distance to the neutral axis (Fig. 8-10c). For $f'_c \leq 4,000$ psi, $\beta_1 = 0.85$; for greater concrete strengths, β_1 is reduced 0.05 for each 1,000 psi in excess of 4,000.

Formulas in the ACI Code based on these assumptions usually contain a factor ϕ to provide for the possibility that small adverse variations in materials, workmanship, and dimensions, while individually within acceptable tolerances, occasionally may combine, and actual capacity may be less than that computed. The coefficient ϕ may be taken as 0.90 for flexure, 0.85 for shear and torsion, 0.75 for spirally reinforced compression members, and 0.70 for tied compression members. Under certain conditions of load (as the value of the axial load approaches zero) and geometry, the ϕ value for compression members may increase linearly to a maximum value of 0.90.

Crack Control. Because of the risk of large cracks opening up when reinforcement is subjected to high stresses, the ACI Code recommends that designs be based on a steel yield strength f_y no larger than 80 ksi. And when design is based on a yield strength f_y greater than 40 ksi, the cross sections of maximum positive and negative moment shall be proportioned for crack control so that specific limits are satisfied by

$$z = f_s \sqrt[3]{d_c A} \tag{8-6}$$

where f_s = calculated stress, ksi, in reinforcement at service loads

d_c = thickness of concrete cover, in., measured from the extreme tension surface to the center of the bar closest to that surface

A = effective tension area of concrete, sq in. per bar. This area should be taken as that surrounding the main tension reinforcement and having the same centroid as that reinforcement multiplied by the ratio of the area of the largest bar used to the total area of the tension reinforcement

These limits are $z \leq 175$ kips per in. for interior exposures and $z \leq 145$ kips per in. for exterior exposures. These correspond to limiting crack widths of 0.016 and 0.013 in., respectively, at the extreme tension edge under service loads. In Eq. (8-6), f_s should be computed by dividing the bending moment by the product of the steel area and the internal moment arm, but f_s may be taken as 60% of the steel yield strength without computation.

Required Strength. For combinations of loads, the ACI Code requires that a structure and its members should have the following ultimate strengths (capacity to resist design loads and their related internal moments and forces):

Wind and earthquake loads not applied:

$$U = 1.4D + 1.7L \tag{8-7}$$

where D = effect of basic load consisting of dead load plus volume change (shrinkage, temperature)

L = effect of live load plus impact

When wind loads are applied, the largest of Eq. (8-7) and Eq. (8-8a and b) determines the
required strength.

$$U = 0.75 \, (1.4D + 1.7L + 1.7W) \qquad (8\text{-}8a)$$

$$U = 0.9D + 1.3W \qquad (8\text{-}8b)$$

where W = effect of wind load.

If the structure will be subjected to earthquake forces E, substitute $1.1E$ for W in Eq. (8-8).

Where the effects of differential settlement, creep, shrinkage, or temperature change may be
critical to the structure, they should be included with the dead load D, and the strength should
be at least equal to

$$U = 0.75 \, (1.4D + 1.7L) \qquad (8\text{-}9)$$

For ultimate-strength loads (load-factor method) for bridges, see Art. 17-4.

Though structures may be designed by ultimate-strength theory, it is not anticipated that
service loads will be substantially exceeded. Hence, deflections that will be of concern to the
designer are those that occur under service loads. These deflections may be computed by
working-stress theory. (See Art. 8-21.)

Deep Members. Due to the nonlinearity of strain distribution and the possibility of lateral
buckling, deep flexural members must be given special consideration. The ACI Code considers
members with overall depth-to-span ratios greater than $^2/_5$ for continuous spans ($^4/_5$ for simple
spans) as deep members. The ACI Code provides special shear design requirements and
minimum requirements for both horizontal and vertical reinforcement for such members. See
Sec. 10-7, ACI 318-71.

(G. Winter and A. H. Nilson, "Design of Concrete Structures," McGraw-Hill Book Company,
New York; P. F. Rice and E. S. Hoffman, "Structural Design Guide to the ACI Building Code,"
Van Nostrand Reinhold Company, New York.)

8-21. Working-Stress Theory for Reinforced Concrete Beams. Stress distribution in a
reinforced-concrete beam under service loads is different from that at ultimate strength (Art.
8-20). Knowledge of this stress distribution is desirable for many reasons, including the requirements of some design codes that specified working stresses in steel and concrete not be
exceeded.

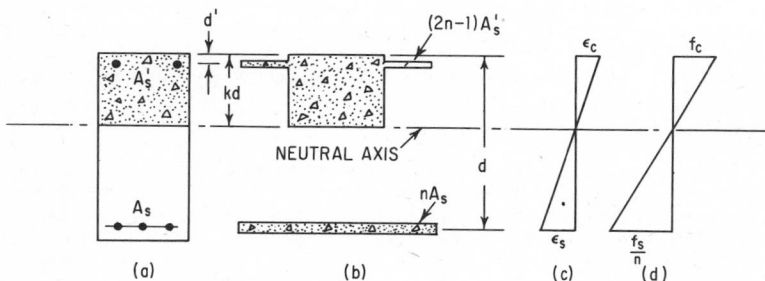

Fig. 8-11. Typical cross section (a) of a reinforced concrete beam may be treated in design as an all-concrete
(transformed) section (b). Strains (c) and stresses (d) are assumed to vary linearly.

Working stresses in reinforced-concrete beams are computed from the following assumptions:
Longitudinal stresses and strains vary with distance from the neutral axis (Fig. 3-11c and d).
The concrete develops no tensile stress.

Except in anchorage zones, strain in reinforcing steel equals strain in adjoining concrete.
But because of creep, strain in compressive steel in beams may be taken as half that in the
adjoining concrete.

The modular ratio $n = E_s/E_c$ is constant. E_s is the modulus of elasticity of the reinforcing
steel, and E_c of the concrete.

The following allowable stresses may be used for flexure:

	Buildings	Bridges
Compression in extreme compression surface	$0.45f'_c$	$0.4f'_c$
Tension in reinforcement		
Grade 40 or 50 steel	.20 ksi	20 ksi
Grade 60 or higher yield strength	.24 ksi	24 ksi

where f'_c is the 28-day compressive strength of the concrete.

For other than such flexural stresses, allowable or maximum stresses to be used in design are stated as a percentage of the values given for ultimate-strength design. For example, for service loads:

Type of Member and Stress	Allowable stresses or Capacity, % of Ultimate
Compression members, walls	.40
Shear or tension in beams, joists, walls, one-way slabs	.55
Shear or tension in two-way slabs, footings	.50
Bearing in concrete	.35

Allowable stresses may be increased one-third when wind or earthquake forces are combined with other loads, but the capacity of the resulting section should not be less than that required for dead plus live loads.

Other equivalency factors are also given in terms of ultimate strength values. Thus, the predominant design procedure is the ultimate-strength method, but for reasons of background and historical significance and also because the working-stress design method is sometimes preferred for bridges and certain foundation and retaining-wall design, examples of working-stress design procedure are presented in Arts. 8-24, 8-29, and 8-31.

Transformed Section. According to working-stress theory for reinforced-concrete beams (Art. 8-21), strains in reinforcing steel and adjoining concrete are equal. Hence f_s, the stress in the steel, is n times f_c, the stress in the concrete, where n is the ratio of modulus of elasticity of the steel E_s to that of the concrete E_c. The total force acting on the steel then equals $(nA_s)f_c$. This indicates that the steel area can be replaced in stress calculations by a concrete area n times as large.

The transformed section of a concrete beam is one in which the reinforcing has been replaced by an equivalent area of concrete (Fig. 8-11b). (In doubly reinforced beams and slabs, an effective modular ratio of $2n$ should be used to transform the compression reinforcement, to account for the effects of creep and nonlinearity of the stress-strain diagram for concrete. But the computed stress should not exceed the allowable tensile stress.) Since stresses and strains are assumed to vary with distance from the neutral axis, conventional elastic theory for homogeneous beams holds for the transformed section. Section properties, such as location of neutral axis, moment of inertia, and section modulus S, can be computed in the usual way, and stresses can be found from the flexure formula $f = M/S$, where M is the bending moment.

(G. Winter and A. H. Nilson, "Design of Concrete Structures," McGraw-Hill Book Company, New York; P. Rice and E. S. Hoffman, "Structural Design Guide to the ACI Building Code," Van Nostrand Reinhold Company, New York.)

8-22. Deflection Computations and Criteria. The assumptions of working-stress theory (Art. 8-21) may also be used for computing deflections under service loads; that is, elastic-theory deflection formulas may be used for reinforced concrete beams. In these formulas, the effective moment of inertia I_e should be taken as not greater than I_g, the moment of inertia of the gross concrete section, or as given by Eq. (8-10).

$$I_e = \left(\frac{M_{cr}}{M_a}\right)^3 I_g + \left[1 - \left(\frac{M_{cr}}{M_a}\right)^3\right] I_{cr} \qquad (8\text{-}10)$$

where M_{cr} = cracking moment
M_a = moment for which deflection is being computed
I_{cr} = moment of inertia of cracked concrete section

If y_t is taken as the distance from the centroidal axis of the gross section, neglecting the reinforcement, to the extreme surface in tension, the cracking moment may be computed from

$$M_{cr} = \frac{f_r I_g}{y_t} \qquad (8\text{-}11)$$

with the modulus of rupture of the concrete $f_r = 7.5\sqrt{f'_c}$. Equation (8-10) takes into account the variation of the moment of inertia of a concrete section based on whether the section is cracked or uncracked. The modulus of elasticity of the concrete E_c may be computed from Eq. (8-2) in Art. 8-1.

The deflections thus calculated are those assumed to occur immediately on application of load. Additional long-time deflections can be estimated by multiplying the immediate deflection by 2 when there is no compression reinforcement or by $2 - 1.2A'_s/A_s \geq 0.6$, where A'_s is the area of compression reinforcement and A_s is the area of tension reinforcement.

Deflection Limitations. The ACI Code recommends the following limits on deflections in buildings:

For roofs not supporting plastered ceilings or not attached to nonstructural elements, maximum immediate deflection under live load should not exceed $L/180$, where L is the span of beam or slab.

For floors not supporting partitions or not attached to nonstructural elements, the maximum immediate deflection under live load should not exceed $L/360$.

For a floor or roof construction intended to support or to be attached to partitions or other construction likely to be damaged by large deflections of the support, the allowable limit for the sum of immediate deflection due to live loads and the additional deflection due to shrinkage and creep under all sustained loads should not exceed $L/480$. If the construction is not likely to be damaged by large deflections, the deflection limit may be increased to $L/240$. But tolerances should be established and adequate measures should be taken to prevent damage to supported or nonstructural elements resulting from the deflections of structural members.

8-23. Ultimate-Strength Design of Rectangular Beams with Tension Reinforcement Only. (See also Art. 8-20.) Generally, the area A_s of tension reinforcement (Fig. 8-10) in a reinforced-concrete beam is represented by the ratio $\rho = A_s/bd$, where b is the beam width and d the distance from extreme compression surface to the centroid of tension reinforcement. At ultimate strength, the steel at a critical section of the beam will be at its yield strength f_y if the concrete does not fail in compression first. Total tension in the steel then will be $A_s f_y = \rho f_y bd$. It will be opposed, according to Fig. 8-10, by an equal compressive force, $0.85f'_c ba = 0.85f'_c b\beta_1 c$, where f'_c is the 28-day strength of the concrete, a the depth of the equivalent rectangular stress distribution, c the distance from the extreme compression surface to the neutral axis, and β_1 a constant (see Art. 8-20). Equating the compression and tension at the critical section yields

$$c = \frac{\rho f_y}{0.85\beta_1 f'_c} d \qquad (8\text{-}12)$$

The criterion for compression failure is that the maximum strain in the concrete equals 0.003 in. per in. In that case

$$c = \frac{0.003}{f_s/E_s + 0.003} d \qquad (8\text{-}13)$$

where f_s = steel stress, psi
E_s = modulus of elasticity of the steel = 29,000,000 psi

Table 8-8 lists the nominal diameters, weights, and cross-sectional areas of standard steel reinforcing bars.

Balanced Reinforcing. Under balanced conditions, the concrete will reach its maximum strain of 0.003 when the steel reaches its yield strength f_y. Then, c as given by Eq. (8-12) will equal c as given by Eq. (8-13), since c determines the location of the neutral axis. This determines the steel ratio for balanced conditions:

$$\rho_b = \frac{0.85\beta_1 f'_c}{f_y} \frac{87,000}{87,000 + f_y} \qquad (8\text{-}14)$$

Maximum Reinforcing. All structures are designed to collapse not suddenly but by gradual deformation when overloaded. To achieve this end in concrete, the reinforcement should yield before the concrete crushes. This will occur if the quantity of tensile reinforcement is less than the critical percentage determined by ultimate-strength theory [Eq. (8-14)]. The ACI Code, to avoid compression failures, limits the steel ratio ρ to a maximum of $0.75\rho_b$. The Code also requires that ρ for positive-moment reinforcement be at least $200/f_y$.

Table 8-8. Areas of Groups of Standard Bars, Sq In.

Bar No.	Diam, in.	Weight, lb per ft	Number of bars								
			1	2	3	4	5	6	7	8	9
2	0.250	0.167	0.05	0.10	0.15	0.20	0.25	0.30	0.35	0.40	0.45
3	0.375	0.376	0.11	0.22	0.33	0.44	0.55	0.66	0.77	0.88	0.99
4	0.500	0.668	0.20	0.39	0.58	0.78	0.98	1.18	1.37	1.57	1.77
5	0.625	1.043	0.31	0.61	0.91	1.23	1.53	1.84	2.15	2.45	2.76
6	0.750	1.502	0.44	0.88	1.32	1.77	2.21	2.65	3.09	3.53	3.98
7	0.875	2.044	0.60	1.20	1.80	2.41	3.01	3.61	4.21	4.81	5.41
8	1.000	2.670	0.79	1.57	2.35	3.14	3.93	4.71	5.50	6.28	7.07
9	1.128	3.400	1.00	2.00	3.00	4.00	5.00	6.00	7.00	8.00	9.00
10	1.270	4.303	1.27	2.53	3.79	5.06	6.33	7.59	8.86	10.12	11.39
11	1.410	5.313	1.56	3.12	4.68	6.25	7.81	9.37	10.94	12.50	14.06
14	1.693	7.650	2.25	4.50	6.75	9.00	11.25	13.50	15.75	18.00	20.25
18	2.257	13.600	4.00	8.00	12.00	16.00	20.00	24.00	28.00	32.00	36.00

Moment Capacity.   For such underreinforced beams, the bending-moment capacity at ultimate strength is

$$M_u = 0.90[bd^2 f'_c \omega(1 - 0.59\omega)] = 0.90\left[A_s f_y\left(d - \frac{a}{2}\right)\right] \qquad (8\text{-}15)$$

where $\omega = \rho f_y/f'_c$
$a = A_s f_y/0.85 f'_c b$

Shear and Torsion Reinforcement.   The nominal ultimate design shear stress, as a measure of diagonal tension, under an ultimate design shear V_u should be computed from

$$v_u = \frac{V_u}{\phi b_w d} \qquad (8\text{-}16)$$

where ϕ = capacity reduction factor (0.85 for shear and torsion)
b_w = width of beam web

Except for brackets and other short cantilevers, the section for maximum shear may be taken at a distance equal to d from the face of the support.   The shear stress v_c carried by the concrete alone should not exceed $2\sqrt{f'_c}$.   (As an alternative, the maximum for v_c may be taken as

$$v_c = 1.9\sqrt{f'_c} + 2{,}500\rho_w\,\frac{V_u d}{M_u} \le 3.5\sqrt{f'_c} \qquad (8\text{-}17)$$

where V_u and M_u are the shear and bending moment, respectively, at the section considered, but M_u should not be less than $V_u d$.)

When v_u is larger than v_c, the excess shear will have to be resisted by web reinforcement.   In general, this reinforcement should be stirrups perpendicular to the axis of the member (Fig. 8-12) or welded-wire fabric with wires perpendicular to the axis of the member.   In members without prestressing, however, the stirrups may be inclined, as long as the angle is at least 45°

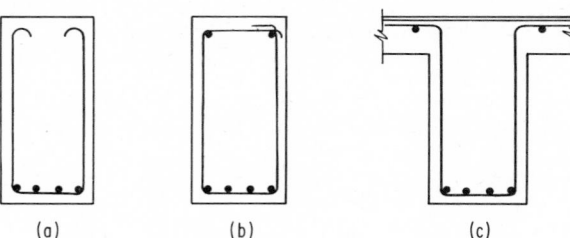

(a)   (b)   (c)

Fig. 8-12.   Typical stirrups in a concrete beam.

with the axis of the member.   As an alternative, longitudinal bars may be bent up at an angle of 30° or more with the axis, or spirals may be used.   Spacing should be such that every 45° line, representing a potential crack and extending from middepth $d/2$ to the longitudinal tension bars, should be crossed by at least one line of reinforcing.

The area of steel required in vertical stirrups, sq in. per stirrup, with a spacing s is

$$A_v = \frac{(v_u - v_c)b_w s}{f_y} \tag{8-18a}$$

where f_y = yield strength of the shear reinforcement. A_v is the area of the stirrups cut by a horizontal plane.   The value of $v_u - v_c$ should not exceed $8\sqrt{f_c'}$ in sections with web reinforcement, nor should f_y exceed 60 ksi.   Where shear reinforcement is required and is placed perpendicular to the axis of the member, it should not be spaced farther apart than $0.5d$, nor more than 24 in. c to c.   When $v_u - v_c$ exceeds $4\sqrt{f_c'}$, however, the maximum spacing should be limited to $0.25d$.

Alternatively, for practical design, Eq. (8-18a) can be transformed into Eq. (8-18b) to indicate the stirrup spacing s for the design shear V_u, stirrup area A_v, and geometry of the member b_w and d:

$$s = \frac{A_v \phi f_y d}{V_u - 2\phi \sqrt{f_c'}\, b_w d} \tag{8-18b}$$

The area required when a single bar or a single group of parallel bars are all bent up at the same distance from the support at angle α with the longitudinal axis of the member is

$$A_v = \frac{(v_u - v_c)b_w d}{f_y \sin \alpha} \tag{8-18c}$$

A_v is the area cut by a plane normal to the axis of the bars.   The area required when a series of such bars are bent up at different distances from the support or when inclined stirrups are used is

$$A_v = \frac{(v_u - v_c)b_w d}{f_y (\sin \alpha + \cos \alpha)} \tag{8-18d}$$

A minimum area of shear reinforcement is required in all members, except slabs, footings, and joists or where v_u is less than $0.5v_c$.   When the nominal torsion stress v_{tu} [Eq. (8-21)] does not exceed $1.5\sqrt{f_c'}$, this area is given by

$$A_v = 50 \frac{b_w s}{f_y} \tag{8-19}$$

But when the nominal torsional stress v_{tu} exceeds $1.5\sqrt{f_c'}$ and where web reinforcement is required, either nominally or by calculation, the minimum area of closed stirrups required is

$$A_v + 2A_t = \frac{50\, b_w s}{f_y} \tag{8-20}$$

where A_t is the area of one leg of a closed stirrup resisting torsion within a distance s.

Shear or torsion reinforcement should extend the full depth d of the member and should be adequately anchored at both ends to develop the design yield strength of the reinforcement. While shear reinforcement may consist of stirrups (Fig. 8-12), bent-up longitudinal bars, spirals, or welded-wire fabric, torsion reinforcement should consist of closed ties, closed stirrups, or spirals—all combined with longitudinal bars.   Closed ties or stirrups may be formed either in one piece by overlapping standard tie end hooks around a longitudinal bar (Fig. 8-12b) or in two pieces spliced as a Class C splice or adequately embedded.   Pairs of U stirrups placed so as to form a closed unit should be lapped at least $1.7l_d$, where l_d is the development length (Art. 8-15).

Torsion effects should be considered whenever the nominal torsion stress v_{tu} exceeds $1.5\sqrt{f_c'}$ when computed from

$$v_{tu} = \frac{3T_u}{\phi \Sigma x^2 y} \tag{8-21}$$

where ϕ = capacity reduction factor = 0.85

$\quad T_u$ = ultimate design torsional moment

$\quad \Sigma x^2 y$ = sum for the component rectangles of the section of the product of the square of the shorter side and the longer side of each rectangle (where a T section applies, the overhanging flange width used in design should not exceed three times the flange thickness)

The torsion stress v_{tc} carried by the concrete alone should not exceed

$$v_{tc} = \frac{2.4\sqrt{f_c'}}{\sqrt{1 + (1.2v_u/v_{tu})^2}} \tag{8-22}$$

Torsion reinforcement should be provided in addition to that required for flexure, shear, and axial forces. The requirements for torsion reinforcement may be combined with those for other forces if the area provided equals or exceeds the sum of the individual required areas and spacing of reinforcement meets the most restrictive of the spacing requirements.

Spacing of closed stirrups for torsion should be computed from

$$s = \frac{3A_t \alpha_t x_1 y_1 f_y}{(v_{tu} - v_{tc})\Sigma x^2 y} \tag{8-23}$$

where A_t = area of one leg of a closed stirrup
$\alpha_t = 0.66 + 0.33y_1/x_1$ but not more than 1.50
f_y = yield strength of the torsion reinforcement
x_1 = shorter dimension c to c of the legs of a closed stirrup
y_1 = longer dimension c to c of the legs of a closed stirrup

The spacing of closed stirrups, however, should not exceed $(x_1 + y_1)/4$ nor 12 in. Torsion reinforcement should be provided over at least a distance of $d + b$ beyond the point where it is theoretically required, where b is the beam width.

At least one longitudinal bar should be placed in each corner of the stirrups. Size of longitudinal bars should be at least No. 3, and their spacing around the perimeters of the stirrups should not exceed 12 in. Longitudinal bars larger than No. 3 are required if indicated by the larger of the values of A_l computed from Eqs. (8-24) and (8-25).

$$A_l = 2A_t \frac{x_1 + y_1}{s} \tag{8-24}$$

$$A_l = \left[\frac{400xs}{f_y}\left(\frac{v_{tu}}{v_{tu} + v_u}\right) - 2A_t\right]\left(\frac{x_1 + y_1}{s}\right) \tag{8-25}$$

In Eq. (8-25), $50b_w s/f_y$ may be substituted for $2A_t$.

Development of Reinforcement. To prevent bond failure or splitting, the calculated stress in any bar at any section must be developed on each side of the section by adequate embedment length, end anchorage, or hooks. The critical sections for development of reinforcement in flexural members are at points of maximum stress and at points within the span where adjacent reinforcement terminates. See Art. 8-25.

At least one-third of the positive-moment reinforcement in simple beams and one-fourth of the positive-moment reinforcement in continuous beams should extend along the same face of the member into the support, and for beams at least 6 in. into the support. At simple supports and at points of inflection, the diameter of the reinforcement should be limited to a diameter such that the development length l_d defined in Art. 8-15 satisfies

$$l_d = \frac{M_t}{V_u} + l_a \tag{8-26}$$

where M_t = computed flexural strength with all the reinforcing steel at the section stressed to f_y
V_u = applied shear at the section
l_a = additional embedment length beyond the inflection point or the center of the support

Negative-moment reinforcement should have an embedment length into the span to develop the calculated tension in the bar, or a length equal to the effective depth of the member, or 12 bar diameters, whichever is greatest. At least one-third of the total negative reinforcement should have an embedment length beyond the point of inflection not less than the effective depth of the member, or 12 bar diameters, or one-sixteenth of the clear span, whichever is greatest.

Hooks on Bars. Standard hooks are considered to develop a tensile stress in bar reinforcement of

$$f_h = \xi \sqrt{f_c'} \tag{8-27}$$

where ξ has the values given in Table 8-9 for $f_y = 60$ ksi. For $f_y = 40$ ksi, $\xi = 360$ for No. 11 and smaller bars, 330 for No. 14 bars, and 220 for No. 18 bars. The given values may be increased 30% where enclosure is provided perpendicular to the plane of the hook.

Table 8-9. ξ for $f_y = 60$ ksi

Bar size No.	Top bars	Other bars
3, 4, 5	540	540
6	450	540
7, 8, 9	360	540
10	360	480
11	360	420
14	330	330
18	220	220

An equivalent embedment length l_e for hooks may be computed from Eq. (8-4) by substituting l_e for l_d and f_h for f_y. Also, the development length may consist of a combination of embedment length plus the equivalent embedment length of a hook.

Hooks should not be considered effective in adding to the compressive resistance of reinforcement. Thus, hooks should not be used on footing dowels. Instead, when depth of footing is less than that required by large-size bars, the designer should substitute smaller-diameter bars with equivalent area and lesser embedment length. It may be possible sometimes to increase the footing depth where large-diameter dowel reinforcement is used, so that footing dowels can have the proper embedment length. Footing dowels need only transfer the excess load above that transmitted in bearing and therefore may be bars with areas different from those required for compression design for the first column lift.

(P. F. Rice and E. S. Hoffman, "Structural Design Guide to the ACI Building Code," Van Nostrand Reinhold Company, New York; "CRSI Handbook," Concrete Reinforcing Steel Institute, Chicago, Ill.; ACI SP-17, Design Handbook in Accordance with the Strength Design Method of ACI 318-71, American Concrete Institute, Detroit, Mich.; P. Rogers, "Reinforced Concrete Design for Buildings," Van Nostrand Reinhold Company, New York.)

8-24. Working-Stress Design of Rectangular Beams with Tension Reinforcement Only. (See also Art. 8-21.) From the assumption that stress varies across a beam section with the distance from the neutral axis, it follows that (see Fig. 8-11d)

$$\frac{nf_c}{f_s} = \frac{k}{1 - k} \tag{8-28}$$

where n = modular ratio E_s/E_c
E_s = modulus of elasticity of steel reinforcement, psi
E_c = modulus of elasticity of concrete, psi
f_c = compressive stress in extreme surface of concrete, psi
f_s = stress in steel, psi
kd = distance from extreme compression surface to neutral axis, in.
d = distance from extreme compression to centroid of reinforcement, in.

When the steel ratio $\rho = A_s/bd$, where A_s = area of tension reinforcement, sq in., and b = beam width, in., is known, k can be computed from

$$k = \sqrt{2n\rho + (n\rho)^2} - n\rho \tag{8-29}$$

Wherever positive-moment steel is required, ρ should be at least $200/f_y$, where f_y is the steel yield stress. The distance jd between the centroid of compression and the centroid of tension, in., can be obtained from

$$j = 1 - \frac{k}{3} \tag{8-30}$$

The moment resistance of the concrete, in.-lb, is

$$M_c = \tfrac{1}{2} f_c k j b d^2 = K_c b d^2 \qquad (8\text{-}31)$$

where $K_c = \tfrac{1}{2} f_c k j$. The moment resistance of the steel is

$$M_s = f_s A_s j d = f_s \rho j b d^2 = K_s b d^2 \qquad (8\text{-}32)$$

where $K_s = f_s \rho j$. Allowable stresses are given in Art. 8-21. Table 8-8 lists nominal diameters, weights, and cross-sectional areas of standard steel reinforcing bars.

Shear. The nominal unit shear stress acting on a section with shear V is

$$v = \frac{V}{bd} \qquad (8\text{-}33)$$

Allowable shear stresses are 55% of those for ultimate-strength design (Art. 8-23). Otherwise, designs for shear by the working-stress and ultimate-strength methods are the same. Except for brackets and other short cantilevers, the section for maximum shear may be taken at a distance d from the face of the support. In working-stress design, the shear stress v_c carried by the concrete alone should not exceed $1.1\sqrt{f_c'}$. (As an alternative, the maximum for v_c may be taken as $\sqrt{f_c'} + 1{,}375\rho\, Vd/M$, with a maximum of $1.93\sqrt{f_c'}$. M is the bending moment at the section but should not be less than Vd.)

At cross sections where the torsional stress v_t exceeds $0.825\sqrt{f_c'}$, v_c should not exceed

$$v_c = \frac{1.1\sqrt{f_c'}}{\sqrt{1 + (v_t/1.2v)^2}} \qquad (8\text{-}34)$$

The excess shear $v - v_c$ should not exceed $4.4\sqrt{f_c'}$ in sections with web reinforcement. Stirrups and bent bars should be capable of resisting the excess shear $V' = V - v_c bd$.

The area required in the legs of a vertical stirrup, sq in., is

$$A_v = \frac{V's}{f_v d} \qquad (8\text{-}35)$$

where s = spacing of stirrups, in.
f_v = allowable stress in stirrup steel, psi (see Art. 8-21)

For a single bent bar or a single group of parallel bars all bent at an angle α with the longitudinal axis at the same distance from the support, the required area is

$$A_v = \frac{V'}{f_v d \sin \alpha} \qquad (8\text{-}36)$$

For inclined stirrups and groups of bars bent up at different distances from the support, the required area is

$$A_v = \frac{V's}{f_v d (\sin \alpha + \cos \alpha)} \qquad (8\text{-}37)$$

Where shear reinforcing is required and the nominal torsion stress does not exceed $0.825 \sqrt{f_c'}$, the minimum area of shear reinforcement provided should be that given by Eq. (8-19).

Torsion. Torsion effects should be considered whenever the nominal torsion stress v_t exceeds $0.825\sqrt{f_c'}$ when computed from

$$v_t = \frac{3T}{\sum x^2 y} \qquad (8\text{-}38)$$

where T = torsional moment caused by service loads
$\sum x^2 y$ = sum for the component rectangles of the section of the product of the square of the shorter side and the longer side of each rectangle

The allowable torsion stress on the concrete is 55% of that computed from Eq. (8-22). Spacing of closed stirrups for torsion should be computed from

$$s = \frac{3A_t \alpha_t x_1 y_1 f_v}{(v_t - v_{tc}) \sum x^2 y} \qquad (8\text{-}39)$$

where A_t = area of one leg of a closed stirrup
$\alpha_t = 0.66 + 0.33 y_1/x_1$ but not more than 1.50

v_{tc} = allowable torsion stress on the concrete

x_1 = shorter dimension c to c of the legs of a closed stirrup

y_1 = longer dimension c to c of the legs of a closed stirrup

Development of Reinforcement. To prevent bond failure or splitting, the calculated stress in reinforcement at any section should be developed on each side of that section by adequate embedment length, end anchorage, or, for tension only, hooks. Requirements are the same as those given for ultimate-strength design in Art. 8-23. Embedment length required at simple supports and inflection points can be computed from Eq. (8-26) by substituting double the computed shears for V_u. In computation of M_t, the moment arm $d - a/2$ may be taken as $0.85d$ (Fig. 8-10). See also Art. 8-25.

8-25. Bar Cutoffs and Bend Points. It is common practice to stop or bend main reinforcement in beams and slabs where it is no longer required. But tensile steel should never be discontinued exactly at the theoretical cutoff or bend points. It is necessary to resist tensile forces in the reinforcement through embedment beyond those points.

All reinforcement should extend beyond the point at which it is no longer needed to resist flexure for a distance equal to the effective depth of the member or 12 bar diameters, whichever is greater. Lesser extensions, however, may be used at supports of a simple span and at the free end of a cantilever. See Art. 8-23 for embedment requirements at simple supports and inflection points and for termination of negative-moment bars. Continuing reinforcement should have an embedment length beyond the point where bent or terminated reinforcement is no longer required to resist flexure. The embedment should be at least as long as the development length l_d defined in Art. 8-15.

Flexural reinforcement should not be terminated in a tension zone unless one of the following conditions is satisfied:

1. Shear is less than two-thirds that normally permitted, including allowance for shear reinforcement, if any.

2. Continuing bars provide double the area required for flexure at the cutoff, and the shear does not exceed three-quarters of that permitted.

3. Stirrups in excess of those normally required are provided each way from the cutoff for a distance equal to 75% of the effective depth of the member. Area and spacing of the excess stirrups should be such that

$$A_v \geq 60 \frac{b_w s}{f_y} \tag{8-40}$$

where A_v = stirrup cross-sectional area, sq in.

b_w = web width, in.

s = stirrup spacing, in.

f_y = yield strength of stirrup steel, psi

Stirrup spacing s should not exceed $d/8\beta_b$, where β_b is the ratio of the area of bars cut off to the total area of bars at the section and d is the effective depth of the member.

The location of theoretical cutoffs or bend points may usually be determined from bending moments, since the steel stresses are approximately proportional to them. The bars generally are discontinued in groups or pairs. So, for example, if one-third the bars are to be bent up, the theoretical bend-up point lies at the section where the bending moment is two-thirds the maximum moment. The point may be found analytically or graphically.

(G. Winter and A. H. Nilson, "Design of Concrete Structures," McGraw-Hill Book Company, New York; P. F. Rice and E. S. Hoffman, "Structural Design Guide to the ACI Building Code," Van Nostrand Reinhold Company, New York; ACI 315, "Manual of Standard Practice for Detailing Reinforced Concrete Structures," American Concrete Institute, Detroit, Mich.)

8-26. One-Way Slabs. If a slab supported on beams or walls spans a distance in one direction more than twice that in the perpendicular direction, so much of the load is carried on the short span that the slab may reasonably be assumed to be carrying all the load in that direction. Such a slab is called a one-way slab.

Generally, a one-way slab is designed by selecting a 12-in.-wide strip parallel to the short direction and treating it as a rectangular beam. Reinforcing steel usually is spaced uniformly along both spans (Table 8-10). In addition to the main reinforcing in the short span, steel should be provided in the long direction to distribute concentrated loads and resist shrinkage and thermal stresses. The bars or wires should not be spaced farther apart than 18 in. or five times the slab thickness.

For shrinkage and temperature stresses, ACI 318, Building Code Requirements for Reinforced Concrete, requires the following minimum areas of steel, sq in. per ft: deformed bars with

Table 8-10. Areas of Bars in Slabs, Sq In. per Ft of Slab

Spacing, in.	Bar No.								
	3	4	5	6	7	8	9	10	11
3	0.44	0.78	1.23	1.77	2.40	3.14	4.00	5.06	6.25
3½	0.38	0.67	1.05	1.51	2.06	2.69	3.43	4.34	5.36
4	0.33	0.59	0.92	1.32	1.80	2.36	3.00	3.80	4.68
4½	0.29	0.52	0.82	1.18	1.60	2.09	2.67	3.37	4.17
5	0.26	0.47	0.74	1.06	1.44	1.88	2.40	3.04	3.75
5½	0.24	0.43	0.67	0.96	1.31	1.71	2.18	2.76	3.41
6	0.22	0.39	0.61	0.88	1.20	1.57	2.00	2.53	3.12
6½	0.20	0.36	0.57	0.82	1.11	1.45	1.85	2.34	2.89
7	0.19	0.34	0.53	0.76	1.03	1.35	1.71	2.17	2.68
7½	0.18	0.31	0.49	0.71	0.96	1.26	1.60	2.02	2.50
8	0.17	0.29	0.46	0.66	0.90	1.18	1.50	1.89	2.34
9	0.15	0.26	0.41	0.59	0.80	1.05	1.33	1.69	2.08
10	0.13	0.24	0.37	0.53	0.72	0.94	1.20	1.52	1.87
12	0.11	0.20	0.31	0.44	0.60	0.79	1.00	1.27	1.56

yield strength less than 60,000 psi, 0.024; deformed bars with 60,000 psi yield strength or welded-wire fabric with wires not more than 12 in. apart, 0.0216. For highway bridge slabs, Standard Specifications for Highway Bridges (American Association of State Highway and Transportation Officials) requires reinforcing steel in the bottoms of all slabs transverse to the main reinforcement for lateral distribution of wheel loads. The area of the distribution steel should be at least the following percentages of the main steel required for positive moment, where S is the effective span, ft (Art. 8-27): When the main steel is parallel to traffic, $100/\sqrt{S}$, with a maximum of 50%; when the main steel is perpendicular to traffic, $200/\sqrt{S}$, with a maximum of 67%.

To control deflections, the ACI Code sets limitations on slab thickness unless deflections are computed and determined to be acceptable (Art. 8-22). Otherwise, thickness of one-way slabs must be at least $L/20$ for simply supported slabs; $L/24$ for slabs with one end continuous; $L/28$ for slabs with both ends continuous; and $L/10$ for cantilevers; where L is the span, in.

8-27. Design Spans for Beams and Slabs. ACI 318, Building Code Requirements for Reinforced Concrete, specifies that the span of members not integral with supports should be taken as the clear span plus the depth of the member but not greater than the distance center of supports. For analysis of continuous frames, spans should be taken from center to center of supports for determination of bending moments in beams and girders, but moments at the faces of supports may be used in the design of the members. Solid or ribbed slabs integral with supports and with clear spans up to 10 ft may be designed for the clear span.

Standard Specifications for Highway Bridges (American Association of State Highway and Transportation Officials) has the same requirements as the ACI Code for spans of simply supported beams and slabs. For slabs continuous over more than two supports, the effective span is the clear span for slabs monolithic with beams or walls (without haunches); the distance between stringer-flange edges plus half the stringer-flange width for slabs supported on steel stringers; clear span plus half the stringer thickness for slabs supported on timber stringers. For rigid frames, the span should be taken as the distance between centers of bearings at the top of the footings. The span of continuous beams should be the clear distance between faces of supports.

Where fillets or haunches make an angle of 45° or more with the axis of a continuous or restrained slab and are built integral with the slab and support, AASHTO requires that the span be measured from the section where the combined depth of the slab and fillet is at least 1.5 times the thickness of slab. The moments at the ends of this span should be used in the slab design, but no portion of the fillet should be considered as adding to the effective depth of the slab.

8-28. Rectangular Beams with Compression Bars—Ultimate-Strength Design. (See also Arts. 8-20 and 8-23.) The steel ratio ρ_b for balanced conditions at ultimate strength of a rectangular beam is given by Eq. (8-14). When the tensile steel ratio ρ exceeds $0.75\rho_b$, compression reinforcement should be used. When ρ is equal to or less than $0.75\rho_b$, the strength of the beam may be approximated by Eq. (8-15), disregarding any compression bars that may be present, since the strength of the beam will usually be controlled by yielding of the tensile steel.

The bending-moment capacity of a rectangular beam with both tension and compression steel is

$$M_u = 0.90 \left[(A_s - A_s')f_y \left(d - \frac{a}{2} \right) + A_s' f_y (d - d') \right] \tag{8-41}$$

where a = depth of equivalent rectangular compressive stress distribution
    = $(A_s - A_s')f_y/f_c'b$
b = width of beam, in.
d = distance from extreme compression surface to centroid of tensile steel, in.
d' = distance from extreme compression surface to centroid of compressive steel, in.
A_s = area of tensile steel, sq in.
A_s' = area of compressive steel, sq in.
f_y = yield strength of steel, psi
f_c' = 28-day strength of the concrete, psi

Equation (8-41) is valid only when the compressive steel reaches f_y. This occurs when

$$(\rho - \rho') \geq 0.85\beta_1 \frac{f_c'd'}{f_y d} \frac{87,000}{87,000 - f_y} \tag{8-42}$$

where $\rho = A_s/bd$, $\rho' = A_s'/bd$, and β_1 is a constant defined in Art. 8-20. When $\rho - \rho'$ is less than the right-hand side of Eq. (8-42), calculate the moment capacity from Eq. (8-15) or from an analysis based on the assumptions of Art. 8-20. ACI 318, Building Code Requirements for Reinforced Concrete (American Concrete Institute) requires also that $\rho - \rho'$ not exceed $0.75\rho_b$ to avoid brittle failure of the concrete.

Compressive steel should be anchored by ties or stirrups at least $\frac{3}{8}$ in. in diameter and spaced no more than 16 bar diameters or 48 tie diameters apart. At least one tie within the required spacing, throughout the length of beam where compressive reinforcement is required, should extend completely around all longitudinal bars.

Design for shear and development lengths of reinforcement is the same as for beams with tension reinforcement only (Art. 8-23).

8-29. Rectangular Beams with Compression Bars—Working-Stress Design. (See also Arts. 8-21, 8-22, and 8-24.) The following formulas, based on the linear variation of stress and strain with distance from the neutral axis (Fig. 8-11), may be used in design:

$$k = \frac{1}{1 + f_s/nf_c} \tag{8-43}$$

where f_s = stress in tensile steel, psi
f_c = stress in extreme compression surface, psi
n = modular ratio, E_s/E_c

$$f_s' = \frac{kd - d'}{d - kd} 2f_s \tag{8-44}$$

where f_s' = stress in compressive steel, psi
d = distance from extreme compression surface to centroid of tensile steel, in.
d' = distance from extreme compression surface to centroid of compressive steel, in.

The factor 2 is incorporated in Eq. (8-44) in accordance with ACI 318, Building Code Requirements for Reinforced Concrete, to account for the effects of creep and nonlinearity of the stress-strain diagram for concrete. But f_s' should not exceed the allowable tensile stress for the steel.

Since total compressive force equals total tensile force on a section,

$$C = C_c + C_s' = T \tag{8-45}$$

where C = total compression on a beam cross section, lb
C_c = total compression on the concrete, lb, at the section
C_s' = force acting on the compressive steel, lb
T = force acting on the tensile steel, lb

$$\frac{f_s}{f_c} = \frac{k}{2[\rho - \rho'(kd - d')/(d - kd)]} \tag{8-46}$$

where $\rho = A_s/bd$ and $\rho' = A_s'/bd$

For reviewing a design, the following formulas may be used:

$$k = \sqrt{2n\ \left(\rho + \rho'\ \frac{d'}{d}\right) + n^2(\rho + \rho')^2} - n(\rho + \rho') \tag{8-47}$$

$$\bar{z} = \frac{(k^3 d/3) + 2n\rho' d'[k - (d'/d)]}{k^2 + 2n\rho'[k - (d'/d)]} \tag{8-48}$$

$$jd = d - \bar{z} \tag{8-49}$$

where jd is the distance between the centroid of compression and the centroid of the tensile steel. The moment resistance of the tensile steel is

$$M_s = Tjd = A_s f_s jd \tag{8-50a}$$

$$f_s = \frac{M}{A_s jd} \tag{8-50b}$$

where M is the bending moment at the section of beam under consideration. The moment resistance in compression is

$$M_c = \frac{1}{2} f_c jbd^2 \left[k + 2n\rho' \left(1 - \frac{d'}{kd}\right)\right] \tag{8-51a}$$

$$f_c = \frac{2M}{jbd^2\{k + 2n\rho'[1 - (d'/kd)]\}} \tag{8-51b}$$

Solution of the preceding equations often is facilitated by tables and charts. Many designers, however, prefer the following approximate formulas:

$$M_1 = \frac{1}{2} f_c bkd \left(d - \frac{kd}{3}\right) \tag{8-52}$$

$$M_s' = M - M_1 = 2f_s' A_s'(d - d') \tag{8-53}$$

where M = bending moment

M_s' = moment-resisting capacity of the compressive steel

M_1 = moment-resisting capacity of the concrete

For determination of shear, see Art. 8-24. Compressive steel should be anchored by ties or stirrups at least No. 3 in size and spaced not more than 16 bar diameters or 48 tie diameters apart. At least one tie within the required spacing, throughout the length of the beam where compressive reinforcement is required, should extend completely around all longitudinal bars.

8-30. I and T Beams—Ultimate-Strength Design. (See also Arts. 8-20 and 8-23.) A reinforced-concrete beam may be shaped in cross section like a T, or it may be composed of a slab and integral rectangular beam that, in effect, act as a T beam. According to ACI 318, Building Code Requirements for Reinforced Concrete (American Concrete Institute), and Standard Specifications for Highway Bridges (American Association of State Highway and Transportation Officials), when the slab forms the compression flange, its effective width b should not exceed one-fourth the beam span, and it should not be greater than the distance center to center of beams. In addition, the ACI Code requires that the overhanging width on either side of the beam web should not exceed eight times the slab thickness. The AASHTO specifications more conservatively limit the effective width to twelve times the slab thickness plus the beam width. For beams with a flange on only one side, the effective overhanging flange width should not exceed one-twelfth the beam span, or six times the slab thickness, or half the clear distance to the next beam.

Two cases may occur in the design of T and I beams: the neutral axis lies in the compression flange (Fig. 8-13a and b) or in the web (Fig. 8-13c and d). For negative moment, a T beam should be designed as a rectangular beam with width b equal to that of the stem.

When the neutral axis lies in the flange, the member may be designed as a rectangular beam, with effective width b and depth d, by Eq. (8-15). For that condition, the flange thickness t will be greater than the distance from the extreme compression surface to the neutral axis,

$$c = \frac{1.18\omega d}{\beta_1} \tag{8-54}$$

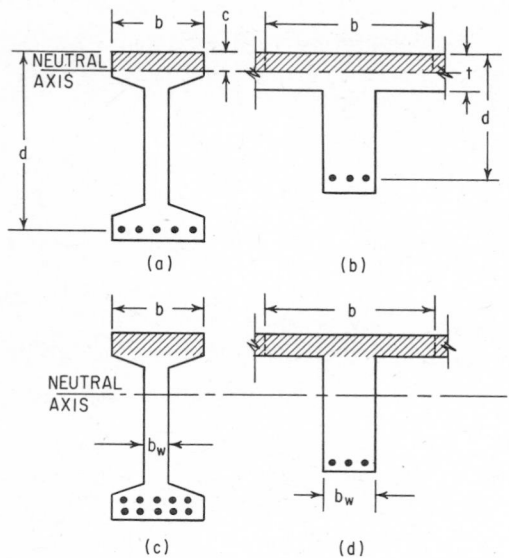

Fig. 8-13. I and T beams with neutral axis in the flange (*a*) and (*b*) are designed as rectangular beams. When the neutral axis lies in the web (*c*) and (*d*), the usual practice is to ignore the compression in the web in design.

where $\beta_1 =$ constant defined in Art. 8-20
  $\omega = A_s f_y/bdf'_c$
  $A_s =$ area of tensile steel, sq in.
  $f_y =$ yield strength of the steel, psi
  $f'_c =$ 28-day strength of the concrete
When the neutral axis lies in the web, the ultimate moment should not exceed

$$M_u = 0.90 \left[(A_s - A_{sf})f_y \left(d - \frac{a}{2} \right) + A_{sf}f_y \left(d - \frac{t}{2} \right) \right] \tag{8-55}$$

where $A_{sf} =$ area of tensile steel required to develop the compressive strength of the overhanging flange, sq in. $= 0.85(b - b_w) \, tf'_c/f_y$
  $b_w =$ width of beam web or stem, in.
  $a =$ depth of equivalent rectangular compressive stress distribution, in.
   $= (A_s - A_{sf})f_y/0.85f'_c b_w$
The quantity $\rho_w - \rho_f$ should not exceed $0.75\rho_b$, where ρ_b is the steel ratio for balanced conditions [Eq. (8-14)], $\rho_w = A_s/b_w d$, and $\rho_f = A_{sf}/b_w d$.
  For determination of ultimate shear, see Art. 8-23. Note, however, that the web or stem width b_w should be used instead of b in these calculations.
  8-31. I and T Beams—Working-Stress Design. (See also Arts. 8-21, 8-22, and 8-24.) For T beams, effective width of compression flange is determined by the same rules as for ultimate-strength design (Art. 8-30). Also, for working-stress design, two cases may occur: the neutral axis may lie in the flange (Fig. 8-13*a* and *b*) or in the web (Fig. 8-13*c* and *d*). (For negative moment, a T beam should be designed as a rectangular beam with width b equal to that of the stem.)
  If the neutral axis lies in the flange, a T or I beam may be designed as a rectangular beam with effective width b. If the neutral axis lies in the web or stem, an I or T beam may be designed by the following formulas, which ignore the compression in the stem, as is customary:

$$k = \frac{1}{1 + f_s/nf_c} \tag{8-56}$$

where $kd =$ distance from extreme compression surface to neutral axis, in.
  $d =$ distance from extreme compression surface to centroid of tensile steel, in.
  $f_s =$ stress in tensile steel, psi
  $f_c =$ stress in concrete at extreme compression surface, psi
  $n =$ modular ratio $= E_s/E_c$

Since the total compressive force C equals the total tension T,

$$C = \frac{1}{2} f_c (2kd - t) \frac{bt}{kd} = T = A_s f_s \tag{8-57}$$

$$kd = \frac{2ndA_s + bt^2}{2nA_s + 2bt} \tag{8-58}$$

where A_s = area of tensile steel, sq in.
$\quad\quad t$ = flange thickness, in.
The distance between the centroid of the area in compression and the centroid of the tensile steel is

$$jd = d - \bar{z} \tag{8-59}$$

$$z = \frac{t(3kd - 2t)}{3(2kd - t)} \tag{8-60}$$

The moment resistance of the steel is

$$M_s = Tjd = A_s f_s jd \tag{8-61}$$

The moment resistance of the concrete is

$$M_c = Cjd = \frac{f_c btjd}{2kd} (2kd - t) \tag{8-62a}$$

In design, M_s and M_c can be approximated by

$$M_s = A_s f_s \left(d - \frac{t}{2} \right) \tag{8-61b}$$

$$M_c = \frac{1}{2} f_c bt \left(d - \frac{t}{2} \right) \tag{8-62b}$$

derived by substituting $d - t/2$ for jd and $f_c/2$ for $f_c(1 - t/2kd)$, the average compressive stress on the section.

For determination of shear, see Art. 8-24. Note, however, that the web or stem width b_w should be used instead of b in these calculations.

8-32. Torsion in Reinforced Concrete Members.   Under twisting or torsional loads, a member develops normal (warping) and shear stresses. The warping, normal stresses help greatly in resisting torsion. But there are no accurate ways of computing this added resistance.

The maximum shears at any point are accompanied by equal tensile stresses on planes bisecting the angles between the planes of maximum shears. The unit shear stresses may be computed from Eqs. (6-92), (6-95), (6-98), and (6-99).

As for ordinary shear, reinforcement should be incorporated to resist the diagonal tension in excess of the tensile capacity of the concrete. If web reinforcement is required for vertical shear in a horizontal beam subjected to both flexure and torsion, additional web reinforcement should be included to take care of the full torsional shear.

Design for torsion in combination with shear and requirements for torsion reinforcement are discussed in Art. 8-23.

8-33. Two-Way Slabs.   When a rectangular reinforced concrete slab is supported on all four sides, reinforcement placed perpendicular to the sides may be assumed to be effective in the two directions if the ratio of the long sides to the short sides is less than about 2 to 1. [Standard Specifications for Highway Bridges (American Association of State Highway and Transportation Officials) requires that the slab be designed as a one-way slab if the ratio is more than 1.5 to 1.] In effect, a two-way slab distributes part of the load on it in the long direction and usually a much larger part in the short direction. For a symmetrically supported square slab, however, distribution is the same in the two directions for symmetrical loading.

Because precise determination of reactions and moments for two-way slabs with various edge conditions is complex and tedious, most codes offer empirical formulas to simplify the calculation.

According to the AASHTO specifications, the proportion of the load carried by the short span of the slab should be assumed as follows: For load uniformly distributed:

$$p = \frac{B^4}{A^4 + B^4} \tag{8-63}$$

For load concentrated at the center of the slab:

$$p = \frac{B^3}{A^3 + B^3} \qquad (8\text{-}64)$$

where A = length of short span of slab

B = length of long span of slab

Moments obtained with Eqs. (8-63) and (8-64) should be used in designing the center half of the slab in the short and long directions. Reinforcing steel in the outer quarters in both directions may be reduced to 50% of that required in the center half.

The reactions of the slab on supporting beams and walls are not constant along the sides. This should be taken into account in the design of the supports. (One method is to use a triangular distribution on the short sides and a trapezoidal distribution on the long sides. The legs of the triangles and the trapezoids usually are assumed to make a 45° angle with the slab edges.)

ACI 318-71, Building Code Requirements for Reinforced Concrete (American Concrete Institute), considers the design of two-way slabs, supported on all four sides, to be subject to the same fundamental principles as design of any slab system (flat slabs, flat plates, waffle slabs) reinforced for flexure in more than one direction. Article 8-35 may also be applied in design of two-way slabs.

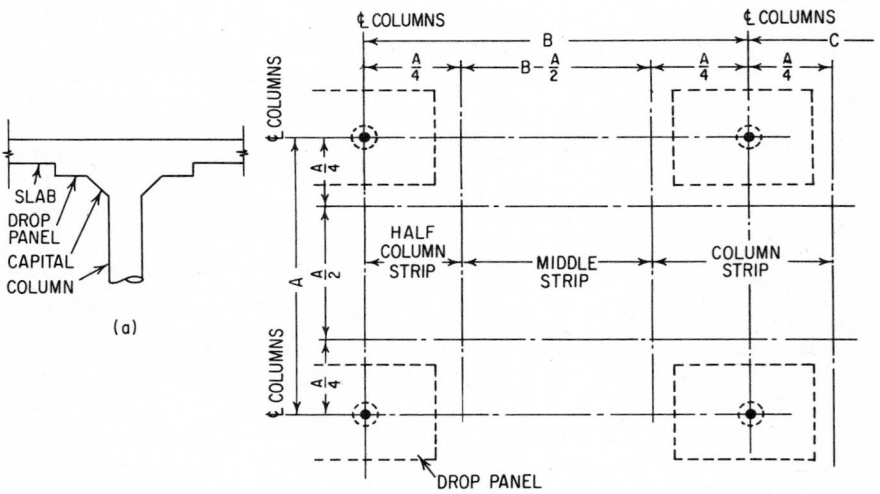

Fig. 8-14. Flat slab, with drop panels at supports (a), is divided into column and middle strips (b) for design purposes.

8-34. Flat-Slab Construction. Slabs supported directly on columns, without beams or girders, are classified as flat slabs. Generally, the columns flare out at the top in capitals (Fig. 8-14a). But only the portion of the inverted truncated cone thus formed that lies inside a 90° vertex angle is considered effective in resisting stress. Sometimes, the capital for an exterior column is a bracket on the inner face.

To reduce the shear stresses in the region of the columns and the amount of steel needed for negative bending moments, especially when the live load exceeds 150 psf, a rectangular drop panel, or thicker slab, is formed over the columns (Fig. 8-14a). For the full effective depth of the drop to be used in determination of negative-moment reinforcement, ACI 318, Building Code Requirements for Reinforced Concrete (American Concrete Institute), specifies that a drop panel should extend in each direction from the center of support a distance equal to at least one-sixth the span in that direction. The difference in thickness between the drop panel and slab should be at least one-fourth the slab thickness but, for determining reinforcement, should not be taken as more than one-fourth the distance from the edge of the drop panel to the edge of the column or capital. For such flat slabs, requirements for minimum thickness may be reduced 10% but not to less than 4 in.

The slab may be solid, hollow, or waffle. A waffle slab usually is the most economical type for long spans, though formwork may be more expensive than for a solid slab. A waffle slab omits much of the concrete that would be in tension and thus not considered effective in resisting stresses. To control deflection, the ACI Code establishes minimum thicknesses for slabs (Table 8-11).

Table 8-11. Minimum Slab Thickness for Two-Way Slabs*

1. Slabs with beams along one or more edges†:

$$h = \frac{l_n(800 + 0.005f_y)}{36,000 + 5,000\beta \left[\alpha_m - 0.5(1 - \beta_s)\left(1 + \frac{1}{\beta}\right)\right]} \qquad (8\text{-}65)$$

2. Edge slabs without beams ‡:

$$h = \frac{l_n(800 + 0.005f_y)}{36,000 + 5,000\beta(1 + \beta_s)} \qquad (8\text{-}66)$$

3. Interior slabs without beams §:

$$h = \frac{l_n(800 + 0.005f_y)}{36,000} \qquad (8\text{-}67)$$

where h = slab thickness, in.
l_n = length of clear span in long direction, in.
f_y = yield strength of reinforcement, psi
β = ratio of clear span in long direction to clear span in short direction
β_s = ratio of length of continuous edges to total perimeter of a slab panel
α = ratio of flexural stiffness EI of beam section to flexural stiffness EI of a width of slab bounded laterally by the center line of the adjacent panel, if any, on each side of the beam
α_m = average value of α for all beams on the edges of a panel

*Thicknesses less than those in the table may be used if computations prove deflection criteria (Art. 8-22) will not be exceeded.
†Thickness should not be less than that given by Eq. (8-66) but need not be more than that given by Eq. (8-67). Slabs having beams on all four edges with $\alpha_m \geq 2$ need not be thicker than 3½ in. Unless a beam with $\alpha > 0.80$ is provided at discontinuous edges, thickness required by Eq. (8-65) should be increased 10%.
‡For slabs with drop panels, thickness required by Eq. (8-66) may be reduced 10% but not to less than 4 in. Without drop panels, minimum thickness is 5 in. Unless an edge beam with $\alpha > 0.80$ is provided at discontinuous edges, thickness required for Eq. (8-66) should be increased 10%.
§For slabs with drop panels, thickness required by Eq. (8-67) may be reduced 10% but not to less than 4 in. Without drop panels, minimum thickness is 5 in.

In general, flat slabs are more economical than beam-and-girder construction. They yield a lower building for the same number of stories. Formwork is simpler. Fire resistance is greater, because of fewer sharp corners where spalling may occur. And there is less obstruction to light with flat slabs. The design procedure is similar to that for flat plates and is described in Art. 8-35.

8-35. Flat-Plate Construction. Flat slabs with constant thickness between supports are called flat plates. Generally, capitals are omitted from the columns.

Exact analysis or design of flat slabs or flat plates is very complex. It is common practice to use approximate methods. The ACI Code presents two such methods—direct design and equivalent-frame methods.

In both methods, a flat slab is considered to consist of strips parallel to column lines in two perpendicular directions. In each direction, a **column strip** spans between columns and has a width of one-fourth the shorter of the two perpendicular spans on each side of the column center line. The portion of a slab between parallel column strips in each panel is called the **middle strip** (Fig. 8-14b).

Direct Design Method. This may be used when all the following conditions exist:

The slab has three or more bays in each direction.
Ratio of length to width of panel is 2 or less.
Ratio of live to dead load is 3 or less.
Columns form an approximately rectangular grid (10% maximum offset).
Successive spans in each direction do not differ by more than one-third of the longer span.
When a panel is supported by beams on all sides, the relative stiffness of the beams satisfies

$$0.2 \leq \frac{\alpha_1}{\alpha_2} \left(\frac{l_2}{l_1}\right)^2 \leq 5 \tag{8-68}$$

where $\alpha_1 = \alpha$ in the direction of l_1

$\alpha_2 = \alpha$ in the direction of l_2

$l_1 = $ span in direction moments are being determined, c to c supports

$l_2 = $ span perpendicular to l_1, c to c supports

$\alpha = $ ratio of flexural stiffness $E_{cb}I_b$ of beam section to flexural stiffness $E_{cs}I_s$ of a width of slab bounded laterally by the center line of the adjacent panel, if any, on each side of the beam

$E_{cb} = $ modulus of elasticity of beam concrete

$E_{cs} = $ modulus of elasticity of slab concrete

$I_b = $ moment of inertia about centroidal axis of gross section of beam, including that portion of the slab on each side of the beam that extends a distance equal to the projection of the beam above or below the slab, whichever is greater, but not more than four times the slab thickness

$I_s = $ moment of inertia about centroidal axis of gross section of slab $= h^3/12$ times slab width specified in definition of α (Table 8-11)

$h = $ overall thickness of slab

The basic equation used in direct design is the total static design moment in a strip bounded laterally by the center line of the panel on each side of the center line of the supports:

$$M_o = \frac{wl_2l_n^2}{8} \tag{8-69}$$

where $w = $ uniform design load per unit of slab area

$l_n = $ clear span in direction moments are being determined.

The strip, with width l_2, should be designed for bending moments for which the sum in each span of the absolute values of the positive and average negative moments equals or exceeds M_o.

Interior Panels.   Following is the procedure for direct design of an interior panel of a flat slab (or flat plate or two-way beam-and-slab construction):

Step 1.   Determine the minimum allowable and practical slab thickness from whichever of Eqs. (8-65) to (8-67) is appropriate.

Step 2.   Determine the design ultimate load from Eq. (8-7), $U = 1.4D + 1.7L$, where D represents the moments and shears caused by dead load and L, those caused by live load. (This assumes that horizontal loads are taken by shear walls or other vertical elements.)

Step 3.   Determine and check column size. To take into account the effect of pattern loading when β_a, the ratio of dead to live load (without load factors), is less than 2, one of the following two conditions should be satisfied:

1. The sum of the flexural stiffnesses of the columns above and below the slab ΣK_c should be such that

$$\alpha_c = \frac{\Sigma K_c}{\Sigma(K_s + K_b)} \geq \alpha_{min} \tag{8-70}$$

where $K_c = $ flexural stiffness of column $= E_{cc}I_c$

$E_{cc} = $ modulus of elasticity of column concrete

$I_c = $ moment of inertia about centroidal axis of gross section of column

$K_s = E_{cs}I_s$

$K_b = E_{cb}I_b$

$\alpha_{min} = $ minimum value of α_c as given in Table 8-12

2. If the columns do not satisfy condition 1, the design positive moments in the panels should be multiplied by the coefficient

$$\delta_s = 1 + \frac{2 - \beta_a}{4 + \beta_a}\left(1 - \frac{\alpha_c}{\alpha_{min}}\right) \tag{8-71}$$

Preferably, the column size should be increased so that α_c is greater than α_{min} to minimize column moments.

Step 4.   Determine M_o from Eq. (8-69).

Step 5.   For an interior span, distribute M_o as follows:

Negative design moment $= 0.65M_o$

Positive design moment $= 0.35M_o$

Table 8-12. Minimum Column-Slab Stiffness Ratios α_{min}

Ratio of dead to live load β_a	Span ratio l_2/l_1	$\alpha = K_b/K_s$				
		0	0.5	1.0	2.0	4.0
2.0	0.5–2.0	0	0	0	0	0
1.0	0.50	0.6	0	0	0	0
	0.80	0.7	0	0	0	0
	1.00	0.7	0.1	0	0	0
	1.25	0.8	0.4	0	0	0
	2.00	1.2	0.5	0.2	0	0
0.5	0.50	1.3	0.3	0	0	0
	0.80	1.5	0.5	0.2	0	0
	1.00	1.6	0.6	0.2	0	0
	1.25	1.9	1.0	0.5	0	0
	2.00	4.9	1.6	0.8	0.3	0
0.33	0.50	1.8	0.5	0.1	0	0
	0.80	2.0	0.9	0.3	0	0
	1.00	2.3	0.9	0.4	0	0
	1.25	2.8	1.5	0.8	0.2	0
	2.00	13.0	2.6	1.2	0.5	0.3

The negative-moment section should be designed to resist the larger of the two interior negative design moments determined for the spans framing into a common support.

Step 6. Proportion design moments and shears in column and middle strips as follows:

1. Column Strip. The interior negative design moment should be determined in accordance with Table 8-13. Values not given may be obtained by linear interpolation.

Table 8-13. Percent of Interior Negative Design Moment in Column Strips

$\alpha_1 l_2/l_1$	Span ratio l_2/l_1		
	0.5	1.0	2.0
0	75	75	75
1 or more	90	75	45

The positive design moment should be determined in accordance with Table 8-14. Values not given may be obtained by linear interpolation.

When there is a beam between columns in the direction of the span in which moments are being considered, the beam should be proportioned to resist 85% of the column strip moment if $\alpha_1 l_2/l_1$ is greater than 1.0. For values of $\alpha_1 l_2/l_1$ between 1.0 and zero, the proportion of moment resisted by the beam may be obtained by linear interpolation between 85 and 0%. The slab in the column strip should be proportioned to resist that portion of the design moment not resisted by the beam.

Table 8-14. Percent of Positive Design Moment in Column Strips

$\alpha_1 l_2/l_1$	Span ratio l_2/l_1		
	0.5	1.0	2.0
0	60	60	60
1.0 or more	90	75	45

2. *Middle Strip.* The interior negative or positive design moment assigned to a middle strip is that portion of the design moments not resisted by the column strips bounding it. Thus, each middle strip should be proportioned to resist the sum of the negative moment not taken by the column strip along one side and the negative moment not resisted by the column strip on the other side, and similarly, the sum of the positive moments.

3. *Moment Redistribution.* A design moment may be modified by 10% if the total static design moment for the panel in the direction considered is not less than that required by Eq. (8-69).

Step 7. Walls and columns built integrally with the slab should be designed to resist the moments due to loads on the slab system.

Exterior Panels. The ACI Code lists design criteria for exterior panels for a wide range of support conditions. These criteria require determination of the relative flexural stiffness of supports at edges, including torsional resistance.

Equivalent-frame Method. The slab is initially divided into a series of bents, or equivalent frames, on column lines taken longitudinally and transversely through the building. Each frame consists of a row of equivalent columns and slab-beam strips, bounded laterally by the center line of the panel on each side of the column line under investigation. Each such frame may be analyzed in its entirety. Or for vertical loads, each floor may be analyzed, with columns, above and below, assumed fixed at floors above and below. For purposes of computation, the slab-beam may be assumed fixed at any support two panels away from the support where the bending moment is being determined. The moments thus determined may be distributed to the column strips, middle strips, and beams as previously described for the direct design method if Eq. (8-68) is satisfied.

The critical section for negative moment in both the column and middle strips should be taken at the face of supports, but in no case farther than $0.175l_1$ from the center of the column, where l_1 is the span c to c of supports.

Note that where slabs designed by the equivalent-frame method meet the criteria of the direct design method, the computed moments in any span may be reduced in a proportion such that the sum of the absolute values of the positive and average negative bending moments used in design does not exceed M_o given by Eq. (8-69).

Determination of reinforcement, based on the bending moments at critical sections, is the same as described for rectangular beams (Art. 8-23). Requirements for minimum reinforcement should be respected.

The equivalent-frame method attempts to represent the effects of torsional stiffness of the three-dimensional slab system by defining and using the flexural stiffness of the slab-beam-column system in geometric terms applicable to a two-dimensional analysis. The ACI Code assigns a finite moment of inertia to the slab-beam from center to face of column equal to the moment of inertia of the slab-beam at the face of the column divided by $(1 - c_2/l_2)^2$, where c_2 is the dimension of column, capital, or bracket in the direction of l_2. This assigned I represents the flexibility of the slab on the sides of the column. This simulates additional stiffness in the area of the slab-column and is reflected by the change in the coefficients used to determine fixed-end moments, stiffness factors, and carry-over factors for slabs. The ACI Code also modifies the column flexural stiffness to account for the torsional flexibility of the slab. The part of the slab providing the torsional restraint is transverse to the direction in which moments are being determined for the width of the column and extends to the bounding lateral panel center lines on each side of the column.

The ACI Code also provides formulas for determining the stiffness of the equivalent column K_{ec} that give its flexibility (inverse of stiffness) as the sum of the flexibilities of the columns above and below the slab-beam and the flexibility of the slab torsional member. The stiffness factors for the slab and column, incorporating factors for flexural and torsional effects, are then used to determine the relative stiffness of the elements at a slab-column joint. Once the geometric properties of the equivalent frame are known, the moment distribution may be applied to determine the bending moments at the critical sections.

Shear. Slabs should also be investigated for shear, either beam-type or punching shear.

For beam-type shear, the slab is considered as a thin, wide rectangular beam. The critical section for diagonal tension should be taken at a distance from the face of the column or capital equal to the effective depth d of the slab. The critical section extends across the full width b of the slab. Across this section, the nominal shear stress v_u on the unreinforced concrete should not exceed the ultimate capacity $2\sqrt{f_c'}$ or the allowable working stress $1.1\sqrt{f_c'}$ (Arts. 8-23 and 8-24).

Punching shear may occur along several sections extending completely around the support, for example, around the face of the column or column capital or around the drop panel. These critical sections occur at a distance $d/2$ from the faces of the supports, where d is the effective depth of the slab or drop panel. Along these sections, the nominal shear stress is given by

$$v_u = \frac{V_u}{\phi b_o d} \tag{8-72}$$

where b_o = perimeter of critical section
ϕ = capacity reduction factor = 0.85
and v_u should not exceed an ultimate capacity of $4\sqrt{f_c'}$ or an allowable working stress of $2.2\sqrt{f_c'}$. But if shear reinforcement is provided, the allowable shear stress may be increased a maximum of 50% if shear reinforcement consisting of bars is used and increased a maximum of 75% if shearhead reinforcement consisting of steel shapes is used.

Shear reinforcement for slabs generally consists of bent bars and is designed in accordance with the provisions for beams (Art. 8-23), with the allowable shear stress in the concrete at critical sections taken as $2\sqrt{f_c'}$ at ultimate strength. Extreme care should be taken to assure that shear reinforcement is accurately placed and properly anchored, especially in thin slabs.

The ACI Code also includes instructions for design of steel shear heads. Because of the cost of steel shear-head reinforcement, however, it is preferable to either thicken the slab or design concrete beams to support heavy loads.

Column Moments. Another important consideration in design of two-way slab systems is the transfer of moments to columns. This is generally a critical condition at edge columns, where the unbalanced slab moment is very high due to the one-sided panel.

The unbalanced slab moment is considered to be transferred to the column partly by flexure across a critical section, which is $d/2$ from the periphery of the column, and partly by eccentric shear forces acting about the centroid of the critical section.

That portion of unbalanced slab moment M_u transferred by the eccentricity of the shear is given by

$$M_{ve} = \left(1 - \frac{1}{1 + \frac{2}{3}\sqrt{\frac{c_1 + d}{c_2 + d}}} \right) M_u \tag{8-73}$$

where $c_2 + d$ = width of the edges of the critical section resisting the moment
$c_1 + d$ = width of the edge of the critical section perpendicular to $c_2 + d$.
As the width of the critical section resisting moment increases (rectangular column), that portion of the unbalanced moment transferred by flexure also increases. The maximum shear stress, which is determined by combining the vertical load and that portion of shear due to the unbalanced moment being transferred, should not exceed $4\sqrt{f_c'}$. Where the allowable shear stress is exceeded, considerations for shear design should be the same as discussed previously in this article.

For that portion of the unbalanced moment transferred to the column by flexure, it is accepted practice to concentrate or add reinforcement across the critical slab width, determined as the sum of the column width plus the thickness of the slab.

(G. Winter and A. H. Nilson, "Design of Concrete Structures," McGraw-Hill Book Company, New York; P. F. Rice and E. S. Hoffman, "Structural Design Guide to the ACI Building Code," Van Nostrand Reinhold Company, New York; P. Rogers, "Reinforced Concrete Design for Buildings," Van Nostrand Reinhold Company, New York; "CRSI Handbook" and "Two-way Slab Design Supplements," Concrete Reinforcing Steel Institute, Chicago, Ill.)

8-36. Brackets and Corbels. Brackets and corbels are members having a ratio of shear span to depth a/d of 1 or less. The shear span a is the distance from the point of load to the face of support.

The depth of a bracket or corbel at its outer edge should not be less than one-half of the required depth d at the support. Reinforcement should consist of main tension bars with area A_s and shear reinforcement with area A_h, consisting of closed ties parallel to the main tension reinforcement. The area of the shear bars should not be less than $0.5A_s$ or more than $1.0A_s$ and should be uniformly distributed within two-thirds of the depth of the bracket adjacent to the main tension bars. Also, the ratio $\rho = A_s/bd$ should not be less than $0.04f_c'/f_y$.

It is good practice to anchor main tension reinforcement bars as close as possible to the outer edge by welding a cross bar or steel angle to them. Also, the bearing area should be kept at least 2 in. in from the outer edge, and the bearing plate should be welded to the main tension reinforcement if horizontal forces are present.

Shear Friction. When the ratio of a/d is 0.5 or less, the bracket design may comply with the ACI 318 Code provisions for shear friction. With this method, a failure crack location is assumed. For example, a corbel may be assumed to fail by cracking along the face of its support. Reinforcement then is provided perpendicular to the crack to prevent failure. Due to the rough surface at a crack, friction develops under the tension in the reinforcement and holds together the two sections on opposite sides of the crack. An equal compression load develops in the concrete at the confined crack.

The shear stress at the crack, at the face of the column or bracket support, is limited to $0.2f'_c$, or 800 psi maximum.

The area of shear-friction reinforcement A_{vf} required in addition to reinforcement provided to take the direct tension due to temperature changes or shrinkage should be computed from

$$A_{vf} = \frac{V_u}{\phi f_y \mu} \tag{8-74}$$

where V_u is the design shear at the section; f_y is the reinforcement yield strength, but not more than 60 ksi; and μ, the coefficient of friction, is 1.4 for monolithic concrete, 1.0 for concrete placed against hardened concrete, and 0.7 for concrete placed against structural rolled-steel members. The shear-friction reinforcement should be well distributed across the face of the crack and properly anchored at each side.

Tension Reinforcement. A_s should be adequate at the face of the support to resist the moments due to the vertical load and any horizontal forces. This reinforcement must be properly developed to prevent pullout, by proper anchorage within the support and by a cross bar welded to the bars at the end of the bracket.

8-37. Compression Members and Slenderness Effects. Very short compression members, piers or pedestals, may be unreinforced if the unit compressive stress on the cross-sectional area is less than the allowable bearing stress $0.85\phi f'_c$, where f'_c is the 28-day compressive strength of the concrete and ϕ, the capacity reduction factor, is 0.70. The depth or width of an unreinforced pier or pedestal on soil shall be such that the flexural tensile stress in the concrete does not exceed $5\phi\sqrt{f'_c}$, where $\phi = 0.65$, when computed by the ultimate-strength method. The ratio of height to least dimension should not exceed 3 for unreinforced pedestals. In any event, pedestals must be designed as reinforced columns when loaded beyond the capacity of plain concrete.

In reinforced concrete columns, longitudinal steel bars assist the concrete in carrying the load. Steel ties or spiral wrapping around those bars prevent the bars from buckling outward and spalling the outer concrete shell. Since spirals are more effective, columns with closely spaced spirals are allowed to carry greater loads than comparable columns with ties. Both types of columns may be designed by ultimate-load theory (Art. 8-38) or working stresses (Art. 8-39).

ACI 318, Building Code Requirements for Reinforced Concrete, American Concrete Institute, sets limitations on column geometry and reinforcement. Following are some of the more important.

Reinforcement Cover. For cast-in-place columns, spirals and ties should be protected with a monolithic concrete cover of at least 1½ in. But for severe exposures, the amount of cover should be increased.

Spirals. This type of transverse reinforcement should be at least ⅜ in. in diameter. A spiral may be anchored at each of its ends by 1½ extra turns of the spiral. Splices may be made by welding or by a lap of 48 bar diameters (but at least 12 in.). Spacing (pitch) of spirals should not exceed 3 in. or be less than 1 in. Clear spacing should be at least 1⅓ times the maximum size of coarse aggregate.

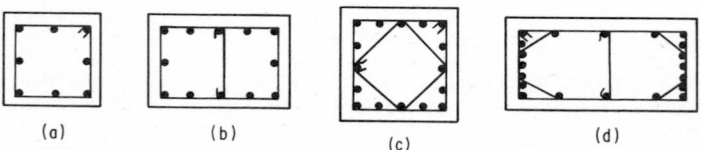

Fig. 8-15. Column ties provide lateral support at corners to alternate reinforcing bars.

A spiral should extend to the level of the lowest horizontal reinforcement in the slab, beam, or drop panel above. Where beams are of different depth or are not present on all sides of a column, ties should extend above the termination of the spiral to the bottom of the shallowest member. In a column with a capital, the spiral should extend to a plane at which the diameter or width of the capital is twice that of the column.

The ratio of the volume of spiral steel to volume of concrete core (out to out of spiral) should be at least

$$\rho_s = 0.45 \left(\frac{A_g}{A_c} - 1\right) \frac{f_c'}{f_y} \qquad (8-75)$$

where A_g = gross area of column
A_c = core area of column measured to outside of spiral
f_y = spiral steel yield strength
f_c' = 28-day compressive strength of concrete

Ties. Lateral ties should be at least ⅜ in. in diameter for No. 10 or smaller bars and ½ in. in diameter for No. 11 and larger bars. Spacing should not exceed 16 bar diameters, 48 tie diameters, or the least dimension of the column. The ties should be so arranged that every corner bar and alternate longitudinal bars will have lateral support provided by the corner of a tie having an included angle of not more than 135° (Fig. 8-15). No bar should be more than 6 in. from such a laterally supported bar. Where bars are located around a circle, a complete circular tie may be used. (For more details, see ACI 315, Manual of Standard Practice for Detailing Reinforced Concrete Structures, American Concrete Institute, Detroit, Mich.)

Minimum Reinforcement. Columns should be reinforced with at least six longitudinal bars in a circular arrangement or with four longitudinal bars in a rectangular arrangement, at least No. 5 in bar size. Area of column reinforcement should not be less than 1% or more than 8% of the gross cross-sectional area of a column.

Excess Concrete. In a column that has a larger cross section than that required by load, the effective area A_g used to determine minimum reinforcement area and load capacity may be reduced proportionately, but not to less than half the total area.

Slenderness Ratios. Building columns generally are relatively short. Thus, an approximate evaluation of slenderness effects can usually be used in design. Slenderness, which is a function of column geometry and bracing, can reduce the load-carrying capacity of compression members by introducing bending stresses and can lead to a buckling failure.

Load-carrying capacity of a column decreases with increase in unsupported length l_u beyond a certain length. In buildings, l_u should be taken as the clear distance between floor slabs, girders, or other members capable of providing lateral support to the column, or as the distance from a floor to a column capital or a haunch, if one is present.

In contrast, load-carrying capacity increases with increase in radius of gyration r of the column cross section. For rectangular columns, r may be taken as 30% of the overall dimension in the direction in which stability is being considered and for circular members as 25% of the diameter.

Also, the greater the resistance offered by a column to sidesway, or drift, because of lateral bracing or restraint against end rotations, the higher the load-carrying capacity. This resistance is represented by application of a factor k to the unsupported length of the column, and kl_u is referred to as the effective length of the column.

The combination of these factors, which is a measure of the slenderness of a column, kl_u/r, is called the slenderness ratio of the column.

The effective length factor k can be determined by analysis. If an analysis is not made, for compression members braced against sidesway, k should be taken as unity. For columns not braced against sidesway, k will be greater than unity; analysis should take into account the effects of cracking and reinforcement on relative stiffness.

ACI Committee 441 has proposed that k should be obtained from the Jackson and Moreland alignment chart, reproduced as Fig. 8-16 (Commentary on ACI 318-71, American Concrete Institute). For determination of k with this chart, a parameter ψ_A must be computed for end A of column AB, and a similar parameter ψ_B must be computed for end B. Each parameter equals the ratio at that end of the column of the sum of EI/l_u for the compression members meeting there to the sum of EI/l for the flexural members meeting there, where EI is the flexural stiffness of a member.

As a guide in judging whether a frame is braced or unbraced, note that the Commentary on ACI 318-71 indicates that a frame may be considered braced if the bracing elements, such as shear walls, shear trusses, or other means resisting lateral movement in a story, have a total

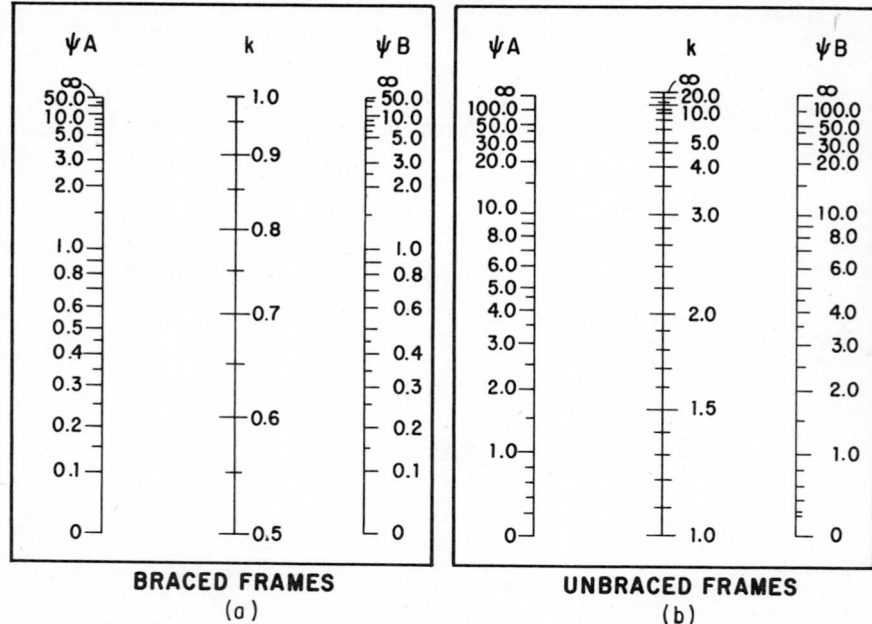

Fig. 8-16. Alignment charts for determination of effective length factor k for columns. ψ is the ratio for each end of a column of $\Sigma EI/l_u$ for the compression members to $\Sigma EI/l$ for the girders.

stiffness at least six times the sum of the stiffnesses of all the columns resisting lateral movement in that story.

The slenderness effect may be neglected under the following conditions:

For columns braced against sidesway, when

$$\frac{kl_u}{r} < 34 - 12 \frac{M_1}{M_2} \tag{8-76}$$

where M_1 = smaller of the two end moments on a column as determined by conventional elastic frame analysis, with positive sign if column is bent in single curvature and negative sign if column is bent in double curvature

M_2 = absolute value of the larger of the two end moments on a column as determined by conventional elastic frame analysis

For columns not braced against sidesway, when

$$\frac{kl_u}{r} < 22 \tag{8-77}$$

Column Design Loads.   Analysis taking into account the influence of axial loads and variable moment of inertia on member stiffness and fixed-end moments, the effects of deflections on moments and forces, and the effects of duration of loads is required for all columns when

$$\frac{kl_u}{r} > 100 \tag{8-78}$$

For columns for which the slenderness ratio lies between 22 and 100, and therefore the slenderness effect on load-carrying capacity must be taken into account, either an elastic analysis can be performed to evaluate the effects of lateral deflections and other effects producing secondary stresses or an approximate method presented in the ACI Code may be used.   In the approximate method, the column is designed for the design axial load P_u and a magnified moment M_c defined by

$$M_c = \delta M_2 \tag{8-79}$$

where δ is the magnification factor, a function of the shape of the deflected column. δ may be determined from

$$\delta = \frac{C_m}{1 - P_u/\phi P_c} \geq 1 \tag{8-80}$$

where C_m = a factor relating the actual moment diagram to that for an equivalent uniform moment

ϕ = capacity reduction factor = 0.75 for spiral-reinforced columns, otherwise 0.70

P_c = critical load for the column

For members braced against sidesway and without transverse loads between supports,

$$C_m = 0.6 + 0.4 \frac{M_1}{M_2} \geq 0.4 \tag{8-81}$$

For other members, $C_m = 1$.

The critical load is given by

$$P_c = \frac{\pi^2 EI}{(kl_u)^2} \tag{8-82}$$

where EI is the flexural stiffness of the column.

The flexural stiffness EI may be computed approximately from

$$EI = \frac{E_c I_g / 2.5}{1 + \beta_d} \tag{8-83}$$

where E_c = modulus of elasticity of the concrete

I_g = moment of inertia about the centroidal axis of gross concrete section, neglecting load reinforcement

β_d = ratio of maximum design dead load to total load moment (always taken positive)

Because a column has different properties, such as stiffness, slenderness ratio, and δ, in different directions, it is necessary to check the strength of a column in each of its two principal directions.

(G. Winter and A. H. Nilson, "Design of Concrete Structures," McGraw-Hill Book Company, New York; P. F. Rice and E. S. Hoffman, "Structural Design Guide to the ACI Building Code," Van Nostrand Reinhold Company, New York.)

8-38. Ultimate-Strength Design of Columns. (See also Arts. 8-20 and 8-37.) At ultimate strength, columns should be capable of sustaining loads as given by Eqs. (8-7) to (8-9), at actual eccentricities. Columns reinforced with spirals should be designed for an eccentricity of at least $0.05h$, where h is the overall thickness of a rectangular section or diameter of a circular section. Tied columns should be sized for a minimum eccentricity of $0.10h$. All columns must be designed for at least a 1-in. eccentricity.

These eccentricities are measured from the plastic centroid. This is the centroid of the resistance to load computed for the assumptions that the concrete is stressed uniformly to $0.85f'_c$ and the steel is stressed uniformly to f_y, where f'_c is the 28-day strength of the concrete and f_y the yield strength of the steel.

Columns are designed under the ACI Code assumptions and requirements pertaining to members subject to combined flexure and axial load. These assumptions are:

1. Loads and stresses are in equilibrium and strains are compatible.

2. Strains in both reinforcing steel and concrete are proportional to distance from the neutral axis.

3. The maximum strain at the extreme concrete compression surface does not exceed 0.003 in. per in.

4. Stress in the reinforcement is E_s times the steel strain, where E_s is the modulus of elasticity of the steel, and the stress does not exceed f_y.

5. Tensile strength of concrete is negligible.

6. The concrete block may be taken as rectangular (Fig. 8-10c), with a concrete stress equal to $0.85f'_c$ extending from the extreme compression surface to a line parallel to the neutral axis and at a distance $a = \beta_1 c$ from the extreme compression surface. $\beta_1 = 0.85$ for concrete strengths up to $f'_c = 4{,}000$ psi and decreases at a rate of 0.05 for each 1,000 psi of strength above 4,000 psi; and c is the distance from the extreme compression surface to the neutral axis.

Strength computed in accordance with these assumptions should be modified by a capacity reduction factor ϕ. It is equal to 0.75 for columns with spiral reinforcement and 0.70 for tied columns.   A larger value of ϕ may be used for small design axial compression loads P_u.   For symmetrically reinforced columns, ϕ generally may be increased when $P_u \leq 0.10f'_cA_g$, where A_g is the gross area of the section.   For unsymmetrically reinforced columns, ϕ may be increased when P_u is less than the smaller of $0.10f'_cA_g$ and the design axial load for balanced conditions P_b.   (In that case, the reinforcement ratio ρ should not exceed $0.75\rho_b$ for balanced conditions, as indicated for beams in Art. 8-23.   Balanced conditions exist at a cross section when tension reinforcement reaches f_y just as the concrete in compression reaches its ultimate strain of 0.003 in. per in.)   When $P_u = 0$, $\phi = 0.90$, the capacity reduction factor for pure bending. ϕ may be assumed to increase linearly to 0.90 as P_u decreases from $0.10f'_cA_g$ or P_b to zero.

The axial-load capacity P_u, lb, of short, rectangular members subject to axial load and bending may be determined from

$$P_u = \phi(0.85f'_cba + A'_sf_y - A_sf_s) \tag{8-84}$$

$$P_ue' = \phi\left[0.85f'_cba\left(d - \frac{a}{2}\right) + A'_sf_y(d - d')\right] \tag{8-85}$$

where e' = eccentricity, in., of axial load at end of member with respect to centroid of tensile reinforcement, calculated by conventional methods of frame analysis
b = width of compression face, in.
a = depth of equivalent rectangular compressive-stress distribution, in.
A'_s = area of compressive reinforcement, sq in.
A_s = area of tension reinforcement, sq in.
d = distance from extreme compression surface to centroid of tensile reinforcement, in.
d' = distance from extreme compression surface to centroid of compression reinforcement, in.
f_s = tensile stress in steel, psi

Equations (8-84) and (8-85) assume that a does not exceed h, that the reinforcement is in one or two faces, each parallel to the axis of bending, and that all reinforcement in any one face is located at about the same distance from the axis of bending.   Whether the compression steel will actually yield at ultimate strength, as assumed in these and the following equations, can be verified by strain compatibility calculations.   That is, the strain in the compression steel, $0.003(c - d')/c$, when the concrete crushes must be larger than the strain when the steel starts to yield, f_y/E_s, where c is the distance, in., from the extreme compression surface to the neutral axis and E_s is the modulus of elasticity of the steel, psi.

The load P_b for balanced conditions can be computed from Eq. (8-84) with $f_s = f_y$ and

$$a = a_b = \beta_1c_b = \frac{87,000\beta_1d}{87,000 + f_y} \tag{8-86}$$

The balanced moment can be obtained from

$$M_b = P_be_b = \phi\left[0.85f'_cba_b\left(d - d'' - \frac{a_b}{2}\right) + A'_sf_y(d - d' - d'') + A_sf_yd''\right] \tag{8-87}$$

where e_b is the eccentricity, in., of the axial load with respect to the plastic centroid, and d'' is the distance, in., from plastic centroid to centroid of tension reinforcement.

When P_u is less than P_b, or the eccentricity e is greater than e_b, tension governs.   In that case, for unequal tension and compression reinforcement, the ultimate strength is

$$P_u = 0.85f'_cbd\phi\left\{\rho'm' - \rho m + \left(1 - \frac{e'}{d}\right)\right.$$
$$\left. + \sqrt{\left(1 - \frac{e'}{d}\right)^2 + 2\left[(\rho m - \rho'm')\frac{e'}{d} + \rho'm'\left(1 - \frac{d'}{d}\right)\right]}\right\} \tag{8-88}$$

where $m = f_y/0.85f'_c$
$m' = m - 1$
$\rho = A_s/bd$
$\rho' = A'_s/bd$

For symmetrical reinforcement in two faces, Eq. (8-88) becomes

$$P_u = 0.85f'_cbd\phi\left\{-\rho + 1 - \frac{e'}{d} + \sqrt{\left(1 - \frac{e'}{d}\right)^2 + 2\rho\left[m'\left(1 - \frac{d'}{d}\right) + \frac{e'}{d}\right]}\right\} \tag{8-89a}$$

For no compression reinforcement, Eq. (8-88) becomes

$$P_u = 0.85f_c'bd\phi\left[-\rho m + 1 - \frac{e'}{d} + \sqrt{\left(1 - \frac{e'}{d}\right)^2 + 2\,\frac{e'\rho m}{d}}\right] \tag{8-89b}$$

When P_u is greater than P_b, or e is less than e_b, compression governs. In that case, the ultimate strength is approximately

$$P_u = P_o - (P_o - P_b)\frac{M_u}{M_b} \tag{8-90a}$$

$$P_u = \frac{P_o}{1 + (P_o/P_b - 1)(e/e_b)} \tag{8-90b}$$

where M_u = moment capacity under combined axial load and bending, in.-lb
P_o = axial load capacity of the member when concentrically loaded, lb

$$P_o = \phi[0.85f_c'(A_g - A_{st}) + A_{st}f_y] \tag{8-91}$$

where A_g = gross area of section, sq in.
A_{st} = total area of longitudinal reinforcement, sq in.
For symmetrical reinforcement in single layers, the ultimate strength when compression governs may be computed from

$$P_u = \phi\left(\frac{A_s'f_y}{\dfrac{e}{d - d'} + 0.5} + \frac{bhf_c'}{\dfrac{3he}{d^2} + 1.18}\right) \tag{8-92}$$

Ultimate strength of short, circular members with bars in a circle may be determined by the theory discussed in Art. 8-20 or from the following:
When tension controls:

$$P_u = 0.85f_c'D^2\phi\left[\sqrt{\left(\frac{0.85e}{D} - 0.38\right)^2 + \frac{\rho_t m D_s}{2.5D}} - \left(\frac{0.85e}{D} - 0.38\right)\right] \tag{8-93}$$

where D = overall diameter of the section, in.
D_s = diameter of the circle through the reinforcement, in.
$\rho_t = A_{st}/A_g$
When compression governs:

$$P_u = \phi\left[\frac{A_{st}f_y}{\dfrac{3e}{D_s} + 1} + \frac{A_g f_c'}{\dfrac{9.6De}{(0.8D + 0.67D_s)^2} + 1.18}\right] \tag{8-94}$$

The eccentricity for the balanced condition is given approximately by

$$e_b = (0.24 + 0.39\rho_t m)D \tag{8-95}$$

Ultimate strength of short, square members with bars in a circle may be computed from the following:
When tension controls:

$$P_u = 0.85bhf_c'\phi\left[\sqrt{\left(\frac{e}{h} - 0.5\right)^2 + 0.67\frac{D_s}{h}\rho_t m} - \left(\frac{e}{h} - 0.5\right)\right] \tag{8-96}$$

When compression governs:

$$P_u = \phi\left[\frac{A_{st}f_y}{\dfrac{3e}{D_s} + 1} + \frac{A_g f_c'}{\dfrac{12he}{(h + 0.67D_s)^2} + 1.18}\right] \tag{8-97}$$

When the slenderness of a column has to be taken into account, the eccentricity should be determined from $e = M_c/P_u$, where M_c is the magnified moment given by Eq. (8-79).

As outside temperatures vary, exposed columns in tall buildings may undergo large changes in length relative to interior columns. The resulting floor warpage may crack partitions unless they are detailed to take the cracking.

(P. F. Rice and E. S. Hoffman, "Structural Design Guide to the ACI Building Code," Van Nostrand Reinhold Company, New York; P. Rogers, "Reinforced Concrete Design for Buildings," Van Nostrand Reinhold Company, New York; "CRSI Handbook," Concrete Reinforcing Steel Institute, Chicago, Ill.; "Design Handbook in Accordance with the Strength Design Method of ACI 318-71," American Concrete Institute, Detroit, Mich.; "Strength Design of Reinforced Concrete Column Sections" (computer program), Portland Cement Association, Old Orchard Road, Skokie, Ill. 60076.)

8-39. Working-Stress Design of Columns. (See also Art. 8-36.) In working-stress design, the capacity of a column is taken as 40% of that determined by the ultimate-strength method (Art. 8-38) with $\phi = 1$. The capacity should be equal to or greater than the service loads on the column.

8-40. Walls. Concrete walls may be classified as non-load-bearing, load-bearing, or shear walls. The last may be load-bearing or non-load-bearing.

Non-load-bearing Walls. These are generally basement, retaining, or facade-type walls that support only their own weight and also resist lateral loads. Such walls are principally designed for flexure. By the ACI Code, design requirements include:

1. Ratio of vertical reinforcement to gross concrete area should be at least 0.0012 for deformed bars No. 5 or smaller, 0.0015 for deformed bars No. 6 and larger, and 0.0012 for welded-wire fabric not larger than ⅝ in. in diameter.

2. Spacing of vertical bars should not exceed three times the wall thickness or 18 in.

3. Lateral or cross ties are not required if the vertical reinforcement is 1% or less of the concrete area, or where the vertical reinforcement is not required as compression reinforcement.

4. Ratio of horizontal reinforcement to gross concrete area should be at least 0.0020 for deformed bars No. 5 or smaller, 0.0025 for deformed bars No. 6 and larger, and 0.0020 for welded-wire fabric not larger than ⅝ in. in diameter.

5. Spacing of horizontal bars should not exceed 3 times the wall thickness or 18 in.

Note that there is no requirement for the reinforcement to be placed at both faces of the wall, but it is good practice to provide nominal reinforcing to control shrinkage in the nonstressed face of foundation walls over 10 or 12 ft in height and also at the faces of walls exposed to view.

Load-bearing Walls. These are subject to axial compression loads in addition to their own weight and, where there is eccentricity of load or lateral loads, also to flexure. Load-bearing walls may be designed in a manner similar to that for columns (Arts. 8-38 and 8-39), but including the preceding design requirements for non-load-bearing walls. As an alternative, load-bearing walls may be designed by an empirical procedure given in the ACI Code when the eccentricity of the resulting compressive load is equal to or less than one-sixth the thickness of the wall.

In the empirical method, the axial capacity of the wall is

$$P_u = 0.55\phi f'_c A_g \left[1 - \left(\frac{l_c}{40h} \right)^2 \right] \qquad (8\text{-}98)$$

where f'_c = 28-day compressive strength of the concrete, psi

A_g = gross area of wall section, sq in.

ϕ = capacity reduction factor = 0.70

l_c = vertical distance between supports, in.

h = overall thickness of wall, in.

The effective length of wall supporting a concentrated load should be taken as the smaller of the distance c to c between loads and the bearing width plus $4h$.

Reinforced bearing walls designed using Eq. (8-98) should have a thickness of at least ¹/₂₅ of the unsupported height or width, whichever is shorter, but not less than 6 in. (See also Art. 15-6.) Also, walls more than 10 in. thick, except for basement walls, should have two layers of reinforcement in each direction, with between one-half and two-thirds of the total steel area in the layer near the exterior face of the wall. Walls should be anchored to the floors, or to the columns, pilasters, or intersecting walls.

If structural analysis indicates adequate strength and stability with less thickness and reinforcement than the ACI Code requires, less may be used.

Walls designed as grade beams should have top and bottom reinforcement as required by the ACI Code for beam design.

Shear Walls. Walls subject to horizontal shear forces in the plane of the wall should, in addition to satisfying flexural requirements, be capable of resisting the shear. The nominal shear stress can be computed from

$$v_u = \frac{V_u}{\phi h d} \qquad (8\text{-}99)$$

where V_u = total design shear force
  ϕ = capacity reduction factor = 0.85
  $d = 0.8 l_w$
  l_w = horizontal length of wall
  h = overall thickness of wall

The shear stress carried by the concrete depends on whether N_u, the design axial load, lb, normal to the wall horizontal cross section and occurring simultaneously with V_u at the section, is a compression or tension force. When N_u is a compression force, v_c may be taken as $2\sqrt{f_c'}$. When N_u is a tension force,

$$v_c = 2 \left(1 - 0.002 \, \frac{N_u}{A_g} \right) \sqrt{f_c'} \qquad (8\text{-}100)$$

When the applied shear stress v_u is less than $0.5 v_c$, reinforcement should be provided as required by the empirical method for bearing walls.

When v_u exceeds $0.5 v_c$, reinforcement should be provided in accordance with Eq. (8-18). Also, the ratio ρ_h of horizontal shear reinforcement to the gross concrete area of the vertical section of the wall should be at least 0.0025. Spacing of horizontal shear bars should not exceed $l_w/5$, $3h$, or 18 in. In addition, the ratio of vertical shear reinforcement area to gross concrete area of the horizontal section of wall need not be greater than that required by Eq. (8-18) but should not be less than

$$\rho_n = 0.0025 + 0.5 \left(2.5 - \frac{h_w'}{l_w} \right) (\rho_h - 0.0025) \qquad (8\text{-}101)$$

where h_w = total height of wall
Spacing of vertical shear reinforcement should not exceed $l_w/3$, $3h$, or 18 in.

In no case should the total design shear stress v_u exceed $10\sqrt{f_c'}$ at any section.

8-41. Composite Columns. A composite column consists of a structural steel shape, pipe, or tube compression member completely encased in concrete, with or without longitudinal reinforcement.

Composite compression members should be designed in accordance with the provisions applicable to ordinary reinforced concrete columns. Loads assigned to the concrete portion of a member must be transferred by direct bearing on the concrete through brackets, plates, reinforcing bars, or other structural shapes that have been welded to the central structural steel compression members prior to placement of the perimeter concrete. The balance of the load should be assigned to the structural steel shape and should be developed by direct connection to the structural shape.

Concrete-filled Steel Columns. When the composite member consists of a steel-encased concrete core, the required thickness of the metal wall for each face of width b of a rectangular

$$t = b \sqrt{\frac{f_y}{3 E_s}} \qquad (8\text{-}102a)$$

and for circular sections of diameter h,

$$t = h \sqrt{\frac{f_y}{8 E_s}} \qquad (8\text{-}102b)$$

where f_y is the yield strength and E_s the modulus of elasticity of the steel.

Steel-Core Columns. When the composite member consists of a spiral-bound concrete encasement around a structural-steel core, the concrete should have a minimum strength of 2,500 psi, and spiral reinforcement should conform to the requirements of Art. 8-37.

When the composite member consists of a laterally tied concrete encasement around a steel core, the concrete should have a minimum strength of 2,500 psi. The lateral ties should completely encase the core. Ties should be No. 3 to No. 5 bar size but should have a diameter of at least $1/50$ the longest side of the cross section. Vertical spacing should not exceed one-half of the least width of the cross section, or 48 tie bar diameters, or 16 longitudinal bar diameters.

The area of vertical reinforcing bars within the ties should not be less than 1% or more than 8% of the net concrete section. In rectangular sections, at least three vertical bars should be placed along each face of the cross section, including one bar in each corner.

The design yield strength of the structural core should not be taken greater than 50 ksi, even though a larger yield strength may be specified.

8-42. Basic Principles of Prestressed Concrete.   Prestressing is the application of permanent forces to a member or structure to counteract the effects of subsequent loading. Applied to concrete, prestressing takes the form of precompression, usually to eliminate disadvantages stemming from the weakness of concrete in tension. The usual procedure is to tension high-strength steel (Art. 8-16) and anchor it to the concrete, which resists the tendency of the stretched steel to shorten and thus is compressed. The amount of prestress used generally is sufficient to prevent cracking or sometimes to avoid tension entirely, under service loads. As a result, the whole concrete cross section is available to resist tension and bending, whereas in reinforced concrete construction, concrete in tension is considered ineffective. Hence, it is particularly advantageous with prestressed concrete to use high-strength concrete. (See also Art. 8-17.)

Prestressed-concrete pipe and tanks are made by wrapping steel wire under high tension around concrete cylinders. Domes are prestressed by wrapping tensioned steel wire around the ring girders. Beams and slabs are prestressed linearly with steel tendons anchored at their ends or bonded to the concrete (Art. 8-17). Piles also are prestressed linearly, usually to counter-act handling stresses.

The final precompression of the concrete is not equal to the initial tension applied to the tendons. There are immediate losses, for example, due to elastic shortening of the concrete, frictional losses from curvature of the tendons, and slip at anchorages. And there are long-time

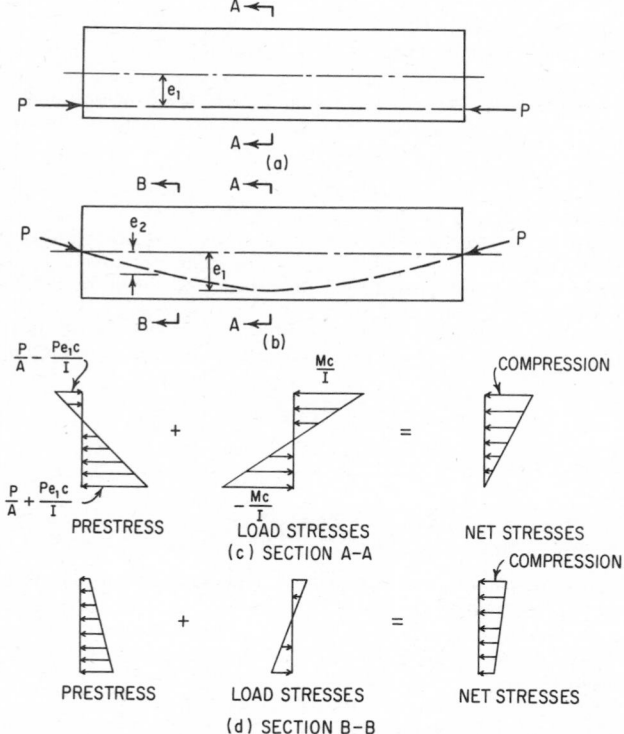

Fig. 8-17. Concrete beams may be prestressed with straight tendons (*a*) or curved (*b*). Stresses at midspan may be the same for both positions (*c*); but with the curved tendons, net stresses may remain compressive away from midspan (*d*), whereas they become tensile near the ends with straight tendons.

losses, such as those due to shrinkage and creep of the concrete and possibly relaxation of the steel. These losses should be computed as accurately as possible, or determined experimentally, or estimated. They should be deducted from the initial prestress to determine the effective prestress to be used in design. One of the reasons that high-tensioned tendons are used for prestressing is to maintain the sum of these losses at a small percentage of the applied prestress.

In determining stresses in prestressed members, the prestressing forces may be treated in the same way as other external loads. If the prestress is large enough to prevent cracking under design loads, elastic theory may be applied to the entire concrete cross section.

For example, consider the simple beam in Fig. 8-17a. Prestress P is applied by a straight tendon at a distance e_1 below the neutral axis. The resulting prestress in the extreme surfaces throughout equals $P/A \pm Pe_1c/I$, where P/A is the average stress on a cross section and Pe_1c/I, the bending stress (+ represents compression, − represents tension), as indicated in Fig. 8-17c. If, now, stresses $\pm Mc/I$ due to downward-acting loads are superimposed at midspan, the net stresses in the extreme surfaces may become zero at the bottom and compressive at the top (Fig. 8-17c). Since the stresses due to loads at the beam ends are zero, however, the prestress is the final stress there. Hence, the top of the beam at the ends will be in tension.

If this is objectionable, the tendons may be draped, or harped, in a vertical curve, as shown in Fig. 8-17b. Stresses at midspan will be substantially the same as before (assuming the horizontal component of P approximately equal to P), and the stress at the ends will be a compression, P/A, since P passes through the centroid of the section there. Between midspan and the ends, the cross sections also are in compression (Fig. 8-17d).

8-43. Losses in Prestress. As pointed out in Art. 8-42, the prestressing force acting on the concrete differs from the initial tension on the tendons by losses that occur immediately and over a long time.

Elastic Shortening of Concrete. In pretensioned members (Art. 8-17), when the tendons are released from fixed abutments and the steel stress is transferred to the concrete by bond, the concrete shortens because of the compressive stress. For axial prestress, the decrease in inches per inch of length may be taken as P_i/AE_c, where P_i is the initial prestress, lb; A the concrete area, sq in.; and E_c the modulus of elasticity of the concrete, psi. Hence, the decrease in unit stress in the tendons equals $P_iE_s/AE_c = nf_c$, where E_s is the modulus of elasticity of the steel, psi; n the modular ratio; and f_c the stress in the concrete, psi.

In posttensioned members, if tendons or cables are tensioned individually, the stress loss in each due to compression of the concrete depends on the order of tensioning. The loss will be greatest for the first tendon or cable tensioned and least for the last one. The total loss may be approximated by assigning half the loss in the first cable to all. As an alternative, the tendons may be brought to the final prestress in steps.

Frictional Losses. In posttensioned members, there may be a loss of prestress where curved tendons rub against their enclosure. For harped tendons, the loss may be computed in terms of a curvature-friction coefficient μ. Losses due to unintentional misalignment may be calculated from a wobble-friction coefficient K (per lin ft). Since the coefficients vary considerably with duct material and construction methods, they should, if possible, be determined experimentally or obtained from the tendon manufacturer. Table 8-15 lists values of K and μ suggested in the Commentary on ACI 318-71, American Concrete Institute, for posttensioned tendons.

Table 8-15. Friction Coefficients For Posttensioned Tendons

Types of tendons and sheathing	Wobble coefficient, K	Curvature coefficient μ
Grouted tendons in metal sheathing:		
Wire tendons	0.0010–0.0015	0.15–0.25
High-strength bars	0.0001–0.0006	0.08–0.30
7-wire strand	0.0005–0.0020	0.15–0.25
Unbonded tendons		
Mastic-coated:		
Wire tendons	0.001–0.002	0.05–0.15
7-wire-strand	0.001–0.002	0.05–0.15
Pregreased:		
Wire tendons	0.0003–0.002	0.05–0.15
7-wire strand	0.0003–0.002	0.05–0.15

With K and μ known or estimated, the friction loss can be computed from

$$P_s = P_x e^{Kl + \mu\alpha} \tag{8-103}$$

where P_s = force in tendon at prestressing jack, lb
  P_x = force in tendon at any point x ft from jack, lb
  $e = 2.718$
  l = length of tendon from jacking point to point x, ft
  α = total angular change of tendon profile from jacking end to point x, radians
When $Kl + \mu\alpha$ does not exceed 0.3, P_s may be obtained from

$$P_s = P_x(1 + Kl + \mu\alpha) \tag{8-104}$$

Slip at Anchorages. For posttensioned members, prestress loss may occur at the anchorages during the anchoring. For example, seating of wedges may permit some shortening of the tendons. If tests of a specific anchorage device indicate a shortening δl, the decrease in unit stress in the steel is $E_s \delta l / l$, where l is the length of the tendon.

Shrinkage of Concrete. Change in length of a member due to concrete shrinkage results over a period of time in prestress loss. This should be determined from test or experience. Generally, the loss is greater for pretensioned members than for posttensioned members, which are prestressed after much of the shrinkage has occurred. Assuming a shrinkage of 0.0002 in. per in. for a pretensioned member, the loss in tension in the tendons will be

$$0.0002E_s = 0.0002 \times 30 \times 10^6 = 6{,}000 \text{ psi}$$

Creep of Concrete. Change in length of concrete under sustained load induces a prestress loss over a period of time. This loss may be several times the elastic shortening. An estimate of the loss may be made with a creep coefficient C_e, equal to the ratio of additional long-time deformation to initial elastic deformation, determined by test. Hence, for axial prestress, the loss in tension in the steel is $C_e n f_c$, where n is the modular ratio and f_c is the prestressing force divided by the concrete area. (Values ranging from 1.5 to 2.0 have been recommended for C_e.)

Relaxation of Steel. Decrease in stress under constant high strain occurs with some steels. For example, for steel tensioned to 60% of ultimate strength, relaxation loss may be 3%. This type of loss may be reduced by temporary overtensioning, stabilizing the strand by artificially accelerating relaxation and thus reducing the loss that will occur later at lower stresses.

Standard Specifications for Highway Bridges (American Association of State Highway and Transportation Officials) permit the sum of the prestress losses in the steel from all causes except friction to be assumed as 35,000 psi for pretensioned members and 25,000 psi for posttensioned members.

8-44. Allowable Stresses in Prestressed Concrete. In setting allowable stresses for prestressed concrete, design codes recognize two loading stages, application of initial stress and loading under service conditions. The codes permit higher stresses for the temporary loads during the initial stage.

Stresses due to the jacking force and those produced in the concrete and steel immediately after prestress transfer or tendon anchorage, before losses due to creep and shrinkage, are considered temporary. Permissible temporary stresses in the concrete are specified as a percentage of f'_{ci}, the compressive strength of the concrete, psi, at time of initial prestress, instead of the usual f'_c, 28-day strength of the concrete. This is done because prestress usually is applied only a few days after casting the concrete. With f_{pu} as the ultimate strength of tendons, the allowable stresses for prestressed concrete, in accordance with ACI 318, Building Code Requirements for Reinforced Concrete (American Concrete Institute), are given in Table 8-16.

Bearing stress on the concrete from anchorages of posttensioned members with adequate reinforcement in the end region should not exceed f'_{ci} or

$$f_{cp} = 0.6 f'_{ci} \sqrt[3]{\frac{A'_b}{A_b}} \tag{8-105}$$

where A_b = bearing area of anchor plate
  A'_b = maximum area of the portion of the anchorage surface that is geometrically similar to and concentric with the area of the anchor plate
A more refined analysis may be applied in the design of the end-anchorage regions of prestressed members to develop the ultimate strength of the tendons. ϕ should be taken as 0.90 for the concrete. Adequate reinforcement should be provided to prevent bursting, horizontal

Table 8-16. Allowable Stresses in Prestressed Concrete Flexural Members

Stresses at transfer or anchoring:

Compression in concrete	$0.60f'_{ci}$
Tension in concrete without auxiliary reinforcement in the tension zone*	$3\sqrt{f'_{ci}}$
Prestress in tendons due to jacking force†	$0.80f_{pu}$
Prestress in tendons immediately after transfer or anchoring	$0.70f_{pu}$
Stresses under service loads:	
Compression in concrete	$0.45f'_c$
Tension in concrete‡	$6\sqrt{f'_c}$
Tension in precompressed concrete where deflections based on transformed cracked section meet limits in Art. 8-22§	$12\sqrt{f'_c}$

*Where the calculated tension stress exceeds this value, reinforcement should be provided to resist the total tension force on the concrete computed on the assumption of an uncracked section.

†But not greater than the maximum value recommended by the manufacturer of the steel or anchorages.

‡Allowable tension in members not exposed to freezing or to a corrosive environment. For members exposed to such conditions, more concrete cover should be provided around the steel than that normally required by the ACI Code (Art. 8-17), and crack-controlling reinforcement should be provided in the tension zone.

§Cover should be increased 50% over that normally required for prestressed concrete (Art. 8-17).

splitting, and spalling. End blocks should be used to distribute end-bearing loads and those due to concentrated prestressing forces.

(J. R. Libby, "Modern Prestressed Concrete," Van Nostrand Reinhold Company, New York.)

8-45. Design of Prestressed-Concrete Beams. (See also Arts. 8-16, 8-17, and 8-42 to 8-44.) This involves selection of shape and dimensions of the concrete portion, type and positioning of tendons, and amount of prestress. After a concrete shape and dimensions have been assumed, determine geometric properties: cross-sectional area, center of gravity, distances of extreme surfaces from the centroid, section moduli, and dead load of member per unit of length. Treat prestressing forces as a system of external forces acting on the concrete.

Compute bending stresses due to dead and live loads. From these, determine the magnitude and location of the prestressing force at points of maximum moment. This force must provide sufficient compression to offset the tensile stresses caused by the bending moments due to loads (Fig. 8-17). But at the same time, it must not create any allowable stresses exceeding those listed in Art. 8-44. Investigation of other sections will guide selection of tendons to be used and determine their position in the beam.

After establishing the tendon profile, prestressing forces, and tendon areas, check critical points along the beam under initial and final conditions, on removal from the forms, and during erection. Check ultimate strength in flexure, diagonal tension, and percentage of prestressing steel. Design anchorages, if required, and diagonal-tension steel. Finally, check camber.

The design may be based on the following assumptions: Strains vary linearly with depth. At cracked sections, the concrete cannot resist tension. Before cracking, stress is proportional to strain. The transformed area of bonded tendons may be included in pretensioned members and in posttensioned members after the tendons have been grouted. Areas of open ducts should be deducted in calculations of section properties before bonding of tendons. The modulus of rupture should be determined from tests, or the cracking stress may be assumed as $7.5\sqrt{f'_c}$, where f'_c is the 28-day strength of the concrete, psi.

Prestressed beams may be designed by ultimate-strength theory (Art. 8-20). Beams for buildings should be capable of supporting the factored loads given by Eqs. (8-6) and (8-8). For bridge beams, the ultimate-load capacity should not be less than

$$U = \frac{1.30}{\phi}\left[D + \frac{5}{3}(L + I)\right] \tag{8-106}$$

where D = effect of dead load
L = effect of design live load
I = effect of impact

$\phi = 1.0$ for factory-produced precast, prestressed members
$= 0.95$ for posttensioned, cast-in-place members
$= 0.90$ for shear

Ultimate Strength in Bending. After cracking, a prestressed beam behaves essentially as an ordinary reinforced concrete beam. For rectangular sections, or flanged sections in which the neutral axis lies within the flange, the design resisting moment, in.-lb, can be computed from

$$M_t = 0.90[A_{ps}f_{ps}d(1 - 0.59\omega_p)] = 0.90\left[A_{ps}f_{ps}\left(d - \frac{a}{2}\right)\right] \tag{8-107}$$

where A_{ps} = area of tendons, sq in.
$\quad f_{ps}$ = calculated stress in tendons at design load, psi
$\quad d$ = distance from extreme compression surface to centroid of prestressing force, in.
$\quad \omega_p = \rho_p f_{ps}/f'_c$, with $\rho_p = A_{ps}/bd$
$\quad b$ = width of compression face, in.
$\quad a$ = depth of equivalent rectangular compressive-stress distribution, in.
$\quad\quad = A_{ps}f_{ps}/0.85f'_c b$

The neutral axis usually will lie within the flange when the flange thickness h_f, in., exceeds $1.4d\rho_p f_{ps}/f'_c$.

For flanged sections with neutral axis outside the flange (usually when h_f is less than $1.4d\rho_p f_{ps}/f'_c$):

$$M_t = 0.90\left[A_{pf}f_{ps}\left(d - \frac{a}{2}\right) + 0.85f'_c(b - b_w)h_f\left(d - \frac{h_f}{2}\right)\right] \tag{8-108}$$

where A_{pf} = area of steel required to develop compressive strength of overhanging flanges, sq in. $= 0.85f'_c(b - b_w)h_f/f_{ps}$
$\quad A_{pw}$ = area of tendons to develop web, sq in. $= A_{ps} - A_{pf}$
$\quad b_w$ = minimum width of web, in.

When information for determination of f_{ps} is not available, and if the effective prestress, after losses, is at least half the ultimate strength of the tendons, the following approximate value should be used for bonded tendons:

$$f_{ps} = f_{pu}\left(1 - \frac{0.5\rho_p f_{pu}}{f'_c}\right) \tag{8-109}$$

where f_{pu} = ultimate strength of the tendons, psi
and for unbonded tendons:

$$f_{ps} = f_{se} + 10,000 + \frac{f'_c}{100\rho_p} \tag{8-110}$$

but not more than $f_{se} + 60,000$, where f_{se} is the effective prestress in tendons, psi, or more than the tendon yield strength.

Nonprestressed reinforcement, in combination with tendons, may be assumed equivalent at ultimate moment to its area times its yield point. But this is permitted only if

$$\frac{\rho_p f_{ps}}{f'_c} + \frac{\rho f_y}{f'_c} \leq 0.3 \tag{8-111}$$

where $\rho = A_s/bd$
$\quad A_s$ = area of unprestressed reinforcement, sq in.
$\quad f_y$ = yield strength of unprestressed reinforcement, psi

The total amount of prestressed and unprestressed reinforcement should be adequate to develop an ultimate load in flexure at least 1.2 times the cracking load calculated for a modulus of rupture of $7.5\sqrt{f'_c}$.

Limitations on Steel Ratios. In calculating M_t from Eqs. (8-107) and (8-108), the amount of prestressing steel should be such that $\rho_p f_{ps}/f'_c$ is not more than 0.30. For flanged sections, ρ_p should be taken as A_{ps}/bd. For larger steel ratios, the ultimate moment should not exceed the following:

For rectangular sections, or flanged sections in which the neutral axis lies within the flange:

$$M_u = 0.225f'_c bd^2 \tag{8-112}$$

For flanged sections with the neutral axis outside the flange:

$$M_u = 0.90 \, [0.25f'_c b_w d^2 + 0.85f'_c(b - b_w)h_f(d - 0.5h_f)] \tag{8-113}$$

Shear. ACI 318, Building Code Requirements for Reinforced Concrete (American Concrete Institute), and Standard Specifications for Highway Bridges (American Association of State Highway and Transportation Officials) require that prestressed beams be designed to resist diagonal tension by ultimate-strength theory.

Shear reinforcement should consist of stirrups or welded-wire fabric. The area of shear reinforcement, sq in., set perpendicular to the beam axis, should not be less than

$$A_v = 50 \, \frac{b_w s}{f_y} \tag{8-114}$$

where s is the reinforcement spacing, in., except when v_u is less than one-half v_c; or when the depth of the member h is less than 10 in. or 2.5 times the thickness of the compression flange, or one-half the width of the web, whichever is largest.

Alternatively, a minimum area

$$A_v = \frac{A_{ps} f_{pu} s}{80 f_y d} \, \sqrt{\frac{d}{b_w}} \tag{8-115}$$

may be used if the effective prestress force is at least equal to 40% of the tensile strength of the flexural reinforcement.

The yield strength of shear reinforcement f_y used in design calculations should not exceed 60,000 psi.

Where shear reinforcement is required, it should be placed perpendicular to the axis of the member and should not be spaced farther apart than $0.75h$, where h is the overall depth of the member, or 24 in. Web reinforcement between the face of support and the section at a distance $h/2$ from it should be the same as the reinforcement required at that section.

The nominal shear stress v_u should be computed from

$$v_u = \frac{V_u}{\phi b_w d} \tag{8-116}$$

where V_u = total applied design shear force
  ϕ = capacity reduction factor = 0.85
  b_w = web width
  d = distance from extreme compression surface to centroid of tendons or $0.80h$, whichever is larger

When v_u exceeds the nominal shear stress v_c carried by the concrete, shear reinforcement must be provided. v_c may be computed from Eq. (8-117) when the effective prestress force is 40% or more of the tensile strength of the flexural reinforcement, but this shear stress should not exceed 5 $\sqrt{f'_c}$.

$$v_c = 0.6\sqrt{f'_c} + 700 \, \frac{V_u d}{M_u} \geq 2\sqrt{f'_c} \tag{8-117}$$

where M_u = design moment at a section occurring simultaneously with the shear V_u at the section
  d = distance from extreme compression surface to centroid of tendons

$V_u d/M_u$ should not be taken greater than 1. For some sections, such as medium- and long-span I-shaped members, Eq. (8-117) may be overconservative, and the following more detailed analysis would be preferable.

The ACI Code requires a more detailed analysis when the prestress force is less than 40% of the tensile strength of the flexural reinforcement. The governing shear stress is the smaller of the values computed for inclined flexure-shear cracking v_{ci} from Eq. (8-118) and web-shear cracking v_{cw} from Eq. (8-119).

$$v_{ci} = 0.6\sqrt{f'_c} + \frac{V_d + V_i M_{cr}/M_{max}}{b_w d} \geq 1.7\sqrt{f'_c} \tag{8-118}$$

$$v_{cw} = 3.5\sqrt{f'_c} + 0.3f_{pc} + \frac{V_p}{b_w d} \tag{8-119}$$

where V_d = shear due to dead load
  V_i = shear occurring simultaneously with M_{max} and produced by external loads

M_{cr} = cracking moment [see Eq. (8-120)]

M_{max} = maximum bending moment due to external design loads

b_w = web width or diameter of circular section

d = distance from extreme compression surface to centroid of prestressing force of 80% of overall depth of beam, whichever is larger

f_{pc} = compressive stress in the concrete occurring, after all prestress losses have taken place, at the center of the cross section resisting the applied loads or at the junction of web and flange when the centroid lies in the flange

V_p = vertical component of effective prestress force at section considered

The cracking moment is given by

$$M_{cr} = \frac{I}{y_t}(6\sqrt{f_c'} + f_{pe} - f_d) \tag{8-120}$$

where I = moment of inertia of section resisting external design loads, in.[4]

y_t = distance from centroidal axis of gross section, neglecting reinforcement, to extreme surface in tension, in.

f_{pe} = compressive stress in concrete due to prestress only, after all losses, occurring at extreme surface of a section at which tension is produced by applied loads, psi

f_d = stress due to dead loads at extreme surface of a section at which tension is produced by applied loads, psi

Alternatively, v_{cw} may be taken as the shear stress corresponding to the design load that induces a principal tensile stress of $4\sqrt{f_c'}$ at the centroidal axis of the member or, when the centroidal axis is in the flange, induces this tensile stress at the intersection of flange and web.

The values of M_{max} and V_i used in Eq. (8-118) should be those resulting from the distribution of loads causing maximum moment to occur at the section.

In a pretensioned beam in which the section at a distance of half the overall beam depth $h/2$ from the face of support is closer to the end of the beam than the transfer length of the tendon, the reduced prestress in the concrete at sections falling within the transfer length should be considered when calculating v_{cw}. The prestress may be assumed to vary linearly along the centroidal axis from zero at the beam end to a maximum at a distance from the beam end equal to the transfer length. This distance may be assumed to be 50 diameters for strand and 100 diameters for single wire.

When $v_u - v_c$ exceeds $4\sqrt{f_c'}$, the maximum spacing of stirrups should be reduced to $0.375h$ but not to more than 12 in. But $v_u - v_c$ should not exceed $8\sqrt{f_c'}$.

Bonded Reinforcement. When prestressing steel is not bonded to the concrete, some bonded reinforcement should be provided in the precompressed tension zone of flexural members. The bonded reinforcement should be distributed uniformly over the tension zone near the extreme tension surface in beams and one-way slabs and should have an area of at least the larger of the values of A_s computed from Eqs. (8-121) and (8-122).

$$A_s = \frac{N_c}{0.5f_y} \tag{8-121}$$

$$A_s = 0.004A \tag{8-122}$$

where N_c = tensile force in the concrete under dead load plus 1.2 times live load

f_y = yield strength of the bonded reinforcement $\leq$ 60,000 psi

A = area of that part of cross section between flexural tension face and center of gravity of gross section

Compression Members. Members subject to axial compression and with an average prestress $f_{se}A_{ps}/A_c$ more than 225 psi or to axial load and bending should be designed by ultimate-strength methods (Arts. 8-37 and 8-38), including effects of prestress, shrinkage, and creep. Reinforcement in columns with an average prestress less than 225 psi should have an area equal to at least 1% of the gross concrete area A_c. For walls subject to an average prestress greater than 225 psi and for which structural analysis shows adequate strength, the minimum reinforcement requirements given in Art. 8-40 may be waived.

Tendons in columns should be enclosed in spirals or closed lateral ties. The spiral should comply with the requirements given in Art. 8-37. Ties should be at least No. 3 bar size, and spacing should not exceed 48 tie diameters or the least dimension of the column.

Ducts for Posttensioning. Tendons for posttensioned members generally are sheathed in ducts before prestress is applied so that the tendons are free to move when tensioned. The tendons may be grouted in the ducts after transfer of prestress to the concrete and thus bonded to the concrete.

Ducts for grouting bonded tendons should be at least ¼ in. larger than the diameter of the posttensioning tendons or large enough to produce an internal area at least twice the gross area of the tendons. The temperature of members at time of grouting should be above 50° F, and members must be maintained at this temperature for at least 48 hr.

Unbonded tendons should be completely coated with suitable material to insure corrosion protection and to protect the tendons against infiltration of cement during casting operations.

Deflections. The immediate deflection of prestressed members may be computed by the usual formulas for elastic deflections. In these formulas, the moment of inertia used should be that of the gross uncracked concrete section. Long-time deflection computations should include effects of the sustained load and effects of creep and shrinkage and relaxation of the steel (Art. 8-22).

(P. F. Rice and E. S. Hoffman, "Structural Design Guide to the ACI Code," Van Nostrand Reinhold Company, New York; J. R. Libby, "Modern Prestressed Concrete," Van Nostrand Reinhold Company, New York; "PCI Design Handbook," Prestressed Concrete Institute, 20 North Wacker Drive, Chicago, Ill. 60606.)

8-46. Concrete Gravity Walls. Generally economical for walls up to about 15 ft high, gravity walls use their own weight to resist lateral forces from earth or other materials (Fig. 8-18a). Such walls usually are sufficiently massive to be unreinforced. In such cases, tensile stresses should not exceed $1.6 \sqrt{f'_c}$, where f'_c, is the 28-day strength of the concrete, as computed by the working-stress method.

Forces acting on gravity walls include their own weight, the weight of the earth on the sloping back and heel, lateral earth pressure, and resultant soil pressure on the base. It is advisable to include a force at the top of the wall to account for frost action, perhaps 700 lb per lin ft. A wall, consequently, may fail by overturning or sliding, overstressing of the concrete, or settlement due to crushing of the soil.

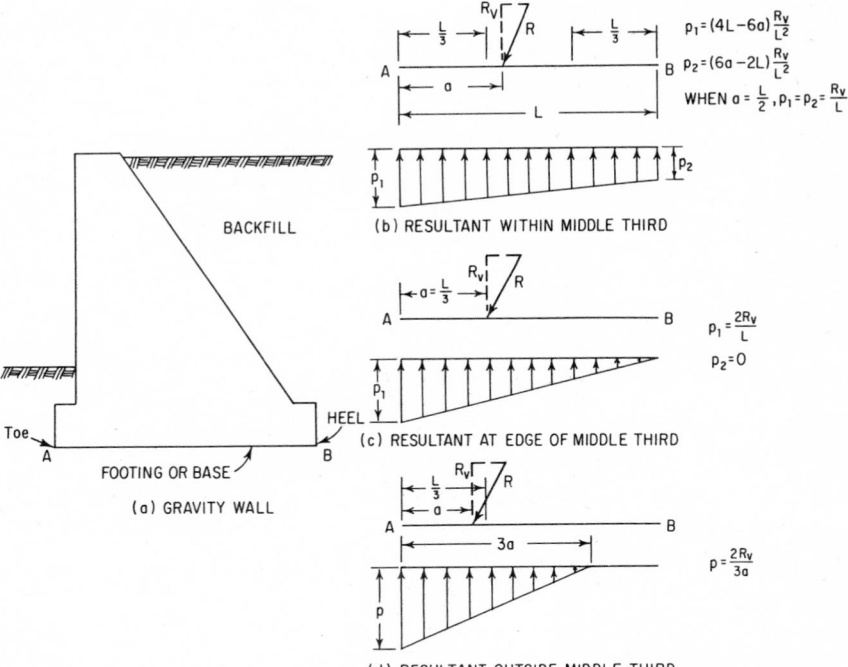

Fig. 8-18. Pressure diagram for base of concrete gravity wall (a) depends on whether the resultant of the forces acting on it lies within the middle third (b), at the edge of the middle third (c), or outside (d).

Design usually starts with selection of a trial shape and dimensions, and this configuration is checked for stability. For convenience, when the wall is of constant height, a 1-ft-long section may be analyzed. Moments are taken about the toe. The sum of the righting moments should be at least 1.5 times the sum of the overturning moments. To prevent sliding

$$\mu R_v \geq 1.5P_h \tag{8-123}$$

where μ = coefficient of sliding friction
R_v = total downward force on the soil, lb
P_h = horizontal component of the earth thrust, lb

Next, the location of the vertical resultant R_v should be found at various sections of the wall, by taking moments about the toe and dividing the sum by R_v. The resultant should act within the middle third of each section if there is to be no tension in the wall.

Finally, the pressure exerted by the base on the soil should be computed, to insure that the allowable pressure will not be exceeded. When the resultant is within the middle third, the pressures, psf, under the ends of the base are given by

$$p = \frac{R_v}{A} \pm \frac{Mc}{I} = \frac{R_v}{A}\left(1 \pm \frac{6e}{L}\right) \tag{8-124}$$

where A = area of base, sq ft
L = width of base, ft
e = distance, parallel to L, from centroid of base to R_v, ft

Figure 8-18b shows the pressure distribution under a 1-ft strip of wall for $e = L/2 - a$, where a is the distance of R_v from the toe. When R_v is exactly $L/3$ from the toe, the pressure at the heel becomes zero (Fig 8-18c). When R_v falls outside the middle third, the pressure vanishes under a zone around the heel, and pressure at the toe is much larger than for the other cases (Fig. 8-18d).

Standard Specifications for Highway Bridges (American Association of State Highway and Transportation Officials) requires that contraction joints be provided at intervals not exceeding 30 ft. Alternate horizontal bars should be cut at these joints for crack control. Expansion joints should be located at intervals of up to 90 ft.

(C. W. Dunham, "Theory and Practice of Reinforced Concrete," McGraw-Hill Book Company, New York.)

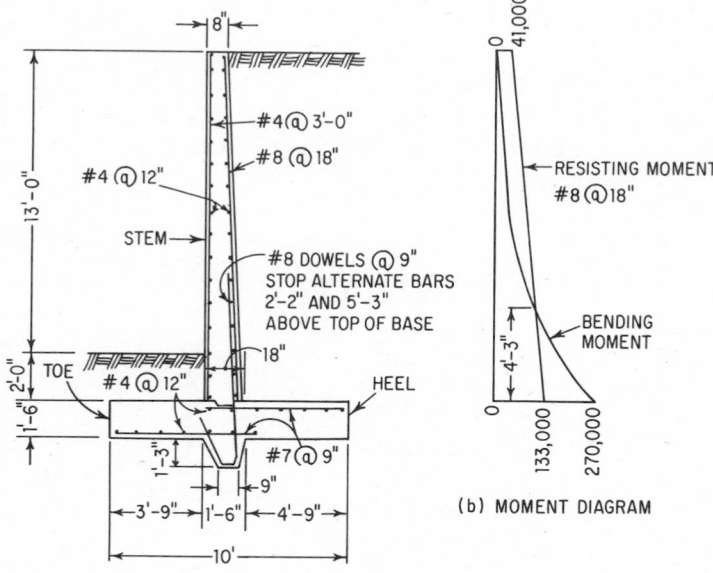

(a) TYPICAL WALL SECTION

Fig. 8-19. Cantilever retaining wall (a) has main reinforcing placed vertically in the stem. Reinforcing requirements may be determined from bending moment diagram (b).

8-47. Cantilever Retaining Walls.   This type of wall resists the lateral thrust of earth pressure through cantilever action of a vertical stem and horizontal base (Fig. 8-19a).   Cantilever walls generally are economical for heights from 10 to 20 ft.   For lower walls, gravity walls may be less costly; for taller walls, counterforts may be less expensive.

Usually, the force acting on the stem is the lateral earth pressure, including the effect of frost action, perhaps 700 lb per lin ft.   The base is loaded by the moment and shear from the stem, upward soil pressure, its own weight, and that of the earth above.   The weight of the soil over the toe, however, may be ignored in computing stresses in the toe, since the earth may not be in place when the wall is first loaded or may erode.   For walls of constant height, it is convenient to design and analyze a 1-ft-long strip.

The stem is designed to resist the bending moments and shear due to the earth thrust.   Then, the size of the base slab is selected to meet requirements for resisting overturning and sliding and to keep the pressure on the soil within the allowable.   If the flat bottom of the slab does not provide sufficient friction [Eq. (8-123)], a key, or lengthwise projection, may be added on the bottom for the purpose.   The key may be reinforced by extending and bending up the dowels between stem and base.

To provide an adequate safety factor against overturning, the sum of the righting moments about the toe should be at least 1.5 times the sum of the overturning moments.   The pressure under the base can be computed, as for gravity walls, from Eq. (8-124).   (See also Fig. 8-18b to d.)

Generally, the stem is made thicker at the bottom than required for shear and balanced design for moment, because of the saving in steel.   Since the moment decreases from bottom to top, the earth side of the wall usually is tapered, and the top is made as thin as convenient concreting will permit (8 to 12 in.).   The main reinforcement is set, in vertical planes, parallel to the sloping face and 3 in. away.   The area of this steel at the bottom can be computed from Eq. (8-32).   Some of the steel may be cut off where it no longer is needed.   Cutoff points may be determined graphically (Fig. 8-19b).   The bending-moment diagram is plotted and the resisting moment of steel not cut off is superimposed.   The intersection of the two curves determines the theoretical cutoff point.   The bars should extend upward beyond this point a distance equal to d or 12 bar diameters.

In addition to the main steel, vertical steel is set in the front face of the wall and horizontal steel in both faces to resist thermal and shrinkage stresses (Art. 8-26).   Standard Specifications for Highway Bridges (American Association of State Highway and Transportation Officials) requires at least ⅛ sq in. of horizontal reinforcement per foot of height.

The heel and toe portions of the base are both designed as cantilevers supported by the stem. The weight of the earth tends to bend the heel down against relatively small resistance from soil pressure under the base.   In contrast, the upward soil pressure tends to bend the toe up.   So for the heel, main steel is placed near the top, and for the toe, near the bottom.   Also, temperature steel is set lengthwise in the bottom.   The area of the main steel may be computed from Eq. (8-32), but the bars should be checked for development length, because of the relatively high shear.

To eliminate the need for diagonal-tension reinforcing, the thickness of the base should be sufficient to hold the shear stress, $v_c = V/bd$, below $1.1 \sqrt{f'_c}$, where f'_c is the 28-day strength of the concrete, as computed by the working-stress method.   The critical section for shear is at a distance d from the face of the stem, where d is the distance from the extreme compression surface to the tensile steel.

The stem is constructed after the base.   A key usually is formed at the top of the base to prevent the stem from sliding.   Also, dowels are left projecting from the base to tie the stem to it, one dowel per stem bar.   The dowels may be extended to serve also as stem reinforcing (Fig. 8-19a).

The AASHTO specifications require that contraction joints be provided at intervals not exceeding 30 ft.   Expansion joints should be located at intervals up to 90 ft.

To relieve the wall of water pressure, weep holes should be formed near the bottom of the stem.   Also, porous pipe and backfill may be set behind the wall to conduct water to the weep holes.

(C. W. Dunham, "Theory and Practice of Reinforced Concrete," McGraw-Hill Book Company, New York; "CRSI Handbook," Concrete Reinforcing Steel Institute, 180 North La Salle St., Chicago, Ill. 60601.)

8-48. Counterfort Retaining Walls.   Counterforts are ties between the vertical stem of a wall and its base (Fig. 8-20a).   Placed on the earth side of the stem, they are essentially wedge-shaped

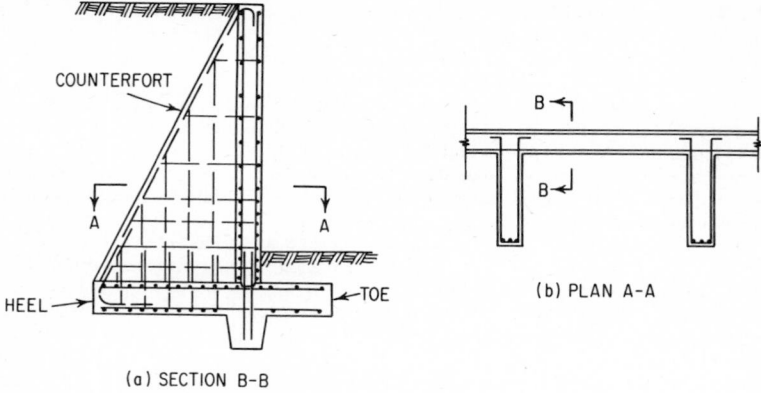

(a) SECTION B-B

(b) PLAN A-A

Fig. 8-20. Counterfort retaining wall.

cantilevers. (Walls with supports on the opposite side are called buttressed retaining walls.) Counterfort walls are economical for heights for which gravity and cantilever walls are not suitable.

Stability design is the same as for gravity walls (Art. 8-46) and cantilever walls (Art. 8-47). But the design is applied to a section of wall center to center of counterforts.

The vertical face resists lateral earth pressure as a continuous slab supported by the counterforts. It also is supported by the base, but an exact analysis of the effects of the three-sided supports would not be worthwhile except for very long walls. Similarly, the heel portion of the base is designed as a continuous slab supported by the counterforts. In turn, the counterforts are subjected to lateral earth pressure on the sloping face and the pull of the vertical stem and base. The toe of the base acts as a cantilever, as in a cantilever wall.

Main reinforcing in the vertical face is horizontal. Since the earth pressure increases with depth, reinforcing area needed also varies with depth. It is customary to design a 1-ft-wide strip of slab spanning between counterforts at the bottom of the wall and at several higher levels. The steel area and spacing for each strip then are held constant between strips. Negative-moment steel should be placed near the earth face of the wall at the counterforts, and positive-moment steel near the opposite face between counterforts (Fig. 8-20b). Concrete cover should be 3 in. over reinforcing throughout the wall. Design requirements are substantially the same as for rectangular beams and one-way slabs, except the thickness is made large enough to eliminate the need for shear reinforcing (Arts. 8-23 to 8-26). The vertical face also incorporates vertical steel, equal to about 0.3 to 1% of the concrete area, for placement purposes and to resist temperature and shrinkage stresses.

In the base, main reinforcing in the heel portion extends lengthwise, whereas that in the toe runs across the width. The heel is subjected to the downward weight of the earth above and its own weight and to the upward pressure of the soil below and the pull of the counterforts. So longitudinal steel should be placed in the top face at the counterforts and near the bottom between counterforts. Main transverse steel should be set near the bottom to resist the cantilever action of the toe.

The counterforts, resisting the lateral earth pressure on the sloping face and the pull of the vertical stem, are designed as T beams. Maximum moment occurs at the bottom. It is resisted by main reinforcing along the sloping face. (The effective depth should be taken as the distance from the outer face of the wall to the steel along a perpendicular to the steel.) At upper levels, main steel not required may be cut off. Some of the steel, however, should be extended and bent down into the vertical face. Also, dowels equal in area to the main steel at the bottom should be hooked into the base to provide anchorage.

Shear unit stress on a horizontal section of a counterfort may be computed from $v_c = V_1/bd$, where b is the thickness of the counterfort and d is the horizontal distance from face of wall to main steel.

$$V_1 = V - \frac{M}{d}(\tan\theta + \tan\phi) \tag{8-125}$$

where V = shear on section

M = bending moment at section

θ = angle earth face of counterfort makes with vertical

ϕ = angle wall face makes with vertical

For a vertical wall face, $\phi = 0$ and $V_1 = V - (M/d) \tan \theta$. The critical section for shear may be taken conservatively at a distance up from the base equal to $d' \sin \theta \cos \theta$, where d' is the depth of counterfort along the top of the base.

Whether or not horizontal web reinforcing is needed to resist the shear, horizontal bars are required to dowel the counterfort to the vertical face (Fig. 8-20b). They should be designed for the full wall reaction. Also, vertical bars are needed in the counterfort to resist the pull of the base. They should be doweled to the base.

The base is concreted first. Vertical bars are left projecting from it to dowel the counterforts and the vertical face. Then, the counterforts and vertical stem are cast together.

8-49. Types of Footings. Footings should be designed to satisfy two objectives: limit total settlement to an acceptable small amount and eliminate differential settlement between parts of a structure as nearly as possible. To limit the amount of settlement, a footing should be constructed on soil with sufficient resistance to deformation, and the load should be spread over a large soil area. The load may be spread horizontally, as is done with spread footings, or vertically, as with friction-pile foundations.

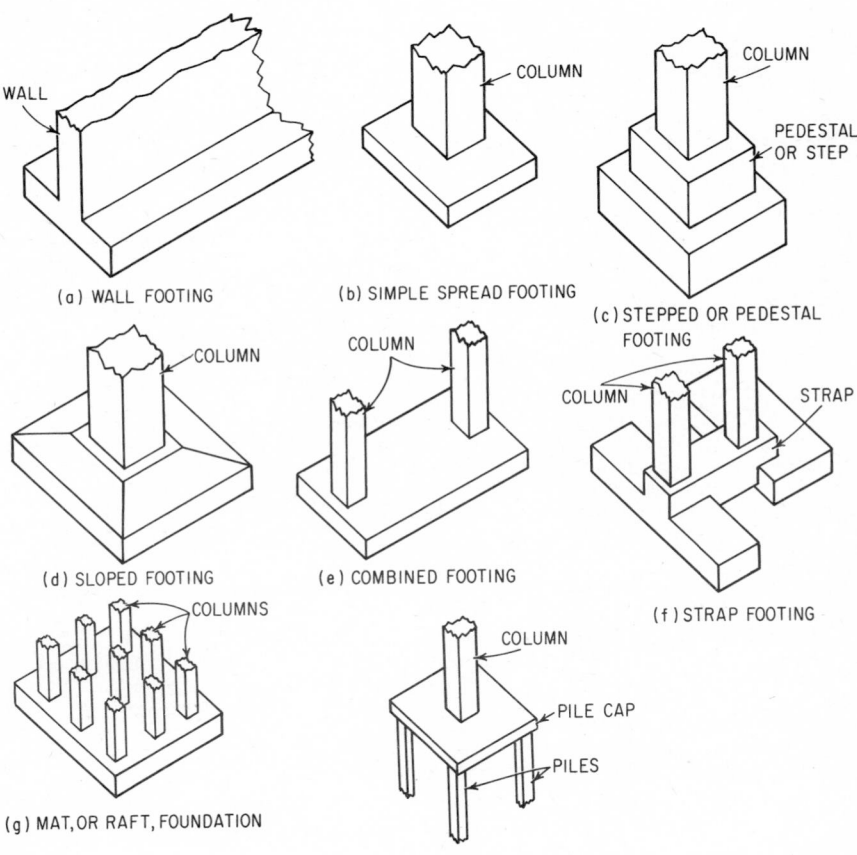

(a) WALL FOOTING

(b) SIMPLE SPREAD FOOTING

(c) STEPPED OR PEDESTAL FOOTING

(d) SLOPED FOOTING

(e) COMBINED FOOTING

(f) STRAP FOOTING

(g) MAT, OR RAFT, FOUNDATION

(h) PILE FOUNDATION

Fig. 8-21. Common types of footings for walls and columns.

There are a wide variety of spread footings. The most commonly used ones are illustrated in Fig. 8-21a to g. A simple pile footing is shown in Fig. 8-21h.

For walls, a spread footing is a slab wider than the wall and extending the length of the wall (Fig. 8-21a). Square or rectangular slabs are used under single columns (Fig. 8-21b to d). When two columns are so close that their footings would merge or nearly touch, a combined footing (Fig. 8-21e) extending under the two should be constructed. When a column footing cannot project in one direction, perhaps because of the proximity of a property line, the footing may be helped out by an adjacent footing with more space; either a combined footing or a strap footing (Fig. 8-21f) may be used under the two.

For structures with heavy loads relative to soil capacity, a mat or raft foundation (Fig. 8-21g) may prove economical. A simple form is a thick, two-way-reinforced concrete slab extending under the entire structure. In effect, it enables the structure to float on the soil, and because of its rigidity, it permits negligible differential settlement. Even greater rigidity can be obtained by building the raft foundation as an inverted beam-and-girder floor, with the girders supporting the columns. Sometimes, also, inverted flat slabs are used as mat foundations.

In general, footings should be so located under walls or columns as to develop uniform pressure below. The pressure under adjacent footings should be as nearly equal as possible, to avoid differential settlement. In the computation of stresses in spread footings, the upward reaction of the soil may be assumed to vary linearly. For pile-cap stresses, the reaction from each pile may be assumed to act at the pile center.

Simple footings act as cantilevers under the downward column or wall loads and upward soil or pile reactions. Therefore, they can be designed as rectangular beams (Arts. 8-23 to 8-26) by working-stress or ultimate-strength theory.

8-50. Stress Transfer to Footings. For a footing to serve its purpose, column stresses must be distributed to it and spread over the soil or to piles, with a safety factor against failure of the footing. Stress in the longitudinal reinforcement of a column should be transferred to its pedestal or footing either by extending the longitudinal steel into the support or by dowels. At least four bars should be extended or four dowels used. In any case, a minimum steel area of 0.5% of the column area should be supplied for load transfer. The stress-transfer bars should project into the base a sufficient compression-embedment distance to transfer the stress in the column bars to the base concrete. Where dowels are used, their total area should be adequate to transfer the compression in excess of that transmitted by the column concrete to the footing in bearing, and the dowel diameter should not exceed the column-bar diameter by more than 0.15 in. If the required dowel length is larger than the footing depth less 3 in., either smaller-diameter bars with equivalent area should be used or a monolithic concrete cap should be added to increase the concrete depth. The dowels, in addition, should provide at least one-quarter of the tension capacity of the column bars on each column face. The dowels should extend into the column a distance equal to that required for compression lapping of column bars (Art. 8-15).

Stress in the column concrete should be considered transferred to the top of the pedestal or footing by bearing. ACI 318, Building Code Requirements for Reinforced Concrete (American Concrete Institute), specifies two bearing stresses:

For a fully loaded area, such as the base of a pedestal, the allowable bearing stress is $0.60f'_c$ (ultimate-strength method), where f'_c is the strength of the concrete.

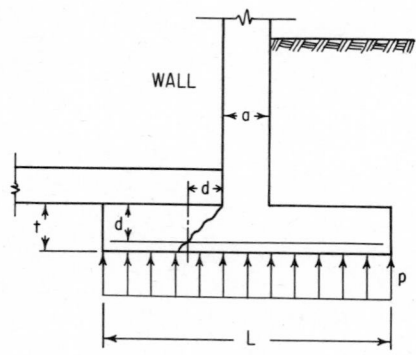

Fig. 8-22. Reinforced concrete wall footing.

If the area A_1, the loaded portion at the top of a pedestal or footing, is less than the area of the top, the allowable pressure may be multiplied by $\sqrt{A_2/A_1}$, but not more than 2, where A_2 is the area of the top that is geometrically similar to and concentric with the loaded area A_1.

For working-stress design, the allowable bearing stress is $0.30f'_c$.

8-51. Wall Footings.   (See also Arts. 8-49 and 8-50.)   The spread footing under a wall (Fig. 8-21a) distributes the wall load horizontally to preclude excessive settlement.   (For retaining-wall footings, see Arts. 8-46 to 8-48.)   The wall should be so located on the footing as to produce uniform bearing pressure on the soil (Fig. 8-22), ignoring the variation due to bending of the footing.   The pressure, psf, is determined by dividing the load per foot by the footing width, ft.

The footing acts as a cantilever on opposite sides of the wall under downward wall loads and upward soil pressure.   For footings supporting concrete walls, the critical section for bending moment is at the face of the wall; for footings under masonry walls, halfway between the middle and edge of the wall.   Hence, for a 1-ft-long strip of symmetrical concrete-wall footing, symmetrically loaded, the maximum moment, ft-lb, is

$$M = \frac{p}{8}(L - a)^2 \tag{8-126}$$

where p = uniform pressure on soil, psf
     L = width of footing, ft
     a = wall thickness, ft

If the footing is sufficiently deep that the tensile bending stress at the bottom, $6M/t^2$, where t is the footing depth, in., does not exceed $1.6\sqrt{f'_c}$, where f'_c is the 28-day concrete strength, psi, the footing need not be reinforced.   If the tensile stress is larger, the footing should be designed as a 12-in.-wide, rectangular, reinforced beam.   Bars should be placed across the width of the footing, 3 in. from the bottom.   Bar development length is measured from the point at which the critical section for moment occurs.   Wall footings also may be designed by ultimate-strength theory.

ACI 318, Building Code Requirements for Reinforced Concrete (American Concrete Institute), requires at least 6 in. of cover over the reinforcement at the edges.   Hence, allowing about 1 in. for the bar diameter, the minimum footing thickness is 10 in.

The critical section for shear is at a distance d from the face of the wall, where d is the distance from the top of the footing to the tensile reinforcement, in.   Since diagonal-tension reinforcement is undesirable, d should be large enough to keep the shear unit stress, $V/12d$, below $1.1\sqrt{f'_c}$, as computed by the working-stress method.

In addition to the main steel, some longitudinal steel also should be placed parallel to the wall to resist shrinkage stresses and facilitate placement of the main steel.

(G. Winter and A. H. Nilson, "Design of Concrete Structures," McGraw-Hill Book Company, New York; C. W. Dunham, "Theory and Practice of Reinforced Concrete," McGraw-Hill Book Company, New York.)

8-52. Single-Column Spread Footings.   (See also Arts. 8-49 and 8-50.)   The spread footing under a column (Fig. 8-21b to d) distributes the column load horizontally to prevent excessive total and differential settlement.   The column should be located on the footing so as to produce uniform bearing pressure on the soil (Fig. 8-23), ignoring the variation due to bending of the footing.   The pressure equals the load divided by the footing area.

Single-column footings usually are square, but they may be made rectangular to satisfy space restrictions or to support elongated columns.

Under the downward load of the column and the upward soil pressure, a footing acts as a cantilever in two perpendicular directions.   For rectangular concrete columns and pedestals, the critical section for bending moment is at the face of the loaded member (ab in Fig. 8-24a).   (For

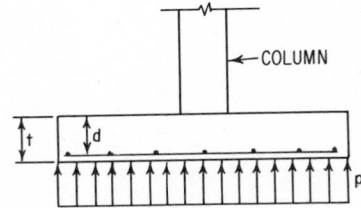

Fig. 8-23. Spread footing for column.

round or octagonal columns or pedestals, the face may be taken as the side of a square with the same area.) For steel baseplates, the critical section for moment is halfway between the face of the column and the edge of the plate.

The bending moment on *ab* is produced by the upward pressure of the soil on the area *abcd*. That part of the footing is designed as a rectangular beam to resist the moment. Another critical section lies along a perpendicular column face and should be similarly designed. If the footing is sufficiently deep that the tensile bending stress at the bottom does not exceed $1.6\sqrt{f_c'}$, where f_c' is the 28-day strength of the concrete, psi, the footing need not be reinforced. If the tensile stress is larger, reinforcement should be placed parallel to both sides of the footing, with the lower layer 3 in. above the bottom of the footing and the upper layer a bar diameter higher. The critical section for anchorage (or bar embedment length) is the same as for moment.

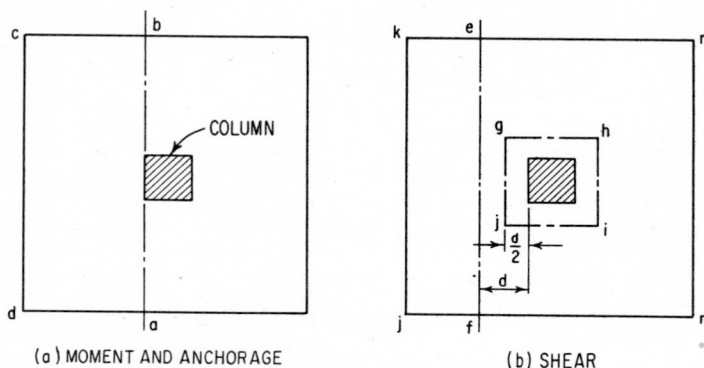

(a) MOMENT AND ANCHORAGE   (b) SHEAR

Fig. 8-24. Critical sections in a column footing.

Ultimate-strength design may be used for footings instead of working-stress design.

In square footings, the steel should be uniformly spaced in each layer. Though the effective depth d is less for the upper layer, thus requiring more steel, it is general practice to compute the required area and spacing for the upper level and repeat them for the lower layer.

In rectangular footings, reinforcement parallel to the long side, with length A, ft, should be uniformly distributed over the width of the footing, B, ft. Bars parallel to the short side should be more closely spaced under the column than near the edges. ACI 318, Building Code Requirements for Reinforced Concrete (American Concrete Institute), recommends that the short bars should be given a constant but closer spacing over a width B centered under the column. The area of steel in this band should equal twice the total steel area required in the short direction divided by $A/B + 1$. The remainder of the reinforcement should be uniformly distributed on opposite sides of the band.

Two types of shear should be investigated, two-way action and beam-type shear. The critical section for beam-type shear lies at a distance d from the face of column or pedestal (*ef* in Fig. 8-24b). The shear equals the total upward pressure on area *efjk*. To eliminate the need for diagonal-tension reinforcing, d should be made large enough that the unit shear stress does not exceed $1.1\sqrt{f_c'}$ ($2\sqrt{f_c'}$ for ultimate-strength design).

The critical section for two-way action (punching shear) is concentric with the column or pedestal. It lies at a distance $d/2$ from the face of the loaded member (*ghij* in Fig. 8-24b). The shear equals the column load less the upward soil pressure on area *ghij*. In this case, d should be large enough that the unit shear stress does not exceed $2\sqrt{f_c'}$ ($4\sqrt{f_c'}$ for ultimate-strength design). Shearhead reinforcement (steel shapes), although generally uneconomical, may be used to obtain a shallow footing. With shearheads, allowable shear stress may be increased 75%.

Footings for columns designed to take moment at the base should be designed against overturning and nonuniform soil pressures. When the moments are about only one axis, the footing may be made rectangular with the long direction perpendicular to that axis, for economy. Design for the long direction is similar to that for retaining-wall bases (Arts. 8-46 to 8-48).

(G. Winter and A. H. Nilson, "Design of Concrete Structures," McGraw-Hill Book Company, New York; C. W. Dunham, "Theory and Practice of Reinforced Concrete," McGraw-Hill.

Book Company, New York; "CRSI Handbook," Concrete Reinforcing Steel Institute, Chicago, Ill.; ACI SP-17, "Design Handbook," American Concrete Institute, Detroit, Mich.)

8-53. Combined Footings. (See also Arts. 8-49 and 8-50.) These are spread footings extended under more than one column (Fig. 8-21e). They may be necessary when two or more columns are so closely spaced that individual footings would interfere with each other. Or they may be desirable when space is restricted for a column footing, such as an exterior member so close to a property line that an individual footing would be so short that it would have excessive eccentric loading. In that case, the footing may be extended under a rear column. If the footing can be continued past that column a sufficient distance, and the exterior column has a lighter load, the combined footing may be made rectangular (Fig. 8-25a). If not, it may be made trapezoidal.

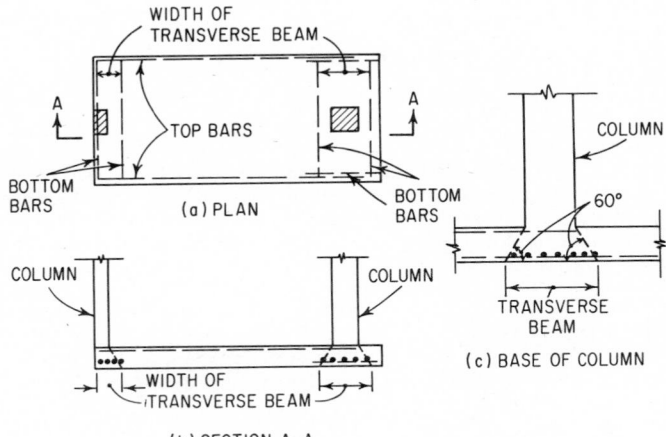

Fig. 8-25. Combined footing.

If possible, the columns should be so placed on the combined footing as to produce a uniform pressure on the soil. Hence, the resultant of the column loads should coincide with the centroid of the footing in plan. This requirement usually determines the length of the footing. The width is computed from the area required to keep the pressure on the soil within the allowable.

In the longitudinal direction, the footing should be designed as a rectangular beam with overhangs. This beam is subjected to the upward pressure of the soil. Hence, the main steel consists of top bars between the columns and bottom bars at the columns where there are overhangs (Fig. 8-25b). Depth of footing may be determined by moment or shear (see Art. 8-52).

The column loads may be assumed distributed to the longitudinal beam by beams of the same depth as the footing but extending in the narrow, or transverse, direction. Centered, if possible, under each column, the transverse member should be designed as a rectangular beam subjected to the downward column load and upward soil pressure under the beam. The width of the beam may be estimated by assuming a 60° distribution of the column load, as indicated in Fig. 8-25c. Main steel in the transverse beam should be placed near the bottom.

Design procedure for a trapezoidal combined footing is similar. But the reinforcing steel in the longitudinal direction is placed fanwise, and alternate bars are cut off as the narrow end is approached.

(G. Winter and A. H. Nilson, "Design of Concrete Structures," McGraw-Hill Book Company, New York; C. W. Dunham, "Theory and Practice of Reinforced Concrete," McGraw-Hill Book Company, New York.)

8-54. Strap Footing. (See also Arts. 8-49 and 8-50.) In Art. 8-53, the design of a combined footing was explained for a column footing in restricted space, such as an exterior column at a property line. As the distance between such a column and a column with adequate space around it increases, the cost of a combined footing rises rapidly. For column spacing more than about 15 ft, a strap footing (Fig. 8-21f) may be more economical. It consists of a separate footing under each column connected by a beam or strap to distribute the column loads (Fig. 8-26a).

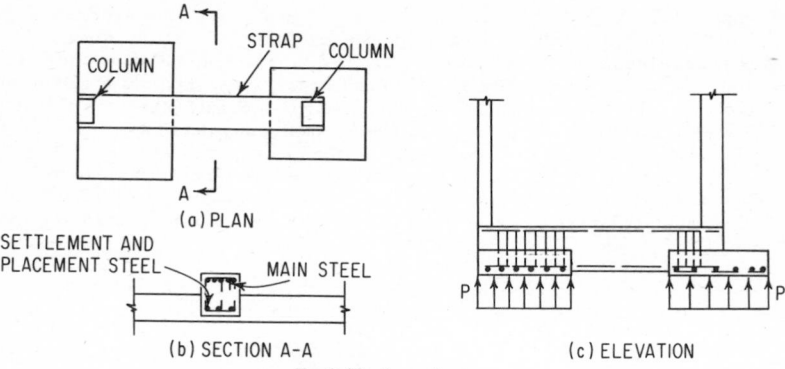

Fig. 8-26. Strap footing.

The footings are sized to produce the same, constant pressure under each (Fig. 8-26c). This requires that the centroid of their areas coincide with the resultant of the column loads. Usually, the strap is raised above the footings so as not to bear on the soil. The sum of the footing areas, therefore, must be large enough for the allowable bearing capacity of the soil not to be exceeded. When these requirements are satisfied, the total net pressure under a footing does not necessarily equal the column load on the footing.

The strap should be designed as a rectangular beam spanning between the columns. The loads on it include its own weight (when it does not rest on the soil) and the upward pressure from the footings. Width of the strap usually is selected arbitrarily as equal to that of the largest column plus 4 to 8 in. so that column forms can be supported on top of the strap. Depth is determined by the maximum bending moment.

The main reinforcing in the strap is placed near the top. Some of the steel can be cut off where not needed. For diagonal tension, stirrups normally will be needed near the columns (Fig. 8-26b). In addition, longitudinal placement steel is set near the bottom of the strap, plus reinforcement to guard against settlement stresses.

The footing under the exterior column may be designed as a wall footing (Art. 8-51). The portions on opposite sides of the strap act as cantilevers under the constant upward pressure of the soil.

The interior footing should be designed as a single-column footing (Art. 8-52). The critical section for punching shear, however, differs from that for a conventional footing. This shear should be computed on a section parallel to the strap and at a distance $d/2$ from the sides and extending around the column at a distance $d/2$ from its faces; d is the effective depth of the footing, the distance from the bottom steel to the top of the footing.

(G. Winter and A. H. Nilson, "Design of Concrete Structures," McGraw-Hill Book Company, New York.)

8-55. Footings on Piles. (See also Arts. 8-49 and 8-50.) When piles are required to support a structure, they are capped with a thick concrete slab, on which the structure rests. The pile cap should be reinforced. ACI 318, Building Code Requirements for Reinforced Concrete (American Concrete Institute), requires that the thickness above the tops of the piles be at least 12 in. The piles should be embedded from 6 to 9 in., preferably the larger amount, into the footing. They should be cut to required elevation before the footing is cast.

Like spread footings, pile footings for walls are continuous, the piles being driven in line under the wall. For a single column or pier, piles are driven in a cluster. Standard Specifications for Highway Bridges (American Association of State Highway and Transportation Officials) requires that piles be spaced at least 2 ft 6 in. center to center. And the distance from the side of a pile to the nearest edge of the footing should be 9 in. or more.

Whenever possible, the piles should be located so as to place their centroid under the resultant of the column load. If this is done, each pile will carry the same load. If the load is eccentric, then the load on a pile may be assumed to vary linearly with distance from an axis through the centroid.

The critical section for bending moment in the footing and embedment length of the reinforcing should be taken as follows: At the face of the column, pedestal, or wall, for footings supporting a concrete column, pedestal, or wall.   Halfway between the middle and edge of the wall, for footings under masonry walls.   Halfway between the face of the column or pedestal and the edge of the metallic base, for footings under steel baseplates.   The moment is produced at the critical section by the upward forces from all the piles lying between the section and the edge of the footing.

For diagonal tension, two types of shear should be investigated, punching shear and beamlike shear, as for single-column spread footings (Art. 8-52).   The ACI Code requires that in computing the external shear on any section through a footing supported on piles, the entire reaction from any pile whose center is located half the pile diameter or more outside the section shall be assumed as producing shear on the section; the reaction from any pile whose center is located half the pile diameter or more inside the section shall be assumed as producing no shear on the section.   For intermediate positions of the pile center, the portion of the pile reaction to be assumed as producing shear on the section shall be based on straight-line interpolation between the full value at half the pile diameter outside the section and zero value at that distance inside the section."

(G. Winter and A. H. Nilson, "Design of Concrete Structures," McGraw-Hill Book Company, New York; C. W. Dunham, "Theory and Practice of Reinforced Concrete," McGraw-Hill Book Company, New York.)

8-56. Considerations in Design of Concrete Rigid Frames, Arches, Folded Plates, and Shells. Analysis of structural frames yields values of internal forces and moments at various sections. Results include bending moments (about two principal axes of each section), concentric normal forces (axial tension or compression), tangential forces (shear), and torsion (bending moment parallel to the section).   In design, critical cross sections are selected and designed to resist the internal forces and moments acting on them.

Geometry of a structural frame and its components has a great bearing on distribution of internal forces and moments and their magnitude.   Thus, the geometry affects economy and aesthetics of a structural system and its components.   Rigid frames, arches, folded plates, and shells are examples of the use of geometry for support of loads at relatively low cost.

Once any of these structures has been analyzed and internal forces and moments on critical cross sections have been determined, design becomes nearly identical with that of cross sections covered in previous articles in this section.   Additional consideration, however, should be given to secondary stresses in detailing the reinforcement.

In practice, most structures and their components are analyzed only for the primary stresses caused by external loads.   But most structural components, including beams, columns, and slabs discussed previously, are subjected to secondary stresses.   They could be due to many causes:

External loads normally not considered during the design, for example, when one side of a building is heated by sun more than the others

Nonhomogeneity of material, such as concrete

Geometry of structural members; for example, deep rather than shallow cross sections

Additional forces and moments due to deformations

Most of the formulas used in everyday structural design are simplified versions of more accurate but complicated mathematical expressions.   The simplified formulas give results only for an approximate stress distribution.   It is common practice to define the difference between the more exact stress distribution and the approximate one as secondary stresses.   Stress concentration, for example, is a secondary stress.   In general, there are no set rules or formulas for predicting secondary stresses and designing for them.

In conventional reinforced concrete structures, secondary stresses are relatively small compared with the primary stresses.   But if secondary stresses are not provided for in design, cracks may develop in the structure.   Usually, these cracks are not serious and are acceptable.   So in view of the difficulty, perhaps impossibility, of predicting the location and magnitude of secondary stresses in most cases, normal practice does not include analysis of structures for secondary stresses.

To protect structures against unpredictable stresses, ACI 318, Building Code Requirements for Reinforced Concrete (American Concrete Institute), specifies minimum reinforcement for beams, columns, and slabs.   Spacing and size of this reinforcement take care of the secondary stresses.   These provisions and some additional reinforcement requirements apply to design of rigid frames, arches, folded plates, and shells.   But these types of structures often have larger secondary stresses than conventional structures, and these stresses are distributed differently from those in beams and columns.   There are no code provisions for designing against these secondary stresses other than the general requirements of elastic behavior, equilibrium checks, and accounting for effects of large deflections, creep, and possible construction defects.   But observations of the behavior of rigid frames, arches, folded plates, and shells, along with more accurate mathematical treatment and analysis, do help to design against secondary stresses.

In the following articles, the more salient considerations in designing these reinforced concrete structures are pointed out.   Engineers, however, should have sufficient experience in design of such structures to take steps to avoid undue cracking of concrete.

One of the most important duties of a structural engineer is to choose an appropriate structural system, for example, to decide whether to span with a simply supported beam, a rigid frame, an arch, a folded plate, or a shell.   The engineer must know the advantages of these structural systems to be able to select a proper structure for a project.

In indeterminate structures such as rigid frames, arches, folded plates, and shells, the sizes and thicknesses of the components of these structures affect the magnitude and distribution of the bending moments and, hence, shears and axial forces.   For example, if the horizontal member of the rigid frame of Fig. 8-27a is made much deeper than the width of the vertical member, that is, the beam is much stiffer than the column, the maximum moment in the beam would be relatively large and that in the column, small.   Conversely, if the vertical member is made much wider than the depth of the horizontal member, that is, the column is much stiffer than the beam, the maximum bending moment in the column would be relatively large.

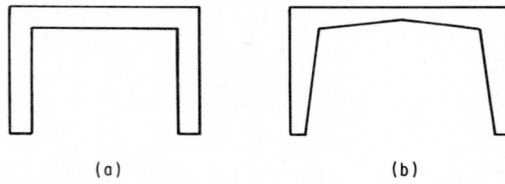

Fig. 8-27. Rigid frames: (a) with prismatic members; (b) with haunched beam.

Similarly, deepening the haunches in the horizontal member of Fig. 8-27b would increase the negative bending moment at the haunches and decrease the positive bending moment at midspan, where the beam is shallow.

Because of the properties described, indeterminate structures are analyzed by first assuming sizes and shapes of components.   After internal forces and moments have been determined, the assumed sections are checked for adequacy.   If the assumed sizes must be adjusted, another analysis is performed with the adjusted sizes.   Then, these are checked for adequacy.   If necessary, the cycle is repeated.

8-57. Concrete Rigid Frames.   (See also Art. 8-56.)   Rigid frame implies a plane structural system consisting of straight members meeting each other at an angle and rigidly connected at the junction.   A rigid connection keeps unchanged the angle between members as the entire frame distorts under load.

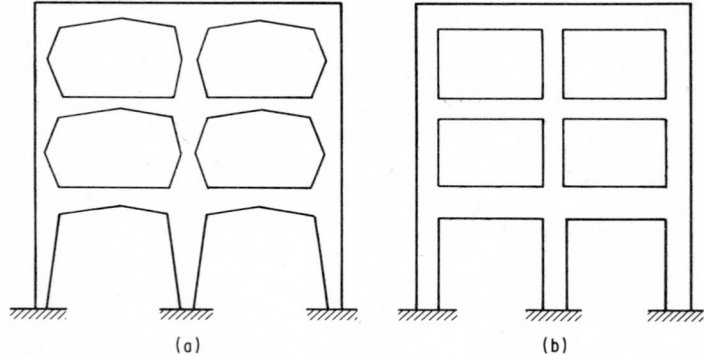

Fig. 8-28. Multistory rigid frames: (a) with haunched members; (b) with prismatic members.

Rigid frames may be one bay long and one tier high (Fig. 8-27a and b), or they may have multiple bays and multiple tiers (Fig. 8-28a and b). They may be built of reinforced concrete or prestressed concrete, cast in place or precast.

Because of continuity between columns and beams, columns in rigid frames participate with the beams in bending and thus in resisting external loads. This participation results in both smaller bending moments and different moment distribution along the beam than in a simply supported beam with the same span and loads. But for these advantages in bending-moment distribution along the beam, the column is penalized. Under vertical loading, for example, it is subjected to bending moments in addition to axial force.

Since the bases of most rigid frames develop horizontal reactions, the beams usually are subjected to a small axial force. Also, the beams and columns are subjected to shear forces.

It is not advisable, in general, to differentiate between beams and columns in a rigid frame, but rather to consider each as an axial member. Find bending moments, shear, and axial forces in each, and design for these.

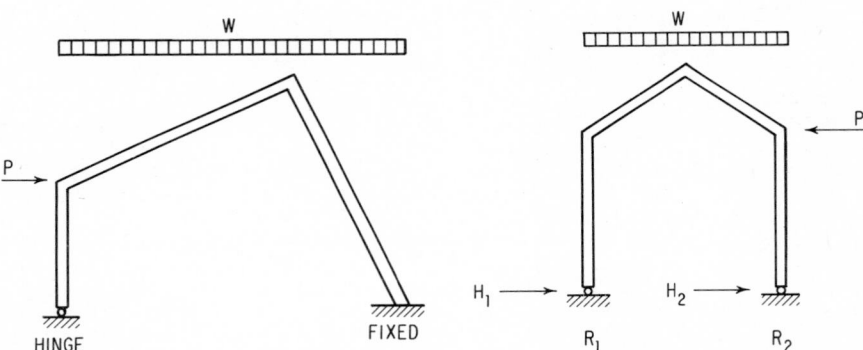

Fig. 8-29. Rigid frame with sloping beam and column.

Fig. 8-30. Gable frame has sloping beams, vertical columns.

Because of continuity between members in a rigid frame, this type of structure is particularly advantageous in resisting wind loads. It does not necessarily have to be subjected to vertical loads only or consist of vertical and horizontal members. Figures 8-29 and 8-30 show examples of rigid frames with sloping members subjected to vertical and lateral loads.

Dimensions of cross sections and the amount of reinforcement in concrete rigid frames are determined by primary stresses due to bending moments, axial forces, and shears, as in beams and columns. In addition, the following require special attention:

    Rigid joints, where members meet, particularly at reentrant corners

    Toes of legs at the foundations

    Exceptionally deep members (Art. 8-20)

Typical details of rigid joints in a reinforced-concrete rigid frame are shown in Fig. 8-31a and b. Ample embedment of bars, at least equal to that recommended by the ACI at supports,

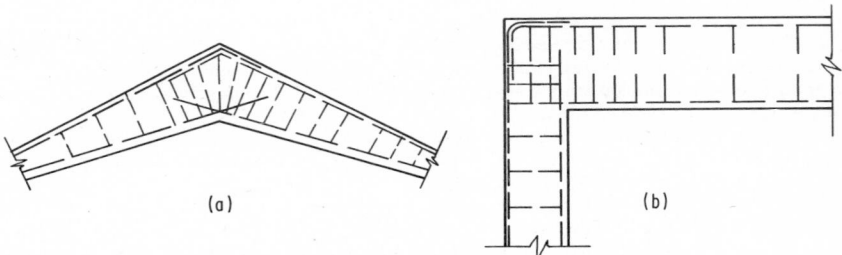

Fig. 8-31. Reinforcing arrangements in joints of rigid frames.

should be provided at all corners, as well as at overlaps (ACI 318, Building Code Requirements for Reinforced Concrete, American Concrete Institute).   No interior or exterior face of a rigid joint should be left without reinforcement.

Note in Fig. 8-31 that reinforcing bars extend without bends past the reentrant corners. Reinforcing never should be bent around a reentrant corner.   When the reinforcement is in tension, it tends to tear concrete at the corner away from the joint.   Furthermore, sufficient stirrups should be provided around all bars that cross a joint.   The amount of stirrups may be computed from the component of tensile force in the reinforcement, but preferably a lower limit should be the minimum size and number of ties required by the ACI for columns.

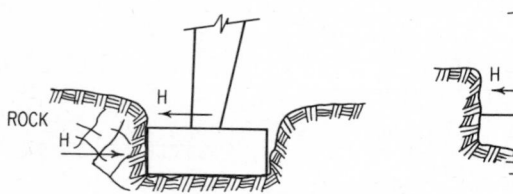

Fig. 8-32. Footing thrust resisted by side bearing.

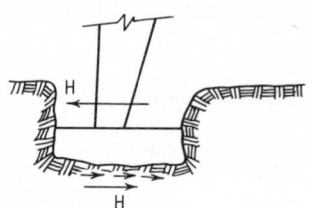

Fig. 8-33. Footing thrust resisted by base friction.

All toes of rigid frames are subjected to horizontal forces, or thrust.   In a hinged rigid frame, an additional axial force (compression or tension) acts on the base, while in a fixed rigid frame, an additional axial force and a bending moment act.

Usually, analysis assumes that the toes of rigid frames do not move relative to each other.   The designer should check this assumption in the design.   If the toes do spread under load, the horizontal thrust, as well as all the internal forces and moments within the frame, will change. The actual internal forces due to movement of the toes should be computed and the frame designed accordingly.   Similarly, if the base is not truly hinged or fixed, but only partly so, the effect of partial fixity on the frame should be taken into account.

The thrust may be resisted by a footing pressing against rock (Fig. 8-32), or by friction of the footing against the soil (Fig. 8-33), or by a tie (Fig. 8-34).   In the cases illustrated by Figs. 8-33 and 8-34, the likelihood of the toes spreading apart is considerable.

If the toe is hinged, the hinge detail could be provided in the field (Fig. 8-35).   Or it could be a prefabricated steel hinge (Fig. 8-36).

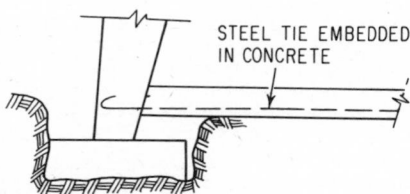

STEEL TIE EMBEDDED IN CONCRETE

Fig. 8-34. Tie between footings takes thrust at rigid-frame base.

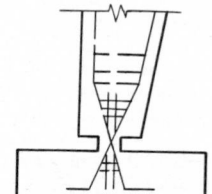

Fig. 8-35. Hinge built with re-inforcing bars at footing.

In a fixed rigid frame, the connection of the toe to the footing (Fig. 8-37) should be strong enough to develop the computed bending moment.   Since this moment is to be transferred to the ground, it is usual to construct a heavy eccentric footing that counterbalances this moment by its weight, as shown in Fig. 8-37.

To obtain an advantageous moment distribution in a frame, a designer might find it desirable to increase the sizes of some of the members of the frame.   For example, for a large-span, low, rigid frame, increasing the width of the vertical legs would reduce positive bending moments in the horizontal members and increase moments in the vertical members.   The vertical members could become stubby, as in Fig. 8-38.   According to the ACI Code, when the ratio of depth d to length L of a continuous member exceeds 0.4, the member becomes a "deep" beam; the bending stresses and resistance to them do not follow the patterns described previously in this section. The designer should provide more than the usual stirrups and distribute reinforcement along the faces of the deep members, as in Fig. 8-38 (see Sec. 10.7, ACI 318-71).

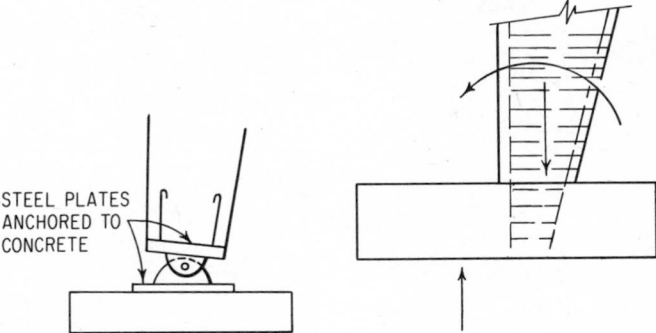

Fig. 8-36. Concrete column with steel hinge at base.

Fig. 8-37. Column base designed to resist bending moment.

Design of precast-concrete rigid frames is identical to that of cast-in-place frames. It is quite common to precast parts of frames between points of counterflexure, or sections where bending moment is small, as shown in Fig. 8-39a. This eliminates the need for a moment connection (often referred to as a continuity conection) at a joint. Only a shear connection is required (Fig. 8-39b). Since some bending moment might occur at the joint due to live loads, wind loads, etc., moment resistance should be provided by grouting longitudinal bars (Fig. 8-39b) or welding steel plates embedded in the precast concrete (Fig. 8-39c).

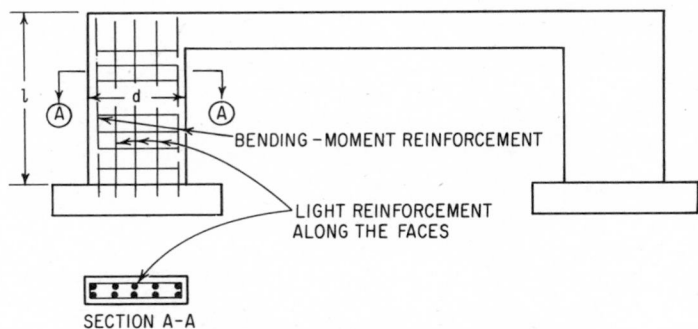

Fig. 8-38. Rigid frame with stubby columns.

Rigid frames also may be prestressed and cast in place or precast. Prestressed, cast-in-place frames are posttensioned. Usually, the prestress is applied to each member with tendons anchored within the member (Fig. 8-40). Although continuous tendons may be more efficient structurally, friction losses due to bending the tendons make application of prestress in the field as intended by the design difficult. Such losses cannot be estimated. Hence, the magnitude of the prestress imparted is uncertain. The rigid joints, though, may be prestressed by individual straight or slightly bent tendons anchored in adjacent members (tendons B in Fig. 8-40).

In selecting the magnitude of the prestressing force in each member, the designer should ascertain that the bending moments at the ends of members meeting at a joint are in equilibrium and that the end rotation there is the same for each member.

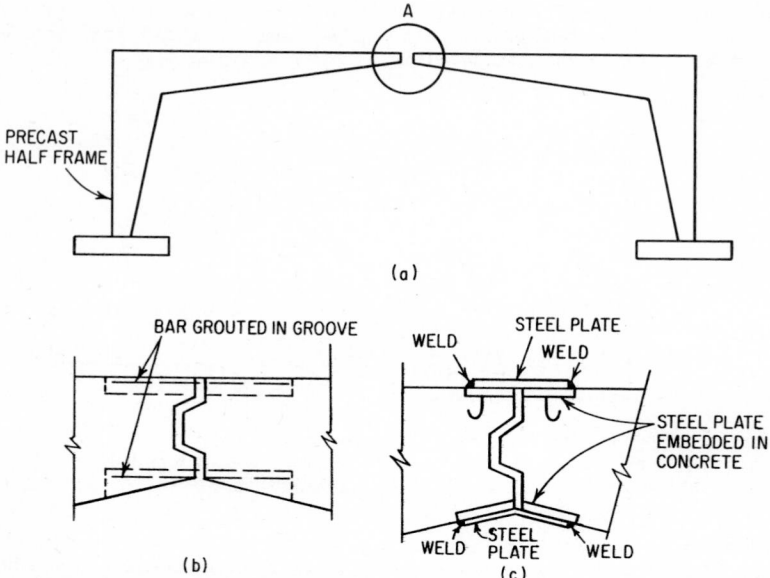

Fig. 8-39. (a) Rigid frame precast in halves. (b) and (c) Typical joints at midspan.

Precast rigid frames may be pretensioned, posttensioned, or both. In prestressed, precast rigid frames, it is common to fabricate the individual members between joints, rather than between points of counterflexure, and connect them rigidly at the joints. The members are con-

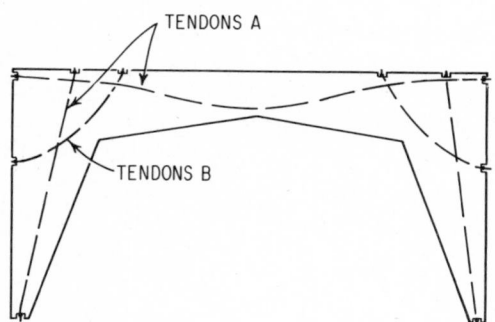

Fig. 8-40. Prestressed-concrete rigid frame.

nected at the rigid joints by grouting reinforcing bars, welding steel inserts, or posttensioning. In all cases, the designer should make sure that the rotations of the ends of all members meeting at a joint are equal.

8-58. Concrete Arches. (See also Art. 8-56.) Structurally, arches are, in many respects, similar to rigid frames (Art. 8-57). An arch may be considered a rigid frame with one curved member instead of a number of straight members (Fig. 8-41). The internal forces in the two structural systems are of the same nature: bending moments, axial forces, and shears. The difference is that bending moments predominate in rigid frames, while arches may be shaped so that axial (compression) force predominates. Nevertheless, general design procedures for arches and rigid frames are identical.

Design of details, however, differs, since arches have no rigid joints above the abutments, and arches, being predominantly subjected to compression, must be provided with more resistance against buckling.

RIGID FRAME

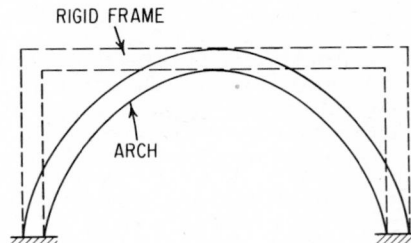

ARCH

Fig. 8-41. Arch replacement for rigid frame.

Precasting of arches is not common because the curvature makes stacking for transportation difficult. Some small-span site-precast arches have been successfully erected.

Prestressing of arch ribs is not very common because the arches are subjected to large compressive forces; thus, prestressing rarely offers advantages. But prestressing of abutments and of connections of a fixed-end arch to abutments, where bending moments are large, could be beneficial in resisting these moments.

(G. Winter and A. H. Nilson, "Design of Concrete Structures," McGraw-Hill Book Company, New York.)

8-59. Concrete Folded Plates. (See also Art. 8-56.) The basic structural advantage of a folded-plate structure (Fig. 8-42) over beams and slabs for a given span is that more material in a

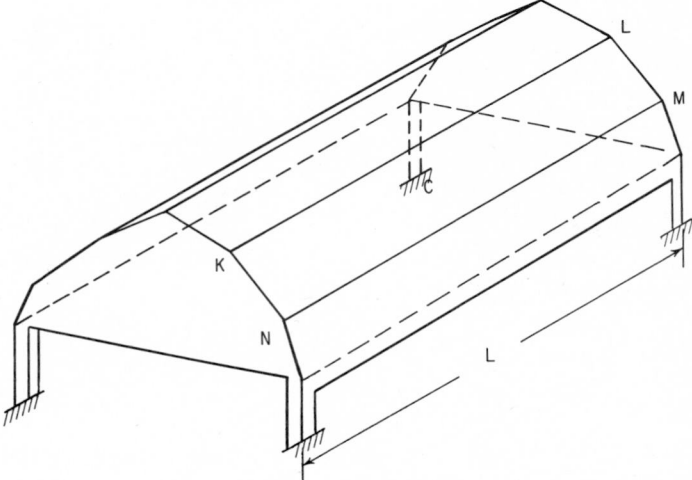

Fig. 8-42. Folded-plate roof.

folded plate carries stresses, and stress distribution may be more uniform. For example, Fig. 8-43a shows cross sections of alternative structural systems of the same span and depth superimposed. One section is for a folded plate; the other, for a system with two solid beams. The stress distribution in the solid beams is shown in Fig. 8-43b. Only the extreme fibers are stressed to the maximum allowable, while the remainder, the largest part of the cross section, is subjected to much smaller stresses. The stresses in the folded plate, as shown in Fig. 8-43c, are more uniformly distributed through the depth D of the structure. Hence, a folded-plate structure needs less material than solid beams and may therefore be more economical.

It should be noted, however, that longitudinal-stress distribution in a folded-plate structure spanning a distance L (Fig. 8-43a) is not given accurately by simple-beam theory; that is, the longitudinal normal stresses are not as shown in Fig. 8-43b. Under vertical loads, one cannot compute the moment of inertia of the folded-plate section in Fig. 8-43a about the centroidal axis and find the stresses from Mc/I. The cross section distorts under load, invalidating the elementary bending theory.   Hence, the result may be more nearly the stress distribution shown in Fig. 8-43c.

These normal stresses are perpendicular to the plane of the folded-plate section (Fig. 8-43a). They and the shear stresses parallel to the section may be assumed uniformly distributed over the thickness of the plates.   The same is true of membrane stresses in shell structures.

Reinforcement in each plate, such as *KLMN* (Fig. 8-42), in the transverse and longitudinal directions, is determined from stresses obtained from analysis.   Typical reinforcement is shown in Fig. 8-44.   The quantity of longitudinal reinforcement is determined by the tensile stresses in each plate.   But reinforcement should not be less than that required by ACI 318, Building Code Requirements for Reinforced Concrete, for minimum quantity in slabs.   In addition, a minimum of temperature reinforcement as required for slabs should be distributed uniformly throughout each plate.

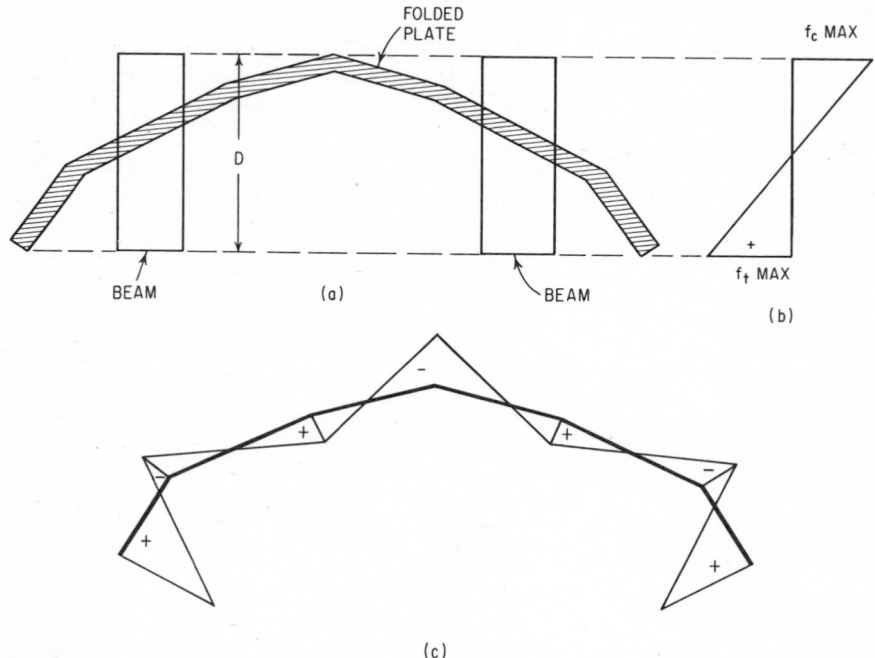

Fig. 8-43. Beams replacing folded plate (*a*) would have stress distribution shown in (*b*).   Longitudinal stresses in folded plate would be distributed as indicated in (*c*).

Transverse reinforcement is determined by the transverse bending in each plate between support points $A, B, C, D, \ldots$ (Fig. 8-44).   This reinforcement should not be less than the temperature reinforcement required by the ACI Code.   Because the regions around plate intersections, such as B and C, are subjected to negative transverse bending moments, negative (top) reinforcement is required there.   This reinforcement, as well as the bottom bars, should be carried far enough past the corner for proper embedment.   Because of the distortions of the section and the uncertainty of the extent of transverse negative moments, it is good practice to carry reinforcement along the top of all plates, as shown for plate CD (Fig. 8-44).   Such top reinforcement also is efficient in resisting shear.

Essentially, Fig. 8-44 represents a cross section of a rigid frame.   The joints between plates have to be maintained rigid to correspond to assumptions made in the analysis.   Thus, these joints should be reinforced as in rigid frames.   When the angle between two plates is large, it is desirable to tie top and bottom reinforcement with ties, as indicated in Fig. 8-44.

If the concrete alone is not sufficient to resist diagonal tension due to shear, reinforcement should be provided for the excess diagonal tension.   Such reinforcement may be inclined, as at

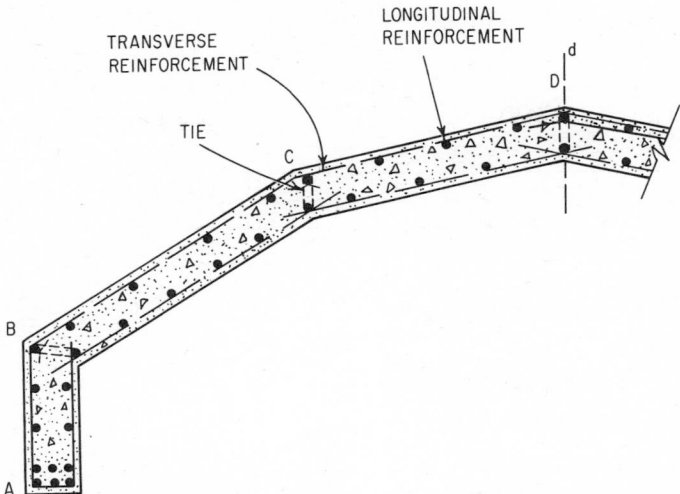

Fig. 8-44. Typical reinforcing arrangement in section of a folded plate.

A in Fig. 8-45, or a grid of longitudinal and transverse bars may be used, as at *B*. In the latter case, the reinforcement will have the pattern indicated in Fig. 8-44. The quantity needed to resist diagonal tension, then, should be added to that required for bending. Both the transverse and longitudinal reinforcement inserted for this purpose preferably should be distributed evenly between the top and bottom faces of the plates.

Elementary analysis of folded plates usually assumes that the cross sections at the supports do not distort. Therefore, it is common practice to provide rigid diaphragms at the ends of folded plates in planes of supports (Fig. 8-46). The diaphragms act as transverse beams, as well as ties, between supports. Hence, they usually have relatively heavy bottom reinforcement. The strains in the end diaphragms should be kept small, to keep the end sections of the folded-plate

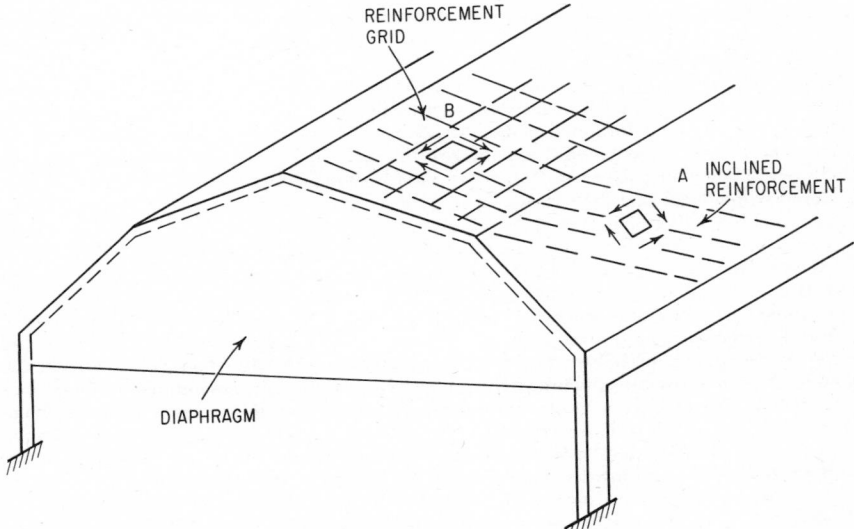

Fig. 8-45. Reinforcement patterns in a folded plate.

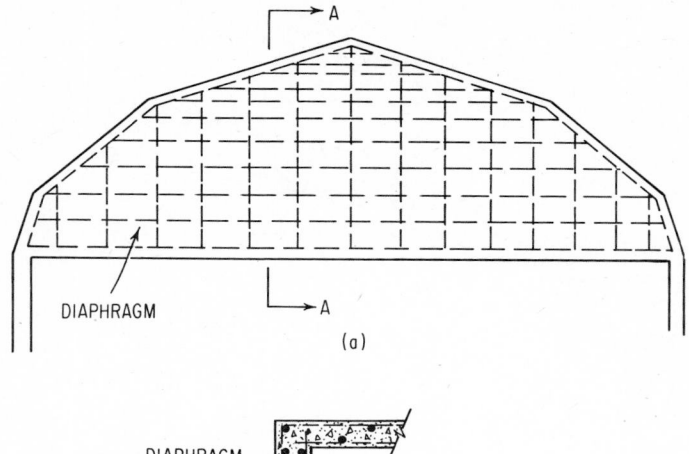

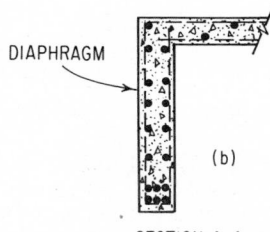

SECTION A-A

Fig. 8-46. Concrete diaphragm for folded plate.

structure from distorting. It is advisable, therefore, that the reinforcement in the diaphragm be evenly distributed throughout each face.

8-60. Concrete Shells. (See also Art. 8-56.) Many forms of concrete shells are used. To be amenable to theoretical analysis, these forms have geometrically expressible surfaces. The ACI Code defines thin shells as curved or folded slabs whose thicknesses are small compared with their other dimensions. In addition, shells are characterized by their three-dimensional load-carrying behavior, which is determined by their geometric shape, their boundary conditions, and the nature of the applied load.

Elastic behavior is usually assumed for shell structural analysis, with suitable assumptions to approximate the three-dimensional behavior of shells.

Stresses usually are determined by membrane theory, but sometimes also by bending theory. Membrane stresses are assumed constant across the shell thickness. The membrane theory for shells neglects bending stresses. Yet, every shell is subjected to bending moments, not only under unsymmetrical loads but also under uniform and symmetrical loads. But accurate determination of bending moments is difficult and sometimes impossible.

While unsymmetrical loads cause bending moments throughout a whole shell, symmetrical loads cause moments mainly at edges and supports. These edge and support moments may be very large. Provision should be made to resist them. If they are not properly provided for, not only would unsightly cracks occur, but the shell may distort, progressively increasing the size of the cracks and causing large deflections. This would render the shell unusable. Therefore, past experience in design, field observations, and knowledge of results of tests on shells are a necessity for design of shell structures, to insure the proper quantity of reinforcement in critical locations, even though the reinforcement is not predicted by theory. Model testing is a helpful tool for shell design, but small-scale models may not predict all the possible stresses in a prototype.

Because of the difficulties in determining stresses accurately, only those forms of shell that have been successfully constructed and tested in the past are usually undertaken for commercial uses. These forms include barrel arches, domes, and hyperbolic paraboloids (Fig. 8-47).

Barrel shells may consist of single transverse spans (Fig. 8-47a) or multiple spans (Fig. 8-48). Analysis yields a different stress distribution for a single barrel shell from that for a multiple one. But design considerations are the same.

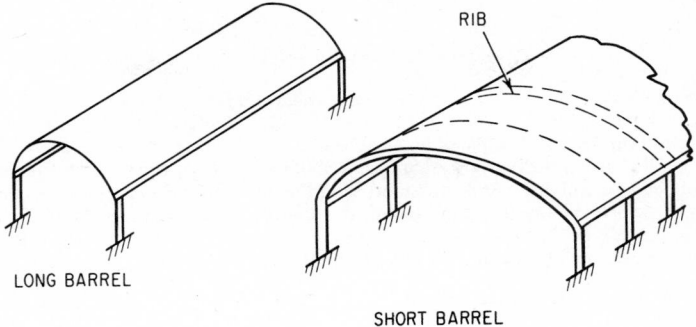

LONG BARREL

RIB

SHORT BARREL

(a)

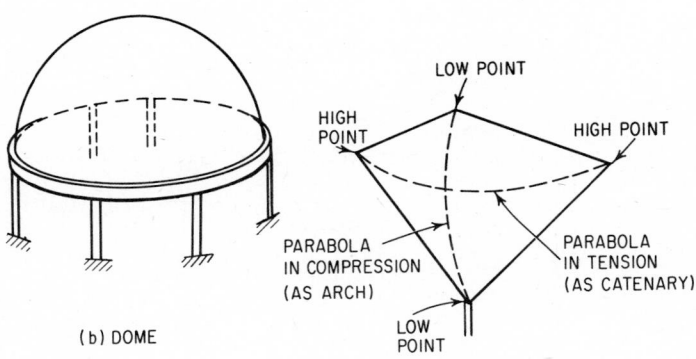

(b) DOME

LOW POINT

HIGH POINT

HIGH POINT

PARABOLA IN COMPRESSION (AS ARCH)

PARABOLA IN TENSION (AS CATENARY)

LOW POINT

(c) HYPERBOLIC PARABOLOID

Fig. 8-47. Common types of concrete shells.

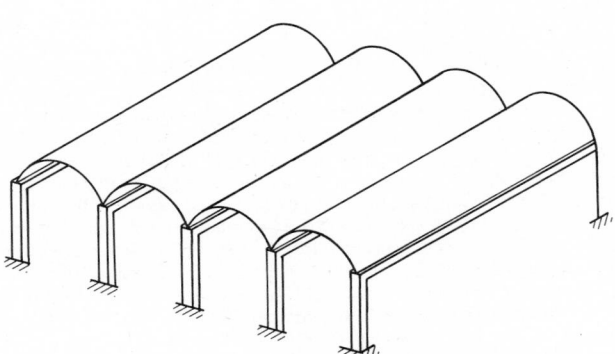

Fig. 8-48. Multiple barrel-arch roof.

Usually, the design stresses in a shell are quite small, requiring little reinforcement. The reinforcement, both circumferential and longitudinal, should not be less than the minimum reinforcement required for slabs by ACI 318, Building Code Requirements for Reinforced Concrete (American Concrete Institute).

Barrel shells usually are relatively thin. Thickness varies from 4 to 6 in. for most parts of shells with spans up to 300 ft transversely and longitudinally. But the shells generally are thickened at edges and supports and stiffened by edge beams. If much time is spent in analysis, including model testing, it is possible to design barrel shells of uniform thickness throughout, without stiffening edge members. But if the more simplified method of analysis (membrane theory) is employed, which is more usual and practical, stiffening edge members should be provided, as shown in Fig. 8-49. These consist of edge beams *AB* and end arch ribs *AA* and *BB*. Instead of an end arch rib, an end diaphragm may be employed (Fig. 8-50).

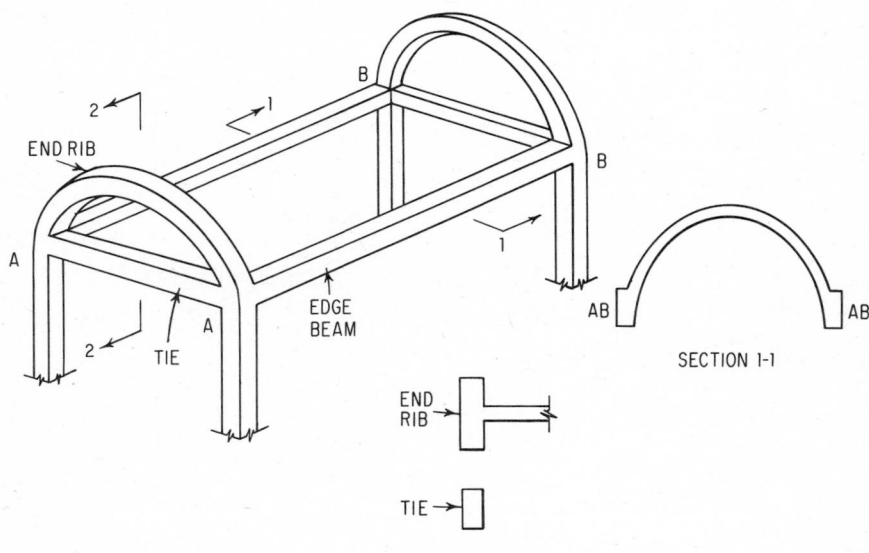

SECTION 2-2
Fig. 8-49. Stiffening members in thin-shell arch roof.

Stresses determined from analysis may be combined to give the principal stresses, or maximum tension and compression, at each point in the shell. It these are plotted on a projection of the shell, the lines of constant stress, or stress trajectories, will be curved. The tensile-stress trajectories generally follow a diagonal pattern near supports and are nearly horizontal around midspan. Reinforcing bars to resist these stresses, therefore, may be draped along the lines of principal stress. This, however, makes field work difficult, because large-diameter bars may have to be bent and extra care is needed in placing them. Hence, main steel usually is placed in a grid pattern, with the greatest concentration along longitudinal edges or valleys. To control temperature and shrinkage cracks, minimum reinforcement should be provided.

Reinforcement may be placed in the shell in one layer (Fig. 8-51a) or two layers (Fig. 8-51b), depending on the stresses, that is, the span and design loads. (Very thin shells, for example, those 3 to 4½ in. thick, may offer space for only a single layer.) Shells with one layer of reinforcement are more likely to crack because of local deformations. Although such cracks may not be structurally detrimental, they could permit rainwater leakage. Hence, shells with one layer of reinforcement should have built-up roofing or other waterproofing applied to the outer surface. In reinforcing small-span shells, two-way wire fabric may be used instead of individual bars.

The area of reinforcement, sq in. per ft width of shell, should not exceed $7.2hf'_c/f_y$ nor $29,000h/f_y$, where h is the overall thickness of the shell, f_y the yield strength of the reinforcement, and f'_c the compressive strength of the concrete. Reinforcement should not be spaced

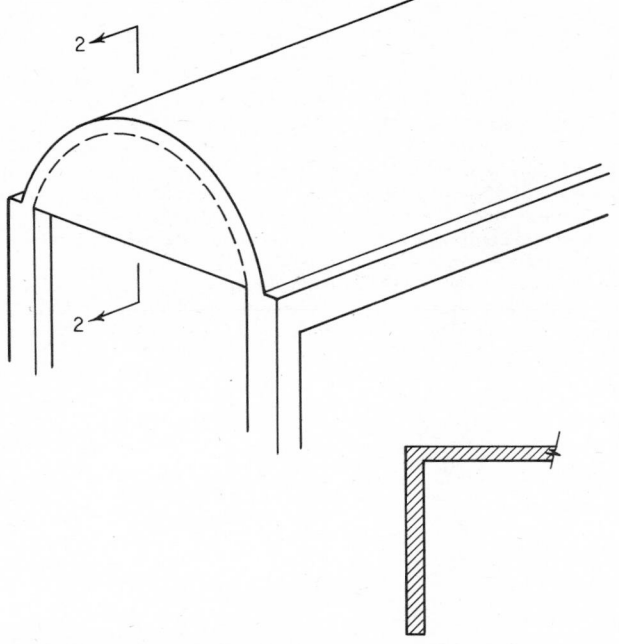

Fig. 8-50. Diaphragm for barrel arch.

SECTION 2-2

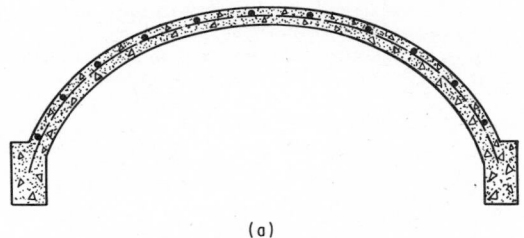

(a)

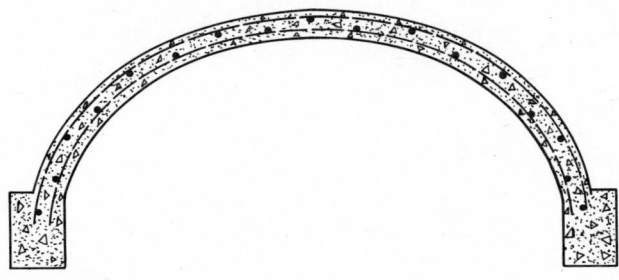

(b)

Fig. 8-51. Arch cross section: (a) single layer of reinforcement; (b) double layer.

farther apart than five times the shell thickness or 18 in.   Where the computed principal tensile stress exceeds $3.6\sqrt{f'_c}$, the reinforcement should not be spaced farther apart than three times the shell thickness.

Minimum specified compressive strength of concrete f'_c should not be less than 3,000 psi, while specified yield strength of reinforcement f_y should not exceed 60,000 psi.

Edge beams of barrel arches behave like ordinary beams under vertical loads, except that additional horizontal shear is applied at the top face at the junction with the shell.   (If these shear stresses are high, reinforcement should be provided to resist them.)   Also, a portion of the shell equal to the flange width permitted for T beams may be assumed to act with the supporting members.   Furthermore, transverse reinforcement from the shell equal to that required for the flange of a T beam should be provided and should be adequately anchored into the edge beam. A typical detail of an edge beam is shown in Fig. 8-52.

Computed stresses in the end arch ribs or diaphragms usually are small.   The minimum amount of reinforcement in a rib should be the minimum specified by the ACI Code for a beam and, in a diaphragm, the minimum specified for a slab.   Longitudinal reinforcement from the

Fig. 8-52.   Edge beam for arch.

shell should be adequately embedded in the ribs.   Because of shear transmission between shell and ribs, the shear stresses should be checked and adequate shear reinforcement provided, if necessary.   Typical reinforcement in end ribs and diaphragms is shown in Fig. 8-53.

High tensile stresses and considerable distortions, particularly in long barrels, usually occur near supports.   If the stresses in those areas are not computed accurately, reinforcement should be increased there by at least 50% over that required by simplified analysis.   The increased quantity of reinforcement should consist of a grid, as shown in Fig. 8-54a.   In arches with very long spans and where stresses are computed more accurately, prestressing of critical areas, as shown in Fig. 8-54b, may be efficient and economical.   But the ratio of steel to concrete in any portion of the tensile zone should be at least 0.0035.

When barrel shells are subjected to heavy concentrated loads, such as in factory roofs or bridges, economy may be achieved by providing interior ribs (Fig. 8-55), rather than increasing

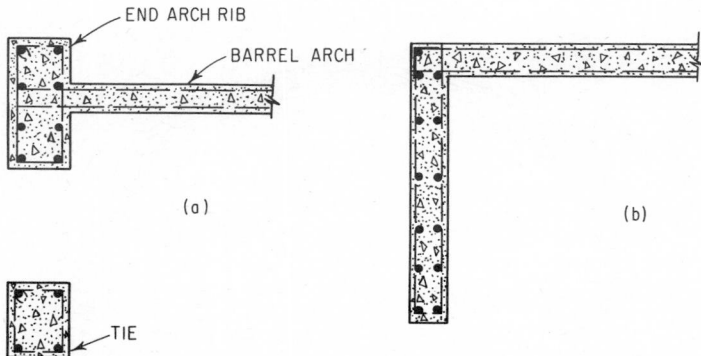

Fig. 8-53.   Reinforcing in end ribs, tie, and diaphragm of arch.

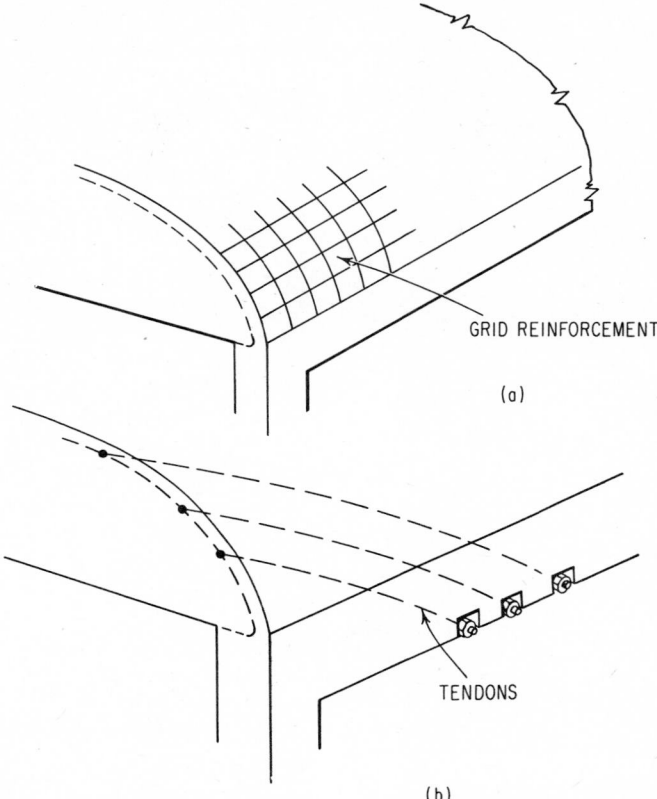

GRID REINFORCEMENT

(a)

TENDONS

(b)

Fig. 8-54. (a) Reinforcement pattern in thin-shell arch. (b) Prestressing of critical zone with tendons.

the thickness throughout the whole shell. Such ribs increase both the strength and stiffness of the shell without increasing the weight very much.

In many cases, only part of a barrel shell may be used. This could occur in end bays of multiple barrels or in interior barrels where large openings are to be provided for windows. Stress distribution in such portions of shells is different from that in whole barrels, but design considerations for edge members and reinforcement placement are the same.

Domes are shells curved in two directions. They are one of the oldest types of construction. At one time, domes were built of large stone pieces, which gave the structures a high ratio of thickness to span; thus, they are excluded from the family of thin shells.

Concrete domes are built relatively thin. It is not uncommon to construct a 6-in.-thick dome spanning 300 ft. Ratio of rise to span usually is in the range of 0.10 to 0.25.

A dome of revolution is subjected mostly to pure membrane stresses under symmetrical, uniform live load. These stresses are compressive in most of the dome and tensile in some other portions, mainly in the circumferential direction. Under unsymmetrical loading, bending moments may occur. Hence, it is common to place reinforcement both in the circumferential direction and perpendicular to it (Fig. 8-56). The reinforcement may be welded-wire fabric or individual bars. It may be placed in one layer (Fig. 8-56b), depending on stresses. Concrete for domes may be cast in forms, as are other more conventional structures, or sprayed.

The critical portion of a dome is its base. Whether the dome is supported continuously there, for example, on a continuous footing, or on isolated supports (Fig. 8-56a), relatively large bending moments and distortions occur in the shell close to the supports. Hence, these regions should be designed to resist the resulting stresses. In domes reinforced with one layer of bars

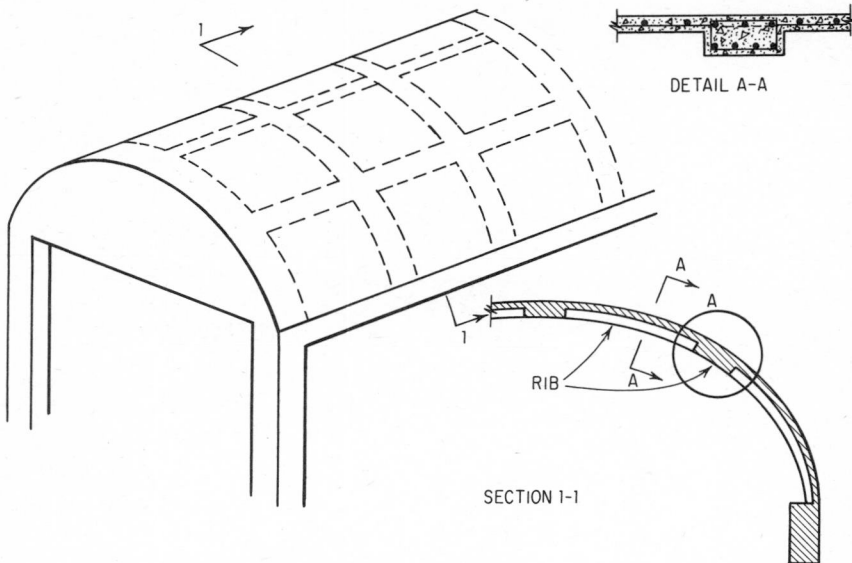

Fig. 8-55. Arch with ribs in longitudinal and transverse directions.

or mesh, it is advisable to provide in the vicinity of the base a double layer of reinforcement (Fig. 8-56b). It also is advisable to thicken the dome close to its base.

The base is subjected to a very large outward-acting radial force, causing large circumferential tension. To resist this force, a concrete ring is constructed at the base (Fig. 8-56). The ring and thickening of the concrete shell in the vicinity of the ring help reduce distortions and cracking of the dome at its base.

Reinforcement of the shell should be properly embedded in the ring (detail A, Fig. 8-56c). The ring should be reinforced or prestressed to resist the circumferential tension. Prestressing is efficient and hence often used. One method of applying prestress is shown in detail A, Fig. 8-56e. Wires are wrapped under tension around the ring and then covered with mortar, for protection against rust and fire. Stirrups should be provided throughout the ring.

Hyperbolic-paraboloid shells, or **hypars,** also are double-curved, but they can be formed with straight boards. Furthermore, since the principal stresses throughout the shell interior consist of equal tension and compression in two perpendicular, constant directions, placement of reinforcement is simple.

Figure 8-57a shows a plan of a hypar supported by two columns at the low points L. The other corners H are the two highest points of the shell. Though strips parallel to LL are in compression and strips parallel to HH in tension, it is customary to place reinforcement in two perpendicular directions parallel to the generatrices of the shell, as shown at section A-A, Fig. 8-57a. The reinforcement should be designed for diagonal tension parallel to the generatrices. Since considerable bending moments may occur in the shell at the columns, this region of the shell usually is made thicker than other portions and requires more reinforcement. The added reinforcement may be placed in the HH and LL directions, as shown at section B-B, Fig. 8-57a.

Shell reinforcement may be placed in one or two layers, depending on the intensity of stresses and distribution of superimposed load. If the superimposed load is irregular and can cause significant bending moments, it is advisable to place the reinforcement in two layers.

As for other types of shell, edges of a hypar are subjected to larger distortions and bending moments than its interior. Therefore, it is desirable to construct edge beams and to thicken the shell in the vicinity of these beams (Fig. 8-57b). A double layer of reinforcement at the edge beams helps reduce cracking of the shell in the vicinity of the beams.

The edge beams are designed as compression or tension members, depending on whether the hypar is supported at the low points, as in Fig. 8-57a, or high points. Prestress in the shell is

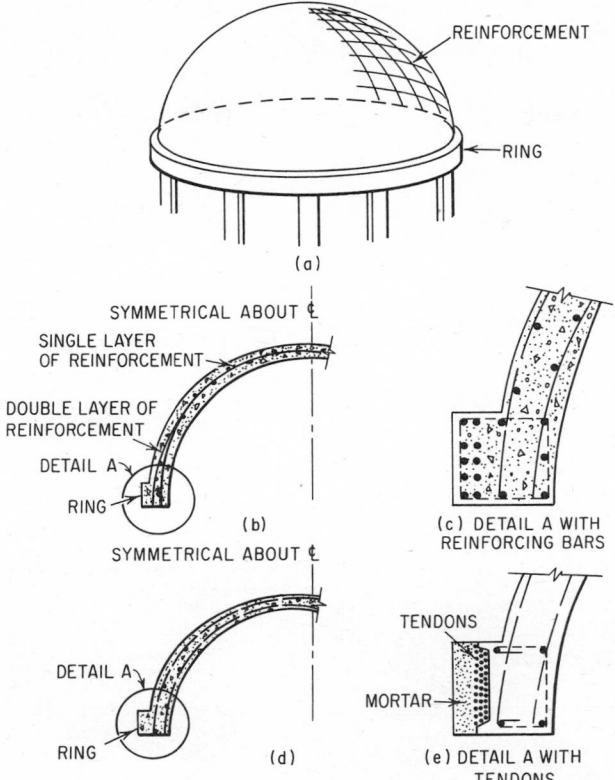

REINFORCEMENT

RING

(a)

SYMMETRICAL ABOUT ℄

SINGLE LAYER
OF REINFORCEMENT

DOUBLE LAYER OF
REINFORCEMENT

DETAIL A

RING

(b)

(c) DETAIL A WITH
REINFORCING BARS

SYMMETRICAL ABOUT ℄

DETAIL A

RING

(d)

TENDONS

MORTAR

(e) DETAIL A WITH
TENDONS

Fig. 8-56. Reinforcing arrangements for dome.

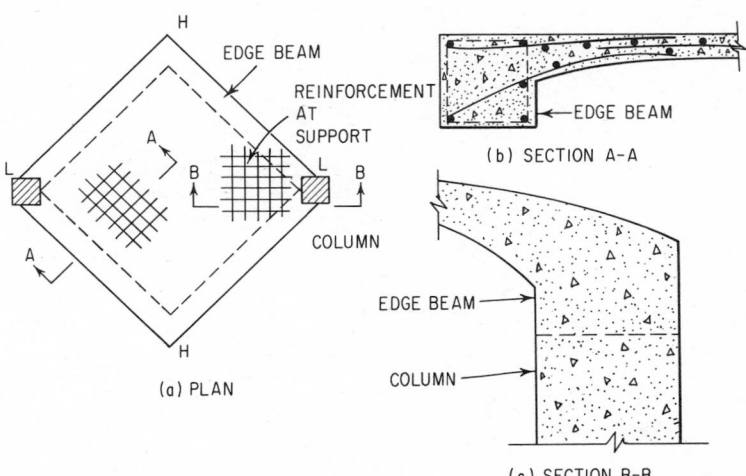

H

EDGE BEAM

REINFORCEMENT
AT
SUPPORT

A

L

B

L

B

COLUMN

A

H

(a) PLAN

EDGE BEAM

(b) SECTION A-A

EDGE BEAM

COLUMN

(c) SECTION B-B

Fig. 8-57. Hyperbolic-paraboloid shell of reinforced concrete. *H* indicates high point; *L*, low point.

most efficient in the vicinity of supports (Fig. 8-58). It also is efficient along the edge beams if supports are at the high points.

Curved shells also may be built with more complex shapes. An interesting example of the versatility of concrete is the undulating shell for the Eastman Kodak pavilion at the 1964–1965 World's Fair in New York (Fig. 8-59). This 300-ft-long shell had a completely arbitrary free form, undefinable geometrically. Thickness of the shell varied from 6 in. at the edges to 14 in., averaging 11 in. The shell had no edge beams. Design, by Lev Zetlin and Associates, Con-

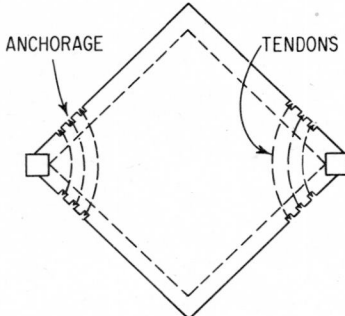

ANCHORAGE   TENDONS

Fig. 8-58. Tendons prestress critical zone of hypar.

Fig. 8-59. Eastman Kodak Pavilion, 1964–1965 World's Fair, New York.

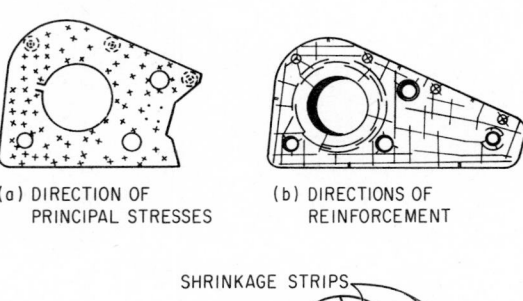

(a) DIRECTION OF
PRINCIPAL STRESSES

(b) DIRECTIONS OF
REINFORCEMENT

SHRINKAGE STRIPS

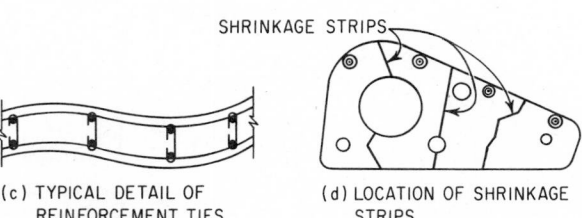

(c) TYPICAL DETAIL OF
REINFORCEMENT TIES

(d) LOCATION OF SHRINKAGE
STRIPS

Fig. 8-60. Details of undulating shell roof shown in Fig. 8-59.

sulting Engineers, was based on theoretical analysis and model tests (Lev Zetlin, "Eastman Kodak Pavilion," *Journal, American Concrete Institute*, October 1964, pp. 1249–1259). Figure 8-60*a* shows the direction of principal stresses and Fig. 8-60*b* the plan of the reinforcement, which was placed in two layers. Since the shell changed curvature several times along a cross section, top and bottom bars were connected with ties (Fig. 8-60*c*) so as not to spall the concrete when the bars tried to straighten under stress. To eliminate shrinkage cracks, three shrinkage strips (Fig. 8-60*d*) were left unconcreted until most of the shrinkage had taken place in adjoining pours.

(G. Winter and A. H. Nilson, "Design of Concrete Structures," McGraw-Hill Book Company, New York; D. P. Billington, "Thin Shell Concrete Structures," McGraw-Hill Book Company, New York.)

Structural-Steel Design and Construction

ROBERT O. DISQUE and FRANK W. STOCKWELL, JR.

**Chief Engineer and Assistant Chief Engineer,
American Institute of Steel Construction, New York, N.Y.**

The many desirable characteristics of structural steels have brought them into widespread use in a large variety of applications. Steel has high strength. It has a very high modulus of elasticity, so that deformations under load are very small. And the modulus is the same in tension and compression. Structural steels also have high ductility. And they have a linear or nearly linear stress-strain relationship up to relatively large stresses. Hence, behavior under working loads can be accurately predicted by elastic theory. Furthermore, structural steels are made under controlled conditions, so that purchasers are assured of uniformly high quality.

Standardization of sections—shapes and plates—has helped make design easy and keep down the cost of structural steels. For tables of properties of these sections, see "Steel Construction Manual," American Institute of Steel Construction, 1221 Avenue of the Americas, New York, N.Y. 10020.

9-1. Characteristics of Structural Steels. (See also Arts. 5-23 to 5-26.) Steels for structural uses are purchased on the basis of stipulated mechanical properties and must be produced with these properties. The most important influence is the chemical composition of each heat of steel. In carbon steels, the elements carbon and manganese have the greatest effects on strength, hardness, ductility, and weldability.

Total reduction in size from ingot to finished product is also important in determining the final mechanical properties. For example, a 1-in.-thick plate rolled from a particular ingot might have a yield point of 64 ksi, while a ¼-in. plate from the same ingot might have a yield point of 71 ksi. Other factors include the amount of residual elements, finishing temperature, and rate of cooling.

Manufacture of steel, therefore, involves meticulous production control. Chemistry is balanced against the other variable factors so that finished products conform to the required mechanical properties.

Yield point F_y is that unit stress, ksi, at which the stress-strain curve exhibits a well-defined increase in strain without increase in stress. Many design rules are based on steel yield points.

Tensile strength or ultimate strength is the largest unit stress the material can achieve in a tensile test.

Modulus of elasticity E is the slope of the stress-strain curve. It is computed by dividing unit stress, ksi, by unit strain, in. per in., at yield stress. For all structural steels, it is usually taken as 29,000 ksi for design calculations.

Ductility is the ability of the material to undergo large inelastic deformations without fracture. It is generally measured by the percent elongation for a specified gage length (usually 8 in.). Structural steel has considerable ductility, which is recognized in many design rules.

Weldability is the ability of steel to be welded without changing its basic mechanical properties. Generally, weldability decreases with increase in carbon and manganese.

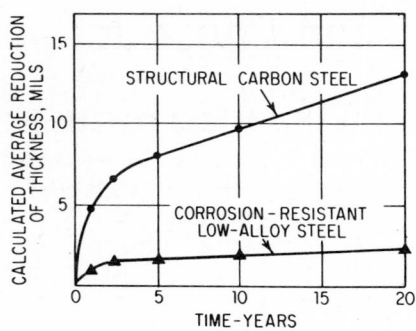

Fig. 9-1. Comparison of loss of thickness due to corrosion of two structural steels exposed in an industrial atmosphere.

Corrosion resistance has no specific index. Some high-strength structural steels are alloyed, generally with copper, to produce high resistance to atmospheric deterioration. These steels develop a tight oxide that inhibits further corrosion. Figure 9-1 compares the rate of reduction of thickness of a typical proprietary "corrosion-resistant" steel with that of ordinary structural steel.

9-2. Available Grades of Structural Steel. Specifications of the American Society for Testing and Materials contain chemical and mechanical requirements for structural steels. The grades of steel are usually referred to by their ASTM specification designation.

A36 steel has a yield point of 36 ksi and a tensile strength of 58 to 80 ksi. Restrictions on maximum carbon content, depending on thickness, insure weldability. A36 is the most commonly used grade.

A529 steel has a yield point of 42 ksi and is limited to smaller-sized shapes and plates. Its carbon content is controlled so that it is weldable. This steel was developed primarily for use in manufactured standard products for which a yield point slightly above 36 ksi could be economically utilized.

A242 is a general specification to cover steels alloyed for corrosion resistance. Minimum yield strength is rated at 50 ksi, but it actually depends on the thickness of the shape or plate and varies from 42 ksi for thick material to 50 ksi for the thinnest material (Tables 9-1 and 9-2). Because several mills produce different steels within the specification, designers should be specific in stating their requirements for corrosion resistance and weldability.

A440, a 50-ksi-yield-strength steel, is applicable primarily to riveted and bolted structures. It can be welded, but only under controlled and proper procedures.

A441 is a high-strength, low-alloy steel for use in riveted, bolted, and welded work. Its minimum yield strength varies from 50 to 42 ksi in the same manner as those of A242 and A440 steels.

A514 is a heat-treated steel with a 100-ksi yield strength. Having a low carbon content, it may be welded by special procedures. This steel is available in plates and bars up to 4 in. thick. For thicknesses of 2½ in. and over, however, the minimum yield strength drops to 90 ksi.

A517, another heat-treated steel, is available in plates with thicknesses up to 2½ in. and minimum yield strength of 100 ksi. It generally is used for bridges rather than for buildings.

A572 includes several grades of high-strength, low-alloy columbium-vanadium steels. Yield strengths range from 42 to 65 ksi. Grades with yield strengths of 55 ksi or more are not approved for welded bridges but may be riveted or bolted.

Table 9-1. Wide-Flange Size Groupings for Tensile Property Classification

Group 1	Group 2	Group 3	Group 4	Group 5
W24 × 55,61	W36 × 135–194	W36 × 230–300	W14 × 219–550	W14 × 605–730
W21 × 44,49	W33 × 118–152	W33 × 200–240		
W18 × 35–60	W30 × 99–210	W14 × 142–211		
W16 × 26–50	W27 × 84–177	W12 × 120–190		
W14 × 22–53	W24 × 68–160			
W12 × 14–58	W21 × 55–142			
W10 × 11.5–45	W18 × 64–114			
W8 × 10–48	W16 × 58–96			
W6 × 8.5–25	W14 × 61–136			
W5 × 16,18.5	W12 × 65–106			
W4 × 13	W10 × 49–112			
	W8 × 58,68			

Table 9-2. Tensile Requirements for High-Strength Steels

ASTM specification	Size grouping	Minimum tensile strength, ksi	Minimum yield strength, ksi	Elongation in 8 in., min, %	Elongation in 2 in., min, %
A242, A440, and A441	1 and 2	70	50	18	
	3	67	46	18	
	4 and 5	63	42	18	21*
A572 Grade 42	1–4†	60	42	20	24
A572 Grade 45	1–4†	60	45	19	22
A572 Grade 50	1–4†	65	50	18	21
A572 Grade 55	1–4†	70	55	17	20
A572 Grade 60	1 and 2	75	60	16	18
A572 Grade 65	1	80	65	15	17
A588	1–4	70	50	18	21*
A588	5	67	46		18

*For shapes over 426 lb per ft, elongation in 2 in. of 18% minimum applies.
†Available in shapes up to 426 lb per ft.

A588 is a high-strength, low-alloy steel with a minimum yield strength of 50 ksi. Intended for use in riveted, bolted, and welded structures, this steel has an atmospheric corrosion resistance four times that of plain carbon steel, A36. A588 steel is available in all shapes and in plate thicknesses up to 8 in., but the yield strength is lower for thick material (Table 9-2).

A709 covers several grades of structural steel intended for use in bridges. Approved in 1974 by ASTM, the specification for these steels was the first in a continuing program to consolidate the specifications applicable to structural steels. Under this specification, grades 36, 50, and 100 are steels with yield strengths of 36, 50, and 100 ksi, respectively. The grade designation is followed by the letter W indicating whether ordinary or high atmospheric corrosion resistance is required. An additional letter T indicates that Charpy V-notch impact tests must be conducted on the steels, and a following numeral, 1, 2, or 3, indicates the testing zone into which the steels to be used fall. The testing zones relate to the lowest ambient temperature expected for the bridge site. The impact requirements under this specification depend on grade and thickness of material and correspond with applicable requirements listed in Table 9-3.

Identification. All shapes and plates used in main components made of steel with a yield strength greater than 36 ksi are marked with the ASTM specification designation for easy identification. This marking originates at the mill and is carried through all fabrication and painting operations. In addition, the fabricator will, if requested, provide an affidavit stating that the structural steel meets the requirements of the grade specified. Furthermore, main tension members in some bridges may be required to meet notch-toughness criteria (Table 9-3).

Table 9-3. Toughness Criteria for A709 Bridge Steels

Grade	Thickness	Minimum average energy and testing temperatures in Charpy V-notch impact tests		
		Zone 1*	Zone 2*	Zone 3*
36T	To 4 in. inclusive	15 ft-lb @ 70° F	15 ft-lb @ 40° F	15 ft-lb @ 10° F
50T†, 50WT†	To 2 in. inclusive	15 ft-lb @ 70° F	15 ft-lb @ 40° F	15 ft-lb @ 10° F
50WT†	Over 2 in. to 4 in.	20 ft-lb @ 70° F	20 ft-lb @ 40° F	20 ft-lb @ 10° F
100T, 100WT	To 2½ in. inclusive	25 ft-lb @ 30° F	25 ft-lb @ 0° F	25 ft-lb @ −30° F
100T, 100WT	Over 2½ in. to 4 in.	35 ft-lb @ 30° F	35 ft-lb @ 0° F	35 ft-lb @ −30° F

*Minimum service temperature: for Zone 1, 0° F and above; for Zone 2, −30° F; for Zone 3, −31° F to −60° F.

† If F_y exceeds 65 ksi, reduce testing temperature by 15° F for each increment (or fraction) of 10 ksi above 65 ksi.

For these, the fabricator may be required to demonstrate a method of tracing these members in order to verify their proper application.

Fasteners. A502 covers two grades of steel for structural rivets in diameters from ½ to 1½ in., inclusive. Grade 1 is a carbon-steel rivet for general purposes. Grade 2 is a carbon-manganese steel rivet, suitable for use with high-strength carbon structural steels and high-strength low-alloy structural steels.

Steels for structural bolts are covered by A307, A325, and A490. A307 specifies the chemical and mechanical requirements for low-carbon-steel common bolts. These bolts are used primarily for low-stress connections or connections of secondary members.

A325 covers the chemical and mechanical requirements of quenched and tempered high-strength bolts, ½ to 1½ in., inclusive, in diameter, made of medium-carbon steel. They are used primarily with A36 steel.

A490 covers the chemical and mechanical requirements of quenched and tempered, alloy, high-strength bolts, ½ to 1½ in., inclusive, in diameter. They are used primarily with higher-strength steels.

9-3. Structural-Steel Shapes. Most structural steel used in building construction is fabricated from rolled shapes. In bridges, greater use is made of plates, since girders spanning over about 90 ft are usually built-up sections.

Many different rolled shapes are available. They are designated as W shapes (wide-flange shapes), M shapes (miscellaneous shapes), S shapes (standard I sections), angles, channels, and bars. The "Steel Construction Manual," American Institute of Steel Construction, lists properties of these shapes.

Wide-flange shapes normally range from a W4 × 13 (4 in. deep weighing 13 lb per lin ft) to a W36 × 300 (36 in. deep weighing 300 lb per lin ft). The heaviest shapes available are the "jumbo" column sections, which range up to W14 × 730.

In general, wide-flange shapes are the most efficient beam section. They have a high proportion of the cross-sectional area in the flanges and thus a high ratio of section modulus to weight. The 14-in. W series includes shapes proportioned for use as column sections; the relatively thick web results in a large area-to-depth ratio.

Since the flange and web of a wide-flange beam do not have the same thickness, their yield points differ. In accordance with design rules for structural steel based on yield point, it is therefore necessary to establish a "design yield point" for each section. In practice, all beams rolled from A36 steel (Art. 9-2) are considered to have a yield point of 36 ksi. Wide-flange shapes, plates, and bars rolled from higher-strength steels are required to have the minimum yield and tensile strength shown in Table 9-2.

Square, rectangular, and round structural tubular members are available with a variety of yield strengths. Suitable for columns because of their symmetry, these members are particularly useful in low buildings and where they are exposed for architectural effect. They may be used in tall buildings if closely spaced.

Connection Material. Connections are normally made with A36 steel. If, however, higher-strength steels are used, the structural size groupings for angles and bars are:

Group 1: thicknesses of ½ in. or less
Group 2: thicknesses exceeding ½ in. but not more than ¾ in.
Group 3: thicknesses exceeding ¾ in.

Structural tees fall into the same group as the wide-flange or standard sections from which they are cut. (A WT7 × 13, for example, designates a tee formed by cutting in half a W14 × 26 and therefore is considered a Group 1 shape, as is that W14.)

9-4. Selecting Structural Steels. The only way to achieve absolute economy in selecting steels would be to compare costs of structures completely erected. This, of course, is impossible. Some guidelines, however, may be useful.

In design of beams or girders, deflection requirements must be examined before steels stronger than A36 are considered. If there are no deflection limitations, a high-strength steel usually is economical.

For a fully braced beam, a good material-cost analysis may be made by comparing directly the cost-yield-strength ratios. This is possible because the section modulus of rolled shapes varies fairly uniformly with weight. If, however, the allowable stress must be reduced because of limited bracing, it may be necessary to design the beam of each grade of steel to determine the most economical one.

Use of high-strength steel in bridge girders often proves economical at points of maximum moment. In fact, a girder often may be designed with two or more grades of steel to reduce costs.

High-strength steels are often used for the lower columns of multistory buildings. With high axial loads, the required section area is nearly a direct function of the yield-strength stress F_y. Table 9-4 gives allowable stress as a function of slenderness ratio Kl/r for various grades of steel. By working with this table and local mill prices, you can determine the most economical grade for a column. For tension members, yield strength divided by price is a good measure of relative cost.

Table 9-4. Ratio of Allowable Stress in Columns of High-Strength Steel to That of A36 Steel

Specified yield strength F_y, ksi	Slenderness ratio Kl/r											
	5	15	25	35	45	55	65	75	85	95	105	115
65	1.80	1.78	1.75	1.72	1.67	1.62	1.55	1.46	1.35	1.22	1.10	1.03
60	1.66	1.65	1.63	1.60	1.56	1.52	1.47	1.40	1.32	1.21	1.10	1.03
55	1.52	1.51	1.50	1.48	1.45	1.42	1.38	1.33	1.27	1.20	1.10	1.03
50	1.39	1.38	1.37	1.35	1.34	1.32	1.29	1.26	1.22	1.17	1.10	1.03
45	1.25	1.24	1.24	1.23	1.22	1.21	1.19	1.17	1.15	1.12	1.08	1.03
42	1.17	1.16	1.16	1.15	1.15	1.14	1.13	1.12	1.10	1.08	1.06	1.03

On a piece-for-piece basis, there is substantially no difference in the cost of fabricating and erecting the different grades. Higher-strength steels, however, may afford an opportunity to reduce the number of members, thus reducing both fabrication and erection costs.

9-5. Tolerances for Structural Shapes. ASTM Specification A6 lists mill tolerances for rolled-steel plates, shapes, sheet piles, and bars. Included are tolerances for rolling, cutting, section area, and weight, ends out of square, camber, and sweep. The "Steel Construction Manual," American Institute of Steel Construction, contains tables for applying these tolerances.

The AISC Specification for the Design, Fabrication and Erection of Structural Steel for Buildings and Code of Standard Practice give fabrication and erection tolerances for structural steel for building work. Figure 9-2 shows permissible tolerances for column erection for a multistory building.

Both mill and fabrication tolerances should be considered in designing and detailing structural steel. A column section, for instance, may have an actual depth up to ½ in. greater or less than the nominal depth. An accumulation of dimensional variations, therefore, would cause serious trouble in the erection of a building with many bays. Provision should be made for such a possibility.

Tolerances for fabrication and erection of bridge girders are usually specified by highway departments.

9-6. Structural-Steel Design Specifications. The basis for design of practically all structural steel for buildings in the United States is the AISC Specification for the Design, Fabrication and Erection of Structural Steel for Buildings (American Institute of Steel Construction, Inc., 1221 Avenue of the Americas, New York, N.Y. 10020). This specification relates allowable working stresses to yield point of material; the same relationships may be used for new steels when they are developed. Revisions may be made annually and are issued as supplements. Design criteria for buildings presented in this section derive from the basic 1969 Specification and supplements adopted in 1970, 1971, and 1974.

Design rules for bridges have been developed by the American Association of State Highway and Transportation Officials (AASHTO), National Press Building, Washington, D.C. 20004 (Standard Specifications for Highway Bridges). They are slightly more conservative than AISC rules, because bridges are subject to dynamic loading and fatigue (see Sec. 17).

In addition, there are other important specifications relating to steel structures:

American Iron and Steel Institute (AISI) publishes rules for the design of structural members cold-formed to shape from sheet or strip and used for load-carrying purposes in buildings (see Sec. 10).

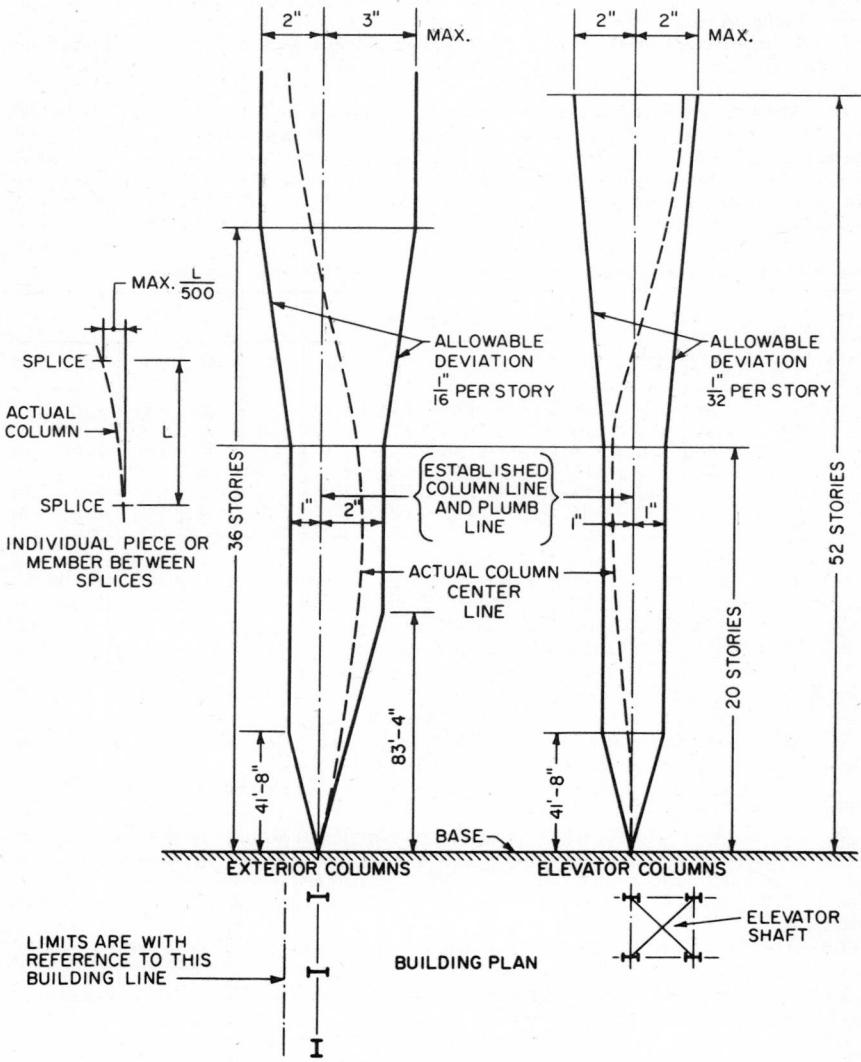

Fig. 9-2. Tolerances for column erection in a multistory building.

American Welding Society (2501 N.W. 7th St., Miami, Fla. 33125) publishes the Structural Welding Code, AWS D1.1, applicable to welding of bridges, buildings, and structures with tubular members.

American Railway Engineering Association publishes rules for design, fabrication, and erection of steel railway bridges (see Sec. 17).

AISC and the Steel Joist Institute (SJI) jointly publish specifications that cover design, manufacture, and use of open-web steel joists (see Sec. 10).

AISC also publishes a Specification for the Design of Architecturally Exposed Structural Steel, which applies to steel structures subject to close inspection by the public.

9-7. Structural-Steel Design Theories.   Structural steel for buildings may be designed by either the elastic or plastic theories.   Design rules for both methods are included in the AISC Specification for the Design, Fabrication and Erection of Structural Steel for Buildings (Art. 9-6).

Elastic-design rules take into account the principles of plastic design, so that for many structures, neither method offers appreciable material saving compared with the other.   Most engineers, however, consider plastic design to be simpler, more direct, and philosophically superior.

In elastic or "allowable-stress" design, a member is selected so that stresses will not exceed a specified value under working or service loads.   Allowable stresses are listed in the AISC Specification.   They are determined by dividing yield stress by a safety factor.

Plastic design is based on the ability of structural steel to deform plastically at and above the yield point.   If the load on a steel structure is increased sufficiently, the material will pass through the elastic range into the plastic range of stress.

The load-carrying capacity of a continuous beam is computed on the assumption that highly stressed areas may yield in bending.   When these highly stressed zones yield, they no longer resist further bending.   As a result, with further increases in load, the additional moment is transferred to less highly stressed areas.   Thus, an overloaded continuous structure will readjust itself to carry the load.

Actually, these plastic hinges will not form under service loads.   Plastic design rules include a safety factor against failure about the same as that for elastic design.

For buildings, plastic design, under the AISC Specification, may be applied only to steels with yield strengths up to 65 ksi, one- and two-story rigid frames, and planar multistory frames (Art. 9-29).

For highway bridges, a similar method, called load-factor design, is acceptable to AASHTO (Art. 9-6) as an alternative to allowable-stress design (Art. 9-30).

9-8. Allowable Tension in Steel.   For buildings, AISC (Art. 9-6) specifies a basic allowable unit tensile stress, ksi, $F_t = 0.60F_y$, where F_y is the yield strength of the steel, ksi (Table 9-5). F_t is subjected to the further limitation that it should not exceed one-half the specified minimum tensile strength F_u of the material.   This limitation, however, does not control for steels for which F_y does not exceed 65 ksi.   On the net section through pinholes in eyebars, pin-connected plates, or built-up members, $F_t = 0.45F_y$.

For bridges, AASHTO (Art. 9-6) specifies allowable tensile stresses as the smaller of $0.55F_y$ or $0.46F_u$ (Table 9-5).

Table 9-5 and subsequent tables apply to two strength levels, $F_y = 36$ ksi and $F_y = 50$ ksi, the ones generally used for construction.

These allowable stresses are applied to the net cross-sectional area of a member.   The net section for a tension member with a chain of holes extending across a part in any diagonal or zigzag line is defined in the AISC Specification as follows: The net width of the part shall be obtained by deducting from the gross width the sum of the diameters of all the holes in the chain, and adding, for each gage space in the chain, the quantity $s^2/4g$, where s = longitudinal spacing (pitch), in., of any two consecutive holes and g = transverse spacing (gage), in., of the same two holes.   The critical net section of the part is obtained from the chain that gives the least net width.

Table 9-5. Allowable Tensile Stresses in Bridge and Building Steels, Ksi

Yield strength	Buildings	Bridges
36	22	20
50	30	27

But the net section taken through a hole shall in no case be considered as more than 85% of the corresponding gross section.

9-9. Allowable Shear in Steel. The AISC Specification for buildings (Art. 9-6) specifies allowable unit shear, ksi, as $F_v = 0.40F_y$, where F_y is the yield point of the steel, ksi. The AASHTO Specification for bridges (Art. 9-6) gives $F_v = 0.33F_y$ (see Table 9-6).

Table 9-6. Allowable Shear Stresses in Bridge and Building Steels, Ksi

Yield strength	Buildings	Bridges
36	14.5	12
50	20	17

The area used to compute shear stress in a rolled beam is defined as the product of the web thickness and the overall beam depth. The webs of all rolled structural shapes are of such thickness that shear is seldom the criterion for design.

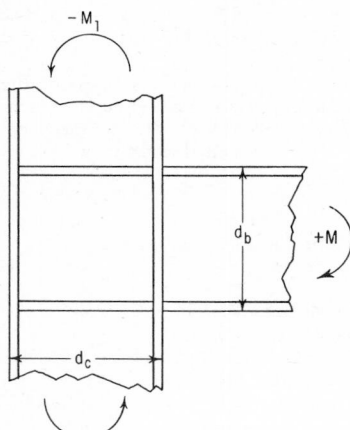

Fig. 9-3. Rigid connection of two steel members with webs in a common plane.

Within the boundaries of a rigid connection of two or more members whose webs lie in a common plane, the web shear stresses are generally high. For buildings, the Commentary on the AISC Specification states that such webs need no reinforcement if the web thickness, in., is greater than that calculated from

$$t_w = \frac{23M}{A_{bc}F_y} \tag{9-1}$$

where M = algebraic sum of clockwise and counterclockwise moments applied on opposite sides of the connection boundary, ft-kips

A_{bc} = planar area of the connection web, sq in. (in Fig. 9-3, $d_b d_c$)

Additional requirements for column web stiffeners are presented in Art. 9-15.

9-10. Allowable Compression in Steel. (See also Art. 9-13.) The allowable compressive unit stress for a column is a function of its slenderness ratio. The slenderness ratio is defined in the AISC Specification for the Design, Fabrication and Erection of Structural Steel for Buildings as Kl/r, where K = effective-length factor, which depends on restraints at top and bottom of the column; l = length of column between supports, in.; and r = radius of gyration of column section, in.

The AISC specification for buildings provides two formulas for computing allowable compressive stress F_a, ksi, for main members. The formula to use depends on the relationship of the largest effective slenderness ratio Kl/r of the cross section of any unbraced segment to a factor C_c defined by Eq. (9-2). See Table 9-7a.

$$C_c = \sqrt{\frac{2\pi^2 E}{F_y}} = \frac{756.6}{\sqrt{F_y}} \tag{9-2}$$

where E = modulus of elasticity of the steel = 29,000 ksi
    F_y = yield stress of the steel, ksi
When Kl/r is less than C_c,

$$F_a = \frac{\left[1 - \dfrac{(Kl/r)^2}{2C_c{}^2}\right] F_y}{\text{F.S.}} \tag{9-3a}$$

where F.S. = safety factor = $\dfrac{5}{3} + \dfrac{3(Kl/r)}{8C_c} - \dfrac{(Kl/r)^3}{8C_c{}^3}$
See Table 9-7b.
When Kl/r exceeds C_c,

$$F_a = \frac{12\pi^2 E}{23(Kl/r)^2} \tag{9-3b}$$

Table 9-7a. Values of C_c

F_y	C_c
36	126.1
50	107.0

Table 9-7b. Allowable Stresses F_a, Ksi, in Steel Building Columns for $Kl/r \le 120$

Kl/r	Yield strength of steel F_y, ksi	
	36	50
10	21.16	29.26
20	20.60	28.30
30	19.94	27.15
40	19.19	25.83
50	18.35	24.35
60	17.43	22.72
70	16.43	20.94
80	15.36	19.01
90	14.20	16.94
100	12.98	14.71
110	11.67	12.34*
120	10.28	10.37*

*From Eq. (9-3b), because $Kl/r > C_c$.

Table 9-7c. Allowable Stresses, Ksi, in Steel Building Columns for $Kl/r > 120$

Kl/r	F_a	F_{as}
130	8.84	9.30
140	7.62	8.47
150	6.64	7.81
160	5.83	7.29
170	5.17	6.89
180	4.61	6.58
190	4.14	6.36
200	3.73	6.22

On the gross section of axially loaded bracing and secondary building members, when l/r exceeds $120\,(K = 1)$, the allowable compressive stress, ksi, is

$$F_{as} = \frac{F_a}{1.6 - l/200r} \tag{9-4}$$

In Eq. (9-4), F_a is obtained from Eq. (9-3a) or (9-3b), depending on the value of C_c (see Table 9-7c).

The effective-length factor K, equal to the ratio of effective column length to actual unbraced length, may be greater or less than 1.0. Theoretical K values for six idealized conditions, in which joint rotation and translation are either fully realized or nonexistent, are tabulated in Fig. 9-4.

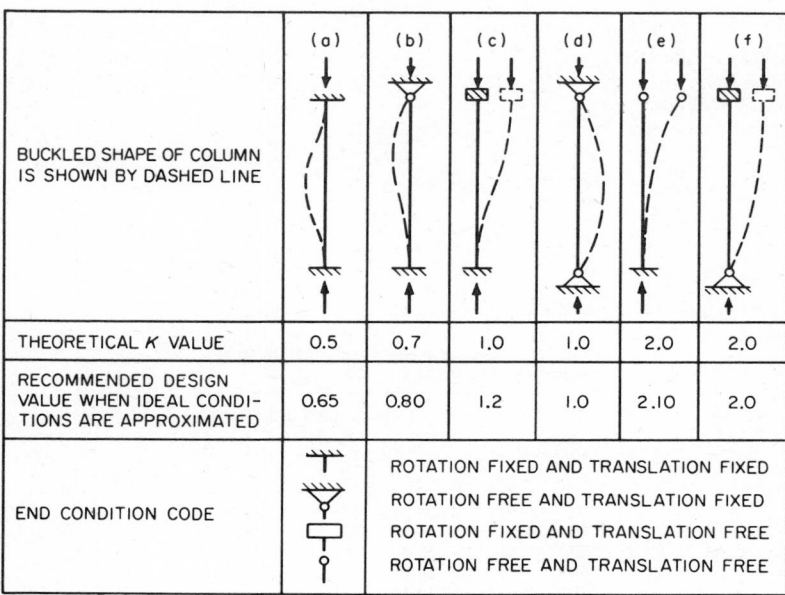

Fig. 9-4. Values of effective-length factor K for columns.

An alternative, and more precise, method of calculating K for an unbraced column uses a nomograph given in the "Commentary on AISC Specification" ("Steel Construction Manual," American Institute of Steel Construction). This method requires calculation of "end-restraint factors" for the top and bottom of the column, to permit K to be determined from the chart.

For bridge design, AASHTO recommends a column formula for each grade of steel (Table 9-8).

Table 9-8. Column Formulas for Bridge Design

		Allowable stress, ksi	
Yield strength, ksi	Maximum l/r*	Riveted or bolted ends	Pinned ends
36	130	$16 - 0.00030\dfrac{l^2}{r^2}$	$16 - 0.00038\dfrac{l^2}{r^2}$
50	125	$22 - 0.00056\dfrac{l^2}{r^2}$	$22 - 0.00074\dfrac{l^2}{r^2}$

*l = length of member, in.; r = least radius of gyration of section, in.

Table 9-9. Allowable Bending Stresses in Braced Beams for Buildings, Ksi

Yield strength, ksi	Compact $(0.66F_y)$	Noncompact $(0.60F_y)$
36	24	22
50	33	30

9-11. Allowable Stresses in Bending. (See also Art. 9-13.) The allowable stress in the compression flange usually governs the load-carrying capacity of steel beams or girders.

In building construction, the maximum fiber stress in bending for laterally supported beams and girders is $F_b = 0.66F_y$ for those sections classified as "compact" (Art. 9-25) and $F_b = 0.60F_y$ for those not conforming to the specified requirements. F_y is the yield strength of the steel, ksi. Table 9-9 lists values of F_b for two grades of steel.

Because continuous steel beams have considerable reserve strength beyond the yield point, a redistribution of moments may be assumed when compact sections are continuous over supports or rigidly framed to columns. In that case, negative gravity-load moments over the supports may be reduced 10%. If this is done, the maximum positive moment in each span should be increased by 10% of the average negative moments at the span ends.

The allowable extreme-fiber stress of $0.60F_y$ applies to laterally supported, unsymmetrical members, except channels, and to noncompact box sections. Compression on extreme fibers of channels should not exceed $0.60F_y$ or the value given by Eq. (9-7).

The allowable fiber stress of $0.66F_y$ for compact members should be reduced to $0.60F_y$ when the compression flange is unsupported for a length, in., exceeding

$$l_{max} = \frac{76.0b_f}{\sqrt{F_y}} \quad \text{or} \quad l_{max} = \frac{20,000}{F_y d/A_f}$$

where b_f = width of compression flange, in.

d = beam depth, in.

A_f = area, sq in., of compression flange

The allowable stress should be reduced even more when l/r_T exceeds certain limits, where l is the unbraced length, in., of the compression flange and r_T is the radius of gyration, in., of a portion of the beam consisting of the compression flange and one-third of the part of the web in compression.

For $\sqrt{102,000C_b/F_y} \le l/r_T \le \sqrt{510,000C_b/F_y}$, use

$$F_b = \left[\frac{2}{3} - \frac{F_y(l/r_T)^2}{1,530,000C_b} \right] F_y \tag{9-5a}$$

For $l/r_T > \sqrt{510,000C_b/F_y}$, use

$$F_b = \frac{170,000C_b}{(l/r_T)^2} \tag{9-6a}$$

where C_b = modifier for moment gradient (discussed later)

When, however, the compression flange is solid and nearly rectangular in cross section, and its area is not less than that of the tension flange, the allowable stress may be taken as

$$F_b = \frac{12,000C_b}{ld/A_f} \tag{9-7}$$

See Table 9-10. When Eq. (9-7) applies (except for channels), F_b should be taken as the larger of the values computed from Eqs. (9-7) and (9-5a) or (9-6a) but not more than $0.60F_y$.

The moment-gradient factor C_b in Eqs. (9-5) to (9-7) may be computed from

$$C_b = 1.75 + 1.05\frac{M_1}{M_2} + 0.3\left(\frac{M_1}{M_2}\right)^2 \le 2.3 \tag{9-8}$$

where M_1 = smaller beam end moment

M_2 = larger beam end moment

Table 9-10. Allowable Bending Stress F_b* Ksi, for Eq. (9-7)

ld/A_f	C_b							
	1.0	1.2	1.4	1.6	1.8	2.0	2.2	2.3
400	30.0	. . .	. . .	. . .	. . .	. . .	. . .	. . .
450	26.7	32.0	. . .	. . .	. . .	. . .	. . .	. . .
500	24.0	28.8	. . .	. . .	. . .	. . .	. . .	. . .
550	21.8	26.2	30.5	. . .	. . .	. . .	. . .	. . .
600	20.0	24.0	28.0	32.0	. . .	. . .	. . .	. . .
650	18.5	22.2	22.8	29.5	. . .	. . .	. . .	. . .
700	17.1	20.6	24.0	27.4	30.9	. . .	. . .	. . .
750	16.0	19.2	22.4	25.6	28.8	. . .	. . .	. . .
800	15.0	18.0	21.0	24.0	27.0	30.0	. . .	. . .
850	14.1	16.9	19.8	22.6	25.4	28.2	31.1	. . .
900	13.3	16.0	18.7	21.3	24.0	26.7	29.3	30.7
950	12.6	15.2	17.7	20.2	22.7	25.3	27.8	29.1
1,000	12.0	14.4	16.8	19.2	21.6	24.0	26.4	27.6
1,100	10.9	13.1	15.3	17.5	19.6	21.8	24.0	25.1
1,200	10.0	12.0	14.0	16.0	18.0	20.0	22.0	23.0
1,300	9.23	11.1	12.9	14.8	16.6	18.5	20.3	21.2
1,400	8.57	10.3	12.0	13.7	15.4	17.1	18.9	19.7
1,500	8.00	9.60	11.2	12.8	14.4	16.0	17.6	18.4
1,600	7.50	9.00	10.5	12.0	13.5	15.0	16.5	17.3

*F_b may not exceed $0.60F_y$. Dots indicate that F_b is greater than 0.6×50 ksi; therefore, use 30 ksi. Values greater than 30 ksi are shown for interpolation purposes only and may not be used as F_b for $F_y = 50$ ksi material.

The algebraic sign of M_1/M_2 is positive for double-curvature bending and negative for single-curvature bending. For braced frames, C_b should be taken as unity for computation of F_{bx} and F_{by} with Eq. (9-13).

Equations (9-5a) and (9-6a) can be simplified by introduction of a new term:

$$Q = \frac{(l/r_T)^2 F_y}{510,000 C_b} \tag{9-9}$$

Now, for $0.2 \le Q \le 1$,

$$F_b = \frac{(2 - Q)F_y}{3} \tag{9-5b}$$

For $Q > 1$,

$$F_b = \frac{F_y}{3Q} \tag{9-6b}$$

As for the preceding equations, when Eq. (9-7) applies (except for channels), F_b should be taken as the largest of the values given by Eqs. (9-7) and (9-5b) or (9-5c), but not more than $0.60F_y$.

For bridge design, AASHTO (Art. 9-6) gives the allowable unit (tensile) stress in bending as $F_b = 0.55F_y$ (Table 9-11). The same stress is permitted for compression when the compression flange is supported laterally for its full length by embedment in concrete or by other means.

Table 9-11. Allowable Bending Stress in Braced Bridge Beams, Ksi

F_y	F_b
36	20
50	27

When the compression flange is partly supported or unsupported in a bridge, the allowable bending stress, ksi, is

$$F_b = \left[1 - \frac{(3l^2/b^2)F_y}{\pi^2 E} \right] 0.55F_y \tag{9-10}$$

where b = flange width, in.

l = length of unsupported compression flange between lateral connections, knee braces, or other points of support, in., or twice the length of a cantilever if this is less than the preceding

For F_y = 36 ksi, l/b should not exceed 36, and for F_y = 50 ksi, l/b should not be more than 30 (Table 9-12). For negative-moment regions of continuous girders, l may be taken as the distance from the interior support to point of dead-load contraflexure if this distance is less than that between lateral connections, knee braces, or other supports. Continuous or cantilever bridge girders may be proportioned for negative moment at interior supports with an allowable unit stress 20% higher than permitted by Eq. (9-10) but in no case exceeding $0.55F_y$.

For each grade of steel, Eq. (9-10) reduces to the expressions given in Table 9-12.

Table 9-12. Allowable Compressive Stress in Flanges of Bridge Beams, Ksi

F_y	Max l/b	F_b
36	36	$20-0.0075(l/b)^2$
50	30	$27-0.0144(l/b)^2$

9-12. Allowable Bearing Stress. In building construction, allowable bearing stress for milled surfaces, including bearing stiffeners, and pins in reamed, drilled, or bored holes, is $F_p = 0.90F_y$, where F_y is the yield strength of the steel, ksi.

For expansion rollers and rockers, the allowable bearing stress, kips per lin in., is

$$F_p = \frac{F_y - 13}{20}0.66d \qquad (9\text{-}11)$$

where d is the diameter, in., of the roller or rocker. When parts in contact have different yield strengths, F_y is the smaller value.

For highway design, AASHTO (Art. 9-6) limits the allowable bearing stress on milled stiffeners and other steel parts in contact to $F_p = 0.80F_y$ (Table 9-13). Allowable bearing stresses on pins are given in Table 9-14.

Table 9-13. Allowable Bearing Stress on Stiffeners of Bridge Girders, Ksi

F_y	F_p
36	29
50	40

Table 9-14. Allowable Bearing Stresses on Pins, Ksi

F_y	Buildings $F_p = 0.90F_y$	Bridges	
		Pins subject to rotation $F_p = 0.40F_y$	Pins not subject to rotation $F_p = 0.80F_y$
36	33	14	29
50	45	20	40

The allowable bearing stress for expansion rollers and rockers used in bridges depends on the yield point in tension F_y of the steel in the roller or the base, whichever is smaller. For diameters up to 25 in., the allowable stress, kips per lin in., is

$$p = \frac{F_y - 13}{20}0.6d \qquad (9\text{-}12a)$$

For diameters from 25 to 125 in.,

$$p = \frac{F_y - 13}{20} \, 3 \, \sqrt{d} \tag{9-12b}$$

where d = diameter of roller or rocker, in.

9-13. Combined Axial Compression or Tension and Bending. The AISC specification for buildings (Art. 9-6) includes three interaction formulas for combined axial compression and bending:

When the ratio of computed axial stress to allowable axial stress $f_a/F_a > 0.15$, both Eqs. (9-13) and (9-14) must be satisfied.

$$\frac{f_a}{F_a} + \frac{C_{mx} f_{bx}}{(1 - f_a/F'_{ex})F_{bx}} + \frac{C_{my} f_{by}}{(1 - f_a/F'_{ey})F_{by}} \le 1 \tag{9-13}$$

$$\frac{f_a}{0.60F_y} + \frac{f_{bx}}{F_{bx}} + \frac{f_{by}}{F_{by}} \le 1 \tag{9-14}$$

When $f_a/F_a \le 0.15$, Eq. (9-15) may be used instead of Eqs. (9-13) and (9-14).

$$\frac{f_a}{F_a} + \frac{f_{bx}}{F_{bx}} + \frac{f_{by}}{F_{by}} \le 1 \tag{9-15}$$

In the preceding equations, subscripts x and y indicate the axis of bending about which the stress occurs, and

F_a = axial stress that would be permitted if axial force alone existed, ksi (see Arts. 9-8 and 9-10)
F_b = compressive bending stress that would be permitted if bending moment alone existed, ksi (see Art. 9-11)
$F'_e = 149{,}000/(Kl_b/r_b)^2$, ksi; as for F_a, F_b, and $0.6F_y$, F'_e may be increased one-third for wind and seismic loads
l_b = actual unbraced length in plane of bending, in.
r_b = radius of gyration about bending axis, in.
K = effective-length factor in plane of bending
f_a = computed axial stress, ksi
f_b = computed compressive bending stress at point under consideration, ksi
C_m = adjustment coefficient

For compression members in frames subject to joint translation (sidesway), $C_m = 0.85$ in Eq. (9-13). For restrained compression members in frames braced against joint translation and not subject to transverse loading between supports in the plane of bending, $C_m = 0.6 - 0.4M_1/M_2$, but not less than 0.4. M_1/M_2 is the ratio of the smaller to larger moment at the ends of that portion of the member unbraced in the plane of bending under consideration. M_1/M_2 is positive when the member is bent in reverse curvature and negative when it is bent in single curvature. For compression members in frames braced against joint translation in the plane of loading and subject to transverse loading between supports, the value of C_m may be determined by rational analysis. But in lieu of such analysis, the following values may be used: For members whose ends are restrained, $C_m = 0.85$. For members whose ends are unrestrained, $C_m = 1.0$.

Building members subject to combined axial tension and bending should satisfy Eq. (9-14), with f_b and F_b, respectively, as the computed and permitted bending tensile stress. But the computed bending compressive stress is limited by Eqs. (9-7) and (9-5a) or (9-6a).

Interaction in bridge design is based on a complicated formula that includes terms for assumed and known eccentricities. The formula incorporates a cosecant function. Solution requires iterative methods and is difficult. An appendix of the AASHTO Specification contains charts to simplify the calculations (Art. 9-6).

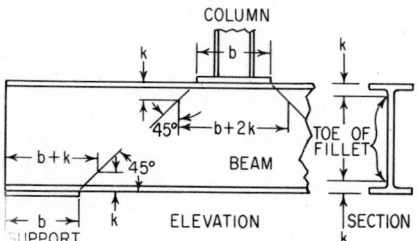

Fig. 9-5. Compressive stress is assumed to spread at 45° angle in web of rolled beam.

9-14. Web Crippling. The possibility of web crippling must be investigated in rolled beams and built-up girders at points of concentrated loads.

The AISC Specification for buildings (Art. 9-6) limits the compressive stress at the toe of the web of a rolled beam due to a concentrated load to $F_a = 0.75F_y$, where F_y is the yield point of the steel, ksi. In calculating the stress area, a 45° distribution should be assumed, as indicated in Fig. 9-5. Bearing stiffeners are required when F_a is exceeded.

For plate girders used in buildings, the sum of the compression stresses resulting from concentrated and distributed loads, bearing directly on or through a flange plate on the compression edge of the web plate, should not exceed the following:

When the flange is restrained against rotation,

$$F_a = \left[5.5 + \frac{4}{(a/h)^2} \right] \frac{10,000}{(h/t)^2} \tag{9-16}$$

When the flange is not restrained against rotation,

$$F_a = \left[2 + \frac{4}{(a/h)^2} \right] \frac{10,000}{(h/t)^2} \tag{9-17}$$

where a = clear distance between transverse stiffeners, in.
h = clear distance between flanges, in.
t = web thickness, in.

The load may be considered distributed over a web length equal to the panel length or girder depth, whichever is less (see also Art. 9-23).

Rolled beams used as bridge girders must be provided with suitable stiffeners at bearings when the unit shear in the web adjacent to the bearing exceeds 75% of the allowable shear for girder webs (Table 9-6).

With welded bridge girders, bearing stiffeners are always required over the end bearings and over the intermediate bearings of continuous girders. Design of these stiffeners is covered in Art. 9-24.

9-15. Restrained Members. When fully restrained beams are welded to the flange of a W-shape column, the column flange and web thicknesses may be such as to require column web stiffeners.

Opposite the beam compression flange, the web is critical and must be stiffened when its thickness is equal to or less than the smaller of

$$t = \frac{C_1 A_f}{t_b + 5k} \tag{9-18}$$

$$t = \frac{d_c \sqrt{F_y}}{180} \tag{9-19}$$

Opposite the beam tension flange, column-flange bending is critical, and the column should be stiffened when the flange thickness is less than

$$t_f = 0.4 \sqrt{C_1 A_f} \tag{9-20}$$

When stiffeners are required by Eqs. (9-18) to (9-20), the stiffener cross-sectional area, sq in., should be at least

$$A_{st} = [C_1 A_f - t (t_b + 5k)] C_2 \tag{9-21}$$

where t = column-web thickness, in.
k = distance from column-flange face to column-web toe of fillet, in.
t_b = beam-flange thickness, in.
t_f = column-flange thickness, in.
A_f = area of beam flange, sq in.
d_c = column-web depth clear of fillets, in.
$C_1 = F_y$ (beam)/F_y (column)
$C_2 = F_y$ (column)/F_y (stiffener)

Stiffeners opposite the compression flange may be fitted to bear on the inside of the column flange, whereas opposite the tension flange, they should be welded.

9-16. Design of Beam Sections for Torsion. Torsional stresses may be induced in steel beams either by unsymmetrical loading or by symmetrical loading on unsymmetrical shapes, such as channels or angles. However, it rarely is necessary to compute torsional stresses, because they generally are much smaller than the concurrent axial or bending stresses.

9-17. Wind and Seismic Stresses.   For buildings, allowable stresses may be increased one-third under wind or seismic forces acting alone or with gravity loads.   The resulting design, however, should not be less than that required for dead and live loads without the increase in allowable stress.   The increased stress is permitted because of the short duration of the load.   Its validity has been justified by many years of satisfactory performance.

For allowable stresses including wind and seismic effects on bridges, see Art. 17-4.

Successful wind or seismic design is dependent on close attention to connection details.   It is good practice to provide as much ductility as practical in such connections so that the fasteners are not overstressed.

9-18. Cyclic-loaded Members.   Fatigue loading in buildings is seldom a design consideration and is usually limited to main supporting members for heavily vibrating machinery, large cranes, moving equipment, etc.   Where fatigue must be considered, a member's strength depends on the applied stress range, the number of cycles of stress changes, and the type of detail.   The AASHTO and AISC Specifications provide for four loading conditions to which a member may be subjected (Table 9-15).

Table 9-15. Loading Conditions, Cycles, for Fatigue

a. FOR BUILDINGS

Loading condition	Number of loading cycles	
	From	To
1	20,000*	100,000
2	100,000	500,000
3	500,000	2,000,000
4	Over 2,000,000	

b. FOR BRIDGES

MAIN (LONGITUDINAL) LOAD-CARRYING MEMBERS

Type of road	Case	ADTT†	Truck loading	Lane loading‡
Freeways, expressways, major highways, and streets	I§ II	2,500 or more Less than 2,500	Over 2,000,000 500,000	500,000 100,000
Other highways and streets not included in Case I or Case II	III	⋯	100,000	100,000

TRANSVERSE MEMBERS AND DETAILS SUBJECTED TO WHEEL LOADS

Type of road	Case	ADTT†	Truck loading
Freeways, expressways, major highways, and streets	I§ II	2,500 or more Less than 2,500	Over 2,000,000 2,000,000
Other highways and streets	III	⋯	500,000

*Approximately 2 applications per day for 25 years.
†Average daily truck traffic.
‡Longitudinal members should also be checked for truck loading.
§This condition corresponds to an extremely heavily traveled artery.

The AASHTO Specification shows 17 different types of details, whereas the AISC Specification lists 27 to which specific stress categories are applied, depending on stress concentrations, discontinuities, and other factors that relate to fatigue strength.   These categories, together with the various loading conditions, define an allowable stress range F_{sr}, as shown in Table 9-16.   The allowable stress range in a tension member is the difference between maximum and minimum stress, ksi.   In a member subjected to stress reversal, the allowable stress range is the numerical

sum of the maximum tension and maximum compression stresses. Fatigue is not a consideration in members that are continuously in compression.

9-19. Stresses for Welds. Allowable stresses for welds are based on the material welded, the
welding electrode, and the joint detail.

Butt Welds. For both buildings and bridges, the allowable stress in butt-welded joints is the
same as in the base metal. The yield and ultimate strengths of electrodes used should exceed
those of the base metal.

Table 9-16. Allowable Stress Range F_{sr}, Ksi, in Buildings and Bridges

Stress* category	Buildings				Bridges			
	Load category				Maximum number of cycles			
	1	2	3	4	100,000	500,000	2,000,000	Over 2,000,000
A†	40	32	24	24	60	36	24	24
B	33	25	17	15	45	27.5	18	16
C	28	21	14	12	32	19	13	10, 12‡
D	24	17	10	9	27	16	10	7
E	17	12	7	6	21	12.5	8	5
F	17	14	11	9	15	12	9	8
G	15	12	9	8	Not Applicable			

*For stress categories, see list of specific details in AASHTO and AISC Specifications.
†For buildings with A514 steels in Category A, use the following values, in order, for load categories 1 to 4:
45, 35, 25, 25.
‡For transverse stiffener welds on girder webs or flanges.

Fillet Welds. The allowable stresses differ for bridges and buildings and are as shown in Tables
9-17a and 9-17b. For bridges (Table 9-17a), the AASHTO Specification (Art. 9-6) relates F_v,
the allowable shear stress, to the yield strength of the material being welded, and the electrodes
are required to have yield and ultimate strengths greater than those of the material being welded.

Plug and Slot Welds. Allowable stresses are given in Table 9-17.

Welds between Different Materials. The allowable weld stress is based on the lower of the
yield strengths of the materials joined.

Excessive-Strength Electrodes. If an electrode of strength higher than that specified for a
particular base metal is used, the weld strength should be computed for the strength of the electrode matching the base metal. For example, referring to Table 9-17b, if E80 electrodes are
used to attach A36 steel, for which E70 electrodes are required, the allowable shear stress in
the weld is 21.0 ksi, corresponding to the E70 electrodes.

Groove Welds. In buildings, for tension or compression applied in any direction to partial- or
complete-penetration groove welds, the allowable stress is the same as that of the base metal if
the strength of the electrode is greater than that of the base metal used. The exception to this
is tension applied normal to the axis of the effective throat of a partial penetration groove weld,
where the allowable stress is that given in Table 9-17b.

Fatigue. Where cyclic loading is a consideration, the allowable stresses may have to be modified according to the welding detail (see Appendixes of the AISC and AASHTO Specifications
for buildings and bridges, respectively).

9-20. Stresses for Rivets and Bolts. For buildings, AISC (Art. 9-6) specifies allowable unit
tension and shear stresses on rivets, bolts, and threaded parts (ksi on area of rivets before driving
or on unthreaded body area of bolts and threaded parts) as shown in Table 9-18. In general,
rivets should not be used in direct tension.

The allowable bearing stress, ksi, on the projected area of rivets and bolts in bearing-type
connections is

$$F_p = 1.35F_y \tag{9-22}$$

where F_y is the yield strength, ksi, of the connected part. (Bearing stress is not restricted in
friction-type connections assembled with A325 and A490 bolts.)

Table 9-17. Allowable Shear Stresses in Fillet and Plug Welds, Ksi

a. FOR BRIDGES

F_y for base metal	Allowable shear stress F_v	
	Fillet welds	Plug welds
36	12.4	12.4
40–50	14.7	12.4
90–100	25.0	12.4

b. FOR BUILDINGS

Electrodes used	Allowable shear stress	Matching Base Metal	
		ASTM designation	F_y
E60	18	A500* Grade A.	33
		A570† Grade D	44
E70	21	A36	36
		A242	42–50
		A441	42–50
		A500* Grade B	42
		A572 Grade 42	42
		to Grade 60	60
		A588	50
E80	24	A572 Grade 65	65
E90	27	A514, over	90
E100	30	2½ in. thick	
E110	33	A514, under	100
		2½ in. thick	

*Structural tubing, carbon steel.
†Sheet and strip, carbon steel.

Table 9-18. Allowable Stresses in Rivets and Bolts, Ksi, in Buildings

Description of fastener	Tension F_t	Shear F_v	
		Friction-type connections	Bearing-type connections
A502, Grade 1, hot-driven rivets	20		15
A502, Grade 2, hot-driven rivets	27		20
A307 bolts	20*		10
Threaded parts of other steels	$0.60F_y$*		$0.30F_y$*
A325 bolts when threading is *not* excluded from shear planes	40†	15	15
A325 bolts when threading is excluded from shear planes	40†	15	22
A490 bolts when threading is *not* excluded from shear planes	54†	20	22.5
A490 when threading is excluded from shear planes	54†	20	32

*Applied to tension stress area = $0.7854 (D - 0.9743/n)^2$, where D is thread diameter, in., and n is number of threads per inch.
†Applied to nominal bolt area.

Friction-type connections are used when it is considered undesirable for the bolt to slip into bearing prior to failure. Bearing-type connections are usually specified in building construction, since both connections have about the same ultimate capacity.

Combined Stresses. In buildings, rivets and bolts subject to combined shear and tension should be so proportioned that the tension stress, ksi, produced by the resultant force does not exceed the following:

For A502, Grade 1, rivets:

$$F_t = 28 - 1.6f_v \leq 20 \tag{9-23a}$$

For A502, Grade 2, rivets:

$$F_t = 38 - 1.6f_v \leq 27 \tag{9-23b}$$

For A307 bolts on tension stress area (Table 9-18):

$$F_t = 28 - 1.6f_v \leq 20 \tag{9-23c}$$

For A325 bolts in bearing-type joints:

$$F_t = 50 - 1.6f_v \leq 40 \tag{9-23d}$$

For A490 bolts in bearing-type joints:

$$F_t = 70 - 1.6f_v \leq 54 \tag{9-23e}$$

where f_v is the shear stress produced by the resultant force. This stress should not exceed the value for shear given in Table 9-18.

For bolts used in friction-type joints, the shear stress allowed in Table 9-18 should be reduced for combined shear and tension so that

For A325 bolts:
$$F_v \leq 15 \left(1 - \frac{f_t A_b}{T_b}\right) \tag{9-23f}$$

For A490 bolts:
$$F_v \leq 20 \left(1 - \frac{f_t A_b}{T_b}\right) \tag{9-23g}$$

where f_t is the tensile stress, ksi, due to applied load; A_b is the area of the bolt, sq in.; and T_b is the specified pretension load, kips, of the bolt.

When wind or seismic loading is combined with gravity loading, the allowable stresses in Table 9-18 may be increased one-third. For combined tension and shear in fasteners under such loadings, the constants of Eq. (9-23) may be increased one-third, but $1.6f_v$ should remain unchanged. For example, with wind or seismic loading, for A325 bolts in bearing-type joints, Eq. (9-23d) becomes:

$$\begin{aligned} F_t &= \frac{4}{3} \times 50 - 1.6f_v \leq \frac{4}{3} \times 40 \\ &= 66.7 - 1.6f_v \leq 53.3 \end{aligned} \tag{9-24}$$

Table 9-19. Allowable Stresses in Rivets and Bolts in Bridges, Ksi

Description of fastener	Tension	Bearing	Shear	
			Friction-type connection	Bearing-type connection
A502, Grade 1, rivets		40		13.5
A502, Grade 2, rivets		40		20
A307 bolts	13.5*	20		11
A325 bolts	36	40	13.5	20†

*Based on area at the root of the thread. A307 bolts should not be used in connections subject to fatigue.
†Threads are excluded from shear planes. Reduce allowable shear by 20% for connected material with F_y less than 42 ksi when the end of the splice material is more than 24 in. from the end of the connected member, as measured along the gage line of the bolts.

Bridge Fasteners. For bridges, AASHTO (Art. 9-6) specifies the working stresses for rivets and bolts listed in Table 9-19. Friction-type connections should be specified where stress reversals may occur or where slippage would be undesirable.

For combined shear and tension in friction-type joints, AASHTO requires that the computed shear stress, ksi, in A325 bolts not exceed

$$F_v = 13.5 - 0.22f_t \qquad (9\text{-}25a)$$

where f_t = computed tension stress due to applied load, ksi.

For combined shear and tension in bearing-type joints, computed stresses in rivets and bolts should satisfy

$$f_v{}^2 + (kf_t)^2 \le F_v{}^2 \qquad (9\text{-}25b)$$

where f_v = computed shear stress, ksi
f_t = computed tensile stress, ksi
F_v = allowable shear stress, ksi
k = 0.75 for rivets and 0.555 for A325 bolts with threads excluded from shear plane

9-21. Composite Construction. In composite construction, steel beams and a concrete slab are connected so that they act together to resist the load on the beam. The slab, in effect, serves as a cover plate. As a result, a lighter steel section may be used.

In building construction, there are two basic methods of composite construction.

Method 1. The steel beam is entirely encased in the concrete. Since the beam is completely braced laterally, the allowable stress in the flanges is $0.66F_y$, where F_y is the yield strength, ksi, of the steel. Assuming the steel to carry the full dead load and the composite section to carry the live load, the maximum unit stress, ksi, in the steel is

$$f_s = \frac{M_D}{S_s} + \frac{M_L}{S_{tr}} \le 0.66F_y \qquad (9\text{-}26)$$

where M_D = dead-load moment, in.-kips
M_L = live-load moment, in.-kips
S_s = section modulus of steel beam, in.^{3}
S_{tr} = section modulus of transformed composite section, in.^{3}

An alternative, shortcut method is permitted by the AISC specification (Art. 9-6). It assumes the steel beam will carry both live and dead loads and compensates for this by permitting a higher stress in the steel:

$$f_s = \frac{M_D + M_L}{S_s} \le 0.76F_y \qquad (9\text{-}27)$$

Method 2. The steel beam is connected to the concrete slab by shear connectors: studs, wire spirals, or channels. Design is based on ultimate load and is independent of use of temporary shores to support the steel until the concrete hardens. The maximum stress in the bottom flange is

$$f_s = \frac{M_D + M_L}{S_{tr}} \le 0.66F_y \qquad (9\text{-}28)$$

To obtain the transformed composite section, treat the concrete above the neutral axis as an equivalent steel area by dividing the concrete area by n, the ratio of modulus of elasticity of steel

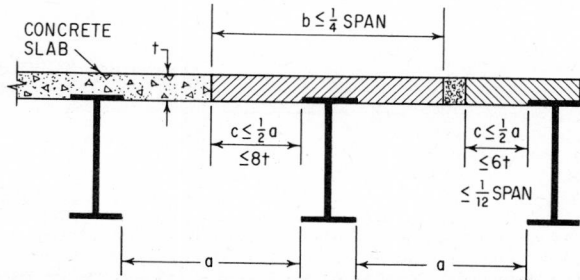

Fig. 9-6. Limitations on effective width of concrete slab in composite steel-concrete beam.

to that of the concrete. In determining the transformed section, the width of the slab should not exceed one-fourth the beam span or, for an interior beam, eight times the slab thickness or half the clear distance to the adjacent beam. For an exterior beam, the effective projection should not be more than one-twelfth the beam span or six times the slab thickness or half the clear distance to the adjacent beam. (See Fig. 9-6.)

For unshored construction, to limit the tension flange to considerably less than yield stress, the section modulus of the transformed section (referred to the tension flange) used in computation of the bottom-flange stress should not exceed

$$S_{tr} = \left(1.35 + 0.35 \frac{M_L}{M_D}\right) S_s \qquad (9\text{-}29)$$

Shear on Connectors. The total horizontal shear to be resisted by the shear connectors is taken as the smaller of the values given by Eqs. (9-30) and (9-31).

$$V_h = \frac{0.85 f'_c A_c}{2} \qquad (9\text{-}30)$$

$$V_h = \frac{A_s F_y}{2} \qquad (9\text{-}31)$$

where V_h = total horizontal shear, kips, between the maximum positive moment and each end of the steel beams (or between the point of maximum positive moment and a point of contraflexure in a continuous beam)

f'_c = specified compressive strength of concrete at 28 days, ksi
A_c = actual area of effective concrete flange, sq in.
A_s = area of steel beam, sq in.

In continuous composite construction, longitudinal reinforcing steel may be considered to act compositely with the steel beam in negative-moment regions. In this case, the total horizontal shear, kips, between an interior support and each adjacent point of contraflexure should be taken as

$$V_h = \frac{A_{sr} F_{yr}}{2} \qquad (9\text{-}32)$$

where A_{sr} = area of longitudinal reinforcement at the support within the effective area, sq in.
F_{yr} = specified minimum yield stress of longitudinal reinforcement, ksi

Number Required. The total number of connectors to resist V_h is computed from V_h/q, where q is the allowable shear for one connector, or one pitch of a spiral, kips. Values of q for connectors in buildings are given in Table 9-20.

Table 9-20. Allowable Shear Loads on Connectors for Composite Construction in Buildings

Type of connector	Allowable horizontal shear load q, kips (applicable only to concrete made with ASTM C33 aggregates)		
	f'_c, ksi		
	3.0	3.5	4.0
½-in. dia × 2-in. hooked or headed stud	5.1	5.5	5.9
⅝-in. dia × 2½-in. hooked or headed stud	8.0	8.6	9.2
¾-in. dia × 3-in. hooked or headed stud	11.5	12.5	13.3
⅞-in. dia × 3½-in. hooked or headed stud	15.6	16.8	18.0
3-in. channel, 4.1 lb	$4.3w$	$4.7w$	$5.0w$
4-in. channel, 5.4 lb	$4.6w$	$5.0w$	$5.3w$
5-in. channel, 6.7 lb.	$4.9w$	$5.3w$	$5.6w$

w = length of channel, in.

Table 9-20 is applicable only to composite construction with concrete made with stone aggregate conforming to ASTM C33.   For lightweight concrete weighing at least 90 pcf and made with rotary-kiln-produced aggregates conforming to ASTM C330, the allowable shear values of Table 9-20 should be reduced by multiplying by the appropriate coefficient of Table 9-21.

Table 9-21. Shear Coefficient for Lightweight Concrete with Aggregates Conforming to ASTM C330

	Air dry unit weight, lb per cu ft						
	90	95	100	105	110	115	120
When $f'_c \le 4$ ksi	0.73	0.76	0.78	0.81	0.83	0.86	0.88
When $f'_c \ge 5$ ksi	0.82	0.85	0.87	0.91	0.93	0.96	0.99

The required number of shear connectors may be spaced uniformly between the sections of maximum and zero moment.   Shear connectors should have at least 1 in. of concrete cover in all directions, and unless studs are located directly over the web, stud diameters may not exceed 2.5 times the beam-flange thickness.

With heavy concentrated loads, the uniform spacing of shear connectors may not be sufficient between a concentrated load and the nearest point of zero moment.   The number of shear connectors in this region should be at least

$$N_2 = \frac{N_1 \left(\dfrac{M\beta}{M_{max}} - 1 \right)}{\beta - 1} \tag{9-33}$$

where M = moment at concentrated load, ft-kips
$M_{\max}$ = maximum moment in span, ft-kips
N_1 = number of shear connectors required between $M_{\max}$ and zero moment
$\beta = S_{tr}/S_s$ or S_{eff}/S_s, as applicable
S_{eff} = effective section modulus for partial composite action, in.[3]

Partial composite construction is used when the number N_1 of shear connectors required would provide a beam considerably stronger than necessary.   In that case, the effective section modulus is used in stress computation instead of the transformed section modulus, and S_{eff} is calculated from Eq. (9-34).

$$S_{eff} = S_s + \frac{V'_h}{V_h} \left(S_{tr} - S_s \right) \tag{9-34}$$

where V'_h = number of shear connectors provided times allowable shear load q of Table 9-20 (times coefficient of Table 9-21, if applicable)

Composite design for highway bridges is based on elastic rather than ultimate-load considerations.   The total stress in the tension flange of the composite section must not exceed the allowable bending stress of the steel.   When shores are used and kept in place until the concrete has attained 75% of its 28-day strength, the stress due to live and dead loads is computed for the composite section.

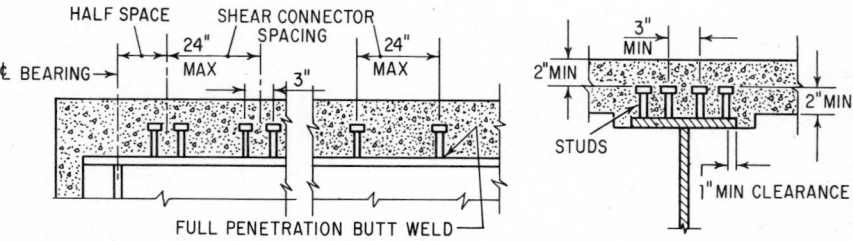

Fig. 9-7. Maximum pitch for stud shear connectors in composite beams.

The effective width of the concrete flange is the same as for the section in building construction (Fig. 9-6), except that for interior bridge beams effective width may be increased from 8 to 12 times slab thickness.

Maximum spacing of shear connectors is 24 in., except over interior supports, where wider spacing may be used. Connectors must have a minimum of 2 in. cover and must project a minimum of 2 in. above the bottom of the slab (Fig. 9-7).

Bending stresses in composite beams in bridges depend on whether or not the members are shored; they are determined as for beams in buildings [see Eqs. (9-26) and (9-28)] except that the stresses in the steel may not exceed $0.55F_y$ [see Eqs. (9-35) and (9-36)].

Unshored:

$$f_s = \frac{M_D}{S_s} + \frac{M_L}{S_{tr}} \leq 0.55F_y \tag{9-35}$$

Shored:

$$f_s = \frac{M_D + M_L}{S_{tr}} \leq 0.55F_y \tag{9-36}$$

Shear Range. Shear connectors in bridges are designed for fatigue and then are checked for ultimate strength. The horizontal-shear range for fatigue is computed from

$$S_r = \frac{V_r Q}{I} \tag{9-37}$$

where S_r = horizontal-shear range at juncture of slab and beam at point under consideration, kips per lin in.

V_r = shear range (difference between minimum and maximum shears at the point) due to live load and impact, kips

Q = static moment of transformed compressive concrete area about neutral axis of the transformed section, in.[3]

I = moment of inertia of transformed section, in.[4]

The transformed area is the actual concrete area divided by n (Table 9-22).

Table 9-22. Ratio of Moduli of Elasticity of Steel and Concrete for Bridges

f'_c for concrete, psi	$n = \dfrac{E_s}{E_c}$
2,000–2,400	15
2,500–2,900	12
3,000–3,900	10
4,000–4,900	8
5,000 or above	6

The allowable range of horizontal shear Z_r, kips, for an individual connector is given by Eq. (9-38) or (9-39), depending on the connector used.

For channels:

$$Z_r = Bw \tag{9-38}$$

where w = channel length, in., in transverse direction on girder flange

B = cyclic variable = 4.0 for 100,000 cycles, 3.0 for 500,000 cycles, 2.4 for 2,000,000 cycles

For welded studs (with height-diameter ratio $H/d \geq 4$):

$$Z_r = \alpha d^2 \tag{9-39}$$

where d = stud diameter, in.

α = cyclic variable = 13.0 for 100,000 cycles, 10.6 for 500,000 cycles, 7.85 for 2,000,000 cycles

Required pitch of shear connectors is determined by dividing the allowable range of horizontal shear of all connectors at one section Z_r, kips, by the horizontal range of shear S_r, kips per lin in.

Number of Connectors. The ultimate strength of the shear connectors is checked by computation of the number of connectors required from

$$N = \frac{P}{\phi S_u} \qquad (9\text{-}40)$$

where N = number of shear connectors between maximum positive moment and end supports or dead-load points of contraflexure, or between maximum negative moment and points of contraflexure

S_u = ultimate shear connector strength, kips [see Eq. (9-42) and Table 9-23]

ϕ = reduction factor = 0.85

P = force in slab, kips

At points of maximum positive moments, P is the smaller of P_1 and P_2, computed from Eqs. (9-41a and b).

$$P_1 = A_s F_y \qquad (9\text{-}41a)$$

$$P_2 = 0.85 f'_c A_c \qquad (9\text{-}41b)$$

where A_c = effective concrete area, sq in.

f'_c = 28-day compressive strength of concrete, ksi

A_s = total area of steel section, sq in.

F_y = steel yield strength, ksi

At points of maximum negative moments, P is equal to P_3, computed from

$$P_3 = A_{sr} F_{yr} \qquad (9\text{-}41c)$$

where A_{sr} = area of longitudinal reinforcing within effective flange, sq in.

F_{yr} = reinforcing steel yield strength, ksi

Ultimate Shear Strength of Connectors, Kips.

For channels:

$$S_u = 17.4(h + t/2) \sqrt{f'_c} \qquad (9\text{-}42a)$$

where h = average channel-flange thickness, in.

t = channel-web thickness, in.

For welded studs ($H/d \ge 4$ in.):

$$S_u = 29.4 d^2 \sqrt{f'_c} \qquad (9\text{-}42b)$$

Table 9-23. Ultimate Strength of Shear Connectors for Composite Construction in Bridges, Kips

Type of Connector	f'_c, ksi			
	2.5	3.0	4.0	5.0
3-in. channel, 4.1 lb	$9.45w$	$10.36w$	$11.96w$	$13.37w$
3-in. channel, 6 lb	$12.03w$	$13.18w$	$15.22w$	$17.01w$
4-in. channel, 5.4 lb	$11.17w$	$12.24w$	$14.13w$	$15.80w$
4-in. channel, 7.25 lb	$12.89w$	$14.72w$	$17.00w$	$19.01w$
5-in. channel, 6.7 lb	$11.17w$	$12.24w$	$14.13w$	$15.80w$
5-in. channel, 9 lb	$12.89w$	$14.72w$	$17.00w$	$19.01w$
6-in. channel, 8.2 lb	$11.17w$	$12.24w$	$14.13w$	$15.80w$
¾-in. dia × 4-in. stud	26.16	28.65	33.09	36.99
⅞-in. dia × 4-in. stud	35.60	39.00	45.03	50.35

w = length of channel, in.

Creep and Shrinkage. AASHTO requires that the effects of creep be considered in the design of composite beams with dead loads acting on the composite section. For such beams, tension, compression, and horizontal shears produced by dead loads acting on the composite section should be computed for n or $3n$, whichever gives the higher stresses.

Shrinkage also should be considered. Resistance of a steel beam to longitudinal contraction of the concrete slab produces shear stresses along the contact surface. Associated with this shear are tensile stresses in the slab and compressive stresses in the steel top flange. These stresses also affect the beam deflection. The magnitude of the shrinkage effect varies within wide limits. It can be qualitatively reduced by appropriate casting sequences, for example, by placing concrete in a checkerboard pattern.

Span-Depth Ratios. In bridges, for composite beams, preferably the ratio of span to steel-beam depth should not exceed 30 and the ratio of span to depth of steel beam plus slab should not exceed 25.

9-22. Plate-Girder Design Criteria for Buildings and Bridges. (See also Arts. 9-23 and 9-24.) In computation of stresses in plate girders, the moment of inertia I, in.^{4}, of the gross cross section generally is used. The bending stress f_b due to bending moment M is computed from $f = Mc/I$, where c is the distance, in., from the neutral axis to the extreme fiber. For determination of tensile stresses in riveted bridge girders, however, the moments of inertia of all holes about the axis of the gross cross section should be deducted. For building girders, no deduction need be made for rivet or bolt holes unless the reduction in flange area, calculated as indicated in Art. 9-8, exceeds 15%; then the excess should be deducted.

In riveted plate girders, the flange angles should form as large a part of the area of the flange as practicable. Side plates should not be used unless the flange angles would have to exceed $\frac{7}{8}$ in. in thickness. The gross area of the compression flange should not be less than that of the tension flange. If several flange plates are used in a flange and the thicknesses are not the same, the plates should decrease in thickness from the flange angles outward. No plate should be thicker than the flange angles. At least one riveted cover plate should extend the full length of the girder, unless the flange is covered with concrete. Any cover plate that is not full length should extend far enough to develop the capacity of the plate beyond the theoretical end, the section where flange stress without that cover plate equals the allowable stress.

In welded plate girders, each flange should consist of a single plate. It may, however, comprise a series of shorter plates joined end to end by full-penetration butt welds. Its thickness may be increased or decreased at a slope of not more than 1 in 2.5 as stress requirements permit.

In bridges, the ratio of compression-flange width to thickness should not exceed 24 or $103/\sqrt{f_b}$, where f_b = computed maximum bending stress, ksi.

The web depth-to-thickness ratio is defined as h/t, where h is the clear distance between flanges, in., and t is the web thickness, in. Several design rules for plate girders depend on this ratio.

9-23. Criteria for Plate Girders in Buildings. (See also Art. 9-22.) For greatest resistance to bending, as much of a plate girder cross section as practicable should be concentrated in the flanges, at the greatest distance from the neutral axis. This might require, however, a web so thin that the girder would fail by web buckling before it reached its bending capacity. To preclude this, the AISC Specification (Art. 9-6) limits the web clear-depth-to-thickness ratio h/t.

Critical Depth-Thickness Ratios. For an unstiffened web, this ratio should not exceed

$$\frac{h}{t} = \frac{14,000}{\sqrt{F_y(F_y + 16.5)}} \qquad (9\text{-}43a)$$

where F_y = yield strength of compression flange, ksi

Larger values of h/t may be used, however, if the web is stiffened at appropriate intervals. For the purpose, vertical angles may be fastened to the web or vertical plates welded to it. These transverse stiffeners are not required, though, when h/t is less than the value computed from Eq. (9-43a) or given in Table 9-24.

Table 9-24. Critical h/t for Plate Girders in Buildings

F_y, ksi	$\dfrac{14,000}{\sqrt{F_y(F_y + 16.5)}}$	$\dfrac{2,000}{\sqrt{F_y}}$
36	322	333
50	243	283

With transverse stiffeners spaced not more than 1.5 times the girder depth apart, the web clear-depth-to-thickness ratio may be as large as

$$\frac{h}{t} = \frac{2,000}{\sqrt{F_y}} \qquad (9\text{-}43b)$$

(See Table 9-24.)   If, however, the web depth-to-thickness ratio h/t exceeds $760/\sqrt{F_b}$, where F_b is the allowable bending stress in the compression flange that would ordinarily apply, this stress should be reduced to F_b', given by Eq. (9-44a).

$$F_b' = F_b \left[1 - 0.0005 \frac{A_w}{A_f} \left(\frac{h}{t} - \frac{760}{\sqrt{F_b}} \right) \right] \qquad (9\text{-}44a)$$

where A_w = web area, sq in.
A_f = area of compression flange, sq in.

In a hybrid girder, where the flange steel has a higher yield strength than the web, the allowable bending stress in top or bottom flange should be taken as the smaller of the values given by Eqs. (9-44a and b).

$$F_b' = F_b \left[\frac{12 + (A_w/A_f)(3\alpha - \alpha^3)}{12 + 2(A_w/A_f)} \right] \qquad (9\text{-}44b)$$

where α = ratio of web yield strength to flange yield strength
This protects against excessive yielding of the lower-strength web in the vicinity of the higher-strength flanges.

Stiffener Spacing.   The shear and allowable shear stress may determine required web area and stiffener spacing.   Equations (9-45) and (9-46) give the allowable web shear F_v, ksi, for any panel of a building girder between transverse stiffeners.

When a factor C_v is less than 1.0, except for hybrid girders,

$$F_v = \frac{F_y}{2.89} \left[C_v + \frac{1 - C_v}{1.15\sqrt{1 + (a/h)^2}} \right] \qquad (9\text{-}45)$$

When C_v is more than 1.0 or when intermediate stiffeners are omitted, and for hybrid girders,

$$F_v = \frac{F_y C_v}{2.89} \leq 0.4F_y \qquad (9\text{-}46)$$

where a = clear distance between stiffeners, in.
h = clear distance between flanges, in.
$C_v = \dfrac{45,000k}{F_y(h/t)^2}$ when C_v is less than 0.8
$\quad = \dfrac{190}{h/t}\sqrt{\dfrac{k}{F_y}}$ when C_v is more than 0.8
t = web thickness, in.
$k = 5.34 + 4/(a/h)^2$ when $a/h > 1$
$\quad = 4 + 5.34/(a/h)^2$ when $a/h < 1$

The average shear stress f_v, ksi, in a panel of a plate girder (web between successive stiffeners) is defined as the largest shear, kips, in the panel divided by the web cross-sectional area, sq in. As f_v approaches F_v given by Eq. (9-45), combined shear and tension become important.   In that case, the tensile stress in the web due to bending in its plane should not exceed $0.6F_y$ or $(0.825 - 0.375f_v/F_v)F_y$, where F_v is given by Eq. (9-45).

Neither h, the clear distance between flanges, nor a, the spacing of stiffeners for an end panel, or for any panel containing large holes and adjacent panels, should exceed $348t/\sqrt{f_v}$.

Intermediate stiffeners, when required, should be spaced so that a/h is less than 3 and less than $[260/(h/t)]^2$.   Such stiffeners are not required when h/t is less than 260 and f_v is less than F_v computed from Eq. (9-46).

An infinite combination of web thicknesses and stiffener spacings is possible with a particular girder.   Figure 9-8, developed for A36 steel, facilitates the trial-and-error process of selecting a suitable combination.   Similar charts can be developed for other steels.

Stiffener Properties.   The required area of intermediate stiffeners is determined by

$$A_{st} = \frac{1 - C_v}{2} \left[\frac{a}{h} - \frac{(a/h)^2}{\sqrt{1 + (a/h)^2}} \right] YDht \qquad (9\text{-}47)$$

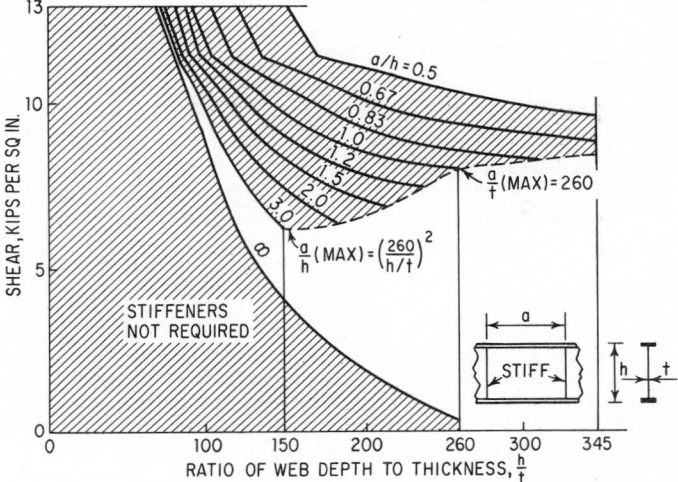

Fig. 9-8. Chart for determining spacing of girder stiffeners of A36 steel.

where A_{st} = gross stiffener area, sq in. (total area, if in pairs)

  Y = ratio of yield point of web steel to yield point of stiffener steel

  D = 1.0 for stiffeners in pairs

  = 1.8 for single-angle stiffeners

  = 2.4 for single-plate stiffeners

If the computed web shear stress f_v is less than F_v computed from Eq. (9-45), A_{st} may be reduced by the ratio f_v/F_v.

The moment of inertia of a stiffener or pair of stiffeners should be at least $(h/50)^4$.

Stiffener Connections. The stiffener-to-web connection should be designed for a shear, kips per lin in. of single stiffener, or pair of stiffeners, of at least

$$f_{vs} = h \sqrt{\left(\frac{F_y}{340}\right)^3}$$
(9-48)

This shear may also be reduced by the ratio f_v/F_v.

Spacing of fasteners connecting stiffeners to the girder web should not exceed 12 in. c to c. If intermittent fillet welds are used, the clear distance between welds should not exceed 10 in. or 16 times the web thickness.

Bearing Stiffeners. These are required on webs where ends of plate girders do not frame into columns or other girders. They may also be needed under concentrated loads and at reaction points. Bearing stiffeners should be designed as columns, assisted by a strip of web. The width of this strip may be taken as 25t at interior stiffeners and 12t at the end of the web. Effective length for l/r (slenderness ratio) should be at least 0.75 of the stiffener length. See Art. 9-14 for prevention of web crippling.

Splices. Butt-welded splices should be complete-penetration groove welds and should develop the full strength of the smaller spliced section. Other types of splices in cross sections of plate girders should develop the strength required by the stresses at the point of splice but not less than 50% of the effective strength of the material spliced.

Flange Connections. Rivets, high-strength bolts, or welds connecting flange to web, or cover plate to flange, should be proportioned to resist the total horizontal shear from bending. The longitudinal spacing of the fasteners, in., may be determined from

$$p = \frac{R}{q}$$
(9-49)

where R = the allowable force, kips, on the rivets, bolts, or welds that serve the length p

  q = horizontal shear, kips per in.

For a rivet or bolt, $R = A_v F_v$, where A_v is the cross-sectional area, sq in., of the fastener and F_v the allowable shear stress, ksi. For a weld, R is the product of the length of weld, in., and allowable stress, kips per in. The horizontal shear may be computed from

$$q = \frac{VQ}{I} \qquad (9\text{-}50a)$$

where V = shear, kips, at point where pitch is to be determined

I = moment of inertia of the section, in.[4]

Q = static moment about the neutral axis of the flange cross-sectional area between the outermost surface and the surface at which the horizontal shear is being computed, in.[3]

Approximately,

$$q = \frac{V}{d} \frac{A}{A_f + A_w/6} \qquad (9\text{-}50b)$$

where d = depth of web, in., for welds between flange and web; distance between centers of gravity of tension and compression flanges, in., for rivets between flange and web; distance back to back of angles, in., for rivets between cover plates and angles

A = area of flange, sq in., for welds, rivets, and bolts between flange and web; area of cover plates only, sq in., for rivets between cover plates and angles

A_f = flange area, sq in.

A_w = web area, sq in.

If the girder supports a uniformly distributed load w, kips per in., on the top flange, the pitch should be determined from

$$p = \frac{R}{\sqrt{q^2 + w^2}} \qquad (9\text{-}51)$$

(See also Art. 9-14.)

Maximum longitudinal spacing permitted in the compression-flange cover plates is 12 in. or the thickness of the thinnest plate times $127/\sqrt{F_y}$ when fasteners are provided on all gage lines at each section or when intermittent welds are provided along the edges of the components. When rivets or bolts are staggered, the maximum spacing on each gage line should not exceed 18 in. or the thickness of the thinnest plate times $190/\sqrt{F_y}$. Maximum spacing in tension-flange cover plates is 12 in. or 24 times the thickness of the thinnest plate. Maximum spacing for connectors between flange angles and web is 24 in.

9-24. Design Criteria for Plate Girders in Bridges. (See also Art. 9-22.) For highway bridges, Table 9-25 gives critical web thicknesses t, in., for two grades of steel as a fraction of h, the clear distance, in., between flanges. When t is larger than the value in column 1, intermediate transverse (vertical) stiffeners are not required. If shear stress is less than the allowable, the web may be thinner. Thus, stiffeners may be omitted if $t \geq h\sqrt{f_v}/237$, where f_v = average unit shear, ksi (vertical shear at section, lb, divided by web cross-sectional area). But t should not be less than $h/150$.

When t lies between the values in columns 1 and 2, transverse intermediate stiffeners are required. Webs thinner than the values in column 2 are permissible if they are reinforced by a longitudinal (horizontal) stiffener. If the computed maximum compressive bending stress f_b, ksi, at a section is less than the allowable bending stress, a longitudinal stiffener is not required if $t \geq h\sqrt{f_b}/727$; but t should not be less than $h/170$. When used, a plate longitudinal stiffener should be attached to the web at a distance $h/5$ below the inner surface of the compression flange. [See also Eqs. (9-68) and (9-69).]

Table 9-25. Minimum Web Thickness, In., for Highway-bridge Plate Girders *

Yield strength, ksi	Without intermediate stiffeners (1)	Transverse stiffeners, no longitudinal stiffeners (2)	Longitudinal stiffener, transverse stiffeners (3)
36	$h/68$	$h/165$	$h/330$
50	$h/58$	$h/140$	$h/280$

*Standard Specifications for Highway Bridges, American Association of State Highway and Transportation Officials.

Webs thinner than the values in column 3 are not permitted, even with transverse stiffeners and one longitudinal stiffener, unless the computed compressive bending stress is less than the allowable. When it is, t may be reduced to $h\sqrt{f_b}/1{,}450$, but it should not be less than $h/340$.

Stiffener Spacing. When transverse stiffeners are required in bridge girders, their spacing should not exceed h or $348t/\sqrt{f_v}$. The first two stiffener spaces at the ends of simply supported girders should not exceed half these values.

Stiffener Properties. Intermediate stiffeners may be a single angle fastened to the web or a single plate welded to the web. But preferably they should be attached in pairs, one on each side of the web. Stiffeners on only one side of the web should be attached to the outstanding leg of the compression flange. At points of concentrated loading, stiffeners should be placed on both sides of the web and designed as bearing stiffeners.

The minimum moment of inertia, in.[4], of a transverse stiffener should be at least

$$I = \frac{a_0 t^3 J}{10.92} \tag{9-52}$$

where $J = 25h^2/a^2 - 20 \geq 5$
h = clear distance between flanges, in.
a = required stiffener spacing, in. = h or $348t/\sqrt{f_v}$, whichever is smaller
f_v = average shear stress, ksi, at stiffener
a_0 = actual stiffener spacing, in.
t = web thickness, in.

For paired stiffeners, the moment of inertia should be taken about the center line of the web; for single stiffeners, about the face in contact with the web.

The width of an intermediate transverse stiffener, plate or outstanding leg of an angle, should be at least 2 in. plus $1/30$ of the depth of the girder and preferably not less than one-fourth the width of the flange. Minimum thickness is $1/16$ of the width. The stiffeners should fit sufficiently tight to exclude water after they are painted.

Bearing Stiffeners. These are required at all concentrated loads, including supports. Such stiffeners should be attached to the web in pairs, one on each side, and they should extend as nearly as practicable to the outer edges of the flanges. If angles are used, they should be proportioned for bearing on the outstanding legs of the flange angles or plates. (No allowance should be made for the portion of the legs fitted to the fillets of flange angles.) The stiffener angles should not be crimped.

Bearing stiffeners should be designed as columns. The allowable unit stress is given in Table 9-8, with $L = h$. For plate stiffeners, the column section should be assumed to consist of the plates and a strip of web. The width of the strip may be taken as 18 times the web thickness t for a pair of plates. For stiffeners consisting of four or more plates, the strip may be taken as the portion of the web enclosed by the plates plus a width of not more than $18t$. Minimum bearing-stiffener thickness is $(b'/12)\sqrt{F_y/33}$, where b' = stiffener width, in.

Bearing stiffeners must be ground to fit against the flange through which they receive their load or attached to the flange with full-penetration groove welds. But welding transversely across the tension flanges should be avoided to prevent creation of a severe fatigue condition.

Termination of Top Flange. Upper corners of through-plate girders, where exposed, should be rounded to a radius consistent with the size of the flange plates and angles and the vertical height of the girder above the roadway. The first flange plate, or a plate of the same width, should be bent around the curve and continued to the bottom of the girder. In a bridge consisting of two or more spans, only the corners at the extreme ends of the bridge need to be rounded, unless the spans have girders of different heights. In such a case, the higher girders should have the top flanges curved down at the ends to meet the top corners of the girders in adjacent spans.

Seating at Supports. Sole plates should be at least $3/4$ in. thick. Ends of girders on masonry should be supported on pedestals so that the bottom flanges will be at least 6 in. above the bridge seat.

Longitudinal Stiffeners. These should be placed with the center of gravity of fasteners $h/5$ from the toe, or inner face, of the compression flange. Moment of inertia, in.[4], should be at least

$$I = ht^3 \left(2.4 \frac{a_o^2}{h^2} - 0.13 \right) \tag{9-53}$$

where a_o = actual distance between transverse stiffeners, in.
t = web thickness, in.

Thickness of stiffener, in., should be at least $b\sqrt{f_b}/71.2$, where b is the stiffener width, in., and f_b the flange compressive bending stress, ksi. The bending stress in the stiffener should not exceed the allowable for the material.

Longitudinal stiffeners usually are placed on one side of the web. They need not be continuous. They may be cut at their intersections with transverse stiffeners.

Splices. These should develop the strength required by the stresses at the splices but not less than 75% of the effective strength of the material spliced. Splices in riveted flanges usually are avoided. In general, not more than one part of a girder should be spliced at the same cross section. Bolted web splices should have plates placed symmetrically on opposite sides of the web. Splice plates for shear should extend the full depth of the girder between flanges. At least two rows of bolts on each side of the joint should fasten the plates to the web.

Rivets, high-strength bolts, or welds connecting flange to web, or cover plate to flange, should be proportioned to resist the total horizontal shear from bending, as described for plate girders in buildings (*Flange Connections* in Art. 9-23). In riveted bridge girders, legs of angles 6 in. or more wide connected to webs should have two lines of rivets. Cover plates over 14 in. wide should have four lines of rivets.

Hybrid Bridge Girders. These may have flanges with larger yield strength than the web and may be composite or noncomposite with a concrete slab, or they may utilize an orthotropic-plate deck as the top flange. With composite or noncomposite girders, the web should have a yield strength at least 35% of the minimum yield strength of the tension flange. In noncomposite girders, both flanges should have the same yield strength. In composite girders, the compression flange may have a yield strength the same as that of the web. In girders with an orthotropic-plate deck, the yield strength of the web should be at least 35% of the bottom-flange yield strength in positive-moment regions and 50% in negative-moment regions.

Computation of bending stresses and allowable stresses is generally the same as for girders with uniform yield strength. Bending stress in the web, however, may exceed the allowable bending stress if the computed flange bending stress does not exceed the allowable stress multiplied by a factor R.

$$R = 1 - \frac{\beta\psi(1-\alpha)^2(3-\psi+\psi\alpha)}{6+\beta\psi(3-\psi)} \qquad (9\text{-}54)$$

where α = ratio of web yield strength to flange yield strength
ψ = distance from outer edge of tension flange or bottom flange of orthotropic deck to neutral axis divided by depth of steel section
β = ratio of web area to area of tension flange or bottom flange of orthotropic-plate bridge

Table 9-26. Dimensional Restrictions on Sections for Buildings*

Maximum projecting-element ratios b/t for compression	
Single-angle struts; double-angle struts with separators	$76/\sqrt{F_y}$
Struts comprising double angles in contact; angles or plates projecting from girders, columns, or other compression members; compression flanges of beams; stiffeners on plate girders	$95/\sqrt{F_y}$
Stem of tees	$127/\sqrt{F_y}$
Flanges of compact W sections	$65/\sqrt{F_y}$
Flanges of compact box sections	$238/\sqrt{F_y}$
Cover plates perforated with access holes	$317/\sqrt{F_y}$
Other uniformly compressed stiffened elements	$253/\sqrt{F_y}$
Maximum web ratios d/t	
Webs of compact W and box sections	$190/\sqrt{F_y}$†
Other uniformly compressed stiffened elements	$253/\sqrt{F_y}$

*AISC Specification for the Design, Fabrication, and Erection of Buildings.
†When a member is subjected to combined axial force and moment, the ratio should not exceed

$$\frac{d}{t} \le \frac{640}{\sqrt{F_y}}\left(1 - 3.74\frac{f_a}{F_a}\right)$$

when $f_a/F_a \le 0.16$, where f_a is the axial unit stress, ksi, and F_a is the allowable axial unit stress, ksi. When $f_a/F_a > 0.16$, d/t should not exceed $257/\sqrt{F_y}$.

9-25. Dimensional Limitations on Steel Sections. Design specifications for structural steel include dimensional limitations on sections to prevent local buckling. AISC and AASHTO (Art. 9-6) stipulate maximum width-thickness (b/t) ratios for projecting elements under compression and depth-thickness ratios (d/t) for webs for various applications.

Table 9-26 lists some of the more important dimensional limitations for buildings. In the table, F_y is the yield strength of the steel, ksi. Dimensional limitations for webs and stiffeners of plate girders are given in Arts. 9-23 and 9-24. (See also Art. 9-29.)

9-26. Deflection Limitations. For buildings, beams and girders supporting plastered ceilings should not deflect under live load more than $1/360$ of the span. To control deflection, fully stressed floor beams and girders should have a minimum depth of $F_y/800$ times the span, where F_y is the steel yield strength, ksi. Depth of fully stressed roof purlins should be at least $F_y/1,000$ times the span, except for flat roofs, for which ponding conditions should be considered (Art. 9-28).

For bridges, simple-span or continuous girders should be designed so that deflection due to live load plus impact should not exceed $1/800$ of the span. For bridges located in urban areas and used in part by pedestrians, however, deflection preferably should not exceed $1/1,000$ of the span. To control deflections, depth of noncomposite girders should be at least $1/25$ of the span. For composite girders, overall depth, including slab thickness, should be at least $1/25$ of the span, and depth of steel girder alone, at least $1/30$ of the span. For continuous girders, the span for these ratios should be taken as the distance between inflection points.

9-27. Vibration Considerations in Buildings. In large open areas of buildings, where there are few partitions or other sources of damping, transient vibrations caused by pedestrian traffic may become annoying. Instead of a complicated vibration analysis, AISC suggests that perceptible vibration may be minimized by keeping depth of steel beams equal to or greater than $1/20$ of the beam span.

9-28. Ponding Considerations in Buildings. Flat roofs on which water may accumulate may require analysis to insure that they are stable under ponding conditions. A flat roof may be considered stable and an analysis need not be made if both Eqs. (9-55) and (9-56) are satisfied.

$$C_p + 0.9C_s \leq 0.25 \tag{9-55}$$

$$I_d \geq 25S^4/10^6 \tag{9-56}$$

where $C_p = 32L_sL_p{}^4/10^7I_p$
 $C_s = 32SL_s{}^4/10^7I_s$
 L_p = length, ft, of primary member or girder
 L_s = length, ft, of secondary member or purlin
 S = spacing, ft, of secondary members
 I_p = moment of inertia of primary member, in.^{4}
 I_s = moment of inertia of secondary member, in.^{4}
 I_d = moment of inertia of steel deck supported on secondary members, in.^{4} per ft
For trusses and other open-web members, I_s should be decreased 15%.

9-29. Plastic-Design Criteria for Buildings. Plastic design (Art. 9-7) provides a method of selecting structural-steel members for load-carrying capacity rather than resistance to a particular stress. The AISC Specification for Design, Fabrication, and Erection of Structural Steel for Buildings (American Institute of Steel Construction) limits the method to steels with specified minimum yield strength up to 65 ksi.

The basic principles of plastic design are incorporated in the elastic-design provisions of the AISC Specification:

   1. A 10% allowable stress increase is permitted for "compact" sections.

   2. A 10% decrease in negative moment under gravity loads with a corresponding increase in positive moment is permitted in continuous construction.

Because of these adjustments in the elastic-design rules, some of the economies inherent in plastic design may be realized in elastic design. Plastic design, however, often requires less design time and is considered by most engineers to be more logical (Arts. 6-91 and 6-92).

Load Factors. Using load factors by which design loads are multiplied, a uniform factor of safety (comparable with that of an elastically designed simple beam) is attained for plastic design for all conditions. These load factors are 1.7 for live and dead loads in combination with 1.3 times wind or earthquake loads.

Width-Thickness Ratios. Plastic design is restricted to sections that meet certain dimensional limitations. These limitations were established to allow a member to yield sufficiently to form a plastic hinge without local buckling. The projecting-flange limits are listed in Table 9-27, where F_y is the steel yield strength, ksi; b_f is the flange width; and t_f the flange thickness.

Table 9-27. Minimum Width-Thickness Ratios for Plastic-Design Beams

F_y, ksi	$b_f/2t_f$
36	8.5
42	8.0
45	7.4
50	7.0
55	6.6
60	6.3
65	6.0

Web Area.   Unless reinforced by diagonal stiffeners or a double plate, the webs of columns, beams, and girders must also be proportioned so that

$$V_u \leq 0.55F_y dt \tag{9-57}$$

where V_u = shear, kips
   d = depth of member, in.
   t = web thickness, in.

Web stiffeners are always required under concentrated loads if coincidental with a plastic hinge.

Connections.   In plastic design, connections also are designed on the ultimate-load basis. Rivets, welds, high-strength bolts, and A307 bolts are proportioned with a load factor of 1.7 applied to elastic unit stresses. When fasteners are used in tension, the minimum proof load is the design criterion.

Depth-Thickness Ratios.   Beam-depth limitation is given by Eq. (9-58) and is similar to that of elastic design of compact sections, except that the load factor of 1.7 has been incorporated.

$$\frac{d}{t} = \frac{412}{\sqrt{F_y}} \left(1 - 1.4\frac{P}{P_y}\right) \leq \frac{257}{\sqrt{F_y}} \tag{9-58}$$

where P = factored axial load, kips
   $P_y = F_y A$, kips
   A = cross-sectional area of member, sq in.

Slenderness Ratios.   The unbraced length of a beam's compression flange is critical and is given by Eqs. (9-59a) and (9-59b), which depend on the moment gradient applied to the beam.

When $+1.0 > M/M_p > -0.5$,

$$\frac{l_c}{r_y} \leq \frac{1,375}{F_y} + 25 \tag{9-59a}$$

When $-0.5 > M/M_p > -1.0$,

$$\frac{l_c}{r_y} \leq \frac{1,375}{F_y} \tag{9-59b}$$

where l_c = length, in., of laterally unsupported compression flange
   r_y = radius of gyration, in., about weak axis
   M = lesser of the end moments at the ends of the unbraced segment, ft-kips
   M_p = plastic moment, ft-kips = $F_y Z_x/12$
   Z_x = plastic section modulus, in.^{3} (See Art. 6-91 and tables for W shapes in AISC "Steel Construction Manual.")

The end moment ratio M/M_p is positive when the unbraced segment is bent in double curvature and negative when the segment is bent in single curvature.

Columns.   Under only axial load, columns are designed for a critical stress that is 1.7 times the allowable elastic axial stress F_a, computed as indicated in Art. 9-10. Thus, the factored axial load should not exceed

$$P_{cr} = 1.7AF_a \tag{9-60}$$

For combined axial load and moment, however, columns should be designed to satisfy Eqs. (9-61a) and (9-61b).

$$\frac{P}{P_{cr}} + \frac{C_m M}{(1 - P/P_e)M_m} \leq 1 \qquad (9\text{-}61a)$$

$$\frac{P}{P_y} + \frac{M}{1.18M_p} \leq 1 \qquad M \leq M_p \qquad (9\text{-}61b)$$

where M = maximum applied moment, ft-kips
$P_e = (23/12)AF_e'$ (see Art. 9-13 for F_e')
C_m = coefficient defined in Art. 9-13
M_m = maximum moment that can be resisted by member without axial load, ft-kips.
For columns braced about the weak axis,

$$M_m = M_p = F_y Z_x \qquad (9\text{-}62a)$$

For columns not braced about the weak axis,

$$M_m = \left[1.07 - \frac{(l/r_y)\sqrt{F_y}}{3,160}\right] M_p \leq M_p \qquad (9\text{-}62b)$$

Bracing for Frames. For plastic-design multistory frames under factored gravity loads (service loads times 1.7) or under factored gravity plus wind loads (service loads times 1.3), unbraced frames may be used if designed to preclude instability, including the effects of axial deformation of columns, without factored column axial loads exceeding $0.75AF_y$. Otherwise, frames should incorporate a vertical bracing system to maintain lateral stability. This vertical system may be used in selected braced bents that must carry not only horizontal loads directly applied to them but also the horizontal loads of unbraced bents. The latter loads may be transmitted through diaphragm action of the floor system.

Determination of Moments. In multistory plastic design, the girders are designed as three-hinged mechanisms. The columns are designed for girder plastic moments distributed to the attached columns plus the moments due to girder shears at the column faces. Additional consideration should be given to moment-end rotation characteristics of the column above and the column below each joint.

("*Plastic Design of Braced Multistory Steel Frames*," American Institute of Steel Construction; R. Disque, "Applied Plastic Design in Steel," Van Nostrand Reinhold Company, New York.)

9-30. Load-Factor Design for Bridges. For bridges, AASHTO (Art. 9-6) permits load-factor design for simple and continuous beam-and-girder structures of moderate length. The design strength of the steel is taken as the specified minimum yield strength F_y, ksi, and the applied loads are factored to provide the appropriate safety factor.

Load Factors. These are classified by groups. Group I is the normal design loading.
For Group I:

$$\text{Load factor} = 1.30\left[D + \frac{5}{3}(L + I)\right] \qquad (9\text{-}63a)$$

where D = dead-load effect (force, moment, stress, etc.)
L = live-load effect corresponding to D
I = impact effect corresponding to D
For bridges with less than H20 loading, Group IA is used to account for infrequent heavy loads. Live load is assumed to occur on only one lane.
For Group IA:

$$\text{Load factor} = 1.30\ [D + 2.2\ (L + I)] \qquad (9\text{-}63b)$$

Groups II and III are load combinations incorporating wind, ice, friction, earthquake, etc. For these groups, design loads equal service loads times 1.3.

The preceding load factors are used with a variety of allowable stresses that depend on section used and design procedure employed.

Bending Strength. For symmetrical beams and girders, there are three general types of members to consider: compact, braced noncompact, and unbraced sections. The maximum strength of each (moment, in.-kips) depends on member dimensions and unbraced length as well as on applied shear and axial load (Table 9-28).

Table 9-28. Design Criteria for Symmetrical Flexural Sections for Load-factor Design of Bridges.

Type of section	Maximum bending strength M_u, in.-kips	Flange minimum thickness t_f, in.	Web minimum thickness t_w, in.	Maximum unbraced length L_b, in.	Maximum shear V, kips
Compact*	$F_y Z$	$\dfrac{b'\sqrt{F_y}}{50.6}$	$\dfrac{d\sqrt{F_y}}{421}$	$\dfrac{221 r_y}{\sqrt{F_y}}$ when $\dfrac{M_2}{M_1} \geq 0.7$ $\dfrac{379 r_y}{\sqrt{F_y}}$ when $\dfrac{M_2}{M_1} < 0.7$ and $\dfrac{20{,}000 A_f}{F_y d}$	$0.55 F_y d t_w$
Braced noncompact*	$F_y S$	$\dfrac{b'\sqrt{F_y}}{69.6}$	$\dfrac{h}{150}$	$\dfrac{20{,}000 A_f}{F_y d}$	$3.5 E t_w^3 / h$ but not more than $0.58 F_y h t_w$
Unbraced†	$F_y S \left[1 - R_u \left(\dfrac{L_b}{b'} \right)^2 \right]$	$\dfrac{b'\sqrt{F_y}}{69.6}$	$\dfrac{h}{150}$		

*Straight-line interpolation between compact and braced noncompact moments may be used for intermediate criteria, except that $t_w \geq d\sqrt{F_y}/421$ should be maintained.

†If $M_2/M_1 < 0.7$, M_u may be increased by 20% but may not exceed $F_y S$.

The maximum strengths given by the formulas in Table 9-28 apply only when the maximum axial stress does not exceed $0.15 F_y A$, where A is the area of the member. Symbols used in Table 9-28 are defined as follows:

F_y = steel yield strength, ksi
Z = plastic section modulus, in.³ (See Art. 6-91.)
S = section modulus, in.³
b' = width of projection of flange, in.
d = depth of section, in.
h = unsupported distance between flanges, in.
M_2/M_1 = ratio of moments at braced points, where $M_1 \geq M_2$
$R_u = \dfrac{3 F_y}{4\pi^2 E} = \dfrac{1}{10{,}600}$ for $F_y = 36$ ksi and $\dfrac{1}{7{,}630}$ for $F_y = 50$ ksi

Transverse Stiffeners. When the shear capacity exceeds the maximum shear given in Table 9-28, transverse stiffeners are required, and the minimum web thickness becomes $h\sqrt{F_y}/1{,}154$. The shear capacity is given by

$$V_u = V_p \left[C + \frac{0.87(1-C)}{\sqrt{1+(d_o/h)^2}} \right] \tag{9-64}$$

where $V_p = 0.58 F_y h t_w$

$C = \dfrac{569 t_w}{h} \sqrt{\dfrac{1 + (h/d_o)^2}{F_y}} - 0.3 \leq 1$

d_o = stiffener spacing, in.

If applied shear V and moment are both high and $V \geq 0.6 V_u$, then the moment should not exceed

$$M = (1.375 - 0.625 V/V_u) M_u \tag{9-65}$$

The stiffeners should have a moment of inertia of at least

$$I = d_o t_w^3 J \tag{9-66}$$

where $J = 2.5(h/d_o)^2 - 2 \geq 0.5$

Stiffener area should be at least

$$A_{st} = \left[0.15 B h t_w (1 - C) \frac{V}{V_u} - 18 t_w^2 \right] Y \tag{9-67}$$

where $Y = F_y$ (beam web)$/F_y$ (stiffener)

$B = 1.0$ for stiffener pairs

$= 1.8$ for single angles

$= 2.4$ for single plates

Longitudinal Stiffeners. With longitudinal stiffeners at $h/5$ below the compression flange, the web may be as thin as $h\sqrt{F_y}/2{,}308$, in. The stiffener moment of inertia should be at least

$$I_{st} = h t_w^3 \left[2.4 \left(\frac{d_o}{h} \right)^2 - 0.13 \right] \tag{9-68}$$

and the stiffener radius of gyration, in., should be at least

$$r_{st} = \frac{d_o \sqrt{F_y}}{727} \tag{9-69}$$

where I_{st} and r_{st} refer to the stiffeners plus a centrally located strip of web not more than $18t_w$ in width.

Unsymmetrical and Other Types. Unsymmetrical beams and girders as well as composite and hybrid sections used as bending members are also covered by AASHTO design rules in a similar manner to the preceding (see AASHTO Standard Specifications for Highway Bridges).

Columns. Compression members designed by load-factor design should have a maximum strength, kips,

$$P_u = 0.85 A_s F_{cr} \tag{9-70}$$

where A_s = gross effective area of column cross section, sq in.

For $KL_c/r \le \sqrt{2\pi^2 E/F_y}$,

$$F_{cr} = F_y \left[1 - \frac{F_y}{4\pi^2 E} \left(\frac{KL_c}{r} \right)^2 \right] \tag{9-71a}$$

For $KL_c/r > \sqrt{2\pi^2 E/F_y}$,

$$F_{cr} = \frac{\pi^2 E}{(KL_c/r)^2} \tag{9-72a}$$

where F_{cr} = buckling stress, ksi

F_y = yield strength of the steel

K = effective length factor in plane of buckling

L_c = length of member between supports, in.

r = radius of gyration in plane of buckling, in.

E = modulus of elasticity of the steel, ksi

Equations (9-71a) and (9-72a) can be simplified by introducing a Q factor as was done in Art. 9-11. Define Q as

$$Q = \left(\frac{KL_c}{r} \right)^2 \frac{F_y}{2\pi^2 E} \tag{9-73}$$

Then, Equations (9-71a) and (9-72a) can be rewritten as follows:

For $Q \le 1.0$.

$$F_{cr} = \left(1 - \frac{Q}{2} \right) F_y \tag{9-71b}$$

For $Q > 1.0$.

$$F_{cr} = \frac{F_y}{2Q} \tag{9-72b}$$

For combined axial load and bending, the interaction equations (9-74a) and (9-74b) must both be satisfied.

$$\frac{P}{0.85 A_s F_{cr}} + \frac{MC}{M_u \left(1 - \dfrac{P}{A_s F_e} \right)} \le 1 \tag{9-74a}$$

$$\frac{P}{0.85 A_s F_y} + \frac{M}{M_p} \le 1 \tag{9-74b}$$

where M_u = maximum strength of member in bending, in.-kips

M = maximum moment, in.-kips

P = maximum axial load, kips

$F_e = \dfrac{\pi^2 E}{(KL_c/r)^2}$

$M_p = F_y Z$

C = equivalent moment factor

= $0.6 + 0.4\,(M_2/M_1) \geq 0.4$

M_1 and M_2 are the bending moments acting at the ends of the member, and $M_1 \geq M_2$. M_2/M_1 is positive for single curvature and negative for double curvature.

9-31. Bracing. It usually is necessary to provide bracing for the main members or secondary members in most buildings and bridges.

In Buildings. There are two general classifications of bracing for building construction: sway bracing for lateral loads and lateral bracing to increase the capacity of individual beams and columns.

Both low-rise and high-rise buildings require sway bracing to provide stability to the structure and to resist lateral loads from wind or seismic forces. This bracing can take the form of diagonal members or X bracing, knee braces, moment connections, and shear walls.

X bracing is probably the most efficient and economical bracing method. Fenestration or architectural considerations, however, often preclude it. This is especially true for high-rise structures.

Knee braces are often used in low-rise industrial buildings. They can provide local support to the column as well as stability for the overall structure.

Moment connections are frequently used in high-rise buildings. They can be welded, riveted, or bolted, or a combination of welds and bolts can be used. End-plate connections, with shop welding and field bolting, are an economical alternative. Figure 9-9 shows examples of various moment connections.

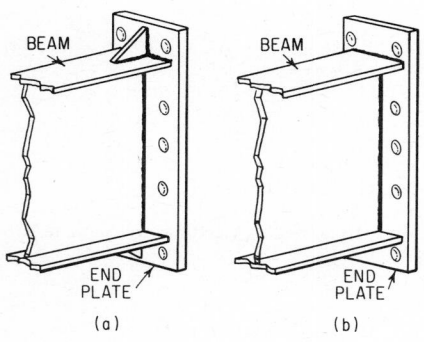

Fig. 9-9. End-plate connections for girders: (a) stiffened moment connection; (b) unstiffened moment connection.

In many cases, moment connections may be used in steel frames to provide continuity and thus reduce the overall steel weight. This type of framing is especially suitable for welded construction; full moment connections made with bolts may be cumbersome and expensive.

In low buildings and the top stories of high buildings, moment connections may be designed to resist lateral forces alone. While the overall steel weight is larger with this type of design, the connections are light and usually inexpensive.

Shear walls are also used to provide lateral bracing in steel-framed buildings. For this purpose, it often is convenient to reinforce the walls normally needed for fire walls, elevator shafts, divisional walls, etc. Sometimes shear walls are used in conjunction with other forms of bracing.

Lateral bracing of columns, arches, beams, and trusses in building construction is used to reduce their critical or effective length, especially of those portions in compression. In floor or roof systems, for instance, it may be economical to provide a strut at midspan of long members to obtain an increase in the allowable stress for the load-carrying members. (See also Arts. 9-10 and 9-11 for effects on allowable stresses of locations of lateral supports.)

Usually, normal floor and roof decks can be relied upon to provide sufficient lateral support to compression chords or flanges to warrant use of the full allowable compressive stress. Examples of cases where it might be prudent to provide supplementary support include purlins framed into beams well below the compression flange or precast-concrete planks inadequately secured to the beams.

In Bridges.   Bracing requirements for highway bridges are given in detail in Standard Specifications for Highway Bridges, American Association of State Highway and Transportation Officials.

Through trusses require top and bottom lateral bracing.   Top lateral bracing should be at least as deep as the top chord.   Portal bracing of the two-plane or box type is required at the end posts and should take the full end reaction of the top-chord lateral system.   In addition, sway bracing at least 5 ft deep is required at each intermediate panel point.

Deck-truss spans and spandrel arches also require top and bottom lateral bracing.   Sway bracing, extending the full depth of the trusses, is required in the plane of the end posts and at all intermediate panel points.   The end sway bracing carries the entire upper lateral stress to the supports through the end posts of the truss.

A special case arises with a half-through truss because top lateral bracing is not possible.   The main truss and the floor beams should be designed for a lateral force of 300 lb per lin ft, applied at the top-chord panel points.   The top chord should be treated as a column with elastic lateral supports at each panel point.   The critical buckling force should be at least 50% greater than the maximum force from dead load, live load, and impact in any panel of the top chord.

Lateral bracing is not usually necessary for deck plate-girder or beam bridges.   Most deck construction is adequate as top bracing, and substantial diaphragms (with depth preferably half the girder depth) or cross frames obviate the necessity of bottom lateral bracing.   Cross frames are required at each end to resist lateral loads.   In spans over 125 ft, lateral bracing should be placed near the bottom flange in at least one-third of the bays.

Through-plate girders should be stiffened against lateral deformation by gusset plates or knee braces attached to the floor beams.   If the unsupported length of the inclined edge of a gusset plate exceeds sixty times the plate thickness, it should be stiffened with angles.

All highway bridges should be provided with cross frames or diaphragms spaced at a maximum of 25 ft.

("Structural Steel Detailing," American Institute of Steel Construction.)

9-32. Mechanical Fasteners.   **Rivets,** made from bar stock, come with one manufactured head (see also Art. 9-2).   Riveted connections usually are made by driving the rivet hot (cherry-red) with pneumatic, hydraulic, or electric hammers, or guns.   Large riveting machines are used for production-shop work.   AASHTO (Art. 9-6) permits cold driving of rivets ⅝ in. in diameter and less.   Larger-size rivets should be heated to a light cherry-red for driving, and all rivets more than ⅞ in. in diameter should be driven mechanically.

ASTM A502, Grade 1, rivets are usually used with A36 steel while A502, Grade 2, is most efficient with high-strength low-alloy steels, such as A242 and A572.   For A588 steel, special rivets with similar corrosion resistance are available and should be used.   See also Art. 9-20.

For economic reasons, bolts or welds generally are preferred to rivets.

Unfinished bolts are used mainly in building construction where slip and vibration are not a factor.   Characterized by a square head and nut, they also are known as machine, common, ordinary, or rough bolts.   They are covered by ASTM A307 and are available in diameters over a wide range (see also Art. 9-2).

A325 bolts are identified by three radial lines on the head and the notation "A325."   The nut is marked with three circumferential marks spaced at 120° or by the number 2.

A490 bolts are marked with "A490" on the head, and the nut is identified by the mark "2H."

Bearing vs. Friction Connections.   Two different types of bolted connections are recognized for bridges and buildings, bearing type and friction type.   Bearing-type connections are allowed higher shear stresses (Art. 9-20).   Thus, they require fewer bolts.   Friction-type connections offer greater resistance to repeated loads and therefore are used when connections are subjected to stress reversal or where slippage would be undesirable.   To qualify for the higher allowable stresses in bearing-type connections, the bolt thread is excluded from the shear plane.   Tests have demonstrated that the ultimate strength of both connections is about the same.   Most building construction is done with bearing-type connections.

Bolt Tightening.   High-strength bolts are tightened by a calibrated wrench or by the "turn-of-the-nut" method.   Calibrated wrenches are powered and have an automatic cutoff set for a predetermined torque.   The "turn-of-the-nut" method requires snugging the plies together and then turning the nut a specified amount, such as one-half or three-quarters of a turn.   Alternatively, a direct tension indicator, such as a load-indicating washer, may be used.   This type of washer has on one side raised surfaces which when compressed to a predetermined height (0.005 in. measured with a feeler gage) indicate attainment of required bolt tension. The Specification for Structural Steel Joints Using A325 or A490 Bolts, adopted by the American Institute of Steel Construction, gives detailed specifications for both tightening methods.

Overtightening a bolt is usually not a serious problem. The bolt works just as well in the plastic as in the elastic range. Excessive overtightening will cause failure, but the operator need only replace the bolt.

Undertightening will result in insufficient friction in a friction-type connection or eventual backing off of the nut and may lead to failure of the connection.

Holes. These generally should be $^1/_{16}$ in. larger than the nominal fastener diameter. Oversize and slotted holes may be used subject to the limitations of Table 9-29.

Symbols used to represent fasteners on drawings are presented in Art. 9-35. ("Structural Steel Detailing," American Institute of Steel Construction.)

9-33. Welded Connections. Welding, a method of joining steel by fusion, is used in both buildings and bridges. It usually requires less connection material than other methods, and in some cases, the quietness of the process is an advantage. No general rules are possible regarding the economics of the various connection methods; each job must be individually analyzed.

Although there are many different welding processes, shielded-arc welding is used almost exclusively in construction. Shielding serves two purposes: it prevents the molten metal from oxidizing and it acts as a flux to cause impurities to float to the surface.

In manual arc welding, an operator maintains an electric arc between a coated electrode and the work. Its advantage lies in its versatility; a good operator can make almost any type of weld. It is used for fitting up as well as for finished work. The coating turns into a gaseous shield, protecting the weld and concentrating the arc for greater penetrative power.

Table 9-29. Oversized and Slotted Hole Limitations for Structural Joints with A325 and A490 Bolts.

Bolt diameter, in.	Maximum hole size, in.		
	Oversize holes*	Short slotted holes†	Long slotted holes†‡
$^5/_8$	$^{13}/_{16}$	$^{11}/_{16} \times ^7/_8$	$^{11}/_{16} \times 1^9/_{16}$
$^3/_4$	$^{15}/_{16}$	$^{13}/_{16} \times 1$	$^{13}/_{16} \times 1^7/_8$
$^7/_8$	$1^1/_{16}$	$^{15}/_{16} \times 1^1/_8$	$^{15}/_{16} \times 2^3/_{16}$
1	$1^1/_4$	$1^1/_{16} \times 1^5/_{16}$	$1^1/_{16} \times 2^1/_2$
$1^1/_8$	$1^7/_{16}$	$1^3/_{16} \times 1^1/_2$	$1^3/_{16} \times 2^{13}/_{16}$
$1^1/_4$	$1^9/_{16}$	$1^5/_{16} \times 1^5/_8$	$1^5/_{16} \times 3^1/_8$
$1^3/_8$	$1^{11}/_{16}$	$1^7/_{16} \times 1^3/_4$	$1^7/_{16} \times 3^7/_{16}$
$1^1/_2$	$1^{13}/_{16}$	$1^9/_{16} \times 1^7/_8$	$1^9/_{16} \times 3^3/_4$

*Not allowed in bearing-type connections.
†In bearing-type connections, slot must be perpendicular to direction of load application.
‡In friction-type connections, increase by one-third the number of bolts required from stress considerations.

Automatic welding, generally the submerged-arc process, is used in the shop, where long lengths of welds in the flat position are required. In this method, the electrode is a base wire (coiled) and the arc is protected by a mound of granular flux fed to the work area by a separate

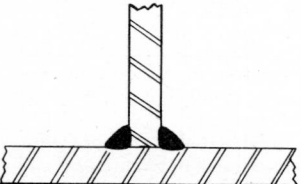

Fig. 9-10. Typical fillet welds.

Fig. 9-11. Typical full-penetration groove weld.

Fig. 9-12. Typical incomplete-penetration groove weld.

flux tube. Most welded bridge girders are fabricated by this method, including the welding of transverse stiffeners.

There are basically two types of welds: fillet welds (Fig. 9-10) and groove welds (Fig. 9-11).

The strength of a fillet weld is based on the shear stress at the throat of the weld. The allowable stress, ksi (Art. 9-19), is multiplied by 0.707 times the nominal leg dimension of the weld to give the allowable shear in kips per linear inch of weld.

The strength of a groove weld depends on the allowable stress of the base metal in tension, compression, bending, or shear. The electrodes should match the base metal (see Table 9-17, Art. 9-19). AISC (Art. 9-6) permits incomplete-penetration groove welds (Fig. 9-12) with a 25% reduction in allowable stress. AASHTO (Art. 9-6) does not recognize incomplete-penetration groove welds for bridges.

("Structural Steel Detailing," American Institute of Steel Construction.)

9-34. Combinations of Fasteners. In new construction, different types of fasteners (rivets, bolts, or welds) are generally not combined to share the same load. The reason for this is that varying amounts of deformation are required to load the different fasteners properly. AISC (Art. 9-6) permits one exception to this rule: Friction-type high-strength bolt connections may be used with welds if the bolts are tightened prior to welding.

In an existing building, a connection may be reinforced for dead load by combining rivets and high-strength bolts, and for dead and live loads, welds may be used. This assumes that no slip is likely beyond that which has already occurred (see also Art. 9-32).

9-35. Fastener Symbols. These are used to denote the type and size of rivets, bolts, and welds on design drawings, as well as on shop and erection drawings. The practice for buildings and bridges is similar.

Figure 9-13 shows the conventional signs for rivets and bolts. Figure 9-14 shows standard welding symbols, as recommended by the American Welding Society.

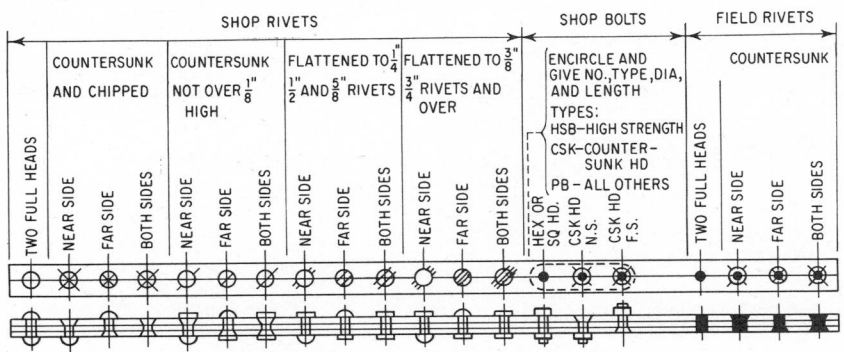

Fig. 9-13. Conventional symbols for rivets and bolts.

9-36. Column Splices. Connections between lengths of a compression member are designed more as an erection device than as stress-carrying elements.

Building columns usually are spliced at every second or third story, about 2 ft above the floor. AISC (Art. 9-6) requires that the connectors and splice material be designed for 50% of the stress in the columns. In addition, they must be proportioned to resist tension that would be developed by lateral forces acting in conjunction with 75% of the calculated dead-load stress and without live load.

Figures 9-15 to 9-17 show typical column splices for riveted, bolted, and welded buildings.

Bridge Splices. AASHTO (Art. 9-6) requires splices (tension, compression, bending, or shear) to be designed for the average of the stress at the point of splice and the strength of the member but not less than 75% of the strength of the member. Splices in riveted members should be located as close as possible to panel points.

In bridges, the ends of columns to be spliced must be milled. In buildings, AISC permits other means of surfacing the end, such as sawing, if the end is accurately finished to a true plane.

("Structural Steel Detailing," American Institute of Steel Construction.)

9-37. Beam Splices. Connections between lengths of a beam or girder are designed as either shear or moment connections, depending on their location and function in the structure.

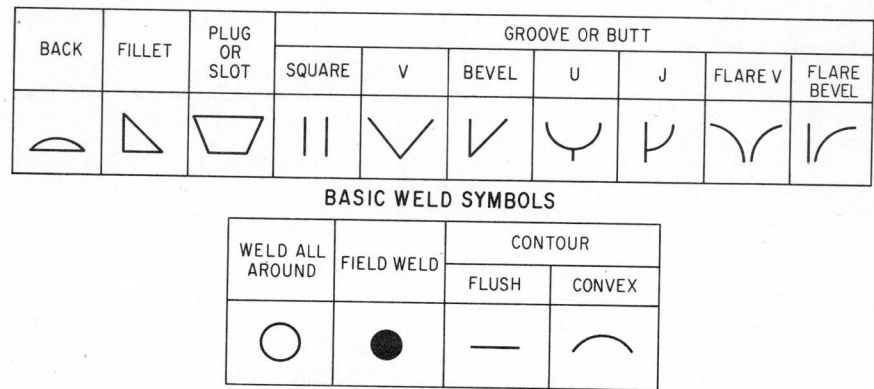

BASIC WELD SYMBOLS

SUPPLEMENTARY WELD SYMBOLS

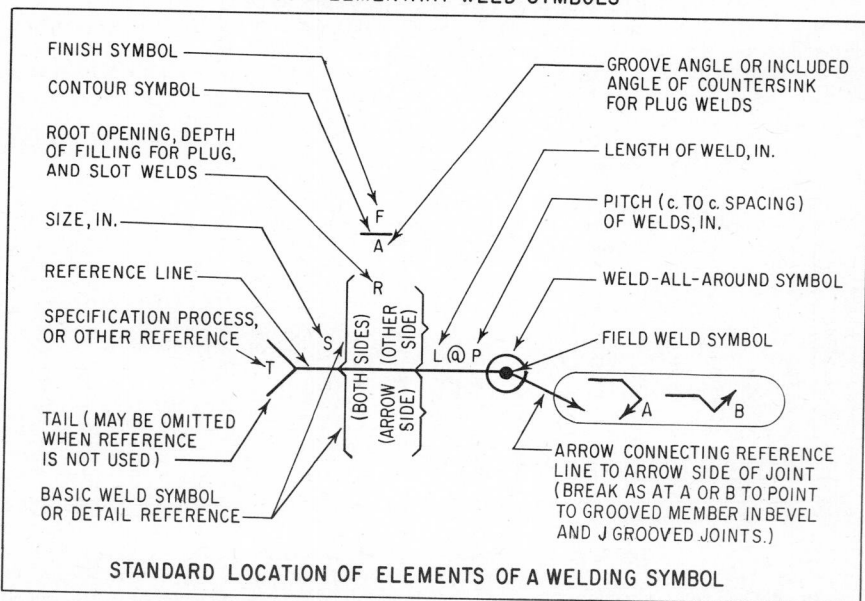

STANDARD LOCATION OF ELEMENTS OF A WELDING SYMBOL

Fig. 9-14. Symbols for welded joints recommended by the American Welding Society. Size, weld symbol, length of weld, and spacing should read in that order from left to right along the reference line, regardless of its orientation or arrow location. The perpendicular leg of symbols for fillet, bevel, J, and flare-bevel groove welds should be at left. Arrow and Other Side welds should be the same size. Symbols apply between abrupt changes in direction of welding unless governed by the "all around" symbol or otherwise dimensioned. When billing of detail material discloses the identity of the far side of a member (such as a stiffened web or truss gusset) with the near side, welding shown for the near side should also be duplicated on the far side.

In cantilever or hung-span construction in buildings, where beams are extended over the tops of columns and spliced, or connected by another beam, it is sometimes possible to use only a shear splice (Fig. 9-18), if no advantage is taken of continuity and alternate span loading is not likely. Otherwise, at least a partial moment splice is required, depending on the loading and span conditions. Splices may be welded or bolted (Fig. 9-19).

Another common beam splice in building construction is the moment connection at the ridge of a rigid frame. A typical connection of this type is shown in Fig. 9-20.

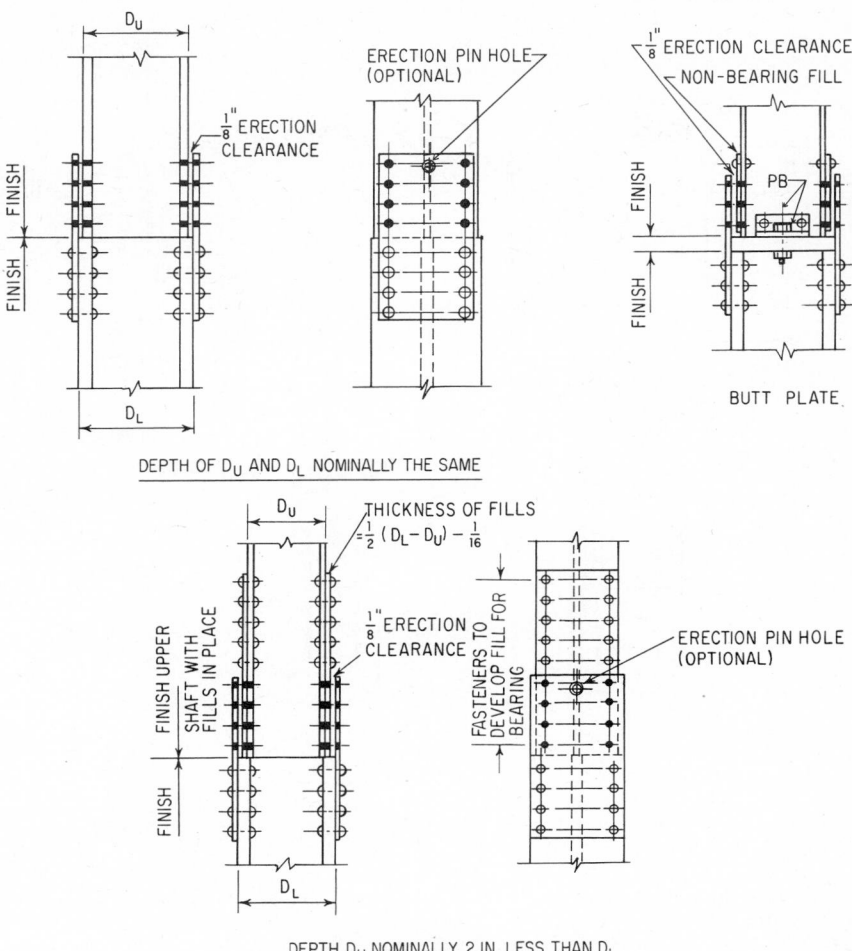

Fig. 9-15. Typical riveted and bolted column splices for buildings. Shims may be used, as required.

For continuous bridges, beam splices are designed for the full moment capacity of the beam or girder and are usually bolted. Figure 9-21 shows such a typical splice for a W36 × 230.

Field-welded splices, though not so common as field-bolted splices, may be an economical alternative.

Special flange splices are always required on welded girders where the flange thickness changes. Care must be taken to insure that the stress flow is uniform. Figure 9-22 shows a typical detail.

("Structural Steel Detailing," American Institute of Steel Construction.)

9-38. Erecting Structural Steel.   Structural steel is erected by either hand-hoisting or power-hoisting devices.

The simplest hand device is the *gin pole* (Fig. 9-23). The pole is usually a sound, straight-grained timber, although metal poles can also be used. The guys, made of steel strands, generally are set at an angle of 45° or less with the pole. The hoisting line may be manila or wire rope. The capacity of a gin pole is determined by the strength of the pole; for instance, an

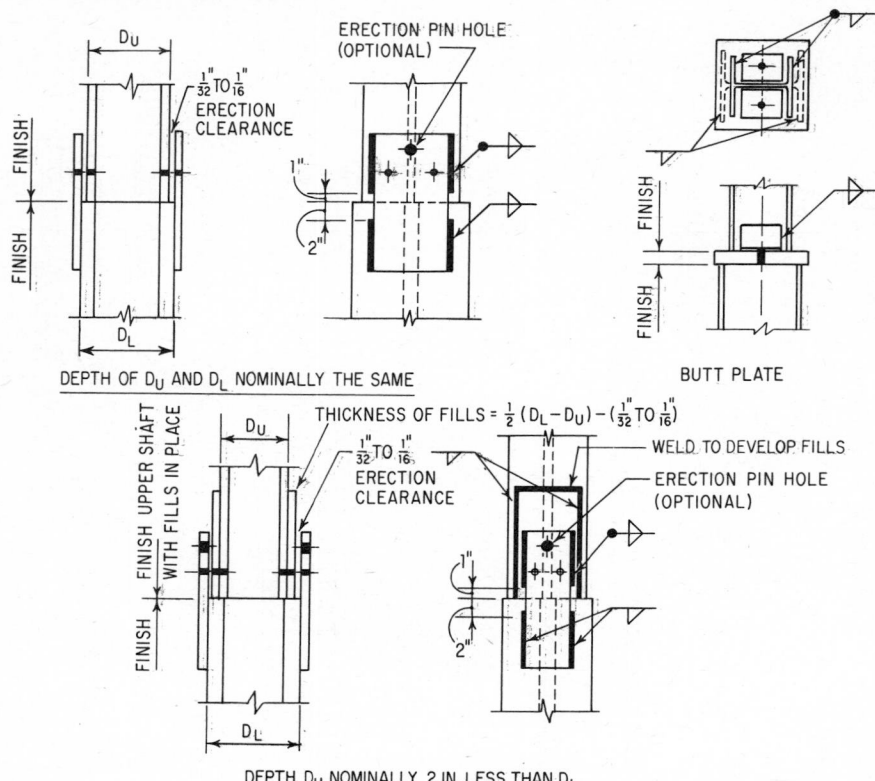

DEPTH OF D_U AND D_L NOMINALLY THE SAME

BUTT PLATE

DEPTH D_U NOMINALLY 2 IN. LESS THAN D_L

Fig. 9-16. Typical welded column splices for buildings. When D_U and D_L are nominally the same and thin fills are required, the shop may attach the splice plate to the upper section and provide field clearance over the lower section.

8×8 in. spruce member, 40 ft high, should carry 3 tons. The capacity increases, of course, with a shorter pole.

There are several variations of gin poles, such as the *A frame* (Fig. 9-24) and the *Dutchman* (Fig. 9-25).

A *stiffleg derrick* consists of a boom, vertical mast, and two inclined braces, or stifflegs (Fig. 9-26). It is provided with a special winch, which is furnished with two hoisting drums to provide separate load and boom lines. It is easily moved from one location to another. A stiffleg derrick is used when a more permanent device is warranted. For instance, after the structural frame of a high building has been completed, a stiffleg may be installed on the roof to hoist building materials, mechanical equipment, etc., to various floors.

A *Chicago boom* is a lifting device that uses the structure being erected to support the boom (Fig. 9-27).

These hoisting devices may be powered by electricity, steam, petroleum, or a combination.

Cranes are powered erection equipment consisting primarily of a rotating cab with a counterweight and a movable boom (Fig. 9-28). Sections of boom may be inserted and removed, and jibs may be added to increase the reach.

Cranes may be mounted on a truck, crawler, or locomotive frame. The truck frame requires firm ground. It is useful on small jobs, where maneuverability and reach are required. Crawler cranes are used on soggy soil or where an irregular or pitched surface exists. Locomotive cranes are used for bridge erection or for jobs where railroad track exists or when it is economical to lay track.

Guy derricks (Fig. 9-29) are advantageous in erecting multistory buildings. These derricks can "jump" themselves from one story to another. The boom temporarily serves as a gin pole to hoist the mast to a higher level. The mast is then secured in place and, acting as a gin pole, hoists the boom into its next position. Slewing (rotating) the derrick may be handled manually or by power.

The *tower or slewing crane* (Fig. 9-30) is more expensive than other types but has important advantages. The control station can be located on the crane or at a distant position that enables the operator to see the load at all times. Also, the equipment can be used to place concrete directly in the forms for floors and roofs, eliminating chutes, hoppers, and barrows.

A variation of the tower crane is the *kangaroo-type tower crane* (Fig. 9-31). The control station is located at the top of the tower and gives the operator a clear view of erection from above. A hydraulic jacking system is built into the fixed mast, and new mast sections are added to increase the height. As the tower gets higher, the mast must be tied into the structural framework for stability.

No general rules can be given regarding the choice of an erection device for a particular job. The main requirement is usually speed of erection, but other factors must be considered, such as the cost of the machine, labor, insurance, and cost of the power.

9-39. Clearances for Erecting Beams and Tolerances. It is the duty of the structural-shop draftsman to detail the steel so that each member may be swung into position without shifting members already in place.

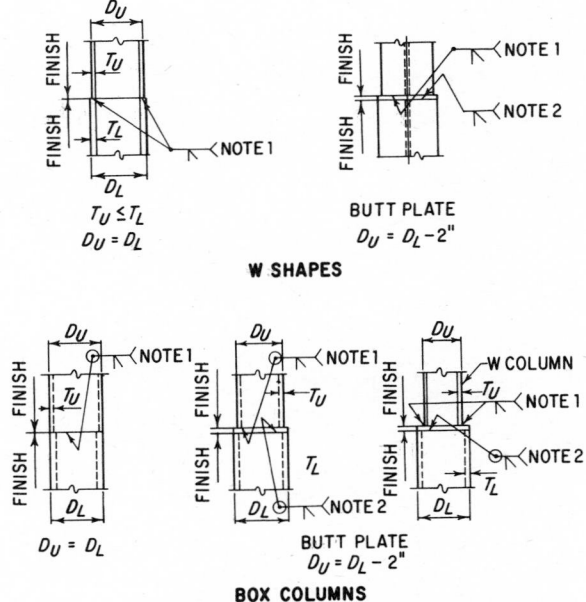

Fig. 9-17. Typical butt-welded details for W shapes and box columns. D_U and D_L relationships indicated are nominal. **NOTE** 1: Weld size is based on T_U; **NOTE** 2: Weld size is based on T_L.

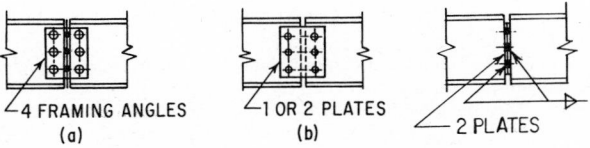

Fig. 9-18. Typical bolted shear splices in beams.

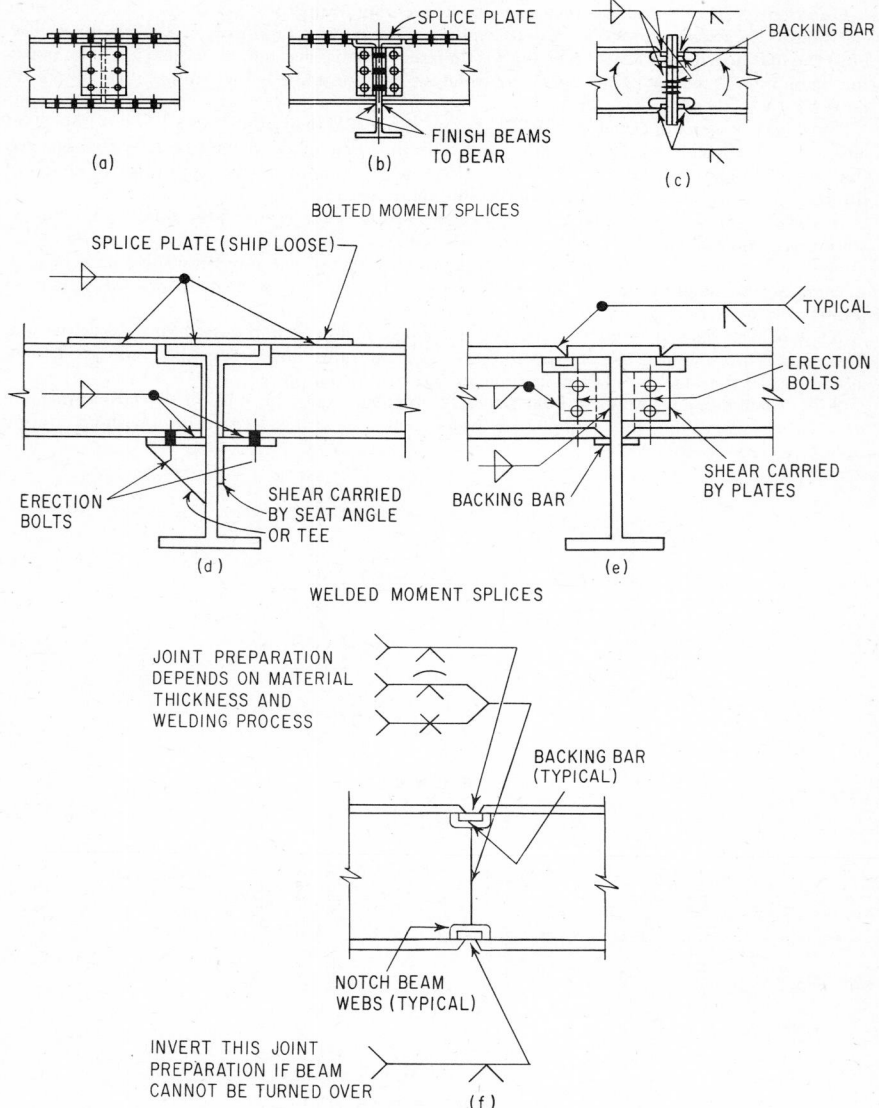

Fig. 9-19. Typical moment splices in beams.

Over the years, experience has resulted in "standard" practices in building work. The following are some examples:

In a framed connection, the total out-to-out distance of beam framing angles is usually ⅛ in. shorter than the face-to-face distance between the columns or other members to which the beam will be connected. Once the beam is in place, it is an easy matter to bend the outstanding legs of the angle, if necessary, to complete the connection. With a relatively short beam, the draftsman may determine that it is impossible to swing the beam into place with only the ⅛ in. clearance. In such cases, it may be necessary to ship the connection angles "loose" for one end of the beam.

Alternatively, it may be advantageous to connect one angle of each end connection to the supporting member and complete the connection after the beam is in place.

The common case of a beam framing into webs of columns must also be carefully considered. The usual practice is to place the beam in the "bosom" of the column by tilting it in the sling as shown in Fig. 9-32. It must, of course, clear any obstacle above. Also, the greatest diagonal distance G must be about ⅛ in. less than the distance between column webs. After the beam is seated, the top angle may be attached.

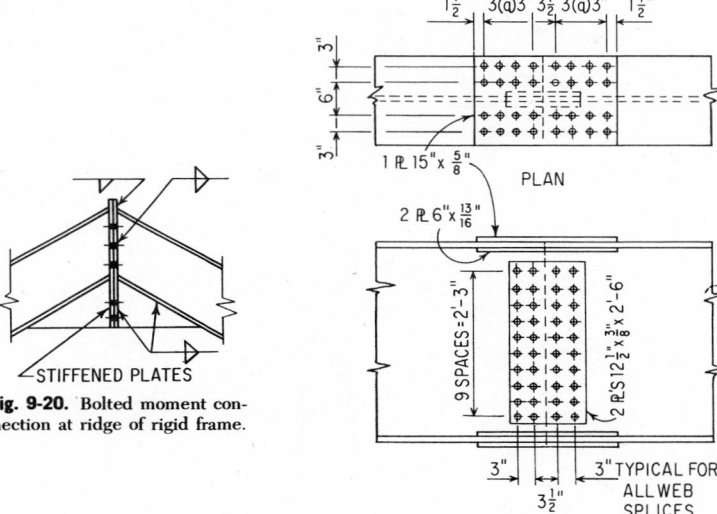

Fig. 9-20. Bolted moment connection at ridge of rigid frame.

Fig. 9-21. Moment splice in continuous bridge beam.

Fig. 9-22. Typical welded-bridge flange splice.

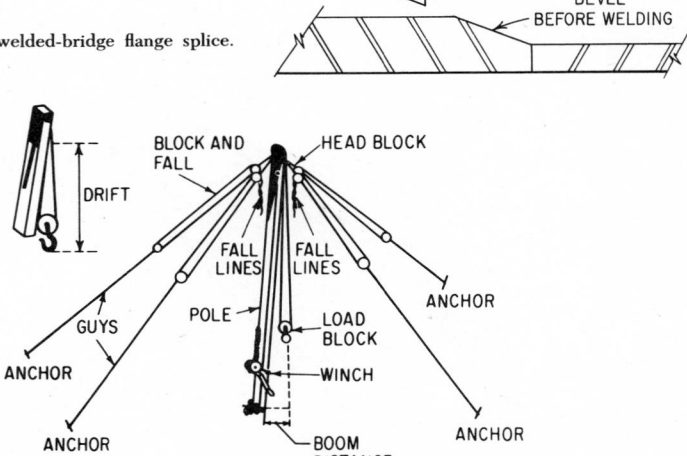

Fig. 9-23. Gin pole.

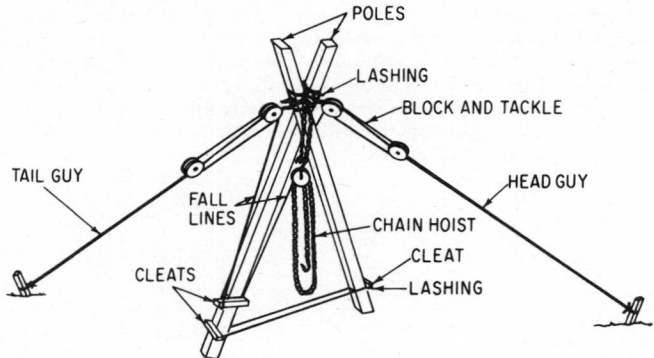

Fig. 9-24. A or shear-leg frame.

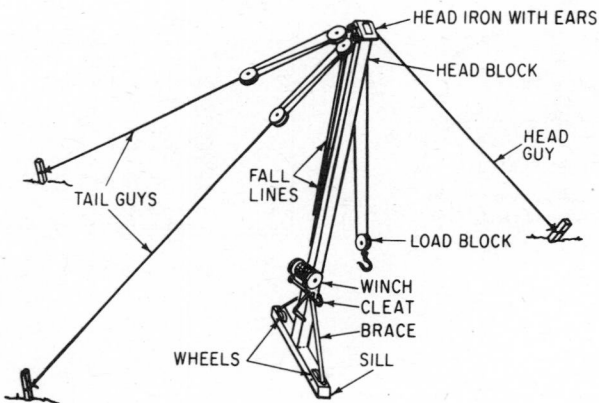

Fig. 9-25. Dutchman.

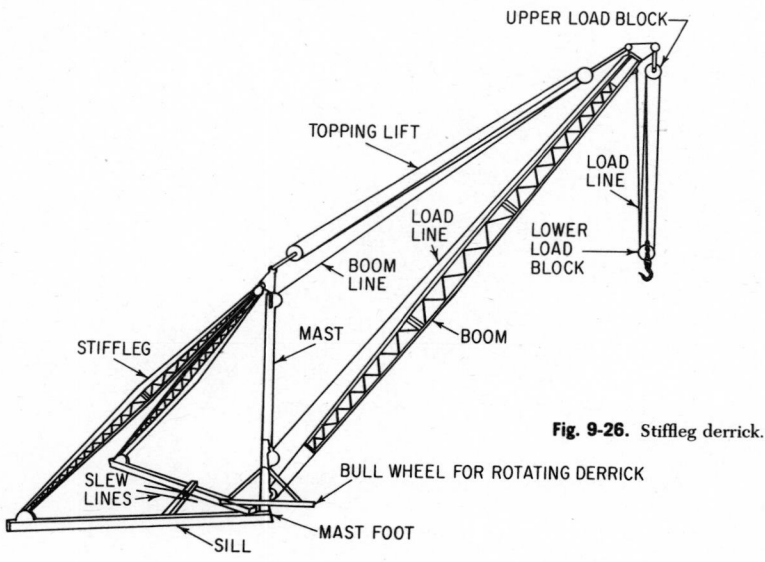

Fig. 9-26. Stiffleg derrick.

It is standard detailing practice to compensate for anticipated mill variations.   The limits for mill tolerances are prescribed in the American Society for Testing and Materials Specification A6, General Requirements for Delivery of Rolled Steel Plates, Shapes, Sheet Piling, and Bars for Structural Use.   For example, wide-flange beams are considered straight, vertically or laterally, if they are within ⅛ in. for each 10 ft of length.   Similarly, columns are straight if the deviation is within ⅛ in. per 10 ft, with a maximum deviation of ⅜ in.

The "Code of Standard Practice" of the American Institute of Steel Construction gives permissible tolerances for the completed frame.   Figure 9-2 summarizes these.   As shown, beams are considered level and aligned if the deviation does not exceed 1:500.   With columns, the 1:500 limitation applies to individual pieces between splices.   The total or cumulative displacement for multistory buildings is also given.   The control is placed on exterior columns or those in elevator shafts.

There are no rules covering tolerances for milled ends of columns.   It is seldom possible to achieve tight bearing over the cross section, and there is little reason for such a requirement.   As the column receives its load, portions of the bearing area may quite possibly become plastic,

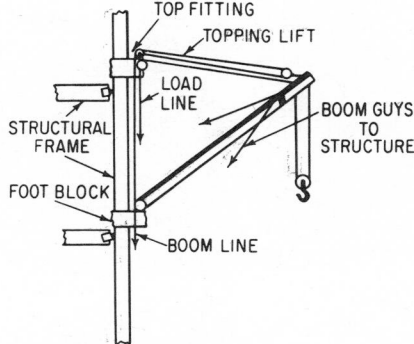

Fig. 9-27. Chicago boom.

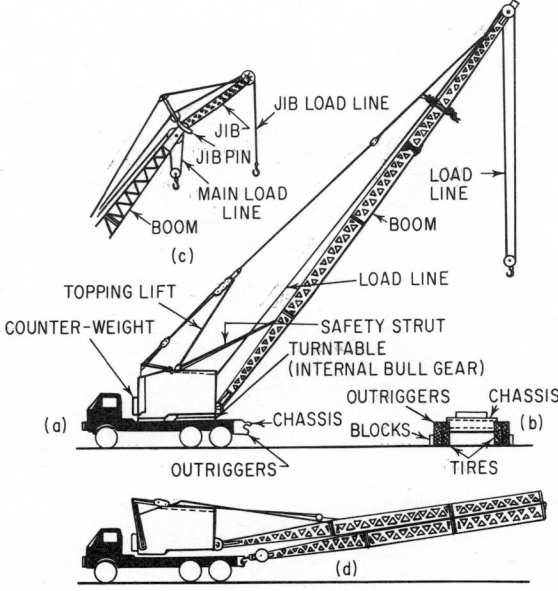

Fig. 9-28. Truck crane.

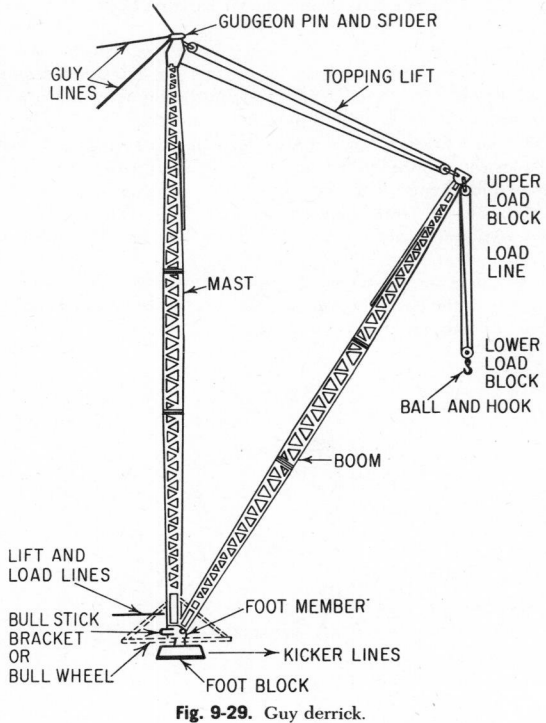

Fig. 9-29. Guy derrick.

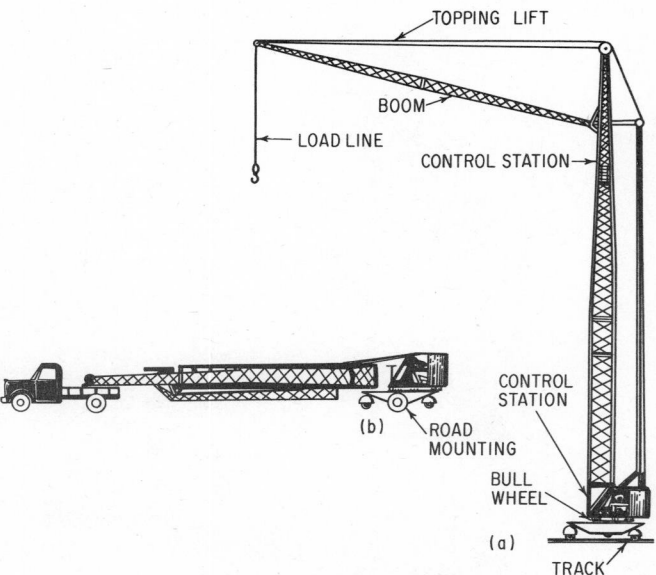

Fig. 9-30. Tower or slewing crane.

resulting in a uniform stress pattern.   No harm is done to the load-carrying capacity of the member.

9-40. Fire Protection of Steel.   Although structural steel does not support combustion and retains its strength to a high degree at temperatures up to 1000°F, the threat of sustained high-temperature fire, in certain types of construction and occupancies, requires that a steel frame be protected with fire-resistive materials.

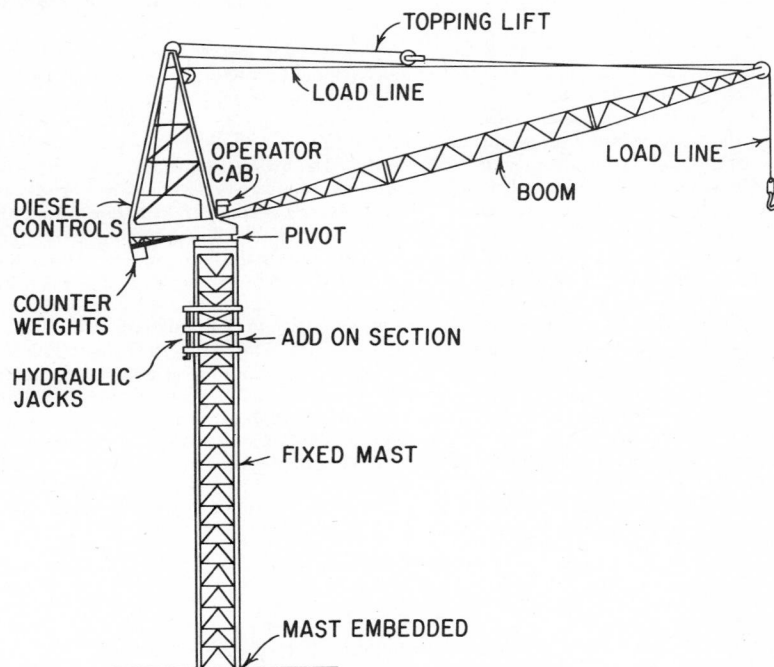

Fig. 9-31. Kangaroo tower crane.

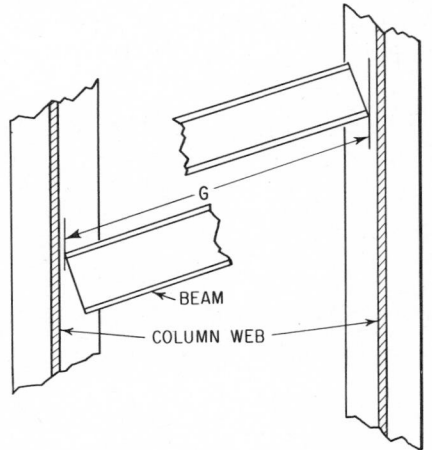

Fig. 9-32. Diagonal distance *G* for beam should be less than the clear distance between column webs to provide erection clearance.

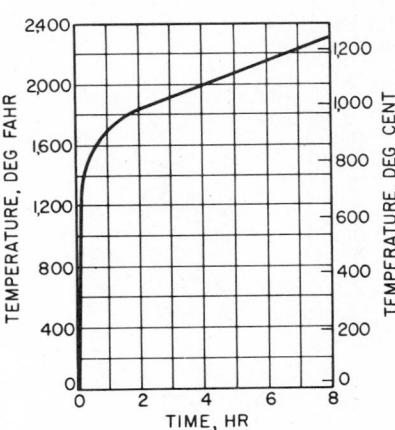

Fig. 9-33. ASTM time-temperature curve for fire test. Air temperature reaches 1000°F in 5 min, 1700°F in 1 hr, and 2000°F in 4 hr.

In many buildings, no protection at all is required because they house little combustible material or they incorporate sprinkler systems. Therefore, "exposed" steel is often used for industrial-type buildings, hangars, auditoriums, stadiums, warehouses, parking garages, billboards, towers, and low stores, schools, and hospitals. Bridges require no fire protection.

The factors that determine fire-protection requirements, if any, are height, floor area, type of occupancy (a measure of combustible contents), availability of fire-fighting apparatus, sprinkler systems, and location in a community (fire zone), which is a measure of hazard to adjoining properties.

Fire Ratings. Based on the above factors, building codes specify minimum fire-resistance requirements. The degree of fire resistance required for any structural component is expressed in terms of its ability to withstand fire exposure in accordance with the requirements of the ASTM standard time-temperature fire test, as shown in Fig. 9-33.

Under the standard fire-test specification of ASTM (E119), each tested assembly is subjected to the standard fire of controlled extent and severity. The fire-resistance rating is expressed as the time, in hours, that the assembly is able to withstand exposure to the standard fire before the first critical point in its behavior is reached. These tests indicate the period of time during which the structural members, such as columns and beams, are capable of maintaining their strength and rigidity when subjected to the standard fire. They also establish the period of time during which floors, roofs, walls, or partitions will prevent fire spread by protecting against the passage of flame, hot gases, and excessive heat.

Strength Changes. In evaluating fire-protection requirements for structural steel, it is useful to consider the effect of heat on its strength. At about 500°F the strength is about 25% larger than that at 65°F. Above that, strength decreases to about normal at 800°F. At 1000°F, compressive strength is about the same as the maximum allowable working stress in columns. Tension and compression members, therefore, are permitted to carry their maximum working stresses if the average temperature in the member does not exceed 1000°F or the maximum at any one point does not exceed 1200°F. (For steels other than carbon or low-alloy, some adjustment may be necessary.)

Coefficient of Expansion. The average coefficient of expansion for structural steel between temperatures of 100 and 1200°F is given by the formula

$$c = (6.1 + 0.0019t) \times 10^{-6} \tag{9-75}$$

where c = coefficient of expansion per °F

t = temperature, °F

Change in Modulus. The modulus of elasticity is about 29,000 ksi at room temperature and decreases linearly to about 25,000 ksi at 900°F. Above that, it decreases more rapidly.

Fire-Protection Methods. Once the required rating has been established for a structural component, there are many ways in which the steel frame may be protected. For columns the most popular fire protection is lightweight plaster (Fig. 9-34). Generally, a vermiculite and perlite plaster thickness of 1 to 1¾ in. affords protection of 3 to 4 hr, depending on construction details.

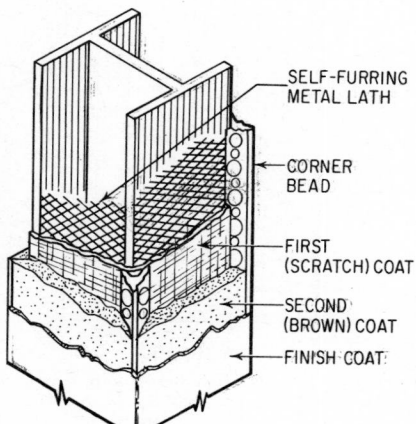

SELF-FURRING
METAL LATH

CORNER
BEAD

FIRST
(SCRATCH) COAT

SECOND
(BROWN) COAT

FINISH COAT

Fig. 9-34. Column fireproofing with lath and plaster.

Concrete, brick, or tile is sometimes used on columns where rough usage is expected. Ordinarily, however, these materials are inefficient because of the large dead weight they add to the structure. Lightweight aggregates would, of course, reduce this.

Beams, girders, and trusses may be fireproofed individually or by a membrane ceiling. Lath and plaster, sprayed mineral fibers, or concrete encasement may be used. As with columns, concrete adds considerably to the weight. The sprayed systems usually require some type of finish for architectural reasons.

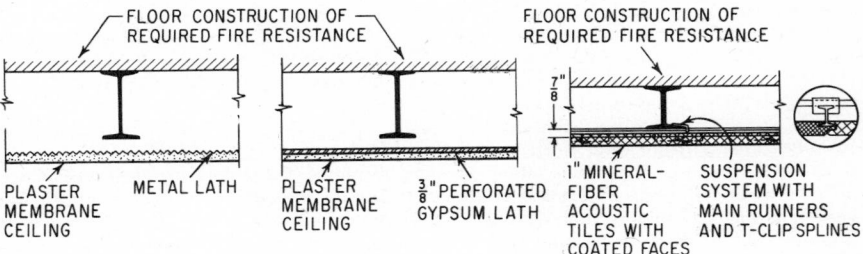

Fig. 9-35. Ceiling-membrane fireproofing.

The membrane ceiling is used quite often to fireproof the entire structural floor system, including beams, girders, and floor deck. For many buildings, a finished ceiling is required. It is therefore logical and economical to employ the ceiling for fire protection. Figure 9-35 illustrates typical installations. As can be seen, the rating depends on the thickness and type of material.

Two alternative methods of fire protection are flame shielding and water-filled columns. These methods are usually used together and are employed where the exposed steel frame is used architecturally.

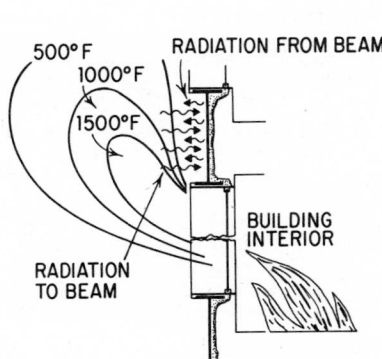

Fig. 9-36. Flame-shielded spandrel girder. (*From "Fire-Resistant Steel-Frame Construction," American Iron and Steel Institute, with permission.*)

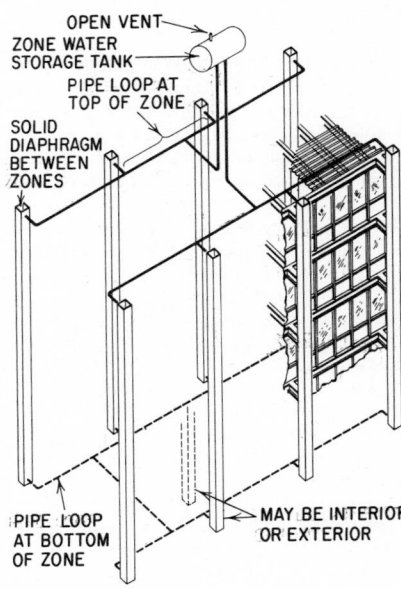

Fig. 9-37. Piping arrangement for typical liquid-filled column fire-protection system. (*From "Fire-Resistant Steel-Frame Construction," American Iron and Steel Institute, with permission.*)

Figure 9-36 illustrates the principle of flame shielding. The spandrel web is exposed on the exterior side and sprayed with fireproofing material on the inside. The shield in this case is the insulated bottom flange, and its extension protects the web from direct contact with the flame. The web is heated by radiation only and will achieve a maximum temperature well below the critical temperature associated with structural failure.

Water-filled columns can be used with flame-shielded spandrels and are an effective fire-resistance system. The hollow columns are filled with water plus antifreeze (in northern climates). The water is stationary until the columns are exposed to fire. Once exposed, heat that penetrates the column walls is absorbed by the water. The heated water rises, causing water in the entire column system to circulate. This takes heated water away from the fire and brings cooler water to the fire-affected columns (Fig. 9-37).

Another alternative in fire protection is intumescent paint. Applied by spray or trowel, this material has achieved a 1-hr rating and is very close to a 2-hr rating. When subjected to heat, it puffs up to form an insulating blanket. It can be processed in many colors and has an excellent architectural finish.

In building construction, it is often necessary to pierce the ceiling for electrical fixtures and air-conditioning ducts. Tests have provided data for the effect of these openings. The rule that has resulted is that ceilings should be continuous except that openings for noncombustible pipes, ducts, and electrical outlets are permissible if they do not exceed 100 sq in. in each 100 sq ft of ceiling area. All openings must be protected with approved fusible-link dampers.

Summaries of established fire-resistance ratings are available from the following organizations:

American Insurance Association, 85 John St., New York, N.Y. 10038.
National Bureau of Standards, Washington, D.C. 20234.
Gypsum Association, 1603 Orrington Ave., Evanston, Ill. 60201.
Metal Lath Association, 221 North LaSalle St., Chicago, Ill. 60601
Perlite Institute, 45 West 45th St., New York, N.Y. 10036.
Vermiculite Institute, 208 South LaSalle St., Chicago, Ill. 60604.
American Iron and Steel Institute, 1000 16th St., Washington, D.C. 20036.
American Institute of Steel Construction, 1221 Avenue of the Americas, New York, N.Y. 10020.

9-41. Corrosion Protection of Steel. Corrosion-resistant structural steels, such as A588 and A242 with copper, ordinarily do not require additional protection against corrosion. The following discussion pertains to other steels.

Steel does not rust except when exposed to atmospheres above a critical relative humidity of about 70%. Serious corrosion occurs at normal temperature only in the presence of both oxygen and water, both of which must be replenished continually. In a hermetically sealed container, corrosion of steel will continue only until either the oxygen or water, or both, are exhausted.

To select a paint system for corrosion prevention, therefore, it is necessary to begin with the function of the structure, its environment, maintenance practices, and appearance requirements. For instance, painting steel that will be concealed by an interior building finish is usually not required. On the other hand, a bridge exposed to severe weather conditions would require a paint system specifically designed for that purpose.

The Steel Structures Painting Council (4400 Fifth Ave., Pittsburgh, Pa. 15213) issues specifications covering practical and economical methods of surface preparation and painting steel structures. The Council also engages in research aimed at reducing or preventing steel corrosion. This material is published in two volumes: I, "Good Painting Practice," and II, "Systems and Specifications."

The SSPC specifications include 13 paint systems. By reference to a specific specification number, it is possible to designate an entire proved paint system, including a specific surface preparation, pretreatment, paint-application method, primer, and intermediate and top coat. Each specification includes a "scope" clause recommending the type of usage for which the system is intended.

In addition to the overall system specification, the SSPC publishes individual specifications for surface preparation and paints. Surface preparations included are solvent, hand tool, power tool, pickling, flame, and several blast techniques.

In developing a paint system, it is extremely important to relate properly the type of paint to the surface preparation. For instance, a slow-drying paint containing oil and rust-inhibitive pigments and one possessing good wetting ability could be applied on steel nominally cleaned.

On the other hand, a fast-drying paint with poor wetting characteristics requires exceptionally good surface cleaning, usually entailing complete removal of mill scale.

In the absence of a specified paint system, for building work the practice described in the American Institute of Steel Construction Specification for the Design, Fabrication, and Erection of Structural Steel for Buildings may be useful. This "nominal" system requires the steel to be brushed by hand or power to remove loose mill scale, loose rust, weld slag, flux deposit, dirt, and foreign matter. Oil and grease are removed by solvent. The shop coat is a commercial-quality paint, usually applied by brushing or spraying to a 2-mil thickness.

The SSPC specifications also include systems for highway bridges. For instance, the scope of Oil Base Paint System 1.01-647 states: "This specification outlines a complete oil-base system for steel bridges and other structural steel surfaces that will be wire brushed, painted and exposed to the weather in moderately corrosive atmospheres . . ."

Standard Specifications for Highway Bridges, American Association of State Highway and Transportation Officials, gives detailed specifications and procedures for the various painting operations but does not outline complete systems. AASHTO specifications for surface preparation include hand cleaning, sandblasting, and flame cleaning. Application procedures are given for brushing and spraying as well as general requirements. In addition, AASHTO specifies various prime, first-field-coat, and finish-coat paints as well as number of coats, color, mixing procedures, weather conditions, and thinning.

Concrete Protection. In other bridge and building construction, steel may be in contact with concrete. According to the "Steel Structures Painting Manual," vol. 1, "Good Painting Practice":

1. Steel that is embedded in concrete for reinforcing should not be painted. Design considerations require strong bond between the reinforcing and the concrete so that the stress is distributed; painting of such steel does not supply sufficient bond. If the concrete is properly made and of sufficient thickness over the metal, the steel will not corrode.

2. Steel that is encased with exposed lightweight concrete that is porous should be painted with at least one coat of good-quality rust-inhibitive primer. When conditions are severe or humidity is high, two or more coats of paint should be applied, since the concrete may accelerate corrosion.

3. When steel is enclosed in concrete of high density or low porosity, and when the concrete is at least 2 to 3 in. thick, painting is not necessary, since the concrete will protect the steel.

4. Steel in partial contact with concrete is generally not painted. This creates an undesirable condition, for water may seep into the crack between the steel and the concrete, causing corrosion. A sufficient volume of rust may build up, spalling the concrete. The only remedy is to chip or leave a groove in the concrete at the edge next to the steel and seal the crack with an alkali-resistant calking compound (such as bituminous cement).

5. Steel should not be encased in concrete that contains cinders, since the acidic condition will cause corrosion of the steel.

Section **10**

Cold-formed Steel Design and Construction

PAUL S. BUKER and DON S. WOLFORD

Development Coordinator, Building Systems, and
Principal Research Structural Engineer,
Armco Steel Corporation, Middletown, Ohio

COLD-FORMED STEEL DESIGN

The introduction of sheet rolling mills in England in 1784 by Henry Cort led to the first cold-formed steel structural application, light-gage corrugated steel sheets for building sheathing. Continuous hot-rolling mills, developed in America in 1923 by John Tytus, led to the present fabricating industry based on coiled strip steel. This is now available in widths up to 90 in. and in coil weights up to 40 tons, hot- or cold-rolled.

Formable, weldable, flat-rolled steel is available in a variety of strengths and in black, galvanized, or aluminum-coated. Thus, fabricators can choose from an assortment of raw materials for producing cold-formed steel products. (In cold forming, bending operations are done at room temperature.) Large quantities of cold-formed sections are most economically produced on multistand roll-forming machines from slit coils of strip steel. Small quantities can still be produced to advantage in presses and bending brakes from sheared blanks of sheet and strip steel. Innumerable cold-formed steel products are now made for building, drainage, road, and construction uses. Design and application of such lightweight-steel products are the principal concern of this section.

10-1. How Cold-formed Shapes Are Made. Cold-formed shapes are relatively thin sections made by bending sheet or strip steel in roll-forming machines, press brakes, or bending brakes. Because of the relative ease and simplicity of the bending operation and the comparatively low cost of forming rolls and dies, the cold-forming process also lends itself well to the manufacture of special shapes for specific architectural purposes and for maximum section stiffness.

Door and window frames, partitions, wall studs, floor joists, sheathing, and moldings are made by cold forming. There are no standard series of cold-formed structural sections, like those for hot-rolled structural shapes, although groups of such sections have been designed for comparison purposes.

Cold-formed shapes cost a little more per pound than hot-rolled sections. They will nevertheless be found more economical under light loading.

10-2. Steel for Cold-formed Shapes. Cold-formed shapes are made from sheet or strip steel, usually from 0.020 to 0.125 in. thick. In thicknesses available (usually 0.060 to ½ in.), hot-rolled steel usually costs less to use. Cold-rolled steel is used in the thinner gages or where the surface finish, mechanical properties, or more uniform thickness resulting from cold reducing are desired. (The commercial distinction between steel plates, sheets, and strip is principally a matter of thickness and width of material.)

Cold-formed shapes may be either black (uncoated) or galvanized. Despite its higher cost, galvanized material is preferable where exposure conditions warrant paying for increased corrosion protection. Uncoated material to be used for structural purposes generally conforms to one of the standard specifications of the American Society for Testing and Materials for structural-quality sheet and strip (A570 and A611). ASTM A446 covers structural-quality galvanized sheets. Steel with a hot-dipped aluminized coating is also available.

The choice of grade of material usually depends on the severity of the forming operation required to make the desired shape. Low-carbon steel has wide usage. Most shapes used for structural purposes in buildings are made from material with a yield point of 33 ksi—Grade C in ASTM A570 for hot-rolled sheets and strip and A611 for cold-rolled sheet. Steel conforming generally to ASTM A606, High-Strength, Low-Alloy, Hot-rolled and Cold-rolled Steel Sheet and Strip with Improved Corrosion Resistance, and A607, Low-Alloy Columbium and/or Vanadium, Hot-rolled and Cold-Rolled Steel Sheet and Strip, is used to achieve lighter weight.

Sheet and strip for cold-formed shapes are usually ordered and furnished in decimal thicknesses.

10-3. Types of Cold-formed Shapes. Some cold-formed shapes used for structural purposes are similar in general configuration to hot-rolled structural shapes. Channels, angles, and Z's can be roll-formed in a single operation from one piece of material. I sections are usually made by welding two channels back to back, or by welding two angles to a channel. All such sections may be made with either plain flanges, as in Fig. 10-1a to d, j, and m, or with flanges stiffened by lips at outer edges, as in Fig. 10-1e to h, k, and n.

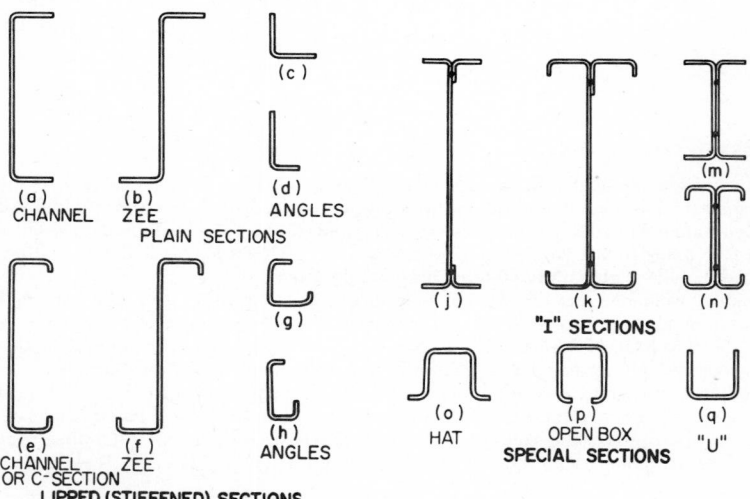

Fig. 10-1. Typical cold-formed steel structural sections.

In addition to these sections, the flexibility of the forming process makes it relatively easy to obtain hat-shaped sections, open box sections, or inverted-U sections (Fig. 10-1o, p, and q). These sections are very stiff in a lateral direction.

The thickness of cold-formed shapes can be assumed to be uniform throughout in computing weights and section properties. The fact that cold-formed sections have corners rounded on both the inside and outside of the bend has only a slight effect on the section properties, and so computations may be based on sharp corners without serious error.

For Grade C material, the inside radius of bends should not be less than three-quarters of the thickness and preferably not less than the thickness. The purpose of this requirement is to avoid cracking during forming, particularly if the bend is parallel to the rolling direction.

10-4. Design Principles for Cold-formed Sections. Starting in 1939, the American Iron and Steel Institute undertook to develop light-gage-steel design data and procedures. Since then, a series of comprehensive test programs with analytical treatments have been performed at Cornell University under the direction of Prof. George Winter. The researches have been directed for AISI by a group of structural specialists as subcommittee members of the Committees of Sheet and Strip Steel Producers. Their efforts led in 1946 to the first AISI Specification for the Design of Light Gage Cold-formed Steel Structural Members (American Iron and Steel Institute, 1000 16th St., N.W., Washington, D.C. 20036). The specification has been revised repeatedly and has been widely adopted in building codes.

The structural behavior of cold-formed shapes conforms to classic principles of structural mechanics, as does that of hot-rolled structural-steel shapes and plates. But section distortion from buckling and other similar effects must be taken into account.

The uniform thickness of most cold-formed sections and the large width-thickness ratios of the various elements comprised in such sections make it possible to assume that structural properties, such as moment of inertia and section modulus, vary directly as the first power of the thickness. (Properties of line elements are given in Supplementary Information on the 1968 edition of the AISI Specification for the Design of Cold-formed Steel Structural Members.)

When wide, thin elements are in axial compression (for example, in beam flanges or column elements), they tend to buckle locally at stresses below the yield point (elastically). (This is not to be confused with overall buckling of long columns or of laterally unsupported beams.) Other factors—such as shear lag, which gives rise to nonuniform stress distribution, and torsional instability, which may be considerably more pronounced in thin, open sections than in thick, closed ones—must sometimes also be considered with thin material.

10-5. Structural Behavior of Flat Compression Elements. In buckling of flat compression elements in beams and columns, the flat-width ratio w/t is an important factor. It is the ratio of width w of a single flat element, exclusive of edge fillets, to the thickness t of the element (Fig. 10-2).

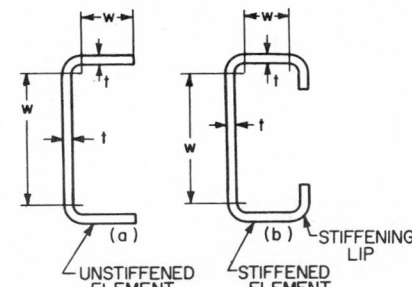

Fig. 10-2. Compression elements.

Flat compression elements of cold-formed structural members are classified as stiffened elements and unstiffened elements. Stiffened compression elements have both edges parallel to the direction of stress stiffened by a web, flange, or stiffening lip. Unstiffened compression elements have only one edge parallel to the direction of stress stiffened. If the sections in Fig. 10-1a to n are used as compression members, the webs are considered stiffened compression elements. But the wide, lipless flange elements and the lips that stiffen the outer edges of the flanges are unstiffened elements. Any section composed of a number of plane elements can be broken down into a combination of stiffened and unstiffened elements.

In order that a compression element may qualify as a stiffened compression element, the minimum moment of inertia of each edge stiffener about its own centroidal axis parallel to the stiffened element must comply with

$$I_{min} = 1.83t^4 \sqrt{\left(\frac{w}{t}\right)^2 - \frac{4,000}{F_y}} \geq 9.2t^4 \qquad (10\text{-}1)$$

where w/t = flat-width ratio of stiffened element

   t = thickness, in.

   F_y = yield strength of the steel, ksi

Where the stiffener consists of a simple lip bent at right angles to the stiffened element, the minimum overall depth d_{min}, in., of such a lip may be determined from

$$d_{min} = 2.8t \sqrt[6]{\left(\frac{w}{t}\right)^2 - \frac{4,000}{F_y}} \geq 4.8t \qquad (10\text{-}2)$$

A simple lip should not be used as an edge stiffener for any element with a flat-width ratio greater than 60.

For unstiffened elements, elastic buckling usually does not have to be considered unless the flat-width ratio exceeds 11 for $F_y \leq 33$ ksi or 8 for $33 < F_y \leq 65$ ksi. For stiffened elements, which buckle differently from unstiffened elements, the flat-width ratio beyond which allowance must be made for local buckling depends on unit stress. The effect of local buckling may be ignored for stiffened elements with flat-width ratios equal to or less than $(w/t)_{lim}$ in Eqs. (10-8) and (10-10).

10-6. Unstiffened Cold-formed Elements Subject to Local Buckling. To design such elements, compute the section properties of the full section in the conventional manner but use reduced allowable compressive stress F_c, ksi. Equations (10-3) to (10-7) give the applicable stresses.

For w/t not greater than $63.3/\sqrt{F_y}$:

$$F_c = 0.60F_y \qquad (10\text{-}3)$$

For w/t greater than $63.3/\sqrt{F_y}$ but not more than $144/\sqrt{F_y}$:

$$F_c = F_y \left(0.767 - 0.00264 \frac{w}{t} \sqrt{F_y}\right) \qquad (10\text{-}4)$$

For w/t greater than $144/\sqrt{F_y}$ but not more than 25:

$$F_c = \frac{8,000}{(w/t)^2} \qquad (10\text{-}5)$$

For angle struts with w/t from 25 to 60:

$$F_c = \frac{8,000}{(w/t)^2} \qquad (10\text{-}6)$$

For all other sections with w/t from 25 to 60:

$$F_c = 19.8 - 0.28 \frac{w}{t} \qquad (10\text{-}7)$$

where w/t = flat-width ratio of unstiffened element

   F_y = yield strength of steel, ksi

Equations (10-3) to (10-7) are plotted in Fig. 10-3 for yield points of 25 to 65 ksi.

10-7. Stiffened Cold-formed Elements Subject to Local Buckling. To design such elements, compute section properties based on an effective width of each stiffened element (Fig. 10-4). Determine the effective width b, in., from Eqs. (10-9) and (10-11).

As shown in Fig. 10-4, that portion of the element not considered to be effective should be considered located symmetrically about the center line of the element.

For determination of safe loads, widths of stiffened compression elements may be considered fully effective ($b = w$) when the flat-width ratio w/t does not exceed

$$\left(\frac{w}{t}\right)_{lim} = \frac{171}{\sqrt{f}} \qquad (10\text{-}8)$$

where f = computed compressive stress for full width, ksi

When w/t exceeds $(w/t)_{lim}$, the effective width b may be calculated from

$$\frac{b}{t} = \frac{253}{\sqrt{f}} \left[1 - \frac{55.3}{(w/t)\sqrt{f}}\right] \qquad (10\text{-}9)$$

where f = actual stress in compression element computed for effective design width, ksi

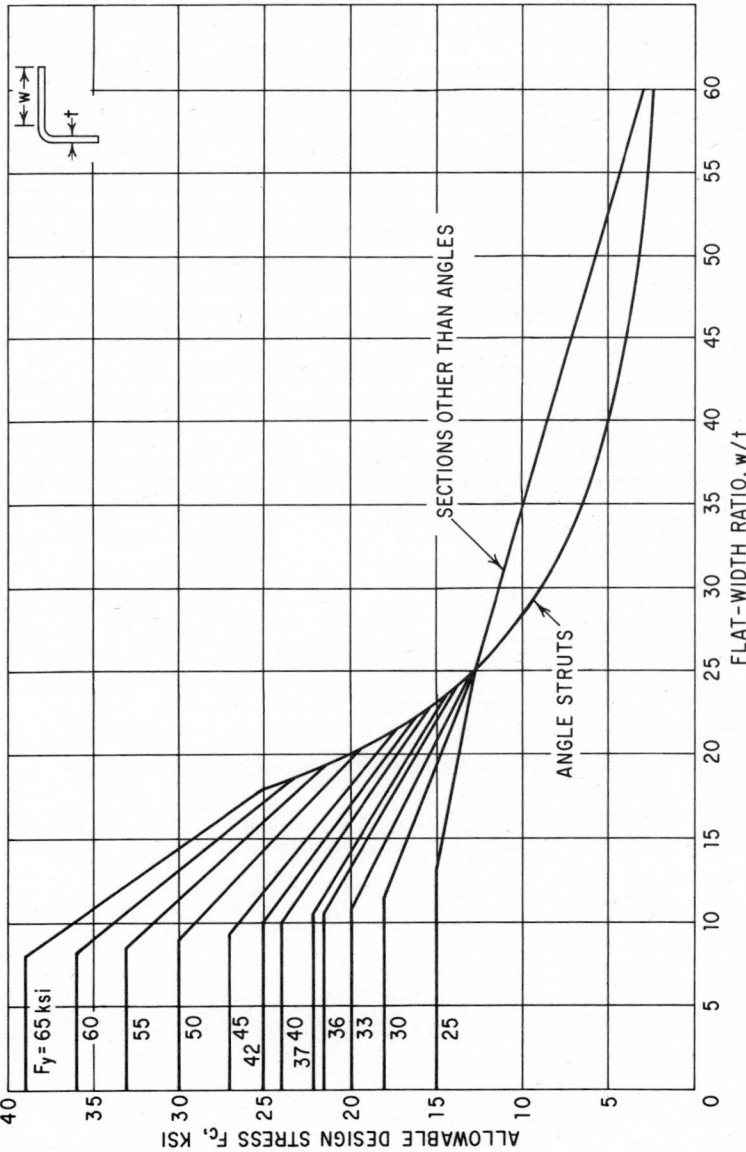

Fig. 10-3. Allowable compression stress in unstiffened steel elements.

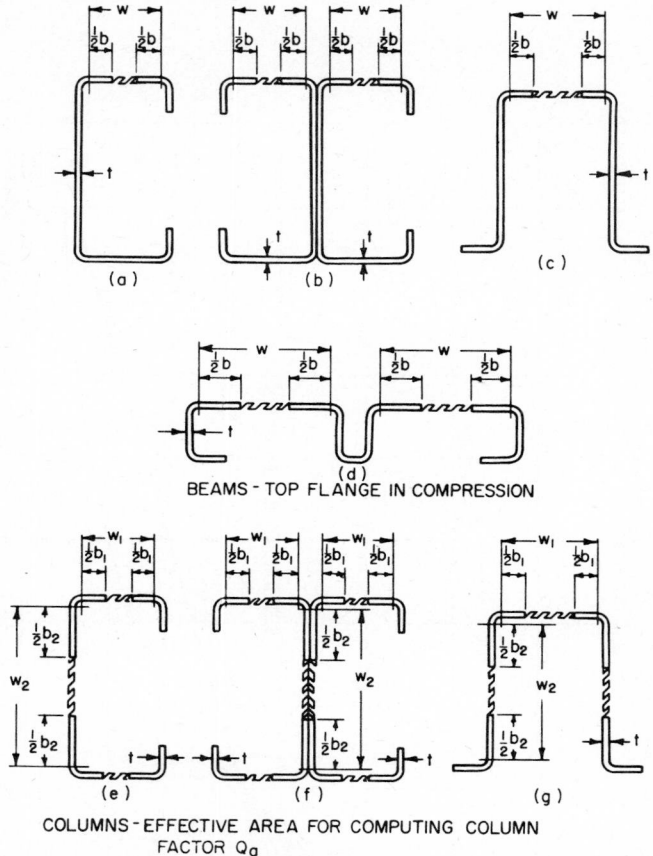

Fig. 10-4. Effective width of stiffened compression elements.

For computation of moment of inertia to be used in deflection calculations or in other calculations involving stiffness, widths may be considered fully effective when w/t does not exceed

$$\left(\frac{w}{t}\right)_{\text{lim}} = \frac{221}{\sqrt{f}} \tag{10-10}$$

where f = computed compressive stress for full width, ksi
When w/t exceeds $(w/t)_{\text{lim}}$, b may be computed from

$$\frac{b}{t} = \frac{326}{\sqrt{f}}\left[1 - \frac{71.3}{(w/t)\sqrt{f}}\right] \tag{10-11}$$

where f is defined as for Eq. (10-9).

The curves of Fig. 10-5 were plotted from Eq. (10-9). They may be used to determine b/t for different values of w/t and unit stresses f.

The effective width b in Eqs. (10-9) and (10-11) is dependent on the actual stress f, which, in turn, is determined by reduced-section properties that are a function of effective width. Employment of successive approximations consequently may be necessary in solving Eqs. (10-9) and (10-11). This can be avoided and the correct values of b/t obtained directly from the formulas when f is known or is held to a specified maximum allowable value (usually 20 ksi for flexural

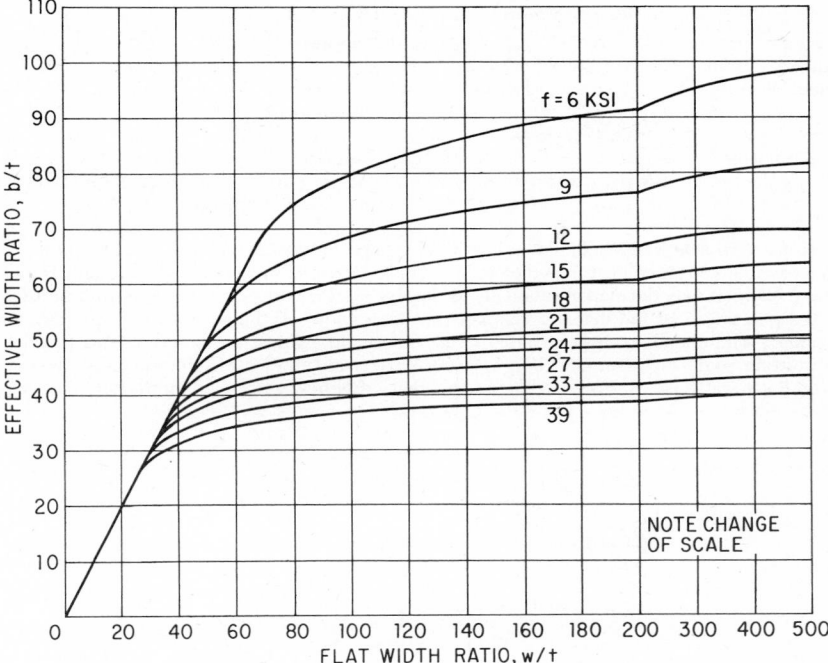

Fig. 10-5. Effective width for safe-load determination where local buckling may occur.

members of Grade C steel). This is true, though, only when the neutral axis of the section is closer to the tension flange than to the compression flange, so that compression controls. The latter condition holds for symmetrical channels, zees, and I sections used as flexural members about their major axes, such as Fig. 10-1e, f, k, and n, or for unsymmetrical channels, zees, and I sections. If w/t of the compression flanges does not exceed about 60, and Grade C steel is used, with an allowable working stress of 20 ksi in beams, the error in basing the effective width of the compression flange on $f = 20$ ksi will generally be negligible. This is so even though the neutral axis is above the geometric center line. For wide, inverted, pan-shaped sections, such as deck and panel sections, a somewhat more accurate determination will frequently prove desirable.

For computation of moment of inertia for deflection or stiffness calculations, properties of the full unreduced section can be used without significant error when w/t of the compression elements does not exceed 60. For greater accuracy, use Equation (10-11).

Example. As an example of effective-width determination, consider the hat section in Fig. 10-6. The section is to be made of Grade C steel and used as a simply supported beam with the top flange in compression, at a basic working stress of 20 ksi, Safe load-carrying capacity is to be computed.

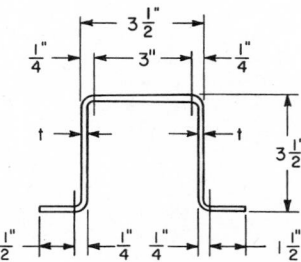

Fig. 10-6. Hat section.

The top flange is a stiffened compression element 3 in. wide. If the thickness is $\frac{1}{16}$ in., then the flat-width ratio is 48 ($>171/\sqrt{f}$), and Eq. (10-9) applies. For this value of w/t and $f = 20$ ksi, Eq. (10-9) or Fig. 10-5 gives b/t as 42. Thus, only 87% of the top flange can be considered effective in this case. The neutral axis of the section will lie below the horizontal center line, and compression will control. Since that stress is limited to 20 ksi, the effective width may be directly determined from Eq. (10-9).

For a wide hat section in which the horizontal centroidal axis is nearer the compression than the tension flange, the stress in the tension flange controls. So determination of unit stress and effective width of the compression flange requires successive approximations. (See also Art. 10-9.)

10-8. Maximum Flat-Width Ratios for Cold-formed Elements. When the flat-width ratio exceeds about 30 for an unstiffened element and 250 for a stiffened element, noticeable buckling of the element may develop at relatively low stresses. Present practice is to permit buckles to develop in the sheet and to take advantage of what is known as the postbuckling strength of the section. The effective-width formulas [Eqs. (10-9) and (10-11)] are based on this practice of permitting some incipient buckling to occur. To avoid intolerable deformations, however, overall flat-width ratios, disregarding intermediate stiffeners and based on the actual thickness of the element, should not exceed the following:

> Stiffened compression element having one longitudinal edge connected to a web or flange, the other to a simple right-angle lip 60
> Stiffened compression element having both edges stiffened by stiffening means other than a simple right-angle lip 90
> Stiffened compression element with both longitudinal edges connected to a web or flange element, such as in a hat, U, or box type of section 500
> Unstiffened compression element 60

10-9. Unit Stresses for Cold-formed Steel. For sheet and strip of Grade C steel with a specified minimum yield point $F_y = 33$ ksi, use a basic allowable stress $F_b = 20$ ksi in tension and bending. For high-strength steel ($F_y = 50$ ksi), a basic stress of 60% of the yield point (corresponding to a safety factor of 1.67 on the yield point) may be used. Basic stress must be reduced for wide unstiffened compression elements, using the equations in Art. 10-6. An increase of $33\frac{1}{3}\%$ in allowable stress is customary for combined wind or earthquake forces and gravity loads.

10-10. Laterally Unsupported Cold-formed Beams. In the relatively infrequent cases in which cold-formed sections used as beams are not laterally supported at frequent intervals, the unit stress must be reduced to avoid failure from lateral instability. The amount of reduction depends on the shape and proportions of the section and the spacing of lateral supports. This is not a difficult obstacle. (For details, see the AISI Specification for the Design of Cold-formed Steel Structural Members, American Iron and Steel Institute.)

Because of the torsional flexibility of light-gage channel and zee sections, their use as beams without lateral support is not recommended. When one flange is connected to a deck or sheathing material, bracing of the other flange may not be needed to prevent twisting of the member, depending on the collateral material and its connections, dimensions of the member and the span, and whether the unbraced flange is in compression.

When laterally unsupported beams must be used, or where lateral buckling of a flexural member is likely to be a problem, consideration should be given to the use of relatively bulky sections which have two webs, such as hat or box sections (Fig. 10-1o and p).

10-11. Web Stresses in Cold-formed Members. The shear stress F_v, ksi, on the gross web section in cold-formed flexural members should not exceed 0.40 of the yield point F_y or Eqs. (10-12a) or (10-12b).

For h/t not greater than $547/\sqrt{F_y}$:

$$F_v = \frac{152\sqrt{F_y}}{h/t} \tag{10-12a}$$

For h/t greater than $547/\sqrt{F_y}$:

$$F_v = \frac{83,200}{(h/t)^2} \qquad (10\text{-}12b)$$

where t = web thickness, in.

h = clear distance between flanges measured along plane of web, in.

In Eqs. (10-12a) and (10-12b), h/t is the ratio of depth between flanges to web thickness. Where the web consists of two sheets, as in the case of two channels fastened back to back to form an I section, each sheet must be considered as a separate web carrying its share of the shear. For Grade C steel, the maximum allowable shear on the gross section of the web is specified as 13,300 psi, except for such increases as may be allowed for combined gravity and wind loading. (See Art. 10-9.)

Use of unstiffened webs in which h/t exceeds 150 is not recommended.

10-12. Cold-formed Steel Columns. When cold-formed sections are used as columns, there usually need be no modification of conventional procedure if the section does not contain any elements that exceed the limits $w/t = 63.3/\sqrt{f_y}$ for unstiffened elements and $w/t = 171/\sqrt{f}$ for stiffened elements, where F_y is the steel yield strength, ksi, and f is the basic design stress, ksi—20 ksi for Grade C steel. Where w/t limits are exceeded, provision must be made against failure by local buckling.

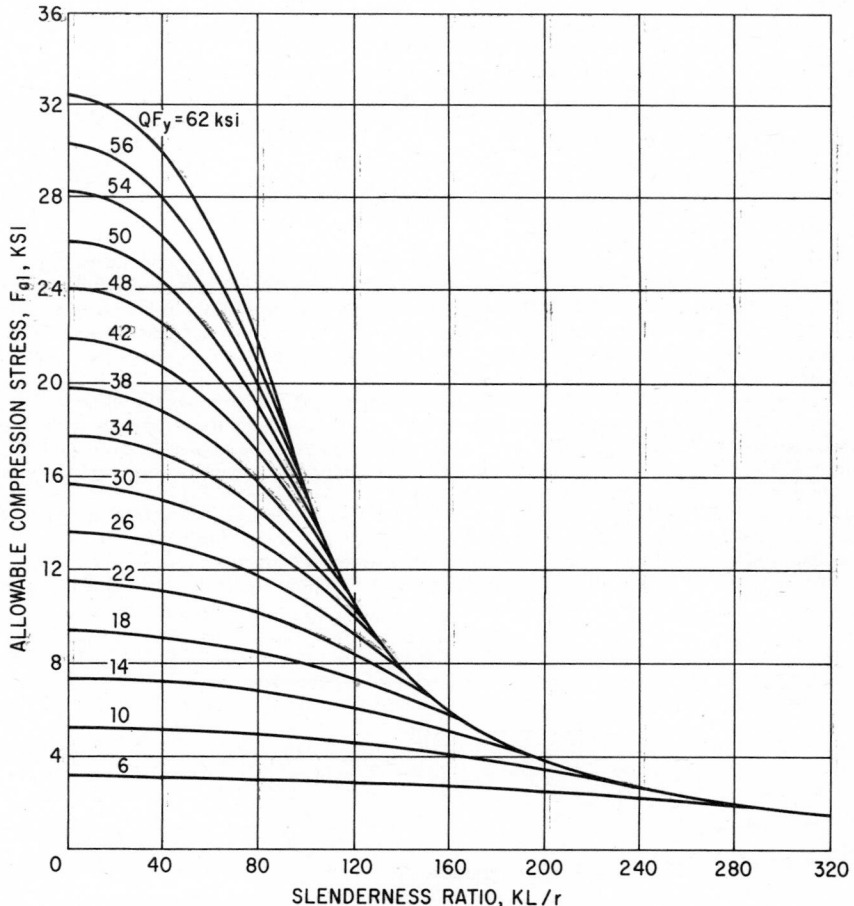

Fig. 10-7. Column design curves for cold-formed shapes.

The column-design formulas recommended by the American Iron and Steel Institute specifications for cold-formed sections consist of a family of Johnson parabolas all tangent to a single Euler curve, generalized for yield point as follows:

For KL/r less than $C_c/\sqrt{Q}$:

$$\frac{P}{A} \leq F_{a1} = 0.522QF_y - \left(\frac{QF_yKL/r}{1,494}\right)^2 \qquad (10\text{-}13a)$$

For KL/r equal to or greater than $C_c/\sqrt{Q}$:

$$\frac{P}{A} \leq F_{a1} = \frac{151,900}{(KL/r)^2} \qquad (10\text{-}13b)$$

where $C_c = \sqrt{2\pi^2 E/F_y}$
    P = total load, kips
    A = full, unreduced cross-sectional area of the member, sq in.
    E = modulus of elasticity of the steel = 29,500 ksi
    F_{a1} = allowable average compression stress under concentric loading, ksi
    K = effective length factor, as for structural steel columns
    L = unbraced length of member, in.
    r = radius of gyration of full, unreduced cross section, in.
    Q = a form and buckling factor determined from Eq. (10-14)

Allowable compression stresses determined by these equations are plotted in Fig. 10-7.

In Eqs. (10-13a) and (10-13b), a safety factor of 1.92 contains an eccentricity factor of 1.25. [Observe that F_y does not appear in Eq. (10-13b).] The point of tangency between these equations is always at a KL/r value equal to C_c.

The form factor and buckling factor Q in Eq. (10-13a), in general, may be evaluated from

$$Q = Q_aQ_s = \frac{A_{eff}}{A}\frac{F_c}{F_{a1}} \qquad (10\text{-}14)$$

For members composed entirely of stiffened elements, Q_a is the ratio of the effective design area A_{eff}, as determined from the effective-design widths of such elements, to the full or gross area of the cross section A. Effective widths may be computed from Eq. (10-9) or (10-11). (In the quantity P/A, A always should be taken as the full cross-sectional area.) The effective design area used in determining Q_a should be based on the basic design stress allowed in tension and bending—20 ksi for Grade C steel.

For members composed entirely of unstiffened elements, Q_s is the ratio of the allowable compression stress F_c, as defined in Art. 10-6, for the weakest element of the cross section (the element having the largest flat-width ratio) to the basic design stress in axial compression F_{a1}. Since F_c can equal but never be greater than F_{a1}, the value of Q_s, used as a buckling factor, can equal but never exceed 1.0.

For members composed of both stiffened and unstiffened elements, the stress on which Q_a is based should be the unit stress F_c used in computing Q_s, and the effective area to be used in computing Q_a should include the full area of all unstiffened elements.

If a section does not contain any element with a flat-width ratio exceeding that for full effectiveness ($w/t = 63.3/\sqrt{F_y}$ for unstiffened elements and $171/\sqrt{f}$ for stiffened elements), $Q = 1.0$ and may be disregarded in Eq. (10-13a).

For treatment of open cross sections that may be subject to torsional-flexural buckling, refer to the AISI Specification for the Design of Cold-formed Steel Structural Members.

For axially loaded bracing and secondary members, the allowable stress, when $L/r > 120$, for concentric loading may be computed from

$$\frac{P}{A} \leq F_{as} = \frac{F_{a1}}{1.3 - L/400r} \qquad (10\text{-}13c)$$

The slenderness ratio L/r should not exceed 200, but during construction a value of 300 may be allowed.

10-13. Combined Axial and Bending Stresses. Combined axial and bending stresses in cold-formed sections can be handled in exactly the same way as in structural steel (Art. 9-13). The interaction criterion to be used is given in the AISI Specification for the Design of Cold-formed Steel Structural Members.

10-14. Welding of Cold-formed Steel.   Welding offers important advantages to fabricators and erectors in joining metal structural components.   Welded joints make possible continuous structures, with economy and speed in fabrication; 100% joint efficiencies are possible.

Conversion to welding of joints initially designed for mechanical fasteners is poor practice. Joints should be specifically designed for welding, to take full advantage of possible savings. Important considerations include the following: The overall assembly should be weldable, welds should be located so that notch effects are minimized, the final appearance of the structure should not suffer from unsightly welds, and welding should not be expected to correct poor fit-up.

Steels bearing protective coatings require special consideration.   Surfaces precoated with paint or plastic are usually damaged by welding.   And coatings may adversely affect weld quality. Metallically coated steels, such as galvanized (zinc-coated), aluminized, and terne-coated (lead-tin alloy), are now successfully welded using procedures tailored for the steel and its coating.

Generally, steel to be welded should be clean and free of oil, grease, paints, scale, etc.   Paint should be applied only after the welding operation.

("Welding Handbook," American Welding Society, 2501 N.W. 7th St., Miami, Fla. 33125; O. W. Blodgett, "Design of Weldments," James F. Lincoln Arc Welding Foundation, Cleveland, Ohio 44117.)

10-15. Arc Welding of Cold-formed Steel.   (See also Art. 10-14.)   Arc welding may be done in the shop and in the field.   Factors favoring arc welding are portability and versatility of equipment and freedom in joint design.   Only one side of a joint need be accessible, and overlap of parts is not required if joint fit-up is good.

Distortion is a problem with light-gage steel weldments.   It can be minimized by avoiding overwelding.   Weld sizes should be matched to service requirements.

Always design joints to minimize shrinking, warping, and twisting.   Jigs and fixtures for holding light-gage work during welding should be used to control distortion.   Tooling designed to serve as heat sinks and to provide restraining forces minimizes distortion.   Directions and amounts of distortion can be predicted and sometimes counteracted by preangling of the parts. Discrete selection of welding sequence can also be used to control distortion.

Groove welds (made by butting the sheet edges together) can be designed for 100% joint efficiency.   Calculation of design stress is usually unnecessary if the weld penetrates 100% of the section.

Stresses in fillet welds should be considered as shear on the throat for any direction of the applied stress.   The dimension of the throat is calculated as 0.707 times the length of the shorter leg of the weld.   For example, a 12-in.-long, ¼-in. fillet weld has a leg dimension of ¼ in., a throat of 0.177 in., and an equivalent area of 2.12 sq in.   For all grades of steel, fillet and plug welds should be proportioned so that the unit stresses do not exceed 13,600 psi in shear on the throat.

Shielded-metal arc welding, also called manual stick electrode, is the most common arc welding process because of its versatility, but it calls for skilled operators.   The welds can be made in any position.   Vertical and overhead welding should be avoided when possible.

Gas-metal arc welding uses special equipment to feed a continuous spool of bare or flux-cored wire into the arc.   A shielding gas such as argon or carbon dioxide is used to protect the arc zone from the contaminating effects of the atmosphere.   The process is relatively fast, and close control can be maintained over the deposit.   The process is not applicable to materials below $\frac{1}{32}$ in. thick but is extensively used for thicker steels.

Gas-tungsten arc welding operates by maintaining an arc between a nonconsumable tungsten electrode and the work.   Filler metal may or may not be added.   Close control over the weld can be maintained.   This process is not widely used for high-production fabrication, except in specialized applications, because of higher cost.

One form of *arc spot welding* is an adaption of gas-metal arc welding wherein a special welding torch and automatic timer are employed.   The welding torch is positioned on the work and a weld is deposited by burning through the top component of the lap joint.   The filler wire provides sufficient metal to fill the hole, thereby fusing the two parts together.   Access to only one side of the joint is necessary.   Field welding by unskilled operators often makes this process desirable.

Another form of arc spot welding utilizes gas-tungsten arc welding.   The heat of the arc melts a spot through one of the sheets and partly through the second.   When the arc is cut off, the pieces fuse.   No filler metal is added.

10-16. Resistance Welding of Cold-formed Steel.   Because of the size of the equipment required, resistance welding is essentially a shop process.   Speed and low cost are factors favoring its selection.

Almost all resistance-welding processes require a lap-type joint. The amount of contacting overlap varies from ⅜ to 1 in., depending on sheet thickness. Access to both sides of the joint is normally required. Adequate clearance for electrodes and welder arms must be provided.

Spot welding is the most common resistance-welding process. The work is held under pressure between two electrodes through which an electric current passes. A weld is formed at the interface between the pieces being joined and consists of a cast-steel nugget. The nugget has a diameter about equal to that of the electrode face and should penetrate about 60 to 80% of each sheet thickness.

In structural design, spot welds can be treated the same way as rivets, but no reduction in net section for holes need be made. Table 10-1 covers some pertinent details for spot-weld design.

Table 10-1. Design Data for Spot and Projection Welding

Thickness t of thinnest outside piece, in.	Min OD of electrode, D, in.	Min contacting overlap, in.	Min weld spacing c to c, in.	Approx dia of fused zone, in.	Min shear strength per weld, lb	Dia of projection, D, in.
SPOT WELDING						
0.021	⅜	⁷/₁₆	⅜	0.13	320	
0.031	⅜	⁷/₁₆	½	0.16	570	
0.040	½	½	¾	0.19	920	
0.050	½	⁹/₁₆	⅞	0.22	1,350	
0.062	½	⅝	1	0.25	1,850	
0.078	⅝	¹¹/₁₆	1¼	0.29	2,700	
0.094	⅝	¾	1½	0.31	3,450	
0.109	⅝	¹³/₁₆	1⅝	0.32	4,150	
0.125	⅞	⅞	1¾	0.33	5,000	
PROJECTION WELDING						
0.125		¹¹/₁₆	⁹/₁₆	0.338	4,800	0.281
0.140		¾	⅝	⁷/₁₆	6,000	0.312
0.156		¹³/₁₆	¹¹/₁₆	½	7,500	0.343
0.171		⅞	¾	⁹/₁₆	8,500	0.375
0.187		¹⁵/₁₆	¹³/₁₆	⁹/₁₆	10,000	0.406

Projection welding is a form of spot welding in which a projection or protuberance is preformed on one of the mating parts. When the parts are brought together in welding, current flows and controls weld formation. Projection welding may be considered for resistance welding of material over ⅛ in. thick. (For maximum spacing of spot welds and fasteners with similar action, such as screws and rivets, consult the AISI Specification for the Design of Cold-formed Steel Structural Members.

("Resistance Welding Manual," Resistance Welder Manufacturer's Association.)

10-17. Bolting Cold-formed Members. Bolting is convenient for field connections in cold-formed steel construction. The distance, in., between bolt centers and from bolt center to edge of sheet, in the line of stress, should not be less than 1½ times the bolt diameter d, in., or less than

$$S = \frac{P}{0.6 F_y t} \tag{10-15}$$

where P = force transmitted by bolt, kips
t = thickness of thinnest connected sheet, in.
F_y = yield strength of steel being connected, ksi.

The designer should set a limit of $2.1 F_y$ for the bearing stress. Maximum allowable tension stress on the net section of a sheet should not exceed

$$F_t = \left(1.0 - 0.9r + 3\frac{rd}{s}\right) 0.6 F_y \tag{10-16}$$

where r = force transmitted by bolts at section considered divided by tension force in member at that section. If r is less than 0.2, it may be taken equal to zero

s = spacing of bolts perpendicular to line of stress, in. For a single bolt, s = width of sheet

d = diameter of bolt, in.

10-18. Tapping Screws for Joining Light-gage Members. Tapping screws are often used for making field joints in light-gage construction, especially in connections which do not carry any calculated gravity load. Such screws are of several types (Fig. 10-8). Tapping screws used for fastening sheet-metal siding and roofing are generally preassembled with Neoprene washers for effective control of leaks, squeaks, cracks, or crazing, depending on the surface of the material. For best results, when Type A sheet-metal screws are specified, screws should be fully threaded to the head to assure maximum hold in sheet metal.

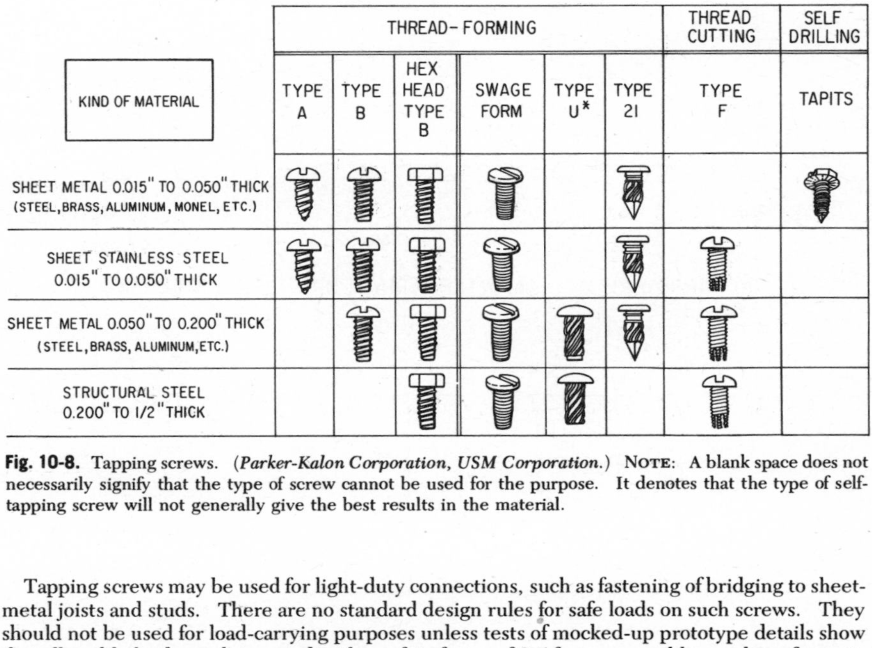

KIND OF MATERIAL	THREAD-FORMING						THREAD CUTTING	SELF DRILLING
	TYPE A	TYPE B	HEX HEAD TYPE B	SWAGE FORM	TYPE U*	TYPE 21	TYPE F	TAPITS
SHEET METAL 0.015" TO 0.050" THICK (STEEL, BRASS, ALUMINUM, MONEL, ETC.)	✦	✦	✦	✦		✦		✦
SHEET STAINLESS STEEL 0.015" TO 0.050" THICK	✦	✦	✦	✦		✦	✦	
SHEET METAL 0.050" TO 0.200" THICK (STEEL, BRASS, ALUMINUM, ETC.)		✦	✦	✦	✦		✦	
STRUCTURAL STEEL 0.200" TO 1/2" THICK			✦	✦	✦		✦	

Fig. 10-8. Tapping screws. (*Parker-Kalon Corporation, USM Corporation.*) NOTE: A blank space does not necessarily signify that the type of screw cannot be used for the purpose. It denotes that the type of self-tapping screw will not generally give the best results in the material.

Tapping screws may be used for light-duty connections, such as fastening of bridging to sheet-metal joists and studs. There are no standard design rules for safe loads on such screws. They should not be used for load-carrying purposes unless tests of mocked-up prototype details show that allowable loads can be carried with a safety factor of 2.5 for a reasonable number of repetitions when repeated or reversed loads are expected. Otherwise, tapping-screw manufacturers' recommendations should be followed explicitly.

STEEL ROOF DECK

Steel roof deck consists of ribbed sheets with nesting or upstanding-seam side joints designed for the support of roof loads between purlins or frames. A typical roof-deck assembly is shown in Fig. 10-9. The Steel Deck Institute, Westchester, Ill. 60153, has developed much useful information on steel roof deck.

10-19. Types of Steel Roof Deck. As a result of the Steel Deck Institute's efforts to improve standardization, steel roof deck can be classified as narrow-rib, intermediate-rib, and wide-rib. All types consist of long, narrow sections with longitudinal ribs at least 1½ in. deep spaced about 6 in. on centers. Other rib dimensions are shown in Fig. 10-10a, b, and c for standard styles. Such steel roof deck is commonly available in 24- and 30-in. covering widths, but sometimes in 18- and 36-in. widths, depending on the manufacturer. Figure 10-10d and e shows full-width executions in cross section. Usual spans, which may be simple, two-span continuous, or three-

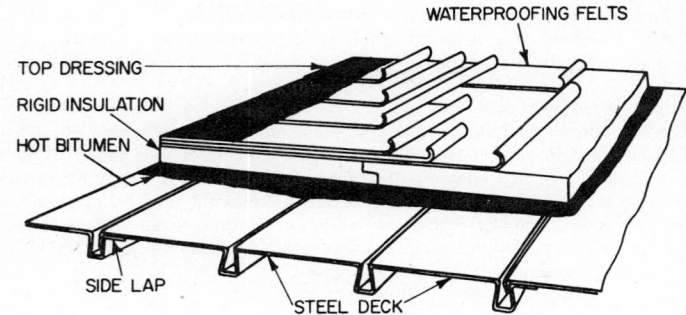

Fig. 10-9. Roof-deck assembly.

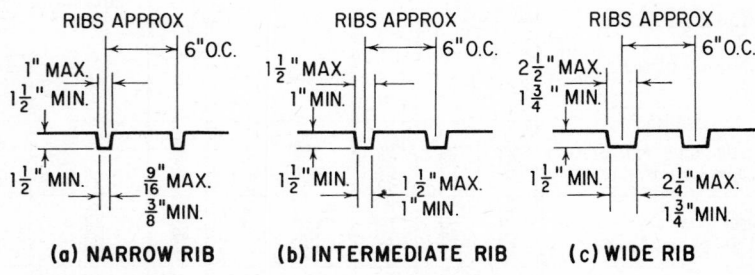

(a) NARROW RIB   **(b) INTERMEDIATE RIB**   **(c) WIDE RIB**

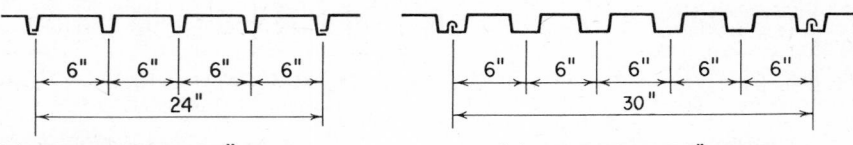

(d) NARROW RIB IN 24" WIDTH WITH NESTED SIDE LAPS   **(e) WIDE RIB IN 30" WIDTH WITH UPSTANDING SEAMS**

Fig. 10-10. Typical ribbed-steel roof-deck sections.

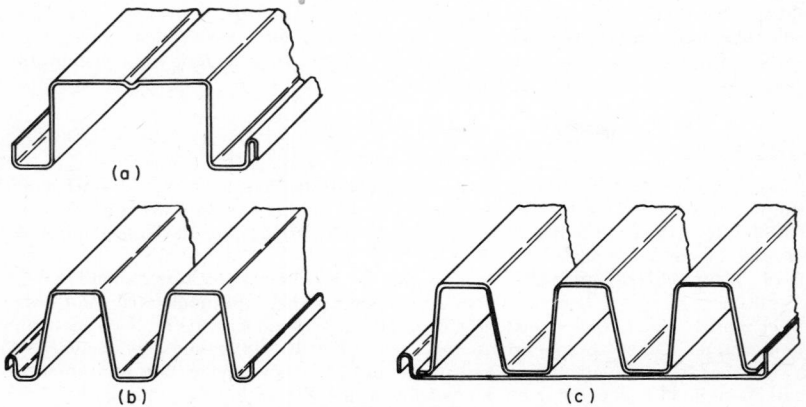

Fig. 10-11. Long-span roof-deck sections.   (*H. H. Robertson Co.*)

span continuous, range from 5 to 10 ft.   The SDI Steel Roof Deck Design Manual gives loading values for various gages, spans, and rib widths.

Some manufacturers make special long-span roof-deck sections (Fig. 10-11).   Cellular floor constructions (Arts. 10-24 and 10-25) may also be used for roofs.

The weight of the steel roof deck shown in Fig. 10-10 varies, depending on rib dimensions and edge details.   For structural design purposes, weights of 3.3, 2.5, and 2.1 psf can be used for 18, 20, and 22 gage (0.048-, 0.036-, and 0.030-in. black steel, respectively) as commonly supplied.

Steel roof deck is usually made of structural-quality sheet or strip, either black or galvanized, ASTM A611, Grade C, or A446, Grade A, respectively.   Both steels have specified minimum yield strengths of 33 ksi.   Black steel is given a shop coat of priming paint by the roof deck manufacturer.   Galvanized steel may or may not be painted; if painted, it should first be bonderized.

The gages of steel commonly used for roof deck are 18 and 20, although most building codes also permit 22-gage steel to be used.   The deep long-span sections of Fig. 10-11 are furnished in heavier gages of black steel, ranging from 18 to 12 (0.048 to 0.105 in. thick).

SDI has made available Recommendations for Site Storage and Erection and also provides standard details for accessories.

10-20. Load-carrying Capacity of Steel Roof Deck.   The Steel Deck Institute has adopted a set of basic design specifications, with limits on rib dimensions, as shown in Fig. 10-10a, b, and c, to foster standardization of steel roof deck.   This also has made possible publication by SDI of allowable uniform loading tables.   These tables are based on section moduli and moments of inertia computed with effective-width procedures stipulated in the AISI Specification for the Design of Cold-formed Steel Structural Members (Art. 10-7).   SDI has banned compression-flange widths otherwise assumed to be effective and also the use of testing to determine vertical load-carrying capacity of steel roof deck.   SDI Basic Design Specifications contain the following provisions:

Moment and Deflection Coefficients.   Where steel roof decks are welded to the supports, a moment coefficient of $1/10$ (applied to WL) shall be used for three or more spans, and a deflection coefficient of $3/384$ (applied to WL^3/EI) shall be used for all except simple spans.   All other steel roof-deck installations shall be designed as simple spans, for which moment and deflection coefficients are $\frac{1}{8}$ and $5/384$, respectively.

Maximum Deflections.   The deflection under live load shall not exceed $1/240$ of the clear span, center to center of supports.   (Suspended ceiling, lighting fixtures, ducts, or other utilities shall not be supported by the roof deck.)

Anchorage.   Steel roof deck shall be anchored to the supporting framework to resist the following gross uplifts:

    45 psf for eave overhang
    30 psf for all other roof areas

The dead load of the roof-deck construction may be deducted from the above uplift forces.

Diaphragm Action.   Steel deck when properly attached to a structural frame becomes a diaphragm capable of resisting in-plane shear forces.   A major SDI steel-deck diaphragm testing program at West Virginia University has led to tentative shear design recommendations that are available on request from SDI.

10-21. Details and Accessories for Steel Roof Deck.   In addition to the use of nesting or upstanding seams, most roof-deck sections are designed so that ends can be lapped shingle fashion.

Special ridge, valley, eave, and cant strips are provided by the roof-deck manufacturers.

Roof decks are commonly arc welded to structural steel with puddle welds at least ½ in. in diameter or with elongated welds of equal perimeter.   Electrodes should be selected and amperage adjusted to fuse all layers of roof deck to steel supporting members without creating blowholes around the welds.

One-inch-long fillet welds should be used to connect lapped edges of roof deck.

Tapping screws are an alternative means of attaching steel roof deck to structural support members, which should be at least $1/16$ in. thick.   All edge ribs and a sufficient number of interior ribs should be connected to supporting frame members at intervals not exceeding 18 in.   When standard steel roof-deck spans are 5 ft or more, adjacent sheets should be fastened together at midspan with either welds or screws.   Details to be used depend on job circumstances and manufacturer's recommendations.

10-22. Insulation of Roof Deck.   Although insulation is not ordinarily supplied by the roof-deck manufacturer, it is standard practice to install ¾- or 1-in.-thick mineral fiberboard between roof deck and roofing.   The SDI Steel Roof Deck Design Manual further recommends:

All steel decks shall be covered with a material of sufficient insulating value as to prevent condensation under normal occupancy conditions. Insulation shall be adequately attached to the steel deck by means of adhesives or mechanical fasteners. Insulation materials shall be protected from the elements at all times during storage and installation.

10-23. Fire Resistance of Roof Deck. The SDI Steel Roof Deck Design Manual contains fire-resistance ratings for steel roof deck construction of up to 2 hr. (See Table 10-2.)

Table 10-2. Fire-Resistance Ratings for Steel Roof Deck Construction

Roof materials	Type of ceiling	U.L. design designation	Fire-resistance rating, hr
MAXIMUM OF 2-HOUR RATING WITH DIRECT APPLIED INSULATION*			
Roof covering, roof insulation, adhesive, vapor barrier, steel roof deck units	2¼-in.-thick sprayed insulation under roof deck	P801	2
	1½-in.-thick sprayed insulation	P703	1½
	1-in.-thick sprayed insulation under roof deck	P701	1
MAXIMUM OF 2-HOUR RATING WITH METAL LATH AND PLASTERED CEILING			
Steel roof deck, steel joist, 6 ft c to c. maximum; insulation, 1 in. minimum; U.L.-listed mineral fiberboard	Suspended ceiling, ⅞ in. lightweight aggregate, gypsum plaster on metal lath	RC1	2
MAXIMUM OF 1-HOUR RATING WITH SUSPENDED ACOUSTICAL CEILING			
Steel roof deck, 1½ in. deep minimum; steel joist, 7 ft. c to c. maximum; insulation, ¾ in. U.L.-listed mineral fiberboard	Suspended ceiling, —⅝ in., U.L.-listed acoustical lay-in boards, and U.L.-listed ceiling grid	RC7	1

*Use 1¼-in.-thick sprayed insulation when in contact with support beams, or ⅞-in.-thick sprayed insulation for support-beam cages.

CELLULAR STEEL FLOOR AND ROOF PANELS

A number of different designs of cellular steel panels for floor and roof construction are on the market. Sections of some of these panels are illustrated in Fig. 10-12. One form of cellular steel floor assembly showing concrete fill, utilities, and ceiling is illustrated in Fig. 10-13.

One of the principal advantages claimed for cellular-type construction is that space is automatically provided to accommodate wiring. Particular emphasis is laid on the electrical-raceway capacity such floors provide, and on the ease with which changes in electrical services can be made. Such panels can also be provided with acoustic metal ceilings (Fig. 10-14).

10-24. Structural Design and Materials for Cellular Decking. Cellular floor and roof steel panels usually are made of 18-gage or heavier steel complying with requirements of Grade C of American Society for Testing and Materials specifications for structural-quality sheet or strip steel. They may be either galvanized or painted.

Structural design of sheet-metal floor and roof panels is usually based on the design principles of Arts. 10-4 to 10-11.

Details of design and installation vary with types of panels and manufacturers. In any particular instance, reference should be made to the manufacturer's recommendations.

10-25. Fire Resistance of Cellular Steel Decking. Any desired degree of fire protection for cellular steel floor and roof assemblies can be obtained with concrete toppings and plaster ceilings or direct-application compounds (sprayed-on fireproofing). Fire-resistance ratings for a considerable number of assemblies are available. (See "Fire-resistant Construction in Modern Steel-framed Buildings," American Institute of Steel Construction; and publications of the

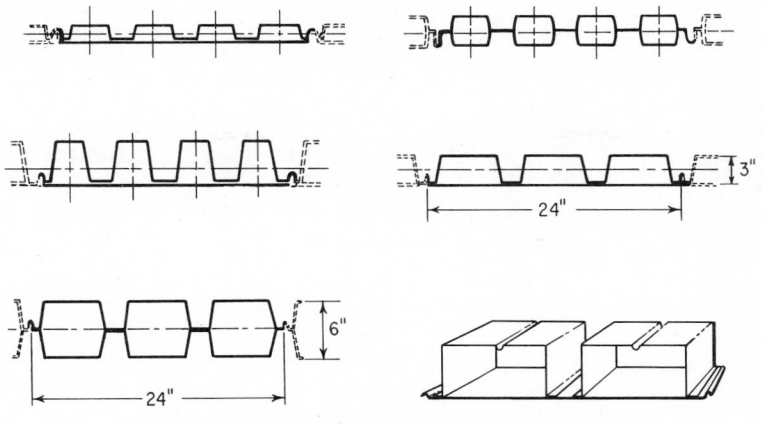

Fig. 10-12. Cellular steel floor sections. (*H. H. Robertson Co.*)

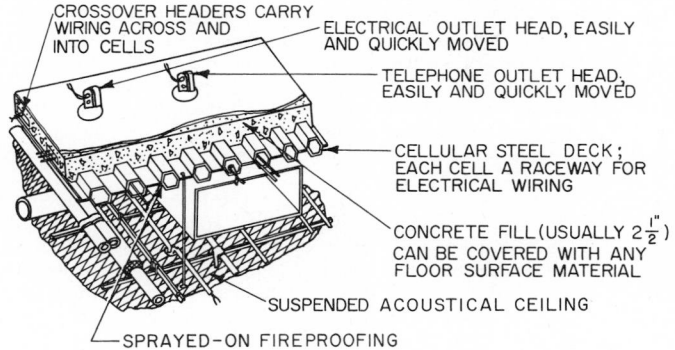

CROSSOVER HEADERS CARRY WIRING ACROSS AND INTO CELLS

ELECTRICAL OUTLET HEAD, EASILY AND QUICKLY MOVED

TELEPHONE OUTLET HEAD, EASILY AND QUICKLY MOVED

CELLULAR STEEL DECK; EACH CELL A RACEWAY FOR ELECTRICAL WIRING

CONCRETE FILL (USUALLY $2\frac{1}{2}''$) CAN BE COVERED WITH ANY FLOOR SURFACE MATERIAL

SUSPENDED ACOUSTICAL CEILING

SPRAYED-ON FIREPROOFING

Fig. 10-13. Cellular steel floor assembly. (*H. H. Robertson Co.*)

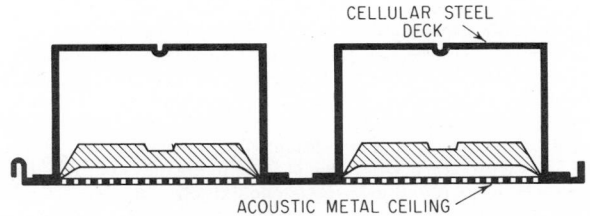

CELLULAR STEEL DECK

ACOUSTIC METAL CEILING

Fig. 10-14. Cellular steel floor and roof panel with acoustic ceiling. (*H. H. Robertson Co.*)

Underwriters' Laboratories, Inc., and other fire-testing agencies and of manufacturers of steel floor and roof units.)

OPEN-WEB STEEL JOISTS

As defined by the Steel Joist Institute, open-web steel joists are load-carrying members suitable for the direct support of floors and roof decks in buildings when these members are designed in accordance with SJI specifications and standard load tables. Those used on spans up to about 60 ft are commonly known as open-web steel joists. Those used on spans of about 25 to 144 ft or more are commonly known as long-span joists. The latter are essentially structural-steel trusses, designed for heavier duty than the open-web joists.

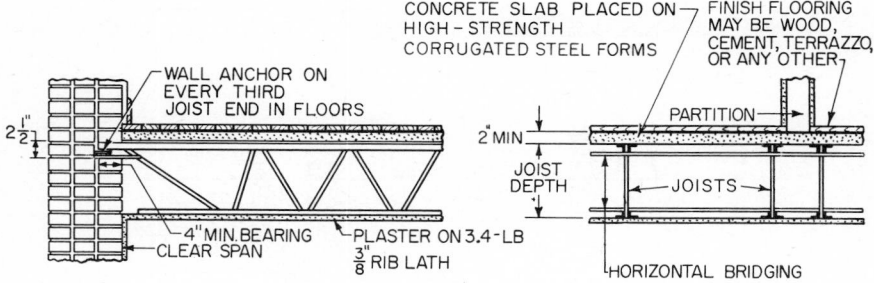

Fig. 10-15. Open-web steel-joist construction.

As usually employed in floor construction, open-web steel joists support on top a slab of concrete, 2 to 2½ in. thick, placed on permanent forms (Fig. 10-15). In addition to light weight, one of the advantages claimed for open-web steel-joist construction is that the open-web system provides space for electrical work, ducts, and piping.

10-26. Joist Fabrication. Standardization under the specifications of the Steel Joist Institute consists of definition of product; specification of materials, design stresses, manufacturing features, accessories, and installation procedures; and handling and erection techniques. Most manufacturers have made uniform certain details, such as end depths, which are desirably standardized for interchangeability. Exact forms of the members, configuration of web systems, and methods of manufacture are matters for the individual manufacturers of these joists. A number of proprietary designs have been developed.

Open-web joists are manufactured in standard depths from 8 to 30 in. in increments of 2 in. and in different weights.

There are two groups, the J series and the H series. The J series is designed with a basic allowable stress of 22,000 psi, contemplating material with a specified minimum yield point of 36,000 psi. The H series is designed with higher allowable stresses, for either high-strength, hot-rolled steel or cold-worked sections that utilize an increase in base-material yield point. A basic allowable stress of 30,000 psi is used for steel with a specified minimum yield point of 50,000 psi.

The safe load capacities of each group are listed in SJI Standard Load Tables. These are available from the Steel Joist Institute, 2001 Jefferson Davis Highway, Arlington, Va. 22202, or the American Institute of Steel Construction.

Open-web steel joists are different in one important respect from fabricated structural-steel framing members commonly used in building construction: The joists usually are manufactured by production-line methods with special equipment designed to produce a uniform product. Component parts generally are joined by either resistance or electric-arc welding. Various joist designs are shown in Fig. 10-16.

10-27. Design of Open-Web Joist Floors. Open-web joists are designed primarily for use under uniformly distributed loading and at substantially uniform spacing. But they can safely carry concentrated loads if proper consideration is given to the effect of such loads. Good practice requires that heavy concentrated loads be applied at joist panel points. The weight of a partition running crosswise to the joists usually is considered satisfactorily distributed by the floor slab and is assumed not to cause local bending in the top chords of the joists. Even so, joists must be selected to resist the bending moments, shears, and end reactions due to such loads.

The method of selecting joist sizes for any floor depends on whether or not the effect of any cross partitions or other concentrated loads must be considered. Under uniform loading only, joist sizes and spacings are most conveniently selected from a table of safe loads. Where concentrated or nonuniform loads exist, calculate bending moments, end reactions, and shears, and select joists accordingly.

The chord section and web details are different for different joist designs made by different manufacturers. Information relating to the size and properties of the members may be obtained from manufacturers' catalogs.

Open-web steel-joist specifications require that for joists used in floors, the clear span should not exceed 20 times the depth of the joist. For roofs, spans up to 24 times the depth of the joist are permitted.

("Standard Specifications and Load Tables," Steel Joist Institute, 2001 Jefferson Davis Highway, Arlington, Va. 22202.)

10-28. Construction Details for Open-Web Steel Joists.   It is essential that bridging be installed between joists as soon as possible after the joists have been placed and before application of construction loads.   The most commonly used type of bridging is a continuous horizontal bracing composed of rods fastened to the top and bottom chords of the joists.   The attachment of the floor or roof deck must provide lateral support for design loads.

It is important that masonry anchors be used on wall-bearing joists.   Where the joists rest on steel beams, the joists should be welded, bolted, or clipped to the beams.

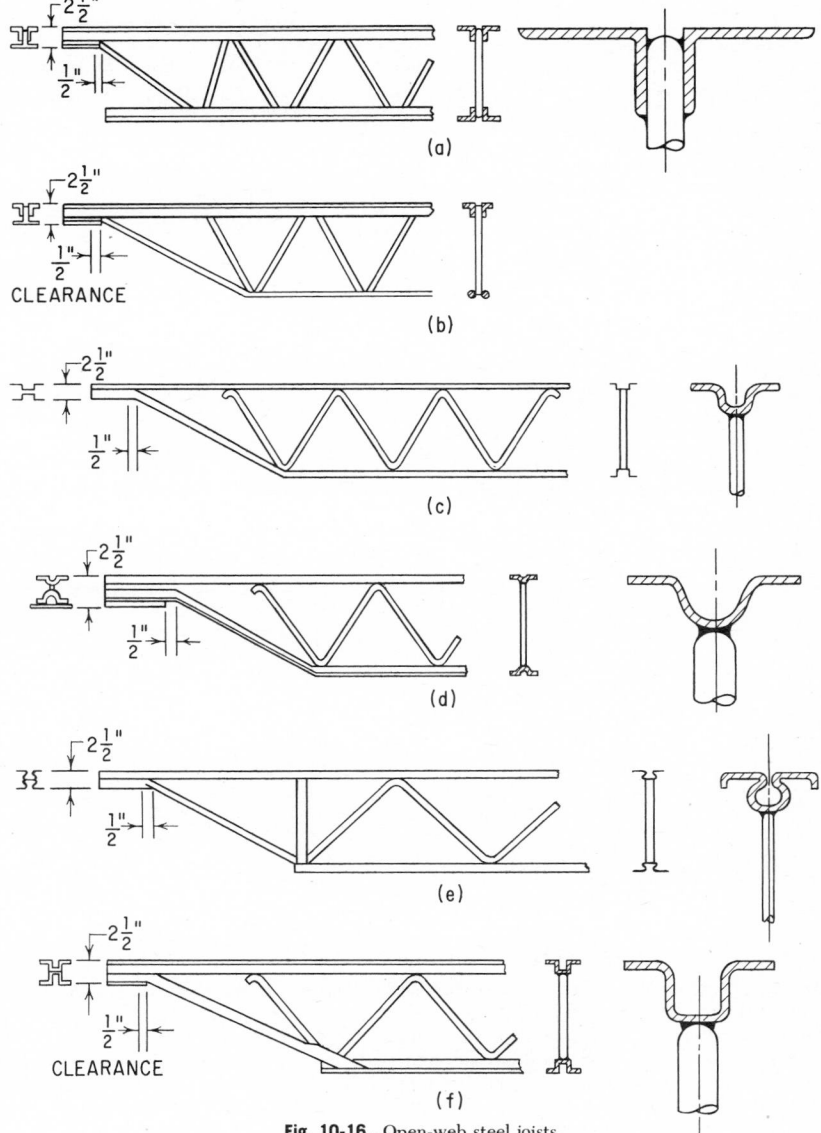

Fig. 10-16. Open-web steel joists.

Forms for the concrete slab usually consist of corrugated steel sheets. Expanded-metal rib lath or paper-backed, welded-wire fabric sometimes is used. Corrugated sheets can be fastened with self-tapping screws or welded to the joists, with a bent washer to reinforce the weld and anchor the slab.

When the usual cast-in-place concrete floor slab is used, it is customary to install reinforcing bars in two perpendicular directions or welded-wire fabric. No other reinforcement is usually considered necessary.

Table 10-3. Fire-Resistance Ratings with Steel Joists*

FLOOR ASSEMBLIES (LISTED)

1- or 1½-hr Fire Resistance
  2-in. reinforced concrete slab, ½-in. for 1 hr and ⅝-in. for 1½ hr rating acoustical tile ceiling, concealed ceiling grid†
  2-in. reinforced concrete slab, ½-in. acoustical board ceiling, listed exposed ceiling grid†
  2-in. reinforced concrete slab, ½-in. gypsumboard ceiling‡
2-Hr Fire Resistance
  2½-in. reinforced concrete slab, ⅝-in. acoustical tile ceiling, listed concealed ceiling grid†
  2½-in. reinforced concrete slab, ½-in. acoustical board ceiling, listed exposed ceiling grid†
  2-in. reinforced concrete slab, ⅝-in. gypsumboard ceiling‡
  2½-in. reinforced concrete slab, ½-in. gypsumboard ceiling‡
3-Hr Fire Resistance
  2½-in. reinforced concrete slab, ¾-in. acoustical ceiling tiles, concealed ceiling grid†
  2½-in. reinforced concrete slab, ⅝-in. gypsumboard ceiling‡
4-Hr Fire Resistance
  2½-in. reinforced concrete slab, ¾-in. plaster-on-metal-lath ceiling‡

ROOF ASSEMBLIES

1-Hr Fire Resistance
  Built-up roofing on 2-in. structural wood-fiber units on steel joists, listed ¾-in. acoustical ceiling tiles in concealed ceiling grid†
Built-up roofing and insulation on 0.018-in.-thick or heavier steel deck on steel joists, listed ⅝-in. acoustical ceiling boards, listed exposed ceiling grid†
Built-up roofing over 2-in. vermiculite concrete on forms supported on joists, listed ½-in. acoustical ceiling boards, listed exposed ceiling grid†
2-Hr Fire Resistance
  Built-up roofing on 2-in. listed gypsum building units on steel joists, listed ⅝-in. acoustical ceiling boards, listed exposed ceiling grid†
Built-up roofing on 0.030-in.-thick or heavier steel deck on steel joists, suspended ⅞-in. metal lath and plaster ceiling

---

*Recommended by Steel Joist Institute. Listed means Underwriters' Laboratories listed or Factory Mutual approved, as appropriate.
†Suspended from joists.
‡Fastened to joists.

Almost any desired degree of fire resistance can be obtained in steel-joist construction with proper protection. The ratings shown in Table 10-3 were given in the 1975 edition of Standard Specifications and Load Tables for Open-Web Steel Joists, Steel Joist Institute, 2001 Jefferson Davis Highway, Arlington, Va. 22202. See also SJI Technical Digest No. 4, "Design of Fire-Resistive Assemblies with Steel Joists."

PRE-ENGINEERED STEEL BUILDINGS

10-29. Characteristics of Pre-engineered Steel Buildings. These structures may be selected from a catalog fully designed and supplied with all structural and covering material, with all components and fasteners. Such buildings eliminate the need for engineers and architects to design and detail both the structure and the required accessories and openings, as would be done for conventional buildings with components from many individual suppliers. Available with floor area of up to 1,000,000 sq ft, pre-engineered buildings readily meet requirements for single-story structures, especially for industrial plants and commercial buildings (Fig. 10-17).

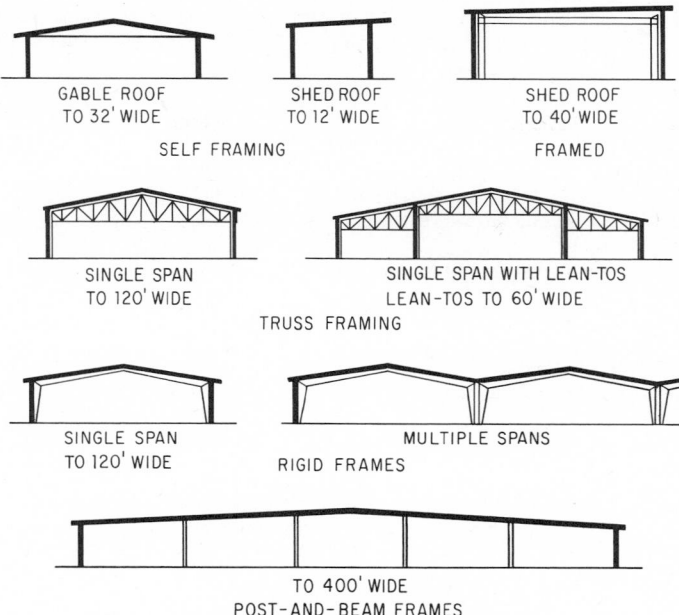

Fig. 10-17. Principal framing systems for pre-engineered buildings.

Pre-engineered buildings may be provided with custom architectural accents. Exterior wall panels are available with durable factory-applied colors.

Pre-engineered buildings make extensive use of cold-formed structural members. These parts lend themselves to mass production, and their design can be more accurately fitted to the specific structural requirement. For instance, a roof purlin can be designed with the depth, moment of inertia, section modulus, and gage required to carry the load, as opposed to picking the next higher size standard hot-rolled shape, with more weight than required. Also, because this purlin is used on thousands of buildings, the quantity justifies investment in automated equipment for forming and punching. This equipment is flexible enough to permit a change of gage or depth of section to produce similar purlins for other loadings.

The engineers designing a line of pre-engineered buildings can, because of the repeated use of the design, justify spending additional design time in refining and optimizing the design. Most pre-engineered buildings are designed with the aid of electronic computers. Their programs are specifically tailored for the product. A rerun of a problem to eliminate a few pounds of steel is justified, since the design will be reused many times during the life of that model.

10-30. Structural Design of Pre-engineered Buildings. The buildings are designed for loading criteria in such a way that any building may be specified to meet the geographical requirements of any location. Combinations of dead load and wind load conform with requirements of several model building codes (Uniform Building Code, Southern Standard Building Code, and Basic Building Code).

The Metal Building Manufacturers Association, an organization of companies producing pre-engineered metal buildings, has established standards of design ("Recommended Design Practices Manual," MBMA, 2130 Keith Building, Cleveland, Ohio 44115). These standards discuss methods of load application and minimum loadings where load requirements are not established by local codes.

Because of the wide variation and lack of definition in some building codes, load application is well covered in the MBMA design manual. It follows the criteria in the U.S. Navy Technical Publication Navdocks DM-2. Treatment of wind loads is based on an extensive testing and research program (N. Chien, Y. Feng, H. J. Wang, and T. T. Siao, "Wind-tunnel Studies of Pressure Distribution on Elementary Building Forms," State University of Iowa, Iowa City).

This testing, performed in a wind tunnel on small plastic models of different shapes and proportions, led to a rational approach in which pressure on a windward wall was related to the height-to-width ratio of the building.

Wind causes suction on the windward slope of a roof for all height-to-width ratios and roof slopes less than 20°. Some combinations of steeper roof slopes and large height-to-width ratios have positive wind pressure on the windward roof slope.

The leeward roof and the leeward wall always have external suction on them when the wind blows.

Air leakage through doors and windows left open or broken by wind-borne debris may modify the wind forces by increasing the internal pressure from windward openings or internal suction from leeward openings.

The Metal Building Manufacturers Association, in its Recommended Design Practices Manual, recommends:

The following combinations of loads where they apply shall be considered in the design of all members of the structure.
Dead load plus live load.
Dead load plus wind load.
Dead load plus one-half wind load plus live load.
Dead load plus wind load plus one-half live load uniformly distributed over full span.
Dead load plus crane load plus one-half wind load or one-half live load, whichever is critical.
For truss-type buildings, dead load plus wind load plus full live load uniformly distributed over the leeward one-half span.
For truss-type buildings, dead load plus one-half live load distributed over either half of the span.
Dead load plus seismic force. (This combination applies only where seismic design is required.)

Each member should be designed to withstand the stresses resulting from the combinations of loads that produce an actual stress with the maximum percentage of allowable stress in that member.

Allowable stresses for combinations including wind or earthquake load may be increased by 33⅓%, provided the member thus required is not smaller than that required for the combination of dead load, live load, and impact (if any).

Because of the low mass of pre-engineered steel buildings, seismic forces are seldom critical design requirements. Owing to the wide range of requirements for cranes, their design is not included in any standard line of buildings. But the criterion for cranes is used for special designs.

Standard design specifications for pre-engineered steel buildings are:

Structural Steel—Specification for Design, Fabrication, and Erection of Structural Steel for Buildings, American Institute of Steel Construction.

Light-gage Steel—Specification for the Design of Cold-formed Steel Structural Members, American Iron and Steel Institute.

Welding—Structural Welding Code, D1.1, American Welding Society.

STRUCTURAL DESIGN OF CORRUGATED STEEL PIPE

10-31. Corrugated Steel Pipe. Corrugated steel pipe was first developed and used for culvert drainage in 1896. It is now produced in full-round diameters from 6 in. in diameter and 16 gage to 120 in. in diameter and 8 gage. Heights of cover up to 100 ft are permissible with highway or railway loadings.

(a)   (b)

Fig. 10-18. Corrugated steel structures. (*a*) Riveted pipe-arch. (*b*) Helical pipe.

Riveted corrugated pipe (Fig. 10-18a shows pipe-arch shape) is produced by riveting together circular corrugated sheets to form a tube. The corrugations are annular.

Helically corrugated pipe (Fig. 10-18b) is manufactured by spirally forming a continuously corrugated strip into a tube with a locked or welded seam joining abutting edges. This pipe is stronger in ring compression because of the elimination of the longitudinal riveted joints. Also, the seam is more watertight than the lap joints of riveted pipe.

Besides being supplied in full-round shapes, both types of pipe are also available in a pipe-arch shape. This configuration, with a low and wide waterway area in the invert, is beneficial for low headroom conditions. It provides adequate flow capacity without raising the grade.

Corrugated steel pipe and pipe arch is produced with a variety of coatings to resist corrosion and erosion.

The zinc coating provided on these structures is adequate protection under normal drainage conditions with no particular corrosion hazard. Additional coatings or pavings may be specified for placing over the galvanizing.

Asbestos-bonded steel has a coating in which a layer of asbestos fiber is embedded in molten zinc and then saturated with bituminous material. This provides protection for extreme corrosion conditions. Asbestos-bonded steel is available in riveted pipe only. Helical corrugated structures may be protected with a hot-dip coating of bituminous material for severe soil or effluent conditions.

For erosive hazards, a paved invert of bituminous material can be applied to give additional protection to the bottom of the pipe. And for improved flow, these drainage conduits may also be specified with a full interior paving of bituminous material.

Normally, pipe-arch structures are supplied in a choice of span-and-rise combinations that have a periphery equal to that available with full-round corrugated pipe.

10-32. Structural Plate Pipe. To extend the diameter or span-and-rise dimensions of corrugated steel structures beyond that (120 in.) available with factory-fabricated drainage conduits, structural plate pipe and other shapes may be used. These are made of heavier gages of steel and are composed of curved and corrugated steel plates that are bolted together at the installation site. Their shapes include full-round, elliptical, pipe-arch, arch, and horseshoe or underpass shapes. Applications include storm drainage, stream enclosures, vehicular and pedestrain underpasses, and small bridges.

Such structures are field-assembled with curved and corrugated steel plates that may be 10 or 12 ft long (Fig. 10-19). The wall section of the structures has 2-in.-deep corrugations, 6 in. c to c. Thickness ranges from 12 gage (0.1046 in.) to 1 gage (0.2758 in.). Each of the plates is

Fig. 10-19. Structural plate pipe is shown being bolted together at right. Completely assembled structural plate pipe arch is shown at left.

punched for field bolting with special high-strength bolts supplied with each structure. The number of bolts used can be varied to meet the ring-compression stress.

Circular pipes are available in diameters ranging from 5 to 24 ft, with structures of other configurations available in a similar approximate size range. Special end plates can be supplied to fit a skew or bevel, or a combination of both.

Plates of all structures are hot-dip galvanized. They are normally shipped in bundles for handling convenience. Instructions for assembly are also provided.

10-33. Design of Culverts. Formerly, design of corrugated steel structures was based on observations of how such pipes performed structurally under service conditions. From these observations, data were tabulated and gage tables established. As larger pipes were built and installed and experience was gained, these gage tables were revised and enlarged.

The following is the design procedure for corrugated steel structures recommended in the "Handbook of Steel Drainage and Highway Construction Products" (American Iron and Steel Institute, 1000 16th St., N.W., Washington, D.C. 20036).

1. Backfill Density. Select a percent compaction of pipe backfill for design. The value chosen should reflect the importance and size of the structure and the quality that can reasonably be expected. The recommended value for routine use is 85%. This value will usually apply to ordinary installations for which most specifications will call for compaction to 90%. But for more important structures in higher-fill situations, consideration must be given to selecting higher-quality backfill and requiring this quality for construction.

2. Design Pressure. When the height of cover is equal to or greater than the span or diameter of the structure, enter the load-factor chart (Fig. 10-20) to determine the percentage of the total load acting on the steel. For routine use, the 85% soil compaction will provide a load factor $K = 0.86$. The total load is multiplied by K to obtain the design pressure P_v acting on the steel. If the height of cover is less than one pipe diameter, the total load TL is assumed to act on the pipe, and $TL = P_v$; that is,

$$P_v = DL + LL + I \qquad H < S \qquad (10\text{-}17a)$$

When the height of cover is equal to or greater than one pipe diameter,

$$P_v = K(DL + LL + I) \qquad H \geq S \qquad (10\text{-}17b)$$

where P_v = Design pressure, ksf
K = load factor
DL = dead load, ksf
LL = live load, ksf
I = impact, ksf
H = height of cover, ft
S = span or pipe diameter, ft

3. Ring Compression. The compressive thrust C, kips per ft, on the conduit wall equals the radial pressure P_v, ksf, acting on the wall multiplied by the wall radius R, ft, or

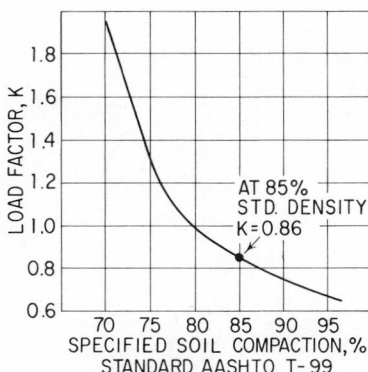

Fig. 10-20. Load factors for corrugated steel pipe are plotted as a function of specified compaction of pipe backfill.

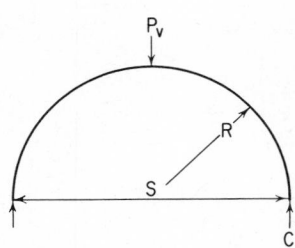

Fig. 10-21. Pressure on curved conduit.

$C = P_v R$. This thrust, called ring compression, is the force carried by the steel. The ring compression is an axial load acting tangentially to the conduit wall (Fig. 10-21). For conventional structures in which the top arc approaches a semicircle, it is convenient to substitute half the span for the wall radius. Then

$$C = P_v \frac{S}{2} \tag{10-18}$$

4. *Allowable Wall Stress.* The ultimate compression in the pipe wall is expressed by Eqs. (10-19). Equation (10-19a) sets the ultimate wall stress equal to the specified minimum yield point of the steel and applies to the zone of wall crushing or yielding. Equation (10-19b) applies to the interaction zone of yielding and ring buckling. And Eq. (10-19c) applies to the ring buckling zone.

When the ratio D/r of pipe diameter or span D, in., to radius of gyration r, in., of the pipe cross section does not exceed 294, the ultimate wall stress may be taken as equal to the steel yield strength:

$$F_b = F_y = 33 \text{ ksi} \tag{10-19a}$$

When D/r exceeds 294 but not 500, the ultimate wall stress, ksi, is given by

$$F_b = 40 - 0.000081 \left(\frac{D}{r}\right)^2 \tag{10-19b}$$

When D/r is more than 500,

$$F_b = \frac{4.93 \times 10^6}{(D/r)^2} \tag{10-19c}$$

A safety factor of 2 is applied to the ultimate wall stress to obtain the design stress F_c, ksi.

$$F_c = \frac{F_b}{2} \tag{10-20}$$

5. *Wall Thickness.* Required wall area A is computed from the calculated compression C in the pipe wall and the allowable stress F_c.

$$A = \frac{C}{F_c} \tag{10-21}$$

From Table 10-4, select the wall thickness that provides the required area with the same corrugation used for selection of the allowable stress.

Table 10-4. Moments of Inertia and Cross-Sectional Areas of Corrugated Steel Sheets and Plates for Underground Conduits*

Corrugation Pitch × Depth, In.	Specified thickness, in.											
	0.034	0.040	0.052	0.064	0.079	0.109	0.138	0.168	0.188	0.218	0.249	0.280
	MOMENT OF INERTIA, I, IN. PER FT OF WIDTH											
1½ × ¼	0.0025	0.0030	0.0041	0.0053	0.0068	0.0103	0.0145	0.0196				
2 × ½	0.0118	0.0137	0.0184	0.0233	0.0295	0.0425	0.0566	0.0719				
2⅔ × ½	0.0112	0.0135	0.0180	0.0227	0.0287	0.0411	0.0544	0.0687				
3 × 1	0.0514	0.0618	0.0827	0.1039	0.1306	0.1855	0.2421	0.3010				
6 × 2						0.725	0.938	1.154	1.296	1.523	1.754	1.990
	CROSS SECTIONAL WALL AREA, SQ IN. PER FT OF WIDTH											
1½ × ¼	0.3801	0.456	0.608	0.761	0.950	1.331	1.712	2.093				
2 × ½	0.4086	0.489	0.652	0.815	1.019	1.428	1.838	2.249				
2⅔ × ½	0.3873	0.465	0.619	0.775	0.968	1.356	1.744	2.133				
3 × 1	0.4445	0.534	0.711	0.890	1.113	1.560	2.008	2.458				
6 × 2						1.556	2.003	2.449	2.739	3.199	3.658	4.119

*Corrugation dimensions are nominal, subject to manufacturing tolerances.

6. Check Handling Stiffness. Minimum pipe stiffness requirements for practical handling and installation, without undue care or bracing, have been established through experience. The resulting flexibility factor *FF* limits the size of each combination of corrugation pitch and metal thickness.

$$FF = \frac{D^2}{EI} \tag{10-22}$$

where E = modulus of elasticity, ksi, of the steel = 30,000 ksi
$\quad\quad I$ = moment of inertia of wall, in.[4] per in.

The following maximum values of *FF* are recommended for ordinary installations:

$FF = 0.0433$ for factory-made pipe less than 120 in. in diameter and made with riveted, welded, or helical seams

$FF = 0.0200$ for field-assembled pipe over 120 in. in diameter or made with bolted seams

Higher values can be used with special care or where experience indicates. Trench condition, as in sewer design, can be one such case. Use of aluminum pipe is another. For example, the flexibility factor permitted for aluminum pipe in some national specifications is more than twice that recommended here for steel. This has come about because aluminum has only one-third the stiffness of steel, the modulus for aluminum being about 10,000 ksi versus 30,000 ksi for steel. Where a high degree of flexibility is acceptable for aluminum, it will be equally acceptable for steel.

7. Check Bolted Seams. Standard factory-made pipe seams are satisfactory for all designs within the maximum allowable wall stress of 16.5 ksi. Seams bolted in the shop or field, however, will continue to be evaluated on the basis of test values for uncurved, unsupported columns. A bolted seam (standard for structural plate) must have a test strength of twice the design load in the pipe wall.

Table 10-5. Bolted Seam Design Data

Thickness, in.	Structural plate pipe 6 × 2 in. corrugations (four ¾-in. bolts per ft)		Corrugated steel pipe 3 × 1 in. corrugations (eight ½-in. bolts per ft)	
	Allowable strength (½ ultimate), kips per ft	Corresponding wall stress, ksi	Allowable strength (½ ultimate), kips per ft	Corresponding wall stress, ksi
0.064			14.4	16.2
0.079			17.9	15.8
0.109	21	13.5	26.5	17.0
0.138	31	15.5	31.9	15.9
0.168	40	16.5	35.4	14.4
0.188	46	17.0		
0.218	56	17.5		
0.249	66	18.1		
0.280	72	17.5		

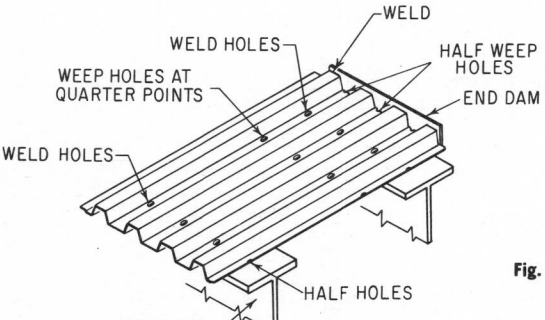

Fig. 10-22. Lightweight-steel bridge plank.

Table 10-5 lists the allowable design values (one-half the ultimate) of bolted joints for 6 × 2 in. and 3 × 1 in. corrugations tested as unsupported short columns. For convenience, the wall stress that corresponds to the allowable joint strength is also shown.

OTHER TYPES OF LIGHTWEIGHT-STEEL CONSTRUCTION

10-34. Lightweight-Steel Bridge Decking. This trapezoidal-corrugated plank, 2 in. deep by 18 or 24 in. wide, welded to steel (Fig. 10-22) or lagged to wood stringers, gives a strong, secure base for a smooth bituminous traffic surface. It may be used for replacement of old wood decks and for new construction.

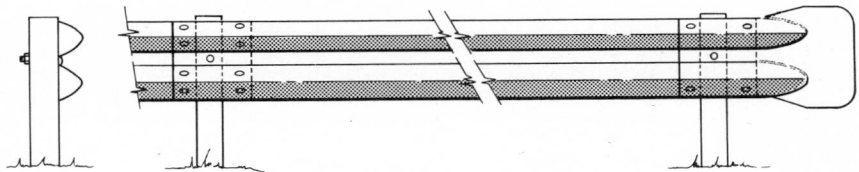

Fig. 10-23. Beam-type guardrail of steel.

10-35. Beam-type Guardrail. (See also Art. 16-30.) The beam-type guardrail in Fig. 10-23 has the flexibility necessary to absorb impact as well as the beam strength to prevent pocketing of a car against a post. Standard post spacing is 12½ ft. The rail is anchored with one bolt to each post, and there are eight bolts in the rail splice to assure continuous-beam strength. Available lengths are 12½ and 25 ft. Standard weight is 12-gage steel; heavy-duty is 10 gage. The guardrail is furnished galvanized or as prime-painted steel.

10-36. Bin-type Retaining Wall. A bin-type retaining wall (Fig. 10-24) is a series of closed-face bins, which when backfilled transform the soil mass into an economical retaining wall. The flexibility of steel allows for adjustments due to uneven ground settlement. There are standard designs for these walls with vertical or battered face, heights to 30 ft, and various conditions of surcharge.

Fig. 10-24. Bin-type retaining wall of light-gage steel.

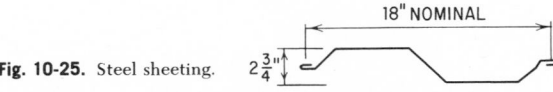

Fig. 10-25. Steel sheeting.

10-37. Lightweight-Steel Sheeting.   Corrugated sheeting has beam strength to support earth pressure on walls of trenches and excavations, and column strength for driving.   The sheeting presents a small end cross section for easy driving (Fig. 10-25).   Physical properties of the sheeting shown in Fig. 10-25 are listed in Table 10-6.

Table 10-6. Physical Properties of Corrugated Steel Sheeting (Fig. 10-25)*

| Thickness | | Cross-sectional area | | Moment of inertia | | Section modulus | | Weight | |
Gage	In.	Sq in. per section	Sq in. per lin ft of wall	In.^{4} per section	In.^{4} per lin ft of wall	In.^{3} per section	In.^{3} per lin ft of wall	Lb per lin ft	Lb per sq ft
12	0.1046	2.43	1.62	2.23	1.49	1.62	1.08	8.48	5.65
10	0.1345	3.13	2.09	2.83	1.92	2.06	1.37	10.90	7.27
8	0.1644	3.82	2.55	3.67	2.44	2.67	1.78	13.32	8.88

*Based on data given in the AISI "Handbook of Steel Drainage and Highway Construction Products."

Wood Design and Construction

MAURICE J. RHUDE

President, Sentinel Structures, Inc., Peshtigo, Wis.

Wood is remarkable for its beauty, versatility, strength, durability, and workability. It possesses a high strength-to-weight ratio. It has flexibility. It performs well at low temperatures. It withstands substantial overloads for short periods. It has low electrical and thermal conductance. It resists the deteriorating action of many chemicals that are extremely corrosive to other building materials. There are few materials that cost less per pound than wood.

As a consequence of its origin, wood as a building material has inherent characteristics with which users should be familiar. For example, although cut simultaneously from trees growing side by side in a forest, two boards of the same species and size most likely do not have the same strength. The task of describing this nonhomogeneous material, with its variable biological nature, is not easy. But it can be described accurately, and much better than was possible in the past. For research has provided much useful information on wood properties and behavior in structures.

Research has shown, for example, that a compression grade cannot be used, without modification, for the tension side of a deep bending member. Also, a bending grade cannot be used, unless modified, for the tension side of a deep bending member, or for a tension member. Experience indicates that typical growth characteristics are more detrimental to tensile strength than to compressive strength. Furthermore, research has made possible better estimates of engineering qualities of wood. No longer is it necessary to use only visual inspection, keyed to averages, for estimating the engineering qualities of a piece of wood. With a better understanding of wood now possible, the availability of sound structural design criteria, and development of economical manufacturing processes, greater and more efficient use is being made of wood for structural purposes.

Improvements in adhesives also have contributed to the betterment of wood construction. In particular, the laminating process, employing adhesives to build up thin boards into deep timbers, improves on nature. Not only are stronger structural members thus made available, but also higher grades of lumber can be placed in regions of greatest stress, and lower grades in regions of lower stress, for overall economy. Despite variations in strength of wood, lumber can be transformed into glued-laminated timbers of predictable strength and with very little variability in strength.

11-1. Basic Characteristics and How to Use Them. (See also Art. 5-31.) Wood differs in several significant ways from other building materials. Its cellular structure is responsible, to a considerable degree, for this. Because of this structure, structural properties depend on orientation. While most structural materials are essentially isotropic, with nearly equal properties in all directions, wood has three principal grain directions—longitudinal, radial, and tangential. (Loading in the longitudinal direction is referred to as parallel to the grain, whereas transverse loading is considered across the grain.) Parallel to the grain, wood possesses high strength and stiffness. Across the grain, strength is much lower. (In tension, wood stressed parallel to the grain is 25 to 40 times stronger than when stressed across the grain. In compression, wood loaded parallel to the grain is 6 to 10 times stronger than when loaded perpendicular to the grain.) Furthermore, a wood member has three moduli of elasticity, with a ratio of largest to smallest as large as 150:1.

Wood undergoes dimensional changes from causes different from those for dimensional changes in most other structural materials. For instance, thermal expansion of wood is so small as to be unimportant in ordinary usage. Significant dimensional changes, however, occur because of gain or loss in moisture. Swelling and shrinkage from this cause vary in the three grain directions; size changes about 6 to 16% tangentially, 3 to 7% radially, but only 0.1 to 0.3% longitudinally.

Wood offers numerous advantages nevertheless in construction applications—beauty, versatility, durability, workability, low cost per pound, high strength-to-weight ratio, good electrical insulation, low thermal conductance, and excellent strength at low temperatures. It is resistant to many chemicals that are highly corrosive to other materials. It has high shock-absorption capacity. It can withstand large overloads of short time duration. It has good wearing qualities, particularly on its end grain. It can be bent easily to sharp curvature. A wide range of finishes can be applied for decorative or protective purposes. Wood can be used in both wet and dry applications. Preservative treatments are available for use when necessary, as are fire retardants. Also, there is a choice of a wide range of species with a wide range of properties.

In addition, a wide variety of wood framing systems is available. The intended use of a structure, geographical location, configuration required, cost, and many other factors determine the framing system to be used for a particular project.

Design Recommendations. The following recommendations aim at achieving economical designs with wood framing:

Use standard sizes and grades of lumber. Consider using standardized structural components, whether lumber, stock glued beams, or complex framing designed for structural adequacy, efficiency, and economy.

Use standard details wherever possible. Avoid specially designed and manufactured connecting hardware.

Use as simple and as few joints as possible. Place splices, when required, in areas of lowest stress. Do not locate splices where bending moments are large, and thus avoid design, erection, and fabrication difficulties.

Avoid unnecessary variations in cross section of members along their length.

Use identical member designs repeatedly throughout a structure, whenever practicable. Keep the number of different arrangements to a minimum.

Consider using roof profiles that favorably influence the type and amount of load on the structure.

Specify allowable design stresses rather than the lumber grade or combination of grades to be used.

Select an adhesive suitable for the service conditions, but do not overspecify. For example, waterproof resin adhesives need not be used where less expensive water-resistant adhesives will do the job.

Use lumber treated with preservatives where service conditions dictate. Such treatment need not be used where decay hazards do not exist. Fire-retardant treatments may be used to meet a specific flame-spread rating for interior finish, but are not necessary for large-cross-section members that are widely spaced and already a low fire risk.

Instead of long, simple spans, consider using continuous or suspended spans or simple spans with overhangs.

Select an appearance grade best suited to the project. Do not specify premium appearance grade for all members if it is not required.

Table 11-1 may be used as a guide to economical span ranges for roof and floor framing in buildings.

Table 11-1. Economical Span Range for Framing Members

Framing member	Economical span range, ft	Usual spacing, ft
Roof beams (generally used where a flat or low-pitched roof is desired):		
Simple span:		
Constant depth		
Solid-sawn	0–40	4–20
Glued-laminated	20–100	8–24
Tapered	25–100	8–24
Double tapered (pitched beams)	25–100	8–24
Curved beams	25–100	8–24
Simple beam with overhangs (usually more economical than simple span when span is over 40 ft):		
Solid-sawn	24	4–20
Glued-laminated	10–90	8–24
Continuous span:		
Solid-sawn	10–50	4–20
Glued-laminated	10–50	8–24
Arches (three-hinged for relatively high-rise applications and two-hinged for relatively low-rise applications):		
Three-hinged:		
Gothic	40–90	8–24
Tudor	30–120	8–24
A-frame	20–160	8–24
Three-centered	40–250	8–24
Parabolic	40–250	8–24
Radial	40–250	8–24
Two-hinged:		
Radial	50–200	8–24
Parabolic	50–200	8–24
Trusses (provide openings for passage of wires, piping, etc.):		
Flat or parallel chord	50–150	12–20
Triangular or pitched	50–90	12–20
Bowstring	50–200	14–24
Tied arches (where no ceiling is desired and where a long, clear span is desired with low rise):		
Tied segment	50–100	8–20
Buttressed segment	50–200	14–24
Domes	50–350	8–24
Simple-span floor beams:		
Solid-sawn	6–20	4–12
Glued-laminated	6–40	4–16
Continuous floor beams	25–40	4–16
Roof sheathing and decking:		
1-in. sheathing	1–4	
2-in. sheathing	6–10	
3-in. roof deck	8–15	
4-in. roof deck	12–20	
Plywood sheathing	1–4	
Sheathing on roof joists	1.33–2	
Plank floor decking (floor and ceiling in one):		
Edge to edge	4–16	
Wide face to wide face	4–16	

11-2. Standard Sizes of Lumber and Timber.   Details regarding dressed sizes of various species of wood are given in the grading rules of agencies that formulate and maintain such rules. Dressed sizes in Table 11-2 are from the American Softwood Lumber Standard, Voluntary Product Standard PS20-70.   These sizes are generally available, but it is good practice to consult suppliers before specifying sizes not commonly used to find out what sizes are on hand or can be readily secured.

Table 11-2. Nominal and Minimum Dressed Sizes of Boards, Dimension, and Timbers

Item	Thickness, in.			Face width, in.		
	Nominal	Minimum dressed		Nominal	Minimum dressed	
		Dry*	Green†		Dry*	Green†
Boards	1	¾	$^{25}/_{32}$	2	1½	$1^9/_{16}$
	1¼	1	$1^1/_{32}$	3	2½	$2^9/_{16}$
	1½	1¼	$1^9/_{16}$	4	3½	$3^9/_{16}$
				5	4½	4⅝
				6	5½	5⅝
				7	6½	6⅝
				8	7¼	7½
				9	8¼	8½
				10	9¼	9½
				11	10¼	10½
				12	11¼	11½
				14	13¼	13½
				16	15¼	15½
Dimension	2	1½	$1^9/_{16}$	2	1½	$1^9/_{16}$
	2½	2	$2^1/_{16}$	3	2½	$2^9/_{16}$
	3	2½	$2^9/_{16}$	4	3½	$3^9/_{16}$
	3½	3	$3^1/_{16}$	5	4½	4⅝
				6	5½	5⅝
				8	7¼	7½
				10	9¼	9½
				12	11¼	11½
				14	13¼	13½
				16	15¼	15½
	4	3½	$3^9/_{16}$	2	1½	$1^9/_{16}$
	4½	4	$4^1/_{16}$	3	2½	$2^9/_{16}$
				4	3½	$3^9/_{16}$
				5	4½	4⅝
				6	5½	5⅝
				8	7¼	7½
				10	9¼	9½
				12	11¼	11½
				14		13½
				16		15½
Timbers	5 and thicker		½ in. less	5 and wider		½ in. less

*Dry lumber is defined as lumber seasoned to a moisture content of 19% or less.
†Green lumber is defined as lumber having a moisture content in excess of 19%.

11-3. Sectional Properties of Lumber and Timber.   Table 11-3 lists properties of sections of solid-sawn lumber and timber.

Table 11-3. Properties of Sections of Solid-Sawn Wood

Nominal size, in.	Standard dressed size, in. (S4S)	Area of section sq in.	Moment of inertia $in.^4$	Section modulus, $in.^3$	Weight, lb per lin ft, of piece when weight of wood, lb per cu ft, equals					
					25	30	35	40	45	50
1 × 3	¾ × 2½	1.875	0.977	0.781	0.326	0.391	0.456	0.521	0.586	0.651
1 × 4	¾ × 3½	2.625	2.680	1.531	0.456	0.547	0.638	0.729	0.820	0.911
1 × 6	¾ × 5½	4.125	10.398	3.781	0.716	0.859	1.003	1.146	1.289	1.432
1 × 8	¾ × 7¼	5.438	23.817	6.570	0.944	1.133	1.322	1.510	1.699	1.888
1 × 10	¾ × 9¼	6.938	49.466	10.695	1.204	1.445	1.686	1.927	2.168	2.409
1 × 12	¾ × 11¼	8.438	88.989	15.820	1.465	1.758	2.051	2.344	2.637	2.930
2 × 3*	1½ × 2½	3.750	1.953	1.563	0.651	0.781	0.911	1.042	1.172	1.302
2 × 4	1½ × 3½	5.250	5.359	3.063	0.911	1.094	1.276	1.458	1.641	1.823
2 × 6	1½ × 5½	8.250	20.797	7.563	1.432	1.719	2.005	2.292	2.578	2.865
2 × 8	1½ × 7¼	10.875	47.635	13.141	1.888	2.266	2.643	3.021	3.398	3.776
2 × 10	1½ × 9¼	13.875	98.932	21.391	2.409	2.891	3.372	3.854	4.336	4.818
2 × 12	1½ × 11¼	16.875	177.979	31.641	2.930	3.516	4.102	4.688	5.273	5.859
2 × 14	1½ × 13¼	19.875	290.775	43.891	3.451	4.141	4.831	5.521	6.211	6.901
3 × 1	2½ × ¾	1.875	0.088	0.234	0.326	0.391	0.456	0.521	0.586	0.651
3 × 2	2½ × 1½	3.750	0.703	0.938	0.651	0.781	0.911	1.042	1.172	1.302
3 × 4	2½ × 3½	8.750	8.932	5.104	1.519	1.823	2.127	2.431	2.734	3.038
3 × 6	2½ × 5½	13.750	34.661	12.604	2.387	2.865	3.342	3.819	4.297	4.774
3 × 8	2½ × 7¼	18.125	79.391	21.901	3.147	3.776	4.405	5.035	5.664	6.293
3 × 10	2½ × 9¼	23.125	164.886	35.651	4.015	4.818	5.621	6.424	7.227	8.030
3 × 12	2½ × 11¼	28.125	296.631	52.734	4.883	5.859	6.836	7.813	8.789	9.766
3 × 14	2½ × 13¼	33.125	484.625	73.151	5.751	6.901	8.051	9.201	10.352	11.502
3 × 16	2½ × 15¼	38.125	738.870	96.901	6.619	7.943	9.266	10.590	11.914	13.238
4 × 1	3½ × ¾	2.625	0.123	0.328	0.456	0.547	0.638	0.729	0.820	0.911
4 × 2	3½ × 1½	5.250	0.984	1.313	0.911	1.094	1.276	1.458	1.641	1.823
4 × 3	3½ × 2½	8.750	4.557	3.646	1.519	1.823	2.127	2.431	2.734	3.038
4 × 4	3½ × 3½	12.250	12.505	7.146	2.127	2.552	2.977	3.403	3.828	4.253
4 × 6	3½ × 5½	19.250	48.526	17.646	3.342	4.010	4.679	5.347	6.016	6.684
4 × 8	3½ × 7¼	25.375	111.148	30.661	4.405	5.286	6.168	7.049	7.930	8.811
4 × 10	3½ × 9¼	32.375	230.840	49.911	5.621	6.745	7.869	8.933	10.117	11.241
4 × 12	3½ × 11¼	39.375	415.283	73.828	6.836	8.203	9.570	10.938	12.305	13.672
4 × 14	3½ × 13½	47.250	717.609	106.313	8.203	9.844	11.484	13.125	14.766	16.406
4 × 16	3½ × 15½	54.250	1,086.130	140.146	9.418	11.302	13.186	15.069	16.953	18.837
6 × 1	5½ × ¾	4.125	0.193	0.516	0.716	0.859	1.003	1.146	1.289	1.432
6 × 2	5½ × 1½	8.250	1.547	2.063	1.432	1.719	2.005	2.292	2.578	2.865
6 × 3	5½ × 2½	13.750	7.161	5.729	2.387	2.865	3.342	3.819	4.297	4.774
6 × 4	5½ × 3½	19.250	19.651	11.229	3.342	4.010	4.679	5.347	6.016	6.684
6 × 6	5½ × 5½	30.250	76.255	27.729	5.252	6.302	7.352	8.403	9.453	10.503
6 × 8	5½ × 7½	41.250	193.359	51.563	7.161	8.594	10.026	11.458	12.891	14.323
6 × 10	5½ × 9½	52.250	392.963	82.729	9.071	10.885	12.700	14.514	16.328	18.142
6 × 12	5½ × 11½	63.250	697.068	121.229	10.981	13.177	15.373	17.569	19.766	21.962
6 × 14	5½ × 13½	74.250	1,127.672	167.063	12.891	15.469	18.047	20.625	23.203	25.781
6 × 16	5½ × 15½	85.250	1,706.776	220.229	14.800	17.760	20.720	23.681	26.641	29.601
6 × 18	5½ × 17½	96.250	2,456.380	280.729	16.710	20.052	23.394	26.736	30.078	33.420
6 × 20	5½ × 19½	107.250	3,398.484	348.563	18.620	22.344	26.068	29.792	33.516	37.240
6 × 22	5½ × 21½	118.250	4,555.086	423.729	20.530	24.635	28.741	32.847	36.953	41.059
6 × 24	5½ × 23½	129.250	5,948.191	506.229	22.439	26.927	31.415	35.903	40.391	44.878
8 × 1	7¼ × ¾	5.438	0.255	0.680	0.944	1.133	1.322	1.510	1.699	1.888
8 × 2	7¼ × 1½	10.875	2.039	2.719	1.888	2.266	2.643	3.021	3.398	3.776
8 × 3	7¼ × 2½	18.125	9.440	7.552	3.147	3.776	4.405	5.035	5.664	6.293
8 × 4	7¼ × 3½	25.375	25.904	14.802	4.405	5.286	6.168	7.049	7.930	8.811
8 × 6	7½ × 5½	41.250	103.984	37.813	7.161	8.594	10.026	11.458	12.891	14.323
8 × 8	7½ × 7½	56.250	263.672	70.313	9.766	11.719	13.672	15.625	17.578	19.531
8 × 10	7½ × 9½	71.250	535.859	112.813	12.370	14.844	17.318	19.792	22.266	24.740
8 × 12	7½ × 11½	86.250	950.547	165.313	14.974	17.969	20.964	23.958	26.953	29.948
8 × 14	7½ × 13½	101.250	1,537.734	227.813	17.578	21.094	24.609	28.125	31.641	35.156
8 × 16	7½ × 15½	116.250	2,327.422	300.313	20.182	24.219	28.255	32.292	36.328	40.365
8 × 18	7½ × 17½	131.250	3,349.609	382.813	22.786	27.344	31.901	36.458	41.016	45.573
8 × 20	7½ × 19½	146.250	4,634.297	475.313	25.391	30.469	35.547	40.625	45.703	50.781
8 × 22	7½ × 21½	161.250	6,211.484	577.813	27.995	33.594	39.193	44.792	50.391	55.990
8 × 24	7½ × 23½	176.250	8,111.172	690.313	30.599	36.719	42.839	48.958	55.078	61.198
10 × 1	9¼ × ¾	6.938	0.325	0.867	1.204	1.445	1.686	1.927	2.168	2.409
10 × 2	9¼ × 1½	13.875	2.602	3.469	2.409	2.891	3.372	3.854	4.336	4.818
10 × 3	9¼ × 2½	23.125	12.044	9.635	4.015	4.818	5.621	6.424	7.227	8.030
10 × 4	9¼ × 3½	32.375	33.049	18.885	5.621	6.745	7.869	8.993	10.117	11.241
10 × 6	9½ × 5½	52.250	131.714	47.896	9.071	10.885	12.700	14.514	16.328	18.142
10 × 8	9½ × 7½	71.250	333.984	89.063	12.370	14.844	17.318	19.792	22.266	24.740
10 × 10	9½ × 9½	90.250	678.755	142.896	15.668	18.802	21.936	25.069	28.203	31.337
10 × 12	9½ × 11½	109.250	1,204.026	209.396	18.967	22.760	26.554	30.347	34.141	37.934
10 × 14	9½ × 13½	128.250	1,947.797	288.563	22.266	26.719	31.172	35.625	40.078	44.531
10 × 16	9½ × 15½	147.250	2,948.068	380.396	25.564	30.677	35.790	40.903	46.016	51.128
10 × 18	9½ × 17½	166.250	4,242.836	484.896	28.863	34.635	40.408	46.181	51.953	57.726
10 × 20	9½ × 19½	185.250	5,870.109	602.063	32.161	38.594	45.026	51.458	57.891	64.323
10 × 22	9½ × 21½	204.250	7,867.879	731.896	35.460	42.552	49.644	56.736	63.828	70.920
10 × 24	9½ × 23½	223.250	10,274.148	874.396	38.759	46.510	54.262	62.014	69.766	77.517
12 × 1	11¼ × ¾	8.438	0.396	1.055	1.465	1.758	2.051	2.344	2.637	2.930
12 × 2	11¼ × 1½	16.875	3.164	4.219	2.930	3.516	4.102	4.688	5.273	5.859
12 × 3	11¼ × 2½	28.125	14.648	11.719	4.883	5.859	6.836	7.813	8.789	9.766
12 × 4	11¼ × 3½	39.375	40.195	22.969	6.836	8.203	9.570	10.938	12.305	13.672
12 × 6	11½ × 5½	63.250	159.443	57.979	10.981	13.177	15.373	17.569	19.766	21.962
12 × 8	11½ × 7½	86.250	404.297	107.813	14.974	17.969	20.964	23.958	26.953	29.948
12 × 10	11½ × 9½	109.250	821.651	172.979	18.967	22.760	26.554	30.347	34.141	37.934
12 × 12	11½ × 11½	132.250	1,457.505	253.479	22.960	27.552	32.144	36.736	41.328	45.920
12 × 14	11½ × 13½	155.250	2,357.859	349.313	26.953	32.344	37.734	43.125	48.516	53.906

Table 11-3. Properties of Sections of Solid-Sawn Wood *(Continued)*

Nominal size, in.	Standard dressed size, in. (S4S)	Area of section, sq in.	Moment of inertia $in.^4$	Section modulus, $in.^3$	Weight, lb per lin ft, of piece when weight of wood, lb per cu ft, equals 25	30	35	40	45	50
12 × 16	11½ × 15½	178.250	3,568.713	460.479	30.946	37.135	43.325	49.514	55.703	61.892
12 × 18	11½ × 17½	201.250	5,136.066	586.979	34.939	41.927	48.915	55.903	62.891	69.878
12 × 20	11½ × 19½	224.250	7,105.922	728.813	38.932	46.719	54.505	62.292	70.078	77.865
12 × 22	11½ × 21½	247.250	9,524.273	885.979	42.925	51.510	60.095	68.681	77.266	85.851
12 × 24	11½ × 23½	270.250	12,437.129	1,058.479	46.918	56.302	65.686	75.069	84.453	93.837
14 × 2	13¼ × 1½	19.875	3.727	4.969	3.451	4.141	4.831	5.521	6.211	6.901
14 × 3	13¼ × 2½	33.125	17.253	13.802	5.751	6.901	8.051	9.201	10.352	11.502
14 × 4	13¼ × 3½	46.375	47.34	27.052	8.047	9.657	11.266	12.877	14.485	16.094
14 × 6	13½ × 5½	74.250	187.172	68.063	12.891	15.469	18.047	20.625	23.203	25.781
14 × 8	13½ × 7½	101.250	474.609	126.563	17.578	21.094	24.609	28.125	31.641	35.156
14 × 10	13½ × 9½	128.250	964.547	203.063	22.266	26.719	31.172	35.625	40.078	44.531
14 × 12	13½ × 11½	155.250	1,710.984	297.563	26.953	32.344	37.734	43.125	48.516	53.906
14 × 16	13½ × 15½	209.250	4,189.359	540.563	36.328	43.594	50.859	58.125	65.391	72.656
14 × 18	13½ × 17½	236.250	6,029.297	689.063	41.016	49.219	57.422	65.625	73.828	82.031
14 × 20	13½ × 19½	263.250	8,341.734	855.563	45.703	54.844	63.984	73.125	82.266	91.406
14 × 22	13½ × 21½	290.250	11,180.672	1,040.063	50.391	60.469	70.547	80.625	90.703	100.781
14 × 24	13½ × 23½	317.250	14,600.109	1,242.563	55.078	66.094	77.109	88.125	99.141	110.156
16 × 3	15¼ × 2½	38.125	19.857	15.885	6.619	7.944	9.267	10.592	11.915	13.240
16 × 4	15¼ × 3½	53.375	54.487	31.135	9.267	11.121	12.975	14.828	16.682	18.536
16 × 6	15½ × 5½	85.250	214.901	78.146	14.800	17.760	20.720	23.681	26.641	29.601
16 × 8	15½ × 7½	116.250	544.922	145.313	20.182	24.219	28.255	32.292	36.328	40.365
16 × 10	15½ × 9½	147.250	1,107.443	233.146	25.564	30.677	35.790	40.903	46.016	51.128
16 × 12	15½ × 11½	178.250	1,964.463	341.646	30.946	37.135	43.325	49.514	55.703	61.892
16 × 14	15½ × 13½	209.250	3,177.984	470.813	36.328	43.594	50.859	58.125	65.391	72.656
16 × 16	15½ × 15½	240.250	4,810.004	620.646	41.710	50.052	58.394	66.736	75.078	83.420
16 × 18	15½ × 17½	271.250	6,922.523	791.146	47.092	56.510	65.929	75.347	84.766	94.184
16 × 20	15½ × 19½	302.250	9,577.547	982.313	52.474	62.969	73.464	83.958	94.453	104.948
16 × 22	15½ × 21½	333.250	12,837.066	1,194.146	57.856	69.427	80.998	92.569	104.141	115.712
16 × 24	15½ × 23½	364.250	16,763.086	1,426.646	63.238	75.885	88.533	101.181	113.828	126.476
18 × 6	17½ × 5½	96.250	242.630	88.229	16.710	20.052	23.394	26.736	30.078	33.420
18 × 8	17½ × 7½	131.250	615.234	164.063	22.786	27.344	31.901	36.458	41.016	45.573
18 × 10	17½ × 9½	166.250	1,250.338	263.229	28.863	34.635	40.408	46.181	51.953	57.726
18 × 12	17½ × 11½	201.250	2,217.943	385.729	34.939	41.927	48.915	55.903	62.891	69.878
18 × 14	17½ × 13½	236.250	3,588.047	531.563	41.016	49.219	57.422	65.625	73.828	82.031
18 × 16	17½ × 15½	271.250	5,430.648	700.729	47.092	56.510	65.929	75.347	84.766	94.184
18 × 18	17½ × 17½	306.250	7,815.754	893.229	53.168	63.802	74.436	85.069	95.703	106.337
18 × 20	17½ × 19½	341.250	10,813.359	1,109.063	59.245	71.094	82.943	94.792	106.641	118.490
18 × 22	17½ × 21½	376.250	14,493.461	1,348.229	65.321	78.385	91.450	104.514	117.578	130.642
18 × 24	17½ × 23½	411.250	18,926.066	1,610.729	71.398	85.677	99.957	114.236	128.516	142.795
20 × 6	19½ × 5½	107.250	270.359	98.313	18.620	22.344	26.068	29.792	33.516	37.240
20 × 8	19½ × 7½	146.250	685.547	182.813	25.391	30.469	35.547	40.625	45.703	50.781
20 × 10	19½ × 9½	185.250	1,393.234	293.313	32.161	38.594	45.026	51.458	57.891	64.323
20 × 12	19½ × 11½	224.250	2,471.422	429.813	38.932	46.719	54.505	62.292	70.078	77.865
20 × 14	19½ × 13½	263.250	3,998.109	592.313	45.703	54.844	63.984	73.125	82.266	91.406
20 × 16	19½ × 15½	302.250	6,051.297	780.813	52.474	62.969	73.464	83.958	94.453	104.948
20 × 18	19½ × 17½	341.250	8,708.984	995.313	59.245	71.094	82.943	94.792	106.641	118.490
20 × 20	19½ × 19½	380.250	12,049.172	1,235.813	66.016	79.219	92.422	105.625	118.828	132.031
20 × 22	19½ × 21½	419.250	16,149.859	1,502.313	72.786	87.344	101.901	116.458	131.016	145.573
20 × 24	19½ × 23½	458.250	21,089.047	1,794.813	79.557	95.469	111.380	127.292	243.203	159.115
22 × 6	21½ × 5½	118.250	298.088	108.396	20.530	24.635	28.741	32.847	36.953	41.059
22 × 8	21½ × 7½	161.250	755.859	201.563	27.995	33.594	39.193	44.792	50.391	55.990
22 × 10	21½ × 9½	204.250	1,536.130	323.396	35.460	42.552	49.644	56.736	63.828	70.920
22 × 12	21½ × 11½	247.250	2,724.901	473.896	42.925	51.510	60.095	68.681	77.266	85.851
22 × 14	21½ × 13½	290.250	4,408.172	653.063	50.391	60.469	70.547	80.625	90.703	100.781
22 × 16	21½ × 15½	333.250	6,671.941	860.896	57.856	69.427	80.998	92.569	104.141	115.712
22 × 18	21½ × 17½	376.250	9,602.211	1,097.396	65.321	78.385	91.450	104.514	117.578	130.642
22 × 20	21½ × 19½	419.250	13,284.984	1,362.563	72.786	87.344	101.901	116.458	131.016	145.573
22 × 22	21½ × 21½	462.250	17,806.254	1,656.396	80.252	96.302	112.352	128.403	144.453	160.503
22 × 24	21½ × 23½	505.250	23,252.023	1,978.896	87.717	105.260	122.804	140.347	157.891	175.434
24 × 6	23½ × 5½	129.250	325.818	118.479	22.439	26.927	31.415	35.903	40.391	44.878
24 × 8	23½ × 7½	176.250	826.172	220.313	30.599	36.719	42.839	48.958	55.078	61.198
24 × 10	23½ × 9½	223.250	1,679.026	353.479	38.759	46.510	54.262	62.014	69.766	77.517
24 × 12	23½ × 11½	270.250	2,978.380	517.979	46.918	56.302	65.686	75.069	84.453	93.837
24 × 14	23½ × 13½	317.250	4,818.234	713.813	55.078	66.094	77.109	88.125	99.141	110.156
24 × 16	23½ × 15½	364.250	7,292.586	940.979	63.238	75.885	88.533	101.181	113.828	126.476
24 × 18	23½ × 17½	411.250	10,495.441	1,199.479	71.398	85.677	99.957	114.236	128.516	142.795
24 × 20	23½ × 19½	458.250	14,520.797	1,489.313	79.557	95.469	111.380	127.292	143.203	159.115
24 × 22	23½ × 21½	505.250	19,462.648	1,810.479	87.717	105.260	122.804	140.347	157.891	175.434
24 × 24	23½ × 23½	552.250	25,415.004	2,162.979	95.877	115.052	134.227	153.403	172.578	191.753

* For lumber surfaced $1^5/_8$ in. thick, instead of $1^1/_2$ in., the area, moment of inertia, and section modulus may be increased 8.33%.

11-4. Standard Sizes of Glued-Laminated Timber. Standard finished sizes of structural glued-laminated timber should be used to the extent that conditions permit. These standard finished sizes are based on lumber sizes given in Voluntary Product Standard PS20-70. Other finished sizes may be used to meet the size requirements of a design, or to meet other special requirements.

Nominal 2-in.-thick lumber, surfaced to 1½ in. before gluing, is used to laminate straight members and curved members with radii of curvature within the bending-radius limitations for the species. (Formerly, a net thickness of 1⅝ in. was common for nominal 2-in. lumber and may be used, depending on its availability.) Nominal 1-in.-thick lumber, surfaced to ¾ in. before gluing, may be used for laminating curved members when the bending radius is too short to permit use of nominal 2-in.-thick laminations, if the bending-radius limitations for the species are observed. Other limitation thicknesses may be used to meet special curving requirements.

11-5. Sectional Properties of Glued-Laminated Timber. Table 11-4 lists properties of sections of glued-laminated timber.

11-6. Basic and Allowable Stresses for Timber. Testing of a species to determine average strength properties should be carried out from either of two viewpoints:

1. Tests should be made on specimens of large size containing defects. Practically all structural uses involve members of this character.

2. Tests should be made on small, clear specimens to provide fundamental data. Factors to account for the influence of various characteristics may be applied to establish the strength of structural members.

Tests made in accordance with the first viewpoint have the disadvantage that the results apply only to the particular combination of characteristics existing in the test specimens. To determine the strength corresponding to other combinations requires additional tests; thus, an endless testing program is necessary. The second viewpoint permits establishment of fundamental strength properties for each species, and application of general rules to cover the specific conditions involved in a particular case.

It is this second viewpoint that has been generally accepted. When a species has been adequately investigated under this concept, there should be no need for further tests on that species unless new conditions arise. ASTM Standard D143, "Standard Methods of Testing Small, Clear Specimens of Timber," gives the procedure for determination of fundamental data on wood species.

Basic stresses are essentially unit stresses applicable to clear and straight-grained defect-free material. These stresses, derived from the results of tests on small, clear specimens of green wood, include an adjustment for variability of material, length of loading period, and factor of safety. They are considerably less than the average for the species. They require only an adjustment for grade to become allowable unit stresses.

Allowable unit stresses are computed for a particular grade by reducing the basic stress according to the limitations on defects for that grade. The basic stress is multiplied by a strength ratio to obtain an allowable stress. This strength ratio represents that proportion of the strength of a defect-free piece that remains after taking into account the effect of strength-reducing features.

The principal factors entering into the establishment of allowable unit stress for each species include inherent strength of wood, reduction in strength due to natural growth characteristics permitted in the grade, effect of long-time loading, variability of individual species, possibility of some slight overloading, characteristics of the species, size of member and related influence of seasoning, and factor of safety. The effect of these factors is a strength value for practical-use conditions lower than the average value taken from tests on small, clear specimens.

Basic stresses for laminated timbers under wet service conditions are the same as for solid timbers; i.e., these stresses are based on the strength of wood in the green condition. When moisture content in a member will be low throughout its service, a second set of higher basic stresses, based on the higher strength of dry material, may be used. Technical Bulletin 479, U.S. Department of Agriculture, "Strength and Related Properties of Woods Grown in the United States," presents test results on small, clear, and straight-grained wood species in the green state and in the 12%-moisture-content, air-dry condition. Technical Bulletin 1069, U.S. Department of Agriculture, "Fabrication and Design of Glued-Laminated Structural Members," gives the basic stresses for clear, solid-sawn members and glued-laminated timbers in the wet condition, and basic stresses for clear, glued-laminated timbers in the dry condition.

Allowable unit stresses for commercial species of lumber, widely accepted for wood construction in economical and efficient designs, are given in "National Design Specification for Stress-grade Lumber and Its Fastenings," National Forest Products Association. Allowable unit stresses for glued-laminated timber are given in the AITC "Timber Construction Manual," John Wiley & Sons, Inc., and in the AITC "Standard Specifications for Structural Glued-Laminated Timber," AITC 117, American Institute of Timber Construction, 333 W. Hampden Ave., Englewood, Colo. 80110.

See also Arts. 11-9 to 11-12 and 11-14.

11-7. Structural Grading of Wood. Strength properties of wood are intimately related to moisture content and specific gravity. Therefore, data on strength properties unaccompanied by corresponding data on these physical properties would be of little value.

Table 11-4. Properties of Sections of Glued-Laminated Timber*

2¼-in. Width

No. of laminations 1½-in.	No. of laminations ¾-in.	d	C_F	A	S	I	Vol.
2	4	3.00	1.00	6.8	3.4	5.1	0.05
	5	3.75	1.00	8.4	5.3	9.9	0.06
3	6	4.50	1.00	10.1	7.6	17.1	0.07
	7	5.25	1.00	11.8	10.3	27.1	0.08
4	8	6.00	1.00	13.5	13.5	40.5	0.09
	9	6.75	1.00	15.2	17.1	57.7	0.11
5	10	7.50	1.00	16.9	21.1	79.1	0.12
	11	8.25	1.00	18.6	25.5	105.3	0.13
6	12	9.00	1.00	20.2	30.4	136.7	0.14
	13	9.75	1.00	21.9	35.6	173.8	0.15
7	14	10.50	1.00	23.6	41.3	217.0	0.16
	15	11.25	1.00	25.3	47.5	267.0	0.18
8	16	12.00	1.00	27.0	54.0	324.0	0.19
	17	12.75	0.99	28.7	61.0	388.6	0.20
9	18	13.50	0.99	30.4	68.3	461.3	0.21
	19	14.25	0.98	32.1	76.1	542.6	0.22
10	20	15.00	0.98	33.8	84.4	632.8	0.23

3⅛-in. Width

No. of laminations 1½-in.	No. of laminations ¾-in.	d	C_F	A	S	I	Vol.
2	4	3.00	1.00	9.4	4.7	7.0	0.06
	5	3.75	1.00	11.7	7.3	13.7	0.08
3	6	4.50	1.00	14.1	10.5	23.7	0.10
	7	5.25	1.00	16.4	14.4	37.7	0.11
4	8	6.00	1.00	18.8	18.8	56.3	0.13
	9	6.75	1.00	21.1	23.7	80.1	0.15
5	10	7.50	1.00	23.4	29.3	109.9	0.16
	11	8.25	1.00	25.8	35.4	146.2	0.18
6	12	9.00	1.00	28.1	42.2	189.8	0.20
	13	9.75	1.00	30.5	49.5	241.4	0.21
7	14	10.50	1.00	32.8	57.4	301.5	0.23
	15	11.25	1.00	35.2	65.9	370.8	0.24
8	16	12.00	1.00	37.5	75.0	450.0	0.26
	17	12.75	0.99	39.8	84.7	539.8	0.28
9	18	13.50	0.99	42.2	94.9	640.7	0.29
	19	14.25	0.98	44.5	105.8	753.6	0.31
10	20	15.00	0.98	46.9	117.2	878.9	0.33
	21	15.75	0.97	49.2	129.2	1,017.4	0.34
11	22	16.50	0.97	51.6	141.8	1,169.8	0.36
	23	17.25	0.96	53.9	155.0	1,336.7	0.37
12	24	18.00	0.96	56.3	168.8	1,518.8	0.39
	25	18.75	0.95	58.6	183.1	1,716.6	0.41
13	26	19.50	0.95	60.9	198.0	1,931.0	0.42
	27	20.25	0.94	63.3	213.6	2,162.4	0.44

6¾-in. Width

No. of laminations 1½-in.	No. of laminations ¾-in.	d	C_F	A	S	I	Vol.
4	8	6.00	1.00	40.5	40.5	121.5	0.28
	9	6.75	1.00	45.6	51.3	173.0	0.32
5	10	7.50	1.00	50.6	63.3	237.3	0.35
	11	8.25	1.00	55.7	76.6	315.9	0.39
6	12	9.00	1.00	60.8	91.1	410.1	0.42
	13	9.75	1.00	65.8	106.9	521.0	0.46
7	14	10.50	1.00	70.9	124.0	651.2	0.49
	15	11.25	1.00	75.9	142.4	800.9	0.53
8	16	12.00	1.00	81.0	162.0	972.0	0.56
	17	12.75	0.99	86.1	182.9	1,165.9	0.60
9	18	13.50	0.99	91.1	205.0	1,384.0	0.63
	19	14.25	0.98	96.2	228.4	1,627.7	0.67
10	20	15.00	0.98	101.3	253.1	1,898.4	0.70
	21	15.75	0.97	106.3	279.1	2,197.7	0.74
11	22	16.50	0.97	111.4	306.3	2,526.8	0.77
	23	17.25	0.96	116.4	334.8	2,887.3	0.81
12	24	18.00	0.96	121.5	364.5	3,280.5	0.84
	25	18.75	0.95	126.6	395.5	3,707.9	0.88
13	26	19.50	0.95	131.6	427.8	4,170.0	0.91
	27	20.25	0.94	136.7	461.3	4,670.0	0.95
14	28	21.00	0.94	141.8	496.1	5,209.3	0.98
	29	21.75	0.94	146.8	532.2	5,787.6	1.02
15	30	22.50	0.93	151.9	569.5	6,407.2	1.05
	31	23.25	0.93	156.9	608.1	7,069.5	1.09
16	32	24.00	0.93	162.0	648.0	7,776.0	1.12
	33	24.75	0.92	167.1	689.1	8,528.0	1.16
17	34	25.50	0.92	172.1	731.5	9,327.0	1.20
	35	26.25	0.92	177.2	775.2	10,174.4	1.23
18	36	27.00	0.91	182.3	820.1	11,071.7	1.27
	37	27.75	0.91	187.3	866.3	12,020.2	1.30
19	38	28.50	0.91	192.4	913.8	13,021.4	1.34
	39	29.25	0.91	197.4	962.5	14,076.7	1.37
20	40	30.00	0.90	202.5	1,012.5	15,187.5	1.41
	41	30.75	0.90	207.6	1,063.8	16,355.3	1.44
21	42	31.50	0.90	212.6	1,116.3	17,581.4	1.48
	43	32.25	0.90	217.7	1,170.1	18,867.4	1.51
22	44	33.00	0.89	222.8	1,225.1	20,214.6	1.55
	45	33.75	0.89	227.8	1,281.4	21,624.4	1.58
23	46	34.50	0.89	232.9	1,339.0	23,098.3	1.62
	47	35.25	0.89	237.9	1,397.9	24,637.7	1.65
24	48	36.00	0.88	243.0	1,458.0	26,244.0	1.69
	49	36.75	0.88	248.1	1,519.4	27,918.7	1.72
25	50	37.50	0.88	253.1	1,582.0	29,663.1	1.76
	51	38.25	0.88	258.2	1,645.9	31,478.7	1.79

8¾-in. Width

No. of laminations 1½-in.	No. of laminations ¾-in.	d	C_F	A	S	I	Vol.
6	12	9.00	1.00	78.8	118.1	531.6	0.55
	13	9.75	1.00	85.3	138.6	675.8	0.59
7	14	10.50	1.00	91.9	160.8	844.1	0.64
	15	11.25	1.00	98.4	184.6	1,038.2	0.68
8	16	12.00	1.00	105.0	210.0	1,260.0	0.73
	17	12.75	0.99	111.6	237.1	1,511.3	0.77
9	18	13.50	0.99	118.1	265.8	1,794.0	0.82
	19	14.25	0.98	124.7	296.1	2,109.9	0.87
10	20	15.00	0.98	131.3	328.1	2,460.9	0.91
	21	15.75	0.97	137.8	361.8	2,848.8	0.96
11	22	16.50	0.97	144.4	397.0	3,275.5	1.00
	23	17.25	0.96	150.9	433.9	3,742.8	1.05
12	24	18.00	0.96	157.5	472.5	4,252.5	1.09
	25	18.75	0.95	164.1	512.7	4,806.5	1.14
13	26	19.50	0.95	170.6	554.5	5,406.7	1.18
	27	20.25	0.94	177.2	598.0	6,054.8	1.23
14	28	21.00	0.94	183.8	643.1	6,752.8	1.28
	29	21.75	0.94	190.3	689.9	7,502.5	1.32
15	30	22.50	0.93	196.9	738.3	8,305.7	1.37
	31	23.25	0.93	203.4	788.3	9,164.2	1.41
16	32	24.00	0.93	210.0	840.0	10,080.0	1.46
	33	24.75	0.92	216.6	893.3	11,054.8	1.50
17	34	25.50	0.92	223.1	948.3	12,090.6	1.55
	35	26.25	0.92	229.7	1,004.9	13,189.1	1.59
18	36	27.00	0.91	236.3	1,063.1	14,352.2	1.64
	37	27.75	0.91	242.8	1,123.0	15,581.7	1.69
19	38	28.50	0.91	249.4	1,184.5	16,879.6	1.73
	39	29.25	0.91	255.9	1,247.7	18,247.5	1.78
20	40	30.00	0.90	262.5	1,312.5	19,687.5	1.82
	41	30.75	0.90	269.1	1,378.9	21,201.3	1.87
21	42	31.50	0.90	275.6	1,447.0	22,790.7	1.91
	43	32.25	0.90	282.2	1,516.8	24,457.7	1.96
22	44	33.00	0.89	288.8	1,588.1	26,204.1	2.00
	45	33.75	0.89	295.3	1,661.1	28,031.6	2.05
23	46	34.50	0.89	301.9	1,735.8	29,942.2	2.10
	47	35.25	0.89	308.4	1,812.1	31,937.7	2.14
24	48	36.00	0.88	315.0	1,890.0	34,020.0	2.19
	49	36.75	0.88	321.6	1,969.6	36,190.9	2.23
25	50	37.50	0.88	328.1	2,050.8	38,452.2	2.28
	51	38.25	0.88	334.7	2,133.6	40,805.7	2.32
26	52	39.00	0.88	341.3	2,218.1	43,253.4	2.37
	53	39.75	0.88	347.8	2,304.3	45,797.1	2.42
27	54	40.50	0.87	354.4	2,392.2	48,438.6	2.46
	55	41.25	0.87	360.9	2,481.4	51,179.8	2.51

5⅛-in. Width

Cross-section with neutral axis X–X, width b and depth d.

Table (upper-left band)

	d		Area	S	I	
14	28	21.00	0.94	65.6	229.7	2,411.7	0.46
	29	21.75	0.94	68.0	246.4	2,679.5	0.47
15	30	22.50	0.93	70.3	263.7	2,966.3	0.49
	31	23.25	0.93	72.7	281.5	3,272.9	0.50
16	32	24.00	0.93	75.0	300.0	3,600.0	0.52

Main table, 5⅛-in. Width

	d		Area	S	I	
3	6	4.50	1.00	23.1	17.3	38.9	0.16
	7	5.25	1.00	26.9	23.5	61.8	0.19
4	8	6.00	1.00	30.8	30.8	92.3	0.21
	9	6.75	1.00	34.6	38.9	131.8	0.24
5	10	7.50	1.00	38.4	48.0	180.2	0.27
	11	8.25	1.00	42.3	58.1	239.8	0.29
6	12	9.00	1.00	46.1	69.2	311.3	0.32
	13	9.75	1.00	50.0	81.2	395.8	0.35
7	14	10.50	1.00	53.8	94.2	494.4	0.37
	15	11.25	1.00	57.7	108.1	608.1	0.40
8	16	12.00	1.00	61.5	123.0	738.0	0.43
	17	12.75	0.99	65.3	138.9	885.2	0.45
9	18	13.50	0.99	69.2	155.7	1,050.8	0.48
	19	14.25	0.98	73.0	173.4	1,235.8	0.51
10	20	15.00	0.98	76.9	192.0	1,441.4	0.53
	21	15.75	0.97	80.7	211.9	1,668.6	0.56
11	22	16.50	0.97	84.6	232.5	1,918.5	0.59
	23	17.25	0.96	88.4	254.2	2,192.2	0.61
12	24	18.00	0.96	92.3	276.8	2,490.8	0.64
	25	18.75	0.95	96.1	300.3	2,815.2	0.67
13	26	19.50	0.95	99.9	324.8	3,166.8	0.69
	27	20.25	0.94	103.8	350.3	3,546.4	0.72
14	28	21.00	0.94	107.6	376.7	3,955.2	0.75
	29	21.75	0.94	111.5	404.1	4,394.3	0.77
15	30	22.50	0.93	115.3	432.4	4,864.7	0.80
	31	23.25	0.93	119.2	461.7	5,367.6	0.83
16	32	24.00	0.92	123.0	492.0	5,904.0	0.85
	33	24.75	0.92	126.8	523.2	6,475.0	0.88
17	34	25.50	0.92	130.7	555.4	7,081.6	0.91
	35	26.25	0.92	134.5	588.6	7,725.0	0.93
18	36	27.00	0.91	138.4	622.7	8,406.3	0.96
	37	27.75	0.91	142.2	657.8	9,126.4	0.99
19	38	28.50	0.91	146.1	693.8	9,886.6	1.01
	39	29.25	0.90	149.9	730.8	10,687.8	1.04
20	40	30.00	0.90	153.8	768.8	11,531.8	1.07
	41	30.75	0.90	157.6	807.7	12,417.9	1.09
21	42	31.50	0.90	161.4	847.5	13,348.9	1.12
	43	32.25	0.89	165.3	888.4	14,325.2	1.15
22	44	33.00	0.89	169.1	930.2	15,348.1	1.17
	45	33.75	0.89	173.0	972.9	16,418.5	1.20
23	46	34.50	0.89	176.8	1,016.7	17,537.6	1.23
	47	35.25	0.89	180.7	1,061.4	18,706.4	1.25
24	48	36.00	0.88	184.5	1,107.0	19,926.0	1.28

Table (middle band)

	d		Area	S	I	
26	52	39.00	0.88	263.3	1,711.1	33,366.9	1.83
	53	39.75	0.88	268.3	1,777.6	35,329.2	1.86
27	54	40.50	0.87	273.4	1,845.3	37,367.0	1.90
	55	41.25	0.87	278.4	1,914.3	39,481.6	1.93
28	56	42.00	0.87	283.5	1,984.5	41,674.5	1.97
	57	42.75	0.87	288.6	2,056.0	43,947.2	2.00
29	58	43.50	0.87	293.6	2,128.8	46,301.0	2.04
	59	44.25	0.86	298.7	2,202.8	48,737.4	2.07
30	60	45.00	0.86	303.8	2,278.1	51,257.8	2.11
	61	45.75	0.86	308.8	2,354.7	53,863.7	2.14
31	62	46.50	0.86	313.9	2,432.5	56,556.4	2.18
	63	47.25	0.86	318.9	2,511.6	59,337.3	2.21
32	64	48.00	0.86	324.0	2,592.0	62,208.0	2.25

Table (upper band)

	d		Area	S	I	
28	56	42.00	0.87	367.5	2,572.5	54,022.5	2.60
	57	42.75	0.87	374.1	2,665.2	56,968.6	2.55
29	58	43.50	0.87	380.6	2,759.5	60,019.8	2.64
	59	44.25	0.87	387.2	2,855.5	63,178.1	2.69
30	60	45.00	0.86	393.8	2,953.1	66,445.3	2.73
	61	45.75	0.86	400.3	3,052.4	69,823.3	2.78
31	62	46.50	0.86	406.9	3,153.3	73,313.8	2.83
	63	47.25	0.86	413.4	3,255.8	76,918.8	2.87
32	64	48.00	0.86	420.0	3,360.0	80,640.0	2.92
	65	48.75	0.85	426.6	3,465.8	84,479.4	2.96
33	66	49.50	0.85	433.1	3,573.3	88,438.7	3.01
	67	50.25	0.85	439.7	3,682.4	92,519.9	3.05
34	68	51.00	0.85	446.3	3,793.1	96,724.7	3.10
	69	51.75	0.85	452.8	3,905.5	101,055.0	3.14
35	70	52.50	0.85	459.4	4,019.5	105,512.7	3.19
	71	53.25	0.85	465.9	4,135.2	110,099.6	3.24
36	72	54.00	0.85	472.5	4,252.5	114,817.5	3.28
	73	54.75	0.84	479.1	4,371.4	119,668.3	3.33
37	74	55.50	0.84	485.6	4,492.0	124,653.9	3.37
	75	56.25	0.84	492.2	4,614.3	129,776.0	3.42
38	76	57.00	0.84	498.8	4,738.1	135,036.6	3.46
	77	57.75	0.84	505.3	4,863.6	140,437.4	3.51
39	78	58.50	0.84	511.9	4,990.8	145,980.4	3.55
	79	59.25	0.84	518.4	5,119.6	151,667.3	3.60
40	80	60.00	0.84	525.0	5,250.0	157,500.0	3.65
	81	60.75	0.84	531.6	5,382.1	163,480.4	3.69
41	82	61.50	0.84	538.1	5,515.8	169,610.3	3.74
	83	62.25	0.83	544.7	5,651.1	175,891.5	3.78
42	84	63.00	0.83	551.3	5,788.1	182,326.0	3.83

Table 11-4.   Properties of Sections of Glued-Laminated Timber* (Continued)

10¾-in. Width

No. of laminations 1½-in.	No. of laminations ¾-in.	d	C_F	A	S	I	Vol.
7	14	10.50	1.00	112.9	197.5	1,037.0	0.78
	15	11.25	1.00	120.9	226.8	1,275.5	0.84
8	16	12.00	1.00	129.0	258.0	1,548.0	0.90
	17	12.75	0.99	137.1	291.3	1,856.8	0.95
9	18	13.50	0.99	145.1	326.5	2,204.1	1.01
	19	14.25	0.98	153.2	363.8	2,592.2	1.06
10	20	15.00	0.98	161.3	403.1	3,023.4	1.12
	21	15.75	0.97	169.3	444.1	3,500.0	1.18
11	22	16.50	0.97	177.4	487.8	4,024.2	1.23
	23	17.25	0.96	185.4	533.1	4,598.3	1.29
12	24	18.00	0.96	193.5	580.5	5,224.5	1.34
	25	18.75	0.95	201.6	629.9	5,905.2	1.40
13	26	19.50	0.95	209.6	681.3	6,642.5	1.46
	27	20.25	0.94	217.7	734.7	7,438.8	1.51
14	28	21.00	0.94	225.8	790.1	8,296.3	1.57
	29	21.75	0.94	233.8	847.6	9,217.3	1.62
15	30	22.50	0.93	241.9	907.0	10,204.1	1.68
	31	23.25	0.93	249.9	968.5	11,258.9	1.74
16	32	24.00	0.93	258.0	1,032.0	12,384.0	1.79
	33	24.75	0.92	266.1	1,097.5	13,581.7	1.85
17	34	25.50	0.92	274.1	1,165.0	14,854.1	1.90
	35	26.25	0.92	282.2	1,234.6	16,203.7	1.96
18	36	27.00	0.91	290.3	1,306.1	17,632.7	2.02
	37	27.75	0.91	298.3	1,379.7	19,143.3	2.07
19	38	28.50	0.91	306.4	1,455.3	20,737.8	2.13
	39	29.25	0.91	314.4	1,532.9	22,418.4	2.18
20	40	30.00	0.90	322.5	1,612.5	24,187.5	2.24
	41	30.75	0.90	330.6	1,694.1	26,047.3	2.30
21	42	31.50	0.90	338.6	1,777.8	28,000.1	2.35
	43	32.25	0.90	346.7	1,863.4	30,048.1	2.41
22	44	33.00	0.89	354.8	1,951.1	32,193.6	2.46
	45	33.75	0.89	362.8	2,040.8	34,438.8	2.52
23	46	34.50	0.89	370.9	2,132.5	36,786.2	2.58
	47	35.25	0.89	378.9	2,226.3	39,237.8	2.63
24	48	36.00	0.88	387.0	2,322.0	41,796.0	2.69
	49	36.75	0.88	395.1	2,419.8	44,463.1	2.74
25	50	37.50	0.88	403.1	2,519.5	47,241.2	2.80
	51	38.25	0.88	411.2	2,621.3	50,132.8	2.86
26	52	39.00	0.88	419.3	2,725.1	53,139.9	2.91
	53	39.75	0.87	427.3	2,830.9	56,265.0	2.97
27	54	40.50	0.87	435.4	2,938.8	59,510.3	3.02
	55	41.25	0.87	443.4	3,048.6	62,878.1	3.08
28	56	42.00	0.87	451.5	3,160.5	66,370.5	3.14
	57	42.75	0.87	459.6	3,274.4	69,989.9	3.19
29	58	43.50	0.87	467.6	3,390.3	73,738.6	3.25
	59	44.25	0.86	475.7	3,508.2	77,618.8	3.30
30	60	45.00	0.86	483.8	3,628.1	81,632.8	3.36
	61	45.75	0.86	491.8	3,750.1	85,782.9	3.42
31	62	46.50	0.86	499.9	3,874.0	90,071.2	3.47
	63	47.25	0.86	507.9	4,000.0	94,500.2	3.53
32	64	48.00	0.86	516.0	4,128.0	99,072.0	3.58

12¼-in. Width

No. of laminations 1½-in.	No. of laminations ¾-in.	d	C_F	A	S	I	Vol.
8	16	12.00	1.00	147.0	294.0	1,764.0	1.02
	17	12.75	0.99	156.2	331.9	2,115.8	1.08
9	18	13.50	0.99	165.4	372.1	2,511.6	1.15
	19	14.25	0.98	174.6	414.6	2,953.8	1.21
10	20	15.00	0.98	183.8	459.4	3,445.3	1.28
	21	15.75	0.97	192.9	506.4	3,988.6	1.34
11	22	16.50	0.97	202.1	555.8	4,585.1	1.40
	23	17.25	0.96	211.3	607.5	5,239.7	1.47
12	24	18.00	0.96	220.5	661.5	5,953.5	1.53
	25	18.75	0.95	229.7	717.7	6,728.9	1.60
13	26	19.50	0.95	238.9	776.3	7,569.4	1.66
	27	20.25	0.94	248.1	837.2	8,476.5	1.72
14	28	21.00	0.94	257.2	900.4	9,453.9	1.79
	29	21.75	0.94	266.4	965.8	10,503.1	1.85
15	30	22.50	0.93	275.6	1,033.6	11,627.9	1.91
	31	23.25	0.93	284.8	1,103.6	12,829.5	1.97
16	32	24.00	0.93	294.0	1,176.0	14,112.0	2.04
	33	24.75	0.92	303.2	1,250.6	15,476.3	2.10
17	34	25.50	0.92	312.4	1,327.6	16,926.8	2.17
	35	26.25	0.92	321.6	1,406.8	18,464.1	2.23
18	36	27.00	0.91	330.8	1,488.4	20,093.1	2.30
	37	27.75	0.91	339.9	1,572.2	21,813.7	2.36
19	38	28.50	0.91	349.1	1,658.3	23,631.4	2.42
	39	29.25	0.91	358.3	1,746.7	25,545.7	2.49
20	40	30.00	0.90	367.5	1,837.5	27,562.5	2.55
	41	30.75	0.90	376.7	1,930.5	29,680.8	2.62
21	42	31.50	0.90	385.9	2,025.8	31,907.0	2.68
	43	32.25	0.90	395.1	2,123.4	34,239.7	2.74
22	44	33.00	0.89	404.2	2,223.4	36,685.7	2.81
	45	33.75	0.89	413.4	2,325.6	39,244.1	2.87
23	46	34.50	0.89	422.6	2,430.1	41,919.1	2.93
	47	35.25	0.89	431.8	2,536.9	44,712.7	3.00
24	48	36.00	0.88	441.0	2,646.0	47,628.0	3.06
	49	36.75	0.88	450.2	2,757.4	50,667.0	3.13
25	50	37.50	0.88	459.4	2,871.1	53,833.0	3.19
	51	38.25	0.88	468.6	2,987.1	57,127.8	3.25
26	52	39.00	0.88	477.8	3,105.4	60,554.8	3.32
	53	39.75	0.87	486.9	3,226.0	64,115.8	3.38
27	54	40.50	0.87	496.1	3,348.8	67,814.1	3.45
	55	41.25	0.87	505.3	3,474.0	71,651.5	3.51
28	56	42.00	0.87	514.5	3,601.5	75,631.5	3.57
	57	42.75	0.87	523.7	3,731.3	79,755.7	3.64
29	58	43.50	0.87	532.9	3,863.3	84,027.7	3.70
	59	44.25	0.86	542.1	3,997.1	88,449.1	3.76
30	60	45.00	0.86	551.2	4,134.4	93,024.2	3.83
	61	45.75	0.86	560.4	4,273.3	97,752.2	3.89
31	62	46.50	0.86	569.6	4,414.6	102,639.3	3.96
	63	47.25	0.86	578.8	4,558.1	107,685.9	4.02
32	64	48.00	0.86	588.0	4,704.0	112,896.0	4.08
	65	48.75	0.86	597.2	4,852.1	118,270.6	4.15
33	66	49.50	0.85	606.4	5,002.6	123,814.2	4.21

14¼-in. Width

No. of laminations 1½-in.	No. of laminations ¾-in.	d	C_F	A	S	I	Vol.
9	18	13.50	0.99	192.4	432.8	2,921.7	1.34
	19	14.25	0.98	203.1	482.3	3,436.2	1.41
10	20	15.00	0.98	213.8	534.4	4,007.8	1.48
	21	15.75	0.97	224.4	589.1	4,639.5	1.56
11	22	16.50	0.97	235.1	646.6	5,334.4	1.63
	23	17.25	0.96	245.8	706.7	6,095.4	1.71
12	24	18.00	0.96	256.5	769.5	6,925.5	1.78
	25	18.75	0.95	267.2	835.0	7,827.8	1.86
13	26	19.50	0.95	277.9	903.1	8,805.2	1.93
	27	20.25	0.94	288.6	973.9	9,860.7	2.00
14	28	21.00	0.94	299.3	1,047.4	10,997.4	2.08
	29	21.75	0.94	309.9	1,123.5	12,218.3	2.15
15	30	22.50	0.93	320.6	1,202.3	13,526.4	2.23
	31	23.25	0.93	331.3	1,283.8	14,924.6	2.30
16	32	24.00	0.93	342.0	1,368.0	16,416.0	2.38
	33	24.75	0.92	352.7	1,454.8	18,003.6	2.45
17	34	25.50	0.92	363.4	1,544.3	19,690.4	2.52
	35	26.25	0.92	374.1	1,636.5	21,479.4	2.60
18	36	27.00	0.91	384.8	1,731.4	23,373.6	2.67
	37	27.75	0.91	395.4	1,829.0	25,376.0	2.75
19	38	28.50	0.91	406.1	1,929.1	27,489.6	2.82
	39	29.25	0.91	416.8	2,032.0	29,717.4	2.89
20	40	30.00	0.90	427.5	2,137.5	32,062.5	2.97
	41	30.75	0.90	438.2	2,245.7	34,527.8	3.04
21	42	31.50	0.90	448.9	2,356.6	37,116.4	3.12
	43	32.25	0.90	459.6	2,470.1	39,831.1	3.19
22	44	33.00	0.89	470.3	2,586.4	42,675.2	3.27
	45	33.75	0.89	480.9	2,705.3	45,651.5	3.34
23	46	34.50	0.89	491.6	2,826.8	48,763.1	3.41
	47	35.25	0.89	502.3	2,951.1	52,012.9	3.49
24	48	36.00	0.88	513.0	3,078.0	55,404.0	3.56
	49	36.75	0.88	523.7	3,207.6	58,939.4	3.64
25	50	37.50	0.88	534.4	3,339.8	62,622.1	3.71
	51	38.25	0.88	545.1	3,474.8	66,455.0	3.79
26	52	39.00	0.88	555.8	3,612.4	70,441.3	3.86
	53	39.75	0.87	566.4	3,752.6	74,583.9	3.93
27	54	40.50	0.87	577.1	3,895.6	78,885.8	4.01
	55	41.25	0.87	587.8	4,041.2	83,350.0	4.08
28	56	42.00	0.87	598.5	4,189.5	87,979.5	4.16
	57	42.75	0.87	609.2	4,340.5	92,777.4	4.23
29	58	43.50	0.87	619.9	4,494.1	97,746.5	4.30
	59	44.25	0.86	630.6	4,650.4	102,890.1	4.38
30	60	45.00	0.86	641.3	4,809.4	108,211.0	4.45
	61	45.75	0.86	651.9	4,971.0	113,712.2	4.53
31	62	46.50	0.86	662.6	5,135.3	119,396.7	4.60
	63	47.25	0.86	673.3	5,302.3	125,267.4	4.68
32	64	48.00	0.86	684.0	5,472.0	131,328.0	4.75
	65	48.75	0.86	694.7	5,644.3	137,580.7	4.82
33	66	49.50	0.85	705.4	5,819.3	144,028.8	4.90
	67	50.25	0.85	716.1	5,997.0	150,675.2	4.97
34	68	51.00	0.85	726.8	6,177.4	157,523.0	5.05

Nominal width 10.75 in.

No.	Lam.	d	C_F	A	S	I	Vol.
67	34	50.25	0.85	540.2	4,524.1	113,667.3	3.75
68	34	51.00	0.85	548.3	4,660.1	118,833.2	3.81
69	35	51.75	0.85	556.3	4,798.2	124,153.3	3.86
70	35	52.50	0.85	564.4	4,938.5	129,629.9	3.92
71	36	53.25	0.85	572.4	5,080.4	135,265.2	3.98
72	36	54.00	0.85	580.5	5,224.5	141,061.5	4.03
73	37	54.75	0.84	588.6	5,370.6	147,021.1	4.09
74	37	55.50	0.84	596.6	5,518.8	153,146.2	4.14
75	38	56.25	0.84	604.7	5,668.9	159,439.1	4.20
76	38	57.00	0.84	612.8	5,821.1	165,902.1	4.26
77	39	57.75	0.84	620.8	5,975.2	172,537.4	4.31
78	39	58.50	0.84	628.9	6,131.5	179,347.3	4.37
79	40	59.25	0.84	636.9	6,289.8	186,334.1	4.42
80	40	60.00	0.84	645.0	6,450.0	193,500.0	4.48
81	41	60.75	0.83	653.1	6,612.3	200,847.4	4.54
82	41	61.50	0.83	661.1	6,776.5	208,378.4	4.59
83	42	62.25	0.83	669.2	6,942.8	216,095.3	4.65
84	42	63.00	0.83	677.3	7,111.1	224,000.5	4.70
85	43	63.75	0.83	685.3	7,281.4	232,096.1	4.76
86	43	64.50	0.83	693.4	7,453.8	240,384.5	4.82
87	44	65.25	0.83	701.4	7,628.1	248,867.9	4.87
88	44	66.00	0.83	709.5	7,804.5	257,548.5	4.93
89	45	66.75	0.83	717.6	7,982.9	266,428.8	4.98
90	45	67.50	0.83	725.6	8,163.3	275,510.8	5.04
91	46	68.25	0.82	733.7	8,345.7	284,796.9	5.10
92	46	69.00	0.82	741.8	8,530.1	294,289.3	5.15
93	47	69.75	0.82	749.8	8,716.6	303,990.5	5.21
94	47	70.50	0.82	757.9	8,905.0	313,902.4	5.26
95	48	71.25	0.82	765.9	9,095.5	324,027.5	5.32
96	48	72.00	0.82	774.0	9,288.0	334,368.0	5.38
97	49	72.75	0.82	782.1	9,482.5	344,926.3	5.43
98	49	73.50	0.82	790.1	9,679.0	355,704.5	5.49
99	50	74.25	0.82	798.2	9,877.6	366,704.8	5.54
100	50	75.00	0.82	806.3	10,078.1	377,929.7	5.60

Nominal width 12.25 in.

No.	Lam.	d	C_F	A	S	I	Vol.
69	35	51.75	0.85	633.9	5,467.2	141,476.6	4.40
70	35	52.50	0.85	643.1	5,627.3	147,717.8	4.47
71	36	53.25	0.85	652.3	5,789.3	154,138.9	4.53
72	36	54.00	0.85	661.5	5,953.5	160,744.5	4.59
73	37	54.75	0.84	670.7	6,120.0	167,535.1	4.66
74	37	55.50	0.84	679.9	6,288.8	174,515.4	4.72
75	38	56.25	0.84	689.1	6,459.9	181,685.8	4.79
76	38	57.00	0.84	698.2	6,634.4	189,051.2	4.85
77	39	57.75	0.84	707.4	6,809.1	196,611.7	4.91
78	39	58.50	0.84	716.6	6,987.1	204,372.5	4.98
79	40	59.25	0.84	725.8	7,167.4	212,333.5	5.04
80	40	60.00	0.84	735.0	7,350.0	220,500.0	5.10
81	41	60.75	0.83	744.2	7,534.9	228,871.8	5.17
82	41	61.50	0.83	753.4	7,722.1	237,454.4	5.23
83	42	62.25	0.83	762.6	7,911.6	246,247.3	5.30
84	42	63.00	0.83	771.8	8,103.4	255,256.3	5.36
85	43	63.75	0.83	780.9	8,297.4	264,480.7	5.42
86	43	64.50	0.83	790.1	8,493.8	273,926.5	5.48
87	44	65.25	0.83	799.3	8,692.5	283,592.7	5.55
88	44	66.00	0.83	808.5	8,893.5	293,485.5	5.61
89	45	66.75	0.83	817.7	9,096.1	303,603.8	5.68
90	45	67.50	0.83	826.9	9,302.3	313,954.1	5.74
91	46	68.25	0.82	836.1	9,510.2	324,534.9	5.81
92	46	69.00	0.82	845.2	9,720.4	335,352.9	5.87
93	47	69.75	0.82	854.4	9,932.8	346,406.5	5.93
94	47	70.50	0.82	863.6	10,147.6	357,702.7	6.00
95	48	71.25	0.82	872.8	10,364.6	369,239.4	6.06
96	48	72.00	0.82	882.0	10,584.0	381,024.0	6.12
97	49	72.75	0.82	891.2	10,805.6	393,054.2	6.19
98	49	73.50	0.82	900.4	11,029.6	405,337.6	6.25
99	50	74.25	0.82	909.6	11,255.8	417,871.5	6.32
100	50	75.00	0.82	918.8	11,484.4	430,664.1	6.38
101	51	75.75	0.81	927.9	11,715.2	443,712.2	6.44
102	51	76.50	0.81	937.1	11,948.3	457,024.1	6.51
103	52	77.25	0.81	946.3	12,183.7	470,596.7	6.57
104	52	78.00	0.81	955.5	12,421.5	484,438.5	6.64
105	53	78.75	0.81	964.7	12,661.5	498,545.9	6.70
106	53	79.50	0.81	973.9	12,903.8	512,927.8	6.76
107	54	80.25	0.81	983.1	13,148.4	527,580.3	6.83
108	54	81.00	0.81	992.2	13,395.4	542,512.7	6.89
109	55	81.75	0.81	1,001.4	13,644.5	557,720.6	6.95
110	55	82.50	0.81	1,010.6	13,896.1	573,213.9	7.02
111	56	83.25	0.81	1,019.8	14,149.9	588,987.6	7.08
112	56	84.00	0.81	1,029.0	14,406.0	605,052.0	7.15

Nominal width 14.25 in.

No.	Lam.	d	C_F	A	S	I	Vol.
71	36	53.25	0.85	758.8	6,754.5	179,305.0	5.27
72	36	54.00	0.85	769.5	6,925.5	186,988.5	5.34
73	37	54.75	0.84	780.2	7,119.2	194,888.4	5.42
74	37	55.50	0.84	790.9	7,315.6	203,007.7	5.49
75	38	56.25	0.84	801.6	7,514.6	211,349.5	5.57
76	38	57.00	0.84	812.3	7,716.4	219,916.7	5.64
77	39	57.75	0.84	822.9	7,920.8	228,712.4	5.71
78	39	58.50	0.84	833.6	8,127.8	237,739.5	5.79
79	40	59.25	0.84	844.3	8,337.6	247,001.0	5.86
80	40	60.00	0.84	855.0	8,550.0	256,500.0	5.94
81	41	60.75	0.83	865.7	8,765.1	266,239.5	6.01
82	41	61.50	0.83	876.4	8,982.8	276,222.5	6.09
83	42	62.25	0.83	887.1	9,203.3	286,451.9	6.16
84	42	63.00	0.83	897.8	9,426.4	296,930.8	6.23
85	43	63.75	0.83	908.4	9,652.1	307,662.3	6.31
86	43	64.50	0.83	919.1	9,880.6	318,649.2	6.38
87	44	65.25	0.83	929.8	10,111.7	329,894.6	6.46
88	44	66.00	0.83	940.5	10,345.5	341,401.5	6.53
89	45	66.75	0.83	951.2	10,582.0	353,173.0	6.61
90	45	67.50	0.83	961.9	10,821.1	365,212.0	6.68
91	46	68.25	0.82	972.6	11,062.9	377,521.5	6.75
92	46	69.00	0.82	983.3	11,307.4	390,104.5	6.83
93	47	69.75	0.82	993.9	11,554.5	402,964.0	6.90
94	47	70.50	0.82	1,004.6	11,804.3	416,103.1	6.98
95	48	71.25	0.82	1,015.3	12,056.8	429,524.8	7.05
96	48	72.00	0.82	1,026.0	12,312.0	443,232.0	7.12
97	49	72.75	0.82	1,036.7	12,569.8	457,227.8	7.20
98	49	73.50	0.82	1,047.4	12,830.3	471,515.1	7.27
99	50	74.25	0.82	1,058.1	13,093.5	486,097.1	7.35
100	50	75.00	0.82	1,068.8	13,359.4	500,976.6	7.42
101	51	75.75	0.81	1,079.4	13,627.9	516,156.7	7.50
102	51	76.50	0.81	1,090.1	13,899.1	531,640.5	7.57
103	52	77.25	0.81	1,100.8	14,173.0	547,430.7	7.64
104	52	78.00	0.81	1,111.5	14,449.5	563,530.6	7.72
105	53	78.75	0.81	1,122.2	14,728.7	579,943.1	7.79
106	53	79.50	0.81	1,132.9	15,010.6	596,671.2	7.87
107	54	80.25	0.81	1,143.6	15,295.1	613,718.0	7.94
108	54	81.00	0.81	1,154.3	15,582.4	631,086.2	8.02
109	55	81.75	0.81	1,164.9	15,872.3	648,779.2	8.09
110	55	82.50	0.81	1,175.6	16,164.8	666,799.8	8.16
111	56	83.25	0.81	1,186.3	16,460.1	685,151.2	8.24
112	56	84.00	0.81	1,197.0	16,758.0	703,836.1	8.31
113	57	84.75	0.80	1,207.7	17,058.6	722,857.7	8.39
114	57	85.50	0.80	1,218.4	17,361.8	742,218.8	8.46
115	58	86.25	0.80	1,229.1	17,667.8	761,922.8	8.54
116	58	87.00	0.80	1,239.8	17,976.4	781,972.3	8.61
117	59	87.75	0.80	1,250.4	18,287.6	802,370.6	8.68
118	59	88.50	0.80	1,261.1	18,601.6	823,120.6	8.76
119	60	89.25	0.80	1,271.8	18,918.2	844,225.2	8.83
120	60	90.00	0.80	1,282.5	19,237.5	865,687.6	8.91
121	61	90.75	0.80	1,293.2	19,559.5	887,510.6	8.98
122	61	91.50	0.80	1,303.9	19,884.1	909,697.3	9.05
123	62	92.25	0.80	1,314.6	20,211.4	932,250.8	9.13
124	62	93.00	0.80	1,325.3	20,541.4	955,174.0	9.20
125	63	93.75	0.80	1,335.9	20,874.0	978,470.0	9.28
126	63	94.50	0.80	1,346.6	21,209.3	1,002,141.6	9.35
127	64	95.25	0.79	1,357.3	21,547.3	1,026,192.0	9.43
128	64	96.00	0.79	1,368.0	21,888.0	1,050,624.2	9.50

* d = section depth, in.; C_F = size factor; A = cross-sectional area, sq in.; S = section modulus, in.^{3}; I = moment of inertia, in.^{4}; Vol. = volume, cu ft per lin ft.

The strength of wood is actually affected by many other factors, such as rate of loading, duration of load, temperature, direction of grain, and position of growth rings. Strength is also influenced by such inherent growth characteristics as knots, cross grain, shakes, and checks.

Analysis and integration of available data have yielded a comprehensive set of simple principles for grading structural timber (Tentative Methods for Establishing Structural Grades of Lumber, ASTM D245).

The same characteristics, such as knots and cross grain, that reduce the strength of solid timber also affect the strength of laminated members. However, there are additional factors, peculiar to laminated wood, that must be considered: Effect on strength of bending members is less from knots located at the neutral plane of the beam, a region of low stress. Strength of a bending member with low-grade laminations can be improved by substituting a few high-grade laminations at the top and bottom of the member. Dispersement of knots in laminated members has a beneficial effect on strength. With sufficient knowledge of the occurrence of knots within a grade, mathematical estimates of this effect may be established for members containing various numbers of laminations.

Allowable design stresses taking these factors into account are higher than for solid timbers of comparable grade. But cross-grain limitations must be more restrictive than for solid timbers, to justify these higher allowable stresses.

11-8. Weight and Specific Gravity of Commercial Lumber Species. Specific gravity is a reliable indicator of fiber content. Also, specific gravity and the strength and stiffness of solid wood or laminated products are interrelated. See Table 11-5 for weights and specific gravities of several commercial lumber species.

11-9. Moisture Content of Wood. Wood is unlike most other structural materials in regard to the causes of its dimensional changes. These are primarily from gain or loss of moisture, not change in temperature. It is for this reason that expansion joints are seldom required for wood structures to permit movement with temperature changes. It partly accounts for the fact that wood structures can withstand extreme temperatures without collapse.

A newly felled tree is green (contains moisture). In removing the greater part of this water, seasoning first allows free water to leave the cavities in the wood. A point is reached where these cavities contain only air, and the cell walls still are full of moisture. The moisture content at which this occurs, the fiber-saturation point, varies from 25 to 30% of the weight of the oven-dry wood.

During removal of the free water, the wood remains constant in size and in most properties (weight decreases). Once the fiber-saturation point has been passed, shrinkage of the wood begins as the cell walls lose water. Shrinkage continues nearly linearly down to zero moisture content (Table 11-6). (There are, however, complicating factors, such as the effects of timber size and relative rates of moisture movement in three directions, longitudinal, radial, and tangential to the growth rings.) Eventually, the wood assumes a condition of equilibrium, with the final moisture content dependent on the relative humidity and temperature of the ambient air. Wood swells when it absorbs moisture, up to the fiber-saturation point. The relationship of wood moisture content, temperature, and relative humidity can actually define an environment (Fig. 11-1).

This explanation has been simplified. Outdoors, rain, frost, wind, and sun can act directly on the wood. Within buildings, poor environmental conditions may be created for wood by localized heating, cooling, or ventilation. The conditions of service must be sufficiently well known to be specifiable. Then, the proper design stress can be assigned to wood and the most suitable adhesive selected.

Dry Condition of Use. *Design Stresses.* Allowable unit stresses for dry condition of use are applicable for normal loading when the wood moisture content in service is less than 16%, as in most covered structures.

Adhesive Selection. Dry-use adhesives are those that perform satisfactorily when the moisture-content of wood does not exceed 16% for repeated or prolonged periods of service, and are to be used only when these conditions exist.

Wet Condition of Use. *Design Stresses.* Allowable unit stresses for wet condition of use are applicable for normal loading when the moisture content in service is 16% or more. This may occur in members not covered or in covered locations of high relative humidity.

Adhesive Selection. Wet-use adhesives will perform satisfactorily for all conditions, including exposure to weather, marine use, and where pressure treatments are used, whether before or after gluing. Such adhesives are required when the moisture content exceeds 16% for repeated or prolonged periods of service.

Table 11-5. Weights and Specific Gravities of Commercial Lumber Species

Species	Specific gravity based on oven-dry weight and volume at 12% moisture content	Weight, lb per cu ft			Moisture content when green (avg), %	Specific gravity based on oven-dry weight and volume when green	Weight when green, lb per cu ft
		At 12% moisture content	At 20% moisture content	Adjusting factor for each 1% change in moisture content			
Softwoods:							
Cedar:							
Alaska	0.44	31.1	32.4	0.170	38	0.42	35.5
Incense	0.37	25.0	26.4	0.183	108	0.35	42.5
Port Orford	0.42	29.6	31.0	0.175	43	0.40	35.0
Western red	0.33	23.0	24.1	0.137	37	0.31	26.4
Cypress, southern	0.46	32.1	33.4	0.167	91	0.42	45.3
Douglas fir:							
Coast region	0.48	33.8	35.2	0.170	38	0.45	38.2
Inland region	0.44	31.4	32.5	0.137	48	0.41	36.3
Rocky Mountain	0.43	30.0	31.4	0.179	38	0.40	34.6
Fir, white	0.37	26.3	27.3	0.129	115	0.35	39.6
Hemlock:							
Eastern	0.40	28.6	29.8	0.150	111	0.38	43.4
Western	0.42	29.2	30.2	0.129	74	0.38	37.2
Larch, western	0.55	38.9	40.2	0.170	58	0.51	46.7
Pine:							
Eastern white	0.35	24.9	26.2	0.167	73	0.34	35.1
Lodgepole	0.41	28.8	29.9	0.142	65	0.38	36.3
Norway	0.44	31.0	32.1	0.142	92	0.41	42.3
Ponderosa	0.40	28.1	29.4	0.162	91	0.38	40.9
Southern shortleaf	0.51	35.2	36.5	0.154	81	0.46	45.9
Southern longleaf	0.58	41.1	42.5	0.179	63	0.54	50.2
Sugar	0.36	25.5	26.8	0.162	137	0.35	45.8
Western white	0.38	27.6	28.6	0.129	54	9.36	33.0
Redwood	0.40	28.1	29.5	0.175	112	0.38	45.6
Spruce:							
Engelmann	0.34	23.7	24.7	0.129	80	0.32	32.5
Sitka	0.40	27.7	28.8	0.145	42	0.37	32.0
White	0.40	29.1	29.9	0.104	50	0.37	33.0
Hardwoods:							
Ash, white	0.60	42.2	43.6	0.175	42	0.55	47.4
Beech, American	0.64	43.8	45.1	0.162	54	0.56	50.6
Birch:							
Sweet	0.65	46.7	48.1	0.175	53	0.60	53.8
Yellow	0.62	43.0	44.1	0.142	67	0.55	50.8
Elm, rock	0.63	43.6	45.2	0.208	48	0.57	50.9
Gum	0.52	36.0	37.1	0.133	115	0.46	49.7
Hickory:							
Pecan	0.66	45.9	47.6	0.212	63	0.60	56.7
Shagbark	0.72	50.8	51.8	0.129	60	0.64	57.0
Maple, sugar	0.63	44.0	45.3	0.154	58	0.56	51.1
Oak:							
Red	0.63	43.2	44.7	0.187	80	0.56	56.0
White	0.68	46.3	47.6	0.167	68	0.60	55.6
Poplar, yellow	0.42	29.8	31.0	0.150	83	0.40	40.5

11-10. Checking in Timbers. Separation of grain, or checking, is the result of rapid lowering of surface moisture content combined with a difference in moisture content between inner and outer portions of the piece. As wood loses moisture to the surrounding atmosphere, the outer cells of the member lose at a more rapid rate than the inner cells. As the outer cells try to shrink,

Table 11-6. Shrinkage Values of Wood Based on Dimensions When Green

Species	Dried to 20% MC*			Dried to 6% MC†			Dried to 0% MC		
	Radial, %	Tangential, %	Volumetric, %	Radial, %	Tangential, %	Volumetric, %	Radial, %	Tangential, %	Volumetric, %
Softwoods:‡									
Cedar:									
Alaska	0.9	2.0	3.1	2.2	4.8	7.4	2.8	6.0	9.2
Incense	1.1	1.7	2.5	2.6	4.2	6.1	3.3	5.2	7.6
Port Orford	1.5	2.3	3.4	3.7	5.5	8.1	4.6	6.9	10.1
Western red	0.8	1.7	2.3	1.9	4.0	5.4	2.4	5.0	6.8
Cypress, southern	1.3	2.1	3.5	3.0	5.0	8.4	3.8	6.2	10.5
Douglas fir:									
Coast region	1.7	2.6	3.9	4.0	6.2	9.4	5.0	7.8	11.8
Inland region	1.4	2.5	3.6	3.3	6.1	8.7	4.1	7.6	10.9
Rocky Mountain	1.2	2.1	3.5	2.9	5.0	8.5	3.6	6.2	10.6
Fir, white	1.1	2.4	3.3	2.6	5.7	7.8	3.2	7.1	9.8
Hemlock:									
Eastern	1.0	2.3	3.2	2.4	5.4	7.8	3.0	6.8	9.7
Western	1.4	2.6	4.0	3.4	6.3	9.5	4.3	7.9	11.9
Larch, western	1.4	2.7	4.4	3.4	6.5	10.6	4.2	8.1	13.2
Pine:									
Eastern white	0.8	2.0	2.7	1.8	4.8	6.6	2.3	6.0	8.2
Lodgepole	1.5	2.2	3.8	3.6	5.4	9.2	4.5	6.7	11.5
Norway	1.5	2.4	3.8	3.7	5.8	9.2	4.6	7.2	11.5
Ponderosa	1.3	2.1	3.2	3.1	5.0	7.7	3.9	6.3	9.6
Southern (avg.)	1.6	2.6	4.1	4.0	6.1	9.8	5.0	7.6	12.2
Sugar	1.0	1.9	2.6	2.3	4.5	6.3	2.9	5.6	7.9
Western white	1.4	2.5	3.9	3.3	5.9	9.4	4.1	7.4	11.8
Redwood (old growth)	0.9	1.5	2.3	2.1	3.5	5.4	2.6	4.4	6.8
Spruce:									
Engelmann	1.1	2.2	3.5	2.7	5.3	8.3	3.4	6.6	10.4
Sitka	1.4	2.5	3.8	3.4	6.0	9.2	4.3	7.5	11.5
Hardwoods:‡									
Ash, white	1.6	2.6	4.5	3.8	6.2	10.7	4.8	7.8	13.4
Beech, American	1.7	3.7	5.4	4.1	8.8	13.0	5.1	11.0	16.3
Birch:									
Sweet	2.2	2.8	5.2	5.2	6.8	12.5	6.5	8.5	15.6
Yellow	2.4	3.1	5.6	5.8	7.4	13.4	7.2	9.2	16.7
Elm, rock	1.6	2.7	4.7	3.8	6.5	11.3	4.8	8.1	14.1
Gum, red	1.7	3.3	5.0	4.2	7.9	12.0	5.2	9.9	15.0
Hickory:									
Pecan§	1.6	3.0	4.5	3.9	7.1	10.9	4.9	8.9	13.6
True	2.5	3.8	6.0	6.0	9.0	14.3	7.5	11.3	17.9
Maple, hard	1.6	3.2	5.0	3.9	7.6	11.9	4.9	9.5	14.9
Oak:									
Red	1.3	2.7	4.5	3.2	6.6	10.8	4.0	8.2	13.5
White	1.8	3.0	5.3	4.2	7.2	12.6	5.3	9.0	15.8
Poplar, yellow	1.3	2.4	4.1	3.2	5.7	9.8	4.0	7.1	12.3

*MC = moisture content, as a percent of weight of oven-dry wood. These shrinkage values have been taken as one-third the shrinkage to the oven-dry condition as given in the last three columns of this table.

†These shrinkage values have been taken as four-fifths of the shrinkage to the oven-dry condition as given in the last three columns of this table.

‡The total longitudinal shrinkage of normal species from fiber saturation to oven-dry condition is minor. It usually ranges from 0.17 to 0.3% of the green dimension.

§Average of butternut hickory, nutmeg hickory, water hickory, and pecan.

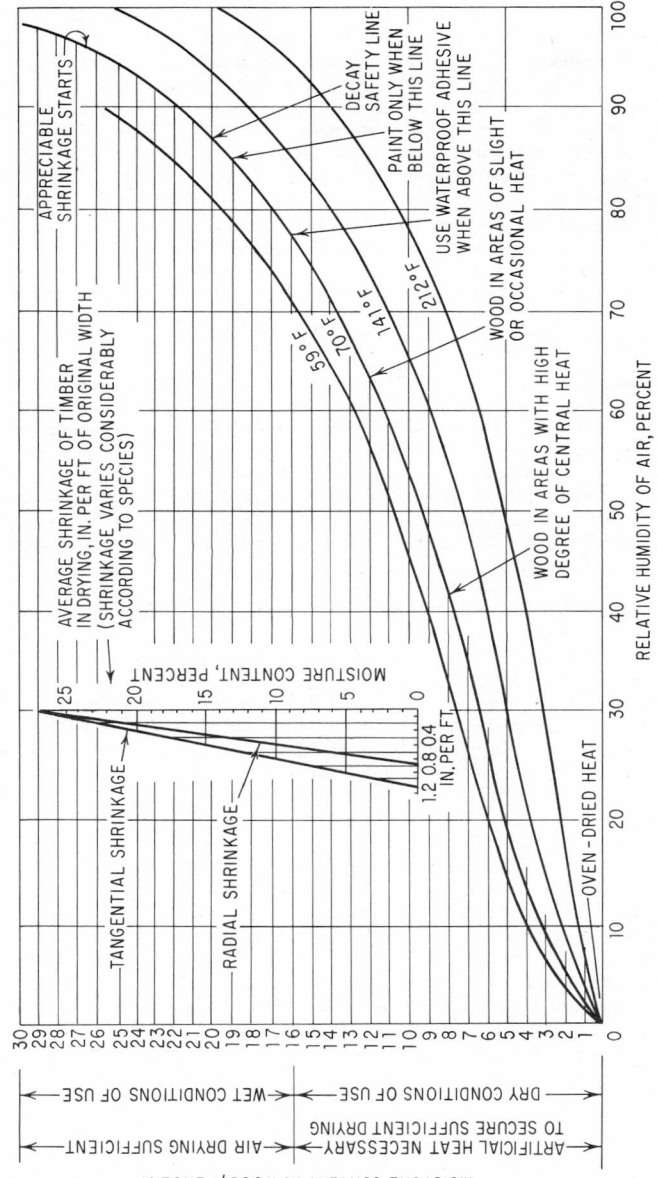

Fig. 11-1. Approximate relationship of wood equilibrium moisture content and temperature and relative humidity is shown by the curves. The triangular diagram indicates the effect of wood moisture content on shrinkage.

they are restrained by the inner portion of the member. The more rapid the drying, the greater will be the differential in shrinkage between outer and inner fibers, and the greater the shrinkage stresses. Splits may develop.

Checks affect the horizontal shear strength of timber. A large reduction factor is applied to test values in establishing allowable unit stresses, in recognition of stress concentrations at the ends of checks. Allowable unit stresses for horizontal shear are adjusted for the amount of checking permissible in the various stress grades at the time of the grading. Since strength properties of wood increase with dryness, checks may enlarge with increasing dryness after shipment without appreciably reducing shear strength.

Cross-grain checks and splits that tend to run out the side of a piece, or excessive checks and splits that tend to enter connection areas, may be serious and may require servicing. Provisions for controlling the effects of checking in connection areas may be incorporated in design details.

To avoid excessive splitting between rows of bolts due to shrinkage during seasoning of solid-sawn timbers, the rows should not be spaced more than 5 in. apart, or a saw kerf, terminating in a bored hole, should be provided between the lines of bolts. Whenever possible, maximum end distances for connections should be specified to minimize the effect of checks running into the joint area. Some designers require stitch bolts in members, with multiple connections loaded at an angle to the grain. Stitch bolts, kept tight, will reinforce pieces where checking is excessive.

One of the principal advantages of glued-laminated timber construction is relative freedom from checking. Seasoning checks may, however, occur in laminated members for the same reasons that they exist in solid-sawn members. When laminated members are glued within the range of moisture contents set in AITC 103, "Standard for Structural Glued-Laminated Timber," they will approximate the moisture content in normal-use conditions, thereby minimizing checking. Moisture content of the lumber at the time of gluing is thus of great importance to the control of checking in service. However, rapid changes in moisture content of large wood sections after gluing will result in shrinkage or swelling of the wood, and during shrinking, checking may develop in both glued joints and wood.

Differentials in shrinkage rates of individual laminations tend to concentrate shrinkage stresses at or near the glue line. For this reason, when checking occurs, it is usually at or near glue lines. The presence of wood-fiber separation indicates adequate glue bonds, and not delamination.

In general, checks have very little effect on the strength of glued-laminated members. Laminations in such members are thin enough to season readily in kiln drying without developing checks. Since checks lie in a radial plane, and the majority of laminations are essentially flat grain, checks are so positioned in horizontally laminated members that they will not materially affect shear strength. When members are designed with laminations vertical (with wide face parallel to the direction of load application), and when checks may affect the shear strength, the effect of checks may be evaluated in the same manner as for checks in solid-sawn members.

Seasoning checks in bending members affect only the horizontal shear strength. They are usually not of structural importance unless the checks are significant in depth and occur in the midheight of the member near the support, and then only if shear governs the design of the members. The reduction in shear strength is nearly directly proportional to the ratio of depth of check to width of beam. Checks in columns are not of structural importance unless the check develops into a split, thereby increasing the slenderness ratio of the columns.

Minor checking may be disregarded, since there is an ample factor of safety in allowable unit stresses. The final decision as to whether shrinkage checks are detrimental to the strength requirements of any particular design or structural member should be made by a competent engineer experienced in timber construction.

11-11. Allowable Unit Stresses and Modifications for Stress-grade Lumber. (See also Arts. 11-6 to 11-10.) A wide range of allowable unit stresses for many species is given in the "National Design Specification for Stress-grade Lumber and Its Fastenings" (NDS). They apply to solid-sawn lumber under normal duration of load and under continuously dry service conditions, such as those in most covered structures. For continuously wet-use service conditions, apply the appropriate factor from Table 11-7. NDS may be obtained from the National Forest Products Association, 619 Massachusetts Ave., N.W., Washington, D.C. 20036.

The modifications for duration of load, temperature, treatment, and size are the same as for glued-laminated timber (Art. 11-12).

11-12. Allowable Unit Stresses and Modifications for Structural Glued-Laminated Timber. (See also Arts. 11-6 to 11-10.) The allowable unit stresses given in the AITC "Timber Construction Manual" and "Standard Specification for Structural Glued-Laminated Timber," AITC 117,

Table 11-7. Allowable-Stress Modification Factors for Moisture in Solid-Sawn Lumber

| Service condition | Allowable unit stress | | | | Modulus of elasticity E |
	Extreme fiber in bending F_b or tension parallel to grain F_t	Compression parallel to grain F_c	Compression perpendicular to grain F_c	Horizontal shear F_v	
At or above fiber-saturation point, or continuously submerged	1.00	0.90	0.67	1.00	0.91

are for normal conditions of loading, assuming uniform loading, horizontal laminating, and a 12-in.-deep member with a span-to-depth ratio of 21:1. Tables are given for dry-use and for wet-use conditions. AITC specifications may be obtained from the American Institute for Timber Construction, 333 W. Hampden Ave., Englewood, Colo. 80110.

Allowable unit stresses for dry-use conditions are applicable when the moisture content in service is less than 16%, as in most covered structures. Allowable unit stresses for wet-use conditions are applicable when the moisture content in service is 16% or more, as may occur in exterior or submerged construction and in some structures housing wet processes or otherwise having constantly high relative humidities.

Allowable stresses for vertically laminated members made of combination grades of lumber are the weighted average of the lumber grades.

Requirements for limiting cross grain, type, and location of end joints, and certain manufacturing and other requirements, must be met for these allowable-unit-stress combinations to apply.

Species other than those referenced in these specifications may be used if allowable unit stresses are established for them in accordance with the provisions of U.S. Product Standard PS56-73.

For laminated bending members 16¼ in. or more in depth, the outermost tension-side laminations representing 5% of the total depth of the member should have additional grade restrictions, as described in AITC 117, "Standard Specification for Structural Glued-Laminated Timber of Douglas Fir, Southern Pine, Western Larch, and California Redwood."

Use of hardwoods in glued-laminated timbers is covered by AITC 119, "Standard Specification for Hardwood Glued-Laminated Timber."

Lumber that is E-rated, as well as visually graded and then positioned in laminated timber, is covered by AITC 120, "Standard Specification for Structural Glued-Laminated Timber Using E-rated and Visually Graded Lumber of Douglas Fir, Southern Pine, Hemlock, and Lodgepole Pine." The lumber is sorted into E grades by measuring the modulus of elasticity of each piece and positioning the lumber in the member according to stiffness, but still grading visually so as to meet the visual grade requirements.

Modifications for Duration of Load. Wood can absorb overloads of considerable magnitude for short periods; thus, allowable unit stresses are adjusted accordingly. The elastic limit and ultimate strength are higher under short-time loading. Wood members under continuous loading for years will fail at loads one-half to three-fourths as great as would be required to produce failure in a static-bending test when the maximum load is reached in a few minutes.

Normal load duration contemplates fully stressing a member to the allowable unit stress by the application of the full design load for a duration of about 10 years (either continuously or cumulatively).

When a member is fully stressed by maximum design loads for long-term loading conditions (greater than 10 years, either continuously or cumulatively), the allowable unit stresses are reduced to 90% of the tabulated values.

When duration of full design load (either continuously or cumulatively) does not exceed 10 years, tabulated allowable unit stresses can be increased as follows: 15% for 2-month duration, as for snow; 25% for 7-day duration; 33⅓% for wind or earthquake; and 100% for impact. These increases are not cumulative. The allowable unit stress for normal loading may be used without regard to impact if the stress induced by impact does not exceed the allowable unit stress for normal loading. These adjustments do not apply to modulus of elasticity, except when it is used to determine allowable unit loads for columns.

Modifications for Temperature. Tests show that wood increases in strength as temperature is lowered below normal. Tests conducted at about $-300°$ F indicate that the important strength properties of dry wood in bending and compression, including stiffness and shock resistance, are much higher at extremely low temperatures.

Some reduction of the allowable unit stresses may be necessary for members subjected to elevated temperatures for repeated or prolonged periods. This adjustment is especially desirable where high temperature is associated with high moisture content.

Temperature effect on strength is immediate. Its magnitude depends on the moisture content of the wood and, when temperature is raised, on the duration of exposure.

Between 0 and 70°F, the static strength of dry wood (12% moisture content) roughly increases from its strength at 70°F about ⅓ to½% for each 1°F decrease in temperature. Between 70 and 150°F, the strength decreases at about the same rate for each 1°F increase in temperature. The change is greater for higher wood moisture content.

After exposure to temperatures not much above normal for a short time under ordinary atmospheric conditions, the wood, when temperature is reduced to normal, may recover essentially all its original strength. Experiments indicate that air-dry wood can probably be exposed to temperatures up to nearly 150°F for a year or more without a significant permanent loss in most strength properties. But its strength while at such temperatures will be temporarily lower than at normal temperature.

When wood is exposed to temperatures of 150°F or more for extended periods of time, it will be permanently weakened. The nonrecoverable strength loss depends on a number of factors, including moisture content and temperature of the wood, heating medium, and time of exposure. To some extent, the loss depends on the species and size of the piece.

Glued-laminated members are normally cured at temperatures of less than 150°F. Therefore, no reduction in allowable unit stresses due to temperature effect is necessary for curing.

Adhesives used under standard specifications for structural glued-laminated members, for example, casein, resorcinol-resin, phenol-resin, and melamine-resin adhesives, are not affected substantially by temperatures up to those that char wood. Use of adhesives that deteriorate at high temperatures is not permitted by standard specifications for structural glued-laminated timber. Low temperatures appear to have no significant effect on the strength of glued joints.

Modifications for Pressure-applied Treatments. The allowable stresses given for wood also apply to wood treated with a preservative when this treatment is in accordance with American Wood Preservers Association (AWPA) standard specifications. These limit pressure and temperature. Investigations have indicated that, in general, any weakening of timber as a result of preservative treatment is caused almost entirely by subjecting the wood to temperatures and pressures above the AWPA limits.

Highly acidic salts, such as zinc chloride, tend to hydrolyze wood if they are present in appreciable concentrations. Fortunately, the concentrations used in wood preservative treatments are sufficiently small that strength properties other than impact resistance are not greatly affected under normal service conditions. A significant loss in impact strength may occur, however, if higher concentrations are used.

None of the other common salt preservatives is likely to form solutions as highly acidic as those of zinc chloride. So, in most cases, their effect on the strength of the wood can be disregarded.

In wood treated with highly acidic salts, such as zinc chloride, moisture is the controlling factor in corrosion of fastenings. Therefore, wood treated with highly acidic salts is not recommended for use under highly humid conditions.

The effects on strength of all treatments, preservative and fire-retardant, should be investigated, to assure that adjustments in allowable unit stresses are made when required. ("Manual of Recommended Practice," American Wood Preservers Association.)

Modifications for Size Factor. When the depth of a rectangular beam exceeds 12 in., the tabulated unit stress in bending F_b should be reduced by multiplication by a size factor C_F.

$$C_F = \left(\frac{12}{d}\right)^{1/9} \tag{11-1}$$

where d = depth of member, in.

Table 11-4 lists values of C_F for various cross-sectional sizes.

Equation (11-1) applies to bending members satisfying the following basic assumptions: simply supported beam, uniformly distributed load, and span-depth ratio L/d = 21. C_F may thus be applied with reasonable accuracy to beams usually used in buildings. Where greater accuracy is required for other loading conditions or span-depth ratios, the percentage changes given in Table 11-8 may be applied directly to the size factor given by Eq. (11-1). Straight-line interpolation may be used for L/d ratios other than those listed in the table.

Table 11-8. Change in Size Factor C_F

Span-depth ratio	Change, %	Loading condition for simply supported beams	Change, %
7	+6.2	Single concentrated load	+7.8
14	+2.3	Uniform load	0
21	0	Third-point load	−3.2
28	−1.6		
35	−2.8		

Modifications for Radial Tension or Compression. The radial stress induced by a bending moment in a member of constant cross section may be computed from

$$f_r = \frac{3M}{2Rbd} \qquad (11\text{-}2)$$

where M = bending moment, in.-lb
R = radius of curvature at centerline of member, in.
b = width of cross section, in.
d = depth of cross section, in.

Equation (11-2) can also be used to estimate the stresses in a member with varying cross section. Information on more exact procedures for calculating radial stresses in curved members with varying cross section can be obtained from the American Institute of Timber Construction.

When M is in the direction tending to decrease curvature (increase the radius), tensile stresses occur across the grain. For this condition, the allowable tensile stress across the grain is limited to one-third the allowable unit stress in horizontal shear for southern pine for all load conditions, and for Douglas fir and larch for wind or earthquake loadings. The limit is 15 psi for Douglas fir and larch for other types of loading. These values are subject to modification for duration of load. If these values are exceeded, mechanical reinforcement sufficient to resist all radial tensile stresses is required.

When M is in the direction tending to increase curvature (decrease the radius), the stress is compressive across the grain. For this condition, the allowable stress is limited to that for compression perpendicular to grain for all species covered by AITC 117.

Modifications for Curvature Factor. For the curved portion of members, the allowable unit stress in bending should be modified by multiplication by the following curvature factor:

$$C_c = 1 - 2{,}000 \left(\frac{t}{R} \right)^2 \qquad (11\text{-}3)$$

where t = thickness of lamination, in.
R = radius of curvature of lamination, in.

t/R should not exceed $1/100$ for hardwoods and southern pine, or $1/125$ for softwoods other than southern pine. The curvature factor should not be applied to stress in the straight portion of an assembly, regardless of curvature elsewhere.

The recommended minimum radii of curvature for curved, structural glued-laminated members are 9 ft 4 in. for ¾-in. laminations, and 27 ft 6 in. for 1½-in. laminations. Other radii of curvature may be used with these thicknesses, and other radius-thickness combinations may be used.

Certain species can be bent to sharper radii, but the designer should determine the availability of such sharply curved members before specifying them.

Modifications for Lateral Stability. The tabulated allowable bending unit stresses are applicable to members that are adequately braced. When deep and slender members not adequately braced are used, allowable bending unit stresses must be reduced. For the purpose, a slenderness factor should be applied, as indicated in the AITC "Timber Construction Manual."

The reduction in bending stresses determined by applying the slenderness factor should not be combined with a reduction in stress due to the application of the size factor. In no case may the allowable bending unit stress used in design exceed the stress determined by applying a size factor.

11-13. Lateral Support of Wood Framing. To prevent beams and compression members from buckling, they may have to be braced laterally. Need for such bracing and required spacing depend on the unsupported length and cross-sectional dimensions of members.

When buckling occurs, a member deflects in the direction of its least dimension b, unless prevented by bracing. (In a beam, b usually is taken as the width.) But if bracing precludes buckling in that direction, deflection can occur in the direction of the perpendicular dimension d. Thus, it is logical that unsupported length L, b, and d play important roles in rules for lateral support, or in formulas for reducing allowable stresses for buckling.

For glued-laminated beams, design for lateral stability is based on a function of Ld/b^2. For solid-sawn beams of rectangular cross section, maximum depth-width ratios should satisfy the approximate rules, based on nominal dimensions, summarized in Table 11-9. When the beams are adequately braced laterally, the depth of the member below the brace may be taken as the width. For glued-laminated arches, maximum depth-width ratios should satisfy the approximate rules, based on actual dimensions, in Table 11-10.

Table 11-9. Approximate Lateral-Support Rules for Sawn Beams*

Depth-width ratio (nominal dimensions)	Rule
2 or less	No lateral support required
3	Hold ends in position
4	Hold ends in position and member in line, e.g., with purlins or sag rods
5	Hold ends in position and compression edge in line, e.g., with direct connection of sheathing, decking, or joists
6	Hold ends in position and compression edge in line, as for 5 to 1, and provide adequate bridging or blocking at intervals not exceeding 6 times the depth
7	Hold ends in position and both edges firmly in line.

If a beam is subject to both flexure and compression parallel to grain, the ratio may be as much as 5:1 if one edge is held firmly in line, e.g., by rafters (or roof joists) and diagonal sheathing. If the dead load is sufficient to induce tension on the underside of the rafters, the ratio for the beam may be 6:1.

*"Wood Handbook," Agricultural Handbook 72, U.S. Forest Service, Forest Products Laboratory, Madison, Wis., 1955.

Table 11-10. Approximate Lateral-Support Rules for Glued-Laminated Arches

Depth-width ratio (actual dimensions)	Rule
5 or less	No lateral support required
6	Brace one edge at frequent intervals

Where joists frame into arches or the compression chords of trusses and provide adequate lateral bracing, the depth, rather than the width, of arch or truss chord may be taken as b. The joists should be erected with upper edges at least ½ in. above the supporting member, but low enough to provide adequate lateral support. The depth of arch or compression chord also may be taken as b where joists or planks are placed on top of the arch or chord and securely fastened to them and blocking is firmly attached between the joists.

For glued-laminated beams, no lateral support is required when the depth does not exceed the width. In that case also, the allowable bending stress does not have to be adjusted for lateral instability. Similarly, if continuous support prevents lateral movement of the compression flange, lateral buckling cannot occur and the allowable stress need not be reduced.

When the depth of a glued-laminated beam exceeds the width, bracing must be provided at supports. This bracing must be so placed as to prevent rotation of the beam in a plane perpendicular to its longitudinal axis. And unless the compression flange is braced at sufficiently close intervals between the supports, the allowable stress should be adjusted for lateral buckling. All other modifications of the stresses, except for size factor, should be made.

When the buckling factor

$$C = \frac{L_e d}{b^2} \tag{11-4}$$

where L_e is the effective length, in., and b and d also are in in., does not exceed 100, the allowable bending stress does not have to be adjusted for lateral buckling. Such beams are classified as short beams.

For intermediate beams, for which C exceeds 100 but is less than

$$K = \frac{3E}{5F_b} \tag{11-5}$$

the allowable bending stress should be determined from

$$F_b' = F_b \left[1 - \frac{1}{3} \left(\frac{C}{K} \right)^2 \right] \tag{11-6}$$

where E = modulus of elasticity, psi
    F_b = allowable bending stress, psi, adjusted, except for size factor

For long beams, for which C exceeds K but is less than 2,500, the allowable bending stress should be determined from

$$F_b' = \frac{0.40E}{C} \tag{11-7}$$

In no case should C exceed 2,500.

The effective length L_e for Eq. (11-4) is given in terms of unsupported length of beam in Table 11-11. Unsupported length is the distance between supports or the length of a cantilever

Table 11-11. Ratio of Effective Length to Unsupported Length of Glued-Laminated Beams

Simple beam, load concentrated at center	1.61
Simple beam, uniformly distributed load	1.92
Simple beam, equal end moments	1.84
Cantilever beam, load concentrated at unsupported end	1.69
Cantilever beam, uniformly distributed load	1.06
Simple or cantilever beam, any load (conservative value)	1.92

when the beam is laterally braced at the supports to prevent rotation and adequate bracing is not installed elsewhere in the span. When both rotational and lateral displacement are also prevented at intermediate points, the unsupported length may be taken as the distance between points of lateral support. If the compression edge is supported throughout the length of the beam and adequate bracing is installed at the supports, the unsupported length is zero.

Acceptable methods of providing adequate bracing at supports include anchoring the bottom of a beam to a pilaster and the top of the beam to a parapet; for a wall-bearing roof beam, fastening the roof diaphragm to the supporting wall or installing a girt between beams at the top of the wall; for beams on wood columns, providing rod bracing.

For continuous lateral support of a compression flange, composite action is essential between deck elements, so that sheathing or deck acts as a diaphragm. One example is a plywood deck with edge nailing. With plank decking, nails attaching the plank to the beams must form couples, to resist rotation. In addition, the planks must be nailed to each other, for diaphragm action. Adequate lateral support is not provided when only one nail is used per plank and no nails are used between planks.

(American Institute of Timber Construction, "Timber Construction Manual," John Wiley & Sons, Inc., New York; R. F. Hooley and B. Madsen, "Lateral Stability of Glued-Laminated Beams," *Journal of the Structural Division*, No. ST3, June, 1964, American Society of Civil Engineers.)

11-14. Combined Stresses in Timber. Allowable unit stresses given in the "National Design Specification for Stress-grade Lumber and Its Fastenings" apply directly to bending, horizontal shear, tension parallel or perpendicular to grain, and compression parallel or perpendicular to grain. For combined axial stress P/A and bending stress M/S, stresses are limited by

$$\frac{P/A}{F} + \frac{M/S}{F_b} \leq 1 \qquad (11\text{-}8)$$

where F = allowable tension stress parallel to grain or allowable compression stress parallel to grain adjusted for unsupported length, psi

F_b = allowable bending stress, psi

When a bending stress f_x, shear stress f_{xy}, and compression or tension stress f_y perpendicular to the grain exist simultaneously, the stresses should satisfy the Norris interaction formula:

$$\frac{f_x{}^2}{F_x{}^2} + \frac{f_y{}^2}{F_y{}^2} + \frac{f_{xy}{}^2}{F_{xy}{}^2} \leq 1 \qquad (11\text{-}9)$$

where F_x = allowable bending stress F_b modified for duration of loading but not for size factor

F_y = allowable stress in compression or tension perpendicular to grain $F_{c\perp}$ or $F_{b\perp}$ modified for duration of loading

F_{xy} = allowable horizontal shear stress F_v modified for duration of loading

The usual design formulas do not apply to sharply curved glued-laminated beams. Nor do they hold where laminations of more than one species, each with markedly different modulus of elasticity, are comprised in the same member.

11-15. Effect of Shrinkage or Swelling on Shape of Curved Members. Wood shrinks or swells across the grain but has practically no dimensional change along the grain. Radial swelling causes a decrease in the angle between the ends of a curved member; radial shrinkage causes an increase in this angle.

Such effects may be of great importance in three-hinged arches that become horizontal, or nearly so, at the crest of a roof. Shrinkage, increasing the relative end rotations, may cause a depression at the crest and create drainage problems. For such arches, therefore, consideration must be given to moisture content of the member at time of fabrication and in service, and to the change in end angles that results from change in moisture content and shrinkage across the grain.

11-16. Fabrication of Structural Timber. Fabrication consists of boring, cutting, sawing, trimming, dapping, routing, planing and otherwise shaping, framing, and furnishing wood units, sawn or laminated, including plywood, to fit them for particular places in a final structure. Whether fabrication is performed in shop or field, the product must exhibit a high quality of workmanship.

Jigs, patterns, templates, stops, or other suitable means should be used for all complicated and multiple assemblies to insure accuracy, uniformity, and control of all dimensions. All tolerances in cutting, drilling, and framing must comply with good practice in the industry and applicable specifications and controls. At the time of fabrication, tolerances must not exceed those listed below unless they are not critical and not required for proper performance. Specific jobs, however, may require closer tolerances.

Location of Fastenings. Spacing and location of all fastenings within a joint should be in accordance with the shop drawings and specifications with a maximum permissible tolerance of $\pm 1/16$ in. The fabrication of members assembled at any joint should be such that the fastenings are properly fitted.

Bolt-Hole Sizes. Bolt holes in all fabricated structural timber, when loaded as a structural joint, should be $1/16$ in. larger in diameter than bolt diameter for $1/2$-in. and larger-diameter bolts, and $1/32$ in. larger for smaller-diameter bolts. Larger clearances may be required for other bolts, such as anchor bolts and tension rods.

Holes and Grooves. Holes for stress-carrying bolts, connector grooves, and connector daps must be smooth and true within $1/16$ in. per 12 in. of depth. The width of a split-ring connector groove should be within $+0.02$ in. of and not less than the thickness of the corresponding cross section of the ring. The shape of ring grooves must conform generally to the cross-sectional shape of the ring. Departure from these requirements may be allowed when supported by test data. Drills and other cutting tools should be set to conform to the size, shape, and depth of holes, grooves, daps, etc., specified in the National Design Specification for Stress-grade Lumber and Its Fastenings, National Forest Products Association (formerly National Lumber Manufacturers Association).

Lengths. Members should be cut within $\pm 1/16$ in. of the indicated dimension when they are up to 20 ft long and $\pm 1/16$ in. per 20 ft of specified length when they are over 20 ft long. Where length dimensions are not specified or critical, these tolerances may be waived.

End Cuts. Unless otherwise specified, all trimmed square ends should be square within $^{1}/_{16}$ in. per ft of depth and width.   Square or sloped ends to be loaded in compression should be cut to provide contact over substantially the complete surface.

11-17. Fabrication of Glued-Laminated Timbers. Structural glued-laminated timber is made by bonding layers of lumber together with adhesive so that the grain direction of all laminations is essentially parallel.   Narrow boards may be edge-glued; short boards, end-glued; and the resultant wide and long laminations then face-glued into large, shop-grown timbers.

Recommended practice calls for lumber of nominal 1- and 2-in. thicknesses for laminating. The lumber is dressed to ¾- and 1½-in. thicknesses, respectively, just prior to gluing.   The thinner laminations are generally used in curved members.

Depth of constant-depth members normally is a multiple of the thickness of the lamination stock used.   Depth of variable-depth members, due to tapering or special assembly techniques, may not be exact multiples of these lamination thicknesses.

Industry-standard finished widths correspond to the following nominal widths after allowance for drying and surfacing of nominal lumber widths:

Nominal width of stock, in	3	4	6	8	10	12	14	16
Standard member finished width, in.	2¼	3⅛	5⅛	6¾	8¾	10¾	12¼	14¼

Standard widths are most economical, since they represent the maximum width of board normally obtained from the lumber stock used in laminating.

When members wider than the stock available are required, laminations may consist of two boards side by side.   These edge joints must be staggered, vertically in horizontally laminated beams (load acting normal to wide faces of laminations) and horizontally in vertically laminated beams (load acting normal to the edge of laminations).   In horizontally laminated beams, edge joints need not be edge-glued.   But edge gluing is required in vertically laminated beams.

Edge and face gluings are the simplest to make, end gluings the most difficult.   Ends are also the most difficult surfaces to machine.   Scarfs or finger joints generally are used to avoid end gluing.

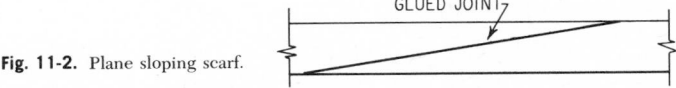

Fig. 11-2. Plane sloping scarf.

A plane sloping scarf (Fig. 11-2), in which the tapered surfaces of laminations are glued together, can develop 85 to 90% of the strength of an unscarfed, clear, straight-grained control specimen.   Finger joints (Fig. 11-3) are less wasteful of lumber.   Quality can be adequately controlled in machine cutting and in high-frequency gluing.   A combination of thin tip, flat slope on the side of the individual fingers, and a narrow pitch is desired.   The length of fingers should be kept short for savings of lumber but long for maximum strength.   Obviously a satisfactory compromise must be selected.

The usefulness of structural glued-laminated timbers is determined by the lumber used and glue joint produced.   Certain combinations of adhesive, treatment, and wood species do not produce the same quality of glue bond as other combinations, though the same gluing procedures are used.   Thus, a combination must be supported by adequate experience with a laminator's gluing procedure (see also Art. 11-18).

The only adhesives currently recommended for wet-use and preservative-treated lumber, whether gluing is done before or after treatment, are the resorcinol and phenol-resorcinol resins.   The prime adhesive for dry-use structural laminating is casein.   Melamine and melamine-urea blends are used in smaller amounts for high-frequency curing of end gluings.

Glued joints are cured with heat by several methods.   R. F. (high-frequency) curing of glue lines is used for end joints and for limited-size members where there are repetitive gluings of the same cross section.   Low-voltage resistance heating, where current is passed through a strip of metal to raise the temperature of a glue line, is used for attaching thin facing pieces.   The metal may be left in the glue line as an integral part of the completed member.   Printed electric circuits, in conjunction with adhesive films, and adhesive films, impregnated on paper or on each side of a metal conductor placed in the glue line, are other alternatives.   Also, an aluminum-

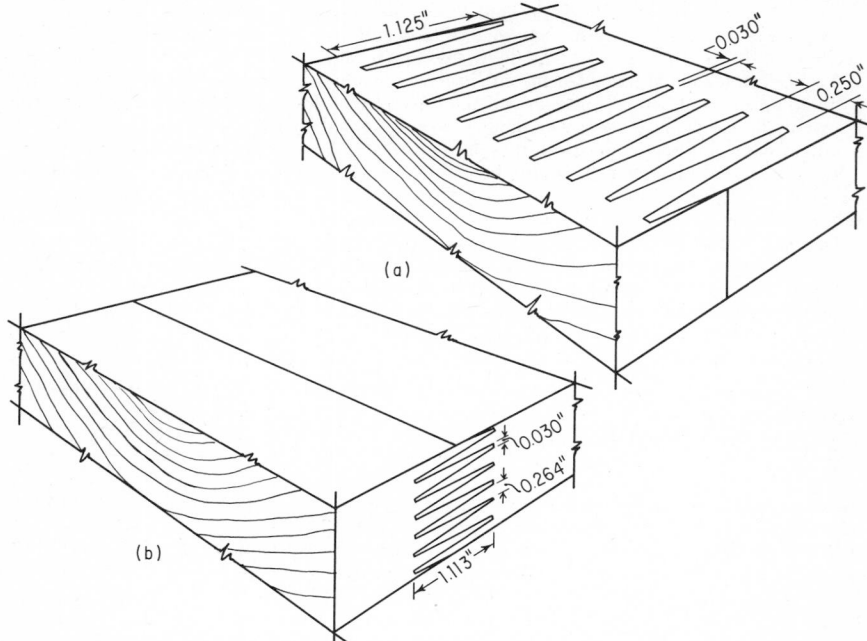

Fig. 11-3. Finger joint: (*a*) fingers formed by cuts perpendicular to the wide face of the board; (*b*) fingers formed by cuts perpendicular to the edges.

foil system, with the aluminum foil as the electrical conductor, faced with dry-film phenolic resin, may be used for resistance heating.

Preheating the wood to insure reactivity of the applied adhesive has limited application in structural laminating. The method requires adhesive application as a wet or dry film simultaneously to all laminations and then rapid handling of multiple laminations.

Curing the adhesive at room temperature has many advantages. Since wood is an excellent insulator, a long time is required for elevated ambient temperature to reach inner glue lines of a large assembly. With room-temperature curing, equipment needed to heat the glue line is not required, and the possibility of injury to the wood from high temperatures is avoided.

(AITC 103, Standard for Structural Glued-Laminated Timber, American Institute of Timber Construction, Washington.)

11-18. Preservative Treatments for Wood. Wood-destroying fungi must have air, suitable moisture, and favorable temperatures to develop and grow in wood. Submerge wood permanently and totally in water, or keep the moisture content below 18 to 20%, or hold temperatures below 40°F or above 110°F, and wood remains permanently sound. If wood moisture content is kept below the fiber-saturation point (25 to 30%) when the wood is untreated, decay is greatly retarded. Below 18 to 20% moisture content, decay is completely inhibited.

If wood cannot be kept dry, a wood preservative, properly applied, must be used. The following can serve as a guide to determine if treatment is necessary.

Wood members are permanent without treatment if located in enclosed buildings where good roof coverage, proper roof maintenance, good joint details, adequate flashing, good ventilation, and a well-drained site assure moisture content of the wood continuously below 20%. Also, in arid or semiarid regions where climatic conditions are such that the equilibrium moisture content seldom exceeds 20%, and then only for short periods, wood members are permanent without treatment.

Where wood is in contact with the ground or with water, where there is air and the wood may be alternately wet and dry, a preservative treatment, applied by a pressure process, is necessary to obtain an adequate service life. In enclosed buildings where moisture given off by wet-process operations maintains equilibrium moisture contents in the wood above 20%, wood structural members must be treated with a preservative. So must wood exposed outdoors without protective roof covering and where the wood moisture content can go above 18 to 20% for repeated or prolonged periods.

Table 11-12. Recommended Minimum Retentions of Preservatives, Lb per Cu Ft*

Preservatives	Sawn and laminated timbers		Laminations		Sawn and laminated timbers		Laminations	
	Western woods†	Southern pine	Western woods†	Southern pine	Western woods†	Southern pine	Western woods†	Southern pine
	Coastal waters				General use			
Creosote or creosote solutions:								
Coal-tar creosote	12	20	15	25	8‡	8	10	10
Creosote–coal-tar solution	Not recommended	20	Not recommended	25	8‡	8	10	10
Creosote-petroleum solution	Not recommended				8‡	8	10	10
Oil-borne chemicals:								
Pentachlorophenol, (5% in specified petroleum oil)	Not recommended				0.4§	0.4	0.5	0.5
Penta (water-repellent moderate decay hazard)	Not recommended				0.2	0.2	0.25	0.25
	Ground contact				Above ground			
Water-borne inorganic salts:								
Chromated zinc arsenate (Boliden salt)	1.00	1.00	1.25	1.25	0.50	0.50	0.625	0.625
Acid copper chromate (Celcure)	1.00	1.00	1.25	1.25	0.50	0.50	0.625	0.625
Ammoniacal copper arsenite (Chemonite)	0.50	0.50	0.625	0.625	0.30	0.30	0.375	0.375
Chromated zinc chloride	1.00	1.00	1.25	1.25	0.75	0.75	0.94	0.94
Copperized chromated zinc chloride	1.00	1.00	1.25	1.25	0.75	0.75	0.94	0.94
Chromated copper arsenite (Greensalt, Erdalith)	0.75	0.75	0.94	0.94	0.35	0.35	0.44	0.44
Fluor chrome arsenate phenol (Tanalith, Wolman salt)	0.50	0.50	0.625	0.625	0.35	0.35	0.44	0.44
Fluor chrome arsenate Phenol (Osmossar, Osmosalt)	0.50	0.50	0.625	0.625	0.35	0.35	0.44	0.44

*AITC 109, Treating Standard for Structural Timber Framing, American Institute of Timber Construction.
†Douglas fir, western hemlock, western larch.
‡10 lb for timber less than 5 in. thick.
§0.5 lb for timber less than 5 in. thick.

Where wood structural members are subject to condensation by being in contact with masonry, preservative treatment is necessary.

To obtain preservative-treated glued-laminated timber, lumber may be treated before gluing and the members then glued to the desired size and shape. Or the already-glued and machined members may be treated with certain treatments. When laminated members do not lend themselves to treatment because of size and shape, gluing of treated laminations is the only method of producing adequately treated members.

There are problems in gluing some treated woods. Certain combinations of adhesive, treatment, and wood species are compatible; other combinations are not. All adhesives of the same type do not produce bonds of equal quality for a particular wood species and preservative. The bonding of treated wood is dependent on the concentration of preservative on the surface at the time of gluing and on the chemical effects of the preservative on the adhesive. In general, longer curing times or higher curing temperatures, and modifications in assembly times, are needed for treated than for untreated wood to obtain comparable adhesive bonds (see also Art. 11-17).

Each type of preservative and method of treatment has certain advantages. The preservative to be used depends on the service expected of the member for the specific conditions of exposure. The minimum retentions shown in Table 11-12 may be increased where severe climatic or exposure conditions are involved.

Creosote and creosote solutions have low volatility. They are practically insoluble in water and thus are most suitable for severe exposure, contact with ground or water, and where painting is not a requirement or a creosote odor is not objectionable.

Oil-borne chemicals are organic compounds dissolved in a suitable petroleum carrier oil and are suitable for outdoor exposure or where leaching may be a factor, or where painting is not required. Depending on the type of oil used, they may result in relatively clean surfaces. While there is a slight odor from such treatment, it is usually not objectionable.

Water-borne inorganic salts are dissolved in water or aqua ammonia, which evaporates after treatment and leaves the chemicals in the wood. The strength of solutions varies to provide net retention of dry salt required. These salts are suitable where clean and odorless surfaces are required. The surfaces are paintable after proper seasoning.

When treating before gluing is required, water-borne salts, or oil-borne chemicals in mineral spirits, or AWPA P9 volatile solvent are recommended. When treatment before gluing is not required or desired, creosote, creosote solutions, or oil-borne chemicals are recommended.

11-19. Resistance of Wood to Chemical Attack. Wood is superior to many building materials in resistance to mild acids, particularly at ordinary temperatures. It has excellent resistance to most organic acids, notably acetic. However, wood is seldom used in contact with solutions that are more than weakly alkaline. Oxidizing chemicals and solutions of iron salts, in combination with damp conditions, should be avoided.

Wood is composed of roughly 50 to 70% cellulose, 25 to 30% lignin, and 5% extractives with less than 2% protein. Acids such as acetic, formic, lactic, and boric do not ionize sufficiently at room temperature to attack cellulose and thus do not harm wood.

When the pH of aqueous solutions of weak acids is 2 or more, the rate of hydrolysis of cellulose is small and is dependent on the temperature. A rough approximation of this temperature effect is that for every 20°F increase, the rate of hydrolysis doubles. Acids with pH values above 2 or bases with pH below 10 have little weakening effect on wood at room temperature if the duration of exposure is moderate.

11-20. Designing for Fire Safety. Maximum protection of the occupants of a building and of the property itself can be achieved in timber design by taking advantage of the fire-endurance properties of wood in large cross sections and by close attention to details that make a building fire-safe. Building materials alone, building features alone, or detection and fire-extinguishing equipment alone cannot provide maximum safety from fire in buildings. A proper combination of these three will provide the necessary degree of protection for the occupants and the property.

The following should be investigated:

Degree of protection needed, as dictated by occupancy or operations taking place.

Number, size, type (such as direct to the outside), and accessibility of exitways (particularly stairways), and their distance from each other.

Installation of automatic alarm and sprinkler systems.

Separation of areas in which hazardous processes or operations take place, such as boiler rooms and workshops.

Enclosure of stairwells and use of self-closing fire doors.

Fire stopping and elimination, or proper protection of concealed spaces.

Interior finishes to assure surfaces that will not spread flame at hazardous rates.

Roof venting equipment or provision of draft curtains where walls might interfere with production operations.

When exposed to fire, wood forms a self-insulating surface layer of char, which provides its own fire protection. Even though the surface chars, the undamaged wood beneath retains its strength and will support loads in accordance with the capacity of the uncharred section. Heavy-timber members have often retained their structural integrity through long periods of fire exposure and remained serviceable after refinishing of the charred surfaces. This fire endurance and excellent performance of heavy timber are attributable to the size of the wood members and to the slow rate at which the charring penetrates.

The structural framing of a building, which is the criterion for classifying a building as combustible or noncombustible, has little to do with the hazard from fire to the building occupants. Most fires start in the building contents and create conditions that render the inside of the structure uninhabitable long before the structural framing becomes involved in the fire. Thus, whether the building be of combustible or noncombustible classification has little bearing on the potential hazard to the occupants. However, once the fire starts in the contents, the material of which the building is constructed can be of significant help in facilitating evacuation, fire fighting, and property protection.

The most important protection factors for occupants, fire fighters, and the property, as well as adjacent exposed property, are prompt detection of the fire, immediate alarm, and rapid extinguishment of the fire. Fire fighters do not fear fires in buildings of heavy-timber construction as they do those in buildings of many other types of construction. They need not fear sudden collapse without warning; they usually have adequate time, because of the slow-burning characteristics of the timber, to ventilate the building and fight the fire from within the building or on top.

With size of member of particular importance to fire endurance of wood members, building codes specify minimum dimensions for structural members and classify buildings with wood framing as heavy-timber construction, ordinary construction, or wood-frame construction.

Heavy-timber construction is that type in which fire resistance is attained by placing limitations on the minimum size, thickness, or composition of all load-carrying wood members; by avoidance of concealed spaces under floors and roofs; by use of approved fastenings, construction details, and adhesives; and by providing the required degree of fire resistance in exterior and interior walls. (See AITC 108, Heavy Timber Construction, American Institute of Timber Construction.)

Ordinary construction has exterior masonry walls and wood-framing members of sizes smaller than heavy-timber sizes.

Wood-frame construction has wood-framed walls and structural framing of sizes smaller than heavy-timber sizes.

Depending on the occupancy of a building or hazard of operations within it, a building of frame or ordinary construction may have its members covered with fire-resistive coverings. The interior finish on exposed surfaces of rooms, corridors, and stairways is important from the standpoint of its tendency to ignite, flame, and spread fire from one location to another. The fact that wood is combustible does not mean that it will spread flame at a hazardous rate. Most codes exclude the exposed wood surfaces of heavy-timber structural members from flame-spread requirements because such wood is difficult to ignite and, even with an external source of heat, such as burning contents, is resistant to spread of flame.

Fire-retardant chemicals may be impregnated in wood with recommended retentions to lower the rate of surface flame spread and make the wood self-extinguishing if the external source of heat is removed. After proper surface preparation, the surface is paintable. Such treatments are accepted under several specifications, including Federal and military. They are recommended only for interior or dry-use service conditions or locations protected against leaching. These treatments are sometimes used to meet a specific flame-spread rating for interior finish or as an alternate to noncombustible secondary members and decking meeting the requirements of Underwriters' Laboratories, Inc., NM 501 or NM 502, nonmetallic roof-deck assemblies in otherwise heavy-timber construction.

11-21. Mechanical Fastenings. Various kinds of mechanical fastenings are used in wood construction. The commonest are nails, spikes, screws, lags, bolts, and timber connectors, such as shear plates and split rings. Joint-design data have been established by experience and tests because determination of stress distribution in wood and metal fasteners is complicated.

Allowable loads or stresses and methods of design for bolts, connectors, and other fasteners used in one-piece sawn members also are applicable to laminated members.

Problems can arise, however, if a deep-arch base section is bolted to the shoe attached to the foundation by widely separated bolts. A decrease in wood moisture content and shrinkage will set up considerable tensile stress perpendicular to grain, and splitting may occur. If the moisture content at erection is the same as that to be reached in service, or if the bolt holes in the shoe are slotted to permit bolt movement, the tendency to split will be reduced.

Fasteners subject to corrosion or chemical attack should be protected by painting, galvanizing, or plating. In highly corrosive atmospheres, such as in chemical plants, metal fasteners and connections should be galvanized or made of stainless steel. Consideration may be given to covering connections with hot tar or pitch. In such extreme conditions, lumber should be at or below equilibrium moisture content at fabrication, to reduce subsequent shrinkage, which could open avenues of attack for the corrosive atmosphere.

Iron salts are frequently very acidic and show hydrolytic action on wood in the presence of free water. This accounts for softening and discoloration of wood observed around corroded nails. This action is especially pronounced in acidic woods, such as oak, and in woods containing considerable tannin and related compounds, such as redwood. It can be eliminated, however, by using zinc-coated, aluminum, or copper nails.

Nails and Spikes. Common wire nails and spikes conform to the minimum sizes in Table 11-13.

Table 11-13. Nail and Spike Dimensions

Pennyweight	Length, in.	Wire dia, in.
Nails:		
6d	2	0.113
8d	2½	0.131
10d	3	0.148
12d	3¼	0.148
16d	3½	0.162
20d	4	0.192
30d	4½	0.207
40d	5	0.225
50d	5½	0.244
60d	6	0.263
Spikes:		
10d	3	0.192
12d	3¼	0.192
16d	3½	0.207
20d	4	0.225
30d	4½	0.244
40d	5	0.263
50d	5½	0.283
60d	6	0.283
5/16	7	0.312
3/8	8½	0.375

Hardened deformed-shank nails and spikes are made of high-carbon-steel wire and are headed, pointed, annularly or helically threaded, and heat-treated and tempered, to provide greater strength than common wire nails and spikes. But the same loads as given for common wire nails and spikes or the corresponding lengths are used with a few exceptions.

Nails should not be driven closer together than half their length, unless driven in prebored holes. Nor should nails be closer to an edge than one-quarter their length. When one structural member is joined to another, penetration of nails into the second or farther timber should be at least half the length of the nails. Holes for nails, when necessary to prevent splitting, should be bored with a diameter less than that of the nail. If this is done, the same allowable load as for the same-size fastener with a bored hole applies in both withdrawal and lateral resistance.

Nails or spikes should not be loaded in withdrawal from the end grain of wood. Also, nails inserted parallel to the grain should not be used to resist tensile stresses parallel to the grain.

Safe lateral loads and safe resistance to withdrawal of common wire nails are given in Table 11-14 and of spikes in Table 11-15. The tables provide a safety factor of about 6 against failure. Joint slippage would be objectionable long before ultimate load is reached.

For lateral loads, if a nail or spike penetrates the piece receiving the point for a distance less than 11 times the nail or spike diameter, the allowable load is determined by straight-line interpolation between full load at 11 diameters, as given in the tables, and zero load for zero penetration. But the minimum penetration in the piece receiving the point must be at least 3⅔ diameters. Allowable lateral loads for nails or spikes driven into end grain are two-thirds the

Table 11-14. Strength of Wire Nails

	6d	8d	10d	12d	16d	20d	30d	40d	50d	60d
Size of nail, pennyweight	6d	8d	10d	12d	16d	20d	30d	40d	50d	60d
Length of nail, in.	2	2½	3	3¼	3½	4	4½	5	5½	6
Safe lateral strength, lb (inserted perpendicular to the grain of the wood, penetrating 11 diameters), Douglas fir or southern pine	63	78	94	94	107	139	154	176	202	223
Safe resistance to withdrawal, lb per lin in. of penetration into the member receiving the point (inserted perpendicular to the grain of the wood):										
Douglas fir	29	34	38	38	42	49	53	58	63	68
Southern pine	42	48	55	55	60	71	76	83	90	97

Table 11-15. Strength of Spikes

	10d	12d	16d	20d	30d	40d	50d	60d	5/16 in.	⅜ in.
Size of spike, pennyweight	10d	12d	16d	20d	30d	40d	50d	60d	5/16 in.	⅜ in.
Length of spike, in.	3	3¼	3½	4	4½	5	5½	6	7	8½
Safe lateral strength, lb (inserted perpendicular to the grain of the wood, penetrating 11 diameters), Douglas fir or southern pine	139	139	155	176	202	223	248	248	289	380
Safe resistance to withdrawal, lb per lin in. of penetration into the member receiving the point (inserted perpendicular to the grain of the wood):										
Douglas fir	49	49	53	58	63	68	73	73	80	96
Southern pine	71	71	76	83	90	97	104	104	115	138

tabulated values. If a nail or spike is driven into unseasoned wood that will remain wet or will be loaded before seasoning, the allowable load is 75% of the tabulated values, except that for hardened, deformed-shank nails the full load may be used. Where properly designed metal side plates are used, the tabulated allowable loads may be increased 25%.

Wood Screws. The common types of wood screws have flat, oval, or round heads. The flat-head screw is commonly used if a flush surface is desired. Oval and round-headed screws are used for appearance or when countersinking is objectionable.

Wood screws should not be loaded in withdrawal from end grain. They should be inserted perpendicular to the grain by turning into predrilled holes and should not be started or driven with a hammer. Spacings, end distances, and side distances must be such as to prevent splitting.

Table 11-16 gives the allowable loads for lateral resistance at any angle of load to grain, when the wood screw is inserted perpendicular to the grain (into the side grain of main member) and a wood side piece is used. Embedment must be seven times the shank diameter into the member holding the point. For less penetration, reduce loads in proportion; penetration, however, must not be less than four times the shank diameter.

Table 11-16 also gives allowable withdrawal loads in pounds per inch of penetration of the threaded portion of a screw into the member holding the point when the screw is inserted perpendicular to the grain of the wood.

When metal side plates rather than wood side pieces are used, the allowable lateral load, at any angle of load to grain, may be increased 25%. Allowable lateral loads when loads act perpendicular to the grain and the screw is inserted into end grain are two-thirds of those shown.

Table 11-16. Strength of Wood Screws

Gage of screw	6	7	8	9	10	12	14	16	18	20	24
Dia of screw, in.	0.138	0 151	0.164	0.177	0.190	0.216	0.242	0.268	0.294	0.320	0.372
Allowable lateral load, lb (normal duration): Douglas fir or southern pine	75	90	106	124	143	185	232	284	342	406	548
Allowable withdrawal load, lb per in. of penetration (normal duration):											
Douglas fir	102	112	121	131	141	160	179	199	218	237	276
Southern pine	137	150	163	176	189	214	240	266	292	317	369

For Douglas fir and southern pine, the lead hole for a screw loaded in withdrawal should have a diameter of about 70% of the root diameter of the screw. For lateral resistance, the part of the hole receiving the shank should be about seven-eighths the diameter of the shank and that for the threaded portion should be about seven-eighths the diameter of the screw at the root of the thread.

The loads given in Table 11-16 are for wood screws in seasoned lumber, for joints used indoors or in a location always dry. When joints are exposed to the weather, use 75%, and where joints are always wet, 67% of the tabulated loads. For lumber pressure-impregnated with fire-retardant chemicals, use 75% of the tabulated loads.

Lag Screws or Lag Bolts. Lag screws are commonly used because of their convenience, particularly where it would be difficult to fasten a bolt or where a nut on the surface would be objectionable. They range, usually, from about 0.2 to 1.0 in. in diameter and from 1 to 16 in. in length. The threaded portion ranges from ¾ in. for 1- and 1¼-in.-long lag screws to half the length for all lengths greater than 10 in.

Lag screws, like wood screws, require prebored holes of the proper size. The lead hole for the shank should have the same diameter as the shank. The lead-hole diameter for the threaded portion varies with the density of the wood species. For Douglas fir and southern pine, the hole for the threaded portion should be 60 to 75% of the shank diameter. The smaller percentage applies to lag screws of smaller diameters. Lead holes slightly larger than those recommended for maximum efficiency should be used with lag screws of excessive length.

In determining withdrawal resistance, the allowable tensile strength of the lag screw at the net or root section should not be exceeded. Penetration of the threaded portion to a distance of about seven times the shank diameter in the denser species and ten to twelve times the shank diameter in the less dense species will develop approximately the ultimate tensile strength of a lag screw.

The resistance of a lag screw to withdrawal from end grain is about three-quarters that from side grain.

Table 11-17 gives the allowable normal-duration lateral and withdrawal loads for Douglas fir and southern pine. When lag screws are used with metal plates, the allowable lateral loads parallel to the grain are 25% higher than with wood side plates. No increase is allowed for load applied perpendicular to the grain.

Lag screws should preferably not be driven into end grain because splitting may develop under lateral load. If lag screws are so used, however, the allowable loads should be taken as two-thirds the lateral resistance of lag screws in side grain with loads acting perpendicular to the grain.

Spacings, edge and end distances, and net section for lag-screw joints should be the same as those for joints with bolts of a diameter equal to the shank diameter of the lag screw.

For more than one lag screw the total allowable load equals the sum of the loads permitted for each lag screw, provided that spacings, end distances, and edge distances are sufficient to develop the full strength of each lag screw.

The allowable loads in Table 11-17 are for lag screws in lumber seasoned to a moisture content about equal to that which it will eventually have in service. For lumber installed unseasoned and seasoned in place, the full allowable lag-screw loads may be used for a joint having a single lag screw and loaded parallel or perpendicular to grain; or a single row of lag screws loaded parallel to grain; or multiple rows of lag screws loaded parallel to grain with separate splice plates

Table 11-17. Strength of Lag Screws

Dia of lag screw, in.	1/4	5/16	3/8	7/16	1/2	9/16	5/8	3/4	7/8	1	1 1/8	1 1/4
Allowable withdrawal load, lb per in. of penetration of the threaded part into the member holding the point (normal duration):												
Douglas fir	232	274	313	352	389	425	460	528	593	655	716	774
Southern pine	289	341	391	439	485	529	573	657	738	815	891	964

Allowable lateral load, lb per lag screw, in single shear with 1/2-in. metal side plates (Douglas fir or southern pine) for normal duration

Length of lag screw in main members, in.	1/4 Parallel to grain	1/4 Perpendicular to grain	5/16 Parallel to grain	5/16 Perpendicular to grain	3/8 Parallel to grain	3/8 Perpendicular to grain	7/16 Parallel to grain	7/16 Perpendicular to grain	1/2 Parallel to grain	1/2 Perpendicular to grain	9/16 Parallel to grain	9/16 Perpendicular to grain	5/8 Parallel to grain	5/8 Perpendicular to grain	3/4 Parallel to grain	3/4 Perpendicular to grain	7/8 Parallel to grain	7/8 Perpendicular to grain	1 Parallel to grain	1 Perpendicular to grain	1 1/8 Parallel to grain	1 1/8 Perpendicular to grain	1 1/4 Parallel to grain	1 1/4 Perpendicular to grain
3	210	160	265	180	320	245	370	210	415	215	455	225	490	235										
4	235	185	355	240	480	290	575	320	625	325	680	340	740	355										
5			375	255	535	325	710	405	850	440	930	460	1,005	480	1,190	525								
6			400	270	545	330	735	415	945	490	1,095	540	1,250	600	1,480	650								
7					555	340	750	425	970	505	1,210	600	1,460	700	2,030	890								
8							760	430	985	510	1,240	615	1,500	720	2,130	935	2,720	1,130						
9									990	515	1,250	620	1,510	725	2,160	950	2,880	1,200						
10													1,540	740	2,190	965	2,960	1,230	3,710	1,485				
11															2,220	970	2,990	1,240	3,880	1,550				
12																	3,000	1,250	3,900	1,560	4,900	1,960		
13																	3,030	1,260	3,930	1,570	4,920	1,970		
14																			3,950	1,570	4,950	1,980	6,060	2,420
15																			3,960	1,580	4,980	1,990	6,110	2,450
16																			3,960	1,580	5,000	2,000	6,150	2,460

for each row. For other types of lag-screw joints, the allowable loads are 40% of the tabulated loads. For lumber partly seasoned when fabricated, proportionate intermediate loads between 100 and 40% may be used.

For lumber pressure-impregnated with fire-retardant chemicals, allowable loads for lag screws are the same as those for unseasoned lumber. Where joints are to be exposed to weather, use 75%, and where joints are always wet, 67% of the tabulated allowable loads.

Bolts. Standard machine bolts with square heads and nuts are used extensively in wood construction. Spiral-shaped dowels are also used at times to hold two pieces of wood together; they are used to resist checking and splitting in railroad ties and other solid-sawn timbers.

Holes for bolts should always be prebored and should have a diameter that permits the bolt to be driven easily (Art. 11-16). Careful centering of holes in main members and splice plates is necessary.

Center-to-center distance along the grain between bolts acting parallel to the grain should be at least four times the bolt diameter. For a tension joint, distance from end of wood to center of nearest bolt should be at least seven times the bolt diameter for softwoods and five times for hardwoods. For compression joints, the end distance should be at least four times the bolt diameter for both softwoods and hardwoods. If closer spacings are used, the loads allowed should be reduced proportionally.

Also for bolts acting parallel to the grain, the distance from the center of a bolt to the edge of the wood should be at least 1.5 times the bolt diameter. Usually, however, the edge distance is set at half the distance between bolt rows. In any event, the area at the critical section through the joint must be sufficient to keep unit stresses in the wood within the allowable.

The critical section is that section at right angles to the direction of the load that gives maximum stress in the member over the net area remaining after bolt holes at the section are deducted. For parallel-to-grain loads, the net area at a critical section should be at least 100% for hardwoods and 80% for softwoods of the total area in bearing under all the bolts in the joint.

For parallel- or perpendicular-to-grain loads, spacing between rows paralleling a member should not exceed 5 in. unless separate splice plates are used for each row.

For bolts bearing perpendicular to the grain, center-to-center spacing across the grain should be at least four times the bolt diameter if wood side plates are used. But if the design load is less than the bolt bearing capacity of the plates, the spacing may be reduced proportionally. For metal side plates, the spacing only need be sufficient to permit tightening of the nuts. Distance from edge of wood to center of bolt should be at least four diameters. So should the distance between center of bolt and edge toward which the load is acting. The edge distance at the opposite edge is relatively unimportant.

A load applied to only one end of a bolt perpendicular to its axis may be taken as half the symmetrical two-end load.

Bearing Stresses. Basic bolt bearing stresses for calculating allowable loads parallel and perpendicular to the grain are the same for Douglas fir and southern pine:

> Parallel to grain . . . . 1,450 psi
> Perpendicular to grain . . .   320 psi

These basic stresses are for permanent loads. For short-duration loads, basic stresses may be modified as noted in Art. 11-12. Allowable stresses for aligned bolts satisfying the spacing requirements previously given and with the load applied, parallel or perpendicular to grain, through metal plates to both ends of the bolt are calculated by multiplying the appropriate basic stress by a factor r (Table 11-18). For loads perpendicular to grain, an additional diameter factor v should be applied to the basic stresses (Table 11-19).

Bolt Groups. When bolts are properly spaced and aligned, the allowable load on a group of bolts may be taken as the sum of the individual load capacities.

Allowable Loads. To determine allowable bolt load parallel to grain: Select the basic bearing stress parallel to grain for the wood species. Adjust this stress for the service condition if the joint is to be used in other than a dry, inside location. Call the adjusted stress S_1. Calculate the bolt length-diameter ratio L/D. For this value, select r from Table 11-18. Multiply S_1 by r to obtain the allowable unit stress S_2. Finally, multiply S_2 by the projected area of the bolt to obtain the allowable load for the bolt.

To determine allowable bolt load perpendicular to grain: Select the basic bearing stress perpendicular to grain for the wood species. Adjust it for the service condition if the joint is used in other than a dry, inside location. Call this stress S_1. Calculate the bolt length-diameter ratio L/D. For this value, select r from Table 11-18; then, pick v corresponding to the bolt diameter from Table 11-19. Multiply S_1 by r and v to obtain the allowable unit stress S_2. Finally,

Table 11-18. Factor _r_ for Adjusting Basic Stresses for Bolts
(Used in Calculating Allowable Bearing Stresses for Common Bolts When Load Is Applied through Metal Splice Plates)

Ratio of bolt length to diameter _L/D_	Bearing stress, % of basic stress for	
	Bolts bearing parallel to grain* when basic stress is 1,300 to 2,000 psi	Bolts bearing perpendicular to grain† when basic stress is 300 to 350 psi
1	100.0	100.0
2	100.0	100.0
3	99.0	100.0
4	92.5	100.0
5	80.0	100.0
6	67.2	100.0
7	57.6	97.3
8	50.4	88.1
9	44.8	76.7
10	40.3	67.2
11	36.6	59.3
12	33.6	52.0
13	31.0	45.9

*For wood splice plates, each of which is half the thickness of the main member, the allowable load should be taken as four-fifths that computed for metal splice plates.

†No reduction need be made when wood splice plates are used, except that the allowable load perpendicular to the grain should never exceed the allowable load parallel to the grain for any given size and quality of bolt and timber.

Table 11-19. Factor _v_ for Adjusting Basic Stresses for Bolts
(For Loads Acting Perpendicular to the Grain and Applied through Metal Plates)

Bolt dia, in.	Dia factor _v_
¼	2.50
⅜	1.95
½	1.68
⅝	1.52
¾	1.41
⅞	1.33
1	1.27
1¼	1.19
1½	1.14
1¾	1.10
2	1.07
3 or over	1.00

multiply S_2 by the projected area of the bolt to obtain the allowable load for the bolt loaded through wood or metal plates.

Timber Connectors. These are metal devices used with bolts for producing joints with fewer bolts without reduction in strength. Several types of connectors are available. Usually, they are either steel rings that are placed in grooves in adjoining members to prevent relative movement or metal plates embedded in the faces of adjoining timbers. The purpose of the bolts used with these connectors is to prevent the timbers from separating. The load is transmitted across the joint through the connectors.

Split rings are the most efficient device for joining wood to wood. They are placed in circular grooves cut by a hand tool in the contact surfaces. About half the depth of each ring is in each of the two members in contact. A bolt hole is drilled through the center of the core encircled by the groove.

A split ring has a tongue-and-groove split to permit simultaneous bearing of the inner surface of the ring against the core and the outer surface of the ring against the outer wall of the groove. The ring is beveled for ease of assembly. Rings are manufactured in 2½- and 4-in. diameters.

Shear plates are intended for wood-to-steel connections. But when used in pairs, they may be used for wood-to-wood connections, replacing split rings. Set with one plate in each member at the contact surface, they enable the members to slide easily into position during fabrication of the joint, thus reducing the labor needed to make the connection. Shear plates are placed in precut daps and are completely embedded in the timber, flush with the surface. As with split rings, the role of the bolt through each plate is to prevent the components of the joint from separating; loads are transmitted across the joint through the plates. They come in 2⅝- and 4-in. diameters.

Shear plates are useful in demountable structures. They may be installed in the members immediately after fabrication and held in position by nails.

Toothed rings and *spike grids* sometimes are used for special applications. But split rings and shear plates are the prime connectors for timber construction.

Safe long-time working loads for split rings and shear plates have been derived from tests of full-scale joints and other basic information on strength and behavior of timber. Particular consideration has been given to the effects of short-time loading and allowances for variability in timber quality.

Working loads acting parallel to the grain of the wood were derived by applying a reduction factor to the ultimate loads observed in tests. A reduction factor of 4 kept working loads for split rings and shear plates within five-eighths of the test loads at the proportional limit.

Allowable Loads. Table 11-20 gives safe working loads for split rings and bolts. Table 11-21 gives safe working loads for shear plates and bolts.

Table 11-20. Allowable Loads for One Split Ring and Bolt in Single Shear (Normal Loading)

Split-ring dia, in.	Bolt dia, in.	No. of faces of piece with con-nectors on same bolt	Thickness (net) of lumber, in.	Loaded parallel to grain		Loaded perpendicular to grain		
				Min edge dis-tance, in.	Allow-able load per con-nector and bolt, lb	Min edge distance, in.		Allow-able load per con-nector and bolt, lb
						Un-loaded edge	Loaded edge	
2½	½	1	1 min	1¾	2,270	1¾	1¾	1,350
							2¾	1,620
			1⅝ and thicker	1¾	2,730	1¾	1¾	1,620
							2¾	1,940
		2	1⅝ min	1¾	2,270	1¾	1¾	1,350
							2¾	1,620
			2 and thicker	1¾	2,730	1¾	1¾	1,620
							2¾	1,940
4	¾	1	1 min	2¾	3,510	2¾	2¾	2,030
							3¾	2,440
			1⅝ and thicker	2¾	5,260	2¾	2¾	3,050
							3¾	3,660
		2	1⅝ min	2¾	3,690	2¾	2¾	2,140
							3¾	2,570
			2	2¾	4,250	2¾	2¾	2,470
							3¾	2,960
			2⅝	2¾	5,160	2¾	2¾	3,000
							3¾	3,600
			3 and thicker	2¾	5,260	2¾	2¾	3,050
							3¾	3,660

In design for wind forces acting alone or with dead and live loads, the safe working loads for connectors may be increased by one-third. But the number and size of connectors should not be less than those required for dead and live loads alone.

Table 11-21. Allowable Loads for One Shear Plate and Bolt in Single Shear
(Normal Load Duration and Wood Side Plates*)

Shear-plate dia, in.	Bolt dia, in.	No. of faces of piece with connectors on same bolt	Thickness (net) of lumber, in.	Loaded parallel to grain		Loaded perpendicular to grain		
				Min edge distance, in.	Allowable load per connector and bolt, lb	Edge distance, in.		Allowable load per connector and bolt, lb
						Unloaded edge	Loaded edge	
2⅝	¾	1	1½ min	1¾	2,760	1¾	1¾ min	1,550
							2¾ or more	1,860
		2	1½ min	1¾	2,080	1¾	1¾ min	1,210
							2¾ or more	1,450
			2	1¾	2,730	1¾	1¾ min	1,590
							2¾ or more	1,910
			2½ or more	1¾	2,860	1¾	1¾ min	1,660
							2¾ or more	1,990
4	¾ or ⅞	1	1½ min	2¾	3,750	2¾	2¾ min	2,180
							3¾ or more	2,620
			1¾ or more	2¾	4,360	2¾	2¾ min	2,530
							3¾ or more	3,040
		2	1¾ min	2¾	2,910	2¾	2¾ min	1,680
							3¾ or more	2,020
			2	2¾	3,240	2¾	2¾ min	1,880
							3¾ or more	2,260
			2½	2¾	3,690	2¾	2¾ min	2,140
							3¾ or more	2,550
			3	2¾	4,140	2¾	2¾ min	2,400
							3¾ or more	2,880
			3½ or more	2¾	4,320	2¾	2¾ min	2,500
							3¾ or more	3,000

*Tabulated loads also apply to metal side plates, except that for 4-in. shear plates the parallel-to-grain (not perpendicular) loads for wood side plates shall be increased 11%; but loads shall not exceed: for all loadings, except wind, 2900 lb for 2⅝-in. shear plates; 4,970 and 6,760 lb for 4-in. shear plates with ¾- and ⅞-in. bolts, respectively; or for wind loading, 3,870, 6,630, and 9,020 lb, respectively. If bolt threads are in bearing on the shear plate, reduce the preceding values by one-ninth.

Metal side plates, when used, shall be designed in accordance with accepted metal practices. For A36 steel, the following unit stresses, psi, are suggested for all loadings except wind: net section in tension, 22,000; shear, 14,500; bearing, 33,000. For wind, these values may be increased one-third. If bolt threads are in bearing, reduce the preceding shear and bearing values by one-ninth.

Impact may be disregarded up to the following percentages of the static effect of the live load producing the impact:

Connector	Impact Allowance, %
Split ring, any size, bearing in any direction	100
Shear plate, any size, bearing parallel to grain	66⅔
Shear plate, any size, bearing perpendicular to grain	100

One-half of any impact load that remains after disregarding the percentages indicated should be included with the dead and live loads in designing the joint.

The above procedures for increasing the allowable loads on connectors for suddenly applied and short-duration loads do not reduce the actual factor of safety of the joint. They are recommended because of the favorable behavior of wood under such forces. Different values are allowed for different types and sizes of connectors and directions of bearing because of variations in the extent to which distortion of the metal, as well as the strength of the wood, affects the ultimate strength of the joint.

The tabulated loads apply to seasoned timbers used where they will remain dry. If the timbers will be more or less continuously damp or wet in use, two-thirds of the tabulated values should be used.

Safe working loads for split rings and shear plates for angles between 0° (parallel to grain) and 90° (perpendicular to grain) may be obtained from the *Scholten nomograph* (Fig. 11-4). This determines the bearing strength of wood at various angles to the grain. The chart is a graphical solution of the Hankinson formula:

$$N = \frac{PQ}{P \sin^2\theta + Q \cos^2\theta}$$

$$(11\text{-}10)$$

where N, P, and Q are, respectively, the allowable load, lb, or stress, psi, at inclination θ with the direction of grain, parallel to grain, and perpendicular to grain.

For example, given $P = 6,000$ lb and $Q = 2,000$ lb, find the allowable load at an angle of 40° with the grain. Connect 6 on the 0° line (point a) with the intersection of a vertical line through 2 and the 90° line (point b). $N = 3,280$ lb is found directly below the intersection of line ab with the 40° line.

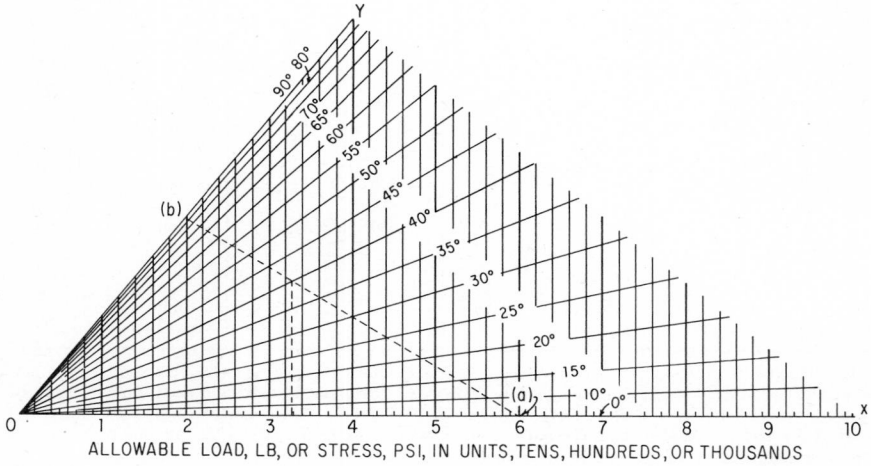

Fig. 11-4. Scholten nomograph for determining allowable bolt load on or bearing stress in wood at various angles to the grain.

Safe working loads are based on the assumption that the wood at the joint is clear and relatively free from checks, shakes, and splits. If knots are present in the longitudinal projection of the net section within a distance from the critical section of half the diameter of the connector, the area of the knot should be subtracted from the area of the critical section. It is assumed that slope of the grain at the joint does not exceed 1 in 10.

The stress, whether tension or compression, in the net area, the area remaining at the critical section after subtracting the projected area of the connectors and the bolt from the full cross-sectional area of the member, should not exceed the safe stress of clear wood in compression parallel to the grain.

Minimum Thickness. Tables 11-20 and 11-21 list the least thickness of member that should be used with the various sizes of connectors. The loads listed for the greatest thickness of member with each type and size of connector unit are the maximums to be used for all thicker material. Loads for members with thicknesses between those listed may be obtained by interpolation.

Connector Groups. For connectors in a multiple joint, there are several rules to follow based on observed behavior of single-connector joints tested with variables that simulate those in a multiple joint:

When two or more connectors in the same face of a member are in line at right angles to the grain of the member and are bearing parallel to the grain, the clear distance between the connectors should not be less than ½ in.

When two or more connectors act perpendicular to the grain and are on a line at right angles to the length of the member, rules for width of member and edge distances used with one connector are applicable to width and edge distance for multiple connectors. The clear distance between the connectors should equal the clear distance between the edge of the timber toward which the load is acting and the connector nearest this edge.

In a joint with two or more connectors on a line parallel to the grain and with the load acting perpendicular to the grain, the clear spacing between adjacent connectors should not be less than 1 in. The total load used should equal the full load of one connector plus one-third this amount for each additional connector. In a joint of this type, somewhat more favorable results are obtained in tests if the connectors are staggered so that they do not act along the same line with respect to the grain of the transverse member.

Placement of connectors in joints with members at right angles to each other is subject to the limitations of either member. Since rules for alignment, spacing, and edge and end distance of connectors for all conceivable directions of applied load would be complicated, designers must rely on a sense of proportion and adequacy in applying the above rules to conditions of loading outside the specific limitations mentioned.

Anchor Bolts. To attach columns or arch bases to concrete foundations, anchor bolts are embedded in the concrete, with sufficient projection to permit placement of angles or shores bolted to the wood. Sometimes, instead of anchor bolts, steel straps are embedded in the concrete with a portion projecting above for bolt attachment to the wood members. Table 11-22 lists allowable shear loads for anchor bolts and steel base shoes.

Table 11-22. Anchor-Bolt Load Values for Steel Base Shoes

Bolt dia, in.	Allowable shear load,* lb	Embedment in concrete, in.
½	1,500	4
⅝	2,200	4
¾	3,000	5
⅞	3,500	6
1	4,000	8
1¼	5,000	8
1½	7,000	9
1¾	9,000	9
2	11,000	10
2¼	13,500	10
2½	16,000	12
2¾	19,000	12
3	22,000	14

*Shear load based on 3,000-psi concrete and bolt yield strength of 50,000 psi.

Washers. Bolt heads and nuts bearing on wood require metal washers to protect the wood and to distribute the pressure across the surface of the wood. Washers may be cast, malleable, cut, round-plate, or square-plate. When subjected to salt air or salt water, they should be galvanized or given some type of effective coating. Ordinarily, washers are dipped in red lead and oil prior to installation.

Setscrews should never be used against a wood surface. It may be possible, with the aid of proper washers, to spread the load of the setscrew over sufficient surface area of the wood that the compression strength perpendicular to grain is not exceeded.

Table 11-23 gives washer sizes capable of developing the capacity of A307 bolts.

Tie Rods. To resist the horizontal thrust of arches not buttressed, tie rods are required. The tie rods may be installed at ceiling height or below the floor. Table 11-24 gives the allowable loads on bars used as tie rods.

Hangers. Standard and special hangers are used extensively in timber construction. Stock hangers are available from a number of manufacturers. But by far the greater number of hangers are of special design. Where appearance is of prime importance, concealed hangers are frequently selected.

(National Design Specification for Stress-grade Lumber and Its Fastenings, National Forest Products Association, Washington; "Design Manual for TECO Timber Connector Construction," Timber Engineering Co., 1619 Massachusetts Ave., N.W., Washington, D.C. 20036; American Institute of Timber Construction "Timber Construction Manual," John Wiley & Sons, Inc., New York.)

Table 11-23. Allowable Loads for Bolts and Washers

Rod or bolt		Plate washers†		Cut washers			
Dia, in.	Tensile capacity,* lb	Side of square, in.	Thickness, in.	Outside dia, in.	Hole dia, in.	Thickness, in.	Max load,‡ lb
⅜	1,550	1¾	³/₁₆	1	⁷/₁₆	⁵/₆₄	290
⁷/₁₆	2,100	2⅛	¼	1¼	½	⁵/₆₄	460
½	2,750	2⅜	¼	1⅜	⁹/₁₆	⁷/₆₄	550
⅝	4,300	3	⁵/₁₆	1¾	¹¹/₁₆	⁹/₆₄	910
¾	6,190	3¾	⅜	2	¹³/₁₆	⁵/₃₂	1,100
⅞	8,420	4⅜	½	2¼	¹⁵/₁₆	¹¹/₆₄	1,100
1	11,000	5	½	2½	1 ¹/₁₆	¹¹/₆₄	1,800
1⅛	13,920	5⅝	⅝	2¾	1 ¼	¹¹/₆₄	2,100
1¼	17,180	6⅜	¾	3	1 ⅜	¹¹/₆₄	2,500
1⅜	20,800	7	¾	3¼	1 ½	³/₁₆	2,900
1½	24,700	7¾	⅞	3½	1 ⅝	³/₁₆	3,400
1⅝	29,000	8⅜	⅞	3¾	1 ¾	³/₁₆	3,900
1¾	33,700	9	1	4	1 ⅞	³/₁₆	4,400
1⅞	38,700	9¾	1	4¼	2	³/₁₆	5,000
2	44,000	10¼	1⅛	4½	2 ⅛	³/₁₆	5,500
2⅛	49,700	11	1⅛				
2¼	55,700	11¾	1¼	4¾	2 ⅜	⁷/₃₂	6,000
2⅜	62,000	12½	1¼				
2½	68,700	13	1⅜	5	2 ⅝	¹⁵/₆₄	6,500
2⅝	75,800	13¾	1½				
2¾	83,200	14½	1½	5¼	2 ⅞	¼	6,800
2⅞	90,900	15¼	1⅝				
3	99,000	15¾	1⅝	5½	3 ⅛	⁹/₃₂	7,200

*Based on ASTM A36 steel and A307 bolts.
†Size required to develop capacity of rod or bolt in accordance with note ‡.
‡Based on allowable strength of wood in compression perpendicular to grain of 450 psi.

11-22. Glued Fastenings. Glued joints are generally between two pieces of wood where the grain directions are parallel (as between the laminations of a beam or arch). Or such joints may be between solid-sawn or laminated timber and plywood, where the face grain of the plywood is either parallel or at right angles to the grain direction of the timber.

It is only in special cases that lumber may be glued with the grain direction of adjacent pieces at an angle. When the angle is large, dimensional changes from changes in wood moisture content set up large stresses in the glued joint. Consequently, the strength of the joint may be considerably reduced over a period of time. Exact data are not available, however, on the magnitude of this expected strength reduction.

In joints connected with plywood gusset plates, this shrinkage differential is minimized because plywood swells and shrinks much less than does solid wood.

Table 11-24. Allowable Tension on Round and Square Upset Bars

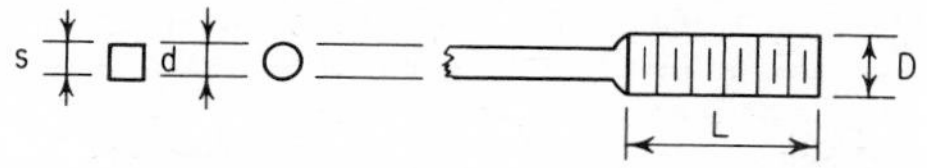

UNC and 4UN Class 2A Thread

Dia *d* or side *s*, in.	Round bars			Square bars		
	Capacity,* lb	Upset		Capacity,* lb	Upset	
		Dia *D*, in.	Length *L*, in.		Dia. *D*, in.	Length *L*, in.
¾	9.700	1	4	12,400	1⅛	4
⅞	13,200	1⅛	4	16,900	1¼	4
1	17,300	1⅜	4	22,000	1½	4
1⅛	21,800	1½	4	27,800	1¾	4
1¼	27,000	1¾	4	34,400	2	4½
1⅜	32,700	1¾	4	41,600	2	4½
1½	38,900	2	4½	49,500	2¼	5
1⅝	45,600	2¼	5	58,100	2½	5½
1¾	53,000	2¼	5	67,400	2½	5½
1⅞	60,800	2⅛	5½	77.300	2¾	5½
2	69,100	2½	5½	88,000	2¾	5½
2⅛	78,000	2¾	5½	99,300	3	6
2¼	87,500	2¾	5½	111,400	3¼	6½
2⅜	97,500	3	6	124,100	3¼	6½
2½	108,000	3¼	6½	137,500	3½	7
2⅝	119,100	3¼	6½	151,600	3¾	7

*Based on ASTM A36 steel.

Glued joints can be made between end-grain surfaces. But they are seldom strong enough to meet the requirements of even ordinary service. Seldom is it possible to develop more than 25% of the tensile strength of the wood in such butt joints. It is for this reason that plane sloping scarfs of relatively flat slope (Fig. 11-2) or finger joints with thin tips and flat slope on the sides of the individual fingers (Fig. 11-3) are used to develop a high proportion of the strength of the wood.

Joints of end grain to side grain are also difficult to glue properly. When subjected to severe stresses as a result of unequal dimensional changes in the members due to changes in moisture content, joints suffer from severely reduced strength.

For the above reasons, joints between end-grain surfaces and between end-grain and side-grain surfaces should not be used if the joints are expected to carry load.

For joints made with wood of different species, the allowable shear stress for parallel-grain bonding is equal to the allowable shear stress parallel to the grain for the weaker species in the joint. This assumes uniform stress distribution in the joint. When grain direction is not parallel, the allowable shear stress on the glued area between the two pieces may be estimated from the Scholten nomograph (Fig. 11-4).

Adhesives used for fabricating structural glued-laminated timbers (Art. 11-12) also are satisfactory for other glued joints. In selecting an adhesive, consideration should be given to wood moisture content.

[AITC 103, Standard for Structural Glued-Laminated Timber, American Institute of Timber Construction, Englewood, Colo.; Federal Specification MMM-A-125, Adhesive, Casein-type, Water- and Mold-resistant, General Services Administration, Washington, D.C. 20405; Military Specification MIL-A-397B, Adhesive, Room-Temperature and Intermediate-Temperature Setting Resin (Phenol, Resorcinol, and Melamine Base), and Military Specification MIL A-5534A, Adhesive, High-temperature Setting Resin (Phenol, Melamine, and Resorcinol Base), U.S. Naval Supply Depot, Philadelphia, Pa. 19120.]

11-23. Wood Columns. A wood compression member may be a solid piece of timber (Fig. 11-5a) or laminated (Fig. 11-5b) or built up of spaced members (Fig. 11-5c). The latter are comprised of two or more wood compression members with parallel longitudinal axes. The members are separated at ends and midpoints by blocking, and joined to the end blocking with connectors with adequate shear resistance.

Columns with a ratio of unsupported length L, in., to least dimension d, in., less than 11 fail by crushing. The allowable concentric load for such members equals the cross-sectional area, sq in., times F_c, the allowable compression stress, psi, parallel to grain for the species, adjusted for service conditions and duration of load.

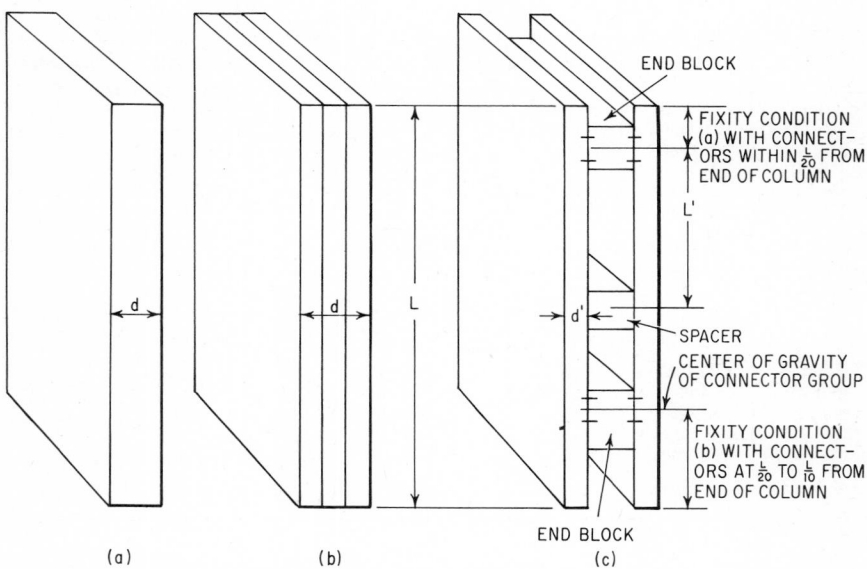

Fig. 11-5. Behavior of wood columns depends on the ratio of length L or L' to least dimension d or d'. (a) Solid-sawn timber column; (b) glued-laminated column; (c) spaced column.

When the slenderness ratio L/d exceeds 11, wood columns generally fail by buckling. In that case, the allowable stress is determined from formulas that yield values less than F_c. The computed allowable stress must be adjusted for duration of load.

For rectangular solid-sawn or glued-laminated columns, the allowable unit stress may not exceed F_c or

$$F'_c = \frac{0.30E}{(L/d)^2} \qquad (11\text{-}11)$$

where E = modulus of elasticity, psi, adjusted for duration of load. In no case may L/d exceed 50. The formula was derived for pin-end conditions but may also be used for square-cut ends.

For round columns, the allowable stress may not exceed that for a square column of the same cross-sectional area or

$$F'_c = \frac{3.619E}{(L/r)^2} \qquad (11\text{-}12)$$

where r = radius of gyration of cross section, in.

For tapered columns, d may be taken as the sum of the least dimension plus one-third the difference between this and the maximum thickness parallel to this dimension.

For spaced columns, the allowable load is the smallest computed with F_c, F'_c calculated from Eq. (11-11) with the overall dimensions of the column, and F''_c determined from Eq. (11-13a) or (11-13b) with the dimensions of the individual members. For the individual members, L/d' may not exceed 80 nor may L'/d' exceed 40, where L' is the spacing of the blocking and d' the least dimension.   End blocks are required to insure spaced-column action when L/d' exceeds $\sqrt{0.30E/F_c}$.   For smaller L/d', the individual members should be designed for allowable stresses computed from Eq. (11-11).   Allowable stresses for spaced-column action depend on the fixity condition (Fig. 11-5c).   For fixity condition (a),

$$F''_c = \frac{0.75E}{(L/d')^2}$$ (11-13a)

For fixity condition (b),

$$F''_c = \frac{0.90E}{(L/d')^2}$$ (11-13b)

Each member of a spaced column should be designed separately on the basis of its L/d'.   The allowable load on each equals its cross-sectional area, sq in., times its allowable stress, psi, adjusted for duration of load.   The sum of the allowable loads on the individual members should equal or exceed the total load on the column.

When a single spacer block is placed in the middle tenth of a spaced column, connectors are not required.   But they should be used for multiple spacer blocks.   The distance between two adjacent blocks may not exceed half the distance between centers of connectors in the end blocks in opposite ends of the column.

For all types of columns, the distance between adequate bracing, including beams and struts, should be used as L in determining L/d.   The largest L/d for the column or any component, whether it be computed for a major or minor axis, should be used in calculating the allowable unit stress.

For combined axial and bending stress, P/A divided by the allowable compression stress plus M/S divided by the allowable bending stress must not exceed unity. P is the axial load, lb; A the area of the column, sq in.; M the bending moment, in.-lb; and S the section modulus, in.³   Bending caused by transverse loads, wind loads, or eccentric loads, or any combination of these, should be taken into account.

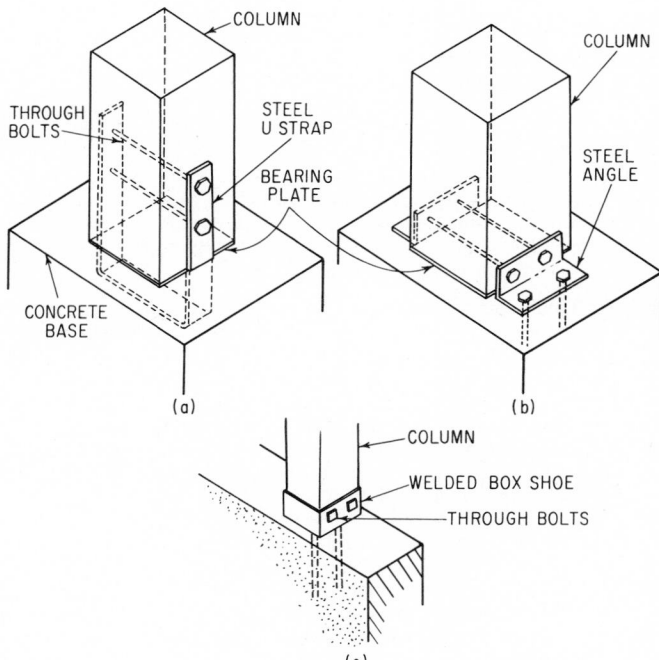

Fig. 11-6. Typical anchorages of wood column to base. (a) Wood column anchored to concrete base with U strap; (b) anchorage with steel angles; (c) with a welded box shoe.

The critical section of columns supporting trusses frequently exists at the connection of knee brace to column. Where no knee brace is used, or the column supports a beam, the critical section for moment usually occurs at the bottom of truss or beam. Then, a rigid connection must be provided to resist moment, or adequate diagonal bracing must be provided to carry wind loads into a support.

Figure 11-6 shows typical column base anchorages and Fig. 11-7 typical beam-to-column connections (AITC 104, Typical Construction Details, American Institute of Timber Construction).

(American Institute of Timber Construction, "Timber Construction Manual," John Wiley & Sons, Inc., New York.)

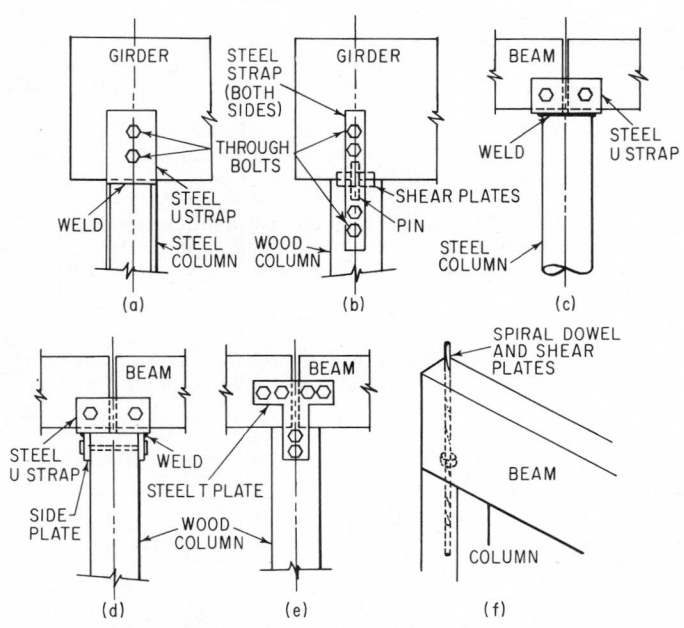

Fig. 11-7. Typical wood beam and girder connections to columns. (*a*) Girder to steel column; (*b*) girder to wood column; (*c*) beam to pipe column; (*d*) beam to wood column, with steel strap welded to steel side plates; (*e*) beam to wood column, with a T plate; (*f*) with spiral dowel and shear plates.

11-24. Design of Timber Joists.   Joists are relatively narrow beams, usually spaced 12 to 24 in. c to c. They generally are topped with sheathing and braced with diaphragms or cross bridging at intervals up to 10 ft. For joist spacings of 16 to 24 in. c to c, 1-in. sheathing usually is required. For spacings over 24 in., 2 in. or more of wood decking is necessary.

Standard beam formulas for bending, shear, and deflection may be used to determine joist sizes. Connections shown in Figs. 11-7 and 11-8 may be used for joists.

11-25. Design of Timber Beams.   Standard beam formulas for bending, shear, and deflection may be used to determine beam sizes. Ordinarily, deflection governs design; but for short, heavily loaded beams, shear is likely to control.

Figure 11-9 shows the types of beams commonly produced in timber. Straight and single- and double-tapered straight beams can be furnished solid-sawn or glued-laminated. The curved members can be furnished only glued-laminated. Beam names describe the top and bottom surfaces of the beam. The first part describes the top surface, the word following the hyphen the bottom. Sawn surfaces on the tension side of a beam should be avoided.

Table 11-25 gives the load-carrying capacity for various cross-sectional sizes of glued-laminated, simply supported beams.

Example.   Design a straight, glued-laminated beam, simply supported and uniformly loaded: span, 28 ft; spacing, 9 ft c to c; live load, 30 psf; dead load, 5 psf for deck and 7.5 psf for roofing.

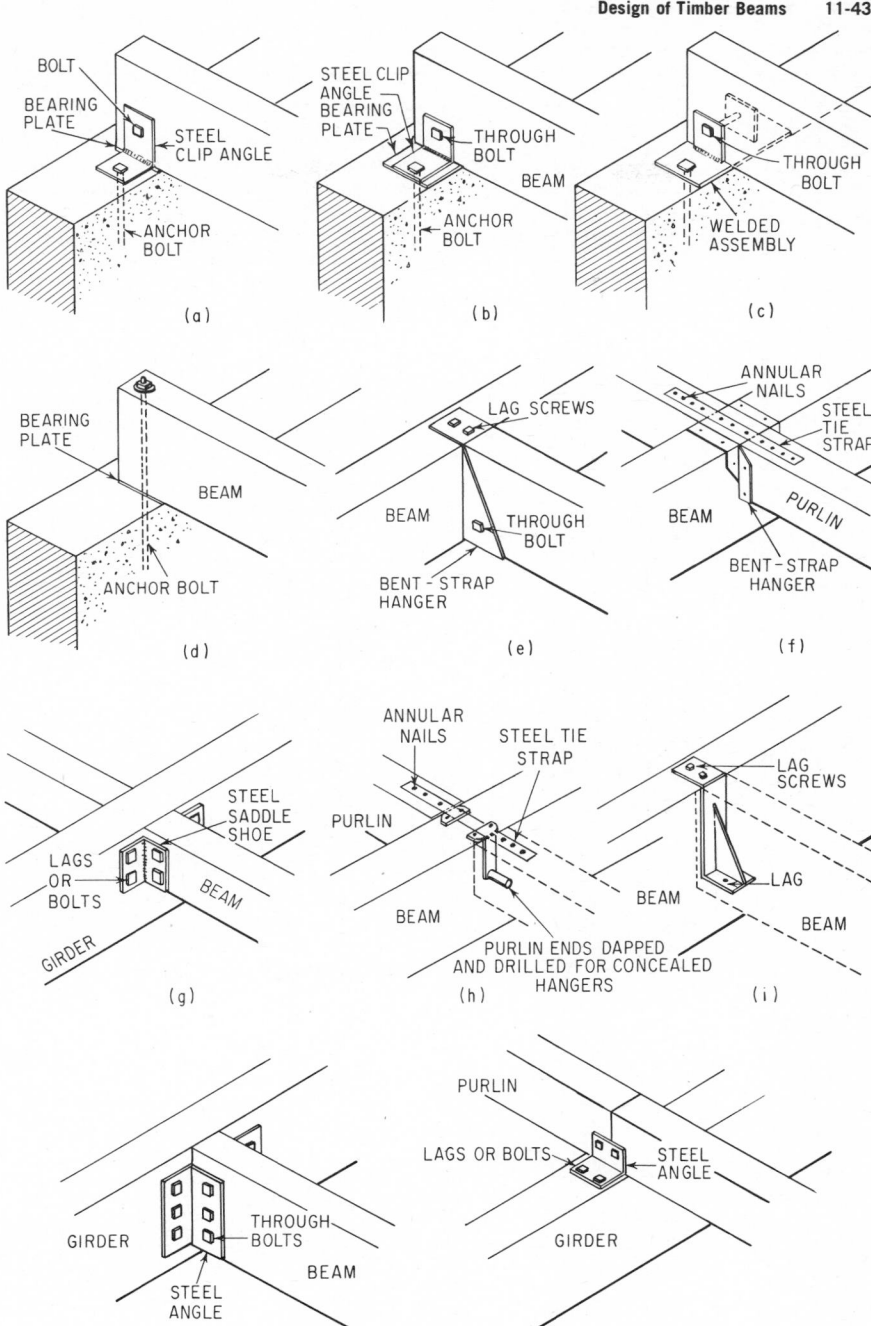

Fig. 11-8. Beam connections. (*a*) and (*b*) Wood beam anchored on wall with steel angles; (*c*) with welded assembly. (*d*) Beam anchored directly with bolt. (*e*) Beam supported on girder with bent-strap hanger. (*f*) Similar support for purlins. (*g*) Saddle connects beam to girder (suitable for one-sided connection). (*h*) and (*i*) Connections with concealed hangers. (*j*) and (*k*) Connections with steel angles.

Table 11-25. Load-Carrying Capacity of Simple-Span Laminated Beams*

Span, ft	Spacing, ft	Roof beam total load-carrying capacity						Floor beams total load
		30 psf	35 psf	40 psf	45 psf	50 psf	55 psf	50 psf
8	4	3⅛ × 4½	3⅛ × 4½	3⅛ × 6	3⅛ × 6	3⅛ × 6	3⅛ × 6	3⅛ × 6
	6	3⅛ × 4½	3⅛ × 4½	3⅛ × 6	3⅛ × 6	3⅛ × 6	3⅛ × 6	3⅛ × 6
	8	3⅛ × 4½	3⅛ × 4½	3⅛ × 6	3⅛ × 6	3⅛ × 6	3⅛ × 6	3⅛ × 7½
10	4	3⅛ × 4½	3⅛ × 4½	3⅛ × 6	3⅛ × 6	3⅛ × 6	3⅛ × 6	3⅛ × 7½
	6	3⅛ × 4½	3⅛ × 6	3⅛ × 6	3⅛ × 6	3⅛ × 6	3⅛ × 7½	3⅛ × 7½
	8	3⅛ × 6	3⅛ × 6	3⅛ × 7½	3⅛ × 7½	3⅛ × 7½	3⅛ × 7½	3⅛ × 9
	10	3⅛ × 6	3⅛ × 7½	3⅛ × 7½	3⅛ × 7½	3⅛ × 7½	3⅛ × 9	3⅛ × 9
12	6	3⅛ × 6	3⅛ × 6	3⅛ × 7½	3⅛ × 7½	3⅛ × 7½	3⅛ × 7½	3⅛ × 9
	8	3⅛ × 6	3⅛ × 7½	3⅛ × 9	3⅛ × 9	3⅛ × 9	3⅛ × 9	3⅛ × 10½
	10	3⅛ × 7½	3⅛ × 7½	3⅛ × 9	3⅛ × 9	3⅛ × 9	3⅛ × 10½	3⅛ × 10½
	12	3⅛ × 7½	3⅛ × 9	3⅛ × 9	3⅛ × 9	3⅛ × 10½	3⅛ × 10½	3⅛ × 12
14	8	3⅛ × 7½	3½ × 9	3⅛ × 9	3⅛ × 9	3⅛ × 10½	3⅛ × 10½	3⅛ × 12
	10	3⅛ × 9	3⅛ × 9	3⅛ × 10½	3⅛ × 10½	3⅛ × 10½	3⅛ × 12	3⅛ × 12
	12	3⅛ × 9	3⅛ × 10½	3⅛ × 10½	3⅛ × 10½	3⅛ × 12	3⅛ × 12	3⅛ × 13½
	14	3⅛ × 10½	3⅛ × 10½	3⅛ × 12	3⅛ × 12	3⅛ × 12	3⅛ × 13½	3⅛ × 13½
16	8	3⅛ × 9	3⅛ × 9	3⅛ × 10½	3⅛ × 10½	3⅛ × 12	3⅛ × 12	3⅛ × 13½
	12	3⅛ × 10½	3⅛ × 12	3⅛ × 12	3⅛ × 12	3⅛ × 13½	3⅛ × 13½	3⅛ × 15
	14	3⅛ × 12	3⅛ × 12	3⅛ × 13½	3⅛ × 13½	3⅛ × 15	3⅛ × 15	3⅛ × 15
	16	3⅛ × 12	3⅛ × 13½	3⅛ × 13½	3⅛ × 15	3⅛ × 15	3⅛ × 16½	3⅛ × 15
18	8	3⅛ × 9	3⅛ × 10½	3⅛ × 12	3⅛ × 12	3⅛ × 12	3⅛ × 13½	3⅛ × 15
	12	3⅛ × 12	3⅛ × 12	3⅛ × 13½	3⅛ × 13½	3⅛ × 15	3⅛ × 16½	3⅛ × 16½
	16	3⅛ × 13½	3⅛ × 15	3⅛ × 15	3⅛ × 16½	5⅛ × 13½	5⅛ × 13½	5⅛ × 15
	18	3⅛ × 15	3⅛ × 15	3⅛ × 16½	3⅛ × 18	5⅛ × 15	5⅛ × 15	5⅛ × 15
20	8	3⅛ × 12	3⅛ × 12	3⅛ × 13½	3⅛ × 13½	3⅛ × 13½	3⅛ × 15	3⅛ × 16½
	12	3⅛ × 13½	3⅛ × 13½	3⅛ × 15	3⅛ × 16½	3⅛ × 16½	5⅛ × 13½	5⅛ × 15
	16	3⅛ × 15	3⅛ × 16½	3⅛ × 18	3⅛ × 18	5⅛ × 15	5⅛ × 16½	5⅛ × 18
	18	3⅛ × 16½	3⅛ × 16½	3⅛ × 18	5⅛ × 15	5⅛ × 16½	5⅛ × 16½	5⅛ × 18
22	8	3⅛ × 13½	3⅛ × 13½	3⅛ × 13½	3⅛ × 15	3⅛ × 15	3⅛ × 16½	5⅛ × 15
	12	3⅛ × 15	3⅛ × 15	3⅛ × 16½	3⅛ × 18	3⅛ × 18	5⅛ × 15	5⅛ × 16½
	16	3⅛ × 16½	3⅛ × 18	5⅛ × 15	5⅛ × 16½	5⅛ × 16½	5⅛ × 18	5⅛ × 19½
	18	3⅛ × 18	5⅛ × 15	5⅛ × 16½	5⅛ × 16½	5⅛ × 18	5⅛ × 18	5⅛ × 19½
24	8	3⅛ × 13½	3⅛ × 15	3⅛ × 15	3⅛ × 16½	3⅛ × 16½	3⅛ × 18	5⅛ × 16½
	12	3⅛ × 16½	3⅛ × 16½	3⅛ × 18	5⅛ × 15	5⅛ × 16½	5⅛ × 16½	5⅛ × 18
	16	3⅛ × 18	5⅛ × 16½	5⅛ × 16½	5⅛ × 18	5⅛ × 18	5⅛ × 19½	5⅛ × 21
	18	5⅛ × 15	5⅛ × 16½	5⅛ × 18	5⅛ × 18	5⅛ × 19½	5⅛ × 21	5⅛ × 21
26	8	3⅛ × 15	3⅛ × 16½	3⅛ × 16½	3⅛ × 16½	3⅛ × 18	5⅛ × 16½	5⅛ × 18
	12	3⅛ × 18	3⅛ × 18	5⅛ × 16½	5⅛ × 16½	5⅛ × 18	5⅛ × 18	5⅛ × 19½
	16	5⅛ × 16½	5⅛ × 16½	5⅛ × 18	5⅛ × 18	5⅛ × 19½	5⅛ × 21	5⅛ × 22½
	18	5⅛ × 16½	5⅛ × 18	5⅛ × 18	5⅛ × 19½	5⅛ × 21	5⅛ × 21	5⅛ × 22½
28	8	3⅛ × 16½	3⅛ × 16½	3⅛ × 16½	3⅛ × 18	5⅛ × 16½	5⅛ × 16½	5⅛ × 19½
	12	3⅛ × 18	5⅛ × 16½	5⅛ × 18	5⅛ × 18	5⅛ × 18	5⅛ × 19½	5⅛ × 21
	16	5⅛ × 18	5⅛ × 18	5⅛ × 19½	5⅛ × 19½	5⅛ × 21	5⅛ × 22½	5⅛ × 24
	18	5⅛ × 18	5⅛ × 19½	5⅛ × 19½	5⅛ × 21	5⅛ × 22½	5⅛ × 24	5⅛ × 24
30	8	3⅛ × 18	3⅛ × 18	5⅛ × 16½	5⅛ × 16½	5⅛ × 18	5⅛ × 18	5⅛ × 21
	12	5⅛ × 16½	5⅛ × 18	5⅛ × 18	5⅛ × 19½	5⅛ × 19½	5⅛ × 21	5⅛ × 22½
	16	5⅛ × 18	5⅛ × 19½	5⅛ × 21	5⅛ × 21	5⅛ × 22½	5⅛ × 24	5⅛ × 25½
	18	5⅛ × 19½	5⅛ × 21	5⅛ × 21	5⅛ × 22½	5⅛ × 24	5⅛ × 25½	5⅛ × 27
32	8	3⅛ × 18	5⅛ × 16½	5⅛ × 18	5⅛ × 18	5⅛ × 18	5⅛ × 19½	5⅛ × 21
	12	5⅛ × 18	5⅛ × 19½	5⅛ × 19½	5⅛ × 21	5⅛ × 21	5⅛ × 22½	5⅛ × 24
	16	5⅛ × 19½	5⅛ × 21	5⅛ × 22½	5⅛ × 22½	5⅛ × 24	5⅛ × 25½	5⅛ × 27
	18	5⅛ × 21	5⅛ × 21	5⅛ × 22½	5⅛ × 24	5⅛ × 25½	5⅛ × 27	5⅛ × 28½
34	8	5⅛ × 16½	5⅛ × 18	5⅛ × 18	5⅛ × 19½	5⅛ × 19½	5⅛ × 21	5⅛ × 22½
	12	5⅛ × 19½	5⅛ × 19½	5⅛ × 21	5⅛ × 21	5⅛ × 22½	5⅛ × 24	5⅛ × 25½
	16	5⅛ × 21	5⅛ × 22½	5⅛ × 22½	5⅛ × 24	5⅛ × 25½	5⅛ × 27	5⅛ × 28½
	18	5⅛ × 22½	5⅛ × 22½	5⅛ × 24	5⅛ × 25½	5⅛ × 27	5⅛ × 28½	5⅛ × 28½
36	12	5⅛ × 19½	5⅛ × 21	5⅛ × 22½	5⅛ × 22½	5⅛ × 24	5⅛ × 25½	6¾ × 25½
	16	5⅛ × 22½	5⅛ × 24	5⅛ × 24	5⅛ × 25½	5⅛ × 27	5⅛ × 28½	6¾ × 27
	18	5⅛ × 22½	5⅛ × 24	5⅛ × 25½	5⅛ × 28½	5⅛ × 30	6¾ × 27	6¾ × 28½
	20	5⅛ × 24	5⅛ × 25½	5⅛ × 27	5⅛ × 30	6¾ × 27	6¾ × 28½	6¾ × 30
38	12	5⅛ × 21	5⅛ × 22½	5⅛ × 24	5⅛ × 24	5⅛ × 25½	5⅛ × 27	6¾ × 27
	16	5⅛ × 24	5⅛ × 24	5⅛ × 25½	5⅛ × 27	5⅛ × 28½	5⅛ × 30	6¾ × 28½
	18	5⅛ × 24	5⅛ × 25½	5⅛ × 27	5⅛ × 30	6¾ × 27	6¾ × 28½	6¾ × 30
	20	5⅛ × 25½	5⅛ × 27	5⅛ × 28½	6¾ × 27	6¾ × 28½	6¾ × 30	6¾ × 31½

Table 11-25. Load-Carrying Capacity of Simple-Span Laminated Beam * *(Continued)*

Span, ft	Spacing, ft	Roof beams total load-carrying capacity						Floor beams total load
		30 psf	35 psf	40 psf	45 psf	50 psf	55 psf	50 psf
40	12	5⅛ × 22½	5⅛ × 24	5⅛ × 24	5⅛ × 25½	5⅛ × 27	6¾ × 25½	6¾ × 28½
	16	5⅛ × 24	5⅛ × 25½	5⅛ × 27	5⅛ × 28½	6¾ × 27	6¾ × 28½	6¾ × 31½
	18	5⅛ × 25½	5⅛ × 27	5⅛ × 28½	6¾ × 27	6¾ × 28½	6¾ × 30	6¾ × 31½
	20	5⅛ × 27	5⅛ × 28½	6¾ × 27	6¾ × 28½	6¾ × 30	6¾ × 31½	6¾ × 33
42	12	5⅛ × 24	5⅛ × 24	5⅛ × 25½	5⅛ × 27	6¾ × 25½	6¾ × 25½	6¾ × 30
	16	5⅛ × 25½	5⅛ × 27	5⅛ × 28½	5⅛ × 30	6¾ × 28½	6¾ × 30	6¾ × 33
	18	5⅛ × 27	5⅛ × 28½	5⅛ × 30	6¾ × 28½	6¾ × 30	6¾ × 31½	6¾ × 33
	20	5⅛ × 28½	5⅛ × 30	6¾ × 28½	6¾ × 30	6¾ × 31½	6¾ × 33	6¾ × 34½
44	12	5⅛ × 24	5⅛ × 25½	5⅛ × 27	5⅛ × 27	6¾ × 25½	6¾ × 27	6¾ × 31½
	16	5⅛ × 27	5⅛ × 28½	5⅛ × 30	6¾ × 28½	6¾ × 30	6⅜ × 31½	6¾ × 33
	18	5⅛ × 28½	5⅛ × 30	6¾ × 28½	6¾ × 30	6¾ × 31½	6¾ × 33	6¾ × 34½
	20	5⅛ × 30	6¾ × 27	6¾ × 30	6¾ × 30	6¾ × 33	6¾ × 34½	6¾ × 36
46	12	5⅛ × 25½	5⅛ × 27	5⅛ × 28½	6¾ × 25½	6¾ × 27	6¾ × 28½	6¾ × 31½
	16	5⅛ × 28½	5⅛ × 30	6¾ × 28½	6¾ × 28½	6¾ × 31½	6¾ × 33	6¾ × 36
	18	5⅛ × 28½	6¾ × 28½	6¾ × 30	6¾ × 31½	6¾ × 33	6¾ × 34½	6¾ × 36
	20	5⅛ × 30	6¾ × 28½	6¾ × 31½	6¾ × 33	6¾ × 34½	6¾ × 36	8¾ × 34½
48	12	5⅛ × 27	5⅛ × 28½	5⅛ × 30	5⅛ × 30	6¾ × 28½	6¾ × 30	6¾ × 33
	16	5⅛ × 30	6¾ × 28½	6¾ × 30	6¾ × 30	6¾ × 31½	6¾ × 34½	6¾ × 37½
	18	5⅛ × 30	6¾ × 30	6¾ × 30	6¾ × 33	6¾ × 34½	6¾ × 36	8¾ × 34½
	20	6¾ × 28½	6¾ × 30	6¾ × 31½	6¾ × 34½	6¾ × 36	6¾ × 37½	8¾ × 36
50	12	5⅛ × 28½	5⅛ × 28½	5⅛ × 30	6¾ × 28½	6¾ × 30	6¾ × 31½	6¾ × 34½
	16	5⅛ × 30	6¾ × 30	6¾ × 30	6¾ × 31½	6¾ × 33	6¾ × 36	8¾ × 34½
	18	6¾ × 28½	6¾ × 30	6¾ × 31½	6¾ × 34½	6¾ × 36	8¾ × 33	8¾ × 36
	20	6¾ × 30	6¾ × 31½	6¾ × 33	6¾ × 36	6¾ × 37½	8¾ × 34½	8¾ × 37½
52	12	5⅛ × 28½	5⅛ × 30	6¾ × 28½	6¾ × 30	6¾ × 31½	6¾ × 31½	6¾ × 36
	16	6¾ × 28½	6¾ × 30	6¾ × 31½	6¾ × 33	6¾ × 34½	6¾ × 37½	8¾ × 36
	18	6¾ × 30	6¾ × 31½	6¾ × 33	6¾ × 34½	6¾ × 37½	6¾ × 39	8¾ × 37½
	20	6¾ × 31½	6¾ × 33	6¾ × 34½	6¾ × 37½	6¾ × 39	8¾ × 36	8¾ × 39
54	12	5⅛ × 30	6¾ × 28½	6¾ × 30	6¾ × 31½	6¾ × 33	6¾ × 33	6¾ × 37½
	16	6¾ × 30	6¾ × 31½	6¾ × 33	6¾ × 34½	6¾ × 36	6¾ × 37½	8¾ × 37½
	18	6¾ × 31½	6¾ × 33	6¾ × 34½	6¾ × 36	6¾ × 39	8¾ × 36	8¾ × 39
	20	6¾ × 33	6¾ × 34½	6¾ × 36	6¾ × 39	8¾ × 36	8¾ × 37½	8¾ × 40½
56	12	6¾ × 28½	6¾ × 30	6¾ × 31½	6¾ × 33	6¾ × 33	6¾ × 34½	8¾ × 36
	16	6¾ × 31½	6¾ × 33	6¾ × 34½	6¾ × 36	6¾ × 37½	8¾ × 34½	8¾ × 39
	18	6¾ × 33	6¾ × 34½	6¾ × 36	6¾ × 37½	8¾ × 34½	8¾ × 37½	8¾ × 40½
	20	6¾ × 33	6¾ × 36	6¾ × 37½	8¾ × 34½	8¾ × 37½	8¾ × 39	8¾ × 42
58	12	6¾ × 30	6¾ × 31½	6¾ × 31½	6¾ × 33	6¾ × 34½	6¾ × 36	8¾ × 37½
	16	6¾ × 31½	6¾ × 34½	6¾ × 36	6¾ × 37½	6¾ × 39	8¾ × 36	8¾ × 40½
	18	6¾ × 33	6¾ × 34½	6¾ × 37½	6¾ × 39	8¾ × 36	8¾ × 39	8¾ × 42
	20	6¾ × 34½	6¾ × 36	6¾ × 39	8¾ × 36	8¾ × 39	8¾ × 40½	8¾ × 43½
60	12	6¾ × 30	6¾ × 31½	6¾ × 33	6¾ × 34½	6¾ × 36	6¾ × 37½	8¾ × 39
	16	6¾ × 33	6¾ × 34½	6¾ × 36	6¾ × 39	8¾ × 36	8¾ × 37½	8¾ × 42
	18	6¾ × 34½	6¾ × 36	6¾ × 39	8¾ × 36	8¾ × 37½	8¾ × 39	8¾ × 43½
	20	6¾ × 36	6¾ × 37½	8¾ × 36	8¾ × 37½	8¾ × 40½	8¾ × 42	8¾ × 45

* This table applies to straight, simply supported, laminated timber beams. Other beam support systems may be employed to meet varying design conditions.

1. Roofs should have a minimum slope of ¼ in. per ft to eliminate water ponding.
2. Beam weight must be subtracted from total load carrying capacity. Floor beams are designed for uniform loads of 40-psf live load and 10-psf dead load.
3. Allowable stresses:
   Bending stress, F_b = 2,400 psi (reduced by size factor).
   Shear stress F_v = 165 psi.
   Modulus of elasticity, E = 1,800,000 psi.
   For roof beams, F_b and F_v were increased 15% for short duration of loading.
4. Deflection limits:
   Roof beams—$\frac{1}{180}$ span for total load.
   Floor beams—$\frac{1}{360}$ span for 40-psf live load only.
   For preliminary design purposes only.
   For more complete design information, see the AITC "Timber Construction Manual."

Allowable bending stress of combination grade is 2,400 psi, with modulus of elasticity E = 1,800,000 psi. Deflection limitation is $L/180$, where L is the span, ft. Assume the beam is laterally supported by the deck throughout its length.

With a 15% increase for short-duration loading, the allowable bending stress F_b becomes 2,760 psi and the allowable horizontal shear F_v, 230 psi.

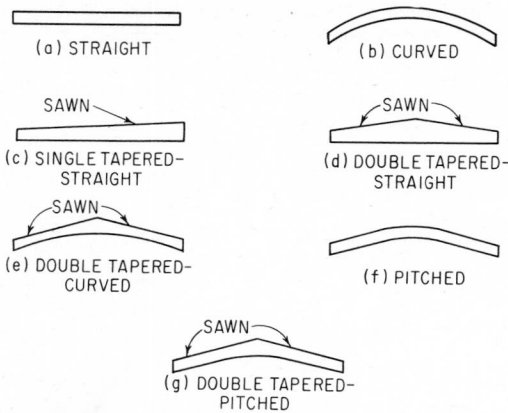

Fig. 11-9. Types of glued-laminated beams.

Assume the beam will weigh 22.5 lb per lin ft, averaging 2.5 psf. Then, the total uniform load comes to 45 psf. So the beam carries $w = 45 \times 9 = 405$ lb per lin ft.

The end shear $V = wL/2$ and the maximum shearing stress $= 3V/2 = 3wL/4$. Hence, the required area, sq in., for horizontal shear is

$$A = \frac{3wL}{4F_v} = \frac{wL}{306.7} = \frac{405 \times 28}{306.7} = 37.0 \text{ sq in.}$$

The required section modulus, in.^{3}, is

$$S = \frac{1.5wL^2}{F_b} = \frac{1.5 \times 405 \times 28^2}{2,760} = 172.6 \text{ in.}^3$$

If $D = 180$, the reciprocal of the deflection limitation, then the maximum deflection equals $5 \times 1,728wL^4/384EI \leq 12L/D$, where I is the moment of inertia of the beam cross section, in.^{4} Hence, to control deflection, the moment of inertia must be at least

$$I = \frac{1.875DwL^3}{E} = \frac{1.875 \times 180 \times 405 \times 28^3}{1,800,000} = 1,688 \text{ in.}^4$$

Assume that the beam will be fabricated with 1½-in. laminations. From Table 11-4, the most economical section satisfying all three criteria is 5⅛ × 16½, with $A = 84.6$, $S = 232.5$, and $I = 1,918.5$. But it has a size factor of 0.97. So the allowable bending stress must be reduced to $2,760 \times 0.97 = 2,677$ psi. And the required section modulus must be increased accordingly to $172.6/0.97 = 178$. Nevertheless, the selected section still is adequate.

Suspended-Span Construction. Cantilever systems may comprise any of the various types and combinations of beam illustrated in Fig. 11-10. Cantilever systems permit longer spans or larger loads for a given size member than do simple-span systems, if member size is not controlled by compression perpendicular to grain at the supports or by horizontal shear. Substantial design economies can be effected by decreasing the depths of the members in the suspended portions of a cantilever system.

For economy, the negative bending moment at the supports of a cantilevered beam should be equal in magnitude to the positive moment.

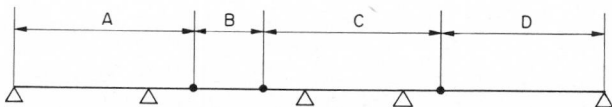

Fig. 11-10. Cantilevered-beam systems. A is a single cantilever; B is a suspended beam; C has a double cantilever; and D is a beam with one end suspended.

Consideration must be given to deflection and camber in cantilevered multiple spans. When possible, roofs should be sloped the equivalent of ¼ in. per ft of horizontal distance between the level of drains and the high point of the roof to eliminate water pockets, or provision should be made to insure that accumulation of water does not produce greater deflection and live loads than anticipated. Unbalanced loading conditions should be investigated for maximum bending moment, deflection, and stability.

(American Institute of Timber Construction, "Timber Construction Manual," John Wiley & Sons, Inc., New York.)

11-26. Deflection and Camber of Timber Beams. The design of many structural systems, particularly those with long spans, is governed by deflection. Strength calculations based on allowable stresses alone may result in excessive deflection. Limitations on deflection increase member stiffness.

Table 11-26 gives recommended deflection limits, as a fraction of the beam span, for timber beams. The limitation applies to live load or total load, whichever governs.

Table 11-26. Recommended Beam Deflection Limitations, In.*
(In Terms of Span l, In.)

Use classification	Live load only	Dead load plus live load
Roof beams:		
Industrial	$l/180$	$l/120$
Commercial and institutional:		
Without plaster ceiling	$l/240$	$l/180$
With plaster ceiling	$l/360$	$l/240$
Floor beams:		
Ordinary usage†	$l/360$	$l/240$
Highway bridge stringers	$l/200$ to $l/300$	
Railway bridge stringers	$l/300$ to $l/400$	

*Camber and Deflection, AITC 102, Appendix B, American Institute of Timber Construction.
†Ordinary usage classification is intended for construction in which walking comfort, minimized plaster cracking, and elimination of objectionable springiness are of prime importance. For special uses, such as beams supporting vibrating machinery or carrying moving loads, more severe limitations may be required.

Glued-laminated beams are cambered by fabricating them with a curvature opposite in direction to that corresponding to deflections under load. Camber does not, however, increase stiffness. Table 11-27 lists recommended minimum cambers for glued-laminated timber beams.

Table 11-27. Recommended Minimum Camber for Glued-Laminated Timber Beams*

Roof beams†	1½ times dead-load deflection
Floor beams‡	1½ times dead-load deflection
Bridge beams: §	
Long span	2 times dead-load deflection
Short span	2 times dead load + ½ applied-load deflection

*Camber and Deflection, AITC 102, Appendix B, American Institute of Timber Construction.
†The minimum camber of 1½ times dead-load deflection will produce a nearly level member under dead load alone after plastic deformation has occurred. Additional camber is usually provided to improve appearance or provide necessary roof drainage (Art. 11-27).
‡The minimum camber of 1½ times dead-load deflection will produce a nearly level member under dead load alone after plastic deformation has occurred. On long spans, a level ceiling may not be desirable because of the optical illusion that the ceiling sags. For warehouse or similar floors where live load may remain for long periods, additional camber should be provided to give a level floor under the permanently applied load.
§Bridge members are normally cambered for dead load only on multiple spans to obtain acceptable riding qualities.

11-27. Minimum Roof Slopes. Flat roofs have collapsed during rainstorms, though they were adequately designed on the basis of allowable stresses and definite deflection limitations. The reason for these collapses was the same, regardless of the structural framing used. The failures were caused by ponding of water as increasing deflections permitted more and more water to collect.

Roof beams should have a continuous upward slope equivalent to ¼ in. per ft between a drain and the high point of a roof, in addition to minimum recommended camber (Table 11-26), to avoid ponding. When flat roofs have insufficient slope for drainage (less than ¼ in. per ft), the stiffness of supporting members should be such that a 5-psf load will cause no more than ½-in. deflection.

Because of ponding, snow loads or water trapped by gravel stops, parapet walls, or ice dams magnify stresses and deflections from existing roof loads by

$$C_p = \frac{1}{1 - W'L^3/\pi^4 EI} \tag{11-14}$$

where C_p = factor for multiplying stresses and deflections under existing loads to determine stresses and deflections under existing loads plus ponding
  W' = weight of 1 in. of water on roof area supported by beam, lb
  L = span of beam, in.
  E = modulus of elasticity of beam material, psi
  I = moment of inertia of beam, in.[4]

(Kuenzi and Bohannan, "Increases in Deflection and Stresses Caused by Ponding of Water on Roofs," Forest Products Laboratory, Madison, Wis.)

11-28. Design of Wood Trusses. Type of truss and arrangement of members may be chosen to suit the shape of the structure, the loads, and the stresses involved. The types most commonly built are bowstring, flat or parallel chord, pitched, triangular or A type, camelback, and scissors (Fig. 11-11). For most construction other than houses, trusses usually are spaced 12 to 20 ft apart. For houses, very light trusses generally are erected 16 to 24 in. c to c.

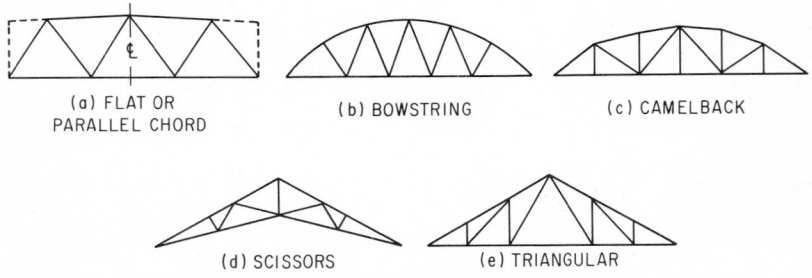

(a) FLAT OR PARALLEL CHORD

(b) BOWSTRING

(c) CAMELBACK

(d) SCISSORS

(e) TRIANGULAR

Fig. 11-11. Types of wood trusses.

Joints are critical in the design of a truss. Use of a specific truss type is often governed by joint considerations.

Chords and webs may be single-leaf (or monochord), double-leaf, or multiple-leaf members. Monochord trusses and trusses with double-leaf chords and single-leaf web system are the commonest arrangements. Web members may be attached to the sides of the chords. Or the web members may be in the same plane as the chords and attached with straps or gussets.

Individual truss members may be solid sawn, glued laminated, or mechanically laminated. Glued-laminated chords and solid-sawn web members are usually used. Steel rods or other steel shapes may be used as members of timber trusses if they meet design and service requirements.

The bowstring truss is by far the most popular. Spans of 100 to 200 ft are common, with single or two-piece top and bottom chords of glued-laminated timber, webs of solid-sawn timber, and metal heel plates, chord splice plates, and web-to-chord connections. This system is light in weight for the loads that it can carry. It can be shop- or field-assembled. Attention to the top chord, bottom chord, and heel connections is of prime importance, since they are the major stress-carrying components. Since the top chord is nearly the shape of an ideal arch, stresses in chords are almost uniform throughout a bowstring truss; web stresses are low under uniformly distributed loads.

Parallel-chord trusses, with slightly sloping top chords and level bottom chords, are used less often, because chord stresses are not uniform along their length and web stresses are high. Hence, different cross sections are required for successive chords, and web members and web-to-chord connections are heavy. Eccentric joints and tension stresses across the grain should be avoided in truss construction whenever possible, but particularly in parallel-chord trusses.

Triangular trusses and the more ornamental camelback and scissors trusses are used for shorter spans. They usually have solid-sawn members for both chords and webs where degree of seasoning of timbers, hardware, and connections are of considerable importance.

For joints, split-ring and shear-plate connectors are generally most economical. Sometimes, when small trusses are field-fabricated, only bolted joints are used. However, grooving tools for connectors can also be used effectively in the field.

Longitudinal sway bracing perpendicular to the plane of the truss is usually provided by solid-sawn X bracing. Lateral wind bracing may be provided by end walls or intermediate walls, or both. The roof system and horizontal bracing should be capable of transferring the wind load to the walls. Knee braces between trusses and columns are often used to provide resistance to lateral loads.

Horizontal framing between trusses consists of struts between trusses at bottom-chord level and diagonal tie rods, often of steel with turnbuckles for adjustment.

Table 11-28 gives typical bowstring-truss dimensions based on uniform loading conditions on the top chord, as normally imposed by roof joists.

For ordinary roof loads and spacing of 16 to 24 ft, vertical X bracing is required every 30 to 40 ft of chord length. This bracing is placed in alternate bays. Horizontal T-strut bracing should

Table 11-28. Bowstring-Truss Dimensions

Span range, ft	No. of panels	Avg truss height*	Avg arc length†
53–57	6	8 ft 7 in.	60 ft 8 in.
58–62		9 ft 3 in.	65 ft 11 in.
63–67		9 ft 11 in.	71 ft 5 in.
68–72	8	10 ft 7 in.	76 ft 5 in.
73–77		11 ft 3 in.	81 ft 8 in.
78–82		11 ft 11 in.	86 ft 10 in.
83–87		12 ft 7 in.	92 ft 1 in.
88–92		13 ft 3 in.	97 ft 4 in.
93–97		13 ft 11½ in.	102 ft 7 in.
98–102		14 ft 7½ in.	107 ft 10 in.
103–107	10	15 ft 5 in.	113 ft 2 in.
108–112		16 ft 1 in.	118 ft 5 in.
113–117		16 ft 9 in.	123 ft 8 in.
118–122		17 ft 5 in.	128 ft 11 in.
123–127		18 ft 1 in.	134 ft 2 in.
128–132	12	18 ft 11 in.	139 ft 6 in.
133–137		19 ft 7 in.	144 ft 9 in.

*The vertical distance from top of truss heel bearing plate to top of chord at midspan.
†Measured along the top of the top chord, wood to wood. Dimensions shown are for the longest of the span range. For smaller spans, lengths are proportionately shorter.

be placed from lower chord to lower chord in the same line as the vertical X bracing for the complete length of the building. If a ceiling is framed into the lower chords, these struts may be omitted.

Joists, spaced 12 to 24 in. c to c, usually rest on the top chords of the trusses and are secured there by toenailing. They may also be placed on ledgers attached to the sides of the upper chords or set in metal hangers, thus lowering the roof line.

Purlins, large-cross-section joists spaced 4 to 8 ft c to c, are often set on top of the top chords, where they are butted end to end and secured to the chords with clip angles. Purlins may also be set between the top chords on metal purlin hangers. Roof sheathing, 1-in. D&M (dressed & matched) or ⅜- to ½-in.-thick plywood, is laid directly on joists. For purlin construction, 2-in. D&M roof sheathing is normally used.

Table 11-29 Preliminary Design Dimensions, In., for Three-Hinged Tudor Arches*

Loading	Roof pitch	Wall hgt, ft	30-ft span					35-ft span					40-ft span					50-ft span				
			Width	Base	Lower tang.	Upper tang.	Crown	Width	Base	Lower tang.	Upper tang.	Crown	Width	Base	Lower tang.	Upper tang.	Crown	Width	Base	Lower tang.	Upper tang.	Crown
Vertical dead + live load = 400 lb per ft	3:12	12	5⅛	7½	11	10¾	7½	5⅛	7½	12	12	7½	5⅛	7½	13¼	13	7½	5⅛	10½	14¾	14½	7½
		14	5⅛	7½	12	12	7½	5⅛	7½	13½	13¼	7½	5⅛	7½	14¾	14½	7½	5⅛	9½	16¾	16¼	7½
		16	5⅛	7½	13¼	13	7½	5⅛	7½	14¾	14½	7½	5⅛	7½	16	16	7½	5⅛	8¾	18¾	17½	7½
		18	5⅛	7½	14¼	14¼	7½	5⅛	7½	16	15¾	7½	5⅛	7½	17½	17¼	7½	5⅛	8	20½	19	7½
		20	5⅛	7½	15¼	15¼	7½	5⅛	7½	17	16¾	7½	5⅛	7½	18¾	18½	7½	5⅛	7½	22	20¾	7½
	4:12	12	5⅛	7½	10¾	10¾	7½	5⅛	7½	11¾	11¾	7½	5⅛	7½	12¾	12½	7½	5⅛	9½	14½	13½	7½
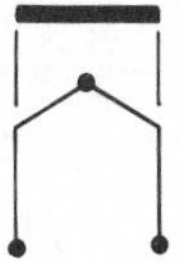		14	5⅛	7½	12	12	7½	5⅛	7½	13¼	13	7½	5⅛	7½	14¼	14¼	7½	5⅛	8¾	16¾	15	7½
		16	5⅛	7½	13¼	13	7½	5⅛	7½	14½	14½	7½	5⅛	7½	15¾	15½	7½	5⅛	8	18½	16½	7½
		18	5⅛	7½	14¼	14	7½	5⅛	7½	15¾	15½	7½	5⅛	7½	17	17	7½	5⅛	7½	20¼	18	7½
		20	5⅛	7½	15¼	15	7½	5⅛	7½	17	16¾	7½	5⅛	7½	18¼	18¼	7½	5⅛	7½	21¾	19½	7½
	6:12	12	5⅛	7½	10½	10½	7½	5⅛	7½	11½	11¼	7½	5⅛	7½	12¼	11½	7½	5⅛	8	14¾	11¾	7½
		14	5⅛	7½	11¾	11¾	7½	5⅛	7½	12¾	12¾	7½	5⅛	7½	13¾	13¼	7½	5⅛	7½	16½	12¾	7½
		16	5⅛	7½	13	12¾	7½	5⅛	7½	14	14	7½	5⅛	7½	15	15	7½	5⅛	7½	18¼	14¼	7½
		18	5⅛	7½	14	14	7½	5⅛	7½	15¼	15¼	7½	5⅛	7½	16½	16¼	7½	5⅛	7½	19¾	16	7½
	8:12	12	5⅛	7½	10¼	10	7½	5⅛	7½	11	10½	7½	5⅛	7½	12	10¼	7½	5⅛	7½	14¼	11½	7⅛
		14	5⅛	7½	11½	11¼	7½	5⅛	7½	12¼	12	7½	5⅛	7½	13½	11¾	7½	5⅛	7½	16	12¼	7½
		16	5⅛	7½	12¾	12½	7½	5⅛	7½	13½	13½	7½	5⅛	7½	14¾	13½	7½	5⅛	7½	17¾	13	7½
		18	5⅛	7½	13¾	13½	7½	5⅛	7½	14¾	14¾	7½	5⅛	7½	16	15	7½	5⅛	7½	19¼	14	7½
Vertical dead + live load = 600 lb per ft	3:12	12	5⅛	7½	12	12	7½	5⅛	8½	13½	13	7½	5⅛	10¾	15	13½	7½	5⅛	15½	17½	16	7½
		14	5⅛	7½	13½	13¼	7½	5⅛	7¾	15½	14¼	7½	5⅛	9¾	17½	14½	7½	5⅛	14	21	16½	7½
		16	5⅛	7½	14¾	14¾	7½	5⅛	7½	17¼	15¼	7½	5⅛	8¾	19¼	15¾	7½	5⅛	12¾	23½	17	7½
		18	5⅛	7½	16	16	7½	5⅛	7½	18¾	16¼	7½	5⅛	8	21¼	16¾	7½	6¾	9	20½	20¼	7½
		20	5⅛	7½	17¼	17	7½	5⅛	7½	20	17½	7½	5⅛	7½	22¾	18	7½	6¾	8¼	22	22	7½
	4:12	12	5⅛	7½	12	12	7½	5⅛	8	13¾	12¼	7½	5⅛	9¾	15¼	12¾	7½	5⅛	14	18	15¼	7½
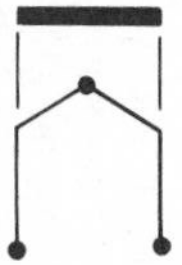		14	5⅛	7½	13½	13¼	7½	5⅛	7½	15½	13¾	7½	5⅛	9	17½	13¾	7½	5⅛	12¾	21	16	7½
		16	5⅛	7½	14¾	14¾	7½	5⅛	7½	17	15	7½	5⅛	8¼	19¼	15¼	7½	5⅛	11¾	23½	16½	7½
		18	5⅛	7½	16	16	7½	5⅛	7½	18½	16¼	7½	5⅛	7½	21	16½	7½	5⅛	10¾	25½	16¾	7½
		20	5⅛	7½	17¼	17	7½	5⅛	7½	20	17	7½	5⅛	7½	22¾	17¼	7½	6¾	7¾	21½	21½	7½
	6:12	12	5⅛	7½	12	11¼	7½	5⅛	7½	13¾	11¼	7½	5⅛	8½	15¼	11	7½	5⅛	11¾	18¼	14½	7¾
		14	5⅛	7½	13½	12¾	7½	5⅛	7½	15½	12½	7½	5⅛	7¾	17¼	12½	7½	5⅛	10¾	20¾	15¼	7½
		16	5⅛	7½	14¾	14	7½	5⅛	7½	17	14	7½	5⅛	7½	19	14	7½	5⅛	10	23	15¾	7½
		18	5⅛	7½	16	15¼	7½	5⅛	7½	18¼	15½	7½	5⅛	7½	20½	15¼	7½	5⅛	9½	25	16½	7½
	8:12	12	5⅛	7½	12	10½	7½	5⅛	7½	13½	10¼	7½	5⅛	7½	15¼	11	7½	5⅛	10	18	14	9½
		14	5⅛	7½	13¼	12¼	7½	5⅛	7½	15¼	11½	7½	5⅛	7½	17	11¾	7½	5⅛	9¼	20¼	15	8
		16	5⅛	7½	14½	13½	7½	5⅛	7½	16½	13¼	7½	5⅛	7½	18½	12¾	7½	5⅛	8¾	22¼	15¾	7½
		18	5⅛	7½	15¾	14¾	7½	5⅛	7½	18	14¼	7½	5⅛	7½	20¼	13¾	7½	5⅛	8¼	24¼	16¼	7½
Vertical dead + live load = 800 lb per ft	3:12	12	5⅛	8¾	13¾	12	7½	5⅛	11¼	15¾	12½	7½	5⅛	14	17¼	13¼	7½	5⅛	20½	20½	18½	12¼
		14	5⅛	7¾	15¾	12¾	7½	5⅛	10	18¼	13¼	7½	5⅛	12¾	20½	13¾	7½	5⅛	18½	24½	19	12¼
		16	5⅛	7½	17½	13¾	7½	5⅛	9¼	20¼	14	7½	5⅛	11½	22¾	14½	7½	6¾	13	22	18¾	12¼
		18	5⅛	7½	19¼	14¼	7½	5⅛	8¼	22¼	14¾	7½	5⅛	10½	25	15¼	7½	6¾	12	24	20	12¼
		20	5⅛	7½	20¾	14¾	7½	5⅛	7¾	24	15½	7½	6¾	7½	21	20¾	7½	6¾	11	25¾	21½	12¼
	4:12	12	5⅛	8	14	11½	7½	5⅛	10½	16	11¾	7½	5⅛	12¾	17¾	13	7½	5⅛	18½	20¾	17½	12
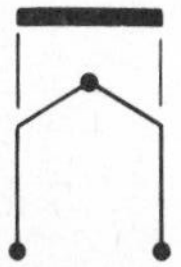		14	5⅛	7½	16	12½	7½	5⅛	9½	18¼	12¾	7½	5⅛	11¾	20½	13½	7½	5⅛	16¾	24¾	18¼	12
		16	5⅛	7½	17½	13½	7½	5⅛	8½	20¼	13¾	7½	5⅛	10¾	22¾	13¾	7½	6¾	12	22	17¾	12
		18	5⅛	7½	19	14½	7½	5⅛	7¾	22	14¾	7½	5⅛	9¾	25	14½	7½	6¾	11	23¾	19¼	12
		20	5⅛	7½	20¾	14¾	7½	5⅛	7½	23¾	15½	7½	6¾	7½	20¾	20½	7½	6¾	10¼	25¼	21	12
	6:12	12	5⅛	7½	14¼	10¾	7½	5⅛	9	16¼	10½	7½	5⅛	11	18	12½	7½	5⅛	15¼	21½	16½	11¼
		14	5⅛	7½	16	11¾	7½	5⅛	8¼	18¼	11¾	7½	5⅛	10	20¼	13¼	7½	5⅛	14¼	24½	17¼	11¼
		16	5⅛	7½	17½	13	7½	5⅛	7½	20	13	7½	5⅛	9¼	22½	13¾	7½	6¾	10¼	21½	16	11¼
		18	5⅛	7½	19	14	7½	5⅛	7½	21¾	14	7½	5⅛	8¾	24½	14¼	7½	6¾	9½	23	17½	11¼
	8:12	12	5⅛	7½	14¼	9¾	7½	5⅛	8	16	10¾	7½	5⅛	9½	17¾	12½	8	5⅛	13¼	21¼	16	11
		14	5⅛	7½	15¾	11¼	7½	5⅛	7½	18	11½	7½	5⅛	9	20	13½	7½	5⅛	12½	24	17	10½
		16	5⅛	7½	17¼	12½	7½	5⅛	7½	19¾	12	7½	5⅛	8¼	22	14	7½	6¾	9	20¾	16	10½
		18	5⅛	7½	18¾	13½	7½	5⅛	7½	21½	13	7½	5⅛	7¾	24	14½	7½	6¾	8¼	22¼	16¼	10½

Vertical dead
+ live load
= 1000 lb per ft

Vertical dead
= 240 lb per ft
Horizontal wind
= 320 lb per ft

Vertical dead
= 320 lb per ft
horizontal wind
= 320 lb per ft

Vertical dead
= 480 lb per ft
horizontal wind
= 320 lb per ft

Table 11-29 Preliminary Design Dimensions, In., for Three-Hinged Tudor Arches* *(Continued)* *ıtinued)*

Loading	Roof pitch	Wall hgt, ft.	60-ft span Width	60-ft span Base	60-ft span Lower tang.	60-ft span Upper tang.	60-ft span Crown	70-ft span Width	70-ft span Base	70-ft span Lower tang.	70-ft span Upper tang.	70-ft span Crown	80-ft span Width	80-ft span Base	80-ft span Lower tang.	80-ft span Upper tang.	80-ft span Crown	90-ft span Width	90-ft span Base	90-ft span Lower tang.	90-ft span Upper tang.	90-ft span Crown
Vertical dead + live load = 400 lb per ft	3:12	12	5⅛	14	16¼	16¾	7½	5⅛	17¾	17¾	20	7½	5⅛	21¾	21¾	23	7½	6¾	20	20	22¾	8
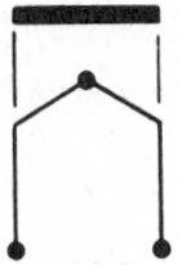		14	5⅛	12¾	19½	17½	7½	5⅛	16¼	22	20¾	7½	5⅛	20	24	24	7½	6¾	18½	21½	24	7½
		16	5⅛	11¾	22	17¾	7½	5⅛	15	25	21¼	7½	6¾	14¼	22½	22	7½	6¾	17¼	24¾	24¾	7½
		18	5⅛	10¾	24	19¼	7½	6¾	10¾	22	21¼	7½	6¾	13¼	24½	22¾	7½	6¾	16	27	25¾	7½
		20	6¾	7¾	22	22	7½	6¾	10	23½	23¼	7½	6¾	12½	26½	23	7½	6¾	15	29¼	26	7½
	4:12	12	5⅛	12½	16¾	15¾	7½	5⅛	15¾	18½	18¾	7¾	5⅛	19¼	20	21½	9½	5⅛	22¾	22¾	24¼	12
		14	5⅛	11½	19½	16½	7½	5⅛	14½	22	19½	7½	5⅛	17¾	24¼	22¾	8¼	6¾	16¼	21½	22¼	8½
		16	5⅛	10½	21¾	17	7½	5⅛	13½	24½	20¼	7½	6¾	12¾	22	20¾	7½	6¾	15¼	24¼	23¼	7½
		18	5⅛	10	23¾	17¾	7½	6¾	9¾	21½	19½	7½	6¾	12	24	21¼	7½	6¾	14¼	26½	24¼	7½
		20	5⅛	9¼	25½	19¼	7½	6¾	9	23	21½	7½	6¾	11¼	25¾	22	7½	6¾	13½	28½	24¾	7½
	6:12	12	5⅛	10¼	16¾	14¼	8¼	5⅛	12¾	18¾	16¾	10½	5⅛	15¼	20½	19	13¼	5⅛	18	22¼	21¼	16
		14	5⅛	9½	19	15¼	7½	5⅛	12	21½	17¾	9	5⅛	14½	23¾	20¼	11½	6¾	13	21	19¾	12
		16	5⅛	9	21	16	7½	5⅛	11¼	23¾	18¾	7¾	6¾	10½	21¼	18¾	8½	6¾	12¼	23¼	21	10½
		18	5⅛	8½	23	16½	7½	6¾	8	20¾	17¼	7½	6¾	9¾	23	19½	7½	6¾	11¾	25¼	22	9¼
	8:12	12	5⅛	8¾	16½	13½	10	5⅛	10¾	18½	15½	12¾	5⅛	12¾	20¼	17¼	16	5⅛	15	22	19¼	19¼
		14	5⅛	8¼	18½	14½	8¾	5⅛	10¼	20¾	16¾	11¼	5⅛	12¼	23	18¾	14	5⅛	14¼	25¼	20¾	17
		16	5⅛	7¾	20½	15½	7½	5⅛	9½	23	17¾	9¾	5⅛	11½	25½	20	12½	6¾	10¼	22¼	19½	12¾
		18	5⅛	7½	22¼	16¼	7½	5⅛	9¼	25¼	18¾	8¾	6¾	8½	22	18½	9¼	6¾	10	24	20½	11½
Vertical dead + live load = 600 lb per ft	3:12	12	5⅛	20½	20½	20¼	7½	5⅛	26¼	26¼	24	12¼	6¾	25	25	24½	12¼	6¾	29¾	29¾	27½	12¼
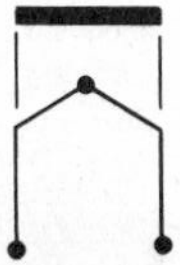		14	5⅛	18¾	24	21	7½	6¾	18½	22	22¼	12¼	6¾	23	23½	26	12¼	6¾	27½	27½	29¼	12¼
		16	6¾	13¼	22	19	7½	6¾	17¼	25	23	12¼	6¾	21¼	27½	26¾	12¼	6¾	25½	30	30¼	12¼
		18	6¾	12¼	24	20¼	7½	6¾	16	27¼	23½	12¼	6¾	19¾	30½	27¼	12¼	6¾	23¾	33½	31	12¼
		20	6¾	11½	25¾	21¾	7½	6¾	14¾	29½	24	12¼	6¾	18½	33	28	12¼	8¾	17½	30	28¼	12¼
	4:12	12	5⅛	18¼	20¼	19¼	7½	5⅛	23¼	23¼	22¾	12	6¾	21¾	21¾	23	12	6¾	25¾	25¾	25½	12¾
		14	5⅛	17	24¼	20	7½	6¾	16½	22	21¼	12	6¾	20¼	24¼	24¼	12	6¾	24	26	27¼	12
		16	6¾	12	21¾	18¼	7½	6¾	15½	24¾	21¾	12	6¾	18¾	27¼	25	12	6¾	22½	29¾	28¼	12
		18	6¾	11¼	23½	18¾	7½	6¾	14½	27	22½	12	6¾	17¾	30	26	12	6¾	21¼	32¾	29¼	12
		20	6¾	10½	25¼	20½	7½	6¾	13½	28¾	23	12	6¾	16¾	32¼	26½	12	8¾	15½	19¼	26¾	12
	6:12	12	5⅛	15¼	21	17½	10¼	5⅛	18¾	23¼	20½	13¼	5⅛	22¾	25½	23½	16¼	6¾	20½	22¼	22¾	17
		14	5⅛	14	24	18½	8¾	6¾	13½	21¾	19¼	11¼	6¾	16¼	23¾	22	12	6¾	19¼	26	24¼	14¾
		16	6¾	10¼	21¼	17¼	7½	6¾	12¾	24	20	11¼	6¾	15½	26½	23	11¼	6¾	18¼	29	25½	13
		18	6¾	9½	22¾	17¾	7½	6¾	12	26	21	11¼	6¾	14¾	28¾	23¾	11¼	6¾	17¼	31½	26¾	11¾
	8:12	12	5⅛	13	20¾	16½	12½	5⅛	16	23¼	19	16	5⅛	19	25½	21¼	19¾	6¾	17	22¼	20½	20¼
		14	5⅛	12¼	23¼	17¾	10¾	6¾	11½	21	18	12	6¾	13¾	23¼	20¼	14¾	6¾	16	25¼	22¼	18
		16	6¾	8¾	20½	16½	8	6¾	11	23	19¼	10½	6¾	13	25½	21½	13¼	6¾	15½	27¾	23¾	16
		18	6¾	8¼	22	17½	7½	6¾	10½	25	20	10½	6¾	12½	27½	22½	11¾	6¾	14¾	30¼	25	14½
Vertical dead + live load = 800 lb per ft	3:12	12	5⅛	27¼	27¼	23	12¼	6¾	26¾	26¾	24¼	12¼	6¾	32½	32½	27¾	12¼	6¾	38¾	38¾	31	12½
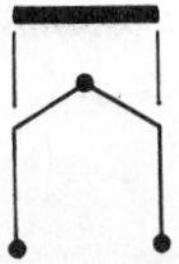		14	6¾	19¼	22½	21¼	12¼	6¾	24½	24½	25¾	12¼	6¾	30	30	29½	12¼	8¾	28¼	28¼	29¾	12¼
		16	6¾	17½	25¾	21¾	12¼	6¾	22½	28¾	26¼	12¼	6¾	27¾	31¾	30¾	12¼	8¾	26¼	29	30¾	12¼
		18	6¾	16¼	28	22¼	12¼	6¾	21	31¾	26¾	12¼	8¾	20¼	29½	27¾	12¼	8¾	24½	32¼	31¾	12¼
		20	6¾	15	30¼	22½	12¼	8¾	15¼	28¼	24¾	12¼	8¾	19	31¾	28½	12¼	8¾	23	34¾	32¼	12¼
	4:12	12	5⅛	24¼	24¼	22	12	6¾	23½	23½	23	12	6¾	28½	28½	26	12½	6¾	33¾	33¾	29¼	15¼
		14	6¾	17¼	22¾	20¼	12	6¾	21¾	25¼	24¼	12	6¾	26½	27½	28	12	6¾	31½	31½	31¼	12¾
		16	6¾	16	25½	21	12	6¾	20¼	28¾	25	12	6¾	24¾	31¾	28¾	12	8¾	23	29	29	12
		18	6¾	14¾	27¾	21½	12	6¾	19	31½	25¾	12	8¾	18¼	29	26½	12	8¾	21¾	31¾	29¾	12
		20	6¾	13¾	29¾	22	12	6¾	17¾	33¾	26¼	12	8¾	17	31	27¼	12	8¾	20½	34	30¾	12
	6:12	12	5⅛	20	24¼	20¼	12	6¾	19	22	21	13	6¾	23	23¾	23¾	16¼	6¾	27	27	26¼	19¾
		14	6¾	14¼	22½	18¾	11¼	6¾	17¾	25¼	22	11¼	6¾	21½	27¾	25	14¼	6¾	25½	30	28	17½
		16	6¾	13½	24¾	19¾	11¼	6¾	16¾	28	23	11¼	6¾	20¼	31	26¼	12½	6¾	24	33¾	29½	15½
		18	6¾	12¾	26¾	20½	11¼	6¾	16	30½	24	11¼	6¾	19¼	33¾	27½	11¼	8¾	17¾	30¼	27¼	11¾
	8:12	12	5⅛	17	24½	19	14¾	6¾	16	22	19¼	16	6¾	19¼	24	21½	19¾	6¾	22½	25¾	23¾	23¾
		14	6¾	12¼	22	18	10¾	6¾	15¼	24½	20¾	14	6¾	18¼	27¼	23¼	17½	6¾	21¼	29½	25¾	21
		16	6¾	11½	24	19	10½	6¾	14½	27	22	12¼	6¾	17¼	30	24¾	15½	6¾	20¼	32¾	27¼	19

Vertical dead + live load = 1000 lb per ft	3:12	12	6¾	26	26	22½	12¼	6¾	32¾	32¾	26½	12¼	6¾	40½	40½	30¾	17	8¾	37¾	37¾	30¾	17
		14	6¾	23½	25	23¾	12¼	6¾	30	30	28½	12¼	6¾	37¼	37¼	32¾	17	8¾	35	35	32¾	17
		16	6¾	21¾	28¾	24¼	12¼	6¾	27¾	32¼	29¼	12¼	8¾	27	29¾	30½	17	8¾	32½	32½	34½	17
		18	6¾	20	31½	24¾	12¼	8¾	20¼	29¾	26½	12¼	8¾	25¼	33	31	17	8¾	30¼	36	35½	17
		20	8¾	14½	28	23½	12¼	8¾	18¾	31¾	27	12¼	8¾	23½	35½	31¾	17	8¾	28½	39	36	17
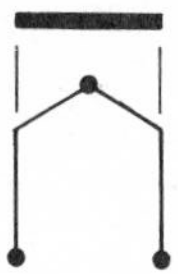	4:12	12	6¾	23	23	21½	12	6¾	29	29	25¼	12	6¾	35¼	35¼	29¼	16¾	8¾	32¾	32¾	28¾	16¾
		14	6¾	21¼	25½	22½	12	6¾	26¾	28	27	12	6¾	33	33	31	16¾	8¾	30½	30½	31	16¾
		16	6¾	19¾	28¾	23¼	12	6¾	25	32¼	27¾	12	8¾	24	29¾	28½	16¾	8¾	28¾	32	32¼	16¾
		18	6¾	18¼	31¼	24	12	8¾	18¼	29¼	25½	12	8¾	22½	32½	29½	16¾	8¾	27	35½	33¼	16¾
		20	6¾	17¼	33¾	24¼	12	8¾	17¼	31¼	26	12	8¾	21¼	35	30¼	16¾	8¾	25½	38¼	34¼	16¾
	6:12	12	6¾	19	22	19¾	11½	6¾	23½	24¼	23¼	14¾	6¾	28½	28½	26¼	18½	6¾	33¼	33¼	29	22¾
		14	6¾	17¾	25¼	21	11¼	6¾	22	28¼	24½	13	6¾	26¾	31	28	16¼	8¾	24¾	28½	27¾	16¾
		16	6¾	16½	28	22	11¼	6¾	20¾	31½	25¾	11¼	8¾	19¾	29	26	15¾	8¾	23½	31½	29¼	15¾
		18	6¾	15¾	30½	22¾	11¼	8¾	15¼	28¼	23¾	11¼	8¾	18¾	31¼	27	15¾	8¾	22¼	34¼	30¼	15¾
	8:12	12	6¾	16¼	22	18¾	14	6¾	20	24½	21½	18	6¾	23¾	26¾	24¼	22¼	6¾	27¾	28¾	26¾	26¾
		14	6¾	15¼	24¾	20¼	12¼	6¾	18¾	27¾	23¼	15¾	6¾	22½	30¾	26	19¾	6¾	26½	33¼	28¾	24
		16	6¾	14¼	27¼	21¼	10¾	6¾	18	30½	24½	14	8¾	16¾	28	24¼	15	8¾	19¾	30½	27	18½
		18	6¾	13¾	29½	22¼	10½	6¾	17	33¼	25¾	12½	8¾	16	30	25¾	14¾	8¾	18¾	32¾	28½	16½

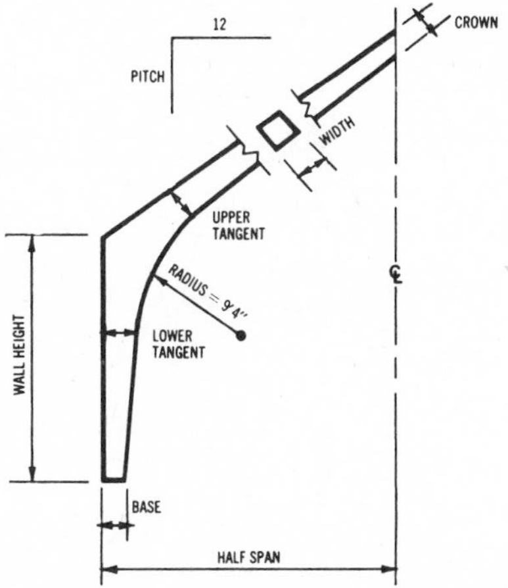

12
PITCH
CROWN
WIDTH
UPPER TANGENT
RADIUS = 9'4"
LOWER TANGENT
WALL HEIGHT
BASE
HALF SPAN

* Typical dimensions for three-hinged tudor arches for preliminary design purposes are given in Table 8-29. Sizes are based on Douglas fir laminated timber, developing an allowable shear stress of 165 psi and with a bending radius of 9 ft 4 in. For southern pine laminated timber, an allowable shear stress of 200 psi and a bending radius of 7 ft 2 in. may be used.

For roof pitches less than 10:12, the critical loading is generally the combined dead and live load on the horizontal projection of the full span. For roof pitches of 10:12 or greater, the critical loading is generally a combination of dead load and horizontal wind load. Sizes shown were determined for a uniformly distributed wind load applied on the vertical projection of the roof arm with a concentrated wind load equal to ½ the total wind load on the wall height acting at the haunch.

In the combined stress analysis, it was assumed that the bending portion of the loading exceeded the axial compression portion. The section sizes are based on the following design criteria:

1. Uniform loading.
2. Radius of curvature at the haunch = 9 ft 4 in.
3. Allowable stresses:
   Bending stress F_b = 2,400 psi (reduced by size factor and curvature factor when applicable).
   Shear stress F_v = 165 psi.
   Compression parallel to grain stress F_c = 1,500 psi (adjusted for $4d$ ratio).
   Modulus of elasticity E = 1,800,000 psi.
   These stresses were increased 15% for short duration of loading and 33⅓% for wind loading when applicable.
4. Deflection limits: $l/180$ for dead plus live load; $l/240$ for dead load only.
5. Vertical arch legs are laterally unsupported with tangent-point depth-to-width ratio not exceeding 5:1. (When vertical arch legs are laterally supported, tangent-point depth-to-ratio not exceeding 6:1 may be used.)

("Design Manual for TECO Timber Connector Construction," Timber Engineering Co., Washington; AITC 102, Appendix A, "Trusses and Bracing," American Institute of Timber Construction, Englewood, Colo.)

11-29. Design of Timber Arches. Arches may be two-hinged, with hinges at each base, or three-hinged, with a hinge at the crown. Figure 11-12 presents typical forms of arches.

Tudor arches are gabled rigid frames with curved haunches. Columns and pitched roof beam on each side of the crown usually are one piece of glued-laminated timber. This type of arch is frequently used in church construction with a high rise. Table 11-29 gives typical dimensions of three-hinged Tudor arches.

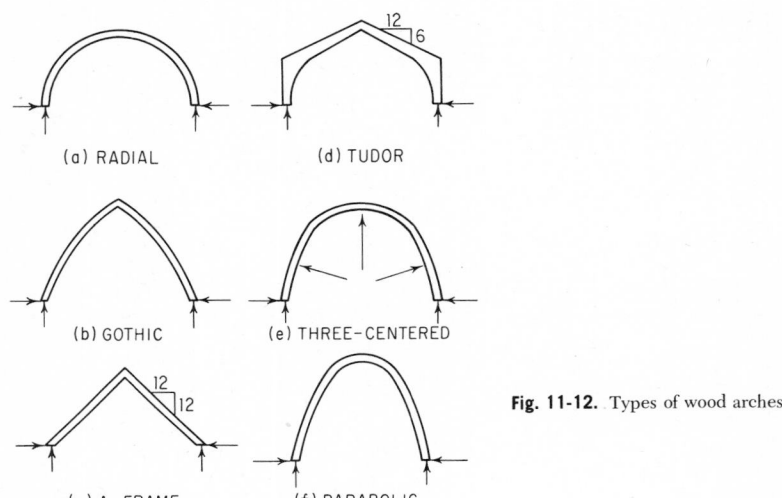

(a) RADIAL (d) TUDOR

(b) GOTHIC (e) THREE-CENTERED

(c) A-FRAME (f) PARABOLIC

Fig. 11-12. Types of wood arches.

Preliminary sizes of key sections of a Tudor arch, for use in a detailed arch analysis, may be determined for uniform vertical loads as follows (for steep roofs, the effect of wind loads must be investigated): Assume an allowable bending stress $F = 2,200$ psi, after adjustment for short-duration loading, curvature, and size factor; horizontal shear stress $F_v = 230$ psi; live plus dead load of 45 psf; span $L = 60$ ft; rise from base to crown $T = 25$ ft; arch spacing = 16 ft; and haunch radius = 7 ft for ¾-in. laminations (Fig. 11-13a).

The total uniform vertical load $w = 45 \times 16 = 720$ lb per lin ft. Compute the horizontal thrust:

$$H = \frac{wL^2}{8T} = \frac{720 \times 60^2}{8 \times 25} = 12,960 \text{ lb}$$

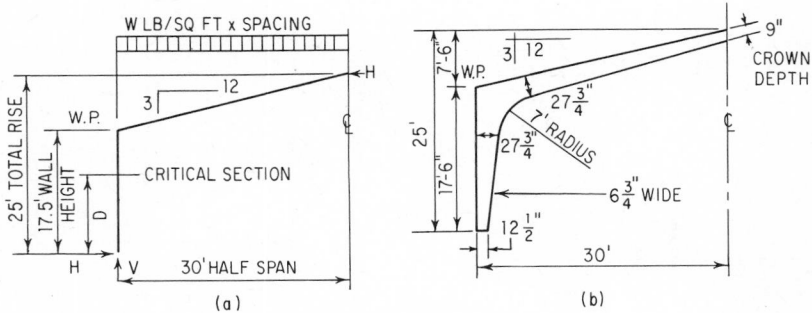

Fig. 11-13. Tudor arch.

Table 11-30. Distance of Critical Section below Working Point (WP in Fig. 11-13)

Pitch	1:12	2:12	3:12	4:12	5:12	6:12	8:12	8:12	9:12
Distance, ft	6.43	5.94	5.47	5.05	4.67	4.32	4.03	3.75	3.5

For this shearing force, determine the base depth d for an assumed width b (from Table 11-29) of 6¾ in.

$$d = \frac{3H}{2F_v b} = \frac{3 \times 12,960}{2 \times 230 \times 6.75} = 12.6$$

Use $d = 12\frac{1}{2}$ in. Next compute the bending moment M at the critical section, which is a distance D from the base hinge. To find D, subtract the dimension given in Table 11-30 for a 3:12 pitch from the wall height: $17.50 - 5.47 = 12.03$ ft.

$$M = 12HD = 12 \times 12,960 \times 12.03 = 1,871,000 \text{ in.-lb}$$

The section modulus required at the critical section then is

$$S = \frac{M}{F_b} = \frac{1,871,000}{2,200} = 850 \text{ in.}^3$$

From Table 11-4, for ¾-in. laminations, select the most economical section with this section modulus: 6¾ × 27, with $S = 866$. Make the upper tangent the same size as the lower. Finally, determine the depth d at the crown. This depth should at least equal the width, and preferably, it should be one-third larger. Hence,

$$d = 1.33 \times 6.75 = 9 \text{ in.}$$

This may be increased for architectural reasons or to equal depth of purlins. Figure 11-13b summarizes the preliminary dimensions of the arch.

Segmented arches are fabricated with overlapping lumber segments, nailed- or glued-laminated. They generally are three-hinged, and they may be tied or buttressed. If an arch is tied, the tie rods, which resist the horizontal thrust (Table 11-31), may be above the ceiling or below grade, and simple connections may be used where the arch is supported on masonry walls, concrete piers, or columns (Fig. 11-14).

Table 11-31. Tie-Rod Capacities Based on Tensile-Stress Area*

Bar no.	Dia of rod, in.	Area of rod, sq in.	Stress area at root of thread, sq in.	Weight, lb per lin ft	Permissible horizontal thrust H			
					Threaded		Unthreaded	
					20,000 psi	32,000 psi	20,000 psi	32,000 psi
4	½	0.1964	0.1416	0.67	2,832	4,531	3,928	6,284
5	⅝	0.3068	0.2256	1.04	4,512	7,219	6,136	9,817
6	¾	0.4418	0.3340	1.50	6,680	10,688	8,836	14,138
7	⅞	0.6013	0.4612	2.04	9,224	14,758	12,026	19,242
8	1	0.7854	0.6051	2.67	12,102	19,363	15,708	25,133
9	1⅛	0.9940	0.7627	3.38	15,254	24,406	19,880	31,808
10	1¼	1.2272	0.9684	4.17	19,368	30,989	24,544	39,270
11	1⅜	1.4849	1.1538	5.05	23,076	36,922	29,698	47,517
12	1½	1.7671	1.4041	6.01	28,082	44,931	35,342	56,547
	1¾	2.4053	1.8983	8.18	37,966	60,746	48,106	76,970
	2	3.1416	2.4971	10.68	49,942	79,907	62,832	100,530
	2½	4.9087	3.9976	16.69	79,952	127,920	98,174	157,080
	3	7.0686	5.9659	24.03	119,320	190,910	141,370	226,200
	3½	9.6211	8.3268	32.71	166,540	266,460	192,420	307,880
	4	12,5660	11.0800	42.73	221,600	354,560	251,320	402,110

*To determine tie-rod size to resist horizontal thrust H, lb, of the tied segment arch: Compute thru..t $H = 0.932wL$, where w = total load, lb per lin ft. Span L = radius. Select rod size with adequate permissible thrust H from above table.

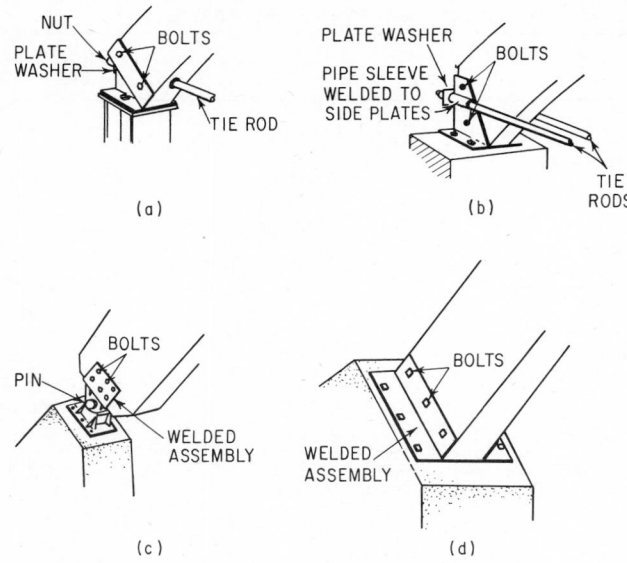

Fig. 11-14. Bases for segmented wood arches: (*a*) and (*b*) tie rod anchored to arch shoe; (*c*) hinge anchorage for large arch; (*d*) welded arch shoe.

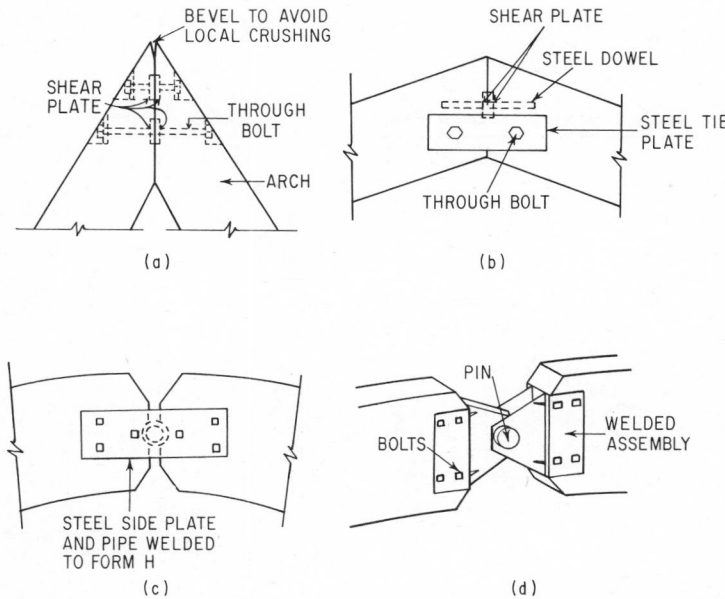

Fig. 11-15. Crown connections for arches. (*a*) For arches with slope 4:12 or greater, connection consists of pairs of back-to-back shear plates with through bolts or threaded rods counterbored into the arch. (*b*) For arches with flatter slopes, shear plates centered on a dowel may be used in conjunction with tie plates and through bolts. (*c*) and (*d*) Hinge at crown.

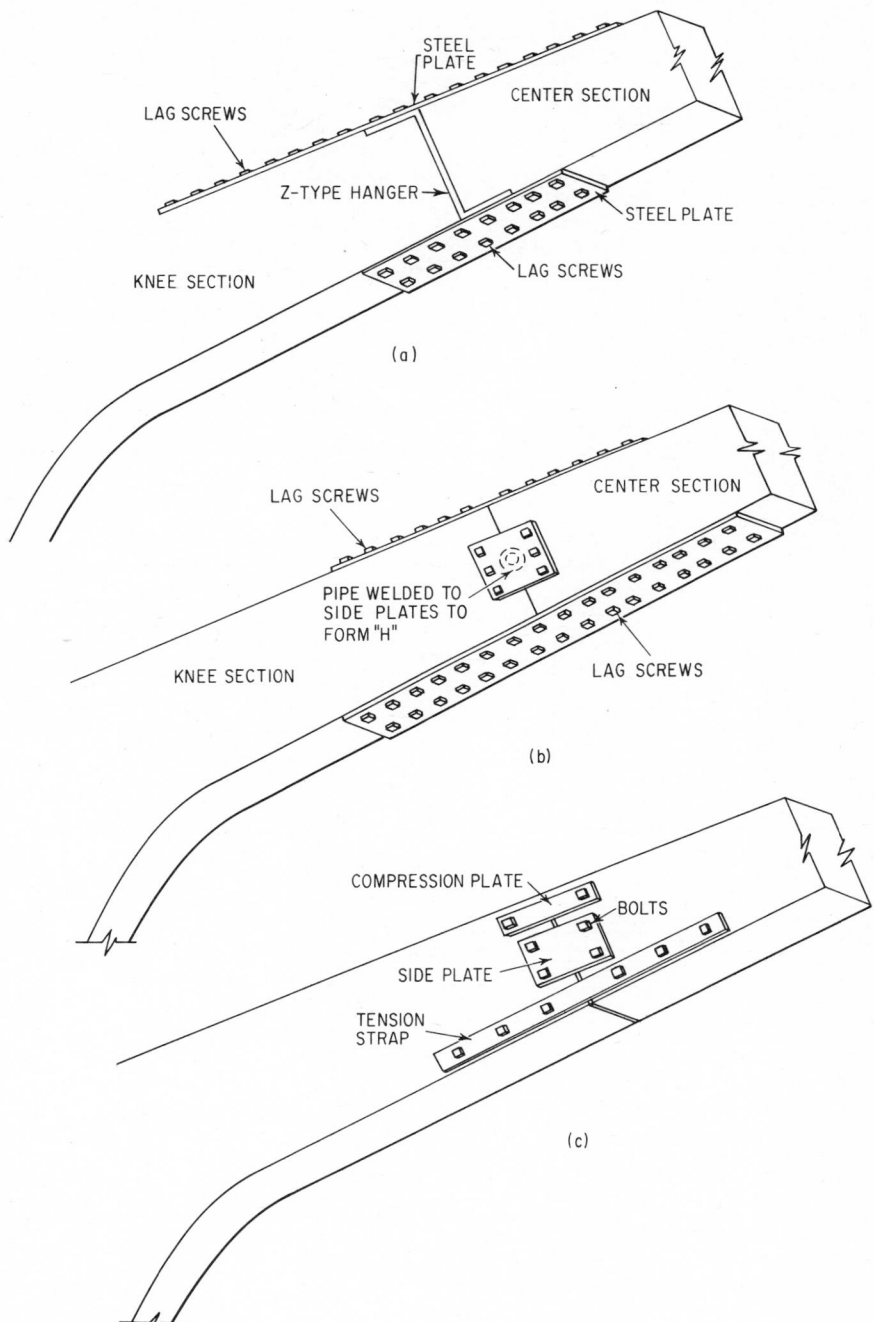

Fig. 11-16. Moment connections for an arch: (*a*) and (*b*) connections with top and bottom steel plates; (*c*) connections with side plates.

Segmented arches are economical because of the ease of fabricating them and simplicity of field erection. Field splice joints are minimized; generally there is only one simple connection, at the crown (Fig. 11-15c). Except for extremely long spans, they are shipped in only two pieces. Erected, they need not be concealed by false ceilings, as may be necessary with trusses. And the cross section is large enough for segmented arches to be classified as heavy-timber construction.

Figure 11-16 shows typical moment connections for segmented arches.

11-30. Timber Decking. Wood decking used for floor and roof construction may consist of solid-sawn planks with nominal thickness of 2, 3, or 4 in. Or it may be panelized or laminated. Panelized decking is made up of splined panels, usually about 2 ft wide.

Mechanically laminated decks are erected by setting square-edged dimension lumber on edge and fastening wide face to wide face. If side nails are used, they should be long enough to penetrate about 2½ lamination thicknesses for load transfer. Where supports are 4 ft c to c or less, side nails should be spaced not more than 30 in. c to c. They should be staggered one-third of the spacing in adjacent laminations. When supports are more than 4 ft c to c, nails should be spaced about 18 in. c to c alternately near top and bottom edges. These nails also should be staggered one-third of the spacing in adjacent laminations. Two side nails should be used at each end of butt-joined pieces. In addition, laminations should be toenailed to supports with nails 4 in. (20d) or longer. When supports are 4 ft c to c or less, alternate laminations should be nailed to alternate supports; for longer spans, alternate laminations at least should be nailed to every support.

For glued-laminated decking, two or more pieces of lumber are laminated into a single decking member, usually with 2- to 4-in. nominal thickness.

Solid-sawn decking usually is fabricated with edges tongued and grooved, shiplap, or groove cut for splines, for transfer of vertical load between pieces. The decking may be end matched,

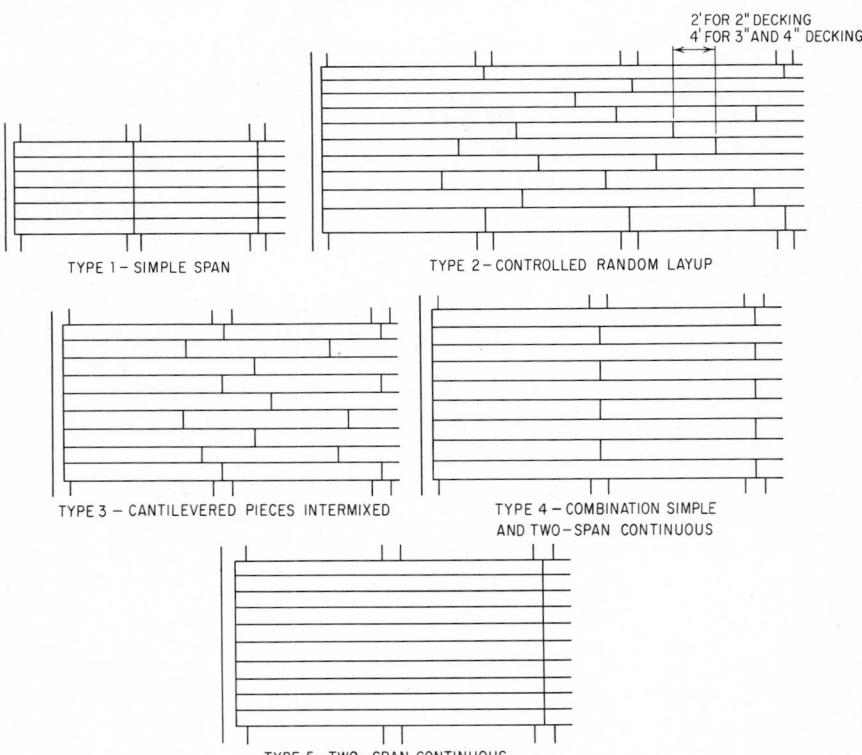

Fig. 11-17. Typical arrangement patterns for heavy-timber decking.

square end, or end grooved for splines. As indicated in Fig. 11-17, the decking may be arranged in various patterns over supports.

In Type 1, the pieces are simply supported. Type 2 has a controlled random layup. Type 3 contains intermixed cantilevers. Type 4 consists of a combination of simple-span and two-span continuous pieces. Type 5 is two-span continuous.

In Types 1, 4, and 5, end joints bear on supports. For this reason, these types are recommended for thin decking, such as 2-in.

Type 3, with intermixed cantilevers, and Type 2, with controlled random layup, are used for deck continuous over three or more spans. These types permit some of the end joints to be located between supports. Hence, provision must be made for stress transfer at those joints. Tongue-and-groove edges, wood splines on each edge of the course, horizontal spikes between courses, and end matching or metal end splines may be used to transfer shear and bending stresses.

In Type 2, the distance between end joints in adjacent courses should be at least 2 ft for 2-in. deck and 4 ft for 3- and 4-in. deck. Joints approximately lined up (within 6 in. of being in line) should be separated by at least two courses. All pieces should rest on at least one support. And not more than one end joint should fall between supports in each course.

In Type 3, every third course is simple span. Pieces in other courses cantilever over supports, and end joints fall at alternate quarter or third points of the spans. Each piece rests on at least one support.

To restrain laterally supporting members of 2-in. deck in Types 2 and 3, the pieces in the first and second courses and in every seventh course should bear on at least two supports. End joints in the first course should not occur on the same supports as end joints in the second course unless some construction, such as plywood overlayment, provides continuity. Nail end distance should be sufficient to develop the lateral nail strength required.

Heavy-timber decking is laid with wide faces bearing on the supports. Each piece must be nailed to each support. Each end at a support should be nailed to it. For 2-in. decking, a 3½-in. (16d) toe and face nail should be used in each 6-in.-wide piece at supports, and three nails for wider pieces. Tongue-and-groove decking generally is also toenailed through the tongue. For 3-in. decking, each piece should be toenailed with one 4-in. (20d) spike and face-nailed with one 5-in. (40d) spike at each support. For 4-in. decking, each piece should be toenailed at each support with one 5-in. (40d) nail and face-nailed there with one 6-in. (60d) spike.

Courses of 3- and 4-in. double tongue-and-groove decking should be spiked to each other with 8½-in. spikes not more than 30 in. apart. One spike should not be more than 10 in. from each end of each piece. The spikes should be driven through predrilled holes. Two-inch decking is not fastened together horizontally with spikes.

Deck design usually is governed by maximum permissible deflection in end spans. But each design should be checked for bending stress. Deflection, in., may be calculated for a 12-in.-wide section from

$$\delta = \frac{wL^5}{K_1 EI} \tag{11-15}$$

where w = uniform load, psf
L = span, ft
E = modulus of elasticity, psi
I = moment of inertia, in.[4]
K_1 = constant, obtained from Table 11-32

Bending stress, psi, for a 12-in.-wide section may be computed from

$$f_b = \frac{wL^2}{K_2 S} \tag{11-16}$$

where S = section modulus, in.[3] Use the full cross section of the deck for Types 1, 4, and 5; use two-thirds for Types 2 and 3
K_2 = constant, obtained from Table 11-32

Table 11-33 gives the maximum allowable, uniformly distributed total load for heavy-timber decking for various spans. The table is based on use of seasoned lumber, normal duration of loading, and loads applied normal to the deck. A limiting deflection ratio of $l/240$ and allowable bending stress of 1,200 psi have been assumed. For allowable bending stresses F_b other than 1,200, multiply tabulated values by $F_b/1,200$. For modulus of elasticity E other than 1,760,000, multiply tabulated values by $E/1,760,000$. For deflection limitations δ_A other than $l/240$, multiply tabulated values by $240\delta_A/l$; for example, for $l/360$, multiply by $^{240}/_{360}$.

(AITC 112, Standard for Heavy Timber Roof Decking, American Institute of Timber Construction, Englewood, Colo.; AITC "Timber Construction Manual," John Wiley & Sons, Inc., New York.)

Table 11-32. Constants for Computing Deck Deflections and Bending Stresses

Deck type	K_1	K_2
1	0.0445	0.667
2	0.0566*	0.555
3	0.0608	0.555
4	0.0631	0.667
5	0.1071	0.667

*For 2-in. deck. Use 0.0671 for 3- and 4-in. deck.

Table 11-33. Allowable Total Uniform Load, Psf, for Heavy-Timber Deck

Nominal deck thickness	Span, ft	When bending governs		When deflection governs				
		$F_b = 1,200$ psi		$E = 1,760,000$ psi; $\delta_A = l/240$*				
		Types 1, 4, 5	Types 2 and 3	Type 1	Type 2	Type 3	Type 4	Type 5
2 in. (dressed thickness 1½ in.) single tongue and groove	7	73	61	39	50	53	55	93
	8	56	47	26	33	35	37	62
	9	44	37	18	24	24	26	43
	10	36	30	13	17	18	19	31
3 in. (dressed thickness 2½ in.) double tongue and groove	11	83	69	46	70	62	66	110
	12	70	58	35	54	44	51	85
	13	60	49	28	42	38	40	67
	14	51	43	22	34	31	32	54
	15	44	37	18	28	25	26	43
4 in. (dressed thickness 3½ in.) double tongue and groove	14	100	83	61	93	83	88	147
	15	87	73	50	75	68	71	120
	16	77	64	41	62	56	58	98
	17	68	56	34	52	46	48	81
	18	61	50	29	44	40	41	70
	19	54	45	24	37	33	35	57
	20	50	41	21	32	29	30	50

*l = span, in.

11-31. Plywood. For construction, plywood is made of softwood layers glued together. The grain of each layer is perpendicular to the grain of adjoining layers. Softwood plywood is manufactured in accordance with U.S. Product Standard PS1-66 for Softwood Plywood—Construction and Industrial. Plywood for exterior use should be made with an adhesive that is fully waterproof and suitable for wet service conditions and severe exposure.

PS1-66 classifies the softwoods according to stiffness:

Group 1. Douglas fir from Washington, Oregon, California, Idaho, Montana, Wyoming, British Columbia, and Alberta; western larch; southern pine (loblolly, longleaf, shortleaf, slash); yellow birch; tan oak.

Group 2. Port Orford cedar; Douglas fir from Nevada, Utah, Colorado, Arizona, and New Mexico; fir (California red, grand, noble, Pacific silver, white); western hemlock; red and white lauan; western white pine; Sitka spruce.

Group 3.   Red alder, Alaska yellow cedar, lodgepole and ponderosa pine, redwood.

Group 4.   Incense and western red cedar, subalpine fir, sugar pine, western poplar, Engelmann spruce.

Group 5.   Balsam fir and balsam poplar.

Construction plywood is available under the standard in two grades made with exterior glue, Structural I and Structural II, and grades C-D, C-C, and Standard. Structural I is made only of Group 1 species. Structural II may be made of Group 1, 2, 3 or of any combination of these species. Structural I and II are suitable for such applications as box beams, gusset plates, stressed-skin panels, and folded-plate roofs. Structural I A-C (face and back of veneer grade A and inner plies of grade C) is commonly used as an exterior type. Subflooring, wall sheathing, and roof decks can be made of Standard grade with exterior glue. Structural I is stiffer than the other grades.

Table 11-34. Plywood Roof Sheathing and Subflooring Recommendations[a]

a. Roof Sheathing Laid with Face Grain Perpendicular to Joists[b]

Panel identification index	Plywood thickness, in.	Maximum span, in.[c]	Unsupported edge—max. length, in.[d]	Allowable roof loads, psf[e] Spacing of supports, in., c to c										
				12	16	20	24	30	32	36	42	48	60	72
12/0	5/16	12	12	100 (130)										
16/0	5/16, 3/8	16	16	130 (170)	55 (75)									
20/0	5/16, 3/8	20	20	...	85 (110)	45 (55)								
24/0	3/8, 1/2	24	24	...	150 (160)	75 (100)	45 (60)							
30/12	5/8	30	26	...	...	145 (165)	85 (110)	40 (55)						
32/16	1/2, 5/8	32	28	...	...	...	90 (105)	45 (60)	40 (50)					
36/16	3/4	36	30	...	...	...	125 (145)	65 (85)	55 (70)	35 (50)				
42/20	5/8, 3/4, 7/8	42	32	...	...	...	...	80 (105)	65 (90)	45 (60)	35 (40)			
48/24	3/4, 7/8	48	36	...	...	...	...	...	105 (115)	75 (90)	55 (55)	40 (40)		
2-4-1	1⅛	72	48	...	...	...	...	...	...	160 (160)	95 (95)	70 (70)	45 (45)	25 (30)
1⅛ in. G 1 and 2	1⅛	72	48	...	...	...	...	...	...	145 (145)	85 (85)	65 (65)	40 (40)	30 (30)
1¼ in. G 3 and 4	1¼	72	48	...	...	...	...	...	...	160 (165)	95 (95)	75 (75)	45 (45)	25 (35)

[a] American Plywood Association.

[b] Applies to Standard, Structural I and II, and C-C grades only. Plywood continuous over two or more spans. For applications where the roofing is to be guaranteed by a performance bond, recommendations may differ somewhat from these values. Contact American Plywood Association for bonded roof recommendations.

Use 2-in. (6d) common, smooth, ring-shank, or spiral-thread nails for ½ in. thick or less, and 2½-in. (8d) common, smooth, ring-shank, or spiral-thread for plywood 1 in. thick or less (if ring-shank or spiral-thread nails have the same diameter as common). Use 2½-in. (8d) ring-shank or spiral-thread or 3-in. (10d) common, smooth-shank nails for 2-4-1, 1⅛-, and 1¼-in. panels. Space nails 6 in. at panel edges and 12 in. at intermediate supports, except that where spans are 48 in. or more, nails shall be 6 in. c to c at all supports.

[c] The spans shall not be exceeded for any load conditions.

[d] Provide adequate blocking, tongued-and-grooved edges, or other suitable edge support such as PlyClips when spans exceed indicated value. Use two PlyClips for 48-in. or greater spans and one for lesser spans.

[e] Uniform load deflection limitation: $^1/_{180}$ the span under live load plus dead load, $^1/_{240}$ under live load only. Allowable live load shown without parentheses and allowable total load shown within parentheses.

Allowable loads were established by laboratory test and calculations assuming evenly distributed loads. Figures shown are not applicable for concentrated loads.

Table 11-34. Plywood Roof Sheathing and Subflooring Recommendations (*Continued*)

    b. Allowable Shears in Plywood Roof Diaphragms for Wind or Seismic Loads, Lb per Ft[f]
    (Plywood and framing assumed already designed for perpendicular loads)

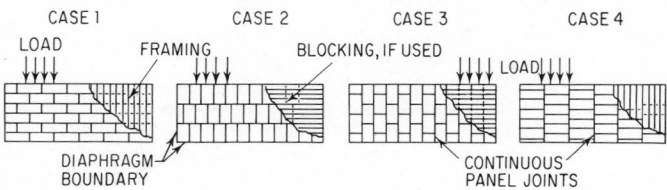

Plywood species and grade	Common nail length, in.	Min nail penetration into framing, in.	Min plywood thickness, in.	Min nominal width of framing member, in.	Blocked diaphragms			Unblocked diaphragms	
					Nail spacing at diaphragm boundaries (all cases) and continuous panel edges parallel to load (cases 3 and 4)[g]			Nails spaced 6 in. max at supported edges[g]	
					4	2½	2	Load perpendicular to unblocked edges and continuous panel joints (case 1)	All other configurations (cases 2, 3, and 4)
					Nail spacing at other plywood panel edges[g]				
					6	4	3		
Structural I	2 (6d)	1¼	5/16 or ¼	2	250	375	420	167	125
				3	280	420	475	187	140
	2½ (8d)	1½	⅜	2	360	530	600	240	180
				3	400	600	675	267	200
	3 (10d)	1⅝	½	2	425	640	730	283	212
				3	480	720	820	320	240
Structural II	2 (6d)	1¼	5/16 or ¼	2	167	250	280	111	84
				3	187	280	317	125	93
	2½ (8d)	1½	⅜	2	240	353	400	160	120
				3	267	400	450	178	133
	3 (10d)	1⅝	½	2	283	427	487	189	141
					320	480	547	213	160

[f] Design for diaphragm stresses depends on direction of continuous panel joints with reference to load, not direction of long dimension of plywood sheet.

[g] Space nails 12 in. c to c along intermediate framing members.

The Standard also classifies plywood made for use as concrete forms in two grades. Plyform (B-B) Class I is limited to Group 1 species on face and back, with limitations on inner plies. Plyform (B-B) Class II permits Group 1, 2, or 3 for face and back, with limitations on inner plies. High-density overlay should be specified for both classes when highly smooth, grain-free concrete surfaces and maximum reuses are required. The bending strength of Plyform Class I is greater than that of Class II. Grades other than Plyform, however, may be used for forms.

Manufacturers stamp plywood to indicate group, grade, and Plyform class. They also include an identification index. This consists of a pair of numbers separated by a slant bar, on unsanded construction grades (Standard, Structural I and II, and C-C). The number on the left indicates the maximum recommended spacing, in., of supports when the panel is used for roof decking. The number on the right gives the maximum recommended spacing, in., of supports when the

Table 11-34. Plywood Roof Sheathing and Subflooring Recommendations *(Continued)*
  c. **Plywood Subflooring and Subfloor Underlayment with Face Grain Perpendicular to Joists**[h]

Panel identification index[i]	Plywood thickness, in.	Maximum span,[j] in.	Nail size and type	Nail spacing, in.	
				Panel edges	Intermediate
30/12	⅝	12[k]	2½-in. (8d) common	6	10
32/16	½, ⅝	16[l]	2½-in. (8d) common[m]	6	10
36/16	¾	16[l]	2½-in. (8d) common	6	10
42/20	⅝, ¾, ⅞	20[l]	2½-in. (8d) common	6	10
48/24	¾, ⅞	24	2½-in. (8d) common	6	10
2-4-1 and					
1⅛-in. Groups 1 and 2	1⅛	48	3-in. (10d) common	6	6
1¼-in. Groups 3 and 4	1¼	48	3-in. (10d) common	6	6

[h] These values apply for Structural I and II, Standard sheathing, and C-C Exterior grades only, for application of 25/32-in. wood-strip flooring or separate underlayment layer. Plywood continuous over two or more spans.

[i] Identification index appears on all panels, except 1⅛- and 1¼-in. panels.
In some nonresidential buildings, special conditions may impose heavy concentrated loads and heavy traffic requiring subfloor constructions in excess of these minimums.

[j] Edges shall be tongued-and-grooved or supported with blocking, unless underlayment is installed or finish floor is 25/32-in. wood strip. Spans limited to values shown because of possible effect of concentrated loads. At indicated maximum spans, panels will support uniform loads of at least 65 psf. For spans of 24 in. or less, panels will support loads of at least 100 psf.

[k] May be 16 in. if 25/32-in. wood-strip flooring is installed at right angles to joists.

[l] May be 24 in. if 25/32-in. wood-strip flooring is installed at right angles to joists.

[m] 2-in. (6d) common nail permitted if plywood is ½ in. thick.

panel is used for subflooring. If a zero is stamped on the right, the panel should not be used for subflooring.

Appearance grades of plywood also are identified by a stamp. The group number indicates the species group of the veneer used on face and back. Letter symbols describe these veneers:

N—Special-order natural-finish veneer. Selected all heartwood or all sapwood. Free of open defects. Allows some repairs.

A—Smooth and paintable. Neatly made repairs permissible. Also used for natural finish in less demanding applications.

B—Solid-surface veneer. Circular repair plugs and tight knots permitted. Can be painted.

C—Minimum veneer permitted in exterior-type plywood. Knotholes up to 1 in. in diameter. (Occasional knotholes ½-in. larger permitted if total width of all knots and knotholes within a specified section does not exceed certain limits.) Limited splits permitted.

C plugged—Improved C veneer with splits limited to ⅛ in. in width and knotholes and bored holes limited to ¼ × ½ in.

D—Used only in interior-type plywood for inner plies and backs. Permits limited splits and knots and knotholes up to 2½ in. (½ in. larger under specified conditions).

Table 11-34 lists allowable loads and nail requirements for plywood roof sheathing. It also gives maximum support spacing and nail requirements for plywood subflooring. Table 11-35 presents similar data for plywood wall sheathing. Table 11-36 gives the allowable uniform load to insure that the deflection of a plywood floor will not exceed $l/180$, where l is the joist spacing, in.

("The Plywood How to Book," "Plywood Construction Guide for Residential Building," "Plywood Construction Systems for Commercial and Industrial Buildings," "Plywood Concrete Forms," "Plywood Design Specification," "Plywood Components," "Plywood Truss Designs," American Plywood Association, 1119 A St., Tacoma, Wash. 98401.)

11-32. Pole Construction. Wood poles are used for various types of construction, including flagpoles, utility poles, and framing for buildings. These employ preservative-treated round poles set into the ground as columns. The ground furnishes vertical and horizontal support and prevents rotation at the base (Table 11-37).

Table 11-35. Plywood Wall Sheathing and Siding Recommendations[a]

a. **Plywood Wall Sheathing**[b]

Panel identification index	Panel thickness, in.	Maximum stud spacing, in.		Nail size[c]	Nail spacing, in.	
		Exterior covering nailed to:			Panel edges (when over framing)	Intermediate (each stud)
		Stud	Sheathing			
12/0, 16/0, 20/0	$5/16$	16	16^d 16	2-in. (6d)	6	12
16/0, 20/0, 24/0	⅜	24	24^d	2-in. (6d)	6	12
24/0, 32/16	½	24	24	2-in. (6d)	6	12

b. **Plywood Siding**

Application	Plywood thickness, in.	Max spacing of supports, in. c to c	Nail length and type[e]	Nail spacing, in.	
				Panel edges	Intermediate
Panel siding[f]	⅜[g]	16	2-in. (6d) casing or siding	6	12
	½	24	2-in. (6d) casing or siding	6	12
	⅝ or thicker	24	2½-in. (8d) casing or siding	6	12
Lap siding	⅜[h]	16	2-in. (6d) casing or siding	One nail per stud along bottom edge	4 at vertical joint; 8 at studs (siding wider than 12 in.)
	½	20	2½-in. (8d) casing or siding		
	⅝	24	2½-in. (8d) casing or siding		
Bevel siding	$9/16$ min butt	16	2-in. (6d) casing or siding	Same as lap siding	
Soffits or ceilings	⅜	24	2-in. (6d) casing or siding	6 (or one nail at each support)	12
	⅝	48			12

[a]American Plywood Association.

[b]The plywood may be installed horizontally or vertically. When plywood sheathing is used, building paper and diagonal wall bracing can be omitted. Plywood should be continuous over two or more spans.

[c]Common, annular, spiral-threaded, or T nails of the same diameter as common nails [0.113 in. for 2-in.-long (6d) nails] may be used. Staples also are permitted, but at reduced spacing.

[d]When sidings such as shingles are nailed only to the sheathing, apply the plywood with the face grain perpendicular to the supports.

[e]Use galvanized, aluminum, or other noncorrosive nails. Nails may be color-coated.

[f]Battens, if used, can be applied with 2½-in.-long (8d) noncorrosive casing nails spaced 12 in. c to c (staggered).

[g]When separate sheathing is applied, ⅜-in. panel siding may be used over supports 24 in. on centers, ¼-in. over supports 16 in. c to c.

[h]When separate sheathing is applied, $5/16$-in. overlaid plywood lap siding may be used over supports 16 in. c to c.

Isolated poles, such as flagpoles or signs, may be designed with lateral bearing values equal to twice these tabulated values.

In buildings, a bracing system can be provided at the top of the poles to reduce bending moments at the base and to distribute loads. Design of buildings supported by poles without bracing requires good knowledge of soil conditions, to eliminate excessive deflection or sidesway.

Bearing values under the base of poles should be checked. For backfilling the holes, well-tamped native soil, sand, or gravel may be satisfactory. But concrete or soil cement is more effective. They can reduce the required depth of embedment and improve bearing capacity by increasing the skin-friction area of the pole. Skin friction is effective in reducing uplift due to wind.

To increase bearing capacity under the base end of poles for buildings, concrete footings often are used. They should be designed to withstand the punching shear of the poles and bending

Table 11-35. Plywood Wall Sheathing and Siding Recommendations (*Continued*)
c. **Allowable Shears in Plywood Walls, lb/lin in.**[i]

Plywood species	Min nominal plywood thickness, in.	Plywood sheathing direct to framing[j]				Plywood sheathing over ½-in. gypsum sheathing[k]				Plywood siding direct to framing[l]		Plywood siding over ½-in. gypsum sheathing[n]	
		Nail spacing, in., at plywood panel edges				Nail spacing, in., at plywood panel edges				Nail spacing, in., at plywood panel edges		Nail spacing, in., at plywood panel edges	
		6	4	2½	2	6	4	2½	2	6	4	6	4
Structural I	5/16	200	300	450	510	200	300	450	510				
	3/8	280	430	640	730	280	430	640	730	160 200[m]	240 300[m]	160 200[o]	240 300[o]
	½, 5/8	340	510	770	870					200	300	200	300
Structural II[p]	5/16	180	270	400	450	180	270	400	450				
	3/8	260	380	570	640	260	380	570	640	140 160[m]	210 240[m]	140 160[o]	210 240[o]
	½, 5/8	310	460	690	770					160	240	160	240

[i] All panel edges should be backed with framing, which should be 2-in. nominal or wider. The plywood may be installed horizontally or vertically. Space nails 12 in. c to c along intermediate framing members.

[j] Smooth, bright, common, or galvanized box nails—use 2-in. (6d) for 5/16-in. plywood, 2½-in. (8d) for 3/8-in., and 3-in. (10d) for ½-in.

[k] Smooth, bright, common, or galvanized box nails—use 2½-in. (8d) for 5/16-in. plywood and 3-in. (10d) for 3/8-in.

[l] Galvanized casing nails—use 2-in. (6d) or 2½-in.[m] (8d) for 3/8-in. plywood and 2½-in. (8d) for ½- and 5/8-in. For siding attached with galvanized box nails, use shears for sheathing with the same size nail. For 3/8- and ½-in. plywood, nails at panel edges should penetrate the full plywood thickness.

[m] Shear value for 2½-in. (8d) nails.

[n] Galvanized casing nails—use 2½-in. (8d) or 3-in.[o](10d) for 3/8-in. plywood and 3-in. (10d) for ½- and 5/8-in. For siding attached with galvanized box nails, use shears for sheathing with the same size nail. For 3/8- and ½-in. plywood, nails at panel edges should penetrate the full plywood thickness.

[o] Shear value for 3-in. (10d) nails.

[p] Shears in this group may be increased 20% when using plywood one or more sizes thicker than minimum.

moments. Thickness of concrete footings should be at least 12 in. Consideration should be given to use of concrete footings even in firm soils, such as hard dry clay, coarse firm sand, or gravel.

Calculation of required depth of embedment in soil of poles subject to lateral loads generally is impractical without many simplifying assumptions. While an approximate analysis can be made, the depth of embedment should be checked by tests or at least against experience in the same type of soil.

Table 11-38 lists standard dimensions of Douglas fir and southern pine poles. Table 11-39 gives allowable stresses for treated poles, and Table 11-40 safe concentric column loads.

11-33. Timber Erection. Erection of timber framing requires experienced crews and adequate lifting equipment to protect life and property and to assure that the framing is properly assembled and not damaged during handling.

Table 11-36. Safe Uniform Floor Load for $l/180$ Deflections *

Joist spacing l, in.	Plywood thickness, in.	Safe load, psf
12	⅜	326
	½	435
	⅝	543
	¾	653
16	⅜	193
	½	327
	⅝	408
	¾	490
24	⅜	57
	½	111
	⅝	217
	¾	311

*Exterior-type Douglas fir, C-C grade, applied with the face grain across supports.

Table 11-37. Allowable Lateral Passive Soil Pressure

Class of material	Allowable stress per foot of depth below natural grade, psf	Max allowable stress, psf
Good: Compact, well-graded sand and gravel, hard clay, well-graded fine and coarse sand (all drained so water will not stand)	400	8,000
Average: Compact fine sand, medium clay, compact sandy loam, loose sand and gravel (all drained so water will not stand)	200	2,500
Poor: Soft clay, clay loam, poorly compacted sand, clays containing large amounts of silt (water stands during wet season)	100	1,500

Table 11-38. Stress Values for Treated Wood Poles *
(Normal Duration of Loading)

Species	Modulus of rupture, extreme fiber in bending, psi	Extreme fiber in bending F_b, psi	Modulus of elasticity E, psi
Cedar, northern white	4,000	1,080	800,000
Cedar, western red	6,000	1,620	1,000,000
Douglas fir	8,000	2,160	1,600,000
Hemlock, western	7,400	2,000	1,400,000
Larch, western	8,400	2,270	1,500,000
Pine, jack	6,600	1,780	1,100,000
Pine, lodgepole	6,600	1,780	1,000,000
Pine, ponderosa	6,000	1,620	1,000,000
Pine, red or Norway	6,600	1,780	1,200,000
Pine, southern	8,000	2,160	1,600,000

*See the latest edition of Standard Specifications and Dimensions for Wood Poles, ANSI 05.1.

Table 11-39. Dimensions of Douglas-Fir and Southern-Pine Poles*

Class		1	2	3	4	5	6	7	9	10
Min circumference at top, in.		27	25	23	21	19	17	15	15	12
Length of pole, ft	Ground-line distance from butt,† ft	Min circumference at 6 ft from butt, in.								
20	4	31.0	29.0	27.0	25.0	23.0	21.0	19.5	17.5	14.0
25	5	33.5	31.5	29.5	27.5	25.5	23.0	21.5	19.5	15.0
30	5½	36.5	34.0	32.0	29.5	27.5	25.0	23.5	20.5	
35	6	39.0	36.5	34.0	31.5	29.0	27.0	25.0		
40	6	41.0	38.5	36.0	33.5	31.0	28.5	26.5		
45	6½	43.0	40.5	37.5	35.0	32.5	30.0	28.0		
50	7	45.0	42.0	39.0	36.5	34.0	31.5	29.0		
55	7½	46.5	43.5	40.5	38.0	35.0	32.5			
60	8	48.0	45.0	42.0	39.0	36.0	33.5			
65	8½	49.5	46.5	43.5	40.5	37.5				
70	9	51.0	48.0	45.0	41.5	38.5				
75	9½	52.5	49.0	46.0	43.0					
80	10	54.0	50.5	47.0	44.0					
85	10½	55.0	51.5	48.0						
90	11	56.0	53.0	49.0						
95	11	57.0	54.0	50.0						
100	11	58.5	55.0	51.0						
105	12	59.5	56.0	52.0						
110	12	60.5	57.0	53.0						
115	12	61.5	58.0							
120	12	62.5	59.0							
125	12	63.5	59.6							

*See the latest edition of Standard Specifications and Dimensions for Wood Poles, ANSI 05.1.
†The figures in this column are intended for use only when a definition of ground line is necessary to apply requirements relating to scars, straightness, etc.

On receipt at the site, each shipment of timber should be checked for tally and evidence of damage. Before erection starts, plan dimensions should be verified in the field. The accuracy and adequacy of abutments, foundations, piers, and anchor bolts should be determined. And the erector must see that all supports and anchors are complete, accessible, and free from obstructions.

Jobsite Storage. If wood members must be stored at the site, they should be placed where they do not create a hazard to other trades or to the members themselves. All framing, and especially glued-laminated members, stored at the site should be set above the ground on appropriate blocking. The members should be separated with strips so that air may circulate around all sides of each member. The top and all sides of each storage pile should be covered with a moisture-resistant covering that provides protection from the elements, dirt, and jobsite debris. (Do not use clear polyethylene films, since wood members may be bleached by sunlight.) Individual wrappings should be slit or punctured on the lower side to permit drainage of water that accumulates inside the wrapping.

Glued-laminated members of Premium and Architectural Appearance (and Industrial Appearance in some cases) are usually shipped with a protective wrapping of water-resistant paper. While this paper does not provide complete freedom from contact with water, experience has shown that protective wrapping is necessary to insure proper appearance after erection. Though used specifically for protection in transit, the paper should remain in place until the roof covering is in place. It may be necessary, however, to remove the paper from isolated areas to make connections from one member to another. If temporarily removed, the paper should be replaced and should remain in position until all the wrapping may be removed.

Table 11-40. Safe Concentric Column Loads on Douglas-Fir and Southern-Pine Poles, Lb *

Top dia, in.	8	7	6	5
ASA pole class	2	3	5	6
Unsupported pole length, ft (above ground line):				
0	65,000	50,000	37,000	25,000
10	65,000	50,000	28,000	14,000
12	65,000	41,000	26,000	12,000
14	52,000	31,000	18,000	9,000
16	40,000	24,000	14,000	7,000
18	33,000	20,000	11,000	6,000
20	28,000	17,000	10,000	5,000
25	19,000	12,000	7,000	
30	14,000	9,000		
35	11,000			

*See the latest edition of Standard Specifications and Dimensions for Wood Poles, ANSI 05.1.

At the site, to prevent surface marring and damage to wood members, the following precautions should be taken:

Lift members or roll them on dollies or rollers out of railroad cars. Unload trucks by hand or crane. Do not dump, drag, or drop members.

During unloading with lifting equipment, use fabric or plastic belts, or other slings that will not mar the wood. If chains or cables are used, provide protective blocking or padding.

Guard against soiling, dirt, footprints, abrasions, or injury to shaped edges or sharp corners.

Equipment. Adequate equipment of proper load-handling capacity, with control for movement and placing of members, should be used for all operations. It should be of such nature as to insure safe and expedient placement of the material. Cranes and other mechanical devices must have sufficient controls that beams, columns, arches, or other elements can be eased into position with precision. Slings, ropes, cables, or other securing devices must not damage the materials being placed.

The erector should determine the weights and balance points of the framing members before lifting begins so that proper equipment and lifting methods may be employed. When long-span timber trusses are raised from a flat to a vertical position preparatory to lifting, stresses entirely different from normal design stresses may be introduced. The magnitude and distribution of these stresses depend on such factors as weight, dimensions, and type of truss. A competent rigger will consider these factors in determining how much suspension and stiffening, if any, is required and where it should be located.

Accessibility. Adequate space should be available at the site for temporary storage of materials from time of delivery to the site to time of erection. Material-handling equipment should have an unobstructed path from jobsite storage to point of erection. Whether erection must proceed from inside the building area or can be done from outside will determine the location of the area required for operation of the equipment. Other trades should leave the erection area clear until all members are in place and are either properly braced by temporary bracing or permanently braced in the building system.

Assembly and Subassembly. Whether these are done in a shop or on the ground or in the air in the field depends on the structural system and the various connections involved.

Care should be taken with match marking on custom materials. Assembly must be in accordance with the approved shop drawings for the materials. Any additional drilling or dapping, as well as the installation of all field connections, must be done in a workmanlike manner.

Trusses are usually shipped partly or completely disassembled. They are assembled on the ground at the site before erection. Arches, which are generally shipped in half sections, may be assembled on the ground or connections may be made after the half arches are in position. When trusses and arches are assembled on the ground at the site, assembly should be on level blocking to permit connections to be properly fitted and securely tightened without damage. End compression joints should be brought into full bearing and compression plates installed where intended.

Prior to erection, the assembly should be checked for prescribed overall dimensions, prescribed camber, and accuracy of anchorage connections. Erection should be planned and executed in such a way that the close fit and neat appearance of joints and the structure as a whole will not be impaired.

Field Welding.   Where field welding is required, the work should be done by a qualified welder in accordance with job plans and specifications, approved shop drawings, and specifications of the American Institute of Steel Construction and the American Welding Society.

Hardware.   All connections should fit snugly in accordance with job plans and specifications and approved shop drawings.   All cutting, framing, and boring should be done in accordance with good shop practices.   Any field cutting, dapping, or drilling should be done in a workmanlike manner with due consideration given to final use and appearance.

Bracing.   Structural elements should be placed to provide restraint or support, or both, to insure that the complete assembly will form a stable structure.   This bracing may extend longitudinally and transversely.   It may comprise sway, cross, vertical, diagonal, and like members that resist wind, earthquake, erection, acceleration, braking, and other forces.   And it may consist of knee braces, cables, rods, struts, ties, shores, diaphragms, rigid frames, and other similar components in combinations.

Bracing may be temporary or permanent.   Permanent bracing, required as an integral part of the completed structure, is shown on the architectural or engineering plans and usually is also referred to in the job specifications.   Temporary construction bracing is required to stabilize or hold in place permanent structural elements during erection until other permanent members that will serve the purpose are fastened in place.   This bracing is the responsibility of the erector, who normally furnishes and erects it.   It should be attached so that children and other casual visitors cannot remove it or prevent it from serving as intended.   Protective corners and other protective devices should be installed to prevent members from being damaged by the bracing.

In timber-truss construction, temporary bracing can be used to plumb trusses during erection and hold them in place until they receive the rafters and roof sheathing.   The major portion of temporary bracing for trusses is left in place, because it is designed to brace the completed structure against lateral forces.

Failures during erection occur occasionally and regardless of construction material used. The blame can usually be placed on insufficient or improperly located temporary erection guys or braces, overloading with construction materials, or an externally applied force sufficient to render temporary erection bracing ineffective.

Structural members of wood must be stiff, as well as strong.   They must also be properly guyed or laterally braced, both during erection and permanently in the completed structure.   Large rectangular cross sections of glued-laminated timber have relatively high lateral strength and resistance to torsional stresses during erection.   However, the erector must never assume that a wood arch, beam, or column cannot buckle during handling or erection.

Specifications often require that:

1. Temporary bracing shall be provided to hold members in position until the structure is complete.

2. Temporary bracing shall be provided to maintain alignment and prevent displacement of all structural members until completion of all walls and decks.

3. The erector should provide adequate temporary bracing and take care not to overload any part of the structure during erection.

While the magnitude of the restraining force that should be provided by a cable guy or brace cannot be precisely determined, general experience indicates that a brace is adequate if it supplies a restraining force equal to 2% of the applied load on a column or of the force in the compression flange of a beam.   It does not take much force to hold a member in line; but once it gets out of alignment, the force then necessary to hold it is substantial.

Section **12**

Surveying

RUSSELL C. BRINKER

Visiting Professor of Civil Engineering, New Mexico State University

12-1. Types of Surveys. Surveying is the science and art of making the measurements necessary to determine the relative positions of points above, on, or beneath the surface of the earth or to establish such points.

Plane surveying neglects the curvature of the earth and is suitable for small areas.

Geodetic surveying takes into account curvature of the earth. It is applicable for large areas, long lines, and precisely locating basic points suitable for controlling other surveys.

Land, boundary, and cadastral surveys usually are closed surveys that establish property corners and land lines. The term "cadastral" is now generally reserved for surveys of the public lands.

Topographic surveys provide the locations of natural and artificial features and elevations used in map making.

Route surveys normally start at a control point and progress to another control point in the most direct manner permitted by field conditions. These surveys are used for railroads, highways, pipelines, etc.

Hydrographic surveys determine the shoreline and depths of lakes, streams, oceans, reservoirs, and other bodies of water.

Construction surveys give locations and elevations (line and grade) for construction work.

Photogrammetric surveys utilize terrestrial or aerial photographs and can be a part of all the types of surveys listed.

12-2. Surveying Organizations. The National Geodetic Survey, Department of Commerce, obtains data for and monuments points to provide basic vertical and horizontal control on the North American Datum of 1927, which extends through Alaska and Central America. Its activities include triangulation, precise leveling, preparation of nautical and aeronautical charts, photogrammetric surveys, tide and current studies, and collection of magnetic data.

The United States Geological Survey, Department of the Interior, is engaged in mapping and remapping the country, among other assignments. It prepares the commonly used $7\frac{1}{2} \times 15$-minute quadrangles.

The Bureau of Land Management, Department of the Interior, replaced the General Land Office established in 1812. It has jurisdiction over the surveys and sales of the public lands, and prepares and publishes the "Manual of Surveying Instructions" for the survey of the public lands of the United States.

12-3. Units of Measurement.   Units of measurement used in past and present surveys are:

For construction work: feet, inches, fractions of inches.

For most surveys: feet, tenths, hundredths, thousandths.

For National Geodetic Survey control surveys: meters, 0.1, 0.01, 0.001 m.

The most-used equivalents are:

1 meter = 39.37 in. (exactly) = 3.2808 ft

1 rod = 1 pole = 1 perch = 16½ ft

1 engineer's chain = 100 ft = 100 links

1 Gunter's chain = 66 ft = 100 Gunter's links (lk) = 4 rods = ¹/₈₀ mile

1 acre = 100,000 sq (Gunter's) links = 43,560 sq ft = 160 sq rods

= 10 sq (Gunter's) chains = 4,046.87 sq m

1 rood = ¼ acre = 40 sq rods (also local unit = 5½ to 8 yd)

1 arpent = about 0.85 acre, or length of side of 1 square arpent (varies)

1 statute mile = 5,280 ft = 1,609.35 m

1 sq mile = 640 acres

1 nautical mile (U.S.) = 6,080.20 ft = 1,853.248 m

1 nautical mile (international) = 6,070.10 ft = 1,852 m

1 fathom = 6 ft

1 cubit = 18 in.

1 vara = 33 in. (Calif.), 33⅓ in. (Texas), varies

1 degree = ¹/₃₆₀ circle = 60 min = 3,600 sec = 0.01745 radian

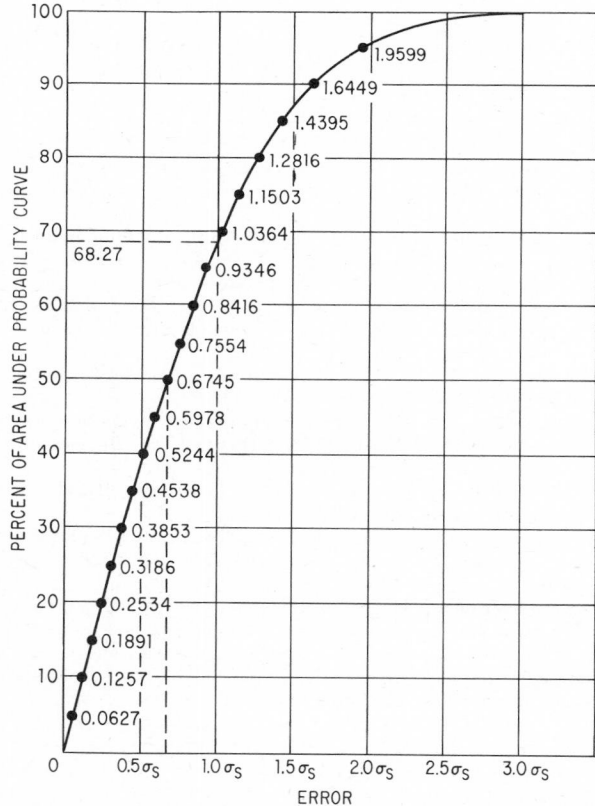

Fig. 12-1. Curve relates error and probability that the error will not be exceeded. (*From Brinker, "Elementary Surveying," Intext Educational Publishers, New York, N.Y.*)

1 radian = 57°17′44.8″ or about 57.30°
1 grad (grade) = $^1/_{400}$ circle = $^1/_{100}$ quadrant = 100 centesimal min
= 10^4 centesimal sec (French)
1 mil = $^1/_{6,400}$ circle = 0.05625°
1 military pace (milpace) = 2½ ft

12-4. Theory of Errors. When a number of measurements of the same quantity have been made, they must be analyzed on the basis of probability and the theory of errors. After all systematic (cumulative) errors and mistakes have been eliminated, random (compensating) errors are investigated to determine the most probable value (**mean**) and other critical values. Formulas for the most common of these values follow:

Standard deviation of a series of observations is

$$\sigma_s = \pm\sqrt{\frac{\Sigma d^2}{n-1}} \tag{12-1}$$

where d = residual (difference from mean) of a single observation
n = number of observations

The **probable error** of a single observation is

$$PE_s = \pm 0.6745\sigma_s \tag{12-2}$$

(The probability that an error within this range will occur is 0.50.)

The probability that an error will lie between two values is given by the ratio of the area of the probability curve included between the values to the total area. Figure 12-1 relates the percent area between plus and minus specific errors and the errors expressed as a constant times standard deviation. For example, the curve shows that the probability that an error will not exceed $1.0\sigma_s$ is 0.6827.

Probable error of a sum is

$$E_{\text{sum}} = \sqrt{E_1^2 + E_2^2 + E_3^2 + \cdots} \tag{12-3}$$

where E_1, E_2, E_3, . . . are probable errors of the separate measurements.

Error of the mean is

$$E_m = \frac{E_{\text{sum}}}{n} = \frac{E_s\sqrt{n}}{n} = \frac{E_s}{\sqrt{n}} \tag{12-4}$$

where E_s = specified error of a single measurement.

Probable error of the mean is

$$PE_m = \frac{PE_s}{\sqrt{n}} = \pm 0.6745\sqrt{\frac{\Sigma d^2}{n(n-1)}} \tag{12-5}$$

12-5. Significant Figures. These are the digits read directly from a measuring device plus one digit that must be estimated and therefore is questionable. For example, a reading of 654.32 ft from a steel tape graduated in tenths of a foot has five significant figures. In multiplying 798.16 by 37.1, the answer cannot have more significant figures than either number used; i.e., three in this case. The same rule applies in division. In addition or subtraction, for example, 73.148 + 6.93 + 482, the answer will have three significant figures, all on the left side of the decimal point.

12-6. Measurement of Distance. Reasonable precisions for different methods of measuring distances are

Pacing (ordinary terrain): $^1/_{50}$ to $^1/_{100}$.

Taping (ordinary steel tape): $^1/_{1,000}$ to $^1/_{10,000}$. (Results can be improved by use of tension apparatus, transit alignment, leveling.)

Base line (invar tape): $^1/_{50,000}$ to $^1/_{1,000,000}$.

Stadia: $^1/_{500}$ to $^1/_{1,000}$ (with special procedures).

Subtense bar: $^1/_{1,000}$ to $^1/_{5,000}$ (for short distances, with a 1-second theodolite, averaging angles taken at both ends).

Tellurometer (uses microwaves): 300 ft to 35 miles, 0.05 ft ± 5 ppm of the distance being measured at long distances. (See also Table 12-15, Art. 12-16.)

Electrotape (uses radio-frequency signals): 100 ft to 30 miles, 1 cm ± $^1/_{300,000}$ of the distance at maximum distances. (See also Table 12-15, Art. 12-16.)

Geodimeter (uses speed of light): 3 miles in daytime and 15 to 20 miles at night, 1 cm ± $^1/_{500,000}$ of the distance at longer distances. (See also Table 12-15, Art. 12-16.)

Slope Corrections. On slope measurements, the horizontal distance $H = L \cos x$, where L = slope distance and x = vertical angle, measured from the horizontal. It is better, however, to use a correction to get slope length from horizontal distance, or vice versa, from the formulas and values in Table 12-1, which also provides corrections for one end of the tape being off line.

Table 12-1. Correction for Slope Distance or Tape off Line

Elevation difference of ends or off-line distance, ft	Correction, ft		Elevation difference of ends or off-line distance, ft	Correction, ft	
	Tape length, ft			Tape length, ft	
	100	300		100	300
0.0	0.000	0.000	1.8	0.016	0.005
0.5	0.001	0.000	2.0	0.020	0.007
0.6	0.002	0.001	3.0	0.045	0.015
0.8	0.003	0.001	4.0	0.080	0.027
0.9	0.004	0.001	5.0	0.125	0.042
1.0	0.005	0.002	6.0	0.180	0.060
1.1	0.006	0.002	7.0	0.245	0.082
1.2	0.007	0.002	8.0	0.320	0.107
1.3	0.008	0.003	9.0	0.405	0.135
1.4	0.010	0.003	10.0	0.500	0.167
1.5	0.011	0.004	15.0	1.131	0.375
1.6	0.013	0.004	20.0	2.020	0.667

NOTE: Values by formula $C = d^2/2L$; for slopes greater than 10%, $C = d^2/2L + d^4/8L^3$, where d is the elevation difference or off-line distance.

Temperature Corrections. Table 12-2 lists temperature corrections for steel tapes. Formulas for other tape corrections, ft, with L as the measured distance, ft, are as follows:
For incorrect tape length,

$$C_t = \frac{(\text{actual tape length} - \text{nominal tape length})L}{\text{nominal tape length}} \qquad (12\text{-}6)$$

For nonstandard tension,

$$C_p = \frac{(\text{applied pull} - \text{standard tension})L}{AE} \qquad (12\text{-}7)$$

where A = cross-sectional area of tape, sq in.
E = modulus of elasticity = 29,000,000 psi for steel
For sag correction between points of support, ft,

$$C = -\frac{w^2 L_s^3}{24P^2} \qquad (12\text{-}8)$$

where w = weight of tape per foot, lb
L_s = unsupported length of tape, ft
P = pull on tape, lb

Sources and Types of Error. There are three sources of error in taping—instrumental, natural, and personal—and nine general types of errors. Table 12-3 lists the types of errors and their sources and classifies them as systematic or accidental.

All errors in Table 12-3 produce, in effect, an incorrect tape length. Therefore, only four basic tape problems exist: *measuring* a line between fixed points with a tape too long or too short, and *laying out* a line from one fixed point with a tape too long or too short. Sketching the conditions provides a foolproof method for determining whether to add or subtract a correction.

In base-line measurements with steel or invar tapes (three or more tapes should be used on different sections of the line), corrections are applied for inclination; temperature; nonstandard length of tape, for both full and partial tape lengths; and reduction to sea level.

Table 12-2. Temperature Corrections for Steel Tapes*

Subtract corrections for these temperatures, °F	Length of line, ft				*Add* corrections for these temperatures, °F
	5,000	1,000	500	100	
68	0.00	0.00	0.00	0.00	68
66	0.06	0.01	0.01	0.00	70
64	0.13	0.03	0.01	0.00	72
62	0.20	0.04	0.02	0.00	74
60	0.26	0.05	0.03	0.01	76
58	0.32	0.06	0.03	0.01	78
56	0.39	0.08	0.04	0.01	80
54	0.46	0.09	0.04	0.01	82
52	0.52	0.10	0.05	0.01	84
50	0.58	0.12	0.06	0.01	86
48	0.65	0.13	0.06	0.01	88
46	0.72	0.14	0.07	0.01	90
44	0.78	0.16	0.08	0.02	92
42	0.84	0.17	0.08	0.02	94
40	0.91	0.18	0.09	0.02	96
38	0.98	0.20	0.10	0.02	98
36	1.04	0.21	0.10	0.02	100
34	1.10	0.22	0.11	0.02	102
32	1.17	0.23	0.12	0.02	104
30	1.24	0.25	0.12	0.02	106
28	1.30	0.26	0.13	0.03	108
26	1.36	0.27	0.14	0.03	110

Example: Given a recorded distance of 8,785.32 ft for a line measured when the average temperature is 80°F. Correction to be added is $0.39 + 3(0.08) + 0.04 + 2(0.01) + 0.01 = 0.70$ ft. Because of rounding off in the table, the total correction of 0.70 is 0.01 ft larger than the value computed directly by formula, $C = 0.0000065(T - 68F)L$.

*By permission of Marvin C. May, University of New Mexico.

Table 12-3. Types, Sources, and Classification of Taping Errors

Type of error	Source*	Classification†	Departure from standard to produce 0.01-ft error for a 100-ft tape
Tape length	I	S	0.01 ft
Temperature	N	S or A	15°F
Tension	P	S or A	15 lb
Sag	N, P	S	7⅜ in. at center as compared with support throughout
Alignment	P	S	1.4 ft at one end, or 8½ in. at midpoint
Tape not level	P	S	1.4 ft
Interpolation	P	A	0.01 ft
Marking	P	A	0.01 ft
Plumbing	P	A	0.01 ft

*I = Instrumental, N = Natural, P = Personal.
†S = Systematic, A = Accidental.

12-7. Leveling.   Differential leveling is the process of determining the difference in elevation of two points. The procedure involves sighting with a level on a ruled rod set on a point of known elevation (backsight or plus sight), then on the rod set on points (or intermediate points) whose elevations are to be determined (foresights). These elevations equal the height of instrument minus the foresight reading. The height of instrument equals the known elevation plus the backsight reading. For accuracy, the sum of backsight and foresight distances should be kept nearly equal.

Table 12-4 shows a typical left-hand page of open-style notes. The right-hand page would contain bench-mark descriptions, sketches, date of survey, names of survey party members, and information on the weather, equipment used, and other necessary remarks.

Table 12-4. Differential Leveling Notes

Station	B.S.	H.I.*	F.S.	Elev.†	Dist.
Differential Leveling—BM Civil to BM Dorm					
BM Civil				100.00	
	4.08	104.08			175
TP 1			0.20	103.88	180
	6.09	109.97			160
BM Dorm			4.32	105.65	155
	10.17		4.52		670
	4.52				
	5.65				
BM Dorm				105.65	
	4.37	110.02			165
TP 2			6.14	103.88	165
	0.93	104.81			170
BM Civil			4.80	100.01	175
	5.30		10.94		675
			5.30		
			5.64		
Elev. Diff. = 5.64 ft					
Loop Closure = 0.01 ft					

*Height of instrument (H.I.) = elevation + backsight (B.S.).
†Elevation = H.I. − foresight (F.S.).

Elevations commonly are taken to 0.01 ft in engineering surveys and to 0.001 m in precise National Geodetic Survey work.

Profile leveling determines the elevations of points at known distances along a line. When these points are plotted, a vertical section through the earth's surface is shown. Elevations are taken at full stations (100 ft, or closer in irregular terrain), at breaks in the ground surface, and at critical points such as bridge abutments and road crossings. Profiles are generally plotted on special paper with an exaggeration of from 5:1 up to 20:1, or even more, so that elevation differentials will show up better. Profiles are needed for route surveys, to select grades and find earthwork quantities. Elevations are usually taken to 0.01 ft on bench marks and 0.1 ft on the ground.

Reciprocal leveling is employed to cross streams, lakes, canyons, and other topographic barriers that prevent keeping backsights and foresights balanced. On each side of the obstruction to be crossed, a plus sight is taken on the near rod and several minus sights on the far rod. The resulting differences in elevation are averaged to eliminate the effects of curvature and refraction, and inadjustment of the instrument. Even though a number of minus sights are taken for averaging, their length may reduce the accuracy of results.

Borrow-pit or cross-section leveling produces elevations at the corners of squares or rectangles whose sides are dependent on the area to be covered, type of terrain, and accuracy desired. For example, sides may be 10, 20, 40, 50, or 100 ft. Layout can be made by tape alone, or transit and tape. Contours can be located readily, topographic features not so well. Quantities of material to be excavated or filled are computed, in cubic yards, by selecting a grade elevation, or final ground elevation, computing elevation differences for the corners, and substituting in

$$Q = \frac{nxA}{108} \tag{12-9}$$

where n = number of times a particular corner enters as part of a division block

x = difference in ground and grade elevation for each corner, ft

A = area of each block, sq ft

Cross-section leveling also is the term applied to the procedure for locating contours or taking elevations on lines at right angles to the center line in a route survey.

Three-wire leveling is a type of differential leveling with three horizontal sighting wires in the level. Upper, middle, and lower wires are read to obtain an average value for the sight, check the precision of reading the individual wires, and secure stadia distances for checking lengths of backsights and foresights. The height of instrument is not needed or computed. The National Geodetic Survey has long used three-wire leveling for its control work, but more general use is now being made of the method.

Grade designates the elevation of the finished surface of an engineering project, and also the rise or fall in 100 ft of horizontal distance; for example, a 4% grade (also called gradient). Note that since the common stadia interval factor is 100, the difference in readings between the middle and upper (or lower) wire represents ½ ft in 100 ft, or a ½% grade.

Types of levels in general use are listed in Table 12-5.

Table 12-5. Types of Levels

Type	Use
Hand level	Rough work. Sights on ordinary level rod limited to about 50 ft because of zero- to 2-power magnification
Engineer's level, Wye or Dumpy	Suitable for ordinary work (third- or fourth-order). Elevations to 0.01 ft without target
Tilting level	Faster, more accurate sighting. Good for third-, second-, or first-order work depending upon refinement
Self-leveling level	Fast, suitable for second-order and third-order work
Precise level	Very sensitive level vials, high magnification, tilting, other features

NOTE: Instruments are arranged in ascending order of cost.

12-8. Vertical Control. The National Geodetic Survey has run many miles of first- and second-order levels to provide vertical control for all types of surveys. Descriptions and elevations of bench marks are furnished free upon request. Allowable closures for the three orders of leveling are as follows:

First order, $C = 0.5 \sqrt{k}$ (Class I)

Second order, $C = 1.0 \sqrt{k}$ (Class I)

Third order, $C = 2.0 \sqrt{k}$

Here C = allowable closure, mm, and k = distance, km. The bench marks are normally set at intervals of ⅝ to 1 mile.

12-9. Magnetic Compass. A magnetic compass consists of a magnetized needle mounted on a pivot at the center of a graduated circle. The compass is now used primarily for retracement purposes and checking, although some surveys not requiring precision are made with a compass; for example, in forestry and geology. American transits have traditionally come with a long compass needle, whereas on European instruments the compass is merely an accessory, and therefore, the instruments can be smaller and lighter.

Table 12-6. Periodic Variations in Declination of Magnetic Needle

Variation	Remarks
Secular	Largest and most important. Produces wide unpredictable swings over a period of years, but records permit comparison of past and present declinations
Daily (diurnal)	Swings about 8 min per day in the U.S. Relatively unimportant
Annual	Periodic swing amounting to less than 1 min of arc; it is unimportant
Irregular	From magnetic storms and other sources. Can pull needle off more than a degree

A small weight is placed on the south end of the needle in the Northern Hemisphere to counteract the dip caused by magnetic lines of force. Since the magnetic poles are not located at the geographic poles, a horizontal angle (declination) results between the axis of the needle and a true meridian. East declination occurs if the needle points east of true north, west declination if the needle points west of true north.

The National Geodetic Survey publishes a world chart every fifth year showing the positions of the agonic line, isogonic lines for each degree, and values for annual variation of the needle. The **agonic line** is a line of zero declination; i.e., a magnetic compass set up on points along this line would point to true north as well as magnetic north. For points along an isogonic line, declination should be constant, barring local attraction.

Table 12-6 lists the periodic variations in the declination of the needle that make it unreliable. In addition, local attraction resulting from power sources, metal objects, etc., may produce considerable error in bearings taken with a compass. If the source of local attraction is fixed and constant, however, angles between bearings are correct, even though the bearings are uniformly distorted.

The Brunton compass or pocket transit has some of the features of a sighting compass, a prismatic compass, a hand level, and a clinometer. It is suitable for some forest, geological, topographical, and preliminary surveys of various kinds.

A common problem today is the conversion of past magnetic bearings based upon the declination of a given date to present bearings with today's declination, or to true bearings. A sketch, such as Fig. 12-2, showing all data will make the answer evident.

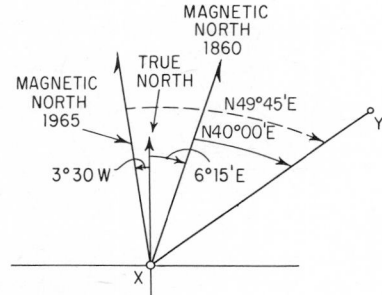

Fig. 12-2. Magnetic bearing of a line *XY* in a past year is found by plotting magnetic north for that year with respect to true north.

12-10. Bearings and Azimuths. The direction of a line is the angle measured from any reference line, such as a magnetic or true meridian. Bearings are angles measured from the north and the south, toward the east or the west. They can never be greater than 90° (Fig. 12-3).

Bearings read in the advancing direction are forward bearings; those in the opposite direction are back bearings. Computed bearings are obtained by using a bearing and applying a direct, deflection, or other angle. Bearings, either magnetic or true, are used in rerunning old surveys, in computations, on maps, and in deed descriptions.

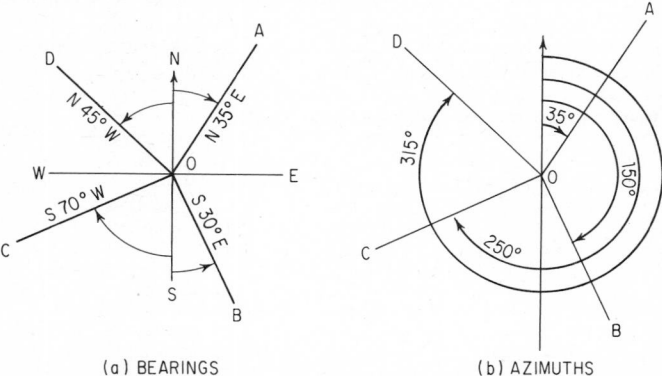

(a) BEARINGS   (b) AZIMUTHS

Fig. 12-3. Direction of lines may be specified by (*a*) bearings or (*b*) azimuths.

An azimuth is a clockwise angle measured from some reference line, usually a meridian. Government surveys use geodetic south as the base of azimuths. Other surveys in the Northern Hemisphere may employ north. Azimuths are advantageous in topographic surveys, plotting, direction problems, and other work where omission of the quadrant letters and a range of angular values from 0 to 360° simplify the work.

12-11. Horizontal Control. All surveys require some kind of control, be it a base line or bench mark, or both. Horizontal control consists of points whose positions are established by traverse, triangulation, or trilateration. The National Geodetic Survey has established control points throughout the country and tabulated azimuths, latitude and longitude, statewide coordinates, and other data for them. Surveys on the statewide coordinate system have increased the number of control points available to all surveyors.

Traverses. For a traverse, the survey follows a line from point to point in succession. The lengths of lines between points and their directions are measured. If the traverse returns to the point of origin, it is called a closed traverse. The United States–Canada boundary, for example, was run by traverse. In contrast, the boundary of a construction site would be surveyed by a closed traverse. Permissible closures for traverses that make a closed loop or connect adjusted positions of equal-order or higher-order control surveys are given in Table 12-7.

Table 12-7. Permissible Traverse Closures *

Traverse order	Max permissible closure after azimuth adjustment	Max azimuth closure at azimuth checkpoint	
		Sec per station	Sec†
First order	1:100,000	1.0	$2\sqrt{N}$
Second order			
Class I	1:50,000	1.5	$3\sqrt{N}$
Class II	1:20,000	2.0	$6\sqrt{N}$
Third order			
Class I	1:10,000	3.0	$10\sqrt{N}$
Class II	1:5,000	8.0	$30\sqrt{N}$

*National Geodetic Survey.
†N = number of stations

Table 12-8. Ratio of Error for Various Surveys

Wasteland	1/500
Ordinary farmland	1/1,000
Small community	1/2,000–1/5,000
Small city	1/5,000–1/10,000
Metropolitan area	1/10,000–1/20,000

Transit-tape traverses provide control for areas of limited size as well as for the final results on property surveys, route surveys, and other work. Stadia traverses are good enough for small-area topographic surveys when tied to higher control. Traverses also can be made with electronic distance-measuring devices (Art. 12-16).

Reasonable ratios of error for different types of property surveys, depending upon land values, are shown in Table 12-8.

Triangulation. In triangulation, points are located at the apexes of triangles, and all angles and one base line are measured. Additional base lines are used when a chain of triangles, quadrilaterals, or central point figures is required (Fig. 12-4). All other sides are computed and adjustments carried from the fixed base lines forward and backward to minimize the corrections. Angles used in computation should exceed 15°, and preferably 30°, to avoid the rapid change in sines for small angles.

Chains of triangles are unsuitable for high-precision work, since they do not permit the rigid adjustments available in quadrilaterals and more complicated figures. Quadrilaterals are advantageous for long, relatively narrow chains; polygons for wide systems and perhaps for large cities, where stations can be set on tops of buildings.

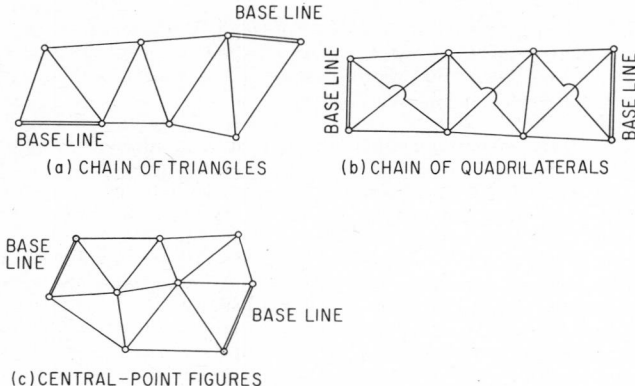

Fig. 12-4. Triangulation chains.

Strength of figure in triangulation is an expression of relative precision possible in the system based upon the route of computation of a triangle side. It is independent of the accuracy of observations and utilizes the number of directions observed, conditions to be satisfied, and rates of changes for the sines of distance angles. Triangulation stations that cannot be occupied require additional computation for reduction to center in obtaining coordinates and other data.

Permissible triangulation closures for the three orders of triangulation specified by the National Geodetic Survey are given in Table 12-9, and specifications for base-line measurements in Table 12-10.

Table 12-9. Triangulation Closures

Specification item	First order	Second order		Third order	
		Class I	Class II	Class I	Class II
Avg triangle closure, sec	1.0	1.2	2.0	3.0	5.0
Max triangle closure, sec	3.0	3.0	5.0	5.0	10.0

Table 12-10. Specifications for Base-line Measurements

Order	Max standard error of base
First	1/1,000,000
Second	
Class I	1/900,000
Class II	1/800,000
Third	
Class I	1/500,000
Class II	1/250,000

Trilateration has replaced triangulation for establishment of control in some cases, such as photogrammetry, since the development of electronic measuring devices. All distances are measured and the angles computed as needed. (See also Art. 12-17.)

12-12. Stadia. Stadia is a method of measuring distances by noting the length of a stadia or level rod intercepted between the upper and lower sighting wires of a transit or level. Most transits and levels have an interval between stadia wires that gives a vertical intercept of 1 ft on a rod 100 ft away. A stadia constant varying from about ¾ to 1¼ ft (usually assumed to be 1 ft) must be added for older-type external-focusing telescopes. The internal-focusing short-length telescopes common today have a stadia constant of only a few tenths of a foot, and so it can be neglected for normal readings taken to the nearest foot.

Figure 12-5 shows stadia relationships for a horizontal sight with the older-type external-focusing telescope. Relationships are comparable for the internal-focusing type.

For horizontal sights, the stadia distance, ft (from instrument spindle to rod), is

$$D = R \frac{f}{i} + C \qquad (12\text{-}10)$$

where R = intercept on the rod between the two sighting wires, ft
  f = focal length of telescope, ft (constant for a specific instrument)
  i = distance between stadia wires, ft
  $C = f + c$
  c = distance from center of spindle to center of objective lens, ft
C is called the stadia constant, although c and C vary slightly.

The value of f/i, the stadia factor, is set by the manufacturer to be about 100, but it is not necessarily 100.00. The value should be checked before use on important work, or when the wires or reticle are damaged and replaced.

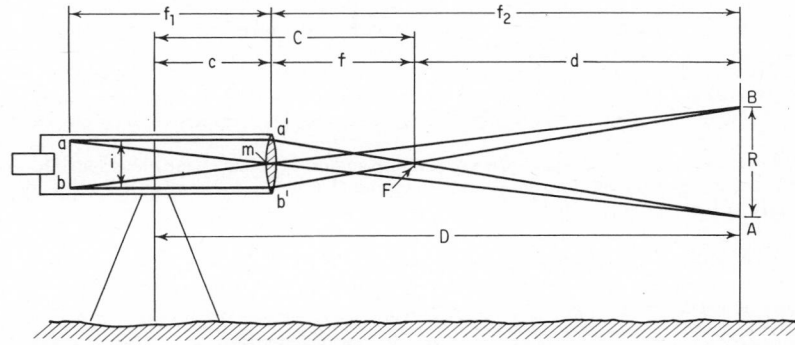

Fig. 12-5. Distance D is measured with an external-focusing telescope by determining the interval R intercepted on a rod AB by two horizontal sighting wires a and b.

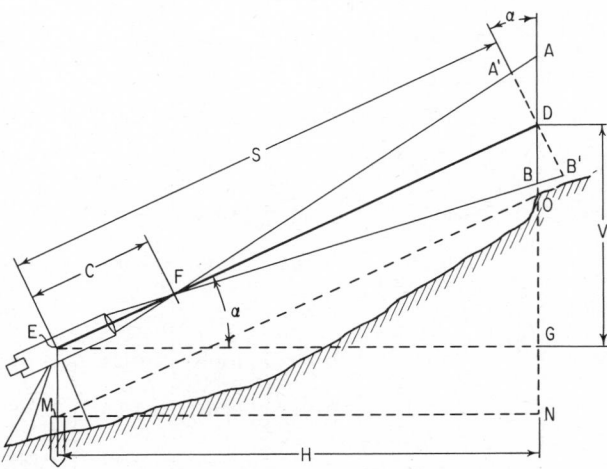

Fig. 12-6. Stadia measurement of vertical and horizontal distances V and H is done by reading rod intercept with telescope and vertical angle α.

For inclined sights (Fig. 12-6) the rod is held vertical, as indicated by a rod level or other means. Reduction to horizontal and vertical distances is made according to formulas, such as

$$H = 100R - 100R \sin^2 \alpha + C \tag{12-11a}$$

$$V = 100R(\tfrac{1}{2} \sin 2\alpha) \tag{12-11b}$$

where H = horizontal distance from instrument to rod, ft
V = vertical distance from instrument to rod, ft
α = vertical angle above or below a level sight

A Beaman arc on transits and alidades simplifies reduction of slope sights. It consists of an H scale and a V scale, both graduated in percent, with spacing based upon the stadia formulas. The H scale gives the correction per 100 ft of slope distance, which is subtracted from $100R + C$ to get the horizontal distance. A V-scale index of 50 for level sights eliminates minus values in determining vertical distance. Readings above 50 are angles of elevation; below 50, angles of depression. Each unit above or below 50 represents 1 ft difference in elevation per 100 ft of sight. By setting the V scale to a whole number, even though the middle wire does not fall on the height of the instrument, you need only mental arithmetic to compute vertical distance. The H scale is read by interpolation, since the value generally is small and falls in the area of wide spaces.

As an illustration, to determine the elevation of a point X from a setup at point Y, elevation X = elevation Y + height of instrument + (arc reading − 50) (rod intercept) − reading of middle wire.

Self-reducing European tachymeters have curved stadia lines engraved on a glass plate, which turns and appears to make the lines move closer or farther apart. A fixed stadia factor of 100 is used for horizontal reduction, but several factors are required for elevation differences, depending upon the slope.

Stadia traverses can be run with direct or azimuth angles. Distances and elevation differences should be averaged for the foresights and backsights. Elevation checks on bench marks are necessary at frequent intervals to maintain reasonable precision.

Table 12-11. Relations between Linear and Angular Errors*

Precision of linear measurements	Allowable angular error	Least reading in angular measurements	Allowable linear error, ft, in				Ratio
			100 ft	500 ft	1,000 ft	5,000 ft	
$\frac{1}{500}$	6'53"	5'	0.145	0.727	1.454	7.272	$\frac{1}{688}$
$\frac{1}{1,000}$	3'26"	1'	0.029	0.145	0.291	1.454	$\frac{1}{3,440}$
$\frac{1}{5,000}$	0'41"	30"	0.015	0.073	0.145	0.727	$\frac{1}{6,880}$
$\frac{1}{10,000}$	0'21"	20"	0.010	0.049	0.097	0.485	$\frac{1}{10,300}$
$\frac{1}{50,000}$	0'04"	10"	0.005	0.024	0.049	0.242	$\frac{1}{20,600}$
$\frac{1}{100,000}$	0'02"	5"	0.002	0.012	0.024	0.121	$\frac{1}{41,200}$
$\frac{1}{1,000,000}$	0'00.2"	2"	0.001	0.005	0.010	0.048	$\frac{1}{103,100}$
		1"		0.002	0.005	0.024	$\frac{1}{206,300}$

The column headers span: "Allowable angular error for given linear precision" covers Precision of linear measurements, Allowable angular error, and Least reading in angular measurements. "Allowable linear error for given angular precision" covers the remaining columns.

*From Brinker, "Elementary Surveying," Intext Educational Publishers, New York, N.Y.

Poor closures in stadia work are usually the result of incorrect rod readings rather than errors in angles. A difference of 1 minute in vertical angle has little effect on the horizontal distance; it produces an elevation difference of less than 0.1 ft for 300-ft sights. Also, for sights of 300 ft or less and stadia readings to the nearest foot, comparable distance-angle precision can be obtained by reading the horizontal angles to the nearest 5 minutes. This can be done by estimation on the scale without using the vernier graduations.

Table 12-11 shows the relation between linear and angular errors, and Table 12-12 the precision of computed values.

12-13. Plane-Table Surveying. This enables maps to be drawn partly or completely in the field as measurements are made. The method is especially suitable for mapping topography. For a plane-table survey, a hard, flat surface that can be adjusted to be level is set up within the area to be mapped. Map paper is fastened to the surface for recording measurements diagrammatically. A surveying instrument, called an alidade, is placed on the plane table and used to sight on a stadia rod and for drawing map lines.

Plane-table surveying has many advantages and disadvantages, as shown by the following list. Although replaced by photogrammetry for most large-area mapping, there still is a place for this time-tested procedure.

Advantages of the plane table:
    Features are mapped while in full view.
    Contours and irregular lines are accurately sketched, with half the number of sights needed for transit stadia.
    No notes need be taken.
    Less total time (field work plus office) is required.
    Setups can be made at desirable points without advance determination by using two- or three-point location.
    Self-reducing alidades and parallel rulers can be used.

Disadvantages of the plane table:
    Plotted control is needed in advance of field work.
    Good weather is needed.
    More field time is required.
    Brush or wooded areas hamper sketching.
    Elbow-height table setups give low sight clearance.
    Computations are needed immediately for plotting.
    Many awkward items must be carried.
    Lengths and angles must be scaled if required later.
    Difficulty in keeping the table level and oriented results from leverage caused by a light pressure on the outer area of the board.
    Considerable experience is essential.

Two basic kinds of table are used: a small traverse table with peep-sight alidade and fixed-leg tripod without leveling head, obviously appropriate only for rough work; and the standard plane-table board, usually 24 by 31 in., set on a tripod having either the National Geodetic Survey four-screw leveling head or the Johnson ball-and-socket head.

Plane tables are oriented by means of a declinator, by backsighting as with a transit, or by resection. Permanent backsights (towers, lone trees, fixed signals) enable the instrumentman to check orientation frequently without interrupting the rodman's movements.

A traverse is run by orienting the table, sighting the next point, drawing a line along the alidade blade and plotting the stadia length, then moving to the plotted point and repeating the process. An average distance and elevation are obtained from the foresight and backsight. Adjusting the vertical-arc index to read zero when the bubble is centered presents an ever-present problem, because the table goes off level.

Topographic details are located by radiation or intersection. Short distances can be measured with a cloth tape for large-scale maps. The intersection method (graphical triangulation) is appropriate for long sights taken from two plane-table stations—or three for checking—to inaccessible points. Elevations of inaccessible points may be determined from vertical angles and scaled map distances.

The stepping method (Fig. 12-7) also is used in rough terrain to save difficult travel for the rodman. To find the elevation of the top of a steep cliff, rod intercept *ab* is taken for a level reading at the base of the cliff, the lower hair set on the ground, and an identifiable spot noted where the upper wire hits the bluff.   The lower hair is moved up to the same spot and the process repeated until the upper hair is near the summit.   A final partial "step" is estimated.   The number of steps times the rod intercept, plus the final estimated number of feet, gives an approximate height of the cliff.

Table 12-12. Precision of Computed Values*

Size of angle and function	Angular error				
	1′	30″	20″	10″	5″
	Precision of computed value using sin or cos				
sin 5° or cos 85°	$\frac{1}{300}$	$\frac{1}{600}$	$\frac{1}{900}$	$\frac{1}{1,800}$	$\frac{1}{3,600}$
10 80	$\frac{1}{610}$	$\frac{1}{1,210}$	$\frac{1}{1,820}$	$\frac{1}{3,640}$	$\frac{1}{7,280}$
20 70	$\frac{1}{1,250}$	$\frac{1}{2,500}$	$\frac{1}{3,750}$	$\frac{1}{7,500}$	$\frac{1}{15,000}$
30 60	$\frac{1}{1,990}$	$\frac{1}{3,970}$	$\frac{1}{5,960}$	$\frac{1}{11,970}$	$\frac{1}{23,940}$
40 50	$\frac{1}{2,890}$	$\frac{1}{5,770}$	$\frac{1}{8,660}$	$\frac{1}{17,310}$	$\frac{1}{34,620}$
50 40	$\frac{1}{4,100}$	$\frac{1}{8,190}$	$\frac{1}{12,290}$	$\frac{1}{24,580}$	$\frac{1}{49,160}$
60 30	$\frac{1}{5,950}$	$\frac{1}{11,900}$	$\frac{1}{17,860}$	$\frac{1}{35,720}$	$\frac{1}{71,440}$
70 20	$\frac{1}{9,450}$	$\frac{1}{18,900}$	$\frac{1}{28,330}$	$\frac{1}{56,670}$	$\frac{1}{113,340}$
80 10	$\frac{1}{19,500}$	$\frac{1}{39,000}$	$\frac{1}{58,500}$	$\frac{1}{117,000}$	$\frac{1}{234,000}$
	Precision of computed value using tan or cot				
tan or cot 5°	$\frac{1}{300}$	$\frac{1}{600}$	$\frac{1}{900}$	$\frac{1}{1,790}$	$\frac{1}{3,580}$
10	$\frac{1}{590}$	$\frac{1}{1,180}$	$\frac{1}{1,760}$	$\frac{1}{3,530}$	$\frac{1}{7,050}$
20	$\frac{1}{1,100}$	$\frac{1}{2,210}$	$\frac{1}{3,310}$	$\frac{1}{6,620}$	$\frac{1}{13,250}$
30	$\frac{1}{1,490}$	$\frac{1}{2,980}$	$\frac{1}{4,470}$	$\frac{1}{8,930}$	$\frac{1}{17,870}$
40	$\frac{1}{1,690}$	$\frac{1}{3,390}$	$\frac{1}{5,080}$	$\frac{1}{10,160}$	$\frac{1}{20,320}$
45	$\frac{1}{1,720}$	$\frac{1}{3,440}$	$\frac{1}{5,160}$	$\frac{1}{10,310}$	$\frac{1}{20,630}$
50	$\frac{1}{1,690}$	$\frac{1}{3,390}$	$\frac{1}{5,080}$	$\frac{1}{10,160}$	$\frac{1}{20,320}$
60	$\frac{1}{1,490}$	$\frac{1}{2,980}$	$\frac{1}{4,470}$	$\frac{1}{8,930}$	$\frac{1}{17,870}$
70	$\frac{1}{1,100}$	$\frac{1}{2,210}$	$\frac{1}{3,310}$	$\frac{1}{6,620}$	$\frac{1}{13,250}$
80	$\frac{1}{590}$	$\frac{1}{1,180}$	$\frac{1}{1,760}$	$\frac{1}{3,530}$	$\frac{1}{7,050}$
85	$\frac{1}{300}$	$\frac{1}{600}$	$\frac{1}{900}$	$\frac{1}{1,790}$	$\frac{1}{3,580}$

*From Brinker, "Elementary Surveying," Intext Educational Publishers, New York, N.Y.

In **resection,** orientation of the table at positions not yet identified on the map is done by either the two- or three-point method.   In two-point location, a direction toward the unselected next point X is plotted as in Fig. 12-8.   After setting up at any selected spot on the projected line bx, the table is oriented using that line.   By sighting to the known plotted point A, which preferably is at an angle of 60 to 90° with bx, the setup location is fixed at point y, the intersection of bx and aA extended.

The three-point location method determines the plane-table position after the table is set up at a place from which three or more prominent, plotted control signals can be seen.   (In the past, navigation and plane-table procedures used directions, but electronic distance-measuring devices solve the problem with lengths.)   Arcs drawn with radii equal to the measured distances and using the plotted signal points as centers give the desired point.   A check is obtained if the three arcs intersect at a unique point.

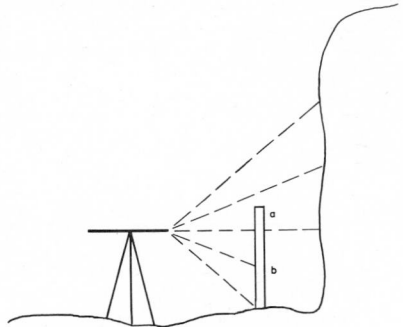

Fig. 12-7. Stepping method determines elevations on steep slope.

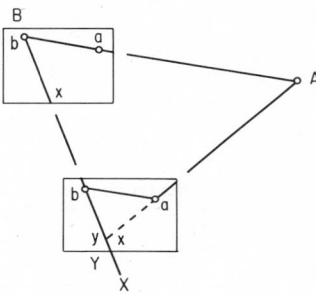

Fig. 12-8. Two-point resection orients plane table at new station when two stations already plotted are visible.

Various solutions are available on the plane table, such as the tracing-paper method, three-arm protractor location, and Lehmann's method.   All give a strong solution of the position (point sought) if it is inside the great triangle (Fig. 12-9) or away from the great circle through the three control points.   An indeterminate solution results if the point sought is on or very close to a great circle through the control points.   The size of the triangle of error in Fig. 12-9 is dependent on how well the table was oriented by estimation to start the process, and the mapping scale.

When lines drawn at the new station to the three control points do not intersect at a point, three simple rules are used to find the point.   (A second, or possibly a third, application of the trial method may be necessary.)

1. The point sought is inside the triangle of error if the station occupied is within the great triangle.

Fig. 12-9. Lehmann method orients plane table at new station when three stations already plotted are visible.

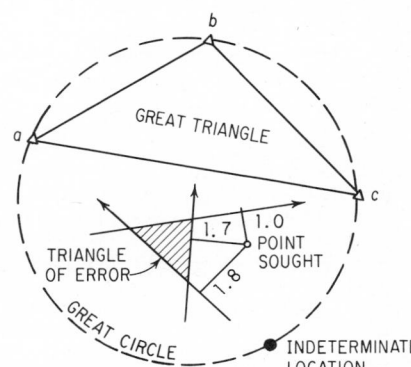

2. The point sought is either to the right or to the left (when facing the control points) of all three resection lines drawn from the signals.

3. The point sought is always distant from each of the three resection lines in proportion to the distances from the respective signals to the plane-table station. In Fig. 12-9, using proportional estimated distances to the three signals, perpendiculars are drawn by trial from the resection lines until they intersect at a unique point, the point sought.

12-14. Topographic Surveys.   Topographic surveys are made to locate natural and artificial (man-made) features for mapping purposes.   By means of conventional symbols, culture (bridges, buildings, boundary lines, etc.), relief, hydrography, vegetation, soil types, and other topographic details are shown for a portion of the earth's surface.

Planimetric (line) maps define natural and cultural features in plan only.   **Hypsometric maps** give elevations by contours, or less definitely by means of hachures, shading, and tinting.

Horizontal and vertical control of a high order are necessary for accurate topographic work. Triangulation, trilateration, traversing, and photogrammetry furnish the skeleton on which the topographic details are hung.   A level net must provide elevations with closures smaller than expected of the topographic traverse and side shots.   For surveys near lake shores or slow-moving streams, the water surface on calm days is a continuous bench mark.

Seven methods are used to locate points in the field, as listed in Table 12-13.   The first four require a "base line" of known length.   An experienced instrumentman selects the simplest method considering both field and office work involved.

Table 12-13. Methods for Locating Points in the Field

Method	Principal Use
1. Two distances	Short taping, details close together, trilateration
2. Two angles	Graphical triangulation, plane table
3. One angle, adjacent distance	Transit and stadia
4. One angle, opposite distance	Special cases
5. One distance, right-angle offset	Route surveys, curved shorelines or boundaries
6. String lines from straddle hubs	Referencing hubs for relocation
7. Two angles at point to be located	Three-point location for plane table, navigation

A **contour** is a line connecting points of equal elevation.   The shoreline of a lake not disturbed by wind, inlet, or outlet water forms a contour.   The vertical distance (elevation) between successive contours is the contour interval.   Intervals commonly used are 1, 2, 5, 10, 20, 25, 40, 50, 80, and 100 ft, depending on the map scale, type of terrain, purpose of the map, and other factors.

Methods of taking topography and pertinent points on the suitability of each for given conditions are given in Table 12-14.

12-15. Photogrammetry.   Photogrammetry is the art and science of obtaining reliable measurements by photography.   It includes the use of terrestrial, aerial, vertical, and oblique photographs and their interpretation.   Some of the advantages of mapping by aerial photographs are rapid coverage of large areas, accessible or inaccessible, and assurance of getting all visible detail.   Note that an aerial photograph is not a map, i.e., an orthographic projection.   Instead, it is a perspective projection and may contain unnecessary details, which tend to overshadow the critical ones.

Aerial cameras commonly employed for topographic mapping are precise instruments with nominal focal lengths of 6, 8¼, or 12 in.   Most use roll film.   Four fiducial marks are printed on each photograph to locate the geometric axes and principal point.   Photographs are made in strips with a side lap (strip overlap) of 25% ± 10%, and a forward overlap (advance) averaging 60% ± 5% to assure that images of ground points appear in at least two and preferably three or more pictures.

Because vertical photographs are perspective views, the scale is not uniform.   Equal-length ground lines at higher elevations and near the edges of the photographs will be longer than those at lower elevations and near the center.   An average scale can be selected as an approximate value.

Table 12-14. Methods of Obtaining Topography

Method	Suitability
Transit and tape	Accurate, but slow and costly. Used where accuracy beyond plotting precision is desired
Transit and stadia	Fast, reasonably accurate for plotting purposes. Contours by direct (trace-contour) method in gently rolling country, or by indirect (controlling-point) system where high, low, and break points are found in rugged terrain, or on uniform slopes, and contours interpolated
Plane table	Plotting and checking in field. Good in cluttered areas of many details. Contours by direct or indirect method. Now generally replaced by photogrammetry for large areas. Used to check photogrammetric maps
Coordinate squares	Better for contours than culture. Elevations at corners and slope changes interpolated for contours. Size of squares dependent on area covered, accuracy desired, and terrain. Best in level to gently rolling country
Offsets from center line or cross sectioning	On route surveys, right-angle offsets taken by eye or prism at full stations and critical points, along with elevations, to get a cross profile and topographic details. Contours by direct or indirect method. Elevations or contours recorded as numerator, and distance out as denominator
Photogrammetry	Fast, cheap, and now commonly used for large areas covering any terrain, where ground can be seen. Basic control by ground methods, some additional control from photographs

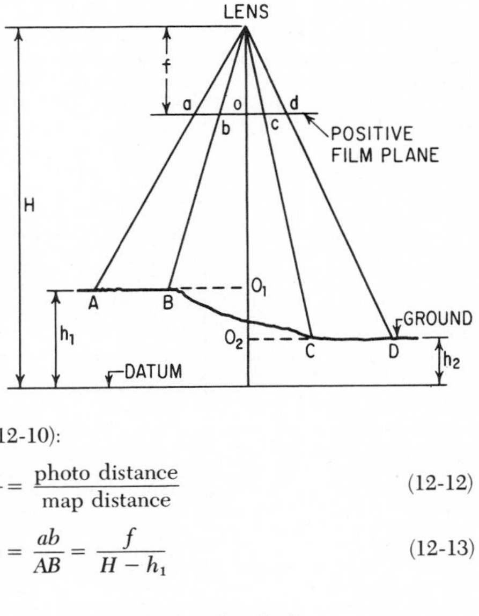

Fig. 12-10. Photographic scale depends on focal length of lens f and height H of airplane. (*From Brinker, "Elementary Surveying," Intext Educational Publishers, New York, N.Y.*)

Scale formulas are as follows (refer to Fig. 12-10):

$$\frac{\text{Photo scale}}{\text{Map scale}} = \frac{\text{photo distance}}{\text{map distance}} \qquad (12\text{-}12)$$

$$\text{Photo scale} = \frac{ab}{AB} = \frac{f}{H - h_1} \qquad (12\text{-}13)$$

where f = focal length of lens, in.

H = flying height of airplane above datum (usually mean sea level), ft

h_1 = elevation of a point, line, or area with respect to the datum, ft

Ground distances can be found from measurements on a photograph by use of photograph coordinates x, y and ground coordinates X, Y (Fig. 12-11). For a line AB with unequal elevations at A and B, length is determined by

$$AB = \sqrt{(X_A - X_B)^2 + (Y_A - Y_B)^2} \qquad (12\text{-}14)$$

where $X_A = x_a(H - h_A)/f$
$Y_A = y_a(H - h_A)/f$
$X_B = x_b(H - h_B)/f$
$Y_B = y_b(H - h_B)/f$

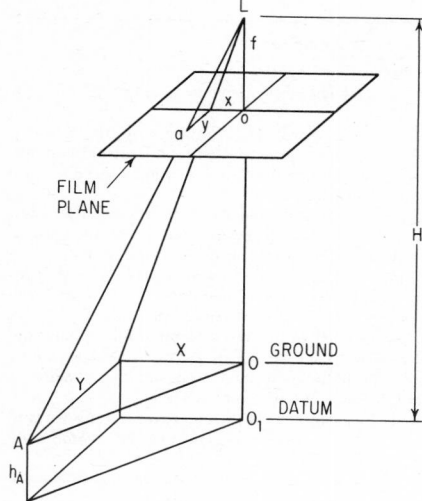

Fig. 12-11. Photograph coordinates x, y are proportional to ground coordinates X, Y when optical axis is vertical. (*From Brinker, "Elementary Surveying," Intext Educational Publishers, New York, N.Y.*)

Average displacements caused by topographic relief on vertical aerial photographs always radiate from the principal point o (Fig. 12-12), which is directly above the nadir point O on the ground when the optical axis is vertical. The displacement d, in., is the distance on a photograph from the image of a ground point to its fictitious image projected on a datum plane (Fig. 12-12). Then

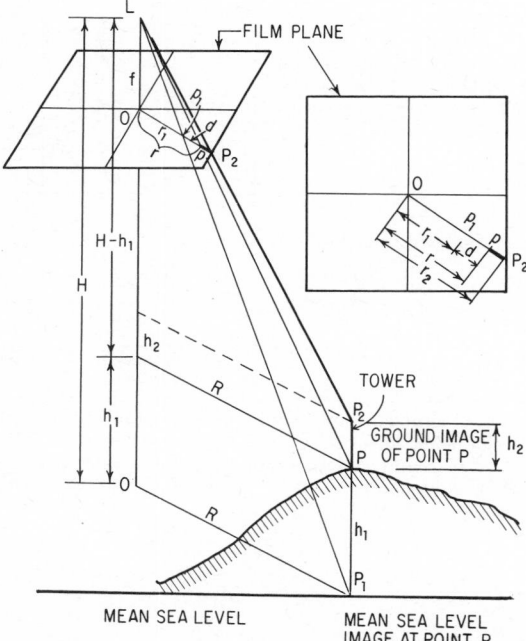

Fig. 12-12. Elevation differences cause topographic-relief displacements. (*From Brinker, "Elementary Surveying," Intext Educational Publishers, New York, N.Y.*)

$$d = r - r_1 \qquad r = \frac{Rf}{H - h_1} \qquad r_1 = \frac{Rf}{H} \tag{12-15}$$

Substituting for r and r_1 in the first equation yields

$$d = \frac{Rf}{H - h_1} - \frac{Rf}{H} = \frac{Rfh_1}{H(H - h_1)} = \frac{rh_1}{H} = \frac{r_1 h_1}{H - h_1} \tag{12-16}$$

where r = radial distance on photograph from principal point to ground image of a point P, in. (or mm)

r_1 = radial distance on photograph from principal point to P_1, the fictitious image position of point P projected to the datum, in. (or mm)

h_1 = height of point P above datum, ft

H = height of airplane above datum, ft

As an example, find the height of a tower on an aerial photograph where the flight altitude above mean sea level is 5,000 ft, ground elevation is 1,000 ft, and measurements give $r_2 = 8.65$ mm and $r = 8.52$ mm (Fig. 12-12).

$$d' = r_2 - r = 0.13 \text{ mm} \qquad h_2 = \frac{d'(H - h_1)}{r_2} = \frac{0.13(5,000 - 1,000)}{8.65} = 60 \text{ ft}$$

Stereoscopic vision is that particular application of binocular vision (simultaneous vision with both eyes) that enables an observer to view two different perspective photographs of an object (such as two photographs taken from different camera stations) and get the mental impression of three dimensions. Thus, a stereoscope permits each eye to see as one a pair of photographs that shows an area from different exposure points and thereby produces a three-dimensional (stereoscopic) image (model).

Parallax is the apparent displacement of the position of a body with respect to a reference point or system caused by a shift in the point of observation. As a result of the forward movement of a camera in flight, positions of all images travel across the focal plane from one exposure to the next, with images of higher elevations moving farther than those at lower levels.

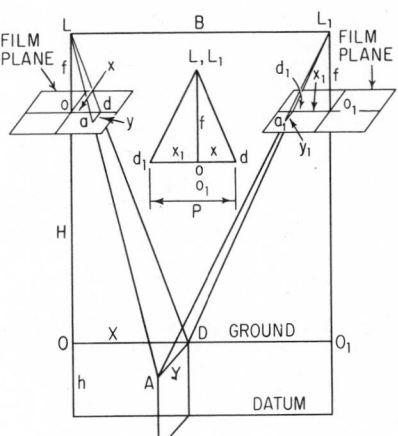

Fig. 12-13. Parallax shifts image of line AD on successive photographs. (*From Brinker, "Elementary Surveying," Intext Educational Publishers, New York, N.Y.*)

Absolute parallax of a point is the total movement of the image of a point in the focal plane between exposures. It is found by locating the principal points of adjacent photographs containing the images of the point (Fig. 12-13); transferring each principal point to the other photograph; connecting each principal and transferred principal point to define the flight line; drawing a line on each photograph through the principal point perpendicular to the flight line; and finally, measuring the x coordinate (parallel to flight line) of the point under study on each photograph.

Absolute parallax of a point, in. (or mm) (observing algebraic signs), is

$$p = x - x_1 \tag{12-17}$$

Also,

$$X = \frac{xB}{p} \qquad Y = \frac{yB}{p} \qquad H - h = \frac{fB}{p} \qquad (12\text{-}18)$$

For untilted photographs,

$$Y = \frac{y_1 B}{p}$$

where X, Y = ground coordinates measured from the plumb point, ft
$\qquad B$ = air base = distance between exposure stations, ft
$\qquad x$, y = photograph coordinates, in. (or mm)
$\qquad H$ = altitude of airplane above datum, ft
$\qquad h$ = elevation of object above datum, ft
$\qquad f$ = focal length of lens, in. (or mm)

Measuring stereoscopes, such as the stereocomparator and contour finder, are satisfactory for small areas. The multiplex, Kelsh plotter, Wild Autograph, and other large plotters usually are preferred for extensive projects. The latter instruments measure differences in parallax by means of a floating dot—actually two dots superimposed on the photographs and mentally fused by the operator to produce the floating dot. The operator places it at apparent ground level in the photograph for contouring or finding spot elevations.

The accuracy of photogrammetric contouring depends on camera precision, type of terrain and ground cover, type of stereoscopic plotter, and experience of the operator.

$$C \text{ factor} = \frac{\text{flying height}}{\text{contour interval}} \qquad (12\text{-}19)$$

is an empirical ratio used to express the efficiency of stereoscopic plotters. Photogrammetrists get C factors of 500 to 1,500, and elevations to the nearest foot and half foot with present equipment.

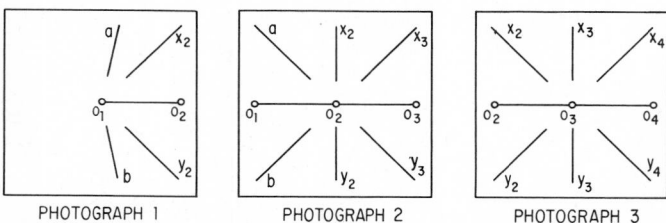

PHOTOGRAPH 1   PHOTOGRAPH 2   PHOTOGRAPH 3

Fig. 12-14. Radial-line plotting of horizontal control points on aerial photographs.

Radial-line plotting is a graphical method of extending horizontal control between fixed ground points on aerial photographs. In Fig. 12-14, points o_1, o_2, and o_3, the principal points in photographs 1, 2, and 3, are located on adjacent photographs. Control points a and b are identified in photograph 1. Additional control points called pass points (x_2 and y_2 in photograph 1, a and b in photograph 2, x_3 and y_3 in photograph 3) are established and transferred to the other photographs. On a sheet of tracing paper or template placed over each photograph, a set of rays is drawn from the principal point through each conjugate principal point, control point, and pass point. The templates are superimposed, as shown in Fig. 12-15, until all rays to each point, such as a or b, provide a single intersection. The map positions of the points then have been fixed.

The method is based upon two fundamental photogrammetric principles:

On truly vertical photographs, image displacements caused by topographic relief radiate from the principal point.

Angles between rays passing through the principal point are equal to the horizontal angles formed by the corresponding lines on the ground.

With positioning of the photographs fixed by the control and pass points, planimetric details can be transferred to a map sheet with an opaque projector or sketchmaster. Either of these instruments can change the photographic scale to fit a map scale, if desired.

12-16. Electronic Surveying Devices and Other Equipment. Article 12-6 notes the basic capabilities of three early electronic distance-measuring instruments—Tellurometer, Electro-

Table 12-15. New Measuring and Alignment Equipment

Instrument	Distance range	Accuracy	Remarks
Tellurometer MA-100	6,600 ft (2 km)	± 0.0065 ft (2 mm) ± 1 ppm (5 mm) at 6,600 ft	Beam width ¼°
Tellurometer MRA 101	300 ft–35 miles (100 m –50 km)	±1.5 cm + 3 ppm	Beam width 6°
Tellurometer CA 1000	150–100,000 ft	±0.05 ft (15 mm) + 5 ppm	Beam width 6°
Electrotape DM 20	100 ft to 30 miles	±1 cm + distance/300,000	
Autotape DM-40	63 miles (100 km)	±50 cm + range/100,000	Display rate 1 per sec
Cubitape DM-60	3–6,800 ft (1–2,000 m)	±0.01 ft (5 mm) + distance/100,000	Measurement in under 15 sec
AGM Laser-Geodimeter 6BL	15 m to 20–25 km	±5 mm + 1 mm/km	Measurement in 3 min
AGM Laser-Geodimeter 8	50 ft–40 miles	±0.016 ft + distance/1,000,000	
AGA Geodimeter 700	0.32 ft–3 miles	±0.016 ft + 1 ppm	Electronic theodolite included
AGA Geodimeter 76	Few feet–2 miles	±0.03 ft + 0.005 ft per mile	Measurement normally 30 sec
Microranger EDM	3 ft to 1 mile	±0.02 ft + 2 ppm	Mounts on theodolite, levels
Wild D-10 Distomat	1–2,000 m	±1 cm	Measurement in 15 sec, mounts on some theodolites
HP 3800A Distance Meter	10,000 ft	±0.01 ft + distance/100,000	Measurement in under 2 min
Akuranger Mark 1	0–4,000 ft	Average error ±0.02 ft	Continuous measurements 1-sec intervals
Laser Ranger	1–6 km (4 miles)	±5 mm + 2 ppm	1 mm least count
Kern DM 1000	Few feet–2 miles	±4 mm	Measurement time 15 sec
Kern MDM 500	1,500 ft	±5 mm + 2 ppm	1 mm least count. Fits on theodolites
Spectra Physics Transit Theodolite Laser	1,000 ft–5 miles	Wide expan. ±25 ft/100 ft	For layout, tunneling, guiding dozer blades, etc.
Rotolite Building Laser		Beam ±⅛ in. in 100 ft	General & building construction
AGL Construction Laser	15–850 ft; 30–1,750 ft		Pipe and tunnel control, etc.
Laser Tracking Level 11A	600 ft	Beam dia ±⅝ in. at 600 ft	Control, check elevations over entire job from one setup

tape, and Geodimeter. Since their introduction, improvements have been made in their efficiency, ease of operation, and range. But in the meantime, many other electronic surveying instruments have become available. Table 12-15 lists some of them, along with their ranges and stated accuracies. Various models are suitable for staking and layout work. Displayed distances to reflectors change, in seconds, as a "rod" is moved on line. Correction factors can be dialed in to eliminate calculations.

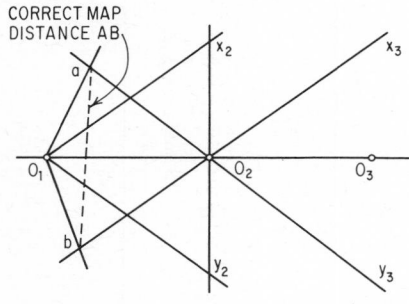

Fig. 12-15. Correct map scale and location of points are obtained with the aid of the radial lines of Fig. 12-14.

Also available are theodolites (1- and 5-sec) for low- to medium-order triangulation. They are less bulky and considerably lighter than older types.

For surveying computations, a programmable calculator is fast and accurate and within the budget of even small surveying offices. For field work, a pocket-size electronic computer is invaluable.

12-17. Trilateration versus Triangulation. Electronic distance-measuring (EDM) devices make precise traverses practical in certain types of terrain and enable trilateration to compete with triangulation (Art. 12-11). Selection of the best method for establishing a control network depends on estimated comparative costs, accuracy required, speed, density of control points, terrain effect on station locations, and convenience (probable weather conditions and day versus night observations), as well as other special factors, such as proper equipment on hand in a particular office. Thus, a traverse, the method most familiar to many surveyors, made with short-range EDM equipment might be preferable for limited-area work. For terrain offering good visibility between appropriate stations from observation towers, the choice of triangulation or trilateration can depend on the need for more substantial towers.

Various field investigations have provided data for comparisons of results of triangulation and trilateration for establishing control. Two of these studies are discussed briefly below. (K. F. Burke, "Why Compare Triangulation and Trilateration?" American Congress on Surveying and Mapping, Annual Meeting, March 1971; P. R. Wolf and S. D. Johnson, "Trilateration with Short-range EDM Equipment and Comparison with Triangulation." American Congress on Surveying and Mapping, Florida Meeting, 1973.)

One study utilized three networks of intermediate-length lines and indicated that for the more complicated geometric figures and network design needed to get sufficient redundancy in pure trilateration, time and effort would be about the same for both methods. Additional measured distances beyond the normal triangulation base lines significantly improved network accuracy. Thus, a combination of observed directions and EDM distances may be better than either individual method.

In the second investigation, for a small-scale network, distance measurements took considerably less time than triangulation observations, and adjustment yielded somewhat higher precision than pure triangulation adjustments with equal degrees of freedom.

High-speed electronic computers facilitated calculations of coordinates and precision based on station error ellipses for either method. Hence, the factors listed previously will govern the choice of method for a specific project.

Earthwork

CHARLES H. SAIN

Consulting Engineer,
Birmingham, Ala.

Earthwork involves movement of a portion of the earth's surface from one location to another and, in its new position, creation of a desired shape and physical condition. Occasionally, the material moved is disposed of as spoil. Because of the wide variety of soils encountered and jobs to be done on them, a wide variety of equipment and methods have been developed for the purpose. This section describes and analyzes them.

13-1. Types of Excavation. A common method of classifying excavation is to associate it with the type of excavated material: topsoil, earth, rock, muck, and unclassified.

Topsoil excavation is removal of the exposed layer of the earth's surface. This task includes stripping of all vegetation. Since the topsoil, or mantle soil, supports growth of trees and other vegetation, this layer contains more moisture than that underneath. So that the lower layer will lose moisture and become easier to handle, it is advantageous to remove the topsoil as soon as possible. When removed, topsoil usually is stockpiled. Later, it is restored on the site for landscaping or to support growth of vegetation to control erosion.

Earth excavation is removal of the layer of soil immediately under the topsoil and on top of rock. Used to construct embankments and foundations, earth usually is easy to move with scrapers or other types of earthmoving equipment.

Rock excavation is removal of a formation that cannot be excavated without systematic drilling and blasting. Any boulder larger than ½ cu yd generally is classified as rock. In contrast, earth is a formation that when plowed and ripped breaks down into small enough pieces to be easily moved, loaded in hauling units, and readily incorporated into an embankment or foundation in relatively thin layers. Rock, when deposited in an embankment, is placed in deep layers, usually exceeding 18 in.

Muck excavation is removal of material that contains an excessive amount of water and undesirable soil. Its consistency is determined by the percentage of water contained. Because of lack of stability under load, muck seldom can be used in an embankment. Removal of water can be accomplished by spreading muck over a large area and letting it dry, or by changing soil characteristics, or by stabilizing it with some other material, thereby reducing the water content.

Unclassified excavation is removal of any combination of topsoil, earth, rock, and muck. Contracting agencies frequently use this classification. It means that earthmoving must be done without regard to the materials encountered.   Much excavation is performed on an unclassified basis because of the difficulty of distinguishing, legally or practically, between earth, muck, and rock.   Unclassified excavation must be carried out to the lines and grades shown on the plans without regard to percentage of moisture and type of material found between the surface and final depth.

Excavation also may be classified in accordance with the purpose of the work, such as stripping, roadway, drainage, bridge, channel, footing, borrow.   In this case, contracting agencies indicate the nature of the excavation for which materials are to be removed.   Designations differ with agencies and locality.   Often, the only reason a certain type of excavation has a particular designation is local custom.

Stripping usually includes removal of all material between the original surface and the top of any material that is acceptable for permanent embankment.

Roadway excavation is that portion of a highway cut that begins where stripping was completed and terminates at the line of finished subgrade or bottom of base course.   Often, however, stripping is made part of roadway excavation.

Drainage excavation or structure excavation is removal of material encountered during installation of drainage structures other than bridges.   Those structures are sometimes referred to as minor drainage structures and include roadway pipe and culverts.   A culvert is usually defined as any structure under a roadway with a clear span less than 20 ft, whereas a bridge is a structure spanning more than 20 ft.   After a pipe or culvert has been installed, backfilling must be done with acceptable material.   This material usually is obtained from some source other than drainage excavation, which generally is not acceptable or workable.   Often, culvert excavation does not include material beyond a specified distance from the end of a culvert.

Bridge excavation is removal of material encountered in digging for footing and abutments. Often, bridge excavation is subdivided into wet, dry, and rock excavation.   The dividing line between wet and dry excavation usually is denoted by specification of a ground elevation, above which material is classified as dry and below which, as wet.   A different elevation may be specified for each foundation.

Channel excavation is relocation of a creek or stream, usually because it flows through a right of way.   A contracting agency will pay for any inlet or outlet ditch needed to route water through a pipe as channel excavation, to the line where culvert excavation starts.

Footing excavation is the digging of a column or wall foundation for a building.   This work usually is done to as neat a line and grade as possible, so that concrete may be cast without forms. While elimination of forms saves money, special equipment and more-than-normal handwork are usually required for this type of excavation.

Borrow excavation is the work done in obtaining material for embankments or fills from a source other than required excavation.   In most instances, obtaining material behind slope lines is classified as borrow, though it commonly is considered as getting material from a source off the site.   Most specifications prohibit borrow until all required excavation has been completed or the need for borrow has been established beyond a reasonable doubt.   In some cases, need for a material not available in required excavation makes borrow necessary.   A borrow pit usually has to be cleared of timber and debris, then stripped of topsoil, before desired material can be excavated.

13-2. Basic Excavating Equipment.   A tractor is the most widely used excavating tool.   Essentially, it is a power source on wheels or tracks (crawler).   Equipped on the front with a **bulldozer,** a steel blade that can be raised and lowered, a tractor can push earth from place to place and shape the ground.   If a scraper is hooked to the drawbar and means of raising, lowering, and dumping are provided, a tractor-drawn scraper results.   Addition of other attachments creates tools suitable for different applications (see also Art. 13-7).

Another basic machine is one that by attachment of different fronts may be converted into a shovel, dragline, clamshell, backhoe, crane, or pile driver.   The basic machine made for a shovel, however, has shorter and narrower tracks than one made for a dragline or clamshell, and more counterweight has to be added to the back.   While a shovel attachment will fit the basic machine made for a dragline or clamshell, the longer tracks will interfere with the shovel (see also Art. 13-4).

Scrapers may be tractor-drawn or self-propelled.   More excavation is moved with self-propelled or rubber-tired scrapers than with scrapers towed and controlled by crawler tractors (see also Art. 13-8).

Trenchers, used for opening trenches and ditches, may be ladder or wheel type. They do most of the pipeline excavation in earth. The ladder type has chains to which are attached buckets that scoop up earth as the chains move. It is adaptable to deep excavation. The wheel type has digging buckets on the circumference of a rotating wheel. The buckets dump excavated material into a conveyor mounted in the center of the wheel. This type of trencher is used mainly for shallow trenches. Neither type is used to any great extent when rock is encountered in trench excavation.

Wheel excavators, used in constructing earth dams or in strip mining, excavate soft or granular materials at very high rates. For example, one excavator with a 28-ft wheel moves 1,500 tons of iron ore per hour. A typical wheel excavator resembles a wheel-type trencher. Buckets mounted on a wheel 12 or more ft in diameter scoop up the earth. They may be 2 or more ft wide, with a capacity of ⅓ cu yd or more, and equipped with a straight cutting edge or teeth. The buckets dump into a hopper, which feeds the earth onto a conveyor belt. The belt moves along a boom, which may be 200 ft or more long, to dump the earth into another hopper. This hopper, in turn, feeds the earth to a stockpile or to earthmoving equipment.

13-3. Selecting Base Equipment. Type of material to be excavated may determine the basic equipment to be used. But length and type of haul road must also be considered. For example, suppose excavation is in earth and best results could be obtained with rubber-tired scrapers, but the haul is over city streets. In this case, this type of equipment probably could not be used because of heavy wheel loads and interference with traffic.

For rock, a front-end loader or a shovel would be the basic rig. For earth, when a haul road can be built, scrapers would be chosen. But if the earth has to be moved several miles over existing streets or highways, the choice would be a front-end loader, shovel, or dragline that would load dump trucks. Whether a shovel or dragline would be used depends on whether the excavation bottom can support a front-end loader or shovel and hauling units. If the bottom is too soft, a dragline would be required. A dragline can sit outside the excavation and load a hauling unit at the same level (loading on top). But when a shovel can be used, it is preferred to a dragline because of greater production.

Therefore, in selecting basic equipment, consider:

Types of material to be excavated.
Types and size of hauling equipment to be used.
Load-supporting ability of original ground.
Load-supporting ability of material to be excavated.
Volume of excavation to be moved.
Volume to be moved per unit of time.
Length of haul.
Type of haul road.

13-4. General Equipment for Excavation and Compaction.

Clearing or Grubbing. Use *tractor* with bulldozer or root rake. *Bulldozer* can fell trees, uproot stumps. *Root rake* piles for burning, makes cleaner pile. For light brush, *brush hog* may be required.

Grubbing. Use low-strength *explosives*, slow detonation speed.

Clearing. Drag *chain* or *chain and heavy ball* between two tractors. Useful for trees that break easily. Tractors equipped with cutter blades can operate on any footing and cut any tree at ground level.

Stripping. *Bulldozers* are limited by length of push or haul but are useful for swampy conditions. *Scrapers* are limited by terrain and support ability of ground; they may be tractor-drawn for short hauls. *Draglines* are limited by depth of stripping, ability to service with hauling units, and space for casting the bucket. They are used where swampy conditions prevent other equipment from being used. *Graders* are limited to use where stripping can be windrowed on final position. Material can be loaded from a windrow by a front-end loader.

Pipe Installation. *Backhoes* are used on firm soil where depth of trench is not excessive; good in rock. *Draglines* are used for deep trenches if the sides can be flattened; they have difficulty digging vertical walls. *Clamshells* are used where sheeting of sides is required and it is necessary to dig between braces and to great depths. They are inefficient in rock. *Bulldozers* are limited to shallow excavation. *Trenching machines* produce vertical or near-vertical walls, and can maintain line and earth grade.

Earth Excavation. *Tractor-drawn scrapers* are limited by length of haul and supporting ability of the soil. Cost gets excessive if haul distance greatly exceeds 1,000 ft. Two-axle, rubber-tired, *self-propelled scrapers* are limited by length of haul, terrain, and supporting ability of the soil; they bounce on long hauls at top speed.   Three-axle, rubber-tired, self-propelled scrapers need maneuvering or working space and are limited by terrain and supporting ability of soil. They are most efficient on long hauls.   Twin-engine, rubber-tired scrapers have few limitations. They are useful in rough terrain and where traction is needed on all wheels.   *Front-end loaders* generally discharge into hauling units if the haul greatly exceeds 100 ft, and they also are limited by digging and dumping ease of excavated material.   *Shovels* are also used to load into hauling units.   Working room must be ample and distance to cast short.   Shovels also have to dig from a face.   *Draglines* may be used where excavation is deep and the material has no supporting ability.   Material should be easy to dig.   Draglines usually load into hauling units.   *Wheel excavators* offer high excavation rate and loading into hauling units with soft or granular soils. *Mobile belt loaders* (Fig. 13-1) give high-production loading into hauling units but are limited

Fig. 13-1. Mobile belt loader.   (*Courtesy of Barber-Greene Co.*)

by working room and supporting capacity of excavation bottom.   Belt loaders are limited to short, infrequent moves.   A wide belt handles some rock excavation.   *Dredges* usually are used where transportation and digging costs are prohibitive if other than water-borne equipment is used.   Water must be available for mixing with the excavated material for pumping through pipes.   Distance to spoil area should not be too great.   *Clamshells* are low producers but are useful in small or deep spaces, where there is no overhead interference with swinging of the boom.   *Gradall*, not a high-production tool, is suitable for dressing or finishing where tolerances are close.   *Scoopers*, hydraulically operated, are high-production equipment, limited by dumping height and to easily dug material.   Production is not so greatly influenced by height of face as for a shovel.

Rock Excavation.   *Shovels* can dig any type of rock broken into pieces that can be easily dug. Limited to digging from a face, shovels are used for high-production loading into hauling units. *Bulldozers* are limited to short movements and easily dug rock.   Sometimes, they are used to dispose of boulders when drilling and blasting are not economical.   *Front-end loaders* are used instead of shovels because of their high production, lower cost of operation, and ease of moving from job to job.   *Backhoes* are used for foundation excavation, trenches, and high production in rough terrain.   They must dig below their tracks.   *Scrapers* are suitable for short movements and rock broken down to small sizes, such as blasted shale.   But tire wear may be greater than in other applications.   *Scoopers* may be used instead of shovels where working space is tight. They are limited by the height of hauling units and to easily dug rock.   *Gradalls* are used for trench and foundation excavation, but material must be well blasted.   *Clamshells* are most suitable for deep foundations or where the reach from machine position to excavation prohibits other equipment from being used.   Rock must be well broken for maximum production.

Compaction.   *Sheepsfoot rollers*, made with feet of various shapes, offer high-speed production.   Compaction depends on unit pressure and speed of roller.   They are not suitable for compacting sand.   They also are limited by depth of layer to be compacted.   *Rubber-tired rollers* are used for granular soils, including shales and rock.   Ranging from very light weight to 200 tons, they may be self-propelled or towed.   Depth of lift compacted depends on weight.   *Vibratory compactors*, towed, self-propelled, or hand-held, are also used for granular soils.   Compaction ability depends on frequency and energy of vibrations.   Depth of lift is not so much a factor as for other types of compactors.   *Grid rollers,* useful in breaking down oversize particles, are limited to shallow lifts of nonsticky material.   They can be towed at any safe, economical speed. *Air tamps* are used to backfill pipe and foundations and for work in areas not accessible to power equipment.   Usually hand-held, they are powered by compressed air imparting reciprocating blows.   They are limited to low production and shallow lifts.   *Paddlefoot rollers,* usually self-propelled, compact from the top of lift down.   They are limited to an average depth (up to 8 in.) of lift in all soils.   A *rubber-tired front-end loader* can be converted to this type of roller by a change of wheels.   *Steel-wheel rollers,* self-propelled, are used where a smooth, sealed surface is desired.   They are limited to shallow depth of lift.

13-5. Power Shovels, Draglines, Clamshells, and Backhoes.   These four machines are made by installing an attachment on a basic machine, which may be mounted on crawler tracks or a trucklike chassis (Art. 13-2).   (See Figs. 13-2 to 13-5.)   When mounted on a trucklike chassis, the machine usually is designed for use as a truck crane, but it also can be used as a shovel or backhoe if mobility is desired and low production is acceptable.   Most backhoes, however, are hydraulic and cannot be converted.

There is not much difference between equipment used as a clamshell and that used as a dragline or crane.   A boom used with a clamshell has two-point sheaves, so that two cables can attach

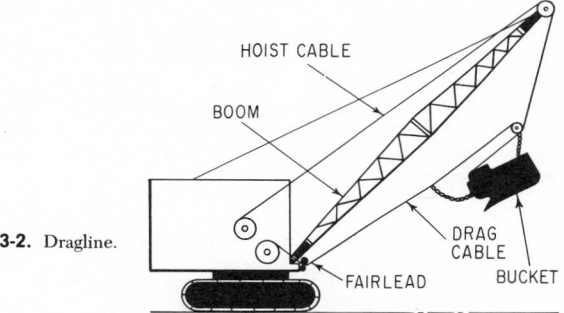

Fig. 13-2. Dragline.

Fig. 13-3. Hydraulic hoe.

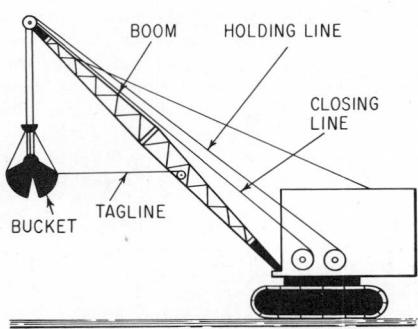

Fig. 13-4. Clamshell.

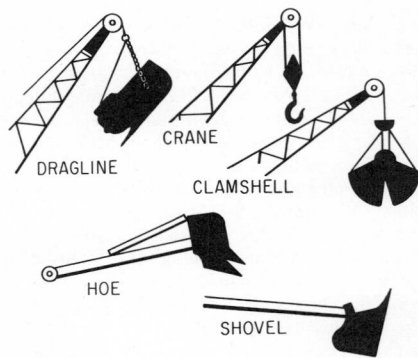

Fig. 13-5. Excavating and crane attachments.

to the bucket.   One cable is used to open and close the bucket and the other to hoist or lift the bucket.   Since the two cables should travel at the same speed, the drums on a clamshell are the same size.   To keep the bucket from spinning and twisting the hoist and closing lines, a tagline extends between the bucket and a spring-loaded reel on the side of the boom (Fig. 13-4).

A dragline has a hoist cable that goes through a point sheave atop the boom and attaches to the bucket.   Another line, the drag cable, goes through the fairlead and attaches to the bucket (Fig. 13-2).   The drum that exerts pull on the drag cable is smaller than the hoist drum because more force is required on the drag cable than on the hoist lines.   Typical performance factors for a dragline are given in Tables 13-1 and 13-6.

Table 13-1. Typical Dragline Calculating Factors
Average Swing Cycle with 110° Swing

Bucket capacity, cu yd	½	1½	2
Time, sec	24	30	33

BUCKET FACTORS

Type of digging	% of rated capacity (approx)
Easy	95–100
Medium	80–90
Medium hard	65–75
Hard	40–65

Table 13-2. Typical Shovel Calculating Factors
Average Swing Cycle with 90° Swing

Bucket capacity, cu yd	½	1	1½	2	2½
Time, sec	20	21	22	23	24

DIPPER FACTORS

Type of digging	% of rated capacity (approx)
Easy	95–100
Medium	85–90
Medium hard	70–80
Hard	50–70

Power shovels are used primarily to load rock into hauling units.   Production depends on type of material to be loaded, overall job efficiency, angle of swing, height of bank or face the shovel digs against, ability of operator, swell of material, slope of ground machine is working on, and whether hauling units are of optimum size and adequate in number.   For highest efficiency, the degree of swing should be held to a minimum.   (Typical performance factors are given in Table 13-2.)   Working the shovel so that a hauling unit can be loaded on each side is desirable so that there is no lost time waiting for a hauling unit to get into position.

Table 13-3 gives estimated hourly production of power shovels.   It is based on bank cubic yards measure, 90° swing, optimum digging depth, grade-level loading, 100% efficiency, 60-min hour, and bucket-fill factor of 1.00 (See Table 13-5).   Table 13-4 indicates the effect on production of depth of cut and angle of swing.

Table 13-3. Estimated Hourly Production of Dipper-type Power Shovel*

Material class	Shovel dipper sizes, cu yd															
	½	¾	1	1¼	1½	2	2½	3	4	4½	5	6	7	8	9	10
Moist loam or sandy clay	115	165	205	250	285	355	405	454	580	635	685	795	895	990	1,075	1,160
Sand and gravel	110	155	200	230	270	330	390	450	555	600	645	740	835	925	1,010	1,100
Common earth	95	135	175	210	240	300	350	405	510	560	605	685	765	845	935	1,025
Clay, tough hard	75	110	145	180	210	265	310	360	450	490	530	605	680	750	840	930
Rock, well blasted	60	95	125	155	180	230	275	320	410	455	500	575	650	720	785	860
Common with rock	50	80	105	130	155	200	245	290	380	420	460	540	615	685	750	820
Clay, wet and sticky	40	70	95	120	145	185	230	270	345	385	420	490	555	620	680	750
Rock, poorly blasted	25	50	75	95	115	160	195	235	305	340	375	440	505	570	630	695

*Caterpillar Tractor Co.

Table 13-4. Correction Factors for Effect of Depth of Cut and Angle of Swing on Power-Shovel Output*

Depth of cut, % of optimum	Angle of swing, deg						
	45	60	75	90	120	150	180
40	0.93	0.89	0.85	0.80	0.72	0.65	0.59
60	1.10	1.03	0.96	0.91	0.81	0.73	0.66
80	1.22	1.12	1.04	0.98	0.86	0.77	0.69
100	1.26	1.16	1.07	1.00	0.88	0.79	0.71
120	1.20	1.11	1.03	0.97	0.86	0.77	0.70
140	1.12	1.04	0.97	0.91	0.81	0.73	0.66
160	1.03	0.96	0.90	0.85	0.75	0.67	0.62

*"Earthmoving Data," Caterpillar Tractor Co.

Table 13-5. Bucket-Fill Factor*

Material	Fill-factor range
Sand and gravel	0.90–1.00
Common earth	0.80–0.90
Hard clay	0.65–0.75
Wet clay	0.50–0.60
Rock, well blasted	0.60–0.75
Rock, poorly blasted	0.40–0.50

*"Earthmoving Data," Caterpillar Tractor Co.

Optimum digging depth is the shortest distance a bucket must travel up a face or bank to obtain its load.   This depth usually is the vertical distance from shipper shaft (dipper-stick pivot shaft) to ground level.   Optimum depth varies with type of material to be excavated, since a lower boom is needed for hard materials than for soft.

Work must be planned to load or move the maximum yardage each shift: Locate the shovel and hauling units for the shortest swing of the shovel. If it is necessary to work high, dig the upper portion first. Move up to the face while a hauling unit is getting into position. Make short moves frequently, instead of less-frequent long moves. Stay close to the face; do not dig at the end of the stick. Lower the dipper only enough to get a full bucket; this cuts down on hoist time. Keep dipper teeth sharp. Have spare cables and dipper teeth readily available near the shovel. Hoist the load no more than necessary to clear the hauling-unit bed. Start the swing when the bucket is full and clear of the bank. Spot the hauling unit under the boom point so that it is not necessary to crowd or retract to dump into the bed (Fig. 13-6). Break rock well for easier digging.

Fig. 13-6. Shovel loads off-highway dump truck. (*Courtesy of Bucyrus Erie.*)

A dragline is more versatile than a shovel. With a dragline, load can be obtained from a greater distance from the machine (reach is greater). Excavation can be done below water and at a long distance above or below the dragline. A larger bucket than the machine's rated capacity can be used if a short boom is installed. It is not uncommon for a machine rated at 2½ cu yd to be loading with a 4-cu yd bucket into hauling units. But weight of bucket and load should not exceed 70% of the tipping load of the machine. (Lifting-crane capacity is based on 75% of actual tipping load. A dragline may approach this if it is on solid footing and is digging good-handling material.)

Since a dragline loads its bucket by pulling it toward the machine, the pit or face slopes from bottom to top toward the dragline. Best production is obtained by removing material in nearly horizontal layers and working from side to side of the excavation. A keyway, or slot, should be cut next to the slope. This keyway should always be slightly lower than the area being taken off in horizontal layers. A good operator fills the bucket as soon as possible, within a distance less than the bucket length. Digging on a slight incline helps fill the bucket. When the bucket is full, it should be nearly under the boom point and should be lifted as drag ceases.

As with shovels, a relatively shallow pit yields the greatest efficiency for draglines. The hauling units should be in the excavation or at the same elevation to which the dragline is digging. So when the bucket is full, it will have a short lift to reach the top of the hauling units. If the pit bottom is soft or for some other reason hauling units cannot be spotted below the machine, then loading on top must be resorted to, with a loss in loading efficiency. Table 13-6 indicates dragline production in cubic yards bank measure per hour. The table is based on suitable depth of cut for maximum effect, no delays, 90° swing, and all materials loaded into hauling units (see also Table 13-1).

Production of clamshells, like that of draglines, depends on radius of operation and lifting capacity. It is general practice to limit the clamshell load, including bucket weight, to 50% of the full power-line pull at the short boom radius.

Types of clamshell bucket are general-purpose, rehandling, and heavy excavating. The rehandling bucket is best for unloading materials from bins or railroad cars or loading materials from stockpiles.   The heavy-excavation bucket is used for extreme service, such as placing rip-rap.   It can be adjusted so that the operation is easy on components, since a clamshell does not demand a tightly adjusted friction band.   The general-purpose bucket is between the rehandling and excavating buckets and can be used with or without teeth.

Table 13-6. Hourly Dragline Handling Capacity, Cu Yd

Class of material	Bucket capacity, cu yd								
	⅜	½	¾	1	1¼	1½	1¾	2	2½
Moist loam or sandy clay	70	95	130	160	195	220	245	265	305
Sand and gravel	65	90	125	155	185	210	235	255	295
Good, common earth	55	75	105	135	165	190	210	230	265
Clay, hard, tough	35	55	90	110	135	160	180	195	230
Clay, wet, sticky	20	30	55	75	95	110	130	145	175

13-6. Tractor Shovels.   Also commonly known as front-end loaders, tractor shovels can be mounted on rubber tires (Fig. 13-7) or crawler tracks (Fig. 13-8).   A crawler is desirable if moving

Fig. 13-7. Rubber-tired tractor shovel and dump truck. (*Courtesy of International Harvester.*)

Fig. 13-8. Crawler-type tractor shovel. (*Courtesy of International Harvester.*)

it from one job to another is no problem, haul distance is short, and type of excavation bottom is not suitable for rubber tires.   Most rubber-tired loaders have four-wheel drive.

Capacity of a tractor shovel depends on unit weight of material to be handled.   So there is a variety of buckets for each loader.   These are of three basic types: hydraulically controlled, gravity dump, and overhead (overshot).   Hydraulically controlled machines are preferable for most operations.   The overhead is desirable where working room for turning is unavailable.

All loaders except overhead use a load, turn, dump cycle.   For best efficiency and reduction of wear on tires or undercarriage, turning should be held to a minimum.

A loader should dig from a relatively low height of bank or face.   Since most loaders are equipped with automatic bucket positions, the height of bank should be adjusted so that it is not higher than necessary to fill the bucket; this is about the same height as the push-arm hinges.

On an average construction job, a front-end loader is a versatile tool.   Attachments are available so that it can be used as a bulldozer, rake, clamshell, log loader, crane, or loader.

13-7. Tractors and Tractor Accessories.   Tractors are the prime movers on any construction job where earth or rock must be moved.   They may be mounted on rubber tires or crawler tracks.   Properly equipped, a tractor usually is the first item moved onto a job and one of the last to finish.

Crawlers are more widely used than rubber-tired tractors.   Crawlers will work on steep, rugged terrain; soft, marshy conditions; and solid rock.   Rubber-tired tractors are suitable for specific projects or uses, such as excavation of earth or sand where track wear would be excessive. Tires and track system are the most expensive parts to maintain.

Basic components of a crawler tractor include engine, radiator, transmission, clutch, steering clutches, final drives, and undercarriage, consisting of tracks, rollers, sprockets, and idlers. Components of a rubber-tired tractor include engine, radiator, transmission, clutch, tires and rear end.   A rubber-tired tractor may have two- or four-wheel drive.   Its travel speed may range from a minimum of 3 mph to a maximum of over 40 mph.   Travel speed of a crawler may range from less than 1 mph to not much more than 8 mph.

A crawler tractor can be equipped with accessories that enable it to perform a wide variety of tasks:

Rear double-drum cable control unit.   Used for pulling a scraper; or cable control for a bulldozer by using only one drum.

Bulldozer, either cable-controlled by rear or front unit, or hydraulically controlled (Fig. 13-9). Several different types of blade are available, such as angle, straight, U, root rake, rock rake, stump dozer, tree dozer, push dozer.

Ripper.   Rear-mounted and hydraulically controlled to provide pressure up or down (Fig. 13-10).

Side boom, a short boom mounted on one side with a counterweight on the opposite side of the tractor.   Cable-operated.   The main use is laying cross-country pipelines (Fig. 13-11).

Tractor crane, a boom with limited swinging radius.

Fig. 13-9. Tractor with bulldozer attachment.   (*Courtesy of International Harvester.*)

Fig. 13-10. Tractor (bulldozer) with ripper. (*Courtesy of International Harvester.*)

Fig. 13-11. Tractor with side boom for pipe laying. (*Courtesy of International Harvester.*)

Pusher block or blade. Used for pushing scrapers, to assist and speed up loading (Fig. 13-12). A pusher block may be attached rigidly to the frame, mounted on an angle bulldozer C frame, or mounted in the center of a bulldozer. Or it can be mounted as a short bulldozer. Though designed especially as a tool for pushing, a pusher block can be used in a limited manner as a bulldozer. One way to absorb the shock when a pusher makes contact with a scraper is with springs. Another is to use a hydraulic accumulator, which eliminates the need for stopping.

Welder. Mounted on the tractor for mobility, welding machines are powered by the tractor engine.

Drills. Often, a tractor serves as the prime mover for a rotary drill. During the drilling, the tractor engine powers the drill-steel rotation, hydraulic pumps, and air compressors. A percussion-type drill and air compressor also can be mounted on a tractor for mobility. Instead of a separate compressor, a piston-type compressor, powered by the tractor engine, can be mounted on the front or rear. Except for very large tractors, horsepower available is sufficient to furnish the required air for only one drill at a time.

13-8. Scrapers. Commonly used for earthmoving, a scraper may be self-propelled or powered by a crawler tractor. A self-propelled scraper may have two or three axles and may be single- or twin-engine. The single engine powers the front wheels (Fig. 13-13). With twin engines, one drives the front wheels; the second, the trailer wheels. Scrapers also can be worked in tandem, that is, two scrapers behind one power unit or tractor.

Essentially, a scraper acts like a scoop. A bowl hung from the frame tilts downward to permit its cutting edge to scrape off a thin layer of earth. As the scraper moves forward, the bowl fills.

Fig. 13-12. Tractor pushing scraper. (*Courtesy of International Harvester.*)

Fig. 13-13. Self-propelled scraper. (*Courtesy of International Harvester.*)

When it is full, it is tilted up and an apron is dropped down over the open end to close the bowl. For discharge in thin layers, the bowl tilts down and an ejector pushes the earth out.

On most scrapers, the bowl and apron are hydraulically operated, with pressure applied to the cutting edge and the apron forceably closed to retain shale, rock, or lumpy material in the bowl. Bowl and apron also may be cable-operated, but with hydraulic pressure on the cutting edge, harder material can be loaded.

Tractor-drawn scrapers are most suitable for short hauls. Maximum economic haul is about 1,000 ft. This type of scraper is useful for stripping topsoil and earthmoving in marshy conditions.

Twin-engine scrapers are suitable for steep grades and swampy conditions. Such a scraper can outperform a crawler tractor-scraper combination on adverse grades and marshy soil. For best performance, a twin-engine scraper should be equipped with the largest-size tires, to obtain the greatest flotation under difficult conditions. Although this machine can obtain a load without a pusher, the scraper can load faster with a pusher, tire wear will be less, and there will be other benefits to offset the higher cost of pushing.

Two-axle self-propelled scrapers are more maneuverable, adaptable to rougher terrain and more difficult conditions, and better for shorter haul distance than three-axle units. The latter are more efficient on long hauls because they move faster on a haul road. Two-axle units bounce at high speeds, even on smooth haul roads. A three-axle unit would be suitable for short hauls where there are no adverse return grades and there is ample maneuvering room, such as sites for airports, railroad classification yards, or industrial buildings.

On jobs with a small amount of rock, scrapers may be competitive with a shovel and rock-type trucks. Scraper tire and cutting-edge costs may exceed normal; but properly evaluated, wear costs may not be excessive. To keep costs within economic limits, cuts must be laid out so that scrapers can load without difficulty. At least half a cut, but preferably the entire cut, should be blasted for its entire length before scrapers start excavation. Better results will be obtained if some earth remains at the end of the cut or in locations where scrapers can complete a load in earth. Broken rock and shale do not "boil" or roll into the scraper; more power is required to force them into the bowl than for earth. When loading is completed in earth, however, rock is forced into the bowl. Hydraulic scrapers can forceably close the apron, thus reducing spillage. But a heaped or full load is very difficult to obtain. So the amount of material moved per trip is less with rock than with earth.

For handling by scrapers, rock has to be broken into small particles efficiently. Blasting holes have to be spaced closer, and more explosives per cubic yard are needed than for shovel-and-truck excavation. Most shales and sandstones can be blasted so that the maximum size can be easily controlled and enough fines produced to facilitate scraper loading. But igneous and metamorphic rocks, when blasted, do not readily produce a material that can be scraper-loaded. They have cleavage planes that form oversize particles and few fines. Experience, observations of rock formation, and comparisons of unit and total cost are necessary to determine whether to use scrapers or a shovel and supporting equipment.

13-9. Formulas for Earthmoving. External forces offer *rolling resistance* to the motion of rubber-tired vehicles, such as tractors and scrapers. The engine has to supply power to overcome this resistance; the greater the resistance, the more power needed to move a load. Rolling resistance depends on the weight on the wheels and the tire penetration into the ground.

$$R = R_f W + R_p p W \qquad (13\text{-}1)$$

where R = rolling resistance, lb
R_f = rolling-resistance factor, lb per ton
W = weight on wheels, tons
R_p = tire-penetration factor, lb per ton per in. penetration
p = tire penetration, in.

R_f usually is taken as 40 lb per ton (or 2% lb per lb) and R_p as 30 lb per ton per in. (1.5% lb per lb per in.). Hence, Eq. (13-1) can be written as

$$R = (2\% + 1.5\%p)W' = R'W' \qquad (13\text{-}2)$$

where W' = weight on wheels, lb
R' = 2% + 1.5%p (see Table 13-7)

Table 13-7. Typical Total Rolling Resistances of Rubber-tired Vehicles

Surface	Lb per ton	Lb per lb
Hard, smooth, stabilized, surfaced roadway without penetration under load, watered, maintained	40	0.020
Firm, smooth roadway, with earth or light surfacing, flexing slightly under load, maintained fairly regularly, watered	65	0.033
Snow:		
Packed	50	0.025
Loose	90	0.045
Earth roadway, rutted, flexing under load, little if any maintenance, no water	100	0.50
Rutted earth roadway, soft under travel, no maintenance, no stabilization	150	0.75
Loose sand or gravel	200	1.00
Soft, muddy, rutted roadway, no maintenance	300–400	1.50–2.00

Additional power is required to overcome rolling resistance on a slope. Grade resistance also is proportional to weight.

$$G = R_g sW \qquad (13\text{-}3)$$

where G = grade resistance, lb
R_g = grade-resistance factor = 20 lb per ton = 1% lb per lb
s = percent grade, positive for uphill motion, negative for downhill
Thus, the total road resistance is the algebraic sum of the rolling and grade resistances, or the total pull, lb, required:

$$T = (R' + R_g s)W' = (2\% + 1.5\% p + 1\% s)W' \qquad (13\text{-}4)$$

In addition, an allowance may have to be made for loss of power with altitude. If so, allow 3% pull loss for each 1,000 ft above 2,500 ft.

Usable pull P depends on the weight W on the drivers:

$$P = fW \qquad (13\text{-}5)$$

where f = coefficient of traction (Table 13-8). (See also Art. 13-12.)

Table 13-8. Approximate Traction Factors*

Traction surface	Traction factors	
	Rubber tires	Tracks
Concrete	0.90	0.45
Clay loam, dry	0.55	0.90
Clay loam, wet	0.45	0.70
Rutted clay loam	0.40	0.70
Loose sand	0.30	0.30
Quarry pit	0.65	0.55
Gravel road (loose not hard)	0.36	0.50
Packed snow	0.20	0.25
Ice	0.12	0.12
Firm earth	0.55	0.90
Loose earth	0.45	0.60
Coal, stockpiled	0.45	0.60

*See also Art. 13-12.

When soils are excavated, they increase in volume, or swell, because of an increase in voids (Table 13-9).

$$V_b = V_L L = \frac{100}{100 + \% \text{ swell}} V_L \qquad (13\text{-}6)$$

where V_b = original volume, cu yd, or bank yards
V_L = loaded volume, cu yd, or loose yards
L = load factor (Tables 13-9 and 13-10)
When soils are compacted, they decrease in volume.

$$V_c = V_b S \qquad (13\text{-}7)$$

where V_c = compacted volume, cu yd
S = shrinkage factor
Bank yards moved by a hauling unit equals weight of load, lb, divided by density of the material in place, lb per bank yard.

13-10. Scraper Production. Production is measured in terms of tons or bank cubic yards of material a machine excavates and discharges, under given job conditions, in 1 hr.

$$\text{Production, bank cu yd per hr} = \text{load, cu yd} \times \text{trips per hr} \qquad (13\text{-}8)$$

$$\text{Trips per hr} = \frac{\text{working min per hr}}{\text{cycle time, min}} \qquad (13\text{-}9)$$

Table 13-9. Load Factors for Earthmoving

Swell, %	Voids, %	Load factor	Swell, %	Voids, %	Load factor
5	4.8	0.952	5.3	5	0.95
10	9.1	0.909	11.1	10	0.90
15	13.0	0.870	17.6	15	0.85
20	16.7	0.833	25.0	20	0.80
25	20.0	0.800	33.3	25	0.75
30	23.1	0.769	42.9	30	0.70
35	25.9	0.741	53.8	35	0.65
40	28.6	0.714	66.7	40	0.60
45	31.0	0.690	81.8	45	0.55
50	33.3	0.667	100.0	50	0.50
55	35.5	0.645			
60	37.5	0.625			
65	39.4	0.606			
70	41.2	0.588			
75	42.9	0.571			
80	44.4	0.556			
85	45.9	0.541			
90	47.4	0.526			
95	48.7	0.513			
100	50.0	0.500			

Table 13-10. Percentage Swell and Load Factors of Materials

Material	Swell, %	Load factor
Cinders	45	0.69
Clay:		
Dry	40	0.72
Wet	40	0.72
Clay and gravel:		
Dry	40	0.72
Wet	40	0.72
Coal, anthracite	35	0.74
Coal, bituminous	35	0.74
Earth, loam:		
Dry	25	0.80
Wet	25	0.80
Gravel:		
Dry	12	0.89
Wet	12	0.89
Gypsum	74	0.57
Hardpan	50	0.67
Limestone	67	0.60
Rock, well blasted	65	0.60
Sand:		
Dry	12	0.89
Wet	12	0.89
Sandstone	54	0.65
Shale and soft rock	65	0.60
Slag, bank	23	0.81
Slate	65	0.60
Traprock	65	0.61

The load, or amount of material a machine carries, can be determined by weighing or estimating the volume. Payload estimating involves determination of the bank cubic yards being carried, whereas the excavated material expands when loaded into the machine. For determination of bank cubic yards from loose volume, the amount of swell or the load factor must be known (Tables 13-9 and 13-10). Then, the conversion can be made by use of Eq. (13-6).

Weighing is the most accurate method of determining the actual load. This is normally done by weighing one wheel or axle at a time with portable scales, adding the wheel or axle weights, and subtracting the weight empty. To reduce error, the machine should be relatively level. Enough loads should be weighed to provide a good average.

$$\text{Bank cu yd} = \frac{\text{weight of load, lb}}{\text{density of material, lb per bank cu yd}} \tag{13-10}$$

For Eq. (13-9), cycle time, the time to complete one round trip, may be measured with a stopwatch. Usually, an average of several complete cycles is taken. Sometimes, additional information is desired, such as load time and wait time, which indicate loading ability and job efficiency. So the watch is kept running continuously and the times for beginning and ending certain phases are recorded. Table 13-11 is an example of a simple time-study form. It can easily be modified to include other segments of the cycle, such as haul time and dump time, if desired. Similar forms can be made for pushers, bulldozers, and other equipment.

Wait time is the time a unit must wait for another machine so that the two can work together, for example, a scraper waiting for a pusher. Delay time is any time other than wait time when a machine is not working, for example, a scraper waiting to cross a road.

Table 13-11. Cycle-Time Observations

Total cycle times (less delays), min	Arrive cut	Wait time	Begin load	Load time	End load	Begin delay	Delay time	End delay
	0.00	0.30	0.30	0.60	0.90			
3.50	3.50	0.30	3.80	0.65	4.45			
4.00	7.50	0.35	7.85	0.70	8.55	9.95	1.00	10.95
4.00	12.50	0.42	12.92	0.68	13.60			

Since cycle time is involved in computation of production [Eq. (13-8)], different types of production may be measured, depending on whether cycle time includes wait or delay time. Measured production includes all waits and delays. Production without delays includes normal wait time but no delay time. For maximum production, wait time is minimized or eliminated and delay time is eliminated. Cycle time may be further altered by using an optimum load time, as determined from a load-growth study (see also Art. 13-13).

Example 13-1. A job study of rubber-tired scrapers yields the following data:
Weight of haul unit empty, 44,800 lb. Average of three weighings of haul unit loaded, 81,970 lb. Density of material to be excavated, 3,140 lb per bank cu yd.

Average wait time, 0.28 min; average delay time, 0.25 min; average load time, 0.65 min; average total cycle time, less delays, 7.50 min.

What will be the production of the haul unit?
The average load will be $81,970 - 44,800$ lb $= 37,170$ lb. This load is equivalent to $37,170/3,140 = 11.8$ bank cu yd. In 60 working minutes per hour, the scraper will make $60/7.50 = 8.0$ trips per hour. Hence, production (less delays) will be $11.8 \times 8.0 = 94$ bank cu yd per hr.

To determine the number of scrapers needed on a job, required production must first be computed.

$$\text{Production required, cu yd per hr} = \frac{\text{quantity, bank cu yd}}{\text{working time, hr}} \tag{13-11}$$

$$\text{No. of scrapers needed} = \frac{\text{production required, cu yd per hr}}{\text{production per unit, cu yd per hr}} \tag{13-12}$$

$$\text{No. of scrapers a pusher will load} = \frac{\text{scraper cycle time, min}}{\text{pusher cycle time, min}} \tag{13-13}$$

For computation of rolling resistance, see Art. 13-9; for traction, see Art. 13-12. ("Earthmoving Data," Caterpillar Tractor Co.)

13-11. Bulldozer Production. Production normally is measured in terms of bank cubic yards dozed per hour. Because of the number of variables involved, determination of bulldozer production is difficult. A simplified method can provide a satisfactory estimate:

Two workers, using a 50-ft tape measure, can determine bulldozer payloads on the job. The bulldozer pushes its load onto a level area, stops, raises the blade while moving slightly forward, then reverses to clear the pile. The workers measure height, width, and length of the pile. To determine the average height, one worker holds the tape vertically at the inside edge of each grouser mark. The second worker, on the other side of the pile, aligns the tape with the top of the pile. To measure average width and length, the two hold the tape horizontally and sight on each end of the pile. Sighting to the nearest tenth of a foot is sufficiently accurate. Multiplication of the dimensions yields the loose volume in cubic feet; division by 27, in cubic yards. Application of a load factor [Eq. (13-6) and Tables 13-9 and 13-10] gives the loads in bank cubic yards.

For computation of rolling resistance, see Art. 13-9; for traction, see Art. 13-12.

13-12. Traction. This normally is measured by the maximum drawbar pull or rim pull, lb, a tractor exerts before the tracks or driving wheels slip and spin. In computing pull requirements for track-type tractors, rolling resistance does not apply to the tractor, only to the trailed unit. Since track-type tractors move on steel wheels rolling on steel "roads," rolling resistance is relatively constant and is accounted for in the drawbar-pull rating.

Traction depends on the weight on driving tracks or wheels, gripping action with the ground, and condition of the ground. The coefficient of traction (Table 13-8) is the ratio of the maximum pull, lb, a tractor exerts when on a specific surface to the total weight on the drivers.

Example 13-2. What usable drawbar pull can a 59,100-lb tractor exert while working on firm earth? On loose earth?

The solution can be obtained with Eq. (13-5) and Table 13-8:

Firm earth: $P = 0.90 \times 59,100 = 53,200$ lb

Loose earth: $P = 0.60 \times 59,100 = 35,500$ lb

If 48,000 lb were required to move a load, then this tractor could move it on firm earth, but on loose earth the tracks would spin.

Example 13-3. What usable rim pull can a rubber-tired tractor-scraper exert while working on firm earth? On loose earth? Assume the weight distribution for the loaded unit as 49,670 lb on the drive wheels and 40,630 lb on the scraper wheels.

The solution can be obtained with Eq. (13-5) and Table 13-8. Use the weight on the drivers only.

Firm earth: $P = 0.55 \times 49,670 = 27,320$ lb

Loose earth: $P = 0.45 \times 49,670 = 22,350$ lb

If 25,000 lb were required to move a load and the engine were sufficiently powerful, the tractor-scraper could move the load on firm earth, but the drivers would slip on loose earth.

Equipment specification sheets show how many pounds pull a machine can exert in a given gear at a given speed. But if the engine works at high altitudes, it cannot produce as much power as it is rated for at sea level because of the decrease in air density. Up to 2,500 ft above sea level, the reduction is insignificant. For each 1,000 ft above 2,500 ft, an engine loses about 3% of its horsepower. But some machines with turbocharged engines operate at altitudes above 2,500 ft without loss of power, so consult service literature on a machine before derating for altitude. ("Earthmoving Data," Caterpillar Tractor Co.)

13-13. How to Estimate Cycle Time and Job Efficiency. Before production on an earthmoving job can be estimated, cycle time for the equipment must be known [Eqs. (13-8) and (13-9)]. Cycle time is the time required to complete one round trip in moving material. Different approaches are used in estimating cycle time for each type of machine.

Track-type Tractor-Scrapers. Cycle time is the sum of fixed times and variable times. Fixed times in scraper work are the number of minutes for loading, turning, dumping, and in bulldozing, for gear shifting. Variable times comprise haul and return times. Experience shows that the fixed times in Table 13-12 are satisfactory for estimating purposes.

Since speeds and distances may vary on haul and return, haul and return times are estimated separately.

$$\text{Variable time, min} = \frac{\text{haul distance, ft}}{88 \times \text{speed, mph}} + \frac{\text{return distance, ft}}{88 \times \text{speed, mph}} \tag{13-14}$$

Haul speed may be obtained from the equipment specification sheet when the drawbar pull required is known.

Table 13-12. Fixed Times for Estimating Cycle Time, Minutes

TRACK-TYPE TRACTOR AND SCRAPER

	Self-loaded		Push loaded 15 cu yd or more
	15 cu yd or more	14 cu yd or less	
Loading	1.5	1.0	1.0
Dumping, turning	1.0	1.0	1.0
Total fixed time	2.5	2.0	2.0

TRACK-TYPE TRACTOR AND BULLDOZER

Shuttle bulldozing using same gear and shifting only forward-reverse lever	0.1
Shuttle bulldozing shifting to higher reverse gear	0.2
Power-shift tractors	0

TRACK-TYPE TRACTOR-SHOVELS
(Fixed time for loading, turning, dumping)

	Manual shift	Power shift
Bank or stockpile loading	0.35	0.25
Excavation	0.61	0.43

WHEEL-TYPE TRACTOR-SHOVELS

Stockpile loading, power shift	0.20

WHEEL-TYPE TRACTOR-SCRAPERS WITH PUSHERS

	5th-gear hauls	4th-gear hauls	3d-gear hauls
Loading	1.0	1.0	1.0
Maneuver and spread	0.5	0.5	0.5
Acceleration and deceleration	1.5	0.8	0.4
Total fixed time	3.0	2.3	1.9

Wheel-type Tractor-Scrapers. The procedure for estimating cycle time for wheel- and track-type tractors is about the same. But for wheel-type tractors time consumed in acceleration and deceleration must be included in the estimate of fixed time. The values given in Table 13-12 may be used for estimating.

To determine the haul speed of a wheel-type tractor-scraper, it is necessary to match the rim pull required (total road resistance) against rim pull available (obtained from equipment specifications) and select a reasonable operating gear (from the specifications). Equation (13-14) may be used to figure variable time. The sum of fixed and variable times gives the estimated cycle time.

Power loss due to altitude is taken into account by dividing the total road-resistance factor [Eq. (13-14)] by a correction factor k. The resulting effective resistance factor then is used to compute travel time.

$$k = 1 - 0.03 \frac{H - 2,500}{1,000} \tag{13-15}$$

where H = altitude above sea level, ft. Travel time can be determined from data supplied by the scraper manufacturer.

Job efficiency depends on many variables, including operator skill, minor repairs and adjustments, delays caused by personnel, and delays caused by job layout. Table 13-13 lists approximate efficiency factors for estimating when job data are unavailable. Production, cu yd per working hour, then equals production, cu yd per 60-min hr, times the efficiency factor.

Table 13-13. Efficiency Factors for Average Job Conditions*

	Min per working hr	Efficiency factor
	DAY OPERATION	
Track-type tractor	50	0.83
Wheel-type tractor	45	0.75
	NIGHT OPERATION	
Track-type tractor	45	0.75
Wheel-type tractor	40	0.67

*These take into consideration only minor delays. No time is included for major overhauls and repairs. Machine availability and weather also should be taken into account.

13-14. Mass Diagram. This is a graph showing the accumulation of cut and fill with distance from a starting point, or origin. Cut usually is considered positive and fill negative. The volume of each is plotted in cubic yards. Distance normally is measured, along the center line of the construction, in stations 100 ft apart, starting with the origin as 0 + 00. Swell factors are applied to the cuts and shrinkage factors to the embankments [Eqs. (13-6) and (13-7)] to obtain bank cubic yards excavated and compacted fill, respectively.

Figure 13-14b shows a mass diagram for the profile in Fig. 13-14a (shrinkage factor of 10% and swell factor of 20% included). Between 0 + 00 and 1 + 00, there is a cut of 2,000 cu yd. This is plotted at 1 + 00. Between 1 + 00 and 2 + 00, there is a cut of 5,000 cu yd, making a total of 7,000 cu yd between 0 + 00 and 2 + 00; 7,000 is plotted at 2 + 00. At 4 + 00, there is a

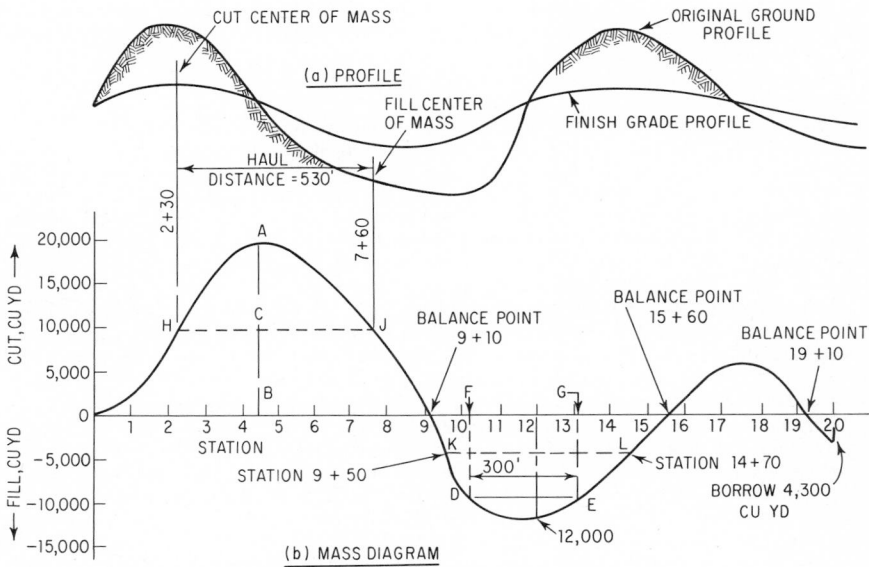

Fig. 13-14. Profile and mass diagram for cut and fill.

total accumulation of 18,000 cu yd of cut.   Between 4 + 00 and 5 + 00, there are 1,000 cu yd of cut and 550 cu yd of embankment (corrected for shrinkage), making a net accumulation of

$$18,000 + 1,000 - 550 = 18,450 \text{ cu yd}$$

From 6 + 00 to 12 + 00, there is mostly embankment, and the accumulation decreases to − 12,000 cu yd.   Cut follows, then some more embankment.   At the end of the construction, 20 + 00, there is a net of − 4,300 cu yd, embankment that must be obtained from borrow.

If a mass curve is horizontal between stations, the implication is that no material has to be moved in that stretch.   Actually, there may be cuts and fills, but they balance.   If work consists of side-hill cuts and fills, the mass diagram tends to flatten, because the cuts can be moved into the fills and not moved from one station to another.   Moving excavation from one side of the center line to the other is called **cross haul.**

The slope of the mass curve increases with volume between stations.   An ascending mass curve indicates cut; a descending diagram, fill.   The curve reaches a maximum where cut ends and fill begins, and a minimum where fill ends and cut begins.

If a mass diagram is intersected by a horizontal line, cuts balance fills between the points of intersection.   If the mass curve loops above the line, cuts will have to be hauled forward (in the direction of increasing stations) for the embankments; if the diagram lies below the horizontal line, the haul will have to be backward.

Haul, station-yards, for a section of earthwork is the product of the amount of excavation, cu yd, and the distance it is moved, stations.   Total haul is the product of total amount of excavation hauled and average haul distance.   The area between the mass diagram and a balancing (horizontal) line equals the haul, station-yards, between the two points cut by that line.   Average haul distance equals the area between the mass diagram and the balancing line divided by the total cut (maximum ordinate) between the points of intersection.

Center of mass of cut and fill can be determined from the mass diagram.   Draw the maximum ordinate between a balancing line and the curve (for example, *BA* in Fig. 13-14*b*).   Then, draw a horizontal line (*HJ*) through the midpoint of that ordinate, and note the stations at the points of intersection with the curve.   The station (*H*) on the increasing portion of the diagram is the center of mass of cut; the station (*J*) on the decreasing portion, the center of fill.   The distance between the stations is the haul distance.

If the mass curve terminates below the horizontal axis, borrow is required.   If the curve ends above the axis, excavation must be wasted.

Free haul is the distance excavation may be moved without an increase in contract price; that is, the unit bid price for excavation applies only to haul distances less than free haul.   Overhaul is haul distance exceeding free haul.   The bid price for overhaul usually is given in terms of dollars per station-yard.

Example 13-4.   For Fig. 13-14, if free haul is 300 ft, determine the overhaul between 9 + 10 and 15 + 60.

Draw horizontal line *DE* with length 300 ft between two points on the mass curve.   Draw ordinates *FD* at *D* and *GE* at *E*.   These vertical lines set the limits of free haul.   Next, the center of mass of cut and fill outside these limits must be found.   To do this, draw a horizontal line through the midpoints of *FD* and *GE* intersecting the mass curve at *K* and *L*.   The center of mass of cut is at *L*, 14 + 70, and of fill, at *K*, 9 + 50.   *KL* = 5.2 stations represents the average haul distance.   Hence, the overhaul equals the product of *DF* = 9,500 cu yd and *KL* less the free-haul distance (5.2 − 3.0), or 20,900 station-yards.

(C. F. Allen, "Railroad Curves and Earthwork," McGraw-Hill Book Company, New York.)

13-15. Rock-Excavation Drilling.   Usually, before rock can be excavated, it must be blasted into pieces small enough for efficient handling by available equipment.   To place explosive charges for this purpose, holes have to be drilled into the rock.   This is done with percussion or rotary drills.   Percussion generally is used for hard rock and small-diameter holes.   Maximum size of bit for percussion drills is about 6 in.   Larger bits may be used on rotary drills (Fig. 13-15), but they rarely exceed 9 in. in diameter.

Normally, percussion drills are mounted on self-propelled crawlers (Fig. 13-16).   Drilling commonly is done with sectional drill steel and carbide-insert bits, both of which have to be rugged.   A bit has first to crush its way into the rock.   Next, the hole must be reamed.   Finally, the cuttings are mixed and blown from the hole by compressed air fed through a hole in the center of the drill steel and discharged through holes in the bit.   Hard rock requires a bit with good crushing or penetrating and reaming ability.   Shales, usually soft, require a bit that mixes material fast.   The bit does not need to have good crushing ability.   For sandstone, the gage will

Fig. 13-15. Tractor-mounted rotary drill with air compressor.

Fig. 13-16. Percussion drill. (*Courtesy of Chicago Pneumatic.*)

usually be destroyed first or the bit will lose its reaming ability.   A bit used in sandstone must have exceptional reaming ability, plus good mixing features.

The best way to determine how well a bit is performing is to inspect the chips.   They should be firm pieces of rock, not dust.   When cuttings are dust, usually the chips are not being blown from the hole until they have been reground several times.   This causes more-than-normal bit wear. Low air pressure also may produce excessive dust.   Pressure at the drill should be at least 90 psi. In computing the pressure, take into account the drop in pressure due to friction in the hose.

Rotary drilling is more suitable for large holes.   With low-cost ammonium nitrate and fuel oil as the explosive, economical production results.   With large holes, spacing can be greater and more cubic yards can be produced per foot of hole.   In determining whether to use large or small holes, the engineer should bear in mind that the amount of explosives is directly proportional to the area of the hole.

In rotary drilling, it is essential to maintain sufficient down pressure, rotation speed, and volume and pressure of air used to blow cuttings out of the hole.   Otherwise, excessive bit wear and low production results.   Down pressure should be at least 5,000 psi per in. of bit diameter. Rotation speed should be the largest possible without regrinding chips before they are blown out of the hole.   Therefore, rotation speed depends on air volume.   Air is blown through the center of the drill steel and discharged through passages in the bit.   Except in extremely deep holes, 40 psi usually is enough pressure to clean holes.

13-16. Explosives for Rock Excavation.   Explosives are used to blast rock into pieces small enough to be handled efficiently by available equipment.   The charges usually are set in holes drilled into the rock (Art. 13-15) and detonated with an electric current.

If the reaction is instantaneous or extremely rapid over the entire mass of the explosive, detonation has occurred.   Deflagration, however, takes place when the reaction particles move away from the unreacted particles or the material burns.   The basic difference between these two reactions is that detonation produces a high-pressure shock wave that is self-propagating throughout the charge.

Several factors contribute to the effectiveness of an explosive charge: confinement, density, most efficient uniform propagation diameter, and critical mass.

Confinement helps the reacted products to contribute to detonation of the unreacted products. If the reacted portions can escape, the reactions will cease.   An air space can be very effective in dampening a reaction.

The denser the mass of the charge, the more effective it will be, up to a point.   For every explosive, there is an optimum density.   Since drilling costs more than explosives per cubic yard of excavation, it is desirable to use as many pounds of explosive per foot of borehole as possible.

The most efficient uniform propagation diameter is the width or length over which the explosive mass will be self-propagating after detonation starts.   This length ranges from very small to about 9 in. for ammonium nitrate.

The self-propagating diameter can be lowered by the overdrive method.   Overdrive is the ability of an explosive to detonate at a rate greater than the self-propagating detonating rate. Suppose, for example, an explosive that detonates at 21,000 fps is set off in contact with another type of explosive that detonates at 12,000 fps.   Then, the slower explosive will detonate at more than 12,000 fps but less than 21,000 fps for a given distance, usually less than 2 ft.

Sensitivity of an explosive is very important from a safety standpoint.   An explosive should be easy to detonate by specific methods, but hard or impossible to set off with normal or careful handling during manufacture, shipment, storage, and preparation for detonation.

Critical mass is that amount of an explosive that must be present for the reaction to change from deflagration to detonation.   This mass is very small for high-order explosives but is about 123 tons for ammonium nitrate.

Explosive manufacturers generally balance the ingredients of their products to get maximum gas volume.   This usually depends on the amount of oxygen available from an unstable oxidizer in the explosive.   A combination of gas ratio and brisance (shattering effect) is called power factor. Explosive ingredients can be combined in many ways to provide almost any power factor.

Rate of detonation is a rough measure of the shattering ability of an explosive.   Mass formations of rock require a rate of at least 12,000 fps.   Maximum detonating rate for commercial explosives is 26,000 fps.

Explosive strength generally is rated in terms of the percent of nitroglycerin or equivalent in explosive power contained in an explosive.   Straight dynamites contain only nitroglycerin and an inert ingredient.   In an ammonia dynamite, some of the nitroglycerin is replaced by other ingredients, such as ammonium nitrate.   Explosive power may be denoted by weight strength or by bulk or cartridge strength. When weight strength is given, an ammonia dynamite will have the same explosive power as a straight dynamite of the same strength. Following are important features of explosives commonly used in construction:

Gelatin Dynamites. Weight strength from 100 to 60%. Detonation rate from 26,200 to 19,700 fps, respectively. Suitable for submarine blasting or for use where considerable water pressure will be encountered. Inflammable. Has high shattering action.

Gelatin Extras. Weight strength from 80 to 30%. Detonation rate from 24,000 to 15,000 fps, respectively. Ammonium nitrate replaces part of nitroglycerin. Gelatin extras have less water resistance than gelatins but can be used satisfactorily except under the most severe conditions.

Extra Dynamites. Weight strength from 60 to 20%. Detonation rate from 12,450 to 8,200 fps. Ammonium nitrate replaces part of nitroglycerin. Extra dynamites can be used in average water conditions if properly wrapped with waterproofing. They usually are called original ammonia dynamites.

Semigelatins. Weight strength from 65 to 40%; bulk strength from 65 to 30%. Detonation speed from 17,700 to 9,850 fps. Higher detonation speeds for larger-diameter cartridges. Can be used instead of gelatins in most blasting uses. Water resistance is adequate for average conditions.

High-Ammonium-Nitrate-Content Dynamites. Weight strength from 68 to 46%; bulk strength from 50 to 20%. Detonation speed from 10,500 to 5,250 fps. Has low water resistance but can be used if fired within a relatively short time of exposure.

Boosters or Primers. Have high density. Detonation speed of 25,000 fps. Used to detonate ammonium nitrates and fuel oil or any non-cap-sensitive explosive, because boosters and primers have a very high detonation pressure.

Detonating Cord. Used as a fuse. Has high-explosive core that detonates at 21,000 fps with sufficient energy to detonate another, less sensitive explosive alongside in a borehole. When strung from top to bottom of a hole, detonating cord will act as a detonating agent throughout the length of the hole.

Ammonium nitrate, for best results, should be mixed with at least 6% fuel oil, by weight. The oil is added for oxygen balancing and to lower the self-propagating diameter. Quantities of fuel oil greatly in excess of 6% have a dampening effect on the explosion. By use of the overdrive method, the rate of detonation for ammonium nitrate and fuel oil will be sufficient to shatter any rock formation encountered. Ammonium nitrate plus 10% booster has a rate of 4,500 to 10,000 fps; when fuel oil is added, the rate increases to 10,000 to 16,500 fps. For overdrive, best results are obtained with at least 5% of a primer with a high detonation rate. The primers should be properly spaced to insure that critical propagation length will not be exceeded and detonation will occur throughout.

Special precautions should be observed when overdrive is used. If free fuel oil is available in the mixture, an ammonia dynamite should not be used as a primer. Fuel oil will desensitize ammonia dynamite, and a partial or complete failure will result. Fuel oil also has an adverse effect on the explosive contained in detonating cord. This, however, can be avoided by use of a plastic coating on the cord.

Table 13-14 gives the approximate amount of ammonium nitrate to use per foot of borehole. The table assumes a density of 47 lb per cu ft for ammonium nitrate and fuel oil.

Ammonium nitrate is soluble in water. It develops some water resistance when mixed with fuel oil. But exposure to water results in loss of efficiency, and detonation becomes difficult.

13-17. Rock-Excavation Blasting. (See also Arts. 13-15 and 13-16.) To secure the desired shape of rock surface after blasting, explosive charges must be placed in boreholes laid out in the proper pattern and of sufficient depth. Before the pattern is chosen, an explosive factor must be selected:

Types of Rock	*Explosive Factor, Lb per Cu Yd*
Shales	0.25–0.75
Sandstone	0.30–0.60
Limestone	0.40–1.00
Granite	1.00–1.50

Next, drill size, burden, and spacing can be selected. Then, the amount of stemming can be determined. Stemming is the top portion of a borehole that contains a tightly tamped backfill, not explosive. Since an explosive exerts equal pressure in all directions, depth of stemming

Table 13-14. Amount of Ammonium Nitrate per Foot of Borehole

Hole dia, in.	Approx. weight, lb per ft	Approx. volume, cu ft per ft
2	1.02	0.0218
2¼	1.29	0.0275
2½	1.59	0.034
3	2.30	0.049
3¼	2.67	0.057
3½	3.00	0.064
4	4.09	0.087
4½	5.17	0.110
5	6.39	0.136
5½	7.75	0.165
6	9.21	0.196
6¼	10.01	0.213
6½	10.81	0.230
6⅞	12.03	0.256
7	12.54	0.267
7¼	13.44	0.286
7⅞	15.79	0.336
8	16.40	0.349
8½	18.51	0.394
9	20.72	0.441
9½	23.12	0.492
10	25.61	0.545
10½	28.24	0.601
11	30.97	0.659
11½	33.88	0.721
12	36.89	0.785

should not exceed the width of burden, the distance from the borehole to the rock face. Burden distance should be less than the hole spacing so that the blasted rock will be thrown in the direction of the burden. Holes should be placed in lines parallel to the rock face, because a rectangular pattern gives better breakage and vibration control. Depth of drill holes is determined by height of face desired and the distance it is necessary to drill below grade so that the bottom can be controlled.

A mathematical check should be made to determine that the explosive factor is correct for the burden and spacing selected. If properly blasted rock is not produced when a drill pattern is tried, a new spacing or burden width should be tried. It is best to vary only one dimension at a time until desired fragmentation is obtained.

Delay caps may be used on the explosive charges for better fragmentation and vibration control. Delay caps permit detonation of explosive charges in different holes at intervals of a few milliseconds. The result is better fragmentation, controlled throw, and less back break, since better displacement is obtained. Table 13-15 gives characteristics of short-period delay caps. Use of regular delays is not recommended because of "hole robbing" and uncontrolled throw.

Presplitting is a technique for producing a reasonably smooth, nonshattered wall, free from loose rock. An objective is to hold maintenance of slopes and ditches to a minimum. Presplit holes are drilled in a single line in a plane that will be the final slope or wall face. Line drilling also may be used, with holes spaced about two times the bit diameter. But for presplitting, the spacing is much greater. Dynamite, evenly spaced on detonating cord, is exploded to break the web between holes. Manufacturers can furnish explosives made for presplitting. When this type of explosive is used, loading of holes is easier, since no detonating cord is required. The resulting saving of labor will usually more than offset additional explosive costs.

Percussion drills commonly are used for drilling presplitting holes. An air track with hydraulic controls is very effective in enabling the driller to move from hole to hole and reset the drill in a minimum time. Number of drills required varies with capacity of loading shovel, width of cut, and spacing of presplit holes.

For presplitting, 40% extra gelatin works satisfactorily. This explosive has a detonation speed that can break the hardest rock formations and is adequate under the most adverse conditions. Speed of detonation should not be less than 15,000 fps for presplitting.

Table 13-15. Characteristics of Millisecond Delay Caps*

Delay period	Nominal firing time, msec	Interval between delay periods, msec
0	12	
SP-1	25	13
SP-2	50	25
SP-3	75	25
SP-4	100	25
SP-5	135	35
SP-6	170	35
SP-7	205	35
SP-8	240	35
SP-9	280	40
SP-10	320	40
SP-11	360	40
SP-12	400	40
SP-13	450	50
SP-14	500	50
SP-15	550	50
SP-16	600	50
SP-17	700	100
SP-18	900	200
SP-19	1,100	200
SP-20	1,300	200
SP-21	1,500	200
SP-22	1,700	200
SP-23	1,950	250
SP-24	2,200	250
SP-25	2,450	250
SP-26	2,700	250
SP-27	2,950	250

*Courtesy of Hercules Powder Co.

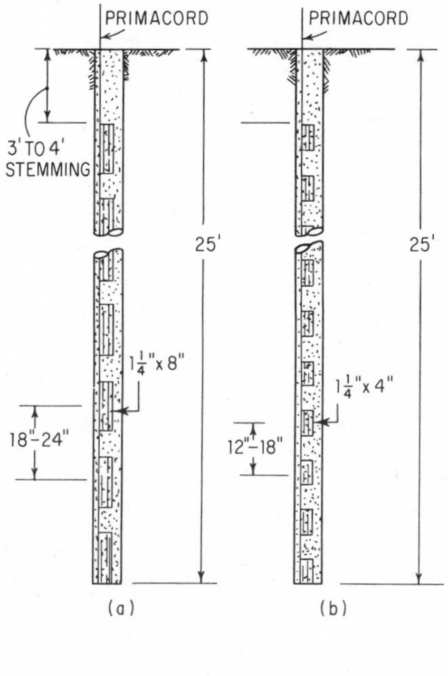

Fig. 13-17. Drill holes loaded with (a) 1¼ × 8 in. cartridges and (b) 1¼ × 4 in. cartridges on detonation cord for presplitting.

Table 13-16. Pounds of 40% Gelatin Extra to Produce 2,500 Sq Ft of Wall by Presplitting

Hole spacing, in.	1¼ × 8 in. cartridges		1¼ × 4 in. cartridges	
	18 in. c to c	24 in. c to c	12 in. c to c	18 in. c to c
18	362	272	272	181
24	270	203	203	135
30	215	161	161	108
36	178	134	134	89
42	152	114	114	76
48	132	99	99	66
54	117	88	88	58
60	105	79	79	52
66	95	71	71	47
72	86	64	64	43

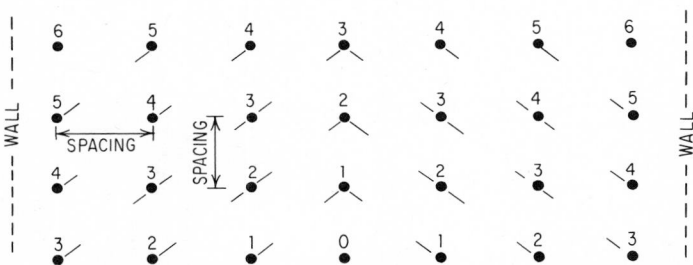

Fig. 13-18. Drill pattern for conventional blasting with holes all of the same diameter. Numbers indicate order of firing with delays.

Table 13-17. Powder Factor for Drill Pattern of Fig. 13-18

Spacing of holes, ft	Burden, cu yd	Powder factor*
FOR 9-IN.-DIA HOLES, 25 FT DEEP, 10 FT LOADED, 207 LB OF AMMONIUM NITRATE		
20 × 18	333	0.62
18 × 16	267	0.78
16 × 14	207	1.00
14 × 12	156	1.33
12 × 10	111	1.87
FOR 6-IN.-DIA HOLES, 25 FT DEEP, 16 FT LOADED, 147 LB OF AMMONIUM NITRATE		
18 × 16	267	0.55
16 × 14	207	0.71
14 × 12	156	0.94
12 × 10	111	1.32
10 × 8	74	1.99
FOR 5-IN.-DIA HOLES, 25 FT DEEP, 17 FT LOADED, 109 LB OF AMMONIUM NITRATE		
16 × 14	207	0.52
14 × 12	156	0.70
12 × 10	111	0.98
10 × 8	74	1.47
8 × 6	44	2.46

*Pounds of ammonium nitrate, density 47 lb per cu ft, per cu yd of burden.

Figure 13-17a shows a presplit hole loaded with 1¼ × 8 in. cartridges spaced 18 to 24 in. apart on primacord; Fig. 13-17b shows 1¼ × 4 in. cartridges on 12- to 18-in spacing. Table 13-16 indicates the number of pounds of 40% gelatin extra required to produce a wall 25 ft high by 100 ft long.

Presplitting should precede the primary blast. Some locations, however, preclude this; for example, a side hill where there would not be sufficient burden in front of the presplit holes. In such a case, presplitting will be accomplished, but the burden in front will be shifted, causing loss of primary blast holes or difficult drilling if the holes were not drilled previously. If a side-hill condition exists, delay caps should be used to insure that presplitting is done before detonation of the primary blast.

Spacing of holes for presplitting varies considerably with material, location, and method of primary blasting. Spacings up to 6 ft have been found adequate where no restrictions are imposed on explosives and primary blasting can be adjusted to obtain correct balance for removal of material within the walls. Obtaining a good wall is the result of balancing primary blasting with as wide a spacing as possible for the type of rock. Use of close spacing of holes without consideration of other factors may be wasteful and not yield best overall results.

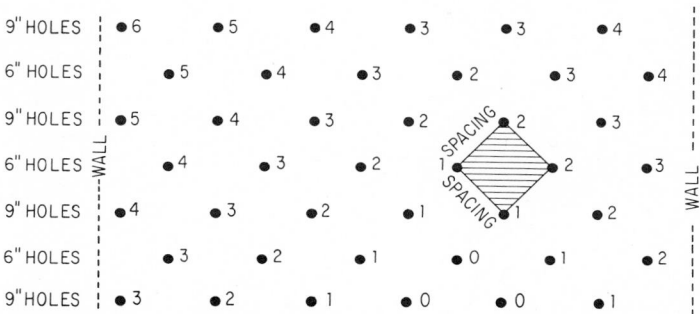

Fig. 13-19. Drill pattern for conventional blasting with two sizes of holes. Numbers indicate order of firing with delays.

Table 13-18. Powder Factor for Drill Pattern of Fig. 13-19

Hole dia, in.	Hole depth, ft	Load depth, ft	Charge	
			Lb	Lb per ft
5	25	17	108.63	6.39
6	25	16	147.36	9.21
9	25	10	207.20	20.72

Spacing, ft	Burden, cu yd	Powder factor*					
		9-in. holes	9- and 6-in. holes	9- and 5-in. holes	6-in. holes	6- and 5-in. holes	5-in. holes
8 × 8	59	3.51	3.00	2.68	2.50	2.17	1.84
10 × 10	93	2.23	1.91	1.70	1.58	1.38	1.17
12 × 12	133	1.56	1.33	1.19	1.11	0.96	0.82
14 × 14	194	1.07	0.91	0.81	0.76	0.66	0.56
16 × 16	237	0.87	0.75	0.67	0.62	0.54	0.46
18 × 18	300	0.69	0.59	0.53	0.49	0.43	0.36
20 × 20	370	0.56	0.48	0.43	0.40	0.35	0.29
22 × 22	448	0.46	0.40	0.35	0.33	0.29	0.24

*Pounds of ammonium nitrate, density 47 lb per cu ft, per cu yd of rock.

Spacing of presplit holes and charges for best results may be determined by trial.   Vary only one variable at a time.   For example, initially drill the holes for 25 ft of wall 18 in. apart and detonate.   Then, for the next 25 ft of wall, drill the holes 24 in. on centers and detonate with the same loading.   Continue increasing the spacing until a maximum is reached.   Next, vary the charge. If too much dynamite is used, the resulting surface between holes will be concave.   Conversely, with insufficient dynamite, the surface will be convex.

In conventional blasting, placing of delays in the primary blast is important.   The more relief that can be given to holes near the wall, the less opportunity for damage to the wall (Figs. 13-18 and 13-19 and Tables 13-17 and 13-18).

The depth of each lift in presplitting is governed by the size of the shovel, but for greatest efficiency, the depth should not exceed the height of the shipper shaft.   Lifts generally average 20 to 25 ft.   The last lift may be deeper to reach grade in one setup.   For efficiency, each lift should be presplit separately.   Drilling speed diminishes rapidly as a 40-ft depth is approached.

When more than one lift is required, the drill has to be set up for successive lifts at least 1 ft away from the face, for clearance (Fig. 13-20).

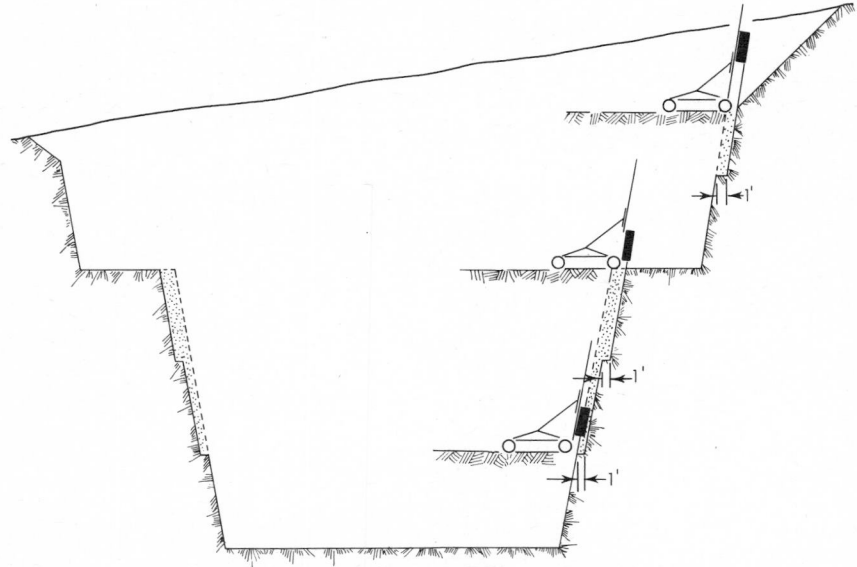

Fig. 13-20. Placement of drill in multilift cut.

Loading of deep holes, particularly if they contain water, can be very difficult.   Stringing sticks of dynamite on a long detonating cord can exceed the structural strength of the cord, breaking it and causing a misfire.

After holes have been drilled, dynamite cartridges are fastened to a detonating cord, usually 50 grain, long enough to reach the bottom of the hole.   Spacing of charges on the cord varies with rock formation and hole spacing.   Charges may be attached with tape or rubber bands.   With rubber bands, spacing is easier to maintain because the charges do not slip so easily.   In a limestone formation with holes drilled at 4-ft intervals, 1¼ × 8 in. charges spaced 18 in. on centers have been found adequate, whereas good results have been obtained in soft shale with a 50% reduction in the charge, to 1¼ × 4 in., and the same hole spacing.   Detonation cord from each hole is attached to a trunk line, which when fired causes each hole to detonate instantaneously.

Stemming can be done in several ways.   In one method, after the charge has been placed in a hole, clean stone chips or sand that will pass a ⅜-in. standard sieve is placed on top.   For best results, the stemming should be worked around the charges by holding the end of the detonation cord in the center of the hole and working it up and down.   Another stemming method is to push newspaper into the hole until it reaches the top charge.   On top of the paper, the hole is stemmed with drill cuttings or other suitable available material.

Contrary to the principles of most blasting procedures, where it is good practice to have as much confinement as possible, some means must be provided in presplitting to allow excess gases to escape. Use of detonation cord and top firing produces best results. Most instant blasting caps have so much delay that breakage occurs in the wall if they are used. To reduce noise and vibration, delay connectors may be used between groups of two or more holes.

The cost of presplitting per cubic yard excavated depends on the distance between walls or volume to be removed per square foot of presplit wall. Presplitting eliminates the need for small-diameter holes for a primary blast, moving of material from behind a pay line, and scaling of slopes. If presplitting is not required and no pay will be received for material excavated behind a pay line set 18 in. beyond the design slope, then, to control excess excavation, small-diameter blast holes would be drilled near the slope at a minimum spacing of 6 ft. These holes would be the same diameter as presplit holes in most cases. Usually, two rows of these holes would be required. The primary blast holes would be at a greater distance from the design slope than for presplitting. When presplitting is used, the spacing of the primary blast holes can be rearranged to produce well-broken rock that will load more easily at less cost.

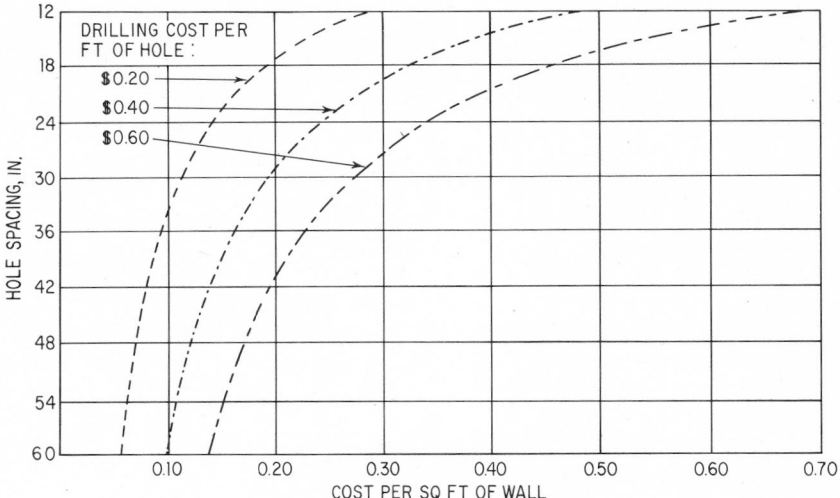

Fig. 13-21. Drilling costs for blasting a wall by presplitting.

A cost comparison between presplitting and conventional blasting should compare the cost of blasting the entire cut without presplitting with the cost of presplitting, rearranging the primary blast, and shooting. Generally, presplitting will cost less. For most formations, this will be true when the ratio of cubic yards excavated to square feet of presplit wall exceeds 1.5:1. Figure 13-21 estimates the cost of producing a wall by presplitting.

13-18. Vibration Control in Blasting. Explosive users should take steps to minimize vibration and noise from blasting and to protect themselves against damage claims.

Before blasting, an explosive user should conduct a survey of nearby structures. Experienced, qualified personnel should make this survey. They should carefully inspect every structure within a preselected distance, at least 500 ft, for cracks, deformation from any cause, and other damage that could be claimed. They should make a written report of all observations, wall by wall, and take pictures of all previous damage. This is known as a pre-blast survey and should be well documented for future use in case of a claim.

Any rock excavation project is a part of some community and has an effect on the surrounding environment. The explosive user can be a good neighbor, enjoying that position, or an undesirable one and suffer the consequences. The decision as to whether the explosive user is an asset or a liability will not be made by people familiar with blasting problems. Quarries and rock excavation projects, therefore, should be operated with the realization that any right to exist will have to be proven by behavior acceptable to the community.

To be a good neighbor, an explosive user must not make noise, create vibrations, or throw projectile rocks. The first and last are easy to control if proper supervision and good guidance are used. If a neighbor does not hear or see the blast, annoyance is greatly diminished.

Noise and throw are best controlled during drilling and loading cycles. No explosives should be loaded closer to the ground surface than the least dimension used for drill-hole spacing. In other words, put explosives in the bottom of holes and use as much stemming as possible; when noise occurs, energy has been wasted. Using larger holes, with resulting wide spacing, will usually produce oversize stone in the top of the shot. This can be controlled by the use of small (satellite) holes drilled to a shallow depth, below the top of the explosives, between large-diameter holes. This is one method used to get explosives evenly distributed.

Extreme care should be exercised with detonating cord. Nothing makes a sharper and more startling noise. When detonating cord is demanded, a low-noise-level cord should be used, and it should be covered with some material that will not contaminate the desired product. Considerable depth of covering is required to control noise. Experience dictates not less than 3 ft for ¼-grain cord.

Knowledge of human habits and how to use the surrounding environment will greatly help to reduce complaints. Set off blasts while people are busy with their daily tasks. Bear in mind that weather conditions affect noise transmission. Blasting during cloudy, overcast weather is like shooting in a room that has a roof. Use other noise- and vibration-producing elements of the surrounding environment for dampening or overriding effects. Scheduling and performing blasts while a freight train passes or while a jet airplane is taking off are examples.

For vibration control, blasting should be controlled with the scaled-distance formula:

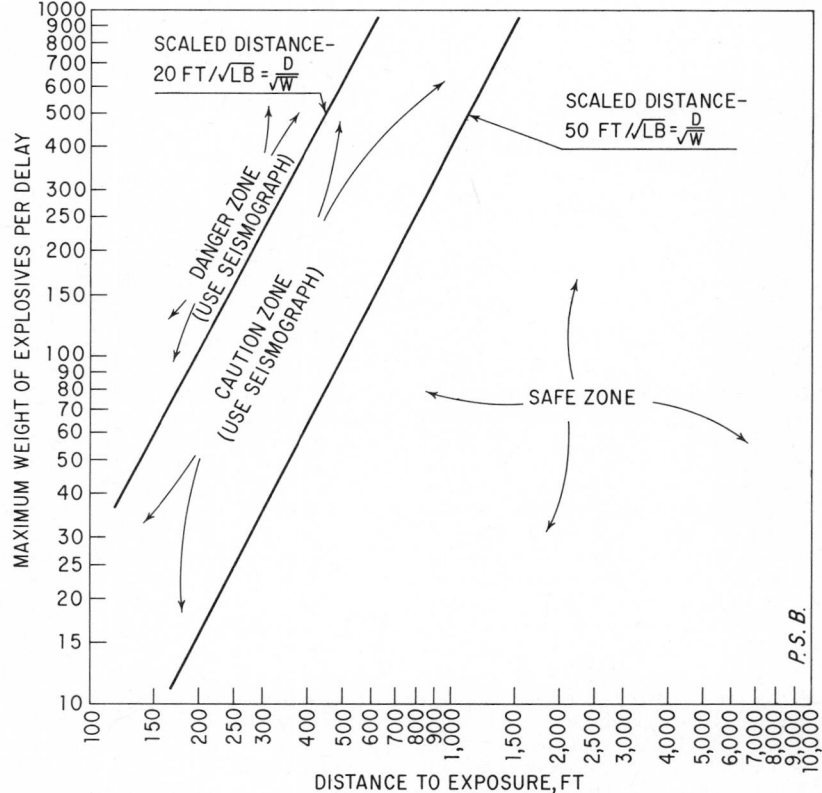

Fig. 13-22. Explosive weight and distance limits for prevention of damage by blasting vibrations.

$$V = H \left(\frac{D}{\sqrt{W}} \right)^{-\beta} \tag{13-16}$$

where V = particle velocity, in. per sec
D = distance to exposure, ft
W = maximum pounds per delay
β = constant (varies for each site)
H = constant (varies for each site)

Distance to exposure, ft, divided by the square root of maximum pounds per delay (Fig. 13-22) is known as **scaled distance.**

Most courts have accepted the fact that a particle velocity not exceeding 2 in. per sec will not damage any part of any structure.

Without specific information about a particular blasting site, the maximum weight of explosives per delay should conform with explosive weight and distance limits to prevent vibration damage. This conforms with a scaled distance of 50 or greater without known facts (Fig. 13-23).

To control vibration, the scaled-distance formula should be applied for each blasting location. If formations vary around the site, each formation will have a different formula, which should be computed. The more blasts used in determining the formula constants, the more accurate the scaled-distance formula becomes. Only two easily determined factors must be known: distance from seismograph and maximum weight of explosive used with any delay. Once a safe scaled distance has been determined, the need to use a seismograph for vibration measurements of future

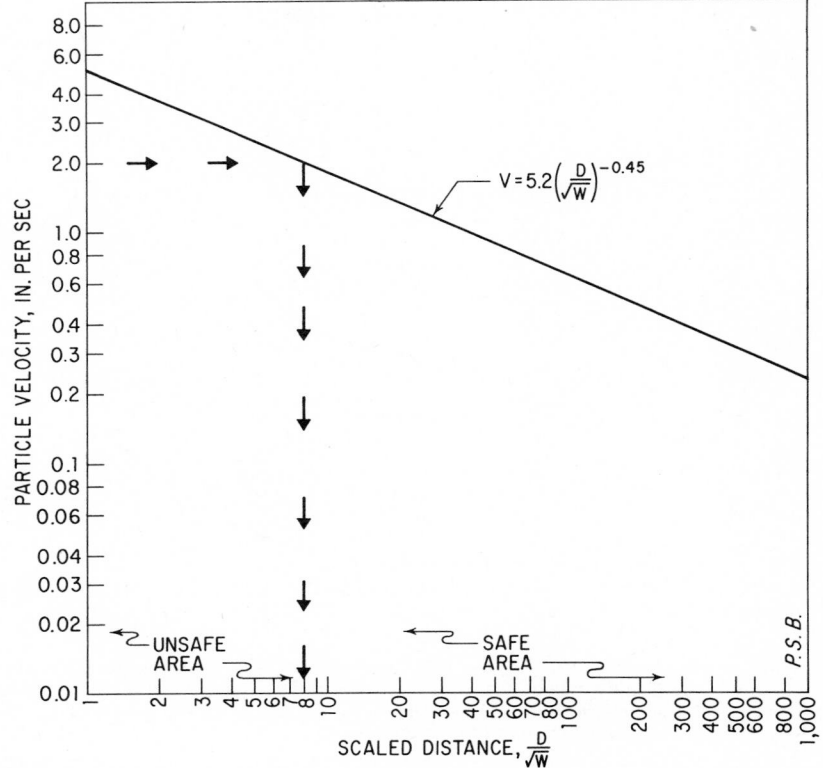

Fig. 13-23. Relationship between particle velocity (vibration) and scaled distance for a specific site, for which $H = 5.2$ and $\beta = 0.45$. For a maximum particle velocity of 2 in. per sec, the scaled distance is 8. Hence, vibration damage is unlikely at scaled distances greater than 8.

blasts is unlikely.   Particle velocity can be computed with actual measured distance and known maximum weight of explosives used with any delay.

There is a direct relationship between particle velocity (vibration) and number of complaints expected from families exposed.   This is shown in Fig. 13-24.

When a complaint is received, it should be handled firmly and expeditiously.   Following are some suggestions:

Assign one person the primary responsibility for handling complaints.   This person should be mature and capable of communicating with complainants who are sincerely upset and afraid of not only property damage but also bodily injury.   A minimum of two employees should, preferably, be detailed, because the primary employee may not always be available.   The primary employee should always be held responsible and should be informed of all complaints.

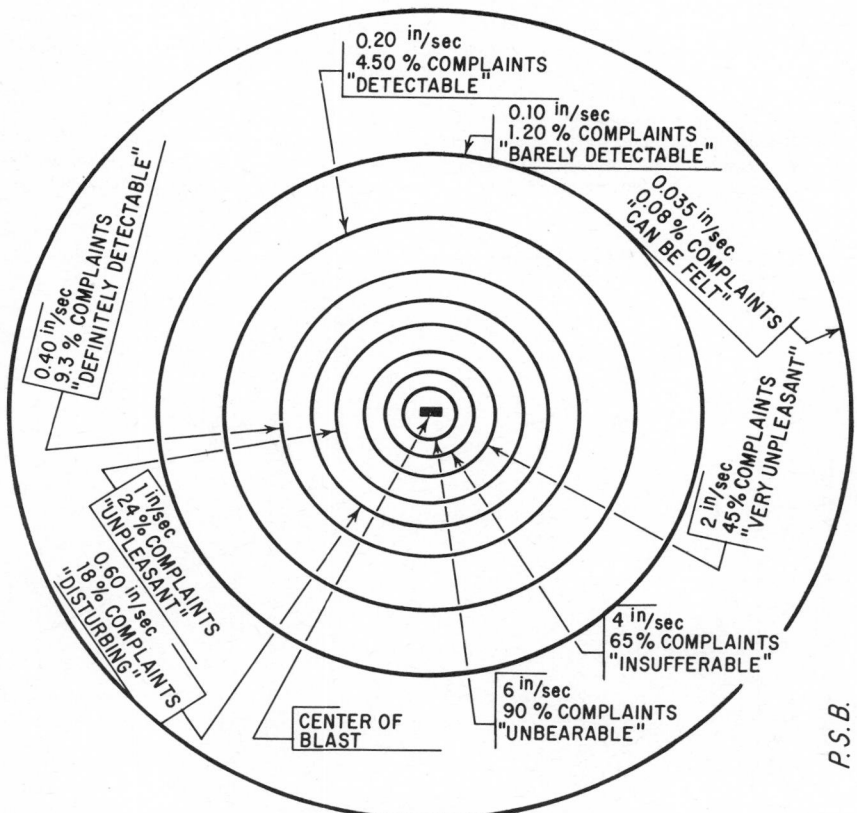

Fig. 13-24. Public reaction to blasting is indicated by the percentage of the total number of families exposed to a specific particle velocity that should be expected to complain, plotted to a logarithmic scale.

Before blasting begins, the public should be advised about whom to contact for any information.   When a complaint is received, record the complainant's name, address, and telephone number.   Ask at what time the blast was felt and heard.   Ask if blast was felt or heard first.   Was the complainant's house included in the preblast survey?

Employees handling complaints should be courteous but firm, never apologizing or saying that less explosives will be used in the future.   Never admit or imply any possible damage until your consultant has advised you of the findings.   A completely informed public will want progress, and that is what your organization owes its success to.

Inform the complainants that a consulting engineer has been retained to design and control your blasting, and this consultant is concerned with nothing but facts. This consultant has been retained to protect the public, to help you do a more efficient job, and to inform the blaster of any potential liability. An independent consultant will know if and where damage may have occurred, probably before the property owner does.

Emphasize that your organization does blasting as a normal operation and has enjoyed success for a considerable length of time, that you have very competent personnel with years of experience, and that you are producing work as efficiently as possible with the least inconvenience to everyone.

People are afraid of noise made by explosives. Noise can be controlled by proper drilling, loading, and stemming. If a shot cannot be seen or heard, your complaints will be few. Remember, it takes only one hole that is not properly tamped to blow out, and then everyone believes the entire shot was not controlled.

Safe blasting is not only demanded and practical, but essential.

13-19. Compaction. This is the process by which soils are densified. It may be done by loading with static weight, striking with an object, vibration, explosives, or rolling. Compaction is used to help eliminate settlement and to make soil more impervious to water. Compaction is costly, and for some embankments, the results cannot be justified, because reduced settlement and other desired benefits are not economical.

For a given soil and given compactive effort, there is an optimum moisture content, expressed in percent of soil dry weight, which gives the greatest degree of compaction. ASTM D698, AASHTO T99, and a modified AASHTO method are widely used for determining moisture content. The modified method may be specified if the soil engineer's investigation indicates that T99 will not yield the desired consolidation. In these tests, soil density of a compacted sample is plotted against percent of moisture in the sample. Maximum density and optimum moisture for the sample can be determined from the resulting curve (Fig. 13-25).

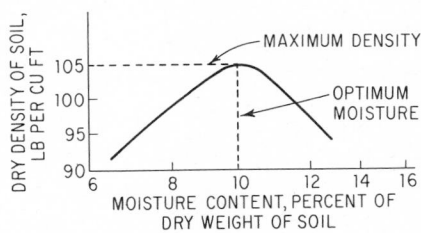

Fig. 13-25. Maximum density graph.

Compaction to be obtained on embankments is expressed as a percent of maximum density. For example, 90% compaction means that the soil in place in the field should have a density of 90% of the maximum obtained in the laboratory. Moisture content should not vary more than 3% above or below optimum. To obtain proper compaction in the field, moisture must be controlled and compactive effort applied to the entire lift.

The calibrated sand method is usually used to determine density in the field. A hole is bored or dug the full depth of a compacted lift. All soil removed from the hole is weighed, and a sample is saved for moisture determination. Then, the hole is filled with dry sand of known density. The weight of sand used to fill the hole is determined and used to compute the volume of the hole. Characteristics of the soil are computed from

$$\text{Volume of soil, cu ft} = \frac{\text{weight of sand filling hole, lb}}{\text{density of sand, lb per cu ft}} \qquad (13\text{-}17)$$

$$\% \text{ moisture} = \frac{100(\text{weight of moist soil} - \text{weight of dry soil})}{\text{weight of dry soil}} \qquad (13\text{-}18)$$

$$\text{Field density, lb per cu ft} = \frac{\text{weight of soil, lb}}{\text{volume of soil, cu ft}} \qquad (13\text{-}19a)$$

$$\text{Dry density} = \frac{\text{field density}}{1 + \% \text{ moisture}/100} \qquad (13\text{-}19b)$$

$$\% \text{ compaction} = \frac{100(\text{dry density})}{\text{max dry density}} \qquad (13\text{-}20)$$

Maximum density is found by plotting a density-moisture curve, similar to Fig. 13-25, and corresponds to optimum moisture. Table 13-19 lists recommended compaction for fills.

Table 13-19. Recommended Compaction of Fills

Dry density, lb per cu ft	Recommended compaction, %
Less than 90	—
90–100	95–100
100–110	95–100
110–120	90–95
120–130	90–95
Over 130	90–95

A mistake commonly made in the field is application of compactive effort when either insufficient or excessive moisture is present in the soil. Under such conditions, it is impossible to obtain recommended compaction no matter how great the effort.

A wide variety of equipment is used to obtain compaction in the field (Table 13-20). Sheepsfoot rollers generally are used on soils that contain high percentages of clay. Vibrating rollers are used on more granular soils.

Table 13-20. Compaction Equipment

Compactor type	Soil best suited for	Max effect in loose life, in.	Density gained in lift*	Max weight, tons
Steel tandem 2–3 axle	Sandy silts, most granular materials, some clay binder	4–8	Average	16
Grid and tamping rollers	Clays, gravels, silts with clay binder	7–12	Nearly uniform	20
Pneumatic small tire	Sandy silts, sandy clays, gravelly sands and clays, few fines	4–8	Uniform to average	12
Pneumatic large tire	All (if economical)	To 24	Average	50
Sheepsfoot	Clays, clay silts, silty clays, gravels with clay binder	7–12	Nearly uniform	20
Vibratory	Sands, sandy silts, silty sands	3–6	Uniform	30
Combinations	All	3–6	Uniform	20

*Density diminishes with depth.

To determine maximum depth of lift, make a test fill. In the process, the most suitable equipment and pressure to be applied, psi of ground contact, also can be determined. Equipment selected should be able to produce desired compaction with four to eight passes. Speed of rolling also can be found. Following are average speeds, mph, under normal conditions:

Grid rollers............................ 12
Sheepsfoot rollers 3
Tamping rollers 10
Pneumatic rollers 8

Compaction production can be computed from

$$\text{Cu yd per hr} = \frac{16WSLFE}{P}$$

$$(13\text{-}21)$$

where W = width of roller, ft
S = roller speed, mph
L = lift thickness, in.
F = ratio of pay cu yd to loose cu yd

E = efficiency factor (allows for time losses, such as those due to turns): 0.90, excellent; 0.80, average; 0.75, poor

P = number of passes

13-20. Dredging. Dredges are used for excavating in or under water. They may be classi-fied as clamshell, suction, bucket, and dipper. In determining equipment capacity, keep in mind that submerged objects weigh less than the same objects will when out of the water.

Clamshell dredges are used where the material to be excavated is rock or hard material that requires blasting or where the spoil area is a considerable distance from the excavation area and hauling is required. Depth of material can be a factor in selection of a clamshell; it has a long reach. A clamshell is a machine that can raise, lower, and close a bucket and swing it on a boom. A machine that is used as a clamshell on land can become a dredge when mounted on a barge. With this type of dredge, production is very low and unit cost of material excavated high.

Suction dredges, available in many sizes, are the most widely used type of dredge. They move soil by suction and pumping through pipes (Fig. 13-26). Size is determined from the amount of material to be moved or production desired. Pump and discharge-line sizes are im-portant in determining production.

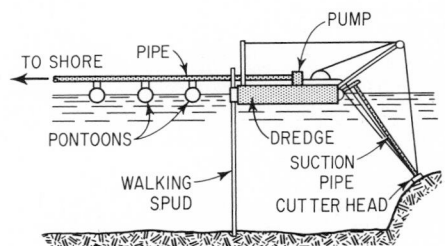

Fig. 13-26. Suction dredge.

The type of material to be excavated determines whether a cutter head will be required on the end of the suction line. A cutter head is used when material has to be loosened and cut up into small enough pieces to get into the suction line and flow through the pipes to discharge. For a free-flowing material, such as sand and gravel, no cutter head would be needed.

Brake horsepower needed for a dredge pump can be computed from

$$hp = \frac{Qhw}{3,960E} \qquad (13\text{-}22)$$

where Q = gallons per minute
h = head, ft
w = specific gravity
E = pump efficiency

Specific gravity depends on the amount of solids to be pumped; for example, it may be taken as 1.3 for 30% solids, 1.35 for 35% solids, etc. The maximum percentage of solids that can be put through a pipe is the plugging limit at lower velocities (12 to 30 fps). Percentage of solids should not be estimated at more than 25%. Dredging capacity at 25% solids is

Pump Size, In.	Production, Cu Yd per Hr
6	80
8	160
10	240
12	425
16	575
18	725
20	900
24	1,300
30	2,100

Since dredge pumps are made with large clearances to pass big particles, an efficiency of 60% would be good.

Maximum length of pipeline that can be used depends on the specific gravity of material to be pumped. If velocities get too low, a booster pump can be installed in the discharge line. Wear on the discharge line depends on abrasiveness of material, velocity, and specific gravity of pumped material.

Bucket dredges are used where depth is not large. They are especially adaptable to trenching under water. A bucket dredge actually is a chain conveyor with buckets attached that dump onto a belt conveyor.

Dipper dredges are comparable with land-type crawler shovels. Their main use is for rock that breaks into large, heavy pieces.

Municipal and Regional Planning

GUSTAV J. REQUARDT,
KENNETH A. McCORD, and FREDERICK R. KNOOP, Jr.*

Partners, Whitman, Requardt and Associates,
Engineers and Consultants, Baltimore, Md.

This is a rapidly changing world. On every continent, population is increasing. Civilization reaches into areas of the globe only recently inaccessible. Independent governments replace colonies.

Changes in the United States since World War II have been caused principally by the following: an exploding population; growth of industry, which enables people to move from rural to urban areas; greater efficiency in production of food, as a result of which less land area is required for raising food; vast increases in means of transportation; new developments in the fields of communications, such as telephone, radio, television, and satellites; better building materials, resulting in better housing at less cost; cybernetics and automation, making possible lower labor unit costs while opening wider fields for employment; increasing use of computers, which reduces working hours in business and industry; better use of leisure, thus creating vast new enterprises in the fields of recreation and amusement; and governmental efforts to improve the lot of the unemployed and poverty-stricken. All these changes have required rapid development of marginal and rural lands for residential and industrial use and redevelopment of large urban areas to put them to better use.

This section deals with planning for the development and better use of land in the United States.

14-1. The Challenge of Land Planning. Population in the continental United States grew from about 76 million in 1900 to more than 200 million persons in 1976. Population is continuing to grow but not as rapidly. Planning for the effects of growth becomes more and more essential.

Population increase plus the great variety of new uses of land and people's insistence on greater access to sun and air make necessary more intensive planning of the development of land than ever before. All around is evidence of previous haphazard planning and layout of land areas: narrow, crooked streets with doglegs and dead ends, block after block of narrow

---

*With minor revisions for the second edition by Frederick S. Merritt, Consulting Engineer.

row houses, lack of provision for storm flow into streamlets and valleys, jumbles of residences mixed in with industry. The objective of planning is the development of rural land and redevelopment of presently ill-used land in cities for the most efficient, durable, pleasant, and economical living, as well as for occupational and recreational use of the people (Art. 14-4).

A tremendous amount of information is needed for the proper development of land areas. Land must be developed for residences; areas must be set aside for industry and utilities; and the whole plan must fit the terrain with the least amount of ground disturbance and with provision for rapid transit and communications (Art. 14-11).

Planners are not done when they have designed the land for pleasant, economical, and durable use. They must, in addition, see that operation and maintenance of the land are also planned. Also, they must evaluate the impact on the environment of making or not making the planned changes and of appropriate alternatives.

Permanently operating systems must be planned for water supply, waste-water disposal, rainwater discharge, transportation, and electricity, telephone, and gas supply. Other necessary services are organized after occupancy but must be provided for in the plan. Those solid refuse wastes, including household garbage, not shredded and disposed of with waste water must be removed from the premises. A convenient location must be chosen for refuse cans at each building and arrangements made for regular collection of refuse and delivery in closed, rubber-tired vehicles to points of disposal. These may be properly operated sanitary landfills or modern incinerators located where they will not be detrimental to the plan.

Lawn cuttings, garden and tree trimmings, and leaves must be removed during the growing seasons. Snow must be cleared from roads and sidewalks in winter. And provision must be made for disposal of old furniture, tree trunks, broken refrigerators, bedsprings, and other bulky materials, such as ancient automobiles abandoned on the streets. The removal of such waste items must be handled promptly; else they will make life burdensome.

Beauty is very important in land planning. And planners know that, with little or no extra cost, the looks of the land, and even of mundane surface utilities, can be made attractive. Grants of funds for development of roadways, public buildings, and urban-renewal areas by government agencies are contingent on the degree of attention to beauty designed into projects by planners.

Water is fundamental in land planning. It is needed for drinking, use in foods, swimming, fishing, fire fighting, boiler feed, and industrial processes. Water is not becoming scarce in the United States. Its distribution, however, is irregular: 40 in. or more of rain per year on the eastern seaboard, only a few inches in the western desert areas, and over 100 in. in the northwestern region. Since water can be transported for considerable distances quite cheaply, no area of the country need fear a lack of water or a prohibitive cost. The United States must, however, preserve its not inexhaustible water. Pollution must be abated; water supply must not be carelessly wasted; and increased means must be found for reuse of water. It remains for the land planner to design projects so that all streams bordering them flow past with no additional load of pollutants and so that no underground supply is damaged.

Fresh air is a godsend to any community. Consider, for example, the discomfort in communities such as Los Angeles, where temperature inversion occurs. (Warm air moves in on top of cooler air and forms a blanket that prevents polluted air from rising and diluting itself with the fresh atmosphere above.)

As population and congestion increase, more pollutants will be released to the atmosphere all over the country. Stringent regulations are necessary to keep existing conditions from becoming worse. The air we breathe and use for many necessary purposes must be kept sensibly free of noxious ingredients, chemicals, and odors.

Planning of the environment poses problems embracing legal, ethical, medical, technological, economic, social, cultural, aesthetic, and political considerations. Teamwork by civil engineers, architects, landscape architects, and planners, who are the design professionals principally concerned with finding solutions to these complex problems, is essential. A private developer or public body, as the owner in need of planning services, must select and engage the prime professional based on demonstrated qualifications and availability. The prime professional will provide specialized services not available in his or her own organization through collaboration with design professionals approved by the owner, and will be responsible for coordinating and directing all the services for which the firm is engaged.

14-2. Planning Trends. The need to plan the environment has existed since primitive times, to provide security from hostile enemies, predatory animals, and nature, and to maintain health and the general welfare of the community. The need to plan is a continuing one, because the requirements of civilization are constantly changing. Primitive camps, which provided for the security of the community, evolved into feudal estates. These offered provision for more extensive human needs, and from the feudal estates, small towns started to grow. Towns further provided for the evolutionary needs of society.

The problem of meeting constant change by replanning an old city was met in London after the fire of 1666, and for planning a new town in Williamsburg, Va., shortly thereafter. Washington, D.C., is an example of a preplanned city. It was laid out by the French engineer P. C. L'Enfant, during the presidency of George Washington.

Formal planning of land has become more necessary with increases in the concentration of populations. While large estates, small towns, and frontier villages were arranged to suit the needs and provide for the interests of only a few individuals, the large metropolitan areas and regions must now provide for the complex interests of present and future populations.

Technological Influences. The capability of using resources in building and the continual changes in this capability are the most decisive elements affecting sound planning. Developments in metallurgy and the variety of new materials that have resulted; progress with portland cement and the ability to construct reinforced concrete; the accumulation of knowledge of the behavior of soils; the means of controlling erosion and of constructing foundations; the development of economical electric power, which provides for the pumping and treating of water supplies and waste water; and the development of air-conditioning and heating equipment are some examples of the technological advances that make the concepts of present-day planning possible.

Changes in the means of transportation alone illustrate the fundamental effect of technology on planning. Prior to the development of railroads, the sites for all major cities were either coastal or on navigable rivers or canals. Railroads provided the transportation that made the location of cities almost anywhere economically feasible and permitted development of the interior of the continent. Automobiles, trucks, and airplanes have added further flexibility in communication and transportation, so that many areas previously deemed unsuitable may now be considered for economic development. Rapid mass transportation has had a major influence on certain areas in the past, and will have still larger influences in many more areas as we progress. Thus, the ability to conceive, design, and successfully use materials to construct and the availability of cheap power and quick communication form part of the engineering base from which planning emerges.

Social and Economic Influences. Sociological changes have developed from technological advances and from the accompanying economic effect on the community. The suitability of plantation communities, large town houses, and estates, with their need for household servants, has diminished, while many people now seek the mass-production economies of small-home building or apartment living to take advantage of their ability to move about freely by motor vehicle. Industrial activities, with the accompanying decrease in the numbers of persons engaged in agriculture, have tended to concentrate populations in urban areas and to advance the economy, and have consequently had a major effect on planning. The changing requirements of planning resulting from economic advances are illustrated by the lack of need today for the company-owned towns that once were prevalent for basic industries, such as mining, textiles, and steel production.

Changing Emphasis. New requirements or more clearly defined needs of communities have always been reflected in the changing of plans. Rural land use has changed from agricultural to industral and residential, while within cities, land has gone from residential to commercial and business uses. With changes in technology and economy, the dispersal of residential, shopping, and industrial activities has grown. Also, the function and nature of downtown areas of large cities have been modified. National concern for renewal of run-down areas has alerted citizens to the desirability of planning, and the availability of Federal funds has expedited rejuvenation of older areas and encouraged movement of certain activities to other areas. Whereas some years ago the approach to residential districts was generally toward subdivision of small or medium-sized tracts of land into individual building lots, the emphasis now is generally on very large tracts that offer a variety of living opportunities, such as cluster villages, town-house apartments, and high-rise apartments, combined with facilities for recreation and leisure-time activities. Retirement communities are still another illustration of the changing emphasis in planning. Furthermore, attention today is more heavily on the preservation of natural beauty and aesthetic design than it was in the past.

14-3. Fundamental Considerations in Planning. Any plan of an area must begin with a recognition of its physical and political dimensions, its population and their mobility, its ecology, and its economic framework and prospects.

The natural resources and physical composition of an area support and limit plans for its development.   The availability of a water supply and the means of discharging waste water and surface drainage (Arts. 14-7 and 14-8), the grade and elevation of the land, and the subsoil characteristics (Art. 14-9) are among the principal elements to be considered in the initial stages.   Also, consideration should be given to the environmental impact of land development.   An appreciation of the economic influence and relative costs of these elements is fundamental to any plan, to achieve successful execution.

A land plan must provide the means to accommodate future needs, which may not now be clearly defined.   At the same time, the plan must be sufficiently practical to be adopted today, and must be economically feasible under present financial circumstances.   It must also make effective use of available resources.

Regional planning, in many cases, depends on adoption of new laws to provide for intergovernmental cooperation.   As population expands, the need for regional planning increases along with the need for municipal planning.

Assignment of land use is a very difficult decision in planning because of the constantly changing technological and economic capabilities of society (see also Arts. 14-5 and 14-11).

14-4. Objectives of Planning. The basic objective of land planning is the creation of an environment in which all of life's processes, including growing, working, and living, can be carried out in the most economic and socially desirable manner, utilizing the natural resources and means available in the region.

By planning, a means of providing for the many necessities of life can be achieved, and a better result can be obtained than can occur by allowing the complicated needs of society to be met by chance.   Planning includes economy as an inherent ingredient because planning aims at efficient utilization of available resources.   In good planning, therefore, the least cost is expended in achieving the desired goals.

One objective of planning must be the adoption of a flexible plan.   Such a plan maintains maximum freedom of choice for individual purchasers and for developers of tracts.   The diversity of individual thought and its expression in the construction of imaginative projects of all kinds must not be limited by unnecessarily rigid controls and restrictions.   The interest that creative designs and personal concepts provide, and their value as an environmental influence, are highly important to a community.

Renewal of run-down areas and prevention of the decay of neighborhoods are a social and economic objective of planning.   They offer great challenge and require cooperation with health, welfare, civic, and philanthropic groups, and involve some of the most complex problems facing society today.

Another fundamental objective of planning is the development of a visionary plan that will be understood and adopted by the public bodies in authority, and can be implemented and controlled effectively.   This objective requires successful public relations, the interest of investors, and the creation of local laws guaranteeing sufficient control to maintain the goals of the plan.

14-5. Master Plans. (See also Arts. 14-3 and 14-4.) A master plan shows by means of maps and written descriptions the use to be made of a regional or municipal area.   The plan also sets the basis on which governing agencies can initiate the necessary controls and actions for its implementation.

Care must be taken in preparation of the plan to preserve flexibility.   Planners should always bear in mind that as changes in desire occur and technological and sociological advances are made, the master plan must be readily adaptable to the changes.

The master plan should clearly show, by appropriate maps, the assignment of land to residential, commercial, industrial, agricultural, educational, recreational, public, and civic uses (Art. 14-11).   Sufficient information must be included to indicate the demands of land use for and the method of providing essential services, such as water, waste water, surface drainage, power, and transportation facilities.   The plan, however, need not go into great detail, such as showing the means of servicing individual residences or structures.   Such details will be designed as the master plan is implemented and individual areas are developed.

Transportation and circulation of traffic are matters of high importance in development of a master plan.   The methods of transporting people throughout the area, whether by individually owned vehicles or public systems, and of moving freight and everyday essentials must be indicated.   Provision for air traffic must also be included if the area is sufficiently large to support air-travel facilities.

The plan should provide for public recreation and safety.   It should include areas for schools, libraries, civic centers, parks, playgrounds, firehouses, police stations, and hospitals.

While the total population is a major factor in the preparation of a plan, the variations in its density and its distribution are of the highest importance.

It is essential that the master plan be prepared so as to satisfy all the needs of the area and the people who will inhabit it, and that these needs be provided for in the most economical way. The master plan must take advantage of all the natural resources at hand.   Every effort must be made to blend nature's treatment of the area into the master plan.

Where water supply and disposal of waste are limiting considerations, the master plan should include suggested restrictions so that proper use is made of the limited services, to insure successful implementation of the plan.   Other controls necessary for successful operation of the plan should be included when applicable.

14-6. Information Needed in Planning.   The information needed for preparation of a master plan includes physical, social, legal, and economic data that establish the pattern of the area under consideration and must be related through the skills and training of the planner.

Physical features of the area to be planned heavily influence the assignment of land for residential, commercial, industrial, agricultural, educational, recreational, public, and civic uses.   Series of maps useful in planning are available from agencies of the Federal and state governments for almost the whole of the United States.   On these maps are shown the general topography, soils, and such physical features as wooded areas, water courses, roads, power-transmission lines, political boundaries, and other existing objects that may affect the general appearance and future use of the area.   Aerial photographs are available for many areas, or they may be ordered for planning use.   The services of land surveyors may be required in obtaining new data or in assembling existing material.

Special care should be taken in locating boundaries.   Inadequate project boundary information will result in poor legal descriptions and inaccuracies in the layout of the plan.

Records of rainfall, stream flows, flood flows, wind direction and velocity, and other pertinent data are published by Federal, state, and municipal agencies.   Existing traffic activity and densities are often available.   Such information is essential to the preparation of the land-use portion of a master plan.   Maps showing present zoning also should be obtained.

Nothing, however, takes the place of the planner's own field reconnaissance of the area.   The planner should be fully acquainted with its physical characteristics and peculiarities, which, in many instances, do not appear on published maps or are not covered by available data.

Consideration must be given the economic aspects of the area.   These play an essential role in determining size, location, and characteristics of the planned industrial and commercial centers.   Census maps, past population data, and estimates of the probable future population growth must be analyzed with the economic factors relating to potential industrial and commercial enterprises to determine the type and locations of housing required in the community under consideration.

The social characteristics of the area affect the location and number of educational, recreational, public, and civic facilities needed to serve the population.

14-7. Significance of Water Supply and Waste Disposal in Planning.   Two controlling considerations in the initial concepts of a master plan are the availability of a potable water supply and means of disposal of domestic and industrial wastes.   Until it can be well established that an existing water supply in a reasonable and economic distance of the area being planned is available, or that one can be developed within the area itself, little or no detailed study need be expended in the preparation of the master plan.   In a similar way, a determination regarding waste disposal into the air or water or on the land must be made at the outset.

Water.   An abundance of aquifers or a great distance from usable surface water suggests the dependence of an area on wells as a source of water.   In such cases, early investigation should be made with test wells to ascertain the productivity of aquifers and quality of the water produced.

For areas within reasonable distance of a surface supply, early investigation of its safe yield and quality is desirable.   Then, planners can make an economic choice of the supply most suited to the contemplated master plan.

For municipal planning and, in some instances, regional planning, the water supply problem may best be solved by reliance on adjacent sanitary districts or other large municipalities.   Such a solution was adopted for the new town of Columbia on 13,000 acres in Howard County, Maryland, between Baltimore and Washington, D.C.   Early in the planning for Columbia, studies were made that proved conclusively that the most economic source of water would be the Baltimore municipal system.   Thus assured of an adequate and safe supply of water for the development of this new town of 100,000 population, other aspects of planning could be confidently undertaken.

Waste Water.   Following a preliminary study of an area to determine land use and the resulting requirements for potable and waste water, early investigations must be made to determine whether economic means of disposal of domestic and industrial wastes are available.   As for water supply, expensive methods of disposing of wastes will be reflected in the economic consideration of the plan.

In many cases, the problem of disposal of domestic and industrial wastes is more difficult than that of providing for an adequate water supply.   The location of many industries depends on the availability of an inexpensive source of water and an easy way to discharge wastes, after moderate treatment, into water courses of sufficient flow to assimilate the wastes without the expense of higher degrees of treatment.   Industries whose waste discharges will be highly acid or caustic or are toxic are separate problems.

As planned, the new town of Columbia is located within two drainage areas, each consisting of about 60 square miles of watershed, with low flows too small to take care of residential, commercial, and industrial liquid wastes after a moderate degree of treatment.   Disposal of waste water originating in this region can be accomplished only by construction of treatment facilities to accomplish over 90% reduction of pollution.   Thus, with knowledge beforehand of the nature of the problem of waste disposal and the approximate expense, the planners were able to prepare the master plan of Columbia from a sound base.   The likelihood of industrial installations utilizing large quantities of water locating in Columbia is remote, in part because of the relatively high cost of water.   Further, the relatively high cost of sewer service, 100% of the water rate, offers additional economic restrictions on the large water user in this area.

Solid-Waste Disposal.   In laying out land for its many uses, planners sometimes put aside for later solution the problem of how to dispose of solid wastes, including garbage, waste paper, and other trash.   Such deferment is likely to be a mistake, for solid wastes accumulate at a rate ranging from a few pounds per person per day for domestic refuse to tons of special wastes by industry and commerce.

The householder often provides a place at the rear entrance for a container for deposit of food scraps, waste paper, food containers, bags, cans, bottles, and other waste items.   Also, newspapers, magazines, tree trimmings, and, periodically, discarded bedsprings, baby carriages, old tires, and excess building materials accumulate and have to be removed regularly.

Generally accepted methods include collection of household wastes several times a week.   Such material is hauled to designated points of discharge where it is placed in open dumps or prepared sanitary landfills, or burned in incinerators, or decomposed in compost plants.

Convenient locations for the disposal method selected must be provided because of the high cost of hauling wastes longer distances.   Often, however, such a site cannot be arbitrarily selected because of objections of nearby residents to close location of a disposal facility.   Politics has a strong effect on rejection of a feasible disposal site.

An incinerator, however, can be built to look like an industrial building, indeed like a school.   It need not have visible stacks.   Gas emissions are controlled so that no floating ash is apparent.   The material furnished the plant is almost fully burned.   The residue ash is inert and makes good fill material with no subsequent subsidence.

Planners should provide for collection of solid wastes by designating locations for refuse disposal.   Otherwise difficulties will be encountered from new land owners and from elected officials over the subsequent location of such facilities.

Air Pollution.   Following a preliminary study of an area for land-use purposes and determination of the probable resulting atmospheric pollution, consideration must be given to the limitations of the surrounding atmosphere in absorbing pollutants.   Stack discharges from refineries, power plants, chemical manufacturers, residences, and many others dissipate particulate matter, dust, and chemicals, which combine to create health hazards and to damage vegetation and materials.   Not only should industry meet Federal and local air-pollution standards for emissions, but also sites for industry should be planned to minimize the possible effects of air pollution on other land uses.   Reference to meteorological data is required for such planning determinations.   Planning of open land in the "downwind" direction from industrial areas is sometimes made use of as an "air shed" to aid in reduction of air pollution.

After determining that economic and adequate water supply and waste disposal can be obtained, planners can study the preliminary land-use plans to locate the water-supply and waste-disposal facilities and decide on the use of lands surrounding these facilities so that they may be properly merged into the master plan.

14-8. Significance of Drainage in Planning.   A matter to be investigated and thoroughly studied during the assignment of land use for a master plan is provision of drainage for rainfall runoff.   Topographic maps are required for such studies.

Natural channels often provide the most economic means for conveyance of surface drainage to points of disposal.   But the quantity of water to be conveyed is the main consideration when choosing between a closed conduit and an open channel.

Rainfall runoff may be estimated by use of the rational formula $Q = CiA$, in which Q = quantity of runoff, cfs; C = average runoff coefficient, depending on slope, soil, land cover, and time of concentration; i = intensity of rainfall, in. per hr; and A = area under consideration, acres. The runoff coefficient must be selected from knowledge of the area and experience and must be appropriate to the types of development contemplated by the master plan.   For areas where the exact land use is known, a precise runoff coefficient can be applied; however, as the drainage structures in the lower reaches of the area are considered, with variable land uses contributing to the runoff, a calculated composite coefficient must be used.

Government agencies at all levels maintain rainfall records covering many years.   From these records, rainfall intensities have been computed corresponding to storms of 1-, 2-, 5-, 10-, 25-, and 100-year frequency.   Rainfall-intensity curves have been developed by governmental agencies and many engineers in private practice, and these curves are available for use in preparing preliminary drainage studies for master planning.   For small local areas, it is common practice to design for rainfall intensities of 10-year frequency.   As larger areas are considered, particularly where open channels are utilized for conveyance, the rainfall intensities for design purposes are those corresponding to 25-year and, in some instances, 50-year frequencies.   For those areas where published data are not available, the experience of an engineer must be relied on to determine the appropriate rainfall intensities for preliminary purposes.

The cost of providing drainage facilities is frequently one of the largest single costs in the orderly development of an area.   The land use selected for each section directly affects the drainage cost.   In many instances where one use of the land could not economically justify construction of closed conduits, such as in single-family residential areas, a different use, such as industrial or commercial development, could easily support the costly construction of closed conduits.

Once the preliminary quantities of runoff and general alignments of the larger drainage facilities have been determined, it is possible to assign preliminary land use to the area.   It may later be necessary to reestablish the preliminary land use after more specific drainage studies have been made in conjunction with other items affecting the cost of development of any land area.

Storm-drainage facilities are generally located along roadways; in some instances, they determine the alignment of streets and roads.   Open water courses, if capable of carrying the estimated flows with little or no improvement to the existing stream channel, are often utilized. Areas adjacent to natural waterways may be assigned to recreational or park-land uses.   Exceptions occur in industrial and commercial areas where the land value may make it economically necessary to enclose the drainage ways and to utilize the area gained thereby for high-density development.

Because of the relatively high cost of drainage facilities for an area, it is mandatory that careful consideration be given to the drainage problems prior to and during the development of land-use maps and master plans.   Inadequate storm drains may flood and damage private property and be the cause of expensive legal proceedings and claims.

14-9. Importance of Surface and Subsurface Features in Planning.   The master plan for regional or municipal areas or for renewal of run-down districts must, of necessity, be prepared so that its implementation can be accomplished with a minimum of expense and yet result in an environment that is aesthetically pleasing and fully useful.   To accomplish these results, planners must give considerable thought to the surface and subsurface features of the area included in the plan.

Surface Features.   A review of existing topographic maps or aerial photographs of the area will acquaint the planner with general surface features, such as woodlands, lakes, location of waterways, and the ground slope.   Such information must be substantiated, and the planner should obtain a broader outlook by visits to the area.

Surface featues such as ridges, knolls, valleys, woodlands, water courses, and lakes should be utilized to as great a degree as possible to achieve the economic development of the master plan and to obtain maximum aesthetic and environmental values at a minimum of cost. Such information as is necessary for the planner to achieve these goals must be obtained in developing the land-use map and in making the master plan.

Subsurface Features. In many localities, geologic surveys have been made and maps prepared that provide invaluable information as to subsurface characteristics. Logs of test wells and producing wells are also available which record extensive information for planning purposes, such as strata of soils, elevation of groundwater, and location of rock.

Three of the most costly subsurface conditions encountered during construction are high water tables, unstable soil, and solid rock close to the surface. Locations of such unfavorable subsurface conditions should be clearly established before preparation of the master plan, and consideration must be given these conditions in establishment of land uses. Where existing information is not available for the area, it is essential that elaborate subsurface investigations be completed prior to development of the land-use map and preparation of the master plan. Soil classification by qualified engineers using ground-testing equipment is required in such investigations.

Occurrence of rock at a shallow depth must be considered in preparation of the master plan; it will have a substantial effect on the construction cost of each element of the plan. Careful selection of land use and the elevations and grades of actual construction are important in reducing the economic disadvantages of developing such areas. In areas having shallow subsurface rock, little or nothing can be done to eliminate increased construction costs. But the astute planner, with a knowledge of such costs, can control and reduce these increased costs by careful studies during the planning period.

Water Table. A water table close to the surface should be of great significance to the planner in preparing the land-use plan. In general, lowering the water table temporarily for construction purposes is relatively expensive. So every effort should be made to keep penetration of construction into the water table to an absolute minimum. In some instances, it is possible, through construction of drainage facilities, to lower the water table permanently either by gravity or by automatic pumps. The decision to do this should be made before preparation of the land-use map so that the general plan can reflect this provision.

Accompanying a high water table generally are unstable soil conditions. These should be explored and considered prior to preparation of the plan. The existence of such material should be known early and taken into account in preparation of the land-use map, to minimize construction and development expenses that would otherwise occur.

Explorations. Subsurface explorations should be made in areas where the preliminary land-use study calls for construction of major structures. Such explorations may indicate unsuitable subsurface conditions requiring expenditure of large sums of money to accommodate the planned structures. Movement of structures instead to another area may accomplish the same overall purpose in land use and materially reduce cost of construction.

In many cases, insufficient data have been assembled concerning subsurface conditions, and too much reliance has been placed on guesswork in preparation of master plans. Adequate planning demands data collected from proper subsurface explorations so that the reasonableness of development costs may be considered.

14-10. Population Studies and Analysis. Following the preliminary study and investigation of water supply, waste disposal, drainage, surface and subsurface features, and the economic potential of an area, all of which support and limit its potential and affect the selection of land uses, a determination of future population over a range of years must be made.

To determine the future population of an area, studies of similar areas and their growth over past years should be considered. Such growth characteristics, however, should not be the only consideration. To utilize the various methods of projecting population from past growth may jeopardize preparation of a master plan.

Population projections based on several land-use "models" of the area, arranged for statistical comparison with the aid of electronic computers, are sometimes used for study purposes. The "models" make allowance for various land-use patterns and densities; availability of transportation and public utilities; economic, zoning, and regulatory factors that mutually influence the size of the estimated population and its distribution at selected future times. Several studies based on differing reasonable assumptions offer a range of future population predictions to which judgment may be applied in making final estimates.

In many instances, the projection of population for master-plan use will be complicated by the existence of adjacent areas, or included areas, that have developed without the benefit of a master plan, resulting in an imbalance between industrial, commercial, residential, and other uses affecting population. In some areas, an unfavorable ratio of population to employment opportunities exists, with accompanying economic problems and low per-capita income. Such conditions cause the planner difficult problems in providing that population with adequate facilities, while attempting at the same time to develop a proper master plan for the entire region or municipality.

14-11. Assignment of Land Use. Assignment of land for various purposes depends on the geographic location of the region or municipality for which a master plan is being prepared. Assignment of land use also depends on the characteristics of atmospheric conditions, topography, soils, vegetation, and drainage of the area. The master plan must also allow for existing development and provide for correction of deficiencies within such existing areas on adoption of the master plan.

Balanced Allocations. Except for areas oriented to a single purpose, such as recreation, a master plan usually must provide for a self-sustaining community and must include balanced allocations of land for industrial, commercial, residential, educational, public, and civic uses. In allotting land for industrial and commercial purposes, planners should consider the types of activities that can utilize the area. Industrial and commercial development play a large part in the character of the area and have a strong influence on preparation of a master plan.

Areas adjacent to port facilities, rail centers, or raw-material deposits will naturally be centers of heavy industry, and allocation of land for residential, commercial, and municipal purposes should be made accordingly. For areas where industry is composed of research and development enterprises whose employees have above-average personal incomes and scientific educational interests, land-use allocations for residential, commercial, and municipal purposes must suit the needs associated with that type of industry.

Once the economic potential has been determined and land allocations made for industry and commerce, the related uses of the area can be established. Then, the land-use map can be drawn to accommodate balanced residential, educational, recreational, public, and civic needs.

Master plans for recreationally oriented areas require little consideration for industrial land use. In many instances, attention must be paid to a combination of a seasonal recreational area and a year-round occupancy area. In such situations, large populations must be accommodated during the recreational season, whereas only small populations are available to support the facilities during other periods.

Families. Establishment of land use for a master plan of a region must take into consideration the appropriate land-use assignment for each individual unit of the region, from the individual family to the region itself.

In preparing the regional land-use map, the family is considered first, since it is the smallest social unit. For general planning purposes, it may be considered that the family consists of 3.6 persons. The needs of this basic unit must be satisfied. The family must be protected from traffic and other nuisances and have adequate fresh air, sunshine, an attractive outlook, privacy, and recreational facilities.

Allowance must be made for groups of families that will, of necessity, live together and will constitute the smallest of the interfamily social groups. Their requirements, although almost identical to the needs of the family unit, call for a focal area where the family groups together can find harmonious and pleasant recreational facilities free from nuisances and with adequate amenities.

Neighborhood. The next larger group or unit that must be considered is the neighborhood. This usually consists of about 500 to 600 families but may in higher-density areas consist of as many as 1,000 families. Planners should consider providing an elementary school for every neighborhood (Art. 14-13). This school, combined with park land, should be the focal point of the people living in the neighborhood.

Community. Grouping of neighborhoods makes up the next larger unit, the community. Education once again is one of the foremost bases for planning of communities. A community consisting of about 10,000 family units will require one high school and two junior high schools, in addition to the elementary schools in each neighborhood (Art. 14-13). Needs of the community, in addition to education and recreation, include a hospital, library, churches, athletic fields, extensive shopping facilities, and small industrial and manufacturing installations. These should be provided for in the plan.

City. The next larger unit in planning is a city or town, which is made up of its communities. A city contains public buildings and civic centers with offices, department stores, theaters, museums, and other facilities to provide for full economic, social, and political activity. The city is the center of the economy and culture for the surrounding communities. It provides a place for people to meet to conduct business and exchange ideas. It serves the population's basic needs for political and social activities. The plan must reflect all this.

The city plan must provide a network of highways so that people can readily be transported and goods and services distributed. The city plan must include parks and other spaces that contribute to the beauty of the area, improve the environment, and contribute to the sense of well-being of the people (Art. 14-14). In addition, the plan should recognize that the city must provide, operate, and maintain the necessary public utilities for health and safety.

Region. The largest unit in planning is the region, which includes several cities and towns with rural areas or a major city with surrounding metro-towns. It is mandatory that the region be planned to provide for its best uses after the many factors and natural resources that affect it have been considered. Ideally, the region should furnish the necessary agricultural land for subsistence, industrial and commercial enterprises for livelihood, and forests and open spaces for recreation and pleasant environment. The region must have a form of government that will avoid overlapping services, provide the people with an equitable tax structure, and prevent unhealthy competition for revenue or power among smaller political subdivisions.

14-12. Residential Districts. Planning for residential districts centers around the neighborhood and community concepts, rather than around the larger regional form (Art. 14-11). Residential districts are dependent on the overall neighborhood plan, which includes the street pattern and location of schools, community facilities, and shopping areas.

For the actual planning of the residential areas, certain general criteria and standards should be applied. Good planning provides for adequate light and clean air, protection of residents against everyday noises, outdoor space for daily needs, and safe surroundings in which to live. Beauty of surroundings, which enriches the quality of life, should not be overlooked. Since there are various types of residential areas, the above criteria may be satisfied in numerous ways.

Building Types and Purposes. Residential buildings consist of single-family separate and semidetached homes, group or town houses, and multifamily apartment structures. In general, families with children prefer one- and two-family houses or garden apartments. Elderly people and families with no children or grown children usually prefer garden or high-rise apartment buildings.

Selection of different residential building types should be based on the community concept, since a neighborhood could consist of only one type of residential building.

A detached single-family dwelling is the most desirable type of housing for obtaining an abundance of air, light, and privacy. Such a single-family home is located on its own lot. The yards surrounding the houses are for the private use of the family. They offer the children freedom to play and an opportunity to make the normal noises of living without abnormal disturbance to the neighbors. When one-family houses are crowded on small lots, however, the long, dark passages between buildings present an unattractive appearance. In such cases, group or town houses, which have more space between building units, are preferable. The space between a detached house and an adjacent one should at least equal the width of the house.

For two-family dwellings, semidetached houses are more desirable than duplex buildings. The former offers greater yard privacy and noise isolation.

In planning for group or town houses, the number of units per group should be restricted to no more than eight to ten, to avoid monotony in appearance and provide occasional access to rear yards.

In planning neighborhoods and communities, it is important that the dwelling units selected provide for the economic cross section of the area being planned. The normal residential cycle starts with the newlywed occupying relatively small quarters. As financial ability and family size increase, movement to a more adequate dwelling, such as a detached home, takes place. After the children are grown, the need for large living space is usually diminished, and an apartment or group or town house is then preferred.

Terrain. Locations for the various types of residential dwellings can be determined after a study of the topography and subsoil conditions within the area. In general, all types of dwellings can be constructed in relatively level areas if the subsoil conditions are satisfactory. For rugged terrain, selective siting, larger lots, split-level construction, and high-density dwellings are more applicable. In many instances, rough terrain that does not lend itself to economical individual-home construction is satisfactory and attractive for apartment buildings with several entrance levels.

Sanitary Systems. Often, the absence of public water and sewage facilities determines the density and type of residential dwellings that may be constructed. In general, where only private on-site sanitary facilities are possible, construction should be limited to detached houses on lots with a minimum area of 20,000 sq ft.

For areas where public water and sewage facilities are available, a one-family detached dwelling may be located on a lot with a minimum area of 6,000 sq ft and a frontage of 60 ft. Where more expensive detached houses are constructed, the area of the lot may be increased to about 20,000 sq ft and the frontage to 100 ft. A two-family duplex house may be located on a minimum of 8,000 sq ft, a net area of 4,000 sq ft being allocated to each family. A lot frontage of 80 ft is appropriate for such cases. Semidetached dwellings also should be located on a lot at least 8,000 sq ft in area, with 4,000 sq ft allocated to each residence. For group homes or town houses, the average lot area per family should be a minimum of 2,400 sq ft. This can be obtained by allowing a 20-ft front for each unit and a 40-ft side yard between each group of eight to ten units.

Densities. Based on the lot sizes suggested, the average family density per acre of net land area (lots only) should be 5 units for one-family detached homes, 10 units for the two-family duplex or semidetached, and 16 units for the group or town-house dwellings. For apartments, recommended density varies with the number of stories in the building. For three-story garden apartments, density is often about 40 apartments per net acre, whereas for a 10-story high-rise apartment building, density may be about 80 units per net acre.

The ideal average density for all residential purposes of different types in a completely planned community is 2.5 dwelling units per gross acre.

14-13. Schools. Education of residents of neighborhoods, communities, cities, and regions has become and undoubtedly will continue to be one of the most important and costly facets of American life. Great care, therefore, must be taken in preparation of a master plan to provide for education in the area being considered.

Preschool. Education of children naturally begins in the home. Through proper planning, the environment of the individual home can afford parents the maximum of educational facilities for preschool children.

Planning must consider nursery schools and kindergartens. In many areas, responsibilities for these facilities are not borne by elected officials, and the residents must look to private or parochial facilities to provide this beginning of formal or group education. The master plan should provide for the requirements of such facilities, and the land-use map should make provision for their proper location in each neighborhood. Where kindergartens are not part of the public-school system, they should be combined with nursery schools. If the kindergarten is part of the public-school system, it should be included in the design of the neighborhood elementary schools.

Elementary Schools. Cost of transportation of students and their safety are major reasons for planning elementary schools on a neighborhood basis and so locating them that public transportation is not required. Based on an average family of 3.6 persons, a neighborhood will contain about 2,000 persons, of whom about 200 will be of elementary-school age. Elementary schools consist of grades 1 through 6. In general, one elementary school consisting of about eight classrooms planned for a maximum of 30 pupils per classroom should therefore be provided for a neighborhood consisting of about 500 to 600 families.

The elementary-school playground should be planned as a neighborhood focal point. It should consist of playground toys, ball fields, basketball courts, and other recreational facilities required by younger school children. Elementary-school sites should be located, if possible, adjacent to park areas, to add to the value of the playground as a neighborhood recreation area (see also Art. 14-14).

To provide adequate space for an elementary school and playground, a site of 20 to 25 acres is required. This area may be reduced to 10 to 15 acres if the site is planned to make use of adjacent park areas.

Junior High and High Schools. Junior high schools and high schools should be considered on a community or district basis. In either event, a population of 10,000 families would constitute the community or district. Based on 3.6 persons per family, the population of such an area would be 36,000 persons.

Junior high schools consist of grades 7 through 9. Two such schools would be required for a community or district with a population of 10,000 families. The 10,000 families will yield about

630 students for each junior high school. With a maximum of 30 pupils per classroom, each school would have about 21 rooms.

High schools consist of grades 10 through 12. One high school is sufficient to serve the community or district. The 10,000 families will yield about 1,000 students, and based on 30 pupils per classroom, they would require a 34-classroom building. Such schools require up to 50 acres of land.

Locations of junior high and high schools should be such that the minimum amount of transportation is needed by the students. This criterion would suggest the location of such facilities in high-density residential areas or in the vicinity of civic centers. Planning should anticipate the use of land around the school to provide outdoor recreational facilities and indoor recreational, social, and cultural facilities for the community or district.

Safety. All schools should be located so that children using the facilities during school or nonschool hours will not have to cross heavily traveled arterial roads or major highways. Sites that minimize street crossings will reduce accidents and the cost of patrolling the crossings during school hours.

Private Schools and Colleges. In preparation of a master plan, the needs of parochial and private educational facilities through high school must be considered. Such facilities influence the number of public-school rooms required, since the combined institutions should be scaled to serve the population.

Campuses for junior colleges and universities must be anticipated for regions and large municipalities. Location and size of such facilities can best be planned after discussion with the responsible educational body.

14-14. Recreation. All areas occupied by people must incorporate suitable places for recreational use, whether the unit under consideration be as small as a residential lot or as large as an entire region. The character and size of recreational facilities are related to the needs of the population being served and the ability of the residents to bear the expense.

Since recreational activities and the demand for facilities for such purposes vary greatly among individuals and groups, planners should know the area and interests of the people and plan recreational facilities so that the basic needs can be fulfilled.

Some areas contain natural recreational features. Communities located on bodies of water, for example, naturally turn to the water for a large portion of their recreation. Since such communities are oriented toward the natural recreational feature, the demand for additional recreational facilities is less than that in areas without such natural features.

School Grounds and Buildings. In a neighborhood, the basic facility available for both outdoor and indoor recreational purposes is the elementary school. Junior high and high schools are similarly available for neighborhood, community, and district use. Use of school grounds requires cooperation between school authorities and the recreation department, which organizes, operates, and supervises use of the playground during nonschool hours. The recreation department sometimes assumes responsibility for programs within the school building. In addition to providing recreational facilities for the children in the neighborhood, the school provides some recreational possibilities for adults, such as softball, tennis, basketball, and badminton. The school building also can be used for adult education and indoor social and cultural activities. All schools generally include an auditorium, gymnasium, library, and some arts and crafts rooms, which can be used for recreation for children and adults in the neighborhood (see also Art. 14-13).

In addition to the active recreational facilities afforded a neighborhood by elementary schools, there are also passive facilities that should be included in each neighborhood, such as a park, which will provide opportunity for less strenuous exercise and exposure of people to fresh air and sunshine. Neighborhood parks should be located near the school. When used in combination with school facilities, they should be planned to fulfill both active and passive recreational needs of all ages.

The recreational needs of communities and larger areas are served in various ways by the facilities of junior high and senior high schools; junior colleges; universities; golf, swimming, tennis, and country clubs; lakes created by water-storage reservoirs for domestic use or power generation; marinas; indoor and outdoor theaters; municipal stadiums and sports centers; riding academies; wildlife preserves; park lands; and other public open spaces.

Playgrounds. It has been found that the smallest playground that can provide acceptable types of equipment and activity space is about 3 acres. This minimal playground will serve about 300 families. Six acres is considered the maximum size for a single playground. It will serve about 1,200 families. The play area within a neighborhood playground should be about 200 sq ft per family.

In selection of a site for a playground, careful attention must be given to the topography of the area. Playgrounds should be located on relatively level, well-drained land. Unless proper drainage is provided, the number of days playgrounds cannot be used after inclement weather will be excessive.

Construction materials for playgrounds must be selected carefully, since the areas are primarily for children. Use of resilient surfaces, rather than concrete, is preferred. Sod is ideal, but its maintenance cost is usually prohibitive. Unsurfaced playgrounds are usually dusty. Their use is limited during inclement weather because of muddy conditions and the slowness of drying.

Park Areas and Locations. Neighborhood parks need be only 2 or 3 acres in size for one- or two-family-dwelling areas. For multifamily-dwelling areas, the park for the same number of persons should be about twice as large.

Parks should be designed to provide pleasant views. They should have walkways and adequate shade and benches for resting and contemplation. Parks may also include some active recreation, such as horseshoe pitching, croquet, and other not-too-strenuous exercise.

If parks are to be used at night, adequate lighting must be provided for safety purposes. The park area should not be divided by streets or thoroughfares and, as in the case of schools, should be located so that children may have ready and safe access.

The desirability of preserving natural woodlands and the suitability of land along streams make these areas desirable choices for park lands.

Parks or passive recreational areas need not be located in one parcel. They may include the land surrounding public buildings and civic centers if such areas contain walks, benches, landscaping, fountains, and other recreational features.

In determining areas of parks, planners should not consider the land devoted to vehicular traffic or street and highway rights of way as park land. The area allocated for park land or open space in planning should be approximately as follows:

Neighborhood: 3 acres per 1,000 population. This area includes school playgrounds, parks, and other recreational areas.

Community: 5 acres per 1,000 population. This area includes school playgrounds, public and private athletic fields, and park land.

City: 10 acres per 1,000 population. This area includes all public open spaces, parks, and playgrounds.

Regional: 20 acres per 1,000 population. This area includes all public open spaces, parks, playgrounds, and recreational areas for hunting, fishing, and wildlife.

14-15. Shopping Centers. Easier transit to outlying districts than to downtown areas combined with an affluent economy has led to dispersal of shopping facilities over wide areas and changes in people's buying habits. Shopping centers of various sizes and shapes are the result. These are now an important part of municipal and regional planning.

Location of shopping centers depends on the distribution of population and its economic level, location of competing facilities, and availability of convenient access. The decision of developers to construct a shopping center is based entirely on economic considerations. Developers weigh the cost of land, cost of interest on borrowed money, and other developmental and annual expenses against income the property can produce. If this comparison is favorable, the project will proceed. If not, it will be deferred until the economic variables indicate a more attractive investment opportunity. In making a decision to construct a shopping center, a developer often needs the advice of market-research specialists, commercial realtors, accountants, engineers, architects, and contractors to collect and analyze important information affecting the decision.

As a rule of thumb, shopping centers are identified as neighborhood, community, and regional, depending on their size and number and types of stores included. A neighborhood center usually serves a population of from 1,000 to 5,000; a community center, from 5,000 to 50,000; and a regional center, large urban areas. The area of sales space required to serve the higher limit of the range of populations indicated is about 30,000 sq ft for a neighborhood center; 100,000 to 250,000 sq ft for a community center, and up to more than 1 million sq ft for a regional center. The minimum land areas required for the lower range of populations indicated are about 5 acres for a neighborhood center, 12 acres for a community center, and 30 acres for a regional center. (Many believe that 50 acres is the minimum adequate land size for a regional center.)

Next in importance to the nature and availability of a buying market in selection of a shopping-center site is its ready accessibility. Shopping centers in outlying areas depend on transportation of customers by private automobile. It is customary, therefore, for shopping centers to be located along main arteries, preferably at the intersection of well-traveled thoroughfares.

A neighborhood center may be successfully operated with access from the local street pattern, while a community center depends on access along thoroughfares, and regional centers require access from one or more major highways. A minimum of several points of entrance and exit is necessary in all types of shopping centers.

Location of entrances and exits with respect to the existing street and highway pattern must be carefully planned to reduce congestion and provide for left-turn movements and storage, acceleration, and deceleration lanes. Traffic signals, channelization, and other traffic-control measures demand study by traffic engineers to achieve the best results.

Facilities Provided. The kinds of stores needed to offer the services required vary with the type of shopping center. A neighborhood center customarily depends on a food market and drugstore as its major, and sometimes only, features. Often, small service shops, such as laundry, bakery, and beauty salon, are included. These stores cater to the daily life and health needs of the neighborhood. The needs of larger population groups are usually met in a community shopping center by a broader variety of service stores, including a junior department store, two or more food markets, drugstore, restaurant, dress and shoe shops, bank, liquor, hardware, and service stores. A regional center includes one or two major department stores, variety store, two or more food markets, drugstore, specialty shops, apparel stores, bank, and sometimes general and professional office space, medical clinics, motel, and recreational facilities.

Building Types and Arrangement. The shape of the land allocated to shopping-center purposes, its topographic features, and its elevation with respect to surrounding access roads control the dimensions and areas of the building groups planned for the tract. A compact, rather than strung-out, land shape is desirable. Land having less than 600 ft in depth from the front roadway to its back property line should be considered marginal for even a neighborhood shopping center.

Buildings may be arranged as a row, or strip of stores facing the street, with parking in front and loading in the rear; in a crescent or U shape; in an L shape; as a hollow square or rectangle with a central court (either open or enclosed); or as small groups clustered in a pattern affording a court or mall arrangement.

The buildings are usually one-story, with partial basement. Department stores, however, are often three-story. The larger food markets, variety shops, and junior department stores frequently range from 20,000 to 100,000 sq ft each in floor area and have width and depth dimensions of 150 to 250 ft. Department stores often have dimensions ranging from 250 to 400 ft with 50,000 to 150,000 sq ft on each floor.

Parking. Within a shopping center, customer parking, traffic control, and access of service vehicles should be carefully planned. In large centers, a roadway that circles the property between the entrance and the stores permits good movement and better use of all the parking land. The width of entrance and exit facilities, the circulation roadway, and the service areas must be adequately planned. Parking fields should be divided by circulation roadways, channelization, landscaping, or other means, for convenience, identity, and appearance. Too many raised curbs for parking controls should, however, be avoided, because of the confusion and danger to drivers, additional maintenance, and difficulties of snow removal.

Individual parking spaces should be a minimum of 9 ft wide, preferably 10 ft with double-divider markings. More vehicles can be accommodated with parking spaces at right angles to the lane from which they are entered than with diagonal parking.

The most remote parking should not be more than 600 ft from the store entrances. It is generally assumed that those parking spaces will be utilized only during periods of maximum shopping activity. During average periods of activity, only the spaces up to about 300 ft from the store entrances will be used.

The number of car parking spaces per unit of store sales area frequently depends on the requirements set in the leases by large national retail chains who are the principal tenants. Parking areas may be three to four times the area of land occupied by stores. With the higher ratio, about 10 car spaces per 1,000 sq ft of store are required. The Urban Land Institute studies show that 5.5 car parking spaces per 1,000 sq ft of store area are adequate for all but the 10 busiest shopping hours annually.

Services. Store service areas are in the rear of strip shopping centers. Such areas may face the entrance roadway of hollow-square- or enclosed-mall-type shopping centers or may be underground, with entrance by tunnel, where terrain and store arrangement offer that possibility. Service areas must be adequately screened from view and afford parking and turning space for tractor trailers.

Water supply, waste disposal, refuse, collection, storm drainage, electrical power, telephone, and gas service are required to serve the needs of shopping centers. Potable water supply must be adequate for normal domestic, lunch counter, restaurant, washing and sprinkling, laundry, and other special uses, and for air conditioning, and fire protection. Shopping centers customarily depend on municipal water systems to meet such requirements. Waste-water pipes must be provided for each store space at an elevation low enough to drain basement fixtures and are customarily connected to a municipal sewerage system.

Storm drainage and the manner of catchment of the surface water must be carefully considered in selecting the grades for paved areas. Parking areas must be positively drained for convenience to foot traffic and economy in maintenance of the paving.

Electric and telephone services into the property and to each store space are customarily run underground for appearance. Transformer stations may be underground or on the surface, concealed from view. Outdoor lighting systems are installed for appearance and safety.

Esthetics. As a result of relatively large building dimensions and relatively low silhouette, in combination with a large number of acres of pavement, shopping centers may be unattractive in appearance. Therefore, they deserve the best architectural, engineering, and landscaping advice. The diversity of the demands of major tenants, the owner, and the public in shopping centers creates a complicated design problem and also demands the best professional talent. The architectural treatment of the buildings, control of tenants' signs, treatment of outdoor lighting, and screening and landscaping of the site are important in the aesthetics of modern shopping centers.

14-16. Industry.   Sites selected by industry for its various operations are those that offer the most economic advantages.   Considerations affecting site selection include source of raw materials, area of sales market, local taxes, availability of labor and transportation, and necessity for special services. With regard to transportation, all modes, including air, sea, rail, and highway, may be involved, and any one of these may limit the number of locations that may be considered. For example, a firm manufacturing specialized electronic equipment for aircraft may need to be situated at an airport so that the output may be factory-installed; an industry importing bulk raw materials and shipping heavy finished products to worldwide users may need to be on navigable waters.

Regarding the necessity for special services, the primary needs of industry are electric power, water supply, and waste disposal, all of which must be considered in planning industrial locations.   Every industry needs to dispose of liquid, solid, and gaseous wastes, and provisions for collecting, treating, and controlling these are a major consideration in planning.   Where municipal sewerage systems exist, industrial wastes are customarily discharged to them, with prior treatment by the industry in some cases.   Where public sewerage systems are not available, private means of treating and disposing of industrial wastes must be provided.

In general, basic industrial demands are met by facilities that mine, process, manufacture, assemble, warehouse, and distribute materials and products. Other industrial buildings include research laboratories, regional or corporate headquarters, and sales offices. These facilities are frequently situated in areas away from the more basic industrial activities.

An individual site occupied by only one firm is common for a large facility and those having specialized needs. For smaller and less demanding operations, however, a number of industries may cluster into an area planned as an industrial park. From the planning standpoint, the industrial park offers the means of grouping industries in an economic and harmonious manner.

Land. The size and shape of land areas allocated for industrial use must vary with the intended purpose. The surface topography and subsurface conditions are major factors to be considered because of their influence on cost of development. It is important to obtain knowledge of subsurface conditions for use as one of the bases for allocating industrial land.

Relatively flat land is desirable. It permits lower site-development costs. But rolling land, if otherwise suitable, can be developed if it is understood that higher development costs will be incurred.

Industrial parks should generally range in size from 100 to 500 acres and should include a generous allowance for future growth. Individual tracts within an industrial park may vary from about 2 acres to more than 40. Industries requiring still larger tracts, however, more frequently locate on individually selected sites, rather than in industrial parks.

Traffic. The layout of industrial parks must provide flexibility to suit varying tract requirements, which are seldom known at the outset. Planners must make provisions for good movement of trucks and vehicles and expansion of utilities to all areas within the park.

Entrance and exit connections to the highway system must be carefully planned to avoid hazards and to provide for control of traffic, especially at peak hours, such as times when shifts change.

Parking. Facilities may be provided in lots for the common use of a number of industries. Frequently, however, parking spaces are located on the tract developed for each plant and are limited to its own employees and visitors. It is common to provide parking space for employees based on 1.5 to 3.0 individuals per car. For all-day parking, somewhat greater walking distances than in the case of shopping centers are acceptable. A maximum distance of about 1,000 ft is considered reasonable.

Buildings. Industrial buildings are commonly one story in height, sometimes with a mezzanine level. Generally, they are attractive in outer appearance, at least in the direction from which they are most usually approached. Modern industrial buildings include complex mechanical and electrical systems for control of operations. The number of employees is frequently low and for many industries may be as few as 5 per acre. The average number of employees may be estimated from assumptions as to the kind of industry expected to occupy the industrial area.

Water Supply. Industrial needs for water involve a variety of water qualities, including highly purified and demineralized water for boiler plants and drug manufacture; potable water for domestic, cafeteria, and certain process work; cooling water, normally taken from a river and returned to it without treatment; salt water for fire or deluge systems; and industrial process water of various qualities for cooling, rinsing, cleaning, and other purposes. The volumetric requirements for the various classes of water range greatly with the type of industry.

Waste Disposal. Industrial wastes vary from simple domestic wastes to those from complex mine, petroleum refining, food processing, paper, drug manufacture, and pickling and plating processes. Some of these have high bacteria count and high oxygen demand. Others are toxic and, therefore, harmful to people and aquatic life. The requirements, therefore, for industrial waste treatment cannot be generalized but must fit the individual need. In planning the waste treatment system for an industrial park where a public sewerage system is not available, planners may have to allow for construction of the plant in stages and provide for its enlargement to meet new needs as the park grows. It may be necessary to exclude some industries whose wastes are highly toxic or caustic.

Storm drainage for industrial sites must be considered in planning to avoid the economic loss resulting from damage to goods and property in periods of heavy rainfall. The grades of individual tracts and roadways must be carefully planned and coordinated to allow for the flow of water to a collection system and to avoid flooding. Extensive drainage systems are necessary for protection of property but may often include some ditches and open channels, instead of pipe, for economy in first cost.

Services. Electric, telephone, and gas utilities must all be laid out for flexibility and extension. Wiring is frequently installed underground, with ground-level transformers, to avoid the unsightly appearance of pole lines.

Industrial employees are sometimes considered isolated from convenience services, especially when in industrial parks. As a result, it is desirable for the larger industrial areas to provide space for stores, banking, bus terminal, post office, restaurants, bars, and, occasionally, motels, and park land. Where park land is provided, it affords a desirable change from the dense industrial environment, a means of recreation and of landscaping. Landscaping and screening are required for all well-planned industrial sites.

14-17. Public Buildings and Civic Centers. For neighborhoods of less than 5,000 population, master plans need not consider public buildings and civic centers, except for a firehouse, police station, school, or recreation facility. The limited needs of small groups for public space are often met by the rental of areas in privately owned commercial and office buildings. Above about 5,000 persons, however, the needs of community, town, county, state, and Federal departments and agencies become increasingly important to the master plan. In recent years, government staffs at all levels have needed increasingly larger facilities, and this trend will undoubtedly continue.

Public buildings usually associated with municipal government, based on services provided, may be divided into two categories: those requiring a single installation, and those requiring multiple installations. Those uses requiring a single installation in small towns may require multiple installations in large cities. Among the services usually requiring a single installation are Board of Education Administrative Headquarters, Town Hall, Department of Public Works, dog pounds, refuse incinerators, offices for the Mayor or Chief Administrator and staff. Among the services usually requiring multiple installations are libraries, schools, fire and police stations, recreation facilities, health centers, water-supply and waste-water treatment facilities, courts, memorial and monumental buildings, and maintenance shops.

The types for state government operations for which space must be allocated frequently require multiple locations within a state but may involve only a single location within a municipality or region.   These include facilities for highway departments, correctional agencies and institutions, departmental and agency administrative offices, unemployment centers, and armories and other military facilities.

Federal departments and agencies, like those of the states, may often require multiple locations within a state but only single locations within a region or municipality.   These typically include post offices, civil defense, military installations, health facilities, and offices of Departments of Commerce, Interior, and Justice.

Master planning of spaces for public buildings should provide primarily for the greatest ease of access by the public.   Both private and mass transportation means must be considered.   In many cases, public buildings can enhance surrounding areas because of their stabilizing influence and permanency.   Provision of landscaped areas and small parks in the vicinity of public buildings, while an expense, can create a desirable environment from which surrounding developments may benefit.

Location of certain public buildings near private financial and legal offices, other businesses, and even shopping areas is mutually beneficial.   The interrelation of the activities of most governmental offices with all the business activities of a community requires convenient access from one to the other for efficient conduct of the people's affairs.

Criteria for proper location of public buildings generally require that they be at the center or hub of the business activity of the area being planned.   Exceptions to this rule, of course, occur for fire and police departments and other local facilities that must be spread throughout a region.

14-18. Transportation of People.   The evolving pattern of land use is the primary determinant of urban travel demand.   Conversely, the supply of transportation influences land use.   Every facet of our economy depends on getting people between their homes and places of their activities.

As in the purchase of other goods and services, selection of means of transportation depends largely on individual comparisons of competitive alternatives, taking into account relative speed, cost, and convenience.   In an affluent society, the influence of cost is not so great as that of relative quickness and comfort of alternative means of transportation.   The tremendous growth in the use of private vehicles attests to this.   More efficiency in movement of people and lower costs to all can be achieved by mass transportation, but it must more nearly compete in convenience and time with private vehicles to be successful.

The investments in land area and in capital cost required by the various means of public transportation are so great that, once established, the systems become very difficult to change. Changes should be based on careful evaluation of expenditures and revenues so that the methods of financing new systems may be the most effective.

Equipment suited to movement of people by public systems includes fixed rail, buses, airplane and helicopter, and combinations of these.   Rail may be at the surface, elevated, or subway, or by rapid intercity train.   The Tokyo-Osaka train in Japan, which travels at speeds of over 100 mph, has proved highly successful, and plans have been made for track and trains offering similar speeds between Washington and Boston.   Vertical takeoff and landing aircraft, magnetic-levitation vehicles, and air-cushion-type vehicles are also receiving attention as possible means of mass transportation.

The broadest choices of types of systems for transportation of people are available to regions and large municipalities.   Smaller municipalities are limited in choice because of the smaller income available to finance the fixed charges and operation of the system.

In planning transportation of people, planners should realize that systems offering the maximum capacities and efficiency lack the flexibility of private vehicles.   For this reason, excellent service, economy, and attractive and comfortable vehicles must be provided by the plan to attract passengers.   Suburban terminals for mass-transit facilities should provide ample parking for those wishing to leave their cars while using the system.   Use of private vehicles becomes less desirable as congestion and parking problems increase.

Buses may operate with mixed traffic, or in special lanes, or on roadways devoted only to mass transit.   Express buses moving in such lanes serving suburban terminals are being used successfully in some areas.

Use of balanced transportation systems that combine automobiles on expressways, outlying parking, feeder buses, and high-speed rail rapid transit, each used to its best advantage, can make a significant contribution to the movement of great numbers of people. One way toward this goal is to design certain roads to the outlying areas as mass-transit routes and to encourage the public to use these by intentional omission of convenient expressways for private vehicles in those directions. In contrast, the system may provide excellent expressways for private vehicles in other directions as a predominant transportation system. Each element of the transportation link is able to do the job for which it is best suited in a balanced system. Such systems are already helping some suburbs grow in a more orderly manner.

Most large municipalities and regions require a combination of all means of transportation available to meet the diverse requirements. Research and experiments are in progress on new types of mass-transit systems that will prove more acceptable to the general public.

14-19. Fire and Police Protection.   Protection of life and property and enforcement of law and order depend on the action of well-organized fire and police departments. The physical facilities required to conduct such operations are rather flexible as to location; few standards can be cited. Both protective bodies, however, have a common interest in the mobilization of forces to meet emergencies, such as large fires, explosions, riots, bad storms, and civil defense. Both agencies also need communications systems, and must operate vehicles, conduct inspections, and maintain records that relate intimately with other municipal departments. For these reasons, the headquarters of fire and police departments are often situated in the central local government offices. Development of an emergency operating center to include fire, police, civil defense, and fallout protection is encouraged by the availability of Federal government participation in the cost of such centers.

Means of fire protection in a community presupposes an adequate water system having the approval of fire insurance underwriters. In addition to their interest in adequate quantities and pressures of water, the underwriters grade and set standards for the kinds and number of pieces of fire-fighting equipment and operating personnel. In planning for fire protection of a community, planners usually attempt to obtain the lowest insurance rates for the property owners and hence consider the requirements of the fire underwriters in achieving this goal.

Most water systems for fire-fighting purposes are integral with the domestic water supply of the community. They include distribution mains with piping no smaller than 6 in. in diameter and with fire hydrants spaced at suitable intervals. Separate high-pressure fire-fighting zones and, for waterfront areas, separate nonpotable water systems with fire boats are sometimes provided.

The kinds of mobile fire-fighting equipment required depend on height and density of buildings and nature of specialized risks to be protected. Petroleum and large electrical installations, for example, require use of specialized methods. In some cases, large industries such as steel mills may maintain private fire-protection companies.

To conduct its operations, a fire department requires facilities for training, for local fire stations, for central headquarters and communications, and often for maintenance garages.

In the conduct of their operations, police departments require physical facilities for training, teletyping and communication, detention, security, patrol, investigation, court, parking, records, and barracks space where food, dining, and medical attention are available. A number of these operations are usually grouped into a central headquarters. But in large areas, some of the operations may require separate locations. For neighborhoods, space must be planned, at least, for operation of security and patrol activities.

14-20. Land-Use Control Methods.   After the many circumstances that affect a plan have been studied and evaluated, a master plan is prepared and presented to the owners or the public and governmental bodies having jurisdiction, with recommendations for adoption. Usually, prior to adoption of a master plan, local legislation requires public hearings. These allow property owners affected both beneficially and adversely to express their views and the public in general to offer constructive criticisms and to approve or disapprove of the plan. Such discussions, in many instances, will tend to improve the plan. A master plan prepared for one area and acclaimed by those that it affects as beneficial may be wholly unacceptable in another area.

To insure approval when a master plan is presented at a public hearing, it is mandatory that the plan take into consideration the local likes and dislikes, traditions, and other factors that will make it more acceptable to those who must approve of the plan and participate in its implementation.

Once the plan has met the approval of the public and the governmental bodies, steps must be taken to insure that it is enforced. Usually, these involve preparation and adoption of zoning ordinances written to support and implement the plan. After the zoning ordinances are adopted, it becomes the responsibility of governmental agencies to see that the only changes permitted are those that correct an error or a hardship in the initial master plan or are necessary to accommodate changed conditions. Changes in zoning for reasons other than these generally fall into the category of political expediency and reduce the effectiveness of the master plan.

14-21. Subdivision Regulations and Building Codes. The means available to carry out effectively the master plan and the zoning adopted as a result of the master plan are subdivision regulations and building codes. In many instances, enactment of a zoning ordinance and adoption of subdivision regulations and building codes are done concurrently to avoid the likelihood of conflicting terms and requirements.

Regulations. Subdivision regulations cause land development to conform to zoning provisions, give the governmental body control of the subdivision of land, and prescribe the procedure for accomplishing the subdivision. The master plan, at best, indicates the general locations recommended for major roadways, parks, open spaces, public building sites, general routes for public utilities, and the various densities of population. The exact location of all facilities, as well as the specific shape of areas, is determined through the subdivision procedure by developing preliminary plans. These indicate the location and shape of units as small as individual lots.

In general, preliminary subdivision plans drawn to a scale of 1 in. = 200 ft or 1 in. = 100 ft are prepared and filed for approval of the governmental agency having jurisdiction. Such plats are drawn to an accuracy compatible with their scale and show the topographic features of the area as well as the uses to be made of the land. The preliminary plan is prepared in accordance with criteria for street and right-of-way widths, block sizes, and other requirements of the subdivision regulations. After the preliminary plan has been approved, final plans are prepared to show horizontal and vertical locations to an accuracy of 0.01 ft. Following approval of the final plans, the subdivision is ready for construction.

Included in the subdivision regulations are engineering criteria that are approved for the area and must be applied. By means of the regulations, the specific requirements of the master plan can be enforced. Subdivision regulations control placement of dwellings and structures on the land, assure adequate air and light in the developed area, and include standards for such improvements as storm drainage, streets, curbs, gutters, and sidewalks.

An illustration of the effectiveness of carefully planned deed restrictions on the maintenance of a neighborhood is found in the Baltimore community of Roland Park, which, since the start of the twentieth century, has continued its dignified appearance and residential atmosphere. The pleasant streets, wide-spaced lots, and restful houses, with little knots of business tucked away in unobtrusive corners, have been based on legal restrictions placed in all deeds at the inception, as follows:

Lots are to be used exclusively for residential purposes with only one dwelling permitted on each lot. No tent, shack or trailer may be maintained on any lot.

Plans for buildings, additions or alterations must be approved by the Roland Park Roads and Maintenance Corp. as being in harmony with the surrounding area and properties. Similarly, no fences or signs are to be erected on any lot unless approved by the Corporation.

No fowl or animals, except domestic pets, may be kept on the premises.

Codes. Building codes are drawn up to protect the health, welfare, and safety of the public. They set minimum standards for structures to be built on the land, as zoned and subdivided. Design criteria for structures and minimum specifications for materials to be incorporated in the construction are included in building codes. The codes are prepared in great detail. They cover the construction of all types of buildings. Through the use of building codes, safe structures are assured and a minimum standard is established that is enforced through inspection.

14-22. Preventing Decay of Neighborhoods. Prevention of decay requires action beforehand to eliminate the harmful results of economic and social forces on residential, commercial, or industrial areas. If the causes of decay are not recognized at the outset, conditions generally deteriorate beyond the limits where preventive measures are effective. Complete renewal of run-down areas will then be necessary. The planner's role is to be alert to changing conditions by recognizing the signs, and to plan and initiate action that will avert decay.

Recently developed communities have a health and vitality that carry them through certain economic cycles and other social forces, because of the investment of interested owners of such property. Older communities, however, from which a substantial proportion of the economic value has been realized, do not always have the stability to withstand the pressures of changing neighborhoods, changes in manufacturing-facility requirements, and other economic and social changes. Methods of measuring residential blight have been developed and are widely accepted. But methods for similar measurements in nonresidential areas have not been perfected.

Indications of impending decay of residential neighborhoods include such conditions as a lowering economic level of the residents; changes in single-family dwellings to multifamily occupancy; an increase in "For Sale" and "For Rent or Lease" signs; subdividing of larger residential lots by construction of additional dwellings on unused portions; and lack of normal maintenance.

Symptoms of decay in industrial and commercial property include reduction of general business activity in that area; lack of maintenance; sale of parts of tracts for other uses; and frequent changes in type of occupancy.

Before a course of action can be developed, a determination must be made of the causes. These may include inadequate transportation as a result of changing transportation modes; changes in technology resulting in different manufacturing-facility requirements that make older plants obsolete; inadequate space for expansion required to meet changing market conditions; and uneconomic relation of the facility to supporting elements, such as suppliers of raw material, conveniences for workers, and amenities.

After the causes have been identified, a program can be formulated to counteract the developing decay by planning the elements needed to stabilize the community. Achievement of these goals is frequently difficult to attain because it requires public and industrial support combined with civic action.

Prevention of decay by building-inspection procedures is not effective in averting or counteracting the causes. But it may be useful in delaying or forestalling further deterioration before complete renewal becomes necessary.

14-23. Renewal of Run-down Areas. Individual properties and neighborhoods are allowed to run down by owners when a large portion of the economic value of the investment has been realized and when economic and social changes affect an area (Art. 14-20). If such changes were always obvious at the outset, and if the remedies for them were easily applied, decay of neighborhoods could be successfully prevented, and run-down areas would not occur. The influence of individual run-down properties is felt in the value of adjoining properties. Thus, these areas spread rapidly to natural barriers.

Planners must decide which properties in a run-down area can be rejuvenated under a new plan and which properties have faded beyond their social and economic purpose to a point where they must be torn down in favor of new structures.

The planning of areas for demolition, restoration, or a combination may be undertaken by private interests in limited areas but usually requires public participation in larger projects. The interest and support of the Federal government in urban renewal have brought much publicity and controversy to this subject.

Renewal of areas is generally a more difficult planning project than planning rural areas for initial development. Planners must have much sympathy and patience for the complexities of cities in developing a renewal plan for run-down areas.

Plans for the renewal of run-down areas must develop general principles from which a variety of land uses may grow. The influences on use of the land through a number of years of development are not fully known at the outset. Much elaboration and evolution must occur after adoption of the general principles, as expressed in a master plan.

Run-down areas must be replanned to integrate into their surroundings, rather than to become an island in the middle of earlier development. Renewal plans that circle an area with parking, for example, tend to isolate the area from the adjoining community.

The difficulty of deciding which existing buildings or groups should be retained sometimes causes an impatient decision to sweep the area clean and start afresh. Often, this decision ignores the economic utility of existing buildings as well as their aesthetic value and stabilizing influence.

Most renewal areas depend for their development on investors who select individual parcels and carry out plans approved by the agency exercising authority. The role of the planner in the initial concept should be the establishment of density, circulation, building mass, and economic uses, leaving to the developer the fitting of the exact facility to the needs as expressed by the market. Controls by a governmental agency are necessary to guarantee to all developers and investors that the basic concepts of the plan will be maintained. Without such controls, the life span of the renewal area is uncertain. The economic loss to the community may be great if the life span should be greatly shortened.

Construction in a large renewal area may easily take 10 to 15 years and presents difficult problems to planners and developers. To avoid the long idleness of property due to advance clearing, it is necessary that a renewal plan be carefully related to the economy of the region and acted on quickly. Construction of new projects on a renewal site that has not been fully cleared, however, raises difficulties with new occupants during the period the adjoining run-down property is still in use.

Large renewal areas sometimes encourage only large developers, thus squeezing out the smaller operators who offer diversity, interest, and a broader representation of the community.

14-24. Financing Capital Improvements. Development of land entails expenditure of large sums of money by private and governmental sources. In all cases, expenses and revenues must be estimated and long- or short-term financing must be arranged.

The policies of local government toward financing land-development improvements have a major influence on the orderly growth of areas and on the nature and timing of projects. The magnitude of development expenses has created serious economic problems for many local governments, limited the facilities available in some areas, and affected the resulting environment in all instances.

Financing of improvements by private owners depends on the availability and cost of money. Such financing is restricted by evaluations of the marketability of the development by financial institutions.

To encourage improved planning and development of areas, the Federal government has enacted many programs for advancing loans and grants to individuals and local governments. These funds are available for technical assistance, research, planning, housing, transportation, utilities, community facilities, and recreational and cultural projects. Many of the programs are available for either development of new areas or rehabilitation of existing areas.

Public financing of capital improvements is accomplished by many methods, some of which are appropriate and some of which are inequitable. The method of financing should be designed to assess the cost of the improvements on persons or areas benefiting. General taxation against real property, special area charges or assessments, front-foot benefit charges, service charges for connection, usage charges based on quantities consumed or discharged, flat-rate charges, and other methods of charging for benefits are used. In many cases, a combination of charges that correspond to the benefits afforded by different parts of utility systems is applicable. The water-distribution main constructed in front of an improved lot or to serve a dwelling unit should be paid, for example, by the owner of the lot or dwelling unit served, whereas the cost of transmission mains and treatment and pumping facilities, which serve a larger area or may indeed serve the entire region, should be paid for by the residents of the region.

In some areas, the developer of land is required to build roads and install drains and water and sewerage facilities, and to convey these without charge to the governmental agency having jurisdiction for ownership, maintenance, and operation. Although this arrangement relieves the municipality of the burden of financing the facilities, it causes the developer to include the cost of the improvements in the prices of the dwelling units. Residents of the area must then bear the resulting greater cost, generally through a mortgage on their property. Individual mortgages command interest rates, in many instances, double those for municipally financed general-obligation bonds. In such cases, the reduction of principal is spread over shorter rather than longer periods. The policy of requiring a developer to donate land for schools, park land, recreational facilities, and other municipal purposes causes the same problem.

Private utility corporations and sanitary districts that operate in regional areas with numerous towns and communities offer desirable means of planning, organizing, and financing capital improvements in many instances.

In the undertaking of large land-development projects by private enterprise, leading financial institutions look with favor on employment of reputable engineers, architects, and planners for preparation of master plans and implementation of the resulting construction. In most cases, the reputation of the prime professional responsible for the master plan is among the most important considerations in the approval of financing.

14-25. Suggested References. The following publications are recommended to readers who wishes to extend their familiarity with the subject. Among the list are engineering and planning texts, government publications, and periodic journals of technical societies which treat in greater depth each of the subjects contained in Sec. 14.   The listed publications also offer many excellent illustrations of the principles that have been presented.

1. ASCE "Urban Planning Guide," M&R 49, American Society of Civil Engineers, New York.

2. Babbitt, H. E., J. J. Doland, and J. L. Cleasby, "Water Supply Engineering," McGraw-Hill Book Company, New York.

3. Banz, G., "Elements of Urban Form," McGraw-Hill Book Company, New York.

4. Bogue, D. J., "Principles of Demography," John Wiley & Sons, Inc., New York.

5. Butler, G. D., "Introduction to Community Recreation," McGraw-Hill Book Company, New York.

6. Claire, W. H., "Handbook on Urban Planning," Van Nostrand Reinhold Company, New York.

7. DeChiara, J., and L. Koppelman, "Planning Design Criteria," Van Nostrand Reinhold Company, New York.

8. Dickey, J. W., "Metropolitan Transportation Planning," McGraw-Hill Book Company, New York.

9. Dober, R. P., "Environmental Design," Van Nostrand Reinhold Company, New York.

10. Eckbo, G., "The Landscape We See," McGraw-Hill Book Company, New York.

11. Fair, G. M., J. C. Geyer, and D. A. Okun, "Water and Wastewater Engineering," John Wiley & Sons, Inc., New York.

12. *Journal of the American Institute of Planners*, Washington, D.C.

13. *Journal of the Environmental Engineering Division*, American Society of Civil Engineers, New York.

14. *Journal of the Urban Planning and Development Division*, American Society of Civil Engineers, New York.

15. Local Climatological Data, Weather Bureau, Department of Commerce, Government Printing Office, Washington, D.C. (available on request from all local weather bureaus).

16. Mayer, A., "The Urgent Future: People, Housing, City, Region," McGraw-Hill Book Company, New York.

17. Metcalf & Eddy, Inc., "Wastewater Engineering," McGraw-Hill Book Company, New York.

18. Rubenstein, H. M., "A Guide to Site and Environmental Planning," John Wiley & Sons, Inc., New York.

19. Rutledge, A., "Anatomy of a Park: The Essentials of Recreation Area Planning and Design," McGraw-Hill Book Company, New York.

20. "Standard Land Use Coding Manual," Government Printing Office, Washington, D.C.

21. "Urban Renewal Manual," Urban Renewal Administration, Government Printing Offiice, Washington, D.C.

22. "U.S. Census of Population—General Population Characteristics," Government Printing Office, Washington, D.C.

23. Whittick, A., "Encyclopedia of Urban Planning," McGraw-Hill Book Company, New York.

Section 15

Building Engineering

FREDERICK S. MERRITT
Consulting Engineer, Syosset, N.Y.

Buildings include a wide range of construction intended for human occupancy or for sheltering machines or stored goods. Civil engineers play an important role in the design and construction of such structures. But sometimes, the civil engineer is only one of many design professionals participating in the planning and design of a building. Therefore, it is necessary that his or her design decisions take into consideration the objectives and needs of the other professionals. For this purpose, civil engineers must be well informed on such subjects as architecture, building layout, lighting, electrical systems, elevators, plumbing, heating, and air conditioning, as well as structural design. To serve this need, this section summarizes briefly the design principles of those fields and lists references for more detailed study.

15-1. Influence of Zoning on Building Design. Localities use zoning to regulate use of land, control types of occupancies and size of buildings, and in various other ways safeguard the public health, safety, and welfare. Zoning regulations supplement building-code requirements.

In selecting land for a building, the local zoning code should be checked to see if the type of occupancy planned—residential, commercial, industrial, school, church—is permitted. If it is not, the possibility of a variance or a code change should be investigated.

For some types of construction—housing, for example—lack of a zoning code may discourage selection of a parcel of land. Uncontrolled land use may permit undesirable neighbors—junkyards or odorous factories—with a resulting deflation in property values. When a zoning code exists, its control over land adjoining the parcels being considered should be examined, to determine whether neighboring occupancies will be desirable.

To insure light and air to adjacent property, zoning codes restrict the height and bulk of buildings. Some codes limit the number of stories; some place a maximum on the height above the street. In some cases codes permit unlimited height but require the buildings to be set back from the base after certain heights are reached, depending on the width of the street, measured between building lines. This type of requirement led in the past to "wedding-cake" architecture, buildings that were made narrower and narrower in steps as they rose, because of frequent setbacks. As an alternative, in which internal space is sacrificed in the interests of aesthetics, a building may satisfy that type of regulation yet not have setbacks if it is erected as a sheer tower occupying only part of its site.

Some codes control height and bulk by establishing a ratio between total floor area permitted and the area of the site. Extra floor area sometimes is allowed if part of the site at or near street level is devoted to a plaza. Thus, designers may shape the building any way they please within the building lines; they may make it tall and thin or short and broad, so long as the total floor area does not exceed that permitted. Codes, however, sometimes also require buildings to be set back at the base minimum distances from lot lines; these regulations should be determined and observed in locating a building on its site.

In addition to being restricted by zoning, building height may be limited by Federal aviation authorities, especially in the vicinity of airports. These regulations should be considered before land is selected for a project and especially before building height is decided upon in the early design stages.

15-2. Building Codes. Localities, and sometimes the states, exercise the police power to safeguard public health, safety, and welfare by controlling building design and construction through building codes. The control generally extends over all phases, including specification of permissible design and construction methods, as well as field inspection to insure compliance.

Codes may be classified as specification or performance-standards type. Specification-type codes are characterized by requirements that list acceptable materials and their minimum sizes for specific applications. Performance-standards codes specify the end result to be obtained in terms of such characteristics as strength, stability, permeability, hardness, and fire resistance. In practice, this type of code generally is supplemented by a catalog of acceptable materials and constructions after tests show they meet code requirements.

Since most communities have their own codes, which may differ from those of adjoining localities, building designers should become familiar with the local building codes for the areas in which their projects will be erected. Even where state codes exist, communities may have the power to set more stringent requirements than the state.

For projects in areas not under the jurisdiction of a state or local building code, building designers should adopt the code of a nearby large city or a model code applicable to the region. Nationally recognized model codes include:

National Building Code—American Insurance Association, 85 John St., New York, N.Y. 10038.

Uniform Building Code—International Conference of Building Officials, 5360 South Workman Mill Road, Whittier, Calif. 90601.

Southern Standard Building Code—Southern Building Code Congress, 1116 Brown-Marx Building, Birmingham, Ala. 35203.

Basic Building Code—Building Officials and Code Administrators International, Inc., 1313 E. 60th St., Chicago, Ill. 60637.

15-3. Fire Protection for Buildings. An important consideration in the design of nearly all buildings is the fire resistance required by building codes and insurance companies. This resistance may require use of incombustible materials, fire-protective coverings, and sprinkler systems, which generally cost more than constructions of lower fire resistance. Also, codes may prohibit use of hazardous materials, for example, materials that may explode or emit excessive smoke or poisonous gases.

Sometimes, the lowest long-run costs for a building are obtained with a higher fire resistance than required by the local building code because of reductions in fire insurance premiums.

Fire-resistant construction aims at withstanding a fire locally for a specific period of time and preventing it from spreading—throughout the level at which it starts, or from story to story, or to adjacent buildings. The objective of sprinkler systems is to extinguish the fire quickly.

Fire ratings are assigned to building components in accordance with their performance in standard fire tests (ASTM E119, Standard Methods of Fire Tests of Building Construction and Material). If a component meets requirements after 1 hour exposure in a standard furnace test, it is given a 1-hr rating; if it withstands the test for 2 hours, it gets a 2-hr rating, etc.

Fire protection for a building and its occupants involves prevention, detection and warning, containment, and extinguishment of fires, and provisions for life safety.

Prevention. Buildings should be designed to minimize the possibility of fire other than in such authorized places as furnaces and fireplaces. Where possible, construction materials—roofing, flooring, ceilings, and sash, for example—and coatings, paints, and curtains should be incombustible. Also, the fuel load from furnishings should be kept small.

Detection and Warning. Buildings should be equipped in every story with devices that can detect fire or smoke and sound an alarm. Such devices can also automatically instigate extinguishment procedures.   There are five general types of detectors.   Each employs a different physical means of operation.

Fixed-temperature detectors indicate the presence of fire when the device reaches a predetermined temperature.

Rate-of-rise detectors function when there is a rapid increase in temperature.

Photoelectric detectors are sensitive to smoke.

Combustion-products detectors recognize combustion products and are designed for very early warning.

Flame detectors respond to light, infrared or ultraviolet, produced by combustion reactions.

Detection should immediately signal a warning so that building occupants who may be endangered, life-safety supervisory personnel, and firefighters may be alerted.

Large buildings, especially those with many occupants, should have an emergency control center, or fire command station, on the ground floor to which detection signals are communicated.   The center should have and control two-way communication with every floor and be able to direct rescuers and fire fighters and transmit instructions to occupants to guide them to safety. The center should also be able to control all electromechanical systems, such as elevators, air conditioning, and fans.   To assist firefighters, controls should be capable of venting, pressurizing, or sealing any zone in a building.

Containment.   Buildings should be so designed that, if fire or smoke should occur, it would be extinguished almost immediately, but in any event, it could not spread much beyond the place of occurrence.   Spread of fire or smoke can be prevented by fire barriers, venting of heat and gases, and dampers.

Barriers.   Large floor areas should be partitioned by fire walls into smaller areas.   Fire doors protecting wall openings should be kept closed.   Plenums, such as the spaces between floor and ceiling or roof and ceiling, should be isolated at frequent intervals by fire stops.   Spandrels should have a high fire rating and should be sufficiently deep at each floor level to prevent flames extending out the windows in one story from igniting materials in the story above.   (The National Building Code recommends a minimum depth of 3 ft.)

Venting systems should be provided to cool fires and to keep heat and smoke from escape routes and refuge areas.   Areas adjacent to a fire should be pressurized to keep out smoke.   To clear smoke, windows should be openable or have smoke ventilation panels.   Alternatively or in addition, an automatically vented smoke shaft should be provided.   Also, the tops of fire towers enclosing elevators or stairs should permit venting of hot gases and smoke.   Emergency ventilation of stairwells and elevator shafts may be assisted by fans.   Fresh make-up air should be provided to keep safe areas habitable.

Automatic fire dampers should be installed in ducts, along with fire or smoke detectors that sample all air passing through.   The dampers should be controlled to seal control zones, prevent smoke from spreading to escape routes and refuge areas, and guide ventilation area air to points where it is needed.

Extinguishment.   Means of fire suppression range from hand-held extinguishers to high-pressure water flows from hoses and sprays from installed sprinkler systems (Art. 15-32).   (For some types of fires, carbon dioxide or chemicals may be necessary instead of water.)   In addition, firefighters have various types of equipment for fire fighting.   Regardless of the means used, life safety and property damage depend primarily on early detection of fire and smoke and rapid application of the appropriate extinguishment method.

To assist firefighters, water must be supplied to them in adequate quantities and at sufficient pressures for fire fighting.   If necessary, storage or pumping facilities must be provided.   An elevated water tank may be used for this purpose (National Fire Protection Association standard 22).   The supply may be augmented by a fire pump (NFPA standard 20).   Pressure should be at least 15 psi at the highest level of sprinklers, while flow at the base of the riser is at least 250 gpm for light-hazard occupancies and 500 gpm for ordinary-hazard occupancies.   (Local building codes usually specify minimum pressures.)

The usual means of manually applying water to interior building fires is with hoses receiving water from standpipes.   These generally are required in buildings more than about 50 ft high and should be so located that any part of a floor is not more than 130 ft from a standpipe outlet valve.   Risers up to 75 ft high may be 4 in. in diameter, but for greater heights, 6 in.   Hose valves usually are 2½ in. in diameter.

Life Safety.   Buildings should provide for safe, easy escape in emergencies, preferably but not necessarily to outdoors at ground level.   In some cases, it may be advisable to instruct occupants to stay in place or to provide refuge areas within the buildings to which occupants may proceed when alerted.   Doors, hallways, and stairs should be adequate in number, size, and location to accommodate the number of occupants that may have to be evacuated in emergencies.   (Requirements are specified in local building codes and in "Life Safety Code," National Fire Protection Association, Boston, Mass.)   In addition, firefighters should be provided with safe access to fires.

In buildings with elevators, the cars should be equipped with controls for emergency use by firefighters and should move automatically to the lobby floor to be available to them if needed. Control wiring should be protected against accidental operation by high temperatures.

Elevators and stairs should be enclosed in fire towers with walls having a 4-hr rating (*fire walls*) and fire-resistant doors that are kept closed.   Building entrances and exits should be especially protected.   (See also Art. 15-17.)

(F. S. Merritt, "Building Construction Handbook," McGraw-Hill Book Company, New York; Factory Mutual System, "Handbook of Industrial Loss Prevention," McGraw-Hill Book Company, New York.)

15-4. Design Loads for Buildings.   All structural components of a building must be designed for the full dead load as a minimum.   When the dead load is not certain, for example, when the location of partitions is not known when the design is made, an allowance should be made.   Some building codes require a uniformly distributed load of 20 psf to be added to the known dead load to allow for partitions not definitely located.   Table 15-1 lists minimum design dead loads for various materials.   In computing dead load, the weight of the member being designed should be included, as well as the weight of the rest of the structure that it has to support.

Live loads for buildings generally are assumed uniformly distributed, except, of course, that the live load transmitted from a beam to a girder is a concentrated load on the girder.   Some codes also require an additional concentrated load, applied at any point in a bay, for garages, machine rooms, and offices.   But such loads as those from a moving crane on crane girders and columns should be treated as moving concentrated loads.   Table 15-2 lists minimum design live loads for various occupancies.

Table 15-1. Minimum Design Dead Loads

	Psf
Walls	
Clay brick	
High-absorption, per 4-in. wythe	34
Medium-absorption, per 4-in. wythe	39
Low-absorption, per 4-in. wythe	46
Sand-lime brick, per 4-in. wythe	38
Concrete brick	
4-in., with heavy aggregate	46
4-in., with light aggregate	33
Concrete block, hollow	
8-in., with heavy aggregate	55
8-in., with light aggregate	35
12-in., with heavy aggregate	85
12-in., with light aggregate	55
Clay tile, load-bearing	
4-in.	24
8-in.	42
12-in.	58
Clay tile, non-load-bearing	
2-in.	11
4-in.	18
8-in.	34
Furring tile	
1½-in.	8
2-in.	10
Glass block, 4-in.	18
Gypsum block, hollow	
2-in.	9.5
4-in.	12.5
6-in.	18.5

Table 15-1. Minimum Design Dead Loads *(Continued)*

Partitions	Psf
Plaster on masonry	
Gypsum, with sand, per in. of thickness	8.5
Gypsum, with lightweight aggregate, per in.	4
Cement, with sand, per in. of thickness	10
Cement, with lightweight aggregate, per in.	5
Plaster, 2-in. solid	20
Metal studs, plastered two sides	18
Wood studs, 2 × 4-in.	
Unplastered	3
Plastered one side	11
Plastered two sides	19

Concrete Slabs	
Stone aggregate, reinforced, per in. of thickness	12.5
Cinder, reinforced, per in. of thickness	9.25
Lightweight aggregate, reinforced, per in. of thickness	9

Insulation	
Cork, per in. of thickness	1.0
Foamed glass, per in. of thickness	0.8
Glass-fiber bats, per in. of thickness	0.5
Polystyrene, per in. of thickness	0.08
Urethane, 2-in.	1.2
Vermiculite, loose fill, per in. of thickness	0.05

Floor Finishes	
Asphalt block, 2-in.	24
Cement, 1-in.	12
Ceramic or quarry tile, 1-in.	12
Hardwood flooring, ⅞-in.	4
Plywood subflooring, ½-in.	1.5
Resilient flooring, such as asphalt tile and linoleum	2
Slate, 1-in.	15
Softwood subflooring per in. of thickness	3
Terrazzo, 1-in.	12
Wood block, 3-in.	4

Floor Fill	
Cinders, no cement, per in. of thickness	5
Cinders, with cement, per in. of thickness	9
Sand, per in. of thickness	8

Waterproofing	
Five-ply membrane	5

Glass	
Single-strength	1.2
Double-strength	1.6
Plate, ⅛-in.	1.6

Wood joists, double wood floor, joist size	Psf	
	12-in. spacing	16-in. spacing
2 × 6	6	5
2 × 8	6	6
2 × 10	7	6
2 × 12	8	7
3 × 6	7	6
3 × 8	8	7
3 × 10	9	8
3 × 12	11	9
3 × 14	12	10

Table 15-1. Minimum Design Dead Loads   *(Continued)*

Ceilings	Psf
Plaster (on tile or concrete)	5
Suspended metal lath and gypsum plaster	10
Suspended metal lath and cement plaster	15

Roof and Wall Coverings	
Asbestos-cement corrugated or shingles	4
Asphalt shingles	2
Composition:	
3-ply ready roofing	1
4-ply felt and gravel	5.5
5-ply felt and gravel	6
Copper or tin	1
Corrugated iron	2
Sheathing (gypsum), ½-in.	2
Sheathing (wood), per in. thickness	3
Slate, $^3/_{16}$-in.	7
Wood shingles	3

Masonry	Pcf
Cast-stone masonry	144
Concrete, stone aggregate, reinforced	150
Ashlar:	
Granite	165
Limestone, crystalline	165
Limestone, oölitic	135
Marble	173
Sandstone	144

Live loads should be placed on a structure to produce maximum stress and deformation in components being designed. For example, in design of a continuous beam for maximum positive moment at midspan, only alternate spans, including the one being designed, should carry full live load. Machine weight should be increased 25% and elevator loads 100% for impact.

When a very large area contributes live load to a member, most codes permit a reduction from that required for a member supporting a small loaded area. For floors, for example, some codes permit a reduction for members supporting 150 sq ft or more of 0.08% per sq ft. But the reduction cannot exceed 60% or 23.1(1 + D/L)%, where D is the dead load per square foot of area supported and L the live load per square foot. And the reduction does not apply to places of public assembly or to live loads exceeding 100 psf.

Snow loads on roofs should be treated as live load and placed to produce maximum stress and deformation. Ordinary roofs should be designed for a live load of at least 20 psf of horizontal projection to provide for sleet and minor snow loads and loads incidental to construction and repair. Where snow loads may exceed 20 psf, the roof should be designed for the maximum anticipated or that required by the local building code or the loads given in Table 15-2d. Roofs used for incidental promenade purpose should be designed for a minimum live load of 60 psf. When used for roof gardens or assembly purposes, they should be designed for 100 psf.

Wind loads vary with location and height of building (Table 15-2c). Buildings should be designed for wind coming from any direction. Wind loads and live loads may act simultaneously, but wind loads need not be combined with seismic loads.

Earthquake forces may be taken as horizontal concentrated loads acting at each floor and roof above the foundation. These forces may act simultaneously in any direction. The sum of the forces, the total shear at the base, kips, should be at least

$$V = ZKCW \tag{15-1}$$

Z provides for variation in design forces with changes in the probability of seismic intensity throughout the country. Z = 0.25 for zone 1; Z = 0.5 for zone 2; and Z = 1 for zone 3 (Fig. 15-1).

W, in general, is the total dead load, kips. For storage and warehouse occupancies, however, W should be taken as the dead load plus 25% of the live load, and there are other cases where designers would find it prudent and realistic to include a portion of the live load in W.

Table 15-2. Minimum Design Live, Wind, and Snow Loads

a. UNIFORMLY DISTRIBUTED LIVE LOADS, PSF, IMPACT INCLUDED[a]

Occupancy or use	Load	Occupancy or use	Load
Assembly spaces:		Marquees	75
Auditoriums[b] with fixed seats	60	Morgue	125
Auditoriums[b] with movable seats	100	Office buildings:	
Ballrooms and dance halls	100	Corridors above first floor	80
Bowling alleys, poolrooms, similar		Files	125
recreational areas	75	Offices	50
Conference and card rooms	50	Penal institutions:	
Dining rooms, restaurants	100	Cell blocks	40
Drill rooms	150	Corridors	100
Grandstand and reviewing-stand seat-	100	Residential:	
ing areas	100	Dormitories:	
Gymnasiums	100	Nonpartitioned	60
Lobbies, first-floor	100	Partitioned	40
Roof gardens, terraces	100	Dwellings, multifamily:	
Skating rinks	100	Apartments	40
Bakeries	150	Corridors	80
Balconies (exterior)	100	Hotels:	
Up to 100 sq ft on one and two-family		Guest rooms, private corridors	40
houses	60	Public corridors	80
Bowling alleys, alleys only	40	Housing, one- and two-family:	
Broadcasting studios	100	First floor	40
Catwalks	30	Storage attics	80
Corridors:		Uninhabitable attics	20
Areas of public assembly, first-floor		Upper floors, habitable attics	30
lobbies	100	Schools:	
Other floors same as occupancy		Classrooms	40
served, except as indicated		Corridors	80
elsewhere in this table		Shops with light equipment	60
Fire escapes:		Stairs and exitways	100
Multifamily housing	40	Handrails, vertical and horizontal	
Others	100	thrust, lb per lin ft	50
Garages:		Storage warehouses:	
Passenger cars	50	Heavy	250
Trucks and buses	c	Light	125
Hospitals:		Stores:	
Operating rooms, laboratories, ser-		Retail:	
vice areas	60	Basement and first floor	100
Patients' rooms, wards, personnel areas	40	Upper floors	75
Kitchens other than domestic	150	Wholesale	100
Laboratories, scientific	100	Telephone equipment rooms	80
Libraries:		Theaters:	
Corridors above first floor	80	Aisles, corridors, lobbies	100
Reading rooms	60	Dressing rooms	40
Stack rooms, books and shelving at 65		Projection rooms	100
pcf, but at least	150	Stage floors	150
Manufacturing and repair areas:		Toilet areas	40
Heavy	250		
Light	125		

[a]For live loads of 100 psf or less on a member carrying a loaded area of 150 sq ft or more, other than roofs, garages, or places of assembly, design live loads may be reduced at the rate of 0.08% per sq ft of supported area, but not more than $23(1 + D/L)\% < 60\%$, where D = dead load, psf, and L = live load, psf, on member. No reductions may be made for live loads exceeding 100 psf, except that design live loads on columns, piers, and walls supporting such loaded areas exceeding 150 sq ft, including storage areas, parking spaces, and places of assembly, may be reduced a maximum of 20%.

[b]Including churches, schools, theaters, courthouses, and lecture halls.

[c]Use American Association of State Highway and Transportation Officials highway lane loadings.

Table 15-2. Minimum Design Live, Wind, and Snow Loads *(Continued)*

b. Concentrated Live Loads[d]

Location	Load, lb
Elevator machine room grating (on 4-sq-in. area)	300
Finish, light floor-plate construction (on 1-sq-in. area)	200
Garages:	
Passenger cars:	
Manual parking (on 20-sq in. area)	2,000
Mechanical parking (no slab), per wheel	1,500
Trucks, buses (on 20-sq in. area), per wheel	16,000
Office floors (on area 2.5 ft square)	2,000
Roof-truss panel point over garage, manufacturing, or storage floors	2,000
Scuttles, skylight ribs, and accessible ceilings (on area 2.5 ft square)	200
Sidewalks (on area 2.5 ft square)	8,000
Stair treads (on 4-sq-in. area at center of tread)	300

c. Wind Pressures for Design of Framing, Vertical Walls, and Windows of Ordinary Rectangular Buildings, Psf[e]

Height zone, ft above curb	Exposures								
	A^f	B^g	C^h	A^f	B^g	C^h	A^f	B^g	C^h
	Coastal areas, N.W. and S.E. United States[i]			Northern and central United States[j]			Other parts of United States[k]		
0–50	20	40	65	15	25	40	15	20	35
51–100	30	50	75	20	35	50	15	25	40
101–300	40	65	85	25	45	60	20	35	45
301–600	65	85	105	40	55	70	35	45	55
Over 600	85	100	120	60	70	80	45	55	65

[d] Use instead of live load, except for roof trusses, if concentrated loads produce greater stresses or deflections. Add impact factor for machinery and moving loads: 100% for elevators, 20% for light machines, 50% for reciprocating machines, 33% for floor or balcony hangers. For craneways, add a vertical force equal to 25% of maximum wheel load; a lateral force equal to 10% of the weight of trolley and lifted load, at the top of each rail; and a longitudinal force equal to 10% of maximum wheel loads, acting at top of rail.

[e] For winds with 50-year recurrence interval. For computation of more exact wind pressures, see ANSI Standard A58.1-1972.

[f] Centers of large cities and very rough, hilly terrain.

[g] Suburban areas, towns, city outskirts, wooded areas, and rolling terrain.

[h] Flat, open country; open flat coastal belts, and grassland.

[i] 110-mph basic wind speed.

[j] 90-mph basic wind speed.

[k] 80-mph basic wind speed.

$C = 0.05/\sqrt[3]{T}$ is called the seismic coefficient. T is the fundamental period of vibration of the structure, sec. For a building as a whole, C need not exceed 0.1. For one- and two-story buildings, it may be taken as 0.1.

T may be computed from

$$T = \frac{0.05h_n}{\sqrt{D}} \qquad (15\text{-}2)$$

where h_n = height, ft, of uppermost level above base

D = plan dimension, ft, in direction parallel to applied forces

More appropriate values of T may be used if they can be substantiated by technical data. Note, however, that C is not sensitive to small changes in T. If the lateral-force-resisting system is a

Table 15-2. Minimum Design Live, Wind, and Snow Loads (*Continued*)

d. SNOW LOADS ON SLOPING ROOFS, PSF[l]

Basic load, psf[m]	Shed roof				Gable roof[n]				
	Roof angle with horizontal, deg								
	0–30	40	50	60	0–20	30	40	50	60
5	5	5	0	0	5	5	5	5	0
10	10	5	5	0	10	10	10	5	5
15	10	10	5	5	10	15	10	10	5
20	15	10	10	5	15	20	15	15	5
30	25	20	10	5	25	30	25	20	10
40	30	25	15	10	30	40	30	25	10
50	40	30	20	10	40	50	40	30	15
60	50	35	25	10	50	60	45	40	15

e. MINIMUM DESIGN LOADS FOR MATERIALS

Material	Load, lb per cu ft	Material	Load, lb per cu ft
Aluminum, cast	165	Gravel, dry	104
Bituminous products:		Gypsum, loose	70
Asphalt	81	Ice	57.2
Petroleum, gasoline	42	Iron, cast	450
Pitch	69	Lead	710
Tar	75	Lime, hydrated, loose	32
Brass, cast	534	Lime, hydrated, compacted	45
Bronze, 8 to 14% tin	509	Magnesium alloys	112
Cement, portland, loose	90	Mortar, hardened:	
Cement, portland, set	183	Cement	130
Cinders, dry, in bulk	45	Lime	110
Coal, anthracite, piled	52	Riprap (not submerged):	
Coal, bituminous or lignite, piled	47	Limestone	83
Coal, peat, dry, piled	23	Sandstone	90
Charcoal	12	Sand, clean and dry	90
Copper	556	Sand, river, dry	106
Cork, compressed	14.4	Silver	656
Earth (not submerged):		Steel	490
Clay, dry	63	Stone, quarried, piled:	
Clay, damp	110	Basalt, granite, gneiss	96
Clay and gravel, dry	100	Limestone, marble, quartz	95
Silt, moist, loose	78	Sandstone	82
Silt, moist, packed	96	Shale, slate	92
Sand and gravel, dry, loose	100	Tin, cast	459
Sand and gravel, dry, packed	110	Water, fresh	62.4
Sand and gravel, wet	120	Water, sea	64
Gold, solid	1,205	Zinc	450

[l]For snow loads of uniform depth with 50-year mean recurrence interval and snow load coefficient of 0.8 adjusted for slopes. For mountainous regions, snow load should be based on analysis of local climate and topography. For computation of more accurate snow loads, see ANSI Standard A58.1-1972.

[m]0–5 psf—Southern and southwestern states, except for mountainous regions.

10 psf—Central and northwestern states.

20–30 psf—Middle Atlantic states and southern Great Lakes area.

40–50 psf—Northern Great Lakes area and New England states.

60–80 psf—Northern New England.

[n]Investigate roof fully loaded and alternatively with only one slope loaded.

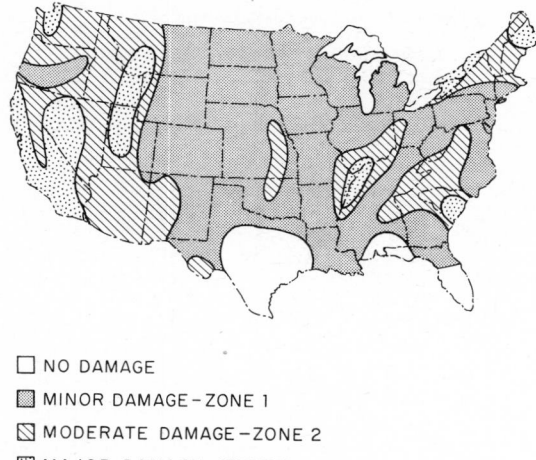

☐ NO DAMAGE

▨ MINOR DAMAGE – ZONE 1

▧ MODERATE DAMAGE – ZONE 2

▦ MAJOR DAMAGE – ZONE 3

Fig. 15-1. Map of the United States showing zones of probable seismic intensity.

moment-resisting space frame designed to withstand all the lateral forces, and is not enclosed or adjoined by more rigid elements, T may be taken as $0.1N$, where N = number of stories above grade.

Space frames, as defined in the Uniform Building Code (Art. 15-2), are three-dimensional structural systems composed of interconnected members, other than bearing walls, and laterally supported to function as a complete, self-contained unit, with or without the aid of horizontal diaphragms or floor bracing systems. Such vertical-load-carrying frames may be considered moment-resisting if the members and joints are capable of resisting design lateral forces by bending moments.

K takes into account the potential for inelastic energy absorption in moment-resisting frames. It also recognizes the redundancy of framing, or second line of defense, present in most complete frames, whether or not they are designed to resist lateral loads. Buildings that do not possess at least a complete vertical-load-carrying space frame are penalized by assignment of a high K. Table 15-3 lists suggested values for K.

Table 15-3. K for Aseismic Design of Buildings and Other Structures*

Framing system	K
Building framing systems other than those listed below	1.00
Building with a box system	1.33
Buildings with a ductile, moment-resisting space frame and shear walls designed so that:	
Frames and shear walls resist the total lateral force in proportion to their rigidities, taking into account interaction of shear walls and frames	
Shear walls acting independently of the moment-resisting space frame resist the total required lateral force	
The ductile, moment-resisting space frame can resist at least 25% of the required lateral force	0.80
Buildings with a ductile, moment-resisting space frame with capacity to resist the total required lateral force	0.67
Elevated tanks plus full contents, on four or more cross-braced legs and not supported by a building†	3.00
Structures other than buildings and those listed in Table 15-4	2.00

*Where prescribed wind loads produce higher stresses, those loads should be used instead of the seismic loads.

†The minimum value of KC in Eq. (15-1) is 0.12, and the maximum value need not exceed 0.25. Framing should be designed for an accidental horizontal torsion due to the lateral shear acting with an eccentricity of 5% of the maximum dimension in plan at the level of the shear. For tanks on the ground or connected to or part of a building, see Table 15-4.

Box systems, for which $K = 1.33$, employ shear walls or braced frames subjected primarily to axial stresses to resist the required lateral forces. Such systems are considered to lack a complete vertical-load-carrying space frame.

Note, however, that a building with shear walls or braced frames may be assigned $K = 0.80$ if it also has a ductile moment-resisting space frame capable of resisting at least 25% of the lateral force. In that case, the shear walls, acting independently of the space frame, must be able to withstand the total required lateral force.

A ductile, moment-resisting space frame of structural steel is a space frame capable of developing plastic hinges at connections of beams and girders to columns. In zones 2 and 3, each such connection should be capable of developing the full plastic capacity of the beam or girder at the connection, unless analysis or tests indicate that a smaller capacity would be adequate. To insure adequate ductility where beam flange area is reduced, for example, by bolt holes, plastic hinges should not be permitted to occur at such locations unless the ratio of ultimate strength to yield strength of the steel exceeds 1.5. In addition, special precautions must be taken to prevent buckling; for example, width-thickness and depth-thickness ratios must satisfy plastic-design criteria. And the effective lengths used in determining the slenderness ratios of columns should ignore any assistance provided by shear walls and braced frames in resisting lateral forces. Furthermore, the capacity of butt welds in tension should be verified by nondestructive testing.

For ductile, moment-resisting space frames of reinforced concrete, the Uniform Building Code places stringent requirements on member sizes, amount and location of reinforcement, concrete strength, anchorage of reinforcement, and beam-column joints.

A building whose sole resistance to lateral forces is provided by a ductile, moment-resisting space frame may be assigned $K = 0.67$. Before assigning a K value of 1.00 or less, however, designers should ascertain that action or failure of more rigid elements surrounding or adjoining the frame will not impair its ability to carry vertical or lateral loads.

Buildings often may qualify for more than one K value, for example, when they have different framing systems in perpendicular directions. Note, however, that if $K = 1.33$ is required in one direction, it also will be required in the other.

For elevated tanks independent of buildings, the relatively high value $K = 3$ is assigned because of their relatively poor performance in earthquakes. This high value is especially desirable for water tanks because of their importance in case of fire after an earthquake.

Structures other than buildings are assigned $K = 2$. This relatively high value reflects a general lack of redundancy, so that little yielding can be tolerated without failure. This value also assumes absence of nonstructural elements that can contribute materially to damping.

Building components should be designed for seismic forces given by

$$F_p = ZC_pW_p \tag{15-3}$$

(This equation has no bearing on determination of lateral forces to be applied to the building as a whole.)

Values of the seismic coefficient C_p are given in Table 15-4. W_p, in general, is the weight of the component. For floors and roofs acting as diaphragms, however, W_p is the tributary load from that story. And for storage and warehouse occupancies, W_p should be taken as the dead load plus 25% of the floor live load.

Distribution of Seismic Loads. Seismic forces are assumed to act at each floor level on vertical planar frames, or bents, extending in the direction of the loads. The seismic loads are distributed in proportion to the tributary weights, or masses carried by the bents. Thus, the seismic design force at any floor or roof level is proportional to that portion of W, used in Eq. (15-1), that is located at or assigned to the level.

Total horizontal shear at any level should be distributed to the bents of the lateral-force distribution system in proportion to their rigidities. The distribution should, however, take into account the rigidities of horizontal bracing and diaphragms (floors and roofs). In lightly loaded structures, for example, diaphragms may be sufficiently flexible to permit independent action of the lateral-force-resisting bents. A strong temblor could cause severe distress in frames and diaphragms if relative rigidities were not properly evaluated.

The total horizontal shear, kips, on a structure is V, given by Eq. (15-1). The Uniform Building Code recommends that the portion of V to be assigned to the top of a multistory structure be computed from

$$F_t = 0.004V \left(\frac{h_n}{D_s}\right)^2 \tag{15-4}$$

where h_n = height, ft, of top of structure above ground level

D_s = plan dimension, ft, of the vertical lateral-force-resisting system in the direction of the seismic forces

F_t need not be more than $0.15V$ and may be taken as zero when $h_n/D_s < 3$. Equation (15-4) recognizes the influence of higher modes of vibration as well as deviations from straight-line deflection patterns, particularly in tall buildings with relatively small dimensions in plan.

The code also recommends that the seismic force F_x to be assigned to any level at a height h_x, ft, above the ground be calculated from

$$F_x = (V - F_t)\ \frac{w_x h_x}{\sum\limits_{i=1}^{n} w_i h_i}$$ (15-5)

where w_x = portion of W located at or assigned to level x

h_x = height, ft, of level x above ground level

w_i = portion of W located at or assigned to level i

h_i = height, ft, of level i above ground level

n = number of levels in structure

Table 15-4. C_p **for Aseismic Design of Components of Buildings and Other Structures**

Components	Direction of force	C_p
Exterior bearing and nonbearing walls, interior bearing walls and partitions, interior nonbearing walls and partitions over 10 ft high, masonry or concrete fences over 6 ft high*	Normal to flat surface	0.20
Cantilever parapet and other cantilever walls, except retaining walls	Normal to flat surface	1.00
Exterior and interior ornamentations and appendages	Any direction	1.00
When connected to or part of a building: towers, tanks, towers and tanks plus contents, chimneys, smokestacks, and penthouses	Any direction	0.20†
When resting on the ground: tank plus effective mass of its contents	Any direction	0.10
Floors and roofs acting as diaphragms‡	Any direction	0.10
Connections for exterior panels or other elements attached to or enclosing the exterior§	Any direction	2.00
Connections for prefabricated structural elements other than walls, with force applied at center of gravity of assemblage¶	Any horizontal direction	0.30

*Deflection of walls or partitions under a force of 5 psf applied perpendicular to them shall not exceed $1/240$ the span for walls with brittle finishes or $1/120$ the span for walls with flexible finishes.

†For any building, when $h_n/D \geq 5$, increase C_p by 50%.

‡Apply a minimum value of C_p of 0.10 to load tributary from that story unless a greater value of C is required for Eq. (15-1).

§Precast, nonbearing, nonshear wall panels or other elements attached to or enclosing the exterior shall accommodate movements of the structure resulting from lateral forces or temperature changes. Connections and panel joints shall allow for a relative movement between stories of at least ¼ in. or twice the story drift under wind or seismic forces, whichever is greater. Connections should have sufficient ductility and rotation capacity to preclude fracture of concrete or brittle failures at or near welds. Movement in the plane of a panel for story drift may be provided by sliding connections with slotted or oversize holes, or by bending of steel.

¶Use dead load plus 25% of the floor live load for W_p in storage and warehouse occupancies.

For one- and two-story buildings, however, a uniform distribution of seismic loading should be used, because of their relatively large stiffness.

Overturning.   The equivalent static lateral forces applied to a building at various levels induce overturning moments. At any level, the overturning moment equals the sum of the products of each force and its height above that level. The overturning moments acting on the base of the structure and in each story are resisted by axial forces in vertical elements and footings.

At any level, the increment in the design overturning moment should be distributed to the resisting elements in the same proportion as the distribution of shears to those elements. Where a vertical resisting element is discontinued, the overturning moment at that level should be carried down as loads to the foundation.

Design Loads. Loads used in the design of a building should be the maximum probable loads to which the structure may be subjected. They should not be less, however, than the loads specified by the local building code. In the absence of a local code, the loads given in this article may be used, or the loads in a model building code, or the loads in "Building Code Requirements for Minimum Design Loads in Buildings and Other Structures," A58.1, American National Standards Institute, 1430 Broadway, New York, N.Y. 10018.

15-5 Modular Measure. This is a dimensioning system for building components and equipment to permit them to be field-assembled without cutting. The basic unit is a 4-in. cube. Thus, buildings may be laid out around a continuous, three-dimensional rectangular grid with 4-in. spacing (Fig. 15-2a).

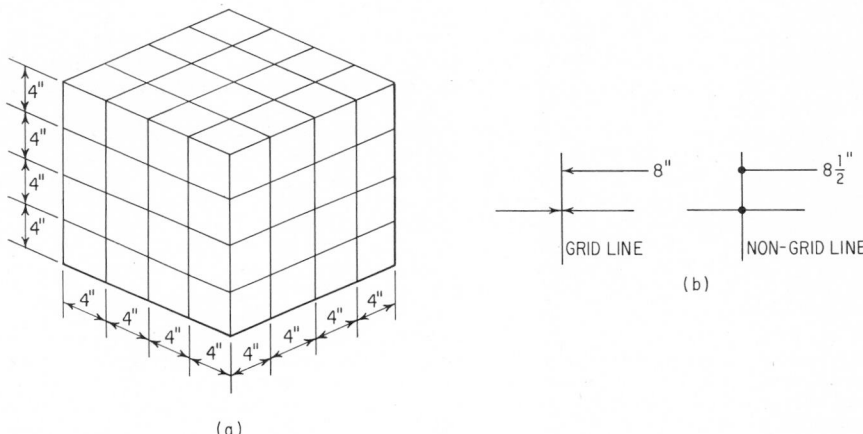

Fig. 15-2. Elements of modular measure.

Manufacturers make many building materials and some installed equipment to correspond to this module. The grid is a convenient tool for drawing assemblies of building products, be they modular or nonmodular.

Modular building products are assigned nominal dimensions corresponding to an even number of modules, though the actual dimensions may be slightly less to allow for joints. Nominal masonry dimensions, for example, equal the dimensions of a unit plus the thickness of one mortar joint. (Standard joint thickness is ⅜ in. for concrete block; ½ in. for clay backup and structural units; ⅜ or ½ in. for brick; and ¼ in. for salt-glazed, clear-glazed, and ceramic-glazed facing units.)

In preparing drawings, the designer can use the grid for both small-scale plans and large-scale details. At scales less than ¾ in. = 1 ft, however, it is not practical to show grid lines at 4-in. spacing. The designer should select a larger planning module that is a multiple of 4 in. For floor plans and elevations, for example, the module may be 2 ft 8 in., 4 ft, 5 ft, 6 ft 4 in., etc. Materials should be shown actual size, or to scale, and located on or related to a grid line by a reference dimension. Dimensions on grid lines are shown by arrows; those not on grid lines, by dots (Fig. 15-2b).

15-6. Structural Systems. Foundations for buildings should be selected and designed in accordance with the principles given in Sec. 7. Basic principles for superstructure design are given in Secs. 6 and 8 to 11.

Buildings may have load-bearing-wall construction, skeleton framing, or a combination of the two. Generally, the engineer's responsibility is to select that type of construction that will serve the owner's total needs most economically. Thus, the most economical construction may not necessarily be the one that requires the least structural materials, or even the one that also has the lowest fabrication and erection costs. Architectural, mechanical, electrical, and other costs that may be affected by the structural system must be considered in any cost comparison.

Because of the large number of variables, which change with time and location, the superiority of one type of construction over the others is difficult to demonstrate, even for a specific building at a given location and time.   Availability of materials and familiarity of contractors with required construction methods, or their willingness to take on a job, are important factors that complicate the selection of a structural system still more.   Consequently, engineers should consider the specific conditions for each building in selecting the structural system.

Also, deciding on the spans to be used is no simple matter.   Foundations, column or wall height, live load, bracing, and provisions for ducts and piping vary with each building and must be taken into account, along with the factors previously mentioned.   It is possible, however, to standardize designs for simple buildings, such as one-story warehouses or factories, and determine the most economical arrangement and spans of structural components.   But such designs should be reviewed and updated periodically, because changing conditions, such as the introduction of new materials, new shapes, or new construction methods and price revisions, could change the economic balance.

Engineers also should bear in mind that the relative economics of a structural system can be improved if it can be made to serve more than just structural purposes.   Money is saved if a facade also carries loads or if a structural slab is both floor and ceiling and also serves as air-conditioning ducts.

Load-bearing walls are such a multipurpose construction.   They serve as facades, separators, or enclosures, and also carry floor and roof loads to the ground.

Load-bearing wood walls frequently are used for one- and two-story houses.   They usually consist of 2 × 4 in. studs spaced 16 in. c to c and set with wide faces perpendicular to the face of the wall.   The walls have top and bottom plates, each consisting of two 2 × 4's.   Unless supported laterally by adequate framing, maximum height of such a wall is 15 ft.   Lumber or plywood sheathes the exterior; plaster or wallboard is placed on the interior.   ("Western Woods Use Book," Western Wood Products Association, Portland, Ore.)

Load-bearing masonry walls have been used for buildings 10 or more stories high.   But unless design is based on rational engineering analysis instead of empirical requirements, thickness required at the base is very large.   Many building codes require plain masonry bearing walls to have a minimum thickness of 12 in. for the top 35 ft and to increase in thickness 4 in. for each successive 35 ft down.   Thus, walls for a 20-story building would have to be about 3 ft thick at the bottom.

Since thickness must be increased from the top down, a natural shape in vertical cross section for load-bearing masonry walls is trapezoidal.   With the widest section at the bottom, such a shape is good for resisting overturning.   In practice, however, the exterior wall face usually is kept plumb and the inside face is stepped where thickness must be increased.

In the past, architects and engineers have tried to create an appearance of strength for such walls.   Hence, they kept openings as small as possible; when wide openings were necessary, they were flanked by wide wall areas, and the designers gave definite expression to the lintel or arch spanning the openings.   In multistory buildings, openings generally were placed one above the other and continuous piers provided from ground to roof.   Windows and doors were set well back from the exterior face to emphasize the thickness.

Much thinner walls can be used with steel-reinforced masonry designed in accordance with "Building Code Requirements for Engineered Brick Masonry," Brick Institute of America, McLean, Va.

("Recommended Practice for Engineered Brick Masonry," Brick Institute of America, McLean, Va.; F. S. Merritt, "Building Construction Handbook," McGraw-Hill Book Company, New York.)

Load-bearing reinforced concrete walls may be much thinner than masonry for a given height. The American Concrete Institute Building Code Requirements for Reinforced Concrete (ACI 318-71) set a minimum thickness of 6 in. for the top 15 ft and require an additional 1 in. for each successive 25 ft down.   So walls for a 20-story building would be 15 in. thick.   But these minimum thicknesses may be waived if structural analysis shows a thinner wall would have adequate strength and stability.

In low buildings, minimum thickness may be governed by the ratio of unsupported wall height or length to thickness, whichever ratio is smaller.   For solid masonry and grouted masonry bearing walls, this ratio should not exceed 20; for hollow masonry or cavity walls, 18; for reinforced concrete and reinforced grouted masonry, 25.   (For cavity walls, thickness is the sum of the nominal thicknesses of inner and outer wythes.)   Usually, bearing-wall thickness must be at least 6 in.; check the local building code.   (See also Art. 15-11.)

Load-bearing walls may be used for the exterior, partitions, wind bracing, and service-core enclosure.   For these purposes, masonry has the disadvantage when used in combination with skeleton framing of being erected more slowly.   Thus, there may be delays in erection of the framing while masonry is being placed to support it.

When **load-bearing partitions** can be placed at relatively short intervals across the width of a building, curtain walls can be used on the exterior along the length of the building.   Such partitions, together with flat-plate reinforced concrete floors (Fig. 15-3), make an efficient structural system for certain types of buildings, such as multistory apartment houses.   In such buildings, also, concrete walls around closets can double as columns.

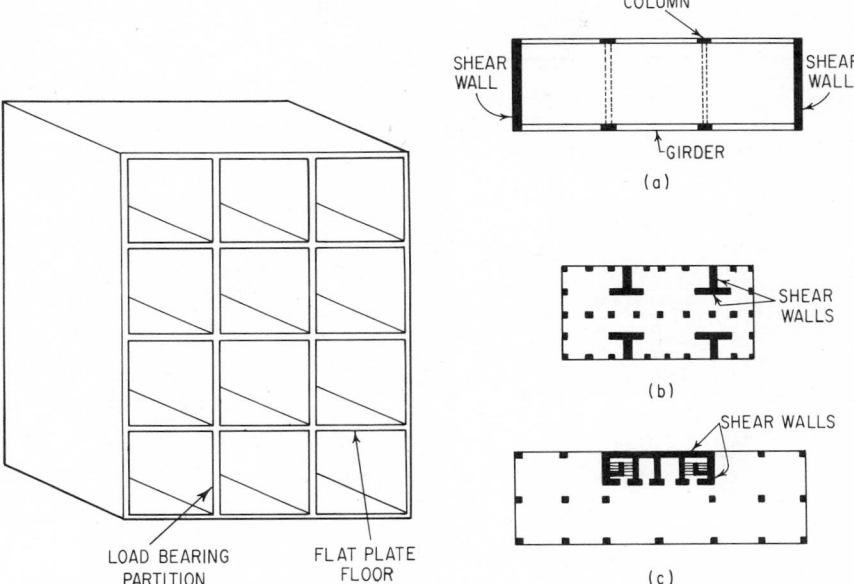

Fig. 15-3. Load-bearing partitions support flat-plate floors in an apartment building.

Fig. 15-4. Arrangements of shear walls for resisting lateral forces.

Load-bearing walls may serve as shear walls.   But unless they are relatively long, bending stresses due to lateral forces acting parallel to the walls may be large.   Thus, the walls, if properly arranged, will resist wind and earthquake forces in shear and bending.   For example, in Fig. 15-4a, shear walls placed at the ends of the building may be designed to resist lateral forces in the narrow direction.   In Fig. 15-4b, perpendicular shear walls can take lateral forces from all directions, since the forces can be resolved into components parallel to the walls.   In Fig. 15-4c, walls enclosing stairs, elevators, toilets, and service rooms (**service core**) may serve as shear walls in perpendicular directions.   For earthquake forces, however, it is desirable to supplement the shear walls with a ductile, moment-resisting space frame, to prevent sudden collapse if a shear wall should fail.

Load-bearing service-core walls can be designed, however, to carry all the loads in a building. In that case, the roof and floors cantilever from the walls (Fig. 15-5a).   When spans are large, cantilevers become uneconomic.   Instead, columns may assist the service-core walls in carrying the vertical loads (Fig. 15-5b).   As an alternative, the outer ends of the floors may be suspended from roof trusses, which are supported on but cantilever beyond the core walls (Fig. 15-5c).   Other possibilities include service cores in pairs with floors supported between them, on girders, trusses, cables, or arches or combinations of these.

Architectural-structural walls represent a type of exterior construction somewhere between load-bearing walls and skeleton framing with curtain walls.   The load-bearing elements in architectural-structural walls are linear, as in skeleton framing, rather than planar, as in load-bearing walls, and their function is clearly expressed architecturally.   Spaces between the structural elements may be screens or curtains, or glass.   The structural elements may lie on

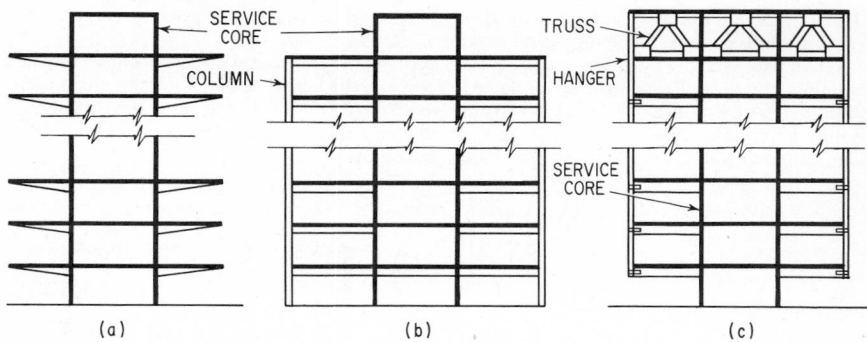

Fig. 15-5. Framing arranged to place all, or nearly all, loads on service-core walls.

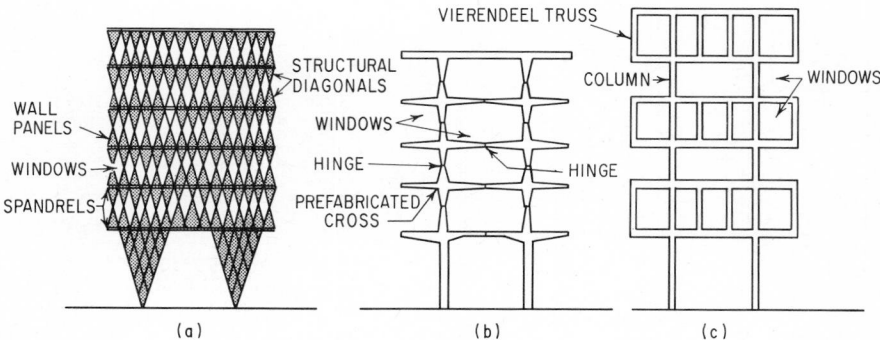

Fig. 15-6. Examples of architectural-structural walls.

diagonal lines or verticals (Fig. 15-6a); they may be cross-shaped, combining columns and spandrels (Fig. 15-6b); they may be horizontal or vertical Vierendeel trusses (Fig. 15-6c); or they may be any other system that is structurally sound.

In **skeleton framing,** columns carry building loads to the foundations. Lateral forces are resisted by the columns and diagonal bracing or by rigid-frame action.

Floor and roof construction are much the same for skeleton and load-bearing construction. One principal component is a horizontal structural slab or deck. The deck underside may serve as or carry a ceiling. The upper surface may serve as or carry a wearing surface for traffic or weatherproofing. The deck may be solid, or it may be hollow to reduce weight, permit pipe and wiring to pass through, and serve as air ducts. When the deck does not transmit its loads directly to columns, as it does in flat-slab and flat-plate construction, other major components of floor and roof systems are trusses, beams, and girders (sometimes also called joists, purlins, or rafters, depending on arrangement and location). These support the deck and transmit the load to the columns.

Flat-plate construction, where the deck has a constant thickness in each bay and transmits the load directly to columns, generally is economical for residential and other lightly loaded structures, where spans are fairly short. It is used for *lift-slab construction,* in which the concrete slabs are cast on the ground, then raised to final position by jacks set on the columns. For longer spans, a waffle or two-way ribbed plate may be used.

Flat-slab reinforced concrete construction may be more suitable for heavier loads. Also transmitting loads directly to columns, it differs from flat-plate in that the slab is thickened in the region around the columns (*drop panels*). Often, too, the columns flare at the top (*capitals*). Waffle construction may be used for longer spans.

Slab-band construction is a variation of flat-plate and flat-slab in which wide, shallow beams are used to support the slab and transmit loads to the columns.

Two-way slabs are another variation; they are supported on girders spanning between columns along the border of each bay. Thus, longer spans and heavier loads can be supported more economically.

Beam-and-girder construction is economical for a wide range of conditions. In one- and two-story houses, wood joists or rafters spaced 16 or 24 in. c to c generally are used on short spans in conjunction with lumber or plywood decking. For other lightly loaded structures, open-web steel joists, light, rolled steel beams, or precast-concrete plank may be used, with wood or concrete floors. For heavier loads and longer spans, one-way ribbed-concrete slabs and girders (metal-pan construction); prestressed-concrete plank, tees, double tees, or girders; reinforced concrete beams and girders; laminated-wood girders; or structural-steel beams and girders, including steel-concrete composite construction, may be more suitable. For still longer spans, as usually is the case in industrial buildings, beams and trusses may be most economical.

Arches and catenary construction are appropriate for very long spans. Usually, they are used to support roofs of hangars, stadiums, auditoriums, railroad terminals, and exhibition halls. Their design must provide a means of resisting the horizontal thrust of their reactions.

Thin-shell construction is suitable for uniform loading where curved surfaces are permissible or desirable. It is economical for very long spans. **Folded-plate construction** often is an economic alternative.

(F. S. Merritt, "Building Construction Handbook," McGraw-Hill Book Company, New York; T. Hamlin, "Forms and Functions of Twentieth Century Architecture," vols. 1 and 2, Columbia University Press, New York; F. S. Merritt, "Structural Steel Designers' Handbook," McGraw-Hill Book Company, New York.)

15-7. Lateral-Force Bracing. No structural system is complete unless it transmits all forces acting on it into adequate support in the ground. Hence, provision must be made in both low and tall buildings to carry into the foundations not only vertical loads but also lateral forces, such as those from wind and earthquake. Also, the possibilities of blast loading and collision with vehicles must be considered. Without adequate provision for resisting lateral forces, buildings may be so unstable that they may collapse during or after construction under loads considerably less than the full design wind or seismic loads.

Low Wood Buildings. In wood-frame houses one and two stories high, plywood or diagonal lumber sheathing may provide adequate resistance to lateral forces if it is properly nailed and glued. With diagonal lumber, each board should be nailed with two nails to every stud it crosses. Plywood ⅝ in. thick should be nailed with 8d common nails, 6 in. c to c; ¼ in. thick, with 6d nails, 3 in. c to c. With other types of sheathing, it is advisable to brace the frame with diagonal studs, especially at end corners of the outside walls and important intermediate corners.

Rigid Frames. Buildings of reinforced concrete beam-and-girder construction generally are designed as rigid frames, capable of taking lateral forces. Except possibly for tall structures subjected to severe earthquakes, rarely does additional provision have to be made for bracing against lateral forces. Tall flat-plate buildings also may be designed as rigid frames to resist wind. If the height-width ratio is large, wind resistance can be improved at relatively low cost by placing wings perpendicular or nearly so to the main portion, so that there are rigid frames with several bays parallel to the directions in which wind-force components may be resolved. Thus, the buildings may be made T-shaped, H-shaped, or cross-shaped in plan, or may have V-shaped wings at the ends. Alternatively, buildings may be curved in plan to improve wind resistance.

Shear Walls. When it is impractical to rely on a moment-resisting space frame to take 100% of the lateral forces, shear walls can be used to take all or part of them. Made of reinforced brick or concrete, such walls should be long enough parallel to the wind that bending stresses are within the allowable for the concrete and steel. As shown in Fig. 15-4, shear walls may be placed parallel to the narrow width of the building and rigid frames used in the longitudinal direction, or perpendicular shear walls may take lateral forces from any direction, or service-core walls may double as shear walls (Art. 15-6). The floors should be designed to act as diaphragms or adequate horizontal bracing should be provided to insure transfer of horizontal forces to the walls. For wind loads, provision must be made to brace exterior walls and transmit the loads from them to the floors. Walls should be adequately anchored to floors and roofs to prevent separation by wind suction or seismic forces.

In areas subjected to severe earthquakes, it is advisable that shear walls be supplemented by ductile, moment-resisting space frames, to prevent sudden collapse if the walls should fail.

Braced Frames. Another method of resisting lateral forces is to use diagonal bracing. Frames that are X-braced generally are stiffer than similar frames relying solely on rigid-frame action.

Roof trusses should be braced against horizontal forces, since the spans usually are long and roof decks are made of light material. Additional horizontal and vertical trusses may be used for the purpose. Also, the framework in the plane of the trusses may be stiffened by inserting knee braces between the columns supporting the trusses and the bottom chord. Purlins carrying the roof deck should be securely fastened to the top chords, which are in compression, to brace them laterally.

Trussed roof bracing may be placed in the plane of the top or bottom chords. Putting it in the plane of the top chords offers the advantages of simpler details, shorter unsupported length of diagonals, and less sagging of bracing, because it can be connected to the purlins at all intersections. Bracing both top and bottom chords with separate truss systems seldom is necessary. But the bottom chord should be braced at frequent intervals, even though it is a tension member, to reduce its unsupported length.

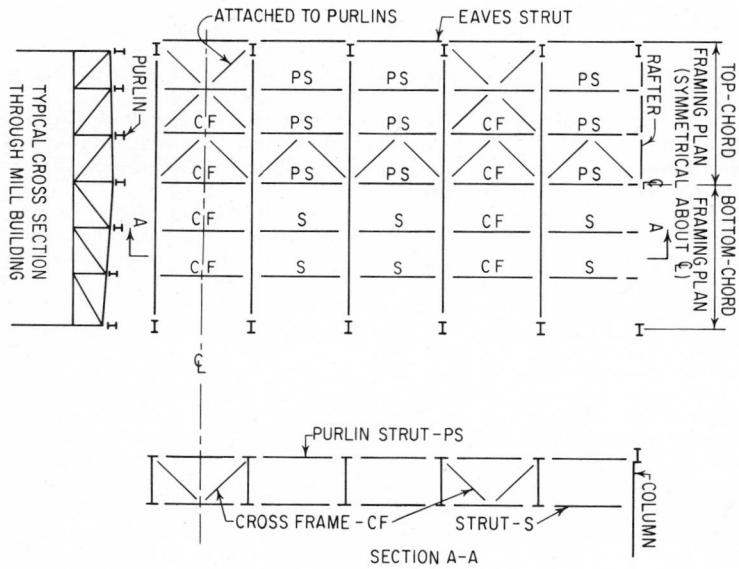

Fig. 15-7. Lateral bracing of roof trusses.

Figure 15-7 illustrates typical bracing for a mill-building roof. Diagonal bracing is placed in the plane of the top chord in three bays, assuming that the purlins will be sufficiently well connected to the trusses to transmit longitudinal forces from the unbraced trusses to the braced bays. Not more than five unbraced bays should be permitted between braced trusses. Struts are shown between lower chords at every panel point, but for a long truss, the struts may be placed at alternate panel points. At corresponding top-chord panel points, the purlins should be capable of carrying compressive forces in addition to vertical loading. The struts between the upper and lower chords should transmit longitudinal forces to the laterally braced bays, where cross frames are placed between the trusses in the plane of the struts, as indicated in Fig. 15-7, to prevent the trusses from tipping over.

Bracing the roof trusses, however, is not enough. The horizontal forces in the roof system must be brought to the ground. The designer must consider the building as a whole.

Figure 15-8 shows a simple bracing system to illustrate the principle. Wind forces on the windward long side of the building are transmitted to the leeward roof truss. This truss carries the loads to the ends of the building, where diagonals in the planes of the ends take the loads to the foundations. Wind on the ends is resisted by bracing in the side walls.

Tall Buildings. Similarly, in designing bracing for a tall building, the designer should consider the building as a whole. For example, lateral forces may be resisted by all the bents (Fig. 15-9a) or only the outer bents (Fig. 15-9b). In the latter case, the building may be designed as a hollow-tube cantilever for the horizontal forces.   The floor and roof systems, then, must be capable of distributing the loads from the windward wall to the side and leeward walls.

For the bents individually, X bracing (Fig. 15-9c) is both efficient and economical.   But it usually is impractical because it interferes with doors, windows, and clearance between floors and ceilings above.   Generally, the only places X bracing can be installed in tall buildings are in walls without openings, such as elevator-shaft and fire-tower walls.   So if X bracing can be used at all, additional bracing that does not interfere with openings must be placed in each bent.

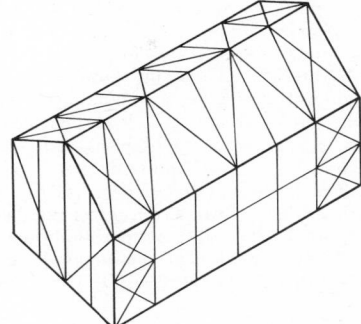

Fig. 15-8. X bracing carries lateral loads from roof to foundations.

There are many alternatives to X bracing.   One is knee bracing between girders and columns (Fig. 15-9d); but the braces may interfere with windows in exterior bents or may be objectionable in interior bents because they are unsightly or reduce floor-to-ceiling clearance.   Portal framing of several types, including haunched, solid-web spandrels (Fig. 15-9e) or trusses, are other alternatives.   At the columns, these members provide sufficient depth for moment resistance, but at a short distance away from the columns, they become shallow enough to clear windows and doors.   In exterior bents, the spandrels can extend from the window head in one

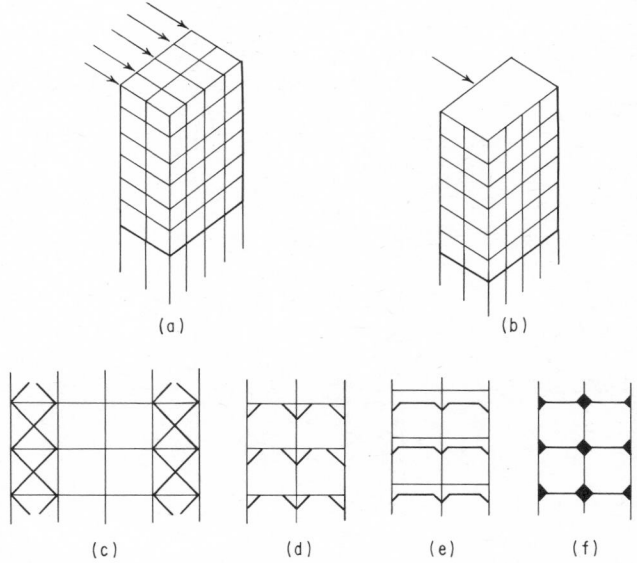

Fig. 15-9. Bracing for high-rise buildings.   (*a*) All transverse bents resist lateral forces; (*b*) building acts as vertical tube; (*c*) bent with X bracing; (*d*) knee bracing between columns and girders; (*e*) haunched spandrels; (*f*) moment-resisting connections between columns and girders.

story to the window sill in the story above.   In interior bents, however, they may have the same disadvantages as knee braces.

Another alternative to diagonal bracing for tall buildings is moment-resisting or wind connections of the bracket type (Fig. 15-9f).   Different types may be used, depending on size of members, magnitude of wind moment, and compactness needed to satisfy floor-to-ceiling clearance.   The minimum type consists of angles attached to the column and to top and bottom girder flanges.   Plates welded to both girder flanges and butt-welded to the column are an alternative.   When greater moment resistance is needed, the angles may be replaced by tees (made by splitting a wide-flange beam at middepth).   Also, the bottom flange may be seated on a beam-stub bracket.

The continuous rigid frames formed with these connections can be analyzed by the methods of Arts. 6-64 to 6-69.   Generally, however, approximate methods are used (Arts. 15-8 and 15-9).

It is noteworthy that for most buildings even the "exact" methods are not exact.   In the first place, the forces are not static loads, but generally dynamic; they are uncertain in intensity, direction, and duration.   Earthquake forces, usually assumed proportional to the weight of the building at each level and to the height above ground, act at the base of the structure, not at each floor level as is assumed in design.   Also, at the beginning of a design, the sizes of members are not known, so the exact resistance to lateral deformation cannot be calculated.   Furthermore, floors, walls, and partitions help resist the lateral forces in a very uncertain way.

(F. S. Merritt, "Building Construction Handbook," McGraw-Hill Book Company, New York; F. S. Merritt, "Structural Steel Designers' Handbook," McGraw-Hill Book Company, New York.)

15-8. Portal Method.   Since, as pointed out in Art. 15-7, an exact analysis of stresses due to lateral forces on a tall building is impractical, most designers prefer a wind-analysis method basis on reasonable assumptions and requiring a minimum of calculations.   One such method is the so-called "portal method."

It is based on the assumptions that points of inflection occur at the midpoints of all members and that exterior columns take half as much shear as do interior columns.   These assumptions enable all moments and shears to be computed by the laws of statics.

Consider, for example, the roof level (Fig. 15-10a) of a tall building.   A wind load of 600 lb is assumed to act at the top line of girders.   To apply the portal method, cut the building frame

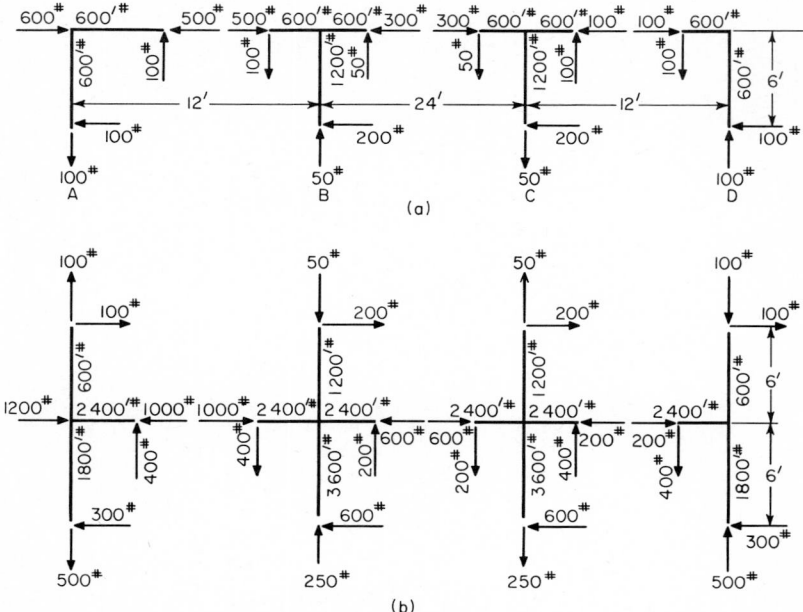

Fig. 15-10. Wind stresses in a tall building computed by the portal method.

along a section through the inflection points of the top-story columns. These points of zero moment are assumed here to be at the column midpoints, 6 ft down from the top of the building. (Some designers prefer to take the top-story inflection points one-third the story height down from the roof girders, because the sum of the stiffnesses of the members at each roof joint is likely to be much less than that at each joint in the story below. Similarly, they assume inflection points in the bottom story to be two-thirds the story height up from the base, because the anchorage tends to fix the base.) Now, let us compute the stresses in the members above the section.

Since the exterior columns take only half as much shear as do the interior columns, 100 lb of the total 600-lb load is apportioned to each of the exterior columns and 200 lb to each of the interior columns. The moments at the top of the columns equal these shears times the distance to the inflection point. The wall end of the end girders carries a moment equal to that in the exterior column. (At the floor below, as indicated in Fig. 15-10b, that end of the end girder carries a moment equal to the sum of the column moments.) Since the inflection point in the girder is at the midpoint, the moment at the inner end of the girder must be the same as the outer end. The moment in the adjoining girder can be found by subtracting the end-girder moment from the column moment, because the sum of the moments at the joint must be zero. (At the floor below, as shown in Fig. 15-10b, the moment in the interior girder is found by subtracting the moment in the end girder from the sum of the column moments.)

Girder shears then can be computed by dividing girder moments by the half span. When these shears have been found, column loads can be easily calculated from the fact that the sum of the vertical loads must be zero, by taking a section around each joint through column and girder inflection points. As a check, it should be noted that the column loads produce a moment that must be equal to the sum of the moments of the wind loads above the section for which the column loads were computed. For the roof level section (Fig. 15-10a), for example, $-50 \times 24 + 100 \times 48 = 600 \times 6$ (see also Art. 15-9).

(C. H. Norris and J. B. Wilbur, "Elementary Structural Analysis," McGraw-Hill Book Company, New York; F. S. Merritt, "Structural Steel Designers' Handbook," McGraw-Hill Book Company, New York.)

15-9. Cantilever Method. (See also Art. 15-8.) This is an approximate method for determining stresses in tall buildings due to lateral forces. Basic assumptions are that inflection points are at the midpoints of all members and that direct stress in a column is proportional to its distance from the center of gravity of all the columns in the bent. The assumptions are sufficient to enable shears and moments in the frame to be determined from the laws of statics.

For multistory buildings with height-to-width ratio of 4 or more, the Spurr modification is recommended ("Welded Tier Buildings," U.S. Steel Corp.). In this method, the moments of inertia of the girders at each level are made proportional to the girder shears when the spans are equal; otherwise, the moments of inertia must also be proportional to the square of the spans.

The results obtained from the cantilever method generally will be different from those obtained from the portal method. In general, neither solution is correct, but the answers provide a reasonable estimate of the resistance to be provided against lateral forces. In buildings over about 25 stories high, the effects of changes in column lengths should be considered in the analysis. (See also *Transactions of the American Society of Civil Engineers,* vol. 105, pp. 1713–1739, 1940; vol. 126, pp. 1124–1198, 1961.)

15-10. Floor Coverings. When concrete is used as the structural deck in a building, it may be left exposed to serve as a wearing surface, depending on the quality of surface and the type of occupancy. This is generally done in warehouses and industrial buildings with heavy moving loads. Some engineers, however, prefer to place a higher-quality topping on the structural concrete slab. The topping may be applied before or after the base slab has hardened. Usually, integral toppings are ½ in. thick, independent toppings about 1 in. ("Surface Treatments for Concrete Floors," IS147T, "Suggested Specifications for Heavy-duty Concrete Floor Topping," IS021B, Portland Cement Association, Skokie, Ill.). In office buildings where electricity and telephone wiring are distributed above the structural slab, a lightweight concrete fill covers the conduit and a floor covering protects the fill.

Wood floors may be made of the hardwoods, maple, beech, birch, oak, or pecan; or of the softwoods, yellow pine, Douglas fir, or western hemlock. The hardwoods are more resistant to wear and indentation. Solid-unit wood blocks are made from two or more units of strip-wood flooring fastened together with metal splines or other suitable devices. The blocks, tongued and grooved, are held in place with nails or an asphalt adhesive. Also, a laminated block is formed with plywood. Average moisture content of wood flooring at time of installation should be 6% in the dry southern states, 10% in damp southern coastal states, and 7% in the rest of the United States ("Moisture Content of Wood in Use," *U.S. Forest Products Laboratory Publication* 1655, Madison, Wis.). Leave at least 1 in. of expansion space at walls and columns.

Asphalt tiles, composed of asbestos fibers, mineral coloring pigments, and inert fillers with asphalt as the binder, are intended for use on rigid subfloors. They may be used on below-grade concrete subject to slight moisture from the ground.

Cork tile is made by baking cork granules with phenolic or other resin binders under pressure. It yields a surface suitable for areas where quiet and comfort are of utmost importance. It is intended for use on rigid subfloors above grade and free of moisture. Cork tile with natural finish should be sanded to level, sealed, and waxed immediately after installation. All cork floors must be maintained with sealers and protective coatings to prevent soiling.

Unbacked vinyl floorings, for use on rigid subfloors above grade, is made of polyvinyl chloride resin as a binder, plasticizers, stabilizers, extenders, inert fillers, and coloring pigments. Resilient under foot, it can withstand heavy loads without indentation but is easily scuffed and scratched unless protected with a floor polish. It is practically unaffected by grease, fat, oils, household cleaners, or solvents.

Vinyl-asbestos tiles are similar in composition to vinyl tiles except for the addition of asbestos fibers. They may be installed on below-grade concrete subject to slight moisture from the ground. The surface is harder than that of vinyl.

Vinyl also may be laminated to various backing materials. With an asbestos backing, the flooring may be used in moist areas.

Rubber flooring generally is intended for use on rigid subfloors above grade. It is resilient and has excellent resistance to permanent deformation under load.

Linoleum is made from drying oils, such as linseed, natural and synthetic resins, a filler, and pigments similar to those used in paints. Usually, it is backed with burlap or rag felt. Since the backing is susceptible to moisture and fungus attack, linoleum should not be used for floors where moisture can reach the backing. Properly maintained, it performs outstandingly on rigid subfloors above grade in residential and commercial buildings.

Since protection from moisture is a prime consideration for most thin floor coverings, moisture within a concrete slab must be brought to a low level before installation of the flooring begins. Moisture barriers should be placed under concrete slabs at or below grade, and a minimum of 30 days drying time should be allowed after concrete placement before installing the flooring. A longer drying time should be allowed for lightweight concrete.

Adhesives. Concrete surfaces to receive adhesive-applied thin flooring should be smooth. A troweled-on underlayment of rubber latex composition or asphalt mastic should be used over rough floors. Usually, the adhesive for asphalt and vinyl-asbestos tiles is an emulsion or cut-back asphalt; for rubber and vinyl above grade, latex; for linoleum, cork, and vinyl backed with felt, linoleum paste; for laminated or solid wood block, hot-melt or cutback asphalt; for vinyl backed with asbestos felt, latex on concrete and linoleum paste on plywood and hardboard. Laminated wood blocks also may be set with a rubber-base adhesive.

Ceramic tiles generally are bonded to the subfloor with portland-cement mortar (see ANSI Standard Specifications for Ceramic Mosaic Tile, Quarry Tile, and Pavers Installed in Portland Cement Mortars, ANS A108.2, A108.3, and A108.5). For areas not subject to heavy traffic, concentrated loads, or excessive amounts of water, organic-adhesive thin setting beds may be used instead. Appearance and resistance to wear make ceramic tiles suitable for use in kitchens and bathrooms.

Ceramic mosaic tile is less than 6 sq in. in area. Paver tiles are larger, usually 3 × 3 to 6 × 6 in. Quarry tile is a denser product, highly resistant to freezing, abrasion, and moisture.

Terrazzo is a mosaic topping composed of two parts marble chips to one part portland cement, sometimes with color pigments, applied to concrete or steel decks. Rubber latex, epoxy, and polyesters are alternative matrix materials. The topping may be precast or cast in place. Sand cushion (floating) terrazzo, at least 3 in. thick, is used where structural movement that might injure the topping is anticipated. When terrazzo is bonded to the under slab, the topping usually is at least 1¾ in. thick; a monolithic topping may be ⅝ in. thick.

Detailed information on installation and maintenance of flooring may be obtained from:

American Plywood Association, 1119 A St., Tacoma, Wash. 98402.

Asphalt and Vinyl Asbestos Tile Institute, 101 Park Ave., New York, N.Y. 10017.

Ceramic Tile Institute, 3415 West 8th St., Los Angeles, Calif. 90005.

Hardwood Plywood Manufacturers Association, P.O. Box 6246, Shirlington Station, Arlington, Va. 22208.

Maple Flooring Manufacturers Assocation, 424 Washington Ave., Oshkosh, Wis. 54901.

National Oak Flooring Manufacturers Association, 814 Sterick Building, Memphis, Tenn. 38103.

National Terrazzo and Mosaic Association, 716 Church St., Alexandria, Va. 22314.

Portland Cement Association, Old Orchard Road, Skokie, Ill. 60076.

Rubber and Vinyl Flooring Council Division, Rubber Manufacturers Association, Inc., 444 Madison Ave., New York, N.Y. 10022.

Tile Contractors Association of America, Inc., 112 North Alfred St., Alexandria, Va. 22314.

Tile Council of America, P.O. Box 326, Princeton, N.J. 08540.

Wood Flooring Institute, 201 North Wells St., Chicago, Ill. 60606.

15-11. Masonry Walls. Different design criteria are applied to masonry walls, depending on whether they are load-bearing or non-load-bearing. Minimum requirements for both types are given in ANSI Standard Building Code Requirements for Masonry, A41.1. See also ANSI Standard Building Code Requirements for Reinforced Masonry, A41.2, and Building Code Requirements for Engineered Brick Masonry, Brick Institute of America.

Following are some of the terms most commonly encountered in masonry construction:

Architectural Terra Cotta. (See Ceramic Veneer.)

Ashlar Masonry. Masonry composed of rectangular units, usually larger in size than brick and properly bonded, having sawed, dressed, or squared beds. It is laid in mortar.

Bonder. (See Header.)

Brick. A rectangular masonry building unit, not less than 75% solid, made from burned clay, shale, or a mixture of these materials.

Buttress. A bonded masonry column built as an integral part of a wall and decreasing in thickness from base to top, though never thinner than the wall. It is used to provide lateral stability to the wall.

Ceramic Veneer. Hard-burned, non-load-bearing, clay building units, glazed or unglazed, plain or ornamental.

Chase. A continuous recess in a wall to receive pipes, ducts, and conduits.

Column. A compression member with width not exceeding four times the thickness, and with height more than three times the least lateral dimension.

Concrete Block. A machine-formed masonry building unit composed of portland cement, aggregates, and water.

Coping. A cap or finish on top of a wall, pier, chimney, or pilaster to prevent penetration of water to masonry below.

Corbel. Successive courses of masonry projecting from the face of a wall to increase its thickness or to form a shelf or ledge.

Course. A continuous horizontal layer of masonry units bonded together.

Cross-sectional Area. Net cross-sectional area of a masonry unit is the gross cross-sectional area minus the area of cores or cellular spaces. Gross cross-sectional area of scored units is determined to the outside of the scoring, but the cross-sectional area of the grooves is not deducted to obtain the net area.

Grout. A mixture of cementitious material, fine aggregate, and sufficient water to produce pouring consistency without segregation of the constituents.

Grouted Masonry. Masonry in which the interior joints are filled by pouring grout into them as the work progresses.

Header (Bonder). A brick or other masonry unit laid flat across a wall with end surface exposed, to bond two wythes.

Height of Wall. Vertical distance from top of wall to foundation wall or other intermediate support.

Hollow Masonry Unit. Masonry with net cross-sectional area in any plane parallel to the bearing surface less than 75% of its gross cross-sectional area measured in the same plane.

Masonry. A built-up construction or combination of masonry units bonded together with mortar or other cementitious material.

Mortar. A plastic mixture of cementitious materials, fine aggregates, and water.

Partition. An interior wall one story or less in height.

Pier. An isolated column of masonry. A bearing wall not bonded at the sides into associated masonry is considered a pier when its horizontal dimension measured at right angles to the thickness does not exceed four times its thickness.

Pilaster. A bonded or keyed column of masonry built as part of a wall and of uniform thickness throughout its height. It serves as a vertical beam, column, or both.

Rubble:

Coursed Rubble. Masonry composed of roughly shaped stones fitting approximately on level beds, well bonded and brought at vertical intervals to continuous level beds or courses.

Random Rubble. Masonry composed of roughly shaped stones, well bonded and brought at irregular intervals vertically to discontinuous but approximately level beds or courses.

Rough or Ordinary Rubble. Masonry composed of irregularly shaped stones laid without regularity of coursing, but well bonded.

Solid Masonry Unit. A masonry unit with net cross-sectional area in every plane parallel to the bearing surface 75% or more of its gross cross-sectional area measured in the same plane.

Veneer. A wythe securely attached to a wall but not considered as sharing load with or adding strength to it.

Wall. Vertical or near-vertical construction for enclosing space or retaining earth or stored materials.

Bearing Wall. A wall that supports any vertical load in addition to its own weight.

Cavity Wall. (See Hollow Wall.)

Curtain Wall. A non-load-bearing exterior wall.

Faced Wall. A wall in which the masonry facing and the backing are of different materials and are so bonded as to exert a common reaction under load.

Hollow Wall. A wall of masonry so arranged as to provide an air space within the wall between the inner and outer wythes. A cavity wall is built of masonry units or plain concrete, or a combination of these materials, so arranged as to provide an air space within the wall, which may be filled with insulation, and in which inner and outer wythes are tied together with metal ties.

Nonbearing Wall. A wall that supports no vertical load other than its own weight.

Party Wall. A wall on an interior lot line used or adapted for joint service between two buildings.

Shear Wall. A wall that resists horizontal forces applied in the plane of the wall.

Spandrel Wall. An exterior curtain wall at the level of the outside floor beams in multistory buildings. It may extend from the head of the window below the floor to the sill of the window above.

Veneered Wall. A wall having a facing of masonry or other material securely attached to a backing, but not so bonded as to exert a common reaction under load.

Wythe. Each continuous vertical section of a wall one masonry unit in thickness.

Materials used in masonry construction should be capable of meeting the requirements of the applicable standard of the American Society for Testing and Materials. For unit masonry, mortar should meet the requirements of ASTM Specifications C270 and C476. Mortars containing lime generally are preferred because of greater workability. Commonly used:

For concrete block, 1 part cement, 1 part lime putty, 5 to 6 parts sand.

For rubble, 1 part cement, 1 to 2 parts lime hydrate or putty, 5 to 7 parts sand.

For brick, 1 part cement, 1 part lime, 6 parts sand.

For setting tile, 1 part cement, ½ part lime, 3 parts sand.

Stresses. In determining stresses, effects of loads should be computed on actual dimensions, not nominal. Stresses should not exceed the allowable stresses given in ANS A41.1 or the local building code. In composite walls, maximum stress should not exceed the allowable stress for the weakest of the combination of units and mortars. To insure desired compressive and tensile strengths, specify a full bed of mortar, with each course well hammered down, and all joints completely filled with mortar.

Stability. For solid or grouted masonry walls or partitions, ratio of unsupported height to nominal thickness, or the ratio of unsupported length to nominal thickness, should not exceed 20. For reinforced grouted masonry, the ratio should not exceed 25. For hollow walls or walls of hollow masonry units, the ratio should be 18 or less. For cavity walls, thickness is the sum of nominal thicknesses for all wythes. Nonbearing masonry partitions should be supported laterally at distances not exceeding 36 times the actual thickness of the partition, including plaster. Cantilever walls and masonry walls in locations exposed to high winds should not be built higher than 10 times their thickness unless adequately braced. Backfill should not be placed against foundation walls until they have been braced to withstand horizontal pressure. Veneers should not be considered part of the wall in computing thickness for strength or stability.

In determining the unsupported length of walls, you may assume existing cross walls, piers, or buttresses as lateral supports if these members are well bonded or anchored to the walls and capable of transmitting the lateral forces to connected structural members or to the ground. In determining the unsupported height of walls, you may consider floors and roofs as lateral supports, if provision is made in the building to transmit the lateral forces to the ground. Ends of floor joists or beams bearing on masonry walls should be securely fastened to the walls. If lateral support of a partition depends on a ceiling, floor, or roof, the top of the partition should have adequate anchorage to transmit the forces. This anchorage may be accomplished with metal anchors or by keying the top of the partition to overhead work. Suspended ceilings may be considered as lateral support if ceilings and anchorages are capable of resisting a horizontal force of 150 lb per lin ft of wall.

Thickness Requirements. Walls should not vary in thickness between lateral supports. When it is necessary to change thickness between floor levels to meet minimum thickness requirements, the greater thickness should be carried up to the next floor level. Where walls of hollow units or bonded hollow walls are decreased in thickness, a course of solid masonry should be interposed between the wall below and the thinner wall above, or else special units or construction should be used to transmit the loads between the walls of different thickness.

Many building codes require that, in general, masonry bearing walls should be at least 12 in. thick for the top 35 ft of their height. Thickness should be increased 4 in. for each successive 35 ft or fraction of this distance, measured down from the top of the wall. Rough or random or coursed rubble stone walls should be 4 in. thicker than this. For other than rubble stone walls, the following exceptions apply to masonry bearing walls:

Stiffened Walls. Where solid masonry bearing walls are braced at distances not greater than 12 ft by masonry cross walls or by reinforced concrete floors, they may be 12 in. thick for the top 70 ft. They should be increased 4 in. in thickness for each successive 70 ft or fraction of that distance.

Top-Story Walls. The top-story bearing wall of a building not over 35 ft high may be made 8 in. thick. But this wall should be no more than 12 ft high and should not be subjected to lateral thrust from the roof construction.

Residential Walls. In dwellings up to three stories high, walls may be 8 in. thick (if not more than 35 ft high), if not subjected to lateral thrust from the roof construction. Such walls in one-story houses and one-story private garages may be 6 in. thick if the height is 9 ft or less or if the height to the peak of a gable does not exceed 15 ft.

Penthouses and Roof Structures. Masonry walls up to 12 ft high above roof level, enclosing stairways, machinery rooms, shafts, or penthouses, may be 8 in. thick. They need not be included in determining the height for meeting thickness requirements for the wall below.

Plain Concrete Walls. Such walls may be 2 in. less in thickness than the minimum requirements for other types of masonry walls but, in general, not less than 8 in.—and not less than 6 in. in one-story dwellings and garages.

Hollow Walls. Cavity or masonry bonded hollow walls should not be more than 35 ft high. In particular, 10-in. cavity walls should be limited to 25 ft in height above supports. The facing and backing of cavity walls should be at least 4 in. thick, and the cavity should not be less than 2 in. or more than 3 in. wide.

Faced Walls. Neither the height of faced (composite) walls nor the distance between lateral supports should exceed that prescribed for masonry of either of the types forming the facing and the backing. Actual (not nominal) thickness of material used for facings should not be less than 2 in. or one-eighth the height of the unit.

Parapet Walls. In general, these should be at least 8 in. thick, and the height should not exceed three times the thickness. The thickness may be less than 8 in., however, if the parapet is reinforced to withstand earthquake and wind forces to which it may be subjected.

Piers. Unsupported height should not exceed 10 times the least dimension. But when structural clay tile or hollow concrete units are used for isolated piers to support beams, unsupported height should not exceed four times the least dimension unless the cellular spaces are filled solidly with concrete or high-strength mortar.

Headers. When headers are used for bonding the facing and backing in solid walls and faced walls, not less than 4% of the wall surface of each face should be composed of headers, which should extend at least 4 in. into the backing. These headers should not be more than 24 in. apart vertically or horizontally. In walls in which a single bonder does not extend through the wall, headers from opposite sides should overlap at least 4 in. Or else, the headers should be covered with another bonder course overlapping headers below at least 4 in.

Ties. Metal ties for bonding should be corrosion-resistant. For bonding facing and backing of solid masonry walls and faced walls, there should be at least one metal tie for each 4½ sq ft of wall area. Ties in alternate courses should be staggered. The maximum vertical distance between ties should not exceed 18 in., and maximum horizontal distance not more than 36 in.

Cavity-wall wythes should be bonded with ³/₁₆-in.-diameter steel rods or metal ties of equivalent stiffness embedded in horizontal joints. There should be at least one metal tie for each 4½ sq ft of wall area, arranged and spaced as in solid walls. Rods bent to rectangular shape should be used with hollow masonry units laid with cells vertical. In other walls, the ends of ties should be bent to 90° angles to provide hooks at least 2 in. long. Additional bonding ties should be installed at all openings. These ties should be spaced not more than 3 ft apart around the perimeter and within 12 in. of the opening.

When two bearing walls intersect and the courses are built up together, the intersections should be bonded by laying in true bond at least half of the units at the intersection. When the courses are carried up separately, the intersecting walls should be regularly toothed or blocked with 8-in. maximum offsets. The joints should be provided with metal anchors having a minimum section of ¼ × 1½ in. with ends bent up at least 2 in. or with cross pins to form an anchorage. Such anchors should be at least 2 ft long and spaced not more than 4 ft apart.

("Recommended Practice for Engineered Brick Masonry," and H. C. Plummer, "Brick and Tile Engineering," Brick Institute of America, McLean, Va.; J. R. Dalzell and F. S. Merritt, "Simplified Masonry Planning and Building," McGraw-Hill Book Company, New York.)

15-12. Glass Block. Masonry walls of glass block may be used to control light that enters a building and to obtain better thermal and acoustic insulation than with ordinary glass panes. These units are hollow, 3⅞ in. thick by 6, 8, or 12 in. square (actual length and height ¼ in. less, for modular dimensioning). Faces of the units may be cut into prisms to direct light, or the block may be treated to diffuse light.

Glass block may serve as nonbearing walls or to fill openings in walls. Block so used should have a minimum thickness of 3½ in. at the joint. Also, surfaces of the block should be treated to permit satisfactory mortar bonding.

For exterior walls, glass-block panels should not have an unsupported area exceeding 144 sq ft. They should not be more than 25 ft long or more than 20 ft high between supports.

For interior walls, glass-block panels should not have an unsupported area of more than 250 sq ft. Neither length nor height should exceed 25 ft.

Exterior panels should be held in place in the wall opening to resist both internal and external wind pressures. The panels should be set in recesses at the jambs so as to provide a bearing surface at least 1 in. wide along the edges. Panels more than 10 ft long also should be recessed at the head. (Some building codes, however, permit anchoring small panels in low buildings with noncorrodible perforated metal strips.) The sides and top, kept free of mortar and filled with resilient material, should permit expansion.

Mortar joints should be from ¼ to ⅜ in. thick. Steel reinforcement should be placed in the horizontal mortar joints at vertical intervals of 2 ft or less. It should extend the full length of the joints. When splices are necessary, the reinforcement should be lapped at least 6 in. It should consist of two parallel longitudinal galvanized-steel wires. They should be No. 9 gage or larger, spaced 2 in. apart, and having welded to them No. 14 or heavier gage cross wires at intervals of up to 8 in.

15-13. Curtain Walls. With skeleton-frame construction, exterior walls need carry no load other than their own weight. Their principal function is to keep wind and weather out of the building—hence the name curtain wall. Nonbearing walls may be supported on the structural frame of the building or projections from it, on supplementary framing (girts, for example) in turn supported on the structural frame, or on the floors.

Curtain walls need not be any thicker than required to serve their principal function. Many industrial buildings are enclosed only with light-gage metal. For structures with certain types of occupancies and for buildings close to others, however, fire resistance is an important characteristic. Fire-resistance requirements in local building codes often govern in determining the thickness and type of material used for curtain walls.

In many types of buildings, it is desirable to have an exterior wall with good insulating properties. Sometimes, a dead air space is used for this purpose. Sometimes, insulating material is incorporated in the wall or erected as a backup.

The exterior surface of a curtain wall should be made of a durable material, capable of lasting as long as the building. Maintenance should be a minimum; initial cost of the wall is not so important as the annual cost (amortized initial cost plus annual maintenance and repair costs).

Wood walls are used on one- and two-story buildings, generally with a wood frame. The frame may be sheathed on its outer sides with gypsum, lumber, or plywood and then a finish applied, or sheathing and siding can be combined in one unit. The exterior finish may be in the form of wood shingles, siding, half timbers, or plywood sheets.

Drop, or novelty, siding—tongued-and-grooved boards—is not considered a good finish for permanent structures. **Lap siding** or **clapboards** are better. These are beveled boards, thinner along one edge than the opposite edge, which are nailed over sheathing and building paper. Usually, narrow boards lap each other 1 in., wide boards more than 2 in. Normally, clapboards are installed with edges horizontal.

Half timbers may be used to form a structural frame of heavy horizontal, vertical, and diagonal members, the spaces between being filled with brick. This type of construction sometimes is imitated by nailing boards in a pattern similar to an ordinary sheathed frame and filling the spaces between boards with stucco.

Plywood for exterior use should be an exterior grade, with plies bonded with permanent waterproof glue. The curtain wall may consist of a single sheet of plywood or of a sandwich of which plywood is a component. Also, plywood may be laminated to another material, such as a light-gage metal, to give it stiffness.

Stucco is an exterior wall finish applied like plaster. It is made of sand, portland cement, lime, and water. Two coats are applied to masonry, three coats on metal lath. The lath should be heavily galvanized. It should weigh 3.4 lb per sq yd when supports are 16 in. c to c and at least 2.5 lb with closely spaced furring. For the first coat, a common mix is 1 part portland cement, 1 part lime putty, and 5 or 6 parts sand. The second, or brown, coat may be based on lime or portland cement. With lime, the mix may be 1 part quicklime putty or hydrated lime putty and 3 parts sand by volume. With cement, the mix may be 1 part portland cement to 3 parts sand, plus lime putty in an amount equal to 15 to 25% of the volume of cement. The finish coat may have the same proportions as the brown coat. The brown coat may be applied as soon as the first, or scratch, coat has hardened, usually in 7 to 10 days. For the finish coat, it may be wise to wait several months to give the building a chance to settle and the base coats to shrink.

Metal or asbestos-cement siding may be used as curtain walls with or without insulating backup. Precautions should be taken to prevent water from penetrating between sheets. With corrugated sheeting, horizontal splices should lap about 4 in. Vertical splices should lap at least 1½ corrugations. Flat sheets may be installed in sash like window glass, or the splices may be covered with battens. Edges of metal sheets may be flanged to interlock and exclude wind and rain. Provision should be made in all cases for expansion and contraction with temperature changes.

In contrast to siding, in which a single material may form the complete wall, metal or glass sometimes is used as a facing and backed up with insulation, fire-resistant material, and an interior finish. The glass usually is tinted and is held in a light frame in the same manner as window glass. Metal panels may be fastened similarly in a light frame, attached to mullions or other secondary framing members, anchored to brackets at each floor level, or connected to the structural frame of the building. The panels may be small and light enough for one person to carry or as big as one or two stories in height, prefabricated with windows. Provisions should be made for expansion and contraction and to prevent penetration of moisture at joints. Flashing and other details should be so arranged that should water penetrate the facing, it will drain to the outside.

Curtain walls also may be built of prefabricated panels consisting of an insulation core sandwiched between a thin, lightweight facing and backing. Such panels may be fastened in place in much the same manner as metal or glass facings. And the same precautions regarding expansion, contraction, and water penetration should be taken.

Metal curtain walls may be custom, commercial, or industrial type. Custom-type walls are those designed for a specific project, generally multistory buildings. Commercial-type walls are those built up of parts standardized by manufacturers. Industrial-type walls comprise ribbed, fluted, or otherwise preformed metal sheets in stock sizes, standard metal sash, and insulation.

Metal curtain walls may be classified according to the methods used for field installation:

Stick Systems (Fig. 15-11a). Walls installed piece by piece. Each principal framing member, with windows and panels, is assembled in place separately (Fig. 15-11b). This type of system involves more parts and field joints than other types and is not so widely used.

Mullion-and-Panel Systems (Fig. 15-11c). Walls in which vertical supporting members (mullions) are erected first, and then wall units, usually incorporating windows (generally

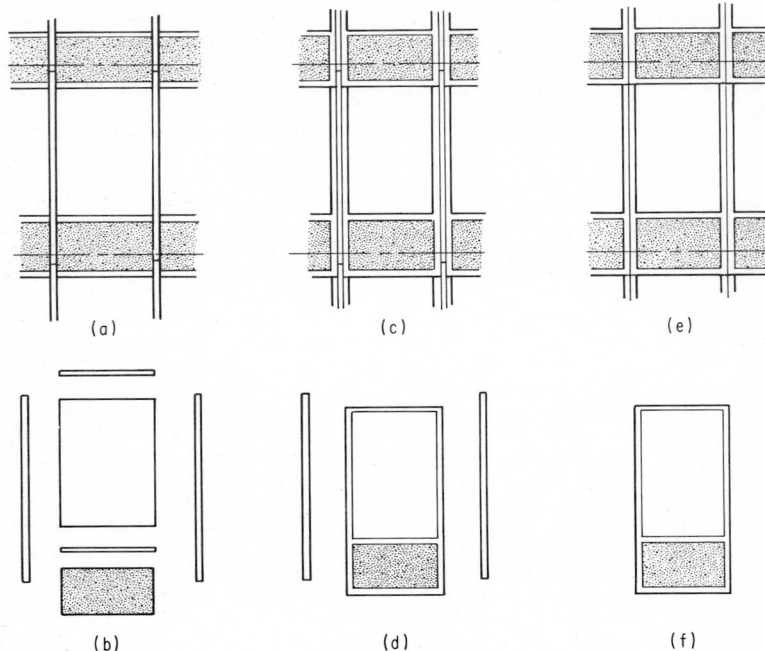

Fig. 15-11. Metal curtain walls. Stick systems (*a*) are installed piece by piece (*b*). Mullion-and-panel systems (*c*) have panels placed between verticals (*d*). Panel systems (*e*) come factory-assembled (*f*).

unglazed), are placed between them (Fig. 15-11*d*). Often, a cover strip is added to cap the vertical joint between units.

Panel Systems (Fig. 15-11*e*). Walls composed of factory-assembled units (generally unglazed) and installed by connecting to anchors on the building frame and to each other (Fig. 15-11*f*). Units may be one or two stories high. This system requires fewer pieces and field joints than the other systems.

When mullions are used, it is customary to provide for horizontal movements of the wall at the mullions and, in multistory buildings, to accommodate vertical movement at each floor, or at alternate floors when two-story-high components are used. Common ways of providing for horizontal movements include use of split mullions, bellows mullions, batten mullions, and elastic structural gaskets. To permit vertical movement, mullions are spliced with a telescoping slip joint. When mullions are not used and wall panels are connected to each other along their vertical edges, the connection is generally made through deep flanges. With the bolts several inches from the face of the wall, movement is permitted by the flexibility of the flanges. Slotted holes are unreliable as a means of accommodating wall movement.

(W. F. Koppes, "Metal Curtain Wall Specifications Manual," National Association of Architectural Metal Manufacturers, 1010 West Lake St., Oak Park, Ill. 60301; "Curtain Wall Handbook," U.S. Gypsum Co., Chicago, Ill. 60606; F. S. Merritt, "Building Construction Handbook," McGraw-Hill Book Company, New York.)

15-14. Partitions. These are walls one story or less in height used to subdivide the interior space in buildings. They may be bearing or nonbearing walls.

Bearing partitions may be built of masonry or of wood or light-gage steel studs. The masonry or studs may be faced with plaster, wallboard, plywood, wood boards, plastic, or other materials that meet functional and architectural requirements.

Nonbearing partitions may be permanently fixed in place, or temporary (movable) so that the walls may be easily shifted when desired. The temporary type includes folding partitions. Since the principal function is to separate space, construction and materials used vary widely. Partitions may be opaque or transparent; they may be louvered or hollow or solid; they may extend from floor to ceiling or only partway; and they may serve additionally as cabinets or closets or as a concealment for piping and electrical conduit. Fire resistance sometimes dictates the type of construction. Acoustic treatment may range from acoustic finishes on the surfaces to use of double walls separated completely by an air space or an insulating material.

Folding partitions, in a sense, are large doors. Depending on size and weight, they may be electrically or manually operated. They may be made of wood, light-gage metal, or synthetic fabric on a light collapsible frame. Provision should be made for framing and supporting them in a manner similar to that for large folding doors (Art. 15-17).

(F. S. Merritt, "Building Construction Handbook," McGraw-Hill Book Company, New York.)

15-15. Windows. Many building codes require that glass areas be equal to at least 10% of the floor area of each room. But for many types of occupancy, it is good practice to provide glass areas in excess of 20%. Windows should be continuous and located as high as possible to lengthen the depth of light penetration. Continuous sash or one large window in a room gives better light distribution than separated narrow windows. Since windows also are used to provide ventilation, the designer sometimes must compromise between locations, sizes, and arrangements of windows that give best lighting or best ventilation.

Window sash and frames generally are made of wood or metal. Fire-resistance requirements of building codes usually dictate use of metal.

White pine, sugar pine, ponderosa pine, fir, redwood, cedar, and cypress are used for exposed wood window parts because of their resistance to shrinkage and warping. A relatively hard wood should be used for the stiles against which a double-hung window slides. Inside parts of wood windows are usually made of the same material as trim. Detailed information can be obtained from the National Woodwork Manufacturers Association, 400 West Madison Ave., Chicago, Ill. 60606.

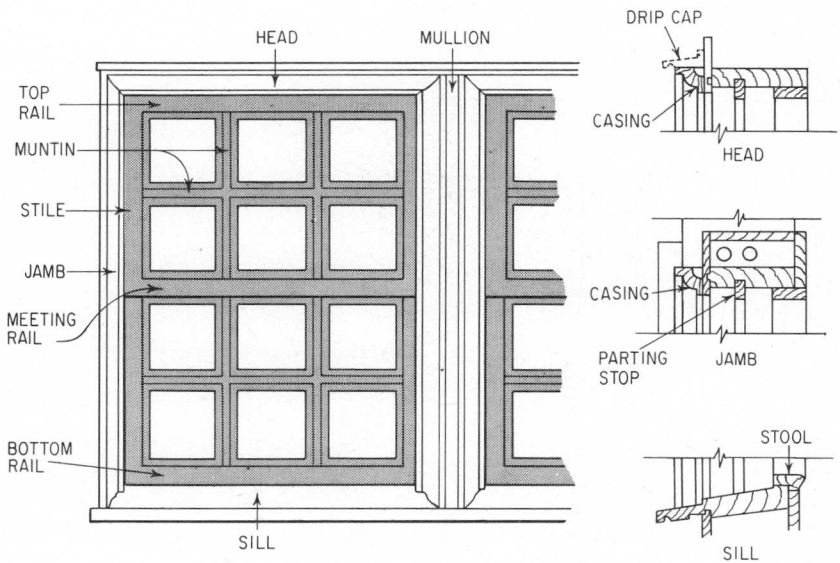

Fig. 15-12. Components of a window.

Components of a typical window are shown in Fig. 15-12. Some commonly used terms are:

Sash. A single assembly of stiles and rails made into a frame for holding glass, with or without dividing bars. It may be supplied glazed or unglazed.

Window. Sash and the glass that fill an opening.

Stiles. Upright, or vertical, border pieces of a sash.

Rails. Cross, or horizontal, members of a sash.

Check rails. Meeting rails sufficiently thicker than the window to fill the opening between the top and bottom sash made by the check strip or parting strip in the frame. They usually are beveled and rabbeted.

Bar. Member that extends the height or width of the opening to be glazed.

Muntin. A short, light bar.

Frame. Wood parts machined and assembled to form an enclosure and support for a window.

Jamb. Part of a frame that surrounds and contacts the window it supports. Side jambs form the vertical sides; head jamb, the top. A rabbeted jamb has a rectangular groove along its edges to receive a window.

Sill. The horizontal bottom part of a frame.

Stool. The part of the sill inside the building.

Pulley stile. A side jamb in which a pulley is installed and along which the sash slides.

Casing. Molding of various widths and thicknesses used to trim window openings.

Parting stop. A molding placed on top of the head casing of a window frame to direct water away from it.

Blind stop. A thin strip of wood machined to fit the exterior vertical edge of the pulley stile or jamb and keep the sash in place.

Dado. A rectangular groove cut across the grain of a frame member.

Jamb liner. A small strip of wood, either surfaced four sides or tongued on one edge, which, when applied to the inside edge of a jamb, increases its width for use in thicker walls.

Steel windows, in general, are made from hot-rolled structural-grade new billet steel. Dimensions usually conform to the specifications of The Steel Window Institute, 18455 Harvest Lane, Brookfield, Wis. 53005. Similar types of windows also are available in aluminum but conforming to the specifications of the Aluminum Window Manufacturers Association, State Highway No. 3 and Seventh Street, Secaucus, N. J.

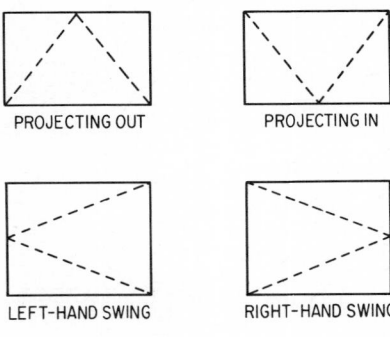

PROJECTING OUT   PROJECTING IN

LEFT-HAND SWING   RIGHT-HAND SWING

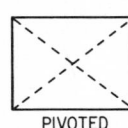

PIVOTED

Fig. 15-13. Symbols for common window types (viewed from outside).

Many types of windows are available (see symbols, Fig. 15-13). Among the more commonly used types are:

Pivoted windows (Fig. 15-14a), an industrial window used where very tight closure is not a necessity. Vents are pivoted about 2 in. above the center. Top swings in. They may be mechanically operated in groups.

Projected windows (Fig. 15-14b), similar to pivoted windows except that the pivot is at the top or bottom. Commercial projected windows are used in commercial and industrial installations where initial cost is a prime consideration. Maximum opening is about 35°. Architectural projected windows are medium-quality, used for commercial, institutional, and industrial buildings. Intermediate projected windows are high-quality, used for schools, hospitals, commercial buildings, and many other structures. Basement and utility windows usually are projected, opening inward, generally with pivot at the bottom. Also, security windows, psychiatric projected, and detention windows generally are bottom-pivoted.

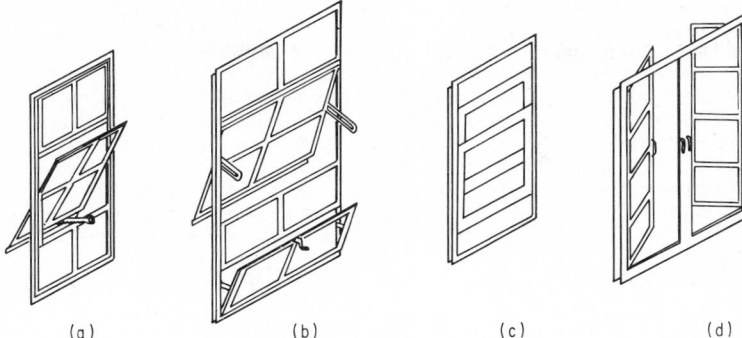

Fig. 15-14. Types of windows: (*a*) pivoted; (*b*) projected; (*c*) double hung; (*d*) casement.

Double-hung windows (Fig. 15-14*c*), comprising a pair of vertically sliding sash. They are used for all types of buildings. Usually, the sash are balanced, to permit easy movement, by weights or other devices in the jambs. Horizontally sliding windows also are available.

Casement windows (Fig. 15-14*d*), consisting of a pair of vertically pivoted sash, generally opening outward. Rotary or lever operators hold the vents at the desired position up to full opening. **Intermediate combination windows** come with casement windows above a vent that projects in.

Picture windows, fixed sash, sometimes with ventilating units.

Weather stripping is used on windows to reduce air leakage around joints. It is made of metal or a compressible resilient material.

Storm sash is another means of reducing heat loss. It is, in effect, a second window installed outside the main window. The objective is to create a dead air space, which offers good thermal insulation, without decreasing visibility appreciably.

Windows may be supplied with or without screens. Storm sash and screens may be obtained as a single unit for some types of windows; for example, for double-hung and horizontally sliding windows.

(F. S. Merritt, "Building Construction Handbook," McGraw-Hill Book Company, New York.)

15-16. Glazing. Window glass in common use includes:

Clear window glass. Used in all types of buildings. Classified by Federal standards according to defects. Grade A is used where appearance is important; Grade B for industrial buildings, low-cost housing, basements, garages, etc. It comes in single strength, $3/32$ in. thick, up to 40×50 in.; double strength, $\frac{1}{8}$ in. thick, up to 60×80 in.; and heavy sheet, $7/32$ in., up to 76×120 in. For appearance's sake, single and double strength should be limited in area to 7 sq ft.

Plate and float glass. Used in store windows, picture windows, and better-grade buildings. Better appearance; no distortion of vision. Thickness ranges from $\frac{1}{8}$ to $\frac{7}{8}$ in.

Processed glass and rolled figure sheet. Types of obscure glass.

Obscure wired glass. Used where resistance to fire or breakage is desired.

Polished wired glass. More expensive than obscure wired glass. Used where clear vision is desired, such as in school or institutional doors.

There are also many special glasses for specific purposes:

Heat-absorbing glass reduces heat, glare, and a large percentage of ultraviolet rays, which bleach colored fabrics. It often is used for comfort and reduction of air-conditioning loads where large areas of glass have a severe sun exposure. Because of differential temperature stresses and expansion induced by heat absorption under severe sun exposure, special attention must be given to edge conditions. Glass with clean-cut edges is particularly desirable, because these affect the edge strength, which, in turn, must resist the central-area expansion. A resilient glazing material should be used.

Corrugated glass, corrugated wire glass, and corrugated plastic panels are used for decorative treatments, diffusing light, or as translucent structural panels with color.

Laminated glass consists of two or more layers of glass laminated together by one or more coatings of a transparent plastic. This construction adds strength. Some types of laminated glass also provide a degree of security, sound isolation, heat absorption, and glare reduction. Where color and privacy are desired, fadeproof opaque colors can be included. When fractured, laminated glass tends to adhere to the inner layer of plastic and, therefore, shatters into small splinters, thus minimizing the hazard of flying glass.

Bullet-resisting glass is made of three or more layers of plate glass laminated under heat and pressure. Thicknesses of this glass vary from ¾ to 3 in. The more common thicknesses are 1³/₁₆ in., to resist medium-powered small arms; 1½ in., to resist high-powered small arms; and 2 in., to resist rifles and submachine guns. (Underwriters Laboratories lists materials having the required properties for various degrees of protection.) Greater thicknesses are used for protection against armor-piercing projectiles. Uses of bullet-resisting glass include cashier windows, bank teller cages, toll-bridge booths, peepholes, and many industrial and military applications. Transparent plastics also are used as bullet-resistant materials, and some of these materials have been tested by the Underwriters' Laboratories. Thicknesses of 1¼ in. or more have met UL standards for resisting medium-powered small arms.

Tempered glass is produced by a process of reheating and sudden cooling that greatly increases strength. All cutting and fabricating must be done before tempering. Doors of ½- and ¾-in.-thick tempered glass are commonly used for commercial buildings. Other uses, with thicknesses from ⅛ to ⅞ in., include backboards for basketball, showcases, balustrades, sterilizing ovens, and windows, doors, and mirrors in institutions. Although tempered glass is 4½ to 5 times as strong as annealed glass of the same thickness, it is breakable, and when broken, disrupts into innumerable small fragments of more-or-less cubical shape.

Tinted and coated glasses are available in several types and for varied uses. As well as decor, these uses can provide for light and heat reflection, lower light transmission, greater safety, sound reduction, reduced glare, and increased privacy.

Transparent mirror glass appears as a mirror when viewed from a brightly lighted side, and is transparent to a viewer on the darker opposite side. This one-way-vision glass is available as a laminate, plate or float, tinted, and in tempered quality.

Plastic window glazing, made of such plastics as acrylic or polycarbonate, is used for urban school buildings and in areas where high vandalism might be anticipated. These plastics have substantially higher impact strength than glass or tempered glass. Allowance should be made in the framing and installation for expansion and contraction of plastics, which may be about eight times as much as that of glass. Note also that the modulus of elasticity (stiffness) of plastics is about one-twentieth that of glass. Standard sash, however, usually will accommodate the additional thickness of plastic and have sufficient rabbet depth.

Suspended glazing utilizes metal clamps bonded to tempered plate glass at the top edge, with vertical glass supports at right angles for resistance to wind pressure. These vertical supports, called stabilizers, have their exposed edges polished. The joints between the large plates and the stabilizers are sealed with a bonding cement. The bottom edge or sill is held in position by a metal channel and sealed with resilient waterproofing. Suspended glazing offers much greater latitude in use of glass and virtually eliminates visual barriers.

Special glazing. A governmental specification Z-97, adopted by many states, requires entrance-way doors and appurtenances glazed with tempered, laminated, or plastic material.

Factory-sealed double glazing is an insulating-glass unit composed of two panes of glass separated by a dehydrated air space. This type of sash is also manufactured with three panes of glass and two air spaces, providing additional insulation against heat flow or sound transmission. Heat loss and heat gain can be substantially reduced by this insulated glass, permitting larger window areas and added indoor comfort. Heat-absorbing glass often is used for the outside pane and a clear plate or float glass for the inside.

Glass Thickness for Wind. Figure 15-15 can be used to determine the nominal thickness of plate, float, or sheet glass for a given glass area, or the maximum area for a given thickness to withstand a specified wind pressure. Based on minimum thickness permitted by Federal Specification DD-G-451c, the wind-load chart provides a safety factor of 2.5. It is intended for rectangular lights with four edges glazed in a stiff, weathertight rabbet.

For example, determine the thickness of a 108 × 120-in. (90-sq-ft) light of polished plate glass to withstand a 20-psf wind load. Since the 20-psf and 90-sq-ft ordinates intersect at the ⅜-in. glass thickness line, the thickness to use is ⅜ in.

The correction factors in Table 15-5 also allow Fig. 15-15 to be used to determine the thickness for certain types of fabricated glass products. The table, however, makes no allowance for the weakening effect of such items as holes, notches, grooves, scratches, abrasion, and welding splatter.

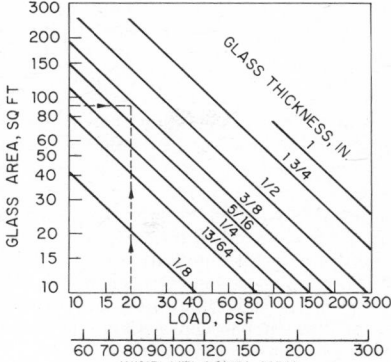

Fig. 15-15. Wind load chart indicates maximum glass area for various nominal thicknesses of plate or sheet glass to withstand specified wind pressures, with design factor of 2.5. (*Pittsburgh Plate Glass Co.*)

Table 15-5. Resistance of Glass to Wind Loads

Product	Factor*
Plate, float, or sheet glass	1.0
Rough rolled plate glass	1.0
Sand-blasted glass	2.5
Laminated glass†	2.0
Sealed double glazing:‡	
Metal edged, up to 30 sq ft	0.667
Metal edged, over 30 sq ft	0.556
Glass edged	0.5
Heat-strengthened glass	0.5
Fully tempered glass	0.25

*Enter Fig. 15-15 with the product of the wind load in psf multiplied by the factor.
†At 70° F or above, for two lights of equal thickness laminated to 0.015-in.-thick vinyl. At 0° F, factor approaches 1.
‡For thickness, use thinner of the two lights, not total thickness.

The appropriate thickness for the fabricated glass product is obtained by multiplying the wind load, psf, by the factor given in Table 15-5. The intersection of the vertical line drawn from the adjusted load and the horizontal line drawn from the glass area indicates the minimum recommended glass thickness.

Glazing Compounds. Glass usually is held in place in sash by putty, glazing compound, rubber, plastic strips, metal clips (with metal sash), or glazing points (with wood sash). Commonly used glazing compounds include vegetable-oil base (skin-forming type), vegetable-oil rubber or nondrying oil blends (polybutene), nondrying oil types—all of which may be applied by gun or knife—butyl rubber or polyisobutylene, applied as tacky tape; polysulfide rubber, applied by gun; Neoprene, applied by gun or as a preformed gasket; and vinyl chloride and copolymer, applied as preformed gaskets. Bedding of glass in glazing compound is desirable, because it furnishes a smooth bearing surface for the glass, prevents rattling, and eliminates voids where moisture can collect. A thin layer of putty or bedding compound is first placed in the rabbet of the sash; then, the glass is pressed into this bed, after which the sash is face-puttied and excess putty removed from the back.

Gaskets. Preformed structural gaskets can be used as an alternative to sash. Gaskets are extruded from rubberlike materials or plastics in a single strip, molded into the shape of the window perimeter, and installed against the glass and window frame. The gasket may fit into a groove or, H-shaped in cross section, grip the glass and a continuous metal fin on the frame. A continuous locking strip of the same material as the gasket is forced into one side of the gasket to make it grip.

(F. S. Merritt, "Building Construction Handbook," McGraw-Hill Book Company, New York.)

15-17. Doors.   Selection of a door depends on more than just its function as a barrier.   Cost, psychological effect, fire resistance, architectural harmony, and ornamental considerations are only a few of the other factors that must be taken into account.

Traffic Flow and Safety.   Openings in walls and partitions must be sized for their primary function of providing entry to or exit from a building or its interior spaces, and doors must be sized and capable of operating so as to prevent or permit such passage, as required by the occupants of the building.   In addition, openings must be adequately sized to serve as an exit under emergency conditions.   In all cases, traffic must be able to flow smoothly through the openings.

To serve these needs, doors must be properly selected for the use to which they are to be put, and properly arranged for maximum efficiency.   In addition, they must be equipped with suitable hardware for the application.

Safety.   Exit doors and doors leading to exit passageways should be so designed and arranged as to be clearly recognizable as such and to be readily accessible at all times.   A door from a room to an exit or to an exit passageway should be the swinging type, installed to swing in the direction of travel to the exit.

Code Limitations on Door Sizes.   To insure smooth, safe traffic flow, building codes generally place maximum and minimum limits on door sizes.   Typical restrictions are as follows:

No single leaf in an exit door should be less than 28 in. wide or more than 48 in. wide.   Minimum nominal width of opening should be at least:

36 in. for single corridor or exit doors.

32 in. for each of a pair of corridor or exit doors with central mullion.

48 in. for a pair of doors with no central mullion.

32 in. for doors to all occupiable and habitable rooms.

44 in. for doors to rooms used by bedridden patients and single doors used by patients in such buildings as hospitals, sanitariums, and nursing homes.

32 in. for toilet-room doors.

Jambs, stops, and door thickness when the door is open should not restrict the required width of opening by more than 3 in. for each 22 in. of width.

Nominal opening height for exit and corridor doors should be at least 6 ft 8 in.   Jambs, stops, sills, and closures should not reduce the clear opening to less than 6 ft 6 in.

Opening Width Determined by Required Capacity.   The width of an opening used as an exit is a measure of the traffic flow that the opening is permitted to accommodate.   Capacities of exits and access facilities generally are measured in units of width of 22 in., and the number of persons per unit of width is determined by the type of occupancy.   Thus, the number of units of exit width for a doorway is found by dividing by 22 the clear width of the doorway when the door is in the open position.   (Projections of stops and hinge stiles may be disregarded.)   Fractions of a unit of width less than 12 in. should not be credited to door capacity.   If, however, 12 in. or more is added to a multiple of 22 in., one-half unit of width can be credited.   Local building codes list capacities in persons per unit of width that may be assumed for various types of occupancy.

Every floor of a building should be provided with exit facilities for its occupant load.   The number of occupants for whom exit facilities must be provided is determined by the actual number of occupants for which the space is designed, or by dividing the net floor area by the net floor area per person specified in the local building code.

Fire and Smokestop Doors.   Building codes require fire-resistant doors in critical locations to prevent passage of fire.   Such doors are required to have a specific minimum fire-resistance rating, and are usually referred to as fire doors.   The codes also may specify that doors in other critical locations be capable of preventing passage of smoke.   Such doors, known as smokestop doors, need not be fire rated.

Fire protection of an opening in a wall or partition depends on the door frame and hardware, as well as on the door.   All these components must be "labeled" or "listed" as suitable for the specific application.   Bear in mind that fire doors are tested as an assembly of these components, and hence only approved assemblies should be specified.

All fire doors should be self-closing or should close automatically when a fire occurs.   In addition, they should be self-latching, so that they remain closed.   Push-pull hardware should not be used.   Exit doors for places of assembly for more than 100 persons usually must be equipped with fire-exit (panic) hardware capable of releasing the door latch when pressure of 15 lb, or less, is applied to the device in the direction of exit.   Combustible materials, such as flammable carpeting, should not be permitted to pass under a fire door.

Fire-door assemblies are rated, in hours, according to their ability to withstand a standard fire test, such as that specified in American Society for Testing and Materials Standard E152. They may be identified as products qualified by tests by a UL label, provided by Underwriters' Laboratories, Inc.; an FM symbol of approval, authorized by Factory Mutual Research Corp.; or a self-certified label, provided by the manufacturer (not accepted by National Fire Protection Association and some code officials).

Openings in walls and partitions that are required to have a minimum fire-resistance rating must have protection with a corresponding fire-resistance rating. Typical requirements are listed in Table 15-6.

Table 15-6. Typical Fire Ratings Required for Doors

Door use	Rating, hr*
Doors in 3- or 4-hr fire barriers	3†
Doors in 2- or 1½-hr fire barriers	1½
Doors in 1-hr fire barriers	¾
Exit doors	1½‡
Doors to stairs and exit passageways	¾
Doors in 1-hr corridors	¾
Other corridor doors	0§
Smokestop doors	0¶

*Self-closing, swinging doors. Normally kept closed.
†Some codes require two 1½-hr opening protectives, with one protective installed on each face of a fire barrier.
‡No rating required for street-floor exit doors with an exterior separation of more than 15 ft, or for exit doors of one- or two-family houses.
§Should be noncombustible or 1¾-in. solid-core wood doors. Some codes do not require self-closing for doors in hospitals, sanitariums, nursing homes, and similar occupancies.
¶Should be metal, metal-covered, or 1¾-in. solid-core wood doors (1⅜-in. in buildings less than three stories high), with 600-sq-in. or larger, clear, wire-glass panels in each door.
Source: Based on New York City Building Code.

This table also gives typical requirements for fire resistance of exit doors, doors to stairs and exit passageways, corridor doors, and smokestop doors.

In addition, some building codes also limit the size of openings in fire barriers. Typical maximum areas, maximum dimensions, and maximum percent of wall length occupied by openings are given in Table 15-7.

Table 15-7. Maximum Sizes of Openings in Fire Barriers

Protection of adjoining areas	Max area, sq ft	Max dimension, ft
Unsprinklered	120*	12†
Sprinklers on both sides	150*	15*
Building fully sprinklered	Unlimited*	Unlimited*

*But not more than 25% of the wall length or 56 sq ft per door if the fire barrier serves as a horizontal exit.
†But not more than 25% of the wall length.
Source: Based on New York City Building Code.

Smokestop doors should be of the construction indicated in the footnote to Table 15-6. They should close openings completely, with only the amount of clearance necessary for proper operation.

["Standard for Fire Doors and Windows," NFPA No. 80; Life Safety Code, NFPA No. 101; "Fire Tests of Door Assemblies," NFPA No. 252, National Fire Protection Association, 60 Batterymarch St., Boston, Mass. 02110.

"Fire Tests of Door Assemblies," Standard UL 10(b); "Fire Door Frames," Standard UL 63; "Building Materials List" (annual, with bimonthly supplements), Underwriters' Laboratories, Inc., 207 East Ohio St., Chicago, Ill. 60611.

"Factory Mutual Approval Guide," Factory Mutual Research Corp., 1151 Boston-Providence Turnpike, Norwood, Mass. 02062.

"Hardware for Labeled Fire Doors," National Builders Hardware Association, 1290 Avenue of the Americas, New York, N.Y. 10019.]

Swinging doors are hung on butts or hinges that permit rotation about a vertical axis at an edge. The door is hinged to and closes against a door frame. The frame consists of two verticals, or jambs, and a horizontal member, the header (Fig. 15-16). *Single-acting doors* can swing 90° or more in only one direction; *double-acting doors* can swing 90° or more in each of two directions.

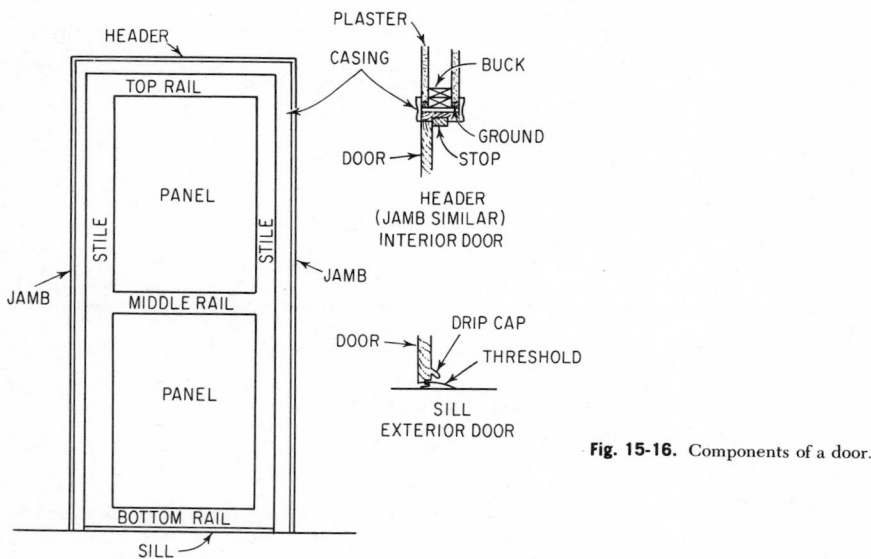

Fig. 15-16. Components of a door.

To stop drafts and passage of light, the jamb about which the door swings has a rebate or projection, extending the full height, against which the door closes. The projection may be integral with the frame, or formed by attaching a stop on the surface of the frame, or inset slightly. With single-acting doors, the opposite jamb also is provided with a stop, against which the door closes.

Door frames for swinging doors generally are fastened to bucks, rough construction members. Joints between the frame and wall are covered with casings, or trim. With metal construction, the trim often is integral with the frame and designed to grip the bucks. The sill, at the bottom of the door, forms a division between the finished floor on one side and that on the other side. The sill generally serves also as a step, since the door opening usually is raised above grade to prevent rain from entering. The top of the sill is sloped to drain water away from the interior. A raised section or separate threshold at the door is an additional barrier to water. A weather strip in the form of a hooked length of metal may be attached to the underside of the door. When the door is closed, the weather strip locks into the threshold to seal out water and reduce air leakage.

Selection of a swinging door involves consideration of the jamb to which the door is hinged and the direction in which it is to open. This relationship is called the *hand* of the door. Hand and hardware for swinging doors are discussed later in this article.

Horizontally sliding doors roll on rails at top or bottom and slide in guides at the opposite edge. Some doors fold or collapse like an accordion, to occupy less space when open. A pocket should be provided in the walls on either or both sides to receive rigid doors; with folding or accordion types, a pocket is optional.

Vertically sliding doors may rise straight up, may rise up and swing in, or may pivot outward to form a canopy. Sometimes, the door may be in two sections, one rising up and the other dropping down. Generally, all types are counterweighted for ease of operation. To exclude weather, either the upper part is recessed into the wall above or the top part of the door extends slightly above the bottom of the wall on the inside.   Similarly, door sides are recessed into the walls or lap them and are held firmly against the inside.   Also, the finished floor is raised a little above outside grade.

Revolving doors are generally selected for entranceways carrying a continuous flow of traffic without a very high peak.   They offer the advantage of keeping interchange of inside and outside air to a relatively small amount compared with other types of doors.   Usually, they are used in combination with swinging doors because of the inability to handle large groups of people in a short time.   Revolving doors consist of four leaves that rotate about a vertical axis inside a cylindrical enclosure with a 4- to 5-ft-wide opening.

Special Doors.   Large-sized doors, such as those for hangars, garages, and craneway openings and for subdividing gymnasiums and auditoriums, often have to be designed individually, with special attention to their supports and controls.   Manufacturers classify such special-purpose doors as horizontal-sliding, vertical-sliding, swing, and top- or horizontal-hinge.

Telescoping doors, a horizontal-sliding type, are used for airplane hangars.   Normally composed of 6 to 20 leaves, they generally are center-parting.   When opened, the doors are stacked in pockets at each end of the opening.   Operation is by electric motor installed in the end pockets and driving an endless chain attached to the tops of the center leaves.

Folding doors are commonly used for subdividing gymnasiums, auditoriums, and cafeterias and for hangars with very wide openings.   This type of door is made up of a series of leaves hinged together in pairs.   The leaves fold outward, and when the door is shut, they are held by automatic folding stays.   Motors that operate the doors usually are installed in mullions adjacent to the center of the opening.   The mullions are connected by cables to the ends of the opening, and when the door is to be opened, the mullions are drawn toward the ends, sweeping the leaves along.   The chief advantage over telescoping types is that only two guide channels are required.

Vertical-sliding doors are advantageous when space is available above and below an opening into which door leaves can be moved.   Usually, the doors are counterweighted, even when motor-operated.

Large swing doors are used when there is insufficient space around openings for sliding doors. Common applications have been for firehouses, where width-of-building clearance is essential, and railway entrances, where doors are interlocked with the signal system.   Common variations include single-swing (solid leaf with vertical hinge on one jamb), double-swing (hinges on both jambs), two-fold (hinge on one jamb and another between folds and leaves), and four-fold (hinges on both jambs and between each pair of folds).   The more folds, the less time required for opening and the smaller the radius needed for swing.

Horizontal-hinge doors are used in craneway entrances to buildings.   Sometimes, horizontal-sliding doors are installed below the crane doors to increase the opening.   If so, the top guides are contained in the bottom of the crane door, so that the sliding door must be opened before the swing door.

Radiation-shielding doors are used as a barrier against harmful radiation and atomic particles across openings for access to "hot" cells, and against similar radioactive-isotope handling arrangements and radiation chambers of high-energy x-ray machines or accelerators.   Usually, the doors must protect not only personnel but also instruments even more sensitive to radiation than people.   Shielding doors usually are much thicker and heavier than ordinary doors, because density is an important factor in barring radiation.   These special-purpose doors are made of steel plates, steel-sheathed lead, or concrete.   To reduce thickness, concrete doors may be of medium-heavy (240 lb per cu ft) or heavy (300 lb per cu ft) concrete, often made with iron-ore aggregate. They usually are operated hydraulically or by electric motor.

Common types of shielding doors include hinged, plug, and overlap.   The hinged type is similar to a bank-vault door.   The plug type, flush with the walls when closed, may roll on floor-mounted tracks or hang from rails.   Overlap doors, surface-mounted, also may roll or hang from rails.   In addition, vertical-lift doors sometimes are used.

Door Materials.   Doors are made of a wide variety of materials.   Wood is used in several forms.   Better-grade doors are made with panels set in a frame or with flush construction.   Paneled doors consist of solid wood or plywood panels held in place by stiles and rails (Fig. 15-16). The joints permit expansion and contraction of the wood with atmospheric changes.   If the rails and stiles are made of a single piece of wood, the paneled door is called solid.   When hardwood or better-quality woods are used, the doors generally are veneered; rails and stiles are made with cores of softwood sandwiched between the desired veneer.   Tempered glass or plastic may be used instead of wood for panels.   Flush doors also may be solid or veneered, or hollow-core.

Metal doors generally are constructed in one of three ways: cast as a single unit or separate frame and panel pieces; metal frame covered with sheet metal; and sheet metal over a wood or other type of insulating core. The heavier metal doors of the swinging type usually are pivoted at top and bottom. Metal-covered doors may be obtained with a wide variety of fire-resistant cores. **A Kalamein door** has a wood core (the wood will not burn as long as the sheet-metal cover prevents oxygen from reaching it).

Doors may be made wholly or partly transparent or translucent. Lights may be made of tempered glass or plastic. Doors made completely of glass are pivoted at top and bottom because the weight makes it difficult to support them with hinges or butts.

Sliding doors of the collapsible accordion type generally consist of wood slats or a light steel frame covered with textile. Plastic coverings frequently are used.

Hand of Doors. Swinging doors are called left-hand doors if, when viewed from the outside, they are hinged to the left-hand jamb and open inward; they are left-hand reverse if they are hinged to the left-hand jamb and open outward. Similarly, they are right-hand and right-hand reverse, respectively, if they open inward and outward when hinged to the right-hand jamb. Since some butts and hinges are handed, the type of door may determine the type of hinge (Fig. 15-17).

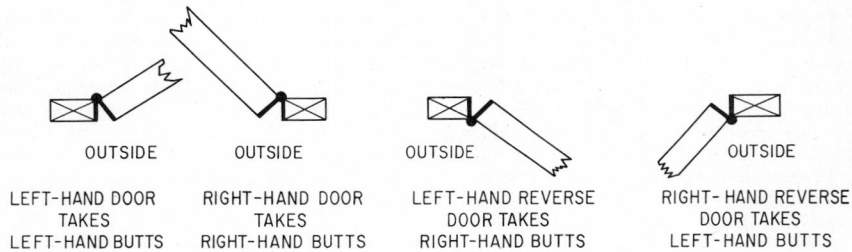

OUTSIDE	OUTSIDE	OUTSIDE	OUTSIDE
LEFT-HAND DOOR TAKES LEFT-HAND BUTTS	RIGHT-HAND DOOR TAKES RIGHT-HAND BUTTS	LEFT-HAND REVERSE DOOR TAKES RIGHT-HAND BUTTS	RIGHT-HAND REVERSE DOOR TAKES LEFT-HAND BUTTS

Fig. 15-17. Direction of opening classifies swinging doors.

Door Hardware. The term **hinge** usually refers to the elongated strap type (Fig. 15-18a and b). It is suitable for mounting on the surface of a door. It consists of two leaves joined by a pin passing through knuckle joints where the leaves fit together.

When the device is to be mounted on the edge of a door, the length of the leaves must be shortened. The leaves thus retain only the portion near the pin, or the butt end of the hinge (Fig. 15-18c to g). Thus, hinges applied to the edge of a door have come to be known as **butts** or **butt hinges.**

Butts usually are mortised into the door edge. The number of butts required per door depends on the size and weight of the door and the conditions of use. In general, use two butts on

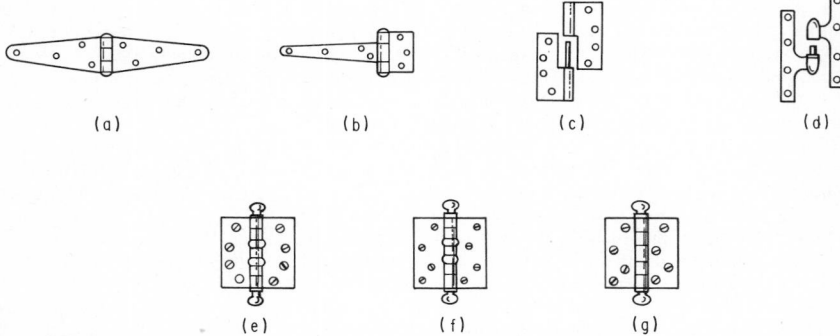

(a)   (b)   (c)   (d)

(e)   (f)   (g)

Fig. 15-18. Typical hinges and butts: (a) heavy strap hinge; (b) heavy tee hinge; (c) loose-joint hinge; (d) olive knuckle hinge; (e) bearing template hinge; (f) bearing nontemplate hinge; (g) plain bearing nontemplate hinge.

doors up to 68 in. high, and three butts on doors 68 to 90 in. high.   The weight and usage of the door also determine whether the butts should be two-bearing or four-bearing.   When the butts are the type that may be mounted on both left-hand and right-hand doors, only half the bearing units available participate in carrying the vertical load.   This should be taken into consideration in selecting butts.

The upper butt may be attached with its top about 5 in. below the rabbet of the head jamb.   The lowest butt may be set with its bottom 10 in. from the finished floor.   A third butt may be installed about midway between the other two.

Bearing butts (Fig. 15-18e and f) or butts with Oilite bearings are used for doors requiring silent operation, or subject to heavy usage, or equipped with a door closer.   Plain bearings (Fig. 15-18g) usually are used for residential doors.

Template hardware, manufactured to close tolerances, is attached to metal jambs and doors with machine screws.   In butts, holes that are template-drilled usually form a crescent pattern (Fig. 15-18c and e).   When the holes are staggered, the butts are nontemplate (Fig. 15-18f and g).

Butts and hinges come with loose or fast pins.   Loose pins are used wherever practicable, because they simplify door hanging.   The fast (or tight) pin is permanently set in the butt at the time of manufacture; thus, a locked door cannot be opened by removing the pins and separating the leaves of the butts.   Also, nonremovable loose pins are available to serve the same purpose.   Another type of loose pin is the nonrising type, which does not have the disadvantage of ordinary loose pins of working upward with repeated movements of the door.

Doors generally are equipped with locks or latches to hold them closed.   **Rim locks or latches** are fastened on the surface of the door.   Those mortised into the edge of the door are called **mortise locks or latches.**   A latch has a beveled locking bolt, which slides into position automatically when the door is closed.   Usually, it is operated by a knob or lever.   In specifying a latch, the hand of the door should be given.

When the locking bolt is rectangular in shape and must be moved in and out by a thumb turn or key, the bolt is called a **dead bolt** and the lock a **dead lock.**   A unit composed of latch bolts and dead bolts is known as a lock.

Unit locks are complete assemblies that can be installed in a standard notch.   Bored-in locks similarly are complete assemblies, but installed in circular holes.   Depending on the arrangement of holes, bored-in locks may be tubular or cylindrical lock sets.

Tubular locks have a horizontal tubular case perpendicular to the door edge.   Another small hole is required normal to the first hole for the locking cylinder.

Cylindrical locks need a relatively large hole perpendicular to the face of the door for the cylindrical case.   Another small hole perpendicular to the door edge accommodates the bolt.

In selecting locks, choose a uniform size, if practicable, for the project.   Then, standard-size cutouts or sinkages can be used throughout.   This will reduce installation costs.   In addition, if changes are made as the job progresses, hardware changes will be simple and special hardware avoided.

(F. S. Merritt, "Building Construction Handbook," McGraw-Hill Book Company, New York; American Institute of Architects, C. G. Ramsey and H. R. Sleeper, "Architectural Graphic Standards," John Wiley & Sons, Inc., New York; "Life Safety Code," National Fire Protection Association, Boston, Mass.; Standards for Wood Doors and Frames, National Woodwork Manufacturers Association, 400 W. Madison Ave., Chicago, Ill. 60606; "Recommended Standard Details, Steel Doors and Frames," Steel Door Institute, 2130 Keith Building, Cleveland, Ohio, 44115; "Entrance Manual," National Association of Architectural Metal Manufacturers, 228 N. LaSalle St., Chicago, Ill. 60601.)

15-18. Roof Coverings.   Success of a roofing installation depends heavily on the roof deck. Roof framing should be sized and spaced to prevent significant deflection of the deck and consequent damage to the roofing.   The deck itself should be smooth, dry, and clean.   Many roof failures have resulted from application of roof coverings to damp decks; pressures developed by the entrapped moisture caused blisters and rupture of the coverings.

Roofing may be single-unit or multiple-unit types.   The single-unit type, which includes built-up roofing of asphalt or coal-tar pitch, sprayed-on products, and flat-seam metal roofing, is suitable for flat roof decks, where water can collect before proceeding slowly to drainage outlets. Multiple-unit coverings, including shingles, tile, slate, and standing-seam metal panels, are used on steep roof decks, where water flows swiftly over each exposed unit to gutters and leaders.

A built-up roof consists of plies of felt mopped with asphalt or pitch.   These form a seamless piece of flexible, waterproofed material, custom-built to conform to the roof deck and to protect all angles formed by the roof deck and projecting surfaces.

Bitumen is a generic term used to indicate either asphalt or coal-tar pitch.

Asphalt is a by-product of the refining processes of petroleum oils.

Coal-tar pitch is a by-product of crude tars derived from the coking of coal. The crude tars are distilled to produce coal-tar pitch. It has a lower melting point than asphalt; hence pitch roofs must be protected by a covering of slag or gravel. Asphalt can be used on steeper slopes than pitch.

Felts, made of wood pulp and rag or of asbestos or glass fibers, aid the bitumens in water shedding and waterproofing. The felts are saturated with bitumen and cemented to the deck and to each other with bitumen. For roof decks that permit nailing, the first two plies of felt are nailed to the deck. For slopes steeper than 2 in. on 12 in., where there is difficulty holding a slag or gravel covering, felt with minerals embedded in the surface may be used for the top, or cap, sheet. Minimum weight of the sheet should be 55 lb per square (100 sq ft). For best results, the cap sheets should be applied over two 15-lb-per-square felts—the first one nailed, the second mopped.

For built-up roofing, five plies of felt are generally used on a wood-sheathed or metal deck and four plies on a cast-in-place concrete deck. "Minimum" specifications require one or two fewer plies and layers of bitumen.

The bituminous roofing may be hot- or cold-process. For cold-process roofing, the bituminous materials, in some cases combined with chemicals, such as polyurethane, are thinned with solvents—for example, kerosene to cut back asphalt, toluene for tar—or emulsified with water. Felts and fabrics differ in the hot and cold processes. Since, in the cold process, cementing occurs on evaporation of solvents or emulsion water, the felts should be of a type that speeds the drying-out process, so an open weave is desirable. Cold-process materials are applied by brush or spray; hot bitumen is mopped in place.

Asphalt roofing materials may be used on slopes up to 4 in. per ft:

> 1 in. per ft . . . . . 19-in. selvage-edge roofing; built-up roofing
> 2 in. per ft . . . . . Mineral-surfaced roll roofing; blind-nailed, square-butt strip shingles with
> 2-ply felt underlay
> 3 in. per ft . . . . . Mineral-surfaced roll roofing, exposed nailed
> 4 in. per ft . . . . . Hexagonal, individual, 3-tab square butt and lock-type shingles

Asphalt shingles are made of asphalt-saturated and coated felt in which is embedded a permanent mineral surfacing. Square-butt strip shingles are slotted at the butts to give the appearance of individual units. Other shapes, such as hexagonal, also are available. The shingles are installed over an underlayment of No. 15 asphalt felt fastened to the roof deck. A starter course at eaves should be cemented to the felt to prevent leakage through nail punctures. The starter strip may be a row of shingles turned upside down or, preferably, mineral-surfaced roll roofing, 18 in. wide in normal-wind areas, 36 in. wide in high-wind areas. The shingles should be nailed with nails at least 1¼ in. long on new work and 1¾ in. long on reroofing.

Wood shingles come in two varieties, machine-sawn and hand-split. Standard exposure is 5 in. but may range from 3½ to 12 in. The shingles are laid in alternate courses, with joints broken relative to courses above and below, starting preferably with a triple layer at the eaves.

Asbestos-cement shingles are composed of 15 to 25% asbestos fiber and 75 to 85% portland cement. Standard types include American-method, multiple-unit, Dutch-lap, and hexagonal. *American-method shingles* have a simulated wood-grain surface. Laid in a rectangular pattern, they resemble wood shingles. *Multiple-unit shingles* are large-size; each covers an area equal to that of two to five standard-size shingles. *Dutch-lap shingles* (also called Scotch lap) are lapped at top and one side. The *hexagonal shingles* actually are nearly square, but laid in a diamond pattern with overlap at top and bottom, they produce a hexagonal effect (*French Method*). Special starters are made for the first course, and specials also are available for hips and ridges.

Slate roofs may employ standard commercial slating or textural (random) slating. For commercial slating, the material is graded at the quarry; for textural slating, the slates are delivered to the job in random sizes. Longer and heavier slates are placed at the eaves, medium-sized at the center, and the smallest at the ridge. Application starts with an undereaves course fastened over a batten, which slopes the first course. Slating nails should be driven level with the slate.

Clay roofing tile comes in two varieties, roll and flat. Roll tiles may be semicircular, reverse curve, pan and cover, or flat shingle. Application is much the same as for slate.

Metal Roofing. Principal metals used in roofing are galvanized iron, terneplate, Monel metal, aluminum, and copper. Sometimes, zinc, lead, cast iron, or stainless steel is used. In all cases, care should be taken to prevent corrosion, especially from galvanic action. For example, a copper roof should not be applied directly over a wood deck; the copper must be insulated from attack by the steel nails in the deck.   Fasteners used with metal roofing should preferably be of the same metal as the roofing.   Also, provision should be made for expansion and contraction.

Common types of metal roof installation that allow for thermal movements include batten-seam, standing-seam, flat-seam, and corrugated-metal.   Batten-seam and standing-seam coverings consist of narrow strips with loose-locked seams that permit lateral motion.   Flat-seam construction consists of small sheets soldered on all edges.   Movement is absorbed by buckling in the center of the small sheets.

Except for corrugated metal, large roof areas should be broken into small units, and no metal sheets should be fastened directly to the deck.   Clips or battens should be used for fastening. With corrugated-metal roofing, the corrugations absorb thermal movements.   Fasteners for corrugated metal include nails, clips, straps, clinch rivets, hook bolts, and welded studs.   Where the sheets are fastened directly to roof framing, lead or Neoprene washers should be used with the fasteners.

Corrugated asbestos-cement sheets are applied much like corrugated metal, but they differ in shape.   Asbestos-cement is molded so that the first and last corrugations point down.   This enables the minimum lap to be one corrugation instead of 1½.   In the step method, alternate courses are started with half sheets.   In the miter method, no effort is made to break joints, and with one corner of each sheet mitered, end and side laps can be of uniform thickness.

Plastics may be applied as a roof covering in liquid, sheet, or rigid form.   Fluid-applied *elastomeric coatings*, such as Neoprene and Hypalon, conform to any shape of surface and expand and contract with it.   Manufacturer's specifications should be followed carefully in all applications. *Plastic roofing sheets* may be made of Neoprene, polyvinyl fluoride, polyisobutylene, or other suitable materials, alone or in combination with other materials; for example, bonded to asbestos felt.   They are applied in the same way as conventional roofing sheets.   *Rigid polyvinyl chloride* is available as flat, corrugated, and ribbed panels.   They can be cut on the job with portable power saws with abrasive cutting wheels.

(American Institute of Architects, "Manual of Built-up Roofing Systems," McGraw-Hill Book Company, New York; F. S. Merritt, "Building Construction Handbook," McGraw-Hill Book Company, New York.)

15-19. Flashing.   At all intersecting surfaces on a building exterior, flashing is necessary to prevent penetration of water through the joints or cracks that might form.   Since thermal movements are likely to occur at the intersections, flashing should be elastic or shaped to permit motion.

Bituminous flashings have the ability to hug tight against building surfaces.   Metal flashings require added protection, such as cap flashing, installed above and covering the top edge of the base flashing.   Plastic flashing sheets have been particularly useful in sealing the junction of vents and pipes with roof decks.

At intersections of flat roofs and walls, at least 6 in. of the base flashing should be fastened to the deck and 8 in. to the wall.   Counterflashing should overlap the base flashing from above at least 4 in. and should penetrate at least 1½ in. into a raggle cut into the mortar line between the nearest row of bricks above the base flashing.

Step flashing should be used at intersections of walls and steep roofs.   For this purpose, short pieces of metal are bent at right angles and one flange sandwiched between every roofing unit at the intersection and the other flange set in contact with the wall.   Each flashing unit should lap the one below at least 2 in.   Counterflashing also should be installed in steps.

Crickets or flashing saddles are needed between chimneys and sloping roofs to guide water away from the intersection.   The saddle is a miniature roof, usually of metal, with a ridge and two valleys, secured to a base step flashing.

Flashing also is required at a number of places in exterior walls; for example, around spandrel beams, coping, sills, at grade, belt courses, water tables, cornices, roof valleys, gables, openings in roofs, and window and door heads.

(American Institute of Architects, C. G. Ramsey and H. R. Sleeper, "Architectural Graphic Standards," John Wiley & Sons, Inc., New York.)

15-20. Waterproofing.   Properly built concrete and masonry walls, whether above or below grade, can keep water out of a building without protective coatings or integral waterproofing. Leakage through masonry walls usually occurs at the joints and results from failure to fill them with mortar and from poor bond between masonry and mortar.   Leakage through concrete walls usually occurs in porous material, or at wall ties, or at intersections with other surfaces.

Permeability is the quality or state of permitting passage of water and water vapor into, through, and from pores and interstices without causing rupture or displacement.

The following terms are listed in decreasing order of permeability:

Pervious or leaky.   Cracks, crevices, leaks, or holes larger than capillary pores, which permit a flow or leakage of water, are present.   The material may or may not contain capillary pores.

Water-resistant.   Capillary pores exist that permit passage of water and water vapor, but there are few or no openings larger than capillaries that permit leakage of significant amounts of water.

Water-repellent.   Not "wetted" by water; hence, not capable of transmitting water by capillary forces alone.   However, the material may allow transmission of water under pressure and may be permeable to water vapor.

Waterproof.   No openings are present that permit leakage or passage of water and water vapor.   The material is impervious to water and water vapor, whether or not under pressure.

Water-repellent admixtures sometimes are added to concrete, with the objective of keeping the surface of voids and capillaries in concrete from being wetted by water.   Since such admixtures usually reduce concrete strength, the amount used is limited, rarely exceeding 0.2% of the weight of cement.   Concrete, however, is not made waterproof by use of an integral water repellent; such concrete is permeable to water vapor, and generally also to water under pressure.

Soaps or salts of fatty acids, such as calcium, aluminum, and ammonium stearates and oleates, added in amounts equal to about 0.2% of the cement weight, may reduce absorption of a concrete about 30%.   Butyl stearate, added as an emulsion at 1% by weight of cement, gives better results, however, with little effect on strength.   Heavy mineral oil (viscosity not less than SAE 60) at 4% by weight of cement and containing 12% stearic acid significantly reduced permeability of concrete in tests under 20-psi water pressure.   Absorption was about 45% of that of the plain concrete and compressive strength 85%.   Asphalt emulsions and coal-tar pitch cut with benzene have been used as admixtures with some effectiveness.

Drainage.   When buildings are partly below grade, surface water should be diverted by grading the ground surface away from the walls and by carrying the runoff from roofs away from the building.   Ground slope should be at least ¼ in. per ft for at least 10 ft from the walls.

Also, groundwater should be drained away from basement walls and floors.   Drain tile should be at least 6 in. in diameter.   Laid in gravel or other porous material at least 6 in. below the basement floor, it should slope continuously to a storm sewer or to a sump with adequate capacity.   Open joints between the tile should be covered with a wire screen or with building paper to prevent fine material from clogging the drain.   Gravel should be placed above the tile to a level well above the footing, and for a distance of at least 12 in. from the wall (Fig. 15-19).

Slabs at Grade.   Concrete floors on ground should have the finished surface at least 6 in. above grade.   (Such floors should not be used where there is danger of flooding or groundwater may be present.)   Subsurface drains at wall-footing level can give added protection from runoff of heavy rains.   To prevent the concrete from absorbing moisture from the ground, a continuous

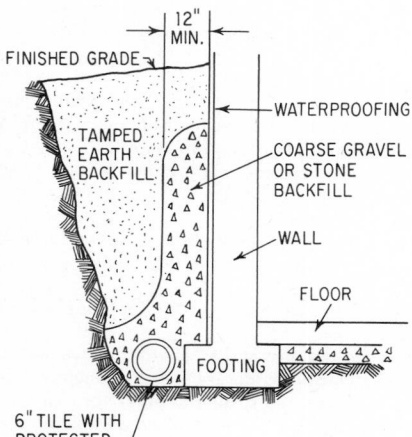

Fig. 15-19. Drainage of basement wall with drain tile along the footing and gravel fill.

waterproof membrane should be placed under the floor and extended up the walls to about 1 in. above the finished floor level.

If there is no danger of water reaching the floor underside, the membrane may be a single layer of 55-lb smooth-surface asphalt roll roofing, or equivalent, with joints lapped and sealed with bituminous mastic. Where a more complete barrier is necessary, use two-ply membrane waterproofing, installed like roofing. The underlying surface should be smooth, for example, a ½-in.-thick grout coat atop granular fill.

Basements. In general, design and construction of concrete basement floors are similar to those of concrete floors on ground. A bituminous-filled joint between walls and floor will prevent leakage into the basement of water that may accumulate occasionally under the slab.

Cast-in-place concrete basement walls should be made of dense concrete, placed in the forms in level layers not more than 18 in. thick, vibrated or carefully tamped, and moist-cured for at least 7 days. Entrance holes for form ties should be sealed with mortar after form removal; if twisted ties are used, they first should be cut back 1.5 in. from the wall faces.

The exterior face of the wall may be given a bituminous coating to increase resistance to passage of water vapor and capillary penetration of water. Cementitious brush-applied paints and grouts and trowel coatings of mortar increase moisture resistance of walls, especially if such coatings contain a water repellent. But if cracks should later develop in the walls, such coatings may be incapable of bridging the openings, and leakage may occur.

Masonry basement walls should be protected against leakage by trowel-applied mortar coats or parging, or by elastic membranes. One trowel coat of mortar may suffice for brick walls, two coats for hollow masonry. The mortar may consist of 1 part portland cement to 3 parts sand by volume. A preparatory coat of portland cement and water of thick-cream consistency should first be scrubbed onto the dampened masonry. Then, the two mortar coats should be applied, at least 1 day apart, each ⅜ in. thick. The coating should be moist-cured for at least 3 days or until the backfill is placed. If resistance to penetration of water vapor is desired, a bituminous coat should be applied to the parging.

Bellows-type water stops should be placed in expansion joints in basement walls. Made of 16-oz copper sheets, they should extend at least 6 in. on each side of a joint.

Bituminous membranes may be used to prevent penetration of water vapor and water under hydrostatic pressure. Continuous in walls and floors below grade, a membrane usually consists of three or more alternate layers of hot, mopped-on asphalt or coal-tar pitch and bituminous-saturated felt or woven cotton fabric. The fabric is stronger and more extensible, but more expensive and more difficult to place than the felt. At least one ply of a membrane should be fabric. Minimum weight of felt should be 13 lb per 100 sq ft; of fabric, 10 oz per sq yd. About 1 gal of primer per 100 sq ft of wall should be applied to the walls before the first mopping of bitumen. Immediately after a membrane is completed, it should be protected by a 1-in. mortar coat or by other facings.

An alternative to a bituminous membrane is a $^{1}/_{16}$- to ⅛-in.-thick layer of butyl rubber secured with a compatible adhesive.

Metal-bellows water stops should be placed in both expansion and contraction joints if there is hydrostatic pressure. The protective facing of the membrane should be disconnected at expansion joints and the space between the membrane and the line of the facing filled with a bituminous cement.

Above-grade Walls. Masonry walls above grade that leak may be made water-resistant with coatings of portland-cement paints, grouts, stuccos, or pneumatically applied mortars. Pigmented organics, including conventional paints, if applied as a continuous coating without pinholes, also may be used. They are decorative but may not be so water-resistant, economical, or durable as cementitious coatings. Leakage through joints in masonry walls can be stopped by either repointing or grouting the joints. Repointing consists of cutting away and replacing the mortar from all joints to a depth of about ⅝ in. Grouting consists of scrubbing a thin coating of grout over the joints. The grout may be composed of equal parts by volume of portland cement and sand passing a No. 30 sieve. Repointing is more effective, but also more expensive.

(F. S. Merritt, "Building Construction Handbook," McGraw-Hill Book Company, New York; ACI Committee 212, "Guide for Use of Admixtures in Concrete," American Concrete Institute, Detroit, Mich.; "Concrete Floor Construction," Portland Cement Association, Skokie, Ill.; ACI Committee 504, "Guide to Joint Sealants for Concrete Structures," American Concrete Institute, Detroit; "Manual for Railway Engineering," American Railway Engineering Association, 59 East Van Buren St., Chicago, Ill. 60605; C. C. Fishburn, D. Watstein, and D. E. Parsons, "Water Permeability of Masonry Walls," C. C. Fishburn, "Water Permeability of Walls Built of

Masonry Units," C. C. Fishburn and D. E. Parsons, "Tests of Cement-Water Paints and Other Waterproofing for Unit-Masonry Walls," C. C. Fishburn, "Effect of Outdoor Exposure on the Water Permeability of Masonry Walls," *National Bureau of Standards BMS Reports* 7, 82, 95, and 76.)

15-21. Stairs.   Principal components of a stairway are:

Flight.   A series of steps extending from floor to floor, or from a floor to an intermediate landing or platform.   Landings are used where turns are necessary or to break up long climbs.

Rise.   Distance from floor to floor.

Run.   Total length of stairs in a horizontal plane, including landings.

Riser.   Vertical face of a step.   Its height generally is taken as the vertical distance between treads.

Tread.   Horizontal face of a step.   Its width usually is taken as the horizontal distance between risers.

Nosing.   Projection of a tread beyond the riser below.

Carriage.   Rough timber supporting the steps of wood stairs.

Stringers.   Inclined members along the sides of a stairway.   The stringer along a wall is called a wall stringer.   Open stringers are those cut to follow the lines of risers and treads.   Closed stringers have parallel top and bottom, and treads and risers are supported along their sides or mortised into them.   In wood stairs, stringers are placed outside the carriage to provide a finish.

Railing.   Protective bar placed at a convenient distance above the stairs for a handhold.

Balustrade.   A railing composed of balusters capped by a handrail.

Handrail.   Protective bar placed at a convenient distance above the stairs for a handhold.

Baluster.   Vertical member supporting the railing.

Newel Post.   Post at which the railing terminates at each floor level.

Angle Post.   Railing support at landings or other breaks in the stairs.   If an angle post projects beyond the bottom of the strings, the ornamental detail formed at the bottom of the post is called the **drop.**

Winders.   Steps with tapered treads in sharply curved stairs.

Headroom.   Minimum clear height from a tread to overhead construction, such as the ceiling of the next floor, ductwork, or piping.

Safety Rules.   Building codes restrict stair dimensions and also control the number of stairways.   This control may be achieved by restricting the horizontal distance from any point on a floor to a stairway or the floor area contributory to a stairway.   In addition, codes usually have special provisions for public buildings and on the maximum capacity of a stairway.

Vertical Clearance.   Minimum vertical distance from the nosing of a tread to overhead construction should never be less than 6 ft 8 in. and preferably not less than 7 ft.   But in general, a person of average height should be able to extend one hand forward and upward without touching an obstruction.

Stair Widths.   Building codes usually specify minimum width of stairs for buildings of various types of occupancy.   But the stairs should be wider than these minimums if necessary to accommodate the number of people who will use them in peak periods and emergencies.   (See also "Life Safety Code," National Fire Protection Association, Boston, Mass.)

Step Sizes.   The most comfortable height of risers is 7 to 7½ in.   Risers less than 6 in. and more than 8 in. high should not be used.   Treads should be 10 to 13 in. wide, exclusive of nosing. Simple formulas generally used to proportion risers and treads include:

1. Product of riser and tread must be between 70 and 75.
2. Riser plus tread must equal 17 to 17.5.
3. Sum of the tread and twice the riser must lie between 24 and 25.5.

In designing stairs, account should be taken of the fact that there always is one less tread than riser per flight.

Railings.   Handrails generally are set 2 ft 6 in. to 2 ft 10 in. above the intersections of tread and risers at the front of the steps.   At landings, railings usually are from 2 ft 10 in. to 3 ft high, though lower railings can be used safely if the parapet is very wide.   Low, wide railings usually are used for monumental stairways.

Emergency Use.   In many types of buildings, exit stairs must be enclosed with walls having a high resistance to fire and with self-closing fire-resistant doors, to prevent spread of smoke and flame (Arts. 15-3 and 15-17).   In public buildings, there should be more than one fire tower, and these should be as far apart as possible.

Materials.   Stairs may be constructed of wood for wood-frame buildings, low nonfireproof buildings, and one- and two-story houses (Fig. 15-20).   They may be built in place or shop-fabricated.

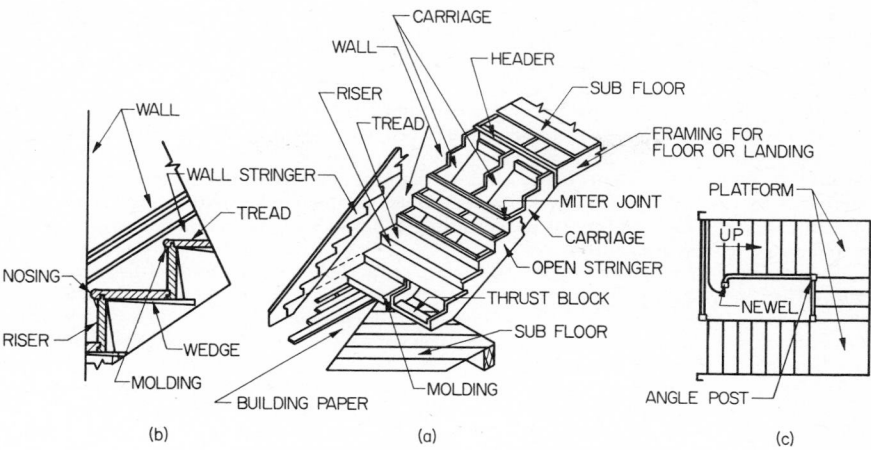

Fig. 15-20. Typical construction for wood stairs.

Press-sheet steel stairs generally are used in fire-resistant buildings. The sheets are formed into risers and subtreads or pans, into which one of several types of treads may be inserted. Treads may be made of stone, concrete, composition, or metal, and usually have a nonslip surface. Stringers generally are channel-shaped.

Concrete stairs may be designed as cantilevered or inclined beams and slabs. The entire stairway may be cast in place as a single unit, or slabs or T beams formed first and the steps built up later. Concrete treads should have metal nosings to protect the edges.

(F. S. Merritt, "Building Construction Handbook," McGraw-Hill Book Company, New York; T. Hamlin, "Forms and Functions of Twentieth Century Architecture," vol. 1, Columbia University Press, New York; American Institute of Architects, C. G. Ramsey and H. R. Sleeper, "Architectural Graphic Standards," John Wiley & Sons, Inc., New York.)

15-22. Escalators. Providing continuous operation without operators, escalators, or powered stairs, are used when it is necessary to move large numbers of people from floor to floor. They have large capacity with low power consumption. Large department stores provide vertical-transportation facilities for one person per hour for every 20 to 25 sq ft of sales area above the entrance floor, and powered stairs generally carry 75 to 90% of the traffic, elevators the rest.

In effect, an escalator is an inclined bridge spanning between floors, with an endless belt to transport passengers. Main components are a steel trussed framework, handrails, and an endless belt with steps. At the upper end are a pair of motor-driven sprocket wheels and a worm-gear driving machine. At the lower end is a matching pair of sprocket wheels. Two precision-made roller chains travel over the sprockets pulling the endless belt. The steps move on an accurately made set of tracks attached to the trusses. Each step is mounted on resilient rollers.

Normally, escalators move at 90 or 120 fpm and are reversible in direction. Slope is standardized at 30°.

For a given speed, width of step determines the capacity of the powered stairs. Standard widths are 32 and 48 in. between handrails, with corresponding capacities at 90 fpm of 5,000 and 8,000 persons per hour. At 120 fpm, a 48-in. escalator can carry as many as 10,000 persons per hour.

Escalators usually are installed in pairs—one for carrying traffic up and the other for moving traffic down. The units may be placed parallel to each other in each story or crisscrossed. The latter generally is preferred for compactness. Fire-protection devices may be incorporated with the stairway installation.

A structural frame should be erected around the stairwell to carry the floor and wellway railing. The stairway should be independent of this frame.

("Life Safety Code," National Fire Protection Association, Boston, Mass.; "Safety Code for Elevators, Dumbwaiters, Escalators, and Moving Walks," A17.1, American National Standards Institute, New York; G. R. Strakosch, "Vertical Transportation: Elevators and Escalators," John Wiley & Sons, Inc., New York.)

15-23. Elevators.   Electric traction elevators are used exclusively in tall buildings.   Hydraulic elevators are usually used for low-rise freight, with lifts up to about 50 ft, but may be used for passenger service in buildings up to six stories high, where they cost less to operate.

Major components of an electric traction installation include the car or cab, hoist wire ropes, driving machine, control equipment, counterweights, shaft, rails, penthouse, and pit.   The car is a cage of light metal supported on a structural frame, to the top of which the ropes are attached.   The ropes raise and lower the car.   They pass over a grooved motor-driven sheave and are fastened to the counterweights.   The elevator machine that drives the sheave consists of an electric motor, brakes, and auxiliary equipment, which are mounted, with the sheave, on a heavy structural frame.   The counterweights, consisting of blocks of cast iron in a frame, are needed to reduce power requirements.

The paths of the counterweights and the car are controlled by separate sets of T-shaped guide rails.   The control and operating machinery may be placed in a penthouse above the shaft or in the basement.   Safety springs or buffers are placed in the pit to bring the car and counterweights to a safe stop if either passes the bottom terminal at normal speed.   Shafts must be enclosed with noncombustible materials of high fire resistance (Art. 15-3).

Elevators and related equipment, such as machinery, signal systems, controls, ropes, and guide rails, are generally supplied and installed by the manufacturer.   The general contractor has to guarantee the dimensions of the shaft and its freedom from encroachments.   The owner's architect or engineer is responsible for the design and construction of components needed for supporting the plant, including buffer supports, machine-room floors, trolley beams, and guide-rail bracket supports.   Magnitudes of loads generally are supplied by the manufacturer with a 100% allowance for impact.

Driving machines may be winding-drum or traction type, depending on whether the ropes are wound on drums on the drive shaft or are powered by a drive sheave.   The traction type is usually used.   It may be double-wrap or single-wrap.   For the double-wrap, to obtain sufficient traction between the ropes and the driving sheave, which has U-shaped or round-seat

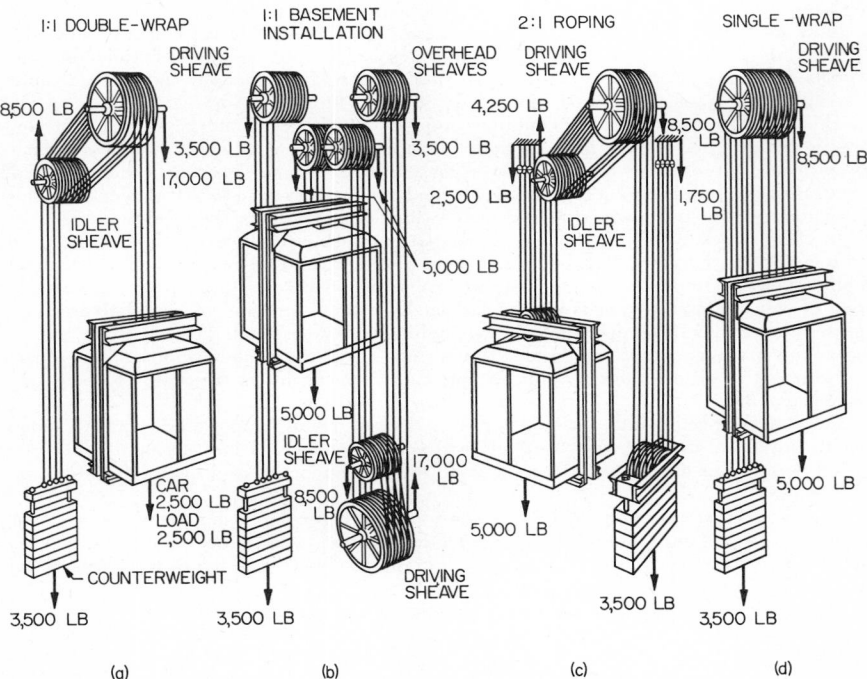

Fig. 15-21. Types of roping for elevators driven by traction machines.

grooves, a secondary or idler sheave is used (Fig. 15-21*a*). In the single-wrap types, the ropes pass over the traction or driving sheave only once, so there is a single wrap, or less, of the ropes on the sheave (Fig. 15-21*d*). The traction sheave has wedge-shaped or undercut grooves for gripping the ropes. For the same weight of car and counterweight, the sheave has half the loading of the double-wrap machine.

In most buildings, driving machines are installed in a penthouse. When a machine must be installed in the basement (Fig. 15-21*b*), the load on the overhead supports is increased, cable length is tripled, and additional sheaves are needed, adding substantially to the cost. When heavy loads are to be handled and speed is not important, a 2:1 roping may be used (Fig. 15-21*c*), in which case the car speed is only half that of the rope. Ends of the rope are anchored to the overhead beams, instead of being attached to the car and counterweights, as for 1:1 roping. With this arrangement, the anchorages carry half the weight of car and counterweights. So the loading on the traction and secondary sheaves is only about half that for the 1:1 machine. Therefore, a less costly motor can be used.

Passenger Elevators. The number of passenger elevators needed to serve a building adequately depends on their capacity, volume of traffic, and interval between cars. Platform sizes should conform to the standards of the National Elevator Industry, Inc.

Traffic is measured by the number of persons handled in 5-min periods. Dividing the peak 5-min traffic flow by the 5-min handling capacity of an elevator gives the minimum number of elevators required. The 5-min handling capacity of an elevator is determined from the round-trip time. Round trip time is composed principally of the time for a full-speed round trip without stops, time for accelerating and decelerating per stop, time for leveling at each stop, time for opening or closing gates and doors, time for passengers to move in and out, reaction time of operator, lost time due to false stops, and standing time at top and bottom floors.

After the number of elevators has been computed on the basis of traffic flow, a check should be made on the **interval,** the average time between elevators leaving the ground floor. Interval is a significant measure of good service.

Fully automatic elevators are used in many tall office and apartment buildings. These systems are capable of adjusting to varying traffic conditions. Since the elevators are operatorless, several safety devices are incorporated in addition to those commonly installed in manually operated systems—an automatic load weigher to prevent overcrowding, buttons in car and starter station to stop the doors from closing and to hold them open, lights to indicate floor stops pressed, two-way loudspeaker system for communication with the starter station, and auxiliary power systems if the primary power and supervisory systems should fail. Safety devices also prevent the doors from closing when a passenger is standing in the doorway. Of course, the elevators cannot move when the doors are open.

Department stores should be served by a coordinated system of moving stairs and elevators (see Art. 15-22). The required capacity of the vertical-transportation system should be based on the transportation or merchandising area and the maximum density to which it is expected to be occupied by shoppers. The transportation area is all the floor space above or below the first floor to which shoppers and employees must be moved. The transportation capacity is the number of persons per hour that the vertical-transportation system can distribute from the main floor to the other merchandising floors. The ratio of the peak transportation capacity to the transportation area is called the density ratio. This ratio is about 1:20 for a busy department store. So the required hourly handling capacity of a combined moving stairs and elevator system equals 5% of the transportation area. The elevator system generally is designed to handle about 10% of the total.

Multivoltage controls normally are used for passenger elevators. Freight elevators may have variable voltage or ac rheostatic. With multivoltage, the hoisting motor is dc operated. A motor-generator set is provided for each elevator, and the speed and direction of motion of the car are controlled by varying the generator field. This type of elevator permits the most accurate stops, the most rapid acceleration and deceleration, and minimum power consumption for an active elevator. Automatic leveling to compensate for rope stretch or other variations from floor level is an inherent part of multivoltage equipment. The ac rheostatic type generally is chosen to keep initial cost down when the elevator is to be used infrequently (less than five trips per hour on a normal business day).

For low-rise elevators, hydraulic equipment may be used to lift the car. It sits atop a plunger, or ram, which operates in a pressure cylinder. Oil serves as the pressure fluid and is supplied through a motor-driven positive-displacement pump, actuated by an electric-hydraulic control system. To raise the car, the pump is started, discharging oil into the pressure cylinder and forcing the ram up.   When the car reaches the desired level, the pump stops.   To lower the car, oil is released from the pressure cylinder and returns to a storage tank.

Capacity of electrohydraulic passenger elevators ranges from 1,000 to 4,000 lb at speeds from 40 to 125 fpm.   With gravity lowering, down speed may be 1.5 to 2 times up speed.   So the average speed for a round trip can be considerably higher than the up speed.   Capacity of standard electrohydraulic freight elevators ranges from 2,000 to 20,000 lb at 20 to 85 fpm, but they can be designed for much greater loads.

(G. R. Strakosch, "Vertical Transportation Elevators and Escalators," John Wiley & Sons, Inc., New York; "Safety Code for Elevators, Dumbwaiters, Escalators, and Moving Walks," American National Standards Institute, New York; F. S. Merritt, "Building Construction Handbook," McGraw-Hill Book Company, New York.)

15-24. Thermal Insulation.   Heat transfer into and out of a building or its parts may be substantially decreased by use of materials that resist heat flow or by a type of construction that achieves the purpose.   Some structural materials, such as wood and lightweight concrete, also have good insulating properties.   But in general, certain nonstructural materials offer greater resistance to heat flow for a given thickness and therefore may be more economical in many applications.

Most insulating materials employ still air as the insulator.   Some, such as cork, cellular glass, and foamed plastics, enclose small particles of air in cells.   Granular materials, such as pumice, vermiculite, and perlite, trap air in relatively large enclosures.   In fibrous materials, thin films of air cling persistently to all surfaces and serve as the heat barrier.   In cavity-wall construction, a dead air space is formed between the wythes.

Reflective insulation involves a different principle.   Metal foil is combined with an air gap to reduce heat flow.   The shiny metal reflects heat, conducts it rapidly away from a heat source, and radiates heat slowly.   An air gap of ¾ to 2 in. on at least one side of the foil acts as a barrier to heat transfer by conduction.   So if heat is radiated to a bright aluminum foil, 95% will be reflected back.   If it receives heat by conduction, it will lose only 5% by radiation from the opposite face.   To prevent condensation troubles, use at least two reflective surfaces separated by a dead air space.   But do not place a foil on the cold side of a construction unless a better vapor barrier is provided close to the warm side.

Heat is transmitted by conduction, convection, and radiation.   All materials conduct heat; but some, such as metals, are excellent conductors, while others, such as cork, are poor conductors. Convection occurs when heat is transmitted by air flow; heat is transferred by conduction from a warm surface to cooler air in contact with it and from warm air to a cooler surface.   Since warm air tends to rise and cool air to fall, the air flow may carry heat from a warm area to a cooler one. Heat transmitted by conduction or convection is proportional to temperature difference.   Radiation, in contrast, is the flow of heat between a warm and cool surface with no material contact.

Heat usually is measured in **British thermal units (Btu)**.   For practical purposes, 1 Btu is the amount of heat required to raise the temperature of 1 lb of water 1 degree Fahrenheit.   Heat flow is measured in terms of **thermal conductivity K**.   This is defined as the number of Btu that will flow in 1 hr through a material 1 ft square and 1 in. thick because of a temperature difference of $1°F$.   Similarly, **thermal conductance C** is defined as the heat flow through a given thickness of 1-ft-square material with a $1°F$ temperature differential.   Note that these basic units do not include the insulating values of air films at the surfaces of the material, but only flow from surface to surface.   **Resistance** is the reciprocal of conductance.

Since building components are built up of several materials, including air spaces and surface films, the **overall conductance U** of a construction is needed in heat-transfer calculations.   This

Table 15-8. Calculation of Overall Conductance of a Wall

Item	K	Thickness, in.	C	R = 1/C
Outside film			6	0.166
Brick	9.2	4	2.30	0.434
Air space			1.10	0.910
Wallboard		½	1.00	1.000
Inside film			1.65	0.606
Total resistance				3.116

Overall conductance $U = 1/3.116 = 0.32$

factor is defined as the number of Btu that will flow in 1 hr through 1 sq ft of the structure from air to air with a temperature differential of 1°F. Values of K, C, and U have been determined experimentally for many materials and types of construction ("Handbook of Fundamentals," American Society of Heating, Refrigerating and Air Conditioning Engineers, 345 E. 47th St., New York, N.Y. 10017).

The thermal conductance of an outside air film in a 15-mph wind is 6.00 Btu per hr; of an inside air film (still air), 1.65 Btu per hr; and of an air space ¾ in. or more wide, 1.10.

When the overall conductance of a construction is not given in a table, it may be computed from tabulated values of conductance of each component and air films. For example, consider a wall composed of 4 in. of brick ($K = 9.2$) and ½-in. wallboard ($C = 1.00$) separated by an air space ($C = 1.10$). The calculations are shown in Table 15-8.

Suppose, now, 1 in. of insulation ($K = 0.25$) were to be incorporated in this wall. The resistance R of the insulation ($1/K$) is 4. Thus, the resistance of the original wall is increased to 3.116 + 4, or 7.116. And the new overall conductance U becomes $1/7.116 = 0.14$.

(T. S. Rogers, "Thermal Design of Buildings," John Wiley & Sons, Inc., New York.)

15-25. Prevention of Condensation. Normally, air contains water vapor, which tends to move from a warm region to a cooler one. The lower the temperature, the less vapor the air can hold. If the air is saturated (100% relative humidity), a temperature drop will cause some of the vapor to condense. The temperature at which this occurs is called the **dew point.**

Since almost all building materials or the joints between them are porous, vapor will permeate them. If the dew point is reached between inner and outer surfaces, the vapor will condense, and the temperature differential will cause more vapor to penetrate, repeating the process. In cold weather, the dew point often occurs within insulation in walls and roofs. If vapor reaches it, the condensation may saturate the insulation, drastically reducing its insulating value. Furthermore, the moisture may rot or rust the structure or stain interior finishes. If temperatures are low, the water may freeze and in expanding, as ice always does, crack the structure.

A simple solution to condensation is to stop the flow of water vapor with a vapor barrier on the warm side. Since the dangers of condensation are greatest in the heating season, vapor barriers should be installed on the interior side of walls and roofs, next to the insulation.

Aluminum foil is a good, economical vapor barrier. Some insulations come equipped with it attached to one side. Other vapor barriers include aluminum paints, plastic paints and films, asphalt paints, rubber-base paints, asphalt, and foil-laminated papers.

The ability of a material to pass vapor is measured by the **perm.** It is defined as a vapor-transmission rate of 1 grain of water vapor through 1 sq ft of material per hour when the vapor-pressure difference equals 1 in. of mercury (7,000 grains = 1 lb). A material with a vapor-transmission rate of 1 perm or less is considered a good vapor barrier. **Rep** is the reciprocal of perm. It measures resistance to vapor transmission.

Since vapor barriers are not likely to be perfect or installed perfectly, some vapor may penetrate to the insulation. Means should be provided to let this vapor escape. Hence, the exterior surface should be as porous as possible or vented and yet prevent rain from penetrating. Cold-side venting may be desirable, even though condensation does not occur in the insulation, because it may occur instead in back of the exterior facing. Whenever the dew point occurs within a material, condensation will not take place until the water vapor encounters the surface of another material with greater resistance to the vapor flow.

Vapor also tends to flow through insulated ceilings into attics and air spaces under a roof. If these spaces are not ventilated with air capable of removing the moisture, it can cause trouble. In general, vent area should total about $1/300$ of the horizontal projection of the roof area. If possible, both high and low vents should be installed to insure air flow.

(T. S. Rogers, "Thermal Design of Buildings," John Wiley & Sons, Inc., New York.)

15-26. Heating. Required capacity of a heating plant is determined mainly by the total heat loss from a building through conduction, radiation, and infiltration. To allow for the temperature pickup usually required in the morning, however, the plant should have a capacity 20% larger than this heat loss. But do not choose too large a unit, because then operating efficiency suffers.

The heat loss depends on the design inside and outside temperatures. (See tables in the "ASHRAE Guide," American Society of Heating, Refrigerating and Air Conditioning Engineers, New York. The design outdoor temperatures are not the lowest ever attained in the region, but a slightly higher recommended value.) The difference between inside and outside temperatures is the temperature gradient. When multiplied by the exposed surface area of a material or construction and its overall thermal conductance U (Art. 15-24), the gradient determines the hourly heat flow in Btu. The sum of these products for all exposed surfaces of walls, windows, roofs, etc., yields the total heat loss through them.

Heat loss through basement floors and walls may be determined from groundwater temperature, which ranges from about 40 to 60°F in the northern sections of the United States and from 60 to 76°F in the southern sections. (For specific areas, see the "ASHRAE Guide.")

Groundwater temp, °F	Basement-floor loss,* Btu per hr per sq ft	Below-grade wall loss, Btu per hr per sq ft
40	3.0	6.0
50	2.0	4.0
60	1.0	2.0

*Based on basement temperature of 70°F.

Heat loss from a floor on grade without edge insulation is about 75 Btu per hr per lin ft of exposed edge in the cold northern sections of the United States, 65 in the temperate zones, and 60 in the warm south. With 1 in. of insulation, these rates drop to 60, 55, and 50; with 2 in., to 50, 45, and 40.

To obtain the heat loss through unheated attics, the equilibrium attic temperature must first be computed by equating heat gain to the attic via the ceiling to heat loss through the roof. The same procedure should be used to obtain the temperature of other unheated spaces, such as cellars and attached garages.

To the heat load for exposed surfaces must be added the load due to cold air infiltrating and warm air leaking out. The amount of leakage depends on crack area, wind velocity, and number of exposures, among other things. To account for leakage, the assumption is made that cold outside air will be heated and pumped into the building to create a static pressure large enough to prevent cold air from infiltrating. The amount of heat q in Btu per hour required to warm up this cold air is given by

$$q = 1.08QT \tag{15-6}$$

where Q = cfm of air to be warmed = $VN/60$
T = temperature rise of air, °F
V = volume of room, cu ft
N = number of air changes per hr

If the heating plant also will be used to produce hot water, the added capacity for this purpose should be determined and added to the heat load.

A warm-air heating system supplies heat to a room by bringing in a quantity of air above room temperature. The amount of heat added by the air must be at least equal to that required to counteract heat losses. Equation (15-6) gives this heat if T is taken as the difference between the temperature of the air leaving the grille and the room temperature, and Q as the cfm of air supplied to the room. In good systems, the discharge temperatures range from 135 to 140° F. Supply grilles should be arranged to blow a curtain of air across exposed walls and windows. The best location is near the floor. Return-air grilles should be installed in the interior, preferably at the ceiling.

Ducts for warm-air systems generally are designed by the equal-friction method. Sizes are calculated to accommodate the design air quantity of the heater with a predetermined friction factor. The pressure loss due to friction should not exceed 0.15 in. of water per 100 ft of duct. Also, starting velocity of the air in main ducts should be kept below 900 fpm in residences; 1,300 fpm in schools, theaters, and public buildings; and 1,800 fpm in industrial buildings. Velocity in branch ducts should be about two-thirds of these, and in branch risers, about one-half. But too low a velocity will require uneconomical, bulky ducts.

Normally, rectangular ducts are used, because dimensions can be easily changed to maintain the required area. Charts in the "ASHRAE Guide" give for round ducts the relationship between duct diameter in inches, air velocity in fpm, and friction in inches of water pressure drop per 100 ft of duct. Table 15-9 is based on data in the Guide. Other tables in the Guide give equivalent sizes of rectangular ducts. Table 15-10 is a shortened version.

In a forced warm-air system, a thermostat calls for heat, starting a heat source. When the air chamber in the heater reaches about 120°F, the fan starts. (If the discharge temperature

Table 15-9. Sizes of Round Ducts for Air Flow*

Friction, in. per 100 ft	0.05		0.10		0.15		0.20		0.25		0.30	
Air flow, cfm	Dia, in.	Veloc-ity, fpm	Dia, in.	Veloc-ity, fpm	Dia, in.	Veloc-ity, fpm	Dia, in.	Veloc-ity, fpm	Dia, in.	Veloc-ity, fpm	Dia, in.	Veloc-ity, fpm
50	5.3	350	4.6	450	4.2	530	3.9	600	3.8	660	3.7	710
100	6.8	420	5.8	550	5.4	640	5.1	720	4.8	780	4.7	850
200	8.7	480	7.6	650	6.9	760	6.6	860	6.3	940	6.1	1,020
300	10.2	540	8.8	730	8.2	850	7.7	960	7.3	1,050	7.1	1,120
400	11.5	580	9.8	770	9.0	920	8.5	1,040	8.2	1,130	7.8	1,200
500	12.4	620	11.8	820	9.8	970	9.3	1,080	8.8	1,160	8.6	1.270
1,000	15.8	730	13.7	970	12.8	1,140	12.0	1,280	11.5	1,400	11.2	1,500
2,000	20.8	870	18.0	1,150	16.6	1,370	15.7	1,520	15.0	1,660	14.5	1,780
3,000	24.0	960	21.0	1,280	19.7	1,500	18.3	1,680	17.5	1,850		
4,000	26.8	1,050	23.4	1,360	21.6	1,600	20.2	1,800				
5,000	29.2	1,100	25.5	1,460	23.7	1,700	22.2	1,900				
10,000	37.8	1,310	33.2	1,770	30.3	2,000						

*Based on data in the "ASHRAE Guide," American Society of Heating, Refrigerating and Air Conditioning Engineers.

Table 15-10. Diameters of Circular Ducts in Inches Equivalent to Rectangular Ducts

Side	4	8	12	18	24	30	36	42	48	60	72	84
3	3.8	5.2	6.2									
4	4.4	6.1	7.3									
5	4.9	6.9	8.3									
6	5.4	7.6	9.2									
7	5.7	8.2	9.9									
12		10.7	13.1									
18		12.9	16.0	19.7								
24		14.6	18.3	22.6	26.2							
30		16.1	20.2	25.2	29.3	32.8						
36		17.4	21.9	27.4	32.0	35.8	39.4					
42		18.5	23.4	29.4	34.4	38.6	42.4	45.9				
48		19.6	24.8	31.2	36.6	41.2	45.2	48.9	52.6			
60		21.4	27.3	34.5	40.4	45.8	50.4	54.6	58.5	65.7		
72		23.1	29.5	37.2	43.8	49.7	54.9	59.6	63.9	71.7	78.8	
84			39.9	46.9	53.2	58.9	64.1	68.8	77.2	84.8	91.9	
96				49.5	56.3	62.4	68.2	73.2	82.6	90.5	97.9	

exceeds 180°F, a safety element in the air chamber shuts off the heat source.) The heat source stops when the indoor temperature reaches the value at which the thermostat is set. But the fan continues to operate until the air cools to below 120°F.

In warm-air perimeter heating, often used with concrete floors on ground, the heater discharges warm air to two or more underfloor radial ducts feeding a perimeter duct. Floor grilles or baseboard grilles are located as in a conventional warm-air heating system, with collars connected to the perimeter duct.

A hot-water heating system consists of a heater or furnace, radiators, piping systems, and circulator. Normally, forced circulation systems are used because they can maintain higher water velocities and therefore require smaller pipes and provide more sensitive control.

Three types of piping systems are in general use. The one-pipe system (Fig. 15-22a) has many disadvantages and is not usually recommended. The two-pipe direct-return system (Fig. 15-22b) provides all radiators with the same supply-water temperature, but the last radiator has more pipe resistance than the first. This can be balanced out by installing orifices in the other radiators to add an equivalent resistance and by sizing the pump for the longest run. In the two-pipe reversed-return system (Fig. 15-22c), the total pipe resistance is about the same for all radiators.

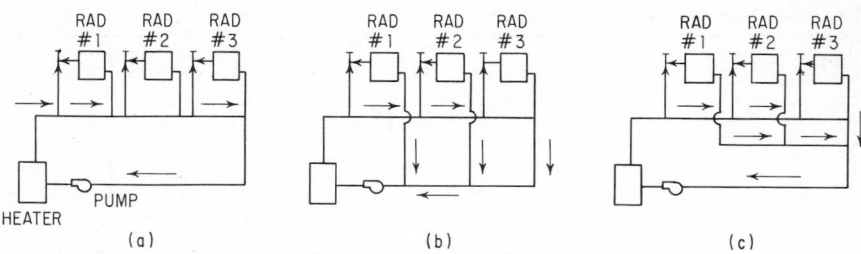

Fig. 15-22. Types of hot-water heating systems: (*a*) one-pipe; (*b*) two-pipe, direct-return; (*c*) two-pipe, reversed-return.

For a hot-water system, supply design temperatures usually are 180°F, with a 20°F drop assumed through the radiators. Thus, the temperature of the return riser would be 160°F. The amount of heat required to offset the 20°F drop, in Btu per hour, is

$$q = 10,000Q \tag{15-7}$$

where Q = flow of water, gpm. Piping may be sized for the required water flow with the aid of friction flow charts and tables showing equivalent pipe lengths for fittings. (See, for example, the "ASHRAE Guide.") Water velocity should be limited to a maximum of 4 fps. Loss of pressure due to friction should be between 0.25 and 0.60 in. of water per foot. The system must be provided with an expansion tank, located at least 3 ft above the highest radiator and in a location where the water will not freeze. The tank should be sized for 6% of the total volume of water in radiators, heaters, and piping. In very tall buildings, to avoid too high a static pressure on the boiler, heat exchangers should be provided in the upper levels.

In a hot-water system, an immersion thermostat in the heater controls the heat source to maintain design heater water temperature (usually about 180°F). When the room thermostat calls for heat, the circulator starts. Thus, an immediate supply of hot water is available for the radiators. For 170°F average water temperature, 1 sq ft of radiation surface emits 150 Btu per hr.

A steam-heating system consists of a boiler or steam generator and a piping system connecting to individual radiators or convectors. In a one-pipe system (Fig. 15-23*a*), the pipe supplying steam to the radiators also is used to return condensate to the boiler. On start-up, the steam must push air out of the pipe and radiators. For the purpose, the radiators are equipped with thermostatic air valves. Orifice size in the air vents must be varied to balance the system; otherwise, radiators at the far end of a pipe run may get steam much later than the near end. Valves in a one-pipe system must be fully open or closed. In a two-pipe system (Fig. 15-23*b*), steam is fed to the radiators through one pipe and the condensate returned through a second pipe. When condensate cools the radiator below 180°F, a trap opens to allow the condensate to return to a collecting tank, from which it is pumped to the boiler. The wet-return system (Fig. 15-23*c*) usually has a smaller pressure head available for pipe loss. It is a self-adjusting system depending on the load. When the condensate collects sufficiently in the return main above boiler level, the pressure will force the condensate into the boiler.

In all cases, the steam-supply pipes must be pitched to remove condensate from the pipe. Where condensate flows against the steam, the pipe may have to be oversized. Pipe capacities for supply risers, runouts, and radiator connections are given in the "ASHRAE Guide." Capacities are expressed in square feet of equivalent direct radiation (EDR).

$$1 \text{ sq ft EDR} = 240 \text{ Btu per hr} \tag{15-8}$$

Where capacities are in pounds per hour, 1 lb per hr = 970 Btu per hr.

A vacuum-heating system is similar to a steam pressure system with a condensate return pump. The vacuum pump pulls noncondensables from the piping and radiators for discharge to the atmosphere, whereas in a steam pressure system thermostatic vents are opened for this purpose.

Unit heaters often are used for large open areas, such as garages, showrooms, stores, and workshops. The units usually consist of a heat source or heat exchanger and an electrically operated fan. Heat may be supplied by steam or electricity or by burning gas. When gas-fired unit heaters are used, however, an outside flue must be provided to dispose of the products of combustion. Sizes of gas piping and burning rates for gas can be obtained from the "ASHRAE Guide." Efficiency of most gas-fired equipment is between 70 and 80%.

Radiant heating, or panel heating, consists of a warm pipe or electric cables embedded in the floor, ceiling, or walls. Joints in ferrous pipe should be welded, whereas those in nonferrous pipe should be soldered, and return bends should be made with a pipe bender instead of with fittings, to avoid joints. All piping should be subjected to a hydrostatic test of at least three times the working pressure, with a minimum of 150 psig. Repairs are costly after construction has been completed. Piping and circuiting are similar to those for a hot-water system with radiators and convectors, except that cooler water is used. But a 20°F temperature drop usually is assumed. Therefore, charts used for the design of hot-water piping systems may be used for panel heating, too. Floor panel temperatures generally are maintained about 85°F or lower and ceiling panel temperatures at 100°F or lower. While it is possible with panel heating to maintain relatively low room air temperatures with comfort, the system should be designed for standard room temperatures to prevent discomfort after the thermostat stops water circulation.

15-27. Air Conditioning.   Required capacity of a cooling plant is determined by the heat transmitted to the conditioned space through the walls, glass, ceiling, floor, etc., and all the heat generated in the space. The total cooling load consists of sensible and latent heat. Sensible heat

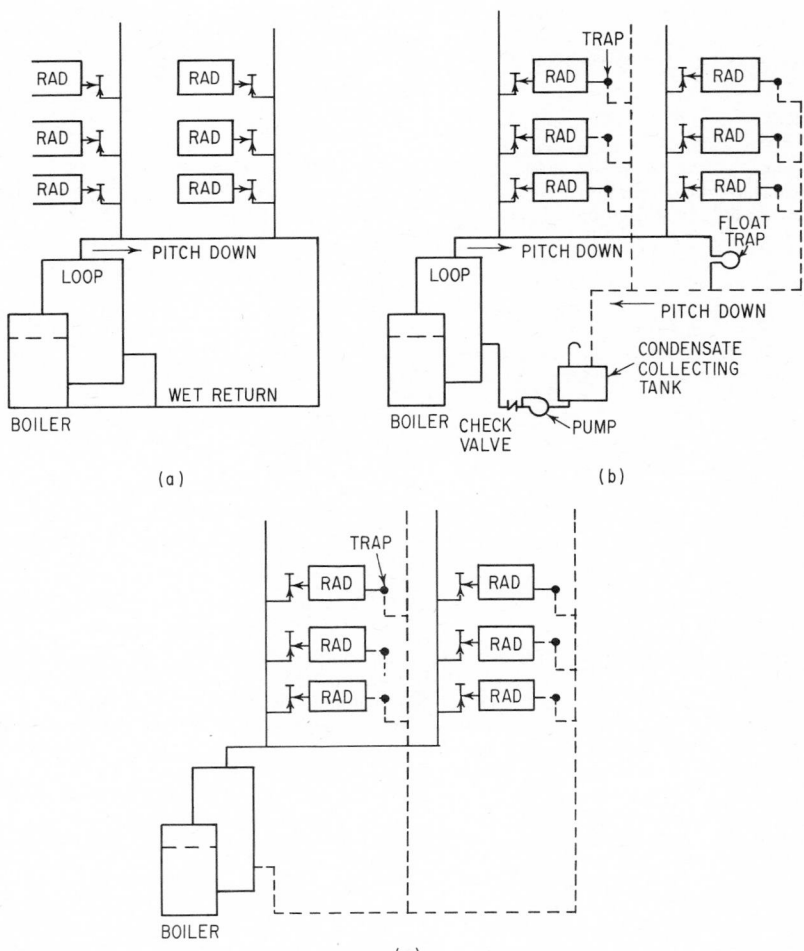

Fig. 15-23. Types of steam-heating systems: (*a*) one-pipe, with condensate returning through the supply pipe; (*b*) two-pipe; (*c*) two-pipe, wet return.

is the part that shows up in the form of a dry-bulb temperature rise. It includes heat transmitted through the building enclosure; radiation from the sun; and heat from lights, people, electrical and gas appliances, and outside air brought into the air-conditioned space. Latent heat is that needed to remove moisture from the air. Usually, the moisture is condensed out on the cooling coils in the cooling unit; 1,050 Btu is required per lb of condensation.

Design conditions for comfort cooling usually are 80°F dry bulb and 50% relative humidity. Design outdoor temperatures are not the highest ever recorded in a region but a slightly lower recommended value. (See tables in the "ASHRAE Guide," American Society of Heating, Refrigerating and Air Conditioning Engineers, New York.) The difference between indoor and outdoor temperatures multiplied by the area of walls, roofs, windows, etc., and the respective overall coefficients of conductance U (Art. 15-24) yields the heat gain through each.

Radiation from the sun through glass and roofs adds substantially to the heat load; the sun effect on walls generally can be neglected. Sun through unshaded window glass can add about 200 Btu per hr per sq ft through windows facing east and west; about three-fourths as much for windows facing northeast and northwest; and half as much for windows facing south. For most roofs, total equivalent temperature difference for calculating heat gain due to the sun is about 50°F. Roof sprays sometimes are used to reduce this load. With a water spray, the equivalent temperature difference may be taken as 18°F.

Heat from electric lights and other electrical appliances can be computed from

$$q = 3.42W \tag{15-9}$$

where q = Btu per hr developed
W = watts of electricity used per hr
For fluorescent lighting, add 25% of the lamp rating for the heat generated in the ballast.

Heat gain from people for various types of activities is given in tables in the "ASHRAE Guide." The sensible heat from outside air brought into a conditioned space can be computed from

$$q_s = 1.08Q(T_o - T_i) \tag{15-10}$$

where q_s = sensible load due to outside air, Btu per hr
Q = cfm of outside air brought into conditioned space
T_o = design dry-bulb temperature of outside air
T_i = design dry-bulb temperature of conditioned space
The latent load due to outside air in Btu per hr is

$$q_l = 0.67Q(G_o - G_i) \tag{15-11}$$

where G_o = moisture content of outside air, grains per lb of air
G_i = moisture content of inside air, grains per lb of air
The moisture content of air at various conditions may be obtained from a psychrometric chart.

The total heat load for sizing a cooling plant also must include heat from fans in the air-conditioning system, which usually ranges from 3½ to 5% of the sensible load, and heat loss from ducts. The load can be converted to tons of refrigeration by

$$\text{Load in tons} = \frac{\text{load in Btu per hr}}{12,000} \tag{15-12}$$

A ton of refrigeration is the amount of cooling that can be done by a ton of ice melting in 24 hr.

Basic Cycle. Figure 15-24a shows the basic air-conditioning cycle of the direct-expansion type. The compressor takes refrigerant gas at a relatively low pressure and compresses it to a higher pressure. The hot gas is passed to a condenser where heat is removed and the refrigerant liquefied. This liquid then is piped to the cooling coil of an air-handling unit and allowed to expand to a lower pressure (suction pressure). The liquid vaporizes or is boiled off by the relatively warm air passing over the coil. The compressor pulls away the vaporized refrigerant to maintain the required low coil pressure with its accompanying low temperature. A system in which the refrigerant chills water, which is circulated to air-handling units for cooling air, is shown in Fig. 15-24b.

Air Quantity. The amount of air in cfm to be handled can be computed from

$$Q = \frac{q_s}{1.08(T_i - T_d)} \tag{15-13}$$

where q_s = total sensible heat load, Btu per hr

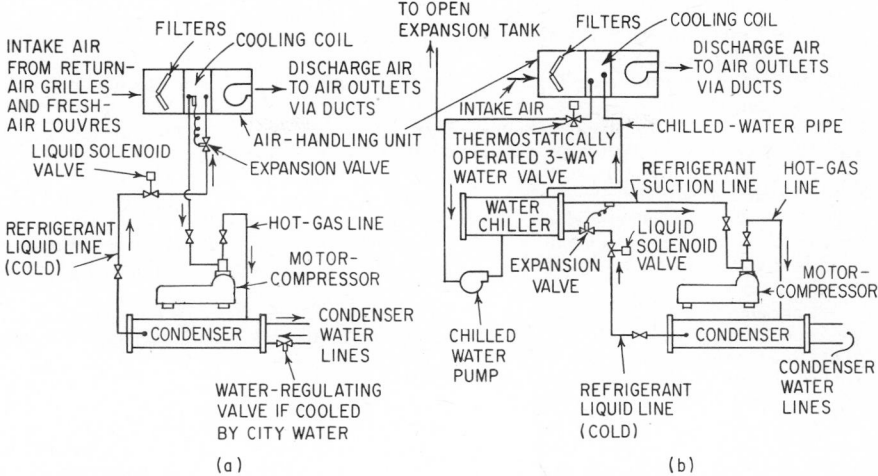

Fig. 15-24. (*a*) Direct-expansion air-conditioning cycle.   (*b*) Chilled-water air-conditioning cycle.

T_i = indoor temperature (dry bulb)

T_d = dry-bulb temperature of air discharged from air-handling unit

T_d should be about 3°F higher than the room dew point, to avoid sweating ducts.

Condensers.   If a water-cooled condenser is used to remove heat from the refrigerant, it may be supplied with city water, and the warm water may be discharged to a sewer.   Or a water tower may be used to cool condenser water, which then can be recirculated to the condenser.   If the wet-bulb temperature is low enough, the condenser and water tower can be replaced by an evaporative condenser.   The capacity of such water savers as towers and evaporative condensers decreases as the wet-bulb temperature increases.   The amount of water in gallons per minute required for condensers is

$$Q = \frac{\text{tons of cooling} \times 30}{\text{water-temperature rise}} \qquad (15\text{-}14)$$

Condensers for small cooling units can be cooled by a fan blowing air over the refrigerant coils.

Zoning.   Multizone air-handling units control the temperature of several zones in a building without a separate air-handling unit for each zone.   When a zone thermostat calls for cooling, the damper motor for that zone opens the cold deck dampers and throttles the warm deck dampers.   Thus, the same unit can provide cooling for one zone while it supplies heat for another zone.

Filters.   The area of the filters in the air-handling units should be large enough so that the air velocity does not exceed 350 fpm for low-velocity filters and 550 fpm for high-velocity filters. Minimum filter area in square feet equals air flow in cfm divided by maximum air velocity across the filters, fpm.   Most filters are the throw-away or cleanable type.   Electrostatic filters usually are used in industrial installations, where a higher percentage of dust removal must be obtained, in combination with regular throw-away or cleanable filters, which remove large particles.

Package Units.   For lower-cost air-conditioning installations, "package" or preassembled units may be used.   They generally operate on the complete cycle shown in Fig. 15-24*a*.   For window units, the condenser, projecting outside the building, is air-cooled, and the same motor usually runs both the fan for the cooling coil and that for the condenser.   Small floor-type units may be air-cooled; larger ones generally are water-cooled.

Absorption chillers use a source of heat to regenerate the refrigerant.   The compressor of the basic air-conditioning cycle (Fig. 15-24*a*) is replaced by an absorber, pump, and generator.   The refrigerant is regenerated by absorption in a weak solution of refrigerant and water, forming a strong solution, which is heated in the generator.   The refrigerant vapor thus is driven out of the solution and brought to the condenser under pressure.   When low-cost steam is available, absorption systems may be more economical to operate than systems with compressors.   In general, steam consumption is about 20 lb per hr per ton of refrigeration.

Ducts for air conditioning may be designed much as for forced warm-air heating systems (Art. 15-26). But the discharge grilles should preferably be installed in or near the ceiling, while the return-air grilles should be near the floor.

High-velocity air-distribution duct systems sometimes are used for large multistory buildings. These systems operate at air velocities well above 3,000 fpm and above 3-in. static pressure. Obvious advantages include smaller ducts and lower buildings, since smaller plenums are needed above hung ceilings. Disadvantages are high power consumption for fans and need for an air-pressure-reducing valve and sound-attenuation box for each air outlet.

Double-duct air-distribution systems comprise a high-pressure warm-air system and cold-air system. Each outlet is mounted in a sound-attenuation box with pressure-reducing valves and branches from the two systems. Room temperature is controlled by a thermostat actuating two motorized volume dampers, one for the cold air and one for the warm air fed into the sound-attenuation box.

A heat pump comprises equipment for using the heat removed by the condenser in a refrigeration cycle for heating the building. The heat absorbed by the refrigerant evaporator is taken from some other heat source, often outdoor air, instead of from the building. When the heat is exhausted outside the building, the heat pump also can be used for cooling. In general, it is economical for regions where the cooling season is substantially longer than the heating season and winter temperatures are not extreme. The colder the outdoor temperature, the lower the heat-pump capacity becomes, unless a constant-temperature heat source, such as warm water from a deep well, is available.

(F. S. Merritt, "Building Construction Handbook," McGraw-Hill Book Company, New York.)

15-28. Ventilation. Natural air movement or air replacement in a room depends on prevailing winds, temperature difference between interior and exterior, height of structure, window openings, etc. For controlled ventilation, a mechanical method of air change is desirable.

Where people are working, the amount of ventilation air required will vary from one air change per hour where no heat or offensive odors are generated to about 60 air changes per hour. Table 15-11 gives the minimum amount of air recommended for various activities.

Table 15-11. Minimum Ventilation Air for Various Activities

Type of occupancy	Ventilation air, cfm per person
Inactive, theaters	5
Light activity, offices	10
Light activity with some odor generation, restaurants	15
Light activity with moderate odor generation, bars	20
Active work, shipping rooms	30
Very active work, gymnasiums	50

The number of air changes per hour equals $60Q/V$, where Q is the air supplied, cfm, and V is the volume of ventilated space, cu ft. If there is less than one air change per hour, the ventilation system will take too long to produce a noticeable effect when first put into operation. Five air changes per hour generally is considered a practical minimum. Air changes above 60 per hour usually will create some discomfort because air velocities will be too high. Toilet and locker-room ventilation generally are covered by local building codes; 50 cfm per water closet and urinal is the usual minimum for toilets and six changes per hour minimum for both toilets and locker rooms.

Removal of heat by ventilation is best done by installing exhaust outlets close to the heat source. Where concentrated sources of heat are present, canopy hoods should be used. When heat is discharged into a room, the amount of ventilation air, cfm, required to remove heat not lost by transmission through enclosures is

$$Q = \frac{q}{1.08(T_i - T_o)} \tag{15-15}$$

where q = heat, Btu per hr, carried away by ventilation air
T_i = indoor temperature to be maintained
T_o = temperature of fresh air (usually outdoor air)

If a gas or moisture in the air is to be diluted, the amount of ventilation air, cfm, required is $Q = X/Y$, where the vapor or gas is formed at the rate of X lb per min and Y is the allowable concentration, lb per cu ft.

15-29. Electric Power for Buildings. Electrical design and construction for buildings are based usually on the National Electrical Code (National Fire Protection Association, 60 Batterymarch St., Boston, Mass. 02110; American Insurance Association, 85 John St., New York, N.Y. 10038). But local building codes may have some more restrictive requirements and should be checked. These codes contain minimum safety standards. Use of these standards does not guarantee adequate performance of an electrical system.

A building's electrical systems operate on electric currents supplied at specified effective voltages. An electric current I, amperes, is the rate at which electricity flows through a circuit. If it always flows in the same direction, it is called a direct current (dc). The current is assumed to flow from a positive to a negative terminal. An alternating current (ac) reverses direction at regular intervals.

Electromotive force or potential difference E, volts, is the force that makes electrons move in the circuit. It is opposed by a resistance R. Ohm's law relates E, I, and R:

$$E = IR \qquad (15\text{-}16)$$

Electric power W, watts or kilowatts (1,000 watts), is the rate of doing electrical work: 746 watts $= 0.746$ kw $= 1$ hp. Direct-current power, watts, is given by

$$W = EI = I^2 R \qquad (15\text{-}17)$$

Phases. In single-phase ac circuits power is the product of voltage, current, and a power factor, which equals 100% only when current and voltage are in phase, that is, pass through zero, maximums, and minimums at the same time.

If current and voltage are represented by a sine curve, one may lead or lag the other by nearly 360°. If, for example, the maximum of a sinusoidal current occurs 60° before the maximum of the voltage, the current leads the voltage by 60° or lags by 300°. In a single-phase ac system, the power factor equals the cosine of the angle between the voltage and current phases. Hence, the closer the phase angle is to 90° or 270°, the smaller the power factor, and the larger the equipment and conductors needed to deliver the required power. Low power factors often may be corrected by installing synchronous motors, or by connecting static condensers across the line.

Inductance L, henrys, makes current lag voltage. **Capacitance** C, farads, makes current lead voltage. Both inductive reactance X_L, ohms, and capacitive reactance X_c, ohms, impede the flow of current. **Impedance** Z, ohms, is the total opposition to the flow of current and equals the vector sum of resistance and reactance:

$$Z^2 = R^2 + (X_L - X_c)^2 \qquad (15\text{-}18)$$

Maximum voltage drop across an impedance equals maximum current times impedance.

Types of Circuits. Basic circuits are either series or parallel types. A series circuit has components connected in sequence. If there is a break in a series circuit, current will not flow; hence, if one lamp goes out, all go out. Parallel, multiple, or shunt circuits, in contrast, have components with common terminals. Voltage across the components is the same, and the current divides among them, in accordance with Ohm's law [Eq. (15-16)]. Parallel circuits generally are used for electrical distribution in buildings, whereas series circuits often are used for street lighting.

Service equipment consists of a circuit breaker or switch and fuses, and their accessories, located near the point of entrance of supply conductors to a building and intended to constitute the main control and means of cutoff of the supply. **Feeders** are the conductors between the service equipment, or the generator switchboard of an isolated plant, and branch-circuit overcurrent-protective devices. A **branch circuit** is the part of the system between the feeder and the load, or current-consuming equipment. Branch circuits deliver current to outlets, points where current is taken for equipment. A **receptacle,** or convenience outlet, permits the circuit to be tapped with a plug and flexible cord.

Electrical Loads. All conductors should be sized for the sum of the loads, in kilowatts, for lighting, motors, and appliances. Since all lights may not be on at the same time, codes generally permit for feeders a reduction in the lighting load by application of a demand factor. Table 15-12 shows the demand factors for sizing feeders permitted by the National Electrical Code. Codes also specify that feeders and branch circuits be sized for a minimum load, in watts per sq ft of floor area, that depends on type of occupancy. But usually, the actual load will exceed these minimums.

Except for household and kitchen appliances, motor and appliance loads are taken at full value. Codes usually list demand factors for reducing the total wattage for household and kitchen appliances. At least two 20-amp, appliance-receptacle circuits should be installed in the kitchen of a dwelling. (No more than two outlets should be connected in a 20-amp appliance circuit.) In addition, there should be at least one 15-amp branch circuit for each 500 sq ft of floor area in the dwelling. Each of these circuits should be limited to six to eight outlets, although codes may permit twelve.

Conductors should not be smaller than No. 12 in branch circuits.

Small installations, such as dwellings, usually are supplied with three-wire service. This consists of a neutral and two power wires with current differing 180° in phase. Tapping across the phase wires yields a single-phase two-wire 230-volt supply. Either phase wire and the neutral yield a single-phase two-wire 115-volt supply. In addition, for safety reasons, a separate ground wire should be provided, since the neutral, though grounded, carries current.

For larger installations, a 120/208-volt three-phase four-wire system usually is used. This consists of a neutral and three power wires carrying current differing 120° in phase. Tapping

Table 15-12. Code Demand Factors for Lighting*

Type of occupancy	Unit load per sq ft, watts	Load to which demand factor applies, watts	Demand factor, %
Armories and auditoriums	1	Total wattage	100
Banks	2	Total wattage	100
Barber shops and beauty parlors	3	Total wattage	100
Churches	1	Total wattage	100
Clubs	2†	Total wattage	100
Courtrooms	2	Total wattage	100
Dwellings (other than hotels)	3†	3,000 or less	100
		Next 117,000	35
		Over 120,000	25
Garages—commercial (storage)	½	Total wattage	100
Hospitals	2	50,000 or less	40‡
Hotels and motels, including apartment houses without			
provisions for cooking by tenants	2†	Over 50,000	20
		20,000 or less	50‡
		Next 80,000	40
		Over 100,000	30
Industrial commercial (loft) buildings	2	Total wattage	100
Lodge rooms	1½	30,000 or less	100
Office buildings	5	Over 30,000	70
		Total wattage	100
Restaurants	2	Total wattage	100
Schools	3	Total wattage	100
Stores	3	12,500 or less	100
Warehouses, storage	¼	Over 12,500	50
In any of above occupancies except single-family			
dwellings and individual appartments of multifamily			
dwellings:			
Assembly halls and auditoriums	1	Total wattage as specified	
Halls, corridors	½	for the specific occupancy	
Closets, storage spaces	¼		

*From "National Electric Code," 1968.

In view of the trend toward higher-intensity lighting systems and increased loads due to more general use of fixed and portable appliances, each installation should be considered with regard to the load likely to be imposed, and the capacity should be increased to ensure safe operation.

Where electric-discharge lighting systems are to be installed, high-power-factor type should be used, or the conductor capacity may need to be increased.

†For general illumination in dwelling occupancies, it is recommended that one 15-amp branch circuit be installed for each 375 sq ft of floor area.

‡For subfeeders to areas in hospitals and hotels where entire lighting is likely to be used at one time, e.g., in operating rooms, ballrooms, dining rooms, a demand factor of 100% shall be used.

across any two phase wires yields a single-phase two-wire 208-volt supply. Any phase wire and the neutral provide a single phase two-wire 120-volt supply. Other combinations yield two- or three-phase 120/208-volt supplies.

No current flows in the neutral when the loads on the system's circuits are balanced. Hence, the system should be so designed that, under full load, the load on each phase leg will be nearly equal.

Current in a conductor may be computed from the following formulas, in which

I = conductor current, amperes

W = power, watts

f = power factor, as a decimal

E_p = voltage between any two phase legs

E_g = voltage between a phase leg and neutral, or ground

Single-phase two-wire circuits:

$$I = \frac{W}{E_p f} \quad \text{or} \quad I = \frac{W}{E_g f} \tag{15-19}$$

Single-phase three-wire (and balanced two-phase three-wire) circuits:

$$I = \frac{W}{2E_g f} \tag{15-20}$$

Three-phase three-wire (and balanced three-phase four-wire) circuits:

$$I = \frac{W}{3E_g f} \tag{15-21}$$

Voltage drop in a circuit may be computed from the following formulas, in which

V_d = voltage drop between any two phase legs, or between phase leg and neutral when only one phase wire is used in the circuit

L = one-way run, ft

c.m. = circular mils (1 c.m. = area of a circle 0.001 in. in diameter)

Single-phase two-wire (and balanced single-phase three-wire) circuits:

$$V_d = \frac{2\,RIL}{\text{c.m.}} \tag{15-22}$$

Balanced two-phase three-wire, three-phase three-wire, and balanced three-phase four-wire circuits:

$$V_d = \frac{\sqrt{3}\,RIL}{\text{c.m.}} \tag{15-23}$$

Equations (15-22) and (15-23) contain a factor R that represents the resistance to direct current, in ohms, of 1 mil-ft of wire. For wires smaller than No. 3, resistance is the same for ac and dc. For wires larger than No. 3 carrying ac, a correction factor should be applied because of the higher resistance.

In design of feeders and branch conductors, voltage drops may range from 1 to 5%. Some codes limit the voltage drop to 2.5% for combined light and power circuits from the service equipment to branch panels. For economy, the greater part of the voltage drop, 1.5 to 2%, may be assigned to the smaller, more numerous feeders, and only 0.5 to 1% to the heavy main feeders. For motor loads only, the maximum voltage drop may be increased to 5%. Of this, 4% can be assigned to feeders.

The general procedure in sizing conductors is to start with the minimum-size wire permitted by code and test it for voltage drop. If this drop is excessive, test a larger size, and repeat until a wire is found for which the voltage drop is within the desired limit.

Fuses and circuit breakers should be incorporated in the circuits to protect motors from overcurrents of long duration, yet permit high, short-duration starting currents to pass. The National Electrical Code allows such overcurrent protective devices to have a higher ampere rating than the allowable current-carrying capacity of the wire. In branch circuits with one motor, conductors should have an allowable current-carrying capacity of at least 125% of the motor full-load current. For feeders supplying several motors, the conductor capacity should be at least 125% of the full-load current of the largest motor plus the sum of the full-load currents of the other motors.

Table 15-13. Recommended Levels of Illumination

Area	Footcandles on Tasks†
Assembly:	
Rough easy seeing	30
Rough difficult seeing	50
Medium	100
Fine	500‡
Extra fine	1,000‡
Cloth products:	
Cloth inspection	2,000‡
Cutting	300‡
Sewing	500‡
Pressing	300‡
Inspection:	
Ordinary	50
Difficult	100
Highly difficult	200‡
Very difficult	500‡
Most difficult	1,000‡
Machine shops	
Rough bench- and machine work	50
Medium bench- and machine work, ordinary automatic machines, rough grinding, medium buffing and polishing	100
Fine bench- and machine work, fine automatic machines, medium grinding, fine buffing and polishing	500‡
Extra-fine bench- and machine work, grinding, fine work	1,000‡
Offices:	
Cartography, designing, detailed drafting	200
Accounting, auditing, tabulating, bookkeeping, business-machine operation, reading poor reproductions, rough layout drafting	150
Regular office work, reading good reproductions, reading or transcribing handwriting in hard pencil or on poor paper, active filing, index references, mail sorting	100
Reading or transcribing handwriting in ink or medium pencil on good-quality paper, intermittent filing	70
Reading high-contrast or well-printed material; tasks and areas not involving critical or prolonged seeing such as conferring, interviewing, inactive files, and washrooms	30
Corridors, elevators, escalators, stairways	20§
Schools:¶	
Reading printed material	30
Reading pencil writing	70
Spirit duplicated material:	
Good	30
Poor	100
Drafting, benchwork	100‡
Lip reading, chalkboards, sewing	150‡
Woodworking:	
Rough sawing and benchwork	30
Sizing, planing, rough sanding, medium-quality machine- and benchwork, gluing, veneering, cooperage	50
Fine bench- and machine work, fine sanding and finishing	100

  * SOURCE: Table 19-21 from W. T. Stuart, Wiring Design—Commercial and Industrial Buildings, in D. G. Fink and J. M. Carroll (eds.), "Standard Handbook for Electrical Engineers," 10th ed., McGraw-Hill Book Co., 1968. Used by permission.

  † Minimum on the task at any time.

  ‡ Obtained with a combination of general lighting plus specialized supplementary lighting. Care should be taken to keep within the recommended brightness ratios. These seeing tasks generally involve the discrimination of fine detail for long periods of time and under conditions of poor contrast. To provide the required illumination, a combination of the general lighting indicated plus specialized supplementary lighting is necessary. The design and installation of the combination system must provide for not only a sufficient amount of light but also the proper direction of light, diffusion, and eye protection. As far as possible, it should eliminate direct and reflected glare as well as objectionable shadows.

  § Or not less than one-fifth the level in adjacent areas.

  ¶ Tasks are listed here, rather than areas.

The part of the wiring system at service switches and main distribution panels connected near these switches consists of heavy cables or buses and large switches that have very low resistance. If a short circuit occurs, very high currents will flow, and ordinary fuses or circuit breakers will not be able to interrupt them before wiring or equipment is damaged. For the purpose, high-interrupting-capacity current-limiting fuses, such as Amp-Traps and Hi-Caps, are needed. Ask the utility company for the interrupting capacity required.

Codes generally require that incoming service in a multiple-occupancy building be controlled near the point of entry by not more than six switches or circuit breakers. Meters, furnished by the utility company, also must be installed near the point of entry. The service switch and metering equipment may be combined in one unit, or the switch may be connected by conduit to a separate meter trough.

(F. Stetka, "NFPA Handbook of the National Electrical Code," McGraw-Hill Book Company, New York; D. G. Fink and J. M. Carroll, "Standard Handbook for Electrical Engineers," McGraw-Hill Book Company, New York; H. Richter, "Practical Electrical Wiring," McGraw-Hill Book Company, New York.)

15-30. Electric Lighting for Buildings. Artificial illumination is installed primarily for seeing, but it also may serve architectural purposes. With electric lighting, room illumination is not limited to window and skylight openings and by the vagaries of sunlight.

A basic lighting unit usually consists of a light source, or lamp, and a luminaire, or fixture, and accessory equipment, such as ballasts required for fluorescent lighting. Either the lamp itself or, more commonly, both lamp and fixture are designed to control light intensity in various directions and brightness. Generally, comfort in seeing is as important as ease in seeing.

Lumen L is the unit of light quantity. Foot-candle, ft-c, is the unit of light intensity and equals the number of lumens per sq ft on an area. Recommended light intensities for various tasks are given in Table 15-13.

Intensity of a light source is measured in candlepower, cp. A light source of 1 cp produces an intensity of 1 ft-c on a surface 1 ft away. (Intensity varies inversely as the square of the distance from the source.)

The amount of lumens to be provided by a lighting system in a uniformly lighted room may be estimated from

$$L = \frac{A \text{ ft-c}}{UM} \tag{15-24}$$

where ft-c = average illumination on a horizontal plane
A = room area, sq ft
M = maintenance factor
U = coefficient of utilization

The maintenance factor, usually ranging from 0.55 to 0.85, allows for such conditions as depreciation of lamp output and dust accumulation. The coefficient of utilization depends on the luminaire, reflective characteristics of ceiling and walls, and a factor called room or cavity ratio, which takes into account shape of room and height of lamps above the floor. The number of luminaires required is obtained by dividing L by the lumens emitted by each luminaire.

Usual practice is to make the distance between fixtures about equal to their height above the floor, with about half this distance between a wall and the first line of fixtures. Do not permit more than 1,500 watts of incandescent lighting or 1,250 watts of fluorescent lighting on each 15-amp fixture circuit.

Lighting adds 3.416 Btu per watt in heat gain to a room. Fluorescent lamps produce more lumens per watt than do incandescent lamps, and so less wattage is required with fluorescent lamps for a given illumination level. Therefore, they generally are preferred for air-conditioned buildings.

("IES Lighting Handbook," Illuminating Engineering Society, 345 E. 47th St., New York, N.Y. 10017; D. G. Fink and J. M. Carroll, "Standard Handbook for Electrical Engineers," McGraw-Hill Book Company, New York; D. Phillips, "Lighting in Architectural Design," McGraw-Hill Book Company, New York; W. Lam, "Perception and Lighting," McGraw-Hill Book Company, New York.)

15-31. Waste Piping. One function of a plumbing system in a building is to remove safely and quickly human, natural, and industrial wastes. The National Plumbing Code, ANSI Standard A40.8 (American Society of Mechanical Engineers, 345 E. 47th St., New York, N.Y. 10017), contains minimum standards for the design of such systems. But local building codes may have more restrictive requirements and should be checked. Table 15-14 lists the minimum number of fixtures required by the National Plumbing Code for various occupancies.

Associated with each fixture is a soil or waste stack, a vent or vent stack, and a trap. Soil stacks conduct wastes from one or more fixtures to a sloped house or building drain at the base of the building. Vents and vent stacks supply fresh air to the plumbing system to dilute gases and balance air pressure. Connected to each drainage pipe, vent stacks (vertical) must extend above the roof. They may have branch vents connected to them. Traps provide a water seal that prevents gases from discharging from the drainage pipes through the fixtures. The house or build-

Table 15-14. Minimum Facilities Recommended by National Plumbing Code, ANSI A40.8-1955[a]

Type of building or occupancy[b]	Water closets		Urinals	Lavatories		Bathtubs or showers	Drinking fountains[c]
Dwelling or apartment houses[d]	1 for each dwelling or apartment unit			1 for each apartment or dwelling unit		1 for each apartment or dwelling unit	
Schools:[e]	Male	Female					
Elementary	1 per 100	1 per 35	1 per 30 male	1 per 60 persons			1 per 75 persons
Secondary	1 per 100	1 per 45	1 per 30 male	1 per 100 persons			1 per 75 persons
Office or public buildings	No. of persons / 1–15 / 16–35 / 36–55 / 56–80 / 81–110 / 111–150 / 1 fixture for each 40 additional persons	No. of fixtures / 1 / 2 / 3 / 4 / 5 / 6	Wherever urinals are provided for men, one water closet less than the number specified may be provided for each urinal installed[b] except that the number of water closets in such cases shall not be reduced to less than ⅔ of the minimum specified	No. of persons / 1–15 / 16–35 / 36–60 / 61–90 / 91–125 / 1 fixture for each 45 additional persons	No. of fixtures / 1 / 2 / 3 / 4 / 5		1 for each 75 persons
Manufacturing warehouses, workshops, loft buildings, foundries, and similar establishments[f]	No. of persons / 1–9 / 10–24 / 25–49 / 50–74 / 75–100 / 1 fixture for each additional 30 employees	No. of fixtures / 1 / 2 / 3 / 4 / 5	Same substitution as above	1–100 persons, 1 fixture for each 10 persons Over 100, 1 for each 15 persons[g,h]		1 shower for each 15 persons exposed to excessive heat or to skin contamination with poisonous, infectious, or irritating material	1 for each 75 persons

[a] The figures shown are based upon one fixture being the minimum required for the number of persons indicated or any fraction thereof.

[b] Building categories not shown in this table will be considered separately by the administrative authority.

[c] Drinking fountains shall not be installed in toilet rooms.

[d] Laundry trays—one single compartment tray for each dwelling unit or 2 compartment trays for each 10 apartments. Kitchen sinks—1 for each dwelling or apartment unit.

[e] This schedule has been adopted (1945) by the National Council on Schoolhouse Construction.

[f] As required by the ANSI Standard Safety Code for Industrial Sanitation in Manufacturing Establishments (ANS Z4.1-1935).

[g] Where there is exposure to skin contamination with poisonous, infectious, or irritating materials, provide 1 lavatory for each 5 persons.

[h] 24 lin in. of wash sink or 18 in. of a circular basin, when provided with water outlets for such space, shall be considered equivalent to 1 lavatory.

Table 15-14. Minimum Facilities Recommended by National Plumbing Code, ANSI A40.8-1955[a] *(Continued)*

Dormitories[i]	Male: 1 for each 10 persons Female: 1 for each 8 persons Over 10 persons, add 1 fixture for each 25 additional males and 1 for each 20 additional females	1 for each 25 men Over 150 persons, add 1 fixture for each additional 50 men	1 for each 12 persons. (Separate dental lavatories should be provided in community toilet rooms. Ration of dental lavatories for each 50 persons is recommended). Add 1 lavatory for each 20 males, 1 for each 15 females.	1 for each 8 persons. In the case of women's dormitories, additional bathtubs should be installed at the ratio of 1 for each 30 females Over 150 persons, add 1 fixture for each 20 persons	1 for each 75 persons

	No of persons	No. of fixtures		No. of persons	No. of fixtures	No. of persons	No. of fixtures		
Theaters auditoriums	1–100 101–200 201–400 Over 400, add 1 fixture for each additional 500 males and 1 for each 300 females	Male 1 2 3	Female 1 2 3	(Male) 1–200 201–400 401–600 Over 600; 1 for each additional 300 males	 1 2 3	1–200 201–400 401–750 Over 750, 1 for each additional 500 persons	1 2 3		1 for each 100 persons

[i] Laundry trays, 1 for each 50 persons. Slop sinks, 1 for each 100 persons.

General. In applying this schedule of facilities, consideration must be given to the fixtures. Conformity purely on a numerical basis may not result in an installation suited to the need of the individual establishment. For example, schools should be provided with toilet facilities on each floor having classrooms.

Temporary workingmen facilities:

1 water closet and 1 urinal for each 30 workmen.

24-in. urinal trough = 1 urinal.   48-in. urinal trough = 2 urinals.

36-in. urinal trough = 2 urinals.   60-in. urinal trough = 3 urinals.   72-in. urinal trough = 4 urinals.

ing drain, located below the lowest fixture, conducts the waste to the house or building sewer, which starts 4 or 5 ft outside the foundation walls. That sewer, in turn, carries the wastes to a public sewer or other main sewer. Generally, a cleanout is required at the upper end of the house drain.

The piping generally is made of cast iron, copper, vitrified clay, steel, wrought iron, brass, plastics, or lead. Codes specify the type of joint to be used with each material.

For convenience, the discharges from fixtures are measured in terms of fixture units, which are used to determine pipe sizes. Tables 15-15 and 15-16 list the number of fixture units assigned to various types of fixtures in the National Plumbing Code, as well as the minimum trap size recommended. Table 15-17 notes the maximum number of fixture units (equivalent to maximum permissible discharge) that may be connected to stacks and horizontal fixture branches of various diameters. Similarly, Table 15-18 gives the maximum number of fixture units that may be connected to building drains and sewers of various diameters. And Table 15-19 gives the diameter of vent and maximum length permitted with various sizes of soil or waste stacks and various fixture units.

The plumbing system also may be required to dispose of rain on roofs, yards, areaways, and exposed floors. Normally, exterior sheet-metal leaders and gutters are not included in the plumbing contract, but interior leaders and storm-water drains are. While storm drains may

Table 15-15. Fixture Units per Fixture or Group *

Fixture type	Fixture-unit value as load factors	Min size of trap, in.	
1 bathroom group consisting of water closet, lavatory, and bathtub or shower stall	Tank water closet, 6 Flush-valve water closet, 8		
Bathtub† (with or without overhead shower)	2		1½
Bathtub†	3		2
Bidet	3	Nominal	1½
Combination sink and tray	3		1½
Combination sink and tray with food-disposal unit	4	Separate traps	1½
Dental unit or cuspidor	1		1¼
Dental lavatory	1		1¼
Drinking fountain	½		1
Dishwasher, domestic	2		1½
Floor drains‡	1		2
Kitchen sink, domestic	2		1½
Kitchen sink, domestic, with food-waste grinder	3		1½
Lavatory¶	1	Small P.O.	1¼
Lavatory¶	2	Large P.O.	1½
Lavatory, barber, beauty parlor	2		1½
Lavatory, surgeon's	2		1½
Laundry tray (1 or 2 compartments)	2		1½
Shower stall, domestic	2		2
Showers (group) per head	3		
Sinks:			
Surgeon's	3		1½
Flushing rim (with valve)	8		3
Service (trap standard)	3		3
Service (P trap)	2		2
Pot, scullery, etc.	4		1½
Urinal, pedestal, siphon jet, blowout	8	Nominal	3
Urinal, wall lip	4		1½
Urinal stall, washout	4		2
Urinal trough (each 2-ft section)	2		1½
Wash sink (circular or multiple) each set of faucets	2	Nominal	1½
Water closet, tank-operated	4	Nominal	3
Water closet, valve-operated	8	Nominal	3

*From National Plumbing Code, ANS A40.8-1955.

†A shower head over a bathtub does not increase the fixture value.

‡Size of floor drain shall be determined by the area of surface water to be drained.

¶Lavatories with 1¼- or 1½-in. trap have the same load value; larger P.O. (plumbing orifice) plugs have greater flow rate.

Table 15-16. Other Fixture-unit Values *

Fixture drain or trap size, in.	Fixture-unit value
1¼ in. and smaller	1
1½	2
2	3
2½	4
3	5
4	6

*From National Plumbing Code, ANS A40.8-1955.

Table 15-17. Horizontal Fixture Branches and Stacks*

Dia of pipe, in.	Max number of fixture units that may be connected to			
	Any horizontal† fixture branch	One stack of 3 stories in height or 3 intervals	More than 3 stories in height	
			Total for stack	Total at one story or branch interval
1¼	1	2	2	1
1½	3	4	8	2
2	6	10	24	6
2½	12	20	42	9
3	20‡	30¶	60¶	16‡
4	160	240	500	90
5	360	540	1,100	200
6	620	960	1,900	350
8	1,400	2,200	3,600	600
10	2,500	3,800	5,600	1,000
12	3,900	6,000	8,400	1,500
15	7,000			

*From National Plumbing Code, ANS A40.8-1955.
†Does not include branches of the building drain.
‡Not over two water closets.
¶Not over six water closets.

Table 15-18. Sizes of Building Drains and Sewers*

Dia of pipe, in.	Max number of fixture units that may be connected to any portion† of the building drain or the building sewer			
	Fall per ft			
	¹/₁₆ in.	⅛ in.	¼ in.	½ in.
2			21	26
2½			24	31
3		20‡	27‡	36‡
4		180	216	250
5		390	480	575
6		700	840	1,000
8	1,400	1,600	1,920	2,300
10	2,500	2,900	3,500	4,200
12	3,900	4,600	5,600	6,700
15	7,000	8,300	10,000	12,000

*From National Plumbing Code, ANSI A40.8-1955.
†Includes branches of the building drain.
‡Not over two water closets.

discharge into sanitary drains, codes generally prohibit use of storm drains for disposing of sewage. Sizes of vertical leaders and horizontal storm drains recommended in the National Plumbing Code are given in Table 15-20.

Table 15-21 lists recommended sizes of semicircular gutters. For maximum rainfalls differing appreciably from 4 in. per hr, apply correction factors given in the National Plumbing Code.

Indirect-waste piping usually is required for the discharge from commercial food-handling equipment and dishwashers, rinse sinks, laundry washers, steam tables, refrigerators, egg boilers, iceboxes, coffee urns, stills, and sterilizers and from units that must be fitted with drip or drainage connections but are not ordinarily regarded as plumbing fixtures. An indirect-waste pipe is not connected directly to the building drains but discharges wastes into a plumbing fixture or receptacle, from which they flow to the drains. An air gap should separate the indirect-

Table 15-19. Size and Length of Vents*

Size of soil or waste stack, in.	Fixture units connected	Dia of vent required, in.								
		1¼	1½	2	2½	3	4	5	6	8
		Max length of vent, ft								
1¼	2	30								
1½	8	50	150							
1½	10	30	100							
2	12	30	75	200						
2	20	26	50	150						
2½	42		30	100	300					
3	10		30	100	200	600				
3	30			60	200	500				
3	60			50	80	400				
4	100			35	100	260	1,000			
4	200			30	90	250	900			
4	500			20	70	180	700			
5	200				35	80	350	1,000		
5	500				30	70	300	900		
5	1,100				20	50	200	700		
6	350				25	50	200	400	1,300	
6	620				15	30	125	300	1,100	
6	960					24	100	250	1,000	
6	1,900					20	70	200	700	
8	600						50	150	500	130
8	1,400						40	100	400	120
8	2,200						30	80	350	110
8	3,600						25	60	250	80
10	1,000							75	125	100
10	2,500							50	100	50
10	3,800							30	80	35
10	5,600							25	60	25

*From National Plumbing Code, ANSI A40.8-1955.

waste pipe from the drains.   The length of the gap should be at least twice the diameter of the drain served.   This requirement is met by a pipe discharging into a vented or trapped floor drain, slop sink, or similar fixture not used for domestic or culinary purposes.   The indirect-waste pipe should be terminated at least 2 in. above the floor level of the fixture.

On completion of the plumbing, the system should be inspected or tested with either air or water.   In a water test, all openings but the highest one are tightly sealed.   The pipes then are filled with water, so that the minimum head is 10 ft, except for the top 10 ft of the system.   In an air test, the system is sealed and subjected to 5-psi pressure.   For a final test, either a strong-smelling smoke or peppermint is used.   With smoke, a pressure of at least 1 in. of water should be maintained on the sealed system for 15 min before inspection begins.   For the peppermint test, 2 oz of oil of peppermint is injected into each line or stack.

(V. T. Manas, "National Plumbing Code Handbook," McGraw-Hill Book Company, New York; H. E. Babbitt, "Plumbing," McGraw-Hill Book Company, New York.)

15-32. Fire-Sprinkler Systems.   Consisting essentially of parallel horizontal pipes installed near ceilings, sprinklers have been very effective in preventing spread of fires in buildings.   The extinguishing agent usually is water, though for some hazards carbon dioxide is used.   The agent, kept under pressure, is discharged from the pipes through sprinklers preset to open when air temperature rises rapidly or reaches a specified level, usually 135 to 160°F.

Common types of sprinkler systems include wet-pipe, dry-pipe, preaction, and deluge.   Pipes in a wet-pipe system contain water at all times and discharge immediately when the sprinklers open.   In a dry-pipe system, air under pressure in the pipes is discharged when the sprinklers open, thus allowing water pressure to open a valve to allow water to flow to the sprinklers.   Such systems are suitable for unheated areas in cold climates.   Preaction systems have pipes containing air, which may not be under pressure.   Heat-responsive devices near sprinklers open water valves when a fire occurs.   The deluge system has sprinklers attached to a piping system.

Table 15-20. Sizes of Vertical Leaders and Horizontal Storm Drains*

VERTICAL LEADERS

Size of leader or conductor,† in.	Max projected roof area, sq ft
2	720
2½	1,300
3	2,200
4	4,600
5	8,650
6	13,500
8	29,000

HORIZONTAL STORM DRAINS

Dia of drain, in.	Max projected roof area for drains of various slopes, sq ft		
	⅛-in. slope	¼-in. slope	½-in. slope
3	822	1,160	1,644
4	1,880	2,650	3,760
5	3,340	4,720	6,680
6	5,350	7,550	10,700
8	11,500	16,300	23,000
10	20,700	29,200	41,400
12	33,300	47,000	66,600
15	59,500	84,000	119,000

*From National Plumbing Code, ANS A40.8-1955.
†The equivalent diameter of square or rectangular leader may be taken as the diameter of that circle which may be inscribed within the cross-sectional area of the leader.

Table 15-21. Size of Gutters*

Dia of gutters,† in.	Max projected roof area for gutters of various slopes, sq ft			
	1/16-in. slope	⅛-in. slope	¼-in. slope	½-in. slope
3	170	240	340	480
4	360	510	720	1,020
5	625	880	1,250	1,770
6	960	1,360	1,920	2,770
7	1,380	1,950	2,760	3,900
8	1,990	2,800	3,980	5,600
10	3,600	5,100	7,200	10,000

*From National Plumbing Code, ANS A40.8-1955.
†Gutters other than semicircular may be used provided they have an equivalent cross-sectional area.

Heat-responsive devices in the same area as the sprinklers open a valve permitting water to discharge from all sprinklers when a fire occurs in the area served.

The sprinkler system should have a water supply of adequate capacity and pressure. For a secondary supply, a motor-driven automatically controlled fire pump supplied from a water main or pressure-storage system of sufficient capacity may be acceptable.

Local fire-prevention authorities and fire underwriters have specific requirements for type, material, size, and spacing of pipes and sprinklers. They should be consulted before a system is designed.

("Sprinkler Systems," American Insurance Association, New York; "National Fire Codes" and "Handbook of Fire Protection," National Fire Protection Association, Boston, Mass.; H. E. Babbitt, "Plumbing," McGraw-Hill Book Company, New York.)

15-33. Hot- and Cold-Water Piping for Buildings. The National Plumbing Code, ANSI Standard A40.8 (American Society of Mechanical Engineers, 345 E. 47th St. New York, N.Y. 10017), and local building codes contain minimum standards for hot- and cold-water piping. In addition, units may be used for which manufacturers have higher requirements, for example, for pressure.

Table 15-22. Rate of Flow and Required Pressure during Flow for Different Fixtures*

Fixture	Flow pressure,† psi	Flow rate, gpm
Ordinary basin faucet	8	3.0
Self-closing basin faucet	12	2.5
Sink faucet, ⅜ in.	10	4.5
Sink faucet, ½ in.	5	4.5
Bathtub faucet	5	6.0
Laundry-tub cock, ½ in.	5	5.0
Shower	12	5.0
Ball cock for closet	15	3.0
Flush valve for closet	10–20	15–40‡
Flush valve for urinal	15	15.0
Garden hose, 50 ft and sill cock	30	5.0

*From National Plumbing Code, ANS A40.8-1955.
†Flow pressure is the pressure in the pipe at the entrance to the particular fixture considered.
‡Wide range due to variation in design and type of flush-valve closets.

Table 15-22 summarizes National Plumbing Code recommendations for water pressure and flow for ordinary plumbing fixtures. Sufficient allowance should be made for pressure loss in the pipe and fittings between the supply source and the fixtures, so that these pressures will be present at the fixtures. If necessary to maintain pressure, booster pumps and gravity and pressure tanks may have to be used. If there is danger of excessive pressure in some parts of the system causing water hammer, an air chamber or approved device must be installed to protect the pipes from pressure surges and to reduce noise.

Pipes and tubes for water distribution may be made of copper, brass, lead, cast iron, wrought iron, steel, or plastic. The local building code should be checked for approved materials and types of joints for each.

The system should be designed so that there is no possibility of backflow at any time. Codes generally require that an air gap—space between fixture outlet and flood-level rim of the receptacle—of specified size for each type of fixture be maintained.

Hot water may be supplied by upfeed or downfeed systems, with unused water returned to the heater. In an upfeed system, fixtures are supplied with hot water by a riser directly from the heater. In a downfeed system, hot water is brought to the highest floor in a supply riser and vented at the top through a vent valve, and the fixtures are supplied by the return riser.

Generally, hot water is delivered at 130 °F. For floor cleaning, slop sinks may be fed 150 °F water.

Heaters. Two types of heaters are in general use, storage and instantaneous. Heat for storage heaters may be supplied by steam, gas, electricity, or hot water. For occupancies with uneven demand for hot water, such as industrial, office, and school buildings, a relatively large storage capacity is needed. But for occupancies with a nearly uniform demand, such as hotels, apartment houses, and hospitals, storage capacity may be smaller, but capacity of the heating coil must be larger. With storage heaters, water is stored in a tank at a specified temperature. With instantaneous heaters, however, water is heated as needed; there is no storage tank. These heaters have V-shaped or straight tubes through which the supply water passes to be heated. Steam or hot water is the usual heating medium.

Cold Water. Required flow at cold-water fixtures usually is measured in terms of fixture units. Table 15-23 gives the demand in fixture units for various fixtures. The load for each type of fix-

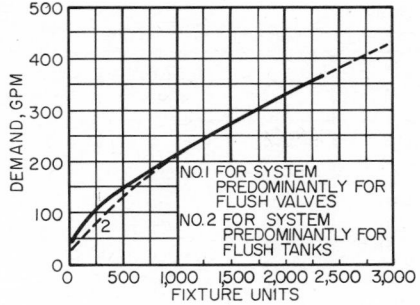

Fig. 15-25. Estimate curves for domestic water-demand load. (*National Plumbing Code.*)

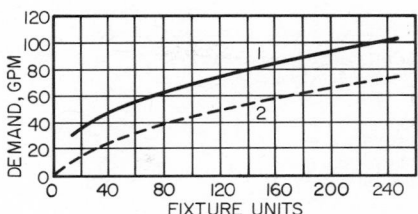

Fig. 15-26. Enlargement of low-demand portion of Fig. 15-25. (*National Plumbing Code.*)

ture is determined by multiplying the number of each to be installed on a riser or in the building by the demand weight in fixture units. Figure 15-25 relates the demand in fixture units to gallons of water per minute. Figure 15-26 is an enlargement of Fig. 15-25 in the range of 0 to 250 units. Hot-water demand per fixture in gallons per hour at 140 °F for various building occupancies is given in Table 15-24.

National Plumbing Code recommendations for minimum sizes for hot- and cold-water fixture-supply pipes are listed in Table 15-25. But required flow and pressure may make larger sizes necessary. Sizes of risers may be computed from the required flow in gallons per minute and the pressure drop permitted between the supply main and the highest fixtures.

Table 15-23. Demand Weight of Fixtures in Fixture Units* †

Fixture or group ‡	Occupancy	Type of supply control	Weight in fixture units¶
Water closet	Public	Flush valve	10
Water closet	Public	Flush tank	5
Pedestal urinal	Public	Flush valve	10
Stall or wall urinal	Public	Flush valve	5
Stall or wall urinal	Public	Flush tank	3
Lavatory	Public	Faucet	2
Bathtub	Public	Faucet	4
Shower head	Public	Mixing valve	4
Service sink	Office, etc.	Faucet	3
Kitchen sink	Hotel or restaurant	Faucet	4
Water closet	Private	Flush valve	6
Water closet	Private	Flush tank	3
Lavatory	Private	Faucet	1
Bathtub	Private	Faucet	2
Shower head	Private	Mixing valve	2
Bathroom group	Private	Flush valve for closet	8
Bathroom group	Private	Flush tank for closet	6
Separate shower	Private	Mixing valve	2
Kitchen sink	Private	Faucet	2
Laundry trays (1–3)	Private	Faucet	3
Combination fixture	Private	Faucet	3

*From National Plumbing Code, ANS A40.8-1955.

†For supply outlets likely to impose continuous demands, estimate continuous supply separately and add to total demand for fixtures.

‡For fixtures not listed, weights may be assumed by comparing the fixtures with a listed one using water in similar quantities and at similar rates.

¶The given weights are for total demand. For fixtures with both hot- and cold-water supplies, the weights for maximum separate demands may be taken as three-fourths the listed demand for supply.

Table 15-24. Hot-Water Demand per Fixture for Various Building Types
(Based on average conditions for types of buildings listed, gallons of water per hour per fixture at 140°F)

Type of fixture	Apartment house	Hospital	Hotel	Industrial plant	Office building
Basins, private lavatories	2	2	2	2	2
Basins, public lavatories	4	6	8	12	6
Showers	75	75	75	225	
Slop sinks	20	20	30	20	15
Dishwashers (per 500 people)	250	250	250	250	250
Pantry sinks	5	10	10		
Demand factor	0.30	0.25	0.25	0.40	0.30
Storage factor	1.25	0.60	0.80	1.00	2.00

Table 15-25. Minimum Sizes for Fixture-Supply Pipes*

Type of fixture or device	Pipe size, in.	Type of fixture or device	Pipe size, in.
Bathtubs	½	Shower (single head)	½
Combination sink and tray	½	Sinks (service, slop)	½
Drinking fountain	⅜	Sinks, flushing rim	¾
Dishwasher (domestic)	½	Urinal (flush tank)	½
Kitchen sink, residential	½	Urinal (direct flush valve)	¾
Kitchen sink, commercial	¾	Water closet (tank type)	⅜
Lavatory	⅜	Water closet (flush valve type)	1
Laundry tray, 1, 2, or 3 com-		Hose bibs	½
partments	½	Wall hydrant	½

*From National Plumbing Code, ANS A40.8-1955.

When the potable-water piping has been completed and before it is put into use, it should be disinfected with chlorine by a procedure approved by the local code.

(V. T. Manas, "National Plumbing Code Handbook," McGraw-Hill Book Company, New York; H. E. Babbitt, "Plumbing," McGraw-Hill Book Company, New York.)

15-34. Acoustics.   As applied to buildings, acoustics involves the creation of conditions for comfortable listening and means for noise control.   At present, acoustics is both an art and a science, for what is comfortable and what is noise depends on judgment and the function of the room to be treated.   A sound one person finds too loud may not bother someone else; what is comfortable in a factory may not be acceptable in a school; the music enjoyed by a high-fidelity fan may be noise to a neighbor trying to sleep.   Noise is unwanted sound.

Sounds are characterized by their pitch, or frequency; intensity, or loudness; and spectral distribution of energy, or sound quality.   An average person can hear from 20 to 20,000 cps (cycles, or vibrations, per second).   High-frequency, or high-pitched, sounds are more annoying to most people than low-pitched sounds of the same intensity.   But high-pitched sounds attenuate faster in air than low-pitched.

Loudness is a subjective evaluation of sound pressure or intensity.   But because human response to loudness varies with frequency, any measure of loudness must, in some way, include frequency as well as pressure or intensity to be of significance in building acoustics.   In addition, changes in human response to loudness depend on the ratio of the intensities of the sound.   In acoustics, the ratio 10:1 is called a bel.   For practical reasons, the unit frequently used, however, is the decibel (dB), equal to 0.1 bel.

Intensity level IL, dB, used as a measure of loudness, is defined by

$$IL = 10 \log_{10} \frac{I}{I_o} \tag{15-25}$$

where I = intensity, measured, watts per sq cm

I_o = reference intensity = 10^{-16} watts per sq cm

Equation (15-25) indicates that zero level corresponds with $I = I_o$, the reference intensity, which, in turn, corresponds with the average threshold of human hearing of sound at about 1,000 Hz (hertz, or cycles per second).

Sound pressure level *SPL*, dB, since intensity varies as the square of pressure, is accordingly defined by

$$SPL = 20 \log_{10} \frac{p}{p_0} \qquad (15\text{-}26)$$

where p = pressure, measured, dynes per sq cm

p_0 = reference pressure = 0.0002 dynes per sq cm

A change in sound level of less than about 3 dB is not likely to be perceptible, but a change of 5 dB will be noticeable. An increase of 10 dB will appear to be twice as great as an increase of 5 dB, and an increase of 20 dB much greater than an increase of 10 dB—not quite proportionately.

Sound levels usually are measured with electronic instruments that respond to sound pressures. Readings on the A scale of such instruments are often used because this scale is adjusted for frequencies that correspond somewhat with the response of the human ear. In such cases, the unit is indicated by dBA.

Table 15-26 presents a comparison of intensity, *SPL*, and common sounds.

Table 15-26. Comparison of Intensity, Sound Pressure Level, and Common Sounds

Relative intensity	SPL, dBA*	Loudness
100,000,000,000,000	140	Jet aircraft and artillery fire
10,000,000,000,000	130	Threshold of pain
1,000,000,000,000	120	Threshold of feeling
100,000,000,000	110	
10,000,000,000	100	Inside propellor plane
1,000,000,000	90	Full symphony or band
100,000,000	80	Inside automobile at high speed
10,000,000	70	Conversation, face-to-face
1,000,000	60	
100,000	50	Inside general office
10,000	40	Inside private office
1,000	30	Inside bedroom
100	20	Inside empty theater
10	10	
1	0	Threshold of hearing

*SPL as measured on A scale of standard sound level meter.

Acoustical analysis and design aim at both sound and vibration control. Sound control is accomplished with barriers and enclosures, acoustically absorbent materials, and other materials properly shaped and assembled. Vibration control is accomplished with energy-absorptive construction, usually with resilient materials, or by damping, involving use of viscoelastic materials.

Table 15-27. *STC* of Various Constructions

Construction	*STC*
¼-in. plate glass	26
¾-in. plywood	28
½-in. gypsumboard, both sides of	
2 × 4 studs	33
¼-in. steel plate	36
6-in. concrete block wall	42
8-in. reinforced concrete wall	51
12-in. concrete block wall	53
Cavity wall, 6-in. concrete block,	
2-in. air space, 6-in. concrete block	56

Effectiveness of a barrier in stopping sound is measured by **sound transmission loss,** the loss in energy level as sound passes through a barrier. The greater the mass of a barrier, the greater the sound transmission loss and the more effective the barrier. Mass and transmission loss, however, are not related linearly. At low frequencies losses tend to be larger, and at other frequencies, smaller, than required for a linear relationship. Table 15-27 lists the performance of various barriers and rating systems for their performance.

The purpose of a barrier with a high sound transmission loss, however, can be defeated if sound can bypass the barrier through openings or by transmission through adjoining construction. Ducts, pipes, and almost any continuous, rigid component of a building can carry sound past a barrier. Therefore, precautions should be taken to prevent such bypassing. Carpet over a resilient pad, for example, is effective in absorbing such sounds as footfalls, heel clicks, and impact of dropped light objects. Openings should be plugged. Vibrations from machines and other equipment can be absorbed by supporting them on springs, elastomeric pads, or other resilient mounts. (For more details, see the "ASHRAE Guide," American Society of Heating, Refrigerating and Air Conditioning Engineers, New York.)

Vibration of barriers resulting from impact of sound or transmission of vibrations from machines can be damped out by proper assembly in any of several ways. One way is to attach to a barrier material with high internal friction or poor connections between particles, or visco-elastic materials, such as asphaltic compounds that are neither completely elastic nor completely viscous. Also, components of the barrier may be connected with a viscoelastic adhesive.

Sound Absorption. Reflection of sound from a surface can be curtailed by covering the surface with an acoustical absorbent, usually lightweight, porous boards, blankets, or panels, which convert the mechanical energy of sound into heat. Exposed surfaces may be smooth or textured, fissured or perforated, or decorated or etched in many ways. Selection of an absorbent usually is based on its absorptive efficiency, appearance, fire resistance, moisture resistance, strength, and maintenance requirements. An absorbent, however, may have poor resistance to sound transmission and should not be used in an attempt to improve the airborne sound isolation of a barrier.

Sound absorption coefficients are used as an indication of the absorptive efficiency of building products. The sound absorption coefficient of a product is the ratio of the energy it will absorb from a sound wave to the total energy impinging. A perfect absorber would be assigned a coefficient of 1. Sound absorption, however, depends on the frequency of the sound. Consequently, coefficients for a product are given for specific frequencies, or sometimes as a composite for a group of frequencies (Table 15-29).

Generally, absorbents are used not only to reduce undesirable sound reflection, such as echoes and flutter, but also to secure desirable reverberations. *Echoes* are distinct reflections. *Flutter* is produced by rapid, repeated, partly distinguishable echoes, such as those that occur between parallel sidewalls of a corridor. *Reverberation* results from very rapid, repeated, jumbled echoes, which produce a continuing indistinct sound that persists after the sound producing the echoes has ceased.

Reverberation within a room can garble speech or distort music. But, properly controlled, reverberation can enhance the sound of music. Good reverberation can be achieved with proper proportioning and shaping of rooms, echo control, and absorption of noise. Generally, acoustical absorbents on room surfaces are desirable to absorb acoustical energy to prevent buildup of undesirable sounds.

Noise reduction NR, dB, achieved through addition of absorbents can be computed from

$$NR = 10 \log_{10} \frac{A_o + A_a}{A_o} \tag{15-27}$$

where A_o = original acoustical absorption present

A_a = added acoustical absorption

Acoustical absorption equals the sum of the products of the absorption coefficient of each material forming the enclosure surface and the corresponding surface area.

Sometimes materials are rated with a *noise reduction coefficient NRC*, the arithmetic average of the sound absorption coefficients of a material at 250, 500, 1,000, and 2,000 Hz.

Reverberation Time. This is the time, in seconds, that a sound pulse within a room takes to decay 60 dB, to one-millionth of its original level. Reverberation time T can be computed from the Sabine formula:

$$T = \frac{0.49V}{A} \tag{15-28}$$

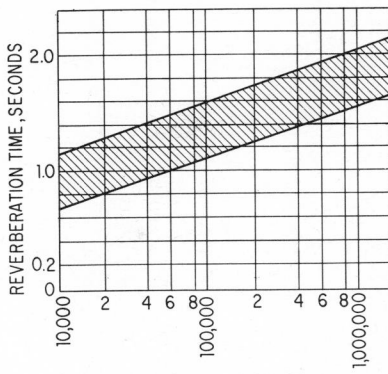

Fig. 15-27. Recommended reverberation time, indicated by shaded area, varies with size of room.

VOLUME OF ROOM, CU FT

where V = volume of room, cu ft
 A = total acoustical absorption in the room
Reverberation times falling within the shaded area in Fig. 15-27 may be considered satisfactory

Table 15-28. Impact Insulation Class of Floor Constructions

Construction	IIC
Oak flooring on ½-in. plywood subfloor,	
2 × 10 joists, ½-in. gypsumboard ceiling	23
With carpet and pad	48
8-in. concrete slab	35
With carpet and pad	57
2½-in. concrete on light metal forms,	
steel bar joists	27
With carpet and pad	50

Table 15-29. Noise Reduction and Sound Absorption Coefficients

Absorbent	Thickness, in.	Density, lb per cu ft	Noise reduction coefficient
Mineral or glass fiber blankets	½–4	½–6	0.45–0.95
Molded or felted tiles, panels, and boards	½–1⅛	8–25	0.45–0.90
Plasters (porous)	⅜–¾	20–30	0.25–0.40
Sprayed-on fibers and binders	⅜–1⅛	15–30	0.25–0.75
Foamed, open-cell plastics, elastomers, etc	½–2	1–3	0.35–0.90
Carpets	Varies with weave, texture, backing, pad, etc.		0.30–0.60
Draperies	Varies with weave, texture, weight, fullness		0.10–0.60

Absorbent	Absorption coefficient per sq ft of floor area at frequencies, Hz:					
	125	250	500	1,000	2,000	4,000
Seated audience	0.60	0.75	0.85	0.95	0.95	0.85
Unoccupied upholstered (fabric) seats	0.50	0.65	0.80	0.90	0.80	0.70

Table 15-30. Guide Criteria for Acoustical Design

Acceptable background levels		Isolation requirements between rooms					
		For sound			For impact		
Space	Background level, dBA	Room	Between and Adjacent area	Sound isolation requirement, STC	Room	Between and Room below	Impact isolation requirement, IIC
Recording studio	25	Hotel bedroom	Hotel bedroom	47	Hotel bedroom	Hotel bedroom	55
Suburban bedroom	30	Hotel bedroom	Corridor	47	Public spaces	Hotel bedroom	60
Theater	30	Hotel bedroom	Exterior	42	Classroom	Classroom	47
Church	35	Normal office	Normal office	33	Music room	Classroom	55
Classroom	35	Executive office	Executive office	42	Music room	Theater	62
Private office	40	Bedroom	Mechanical room	52	Office	Office	47
General office	50	Classroom	Classroom	37			
Dining room	55	Classroom	Corridor	33			
Computer room	70	Theater	Classroom	52			
		Theater	Music rehearsal	57			

under ordinary conditions.   For critical spaces, such as concert halls, radio studios, and auditoriums, advice of an acoustical consultant should be obtained.

Rating Systems.   The American Society for Testing and Materials (ASTM) and the Acoustical and Insulating Materials Association (AIMA) have adopted rating systems for evaluating acoustical performance of materials, such as:

Sound transmission class STC for indicating the insulation against airborne sound of partitions, floor-ceiling assemblies, and other barriers (ASTM E90 and E413).   Table 15-27 lists some typical ratings.

Impact insulation class IIC for indicating the impact insulation of floor-ceiling assemblies (ASTM RM 14-4).   See Table 15-28.

Impact noise rating INR, an alternative measure of impact insulation of floor-ceiling assemblies.   *IIC* ratings can be converted to *INR* by deduction of 51 points.

Sound absorption coefficients for indicating the absorptive efficiency of acoustical absorbents (ASTM C423).   See Table 15-29.

Noise reduction coefficients, an alternative measure of absorptive efficiency (Table 15-29).   AIMA, 205 W. Touhy Ave., Park Ridge, Ill. 60068, publishes tables of coefficients.

Acoustical Criteria.   Though governmental agencies and others have adopted design criteria to achieve desired sound reduction and insulation, the criteria are based on subjective response to acoustical parameters.   Accordingly, small deviations from such criteria may not necessarily result in unsatisfactory performance.   Usually, a tolerance of ±2.5 points from a numerical value is acceptable in practice.   Table 15-30 presents some acoustical criteria that may be used as a guide.

L. L. Beranek, "Music, Acoustics, and Architecture," John Wiley & Sons, Inc., New York; L. L. Beranek, "Noise and Vibration Control," McGraw-Hill Book Company, New York; P. D. Close, "Sound Control and Thermal Insulation of Buildings," Van Nostrand Reinhold Company, New York; M. D. Egan, "Concepts in Architectural Acoustics," McGraw-Hill Book Company, New York; C. M. Harris, "Handbook of Noise Control," McGraw-Hill Book Company, New York; P. M. Morse and K. U. Ingard, "Theoretical Acoustics," McGraw-Hill Book Company, New York; L. F. Yerges, "Sound, Noise and Vibration Control," Van Nostrand Reinhold Company, New York.)

Highway Engineering

RICHARD DUTTENHOEFFER*
Partner

BRUCE E. PODWAL
Associate

AND VIKTORAS A. KIRKYLA
Department Head, Highway-Civil Department

Parsons, Brinckerhoff, Quade & Douglas, New York, N.Y.

Highway engineering is both an art and a science. A well-designed highway should possess internal harmony—motorists should be able to see smooth lines ahead and have a clear vision of the landscape at the sides. The highway also should have external harmony—to the eye of an onlooker, the highway should fit in well with its surroundings. These requirements demand something akin to the vision and imagination of an artist, one who can visualize the three-dimensional aspects of the various combinations of horizontal and vertical curves, of cuts merging smoothly with fills, of side slopes blending with the terrain.

The highway, however, is primarily a transportation medium. It should be built to endure and to provide adequately for safe passage of vehicles. To achieve this objective, the design must adopt certain criteria for strength, safety, and uniformity. Most of these criteria have been developed over many years in the hard school of experience, while some have evolved through research and testing. Thus, certain standard formulas have been established. But these always are subject to modifications, since roads are intimately associated with the earth's surface, which seldom conforms to mathematical concepts.

16-1. Classes of Highways and Financing. Highways run the gamut from *dirt* roads to limited access highways, which to the layman are *superhighways*. Methods of financing their construction are equally varied in character.

*In charge of updating, expanding, and revising Sec. 16, Highway Engineering, prepared for the first edition by Donald R. Goodkind and Walter L. Braybooke.

The dirt road, which is usually sand-clay or gravel, provides access to isolated dwellings in sparsely settled regions. It is generally under the jurisdiction of the township. Construction is financed by local taxation, aided by grants from the county in some instances. As the use of the road increases, it may be improved by adding shoulders and stabilizing the surface with portland cement or asphalt.

The next step up the ladder is the county road. In rural counties, it has one lane of traffic in each direction and often is constructed with asphalt and aggregate surfaces on gravel bases, dry- or water-bound macadam, bituminous macadam, asphaltic concrete, or portland-cement concrete, the choice depending on the volume of traffic and the funds available. It is financed by local taxation plus some state aid. In urban counties, however, one is quite likely to find major highways, some even up to superhighway standards. In such cases, part of the financing may come from bonds issued by the county, part from state aid.

A development of the urban-county condition is a metropolitan district, such as those found in New York, Chicago, Los Angeles, and San Francisco, where the city's geographical boundaries are transcended by the combined needs of the city and its suburbs. One solution to the financing problem in such a case is establishment of an authority empowered to issue revenue bonds and amortize them from tolls.

The authority method of constructing and paying for transportation facilities generally has proved efficient when the existing road network offers little competition to a toll road. The commissioners are relatively free from the political pressures that can affect decisions at city hall or the county courthouse. Also, by utilizing an overall approach rather than a piecemeal program, which local bodies are often forced to follow for financial reasons, a multi-facility authority is able to balance profitable undertakings against those projects that are necessary but not immediately self-supporting. For example, the Port Authority of New York and New Jersey has used the tolls on the George Washington Bridge and the Holland Tunnel to help support the costs of the Goethals Bridge and the Outerbridge Crossing between Staten Island and New Jersey. When the Verrazano-Narrows Bridge replaced the ferry between Staten Island and Brooklyn, those two bridges began to pay for themselves.

Another form of financing is through the sale of general obligation bonds. This method usually involves the raising of funds for large capital expenditures through the sale of bonds that are repaid out of revenue, or through a combination of taxes and tolls. Special legislation is normally required to permit this type of financing since the credit of a governmental body is pledged.

State Roads. In most states, major highways crossing county borders are part of a state system. In a few states, all highways, including township roads, are under the state's jurisdiction. Funds for construction and maintenance of these highways come from highway-user taxes, usually license fees and taxes on gasoline. To keep pace with burgeoning traffic demands, the states' programs become larger annually. Many states have turned to bond issues to obtain sufficient, immediately available funds.

In a number of states, a continuing program to provide adequate highways is handicapped by laws that permit budgeting of state funds for only 1 year ahead. As a result, only a limited number of miles can be constructed each year. This restriction has led to two forms of financing: issuance of bonds and establishment of toll roads, financed by revenue bonds to be amortized by tolls. Often, toll road bonds are backed by the state's credit to make the investment more attractive to investors and to lower the interest rates.

Toll roads, or turnpikes, were one of the earliest classes of highways to be built in both Europe and the United States. However, use of this method of financing had diminished by the late 1930s, only to be revived after World War II. Little highway construction (other than military) had been carried out during the war and during the Depression that preceded the war. Renewal of automobile production and the end of gas rationing after the war made the existing highway systems highly inadequate to accommodate the changed travel habits and increasing volume. The quick answer was turnpikes, and these appeared in many places. The great majority of them have proved to be sound investments and have greatly expanded the network of safe high-speed highways.

Federal Assistance   Federal aid programs for highway construction are provided at several different levels. All projects that receive Federal funds are required to have an environmental assessment in accordance with Section 102 (2) (c) of the National Environmental Policy Act (Public Law 91-190). (See Article 16-18.)

Under most programs, the state-level office of the Federal Highway Administration (FHWA) monitors state operations during highway project location, design, and construction. After the project is completed, state maintenance of the project is reviewed periodically.

The programs include:

1. The regular Federal aid highway program, which has been in existence since 1916, and includes the funding for the Interstate system.

Uses and Use Restrictions.   The funds may be used for planning, surveying, research, engineering, right-of-way acquisition, new construction, reconstruction, repair improvement, roadside beautification, recreation, and rest areas.   All projects in urban areas of more than 50,000 population must be based on a continuing comprehensive planning process.

Some county and local roads and streets are eligible for Federal improvement funds, but only through state highway department initiative and action.

Secondary projects to receive federal aid must be selected in cooperation with local officials.

Also, under the Federal Aid Highway Act of 1968, the Traffic Operations Program to Increase Capacity and Safety (TOPICS) was established.   The main emphasis under this program is aimed at relieving existing traffic problems in urban areas.   The type of improvements made under the TOPICS program include signalization, channelization, pedestrian bridges, grade separations, railroad structures and relocation, lighting, off-street parking, and freeway surveillance and control.

The funding of TOPICS projects is the same as for the federal aid secondary projects.

Assistance Considerations (Formula and Matching Requirements.)   The Formula for Federal financial aid is based on such factors as ratio of population area and intercity mail-route mileage in a single state to the totals for all states, in such proportions as applied by law for the primary, secondary, and urban extension highway programs.   The Interstate formula is based on the cost to complete the remainder of the system.   The normal Federal share is 90 percent for Interstate projects and 70 percent for all other projects.   The Federal share is increased in the case of states with large areas of public lands.   The 70 percent Federal share for all other projects became effective in 1974 for the first time.

Funds also may be used for the repair of highways, roads, and streets which have suffered serious damage as the result of disaster and to replace unsafe bridges.

2. Highway Beautification—Control of Outdoor Advertising and Junkyards, Landscaping and Scenic Enhancement.

Uses and Use Restrictions.   Funds are intended to assist state highway departments (*a*) in beautifying highways and communities by controlling outdoor advertising signs, junkyards, billboards, and displays in areas adjacent to the Interstate and primary highway systems, and (*b*) in landscaping and scenic enhancement.

Application Procedure.   State highway departments should submit a program of desired projects to the state office of FHWA.

Assistance Considerations (Formula and Matching Requirements).   A project grant is based on the estimated cost of removing or screening junkyards on primary highway systems or of removing signs from noncommercial areas.   The Federal share of costs of projects for control of outdoor advertising is generally 75 percent.   Landscaping and scenic enhancement funds are determined by a grant formula based on the total funds apportioned to a state for all Federal aid highway programs with no state matching required.

Other Highway Classes.

Limited-Access Highways.   Limitation or control of access is one of the features that distinguishes the modern highway from those constructed previously.   The first use of the concept of denying the abutting-property owners unrestricted access to the roadway was probably on turnpikes and parkways.   Not being publicly owned and depending on payments from motorists, turnpikes had to bar access except at designated points.   Now, most states have adopted legislation authorizing limited or controlled access on designated highways.   These measures came too late to save some major highways; the heavy build-up of property owners adjacent to these highways, all requiring access, has virtually reduced these roads to main streets with speed limits of 25 mph.

All highways in the Interstate system are designed for limited access.   This is a major reason for their efficiency and for much better safety records than other roads in similar locations.

Parkways.   The term parkway was originally used to designate a road in a national, state, or county park.   Gradually, it came to describe an elongated park formed by a road in the design of which aesthetics was a major consideration.   The modern parkway is a major highway designed to carry a heavy volume of traffic, generally consisting of passenger cars, and with appearance considered at least as important as utility.   Since appearance is also considered in the design of expressways or freeways, parkways differ mainly in their more extensive use of plant material.

Experience has shown that some of the earlier parkways were overlandscaped. Not enough consideration was given, for example, to the growth potential of the planted trees. Also, these were planted too close to both the traveled way and exit and entrance areas, where sight distance is most important. Other hazards associated with having trees directly adjacent to the highway are slippery wet leaves on the pavement in the fall, patches of unexpected icy pavement in the shadow of the foliage, and provision of an object with which a car may collide.

Landscape architects and engineers now utilize natural growth as much as possible instead of importing nursery stock. By use of a process known as selective thinning, appearance of the natural growth is improved and traffic hazards are removed. By this means, landscapers have succeeded in bringing out the native beauty of regions, while at the same time restricting costs. The Garden State Parkway traversing the pine barrens of southern New Jersey is a good example of this.

(K. B. Woods, "Highway Engineering Handbook," McGraw-Hill Book Company, New York; "Catalogue of Federal Domestic Assistance," Executive Office of the President, Office of Management and Budget.)

16-2. Location and Route Surveys. The earliest roads were developed from trails made by either animals or natives of the region. These roads naturally followed the lines of least resistance, avoiding steep grades or treacherous swamps, and circumventing natural obstacles, such as outcroppings of rock. If a river had to be crossed, the trail detoured to a fordable location. To this day, country roads in many areas often follow winding trails. They are interesting to explore if time is no object, but can hardly be classed as efficient transportation facilities.

The first modern road builders were the Romans. Their spreading empire demanded adequate facilities for movement of their legions and the traders who followed. So the Romans constructed straight, stone-paved highways. An abundance of slave labor made it possible for the Roman road builders to take the shortest path between points to be connected. They filled in swamps, cut through hills, bridged or diverted rivers. The driving force was the taskmaster's whip. The product was good and built to last. Some present-day European roads still follow the old Roman highways, and in a few instances, vestiges of the original pavement can be found.

After the Romans came the dark ages, which, from a road-building standpoint, lasted into the Eighteenth Century. Then came such men as MacAdam and Telford, who made big improvements in road construction and whose names are still associated with the pavements of today.

Route location today involves many considerations, not the least of which are user costs, construction cost, community impact, traffic service, environmental impact, and property acquisition.

Construction cost minimization utilizing the cut-equals-fill formula is still followed to some extent, but for modern highways, other considerations often govern the choice of location and alignment. These considerations fall into numerous categories, such as traffic-desire lines, topography, community and environmental considerations, and economics. The alignment that is cheapest to construct is not necessarily the most economical.

Traffic-Desire Lines. Other things being equal, the route of a new highway should conform to the line that the major portion of the traffic would follow if it had a free choice, which is not necessarily the shortest distance between terminals. Traffic may show a preference for a long route that provides superior traffic service to communities along the way, responds to regional land-use planning, permits higher speeds, or bypasses a built-up area.

Topography. This is an important consideration in determining how satisfactory grades and horizontal alignment, suitable bridge crossings, and pleasing appearance may be incorporated into a design. Proper highway location, coordinated with overall plans for community and area development, enhances the land value of the corridor served. Carefully situated, the highway also is a superior facility for moving both local and through traffic.

Land Use. Design criteria for roads in a rural area differ from those for roads in an urban district. In metropolitan locations, design speeds are usually lower because of community restraints, and the problems of serving the higher volume of traffic are more complicated.

Appearance of Highways. This is an important design consideration which warrants a reasonable expenditure of construction funds. Scarred hillsides and untidy embankments are not acceptable by-products of road construction.

Economy. In essence, economy results from the choice of an alignment that, while serving the primary function of the highway, gives the best value for each dollar expended on right-of-way, construction, maintenance, and vehicular operations. Road-user-benefit formulas have been developed to compare the costs of alternative highway locations and their resulting benefits (Art. 16-20).

The cost of right of way must be carefully evaluated, especially in urban areas. The decision as to whether to bypass a built-up area or impact it provides grounds for controversy. The desires of the local community and environmental considerations must be weighed against the benefits to the traveling public.

Information Needed. In locating a new highway, the engineer must collect all pertinent data: aerial photographs, topographic maps (the U.S. Geological Survey and U.S. Coast and Geodetic Survey are especially valuable), county maps, town plans, zoning maps, tax maps, utility company maps, etc. Information from local planners and area inhabitants is useful, particularly in rural districts where official sources of information may not be adequate or even available. Local people can often supply valuable information on floods, swamps, landslides, and other conditions that may affect road construction. The engineer should walk the route if possible, or at least enter it at all accessible points, to gain a feel for the terrain that cannot be equaled by any other procedure. Helicopter reconnaissance is also very useful.

Highway engineers may be greatly aided by aerial photographs and electronic computers. Intelligent use of both not only provides information as to grades, river crossings, and other conditions, but also enables trained soils engineers to obtain information on subsurface soil conditions. By these means, also, a quick and economical evaluation may be made of numerous alternative locations in the search for the most desirable. Aerial photography also offers unique opportunities to make an accurate assessment of environmental factors at an early stage in the planning process. Environmental factors that are particularly amenable to investigation by aerial mapping include ecology, acoustic noise, seismic noise, hydrology, and erosion control ("Photogrammetric Analysis of Urban and Rural Environments," Transportation Research Board, Record No. 452).

Photogrammetry can be used to provide contoured topographic maps. (A scale of 200 ft to 1 in. is commonly used for route-location studies, but heavily built-up areas often require larger-scale mapping.) When the alignment has been decided, more detailed maps to larger scales can be made from lower-altitude photographs. Accurate ground control consisting of baselines, traverses, and elevations established near the highway center line on photo-identifiable objects are required to provide precise control for mapping. The controls should be planned in conjunction with aerial photography with the aid of a highly qualified photogrammetrist to insure production of a usable topographic map.

While common practice is to identify most existing details, both natural and man-made, from the air, supplementary ground surveys are usually needed to locate property lines, below-surface features, and details in heavily wooded areas.

(T. F. Hickerson, "Route Surveys and Design," McGraw-Hill Book Company, New York; P. R. Wolf, "Elements of Photogrammetry," McGraw-Hill Book Company, New York; and D. R. Lueder, "Aerial Photographic Interpretation," McGraw-Hill Book Company, New York.)

16-3. Motor Vehicles: Sizes and Design Loads. Sizes and load classifications of motor vehicles are important in highway design for the following reasons: Lane widths must accommodate

Table 16-1. Design Vehicle Dimensions

Design vehicle	Symbol	Wheelbase, ft	Front overhang, ft	Rear overhang, ft	Overall length, ft	Overall width, ft	Height, ft
Passenger car	P	11	3	5	19	7	—
Single-unit truck	SU	20	4	6	30	8.5	13.5
Single-unit bus	BUS	25	7	8	40	8.5	13.5
Semitrailer combination, intermediate	WB-40	13 + 27 = 40	4	6	50	8.5	13.5
Semitrailer combination, large	WB-50	20 + 30 = 50	3	2	55	8.5	13.5
Semitrailer-full trailer combination	WB-60	9.7 + 20.0 + 9.4* + 20.9 = 60	2	3	65	8.5	13.5

*Distance between rear wheels of front trailer and front wheels of rear trailer.

From "A Policy on Design of Urban Highways and Arterial Streets," American Association of State Highway and Transportation Officials, 1973.

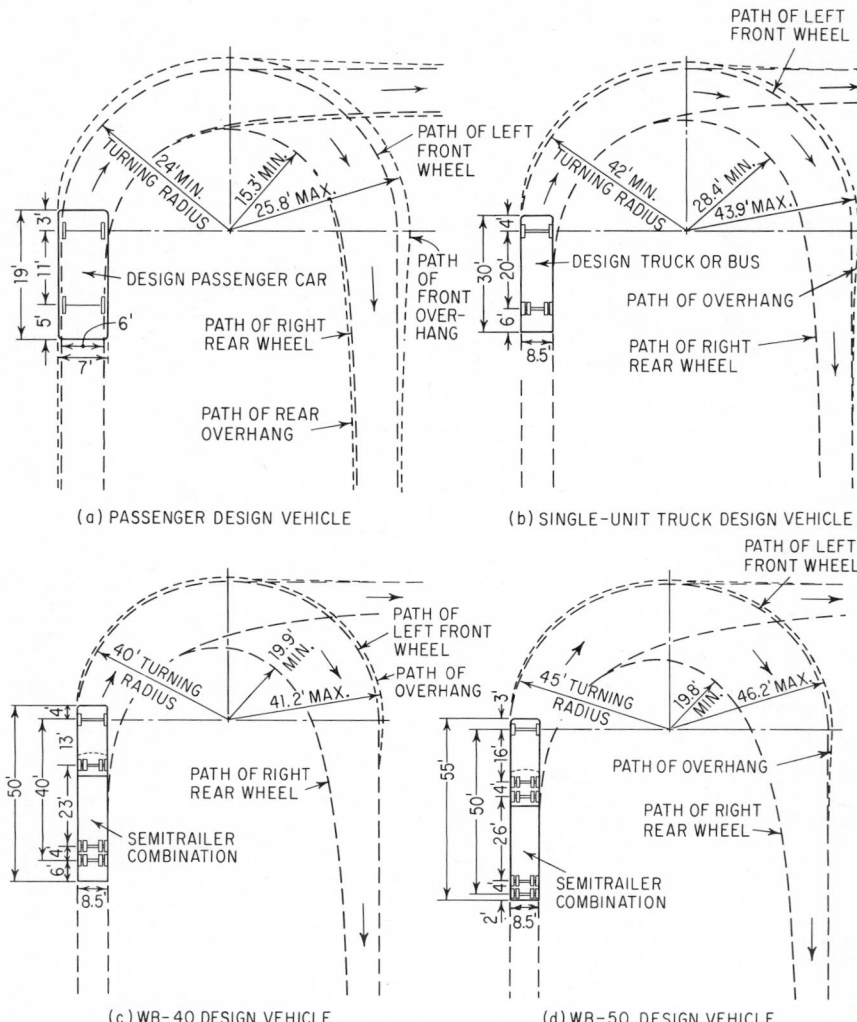

Fig. 16-1. Design vehicles for highways and minimum turning paths. ("*A Policy on Design of Urban Highways and Arterial Streets,*" *American Association of State Highway and Transportation Officials.*)

the widest vehicle (except for occasional overwide vehicles, which have to carry warning signs). Axle loads affect the choice of pavement thickness. Wheel bases influence the choice of minimum radius at intersecting roads; and heights of vehicles affect the decision on vertical clearance at underpasses.

A road designed to carry trucks will accommodate any passenger car. Thus, the design vehicles generally considered are a single-unit truck or bus; semitrailer combination, with tractor; and truck-trailer combination. For a specific road, however, design should be based on the largest vehicle expected, unless the largest vehicle would use the road so infrequently that the added cost of construction is not warranted. In these cases, it is practical to design for a lesser vehicle and allow for occasional lane infringement by the largest of vehicles. Note, however, that constraining elements, such as vertical clearance, must provide for the largest vehicle so that it is not physically restricted from using the road.

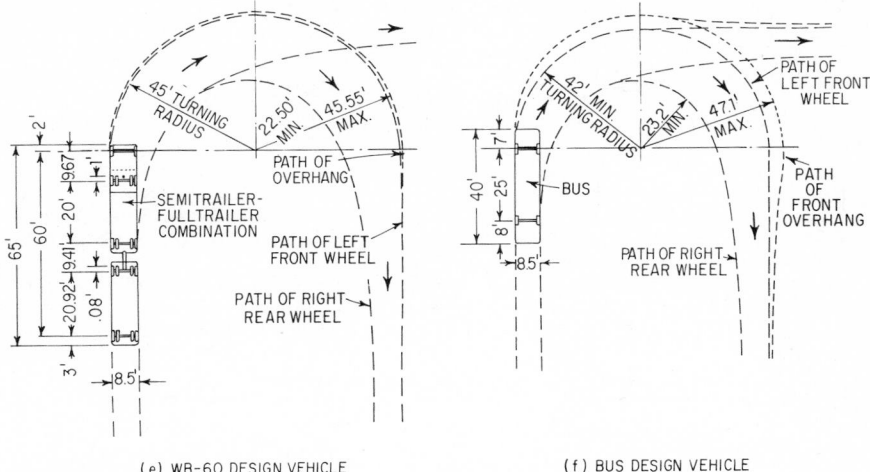

(e) WB-60 DESIGN VEHICLE   (f) BUS DESIGN VEHICLE

Fig. 16-1. *(continued)*

Characteristics of typical design vehicles are given in Table 16-1.   Lengths of semitrailers are listed in Table 16-2.   For design loads, see Fig. 17-3, Art. 17-3.

A nominal 12-ft-wide lane suffices for general use.   The infrequent extra-wide load is allowed to overlap.   Turning paths of typical vehicles are shown in Fig. 16-1.

Nearly all states limit vehicle widths to 8 ft.   Height restrictions generally range from 12.5 to 14 ft.   A 16.5-ft vertical clearance has been provided on certain major highways, mainly for missile carriers.

Table 16-2. Summary of Legally Permissible Lengths of Motor Vehicles*

Vehicle type	Legal length L compared with AASHTO Standards[†]	Number of states[‡]
Single-unit truck	$L > 40$	9
	$L = 40$	20
	$L < 40$	22
Bus	$L > 40$	10
	$L = 40$	38
	$L < 40$	3
Semitrailer	$L > 40$	48
	$L = 40$	1
	$L < 40$	2
Tractor semitrailer	$L > 55$	19
	$L = 55$	31
	$L < 55$	1

*Including Alaska, District of Columbia, and Hawaii.   In states where there are no restrictions on the length of semitrailers the maximum possible length (van bodies) is assumed to be 7 ft less than the permitted tractor-semitrailer combination length.   This 7-ft dimension consists of a bumper-to-rear-of-cab dimension of 4 ft obtainable for cab-over-engine tractor, plus 3 ft of clearance between rear of cab and nose of semitrailer.   Automobile transporter bodies may exceed these lengths when an automobile is carried above the tractor cab, a practice which is permitted in most states.

†American Association of State Highway and Transportation Officials, 1968.

‡From "Summary of Size and Weight Limits and Reciprocity Authority (by Regions) in Effect as of August 1, 1973," American Trucking Association, and "National Cooperative Highway Research Program Report 141," Transportation Research Board.

16-4. Highway Traffic Capacity. The purpose of a highway is to carry traffic; its capacity is a measure of how well it fulfills that purpose. In designing a highway, the engineer should determine whether it will have the capacity to accommodate the predicted traffic volume. But, in general, an exact mathematical answer is not yet obtainable. There are so many intangibles that designers must still, to a great extent, rely on empirical data and good judgment. A fairly accurate figure could be computed for a highway on a tangent, with flat grades, on which vehicles move at a uniform speed and uniform spacing, under ideal weather conditions. Such a situation may occasionally exist, but the predicted traffic volume would need modification for steeper grades, sharper curves, weather conditions, driving habits of the users, composition of the traffic, and other conditions differing from those assumed (Table 16-3). The following discussion should be used as a guide only. For more details, see "Highway Capacity Manual," Transportation Research Board, Washington, D.C.

Capacity is the maximum number of vehicles that can reasonably be expected to pass over a given section of a lane or a roadway in one direction, or both directions if so indicated, during a given time under prevailing roadway and traffic conditions. The period generally used is 1 hr.

Table 16-3. Capacity for Uninterrupted Flow under Ideal Conditions*

Highway type	Capacity (passenger cars per hr)
Multilane	2,000 per lane
Two-lane, two-way	2,000 total both directions
Three-lane, two-way	4,000 total both directions

Adjustments to ideal uninterrupted flow values:
  Roadway factors: Lane width, lateral clearance, shoulders
  Auxiliary lanes: Parking lanes, speed-change lanes, weaving lanes, truck climbing lanes
  Surface conditions: Alignment, grades
  Traffic factors: Trucks (equivalent to 2–3 passenger cars on level terrain, 4 on rolling terrain, 7–8 on mountainous terrain). Buses (equivalent to 2 passenger cars on level terrain, 3–4 on rolling terrain, 5–6 on mountainous terrain)
  Lane distribution, variations in traffic flow (e.g., peak hour)
  Traffic interruptions (e.g., at intersections)

*Adapted from "Highway Capacity Manual," Transportation Research Board.

Prevailing conditions include the physical features of the roadway, nature of the traffic, weather, and visibility.

Level of service for any particular roadway is a function of the volume and composition of the traffic and of the speeds attained. A road can be designed for a certain level of service at a specified volume but will operate at different levels of service as the flow varies. All roadway components, such as ramps, intersections, weaving lanes, should be designed to provide the same level of service as the main roadway.

From the driver's viewpoint, the highest level of service occurs when there is free flow. But this is obtained only when the highway is operating much below capacity (see Table 16-4).

Service volume is the maximum number of vehicles that can pass over a given section of lane or roadway during a specified period (usually 1 hr) while the operating conditions make possible the specified level of service (Table 16-4). For multilane highways, the figure is for one direction only; for other highways, both directions.

Design speed, used when deciding on the geometrics of a proposed highway, is the highest continuous speed for safe driving when that is governed solely by the design features of the highway.

Operating speed is the highest overall speed at which a driver can travel under prevailing conditions without exceeding the safe design speed.

Headway is the time interval between one vehicle and the following vehicle measured from front to front.

Spacing is the distance between successive vehicles, measured front to front.

Volume is the number of vehicles passing over a given section of a lane or roadway in a given time, which can be 1 hr or more (Art. 16-5).

Density is the number of vehicles in a unit length (usually 1 mile) of through-traffic lanes at a given instant.

Critical density occurs when the volume equals the capacity. When the density is greater or less than this, the volume will be less than the capacity.

Studies have shown that the average headway is 1½ sec. (A headway of 1½ sec is equivalent to a spacing of 132 ft at 60 mph and 66 ft at 30 mph.) If this headway were maintained uniformly by all vehicles, it would result in a volume per lane of 2,400 vehicles per hour. This has been

Table 16-4. Levels of Service and Maximum Service Volumes

a. FOR FREEWAYS AND EXPRESSWAYS UNDER UNINTERRUPTED-FLOW CONDITIONS*

| Level of service | Traffic-flow conditions | | Maximum service volume under ideal conditions, assuming 70-mph avg highway speed (total passenger cars per hr, one direction) | | |
	Description	Operating speed, mph	4-lane freeway (2 lanes one direction)	6-lane freeway (3 lanes one direction)	8-lane freeway (4 lanes one direction)
A	Free flow	≦60	1,400	2,400	3,400
B	Stable flow (upper speed range)	≦55	2,000	3,500	5,000
C	Stable flow	≦50	2,300–3,000	3,700–4,800	5,100–6,600
E†	Unstable flow	30–35‡	4,000‡	6,000‡	8,000‡

b. FOR MULTILANE HIGHWAYS, UNDIVIDED OR WITHOUT ACCESS CONTROL, UNDER
UNINTERRUPTED-FLOW CONDITIONS (NORMALLY REPRESENTATIVE OF
RURAL OPERATION)

| Level of service | Traffic-flow conditions | | Maximum service volume under ideal conditions, assuming 70-mph avg highway speed (total passenger cars per hr, one direction) | | |
	Description	Operating speed, mph	4-lane freeway (2 lanes one direction)	6-lane freeway (3 lanes one direction)	Each additional lane
A	Free flow	≦60	1,200	1,800	600
B	Stable flow (upper speed range)	≦55	2,000	3,000	1,000
C	Stable flow	≦45	3,000	4,500	1,500
E†	Unstable flow	30‡	4,000	6,000	2,000

c. FOR TWO-LANE HIGHWAYS UNDER UNINTERRUPTED-FLOW CONDITIONS
(NORMALLY REPRESENTATIVE OF RURAL OPERATION)

| Level of service | Traffic-flow conditions | | Max service volume under ideal conditions, assuming 70-mph avg highway speed (passenger cars, total, both directions, per hr) |
	Description	Operating speed, mph	
A	Free flow	≦60	400
B	Stable flow (upper speed range)	≦50	900
C	Stable flow	≦40	1,400
E†	Unstable flow	30‡	2,000

*Adapted from "Highway Capacity Manual," Transportation Research Board.
†At capacity.
‡Approximately.

observed for short periods on certain freeways under ideal conditions. However, vehicles do not move with uniform spacing but tend to form groups. Observation has shown that generally about two-thirds of the vehicles are likely to be spaced at, or less than, the mean headway. Also, in general, the higher the type of highway, the shorter the headway.

For a highway to provide an acceptable level of service, the service volume must be lower than the roadway capacity. Given a choice of several levels of service below capacity, each related to an operating speed, the engineer should select the type of operation most suitable for the local conditions. In making the choice, the engineer should consider the location (rural or urban); peak-hour conditions, if such exist; to what extent the drivers' freedom to maneuver must be restricted; built-in restrictions or interruptions to free flow; safety; and other significant factors.

Estimating traffic flow at intersections is difficult. There are so many factors involved (Table 16-5) that it is not feasible to make a general statement embracing ideal conditions for uninterrupted flow. For each intersection, conditions on the adjacent portions of the approach roadways must be studied and evaluated.

Table 16-5. Factors Affecting Intersection Capacity and Levels of Service*

Physical and operating conditions:
  Width of approach
  One-way or two-way operation
  Parking conditions
Environmental conditions:
  Load factor
  Peak-hour factor
  Metropolitan area population
  Location within metropolitan area
Traffic characteristics:
  Turning movements
  Trucks and through buses
  Local transit buses
Control measures:
  Traffic signals
  Marking of approach lanes

*"Highway Capacity Manual," Transportation Research Board.

The flow of traffic on a ramp also depends on many variables. Each ramp requires individual study. However, for a well-designed ramp with ample entrance and exit terminals, the lane capacity is about 1,200 passenger vehicles per hour (Art. 16-13). This figure must be reduced if there is a significant proportion of commercial vehicles.

16-5. Collection of Traffic Data. The intelligent collection of traffic data and, especially, its evaluation are the province of a specialist, the traffic engineer. The traffic engineer is asked to predict the traffic volumes to be used in the design of a new facility or improvement of an existing one. The volume figures required are generally for 20 years after completion of project construction.

For data, the traffic engineer is dependent on traffic counts on existing roads, roadside interviews, and polls of potential users. Traffic counts are made by automatic devices on roads serving the areas contiguous to the proposed alignment. For interviews, drivers are halted at predetermined points and asked about such things as the origins and destinations of their journeys and the purpose and frequency of such trips. When this method is used for a city, the survey points are located on roads entering the city (an external cordon) and sometimes supplemented by an internal cordon embracing the central business district. In the polls, a sampling is made of car owners in zones from which the users of the new facility are most likely to come. The polls are made by mail or by home interviews. The questions are framed to elicit such information as routes the people would prefer if they were available (desire lines), their origins and destinations, and frequency and length of trips. The answers are summarized and analyzed, usually with the aid of an electronic computer. Charts can then be drawn (Fig. 16-2) showing the desire lines. Such a chart is useful in determining the most logical alignment corridor for a new expressway, particularly in an urban area.

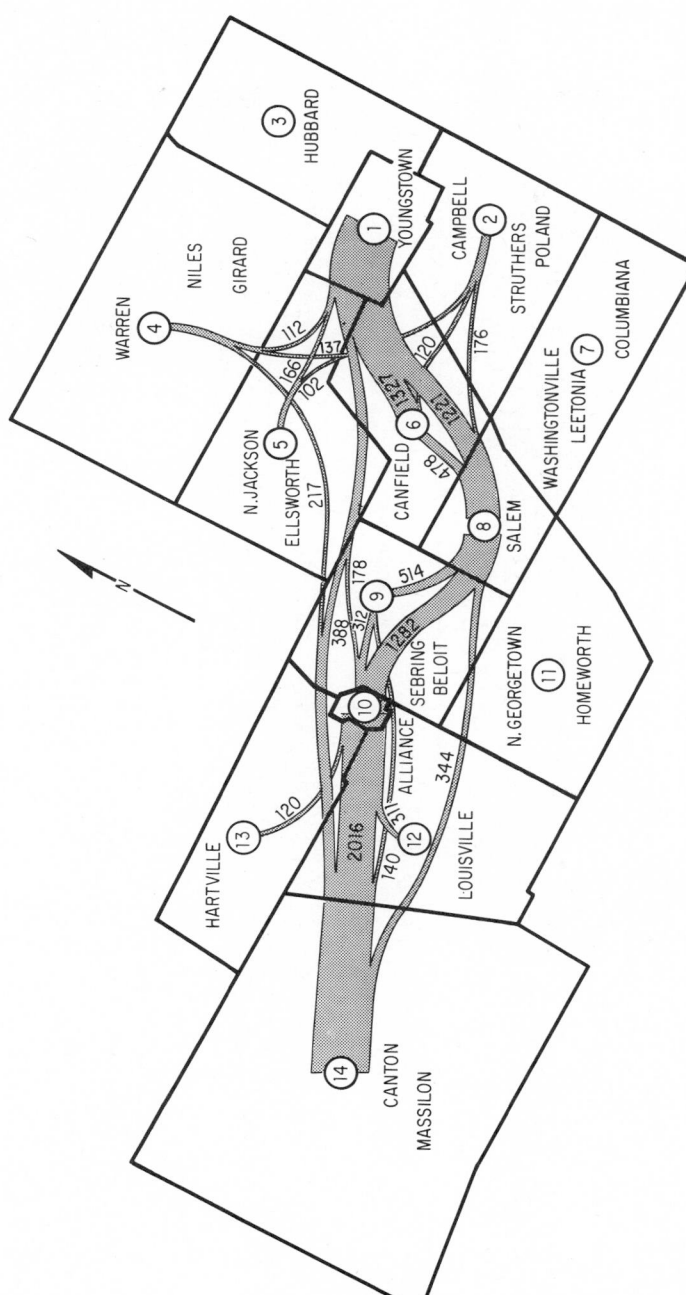

Fig. 16-2. Map indicates with broad lines (desire lines) the routes people prefer. Width of line is proportional to the number of vehicles that would use the route on an average weekday. (*Department of Highways, State of Ohio.*)

Table 16-6. Geometric Design Standards for Interstate and Primary Routes*

Area	Terrain	Design class	Total	Express service	Congestion pt. traffic	Pavement (travel-width)	Travel lane	Parking lanes in addition to travel lanes	Shoulder or clearance lanes	Maximum grade, %	Maximum curve, deg	Min stopping sight distance 44 in. to 4 in.	% mileage with less than min passing sight distance	Speed Design	Speed Running peak hrs.	Truck %	Distance clear of fixed objects from edge of through travel lane	Notes	Design class
Urban	Level or rolling	1	4,400	3,600 800	5,400 1,500	72† 24 & 24	12 12	No prkg 10	12R–6L None	4 7	5 12½	475 275		60 40	35 30	0–10 0–10	30 25	Freeway, full control of access, acceleration lanes, etc., no crossings at grade. 24-ft service roads, parking lane/one side.	1
		2	3,400	2,600 800	3,400 1,500	72† 24 & 24	12 12	No prkg 10	12R–6L None	5 7	5 12½	475 275		60 40	35 30	0–10 0–10	30 25	Expwy, partial control of access, provide structures for cross roads where capacity deficiencies warrant; 24-ft service roads, parking lane one side.	2
		3	3,050	2,400 650	4,200 1,200	48† 22 & 22	12 11	No prkg 10	12R–6L None	4 7	5 12½	475 275		60 40	35 30	0–10 0–10	30 25	Freeway, full control of access, acceleration lanes, etc. 22-ft service roads, parking lane one side.	3
		4	1,600		2,300	48†	12	12		5	8	275		40	30	0–10	25	Urban surface arterial, cross roads, right and left turn lanes, and signals. Parking restricted at intersections.	4
		5	1,100		1,500	48	12	No prkg	2	5	8	275		40	30	0–10	25	Standard arterial st., no service roads, with cross roads and signals. Parking prohibited, 4-ft flush median, 2-ft clearance lanes at curb.	5
		6	900		1,400	48	12	10		5	8	275		40	25	0–10	25	Standard arterial street, parking lanes provided.	6
		7	550 or less		800	24	12	10		5	8	275	30	40	25	0–10	25	Standard arterial street, parking lanes provided.	7

Region	Terrain	No.															No.	Remarks
Rural	Level or rolling	1	3,000	4,000	72†	12	12R–6L	12R–6L	3	3½	600		70	50	0–10	30	1	Freeway, full control of access.
		2	2,000	2,700	48†	12	12R–6L	12R–6L	4	3½	600		70	50	0–10	30	2	Freeway, full control of access.
		3	1,500	2,200	48	12	12R–6L	12R–6L	5	5	475		60	45	10–20	30	3	Control of access not possible.
		4	1,100	1,800	48	12	12	5	5		475		60	45	10–20	30	4	
		5	500	1,300	24	12	12	5	5		475	20	60	45	10	30	5	
		6	300	800	24	12	12	7	7½		350	60	50	40	10	30	6	
		7	200	600	24	12	10	7	7½		350	70	50	40	10	30	7	
		8	70		24	12	6	8	7½		350		50			30	8	
		9	20			12	6	10	12½		275		40			25	9	
	Hilly or mountainous	5M	350	800	24	12	12	7	7½		350	70	50	40	10	30	5M	Provide truck climbing lane on critical grades.
		6M	250	600	24	12	12	8	9		275	80	40	35	10	25	6M	Provide minimum of one passing opportunity per mile.
		7M	125	400	24	12	10	10	10		275	90	40	35	10	25	7M	Provide minimum of one passing opportunity per mile.
		8M	60		24	12	6	10	12½		275		40			25	8M	
		9M	20		24	12	6	10	12½		275		40			25	9M	

General notes:

1. Shoulders indicated by R and L are right and left, respectively, on divided highways in the direction of traffic.

2. The length of passing sight distance shall be determined from Table III-4 of the 1965 AASHTO Rural Policy. Height of object of 4 in. and of eye of 44 in. should be used when computing passing sight distance.

*"Highway Design Manual (Vol. 1) Facilities Design Subdivision," New York State Department of Transportation.

†Divided highway.

Traffic volumes derived from the various surveys are generally expressed as *annual average daily traffic* (AADT) volumes.   Day-by-day volumes fluctuate widely, however.   Hence, the AADT should not be used alone in figuring the required capacity of a highway.

One good criterion for design is the traffic volume for the thirtieth highest hour of a year (30HV).   (There will be only 29 hr in the year during which traffic volume exceeds the 30HV.)   Observations indicate that 30HV and AADT are related.   The ratio 30HV/AADT, however, varies not only with the class of highway but also with other peculiar local conditions.   It has also been found that this ratio generally decreases as the AADT increases ("Highway Capacity Manual," Transportation Research Board).   For preliminary design purposes only, an average value of 0.14 for rural highways and 0.11 for urban highways can be used for 30HV/AADT.

With the design hourly volume established for two-way operation, the volume must now be divided into the fractions going each way.   These proportions vary with the location of the highway, its predominant use, and sometimes the time of day at which the 30HV occurs.   For example, it is possible for a commuter expressway into a city to have two 30HVs, with the major fraction of the traffic in opposite directions.

The final step in making a prediction of future traffic is to increase the design volumes by the growth expected over 20 years, or some other specified period.

Generally, it is not possible to forecast precisely all the large-scale and local behavioral, technological, economic, and social factors that will affect day-by-day and hour-by-hour traffic movements.   Yet, the best possible estimates of probable usage in the design year are necessary to aid in determining whether the project design is rational with respect to probable average or typical needs and to provide a basis for systematic comparison of the effects on traffic of the various alternative project designs under consideration.   Once the probable design-year traffic is established, traffic flow diagrams can be drawn and used as a basis for determining the number of lanes, location of interchanges, and other design features dependent on the traffic patterns and forecasts.

16-6. Geometric Design Standards.   (See also Art. 16-17.)   The design designation of a highway is an expression indicating the major controls or services for which a given highway is designed.   The geometric criteria, as well as other design criteria that govern the design of a highway, are to a great extent dependent on the type of roadway and the projected traffic the highway will service.   The design class of the highway, therefore, is usually based on the forecast of traffic volume 20 years from the time construction ends.

Highway Design Classifications.   For purposes of simplicity, the highway design classifications can be categorized into two major groups: primary and interstate routes, and secondary routes.   These categories can further be subdivided into more distinct classifications on the basis of the projected traffic, location (urban vs. rural), terrain, etc.

Primary and Interstate Routes.   These routes are arterial highways intended for long-distance, high-speed traffic, generally with opposing traffic movements separated by a dividing median and with full control of access.   They are characterized by high design and safety standards including elimination of all grade crossings.

The Interstate Highway System joins principal metropolitan areas, cities, and industrial centers.   In addition, the system serves national defense.   At suitable border points, the system also connects with routes of continental importance.

The primary system consists of certain principal state highways, which are usually through routes between major population centers.

Secondary Routes.   These include state, county, and local roads, which are feeders to the primary and interstate systems.   Because of the generally lower traffic volumes conveyed on these routes, the design standards are not as stringent as for the primary and interstate routes.

Geometric Design Standards—Primary and Interstate Routes.   Table 16-6 illustrates the geometric design standards for various classifications of interstate and primary routes as adopted by the New York State Department of Transportation (NYS DOT).   [Most states have their own design standards, which in some instances vary from the standards established by the American Association of State Highway and Transportation Officials (AASHTO), 341 National Press Building, Washington, D.C. 20004.   Such variations are usually acceptable if they are *better than* the AASHTO standards, thereby yielding a higher grade of design from the standpoint of operations, safety, etc.]   In Table 16-6, NYS DOT has adopted a relationship of height of eye to height of object of 44 in. to 4 in. for determining the minimum stopping sight distance, while AASHTO standards for sight distance are based on a 3.75-ft (45-in.) height of eye and a 6-in. height of object.

Number and Width of Lanes.   Normal width of individual lanes is usually 12 ft.   It is desirable, however, to add 1 or 2 ft to the normal lane width for a lane adjacent to a nonmountable curb.   The number of lanes in each direction is determined on the basis of the design traffic volume.

The usual practical limit to the number of contiguous lanes in one direction is three.   For special situations, however, use of four lanes in one direction is justified.   Where traffic requires more than three lanes in one direction and widening of the road is feasible, dual roadways should be constructed in each direction, with shoulders for each roadway and a divider or, preferably, a mall separating the adjacent pairs.   This dual-dual design, however, raises problems at points where vehicles must move across the divider, as at exits and entrances.   An alternative solution may be an independent parallel highway, if the cost of the necessary right-of-way is not prohibitive.

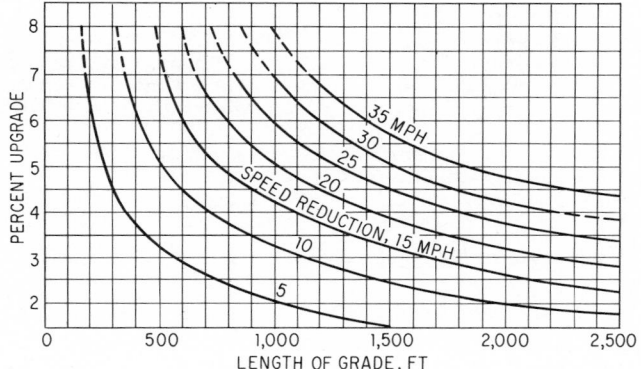

Fig. 16-3. Critical lengths of grade for design.   The curves are based on a heavy truck (400 lb per hp). ("*A Policy on Geometric Design of Rural Highways,*" *American Association of State Highway and Transportation Officials.*)

Table 16-7. Effect of Trucks on Highway Capacity on Grades*

Percentage of trucks during design hour	Minimum design hourly volume suggested for determination of climbing lanes	
	Main highways	Other highways
5	450	500
10	300	400
20	200	300
30	150	200

*"A Policy on Geometric Design of Rural Highways," American Association of State Highway and Transportation Officials.

Where the ratio of trucks to total traffic is large enough to be significant, the efficiency of the road may be reduced by slowing of trucks on long, steep upgrades (Fig. 16-3 and Table 16-7). In such cases, addition of an extra *climbing* lane for truck use is advisable.

Shoulders.   The shoulder is the portion of the roadway contiguous with the traveled way used for accommodation of stopped vehicles, for emergency use, and for lateral support of the base and surface courses of the roadway.   The shoulder pavement should be capable of sustaining the weight of a vehicle without appreciable rutting.

Minimum width of shoulder is 10 ft.   A 12-ft width, however, is desirable, where practical. Where site conditions, such as mountainous terrain, dictate a narrower shoulder width to restrict construction costs, a width of 6 to 8 ft can be used.

Medians.   Highway medians separate opposing streams of traffic.   They can be either raised or flush, as indicated in Fig. 16-4.   Where conditions permit, the desirable minimum width of median is about 36 ft in rural areas and 16 ft in urban areas.

The minimum width of median is generally considered to be about 8 ft (including the inside shoulders).   Use of median barriers should be considered where narrow medians (less than 30 ft) are used, or where possibilities for crossing the median are present.

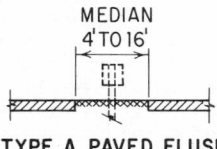

TYPE A. PAVED FLUSH

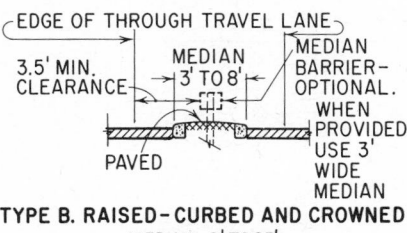

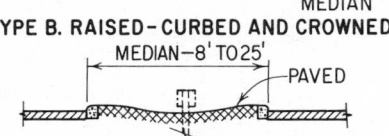

TYPE B. RAISED-CURBED AND CROWNED

TYPE C. RAISED-CURBED AND DEPRESSED

TYPE D. FLUSH—SWALE
WHEN WIDTH OF MEDIAN DOES NOT
EXCEED 36', PAVE AND USE MEDIAN
BARRIER, MAX. SLOPE = 1 : 6

TYPE E. INDEPENDENT ROADWAYS
WITH NATURAL—GROUND MEDIAN

Fig. 16-4. Types of medians, shown in cross section. (*New York State Department of Transportation.*)

Curvature, Superelevation, and Stopping Sight Distances. The relationship between design speeds, curvatures, grades, and stopping sight distances is generally correlated on the basis of criteria established by the American Association of State Highway and Transportation Officials, as outlined in "A Policy on Geometric Design for Rural Highways." Table 16-6 gives, as a sample, the standards adopted by the New York State Department of Transportation for the relationships between design speeds, maximum curvatures, stopping sight distances, and maximum grades for various roadway classifications.

Where superelevation is not required, the normal pavement cross slopes are as indicated in Table 16-8.

Table 16-8. Normal-Pavement Cross Slopes*

Surface type	Range in rate of cross slope	
	In. per ft	Ft per ft
High	⅛ –¼	0.01 –0.02
Intermediate	³/₁₆ –⅜	0.015–0.03
Low	¼ –½	0.02 –0.04

*"Geometric Design Standards for Highways Other than Freeways," American Association of State Highway and Transportation Officials.

Superelevation (banking) of the pavement is required to counteract the excess centrifugal force that is not counteracted by friction between the roadway and the tires. There is a practical limit, however, to the amount of superelevation that can be applied because of snow and ice considerations. If the pavement were fully banked, that is, given the superelevation corresponding to design speed, slow-moving vehicles might slide downward across the pavement when the surface was slippery from rain, snow, or ice. Hence, motorists will have to slow down around some curves.

Maximum superelevation rates vary with geographical location and type of area, urban or rural. These rates also are to a large degree controlled by climatic conditions, such as likelihood of rain, snow, or ice. Typical limits for various design speeds, minimum radii, and superelevation rates and transitions are given in Table 16-9.

Table 16-9. Superelevation e, In. per Ft of Pavement Width, and Spiral Length L, Ft, for Horizontal Curves of Highways*

		V=30 MPH			V=40 MPH			V=50 MPH			V=60 MPH			V=65 MPH			V=70 MPH			V=75 MPH			V=80 MPH		
D_c	R	e	L_s 2 LANE	4 LANE	e	2 LANE	4 LANE	e	2 LANE	4 LANE	e	2 LANE	4 LANE	e	2 LANE	4 LANE	e	2 LANE	4 LANE	e	2 LANE	4 LANE	e	2 LANE	4 LANE
0°15'	22918	NC	0	0	NC	0	0	NC	0	0	NC	0	0	NC	0	0	NC	0	0	NC	0	0	RC	240	240
0°30'	11459'	NC	0	0	NC	0	0	NC	0	0	RC	175	175	RC	190	190	RC	200	200	.022	220	220	.024	240	240
0°45'	7639'	NC	0	0	NC	0	0	RC	150	150	RC	175	175	.025	190	190	.029	200	200	.032	220	220	.036	240	240
1°00'	5730'	NC	0	0	RC	125	125	.021	150	150	.029	175	175	.038	190	200	.038	200	200	.043	220	220	.047	240	240
1°30'	3820'	RC	100	100	.021	125	125	.030	150	150	.040	175	175	.046	190	200	.053	200	240	.060	220	290	.065	240	320
2°00'	2865'	RC	100	100	.027	125	125	.038	150	150	.051	175	210	.057	190	250	.065	200	290	.072	230	340	.076	250	380
2°30'	2292'	.021	100	100	.033	125	125	.046	150	170	.060	175	240	.066	190	290	.072	230	330	.078	250	370	.080	260	400
3°00'	1910'	.025	100	100	.038	125	125	.053	150	190	.067	180	270	.073	210	320	.073	220	330	.080	260	380	.080	260	400
3°30'	1637'	.028	100	100	.043	125	140	.058	150	210	.073	200	300	.077	220	330	.080	240	380	.080	250	380	D Max. = 2.5°		
4°00'	1432'	.052	100	100	.047	125	150	.063	150	230	.077	210	310	.079	230	340	.080	240	360	D Max. = 3.0°					
5°00'	1146'	.038	100	100	.055	125	170	.071	170	260	.080	220	320	.080	230	350	D Max. = 3.5°								
6°00'	955'	.043	100	120	.061	130	190	.077	180	280	.080	220	320	D Max. = 4.5°											
7°00'	819'	.048	100	130	.067	140	210	.079	190	280	D Max. = 5.0°														
8°00'	716'	.052	100	140	.071	150	220	.080	190	290															
9°00'	637'	.056	100	150	.075	160	240	D Max. = 7.5°																	
10°00'	573'	.059	110	160	.077	160	240																		
11°00'	521'	.063	110	170	.079	170	220																		
12°00'	477'	.066	120	180	.080	170	230																		
13°00'	441'	.068	120	180	.080	170	250																		
14°00'	409'	.070	130	190	D Max. = 12.5°																				
16°00'	358'	.074	130	200																					
18°00	318'	.077	140	210																					
20°00	286'	.079	140	210																					
22°00	260'	.080	140	220																					
		.080	140	220																					
		D Max. = 23.0°																							

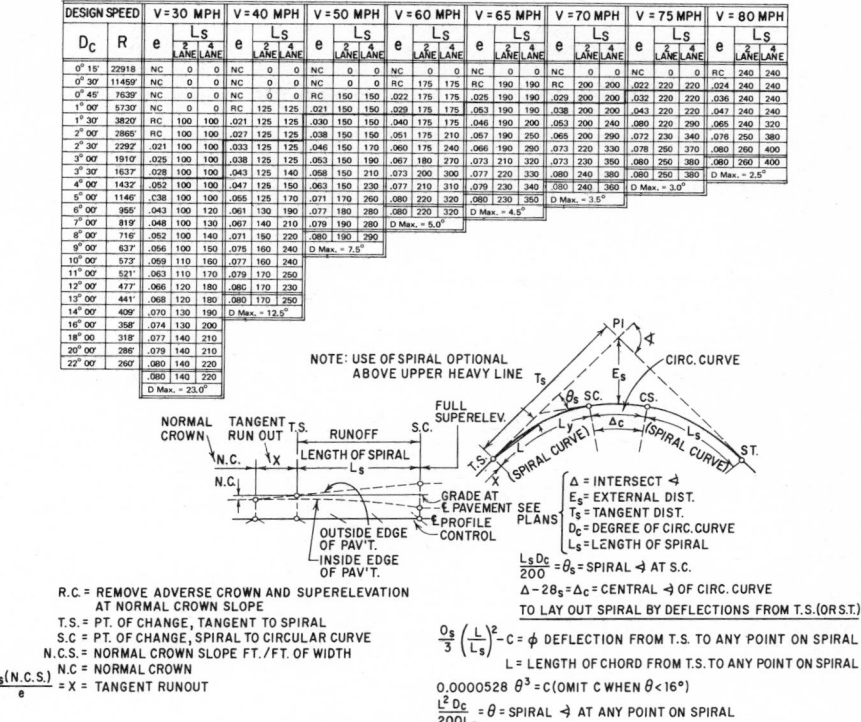

NOTE: USE OF SPIRAL OPTIONAL ABOVE UPPER HEAVY LINE

R.C. = REMOVE ADVERSE CROWN AND SUPERELEVATION AT NORMAL CROWN SLOPE
T.S. = PT. OF CHANGE, TANGENT TO SPIRAL
S.C = PT. OF CHANGE, SPIRAL TO CIRCULAR CURVE
N.C.S. = NORMAL CROWN SLOPE FT./FT. OF WIDTH
N.C = NORMAL CROWN
$\dfrac{L_s(\text{N.C.S.})}{e}$ = X = TANGENT RUNOUT

Δ = INTERSECT ∢
E_s = EXTERNAL DIST.
T_s = TANGENT DIST.
D_c = DEGREE OF CIRC. CURVE
L_s = LENGTH OF SPIRAL

$\dfrac{L_s D_c}{200} = \theta_s$ = SPIRAL ∢ AT S.C.

$\Delta - 2\theta_s = \Delta_c$ = CENTRAL ∢ OF CIRC. CURVE

TO LAY OUT SPIRAL BY DEFLECTIONS FROM T.S.(OR S.T.)

$\dfrac{\theta_s}{3}\left(\dfrac{L}{L_s}\right)^2 - C = \phi$ DEFLECTION FROM T.S. TO ANY POINT ON SPIRAL

L = LENGTH OF CHORD FROM T.S. TO ANY POINT ON SPIRAL

$0.0000528\ \theta^3 = C$ (OMIT C WHEN θ<16°)

$\dfrac{L^2 D_c}{200 L_s} = \theta$ = SPIRAL ∢ AT ANY POINT ON SPIRAL

CURVE SUPERELEVATION AND TRANSITION

*From "Highway Design Manual" (Vol. 1), Facilities Design Subdivision, New York State Department of Transportation.
†C is correction, min, always minus.

When superelevation is used, there must be a transition length in which the change from normal crown section to fully banked section occurs gradually. A good method for providing the transition is to insert a spiral curve between the tangent and the start of the circular curve. The spiral provides a comfortable path for drivers since the radius decreases gradually while the superelevation increases as the driver approaches the sharper curvature. A similar but reversed procedure is followed for the end of the curve. (Compound curves closely simulating the spiral may be substituted for a spiral.) If a spiral is not used, it is good practice to divide the required transition in the proportion of two-thirds on the tangent and one-third on the curve.

Clearances. Where conditions permit, the desirable horizontal clearance from the edge of a through traffic lane to an obstruction should be 30 ft. Where this is not possible, the horizontal clearance from the edges of through traffic lanes to the faces of walls, abutments, or piers should be the usable shoulder width, but not less than 8 ft on the right and 4½ ft on the left. On overpasses, the clearance should be at least 3½ ft between the edges of through traffic lanes and parapets or rails.

Vertical clearances should be at least 14 ft over interstate highways. On certain routes designated by state and Federal agencies, a minimum vertical clearance of 16 ft over interstate highways is required. In each case, an allowance (generally 6 in.) should be added for future resurfacing.

Side Slopes. Cut or fill slopes should be flattened and liberally rounded to blend with the topography and be consistent with the highway class. Safety against rollover accidents, effective erosion control, low-cost maintenance, and adequate drainage of the subgrade are largely dependent on proper shaping of the side slopes.

Slopes should be flattened wherever possible consistent with the availability of excavated material.

Where practical, the side slopes should be 4 (horizontal) on 1 (vertical), or flatter. Side slopes should not be steeper than 2 : 1, except when special stabilization treatments are used or where rock excavation or embankment is encountered.

Right of Way. The highway right of way should be as wide as feasible, consistent with location and cost. In every case, the right of way should, however, be not less than that required for all elements of the design cross section and the appropriate border areas. Such border areas may vary from a few feet in highly developed areas to 100 ft or more in rural areas, where lower land values prevail.

Geometric Design Standards—Secondary Routes. Table 16-10 illustrates geometric design standards for secondary highways of different classifications, as adopted by the New York State Department of Transportation. [The same qualification regarding compatibility with American Association of State Highway and Transportation Officials (AASHTO) standards as given previously for primary and interstate routes applies here.]

Table 16-10. Geometric Criteria for Secondary Roads*

Design class	Forecast volume, design-hour, one-way	Design speed, mph	Width, ft		Min stopping sight distance, 44 in. to 4 in.	Maximum grade, %	Maximum curve, degrees	% mileage with less than min passing sight distance	% trucks	Distance, ft, clear of fixed objects from edge of through travel lanes
			Pavement	Shoulders						
RS5–30		30	24	10	200	10	23.0	90	10	20
RS5–40	240–500	40	24	10	275	10	12.5	80	↑	25
RS5–50		50	24	10	350	7	7.5	60		30
RS5–60		60	24	10	475	5	5.0	20		30
RS6–30		30	22	10	200	10	23.0	90		20
RS6–40	160–240	40	22	10	275	10	12.5	80		25
RS6–50		50	24	10	350	7	7.5	60		30
RS6–60		60	24	10	475	5	5.0	20		30
RS7–30		30	22	8	200	10	23.0	90		20
RS7–40	100–160	40	22	8	275	10	12.5	80		25
RS7–50		50	22	8	350	7	7.5	60		30
RS7–60		60	22	8	475	5	5.0	20		30
RS8–30		30	22	6	200	10	23.0	90		20
RS8–40	60–100	40	22	6	275	10	12.5	80		25
RS8–50		50	22	6	350	7	7.5	60		30
RS8–60		60	22	6	475	5	5.0	20		30
RS9–30		30	22	6	200	10	23.0	90		20
RS9–40	0–60	40	22	6	275	10	12.5	80		25
RS9–50		50	22	6	350	7	7.5	60	↓	30
RS9–60		60	22	6	475	5	5.0	20	10	30

*From "Highway Design Manual" (Vol. 1), Facilities Design Subdivision, New York State Department of Transportation

The following standards are based to a large extent on data given in "Geometric Design Standards for Highways Other than Freeways," "Geometric Design Guide for Local Roads & Streets," "A Policy on Geometric Design of Rural Highways," and "A Policy on Arterial Highways for Urban Areas," American Association of State Highway and Transportation Officials, Washington, D.C.

Number and Width of Lanes.   The number of lanes in each direction is determined on the basis of the design traffic volume and the desired level of service.   Where warranted, special lanes for turning vehicles should be provided.   Parking areas (generally 8 to 10 ft wide) are necessary on certain urban highways and major streets.

Normal lane width of 12 ft is desirable (with a 1- or 2-ft additional offset on curbed sections).   In some instances, however, such as in heavily developed areas where right of way is limited, lane width can be reduced to 11 ft and even, although this is undesirable, to 10 ft.

Shoulders.   Minimum usable width of shoulder should be at least that given in Table 16-10.

Medians.   Where there are more than two lanes, medians are desirable.   The same considerations that were outlined previously for primary routes would generally apply, with variations to suit individual site conditions.

Sight Distance.   For determining stopping sight distance (a vertical-curve factor for divided roadways), assume the height of eye is 3.75 ft and height of object 0.5 ft above grade.   For determining passing sight distance (a vertical-curve factor for undivided roadways), assume the height of eye is 3.75 ft and height of object is 4.5 ft above grade.   Table 16-11 gives values of a coefficient K by which the difference in grades at a change in highway slope may be multiplied to obtain the length, ft of vertical curve that will provide the minimum sight distance (see also Art. 16-7).

For a more in-depth discussion of stopping sight distance criteria, see AASHTO, "A Policy on Design Standards for Stopping Sight Distance."

Table 16-11. Minimum Sight Distances*

Design speed, mph	30	40	50	60	65	70	75	80
Stopping sight distance								
Minimum, ft	200	275	350	475	550	600	675	750
K value† for:								
Crest vertical curve	28	55	85	160	215	255	325	400
Sag vertical curve	35	55	75	105	130	145	160	185
Desirable, ft	200	300	450	650	750	850	950	1,050
K value† for:								
Crest vertical curve	28	65	145	300	400	515	645	780
Sag vertical curve	35	60	100	155	185	215	240	270
Passing sight distance								
Passing distance, ft, 2-lane	1,100	1,500	1,800	2,100	2,300	2,500	2,600	2,700
K value† for:								
Crest vertical curve	365	686	985	1,340	1,605	1,895	2,050	2,210

*From "A Policy on Design Standards for Stopping Sight Distance," American Association of State Highway and Transportation Officials.

†K value is a coefficient by which the algebraic difference in grade may be multiplied to determine the length, ft, of vertical curve that will provide minimum sight distance.

Curvature and Superelevation.   Maximum horizontal curvatures for various design speeds and rates of roadway superelevation are presented in Table 16-12.

Superelevations for rural highways generally should not exceed 0.10 or 0.12 ft per ft.   If snow or ice conditions prevail, superelevations should not exceed 0.06 or 0.08 ft per ft.   For urban highways, superelevation should not exceed 0.06 or 0.08 ft per ft (0.06 ft per ft is usually used as a design maximum).   Table 16-13 gives minimum lengths for superelevation runout in two-lane pavements.

Maximum Grades.   Maximum grades for various design speeds and types of terrain are listed in Table 16-14.

Clearances.   Vertical clearance at underpasses should be at least 14 ft over the entire roadway width, including the shoulders.   An allowance of 6 in. should be added for resurfacing.

Horizontal clearances, both right and left on structures, from edge of traffic lane to face of parapet or railing, should be equal to the width of the entire roadway section, including the usable shoulders.   In special cases, such as on extremely long structures or in tunnels, reduced clearances may be permitted for reasons of economy. At underpasses, the lateral distance from the edge of through traffic lanes to faces of walls, piers, or abutments should be at least the usable shoulder width, but not less than 6 ft on the right and 4.5 ft on the left.

Side Slopes. The basic considerations for desirable and maximum side-slope requirements for secondary roads are the same as those described previously for primary routes, with appropriate adjustments to reflect the roadway classification, economics, site conditions, etc. (Some states have developed their own standards, which relate to a system of highway classifications dependent on anticipated traffic volumes.)

Table 16-12. Maximum Horizontal Curvature for Highways Other than Freeways*

Design speed, mph	Max e †	Total $(e + f)$ †	Min radius, ft	Max degree of curve	Max degree of curve (rounded)
30	0.06	0.22	273	21.0	21.0
40	0.06	0.21	508	11.3	11.5
50	0.06	0.20	833	6.9	7.0
60	0.06	0.19	1,263	4.5	4.5
70	0.06	0.18	1,815	3.2	3.0
30	0.08	0.24	250	22.9	23.0
40	0.08	0.23	464	12.4	12.5
50	0.08	0.22	758	7.6	7.5
60	0.08	0.21	1,143	5.0	5.0
70	0.08	0.20	1,633	3.5	3.5
30	0.10	0.26	231	24.8	25.0
40	0.10	0.25	427	13.4	13.5
50	0.10	0.24	694	8.3	8.5
60	0.10	0.23	1,043	5.5	5.5
70	0.10	0.22	1,485	3.9	4.0
30	0.12	0.28	214	26.7	26.5
40	0.12	0.27	395	14.5	14.5
50	0.12	0.26	641	8.9	9.0
60	0.12	0.25	960	6.0	6.0
70	0.12	0.24	1,361	4.2	4.0

*"Geometric Design Standards for Highways Other than Freeways," American Association of State Highway and Transportation Officials.

† e = rate of roadway superelevation, ft per ft

f = side friction factor (between tires and pavement) for design. It is used to compute R, minimum radius, ft, from $V^2/15(e + f)$, where V = design speed, mph.

Table 16-13. Minimum Length for Superelevation Runout in Two-Lane Pavements*†

Superelevation rate, ft per ft	Length of runout, ft, for design speed, mph, of				
	30	40	50	60	70
0.02	100	125	150	175	200
0.04	100	125	150	175	200
0.06	110	125	150	175	200
0.08	145	170	190	215	240
0.10	180	210	240	270	300
0.12	215	250	290	325	360

*"Geometric Design Standards for Highways Other than Freeways," American Association of State Highway and Transportation Officials.

† For wider pavements, lengths shown in table should be increased as follows: For three lanes in one direction, increase by 20%. For four lanes in one direction, increase by 50%.

Table 16-14. Maximum Grades, %*†

Type of topography	Design speed, mph				
	30	40	50	60	70
Flat	6	5	4	3	3
Rolling	7	6	5	4	4
Mountainous	9	8	7	6	

*"Geometric Design Standards for Highways Other than Freeways," American Association of State Highway and Transportation Officials.
†Short grades (less than 500 ft long) and one-way downgrades may be 1% steeper. For extreme cases, in urban areas and at some underpasses and bridge approaches, steeper grades for relatively short lengths may be considered. For low-volume rural highways, grades may be 2% steeper.

Right of Way. The right-of-way width should be sufficient to accommodate the ultimate planned roadway, including median, shoulders, sidewalks, a utility facility strip in the border areas, and the width necessary for side slopes, except where side slopes are constructed in an easement. The minimum width of right of way for a two-lane local street should be 50 ft and preferably 60 ft. For a four-lane highway, the minimum and desirable widths of the right of way are 60 ft and 80 ft, respectively. These generalized dimensions are exclusive of side-slope considerations.

16-7. Horizontal and Vertical Curves. Horizontal curves are introduced into the roadway alignment primarily to make it conform to the path that a vehicle would normally follow in changing directions. However, unless the radius is sufficiently large, motorists will find themselves constrained to slow down to avoid being compelled by centrifugal force to leave the roadway. If the curve is flat enough, that is, of large enough radius, drivers will not need to reduce speed, since the transverse forces will remain low enough to allow resistive frictional forces to maintain the desired path of travel. The minimum design radius for this result is interrelated with the design speed and frictional forces of the highway and is given in Table 16-15.

Table 16-15. Maximum Curvature for Normal Crown Section*

Design speed, mph	Avg running speed mph	Max degree of curve	Min curve radius, ft	Resulting side friction factor f with adverse crown	
				At design speed	At running speed
30	28	1°21′	4,250	0.026	0.024
40	36	0°48′	7,160	0.027	0.024
50	44	0°32′	10,810	0.027	0.024
60	52	0°23′	14,690	0.028	0.024
65	55	0°20′	17,360	0.028	0.024
70	58	0°18′	19,100	0.029	0.024
75	61	0°16′	21,220	0.030	0.024
80	64	0°14′	24,590	0.029	0.023

*"A Policy on Geometric Design of Rural Highways," American Association of State Highway and Transportation Officials.

Where sharper curvature is used, it becomes necessary to superelevate the pavement so as to counterbalance the centrifugal force. (See Art. 16-6.)

Horizontal highway curves sometimes are rated in degrees. The degree D of a curve is given by the central angle, degrees, subtending an arc of 100 ft along the curve. (Measurement of railroad curves is done with a 100-ft chord instead of the arc.) A curve also may be rated by its radius R, generally in even hundreds of feet.

$$R = \frac{5,729.58}{D}$$ (16-1)

Main components of horizontal curves are indicated in Fig. 16-5.

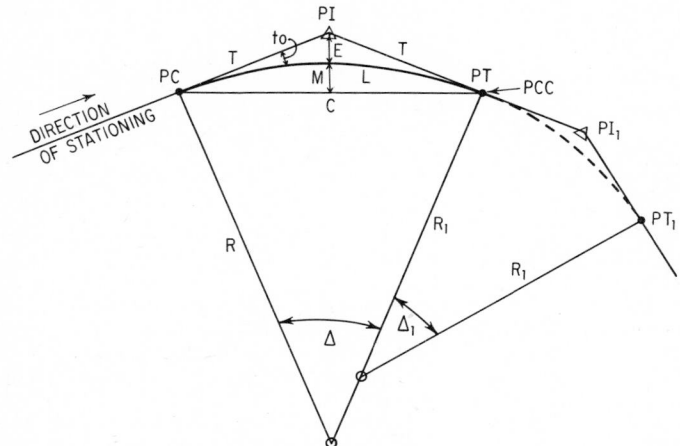

Fig. 16-5. Horizontal curve. *PI* indicates point of intersection of tangents; *PC*, point of curvature, or start of curve; *PT*, point of tangency, or end of curve; *PCC*, point of common curvature (for compound curves); *R*, radius; *T*, tangent length; *L*, length of curve; *C*, length of long chord; *E*, external; *M*, midordinate; Δ, central angle; *to*, tangent offset.

If a curve is too short, especially when combined with a change of grade or cross-section characteristic, it is liable to look kinked.   For small deflection angles, curves should be long enough to avoid the appearance of a kink.   For a central angle of 5°, curves should be at least 500 ft long.   This minimum length should be increased 100 ft for each 1° decrease in the central angle ("A Policy on Geometric Design of Rural Highways," American Association of State Highway and Transportation Officials).

Good design of alignment dictates avoidance of broken-back curves, two horizontal curves in the same direction separated by a short length of tangent.   The desirable minimum length of tangent for expressway design is 2,500 ft.   When conditions do not permit such a length, a compound curve or spiraled transitions are the preferred solutions.   In the design of compound curves, the ratio of the flatter radius to the sharper one should not exceed 1.5.

In the case of reverse curves, a 2,500-ft length for the intervening tangent is not necessary. The only limitation is sufficient distance between the end of one curve and the beginning of the next to accommodate transition spirals or superelevation transition.

One factor sometimes neglected in design of horizontal curves is sight distance.   The standards for expressways, with large minimum radii, wide shoulders, and trees and other natural obstacles excluded for a distance of at least 16 ft from the edge of the traveled way, usually take care of any horizontal sight-distance requirement.   But on other than such primary highways, sight distance must be dealt with.

The criteria for sight distance on vertical curves over crests also should be used for horizontal curves.   The height of the line of vision, however, generally need be considered only for sight across side slopes in cuts.   Obstacles to be checked are those adjacent to the traveled way, such as trees, buildings, retaining walls, bridge piers, abutments, parapets, and side slopes, especially in steep rock cuts.   The sight line to be used in design is the chord of the curve along the center line of the inside lane.

A good method of checking sight distance is to apply a straightedge to a plan of the horizontal curve showing all obstacles.   Sight distances can be read directly at various points along the curve.   In the case of obstacles with heights less than 3.75 ft or of side slopes, the sight distance can be measured across the obstacle.

Vertical Curves.   The grade or profile of a road has a considerable effect on its cost, safety, and efficiency.   Grades are expressed as percentages.   For example, +5.0% signifies a 5-ft rise in a horizontal distance of 100 ft; −5.0%, a 5-ft fall in 100 ft.   A minimum grade of 0% is used only in embankment areas where longitudinal drainage is available outside the pavement and shoulder areas, and across swamps where rolling grades would entail unnecessary height of embankment. But 0.5% is generally the minimum allowable grade where longitudinal drainage is along the pavement and shoulders. Usually, 6% is the maximum permissible, although for low-type roads or for high-type roads in mountainous terrain, higher maximums are sometimes permitted.

Profiles of the existing ground should be drawn to a convenient horizontal scale and to a vertical scale generally ten times larger than the horizontal one so that differences in elevation are more easily discernible. Critical elevations are marked on the profile, such as the elevation of groundwater, required clearance above high water, water courses, rock lines, firm bottom, and elevations of existing roads and railroads. If roads and railroads are to be underpassed or overpassed, clearance elevations at critical points have to be shown. In urban locations, additional critical elevations, such as those of buildings, sidewalks, and driveways, must be considered.

Within the allowable limits for the class of road under design, a grade then may be determined to suit as many of the controlling conditions as possible. If total conformity is impossible, a decision must be made as to which conditions are the most important, and adjustments made for the others; for example, a cross road may be lowered or raised, a stream bed cut down, steps provided to serve an existing building, or deeper cuts made in rock adjacent to a river crossing, thus reducing the height of a proposed bridge and its cost. Other things being equal, the cost of earthwork will be lower when cuts and fills balance, but it is unwise to strain the conditions just to obtain such a balance.

Where the vertical rise from the low point of a grade to the crest exceeds 50 ft, truck climbing lanes may be required. The decision to provide them should take into account the volume of traffic, proportion of trucks, and class of road.

Vertical curves are used to give smooth transitions between tangent-grade changes. These curves must conform to critical elevations, and the curves should provide adequate sight distance over crests and in sags. The rate of change should be sufficient to provide adequate pavement drainage. A K value of 143 is recommended on crests to insure adequate drainage. (K times algebraic difference in grades gives vertical-curve length.) At summits, the safe

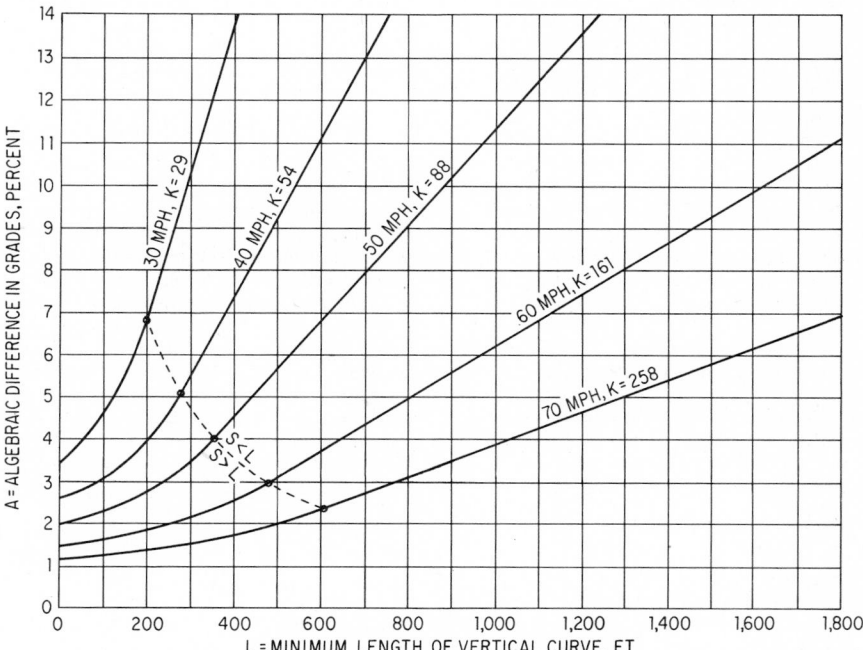

Fig. 16-6. For crest vertical curves, chart gives minimum length of curve to provide required sight distance (height of driver's eye 3.75 ft and of object 0.5 ft above pavement).

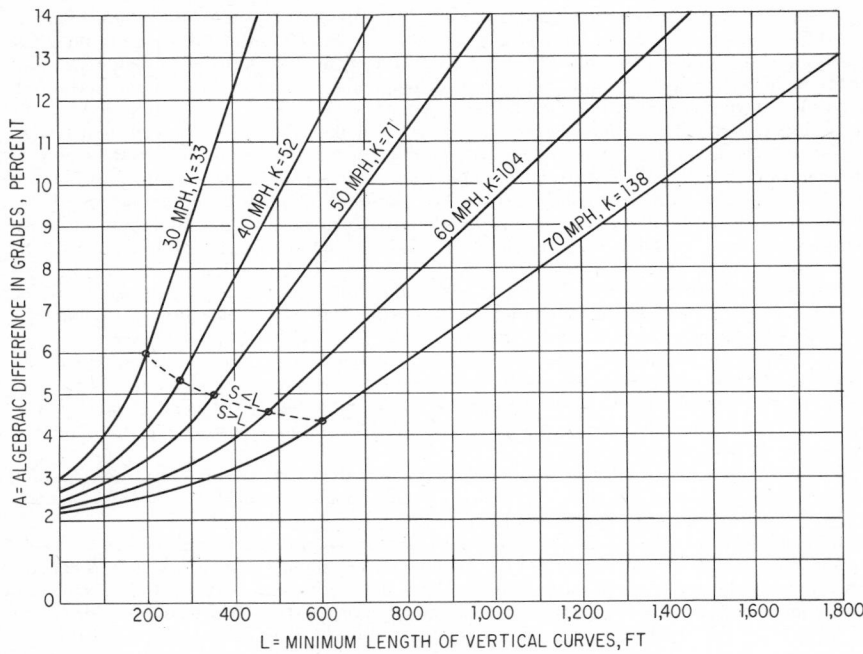

Fig. 16-7. For sag vertical curves, chart gives minimum length of curve to provide required sight distance (height of headlight 2.5 ft above pavement and upward divergence of light beam 1°).

stopping distance governs; at sags, the headlight sight distance governs. Particular care should be taken to provide adequate sight distance at ramp entrances and exits, left-turning locations, and intersections.

Usually, vertical curves are computed as parabolas. But for drawing purposes, circular templates generally are sufficiently accurate.

When vertical curves connect opposing grades, the length depends on the design speed as well as the algebraic difference in grade (Figs. 16-6 and 16-7). Curves should not be longer than necessary if longitudinal drainage will be hindered.

Short, choppy grades should be avoided. Long, smooth-flowing grades are desirable. Grades on which the road disappears from the motorist's vision in a dip ahead and reappears farther on confuse the driver and should be avoided. Vertical curves at sags should be as long as possible, since short curves give a foreshortened effect. Where a vertical curve and a horizontal curve occur at the same position, the horizontal curve should lead the vertical curve.

On divided highways, if the median is sufficiently wide, separate or independent grades for each roadway can prevent headlight-glare hazards. Such construction may also improve the appearance of the highway and lower its costs, when the terrain lends itself to such a design.

For crest vertical curves, the length, ft, may be computed from

$$L = \frac{AS^2}{100(\sqrt{2h_1} + \sqrt{2h_2})^2} \qquad S < L \qquad (16\text{-}2a)$$

$$L = 2S - \frac{200(\sqrt{h_1} + \sqrt{h_2})^2}{A} \qquad S > L \qquad (16\text{-}2b)$$

where A = algebraic difference in grades, %

S = sight distance, ft (Art. 16-6)

h_1 = height of eye above roadway surface, ft

h_2 = height of object above roadway surface, ft

For stopping sight distance, it is general practice to set $h_1 = 3.75$ ft and $h_2 = 0.5$ ft. A simple and useful expression for design control is

$$L = KA \tag{16-3}$$

where K is the length of vertical curve per change in A, percent. There is a positive value of K for each design speed (Fig. 16-6).

For passing sight distance, generally $h_1 = h_2 = 3.75$ ft in Eqs. (16-2). A much longer sight distance is required for passing than for stopping. In most cases, the vertical curve required at a crest becomes too long to be practical. So provision should be made for passing only at places where combinations of alignment and profile do not require the use of crest vertical curves or the algebraic difference in grades is very small.

Length of sag vertical curves is generally related to headlight sight distance, which, for safety, should approximate the stopping sight distance. It is general practice to use 2.5 ft as headlight height and 1° for upward divergence of the light beam. The relationship between K, L, and A is shown in Fig. 16-7. ("A Policy on Geometric Design of Rural Highways," American Association of State Highway and Transportation Officials.)

With grades given and length of vertical curve determined, points on the curve can be located. The tangent offset of the midpoint of the curve equals $AL/800$ (Fig. 16-8). Other offsets vary as the square of their distance from start or end of the curve: $y = (AL/800)x^2/(L/2)^2$. The lowest point on a sag curve and the highest point on a crest curve lie between V and the end of the curve on the flatter gradient. This point is at a distance gL/A from that end, where g is the grade, percent.

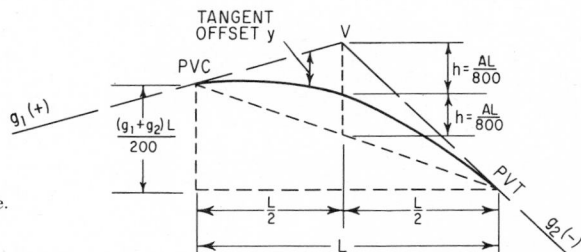

Fig. 16-8. Parabolic vertical curve.

16-8. Cross-sectional Elements of Roads. (See also Art. 16-17.) The cross-sectional elements of a roadway include the traveled way (pavement), shoulders, cross slope (or superelevation), side slopes, and where applicable medians, barriers, guard rails, and longitudinal ditches. In urban areas, such items as retaining walls, curbs, gutters, parking areas, roadway lighting, and sidewalks may also be present.

Each of these individual components is discussed in some detail in other parts of this section. Because of the variety of factors that may influence the design of a roadway, it is difficult to develop any standard guidelines as to the optimum mix of cross-sectional elements that yields the most satisfactory design from the standpoint of roadway operations, safety, etc.

Various Federal, state, city, county, and township agencies have developed their own standards that seem to suit the conditions of the areas under their jurisdiction. These standards generally can serve as a good guide for appropriate design within an individual area. Figures 16-9 and 16-10 illustrate graphically the interrelationship of most of the cross-sectional elements. The typical sections for a two-lane and a four-lane roadway illustrated in these figures are based on the design standards established by the New York State Department of Transportation.

16-9. Earthwork for Highways. Earthwork for highways usually is classified as rock, earth, wet excavation (waterlogged material), unsuitable material, and unclassified excavation (Art. 13-1). Equipment used also is described in Sec. 13.

In roadway areas, rock should be excavated to a depth of at least 1 ft below the normal subgrade on secondary roads; greater depths often will be specified for major roads. Old pavements should be broken up or scarified to a depth of at least 1 ft below subgrade. Stumps and roots should be grubbed out, except sometimes under fills over 4 ft high.

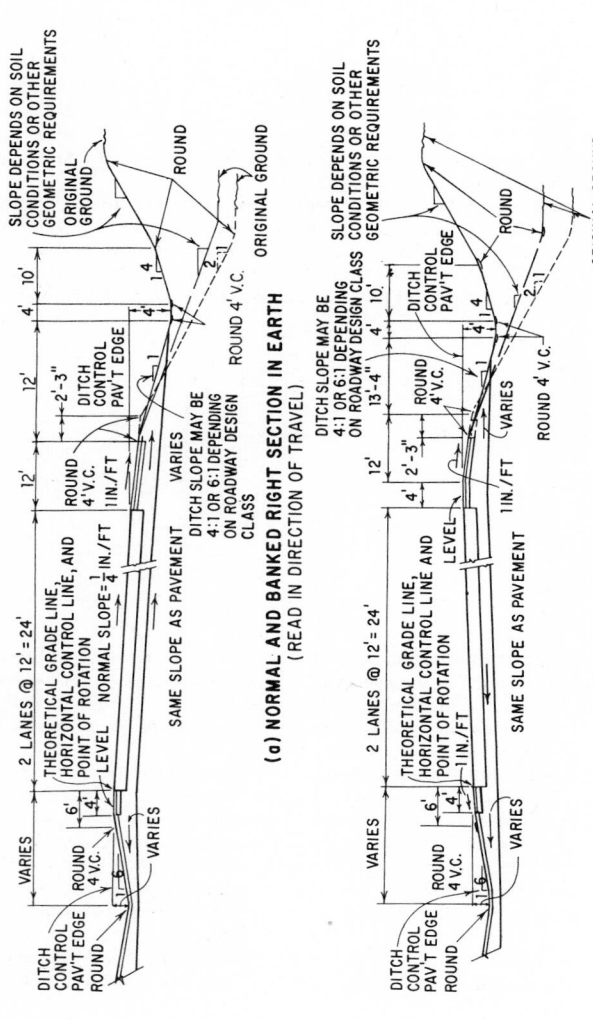

Fig. 16-9. Typical half sections of four-lane roadways.   (*New York State Department of Transportation.*)

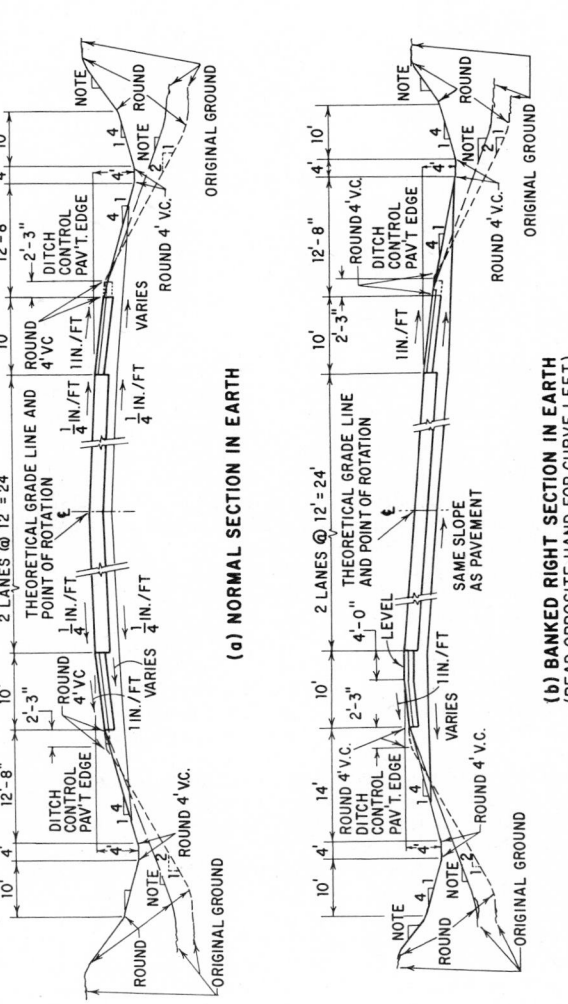

Fig. 16-10. Typical sections of two-lane roadways.   NOTE: Side slope to be used depends on soil conditions and geometric requirements.   *(New York State Department of Transportation.)*

Embankments are built as much as is practicable from the excavated material on the job. Although most unsuitable material has to be hauled away for disposal, wet excavation can often be used to widen and flatten the embankment slopes. There, it can dry out and, after proper treatment, be seeded to provide grass cover. By flattening the slopes, the added material can also eliminate the need for guardrails (Art. 16-17).

When the available excavated material is not sufficient for the embankments, material from outside borrow pits must be brought to the site. Backfill for wet excavation and for the top portions of embankments is preferably made of selected borrow, graded to restrict very-fine-grained soils from these areas.

Embankments are constructed in layers. The specified thickness of each layer varies from state to state, but usually ranges from 6 to 12 in. Compaction of each layer may be accomplished with steel rollers, rubber-wheel rollers, sheepsfoot rollers, etc., depending on the materials being used. In some cases, extra-heavy compactors are employed. These often take the form of a large box mounted on rubber-tired wheels. The box can be filled with soil to achieve the weight specified. In confined spaces, such as those adjacent to bridge abutments and on ramps, small pneumatic or motorized compactors with vibrating plates may be used.

Compaction. The degree of compaction specified is generally higher for the top layers of the embankment than for those below. A 95% compaction requirement means that the material, when compacted, must have a density of 95% of the maximum soil density which is obtainable with the given material brought to the optimum moisture content. The maximum density for the particular fill material may be found from field laboratory tests. Frequent soil tests should be made as compaction proceeds to insure that the minimum specified compaction value is obtained.

When necessary, the fill should be moistened by watering equipment. Water content of the fill material is less critical in granular fills than in fills having fine materials such as silts and clays. Such fills may be rejected when the water content cannot be brought to the specified optimum value because of uncontrollable factors such as wet climate.

Volume changes occur when earth or rock is excavated and when it is placed in embankments (Art. 13-9). Payment for earthmoving should take *swell* and *shrinkage* factors into account. Because of swell, specifications generally limit the sizes of rock pieces and require the voids to be filled with earth to compensate for the swell. Use of large rocks should be restricted to riprap placed against embankments subjected to erosion.

Before embankments are placed, it is advisable, and in shallow fills it is mandatory, to strip off organic topsoil. Sometimes excessive future settlement can be avoided by compacting the existing ground before placing new fill. When new embankment lies against a slope, slippage can be avoided by *benching*, which is done by plowing flat slopes or terracing steep slopes.

Fills on Weak Soils. Placement of embankments on poor foundations, such as swamps, calls for special procedures. Basic treatments include surcharging (overloading), blasting, lateral displacement by overloading, removal and replacement, lightweight embankment, lateral dewatering, sand drains, and construction of the roadway on grade beams supported on pile foundations.

Compression or surcharging can be used when the thickness of the poor material is small compared with the height of the embankment, or when the fill can be allowed to settle for a considerable time (1 year, for example). The rate of application of the fill should be slower than the rate of consolidation of the soft understratum. Vegetation should be removed before placement of fill. This method is particularly effective with silty soils. Organic clays are not easily consolidated by surcharged fills. Careful investigation is needed to determine the optimum method of construction on dry soils.

Dynamiting with controlled charges can produce a semiliquid state in certain types of soil (muck). Embankment may then be end-dumped, displacing the *soup* ahead of the fill. An alternative is to place the embankment first and then explode charges beneath the original ground surface.

Displacement by overloading is useful over an extensive swamp when no property rights will be violated. An overload, roughly equivalent to the expected future live load, displaces the poor material sideways. When check elevations show no further subsidence over a considerable period, the overload can be removed. Care must be exercised to avoid displacement on adjacent properties because of *mud-wave* action.

Removal and replacement are generally economical up to a depth of 15 ft. Backfill must be selected material.

Lightweight embankments are sometimes used where it is possible to keep settlement within tolerable limits by constructing an embankment that is light enough for the *weak* foundation soils to support. This method is particularly useful where lightweight material, such as cinders or expanded slag, is available locally and other soil treatment methods are very expensive or impractical.

Lateral dewatering may be accomplished with underdrains or wellpoints in porous water-bearing soils. Deep side ditches are sometimes used, especially in some southern states.

Sand drains can be used when the depth of poor material is great and consolidation can be achieved by reducing the moisture. In essence, they are sand-filled vertical holes through which the internal water is forced up to a pervious soil layer (sand blanket) at the ground surface. This layer conveys the displaced water into a drainage layer (Fig. 16-11). Spacing of the drains and rate of loading require careful investigations and field controls.

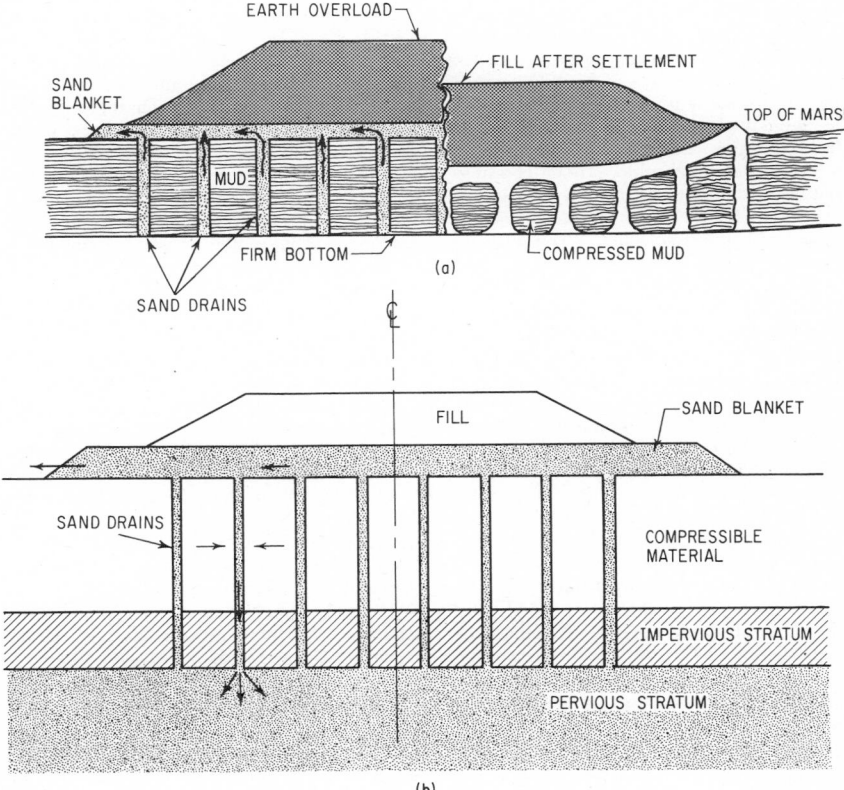

Fig. 16-11. Vertical sand drains used to remove water from marshy ground. (*a*) Mud overlying firm bottom; (*b*) mud overlying an impervious layer.

(See also "Guide Specifications for Highway Construction," American Association of State Highway and Transportation Officials, 341 National Press Building, Washington, D.C. 20004.)

16-10. Highway Drainage. Drainage design is an extremely important element of good highway development.

All highways that encroach on flood plains, large bodies of water, or streams must be designed to permit conveyance of the basic flood without causing significant damage to the highway, the stream, the body of water, or other property.   Without adequate drainage facilities, both surface and subsurface, a road will not endure, no matter how good the pavement.

Surface drainage involves removal of water, from either rain or melting snow, that falls directly on the roadbed, and interception and removal of waters coming to the road from adjacent terrain.

Subsurface drainage is concerned with removal of water from the subgrade and with interception of underground water coming to the subgrade.

Hydraulic Considerations in Highway Planning and Location.   Effects of highway construction on the existing drainage pattern and on the potential flood hazard, as well as effects of floods on the highway, must be assessed in preliminary planning and design stages.   Such an assessment can assist in determining those locations at which construction and maintenance will be unusually expensive or hazardous.   Also, although water laws vary widely throughout the United States and are subject to many different interpretations, statutes usually place responsibility for flood damages upon the person or organization who alters the drainage patterns of a watershed or creates an obstacle to the flow of water in a natural watercourse.

Whenever drainage problems are known to exist or can be identified, drainage and flood easements or other means of avoiding future litigation should be considered.   It is often helpful to document the history and present status of existing conditions or problems and to supplement the record with photographs and a description of field conditions.   Such thoroughness is essential because highway engineers are often blamed for flooding or erosion damages when, in fact, conditions that existed prior to highway construction are the real cause.

Whenever practicable, stream crossings should be made at stable reaches of a stream, but not at meanders that are subject to shifting.   The neck of a horseshoe bend that is subject to overflow is a poor location for a highway because the correct location of relief bridges sometimes varies with the flood stage.   The direction and amount of flood flow at various stages always must be considered when locating bridge openings, to avoid undue scour and erosion, which might result in a complete change in the river channel.

As nearly as practicable, crossings should be made normal to the direction of flow, considering the direction of flood flow if it may be different from that of low water.   Every effort should be made to minimize the number of stream crossings and disturbance of stream beds.

Design of Drainage Facilities.   To properly design any drainage structure, whether it be a simple pipe, a culvert, or a bridge, the engineer should know the quantity of runoff that can be expected to reach the structure.   This quantity will be determined from the heaviest rainfall to be accommodated and depends on the frequency of the storm to be designed for and the watershed characteristics.

In design of drainage facilities to accommodate an active river or stream, stream gage readings and available flood studies can be invaluable.

Storm Frequency.   The first step in the design of a drainage structure is to choose the storm frequency to be used in sizing the waterway opening.   For intense rainfalls, the heavier the storm, the less frequently it is likely to recur.   A rainfall that can be expected once in N years is referred to as an N-year storm, runoff, discharge, or flood.   The N-year or design flood selected should be supported by the following, where applicable: an incremental analysis of estimated construction costs; probable property damage including damage to the highway; potential land growth in terms of development upstream and downstream; cost of traffic delays; availability of alternative routes and emergency supply and evacuation routes; consideration of the potential for loss of life; and budgetary constraints.

Cross drains (small culverts) for defined water courses to be passed under major highways are usually designed to carry a 25-year storm.   For large culverts and bridges on major highways, a 100-year-storm design is advisable.   On other highways, the corresponding design flows for culverts and bridges may range from a 10-year storm for minor water courses on rural roads to a 50-year storm for an important watercourse on more heavily traveled roads.   Special attention is needed when watercourses cross at the low point of a road profile.

In some regions, rainfall records have been kept for years.   Where available, these and records of stream gages are invaluable.   Lacking these, the designer has to use tables, curves, and formulas based on records from similar areas.

Runoff Determination.   Various empirical formulas for obtaining runoff are available, but these should be used with discretion.   One of the more commonly used is the rational formula [Eq. (21-127)].   This formula can give fairly good results in urban areas where conditions lend themselves to a reasonable determination of the time of concentration and runoff coefficient and where watershed areas are usually relatively small in size.   For rural areas and large watershed areas, caution must be exercised in the use of the rational formula because the time of concentration and runoff coefficient are much more difficult to estimate accurately. It is probably safe, however, to apply the rational formula for drainage areas of less than 100 acres or where the runoff is picked up by a number of inlets, thereby creating many small watershed areas.

Land use will have a major effect on peak rates of runoff. The following are recommended for information or possible application, especially when runoff collected in a stream is to be evaluated (see also Art. 21-57):

1. The procedure for estimating peak rates of runoff from small watersheds (up to 25 sq mi in area), as outlined in "Peak Rates of Runoff from Small Watersheds," Hydraulic Design Series No. 2, U.S. Department of Commerce, Bureau of Public Roads, Washington, D.C.

2. U.S. Geological Survey analyses of runoff in various states (available from the local office of the U.S. Geological Survey).

3. U.S. Bureau of Public Roads studies of rainfall and runoff in certain areas of several states. The charts resulting from these studies can be used for other areas where runoff characteristics are similar. The charts, with instructions on their use, are included in various Hydraulic Engineering Circulars prepared by the Hydraulic Branch Bridge Division, Office of Engineering and Operations, Bureau of Public Roads, Washington, D.C.

Structure Capacity. The structure size should be selected to pass the design runoff without flooding the roadway. Also, the capacity should be adequate to limit the amount of backwater upstream of the structure to avoid property damage.

Surface Drainage. Rural highways on embankments are best drained by allowing the runoff from the roadway to flow evenly over the side slopes and spread over adjacent ground. If land use would be damaged by the free flow over adjacent ground, the drainage should be collected in longitudinal ditches at the bottom of slopes and conveyed to a nearby watercourse.

In cuts, roadway drainage should be directed into shallow side ditches, deep enough to drain the pavement subbase and convey the design storm flow. As the volume of water increases, paved gutters or drain pipes are required to replace the ditches. Eventually, the water should be led to the nearest watercourse or swale, which will take it either away from the road or through a cross drain under the highway to an outlet.

Along urban highways and in built-up areas, use of drainage ditches should be avoided where possible because of land-use considerations and the cost of right of way. On embankments, either a curb or an earth berm should be constructed along the outer edge of the roadway, and inlets placed at regular intervals. The inlets should connect to storm sewers that carry the water to points of disposal. In urban areas, it may be necessary to construct considerable lengths of storm sewer to the nearest stream or lake.

Where the median is depressed on a divided highway, runoff from the roadways flows into the median. The water may be carried for some distance by a median ditch. At some point, usually when the depth of flow would reach the pavement subbase or points of convenient outlet, inlets or other drainage structures should be introduced to conduct the median drainage to storm sewers.

Inlets. The design of drop inlets, catch basins, and manholes is well standardized. Some of the larger ones call for special design. Drop inlets (or catch basins) along a storm sewer should not be spaced more than about 300 ft apart. Inlet or catch-basin spacing should be balanced against the volume of water to be intercepted, the shape of the ditch or gutter conveying the water, and the hydraulic capacity of the inlet or catch basin.

Sewers. All changes of direction in a storm sewer should be made at an inlet, catch basin, or manhole. For maintenance purposes, access to the sewer should be provided at distances no greater than about every 500 ft.

For the design of storm sewers, the Manning formula [Eq. (21-33)] may be used if the sewer is to operate under gravity flow and not under pressure. As a storm sewer continues from its beginning to its outlet, pipe sizes should increase in stages to handle the increase in runoff. The storm sewer should connect to an existing storm drainage system, such as a stream or existing storm sewer system adequate to accommodate the design flow.

Outlets. When sewers are connected to streams, investigation of the downstream conditions is necessary to assure that the waterway is adequate and to avoid negative environmental impacts. (See Art. 16-18.)

Improvements to downstream outlets that are required to handle the additional flow or to make the drainage scheme acceptable to local officials should not be overlooked in estimating construction costs of the new facility.

Open Channels. Design of open channels also may be based on the Manning formula. Calculations may be simplified by using the charts contained in a Bureau of Public Roads publication, "Design Charts for Open-Channel Flow, Hydraulic Design Series No. 2," August, 1961.

Side ditches may be V-shaped or trapezoidal, with the latter having the greater capacity for a given depth. V-shaped ditches are often embodied in the typical roadway section. Such ditches often have capacity to spare, because it would not be economical to vary the ditch section to meet conditions at every point along the road, and the normal depth must be sufficient to drain the pavement subbase courses.

On steep grades, erosion becomes a matter for serious consideration. Prevention of erosion can take the form of lining the channel with sod, stone, bituminous or concrete paving; or provision of small check dams at intervals, depending on velocity, type of soil, and depth of flows.

Cross Drains. Cross drains are usually sized to make $H/D = 1.0$ for the design runoff, where H is the height of water at the entrance to the cross drain and D is the diameter or height

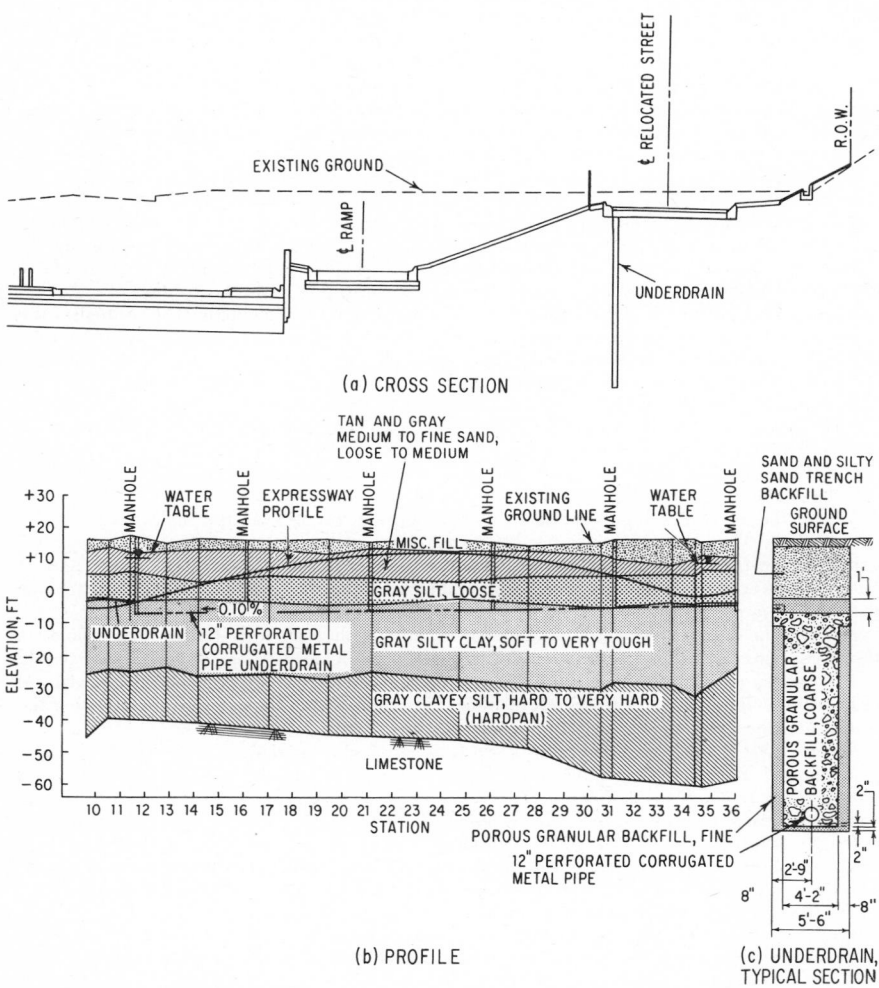

Fig. 16-12. Subsurface drainage system. Underdrains made relatively dry excavation possible during construction of the South Route Expressway, Chicago, despite high water table. Filter material was differentially graded to insure long service without clogging. Trench was capped with impervious fill to prevent intake of surface water. The drains continued to function after construction was completed. (*Highway Report 2-62, Metal Products Division, Armco Steel Corp.*)

of the cross drain.   For the theory of flow in culverts, see Arts. 21-18 and 21-19.   Especially useful also are "Hydraulic Design of Improved Inlets for Culverts," H.E.C. No. 13, August 1972, Federal Highway Administration, and "Practical Guidance for Estimating and Controlling Erosion at Culvert Outfalls," Miscellaneous Paper H-72-5, U.S. Army Engineer Waterways Experiment Station.

In approximations, for estimating purposes, a quick estimate of pipe size can be obtained with the Talbot formula:

$$A = C \sqrt[4]{M^3} \tag{16-4}$$

where A = required waterway, sq ft

$\quad\quad M$ = drainage area, acres

$\quad\quad C$ = coefficient

The Talbot formula was based on a rainfall intensity rate of 4 in. per hr.   The formula can be adjusted for any other desired rainfall intensity by proportion.

Recommended values of C:

> Steep, rocky ground, abrupt slopes 1
> Rough hilly country, moderate slopes . . . . $2/3$
> Uneven valleys, very wide compared with length . . . . $1/2$
> Rolling agricultural country, valley length three to four times width . . . . . $1/3$
> Flat country not affected by accumulated snow or severe floods . . . . $1/5$

Equation (16-4) should be used only for a rough check on required culvert size or for preliminary culvert sizing.   A more accurate analysis should always be used in final culvert design.

Erosion Control.   Provisions for control of erosion and siltation during construction must be made.   Control may require construction of settling basins, weirs, use of jute mesh, straw bales, etc.   The provisions are usually made part of the construction specifications and are implemented according to the construction procedures being employed.

Subsurface Drainage.   Removal of water from below a roadway in swampy areas may require excavation of water-bearing material (wet excavation) and its replacement with selected material that can be drained.   In some instances, recurrence of the wet conditions may necessitate installation of underdrains or deep ditches.

Usually, subsurface drainage is provided through the layer of porous subbase material that is part of the pavement structure.   To insure a positive outlet for the subsurface water, either intercepting underdrains may be installed adjacent to the graded width of the subbase or transverse underdrains may be installed.   Such installations can become quite elaborate, such as the one developed for the South Route Expressway in Chicago (Fig. 16-12).   The section of the Expressway was depressed to a depth of almost 20 ft below the prevailing water table.   The material to be excavated was largely an unstable clay.   Before excavation began, underdrains were placed at each side of the graded width.   The subsurface water was then pumped from these porous drains.   As a result, it was possible to proceed with excavation in relatively dry conditions without extensive sloughing of the side slopes.   After completion of the excavation, drainage continued to flow by gravity in the underdrains to a pumping station.   This system was carefully designed to insure that the water table in adjacent areas was not affected enough to cause subsidence under buildings or on abutting property.

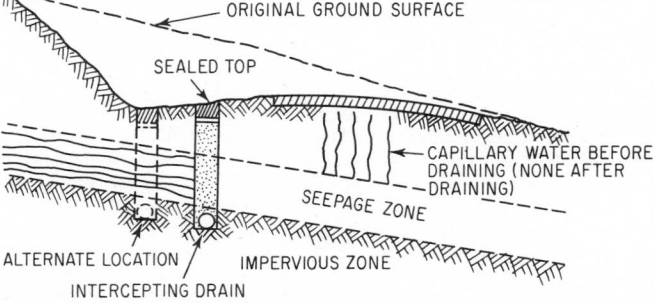

Fig. 16-13.   Drain intercepts source of supply of harmful capillary and free water under a road.   Top of trench was sealed to prevent silting.   (*"Handbook of Drainage and Construction Products,"* Metal Products Division, Armco Steel Corp.)

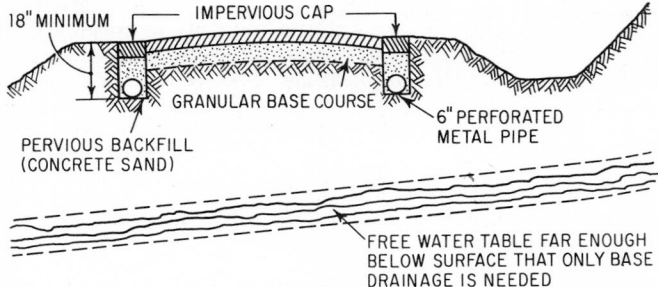

Fig. 16-14. Drains remove surface water that may be trapped when a pervious base is laid over a relatively impervious subgrade. On steep slopes, laterals may be added under the pavement. Longitudinal base drains should be outletted at convenient points, which may be 100 ft or more apart. (*"Handbook of Drainage and Construction Products," Metal Products Division, Armco Steel Corp.*)

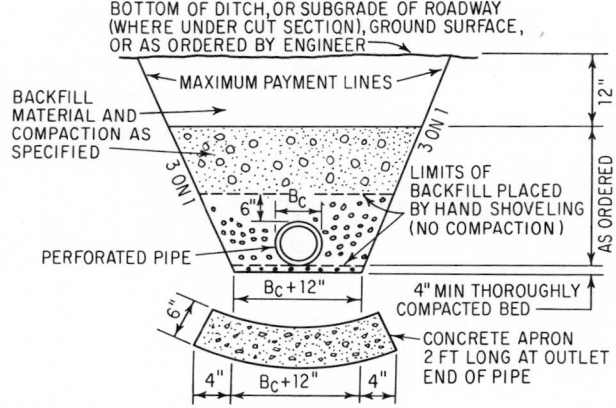

Fig. 16-15. Bedding and backfill details for pipe underdrain. (*New York State Department of Transportation.*)

Examples of subsurface drainage structures are shown in Figs. 16-13 to 16-16. Figure 16-13 shows an intercepting drain installed to cut off the source of supply of harmful capillary and free water under a road. The top of the trench is sealed to prevent silting. In Fig. 16-14, base drains are used to remove surface water that may be trapped when a pervious base is laid over a relatively impervious subgrade. On steep slopes, laterals may be added under the pavement. Figures 16-15 and 16-16 show bedding and backfill details for a pipe underdrain.

("Highway Drainage Guidelines," 3 volumes, American Association of State Highway and Transportation Officials, Washington, D.C.; "Federal-Aid Highway Program Manual," Vol. 6, Chap. 7, Sec. 3, Subsec. 2, Federal Highway Administration.)

16-11. Roadway Pavement Structures. (See also Art. 16-17.) The quality, nature, thickness, and composition to be selected for a roadway structure depend on the volume and type of traffic, cost and availability of materials, climatic and foundation conditions, and whether the pavement is to be constructed in stages over a period of years. The composition of the roadway can range from a stabilized earth surface obtained by reshaping and compacting native soil to a high-quality asphaltic concrete or portland-cement concrete pavement with several layers of base and subbase courses. In the following, the common types of roadway pavements used are listed and the basic considerations in their design and construction are discussed.

Untreated Road Surfaces. In many cases, secondary roads with low traffic volumes can provide satisfactory service with a surface course of untreated soil mixtures consisting of locally available crushed rock or gravel materials. Other locally available materials, such as volcanic cinders, blast-furnace slag, limerock, chert, or shells, may also be used.

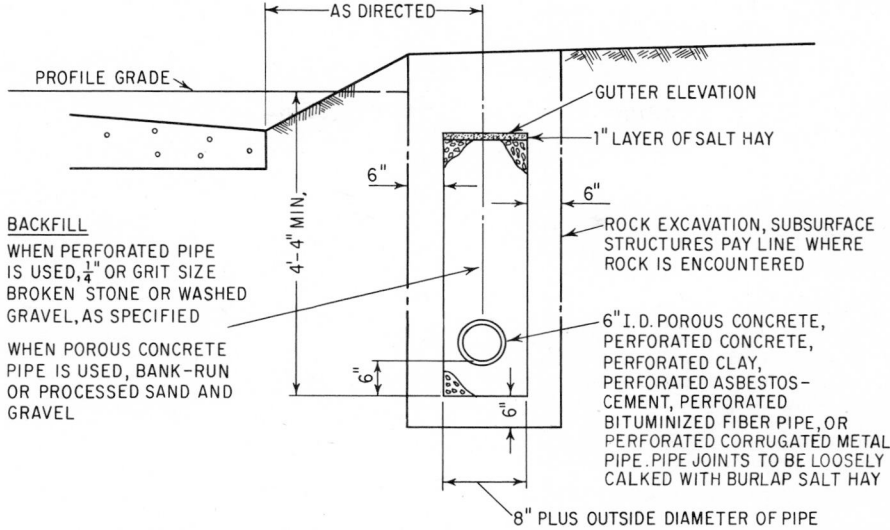

Fig. 16-16. Underdrain detail. (*New Jersey State Department of Transportation.*)

The main characteristics of a good untreated rock or gravel surface include:

1. Ability to withstand abrasion from superimposed traffic loads. This can be achieved with an adequate amount of well-graded coarse aggregate (retained on No. 10 sieve), which will combine with sand to form a tight, water-resistant surface and to provide an interlocking of aggregate to resist shear forces.

2. Ability to support traffic loads without excessive deformation. This can be accomplished by providing enough binding material, such as clay, to cement the coarse aggregate, sand, and silt in a dry or low-moisture condition and yet not so much clay that dislocation of the surface would take place because of expansion caused by high moisture.

3. A surface that drains adequately without permitting excessive infiltration of rain into the subgrade.

4. A surface that allows for some upward percolation of subsurface water to replace moisture lost by evaporation, thereby maintaining a desirable moisture content.

An imbalance in the water content of the surface layer may result in rutting when the roadway is wet or in a dusty condition when inadequate water is present.

Untreated road surfaces may be adaptable to future upgrading. They can provide excellent subgrades for higher-class pavements when traffic volumes and economics warrant such improvement. The relatively low first cost of these road surfaces, however, is offset to some extent by substantial maintenance cost. Maintenance is normally required at least twice a year.

Stabilized Road Surfaces. The term *stabilized road* generally denotes any one of a variety of roadway surfaces composed of a controlled mixture of native soil and additives such as asphalt, portland cement, calcium chloride, and, in some instances, even sand-clay. These mixtures can also serve as an excellent base course for certain types of pavements.

Sand-Clay Roads. A sand-clay surface consists of a controlled mixture of fine and coarse sand (and ideally some fine gravel) with clay and silt. The roadway thickness is generally about 8 in. or more. The controlling factors in successful construction of this type of roadway surface are similar to those described previously for untreated road surfaces.

Stabilization with Calcium Chloride. Calcium chloride is a chemical capable of absorbing moisture from the air and retaining it without becoming liquid. This characteristic makes it an excellent dust palliative as well as a stabilizing agent.

When calcium chloride is used as a stabilizing agent on an existing surface course, the common procedure is to scarify the existing roadway surface and mix about ½ lb per sq yd of calcium chloride per inch of depth. For this process to be successful, adequate moisture must be present. To improve stabilization during dry weather, quantities of water must be added.

For use of calcium chloride as a dust palliative, see Art. 16-19.

Stabilization with Portland Cement. Untreated road surfaces can be stabilized with portland cement if the clay content in the soil is favorable for this type of treatment. Soils that contain less than 35% clay are normally adaptable to this method of stabilization. The required rate of application of cement varies with soil classification and generally ranges from 6 to 12% by volume. The roadway surface to be treated should be scarified for a depth of about 6 in. and the cement applied uniformly to the loose material, then brought to the optimum moisture content and lightly rolled.

A higher quality soil-cement pavement can be constructed by mixing the soils, cement, and water in a central or traveling mixing plant. The mixture then is placed and rolled.

Stabilization with Bituminous Materials. Various asphalt surface treatments can be utilized to stabilize untreated road surfaces.

For use as a dust palliative, liquid asphalt material may be applied at a rate of 0.1 to 0.5 gal per sq yd. Use of bituminous materials as a dust palliative is often referred to as *road oiling.* A dust palliative treatment is frequently used as a preliminary to a progressive improvement of low-type roadways.

As a surface treatment, stabilization with bituminous materials consists of an application of asphalt followed by an application of aggregate. This process can be repeated several times.

Table 16-16. Quantities of Asphalt and Aggregate for Single Surface Treatments and Seal Coats*

Line no.	Nominal size of aggregate, in.	Size	Lbs of aggregate per sq yd†‡	Gallons of asphalt per sq yd†¶	Hot weather (80°F or higher)		Cool weather (up to 80°F)	
					Hard aggregate	Absorbent aggregate	Hard aggregate	Absorbent aggregate
1	¾ to ⅜	6	40–50	0.40–0.50	RC3000, RS2, CRS-1, CRS-2	RC3000, RS2, CRS-1, CRS-2	RC800, RS2, CRS-1, CRS-2	RC800, RS2, CRS-1, CRS-2
2	½ to No. 4	7	25–30	0.30–0.45	RC250, RC800, RS1, RS2, CRS-1, CRS-2	RC250, RC800, RS1, RS2, CRS-1, CRS-2	RC250, RC800, RS1, RS2, CRS-1, CRS-2	RC250, RC800, RS1, RS2, CRS-1, CRS-2
3	⅜ to No. 8	8	20–25	0.20–0.35	RC250, RC800, RS1, RS2, CRS-1, CRS-2	RC250, RC800, RS1, RS2, CRS-1, CRS-2	RC250, RC800, RS1, RS2, CRS-1, CRS-2	RC250, RC800, RS1, RS2, CRS-1, CRS-2
4	No. 4 to No. 16	9	15–20	0.15–0.25	RC250, RC800, RS1, RS2, CRS-1, CRS-2	RC250, RC800, RS1, RS2, CRS-1, CRS-2	RC250, RC800, RS1, RS2, CRS-1, CRS-2	RC250, RC800, RS1, RS2, CRS-1, CRS-2

*These quantities and types of materials may be varied according to local conditions and experience.

Single Surface Treatments. The maximum size aggregate should not be over ½ in. Use line 2. For lighter surface treatments, use line 3 or 4; however, lines 3 and 4 are more for light seal coats.

Double Surface Treatments. The maximum size can be up to ¾ in. First course, use line 1; second course, use line 3 or 4. For lighter double surface treatments, use for first course line 2; for second course, line 3 or 4.

Triple Surface Treatments. The maximum size aggregate is usually ¾ in. The following is recommended: first course, line 1; second course, line 2; third course, line 3 or 4. For most situations, the best probably is lines 1, 2, and 4 for the three courses.

†The lower application rates of asphalt shown in the table should be used for aggregate having gradings on the fine side of the limits specified. The higher application rates should be used for aggregate having gradings on the coarse side of the limits specified.

‡The weight of aggregate shown in the table is based on aggregate with a specific gravity of 2.65. In case the specific gravity of the aggregate used is less than 2.55 or more than 2.75, the amount shown in the table above should be multiplied by the ratio of the bulk specific gravity of the aggregate used to 2.65.

¶Under certain conditions, the grades of MC liquid asphalts may be used satisfactorily.

§In some areas, difficulty in retaining aggregate has been experienced with 200–300 penetration asphalt cements.

From "Surface Treatment Tips," The Asphalt Institute, College Park, Md. 20740.

The results are commonly referred to as double, triple, etc., surface treatments.

Surface treatments are sensitive to certain conditions, among which are the following:

1. Weather conditions free of rain and temperatures above 40°F are highly desirable.

2. The surface to be treated must be dry and well compacted.

3. The quantity and viscosity of the asphalt must be in proper relationship with the temperature, size, and quantity of the aggregate to be used.

4. This type of surface treatment should not be utilized where it will be constructed under heavy high-speed traffic, because this type of usage tends to dislodge the loose aggregate. Once the surface treatment is thoroughly rolled and broomed of loose aggregate, it will provide good service.

Table 16-16 gives some typical surface treatments recommended by The Asphalt Institute. See also the subsequent discussion of bituminous surface treatments.

Waterbound Macadam.   This is one of the oldest types of pavement. Its modern use is chiefly as a base course under a bituminous pavement. In such cases, the waterbound feature may be omitted, to form a *dry-bound macadam.*

A waterbound pavement is usually built in two courses, the lower one about 4 in. thick and the other about 2 in. A typical specification would call for the bottom course stone to pass a 3-in. ring and be retained on a 2-in. one. For the top course, the figures would be 2 in. and 1 in. The stone must pass certain standards of hardness. A good macadam stone is angular and uniform each way, since the stability of the pavement depends largely on the interlocking of the pieces (American Association of State Highway and Transportation Officials M77, Standard Specifications for Crushed Stone and Crushed Slag for Waterbound Surface Course).

Before any stone is placed, the subgrade should be well shaped and adequately drained. The bottom course should be well rolled before the top course is placed. After the top course has been thoroughly rolled, stone chips or stone dust should be spread over it. This generally is done by hand shovels, in successive sweeps. This binder is broomed into the voids, and when these are partly filled, sprinkling is commenced. Alternate applications of binder, water, and rolling continue until a wave of "mortar" appears ahead of the roller. With experienced workers, an excellent pavement, which will shed water and stand up under light traffic in rural areas, can result.

Modern practice has largely eliminated use of this type of base construction because of the advances made in plant equipment and the excellent results that can be obtained with asphalt-concrete or portland-cement-treated bases. Waterbound macadam construction is time-consuming and expensive.

Penetration Macadam.   Penetration macadam has a composition similar to waterbound macadam, except that, instead of screenings or stone dust, asphalt or tar is used for the initial filling of the voids. The stones are then "keyed" in or "choked" with a small-sized stone, and a second application of the bituminous material is made.

A penetration-macadam course is usually from 2 to 3 in. thick. It is placed on a base course about 4 in. thick, similar to the lower course of a waterbound macadam pavement, in which the voids have been filled with screenings. After the base course has been rolled, excess filler should be removed by stiff brooming. Before either course is rolled, it should be made thick enough to produce the specified thickness when compacted. The compacted depth will be about 0.7 of the loose depth, but the shrinkage should be checked for the individual job.

The penetration course should be thoroughly rolled with at least a 10-ton roller before the first application of bituminous material. In the East, this first application is usually from 1.75 to 2.25 gal per sq yd, 3 in. thick. (In California, smaller quantities are used because of the nature of the local asphalt.) The temperature of the oil must be within specified limits (usually 300 to 350°F for asphalt). No application should be made if the air temperature has been below 40°F during the preceding 24 hr. While the poured bitumen is still warm, a small-sized stone (key or choke) should be spread by hand. Excess stone should be broomed off and the surface rolled to insure good keying.

A second application, from 0.5 to 0.75 gal per sq yd, of the bituminous material should then be made. It should be followed by a covering of clean stone chips or pea gravel and final rolling.

Penetration macadam is still used considerably on secondary roads in rural areas. A modified form using less bitumen (1.6 to 2.5 gal per sq yd, according to thickness) is used in some primary roads as an intermediate course between a stone base and bituminous-concrete wearing courses.

Inverted Penetration Macadam.   With the inverted penetration method, the normal penetration process is reversed; that is, the asphalt binder is first sprayed over the prepared surface and then covered with the aggregate. The inverted penetration method can be used for dust control; as a prime coat or tack coat on which a new wearing surface will be constructed; as a surface treatment and armor coat for providing temporary protection for untreated surfaces; and as a seal coat for leveling, strengthening, or otherwise improving existing pavements.

Bituminous Surface Treatments. Asphalt surface treatments can be utilized to serve the following functions:

1. Provide a low-cost, all-weather surface for light-to-medium category highways.
2. Seal an existing roadway surface.
3. Help an overlay course adhere to the previous course.
4. Provide a skid-resistant surface.
5. Rejuvenate existing weathered surfaces.
6. Provide a temporary cover for a new granular base that is not to receive its final cover for an extended period of time.
7. Overlay existing pavements and provide some increase in strength.
8. Serve as a dust palliative.
9. Guide traffic and improve night visibility; for example, through the use of color-contrasted aggregates.

Types of asphalt surface treatments include the following: single surface treatment, multiple surface treatment, seal coat, prime coat, tack coat, dust laying, road oiling, mixed-in-place surface treatment, and plant-mixed surface treatments.

More detailed guidelines for the use and application of various asphalt surface treatments are given in The Asphalt Institute Manual MS-13, "Asphalt Surface Treatments."

Some typical quantities of asphalt and aggregate application for surface treatments are shown in Table 16-16.

Shoulders. Roadway shoulders are utilized to provide an area for disabled vehicles outside the traveled way. They should have adequate structural strength to support the weight of a vehicle without rutting and to resist the intrusion of rainwater and melting snow into the pavement structure.

The type, width, and structural integrity of shoulders are determined to a great extent by the nature and traffic volume of the roadway. For roadways that carry relatively low traffic volumes, stabilized shoulders consisting of a double surface treatment may be adequate. For more heavily traveled roadways, particularly those subject to heavy truck traffic, a more durable shoulder design may be required. A 3-in.-thick asphaltic concrete shoulder supported by adequate subbase material is commonly used on primary highways. The shoulder areas should be different in surface texture and color from the pavement surface to help traffic stay on the roadway.

High-quality Pavements. These are used to support heavy traffic loads on high-traffic-volume roadways. The two basic types of pavements used are bituminous concrete (flexible) and portland cement concrete (rigid). The major criteria for pavement selection and design are outlined in the following discussion. These design considerations and methods are to a large extent based on the American Association of State Highway and Transportation Officials, "Interim Guide for Design of Pavement Structures," 1972.

Type of Pavement Structure. In selection of the type of pavement to be used, the choice between rigid or flexible pavement depends on foundation conditions, local availability of material, relative costs, projected traffic, traffic maintenance during construction and construction practice in the locality, frequency of need to service underground utilities within the pavement area, pavement color, and whether staged construction is being considered. On widening projects and overlays, the existing type of pavement is also a factor.

Highway funding limitations, material scarcities, and the desire for adequate pavement performance intensify the need to optimize pavement design. Several methods have been developed to determine optimal pavement design strategies. An optimal design strategy is defined as: "The design that gives a satisfactory level of performance to the user at a minimum total cost." (See M. I. Darter et al, "Selection of Optimal Pavement Designs Considering Reliability, Performance, Costs," Record No. 485, Transportation Research Board.)

With the use of high-speed computers, possible alternative designs that may satisfy the requirements of a particular pavement situation can be investigated quickly and economically. Generally, several alternative combinations of materials and layer thicknesses can be considered as well as several alternative maintenance or overlay possibilities (including staged construction), thereby permitting a choice between a large number of alternative design strategies. TRB Record No. 485 presents several pavement analysis strategies that have been used successfully.

Foundation Conditions.   A pavement structure is a layered system designed to distribute concentrated traffic loads to the subgrade.   Preparation of the subgrade usually includes at least grading and compaction of the subgrade soils.   Subgrade preparation also may include other means of providing for optimum support of the pavement structure.

The performance of a pavement structure is directly related to the physical properties and condition of the roadbed soils.   The design procedures are based on the assumption that most soils can be adequately represented for pavement design purposes by a soil's support value S for flexible pavements or a modulus of subgrade reaction k for rigid pavements.   Certain soils, however, such as those that are excessively expansive, resilient, frost susceptible, or highly organic, require that additional steps be taken to provide for adequate pavement performance. Other factors related to roadbed soils are nonuniform support resulting from wide variations in soil type or condition; additional densification under traffic of soils that were not adequately compacted during construction; and construction difficulties, particularly those associated with compaction of cohesionless sands and wet, highly plastic clays.

Serviceability Index.   Serviceability of a pavement is defined as the ability to serve high-speed, high-volume automobile and truck traffic.   A procedure developed for periodic rating of serviceability of pavements, known as the Present Serviceability Rating *PSR*, utilizes the mean of individual ratings by a selected panel of experts with long experience in all aspects of highway engineering and pavement design, construction, and performance.   A scale with a value of 5 as the highest index of serviceability and 0 as the lowest is used for the *PSR.*

Selection of a terminal serviceability index p_t is based on the lowest index that will be tolerated before resurfacing or reconstruction becomes necessary.   An index of 2.5 is suggested as a guide for design of major highways and 2.0 for highways with lesser traffic volumes.   For relatively minor highways, if economic considerations dictate that initial capital outlay be kept to a minimum, this may be accomplished by reducing the traffic analysis period rather than by designing for a p_t of less than 2.0.

Design Traffic.   The procedure used in the AASHTO "Interim Guide for Design of Pavement Structures" is to convert the varying axle loads to one design loading and to express the traffic volume as the number of repetitions of the design axle load.   The design loading used is an 18-kip single-axle load.   Thus, traffic is expressed as equivalent 18-kip single-axle loads.

Mixed traffic loads may be converted to equivalent 18-kip single-axle loads with the aid of Tables 16-17 and 16-18.

Prediction of traffic for design purposes must rely on information from past traffic, modified by factors for growth or other expected changes.   Most states accumulate past traffic information in the form of loadometer data in the format of the Federal Highway Administration W4 loadometer tables.   These are tabulations of the number of axles observed within a series of load groups, with each load group usually a 2,000-lb increment.   The tabulations are in a convenient form for conversion, since the number of axles in each load group may be multiplied by an appropriate factor for conversion to equivalent 18-kip single-axle load applications for the load group, and a summation of these for all load groups is the equivalent 18-kip single-axle load application that represents the total traffic for the survey period.   Note that the equations used in the AASHTO Interim Guide are based on the application of a maximum number of loads during a two-year period of road testing.   Extrapolation beyond these total load applications should be used with caution, since additional applications have not been substantiated by road test experience.

Predictions of traffic are made for some convenient period of time.   The traffic analysis period often used is 20 years, which is also a common period used in traffic predictions for capacity design.   However, any period may be used with this design method, because traffic is expressed as daily or total equivalent 18-kip single-axle load applications.   Regardless of the traffic analysis period used, the total equivalent 18-kip single-axle load applications is the total load repetitions that the pavement can be expected to carry from opening of a road to traffic to the time when serviceability of the road is reduced to the selected terminal value; that is, to $p_t = 2.5$ or 2.0.

The equivalent axle loads derived represent the totals for all lanes for both directions of travel. This traffic must be distributed by direction and by lanes for design purposes.   Directional distribution is usually made by assigning 50% of the traffic to each direction, unless special conditions warrant some other distribution.   For lane distribution, 100% of the traffic in each direction is usually assigned to each lane in that direction for purposes of structural design. Some states have developed lane distribution factors for facilities with more than one lane in a given direction.   These factors vary from 80 to 100% of the one-direction traffic for design of each lane when there is a total of four lanes in both directions, and from 60 to 80% of the one-

Table 16-17. Traffic Equivalence Factors, Flexible Pavement[*]

a. SINGLE AXLES, $p_t = 2.0$

Axle load, kips	Structural number SN					
	1	2	3	4	5	6
2	0.0002	0.0002	0.0002	0.0002	0.0002	0.0002
4	0.002	0.003	0.002	0.002	0.002	0.002
6	0.01	0.01	0.01	0.01	0.01	0.01
8	0.03	0.04	0.04	0.03	0.03	0.03
10	0.08	0.08	0.09	0.08	0.08	0.08
12	0.16	0.18	0.19	0.18	0.17	0.17
14	0.32	0.34	0.35	0.35	0.34	0.33
16	0.59	0.60	0.61	0.61	0.60	0.60
18	1.00	1.00	1.00	1.00	1.00	1.00
20	1.61	1.59	1.56	1.55	1.57	1.60
22	2.49	2.44	2.35	2.31	2.35	2.41
24	3.71	3.62	3.43	3.33	3.40	3.51
26	5.36	5.21	4.88	4.68	4.77	4.96
28	7.54	7.31	6.78	6.42	6.52	6.83
30	10.38	10.03	9.24	8.65	8.73	9.17
32	14.00	13.51	12.37	11.46	11.48	12.17
34	18.55	17.87	16.30	14.97	14.87	15.63
36	24.20	23.30	21.16	19.28	19.02	19.93
38	31.14	29.95	27.12	24.55	24.03	25.10
40	39.57	38.02	34.34	30.92	30.04	31.25

b. TANDEM AXLES, $p_t = 2.0$

Axle load, kips	Structural number SN					
	1	2	3	4	6	8
10	0.01	0.01	0.01	0.01	0.01	0.01
12	0.01	0.02	0.02	0.01	0.01	0.01
14	0.02	0.03	0.03	0.03	0.02	0.02
16	0.04	0.05	0.05	0.05	0.04	0.04
18	0.07	0.08	0.08	0.08	0.07	0.07
20	0.10	0.12	0.12	0.12	0.11	0.10
22	0.16	0.17	0.18	0.17	0.16	0.16
24	0.23	0.24	0.26	0.25	0.24	0.23
26	0.32	0.34	0.36	0.35	0.34	0.33
28	0.45	0.46	0.49	0.48	0.47	0.46
30	0.61	0.62	0.65	0.64	0.63	0.62
32	0.81	0.82	0.84	0.84	0.83	0.82
34	1.06	1.07	1.08	1.08	1.08	1.07
36	1.38	1.38	1.38	1.38	1.38	1.38
38	1.76	1.75	1.73	1.72	1.73	1.74
40	2.22	2.19	2.15	2.13	2.16	2.18
42	2.77	2.73	2.64	2.62	2.66	2.70
44	3.42	3.36	3.23	3.18	3.24	3.31
46	4.20	4.11	3.92	3.83	3.91	4.02
48	5.10	4.98	4.72	4.58	4.68	4.83

c. SINGLE AXLES, $p_t = 2.5$

Axle load, kips	Structural number SN					
	1	2	3	4	5	6
2	0.0004	0.0004	0.0003	0.0002	0.0002	0.0002
4	0.003	0.004	0.004	0.003	0.003	0.002

Table 16-17. Traffic Equivalence Factors, Flexible Pavement* (*Continued*)

c. SINGLE AXLES, $p_t = 2.5$

Axle load, kips	Structural number SN					
	1	2	3	4	5	6
6	0.01	0.02	0.02	0.01	0.01	0.01
8	0.03	0.05	0.05	0.04	0.03	0.03
10	0.08	0.10	0.12	0.10	0.09	0.08
12	0.17	0.20	0.23	0.21	0.19	0.18
14	0.33	0.36	0.40	0.39	0.36	0.34
16	0.59	0.61	0.65	0.65	0.62	0.61
18	1.00	1.00	1.00	1.00	1.00	1.00
20	2.61	1.57	1.49	1.47	1.51	1.55
22	2.48	2.38	2.17	2.09	2.18	2.30
24	3.69	3.49	3.09	2.89	3.03	3.27
26	5.33	4.99	4.31	3.91	4.09	4.48
28	7.49	6.98	5.90	5.21	5.39	5.98
30	10.31	9.55	7.94	6.83	6.97	7.79
32	13.90	12.82	10.52	8.85	8.88	9.95
34	18.41	16.94	13.74	11.34	11.18	12.51
36	24.02	22.04	17.73	14.38	13.93	15.50
38	30.90	28.30	22.61	18.06	17.20	18.98
40	39.26	35.89	28.51	22.50	21.08	23.04

d. TANDEM AXLES, $p_t = 2.5$

Axle load, kips	Structural number SN					
	1	2	3	4	5	6
10	0.01	0.01	0.01	0.01	0.01	0.01
12	0.02	0.02	0.02	0.02	0.01	0.01
14	0.03	0.04	0.04	0.03	0.03	0.02
16	0.04	0.07	0.07	0.06	0.05	0.04
18	0.07	0.10	0.11	0.09	0.08	0.07
20	0.11	0.14	0.16	0.14	0.12	0.11
22	0.16	0.20	0.23	0.21	0.18	0.17
24	0.23	0.27	0.31	0.29	0.26	0.24
26	0.33	0.37	0.42	0.40	0.36	0.34
28	0.45	0.49	0.55	0.53	0.50	0.47
30	0.61	0.65	0.70	0.70	0.66	0.63
32	0.81	0.84	0.89	0.89	0.86	0.83
34	1.06	1.08	1.11	1.11	1.09	1.08
36	1.38	1.38	1.38	1.38	1.38	1.38
38	1.75	1.73	1.69	1.68	1.70	1.73
40	2.21	2.16	2.06	2.03	2.08	2.14
42	2.76	2.67	2.49	2.43	2.51	2.61
44	3.41	3.27	2.99	2.88	3.00	3.16
46	4.18	3.98	3.58	3.40	3.55	3.79
48	5.08	4.80	4.25	3.98	4.17	4.49

*From "Interim Guide for Design of Pavement Structures," American Association of State Highway and Transportation Officials, 1972.

direction traffic to one or more of the outer lanes, with lesser values to inner lanes, when there are six or more lanes in both directions.

See also the following discussions of flexible pavements and portland-cement concrete pavements.

Economic Factors. An adequate structural section can be achieved with various combinations of materials. In selection of the appropriate design section, economy, both in first cost and in future maintenance costs, should be a primary consideration.

Table 16-18. Traffic Equivalence Factors, Rigid Pavement*

a. SINGLE AXLES, $p_t = 2.0$

Axle load, kips	Slab thickness D, in.					
	6	7	8	9	10	11
2	0.0002	0.0002	0.0002	0.0002	0.0002	0.0002
4	0.002	0.002	0.002	0.002	0.002	0.002
6	0.01	0.01	0.01	0.01	0.01	0.01
8	0.03	0.03	0.03	0.03	0.03	0.03
10	0.09	0.08	0.08	0.08	0.08	0.08
12	0.19	0.18	0.18	0.18	0.17	0.17
14	0.35	0.35	0.34	0.34	0.34	0.34
16	0.61	0.61	0.60	0.60	0.60	0.60
18	1.00	1.00	1.00	1.00	1.00	1.00
20	1.55	1.56	1.57	1.58	1.58	1.59
22	2.32	2.32	2.35	2.38	2.40	2.41
24	3.37	3.34	3.40	3.47	3.51	3.53
26	4.76	4.69	4.77	4.88	4.97	5.02
28	6.59	6.44	6.52	6.70	6.85	6.94
30	8.92	8.68	8.74	8.98	9.23	9.39
32	11.87	11.49	11.51	11.82	12.17	12.44
34	15.55	15.00	14.95	15.30	15.78	16.18
36	20.07	19.30	19.16	19.53	20.14	20.71
38	25.56	34.54	24.26	24.63	25.36	26.14
40	32.18	30.85	30.41	30.75	31.58	32.57

b. TANDEM AXLES, $p_t = 2.0$

Axle load, kips	Slab thickness D, in.					
	6	7	8	9	10	11
10	0.01	0.01	0.01	0.01	0.01	0.01
12	0.03	0.03	0.03	0.03	0.03	0.03
14	0.05	0.05	0.05	0.05	0.05	0.05
16	0.09	0.08	0.08	0.08	0.08	0.08
18	0.14	0.14	0.13	0.13	0.13	0.13
20	0.22	0.21	0.21	0.20	0.20	0.20
22	0.32	0.31	0.31	0.30	0.30	0.30
24	0.45	0.45	0.44	0.44	0.44	0.44
26	0.63	0.64	0.62	0.62	0.62	0.62
28	0.85	0.85	0.85	0.85	0.85	0.85
30	1.13	1.13	1.14	1.14	1.14	1.14
32	1.48	1.45	1.49	1.50	1.51	1.51
34	1.91	1.90	1.93	1.95	1.96	1.97
36	2.42	2.41	2.45	2.49	2.51	2.52
38	3.04	3.02	3.07	3.13	3.17	3.19
40	3.79	3.74	3.80	3.89	3.95	3.98
42	4.67	4.59	4.66	4.78	4.87	4.93
44	5.72	5.59	5.67	5.82	5.95	6.03
46	6.94	6.76	6.83	7.02	7.20	7.31
48	8.36	8.12	8.17	8.40	8.63	8.79

c. SINGLE AXLES, $p_t = 2.5$

Axle load, kips	Slab thickness D, in.					
	6	7	8	9	10	11
2	0.0002	0.0002	0.0002	0.0002	0.0002	0.0002
4	0.003	0.002	0.002	0.002	0.002	0.002

Table 16-18. Traffic Equivalence Factors, Rigid Pavement* *(Continued)*

c. SINGLE AXLES, $p_t = 2.5$

Axle load, kips	Slab thickness D, in.					
	6	7	8	9	10	11
6	0.01	0.01	0.01	0.01	0.01	0.01
8	0.04	0.04	0.03	0.03	0.03	0.03
10	0.10	0.09	0.08	0.08	0.08	0.08
12	0.20	0.19	0.18	0.18	0.18	0.17
14	0.38	0.36	0.35	0.34	0.34	0.34
16	0.63	0.62	0.61	0.60	0.60	0.60
18	1.00	1.00	1.00	1.00	1.00	1.00
20	1.51	1.52	1.55	1.57	1.58	1.58
22	2.21	2.20	2.28	2.34	2.38	2.40
24	3.16	3.10	3.23	3.36	3.45	3.50
26	4.41	4.26	4.42	4.67	4.85	4.95
28	6.05	5.76	5.92	6.29	6.61	6.81
30	8.16	7.67	7.79	8.28	8.79	9.14
32	10.81	10.06	10.10	10.70	11.43	11.99
34	14.12	13.04	12.94	13.62	14.59	15.43
36	18.20	16.69	16.41	17.12	18.33	19.52
38	23.15	21.14	20.61	21.31	22.74	24.31
40	29.11	26.49	25.65	26.29	27.91	29.90

d. TANDEM AXLES, $p_t = 2.5$

Axle load, kips	Slab thickness D, in.					
	6	7	8	9	10	11
10	0.01	0.01	0.01	0.01	0.01	0.01
12	0.03	0.03	0.03	0.03	0.03	0.03
14	0.06	0.05	0.05	0.05	0.05	0.05
16	0.10	0.09	0.08	0.08	0.08	0.08
18	0.16	0.14	0.14	0.13	0.13	0.13
20	0.23	0.22	0.21	0.21	0.20	0.20
22	0.34	0.32	0.31	0.31	0.30	0.30
24	0.48	0.46	0.45	0.44	0.44	0.44
26	0.64	0.64	0.63	0.62	0.62	0.62
28	0.85	0.85	0.85	0.85	0.85	0.85
30	1.11	1.12	1.13	1.14	1.14	1.14
32	1.43	1.44	1.47	1.49	1.50	1.51
34	1.82	1.82	1.87	1.92	1.95	1.96
36	2.29	2.27	2.35	2.43	2.48	2.51
38	2.85	2.80	2.91	3.04	3.12	3.16
40	3.52	3.42	3.55	3.74	3.87	3.94
42	4.32	4.16	4.30	4.55	4.74	4.86
44	5.26	5.01	5.16	5.48	5.75	5.92
46	6.36	6.01	6.14	6.53	6.90	7.14
48	7.64	7.16	7.27	7.73	8.21	8.55

*From "Interim Guide for Design of Pavement Structures," American Association of State Highway and Transportation Officials, 1972.

Pavement Life. The useful life of a pavement can be defined as that period of time during which the pavement structure is expected to continue to function without any appreciable loss of its support value, while maintaining an acceptable surface condition.

The pavement life (not to be confused with pavement design period) can be extended through the use of various maintenance measures, as well as through planned stage construction. Staged construction consists of the application of successive pavement layers in accordance with a design, taking into account traffic loadings over a predetermined time schedule. This method has many advantages, including improved pavement performance, more accurate analysis of traffic through successive evaluations, and often a more effective utilization of funds.

Flexible Pavements.   A flexible pavement structure may consist of two or more layers.   The layers, beginning at the subgrade and following in order upward, are generally designated as subbase course, base course, and surface course.   The design procedure includes determination of the total thickness of pavement structure as well as the thickness of the individual components, surface, base, and subbase courses.   The procedure should include design of equivalent alternative sections, and selection of structure should be primarily a function of availability of materials and comparative costs.

Subbase Course.   The subbase course is the portion of the flexible pavement structure between the subgrade and the base course.   The subbase usually consists of a compacted layer of granular material, either treated or untreated, or a layer of soil treated with a suitable admixture. In addition to its position in the pavement, it is usually distinguished from the base course material by less stringent specification requirements for strength, aggregate types, and gradation.

The subbase course is usually used to build up the pavement strength economically above that provided by the subgrade soils.   However, the subbase can be omitted if the required pavement structure is relatively thin or if subgrade soils are of high quality, with no moisture problems.   When either is the case, the base course can be constructed directly on the subgrade.

In addition to their major function as a structural portion of the pavement, subbase courses may have additional secondary functions, such as:

1. To prevent intrusion of fine-grained roadbed soils into base courses.   Relatively well-graded materials must be specified if the subbase is intended to serve this purpose.

2. To minimize the damaging effects of frost action.   For this purpose, materials not susceptible to detrimental frost action should be specified.

3. To help prevent accumulation of free water within or below the pavement structure. Relatively free-draining material should be specified if the subbase is intended to serve this purpose, and provision should be made for collecting and removing accumulated water from the subbase.

4. To provide a working platform for construction equipment or for subsequent pavement courses in rock cuts.

Base Course.   The base course is the portion of the flexible pavement structure immediately beneath the surface course.   It is constructed on the subbase course or, if no subbase is used, directly on the subgrade.   It performs its major function as a structural portion of the pavement. The base usually consists of aggregates such as crushed stone, crushed slag, or crushed or uncrushed gravel and sand, or of combinations of these materials.   The aggregates may be used untreated or treated with stabilizing admixtures such as portland cement, asphalt, or lime. Generally, specifications for base-course materials are considerably more stringent than those for subbase materials in requirements for strength, stability, hardness, aggregate types, and gradation.

Requirements in AASHTO Specifications M147 and M75 are typical of specifications for gradation and quality of untreated base aggregates.   However, materials varying in gradation and quality from these specifications have been used in certain areas and have provided satisfactory performance.

A wide variety of materials unsuitable for use as untreated base course have given satisfactory performance when improved by addition of a stabilizing admixture, such as portland cement, asphalt, or lime.   Consideration should be given to the use of such treated materials for base courses whenever it is economically feasible, particularly when suitable untreated materials are in short supply.

Table 16-19 can be used as a guide to establishment of specification requirements for stabilized base courses.   Careful study is required to select the type and amount of admixture to be used for optimum performance and economy.   Use of stabilized base courses can be extremely desirable when traffic must be maintained throughout pavement construction.   Plant-mixed asphalt-concrete bases are often used in these situations.

Surface Course.   In addition to its major function as a structural portion of the pavement, the surface course should be designed to resist the abrasive forces of traffic, limit the amount of surface water that penetrates into the pavement, provide a skid-resistant surface, and furnish a smooth and uniform riding surface.   The surface course also should be durable, able to resist fracture and raveling without becoming unstable under expected traffic and climatic conditions.

Usually constructed on a base course, the surface course of a flexible pavement structure consists of a mixture of mineral aggregates and bituminous materials.   The success of such a

Table 16-19. Some Typical Specification Requirements for Stabilized Base Courses*

Specification	Cement-treated			Bituminous-treated		Lime-treated
	Class A	Class B	Class C	Class 1	Class 2	
Sieve analysis, % passing						
2½ in.	100	100	100			
¾ in.	· · ·	· · ·	75–95			
No. 4	65–100	55–100	25–60			
No. 10	20–45	· · ·	15–45			
No. 40	15–30	25–50	8–30			
No. 200	5–12	5–20	2–15			
Compressive strength, † psi at 7 days	650–1,000	300–650				
Soil support value S	· · ·	· · ·	8.0 min.			10.0 min
Stability						
Hveem Stabilometer				35 min	25 min	
Hubbard-Field				1,200 min	1,000 min	
Marshall-stability				750 min	500 min	
Marshall-flow				16 max	20 max	
Plasticity Index ‡	12 max			6 max	6 max	6 max

*From "Interim Guide for Design of Pavement Structures," American Association of State Highway and Transportation Officials, 1972.

†As determined in unconfined compression tests on cylinders 4 in. in diameter and 4 in. high. Test specimens should contain the same percentage of portland cement, and be compacted to the same density as achieved in construction.

‡Performed on samples prepared in accordance with AASHTO T87; apply to aggregate prior to mixing with the stabilizing admixture, except that in the case of lime-treated base the value applies after mixing.

course depends to a considerable degree on obtaining a mixture with the optimum gradation of aggregate and percent of bituminous binder. Use of a laboratory-tested design procedure or use of a proven specification is essential to insure that a mixture will be satisfactory.

Well-graded aggregates with a maximum size of about ¾ to 1 in. are commonly specified for surface courses for highways. Nevertheless, a wide variety of other gradations—from sand as in *sheet asphalt,* to coarse, open-graded mixtures—has been used and has provided satisfactory performance in specific conditions.

Surface-course asphaltic concrete is usually prepared by plant mixing of heated aggregates, mineral filler, and asphalt cement. Satisfactory performance also has been obtained with plant mixing of cold aggregates and specially formulated asphalt, and also by mixing the composition in place with liquid asphalts or asphalt emulsions.

Construction specifications usually require that before a surface course is placed, liquid bituminous material be applied on untreated aggregate base courses as a prime coat, and on treated base courses and between layers of the surface course as a tack coat.

Minimum Layer Thickness. Design procedures that establish pavement layer thickness also should take into consideration construction requirements for placing the pavement courses. For example, it is impracticable to construct pavement courses in thicknesses less than 1¼ to 1½ times the largest aggregate size of the mixture. Considering aggregate sizes normally used, a guide for the minimum practical thicknesses that can generally be applied is:

> Surface course . . . . 1½ in.
> Base course . . . . . . 3 in.
> Subbase course . . . 4 in.

Flexible Pavement Design Procedure. The design procedure, as outlined in the AASHTO "Interim Guide for Design of Pavement Structures," is based on design equations developed by taking into account the following parameters: terminal serviceability index p_t, equivalent 18-kip single-axle loads, and soil support value S. The last is based on an empirical scale with values from 0 to 10. S = 3.0 represents silty clay roadbed soils used in the AASHTO road tests (a firm and valid point), and S = 10.0 represents a crushed rock base used in the road tests (also a reasonably valid point).

The units of soil support represented by the soil-support scale have no direct relationship to any procedure for testing soils. It is necessary, therefore, to establish a correlation between soil support and some testing procedure for this design method to be utilized properly. Figure 16-17 illustrates the relationship between S and several common methods of establishing subgrade strength.

Regional factor R was included in the design equation to reflect climatic and environmental factors for conditions different from those in the road tests. Based on road-test information, R values that may be used as a guide are:

Subgrade materials frozen to a depth of 5 in. or more ... 0.2 to 1.0
Subgrade materials, dry, summer and fall .... 0.3 to 1.5
Subgrade materials, wet, spring thaw .... 4.0 to 5.0

In general, R should not exceed about 4.0 nor be less than about 0.5 for conditions in the United States. The regional factor may not adjust correctly for special conditions, such as serious frost conditions or other local problems.

Structural number SN is an abstract number expressing the structural strength of pavement required for a given combination of soil-support value, total equivalent 18-kip single-axle loads, terminal serviceability index, and regional factor. The required SN must be converted to actual thicknesses of surfacing, base, and subbase by means of appropriate layer coefficients, which represent the relative strength of the material to be used for each layer, as outlined below.

SN for the entire pavement may be obtained from

$$SN = a_1 D_1 + a_2 D_2 + a_3 D_3 \qquad (16\text{-}5)$$

where a_1, a_2, a_3 = layer coefficients representative of surface, base, and subbase courses, respectively
D_1, D_2, D_3 = actual thickness, in., of surface, base, and subbase courses, respectively

Layer coefficients are assigned to materials used in the pavement structure to convert structural numbers to actual thickness. The layer coefficient for a material expresses the empirical relationship between SN and thickness and is a measure of the relative ability of the

Table 16-20. Structural Layer Coefficients Proposed by AASHTO Committee on Design*

Pavement component	Coefficient
Surface Course	
Road mix (low stability)	0.20
Plant mix (high stability)	0.44†
Sand asphalt	0.40
Base Course	
Sandy gravel	0.07§
Crushed stone	0.14†
Cement-treated (no soil-cement)	
Compressive strength @ 7 days	
650 psi or more‡	0.23§
400 psi to 650 psi	0.20
400 psi or less	0.15
Bituminous-treated	
Coarse-graded	0.34§
Sand asphalt	0.30
Lime-treated	0.15–0.30
Subbase Course	
Sandy gravel	0.11†
Sand or sandy clay	0.05–0.10

*From "Interim Guide for Design of Pavement Structures," American Association of State Highway and Transportation Officials, 1972.
†Established from AASHTO road tests.
‡Compressive strength at 7 days.
§This value has been estimated from AASHTO road tests, but not to the accuracy of those factors marked with †.

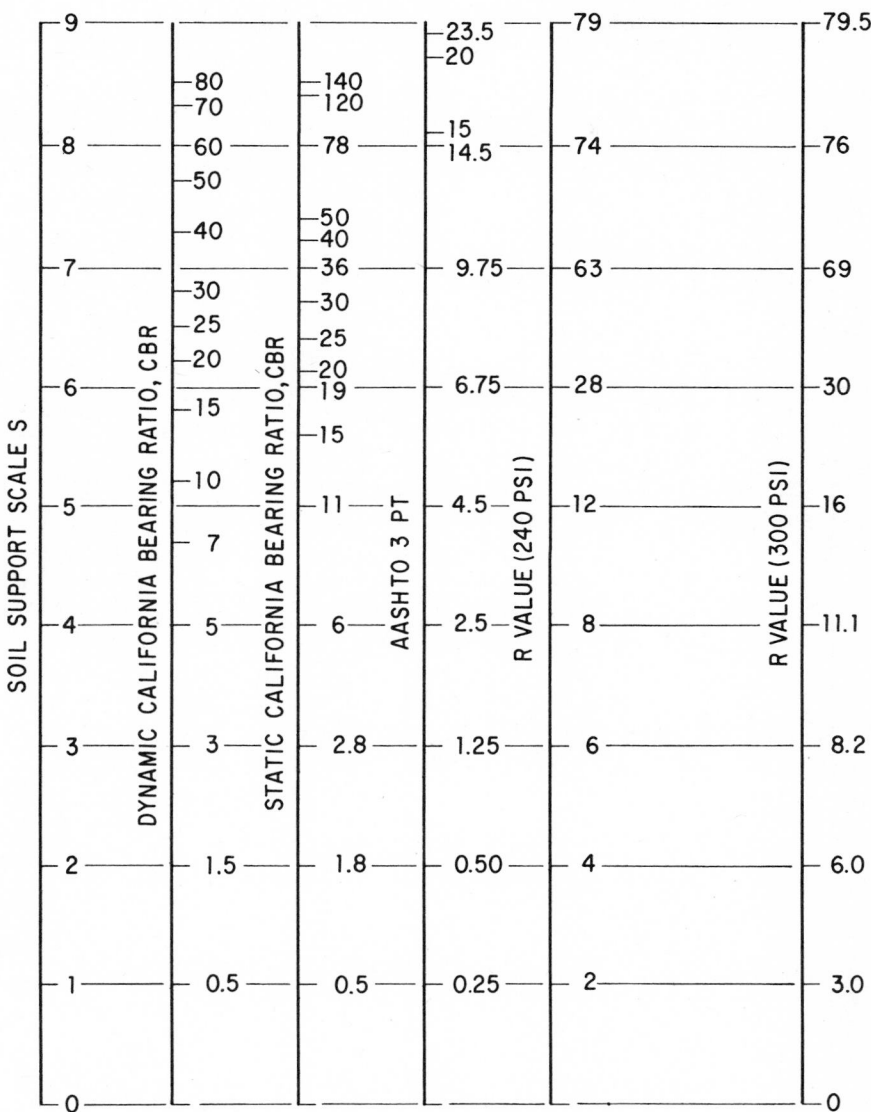

Fig. 16-17. Chart indicates equivalence of soil support values S and soil bearing determinations made by several common methods for evaluating subgrade strengths. ("*Interim Guide for Design of Pavement Structures,*" *American Association of State Highway and Transportation Officials.*)

Table 16-21. Structural Layer Coefficients Used by Various States*

Component	Alabama	Arizona	Delaware	Massachusetts	Minnesota	Montana	Nevada	New Hampshire
Surface Courses								
Plant mix (high stability)	0.44	0.35–0.44	0.35–0.40	0.44	0.315	0.30–0.40	0.30–0.35	0.38
Road mix (low stability)	0.20	0.25–0.38			Plant mix (low stab.) 0.28	0.20	0.17–0.25	0.20
Sand Asphalt	0.40	0.25						0.20
Base Courses								
Untreated	Limestone 0.14 Slag 0.14 Sandstone 0.13 Granite 0.12	Sand & gravel, well graded 0.14 cinders 0.12–0.14 Sandy gravel, mostly sand 0.11–0.13	Waterbound macadam 0.20 Crusher run 0.14 Quarry waste 0.11 Select borrow 0.08	Crushed stone 0.14	Crushed rock (Cl. 5 & 6 gravel) 0.14 Sandy gravel 0.07	Select surf 0.10 Crushed gravel 0.12–0.14	Crushed gravel 0.10–0.12 Crushed rock 0.13–0.16	Crushed gravel 0.10 Bank run gravel 0.07 Crushed stone 0.14
Cement-treated								
650 psi or more	0.23	500 psi + 0.25–0.30	Soil-cement 0.20			400 psi or more 0.20		gravel 0.17
400 to 650 psi	0.20	300–500 psi 0.18–0.25				0.15		
400 psi or less	0.15	less than 300 psi 0.15						
Lime-treated								
Bituminous-treated	Course graded 0.030 Sand 0.25	Sand-gravel 0.25–0.34 Sand 0.20	Asph. stab. 0.10	Black base 0.34 Penetrated crushed stone 0.29	0.175–0.21	0.15–0.20 Plant mix 0.30 Bit. stab. 0.20	Plant mix 0.25–0.34	Bit. conc. 0.34 Gravel 0.24
Subbase	Sand & sandy clay 0.11 Chert, low P.I. 0.10	Sand-gravel, well graded 0.14 Cr. stone or cinders 0.12	Select borrow 0.08	Gravel 0.11 Select material 0.08	Sandy gravel (Cl. 3 & 4 gravel) 0.105 Selected granular	Sand 0.05 Sp. borrow 0.07	Gravel type 1 0.09–0.11 Select material 0.05–0.09	Sand-gravel 0.05

*

Topsoil	Sand & silty clay	(12% minus No. 200)
0.09	0.05–0.10	0.07
Float gravel		
0.09		
Sand & silty clay		
0.05		

Notes:
1. Indiana, Iowa, Montana, New Jersey, Tennessee, and Puerto Rico conform to AASHTO Guides
2. North Carolina conforms to AASHTO Guides, except 0.30 for bituminous-treated base
3. North Dakota conforms to AASHTO Guides, except 0.30 for bituminous aggregate base
4. Maine conforms to AASHTO Guides with some modification. No further information.
5. Maryland substitution values for materials to replace design thickness of asphalt hot-mix are the AASHTO coefficients expressed in equivalent values, in.

*From "Interim Guide for Design of Pavement Structures," American Association of State Highway and Transportation Officials, 1972.

material to function as a structural component of the pavement.   Average values of layer coefficients for the materials used in AASHTO road tests are as follows:

Asphaltic-concrete surface course 0.44
Crushed-stone base course 0.14
Sandy gravel subbase course 0.11

Table 16-20 may be used as a guide to ranges of layer coefficients that were developed by AASHTO.   Table 16-21 lists layer coefficients in use for various materials, reported in a survey of several states.   In most cases, a layer-coefficient value, or a range of values, is assigned on the basis of a description of a material type.   A few states evaluate or measure the coefficient by means of a laboratory test on the material in the pavement structure.

Use of Design Charts.   The design equation for flexible pavements is presented in the form of two nomographs for simplicity of application.   Separate nomographs are presented for a terminal serviceability index $p_t = 2.5$ (Fig. 16-18) and $p_t = 2.0$ (Fig. 16-19).

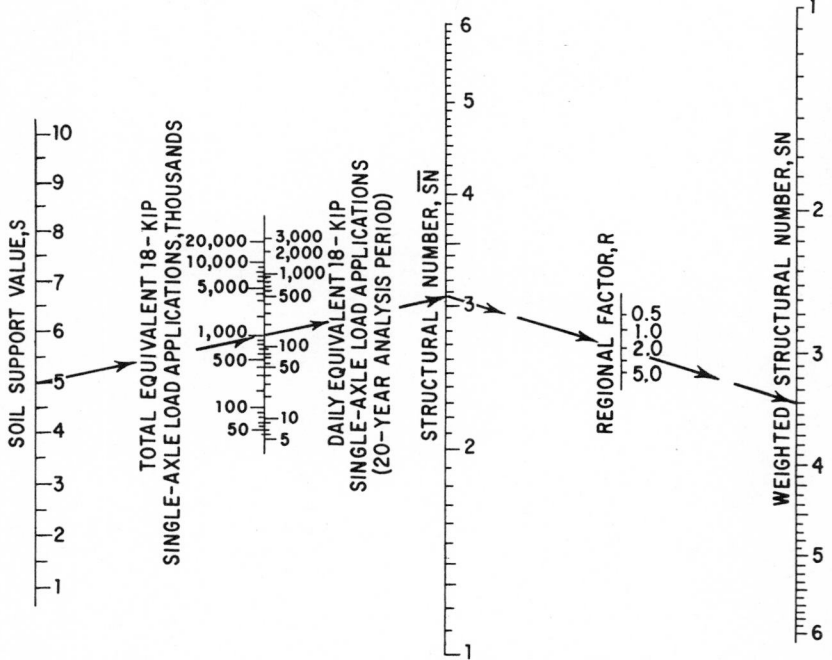

Fig. 16-18. Design chart for flexible pavement, with terminal serviceability index $p_t = 2.5$. *("Interim Guide for Design of Pavement Structures," American Association of State Highway and Transportation Officials.)*

Figure 16-18 is intended for use in design of major highways and assumes that resurfacing or reconstruction will be performed when the level of serviceability reaches 2.5.

Figure 16-19 may be used for other highways where a somewhat lesser level of serviceability (2.0) may be tolerated.   For design of temporary highways or for stage construction, an appropriate traffic analysis period should be used.

Once the decision about the terminal serviceability index p_t has been made and the appropriate design chart has been selected, the following should be determined:

1. Representative values of soil support S for the subgrade soil.

2. The total or daily equivalent 18-kip single-axle loads estimated for the design lane for the traffic analysis period.   Because selection of the traffic equivalence factors to be used to convert mixed traffic to total equivalent 18-kip single-axle loads depends on the structural num-

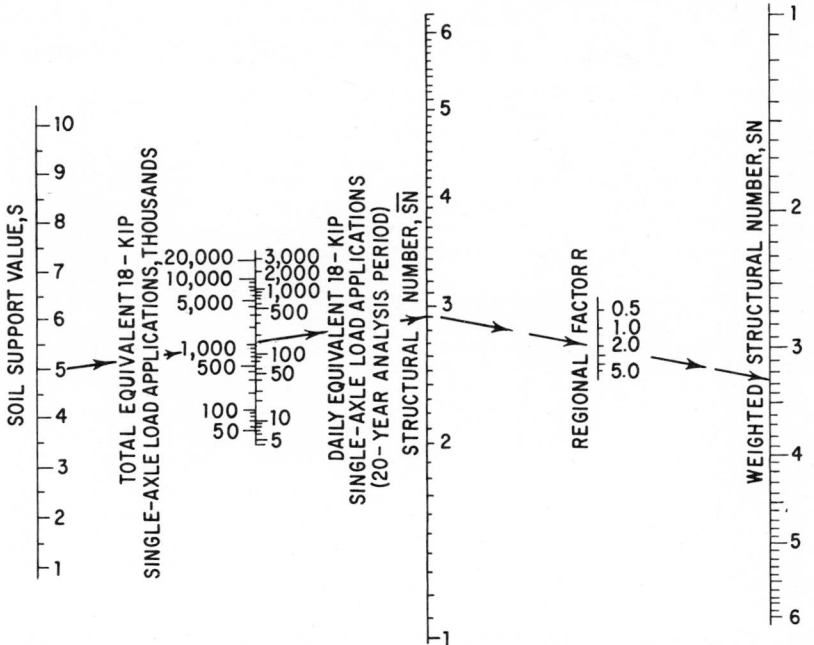

Fig. 16-19. Design chart for flexible pavement, with $p_t = 2.0$. (*"Interim Guide for Design of Pavement Structures," American Association of State Highway and Transportation Officials.*)

ber SN (Table 16-17), this number must be assumed for the initial conversion. The use of an SN of 3 for determination of 18-kip single-axle traffic-equivalence factors will normally give results that are sufficiently accurate for design purposes, even though the final SN determined is substantially different. This assumption will usually result in a conservative estimate of equivalent 18-kip single-axle load applications, but generally the resulting error in SN is not significant.

3. The regional factor R applicable to the region.

The chart (Fig. 16-18 or 16-19) requires two applications of a straightedge for each solution. First, the soil-support value of the subgrade soil (on the left scale) and the total or daily equivalent 18-kip single-axle loads for the traffic analysis period (second scale) are used to solve for the unweighted structural number (center scale). This unweighted structural number is used with the selected regional factor (fourth scale) to solve for the design SN (right scale) applicable to the total pavement structure. Suitable designs are those whose combinations of materials, types, and thicknesses satisfy Eq. (16-5).

If available alternative types of material are to be considered for one or more of the pavement courses, the preceding procedure may be used to prepare the alternative designs of equal total, weighted structural numbers. The resulting alternative designs may then be compared, and the optimum design may be selected on the basis of economics and other applicable considerations.

Full-Depth Asphalt Pavements. A full-depth asphalt pavement is a pavement structure in which asphalt and aggregate mixtures are employed for all courses above the subgrade. There are certain advantages to using this type of pavement construction where availability of material, economic, and construction considerations warrant its use. Some of the advantages of full-depth asphalt pavement are:

1. The construction time is reduced, as compared with a mixed material pavement.
2. It has no permeable granular layers to entrap water and impair performance.
3. The pavement structure is thinner than if untreated granular courses are used.
4. The completed course can be used to serve traffic during construction.

The procedure for designing full-depth asphalt pavement is described in "Thickness Design—Full Depth Asphalt Pavement Structures for Highways and Streets," Manual MS-1, The Asphalt Institute, College Park, Md.

Portland-cement Concrete Pavements. A rigid pavement structure generally consists of two layers, designated as the pavement slab and the subbase course. When the subgrade soils are granular in nature, the subbase course is often omitted.

Subbases. The subbase of a rigid pavement structure consists of one or more compacted layers of granular or stabilized material placed between the subgrade and the rigid slab for the following purposes:

To provide uniform, stable and permanent support; increase the modulus of subgrade reaction k; minimize the damaging effects of frost action, and prevent pumping of fine-grained soils at joints, cracks, and edges of the rigid slab. Also, to reduce cracking and faulting; provide a working platform for construction equipment; provide a means of handling subsurface water, and provide a means of grading a true and uniform surface to support the concrete slab when in rock excavation.

Typical specifications for several types of subbases are shown in Table 16-22.

Table 16-22. Types of Subbases for Rigid Pavements*

Specification	Type A (open-graded)	Type B (dense-graded)	Type C (cement-treated)	Type D (lime-treated)	Type E (bituminous-treated)	Type F (granular)
Sieve analysis % Passing:						
1½ in.	100	100	100			100
¾ in.	60–90	85–100		†	†	
No. 4	35–60	50–80	65–100			65–100
No. 40	10–25	20–35	25–50			25–50
No. 200	0–7	5–12	5–20			0–15

(The minus No. 200 material should be held to a practical minimum)

Specification	Type A (open-graded)	Type B (dense-graded)	Type C (cement-treated)	Type D (lime-treated)	Type E (bituminous-treated)	Type F (granular)
Compressive strength: psi at 28 days			400–750	100		
Stability:						
Hveem Stabilometer					20 min	
Hubbard-Field					1,000 min	
Marshall-stability					500 min	
Marshall-flow					20 max	
Soil constants:						
Liquid limit	25 max	25 max				25 max
Plasticity index‡	N.P.	6 max	10 max§		6 max§	6 max

*From "Interim Guide for Design of Pavement Structures," American Association of State Highway and Transportation Officials, 1972.

†To be determined by complete laboratory analysis, taking into consideration the ability of the stabilized mixture to resist under-slab erosion.

‡As performed on samples prepared in accordance with AASHTO T 87.

§These values apply to the mineral aggregate prior to mixing with the stabilizing agent.

Rigid-Pavement Design Procedure. The design procedure for concrete pavement outlined below was developed on the basis of AASHTO road tests. The specific assumptions and methodology used in developing this design method are given in AASHTO "Interim Guide for Design of Pavement Structures," 1972. This design method is based on the following parameters: terminal serviceability index p_t, equivalent 18-kip single-axle loads, and modulus of subgrade reaction k (Westergaard's modulus of subgrade reaction, referred to as a *gross k* in AASHTO road test reports, which represents the load, psi, on a loaded area divided by the deflection, in., of that area). The scales for k included in design charts (Figs. 16-20 and 16-21) are correlated with values obtained by plate-loading tests performed in accordance with AASHTO T222 with a 30-in. dia plate. The k value may be estimated on the basis of previous experience or by correlation with other tests.

Working Stress in Concrete f_t. The modulus of rupture S_c of concrete at 28 days, as determined by the test procedure in AASHTO T97 with third-point loading, is the basis for determining concrete flexural strengths. If test data are normally available for other than 28-day strengths, the expected 28-day strengths should be obtained from a time-strength correlation

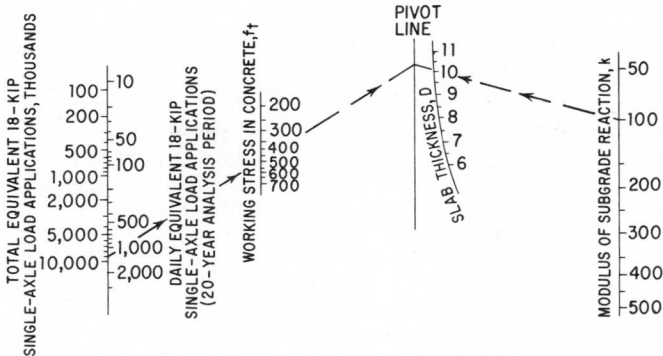

Fig. 16-20. Design chart for rigid pavement, with terminal serviceability index p_t = 2.5. (*"Interim Guide for Design of Pavement Structures,"* American Association of State Highway and Transportation Officials.)

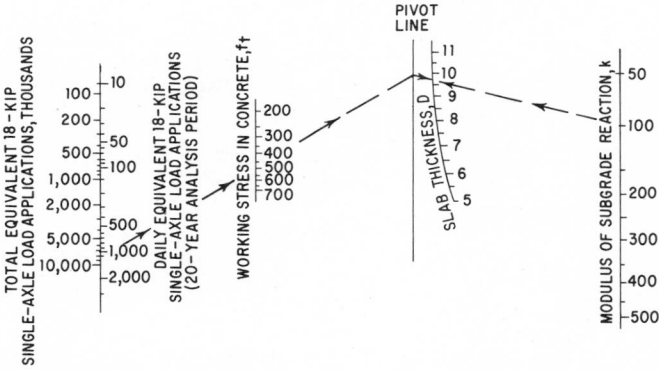

Fig. 16-21. Design chart for rigid pavement, with p_t = 2.0. (*"Interim Guide for Design of Pavement Structures,"* American Association of State Highway and Transportation Officials.)

and the extrapolated or interpolated values used.   Working stress f_t is based on S_c.   The scale included in the design charts (Figs. 16-20 and 16-21) is for working stress, which may be taken as $0.75S_c$.

Pavement Thickness.   Determination of the thickness of the pavement slab is accomplished by use of the design charts (Figs. 16-20 and 16-21).

Select the applicable chart on the basis of the desired terminal serviceability index p_t.   Figure 16-20 is applicable for p_t = 2.5 and Fig. 16-21 for p_t = 2.0.

With a straightedge, draw a line from the estimated equivalent total of daily 18-kip single-axle loads on the left scale through the applicable value of working stress f_t of the concrete on the second scale, to intersect the pivot line.

With a second application of the straightedge, draw a line from the intersection of the pivot line to the applicable value of modulus of subgrade reaction k on the right scale.   The intersection of this line with the second scale from the right is the thickness D, in., of the pavement slab.

Several other design methods for concrete pavements are outlined in the manual "Thickness Design for Concrete Pavements," Portland Cement Association, Old Orchard Road, Skokie, Ill. 60076.

Reinforcement Requirements.   The purpose of distributed steel reinforcement in reinforced concrete pavement is not primarily to prevent cracking, but rather to hold tightly closed any cracks that may form, thus maintaining the pavement as an integral structural unit.   Figure 16-22 shows a reinforced concrete pavement design used by the Virginia Department of High-

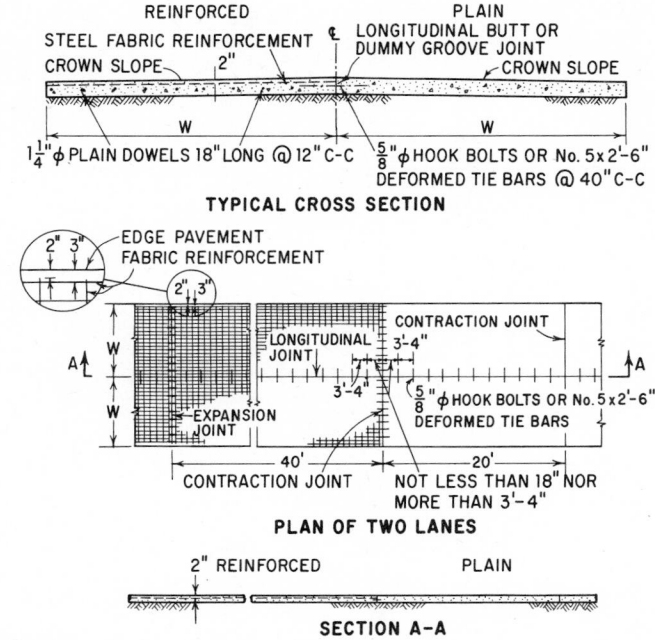

Fig. 16-22. Typical details for reinforced concrete pavement.

NOTES

1. Steel fabric reinforcement: This should consist of members rigidly attached at all joints or points of intersection, except where hinged steel reinforcement may be used instead of rigid sheets. Longitudinal members are 2-gage wire, 6 in. c to c. Transverse members are 4-gage wire, 12 in. c to c.

2. Widths of steel fabric sheets should be 4 in. less than the slab width. Not more than three sheets may be used between contraction and expansion joints.

3. All members should be so cut that projecting ends will extend at least 1 in. but not more than 11 in. from the joints or points of intersection of the fabric members.

4. When lapping of the steel fabric is necessary, the amount of lap should be at least equal to the spacing of the wires parallel to the lap.

5. Contraction joints should be spaced at 40-ft intervals for reinforced concrete pavement and at 20-ft intervals for plain concrete pavement.

6. Longitudinal joints: If the sum of lane widths does not exceed 25 ft, two lanes of the concrete pavement may be constructed simultaneously, if a satisfactory and true longitudinal, dummy groove joint is obtained.

7. The maximum pavement width that may be constructed without a longitudinal joint is 14 ft, except where single lanes of pavement, such as ramps, are to be built, in which case the maximum width may be 16 ft. (*Virginia Department of Highways and Transportation.*)

ways and Transportation. The pavement slab tends to shorten when its temperature drops or its moisture content decreases, which occurs when portland-cement concrete undergoes curing. These contractions are resisted by the subgrade through friction between it and the slab. The resistance to movement should be balanced by the tensile strength of the concrete or the tensile resistance of reinforcing steel. Maximum steel stress will occur at a crack at midlength of a slab. Reinforcement should be designed for the stress developed at this location.

Resistance to movement will vary with the amount of movement, size of slab, rate of temperature change, and characteristics of the subgrade. The maximum resistance is reached when actual sliding of the slab on the subgrade occurs. A coefficient of resistance of 1.5 may be used for design purposes. (This coefficient may vary between 1 and 2.)

The required cross-sectional area of steel A_s, sq in. per ft of slab width, is given by

$$A_s = \frac{FLw}{2f_s} \tag{16-6}$$

where F = coefficient of resistance between slab and subgrade

L = distance, ft, between free transverse joints or between free longitudinal edges

w = weight of pavement slab, lb per sq ft

f_s = allowable working stress in the steel, psi

The formula is used for both longitudinal and transverse steel and is solved graphically in Fig. 16-23, with F assumed to be 1.5 and weight of concrete taken as 150 lb per cu ft.

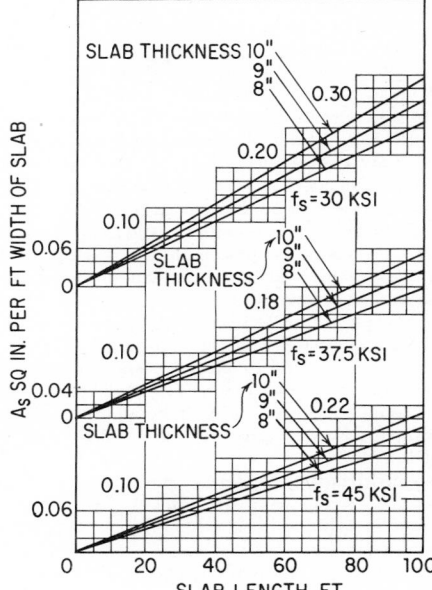

Fig. 16-23. Chart indicates distributed steel reinforcement required for jointed rigid pavements. (*"Interim Guide for Design of Pavement Structures,"* American Association of State Highway and Transportation Officials.)

Continuously Reinforced Pavements. This type of portland-cement concrete pavement is constructed with continuous longitudinal steel reinforcement without the use of intermediate transverse expansion or contraction joints. The pavement is thereby allowed to develop random cracks, which are held together by the steel reinforcement. As a rule, cracking would start within a few days after construction, with full cracking developing within the first few years. The reinforcing steel in continuously reinforced concrete pavements generally consists of deformed bars, bar mats, or deformed-wire mesh.

Of the many factors that influence the size and spacing of cracks, the most important is the percentage of reinforcing steel, often given as a ratio of cross-sectional area of longitudinal steel to cross-sectional area of concrete slab. Current practice calls for reinforcing steel to range from 0.5 to 0.7% of the concrete cross-sectional area. The steel is generally placed at or slightly above middepth of the concrete slab. With construction methods designed for the purpose, transverse reinforcement to serve as tie bars between lanes and hold longitudinal cracks closed is not needed.

Other pertinent factors include bonding area and characteristics of the steel, depth of reinforcement, friction between concrete pavement and base course, the season when construction occurs, strength of concrete, and curing temperature.

At present, there is no single method of determining the thickness of continuously reinforced concrete pavement (CRCP) that is universally accepted. A common practice has been to determine the thickness required for a jointed pavement and then use a lesser thickness for the CRCP. This lesser thickness is arrived at by applying a predetermined ratio to the jointed pavement thickness or by subtracting a specified amount from it. The 1972 AASHTO "Interim Guide" states that the thickness of CRCP "may be less than that obtained from the charts, with the amount of reduction in thickness being based on local experience or other studies."

A discussion in greater depth of design and construction methods for CRCP is given in a Transportation Research Board publication, "Continuously Reinforced Concrete Pavement" (National Cooperative Highway Research Program Synthesis of Highway Practice-16).

The reinforcement layout for CRCP used by the Virginia Department of Highways and Transportation is shown in Fig. 16-24.

Expansion Joints. The primary functions of an expansion joint are preventing development of damaging compressive stresses caused by volume changes in the pavement slab and preventing excessive pressures from being transmitted to adjacent structures.

Frequent expansion joints are not necessary for rigid pavements. Expansion joints should be placed adjacent to structures and at intervals of about 40 to 60 ft. At these locations, expansion joints should be protected with satisfactory load-transfer devices or suitable joint fillers. A ¾- to 1-in. width of joint is generally used. Joint widths of 4 to 5 ft are used for relieving stresses in some locations, such as at bridges and other structures. Consideration can be given to the use of suitable terminal anchorage devices in lieu of expansion joints.

Contraction Joints. The purpose of contraction joints is to provide for an orderly arrangement of the cracking that occurs when concrete undergoes curing. If the joints are properly designed and spaced, a minimum of cracking outside the joints can be expected.

Contraction joints may be sawed in the hardened concrete, formed by plastic inserts, or tooled into the concrete at placement time. Depth of joint should be about one-quarter the thickness of the pavement slab. The design of the joint should be related to the expected joint opening and the elongation of the joint filler used. Adequate load transfer through mechanical means or aggregate interlock should be provided at all joints.

Longitudinal Joints. Longitudinal joints are used to prevent formation of irregular longitudinal cracks and to allow for lane construction. They may be keyed, butted, mechanically formed, or sawed grooves. To keep adjacent lanes from separating and faulting, steel tie bars or connections should be embedded in the concrete transversely to the joint. The depth of formed or sawed grooves should not be less than one-quarter the thickness of the pavement slab.

Load-transfer Devices. Mechanical load-transfer devices used at pavement joints should possess the following attributes:

They should be simple in design, be practical to install, and permit complete encasement in the concrete.

They should distribute the load stresses properly without overstressing the concrete at its contact with the device.

They should offer little restraint to longitudinal movement of the joint at any time.

They should be mechanically stable under the wheel loads and frequencies of loading that will prevail.

They should be resistant to corrosion when used in those geographic locations where corrosion may occur.

A commonly used load-transfer device is the conventional round steel dowel conforming to AASHTO M227, Grade 70 or higher. Minimum design requirements for round dowels are given in Table 16-23a.

Tie Bars. These are either deformed steel bars or connectors used to hold the faces of abutting slabs in firm contact. Tie bars are designed to withstand the maximum tensile forces required to overcome subgrade drag. They are not designed to act as load-transfer devices.

Deformed bars should be made of billet or axle steel, grade 40, conforming to AASHTO M-31 or M-53. The bar sizes, lengths, and spacings for various pavement conditions are given in Table 16-23b.

Alternative Types of Pavements. Several alternative types of pavement construction have been introduced and tested to some extent. In the following, these developments are described briefly.

Porous Pavement. This type of pavement comprises the same material generally used for asphalt pavement, except that the fines (sand) are eliminated from the mix. Consequently, voids are left in the pavement. These allow rainwater to seep through. As a result, porous pavement offers the following advantages:

Allows storm water to percolate into the soil rather than to increase runoff into storm sewers.

Stores polluted storm water and slowly releases it.

Promotes the healthy growth of trees, shrubs, ground cover, and other plantings installed to make a parking area more attractive.

Allows replenishment of underground water supply by percolation of storm water.

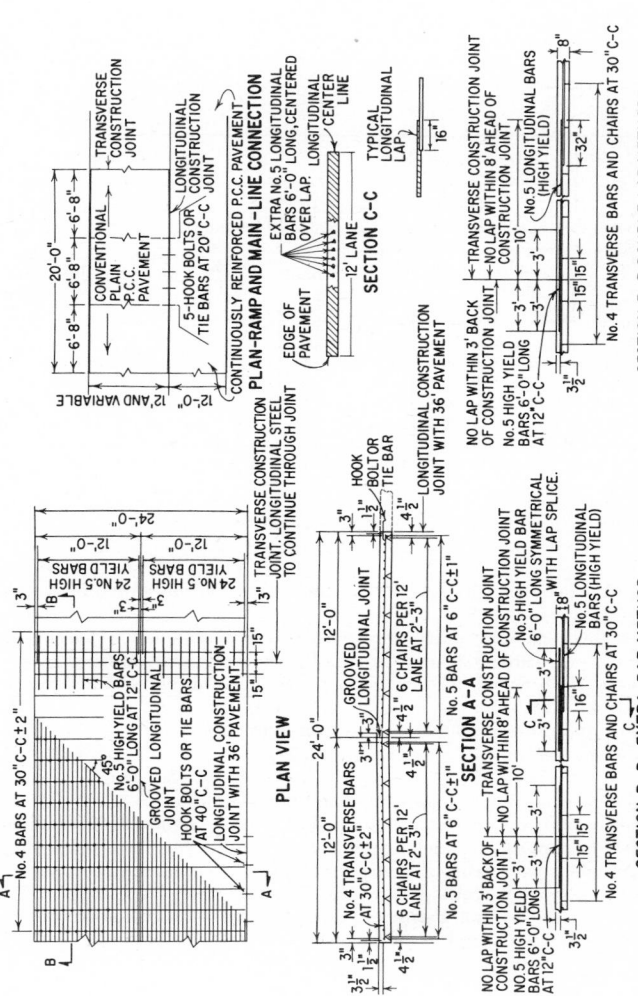

Fig. 16-24. Steel-bar reinforcement for continuously reinforced concrete pavement. End laps of bars should be staggered on an angle of at least 30°, preferably 45°, as viewed in plan. The double-lap requirement of 32 in. (Sec. B-B) and the extra bar method (Sec. B-B) apply only to laps within an area up to 10 ft beyond the construction joint. Lapped bars should be securely fastened to prevent separation during concrete placement. Hook bolts or tie bars should be placed in the same horizontal plane as the No. 4 transverse bars. A 2½-in. minimum clearance should be provided between hook bolts or tie bars and transverse bars. Transverse construction joint bars should be placed in same horizontal plane as No. 5 longitudinal bars, which should be lapped and tied in that plane. For 36-ft-wide pavement, either single 12-ft lanes with two longitudinal construction joints of 12-ft and 24-ft lanes with one longitudinal construction joint and one grooved longitudinal joint should be used. Transverse bars should not extend through longitudinal construction joints but should extend full length (23 ft 9 in.) for grooved longitudinal joints. (This width of pavement is permitted with slip-form paver only.) (*Virginia Department of Highways and Transportation.*)

Table 16-23. Minimum Design Requirements for Dowels and Tie Bars*

a. ROUND DOWELS

Pavement thickness, in.	Dowel dia, in.	Dowel length, in.	Dowel spacing, in.
6	¾	18	12
7	1	18	12
8	1	18	12
9	1¼	18	12
10	1¼	18	12

b. TIE BARS

Type and grade of steel	Working stress, psi	Pavement thickness, in.	½-in.-dia bars				⅝-in.-dia bars			
			Minimum overall length, in. ‡	Maximum spacing, in. †			Minimum overall length, in. ‡	Maximum spacing, in.		
				Lane width 10 ft	Lane width 11 ft	Lane width 12 ft		Lane width 10 ft	Lane width 11 ft	Lane width 12 ft
Grade 40 billet or axle steel	30,000	6	25	48	48	48	30	48	48	48
		7		48	48	45		48	48	48
		8		48	44	40		48	48	48
		9		43	39	35		48	48	48
		10		38	35	32		48	48	48

*From "Interim Guide for Design of Pavement Structures," American Association of State Highway and Transportation Officials, 1972.

†Spacing of tie bars should not exceed 48 in.

‡350 psi assumed for bond stress u. Length includes 3-in. allowance for centering.

Reduces or eliminates the need for a drainage collection system requiring structures, pipes, and sewers.

Prevents wet skidding (hydroplaning) and improves visibility of pavement markings during storms, because of the rapid percolation of water through the porous asphalt surface.

Permits use of urban debris (broken bricks and concrete, ceramic wastes, solidified fly ash, etc.) to form the base course.

Prevents flash flooding and preserves local streams from erosion.

Preserves natural drainage patterns.

There is no basic visual difference between porous and normal, nonpermeable pavement, and porous pavement does not require any specialized equipment for placement. This pavement can be used for a wide range of applications, including highways, local streets, and parking lots.

For a parking lot, for example, a pavement consisting of a 2½-in. surface course of porous bituminous concrete over a 12-in. graded, crushed-stone base has been used successfully. (The 12-in. stone base was used because a subsoil boring indicated that the soil had medium permeability.) The base-course stone was layered in the following order, top to bottom:

Layer	Thickness, in.	Stone size, in.
Top	2	⅜–½
2	3	1½–2
3	2	⅜–½
Bottom	5	1½–2

The smaller stone (⅜ to ½ in.) was used as a binder for the larger stones. The top layer of smaller stone was needed so that the asphaltic-concrete paving machine could lay the surface course with a smoother surface. Cost of pavement for the parking lot was about $11 per sq yd (1972).

("Investigation of Porous Pavements for Urban Runoff Control," Water Pollution Control Research Series, 11034 DUY-03/72, U.S. Environmental Protection Agency.)

Use of Sulfur in Asphalt Pavements. Sulfur has been used successfully as an additive to serve several functions in pavement construction.

Use of Sulfur as a Binder. A process to incorporate sulfur in conventional asphalt paving mixes was developed and field tested by the Research and Development Department of Gulf Oil Canada, Ltd. The tests showed that a significant portion of the asphalt can be replaced by sulfur. The total binder content of the mix, however, is similar to that normally used for mixes with asphalt alone. The basic mix of the binder consists of 50% (by weight) liquid sulfur and 50% asphalt. Use of sulfur as a binder depends on the price of sulfur and price and availability of asphalt.

The principal characteristics of sulfur-asphalt binder construction are briefly as follows:

Elemental sulfur can be dispersed in asphalts to produce a binder for conventional flexible pavement mixes.

The viscosity-temperature characteristics of sulfur-asphalt binders permit better workability under a greater temperature range than for asphalt alone. Pugmill operations and paving procedures can be conducted at lower than conventional temperatures and also over a slightly wider temperature range.

Test specimens of sulfur-asphalt binder mixes demonstrate relatively high stabilities, good flow properties, resistance to water, and good low-temperature and fatigue characteristics.

Sulfur-asphalt binder can be produced in industrial quantities. There is an upper temperature limit for mixing operations, however, because of hydrogen-sulfide formation at higher temperatures.

No additives are required beyond those that might be used in a conventional mix.

Conventional equipment can be used for mix production, material handling, and paving procedures of sulfur-asphalt binder mixes.

Field experiments have demonstrated that sulfur-asphalt binder mixes can be used successfully in rehabilitation applications as well as in full-depth pavement construction.

An economically attractive process with simplicity of construction requirements for the utilization of elemental sulfur in pavements has been demonstrated.

Structural analyses and in-service evaluation of the test roads as well as more comprehensive field experiments are being performed to provide a basis for design and performance correlations.

Sulfur Foam as a Frost-Heave Preventive. A major area of interest in locations subject to permafrost is the use of sulfur foam as protection against frost heave in highway construction. In permafrost regions, sand and gravel are needed to provide a stable load-bearing surface and to provide thermal insulation to protect the underlying permafrost. Typically, gravel depths of 5 ft or more are required. Since gravel is in short supply in many of the northern areas, it often must be transported great distances at substantial cost.

Tests have shown that 7 ft of gravel can be replaced with between 3 and 4 in. of sulfur foam covered with about 3 ft of gravel. Thus, the use of sulfur foam offers an opportunity to minimize gravel requirements as well as to meet certain economic and environmental concerns. The Sulfur Development Institute of Canada has sponsored tests to gain more insight into this method of construction and to establish further possible uses of this material in highway construction.

Recycling of Asphalt-Concrete Pavement. From knowledge gained through recent studies, it has been found that a product comparable to that derived from new materials can be attained through recycling of old pavements. Old asphalt-concrete pavements have been removed and reused in the past as roadway construction but to a limited extent. Often, the old pavement has been used as an aggregate base course. In some instances, it has been run through a mixer and placed as a cold-mix, cold-laid asphalt base or surface course.

Recycling of old asphalt-concrete pavement has been tested in Nevada with specialized equipment with a capacity of 80 tons per hour or more of previously crushed pavement. The recycling process begins with excavation of existing pavement, which is then crushed and screened into piles similar to those of crushed aggregate. The purpose of the screening operation is to allow for the correction of gradation defects that may have existed in the original pavement. The screened material next goes into a heat exchanger in the necessary proportions for correction of any gradation problems. Additional aggregate is added, if necessary. The heated mix is fed into a pugmill, where a special additive is introduced to soften the old asphalt to the desired viscosity. Also, additional new asphalt is added, as necessary, to obtain the optimum asphalt content.

Test results relative to asphalt flow characteristics, bituminous mixture stability, and aggregate gradation show that recycled asphalt pavement can be made to comply with the material specifications of states where asphalt pavement is used. There has been, however, too limited production and testing of this process to make possible realistic cost evaluations on a large scale of the productability and economics of pavement recycling.

On an interstate project in Nevada, the cost of an overlay asphalt pavement on old asphalt-concrete pavement compared with the cost of recycling the old pavement as follows:

The cost per mile-year (cost per mile of roadway per year of life) of overlay (based on an estimated 7-year life) was $5,728 per mile-year, whereas the cost of recycling (based on a 15-year pavement life) was $1,867 per mile-year.

Recycling of old pavements offers the following additional benefits:

Reduces the need for exploring and developing new aggregate sources and conserves existing aggregate sources.

Eliminates the necessity of locating disposal sites for discarded pavements.

Conserves expensive and scarce asphaltic products. Recycled asphalt pavement requires about 75% less asphalt than does virgin material.

Distressed pavements can be recycled in lieu of placement of thin overlays that are especially prone to reflective cracking.

The structural value of a distressed pavement can be increased by recycling a portion of the underlying base aggregates along with the bituminous pavement.

The distressed section of a pavement can be recycled without disturbing any pavement in good condition.

(Robert L. Mendenhall, "Recycling of Asphalt Concrete—Description of Process and Test Sections," Las Vegas Paving Corp., Las Vegas, Nev.)

16-12. Design of Intersections at Grade.   Locations where streams of traffic cross each other at a common elevation are called grade intersections.

Major Design Considerations.   Each intersection, although having many features in common with other intersections, must be treated as an individual project, because there are likely to be differences with significant impact on traffic flow, economics, and the environment.   The major factors that have to be considered in the design of an intersection are:

Traffic Factors.   Design traffic for each movement, including daily and hourly volumes, capacities, turning movements, size and operating characteristics of vehicles, control of movements at points of intersection, vehicle speeds, pedestrian movements, transit operations, accident experience, and storage requirements for traffic-signal-controlled approaches.

Physical Factors.   Topography, improvements and physical requirements for highway and channelization features, adequate sight distance, property restraints, safe location of sidewalks and crosswalks, and accommodation of traffic-control devices.

Economic Factors.   Capital and operating cost of the improvement and the economic effect on abutting businesses where channelization restricts or prohibits certain vehicular movements within the intersectional area.

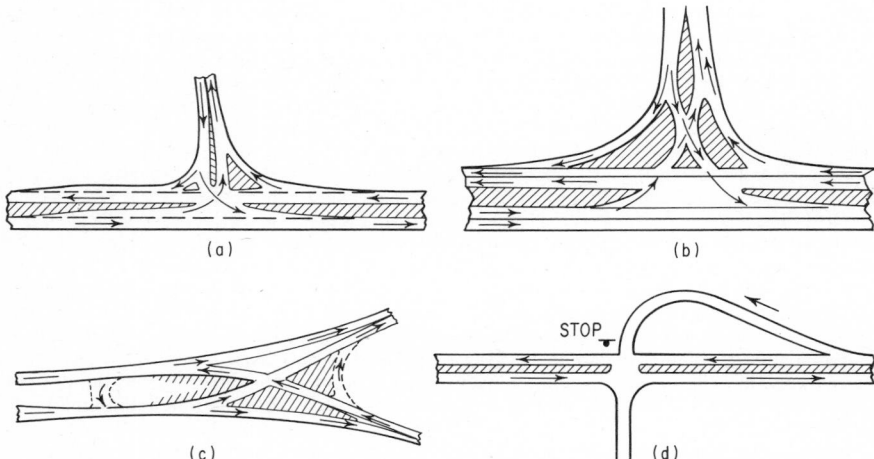

Fig. 16-25. Channelized T intersections at grade (*a*) and (*b*) with median lanes; (*c*) Y intersection; (*d*) "jug handle."

Human Factors. Driving habits, ability of drivers to make decisions, adequate advance warning of intersection, decision and reaction times, and natural paths of movement.

Design Alternatives. The simplest intersection is the meeting of two roads or the crossing of one by the other, in both cases without any widening of either or the posting of any signs to control traffic.

When traffic warrants, a STOP or YIELD sign may be placed on the road with the lower traffic volume.

A further improvement is to widen the pavement on the major road to provide lanes where turning traffic can leave the mainstream. With larger traffic volumes, it is sometimes expedient to separate the traffic entering and leaving the minor road by channelizing the intersection with small islands (Fig. 16-25a). Care should be exercised to limit this treatment so that it does not become unduly confusing.

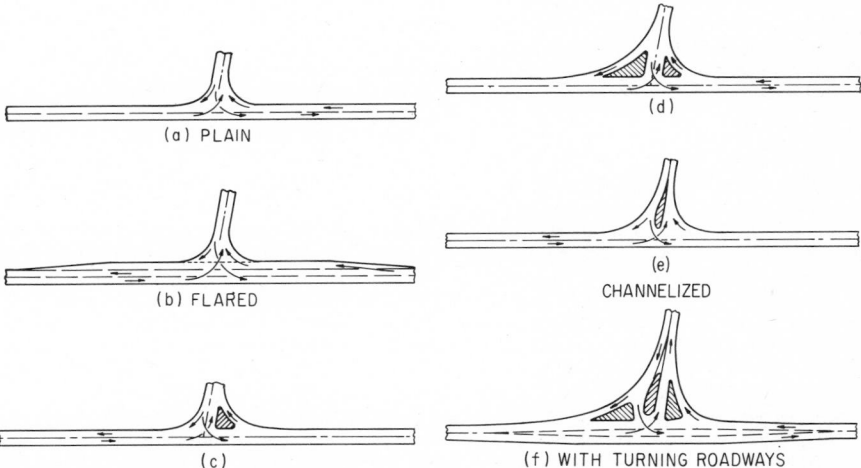

Fig. 16-26. T intersections at grade. (*"A Policy on Geometric Design of Rural Highways,"* American Association of State Highway and Transportation Officials.)

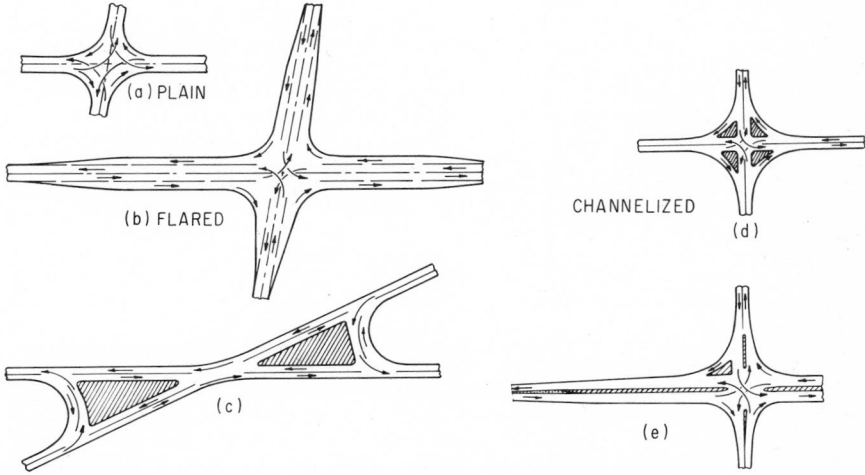

Fig. 16-27. Four-leg intersections at grade. (*"A Policy on Geometric Design of Rural Highways,"* American Association of State Highway and Transportation Officials.)

Greater traffic (a maximum crossing volume of 750 vehicles per hour is a good guide) creates a need for signal lights, especially in urban areas.

When the median in a divided highway is wide enough, a slot can be formed for left and U turns (Fig. 16-25b and c). Otherwise, when right-of-way can be acquired, a loop can be constructed from the right-hand lane to reintersect the highway to permit a right-angle crossing of traffic (Fig. 16-25d). Entrance from the loop is usually controlled by a STOP sign or traffic signal. Because of its shape, this loop is often called a *jug handle*.

Some typical at-grade intersections are shown in Figs. 16-25 to 16-27. For high traffic volumes, bridges are warranted to separate crossing movements, and ramps are installed to interchange vehicles between roads. This construction forms grade-separated intersections, or more commonly, interchanges.

16-13. Design of Interchanges.
An interchange is a system of interconnecting roadways in conjunction with one or more grade separations providing for the movement of traffic between two or more roadways on different levels.

Data Required for Design. The data required are essentially the same as those needed for at-grade intersections (Art. 16-12) and consist generally of traffic volumes; physical site conditions; and geophysical, economic, and environmental factors.

Justification for Interchanges. The major considerations for justifying construction of an interchange are:

Freeway Development. Once the decision has been made to develop a route as a freeway, additional decisions are necessary as to whether each intersecting highway should be terminated, rerouted, or provided with a grade separation or interchange. The chief concern is a safe, uninterrupted flow of traffic on the freeway.

If traffic on any other road must cross the freeway, grade separation is required to eliminate interference with freeway traffic flow. If access from the other road to the freeway is desired, an interchange is required.

Elimination of Bottlenecks or Spot Congestion. Insufficient capacity at an at-grade intersection of heavily traveled highways can result in intolerable congestion on one or all approaches. Inability to provide the essential capacity with an at-grade facility justifies an interchange.

Elimination of Hazard. Some at-grade intersections have a disproportionate share of serious accidents. Unless inexpensive methods of eliminating hazards can be used, a highway grade separation or interchange may be warranted. Accident-prone intersections frequently are found at the junction of comparatively lightly traveled highways in sparsely settled rural areas where speeds are high. In such areas, structures usually can be constructed at little cost, compared with costs for urban areas. Right of way is not expensive, and these lower-cost developments can be justified by elimination of only a few serious accidents.

Road-user Benefits. Road-user costs caused by delays at congested at-grade intersections are large. The annual cost of fuel, tires, oil, repairs, time, accidents, etc., at intersections that require speed changes, stops, and waiting can be well in excess of the annualized cost for interchanges, which permit uninterrupted, safer operations. Interchanges usually require somewhat more total travel distance than direct crossings at grade. The added cost of the extra travel distance, however, is usually offset by the saving realized by reductions in stopping and delay costs. For any type of intersection, the relation of road-user benefits to the cost of improvement indicates whether or not the improvement is warranted on economic grounds. For convenience, the relation is expressed as a ratio of the annual cost benefit to the annualized cost of the improvement. Annual cost benefit is the difference in road-user costs for the existing condition and for the condition after improvement. Annualized cost is the sum of interest and amortization of the cost of the improvement. The larger the ratio, the greater the justification for the improvement. Ratios greater than 1 are necessary for minimum economic justification. Comparison of these ratios for design alternatives is an important factor in deciding which alternative to adopt (Art 16-20).

Preferred Design Criteria for Interchanges. Some basic principles should be considered in the design of an interchange.

Minimum Weaving. Inadequate weaving sections severely reduce speed and capacity and, on high-speed, high-volume highways, cause sharply increased accident rates and congestion. The distance between any entrance and the following exit on limited-access highways should be sufficient to eliminate weaving as a traffic operational restraint. Where acceptable weaving cannot be provided for at the design stage, grade-separated ramp movements should be considered as a substitute.

Ramp Terminals. Locations of exits and entrances to a highway are important. Those affording poor sight distance to drivers or having traffic signs poorly positioned will require too quick a decision by drivers and will be accident-prone interchanges.

In urban areas, where it is necessary to provide frequent access to and from a highway, ramp terminals can be too closely spaced. To avoid this situation, collector-distributor roadways are often introduced. A collector-distributor roadway may be connected to the highway at each end of a series of closely spaced traffic service points. Traffic can leave the highway at one terminal, then use the slower-speed collector-distributor roadway to reach their destinations. Traffic originating from the closely spaced traffic service points use the collector-distributor to gain access to the highway. This arrangement minimizes interference with traffic using the highway while still providing service to locations requiring access to the highway.

Exits and Entrances. Entrances and exits to highways preferably should be on the right of highway traffic. Left-hand entrances and exits are considered undesirable for several reasons.

Decisions and maneuvering take place in the left lanes, which are the higher-speed lanes.

Entering drivers are forced to merge to their right side, where they have reduced visibility and thus more difficulty in making accurate judgments. This problem is greatly magnified when the entering vehicle is a truck.

In view of the preponderance of right-hand exits and entrances, left-side moves tend to confuse and surprise drivers, even with proper signs.

Trucks, which traditionally are restricted to the right-hand freeway lanes, are forced to maneuver across several traffic lanes to reach a left-hand exit or to return to the right lane from a left-hand entrance.

The preceding objections may not be of overriding significance for major forks in which the volumes of traffic are so nearly equal as to make it difficult to favor one movement over the other. Where possible, however, consideration should be given to placing the higher commercial traffic volume in the right-hand fork.

Exits Precede Entrances. Preferably, exits should precede entrances at traffic service locations. One reason is that congestion is reduced if traffic is removed before new traffic is added. Another reason is that main-line weaves of entering and leaving traffic are eliminated.

Single Entrance. Every entrance creates disturbance and friction in the flow of main-line traffic. This results in reduction of main-line speed and capacity. Thus, a reduction in the number of entrances is desirable.

Application of this principle should, however, be tempered by one additional consideration. When traffic volumes on a ramp exceed the capacity of a one-lane entrance, some thought should be given to providing two one-lane entrances rather than one two-lane entrance. This is particularly desirable when the number of through traffic lanes must be maintained, that is, when it is impossible to provide an additional freeway lane downstream from a two-lane entrance. The minimum separation between two successive one-lane entrances is 1,000 ft.

Ramp Design Speeds. Higher ramp design speeds reduce travel time in interchanges and improve ramp terminal operating conditions by reducing the friction caused by speed changes. Ramps should be designed for the desirable design speed, as listed in Table 16-24 or higher.

Table 16-24. Guide Values for Ramp Design Speed*

Highway design speed, mph	30	40	50	60	65	70	75	80
Ramp design speed, mph:								
Desirable	25	35	45	50	55	60	60	65
Minimum	15	20	25	30	30	30	35	40
Corresponding minimum radius, ft:								
Desirable	150	300	550	690	840	1,040	1,040	1,260
Minimum	50	90	150	230	230	230	300	430

*"A Policy for Geometric Design of Rural Highways," American Association of State Highway and Transportation Officials.

Types of Interchanges. Following are descriptions of various types of interchanges.

Three-Leg Interchange. An interchange at an intersection with three intersecting legs consists of one or more highway grade separations and generally one-way roadways for all traffic movements.   When two of the three intersection legs form a through road and the angle of intersection is not acute, the term T interchange applies.   When all three intersection legs are through roads, or the intersection angle of two legs with the third leg is small, the interchange may be considered a Y type.   Regardless of the intersection angle, through-road character, etc., any basic interchange pattern may be adapted to widely varied conditions.   See Fig. 16-28 for examples of three-leg interchanges.

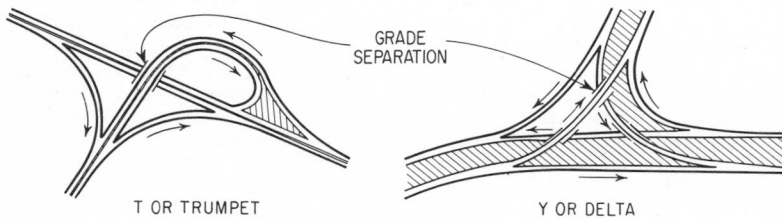

Fig. 16-28. T and Y intersections with grade separation.

Diamond Interchange.   When a heavily traveled highway is crossed by a comparatively lightly traveled one, with a bridge separation, the diamond interchange (Fig. 16-29) is generally a satisfactory solution.   It may also be applicable to a city street crossing an arterial highway, in which case signal control is often introduced at ramp terminals with the city street.

The diamond design is the simplest form of all-movements interchange.   It has four one-way ramps, which can be straight or curved to suit the terrain conditions and generally join the major highway at a flat angle.   The movements at the main-line terminals of the ramps are direct right turns.   Each ramp terminal at the minor road is a **T** or **Y** at-grade intersection, providing for one left and one right turning movement.   Where the traffic volume warrants, the cross street may be divided, with separate lanes provided for the left-turn movements.

Split-Diamond Interchange.   This involves two pairs of ramps.   One pair is provided at each of two parallel, or nearly parallel, but not necessarily consecutive, streets (Fig. 16-30). The arrangement aids in distributing traffic.   Connecting (frontage) roads between the cross streets improve the movements.

The split-diamond interchange works well when the two streets to which the ramps connect are restricted to one-way traffic in opposite directions (Fig. 16-31).   An additional refinement, often used when a freeway traverses an urban area, is construction of a collector-distributor (CD) road along each side of the expressway.   The ramps of the diamond enter and leave

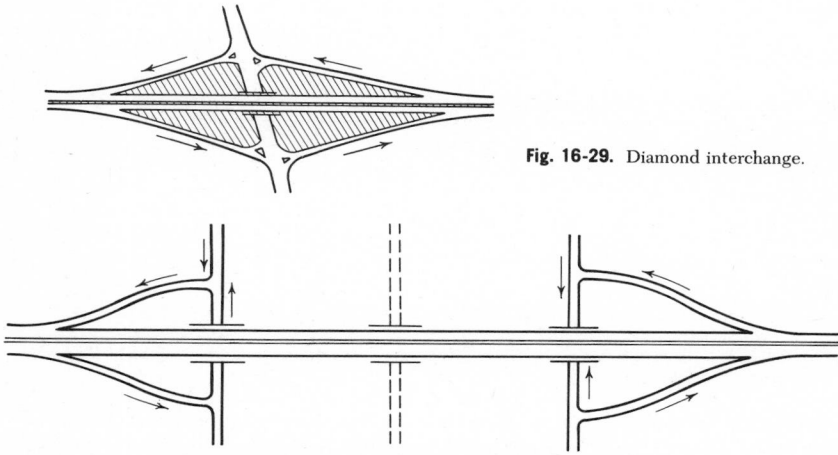

Fig. 16-29. Diamond interchange.

Fig. 16-30. Split-diamond interchange with two-way streets.

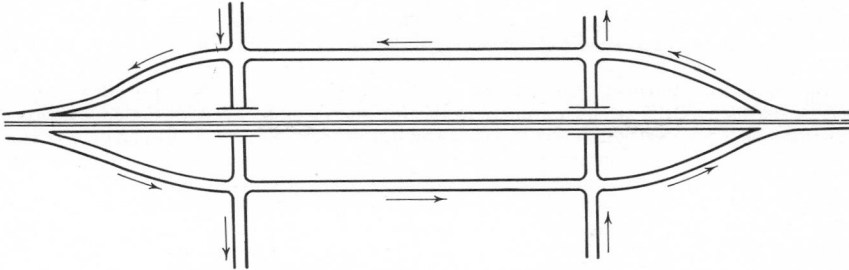

Fig. 16-31. Split-diamond interchange with one-way streets.

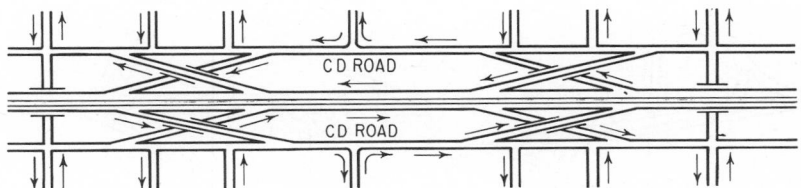

Fig. 16-32. Split-diamond with collector-distributor roads and bridges where ramps cross.

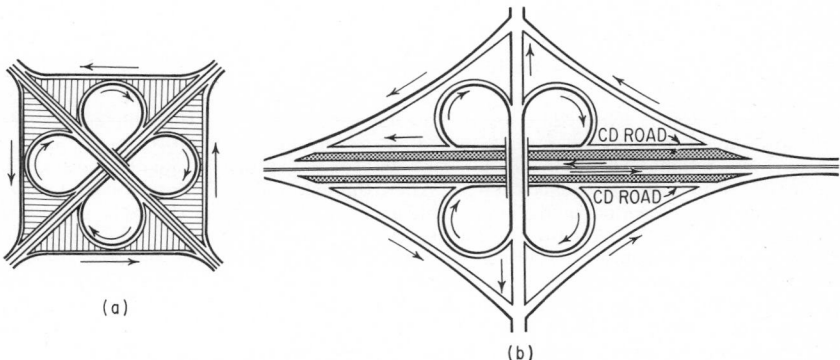

Fig. 16-33. (*a*) Basic cloverleaf. (*b*) Cloverleaf with collector-distributor roads.

along these parallel roads instead of at the cross streets. In confined situations where it is necessary to cross the on-bound and off-bound ramps, a grade-separation bridge is desirable (Fig. 16-32).

Cloverleaf Interchanges. When the first cloverleaf was constructed at Woodbridge, N.J., in 1928, it marked a great innovation in interchange design. This eliminated all left-turning movements, which are a major source of accidents. However, to cope with the speed and volume of modern traffic, the radii of the loop ramps must be increased to such an extent that the cloverleaf often covers a very large area and requires costly right of way (Fig. 16-33*a*). For a design speed of 25 mph, design standards require a loop radius of 150 ft; for 30 mph, that radius becomes 230 ft. Thus, for an increase in speed of only 5 mph, the radius increases more than 50% and the property taken, about 130%. Moreover, since travel time on ramps varies almost directly with length of ramp, the time that might be saved by increased speed is lost over the greater distance that must be traversed.

The cloverleaf requires weaving maneuvers between the entering and leaving traffic on adjoining loops. When this traffic exceeds 1,000 vehicles per hour, there will be serious interference, resulting in a slowing down of the through traffic. Furthermore, because it is seldom practical to provide for more than a single lane on a loop, a ramp can be expected to accommodate no more than 800 vehicles per hour.   These conditions often limit effective use of the cloverleaf. The traffic capacity of a cloverleaf, however, can be increased with collector-distributor (CD) lanes (Fig. 16-33b).

The cloverleaf interchange generally is restricted to rural areas, where it provides adequate service between a freeway and a primary road.   It is not recommended for the intersection of two freeways.   (J. E. Leisch, "Adaptability of Interchange Types on Interstate System," Proceedings, Highway Division, American Society of Civil Engineers, January, 1958.)

Partial Cloverleaf.   Anticipated traffic distribution may not always call for a full cloverleaf. Various modifications can be made, and in some, different ramp arrangements may be introduced (Fig. 16-34).   A major criterion is that the ramps be arranged in such a way that the entrances and exits produce the least impediment to traffic flow on the major highway.

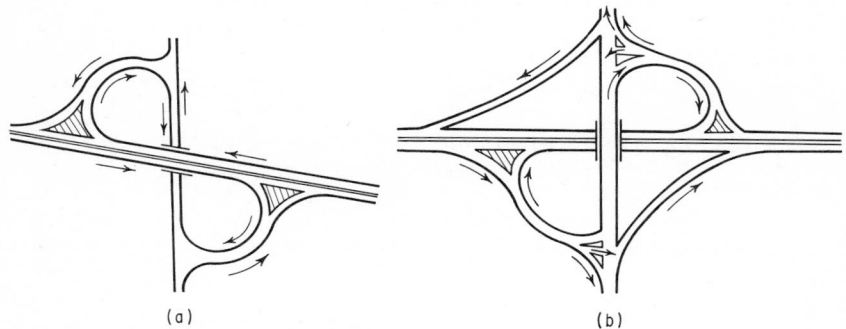

Fig. 16-34. Partial cloverleafs.

This objective can be achieved by applying the following principles.   The ramp arrangement preferably should enable each turning movement to be made by right-turn exits and entrances.   Where right turns are not feasible for both exits and entrances, and either one can be made a right turn, the exit should be chosen.   Where through-traffic volume on a major highway is decidedly greater than that on the intersecting minor road, preference should be given to an arrangement placing the right turns, either exits or entrances, on the major highway, even though this results in a direct left turn off the minor road.   ("A Policy on Geometric Design of Rural Highways," American Association of State Highway and Transportation Officials.)

Directional Interchanges.   An interchange made up only of loops may fail to meet the demands of expressways, with their high speeds and heavy traffic flows.   Thus, the cloverleaf-type interchange is generally relegated to rural locations, where the turning volumes are small compared with the volume of through traffic.   The preferred type of arrangement, the directional interchange, provides direct or semidirect connections between intersecting roads.

This type usually involves several grade separations.   Where several roads intersect, triple-level structures may be required (Fig. 16-35).   Although the individual ramps should satisfy accepted standards for curvature, pavement widths, length of weaving sections, and design of entrances and exits, there are no fixed patterns for this type of interchange.   Each should be studied as a special case and designed to meet terrain conditions and traffic demands. (Fig. 16-35 shows some examples.)

Designs for directional interchanges range from comparatively simple ones to complicated patterns that resemble a tangled maze.   Although the latter may appear to be very complex when viewed from the air, nevertheless, if they are well designed and adequately signed, they should not be confusing to motorists.   Since the turns on a directional interchange are made in the direction the driver is conditioned to expect, the driver is faced with only one decision at a time, and the points for decision are adequately spaced.

In some instances, particularly in rural areas, the traffic pattern does not justify direct connections in more than one or two quadrants.   The left-turning movements in the other quadrants can then be handled satisfactorily by loops.

A matter needing close scrutiny in a directional interchange is weaving.   Entering or leaving traffic must be given a sufficient length of highway to merge with, or cross, the main-line traffic in the same direction.   In most cases, the pavement has to be widened for the weaving section.   The required length of this weaving section is a function of the design speed and

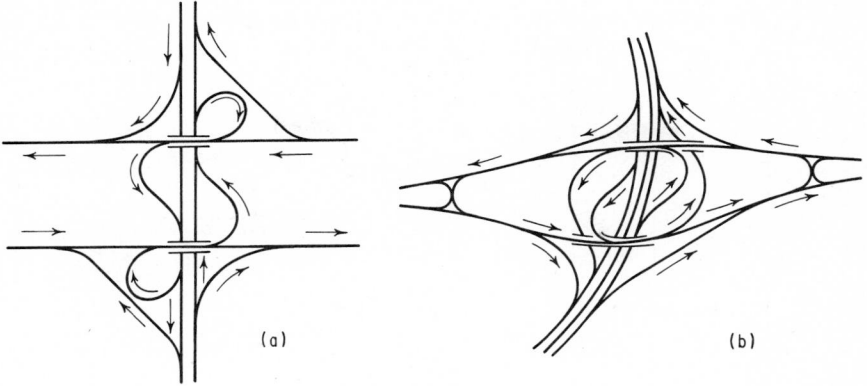

SEMIDIRECT

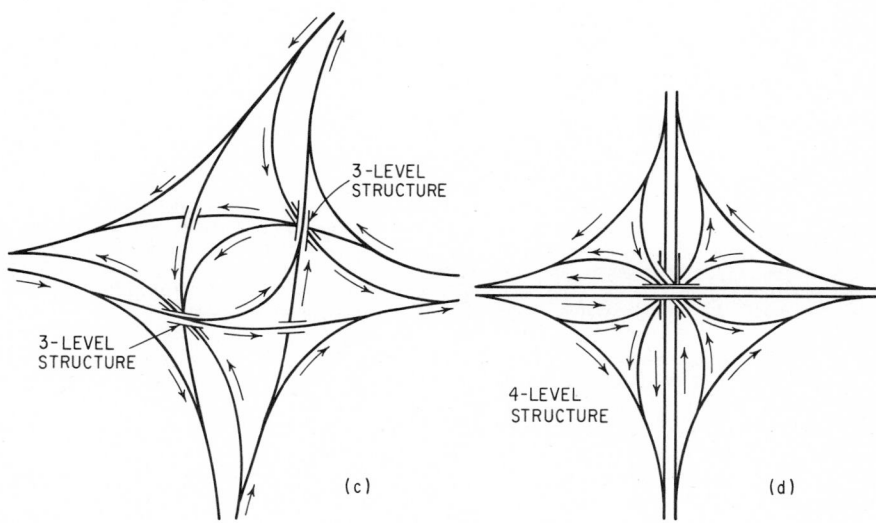

ALL-DIRECTIONAL

Fig. 16-35. Directional interchanges.

number of weaving vehicles per hour (Fig. 16-36). No widening of the pavement is necessary when the length-volume combinations exceed the numbers given in Table 16-25.

Table 16-25. Required Weaving-Section Lengths*

$W_1 + W_2$,† vehicles per hour	Weaving length, ft
500	1,000
1,000	2,300
1,500	4,000
2,000	6,000

*"A Policy on Design of Urban Highways and Arterial Streets," American Association of State Highway and Transportation Officials.

†W_1 = larger weaving movement. W_2 = smaller weaving movement.

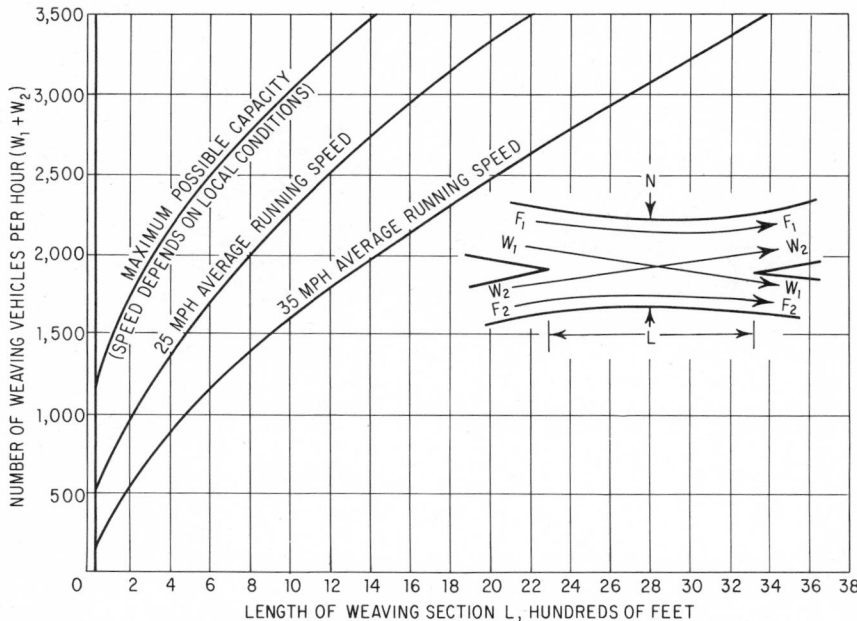

Fig. 16-36. Curves relate length of weaving section to number of weaving vehicles per hour and running speed. Number of lanes N required $= (W_1 + 3W_2 + F_1 + F_2)/C$, where W_1 = vehicles per hour in larger weaving movement, W_2 = vehicles per hour in smaller weaving movement, F_1 and F_2 = vehicles per hour in outer flows, and C = normal uninterrupted flow capacity for approach and exit roadways, vehicles per hour per lane. In general, when an outer flow exceeds 600 passenger cars per hour, the section should be wide enough to provide a separate lane for this movement. (*"A Policy on Design of Urban Highways and Arterial Streets," American Association of State Highway and Transportation Officials.*)

At or near any interchange, motorists should have to make only one decision at a time; for example, a triple fork should be avoided, and the points of decision should be sufficiently spaced on a time scale to avoid undue haste and tension.

Rotary Interchanges. A rotary interchange is often desirable where there are five or more intersection legs and all movements, other than through traffic on the principal highways, can be handled properly on the weaving sections. Rotary interchanges, however, are not suitable where relatively high speeds and traffic volumes are to be maintained on the cross roads.

A rotary interchange with a grade separation is shown in Fig. 16-37. A rotary intersection may be overpassed or underpassed by one of the intersecting highways. With two structures and four diagonal ramps, a complete interchange is provided. The weaving sections on the rotary roadway are a critical design feature of rotary interchanges.

Traffic leaving and entering the principal route does so directly, as on a diamond interchange. Design features, operation, and capacity of the rotary are basically the same as those

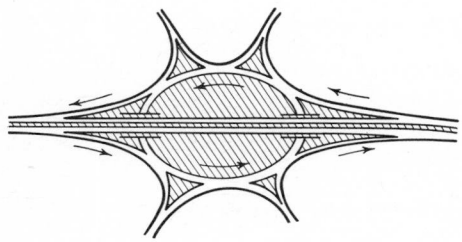

Fig. 16-37. Rotary with grade separation.

of an at-grade rotary.   Normally, a rotary interchange need not occupy any more space than a cloverleaf.

Spacing of Interchanges on Freeways.   In rural areas, spacing of interchanges should be as large as possible, taking into consideration the traffic pattern.   This is particularly necessary for toll turnpikes, where free rides are undesirable and it would be impractical and costly to collect fares at minor entrances or exits.   A 25-mile spacing is not unusual in rural areas.   For both freeways and turnpikes, it is the practice to grade-separate minor cross roads without providing ramps, or to relocate them by connecting them to other roads, or to dead-end them at the expressway.

For freeways in urban areas, however, determination of satisfactory spacing of access is not simple.   Often, traffic demands access at every important cross street.   If the interchanges are too widely spaced, the expressway does not provide the service expected of it.   On the other hand, if the interchanges are too closely spaced, the resulting marginal friction causes loss of efficiency.

Studies of a number of freeways have shown an average interchange spacing of 0.6 mile in urban areas and 1.5 miles in suburban areas (J. E. Leisch, "Spacing of Interchanges on Freeways in Urban Areas," Proceedings, Highway Division, American Society of Civil Engineers, December, 1959).   These distances allow for maneuvering and weaving under most traffic conditions, but an interval of at least 0.75 mile is desirable with the usual urban-traffic volumes.   Traffic on intervening streets should be either detoured to the nearest interchange or accommodated on frontage or collector-distributor roads.

Frontage (Service) Roads.   On limited-access highways, interchanges are usually widely spaced, and cross roads between interchanges pass over or under these highways without access ramps.   When the length of the trip that traffic on such cross roads must take to enter an expressway becomes unreasonable, consideration should be given to provision of a frontage or service road connecting directly to an interchange.

This road becomes particularly necessary in urban areas where heavily traveled streets occur at frequent intervals.   Provision of exit and entrance ramps for an expressway at all streets would reduce the expressway's efficiency to that of an ordinary city thoroughfare.   In such cases, the frontage road functions as a collector-distributor, which, as its name implies, collects traffic leaving the expressway and distributes it to the various streets, and collects street traffic to feed the expressway.   (J. E. Leisch, "Spacing of Interchanges on Freeways in Urban Areas," Proceedings, Highway Division, American Society of Civil Engineers, December, 1959.)

Ramps.   A ramp is a turning roadway connecting two or more legs of an interchange.

In the design of a ramp, efficient accommodation of traffic volume has to be balanced against considerations of topography and cost of right of way.   To conserve land, it may be necessary to locate the ramp so close to the highway that a retaining wall must be constructed.   The cost of such a wall then has to be balanced against the property cost without it.

Ramp Design Speeds.   Usually, steeper grades and sharper curves are permitted on ramps than on the intersecting highways (Tables 16-24 and 16-26).   These modifications in design

Table 16-26. Grades on Ramps, Percent

	Desirable	Maximum
Normal conditions	4	6
Subject to snow or ice	3	5
Used by heavy trucks and buses	3	4

standards are consistent with the lower design speeds for ramps.   Exceptions are the high-speed connectors in directional-type interchanges.   These connector ramps may be designed for speeds approaching those on the highways and consequently must meet correspondingly high geometric standards.

When a ramp connects a high-speed highway with a minor road or a city street, provision should be made for a considerable reduction in speed for traffic leaving the high-speed highway.   An initial speed reduction may be accomplished on a deceleration lane on the main highway.   Next, to allow for continuing deceleration on the ramp, the radii of the curves on the ramp should be reduced in stages.   At the terminal of the ramp, there often is either a STOP sign or signal lights, causing vehicles to come to a halt or proceed slowly.

Ramp Capacity. This generally depends on the design of the entrance. A single-lane ramp may have a capacity of from 600 to 1,200 vehicles per hour, depending on its design and other conditions.

Ramp Shape. This depends on ramp type (for example, loop or diagonal), topography, and relative position of the intersecting roads. Some examples of ramps are shown in Fig. 16-38.

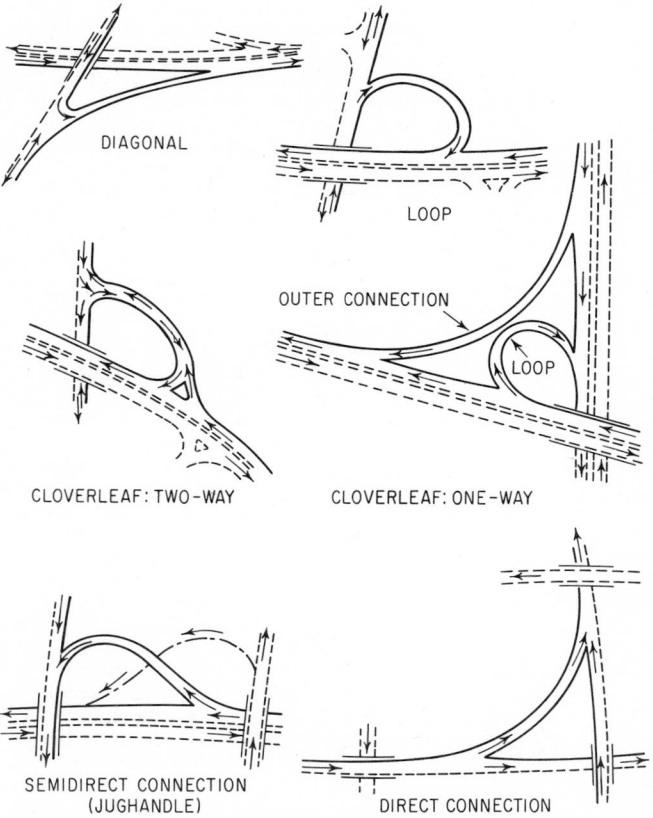

Fig. 16-38. General types of ramps. (*"A Policy on Geometric Design of Rural Highways,"* American Association of State Highway and Transportation Officials.)

Ramp Grades. The portion of a ramp approaching an exit or entrance should be designed to conform with the adjacent pavement on the highway. The grade along the remainder of the ramp is governed by its length. It often is comparatively short, after allowance is made for vertical curves in the areas near the terminals.

The flatter the profile, consistent with economy, the easier will be the change in direction for drivers. Desirable maximum grades are given in Table 16-26.

Sight Distance. For safe driving, it is very important to have adequate sight distance at ramp terminals, especially at an exit to a descending grade on the ramp. An exit to or entrance from an underpassing road should be located at least 200 ft beyond the structure spanning the road.

Ramp Details. Pavement widths, shoulders, and curbs depend on the volume and composition of expected traffic, length of ramps, design speed, etc. Also, the design of the terminals requires careful attention. Examples of standard practice are shown in Figs. 16-39 and 16-40.

Speed-Change Lanes. Except in high-speed direct connections, drivers leaving a highway at an intersection have to reduce speed and drivers entering have to accelerate until their speed

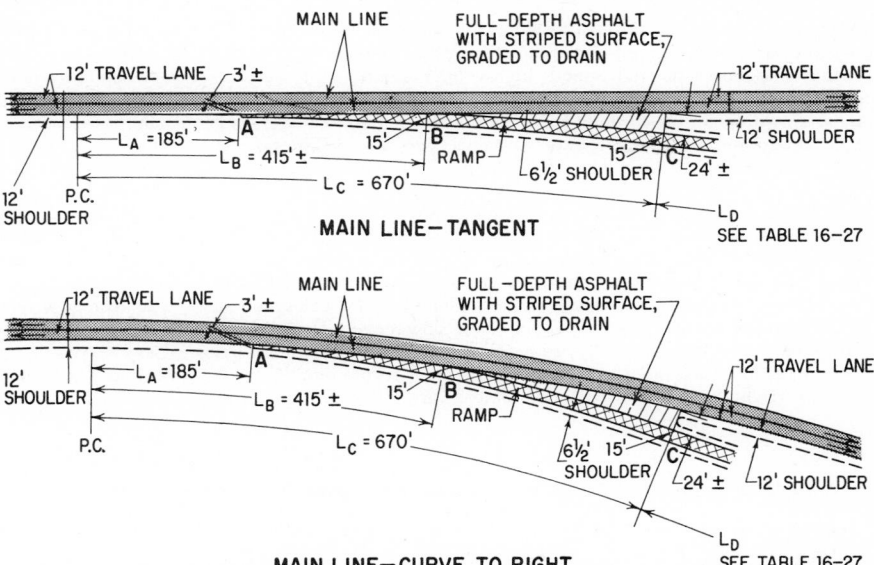

MAIN LINE—TANGENT

SEE TABLE 16-27

MAIN LINE—CURVE TO RIGHT

SEE TABLE 16-27

Fig. 16-39. Typical exit terminals for tangent sections and curves to the right in the direction of traffic. Exit terminals on curves to the left should be avoided. For all cases, the right edge of the ramp should be tangent to the main line. The ramp edge diverges 1° from the main-line curvature; for example, if the main line curves 4° to the right, the ramp curves 5° to the right. L_D is the minimum additional distance required by deceleration before introduction of a sharper ramp curve or stop condition. This design may be used for all exit terminals except those at the end of a combination speed-change lanes. (*New York State Department of Transportation.*)

Table 16-27. Minimum Distance L_D, Ft, Required for Deceleration before Sharp Ramp Curve or Stop*†

Grade, %‡	Highway design speed, mph	Ramp design speed, mph								
		Stop	15	20	25	30	35	40	45	50
		Minimum radius of ramp curve, ft§								
		50	90	150	230	310	430	550	690
+4 to +6	60	0	0	0	0	0	0	0	0	0
	70	0	0	0	0	0	0	0	0	0
+2 to +4	60	0	0	0	0	0	0	0	0	0
	70	55	35	10	10	0	0	0	0	0
−2 to +2	60	15	15	0	0	0	0	0	0	0
	70	115	90	65	65	40	15	0	0	0
−2 to −4	60	115	115	85	55	25	0	0	0	0
	70	235	205	175	175	145	115	25	0	0
−4 to −6	60	190	190	155	125	90	55	0	0	0
	70	325	295	260	260	225	190	90	55	0

*Based on typical exit terminal designs of New York State Department of Transportation.

†L_D should be provided on a smooth, gentle transition from the exit terminal alignment (up to point C in Fig. 16-39) to the usually sharper ramp curvature. The transition may be a tangent or a curve with radius exceeding 700 ft.

‡Average grade between points A and C in Fig. 16-39, in most cases about the same as the adjacent main-line grade.

§Minimum ramp radius should be 230 ft for interstate highways.

conforms to that of the through traffic. (An exception to the latter condition occurs when the entering driver is required by sign to stop or yield before entering.)

It is imperative on high-speed, high-volume roads to provide for such deceleration and acceleration by introducing speed-change lanes. Without a deceleration lane, leaving drivers must slow down in the through-traffic lane, with the risk of a rear-end collision. The acceleration lane permits entering drivers to attain a speed comparable with that of the through traffic before they enter a traffic lane. It also affords them more opportunity to pick a gap in the traffic stream in which to enter. (See Figs. 16-39 and 16-40.)

An acceleration lane should be as long as possible, consistent with traffic volume, design speeds, and steepness of roadway gradient. It should have a width equal to that of a traffic lane for most of its length and then taper to meet the traveled way. If a long taper is used for the entire acceleration lane, the average width should equal that of a regular traffic lane.

There is a practical limit, however, to the length of a deceleration lane. The average driver is not inclined to move over or to commence slowing very far in advance of the intended turn. So deceleration lanes generally are made shorter than acceleration lanes.

Lengths for acceleration and deceleration lanes recommended by the American Association of State Highway and Transportation Officials are given in Table 16-28.

Table 16-28. Design Lengths of Speed-Change Lanes, All Main Highways*
(Flat Grades—2% or Less)†

Design speed of turning roadway curve, mph	Stop con- dition	15	20	25	30	35	40	45	50
Min curve radius, ft		50	90	150	230	310	430	550	690
Design speed of highway, mph	Length of taper, ft	Total length of deceleration lane, including taper, ft							
40	190	325	300	275	250	200			
50	230	425	400	375	350	325	275		
60	270	500	500	475	450	425	400	325	300
65	290	550	550	525	500	475	450	375	325
70	300	600	575	550	550	525	500	425	400	350
75	315	650	625	600	600	575	525	475	450	400
80	330	700	675	650	650	600	575	525	475	450
Design speed of highway, mph	Length of taper, ft	Total length of acceleration lane, including taper, ft							
40	190		325	250	225				
50	230		700	625	600	500	400		
60	270		1,125	1,075	1,000	900	800	600	400
70	300		1,550	1,500	1,400	1,325	1,225	1,000	825	575

NOTE: Uniform 50:1 tapers are recommended where lengths of acceleration lanes exceed 1,300 ft, or where design speeds exceed 70 mph, or elsewhere, if appropriate and space permits.

*"A Policy on Geometric Design of Rural Highways," American Association of State Highway and Transportation Officials.

†Lengths should be adjusted to take into account the effects of roadway grades.

Detours.   Safe and adequate maintenance of traffic during construction is essential.

Stage construction, temporary roads and ramps, temporary structures and approaches, etc., should be considered in design to ascertain that there is a satisfactory method for routing

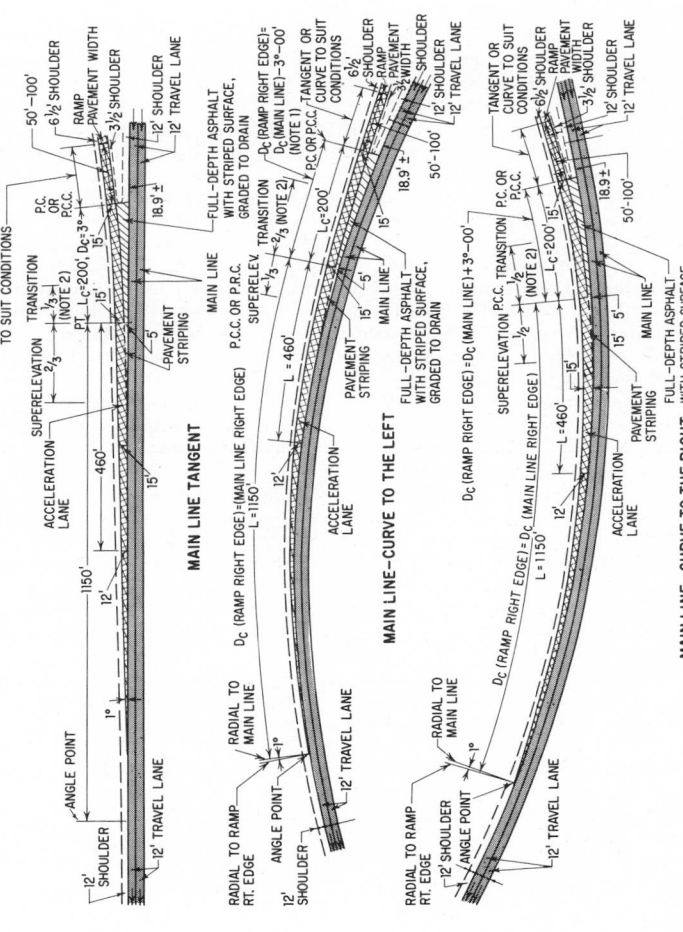

Fig. 16-40. Typical entrance terminals. This design may be used for all entrance terminals except those at the beginning of combination speed-change lanes.

NOTE: On all curves, the right edge of the ramp over Length L_c differs in curvature by 3° from that of the main line. For interstate highways, minimum ramp radius is 230 ft. Superelevation transition from the ramp to the main line should be accomplished in 200 ft. When the main line is on a tangent, one-third of the transition should be completed within L_c. When the main line curves to the left in the direction of traffic, two-thirds of the transition should be completed within L_c. When the main line curves to the right, one-half of the transition should be accomplished in L_c. (*New York State Department of Transportation.*)

traffic during the different phases of construction. Careful consideration also should be given to restriction of construction activities during peak-hour traffic periods and, if possible, during seasonal peak traffic periods. In addition, for work on high-volume traffic routes, consideration should be given to the number of lanes that may be closed at one time. Signs and warning devices should be used in a clear and concise manner to direct drivers for the full length of construction and for all detours.

Generally, detours should be paved when the existing one-way peak hourly volumes exceed 300 and the detour will be used at least 1 month during the construction season. Only in special cases, however, should the detour be paved if the approach roadways are unpaved. Localized weather and traffic conditions may require revision of the preceding criteria.

16-14. Traffic Control Devices. To be effective, a traffic control device should fulfill a need; command attention; convey a clear, simple meaning; command the respect of the highway user; and give adequate time for proper response. Basic considerations to insure meeting these requirements include justification, design, placement, operation, maintenance, and uniformity.

Uniformity and consistency of application, standardization of design, and legibility are essential to obtain maximum benefit. Simply stated, uniformity means treating similar situations in the same way.

Traffic Signs. The three major sign groups are regulatory, warning, and guide signs.

Regulatory signs inform the highway user of certain laws and regulations and include signs regulating movement, speed, stopping, standing, or parking of vehicles and movement of pedestrians. YIELD and STOP signs are used on intersecting roadways to establish which traffic has the right of through movement. These signs are particularly useful in controlling traffic at intersections where control other than a traffic signal is needed. The primary advantage of a YIELD sign over a STOP sign is that it assigns right of way without requiring a complete stop at times when conditions allow safe movement if vehicles slow down to a reasonable speed consistent with highway conditions.

Warning signs are used to alert highway users to physical or operating conditions on or adjacent to the roadway. Warning signs call for caution on the part of motorists and pedestrians.

Guide signs direct motorists to their destination over specified routes and minimize the confusion and potential danger which result when motorists are uncertain of their route. These signs also advise motorists of specific entrances and exits along the highway.

Symbols. As a means of helping drivers recognize a message quickly, day or night, some signs communicate their message largely via nonverbal symbols. These signs are part of a system of uniform, color-coded signs of specified shape that have the same meaning throughout the country.

Variable message signs are signs that display, at the appropriate time, emergency warnings, informational guidance, or special regulations. Such signs may be changed manually, by remote control, or by automatic sensors.

Sign Location. Placement of signs directly over the lanes to which they apply provides added driver safety and convenience, particularly in the communication of directional or route-marker information. Their principal application is on wide and on heavily traveled highways.

Materials. Signs are usually made of light-reflective materials; they are sometimes lighted in areas of high traffic volume.

Delineators. Delineators consist of reflector units, usually of glass or faceted plastic, mounted on supports. They serve as guide markers and are generally intended for use as aids in night driving. Another important advantage of delineators is that they usually remain visible when snow obscures other roadway boundary markers. The delineators at an interchange area are usually of a different color and multiple-mounted to differentiate that area from the normal roadway.

Pavement Markings. Markings are applied for the purposes of regulating and guiding the movement of traffic and promoting safety on highways. Experience has shown that obedience to pavement markings depends on their quality. Therefore, all markings should be properly maintained and reapplied whenever necessary to keep them in a state of good visibility.

The most common method of applying markings is with paint. Pavement markings may also be in the form of plastic or other material attached to or set into the pavement surface. Raised markers, usually less than 1 in. in height, are often set into the pavement at frequent intervals and are quite effective on high-speed roadways. However, their use is limited to locations where there is no chance of snowplows damaging the markers or of pedestrians tripping over them.

Edge marking, usually a painted white stripe between the pavement and the shoulder, has been found effective in guiding drivers on all types of highways without curbs and has become standard practice in most states.

Traffic Signals.   The purposes of traffic signals are to assign right of way at intersections, emphasize a hazardous location, control some types of railroad-highway grade crossings, control travel-lane usage, supplement some signs, and allow safe pedestrian movement. Typical devices include traffic control signals, pedestrian signals, flashing signals, flashing beacons, lane-use control signals, emergency-vehicle signals, and protection at railroad-highway grade crossings.   The basic displays used in traffic control signal operations are red, yellow (or amber), and green in color and have a circular or arrow configuration.

Traffic control signals, however, are not always the best means of controlling traffic at intersections.   For example, unwarranted signal installations can result in excessive delay, disobedience of signal indications, and an increase in accident frequency, especially rear-end collisions.

Analysis and experience have led to the determination of conditions under which a traffic control signal may be justified.   These are included in the "Manual on Uniform Traffic Control Devices for Streets and Highways," U. S. Department of Transportation, Federal Highway Administration, and in various state manuals dealing with traffic control devices.

Principal types of traffic signal control are pretimed, traffic-actuated, and pedestrian-actuated.

Pretimed signals operate on a predetermined, consistent, and regularly repeated cycle and sequence of intervals.

Traffic-actuated control has signal indications that are not of fixed duration, but instead are determined by and, within certain limits, confined to the changing traffic flow as registered by various forms of vehicle and pedestrian detectors.   Typical uses of traffic-actuated signals are for side-road phases or left-turn phases, which are skipped until a demand is registered when a vehicle actuates a detector.

Pedestrian-actuated signals are those signals that let traffic have the right of way until a pedestrian presses a button, which changes a light to halt traffic and allow the pedestrian to cross safely.

Signal Systems.   Great inconvenience and delays often result if independent and non-interrelated operation of adjacent traffic signals is not coordinated.   Major highways in and on approaches to cities and large villages, and through dense urban areas, are particularly suited to the use of signal systems that coordinate movement of traffic.

The most common method of coordinating the operation of signal systems is by a physical interconnection of the signals under the supervision of a master controller, which resynchronizes the various intersectional signal controllers.   The most sophisticated master controllers are computer-operated.

Colored Pavements.   When used for guidance and regulation of traffic, colored pavement surfaces are considered traffic control devices.   They should be used only when they contrast significantly with adjoining paved areas.

Ramp Control.   Highway-entrance-ramp control provides a means for controlling entry of vehicles to limited-access highways.   In certain circumstances, such control has been effective in reducing congestion and improving traffic capacity on the highways.   Control is usually obtained by either ramp closure (completely diverting ramp traffic), ramp metering (requiring drivers to stop and wait before entering the highway when ramp volumes must be restricted), or merge control (a sophisticated ramp-metering system that releases ramp vehicles when the system detects a gap on the highway).

Traffic Surveillance and Control Systems.   These comprise workers and equipment organized to monitor and control traffic flow through a road network in the most efficient possible manner.   These systems include elements, such as vehicle detection loops, helicopters, and closed-circuit TV, suitable for detection of traffic movement that thereby identify the demand for service; elements, such as variable message signs and emergency tow trucks, for traffic control, thereby resolving conflicts that arise; and elements for evaluation and management of the system so as to allow for maintenance and improvement of system operation.

These systems should provide for both normal and abnormal operating conditions, including catastrophes, such as major snowstorms, floods, and transit systems shut down by labor strikes. Statistical analysis may be necessary to predict the frequency of occurrence of such abnormal conditions as more than one lane being blocked at the same time, or the probability of some catastrophe occurring at repeated intervals, for example, predicting that a catastrophe of a particular sort can be expected to occur every 3 yr.

("Manual of Uniform Traffic Control Devices," New York State Department of Transportation; "Manual for Signing and Pavement Marking," American Association of State Highway and Transportation Officials.)

16-15. Highway Lighting. Highway lighting has been effective in improving the safety and comfort of drivers and the safety of pedestrians, facilitating traffic flow, and reducing street crime. But because of the relatively high construction and operation costs for highway lighting, it cannot be justified for all roadways.

In "An Informational Guide for Roadway Lighting," American Association of State Highway and Transportation Officials, several conditions are presented for the purpose of helping to decide which roadways warrant lighting. Separate conditions are given for freeways, highways, streets, interchanges, bridges, and tunnels. Abnormal local conditions, such as frequent occurrence of fog, ice, and snow, could justify modification of the recommendations.

Once the decision to light a roadway has been made, the intensity of illumination may be selected from references giving recommended illumination for different roadway classifications (see, for example, Table 16-29). Note, however, that the recommended illumination values

Table 16-29. Recommended Average Maintained Horizontal Illumination, Foot-candles, for Roadway Lighting*

Roadway classification	Area classification		
	Commercial	Intermediate	Residential
Freeway†	0.6	0.6	0.6
Major and Expressway†	2.0	1.4	1.0
Collector	1.2	0.0	0.6
Local	0.0	0.6	0.4
Alleys	0.6	0.4	0.2

*Adapted from "American National Standard Practice for Roadway Lighting," Illuminating Engineering Society.
†Both main line and ramps.

are meaningful only when designed in conjunction with other elements. The most critical elements are as follows:

Illumination depreciation. The condition just prior to lamp replacement and luminaire washing.

Quality. Relative ability of the light available to provide sufficient contrast differences for quick, accurate, and comfortable recognition of the clues required for seeing tasks.

Uniformity. Relatively even illumination spread on the pavement.

Luminaire mounting heights.

Spacing.

Transverse location of luminaires.

Luminaire selection.

Traffic conflict areas. The values in Table 16-29 are for roadway sections that are straight and level or nearly so. Intersecting, converging, and diverging roadway areas require higher illumination, at least equal to the sum of the values recommended for each roadway.

Border areas. There is value, such as improved depth perception, in illuminating areas that are beyond the roadway proper if the lighting is appropriate to the environment and not objectionable to adjacent property use.

Transition lighting. It is good practice to gradually decrease brightness in the driver's field of view as the vehicle moves from an adequately lighted section of roadway to a darker section.

Generally, highway lighting systems are of one of several types.

Low-level Lighting (Rail Type). Low-mounted, rail-type lighting systems (16-41a) are used most often on bridges, primarily for aesthetic reasons. This type of lighting enhances the low silhouette and appearance of bridges, overpasses, and expressway approaches. Sometimes, since low-mounted lighting equipment is less conspicuous in daytime, it is also appropriate for special conditions, such as areas where conventional overhead-type lighting would tend to detract from the architectural design of the project.

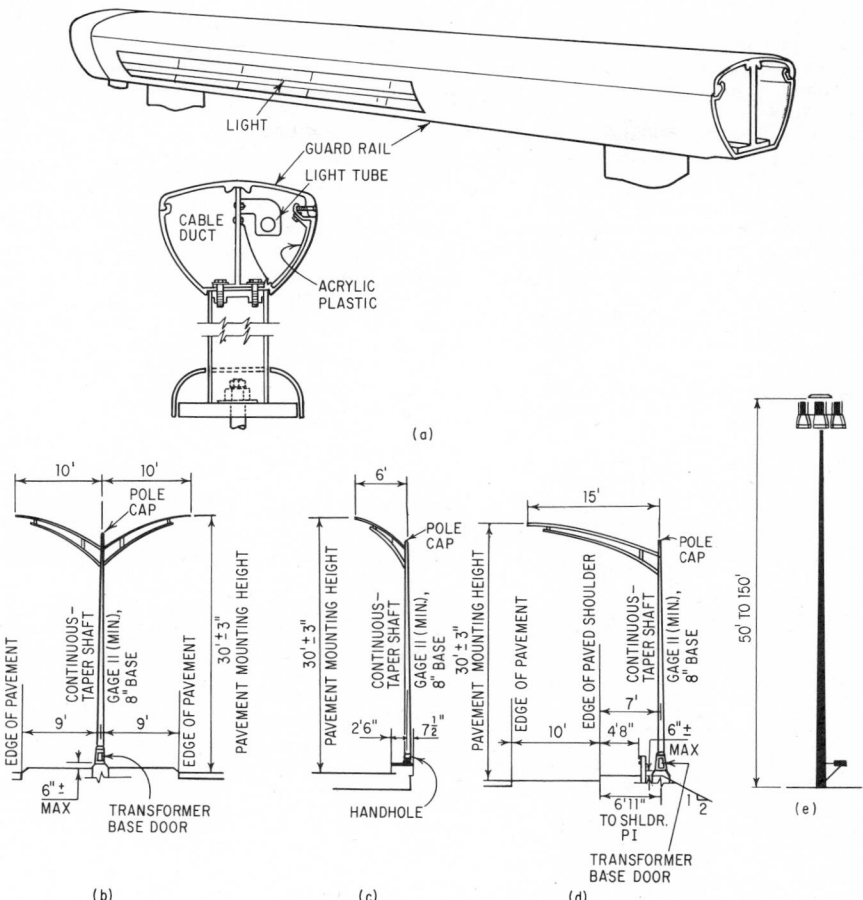

Fig. 16-41. Highway lighting installations.

Low-mounted lighting systems usually incorporate fluorescent luminaires. These may be installed essentially end-to-end in a continuous row along one or both sides of a bridge roadway. A luminaire mounting height in the range of 35½ to 46½ in. is generally used. In most cases, however, the luminaires should be mounted no higher than 40 in., to minimize glare for drivers. The lighting should be continuous, because objectionable flicker occurs for motorists if there are gaps in the light source.

Low-level lighting installations are considerably more costly than the conventional overhead system, both initially and during operating life.

Overhead Lighting (Conventional). Luminaires often are mounted on poles, generally about 30 ft above the pavement, to illuminate roadways (Fig. 16-41b, c, and d). The lamp type commonly used is mercury vapor.

Overhead Lighting (High Mast). A specialized type of lighting, called high mast, is often used to illuminate interchanges, traffic circles, and toll-plaza areas. In this type, luminaires are mounted on tapered steel poles or triangular steel towers that may range in height from 50 to 150 ft (Fig. 16-41e). The luminaires may be lowered to within 3 ft of the ground for inspection and servicing. Hoisting and electric cables can be replaced at ground level, where electrical connections are located for easy maintenance. Lamps may be 1,000-watt mercury vapor, metal halide, or high-pressure sodium vapor.

Tunnels and Underpasses.   For underpasses less than 100 ft long, artificial illumination generally is not necessary in daytime.   At night, lighting should be provided, but it should not exceed twice that on the roadway outside the underpass, unless additional illumination is required in urban areas for pedestrians and policing purposes.   For tunnel lighting, see Art. 20-14.

("An Informational Guide for Roadway Lighting," American Association of State Highway and Transportation Officials; "American National Standard Practice for Roadway Lighting," Illuminating Engineering Society; "Highway Visibility," Special Report 134, Transportation Research Board.)

16-16. Other Highway Appurtenances.   These include curbs, berms, sidewalks, and fences.

Curbs.   The chief functions of a curb are to control drainage and deter vehicles from leaving the roadway or striking bridge parapets.   Especially when reflectorized, curbs also serve to delineate the traveled way.

Barrier curbs may be 6 in. or more in height.   They should be positioned at least 2 ft from the edge of a through lane, or else that lane should be widened, because motorists tend to shy away from an unmountable obstacle.   Since barrier curbs are meant to bar vehicles, curb faces are not battered more than 1 in. in 6 in. (Fig. 16-42).

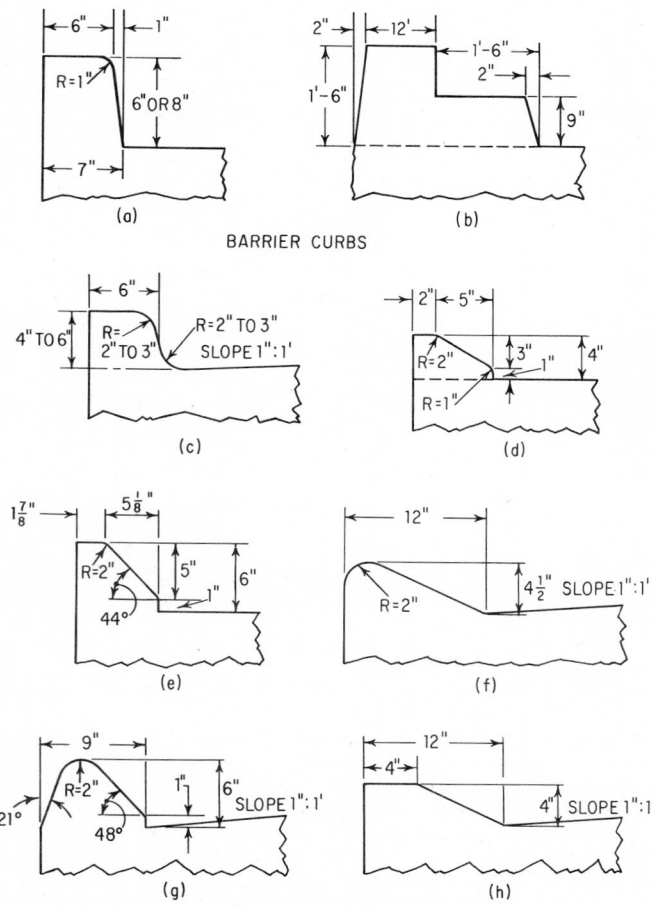

Fig. 16-42. Typical highway curbs. ("*A Policy on Design of Rural Highways,*" American Association of State Highway and Transportation Officials.)

Reflectorized curbs preferably have grooves cut in the face, to accurate angles, so that head-light beams are reflected back to the driver's eyes. Because of their high cost, they are seldom used. Glass beads or *cat's eyes* embedded in the face of the curb also are useful in delineating the edge of the road at night or in fog. A curb faced with a mixture of white cement and white sand is somewhat useful for the same purpose.

Lane dividers are an extreme development of the barrier curb, specially designed to separate opposing traffic on undivided four-lane highways. This type of barrier may be as high as 30 in. and has concave flanks to deflect a colliding vehicle back onto the pavement.

Mountable curbs are low curbs with flat slopes. They are used around islands at channelized intersections, along medians, and at the edge of the traveled way when there is no shoulder and cars that have to stop are expected to use the grass verge. A mountable curb is designed so that a vehicle can cross it with little or no jolt (Fig. 16-42).

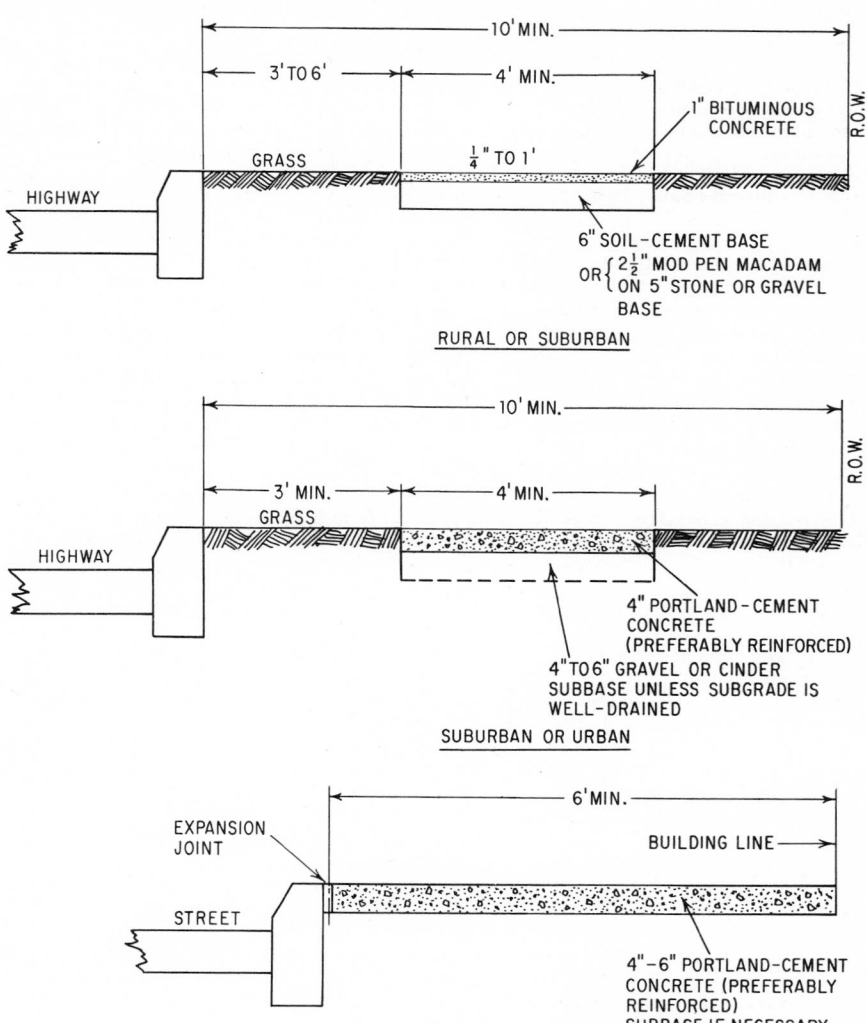

Fig. 16-43. Typical sidewalk cross sections.

Berms are used along rural highways on embankments or around islands to retain the drainage flow in the shoulder and retard side-slope erosion. They can be plain earth, sodded, or paved with a bituminous material of the road- or plant-mix variety.

Sidewalks. Sidewalks are commonly used in urban areas where pedestrian traffic is allowed within the right-of-way. They are also provided along some short stretches of rural highways for the use of schoolchildren.

Although they are generally constructed of concrete, sidewalks may be gravel, cinders, or macadam. It is desirable to have the edge of the sidewalk separated from the curb by a strip of grass to accommodate light poles, fire hydrants, other utilities, and possibly landscaping where it can be safely incorporated into the design. Where there is no curb, the sidewalk should be set back at least 3 ft from the edge of the pavement or shoulder. Typical sidewalks are shown in Fig. 16-43.

("Guide Specifications for Highway Construction," American Association of State Highway and Transportation Officials, Washington, D.C.)

Fences. Fences are provided along highways primarily for access control. Other functions served by fencing include delineation of highway right of way (property lines); on medians, serving as a barrier to prevent indiscriminate crossings of the median by vehicles and pedestrians and to act as a headlight glare screen; and preventing animals from entering the highway.

The most common type of highway fence is the 6-ft chain link fence. In rural areas, a 4-ft *farm fence* is also used frequently.

16-17. Roadway Safety Elements. The geometric design and general location of highways interact with various driver, vehicle, traffic, environmental, and other roadway factors to affect accident frequency, severity, and type. This article identifies those safety-related features that can be incorporated into the design of roadways.

The type and level of traffic demand placed on a highway mandate different types of design standards. High design standards result in lower traffic-accident frequencies.

Access Control. For major highways, the right of owners or occupants of abutting land, and other persons, to access to a highway may be fully or partly controlled by public authority. Full control of access gives preference to through traffic by providing access connections only with selected public roads and by prohibiting crossings at grade and direct private driveway connections. Partly controlled access gives preference to through traffic to some degree, but in addition to access connections with selected public roads, there may be some crossings at grade and some private driveway connections.

Role of Cross-sectional Elements in Safety. The cross-sectional elements of a roadway that have a bearing on its safety include lane width, shoulder width, pavement, cross slope, medians, side slopes and ditches, and clearances to roadside obstructions (Arts. 16-6, 16-8, and 16-11).

Lane Width. Available information on the relationship between lane width and accidents does not justify any change in the present almost universal lane width of 12 to 13 ft. Studies verify this present design practice in that as the width decreases below 11 ft, the accident rate increases.

Shoulders. Studies that have attempted to relate shoulder widths and traffic volumes to accidents have not yielded conclusive information that could serve as a standard for design.

Except for low-volume rural roads, the inclusion of wide, paved shoulders in cross-section design has continued to be the recommended practice.

Pavement. The safety of a pavement surface depends to a large extent on degree of slipperiness, light reflectivity, and riding comfort. A pavement tends to become slippery when:

  Water lubricates the tire pavement-surface contact area.

  The skid resistance of the surface is reduced through wear by traffic.

  High vehicle speeds tend to reduce friction between the tire and the pavement where pavement surfaces are uneven, and this causes a slight bouncing action on the vehicle.

  Skid resistance is reduced by *bleeding* of asphalt caused by excessive asphalt in the pavement mix.

Several methods have been developed to improve skid resistance of pavements. Both portland-cement concrete and bituminous pavements can be given high-skid-resistant surfaces by placement of a surface course with hard, sharp aggregates. Existing pavements that need their skid resistance improved can be given a surface application of a mixture of epoxy resin and abrasives, such as sand, granite, or slag. Sand blasting and acid etching have also been utilized on concrete pavements to expose the surface aggregate, thereby improving skid-resistance characteristics. In concrete pavements, when fine aggregates with a high silica content are exposed, the course provides a longer resistance to wear and polishing than do surface courses made with ordinary sand.

Pavement grooving is another method of reducing wet-pavement skid potential. The patterns and spacing of the pavement grooves have to be carefully established to achieve the desired surface-drainage characteristics without making steering control difficult, especially for light vehicles.

Use of intermittent, abrupt changes in pavement texture (rumble strips) is effective in warning drivers, through noise and vehicle vibration, of upcoming changes in the roadway, such as toll plazas, signals, and intersections. The type, location, and characteristics of this roadway surface treatment should be evaluated in terms of requirements of the individual situation.

Cross Slope. This is a significant component of a road's safety characteristics. Cross slope has to be considered from the standpoint of both operational requirements, such as super-elevation needs, and surface-drainage characteristics, to minimize the possibility of vehicles hydroplaning on flat cross slopes. In ice-susceptible areas, the maximum and minimum cross slopes should be so established as to provide the necessary cross drainage while minimizing the potential for skidding during icy conditions.

Medians. The relationship between the type and width of a median and its safety characteristics depends to a large extent on traffic volumes. Analyses of the frequency, type, and severity of median accidents generally point to a desirable median width of 30 ft or more. When this width cannot be developed because of site limitations or other considerations, a barrier should be placed in the median area to improve safety. Where high traffic volumes warrant, median barriers may also be needed on wider medians. In narrow medians (those about 8 to 10 ft wide), median barriers should have deflection characteristics such that impact deflections will not pose a hazard to oncoming vehicles. Figure 16-44 shows a typical median barrier used by the Virginia Department of Highways and Transportation.

Median openings are sometimes required for maintenance, police, and emergency vehicles. Where interchanges are spaced far apart and the median cannot be easily crossed because of a barrier, it is necessary to provide a median opening. (It is much safer, however, to have all traffic use the nearest interchange to gain access to the opposite roadway if interchange spacing is close enough.) As traffic volume increases, use of the median for crossover becomes more hazardous and more frequent. The number and location of median openings, therefore, have to be carefully considered. Provision should be made for adequate acceleration and deceleration lanes for median openings, to minimize the friction between through traffic and vehicles that have to use the median crossings.

Medians also are useful in reducing or eliminating headlight glare. The glare from the headlights of vehicles traveling in the opposite direction can be annoying and can result in a severe traffic hazard because of loss of visibility. Tests performed to determine the width of separation between opposing streams of traffic necessary to overcome the negative aspects of headlight glare indicate that a minimum separation of about 33 ft is necessary to ameliorate the opposing glare sufficiently to assure comfortable vision for a distance equal to the safe stopping sight distance at 70 mph. To provide an overall comfortable level of glare, the lateral separation should be about 94 ft.

Various types of glare screens have been tested and used successfully. The most commonly used type, where adequate median width is available, consists of screen plantings. Other types of glare screens include slotted fences and other similar devices placed so as to eliminate or minimize the glare of oncoming vehicles.

Side Slopes and Ditches. For reasons of highway safety, road side slopes, from pavement or shoulder to a drainage ditch or channel, should be 6 horizontally on 1 vertically or flatter. For the usual 3-ft height of side slope, the channel bottom lies at least 18 ft away from the pavement or shoulder edge. This clearance adds to the driver's visibility and to the safe maneuver area available for errant vehicles. ("Highway Design and Operational Practices Related to Highway Safety," American Association of State Highway and Transportation Officials.)

Recommended maximum slopes for cut and fill depend on the height of cut and fill and the type of terrain. (Slopes also are kept flat to facilitate mowing of grass.)

Roadside Obstructions. Where site conditions and other design considerations permit, the following measures are recommended from the standpoint of safety:

No roadside obstacles should be located within 30 ft of the edge of the traveled way.

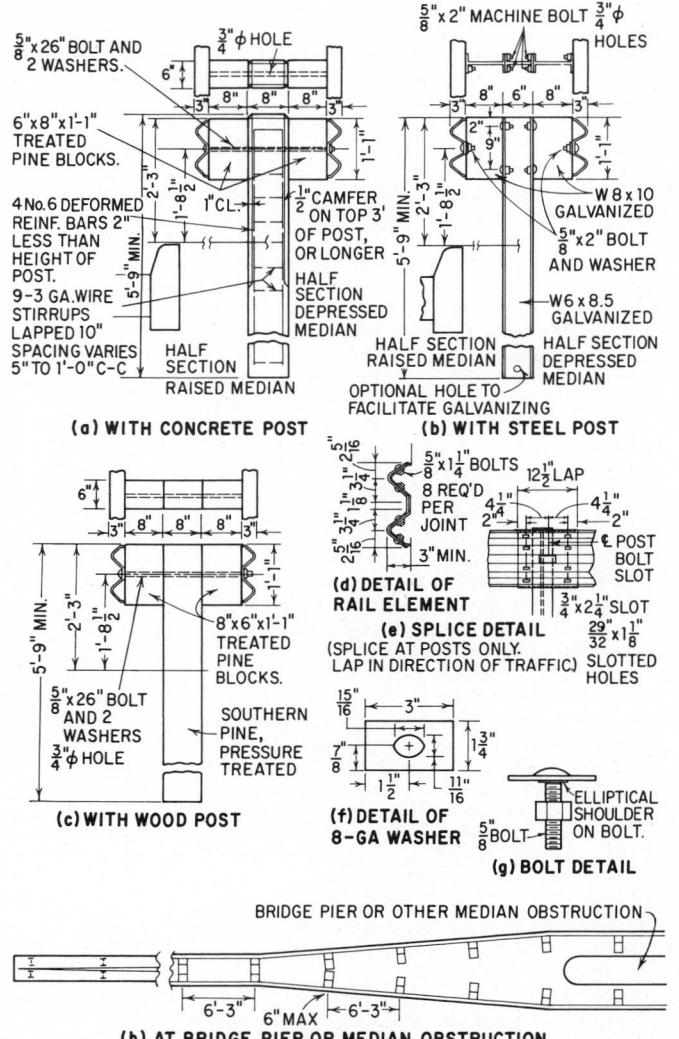

Fig. 16-44. Offset W-beam median barrier. (*Virginia Department of Highways and Transportation.*)

Obstacles that can be moved should be relocated in a safe area.

Breakaway devices should be provided, if practicable, on obstacles that cannot be readily removed.

Through the use of guardrails, median barriers, and impact-attenuation devices, drivers should be protected from obstacles that cannot otherwise be removed or improved.

Cut or fill slopes of 6:1 and flatter should be used wherever possible, practicable, and economically feasible.

Concrete foundations used for signs, light standards, etc., should be flush with the ground surface.

Barriers. Various types of barriers are constructed along roadways to redirect errant vehicles parallel to the traffic flow and to reduce accident severity by preventing vehicles

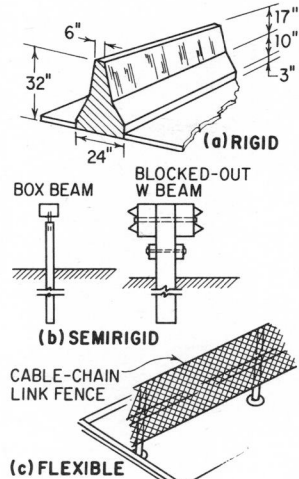

Fig. 16-45. Typical barriers for roadways.

from penetrating dangerous areas. Many different types of barriers have been developed to achieve these results. Figure 16-45 shows commonly used types of roadway barriers.

Current practice is to avoid using barriers wherever possible by eliminating hazards. There are, however, many situations that require the use of barriers. When barriers must be used, it is desirable to use the barrier system that will allow the greatest vehicle deflection on impact that is feasible. This practice will provide the slowest deceleration rate and thereby tend to minimize damage to the vehicle and, even more important, injury to its occupants. Caution must be exercised to avoid selecting a system with so great a deflection that the barrier will deflect on impact into opposing traffic or into a fixed object. Highway barriers can be grouped into three major types: rigid, semirigid, and flexible.

Rigid Barriers. These are intended to redirect vehicles into the normal traffic streams at a low angle of incidence. Rigid barriers permit little or no deflection and therefore inflict a relatively large deceleration rate *g* on vehicles and occupants if struck at any but flat angles. Concrete barriers, such as the one shown in Fig. 16-45a, developed by the New Jersey Department of Transportation and often used as a median barrier, are representative of this type in common use.

In general, rigid barriers are relatively more expensive than other types. They are necessary, however, where large deflections cannot be tolerated, such as in narrow medians.

Semirigid Barriers. This type of barrier includes the commonly used metal guardrails and median barriers of the box beam, W, and three-cable types, attached to heavy posts (Fig. 16-45b).

Flexible Barriers. This type of barrier, including the cable–chain link fence (Fig. 16-45c), developed in California, is designed to permit a large deflection, thereby reducing the deceleration effects experienced by impacting vehicles and their occupants. This type of barrier also has the advantage of presenting a less formidable object for drivers. It has the disadvantage, however, of sometimes causing a vehicle that hits it to spin to a stop, ejecting the occupants.

Guardrails. An intensive testing program carried out by the New York State Department of Transportation to determine deflection characteristics of guardrails and median barriers, including their ability to redirect and contain errant vehicles, indicated that the *strong-beam, weak-post* design was most effective from the standpoint of stopping errant vehicles and redirecting them after impact. The weak-post guardrail, which derives its resistance to vehicle impact almost entirely from bending of the beam-like rail, is applicable where large deflection of a guardrail can be tolerated. The deflection can be controlled to an appreciable degree by selection of beam type and post spacing.

Installation of guardrails is normally warranted by the presence of one or several of the following features along the roadway:

High embankments with steep side slopes.
Sharp curves.

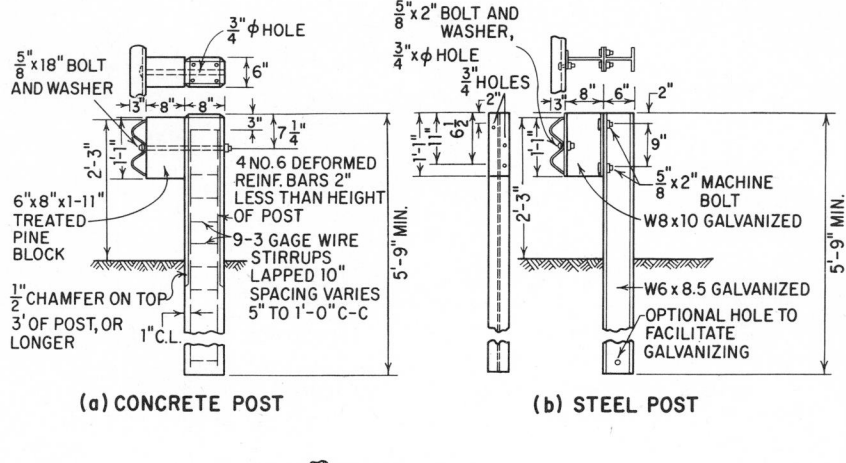

(a) CONCRETE POST   (b) STEEL POST

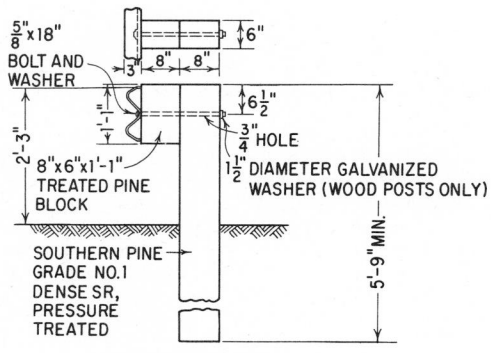

(c) WOOD POST

Fig. 16-46. Typical W-beam guardrails. For details of 8-gage washer, rail element, bolts, and splices, see Fig. 16-44. (*Virginia Department of Highways and Transportation.*)

Obstacles, such as bridges, piers, sign supports, etc., less than 30 ft from the edge of traveled way.

Other nontraversable roadway hazards, such as streams.

The treatment of guardrail terminals is important from the standpoint of safety. Ramped terminals may cause an errant vehicle to tumble and roll. A straight extension, on the other hand, provides the hazard of potential impalement. For these reasons, flaring guardrail ends away from the traffic and anchoring them to the ground are effective in avoiding this hazard.

Figures 16-46 and 16-47 show typical guardrail installations, as well as end treatments, adopted by the Virginia Department of Highways and Transportation.

Impact-attenuation Devices. These are placed in potentially dangerous areas having fixed structures, such as ends of walls, sign supports, and piers. The major purpose of these devices is to reduce injuries and vehicle damage rather than to minimize damage to the barrier. To accomplish this, the devices should be able to decelerate a vehicle at a tolerable rate (in general, not more than 10g to 12g). Several types of impact-attenuation devices have been used successfully.

Hydrocells. This system consists of clusters of water-filled plastic tubes, 6 in. in diameter and about 40 in. high. Each tube is equipped with a vent orifice at the top. The stiffness and impact-attenuation capability of these units can be varied by changing the orifice size, thereby controlling the rate at which the water is expelled on impact. These units, which require special mountings to be effective, can retain a car traveling at 50 mph in about 10 ft, with a deceleration of about 8g.

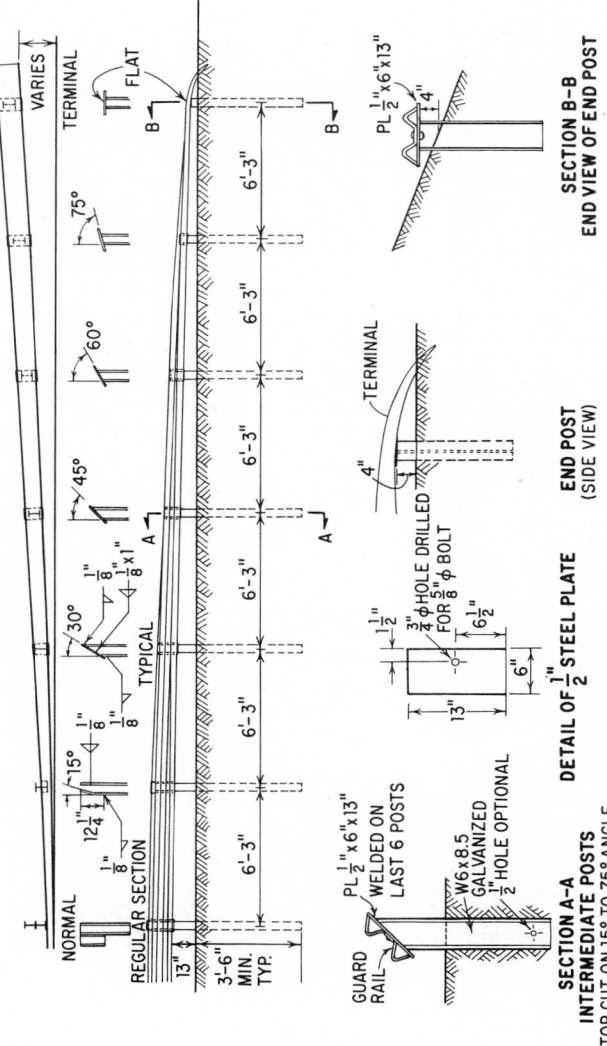

Fig. 16-47. Terminal treatment for W-beam guardrail. For details of 8-gage washer and bolts, see Fig. 16-44. (*Virginia Department of Highways and Transportation.*)

Sand-Filled Modular Units.   These are plastic containers 3 ft high and 3 ft in diameter, partly filled with sand.   They can retain a car traveling at 50 mph in about 30 ft with a deceleration of about 4g to 6g.

Steel Drums.   With portions of their tops cut out, steel drums may be welded together to form a modular system with good impact attenuation.   It can stop vehicles traveling at 60 mph in about 14 ft with deceleration in the range of 8g.

16-18. Community and Environmental Considerations.   The National Environmental Policy Act of 1969 and subsequent related legislation created a Council on Environmental Quality and required the preparation of an environmental impact statement for every federally sponsored action "significantly affecting the quality of the human environment."   Some states have passed similar legislation related to nonfederally sponsored actions.

To develop locally acceptable projects that comply with the appropriate environmental legislation and that can be successfully funded and constructed, traditional engineering skills must be supplemented with input from other disciplines to integrate these projects into the local physical and social environment.   In an assessment of highway impact, the design team should place equal emphasis on ascertaining potential problems and potential opportunities.

Federal Highway Administration Instructional Memorandum 20-4-72, "Guidelines for Consideration of Economic, Social and Environmental Effects," requires those responsible for development of a project to take into consideration, in addition to the need for fast, safe, and efficient transportation, such factors as highway costs, traffic benefits, and public services, including provision for national defense.   The Memorandum also requires study of anticipated economic, social, and environmental effects in seven areas.   Not only must the impact of a proposed project be studied, but also the impact of all alternatives under consideration.   The factors are:

1. Regional and Community Growth.   General plans and proposed land use, total transportation requirements, and status of the planning process require study.   The engineer should consider studying, for example, the direction of urban and suburban growth as a guide and control on future traffic demands.

2. Conservation and Preservation.   The engineer should investigate project impact on soil erosion and sedimentation, the general ecology of the area, and man-made and natural resources, such as park and recreational facilities, wildlife and waterfowl areas, and historic and natural landmarks.

In certain instances, the environmental impact of a highway can extend far beyond the limiting rights of way, in that it may have long-term impact on a nonrenewable natural resource. For example, the construction demands of a highway for select granular material and other earth fill often require large volumes of borrow, or excavations, which can result in alteration of sizable parcels of valuable land.

Identification of easily erodable soils, predictable internal and surface drainage patterns, stream-channel characteristics, and local geologic features are necessary for the basic environmental evaluation.   On occasion, engineers may encounter unique natural areas which, because of unusual ecological conditions, support a rare plant or animal community.   Such areas, on expert evaluation, should be avoided for highway construction if at all possible.

The continuing reduction in the habitats of many forms of wildlife is also reflected in the public concern about environmental impact of highways.   This impact can range from obstruction of trout spawning movements through a culvert to small groundwater-table changes which can contribute to a major ecological impact on large wetland areas.   Special consideration should be given to treatment of wetlands and water courses as wildlife areas.   Stream-channel alterations should be carefully considered to avoid destruction of fish habitats.   Engineers should consider the possibility of recreating suitable fish habitats as part of a project involving stream relocation.

Marshes (fresh, salt, and estuarine), swamps, and bogs are all wetlands that are valued as groundwater recharge basins, wildlife refuges, and open space.   These areas are recognized as producing large quantities of food basic to the ecological *food chain* as well as being the source of many watersheds.   Because of their low human occupancy, they are often regarded as logical areas for highway development.   They are, however, rated high on the environmental scale and should be completely avoided wherever possible.

Note, though, that some controlled-access rural highways with wide rights of way can be designed to provide wildlife habitats for preservation of existing unusual or common species of plants and animals if the presence of the animals does not constitute a safety hazard to the traveling public.

In cases where recreational areas cannot be avoided, engineers should develop detailed plans to minimize highway impact. These plans may involve replacement of the recreational facilities.

3. Public Facilities and Services. Provision should be made for health, religious, and educational facilities, public utilities, fire protection, and other emergency services.

A typical condition to be avoided is locating a highway in such a way that it separates a community facility (school, hospital, etc.) from the neighborhood it is meant to serve.

4. Community Cohesion. Residential and neighborhood character and stability, highway impact on minorities and other specific groups and interests, and effects on local tax base and property values should be evaluated.

5. Displacement of People, Businesses, and Farms. Relocation assistance, availability of adequate replacement housing, and economic activity (employment gains and losses, etc.) should be investigated.

It is important to identify primary and secondary economic impacts of a highway. Often, this identification requires investigation of benefits and costs, both for governmental jurisdictions and for income groups, by assessing the project's impact on the land market, labor market, and fiscal base of the community, and the project's relationship to overall community growth and development. For instance, studies might show that a pocket of unemployment can be eliminated by the improved accessibility brought about by addition of a highway interchange in the area.

Also to be considered is a comparison of the cost of upgrading a highway to take heavy truck traffic with the benefit of having goods delivered more cheaply.

6. Air, Noise, and Water Pollution. Conditions after highway construction should conform with approved air-quality implementation plans, FHWA noise-level standards, and any relevant federal or state water-quality standards.

Air Pollution. Few guidelines are currently available relating highway design to vehicle-emission pollution. In general, air pollution from vehicle emissions can be reduced by avoiding steep grades, conditions requiring acceleration and deceleration, and *closed-in* highway segments with poor air circulation, particularly in urban and suburban areas. Because the last condition may offer a possible solution to noise pollution, maintenance of air-quality and noise standards should be studied concurrently.

Because of rapid changes as a result of laws and improved gasoline engine technology, engineers should be familiar with the latest developments.

The possibility that air pollution may be intensified by unpredictable meteorological events, such as temperature inversions, turbulence, inhomogeneity of the atmosphere, or noncharacteristic wind patterns, should be taken into account.

Air pollution during construction, especially in urban and suburban areas, can be eased by such procedures as dust control, which should be specified in demolition contracts, and converting felled trees into wood chips rather than allowing the trees to be burned.

Noise Pollution. Noise-evaluation techniques and criteria for highways are still in the development stage. In general, while there are techniques for predicting noise levels and evaluating existing noise-abatement barriers, there is little consensus as to the best methods of abatement.

Noise-abatement plantings and barriers continue to be based primarily on empirical data, with generally poor results. Basically, if the adjacent property owners or other segments of the public can see the highway, they can hear it. In many instances, screen plantings have been psychologically successful but have failed as physical noise abatement.

Use of depressed sections, berms or other physical barriers, and smooth-surfaced pavements should be considered for heavily populated areas. Some alignment and grade solutions to noise problems, however, may increase air pollution by entrapping vehicle emission. Furthermore, smooth pavements may increase skidding.

For a highway that passes close to an existing school, recommendations for noise and air-quality control have included air conditioning the school as part of the highway construction so that windows can be kept closed.

The need for noise control during construction should also be considered. Some localities have established guidelines which, in effect, limit the length of time construction equipment can be used as a function of its operating noise and its distance from homes, schools, etc.

Water Pollution. Particular attention should be given to the final disposition of water collected on a project so that physical damage to adjacent properties is avoided and the resultant water is as free of pollutants as possible.

Special attention should also be given to the final disposition of salt and oil from pavement surfaces. In some cases, ditches or closed drainage systems may have to be provided to prevent discharge of these pollutants into sensitive streams or ponds.

The cumulative effects of deicing salts on plants and animals are still not completely known. Nevertheless, engineers should consider the probable long-term environmental impact of pavement and shoulder drainage. The effect of overloading existing combined sewer systems, which might result in untreated sewage bypassing treatment plants, should also be investigated.

During construction, the rapid stabilization of disturbed areas through revegetation is of primary importance. Both temporary and final soil erosion control should be considered by engineers. Note, however, that runoff of excess fertilizer from seeded areas may increase phosphates in adjacent waterways to the point where they become a pollutant.

7. Esthetic and Other Values. Visual quality, such as *view of the road* and *view from the road*, and joint development and multiple use of space should be studied.

By properly relating the project to its surroundings, considering the highway as viewed from adjacent surroundings, the view from the highway itself, and drivers' visual sequence as they drive the highway (Fig. 16-48), designers can help prevent the need for expensive and sometimes ineffective *cosmetic* treatments in later development phases of the project.

If the alignment comes near existing structures of historic importance, visible sites of geologic interest, or some other cultural landmark, engineers should consider the opportunities for multiple use or joint development.

Evaluating the Various Factors. This often leaves the engineer with alternatives that must be compared in a subjective manner. In particular, the objectives of the highway project as viewed by those agencies, groups, and individuals interested in or greatly affected by it should be clearly identified and officially incorporated into the evaluation analysis. The ultimate

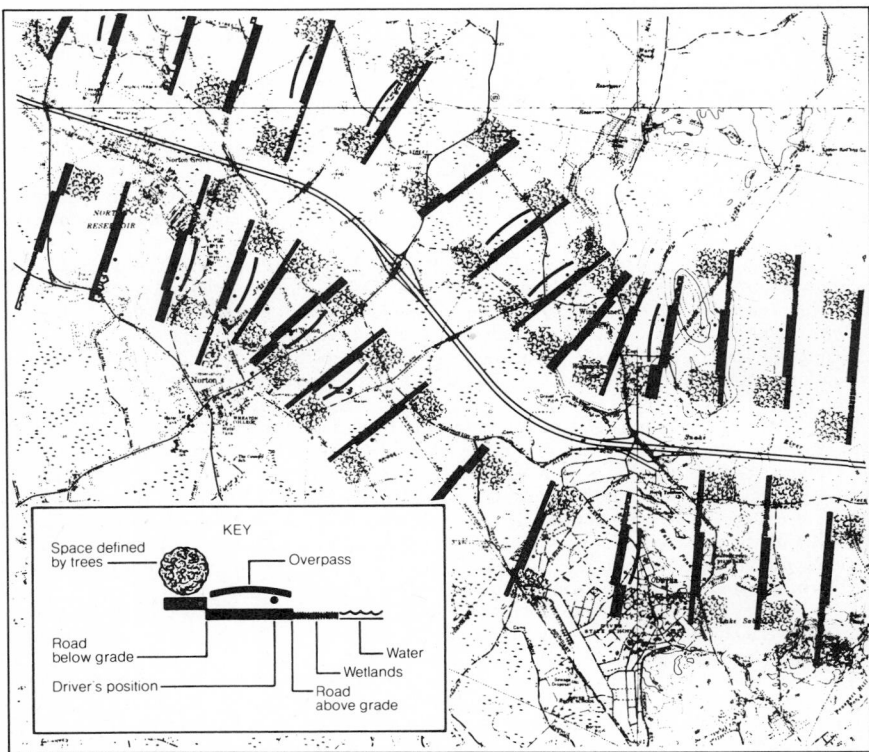

Fig. 16-48. Map of a highway, with appropriate symbols, allows assessment of a driver's visual sequence along the highway.

decision will be facilitated if it is known how each major interest group views the various alternatives and why. This method often shows the relative strengths and weaknesses of alternative solutions, makes the evaluation process systematic and comprehensible, and serves as a mechanism for meaningful community participation. The evaluation can be accomplished with the aid of a planning balance sheet, such as that shown in Table 16-30.

Table 16-30. Sample Planning Balance Sheet*

Interest Groups and Objectives	Weight	Alternative A		Alternative B	
		Rating	Value	Rating	Value
Steps 1 and 2	Step 4	Step 5			
Identify all the interest groups and ask them to define their objectives and concerns with respect to the proposed project.	Request the interest groups to assign weights indicating importance to each of their objectives.	Evaluate and rate the alternatives according to the objectives of the interest groups.			
Step 3 (Optional)		Step 6			
Determine the relative importance of each interest group.		Multiply the weight by the rating to obtain the value for every objective.			
		Step 7			
		Add all the values for each interest group to obtain total comparative values for the alternatives by group. If the groups themselves have been weighted (Step 3), total scores for each alternative can be calculated.			

*From Julie Hetrick Schermer, "Interest Group Impact Assessment in Transportation Planning," *Traffic Quarterly*, January, 1975.

Typical community participation techniques include regular meetings, surveys, interviews, formal reporting to representative major interest groups and citizen advisory committees, and other structured procedures to obtain citizen input to the design process. Contact with the general public can be maintained by use of media, mini-hearings, newsletters, and speakers and by an open-door policy by which the project files and staff are accessible to all.

("Highway Design Manual," New York State Department of Transportation; Special Report 138, "Environmental Considerations in Planning, Design and Construction," Transportation Research Board; G. M. Sturman, "Socioeconomic and Environmental Impact of Highways," Chap. 8, "Environmental Science Policy: An Enunciation of Congressional Concern," M.I.T. Press, Cambridge, Mass.; "Environmental Design and Public Projects," Colorado State University, National Science Foundation, RANN Program Project Report, 1973.)

16-19. Highway Maintenance. For minor roads, the nature of maintenance changed very little over the years, except for increased mechanization. But for major highways, the constant improvement in pavement design and the attention given to appearance have brought about a change of emphasis. Whereas formerly, pavement repairs might have been the major maintenance problem, now the major items in the maintenance budget are likely to be mowing in summer and snow removal in winter.

Early detection and repair of minor defects, however, are still among the most important phases of maintenance work. Cracks and other surface breaks, which in their early stages are almost unnoticeable, may develop, if untended, into major repair jobs after only a few days of heavy traffic. For this reason, frequent close inspection of pavement and other highway structures by competent and experienced personnel is absolutely necessary.

Dirt Roads. Maintenance consists largely of periodic scraping and leveling with a road grader, as well as cleaning out side ditches and culverts. To provide better compaction, blading or dragging should be done as soon after a rain as is practicable, or when surface materials are moist.

Gravel Roads. For untreated gravel roads, a heavy blading of the surface, including side ditches, is called for in early spring and late fall. Badly rutted pavements may need additional gravel. To maintain the pavement in fair driving condition, it should be bladed and shaped at intervals, especially after heavy rains.   Usually, for this purpose, a drag, essentially a frame of heavy timbers, may be pulled over the surface by a truck.

The efficiency of a gravel surface is materially increased by treatment with a dust palliative, such as calcium chloride.   The purposes of dust palliatives are abatement of the dust nuisance to traffic and property adjacent to the highway, provision of additional superficial bond, and conservation of surface material.   Standard practice is to make one application in late spring and a second in midsummer.   The surface is first bladed.   Then, the calcium chloride is applied at a rate of about ¾ to 1 lb per sq yd.   The best time to apply calcium chloride is after a rain, because low humidity inhibits absorption.   When the calcium chloride has been completely absorbed, the surface should be drag-bladed.

Road-mix and Plant-mix Asphalt Surfaces.   A well-constructed pavement of either of these types, with a good base course on a well-drained subgrade, requires comparatively little maintenance if the traffic does not exceed the design loading.   Holes, raveling (progressive disintegration from the surface downward, or edges inward, by the dislodgement of aggregate particles), and ruts should be patched in their early stages with a cold mix of composition similar to that of the original mix.   Before patching is done, loose material should be cleaned out and a prime coat of hot asphaltic oil applied.   Asphalt emulsions are sometimes used as prime coats; cationic emulsions have the advantage of not requiring dry conditions for the job.   With some classes of road or on plant-mix surfaces, the pavement can be renewed by application of a seal coat, followed by application of screenings or sand, or else an overlay of plant-mixed asphaltic concrete.   The patch for a deep hole should be applied in successive thin layers.

Penetration Macadam.   Patching should be done as previously described for asphalt surfaces.   The seal coat should be renewed at intervals depending on the quality of the original construction.   Care should be exercised to keep the seal coat no thicker than necessary, since oil spots (*bleeding*) constitute a traffic hazard.

Bituminous Concrete.   Bituminous concrete surfaces will ravel if they are deficient in asphalt. The raveling can be corrected by applying a seal coat or overlay of plant-mixed asphaltic concrete.

Bituminous concrete, if rich in asphaltic cement, has a tendency to become slippery, because the asphalt works to the surface under the action of weather and traffic (bleeding).   The excess surface asphalt can be removed by planing or infrared heater or can be *blotted up* by spreading heated sand and then brooming.   A seal coat or overlay may then be applied, as warranted by field conditions.   Note that use of seal coats on high-speed, heavily traveled highways should be considered carefully.   Experience indicates that traffic may cause aggregate to be thrown about; an overlay of asphaltic concrete, therefore, is generally more practical in these cases.

Sheet (Sand) Asphalt.   The only maintenance that a surface of well-laid sheet asphalt requires is the filling of minor cracks.   This should preferably be done in cold weather when the cracks are fully dilated.   They should be thoroughly cleaned out, with an air compressor if one is available, and filled with hot asphalt.   Broken areas or holes can be temporarily patched, but they are symptoms of something amiss with the base or drainage.   The cause should be investigated and corrected by removing the surface course and replacing it once the repairs have been made. Repairs to the pavement after it has been cut open for utilities should be carried out by experienced road crews.

Portland-Cement Concrete.   Routine maintenance of this type of pavement consists mainly of filling in cracks and expansion joints.   Unless these are filled with hot asphalt as soon as possible, the edges will begin to break down, and there will be progressive deterioration of the surface adjoining the crack or joint.   A crack may also allow water to enter the subgrade, with a resultant loss of bearing capacity.   The crack should be thoroughly cleaned out with a hard brush or, preferably, an air compressor.   For joints, deteriorated joint filler should be routed out and replaced by a rubber-based asphalt sealer.

A slight vertical movement of one slab in relation to an adjacent one, at a transverse joint, should be corrected with a carefully placed bituminous patch.   Otherwise, the impact of heavy wheels will cause breaking and spalling.

In the past, rock salt or other chemicals used to melt ice have caused considerable scaling or pitting of the concrete surface.   This problem has been corrected to some extent by the use of air-entraining cement and epoxy sealers.   A protective coat of epoxy has been used with some success on bridge decks to prevent the salt from affecting the concrete.

In older pavements laid over inadequate bases or poor subgrades, slabs may settle.   These can be successfully raised several inches by slabjacking.   Holes are drilled through the slab at intervals, and various mixtures (oil and cement, finely ground limestone or sand, fly ash, or portland cement combined with sufficient water to permit pumping) are forced through the holes to the underside of the pavement. The consistency of the mix and the pressure to be used depend on the application. The pressure raises the slab and holds it in place until the grout hardens.

Occasionally, small areas of concrete surface deteriorate and leave shallow depressions. Formerly, the treatment was to scrape out all loose material, spread a neat-cement grout over the surface, and then tamp in a rich, stiff concrete mix. Now, epoxy sometimes is used in making these patches. The epoxy process is also used extensively in repairing bridge decks. Isolated holes are patched with high-early-strength concrete and allowed to harden, and then a coating of epoxy is spread and sanded. This is followed by a thin bituminous-concrete layer. The sanding prevents slippage of the surface coat. The epoxy acts both as an adhesive and as a moisture barrier between the concrete and the resurfacing.

When, after years of service, the concrete surface has deteriorated so badly that routine maintenance measures are unduly costly, the pavement can continue to function usefully as a base for a bituminous-concrete surface course. A course 1 to 1½ in. thick generally suffices.

General Road Maintenance. In addition to pavement maintenance on roads, maintenance operations include shaping and sealing of shoulders, cutting back (or chemically treating) weed and hedge growth, clearing ditches and drains, repairing small structures, removing litter, repainting traffic stripes, etc. For major highways, mowing is a continuous procedure from spring to fall. The grass on narrow medians and along the outer fringes of wide medians should be kept short. The remainder of the grass cover in a median should be kept to a controllable height. These procedures also apply to interchange areas, especially those adjacent to ramp exits and entrances, where unduly high grass not only is unsightly but can interfere with sight distance and can easily become a source of fire during dry periods. An occasional cutting of the grass on the outer portions of the slopes suffices, except along parkways, where esthetics is an important consideration.

Equipment includes large gang-type mowers drawn by tractors, self-propelled mowers with sickle bars, hand-operated motor mowers for narrow places and around guardrails, and in some cases workers with scythes. (A secondary reason for flattening slopes is to permit use of mechanical equipment for mowing.)

With steadily increasing traffic, snow removal has become a major maintenance problem. Motorists expect cleared roadways come what may. To meet this demand, states and municipalities have had to develop organizations comparable to the military. In a typical state program, snow-fighting equipment is distributed to strategic locations in October. Contractors are engaged, on a standby basis, to use both their own equipment and that which is state-owned. With the advent of winter, a headquarters is established, with personnel on duty day and night. Individual engineers or inspectors in the state highway field offices are given the responsibility of notifying headquarters of conditions and acting as liaison with the contractors. When a snowfall begins in an area, the field personnel inform headquarters of its progress. Usually, as the accumulation approaches 3 in., field personnel are told to order equipment into action, although some agencies in heavy snow regions will begin spreading salts and calcium chlorides well before snow accumulation reaches 3 in.

There is an established priority for clearing highways in a district. When a particular highway is hard hit, equipment from less-affected areas is diverted for the emergency. The first step is to clear one lane in each direction. Emergency routes, such as those for fire or hospital vehicles, and school bus routes are usually the first ones cleared. In rural areas, the snow is piled initially on shoulders or in medians, then later pushed over embankments or into rivers, ponds, etc. In urban areas, snow disposal is more difficult. Snow usually is temporarily piled where space is available. As soon as possible, the stored snow is loaded on trucks for dumping into rivers, lakes, or waste areas.

Various types of equipment have been developed for snow removal, but the mainstay is the truck-mounted blade. The truck can be used for other purposes during the remainder of the year and the blade attached as winter nears. Special snow equipment includes a large V-plow and large snow throwers, which pick up the snow and spray it some distance laterally. The use of this equipment, of course, is restricted to open country. Chemicals (generally chlorides) are heavily used to prevent freezing and to melt ice and snow layers on paved surfaces. But concrete deterioration (pavement, curbs, sidewalks, structures, etc.) may be caused by the chemicals.

Maintenance Considerations during Design. The difficulty of mowing narrow medians where there is a risk of workers being struck by passing cars should be considered in the design of highways and is often a good reason for paving those areas with concrete or asphalt.

The difficulty of replacing a nonstandard item incorporated into the construction also should be considered. For example, if a lamp selected for lighting the highway is not an off-the-shelf item, the cost of replacing burnt-out lamps may be prohibitive; the manufacturer may be willing to tool up only for a large number of such lamps. Designers should specify standard items to ease supply and repair operations. If it is necessary to use nonstandard items, designers should specify that an adequate number of spares be furnished to the maintenance shop at the same time as the initial installation. Thus, only one large production run is required to supply original parts plus spares, which will probably lower the unit cost.

Resurfacing of roadways should be planned for in design. The ultimate pavement thickness should be estimated because it is important in establishing clearances beneath underpasses (3 in. of additional clearance is usually satisfactory as a minimum) and in calculating the ultimate dead load on overpasses.

Snow removal should also be planned for during highway design. Since it is easier to push snow off to the side of a roadway than to load and haul it away, sufficient room for storing snow should be provided in the design of a highway. In this regard, the relationship of pavement cross slopes to shoulder cross slopes is very important. If snow is stored on shoulder and median areas, the cross slopes of those areas should be designed to drain the water from melting snow away from the pavement surface to avoid *icing* the pavement during the colder portions of the day. It is also important to keep the low points in the road profile away from shaded areas (caused by bridges and other structures), to help reduce the possibility of icing.

("An Informational Guide for Physical Maintenance," American Association of State Highway and Transportation Officials; K. B. Woods, "Highway Engineering Handbook," McGraw-Hill Book Company, New York.)

16-20. Estimating Costs of Highways. Costs vary widely, depending mostly on the class of highway; its location, which ranges from inexpensive marginal-use land to urban or commercial areas of great value; the topography, which affects costs by requiring heavy or light grading; geological characteristics, which heavily affect cost because rock excavation and extensive treatment of poor soils are expensive; and the frequency and size of cross roads and water or railroad crossings, which are also major causes of high construction cost.

Expenditures for highways have been grouped in major classes: capital outlay, maintenance, administration, highway law enforcement and safety, interest on debt, debt retirement, and intergovernmental payments. Also to be considered are secondary economic impacts (Art. 16-18) and road-user costs and benefits.

Capital outlays are those costs associated with highway improvements, including land acquisition and other right-of-way costs, utility relocations, preliminary and construction engineering, and the construction cost of the highway.

While engineering services have generally become more efficient in recent years through the use of aerial photogrammetry, electronic computers, and other sophisticated tools, costs have tended to increase. This apparent contradiction results from requirements for additional services in the planning stage to satisfy socioeconomic and environmental requirements, including public hearings, and from the need for more complex highway elements than were common in the past, such as curved steel girders, concrete box beams, traffic-actuated signals, and changeable message signs, which require specialized engineering skills. For these reasons, the traditional method of establishing engineering fees—as a function of construction cost—is no longer wholly valid, and fees now are generally established through detailed development of engineering costs and subsequent negotiation of the fee.

Right-of-way costs depend on the width required and the value of the land to be acquired. Rural right of way is less expensive than urban. But certain rural right of way is costly, such as farmland to be split by a highway, land zoned for a special use not allowed elsewhere in the area, or the occasional expensive residential, commercial, or industrial site. Urban right of way usually is expensive in character as well as number of parcels required. Usually, the most costly land areas require the greatest numbers of facilities, such as interchanges and intersections, because they are more intensely developed. Consequential damages, which are damages caused not by the physical impact of the facility but by its effects on the operation of the remaining property, also should be considered.

The most important construction costs are earthwork or grading, pavement, drainage, bridges, and walls. Each varies greatly with conditions encountered, as well as with quantities involved. Grading includes cuts in earth or rock, embankments, and embankment foundations. Costs of pavement, shoulders, and curbs are functions of area needs and the type of pavement required for the class of road. Costs for drainage vary according to the class of highway, the type of collection and disposal system, and the geographical location. Where open ditches are adequate, costs will generally be lowest; while, all other factors being equal, the highest class, with inlets and pipe, will be the most expensive.

Bridge costs vary with the number and size required. A low-class road generally requires bridges only to cross intervening waterways. The highest-class roads require bridges at all important cross roads, all significant waterways, and all railroad crossings. Bridges also are required for ramp crossings at complex interchanges. Walls, while costly, are constructed to avoid other, more costly types of construction and to avoid more expensive property-taking in urban areas.

Most states maintain cost records that can serve as guides for estimating purposes. Every quarter, *Engineering News-Record* publishes a summary of bid prices as reported by the highway departments of representative states. Any attempt to estimate construction costs, however, should take into consideration current trends in construction practice, material prices, and wages; availability of materials; construction capability; construction methods that will be employed; labor union requirements; taxes; and environmental requirements.

A Federal Highway Administration quarterly, "Price Trends for Federal-Aid Highway Construction," presents a composite index based on common excavation, surfacing (portland-cement concrete and bituminous concrete), and structures (reinforcing steel, structural steel, and structural concrete) (Table 16-31). As shown on the graphs in Fig. 16-49, the rise in prices since 1972 has been such that any extrapolations into the future should be made with caution.

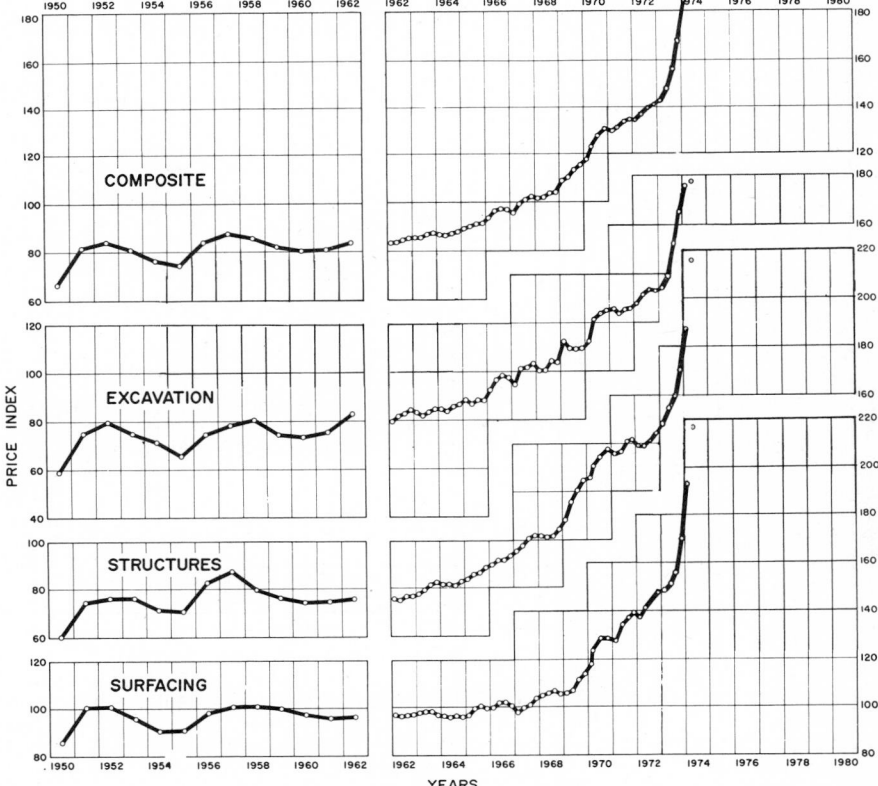

Fig. 16-49. Price trends for federal-aid highway construction. Price index is taken as 100 for the base year 1967. All points from 1950 through 1962 reflect mathematical conversions from a 1957-1959 base to the 1967 base. Beginning with 1962, the points represent three-quarter moving averages for 1967 base quantities and are plotted on the middle quarters of the three-quarter periods. (*Federal Highway Administration.*)

Table 16-31. Price Trends for Federal-Aid Highway Construction*

| Year | Common excavation | | Surfacing | | | | | Structures | | | | | | | Composite index |
| | Average contract price, $ per cu yd | Index | Portland-cement concrete | | Bituminous concrete | | Surfacing Index | Reinforcing steel | | Structural steel | | Structural concrete | | Structures index | |
			Average contract price, $ per sq yd	Index	Average contract price, $ per ton	Index		Average contract price, $ per lb	Index	Average contract price, $ per lb	Index	Average contract price, $ per cu yd	Index		
1967	0.54	100.0	4.43	100.0	6.47	100.0	100.0	0.131	100.0	0.247	100.0	70.30	100.0	100.0	100.0
1968	0.56	102.6	4.79	108.1	6.77	104.7	106.4	0.131	100.5	0.249	100.8	71.81	102.1	101.5	103.4
1969	0.59	108.5	4.87	110.0	7.01	108.4	109.3	0.143	109.6	0.316	128.1	81.34	115.7	118.3	111.8
1970	0.66	121.8	5.42	122.4	8.04	124.3	123.3	0.163	124.9	0.338	137.0	92.73	131.9	132.2	125.6
1971	0.67	123.8	6.06	136.8	8.54	132.1	134.5	0.177	135.3	0.348	141.2	97.02	138.0	138.5	131.7
1972:															
First quarter	0.72	132.4	6.07	137.2	8.53	132.0	134.7	0.185	141.3	0.358	145.0	96.16	136.8	140.0	135.5
Second quarter	0.69	127.5	6.47	146.2	9.05	140.0	143.2	0.176	134.8	0.310	125.7	95.26	135.5	132.5	133.7
Third quarter	0.73	135.0	6.47	146.2	9.50	147.0	146.6	0.177	135.3	0.348	141.1	103.90	147.8	143.6	141.2
Fourth quarter	0.76	140.0	5.93	133.8	10.51	162.6	147.7	0.185	141.6	0.347	140.5	106.75	151.9	146.7	144.4
Average	0.72	133.4	6.25	141.2	9.22	142.6	141.9	0.181	138.2	0.342	138.6	100.17	142.5	140.6	138.2
1973:															
First quarter	0.67	124.7	6.57	148.4	9.85	152.3	150.3	0.181	138.7	0.295	119.4	109.34	155.6	141.9	137.8
Second quarter	0.75	138.0	6.36	143.6	9.90	153.1	148.2	0.193	147.3	0.352	142.5	113.51	161.5	153.4	145.9
Third quarter	0.81	149.5	7.10	160.4	9.61	148.7	154.7	0.212	161.9	0.422	170.9	110.60	157.3	162.1	155.1
Fourth quarter	0.93	172.7	7.43	167.9	10.83	167.5	167.7	0.233	178.0	0.379	153.6	113.51	161.5	162.0	167.8
Average	0.80	147.1	6.87	155.1	9.99	154.5	154.8	0.207	158.0	0.373	151.0	111.83	159.1	156.5	152.4
1974:															
First quarter	0.97	179.7	8.17	184.4	13.28	205.4	194.6	0.281	215.1	0.459	186.0	129.64	184.4	190.2	187.4
Second quarter	0.96	178.0	8.48	191.4	15.77	243.8	216.8	0.342	261.2	0.555	224.9	137.07	195.0	215.4	201.4

*Base for composite index, 1967, involves 1,656,655,000 cu yd of roadway excavation, 79,942,000 sq yd of portland-cement-concrete surfacing with an average thickness of 8.7 in., 51,230,000 tons of bituminous-concrete surfacing, 981,587,000 lb of reinforcing steel for structures, 885,235,000 lb of structural steel, and 5,572,000 cu yd of structural concrete. Index figures are computed from 1967 base quantities and prices. Prices for portland-cement-concrete surfacing reflect adjustments to base period thickness in each state and do not include costs for reinforcing steel and joists. Adapted from "Price Trends for Federal-Aid Highway Construction," Federal Highway Administration.

Costs included under maintenance are of two types: those required to keep the highways in usable condition, such as routine patching and repairs, bridge painting, and other maintenance-of-condition costs; and traffic service costs, such as snow and ice removal, pavement markings, signs, signals, and litter cleanup. Although programs related to street lighting and cleaning, sidewalks, and storm sewers are frequently administered by the municipal government in connection with street programs, expenditures for such programs are generally considered to be for the protection of the health and safety of the public rather than as expenditures for highways.

Classed as administration costs are those for general overhead and for engineering and research not assignable to specific road projects. Included as highway law-enforcement and safety expenditures are activities of state highway patrols, safety education and promotion, driver-training programs, and enforcement of vehicle size and weight limitations. Costs of municipal traffic police also are included where the function is separate from general policing activities.

Intergovernmental payments refer to the actual payment of money from one governmental level to another. Because the expenditures of one governmental agency may become income to another, care must be taken to avoid double counting of income and expense.

Once the need for a highway has been established, an analysis of road-user costs and benefits can assist in choosing between alternatives. Road-user costs and benefits are usually stated in the following terms (from "Road-user Benefit Analyses for Highway Improvements," American Association of State Highway and Transportation Officials).

Road-User Cost. Vehicular operating costs, usually expressed in cents per vehicle-mile, covering all items involved in vehicle ownership and operation. The value of time is included as one of the items of cost.

Road-User Services. Advantages or privileges accruing to the vehicle driver or owner as a result of features of safety, comfort, convenience, etc. In some cases these can be evaluated in cents per vehicle-mile.

Road-User Benefits. The advantages, privileges, or savings that accrue to drivers or owners through the use of one highway facility as compared with the use of another. Benefits are measured in terms of the decrease in road-user costs and the increase in road-user services.

(Robley Winfrey, "Economic Analysis for Highways," Intext Educational Publishers, New York, N.Y.; "Highway Statistics," Federal Highway Administration.)

Bridge Engineering

JOHN J. KOZAK

Chief, Division of Structures
California Department of Transportation, Sacramento, Calif.

and

JOACHIM F. LEPPMANN

Consulting Engineer, Berkeley, Calif.

Bridge engineering covers the planning, design, construction, and operation of structures that carry facilities for movement of humans, animals, or materials over natural or man-made obstacles.

Most of the diagrams used in this section were taken from the "Manual of Bridge Design Practice," State of California Dept. of Highways. The authors express their appreciation for permission to use these illustrations from this comprehensive and authoritative publication.

GENERAL DESIGN CONSIDERATIONS

17-1. Bridge Types. Bridges are of two general types, fixed and movable. They also can be grouped according to the following characteristics:

Supported Facilities: Highway or railway bridges and viaducts, canal bridges and aqueducts, pedestrian or cattle crossings, material-handling bridges, pipeline bridges.

Bridge-Over Facilities or Natural Features: Bridges over highways and over railways; river bridges; bay, lake, and valley crossings.

Basic Geometry: In plan—straight or curved, square or skewed bridges; in elevation—low-level bridges, including causeways and trestles, or high-level bridges.

Structural Systems: Single-span or continuous-beam bridges, single- or multiple-arch bridges, suspension bridges, frame-type bridges.

Construction Materials: Timber, masonry, concrete, and steel bridges.

17-2. Design Specifications.   Designs of highway and railway bridges of concrete or steel often are based on the Standard Specifications for Highway Bridges of the American Association of State Highway and Transportation Officials (AASHTO) and the "Manual for Railway Engineering" of the American Railway Engineering Association (AREA).   Also useful are the "Standard Plans for Highway Bridges," Federal Highway Administration (FHWA), and standard plans issued by various highway administrations and railway companies.

Length, width, elevation, alignment, and angle of intersection of a bridge must satisfy the functional requirements of the supported facilities and the geometric or hydraulic requirements of the bridged-over facilities or natural features.   Figures 17-1 and 17-2 show typical highway and railway clearance diagrams.

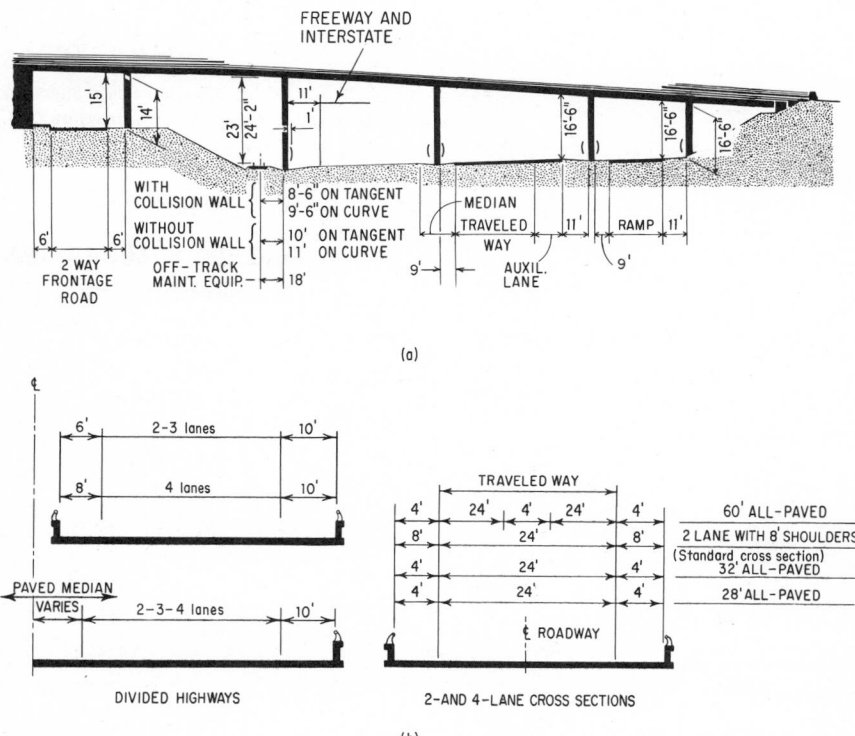

Fig. 17-1. (*a*) Minimum clearances for highway structures.   (*b*) Typical bridge cross sections.   Major long-span bridges may vary from these.

Selection of the structural system and of the construction material and detail dimensions is governed by requirements of structural safety; by economy of fabrication, erection, operation and maintenance; and by aesthetic considerations.

Highway bridge decks should offer comfortable, well-drained riding surfaces.   Longitudinal grades and cross sections are subject to standards similar to those for open highways (Sec. 16). Provisions for roadway lighting and emergency services should be made on long bridges.

Barrier railings should keep vehicles within the roadways and, if necessary, separate vehicular lanes from pedestrians.   Utilities carried on or under bridges should be adequately protected and equipped to accommodate expansion or contraction of the structures.

Most railroads require that the ballast bed be continuous across bridges to facilitate vertical track adjustments.   Long bridges should be equipped with service walkways.

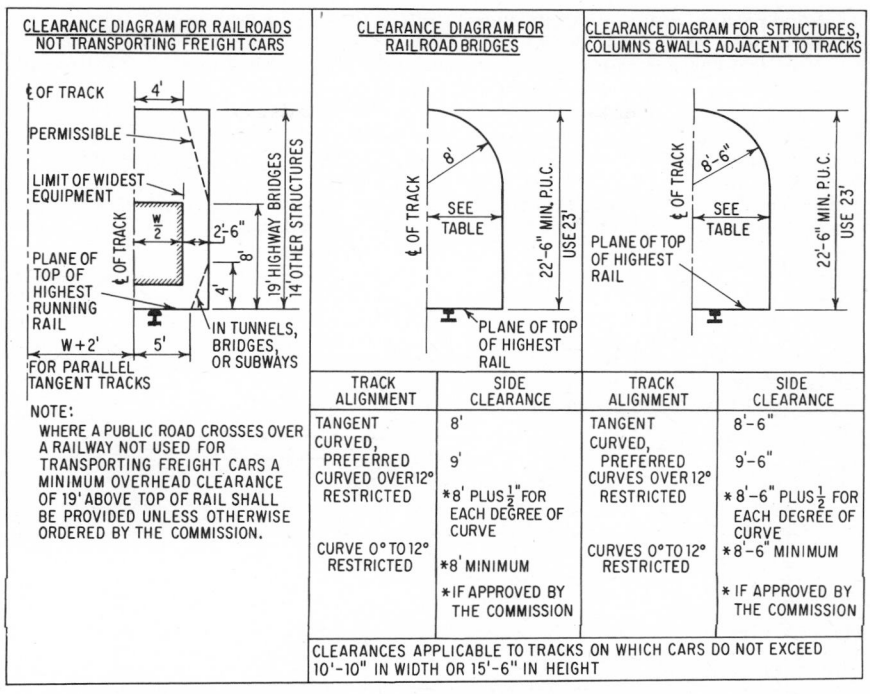

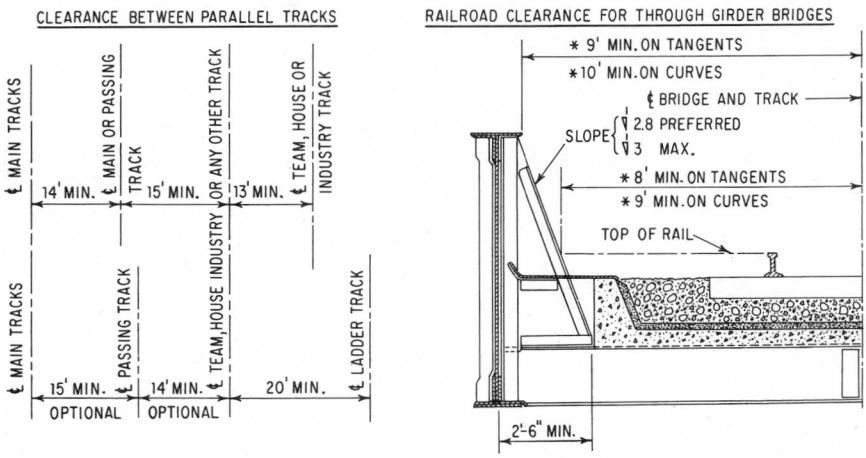

Fig. 17-2. Minimum clearances for railroad structures.

17-3. Design Loads. Bridges must support the following loads without exceeding permissible stresses and deflections:

Dead load D, including permanent utilities.

Live load L and impact I.

Longitudinal forces due to acceleration or deceleration LF and friction F.

Centrifugal forces CF.

Wind pressure acting on the structure W and on the moving load WL.

Earthquake forces EQ.

Earth E, water and ice pressure ICE, stream flow SF, and uplift B acting on the substructure.

Forces resulting from elastic deformations, including rib shortening R.

Forces resulting from thermal deformations T, including shrinkage S.

Highway Bridge Loads. *Vehicular live load* of highway bridges is expressed in terms of design lanes and lane loadings. The number of design lanes depends on the width of the roadway.

Each lane load is represented either by a standard truck with trailer (Fig. 17-3) or, alternatively as a 10-ft-wide uniform load in combination with a concentrated load (Fig. 17-4). As indicated

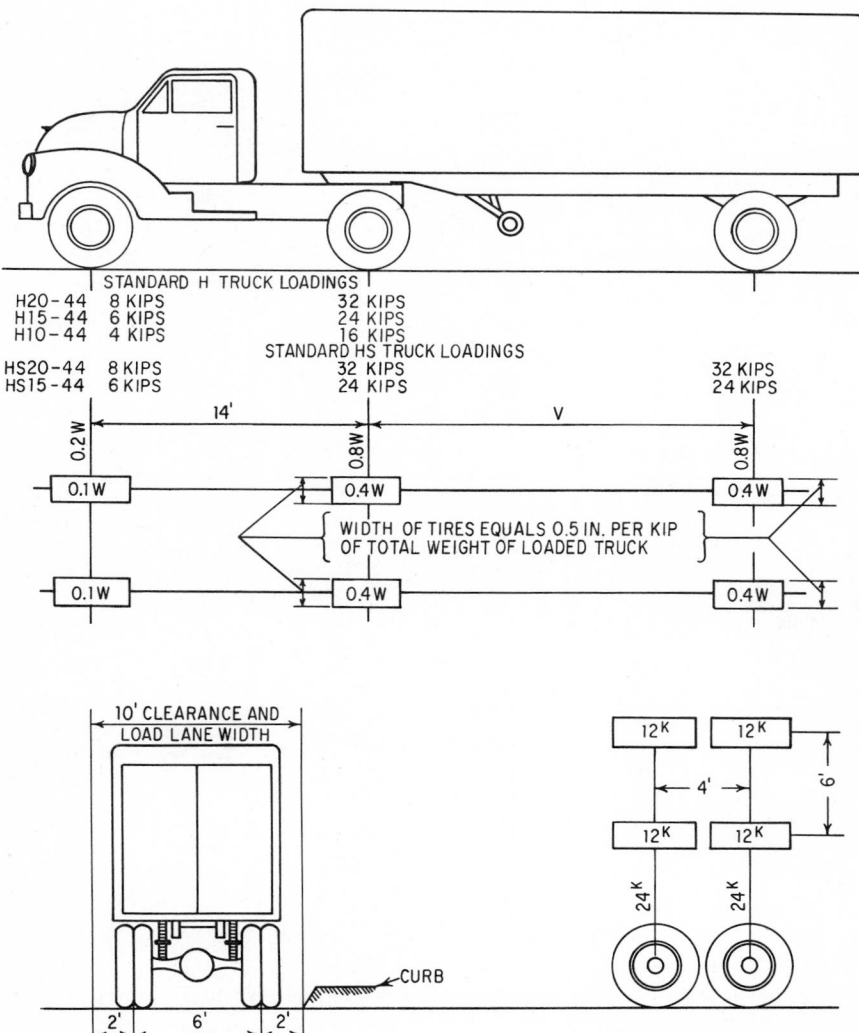

Fig. 17-3. Standard truck loading. For H trucks, W = total weight of truck and load, and for HS trucks, W = combined weight on the first two axles, which is the same weight as for H trucks. V indicates a variable spacing from 14 to 30 ft that should be selected to produce maximum stress.

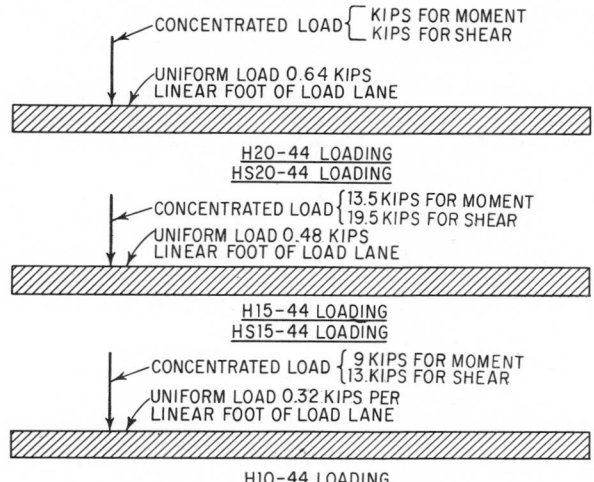

Fig. 17-4. H and HS lane loadings for simply supported spans. For maximum negative moment in continuous spans, an additional concentrated load of equal weight should be placed in one other span for maximum effect. For maximum positive moment, only one concentrated load should be used per lane, but combined with as many spans loaded uniformly as required for maximum effect.

in Fig. 17-3, there are five classes of loading: H20, H15, and H10, which represent a truck with two loaded axles, and HS20 and HS15, which represent a truck and trailer with three loaded axles. These loading designations are followed by a 44, which indicates that the loading standard was adopted in 1944.

In proportioning any member, all lane loads should be assumed to occupy, within their respective lanes, the positions that produce maximum stress in that member. Table 17-1 gives maximum moments, shears, and reactions for one loaded lane. Effects resulting from the simultaneous loading of more than two lanes may be reduced by a loading factor, which is 0.90 for three lanes and 0.75 for four lanes.

In design of steel grid and timber floors for H20 or HS20 loading, one axle load of 24 kips or two axle loads of 16 kips each, spaced 4 ft apart, may be used, whichever produces the greater stress, instead of the 32-kip axle shown in Fig. 17-3. For slab design, the center line of the wheel should be assumed to be 1 ft from the face of the curb.

Sidewalks and their immediate supports should be designed for a uniform live load of 85 psf. The effect of sidewalk live loading on main bridge members should be computed from

$$P = \left(30 + \frac{3,000}{l}\right) \frac{55 - b}{50} \leq 60 \text{ psf} \tag{17-1}$$

where P = sidewalk live load, psf
l = loaded length of sidewalk, ft
b = sidewalk width, ft

Curbs should resist a force of 500 lb per lin ft acting 10 in. above the floor. For design loads for *railings*, see Fig. 17-5.

Impact is expressed as a fraction of live-load stress and is determined by the formula:

$$I = \frac{50}{125 + l} \quad (30\% \text{ maximum}) \tag{17-2}$$

where l = span, ft; or for truck loads on cantilevers, length from moment center to farthermost axle; or for shear due to truck load, length of loaded portion of span. For negative moments in continuous spans, use the average of two adjacent loaded spans. For cantilever shear, use $I = 30\%$. Impact is not figured for abutments, retaining walls, piers, piles (except for steel and concrete piles above ground rigidly framed into the superstructure), foundation pressures and footings, and sidewalk loads.

Table 17-1. Maximum Moments, Shears, and Reactions for Truck Loads on One Lane, Simple Spans*

Span, ft	H15		H20		HS15		HS20	
	Moment†	End shear and end reaction‡	Moment†	End shear and end reaction‡	Moment†	End shear and end reaction‡	Moment†	End shear and end reaction‡
10	60.0§	24.0§	80.0§	32.0§	60.0§	24.0§	80.0§	32.0§
20	120.0§	25.8§	160.0§	34.4§	120.0§	32.2§	160.0§	41.6§
30	185.0§	27.2§	246.6§	36.3§	211.6§	37.2§	282.1§	49.6§
40	259.5§	29.1	346.0§	38.8	337.4§	41.4§	449.8§	55.2§
50	334.2§	31.5	445.6§	42.0	470.9§	43.9§	627.9§	58.5§
60	418.5	33.9	558.0	45.2	604.9§	45.6§	806.5§	60.8§
70	530.3	36.3	707.0	48.4	739.2§	46.8§	985.6§	62.4§
80	654.0	38.7	872.0	51.6	873.7§	47.7§	1,164.9§	63.6§
90	789.8	41.1	1,053.0	54.8	1,008.3§	48.4§	1,344.4§	64.5§
100	937.5	43.5	1,250.0	58.0	1,143.0§	49.0§	1,524.0§	65.3§
110	1,097.3	45.9	1,463.0	61.2	1,277.7§	49.4§	1,703.6§	65.9§
120	1,269.0	48.3	1,692.0	64.4	1,412.5§	49.8§	1,883.3§	66.4§
130	1,452.8	50.7	1,937.0	67.6	1,547.3§	50.7	2,063.1§	67.6
140	1,648.5	53.1	2,198.0	70.8	1,682.1§	53.1	2,242.8§	70.8
150	1,856.3	55.5	2,475.0	74.0	1,856.3	55.5	2,475.1	74.0
160	2,076.0	57.9	2,768.0	77.2	2,076.0	57.9	2,768.0	77.2
170	2,307.8	60.3	3,077.0	80.4	2,307.8	60.3	3,077.0	80.4
180	2,551.5	62.7	3,402.0	83.6	2,551.5	62.7	3,402.0	83.6
190	2,807.3	65.1	3,743.0	86.8	2,807.3	65.1	3,743.0	86.8
200	3,075.0	67.5	4,100.0	90.0	3,075.0	67.5	4,100.0	90.0
220	3,646.5	72.3	4,862.0	96.4	3,646.5	72.3	4,862.0	96.4
240	4,266.0	77.1	5,688.0	102.8	4,266.0	77.1	5,688.0	102.8
260	4,933.5	81.9	6,578.0	109.2	4,933.5	81.9	6,578.0	109.2
280	5,649.0	86.7	7,532.0	115.6	5,649.0	86.7	7,532.0	115.6
300	6,412.5	91.5	8,550.0	122.0	6,412.5	91.5	8,550.0	122.0

*Based on "Standard Specifications for Highway Bridges," American Association of State Highway and Transportation Officials. Impact not included.
†Moments in thousands of ft-lb (ft-kips).
‡Shear and reaction in kips. Concentrated load is considered placed at the support. Loads used are those stipulated for shear.
§Maximum value determined by standard truck loading. Otherwise, standard lane loading governs.

Longitudinal forces on highway bridges should be assumed at 5% of the live load headed in one direction, plus forces resulting from friction in bridge expansion bearings.

Centrifugal forces should be computed as a percentage of design live load

$$C = 6.68S^2/R \qquad (17\text{-}3)$$

where S = design speed, mph
R = radius of curvature, ft

These forces are assumed to act horizontally 6 ft above deck level and perpendicular to the bridge center line.

Wind forces generally are considered as moving loads that may act horizontally in any direction. They apply pressure to the exposed area of the superstructure, as seen in side elevation; to traffic on the bridge, with the center of gravity 6 ft above the deck; and to the exposed areas of the substructure, as seen in lateral or front elevation. Wind loads in Tables 17-2 and 17-3 were derived from "Standard Specifications for Highway Bridges," American Association of State Highway and Transportation Officials. They are based on 100-mph wind velocity. They should be multiplied by $(V/100)^2$ for other design velocities except for Group III loading (Art. 17-4).

In investigating overturning, add to horizontal wind forces acting normal to the longitudinal bridge axis an upward force of 20 psf for the structure without live load or 6 psf when the structure carries live load. This force should be applied to the deck and sidewalk area in plan at the windward quarter point of the transverse superstructure width.

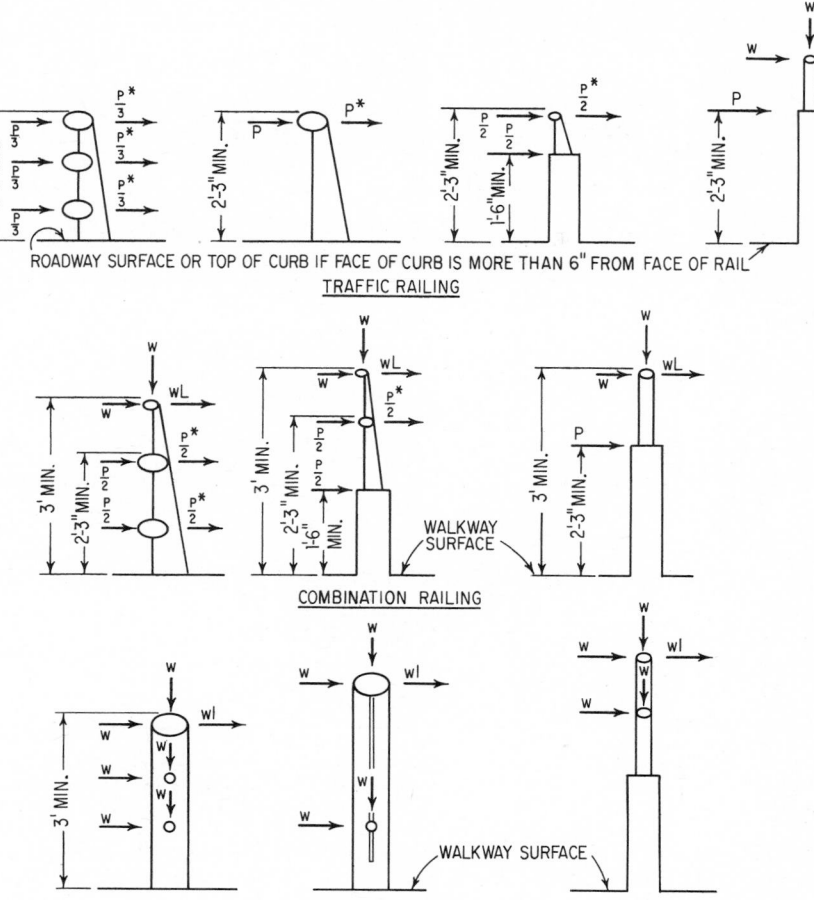

ROADWAY SURFACE OR TOP OF CURB IF FACE OF CURB IS MORE THAN 6" FROM FACE OF RAIL

TRAFFIC RAILING

COMBINATION RAILING

WALKWAY SURFACE

PEDESTRIAN RAILING

* WITH SIMULTANEOUS LONGITUDINAL LOAD OF $\frac{1}{2}$ THIS AMOUNT, DIVIDED AMONG POSTS IN A CONTINUOUS RAIL LENGTH

Fig. 17-5. Design loads for railing. $P = 10,000$ lb; $L =$ post spacing for traffic railing; $w = 50$ lb per lin ft; $l =$ post spacing for pedestrian railing. Rail loads are shown on the left; post loads, on the right. (The shapes of rail members are illustrative only.)

Table 17-2. Wind Loads for Superstructure Design

	Trusses and arches	Beams and girders	Live load
Wind load	75 psf	50 psf	100 lb per lin ft
Minimums:			
On loaded chord	300 lb per lin ft		
On unloaded chord	150 lb per lin ft		
On girders		300 lb per lin ft	

Table 17-3. Wind Loads for Substructure Design

a. Loads Transmitted by Superstructure to Substructure Slab and Girder Bridges
(up to 125-ft Span)

	Transverse	Longitudinal
Wind on superstructure when not carrying live load, psf	50	12
Wind on superstructure when carrying live load, psf	15	4
Wind on live load, lb per lin ft*	100	40

Major and Unusual Structures

Skew angle of wind, deg	No live load on bridge				Live load on bridge				Wind on live load, lb per lin ft*	
	Wind on trusses, psf		Wind on girders, psf		Wind on trusses, psf		Wind on girders, psf			
	Lateral load	Longitudinal load	Lateral load	Longitudinal load	Lateral load	Longitudinal load	Lateral load	Longitudinal load	Lateral load	Longitudinal load
0	75	0	50	0	22.5	0	15	0	100	0
15	70	12	44	6	21	3.6	13.2	1.8	88	12
30	65	28	41	12	19.5	8.4	12.3	3.6	82	24
45	47	41	33	16	14.1	12.3	9.9	4.8	66	32
60	25	50	17	19	7.5	15	5.1	5.7	34	38

b. Loads from Wind acting Directly on the Substructure†

Horizontal wind—no live load on bridge, psf	40
Horizontal wind—live load on bridge, psf	12

*Acting 6 ft of above deck.
†Resolve wind forces acting at a skew into components perpendicular to side and front elevations of the substructure and apply at centers of gravity of exposed areas. These loads act simultaneously with wind loads from superstructure.

Thermal Forces. Provision should be made for expansion and contraction due to temperature variations, and on concrete structures, also for shrinkage. For the continental United States, Table 17-4 covers temperature ranges of most locations and includes the effect of shrinkage on

Table 17-4. Expansion and Contraction of Structures*

Air temp range	Steel		Concrete†	
	Temp rise and fall, °F	Movement per unit length	Temp rise and fall, °F	Movement per unit length
Extreme: 120 °F, certain mountain and desert locations	60	0.00039	40	0.00024
Moderate: 100 °F, interior valleys and most mountain locations	50	0.00033	35	0.00021
Mild: 80 °F, coastal areas, Los Angeles, and San Francisco Bay area	40	0.00026	30	0.00018

*This table was developed for California. For other parts of the United States, the temperature limits given by AASHTO "Standard Specifications for Highway Bridges" should be used.
†Includes shrinkage.

ordinary beam-type concrete structures. The coefficient of thermal expansion for both concrete and steel per degree Fahrenheit is 0.0000065 (approximately 1/150,000). The shrinkage coefficient for concrete arches and rigid frames should be assumed as 0.0002, equivalent to a temperature drop of 31 °F.

Restraint forces, generated by preventing deformations, must be considered in design.

Earthquake (seismic) forces should be assumed to act horizontally at the center of gravity of the structure in the direction that produces maximum stresses in the member or part of the structure under consideration. These forces may be computed from

$$EQ = \phi KCD \qquad (17\text{-}4a)$$

where EQ = total earthquake force, kips, affecting the member or part
D = total dead load supported by the member or part, kips
K = coefficient indicating the energy-absorption characteristics of the structure
= 1.33 for bridges in which a wall with ratio of height to length of 2.5 or less resists horizontal forces acting along the wall
= 1.00 for bridges in which single columns or piers with a ratio of height to length exceeding 2.5 resist the horizontal forces
= 0.67 for bridges in which continuous frames resist the horizontal forces acting along the frames
$C = 0.05/\sqrt[3]{T}$ = coefficient representing the stiffness of the structure
T = natural period of vibration of the structure, sec = $0.32 \sqrt{D_c/P}$
P = force, kips, required to produce a lateral deflection of the structure equal to 1 in.
D_c = "contributory" dead load, kips
ϕ = coefficient related to the regional earthquake probability (California Division of Highways requires $\phi = 2.0$ for structures with spread footings and $\phi = 2.5$ for structures on piles)

The EQ forces calculated from Eq. (17-4a) should not be less than 0.04D. In the structural analysis, special consideration should be given to bridges with large skews or with columns having large differences in stiffness. Careful investigation and accurate analysis also are necessary for structures founded on very poor material, structures adjacent to active faults, and large and high structures.

Restraining features should be provided to limit the displacement of superstructures, such as hinge ties and shear blocks. These should be designed for the force computed from Eq. (17-4b):

$$EQ = 0.25D_c - V \qquad (17\text{-}4b)$$

where V = column shear due to EQ. D_c should be determined from an examination of the entire frame. For example, a simple span with fixed bearings at one end and sliding bearings at the other end will have the weight of the entire superstructure as the contributing dead load for longitudinal forces at the fixed abutment, while one-half of the superstructure dead load will act at each abutment for transverse forces.

For a frame, such as a two-span continuous structure, the full length of the bridge should be used as the contributory length in the longitudinal direction. The resulting force can be reduced by deduction of V.

For hinge restrainers, use 0.25 times D_c of the smaller of the two frames and deduct V.

Stream-flow pressure on a pier should be computed from

$$P = KV^2 \qquad (17\text{-}5)$$

where P = pressure, psf
V = velocity of water, fps
K = 4/3 for square ends, 1/2 for angle ends when angle is 30° or less, and 2/3 for circular piers

Ice pressure should be assumed as 400 psi. The design thickness should be determined locally.

Earth pressure on piers and abutments should be computed by recognized soil-mechanics formulas, but the equivalent fluid pressure should be at least 36 lb per cu ft when it increases stresses and not more than 27 lb per cu ft when it decreases stresses.

Railway Bridge Loads. *Live load* is specified by axle-load diagrams or by the E number of a "Cooper's train," consisting of two locomotives and an indefinite number of freight cars. Table 17-5 shows the typical axle spacing and axle loads for E10 loading and corresponding simple-beam moments, shears, and reactions for spans from 7 to 250 ft. Values of table must be increased proportionally for specified loading other than E10, for example, multiplied by 7.2 for E72 loading. (The American Railway Engineering Association recommended E72 loading for concrete bridges and E80 loading for steel bridges in its "Manual for Railway Engineering," 1972.)

Members receiving load from more than one track should be assumed to be carrying the following proportions of live load: For two tracks, full live load; for three tracks, full live load from two tracks and half from the third track; for four tracks, full live load from two, half from one, and one-fourth from the remaining one.

Table 17-5. Maximum Moments, Shears, and Reactions for Class E10 Engine Loading*

One Track of Two Rails

Axle loads, kips

5.0 10.0 10.0 10.0 10.0 6.5 6.5 6.5 6.5 5.0 10.0 10.0 10.0 10.0 10.0 6.5 6.5 6.5 6.5 12.5 12.5

←8 ft→|←5 ft→|←5 ft→|←5 ft→|←9 ft→|←5 ft→|←6 ft→|←5 ft→| 8 ft |←8 ft→|←5 ft→|←5 ft→|←5 ft→|←9 ft→|←5 ft→|←6 ft→|←5 ft→|←5 ft→| 1 kip per lin ft uniform load or ▨ |←7 ft→|

Span, ft	Max moment, ft-kips	Max shear, kips	Max floor beam reaction, kips†	Equivalent uniform load, kips per ft		
				Moment	Shear	Reaction
10	31.2	16.2	20.0	2.50	3.25	2.00
15	62.5	20.0	27.3	2.22	2.67	1.82
20	103.1	25.0	32.8	2.06	2.50	1.64
25	152.5	28.4	37.8	1.95	2.27	1.51
30	205.2	31.5	43.1	1.82	2.10	1.44
35	261.5	34.6	48.8	1.71	1.98	1.39
40	327.8	37.7	54.0	1.64	1.88	1.35
50	475.5	43.5	64.3	1.52	1.74	1.29
60	649.5	48.8	76.6	1.44	1.63	1.28
70	853.7	55.3	88.5	1.39	1.58	1.26
80	1,080.0	62.1	99.4	1.35	1.55	1.24
90	1,334.7	68.6	109.3	1.32	1.53	1.22
100	1,609.7	75.0	118.6	1.29	1.50	1.19
125	2,497.7	89.7	140.5	1.28	1.44	1.12
150	3,531.0	103.7	162.7	1.25	1.38	1.08
175	4,676.3	117.3	185.8	1.22	1.34	1.06
200	5,939.0	130.5	209.5	1.19	1.31	1.05
250	8,796.3	156.6	257.6	1.13	1.25	1.03

*The standard Class E10 load train consists of two Class E10 engines, coupled front to rear, followed by an indefinite, uniform load of 1 kip per lin ft of track. To obtain the actual design moments, shears, and reactions, the tabulated figures must be multiplied by 7.2 for E72 Loading or by 8.0 for E80 loading.
†From two spans.

Impact of railway loads on steel structures is composed of two components that act vertically on the rails. One is the rolling effect, which increases the load on one rail and decreases the load on the other, each by 10% of the axle load. The second is the vertical effect due to track irregularities, speed, and car impact. This effect acts equally on both rails. Its magnitude depends on type of equipment (steam, electric, or diesel locomotive), type of bridge (rolled beam, girder, truss), and loaded length of structure. For equipment other than steam engines, the impact, expressed as a percentage *I* of live load for steel bridges (without the rolling effect), may be computed from either of the following:

For *l* less than 80 ft,

$$I = 40 - \frac{3l^2}{1,600} \tag{17-6a}$$

For *l* greater than 80 ft,

$$I = 16 + \frac{600}{l - 30} \tag{17-6b}$$

where *l* = span, ft, center to center of supports for stringers, transverse floor beams without stringers, longitudinal girders, and main members of trusses; or length, ft, of the longer adjacent supported stringer, longitudinal beam, girder, or truss for impact in floor beams, floor-beam hangers, subdiagonals of trusses, transverse girders, supports for longitudinal and transverse girders, and viaduct columns.

Members receiving load from more than two tracks should be assumed to take full impact from any two tracks. Members receiving load from two tracks should be assumed to take the full impact from the two tracks when l is less than 175 ft and from only one track when l is greater than 225 ft. For l between 175 and 225 ft, the members should take full impact from one track and a percentage of full impact from the other as given by $450 - 2l$.

Impact on concrete structures, as a percentage of live load, may be computed from

$$I = \frac{100L}{L + D} \tag{17-7}$$

where L is the total live load and D the dead load on the member for which stresses are being computed.

Longitudinal forces should be computed for one track only. They should total either 15% of the entire moving load on the bridge or 25% of the load on the driving axles, whichever is greater.

Centrifugal forces should be computed from Eq. (17-3).

Table 17-6 compares bending moments due to live load plus impact for highway and railway loadings on single-span bridges.

Table 17-6. Moments at Midspan of Highway and Railway Bridges, Ft-kips

Span, ft	Highway One lane of HS20			Railway One track of E60		
	Live load	Impact	Total	Live load	Impact	Total
50	628	180	808	2,853	1,007	3,860
100	1,524	339	1,863	9,660	2,380	12,040
200	4,100	632	4,732	35,634	6,960	42,594

17-4. Proportioning of Members and Sections.   The forces—axial forces N, bending moments M, shears V, and torques M_T—generated by each type of loading that a structure may be subjected to should be computed for all members and relevant sections of the structure in accordance with recognized methods of static analysis (see Sec. 6).

Table 17-7. Allowable Stresses for Loading Combinations*

Group	Combination	% of basic unit stresses
I	$D + L + I + E + B + SF$	100
II	$D + E + B + SF + W$	125
III	Group I + $LF + F$ + 30% $W + WL + CF$	125
IV	Group I + $R + S + T$	125
V	Group II + $R + S + T$	140
VI	Group III + $R + S + T$	140
VII	$D + E + B + SF + EQ$	133⅓
VIII	Group I + ICE	140
IX	Group II + ICE	150

*In frame-type structures, temperature stresses should be included in Group I.

D = dead load
L = live load
I = live-load impact
E = earth pressure
B = buoyancy
W = wind load on structure
WL = wind load on live load
LF = longitudinal force from live load
See also Art. 17-3.

CF = centrifugal force
F = longitudinal force due to friction
R = rib shortening
S = shrinkage
T = temperature
EQ = earthquake
SF = stream flow pressure
ICE = ice pressure

Members and sections then should be proportioned to fulfill either or both of the following conditions:

Working-Stress Design.   The sum of the stresses induced by various combinations of loadings should not exceed the percentage of the basic working stresses for the given materials indicated in Table 17-7.

Groupings other than those listed in Table 17-7 may be specified for spans over 500 ft.   Higher working stresses are also permitted for the operational rating of existing bridges.   In anticipation of such future relaxations of the safety factor, some major bridges have been designed initially with up to 33% higher dead-load stresses than current specifications would permit.

Load-Factor Design.   The totals of the effects (shears, moments, stresses) of the following load groups should not exceed the capacity of the structure, member, or connection:

$$\text{Group I} = 1.3D + \tfrac{5}{3}(L + I) \tag{17-8a}$$

For design loadings less than H20, capacity also must be adequate for

$$\text{Groups IA} = 1.3D + 2.2(L + I) \tag{17-8b}$$

$$\text{Group II} = 1.3(D + W + F + SF + B + S + T) \tag{17-9}$$

(If a structure may be subjected to earthquake forces, use EQ instead of W; and if to ice pressure, use ICE instead of SF.)

$$\text{Group III} = 1.3(D + L + I + CF + 0.3W + WL + F + LF) \tag{17-10}$$

Allowable working stresses and design capacities depend on the quality of the materials of the particular member or connection, on shape or geometry for components or members in compression, and on frequency of loading (fatigue) for connections.   For steel structures, capacities are functions of the ultimate strength F_u, yield strength F_y, or modulus of elasticity E.   For concrete, capacities are functions of the 28-day compressive strength f_c'.

See also Arts. 8-20 to 8-31 and 9-30.

STEEL BRIDGES

Steel is competitive as a construction material for medium spans and is favored for long-span bridges for the following reasons: It has high strength in tension and compression.   It behaves as a nearly perfect elastic material within the usual working ranges.   It has strength reserves beyond the yield point.   The high standards of the fabricating industry guarantee users uniformity of the controlling properties within narrow tolerances.   Connection methods are reliable, and workers skilled in their application are available.

The principal disadvantage of steel in bridge construction, its susceptibility to corrosion, is being increasingly overcome by chemical additives or improved protective coatings.

17-5. Systems Used for Steel Bridges.   The following are typical components of steel bridges. Each may be applied to any of the functional types and structural systems listed in Art. 17-1.

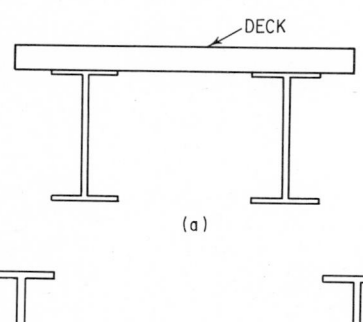

(a)

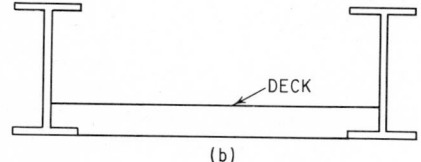

(b)

Fig. 17-6. (a) Deck-type bridge; (b) Through bridge.

Main Support: Rolled beams, plate girders, box girders, or trusses.

Connections: (See also Art. 17-8.) Riveted, high-strength bolted, welded, or combinations.

Materials for Traffic-carrying Deck: Timber stringers and planking, reinforced concrete slab or prestressed-concrete slab, stiffened steel plate (orthotropic deck), or steel grid.

Timber decks are restricted to bridges on roads of minor importance. Plates of corrosion-resistant steel should be used as ballast supports on through plate-girder bridges for railways. For roadway decks of stiffened steel plates, see Art. 17-13.

Deck Framing: Deck resting directly on main members or supported by grids of stringers and floor beams.

Location of Deck: On top of main members: deck spans (Fig. 17-6a); between main members, the underside of the deck framing being flush with that of the main members: through spans (Fig. 17-6b).

17-6. Grades and Permissible Stresses Steel for Bridges. Preferred steel grades, permissible stresses, and standards of details, materials, and workmanship for steel bridges are covered in the "Manual for Railway Engineering," American Railway Engineering Association, and "Standard Specifications for Highway Bridges," American Association of State Highway and Transportation Officials. Properties of the various grades of steel and the testing methods to be used to control them are regulated by specifications of the American Society for Testing and Materials (ASTM). Properties of the structural steels presently preferred in bridge construction are tabulated in Table 17-8.

Table 17-8. Mechanical Properties of Structural Steel for Bridges

Characteristic	ASTM designation	Plates thicknesses, in.	Shape groups*	Min tensile strength, ksi	Yield strength, ksi
Structural carbon steel	A36	To 8 incl.	All	58	36
High-strength low-alloy steel	A242 A440 A441 A588 A572		See Table 9-2		
High-yield-strength, quenched and tempered alloy steel	A514	To 2½ incl. Over 2½ to 4 incl.	Not applicable	110 100	100 90
	A517	To 2½ incl.	Not applicable	115–135	100
Structural steel for bridges	A709 Grade 36	To 8 incl.	To 426 lb per ft Over 426 per ft	58–80 58 min	36 36
		Over 8	. . . .	58 min	32
	Grade 50	To 2 incl.	1,2,3,4	65 min	50
	Grade 50W	To 4 incl.	All	70 min	50
	Grades 100 and 100W	To 2½ incl.	Not applicable	100–130	100
		Over 2½ to 4 incl.	Not applicable	100–130	90

*See Table 9-1.

Dimensions and geometric properties of commercially available rolled plates and shapes are tabulated in the "Steel Construction Manual," American Institute of Steel Construction (AISC), and in manuals issued by the major steel producers.

All members and parts of steel bridges should be designed in accordance with the recognized rules of elastic analysis. They should be so proportioned that all stresses remain within permissible limits.

The permissible tensile stresses are obtained by applying to the yield strength of the given grade of a steel a safety factor of 1.8. Compression stresses are subject to further reduction to compensate for slenderness of member or element, as well as unintentional and calculated eccentricities. The basic stresses are compiled in Sec. 9 and summarized for convenience in Table 17-9. (See also Art. 17-4.)

Table 17-9. Permissible Stresses for Structural Steel for Highway Bridges, Ksi

Axial tension on net section of members with holes: The smaller of

$$F_a = 0.55F_y$$

$$F_a = 0.46F_u$$

where F_y = minimum yield strength, ksi
$\quad\;\, F_u$ = minimum tensile strength, ksi

Axial tension in members without holes
Tension in extreme surfaces of rolled shapes, girders, and built-up sections subject to bending, on net section
Axial compression in stiffeners of plate girders, on gross section
Compression on gross section of splice material

$$F = 0.55F_y$$

Compression in extreme surfaces of rolled shapes, girders, and built-up sections subject to bending (on gross section) when compression flange is
(a) supported laterally for its full length by embedment in concrete

$$F_b = 0.55F_y$$

(b) partly supported or unsupported with ratio of unsupported length L_u to flange width b_f not more than about $215/\sqrt{F_y}$

$$F_b = 0.55F_y\left[1 - \frac{(L_u/b_f)^2 F_y}{95,400}\right]$$

Compression in concentrically loaded columns with slenderness ratios not exceeding

Max L'/r	130	125	120	115	110	90	85
For F_y	36	42–50	55	60	65	90	100

where L' = length of member, in.
$\quad\;\, r$ = least radius of gyration, in.
(a) with riveted or bolted ends

$$F_a = 0.44f_y\left[1 - \frac{(L'/r)^2 F_y}{2,000,000}\right]$$

(b) with pinned ends

$$F_a = 0.44f_y\left[1 - \frac{(L'/r)^2 F_y}{1,500,000}\right]$$

Shear in girder webs

$$F_v = 0.33F_y$$

Bearing on milled stiffeners and other steel parts in contact except rivets and bolts
Stress in extreme surface of pins
Bearing on pins not subject to rotation

$$F = 0.80F_y$$

Shear in pins

$$F_v = 0.40F_y$$

Bearing on pins subject to rotation (as for rockers and hinges)

$$F_p = 0.40F_y$$

Bearing on power-driven rivets and high-strength bolts

$$F_p = 1.22F_y$$

but not more than the allowable bearing on the fasteners

Members, connection material, and fasteners subject to repeated variations or reversals of stress should be proportioned for the allowable stress ranges in Art. 9-18.

17-7. Other Design Limitations. *Depth ratios, slenderness ratios, deflections.* AASHTO and AREA specifications restrict the depth-to-span ratios D/L of bridge structures and the slenderness ratios l/r of individual truss or bracing members to the values in Table 17-10, where D = depth of construction, ft

$\quad L$ = span, ft, c to c bearings for simple spans or distance between points of contraflexure for continuous spans

$\quad l$ = unsupported length of member, in.

$\quad r$ = radius of gyration, in.

These are minimum values. Preferred values are higher.

Table 17-10. Dimensional Limitations for Bridge Members

	AASHTO	AREA
Min depth-span ratios:		
For rolled beams	$^1/_{25}$	$^1/_{15}$
For girders	$^1/_{25}$	$^1/_{12}$
For trusses	$^1/_{10}$	$^1/_{10}$
Max slenderness ratios:		
For main members in compression	120	100
For bracing members in compression	140	120
For main members in tension	200	200
For bracing members in tension	240	200

For plate-girder design criteria, see Arts. 9-22 and 9-24.

Both specifications limit the elastic deflections of bridges under design live load plus impact to $^1/_{800}$ of the span, measured c to c bearings except that $^1/_{1,000}$ may be used for bridges used by pedestrians.

$^1/_{300}$ of the length of cantilever arms.

Deflection calculations should be based on the gross sections of girders or truss members. Anticipated dead-load deflections must be compensated by adequate camber in the fabrication of steel structures.

Splices. Shop, assembly-yard, or erection splices must be provided for units whose overall length exceeds available rolled lengths of plates and shapes or the clearances of available shipping facilities. Splices also must be provided when total weight exceeds the capacity of available erection equipment.

Accessibility. All parts should be accessible and adequately spaced for fabrication, assembly, and maintenance. Closed box girders and box-type sections should be equipped with handholes or manholes.

On long and high bridges, installation of permanent maintenance travelers may be justified.

17-8. Steel Connections in Bridges. *Riveted Connections.* Rivets function as shear connectors and clamping devices, the heads preventing the parts connected from falling apart. Rivet diameters most commonly used on bridges are ¾, ⅞, 1, and occasionally 1⅛ in. Rivet holes are either punched, or subpunched and reamed, or drilled full size. They are usually made $^1/_{16}$ in. larger than the nominal rivet diameter.

The most commonly used rivet steel is ASTM A502, Grade 1. Less common are A502, Grade 2 high-strength structural rivets. Allowable stresses are given in Art. 9-20.

Limitations on rivet spacing and edge distances are spelled out in design specifications. Minimum pitch is controlled by fabrication requirements. Edge distances are controlled by the size of the rivet head, curvature of rolled edge, and rolling tolerances. The maximum pitch of stitch rivets in compression members is limited to prevent buckling of the individual plates, whereas spacing of rivets along free edges of the material is restricted to assure sealing of the joint.

Connections with High-strength Bolts. The parts may be clamped together by bolts of quenched and tempered steel, ASTM A325. The nuts are tightened to specified amounts. Setting of the bolts requires less, preparation of the faying surfaces more, manpower than does riveting.

Details and workmanship are covered by the Specifications for Structural Joints Using ASTM A325 and A490 Bolts, approved by the Research Council on Riveted and Bolted Structural Joints of the Engineering Foundation. Permissible working stresses are given in Art. 9-20.

Welded Connections. In welding, the parts to be connected are fused at high temperatures, usually with addition of suitable metallic material. The "Structural Welding Code," AWS D1.1, American Welding Society, regulates application of the various types and sizes of welds, permissible stresses in the weld and parent metal, permissible edge configurations, kinds and sizes of electrodes, details of workmanship, and qualification of welding procedures and welders. (For allowable welding stresses, see AASHTO Bridge Specifications.)

Welded construction has the following advantages over riveted construction that make it attractive to bridge designers: savings of steel due to elimination of holes that weaken the effective section in tension, and to omission of additional splice material; smoother appearance, easier maintenance, easier repairs; less noise during erection. These advantages are partly offset by some disadvantages in fabrication: restriction in choice of steel to weldable steels; greater shop-space requirements; and necessity of extensive, often costly, inspection. To these must be added some structural shortcomings: distortions that result from differential cooling of the weld and its surrounding metal, and, consequently, formation of hidden, or "locked-up," stresses, which are quantitatively difficult to assess; brittleness of the weld or its surroundings at low temperatures.

For the last reason, welded construction requires special controls for bridges in cold climates. It is still not often used in railway bridges. When welding is used for field splices of bridges under conditions that are less favorable for effective controls than sheltered shops, the splices should receive thorough inspection.

Many designers favor the combination of shop welding with high-strength bolted field connections.

Pin Connections. Hinges between members subject to relative rotation are usually formed with pins, machined steel cylinders. They are held in either semicircular machined recesses or smoothly fitting holes in the connected members.

For fixation of the direction of the pin axis, pins up to 10-in. diameter have threaded ends for recessed nuts, which bear against tne connected members. Pins over 10-in. diameter are held by recessed caps. These, in turn, are held either by tap bolts or by a rod that runs axially through a hole in the pin itself and is threaded and secured by nuts at its ends.

Pins are designed for bending and shear and for bearing against the connected members. (For stresses, see Art. 9-19.)

17-9. Bridge Bearings. Bearings are structural assemblies installed to secure the safe transfer of all reactions from the superstructure to the substructure. They must fulfill two basic requirements: They must spread the reactions over adequate areas of the substructure, and they must be capable of adapting to elastic, thermal, and other deformations of the superstructure without generating harmful restraining forces.

Generally, bearings are classified as fixed or expansion, and as metal or elastomeric.

Fixed bearings adapt only to angular deflections. They must be designed to resist both vertical and horizontal components of reactions.

Expansion bearings adapt to both angular deflections and longitudinal movements of the superstructure. Except for friction, they resist only those components of the superstructure reactions perpendicular to these movements.

In both types of bearings, provision must be made for the safe transfer of all forces transverse to the direction of the span.

Metal bearings are preferably of structural steel, cast steel, or cast iron. Their basic components are an upper unit, which is bolted to the superstructure, and a lower unit (shoe or pedestal), which is anchored to the substructure. Inserted between these, if required, are elements for centering and for adaptation to angular deflections and, in the case of expansion bearings, to longitudinal movements of the superstructure.

According to AASHTO Standard Specifications for Highway Bridges, no provision for angular deflections need be made when spans are less than 50 ft. Bearings then may consist of two plane steel plates in contact with each other. However, for better centering and maintenance, a *bearing bar* of rectangular cross section with chamfered or rounded top may be welded to the lower plate, and a *keeper plate*, cut out to fit over the bearing bar, may be welded to the upper plate.

For steel bearings of spans over 50 ft, AASHTO Standard Specifications for Highway Bridges requires curved plates, hinges, or pins (Fig. 17-7). For expansion bearings in addition, sliding plates, rockers (Fig. 17-7a) or rollers (Fig. 17-8b), or elastomeric pads are required. The top and bottom of each bearing must be held together in an effective way without obstructing the required movements.

Heavier bearings are made of cast steel or built up from structural steel by riveting, bolting, or welding (Figs. 17-7 and 17-8). Such bearings have machined contact surfaces without or with centering pins. Weldments must be stress-relieved before machining.

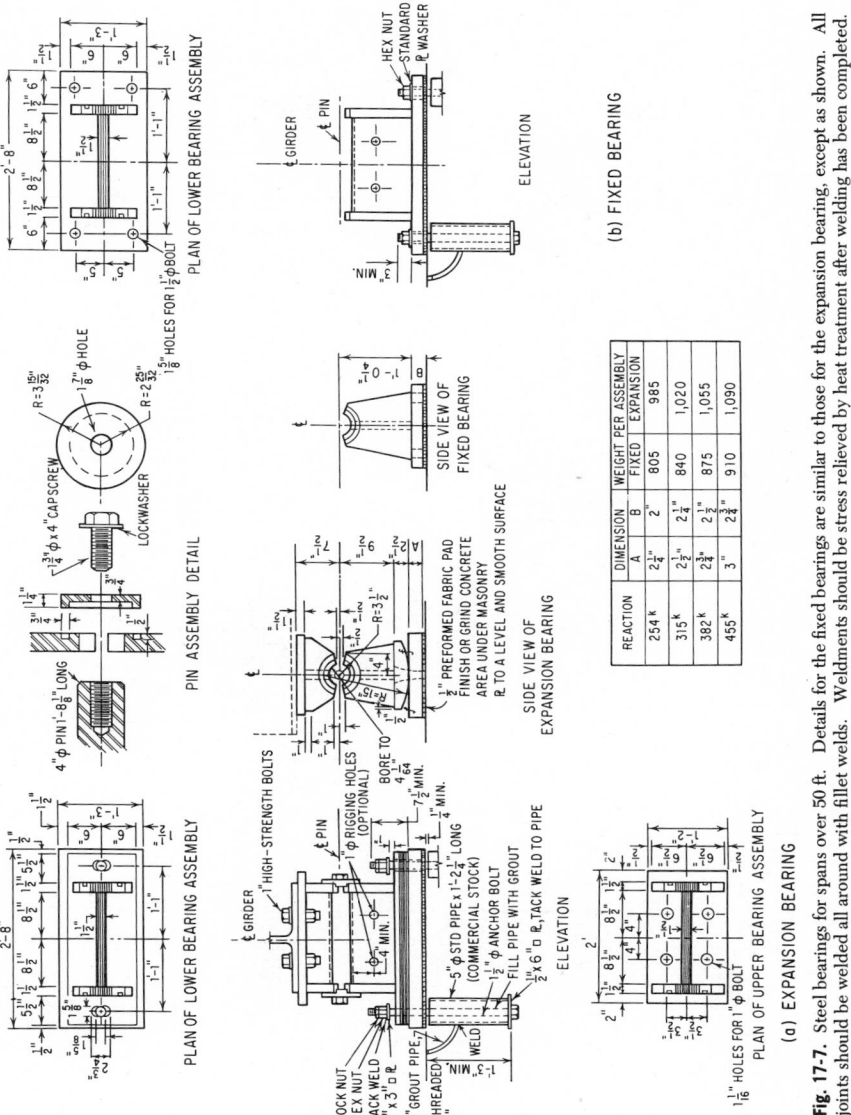

Fig. 17-7. Steel bearings for spans over 50 ft. Details for the fixed bearings are similar to those for the expansion bearing, except as shown. All joints should be welded all around with fillet welds. Weldments should be stress relieved by heat treatment after welding has been completed.

The plan dimensions of bearings are governed by the permissible bearing pressures on the bridge seat. The allowable concrete stresses are:

Under properly hinged bearings, not subject to high edge pressures .... 1,000 psi
Under bearing plates and nonhinged shoes .... 700 psi

These stresses must be reduced by 25% if the concrete edges project less than 3 in. beyond the edge of the base plate.

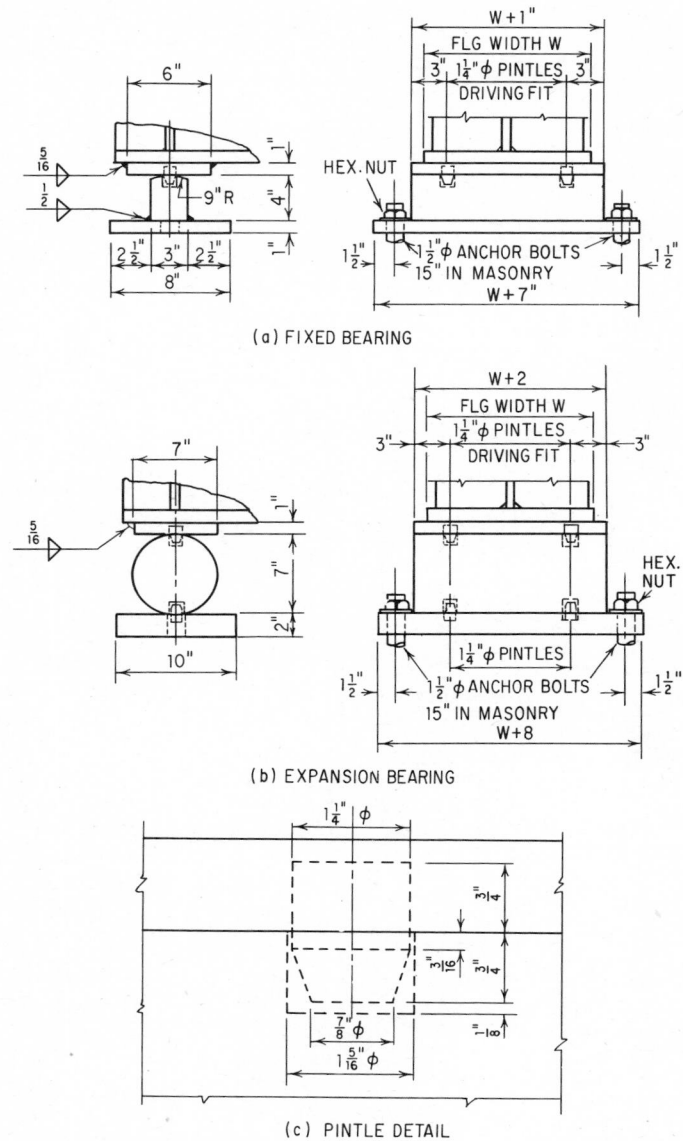

(a) FIXED BEARING

(b) EXPANSION BEARING

(c) PINTLE DETAIL

Fig. 17-8. Bearings for composite welded girders on spans of 90 to 180 feet.

Permissible stresses for proportioning of pins and expansion rollers and rockers are given in Art. 9-2.

Plate thicknesses and other vertical dimensions must be proportioned to resist the flexural stresses that result from the spreading of the load from the upper lines of contact to the bearing areas. For anchorage of base plates to the substructure, AASHTO specifications give the minimum requirements listed in Table 17-11. The mass of masonry engaged by anchor bolts in tension must be sufficient to resist 1.5 times the calculated uplift.

Table 17-11. Requirements for Anchorage of Base Plates

Type and spans of superstructure	Bolts		Embedment in concrete, in.
	Diameter, in.	Number	
Outer beam of I-beam spans	1	2	10
Truss and girder spans:			
To 50 ft	1	2	10
51–100 ft	1¼	2	12
101–150 ft	1½	2	15
Over 150 ft	1½	4	15

Elastomeric bearings are either plain bearings, consisting of elastomer only, or laminated bearings, consisting of layers of elastomer, restrained at their interfaces by bonded laminas. The elastomer may be either natural rubber (polyisoprene) or synthetic rubber (chloroprene, neoprene). Laminas should be rolled, mild-steel sheets.

The bearing pressure of elastomeric bearings should not exceed the following values: For dead load alone: 500 psi; for dead load plus live load: 800 psi. If the pressure from dead load minus uplift drops below 200 psi, restraints are required to prevent lateral crawling.

The capacity of an elastomeric bearing to absorb angular deflections and longitudinal movements of the superstructure is a function of its thickness (or of the sum of the thicknesses of its rubber elements between steel laminas), of its shape factor (area of the loaded face divided by the sum of the side areas free to bulge), and of the properties of the elastomer.

AASHTO Specifications limit the overall thickness of a plain elastomeric bearing to one-fifth its length or width, of a laminated bearing to one-third its length (in the direction of the bridge) or one-half its width, in both cases whichever is smaller. The thickness should be at least twice the sum of the positive and negative movements expected because of temperature changes.

17-10. Rolled-Beam Bridges. The simplest steel bridges consist of rolled I beams or wide-flange beams, which either support the traffic-carrying deck or are fully embedded in it. Rolled beams serve also as floor beams and stringers for decks of plate-girder and truss bridges.

Reductions in steel weight may be obtained, but with greater labor costs, by adding cover plates in the area of maximum moments, by providing continuity over several spans, by utilizing the deck in composite action, or by a combination of these measures. The principles of design and details are essentially identical with those of plate girders (Art. 17-11).

Standard designs by the Federal Highway Administration uses shapes, up to W36 × 245, up to the following maximum spans:

Simple spans, without cover plates . . . . 70 ft
Simple spans, composite and with welded cover plates . . . . 90 ft
Continuous spans with welded cover plates and high-strength bolted splices . . . . 80-100-80 ft

For design of a concrete deck slab, see Art. 17-18.

17-11. Plate-Girder Bridges. The term plate girder applies to structural elements of I-shaped cross section that are either riveted together from plates and angles or welded from plates alone. Plate girders are used as primary supporting elements in many structural systems: as simple beams on abutments or, with overhanging ends, on piers; as continuous or hinged multispan beams; as stiffening girders of arches and suspension bridges, and in frame-type bridges. They also serve as floor beams and stringers on these and other bridge systems.

Their prevalent application on highway and railway bridges is in the form of *deck plate girders* in combination with concrete decks (Fig. 17-9). (For design of concrete deck slabs, see Art. 17-18.) For girders with steel decks (orthotropic decks), see Art 17-13. Girders with track ties mounted directly on the top flanges, *open-deck girders*, are used on branch railways and industrial spurs. *Through plate girders* (Fig. 17-10) are now practically restricted to railway bridges where allowable structure depth is limited.

The two or more girders supporting each span must be braced against each other to provide stability against overturning and flange buckling, to resist transverse forces (wind, earthquake, centrifugal), and to distribute concentrated heavy loads. On *deck girders*, this is done by systems of diagonals in the planes of the top and bottom flanges (Fig. 17-9c) and by transverse bracing in vertical planes. The top lateral system can be omitted if the deck and its connections to the girder are designed to take its place. Bottom lateral systems are required for deck plate

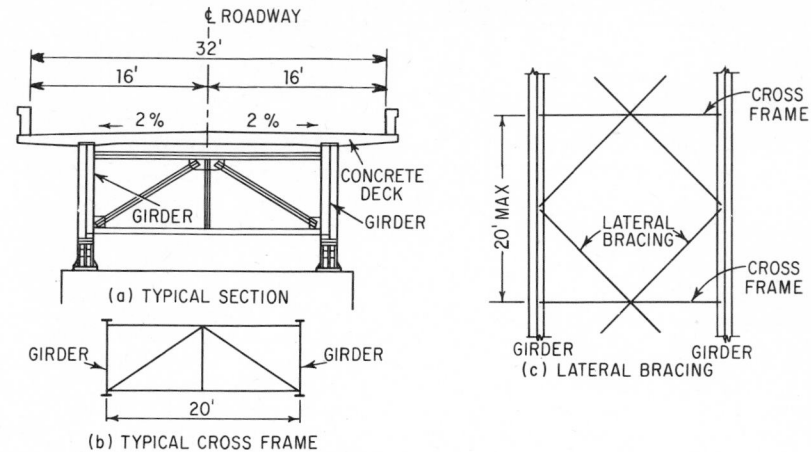

Fig. 17-9. Two-lane, deck-girder highway bridge.

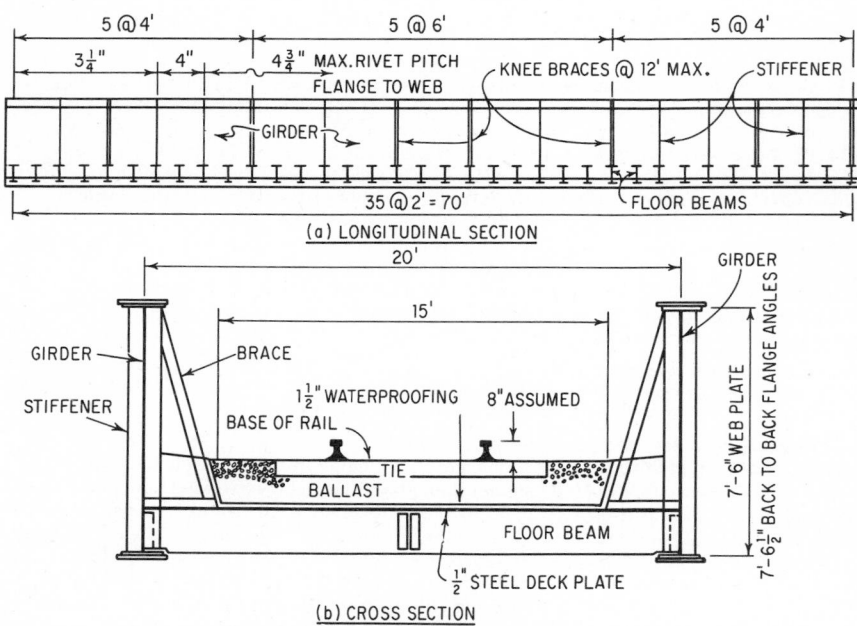

Fig. 17-10. Single-track, through-girder railway bridge.

girders with spans greater than 125 ft.   Transverse bracing should be installed over each bearing and, on bridges of over 40-ft span, at intermediate locations not over 25 ft apart.   This bracing may consist either of full-depth cross frames (Fig. 17-9*b*) or of solid diaphragms of not less than one-third and preferably half the girder depth.   On *through girder spans,* since there can be no provision for top lateral and transverse bracing systems, the top flanges must be braced against the floor system by heavy gusset plates or knee braces (Fig. 17-10*b*).

Standard designs for two-lane welded deck-girder bridges have been prepared by the Federal Highway Administration for spans from 90 to 200 ft.   Figure 17-11 shows typical welded-girder details, and Figs. 17-12 and 17-13, typical riveted-girder details.

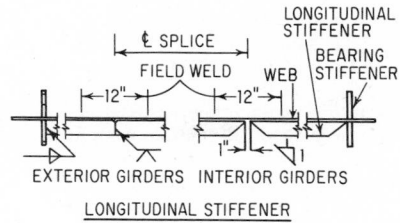

LONGITUDINAL STIFFENER

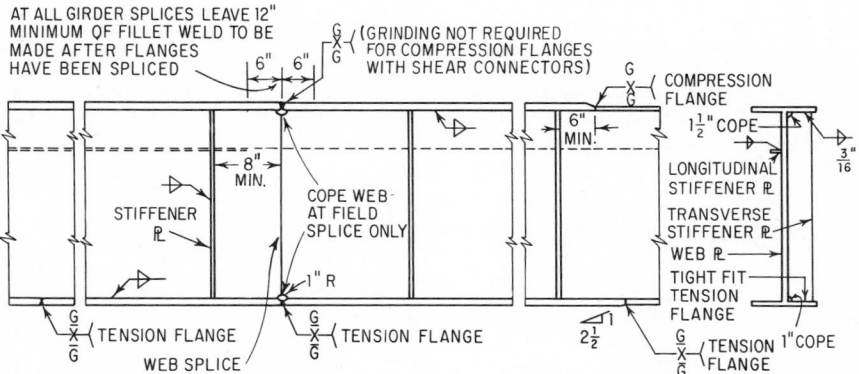

Fig. 17-11. Typical welded-girder details. Top and bottom flange splices may be offset. Web splices should be full-penetration butt welds. Where weld areas on exterior girders are visible from a traveled way, the areas should be ground flush.

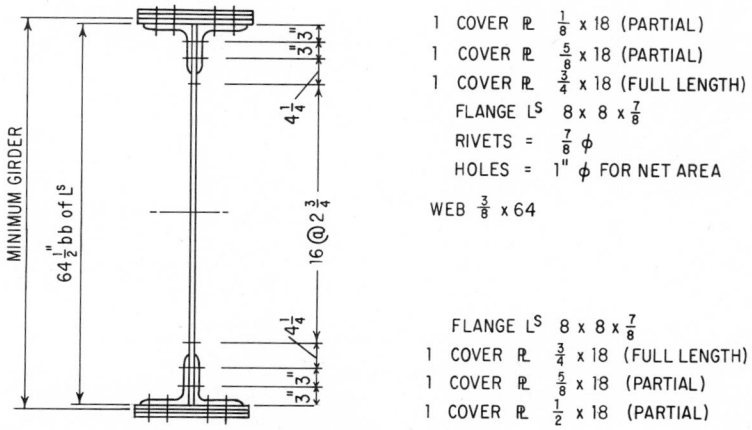

1 COVER ℙ $\frac{1}{8}$ x 18 (PARTIAL)

1 COVER ℙ $\frac{5}{8}$ x 18 (PARTIAL)

1 COVER ℙ $\frac{3}{4}$ x 18 (FULL LENGTH)

FLANGE Lˢ 8 x 8 x $\frac{7}{8}$

RIVETS = $\frac{7}{8}$ ϕ

HOLES = 1" ϕ FOR NET AREA

WEB $\frac{3}{8}$ x 64

FLANGE Lˢ 8 x 8 x $\frac{7}{8}$

1 COVER ℙ $\frac{3}{4}$ x 18 (FULL LENGTH)

1 COVER ℙ $\frac{5}{8}$ x 18 (PARTIAL)

1 COVER ℙ $\frac{1}{2}$ x 18 (PARTIAL)

Fig. 17-12. Typical riveted-girder cross section.

Flexural Analysis. The extreme fiber stresses in flexure of rolled beams and girders are computed under the assumption of linear stress distribution across each section from the equations

$$f_c = \frac{Me_c}{I_{\text{gross}}} \tag{17-11a}$$

$$f_t = \frac{Me_t}{I_{\text{net}}} \tag{17-11b}$$

where M = maximum bending moment, in.-kips

e_c and e_t = distances of the extreme fibers in compression and tension from the neutral axis, in.

I_{gross} = moment of inertia of gross section, in.^{4}

I_{net} = moment of inertia of net section, in.^{4}, referred to the neutral axis of the gross section

f_c and f_t = extreme fiber stresses in compression and tension, ksi

The web shear stresses are computed from

$$v_s = \frac{V}{Dt} \qquad\qquad (17\text{-}12)$$

where V = maximum static shear, kips

D = depth of web, in.

t = thickness of web plate, in.

v_s = average web shear on gross section, ksi

For permissible stresses, see Sec. 9.

In riveted design, the web thickness and flange angle sizes are preferably constant along the girder. The section modulus is adapted to the moment variations by successive addition of flange cover plates to a maximum of three per flange, with thicknesses not exceeding that of the angles.

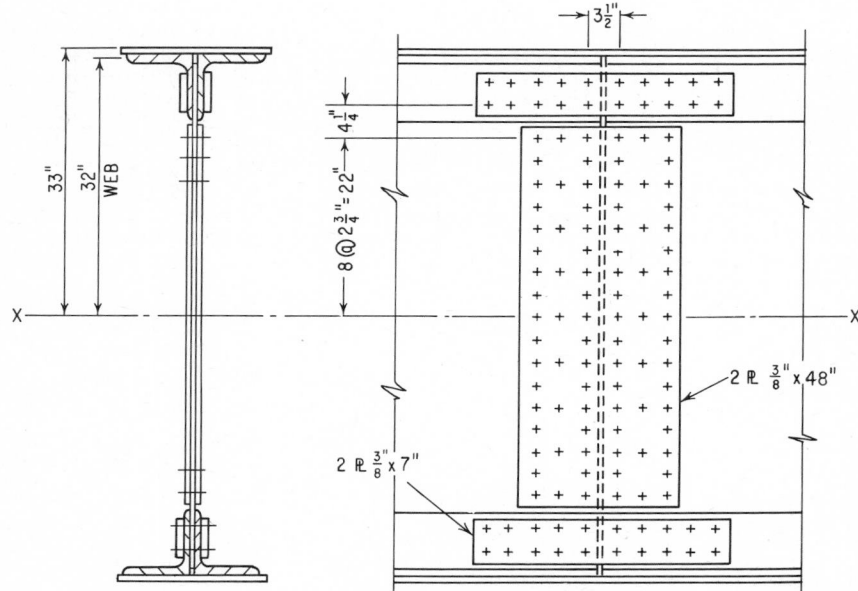

Fig. 17-13. Full splice in riveted plate girder.

In welded design, variations in moment resistance are obtained by using flange plates of different thicknesses, widths, or steel grades, butt-welded to each other in succession. Stacking of plates, as in riveted design, is not recommended. Web thickness, too, may be varied.

Girder webs should be protected against buckling by transverse and, in the case of deep webs, longitudinal stiffeners. Transverse bearing stiffeners are required to transfer end reactions from the web into the bearings and to introduce concentrated loads into the web. Intermediate and longitudinal stiffeners are required if the girder depth-to-thickness ratios exceed critical values (see Art. 9-24).

Stiffeners may be plain plates, angles, or T sections. Transverse stiffeners should, preferably, be in pairs, although single stiffeners are allowed. The AASHTO specifications contain restrictions on width-to-thickness ratios and minimum widths of plate stiffeners (Art. 9-24).

In riveted design, intermediate transverse stiffeners may be crimped at the flange angles, but bearing stiffeners must be straight. Hence, filler plates of the same thickness as the flange angles are required.   No such problem exists with welded girders.   Transverse intermediate stiffeners on only one side of the web should be riveted or welded to the compression flange.

Web-to-flange connections should be capable of carrying the stress flow from web to flange at every section of the girder.   At an unloaded point, the stress flow equals the horizontal shear per linear inch.   Where a wheel load may act, for example, at upper flange-to-web connections of deck girders, the stress flow is the vectorial sum of the horizontal shear per inch and the wheel load (assumed distributed over a web length equal to twice the deck thickness).

Similarly, connection rivets or welds between flange cover plates and between cover plates and flange angles should be capable of carrying the stress flow (see Art. 9-24).

Rivets or welds connecting bearing stiffeners to the web must be designed for the full bearing reaction.

Space restrictions in the shop, clearance restrictions in transportation, and erection considerations may require dividing long girders into shorter sections, which are then joined (spliced) in the field.   Individual segments, plates or angles, must be spliced either in the shop or in the field if they exceed in length the sizes produced by the rolling mills or if shapes are changed in thickness to meet stress requirements.

Specifications require splices to be designed for the average between the stress due to design loads and the capacity of the unspliced segment, but for not less than 75% of the latter.   In riveted design, material may have to be added at each splice to satisfy this requirement.   Each splice element must be connected by a sufficient number of rivets to develop its full strength. Whenever it is possible to do so, splices of individual segments should be staggered.   No splices should be located in the vicinity of the highest-stressed parts of the girder, for example, at midspan of simple-beam spans, or over the bearings on continuous beams.

(F. S. Merritt, "Structural Steel Designers' Handbook," McGraw-Hill Book Company, New York.)

17-12. Composite-Girder Bridges.   Installation of appropriately designed shear connectors between the top flange of girders or beams and the concrete deck allows use of the deck as part of the top flange (equivalent cover plate).   The resulting increase in effective depth of the total section and possible reductions of the top-flange steel usually allow some savings in steel compared with the noncomposite steel section.   The overall economy depends on the cost of the shear connectors and of any other additions to the girder or the deck that may be required and on possible limitations in effectiveness of the composite section as such.

In areas of negative moment, composite effect may be assumed only if the calculated tensile stresses in the deck are either taken up fully by reinforcing steel or compensated by prestressing. The latter method requires special precautions to assure slipping of the deck on the girder during the prestressing operation but rigidity of connection after completion.

If the steel girder is not shored up while the deck concrete is placed, computation of dead-load stresses must be based on the steel section alone.

Shear connectors should be capable of resisting all forces tending to separate the abutting concrete and steel surfaces, both horizontally and vertically.   Connectors should not obstruct placement and thorough compaction of the concrete.   Their installation should not harm the structural steel.

The types of shear connectors presently preferred are channels, or welded studs.   Channels should be placed on beam flanges normal to the web and with the channel flanges pointing toward the girder bearings.

For stress calculations, see Arts. 9-21 and 17-18.

(F. S. Merritt, "Structural Steel Designers' Handbook," McGraw-Hill Book Company, New York.)

17-13. Orthotropic-Deck Bridges.   An orthotropic deck is, essentially, a continuous, flat steel plate, with stiffeners (ribs) welded to its underside in a parallel or rectangular pattern.   The term *orthotropic* is shortened from *orthogonal anisotropic,* referring to the mathematical theory used for the flexural analysis of such decks.

When used on steel bridges, orthotropic decks are usually joined quasi-monolithically, by welding or high-strength bolting, to the main girders and floor beams.   They then have a dual function as roadway and as structural top flange.

The combination of plate or box girders with orthotropic decks allows the design of bridges of considerable slenderness and of nearly twice the span reached by girders with concrete decks. The most widespread application of orthotropic decks is on continuous, two- to five-span girders on low-level river crossings in metropolitan areas, where approaches must be kept short and grades low.   This construction has been used for main spans up to 1,100 ft with bracing by overhead cables and up to 856 ft without such bracing.   There also are some spectacular high-level orthotropic girder bridges and some arch and suspension bridges with orthotropic stiffening girders.   On some of the latter, girders and deck have been combined in a single lens-shaped box section that has great stiffness and low aerodynamic resistance.

Box Girders.   Single-web or box girders are used for orthotropic bridges.   Box girders are preferred if structure depth is restricted.   Their inherent stiffness makes it possible to reduce, or to omit, unsightly transverse bracing systems.   In cross section, they usually are rectangular, occasionally trapezoidal.   Minimum dimensions of box girders are controlled by considerations of accessibility and ease of fabrication.

Wide decks are supported by either single box girders or twin boxes.   Wide single boxes have been built with multiple webs or secondary interior trusses.   Overhanging floor beams sometimes are supported by diagonal struts.

Depth-Span Ratios.   Girder soffits are parallel to the deck, tapered, or curved.   Parallel flanges, sometimes with tapered side spans, generally are used on unbraced girders with depth-to-main-span ratios as low as 1:70.   Parallel-flange unbraced girders are practically restricted to high-level structures with unrestricted clearance.   Unbraced low-level girders usually are designed with curved soffits with minimum depth-to-main-span ratios of about 1:25 over the main piers and 1:50 at the shallowest section.

Cable Systems.   Cable-braced girder systems have either two main girders or one, with two towers or one for each girder.   The cables are curved if the girders are suspended at each floor beam; otherwise they are straight.   In the latter (more usual) case, the cables either are arranged in parallel tiers or converge toward the tower or towers.

Each cable adds one degree of statical indeterminacy to a system.   To make the actual conditions conform to design assumptions, the cable length must be adjustable either at the anchorages to the girders or at the saddles on the towers.

Steel Grades.   The steel most commonly used for orthotropic plates is weldable high-strength, low-alloy structural steel A441.   The minimum plate thickness is seldom less than $7/16$ in. (10 mm), to avoid excessive deflections under heavy wheel loads.   The maximum thickness seldom exceeds ¾ in. because of the decrease in permissible working stresses of high-strength low-alloy steel and the increase of fillet- and butt-weld sizes for plates of greater thickness.

Floor Beams.   If, as in most practical cases, the deck spans transversely between main girders, transverse ribs are replaced by the floor beams.   These are then built up of inverted T sections, with the deck plate acting as top flange.   Floor-beam spacings are preferably kept constant on any given structure.   They range from less than 5 ft to over 15 ft.   Longer spacings have been suggested for greater economy.

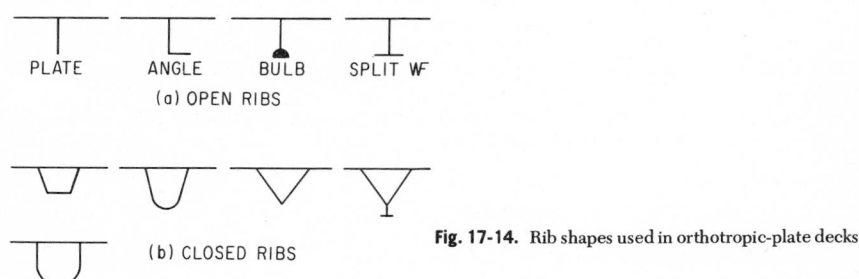

Fig. 17-14.   Rib shapes used in orthotropic-plate decks.

Ribs.   These are either open (Fig. 17-14a) or closed (Fig. 17-14b).   The spacing of open ribs is seldom less than 12 in. or more than 15 in.   The lower limit is determined by accessibility for fabrication and maintenance, the upper by considerations of deck-plate stiffness.   To reduce deformations of the surfacing material under concentrated traffic loads, some specifications require the plate thickness to be not less than $1/25$ of the spacing between open ribs or between the weld lines of closed ribs.

Usually, the longitudinal ribs are made continuous through slots or cutouts in the floor-beam webs to avoid a multitude of butt welds.   Rib splices can then be coordinated with the transverse deck splices.

Closed ribs, because of their greater torsional rigidity, give better load distribution and, other things being equal, require less steel and less welding than open ribs. Disadvantages of closed ribs are their inaccessibility for inspection and maintenance and more complicated splicing details. There have also been some difficulties in defining the weld between closed ribs and deck plate.

Fabrication. Orthotropic decks are fabricated in the shop in as large panels as transportation and erection facilities permit. Deck-plate panels are fabricated by butt-welding available rolled plates. Ribs and floor beams are fillet-welded to the deck plate in upside-down position. Then, the deck is welded to the girder webs.

It is important to schedule all welding sequences to minimize distortion and locked-up stresses. The most effective method has been to fit up all components of a panel—deck plate, ribs, and floor beams—before starting any welding, then to place the fillet welds from rib to rib and from floor beam to floor beam, starting from the panel center and uniformly proceeding toward the edges. Since this sequence practically requires manual welding throughout, American fabricators prefer to join the ribs to the deck by automatic fillet welding before assembly with the floor beams. After slipping the floor-beam webs over the ribs, the fabricators weld manually only the beam webs to the deck. This method requires careful preevaluation of rib distortions, wider floor-beam slots, and consequently more substantial or only one-sided rib-to-floor beam welds.

Analysis. Stresses in orthotropic decks are considered as resulting from a superposition of four static systems:

System I consists of the deck plate considered as an isotropic plate elastically supported by the ribs (Fig. 17-15a). The deck is subject to bending from wheel loads between the ribs.

System II combines the deck plate, as transverse element, and the ribs, as longitudinal elements. The ribs are continuous over, and elastically supported by, the floor beams (Fig. 17-15b). The orthotropic analysis furnishes the distribution of concentrated (wheel) loads to the ribs, their flexural and torsional stresses, and thereby the axial and torsional stresses of the deck plate as their top flange.

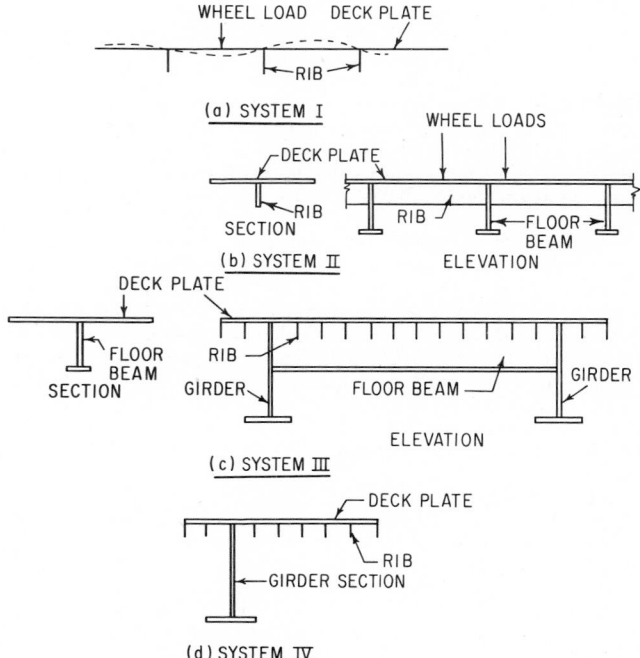

Fig. 17-15. Orthotropic deck may be considered to consist of four systems: (a) Deck plate supported on ribs; (b) rib-deck T beam spanning between floor beams; (c) floor beam with deck plate as top flange, supported on girders; (d) girder with deck plate as top flange.

System III combines the ribs with the floor beams and is treated either as an orthotropic system or as a grid (Fig. 17-15c). Analysis of this system furnishes the flexural stresses of the floor beams, including the stresses the deck plate receives as their top flange.

System IV comprises the main girders with the orthotropic deck as top flange (Fig. 17-15d). Axial stresses in the deck plate and ribs and shear stresses in the deck plate are obtained from the flexural and torsional analysis of the main girders by conventional methods.

Theoretically, the deck plate should be designed for the maximum principal stresses that may result from the simultaneous effect of all four systems. Practically, because of the rare coincidence of the maxima from all systems and in view of the great inherent strength reserve of the deck as a membrane (second-order stresses), a design is generally satisfactory if the stresses from any one system do not exceed 100% of the ordinarily permissible working stresses and 125% from a combination of any two systems.

In the design of long-span girder bridges, special attention must be given to buckling stability of deep webs and of the deck. Also, consideration should be given to conditions that may arise at intermediate stages of construction.

Steel-Deck Surfacing. All traffic-carrying steel decks require a covering of some nonmetallic material to protect them from accidental damage, distribute wheel loads, compensate for surface irregularities, and provide a nonskid, plane riding surface. To be effective, the surfacing must adhere firmly to the base and resist wear and distortion from traffic under all conditions. Problems arise because of the elastic and thermal properties of the steel plate, its sensitivity to corrosion, the presence of bolted deck splices, and the difficulties of replacement or repair under traffic.

The surfacing material usually is asphaltic. Strength is provided by the asphalt itself (mastic-type pavements) or by mineral aggregate (asphalt-concrete pavement). The usefulness of mastic-type pavements is restricted to a limited temperature range, below which they become brittle and above which they may flow. The effectiveness of the mineral aggregate of asphalt concrete depends on careful grading and adequate compaction, which, on steel decks, sometimes is difficult to obtain. Asphalt properties may be improved by admixtures of highly adhesive or ductile chemicals of various plastics families.

("Design Manual for Orthotropic Steel Plate Deck Bridges," American Institute of Steel Construction, 1221 Avenue of the Americas, New York, N.Y. 10020; F. S. Merritt, "Structural Steel Designers' Handbook," McGraw-Hill Book Company, New York.)

17-14. Truss Bridges. Trusses are lattices formed of straight members in triangular patterns. While truss-type construction is applicable to practically every static system, the term is restricted, in this article, to beam-type structures: simple spans and continuous and hinged (cantilever) structures. For other applications see Arts. 17-15 and 17-16. For typical single span bridge truss configurations, see Fig. 6-48. For the stress analysis of bridge trusses, see Arts. 6-44 through 6-51.

Truss bridges require more field labor than comparable plate girders. Also, trusses are more costly to maintain because of the more complicated makeup of members and poor accessibility of the exposed steel surfaces. For these reasons, and as a result of changing aesthetic preferences, use of trusses is increasingly restricted to long-span bridges on which the relatively low weight and consequent easier handling of the individual members are decisive advantages.

The superstructure of a typical truss bridge is composed of two main trusses, the floor system, a top lateral system, a bottom lateral system, cross frames, and bearing assemblies.

Decks for highway truss bridges are usually concrete slabs on steel framing. On long-span railway bridges, the tracks are sometimes mounted directly on steel stringers, although continuity of the track ballast across the deck is usually preferred. Orthotropic decks are rarely used on truss bridges.

Most truss bridges have the deck located between the main trusses, with the floor beams framed into the truss posts. As an alternative, the deck framing may be stacked on top of the top chord. *Deck trusses* have the deck at or above top-chord level (Fig. 17-16); *through trusses*, near the bottom chord (Fig. 17-17). Through trusses whose depth is insufficient for the installation of a top lateral system are referred to as *half through trusses* or *pony trusses*.

Figure 17-17 illustrates a typical cantilever truss bridge. The main span comprises a suspended span and two cantilever arms. The side, or anchor, arms counterbalance the cantilever arms.

Sections of truss members are selected to insure effective use of material, simple details for connections, and accessibility in fabrication, erection, and maintenance. Preferably, they should be symmetrical.

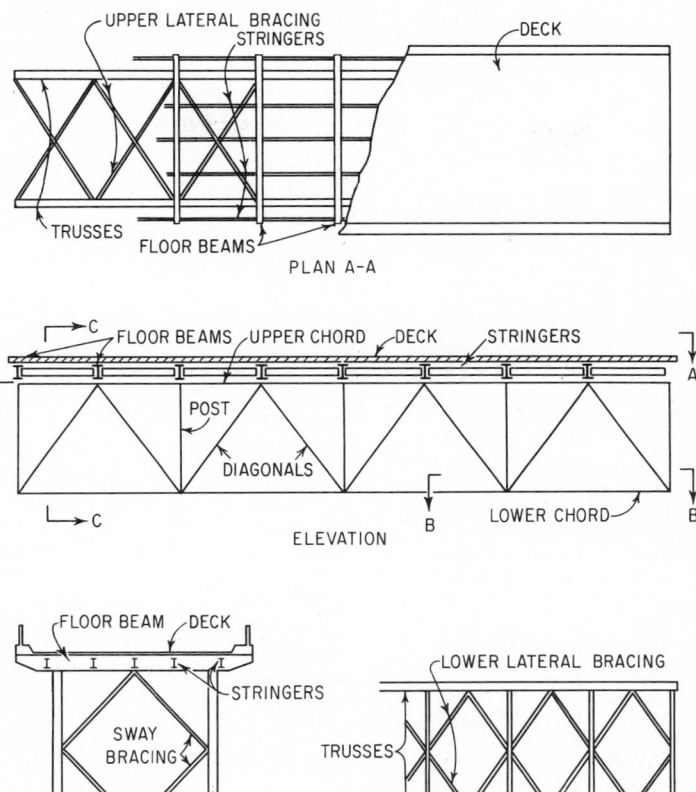

Fig. 17-16. Deck truss bridge.

In riveted design, the members are formed of channels or angles and plates, which are combined into open or half-open sections. Open sides are braced by lacing bars, stay plates, or perforated cover plates. Welded truss members are formed of plates. Figure 17-18 shows typical truss-member sections.

For slenderness restrictions of truss members, see Art. 17-7. For permissible stresses, see Table 17-9. The design strength of tensile members is controlled by their net section, that is, by the section area that remains after deduction of rivet or bolt holes. In shop-welded field-bolted construction, it is sometimes economical to build up tensile members by butt-welding three sections of different thickness or steel grades. Thicker plates or higher-strength steel is used for the end sections to compensate for the section loss at the holes.

The permissible stress of compression members depends on the slenderness ratio (see Art. 9-10). Design specifications also impose restrictions on the width-to-thickness ratios of webs and cover plates to prevent local buckling.

The magnitude of stress variation is restricted for members subject to stress reversal during passage of a moving load (Art. 9-18).

All built-up members must be stiffened by diaphragms in strategic locations to secure their squareness. Accessibility of all members and connections for fabrication and maintenance should be a primary design consideration.

Whenever possible, each web member should be fabricated in one piece reaching from the top to the bottom chord. The shop length of chord members may extend over several panels.

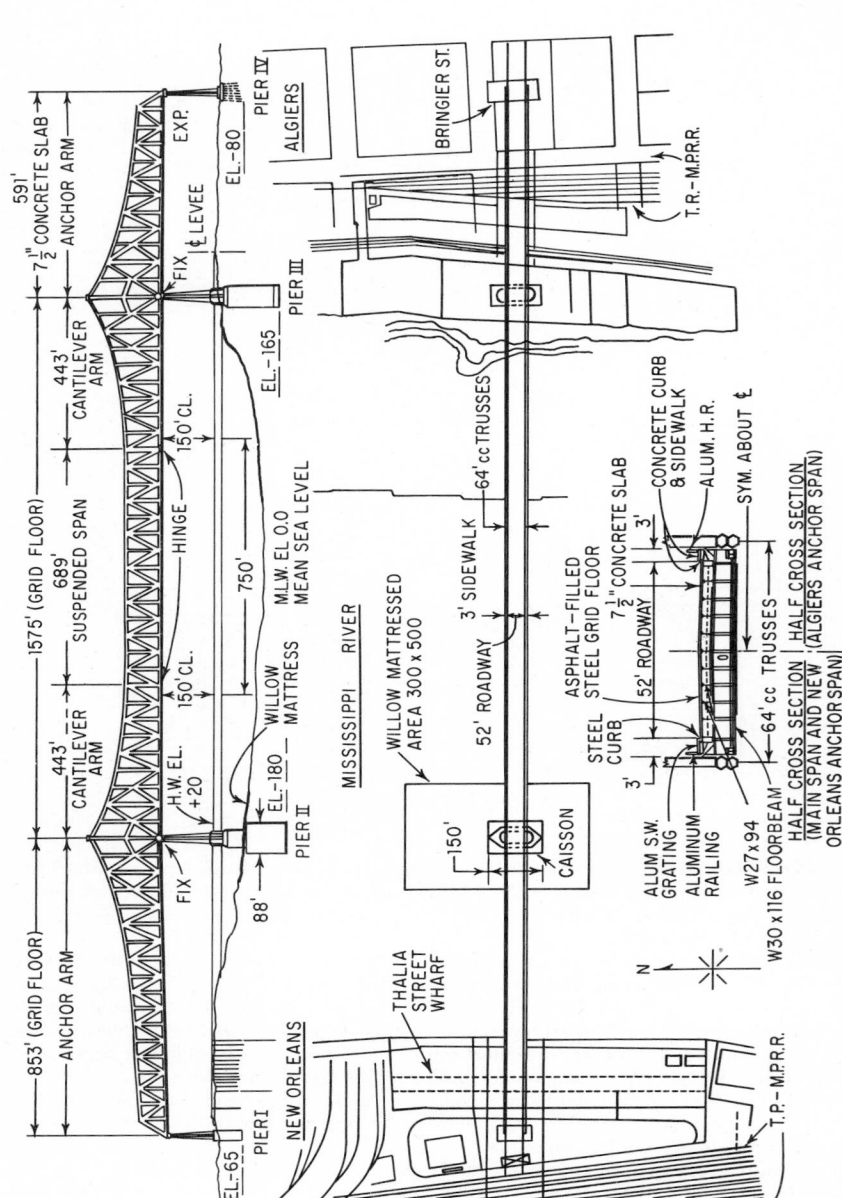

Fig. 17-17. Typical cantilever truss bridge.

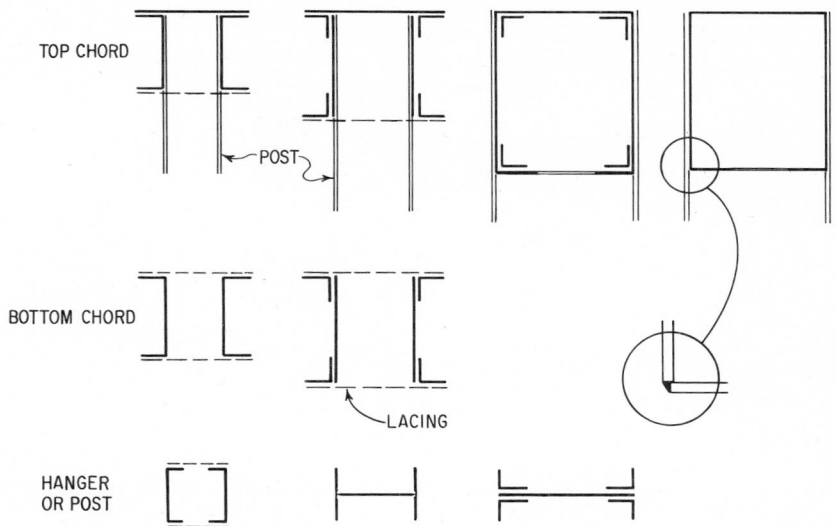

Fig. 17-18. Typical sections used in steel bridge trusses.

Chord splices should be located near joints and may be incorporated into the gusset plates of a joint.

In most trusses, members are joined by riveting, bolting, or welding with gusset plates. Pin connections, which were used frequently in earlier truss bridges, are now the exception. As a rule, the center lines or center-of-gravity lines of all members converging at a joint intersect in a single point (Figs. 17-19 and 17-20).

Stresses in truss members and connections are divided into primary and secondary stresses. Primary stresses are the axial stresses in the members of an idealized truss all of whose joints are made with frictionless pins and all of whose loads are applied at pin centers (Arts. 6-44 to 6-51). Secondary stresses are the stresses resulting from the incorrectness of these assumptions. Somewhat higher stresses are allowed when secondary stresses are considered. (Some specifications require computation of the flexural stresses in compression members caused by their own weight as primary stresses.) Under ordinary conditions, secondary stresses must be computed only for members whose depth is more than one-tenth of their length.

(F. S. Merritt, "Structural Steel Designers' Handbook," McGraw-Hill Book Company, New York.)

17-15. Suspension Bridges. These "hold," as the late D. B. Steinman, noted bridge designer, put it, "the supremacy among all bridges," in magnitude as well as in aesthetic appeal. They are presently the exclusive bridge type for spans over 1,800 ft, and they compete with other systems on shorter spans.

The basic structural system consists of flexible main cables (or, occasionally, eyebar chains) and, suspended from them, stiffening girders or trusses (collectively referred to as "stiffening beams"), which carry the deck framing. The vehicular traffic lanes are, as a rule, accommodated between the main supporting systems. Sidewalks may lie between the main systems or cantilever out on both sides.

Stiffening Beams. The purposes of the stiffening beams are to distribute concentrated loads, reduce local deflections, act as chords for the lateral system, and secure the aerodynamic stability of the structure. Spacing of the stiffening beams is controlled by the roadway width but is seldom less than $1/50$ the span.

Stiffening beams may be either plate girders, box girders, or trusses, the last being preferred because of their smaller air resistance. On major bridges, their depth is at least $1/180$ of the

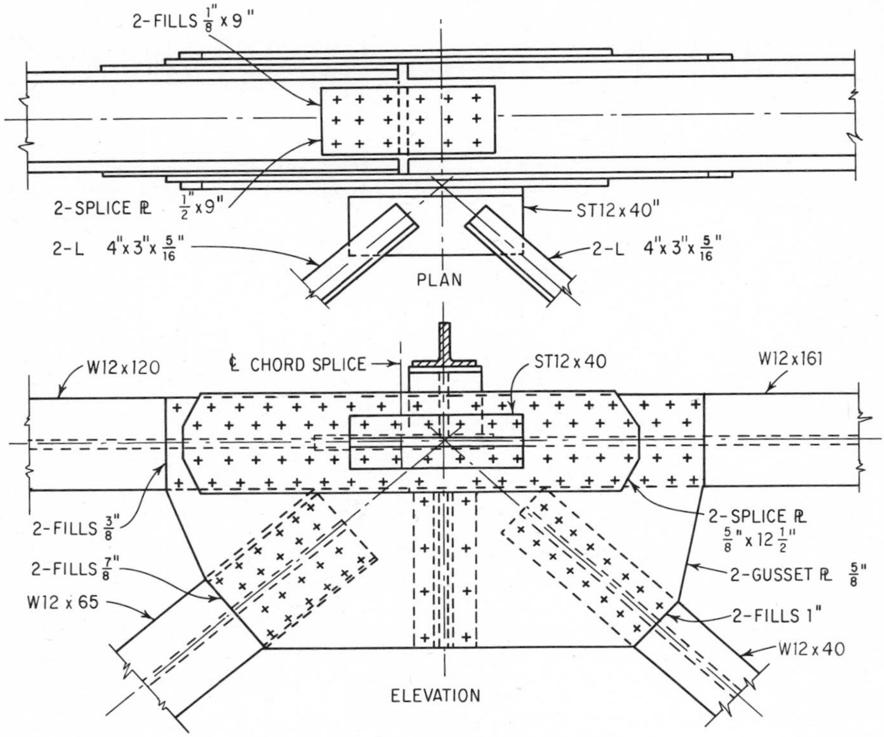

2-FILLS $\frac{1}{8}"$ x 9"

2-SPLICE ℝ $\frac{1}{2}"$ x 9"

2-L   4"x 3"x $\frac{5}{16}"$

ST 12 x 40"

2-L   4"x 3"x $\frac{5}{16}"$

PLAN

W 12 x 120

₵ CHORD SPLICE

ST 12 x 40

W 12 x 161

2-FILLS $\frac{3}{8}"$

2-FILLS $\frac{7}{8}"$

W 12 x 65

2-SPLICE ℝ $\frac{5}{8}"$ x 12 $\frac{1}{2}"$

2-GUSSET ℝ $\frac{5}{8}"$

2-FILLS 1"

W 12 x 40

ELEVATION

Fig. 17-19. Upper-chord joint in bridge truss.

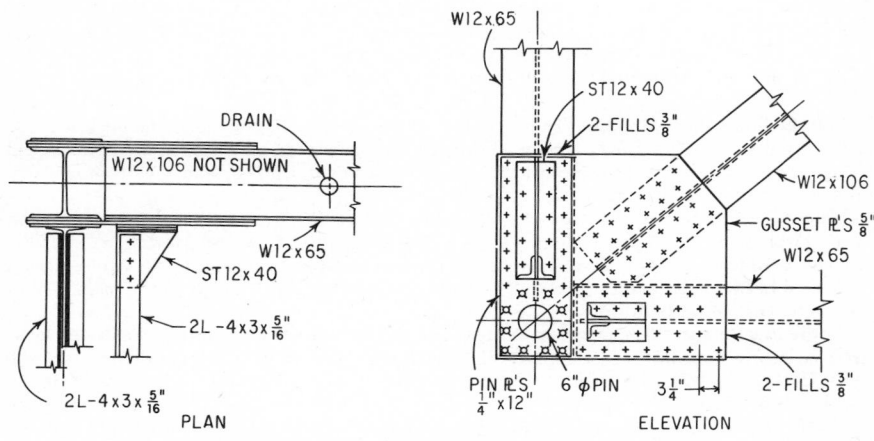

DRAIN

W 12 x 106 NOT SHOWN

W 12 x 65

ST 12 x 40

2 L -4 x 3 x $\frac{5}{16}"$

2 L -4 x 3 x $\frac{5}{16}"$

PLAN

W 12 x 65

ST 12 x 40

2-FILLS $\frac{3}{8}"$

W 12 x 106

GUSSET ℝ'S $\frac{5}{8}"$

W 12 x 65

2-FILLS $\frac{3}{8}"$

PIN ℝ'S
$\frac{1}{4}"$ x 12"

6" ϕ PIN

3 $\frac{1}{4}"$

ELEVATION

Fig. 17-20. Bridge-truss lower-chord pin joint at support.

main span. Stiffening trusses of variable depth have been built. Part of the top chord may be formed by the main supporting eyebar chain. Panel lengths may be equal to, or twice, or one-half the floor-beam spacing, so the truss diagonals will be close to 45°.

Anchorages.   The main cables are anchored in massive concrete blocks or, where rock sub-grade is capable of resisting cable tension, in concrete-filled tunnels. Or the main cables are connected to the ends of the stiffening girders, which then are subjected to longitudinal compression equal to the horizontal component of the cable tension.

Continuity.   Single-span suspension bridges are rare in engineering projects. They may occur in crossings of narrow gorges where the rock on both sides provides a reliable foundation for high-level cable anchorages.

The overwhelming majority of suspension bridges have main cables draped over two towers. Such bridges consist, thus, of a main span and two side spans. Preferred ratios of side span to main span are 1:4 to 1:2. Ratios of cable sag to main span are preferably in the range of 1:9 to 1:11, seldom less than 1:12.

If the side spans are short enough, the main cables may drop directly from the tower tops to the anchorages. In that case, the deck is carried to the abutments on independent, single-span plate girders or trusses. Otherwise, the suspension system is extended over both side spans to the next piers. There, the cables are deflected to the anchorages. The first system allows the designer some latitude in alignment, for example, curved roadways. The second requires straight side spans, in line with the main span. It is the common system for all those suspension bridges that are links in a chain of multiple-span crossings.

When side spans are not suspended, the stiffening beam is, of course, restricted to the main span. When side spans are suspended, the stiffening beams of the three spans may be continuous or discontinuous at the towers. Continuity of stiffening beams is required in self-anchored suspension bridges, where the cable ends are anchored to the stiffening beams.

Cable Systems.   The suspenders between main cables and stiffening beams are usually equally spaced and vertical. Sometimes, for greater aerodynamic stability, the suspenders are interwoven with diagonals that originate at the towers. Zigzag suspender systems have also been used.

Main cables, suspenders, and stiffening beams (girders or trusses) are usually arranged in vertical planes, symmetrical with the longitudinal bridge axis. Bridges with inward- or outward-sloping cables and suspenders and with offset stiffening beams are rare.

Three-dimensional stability is provided by top and bottom lateral systems and transverse frames, similar to those in ordinary girder and truss bridges. Rigid roadway decks may take the place of either or both lateral systems.

In the United States, the main cables are usually made up of 6-gage galvanized bridge wire of 220 to 225 ksi ultimate and 82 to maximum 90 ksi working stress. The wires are placed either parallel or in strands, and compacted and wrapped with No. 9 wire. In Europe, strands containing elaborately shaped heat-treated cast steel wires are sometimes used. Strands must be prestretched. They have a lower and less reliable modulus of elasticity than parallel wires. The heaviest cables, those of the Golden Gate Bridge, are about 36 in. in diameter. Twin cables are used if larger sections are required.

Suspenders may be eyebars, rods, single steel ropes, or pairs of ropes slung over the main cable. Connections to the main cable are made with cable bands. These are cast-steel whose inner faces are molded to fit the main cable. The bands are clamped together with high-strength bolts.

Floor System.   In the design of the floor system, reduction of dead load and of resistance to vertical air currents should be the governing considerations. The deck is usually lightweight concrete or steel grating partly filled with concrete. Expansion joists should be provided every 100 to 120 ft to prevent mutual interference of deck and main structure. Stringers should be made composite with the deck for greater strength and stiffness. Floor beams may be plate girders or trusses, depending on available clearance. With trusses, wind resistance is less.

As an alternative to conventional floor framing, orthotropic decks may be used advantageously (Art. 17-13).

Towers.   The towers may be portal type, or multistory, or diagonally braced frames (Fig. 17-21). They may be of cellular construction, made of steel plates and shapes, or steel lattices, or of reinforced concrete. The substructure below the "spray" line is concrete. The base of steel towers is usually fixed, but it may be hinged. Hinged towers offer some erection difficulties. The cable saddles at the top of fixed towers are sometimes placed on rollers to reduce the effect on the towers of unbalanced cable deflections. Cable bents can be considered as

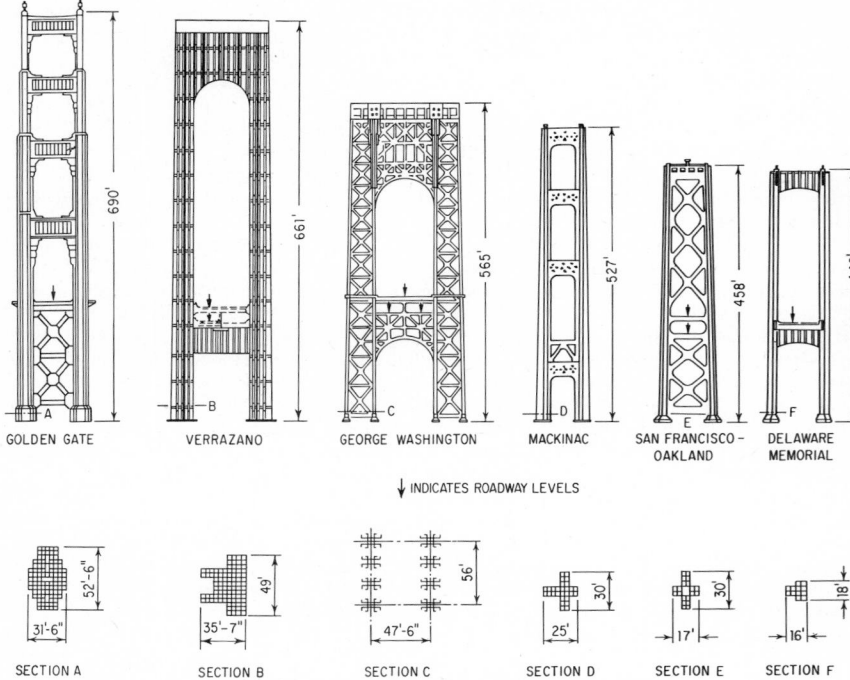

GOLDEN GATE   VERRAZANO   GEORGE WASHINGTON   MACKINAC   SAN FRANCISCO –   DELAWARE
OAKLAND   MEMORIAL

↓ INDICATES ROADWAY LEVELS

SECTION A   SECTION B   SECTION C   SECTION D   SECTION E   SECTION F

Fig. 17-21. Types of towers used for large suspension bridges.

short towers, either fixed or hinged, whose axis coincides with the bisector of the angle formed by the cable.

Analysis.   For gravity loads, the three elements of a suspension bridge in a vertical plane—the main cable or chain, the suspenders, and the stiffening beam—are considered as a single system.   The system of discrete suspenders often is idealized as one of continuous suspension.

The stiffening beam is assumed stressless under dead load, a condition approximated by appropriate methods of erection.   Moments and shears are produced by that part of the live load that is not taken up by the main cable through the suspenders.   Also, moments and shears result from changes in cable length and sag due to temperature variations or unbalanced load-

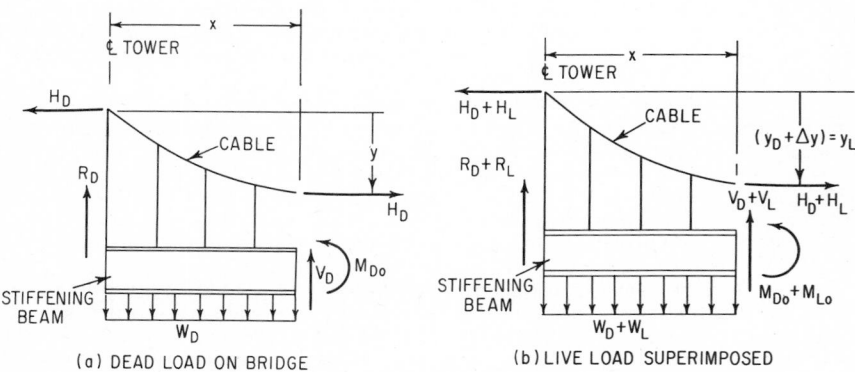

(a) DEAD LOAD ON BRIDGE   (b) LIVE LOAD SUPERIMPOSED

Fig. 17-22. Stresses in cable and stiffening beam of a suspension bridge.

ings of adjacent spans. Deflections of the stiffening beam are strictly elastic; that is, neglecting the effect of shear, the curvature at any section of the elastic line of the loaded beam is proportional to the bending moment divided by the moment of inertia of that section.

The suspenders are subject to tension only. Their elongation under live load is usually neglected in the analysis.

The main cable, too, is assumed to have no flexural stiffness and to be subject to axial tension only. Its shape is that of a funicular polygon of the applied forces (which include the dead weight of the cable). The pole distance H, lb, which is the horizontal component of the cable tension, is constant for a given loading and a given sag. The shape of the cable under given loads, that is, its ordinate y, ft, and slope $\tan \alpha$ at any point with abscissa x, ft, can be expressed in terms of the moment M_o, ft-kips, and shear V, kips, that a simple beam of the same span L, ft, as the cable would have under the same load (Fig. 17-22).

$$y = \frac{M_o}{H} \qquad \tan \alpha = \frac{V}{H} \tag{17-13}$$

In the special case of a uniform load w, kips per lin ft,

$$H = \frac{wL^2}{8f} \tag{17-14}$$

$$y = \frac{wx(L - x)}{2H} \quad \text{or} \quad y = \frac{4fx(L - x)}{L^2} \tag{17-15}$$

where f = cable sag, ft.

The shape of the cable under its own weight without suspended load would be a catenary; under full dead load, the cable shape is usually closer to a parabola. The difference is small.

Concentrated or sectionally uniform live load superimposed on the dead load subjects the cable to additional strain and causes it to adjust its shape to the changed load configuration. The resulting deformations are not exactly proportional to the additional loading; their magnitude is influenced by the already existing dead-load stresses.

If M_o is the bending moment of the stiffening beam under the applied load but without cooperation of the cable, the beam moment M with cooperation of the cable will be

$$M = M_o - Hy \tag{17-16}$$

More specifically, using subscripts D and L, respectively, for dead and live load and considering that

$$y_L = y_D + \Delta y \tag{17-17}$$

one gets the following expression for the dead- plus live-load bending moment of the beam (see Fig. 17-22b):

$$M = M_D + M_L = M_{D0} + M_{L0} - (H_D + H_L)(y_D + \Delta y)$$

But, since $M_D = M_{D0} - H_D y_D = 0$, because the stiffening beam has no bending moment under dead load (ideally),

$$M = M_{L0} - (H_D + H_L) \, \Delta y - H_L y_D \tag{17-18}$$

This is the basic equation of the cable-beam system.

In this equation, M_{L0}, H_D, and y_D are given. H_L and Δy must be so determined that the conditions of static equilibrium of all forces and geometric compatibility of all deformations are satisfied throughout the system.

The mathematically exact solution of the problem is known as the deflection theory. A less exact, older theory is known as the elastic theory. Besides these, there are several approximate methods based on observed regularities in the behavior of suspension bridges, which are sufficiently accurate to serve for preliminary design.

Wind Resistance. Wind acting on the main cables and on part of the suspenders is carried to the towers by the cables. Wind acting on the deck, stiffening beams, and live load is resisted mainly by the lateral bracing system and slightly by the cables, because of the gravity component resulting from any elastic lateral deflection of the main supporting system.

Oscillations of the structure may be generated by live load, earthquake, or wind. Live-load vibrations are insignificant in major bridges. Earthquake load seldom governs the design (N. C. Raab and H. C. Wood, "Earthquake Stresses in the San Francisco–Oakland Bay Bridge,"

Transactions of the American Society of Civil Engineers, vol. 106, 1941). Oscillations due to wind, however, can become dangerous if excessive amplitudes build up, that is, if the exciting impulses approach the natural frequency of the structure. Oscillating wind forces are caused by eddies, which may be generated outside the structure or by the structure itself, especially on the lee side of large plates. Oscillations of the structure may be purely flexural, or purely torsional, or coupled (flutter), the last two being the more dangerous.

Methods used to predict the aerodynamic behavior of suspension bridges include:

Mathematical analysis of the natural frequency of the structure in flexure and torsion [F. Bleich, C. B. McCullogh, R. Rosecrans, and G. S. Vincent, "Mathematical Theory of Vibration in Suspension Bridges," Government Printing Office, Washington, D.C.; A. G. Pugsley, "Theory of Suspension Bridges," Edward Arnold (Publishers) Ltd., London].

Wind-tunnel tests on scale models of the entire structure or of typical sections ("Aerodynamic Stability of Suspension Bridges with Special Reference to the Tacoma Narrows Bridge," *University of Washington Engineering Experiment Station Bulletin* 116).

Application of Steinman's criteria (these are controversial) (D. B. Steinman, "Rigidity and Aerodynamic Stability of Suspension Bridges," with discussion, *Transactions of the American Society of Civil Engineers,* vol. 110, 1945).

To prevent large or annoying oscillations, reduce air resistance, use trusses instead of girders for stiffening beams, and provide openings in the deck (for example, use a steel grid). Also, increase the rigidity of the structure, make the stiffening beams continuous past the towers, and use diagonal stays. ("Aerodynamic Stability of Suspension Bridges," 1952 Report of Advisory Board, *Transactions of the American Society of Civil Engineers,* vol. 120, 1955.)

Tower Stresses.   The towers must resist the forces imposed on them by the main cables in addition to the gravity and wind loads acting directly.

The following forces must be considered: The vertical components of the main cables in main and side spans under dead load, live load, temperature change, and wind. Wind acting on the main cables, both parallel and transverse to the bridge axis. Reactions to longitudinal cable movements due to unbalanced loading. These reactions will develop unless the movements are taken up by hinges or friction-free roller nests. Theoretically, the magnitude of these movements will be affected by the flexural resistance Q of the towers; but this effect, being comparatively small, is usually neglected.

Movement of the tower top generates bending moments. These increase from the top to the bottom at the rate of

$$M_x = Vy + Qx \qquad (17\text{-}19)$$

where V is the vertical cable reaction, x the distance below the top, y the horizontal deflection at x, and Q the horizontal resistance at the top. The magnitude of Q is such that the total deflection equals the longitudinal cable movement. It is found by solving the differential equation for the elastic curve of the tower axis. Thus,

$$y = A \sin cx + B \cos cx - \frac{Q}{V} x = \frac{Q}{V} \left(\frac{\sin cx}{c \cos cL} - x \right) \qquad (17\text{-}20)$$

in which $c = \sqrt{V/EI}$, I = moment of inertia, and E = modulus of elasticity of tower, if the towers have constant cross sections. The bending moment at x is

$$M_x = \frac{Q}{c} \frac{\sin cx}{\cos cL} \qquad (17\text{-}21)$$

where L = height of tower.

If, as is usual, the tower cross section varies in several steps, the coefficients A and B in Eq. (17-20) differ from section to section. They are found from the continuity conditions at each step.

Anchorages and footings should be designed for adequate safety against uplift, tipping, and sliding under any possible combination of acting forces.

(S. Hardesty and H. E. Wessman, "Preliminary Design of Suspension Bridges," *Transactions of the American Society of Civil Engineers,* vol. 104, 1939; R. J. Atkinson and R. V. Southard, "On the Problem of Stiffened Suspension Bridges and Its Treatment by Relaxation Methods," *Proceedings of the Institute of Civil Engineers,* 1939; C. D. Crosthwaite, "The Corrected Theory of the Stiffened Suspension Bridge," *Proceedings of the Institute of Civil Engineers,* 1946; Ling-Hi Tsien, "A Simplified Method of Analyzing Suspension Bridges,"

Transactions of the American Society of Civil Engineers, vol. 114, 1947; F. S. Merritt, "Structural Steel Designers' Handbook," McGraw-Hill Book Company, New York.)

17-16. Steel Arch Bridges.   A typical arch bridge consists of two or (rarely) more parallel arches or series of arches, plus necessary lateral bracing and end bearings, and columns or hangers for supporting the deck framing.   Types of arches correspond roughly to positions of the deck relative to the arch ribs.

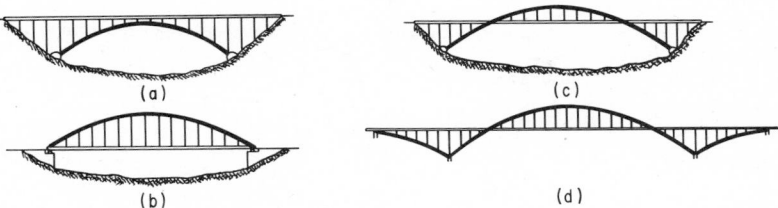

Fig. 17-23. Basic types of steel arch bridges. (*a*) Open spandrel arch; (*b*) tied arch; (*c*) arch with deck at intermediate level; (*d*) multiple-arch bridge.

Bridges with decks above the arches and clear space underneath (Fig. 17-23*a*) are designed as open spandrel arches on thrust-resisting abutments.   Given enough underclearance and adequate foundations, this type is usually the most economical.   Often, it is competitive in cost with other bridge systems.

Bridges with decks near the level of the arch bearings (Fig. 17-23*b*) are usually designed as tied arches; that is, tie bars take the arch thrust.   End bearings and abutments are similar to those for girder or truss bridges.   Tied arches compete in cost with through trusses in locations where underclearances are restricted.   Arches sometimes are preferred for aesthetic reasons. Unsightly overhead laterals can be avoided by using arches with sufficiently high moment of inertia to resist buckling.

Bridges with decks at an intermediate level (Fig. 17-23*c*) may be tied, or may rest on thrust-resisting abutments, or may be combined structurally with side spans that alleviate the thrust of the main span on the main piers (Fig. 17-23*d*).   Intermediate deck positions are used for long, high-rising spans on low piers.

Spans of multiple-arch bridges are usually structurally separated at the piers.   But such bridges may also be designed as continuous structures.

Hinges.   Whether or not hinges are required for arch bridges depends on foundation conditions.   Abutment movements may sharply increase rib stresses.   Fully restrained arches are more sensitive to small abutment movements (and temperature variations) than hinged arches. Flat arches are more sensitive than high arches.   If foundations are not fully reliable, hinged bearings should be used.

Complete independence from small abutment movements is achieved by installing a third hinge, usually at the crown.   This hinge may be either permanent or temporary during erection, to be locked after all dead-load deformations have been accounted for.

Arch Analysis.   The elementary analysis of steel arches is based on the elastic, or first-order, theory, which assumes that the geometric shape of the center line remains constant, irrespective of the imposed load.   This assumption is never mathematically correct.   The effects of deviations caused by overall flattening of the arch due to elastic rib shortening, elastic or inelastic displacements of the abutment, and local deformation due to live-load concentrations increase with initial flatness of the arch.   An effort is usually made to eliminate the dead-load part of the effect of rib shortening and abutment yielding during erection by jacking the legs of an arch toward each other or the crown section apart before final closure.   Arches subject to substantial deformation must be checked by the second-order, or deflection, theory.

For heavy moving loads, it is sometimes advantageous to assign the flexural resistance of the system to special stiffening girders or trusses, analogous to those of suspension bridges (Art. 17-15).   The arches themselves are then subject, essentially, to axial stresses only and can be designed as slender as buckling considerations permit.

Arch Design.   In general, steel arches must be designed for combined stresses due to axial loads and bending.

The height-to-span ratio used for steel arches varies within wide limits. Minimum values are around 1:10 for tied arches, 1:16 for open spandrel arches.

In cross section, steel arches may be I-shaped, box-shaped, or tubular. Or they may be designed as space trusses.

Deck Construction. The roadway deck of steel arch bridges is usually of reinforced concrete, often of lightweight concrete, on a framing of steel floor beams and stringers. To avoid undesirable cooperation with the primary steel structure, concrete decks either are provided with appropriately spaced expansion joints or are prestressed. Orthotropic decks that combine the functions of traffic deck, tie bar, stiffening girder, and lateral diaphragm have been used on some major arch bridges.

(F. S. Merritt, "Structural Steel Designers' Handbook," McGraw-Hill Book Company, New York.)

CONCRETE BRIDGES

Reinforced concrete is used extensively in highway bridges because of its economy in short and medium spans, durability, low maintenance costs, and easy adaptability to horizontal and vertical curvature. The principal types of cast-in-place supporting elements are the longitudinally reinforced slab, T beam or girder, and cellular or box girder. Precast construction, usually prestressed, often employs an I-beam or box-girder cross section. In long-span construction, posttensioned box girders often are used.

17-17. Slab Bridges. Concrete slab bridges, longitudinally reinforced, may be simply supported on piers or abutments, monolithic with wall supports, or continuous over supports.

Design Span. For simple spans, the span is the distance center to center of supports but need not exceed the clear span plus slab thickness. For slabs monolithic with walls (without haunches), use the clear span. For slabs on steel or timber stringers, use the clear span plus half the stringer width.

Load Distribution. In design, usually a 1-ft-wide longitudinal, typical strip is selected and its thickness and reinforcing determined for the appropriate HS loading. Wheel loads may be assumed distributed over a width, ft,

$$E = 4 + 0.06S \leq 7 \tag{17-22}$$

where S = span, ft. Lane loads should be distributed over a width of $2E$.

For simple spans, the maximum live-load moment, ft-kips, per foot width of slab, without impact, for HS20 loading is closely approximated by

$$M = 0.9S \qquad S \leq 50 \text{ ft} \tag{17-23a}$$

$$M = 1.30S - 20 \qquad 50 > S < 100 \tag{17-23b}$$

For HS15 loading, use three-quarters of the value given by Eqs. (17-23).

For longitudinally reinforced cantilever slabs, wheel loads should be distributed over a width, ft,

$$E = 0.35X + 3.2 \leq 7 \text{ ft} \tag{17-24}$$

where X = distance from load to point of support, ft.

The moment, ft-kips per foot width of slab, is

$$M = \frac{P}{E} X \tag{17-25}$$

where P = 16 kips for H20 loading and 12 kips for H15.

Reinforcement. Slabs should also be reinforced transversely to distribute the live loads laterally. The amount should be at least the following percentage of the main reinforcing steel required for positive moment: $100/\sqrt{S}$; but it need not exceed 50%.

The slab should be strengthened at all unsupported edges. In the longitudinal direction, strengthening may consist of a slab section additionally reinforced, a beam integral with and deeper than the slab, or an integral reinforced section of slab and curb. These should be designed to resist a live-load moment, ft-kips, of 1.6S for H20 loading and 1.2S for H15 loading on simply supported spans. Values for continuous spans may be reduced 20%. Greater reductions are permissible if justified by more exact analysis.

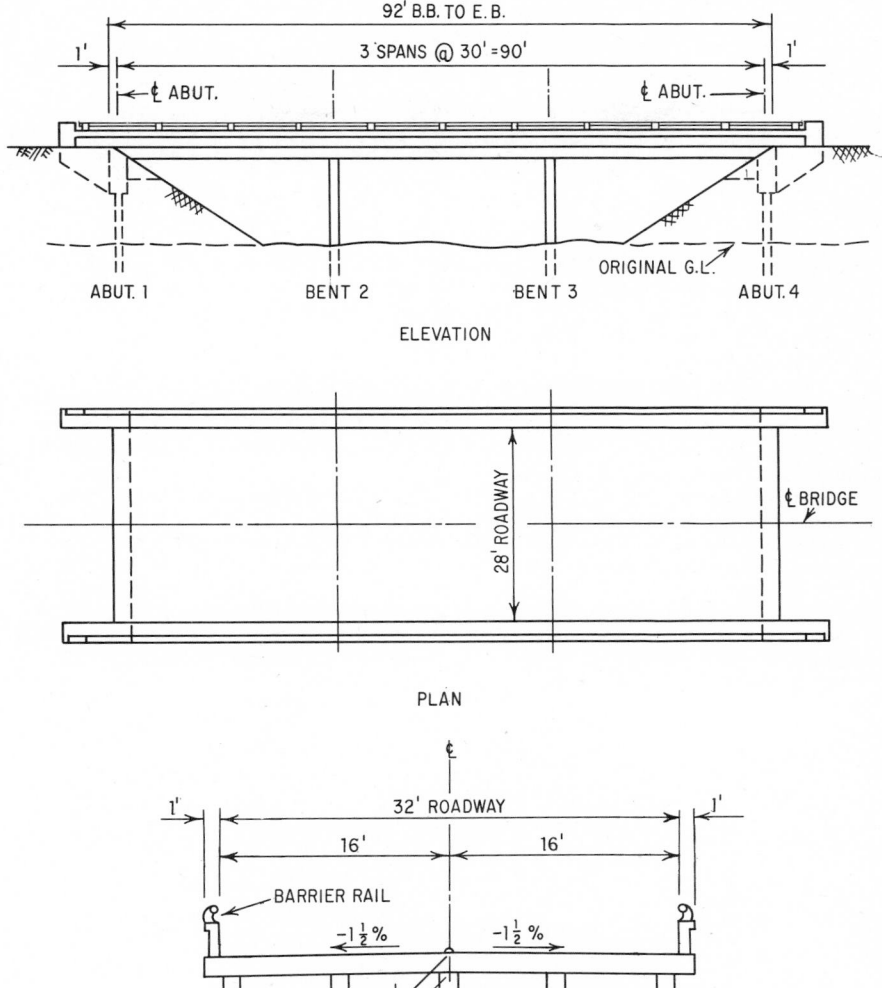

Fig. 17-24. Three-span concrete-slab bridge.

At bridge ends and intermediate points where continuity of the slab is broken, the edges should be supported by diaphragms or other suitable means. The diaphragms should be designed to resist the full moment and shear produced by wheel loads that can pass over them.

Design Procedure. The following procedure may be used for design of a typical longitudinally reinforced concrete slab bridge (Fig. 17-24):

Step 1. Determine the live-load distribution (effective width). For the three-span, 90-ft-long bridge in Fig. 17-24, $S = 30$ ft and

$$E = 4 + 0.06 \times 30 = 5.8 \text{ ft}$$

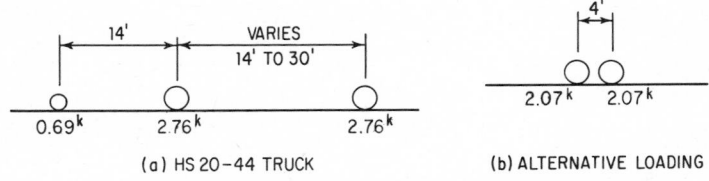

(a) HS 20–44 TRUCK   (b) ALTERNATIVE LOADING

Fig. 17-25. Wheel loads per foot width of slab for bridge of Fig. 17-24.

The distributed load for a 4-kip front wheel then is 4/5.8, or 0.69 kips, and for a 16-kip rear or trailer wheel load, 16/5.8, or 2.76 kips, per ft of slab width. For an alternative 12-kip wheel load, the distributed load is 12/5.8, or 2.07 kips per ft of slab width (see Fig. 17-25).

Step 2. Determine moment-distribution constants.

Step 3. Make up a table of distributed unit moments.

Step 4. Distribute the uniform-load fixed-end moment coefficients.

Step 5. Assume a slab depth.

Step 6. Distribute dead-load moments for the assumed slab depth.

Step 7. Determine live-load moment at point of maximum moment. (This is done at this stage to get a check on the assumed slab depth.)

Step 8. Combine dead-load, live-load, and impact moments at point of maximum moment. Compare the required slab depth with the assumed depth.

Step 9. Adjust the slab depth, if necessary. If the required depth differs from the assumed depth of step 5, the dead-load moments should be revised and step 8 repeated. Usu-

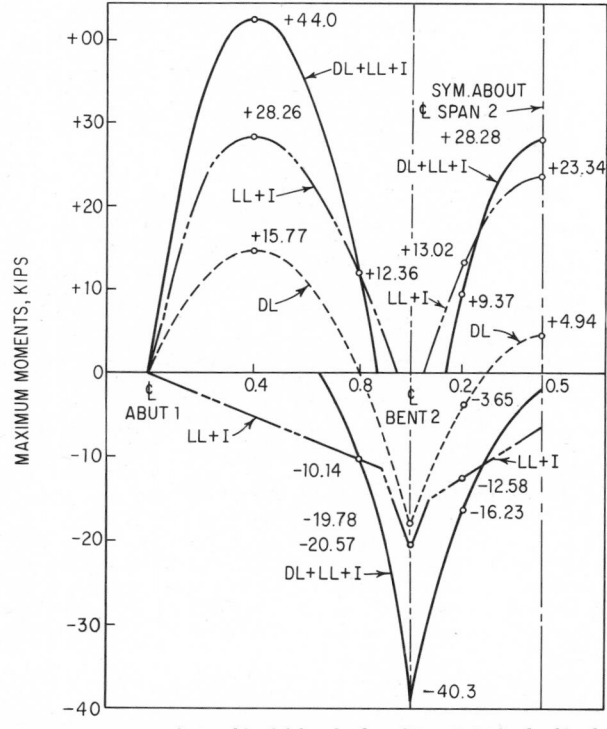

Fig. 17-26. Maximum moments per foot width of slab in bridge of Fig. 17-24 for dead load plus live load plus impact.

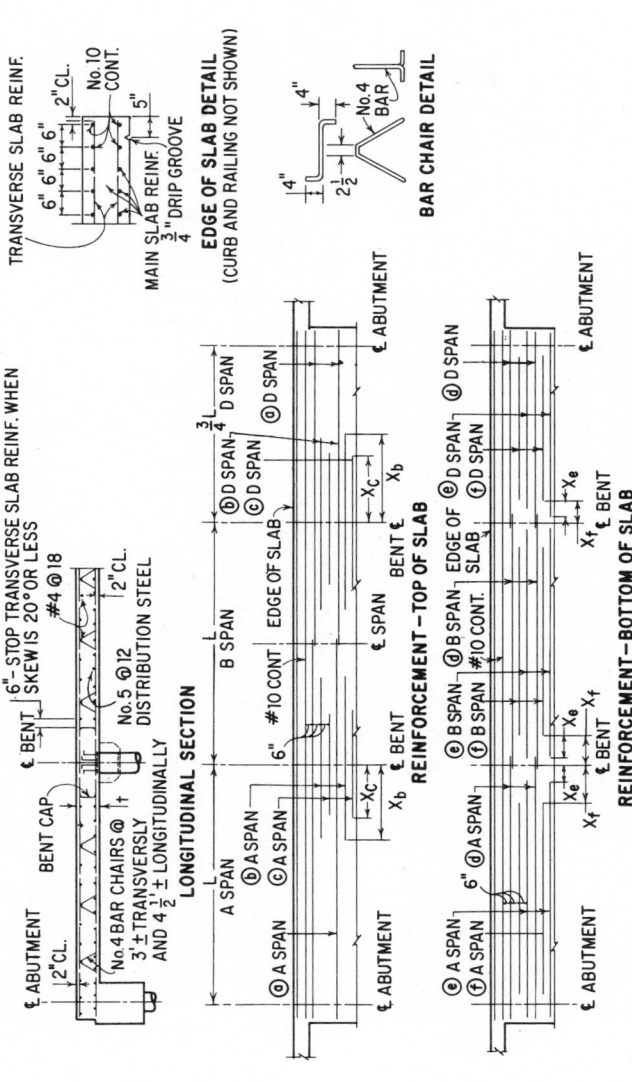

Fig. 17-27. Reinforcing bar layout for typical three-span concrete slab bridge.

Table 17-12. Slab Thicknesses and Reinforcing for Continuous Slab Bridges*

Length of span, ft	16			18			20			22			24		
Type of span	A	B	D	A	B	D	A	B	D	A	B	D	A	B	D
Reinforcement:†															
Top of slab															
a bars Size	No. 7	No. 6	No. 6	No. 7	No. 6	No. 6	No. 7	No. 6	No. 6	No. 7	No. 6	No. 6	No. 8	No. 6	No. 6
Length	27'0"	18'0"	22'0"	30'0"	20'0"	25'0"	33'0"	22'0"	27'0"	36'0"	24'0"	30'0"	39'0"	26'0"	32'0"
b bars Size	No. 6		No. 6	No. 7		No. 6	No. 7		No. 6	No. 7		No. 7	No. 8		No. 7
Length	12'0"		12'0"	12'0"		12'0"	13'0"		12'0"	15'0"		12'0"	14'0"		13'0"
X_b	6'0"		6'0"	6'0"		6'0"	6'6"		6'0"	7'6"		6'0"	7'0"		6'6"
c bars Size	No. 6		No. 6	No. 6		No. 6	No. 7		No. 6	No. 8		No. 6	No. 7		No. 7
Length	6'0"		6'0"	6'0"		6'0"	7'0"		6'0"	8'0"		7'0"	10'0"		7'0"
X_c	3'0"		3'0"	3'0"		3'0"	3'6"		3'0"	4'0"		3'6"	5'0"		3'6"
Bottom of slab															
d bars Size	No. 8		No. 6	No. 8		No. 6	No. 8		No. 7	No. 9		No. 7	No. 9		No. 7
Length	20'0"		14'0"	22'0"		16'0"	24'0"		17'0"	27'0"		19'0"	29'0"		20'0"
e bars Size	No. 6	No. 6	No. 6	No. 8	No. 6	No. 6	No. 9	No. 7	No. 6	No. 9	No. 8	No. 7	No. 10	No. 8	No. 7
Length	14'0"	13'0"	11'0"	15'0"	15'0"	12'0"	18'0"	16'0"	11'0"	18'0"	17'0"	14'0"	21'0"	18'0"	15'0"
X_e	1'6"	1'6"	1'0"	2'6"	1'6"	1'6"	2'0"	2'0"	3'6"	4'0"	2'6"	2'6"	3'0"	3'0"	3'0"
f bars Size	No. 6	No. 6	No. 6	No. 7	No. 7	No. 6	No. 7	No. 7	No. 6	No. 7	No. 7	No. 6	No. 7	No. 6	No. 6
Length	10'0"	8'0"	8'0"	11'0"	11'0"	9'0"	11'0"	13'0"	9'0"	10'0"	14'0"	9'0"	12'0"	16'0"	10'0"
X_f	4'0"	3'0"	3'0"	4'6"	3'6"	3'6"	5'6"	3'6"	4'6"	8'0"	4'0"	5'6"	8'0"	4'0"	6'0"
Quantities per ft of slab width:															
Concrete, cu ft	14.9	14.1	11.4	18.2	17.3	15.1	21.0	20.0	16.0	24.0	23.0	18.2	28.1	27.0	21.4
Steel, lb	135	115	85	155	135	100	180	160	115	200	130	125	260	215	140
Deflection at midspan Δ, ft:															
ABA spans				0.02'			0.02'			0.03'			0.03'		
DBD spans	Negligible deflection →												0.02		
t = thickness of slab, in. ‡		10½"			11½"			12"			12½"			13½"	

Bar Splice Length, In.

*For HS20 loading on any width roadway. A span is end span with same length as interior span *B* (see Fig. 17-27). *D* spans are three-fourths the length of span indicated in the heading. Design stresses: $f_c = 0.4f'_c = 1.3$ ksi; Grade 60 reinforcing bars, $f_s = 24$ ksi.

†Splices in top main bars to be located near center of span. Splices in bottom main bars to be located near bent. Spacing of all transverse bars is measured along center line of roadway.

Skew 0° to 20°: Place all transverse bars parallel to bent.
Skew over 20°: Place transverse slab bars perpendicular to center line of bridge.
†Add ½ in. for "marine environment" and adjust concrete quantity.

Bar Size No.	4	5	6	7	8	9	10	11
All bars, except top bars in spans over 26 ft	18	23	27	32	36	41	51	57
Top bars in spans over 26 ft	26	32	38	45	51	58	72	80

Table 17-12. Slab Thicknesses and Reinforcing for Continuous Slab Bridges* *(Continued)*

Length of span, ft	26			28			30			32			34		
Type of span	A	B	D	A	B	D	A	B	D	A	B	D	A	B	D
Reinforcement:†															
Top of slab															
a bars															
Size	No. 7	No. 7	No. 7	No. 9	No. 7	No. 7	No. 10	No. 7	No. 7	No. 10	No. 7	No. 7	No. 10	No. 8	No. 8
Length	42'0"	28'0"	35'0"	46'0"	30'0"	37'0"	49'0"	32'0"	40'0"	52'0"	34'0"	40'0"	55'0"	37'0"	45'0"
b bars															
Size	No. 9	No. 8	No. 7	No. 8	No. 8	No. 7	No. 8	No. 8	No. 7	No. 9	No. 8	No. 8	No. 9	No. 8	No. 8
Length	17'0"	19'0"	14'0"	13'0"	20'0"	14'0"	11'0"	21'0"	14'0"	13'0"	24'0"	14'0"	15'0"	22'0"	14'0"
X_b	8'6"	3'6"	7'0"	6'6"	4'0"	7'0"	5'6"	4'6"	7'0"	6'6"	4'0"	7'0"	7'6"	6'0"	7'0"
c bars															
Size	No. 9	No. 7	No. 7	No. 7	No. 7	No. 7	No. 7	No. 7	No. 8	No. 7	No. 8	No. 8	No. 9	No. 7	No. 8
Length	7'0"	15'0"	7'0"	8'0"	16'0"	7'0"	10'0"	17'0"	8'0"	9'0"	19'0"	8'0"	10'0"	17'0"	7'0"
X_c	3'6"	4'6"	3'6"	4'0"	6'0"	3'6"	5'0"	6'6"	4'0"	4'6"	6'6"	4'0"	5'6"	8'6"	3'6"
Bottom of slab															
d bars															
Size	No. 9		No. 7	No. 9		No. 7	No. 9		No. 7	No. 9		No. 7	No. 10		No. 8
Length	31'0"		22'0"	33'6"		23'0"	35'0"		25'0"	37'0"		26'0"	39'0"		28'0"
e bars															
Size	No. 9		No. 7	No. 9		No. 7	No. 9		No. 7	No. 11		No. 7	No. 10		No. 7
Length	22'0"		16'0"	23'0"		17'0"	25'0"		18'0"	27'0"		18'0"	26'0"		19'0"
X_e	4'0"		3'6"	4'6"		4'0"	4'6"		4'6"	4'6"		4'6"	7'0"		6'0"
f bars															
Size	No. 9		No. 7	No. 9		No. 7	No. 9		No. 7	No. 7		No. 7	No. 8		No. 7
Length	17'0"		13'0"	18'0"		13'0"	19'0"		14'0"	15'0"		14'0"	17'0"		15'0"
X_f	7'0"		5'0"	8'0"		6'0"	8'6"		6'0"	11'6"		6'0"	11'6"		8'0"
Quantities per ft of slab width:															
Concrete, cu ft	31.5	30.3	23.9	36.2	35.0	27.5	41.3	40.0	31.4	45.4	44.0	33.0	49.5	48.1	37.5
Steel, lb	290	225	155	310	255	175	340	285	190	370	315	210	400	345	220
Deflection at midspan Δ, ft:															
ABA spans	0.04'			0.05'			0.06'			0.07'			0.08'		
DBA spans		0.02'			0.02'			0.03'	0.01'		0.03'	0.01'		0.04'	0.02'
t = thickness of slab, in. †	14"			15"			16"			16½"			17"		

Table 17-12. Slab Thicknesses and Reinforcing for Continuous Slab Bridges* (Continued)

Length of span, ft	36			38			40			42			44		
Type of span	A	B	D	A	B	D	A	B	D	A	B	D	A	B	D
Reinforcement:															
Top of slab															
a bars															
Size	No. 10		No. 8	No. 10		No. 8	No. 10		No. 8	No. 10		No. 8	No. 11		No. 8
Length	58'0"		48'0"	67'0"		50'0"	70'0"		53'0"	73'0"		55'0"	76'0"		58'0"
b bars															
Size	No. 10		No. 8	No. 8		No. 9	No. 10		No. 8	No. 11		No. 8	No. 11		No. 10
Length	17'0"		15'0"	17'0"		16'0"	19'0"		17'0"	19'0"		19'0"	21'0"		21'0"
X_b	8'6"		7'6"	8'6"		8'0"	9'6"		8'6"	9'6"		9'6"	10'6"		10'6"
c bars															
Size	No. 9		No. 9	No. 11		No. 9	No. 10		No. 10	No. 10		No. 11	No. 11		No. 10
Length	8'0"		11'0"	14'0"		9'0"	11'0"		12'0"	10'0"		15'0"	12'0"		11'0"
X_c	4'0"		5'6"	7'0"		4'6"	5'6"		6'0"	5'0"		7'6"	6'0"		5'6"
Bottom of slab															
d bars															
Size	No. 11	No. 8	No. 8	No. 10	No. 8	No. 8	No. 10	No. 8	No. 8	No. 11	No. 9	No. 8	No. 11	No. 10	No. 8
Length	41'0"	39'0"	30'0"	43'0"	41'0"	31'0"	45'0"	43'0"	33'0"	47'0"	46'0"	34'0"	49'0"	48'0"	36'0"
e bars															
Size	No. 11	No. 8	No. 7	No. 11	No. 8	No. 8	No. 10	No. 9	No. 8	No. 11	No. 8	No. 8	No. 11	No. 8	No. 8
Length	29'0"	26'0"	20'0"	30'0"	27'0"	22'0"	32'0"	27'0"	22'0"	33'0"	28'0"	25'0"	34'0"	26'0"	25'0"
X_e	6'6"	5'0"	6'0"	7'0"	5'6"	6'0"	8'0"	6'6"	5'6"	8'6"	7'0"	6'0"	9'0"	9'0"	7'0"
f bars															
Size	No. 8	No. 8	No. 7	No. 8	No. 8	No. 7	No. 10	No. 8	No. 7	No. 10	No. 8	No. 8	No. 10	No. 7	No. 8
Length	18'0"	19'0"	15'0"	19'0"	20'0"	15'0"	24'0"	22'0"	15'0"	23'0"	20'0"	18'0"	24'0"	17'0"	19'0"
X_f	12'6"	8'6"	9'0"	14'6"	9'0"	9'6"	12'0"	9'0"	10'6"	13'6"	11'0"	10'0"	14'0"	13'6"	10'6"
Quantities per ft of slab width:															
Concrete, cu ft	54.0	52.5	40.8	58.5	57.0	44.3	64.9	63.3	49.0	69.8	68.2	52.8	75.0	73.3	56.7
Steel, lb	430	375	250	480	415	280	550	435	290	590	485	320	635	535	340
Deflection at midspan Δ, ft:															
ABA spans	0.10'			0.11'	0.01'		0.13'	0.01'		0.15'	0.01'		0.16'	0.01'	
DBD spans		0.05'	0.02'		0.05'	0.02'		0.06'	0.02'		0.07'	0.03'		0.08'	0.03'
t = thickness of slab, in. ‡	17½"			18"			19"			19½"					

ally, the second assumption is sufficient to yield the proper slab depth. Steps 2 through 9 follow conventional structural theory.

Step 10. Place live loads for maximum moments at other points on the structure to obtain intermediate values for drawing envelope curves of maximum moment.

Step 11. Draw the envelope curves (Fig. 17-26). Determine the sizes and points of cutoff for reinforcing bars.

Step 12. Determine distribution steel. Figure 17-27 shows the reinforcing-bar layout.

Step 13. Determine the number of piles required at each bent.

Table 17-12 lists slab thicknesses and reinforcing requirements for typical three-span slab bridges designed for HS20 loading. In Fig. 17-27 and Table 17-12, an A span is an end span with the same length as the interior span B; a D end span has three-fourths the length of B, which gives better moment balance. Data for B spans also can be used for interior spans of slabs continuous over more than three spans, while the data for A and D can be used for the end spans.

("Bridge Planning and Design Manual," Bridge Department, Division of Highways, State of California Department of Public Works.)

17-18. T-Beam Bridges. Widely used in highway construction, this type of bridge consists of a concrete slab supported on, and integral with, girders. It is especially economical in the 50- to 80-ft range. Where falsework is prohibited, because of traffic conditions or clearance limitations, precast construction of reinforced or prestressed concrete may be used. But adequate bond and shear resistance must be provided at the junction of slab and girders to justify the assumption that they are integral:

Since the girders parallel traffic, main reinforcing in the slab is perpendicular to traffic. For simply supported slabs, the span should be the distance center to center of supports but need not exceed the clear distance plus thickness of slabs. For slabs continuous over more than two girders, the clear distance may be taken as the span.

Bending Moments. The live-load moment, ft-kips, for HS20 loading on simply supported slab spans is given by

$$M = 0.5(S + 2) \qquad (17\text{-}26)$$

where S = span, ft.

For slabs continuous over three or more supports, multiply M in Eq. (17-26) by 0.8 for both positive and negative moment. For HS15 loading, multiply M by ¾.

Reinforcement also should be placed in the slab parallel to traffic to distribute concentrated live loads. The amount should be the following percentage of the main reinforcing steel required for positive moment: $220/\sqrt{S}$, but need not exceed 67%.

Where a slab cantilevers over a girder, the wheel load should be distributed over a distance, ft, parallel to the girder of

$$E = 0.8X + 3.75 \qquad (17\text{-}27)$$

where X = distance, ft, from load to point of support. The moment, ft-kips per ft of slab parallel to girder, is

$$M = \frac{P}{E} X \qquad (17\text{-}28)$$

where P = 16 kips for H20 loading and 12 kips for H15.

Equations (17-26) to (17-28) apply also to concrete slabs supported on steel girders, including composite construction.

Slab Design. In design of the slabs, a 1-ft-wide strip is selected and its thickness and reinforcing determined. The dead-load moments, ft-kips, positive and negative, can be assumed to be $wS^2/10$, where w is the dead load, ksf. Live-load moments are given by Eq. (17-26) with a 20% reduction for continuity. Impact is a maximum of 30%. With these values, standard charts can be developed for design of slabs on steel and concrete girders. An example is illustrated by Table 17-13 and Fig. 17-28. A typical slab-reinforcement layout is shown in Fig. 17-29.

T-Beam Design. The structure shown in Fig. 17-30 is a typical four-span grade-separation structure. The structural frame assumed for analysis is shown in Fig. 17-31. The hinged condition at the base of the columns is probably close to the condition existing in most structures of this type. Completely fixed column bases are difficult to attain because of footing rotation. In addition, economy in footing design results from the uniform footing pressures under a hinged column base.

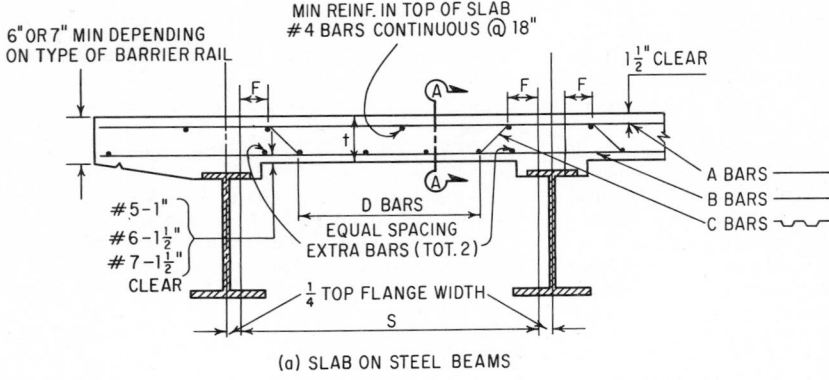

(a) SLAB ON STEEL BEAMS

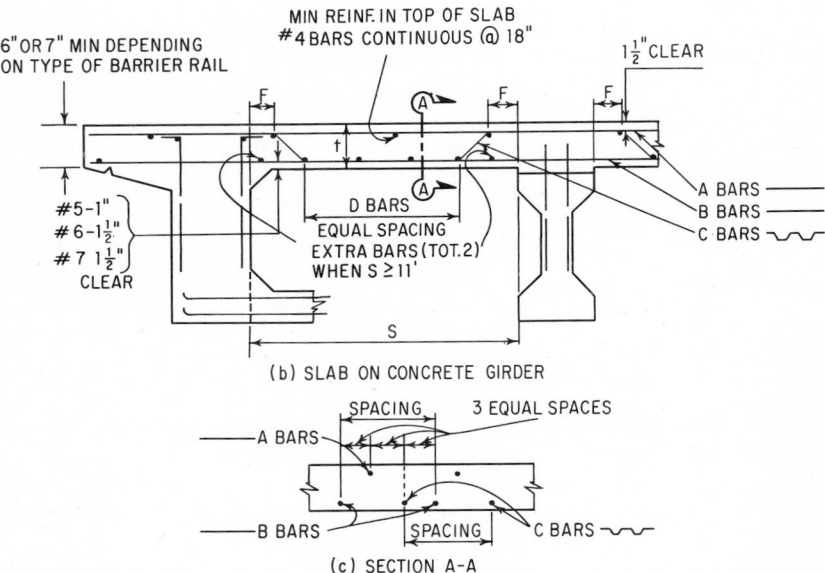

(b) SLAB ON CONCRETE GIRDER

(c) SECTION A-A

Fig. 17-28. Transverse reinforcing for concrete slabs on steel beams or concrete girders. (See also Table 17-13.) For bridges skewed 20° or less, place the reinforcing parallel to the abutments and bents. Space it along the center line of the structure. For skews greater than 20°, place the reinforcing perpendicular to the girders. Use *A* and *B* bars at acute corners at half the given spacing. Add ½ in. to slab thickness above 4,000-ft elevation.

Ratios of beam depths to spans used in continuous T-beam bridges generally range from 0.065 to 0.075. An economical depth usually results when a small amount of compressive reinforcement is required at the interior supports.

Girder spacing ranges from about 7 to 9 ft. Usually, a deck slab overhang of about 2 ft 6 in. is economical.

When the slab is made integral with the girder, its effective width in design may not exceed distance c to c of girders, one-fourth the girder span, or twelve times the least thickness of slab, plus girder web width. For exterior girders, however, effective overhang width may not exceed half the clear distance to the next girder, one-twelfth the girder span, or six times the slab thickness.

Table 17-13. Design Data for Concrete Slabs on Girders with H20 Loading*

Effective span S	t slab thickness, in.	F	A, B, C bars Size No.	A, B, C bars Spacing, in.	D bars, No. of No. 5 bars	Effective span S	t slab thickness, in.	F	A, B, C bars Size No.	A, B, C bars Spacing, in.	D bars, No. of No. 5 bars
4′0″	6	6″	5	13	4	12′0″	8⅝	1′7″	6	11	16
4′3″	6	7″	5	13	4	12′3″	8⅝	1′7″	6	11	16
4′6″	6	7″	5	13	4	12′6″	8¾	1′7″	6	11	16
4′9″	6	7″	5	12	5	12′9″	8⅞	1′8″	6	11	17
5′0″	6	8″	5	12	5	13′0″	8⅞	1′8″	6	10	17
5′3″	6	8″	5	12	5	13′3″	9	1′9″	6	10	17
5′6″	6	9″	5	11	5	13′6″	9	1′9″	6	10	18
5′9″	6	9″	5	11	6	13′9″	9⅛	1′9″	6	10	18
6′0″	6⅛	9″	5	11	6	14′0″	9¼	1′10″	6	10	18
6′3″	6¼	10″	5	11	6	14′3″	9⅜	1′10″	6	10	19
6′6″	6¼	10″	5	11	7	14′6″	9½	1′11″	7	14	19
6′9″	6¼	11″	5	10	7	14′9″	9⅝	1′11″	7	14	20
7′0″	6⅜	11″	5	10	7	15′0″	9¾	1′11″	7	14	20
7′3″	6½	11″	5	10	8	15′3″	9¾	2′0″	7	13	20
7′6″	6⅝	1′0″	5	10	8	15′6″	9¾	2′0″	7	13	21
7′9″	6¾	1′0″	5	10	8	15′9″	9⅞	2′1″	7	13	21
8′0″	7	1′1″	5	10	9	16′0″	10	2′1″	7	13	22
8′3″	7⅜	1′1″	6	13	9	16′3″	10	2′1″	7	13	22
8′6″	7½	1′1″	6	13	10	16′6″	10⅛	2′2″	7	13	22
8′9″	7½	1′2″	6	13	10	16′9″	10¼	2′2″	7	12	23
9′0″	7⅝	1′2″	6	13	11	17′0″	10¼	2′3″	7	12	23
9′3″	7¾	1′2″	6	13	11	17′3″	10⅜	2′3″	7	12	24
9′6″	7¾	1′3″	6	13	11	17′6″	10½	2′3″	7	12	24
9′9″	7⅞	1′3″	6	13	12	17′9″	10½	2′4″	7	12	24
10′0″	8	1′4″	6	12	12	18′0″	10⅝	2′4″	7	12	25
10′3″	8	1′4″	6	12	13	18′3″	10¾	2′4″	7	11	25
10′6″	8⅛	1′4″	6	12	13	18′6″	10¾	2′5″	7	11	26
10′9″	8⅛	1′5″	6	12	14	18′9″	10⅞	2′5″	7	11	26
11′0″	8¼	1′5″	6	12	14	19′0″	11	2′6″	7	11	26
11′3″	8⅜	1′6″	6	12	14	19′3″	11⅛	2′6″	7	11	27
11′6″	8½	1′6″	6	11	15	19′6″	11¼	2′6″	7	11	27
11′9″	8½	1′6″	6	11	15	19′9″	11¼	2′7″	7	11	28
						20′0″	11⅜	2′7″	7	11	28

*Design based on decks having three or more girders. (Add ½ in. to slab thickness above 4,000-ft elevation.) Design stresses: $f_c = 1{,}200$ psi; $f_s = 20{,}000$ psi; $n = 10$. Dead-load moment $= wS^2/10$. Impact factor = 30% for all spans. Moment for live load plus impact, with 80% continuity factor, ft-kips = $0.52(S + 2)$.

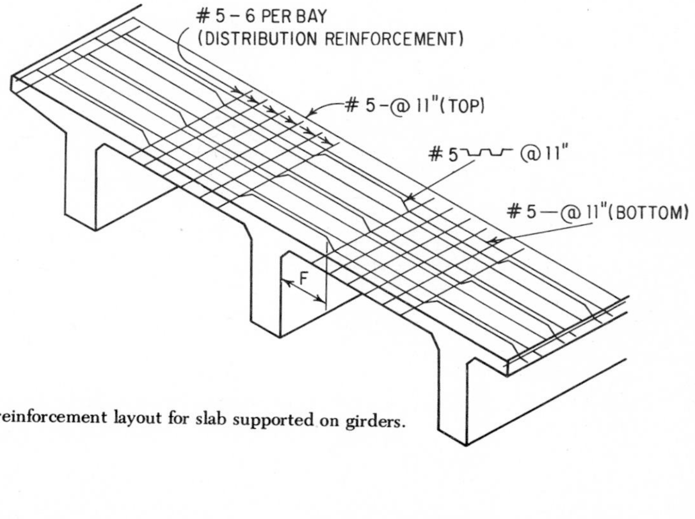

Fig. 17-29. Typical reinforcement layout for slab supported on girders.

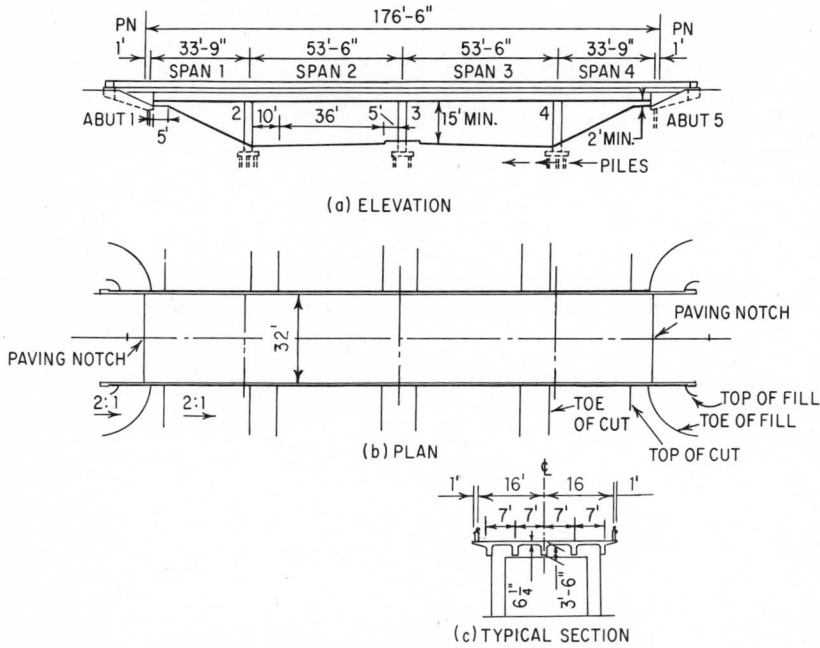

Fig. 17-30. Four-span reinforced concrete T-beam bridge.

Fig. 17-31. Assumed support conditions for bridge of Fig. 17-30.

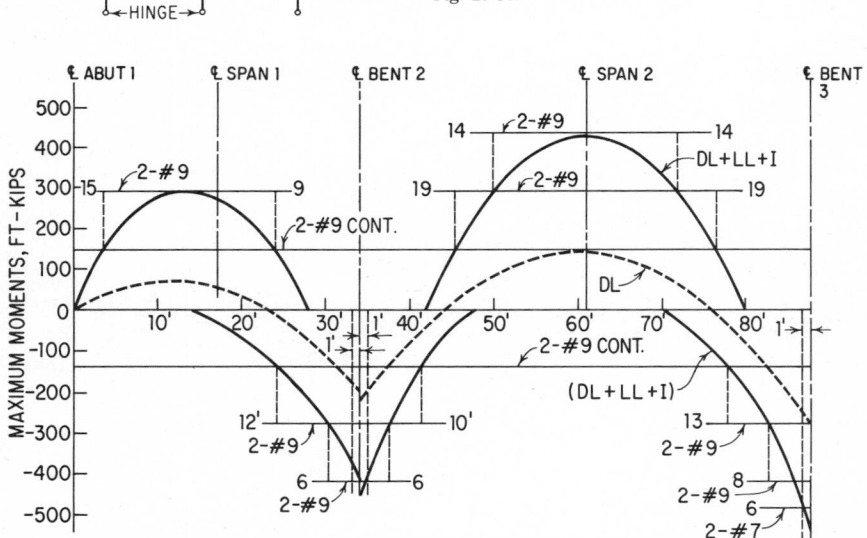

Fig. 17-32. Reinforcing for T beams of Fig. 17-30 is determined from curves of maximum moment. Numbers at ends of bars are distance, ft, from center line of span or bent.

For concrete girder design, curves of maximum moments for dead load plus live load plus impact may be developed to determine reinforcement. For live-load moments, truck loadings are moved across the bridge. As they move, they generate changing moments, shears, and reactions. It is necessary to accumulate maximum combinations of moments to provide an adequate design. For heavy moving loads, extensive investigation is necessary to find the maximum stresses in continuous structures.

Figure 17-32 shows curves of maximum moments consisting of dead load plus live load plus impact combinations that are maximum along the span. From these curves, reinforcing steel amounts and lengths may be determined by plotting the moments developed by bars. Maximum-shear requirements are derived theoretically by a point-to-point study of variations. Usually, a straight line between center line and end maximums is adequate. Figure 17-33 shows curves of maximum shears. Figure 17-34 shows the girder reinforcement layout.

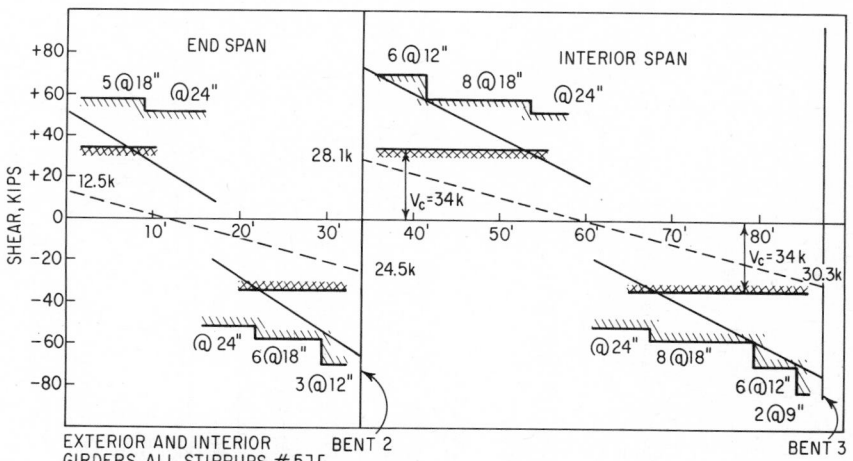

Fig. 17-33. Curves of maximum shear for T beams of Fig. 17-30.

Design of intermediate supports or bents varies widely, according to the designer's preference. A simple two-column bent is shown in Fig. 17-30. But considerable shape variations in column cross section and elevation are possible.

Abutments are usually of the L type or a monolithic end diaphragm supported on piles. ("Bridge Planning and Design Manual," Bridge Department, Division of Highways, State of California Department of Public Works.)

17-19. Box-Girder Bridges. Box or hollow concrete girders are favored by many designers because of the smooth plane of the bottom surface, uncluttered by lines of individual girders. Provision of space in the open cells for utilities is both a structural and an aesthetic advantage. Utilities are supported by the bottom slab, and access can be made available for inspection and repair of utilities.

For sites where structure depth is not severely limited, box girders and T beams have been about equal in price in the 80-ft span range. For shorter spans, T beams usually are cheaper, and for longer spans, box girders. While these cost relations hold in general, box girders have, in some instances, been economical for spans as short as 50 ft when structure depth was restricted.

Girder Design. Structural analysis is usually based on two typical segments, interior and exterior girders (Fig. 17-35). An argument could be made for analyzing the entire cross section as a unit because of its inherent transverse stiffness. Requirements in Standard Specifications for Highway Bridges, American Association of State Highway and Transportation Officials, however, are based on live-load distributions for individual girders, and so design usually is based on the assumption that a box-girder bridge is composed of separate girders.

Effective width of slab as top flange of an interior girder may be taken as the smallest of the distance c to c of girders, one-fourth the girder span, and twelve times the least thickness of

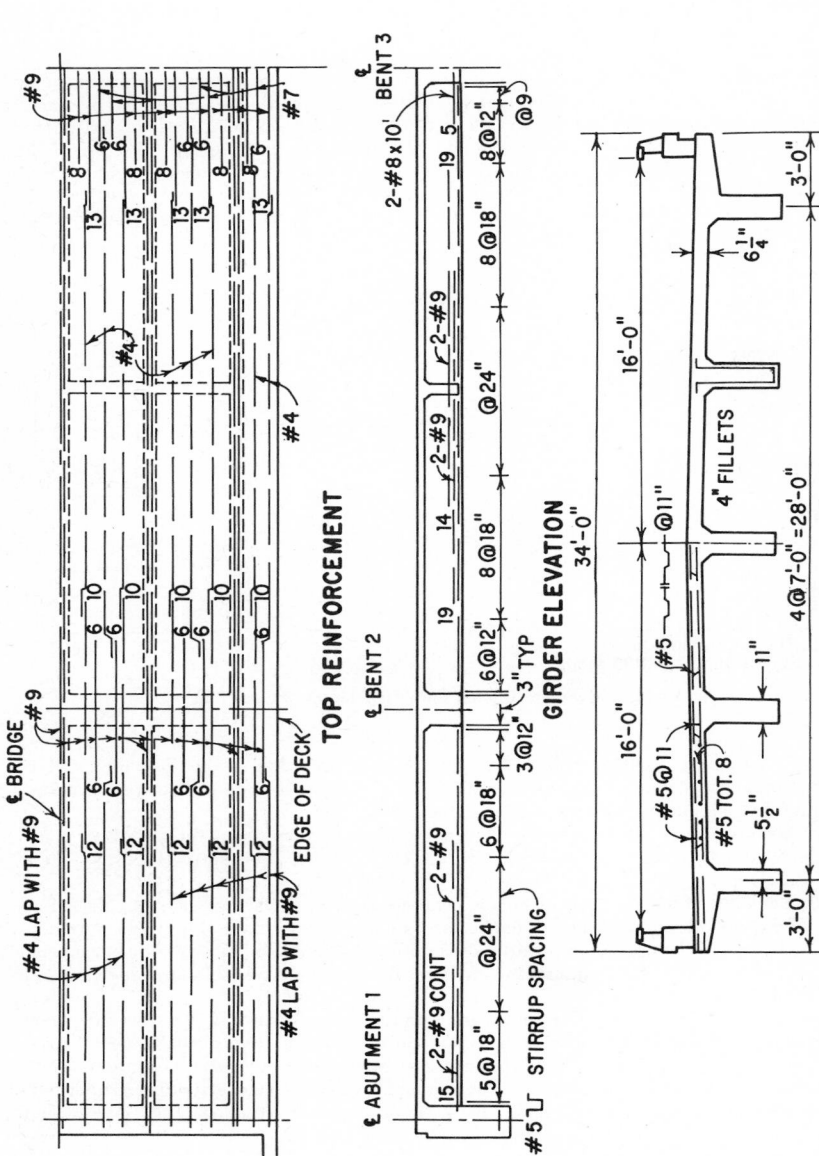

Fig. 17-34. Reinforcement layout for T beams of Fig. 17-30. Reinforcement is symmetrical about center lines of bridge and Bent 3. Numbers at ends of bars indicate distance, ft, from center line of bent or span.

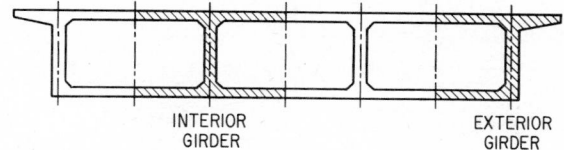

INTERIOR   EXTERIOR
GIRDER   GIRDER

Fig. 17-35. Typical design sections (crosshatched) for box-girder bridges.

slab, plus girder web width. Effective overhang width for an exterior girder may be taken as the smallest of half the clear distance to the next girder, one-twelfth the girder span, and six times the least thickness of the slab.

Usual depth-to-span ratio for continuous spans is 0.055. This may be reduced to about 0.048 with balanced spans, at some sacrifice in economy and increase in deflections. Simple spans usually require a minimum depth-to-span ratio of 0.065.

A typical concrete box-girder highway bridge is illustrated in Fig. 17-36. Spacing of webs could be either 7 ft 4 in. or 9 ft 4 in. The wider spacing is chosen to eliminate one web. Minimum web thickness is determined by shear but generally is at least 8 in. Changes should be gradual, spread over a distance at least twelve times the difference in web thickness.

Top-slab design follows the procedure described for T-beam bridges in Art. 17-18. Bottom-slab thickness and secondary reinforcement are usually controlled by specification minimums. AASHTO specifications require that slab thickness be at least one-sixteenth the clear distance between girders but not less than 6 in. for the top slab and 5½ in. for the bottom slab. Fillets should be provided at the intersections of all surfaces within the cells.

Minimum flange reinforcement parallel to the girder should be 0.6% of the flange area. This steel may be distributed at top and bottom or placed in a single layer at the center of the slab. Spacing should not exceed 18 in. Minimum flange reinforcing normal to the girder should be 0.5% and similarly distributed. Bottom-flange bars should be bent up into the exterior-girder webs at least ten diameters. The top-flange bars should extend to the exterior face of all exterior girders. At least one-third of these bars should be anchored with 90° bends or, where the flange projects beyond the girder sufficiently, extended far enough to develop bar strength in bond.

When the top slab is placed after the web walls have set, at least 10% of the negative-moment reinforcing should be placed in the web. The bars should extend a distance of at least one-fourth the span each side of the intermediate supports of continuous spans, one-fifth the span from the restrained ends of continuous spans, and the entire length of cantilevers. In any event, the web should have reinforcing placed horizontally in both faces, to prevent temperature and shrinkage cracks. The bars should be spaced not more than 2 ft c to c. Total area of this steel should be at least ⅛ sq in. per ft of web height.

Analysis of the structure in Fig. 17-36 for dead loads follows conventional moment-distribution procedure. Assumed end conditions are shown in Fig. 17-37a.

Live loads, positioned to produce maximum negative moments in the girders over Pier 2, are shown in Fig. 17-37b to d. Similar loadings should be applied to find maximum positive and negative moments at other critical points. Moments should be distributed and points plotted on a maximum-moment diagram (for dead load plus live load plus impact), as shown in Fig. 17-38. Layout of main girder reinforcement follows directly from this diagram. Figure 17-39 shows a typical layout.

("Bridge Planning and Design Manual," Bridge Department, Division of Highways, State of California Department of Public Works.)

17-20. Prestressed-Concrete Bridges. In prestressed-concrete construction, concrete is subjected to permanent compressive stresses of such magnitude that little or no tension develops when design loading is applied (Art. 8-42). Prestressing allows considerably better utilization of concrete than conventional reinforcement. It results in an overall dead-load reduction, which makes long spans possible with concrete, sometimes competitive in cost with those of steel. Prestressed concrete, however, requires greater sophistication in design, higher quality of materials (both concrete and steel), and greater refinement and controls in fabrication than does reinforced concrete.

Depending on the methods and sequence of fabrication, prestressed concrete may be precast, pretensioned; precast, posttensioned; cast-in-place posttensioned; composite; or partly prestressed.

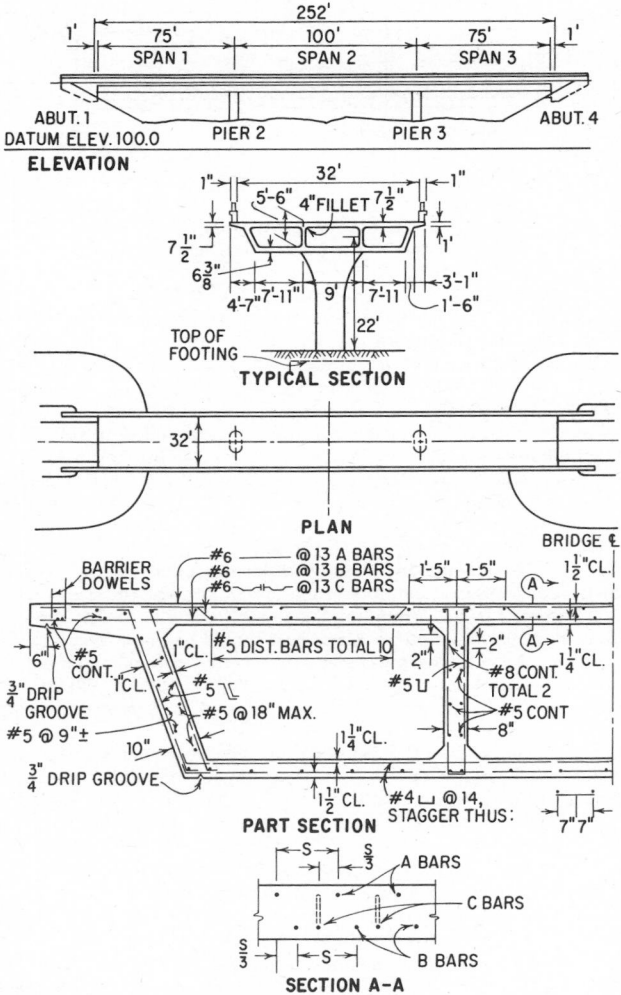

Fig. 17-36. Three-span, reinforced concrete box-girder bridge. For more reinforcing details, see Fig. 17-39.

In precast-beam bridges, the primary structure consists of precast-concrete units, usually I beams, channels, T beams, or box girders. They may be either pretensioned or posttensioned. Precast slabs may be solid or hollow. Precast I beams (Fig. 17-40) may be combined with fully or partly cast-in-place decks. This construction has the advantage that the deck can be shaped closely to the desired specifications. Precast slabs, incorporated in the deck, may be used in lieu of removable deck forms where accessibility is poor, for example, on over-water trestles or causeways. Precast T beams (Fig. 17-41) offer no advantage over the easier to fabricate, more compact I sections. Alignment of the flanges of T sections often is difficult. And as with I beams, the flanges must be connected with cast-in-place concrete. Precast box sections may be placed side by side to form a bridge span. If desired, they may be post-tensioned transversely.

Precast beams mainly are used for spans up to about 90 ft where erection of conventional falsework is not feasible or desirable. Such beams are particularly economical if conditions

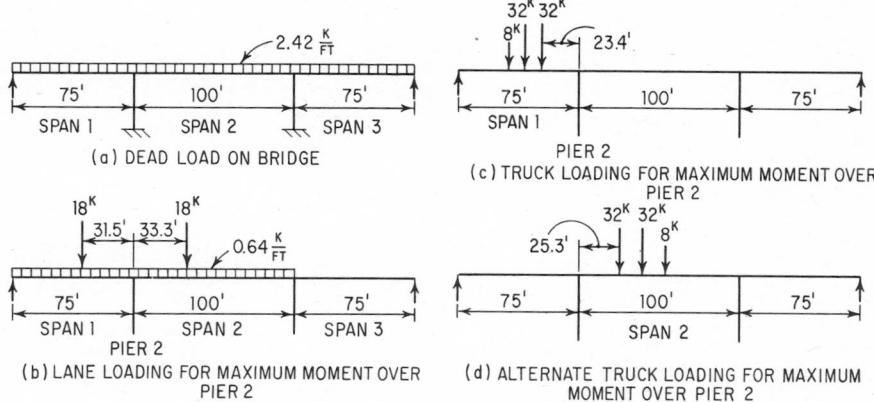

Fig. 17-37. Loading patterns for maximum stresses in a box-girder bridge.

are favorable for mass fabrication; for example, in multispan viaducts or causeways or in the vicinity of centralized fabrication plants. Longer spans are possible but require increasingly heavy handling equipment.

Standard designs for precast, prestressed girders have been developed by the Federal Highway Administration and state highway departments.

Cast-in-place prestressed concrete often is used for low-level bridges, where ground conditions favor erection of conventional falsework. Typical cross sections are essentially similar to those used for conventionally reinforced sections, except that, in general, prestressing permits thinner-walled designs.

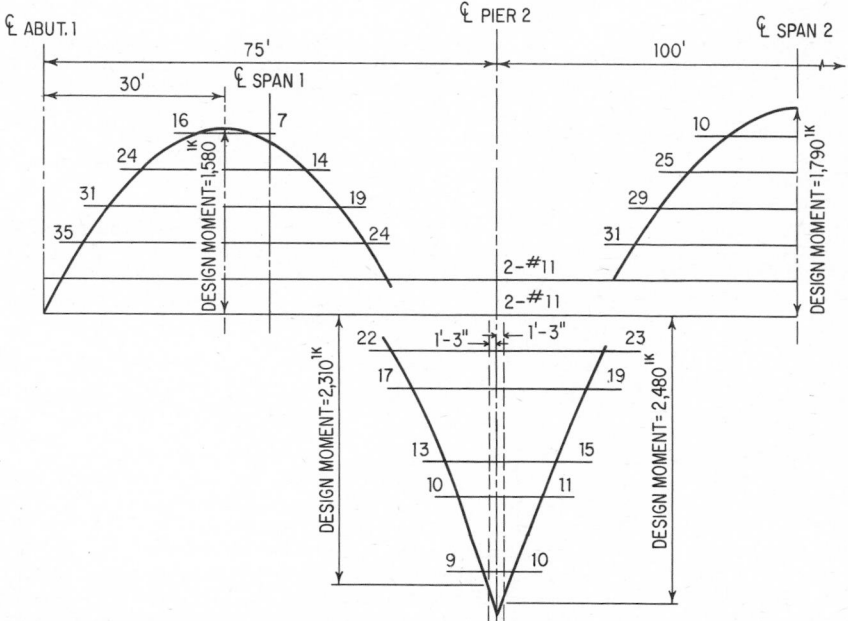

Fig. 17-38. Reinforcing for box girder determined from curves of maximum moment. Numbers at ends of bars indicate distance, ft, from center line of pier or span.

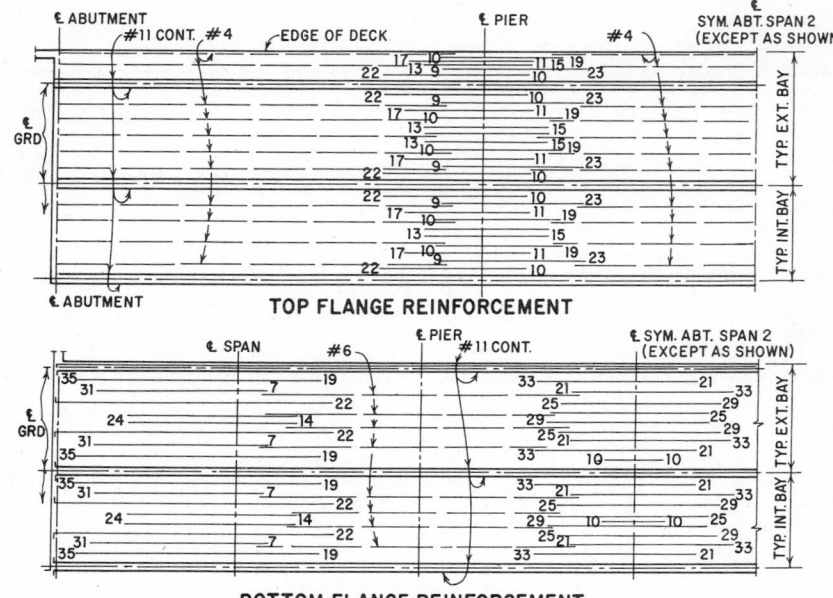

Fig. 17-39. Reinforcing layout for the box-girder bridge of Fig. 17-36. All bars are No. 11 except as noted. Numbers at ends of bars indicate distance, ft, from center line of pier for top reinforcement and center line of span for bottom bars. Design stresses for HS20 loading: $f_c = 1.3$ ksi, except 1.2 ksi for transverse deck slabs; $f_s = 24$ ksi, except 20 ksi for transverse deck slabs and stirrups.

For fully cast-in-place single-span bridges, posttensioning differs only quantitatively from that for precast elements. In design of multispan continuous bridges, the following must be considered: Frictional prestress losses depend on the draping pattern of the ducts. To reduce potential losses and increase the reliability of effective prestress, avoid continuously waving tendon patterns. Instead, use discontinuous simple patterns. Another method is to place tendons, usually bundles of cables, in the hollows of box girders and to bend the tendons at lubricated, accessible bearings.

The cantilever system developed for long-span high-level truss bridges continuous over at least two spans has been successfully adapted to prestressed concrete. Starting from each main pier, segments of superstructure are added symmetrically on opposite sides of the piers. Each segment is clamped to those already erected with tensioned rods or cables. (The segments may be precast or cast in place in forms supported by the previously erected parts of the structure.) The cantilever arms may be connected when they reach the center, or they may be stopped an appropriate distance from the center to support a simple (suspended or hung) span. Among the advantages of cantilever construction are its clear stress pattern (only negative bending moments in the cantilever arms and only positive moments in the suspended spans), full utilization of the individual tendons, and possibility of progressive adjustment of camber to observed deflections.

Posttensioning makes possible widening or strengthening or other remodeling of existing concrete structures. For example, Fig. 17-42 shows a cross section through a double-deck viaduct. The row of columns under the upper deck had to be removed, while capacity had to be increased from H15 to HS20 loading. No interference with upper-deck traffic and a minimum of interference with lower-deck traffic were permitted. This objective was accomplished by reinforcing each floor beam with precast units incorporating preformed ducts for tendons. Then, the entire upper deck was prestressed transversely. This permitted the beams to span the full width of the bridge and to carry the heavier loading. Similar remodeling has been done with cast-in-place concrete.

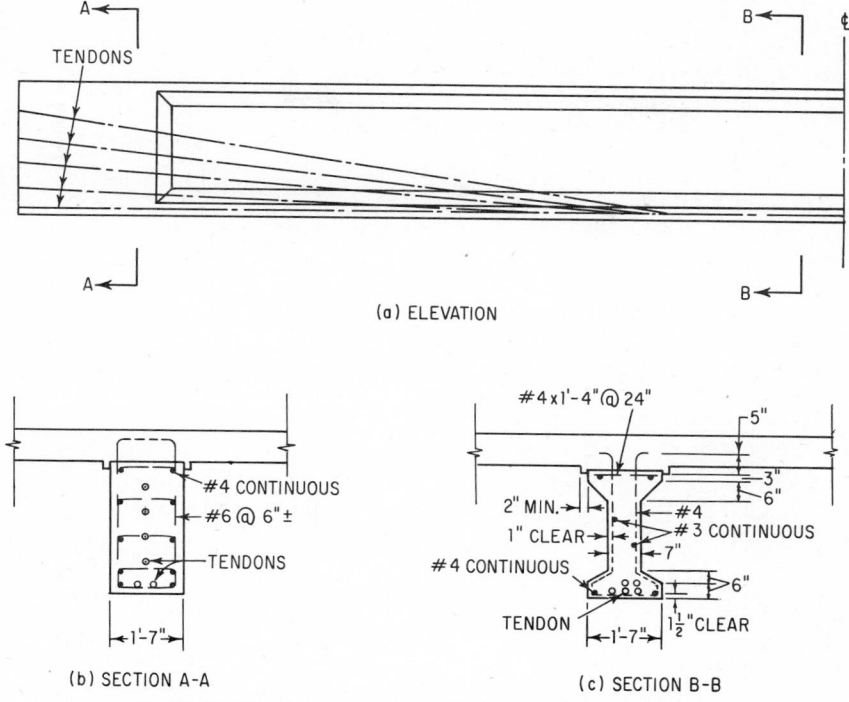

(a) ELEVATION

(b) SECTION A-A   (c) SECTION B-B

Fig. 17-40. Typical precast, prestressed I beam used in highway bridges.

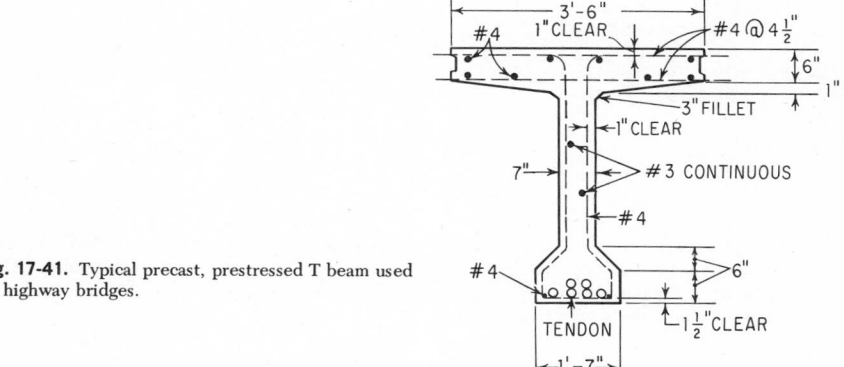

Fig. 17-41. Typical precast, prestressed T beam used
in highway bridges.

Determination of stresses in prestressed bridges is similar to that for other structures. In
analysis of statically indeterminate systems, however, the deformations caused by prestressing
must be taken into account (see also Arts. 8-42 to 8-45).

17-21. Bridge Piers and Abutments. Bridge piers are the intermediate supports of the
superstructure of bridges with two or more openings. Abutments are the end supports and,
usually, have the additional function of retaining earth fill for the bridge approaches.

The minimum height of piers and abutments is governed by requirements of accessibility
for maintenance of the superstructure, including bearings; of protection against spray for

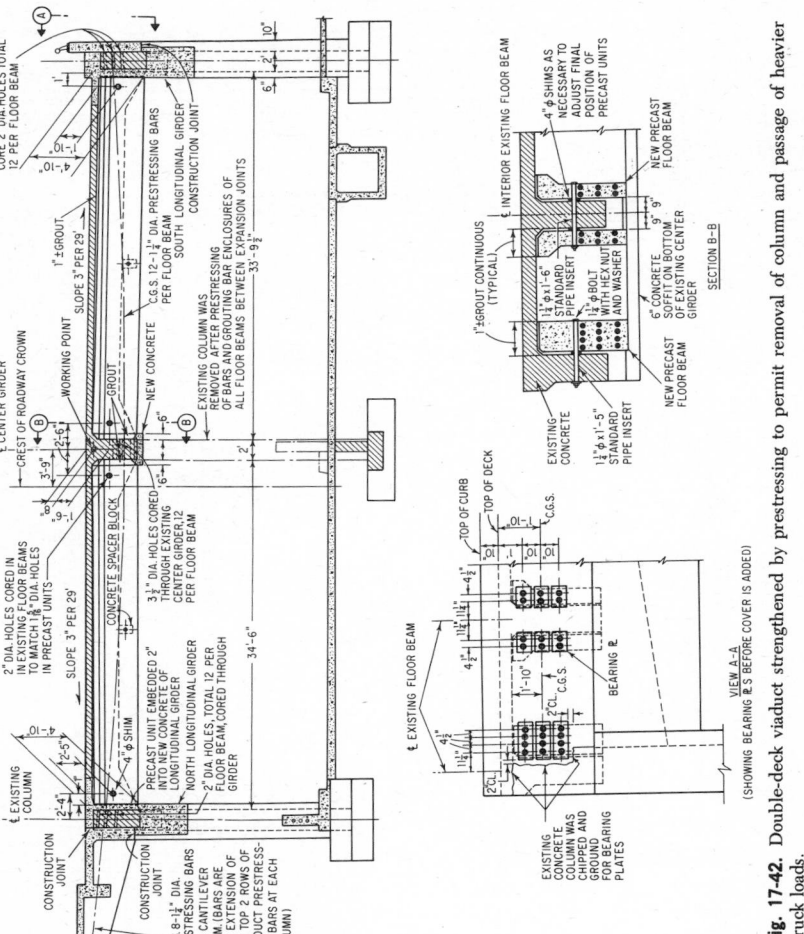

Fig. 17-42. Double-deck viaduct strengthened by prestressing to permit removal of column and passage of heavier truck loads.

bridges over water; and of vertical clearance requirements for bridges over traveled ways. There is no upper limit for pier heights, except that imposed by economic considerations. One of the piers of the Europa Bridge, which carries an international freeway in Austria, for instance, soars to 492 ft above the ground surface of the valley.

The top surface of piers must have adequate length and width to accommodate the bridge bearings of the superstructure. On abutments, added width is required for the back wall (curtain wall or bulkhead), which retains approach fill and protects the end section of the superstructure. In designing the aboveground sections of piers, restrictions resulting from lateral-clearance requirements of adjacent traveled ways and visibility needs may have to be taken into account. Length and width at the base level are controlled by stability, stress limitations in the pier shaft, and foundation design. For design of spread footings, see Arts. 7-6 and 7-7. For pile foundations, see Art. 7-8 and for caisson foundations, Art. 7-11.

For stress and stability analyses, the reactions from loadings (dead and live, but not impact) acting on the superstructure should be combined with those acting directly on the substructure. Longitudinal reactions depend on the type of bearing, whether fixed or expansion.

Piers. A number of basic pier shapes have been developed to meet the widely varying requirements. Enumerated below are some of the more common types and their preferred uses.

Trestle-type piers are preferred on low-level "causeways" carried over shallow waters or seasonally flooded land on concrete slab or beam-and-slab superstructures. Each pier or bent consists of two or more bearing piles, usually all driven in the same plane, and a prismatic cap into which the piles are framed. Both cap and piles may be of timber or, for more permanent construction, of precast conventionally reinforced or prestressed concrete.

Wall-type concrete piers on spread footings are generally used as supports for two-lane overcrossings over divided highways. Given adequate longitudinal support of the super-structure, these piers may be designed as pendulum walls, with joints at top and bottom; otherwise, as cantilever walls.

T-shaped piers on spread footings, with or without bearing piles, may be used to advantage as supports of twin girders. The girders are seated on bearings at both tips of the cross beam atop the pier stem. T-shaped piers have been built either entirely of reinforced concrete or of reinforced concrete in various combinations with structural steel.

Single-shaft piers of rectangular or circular cross section on spread footings may be used to support box girders, with built-in diaphragms acting as cross beams.

Portal frames may be used as piers under heavy steel girders, with bearings located directly over the portal legs (columns). When more than two girders are to be supported, the designer may choose to strengthen the portal cap beam or to add more columns. Preferably, all legs of each portal frame should rest on a common base plate. If instead separate footings are used, as, for instance, on separate pile clusters, adequate tie bars must be used to prevent unintended spreading.

Massive masonry piers have been built since antiquity for multiple-arch river bridges, high-level aqueducts, and, more recently, viaducts. In the twentieth century, their place has been taken by massive concrete construction, with or without natural stone facing. Where reduction of dead load is of the essence, hollow piers, often of heavily reinforced concrete, may be used.

Steel towers on concrete pedestals may be used for high bridge piers. They may be designed either as thin-membered, special trellis or as closed box portals, or combinations of these.

Very tall piers, when used, are usually constructed of reinforced or prestressed concrete, either solid or cellular in design.

Bridge abutments, basically, are piers with flanking (wing) walls. Abutments for short-span concrete bridges, such as T-beam or slab-type highway overcrossings, are frequently simple concrete trestles built monolithically with the superstructure (see Figs. 17-24 and 17-30). Abutments for steel bridges and for long-span concrete bridges that are subject to substantial end rotation and longitudinal movements should be designed as separate structures that provide a level area for the bridge bearings (bridge seat) and a back wall (curtain wall or bulkhead). The wall (stem) below the bridge seat of such abutments may be of solid concrete or thin-walled reinforced concrete construction, with or without counterfort walls; but on rare occasions, masonry is used.

Sidewalls, which retain approach fill, should have adequate length to prevent erosion and undesired spill of the backfill. They may be built either monolithically with the abutment stem and backwall, in which case they are designed as cantilevers subject to two-way bending, or as self-supporting retaining walls on independent footings. Sidewalls may be arranged in a straight line with the abutment face, or parallel to the bridge axis, or at any intermediate angle to the abutment face that may suit local conditions. Given adequate foundation conditions, the parallel-to-bridge-axis arrangement (U-shaped abutment) is often preferred because of its inherent stability.

For design of abutment footings, see Arts. 7-6 through 7-8. Abutments must be safe against overturning about the toe of the footing, against sliding on the footing, and against crushing of the underlying soil or overloading of piles. In earth-pressure computations, the vehicular load on highways may be taken into account in the form of an equivalent layer of soil of 2-ft thickness. Live loads from railroads may be assumed to be 0.5 kips per sq ft over a 14-ft-wide strip for each track.

In computations of internal stresses and stability, the weight of the fill material over an inclined or stepped rear face and over reinforced concrete spread footings should be considered as fully effective. No earth pressure, however, should be assumed from the earth prism in front of the wall. Buoyancy should be taken into account if it may occur.

Section **18**

Airport Engineering

HERBERT H. HOWELL

Airport Consultant, St. Louis, Mo.

Airport engineering involves design and construction of a wide variety of facilities for the landing, takeoff, movement on the ground, and parking of airplanes; maintenance and repair of airplanes; fuel storage; and handling of passengers, baggage, and freight. Thus, at a typical airport, there will be terminal buildings and hangars; pavements for airplane runways, taxiways, and aprons; roads, bridges, and tunnels for automobiles and walks for pedestrians; automobile parking areas; drainage structures; and underground storage tanks. Airport engineers have the responsibility of determining the size and arrangement of these facilities for safe, efficient, low-cost functioning of an airport.

18-1. Classes of Airports. There are two categories of airports in the United States: civil and military. Civil airports serve the scheduled airlines and all phases of general aviation. They are developed through the local initiative of individual communities, with some assistance from state and Federal sources. Military airports serve as bases for Air Force, Army, Navy, and Marine aviation and are developed, as needed, through the Department of Defense.

Civil airports may be further classified as air carrier airports (those that serve the scheduled airlines) and general aviation airports (those that serve business and executive flying, air-taxi operations, commercial and industrial aviation, and student instruction). While all air carrier airports accommodate considerable general aviation activity, the general aviation airports are usually not of a size sufficient to accommodate scheduled aviation. In each instance, the size and type of facility are determined by the existing and anticipated types and volume of air traffic that the facility will serve.

Military airports serve only the nation's defense needs. Only in rare instances is civil aviation activity permitted. There are, however, limited military facilities for Reserve and National Guard purposes at some civil airports. Military development is under the Corps of Engineers, United States Army, or the Facilities Engineering Command, United States Navy. Rigid adherence to standards and specifications for military airports is maintained.

18-2. Federal Aviation Administration. All airport work must be carefully coordinated with the Federal Aviation Administration (FAA) and receive air-space clearance to insure its compatibility with the total airport and air-space system. The Airport Development Aid Program, administered by the FAA, may provide funds for a major part of the development of landing areas. The FAA maintains national airport standards; offers advice on airport planning, design, and construction matters; maintains a national airport systems plan; certificates airports for operation; and conducts a compliance program to insure adherence to regulations and requirements. The FAA operates through conveniently located district offices. Liaison should be effected with the appropriate FAA office to insure full consideration of FAA policies and procedures.

18-3. Airport Master Plans. In the event that a full master planning study has not been made for an existing or proposed airport, such a study might well precede the planning of an improvement to that airport. If a master plan study has been undertaken, it can be used as the basis for further planning, or it can be reappraised. The master plan presents the planner's conception of the ultimate development of a specific airport, together with priority phasing, cost estimates, and financial plan. The master plan should be reevaluated periodically to maintain its validity.

18-4. National Airport Systems Plan. Through constant research, the Federal Aviation Administration, Department of Transportation, has developed criteria for determining the aeronautical potential of a community and for translating that potential into airport requirements. The overall airport needs of a community are summarized in the National Airport Systems Plan, published by the FAA. For existing and proposed airports, the plan shows the type of activity that is forecast and the general facilities required to accommodate it. A brief text spells out broad items of recommended development.

Communities are certificated for airline service by the Civil Aeronautics Board. Exceptions are intrastate carriers, which are subject to state jurisdiction, and commuter air carriers, which operate under special CAB authority. The number of enplaned airline passengers gives an indication of the air carrier potential of a community.

The number of based aircraft at an airport is an indication of the general aviation potential. At air carrier airports, the requirements for facilities to serve scheduled operations are greater than those for general aviation. Consequently, the overall needs of general aviation are usually met at airports that are developed to serve scheduled activity. Thus, the requirements of general aviation become a controlling factor only at airports that are not served by, or built for, the scheduled airline service.

The National Airport Classification System identifies the broad, functional mission of each airport in the National Airport Systems Plan by relating it to the Primary, Secondary, or Feeder system (Table 18-1).

Table 18-1. National Airport Classification System—Aeronautical Activity Levels for Functional-Role Airport Classification System *

Airport category	NASP† Codes	Public service level (annual enplaned passengers)	Aeronautical operational density (annual aircraft operations)
Primary system		More than 1,000,000	
High density	P1		More than 350,000
Medium density	P2		250,000 to 350,000
Low density	P3		Less than 250,000
Secondary system		50,000 to 1,000,000	
High density	S1		More than 250,000
Medium density	S2		100,000 to 250,000
Low density	S3		Less than 100,000
Feeder system		Less than 50,000	
High density	F1		More than 100,000
Medium density	F2		20,000 to 100,000
Low density	F3		Less than 20,000

*Federal Aviation Administration.
†National Airport Systems Plan

18-5. National Airport Standards. The Federal Aviation Administration has developed standards for nationwide application in design and construction of airports. Through the use of these standards, each local airport is made compatible with other airports and to fit in with the national system of airports. Although these standards are widely accepted, their use by local communities is not mandatory unless Federal funds are involved in the development. There is atitude that permits deviation from the standards where justified.

Table 18-2. Federal Aviation Administration Standards for Physical Characteristics of Airports

Item	Airports for general aviation[a]					Air carrier airports[b]			
	Basic utility stage I	Basic utility stage II	General utility	Basic transport	General transport	Design group I	Design group II	Design group III	Design group IV
Length of runway, ft[c]	2,000	2,500	3,000	4,500	5,750	6,500	8,000	10,000	12,000
Width of runway, ft	50	60	75	100[d]	100[d]	150	150	150[e]	150[e]
Width of landing strip, ft[f]	100	120	150	300[g]	300[g]	500	500	500	500
Width of taxiway, ft	20	30	40	40	40	50	75	100	125
Minimum distance between:									
Center lines of parallel runways, ft	300	300	500	700	700	700[h]	700[h]	700[h]	700[h]
Center line of runway and center line of taxiway, ft	150	150	200	200[i]	300[i]	400[j]	400[j]	600	1,000
Center line of runway and aircraft parking area, ft	225	225	275	300[k]	475[k]	650	650	650	650
Center line of taxiway and aircraft parking apron, ft	75	75	75	100[l]	175[l]	250	250	250	250
Center lines of parallel taxiways, ft	100	100	100	150[m]	200[m]	200[m]	300	300	400
Center line of runway to building line or obstruction, ft	250	250	300	300[n]	350[n]	750	750	750	750
Center line of taxiway to obstruction, ft	75	75	75	100[o]	105[o]	105[o]	135[o]	180[o]	235
Maximum runway grades, %:									
Effective[p]	2.0	2.0	2.0	1.0	1.0	1.0	1.0	1.0	1.0
Longitudinal[q]	2.0	2.0	2.0	1.5	1.5	1.5	1.5	1.5	1.5
Transverse[r]	1.5	1.5	1.5	1.5	1.5	1.5	1.5	1.5	1.5

[a] "Utility Airports," FAAA, November, 1968, and "General Aviation Airports—Basic and General Transport," FAA, July, 1969.

[b] AC 150/5330-2A, "Runway Widths for Airline Airports," and AC 150/5335-1A, "Airports Served by Air Carriers—Taxiways."

[c] Runway lengths for general aviation airports are minimum and are to be corrected for local conditions. Runway lengths for air carrier airports are general lengths for airports in each group; actual length is to be determined on the basis of specific aircraft requirements.

[d] 150-ft width for instrument runways.

[e] 200 ft for instrument runways.

[f] Landing strips extend 200 ft beyond pavement at each end of runways, except for utility airports where strips extend 100 ft beyond pavement.

[g] 500 ft for instrument-runway landing strips.

[h] 5,000 ft required for simultaneous landings; consult FAA.

[i] 400 ft for instrument runways.

[j] 600 ft for Category II (low minimum weather) operations.

[k] 650 ft for instrument runways.

[l] 250 ft for instrument runways.

[m] 300 ft for instrument runways.

[n] 750 ft for instrument runways.

[o] 200 ft for instrument runways.

[p] Maximum effective gradient is obtained by dividing the maximum difference in runway center-line elevation by the total length of the runway (Art. 18-7).

[q] Taxiway grades should be held to the same maximum grades as runways.

[r] Gradient shown is for pavement. To improve runoff, shoulder slopes may be increased to 5.0% for a distance of 10 ft from the edge of the pavement, then to 2.0% maximum (3% for utility airports).

Physical characteristics set by national standards for landing areas are summarized in Table 18-2. These are the minimum requirements that the FAA considers acceptable for safe operation. The standards are published by the FAA in a series of advisory circulars. Information as to how to obtain them is available through FAA airports district offices.

A **landing strip** is a graded strip, usually turfed. A **runway** is a paved strip located in the central portion of the landing strip and provided specifically for landings and takeoffs. A **taxiway** is a strip (usually paved) connecting runways with one another and with the aircraft parking apron. **Parallel runways** are two runways laid out in the same direction.

18-6. Runway Lengths. (See also Arts. 18-5.). The runway is the most important part of an airport. It must be of a length and design adequate to accommodate the aircraft it is to serve.

To determine the length required for a given location, the engineer should take into account the takeoff and landing characteristics of the most critical aircraft expected to make regular use of the airport. These characteristics must be studied in the light of the distances to be flown from the site, as well as its elevation, gradient, and temperature characteristics. The length chosen should be thoroughly reviewed and validated.

The FAA issues advisory circulars from time to time, giving performance data on aircraft that supplement its airport engineering data. The safe runway length for nontransport aircraft is based on takeoff and landing over an obstruction. The safe runway length for transport aircraft is based on Federal Aviation Regulations (Part 25), which specify three requirements for civil air transports, each of which must be met:

1. The runway length should be sufficient for the airplane to accelerate to the point of takeoff and then, in case of failure of the critical engine, be braked and brought to a stop within the limits of the runway (or usable landing strip).

2. If failure of the critical engine occurs at point of takeoff, the airplane should be capable of takeoff on the operating engine (or engines). Aircraft powered by reciprocating engines should be able to clear the end of the runway at an elevation of 50 ft and those powered by turbine engines at an elevation of 35 ft.

3. In landing, the airplane should clear the end of the runway by 50 ft and be landed and brought to a stop within 60% of the available runway length.

Runway requirement data published for transport aircraft usually have these factors included so that no further computation is required, except for effective gradient (Art. 18-7).

The normal landing requirements of jet aircraft establish runway lengths that are valid only for normal instrument conditions. For a jet airliner to land at lower weather minimums, the runway should provide more than the normally required landing length. In most instances, this additional requirement will still be less than the required takeoff length.

The length is required for instrument operations below 2,400 ft RVR (runway visual range) and down to 1,200 ft RVR, the equivalent of a 100-ft ceiling and 0.25-mile visibility. With electronic and visual landing aids of greater integrity, weather minimums may be lowered. All-weather operations are the ultimate goal. The corrected landing length should be checked against required takeoff length to obtain an adequate length if lower RVR operations can be forecast.

It might be possible to lower runway-length requirements by taking advantage of clearways and stopways, which permit partial runway credit to be given to areas at the ends of runways that are free from obstructions and capable of supporting aircraft loads. Usually, the real estate requirement and the cost of stabilization make this application uneconomical. Runway thresholds may be displaced to provide better clearance over obstructions, but credit may be taken for the displaced area for takeoff away from the obstruction and for landings involving an approach from the opposite end.

18-7. Runway Grades. Aircraft performance is influenced by the gradient of the runway. Ascending grades increase power required for takeoff. Descending grades increase braking distance. Not only is the gradient at any point along the runway of concern, but also the effective gradient of the overall runway. Other factors that influence grades are the sight distance and the transverse slopes of graded areas.

Longitudinal grades for air carrier airports should not exceed 1.50% at any point on the runway profile, but a 2.0% maximum is allowable for utility airports (Table 18-2).

Runway length determined for the critical aircraft at the elevation and mean temperature of the airport site is further increased at the rate of 20% for each 1.0% of effective gradient.

Longitudinal grade changes should generally be avoided. If changes are necessary, they should be in accordance with Table 18-3, which shows maximum grade changes and minimum length of vertical curves.

Minimum runway sight distances are necessary to permit safe visual aircraft operations. At noncontrolled airports, runway grade changes should be such that there will be an unobstructed line of sight from any point 5 ft above the center line of the runway to any other point 5 ft above

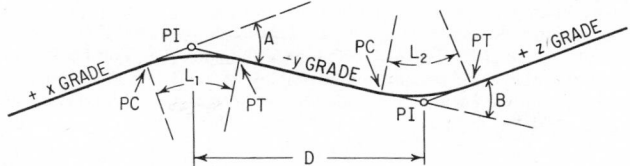

Fig. 18-1. Vertical profile along runway center line shows changes in longitudinal grades. (*Federal Aviation Administration.*)

Table 18-3. Vertical Curve Data and Maximum Grade Changes for Runways

	Utility airports	Air carrier airports
Maximum gradient at ends of runway, such as x grade or z grade (Fig. 18-1)	0 to 2.0%	0 to 1.5%, first and last quarter of runway length
Maximum gradient in middle portion of runway, such as y grade (Fig. 18-1)	0 to 2.0%	0 to 1.5%
Maximum grade change, such as A or B (Fig. 18-1)	2.0%	1.5%
Minimum length of vertical curve L_1 or L_2 (Fig. 18-1) for each 1.0% grade change	300 ft*	1,000 ft
Minimum distance between points of intersection for vertical curves, D (Fig. 18-1)	$250(A + B)$ ft†	$1,000 (A + B)$ ft†
Maximum effective gradient	2.0%‡	1.0%‡

*Vertical curves not required at utility airports for grade changes less than 0.4%.
†$A\%$ and $B\%$ are successive changes in grade.
‡Effective runway gradient is the maximum difference in runway center-line elevations divided by the runway length.

the runway. If the airport has a 24-hr control tower, adherence to runway longitudinal gradient standards will provide an adequate line of sight.

A graded safety area 200 ft in length is required at each end of the runway. An additional 800 ft of extended runway safety area is required for air carrier airports. This area should be 500 ft in width. The maximum grade change in the extended runway safety area is 3% per 100 ft, and the maximum gradient should not exceed 15%.

Transverse grades on runways should not exceed 1.5% (Table 18-2). The unpaved shoulders may be at a steeper gradient to improve runoff. The first 10 ft of shoulder adjacent to the pavement may be as steep as 5.0%, and the transverse grade past this 10-ft distance may be 2.0%. For utility airports and airports with runways less than 3,200 ft, the transverse grade may be increased to 3.0%.

Graded shoulders should be built 1.50 in. below the adjacent pavement edge to preclude future turf from developing a gutter, holding water at the edge of the pavement.

18-8. Runway Numbering System. The runways at each airport are designated by numbers related to azimuth, measured clockwise from magnetic north. For simplicity the numbers are expressed in 10° units of azimuth.

For example, if a runway has an azimuth measured from magnetic south of 32°, the southerly end is numbered 21, since $(32° + 180°)/10° = 21.2$. The other end is numbered 3, since $32°/10° = 3.2$. The runway is referred to as 3-21.

The object of the system is to have the number facing a landing airplane correspond (in 10° units) to the compass course of the airplane. Where there are parallel runways, the runway on the right of the landing airplane is designated with an R (right); the other is designated L (left). For example, if there were a runway parallel to 3-21, the runway would be 3R-21L or 3L-21R.

18-9. Obstruction Criteria. The Federal Aviation Administration has established standards for determining obstructions to airports in Part 77 of the Federal Aviation Regulations. These standards set up *civil imaginary surfaces* (Fig. 18-2 and Table 18-4). Objects that extend above these surfaces are considered obstructions and should be removed or marked and lighted, depending on the nature of the obstruction and the feasibility of its removal.

The **airport reference point** is a centrally located point that defines the geographic location of the airport.   The **primary surface** corresponds to a **landing;** it is a surface longitudinally centered on a runway and extending 200 ft beyond each end of the runway.   The **horizontal surface** is a horizontal plane 150 ft above the established airport elevation (the highest point of the landing area).   It is bounded by a **conical surface,** which has a width of 4,000 ft and rises on a 20:1 slope.

Approach surfaces are longitudinally centered on extended runway center lines.   The dimensions and slopes vary, depending on the nature of the runway involved (Table 18-4).   From the sides of the approach surfaces, **transitional surfaces** extend outward at 7:1 until they intersect the horizontal or conical surfaces.   The transitional surface at each end of a precision instrument runway extends beyond the conical surface for the remaining length of the approach surface and has a width of 5,000 ft.

All feasible steps should be taken to insure adequate protection of airports from obstructions above these imaginary surfaces.

18-10. Runway Clear Zones.   Clear zones are land areas that lie directly beneath the inner portions of runway approach surfaces (Fig. 18-3 and Table 18-5).   The standard configurations of clear zones conform to the inner dimensions of approach surfaces.   The length is determined by the horizontal distance required for the approach surface to reach a height of 50 ft above the terrain or 50 ft above the elevation of the end of the runway, whichever is shorter.

The airport should control sufficient property in the clear-zone areas to provide for the unobstructed passage of aircraft, landing or taking off.   All obstructions should be cleared and future obstructions should be prevented.   It is desirable to clear these areas completely, but grading

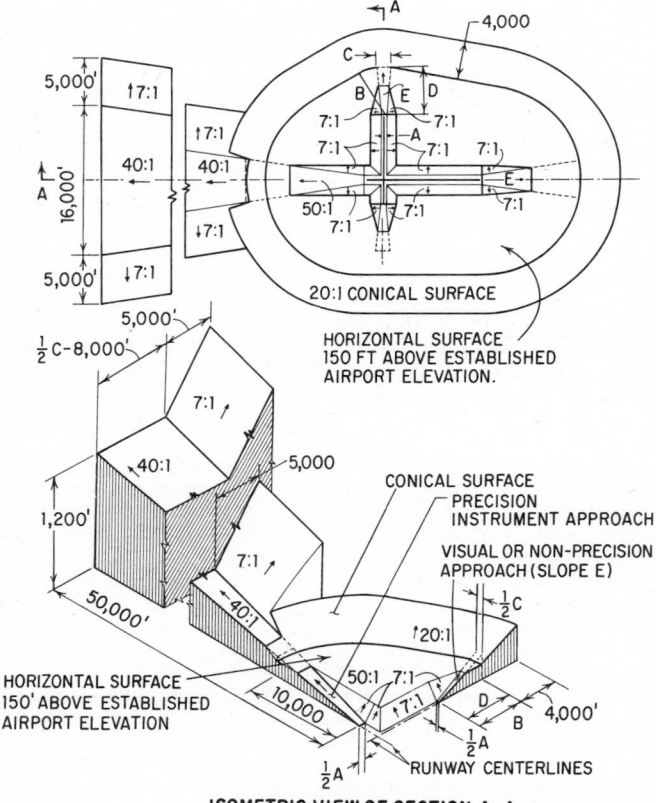

ISOMETRIC VIEW OF SECTION A–A

Fig. 18-2. Airport imaginary surfaces for determining obstructions. (*Federal Aviation Administration.*)

Table 18-4. Criteria for Airport Imaginary Surfaces for Determining Obstructions*

Dimension	Item	Dimensional standards, ft (see Fig. 18-2)					
		Visual runway		Non-precision instrument runway			Precision instrument runway
					Runways larger than utility		
		Utility runways	Runways larger than utility	Utility runways	Visibility minimums greater than ¾ mile	Visibility minimums as low as ¾ mile	
A	Width of primary surface and width of approach surface at inner end	250	500	500	500	1,000	1,000
B	Radius of horizontal surface	5,000	5,000	5,000	10,000	10,000	10,000
C	Approach surface width at end	1,250	1,500	2,000	3,500	4,000	16,000
D	Approach surface length	5,000	5,000	5,000	10,000	10,000	†
E	Approach slope	20:1	20:1	20:1	34:1	34:1	†

*Federal Aviation Administration
†Precision instrument approach slope is 50:1 for inner 10,000 ft and 40:1 for an additional 40,000 ft.

is not necessary. Ownership is desirable, but zoning or easements might give the desired protection.

18-11. Environmental Impact. Airport development is subject to state and Federal regulations that require careful consideration of environmental, ecological, and sociological matters in planning and construction. It is likely that the planning for any proposed airport development will require preparation of a Draft Environmental Impact Statement. Such a statement should include, among other things, a description of the project and a discussion of its purpose, impact on the natural environment, impact on the human environment, alternatives to the proposed development, unavoidable adverse environmental impact, short-term effects, long-term impact, irreversible or irretrievable commitments of resources, and long-term benefits. The statement should be carefully prepared, thorough and complete, unbiased, and clear, in order that its review by many public bodies and agencies not be unduly protracted.

18-12. Reconnaissance for Airport Site Selection. Before investigating possible sites in detail for an airport, the engineer should assemble certain background data. These include U.S. Geological Survey topographic maps, aerial photographs in stereopairs for studying relief and culture, available soils maps and analyses, and overall development plans of the area. Data on winds and weather should be obtained from the most reliable sources possible. It is desirable to get complete weather information for a period of at least 5 years.

The engineer should establish liaison with appropriate representatives of the Federal Aviation Administration, the state aviation agency, local and area planning groups, and the aeronautical interests that can be expected to use the airport. Finally, there must be evaluations, projections, and studies to develop forecasts of the volumes and types of anticipated activity and to establish the general size, character, and scope of the airport. With such information, a reconnaissance of the area can be made and the most likely sites identified for further study (see also Arts. 18-13 to 18-15).

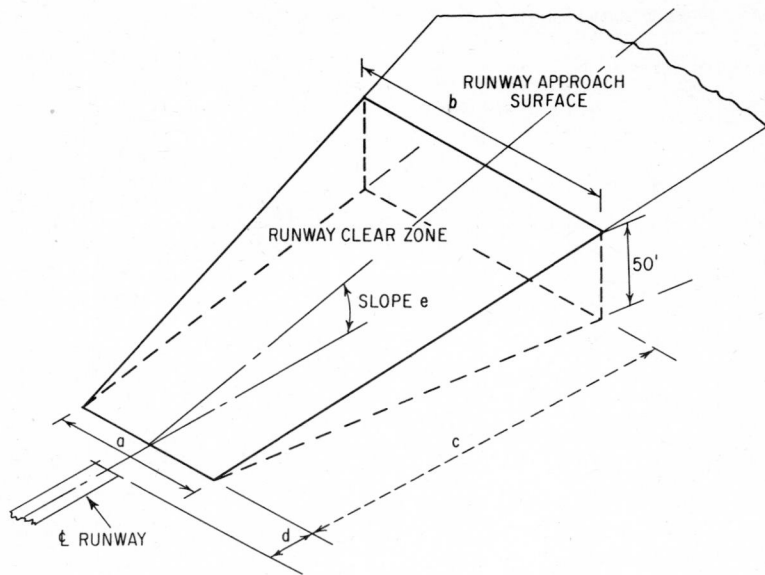

Fig. 18-3. Runway clear zones. (*Federal Aviation Administration.*)

Table 18-5. Maximum Dimensions of Runway Clear Zones*

Type of runway	Distance, ft (see Fig. 18-3)				Slope
	a	*b*	*c*	*d*	*e*
Visual runways utility	250	450	1,000	200	20:1
Larger than utility	500	700	1,000	200	20:1
Non-precision instrument runways utility	500	800	1,000	200	20:1
Larger than utility, visibility minimums greater than ¾ mile	500	1,010	1,700	200	34:1
Larger than utility, visibility minimums as low as ¾ mile	1,000	1,510	1,700	200	34:1
Precision instrument runways	1,000	1,750	2,500	200	50:1

*Federal Aviation Administration.

18-13. Physical Site Characteristics.   Selection of an airport site is influenced by a number of physical factors.   These can affect the utility of the airport and the economy of its development.

Adequate area must be provided to accommodate an airport of the type required and oriented for prevailing winds.   The area is determined by the runway length and runway layout, and by terminal-area requirements.   A small airport may be located on 50 to 100 acres.   A large international airport may cover as much as 15,000 to 40,000 acres.

Possibility for expansion should be insured by the selection of a site that is not hemmed in by built-up property, railroad yards, mountains, rivers, harbors, or other features that prohibit enlargement except at excessive cost.   Although initial acquisition should include all land needed for ultimate development, there should be ample undeveloped land available adjacent to the site. This land should be protected by zoning against uncontrolled growth of industrial or residential property that will block runway extensions or terminal-area expansion.

Terrain should be relatively flat to avoid excessive grading costs.   Elevated sites are preferable to those in lowlands, because they are usually free from obstructions in approach zones, less subject to fog and erratic winds, and easy to drain.

Soils should be studied and evaluated for their effect on grading, drainage, and pavements. The nature of the soil influences the cost of construction. Ideally, the site should be cleared ground that is easily drained and has sandy or gravelly soil that offers a satisfactory foundation for runway pavements without excessively thick subbases and costly subdrainage systems.

Drainage characteristics of the site should be investigated to ascertain the possibility of floods and the existence of high water tables. Natural drainage is most desirable. The ability to dispose of storm water should be evaluated.

Air approaches to the proposed airport should be free of obstacles, such as mountains, hills, tall buildings, transmission lines, chimneys, and towers. (R. Horonjeff, "Planning and Design of Airports," McGraw-Hill Book Company, New York.)

18-14. General Site Characteristics. In addition to the physical characteristics of an airport site (Art. 18-13), there are factors of a more general nature that require consideration.

Accessibility to the community is essential, to preserve the speed advantage of air transportation. In general, accessibility is measured in time rather than distance. Sites near modern express highways are to be sought, and those bounded by traffic-congested streets avoided. On the other hand, the site should not be so remote from the community as to require excessive transportation time.

Availability of utilities, such as electric power, gas, telephone, water, sewers, and public transportation, is an important factor to be investigated. If these utilities are not available, the cost of providing them must be considered.

Control of the site and its surroundings by zoning should be investigated to insure protection of aerial approaches and possibility of expansion. If the airport is located outside the community to be served, the means of guaranteeing proper control should be determined.

Compatibility with local and area planning is an important characteristic. It should be explored so that the airport and the area can develop without one interfering with the other. The effect on land values and tax assessments may be adverse or beneficial depending upon the nature of the site. If the airport is located near residential property, the value of that property could be affected, because of the commercial nature of some types of airports. If located in an undeveloped area, the airport will increase the value of adjacent land for industrial sites and for other uses related to the airport. The possible impact of aircraft noise should be assessed.

Spacing of the airports is a consideration, since airports should not be located so that air-traffic patterns interfere. Approval of the Federal Aviation Administration is necessary to insure airspace compatibility. This approval should be obtained before a final commitment is made for a specific airport site.

18-15. Site Evaluation. Having identified the most likely sites in an area, the engineer should review them on the basis of physical and general characteristics (Arts. 18-12 to 18-14). It is not likely that any one site will possess all the desirable characteristics. Thus, it is necessary to evaluate the good and bad features of each site to make the best selection.

Preliminary runway patterns should be tested, approaches checked, real estate evaluated, and construction costs analyzed. The more promising sites can be evaluated in the field, and specific soil and topographic data developed. Before final selection is made, the engineer should ascertain that the most favored site will receive Federal Aviation Agency air-space clearance, that an acceptable master plan can be developed for that site, and that it offers maximum compatibility with area planning.

18-16. Runway Layout. Choice of runway pattern is influenced by the necessity of obtaining clear approaches, the desirability of providing maximum wind coverage, and the necessity for fitting the layout to the topography so as to secure low grading and drainage costs. Shape and location of the terminal area also influence the layout. Furthermore, short and direct taxiing distances are desired between runways and the airport terminal.

The number of runways will depend upon wind coverage and traffic volume to be handled. To increase capacity, the layout should permit simultaneous use of two or more runways.

Orientation of the runways depends upon obstacle clearance requirements and prevailing wind directions. The instrument runway should, if possible, be aligned with the winds that prevail during instrument-flying conditions. Ideally, runway approaches should, if possible, be over sparsely settled or nonresidential areas where the public will be the least inconvenienced by aircraft operations.

18-17. Wind Coverage. The Federal Aviation Administration specifies that runways should be oriented so that aircraft may be landed at least 95% of the time with cross-wind components not to exceed 15 mph. This is considered the average maximum cross-wind component that can be safely accepted by light- and medium-weight aircraft. Larger transport aircraft can be landed safely at greater cross-wind components, but since most airports are used by light airplanes as well as transports, compliance with a 15-mph component is recommended, wherever practical.

The trend is toward one- or possibly two-directional layouts. In some localities, where the prevailing winds are consistently in one direction or the reverse, a single runway will meet FAA requirements. One-runway layouts are sometimes adopted when wind-coverage requirements are not fully met but the approaches are excellent and other factors are satisfied.

18-18. Wind Rose. To determine the orientation of a runway that will offer the greatest wind coverage, a wind rose may be used. A simple type consists of bars radiating in several compass directions, each representing, to scale, the percent of time that the wind blows from the direction in which that bar points.

For mathematical computation of wind coverage on the basis of cross-wind component, a wind rose similar to that shown in Fig. 18-4 is helpful. This wind rose gives the percentage of time the wind blows in specified speed ranges as well as in specified directions. The small numbers on the diagram represent the percentages of time the wind blows from the several compass directions between specified velocities. For the wind rose in Fig. 18-4, the percentages of winds were known for velocity ranges of 0 to 4 mph (calms), 4 to 15 mph, 16 to 31 mph, 32 to 47 mph, and over 47 mph. Winds over 47 mph accounted for less than 0.1% and were neglected.

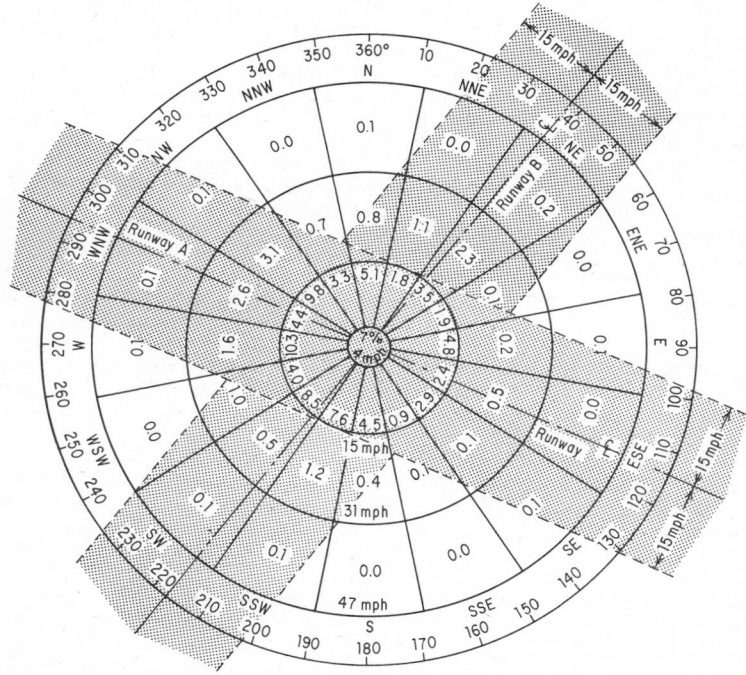

Fig. 18-4. Template aids determination of wind coverage for cross-wind component of 15 mph. (*From L. C. Urquhart, "Civil Engineering Handbook," 4th ed., McGraw-Hill Book Company, New York.*)

This wind rose may be used to determine the maximum wind coverage for a one-, two-, or three-directional runway layout. It may also be used to check the wind coverage for a layout adopted after a study of obstacles in approaches and other factors.

For finding the maximum wind coverage possible for a given runway, a transparent template is made. On it are drawn the runway center line and parallel lines representing the limits of 15-mph cross-wind components on each side of the center line. This template is then superimposed on the wind rose with the center line passing through the center of the rose. Next, the template is rotated until a direction is found in which the greatest percentage of wind is included within the 30-mph-wide band.

If the layout has more than one runway, templates are plotted for each runway and shifted about the center of the wind rose until the direction for each runway is found such that the total percentage of wind coverage by all runways is a maximum.

With Fig. 18-4, for example, a two-runway layout is to be checked for wind coverage; first for Runway A alone, and then for both Runways A and B. The runway center lines are plotted on the wind rose in their proper compass directions. Lines are drawn parallel to each center line, to represent, to the scale of the wind rose, the limits of all cross-wind components of 15 mph. For simplicity, the percentage of winds not covered is computed and deducted from 100. The percentages and fractions of percentages outside the limits of coverage (dashed lines in Fig. 18-4) for Runway A are as follows: in directions NW to E, $0.4 \times 0.1 + 0.0 + 0.6 \times 0.7 + 0.1 \times 0.9 \times 0.8 + 0.0 + 1.1 + 0.2 + 2.3 + 0.0 + 0.8 \times 0.1 + 0.6 \times 0.1 + 0.1 \times 0.2$; from SE to W, $0.4 \times 0.1 + 0.0 + 0.5 \times 0.1 + 0.0 + 0.9 \times 0.4 + 0.1 + 1.2 + 0.1 + 0.9 \times 0.5 + 0.0 + 0.6 \times 1.0 + 0.6 \times 0.1 + 0.1 \times 1.6 = 8.16$ or 91.84% coverage. The addition of Runway B will add the following coverage: from N to ENE, $0.5 \times 0.8 + 0.0 + 1.1 + 0.2 + 2.3 + 0.0 + 0.6 \times 0.1$ and from S to WSW $0.5 \times 0.4 + 0.8 \times 0.1 + 1.2 + 0.1 + 0.9 \times 0.5 + 0.0 + 0.4 \times 1.0 = 6.49$, giving total coverage for two runways of 98.33%.

The analysis may be refined by using more wind-velocity groups if they are available. It may also be applied for other cross-wind components.

The wind rose usually employed for study purposes is plotted on an annual basis. In locations where the wind distribution varies during the year, roses should be plotted for the different seasons and the fluctuations taken into account in design, particularly if the airport is used mostly in certain seasons.

For selecting the instrument-runway orientation, a wind rose for low-visibility conditions is useful and can be developed from special studies undertaken by the U.S. Weather Bureau.

(R. Horonjeff, "Planning and Design of Airports," McGraw-Hill Book Company, New York.)

18-19. Clearance of Obstructions. To test approach zones (Art. 18-9) for clearance of obstructions, a topographic map of the airport site and its environs is required for a radius of at least 5 miles from the airport boundary. A convenient test method is to prepare a transparent template showing the extension of the runway center line, the limits of the runway approach surface, and contour lines representing the elevations of the sloping runway approach surface and 7:1 transition surface. For an instrument runway approach, the transparent template (Fig. 18-5) is fitted to the end of each runway, and the ground-surface contours are compared with those of the runway approach surface. Any high places or man-made features on the ground that will protrude into the runway approach surface are noted. The runway layout is adjusted, if necessary, to avoid obstacles with a minimum sacrifice of wind coverage.

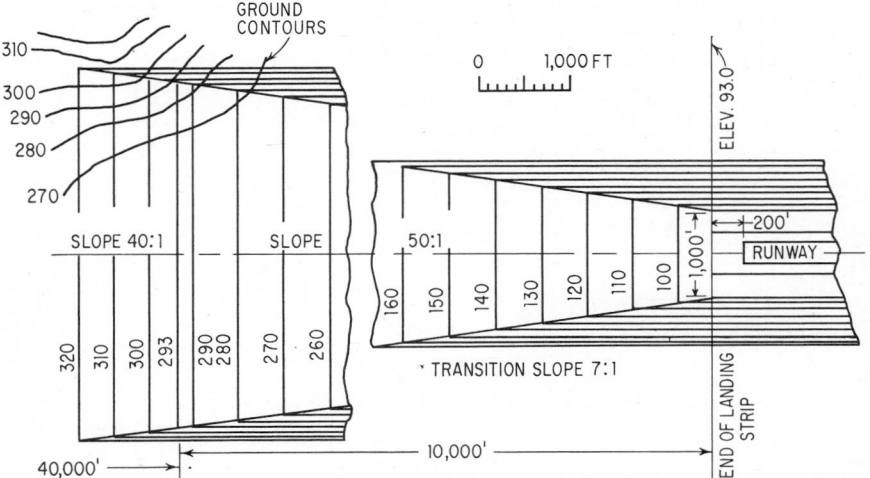

Fig. 18-5. Template for checking approach-zone clearance for instrument runways. Similar templates can be developed for noninstrument runways. (*From L. C. Urquhart, "Civil Engineering Handbook," 4th ed., McGraw-Hill Book Company, New York.*)

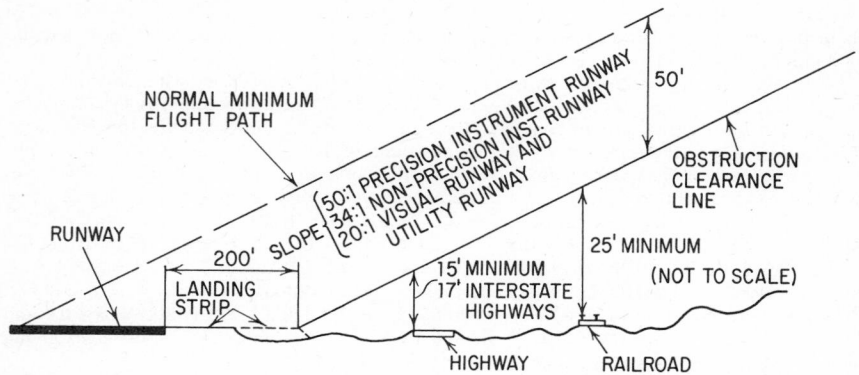

Fig. 18-6. Vertical profile along extended center line of runway shows minimum clearance required by Federal Aviation Administration over highways and railroads.

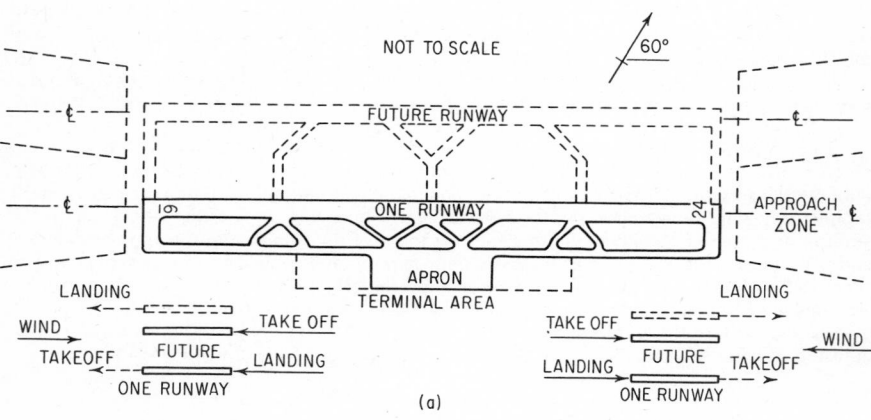

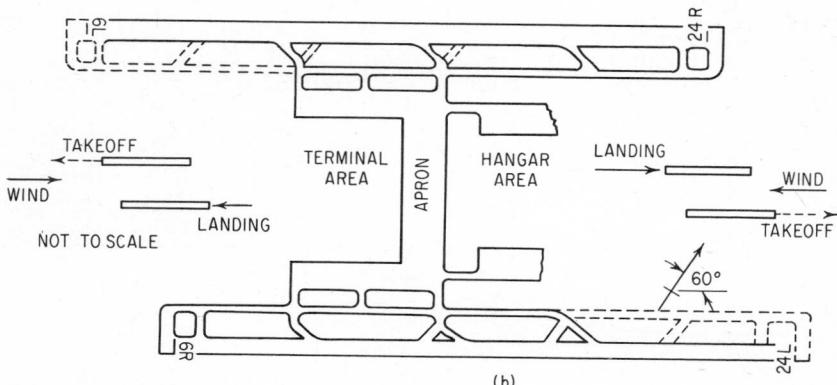

Fig. 18-7. Plan of simple runway layouts. (*a*) Single runway with future parallel runway. (*b*) Two parallel runways. (*From L. C. Urquhart, "Civil Engineering Handbook," 4th ed., McGraw-Hill Book Company, New York.*)

The horizontal surface clearances, 150 ft above the airport (Art. 18-9), are examined in a similar manner. All obstructions above the horizontal surface are spotted. Measures should be taken to remove as many obstructions as possible, and to mark and light those that cannot be removed.

Detailed plans should be made of critical areas in approach zones. The plans should show heights of trees, poles, buildings, etc., that come near the runway approach surface. Steps should then be taken to obtain control of these areas by easement or purchase, so that the obstructions may be removed. Clearances for railroads and highways are shown in Fig. 18-6.

18-20. Runway Configurations. The simplest layout is a single runway with parallel taxiway and centrally located terminal area as shown by full lines in Fig. 18-7a. Two directions of operation are possible, 6-24 or 24-6 (Art. 18-8). Only one landing or takeoff can be made at a time.

Under these conditions, the capacity of the runway is about 50 movements per hour (including both landings and takeoffs). When more capacity is needed, a second parallel runway may be built as shown by dashed lines in Fig. 18-7a.

In this design, the original runway can be used for takeoffs, while the "future" runway is used for landings. The capacity under visual flight rules will be raised to about 70 movements per hour. Landing traffic will have to cross the takeoff runway under control from the tower.

Figure 18-7b shows parallel runways 5,000 ft apart. The terminal area lies between the runways. This arrangement has definite operational advantages over the layout in Fig. 18-7a. Taxiways do not cross runways, the terminal area is centrally located with ample room for expansion, and the wide separation of runway approaches will increase capacity under conditions of low visibility, since the 5,000-ft separation is adequate for simultaneous operations. But the layout in Fig. 18-7b requires a larger area than the one in Fig. 18-7a. The two parallel runways, however, need not be opposite each other. Increasing the offset from the terminal area will decrease taxiing distance but may increase land and construction costs.

Taxiways may be extended to the runway ends to provide exits for incompleted takeoffs, to facilitate landings and takeoffs on the same runway, and to permit simultaneous use of both runways for takeoff or for landing. During peak-hour operations, arrivals and departures are not usually equal, so that simultaneous use of both runways for the same type of operation is often desirable.

In Fig. 18-8 an open V-type layout is shown. This layout gives four directions of wind coverage and also allows simultaneous operation of runways in most directions when wind velocities are not unusually high. The traffic diagrams indicate a separation of landings and takeoffs in three

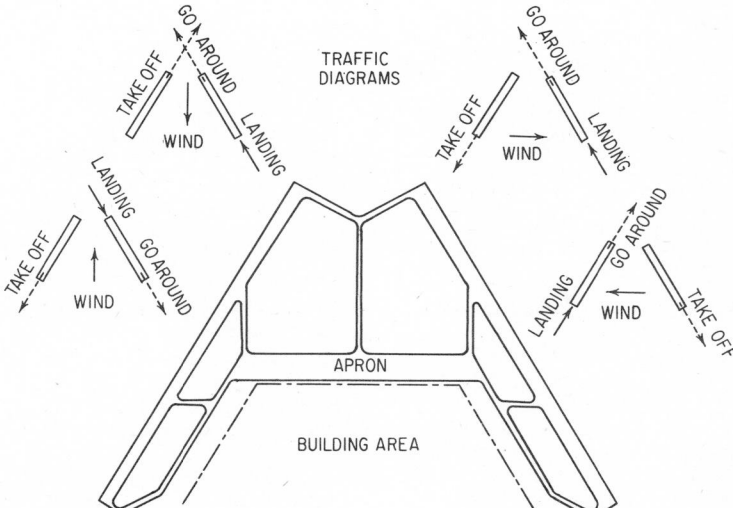

Fig. 18-8. V-type runway layout permits two-directional operation of aircraft. (*Federal Aviation Administration.*)

or four wind directions. In the one situation where the landing go-around path intersects the takeoff path, the landings and takeoffs will have to be rigidly coordinated.

The V shape permits a centrally located terminal area with room for expansion. In some designs the angle of the V is made about 90°.

When additional capacity is required, the designs in Figs. 18-7*b* and 18-8 may be expanded by building a runway parallel to each of the original runways but 1,000 to 3,500 ft farther out. Two runways would then be available for landings and two for takeoffs at all times for the layout in Fig. 18-7*b*, and for most of the time for the layout in Fig. 18-8. The greatest capacity can be obtained from the two sets of parallels for the configuration shown in Fig. 18-7*b* with a third runway at a divergent angle on each side.

Most existing airports have intersecting runways. At some locations, it is impractical to build nonintersecting runways. When winds are not critical, the capacity of these designs can be improved over single-runway operations by using one runway for takeoffs and another for landings. The movements are alternated under rigid coordination from the traffic-control tower.

Airport capacity is reduced under instrument-landing procedures, and delays to landings occur. Improvements in air-traffic control, however, have increased landing rates in overcast weather so they nearly equal those in good weather.

18-21. Taxiway Systems. Taxiways are laid out to connect the terminal area with ends of runways for takeoffs and to tap the runways at several points to provide exits for landing aircraft. Landings usually do not require the full length of the runway.

To clear a runway of landing planes as rapidly as possible, easy turns are introduced at exit taxiways (Fig. 18-7). Even faster aircraft exits are obtained when the runway is equipped with the taxiway illustrated in Fig. 18-9. This is designed for runway exit at 60 mph. These exit taxiways best serve a variety of aircraft when placed about 2,500, 4,000, and 6,000 ft from the runway threshold.

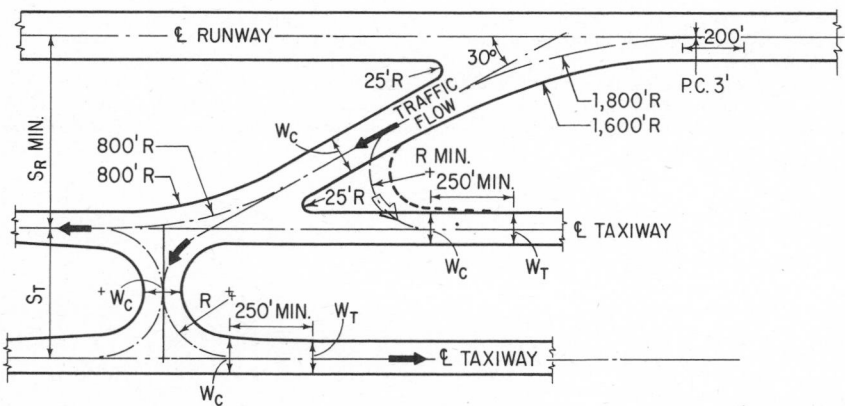

Fig. 18-9. Angled-exit taxiway design with dual parallel and crossover. (*Federal Aviation Administration.*)

Where there is a taxiway parallel to the runway, the exit taxiways can lead into the parallel taxiway with a reverse curve that permits the maintenance of high-speed taxi operations. When applied bidirectionally to the same runway, the effect can be that in Fig. 18-7*a*. At the ends of a runway, the taxiways join the runway at about 90° to give the pilot a view of the runway and its extension in both directions. Additional pavement is added to make room for waiting airplanes and to allow one airplane to pass another in the takeoff sequence. Taxiway widths and clearances are given in Table 18-6.

18-22. Airport Layout Plan. Every airport should have a layout plan showing ultimate development, even though construction is to be in stages. Such a plan is desirable to insure an orderly development and an economical and functionally sound airport. All major components should be worked out in advance.

Table 18-6. FAA Minimum Dimensional Taxiway Criteria for Air Carrier Airports

Design item	Symbol	Dimensional criteria, ft—airplane design group			
		Group I B-727-100 B-737-200 DC-9-30 BAC-1-11	Group II DC-8-63 B-707-300 B-727-200	Group III B-747 L-500 L-1011 DC-10	Group IV Anticipated future aircraft
1. Taxiway structural-pavement width on tangents	W_T	50	75	100	125
2. Taxiway structural-pavement width on curves	W_C	65	90	115	140
3. Taxiway shoulder width		20	25	35	40
4. Safety area width		110	150	220	310
5. Taxiway and apron taxiway obstacle-free area width		210	270	360	470
6. Terminal taxi-lane obstacle-free area width		160	210	290	390
7. Separation distance c to c of taxiways	S_T	200	300	300	400
8. Separation distance c to c between taxiway and runway*	S_R	400	400	600	1,000
9. Radius of taxiway center line curves	R	100	150	150	200

*Where Category II operation is anticipated, use at least 600 ft.

The airport layout plan is the basic element of the airport's master plan and shows all existing and proposed facilities, property lines, topography, utilities, airport approach surfaces and clear zones, in addition to the ultimate runway and taxiway layout. The ultimate plan will provide a basis for acquiring ample land and for determining zoning required to protect future approaches. The plan should be flexible enough to permit modifications between stages of construction to meet the changing demands of air transportation.

18-23. Airport Zoning. In the planning of any airport, it is important that all existing obstructions to air navigation be cleared or marked and lighted, and that future obstructions be prevented. Where legally possible, steps should be taken to adopt appropriate airport zoning legislation to prevent the establishment of obstructions to air navigation. Ideally, the zoning will be developed concurrently with the layout plan. If comprehensive zoning is in force or can be instituted, height restrictions and land use can both be incorporated.

18-24. Airport Construction Plans. Construction plans for airports should include a location and site plan, airport layout plan, airport protection and zoning plan, clearing plan, borings and soils-exploration plot, grading and drainage plan, runway and taxiway profiles, access-road plans and profiles, drainage-line profiles, pavement cross sections, drainage structures, lighting and conduit plan, turfing plan, and summary of construction quantities. Plans are also required for development of terminal area, and parking lots, and for construction of terminal buildings.

18-25. Terminal Concepts. Transition of passengers from ground to air occurs in the terminal area. Various methods are used to accommodate and transfer the public and its goods, arriving either by air or by ground, and to provide for the parking, servicing, and storage of aircraft and vehicles used in ground transportation. The degree of development in the terminal area varies with the volume of airport operations, the type of traffic using the airport, the number of people to be served, and the manner in which they are to be accommodated.

The concept of a very small airport might involve only a hangar with simple office facilities, adequate for limited aeronautical activity. At larger airline terminals, demands are greater. The concept can involve bilevel terminal operations, auto parking in buildings, and elaborate passenger-loading devices. Various concepts of terminal systems are shown in Figs. 18-10 and 18-11.

Frontal layout of facilities is usual at airports of low activity. In the small-airport layout (Fig. 18-10a) the facilities required to serve a moderate volume of general aviation are in a row along the boundary road. At many small airports, the terminal is eliminated and its functions housed in a lean-to of a service hangar. At such an airport, the terminal (or lean-to) would usually have a waiting room, rest rooms, office for the airport manager or flight service operator, and perhaps a restaurant, snack bar, or vending machines.

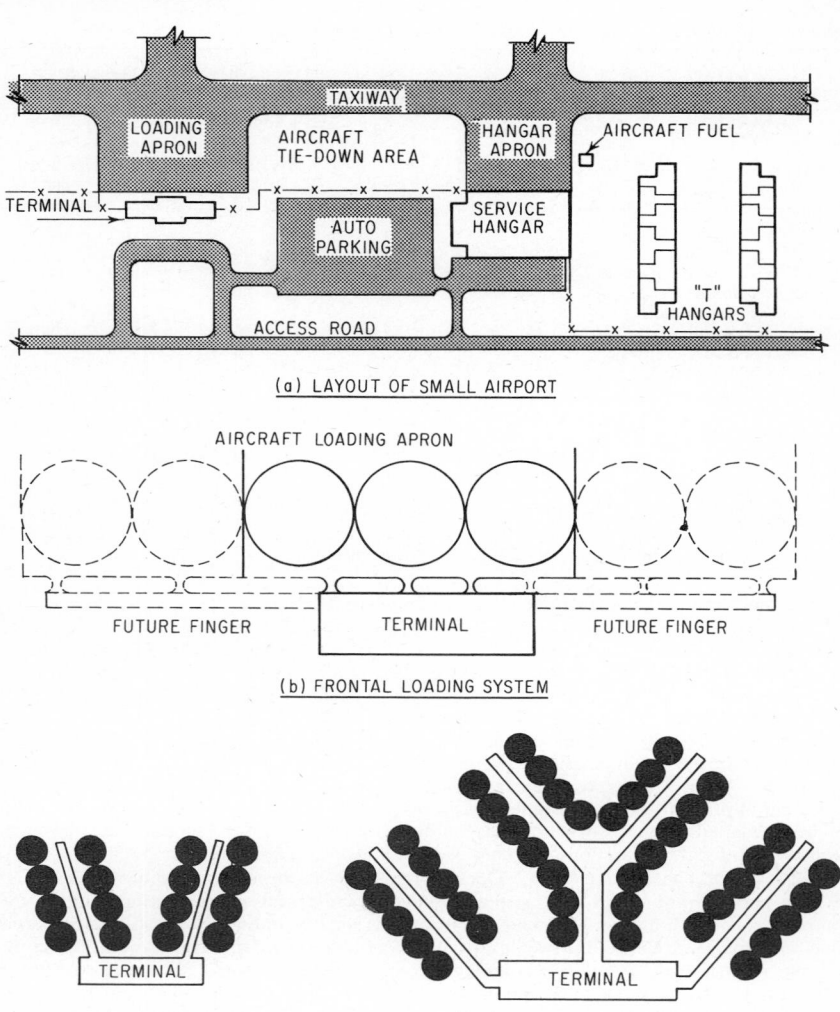

Fig. 18-10. Simple terminal systems. Fingers are added to increase parking capacity for aircraft.

At airline terminals of low activity, a frontal loading system as shown in Fig. 18-10*b* is usually preferred. Expansion possibilities are indicated. As the fingers are extended, however, the passenger walking distances increase. Likewise, the finger structure becomes less economical, since loading positions are on only one side.

Finger systems project onto the parking apron and permit aircraft to park closer to the terminal. This arrangement reduces structural cost since loading is accomplished on both sides.

The finger system shown in Fig. 18-10*c* is a simple solution for a feeder airport. Walking distances to the extreme end positions, however, could be rather long.

A more elaborate finger layout is the split-finger system (Fig. 18-10*d*). Here the passenger walking distances become quite long. A passenger transferring from the end-loading position of one finger to the end-loading position of another finger would walk more than ½ mile, assuming the aircraft parking positions to be 200 ft in diameter. Walking distances are inevitably long at centralized terminals that serve large numbers of gates, unless mechanical transfer of passengers is employed.

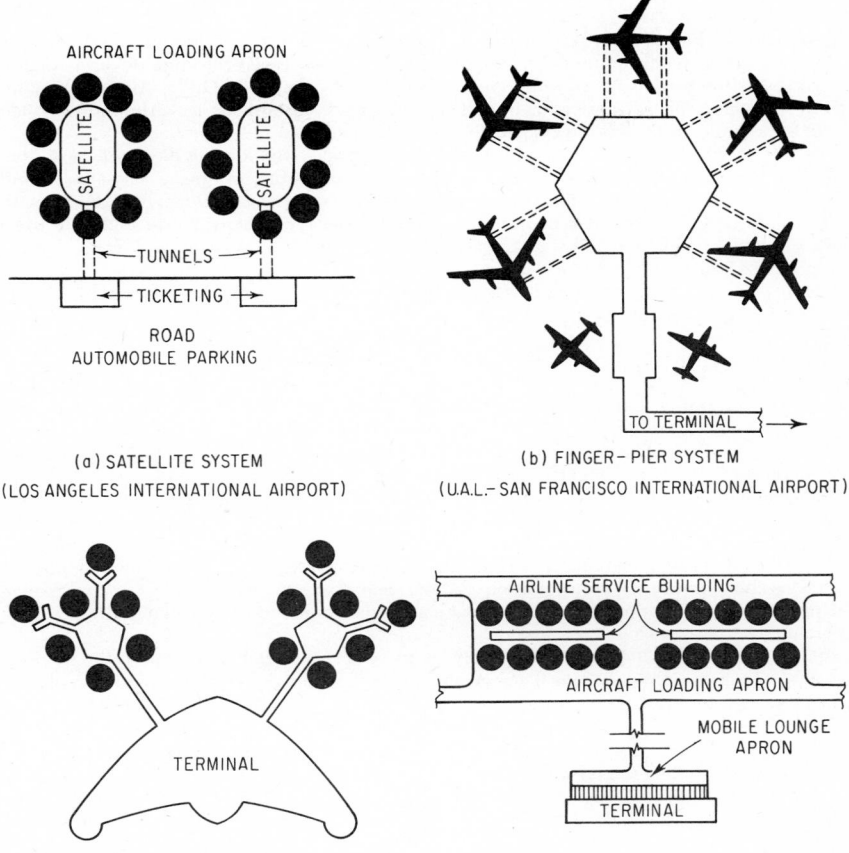

Fig. 18-11. Terminal systems used at some international airports.

Unit terminals concentrate aircraft parking positions and minimize passenger walking distance except where interunit transfers are required. The movement from one unit terminal to another can involve excessive time and distance. Unit terminals are generally designed so that each unit is a self-contained entity.

Satellite terminals also concentrate aircraft parking positions in an effort to minimize walking distances. The satellites shown in Fig. 18-11a are fed by tunnels from the ticketing area and provide a number of aircraft parking positions without excessive walking distances.

In Fig. 18-11b the satellite is a pier at the end of a finger and concentrates parking positions, with a resultant saving in walking distance. The terminal layout shown in Fig. 18-11c has two piers to serve 14 loading positions with relatively short walking distances. The pier-satellite approach offers minimum passenger walking distances for a large number of gate positions.

Remote parking of aircraft minimizes walking distances by using a vehicle to transport passengers from terminal to airplane. At some European airports, buses of varying design accomplish the transfer.

In an elaborate scheme under the remote concept, a mobile lounge moves passengers to and from aircraft parked some distance from the terminal (Fig. 18-11d). The mobile lounge is in use at Dulles International Airport, Washington, D.C. Parked at the terminal, it serves as a waiting room. At flight time, it is driven to the airplane and passengers move directly into the aircraft.

Passenger-loading devices permit weatherproof transfer from terminal to aircraft, usually with no change in level required. The *mobile lounge* is one type of loading device. Figure 18-12*a* shows the terminal end of the type of lounge vehicle used at Dulles Airport. Figure 18-12*b* shows passengers transferring from lounge to aircraft. Local-service airlines using small transport aircraft can be accommodated directly at the terminal.

The *telescoping gangplank,* shown in Fig. 18-12*c* and *d,* is the loading device in most general use. The telescoping passage has a swivel connection at the terminal. The aircraft end rides on a full-swivel gear that is electric-powered. Parallel parking is shown in Fig. 18-12*c,* with gangplanks serving front and rear doors of the aircraft. Airplanes that park at an angle may use a single gangplank, as shown in Fig. 18-12*d.* In both instances the aircraft can taxi into and out of gate positions.

The *nose loading devices* shown in Fig. 18-12*e* and *f* permit aircraft to taxi into the parking position, but aircraft must be towed away from the gate. Deplaning the passengers is faster with nose loading devices, but the departure from the gate is slower.

In Fig. 18-12*e,* an adjustable transfer device suspended from a canopy on the outside of the terminal moves only a few feet to the aircraft door. The nose loading device in Fig. 18-12*f* is pivoted at the terminal and supported on fixed, powered wheels at the aircraft end. When not in use, the device is stored against the wall of the terminal and swings into position to connect to the doorway of the aircraft. Experience has shown that aircraft can be precisely taxied into parking positions so that elaborate adjustments are not required.

The canopy device (Fig. 18-12*e*) is the least expensive, but the swivel type (Fig. 18-12*f*) can serve a greater variety of aircraft since it can serve a wider span of aircraft heights because of the longer ramp.

Other types of passenger-transfer devices include moving sidewalks in fingers and other places where feasible, horizontal transportation systems that connect unit terminals and satellites, and similar systems that can serve individual loading positions, to keep walking distances at a minimum.

18-26. Terminal-Building Layout. The key feature of any terminal-area layout is the terminal building. In size, it can be small for airports with low activity, or large and complex at primary-system terminals.

The terminal should be planned to serve the number of peak-hour passengers forecast for 10 years in the future. Flexibility and expandability are paramount requirements.

The terminal building should provide a smooth flow of passengers from parking lot to aircraft. The passenger should be able to park, or get out of a taxi, bus, or limousine, at a point near the ticket counter. Baggage is checked at this point. Then the passenger proceeds to the aircraft via a waiting room where rest rooms, telephones, concessions, and restaurant facilities should be available. At the loading position, there should be a hold room where the passenger may be processed for boarding the scheduled flight.

Deplaning passengers go directly from the aircraft to the baggage-pickup area, then proceed to taxicab, bus, limousine, or parked automobile. Automobile rental counters should be near the baggage-pickup areas, and there should be telephones and rest rooms nearby.

Visitors should be provided with observation decks. The need for concession, restaurant, and office space will vary at each location. A greater variety of concession potential will obviously develop at the larger airports.

Airline facilities will include ticket counters, ticket offices, baggage-processing areas (with baggage usually mechanically conveyed from the ticket counters), and operational space at the loading position. Inbound baggage should be available to the passenger at a convenient location, either placed by hand on a claim counter or mechanically conveyed by belt with spacers, diverters, or carousels for delivery to the arriving passenger.

At small airline terminals, the entire operation is at a single level. Larger terminals tend to have elevated roadways so that departing passengers enter the terminal at the second level and enter the aircraft at the same general elevation by means of a loading device. Deplaning passengers leave the aircraft at the second level, and escalate down to the ground floor for baggage pickup and ground transportation. There are many variations of this scheme, but the pattern is the same.

The accommodation of Federal Aviation Administration air-traffic-control quarters, as well as weather facilities, will vary from one location to another. There is a trend toward locating these government facilities in separate structures, away from the terminal but nearer to the general

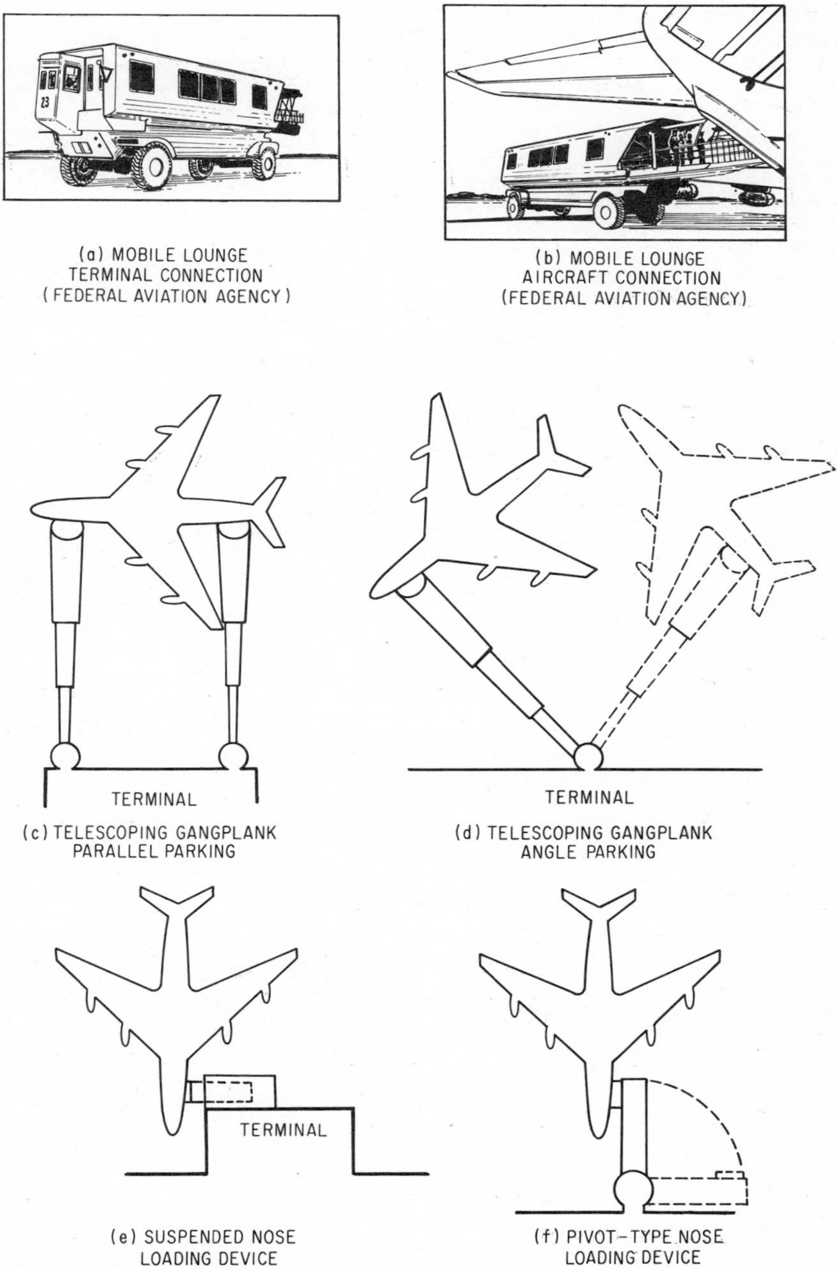

(a) MOBILE LOUNGE
TERMINAL CONNECTION
(FEDERAL AVIATION AGENCY)

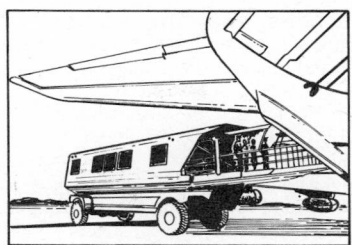

(b) MOBILE LOUNGE
AIRCRAFT CONNECTION
(FEDERAL AVIATION AGENCY)

TERMINAL

(c) TELESCOPING GANGPLANK
PARALLEL PARKING

TERMINAL

(d) TELESCOPING GANGPLANK
ANGLE PARKING

TERMINAL

(e) SUSPENDED NOSE
LOADING DEVICE

(f) PIVOT—TYPE NOSE
LOADING DEVICE

Fig. 18-12. Passenger-loading devices for aircraft.

aviation activity. There is no fixed pattern. The need for such space in the terminal building varies from location to location.

18-27. Access Roads. In preparing a terminal-area layout, the engineer should recognize the importance of vehicular access. The area should be located so that full advantage is gained from freeways and other roads, planned or existing, that will expedite ground transportation to the airport.

Within the airport, the access-roadway system should provide a connection between the terminal area and the best routes to town. The system also should include roads for inter-communication between the separate facilities. Separation of passenger and commercial traffic is desirable, as well as separation of patron, spectator, and employee traffic.

18-28. Automobile Parking Areas. Ample parking facilities are required for airport patrons, passengers, employees, and spectators. Public parking should be developed as near the airline terminal as feasible, to minimize walking distance. Most visitors will come on Sundays and when special events occur at the airport.

The parking lot should be designed to handle overflow traffic, or a supplemental lot should be developed for intermittent use. A design criterion of 150 parked automobiles per acre may be used when estimating the size of parking lot required.

To minimize walking distances, some airports have multilevel parking structures adjoining the terminal. Employee parking facilities are usually separate and more distant.

At busy terminals, temporary storage areas will be needed to park taxicabs, buses, and limousines waiting for turns or for scheduling. Parking might be required for service vehicles such as fuel trucks. There should be adequate truck-parking areas at the terminal for delivery of commodities and supplies.

18-29. Aprons. The apron or "ramp" adjacent to the terminal is used for loading and unloading airplanes, fueling, and minor servicing and checkup. The dimensions of the apron depend on the number of loading positions required and the size and turning characteristics of aircraft. The number of spaces depends on the time of occupancy per aircraft, the time being longer at terminal airports than at en-route stops.

In most instances, airlines desire exclusive use of apron positions, because of the complex equipment required to service transport aircraft. The resultant need is for a greater number of loading positions than would be required if positions were shared.

In determining area requirements for aprons, various methods of aircraft positioning may be explored. The size of airline loading aprons depends upon the number and size of aircraft to be accommodated, as determined from a forecast of peak-hour aircraft movements. Aircraft loading positions are designated by circles of varying diameters, depending on wing span, length, and turning radius of the aircraft that will use the airport.

Provision of underground facilities in the apron is a requirement at some airports. At others, services such as fuel, air, power, and telephone are available at the edge of the apron or from the terminal building. Grounding connections should be provided.

18-30. Hangars. The size of hangars depends upon the dimensions and numbers of aircraft to be serviced. Airports for general aviation usually have one or more service hangars that hold several aircraft, for which repair and maintenance operations are conducted. These hangars are supplemented by nests of T hangars, which provide individual stalls for aircraft storage. At larger airports, the trend is toward cantilever hangars, capable of accommodating the largest aircraft.

Table 18-7 gives the gross weight, wing span, length, and height of typical aircraft. Jet transport aircraft are being produced in short-, medium-, and long-haul versions. The larger jets are becoming even larger, with fuselage increases in excess of 30 ft and with gross weights exceeding 350,000 lb. Supersonic aircraft will exceed the length and weight of the stretched-out jet aircraft. One model is projected to have a length of about 300 ft, a wing span of about 120 ft, and a gross weight of nearly 500,000 lb. Hangars to serve such aircraft must have built-in flexibility.

18-31. Cargo Buildings. At many airports, air cargo is handled through the terminal building. Where separate cargo facilities have been developed, they usually have been located adjacent to terminal areas. Size and type of cargo facilities vary, depending on local need. Most are long, low structures with truck docks on one side and aircraft parking on the other. The roadway level on the truck side should be depressed to provide a truck-high floor for easy loading and unloading.

These separate cargo buildings not only provide facilities to load cargo directly into aircraft on the adjacent apron but also contain facilities for sorting out small freight shipments to be taken to the terminal area on small carts and placed aboard passenger aircraft.

Table 18-7. Physical Data for Selected Aircraft

Name and model	Gross weight, lb	Wing span	Length	Height
Single Engine, Prop.				
Beech Bonanza	3,125	33 ft 5 in.	25 ft 2 in.	7 ft 7 in.
Cessna 210	2,900	36 ft 7 in.	27 ft 9 in.	8 ft 8 in.
Piper Comanche	2,900	36 ft 0 in.	24 ft 11 in.	7 ft 3 in.
Multi-Engine, Prop.				
Aero Commander	8,000	49 ft 0 in.	35 ft 1 in.	14 ft 6 in.
Beech Baron	4,880	37 ft 10 in.	26 ft 8 in.	9 ft 7 in.
Cessna 310	4,830	36 ft 0 in.	29 ft 7 in.	9 ft 11 in.
Piper Aztec	4,800	37 ft 0 in.	27 ft 7 in.	10 ft 3 in.
Executive Jets				
Lockheed Jetstar	35,000	54 ft 5 in.	60 ft 5 in.	20 ft 5 in.
Grumman Gulfstream II	51,340	68 ft 10 in.	79 ft 11 in.	24 ft 6 in.
Learjet 25	13,300	35 ft 7 in.	47 ft 7 in.	12 ft 7 in.
Rockwell Sabreliner	17,500	44 ft 5 in.	43 ft 9 in.	16 ft 0 in.
Airline Transports				
B-737-200	100,800	93 ft 0 in.	100 ft 0 in.	36 ft 9 in.
B-727-200	173,000	108 ft 0 in.	153 ft 2 in.	34 ft 0 in.
DC-9-30	109,000	93 ft 4 in.	107 ft 0 in.	27 ft 6 in.
B-707-300	316,000	142 ft 5 in.	152 ft 11 in.	42 ft 2 in.
DC-8-63	358,000	148 ft 5 in.	187 ft 5 in.	43 ft 0 in.
B-747	775,000	195 ft 8 in.	229 ft 2 in.	64 ft 8 in.
L-1011	432,000	155 ft 4 in.	178 ft 8 in.	55 ft 10 in.
DC-10-30	555,000	161 ft 4 in.	181 ft 11 in.	59 ft 7 in.

At smaller airports, the cargo is handled at the terminal building and is carried only by passenger planes.

18-32. Service Buildings. At most airports, some form of fire-crash-rescue facilities is required. They should be provided at a location having ready access to all parts of the airport. Other buildings that might be required are heating plant, utility buildings, maintenance buildings, equipment-storage buildings, electrical equipment, and transformer vaults.

18-33. Soils Investigations for Airports. A thorough analysis of the soils on an airport site is required for planning of grading and drainage and subdrainage systems, and for designing pavements and base courses. Soils testing is also required for the control of compaction of fills and base courses so that there will be no detrimental settlement under heavy airplane loads.

Procedures for soil sampling and testing are much the same as for highways. Samples should be taken at 200-ft intervals along the center lines of planned runways and taxiways, one boring per 10,000 sq ft for other areas of pavement. Borrow areas should be tested sufficiently to define the borrow material clearly. Results of such tests are plotted on soils profiles or on a boring plan. This plan shows locations of borings with respect to proposed runway layout and individual profiles of the soil layers at each location, with a description of each soil type.

Soil classifications for airports have been adopted by the Federal Aviation Administration. These are useful for correlating soil types and conditions with pavement and base-thickness requirements (Table 18-8). The classifications are supplemented by a chart (Fig. 18-13) to aid in classifying fine-grained soils. In the FAA classification, the soils are arranged in order of decreasing desirability as subgrades. ("Airport Paving," Federal Aviation Administration.)

18-34. Airport Grading. The surface of an airport should be relatively smooth but well drained. Few natural sites provide these ideals; hence, proper grading is important. Grading plans and drainage plans must be carefully coordinated.

The grading plans should consist of runway and taxiway center-line profiles, cross sections showing areas of cut and fill, and a topographic map showing initial and final contours. This latter map becomes the basis of the drainage-layout plan.

Cross sections of runways and taxiways should slope transversely each way from the center line to provide for surface drainage. Paved surfaces should slope 1.5%, and graded areas of the landing strip 1.5 to 2.0%.

Side slopes of cuts and fills should be as flat as possible. In cuts, the sides should not encroach on a lateral clearance ratio of 7:1 measured normal to the edge of the landing strip.

Table 18-8. Federal Aviation Administration Classification of Soils for Airport Construction*

| Soil group† | Material retained on No. 10 sieve, percent‡ | Mechanical analysis | | | Liquid limit | Plasticity index | Subgrade class | | |
| | | Material finer than No. 10 sieve, percent | | | | | Good drainage§ | Poor drainage§ | |
		Coarse sand passing No. 10 retained on No. 40	Fine sand passing No. 10 retained on No. 200	Combined silt and clay passing No. 200			No frost or frost§	No frost	Frost§
E-1	0–45	40+	60–	15–	25–	6–	Fa or Ra	Fa or Ra	F1 or Ra
E-2	0–45	15+	85–	25–	25–	6–	Fa or Ra	F1 or Ra	F2 or Rb
E-3	0–45	—	—	25–	25–	6–	F1 or Ra	F2 or Rb	F3 or Rb
E-4	0–45	—	—	35–	35–	10–	F1 or Ra	F2 or Rb	F4 or Rb
E-5	0–55	—	—	45–	40–	15–	—	F3 or Rb	F5 or Rb
E-6	0–55	—	—	45+	40–	10–	—	F4 or Rc	F6 or Rc
E-7	0–55	—	—	45+	50–	10–30	—	F5 or Rc	F7 or Rc
E-8	0–55	—	—	45+	60–	15–40	—	F6 or Rc	F8 or Rd
E-9	0–55	—	—	45+	40+	30–	—	F7 or Rd	F9 or Rd
E-10	0–55	—	—	45+	70–	20–50	—	F8 or Rd	F10 or Rd
E-11	0–55	—	—	45+	80–	30+	—	F9 or Re	F10 or Re
E-12	0–55	—	—	45+	80+	—	—	F10 or Re	F10 or Re
E-13	Muck and peat—field examination						Not suitable for subgrade		

*From "Airport Paving," Federal Aviation Administration, September, 1971.

†Soils E-1 to E-4 are classed as granular, and E-5 to E-12 as fine-grained.

‡Determination of sand, silt, and clay fractions is made on that portion of the sample passing a No. 10 sieve. If percentage of material retained on No. 10 sieve exceeds that shown, the classification may be raised if such material is sound and fairly well graded.

§See Art. 18-37. The prefixes R and F indicate subgrade classes for rigid and flexible pavements.

Properly designed grades can develop low areas that may be used for temporary ponding of storm runoff in the interest of a more economical storm-sewer system. Typical cross sections of runways are shown in Fig. 18-14. ("Airport Paving," Federal Aviation Administration.)

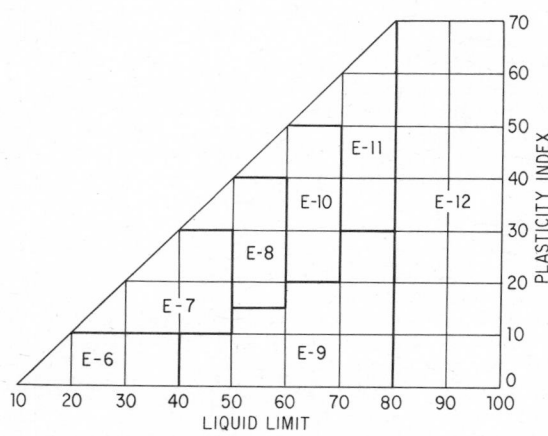

Fig. 18-13. Classification chart for fine-grained soils. (*"Airport Paving," Federal Aviation Administration.*)

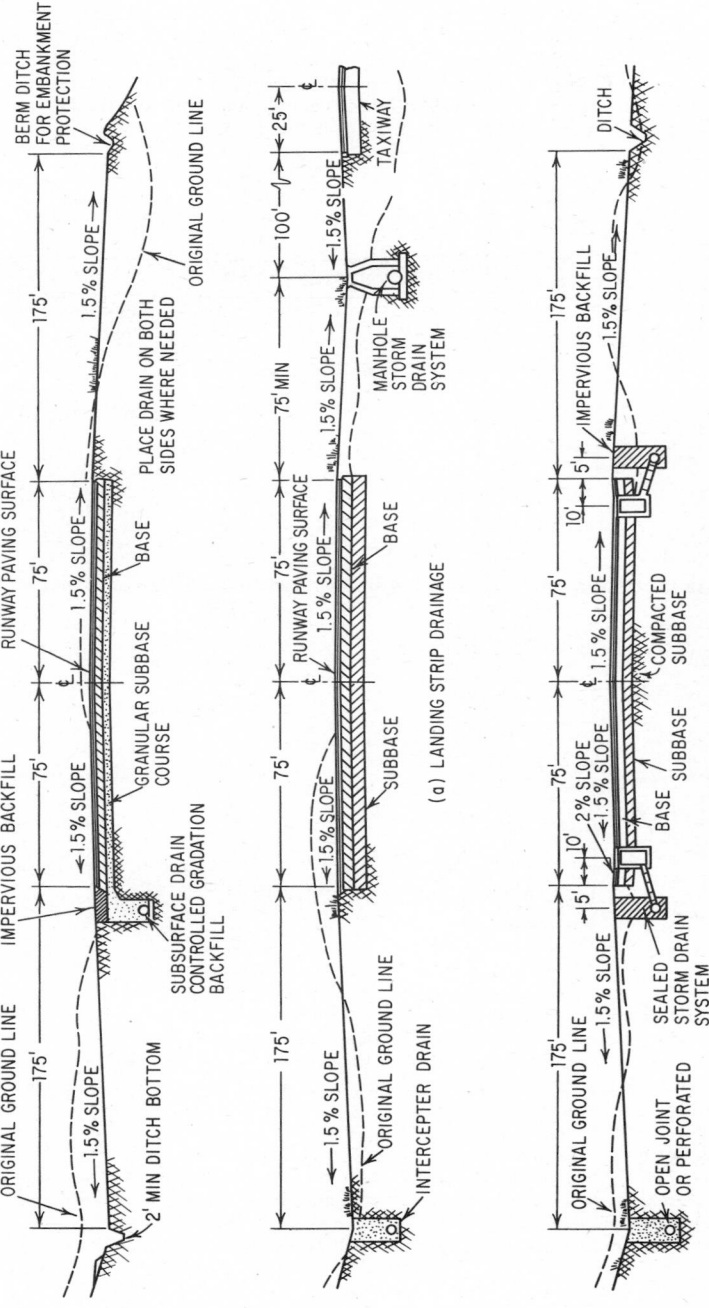

Fig. 18-14. Typical runway cross sections. (*Federal Aviation Administration.*)

18-35. Airport Drainage. With proper grading the surface runoff is drained into collector sewers or ditches. Runoff is usually collected along the edges of landing strips in shallow ditches leading to inlets piped to storm sewers (Fig. 18-14a). At some of the larger airports, with wide paved runways, the surface water is also collected along the edge of the runways (Fig. 18-14b), particularly in northern climates, where snowbanks along the edges of the runway block drainage across the landing strip. Surface drainage inlets may be placed just outside the edges of runways, or they may be set in a shallow depression built in the outer edge of the pavement (Fig. 18-15). Inlets are usually spaced from 200 to 300 ft apart along the runway or taxiway.

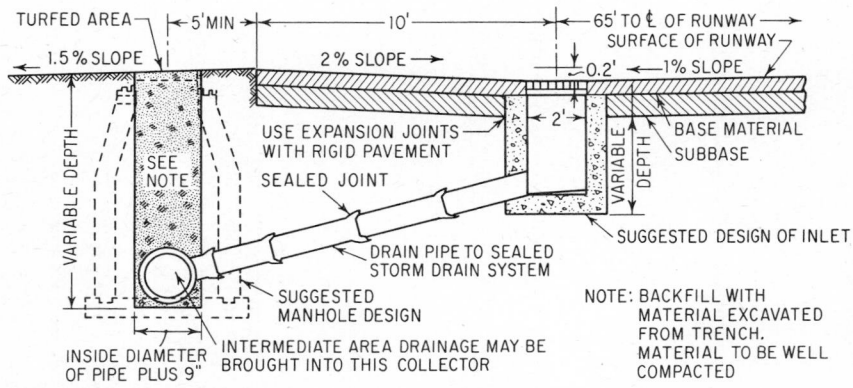

Fig. 18-15. Drainage inlet at outer edge of runway. (*Federal Aviation Administration.*)

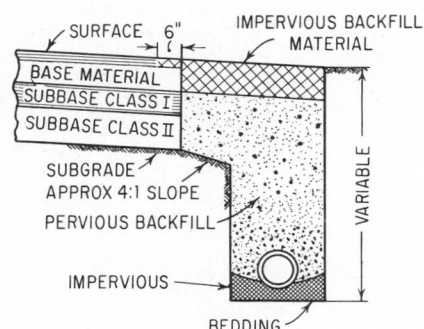

Fig. 18-16. Combined interceptor and base drain. (*Federal Aviation Administration.*)

Subsurface drainage is obtained by the use of interceptor drains and pervious base-course layers, in much the same way that highways are drained. Some of the smaller, turfed fields are drained by a network of subdrains covering the entire area. At airports with paved runways, subdrains are usually placed along the edges of the runways where soil conditions indicate that drainage is needed to lower the groundwater level. A combined interceptor and base drain is often used (Fig. 18-16).

Surface drainage is accomplished by the collection of surface water into inlets. A system of underground pipes is required to carry runoff from inlets and subdrains to outlets into waterways. In low areas, surface waters are sometimes drained into ditches or canals running around the perimeter of the airport.

For design of the drainage system, a topographic map is required. Upon it is plotted the proposed layout of runways, taxiways, aprons, and the terminal plan. The proposed surface grades of these features are shown by contours of small interval: 0.1 or 0.2 ft for paved areas and 0.5 or 1.0 ft for turfed areas. Inlet locations and subdrains are plotted, and storm-drain lines laid out to collect the discharge from them. The system should be as direct as possible

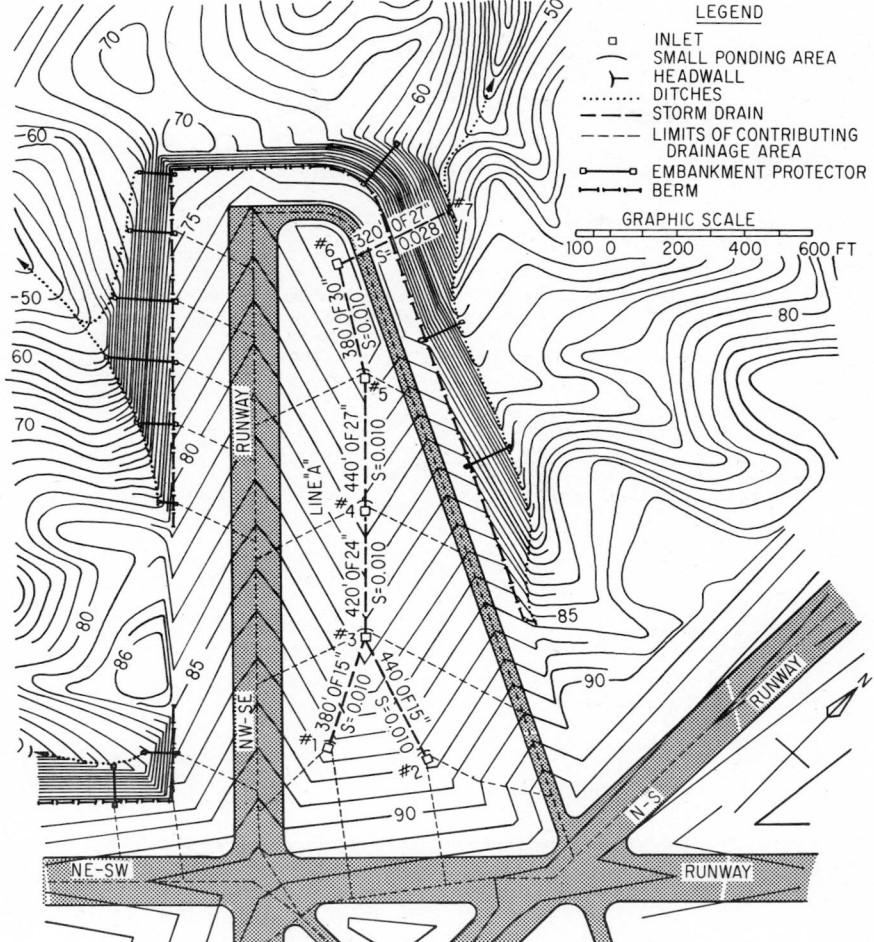

Fig. 18-17. Plan of portion of an airport drainage system. (*Federal Aviation Administration.*)

to avoid excessive lengths of pipe; frequent changes in pipe size should also be avoided. Crossings of pipes under runways should be held to a minimum.

Figure 18-17 shows a portion of an airport drainage system. The pipe sizes are computed to accommodate the discharge from the design storm, which may be taken as that expected once in every 2 to 10 years, depending on how serious an effect an occasional flooding may have on airplane operations. In some designs, a certain amount of ponding is permitted in areas outside the runways.

The rational method (Art. 21-57) of calculating runoff is universally used in airport-drainage design.

The engineer should prepare studies of intersections to insure good drainage. Center-line grades are held constant, and the grades of the outer portion of the runway or taxiway warped or adjusted so that there will be no abrupt changes in grade in the path of airplanes. The surface should have sufficient slope to drain properly. Intersection studies should be made at a scale of 1 in. equals 50 ft. A contour interval of 0.10 ft will permit positive surface drainage to be designed. The studies will also be useful in establishing pavement grades. ("Airport Drainage," Federal Aviation Administration.)

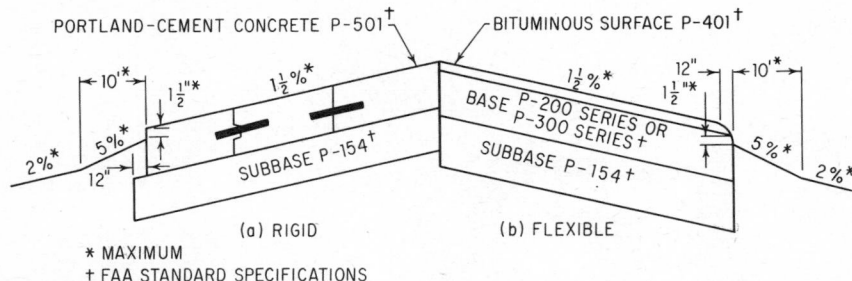

Fig. 18-18. Runway cross section shows typical portland-cement-concrete and bituminous pavement construction. (*"Airport Paving," Federal Aviation Administration.*)

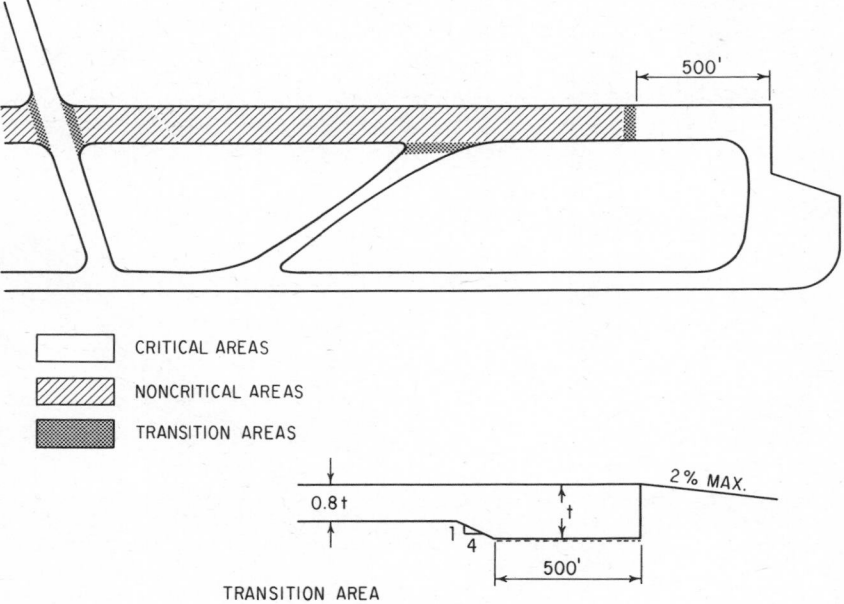

Fig. 18-19. Critical areas of airport pavement; t = total thickness of flexible pavement or concrete thickness for rigid pavement. (*"Airport Paving," Federal Aviation Administration.*)

18-36. Types of Airport Pavements. Airport pavements are constructed to support the loads imposed by aircraft using the airport and to produce a smooth, all-weather surface. Pavements are divided into two general types: *flexible* and *rigid*. Properly designed and constructed, either type will provide a satisfactory airport pavement. Specific types have, however, proved to be beneficial in specific applications: rigid pavements are recommended for areas subjected to appreciable fuel spillage at aircraft gate positions or maintenance positions on the apron; a low-cost flexible pavement is adequate to stabilize an area subject to jet-blast erosion.

The "Airport Paving Manual," published by the Federal Aviation Administration, is the usually accepted guide for design of civil airport pavements. It contains methods and requirements to be used in designing projects involving Federal funds.

Subgrade is the foundation for airport pavements. Its bearing capacity affects the thickness required in flexible and rigid pavements. Depth of frost penetration and influence of drainage conditions can affect the supporting value of the subgrade. Through selective grading, it might be economical to replace inferior subgrade material with superior material so as to reduce the subbase thickness requirement. Subgrades should be thoroughly compacted to provide the highest possible bearing capacity.

Subbase is a granular material placed on the compacted subgrade. It usually is required under flexible and rigid pavements, except for the better soils groups. Thorough compaction is mandatory.

Figure 18-18 shows cross sections of typical runway pavements. The transverse slope of pavements usually is 1.50%, to minimize water ponding on the surface. (Figure 18-18 refers to Federal Aviation Administration Standard Specifications for Construction of Airports, which covers many items of airport development.) The steep slopes at runway edges are for quick removal of rainfall.

Critical areas are those requiring the thickest pavement. They include the ends of runways, all taxiways, and aprons (Fig. 18-19). These are the areas subject to the most adverse aircraft loadings. Pavement thickness in noncritical areas may be reduced from the thickness required in critical areas. ("Airport Paving," Federal Aviation Administration.)

18-37. Flexible Pavements. These consist of a bituminous surface course, a base course of suitable material, and usually a granular subbase course (Fig. 18-18). Design of flexible pavements is based on the results of subgrade soils tests. The Federal Aviation Agency has developed a relationship between soil classes and thickness of surface course, base course, and subbase course required for various gross weights of aircraft, based on different conditions of drainage and frost action (Table 18-8).

Design curves for flexible pavements are shown in Fig. 18-20 (single-wheel landing gear), Fig. 18-21 (dual-wheel landing gear), and Fig. 18-22 (dual-tandem landing gear). The F numbers refer to those given for soil classes in Table 18-8.

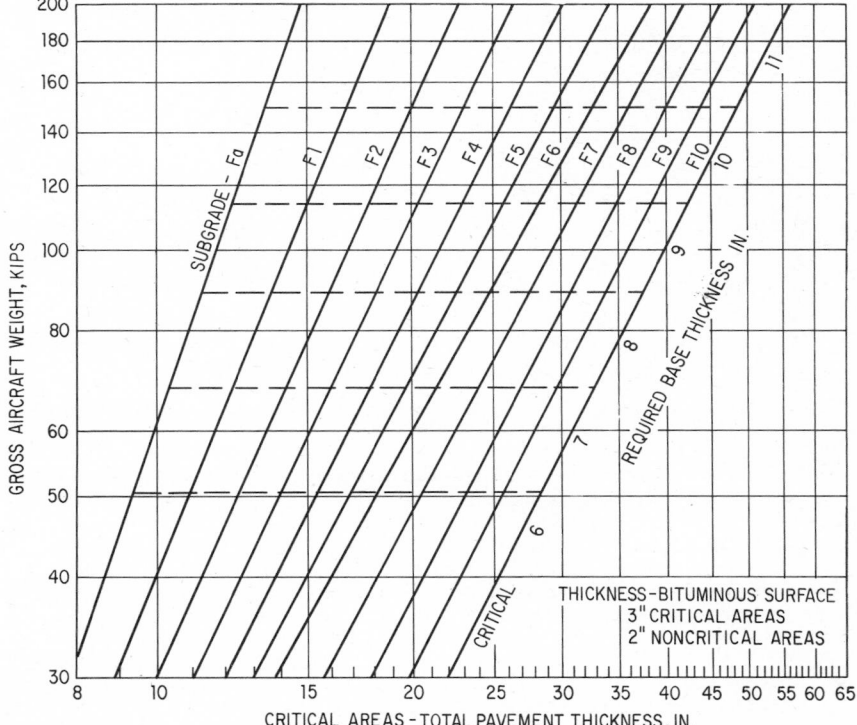

Fig. 18-20. Flexible-pavement thickness for single-wheel landing gear is given by chart. (*"Airport Paving,"* *Federal Aviation Administration.*)

Good drainage implies that surface water will be removed rapidly, groundwater level is low, and there will be no accumulation of water in the soil by either percolation or capillarity. Poor drainage indicates conditions where the subgrade may become unstable because of saturation.

The frost classification applies when the depth of frost penetration for a particular site is greater than the anticipated total pavement thickness as determined for no frost and the appropriate drainage condition.

Before the design curves are used, the weight and landing-gear characteristics of the critical aircraft expected at the airport should be determined. Then, the chart containing the appropriate design curves (Fig. 18-20, 18-21, or 18-22) is entered on the left side, at the gross-weight

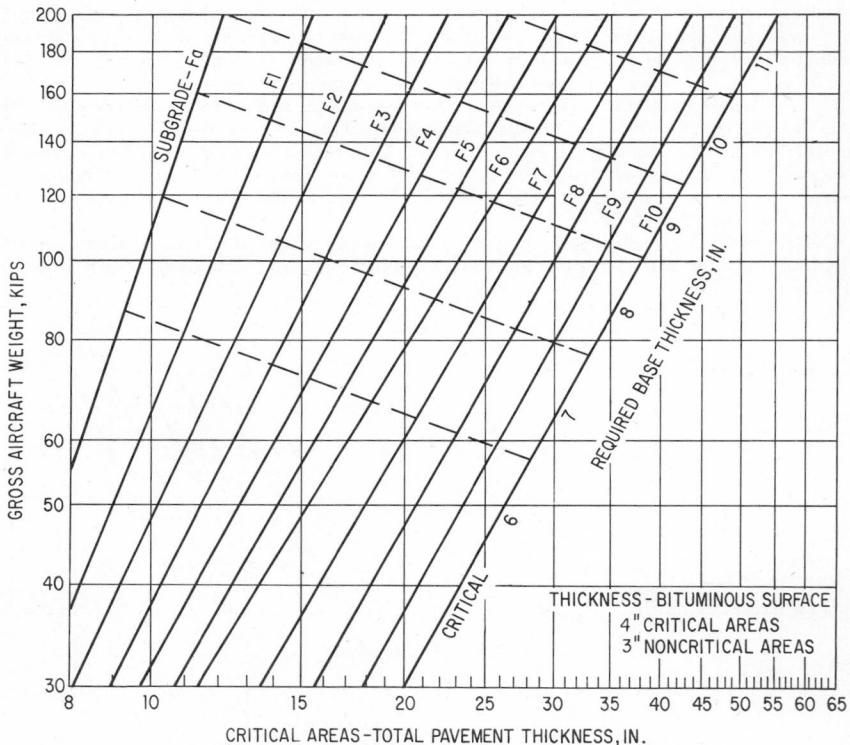

Fig. 18-21. Flexible-pavement thickness for dual-wheel landing gear is indicated by chart. (*"Airport Paving," Federal Aviation Administration.*)

index for the critical aircraft. A horizontal line to the intersection with the appropriate subgrade classification yields total pavement thickness. The point of intersection will also indicate base thickness. The charts indicate a thickness of 2 to 3 in. of surface course in noncritical areas and 3 to 4 in. in critical areas. With total thickness, surface-course, and base-course thicknesses known, the subbase thickness is readily obtained.

Surface-course requirements are to protect the base from surface water and to provide a smooth running surface for aircraft operations. The FAA recommends a dense-graded, hot-laid bituminous concrete produced in a central mixing plant for the wearing course of flexible pavements.

Base-course materials include a wide variety, to take maximum advantage of local materials and construction practices. When high-quality aggregates are used, asphalt or portland-cement treatments produce bases that are more effective than untreated bases. Accordingly, the FAA credits 1.0 in. of certain treated base materials as being equivalent to 1.5 in. of untreated base material.

Subbase is usually an integral part of the flexible-pavement structure. It is protected by the base and surface courses, and so the material requirements are not so strict as for the base course.

Pavements for light aircraft do not need to be so thick as for heavy aircraft. At airports that will not be required to accommodate aircraft in excess of 30,000 lb gross weight, the design curves in Fig. 18-23 should be used. The procedure is the same as with the design curves for heavier aircraft, except there is no reduction for noncritical areas.

A greater variety of local materials may be incorporated. Although a plant-mixed, dense-graded surface course with a minimum thickness of 1 in. is preferred, a double bituminous surface treatment will provide a satisfactory wearing course in about 1 in. thickness. Seal coats should not be used in lieu of a wearing surface course, except as a temporary measure, to be followed by a wearing course as soon as possible. A subbase course should not be introduced until the base requirement exceeds 6 in.

("Airport Paving" and Standard Specifications for Construction of Airports, Federal Aviation Administration.)

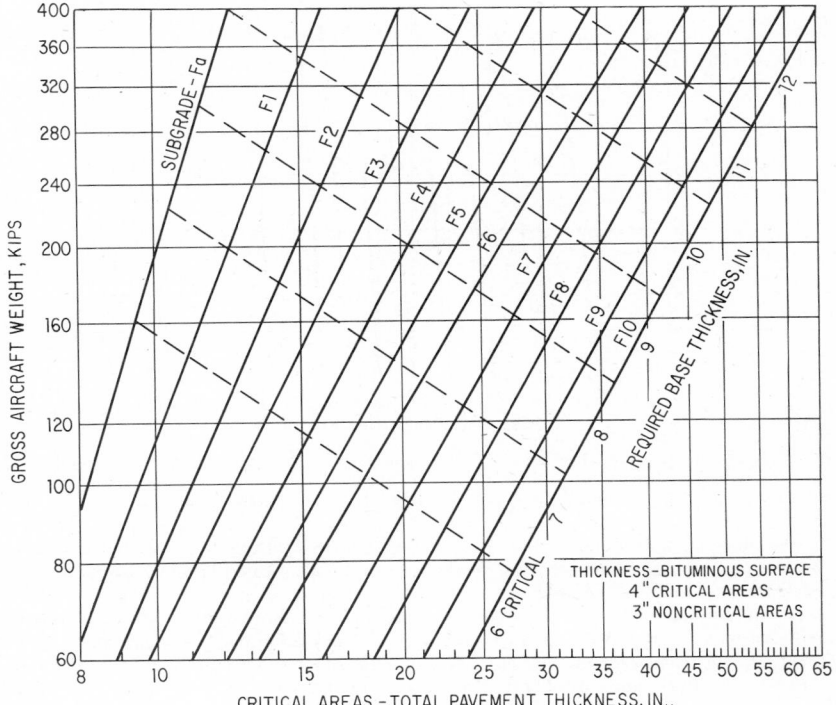

Fig. 18-22. Flexible-pavement thickness for dual-tandem gear is determined by design curves. (*"Airport Paving," Federal Aviation Administration.*)

18-38. Rigid Pavements. These are made of portland-cement concrete, usually placed on a suitable subbase course, which rests on a compacted subgrade (Fig. 18-18). Rigid pavements are designed independently of the subgrade classification, which affects the thickness of subbase. The design curves shown in Fig. 18-24 are used to determine total thickness.

The gross weight of the critical aircraft, as shown on the left scale of the upper curves in Fig. 18-24, is used as a point of entry. A horizontal line through it intersects with the appropriate landing-gear curve, to give the thickness of pavement. From this point of intersection, a vertical downward projection to the appropriate subgrade classification in the lower curves indicates subbase thickness. For noncritical areas, thickness of pavement is reduced 20% (with a minimum thickness of 6 in.), but subbase thickness is not reduced.

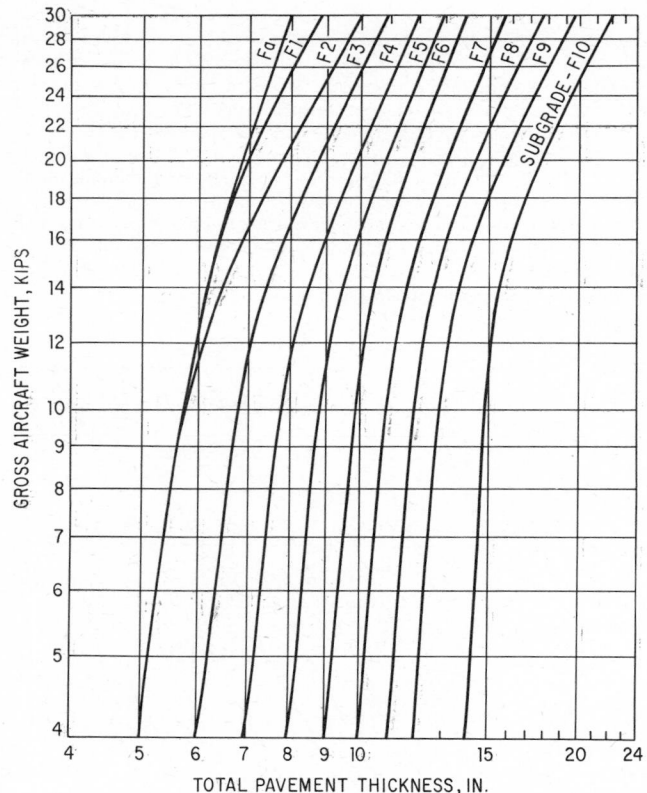

Fig. 18-23. Flexible-pavement thickness for light aircraft may be obtained from the above design curves. (*"Airport Paving," Federal Aviation Administration.*) NOTE: The Fa curve fixes the required base-plus surface-course thickness.   1 in. minimum surface thickness assumed for Fa curve.

The subbase thickness should be increased, as required, where the indicated thickness of pavement and subbase is not adequate to prevent frost heave.   In arid regions the subbase can, in certain instances, be decreased.

Joints and reinforcing used in airport concrete are similar to those used in highway pavements, except that wider slabs and larger dowels are used for thick pavement.   Longitudinal joints are of the dummy-groove or tongue-and-groove types.   These are spaced up to 12.5 ft c to c for slabs less than 10 in. thick, and up to 25 ft apart for thicker slabs.

Longitudinal expansion joints are advisable at runway and taxiway intersections and next to structures.   Transverse contraction or warping joints are spaced 15 to 25 ft apart in unreinforced pavement and 45 to 75 ft apart when distributed reinforcing is used.   Transverse expansion joints are not generally used except at intersections.

Dowels are used across expansion joints, and also across contraction joints in some designs. Tie bars or bonded reinforcing are carried across "hinged" or warping joints.

Construction joints between runs of paving are tongue-and-groove or butt types, with dowels. Diameters of dowels vary from ¾ in. for 6- to 7-in.-thick slabs to 1¼ in. for 11- to 12-in. slabs. Standard length of dowel is 18 to 20 in. and spacing 12 in. c to c.

Reinforcing of the pavement is desirable to control cracks.   Its installation should follow the latest design and construction practices.

Pavements for light aircraft should have a minimum thickness of 6 in.   This will accommodate aircraft up to 30,000 lb gross weight.

("Airport Paving" and Standard Specifications for Construction of Airports, Federal Aviation Administration.)

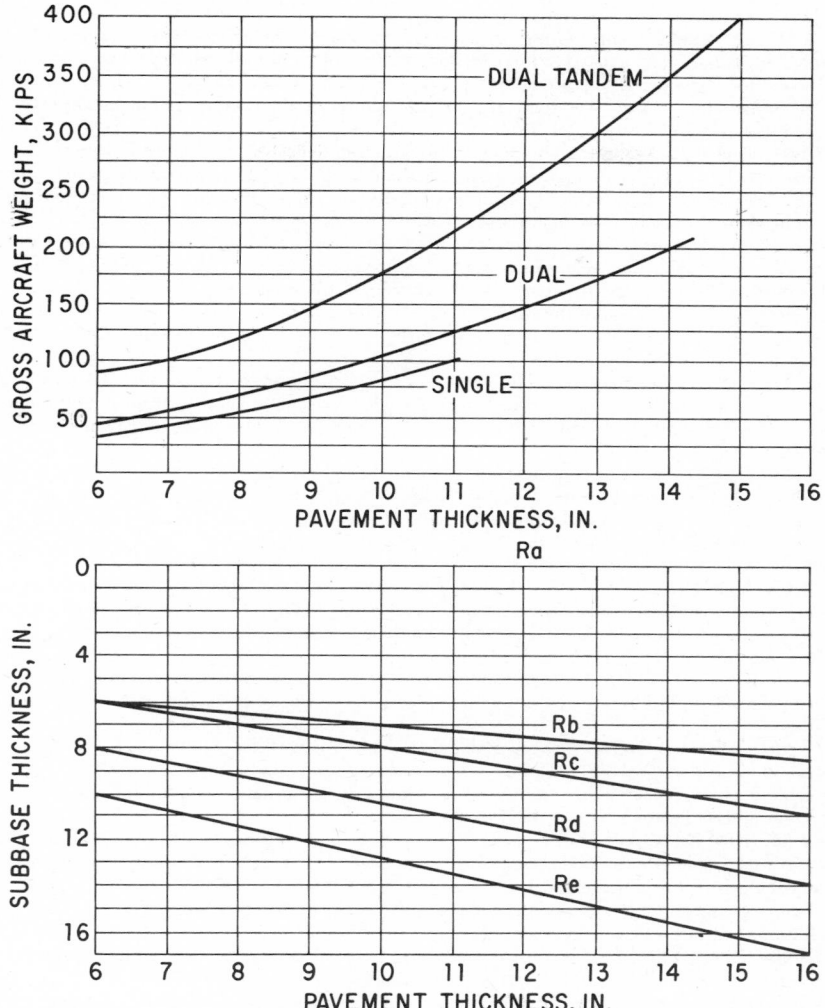

Fig. 18-24. Rigid-pavement thickness for critical areas. *("Airport Paving," Federal Aviation Administration.)*

18-39. Unpaved Surfaces.   Some airports do not require paved surfaces, because of a low volume of traffic and use by only light aircraft.   In some instances, turf surfaces are used for landings and takeoffs at small airports, and on the unpaved areas of landing strips at larger airports.

A tough, thickly matted grass is required in these areas.   The type of grass to use depends on soil characteristics and climate at the site.   If tests show the soil to be deficient in nutrient elements, these may be supplied by appropriate fertilizers.   When a fertile topsoil must be removed during grading operations, it should be stockpiled and later spread on the areas to be turfed.   A vegetative cover is also desirable on embankment, cut slopes, and other interior areas of the airport to prevent dusting and erosion.

Where turf is not adequate by itself, it may be possible to add to stability by adding aggregate to the soil prior to the development of the turf.   This will permit the soil to retain sufficient moisture to promote the growth of grass, yet provide a surface that will not become too soft in wet weather.   The required thickness of the turf-aggregate stabilization can be determined from the curves in Fig. 18-23.

("Airport Paving," Federal Aviation Administration.)

18-40. Soil Stabilization. Granular material, portland cement, tar, cut-back asphalt, or emulsified asphalt may be used to improve the qualities of a soil so that it can serve as a base or subbase. Such stabilized soils are not intended to serve as a surface course. A separate wearing surface must be provided. The same general procedures are followed in stabilizing airport soils as are followed in highway practice.

18-41. Pavement Overlays. The purpose of pavement overlays is to restore a pavement. An overlay may be applied over a pavement that no longer can be maintained satisfactorily. Or an overlay may be used to increase the load-bearing qualities of a satisfactory pavement that must accommodate aircraft heavier than those for which the pavement was designed.

Flexible overlays involve a combination of a base course and a bituminous surface course. *Rigid overlays* involve the use of a layer of portland-cement concrete. *Bituminous overlays* consist entirely of bituminous concrete. In each instance, the qualities of the existing pavement must be fully ascertained and the overlay designed to make the resultant pavement capable of handling the required traffic, following procedures outlined in "Airport Paving," Federal Aviation Administration.

18-42. Basic Airport Lighting. The function of airport lighting is to provide illumination to keep the facilities available around the clock. Lighting, usually kept on from dusk to dawn, assists in location and identification of the airport, outlines the usable areas, and furnishes guidance to moving aircraft.

Basic lighting consists of beacons, lighted wind indicator, runway or strip lights, and such obstruction lights as are required. Figure 18-25 illustrates the basic lighting at a small airport. Airport-lighting equipment and systems are subject to considerable modification in concept and design. The latest Federal Aviation Administration recommendations and practices should be followed (see also Arts. 18-43 to 18-50).

18-43. Airport Beacon. This is a double-end, rotating light situated on or near the airport and visible from considerable distances. Appropriate color coding of the two beacon lenses will identify the airport as an unlighted or unattended facility (both lenses clear); or equipped with lights, burning or readily available (clear-green).

The beacon may be placed atop a structure or on a standard beacon tower. The beams of the beacon are set slightly above the horizontal and should clear all trees and obstructions in the vicinity.

18-44. Obstruction Lights. These red lights mark objects that penetrate approach, horizontal, or conical surfaces (Fig. 18-2). Both steady-burning and flashing obstruction lights are available for use according to requirements. The positions of lights will depend upon the obstruction and its location with respect to the airport.

18-45. Wind Indicator. Wind information is required at all times to permit aircraft to select the most favorable runway or landing strip for takeoff or landing. The simplest indicator is a wind cone, a free-swinging cloth cylinder which gives information as to wind direction and velocity. At larger airports, landing information is furnished by a wind tee. The cone and tee should be illuminated to provide information during hours of darkness.

18-46. Runway Lights. These are low, elevated lights used to outline the edges of paved runways or to define unpaved landing strips. The smallest airports have lights mounted on driven stakes. At larger airports, the lights are mounted on heavy bases or small vaults of metal or concrete. The vault contains the isolating transformer for each fixture; otherwise, the transformer is buried alongside the runway light. The tops of bases and vaults are flush with the airport surface.

The lights are spaced 200 ft apart longitudinally and are usually 10 ft off the edge of the pavement (Fig. 18-25). They are fed from underground cables, either direct-burial or in ducts.

Medium-intensity runway lights are used on noninstrument runways and are adequate for contact operations. The intensity is controlled through a five-step regulator so that minimum intensity can be used in good weather. The lights have a Fresnel lens for optimum light distribution.

High-intensity runway lights are used on runways equipped for instrument landings or designated as instrument-landing runways by the Federal Aviation Administration. These lights concentrate powerful beams down the longitudinal axis of the runway, in both directions. Intensity is controlled so that there is adequate guidance without undue glare.

Threshold Lights. The effective end of each runway is indicated by runway lights (medium- or high-intensity) with green lenses. These lights are placed athwart the runway to mark its actual end, or outboarded beyond the edges of the runway for displaced thresholds. It is usual

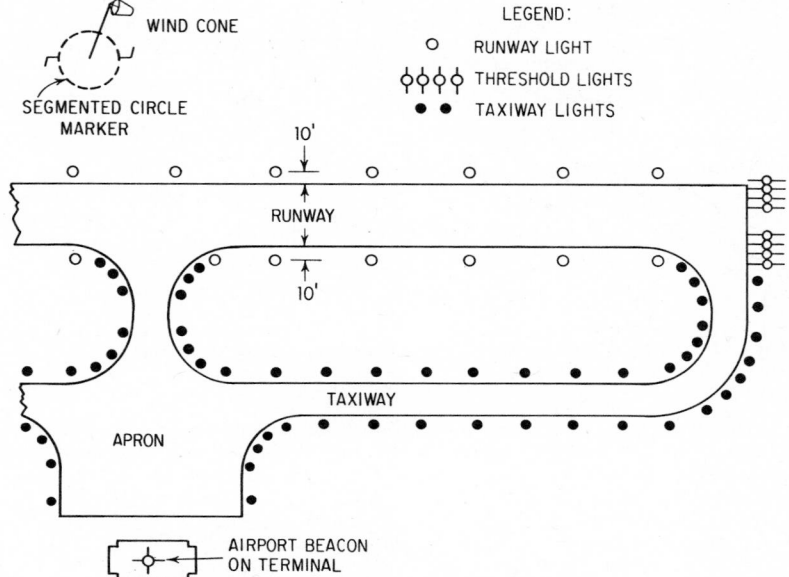

Fig. 18-25. Basic airport-lighting layout.

to place threshold lights at the actual runway ends (Fig. 18-25), except where clearance of obstructions in the approach dictates locating the threshold inward from the actual end.

Taxiway Lights. An airport with paved taxiways should have guidance lights if there is significant traffic at night. Taxiway lights are similar to medium-intensity runway edge lights, except that they are equipped with blue lenses. They are placed along the edges of taxiways to outline usable paved areas (Fig. 18-25). The longitudinal spacing varies with the taxiway configuration.

Taxiway Guidance Signs. These are internally illuminated directional indicators placed low above the ground surface. They give abbreviated guidance to the ends of runways, terminal aprons, hangar areas, and other airport locations. Need for them depends upon the volume of traffic and the complexity of airport layout and development.

In-runway Lights. Use of precision approach facilities to achieve lower weather minimums requires extensive electronic equipment in the aircraft, improved FAA navigation aids on the airport, and high-intensity runway lights, plus "in-runway" lighting. The last consists of center-line runway lights and touchdown-zone lighting. The lighting of the center line of taxiway turn-offs is desirable, since it assists aircraft in clearing the runway during inclement weather.

Taxiway-turnoff lighting (Fig. 18-26) consists of lights installed, for relatively high-speed performance, along the center of a turnoff taxiway to indicate the exit path. The lights are spaced 50 ft apart. The fixtures are similar to those used in the runway center line.

Runway center-line lighting (Fig. 18-27) consists of fixtures installed at uniform intervals along the center line of a runway to give a continuous lighting reference from threshold to threshold. The lights are spaced at 50-ft intervals. Fixtures are installed in shallow holes drilled into the pavement. The lights are fed by cables installed in ¼-in. slots sawed 1 in. deep. Isolating transformers are located at the sides of the runway.

Touchdown-zone lighting (Fig. 18-27) consists of 30 rows of transverse light bars at 100-ft intervals. Each row contains two bars. Set 30 ft on each side of the runway center line, each bar consists of three lights 5 ft apart, flush with the surface of the pavement, and aligned normal to the axis of the runway.

The fixtures are high-intensity lights installed in the pavement and fed through ducts or cemented into shallow holes drilled into the pavement and fed through cable installed in sawed joints. A variety of fixtures is available.

18-47. Airport-Lighting Control. All airport lights should be controlled from a single panel, readily accessible to an operator. At small airports, a regulator assembly with controls built into

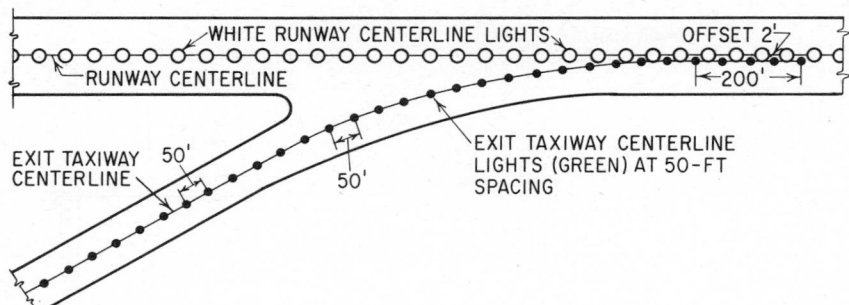

Fig. 18-26. Long-radius taxiway-turnoff lighting. (See Fig. 18-27 for location and spacing of runway centerline lights.) A longitudinal tolerance may be necessary to avoid rigid-pavement joints. (*Federal Aviation Administration.*)

the same cabinet provides a simple solution for basic lighting. Automatic controls (photoelectric or astronomic time switches) may be used where it is not feasible to have an operator on duty, or for remote beacons, obstruction lights, or other equipment, where direct control lines would not be economically feasible.

At airports with more complex installations, relay control equipment is placed in a transformer vault and lights are remotely controlled from the airport traffic-control tower or other central source. The remote-control source should have an adequate control panel, usually mounted on the control-tower console, and should contain circuit-control and brightness-control switches.

18-48. Power Supply. Provision of electric power to an airport for general purposes, as well as for airport lighting, requires a determination of power requirements and study of power availability. Usually, a second source of power is desirable to insure reliability of the lighting system. The overall reliability of power from commercial sources will determine the possible need for standby service or equipment.

18-49. Airport Marking. In addition to airport lighting, marking of facilities assists in guidance day and night and enhances operations in periods of restricted visibility. Federal Aviation Administration national standards should be followed.

The basic marking at an airport consists of a segmented circle marker (Fig. 18-25) and a wind indicator. The segmented circle marker is placed just outside the usable landing area. It identifies an airport and provides a central location for such indicators as exist at that airport.

The marker is a broken circle 100 ft in diameter. At the center is a conventional wind cone. A tee, however, may be used as a landing direction indicator.

Radial extensions beyond the 100-ft circle show the orientation of landing strips or runways. Extensions of the radials to the left or right indicate the airport traffic pattern.

Obstructions should be day-marked for maximum visibility. Other marking includes the numbering and striping of runways for normal identification, striping of taxiways, marking of unusable areas, and special runway marking to facilitate operations during low-visibility weather.

("Airport Design," Federal Aviation Administration.)

18-50. Electrical Ducts. In the preparation of a master plan for an airport, provision of electrical ducts for all cable crossings under paved areas should be carefully studied. The various systems of lighting should be laid out in sufficient detail to permit cable runs to be determined. All lighting that can be contemplated as an ultimate requirement should be studied.

The Federal Aviation Administration should be requested to furnish details of all installations that it might make so that its cable requirements can be incorporated into the duct plan.

When runways, taxiways, or aprons are paved, care should be taken to insure that adequate electrical ducts are provided to preclude costly jacking or cutting of pavements at some future date. A number of spare ducts might well be provided in all instances.

18-51. Fuel Systems. Regardless of the volume of traffic at an airport, some system of supplying fuel to aircraft must be provided. The simplest system at a small airport is an underground tank and an elevated dispenser, not unlike a regular service-station pump. Usually provided by a petroleum company, this system requires aircraft to taxi to a central location for service (Fig. 18-10*a*).

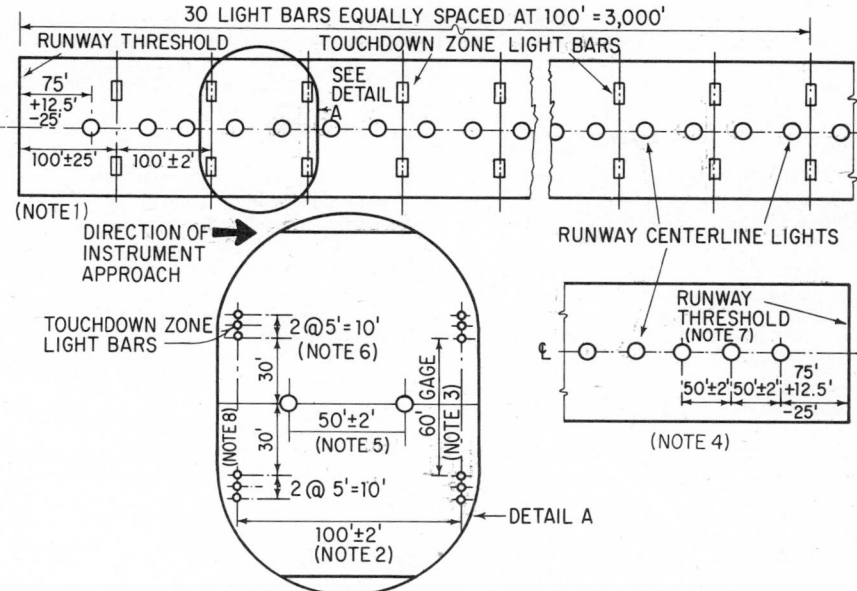

Fig. 18-27. Lighting layout for runway touchdown zone and center-line runway lights. NOTES: (1) In case of unusual joint location in concrete pavement, the first pair of light bars may be located 75 to 125 ft from threshold. (2) Longitudinal tolerance should not exceed 2 ft. (3) Gage may be reduced to 55 ft where dictated by construction requirements. (4) Longitudinal installation tolerance for individual lights should not exceed 2 ft. (5) Center-line lights need not be aligned with transverse light bars. (6) Maximum uniform spacing of lights is 5 ft c to c. (7) Center-line lights may be located up to 2 ft from runway center line to avoid joints. (8) Corresponding pairs of transverse light bars should lie along a line at right angles with runway center line. (*Federal Aviation Administration.*)

Generally, a single grade of fuel is available. Each additional grade and type of fuel requires a separate installation.

Airports with a medium volume of traffic normally use fuel-truck dispensers. These are serviced from local bulk stations if traffic is low and from airport storage if there is a sufficient volume of traffic to warrant. The busiest primary-system airports require fuel in such quantities that supply, storage, and distribution become special and complex problems.

18-52. Fuel Supply. Depending on overall fuel requirements coupled with local conditions, fuel will be supplied to the airport by truck delivery from local sources; tank truck, rail, or barge deliveries from refineries or bulk-storage sources; direct pipeline delivery; or various combinations of these. The heavier the volume of traffic, the more varied will be the types of fuel required.

Even at some airports where large quantities of certain varieties of fuels come by pipeline, the demand for other varieties is so low that truck delivery is employed for them. It is necessary to make a forecast of the demand for various grades of fuel, to determine long-range sources of availability, and to study all possible methods of delivery as a prerequisite to the design of an airport fuel system.

18-53. Fuel Storage. The bulk-storage system at the airport should provide for each type of fuel to be handled. Normal practice is also to maintain brand segregation. The capacity for each type should be adequate to accommodate fueling requirements for several days.

Delivery provisions should be flexible. There should be truck stands so that fuel can be unloaded from trucks and pumped into storage. Even where trucks are not the major source of supply to the airport, truck stands should be adequate to supply the entire fuel demand in an emergency, with pumping capacity sized accordingly. The same pumping capacity can be used for rail or barge supply.

Pipeline delivery will normally not require pumping capacity, inasmuch as fuel can be transferred under pressure to storage tanks. A waste tank should be provided for changing types of fuel without affecting type integrity.

The storage tanks should be interconnected to provide for interchange or transfer of fuel within the storage area. Tanks should be adequate to handle modern jet fuels. Usually, inert-gas explosion suppression is provided, or the tanks have floating roofs. The storage system should be capable of easy expansion or modification.

18-54. Fuel Transfer. Fuel is pumped from storage tanks through filter separators to pits, hydrants, or truck-loading stands, either directly or through satellite storage areas. If the distance is great, the size and number of transfer pipes may be reduced by introducing one or more satellite storage areas.

It is usual to have a separate satellite area for each user or group of users. Pumps take fuel from satellite storage through filter separators and to pits, hydrants, or truck stands.

18-55. Fuel Delivery. Trucks and pits are used for low-capacity delivery of fuel. High-capacity fuel delivery is accomplished by hydrants and hose carts.

Truck stands serve as loading points for fuel trucks, which deliver from the trucks direct into aircraft fuel tanks, through filter separators.

Pits contain booster pumps, filter separators, and coiled hose to deliver fuel direct into aircraft tanks, similar to truck delivery.

Hydrants provide for quick connection to hose carts. These are powered vehicles equipped with filter separators and pressure regulators to deliver fuel at high rates, under pressure, through underwing loading.

18-56. Air-Traffic Control. Airports are developed through the initiative of local communities, but the control of air traffic is a function of the Federal government. It is usual for air-traffic-control facilities to be installed and operated wholly with Federal funds.

Some auxiliary facilities, such as high-intensity runway lights and in-runway lighting, are the responsibility of the local community owning the airport. Facilities that furnish guidance along the airways and assist in the transition from airway to airport are usually installed without local participation. The Federal Aviation Administration has criteria based on volumes of traffic that are used to locate specific control facilities at an airport. Articles 18-57 to 18-59 provide general information concerning location and installation, but the FAA should be contacted for the latest revisions.

("Airport Design Requirements for Terminal Navigation Aids," Federal Aviation Administration.)

18-57. Instrument Landing System (ILS). The ILS is an electronic facility that furnishes three-dimensional information in the final portion of an airport approach, to permit an aircraft to land safely in inclement weather. The system consists of localizer, glide path, outer marker, and middle marker.

Localizer equipment directs an electronic course down the projected center line of the runway for lateral guidance. The equipment is normally installed 1,000 ft beyond the end of the runway opposite the approach direction. The area between the end of the runway and the localizer should be smooth, and within a circular area 500 ft from the localizer there should be no trees, buildings, roads, or fences.

Glide-path direction is transmitted from equipment located 400 to 600 ft off the center line of the runway and 750 to 1,250 ft in from the approach end. A smooth area is necessary for a considerable distance in front of the glide-path unit to insure the stability and accuracy of the electronic emissions.

Outer-marker equipment is located 4 to 7 miles from the airport on the projected center line of runway. The signals from the outer marker indicate distance from the runway end.

Middle-marker signals indicate a point about 3,500 ft from the runway end.

("Airport Design Requirements for Terminal Navigation Aids," Federal Aviation Administration.)

18-58. Approach-Light Systems. This is a system of high-intensity lights that extend outward from the approach end along the projected center line of the runway. They provide visual reference to the instrument runway during the transition from instrument flight to visual flight.

The system consists of horizontal 12-ft bars of high-intensity lights spaced 100 ft apart longitudinally for a distance of 3,000 ft (1,400 ft at small airports). Each bar contains, in addition, a condenser discharge light. These flash in sequence toward the runway.

An area 400 × 3,200 ft is desirable for the installation of the approach-light system. The lights are placed on piers or towers as required to provide a uniform light line at a slope not exceeding 2% upward from the end of the runway or a slope of 1% downward.

Runway-End Identifier Lights.   This system consists of a pair of synchronized flashing lights. One is located on each side of the runway-landing threshold facing into the approach area.   The lights are placed 40 ft outward from the runway edge lights.   The purpose of the flashing lights is to provide rapid and positive identification of the approach end of a particular runway.

Visual-Approach Slope Indicator.   This is a system of visual-approach indication, designed to provide visually the same information that a glide-slope unit provides electronically.   Downwind lights are placed 600 ft in from the runway threshold.   Upwind lights are 1,200 ft in from the threshold.   The lights are placed 50 ft from the runway edges.

The light units have beams elevated so that a specific approach slope is indicated through the proper alignment of the upwind and downwind light bars.   The approach slope may be set to clear a specific obstruction or to enhance noise-abatement procedures.

("Airport Design Requirements for Terminal Navigation Aids," Federal Aviation Administration.)

18-59. Other Airport Traffic Controls.   **Surveillance radar** controls traffic within a considerable distance from the airport, about the same range as covered by the airport traffic-control tower. No unusual siting problems are involved.

Precision-approach radar system is used to monitor or control traffic approaching the instrument runway.   It is located alongside the instrument runway 400 to 750 ft from the center line.

Terminal VOR omnirange is a terminal facility similar to the standard VOR (visual omnirange) navigation device.   When it is sited on an airport, there should be a clearance of 1,200 ft in all directions to insure true azimuth course indication.

Transmissometer is a device that furnishes visibility-measurement information for the runway touchdown area.   The installation is located slightly more than 400 ft from the center line of the instrument landing system runway.

Airport traffic-control towers are provided by the Federal Aviation Administration at those locations where new towers are required.   The control tower should be located at a point from which all portions of the runways, taxiways, and aprons are visible.   Requirements for each airport will vary; hence, they should be checked with the FAA.

Airport surface-detection equipment comprises a radar system that permits the observation of aircraft ground traffic on the airport.   It is usually located on top of airport traffic-control towers.

("Airport Design Requirements for Terminal Navigation Aids," Federal Aviation Administration.)

18-60. Heliports.   Helicopters in civil use vary in size, number of rotors, number of engines, and overall weight.   Small helicopters usually employ a single rotor for lift and a vertical tail rotor for lateral control.   Large civil helicopters have a single main rotor and vertical tail rotor or two main rotors located in tandem along the longitudinal axis of the helicopter.   There are other potential configurations, including intermeshing main rotors placed normal to the main axis, and various models of vertical-lift devices and convertible aircraft that can take off vertically and, through variable aircraft geometry, fly horizontally at speeds greater than those possible for helicopters.

Small helicopters (up to 4 passengers) generally weigh about 3,000 lb, are 30 to 40 ft long, 9 to 10 ft high, and have rotor diameters up to 35 ft.   Larger helicopters in general use weigh up to 20,000 lb, carry as many as 30 passengers, are 65 to 85 ft long, up to 17 ft high, and have rotor diameters up to 55 ft.   Most small helicopters use a skid-type landing gear, but large vehicles use wheel landing gear with a three- or four-wheel configuration.

Helicopters rise vertically a few feet above the heliport surface when taking off.   Then they accelerate upward and forward on a sloping path to climb-out speed and continue to en-route altitude.

Landing involves an approach on a sloping path to a hovering position a few feet above the heliport surface.   Then, the craft descends vertically to a selected landing point.   Sideward flight may be performed easily during the landing maneuver, so the helicopter will land in a precise position.   Ability to operate vertically permits the helicopter to land and take off using areas only slightly larger than its own dimensions.

Heliports are classified by size, use, and support facilities as follows: Class I—private; Class II—public (small); Class III—public (large).   They are further subclassified in accordance with their available support facilities, as follows: Subclass A—minimum support facilities; no shelter,

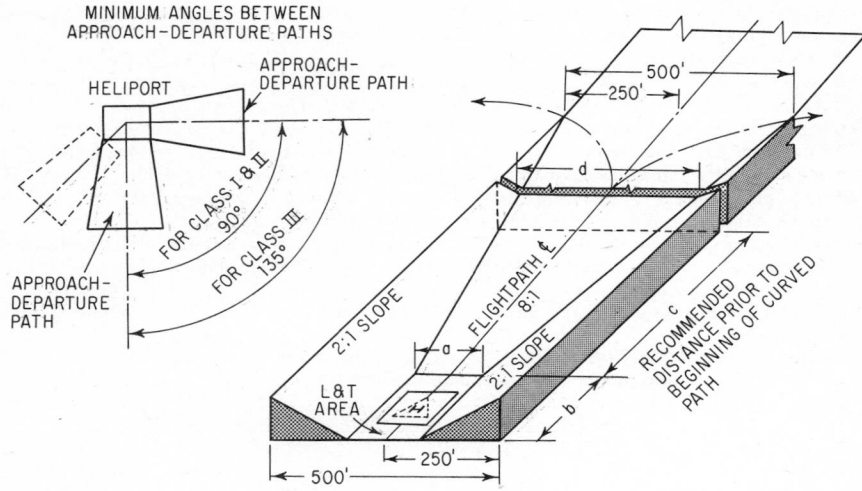

Fig. 18-28. Standards for heliports and approaches. (*"Heliport Design Guide," Federal Aviation Administration.*)

Table 18-9.. Heliport Standards*

Heliport class	Min dimensions (see Fig. 18-28)				Min angle between approach-departure paths, deg
	a	*b*	*c*, ft	*d*, ft	
I—private	1.5	1.5	300	200	90
II—small public	1.5	2.0	300	300	90
III—large public	1.5†	2.0†	400	300	135

Dimensions *a* and *b* (1) are expressed as multiples of overall helicopter length, (2) may be increased or decreased upon evaluation of the site by FAA.

*Federal Aviation Administration, "Heliport Design Guide."

†For scheduled airline operations, other factors related to a specific site would need to be considered.

maintenance, or fueling (a helistop); Subclass B—limited support facilities, no maintenance or fueling; Subclass C—complete support facilities.

Physical characteristics of heliports vary slightly with the various classes. Each heliport should be designed for the largest helicopter expected to land there. The basic heliport dimensions are determined from the overall length of that helicopter. Table 18-9 and Fig. 18-28 give the various standards and minimum dimensions.

Landing and takeoff area lengths are 1.5 times the helicopter length for Class I heliports and 2.0 times the helicopter length for Class II and Class III heliports. The width of the landing and takeoff area is 1.5 times the helicopter length for all classes of heliports. A circular heliport should have a diameter equal to the basic heliport lengths given above.

These lengths are for elevations up to 1,000 ft above sea level. They should be increased for elevation at the rate of 1.5% per 100 ft of elevation above 1,000 ft. Thus, for a heliport at 3,000-ft elevation, the minimum length would be increased 30%.

The dimensions can be adjusted, where fully justified, after a thorough evaluation. An example might be a heliport on a waterfront pier, surrounded on three sides by unobstructed approaches over water. If the length and width requirements are combined with helicopter lengths, the following minimum dimensions of landing and takeoff areas appear feasible for general use:

Class I—42 × 42 ft.
Class II—65 × 80 ft.
Class III—120 × 160 ft.

Approach-departure areas are laid out to offer the best lines of flight. It generally is necessary to have at least two flight paths, usually 180° apart. The paths may be as little as 90° apart for Class I heliports and 135° apart for Class II and Class III heliports.

Curved paths are practicable but should be used with a minimum straightaway approach-departure length of 300 ft (200 ft for Class I). The center-line radius of a curved path will vary, depending on local conditions and type of helicopter used. In general, however, the radius of the curved path should be at least 300 ft.

The approach-departure path has the same width as the contiguous edge of the landing and takeoff area and flares uniformly on each side of the center line to a width of 500 ft at the en-route altitude. The slope of the path is 1 ft vertical for each 8 ft longitudinally (8:1). Objects that extend above this sloping plane are obstructions.

Transition areas are surfaces along the lateral boundaries of the landing and takeoff area and the approach-departure areas. The surfaces, or "side slopes," extend outward and upward from the edges of the heliport and approach-departure areas for a distance of 250 ft from the center line. The slope is 2:1 upward from the edge of the landing and takeoff area or from the edge of the sloping approach-departure plane.

Peripheral areas surround the landing and takeoff area. The peripheral areas should have a width not less than one-fourth of the overall helicopter length. This area should be kept free of obstructions. For ground heliports, a fence or safety barrier should be provided at the boundary of the peripheral area. (See also Arts. 18-61 and 18-62.)

("Heliport Design Guide," Federal Aviation Administration.)

18-61. Heliport Site Selection. A heliport may be sited on the ground or on top of a building. For greatest utility to helicopters, the site should be as close as possible to the locale it serves. It should provide operational safety, have clear approaches, and be compatible with air traffic in the vicinity. It should fit in with area planning and not have an adverse impact on the community.

18-62. Heliport Layout and Design. (See also Art. 18-60.) The small heliport may consist of only a fenced area containing an unsurfaced landing and takeoff area (Fig. 18-29) or of an elaborate facility with a paved landing area, parking and service aprons, heliport terminal, and automobile parking (Fig. 18-30).

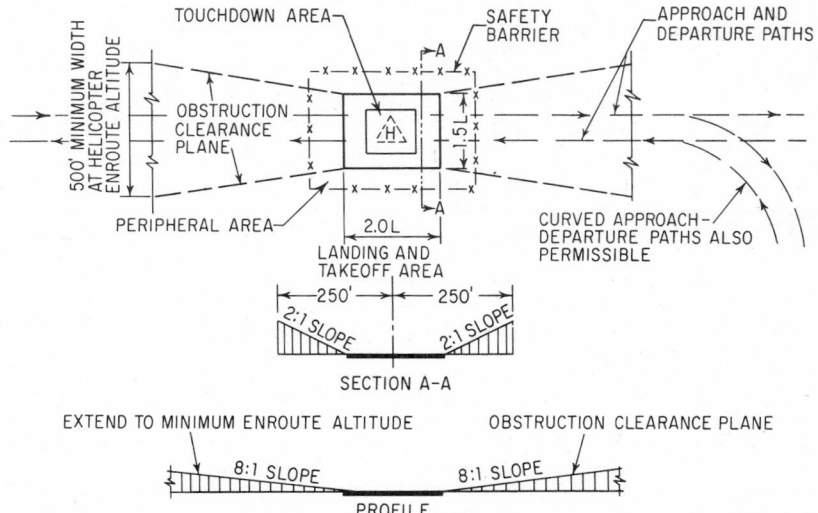

Fig. 18-29. Layout for small heliport. *L* = overall length of helicopter. *("Heliport Design Guide," Federal Aviation Administration.)*

Standard grading and drainage practices should be employed. The rotor downwash of helicopter operations usually requires a stabilized landing and takeoff area at a minimum. A paved touchdown pad is desirable.

Ground locations for heliports usually permit less expensive construction than rooftop sites but are seldom available in congested areas. Rooftop locations usually have advantages of accessibility and clear approaches to counter the disadvantages of limited space, difficulty of locating emergency-landing areas, and the probable need to strengthen the structure. It is necessary to consider wind effects as well as local building codes, zoning, and fire regulations.

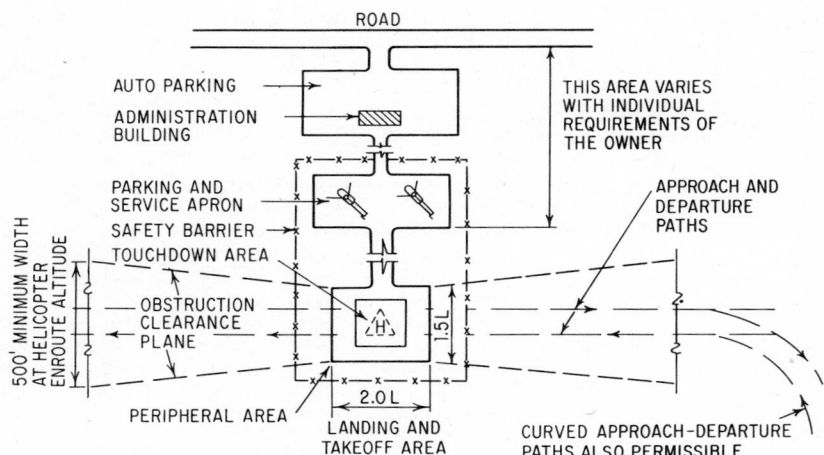

Fig. 18-30. Large-heliport layout. L = overall length of helicopter. (*"Heliport Design Guide," Federal Aviation Administration.*)

If the structure requires reinforcing, a load-distribution pad might be satisfactory. The pad need not be so large as the landing and takeoff area, but the full area should be a clear area. The pad can be as small as 20×20 ft for smaller helicopters, up to 50×50 ft for larger vehicles. The rooftop heliport should be of sufficient strength that it will not fail under unusual, high-impact landings. The landing surface should be designed for a concentrated load equal to 75% of the gross weight of the helicopter on any 1 sq ft of the surface. Fire-resistive materials are recommended.

Wind conditions might require baffles to eliminate turbulence across the surface of the heliport. Also, a safety device should be provided around elevated touchdown areas or landing pads. This should extend outward from the touchdown area.

Standard heliport markers are shown in Fig. 18-31. A marker is placed in the approximate center of the touchdown area.

The touchdown area should be marked with a border at least 1 ft wide. The boundary of the landing and takeoff area should be made conspicuous by low markers spaced 25 ft apart. There should be a wind indicator adjacent to the landing and takeoff area, located to provide true wind information.

Obstructions should be marked and lighted. Yellow boundary lights may be used to outline the landing and takeoff area. Floodlighting will be effective. One method is to place low Fresnel-lens lights around the landing and takeoff area, with a sharp cutoff that will not bother the pilot. Where loading aprons are provided, they should be adequately lighted if night operations are contemplated.

("Heliport Design Guide," Federal Aviation Administration.)

18-63. STOL Ports. There is a great potential for STOL (Short Take-Off and Landing) aircraft in short-haul transportation, serving stage distances of up to 500 miles. There is considerable advantage for city-center-to-city-center and intra-city air-passenger carriers that can provide better service to passengers and relieve both air-space and ground congestion at large airports.

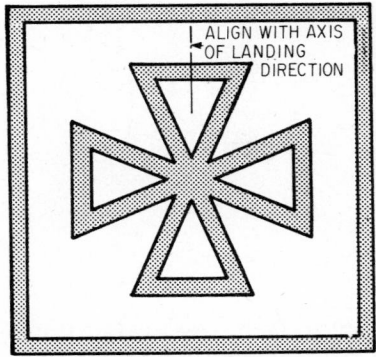

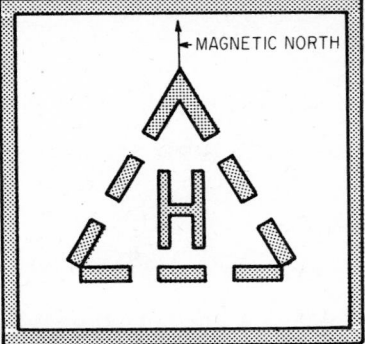

(a) STANDARD HELIPORT MARKER   (b) MARKER FOR PRIVATE-USE HELIPORTS
(BORDER AND SYMBOLS IN WHITE. IF BACKGROUND IS LIGHT COLOR, EDGE SYMBOLS IN BLACK.)

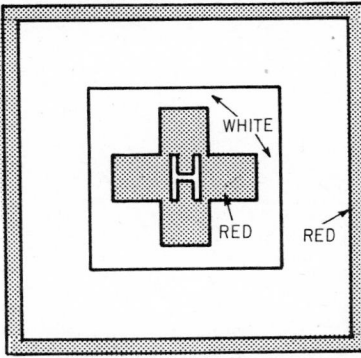

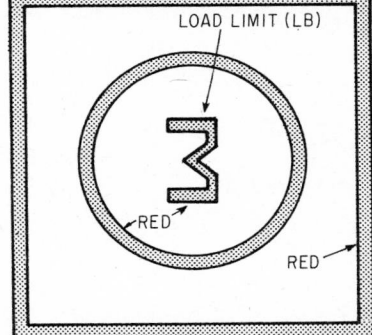

(c) MARKER FOR HOSPITAL HELIPORT   (d) EMERGENCY MARKER FOR ELEVATED
                                          HELIPORTS

Fig. 18-31. Heliport markers. *("Heliport Design Guide," Federal Aviation Administration.)*

Criteria for STOL ports are tentative and subject to change as evaluation of proposed STOL aircraft and operational experience dictate. Significant future changes will be incorporated in revisions to the Federal Aviation Administration publication, "Planning and Design Criteria for Metropolitan STOL Ports." The STOL vehicle promises shorter runways, steeper approach paths, lesser real estate requirements, and the prospect of in-town airport locations.

Rail-Transportation Engineering

G. M. MAGEE

Railroad Engineering Consultant, Retired Assistant Vice President, Research Department, Association of American Railroads, Chicago, Ill.

Rail transportation will become increasingly important as a transportation mode in the future because of substantial growth in population and gross national product. In addition, rail transportation is the most effective way to handle increased transportation demands with low power requirements, a low land requirement, little air pollution, and few accidents involving fatalities and injuries.

Rail transportation is considered here as a system in which vehicles are supported and guided by rails or other guideways. Rail-transportation engineering deals with the need, planning, selection, design, and construction of such systems for movement of passengers and freight. It involves roadbed, track, bridges, trestles, culverts, yards, terminals, stations, office buildings, signals and communications, and protection devices. Engineers also may be responsible for maintenance of way and structures. And they must be familiar with motive power, railway cars, and other equipment.

19-1. Glossary. Following are terms commonly encountered in rail-transportation engineering:

Alignment. Horizontal location of a railroad as described by tangents and curves.

Apron, Car Ferry. Bridge structure supporting tracks and connecting the car deck of a car ferry to land. The apron is hinged at the shore end so that it is free to move vertically at the outboard end to accommodate varying elevations of the ferry.

Apron, Track. Railroad track along the waterfront edge of a pier or wharf for direct transfer of cargo between ship and car.

Ballast. Selected material, such as crushed stone, placed on the roadbed to hold the track in line and surface.

Batter. The deformation of the surface of the head of a rail at the end.

Branch Line. Secondary line or lines of a railway.

Branding. Identification markings hot-rolled in raised figures and letters on a rail web indicating weight of rail and section number, type of rail, kind of steel, name of manufacturer and mill, and year and month rolled.

Car, Heavy-duty. A motorcar weighing more than 1,400 lb designed for hauling trailers and such other equipment as ballast diskers and weed mowers. It also is used for hump-yard service. Seats and deck sometimes are lengthened to accommodate many people. Engines ranging from 12 to 30 hp may propel the car.

Car, Heavy-section. A motorcar weighing from 1,200 to 1,400 lb. with seating capacity of 10 to 12 people. It is propelled by 8- to 12-hp engines.

Car, Light-inspection. A motorcar with a capacity of 650 to 800 lb designed to carry one or two people and tools. Weight ranges from 400 to 600 lb, the lighter car for use by one person.

Car, Light Rail (Trolley Car). A self-propelled vehicle operating on rails, generally in streets, and drawing electric power from overhead or underground conductors.

Car, Light-section. A motorcar weighing from 750 to 900 lb, with seating capacity of four to six people. It can be handled on or off the track by two people. It usually is propelled by 4- to 6-hp engines.

Car, Party-inspection. Any car with a capacity of four or more people used exclusively for inspection.

Car, Push. A four-wheeled railway work car designed to be pushed by hand or towed by a motorcar. It is used to transport materials too heavy to be carried on a motorcar.

Car, Standard-section. A motorcar weighing from 900 to 1,200 lb, with a seating capacity of six to eight people and a total load capacity of 2,500 lb. It usually is propelled by a 6- to 8-hp engine. It generally is used on standard main-line sections.

Car, Track. Any car or machine operated on track, such as a motorcar, handcar, or trailer.

Car, Trailer. A four-wheeled railway work car similar to a push car but equipped with a seat, foot boards, safety rails, and brakes. It is used with a motorcar for transporting people. It may be converted into a push car by removing seat, foot boards, and railings.

Car Retarder. Braking device, usually power-operated, built into a railway track to reduce the speed of cars. Brake shoes, when set in braking position, press against the sides of the lower portion of the car wheels.

Compromise Joint. Joint bars for connecting rails of different fishing height and section, or rails of the same section but with different joint drillings.

Cradle. Structure riding on an inclined track on a riverbank and having a horizontal deck with a track on it for transfer of railroad cars to and from boats at different water elevations.

Crib. Space between two successive ties.

Crossing (Track). Construction used where one track crosses another at grade; it consists of four connected frogs.

Crossing, Bolted Rail. A crossing in which all the running surfaces are of rolled rail. The parts are held together with bolts.

Crossing, Manganese-Steel Insert. A crossing in which a manganese-steel casting is inserted at each of the four intersections. Fitted into rolled rails, the casting forms the points and wings of the crossing frogs.

Crossing, Solid Manganese-Steel. A crossing in which the frogs are of the solid manganese-steel type.

Crossing, Movable-Point. A crossing of small angle in which each of the two center frogs consists essentially of a knuckle rail and two opposed movable center points with the necessary fixtures.

Crossing, Single-Rail. A crossing in which the connections between the end frogs and the center frogs consist of running rails only.

Crossing, Two-Rail. A crossing in which the connections between the end frogs and the center frogs consist of running rails and guardrails.

Crossing, Three-Rail. A crossing in which the connections between the end frogs and the center frogs consist of running rails, guardrails, and easer rails.

Crossing Plates. Plates interposed between a crossing and the ties or other timbers to protect the ties and to support the crossing better by distributing loads over larger areas.

Crossover. Two turnouts with the track between the frogs arranged to form a continuous passage between two nearby and generally parallel tracks (Fig. 19-25).

Crossover, Double. Two crossovers that intersect between the connected tracks.

Curve, Compound. A continuous change in alignment effected with two or more contiguous, simple curves of different radii but with a common tangent at each junction (Fig. 19-13).

Curve, Degree of. (See Degree of Curve.)

Curve, Easement. A curve whose radius varies to provide gradual transition between a tangent and a simple curve or between two simple curves of different radii (Fig. 19-16).

Curve, Lead.   Curve between switch and frog in a turnout (Fig. 19-25).

Curve, Reverse.   Curve formed by two contiguous, simple curves with a common tangent but with centers of curvature on opposite sides of the tangent (Fig. 19-14).

Curve, Simple.   A continuous change in alignment effected with an arc of constant radius and fixed center (Fig. 19-12).

Curve, Spiral.   (See Curve, Easement.)

Curve, Vertical.   An easement curve connecting intersecting grade (sloped) lines (Fig. 19-17).

Degree of Curve.   Angle subtended at the center of a simple curve by a 100-ft chord.

Derail.   A track structure for derailing rolling stock in an emergency.

Easer.   (See Rail, Easer.)

Elevation of Curves (Superelevation).   Height of outer rail above inner rail along a curve.

Fishing Space.   Space between head and base of a rail occupied by a joint bar (Fig. 19-21).

Flangeway.   Open way through a track structure that provides a passageway for wheel flanges (Fig. 19-26).

Flare.   A tapered widening of the flangeway at the end of a guard line of a track structure.   A flare may be at the end of a guardrail or at the end of a frog or crossing wing rail (Fig. 19-26).

Foot Guard.   Filler for space between converging rails to prevent a foot from being accidentally wedged between the rails.

Frog.   A track structure at the intersection of two running rails to provide support for wheels and passageways for their flanges, thus permitting wheels on either rail to cross the other (Fig. 19-26).

Frog, Bolted Rigid.   A frog built of rolled rails with fillers between, held together with bolts.

Frog, Center.   Either of the two frogs at the opposite ends of the short diagonal of a crossing.

Frog, Clamp.   A frog built mainly of rolled rails, with fillers between the rails, and held together by clamps.

Frog, End.   Either of the two frogs at the opposite ends of the long diagonal of a crossing.

Frog, Flange.   (See Frog, Self-guarded.)

Frog, Rail-bound Manganese-Steel.   A frog consisting essentially of a manganese-steel body casting fitted into and between rolled rails and held together with bolts (Fig. 19-26a).

Frog, Self-guarded (Flange Frog).   A frog with guides or flanges above its running surface to contact the tread rims of wheels to guide their flanges safely past the point of the frog (Fig. 19-26b).

Frog, Solid Manganese-Steel.   A frog consisting essentially of a single manganese-steel casting (Fig. 19-26b).

Frog, Spring-Rail.   A frog with a movable wing rail normally held against the point rail by springs.   The rails thus form an unbroken running surface for wheels on one track, whereas the flanges of the wheels on the other track force the movable wing rail away from the point rail to provide a passageway.   Viewed from the toe end toward the point, a right-hand frog has the movable wing rail on the right-hand side.

Frog Angle.   The angle formed by the intersecting gage lines of a frog.

Frog Number.   Half the cotangent of half the frog angle.

Frog Point.   That part of a frog lying between the extensions of the gage lines from their intersection toward the heel end (part farthest from the switch).   The theoretical point is the intersection of the gage lines.   The half-inch point is located at a distance from the theoretical point toward the heel equal, in inches, to half the frog number and at which the spread between the gage lines is ½ in.   Usually, measurements are made from the half-inch frog point.

Gage (Track).   Distance between gage lines (Fig. 19-19).   (Standard gage is 4 ft 8½ in.)

Gage (Track Tool).   A device by which the gage of a track is established or measured.

Gage Line.   A line ⅝ in. below the top of the center line of head of running rail or corresponding location of tread portion of other track structures along that side nearer the center of track.

Gagging.   Work done on a rail at a straightening press with a steel "gag" or tool for taking out a bend.

Grade Line.   Line on profile representing tops of embankments and bottoms of cuts ready to receive ballast.   This line is the intersection of the plane of the roadbed with a vertical plane through the center line.

Guard, Stock.   A barrier between and along track rails to prevent passage of livestock on or along the track.

Guard Check Gage.   Distance between guard and gage lines, measured perpendicular to gage lines across the track.

Guard Face Gage.   Distance between guard lines, measured perpendicular to gage line across the track.

Guard Line. A line along that side of the flangeway nearer the center of track and at the same elevation as the gage line.

Guard Timber. A longitudinal timber placed outside the track rail to maintain tie spacing.

Guardrail. A rail or other structure parallel to the running rails of a track used to prevent wheels from being derailed, or to hold wheels in correct alignment to prevent their flanges from striking the points of turnouts, crossing frogs, or switches. Also, a guardrail is a rail or other structure laid parallel to the running rails of a track to keep derailed wheels adjacent to running rails.

Guardrail, Frog. A rail or other device to guide a wheel flange so that it is kept clear of the point of a frog.

Guardrail, Inner. A longitudinal member, usually a metal rail, secured on top of the ties inside the track rail to guide derailed wheels.

Guardrail, One-piece. A guardrail consisting of a single component so designed that no auxiliary parts or fastenings other than spikes are required for its installation.

Joint, Compromise. See Compromise Joint.

Joint, Rail. Splice uniting abutting ends of contiguous rails. An insulated joint arrests the flow of electric current from rail to rail with insulation between rail ends and other metal parts connecting them.

Joint Bar. A stiff steel member commonly used (in pairs) to join rail ends and to hold them firmly, evenly, and accurately in surface and gage-side alignment (Fig. 19-21).

Joint Gap. Distance between ends of contiguous rails in track, measured on the outside of the head ⅝ in. below top of rail.

Lead. Distance between actual point of a switch and half-inch point of a frog. The actual lead is measured along the line of the parent track (Fig. 19-25). The curved lead is measured to the half-inch point of the frog but along the outside gage line of the turnout. The theoretical lead is the distance from the theoretical point of a uniform turnout curve to the theoretical point of the frog, measured along the line of the parent track.

Nosing. A transverse, horizontal motion of a locomotive that exerts a lateral force on the supporting structure.

Out of Face (Trackwork). Work, such as tie replacement, that proceeds completely and continuously over a given piece of track as distinguished from work at disconnected points.

Rail (Track). A rolled steel shape, commonly a T section, laid end to end, on crossties or other suitable supports, to form a track for railway rolling stock (Fig. 19-20).

Rail, Closure. Rail between the parts of any special trackwork layout, such as the rail between switch and frog in a turnout (sometimes called lead or connecting rail); also the rail connecting the frogs of a crossing or of adjacent crossings but not a part of the crossings (Fig. 19-25).

Rail, Compromise. Relatively short rail with two ends of different section to correspond with the ends of rails to be joined. It provides the transition between rails of different section.

Rail, Easer. A rail that provides a bearing for the portion of hollowed-out treads of worn wheels that overhangs a running rail. Sloped at the ends, an easer is laid with its head along the outside of and close to the head of the running rail.

Rail, Guard. (See Guardrail.)

Rail, Knuckle. A bent rail or equivalent structure forming an obtuse point at a movable-point crossing or slip switch. When set for traffic, the movable points of the crossing or switch rest against the obtuse point.

Rail, Reinforcing. A bent rail placed with its head outside of and close to the head of a knuckle rail to strengthen it and act as an easer rail; or a piece of rail similarly applied to a movable center point.

Rail, Running. Rail or surface on which the tread of a wheel bears.

Rail, Safety. Railing on a motorcar or trailer to serve as a handhold for occupants for safety.

Rail, Stock. Running rail against which the switch rail operates.

Rail, Switch (Switch Point or Switch-Point Rail). Tapered rail of a split switch (Fig. 19-27).

Rail, Welded. Two or more rails welded together to form a length less than 400 ft. When the length is 400 ft or more, the result is called a continuous welded rail.

Retarder, Car. See Car Retarder.

Retarder, Insert. A braking device without external power, built into a railway track to reduce the speed of cars with brake shoes against the sides of the lower portions of wheels. Sometimes, means are provided to open the retarder to nullify its braking effect.

Right of Way. Lands or rights used or held for railroad operation.

Shoulder. That portion of the ballast between the end of the tie and the toe of the ballast slope.

Siding. Track, auxiliary to the main track, used to permit trains to pass.

Spot Board. A sighting board placed above and across the track at a proposed elevation for the rails to indicate the new surface and insure its uniformity.

Stamping. Figures and letters indented, after hot sawing, in the center of the rail web, parallel with the direction of rolling, to indicate the serial heat number, ingot number as cast or rolled, and position of each rail relative to top of ingot.

Station, Loop. A form of through station in which the station track layout embraces a loop or part of a circle. Trains move in one direction only and turn relative to the station.

Station, Stub. Station with tracks connected at one end only.

Station, Through. Station with tracks connected at both ends.

Stock Pass. A culvert or bridge opening under a track primarily for passage of livestock.

Subballast. Material of superior character spread on the finished subgrade of a roadbed below the topballast to provide good drainage, prevent frost upheaval, and distribute the load over the roadbed (Fig. 19-9).

Subgrade. Finished surface of roadbed below ballast and track.

Surface, Running (Tread). Top part of structures on which the treads of wheels bear.

Switch. A track structure for diverting rolling stock from one track to another (Fig. 19-27).

Switch, Slip. A combination of a crossing with left- and right-hand switches and curves between them within the limits of the crossing connecting the two intersecting tracks on both sides of the crossing without separate turnout frogs. A single slip switch combines a crossing with one right-hand and one left-hand switch; a double slip switch, with two right-hand and two left-hand switches.

Switch, Split. A switch consisting essentially of two movable-point rails with necessary fixtures (Fig. 19-27).

Switch, Spring. A switch with an operating mechanism incorporating a spring device to return the movable points automatically to their original or normal position. This action takes place after the points have been shifted by the flanges of trailing wheels passing along the track other than that for which the points are set for facing movements.

Switch Angle. Angle between the gage lines of a stock rail and the switch rail at its point.

Switch Detector Bar. Strip of metal, alongside the track rail, connected with the throwing mechanism of a switch to prevent moving of the switch under trains.

Switch Heel. End of a switch rail farther from its point and nearer the frog. Heel spread is distance at the heel between gage lines of stock and switch rails (standardized at 6¼ in. for straight switches).

Switch Point. (See also Rail, Switch.) Theoretically, the intersection of the gage line of the switch rail, extended, and the gage line of the stock rail. The actual point is that end of the switch rail farther from the frog; the point where the spread between the gage lines of stock and switch rails is sufficient for a practicable switch point (Fig. 19-27).

Switch Stand. Device for manual operation of switches or movable center points.

Switch Throw. Distance through which points of switch rails are moved sideways (standardized at 4¾ in.). It is measured along the center line of the No. 1 switch rod or head rod.

Tangent. Straight rails or track; specifically, straight track contiguous with a curve.

Tie, Cross. The transverse member of the track structure to which rails are fastened to provide proper gage and to cushion and distribute traffic loads (Fig. 19-19). An adzed tie has plate-bearing areas on top made plane and smooth by machine. A bored tie has machine-made holes for spikes. A grooved tie has depressions machine-gouged across its top into which ribs on the bottom of a tie plate fit.

Tie, Heart. A tie with sapwood no wider than one-fourth the width of the tie top between 20 and 40 in. from midlength.

Tie, Sap. A tie with sapwood wider than one-fourth the width of the tie top between 20 and 40 in. from midlength.

Tie, Slabbed (Pole Tie, Round Tie). A tie sawed on ends, top, and bottom only.

Tie, Substitute. A tie of any material other than wood or a combination of any material and wood.

Tie, Switch. A tie that functions as a crosstie but is longer and also supports a crossover or turnout.

Tie Plate. Plate interposed between a tie and rail or other track structure (Fig. 19-19).

Topballast. Material of superior character spread over a subballast to support the track structure, distribute the load; and provide good drainage (Fig. 19-9).

Track. Assembly of rails, ties, and fastenings over which cars, locomotives, and trains move.

Track, Body.   Each of the parallel tracks of a yard on which cars are moved or stored.

Track, Connecting.   Two turnouts with the track between the frogs arranged to form a continuous passage between one track and another intersecting or oblique track or another remote, parallel track.

Track, Crossover.   (See Crossover.)

Track, Drill.   A track connecting with a ladder track and over which locomotives and cars pass in switching.

Track, House.   A track alongside or entering a freight house and used for cars receiving or delivering freight.

Track, Ladder.   Track connecting the body tracks of a yard.

Track, Lead.   An extended track connecting either end of a yard with the main track.

Track, Main.   Track extending through yards and between stations and on which trains are operated by timetable or train order, or both, or the use of which is governed by block signals.

Track, Rider.   A track in a hump yard on which a conveyance is operated for returning car riders to the summit of the hump.

Track, Running.   A track reserved for movement through a yard.

Track, Side.   A track auxiliary to the main track for use other than as a siding.

Track, Special.   All rails, track structures, and fittings other than plain unguarded track that is neither curved nor fabricated before laying.

Track, Spur.   A stub track diverging from another track.

Track, Stub.   Track connected with another track only at one end.

Track, Team.   Track on which cars are placed for transfer of freight between cars and highway vehicles.

Track, Transfer.   A track so located with respect to other tracks and transferring facilities as to facilitate transfer of lading from one car to another.

Track, Wye.   Triangular arrangement of tracks on which cars, locomotives, and trains may be turned.

Track Bolt.   A buttonhead bolt with oval neck and threaded nut for fastening rails and joint bars.

Tread.   Top surface of a railhead that contacts the wheels.

Turnout.   Arrangement of a switch and frog for diverting rolling stock from one track to another (Fig. 19-25).

Turnout Number.   Number corresponding to the frog number of the frog used in the turnout.

Wye.   (See Track, Wye.)

Yard.   System of tracks for such purposes as making up trains, storing cars, and sorting cars and over which movements not authorized by timetable or train order may be made, subject to prescribed signals and rules or special instructions.

Yard, Flat.   A yard in which car movements are accomplished by locomotive, without material aid from gravity.

Yard, Gravity.   Yard in which car classification is accomplished by locomotive, with material aid from gravity.

Yard, Hump.   Yard in which car classification is accomplished by pushing the cars over a summit, beyond which they run by gravity.

Yard, Retarder.   Hump yard equipped with retarders to control car speed during descent to classification tracks.

Yard, Sorting.   Yard in which cars are classified in greater detail after they have passed through a classification yard.

("Manual for Railway Engineering," American Railway Engineering Association, 59 E. Van Buren St., Chicago, Ill. 60605.)

19-2. Rail-Transportation Systems.   There are three principal types of rail-transportation systems: intercity passenger and freight, commuter, and rapid transit.   Outstanding attributes of each are safety, low energy requirements (a rolling resistance of 3 to 8 lb per ton for steel wheels on steel rails), ability to handle 1000 passengers or 10,000 tons of freight (or more) with one train, a minimum amount of land required for right of way, dependability of service under all weather conditions, and little atmospheric pollution.   A fourth type of rail-transportation system receiving increasing attention is personal rapid transit, which has the objective of taking passengers from one station on the line to any other station on the line with a minimum of waiting time for a car and no intervening stops.

Intercity passenger and freight systems are provided by many railways, but some provide only freight service.   Characteristic engineering requirements for a satisfactory passenger service include cars having trucks equipped with very long travel springs, snubbers, cross stabilizers, air conditioning, good lighting, attractive decor, comfortable and roomy seats, clean and adequate toilet facilities, convenient baggage storage, good dining-car service reasonably priced, and vista-dome lounge cars (except for overnight service). Departure times, speed, on-time arrivals, and low fares are also important factors.

Over 90% of intercity passenger travel in the United States is by private automobile and only about 1% by railroad. To relieve the railroads of huge deficits, incurred for intercity passenger service, Congress established a quasi-public corporation—the National Railroad Passenger Corporation, usually called Amtrak—to operate a basic national rail passenger system. When Amtrak commenced operations on May 1, 1971, it operated about 60% of the intercity passenger trains which had existed before its creation. Its basic system was, in large measure, dictated by the Department of Transportation, and it was heavily subsidized by the U.S. government. All railroads operating intercity rail passenger service were given the option of continuing to operate their trains apart from Amtrak or joining Amtrak by contract. On payment of entry fees of considerable amounts, those that joined Amtrak were relieved of all responsibility for provision of intercity rail passenger service. Twenty railroads signed a contract with Amtrak, and three railroads decided to continue to operate their trains independently of Amtrak.

For successful intercity freight service, rates are usually most important. Ease of loading and unloading cargo, time in transit, freedom of lading from damage, and on-time delivery are also important. Good engineering and operation are required to provide profitable freight service. Many types of specialized cars have been developed to meet shippers' needs, and these have been very effective in attracting or holding freight traffic for the railroads.

Commuter systems usually provide short-haul passenger service between a large city and its suburbs and operate as part of a larger rail system. Peak periods for transportation of workers occur during early morning and late afternoon. But some service must be provided throughout the day. Important requirements are reliability, minimum travel time, convenience, comfort, and economy.

Automobile travel competes with commuter service, so it is important that engineers design a commuter system that will attract the maximum possible volume of travel. Attractions are trains at frequent intervals, protection from inclement weather, possible saving in travel time, and economy. Use of double-deck, stainless-steel commuter cars, with air conditioning, good lighting, and comfortable seats; on-time performance and frequent scheduling; and push-pull operation have resulted in substantial increase in commuter travel, even with some increase in fares.

However, even with good commuter service, few railroads have been able to operate this service profitably. As a result, some states have subsidized commuter operation where it was considered advantageous to do so. Also, the U.S. Department of Transportation (DOT) has

Fig. 19-1. Garrett AiResearch linear-induction-motor (LIM) research vehicle built for the U.S. Department of Transportation for testing near Pueblo, Col.

recommended, and the Federal government has made, grants for the purchase of new equipment for some commuter lines.   DOT is also conducting extensive research to develop new concepts and improvements for rail systems.   Figure 19-1 illustrates one research vehicle built for DOT. The justification for such state and Federal aid has been saving of money and land that would otherwise be used for additional expressways, relief of automobile congestion in cities, reduction in the amount of parking space required in cities, fewer automobile accidents, and less noise and air pollution.

Rapid-transit systems are primarily intracity, although some provide service to nearby suburbs.   Characteristic requirements are frequent and dependable service, quick loading and unloading, light weight for rapid acceleration and deceleration, low fares, and a degree of comfort consistent with the other requirements.   During rush hours, standing is general practice for a portion of the run.   In congested areas, the trackage is in subways or elevated.

With population growth, it becomes desirable to extend or add to the rapid-transit system in some cities, and in other cities which have no rapid transit system, to study the desirability of providing a rapid-transit rail system or some other type of system to provide adequate transportation for the increased population.   The advantages of rail rapid transit are much the same as those given previously for commuter lines.   Although the likelihood of a rail rapid-transit system being self-supporting is not good, few cities are able to provide, on existing streets and highways, bus service that is self-supporting.

One principal difference between commuter systems and rapid transit is that rapid transit involves new construction in most cases.   Therefore, studies should be made to determine location and station spacing which will be most compatible with feeder buses at stations and will also be most convenient for the maximum number of people.

Rapid transit is subsidized through the DOT Urban Mass Transit Administration (UMTA). This agency is also sponsoring much research toward improving rapid transit components and developing new concepts.

Fig. 19-2. Model of a station and steel duo-rail-wheel, personal-rapid-transit system proposed for use along the "Strip" between central Las Vegas and the airport.   Operation is under fully automatic train control. (*Aerial Transit Systems of Nevada, Inc.*)

Personal rapid transit (PRT) is a system that provides passengers with individualized service. The system has a siding at each station into which a PRT car can be switched and stopped for loading and unloading passengers by computer control.   The PRT cars are relatively small (Fig. 19-2), usually with a capacity of 4 to 6 persons.   They are electrically operated.   A passenger can call a PRT car to a station by pushing a button or dialing.   After boarding, the passenger can designate the station to which he or she wants to travel by pushing a button or dialing, and the car will proceed to that station without stopping at any intervening station.   The operation is completely automatic.   Interference with other cars operating on the same line is prevented by computer scheduling.   Ticket selling and collecting are also automatic, using computer-controlled vending machines and turnstiles.

Vehicles for rail transportation predominantly use steel wheels on steel rails because of the low rolling resistance and heavy weight that can be supported on a single wheel. A few rapid-transit systems utilize vehicles with rubber tires that run on concrete beams or "rails" and are self-guided. Disadvantages of such vehicles are the higher rolling resistance, greater operating cost, and lower weight-supporting capability. Measurements of noise level and of vertical and lateral accelerations indicate no quieter or more comfortable ride with the rubber-tired systems than with well-designed steel-wheel-on-steel-rail systems. As for acceleration and deceleration, the coefficient of adhesion between steel wheel and steel rail is high enough to permit as fast an acceleration and deceleration as passengers can tolerate with comfort. Therefore, the higher coefficient of friction of the rubber tire on the concrete rail is of no real benefit. And if the rubber-on-concrete system is exposed to snow, or if ice can accumulate on the concrete rails, the coefficient of adhesion can become less than that for steel wheels on steel rails, particularly if the latter are sanded.

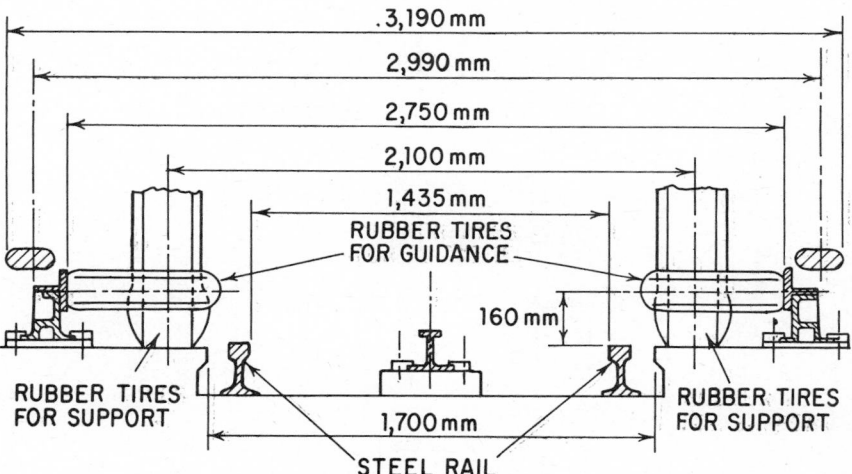

Fig. 19-3. Illustration of guideway for Paris Metro rubber-tired rapid-transit vehicles. Steel rails and wheels are required for guidance through turnouts.

Another serious disadvantage of the rubber-tired system occurs in operation through turnouts. In one system (Fig. 19-3), the rubber tires are spaced far enough apart to permit a regular rail track structure with switch points and frog to be placed at the turnout location. The vehicle has two steel wheels and an axle at each end. As the vehicle approaches the turnout, the concrete rails are ramped downward so that the vehicle is supported on the steel wheels through the turnout, after which the concrete rails are ramped up to support the vehicle again. In another system (Fig. 19-4), one end of the steel guide beam is switched back and forth to line up with the desired track. Vehicles must be operated at slow speed through the turnouts on either of these types of rubber-tired systems.

Method of traction for rail passenger and freight intercity systems is primarily diesel-electric locomotives. Where traffic density warrants, electric locomotives are used with an overhead catenary or a third rail. Most commuter systems are powered by diesel-electric locomotives with push-pull controls in some of the cars, so that the train does not have to be turned around at each terminal of the run. Several commuter systems are electrified, and each car has its own motor drive so that a separate locomotive is not required. All rapid-transit systems are electrified, and each car has a driving motor for each axle to give sufficient adhesion for the rapid acceleration and deceleration required. Personal rapid-transit systems are also electrified. Much research is being sponsored by UMTA on this type of transportation. This includes the following guided minivehicle systems: suspended monorail, air levitation on a concrete guideway, rubber tires on an aluminum guideway, and rubber tires on a concrete guideway. No doubt other systems will be developed.

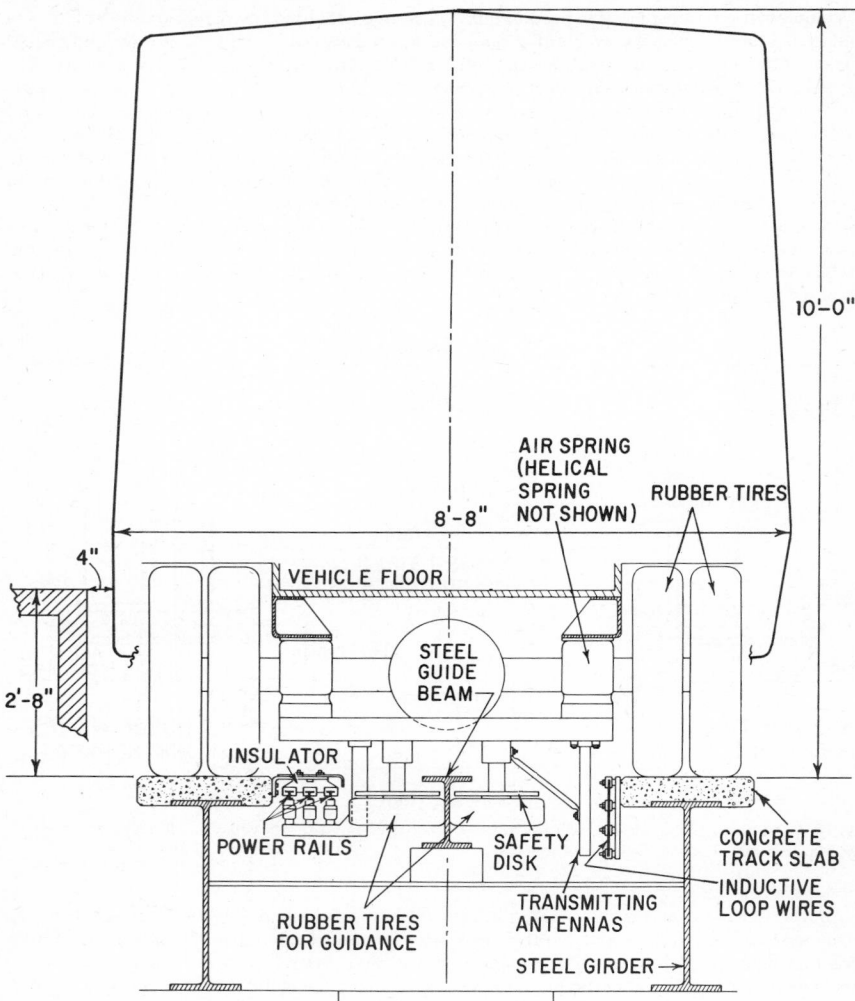

Fig. 19-4. Cross section of guideway and vehicle for Westinghouse "Transit Expressway" system.

Since switching is such an important part of a personal rapid transit system, the conventional duo-rail wheel system has an important advantage that will be difficult but not impossible to overcome. For example, the personal-rapid-transit car in Fig. 19-5 is supported on four rubber-tired wheels and guided in the guideway by another four rubber-tired wheels. The guide wheels may be computer-controlled to make the vehicle follow either the left or right guiding surface. Thus, at a station, the car may be made to pass by directing the guide wheels to follow one guiding surface, or the car may be made to turn into the station track for a stop by directing the guide wheels to follow the other guiding surface. Thus, no moving parts are needed in the guideway to make a car bypass or stop at a station.

Supports for any type of system can be wheels (steel or rubber-tired), air-cushion levitation (Fig. 19-6), or magnetic levitation (Fig. 19-7). Since either type of levitation is costly and complicated, there must be overriding advantages to justify this expense if such a system is to be used.

Power for a transportation system can be diesel-electric, electric, gas-turbine electric, gas-turbine hydraulic, jet propulsion, linear induction motor (Figs. 19-6 and 19-7), or pneumatic.

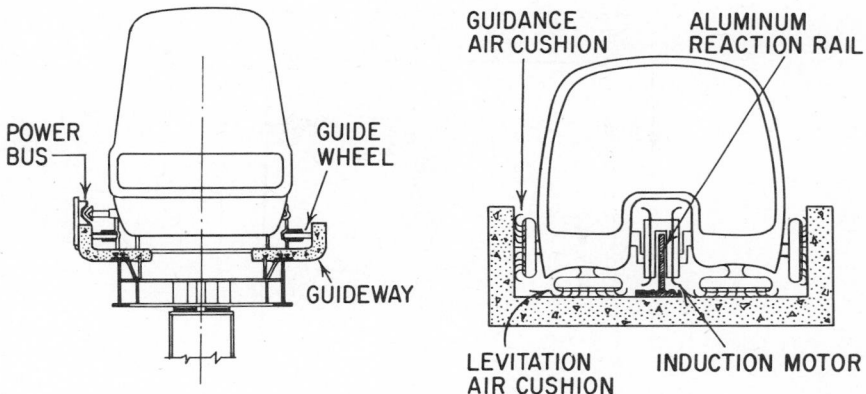

Fig. 19-5 left. Outline diagram of the Alden StaRRcar personal-rapid-transit system. Vehicles may have a 10- or 20-passenger capacity.

Fig. 19-6 right. Schematic illustrating the principle of the tracked air-cushion vehicle with vertical reaction rail for linear induction motor.

The costs and characteristics of each must be taken into account in selection of the type of propulsion for any given transportation system. Much experience has been had with the diesel-electric and electric motor drives. Some experience has been had with the gas-turbine electric motor and gas-turbine hydraulic drive. This experience shows that it is difficult to compete with the diesel-electric or the electric motor drive. So far, the efficiency of the turbo-electric or turbo-hydraulic drive has not been brought up to that of the other two.

For speeds over 100 mph, the electric motor drive has the advantage over the diesel-electric because the electric drive does not have to pull the weight of the electric generating plant and because, for short periods of time, it can draw a great deal of power from the catenary, whereas the diesel-electric has a fixed maximum power. But at speeds above about 150 mph, catenary and rail-wheel adhesion become a problem.

A speed of over 200 mph has been attained in trial runs with an electric-motor-powered vehicle supported on steel wheels on conventional track. However, to attain speeds of 200 to 300 mph regularly, vehicles may have to be powered by a linear induction motor or turbine jet. The latter, however, may be objectionable because of the noise level, and the former poses the difficulty of keeping the track reactor in accurate line and surface for such high speeds, as well as maintaining it free from windblown debris, sand, snow, and ice. At such high speeds, the power required to overcome air drag is considerable. In addition, it would probably be necessary to go to air-

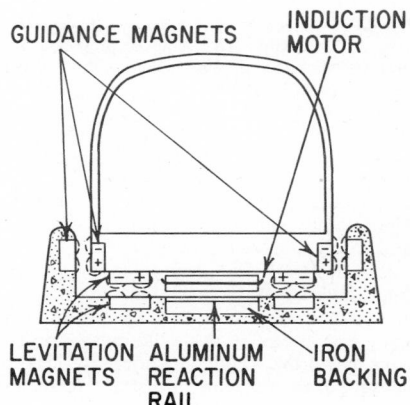

Fig. 19-7. Schematic illustrating the principle of the magnetic levitation vehicle with horizontal reaction rail for linear induction motor.

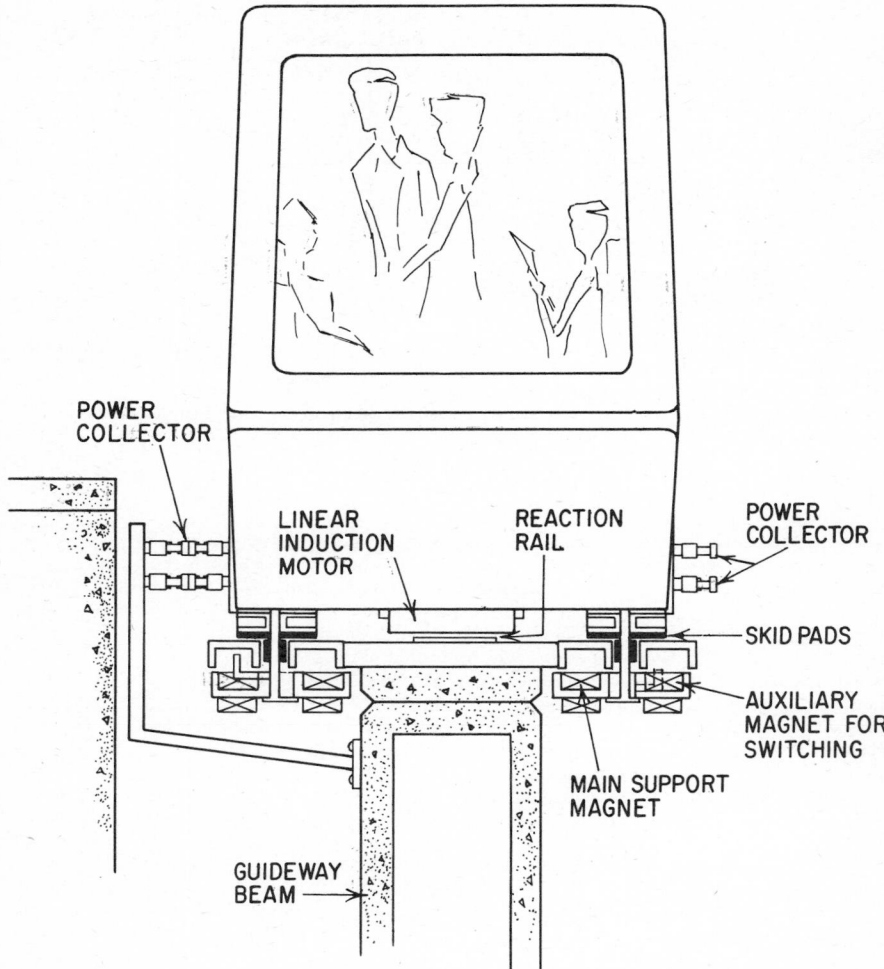

POWER
COLLECTOR

LINEAR
INDUCTION
MOTOR

REACTION
RAIL

POWER
COLLECTOR

SKID PADS

AUXILIARY
MAGNET FOR
SWITCHING

MAIN SUPPORT
MAGNET

GUIDEWAY
BEAM

Fig. 19-8. Schematic of an intermediate-capacity rapid-transit system proposed by Krauss-Maffei A. G., Munich, Germany. The vehicles, propelled by a horizontal linear induction motor, are suspended by magnets while electric sensors control the width of the suspension gap. Automatic train operation·is used with computer control.

cushion (Fig. 19-6) or magnetic levitation (Figs. 19-7 and 19-8), which would require still more power.

Levitation Systems.   Much experimental and developmental work is being done in the United States and other countries on levitation of vehicles by an air cushion or by magnetism with linear-induction-motor propulsions.   There seems to be little likelihood that these systems will be economical for commuter or rapid transit service at relatively low speeds with frequent stops, because of the much higher cost of these systems compared with the conventional steel duo-rail and wheel systems or rubber-tire-supported systems.   However, for intercity passenger service at distances up to 300 miles, particularly in high-population corridor areas, such systems may prove to be more attractive for travel than the duo-rail system and may be desirable because of the present congestion, delay, and travel and waiting time to and from airports for air travel.

Krauss-Maffei states that a high-speed levitation system must offer these advantages: reduced travel time, comfort, safety, punctuality, competitive fares, minimum disruption of the environment, compatibility with other transit systems, and a minimum chance of failure. To this should be added: operable in all weather conditions.

The air-cushion support system is not favored by many engineers working in the levitation field because of its high noise level, its high power requirements, the weight of the air-cushion fans and motors, and the lack of an efficient design for track switches. Levitation requires about ten times as much power as magnetic levitation.

Several systems with magnetic levitation are being tried. One uses magnetic attraction with solid-state sensors to maintain a constant air gap (10 to 20 mm). This requires a very smooth track and an elaborate control system. In another system, magnetic repulsion develops a relatively high magnetic drag, which is equivalent to about 200-mph headwind. Another approach is the superconductive induction system. Cryogenics are used to increase the power of the magnets.

It may be possible to operate with a rail-car gap of as much as 15 to 30 cm with this system. This large a gap would greatly simplify the problem of maintaining the necessary accuracy in trackway line and surface. This approach produces suspension by a repulsive-force interaction between the vehicle magnet and the induced current. Still another approach is to combine the repulsive-magnetic levitation force and the linear-induction traction force by having two angle reaction rails to provide levitation, lateral guidance, and traction. Which of the above systems will be most attractive and practical will depend upon cost, test performance, and passenger acceptance.

(See R. D. Thornton, "Flying Low with Maglev," *IEEE Spectrum,* April, 1973; Gerhart W. Heumann, "German High-Speed Railroads," *Machine Design,* Sept. 6, 1973; Joseph Hanlon, "Magnets Boost High-Speed Trains," *New Scientist,* Feb. 15, 1973; F. N. Houser, "Moving People: Which Is the Shape of Tomorrow?," *Railway Age,* Aug. 13, 1973; Henry H. Kolm and Richard D. Thornton, "Electromagnetic Flight," *Scientific American,* Oct. 1973.)

From the preceding general discussion of the characteristics of different systems, it is possible for an engineer to select one or possibly two systems that look most promising to meet a given set of transportation requirements. More specific details of each system will be given in the following pages, from which it will be possible to make a final selection and recommendation of the type of system to be used.

19-3. Cost-Benefit Analyses of Transportation Systems. In the U.S., new construction of intercity rail and freight systems consists mostly of line changes, grade revisions, and trackage to serve new industries, mines, and quarries. The justification for line changes is primarily reduction of curvature to permit higher speeds or shortening the line to reduce running time, in order to compete better for passenger and freight business. The justification for grade reductions is to permit longer trains to be hauled with one crew or to eliminate the cost of helper engines. The benefits obtained by these measures may be determined from data given in Arts. 19-16 and 19-17. The cost of the line changes or grade revisions should be estimated from the cost of right of way required and the cost of building the new line. The benefits of new line construction for industries, mines, and quarries should be based on the added revenue the new lines may be expected to produce, balanced against the cost of construction, maintenance, and taxes for the added trackage.

For contemplated rapid-transit systems, the cost-benefit analysis is more involved. There are a number of quantifiable benefits that can be included in the analysis. There are also a number of nonquantifiable benefits that should be taken into consideration in making the final decision.

Quantifiable benefits are those which produce a net economic gain and are directly attributable to the rapid-transit system. These include land cost savings, increase in land values, saving of rider time, reduced auto operating and parking costs, reduced congestion for auto traffic, reduced congestion for pedestrians in the business district, reduced need for a second or third car for some families, insurance cost savings, and transportation cost savings.

The cost-benefit analysis should be made for a reasonable period of time in the future, and should include projected population growth, amortization of equipment at the interest rate that must be paid over a period of 25 to 30 years (some equipment has been used longer but should have been replaced because of obsolescence), and interest on the cost of rapid-transit fixed facilities (roadway, stations, and shops). It is assumed that maintenance and operating costs are included in the quantifiable transportation cost savings.

Nonquantifiable benefits include increased regional growth; development of community centers; benefits to minority wage earners; attraction of new industrial development; added employment in the construction, maintenance, and operation of the system; adequate transportation for the young and aged; increased accessibility to educational, institutional, and recreational facilities; reduction in air pollution; and reduction in the total energy required.

A cost-benefit analysis for a personal rapid-transit system would be along the same lines as one for a rapid-transit system but would probably be far less extensive and have a more specific objective or purpose.

19-4. Route Selection.   As stated in Art. 19-3, new construction for intercity passenger and freight systems usually involves line changes, grade revisions, or trackage to new industries.   The only consideration in route selection is to obtain the desired objective at lowest cost with minimum environmental damage.   Since grading and bridge structures will probably be the only items that can be varied, use should be made of existing government topographical and geological maps to the extent they will suffice.   If a considerable amount of trackage is involved, it will probably be desirable to have aerial contour maps made (photogrammetry), first on a large scale to lay out one or more possible routes and then at small scale along each route to arrive at estimated grading quantities.

Maximum grade and rate of curvature must be established before a location can be chosen. Grade is expressed as the ratio of rise to distance in percent (a 1% grade rises 1 ft per 100 ft).   Rate of curvature is the central angle, degrees, subtended by a 100-ft chord.   It is desirable that both grade and rate of curvature be kept to a minimum, but almost always a lower grade and rate of curvature means increased construction cost and sometimes a longer time.   Studies should be made of several routes having different grades and rates of curvature, taking into account the annual carrying charges on construction cost and the estimated costs for the anticipated train operation.   From these studies, the grade and curvature may be selected to give minimum costs. A calculation of running time should be made and considered in making this decision.

Track gage must be decided upon early.   Standard gage for railway track in the United States (and many other countries) is 4 ft 8½ in.   It is measured between the inner sides of the heads of the two rails of the track at a distance ⅝ in. below the top of the rails.   This gage should be used if equipment is to be interchanged with other railroads having standard gage.   Locomotives, cars, and mechanized work equipment are commonly manufactured for this gage.

A roadway cross section must also be adopted.   A minimum width of roadway crown of 22 ft is recommended.   For sidings or multiple track, a minimum distance between track centers of 14 ft is recommended.   On fills, the side slopes should be at least 1 on 1½ in earth, 1 on ½ in loose rock, and 1 on ¼ in solid rock (Fig. 19-9).

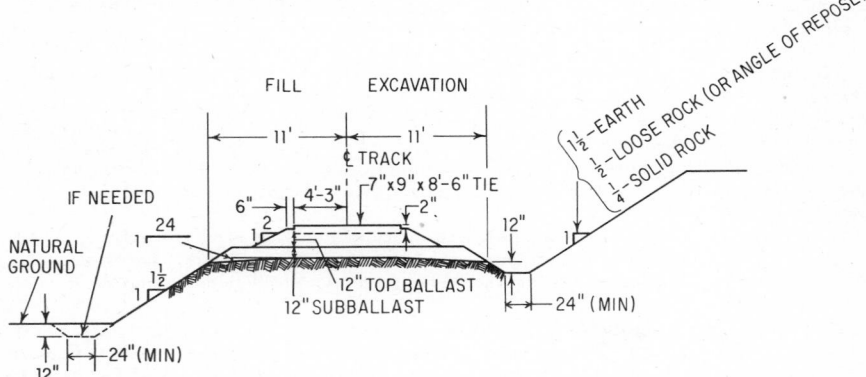

Fig. 19-9. Typical roadbed and ballast cross section for straight single track; topballast, about 3,700 cu yd per mile; subballast, about 4,900 cu yd per mile (includes 15% shrinkage).

Final location of the line should be made with transit and level after the general location has been established.   The transit location should be referenced to land lines so the necessary right of way can be acquired.   Elevations should be taken along the center line of the located track. They also should be taken at right angles on both sides every 100 ft, or oftener if the topography warrants, to permit calculation of quantities of excavation and fill.   Center-line stakes should be driven to show alignment at 100-ft stations on straight track and every 50 ft on curves.   Slope stakes should be placed as required.   If spirals are to be used (as will generally be the case), these need not be staked out during location, but the curves should be offset the proper distance from the tangents to provide for the spirals.

Commuter and Rapid-Transit Lines.   Route selection for these shorter lines will be determined by a number of factors.   Since these systems are to serve people, a route that will be closest to the largest number of people is to be preferred.   This may be done by taking into account the following:

Service to existing land use, which includes major employment areas; residential areas; institutions (hospitals, schools, churches, recreation, and other public facilities); and sports, zoo, parks, and other cultural and recreational areas.

Availability of right of way, an important factor in the cost.   Alternative alignments that make use of existing right of way, vacant undeveloped land, and publicly owned land and streets will minimize acquisition costs and relocation of homes and businesses.

Current plans and proposals for public and private projects that are contemplated for the future.

For rapid-transit lines, it is possible to use steeper grades than for intercity passenger and freight lines, although here again minimum practical grades will afford operating economies.   For horizontal curves, the degree of curvature should be kept to the minimum practicable.   The maximum curvature permitted will depend upon the desired running speed and the amount of superelevation provided.   Consideration should also be given to the length of car to be operated in a subway, because the sharper the curve, the greater the width required (due to overhang) and the higher the cost of the tunnel construction.

For one rapid-transit system, the following track standards have been established:

Tangent: desired minimum length, 500 ft and absolute minimum, 75 ft; extension at stations, 100 ft beyond length of platform.

Curvature: desirable minimum radius for main-line track, 1,000 ft; for yard tracks, 250 ft; minimum track radius for track in circular tunnels, 1,000 ft.

Spiral or transition curves should be used between tangent and curvature of 1° or more, except in yard or slow-speed trackage; also should be used between compound curves.

Grades: maximum between stations, 3.5%; through stations and at terminal storage tracks, 0.3%; other storage tracks and yard areas, level; minimum length of constant-profile grade, 500 ft.

Vertical curves should be used between changes in grade; minimum curve length, 100 times the algebraic difference of the grades being connected, but not less than 200 ft; where vertical and horizontal curves are combined and an unbalanced superelevation in excess of 1 in. is present, the length should be doubled.

Reversed curves should not be used without incorporating the minimum tangent length or length required for the two runoffs of elevation, whichever is greater.

Superelevation should be the equilibrium for the speed permitted, with a maximum of 4 in.; an unbalanced elevation of up to 2½ in. is permitted for speeds requiring more than 4-in. elevation.

Right of Way.   For intercity passenger and freight lines, the right of way needed should be determined to accommodate the number of tracks and the slope for the cuts, fills, and borrow pits.   Unless the line is being located in a densely populated area or land cost is very high, a minimum right-of-way width of 50 ft each side of the track should be obtained.   Allowance should also be made for any stations or yard facilities that may be required.

Location of receiving, classification, and departure yards is governed primarily by operating requirements.   At one time, yards were provided between divisions having different ruling grades.   With diesel power and the need to avoid delay due to switching, it is generally preferable to handle the same train from origin to destination, adding or taking off diesel units at intermediate points if conditions justify.

For rapid-transit systems in densely populated areas and through business districts, the tracks will necessarily be elevated or in subway and wherever possible will be located over or under existing streets.   Because of esthetic considerations, to avoid blocking of light, and for a minimum of noise, a location in subway is to be preferred, unless the underlying ground condition is such as to make subway construction impractical.

19-5. Track Location.   Location of the tracks on prepared roadbed—including cut, fill, and sidehill cut and fill—or natural ground surface is the most economical and is to be preferred where practicable.   However, in some instances, other locations are more desirable for reasons that are more important than the first cost.   This applies particularly to rapid-transit systems that are to be constructed.

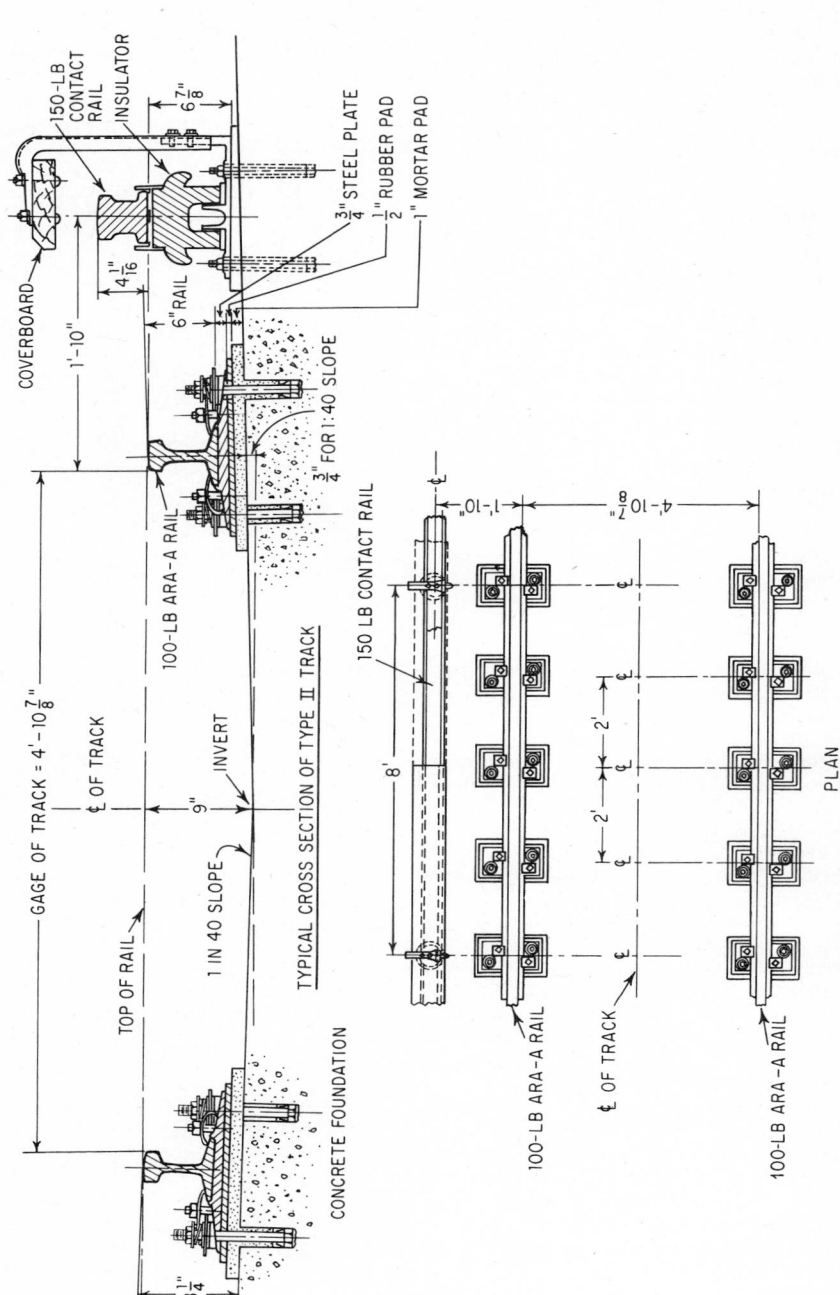

Fig. 19-10. Tangent track construction used in subways of the Toronto Transit Commission.

In some instances, a city has provided space for trackage in the median strip of expressways when they were constructed, anticipating the construction of rapid-transit trackage at a later date when population growth would require it. In this case, the track location has already been decided on. Otherwise, trackage for new rapid-transit systems should be constructed in open roadbed wherever practicable.

In residential areas, the trackage should be elevated or placed in open cuts to avoid street grade crossings. The choice between the two is largely a matter of which costs the least from an overall standpoint of first cost and maintenance. Open cuts will probably require reinforced concrete retaining walls on each side of the trackage with a chain link fence and barbed-wire outriggers on top of each retaining wall to prevent children or others from falling into the cut. This could be avoided by use of a tunnel, but a tunnel is more costly to construct and maintain.

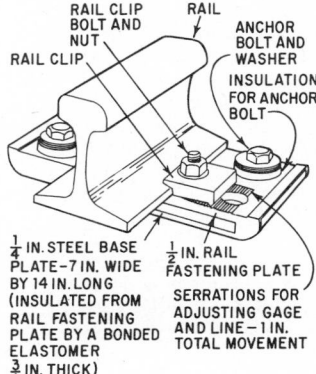

Fig. 19-11. Landis direct-fixation rail fastener.

Elevated trackage is in most cases preferable to cuts. It has its disadvantages, principally in terms of esthetics, and the effects of noise on nearby residents. However, in modern elevated-track construction, the track support is of either reinforced concrete or prestressed concrete, or a combination of the two, and of pleasing appearance. The elevated construction may have a ballast deck so that the trackage can be supported on ballast, which does much to reduce the noise, or else the rails may be supported directly on the concrete floor, in which case special fastenings will be used with resilient pads between the rail and the deck to reduce the noise level. Figure 19-10, for example, shows the rail fasteners used in tangent track construction in subways of the Toronto Transit Commission. A rubber pad is inserted under the steel rail plate for noise and vibration attenuation. Figure 19-11 shows the Landis direct-fixation rail fastener, developed for use on the San Francisco Bay Area Rapid Transit system and later used for other installations. This device for directly affixing rail to a rigid support structure incorporates a shear pad consisting of a ½-in.-thick steel rail fastening plate and a ¼-in.-thick steel base plate bonded on opposite sides of a ¾-in.-thick elastomeric pad. The base plate is bolted directly to the supporting structure. The elastomeric pad not only insulates the rail plate from the base plate but also permits an elastic deflection of about ¼ in., for attenuating noise and vibration.

A noise level of 70 to 75 dB(A) is comparable to noise frequently encountered in residential areas, and 70 to 80 dB(A) to usual noise in commercial and retail districts. In residential areas, tracks should not be closer to homes than 100 to 120 ft for elevated structures or ballast on grade or on fill. For speeds of 50 mph or above, a sound barrier should be placed between the track and any house within 120 ft. In commercial and retail areas, trackage can be as close to buildings as 30 ft if a sound barrier is provided. The sound barrier may be a vertical wall extending from the ground to 10 in. above the bottom of the car side skirt with 8 to 10 in. clearance, lined on the inside with acoustic material. (Most of the noise from the rapid-transit car comes from the trucks and wheel impacts.) Such a sound barrier will lower the noise level about 12 dB(A). Use of continuous welded rail with the running surface periodically ground smooth with a grinding car or a train of such cars will also be helpful in reducing the noise level. It is also desirable to keep the car wheels ground smooth to reduce impact noise.

In the main business district, rapid transit should be placed in subway, if that is at all practicable. This eliminates the problem of train noise and does not require space in the street for the supports of an elevated structure. Also, there is no diminution of daylight at street level. Here again, use of resilient rail supports (Figs. 19-10 and 19-11), continuous welded rail periodically ground, and wheels kept ground smooth will minimize transmission of train vibrations through the ground to nearby buildings.

For intercity passenger and freight lines, the location of passing tracks and yard tracks should be taken into account in establishing the grade line. If the line is for a single track, the time it takes for train *A* to go from one passing track to the next and meet train *B* and then for train *B* to get to the first passing track determines the capacity of the railway in trains per day. Thus, passing tracks spaced close together afford larger line capacity than those spaced far apart.

The sidings should be long enough for the maximum length of train to be clear of the main line. If centralized traffic control is used, even longer passing tracks are desirable to permit passing without stopping the trains.

Passenger Capacities.   Commuter and rapid-transit lines will be double-track in most cases. It is necessary to have crossovers suitably located to permit use of only one track at slack periods so that track repairs can be made, a disabled car or train bypassed, third rail or trolley repairs made, or for other reasons. However, the addition of crossovers cannot be expected to add much to traffic capacity of such a double-track line. Addition of a third and fourth track would be the most effective way to increase capacity, if that were needed.

The capacity of a double-track, steel, duo-rail system is normally about 40,000 passengers per track per hour. This is based on 10-car trains, with 300 passengers per car, operating at 5-minute intervals. The determining factor is the time required for a train to come into a station, unload and load passengers, and depart from the station.

If capacities of this amount or more are contemplated, station design should be planned accordingly. Capacities in excess of 40,000 per track per hour are possible if the stations can be designed to handle passengers at the proposed rate. Normally, all the passengers for one train would not unload or load at a single station. The exceptions to this are stations that serve a baseball field, football stadium, or similar facility from which large numbers of passengers may be discharged in a short time, and some emergency that requires many passengers to leave a station swiftly.

19-6. Station Location and Characteristics.   Station locations have already been established in the United States for passenger trains and most commuter service. Commuter stations are located in suburbs or city areas a few miles apart, and where local or bus transportation generates a large enough volume of passengers to justify a train stop. Usually, not all commuter trains will stop at all the stations, but a schedule will be established to provide reasonably frequent service, particularly in morning and evening rush hours, at stations with a relatively low traffic volume.

Rapid Transit.   In planning new systems, several factors should be considered in deciding on station locations. These are:

Physical constraints: available space for the station, space for parking, space for bus and automobile circulation.

Accessibility: convenient location within network of freeways and arterial and feeder bus routes.

Service potential: number of persons, households, students, and jobs of various types located within 700 ft, 1,500 ft, and 3,000 ft of each station. Most people living or working within 1,500 ft of a station will walk to or from it. In outlying, low-density areas, automobiles and feeder bus lines will expand the service area of a station.

Convenience to major institutions and centers: schools, hospitals, recreational areas (including sports facilities), and major industrial and commercial concentrations located within 700 ft of each station.

Development opportunities: joint development potential of vacant or deteriorated structures within 700 ft of each station.

Impact on neighborhood: localized traffic congestion, reinforcement of community centers and boundaries, and conformance with local development plans.

Projected ridership: number of riders coming to and from each station, projected for 25 years ahead.

From the above, it will be possible to determine the location (tentative) for each station which will attract the maximum number of riders and give the best service. Stations should be placed closer together in areas expecting the greatest number of riders, not only to give better service but also to avoid undue congestion within and outside the station.

Station platforms should be as long as the longest train that will be operated. For passenger and commuter trains, the platforms or paving for loading and unloading are generally outside of the track or two tracks. Also, most existing platforms are at top-of-rail level and are 6 ft or more in width. However, in new construction for commuter service and rapid-transit service, it is preferable to have the platforms between tracks at a height of 42 in. above top of rail. The platform width should in no case be less than 10 ft and should provide 8 sq ft of occupancy space per person for maximum assembly crowds. In one subway system, a platform between tracks about 30 ft wide is provided.

Station construction should comply with the applicable building codes. Suitable lighting, satisfactory noise level, comfortable air conditioning, pleasant appearance from the standpoint of both decor and cleanliness, control of wind and odors, and clear circulation (by appropriate directional signs if needed) should be provided. Signs at street level, illuminated at night, should clearly indicate to a stranger where the rapid transit entrance is located. Steps down to the station should also be illuminated when required for safety. Steps should be kept clear or sanded where exposed to the weather, and handholds should be provided on both sides. All walking surfaces on stairs and in passageways should be kept dry and covered with a suitable non-skid material. Small sidewall depression "trenches" parallel to the side walls and suitably drained should be provided to accomplish this.

Other important criteria, in addition to all appropriate safety measures, are traffic-handling capability, consistently available information, and orientation. Maps of the system showing all lines and station stops should be placed conspicuously, with the particular station clearly designated thereon. Stations should include a free area and a service area. It is desirable that provision be made for concessions, if any, in the free area and that necessary facilities, such as electric outlets, water supply, etc., be provided at a suitable location for the concessions. Automatic coin-operated dispensers should be located in the free area, but both these and concessions should be located so as not to interfere with free circulation of passengers to and from the trains.

A minimum clearance of at least 8 ft should be provided throughout the station and platform. Adequate space should be provided at ticket facilities to allow for a line of ticket purchasers without interference with the normal flow of passengers.

Passageways and stairs should be located to give a balanced train loading and unloading—usually at each end of the platform, or at the platform midlength, or both. An island-type platform between two tracks is preferable to a platform on each side of the two tracks.

Escalators should be provided to carry passengers up whenever the stair height exceeds 12 ft and to carry them down when it exceeds 24 ft.

To determine the space required for exit and entrance doors, which should open out or revolve, the maximum number of passengers that will pass in any 15-min peak period should be estimated. Enough space should then be allowed to clear the platform under normal conditions within the headway time between trains. For emergency evacuation of a train, provision should be made to clear the platform in 4 min. To do this, the designer may assume a crush capacity of 25 passengers per minute per foot width of passageways, 20 passengers per minute per foot width of stairways, and 100 passengers per minute for each 48-in. escalator. Enough supplemental stairway width should be provided to permit evacuation if the escalator should become inoperable.

Between subway stations, tracks will be separated by the tunnel walls or a concrete wall. Walkways with a minimum width of 2 ft should be provided on one side of all line sections of tunnels and subways, and for high-speed train operation, a hand rail should be placed on the wall 3 ft above the walkway floor. Walkways should be placed on adjacent sides of the wall to permit cross connection between the walkways. Crossways should be placed not more than 1,000 ft apart for workers and emergency evacuation of passengers. ("Guidelines and Principles for Design of Rapid Transit Facilities," Institute of Rapid Transit, 1612 K Street, N.W., Washington, D.C. 20006.)

Telephones, toilets, storage lockers, service rooms, and toilets for station personnel should be provided at stations as warranted.

Lighting is of great importance for safety and the security of passengers. Table 19-1 may serve as a guide for minimum illumination levels at different locations.

Subway Ventilation Objectives are to:

Provide a comfortable environment for patrons and staff.

Provide, in the event of fire, control and removal of smoke and a supply of fresh air for evacuation of passengers and for fire-fighting personnel.

Provide for the removal of heat generated by normal train operation.

Provide control of condensate and haze, and removal of objectionable or hazardous odors and gases.

The piston action of trains will provide a considerable amount of ventilation if suitable vent shafts are provided. If necessary, supplemental mechanical ventilation must be provided.

Table 19-1. Recommended Minimum Illumination Levels in Passenger Stations*

Locations	Illumination, foot-candles
Platform, subway	20
Platform, under canopy, surface and aerial	15
Uncovered platform ends, surface	5
Mezzanine	20
Ticketing area, turnstile	30
Passageways	20
Stairs and escalators	25
Fare-collection kiosk	100
Concessions and vending machine areas	30
Elevator (interior)	20
Above-ground entry to subway (day)	30
(night)	10
Washrooms	30
Service and utility rooms	15
Electrical, mechanical, and train-control equipment rooms	20
Storage areas	5

*From "Guidelines and Principles for Design of Rapid Transit Facilities," Institute of Rapid Transit.

Maximum piston-type ventilation can be obtained by having the tunnel or subway section as near the size of the train cross section as clearance requirements will permit and having a separate tunnel or subway for each track.   Fan shafts must be located relative to the vent shafts and stations in such a way as to insure that all sections of the subway and stations can be purged under emergency conditions.   The ventilation rate should satisfy the purge rate.   A minimum air velocity of 4 ft per sec is recommended for determining the sizes of fans and appurtenances.   Vent and other shaft openings on the surface should be located to draw in unpolluted air and protected by gratings or screens.   Acoustic treatment of the shafts should be provided if needed.   Fans for emergency ventilation should be connected to power feeders from two separate sources and should be operable through remote controls located at a control station.   ("Subway Environmental Design Handbook," Vol. I, Urban Mass Transportation Administration, Washington, D.C. 20590.)

Security and communications are of utmost importance for the safety and comfort of passengers and for train operation.   Security can best be provided by closed-circuit television cameras suitably located at strategic locations in the station, passageways, and platforms.   These instruments should be monitored at each station at which there is an agent and at a central control station.   An alternative for trains would be to have the monitors for the cars where the train attendant can see them and to provide the attendant with a means of communicating with the nearest agent and the central control office.   Telephone communication between each station and the central control office should be provided.   Portable radios should be considered for use between security personnel and station agents and the central control office.   This type of radio is not effective in subway or tunnels, so its provision to train attendants will depend on how much of the line is open or aerial structure.   Or a special antenna line can be placed through the subway and tunnel sections to enable portable radio communication.

Fare collection should generally be accomplished on the trains in commuter service but at the stations in rapid transit.   For this purpose, station personnel can be augmented by turnstiles, either coin-operated or using coded tickets or some other suitable method.   In some commuter and rapid-transit service, coin (or currency and coin) vending machines are used for selling coded tickets, and computer-controlled turnstiles are used for collecting and monitoring them.   One system has experienced about 150 failures a month with 43 changemakers and well under the one-failure-per-machine-per-month guarantee for its computerized turnstiles.

19-7. Passenger Terminals.   These comprise all the facilities required for handling passengers, baggage, mail, and express and for servicing, repairing, and storing cars and locomotives.   The facilities include parking space for automobiles; entrance, standing, loading, and exit areas for taxicabs; ticket windows; waiting and rest rooms; newsstands and concessions; lunch counter and dining room; cocktail lounge; drinking fountain; telephone and telegraph booths; baggage carts, lockers, and check room; means for handling mail, express, and baggage; concourse to trains; and stairways, moving stairways, and elevators to trains, if necessary.

Tracks should be on 20-ft centers, with paved platforms between. The platforms should be covered with a suitable type of roof. A minimum width of passage for passengers of 6 ft should be allowed on platforms, stairways, and ramps. The incline of ramps should not exceed 7%. Platforms at service points for trains passing through should have water hydrants, electrical outlets, steam connections, and brake shoes available. Handling of baggage, mail, and express requires separate platforms or wide platforms so that trucks may pass without interfering with passengers; ramps or conveyor belts; and adequate space for sorting and transferring to other trains or trucks.

Servicing of locomotives and cars requires a coach yard, preferably with a mechanical washer to wash all equipment as it enters the yard. Tracks in this yard should be level and on 20-ft centers. They should have concrete platforms between them. An inspection pit 36 in. wide and 38 in. below top of rail is desirable for some of the trackage. Preferably, this area should be covered over to facilitate work in bad weather. Jacking pads and wheel drop pits should be provided. Other facilities needed are water hydrants meeting U.S. Public Health Service requirements; hot water; low-pressure air connections for cleaning; high-pressure connections for air brakes; electrical service outlets, including 220-volt ac for air-conditioning equipment; steam supply lines; adequate lighting for night operation; a convenient supply of brake shoes and mounted car wheels; car pullers; commissary facilities for dining cars; service building providing offices, toilet, wash, locker, and lunch rooms; storehouse; repair shops; refuse disposal; fire protection; bottling plant for refilling gas cylinders; and fuel oil and sand supply for locomotives.

The extent to which all these facilities should be provided depends on the number of trains and passengers to be handled during peak periods, with an allowance for train delays. Detailed recommendations related to the number of passengers handled may be found in "Manual for Railway Engineering," American Railway Engineering Association, Chicago, Ill.

Commuter Terminals. These should provide most of the facilities listed for passenger terminals, except for a lunch counter, dining room, and cocktail lounge, which would depend on the location, passenger volume, and demand. Also, baggage carts and check rooms would probably not be needed, nor would means for handling mail, baggage, and express.

Rapid-Transit Terminals. Requirements for stations were given in Art. 19-6. Terminal facilities should be provided at the end of each line for car storage, and a shop should be available for emergency repairs. A shop for overall and scheduled maintenance of cars should be provided at the location most suitable from the standpoint of accessibility in operation, land availability, environmental factors, etc.

For storage tracks, the length required may be determined by calculating the length of the number of cars required for peak movements plus 10% for spare units to replace cars out of service for repairs. Storage tracks should be level, and lead and other tracks on a grade of not more than 0.3%. Curves should not be less than 200-ft radius, but should be longer if the car units are designed to require a longer radius.

One or more shops should be provided at the chosen shop location for repair of electrical, electronic, hydraulic, pneumatic, control, and undercar equipment (including drop pits); for mechanical repairs; for wheel grinding; and for painting and seat repair. An automatic car washer should be provided. It is desirable that all the above work areas, except the car washer, be located under cover; work should be scheduled on an assembly-line basis; all work should be automated; and all workers should be provided with power tools to the extent such tools are available.

Storage and shop areas should be surrounded by a suitable fence to prevent trespassers from entering, for safety and to avoid pilferage. A watchman or automatically operated gates should guard the entrance tracks and driveway to the storage and shop area. A 7-ft-high chain link fence with barbed-wire outriggers inside the right of way is well suited for this purpose.

19-8. Freight Terminals.

On most railways, one or more yard facilities are required. Such a facility should have a receiving yard, classification yard, hold and repair tracks, engine house, and departure yard ("Manual for Railway Engineering," American Railway Engineering Association).

A yard consists of a series of parallel tracks, called body tracks, on which cars are placed, and a ladder track at each end. A turnout connects each body track to a ladder track. Thus, the ladder track is a means of placing cars on or removing them from each body track.

The receiving yard should be conveniently accessible from the main line. Its tracks should be long enough to hold the longest train without doubling it into two tracks. The number of receiving tracks required depends on the spacing of train arrivals and time required for classification. A spacing of 18 ft should be provided between parallel ladder tracks, 15 ft between a ladder track and any parallel track, and not less than 13 ft, preferably 14 ft, between body tracks. A gradient of not more than 0.15% is desirable to prevent the cars from rolling without setting the brakes. The classification yard may be a flat yard if the number of trains and amount of switching are relatively small. A gravity or hump yard should be used otherwise.

A *hump yard* utilizes gravity to expedite switching of cars. The train of cars is pushed up an incline to a hump, at which point one or more cars are successively uncoupled while moving and allowed to roll down the incline from the hump into the classification yard. The height of the hump must be sufficient to impart enough velocity to overcome the rolling resistance of each car to the farthest point in the yard. Thus, if the distance from the hump to the farthest point is 3,000 ft and the rolling resistance of the slowest-rolling car under adverse weather conditions is 10 lb per ton, equivalent to a 0.50% grade, then a minimum hump height of 15 ft would be required. Another requirement is that the decline from the hump be steep enough and long enough to separate the cars sufficiently to permit operation of switches and to clear the switches ahead of the following car. Usually, the hump height is from 16 to 20 ft. Two or three sets of retarders are provided for controlling the speed of the cars into the classification tracks. The retarders are set so that each car will roll the desired distance and couple to a standing car without undesirable impact (up to 4 mph).

Cars usually have journal-box lids opened and oil is added as they approach the hump. Humping speed is about 1 mph. In a fully automated, or so-called "push-button," yard, the operator pushes a button numbered to correspond to the track number into which a car is to go. When the car is uncoupled, it rolls down the hump and is weighed, if desired, on an electronic, un-coupled-in-motion, track scale. Also, the car's rolling resistance is measured by determining change in speed over a given length of track. This information goes to a control computer.

The approximate wheel load is measured by a track device. This reading goes to the computer, to limit the amount of retardation so that the wheels will not rise out of the retarder. The car speed is measured as it approaches each retarder, and this information goes to the computer. When the operator pushed the button for the track number, the computer was fed the total rolling resistance to the farthest point in that track. A wheel trip on each track corrects this value for the distance taken up by the number of cars that have already been placed in that track. From all these data, the control computer determines the speed the car should have as it leaves the last retarder to roll to the desired point and retards the car to that speed. Usually, a radar device is used to measure car speed. Retarders are pneumatically powered but electrically controlled. Switches are electrically set by the computer for the track number punched for the car.

Various grades have been tried for classification tracks. Good results are usually obtained with a grade of about 0.12%.

The departure yard should be long enough to accommodate the longest train. It should be level, if possible. If the grade is adverse to the direction of starting, it should be at least 20% less than the ruling grade over which the train will operate. Track spacing should be the same as for the receiving yard.

Car-repair tracks should be provided to accommodate the number of cars to be repaired and the repair time. These tracks can be alternately spaced 18 ft and a width sufficient to accommodate mechanical equipment. A paved driveway of rail height should be provided between tracks. It is desirable to have a car-repair building, with the required number of tracks, through which cars can be moved by cables and "rabbits." This provides more efficient working conditions and mechanical equipment for repair work at minimum cost and delay.

An engine house usually is required for servicing and storage of road and yard engines. For most efficient servicing, the engine house should have two levels, one at track level, and the other lower to permit workers to get under locomotives.

Other terminal facilities that may be required are:

Team tracks having an adjoining paved area for loading from trucks into the cars.

Stub tracks with a ramp at the ends. Narrow, car-floor-height platforms, with electric power outlets, should be placed between the tracks for loading and unloading "piggyback" or truck trailers on flatcars.

Elevating-type inclines for end loading of automobiles on auto-rack cars.

House tracks at a freight-station building for less-than-carload shipments.

Docks for loading cars or contents on boats.

Car dumpers that turn upside down and empty hopper-type cars.

Stockpiling facilities for coal and ore. Storage bins and elevators for grain.

Provision of overhead cranes, wheel grinders, wheel drop pits, etc., depends on the extent to which repairs will be made at the particular facility. Servicing facilities must also supply fuel oil, water, and sand.

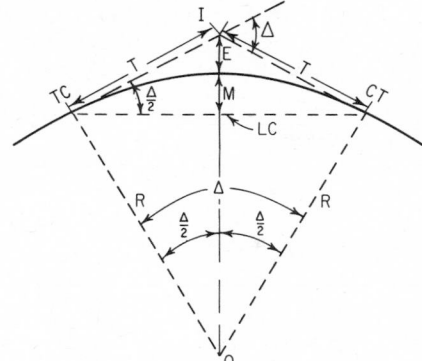

Fig. 19-12. Simple curve.

19-9. Horizontal Curves.   These include simple, compound, and reverse curves; superelevation required for such curves; and spiral curves as a means of introducing the superelevation on a gradual and uniform basis.

Simple Curves.   A simple curve has a constant radius throughout.   The degree of curvature generally is measured by the central angle subtended by a 100-ft-long chord.   Radius R, ft, and degree of curve D are related by

$$R = \frac{50}{\sin(D/2)} \tag{19-1}$$

For curves up to $7°$, length measured along the curve is practically the same as that measured with 100-ft chords.   Hence, the radius R of a curve is given approximately by

$$R = \frac{36,000}{2\pi D} = \frac{5,730}{D} \qquad D < 7 \tag{19-2}$$

For curves of more than $7°$, the error in radius increases with the degree of curve.

In the location or staking of the center line of a simple curve, the tangents (to its ends) should be extended, if possible, to an intersection I and the intersection angle Δ measured by transit (Fig. 19-12).   The tangent distance T from the point of curve, $T.C.$, to I and from the end of curve, $C.T.$, to I may be determined from

$$T = R \tan \frac{\Delta}{2} \tag{19-3}$$

Length of curve, ft, from $T.C.$ to $C.T.$ is given approximately by

$$L = \frac{100\Delta}{D} \tag{19-4}$$

where Δ and D are in degrees.

Stakes should be driven and tacked to mark the $T.\ C.$ and $C.T.$ This can be done by setting the transit at I and sighting along each tangent.   The transit then should be moved to the $T.C.$, sighted on I, and $\Delta/2$ turned for a check on $C.T.$ Next, stakes should be set every 50 ft for flat curves; the measurement should be made with 100-ft chords for curves over $7°$.   It is good practice to mark stations (100-ft intervals) around the curve and to set a stake at each station and at plus 50.

The transit deflection, degrees (angle between tangent and line from $T.C.$ to point on the curve), for each stake equals

$$\alpha = \frac{LD}{200} \tag{19-5}$$

where L = length of curve, ft
        D = degree of curve

Suppose, for example, a stake is to be set and tacked at $1,108 + 50$ when the $T.C.$ comes at $1,108 + 10.5$ and the degree of curve is $2°30'$.   A length of $50 - 10.5 = 39.5$ ft should be taped from the $T.C.$ and a deflection angle of $39.5 \times 2.5 \times 60/200 = 30$ min turned with the transit, to set the stake at $1,108 + 50$.   For each succeeding stake, at 50-ft intervals, the increment of deflection is $50 \times 2.5 \times 60/200 = 37.5$ min.

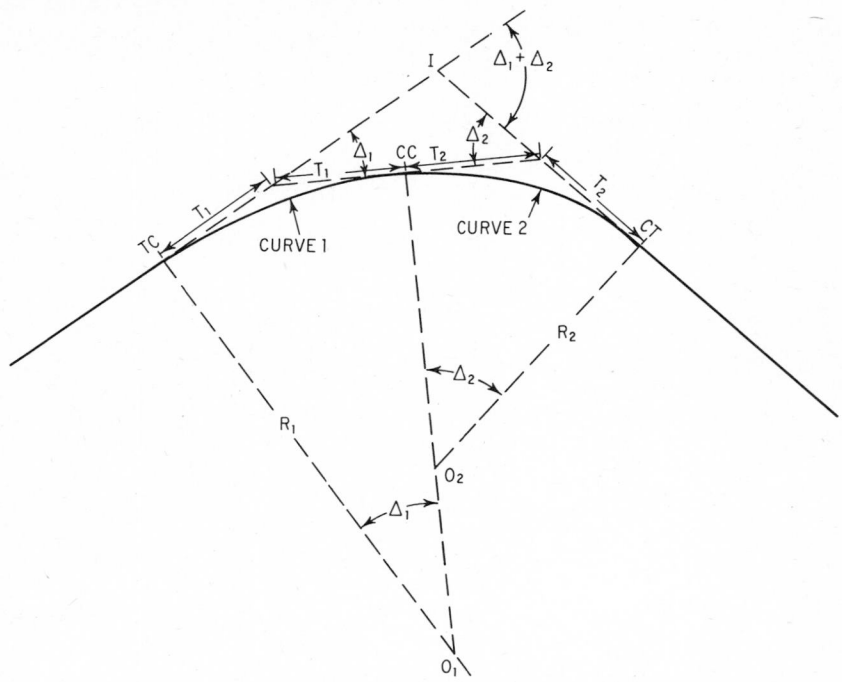

Fig. 19-13. Compound curve.

A railroad surveying handbook with tables of radii, functions of angles, functions of a 1° curve, and logarithms is useful for staking curves. (See also Spirals.)

(C. F. Allen, "Railroad Curves and Earthwork," McGraw-Hill Book Company, New York; T. F. Hickerson, "Route Location and Design," McGraw-Hill Book Company, New York.)

Compound and Reverse Curves.   A compound curve comprises two or more simple curves, each successive curve having a common tangent with the preceding curve (Fig. 19-13). The point of curve, $T.C.$, and end of curve, $C.T.$, are staked as for a simple curve, although calculation of the tangent distances is more involved. The transit should be moved to the beginning point of each simple curve to stake it. The degree and central angle for each simple curve of the com-

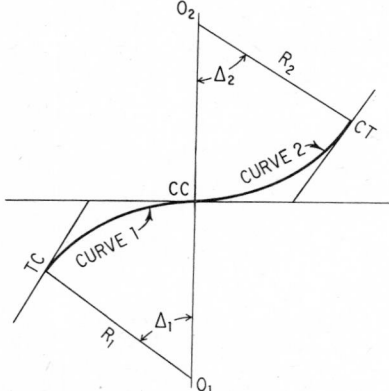

Fig. 19-14. Reverse curve.

pound curve have to be known or decided on in advance. Compound curves should be avoided, but they may be used where excessive excavation or fixed objects that must be cleared justify or require such a curve. (See also Spirals.)

A reverse curve (Fig. 19-14) is a combination of two simple curves with centers on opposite sides of a common tangent. Reverse curves are acceptable in slow-speed passing and yard tracks but should never be used in main line. A short tangent, at least 100 ft long, but preferably more, should be placed between curves of opposite direction in main line. (See also Spirals.)

Elevation of Curves. Elevation of the outer rail of a curve relative to the inner rail is desirable on main-line track. The amount of elevation depends on degree of curvature and desired operating speed around the curve. However, the amount of elevation is usually limited to 6 in. to prevent undue tilting of the train if it stops on the curve. For sharp curves, it may be necessary to restrict train speed so that it will not exceed by too much the speed for which the curve is elevated.

The amount of elevation to be provided on a curve, up to the 6-in. maximum, is a matter of judgment, subject to change from service experience. Usually, and on single-track lines particularly, not all trains will go around a given curve at the same speed. If too little elevation is provided for the predominating traffic and speed, the outer rail will show excessive wear on the gage side from the wheel flanges. If too much elevation is provided, the inner rail will show excessive flow of the top of the railhead toward the gage and field sides and sometimes surface corrugation.

Equilibrium speed is the speed at which outward centrifugal force from curvature is just balanced by the inward component of car weight resulting from elevation of the curve. For a given degree of curve and elevation

$$V = \sqrt{\frac{E}{0.0007D}}$$ (19-6)

where V = equilibrium speed, mph
E = elevation of outer rail, in.
D = degree of curve

Figure 19-15 shows the equilibrium speed for different degrees of curve and amounts of elevation for standard-gage (4 ft 8½ in.) track.

Permissible speed somewhat in excess of equilibrium speed will not cause discomfort to passengers or other undesirable effects. This permissible speed may be obtained readily from Fig. 19-15 by adding 3 in. to the actual elevation of curve. For example, for a 3° curve with

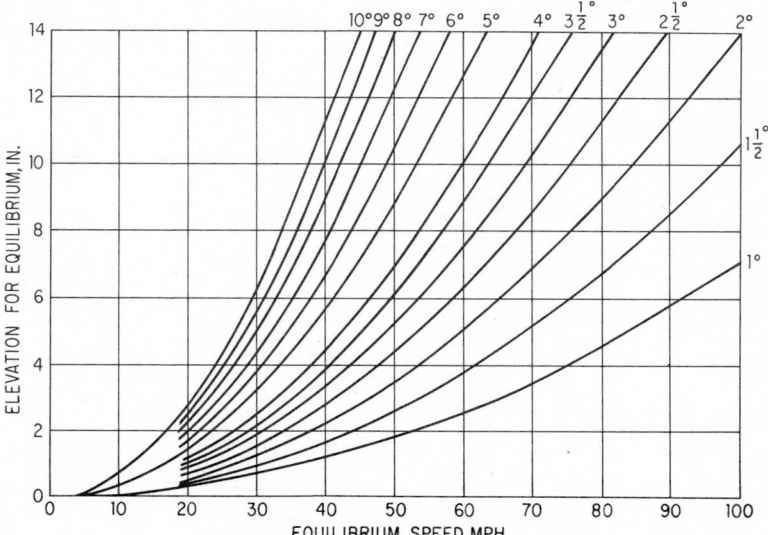

Fig. 19-15. Chart gives required elevation of outer rail above inner rail of standard-gage track on curves for various speeds and curvatures.

5-in. elevation, the equilibrium speed is 49 mph. The permissible speed, however, is 62 mph (equilibrium speed for 8-in. elevation). Thus, the permissible speed will have a deficiency in elevation of 3 in. This is acceptable for the type of equipment in general use in the United States. For passenger cars having antiroll devices, a somewhat higher deficiency is permissible (*Proceedings, American Railway Engineering Association,* vol. 56, p. 125). For some types of freight cars having a very high center of gravity (over 96 in. above top of rail) a somewhat smaller deficiency may be desirable to guard against derailment.

Spirals.   A transition curve or spiral should be placed between tangents and each end of a simple curve and between the simple curves of a compound curve. A spiral increases in curvature gradually, thus avoiding an abrupt change in the rate of lateral displacement of cars. It also provides a means of gradually elevating the high rail in proper relation to the degree of curvature.

Several forms of spiral may be used. The one generally used in the United States increases degree of curvature with length.

$$d = ks \tag{19-7}$$

where d = degree of curvature at any point
k = increase in degree of curvature per 100-ft station
s = length in 100-ft stations from the beginning of spiral to any point
The central angle δ, degrees, from the beginning of spiral, $T.S.$ (Fig. 19-16), and the deflection a, degrees, from the tangent at $T.S.$ vary as the square of the length.

$$\delta = \tfrac{1}{2}ks^2 \tag{19-8}$$

$$a = \tfrac{1}{6}ks^2 \tag{19-9}$$

Also, the offset of the spiral, ft, from either the tangent or the circular curve varies as the cube of the distance. Other key elements shown in Fig. 19-16 may be computed from

$$X_o = S(50 - 0.000508\Delta^2) \tag{19-10}$$

$$T_s = X_o + (R + O) \tan \frac{I}{2} \tag{19-11}$$

$$E_s = O + (R + O) \left(\sec \frac{I}{2} - 1\right) \tag{19-12}$$

$$O = 0.1454\Delta S \tag{19-13}$$

where S = total length of spiral in 100-ft stations
Δ = total central angle of spiral, deg

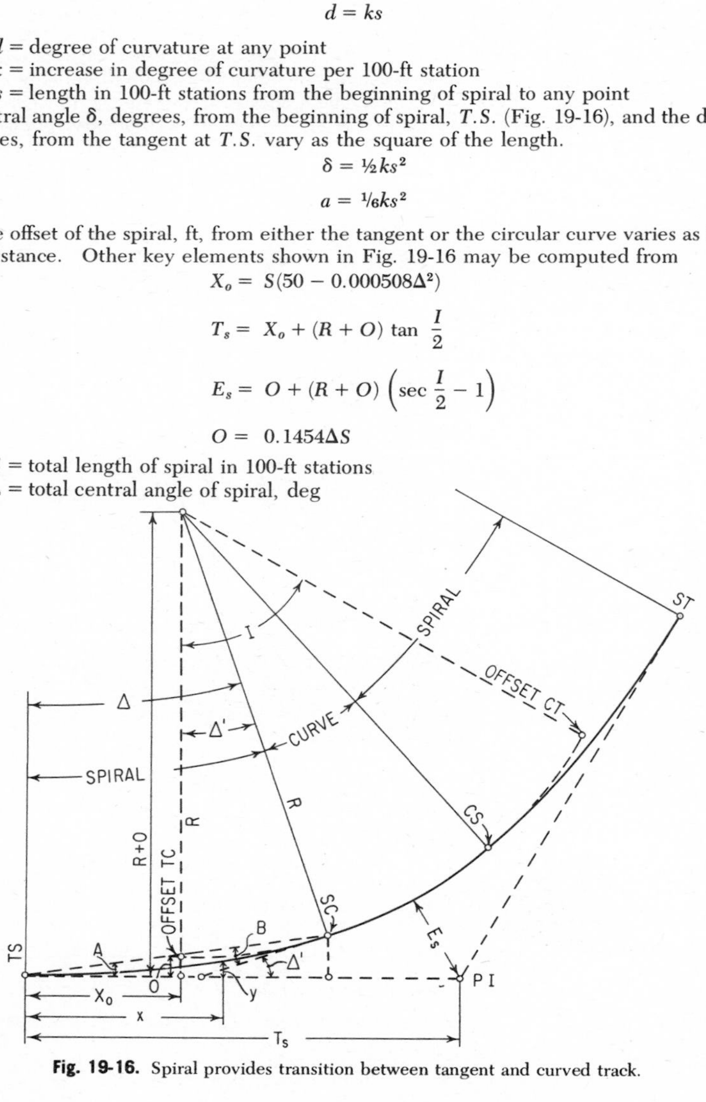

Fig. 19-16. Spiral provides transition between tangent and curved track.

R = radius of circular curve, ft

O = offset, ft, from tangent to circular curve extended at midlength of spiral

The deflection from the tangent to the end of the spiral, $S.C.$, with the transit set at $T.S.$ is one-third of Δ. When the transit is set at $S.C.$ and a backsight is taken on $T.S.$, a deflection of $\frac{2}{3}\Delta$ must be turned off to put the line of sight tangent to the circular curve. The deflections for the circular curve then should be turned from this tangent.

Stakes on the spiral should be set every 50 ft, as for a simple curve. Deflections may be calculated to place the stakes on even stations and plus 50. Or if preferred to simplify calculation of deflections, the spiral may be divided into segments of equal length, say 10. Then, the deflection may be computed for the first segment, multiplied by 4 (square of 2) to obtain the second deflection, by 9 (square of 3) for the third, by 16 for the fourth, etc.

To set the stakes for the spiral at a distance s with the transit at $S.C.$, subtract the deflection computed from Eq. (19-9) from the deflection computed for the same length of the circular curve extended. This deflection is then turned from the tangent at $S.C.$ to locate each stake.

The length of spiral should be such as to give passengers a time interval to adjust to the un-balanced centrifugal force without feeling a jerk upon entering or leaving the curve. Also, the rate of change of elevation should be sufficiently gradual to prevent undue twisting of the car body. The desired minimum length of spiral, ft, is the greater of the lengths determined from

$$L = 1.63 E_u V \qquad (19\text{-}14a)$$

$$L = 62 E_a \qquad (19\text{-}14b)$$

where V = maximum train speed on the curve, mph

E_u = unbalanced elevation (deficiency), in.

E_a = elevation of outer rail, in.

("Manual for Railway Engineering," American Railway Engineering Association).

19-10. Vertical Curves. At changes in grade on main line, a vertical curve should be provided of sufficient length to prevent excessive slack action in long freight trains or a sensation

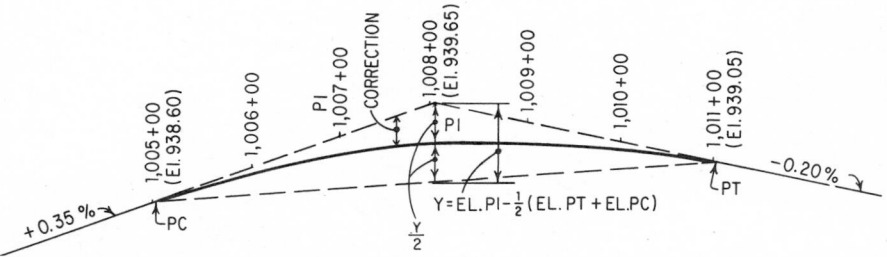

Fig. 19-17. Parabolic vertical curve connects two grades at a summit.

Table 19-2. Offsets from Tangent for Vertical Curve

$$\text{Length of curve} = \frac{+0.35 - (-0.20)}{0.10} = 5.5 \text{ stations}$$

Use 6 stations, or 600-ft vertical curve.

Offset at $P.I.$ = ½[939.65 − ½(938.60 + 939.05)] = 0.41 ft

Station	Grade elevation	Offset, ft*	Vertical-curve elevation
P.C. 1,005 + 00	938.60	0.00	938.60
1,006 + 00	938.95	0.05	938.90
1,007 + 00	939.30	0.18	939.12
P.I. 1,008 + 00	939.65	0.41	939.24
1,009 + 00	939.45	0.18	939.27
1,010 + 00	939.25	0.05	939.20
P.T. 1,011 + 00	939.05	0.00	939.05

*Offset $P.C.$ to $P.I.$ (Fig. 19-17) varies as the square of the distance from $P.C.$ Offset from $P.T.$ to $P.I.$ varies as the square of the distance from $P.T.$

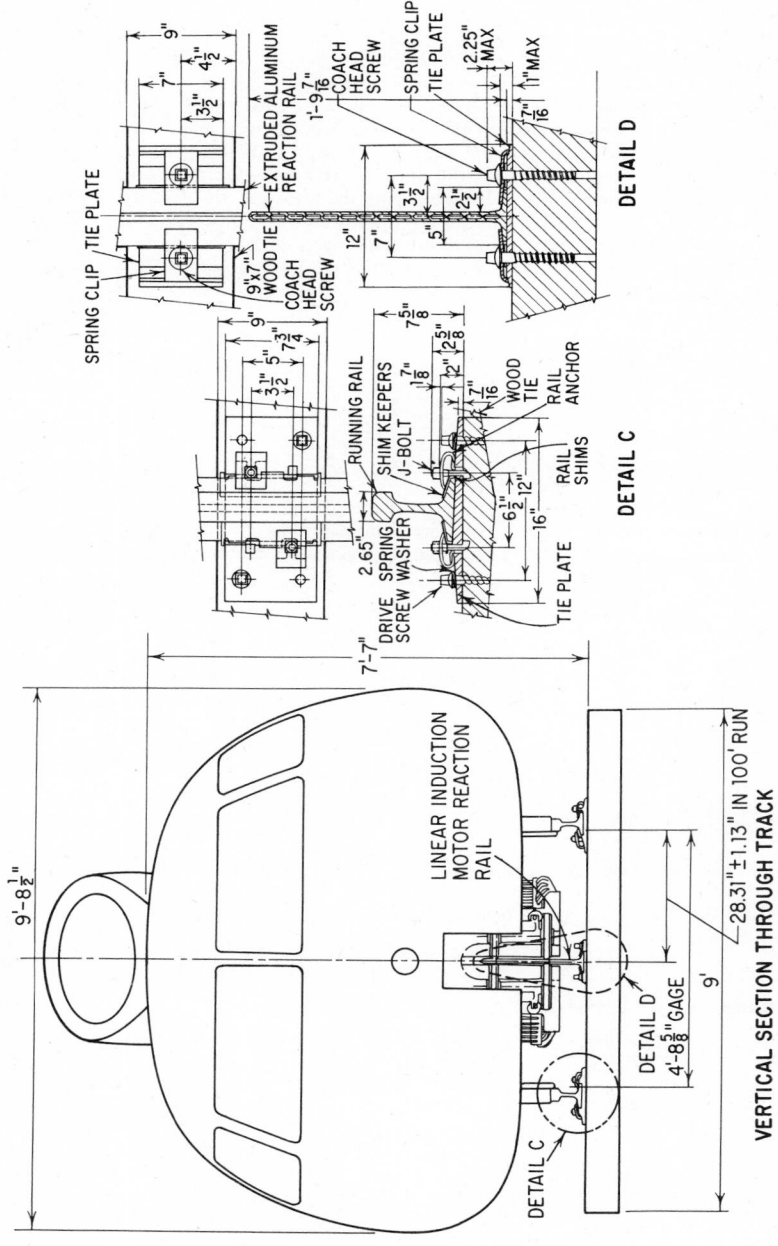

VERTICAL SECTION THROUGH TRACK

Fig. 19-18. Linear-induction-motor propulsion developed for a U.S. Department of Transportation test vehicle by Garrett AiResearch Manufacturing Company.

of discomfort to passengers at maximum speed.   Experience has shown that the rate of change in grade per 100-ft station on vertical curves should not exceed 0.10% on summits or 0.05% in sags.   Thus, if the grade changes from 0.20% descending to 0.20% ascending in a sag, the total change in grade is 0.40%, and a vertical curve 0.40 × 100/0.05 = 800 ft long should be provided. If a similar change in grade occurs on a summit, the length of vertical curve should be 400 ft.

Ordinarily, the length of vertical curve determined in this manner will not be an even number of stations.   It is simpler and satisfactory to use a vertical curve of the next even number of stations; i.e., if the calculated length is 7.2 stations, use a vertical curve of 8 stations.

The form of the vertical curve is parabolic in a vertical plane.   First, determine the elevations at the beginning and end of the vertical curve.   Add these and divide by 2 to obtain the average. Determine the difference between this average and the elevation at the intersection of the two grades.   One-half this difference is the offset from tangent, or correction, to be made at the middle of the vertical curve (Fig. 19-17).   The correction at other points varies as the square of the ratio of the distance from the nearest end of the vertical curve to half the length of the curve.   Table 19-2 illustrates the method of calculating a vertical curve on a summit.

(C. F. Allen, "Railroad Curves and Earthwork," McGraw-Hill Book Company, New York; T. F. Hickerson, "Route Location and Design," McGraw-Hill Book Company, New York.)

19-11. Track Construction.   There are several different types of track construction used, depending on the type of rail-transportation service and the physical characteristics of the environment.   These are as follows:

Duo Rail

Two lines of parallel, steel running rails supported on tie plates, ties, and ballast (Fig. 19-19) for diesel-electric- or electric-powered vehicles.

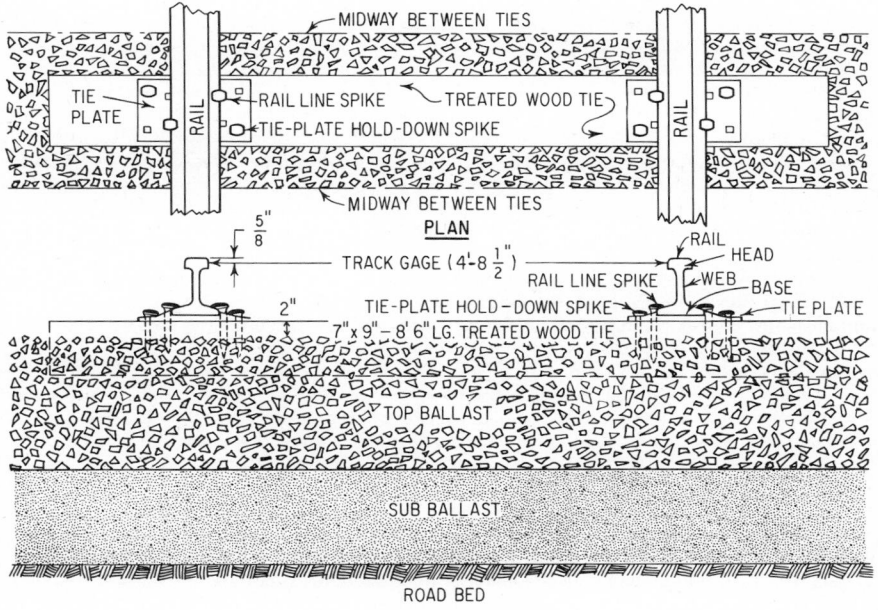

Fig. 19-19.   Typical standard-gage, duo-rail track on tangent.

Same construction except with a vertical or horizontal aluminum reaction rail installed between the steel running rails for linear-induction-motor-powered vehicles (Fig. 19-18).

Two parallel lines of steel or concrete beams (Fig. 19-4) to provide support and guidance for electric-powered rubber-tired vehicles.

Two parallel lines of suitably designed rails for the support of levitation-type vehicles, either air-cushion or magnetic (tracked air-cushion or Maglev vehicles), with linear-induction or turbo-jet motor power for traction.

MONORAIL

One line of a suitable type rail and rail support, with the vehicles supported above or suspended below the monorail and electric-powered.

A monorail supported on its under side may be used for elevated construction and in subway, but if used at ground level, it must have a grade separation at all highway and street crossings. A suspended monorail may be used for elevated construction. Enough clearance should be provided below the vehicle bottom for street and highway crossings. Its use in tunnel and subway construction would require that it be supported by the top of the opening, resulting in a high and costly opening. (The economics of a system in tunnel or subway construction depends a great deal on the area of the required opening.) Also, a monorail system has disadvantages in switching, weight support, construction cost, and ride quality.

Power for Vehicles.   Electric power is well suited to on-ground, elevated, or underground use.   Diesel-electric, turbo-jet, and gas-turbine-electric sources are not well suited for tunnels or subways because of noise and ventilation problems.   The tracked air-cushion vehicle presents the same problems.   The magnetic levitation system does not have these problems and can be used for all conditions.

Steel Duo-rail Tie-and-Ballast System.   This system (Fig. 19-19), which is used for the bulk of the railway track miles in the United States, and its construction will be covered in detail. Other systems either use some of the same components or are proprietary, in which case the details of construction can best be obtained from the owners.

The duo-rail systems with steel wheels on steel rails, with ties and ballast, or rubber-tired wheels on steel or concrete beams (Figs. 19-3 to 19-5) can be used for on-ground or elevated-track construction.   However, in subway construction, the duo-rail steel-wheel-on-steel-rail system has the rails fastened to an insulated tie plate, which is bolted to the invert floor but separated by an insulating and cushioning pad and insulating thimbles and washers for the fastening bolts (Figs. 19-10 and 19-11).   This fastening is more economical than the provision of additional tunnel or subway height to provide for tie and ballast depths, even if a depth of only 6 in. of ballast is used under the tie.   The duo-rail system with rubber tires on steel or concrete beams has some disadvantage in first cost, because a larger-diameter tire must be used than for the steel wheel, thus resulting in increased height of tunnel or subway opening.   This, however, is to a large extent offset by the narrower width of the opening required for the vehicles used in this system.

Rail serves three functions.   It must resist contact pressure from wheels and therefore must be hard.   It must be capable of distributing the wheel load over several ties along the track and thus must be stiff.   And it must be able to do this without breaking from repeated loadings. So it must have flexural and fatigue strength (see also Art. 19-12).

Tie plates protect the ties from abrasive longitudinal and lateral movements of the rail base. The plates provide a larger bearing surface on the ties than does the rail base, thus reducing the intensity of bearing pressure.   And the plates make spikes more effective in holding the rail to gage (see also Art. 19-12).   Cut spikes are used in the United States for rail and tie-plate fastening (see also Art. 19-12).   Screw spikes were tried extensively from 1910 to 1920 with unsatisfactory results.

Ties distribute the wheel load across the track.   Ballast distributes the tie load more uniformly on the subgrade between ties and also extends the distribution across the track.   Ballast also must withstand the bearing pressure from ties without displacement and must provide drainage.

Ballast.   Type and gradation of the material to be used for ballast and the cross section are important with respect to the cost of maintaining line and surface, which must be balanced against the original cost.   In new track construction, best results can usually be obtained by placing a layer of subballast on top of the roadway and supporting the track structure, including the topballast, on this layer.   The subballast should be of small particles of a material that will not disintegrate (American Society for Testing and Materials Specification D1241).   Its purpose is to provide drainage and keep the subgrade from penetrating up into the topballast while wet and under pressure.   Stone or slag screenings, chat (residue after extracting ore from rock), and sand make acceptable subballast.   Subballast should be placed in layers and thoroughly compacted.

The topballast may be of hard rock crushed to suitable size; crushed blast-furnace or properly processed open-hearth slag; or crushed gravel, if there is a sufficient quantity of angular material to prevent rolling. Individual railroads have different preferences for size of ballast. Good results will generally be obtained with a ballast size and gradation of which 100% will pass a sieve with 1½-in.-square openings; 90 to 100%, 1-in.; 40 to 75%, ¾-in.; 15 to 35%, ½-in.; 0 to 15%, ⅜-in.; and 0 to 5%, No. 4. For complete specifications for ballast materials, see "Manual for Railway Engineering," American Railway Engineering Association.

A recommended ballast section is shown in Fig. 19-9. A 12-in. depth of topballast below the bottom of the ties and a 12-in. depth of subballast will generally provide good track support for heavy loading and traffic (AREA Manual). As the roadbed becomes further compacted by traffic, it will be necessary to add additional ballast to resurface the track from time to time. After several years of service, the depth of ballast under the ties will probably be considerably increased.

Ties. Most ties in railway track in the United States are of treated wood, mostly oak, gum, pine, or fir. For main-line track, the most common size is 7 × 9 in. by 8 ft 6 in. long. Smaller sizes are used for yard tracks, such as 6 × 8 in. by 8 ft long. Ties are sawed, rather than hewn. They consist mostly of heartwood. This part of the tree is less desirable for lumber but more desirable for ties. Generally, ties containing the following will not be accepted by purchasers: decay; a hole more than 3 in. deep and ½ in. in diameter when between 20 and 40 in. from midlength, or more than one-fourth the width of the surface on which it appears when outside the sections of the tie between 20 and 40 in. from its middle; a knot having an average diameter in excess of one-fourth the width of the surface on which it appears, except when outside the same 20- to 40-in. zone; a shake larger than one-third the tie width; a split more than 5 in. long; and a slant in grain in excess of 1 in 15 (AREA Manual).

When ties are received at the treating plant for seasoning, some railroads apply antisplitting devices, such as fluted dowels or S or C irons, to some or all ties. Ties should be stacked for seasoning, allowing space for ventilation and for handling with a forklift. Two old or treated ties should be placed on the ground; then nine ties placed on top of these supported at their ends; then a tie crosswise at one end; then nine ties sloping down; then a tie crosswise at the other end; then nine more ties, and so on. The stack can be built to the height the forklift can handle.

The purpose of seasoning is to remove sufficient moisture from the wood to permit addition of a preservative. The stacking should provide ventilation so this can be effected before decay starts, but not so rapidly as to cause excessive checking and splitting. Before the seasoned ties are treated with preservatives, they should be adzed for the tie plates and bored for the track spikes.

The treating processes most generally used for ties are the Lowry or Rueping. Some railways use a creosote-coal tar solution (varying from 80% cresote and 20% coal tar to 50% of each). Others use a creosote-petroleum solution containing not less than 50% creosote. The minimum desirable retention of preservative, lb per cu ft, is as follows: oak, 6; fir, 8 or refusal; gum, 9; and pine 8 (AREA Manual).

Substitute ties have been used experimentally in years past. These were made of steel, concrete, or concrete and steel. None of these has been successful in the United States. Starting abroad about 1950, and in the United States in 1960, prestressed-concrete ties were installed. They hold more promise for successful use than did the earlier steel and concrete ties. With present-day automation procedures, prestressed-concrete ties and fastenings are produced that can be used to construct new trackage at a cost competitive with that of treated wood ties.

On most existing track, ties are renewed only as required or on a spot renewal basis. Prestressed-concrete ties should be placed out of face to give best results. Therefore, they are less economical for tie renewals in existing track.

Tie spacing has been determined, by experience over many years, to afford the most economical track maintenance. Many railroads space ties in main track to provide 24 per 39-ft rail length, or a spacing center to center of ties of 19½ in. Some roads use 23 ties per 39-ft rail (20.3-in. spacing) and some 22 per 39-ft rail (21.3-in. spacing). Preliminary specifications for the design of concrete ties and fastenings with spacing taken into consideration are given in the *American Railway Engineering Association Proceedings*, Vol. 77, pp. 193–236.

Drainage. Suitable drainage openings must be provided where the track construction crosses waterways. The principal factors affecting the required size of a waterway opening are the area of watershed, slope and characteristics of ground within the watershed, and maximum intensity of rainfall that may be expected within a given period of time. The following formula will provide sufficient waterway opening for maximum flood conditions:

$$A = 1,000 \sqrt{M} \qquad (19\text{-}15)$$

where A = area of opening, sq ft

M = square miles of watershed

Flow with A should be attained with a height of water that can be tolerated. If the area given by this formula exceeds what can be provided at an acceptable cost, more detailed study can be given to determine if a smaller area can be justified by the conditions of the watershed and intensity of rainfall (see "Manual for Railway Engineering," American Railway Engineering Association).

The type of waterway opening provided may be galvanized steel or concrete pipe; concrete box; or timber, concrete, or steel bridge. Culvert pipe and box should have headwalls to prevent water erosion of fill. Abutments of bridges should have suitable headwalls to contain fill and prevent water erosion.

Fencing. Right of way should be fenced if it is desired to keep off trespassers, livestock, or poultry. Posts should be not more than 16 ft 6 in. apart. The fencing should be galvanized woven wire of No. 9 gage, or galvanized steel ribbon, smooth round, or barbed wire. The type and height of fencing are dictated by the local conditions and statutory requirements. (For details of fencing, see Sec. 1-6, AREA Manual.)

19-12. Rails and Rail Accessories. (See also Art. 19-11.) To provide flexural stiffness and strength, rail is shaped in section somewhat like an I beam. But the head is made narrower and deeper than the flange of an ordinary I beam to resist the contact pressure and wear from flanged wheels better. Table 19-3 and Fig. 19-20 show the principal dimensions and physical properties of sections that have been rolled in substantial rail tonnage or are being rolled today in the United States. The heavier sections are used for heavy traffic and high-speed lines.

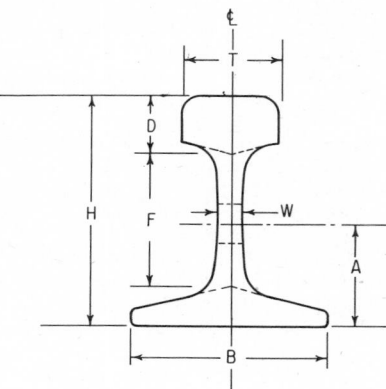

Fig. 19-20. Principal rail dimensions.

The standard length of rail in the United States and Canada is 39 ft.

The branding rolled in raised letters on one side of the rail web gives the weight of rail in pounds per yard, the section number, the mill, the year and month rolled, and the method of manufacture. A typical branding is as follows:

115	RE	CC	Manufacturer	1977	IIIII
(Weight or section number)	(Type)	(If controlled cooled)	(Mill brand)	(Year rolled)	(Month rolled)

On the opposite side of the web, the rail is hot-stamped to show the heat number, rail letter (position in the ingot), and ingot number.

Rail specifications may be found in ASTM Standards and the "Manual for Railway Engineering," American Railway Engineering Association, both being substantially the same. The chemical composition varies somewhat with the weight of rail (Table 19-4).

Table 19-4. Chemical Composition of Rails

Nominal weight of rail, lb per yd	70–80	81–90	91–120	121 and over
Carbon, %	0.55–0.68	0.64–0.77	0.67–0.80	0.69–0.82
Manganese, %	0.60–0.90	0.60–0.90	0.70–1.00	0.70–1.00
Phosphorus, max, %	0.04	0.04	0.04	0.04
Silicon, %	0.10–0.23	0.10–0.23	0.10–0.23	0.10–0.23

Table 19-3. Physical Properties of Rail Sections

Rail section	Weight — Lb per yd Nominal	Weight — Lb per yd Calculated	Net tons per mile	Dimensions, in. (see Fig. 19-20) Height H	Base width B	Fishing F	Max head width T	Head depth D	Min web thickness W	Base to center of both holes, A	Cross-sectional area, sq in.	Moment of inertia, in.^{4}	Section modulus Head, in.^{3}	Base, in.^{3}
AREA (RE)	140	140.6	247.4	$7^5/_{16}$	6	$4^1/_{16}$	3	$2^1/_{16}$	$3/_4$	3	13.8	96.8	24.6	28.7
AREA (RE)	136	136.2	239.7	$7^5/_{16}$	6	$4^3/_{16}$	$2^{15}/_{16}$	$1^{15}/_{16}$	$11/_{16}$	$3^3/_{32}$	13.35	94.9	23.9	28.3
NYC	136	136.3	239.4	$7^9/_{32}$	$6^1/_4$	$4^5/_{32}$	$2^{15}/_{16}$	$1^{31}/_{32}$	$11/_{16}$		13.36	93.9	23.9	28.1
AREA (RE)	132	132.1	232.4	$7^7/_8$	6	$4^3/_{16}$	3	$1^3/_4$	$21/_{32}$	$3^3/_{32}$	12.95	88.2	22.5	27.6
CB	122	122.5	215.6	$6^{25}/_{32}$	6	$3^{19}/_{32}$	$2^{15}/_{16}$	$1^{15}/_{16}$	$21/_{32}$		12.01	74.0	20.6	23.3
CF&I	119	118.8	208.1	$6^{13}/_{16}$	$5^1/_2$	$3^{13}/_{16}$	$2^{21}/_{32}$	$1^7/_8$	$5/_8$		11.65	71.4	19.4	22.9
AREA (RE)	115	114.7	201.9	$6^5/_8$	$5^1/_2$	$3^{13}/_{16}$	$2^{23}/_{32}$	$1^{11}/_{16}$	$5/_8$	$2^7/_8$	11.25	65.6	18.0	22.0
CF&I	106	106.6	187.6	$6^3/_{16}$	$5^1/_2$	$3^3/_8$	$2^{21}/_{32}$	$1^3/_4$	$19/_{32}$		10.45	53.6	16.1	18.8
AREA (RE)	100	101.5	178.6	6	$5^5/_8$	$3^9/_{32}$	$2^{11}/_{16}$	$1^{21}/_{32}$	$9/_{16}$	$2^{45}/_{64}$	9.95	49.0	15.1	17.8
ARA-A (RA-A)	100	100.4	175.6	6	$5^1/_2$	$3^3/_8$	$2^3/_4$	$1^9/_{16}$	$9/_{16}$		9.84	48.9	15.0	17.8
ARA-B (RA-B)	100	100.5	176.9	$5^{41}/_{64}$	$5^9/_{64}$	$2^{55}/_{64}$	$2^{21}/_{32}$	$1^{45}/_{64}$	$9/_{16}$		9.85	41.3	13.7	15.7
ASCE	100	100.4	175.6	$5^3/_4$	$5^3/_4$	$3^5/_{64}$	$2^3/_4$	$1^{45}/_{64}$	$9/_{16}$		9.84	44.0	14.6	16.1
ARA-A (RA-A)	90	90.0	158.4	$5^5/_8$	$5^1/_4$	$3^5/_{32}$	$2^9/_{16}$	$1^{15}/_{32}$	$9/_{16}$	$2^{37}/_{64}$	8.82	38.7	12.6	15.2
ARA-B (RA-B)	90	90.5	159.3	$5^{17}/_{64}$	$4^{49}/_{64}$	$2^7/_8$	$2^9/_{16}$	$1^{39}/_{64}$	$9/_{16}$		8.87	32.3	11.5	13.2
ASCE	90	90.1	158.6	$5^5/_8$	$5^3/_8$	$2^{55}/_{64}$	$2^5/_8$	$1^{19}/_{32}$	$9/_{16}$		8.83	34.4	12.2	13.5
ASCE	85	85.0	149.6	$5^3/_{16}$	$5^3/_{16}$	$2^7/_8$	$2^9/_{16}$	$1^{35}/_{64}$	$35/_{64}$		8.33	30.1	11.1	12.2
ASCE	80	80.2	141.2	5	5	$2^7/_8$	$2^1/_2$	$1^1/_2$	$17/_{32}$		7.86	26.4	10.1	11.1
ASCE	75	74.8	131.7	$4^{13}/_{16}$	$4^{13}/_{16}$	$2^{35}/_{64}$	$2^{15}/_{32}$	$1^{27}/_{64}$	$31/_{64}$		7.33	22.9	9.1	9.9
ASCE	60	60.5	106.5	$4^1/_4$	$4^1/_4$	$2^{17}/_{64}$	$2^3/_8$	$1^7/_{32}$			5.93	14.6	6.6	7.1

Sulfur is not specified because it produces hot shortness during rolling, which requires that it be kept within acceptable limits. The Brinell hardness of rail is not specified, but it normally ranges from 240 to 270. The only physical test required in the specifications is the drop test, which is made on specimens of rail from the top portion of the second, middle, and last full ingots of each heat. The purpose of this test is to determine ductility and resistance to impact, which normally is least at the top of the ingot because of segregation of impurities in solidification.

Control cooling of rail (retarding the cooling rate under controlled conditions) is effective in preventing shatter cracks. These may lead to development of transverse fissures in service. So control cooling is included in rail specifications, except when rails are made from vacuum degassed steel.

On curves, many railroads use fully heat-treated rail or rail with the top part of the head heat-treated to withstand better the flange wear that occurs on the high rail of curves and the flow and corrugation that occur on the low rail.

Rail Joints.   Rail-joint bars are used to join abutting rails together; or rail is butt-welded into long lengths before it is laid in track, and then the welded strings are joined with rail-joint bars or thermite welds.

Most railways use two 36-in. joint bars with six bolts and spring washers per rail joint (Fig. 19-21). There are 271 rail joints per mile of jointed track.

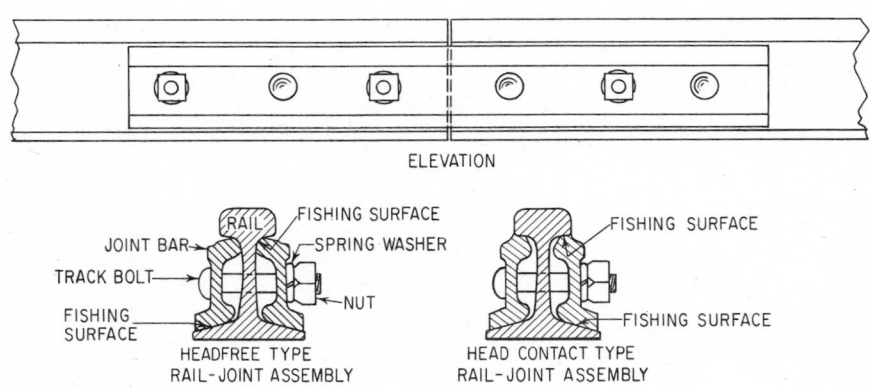

ELEVATION

CROSS SECTION

Fig. 19-21. Six-hole rail joint.

In years past, joint bars were shaped somewhat like an angle in cross section and were called angle bars. Since about 1930, most joint bars have been shaped more like an I beam and are called joint bars or sometimes short-toe joint bars, to distinguish them from the long-toe angle bar. Headfree bars fit into the upper fillet between the web and head. Take-up for contact-surface (fishing-surface) wear is provided in the base. Head contact bars have a slope on both the head and the base to match the fishing surfaces of the rail, and takeup for wear is provided at both head and base. Results with both types of bars have been equal in service tests. Section 4-1, "Manual for Railway Engineering," American Railway Engineering Association, gives designs of joint bars for 90 RA-A, 100 RE, 115 RE, 132 RE, 136 RE, and 140 RE rail. Steel companies that roll joint bars can furnish design drawings of bars they are equipped to roll.

Most rail-joint bars are made of oil-quenched carbon steel, manufactured in accordance with specifications given in the AREA Manual, Sec. 4-2, or ASTM Standards. Carbon is specified at 0.35 to 0.60%; manganese, not over 1.20%; and phosphorus, not over 0.04%. Tensile strength of 100,000 psi, yield point of 70,000 psi, 12% elongation in 2 in., and 25% reduction of area are minimum requirements. A bend test is also required. Brinell hardness is not specified but usually varies from 225 to 275.

Rail-joint bars are punched with alternate oval and circular holes. Hence, the bars can be used on either side of the rail and always have an oval and circular hole match for the track bolt. The recommended dimensions for a 1-in.-nominal-diameter track bolt are $1^{1}/_{16}$-in.-diameter circular hole and $1^{1}/_{16} \times 1^{13}/_{32}$-in. oval hole. Bar punching is spaced 6-6-7⅛-6-6 in. (AREA Manual).

It is important that the bars be straight or cambered in the least harmful direction. A camber of $1/16$ in. in either direction in the horizontal plane is acceptable. But in the vertical plane, the bar may not be low or more than $1/16$ in. high at midlength.

Track bolts are used for bolting a pair of joint bars in position. Most railways purchase heat-treated carbon-steel track bolts and carbon-steel nuts in accordance with specifications in the AREA Manual, Sec. 4-2, or ASTM Standards. Track bolts have a forged button-type head with either an oval or elliptic neck to prevent turning in the joint bar. The threads are rolled. Most railways specify a Class 2 or finger-free fit. The bolt-and-nut design is in accordance with American National Standards Institute Standard B18.2. A minimum carbon of 0.30%, maximum phosphorus of 0.04%, and maximum sulfur of 0.06% are specified. Tensile strength of 110,000 psi, yield point of 80,000 psi, 12% elongation in 2 in., and 25% reduction in area are minimum requirements. A bend test is specified. So is the minimum tension load that the bolt with nut fully engaged must withstand without stripping the nut or breaking the bolt. For the 1-in.-nominal-diameter bolt this is 66,560 lb; $1\frac{1}{16}$-in., 76,360 lb; and $1\frac{1}{8}$-in., 83,900 lb. A bolt tension in track of 15,000 to 25,000 lb is recommended, for which the 1-in.-diameter bolt is adequate. However, the larger-diameter bolts have some value in resisting bending from the contraction force in the rail in cold weather.

Spring washers are used to maintain bolt tension and reduce the amount of bolt tightening required. Tests have shown that track bolts become loose because of fishing-surface wear, which permits the joint bars to move closer together, not because of vibration. Specifications for spring washers in the AREA Manual, Sec. 4-2, require that, with a release of 0.03 in. from an initial compression of 20,000 lb, spring washers will maintain a reactive force of at least 5,000 lb. This amount of release is adequate for the fishing wear that occurs in a year's service, regardless of traffic, and a bolt tension of 5,000 lb is sufficient to insure proper functioning of the rail joint.

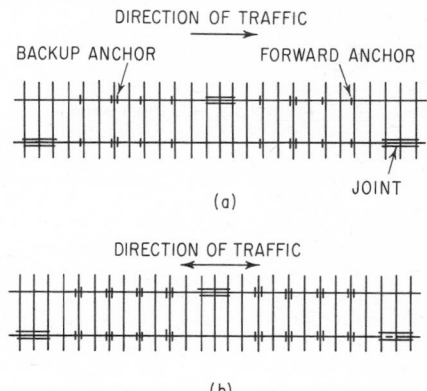

Fig. 19-22. Methods of anchoring track. (*a*) For main track carrying traffic essentially in one direction. Under average conditions, with any type of ballast, use eight forward anchors and two backup anchors per 39-ft rail length. (*b*) For main track carrying traffic in both directions. Under average conditions, with any type of ballast, use eight anchors per 39-ft rail length to resist movement in each direction, a total of 16.

Rail Anchors. A rail anchor is a device used to restrain lengthwise movement of rail. There are many different types in use. Most types engage the rail base by a spring clamping action and bear against the side of the tie or tie plate to restrain rail movement. The anchor should have sufficient holding power to move the tie in the ballast rather than permit the rail to slip through the anchor. Figure 19-22 shows a good method for anchoring track with this type of anchor (AREA Manual).

Tie Plates. (See also Art. 19-11.) A rolled-steel plate is used between rail and tie to reduce tie abrasion and hold the rail to gage better. The trend has been toward larger tie plates and use of double shoulders, instead of just one shoulder to restrain the outer edge of the rail base. The plates, rolled to the desired cross section, are sheared to a width generally of $7\frac{3}{4}$ in. A cant of 1:40 is provided in the rail seat to incline the rail slightly inward. Tie plates having a length of 12 or 13 in. are commonly used in the United States for rails having a base width of $5\frac{1}{2}$ in. and a length of 13 or 14 in. for rails with a base width of 6 in. A greater length of tie plate is provided on the field side of the rail than on the gage side (from $\frac{1}{2}$ to $\frac{3}{4}$ in.) to resist the outward lateral forces on the rail on curves better.

Generally, tie plates have four holes, $\frac{3}{4}$ in. square, punched through the shoulders for spikes to hold the rail to line. The plates also have four holes, $11/16$ in. square, punched near the corners for tie-plate fastening or hold-down spikes (Fig. 19-19). On tangent track, it is usual practice to use for each tie plate two line spikes in staggered holes. Sometimes, two hold-down spikes are used in oppositely staggered holes. On curves, two line and two hold-down spikes are used per plate. On curves of 6° and over with heavy traffic density, an additional line spike is used at the inner edge of the rail base.

Tie plates are made by various processes, generally by open-hearth or basic-oxygen. For steel plates made by these two processes, carbon varies from a minimum of 0.15% for low-carbon plates to a maximum of 0.82% for high-carbon plates. Low-carbon plates may be cold-worked; high-carbon plates must be hot-worked. A copper content of 0.20% is sometimes specified to provide corrosion resistance. Designs and specifications may be found in ASTM Standards and the AREA Manual.

Track Spikes. (See also Art. 19-11.) In the United States cut spikes are used to fasten rails to ties. They are formed with a wedge-shaped point to cut the tie fibers and prevent splitting. The head is rounded on top to facilitate driving; it is oval in shape and eccentric on the shank to provide a length of $^{11}/_{16}$ in. to engage the top of the rail base.

Line spikes, for holding rails to gage, are commonly $^5\!/_8$ in. square and 6 in. long under the head. Hold-down spikes, for fastening tie plates to ties, are commonly $^9/_{16}$ in. square and 5½ in. long under the head. A copper content of 0.20% is sometimes specified to give corrosion resistance. For design and specifications, see ASTM Standards and the AREA Manual.

Continuous Welded Rail. Most railways in the United States butt-weld new rail laid. Much-used rail is cropped to remove worn and battered ends and bolt holes and then is butt-welded before it is laid in track.

Railways in the United States use either oxyacetylene-pressure butt welds or electric-flash-pressure butt welds for continuous welded rail. General practice is to join, at a welding station, a number of rails, usually about 40, into a string about ¼ mile long. Then, the strings are transported on a train of special cars. The cars are equipped with rollers for supporting and handling 11 to 13 strings of rail in each of one to three tiers. When the strings arrive at the track location where they are to be laid, one end of each of two strings is fastened to the rail in track with a cable. Then, the train is pulled out from under these two strings, which are thus lowered on the ties for later positioning.

Expansion and contraction of continuous welded rail are prevented by the rail joints and the rail fastenings or anchors. The restraint stresses the rail. A tensile stress of 195 psi is produced in the rail by a 1°F temperature drop. For example, if continuous welded rail is laid at 70°F and the rail temperature drops to −30°F, a tensile stress of 19,500 psi develops in the rail because it is restrained from shortening.

When the rail temperature increases above the laying temperature, there is no movement of the ends of the long strings. The rail ends are in solid contact and can thus develop the restraint needed to prevent rail lengthening. When the temperature decreases from the laying temperature, if rail joints are used to connect the long strings the restraint from rail-joint friction may not be enough to restrain the rail fully. There may be some rail-end and tie movement. Many railroads anchor every tie for five rail lengths at each end of the welded strings to reduce the amount of end movement and joint gap. For the remainder of the welded rail, only every second or third tie is anchored. Where the strings are connected by field welds, the additional anchors are used only at insulated joints. Some types of insulated joints have enough pull-apart resistance so the additional anchors are not required.

An effort is made to lay continuous welded rail at about a mean temperature. This is not always practical. So it may be desirable to adjust the rail length later if difficulty with track buckling or joint pull-aparts occurs.

Stress and Strain in Rails. *Rail stresses and depressions* for unusually heavy loads may be computed by considering a rail as a continuous beam on an elastic support (*American Railway Engineering Association Proceedings,* vol. 19, pp. 878–896). With the tie spacings in general use, the assumption that rail is continuously supported will not cause significant error. The modulus of elasticity of rail support u is the uniform load, lb per lin in. of rail, required to depress the rail 1 in. It is further assumed that the pressure, lb per in., of the rail on its support at any point is

$$p = uy \tag{19-16}$$

where y = rail depression, in. Another significant term is the distance X_1, in., from point of application of a wheel load to the point where the bending moment caused by that load becomes zero and then reverses in direction.

$$X_1 = \frac{\pi}{4} \sqrt[4]{\frac{4EI}{u}} \tag{19-17}$$

where E = modulus of elasticity of rail steel (30,000,000 psi)

I = moment of inertia of rail, in.[4]

For a single wheel load, the bending moment and rail depression along a rail may be determined in terms of M_o and Y_o from Fig. 19-23.

$$M_o = 0.318PX_1 \tag{19-18}$$

$$Y_o = -0.393 \frac{P}{uX_1} \tag{19-19}$$

where P = wheel load, lb

M_o = bending moment due to the wheel load, in.-lb

Y_o = rail depression under wheel load, in.

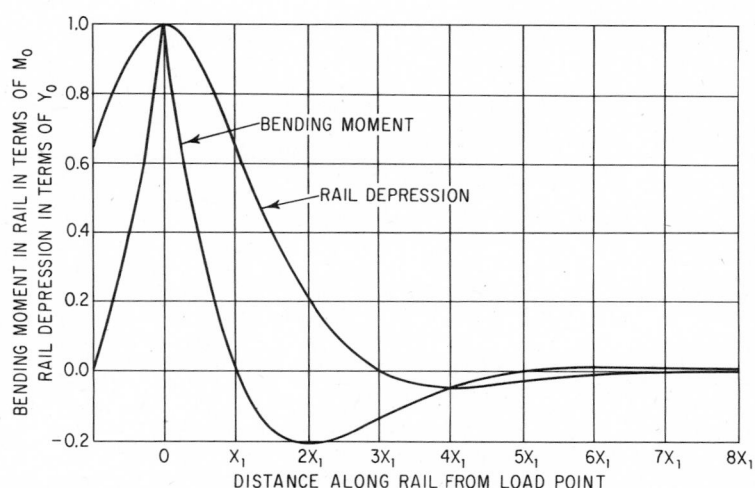

Fig. 19-23. Diagram for calculating rail bending moment and depression under a single wheel load. (*ASCE-AREA Special Committee on Stresses in Railroad Track.*)

Since there is always more than one wheel load, the master diagram may be used to determine the moment and depression at any point in the rail for all wheels by taking one wheel at a time and combining the effects algebraically. The maximum flexural stress in the rail base at this point may then be determined by dividing the total bending moment by the section modulus of the rail for the base. The tie load or reaction can be determined by calculating the average rail depression for the tie spacing and multiplying by the tie spacing and modulus u.

The value of u must be determined by actual measurement in track. This value ranges from 500 for track with little ballast and poorly compacted roadbed to 2,000 or more on track with adequate ballast and well-compacted roadbed. The value of u is not critical in calculating rail stresses but is significant for rail depression.

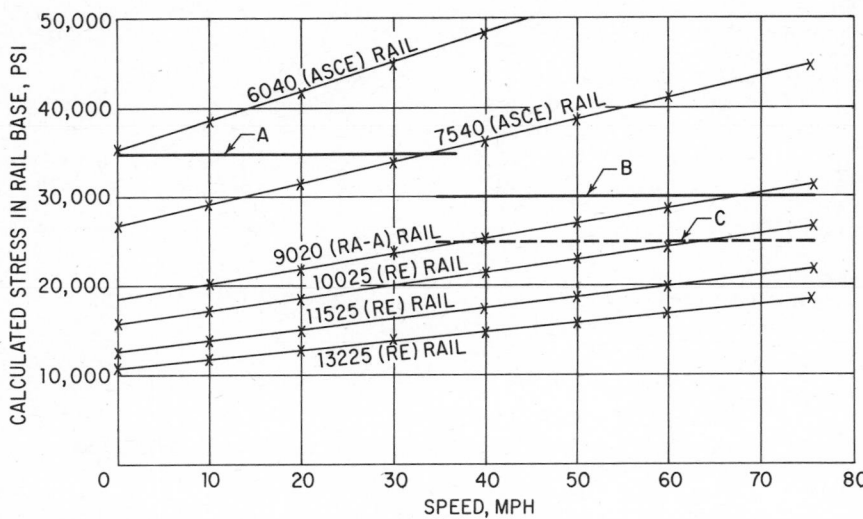

Fig. 19-24. Calculated rail stresses produced by a typical 100-ton-capacity hopper car (gross load of 263,000 lb). Recommended working stresses: (*A*) Jointed rail on branch line with speeds under 35 mph; (*B*) jointed rail on main line; (*C*) continuous welded rail on main line.

Table 19-5. Turnout and Crossover Data for Straight Split Switches*

1	2		3		Closure distance				Lead curve					Gage line offsets					
					4		5		6	7			8		9		10		11
Frog No.	Length of switch rail		Actual lead		Straight closure rail		Curved closure rail		Radius of center line, ft	Degree of curve									
	Ft	In.	Ft	In.	Ft	In.	Ft	In.		Deg	Min	Sec	Ft	In.	Ft	In.	Ft	In.	In.
5	11	0	42	6½	28	0	28	4	177.80	32	39	56	18	0	25	0	32	0	11¹³/₁₆
6	11	0	47	6	32	9	33	0	258.57	22	17	58	19	2¼	27	4½	35	6¾	12⅜
7	16	6	62	1	40	10½	41	1¼	365.59	15	43	16	26	2¼	35	10½	45	6¾	11⅜
8	16	6	68	0	46	5	46	7½	487.28	11	46	44	27	7¼	38	8½	49	9¾	11⅞
9	16	6	72	3½	49	5	49	7¼	615.12	9	19	30	28	10¼	41	2½	53	6¾	12⁵/₁₆
10	16	6	78	9	55	10	56	0	779.39	7	21	24	29	11¾	43	5½	56	11¼	12¼
11	22	0	91	10¼	62	10¼	63	0	927.27	6	10	56	37	8½	53	5	69	1½	12¼
12	22	0	96	8	66	10½	67	0	1,104.63	5	11	20	38	8½	55	5	72	1½	12⁷/₁₆
14	22	0	107	0¾	76	5¼	76	6¾	1,581.20	3	37	28	41	1¼	60	2½	79	3¾	12⅞
15	30	0	126	4½	86	11½	87	0¾	1,720.77	3	19	48	51	9	73	6	95	3	12⅛
16	30	0	131	4	91	11	92	0	2,007.12	2	51	18	53	0	76	0	99	0	12⁷/₁₆
18	30	0	140	11½	99	11	100	0	2,578.79	2	13	20	55	0	80	0	105	0	12¾
20	30	0	151	11½	110	11	111	0	3,289.29	1	44	32	57	9	85	6	113	3	13¹/₁₆

*Adapted from AREA Trackwork Plans. Comfortable speed added. Column numbers refer to dimensions in Fig. 19-25.

Calculated for turnouts from straight track for 4-ft 8½-in. gage.

Turnouts and crossovers recommended: for main-line high-speed movements, No. 16 or No. 20; for main-line slow-speed movements, No. 12 or No. 10; for yards and sidings to meet general conditions, No. 8.

There is no established impact effect or permissible working stress in rail, because of variability of conditions on different railways. The following may be used as a guide: Multiply the stress for static loads by a percent impact factor of $33V/D$, where V is the speed, mph, and D the wheel diameter, in. Thus, with a 36-in.-diameter wheel at 60 mph, the impact factor is 55%. A flexural stress at the extreme fiber of the base in jointed track of 35,000 psi is permissible at speeds below 35 mph, or 30,000 psi at higher speeds; in continuous welded rail, 25,000 psi.

Figure 19-24 shows the bending stresses calculated by this method for a typical 100-ton freight car with four-wheel trucks. An approximate value of stress for other weights may be determined by multiplying the values shown by the ratio of the wheel weights on the rail.

19-13. Turnouts and Crossings. A turnout provides the means for trains to be directed from one track into another. A turnout is made up of a pair of switch points with accessories, a frog, a pair of guardrails, and a set of turnout ties (Fig. 19-25).

A **frog** is a special unit of trackwork that permits two rails to cross. It is designated by number and type.

The frog number is the ratio of the distance from the intersection of two gage lines to the spread, or distance between gage lines, at that distance. The number also is given by half the cotangent of half the frog angle. The frog number determines the frog angle, the degree of turnout curvature, and the lead, or distance from the point of switch to the point of frog. Table 19-5

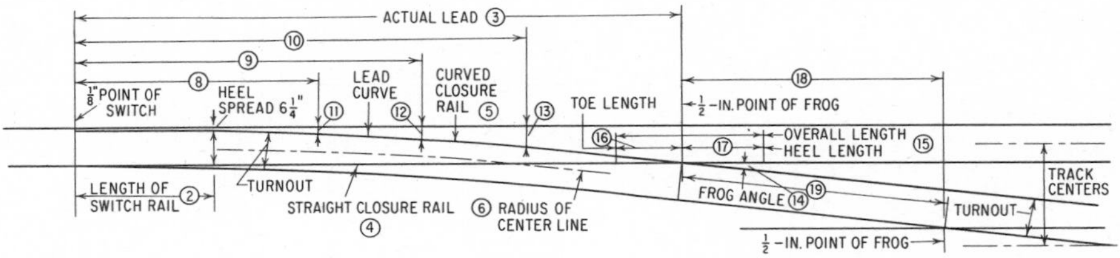

Fig. 19-25. Crossover consists of two turnouts and a crossover (connecting) track. Numbers indicate dimensions given in Table 19-5.

			Properties of frogs									Crossover data								
12	13		14 Frog angle			15 Overall length		16 Toe length		17 Heel length		18 Straight track, 13-ft track centers		19 Crossover track, 13-ft track centers		For change of 12 in. in track centers: Straight track		For change of 12 in. in track centers: Cross-over track		Comfortable speed, mph
In.	Ft	In.	Deg	Min	Sec	Ft	In.	Ft	In.	Ft	In.	Ft	In.	Ft	In.	Ft	In.	Ft	In.	
20⅝	2	8⅞	11	25	16	9	0	3	6½	5	5½	16	10 5/16	18	1⅞	4	11 7/16	5	0⅝	12
21⅝	2	10	9	31	38	10	0	3	9	6	3	20	5½	21	6½	5	11½	6	0½	13
19 9/16	2	6⅞	8	10	16	12	0	4	8½	7	3½	24	0⅜	24	11⅝	6	11 9/16	7	0 7/16	17
20 9/16	2	8 5/16	7	9	10	13	0	5	1	7	11	27	7⅛	28	4⅞	7	11⅝	8	0⅜	19
21⅜	2	9 7/16	6	21	35	16	0	6	4½	9	7½	31	1⅝	31	10⅜	8	11 11/16	9	0 5/16	21
21	2	8⅝	5	43	29	16	6	6	5	10	1	34	8⅛	35	3⅞	9	11 11/16	10	0 5/16	24
21⅜	2	9¾	5	12	18	18	8½	7	0	11	8½	38	2½	38	9½	10	11¾	11	0¼	26
21⅝	2	9⅞	4	46	19	20	4	7	9½	12	6½	41	8¾	42	3¼	11	11¾	12	0¼	28
22 5/16	2	10½	4	5	27	23	7	8	7½	14	11½	48	9¼	49	2 3/16	13	11 13/16	14	0¼	34
21¼	2	9¾	3	49	6	24	4½	9	5	14	11½	52	3 7/16	52	8⅝	14	11 13/16	15	0 3/16	35
21 13/16	2	10 5/16	3	34	47	26	0	9	5	16	7	55	9⅝	56	2½	15	11 13/16	16	0 3/16	38
22⅛	2	10 7/16	3	10	56	29	3	11	0½	18	2½	62	9⅞	63	2 3/16	17	11 13/16	18	0 3/16	40
22 11/16	2	11 3/16	2	51	51	30	10½	11	0½	19	10	69	10	70	2	19	11⅞	20	0⅛	40

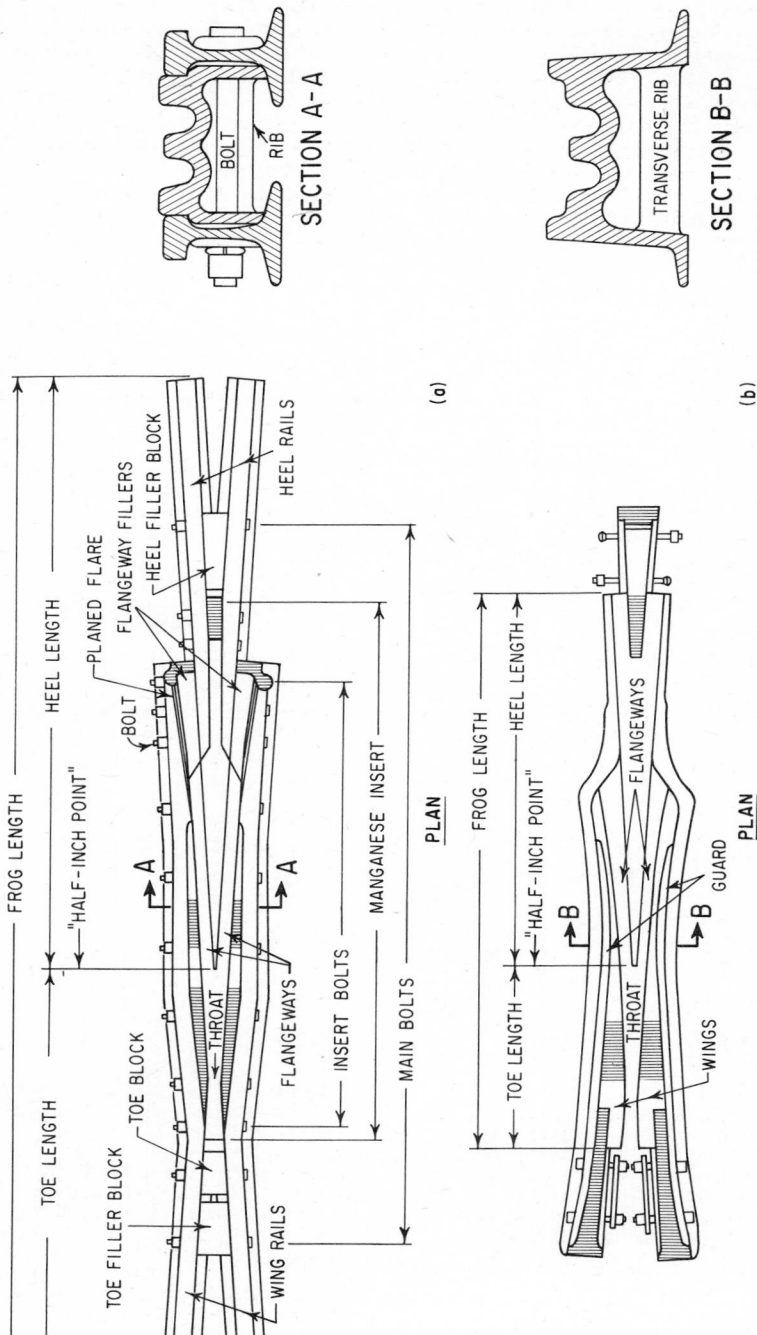

Fig. 19-26. Frogs used where rails intersect. (*a*) Rail-bound manganese-steel frog for main line. (*b*) Solid manganese-steel self-guarded frog for yard tracks.

gives these data for frog numbers from 5 to 20. Since speed is limited by curvature, the frogs with sharper turnout, the smaller-numbered ones, are used in yard tracks where speed is slow. The larger-numbered frogs are used in main-line locations to permit desired speed to the extent practical.

Frogs are either rigid or spring-rail types. Rigid frogs are of bolted rail, rail-bound manganese-steel, or solid manganese-steel construction. In a bolted-rail rigid frog, components are made from regular rolled rail, planed or machined as required. The assembly is held together by bolts through the rail webs, with the components separated by filler blocks to form the flangeway. A rail-bound manganese-steel frog (Fig. 19-26a) includes a cast insert of Hadfield manganese steel, which forms the point and wings, the locations most subject to impact, batter, and wear. (Hadfield manganese steel is a high-manganese alloy, which when properly heat-treated increases in hardness with cold working. So it is especially well suited to resist batter at frog corners.) The insert is supported by bent sections of rail, and the assembly is fastened together with bolts through the binding rails and the insert. A solid manganese-steel rigid frog (Fig. 19-26b) is made entirely of cast Hadfield manganese steel. It usually is self-guarded to save the cost of separate guardrails. The frog is joined to the two running rails at toe and heel by regular rail-joint bars and connecting bolts.

A spring-rail frog is made of machined rail sections. One side of the frog has a regular flangeway like a bolted-rail rigid frog. This is placed in the main running track. The other, or turnout, side, has a spring wing rail, which normally is held against the side of the frog point. Wheels passing through the turnout side force the spring rail out, against spring resistance, to provide a flangeway. The spring-rail frog provides a continuous running surface with a minimum of impact for the main running track.

Spring-type frogs are not recommended where there are many movements through the turnout side requiring frequent opening of the spring rail or on the outside of curves. Bolted-rail frogs cost the least. But they do not last as long and require more maintenance than the rail-bound or solid manganese. Self-guarded solid manganese frogs are used mostly for the smaller-numbered frogs in yard tracks where speeds are relatively slow.

A **guardrail** is fastened to each rail directly opposite the frog point. The purpose is to contact the back of each passing wheel and prevent the flange of its mating wheel on the axle from going down the wrong side of the frog point. Guardrails are of rail or cast-manganese-steel construction. The ends are flared inwardly of the track to engage the back of the wheel flanges and guide the pair of wheels on each axle into proper lateral position in the track. It is important that the guardrails are long enough and properly positioned to insure that the wheels are guarded past the frog point. It is also important that the guard check gage (distance between guard and gage lines) be maintained at not less than 4 ft 6⅝ in. (for standard gage). Guardrails are not required with self-guarded frogs.

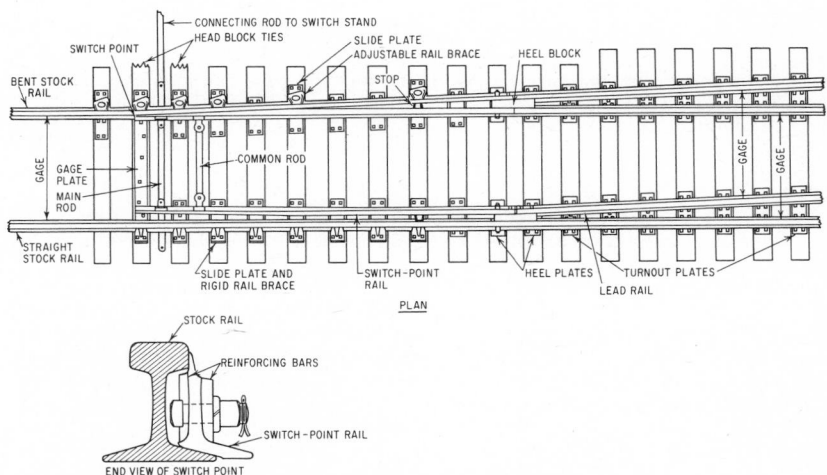

Fig. 19-27. Left-hand, straight, split switch.

A **switch** consists of a pair of switch points, a set of switch slide plates with braces, main and connecting rods, and a manually or power-actuated switch stand (Fig. 19-27). The switch-point rails are planed from regular rolled rail and reinforced on each side of the web with steel straps riveted in place. Heel blocks are used to join each switch-point rail to the adjoining lead rail, and both are fastened to the other running rail.

The heel spread (distance between the two gage lines) is 6¼ in. So the switch angle is fixed by this distance and the length of the switch-point rail. A short switch point and large angle are satisfactory for slow-speed operation. For example, a 16-ft 6-in. length of point is satisfactory for a No. 8 turnout. For a high-speed turnout, such as a No. 20, 30-ft points are used.

Usually, switch points are made straight. But for high speeds, the switch points are sometimes curved and 39 ft long for No. 18 and 20 turnouts. Comfortable operating speeds through turnouts are shown in Table 19-5.

Switch ties must be provided for turnouts. These are usually spaced on about 20-in. centers. Two long ties must be provided at the switch point for the switch stand. Each tie thereafter is made long enough to extend from each outer rail base the same distance as on regular track. Whenever the switch tie becomes as long as twice the length of a regular tie, then the switch ties are discontinued and regular ties used.

A **crossing** of two tracks requires four crossing frogs, frog plates, and crossing ties. Crossing frogs are made of bolted rails with either regular control-cooled rail or heat-treated rail; of rail-bound manganese-steel castings; or of all manganese-steel castings. Each running rail has a guardrail with a 1⅞-in.-wide flangeway between. To insure that such guardrails are effective in preventing the wheel flanges from entering the wrong side of the point, crossings should not be made with an angle of less than 9°36′ on tangents. (For curves, see American Railway Engineering Association Trackwork Plan No. 820.)

It is desirable to locate crossings on tangent on both intersecting tracks. But when this is not practical, crossings can be made to fit any condition of curvature.

Trackwork Plans. Details and specifications of all material required for turnouts and crossings are given in the American Railway Engineering Association Trackwork Plans. In ordering trackwork material from any manufacturer of trackwork material in the United States, it is only necessary to specify the AREA Plan Number and specifications. In ordering crossings, the intersection angle should be specified, the curvature, if any, and the rail size.

19-14. Culverts, Trestles, and Bridges. **Culverts** provide waterway openings under tracks. Usually, culverts consist of galvanized corrugated pipe or arches, reinforced concrete pipe, or reinforced concrete rigid-frame boxes. They are cheaper to install and maintain than other types of opening.

Care must be exercised in placing the fill on the sides and over the larger-sized culverts because side pressure against the culvert is a large factor in its ability to support vertical pressure. Metal culverts of up to 180 sq ft and reinforced concrete culverts of up to 300 sq ft in opening area are in use.

Trestles often are built of treated-timber stringers supported on capped and braced treated-timber piles. Trestles have either an open deck or ballasted deck. Ballasted decks are more expensive in first cost but require less work to keep the track in line and surface and offer less of a fire hazard. Treated-timber trestles are economical, have a life of 40 years or more, and require no painting.

Trestles are also constructed of steel or concrete piles, either reinforced or prestressed, with a concrete cap supporting steel or concrete stringers. Steel or concrete trestles usually have a ballasted deck.

Bridges generally are built of steel, reinforced concrete, or prestressed concrete. Usually, the abutments and piers are of reinforced concrete. For steel bridges, rolled beams are generally used for spans up to 50 ft, plate girders of riveted or welded construction for spans up to about 140 ft, and trusses, either through or deck type, for longer spans; but the deck-type spans are preferred. On steel bridges, either open or ballasted decks are used on the rolled-beam and plate-girder spans, whereas open decks are generally used on truss spans. Reinforced concrete slabs are used for spans up to 50 ft, and prestressed-concrete beams on spans up to 100 ft.

"Manual for Railway Engineering," American Railway Engineering Association, gives recommended designs and specifications for construction of all types of bridges, trestles, and culverts. These include recommendations for live load in terms of Cooper's E loading, allowance for impact effects, and permissible design stresses. (See also Sec. 17.)

Grade-crossing Elimination. Grade separations to avoid crossings of highways at grade are either underpasses or overpasses of the railway. For underpasses, the bridge carries the railway and must be designed and constructed to carry railway loadings. For overpasses, the bridge carries highway traffic and must be designed and constructed to carry highway loadings. Adequate clearance should be provided. In general, vertical clearance is 23 ft for railway traffic at overpasses and 14 to 16 ft for highway traffic at underpasses, but state requirements govern.

It is highly desirable to eliminate highway grade crossings to the extent practicable because of the accidents, injuries, and fatalities that may occur at such crossings, even with signal protection. Elimination of such crossings is a necessity for automatic train operation.

19-15. Vehicles and Propulsion. Different types of vehicles and methods of propulsion are used for the different types of rail-transportation service.

Passenger Cars. Types of passenger cars include mail, express, baggage, coach, diner, lounge, vista dome, parlor, sleeping, and combinations. Dimensions of passenger cars proposed by the American Railway Car Institute and adopted as recommended practice by the Association of American Railroads Mechanical Division are: coupled length, 85 ft; width, 10 ft; height, 13 ft 6 in. (vista dome, 16 ft 6 in.); and truck centers, 59 ft 6 in. Weight of these cars empty ranges from 125,000 to 160,000 lb. Seating capacity ranges from 60 to 80 in coaches and is about 48 in dining cars. A sleeping car has 11 sections, or 22 roomettes, or 11 bedrooms, or some combination of these accommodations. Duplex roomettes and bedrooms are at two levels, and thus more can be placed in a car. A slumber coach has 24 single roomettes and 8 double roomettes. ("Car Builders' Cyclopedia," Simmons-Boardman Publishing Corporation, New York.)

Passenger cars must be constructed to meet requirements of the Railway Mail Service and the AAR for safety and interchange. (These may be obtained from the executive vice-chairman, AAR Mechanical Division, 1920 L St., N.W., Washington, D.C. 20036.) Four-wheel trucks are generally used with 36-in.-diameter wrought-steel wheels, roller bearings, helical springs with about 12-in. travel, snubbers, cross stabilizers, swing hangers, and load equalizers. Passenger-type air-brake equipment and steam and air signal lines are provided. Passenger-carrying cars are air-conditioned and electrically lighted with power from a propane- or diesel-engine generator.

Passenger cars are designed to negotiate a curve of 250 ft minimum radius when coupled together.

Freight Cars. General types of freight cars include flat, box, stock, tank, hopper, covered hopper, gondola, refrigerator, and caboose. Some special types of freight cars include trailer on flat, auto rack, auto pack, container on flat, steel sheet, steel coil, and Hy-Cube. The point of special cars is improved service; for example, auto-pack cars completely enclose the automobile and prevent the damage and pilferage that frequently occur with open auto-rack cars. The **Auto Train** combines auto-rack cars to transport the automobiles of passengers with conventional passenger cars of different types to transport the passengers. The coupled length of freight cars ranges from 24 ft for ore hopper cars to 94 ft for Hy-Cube boxcars.

For freight cars to be freely interchanged in the United States, Canada, and Mexico, many components must have Association of American Railroads approval. These components include couplers, draft gear, center sill, air-brake system, wheels, axles, bearings, truck side frames, springs, snubbers, bolsters, and side bearings. The total rail weight that is permitted is determined by the journal size. For a car having four axles:

Journal Size, In.	Rail Weight, Lb
5 × 9	142,000
5½ × 10	177,000
6 × 11	220,000
6½ × 12	263,000
7 × 12	315,000

Width and height of freight cars must come within Plate B (Fig. 19-28) for unrestricted interchange and Plate C for interchange on most roads, as given in AAR Mechanical Division Specifications for Design, Fabrication, and Construction of Freight Cars. The dimensions of Plate B for width must be reduced for cars having truck centers in excess of 41 ft 3 in.

Freight cars of 45-ft coupled length can be operated around 45° curves coupled together. Boxcars of 94-ft coupled length coupled to a short car can be operated around a curve of about 20°. Curve negotiability depends upon clearance of the car corners and the free angling of the couplers in the draft gear pockets. ("Car Builders' Cyclopedia," Simmons-Boardman Publishing Corporation, New York.)

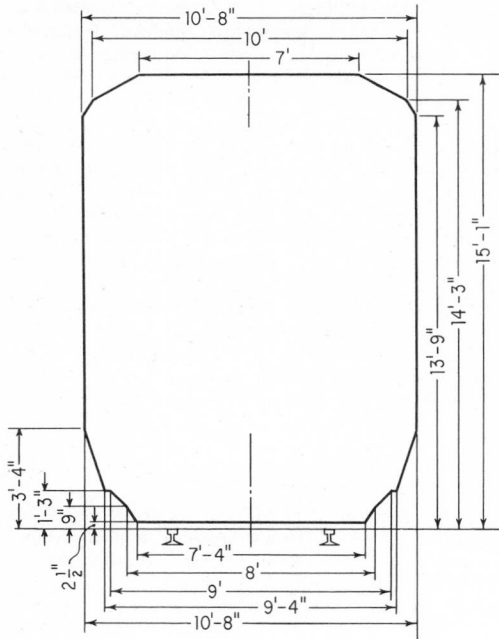

Fig. 19-28. Plate B clearance diagram for freight cars for unrestricted interchange service. Cars may be constructed to an extreme width of 10 ft 8 in. and to the other limits of this diagram when truck centers do not exceed 41 ft 3 in. With truck centers of 41 ft 3 in., the swingout at ends of car should not exceed the swingout at center of car on a 13° curve. A car to these dimensions is defined as the base car. When truck centers exceed 41 ft 3 in., the car width should be reduced to compensate for the increased swingout at the center or ends of the car on a 13° curve, so that the extreme width of car does not project beyond the center of track more than the base car. The 2½-in. clearance above top of rail is an absolute minimum. (*AAR Mechanical Division.*)

Commuter Cars.   Several types of commuter cars are in use.   One type is designed for push-pull operation by a separate locomotive.   It is of semi-monocoque design of aluminum with high-strength steel underframe, 85 ft long, 10 ft 6 in. wide, and 12 ft 8 in. high above top of rail. Truck centers are 59 ft 6 in., and wheel base is 8 ft 6 in.   Trucks are inboard bearing, air suspension, with 32-in.-diameter wheels, eight composition brake shoes, and electropneumatic braking.   Weight is 74,000 lb.   Seating capacity is 104.   There are two 33-in.-wide doors on each side near the car ends.   Loading is from low platform level.

Another type is the bilevel or gallery-type push-pull car.   A typical car of this type is 85 ft long, 10 ft wide, and 15 ft 10 in. high.   It seats 161 passengers in a trailer car, 155 in a cab car.   The cab car weighs 128,500 lb; the trailer car, 123,400 lb.   These cars are pulled or pushed in the train by a diesel-electric locomotive.   A cab car with controls for the engineer is located at one end, the diesel-electric unit at the other end of the train.   As many trailer cars as needed are coupled between them.   These cars have four-wheel trucks 59 ft 6 in. on centers.   Double or triple doors at midlength of the cars expedite loading and unloading at low platform level.

A self-propelled rail diesel car is used to some extent in commuter service.   A typical car is 85 ft long, 10 ft wide, and 14 ft 7 in. high.   It has two four-wheel trucks 59 ft 6 in. on centers.   The weight is 112,800 lb and seating capacity 89.   A 550-hp diesel engine with electric drive powers the car.   These cars are also used for mail, express, and passenger service on lines having light traffic.   Loading is from low platform level.

The fourth type of commuter car is the electric multiple-unit (MU) coach.   This type is used only for high traffic density.   One design of MU car is 85 ft long, 10 ft wide, and 12 ft 6 in. high. It has 59-ft 6-in. truck centers; trucks have two axles spaced 8 ft 6 in. on centers.   The car weighs 105,600 lb and seats 122 passengers.   Generally, several units are used in one train, but each has its own catenary trolley and is powered by four 156-hp motors.   Loading is from floor level.

A fifth type is a double-deck, multiple-unit car with cab controls at opposite ends of adjoining cars.   This type is 85 ft long, 10 ft 5¾ in. wide, and 15 ft 10 in. above top of rail.   Weight is 134,000 lb.   Seating capacity is 156, equally divided between the double doors on each side near car midlength.   It also has a single door on one side at the cab end.   It operates from a 1,500-volt dc catenary system.   The pantograph that collects the current for each car is located in a roof offset at the cab end, measuring 1 ft 10 in. deep and 10 ft 4⁹/₁₆ in. long.   Loading is from floor level.

A sixth type of commuter car is constructed to operate off the third rail in electrified territory and from its own power supply in nonelectrified trackage. These cars are built as pairs with one power source. Each car is 85 ft long, weighs 140,000 lb, and seats 240 in a "married pair" of cars. Loading is from either floor or ground level. Power is supplied by two 550-hp gas turbine–electric generator units, mounted directly under the roof for easy maintenance. The two gas turbines drive alternators providing three-phase power at 420 Hz, 277 to 480 volts. The rectified output is transmitted to a dc-dc chopper circuit that controls separately excited traction motors. The choppers (solid-state electronic switching devices) are advanced means of controlling dc traction-motor input power to provide smooth, efficient, jerkless acceleration for passenger comfort.

Electrically self-propelled commuter and rapid-transit cars may store energy developed by regenerative braking in storage batteries or in a high-speed flywheel for later use in train acceleration. This reserve energy supply could be used to operate the cars to the next station in the event of a power failure and, with storage batteries, to move cars in and out of yard and shop tracks and thus eliminate the need to electrify this trackage (resulting in less cost and greater safety).

All the types of commuter cars described previously have tinted glass windows, are air conditioned, and have comfortable seats, attractive decor, good lighting, racks for luggage or apparel, and toilets.

Rapid-Transit Cars. Essential characteristics of rapid-transit cars are rapid acceleration and deceleration, quick entrance and exit, maximum seating capacity, and passenger comfort. These are provided, respectively, by high-horsepower motors, a combination of dynamic and air brakes, and lightweight construction; several doors per car; loading and unloading at floor level; seats and arrangement designed for best space utilization; and padded upholstered seats, air conditioning, good lighting, and attractive decor. Table 19-6 gives comparable car data for several rapid-transit systems. Cars can carry up to 350 passengers. Seats provided range from 56 to 83.

The Bay Area Rapid Transit (BART) cars are a good example of car design which offers excellent service, comfort, and safety. A cars have one slanted end with a cab for a single attendant for train control (when needed), automatic train operation sensors, and a communications system. A cars are placed with the slanted end at the front and rear of the train, an arrangement that gives a pleasing, streamlined appearance. As many B cars as needed, up to 8, are placed between the two A cars. Vinyl-padded double seats are placed on each side of a middle aisle. The floors are carpeted; smoking is not permitted. The car interior is made of simple, durable, and fire-resistant construction and designed for ease of cleaning. No painting is required, and advertising signs are not used. Lighting fixtures use focusing lenses and provide 30 to 35 foot-candles at reading height, 20 foot-candles at floor level. At two locations in each car, a small push-to-talk intercom set permits passengers to report emergencies or seek information from the attendant. Either the attendant or the central office can make announcements to passengers from speakers in each car. A large enclosed passageway between cars with biparting doors and large panes of glass allows passengers to see seats in adjoining cars. This also facilitates observation of two cars by the attendant during night hours.

Each car has its own air-conditioning system. It provides draft-free, uniform air distribution with fresh air infusion, 12-ton refrigeration, 30-kw heating, and humidity control to below 60% relative humidity.

Automatic train control and cab signals are provided, but the attendant can override the train control in an emergency. Automatic couplers complete 24 electrical circuits throughout the train.

Wheels are designed for light weight and noise reduction. They have AAR wrought-steel, heat-treated rims and aluminum hubs. The car support and trucks include level-controlled air bellows, rubber "doughnuts" around the journal roller bearings, and hydraulic shock absorbers.

A dc chopper is used to control the 450-volt direct current to each motor to give smooth starting and stopping. An automatic car identification system is used, with color-coded labels on each car. Scanners are located on yard leads to record miles run for maintenance purposes and also to determine the location of each car.

Communication between trains and central control is by radio, using a line antenna through subway sections.

A more detailed description of the BART system is given in Modern Railroads Rapid Transit, February, 1972; and "The Bay Area Rapid Transit Vehicle System," by L. A. Irvin and J. R. Asmus, Paper 680544 Society of Automotive Engineers.

Table 19-6. Characteristics of Some Rapid Transit Cars

	Bay Area Rapid Transit District (BART)	Cleveland Transit System	New York City Transit Authority	Southeastern Pennsylvania Transit Authority (SEPTA)	Toronto Transit Commission	Washington Metropolitan Area Transit Authority (WMATA)
Capacity:						
Seats per car	72	80	72[c]	56	83	80
Maximum passenger design	216	—	350	202	300	240
Length over coupler faces	75 ft[a]	70 ft 0 in.	75 ft 0 in.	55 ft 4 in.	74 ft 9⅛ in.	75 ft 0 in.
Height:						
Overall	10 ft 6 in.	12 ft 0 in.	12 ft 1½ in.	12 ft 10 in.	11 ft 11½ in.	10 ft 10½ in.
Headroom	6 ft 9 in.	7 ft 2 in.	6 ft 8⅜ in.[d]	7 ft 4 in.	6 ft 11 in.	6 ft 10 in.
Floor to top of rail	3 ft 3 in.	3 ft 6 in.	3 ft 10⅜ in.	3 ft 10 in.	3 ft 7½ in.	3 ft 4 in.
Width, maximum	10 ft 6 in.	10 ft 5 in.	10 ft 0 in.	9 ft 1 in.	10 ft 4 in.	10 ft 1¾ in.
Weight, total less passengers	56,500 lb[b]	64,000 lb	87,000 lb[e]	48,760 lb	55,500 lb	72,000 lb
Trucks:						
Truck center distance	50 ft 0 in.	49 ft 6 in.	54 ft 0 in.	38 ft 0 in.	54 ft 0 in.	52 ft 0 in.
Wheel diameter	30 in.	28 in.	34 in.	28 in.	28 in.	28 in.
Track gage	5 ft 6 in.	4 ft 8½ in.	4 ft 8½ in.	5 ft 2¼ in.	4 ft 10 in.	4 ft 8½ in.
Wheelbase	7 ft 0 in.	6 ft 6 in.	6 ft 10 in.	6 ft 8 in.	6 ft 10 in.	7 ft 3 in.
Minimum radius horizontal curve	500 ft	120 ft	145 ft	140 ft	250 ft[f]	250 ft[h]
Minimum radius vertical curve	1,670 ft	2,000 ft	2,000 ft	3,000 ft	2,000 ft	[i]
Number of motors	4	4	4	4	4	4
Horsepower per motor	150	100	115	100	125	160
Performance:						
Balancing speed, mph	80	55	80	55	55	75[j]
Initial acceleration rate, mph per sec	3.0	2.75	2.5	3.0	2.5[g]	3.0
Service braking rate, mph per sec	3.0	3.0	3.0	2.75	2.8	3.0[k]
Emergency braking rate, mph per sec	3.3	3.5	3.2	3.0	3.0	3.2[k]
Dynamic brake range	80–4	55–1	70–15	55–1	50–10	15 fade out
Doors:						
Number per side	2	2	4	3	4	3
Height	6 ft 4 in.	6 ft 3 in.	6 ft 3 in.	6 ft 3 in.	6 ft 5¼ in.	6 ft 4 in.
Width	4 ft 6 in.	4 ft 2 in.	4 ft 2 in.	4 ft 1 in.	3 ft 9 in.	4 ft 2 in.
Minimum number of cars per train	2	1	4	2	2	2
Maximum number of cars per train	10	4	8	10	6	8

[a] For A cars; B cars = 70 ft.
[b] For A cars; B cars = 55,000 lb.
[c] For A cars; B cars = 76.
[d] Low ceiling; 7 ft 2⅜ in. for high ceiling.
[e] For A cars; B cars = 84,000 lb.
[f] Minimum desirable for main-line box structure and circular tunnels = 1,000 ft.
[g] High rate; low rate = 1.9.
[h] For yard track; main line = 500 ft.
[i] Parabolic, min. length = $(G_1 - G_2)$ 100 ft, but not less than 200 ft.
[j] On 1% grade.
[k] Below 50 mph.

Railway Service Cars. These cars are nonrevenue and maintenance-of-way equipment cars for various purposes. They include air-dump cars for roadway construction and maintenance; clearance-measuring cars; dynamometer cars for studying efficiency of train operation; instruction, test, and air-brake cars; rail cars for transporting long strings of welded rail; snow-handling cars, including blade and rotary snowplows; scale-test cars; spreader cars for ballast and ditch cleaning; and wrecking cars, including the "big hook" of up to 250 tons capacity, tools and materials, storage, and living quarters for the wrecking crew. ("Car Builders' Cyclopedia," Simmons-Boardman Publishing Corporation, New York.)

Diesel-Electric Locomotives. In the United States, almost all freight and passenger trains are moved and switching operations done with diesel-electric locomotives. Less than 1.5% of locomotives are electric; most of the remainder are diesel-electric. There are a few locomotives of other types in use, such as diesel-hydraulic and gas turbine–electric.

Steam locomotives had an overall efficiency of only about 5%. They caused many train delays for water, fuel, and other servicing. The diesel-electric locomotive has an overall efficiency of about 25%. It can be operated in multiple units by one engineman to afford the horsepower and tractive effort required. It requires relatively few stops for fuel and water. And it has excellent starting characteristics because all the weight is on the driving wheels.

An approximation of the drawbar pull of a diesel-electric locomotive may be obtained by dividing its weight by 4 (coefficient of adhesion of 0.25). Figure 19-29 gives a tractive effort–speed curve and dynamic braking-speed curve for a typical diesel-electric road locomotive. For road service, it is general practice to operate a number of locomotive units coupled together to give the tractive effort and horsepower required.

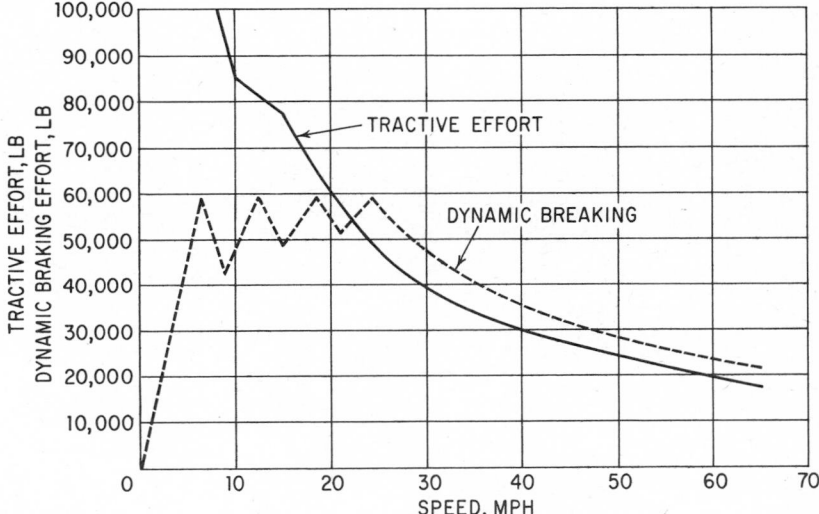

Fig. 19-29. Curves relate speed of a typical diesel-electric locomotive (3,600 hp) with tractive effort and dynamic braking effort. Locomotive characteristics: weight fully loaded, 368,000 lb; coupled length, 68 ft 10 in.; height (over cooling fans), 15 ft 7³/₁₆ in.; width, 10 ft 3¼ in.; two six-wheel trucks; distance between truck pivot points, 43 ft 6 in.; overall wheel base, 56 ft 6¾ in.; axles spaced 6 ft 7⅝ in.; 6 ft 11¾ in.; 29 ft 4 in.; 6 ft 11¾ in., and 6 ft 7⅝ in.; wheel diameter, 40 in.; fuel consumption, about 0.4 lb per hp per hour. (*Courtesy Electro-Motive Division, General Motors Corp., La Grange, Ill.*)

Electric Locomotives. An electric locomotive has good efficiency. But the electric power required is usually generated in a coal-fired steam plant, there is some power loss in the line transmission, and the catenary system represents a considerable investment and maintenance expense. Electric locomotives, in general, are economical only on lines having fast and frequent train schedules. These locomotives have the advantages of being able to develop a high horsepower at high speed and of requiring less maintenance than diesel-electric.

Power Requirements. These vary with the type of service. For intercity passenger and freight trains, the power to pull the trains up grades and make the scheduled time is of paramount importance. For commuter and rapid-transit service, an important factor in the power requirement is the need to accelerate quickly. For personal rapid transit, the speed is relatively slow, but power must be adequate to accelerate quickly to the desired speed. For all types of service, power must be adequate to overcome grade, curve, and rolling resistance; requirements for running time, frequency of service, and operating costs must all be considered.

Grade resistance, offered by an ascending grade, equals 20 times the percent grade per ton of train. Thus, on a 1.5% grade, the grade resistance is 30 lb per ton; on a 1.0% grade, 20 lb per ton; and on a 0.5% grade, 10 lb per ton. On a descending grade, the same forces accelerate the train, which must be controlled by braking.

Curve resistance is the added resistance required to guide and slip the wheels in negotiating a curve. Curve resistance is generally considered equivalent to 0.04% grade per degree of curvature. Thus, the curve resistance on a 4° curve would be $0.04 \times 20 \times 4 = 3.2$ lb per ton of train. It is customary on ruling grades to compensate for curvature by reducing the grade for the length of the curve. Thus, if the ruling grade on a line is 0.5%, compensated for curvature, no consideration need be given to curve resistance in calculating power requirements because it is already included in the grade. Use of rail lubricators reduces curve resistance by about one-half. Curve resistance tends to retard the train on descending grades.

Rolling or train resistance is the resistance to train movement on level tangent track. Train resistance is affected by speed, weight on the axle, and characteristics of the track. This last factor is usually neglected because it is relatively small. Starting resistance is less with roller bearings, but after a train starts, the train resistance is about the same for roller and solid bearings. For example, the starting resistance for a car with solid bearings might be as much as 20 lb per ton, but train resistance becomes 5 lb per ton as soon as the car is in motion. The same car on roller bearings would have the same starting resistance as when moving at slow speed, 5 lb per ton.

There are several formulas for calculating train resistance. The Davis formulas (W. J. Davis, Jr., "The Tractive Resistance of Electric Locomotives and Cars," *General Electric Review*, October, 1926) are representative of results found by several investigators. According to the AREA "Manual for Railway Engineering," the Davis formulas have given satisfactory results for speeds between 5 and 40 mph. However, the increased dimensions and heavier loading of freight cars, the much higher operating speed of freight trains, and changes in types of cars since the formulas were developed have made it desirable to modify the constants in the Davis equation. Recent tests have shown improved results with the following modified Davis formula:

$$R = 0.6 + \frac{20}{W} + 0.01V + \frac{KV^2}{WN} \tag{19-20}$$

where R = resistance, lb per ton
W = weight per axle, tons
N = number of axles per car
V = speed, mph
K = air resistance coefficient
= 0.07 for conventional freight-train equipment
= 0.16 for trailer on flat car (piggy-back)
= 0.0935 for containers on flat cars

The last term in this equation, KV^2/WN, represents the air drag due to train speed. At high speeds, this becomes a major factor in train resistance, and it is necessary to take into account the cross-sectional area of the car, aerodynamic properties of the car design, air density, and wind velocity and direction.

For a detailed treatment of this subject for high-speed passenger service, see J. L. Koffman, "Tractive Resistance of Multi-unit and Locomotive-hauled Passenger Trains," Rail Engineering International, April–May, 1973. The author suggests the following formula as representative of modern passenger-train equipment on British and Continental railways:

$$R = 1.5W + (5.5 + n - 2)\left(\frac{V}{10}\right)^2 \tag{19-21}$$

where R = total tractive resistance of a conventional passenger train, kg
W = total weight of train, metric tons
n = number of coaches in the train
V = train speed, km per hr

and the effective locomotive and coach frontal area is taken to be 10 sq m. Equation (19-21) assumes an air drag coefficient of 0.6 based on experimental data on train speeds up to 100 mph. With some car designs where little consideration was given to aerodynamic properties, the air-drag coefficient was found to be as much as 1.85 for an eight-car train. On the other hand, it was found to be as little as 0.97 for a 249.5-m-long, 10-coach Tokaido train, for which extensive model tests were made in a wind tunnel to obtain good aerodynamic performance.

In design of vehicles to be operated at speeds over 100 mph, it is highly important that aerodynamic performance be considered because air drag causes most of the rolling resistance at these high speeds and increases as the square of the speed.

19-16. Calculating Running Time and Fuel Consumption. Running time and fuel consumption are useful data in comparing the relative desirability of various lines in new construction or in revisions of existing line. Running time may be calculated by the velocity-profile method.

In this method, accelerating force, lb per ton, is computed by subtracting from the drawbar-pull characteristics of the locomotive the train resistance on level track. The computation is repeated for 5-mph increments from starting to maximum permitted operating speed. Since grade resistance is 20 lb per ton (see Art. 19-15), the accelerating force may be converted into an equivalent grade by dividing by 20. The actual profile of the line is plotted on a graph showing elevations vs. distance. On the same graph, for each increment of speed, the equivalent grade is plotted between points for which the vertical difference between the actual and equivalent grades equals the velocity head. The velocity head, ft, for any speed is

$$VH = 0.0355V^2 \qquad (19\text{-}22)$$

where V = speed, mph. This formula expresses the kinetic energy of a train due to its velocity and the rotating energy in its wheels as equivalent potential energy due to height. The same procedure is applied for braking to reduce speed or stop. The series of lines representing equivalent grades is the velocity profile. (Detailed instructions for train performance calculations are given in Sec. 16-2.2, "Manual for Railway Engineering," American Railway Engineering Association.) After the velocity profile has been completed for the line, the running time is found by summing the time required to travel each increment of distance at the average speed for the increment. A computer may be used to facilitate and expedite the calculations required.

The time that a locomotive will be working at full capacity, part capacity, or drifting can be determined from the velocity profile. Multiplying each period of time by the corresponding rate at which fuel is used by a particular locomotive yields the fuel consumption. Another method that may be used to calculate fuel consumption is to first figure the total work done. This consists of the work done in overcoming rolling resistance, plus the resistance of gravity on ascending grades, plus the resistance due to curvature. From this sum should be subtracted the energy of gravity on descending grades, but the loss of energy (velocity head) due to application of brakes should be added to give total work.

The total work done, ft-lb, may be converted to gallons of diesel fuel by multiplying by 4 (efficiency of 25%) and dividing by 90 million (ft-lb of energy per gallon of diesel fuel).

A simpler method that will be sufficiently accurate for most purposes is to approximate a condensed profile of the line with a series of long grades, calculate the speed at which the locomotive can handle the train over each grade, obtain the time over each grade by dividing the distance by the speed, total these, and add an arbitrary 5 to 10 min for each stop and start, depending upon the length of the train. This will give the approximate running time. The fuel consumption can be determined as in the velocity-profile method or from total work done.

19-17. Train Tonnage. The maximum tonnage that can be hauled over a line with a given locomotive is determined by the ruling gradient. However, the locomotive may not be able to handle this much tonnage at a high enough sustained speed to meet competitive traffic requirements or to avoid train-crew overtime. With diesel-electric locomotives, any number of units can be coupled together. But if they are placed at the head end of the train, trouble with broken couplers may be encountered if drawbar pull exceeds 200,000 lb.

If a train is made very long, for example, 200 cars, difficulty may be experienced from slack run-in, from excessive delays for replacement of broken couplers or setting out cars that have developed hot boxes, or from air-brake operation in very cold weather. A train of 100 cars is quite common in the United States. Occasionally, railroads operate trains with as many as 250 cars. Diesel-electric locomotives are sometimes added near midlength of a train and as pushers at the end on steep grades.

Generally, the more tonnage in a train, the lower the operating cost. Thus, train tonnage is a matter of economy, practicality of operation, and meeting competitive traffic requirements for speed and frequency of service.

Since train resistance varies with car weight and number of cars, a locomotive cannot handle so much tonnage in a train of empty cars as in one of loaded cars. Also, a locomotive cannot handle so much tonnage in cold weather as in warm weather. As a convenient means of compensating for these two factors, use may be made of the data shown in Table 19-7.

The adjusted tonnage rating may be considered as the sum of the weight of cars and contents, tons, and an adjustment, tons. For temperatures above 35°F, this sum is called the adjusted tonnage A rating. The adjustment for computing the adjusted tonnage A rating is obtained by multiplying the adjustment factor given in Table 19-7 for a specific ruling grade by the number of cars in a train. For temperatures below 35°F, a percentage of the A rating is used, as indicated in Table 19-7. The adjusted tonnage rating is independent of the number of cars in a train.

Table 19-7. Data for Calculating Adjusted Tonnage Ratings

Ruling grade, %	Adjustment factor, tons per car	% of A rating		
		B, 20–35° F	C, 0–20° F	D, below 0° F
0.1	29	84	70	57
0.2	20	89	78	66
0.3	15	91	82	72
0.4	12	93	85	76
0.5	10	94	87	79
0.7	8	95	90	84
1.0	5	97	93	87
1.5	4	98	95	91
2.0	3	98	96	93
2.5	2	99	97	94
3.0	2	99	97	95

19-18. Train Control.   There are many methods for controlling the movement of trains on tracks, depending upon the number of tracks and the characteristics of the traffic. The objective is to move the trains to conform to desired schedules between departure and destination points with safety the paramount consideration.

Train orders and time schedules are used where only a few trains move over a line per day. Passenger train speeds are limited to 60 mph and freight train speeds to 50 mph by the Federal Railroad Administration where this method of train control is used.

The manual-block system provides a safer operation. Operators stationed between blocks of track do not permit a train to enter the next block until notified by the operator at the other end of block that it is clear. Although safe, this method gives low track capacity, slow schedules, and high cost for block operators. With the manual-block system, speeds up to 79 mph are permitted by FRA order.

The automatic block signal system provides for successive blocks of tracks to be separated electrically by insulated rail joints at both ends. Unless the rail is continuously welded, rail bonds are used at each bolted rail joint to insure continuity of the electric circuit between rail ends. A three-position signal aspect is connected into the electric circuit for each block and for adjoining blocks. Many different types of signal aspects are used. A common signal aspect is a green light to show an approaching train that two blocks ahead are clear, a yellow light to show that the second block ahead is occupied, and a red light to show that the next block ahead is occupied. The block length should not be less than the service braking distance required for the train speed. A block length of 1 mile is frequently used. Speeds up to 79 mph are permissible.

Automatic train control is provided by a wayside inductor located in advance of each block circuit over which the locomotive receiver passes. This receiver is mounted on the locomotive journal box to have 1½-in. clearance with the wayside inductor. An electric circuit is provided so that when a locomotive passes a restrictive signal, the engineman must acknowledge awareness by actuating a contactor. Otherwise, the train brakes are automatically applied. With this system, train speed is not limited by FRA order to 79 mph.

Coded control has the advantage of using one pair of line wires to transmit the signals from blocks ahead instead of requiring many line wires for this purpose. An interruption of the dc voltage is used for different signal indications. For example, on track with one-direction movement, 180 interruptions per minute operates the "proceed" signal; 120, the "approach-medium"; 75, the "approach"; and no code, the "restrictive." Additional signals may be transmitted by combinations of reversed polarity.

Another advantage of coded control is that the code-following track relay must pick up with each pulse. Therefore, the train shunt need only be enough to reduce the track current at the relay below the pickup value, rather than below the dropout value. This permits higher track voltage to be used, and for given ballast resistance conditions, track circuits can be made twice as long.

Continuous cab signals are provided by using alternating current for the track circuits instead of direct current and placing inductive receivers in front of the leading wheels on the locomotive. Thus, the signal passing through the rails is transmitted by the receivers to give signals in the locomotive cab. Any change in signal aspect is immediately visible, whether or not the wayside signal is in sight of the engineman. With this system, wayside signals are not actually required. Coded control can be used with this system by interrupting the ac voltage in the same manner as for dc voltage. With continuous cab signals, train speed may exceed 79 mph.

Interlocking is usually provided at railroad grade crossings and at some turnouts. At crossings without interlocking, each train must first stop at the crossing and then proceed if the crossing is clear.

Mechanical interlocking operated by a towerman permits giving the right of way to one train, holding any on the track being crossed. Signal aspects are operated by levers and long pipe connectors, as are derails on each track.

Electric interlocking permits the operator to actuate signal aspects and derails electrically. The operator can also unlock and throw switches by electric control for crossovers or connecting tracks. Switches are thrown by electropneumatic or electric-motor switch machines. Safety features are provided to prevent an operator from lining up signals, derails, and switches unless the track is clear for such movements.

For very complicated crossings involving many tracks and train movements, route interlocking is used. It is only necessary for the operator to push a button for the point where a train will enter the interlocking and another button for the point where the train is to leave. The best available route will then be automatically selected and lined up for the train. On simple crossings, train movements through an interlocking can be controlled automatically by an electric signal system.

Overlap and absolute permissive block signaling is required to avoid collision of trains moving in opposing directions on the same track. With automatic block signals giving indications for only two blocks ahead, opposing trains could pass a clear signal simultaneously and find the next signal at stop but be unable to do so in time. This situation can be prevented by overlapping the advance blocks so that the stop aspect is displayed more than one block in advance of a train. With the absolute permissive block system, relays can be used to extend the blocks in advance but provide the normal block indications for following trains, thus expediting their movement. Where opposing trains are operated on the same track with absolute permissive block, the block control is extended far enough in advance to include a passing track or crossover so that the trains can pass.

Centralized traffic control (CTC) is officially designated by the FRA as the "traffic-control system." It is defined as "a block system under which train movements are authorized by block signals whose indications supersede the superiority of trains for both opposing and following movements on the same track."

With CTC, one operator directs the movement of all trains and usually of all switches and derails on the trackage under his or her control. For low-traffic-density lines, sometimes the switches are manually thrown by the trainman in accordance with a signal aspect at the switch. A panelboard shows the operator a diagrammatic layout of the trackage, with all turnouts identified and signal aspects at turnouts shown. Lights identify the location of all trains. Signals and switches at the ends of passing tracks are arranged as route-type interlocking. Automatic block signaling controls movement between passing tracks. With only two line wires along the track and different frequencies for transmitting coded data, it is possible for one operator to control train movement over several hundred miles of trackage. The actual operation is performed on a small control board in front of the operator, who merely pushes buttons or turns small switches to send out the directing signal. Acknowledgment is indicated on the panelboard by a signal automatically sent back when the action has been completed.

Automatic train operation is the capability for complete scheduling and operation of trains, including starting and stopping, opening and closing doors, etc., by computer command. Theoretically, a train attendant is not required. Actually, it is usually considered desirable if passenger traffic is involved to have an attendant on each train who can take over in an emergency and override the automatic operation with manual operation. For manual override, automatic block signals or continuous cab signals are required. Also, presence of a train attendant may give the passengers a sense of security.

At the control center, one or more computers control the operation of each train according to schedule but have the capability to automatically change the operation as required by any delays that may occur. One or more "dispatchers" are provided to observe the control board showing the position of all trains and to take over manual operation in an emergency. Experience has shown that automatic train operation will get trains over a route in less time and with more comfort to passengers than can be obtained with manual operation.

The systems and procedures for train control to secure maximum performance have become so sophisticated that specialists in the field should be consulted for selection and design of a system for any given conditions.

Grade-crossing warning is an important signal function in train operation. Block-signal track circuits may be used to actuate flasher lights or crossing gates automatically to warn vehicles of approaching trains at highway grade crossings. Crossing gates are advantageous at crossings of two or more tracks, because of the danger that a motorist will drive on the crossing after a train has cleared without waiting to see if a train may be approaching on another track.

Audio-frequency overlay circuits have been developed to actuate grade-crossing protection without the need for insulated rail joints.

Slide fences are frequently used at locations where falling rocks obstruct the track. These fences are drawn tight by spring tension at one end. The pressure of a rock at any point on the fence will cause end movement, which breaks an electric circuit, causes a control relay to become deenergized, and sets the block signal to a stop position.

19-19. Communications. Many types of communication are available to enhance the safety and performance of train operation and give passengers a feeling of security. These include simple communication by means of a train whistle or warning bell, wayside or cab signals, telegraph, telephone, radio, microwave, and electronic direct circuit or inductance. Many of these have been discussed in Arts. 19-6, 19-15, and 19-18. Specialists in the communications field should be consulted on selection and design of the most suitable communication system for a given rail transportation operation.

Automatic car identification is a development for which the full potential has not yet been realized. This is a system whereby the identifying letters of the car owner, the car number, and the general type of the equipment can be picked up by a scanner located at any desired location and transmitted to a teletype, memory storage of a computer, or any desired location. Truck trailers or individual containers on a flat car can also be identified as being on the particular car. For identification purposes, a reflective color-coded label is placed on each side of each piece of equipment to be identified. The scanner is actuated by the reflection from a light beam moving at a high rate of speed in the vertical direction. To be effective, it was necessary that one system be adopted by all railroads in the United States, Canada, and Mexico where cars are interchanged. The Association of American Railroads, after extensive tests, selected the color-coded label system described (ACI) and made its use mandatory on equipment in interchange service. Although intended primarily for freight cars, many roads are also using ACI on their locomotives and on passenger, commuter, and work equipment. Also, some rapid-transit systems use ACI.

19-20. Maintenance of Way. After a railway is completed, continual maintenance is required to keep it in condition for operation. Mechanized equipment is used for this to a major extent in the United States.

On-track equipment, such as Jordan spreaders and shovels, and off-track equipment, such as bulldozers, shovels, and draglines, may be used for ditching. Side-dump cars may be used for transporting material from ditches in cuts for bank widening on fills. Chemical weed killers or track burners may be used to keep the ballast section clear of vegetation. Chemical weed and brush killers or power mowers or cutters may be used to control undesirable weeds and brush on the right of way. Ballast-cleaning equipment is available for cleaning ballast between ties, in the ballast shoulders, and under the ties.

When rail is replaced, mechanized equipment is available for handling almost every item of work. This equipment includes spike pullers, power wrenches for track bolts, rail cranes, tie-adzing machines, ballast sweepers, creosote applicators, tie-plug drivers, spike drivers, and rail-anchor applicators.

Rail-defect detection is an important factor in safe operation of railroads. Rail defects develop from service use and are classified as transverse fissure, compound fissure, detail fracture, engine-burn fracture, horizontal split head, vertical split head, crushed head, piped rail, split web, head and web separation, bolt-hole crack, broken base, and damaged rail. However, because of improvements in rail design, manufacture, and maintenance practices, the number of rail defects that develop is remarkably small. Most of the rail defects that do develop are in the head or web within the joint-bar area. Rail-defect-detection equipment is available with which such defects can generally be detected. It makes possible removal of defective rail before a service failure occurs.

Rail-defect-detector cars are available that travel over the track at testing speeds of 6 to 15 mph. By utilizing electrical magnetism or ultrasonic waves, equipment on board is able to locate internal defects in the railhead. Ultrasonic equipment is used to detect defects in rail webs, particularly at rail joints, inside road crossings and other paved areas, and in frogs. Most rail in mainline track in the United States is tested once a year or oftener with detector cars.

Rail life is determined by several factors, including head wear, rail-end batter, fishing-surface wear, development of rail failures, surface condition as affected by engine burns or corrugation, etc. Fishing-surface wear occurs at surfaces where rail-joint bars bear, because of slippage of rail ends. For making economic studies, rail life in terms of traffic carried may be determined from

$$T = KWD^{0.565} \qquad (19\text{-}23)$$

where T = rail life on main-line tangent track, million gross tons of traffic carried

K = constant representative of conditions of track-maintenance standards, characteristics of traffic, etc.

W = weight of rail, lb per yd

D = traffic density, million gross tons per year

The term gross tons means the total weight of the locomotives and cars and their lading, in short tons. If the K factor is not known from past experience on a railway, a factor of 0.545 may be used. It is representative of past experience on a number of railways. For life on curves, use the percentage of life on tangents [Eq. (19-23)] given in Table 19-8.

Table 19-8. Life of Rail on Curves*

Degree of curve	% of life on tangents	
	Without oilers	With oilers
0	100	100
1	87	100
2	73	89
3	60	79
4	48	70
5	38	62
6	30	55
7	22	49
8	16	44
9	12	40
10	10	37

NOTE: Oilers may be either track lubricators or locomotive-flange lubricators.

Track lubricators reduce flange wear of outer rails and curve resistance to train movement. The devices are fastened to the rails at curves to apply lubricant to the flange of each passing wheel. Generally, a track lubricator consists of a reservoir containing a suitable type of grease, a plunger activated by each passing wheel to pump a small quantity of grease into the applicator, and the applicator. The last is a steel member, several feet long, placed against the gage side of the rail. This member contains small holes through which the grease is pumped to contact wheel flanges. Several types of lubricators are available. Manufacturer's instructions should be followed with respect to location and type of lubricant.

Some of the lubricant may get on top of the rail, lowering traction of locomotive wheels. To prevent loss of traction on both rails at the same time, separate lubricators should be used for each rail, and they should not be placed opposite each other.

Tie Renewals.   In the United States, only those ties that require renewal are replaced each year—the spot renewal method. The out-of-face renewal method, in which consecutive ties are replaced, is used abroad.

Some railways defer renewals for 2 or 3 years on a section of track to be resurfaced and then renew these ties and any ties that would have to be renewed in the next 2 or 3 years. This is done because ties can be renewed at less cost when track is being resurfaced. Mechanized equipment is available for pulling spikes, lifting up the rail, pushing out the old ties, pulling in the new ties, driving spikes, and tamping the ballast under the ties.

Track Maintenance.   Maintenance of track structure includes, in addition to tie renewals, periodic tightening of track bolts, raising the track to correct variations in surface and cross level, lining the track to correct deviations from alignment, and adding ballast and dressing the ballast section. Mechanized equipment generally used for this work includes power wrenches, rail-joint oilers, power jacks and tampers with self-contained sighting device, multiple-unit tie tampers, track liners with self-contained sighting device, and ballast spreaders.

String Lining. This is a convenient and satisfactory method of checking alignment of curves and spirals. First, the curve is marked off in 31-ft stations. Then, the midordinate to a 62-ft string line or chord is measured at each station. The midordinates are indicative of the degree of curvature, can be used to provide uniform curvature around the curve, and serve also to align the spirals (Sec. 5-3, "Manual for Railway Engineering," American Railway Engineering Association). A specially designed string-lining computing machine is available for this work.

Structures Maintenance. A detailed inspection should be made of all bridges, trestles, and culverts each year. Work not of an emergency nature, such as cleaning and painting, repairing concrete deterioration, and replacement of any parts, should be scheduled. The supervisory forces should note the condition of bridges and trestles at every opportunity. Any locations where scour at the footings occurs should be inspected frequently. Patrolling and inspection of bridge structures may be required during storms.

Mechanized Work Equipment. The rail-transportation engineer will find it helpful to have available a copy of the "Pocket List of Railroad Officials." This is published quarterly by The Railway Equipment and Publication Company, 424 West 33d Street, New York, N.Y. 10001. It contains an alphabetical listing of all railroad equipment and components with the names and addresses of companies that manufacture or sell these products.

Track Safety Standards. Legislation was passed by the U.S. Congress in 1970 requiring the Federal Railroad Administration of the Department of Transportation to establish track safety standards and to see that the railroads complied with them. These standards established by the FRA prescribe minimum standards for the safe operation of trains with regard to drainage, vegetation, ballast, rail defects (including end mismatch and end batter), welds, joints, tie plates, spikes, shims, switches, frogs, track appliances, deviations and variability in track geometry (runoff of elevation, alignment, cross level, and surface), maximum elevation on curves, maximum unbalance on curves, and track inspection. The standards have been established for different classes of track, classification being determined by a range of permitted operating speed. These standards may be revised as the FRA considers necessary or desirable. Copies may be obtained from the Chief of Operations Branch, Office of Safety, Federal Railroad Administration, Department of Transportation, 2100 Second Street, S.W., Washington, D.C. 20032.

Tunnel Engineering

JOHN O. BICKEL

Associate Consultant to Parsons, Brinckerhoff, Quade & Douglas,
Consulting Engineers, New York, N.Y.

Tunnel engineering makes possible many vital underwater and underground facilities. Unique design and construction techniques are involved because of the necessity of protecting the constructors and users of these facilities from alien environments. These facilities must be built to exclude the materials through which they pass, including water. Often, they have to withstand high pressures. And when used for transportation or human occupancy, tunnels must provide adequate lighting and a safe atmosphere, with means for removing pollutants.

20-1. Glossary

Adit. A short, transverse tunnel between parallel tunnels or to the face of the slope in a side-hill tunnel.

Air Lock. A compartment in which air pressure can be made equal to that of the compressed air used in shield tunneling as well as to that of the outside air, to permit passage of workers or material.

Bench. Top of part of a tunnel section, with horizontal or nearly horizontal upper surface, temporarily left unexcavated.

Blowout. A sudden loss of a large amount of compressed air at the top of a tunnel shield.

Breast Boards. Timber planks to hold the face of tunnel excavation in loose soil.

Dry Packing. Filling a void with a stiff mortar, placed in small increments, each rammed into place.

Evase Stack. An air-exhaust stack with a cross section increasing in the direction of air flow at a rate to regain pressure.

Face. The surface at the head of a tunnel excavation.

Grommet. A ring of compressible material inserted under the head and nut of a bolt connecting tunnel liners to seal the bolt hole.

Heading. A small tunnel, or tunnels, excavated within a large tunnel cross section which will be enlarged to the full section.

Jumbo. A frame that rolls on tracks or rubber wheels and carries drills for excavation of rock tunnels.

Lagging. Timber planks or steel plates inserted above tunnel-supporting ribs to hold back rocks or soil.

Liner Plate.   A steel segment to support the interior of a tunnel excavation.

Mucking.   Removal of excavated or blasted material from face of tunnel.

Pilot Tunnel.   A small tunnel excavated over part or the entire length to explore geological conditions and assist in final excavation.

Pioneer Bore.   (See Pilot Tunnel.)

Poling Boards.   Timber planks driven into soft soil, over timber supports, to hold back material during excavation.

Scaling.   Removal of loose rocks from tunnel surface after blasting.

Shield.   A steel cylinder of diameter equal to that of the tunnel, for excavation of tunnels in soft material.

Spiling.   (See Poling Boards.)

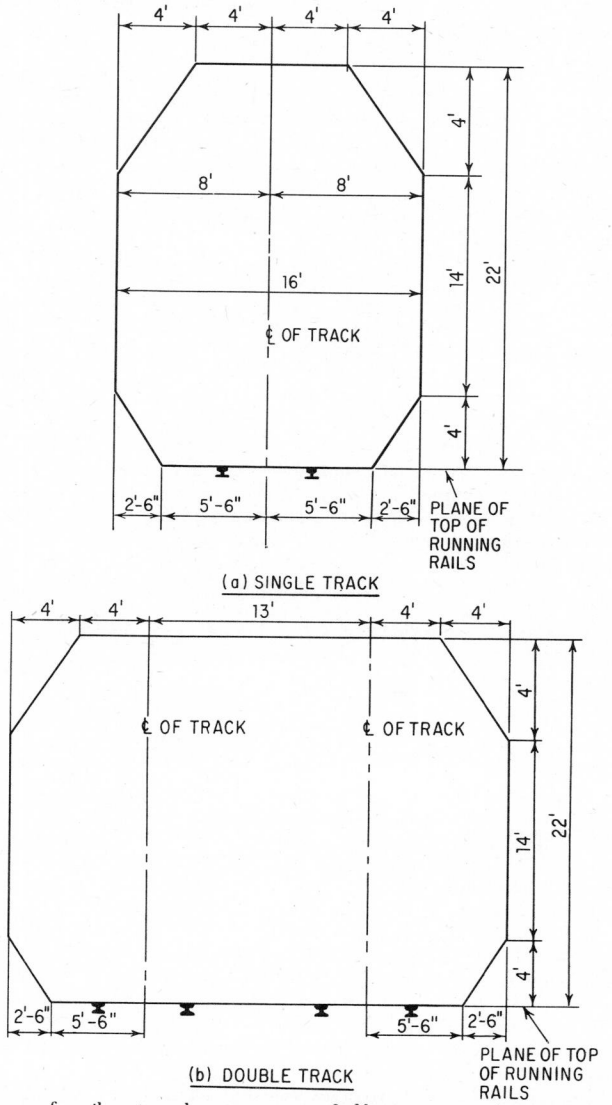

Fig. 20-1. Clearances for railway tunnels on tangent specified by American Railway Engineering Association.

20-2. Clearances for Railroad Tunnels.   Individual railroads have different standards to suit their equipment.   But on tangent tracks, clearances for single- and double-track tunnels should not be less than those shown in Fig. 20-1.   (Clearances shown are essentially those for bridges, because these clearances are more universally applicable than the tunnel diagrams in the "AREA Manual," American Railway Engineering Association, 59 E. Van Buren St., Chicago, Ill. 60605.)

On curved tracks, the clearances should be increased to allow for overhang and tilting of an 85-ft-long car, 60 ft c to c of trucks, and a height of 15 ft 1 in. above top of rail.   (Distance from top of rails to top of ties should be taken as 8 in.)

The track should be superelevated at curves according to AREA standards.

Clearances for pantograph, third-rail, or catenary construction should conform to diagrams published by the Electrical Section, Engineering Division of the Association of American Railroads.

The latest clearance standards of AREA should be checked for new construction.   Local legal requirements should govern if they exceed these standards.

Circular tunnels should be fitted to the clearance diagrams with such modifications as may be permissible.

20-3. Alignment and Grades for Railroad Tunnels.   Straight alignments and grades as low as possible, yet providing good drainage, are desirable for train operation.   But overall construction costs must be taken into account.

Grades in curved tunnels should be compensated for curvature, as is done for open lines.   In general, maximum grades in tunnels should not exceed about 75% of the ruling grade of the line.   This grade should be extended about 3,000 ft below and 1,000 ft above the tunnel.

Short (under 2,500 ft), unventilated tunnels should have a constant grade throughout.   Long, ventilated tunnels may require a high point near the center for better drainage during construction if work starts from two headings.

Radii of curves and superelevation of tracks are governed by maximum train speeds (Art. 19-9).

20-4. Clearances for Rapid-Transit Tunnels.   There are no general standards for clearances in rapid-transit tunnels.   Requirements vary with size of rolling stock used in the system.

Figure 20-2 shows the normal-clearance diagram of the New York City Independent Subway System.   Figure 20-3 gives the clearances established for the San Francisco Bay Area Rapid Transit System, which has cars 10 ft wide and 75 ft long on a 5-ft 6-in. gage track.   The clearances allow not only for overhang of cars, tilting due to superelevation, and sway, but also for a broken spring or defective car suspension.

20-5. Alignment and Grades for Rapid-Transit Tunnels.   Radii of curvature and limiting grades are governed by operating requirements.   The New York City Independent Subway has a 350-ft minimum radius, with transition curves for radii below 2,300 ft.   Maximum grades for this system are 3% between stations and 1.5% for turnouts and crossovers.   The San Francisco system is designed for train speeds of 80 mph.   Relation of speed to radius and superelevation of track for horizontal curves is determined by

$$E = \frac{4.65V^2}{R} - U \tag{20-1}$$

where E = superelevation, in.
    R = radius, ft
    V = train speed, mph
    U = unbalanced superelevation, which should not exceed 2¾ in. optimum or 4 in. as an absolute maximum

For 80-mph design speed, the radius with an optimum superelevation would be 5,000 ft.   For a maximum permissible superelevation of 8¼ in., a minimum radius of 3,600 ft would be required.   The absolute minimum radius for yards and turnouts is 500 ft.   Maximum line grade is 3.0% and 1.0% in stations.   To insure good drainage, grade should preferably be not less than 0.50%.

20-6. Clearances for Highway Tunnels.   The American Association of State Highway and Transportation Officials has established standard horizontal and vertical clearances for various classes of highways.   These have been modified and expanded for the Interstate Highway System under the jurisdiction of the Bureau of Public Roads (Fig. 20-4).

For specially designated parts of the Interstate Highway System, called Missile Routes, a 16-ft vertical clearance is required.

Since construction costs of tunnels are high and vehicle speeds generally limited to 40 mph, the clearance requirements are usually somewhat reduced.   A width of 21 ft between curbs for

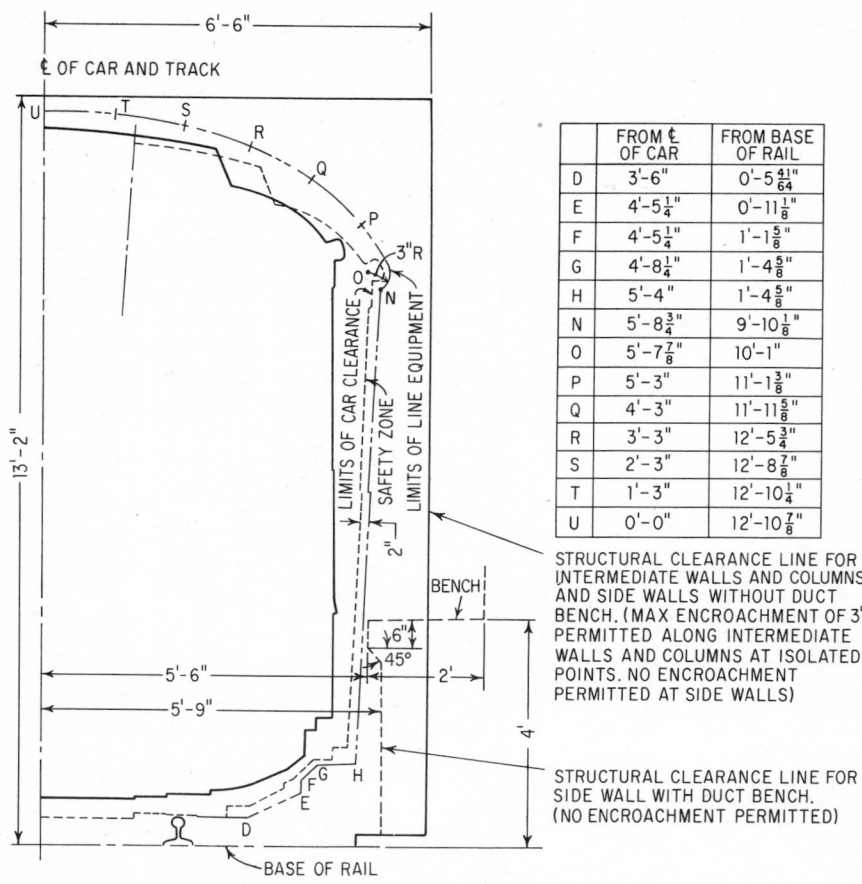

	FROM ℄ OF CAR	FROM BASE OF RAIL
D	$3'-6''$	$0'-5\frac{41}{64}''$
E	$4'-5\frac{1}{4}''$	$0'-11\frac{1}{8}''$
F	$4'-5\frac{1}{4}''$	$1'-1\frac{5}{8}''$
G	$4'-8\frac{1}{4}''$	$1'-4\frac{5}{8}''$
H	$5'-4''$	$1'-4\frac{5}{8}''$
N	$5'-8\frac{3}{4}''$	$9'-10\frac{1}{8}''$
O	$5'-7\frac{7}{8}''$	$10'-1''$
P	$5'-3''$	$11'-1\frac{3}{8}''$
Q	$4'-3''$	$11'-11\frac{5}{8}''$
R	$3'-3''$	$12'-5\frac{3}{4}''$
S	$2'-3''$	$12'-8\frac{7}{8}''$
T	$1'-3''$	$12'-10\frac{1}{4}''$
U	$0'-0''$	$12'-10\frac{7}{8}''$

STRUCTURAL CLEARANCE LINE FOR INTERMEDIATE WALLS AND COLUMNS AND SIDE WALLS WITHOUT DUCT BENCH. (MAX ENCROACHMENT OF 3" PERMITTED ALONG INTERMEDIATE WALLS AND COLUMNS AT ISOLATED POINTS. NO ENCROACHMENT PERMITTED AT SIDE WALLS)

STRUCTURAL CLEARANCE LINE FOR SIDE WALL WITH DUCT BENCH. (NO ENCROACHMENT PERMITTED)

Fig. 20-2. Clearance diagram for Independent Lines of the New York Subway System.

two-lane one-directional traffic, used in many existing tunnels, has proved adequate. For two-way traffic 23 ft between curbs is sufficient. Most recent tunnels have a minimum clearance of 14 ft above the roadway, whereas earlier clearances were 13 ft or less. Usually, a ledge 18 in. wide is placed between curb and tunnel wall on one side and a raised service walk of 2 ft 6 in. wide is set back 15 in. from the curb on the other side. Also, space has to be provided for air ducts, to meet ventilation requirements.

20-7. Alignment and Grades for Highway Tunnels. For tunnels under navigable water carrying heavy traffic, upgrades are generally limited to 3.5%; downgrades of 4% are acceptable. For lighter traffic volumes, grades up to 5% have been used for economy's sake. Between governing navigation clearances, grades are reduced to a minimum adequate for drainage, preferably not less than 1%. For long rock tunnels with two-way traffic, a maximum grade of 3% is desirable to maintain reasonable truck speeds.

Because of low tunnel speed limits, radii of curvature can be reduced to 1,500 ft or less (as little as 900 ft has been used). Short radii require superelevation and some widening of roadway to provide for overhang.

20-8. Pavements and Equipment for Highway Tunnels. Roadway base is a reinforced concrete slab. On this is placed a renewable pavement. Many tunnels have pavements of specially burned bricks. Well-designed bitumastic concrete has given good service and has good riding qualities.

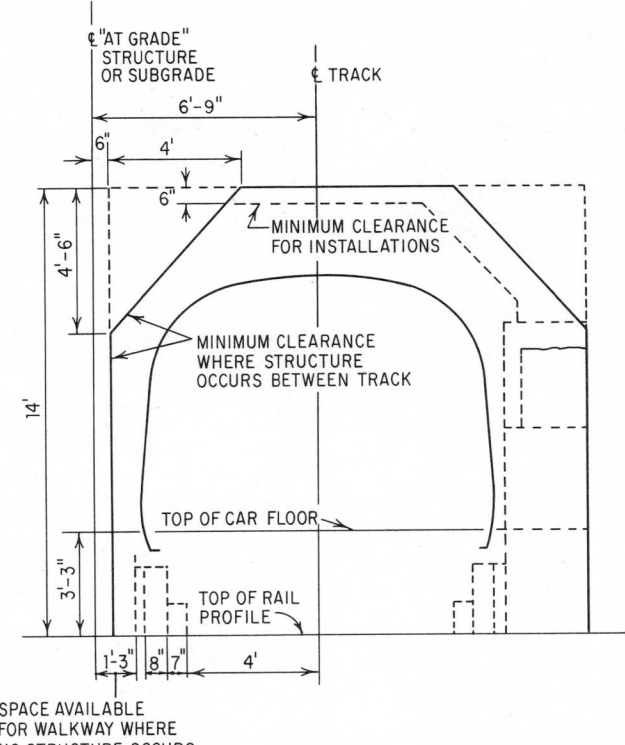

Fig. 20-3. Clearance diagram for San Francisco Bay Area Rapid Transit System.

Average daily traffic capacity of a two-lane two-directional tunnel is about 20,000 vehicles, with a maximum of 1,200 to 1,500 vehicles per lane per hour. For single-direction traffic in both lanes, capacities are 10 to 15% higher.

Red-amber-green traffic lights are installed at about 1,000-ft intervals, or at such spacing that the driver always sees at least one light. Telephones are placed in recesses about 500 ft apart for service and emergency calls.

Most tunnels, particularly those under water, are equipped with fire mains and hose outlets every 300 ft. Booster pumps in ventilation buildings raise supply pressure to 120 psi for use of foam. Fire extinguishers are mounted in recesses of hose outlets. Fire-alarm stations are at the same locations. Emergency trucks with heavy hoists, fire hose, foam equipment, and emergency tools are kept in readiness at each portal. Light tow cars for removing passenger cars are available.

20-9. Preliminary Investigations. Surveys should be made to establish all topographical features and locate all surface and subsurface structures that may be affected by the tunnel

Fig. 20-4. Clearance diagram for interstate highway tunnels.

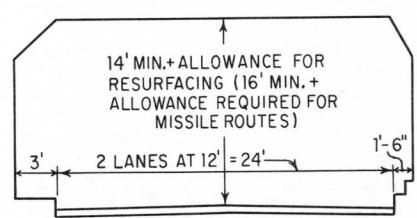

construction.   For underwater tunnels, soundings should be made to plot the bottom of the water.   (Accurate sonic fathometers are available for this.)

Knowledge of geological conditions is useful for all tunnel construction but is of primary importance for rock tunnels.   Explorations by borings for soft ground and underwater tunnels are readily made to the extent necessary.   For rock tunnels, particularly long ones, however, possibilities for borings are often limited.   A thorough investigation should be made by a geologist familiar with the area.   This study should be based on a careful surface investigation and examination of all available records, including records of other construction in the vicinity, such as previous tunnels, mines, quarries, open cuts, shafts, and borings.   The geologist should prepare a comprehensive report for the guidance of designers and contractors.

For soft ground and underwater tunnels, borings should be made at regular intervals.   They should be spaced 500 to 1,000 ft apart, depending on local conditions.   Closer spacing should be used in areas of special construction, such as ventilation buildings, portals, and cut-and-cover sections.   Spoon samples should be taken for soil classification, and undisturbed samples, where possible, for laboratory testing.   Samples not needed in the laboratory, boring logs, and laboratory reports should be preserved for inspection by contractors.   Density, shear and compressive strength, and plasticity of soils are of special interest.

All borings should be carried below tunnel invert.   For compressed-air shield tunnels, borings should be located outside the tunnel cross section.

For rock tunnels as many borings as practicable should be made.   Holes may be inclined, to cut as many layers as possible.   Holes should be carried below the invert and may be staggered on either side of the center line, but preferably outside the tunnel cross section to prevent annoying water leaks.   Where formations striking across the tunnel have steep dips, horizontal borings may give more information; these have been made up to 2,000 ft in length.   All rock cores should be carefully cataloged and preserved for future inspection by contractors.

Groundwater levels should be logged in all borings.   Presence of any noxious, explosive, or other gases should be noted.

Where lowering of groundwater may be employed during construction of cut-and-cover or shield tunnels on land, the permeability of the ground should be tested by pumping tests in deep wells at selected locations.   Rate of pumping and drawdown checked in observation wells at various distances should be recorded, as well as recovery of the water level after stopping the pumps.

Geophysical exploration to determine elevations of distinctive layers of soil or rock surfaces, density, and elastic constants of soil may be used for preliminary investigations.   The findings should be verified by a complete boring program before final design and construction.

20-10. Ventilation Requirements for Railroad and Rapid-Transit Tunnels.   Short tunnels (up to about 2,500 ft) generally have no forced ventilation.   Longer tunnels for steam and diesel trains need some ventilation to purge smoke and exhaust gases.   Tunnels for electric traction are adequately self-ventilated by piston action but may require emergency exhaust ventilation.

A ventilation system dilutes and purges smoke and combustion and exhaust gases.   Its capacity must be adequate to prevent irritating smoke or gas concentrations while a train passes through and to clear the air between train passages.   Diesel exhaust is less irritating than smoke from steam engines, but its content of nitrogen oxides may form corrosive acid in lungs when inhaled for long periods.   The following systems are used by American railroads:

Injecting a stream of air at high velocity in the direction of train movement to keep smoke ahead of the train.

Injecting a high-speed high-volume air stream from the opposite end against the train motion to dilute smoke and clear the tunnel.

Addition of portal doors with the first injection system, to increase efficiency and prevent backflow in case of a stalled train.   Doors are interlocked with signal systems (Moffat Tunnel).

Because of absence of smoke or exhaust gas when electric traction is used, ventilation by piston action of trains is adequate for tunnels for electric trains except under emergency conditions.   Auxiliary exhaust fans should be installed to remove smoke in case of fire and to draw fresh air into the tunnel from the stations or portals.   Fans may be installed in exhaust shafts between stations or in separate ventilation buildings in long underwater tunnels equipped with exhaust ducts.   Air velocity of 1.5 to 2.0 fps around stalled trains is adequate.

High-speed rapid-transit tunnels require air-relief shafts ahead of stations to dissipate the air stream caused by piston action, to prevent air blasts entering the stations.   ("Subway Environmental Design Handbook," Vol. I, Urban Mass Transportation Administration, Washington, D.C. 20590.)

20-11. Ventilation Requirements for Highway Tunnels.   Exhaust gases of internal-combustion engines contain deadly carbon monoxide from gasoline engines and irritating smoke and oil vapors.   Diesel engines may also produce dangerous nitrogen oxides and aldehyde.   The components of exhaust gases vary over a wide range, depending upon adjustment of carburetors and fuel injectors and general engine maintenance.

All highway tunnels, except those on straight grades and less than about 1,000 ft long, require forced ventilation.

Headache is the most reliable indicator of carbon monoxide poisoning.   The tests showed that no one had this symptom to any degree after exposure to a concentration of 4 parts carbon monoxide to 10,000 parts of air for 1 hr.   This has been set as the maximum permissible in a tunnel.

The Federal government or health authorities of states may place more severe restrictions on permissible CO content.   New York State has proposed limits as low as 75 ppm, with 150 ppm for short periods.   With new standards limiting contaminants in exhaust gases, however, it may eventually be possible to meet the CO limitations without extensive increase in ventilation. Engineers should check current rules at time of design.

Haze from exhaust gases, reducing visibility in tunnels, is another controlling factor.   In practice, the CO content is kept at 2.0 to 2.5 parts in 10,000, or less, insuring adequate dilution of irritating parts of exhaust gases and maintenance of visibility.

For preliminary estimates, the following air quantities, cfm per lane, may be estimated for a highway tunnel, at low elevation, carrying about 75% passenger cars and 25% trucks:

On level grade	115
On 3½% downgrade	75
On 3½% upgrade	150

Use of lower quantities should be approached with caution, since ventilation may permanently control the capacity of the tunnel.   Natural draft, because of differences of barometric pressure often prevailing on two sides of large mountain ranges, may influence the ventilation and should be considered.

At higher elevations, CO increases rapidly, both in percent of exhaust gas and in absolute quantity.   Tests made in Switzerland with modern automobiles, trucks, and buses show a considerable reduction in CO emission from the quantities in Table 20-1.   These led to the base values in Table 20-1 for CO per vehicle, cfm, at a speed of 30 mph at sea level:

Table 20-1. CO Produced per Vehicle at 30 Mph at Sea Level

Vehicle	Cfm
Passenger cars	0.792
Trucks:	
5-ton	1.35
10-ton	3.10
15-ton	4.63

These are to be multiplied by factors for higher elevations (Fig. 20-5) and for grade and increased traffic density at reduced speed (Fig. 20-6).   ("Die Luftung der Autotunnel," *Publication* 10, Institute for Highway Construction, Swiss Institute of Technology, Zurich.)

In 1964 and 1965, the Colorado State Highway Department, Bureau of Public Roads, and Public Health Service made tests to determine the effect of elevation on the CO production of passenger cars and its physiological influence.   The road tests gave the CO values in Tables 20-2 and 20-3.   The values in the tables are comparable with, although slightly lower than, those obtained from Figs. 20-5 and 20-6.

As a result of the Colorado tests, the following criteria were proposed for ventilating the two 7,800-ft Straight Creek Highway tunnels through the Continental Divide, which have an average grade of 1.68% and are at 11,000-ft elevation:

CO per vehicle at a speed of 40 mph:

Upgrade: 4.86 cfm (air density of 0.055 lb per cu ft).

Downgrade: 1.82 cfm.

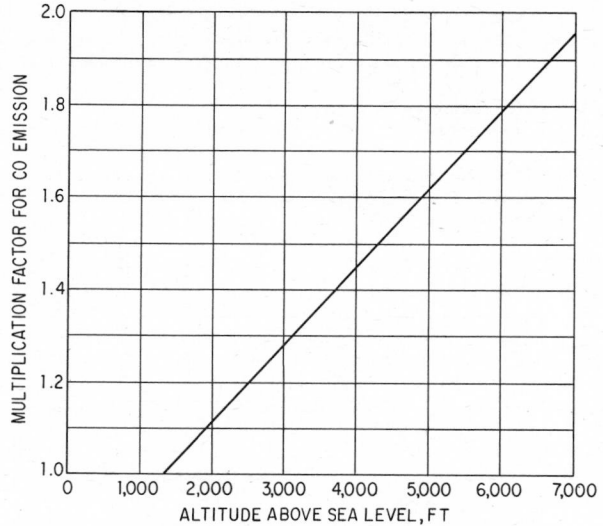

Fig. 20-5. Carbon monoxide emitted from vehicles increases with altitude.

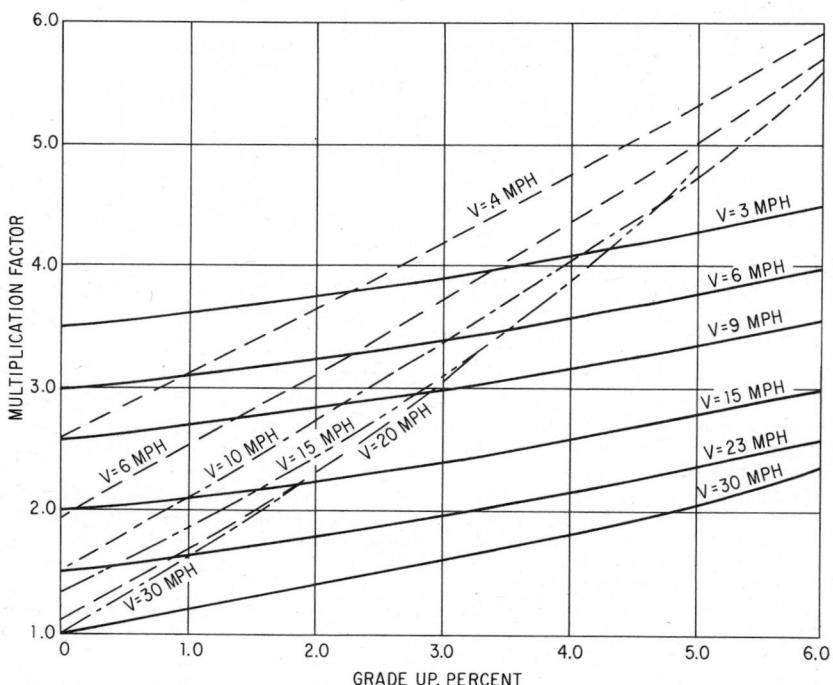

Fig. 20-6. Chart gives the increase in carbon monoxide emitted by vehicles with increase in speed and grade. Solid lines are for passenger cars; broken lines, for buses and trucks.

Table 20-2. CO Produced per Vehicle at 40 Mph at Different Grades and Elevations, Cfm

Grade, %	Altitude		
	5,000 ft	8,000 ft	10,000 ft
−3	0.063	0.070	0.090
−2	0.069	0.073	0.088
−1	0.073	0.080	0.105
0	0.078	0.095	0.137
+1	0.082	0.113	0.185
+2	0.085	0.140	0.250
+3	0.087	0.154	0.290

Table 20-3. CO Produced per Vehicle on Grade of +1.68% at Different Speeds and Elevations, Cfm

Altitude, ft	30 mph	40 mph
5,000	0.075	0.083
6,000	0.076	0.087
7,000	0.085	0.102
8,000	0.100	0.126
9,000	0.120	0.160
10,000	0.150	0.206
11,000	0.183	0.267

Permissible CO concentration in tunnel: 150 ppm for a maximum of 30 min, with an hourly average not exceeding 100 ppm. ("Ventilation Requirements for the Straight Creek Highway Tunnels, Colorado," Report to the State of Colorado Department of Highways by Tippets-Abbett-McCarthy-Stratton, 1965.)

20-12. Types of Ventilation Systems for Highway Tunnels. *Natural Ventilation.* In straight tunnels up to about 1,000 ft in length, natural air flow is usually sufficient, particularly with traffic in one direction. If a tunnel is exposed to heavy traffic congestion at times, installation of exhaust fans in a shaft or adit near the center for emergency ventilation is advisable if the length exceeds 500 ft.

Longitudinal System. For short tunnels, with tubes carrying single-direction traffic, air is injected at high speed at one end in the direction of traffic. Air flow is assisted by drag effect. This system has been used successfully in underwater tunnels in Europe up to about 2,000 ft in length.

Another system mounts fans under the ceiling to push air along the tunnel. This system eliminates a ventilation building but involves maintenance of many pieces of equipment.

Semitransverse System. Fresh air is supplied through a duct for the entire length of a tunnel. Vitiated air leaves through portals. The system has been satisfactory in subaqueous tunnels up to about 3,500 ft long, with air supplied at roadway level, and in straight rock tunnels up to 8,000 ft in length, with air supplied through ceiling ports. Some of these rock tunnels have auxiliary exhaust fans, or duct arrangements permitting regular fans to draw air from the tunnels, to remove smoke in case of fire.

Pure exhaust systems have been used. These draw air through upper and lower air ducts from the center of a tunnel, while fresh air flows in through the portals. These systems are effective in removing smoke in case of fire and letting fire-fighting equipment enter with a fresh-air stream. This type of system, however, may become somewhat unbalanced because of wind pressure in one portal producing a greater air flow from that end and reducing ventilation in the opposite half of the tunnel. This effect can be partly overcome by operating fans at maximum capacity.

Transverse System. Fresh air is supplied at roadway level, and vitiated air is exhausted through ports in or near the ceiling. Fresh-air ducts are below the roadway in circular tunnels and exhaust ducts above the ceiling. In rectangular tunnels, air ducts may be below the roadway, above the ceiling, or on the sides. This is the most effective system and is used in all subaqueous

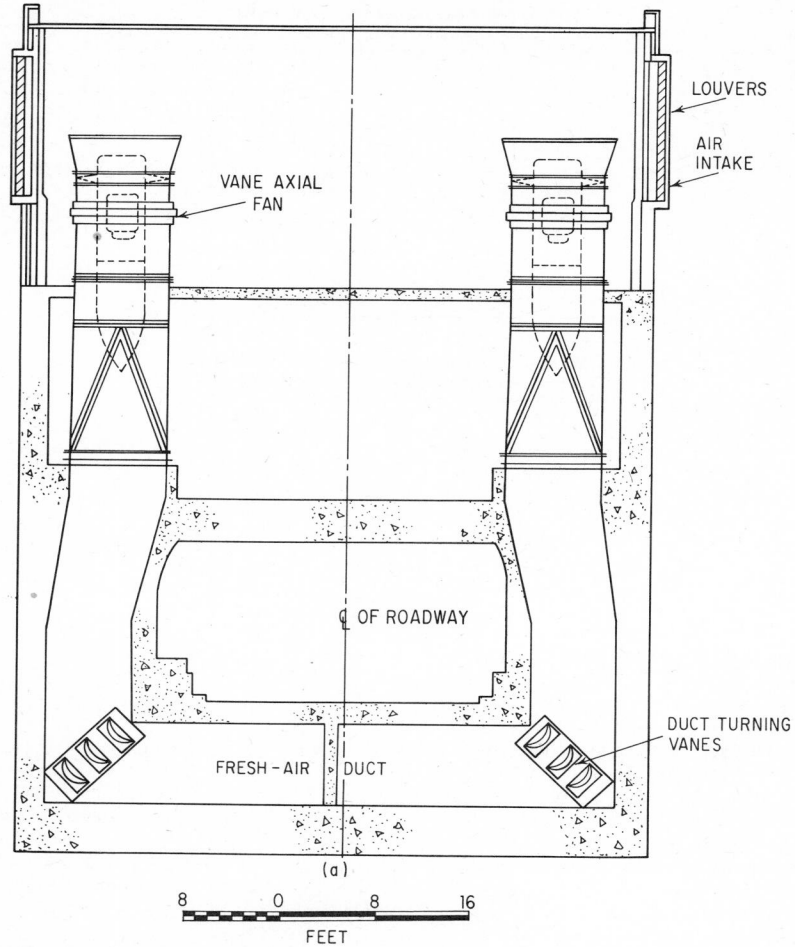

LOUVERS

AIR
INTAKE

VANE AXIAL
FAN

℄ OF ROADWAY

DUCT TURNING
VANES

FRESH–AIR ▷ DUCT

(a)

8   0   8   16

FEET

Fig. 20-7. Sections through Hampton Roads Tunnel Ventilation Building. (a) Fresh-air supply system.

tunnels over 3,500 ft long. It is advisable for heavily traveled straight tunnels exceeding about 5,000 ft in length. Supply and exhaust fans are usually installed at both ends for maximum efficiency; but for short tunnels, 3,500 to 4,000 ft, supply and exhaust from one end may be most economical. Comparative estimates of construction and operating costs should be made.

20-13. Elements of Tunnel Ventilation Systems. (See also Art. 20-12.)

Ventilation Buildings (Figs. 20-7 and 20-8). Fans, electrical transformers and switchgear, control board, and auxiliary equipment are housed in ventilation buildings. In short and medium-length tunnels, one building at either portal is sufficient. Longer tunnels should have a building at each portal. A few of the longest have three or four buildings. For underwater tunnels, ventilation buildings may be at the water's edge, each building controlling a land and a river section of the tunnel.

Fresh air is taken in through large louver areas in the walls of the building. The louvers should be protected by bird screens. Louvers are usually aluminum and arranged for shedding water. Adequate drains should be provided in the fan room to remove rainwater, which may blow in

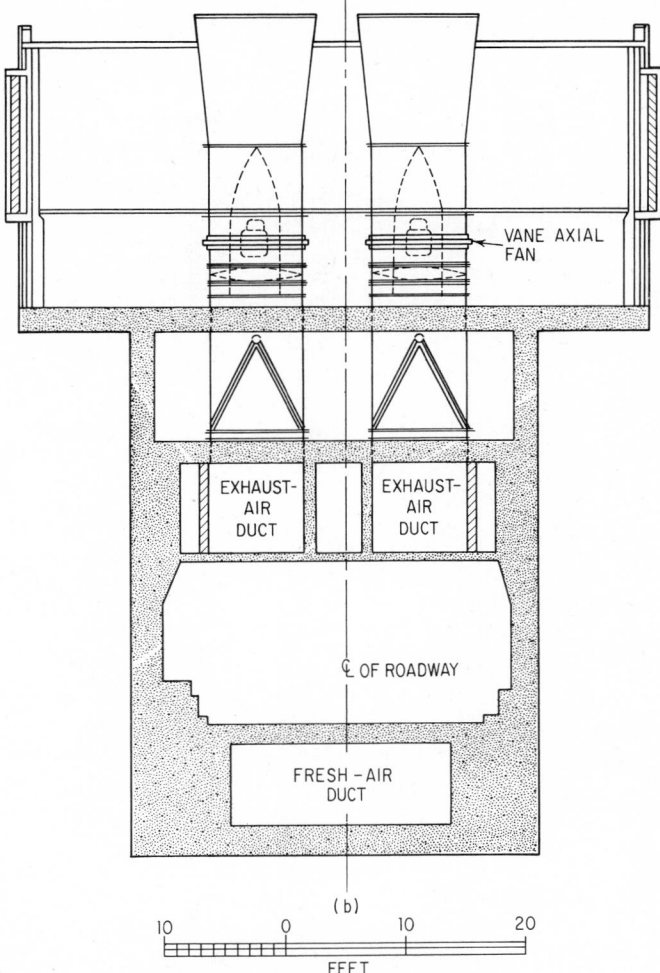

(b)

FEET

Fig. 20-7. (*continued*) (*b*) Exhaust-air system.

through the louvers. Vitiated air is discharged through vertical, tapered stacks, which also should be covered by screens.

Ducts are usually of constant area throughout their length. Concrete surfaces should be smooth for minimum friction. Obstructions, such as ceiling hangers, should be streamlined or at least rounded. Turns in ducts and shafts leading to the tunnel should be equipped with noncorrosive turning vanes for smooth air flow.

Flues spaced about 15 ft apart, extended from the ducts, supply fresh air slightly above roadway level. Ceiling ports are slanted at 45° in the direction of air flow in the ducts. All air openings should be adjusted in size to balance the air flow over the length of the tunnel.

Fans. Two types of fans are available: centrifugal fans, used in all tunnels up to about 1938, and vane axial fans, a later development. Centrifugal fans have backward-curved blades and are nonoverloading. The efficiencies of well-designed fans of either type are about the same. For underwater tunnels, with vertical air shafts in the ventilation buildings, the vane axial fans require considerably less space and avoid the efficiency loss through the fan chambers usually

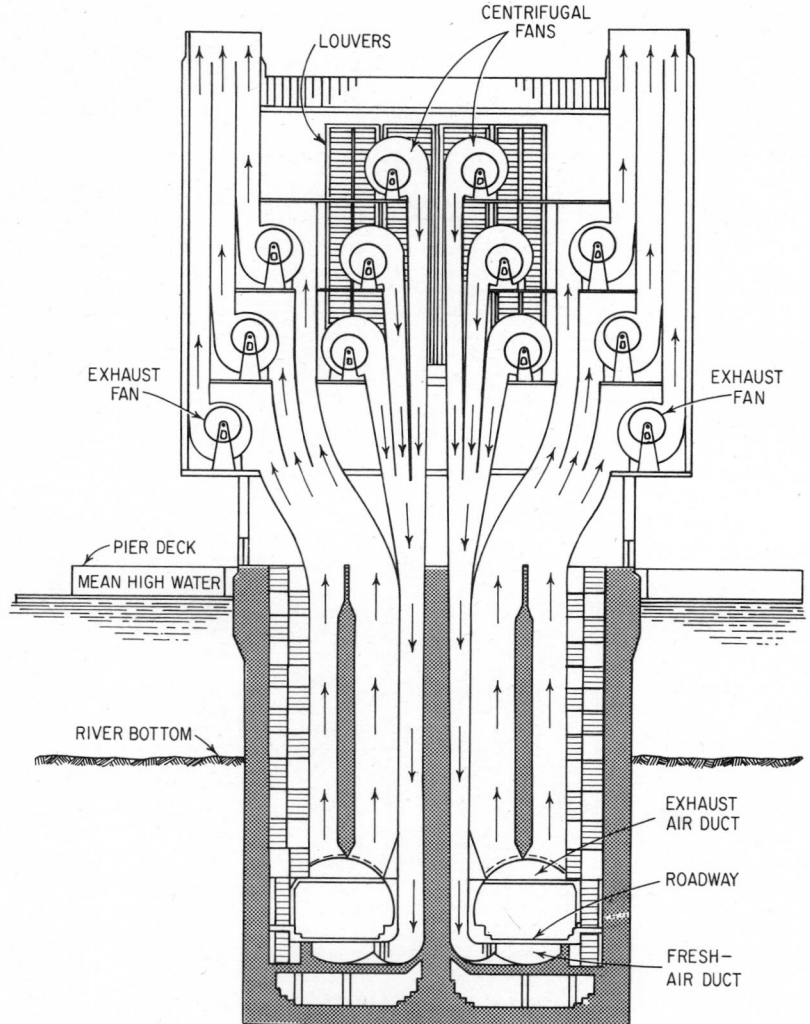

Fig. 20-8. Section through ventilation building of the Holland Tunnel.

associated with centrifugal fans. Their noise level at maximum speed is somewhat higher because of greater tip speed. In sensitive surroundings, the noise from supply and exhaust fans can be dampened by sound baffles.

Centrifugal fans are operated by squirrel-cage motors through chain or multiple V-belt drives. The latter eliminate lubrication problems and wear on a multiplicity of parts (inherent in chain drives), give excellent service, and can be easily replaced. Chains are enclosed in solid housings, belts protected by wire guards.

Vane axial fans may have external drives or motors built into the hub of the impellers. They may be provided with blades adjustable while running, to vary output, a refinement not usually required.

For flexibility, the load is divided between several fans—at least two, sometimes as many as six—for each system. Four is a good number for demands exceeding about 600,000 cfm.

To further adjust supply to variable demand, fan motors are equipped with two-speed windings. Three speeds with two motors have been used in earlier installations but are not necessary with an adequate number of fans. Spare fans may be provided as protection against breakdown, or total fan capacity may be increased by 10 or 15%. With good maintenance, fans are seldom out of commission, and the extra capacity of the system is sufficient to maintain acceptable conditions for limited periods with one unit out of service.

To protect the exhaust fans in case of a serious fire in the tunnel, automatic deluge sprinkler systems should be installed to cool the exhaust air.

Dampers. All fans should be equipped with dampers to prevent short circuiting of air. Their operating motors should be interlocked with the control of the fan motors for automatic opening and closing. Trapdoor-type or multileaf dampers are in use. The latter take less space and time to operate.

Fan Control. In short, unattended tunnels, fans are controlled automatically with carbon monoxide analyzers. Larger tunnels with heavy traffic should have operators stationed in the control room. They operate the fans to control conditions in the tunnel. At least two independent sources of electric power must be available, usually through feeders from different parts of the utility system. If these are not available, a diesel-engine emergency generator sufficient for minimum requirements should be installed.

Carbon Monoxide Analyzers. These take continuous air samples from the tunnel and analyze them for CO content. The results are visually indicated and also recorded on paper tape, with time gradations. The recorders are mounted on the face of the control board, to guide the operator in selection of number of fans and speed necessary. If the CO content reaches 3 parts per 10,000, the recorder sets off an alarm in the control room.

In a longitudinal or semitransverse supply system, air samples are taken from the tunnel proper at points of maximum concentration. In transverse systems, the samples may be taken from the exhaust ducts.

Haze Control. To measure visibility in tunnels affected by haze from exhaust gases, instruments have been developed that give a reliable indication without excessive maintenance. Equipment manufactured for the Port of New York Authority uses the scattering of ultraviolet light by dust particles. European instruments protect the optics by recessing them in tubes through which filtered air is exhausted. Another type of instrument compares the intensities of two branches of a split light beam passing through the same optics, one going through a tube filled with clean air, the other through tunnel air.

Ventilation Power Requirements. These are determined by the following losses:

Supply system: intake losses through louvers, which should not exceed 0.20 in. of water; losses in ducts, bends, and elbows leading to the tunnel; and losses in tunnel air ducts and supply flues.

Exhaust system: losses in air ports leading to the exhaust ducts; losses in tunnel exhaust ducts and in ducts, bends, and elbows leading to fans; losses in fan chamber (if used); and losses or recovery in discharge stacks.

The following equations for pressure losses in the tunnel air ducts of uniform cross section, with uniform air discharge to and air removal from tunnel, are based on scale-model tests and air-flow studies (A. C. Davis, "Development of the Ventilation System of the Holland Tunnel," *Heating, Piping and Air Conditioning Journal*, October, 1930). The equations have proved satisfactory for many tunnel ventilation systems. The coefficient of friction for air ducts may be determined from

$$f = 0.0035 + \frac{0.01433}{m^2 v^2} \tag{20-2}$$

where m = mean hydraulic radius, ft = duct area, sq ft, divided by perimeter, ft
v = velocity of air, fps = Q'/A
Q' = quantity of air, cfs
A = area of duct, sq ft

The pressure drop in straight air ducts, in. of water, is given by

$$H = \frac{12y}{D} \frac{v_1^2}{2g} \frac{fL}{m} \tag{20-3}$$

where y = density of dry air, lb per cu ft = 0.075 at 70° F
D = density of water, lb per cu ft = 62.5 at 70° F
v_1 = velocity of air entering duct, fps

$g = 32.2$ ft per sec^2

L = total length of constant-section duct, ft

The pressure loss, in. of water, in the tunnel supply duct may be computed from

$$TP = \frac{12y}{D} \left\{ \frac{v_1{}^2}{2g} \left[\frac{0.0035LZ^3}{3m} - \frac{1}{2}(1 - K)Z^2 \right] + \frac{0.01433LZ}{2gm^3} \right\} + P_o \qquad (20\text{-}4)$$

where $Z = (L - X)/L$

X = distance from duct entrance to section considered, ft

K = turbulence factor = 0.615

P_o = pressure at bulkhead, in. of water

The suction head, in. of water, in the exhaust duct may be obtained from

$$SP = \frac{12y}{D} \left\{ \frac{v_1{}^2}{2g} \left[\frac{0.0035LZ^3}{(3 + C)m} + \frac{3Z^2}{2 + C} \right] + \frac{0.01433LZ}{2gm^3(1 + C)} \right\} + P'_o \qquad (20\text{-}5)$$

where C = turbulence factor for entrance ports. For air flow in excess of 200 cfm per ft of tunnel, $C = 0.25$; for lower rates of flow, $C = 0.20$. (Air ports at an angle of 45° with the air flow in the duct cause the least amount of turbulence.)

P'_o = suction head through 45° ports; ordinarily not less than 0.03 in. of water

For pressure losses in elbows and effects of sudden changes in cross section, see fan manufacturers' publications or reference books, such as "Heating, Ventilating, Air Conditioning Guide," American Society of Heating, Refrigerating, and Air-Conditioning Engineers.

The static pressure required to discharge supply air to the roadway depends upon the size, shape, and length of flues, but it generally does not exceed 0.25 in. of water.

The losses in plenum chambers of centrifugal exhaust fans may equal the entire velocity head. Discharge stacks from the fans are of the evasé type, with gradually enlarged area toward the top to regain some of the pressure loss.

The fan horsepower is

$$\text{hp} = \frac{62.4PQ}{396,000} \qquad (20\text{-}6)$$

where P = total pressure, in. of water

Q = quantity of air, cfm

20-14. Tunnel Lighting. Since locomotives are equipped with strong headlights, railway tunnels are generally not lighted. Subaqueous tunnels and other tunnels on electrified lines, particularly in cities, are equipped with a nominal amount of lights, especially in refuge niches.

Rapid-transit tunnels are lighted sufficiently to make obstructions on tracks visible and to facilitate maintenance work. The lights are shielded to prevent glare in the motorman's eyes. Fluorescent lights spaced 25 ft apart were installed in the tunnels of the San Francisco Bay Area Rapid Transit System for emergency use. Normally, only some of the lights are in service, but in emergencies they are all switched on from control points in the stations.

For highway tunnels, an important feature of the lighting system is the transition from bright daylight to the reduced lighting level in the interior. The American Society of Illuminating Engineers recommends the following average intensities, measured at pavement level, for daytime lighting:

At entering portal: 75 footcandles.

Tunnel interior: 5 footcandles.

The transition from portal lighting to interior should be made in not less than three steps, extending over a total distance of 700 ft for 30-mph speed and 1,400 ft for 60-mph.

At night, the lighting is uniformly reduced to about 3 to 5 footcandles, depending on the degree of illumination of the approach roads.

Old highway tunnels have incandescent lights mounted in recessed boxes in the side walls below the ceiling or set flush in the ceiling slab. Intensities, particularly at portals, are much lower than those now recommended. Improvements have been made by installing stronger lamps and by replacing the diffusing glass covers with distributing lenses. The intermittent lights may cause a somewhat disturbing flicker reflection on the hoods of automobiles.

Higher intensities can be obtained with mercury lights installed in the tunnel ceilings. The similarity of their color to daylight is another advantage, as is their longer life.

By far the most satisfactory lighting is achieved with continuous rows of fluorescent lights, mounted either on the side walls below the ceiling or directly on the ceiling. Their surface brightness is relatively low, preventing glare, and the flicker effect is avoided. Their color comes closest to daylight, and their intensity can be varied over a wide range by adjusting the electric current. The uniformity of the light source can therefore be maintained even for the reduced lighting at night. Because of the continuous service, the lamp life is very long, usually in excess of 15,000 hr.

20-15. Tunnel Drainage. Most tunnels through hills and mountains have water problems. Surface water penetrates through fissures and percolates through permeable soils. Attempts to seal off the rock by grouting, with either cement or chemicals, usually are not completely successful. Nor are concrete linings completely watertight. Water may find its way through shrinkage cracks in the linings into the interior of tunnels. There, it may freeze in cold weather and produce an unsightly appearance, objectionable in highway tunnels. Provision must be made to drain water from tunnels.

Fire fighting, washing of tunnel interiors, and flushing of pavements also introduce water that must be drained.

While cut-and-cover tunnels can be waterproofed, this is not possible with bored tunnels. If the water problem is not serious, the most economical solution is to seal cracks in the lining that leak. With good concrete control, the number of these should be small.

If water appears in considerable quantity during tunneling operations, tight steel lagging over the tunnel supports and grouting may prevent leakage. In serious cases, it may be necessary to dry-pack between the rock and the tunnel lagging to drain water. This is a slow, costly method requiring much manual labor. Dry pack behind the side walls can easily be placed and is effective in preventing the build-up of a hydrostatic head behind the lining (see also Art. 20-19). Longitudinal drain pipes should be installed behind the base of the side walls, with the laterals at regular intervals leading to the main drain. This is a large drain installed under the roadway, for roadway drainage.

In highway tunnels, catch basins should be installed at regular intervals along the curbs, with cross connections to the main drain. The latter should be of generous size, in longer tunnels preferably large enough to provide crawl space.

Leakage in well-constructed underwater tunnels, either shield-driven or sunken-tube type, is usually minor. It can be controlled by calking of joints in segmental liners or by installation of metal chases in concrete lining where leaks appear. Main sources of water are washing of tunnel interior, fire fighting, drippings from vehicles, and rain collected in open approaches.

Continuous open gutters recessed into the curbs have been used in many subaqueous tunnels. The gutters lead water to a low point, where it is collected in a sump. Drainage inlets, spaced about 50 ft apart along each curb and connected to longitudinal drain lines embedded in the concrete below the curbs, are desirable because they prevent propagation of fire by burning fuel in case of a serious accident. Drain lines should be at least 8 in. in diameter, of cast iron or heavy-wall asbestos cement. They should be equipped with cleanouts, accessible from the lower air duct, every 500 ft.

In straight, open approaches, transverse interceptors about 300 ft apart are most effective in preventing water from entering a tunnel. They are 18 in. wide, extend from curb to curb, and are covered with heavy, cast-iron gratings, with slots parallel to the center line of the roadway. An interceptor is placed in front of the tunnel portal and another about 10 ft inside.

In curved, superelevated approaches, drainage inlets should be installed at regular intervals along the low curb.

All drainage from open approaches should be collected inside the portals in sumps below the roadway. Each sump should be divided into a settling basin and a suction chamber. Easy access must be provided for cleaning out sediments. Three electrically driven, large-clearance drainage pumps should be installed, one as a standby. Alternating automatic controls rotate the pumps in service. High-water-level alarm circuits should be extended to the control room. Sump and pump capacity, with two pumps operating, should be designed for maximum, short-duration rainfall for the locality. An intensity of 4 in. per hr, based on a 15-min downpour at a rate of 8 in. per hr, is ample for most areas.

A smaller sump should be located at the low point of the tunnel, usually in part of the lower air duct. This sump also should be divided into a settling and a suction chamber. Two automatically controlled drainage pumps, with a capacity of 250 gpm each, are usually adequate. Their discharge should be carried to one of the portal sumps.

20-16. Water Tunnels. These may be diversion or intake tunnels for hydropower plants, or aqueducts bringing water to city and municipal distribution systems.

Diversion tunnels carry river water around dam sites during construction. They are designed to carry the maximum expected runoff during this period. They may also serve to discharge excess water after the reservoir has been filled, or be converted to intake tunnels to a powerhouse located in the side of the valley below the dam. If they are not needed after completion of the project, the diversion tunnels are closed with concrete plugs. Extensive diversion tunnels have also been built to collect water from several watersheds for a central power plant.

Intake tunnels bring water from reservoirs to turbines or the heads of penstocks. The tunnels are mostly in rock and operate under a positive hydrostatic head. In pervious and fissured ground, they are lined with reinforced concrete; in sound rock, a sprayed-concrete lining may be adequate to provide a smooth surface.

Many miles of aqueduct tunnels have been built for municipal or area water-distribution systems. These tunnels are, for the most part, in rock but may also contain stretches of soft-ground tunneling. They may be under large hydrostatic pressure, such as the New York City aqueduct, which crosses the Hudson River 600 ft below sea level.

Tunnels with small or no interior pressure generally have a horseshoe section; pressure tunnels are circular. Lining is concrete, 6 to 36 in. thick, depending on size, pressure, and nature of rock. Grade tunnels may be lined with plain concrete, pressure tunnels with reinforced concrete. Diameters range from 7 ft for small aqueducts to 50 ft for Hoover Dam diversion tunnels. In very sound rock, sprayed-concrete lining has been used. Parts of the Colorado River aqueduct are lined with continuous steel shells against concrete backing, and the inside is protected by 2 in. of reinforced sprayed concrete.

To expedite construction, long tunnels are subdivided into several headings by shafts or adits, about 2 to 5 miles apart.

20-17. Sewer and Drainage Tunnels.   Large cities require many miles of tunnels to carry off storm runoff and to conduct sewerage to treatment plants. These are built in a great variety of soils. Some are constructed as box culverts by the cut-and-cover method, but most are tunneled with shields and compressed air. Size varies from about 7 to 15 ft. Drainage tunnels for storm water are usually less extensive since they can discharge into nearest open waters.

The cross section of sewers and drainage tunnels usually is horseshoe or circular, with concrete lining. Quality of concrete is of special importance, to resist the detrimental effect of sewage. Generally, they are grade tunnels, except for siphons under rivers, which are under pressure. A circular or egg-shaped section maintains velocity at low flow to prevent excessive settling of solids.

Alignment is dictated by location of treatment plants, soil conditions, and the street plan of the city. Continuous grades should be maintained except for siphons. A minimum grade should be maintained for gravity flow.

20-18. Cut-and-Cover Tunnels.   Shallow-depth tunnels, such as rapid-transit lines under city streets, underpasses, land sections of underwater tunnels, and end sections of tunnels through hills, are built by cut-and-cover methods. Depth of invert on subways and underpasses usually does not exceed 35 to 40 ft. For connections to subaqueous tunnels, cuts up to 75 ft have been used under special circumstances, and depths to 60 ft are not uncommon.

Where space and depth of excavation permit, open slopes are used in sufficiently firm materials. Groundwater may be lowered, as needed, by tiers of wellpoints. In confined areas, soft material, or greater depth, the trench should be protected by bulkheads. These may be constructed as:

Steel sheetpile walls, for depths to about 30 to 40 ft, supported by wales and cross bracing. The walls keep loss of ground to a minimum.

Soldier piles, made of steel H beams and wood lagging. These are used for greater depth. Lagging must be blocked tight against the earth to control loss of ground. Soldier piles may be combined with sheetpiles, instead of wood lagging, if tight bulkheads are required. Wales and cross bracing support the walls.

Concrete walls built in bentonite-slurry trenches have been used to prevent loss of ground and eliminate or reduce groundwater lowering. Sections of trenches about 20 ft long and 1.5 to 3.0 ft wide are excavated. The trenches are kept filled with bentonite slurry. Then, reinforcing cages are lowered into them, and concrete is placed to fill the trenches, displacing the slurry. Key sections are formed at the ends of the trenches. The walls serve as part of the final structure or as impervious bulkheads.

California wall, a combination of soldier piles and slurry wall, has been used on some stations of the Bay Area Rapid Transit System in San Francisco. Large wide-flange beams are inserted in slurry-filled bored holes, the space between the beams is excavated under slurry, and excavation and pipe holes are filled with concrete. Care must be used in excavation to have the concrete solidly keyed into the space between the flanges. The steel piles in the composite wall act as reinforcing and permit easy attachment of interior bracing.

Trench bulkheads are designed much like cofferdams for foundation excavations.

Trenches may be dewatered by wellpoints or deep wells, depending on depth, permeability of soil, and amount of groundwater.

Where loss of ground or consolidation of relatively loose, granular soils may cause settlement of adjacent buildings, a careful study must be made of the underpinning for these structures.

Subway structures may be reinforced concrete boxes or steel columns and beams with jack arches or reinforced concrete deck slabs and walls. Concrete inverts support tracks and columns. Other types of tunnels are of reinforced concrete box construction, designed as frames for minimum height, or horseshoe section if height permits. Design loads include weight of overburden, horizontal earth pressure, and hydrostatic loads if below water level. Weight of submerged structures must be adequate to prevent flotation.

Tunnels in dry soil need no waterproofing on base and walls, but roof slabs should have at least minimum waterproofing. Tunnels below groundwater level should be waterproofed all around.

Membrane waterproofing consists of layers of cotton or glass-fiber fabric saturated with and bedded in hot asphalt. A minimum of two plies should be used for depths to about 10 ft and a maximum of four plies for depths to 70 ft. Fabric should be laid with lapped joints, offset in the layers. On vertical walls, membranes should be protected against sun in hot weather to prevent peeling. On the bottom, membranes should be laid on a 6-in.-thick concrete leveling slab. Waterproofing should be protected against mechanical damage by asphalt plank, brick, or a layer of concrete.

To save on excavation width, waterproofing for walls may be applied to trench bulkheads and concrete placed against it.

Brick in mastic is more reliable but more expensive than membrane waterproofing and is often used for vehicular tunnels. Construction starts with a four-ply membrane applied on a bottom leveling slab. A single layer of brick is laid over this on hot asphaltic mastic. The mastic also should fill all joints and cover the brick. The concrete base slab goes on top.

Walls should be covered with a two-ply membrane. A protection course of concrete planks should be erected in lifts to allow space for two layers of brick, laid with all joints filled with hot mastic. Lifts are limited in height to 3 ft. Roofs should be covered with a three-ply membrane and a single layer of brick in hot mastic.

The mastic should be at least one-third asphalt and the rest clean concrete sand and lime dust, or sand and cement, thoroughly mixed mechanically at a temperature not exceeding 350° F. Brick should be dry and heated to at least 110° F.

Waterproofing with hot asphalt should be done only in dry weather and on dry surfaces.

Multi-ply membrane waterproofing with cold emulsified asphalt instead of hot asphalt has been used with good results, primarily in Canada. The surface of the structure need not be dry, but the temperature must be above freezing.

Membranes of rubber or rubberlike synthetic sheets also may be used. Coatings of epoxy compounds, usually mixed with coal tar, may provide adequate waterproofing in certain cases, but their use may be limited by their relatively high cost.

20-19. Rock Tunneling. For rock excavation, the most important geological conditions to be anticipated are the presence of faults, usually involving areas of badly fractured rock; direction and degree of stratification; fissures and seams; presence of water, which may be cold or hot or contain corrosive or irritating ingredients; pockets of explosive or toxic gas; and rock strain. The petrography is of lesser importance unless the rock is highly abrasive, causing excessive wear of drills.

Too much information can never be provided for the engineer, to produce a realistic design, and for the contractors, to prepare sound bids. Even at best, unforeseen difficulties must be expected.

In addition to geological surveys and borings (Art. 20-9), engineers may use electric-resistivity measurements and gamma-ray absorption for information on depth and characteristics of rock formations. Information also may be obtained from the U.S. Geological Survey, which has extended its scope and geophysical studies beyond the mining field. Where geological conditions are particularly hard to evaluate or are especially severe, exploratory pilot tunnels, about 10 × 10 ft, may be driven part way from each end or for the entire length of a tunnel, prior to final design and advertising of construction. In these pilot tunnels, internal rock stresses can be measured by pressure cells and strain gages inserted in transverse drill holes, and the nature of the rock, foliation, blockiness, and pressure of faults and water can be inspected.

Headings. In the past, when mucking was done by hand loading into mine cars, and drill equipment was cumbersome, excavation was advanced in drifts or headings. In weak rock or for very wide tunnels, this method is still used. A top heading may be advanced first. This permits installation of crown supports if needed. The rest is excavated by benching down from the top heading. These different levels make transportation of excavated material inconvenient. In wide tunnels, side headings may be advanced, in which the legs of steel sets (supports for side walls and roof) are placed, where necessary. The headings are followed by a top heading and erection of the arch supports. The remaining block can be attacked from the face or from the side drifts.

A bottom heading or pilot tunnel may be used instead. Enlargement proceeds at several places along the heading simultaneously. The pilot tunnel has to be large enough to allow in and out traffic and should be timbered to protect it.

In very long tunnels, a parallel heading, 40 ft or more from the tunnel axis, expedites excavation by providing access to several working faces through cross drifts. From this pilot tunnel, transverse headings are driven at several points to the main tunnel axis, from which tunnel excavation can proceed in both directions. The parallel heading carries all traffic to the different faces and serves as a drainage and ventilation tunnel. This method was used in the 12-mile Simplon tunnel, where the parallel drift was later enlarged to a full-sized single-track railroad clearance, and in the Moffat and New Cascade tunnels.

A center heading may also be used in large rock tunnels. From it, the section is enlarged to full size by radial drilling.

Full-Face Tunneling. To save time and labor, full-face rock excavation is used wherever feasible, for efficient mechanization of the operation. Large track or rubber-wheel-mounted jumbo frames carry high-speed drills. Mucking (removal of excavation) is done by large, mechanized loaders. Muck is carried in diesel trucks, where permissible, or in trains of large mine cars pulled by battery-powered locomotives if laws prohibit use of internal-combustion engines.

Excavation Limits. Contract plans prescribe excavation profiles. An inner A line is the minimum theoretical section to be excavated; to this is added a tolerance, usually 6 in. to the B line or payment line. Any overbreak beyond this is at the contractor's risk and has to be filled at the contractor's expense.

Blasting. Drilling pattern and blasting charges are governed by the rock characteristics, fragmentation desired for mucking, and external conditions, such as proximity of sensitive structures. The procedure should be worked out by an experienced blasting expert and may have to be modified during construction. The center group of holes, fired first, are drilled convergent, so that a conical shape is blasted. Blasting proceeds toward the periphery with short-time delays. A 6- or 8-in.-diameter center, or "burn" hole, without charge, acts as a relief opening, improving blasting effect. Rounds are usually about 10 ft deep but may be more or less, depending on the rock. Line drilling, a ring of straight holes, fairly closely spaced around the periphery, is used if as smooth a section as possible is desired.

Temporary Supports. Practically all rock tunnels need some temporary supports. Timber may be used in pilot tunnels and small headings. For larger tunnel cross sections, steel sets are more economical because of their strength and ease of installation. These are made of I beams cold-rolled into shape. For small tunnels with circular arches, the sets may be continuous frames. In larger tunnels or for flat arches, the sets consist of separate posts and arches (Fig. 20-9). Where roof supports only are necessary, the arches may be supported on plates resting on rock ledges. Steel sections are usually uniform for the entire tunnel, and spacing of sets is varied according to rock loads. Normal spacing is 4 ft, but spacing may be reduced to 2 ft or increased to as much as 6 ft.

The sets should be erected as soon as scaling of loose rock has been completed. Blocking should immediately be wedged between the steel and the rock surface at 3- to 5-ft intervals to prevent rock movement from starting. The steel frames should allow space at the crown, between the lower flange and the concrete surface, for a pipe for placing concrete.

Timber or steel lagging should be placed between the sets. The amount of lagging depends on rock conditions. Lagging may be practically solid, or there may be gaps of various widths between the sheets, as required by circumstances. Badly fragmented rock may require metal panning between sets if water is present. The pans are made of interlocking channels. The space between pans and rock should be dry-packed to allow water to run off into the drainage system.

The concentrated loads on the sets at blocking points produce bending moments in the frames. Table 20-4 presents formulas for loads on supports in rock tunnels (R. V. Proctor and T. L. White, "Rock Tunnels and Steel Supports," Commercial Shearing and Stamping Co., Youngstown, Ohio).

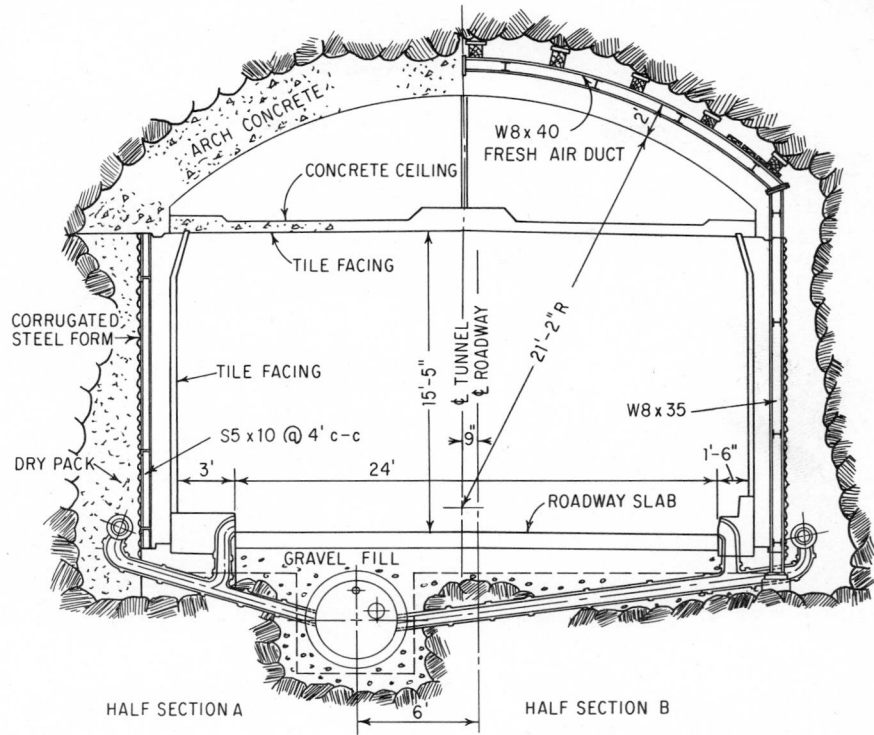

Fig. 20-9. Typical cross section through the Lehigh Tunnel on the Pennsylvania Turnpike Extension.   Half section *A* shows the concrete lining in place.   Half section *B* shows the bracing or sets.

Table 20-4. Load H_p in Feet of Rock on Support in Tunnel*

Rock condition	H_p, ft	Remarks
1. Hard and intact	Zero	Light lining or rock bolts only if spalling or popping occurs
2. Hard, stratified, or schistose	0 to 0.5B	Light supports. Load may change erratically from point to point
3. Massive, moderately jointed	0 to 0.25B	
4. Moderately blocky and seamy†	0.35($B + H_t$) to 1.10($B + H_t$)	No side pressure
5. Very blocky and seamy	0.35($B + H_t$) to 1.10($B + H_t$)	Little or no side pressure
6. Completely crushed, chemically intact†	1.10($B + H_t$)	Considerable side pressure. Requires continuous support of lower ends of ribs or circular ribs
7. Squeezing rock, moderate depth	1.10($B + H_t$) to 2.10($B + H_t$)	Heavy side pressure; invert struts required. Circular ribs recommended
8. Squeezing rock, great depth	2.10($B + H_t$) to 4.50($B + H_t$)	Same as for Type 7
9. Swelling rock	Up to 250 ft, regardless of value ($B + H_t$)	Circular ribs. In extreme cases, use yielding supports

*If depth of rock over tunnel is more than 1.5($B + H_t$), where B is width and H_t is height of tunnel.   From R. V. Proctor and T. L. White, "Rock Tunnels and Steel Supports," Commercial Shearing & Stamping Co., Youngstown, Ohio.

†If roof of tunnel is permanently above the water table, values for Types 4 and 6 can be reduced by 50%.

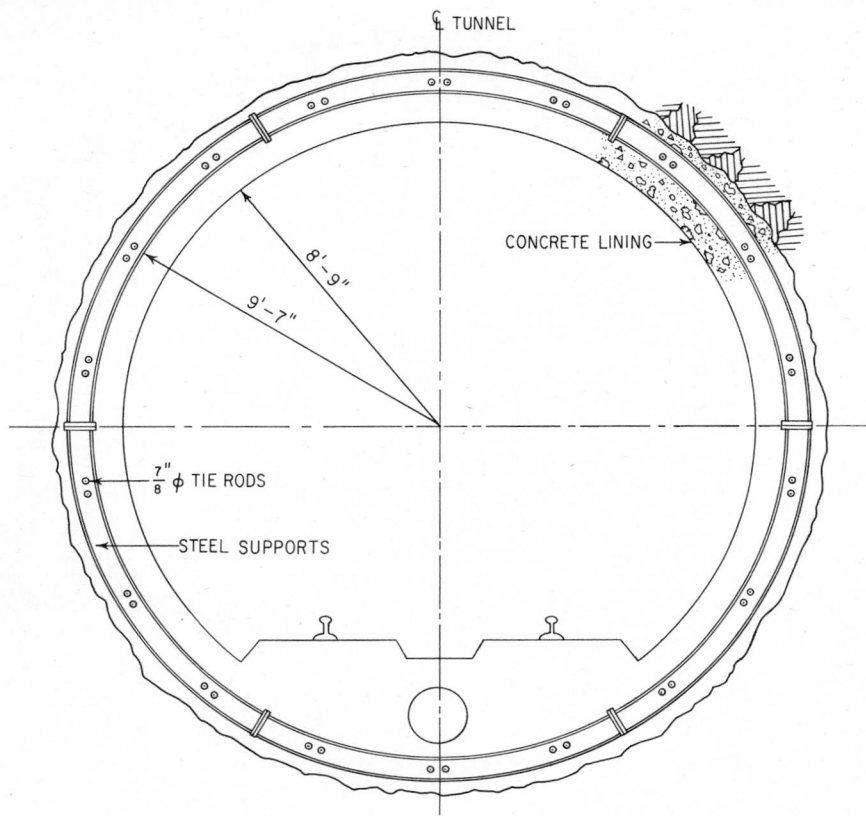

Fig. 20-10. Typical section through Berkeley Hill Rock Tunnel (heavily faulted rock) for the San Francisco Bay Area Rapid Transit.

Through badly faulted rock or pressure areas, circular tunnel sections and ring supports are preferable, particularly in seismic areas (Fig. 20-10).

Roof bolts should be used to hold horizontally bedded rock, prevent highly stressed rock from "popping," or secure sound but somewhat blocky rock. Bolts usually are 1 in. in diameter, 8 ft long. They can be coupled for greater length. Their length must be sufficient to provide grip in solid rock beyond the probable break line. For anchoring in solid rock, the end of each bolt may be split and a wedge inserted. The wedge spreads the "fishtail" when the bolt is driven home, so that the end will grip the rock. A more effective anchorage is provided by one or more conical, split sleeves, which expand when the bolt is threaded in. Sometimes, corrugated reinforcing steel bars, grouted into the bore holes, may be used. They provide adequate holding power. Regardless of the anchorage method, however, the bolts should be tested for pull-out force. For holding the rock in place, a square steel plate is slipped over the exposed, threaded end of the bolt, and a nut is securely tightened against the plate.

Another type of bolt has a perforated sleeve, which is placed in a hole in the rock and filled with grout. As the bolt is pushed into the hole, the grout is squeezed through the perforations and against the rock. Bond between bolt, grout, and rock provides the holding force.

Use of sprayed concrete (gunite or shotcrete) as preliminary tunnel support for rock tunnels was developed in Europe and has also been successful in North America (Canadian National Railroad tunnel in Vancouver). As soon as possible after blasting, while mucking is going on, a layer of concrete is sprayed on the roof. The concrete is made with a well-graded aggregate, up to ¾-in. size, which is dry-mixed with cement and an accelerating agent. The mixture is ejected through a nozzle under pressure by special pumps. Mixing water is added at the nozzle. Initial set takes place in about 30 to 120 sec, final set in 12 min.

Thickness of the initial layer may vary from 2 to 4 in., depending on rock conditions. Additional layers may be sprayed on as needed.

The nozzle may be held directly by an operator or may be attached to a boom manipulated by a worker stationed under the protective roof of the jumbo. Automatic application has been successful in a machine-bored tunnel (Heitersberg Tunnel in Switzerland).

Shotcrete is sprayed on the sidewalls after completion of mucking. Heavy water inflow must be intercepted and drained through inserts in the shotcrete. Well-trained operators and careful supervision and control are essential for good results. Properly executed, the method can be used successfully for fractured rock.

Strength of concrete in place reaches 200 to 250 psi in 2 hr, 1,400 to 1,500 psi in 12 hr. The ultimate compressive strength of 4,000 to 5,500 psi is about 15% less than that of the same concrete without accelerator.

Leakage. Most rock tunnels have water problems. They may range from seepage through fissures or faults to heavy flow from large water pockets or water channels through fractured formations.

Seepage can be controlled to some extent by cement grouting, but complete stoppage is difficult. Water excluded at one place often finds its way to formerly dry cracks. Panning and dry packing in the arch is the best way to carry off seepage that grouting does not stop, but this is costly. It often is more economical to seal shrinkage cracks in the concrete lining that leak or to insert copper drainage channels in chases along the cracks. Dry packing behind the walls is easier and can be done mechanically. Water will flow through the dry pack into open drain lines installed in the bottom and connected to the tunnel drainage system (see also Art. 20-15).

If a heavy flow of water comes through a drill hole, it indicates a water-bearing fault or seam. The flow may be stopped by drilling additional holes and injecting cement grout. Some holes should be slanted to reach beyond the periphery. If dense sand or rock flour in the fault prevents proper pentration of cement grout, chemical grouting may give satisfactory results. In special cases, it may be necessary to drill a pilot hole well ahead of the face to detect severe water conditions, especially substantial quantities under heavy pressure. This must be done for rock tunneling under deep bodies of water.

(R. Hammond, "Tunnel Engineering," Heywood & Co., London.)

20-20. Tunnels in Firm Materials. Materials, other than rock, that may be encountered in tunneling are sands of various densities and grain sizes; sands mixed with silt or clay; clays, either pure or containing silt or sand, and varying from relatively plastic with high water content to firm and dry; and alluvial mixtures of sand and gravel or glacial till. If not subject to hydrostatic pressure of free water, these materials may be excavated by mining. Temporary support is given by timber or steel framing in headings whose size and number depend on local conditions.

Mining of headings in all these materials requires the driving of poling boards, supported by cross timbers and posts to hold the roof. As excavation is advanced on a face as steep as the

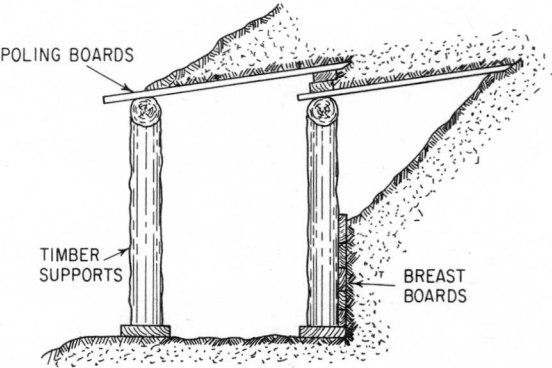

Fig. 20-11. Timber bents support poling boards in basic earth mining.

material will stand, these boards are driven further, with the rear supported by the frame, the front by the soil. A new support is set under the forward end of the poling boards and the process repeated. The sides of the heading are held by boards supported by the posts, as required. Figure 20-11 illustrates the basic procedure for this type of excavation.

Steel supports are often used instead of timber, particularly for large headings. Steel lances, made of small wide-flange beams with wedge-shaped points, may be used instead of wood poling boards. The lances are long enough to be supported on two frames and driven by jacks or air hammers into the soft face for a distance equal to the support spacing.

In loose soil or running sand, the face is supported by breast boards. A shallow slot about 2 ft deep and one or two poling boards wide is excavated in the top of the face, and a short vertical breast board is placed immediately, to hold the face and support the forward end of the poling. After this slot has been excavated across the heading and all vertical breast boards set, a cap is installed, supported by short posts. The rest of the face may then be excavated downward and held by horizontal breast boards (see Fig. 20-12).

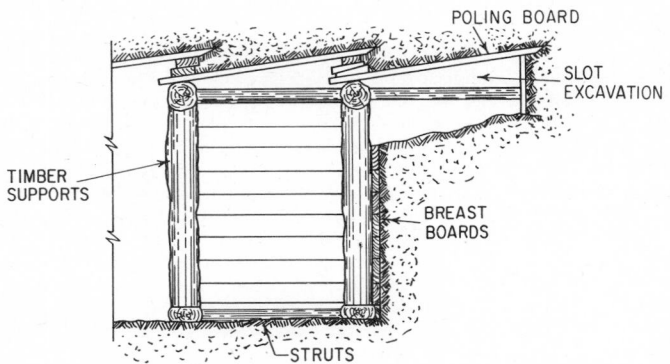

Fig. 20-12. Mining in running ground requires breast boards.

The size of the heading should be as large as soil characteristics allow, but not less than 5 ft wide by 7 ft high. Steel bents shaped to the tunnel arch are preferable to timber framing, if economical, considering both price and speed of operation. Poling may be timber or steel.

Steel liner plates are available in various shapes and sizes. They may be used to support the ground if a limited excavated area of the roof or arch will stand long enough for insertion of the liner plates, starting at the top of the arch and working down. The flange of each plate is bolted to the previously erected liner.

In small tunnels, ribbed or corrugated liner plates may give adequate support. In large tunnels or under heavier loads, the plates are backed up by steel ribs, against which they are blocked. Liner plates without flanges may also be used as lagging or poling.

To prevent settlement or unbalanced load, all voids behind the liner plates should be filled by injection of pea gravel or cement grout.

Small tunnels may consist of a single heading. For large tunnels, various combinations of headings are used. Some of these are known by the country of their origin, as American, Austrian, Belgian, English, German, or Italian methods, but are used in many variations. Orginally, the methods required wood supports, but now steel supports are favored, where economical.

American Method. Excavation starts with a top heading at the tunnel crown, which is supported by poling, posts, and caps. Next, the excavation is widened between two bents and the top arch segments adjoining the crown are set, supported by extra posts or struts. The excavation then is benched down along the sides, and another segment of ribs is set on each side. These are doweled to the upper part and supported by temporary sills. This process is repeated to the invert sill. The bench finally is excavated to full section. Ground between ribs is held by lagging, and voids are packed. This method is suitable in reasonably firm material.

Austrian System. A full-height center heading is advanced. It either starts with a top heading and is cut down to the invert in short lengths, or starts as separate bottom and top headings.

In the latter case, the core between the two is excavated for short distances, and the short posts replaced with long ones. The arch section is widened in short lengths. It is held by segmental arch ribs and longitudinal poling boards. The arch ribs are supported by struts from the center-cut framing and sills at the spring line. The rest of the excavation is advanced to full face in short increments, and posts are set to support the sills. This method is suitable for reasonably stable soil.

Belgian Method.   In firm ground, the upper half of the tunnel is excavated, starting with a center heading from the crown to the spring line. This is widened to both sides, the ground being held by transverse polings. These are supported by longitudinal timbers, in turn supported by struts extending fanlike from a sill in the center heading. Next, a center cut is excavated to the invert, leaving benches to support the arch of the tunnel lining. Slots are cut into the benches at intervals to underpin the arches. The rest of the bench then is removed to complete the side walls, after which the invert is concreted. The excavation may be advanced a considerable distance before following with the tunnel lining.

English Method.   The entire face is excavated a short distance, usually not exceeding 20 ft, ahead of the permanent tunnel lining. Excavation starts with a top heading. In it, two roof timbers or crown bars are set, with one end of each supported on the completed tunnel lining and the forward end resting on posts. Transverse polings are driven over the crown bars. Then, the cut is widened to the end of the polings. Next, vertical breast boards are set below them across the face and held by additional timbers. After that, side bars and polings are placed to permit widening of the excavation. This sequence is repeated until the invert is reached, the face being held by breast boards strutted against the completed lining. The lining is then extended to the end of the excavation and the cycle repeated. To control alignment and assist in drainage, a pilot heading at invert level is sometimes constructed first. This also permits full excavation at several faces.

German Method.   Two bottom headings are advanced, one at each side wall. In these the side walls are built to the roof of the headings. Over these, two more headings are driven and the walls brought up. A top center heading is added and widened until it meets the side headings, the ground over the arch being supported by longitudinal timbers and transverse polings. After the lining arch is completed, the remainder of the ground is removed.

Italian Method.   This was developed for very soft ground, in which only small areas are opened. It is very costly and has been superseded by the shield method, which is used exclusively in this type of ground.

(R. Hammond, "Tunnel Engineering," Heywood & Co., London.)

20-21. Shield Tunneling in Free Air.   Shield tunneling is used generally today in noncohesive, soft ground composed of loose sand, gravel, or silt and in all types of clay, or in mixtures of any of these. It is indispensable for tunneling in these materials below the water table.

The first shield for tunnel construction was designed and patented in 1818 by the English engineer Marc Isambard Brunel. It contained most of the features of today's shields but was not used then with compressed air. The latter idea was patented in 1830 by Lord Cochrane, but he never actually used it.

The shield is a cylinder made of welded steel plate (Fig. 20-13). It has a diameter slightly larger than the outside of the tunnel lining. The plate is stiffened by two interior ring girders, the first one installed a short distance behind the cutting edge. Depending on the diameter and loads, the girders are braced by horizontal and vertical steel struts. The cutting edge is beveled and reinforced by welded steel plates to a thickness of up to 3 or 4 in. For loose ground, the upper half of the shield is extended forward 12 to 18 in. to form a protective hood.

The tail of the shield overlaps slightly the end of the finished lining and provides space for at least one liner ring, and for underwater tunnels is usually long enough to accommodate two rings. The inside of the tail clears the lining by about 1 in. all around. For working in soft clay, the front of the shield may be closed by a steel bulkhead with door-equipped openings through which material is excavated. Soft clay may be extruded through the openings while advancing the shield.

Working platforms that can be advanced and retracted by hydraulic jacks are mounted on the shield bracing. They give access to all parts of the face and, by keeping in contact with it, support it if necessary during shoving. Additional breasting jacks can be mounted in the bracing to hold breast boards against the face if it needs extensive support.

Shield Advancement.   Hydraulic jacks for advancing the shield are set on the webs of the ring girders close to the periphery of the shield. The jacks are evenly spaced around the perimeter and exert pressure against the forward ring girder, which is stiffened by brackets welded to the

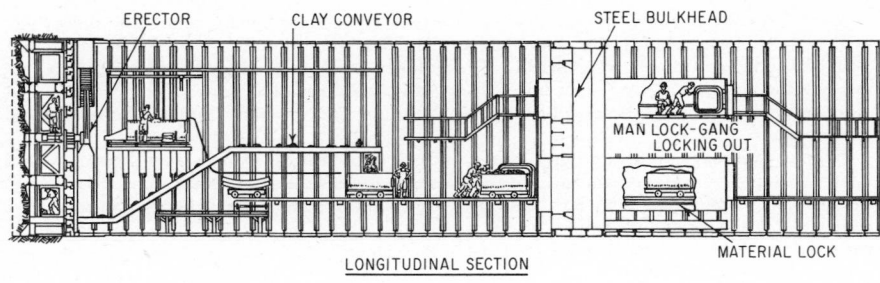

LONGITUDINAL SECTION

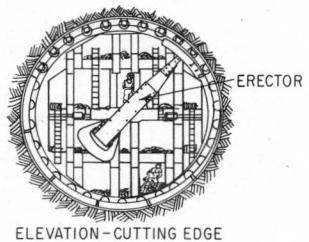

ELEVATION – CUTTING EDGE

Fig. 20-13. Sections through a conventional shield (used in 1930 for the Detroit–Windsor, Ontario, Tunnel).

skin of the cutting edge.   Jack plungers are equipped with shoes bearing against the tunnel lining.   The stroke of the jacks is slightly more than the width of a liner ring.

A rotating erector arm is mounted inside the tail to pick up and place liner segments (Fig. 20-13).   Hydraulic pumps mounted behind the shield supply 5,000- to 6,000-psi pressure to the jacks, erector arm motor, and other hydraulically operated equipment.   Control valves for these devices are mounted on a panel in the shield.

The method of operation, excavation, and speed of advance vary greatly according to the type of soil.   In sand and gravel, the face usually has to be held by breast boards, which are braced by telescoping struts, breasting jacks, or the working platforms.   The breasting may have to be carried down to the invert of the face, which is excavated to the cutting edge of the hood.   If compressed air is used, the breasting may be carried part way down to where the air pressure balances the hydrostatic head, the lower part of the face taking its natural slope.   In firm materials, silty sand, or stiff clay, the full face may be excavated without breasting.   Average progress for the 31-ft-diameter Queens Midtown Tunnel, New York City, in these materials was between 7 and 8 ft in 24 hr.

Shields are not well suited for rock tunneling, but rock or mixed faces, partly rock and partly soil, may be encountered in parts of soft-ground tunnels.   If the rock is high enough, a bottom heading may be excavated ahead of the shield and a concrete cradle placed, with steel rails embedded, to exact line and grade to support the shield as it advances.   A similar bottom heading may be used in a full rock face if the full cross section cannot be excavated.   Then, the rock may be blasted around the periphery of the rest of the cutting edge to permit advancing the shield.   Progress in mixed face in the Queens Midtown Tunnel averaged about 3 to 4 ft per 24-hr day.

Best progress is made in plastic material through which the shield may be shoved blind, that is, without taking any soil into the inside, the volume being displaced by compressing or heaving of the surrounding material.   To counteract the tendency of the tunnel to heave behind the shield, because of buoyancy, enough soil may be admitted through small openings in the face bulkhead and left in the invert to balance the forces until the interior lining is placed.   This method is called shoving half-blind.   In the first tube of the Lincoln Tunnel, New York City, about 20% of the material was taken in.   If displacement or heaving of the soil may cause disturbance of adjacent structures, such as buildings or another tube nearby already in place, the openings should be adjusted to admit nearly all the displaced material.   This was done in the second and third tubes of the Lincoln Tunnel, through openings aggregating 5 to 20% of the face area. Average daily progress was about 30 ft.

Shields are usually started from shafts sunk to the invert grade. These shafts may be specially constructed for this or may later be part of a ventilation building. An opening is provided in the shaft wall to fit the shield and is closed by a timber bulkhead during sinking operations. The shield is erected on a concrete cradle at the base of the shaft. The opposite shaft wall forms the abutment for the jacking forces. A few rings are erected behind the shield, which is advanced through the opening after removal of the bulkhead.

The shield is steered by varying the pressure of the shoving jacks around the periphery. On large tunnels, the total jacking force may be 3,000 to 5,000 tons. If the shield has a tendency to rise, more pressure is applied at the top than at the bottom. Similar corrections are made for other directions.

If the soil is relatively loose, it is excavated at the face by hand tools. In hard-packed silty sands or very stiff clay, air spades are used. Relatively soft clay may be cut by clay knives. The muck is shoveled by hand on a short conveyor in the shield. From there, it is discharged on a loading conveyor mounted on a movable carriage behind the shield. The loading conveyor dumps it into mine cars, usually of about 4 cu yd capacity in large tunnels. The muck trains are rolled back through the tunnels to an access shaft. The individual cars are hoisted up the access shaft and dumped into hoppers for charging into trucks.

Tunnel Linings. Except in very stiff or compact soils, segmental ring liners are used in shield tunnels. These may be of cast iron, steel, or precast concrete. The segments are brought in by mine cars, unloaded by hoists mounted on the conveyor carriage, and deposited within reach of the erector arm. This is a telescoping, counterweighted arm pivoted on the center line of the tunnel for full rotation by a hydraulic motor. A gripper at its outer end engages lugs or bars in the segments and places these, starting at the bottom. A short, tapered segment forms the key (see also Art. 20-23).

In stiff soils, steel ribs, usually 4-in. H beams, and wood lagging may be used as primary lining. The ribs are usually spaced 4 ft c to c and are erected in the tail of the shield. Precut and dressed wood lagging is placed solidly around the circumference between the flanges of the ribs. This lagging also transfers the jacking forces to the tunnel lining. Precast-concrete lagging has also been used successfully.

Packing. Since the shield has a larger diameter than the lining, a void exists around the liner rings. This may permit a cave-in and cause settlement. The usual practice when segmental liners are used is to inject pea gravel into this void through grout holes in the liners immediately after the shield has been advanced. Cement grout is later injected into the gravel to solidify it. In a section of the Victoria line of the London subway in deep, very stiff clay, an articulated cast-iron lining was installed and expanded against the clay behind the shield. The adjacent rings were pressed into contact by the jacking forces but were not bolted. Expansion of steel ribs with wood lagging has also been used to achieve tight fit against the soil.

Semicircular or semielliptical shields have been used as temporary supports for the roof or arch of excavations, mostly in dry or dewatered soils, for example, for tunnels at shallow depth where open-cut operations are prohibited by circumstances. They are advanced in a manner similar to that for circular shields.

20-22. Compressed-Air Tunneling. After about 40 years of use for keeping water out of caissons, compressed air was applied for the first time to tunnel construction in 1879. Then, a heading 18 × 16 ft was driven from a shaft in New Jersey under the Hudson River for a railroad tunnel. Strangely, no shield was used. After 3 years and severe flooding, the project was suspended. It was resumed with a shield in 1889. After interruptions due to financial troubles, the tunnel was completed in 1905.

While tunnel shields in free air are effective in naturally dry soil or ground that can be dewatered (Art. 20-21), compressed air is needed while tunneling below the water table, particularly in subaqueous tunnels. The air pressure counteracts the hydrostatic head. Also the pressure reduces the water content of the soil at the face, making it more stable and safer to excavate.

Air Pressures. Theoretically, the air pressure required to balance the hydrostatic head is 0.43 psi for fresh water and 0.44 psi for seawater per foot of depth. Actually, the pressure depends on the properties of the soil, as well as on the method of excavation. In open material, such as pervious coarse sand and gravel, the full air pressure would be required, whereas in impervious soils, such as stiff clay, no pressure at all may be needed. A careful analysis of the soil at regular intervals along the alignment is needed to estimate the maximum air pressure and air quantities required. Closed shields for tunnels in the Hudson River silt operated with as little as 16-psi air pressure in depths up to 100 ft. In the sand and gravel under the East River in New York, the hydrostatic head was balanced for about one-quarter or one-third the diameter above the invert. To reduce loss of air at the top of the face, breast boards were plastered with clay.

Blowout Prevention.   With the air pressure balancing the head at the bottom of the face, there is an excess of pressure at the top. If the weight of cover over the tunnel is insufficient to hold the excess air pressure safely, a heavy clay blanket may be placed on the river bottom over the tunnel heading to prevent a blowout at the top of the face. If the air pressure equals the water pressure at the invert, the excess pressure at the top of a 30-ft-diameter face would be 13 psi in seawater, or 1,870 psf. For a 10-ft natural cover of 50-lb-per-cu-ft material, the blanket would have to make up the deficiency of 1,370 psf. At 60 lb per cu ft submerged weight, 23 ft of clay would be required. Navigation requirements may make it necessary to remove the blanket after completion of the tunnel. Clay for this blanket should be relatively soft so that it will readily coalesce into an impervious layer.

Bulkheads.   When the shield is started from a shaft, an airtight deck is built above the tunnel, to hold the pressure until the shield is advanced some distance. An airtight bulkhead is then built into the tunnel a sufficient distance behind the shield to provide space for the loading conveyor and a few mine cars. To keep the volume filled with compressed air within reasonable limits, and to comply with safety laws, new bulkheads are built as the tunnel advances, and the old bulkheads removed. Usually, regulations permit a maximum distance of 1,000 ft between the face and the bulkhead, which may be constructed of steel or concrete.

Air Locks.   Worker and material locks are built into the airtight bulkhead. The locks are airtight cylindrical enclosures with gasketed doors (Fig. 20-13). Compressed air is admitted to the lock from the high-pressure side or from the compressed-air line and is exhausted through a connection to the free-air side. Valves of these connections are controlled from the inside in worker locks and generally from the outside in material locks. The door at the high-pressure side opens from the lock into the tunnel; the door at the free-air end opens into the lock chamber. Doors are held tight by the air pressure and cannot be opened until pressures on both sides are equalized. Pressure gages are provided in the locks as well as in the tunnel.

The material lock is at the level of the mine-car track. The lock should be large enough to accommodate several mine cars.

Table 20-5. New York State Regulations for Pressures, Working Hours, and Decompression Time for Work in Compressed Air

Maximum working pressure: 50 psi (may be exceeded in emergencies). Not more than two shifts worked in 24 hr with an interval in fresh air between them.

Gage pressure, psi	Hours of work		
	First shift	Min intervals	Second shift
0–18	4	½	4
18–26	3	1	3
26–33	2	2	2
33–38	1½	3	1½
38–43	1	4	1
43–48	¾	5	¾
48–50	½	6	½

Decompression: Drop to half the maximum gage pressure at a rate of 5 psi per min. Then, complete decompression at a uniform rate so that the total time shall not be less than the maximum pressure multiplied by the rates given below:

Decompression	
Gage Pressure, Psi	Rate, Psi per Min
0–15	3
15–20	2
20–30	1½
30 or over	1

The worker lock is at a higher level and is from 5 to 7 ft in diameter.   This lock is equipped with benches for workers to sit down.   In large tunnels, two sets of locks may be used to speed up operations.   If there is danger of rapid flooding, an extra worker lock may be placed as high as possible, and a hanging safety walk extended at this level from the lock to the shield.   A safety screen placed in the upper part of the tunnel near the heading will trap air above this safety walk in case of flooding and permit workers to escape.   Some safety laws require installation of two worker locks.

Safety and Health.   For all compressed-air work a well-equipped first-aid station and decompression chamber are required, staffed by a trained attendant at all times.   A physician must be available on short notice for emergency calls.

Most states and countries have laws regulating the working hours and locking rates for compressed-air work.   The law of New York State, summarized in Table 20-5, is generally accepted as a standard in many parts of the United States, but local rules should be checked. Limits of union contracts, however, are sometimes more stringent than legal requirements.

The amount of air required for compressed-air tunneling depends on so many variables that exact rules cannot be given.   To determine the size of the compressor plant for a given job requires a great deal of judgment by the engineer, based upon past experience.   Low-pressure machines are installed for the tunnel air and high-pressure units for air tools.   Adequate standby capacity must be provided by using a number of compressors.   High-pressure air may be used as an emergency tunnel supply by interconnecting the compressors through reducing valves.

Shieldless Tunneling.   Some tunnels have been built in water-bearing ground by using compressed air in conjunction with liner plates, without a shield.   Considerable lengths of 7- to 12-ft-

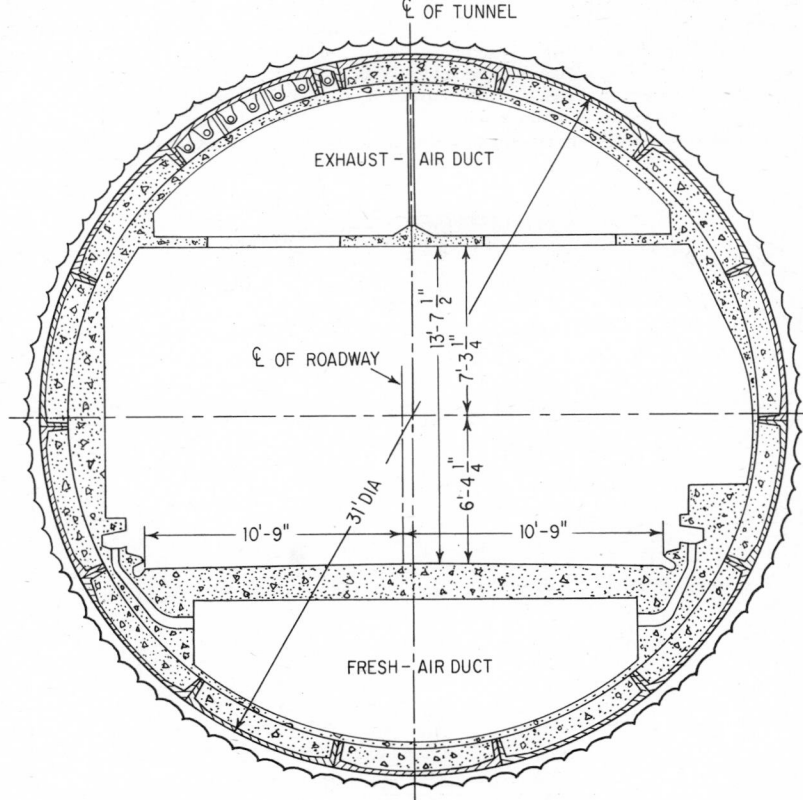

Fig. 20-14. Section shows typical segmental cast-iron lining for a tunnel (Lincoln Tunnel).

diameter interceptor sewers in New York were constructed this way. A few miles of Chicago subway were built in soft clay with steel ribs and liner plates under compressed air.

20-23. Linings for Rock and Firm-Ground Tunnels. Tunnels in very sound rock, not affected by exposure to air, humidity, or freezing, and where appearance is immaterial, are left unlined. This is the case with many railroad tunnels.

Where rock is structurally sound but may deteriorate through contact with atmospheric conditions, it can be protected by coating with sprayed concrete, reinforced with wire fabric or unreinforced. Such a lining may also be used in water tunnels in good rock to provide a smooth surface, reducing the friction factor and turbulence.

Most tunnels in rock, and all tunnels in softer ground, require a solid lining. Highway tunnels of any importance are always lined for appearance and better lighting conditions. Stone or brick masonry has been used to a great extent in the past, but currently concrete is preferred. The thickness of the permanent concrete lining is determined by the size of the tunnel, loading conditions, and the minimum required to embed the steel ribs of the primary lining.

Where the concrete lining is exposed to compression stresses only, it may be without reinforcing. In most cases, reinforcing steel will be required to withstand tension and bending stresses.

The lining is placed in sections 20 to 30 ft in length. Segmental steel forms are universally used. They must be properly braced to support the weight of the fresh concrete. The walls are usually concreted first, up to the spring line. Next come the arch pours. It is important that the space between the forms and the rock or soil surface be completely filled. Grout pipes should be inserted in the arch concrete to permit filling any voids with sand-and-cement grout.

Concrete is placed through ports in the steel lining or pumped through a pipe introduced in the crown, a so-called slick line. Placement starts at the back of the pour, and the pipe is withdrawn slowly. A combination of both methods may be used. Concrete is either pumped or injected by slugs of compressed air. Admixtures are added to get an easily placed mix with low water content and to reduce concrete shrinkage. If there is leakage of water, it usually occurs at shrinkage cracks, which may be sealed with a plastic compound. Or the water may be carried off by copper drainage channels installed in chases cut in the concrete.

Footings for side walls in rock tunnels are cut into the rock below grade. They give adequate stability unless squeezing ground is encountered, in which case a concrete invert lining is placed.

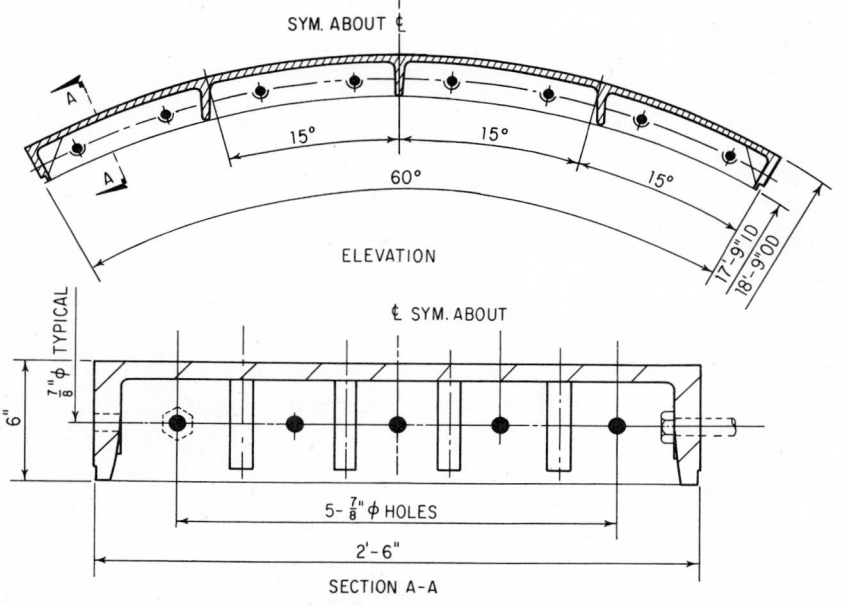

Fig. 20-15. Typical liner segments used for subway tunnels.

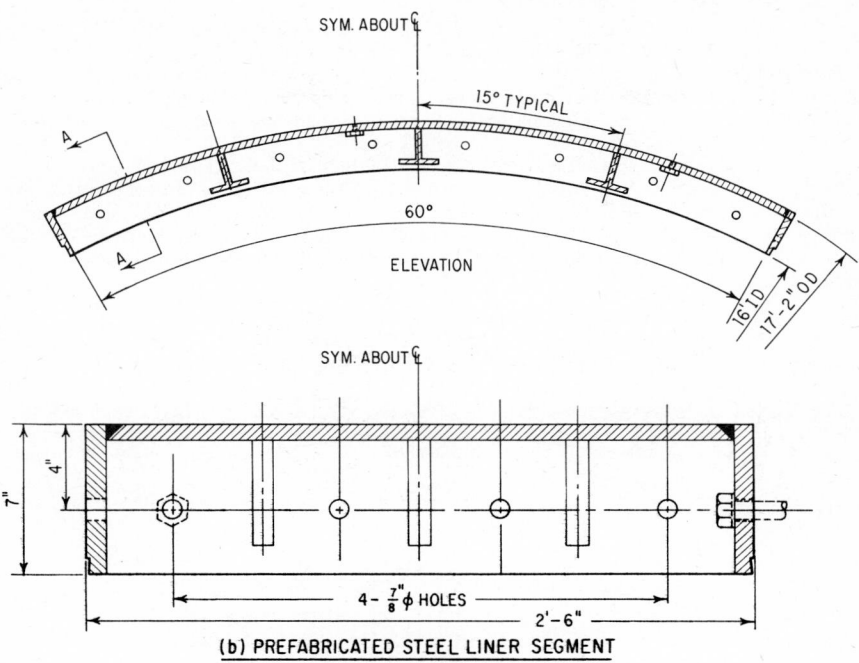

(b) PREFABRICATED STEEL LINER SEGMENT

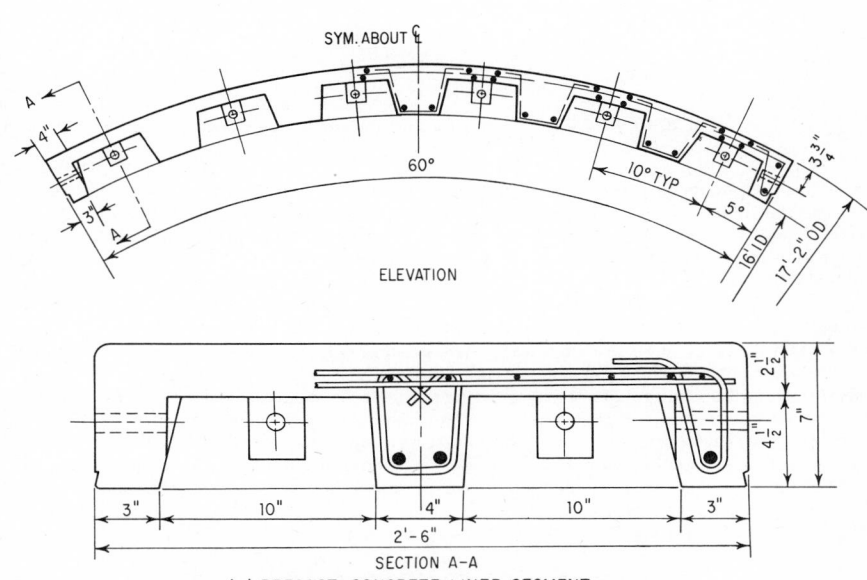

SECTION A-A
(c) PRECAST–CONCRETE LINER SEGMENT

Fig. 20-15. (*continued*)

In soft ground, a concrete slab is placed, to serve as pavement in highway tunnels. If heavy side pressure exists, this slab may have to be made heavier to prevent buckling.

Primary linings in shield tunnels are generally of the segmental type for ease of erection (Fig. 20-14). Attempts have been made, however, to develop a method for placing a continuous concrete lining behind a tunneling machine. Segments usually are of cast iron, welded steel, pressed steel, or precast concrete.

Cast iron has been widely used since the early days of shield tunnels. The segments are made as long as convenient handling permits, usually 6 to 7 ft. The width of the rings depends on the distance the face can be safely excavated ahead of the shield, weight to be handled, and foundry practice. The wider the rings, the longer the tail of the shield and hence the more difficult the steering of the shield. Early tunnels had 18-in.-wide rings. Recent tunnels have gone to 30 or 32 in. (Fig. 20-16).

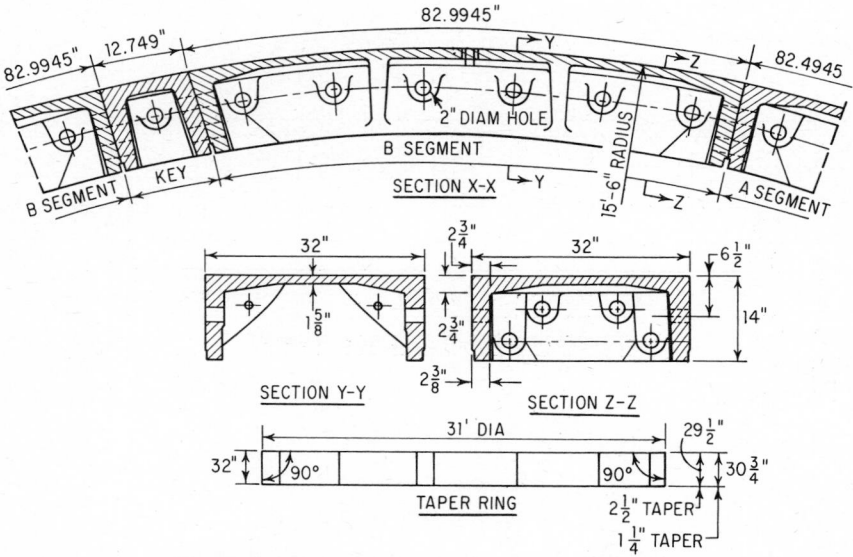

Fig. 20-16. Cast-iron liner ring used for Queens Midtown Tunnel.

Cast-iron segments have machined flanges on all sides (Fig. 20-15a). They are connected by high-strength bolts. Longitudinal joints are offset in successive rings.

The flanges have recesses along their matching edges for calking. These grooves are filled with lead or impregnated asbestos calking strips, pounded in manually. Tests have been made with synthetic sealers, such as silicone rubber and polysulfides, which can be injected into the grooves by calking guns. These compounds adhere to the metal sufficiently to form an effective seal under pressures usually encountered in underwater tunnels.

Each cast-iron segment is provided with a 2-in. grout plug for injection of pea gravel and grout into the space between the lining and the soil. Bolt holes are sealed with grommets of impregnated fabric or plastic grommets, the latter being particularly effective. Bolts are tightened with hydraulic or pneumatic wrenches where possible, otherwise with hand wrenches.

Welded steel segments, similar in shape to cast-iron segments, have been used for economic reasons in some subaqueous tunnels. They were welded in jigs to tolerances as close as practicable, but flanges were not machined and no calking grooves were provided. Difficulties were experienced in making them watertight with gaskets. An improved design includes calking grooves and fabrication tolerances similar to those for cast iron (Fig. 20-15b).

Flanged precast-concrete segments have been proposed for tunnels below the water table (Fig. 20-15c). They are shaped similar to cast-iron segments and cast in steel forms. Grinding flanges to close tolerances and forming of calking grooves would make them suitable for watertight construction.   No installations have yet been made, but prototype rings were tested for possible use in the upper subway level of the San Francisco Bay Area Rapid Transit System.

Heavy, interlocking concrete blocks have been used successfully in relatively dry or impervious soil.   They present difficulties when exposed to water pressure due to leakage.

Except where steel rings and lagging or concrete blocks are used as primary lining, no secondary concrete lining is used, unless required for appearance and interior finish of highway tunnels. In this case, a concrete lining of the minimum thickness practicable is placed.   When the tunnel is to be faced with tile, provision should be made for attaching it.   (To facilitate maintenance and improve lighting, walls and ceilings of highway tunnels are usually finished with ceramic tiles.)   To provide good adherence of the scratch coat, scoring wires may be welded longitudinally on the steel forms for the lining to provide a rough concrete surface.   Coating of smooth concrete surfaces with epoxy compound may result in satisfactory finishes at less cost.

20-24. Design of Tunnel Linings.   (See also Art. 20-23.)   A liner ring is statically indeterminate.   Stresses in it may be computed by the methods of Arts. 6-78 and 6-79 after the ring is made statically determinate by a cut at the top and one end is fixed (Fig. 20-17).

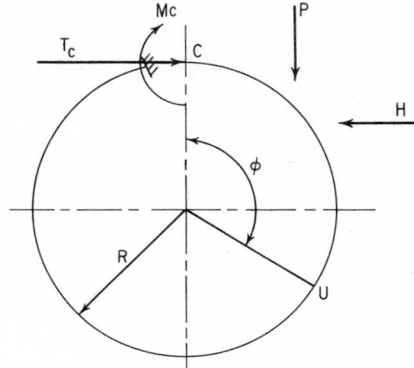

Fig. 20-17. Liner ring may be analyzed by assuming it cut at crown C.

For a circular ring of constant cross section symmetrically loaded, the thrust at the crown C is

$$T_c = \frac{2}{\pi R} \int_0^\pi M \cos \phi \, d\phi \tag{20-7}$$

The vertical shear at the crown is zero.   And the moment is

$$M_c = -RT_c - \frac{1}{\pi} \int_0^\pi M \, d\phi \tag{20-8}$$

where R = radius of the ring

M = bending moment at any point U due to the loads on CU

ϕ = angle between U and crown C

With the thrust and moment at the crown known, the stresses at any point on the ring can be computed, as for an arch.

Loads on a lining include its own weight and internal loads, weight of soil above the tunnel (submerged soil for tunnels below water level), reaction due to vertical loads, uniform horizontal pressure due to soil and water above the crown, and triangular horizontal pressure due to soil and water below the crown.

Because of flexibility, tunnel liner rings can offer only limited resistance to bending produced by unbalanced vertical and horizontal forces.   The lining and soil will distort together until a state of equilibrium is obtained.   So the lining may have to be temporarily braced with tie rods when it leaves the shield until the final loading conditions and passive pressures have been developed.   In certain soft materials, when shields were shoved blind (without material being excavated), initial horizontal pressures exceeded the vertical loads, so that the vertical diameter lengthened temporarily. Ultimately, the section reverted to approximately its initial circular configuration.

Magnitude of loads on tunnel liners depends on types of soil, depth below surface, loads from adjacent foundations, and surface loads. These will require careful analysis, in which observations made on previous tunnels in similar materials will be most helpful.

20-25. Machine Tunneling. To reduce costs and increase the speed of the ever-increasing amount of tunnel construction, a number of tunnel-boring machines for rock and soft ground have been developed. Machines for noncohesive soils have been built, but so far they have had limited success. Universal machines for mixed ground of rock and soft material have not yet been developed to a satisfactory degree.

Rock-boring Machines. Rock-boring machines consist of a rotating head, either solid or with spokes, on which are mounted cutting tools suitable for the type of rock. The machines are mounted on large frames, which carry the driving machinery and auxiliaries, including a series of hydraulic jacks to exert heavy pressure against the face. Chisel cutters serve for soft rock, disk cutters break harder rock by wedge action, and toothed roller cutters with tungsten carbide inserts cut the hardest rocks. The upper limit for present tools is rock of about 35,000-psi compressive strength, although some manufacturers claim success in granite up to 45,000 psi. A critical factor in evaluating production is the amount of down time for maintenance and replacement of cutters and their cost.

Successful rock-boring machines are made by a number of manufacturers:

Robbins Company, Seattle, Wash., designed the first large machines for the 30-ft-diameter tunnels of the Oahe Hydro Project in Wisconsin, in relatively soft shale. Progress was 8.5 to 12 ft per hr of actual working time. So far, this company has built the most machines, the largest of which is the 36-ft 8-in.-diameter machine for the Mangla Dam tunnels in West Pakistan.

Jarva Inc., California, has built machines for tunnels 8 to 21 ft in diameter in types suitable for conditions ranging from soft material to hard rock..

Lawrence Division of Ingersoll-Rand, Oregon, produces a machine that utilizes a pilot hole drilled about 11 ft ahead of the face, into which an expandable anchor is extended to pull the machine forward. Several of these have been used in hard rock.

Hughes Tool Company has built several machines, some together with Mitsubishi for use in Japan.

Wirth Company, Germany, has built a number of machines, one a tandem unit enlarging an initial 11-ft pilot tunnel in two steps, to 22 and 34 ft.

Atlas Copco builds a machine with several toothed wheels mounted on a cutter head with their axes slightly inclined. Head and wheels rotate independently. A variant drives smaller, noncircular tunnels.

McAlpin-Greenside's group builds machines adaptable to various diameters with 85% of the unit remaining unchanged.

Soft-ground-boring Machines. Soft-ground tunneling machines usually operate a rotating cutter head in a shield. In very firm, nonraveling soil, however, a backhoe type is most effective. Where the face needs support, use is made of a solid cutter head on which chisels or drag cutters are mounted.

A number of boring machines have been used for sewer tunnels up to 23 ft in diameter in the stiff clay below Chicago and Detroit. These machines have cutter heads with a slightly conical center from which spokes extend to a cylindrical rim. Cutting knives are set on center and spokes. In very firm clay, the machines have operated without a shield; otherwise, they operate in a conventional shield.

The Toronto Mole of Sir Robert McAlpine & Sons, Ltd., London, operated on a principle similar to that of the Chicago units.

Calweld, California, builds a drum-type machine with cutters on a spoked wheel, divided into four segments that oscillate independently through 30°.

Drum diggers used for the Victoria line of the London subway consisted of a 14-ft-diameter shield within which rotated a 7-ft 6-in. drum which carried six arms, each with eight replaceable knives.

K. M. Tunneling Machines Ltd. builds machines for soft, cohesive soils, from clay to shale and sandstone. Inside a conventional shield rotates a drum with a cutting head to which are attached cutters, usually of the pick-type drag variety.

Flowing Ground. Attempts have been made to develop a boring machine with a closed space located in front of a solid cutting head and pressurized by air or a fluid to support the face. So far, the only successful one was that used in soft silt and sand 33 ft below Tokyo Bay, where salt water was used to pressurize the face.   The machine advanced an average of 13 ft per day in the 23-ft-diameter tunnel.

20-26. Sunken-Tube Tunnels.   Where conditions are suitable, sunken-tube tunnels are the most economical solution for subaqueous crossings.   Tunnel sections are fabricated in convenient lengths on shipways and outfitting piers, in drydocks or improvised floodable basins. They are towed to the site, where a trench has been dredged in the bottom of the waterway, and sunk into position (Fig. 20-18).

Soil conditions must permit dredging a deep enough trench with reasonable side slopes and its maintenance until the tubes are sunk and backfilled.   Since there are limits to the distance to which tubes can be floated into trenches extended into the shores, the tunnel has to be continued by the cut-and-cover method after the end tubes have been backfilled.

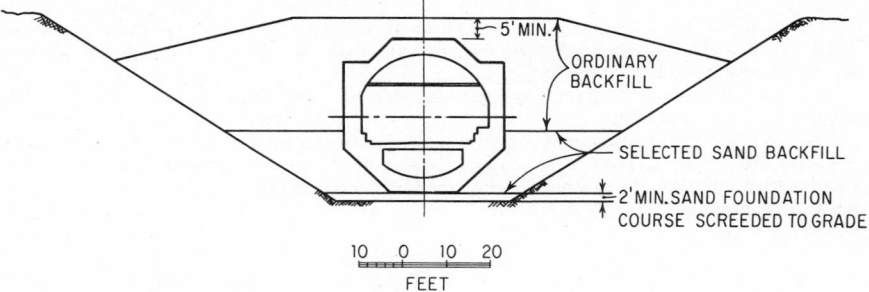

Fig. 20-18. Sunken-tube tunnels are set in a trench, which is then backfilled.

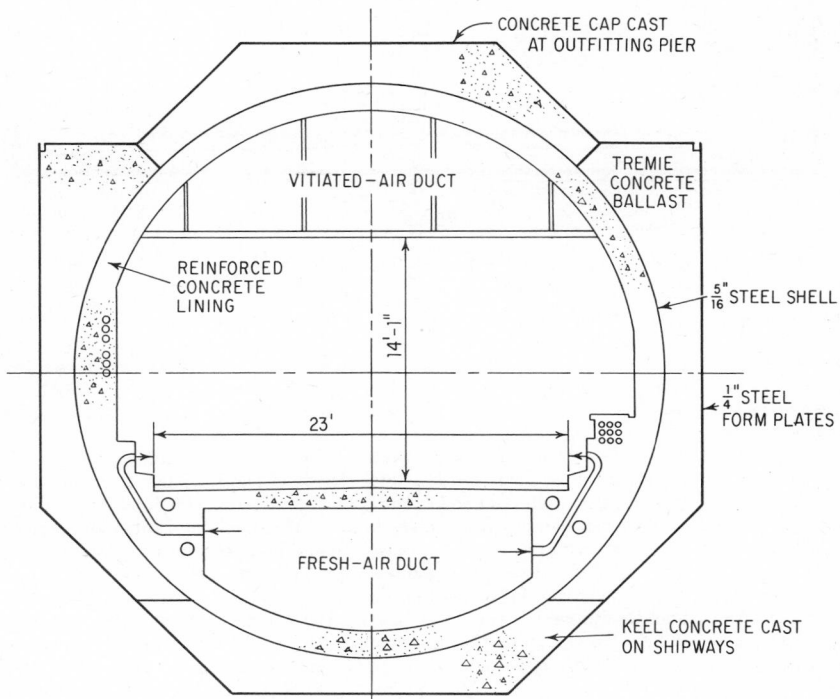

Fig. 20-19. Cylindrical sunken-tube tunnel (Hampton Roads Tunnel).

Figure 20-19 shows a two-lane tunnel with a circular interior section. Its basic element is a circular steel tube made of ⁵/₁₆-in. plate, 31 ft in diameter and about 300 ft long. Exterior diaphragms of roughly octagonal configuration, spaced about 15 ft apart, and longitudinal ribs of flat bars and T sections stiffen the shell. Outside form plates were attached to the diaphragms and supported by angle struts extended from the shell stiffeners.

The tubes were erected on shipways. All welds were tested for watertightness with a compressed-air stream and soap solution. Before the ends were closed with welded watertight bulkheads, the reinforcing steel for the interior concrete lining was placed. The keel concrete in the space between the outside form plates and the bottom of the shell was cast before launching. The concrete lining and roadway slab were cast, while the tube was floating at a fitting-out pier, by pumping concrete through hatches in the top of the shell into segmental steel forms. Pouring sequences were regulated to control increments of water pressure on the shell plate and longitudinal bending moments. The hatches were closed with welded plates, and the exterior concrete cap was cast and enough tremie concrete placed in the side pockets to reduce freeboard to about 1 ft. The tube then was ready for towing. At the site, tremie concrete was added to the side pockets to sink the section.

Two such units were combined into a single structure for the four-lane tunnel under the Potapsco River in Baltimore.

The Posey Tunnel in Alameda, Calif., was built as a circular reinforced concrete section without steel shell. The sections were constructed in a dry basin. A multiple membrane waterproofing was applied to the outside and protected by wood lagging. The tubes were floated into position on pontoons.

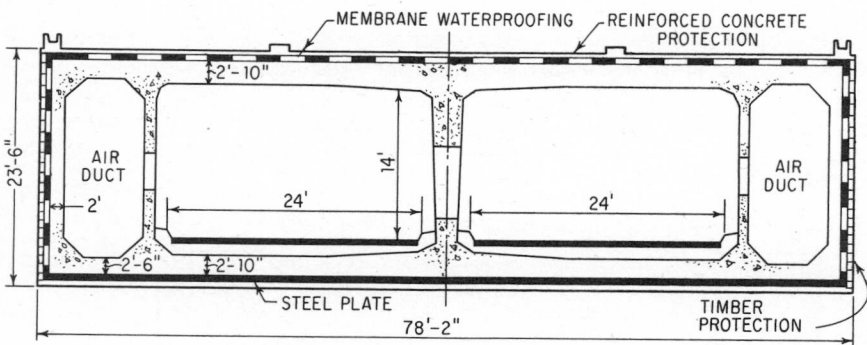

Fig. 20-20. Box-section sunken-tube tunnel (Deas Island Tunnel).

Rectangular concrete sections have been used for a number of four-lane highway tunnels (Fig. 20-20). The tubes are divided into two two-lane compartments, with air ducts either on the side or below the roadway. The sections were built in drydocks. They were floated on pontoons, or towed to the site buoyant and ballasted for sinking. In most cases, watertightness is assured by an exterior enclosure of welded steel plate, protected by a 4- to 6-in.-thick course of concrete.

Figure 20-21 shows the cross section of a double-track rapid-transit tunnel under San Francisco Bay. The tunnel comprises 57 sections, with a total length of about 19,000 ft. The track spaces are separated by an exhaust-air duct and service passage. The section consists of a ⅜-in. exterior steel shell and interior reinforced concrete lining. The steel shell provides security against serious earthquake stresses produced by transverse and longitudinal shear waves. The shell is protected against corrosion by a coating and a cathodic-protection system.

There are two basic systems in use for supporting tunnel tubes on line and grade. In the first (Fig. 20-18), the trench is dredged at least 2 ft below subgrade and a foundation course of well-graded sand and fine gravel is placed in the bottom. This is screeded to exact grade by a heavy grid of steel beams. The grid is suspended from a carriage rolling on tracks set parallel to grade on a screeding barge assembly. This operation also compacts the foundation course. An accuracy of ±0.1 ft of grade can be achieved.

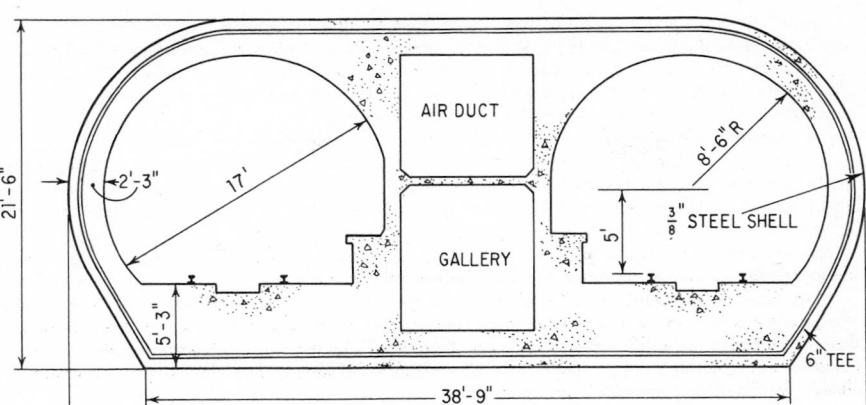

Fig. 20-21. Trans-bay Tube of the San Francisco Rapid Transit System.

In the second method, the trench is also dredged 2 ft below subgrade and the tube is set on two light pile bents, driven to correct grade. Sand is then flushed in under the section. One method for doing this, invented by the Danish firm of Christiani & Nielson, is particularly effective. A sand slurry is injected through a movable nozzle, and the surplus water is pumped off by another nozzle, the rolling motion depositing the sand in a compact layer.

Instead of pile bents, sections have been temporarily supported by jacks, penetrating through the base of the section. The jacks bear on concrete blocks loosely attached to the underside of the tube. By adjustment of the jacks, the section is brought to exact grade. Then, the sand foundation course is flushed in. The inside of the tube must be accessible through stacks projecting above the water.

Sinking of the tubes is regulated by winches on special barges, or by floating cranes, from which they are suspended. Alignment is controlled by instruments set on fixed points and sighting on targets mounted on temporary towers attached to the ends of the sections. For steel-shell tubes, sections are connected with short lengths of shell, which project beyond the end bulkheads. The gap between the ends is covered by hood plates extended from the lower and upper half of the shell extensions. Also, form plates are inserted into guides on the vertical edges of the bulkheads. The space around the joint is filled with tremie concrete as a preliminary seal. The inside of the joint is drained, and closure plates are welded to interior ribs of the shell extensions. Finally, the concrete lining is completed.

For some box sections, rubber gaskets have been used as preliminary seals. The gaskets are mounted on one end of each tube. Jacks pull the tubes in contact to provide an initial seal. The joint is then drained, activating the full hydrostatic pressure on the opposite end of the tube. The pressure compresses the gaskets completely, providing a secure seal. Then, the bulkheads between the connected tubes can be opened and the joint completed from the inside.

For a depth of about 10 ft, the trench is backfilled with well-graded sand to lock the tubes securely into place. Ordinary backfill is placed to a depth of at least 5 ft over the top of the tunnel. If any part of the tunnel projects above the natural bottom, dikes should be built at least 50 ft away on both sides to a height of 5 ft above the tube. The space between the dikes should be filled with backfill, covered with a stone blanket to prevent scour.

Tube sections are designed as rigid structures for the combined hydrostatic and earth pressures acting on the top and sides, as are tunnel linings (Art. 20-24). Although the usual tube section does not exactly conform to a cylinder, design as a cylinder is a sufficiently close approximation, especially since the thickness of concrete is largely determined by weight to overcome buoyancy. Rectangular sections, however, should be designed as rigid frames. Prestressed concrete may be advantageous under certain conditions, particularly for box sections, but its application is limited because of the heavy sections required for weight.

20-27. Shafts. In tunnel work, shafts serve as starting points for excavation in rock or firm material or for shields. For long tunnels, such as aqueducts, several shafts are used to divide construction into shorter sections that can be worked simultaneously.   For vehicular tunnels, especially subaqueous shield tunnels, the shafts are used as bases for ventilation buildings.

Timber shafts are mined and braced in the same manner as tunnels in similar material.   Usually, poling boards 5 to 6 ft long are driven into the ground and braced at regular intervals by rectangular timber frames.   Then, the soil is excavated to the ends of the polings and a new frame installed at this level.

A relatively shallow shaft may be started oversize with sheeting 10 to 20 ft long driven vertically on the outside of the frame bracing.   Intermediate frames are installed as the excavation proceeds.   At the bottom of the tier of sheeting, the sides are stepped in to make room for the next tier of vertical sheeting.

In rock shafts, timbering is used to prevent loose rocks from falling off the walls.   Its placement usually lags an appreciable distance behind the excavation.

Steel liner plates alone, or in combination with horizontal ribs, may be used in soft ground where excavation can be made in increments equal to the width of the liner plates.   H beams driven vertically as soldier piles, with wood or steel lagging and horizontal bracing, may be used for rectangular shafts.   Enclosures of vertical steel sheetpiling, for round or rectangular shafts, are suitable for water-bearing ground.

Where ground conditions are poor and water-bearing, shafts may be constructed with a caisson (hollow box), with compressed air as needed to exclude water.   Gravity pulls the caisson down as excavation proceeds.   Since its weight is relatively small, the caisson may have to be temporarily ballasted or jetted for sinking.   The depth to which a caisson may be sunk is limited by the high cost of compressed-air work, which results from the short working hours permitted under high pressure.

Open-bottom shafts with heavy walls, often circular or subdivided into compartments, may be built on the ground and sunk by excavating the ground underneath.   In dry soil, the excavation may be done directly; if water is present, clamshell buckets and high-pressure jets may be used to loosen the soil and remove it.   On reaching the proper depth, the bottom of the shaft is closed by tremie concrete.

As an alternative method for shaft construction, water-bearing ground may be frozen in a circular ring around the shaft location and the excavation made in the dry.   Closed-end pipes are driven vertically into the ground around the periphery, and open-end smaller pipes inserted into them.   A refrigerant, usually brine, is circulated at temperatures as low as $-30°$ F from the interconnected inner pipes into the larger ones and from them returned to the refrigeration plant. Several months may be required to freeze a deep ring solidly.   The ventilation shaft of the Scheldt River Tunnel in Antwerp was built in this manner, as were a number of mine shafts in Germany and France.

Section **21**

Water Engineering

SAMUEL B. NELSON

Former Director of Public Works, State of California; Retired General Manager
and Chief Engineer, Department of Water and Power, City of Los Angeles;
Former General Manager, Southern California Rapid Transit District; Vice
President, Daniel, Mann, Johnson & Mendenhall; Vice Chairman, Board of
Directors, Metropolitan Water District of Southern California; and Member,
California Water Commission

Water engineering includes the application of fluid mechanics, hydraulics, hydrology, and water-supply theories. Fluid mechanics describes the behavior of water under various static and dynamic conditions. This theory, in general, has been developed for an ideal liquid, a frictionless, inelastic liquid whose particles follow smooth flow paths. Since water only approaches an ideal liquid, empirical coefficients and formulas are used to describe more accurately the behavior of water. These empiricisms are intended to compensate for all neglected and unknown factors.

The relatively high degree of dependence on empiricism, however, does not minimize the importance of an understanding of the basic theory. Since major hydraulic problems are seldom identical to the experiments from which the empirical coefficients were derived, the application of fundamentals is frequently the only means available for attacking problems.

21-1. Dimensions and Units. A list of symbols and their dimensions used in this section is given in Table 21-1. Table 21-2 lists conversion factors for commonly used quantities, including the basic equivalents between the English and metric systems.

FLUID MECHANICS

21-2. Properties of Fluids. **Specific weight** or **unit weight** w is defined as weight per unit volume. The specific weight of water varies from 62.42 lb per cu ft at 32° F to 62.22 lb per cu ft at 80° F but is commonly taken as 62.4 lb per cu ft for the majority of engineering problems. The specific weight of sea water is about 64.0 lb per cu ft.

Density, ρ is defined as mass per unit volume and is significant in all flow problems where acceleration is important. It is obtained by dividing the specific weight w by the acceleration due to gravity g. The variation of g with latitude and altitude is small enough to warrant the assumption that its value is constant at 32.2 ft per sec per sec in the solution of all hydraulics problems.

The **specific gravity** of water is the ratio of its density at some temperature to that of pure water at 68.2°F (20°C).

Modulus of elasticity E of a fluid is defined as the change in pressure intensity divided by the corresponding change in volume per unit volume. Its value for water is about 300,000 psi, varying slightly with temperature. The modulus of elasticity of water is large enough to permit

Table 21-1. Symbols, Dimensions, and Units Used in Water Engineering

Symbol	Terminology	Dimensions	Units
A	Area	L^2	ft^2
C	Chezy roughness coefficient	$L^{1/2}/T$	ft$^{1/2}$ per sec
C_1	Hazen-Williams roughness coefficient	$L^{0.37}/T$	ft$^{0.37}$ per sec
d	Depth	L	ft
d_c	Critical depth	L	ft
D	Diameter	L	ft
E	Modulus of elasticity	F/L^2	psi
F	Force	F	lb
g	Acceleration due to gravity	L/T^2	ft per sec^2
H	Total head, head on weir	L	ft
h	Head or height	L	ft
h_f	Head loss due to friction	L	ft
L	Length	L	ft
M	Mass	FT^2/L	lb-sec^2 per ft
n	Manning's roughness coefficient	$T/L^{1/3}$	sec per ft$^{1/3}$
P	Perimeter, weir height	L	ft
P	Force due to pressure	F	lb
p	Pressure	F/L^2	psf
Q	Flow rate	L^3/T	cfs
q	Unit flow rate	$L^3/T/L$	ft^3 per sec per ft
r	Radius	L	ft
R	Hydraulic radius	L	ft
T	Time	T	sec
t	Time, thickness	T, L	sec, ft
V	Velocity	L/T	fps
W	Weight	F	lb
w	Specific weight	F/L^3	lb per ft^3
y	Depth in open channel, distance from solid boundary	L	ft
Z	Height above datum	L	ft
ϵ	Size of roughness	L	ft
μ	Viscosity	FT/L^2	lb-sec per ft^2
ν	Kinematic viscosity	L^2/T	ft^2 per sec
ρ	Density	FT^2/L^4	lb-sec^2 per ft^4
σ	Surface tension	F/L	lb per ft
τ	Shear stress	F/L^2	psi

SYMBOLS FOR DIMENSIONLESS QUANTITIES

Symbol	Quantity
C	Weir coefficient, coefficient of discharge
C_c	Coefficient of contraction
C_v	Coefficient of velocity
F	Froude number
f	Darcy-Weisbach friction factor
K	Head-loss coefficient
R	Reynolds number
S	Friction slope—slope of energy grade line
S_c	Critical slope
η	Efficiency
Sp. gr.	Specific gravity

the assumption that it is incompressible for all hydraulics problems except those involving water hammer (Art. 21-13).

Surface tension and **capillarity** are a result of the molecular forces of liquid molecules. **Surface tension** σ is due to the cohesive forces between liquid molecules. It shows up as the apparent skin that forms when a free liquid surface is in contact with another fluid. It is expressed as the force in the liquid surface normal to a line of unit length drawn in the surface. Surface tension decreases with increasing temperature and is also dependent on the fluid with which the liquid surface is in contact. The surface tension of water at 70° F in contact with air is 0.00498 lb per ft.

Capillarity is due to both the cohesive forces between liquid molecules and the adhesive forces of liquid molecules. It shows up as the difference in liquid surface elevations between the inside and outside of a small tube that has one end submerged in the liquid. Since the adhesive forces of water molecules are greater than the cohesive forces between water molecules, water wets a surface and rises in a small tube as shown in Fig. 21-1. Capillarity is commonly expressed as the height of this rise. In equation form,

$$h = \frac{2\sigma \cos \theta}{(w_1 - w_2)r} \tag{21-1}$$

where h = capillary rise, ft
σ = surface tension, lb per ft
w_1 and w_2 = specific weights of the fluids below and above the meniscus, respectively, lb per cu ft
θ = angle of contact
r = radius of capillary tube, ft

Capillarity, like surface tension, decreases with increasing temperature. Its temperature variation, however, is small and insignificant in most problems.

Surface tension and capillarity, although negligible in many water engineering problems, are significant in others, such as capillary rise and flow of liquids in narrow spaces, formation of spray from water jets, interpretation of the results obtained on small models, and freezing damage to concrete.

Table 21-2. Conversion Table for Commonly Used Quantities

Area	Discharge
1 acre = 43,560 sq ft	1 cfs = 449 gpm = 646,000 gpd
1 sq mile = 640 acres	1 cfs = 1.98 acre-ft per day = 724 acre-ft per year
Volume	1 cfs = 50 miner's inches in Idaho, Kansas, Nebraska, New Mexico, North Dakota, and South Dakota
1 cu ft = 7.4805 gal	1 cfs = 40 miner's inches in Arizona, California, Montana, and Oregon
1 acre-ft = 325,850 gal	1 mgd = 3.07 acre-ft per day = 1,120 acre-ft per year
1 mg = 3.0689 acre-ft	1 mgd = 1.55 cfs = 694 gpm
Power	1 million acre-ft per year = 1,380 cfs
1 hp = 550 ft-lb per sec	
1 hp = 0.746 kw	
1 hp = 6,535 kwhr per year	
Weight of water	**Pressure**
1 cu ft weighs 62.4 lb	1 psi = 2.31 ft of water
1 gal weighs 8.34 lb	= 51.7 mm of mercury
	1 in. of mercury = 1.13 ft of water
	1 ft of water = 0.433 psi

Metric equivalents
Length: 1 foot = 0.3048 meter

Area: 1 acre = 4,046.9 sq meters

Volume: 1 gal = 3.7854 liters
1 cu meter = 264.17 gal
Weight: 1 pound = 0.4536 kilogram

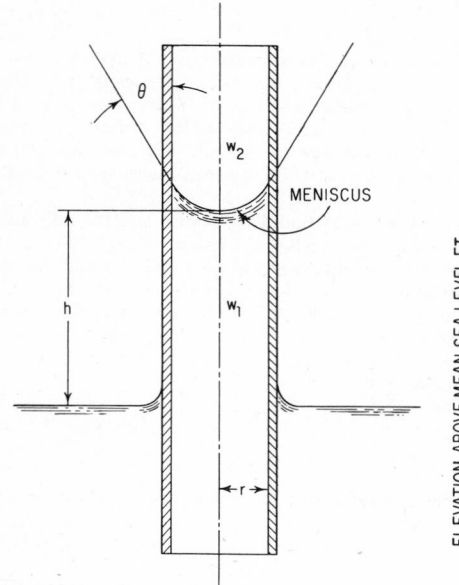

Fig. 21-1. Capillary action raises water in a small tube. Meniscus, or liquid surface, is concave upward.

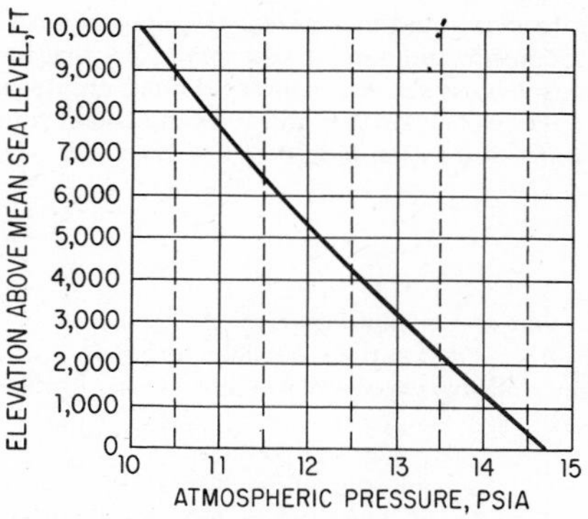

Fig. 21-2. Atmospheric pressure varies with elevation. The curve is based on the ICAO standard atmosphere.

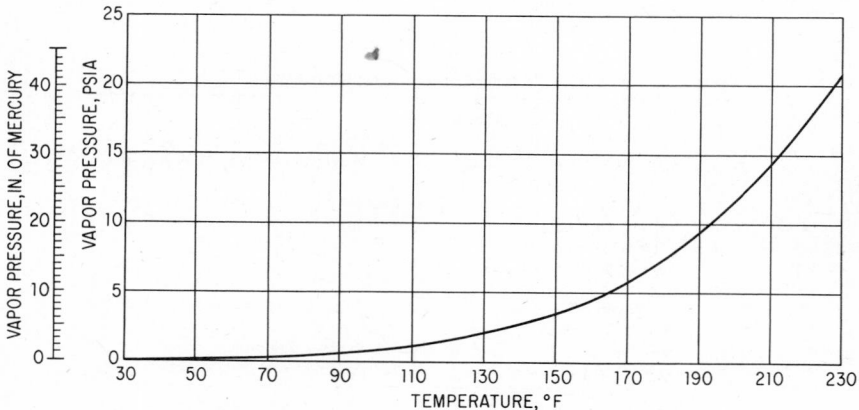

Fig. 21-3. Vapor pressure of water depends on temperature.

Atmospheric pressure is the pressure due to the weight of the air above the earth's surface. Its value at sea level is 2,116 psf or 14.7 psi. The variation in atmospheric pressure with elevation from sea level to 10,000 ft is shown in Fig. 21-2. **Gage pressure,** *psi,* is pressure above or below atmospheric. **Absolute pressure,** *psia,* is the total pressure including atmospheric pressure. Thus, at sea level, a gage pressure of 10 psi is equivalent to 24.7 psia.

Vapor pressure is the partial pressure caused by the formation of vapor at the free surface of a liquid. When the liquid is in a closed container, the partial pressure due to the molecules leaving the surface increases until the rates at which the molecules leave and reenter the liquid are equal. The vapor pressure at this equilibrium condition is called the saturation pressure. Vapor pressure increases with increasing temperature as shown in Fig. 21-3.

Cavitation occurs in flowing liquids at pressures below the vapor pressure of the liquid. Cavitation is a major problem in the design of pumps and turbines, since it causes mechanical vibrations, pitting, and loss of efficiency through gradual destruction of the impeller. The cavitation phenomenon may be described as follows:

Because of low pressures, portions of the liquid vaporize, with subsequent formation of vapor cavities. As these cavities are carried a short distance downstream, abrupt pressure increases force them to collapse, or implode. The implosion and ensuing inrush of liquid produce regions of very high pressure, which extend into the pores of the metal. (Pressures as high as 350,000 psi have been measured in the collapse of vapor cavities by the Fluid Mechanics Laboratory at Stanford University.) Since these vapor cavities form and collapse at very high frequencies, weakening of the metal results as fatigue develops, and pitting appears.

Cavitation may be prevented by designing pumps and turbines so that the pressure in the liquid at all points is always above its vapor pressure.

Viscosity μ of a fluid, also called the **coefficient of viscosity, absolute viscosity,** or **dynamic viscosity,** is a measure of its resistance to flow. It is expressed as the ratio of the tangential shearing stresses between flow layers to the rate of change of velocity with depth:

$$\mu = \frac{\tau}{dV/dy} \qquad (21\text{-}2)$$

where τ = shearing stress, psf
V = velocity, fps
y = depth, ft

Viscosity decreases as temperature increases but may be assumed to be independent of changes in pressure for the majority of engineering problems. Water at 70° F has a viscosity of 0.00002050 lb-sec per sq ft.

Kinematic viscosity ν is defined as viscosity μ divided by density ρ. It is so named because its units, sq ft per sec, are a combination of the kinematic units of length and time. Water at 70° F has a kinematic viscosity of 0.00001059 sq ft per sec.

In hydraulics, viscosity is most frequently encountered in the calculation of Reynolds number (Art. 21-8) to determine whether laminar, transitional, or completely turbulent flow exists.

21-3. Fluid Pressure. **Pressure** or **intensity of pressure** p is the force per unit area acting on any real or imaginary surface within a fluid. Fluid pressure acts normal to the surface at all points. At any depth, the pressure acts equally in all directions. This results from the inability of a fluid to transmit shear when at rest. Liquid pressure and gas pressure differ in that the variation of pressure with depth is linear for a liquid and nonlinear for a gas.

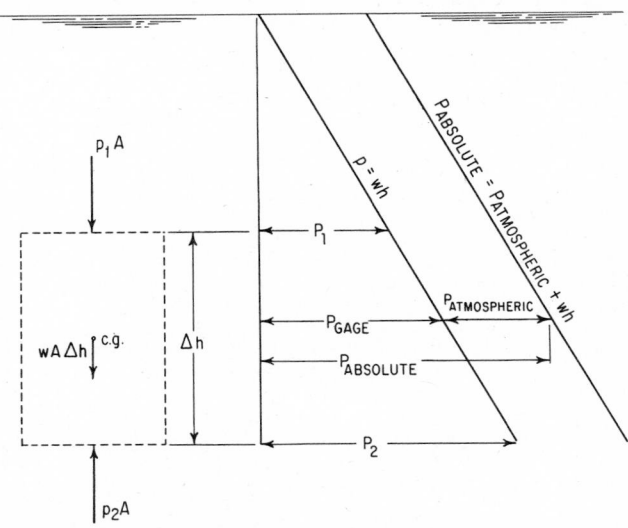

Fig. 21-4. Hydrostatic pressure varies linearly with depth.

Hydrostatic pressure is the pressure due to depth. It may be derived by considering a submerged rectangular prism of water of height Δh, ft, and cross-sectional area A, sq ft, as shown in Fig. 21-4. The boundaries of this prism are imaginary. Since the prism is at rest, the summation of all forces in both the vertical and horizontal directions must be zero. Let w equal the specific weight of the liquid, lb per cu ft. Then, the forces acting in the vertical direction are the weight of the prism $wA\,\Delta h$, the force due to pressure p_1, psf, on the top surface, and the force due to pressure p_2, psf, on the bottom surface. Summing these vertical forces and setting the total equal to zero yields

$$p_2 A - wA\,\Delta h - p_1 A = 0 \qquad (21\text{-}3a)$$

$$p_2 = w\,\Delta h + p_1 \qquad (21\text{-}3b)$$

For the special case where the top of the prism coincides with the water surface, p_1 is atmospheric pressure. Since most hydraulics problems involve gage pressure, p_1 is zero (gage pressure is zero at atmospheric pressure). Taking Δh to be h, the depth below the water surface, ft, then p_2 is p, the pressure, psf, at depth h. Equation (21-3b) then becomes

$$p = wh \qquad h = \frac{p}{w} \qquad (21\text{-}4)$$

Equation (21-4) gives the depth of water h of specific weight w required to produce a gage pressure p. By adding atmospheric pressure p_a to Eq. (21-4), absolute pressure p_{ab} is obtained as shown in Fig. 21-4. Thus,

$$p_{ab} = wh + p_a \qquad (21\text{-}5)$$

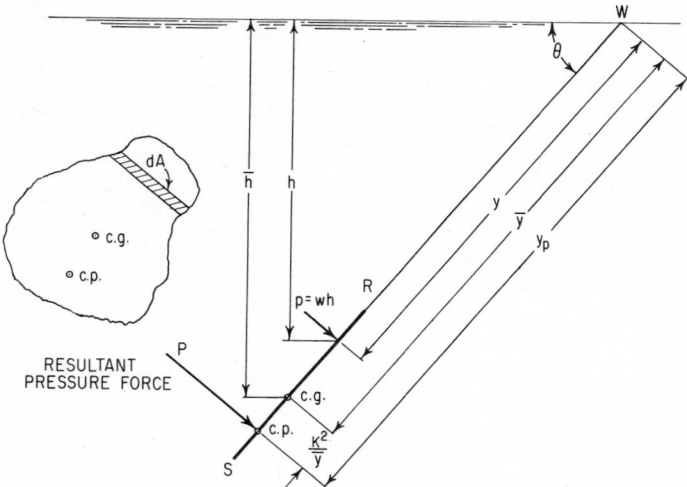

Fig. 21-5. Total pressure on a submerged plane surface depends on pressure at center of gravity (c.g.) but acts at a point (c.p.) that is below the c.g.

Pressure on submerged plane surfaces is important in the design of dams, tanks, and outlet works, such as sluice gates. For horizontal surfaces, the pressure-force determination is a simple matter since the pressure is constant. For determination of the pressure force on inclined or vertical surfaces, however, the summation concepts of integral calculus must be used.

Figure 21-5 represents any submerged plane surface of negligible thickness inclined at an angle θ with the horizontal. The resultant pressure force P, lb, acting on the surface is equal to $\int p\, dA$. Since $p = wh$ and $h = y \sin \theta$, where w is the specific weight of water, lb per cu ft,

$$P = w \int y \sin \theta \, dA \qquad (21\text{-}6)$$

Equation (21-6) can be simplified by setting $\int y \, dA = \bar{y}A$, where A is the area of the submerged surface, sq ft; and $\bar{y} \sin \theta = \bar{h}$, the depth of the centroid, ft. Therefore,

$$P = w\bar{h}A = p_{cg}A \tag{21-7}$$

where p_{cg} is the pressure at the centroid, psf.

The point on the submerged surface at which the resultant pressure force acts is called the **center of pressure** $c.p.$ It is below the center of gravity because the pressure intensity increases with depth. The location of the center of pressure, represented by the length y_p, is calculated by summing the moments of the incremental forces about an axis in the water surface through point W (Fig. 21-5). Thus, $Py_p = \int y \, dP$. Since $dP = wy \sin \theta \, dA$ and $P = w\int y \sin \theta \, dA$,

$$y_p = \frac{\int y^2 \, dA}{\int y \, dA} \tag{21-8}$$

The quantity $\int y^2 \, dA$ is the moment of inertia of the area about the axis through W. It also equals $AK^2 + A\bar{y}^2$, where K is the radius of gyration, ft, of the surface about its centroidal axis. The denominator of Eq. (21-8) equals $\bar{y}A$. Hence

$$y_p = \bar{y} + \frac{K^2}{\bar{y}} \tag{21-9}$$

and $K^2/\bar{y}$ is the distance between the centroid and center of pressure.

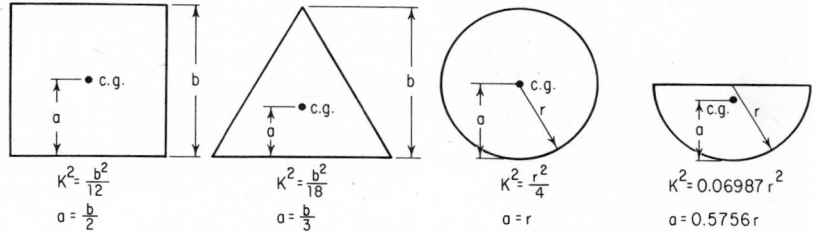

Fig. 21-6. Radius of gyration and location of center of gravity (c.g.) of common shapes.

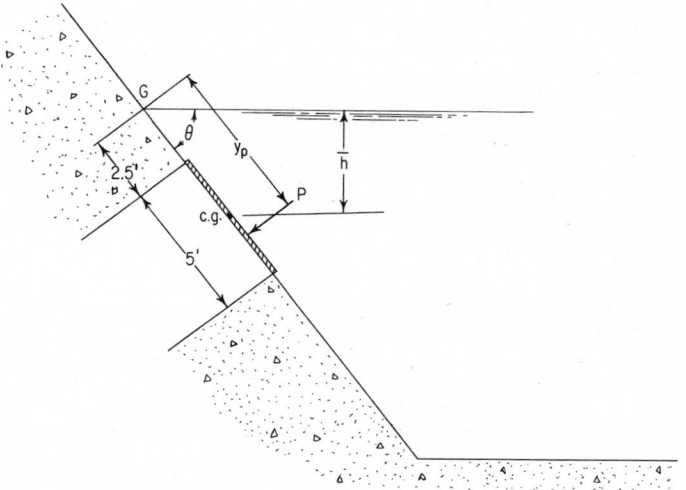

Fig. 21-7. Sluice gate (crosshatched) is subjected to hydrostatic pressure. (See Example 21-1.)

Values of K^2 for some common shapes are given in Fig. 21-6 (see also Fig. 6-29). For areas for which radius of gyration has not been determined, y_p may be calculated directly from Eq. (21-8).

The horizontal location of the center of pressure may be determined as follows: It lies on the vertical axis of symmetry for surfaces symmetrical about the vertical. It lies on the locus of the midpoints of horizontal lines located on the submerged surface, if that locus is a straight line. Otherwise, the horizontal location may be found by taking moments about an axis perpendicular to the one through W in Fig. 21-5 and lying in the plane of the submerged surface.

Example 21-1.   Determine the magnitude and point of action of the resultant pressure force on a 5-ft-square sluice gate inclined at an angle θ of 53.2° to the horizontal (Fig. 21-7).

From Eq. (21-7), the total force $P = w\bar{h}A$, with

$$\bar{h} = [2.5 + \tfrac{1}{2}(5)] \sin 53.2° = 5.0 \times 0.8 = 4.0 \text{ ft}$$
$$A = 5 \times 5 = 25 \text{ sq ft}$$

Thus, $P = 62.4 \times 4 \times 25 = 6{,}240$ lb. From Eq. (21-9), its point of action is a distance $y_p = \bar{y} + K^2/\bar{y}$ from point G, and $\bar{y} = 2.5 + \tfrac{1}{2}(5.0) = 5.0$ ft. Also $K^2 = b^2/12 = 5^2/12 = 2.08$. Therefore, $y_p = 5.0 + 2.08/5 = 5.0 + 0.42 = 5.42$ ft.

Pressure on Submerged Curved Surfaces.   The resultant pressure force on submerged curved surfaces cannot be calculated from the equations developed for the pressure force on submerged plane surfaces because of the variation in direction of the pressure force. The

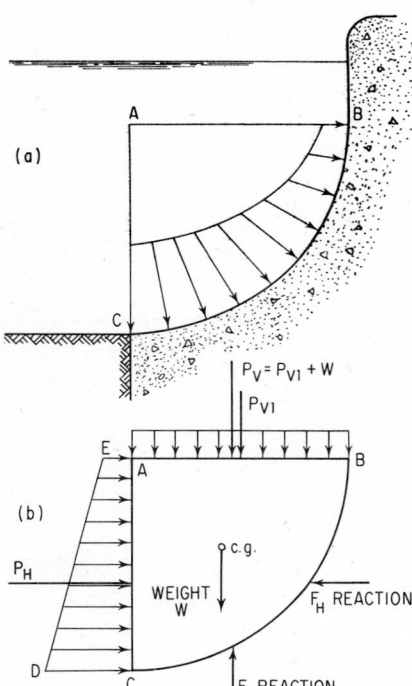

Fig. 21-8. Hydrostatic pressure on a submerged curved surface. (a) General configuration showing pressure variation. (b) Free-body diagram.

resultant pressure force can be calculated, however, by determining its horizontal and vertical components and combining them vectorially.

A typical configuration of pressure on a submerged curved surface is shown in Fig. 21-8. Consider ABC to be a 1-ft-thick prism and analyze it as a free body by the principles of statics. Note:

1. The horizontal component P_H of the resultant pressure force has a magnitude equal to the pressure force on the vertical projection AC of the curved surface and acts at the centroid of pressure diagram $ACDE$.

2. The vertical component P_V of the resultant pressure force has a magnitude equal to the sum of the pressure force on the horizontal projection AB of the curved surface and the weight of the water vertically above ABC. The horizontal location of the vertical component is calculated by taking moments of the two vertical forces about point C.

When water is below the curved surface, such as for a taintor gate (Fig. 21-9), the vertical component P_V of the resultant pressure force has a magnitude equal to the weight of the imaginary volume of water vertically above the surface. P_V acts upward through the center of gravity of this imaginary volume.

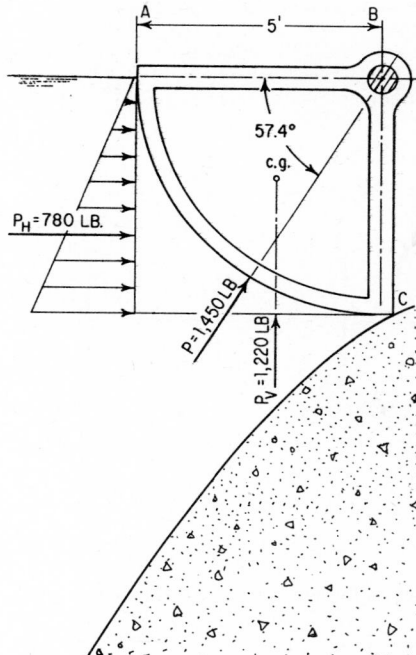

Fig. 21-9. Taintor gate has submerged curved surface under pressure. Vertical component of pressure acts upward. (See Example 21-2.)

Example 21-2. Calculate the magnitude and direction of the resultant pressure on a 1-ft-wide strip of the semicircular taintor gate in Fig. 21-9.

The magnitude of the horizontal component P_H of the resultant pressure force equals the pressure force on the vertical projection of the taintor gate. From Eq. (21-7), $P_H = w\bar{h}A = 62.4 \times 2.5 \times 5 = 780$ lb.

The magnitude of the vertical component of the resultant pressure force equals the weight of the imaginary volume of water in the prism ABC above the curved surface. The volume of this prism is $\pi R^2/4 = 3.14 \times 25/4 = 19.6$ cu ft. So the weight of the water is $19.6w = 19.6 \times 62.4 = 1,220$ lb $= P_V$.

The magnitude of the resultant pressure force equals

$$P = \sqrt{P_H{}^2 + P_V{}^2} = \sqrt{780^2 + 1,220^2} = 1,450 \text{ lb}$$

The tangent of the angle the resultant pressure force makes with the horizontal $= P_V/P_H = 1,220/780 = 1.564$. The corresponding angle is $57.4°$.

The positions of the horizontal and vertical components of the resultant pressure force are not required to find the point of action of the resultant. Its angle with the horizontal is known, and for a constant-radius surface, the resultant must act perpendicular to the surface.

21-4. Submerged and Floating Bodies. The principles of buoyancy govern the behavior of submerged and floating bodies and are important in determining the stability and draft of cargo vessels.

The buoyant force acting on a submerged body equals the weight of the volume of liquid displaced.

A floating body displaces a volume of liquid equal to its weight.

The buoyant force acts vertically through the center of buoyancy c.b., which is located at the center of gravity of the volume of liquid displaced.

For a body to be in equilibrium, whether floating or submerged, the center of buoyancy and center of gravity must be on the same vertical line *AB* (Fig. 21-10*a*). The stability of a ship, its tendency not to overturn when it is in a nonequilibrium position, is indicated by the *meta-center*. It is the point at which a vertical line through the center of buoyancy intersects the rotated position of the line through the centers of gravity and buoyancy for the equilibrium condition *A'B'* (Fig. 21-10*b*). The ship is stable only if the metacenter is above the center of gravity, since the resulting moment for this condition tends to right the ship.

The distance between the ship's metacenter and center of gravity is called the *metacentric height* and is designated by y_m in Fig. 21-10*b*. Given in feet by Eq. (21-10), y_m is a measure of

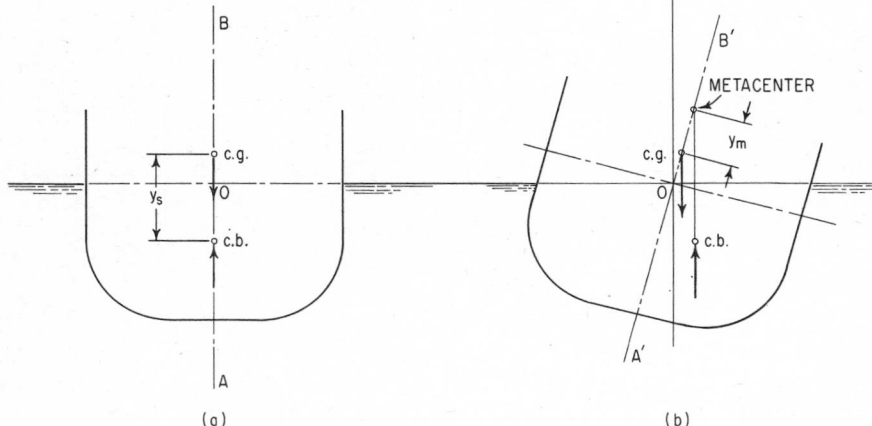

Fig. 21-10. Stability of a ship depends on the location of its metacenter relative to its center of gravity (c.g.).

the degree of stability or instability of a ship, since the magnitudes of moments produced in a roll are directly proportional to this distance.

$$y_m = \frac{I}{V} \pm y_s \qquad (21\text{-}10)$$

where I = moment of inertia of the ship's cross section at waterline about the longitudinal axis through 0, ft^4

V = volume of displaced liquid, cu ft

y_s = distance, ft, between the centers of buoyancy and gravity when the ship is in equilibrium

The negative sign should be used when the center of gravity is above the center of buoyancy.

21-5. Manometers. A manometer is a device for measuring pressure. It consists of a tube containing a column of one or two liquids that balances the unknown pressure. The basis for the calculation of this unknown pressure is provided by the height of the liquid column. All manometer problems may be solved with Eq. (21-4), $p = wh$. Manometers indicate h, the *pressure head*, or the difference in head.

The primary application of manometers is measurement of relatively low pressures, for which aneroid and Bourdon gages are not sufficiently accurate because of their inherent mechanical limitations. However, manometers may also be used in precise measurement of high pressures by arranging several U-tube manometers in series (Fig. 21-12*c*). Manometers are used for both static and flow applications, although the latter is most common.

Three basic types are used. These are shown in Fig. 21-11: piezometer, U-tube manometer, and differential manometer. The following is a brief discussion of the basic types.

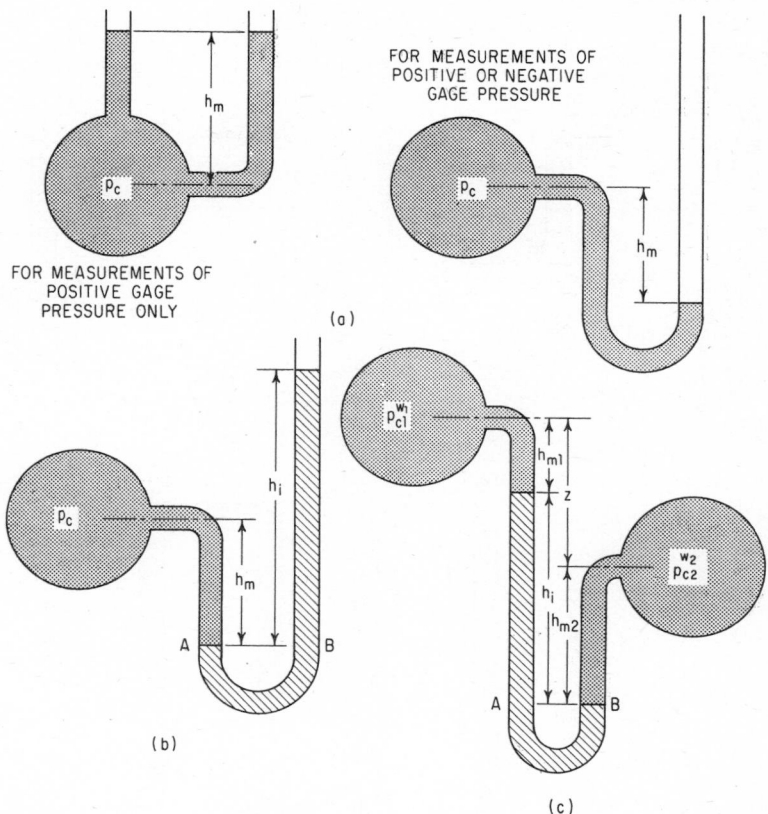

FOR MEASUREMENTS OF
POSITIVE OR NEGATIVE
GAGE PRESSURE

FOR MEASUREMENTS OF
POSITIVE GAGE
PRESSURE ONLY

(a)

(b)

(c)

Fig. 21-11. Basic types of manometers.   (a) Piezometers; (b) U-tube manometer; (c) differential manometer.

The **piezometer** (Fig. 21-11a) consists of a tube with one end tapped flush with the wall of the container in which the pressure is to be measured and the other end open to the atmosphere. The only liquid it contains is the one whose pressure is being measured (the **metered liquid**). The piezometer is a sensitive gage, but it is limited to the measurement of relatively small pressures, usually heads of 5 ft of water or less.   Larger pressures would create an impractically high column of liquid.

Example 21-3.   The gage pressure p_c in the pipe of Fig. 21-11a is 2.17 psi.   The liquid is water with $w = 62.4$ lb per cu ft.   What is h_m?

$$h_m = \frac{p_c}{w} = \frac{2.17 \times 144}{62.4} = 5.0 \text{ ft}$$

For pressures greater than 5 ft of water, the **U-tube manometer** (Fig. 21-11b) is used.   It is similar to the piezometer except that it contains an **indicating liquid** with a specific gravity usually much larger than that of the metered liquid.   The only other criteria are that the indicating liquid should have a good meniscus and be immiscible with the metered liquid.

The U-tube manometer is used when pressures are either too high or too low for the piezometer.   High pressures can be measured by arranging U-tube manometers in series (Fig. 21-12c).   Very low pressures, including negative gage pressures, can be measured if the bottom of the U tube extends below the center line of the container of the metered liquid.   The commonest use of the U-tube manometer is measurement of the pressures of flowing water.   In this application, the usual indicating liquid is mercury.

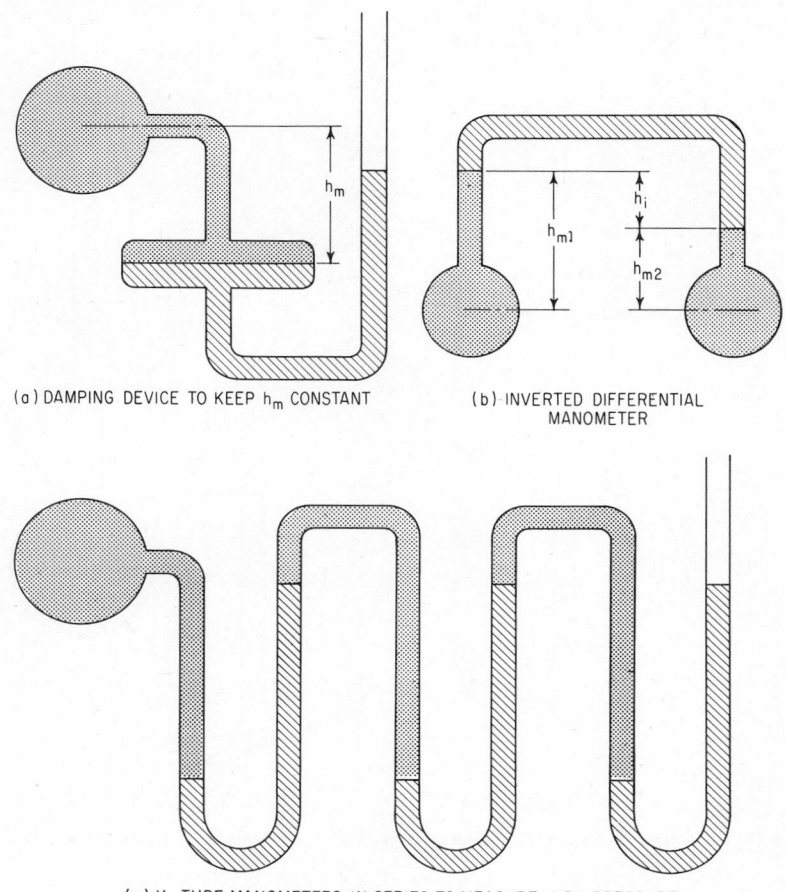

(a) DAMPING DEVICE TO KEEP h_m CONSTANT   (b) INVERTED DIFFERENTIAL
                                                        MANOMETER

(c) U-TUBE MANOMETERS IN SERIES TO MEASURE HIGH PRESSURE

Fig. 21-12. Manometers may contain a sump, as in (a), to damp flow disturbances; they may be shaped like an inverted U (b) for measuring pressures with low-specific-gravity liquids; and they may be connected in series (c) for measuring high pressures.

A movable scale, as opposed to a fixed scale, facilitates reading the U-tube manometer. The scale is positioned between the two vertical legs and moved to adjust for the variation in distance h_m from the center line of the pressure vessel to the indicating liquid. This zero adjustment enables a direct reading of the heights h_i and h_m of the liquid columns. The scale may be calibrated in any convenient units, such as ft of water or psi.

The **differential manometer** (Fig. 21-11c) is identical to the U-tube manometer but measures the difference in pressure between two points. (It does not indicate the pressure at either of the points.) The differential manometer may have either the standard U-tube configuration or an inverted U-tube configuration, depending on the comparative specific gravities of the indicating and metered liquids. The inverted U-tube configuration (Fig. 21-12b) is used when the indicating liquid has a lower specific gravity than the metered liquid.

Example 21-4. A differential manometer (Fig. 21-11c) is measuring the difference in pressure between two water pipes. The indicating liquid is mercury (specific gravity = 13.6), h_i is 2.25 ft, h_{m1} is 9 in., and z is 1.0 ft. What is the pressure differential between the two pipes?

$$h_{m2} = h_i + h_{m1} - z = 2.25 + 0.75 - 1.0 = 2.0 \text{ ft}$$

The pressure at B, psf, is

$$P_B = p_{c2} + w_m h_{m2} = p_{c2} + 62.4 \times 2.0 = p_{c2} + 125$$

The pressure at A, psf, is

$$P_A = p_{c1} + w_m h_{m1} + w_i h_i$$
$$= p_{c1} + 62.4 \times 0.75 + 13.6 \times 62.4 \times 2.25 = p_{c1} + 1{,}957$$

Since the pressure at A must equal that at B,

$$p_{c2} + 125 = p_{c1} + 1{,}957$$
$$p_{c2} - p_{c1} = 1{,}832 \text{ psf} = 12.7 \text{ psi}$$

When small pressure differences in water are measured, if the specific gravity of the indicating liquid is between 1.0 and 2.0 and the points at which the pressure is being measured are at the same level, the actual pressure difference, when expressed in feet of water, is magnified by the differential manometer. For example, if the actual pressure difference is 0.50 ft of water and the indicating liquid has a specific gravity of 1.40, the magnification will be 2.5; that is, the height of the liquid column h_i will be 1.25 ft of water. The closer the specific gravities of the metered and indicating liquids, the greater are the magnification and sensitivity. This is true only up to a magnification of about 5. Above 5, the increased sensitivity may be deceptive because the meniscus between the two liquids becomes poorly defined and sluggish in movement.

Many factors affect the accuracy of manometers. Most of them, however, may be neglected in the majority of hydraulics applications, since they are significant only in precise reading of manometers, such as might be required in laboratories. One factor, however, is significant. This is the existence of surges in the manometer caused by the pulsations and disturbances in the flow of water resulting from turbulence. These surges make reading of the manometer difficult. They may be reduced or eliminated by installing a large-diameter section, or sump, in the manometer, as shown in Fig. 21-12a. This sump will act to damp the pulsations and keep the distance from the center line of the conduit to the indicating liquid essentially at a constant value.

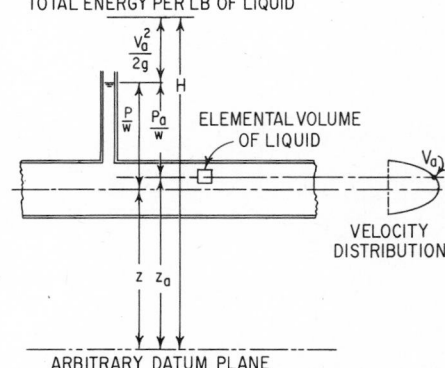

TOTAL ENERGY PER LB OF LIQUID

$\dfrac{V_0^2}{2g}$

H

$\dfrac{P}{w}$ $\dfrac{P_0}{w}$ ELEMENTAL VOLUME OF LIQUID

V_0

VELOCITY DISTRIBUTION

z z_0

ARBITRARY DATUM PLANE

Fig. 21-13. Energy in a liquid depends on elevation, velocity, and pressure.

21-6. Fundamentals of Fluid Flow.

For fluid energy, the law of conservation of energy is represented by the **Bernoulli equation**:

$$Z_1 + \frac{p_1}{w} + \frac{V_1^2}{2g} = Z_2 + \frac{p_2}{w} + \frac{V_2^2}{2g} \tag{21-11}$$

where Z_1 = elevation, ft, at any point 1 of a flowing fluid above an arbitrary datum
Z_2 = elevation, ft, at a downstream point in the fluid above the same datum
p_1 = pressure at 1, psf
p_2 = pressure at 2, psf
w = specific weight of fluid, lb per cu ft
V_1 = velocity of fluid at 1, fps
V_2 = velocity of fluid at 2, fps
g = acceleration due to gravity, 32.2 ft per sec per sec

The left side of the equation sums the total energy per unit weight of fluid at 1, and the right side, the total energy per unit weight at 2. Equation (21-11) applies only to an ideal fluid. Its practical use requires a term to account for the decrease in total head, ft, through friction. This term h_f, when added to the downstream side of Eq. (21-11), yields the form of the Bernoulli equation most frequently used:

$$Z_1 + \frac{p_1}{w} + \frac{V_1{}^2}{2g} = Z_2 + \frac{p_2}{w} + \frac{V_2{}^2}{2g} + h_f \qquad (21\text{-}12)$$

The energy contained in an elemental volume of fluid, thus, is a function of its elevation, velocity, and pressure (Fig. 21-13). The energy due to elevation is the potential energy and equals WZ_a, where W is the weight, lb, of the fluid in the elemental volume and Z_a is its elevation, ft, above some arbitrary datum. The energy due to velocity is the kinetic energy. It equals $WV_a{}^2/2g$, where V_a is the velocity, fps. The pressure energy equals Wp_a/w, where p_a is the pressure, psf, and w is the specific weight of the fluid, lb per cu ft. The total energy, ft-lb, in the elemental volume of fluid is

$$E = WZ_a + \frac{Wp_a}{w} + \frac{WV_a{}^2}{2g} \qquad (21\text{-}13)$$

Dividing both sides of the equation by W yields the energy per unit weight of flowing fluid, or the *total head*, ft:

$$H = Z_a + \frac{p_a}{w} + \frac{V_a{}^2}{2g} \qquad (21\text{-}14)$$

p_a/w is called **pressure head;** $V_a{}^2/2g$, **velocity head.**

As indicated in Fig. 21-13, $Z + p/w$ is constant for any point in a cross section normal to the flow through a pipe or channel. Kinetic energy at the section, however, varies with velocity.

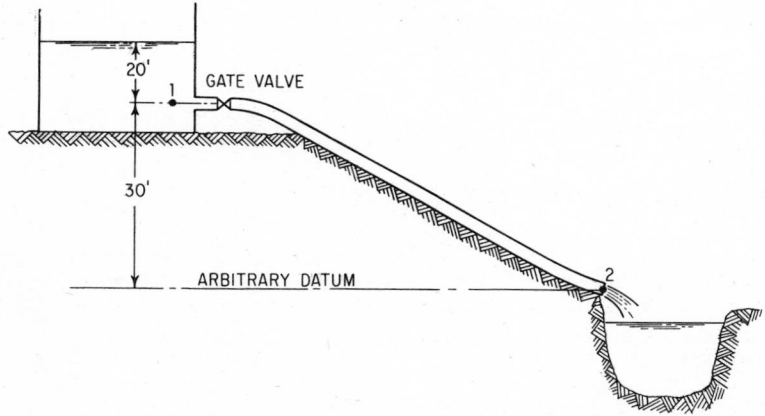

Fig. 21-14. Flow from an elevated reservoir—application of the Bernoulli equation. (See Example 21-5.)

Usually, $Z + p/w$ at the midpoint and the average velocity at a section are assumed when the Bernoulli equation is applied to flow across the section or when total head is to be determined. Average velocity, fps = Q/A, where Q is the quantity of flow, cfs, across the area of the section A, sq ft.

Example 21-5.   Determine the energy loss between points 1 and 2 in the 24-in.-diameter pipe in Fig. 21-14. The pipe carries water flowing at 31.4 cfs.

Average velocity in the pipe = $Q/A = 31.4/3.14 = 10$ fps. Select point 1 far enough from the reservoir outlet that V_1 can be assumed to be 0. Since the datum plane passes through point 2, $Z_2 = 0$. Also, since the pipe has free discharge, $p_2 = 0$. Thus, substitution in Eq. (21-12) yields

$$30 + 20 + 0 = 0 + 0 + \frac{10^2}{64.4} + h_f$$

where h_f is the friction loss, ft. Hence, $h_f = 50 - 1.55 = 48.45$ ft.

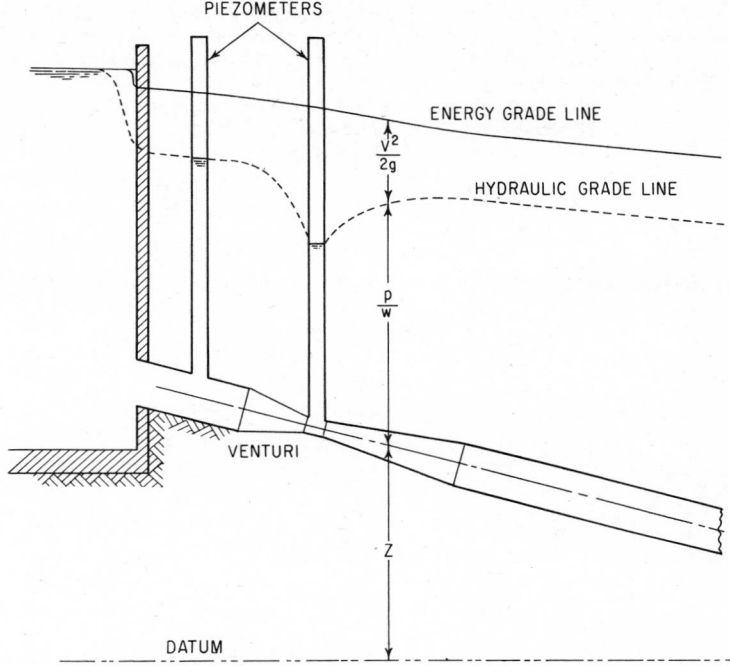

Fig. 21-15. Energy grade line and hydraulic grade line indicate variations in energy and pressure head, respectively, in a liquid as it flows along a pipe or channel.

Note that in this example h_f includes minor losses due to the pipe entrance, gate valve, and any bends.

The Bernoulli equation and the variation of pressure may be represented graphically, respectively, by energy and hydraulic grade lines (Fig. 21-15). The *energy grade line,* sometimes called the total head line, shows the decrease in total energy per unit weight H in the direction of flow. The slope of the energy grade line is called the **energy gradient** or friction slope. The **hydraulic grade line** lies a distance $V^2/2g$ below the energy grade line and shows the variation of velocity or pressure in the direction of flow. The slope of the hydraulic grade line is termed the **hydraulic gradient.** In open-channel flow, the hydraulic grade line coincides with the water surface, while in pressure flow, it represents the height to which water would rise in a piezometer (see also Example 21-7, Art. 21-9).

Momentum is a fundamental concept that must be considered in the design of essentially all waterworks facilities involving flow. A change in momentum, which may result from a change in either velocity, direction, or magnitude of flow, is equal to the impulse, the force F acting on the fluid times the period of time dt over which it acts. Dividing the total change in momentum by the time interval over which the change occurs gives the momentum equation, or impulse-momentum equation:

$$F_x = \rho Q \,\Delta V_x \qquad\qquad (21\text{-}15)$$

where F_x = summation of all forces in the X direction per unit time causing a change in momentum in the X direction, lb

ρ = density of flowing fluid, lb-sec²/ft⁴ (specific weight divided by g)

Q = flow rate, cfs

ΔV_x = change in velocity in the X direction, fps

Similar equations may be written for the Y and Z directions. The impulse-momentum equation often is used in conjunction with the Bernoulli equation [Eq. (21-11) or (21-12)] but may be used separately.

Example 21-6. Calculate the resultant force on the reducer elbow in Fig. 21-16. The pipe center line lies in a horizontal plane. The pipe reduces from 48 in. in diameter to 16 in. The pressure at the upstream side of the reducer bend (point 1) is 100 psi, and the water flow is 100 cfs. (Neglect friction loss at the bend.)

Velocity at points 1 and 2 is found by dividing $Q = 100$ cfs by the respective areas: $V_1 = 100 \times 4/4^2\pi = 7.96$ fps and $V_2 = 100 \times 4/1.33^2\pi = 71.5$ fps.

With p_1 known, the Bernoulli equation yields the pressure p_2 at 2:

$$0 + 100 \times \frac{144}{62.4} + \frac{7.96^2}{2 \times 32.2} = 0 + \frac{p_2}{62.4} + \frac{71.5^2}{2 \times 32.2}$$

$$p_2 = 9,500 \text{ psf}$$

The total pressure force at 1 is $P_1 = p_1 A_1 = 181,000$ lb, and at 2, $P_2 = p_2 A_2 \doteq 13,200$ lb.

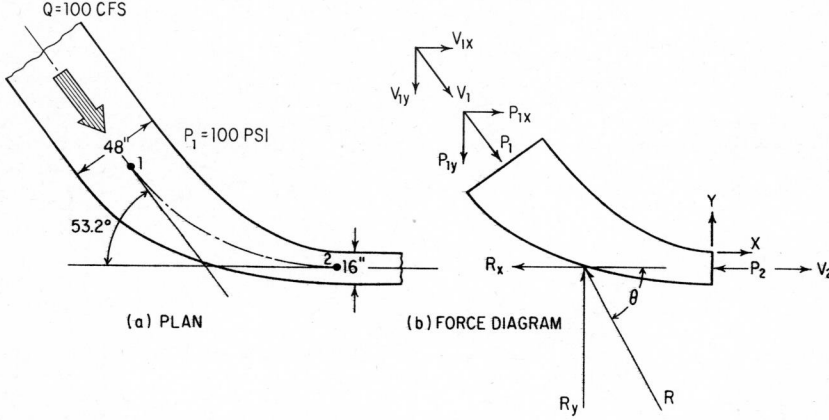

Fig. 21-16. Flow induces forces in a pipe at bends and at changes in size of section—application of momentum equation. (See Example 21-6.)

Let R be the force, lb, exerted by the pipe on the fluid (equal and opposite in direction to the force against the pipe, which is to be determined). Then, the force F changing the momentum of the fluid equals the vector sum $P_1 - P_2 + R$. To find F, apply Eq. (21-15) first in the X direction, then in the Y direction, and determine the resultant of the forces:

In the X direction, since $\Delta V_x = -(7.96 \sin 53.2° - 71.5) = 65.1$ and the density $\rho = 62.4/32.2 = 1.94$.

$$F_x = 181,000 \cos 53.2° - 13,200 + R_x = 1.94 \times 100 \times 65.1$$

$$R_x = -82,600 \text{ lb}$$

In the Y direction, since $\Delta V_y = -(-7.96 \cos 53.2° - 0) = 4.78$,

$$F_y = -181,000 \sin 53.2° + R_y = 1.94 \times 100 \times 4.78$$

$$R_y = 145,700 \text{ lb}$$

The resultant $R = \sqrt{R_x^2 + R_y^2} = 167,500$ lb. It acts at an angle θ with the horizontal such that $\tan \theta = 145,700/82,600$; so $\theta = 60.5°$. The force against the pipe acts in the opposite direction.

21-7. Hydraulic Models. A model is a system whose operation can be used to predict the characteristics of a similar system, or prototype, usually more complex or built to a much larger scale. A knowledge of the laws governing the phenomena under investigation is necessary if the model study is to yield accurate quantitative results.

Forces acting on the model should be proportional to forces on the prototype. The four forces usually considered in hydraulic models are inertia, gravity, viscosity, and surface tension. Because of the laws governing these forces and because the model and prototype are normally not the same size, it is usually not possible to have all four forces in the model in the same proportions as they are in the prototype. It is, however, a simple procedure to have two predominant forces in the same proportion. In most models, the fact that two of the four forces are not in the same proportion as they are in the prototype does not introduce serious error. The inertial force, which is always a predominant force, and one other force are made proportional.

Ratios of the forces of gravity, viscosity, and surface tension to the force of inertia are designated, respectively, Froude number, Reynolds number, and Weber number. Equating the Froude number of the model and the Froude number of the prototype insures that the gravitational and inertial forces are in the same proportion. Similarly, equating the Reynolds numbers of the model and prototype insures that the viscous and inertial forces will be in the same proportion. And equating the Weber numbers insures proportionality of surface tension and inertial forces.

The **Froude number** is

$$F = \frac{V}{\sqrt{Lg}} \tag{21-16}$$

where F = Froude number (dimensionless)
V = velocity of fluid, fps
L = linear dimension (characteristic, such as depth or diameter), ft
g = acceleration due to gravity, 32.2 ft per sec²

For hydraulic structures, such as spillways and weirs, where there is a rapidly changing water-surface profile, the two predominant forces are inertia and gravity. Therefore, the Froude numbers of the model and prototype are equated:

$$F_m = F_p \qquad \frac{V_m}{\sqrt{L_m g}} = \frac{V_p}{\sqrt{L_p g}} \tag{21-17a}$$

where subscript m applies to the model and p to the prototype. Squaring both sides of Eq. (21-17a) and grouping like terms yields

$$\frac{V_m^2}{V_p^2} = \frac{L_m}{L_p} \tag{21-17b}$$

Let $V_r = V_m/V_p$ and $L_r = L_m/L_p$. Then

$$V_r^2 = L_r \qquad V_r = L_r^{1/2} \tag{21-18}$$

The subscript r indicates ratio of quantity in model to that in prototype.

If the ratios of all the physical dimensions of a model to all the corresponding physical dimensions of the prototype are equal to the length ratio, the model is termed a **true model**. In a true model where the Froude number is the governing design criterion, the length ratio is the only variable. Once the length ratio has been set, all the physical dimensions of the model are fixed. The discharge ratio is determined as follows:

$$Q_r = V_r A_r \tag{21-19a}$$

Since $V_r = L_r^{1/2}$ and A_r = area ratio = L_r^2,

$$Q_r = V_r A_r = L_r^{5/2} \tag{21-19b}$$

By this method all the necessary characteristics of a spillway or weir model can be determined.

The **Reynolds number** is

$$R = \frac{VL}{\nu} \tag{21-20}$$

R is dimensionless, and ν is the kinematic viscosity of fluid, sq ft per sec. The Reynolds numbers of model and prototype are equated when the viscous and inertial forces are predominant. Viscous forces are usually predominant when flow occurs in a closed system, such as pipe flow where there is no free surface. The following relations are obtained by equating Reynolds numbers of the model and prototype:

$$\frac{V_m L_m}{\nu_m} = \frac{V_p L_p}{\nu_p} \tag{21-21a}$$

$$V_r = \frac{\nu_r}{L_r} \tag{21-21b}$$

The variable factors that fix the design of a true model when the Reynolds number governs are the length ratio and the viscosity ratio.

The **Weber number** is

$$W = \frac{V^2 L \rho}{\sigma} \qquad (21\text{-}22)$$

where ρ = density of fluid, lb-sec^2 ft^4 (specific weight divided by g)

σ = surface tension of fluid, psf

The Weber numbers of model and prototype are equated in certain types of wave studies, the formation of drops and air bubbles, entrainment of air in flowing water, and other phenomena where surface tension and inertial forces are predominant. The velocity ratio is determined as follows:

$$\frac{V_m{}^2 L_m \rho_m}{\sigma_m} = \frac{V_p{}^2 L_p \rho_p}{\sigma_p} \qquad (21\text{-}23a)$$

$$V_r{}^2 = \frac{\sigma_r}{\rho_r L_r} \qquad (21\text{-}23b)$$

The fluid properties and the length ratio fix the design of a model governed by the Weber number.

In some cases, such as a morning-glory spillway, inertial, viscous, and gravity forces all have an important effect on the flow. In these cases it is usually not possible to have both the Reynolds and Froude numbers of the model and prototype equal. The solution to this type of problem is mostly empirical and may consist of an attempt to evaluate the effects of viscosity and gravity separately.

For the flow of water in open channels and rivers where the friction slope is relatively flat, model designs are often based on the Manning equation. The relations between the model and prototype are determined as follows:

$$\frac{V_m}{V_p} = \frac{(1.486/n_m) R_m{}^{2/3} S_m{}^{1/2}}{(1.486/n_p) R_p{}^{2/3} S_p{}^{1/2}} \qquad (21\text{-}24)$$

where n = Manning roughness coefficient ($T/L^{1/3}$, T representing time)

R = hydraulic radius (L)

S = loss of head due to friction per unit length of conduit (dimensionless)

= slope of energy gradient

For true models, $S_r = 1$, $R_r = L_r$. Hence,

$$V_r = \frac{L_r{}^{2/3}}{n_r} \qquad (21\text{-}25)$$

In models of rivers and channels, it is necessary for the flow to be turbulent. The U.S. Waterways Experiment Station has determined that flow will be turbulent if

$$\frac{VR}{v} \geqq 4,000 \qquad (21\text{-}26)$$

where V = mean velocity, fps

R = hydraulic radius, ft

v = kinematic viscosity, sq ft per sec

If the model is to be a true model, it may have to be uneconomically large for the flow to be turbulent. Another problem also encountered in true models is surface tension. In a true model of a wide river where the depth may be only a fraction of an inch, the surface tension will distort the flow to such an extent that the model may be useless. To overcome the effect of surface tension and to get turbulent flow, the depth scale is often made much larger than the length scale. This type of model is called a **distorted model.**

The relations between a distorted model of a channel and a prototype are determined in the same manner as was Eq. (21-24). The only difference is that the slope ratio S_r equals the depth ratio d_r and the hydraulic-radius ratio is a function of the width ratio and depth ratio.

One type of model, called a movable-bed model, is used to study erosion and transportation of silt in riverbeds. Because the laws governing the transportation of material are not fully understood, movable-bed models are built largely on the basis of experience and give only qualitative results.

(J.E.A. John and W.L. Haberman, "Introduction to Fluid Mechanics," Prentice-Hall, Inc., Englewood Cliffs, N.J.)

PIPE FLOW

The term pipe flow as used in this section refers to flow in a circular closed conduit entirely filled with fluid. For closed conduits other than circular, reasonably good results are obtained in the turbulent range with standard pipe-flow formulas if the diameter is replaced by four times the hydraulic radius. But when there is severe deviation from a circular cross section, as in annular passages, this method gives values that are much too low. (J. F. Walker, G. A. Whan, and R. R. Rothfus, "Fluid Friction in Noncircular Ducts," *Journal of the American Institute of Chemical Engineers,* vol. 3, 1957.)

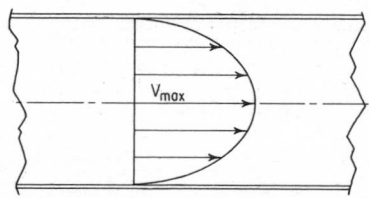

Fig. 21-17. Velocity distribution for laminar flow in a circular pipe is parabolic. Maximum velocity is twice average velocity.

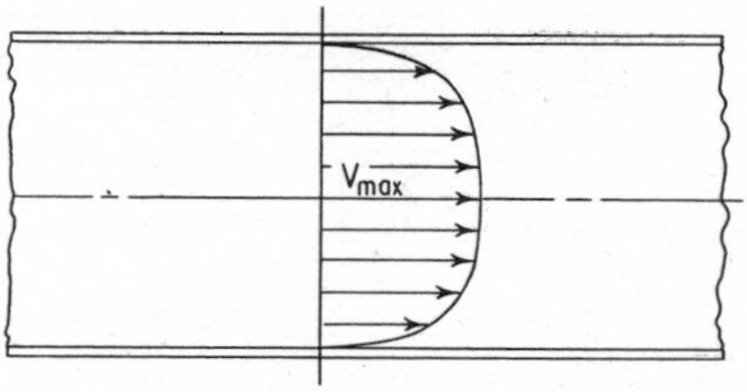

Fig. 21-18. Velocity distribution for turbulent flow in a circular pipe is more nearly uniform than that for laminar flow.

21-8. Laminar Flow. In laminar flow, fluid particles move in parallel layers in one direction. The parabolic velocity distribution in laminar flow, shown in Fig. 21-17, creates a shearing stress $\tau = \mu \, dV/dy$, where dV/dy is the rate of change of velocity with depth and μ is the coefficient of viscosity (see Viscosity, Art. 21-2). As this shearing stress increases, the viscous forces become unable to damp out disturbances, and turbulent flow results. The region of change is dependent on the fluid's velocity, density, and viscosity and on the size of the conduit.

A dimensionless parameter called the Reynolds number has been found to be a reliable criterion for the determination of laminar or turbulent flow. It is the ratio of inertial forces to viscous forces, and is given by

$$\mathbf{R} = \frac{VD\rho}{\mu} = \frac{VD}{\nu} \tag{21-27}$$

where V = fluid velocity, fps
$\quad D$ = pipe diameter, ft
$\quad \rho$ = density of fluid, lb-sec^2/ft^4 (specific weight divided by g, 32.2 ft per sec^2)
$\quad \mu$ = viscosity of fluid, lb-sec per sq ft
$\quad \nu = \mu/\rho$ = kinematic viscosity, sq ft per sec
For a Reynolds number less than 2,000, flow is laminar in circular pipes. When the Reynolds number is greater than 2,000, laminar flow is unstable; a disturbance will probably be magnified, causing the flow to become turbulent.

In laminar flow, the following equation for head loss due to friction can be developed by considering the forces acting on a cylinder of fluid in a pipe:

$$h_f = \frac{32\mu LV}{D^2\rho g} = \frac{32\mu LV}{D^2 w} \tag{21-28}$$

where h_f = head loss due to friction, ft
$\quad L$ = length of pipe section considered, ft
$\quad g$ = acceleration due to gravity, 32.2 ft per sec^2
$\quad w$ = specific weight of fluid, lb per cu ft
Substitution of the Reynolds number yields

$$h_f = \frac{64}{\mathbf{R}} \frac{L}{D} \frac{V^2}{2g} \tag{21-29}$$

For laminar flow, Eq. (21-29) is identical to the Darcy-Weisbach formula Eq. (21-30), since in laminar flow the friction factor $f = 64/\mathbf{R}$.

(H. W. King and E. F. Brater, "Handbook of Hydraulics," McGraw-Hill Book Company, New York.)

21-9. Turbulent Flow.   In turbulent flow, the inertial forces are so great that viscous forces cannot dampen out disturbances caused primarily by surface roughness.   These disturbances create eddies, which have both a rotational and a translational velocity.   The translation of these eddies is a mixing action that effects an interchange of momentum across the cross section of the conduit.   As a result, the velocity distribution is more uniform, as shown in Fig. 21-18, than for laminar flow (Fig. 21-17).

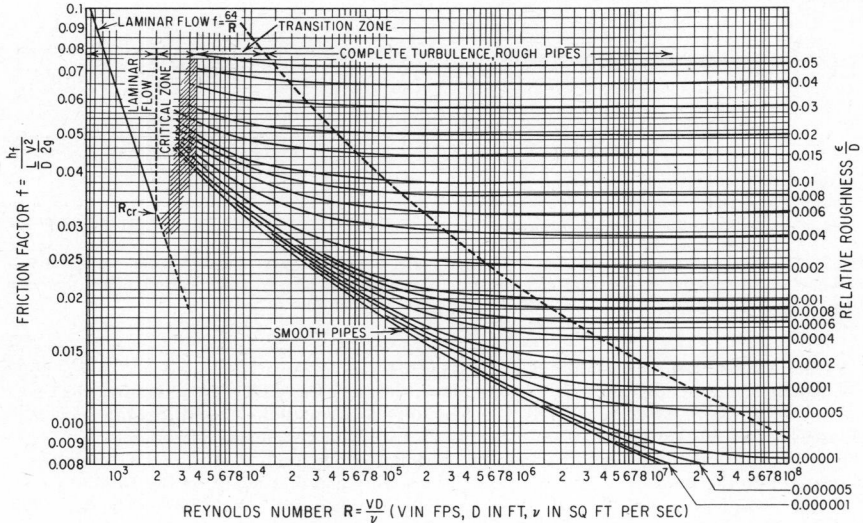

Fig. 21-19.   Chart gives friction factors for flow in pipes.

For a Reynolds number greater than 2,000 but to the left of the dashed line in Fig. 21-19, there is a transition from laminar to turbulent flow.   In this region, there is a laminar film at the boundaries that covers some of the smaller roughness projections.   This explains why the friction loss in this region has both laminar and turbulent characteristics.   As the Reynolds number increases, this laminar boundary layer decreases in thickness until, at completely turbulent flow, it no longer covers any of the roughness projections.   To the right of the dashed line in Fig. 21-19, the flow is completely turbulent, and viscous forces do not affect the friction loss.

Because of the random nature of turbulent flow, it is not practical to treat it analytically. Therefore, formulas for head loss and flow in the turbulent regions have been developed through experimental and statistical means.   Experimentation in turbulent flow has shown that:

The head loss varies directly as the length of the pipe.

The head loss varies almost as the square of the velocity.

The head loss varies almost inversely as the diameter.

The head loss depends upon the surface roughness of the pipe wall.

The head loss depends upon the fluid's density and viscosity.

The head loss is independent of the pressure.

The Darcy-Weisbach formula, one of the most widely used equations for pipe flow, satisfies the above conditions and is valid for laminar or turbulent flow in all fluids.

$$h_f = f \, \frac{L}{D} \frac{V^2}{2g} \tag{21-30}$$

where h_f = head loss due to friction, ft
  f = friction factor (see Fig. 21-19)
  L = length of pipe, ft
  D = diameter of pipe, ft
  V = velocity of fluid, fps
  g = acceleration due to gravity, 32.2 ft per sec²
It employs the Moody diagram (Fig. 21-19) for evaluating the friction factor f. (L. F. Moody, "Friction Factors for Pipe Flow," *Transactions of the American Society of Mechanical Engineers,* November, 1944.)

Because Eq. (21-30) is dimensionally homogeneous, it can be used with any consistent set of units without changing the value of the friction factor.

Roughness values ϵ (ft) for use with the Moody diagram to determine the Darcy-Weisbach friction factor f are given in Table 21-3.

Table 21-3. Typical Values of Roughness for Use in the Moody Diagram (Fig. 21-19) to Determine f

	ϵ, ft
Steel pipe:	
Severe tuberculation and incrustation	0.03 –0.008
General tuberculation	0.008 –0.003
Heavy brush-coat asphalts, enamels, and tars	0.003 –0.001
Light rust	0.001 –0.0005
New smooth pipe, centrifugally applied enamels	0.0002–0.00003
Hot-dipped asphalt; centrifugally applied concrete linings	0.0005–0.0002
Steel-formed concrete pipe, good workmanship	0.0005–0.0002
New cast-iron pipe	0.00085

The following formulas were derived for head loss in waterworks design and give good results for water-transmission and -distribution problems. They contain a factor that depends on the surface roughness of the pipe material. The accuracy of these formulas is greatly affected by the selection of the roughness factor, which requires experience in its choice.

The Chezy formula holds for head loss in conduits and gives reasonably good results for high Reynolds numbers:

$$V = C \sqrt{RS} \tag{21-31}$$

where V = velocity, fps
  C = coefficient, dependent on surface roughness of conduit
  S = slope of energy grade line or head loss due to friction, ft per ft of conduit
  R = hydraulic radius, ft
Hydraulic radius of a conduit is the cross-sectional area of the fluid in it divided by the perimeter of the wetted section.

Manning's formula: Through experimentation, Manning concluded that the C in the Chezy equation [Eq. (21-31)] should vary as $R^{1/6}$:

$$C = \frac{1.486R^{1/6}}{n} \tag{21-32}$$

where n = coefficient, dependent on surface roughness. Substitution into Eq. (21-31) gives

$$V = \frac{1.486}{n} R^{2/3} S^{1/2} \tag{21-33a}$$

Upon substitution of $D/4$, where D is the pipe diameter, for the hydraulic radius of the pipe, the following equations are obtained for pipes flowing full:

$$V = \frac{0.590}{n} D^{2/3} S^{1/2} \tag{21-33b}$$

$$Q = \frac{0.463}{n} D^{8/3} S^{1/2} \tag{21-33c}$$

$$h_f = 4.66n^2 \frac{LQ^2}{D^{16/3}} \tag{21-33d}$$

$$D = \left(\frac{2.159Qn}{S^{1/2}}\right)^{3/8} \tag{21-33e}$$

where Q = flow, cfs.

Tables 21-4 and 21-11 (p. 21-46) gives values of n for the foot-pound-second system.   See also tables in Art. 22-8 for velocity and flow at various slopes.

Table 21-4. Values of n for Pipes, to Be Used with the Manning Formula

Material of pipe	Variation		Use in designing	
	From	To	From	To
Clean cast iron	0.011	0.015	0.013	0.015
Dirty or tuberculated cast iron	0.015	0.035		
Riveted steel	0.013	0.017	0.015	0.017
Welded steel	0.010	0.013	0.012	0.013
Galvanized iron	0.012	0.017	0.015	0.017
Wood stave	0.010	0.014	0.012	0.013
Concrete	0.010	0.017		
Good workmanship			0.012	0.014
Poor workmanship		—	0.016	0.017

The Hazen-Williams formula is one of the most widely used formulas for pipe-flow problems of water utilities, although it was developed for both open channels and pipe flow:

$$V = 1.318C_1 R^{0.63} S^{0.54} \tag{21-34a}$$

For pipes flowing full:

$$V = 0.55C_1 D^{0.63} S^{0.54} \tag{21-34b}$$

$$Q = 0.432C_1 D^{2.63} S^{0.54} \tag{21-34c}$$

$$h_f = \frac{4.727}{D^{4.87}} L \left(\frac{Q}{C_1}\right)^{1.85} \tag{21-34d}$$

$$D = \frac{1.376}{S^{0.205}} \left(\frac{Q}{C_1}\right)^{0.38} \tag{21-34e}$$

where V = velocity, fps
C_1 = coefficient, dependent on surface roughness
R = hydraulic radius, ft
S = head loss due to friction, ft per ft of pipe
D = diameter of pipe, ft
L = length of pipe, ft
Q = discharge cfs
h_f = friction loss, ft

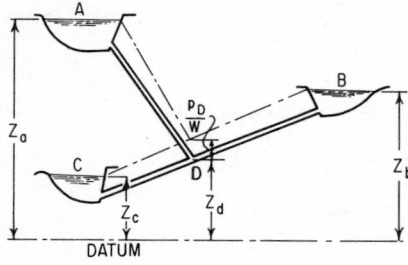

Fig. 21-20. Flow between three reservoirs. (See Example 21-7.)

Table 21-5. Values of C_1 in Hazen and Williams Formula

Type of pipe	C_1
Cast iron:	
New	All sizes, 130
5 years old	All sizes up to 24 in., 120
	24 in. and over, 115
10 years old	12 in., 110
	4 in., 105
	30 in. and over, 85
40 years old	16 in., 80
	4 in., 65
Welded steel	Values the same as for cast-iron pipe, 5 years older
Riveted steel	Values the same as for cast-iron pipe, 10 years older
Wood stave	Average value, regardless of age, 120
Concrete or concrete-lined	Large sizes, good workmanship, steel forms, 140
	Large sizes, good workmanship, wood forms, 120
	Centrifugally spun, 135
Vitrified	In good condition, 110

The C_1 terms in Table 21-5 are in the foot-pound-second system. (See Fig. 21-84, p. 21-101.)

The problem of flow in branching pipes illustrates the use of friction-loss equations and the hydraulic-grade-line concept.

Example 21-7. Figure 21-20 shows a typical three-reservoir problem. The elevations of the hydraulic grade lines for the three pipes are equal at point D. The Hazen-Williams equation for friction loss [Eq. (21-34d)] can be written for each pipe meeting at D. With the continuity equation for quantity of flow, there are as many equations as there are unknowns:

$$Z_a = Z_d + \frac{p_D}{w} + \frac{4.727L_A}{D_A{}^{4.87}}\left(\frac{Q_A}{C_A}\right)^{1.85} \tag{21-35a}$$

$$Z_{tb} = Z_d + \frac{p_D}{w} + \frac{4.727L_B}{D_B{}^{4.87}}\left(\frac{Q_B}{C_B}\right)^{1.85} \tag{21-35b}$$

$$Z_c = Z_d + \frac{p_D}{w} - \frac{4.727L_C}{D_C{}^{4.87}}\left(\frac{Q_C}{C_C}\right)^{1.85} \tag{21-35c}$$

$$Q_A + Q_B = Q_C \tag{21-36}$$

where p_D = pressure at D

w = unit weight of liquid

With the elevations Z of the three reservoirs and the pipe intersection known, the easiest way to solve these equations is by trying different values of p_D/w in Eqs. (21-35) and substituting the values obtained for Q into Eq. (21-36) for a check. If the value of $Z_d + p_D/w$ becomes greater than Z_b, the sign of the friction-loss term is negative instead of positive. This would indicate water is flowing from reservoir A into reservoirs B and C.

21-10. Minor Losses in Pipes. These are losses occurring in contractions, bends, enlargements, and valves and other pipe fittings. These losses can usually be neglected if the length of the pipeline is greater than 1,500 times the pipe's diameter. However, in short pipelines, because these losses may exceed the friction losses, minor losses must be considered.

Sudden Enlargement. The following equation for the head loss, ft, across a sudden enlargement of pipe diameter has been determined analytically and agrees well with experimental results:

$$h_L = \frac{(V_1 - V_2)^2}{2g} \tag{21-37}$$

where V_1 = velocity before enlargement, fps

V_2 = velocity after enlargement, fps

g = 32.2 ft per sec^2

It was derived by applying the Bernoulli equation and the momentum equation across an enlargement.

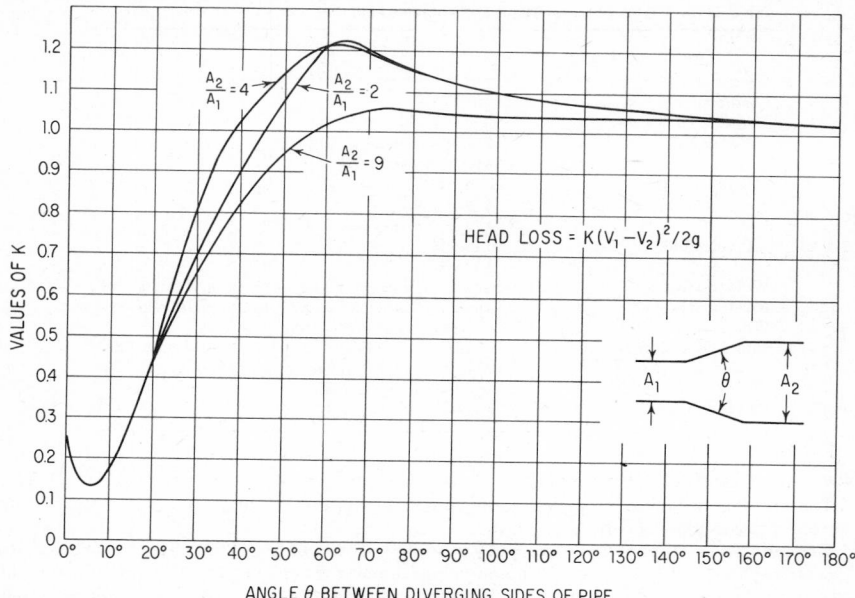

Fig. 21-21. Head loss coefficients for a pipe with diverging sides depend on the angle of divergence.

Another equation for the head loss caused by sudden enlargements was determined experimentally by Archer. This equation gives slightly better agreement with experimental results than Eq. (21-37):

$$h_L = \frac{1.1(V_1 - V_2)^{1.92}}{2g} \qquad (21\text{-}38)$$

A special application of Eq. (21-37) or (21-38) is the discharge from a pipe into a reservoir. The water in the reservoir has no velocity, so a full velocity head is lost.

Gradual Enlargement. The equation for the head loss due to a gradual conical enlargement of a pipe takes the following form:

$$h_L = \frac{K(V_1 - V_2)^2}{2g} \qquad (21\text{-}39)$$

where K = loss coefficient (see Fig. 21-21)

Since the experimental data available on gradual enlargements are limited and inconclusive, the values of K in Fig. 21-21 are approximate. (A. H. Gibson, "Hydraulics and Its Applications," Constable & Co., Ltd., London.)

Sudden Contraction. The following equation for the head loss across a sudden contraction of a pipe was determined by the same type of analytical studies as Eq. (21-37):

$$h_L = \left(\frac{1}{C_c} - 1\right)^2 \frac{V^2}{2g} \qquad (21\text{-}40)$$

where C_c = coefficient of contraction (see Table 21-6)

     V = velocity in smaller-diameter pipe, fps

This equation gives best results when the head loss is greater than 1 ft. Table 21-6 gives C_c values for sudden contractions, determined by Julius Weisbach ("Die Experiments-Hydraulik").

Another formula for determining the loss of head caused by a sudden contraction, determined experimentally by Brightmore, is

$$h_L = \frac{0.7(V_1 - V_2)^2}{2g} \qquad (21\text{-}41)$$

This equation gives best results if the head loss is less than 1 ft.

Table 21-6. C_c **for Contractions in Pipe Area from** A_1 **to** A_2

A_2/A_1	0.1	0.2	0.3	0.4	0.5	0.6	0.7	0.8	0.9	1.0
C_c	0.62	0.63	0.64	0.66	0.68	0.71	0.76	0.81	0.89	1.0

A special case of sudden contraction is the entrance loss for pipes. Some typical values of the loss coefficient K in $h_L = KV^2/2g$, where V is the velocity in the pipe, are presented in Table 21-7.

Table 21-7. Coefficients for Entrance Losses

Pipe projecting into reservoir	$K = 0.80$
Sharp-cornered entrance	$K = 0.50$
Bellmouth entrance	$K = 0.05$
Slightly rounded entrance	$K = 0.25$

Bends and Standard Fitting Losses. The head loss that occurs in pipe fittings, such as valves and elbows, and at bends is given by

$$h_L = \frac{KV^2}{2g} \qquad (21\text{-}42)$$

Table 21-8 gives some typical K values for these losses.

Table 21-8. Coefficients for Fitting Losses and Losses at Bends

Fitting	K
Globe valve, fully open	10.0
Angle valve, fully open	5.0
Swing check valve, fully open	2.5
Gate valve, fully open	0.2
Closed-return bend	2.2
Short-radius elbow $(r/D \approx 1.0)$*	0.9
Long-radius elbow $(r/D \approx 1.5)$	0.6
45° elbow	0.4

*r = radius of bend; D = pipe diameter.

The values in Table 21-8 are only approximate. K values vary not only for different sizes of fitting, but also with different manufacturers. For these reasons, manufacturers' data are the best source for loss coefficients.

Experimental data available on bend losses cover a rather narrow range of laboratory experiments utilizing small-diameter pipes and do not give conclusive results. The data indicate the losses vary with surface roughness, Reynolds number, ratio of radius of bend r to pipe diameter D, and angle of bend. The data are in agreement that the head loss, not including friction loss, decreases sharply as the r/D ratio increases from zero to around 4 or 5. When r/D increases above 4 or 5, there is disagreement. Some experiments indicate that the head loss, not including friction loss in the bend, increases significantly with an increasing r/D. Some recent work by Ito, on smooth pipes, indicates that this increase is very slight and that above an r/D of 4, the bend loss essentially remains constant. (H. Ito, "Pressure Losses in Smooth Pipe Bends," *Transactions of the American Society of Civil Engineers,* series D, vol. 82, no. 1, 1960.)

Because experiments have produced such widely varying data, bend-loss coefficients give only an approximation of losses to be expected. Figure 21-22 gives values of K for 90° bends for use with Eq. (21-42). (K. H. Beij, "Pressure Losses for Fluid Flow in 90° Pipe Bends," *Journal of Research, National Bureau of Standards,* vol. 21, July, 1938.)

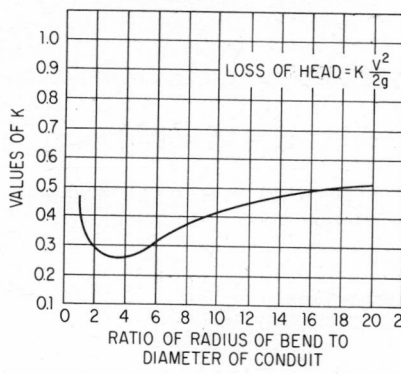

Fig. 21-22. Recommended values of head-loss coefficient K for 90° bends in closed conduits.

To obtain losses in bends other than 90°, Hinds suggested the following formula to adjust the K values given in Fig. 21-22:

$$K' = K \sqrt{\frac{\Delta}{90}} \qquad (21\text{-}43)$$

where Δ = the deflection angle, deg.

The K' value may be used in place of K in Eq. (21-42).

21-11. Orifices. An orifice is an opening with a closed perimeter through which water flows. Orifices may have any shape, although they are usually round, square, or rectangular.

Discharge through an orifice may be calculated from

$$Q = Ca \sqrt{2gh} \qquad (21\text{-}44)$$

where Q = discharge, cfs
C = coefficient of discharge
a = area of orifice, sq ft
g = acceleration due to gravity, ft per sec²
h = head on the horizontal center line of the orifice, ft

Coefficients of discharge C are given in Table 21-9 for low velocity of approach. If this velocity is significant, then its effect should be taken into account. This equation is applicable for any head for which the coefficient of discharge is known. For low heads, measuring the head from the center line of the orifice is not theoretically correct; however, this error is corrected by the C values.

Table 21-9. Smith's* Coefficients of Discharge for Circular and Square Orifices with Full Contraction

Dia of circular orifices, ft				Head, ft	Side of square orifices, ft			
0.02	0.04	0.01	1.0		0.02	0.04	0.1	1.0
	0.637	0.618		0.4		0.643	0.621	
0.655	0.630	0.613		0.6	0.660	0.636	0.617	
0.648	0.626	0.610	0.590	0.8	0.652	0.631	0.615	0.597
0.644	0.623	0.608	0.591	1	0.648	0.628	0.613	0.599
0.637	0.618	0.605	0.593	1.5	0.641	0.622	0.610	0.601
0.632	0.614	0.604	0.595	2	0.637	0.619	0.608	0.602
0.629	0.612	0.603	0.596	2.5	0.634	0.617	0.607	0.602
0.627	0.611	0.603	0.597	3	0.632	0.616	0.607	0.603
0.623	0.609	0.602	0.596	4	0.628	0.614	0.606	0.602
0.618	0.607	0.600	0.596	6	0.623	0.612	0.605	0.602
0.614	0.605	0.600	0.596	8	0.619	0.610	0.605	0.602
0.611	0.603	0.598	0.595	10	0.616	0.608	0.604	0.601
0.601	0.599	0.596	0.594	20	0.606	0.604	0.602	0.600
0.596	0.595	0.594	0.593	50	0.602	0.601	0.600	0.599
0.593	0.592	0.592	0.592	100	0.599	0.598	0.598	0.598

*Hamilton Smith, Jr., "Hydraulics," 1886.

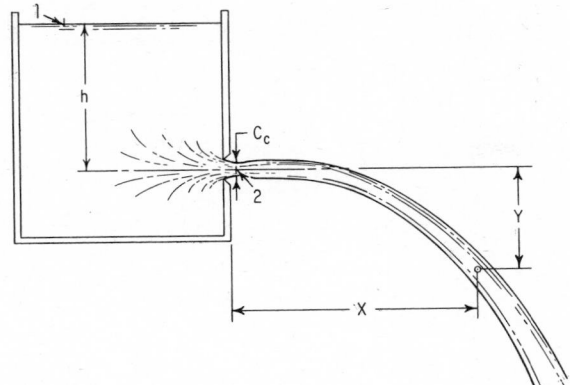

Fig. 21-23. Fluid jet takes parabolic path.

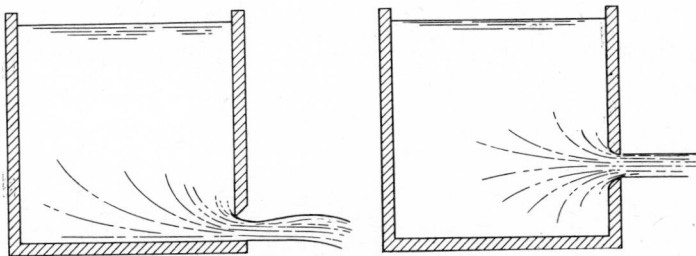

Fig. 21-24. (*a*) Sharp-edged orifice with partly suppressed contraction. (*b*) Round-edged orifice 'has no contraction.

The coefficient of discharge C is the product of the coefficient of velocity C_v and the coefficient of contraction C_c. The coefficient of velocity is the ratio obtained by dividing the actual velocity at the *vena contracta* (contraction of the jet discharged) by the theoretical velocity. The theoretical velocity may be calculated by writing Bernoulli's equation for points 1 and 2 in Fig. 21-23.

$$\frac{V_1^2}{2g} + \frac{p_1}{w} + Z_1 = \frac{V_2^2}{2g} + \frac{p_2}{w} + Z_2 \tag{21-45}$$

With the reference plane through point 2, $Z_1 = h$, $V_1 = 0$, $p_1/w = p_2/w = 0$, and $Z_2 = 0$, and Eq. (21-45) becomes

$$V_2 = \sqrt{2gh} \tag{21-46}$$

The actual velocity, determined experimentally, is less than the theoretical velocity because of the energy loss from point 1 to point 2. Typical values of C_v range from 0.94 to 0.99

The coefficient of contraction C_c is the ratio of the smallest area of the jet, the vena contracta, to the area of the orifice. Contraction of a fluid jet will occur if the orifice is square-edged and is so located that some of the fluid approaches the orifice at an angle to the direction of flow through the orifice. This fluid has a momentum component perpendicular to the axis of the jet which causes the jet to contract. Typical values of the coefficient of contraction range from 0.61 to 0.67.

If the water entering the orifice does not have this momentum, the contraction is completely suppressed. Figure 21-24a is an example of a partly suppressed contraction; no contraction occurs at the bottom of the jet. In Fig. 21-24b, the edges of the orifice have been rounded to reduce or eliminate the contraction. With a partly suppressed orifice, the increased area of jet caused by suppressing the contraction on one side is partly offset, because more water at a higher velocity enters on the other sides. The result is a slightly greater coefficient of contraction.

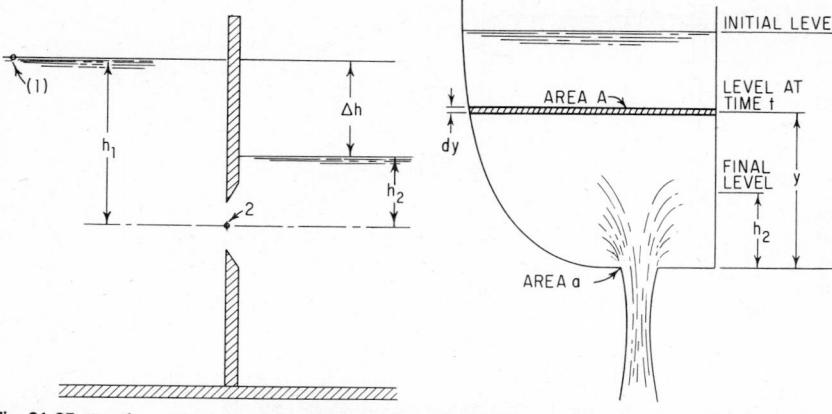

Fig. 21-25. Discharge through a submerged orifice.

Fig. 21-26. Discharge from a reservoir under falling head.

Submerged Orifices. Flow through a submerged orifice may be computed by applying Bernoulli's equation to points 1 and 2 in Fig. 21-25.

$$V_2 = \sqrt{2g\left(h_1 - h_2 + \frac{V_1^2}{2g} - h_L\right)} \tag{21-47}$$

where h_L = losses in head, ft, between 1 and 2

Assuming $V_1 \approx 0$, setting $h_1 - h_2 = \Delta h$, and using a coefficient of discharge C to account for losses, Eq. (21-48) is obtained.

$$Q = Ca\sqrt{2g\,\Delta h} \tag{21-48}$$

Values of C for submerged orifices do not differ greatly from those for nonsubmerged orifices. (For table of values of coefficients of discharge for submerged orifices, see H. W. King and E. F. Brater, "Handbook of Hydraulics," McGraw-Hill Book Company, New York.)

Discharge under Falling Head. The flow from a reservoir or vessel when the inflow is less than the outflow represents a condition of falling head. The time required for a certain quantity of water to flow from a reservoir can be calculated by equating the volume of water that flows through the orifice or pipe in time dt to the volume decrease in the reservoir (Fig. 21-26):

$$Ca\sqrt{2gy}\,dt = A\,dy \tag{21-49}$$

Solving for dt yields

$$dt = \frac{A\,dy}{Ca\sqrt{2gy}} \tag{21-50}$$

where a = area of orifice, sq ft
A = area of reservoir, sq ft
y = head on orifice at time t, ft
C = coefficient of discharge
g = acceleration due to gravity, 32.2 ft per sec²

Expressing the area as a function of y $[A = F(y)]$, and summing from time zero, when $y = h_1$, to time t, when $y = h_2$, Eq. (21-50) becomes

$$t = \int_{h_2}^{h_1} \frac{F(y)\,dy}{Ca\sqrt{2gy}} \tag{21-51}$$

If the area of the reservoir is constant as y varies, Eq. (21-41) upon integration becomes

$$t = \frac{2A}{Ca\sqrt{2g}}(\sqrt{h_1} - \sqrt{h_2}) \tag{21-52}$$

where h_1 = head at the start, ft
  h_2 = head at the end, ft
  t = time interval for head to fall from h_1 to h_2, sec

Fluid Jets. Where the effect of air resistance is small, a fluid discharged through an orifice into the air will follow the path of a projectile. The initial velocity of the jet is

$$V_0 = C_v \sqrt{2gh} \tag{21-53}$$

where h = head on the center line of the orifice, ft
  C_v = coefficient of velocity

The direction of the initial velocity depends on the orientation of the surface in which the orifice is located. For simplicity, the following equations were determined assuming the orifice is located in a vertical surface (Fig. 21-23). The velocity of the jet in the X direction (horizontal) remains constant.

$$V_x = V_0 = C_v \sqrt{2gh} \tag{21-54}$$

The velocity in the Y direction is initially zero and thereafter a function of time and the acceleration of gravity:

$$V_y = gt \tag{21-55}$$

The X coordinate at time t is

$$X = V_x t = t C_v \sqrt{2gh} \tag{21-56}$$

The Y coordinate is

$$Y = V_{avg} t = \frac{gt^2}{2} \tag{21-57}$$

where V_{avg} = average velocity over period of time t. The equation for the path of the jet [Eq. (21-58)], obtained by solving Eq. (21-57) for t and substituting in Eq. (21-56), is that for a parabola:

$$X^2 = C_v^2 4hY \tag{21-58}$$

Equation (21-58) can be used to determine C_v experimentally. Rearranging Eq. (21-58) gives

$$C_v = \sqrt{\frac{X^2}{4hY}} \tag{21-59}$$

The X and Y coordinates can be measured in a laboratory and C_v calculated from Eq. (21-59).

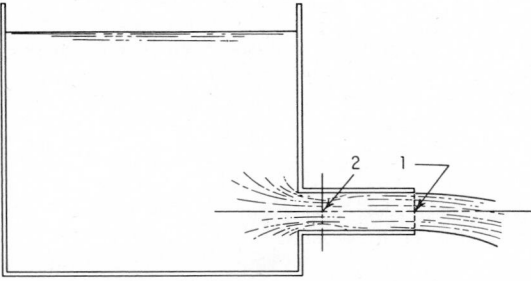

Fig. 21-27. Flow through a tube with a sharp-edged inlet.

Short Tubes. When water flows from a reservoir into a pipe or tube with a sharp leading edge, the same type of contraction occurs as for a sharp-edged orifice. In the tube or pipe, however, the water contracts and then expands to fill the tube. If the tube is discharging at atmospheric pressure, a partial vacuum is created at the contraction, as can be seen by applying the Bernoulli equation across points 1 and 2 in Fig. 21-27. This reduced pressure causes the flow through a short tube to be greater than that through a sharp-edged orifice of the same dimensions. If the head on the tube is greater than 50 ft and the tube is short, the water will shoot through the tube without filling it. When this happens, the tube acts as a sharp-edged orifice.

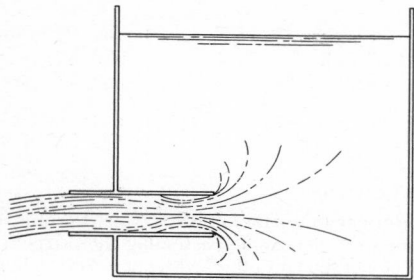

Fig. 21-28. Flow through a reentrant tube resembles that through a flush tube (Fig. 21-27), but head loss is greater.

For a short tube flowing full, the coefficient of contraction $C_c = 1.00$ and the coefficient of velocity $C_v = 0.82$. Therefore, the coefficient of discharge $C = 0.82$. Solving for head loss as a proportion of final velocity head, a K value for Eq. (21-42) of 0.5 is obtained as follows: The theoretical velocity head with no loss is $V_T^2/2g$. Actual velocity head is $V_a^2/2g = (0.82V_T)^2/2g = 0.67V_T^2/2g$. Head loss $h_L = 1.00V_T^2/2g - 0.67V_T^2/2g = 0.33V_T^2/2g$. From $h_L = KV_a^2/2g$, where $V_a^2/2g$ is the actual velocity head, $K = 2gh_L/V_a^2 = (0.33V_T^2 \times 2g)/(2g \times 0.67V_T^2) = 0.5$.

For a reentrant tube projecting into a reservoir (Fig. 21-28), the coefficients of velocity and discharge equal 0.75, and the loss coefficient K equals 0.80.

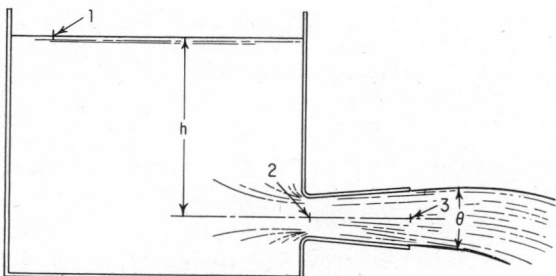

Fig. 21-29. Diverging conical tube increases flow through an orifice.

Diverging Conical Tubes. This type of tube can greatly increase the flow through an orifice by reducing the pressure at the orifice below atmospheric. Equation (21-60) for the pressure at the entrance to the tube is obtained by writing the Bernoulli equation for points 1 and 3 and points 1 and 2 in Fig. 21-29.

$$p_2 = wh\left[1 - \left(\frac{a_3}{a_2}\right)^2\right] \qquad (21-60)$$

where p_2 = gage pressure at tube entrance, psf
w = unit weight of water, lb per cu ft
h = head on center line of orifice, ft
a_2 = area of the smallest part of jet (vena contracta, if one exists), sq ft
a_3 = area of discharge end of tube, sq ft
The discharge is also calculated by writing the Bernoulli equation for points 1 and 3 in Fig. 21-29.

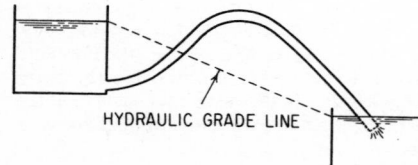

HYDRAULIC GRADE LINE

Fig. 21-30. Siphon is pipe that rises above hydraulic grade line.

For this analysis to be valid, the tube must flow full, and the pressure in the throat of the tube must not fall to the vapor pressure of water.   Experiments by Venturi show the most efficient angle θ to be around 5°.

21-12. Siphons.   A siphon is a closed conduit that rises above the hydraulic grade line and in which the pressure at some point is below atmospheric (Fig. 21-30).   The commonest use of a siphon is the siphon spillway.

Flow through a siphon can be calculated by writing the Bernoulli equation for the entrance and exit.   But the pressure in the siphon must be checked to be sure it does not fall to the vapor pressure of water.   This is accomplished by writing the Bernoulli equation across a point of known pressure and a point where the elevation head or the velocity head is a maximum in the conduit.   If the pressure were to fall to the vapor pressure, vaporization would decrease or totally stop the flow.

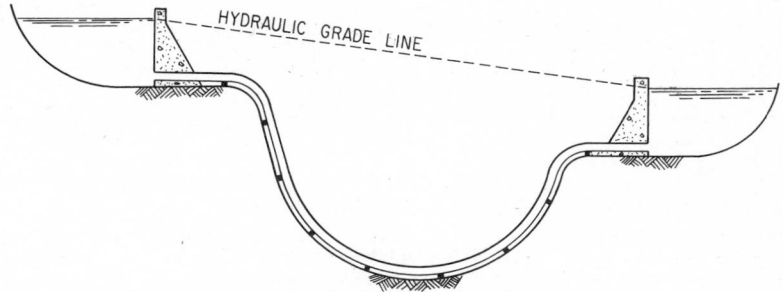

Fig. 21-31. Sag pipe connects two reservoirs.

The pipe shown in Fig. 21-31 is also commonly called a siphon or inverted siphon.   This is a misnomer since the pressure at all points in the pipe is above atmospheric.   The American Society of Civil Engineers recommends that the inverted siphon be called a **sag pipe** to avoid the false impression that it acts as a siphon.

21-13. Water Hammer.   Water hammer is a change in pressure, either above or below the normal pressure, caused by a variation of the flow rate in a pipe.   Every time the flow rate is changed, either increased or decreased, it causes water hammer.   However, the stresses are not critical in small-diameter pipes with flows at low velocities.

The water flowing in a pipe has momentum equal to the mass of the water times its velocity.   When a valve is closed, this momentum drops to zero.   The change causes a pressure rise, which begins at the valve and is transmitted up the pipe.   The pressure at the valve will rise until it is high enough to overcome the momentum of the water and bring the water to a stop.   This pressure buildup travels the full length of the pipe to the reservoir (Fig. 21-32).

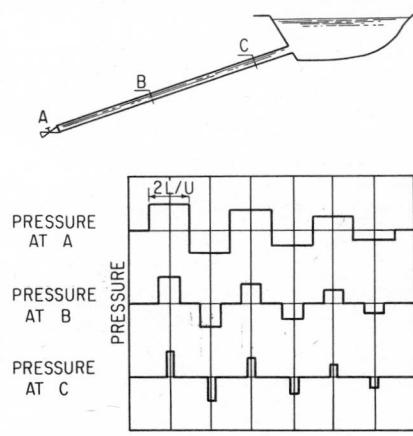

Fig. 21-32. Variation of pressure with time along a penstock, for water hammer caused by instantaneous closure of a valve.

At the instant the pressure wave reaches the reservoir, the water in the pipe is motionless, but at a pressure much higher than normal. The differential pressure between the pipe and the reservoir then causes the water in the pipe to rush back into the reservoir. As the water flows into the reservoir, the pressure in the pipe falls.

At the instant the pressure at the valve reaches normal, the water has attained considerable momentum up the pipe. As the water flows away from the closed valve, the pressure at the valve drops until differential pressure again brings the water to a stop. This pressure drop begins at the valve and continues up the pipe until it reaches the reservoir.

The pressure in the pipe is now below normal, so water from the reservoir rushes into the pipe. This cycle repeats over and over until friction damps these oscillations. Because of the high velocity of the pressure waves, each cycle may take only a fraction of a second.

The equation for the velocity of a wave in a pipe is

$$U = \sqrt{\frac{E}{\rho}} \sqrt{\frac{1}{1 + ED/E_p t}} \tag{21-61}$$

where U = velocity of a pressure wave along a pipe, fps
E = modulus of elasticity of water, 43.2×10^6 psf
ρ = density of water, 1.94 lb-sec^2 per ft^4 (specific weight divided by acceleration due to gravity)
D = diameter of pipe, ft
E_p = modulus of elasticity of pipe material, psf
t = thickness of pipe wall, ft

Instantaneous Closure. The magnitude of the pressure change that results when flow is varied depends on the rate of change of flow and the length of the pipeline. Any gradual movement of a valve that is made in less time than it takes for a pressure wave to travel from the valve to the reservoir and be reflected back to the valve produces the same pressure change as an instantaneous movement. Instantaneous closure:

$$T < \frac{2L}{U} \tag{21-62}$$

where L = length of pipe from reservoir to valve, ft
T = time required to change setting of valve, sec

A plot of pressure vs. time for various points along a pipe is shown in Fig. 21-32 for the instantaneous closure of a valve. Equation (21-63a) for the pressure rise or fall caused by adjusting a valve was derived by equating the momentum of the water in the pipe to the force impulse required to bring the water to a stop.

$$\Delta p = \rho U \, \Delta V \tag{21-63a}$$

In terms of pressure head, Eq. (21-63a) becomes

$$\Delta h = \frac{U \, \Delta V}{g} \tag{21-63b}$$

where Δp = pressure change from normal due to instantaneous change of valve setting, psf
Δh = head change from normal due to instantaneous change of valve setting, ft
ΔV = change in the velocity of the water caused by adjusting a valve, fps

If the closing or opening of a valve is instantaneous, the pressure change can be calculated in one step from Eq. (21-63).

Gradual Closure. The following method of determining the pressure change due to gradual closure of a valve gives a quick, approximate solution. The pressure rise or head change is assumed to be in direct proportion to the closure time:

$$\Delta h_g = \frac{t_i \, \Delta h}{T} = \frac{2L \, \Delta V}{Tg} \tag{21-64}$$

where Δh_g = head change due to gradual closure, ft
t_i = time for wave to travel from the valve to the reservoir and be reflected back to the valve, sec
T = actual closure time of valve, sec
Δh = head rise due to instantaneous closure, ft
L = length of pipeline, ft
ΔV = change in velocity of water due to instantaneous closure, fps
g = acceleration due to gravity, 32.2 ft per sec^2

Arithmetic integration is a more exact method for finding the pressure change due to gradual movement of a valve, and it can be readily programmed for a computer. Integration is a direct means of studying every physical element of the process of water hammer. The valve is assumed to close in a series of small movements, each causing an individual pressure wave. The magnitude of these pressure waves is given by Eq. (21-63). The individual pressure waves are totaled to give the pressure at any desired point for a certain time.

The first step in this method is to choose the time interval for each incremental movement of the valve. (It is convenient to make the time interval some submultiple of L/U, such as L/aU, where a equals any whole integer, so that the pressure waves reflected at the reservoir will be superimposed upon the new waves being formed at the valve. The wave formed at the valve will be opposite in sign to the wave reflected from the reservoir, and so there will be a tendency for the waves to cancel out.) Assuming a valve is fully open and requires T sec for closing, the number of incremental closing movements required is $T/\Delta t$, where Δt, the increment of time, equals L/aU.

Once the time interval has been determined, an estimate of the velocity change ΔV during each time interval must be made, to apply Eq. (21-63). A rough estimate for the velocity following the incremental change is $V_n = V_o(A_n/A_o)$, where V_n is the velocity following a certain incremental movement, V_o the original velocity, A_n the area of the valve opening after the corresponding incremental movement, and A_o the original area of the valve opening.

The change in head can now be calculated with Eq. (21-63). With the head known, the estimated velocity V_n can be checked by the following equation:

$$V_n = \frac{V_o A_n}{A_o} \sqrt{\frac{H_o + \Sigma \Delta h}{H_o}} \qquad (21\text{-}65)$$

where H_o = head at valve before any movement of the valve, ft
$H_o + \Sigma\, \Delta h$ = total pressure at the valve after a particular movement. This includes the pressure change caused by the valve movement plus the effect of waves reflected from the reservoir, ft
A_n = area of the valve opening after n incremental closings. This area can be determined from the closure characteristics of a valve or by assuming its characteristics, sq ft

If the velocity obtained from Eq. (21-65) differs greatly from the estimated velocity, then that obtained from Eq. (21-65) should be used to recalculate Δh.

Two other widely used methods for solution of water-hammer problems are the Angus graphical method and the Allievi chart method. The Angus graphical method is simple yet can handle most complex problems that are encountered. The Allievi chart method is widely used; however, it has a relatively limited range of application. (See C. V. Davis and K. E. Sorensen, "Handbook of Applied Hydraulics," McGraw-Hill Book Company, New York.)

Example 21-8.   The following problem illustrates the use of the preceding methods and compares the results: Steel penstock, length = 3,000 ft, diameter = 10 ft, area = 78.5 sq ft, initial velocity = 10 fps, penstock thickness = 1 in., head at turbine with valve open = 1,000 ft, and modulus of elasticity of steel = 43.2×10^8 psf.

(For penstocks as shown in Fig. 21-32, thickness and diameter normally vary with head. Thus, the velocity of the pressure waves is different in each section of the penstock. Separate calculations for the velocity of the pressure wave should be made for each thickness and diameter of penstock to obtain the time required for a wave to travel to the reservoir and back to the valve.)

Velocity of pressure wave, from Eq. (21-61), is

$$U = \sqrt{\frac{E}{\rho}} \sqrt{\frac{1}{1 + ED/E_p t}}$$

$$= \sqrt{\frac{43.2 \times 10^6}{1.94}} \sqrt{\frac{1}{1 + 43.2 \times 10^6 \times 10 \times 12/(1 \times 43.2 \times 10^8)}}$$

$$= 3{,}180 \text{ fps}$$

The time required for the wave to travel to the reservoir and be reflected back to the valve = $2L/U = 6{,}000/3{,}180 \doteq 1.90$ sec.

If closure time T of the valve is less than 1.90 sec, the closure is instantaneous, and the pressure rise, from Eq. (21-63), is

$$\Delta h = \frac{U\, \Delta V}{g} = \frac{3{,}180 \times 10}{32.2} = 990 \text{ ft}$$

Assuming $T = 4.75$ sec, approximate equation (21-64) gives the following result:

$$\Delta h_g = \frac{t_i \Delta h}{T} = \frac{1.90 \times 990}{4.75} = 396 \text{ ft}$$

By arithmetic integration, the solution requires that the variation of the area of the valve with time be determined. For this example, the assumption will be made that the area varies linearly with time.

The next step is to choose a time interval: $t = 2L/2U = 0.95$ sec. (This time interval is too large for exact calculations, but it illustrates the procedure.) Hence, valve closure comprises $4.75/0.95 = 5$ increments. A table can now be set up and computations made (Table 21-10).

Assuming ΔV for the first incremental closing is 1.5 fps, Δh is calculated as follows:

$$\Delta h = \frac{U \Delta V}{g} = \frac{3,180 \times 1.5}{32.2} = 148 \text{ ft}$$

Now check to see if the assumed ΔV was correct.

$$V_1 = \frac{V_o \times A_n}{A_o} \sqrt{\frac{H_o + \Sigma \Delta h}{H_o}}$$

$$= 10 \times 0.8 \sqrt{\frac{1,000 + 148}{1,000}} = 8.54 \text{ fps}$$

$$\Delta V = V_o - V_1 = 10 - 8.54 = 1.46 \text{ fps}$$

Recalculate Δh:

$$\Delta h = \frac{3,180 \times 1.46}{32.2} = 144 \text{ ft}$$

This procedure for determining Δh is repeated for each incremental movement of the valve. Results are posted in Table 21-10. Column 7 is a total of all the wave pressures at the valve due to reflection. The waves for which heads are given in column 6 are reflected back from the reservoir and reach the valve $2L/U$ sec after they were formed. The wave pressures were Δh when formed but are $-\Delta h$ when reflected back, the difference being $-2\Delta h$. The waves for which values are given in column 7 also are reflected back to the valve every $2L/U$ sec and must be included. For example, for the fifth increment ($t = T = 4.25$), the reflected pressure is -2×234 from the third increment plus $+288$, a total of -180 ft. Note that the waves continue after the valve is closed.

Table 21-10. Calculation of Pressure Due to Gradual Closing of Valve

Time, sec	Interval Δt_i, sec	ΔV, fps	V, fps	$Q = 78.5V$, cfs	Δh at valve, ft	$2\Delta h$ reflected from reservoir, ft	Total head $H + h_o$, ft	Valve area, A_n/A_o	Change in head from static, Col. (8) minus 1,000 ft
(1)	(2)	(3)	(4)	(5)	(6)	(7)	(8)	(9)	(10)
0			10.00	785			1,000	1.00	0
0.95	0.95	1.46	8.54	670	144		1,144	0.80	144
1.90	0.95	1.67	6.87	539	165		1,309	0.60	309
2.85	0.95	2.37	4.50	353	234	−288	1,255	0.40	255
3.80	0.95	2.36	2.14	168	233	−330	1,158	0.20	158
4.75	0.95	2.14	0	0	211	−180	1,189	0.00	189
5.70	0.95	Valve closed at time $t = 4.75$				−136	1,053		53
6.65	0.95					−242	811		−189
7.60						+136	947		−53
8.55						+242	1,189		189
9.50						−136	1,053		53

For further information, see C. V. Davis and K. E. Sorensen, "Handbook of Applied Hydraulics," McGraw-Hill Book Company, New York.

Surge Tanks. It is uneconomical to design long pipelines for pressures created by water hammer or to operate a valve slowly enough to reduce these pressures. Uusually, a surge tank is installed close to valves at the end of long conduits. A surge tank is a tank containing water connected to the conduit; the water column, in effect, floats on the line.

When a valve is suddenly closed, the water in the line rushes into the surge tank. The water level in the tank rises until the increased pressure in the surge tank overcomes the momentum of the water. When a valve is suddenly opened, the surge tank supplies water to the line when the pressure drops. The section of pipe (Fig. 21-33) between the surge tank and the valve must still be designed for water hammer; however, the closure time to reduce the pressures for this section will be only a fraction of the time required without the surge tank.

Although a surge tank is one of the most commonly used devices to prevent water hammer, it is by no means the only device. Various types of relief valves and air chambers are widely used on small-diameter lines, where the pressure of water hammer may be relieved by the release of a relatively small quantity of water.

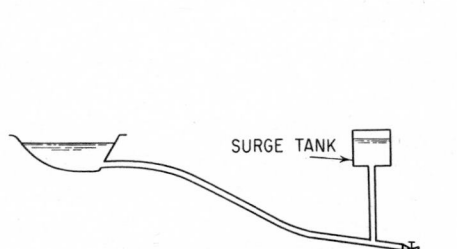

Fig. 21-33. Surge tank is placed near valve on penstock to prevent water hammer.

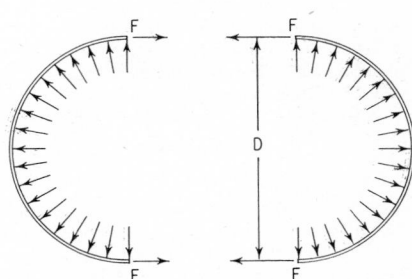

Fig. 21-34. Internal pressure in a pipe produces hoop tension.

PIPE STRESSES

21-14. Stresses Perpendicular to the Longitudinal Axis.
The stresses acting perpendicular to the longitudinal axis of a pipe are caused by either internal or external pressures on the pipe walls.

Internal pressure creates a stress commonly termed hoop tension. It may be calculated by taking a free-body diagram of a 1-in.-long strip of pipe cut by a vertical plane through the longitudinal axis (Fig. 21-34). The forces in the vertical direction cancel out. The sum of the forces in the horizontal direction is

$$pD = 2F \tag{21-66}$$

where p = internal pressure, psi
D = outside diameter of pipe, in.
F = force acting on each cut edge of pipe, lb

Hence, the stress, psi, on the pipe material is

$$f = \frac{F}{A} = \frac{pD}{2t} \tag{21-67}$$

where A = area of cut edge of pipe, sq in.
t = thickness of pipe wall, in.

From the derivation of Eq. (21-67), it would appear that the diameter used for calculations should be the inside diameter. However, Eq. (21-67) is not theoretically exact and gives stresses slightly lower than those actually developed. For this reason the outside diameter often is used (see also Art. 6-10).

Equation (21-67) is exact for all practical purposes when D/t is equal to or greater than 50. If D/t is less than 10, this equation will usually be quite conservative and therefore will yield an uneconomical design. For steel pipes, Eq. (21-67) gives directly the thickness required to resist internal pressure.

For concrete pipes, this analysis is approximate, however, since concrete cannot resist large tensile stresses. The force F must be carried by steel reinforcing. The internal diameter is used in Eq. (21-67) for concrete pipe.

When a pipe has external pressure acting on it, the analysis is much more complex, because the pipe material no longer acts in direct tension. The external pressure creates bending and compressive stresses that cause buckling problems. Equation (21-68) gives the thickness required for an empty steel pipe to resist buckling under uniform external pressure. (S. Timoshenko, "Strength of Materials," Van Nostrand Reinhold Company, New York. For calculation of pressure due to soil loading, see Art. 7-14.)

$$t = D \sqrt[3]{\frac{6p}{E}} \tag{21-68}$$

where t = shell thickness, in.
  D = diameter of pipe, ft
  p = uniform external pressure, psf
  E = modulus of elasticity, psi

21-15. Stresses Parallel to the Longitudinal Axis. If a pipe is supported on piers, it acts like a beam. The stresses created can be calculated from the bending moment and shear equations for a continuous circular hollow beam. This stress is usually not critical in high-head pipes. However, thin-walled steel pipes usually require stiffening to prevent buckling and excessive deflection from the concentrated loads.

21-16. Temperature Expansion. If a pipe is subject to a wide range of temperatures, the stress due to temperature variation must be designed for or expansion joints provided. The stress, psi, due to a temperature change is

$$f = cE\Delta T \tag{21-69}$$

where E = modulus of elasticity of pipe material, psi
  ΔT = temperature change from installation temperature
  c = coefficient of thermal expansion of pipe material
The movement that should be allowed for, if expansion joints are to be used, is

$$\Delta L = Lc\Delta T \tag{21-70}$$

where ΔL = movement in length L of pipe
  L = length between expansion joints

21-17. Forces Due to Pipe Bends. It is common practice to use thrust blocks in pipe bends to take the forces on the pipe caused by the momentum change and the unbalanced internal pressure of the water.

In all bends, there will be a slight loss of head due to turbulence and friction. This loss will cause a pressure change across the bend, but it is usually small enough to be neglected. When there is a change in the cross-sectional area of the pipe, there will be an additional pressure

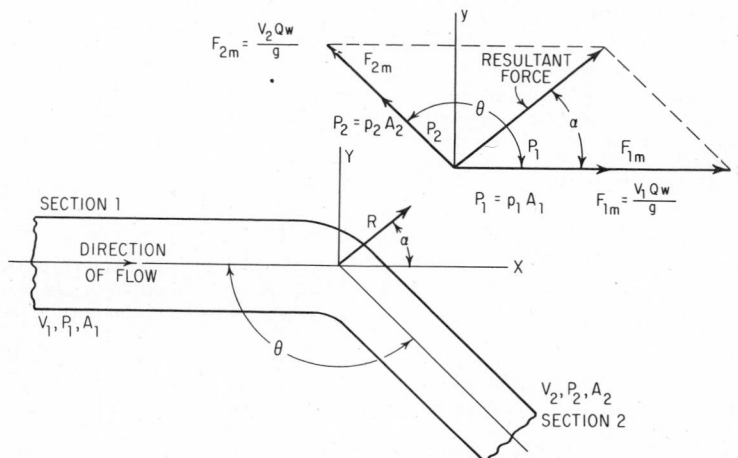

Fig. 21-35. Forces produced by flow at a pipe bend and change in diameter.

change that can be calculated with the Bernoulli equation (see Example 6, Art. 21-6). In this case, the pressure differential may be large and must be considered.

The force diagram in Fig. 21-35 is a convenient method for finding the resultant force on a bend. The forces can be resolved into X and Y components to find the magnitude and direction of the resultant force on the pipe.

In Fig. 21-35:

V_1 = velocity before a change in size of pipe, fps
V_2 = velocity after a change in size of pipe, fps
p_1 = pressure before a bend or size change in pipe, psf
p_2 = pressure after a bend or size change in pipe, psf
A_1 = area before a size change in the pipe, sq ft
A_2 = area after a size change in the pipe, sq ft
F_{2m} = momentum of water in section 2 = $V_2 Qw/g$
F_{1m} = momentum of water in section 1 = $V_1 Qw/g$
P_2 = pressure of water in section 2 times area of section 2 = $p_2 A_2$
P_1 = pressure of water in section 1 times area of section 1 = $p_1 A_1$
w = unit weight of liquid, lb per cu ft
Q = discharge, cfs

If the pressure loss in the bend is neglected and there is no change in magnitude of velocity around the bend, Eqs. (21-71) and (21-72) give a quick solution.

$$R_F = 2A \left(w \frac{V^2}{g} + p \right) \cos \frac{\theta}{2} \qquad (21\text{-}71)$$

$$\alpha = \frac{\theta}{2} \qquad (21\text{-}72)$$

where R_F = resultant force on the bend, lb
α = angle R_F makes with F_{1m}
p = pressure, psf
w = unit weight of water, 62.4 lb per cu ft
V = velocity of flow, fps
g = acceleration due to gravity, 32.2 ft per sec^2
A = area of pipe, sq ft
θ = angle between the pipes ($0° \leq \theta \leq 180°$)

Although thrust blocks are normally used to take the force on bends, in many cases the pipe material takes this force. The stress caused by this force is directly additive to other stresses along the longitudinal axis of the pipe. In small pipes, the force caused by bends can easily be carried by the pipe material; however, the joints must also be able to take these forces.

CULVERTS

A culvert is a closed conduit for the passage of surface drainage under a highway, a railroad, canal, or other embankment. The slope of a culvert and its inlet and outlet conditions are usually determined by the topography of the site. Because of the many combinations obtained by varying the entrance conditions, exit conditions, and slope, no single formula can be given that will apply to all culvert problems.

The basic method for determining discharge through a culvert is by applying the Bernoulli equation between a point just outside the entrance and a point somewhere downstream. An understanding of uniform and nonuniform flow is necessary to understand culvert flow fully. However, an exact theoretical analysis, involving detailed calculation of drawdown and backwater curves, is usually unwarranted because of the relatively low accuracy attainable in determining runoff. Neglecting drawdown and backwater curves does not seriously affect the accuracy but greatly simplifies the calculations.

21-18. Culverts on Critical Slope or Steeper.   Entrance Submerged or Unsubmerged but Free Exit. If a culvert is on critical slope or steeper, that is, the normal depth is equal to or less than the critical depth (Art. 21-23), the discharge will be entirely dependent on the entrance conditions (Fig. 21-36). Increasing the slope of the culvert past critical slope (the slope just sufficient to maintain flow at critical depth) will decrease the depth of flow downstream from the entrance. But the increased slope will not increase the amount of water entering the culvert, since the entrance depth will remain at critical.

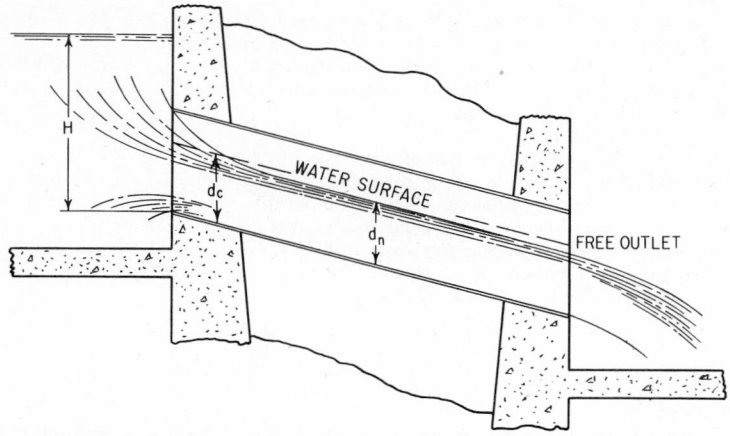

Fig. 21-36. Flow through a culvert with free discharge. Normal depth d_n is less than critical depth d_c; slope is greater than critical. Discharge depends on type of inlet and head H.

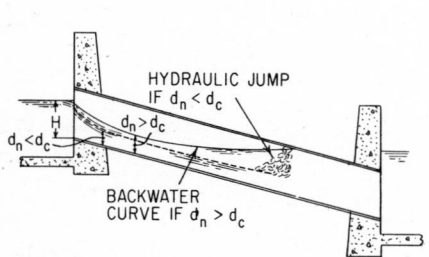

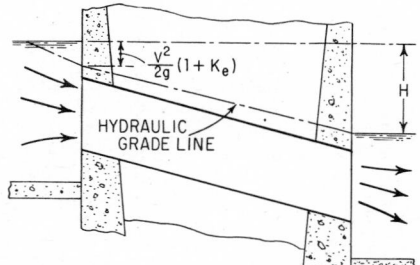

Fig. 21-37. Flow through a culvert with entrance unsubmerged but exit submerged. When slope is less than critical, open-channel flow takes place, and $d_n > d_c$. When slope exceeds critical, flow depends on inlet conditions, and $d_n < d_{dc}$.

Fig. 21-38. With entrance and exit submerged, normal pipe flow occurs in a culvert, and discharge is independent of slope. The fluid flows under pressure. Discharge may be determined from Bernoulli and Manning equations.

The discharge is given by the equation for flow through an orifice if the entrance is submerged, or by the equation for flow over a weir if the entrance is not submerged. Coefficients of discharge for weirs and orifices give good results, but they do not cover the entire range of entry conditions encountered in culvert problems. For this reason, charts and nomographs have been developed and are used almost exclusively in design. ("Handbook of Concrete Culvert Pipe Hydraulics," Portland Cement Association.)

Entrance Unsubmerged but Exit Submerged. In this case, the submergence of the exit will cause a hydraulic jump to occur in the culvert (Fig. 21-37). The jump will not affect the culvert discharge, and the control will still be at the inlet.

Entrance and Exit Submerged. When both the exit and entrance are submerged (Fig. 21-38), the culvert flows full, and the discharge is independent of the slope. This is normal pipe flow and is easily solved by using the Manning or Darcy-Weisbach formula for friction loss [Eq. (21-33d) or (21-30)]. From the Bernoulli equation for the entrance and exit, and the Manning equation for friction loss, the following equation is obtained:

$$H = (1 + K_e)\frac{V^2}{2g} + \frac{V^2 n^2 L}{2.21 R^{4/3}} \tag{21-73a}$$

$$V = \sqrt{\frac{H}{(1 + K_e/2g) + (n^2 L/2.21 R^{4/3})}} \tag{21-73b}$$

where H = elevation difference between headwater and tailwater, ft
  V = velocity in culvert, fps
  g = acceleration due to gravity, 32.2 ft per sec²
  K_e = entrance-loss coefficient (Art. 21-20)
  n = Manning's roughness coefficient
  L = length of culvert, ft
  R = hydraulic radius of culvert, ft
This equation can be solved directly since the velocity is the only unknown.

21-19. Culverts with Slope Less than Critical.   Critical slope is the slope just sufficient to maintain flow at critical depth (Art. 21-23).

Entrance Submerged or Unsubmerged but Free Exit.   For these conditions, depending on the head, the flow can be either pressure or open-channel.

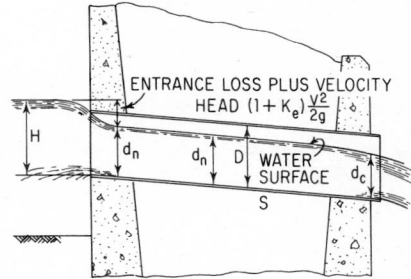

Fig. 21-39. Flow through culvert with free discharge and normal depth d_n greater than critical depth d_c when entrance is unsubmerged or slightly submerged. Open-channel flow occurs and discharge depends on head H, loss at entrance, and slope of culvert.

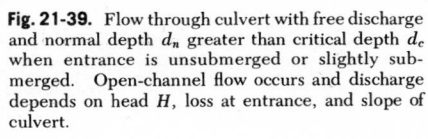

The discharge, for the open-channel condition (Fig. 21-39), is obtained by writing the Bernoulli equation for a point just outside the entrance and a point a short distance downstream from the entrance. Thus,

$$H = K_e \frac{V^2}{2g} + \frac{V^2}{2g} + d_n \tag{21-74}$$

The velocity can be determined from the Manning equation:

$$V^2 = \frac{2.2SR^{4/3}}{n^2} \tag{21-75}$$

Substituting this into Eq. (21-74) yields

$$H = (1 + K_e)\frac{2.2}{2gn^2}SR^{4/3} + d_n \tag{21-76}$$

where H = head on entrance measured from bottom of culvert, ft
  K_e = entrance-loss coefficient (Art. 21-20)
  S = slope of energy grade line, which for culverts is assumed to equal the slope of the bottom of the culvert
  R = hydraulic radius of culvert, ft
  d_n = normal depth of flow, ft
To solve Eq. (21-76), it is necessary to try different values of d_n and corresponding values of R until a value is found that satisfies the equation. If the head on a culvert is high, a value of d_n less than the culvert diameter will not satisfy Eq. (21-76). This means the flow is under pressure (Fig. 21-40), and discharge is given by Eq. (21-73).

When the depth of the water is slightly below the top of the culvert, there is a range of unstable flow fluctuating between pressure and open-channel. If this condition exists, it is good practice to check the discharge for both pressure flow and open-channel flow. The condition that gives the lesser discharge should be assumed to exist.

Short Culvert with Free Exit.   When a culvert on a slope less than critical has a free exit, there will be drawdown of the water surface at the exit and for some distance upstream. The magnitude of the drawdown depends on the friction slope of the culvert and the difference between the critical and normal depths. If the friction slope approaches critical, the difference between normal depth and critical depth is small (Fig. 21-39), and the drawdown will not

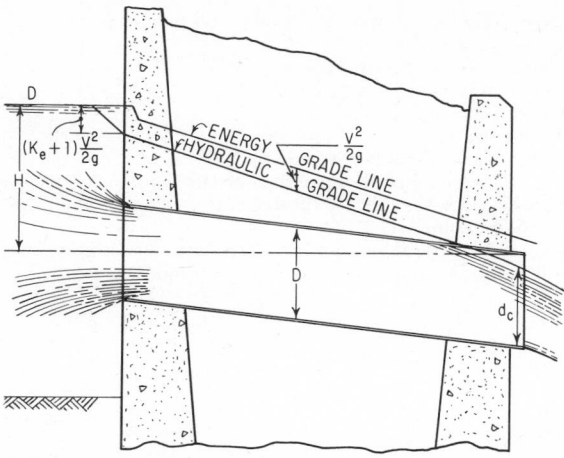

$(K_e+1)\dfrac{v^2}{2g}$

ENERGY GRADE LINE

HYDRAULIC GRADE LINE

$\dfrac{v^2}{2g}$

D

H

D

d_c

Fig. 21-40. Flow through culvert with free discharge and normal depth d_n greater than critical depth d_c when entrance is deeply submerged. The culvert flows full. Discharge is given by equations for pipe flow.

extend for any significant distance upstream. When the friction slope is flat, there will be a large difference between normal and critical depth. The effect of the drawdown will extend a greater distance upstream and may reach the entrance of a short culvert (Fig. 21-41). This drawdown of the water level in the entrance of the culvert will increase the discharge, causing it to be about the same as for a culvert on a slope steeper than critical (Art. 21-18). Most culverts, however, are on too steep a slope for the backwater to have any effect for an appreciable distance upstream.

Entrance Unsubmerged but Exit Submerged. If the level of submergence of the exit is well below the bottom of the entrance (Fig. 21-37), the backwater from the submergence will not extend to the entrance. The discharge for this case will be given by Eq. (21-76).

If the level of submergence of the exit is close to the level of the entrance, it may be assumed that the backwater will cause the culvert to flow full and a pipe flow condition will result. The discharge for this case is given by Eq. (21-73).

When the level of submergence falls between these two cases and the project does not warrant a trial approach with backwater curves, it is good practice to assume the condition that gives the lesser discharge.

21-20. Entrance Losses for Culverts. (See also Arts. 21-18 and 21-19.) The following are coefficients of entrance loss K_e for some typical entrance conditions:

Sharp-edged projecting inlet 0.9
Flush inlet, square edge 0.5
Concrete pipe, groove or bell, projecting 0.15
Concrete pipe, groove or bell, flush 0.10
Well-rounded entrance 0.08

The above values are for culverts flowing full. When the entrance is not submerged, the coefficients are usually somewhat lower. But because of the many unknowns entering into culvert problems, the values tabulated can be used for submerged or unsubmerged cases without sacrificing accuracy.

Example 21-9. *Given:* Maximum head above the top of the culvert = 5 ft, slope = 0.01, length = 300 ft, discharge Q = 40 cfs, n = 0.013, and free exit. *Find:* size of culvert.

Procedure: First assume a trial culvert; then investigate the assumed section to find its discharge. Assume a 2×2 ft concrete box section. Calculate Q, assuming entrance control, with Eq. (21-44) for discharge through an orifice. The coefficient of discharge C for a 2-ft-square orifice is about 0.6. Head h on center line of entrance = $5 + \frac{1}{2} \times 2 = 6$ ft. Entrance area $a = 2 \times 2 = 4$ sq ft.

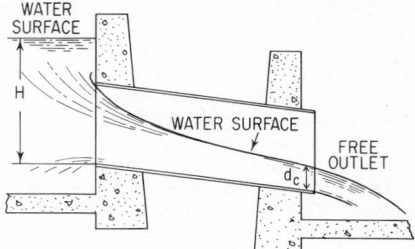

Fig. 21-41. Drawdown of water surface at free exit of a short culvert with slope less than critical affects entrance depth and controls discharge.

$$Q = Ca \sqrt{2gh} = 0.6 \times 4\sqrt{64.4 \times 6} = 47.2 \text{ cfs}$$

For entrance control, the flow must be supercritical and d_n must be less than 2 ft. First find d_n.

For the purpose of calculating the hydraulic radius, assume the depth is slightly less than 2 ft, since this will give the maximum possible value of the hydraulic radius for this culvert.

$$R = \frac{\text{area of flow}}{\text{wetted perimeter}} = \frac{2 \times 2}{6} = 0.67 \text{ ft}$$

Application of Eq. (21-33a) gives

$$V = \frac{1.486}{n} R^{2/3} S^{1/2} = \frac{1.486}{0.013} \times 0.67^{2/3} \times 0.01^{1/2} = 8.76 \text{ fps}$$

$$d_n = \frac{Q}{V \times \text{width}} = \frac{47.2}{8.76 \times 2} = 2.69 \text{ ft}$$

Since d_n is greater than the culvert depth, the flow is under pressure, and the entrance will not control.

Since the culvert is under pressure, Eq. (21-73a) applies. But

$$H = 5 + 0.01 \times 300 = 8 \text{ ft}$$

(see Fig. 21-40). The hydraulic radius for pipe flow is $R = 2^2/8 = \frac{1}{2}$. Substitution in Eq. (21-73a) yields

$$8 = \frac{1.5V^2}{2g} + 0.0575V^2 = 0.0808V^2$$

$$V = \sqrt{8/0.0808} = 9.95 \text{ fps}$$

$$Q = Va = 9.95 \times 4 = 39.8 \text{ cfs}$$

Since the discharge of the assumed culvert section under the allowable head equals the maximum expected runoff, the assumed culvert would be satisfactory.

OPEN-CHANNEL FLOW

21-21. Basic Elements of Open Channels.   Free surface flow, or open-channel flow, includes all cases of flow in which the liquid surface is open to the atmosphere. Thus, flow in a pipe is open-channel flow if the pipe is only partly full.

A **uniform channel** is one of constant cross section. It has **uniform flow** if the grade, or slope, of the water surface is the same as that of the channel. Hence, depth of flow is constant throughout. **Steady flow** in a channel occurs if the depth at any location remains constant with time.

The **discharge** Q at any section is defined as the volume of water passing that section per unit of time. It is expressed in cubic feet per second, cfs, and is given by

$$Q = VA \tag{21-77}$$

where V = *average velocity*, fps
A = cross-sectional *area* of flow, sq ft

When the discharge is constant, the flow is said to be **continuous** and, therefore,

$$Q = V_1 A_1 = V_2 A_2 = \cdots \qquad (21\text{-}78)$$

where the subscripts designate different channel sections. Equation (21-78) is known as the continuity equation for continuous steady flow.

In a uniform channel, **varied flow** occurs if the longitudinal water surface profile is not parallel with the channel bottom. Varied flow exists within the limits of backwater curves, within a hydraulic jump, and within a channel of changing slope or discharge.

Depth of flow d is taken as the vertical distance, ft, from the bottom of a channel to the water surface. The **water perimeter** is the length, ft, of a line bounding the cross-sectional area of flow, minus the free surface width. The **hydraulic radius** R equals the area of flow divided by its wetted perimeter. The **average velocity** of flow V is defined as the discharge divided by the area of flow,

$$V = \frac{Q}{A} \qquad (21\text{-}79)$$

The velocity head H_V, ft, is generally given by

$$H_V = \frac{V^2}{2g} \qquad (21\text{-}80)$$

where V = average velocity from Eq. (21-79)

  g = acceleration due to gravity, 32.2 ft per sec²

Velocity heads of individual filaments of flow vary considerably above and below the velocity head based on the average velocity. Since these velocities are squared in head and energy computations, the average of the velocity heads will be greater than the average-velocity head. The **true velocity head** may be expressed as

$$H_{Va} = \alpha \frac{V^2}{2g} \qquad (21\text{-}81)$$

where α is an empirical coefficient that represents the degree of turbulence. Experimental data indicate that α may vary from about 1.03 to 1.36 for prismatic channels. It is, however, normally taken as 1.00 for practical hydraulic work and is evaluated only for precise investigations of energy loss.

The total energy per pound of water relative to the bottom of the channel at a vertical section is called the **specific energy head** H_e. It is composed of the depth of flow at any point, plus the velocity head at the point. It is expressed in feet as

$$H_e = d + \frac{V^2}{2g} \qquad (21\text{-}82)$$

A longitudinal profile of the elevation of the specific energy head is called the **energy grade line**, or the **total-head line**. A longitudinal profile of the water surface is called the **hydraulic**

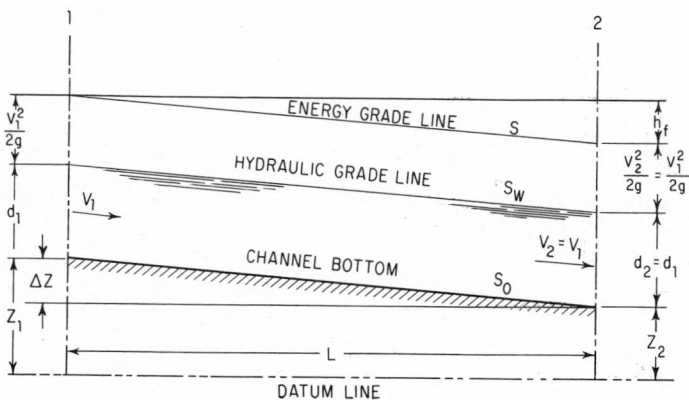

Fig. 21-42. Flow characteristics for uniform open-channel flow.

grade line. The vertical distance between these profiles at any point equals the velocity head at that point.

Figure 21-42 shows a section of uniform open channel for which the slopes of the water surface S_W and of the energy grade line S equal the slope of the channel bottom S_o.

Loss of head due to friction h_f in channel length L equals the drop in elevation of the channel ΔZ in the same distance.

21-22. Normal Depth of Flow. The depth of equilibrium flow that exists in the channel of Fig. 21-42 is called the normal depth d_n. This depth is unique for specific discharge and channel conditions. It may be computed by a trial-and-error process when the channel shape, slope, roughness, and discharge are known. A form of the Manning equation has been suggested for this calculation. (V. T. Chow, "Open-Channel Hydraulics," Mc-Graw-Hill Book Company, New York.)

$$AR^{2/3} = \frac{Qn}{1.486S^{1/2}} \tag{21-83}$$

where A = area of flow, sq ft
R = hydraulic radius, ft
Q = amount of flow or discharge, cfs
n = Manning's roughness coefficient
S = slope of energy grade line or loss of head, ft, due to friction per lin ft of channel

$AR^{2/3}$ is referred to as a section factor. Determination of d_n for uniform channels is simplified by use of tables that relate d_n to the bottom width of a rectangular or trapezoidal channel, or to the diameter of a circular channel. (See, for example, H. W. King and E. F. Brater, "Handbook of Hydraulics," McGraw-Hill Book Company, New York.)

In a prismatic channel of gradually increasing slope, normal depth decreases downstream, as shown in Fig. 21-43, and specific energy first decreases and then increases as shown in Fig. 21-44.

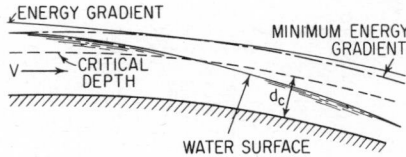

Fig. 21-43. Prismatic channel with gradually increasing bottom slope. Normal depth decreases downstream as slope increases.

The specific energy is high initially where the channel is relatively flat, because of the large normal depth (Fig. 21-43). As the depth decreases downstream, the specific energy also decreases. It reaches a minimum at the point where the flow satisfies the equation

$$\frac{A^3}{T} = \frac{Q^2}{g} \tag{21-84}$$

in which T is the top width of the channel, ft. For a rectangular channel, Eq. (21-84) reduces to

$$\frac{d}{2} = \frac{V^2}{2g} \tag{21-85}$$

where $V = Q/A$ = mean velocity of flow, cfs
d = depth of flow, ft
This indicates that the specific energy is a minimum where the normal depth equals twice the velocity head. As the depth continues to decrease in the downstream direction, the specific energy increases again because of the higher velocity head (Fig. 21-44).

21-23. Critical Depth. The depth of flow that satisfies Eq. (21-84) is called the *critical depth* d_c. For a given value of specific energy, the critical depth gives the greatest discharge, or, conversely, for a given discharge, the specific energy is a minimum for the critical depth (Fig. 21-44).

In the section of mild slope upstream from the critical-depth point in Fig. 21-43, the depth is greater than critical. The flow there is called *subcritical flow*, indicating that the velocity is less than that at critical depth. In the section of steeper slope below the critical-depth point, the depth is below critical. The velocity there exceeds that at critical depth, and flow is *supercritical*.

Critical depth may be computed for a uniform channel once the discharge is known. Determination of this depth is independent of the channel slope and roughness, since critical depth simply represents a depth for which the specific energy head is a minimum. Critical depth may be calculated by trial and error with Eq. (21-84), or it may be found directly from tables (H. W. King and E. F. Brater, "Handbook of Hydraulics," McGraw-Hill Book Company, New York). For rectangular channels, Eq. (21-84) may be reduced to

$$d_c = \sqrt[3]{\frac{Q^2}{b^2 g}}$$

(21-86)

where d_c = critical depth, ft

Q = quantity of flow or discharge, cfs

b = width of channel, ft

Critical slope is the slope of the channel bed that will maintain flow at critical depth. Such slopes should be avoided in channel design because flow near critical depth tends to be unstable and exhibits turbulence and water-surface undulations.

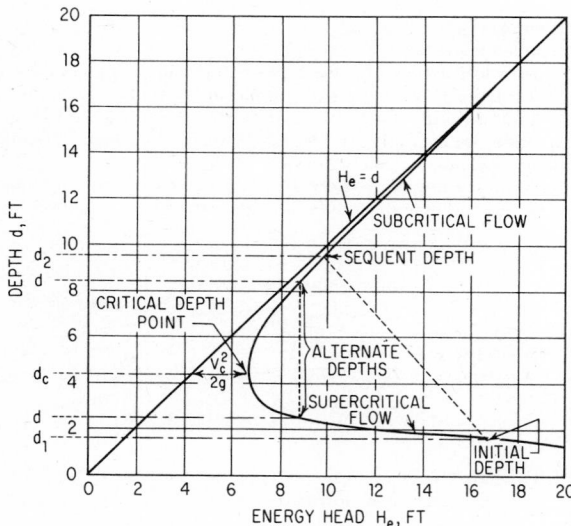

Fig. 21-44. Specific energy head H_e changes with depth for constant discharge in a rectangular channel of changing slope and is a minimum at critical depth.

Critical depth, once calculated, should be plotted for the full length of a uniform channel, regardless of slope, to determine whether the normal depth at any section is subcritical or supercritical. [As indicated by Eq. (21-85), if the velocity head is less than half the depth in a rectangular channel, flow is subcritical, but if velocity head exceeds half the depth, flow is supercritical.] If channel configuration is such that the normal depth must go from below to above critical, a hydraulic jump will occur, along with a high loss of energy. Critical depth will change if the channel cross section changes, and so the possibility of a hydraulic jump in the vicinity of a transition should be investigated.

For every depth greater than critical depth, there is a corresponding depth less than critical that has an identical value of specific energy (Fig. 21-44). These depths of equal energy are called *alternate depths*. The fact that the energy is the same for alternate depths does not mean that the flow may switch from one alternate depth to the other and back again. Flow will always seek to attain the normal depth in a uniform channel and will maintain that depth unless an obstruction is met.

It can be seen from Fig. 21-44 that any obstruction to flow that causes a reduction in total head causes subcritical flow to experience a drop in depth and supercritical flow, an increase in depth.

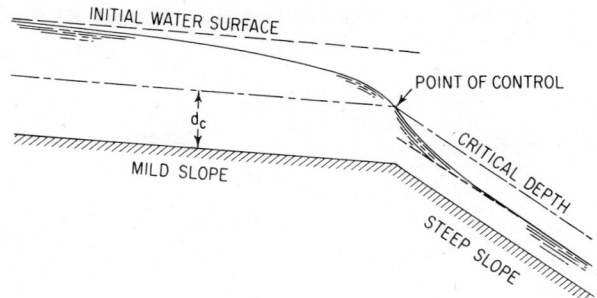

Fig. 21-45. Change in flow stage from subcritical to supercritical occurs gradually.

If supercritical flow exists momentarily on a flat slope, because of a sudden grade change in the channel (Fig. 21-53b), the depth will increase suddenly from the depth below critical to a depth above critical in a hydraulic jump. The depth following the jump will not be the alternate depth, however. There has been a loss of energy in making the jump. The new depth is said to be sequent to the initial depth, indicating an irreversible occurrence. There is no similar phenomenon that allows a sudden change in depth from subcritical flow to supercritical flow with a corresponding gain in energy. Such a change occurs gradually, without turbulence, as indicated in Fig. 21-45.

21-24. Manning's Equation for Open Channels. One of the more popular of the numerous equations developed for determination of discharge in an open channel is Manning's variation of the Chezy formula,

$$V = C \sqrt{RS} \tag{21-87}$$

where R = hydraulic radius, ft
V = mean velocity of flow, fps
S = slope of energy grade line or loss of head due to friction per lin ft of channel
C = Chezy roughness coefficient
Manning proposed

$$C = \frac{1.486 R^{1/6}}{n} \tag{21-88}$$

where n is the coefficient of roughness in the earlier Ganguillet-Kutter formula (see also Art. 21-25). When Manning's C is used in the Chezy formula, the familiar Manning equation results:

$$V = \frac{1.486}{n} R^{2/3} S^{1/2} \tag{21-89}$$

Since the discharge $Q = VA$, Eq. (21-89) may be written

$$Q = \frac{1.486}{n} AR^{2/3} S^{1/2} \tag{21-90}$$

where A = area of flow, sq ft
Q = quantity of flow, cfs

21-25. Roughness Coefficient for Open Channels. (See also Art. 21-24.) Values of the roughness coefficient n have been determined for a wide range of natural and artificial channel construction materials. Excerpts from a table of these coefficients taken from V. T. Chow, "Open-Channel Hydraulics," McGraw-Hill Book Company, New York, are presented in Table 21-11. Dr. Chow compiled data for his table from work by R. E. Horton and from technical bulletins published by the U. S. Department of Agriculture.

Channel roughness does not remain constant with time or even with depth of flow. An unlined channel excavated in earth may have one n value when first put in service and another when overgrown with weeds and brush. If an unlined channel is to have a reasonably constant n value over its useful lifetime, a continuing maintenance program must be provided.

Table 21-11. Values of the Roughness Coefficient n for Use in the Manning Equation

	Min	Avg	Max
A. Open-channel flow in closed conduits			
1. Corrugated-metal storm drain	0.021	0.024	0.030
2. Cement-mortar surface	0.011	0.013	0.015
3. Concrete (unfinished)			
a. Steel form	0.012	0.013	0.014
b. Smooth wood form	0.012	0.014	0.016
c. Rough wood form	0.015	0.017	0.020
B. Lined channels			
1. Metal			
a. Smooth steel (unpainted)	0.011	0.012	0.014
b. Corrugated	0.021	0.025	0.030
2. Wood			
a. Planed, untreated	0.010	0.012	0.014
3. Concrete			
a. Float finish	0.013	0.015	0.016
b. Gunite, good section	0.016	0.019	0.023
c. Gunite, wavy section	0.018	0.022	0.025
4. Masonry			
a. Cemented rubble	0.017	0.025	0.030
b. Dry rubble	0.023	0.032	0.035
5. Asphalt			
a. Smooth	0.013	0.013	
b. Rough	0.016	0.016	
C. Unlined channels			
1. Excavated earth, straight and uniform			
a. Clean, after weathering	0.018	0.022	0.025
b. With short grass, few weeds	0.022	0.027	0.033
c. Dense weeds, high as flow depth	0.050	0.080	0.120
d. Dense brush, high stage	0.080	0.100	0.140
2. Dredged earth			
a. No vegetation	0.025	0.028	0.033
b. Light brush on banks	0.035	0.050	0.060
3. Rock cuts			
a. Smooth and uniform	0.025	0.035	0.040
b. Jagged and irregular	0.035	0.040	0.050

Shallow flow in an unlined channel will result in an increase in the effective n value if the channel bottom is covered with large boulders or ridges of silt, since these projections would then have a larger influence on the flow than for deep flow. A deeper-than-normal flow will also result in an increase in the effective n value if there is a dense growth of brush along the banks within the path of flow. When channel banks are overtopped during a flood, the effective n value increases as the flow spills into heavy growth bordering the channel. The roughness of a lined channel experiences change with age, because of both deterioration of the surface and accumulation of foreign matter; therefore, the average n values given in Table 21-11 are recommended only for well-maintained channels. (See also Art. 21-9 and Table 21-4.)

21-26. Water-Surface Profiles for Gradually Varied Flow. Examples of various surface curves possible with gradually varied flow are shown in Fig. 21-46. These surface profiles represent backwater curves that form under the conditions illustrated in examples (a) through (r).

These curves are divided into five groups according to the slope of the channel in which they appear (Art. 21-23). Each group is labeled with a letter descriptive of the slope: M for mild (subcritical), S for steep (supercritical), C for critical, H for horizontal, and A for adverse. The two dashed lines that appear in the left-hand figure for each of these classes are the *normal-depth line* N.D.L. and the *critical-depth line* C.D.L. The N.D.L. and C.D.L. are identical for a channel of critical slope, and the N.D.L. is replaced by a horizontal line, at an arbitrary elevation, for the channels of horizontal or adverse slope.

There are three types of surface-profile curves possible in channels of mild or steep slope, and two types for channels of critical, horizontal, and adverse slope.

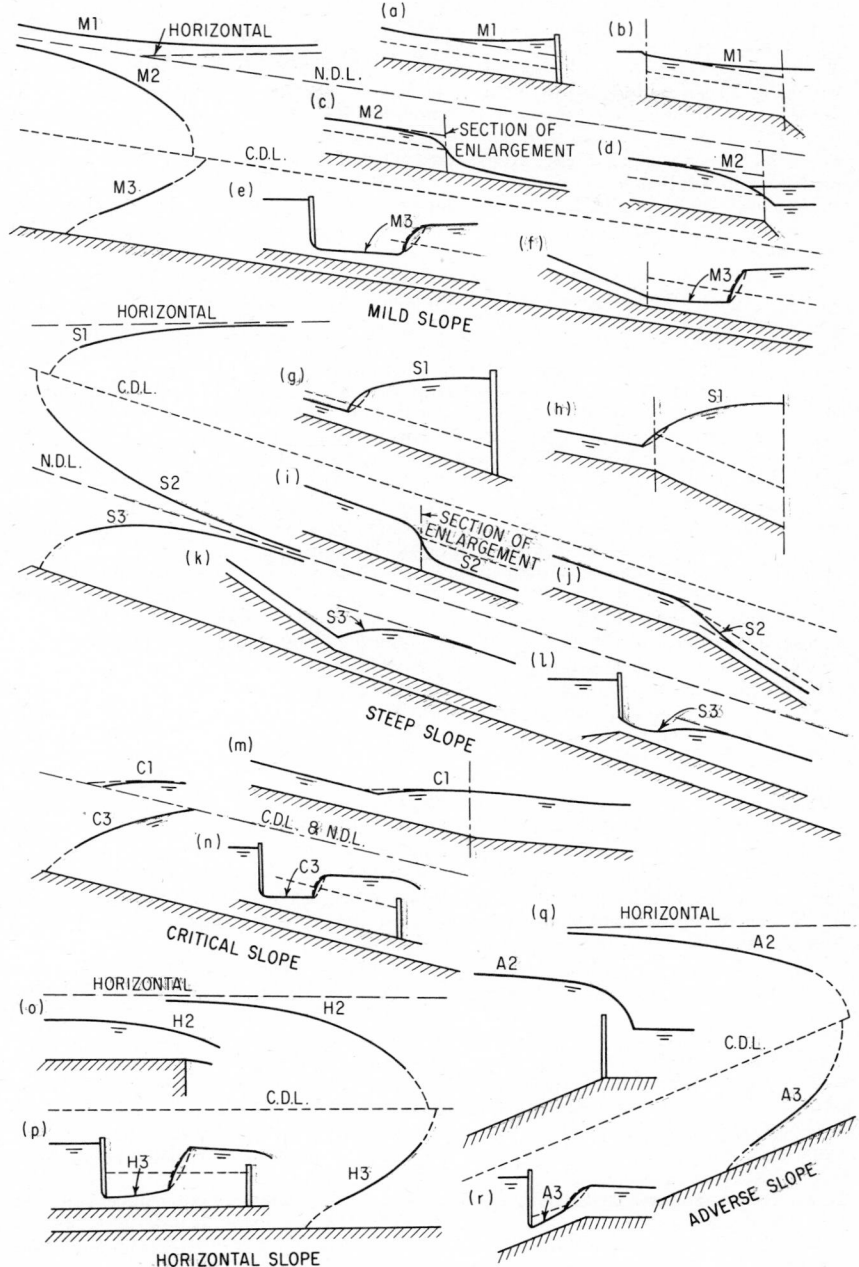

Fig. 21-46. Typical flow profiles for channels with various slopes. N.D.L. indicates normal-depth line; C.D.L., critical-depth line.

The M1 curve is the familiar surface profile from which all backwater curves derive their name, and is the most important from a practical point of view. It forms above the normal-depth line. It occurs when water is backed up a stream by high water in the downstream channel, as shown in Fig. 21-46a and b.

The M2 curve forms between the normal- and critical-depth lines. It occurs under conditions shown in Fig. 21-46c and d, corresponding to an increase in channel width or slope.

The M3 curve forms between the channel bottom and the critical-depth line. It terminates in a hydraulic jump, except where a drop-off in the channel occurs before a jump can form. Examples of the M3 curve appear in Fig. 21-46e and f (a partly opened sluice gate and a decrease in channel slope, respectively).

The S1 curve begins at a hydraulic jump and extends downstream, becoming tangent to a horizontal line (Fig. 21-46g and h) under channel conditions corresponding to those for Fig. 21-46a and b.

The S2 curve, commonly called a drawdown curve, extends downstream from the critical depth and becomes tangent to the normal-depth line under conditions corresponding to those for Fig. 21-46i and j.

The S3 curve is of the transitional type. It forms between two normal depths of less than critical depth under conditions corresponding to those for Fig. 21-46k and l.

Examples in Fig. 21-46m through r show conditions for the formation of C, H, and A profiles.

The curves in Fig. 21-46 approach the normal-depth line asymptotically and terminate abruptly in a vertical line as they approach the critical depth. The curves that approach the bottom intersect it at a definite angle but are imaginary near the bottom since velocity would have to be infinite to satisfy Eq. (21-77) if the depth were zero. The curves are shown dotted near the critical-depth line as a reminder that this portion of the curve does not possess the same degree of accuracy as the rest of the curve because of neglect of vertical components of velocity in the calculations. These curves either start or end at what is called a point of control.

A **point of control** is a physical location in a prismatic channel at which the depth of steady flow may readily be determined. This depth is usually different from the normal depth for the channel because of a grade change, gate, weir, dam, free overfall, or other feature at that location that causes a backwater curve to form. Calculations for the length and shape of the surface profile of a backwater curve start at this known depth and location and proceed either up or downstream depending on the type of flow. For subcritical flow conditions, the curve proceeds upstream from the point of control in a true backwater curve. The surface curve that occurs under supercritical flow conditions proceeds downstream from the point of control and might better be called a downwater curve.

The point of control is always at the downstream end of a backwater curve in subcritical flow and at the upstream end for supercritical flow. This may be explained as follows: A backwater curve may be thought of as being the result of some disruption of uniform flow that causes a wave of disturbance in the channel. The wave travels at a speed, known as its **celerity**, which always equals the critical velocity for the channel. If a disturbance wave attempts to move upstream against supercritical flow (flow moving at a speed greater than critical), it will be swept downstream by the flow and have no effect on conditions upstream. A disturbance wave is held steady by critical flow and moves upstream in subcritical flow.

When a hydraulic jump occurs on a mild slope and is followed by a free overfall (Fig. 21-52), backwater curves form both before and after the jump. The point of control for the curve in the supercritical region above the jump will be located at the vena contracta that forms just below the sluice gate. The point of control for the backwater curve in the subcritical region below the jump is at the free overfall where critical depth occurs. Computations for these backwater curves are carried toward the jump from their respective points of control and are extended across the jump to aid in determining its exact location. But a backwater curve cannot be calculated through a hydraulic jump from either direction. The surface profiles involved terminate abruptly in a vertical line as they approach the critical depth, and a hydraulic jump always occurs across critical depth.

21-27. Backwater-Curve Computations. The solution of a backwater curve involves computation of a gradually varied flow profile. Methods of solution available include the graphical-integration method, direction-integration method, and step method. Explanations of both the graphical- and direct-integration methods may be found in V. T. Chow, "Open-Channel Hydraulics," McGraw-Hill Book Company, New York.

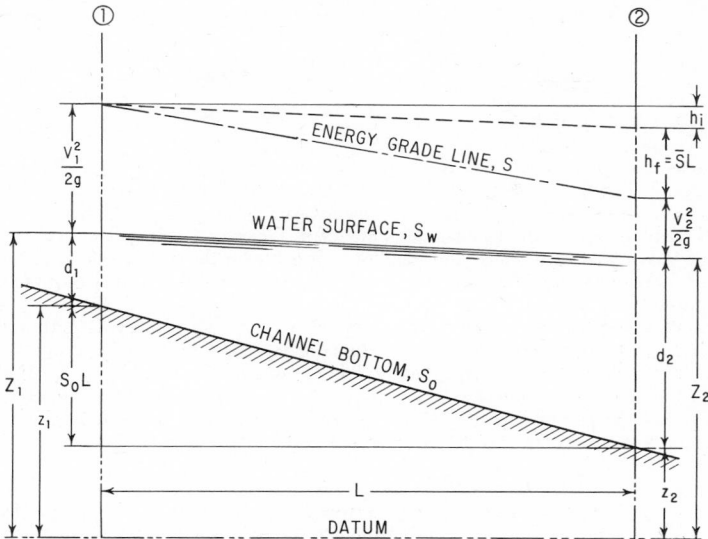

Fig. 21-47. Channel with constant discharge gradually varying in area.

Two variations of the step method include the direct or uniform method and the standard method. They are simple and widely used.

For step-method computations, the channel is divided into short lengths, or reaches, with relatively small variation. In a series of steps starting from a point of control, each reach is solved in succession. Step methods have been developed for channels with uniform or varying cross sections.

Direct step method of backwater computation involves solving for an unknown length of channel between two known depths. The procedure is applicable only to uniform prismatic channels with gradually varying area of flow.

For the section of channel in Fig. 21-47, Bernoulli's equation for the reach between sections 1 and 2 is

$$S_o L + d_1 + \frac{V_1^2}{2g} = d_2 + \frac{V_2^2}{2g} + \bar{S}L \qquad (21\text{-}91)$$

where V_1 and V_2 = mean velocities of flow at sections 1 and 2, fps
d_1 and d_2 = depths of flow at sections 1 and 2, ft
g = acceleration due to gravity, 32.2 ft per sec²
$\bar{S}$ = average head loss due to friction, ft per ft of channel
S_o = slope of channel bottom
L = length of channel between sections 1 and 2, ft

Note that $S_o L = \Delta z$, the change in elevation, ft, of the channel bottom between sections 1 and 2, and $\bar{S}L = h_f$, the head loss, ft, due to friction in the same reach. (For uniform, prismatic channels, h_i, the eddy loss, is negligible and can be ignored.) $\bar{S}$ equals the slope calculated for the average depth in the reach but may be approximated by the average of the values of friction slope S for the depths at sections 1 and 2.

Solving Eq. (21-91) for L gives

$$L = \frac{\left(d_2 + \dfrac{V_2^2}{2g}\right) - \left(d_1 + \dfrac{V_1^2}{2g}\right)}{S_o - \bar{S}} = \frac{H_{e2} - H_{e1}}{S_o - \bar{S}} \qquad (21\text{-}92)$$

where H_{e1} and H_{e2} are the specific energy heads for sections 1 and 2, respectively, as given by

Eq. (21-82). The friction slope S at any point may be computed by the Manning equation, rearranged as follows:

$$S = \frac{n^2 V^2}{2.21 R^{4/3}} \tag{21-93}$$

where R = hydraulic radius, ft
   n = roughness coefficient (Art. 21-25)

Note that the slope S used in the Manning equation is the slope of the energy grade line and not of the channel bottom. Note also that the roughness coefficient n is squared in Eq. (21-93), and its value must therefore be chosen with special care to avoid an exaggerated error in the computed friction slope. The smaller the value of n, the longer will be the backwater curve profile, and vice versa. Therefore, the smallest n possible for the prevailing conditions should be selected for computation of a backwater curve if knowledge of the longest possible flow profile is required.

The first step in using the direct step method involves choosing a series of depths for the end points of each reach. These depths will range from the depth at the point of control to the ending depth for the backwater curve. This ending depth is often the normal depth for the channel (Art. 21-22) but may be some intermediate depth, such as for a curve preceding a hydraulic jump. Depths should be chosen so that the velocity change across a reach does not exceed 20% of the velocity at the beginning of the reach. Also, the change in depth between sections should never exceed 1 ft.

The specific energy head H_e should be computed for the chosen depth at each of the various sections and the change in specific energy between sections determined. Next, the friction slope S should be computed at each section from Eq. (21-93). The average of two sections gives the friction slope $\bar{S}$ between sections. Finally, the difference between $\bar{S}$ and slope of channel bottom S_o should be computed and the length of reach determined from Eq. (21-92).

These computations can be handled most conveniently in a table. The table should be arranged with separate columns for results of calculations and separate rows for each of the chosen depths (Table 21-12a).

Column 1 in Table 21-12a lists d, depth of flow, ft, arbitrarily assigned. In column 2 are areas of flow, sq ft, corresponding to depth in column 1. Column 3 gives wetted perimeter, ft, corresponding to depth in column 1. For column 4, hydraulic radius R, ft, equals column 2 divided by column 3. In column 6, V is the mean velocity, fps, or discharge Q divided by column 2. The values in column 6 are used to compute the velocity head, ft, in column 7, which may be adjusted for turbulence if α is known [see Eq. (21-81)].

Column 8 gives specific energy H_e, ft, the sum of column 7 and column 1 [Eq. (21-82)]. Column 9 is the value in column 8 for a section minus that in the previous section ($H_{e2} - H_{e1}$).

Table 21-12a. Direct-Step Backwater-Curve Calculations for Example 21-10

Section	d, ft (1)	A, sq ft (2)	W.P., ft (3)	R, ft (4)	$R^{4/3}$ (5)	V, fps (6)	$V^2/2g$, ft (7)
1	3.86	38.6	17.72	2.18	2.83	11.14	1.93
2	4.30	43.0	18.60	2.31	3.05	10.00	1.55
3	4.75	47.5	19.50	2.44	3.28	9.05	1.27
4	4.94	49.4	19.88	2.48	3.36	8.70	1.18

Section	H_e, ft (8)	ΔH_e, ft (9)	S (10)	$\bar{S}$ (11)	$S_o - \bar{S}$ (12)	L, ft (13)	ΣL, ft (14)
1	5.79		0.00389				
		−0.06		0.00340	−0.00140	43	43
2	5.85		0.00291				
		−0.17		0.00256	−0.00056	304	347
3	6.02		0.00222				
		−0.10		0.00211	−0.00011	909	1,256
4	6.12		0.00200				

Column 10 lists friction slope S computed from Eq. (21-93), with known value of n, V as given in column 6, and $R^{4/3}$ from column 5. In column 11 is $\bar{S}$, the average friction between steps. It equals the arithmetic mean of the slope for a section in column 10 and the one computed for the previous section. Column 12 is obtained by subtracting column 11 from the known bottom slope. Column 13 gives length of reach, ft, between the consecutive sections, column 9 divided by column 12 [Eq. (21-92)]. Finally, column 14 lists distance from the section under consideration to the point of control, equal to the cumulative sum of the values in column 13.

Example 21-10. Direct Step Method for Computing Backwater Curves. A 10-ft-wide rectangular channel with a slope $S_o = 0.0020$ and roughness factor $n = 0.014$ carries a discharge $Q = 430$ cfs. The channel ends in a free overfall as shown in Fig. 21-48. Determine the water-surface profile and the distance from the free overfall to the location where the backwater curve joins the normal-depth line.

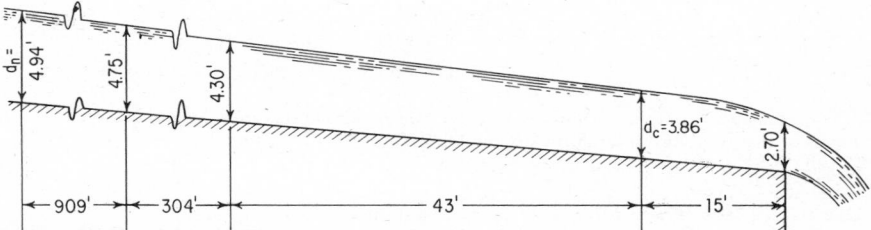

Fig. 21-48. Backwater curve for rectangular channel with free overfall (right) in Example 21-10.

Solution. Critical depth d_c and normal depth d_n are found from Eqs. (21-84) and (21-83) to be 3.86 and 4.94 ft, respectively. Since the normal depth is greater than the critical depth, the slope is mild and flow is subcritical. Calculations for the backwater curve progress upstream from the point of control, which, in this case, is the critical depth (see Table 21-12a). Critical depth may be assumed to occur from $3d_c$ to $4d_c$, about 15 ft, upstream of the overfall (Art. 21-32). The water depth at the overfall is about $0.7d_c = 0.7 \times 3.86 = 2.70$ ft. The water-surface profile is of the M2 type, as shown in Fig. 21-48. The total distance from the overfall to the location where the curve joins the normal-depth line equals 1,256 ft plus 15 ft, or 1,271 ft.

Standard step method allows computation of backwater curves in both nonprismatic natural channels and nonuniform artificial channels, as well as in uniform channels. This method involves solving for the depth of flow at various locations along a channel with Bernoulli's energy equation and a known length of reach.

A surface profile is determined in the following manner: The channel is examined for changes in cross section, grade, or roughness, and the locations of these changes are given station numbers. Stations are also established between these locations such that the velocity change between any two consecutive stations is not greater than 20% of the velocity at the former station. Data concerning the hydraulic elements of the channel are collected at each of these stations. Computation of the surface curve is then made in steps, starting from the point of control and progressing from station to station—in an upstream direction for subcritical flow and downstream for supercritical flow. The length of reach in each step is given by the stationing, and the depth of flow is determined by trial and error.

Nonprismatic channels do not have well-defined points of control to aid in determining the starting depth for a backwater curve. Therefore, the water-surface elevation at the beginning must be determined as follows: The step computations are started at a point in the channel some distance upstream or downstream from the desired starting point, depending on whether flow is supercritical or subcritical, respectively. Then, computations progress toward the initial section. Since this step method is a converging process, this procedure produces the true depth for the initial section within a relatively few steps.

The energy balance used in the standard step method is shown graphically in Fig. 21-47, in which the position of the water surface at section 1 is Z_1 and at section 2, Z_2, referred to a horizontal datum. Writing Bernoulli's equation [Eq. (21-11)] for sections 1 and 2 yields

$$Z_1 + \frac{V_1^2}{2g} = Z_2 + \frac{V_2^2}{2g} + h_f + h_i \qquad (21\text{-}94)$$

where V_1 and V_2 are the mean velocities, fps, at sections 1 and 2; the friction loss ft, in the reach ($\bar{S}L$) is denoted by h_f; and the term h_i is added to account for eddy loss, ft.

Eddy loss, sometimes called **impact loss,** is a head loss caused by flow running contrary to the main current because of irregularities in the channel. No rational method is available for determination of eddy loss, and it is therefore often accounted for, in natural channels, by a slight increase in Manning's n. Eddy loss depends mainly on a change in velocity head. For lined channels, it has been expressed as a coefficient k to be applied as follows:

$$h_i = k\left(\frac{V_1^2}{2g} - \frac{V_2^2}{2g}\right) = k\left(\Delta\frac{V^2}{2g}\right) \tag{21-95}$$

The coefficient k is 0.2 for diverging reaches, from 0 to 0.1 for converging reaches, and about 0.5 for abrupt expansions and contractions.

The total head at any section of the channel is given by

$$H = Z + \frac{V^2}{2g} \tag{21-96}$$

where Z equals the elevation of the channel bottom above the given datum plus the depth of flow d at that section.

The standard step method is most easily used if the computations are arranged in tabular form, similar to that used for the uniform step method (Table 21-12a). One form is shown in Table 21-12b.

Table 21-12b. Computation of a Flow Profile by the Standard Step Method

Station (1)	Z (2)	d (3)	A (4)	V (5)	V²/2g (6)	H (7)	R (8)	R⁴/³ (9)	S (10)	$\bar{S}$ (11)	L (12)	h_f (13)	h_i (14)	H (15)

In column 1 of the table, each section is identified by a station number, such as Station 2 + 80 (280 ft from initial station). Column 2 gives water-surface elevation, ft, at the station. The first entry in this column is the known elevation of the water surface at the initial section. Subsequent entries are trial values, to be verified or rejected by the computations made in the remaining columns of the table.

Column 3 lists depth of flow, ft. This depth corresponds to the elevation in column 2, adjusted for S_oL (see Fig. 21-47). Column 4 shows area of flow, sq ft, corresponding to the depth in column 3. In column 5 is mean velocity, fps, equal to the given discharge Q divided by column 4. It yields velocity head, ft, for column 6. Addition of columns 6 and 2 then produces total head, ft, for column 7 [Eq. (21-96)].

Column 8 contains hydraulic radius R, ft, corresponding to the depth in column 3. In column 10 is friction slope S computed from Eq. (21-93), with V from column 5, $R^{4/3}$ from column 9, and n from Table 21-11. Column 11 lists $\bar{S}$, average friction slope for the reach. It equals the mean of the friction slope for a section in column 10 and that for the previous section. Column 12 gives length of reach, ft, between sections. These values are the differences in station numbers for the reach (column 1). Column 13 contains friction loss h_f in the reach, column 11 times column 12. For column 14, which lists eddy loss h_i in the reach, ft, a coefficient k is multiplied by the result obtained by subtracting the value for a section in column 6 from that for the previous section.

Finally, column 15 gives total head H, ft. This is obtained from Eq. (21-94), which after substitution of H from Eq. (21-96) becomes

$$H_1 = H_2 + h_f + h_i \tag{21-97}$$

where H_1 and H_2 equal the total head of sections 1 and 2, respectively. The value of total head computed from Eq. (21-97) must agree with the value of total head given in column 7 or the assumed water-surface elevation in column 2 is incorrect. Agreement is assumed if the two values of total head are within 0.1 ft in elevation. If the two values of total head do not agree (column 15 ≠ column 7), a new water-surface elevation must be assumed for column 2 and the computations repeated until agreement is obtained. The value that finally leads to agreement gives the correct water-surface elevation. This value should be underlined to indicate its acceptance, and the computations may then proceed for the next step.

Additional columns may be added to the table to give such incidental information as the invert elevation, bottom width of the channel, wetted perimeter, and change in velocity head between sections.

Backwater curves for natural river or stream channels (irregularly shaped channels) are calculated in a manner similar to that described for regularly shaped channels. However, some account must be taken of the varying channel roughness and the differences in velocity and capacity in the main channel and the overbank or flood plain portions of the stream channel. The most expeditious way of determining the backwater curves is to plot the channel cross section to a scale convenient for measurement of lengths and areas; subdivide the cross section into main channels and flood-plain areas; and determine the discharge, velocity, and friction slope for each subarea at selected water surface elevations. Utilizing the above data, determine the total discharge (the sum of the subarea discharges), the mean velocity (the total discharge divided by the total area), and α (the energy coefficient or coriolis coefficient to be applied to the velocity head). (See V. T. Chow, "Open-Channel Hydraulics," McGraw-Hill Book Company, New York.)

The backwater curve is usually started by assuming normal depth at a point some distance downstream from the start of the reach under analysis. Several intermediate cross sections should be taken between the point where normal depth is assumed and the start of the reach for which a detailed water-surface profile is required. The purpose of this is to allow the intermediate sections to "dampen out" any minor errors in the assumed starting water-surface elevation.

The accuracy or validity of the water-surface profile is contingent on an accurate evaluation of the channel roughness and judicious selection of cross-section location. A greater number of cross sections generally enhances the validity of the water-surface profile; however, because of the extensive calculations involved with each cross section, their number should be limited to as few as accuracy permits.

The effect of bridges, approach roadways, bridge piers, and culverts can be determined using procedures outlined in V. T. Chow, "Open-Channel Hydraulics," McGraw-Hill Book Company, New York, and J. N. Bradley, "Hydraulics of Bridge Waterways," Hydraulic Design Series No. 1, 2d ed., U.S. Department of Transportation, Federal Highway Administration, Bureau of Public Roads, 1970.

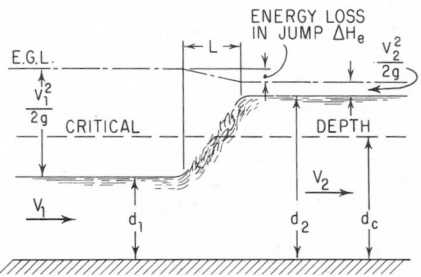

Fig. 21-49. Hydraulic jump.

21-28. Hydraulic Jump. (See also Art. 21-23.) This is an abrupt increase in depth of rapidly flowing water (Fig. 21-49). Flow at the jump changes from a supercritical to a subcritical stage with an accompanying loss of kinetic energy.

A hydraulic jump is the only means by which the depth of flow can change from less than critical to greater than critical in a uniform channel. A jump will occur either where supercritical flow exists in a channel of subcritical slope, as shown in Figs. 21-52 and 21-53b, or where a steep channel enters a reservoir. The first condition is met in a mild channel downstream from a sluice gate or ogee overflow spillway, or at an abrupt change in channel slope from steep to mild. The second condition occurs where flow in a steep channel is blocked by an overflow weir, a gate, or other obstruction.

A hydraulic jump can be either stationary or moving, depending on whether the flow is steady or unsteady, respectively.

Depth at the jump is not discontinuous. The change in depth occurs over a finite distance, known as the length of jump. The upstream surface of the jump, known as the roller, is a turbulent mass of water, which is continually tumbling erratically against the rapidly flowing sheet below.

The depth before a jump is the initial depth, and the depth after a jump is the sequent depth. The specific energy for the sequent depth is less than that for the initial depth because of the energy dissipation within the jump. (Initial and sequent depths should not be confused with the depths of equal energy, or alternate depths.)

According to Newton's second law of motion, the rate of loss of momentum at the jump must equal the unbalanced pressure force acting on the moving water and tending to retard its motion. This unbalanced force equals the difference between the hydrostatic forces corresponding to the depths before and after the jump. For rectangular channels, this resultant pressure force is

$$F = \frac{d_2^2 w}{2} - \frac{d_1^2 w}{2} \qquad (21\text{-}98)$$

where d_1 = depth before the jump, ft
d_2 = depth after the jump, ft
w = unit weight of water, lb per cu ft
The rate of change of momentum at the jump per foot width of channel equals

$$F = \frac{MV_1 - MV_2}{t} = \frac{qw}{g}(V_1 - V_2) \qquad (21\text{-}99)$$

where M = mass of water, lb-sec^2 per ft
V_1 = velocity at depth d_1, fps
V_2 = velocity at depth d_2, fps
q = discharge per foot width of rectangular channel, cfs
t = unit of time, sec
g = acceleration due to gravity, 32.2 ft per sec^2
Equating the values of F in Eqs. (21-98) and (21-99), and substituting $V_1 d_1$ for q and $V_1 d_1/d_2$ for V_2, the reduced equation for rectangular channels becomes

$$V_1^2 = \frac{gd_2}{2d_1}(d_2 + d_1) \qquad (21\text{-}100)$$

Equation (21-100) may then be solved for the sequent depth:

$$d_2 = \frac{-d_1}{2} + \sqrt{\frac{2V_1^2 d_1}{g} + \frac{d_1^2}{4}} \qquad (21\text{-}101)$$

If $V_2 d_2/d_1$ is substituted for V_1 in Eq. (21-100),

$$d_1 = \frac{-d_2}{2} + \sqrt{\frac{2V_2^2 d_2}{g} + \frac{d_2^2}{4}} \qquad (21\text{-}102)$$

Equation (21-102) may be used in determining the position of the jump where V_2 and d_2 are known. Relationships may be derived similarly for channels of any cross section.

The head loss in a jump equals the difference in specific-energy head before and after the jump. This difference (Fig. 21-49) is given by

$$\Delta H_e = H_{e1} - H_{e2} = \frac{(d_2 - d_1)^3}{4 d_1 d_2} \qquad (21\text{-}103)$$

where H_{e1} = specific-energy head of stream before jump, ft
H_{e2} = specific-energy head of stream after jump, ft
The specific energy for free-surface flow is given by Eq. (21-82).

The depths before and after a hydraulic jump may be related to the critical depth by the equation

$$d_1 d_2 \frac{d_1 + d_2}{2} = \frac{q^2}{g} = d_c^3 \qquad (21\text{-}104)$$

where q = discharge, cfs per ft of channel width
d_c = critical depth for the channel, ft
It may be seen from this equation that if $d_1 = d_c$ then d_2 must also equal d_c.

21-29. Jump in Horizontal Rectangular Channels. The form of a hydraulic jump in a horizontal rectangular channel may be of several distinct types, depending upon the Froude number of the incoming flow $\mathbf{F} = V/(gL)^{1/2}$ [Eq. (21-16)], where L is a characteristic length, ft; V is the mean velocity, fps; and g = acceleration due to gravity, ft per sec^2. For open-channel flow, the characteristic length for the Froude number is made equal to the **hydraulic depth** d_h.

Hydraulic depth is defined as

$$d_h = \frac{A}{T} \qquad\qquad (21\text{-}105)$$

where A = area of flow, sq ft

T = width of free surface, ft

For rectangular channels, hydraulic depth equals depth of flow.

Various forms of hydraulic jump, and their relation to the Froude number of the approaching flow F_1, were classified by the U.S. Bureau of Reclamation and are presented in Fig. 21-50. (V. T. Chow, "Open-Channel Hydraulics," McGraw-Hill Book Company, New York.)

For $F_1 = 1$, the flow is critical and there is no jump.

For $F_1 = 1$ to 1.7, there are undulations on the surface. The jump is called an undular jump.

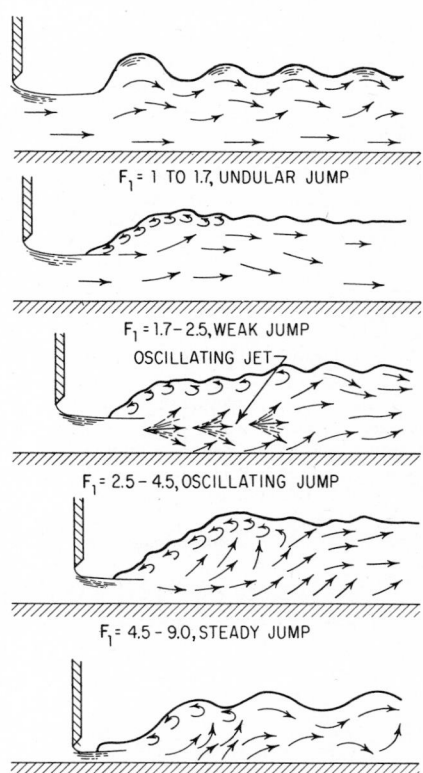

F$_1$ = 1 TO 1.7, UNDULAR JUMP

F$_1$ = 1.7 – 2.5, WEAK JUMP

F$_1$ = 2.5 – 4.5, OSCILLATING JUMP

F$_1$ = 4.5 – 9.0, STEADY JUMP

Fig. 21-50. Type of hydraulic jump depends on Froude number. (*After V. T. Chow, "Open-Channel Hydraulics," McGraw-Hill Book Company, New York.*)

F$_1$ = LARGER THAN 9.0, STRONG JUMP

For $F_1 = 1.7$ to 2.5, a series of small rollers develop on the surface of the jump, but the downstream water surface remains smooth. The velocity throughout is fairly uniform and the energy loss is low. This jump may be called a weak jump.

For $F_1 = 2.5$ to 4.5, there is an oscillating jet entering the jump. The jet moves from the channel bottom to the surface and back again with no set period. Each oscillation produces a large wave of irregular period, which, very commonly in canals, can travel for miles, doing extensive damage to earth banks and riprap surfaces. This jump may be called an oscillating jump.

For $F_1 = 4.5$ to 9.0, the downstream extremity of the surface roller and the point at which the high-velocity jet tends to leave the flow occur at practically the same vertical section. The action and position of this jump are least sensitive to variation in tailwater depth.   The jump is well balanced, and the performance is at its best.   The energy dissipation ranges from 45 to 70%. This jump may be called a steady jump.

For $F_1 = 9.0$ and larger, the high-velocity jet grabs intermittent slugs of water rolling down the front face of the jump, generating waves downstream and causing a rough surface.   The jump action is rough but effective, and the energy dissipation may reach 85%.   This jump may be called a strong jump.

Note that the ranges of the Froude number given for the various types of jump are not clear-cut but overlap to a certain extent, depending on local conditions.

21-30. Hydraulic Jump as an Energy Dissipator.   A hydraulic jump is a useful means for dissipating excess energy in supercritical flow (Art. 21-23).   A jump may be used to prevent erosion below an overflow spillway, chute, or sluice gate, by quickly reducing the velocity of the flow over a paved apron.   A special section of channel built to contain a hydraulic jump is known as a **stilling basin.**

If a hydraulic jump is to function ideally as an energy dissipator, below a spillway, for example, the elevation of the water surface after the jump must coincide with the normal tailwater elevation for every discharge.   If the tailwater is too low, the high-velocity flow will continue downstream for some distance before the jump can occur.   If the tailwater is too high, the jump will be drowned out, and there will be a much smaller dissipation of total head.   In either case, dangerous erosion is likely to occur for a considerable distance downstream.

The ideal condition is to have the *sequent-depth curve,* which gives discharge vs. depth after the jump, coincide exactly with the *tailwater-rating curve.*   The tailwater-rating curve gives normal depths in the discharge channel for the range of flows to be expected.   Changes in the spillway design that can be made to alter the tailwater-rating curve involve changing the crest length, changing the apron elevation, and sloping the apron.

Accessories, such as chute blocks and baffle blocks, are usually installed in a stilling basin to control the jump.   The main purpose of these accessories is to shorten the range within which the jump will take place, not only to force the jump to occur within the basin but also to reduce the size and, therefore, the cost of the basin.   Controls within a stilling basin have additional advantages in that they improve the dissipation function of the basin and stabilize the jump action.

21-31. Length of Hydraulic Jump.   The length of a hydraulic jump L may be defined as the horizontal distance from the upstream edge of the roller to a point on the raised surface immediately downstream from cessation of the violent turbulence.   This length (Fig. 21-49) defies accurate mathematical expression, in part because of the nonuniform velocity distribution within the jump.   But it has been determined experimentally.   The experimental results may be summarized conveniently by plotting the Froude number of the upstream flow F_1 against a dimensionless ratio of jump length to downstream depth L/d_2.   The resulting curve (Fig. 21-51)

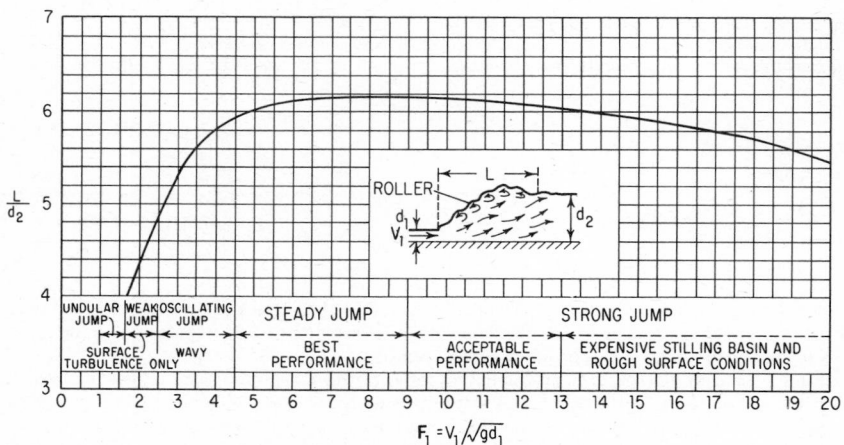

Fig. 21-51. Length of hydraulic jump in horizontal channel depends on sequent depth d_2 and Froude number of approaching flow. (*From V. T. Chow, "Open-Channel Hydraulics," McGraw-Hill Book Company, New York.*)

has a flat portion in the range of steady jumps. The curve thus minimizes the effect of any errors made in calculation of the Froude number in the range where this information is most frequently needed. The curve, prepared by V. T. Chow from data gathered by the U.S. Bureau of Reclamation, was developed for jumps in rectangular channels. But it will give approximate results for jumps formed in trapezoidal channels.

For other than rectangular channels, the depth d_1 used in the equation for Froude number is the hydraulic depth given by Eq. (21-105).

21-32. Location of a Hydraulic Jump. It is important to know where a hydraulic jump will form, since the turbulent energy released in a jump can cause extensive scour in an unlined channel or destruction of paving in a thinly lined channel. Special reinforced sections of channel must be built to withstand the pounding and vibration of a jump and to provide extra freeboard for the added depth at the jump. These features are expensive to build; therefore, a great savings can be realized if their use is restricted to a limited area through a knowledge of the jump location.

The precision with which the location is predicted depends on the accuracy with which the friction losses and length of jump are estimated, and on whether the discharge is as assumed. The method of prediction used for rectangular channels is illustrated in Fig. 21-52.

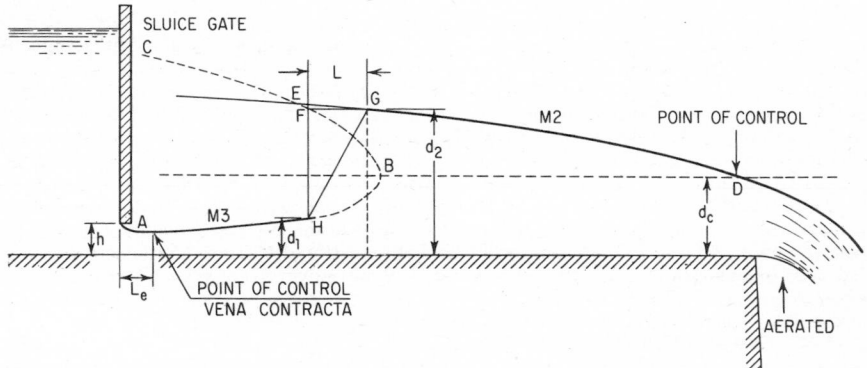

Fig. 21-52. Graphical method for locating hydraulic jump below a sluice gate.

The water-surface profiles of the flow approaching and leaving the jump, curves AB and ED in Fig. 21-52, are type M3 and M2 backwater curves, respectively (Fig. 21-46e and c).

Backwater curve ED has as its point of control the critical depth d_c, which occurs near the channel drop-off. Critical depth does not exist exactly at the edge, as theory would indicate, but instead occurs a short distance upstream. The distance is small (from three to four times d_c) and can be ignored for most problems. The actual depth at the brink is 71.5% of critical depth, but it is normally assumed to be $0.7d_c$ for simplicity.

The point of control for backwater curve AB is taken as the depth at the vena contracta, which forms just downstream from the sluice gate. The distance from the gate to the vena contracta L_e is nearly equal to the size of gate opening h. The amount of contraction varies with both the head on the gate and the gate opening. Depth at the contraction ranges from 50 to over 90% of h. The depth of flow at the vena contracta may be taken as $0.75h$ in the absence of better information.

Jump location is determined as follows: The backwater curves AB and ED are computed in their respective directions until they overlap, using the step methods of Art. 21-27. With values of d_2 obtained from Eq. (21-101), CB, the curve of depths sequent to curve AB, is plotted through the area where it crosses curve ED. A horizontal intercept FG, equal in length to L, the computed length of jump, is then fitted between the curves CB and ED. The jump may be expected to form between the points H and G, since all requirements for the formation of a jump are satisfied at this location.

If the downstream depth is increased because of an obstruction, the jump moves upstream and may eventually be drowned out in front of the sluice gate. Conversely, if the downstream depth is lowered, the jump moves to a new location downstream.

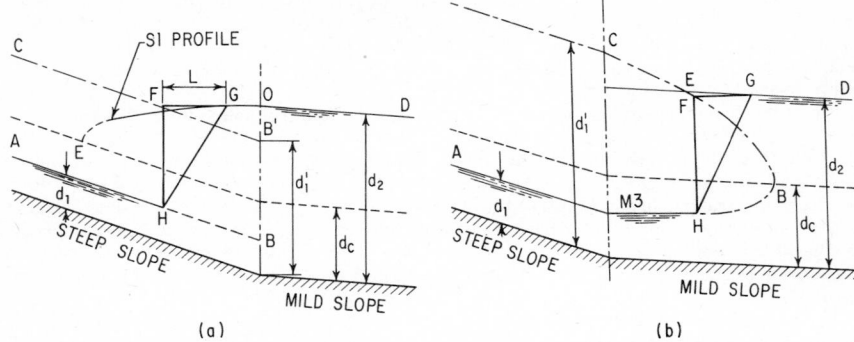

Fig. 21-53. Hydraulic jump may occur above (*a*), or below (*b*), or at a change in bottom slope of a channel.

When the slope of a channel has an abrupt change from steeper than critical (Art. 21-23) to mild, a jump forms that may be located either above or below the grade change. The position of the jump depends on whether the downstream depth d_2 is greater than, less than, or equal to the depth d'_1 sequent to the upstream depth d_1. Two possible positions are shown in Fig. 21-53.

It is assumed, for simplicity, that flow is uniform, except in the reach between the jump and the grade break. If the downstream depth d_2 is greater than the upstream sequent depth d'_1, computed from Eq. (21-101) with d_1 given, the jump occurs in the steep region, as shown in Fig. 21-53*a*. The surface curve *EO* is of the S1 type. It is asymptotic to a horizontal line at *O*. Line *CB'* is a plot of the depth d'_1 sequent to the depth of approach line *AB*. The jump location is found by producing a horizontal intercept *FG*, equal to the computed length of the jump, between lines *CB'* and *EO*. A jump will form between *H* and *G*, since all requirements are satisfied for this location. As depth d_2 is lowered, the jump moves downstream to a new position, as shown in Fig. 21-53*b*. If d_2 is less than d'_1, computed from Eq. (21-102), the jump will form in the mild channel and can be located as described for Fig. 21-52.

21-33. Flow at Entrance to a Steep Channel. The discharge Q, cfs, in a channel leaving a reservoir is a function of the total head H, ft, on the channel entrance, the entrance loss, ft, and the slope of the channel. If the channel has a slope steeper than the critical slope (Art. 21-23), the flow passes through critical depth at the entrance, and discharge is at a maximum. If the channel entrance is rectangular in cross section, the critical depth $d_c = \frac{2}{3}H_e$ [according to Eqs. (21-82) and (21-85)], where H_e is the specific energy head, ft, in the reservoir and datum is the elevation of the lip of the channel (Fig. 21-54*a*).

From $Q = AV$, with the area of flow $A = bd_c = \frac{2}{3}bH_e$ and the velocity

$$V = \sqrt{2g(H_e - d_c)} = \sqrt{\frac{64.4H_e}{3}}$$

the discharge for rectangular channels, ignoring entrance loss, is

$$Q = 3.087bH_e^{3/2} \tag{21-106}$$

where b is the channel width, ft

If the entrance loss must be considered, or if the channel entrance is other than rectangular, the inlet depth must be solved for by trial and error, since the discharge is unknown. The procedure for finding the correct discharge is as follows:

A trial discharge is chosen. Then, the critical depth for the given shape of channel entrance is determined from appropriate tables, such as those in H. W. King and E. F. Brater, "Handbook of Hydraulics," McGraw-Hill Book Company, New York. Adding d_c to its associated velocity head gives the specific energy in the channel entrance, to which the resulting entrance loss is added. This sum then is compared with the specific energy of the reservoir water, which equals the depth of water above datum plus the velocity head of flow toward the channel. (This velocity head is normally so small that it may be taken as zero for most problems.) If the specific energy computed for the depth of water in the reservoir equals the sum of specific energy and entrance loss determined for the channel entrance, then the assumed discharge is correct; if not, a new discharge is assumed, and the computations are continued until a balance is reached.

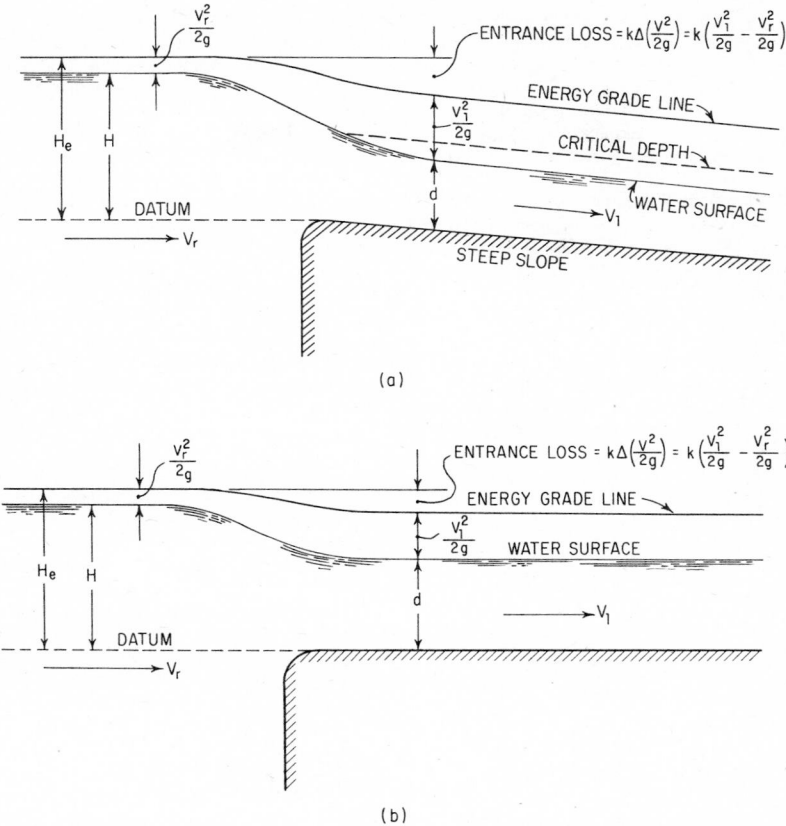

Fig. 21-54. Entrance to (*a*) steep channel, (*b*) mild-slope channel from a reservoir.

A first trial discharge may be found from $Q = A \sqrt{2g(H_e - d)}$, where $(H_e - d)$ gives the actual head producing flow (Fig. 21-54). A reasonable value for the depth d would be $\frac{2}{3}H_e$ for steep channels and an even greater percentage of H_e for mild channels.

The entrance loss equals the product of an empirical constant k and the change in velocity head ΔH_v at the entrance. If the velocity in the reservoir is assumed to be zero, then the entrance loss is $k(V_1^2/2g)$, where V_1 is the velocity computed for the channel entrance. Safe design values for this coefficient vary from about 0.1 for a well-rounded entrance to slightly over 0.3 for one with squared ends.

21-34. Flow at Entrance to a Channel of Mild Slope. When water flows from a reservoir into a channel with slope less than the critical slope (Art. 21-23), the depth of flow at the channel entrance equals the normal depth for the channel (Art. 21-22). The entrance depth and discharge are dependent on each other. The discharge that results from a given head is that for which flow enters the channel without forming either a backwater or drawdown curve within the entrance. This requirement necessitates the formation of normal depth d, since only at this equilibrium depth is there no tendency to change the discharge or to form backwater curves. (In Fig. 21-54*b*, d is normal depth.)

A solution for discharge at entrance to a channel of mild slope is found as follows: A trial discharge, cfs, is estimated from $Q = A \sqrt{2g(H_e - d)}$, where $H_e - d$ is the actual head, ft, producing flow. H_e is the specific energy head, ft, of the reservoir water relative to datum at lip of channel; A is the cross-sectional area of flow, sq ft; and g is acceleration due to gravity, 32.2 ft per sec². The normal depth of the channel is determined for this discharge from Eq. (21-83). (Tables in

H. W. King and E. F. Brater, "Handbook of Hydraulics," McGraw-Hill Book Company, New York, may be used to advantage for this calculation.) The velocity head is computed for this depth-discharge combination, and an entrance-loss calculation is made (see Art. 21-33). The sum of the specific energy of flow in the channel entrance and the entrance loss must equal the specific energy of the water in the reservoir for an energy balance to exist between those points (Fig. 21-54*b*). If the trial discharge gives this balance of energy, then the discharge is correct; if not, a new discharge is chosen, and the calculations continued until a satisfactory balance is obtained.

21-35. Channel Section of Greatest Efficiency.   If a channel of any shape is to reach its greatest hydraulic efficiency, it must have the shortest possible wetted perimeter for a given cross-sectional area. The resulting shape gives the greatest hydraulic radius and therefore the greatest capacity for that area. This can be seen from the Manning equation for discharge [Eq. (21-83)], in which Q is a direct function of hydraulic radius to the two-thirds power.

The most efficient of all possible open-channel cross sections is the semicircle. There are practical objections to the use of this shape, because of the difficulty of construction, but it finds some use in metal flumes where sections can be preformed. The most efficient of all trapezoidal sections is the half hexagon, which is used extensively for large water-supply channels. The rectangular section with the greatest efficiency has a depth of flow equal to one-half the width. This shape is often used for box culverts and for small drainage ditches.

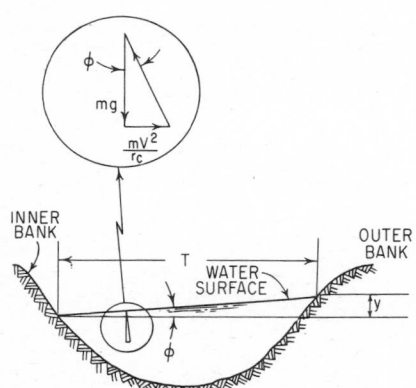

Fig. 21-55. Water-surface profile at a bend in a channel with subcritical flow.

21-36. Subcritical Flow around Bends in Channels.   Because of the inability of liquids to resist shearing stress, the free surface of steady uniform flow is always normal to the resultant of the forces acting on the water. Water in a reservoir has a horizontal surface since the only force acting on it is the force of gravity.

Water reacts in accordance with Newton's first law of motion: It flows in a straight line unless deflected from its path by an outside force. When water is forced to flow in a curved path, its surface assumes a position normal to the resultant of the forces of gravity and radial acceleration. The force due to radial acceleration equals the force required to turn the water from a straight-line path, or mV^2/r_c for m, a unit mass of water, where V is its average velocity, fps, and r_c the radius of curvature, ft, of the center line of the channel.

The water surface makes an angle ϕ with the horizontal such that

$$\tan \phi = \frac{V^2}{r_c g} \tag{21-107}$$

The theoretical difference y, ft, in water-surface level between the inside and outside banks of a curve (Fig. 21-55) is found by multiplying $\tan \phi$ by the top width of the channel T, ft. Thus,

$$y = \frac{V^2 T}{r_c g} \tag{21-108}$$

where the radius of curvature r_c of the center of the channel is assumed to represent the average curvature of flow. This equation gives values of y that are smaller than those actually encountered because of the use of average values of velocity and radius, rather than empirically derived values more representative of actual conditions. The error will not be great, however, if the depth of flow is well above critical (Art. 21-23). In this range, the true value of y would be only a few inches.

The difference in surface elevation found from Eq. (21-108), though it involves some drop in surface elevation on the inside of the curve, does not allow a savings of freeboard height on the inside bank. The water surface there is wavy and thus needs a freeboard height at least equal to that of a straight channel.

The top layer of flow in a channel has a higher velocity than flow near the bottom because of the retarding effect of friction along the floor of the channel. A greater force is required to deflect the high-velocity flow. Therefore, when a stream enters a curve, the higher-velocity flow moves to the outside of the bend. If the bend continues long enough, all the high-velocity water will move against the outer bank and may cause extensive scour unless special bank protection is provided.

Since the higher-velocity flow is pressed directly against the bank, an increase in friction loss results. This increased loss may be accounted for in calculations by assuming an increased value of the roughness coefficient n within the curve. Scobey suggests that the value of n be increased by 0.001 for each 20° of curvature in 100 ft of flume. His values have not been evaluated completely, however, and should be used with discretion. (F. C. Scobey, "The Flow of Water in Flumes," *U.S. Department of Agriculture, Technical Bulletin* 393.)

21-37. Supercritical Flow around Bends. When water, traveling at a velocity greater than critical (Art. 21-23), flows around a bend in a channel, a series of standing waves are produced. Two waves form at the start of the curve. One is a positive wave, of greater-than-average surface elevation, which starts at the outside wall and extends across the channel on the line AME (Fig. 21-56). The second is a negative wave, with a surface elevation of less-than-average height,

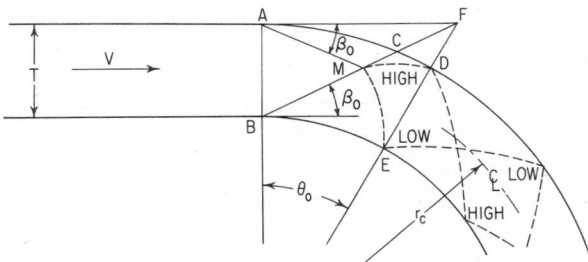

Fig. 21-56. Plan view of supercritical flow around a bend in a channel.

which starts at the inside wall and extends across the channel on the line BMD. These waves cross at M, are reflected from opposite channel walls at D and E, recross as shown, and continue crossing and recrossing.

The two waves at the entrance form at an angle with the approach channel known as the wave angle β_o. This angle may be determined from the equation

$$\sin \beta_o = \frac{1}{\mathbf{F}_1} \tag{21-109}$$

where $\mathbf{F}_1$ represents the Froude number of flow in the approach channel [Eq. (21-16)].

The distance from the beginning of the curve to the first wave peak on the outside bank is determined by the central angle θ_o. This angle may be found from

$$\tan \theta_o = \frac{T}{(r_c + T/2) \tan \beta_o} \tag{21-110}$$

where T is the normal top width of channel and r_c is the radius of curvature of the center of channel. The depths along the banks at an angle $\theta < \theta_o$ are given by

$$d = \frac{V^2}{g} \sin^2 \left(\beta_o \pm \frac{\theta}{2} \right) \tag{21-111}$$

where the positive sign gives depths along the outside wall and the negative sign, depths along the inside wall. The depth of maximum height for the first positive wave is obtained by substituting the value of θ_o found from Eq. (21-110) for θ in Eq. (21-111).

Prevention of standing waves in existing rectangular channels may be accomplished by installation of diagonal sills at the beginning and end of the curve. The purpose of the sills is to introduce a counterdisturbance of the right magnitude, phase, and shape to neutralize the undesirable oscillations that normally form at the change of curvature. The details of sill design have been determined experimentally.

Good flow conditions may be insured in new projects with supercritical flow in rectangular channels by providing transition curves or by banking the channel bottom. Circular transition curves aid in wave control by setting up counterdisturbances in the flow similar to those provided by diagonal sills. A transition curve should have a radius of curvature twice the radius of the central curve. It should curve in the same direction and have a central angle given, with sufficient accuracy, by

$$\tan \theta_t = \frac{T}{2r_c \tan \beta_o} \tag{21-112}$$

Transition curves should be used at both the beginning and end of a curve to prevent disturbances downstream.

Banking the channel bottom is the most effective method of wave control. It permits equilibrium conditions to be set up without introduction of a counterdisturbance. The cross slope required for equilibrium is the same as the surface slope found for subcritical flow around a bend (Fig. 21-55). The angle ϕ that the bottom makes with the horizontal is found from the equation

$$\tan \phi = \frac{V^2}{r_c g} \tag{21-113}$$

21-38. Transitions in Open Channels. A transition is a structure placed between two open channels of different shape or cross-sectional area to produce a smooth, low-head-loss transfer of flow. The major problems associated with design of a transition lie in locating the invert and in determining the various cross-sectional areas so that the flow is in accord with the assumptions made in locating the invert. Many variables, such as flow-rate changes, wall roughness, and channel shape and slope, must be taken into account in design of a smooth-flow transition.

When proceeding downstream through a transition, the flow may remain subcritical or supercritical (Art. 21-23), change from subcritical to supercritical, or change from supercritical to subcritical. The latter flow possibility may produce a hydraulic jump.

Special care must be exercised in the design if the depth in either of the two channels connected is near the critical depth. In this range, a small change in energy head within the transition may cause the depth of flow to change to its alternate depth. A flow that switches to its subcritical alternate depth may overflow the channel. A flow that changes to its supercritical alternate depth may cause excessive channel scour. The relationship of flow depth to energy head can be shown on a plot such as Fig. 21-44.

To place a transition properly between two open channels, it is necessary to determine the design flow and to calculate normal and critical depths for each channel section. Maximum flow is usually selected as the design flow. Normal depth for each section is used for the design depth. After the design has been completed for maximum flow, hydraulic calculations should be made to check the suitability of the structure for lower flows.

The transition length that produces a smooth-flowing, low-head-loss structure is obtained for an angle of about 12.5° between the channel axis and the lines of intersection of the water surface with the channel sides, as shown in Fig. 21-57. The length of the transition L_t is then given by

$$L_t = \frac{\frac{1}{2}(T_2 - T_1)}{\tan 12.5°} \tag{21-114}$$

where T_2 and T_1 are the top widths of sections 2 and 1, respectively.

In design of an inlet-type transition structure, the water-surface level of the downstream channel must be set below the water-surface level of the upstream channel by at least the sum of the increase in velocity head, plus any transition and friction losses. The transition loss, ft, is given by $K(\Delta V^2/2g)$, where K, the loss factor, equals about 0.1 for an inlet-type structure; ΔV is the velocity change, fps; and $g = 32.2$ ft per sec². The total drop in water surface y_d across the inlet-type transition is then $1.1[\Delta(V^2/2g)]$, if friction is ignored.

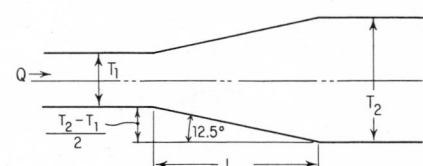

Fig. 21-57. Plan view of transition between two open channels with different widths.

For outlet-type structures, the average velocity decreases, and part of the loss in velocity head is recovered as added depth. The rise of the water surface for an outlet structure equals the decrease in velocity head minus the outlet and friction losses. The outlet loss factor is normally 0.2 for well-designed transitions. If friction is ignored, the total rise in water surface y_r across the outlet structure is $0.8[\Delta(V^2/2g)]$.

Many well-designed transitions have a reverse parabolic water-surface curve that is tangent to the water surfaces in each channel (Fig. 21-58). After such a water-surface profile is chosen, depths and cross-sectional areas are selected at points along the transition to produce this smooth curve. Straight, angular walls usually will not produce a smooth parabolic water surface. Therefore, a transition with a curved bottom or sides has to be designed.

The total transition length L_t is split into an even number of sections of equal length x. For Fig. 21-58, six equal lengths of 10 ft each are used, for an assumed drop in water surface y_d of 1 ft. It is assumed that the water surface will follow parabola AC for the length $L_t/2$ to produce a water-surface drop of $y_d/2$, and that the other half of the surface drop takes place along the parabola CB. The water-surface profile can be determined from the general equation for a parabola, $y = ax^2$, where y is the vertical drop in the distance x, measured from A or B.

The surface drops at sections 1 and 2 are found as follows: At the midpoint of the transition, $y_3 = ax^2 = y_d/2 = 0.5 = a(30)^2$, from which $a = 0.000556$. Then, $y_1 = ax_1^2 = 0.000556(10)^2 = 0.056$ ft and $y_2 = ax_2^2 = 0.000556(20)^2 = 0.222$ ft.

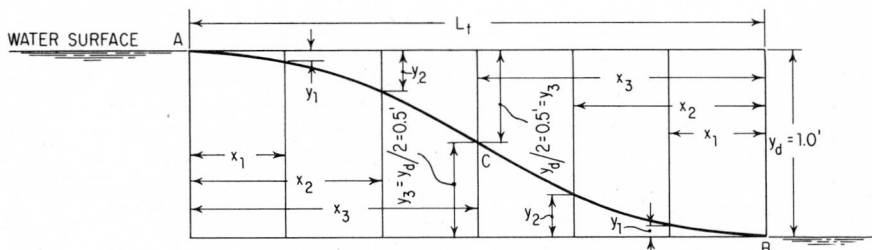

Fig. 21-58. Profile of reverse parabolic water-surface curve for well-designed transitions.

21-39. Types of Weirs.   A weir is a barrier in an open channel over which water flows. The edge or surface over which the water flows is called the *crest*. The overflowing sheet of water is the *nappe*.

If the nappe discharges into the air, the weir has *free discharge*. If the discharge is partly under water, the weir is *submerged* or drowned.

A weir with a sharp upstream corner or edge such that the water springs clear of the crest is a *sharp-crested weir* (Fig. 21-59). All other weirs are classed as *weirs not sharp-crested*. Sharp-crested weirs are classified according to the shape of the weir opening, such as rectangular weirs, triangular or V-notch weirs, trapezoidal weirs, and parabolic weirs. Weirs not sharp-crested are classified according to the shape of their cross section, such as broad-crested weirs, triangular weirs, and, as shown in Fig. 21-60, trapezoidal weirs.

The channel leading up to a weir is the *channel of approach*. The mean velocity in this channel is the *velocity of approach*. The depth of water producing the discharge is the *head*.

Sharp-crested weirs are useful only as a means of measuring flowing water. In contrast, weirs not sharp-crested are commonly incorporated into hydraulic structures as control or regulation devices, with measurement of flow as their secondary function.

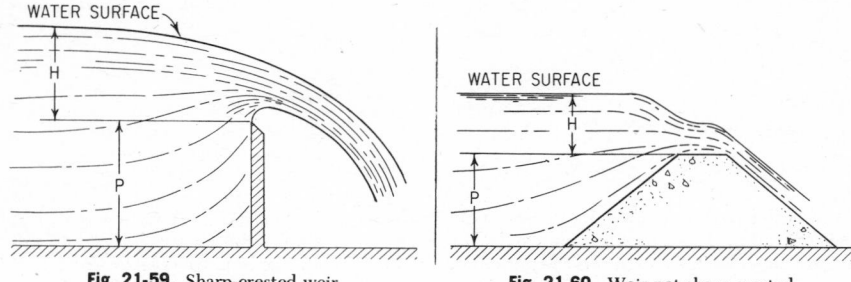

Fig. 21-59. Sharp-crested weir.   **Fig. 21-60.** Weir not sharp-crested.

21-40. Sharp-crested Weirs.   (See also Art. 21-39.)   Discharge over a rectangular sharp-crested weir is given by

$$Q = CLH^{3/2} \qquad (21-115)$$

where Q = discharge, cfs
$\quad\;\; C$ = discharge coefficient
$\quad\;\; L$ = effective length of crest, ft
$\quad\;\; H$ = measured head = depth of flow above elevation of crest, ft
The head should be measured at least $2.5H$ upstream from the weir, to be beyond the drop in the water surface (surface contraction) near the weir.

Numerous equations have been developed for finding the discharge coefficient C. One such equation, which applies only when the nappe is fully ventilated, was developed by Rehbock and simplified by Chow:

$$C = 3.27 + 0.40\frac{H}{P} \qquad (21-116)$$

where P is the height of the weir above the channel bottom (Fig. 21-59) (V. T. Chow, "Open-Channel Hydraulics," McGraw-Hill Book Company, New York).

The height of weir P must be at least $2.5H$ for a complete crest contraction to form. If P is less than $2.5H$, the crest contraction is reduced and is said to be partly suppressed. Equation (21-116) corrects for the effects of friction, contraction of the nappe, unequal velocities in the channel of approach, and partial suppression of the crest contraction and includes a correction for the velocity of approach and the associated velocity head.

To be fully ventilated, a nappe must have its lower surface subjected to full atmospheric pressure. A partial vacuum below the nappe can result through removal of air by the overflowing jet if there is restricted ventilation at the sides of the weir. This lack of ventilation causes increased discharge and a fluctuation and shape change of the nappe. The resulting unsteady condition is very objectionable when the weir is used as a measuring device. At very low heads, the nappe has a tendency to adhere to the downstream face of a rectangular weir even when means for ventilation are provided. A weir operating under such conditions could not be expected to have the same relationship between head and discharge as would a fully ventilated nappe. A V-notch weir should be used for measurement of flow at very low heads if accuracy of measurement is required.

End contractions occur when the weir opening does not extend the full width of the approach channel. Water flowing near the walls must move toward the center of the channel to pass over the weir, thus causing a contraction of the flow. The nappe continues to contract as it passes over the crest. So below the crest, the nappe has a minimum width less than the crest length.

The effective length L, ft, of a contracted-width weir is given by

$$L = L' - 0.1NH \qquad (21-117)$$

where L' = measured length of crest, ft
$\quad\;\; N$ = number of end contractions
$\quad\;\; H$ = measured head, ft
If flow contraction occurs at both ends of a weir, there are two end contractions and $N = 2$. If the weir crest extends to one channel wall but not the other, then there is one end contraction and $N = 1$. The effective crest length of a full-width weir is taken as its measured length. Such a weir is said to have its contractions suppressed.

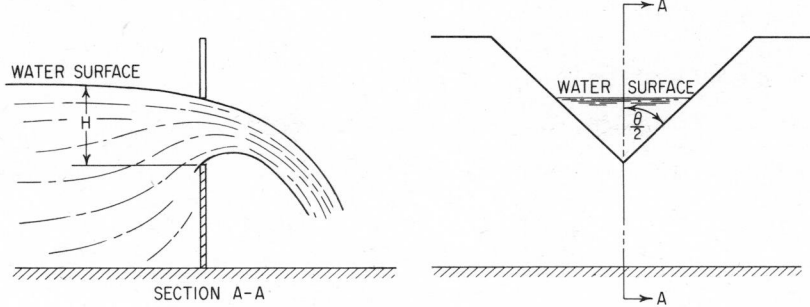

Fig. 21-61. V-notch weir.

21-41. Triangular or V-notch Sharp-crested Weirs.   The triangular or V-notch weir (Fig. 21-61) has a distinct advantage over a rectangular sharp-crested weir (Art. 21-40) when low discharges are to be measured.   Flow over a V-notch weir starts at a point, and both discharge and width of flow increase as a function of depth.   This has the effect of spreading out the low-discharge end of the depth-discharge curve and therefore allows more accurate determination of discharge in this region.

Discharge is given by

$$Q = C_1 H^{5/2} \tan\frac{\theta}{2} \tag{21-118}$$

where θ = notch angle, deg
H = measured head, ft
C_1 = discharge coefficient
The head H is measured from the notch elevation to the water-surface elevation at a distance $2.5H$ upstream from the weir.   Values of the discharge coefficient were derived experimentally by Lenz, who developed a procedure for including the effect of viscosity and surface tension, as well as the effect of contraction and velocity of approach (A. T. Lenz, "Viscosity and Surface Tension Effects on V-notch Weir Coefficients," *Transactions of the American Society of Civil Engineers,* vol. 69, 1943).   His values were summarized by Brater, who presented the data in the form of curves (Fig. 21-62) (E. F. Brater and H. W. King, "Handbook of Hydraulics," McGraw-Hill Book Company, New York).

A V-notch weir tends to concentrate or focus the overflowing nappe, causing it to spring clear of the downstream face for even the smallest flows.   This characteristic prevents a change in the head-discharge relationship at low flows and adds materially to the reliability of the weir.

21-42. Trapezoidal Sharp-crested Weirs.   The discharge from a trapezoidal weir (Fig. 21-63) is assumed to be the same as that from a rectangular weir and a triangular weir in combination.

$$Q = C_2 L H^{3/2} + C_3 Z H^{5/2} \tag{21-119}$$

Fig. 21-62. Chart gives discharge coefficients for sharp-crested V-notch weirs.   Coefficients depend on head and notch angle.

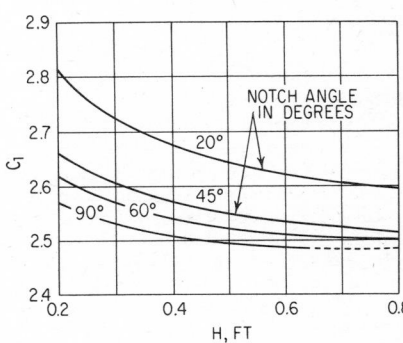

where Q = discharge, cfs
    L = length of notch at bottom, ft
    H = head, measured from notch bottom, ft
    Z = b/H [substituted for tan $(\theta/2)$ in Eq. (21-118)]
    b = half the difference between lengths of notch at top and bottom, ft
No data are available for determination of coefficients C_2 and C_3. These must be determined experimentally for each installation.

21-43. Submerged Sharp-crested Weirs.
The discharge over a submerged sharp-crested

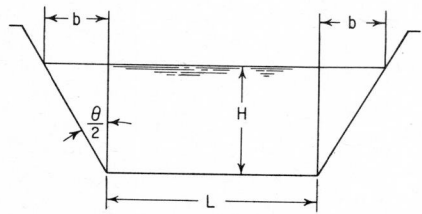

Fig. 21-63. Trapezoidal sharp-crested weir.

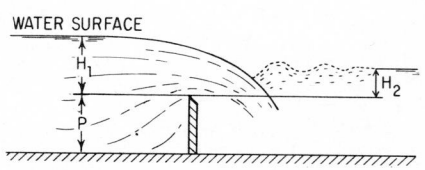

Fig. 21-64. Submerged sharp-crested weir.

weir (Fig. 21-64) is affected not only by the head on the upstream side H_1 but also by the head downstream H_2. Discharge also is influenced to some extent by the height P of the weir crest above the floor of the channel.

The discharge Q_s, cfs, for a submerged weir is related to the free or unsubmerged discharge Q, cfs, for that weir by a function of H_2/H_1. Villemonte expressed this relationship by the equation

$$\frac{Q_s}{Q} = \left[1 - \left(\frac{H_2}{H_1} \right)^n \right]^{0.385} \tag{21-120}$$

where n is the exponent of H in the equation for free discharge for the shape of weir used. (The value of n is $3/2$ for a rectangular sharp-crested weir and $5/2$ for a triangular weir.) To use the Villemonte equation, first compute the rate of flow Q for the weir when not submerged, and then, using this rate and the required depths, solve for the submerged discharge Q_s. (J. R. Villemonte, "Submerged-Weir Discharge Studies," *Engineering News-Record*, Dec. 25, 1947, p. 866.)

Equation (21-120) may be used to compute the discharge for a submerged sharp-crested weir of any shape simply by changing the value of n. The maximum deviation from the Villemonte equation for all test results was found to be 5%. Where great accuracy is essential, it is recommended that the weir be tested in a laboratory under conditions comparable with those at its point of intended use.

21-44. Weirs Not Sharp-crested.
These are sturdy, heavily constructed devices, normally an integral part of hydraulic projects (Fig. 21-60). Typically, a weir not sharp-crested appears as the crest section for an overflow dam or the entrance section for a spillway or channel. Such a weir can be used for discharge measurement, but its purpose is normally one of control and regulation.

The discharge over a weir not sharp-crested is given by

$$Q = CLH_t{}^{3/2} \tag{21-121}$$

where Q = discharge, cfs
    C = coefficient of discharge
    L = effective length of crest, ft
    H_t = total head on crest including the velocity head of approach, ft

The head of water producing discharge over a weir is the total of measured head H and velocity head of approach H_v. The velocity head of approach is accounted for by the discharge coefficient for sharp-crested weirs but must be considered separately for weirs not sharp-crested. Thus, for such weirs, Eq. (21-115) is rewritten in the form

$$Q = CL \left(H + \frac{V^2}{2g} \right)^{3/2} \tag{21-122}$$

where H = measured head, ft

V = velocity of approach, fps

$V^2/2g = H_v$, the velocity head of approach, ft, neglecting the degree of turbulence given by Eq. (21-81)

g = acceleration due to gravity, 32.2 ft per sec²

Since velocity and discharge are dependent on each other in this equation, and both are unknown, discharge must be found by a series of approximations. This may be done as follows: First, compute a trial discharge from the measured head, neglecting the velocity head. Then, using this discharge, compute the velocity of approach, velocity head, and finally total head. From this total head, compute the first corrected discharge. This corrected discharge will be sufficiently accurate if the velocity of approach is small. But the process should be repeated, starting with the corrected discharge, where approach velocities are high.

The discharge coefficient C must be determined empirically for weirs not sharp-crested. If a weir of untested shape is to be constructed, it must be calibrated in place or a model study made to determine its head-discharge relationship. The problem of establishing a fixed relation between head and discharge is complicated by the fact that the nappe may assume a variety of shapes in passing over the weir. For each change of nappe shape, there is a corresponding change in the relation between head and discharge. The effect is most critical for low heads. A nappe undergoes several changes in succession as the head varies, and the successive shapes that appear with an increasing stage may differ from those pertaining to similar stages with decreasing head. Care must be exercised when using these weirs for flow measurement, therefore, to assure that the conditions are similar to those at the time of calibration. (E. F. Brater and H. W. King, "Handbook of Hydraulics," McGraw-Hill Book Company, New York.)

Large weirs not sharp-crested often have piers on their crest to support control gates or a roadway. These piers reduce the effective length of crest by more than the sum of their individual widths because of the formation of flow contractions at each pier. The effective crest length for a weir not sharp-crested is given by

$$L = L' - 2(NK_p + K_a)H_t \qquad (21\text{-}123)$$

where L = effective crest length, ft

L' = net crest length, ft = measured length minus width of all piers

N = number of piers

K_p = pier-contraction coefficient

K_a = abutment-contraction coefficient

H_t = total head on crest including velocity head of approach, ft

(U.S. Department of the Interior, "Design of Small Dams," Government Printing Office, Washington, D.C.)

The pier-contraction coefficient K_p is affected by the shape and location of the pier nose, thickness of pier, head in relation to design head, and approach velocity. For conditions of design head H, the average pier-contraction coefficients are as shown in Table 21-13.

Table 21-13. Pier-Contraction Coefficients

Condition	K_p
Square-nosed piers with corners rounded on a radius equal to about 0.1 of the pier thickness	0.02
Round-nosed piers	0.01
Pointed-nosed piers	0

The abutment-contraction coefficient K_a is affected by the shape of the abutment, the angle between the upstream approach wall and the axis of flow, the head in relation to the design head, and the approach velocity. For conditions of design head H_d, average coefficients may be assumed as shown in Table 21-14.

21-45. Submergence of Weirs Not Sharp-crested. Spillways and other weirs not sharp-crested are submerged when their tailwater level is high enough to affect their discharge. Because of the surface disturbance produced in the vicinity of the crest, such a spillway or weir is unsatisfactory for accurate flow measurement.

Table 21-14. Abutment-contraction Coefficients

Condition	K_a
Square abutment with headwall at 90° to direction of flow	0.20
Rounded abutments with headwall at 90° to direction of flow when $0.5H_d \geq r^* \geq 0.15H_d$	0.10
Rounded abutments where $r^* > 0.5H_d$ and headwall is placed not more than 45° to direction of flow	0

$*r$ = radius of abutment rounding.

Approximate values of discharge may be found by applying the following rules proposed by E. F. Brater: "(1) If the depth of submergence is not greater than 0.2 of the head, ignore the submergence and treat the weir as though it had free discharge. (2) For narrow weirs having a sharp upstream leading edge, use a submerged-weir formula for sharp-crested weirs. (3) Broad-crested weirs are not affected by submergence up to approximately 0.66 of the head. (4) For weirs with narrow rounded crests, increase discharge obtained by a formula for submerged sharp-crested weirs by 10% or more. Of the above rules, 1, 2, and 3 probably apply quite accurately, while 4 is simply a rough approximation." (L. C. Urquhart, "Civil Engineering Handbook," McGraw-Hill Book Company, New York.)

21-46. The Ogee-crested Weir. The ogee-crested weir was developed in an attempt to produce a weir that would not have the undesirable nappe variation normally associated with weirs not sharp-crested. A shape was needed that would force the nappe to assume a single path for any discharge, thus making the weir consistent for flow measurement. The ogee-crested weir (Fig. 21-65) has such a shape. Its crest profile conforms closely to the profile of the lower surface of a ventilated nappe flowing over a rectangular sharp-crested weir.

The shape of this nappe, and therefore of an ogee crest, depends on the head producing the discharge. Consequently, an ogee crest is designed for a single total head, called the design head H_d. When an ogee weir is discharging at the design head, the flow glides over the crest with no interference from the boundary surface and attains near-maximum discharge efficiency.

For flow at heads lower than the design head, the nappe is supported by the crest and pressure develops on the crest that is above atmospheric but less than hydrostatic. This crest pressure reduces the discharge below that for ideal flow. (Ideal flow is flow over a fully ventilated sharp-crested weir under the same head H.)

When the weir is discharging at heads greater than the design head, the pressure on the crest is less than atmospheric, and the discharge increases over that for ideal flow. The pressure may become so low that separation in flow will occur; however, according to Chow, the design head may be safely exceeded by at least 50% before harmful cavitation develops (V. T. Chow, "Open-Channel Hydraulics," McGraw-Hill Book Company, New York).

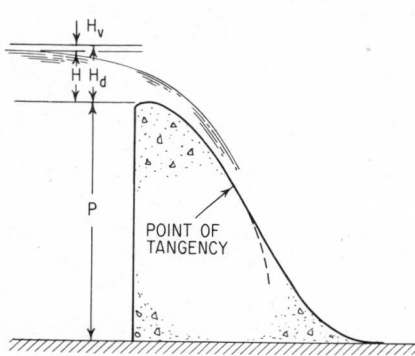

Fig. 21-65. Ogee-crested weir with vertical upstream face.

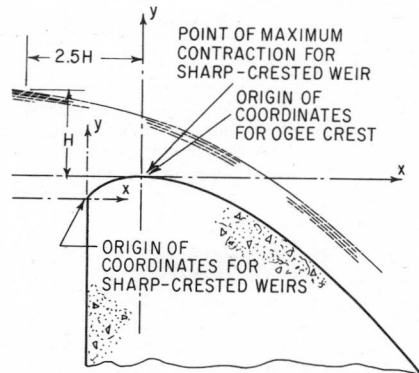

Fig. 21-66. Location of origin of coordinates for sharp-crested and ogee-crested weirs.

The measured head H on an ogee-crested weir is taken as the distance from the highest point of the crest to the level of the water surface at a distance $2.5H$ upstream. This depth coincides with the depth measured between the upstream water level and the bottom of the nappe, at the point of maximum contraction, for a sharp-crested weir. This relationship is shown in Fig. 21-66.

Discharge coefficients for ogee-crested weirs are therefore determined from sharp-crested-weir coefficients after an adjustment for this difference in head. These coefficients are a function of the approach velocity, which varies with the ratio of height of weir P to actual total head H_t where discharge is given by Eq. (21-122). Figure 21-67 for an ogee weir with a vertical upstream face gives coefficient C_d for discharge at design head H_d. (U.S. Department of the

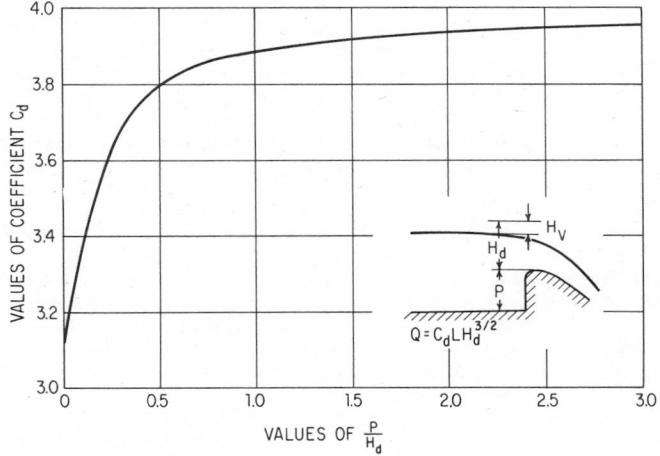

Fig. 21-67. Chart gives discharge coefficients at design head H_d for vertical-faced ogee-crested weirs. (*From "Design of Small Dams," U.S. Department of Interior.*)

Interior, "Design of Small Dams," Government Printing Office, Washington, D.C. This manual and V. T. Chow, "Open-Channel Hydraulics," McGraw-Hill Book Company, New York, present methods for determination of the shape of an ogee crest profile.) When the weir is discharging at other than the design head, the flow differs from ideal, and the discharge coefficient changes from that given in Fig. 21-67.

Figure 21-68 gives values of the discharge coefficient C as a function of the ratio H_t/H_d, where H_t is the actual head being considered and H_d is the design head.

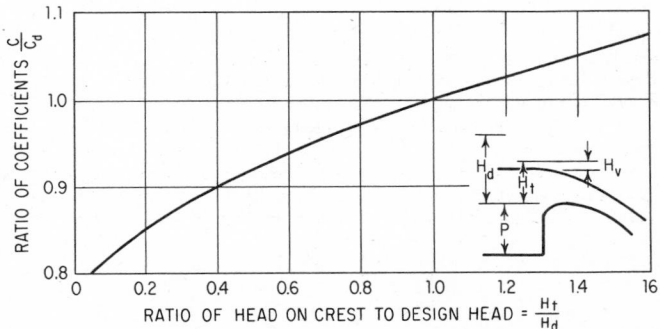

Fig. 21-68. Chart gives discharge coefficients for vertical-faced ogee-crested weirs at head H_t other than design head. (*From "Design of Small Dams," U. S. Department of Interior.*)

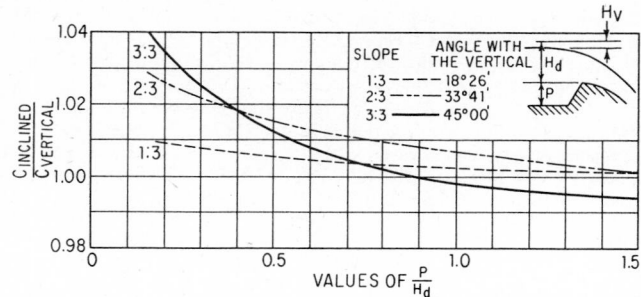

Fig. 21-69. Chart gives discharge coefficients at design head H_d for ogee-crested weirs with sloping upstream face. (*From "Design of Small Dams," U.S. Department of Interior.*)

If an ogee weir has a sloping upstream face, there is a tendency for an increase in discharge over that for a weir with a vertical face.   Figure 21-69 shows the ratio of the coefficient for an ogee weir with a sloping face to the coefficient for a weir with a vertical upstream face.   The coefficient of discharge for an ogee weir with a sloping upstream face, if flow is at other than the design head, is determined from Fig. 21-67 and is then corrected for head and slope with Figs. 21-68 and 21-69.

21-47. Broad-crested Weir.   This is a weir with a horizontal or nearly horizontal crest.   The crest must be sufficiently long in the direction of flow that the nappe is supported and hydrostatic pressure developed on the crest for at least a short distance.   A broad-crested weir is nearly rectangular in cross section.   Unless otherwise noted, it will be assumed to have vertical faces, a plane horizontal crest, and sharp right-angled edges.

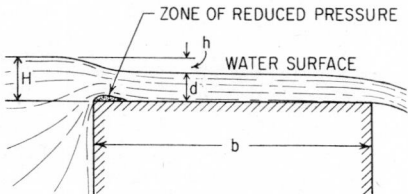

Fig. 21-70.   Broad-crested weir.

Figure 21-70 shows a broad-crested weir that, because of its sharp upstream edge, has contraction of the nappe.   This causes a zone of reduced pressure at the leading edge.   When the head H on a broad-crested weir reaches one to two times its breadth b, the nappe springs free, and the weir acts as a sharp-crested weir.

Discharge over a broad-crested weir is given by Eq. (21-115), since the velocity of approach was ignored in experiments performed to determine the coefficient of discharge.   These coefficients probably apply more accurately, therefore, where the velocity of approach is not high.   Values of the discharge coefficient, compiled by King, appear in Table 21-15 (H. W. King and E. F. Brater, "Handbook of Hydraulics," McGraw-Hill Book Company, New York).

21-48. Weirs of Irregular Section.   This group includes those weirs whose cross section deviates from typical broad-crested weirs or ogee-crested weirs.   Weirs of irregular section, fairly common in waterworks projects, are used as spillways and control structures.   Experimental data are available on the more common shapes.   (See, for example, H. W. King and E. F. Brater, "Handbook of Hydraulics," McGraw-Hill Book Company, New York.)

21-49. Sediment Transfer and Deposition in Open Channels.   According to data gathered by B. J. Witzig in 1943, 64% of all reservoirs in the United States have a useful life of less than 100 years because of loss of storage through the accumulation of silt.   Sediment causes a hazard in navigable channels and harbors and an increase in frequency of flooding due to aggradation of rivers and flood channels.   Silting of arable land by flooding rivers destroys fertility when the silt originates from bank or gully erosion rather than from surface, or soil, erosion.   The cost of operating irrigation systems is increased by the need for frequent dredging.   Water-supply facilities have increased costs because of the necessity of providing desilting works and because

Table 21-15. Values of C in $Q = CLH^{3/2}$ for Broad-crested Weirs

Measured head H ft	Breadth of crest of weir, ft										
	0.50	0.75	1.00	1.50	2.00	2.50	3.00	4.00	5.00	10.00	15.00
0.2	2.80	2.75	2.69	2.62	2.54	2.48	2.44	2.38	2.34	2.49	2.68
0.4	2.92	2.80	2.72	2.64	2.61	2.60	2.58	2.54	2.50	2.56	2.70
0.6	3.08	2.89	2.75	2.64	2.61	2.60	2.68	2.69	2.70	2.70	2.70
0.8	3.30	3.04	2.85	2.68	2.60	2.60	2.67	2.68	2.68	2.69	2.64
1.0	3.32	3.14	2.98	2.75	2.66	2.64	2.65	2.67	2.68	2.68	2.63
1.2	3.32	3.20	3.08	2.86	2.70	2.65	2.64	2.67	2.66	2.69	2.64
1.4	3.32	3.26	3.20	2.92	2.77	2.68	2.64	2.65	2.65	2.67	2.64
1.6	3.32	3.29	3.28	3.07	2.89	2.75	2.68	2.66	2.65	2.64	2.63
1.8	3.32	3.32	3.31	3.07	2.88	2.74	2.68	2.66	2.65	2.64	2.63
2.0	3.32	3.31	3.30	3.03	2.85	2.76	2.72	2.68	2.65	2.64	2.63
2.5	3.32	3.32	3.31	3.28	3.07	2.89	2.81	2.72	2.67	2.64	2.63
3.0	3.32	3.32	3.32	3.32	3.20	3.05	2.92	2.73	2.66	2.64	2.63
3.5	3.32	3.32	3.32	3.32	3.32	3.19	2.97	2.76	2.68	2.64	2.63
4.0	3.32	3.32	3.32	3.32	3.32	3.32	3.07	2.79	2.70	2.64	2.63
4.5	3.32	3.32	3.32	3.32	3.32	3.32	3.32	1.88	2.74	2.64	2.63
5.0	3.32	3.32	3.32	3.32	3.32	3.32	3.32	3.07	2.79	2.64	2.63
5.5	3.32	3.32	3.32	3.32	3.32	3.32	3.32	3.32	2.88	2.64	2.63

of the wear on mechanical equipment, such as gates, valves, and turbines. (B. J. Witzig, "Sedimentation in Reservoirs," Paper 2227, *Transactions of the American Society of Civil Engineers,* vol. 109, pp. 1047–1067, 1944.)

Deposition of silt results when the transporting forces of a river are dissipated as the river enters a body of still water, such as a reservoir. Heavier silt sizes, those forming the bed load, are deposited in a delta as the river enters calm water. The smaller silt sizes, those carried in suspension, travel farther into the reservoir before deposition.

This incoming water, with its load of suspended silt, has a specific gravity greater than that of the clear water in the reservoir and may form a *density current,* rather than mixing with the clear water immediately. A density current, once formed, quickly moves to the bottom and flows in a dense cloud down the slopes of the reservoir until it is blocked by the dam. The dense flow then spreads out in this deeper area, where the stilling effect of the basin eventually causes deposition of the sediment. Deposits of fine sediment form about one-third of the volume of silt deposits in a reservoir. Much or all of this fine sediment is transported to its final location by density currents. The visible delta formed by the coarse sediments frequently distracts attention from the unseen bottom deposits of fine sediment, which are often of equal consequence.

Most reservoirs trap from 70 to almost 100% of the incoming sediment, depending on whether the reservoir is used for flood control or for storage. Flood-control reservoirs are normally emptied shortly after a storm, and so the suspended materials are carried out with the water before settling can occur. This procedure results in the reduction of new deposits by almost 30% after each storm. Storage reservoirs, on the other hand, normally retain any inflow long enough for settlement of all suspended matter to occur. Their discharges are regulated to allow generation of power or to produce a uniform flow downstream with no thought to the venting of silt-laden storm flows.

The greater part of the annual suspended silt load in a stream may be carried in a relatively short time. The stream runs comparatively clear during the remainder of the year.

Venting of much of the annual suspended silt load is feasible through the use of density currents. These currents are stable, once formed, and often extend to the reservoir outlet. If density currents are observed and their time of arrival at the outlet determined, appropriate gates can be opened and much of the fine sediment entering a storage reservoir can be vented before it has time to form permanent deposits. This venting operation can extend the life of a reservoir by many years.

There are numerous phenomena that can destroy a reservoir, such as loss of storage capacity by landslide, and loss of the dam by earthquake, landslide, overtopping, or failure of materials. The commonest manner of destruction, however, is through loss of storage by deposition of silt.

Redemption of reservoir capacity lost through silting is almost always economically unfeasible because of the wide distribution of deposits in a reservoir and the large quantity present. The most practicable means of avoiding a loss in reservoir capacity are to prevent formation of permanent deposits by taking advantage of density currents and to control rate of sediment production from eroding areas. When neither can be done, sufficient storage space must be provided in the design of the reservoir to compensate for depletion by silting during a reasonable economic lifetime.

Sediment production and its transportation to reservoirs or navigable waters cannot be prevented at costs proportionate to the resulting benefits. However, nature may be economically improved on at times through a program of erosion control, reducing sediment production to less than that normally found under virgin conditions.

The deposits of silt that form in a storage reservoir are categorized into two distinct types: *Delta deposits,* formed from the *bed load,* are coarse-grained, with an in-place weight of about 80 lb per cu ft. Deposits produced from the *suspended load* are fine-grained, with an average weight of about 30 lb per cu ft. They constitute about one-sixth of the total weight of sediment delivered but account for about one-third of the volume of all deposits in a storage reservoir because of their low density. If sediment deposits are periodically above water, because of fluctuations in the reservoir water level, their density increases and the volume ratios given above for continued submergence no longer apply.

21-50. Erosion Control. The various methods used in erosion control are collectively called *upstream engineering.* They consist of reforestation, check-dam construction, planting of burned-over areas, contour plowing, and regulation of crop and grazing practices. Also included are measures for proper treatment of high embankments and cuts, and stabilization of stream banks by planting or by revetment construction.

One phase of reforestation that may be applied near a reservoir is planting of vegetation screens. Such screens, planted on the flats adjacent to the normal stream channel at the head of a reservoir, reduce the velocity of silt-laden storm inflows that inundate these areas. This stilling action causes extensive deposition to occur before the silt reaches the main cavity of the reservoir. Use of vegetation screens, debris barriers, or desilting basins above a reservoir should be planned with future development in mind. For instance, if the dam is raised at a later date, the accumulated silt in this area would detract from the added storage that might otherwise have been obtained.

21-51. Prediction of Sediment-Delivery Rate. Two methods of approach are available for predicting the rate of sediment accumulation in a reservoir. They both involve predicting the rate of sediment delivery.

One approach depends on historical records of the silting rate for existing reservoirs and is purely empirical. By this method, the silting records of a reservoir may be used to predict either the silting rate for that reservoir or the probable pattern of silt accumulation for a proposed reservoir in a similar area. This method allows transposition of data from one watershed to another because the measured annual sediment accumulation of a reservoir is expressed as a *rate of sediment delivery* per unit area of its watershed. Of course, the rate is not uniform during the year, or from year to year, because of variations in rainfall, but it should average the computed annual amount over the life of the project. The annual silt accumulation in a reservoir is determined by surveying exposed deltas and by taking depth soundings. The resulting volume is adjusted to account for any silt loss through sluice gates or over the spillway and is then expressed as silt delivery per square mile of drainage area. This silt-delivery figure is further adjusted for rainfall and runoff conditions, to give a figure that could reasonably be expected during a year of average rainfall. If this adjusted figure is to be transposed to a neighboring drainage basin, adjustments should be made to account for both soil cover and rainfall differences between the basins. (For a discussion of the factors upon which this adjustment is based, see Art. 21-55.)

Silt-delivery figures do not give total silt production for an area, because part of the silt produced in a basin is deposited on floodplains and in channels before it reaches a reservoir. The difference between the amount of silt produced and that delivered increases as the size of the drainage area increases, because of the increased chance that the silt will be deposited before it reaches the reservoir. If, therefore, a silt-delivery figure is to be transposed to a basin of different size, an adjustment should be made to account for this discrepancy as well. The information for this adjustment can come only from field reconnaissance of the two areas to determine differences that might account for a variation in deposition of silt along the water courses.

The second general method of calculating sediment-delivery rate involves determining the rate of sediment transport as a function of stream discharge and density of suspended silt. The total sediment inflow for the year is then computed from these relationships and the recorded stream-discharge data.

The total quantity of sediment carried by a river is assumed to be transported either as suspended load or as bed load (Art. 21-49). The division is based on particle size but depends on velocity of flow as well, so that there is no sharp line of demarcation between the two classes. According to Witzig, about 80% of the volume of all sediment is produced by stream-bank erosion. The remaining 20% is produced by land-surface erosion. Constant erosion of the stream banks keeps the stream bed well supplied with the coarse silt that travels as bed load. The fine silt that travels in suspension is produced in small amounts by stream-bank erosion. But for the most part, this silt comes from land-surface erosion, which generally occurs only during a storm.

The total quantity of sediment in suspension is not necessarily related directly to discharge at all times. The quantity is affected by seasonal variations in the supply and source of fine sediment and by distribution of rainfall and runoff from the watershed. Therefore, measurement of the sediment load for a given discharge does not necessarily indicate the amount that may be carried by an equal discharge at another time.

The bed load consists of the silt particles too large to be held in suspension. This size range includes particles of coarse sand, gravel, and boulders. The bed-load particles are moved by rolling along the bed of the stream. Some of the finer bed-load particles are moved in a series of steps or jumps representing a transition between transportation as bed load and suspended load.

The quantity of bed load is considered to be a constant function of the discharge, because the sediment supply for the bed-load forces is always available in all but lined channels. An accepted formula for the quantity of sediment transported as bed load is the Schoklitsch formula:

$$G_b = \frac{86.7}{D_g^{1/2}} S^{3/2}(Q_i - bq_o) \qquad (21\text{-}124)$$

where G_b = total bed load, lb per sec
D_g = effective grain diameter, in.
S = slope of energy gradient
Q_i = total instantaneous discharge, cfs
b = width of river, ft
q_o = critical discharge, cfs per ft of river width
$= (0.00532/S^{4/3})D_g$

An approximate solution for bed load by the Schoklitsch formula can be made by determining or assuming mean values of slope, discharge, and a single grain size representative of the bed-load sediment. A mean grain size of 0.04 in. in diameter (about 1 mm) is reasonable for a river with a slope of about 1.0 ft per mile.

The size of grains moving on the bed of a river depends on velocity of flow, which varies with both slope and discharge. Therefore, the mean grain size changes as the flow increases during a storm or as the river changes slope along its course. It is obvious that considerable error could result from the use of Eq. (21-124) if it is necessary to guess at a mean grain diameter in the absence of carefully collected field data. Frequently, however, if insufficient data or lack of money prevent more thorough investigations, this short cut can give results of sufficient accuracy.

Numerous formulas have been developed to represent the condition of flow involved in transportation of suspended sediment. These formulas express the degree of turbulent energy involved in suspension of the sediment and the mode of transfer of this energy to the silt and other fluid particles. The formulas require a number of empirical constants but are based on a sound physical and rational foundation. They require information as to the sediment composition by grain size, the actual quantity of silt in suspension at a given depth, and the stream velocity. (See H. A. Einstein, "The Bed-Load Function for Sediment Transportation in Open-Channel Flows," U.S. Department of Agriculture.)

An approximate determination of suspended load may be made without using these complicated formulas. The weight of suspended sediment transported by a river in an average year normally equals about 20% of the weight transported as bed load. The total weight of material annually moved by a river is therefore equal to 120% of the weight of material transported as bed load during the year as computed from Eq. (21-124).

(W. H. Graf, "Hydraulics of Sediment Transport," McGraw-Hill Book Company, New York.)

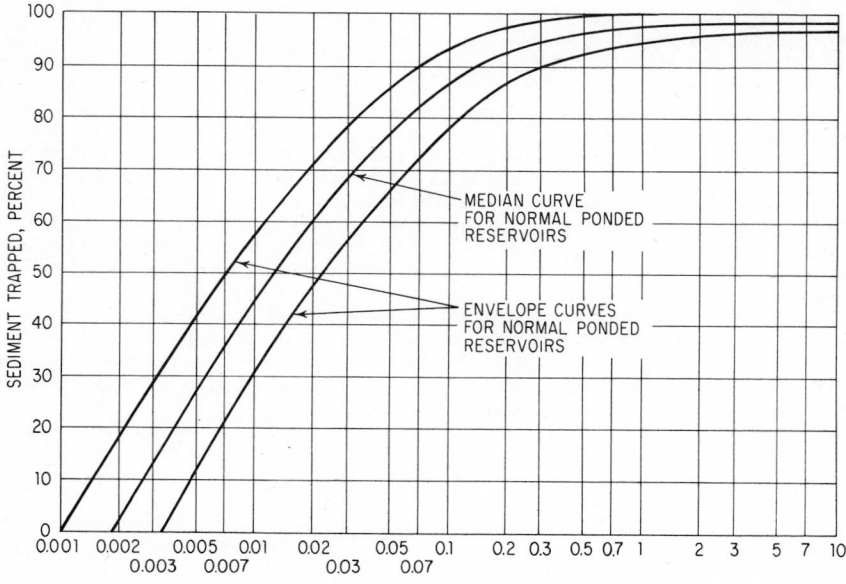

Fig. 21-71. Chart indicates percent of incoming sediment trapped in reservoirs.

21-52. Reservoir Trap Efficiency. The methods of Art. 21-51 for determining the quantities of sediment delivered to a reservoir require knowledge of the trap efficiency of the reservoir before the percentage of the incoming silt that will remain to reduce storage can be determined. Studies of trap efficiency were made by G. M. Brune, who developed a curve to express the relationship between trap efficiency and what he called the *capacity-inflow ratio* for a reservoir (Fig. 21-71) (G. M. Brune, "Trap Efficiency of Reservoirs," *Transactions of the American Geophysical Union,* vol. 34, no. 3, June, 1953).

The higher the capacity-inflow ratio, acre-feet of storage per acre-foot of annual inflow, the greater will be the percentage of sediment trapped in a reservoir. For any given storage reservoir, the trap efficiency decreases with time, since the capacity-inflow ratio decreases as sediment builds up. The rate of silting of a storage reservoir decreases when the capacity is reduced to an amount such that some spillage of silt-laden water occurs with each major storm. This rate decrease occurs because an increasing percentage of the annual suspended silt load is vented before sedimentation can occur.

HYDROLOGY

21-53. Precipitation. The primary concern with precipitation in water engineering is forecasting it. The means for doing so are based on either current or past data, or a combination of the two.

Current data, in the form of synoptic weather charts, are published daily by the U.S. Weather Bureau. These charts summarize the various meteorological factors, such as wind, temperature, and pressure, through whose interaction precipitation is produced.

Past data are primarily in the form of rainfall records for a standard period, such as an hour, day, or year. They provide the major source of data for determination of the recurrence interval for storms of a definite magnitude and the magnitude of storms in a definite recurrence interval.

Rainfall records are obtained from rain gages, which are of two types. The first type is a recording or automatic gage. It continually records, by ink pen and revolving drum, the variation in rainfall intensity, as well as the total rainfall volume. The second type is a nonrecording gage. It measures only the total rain volume that fell during the period between observations. The standard observation time for nonrecording gages for the U.S. Weather Bureau is 24 hr.

Corrections must be made to rain-gage records to account for the mean precipitation over the entire drainage basin, for hourly rainfall rates when only daily volumes are given, and for errors arising out of the location of the gage.   Most of the methods used in runoff determinations are based on the assumption that rainfall is uniform over the entire drainage basin.   This necessitates development of a correction factor to balance out the rainfall variation caused by various topographical features in the watershed.   Rain gages tend to give rainfall volumes that are too small. This error is caused by the movement of wind around the gage, and it increases as wind velocity increases.   This "windage" error is much more pronounced when the rain gage is near the top or bottom of a cliff or near other big obstructions.   Care must be exercised in placement of rain gages to insure accuracy.

The probable maximum precipitation is the greatest rainfall intensity or volume that could ever be expected to occur in a specific drainage basin.   This rainfall magnitude is frequently used as the design storm for major hydraulic structures to serve the basin when the rainfall records are short and extrapolation to the desired design-storm frequency could be grossly inaccurate. The magnitude of probable maximum precipitation is based on simultaneous occurrence of the maximum values of the meteorological factors that combine to form precipitation.   The two most important factors are wind and air-mass moisture content.   An idea of the magnitude of the probable maximum precipitation can also be obtained by transposing the greatest rainfall that has occurred in a meteorologically homogeneous region.   For a method for determining the probable maximum precipitation, see D. M. Hershfield, "Estimating the Probable Maximum Precipitation," *Journal of the Hydraulics Division*, American Society of Civil Engineers, vol. 87, no. HY 5, September, 1961.

Not all rain reaches the ground.   A portion may evaporate as it falls, while another portion may be caught on leaves, branches, and other vegetation surfaces.   This phenomenon, called **interception**, is a loss from a runoff standpoint, since the rain evaporates and never reaches the ground.   Interception may be significant for small-intensity storms occurring with little or no wind over an area with heavy vegetation growth.

21-54. Evaporation and Transpiration.   These are processes by which moisture is returned to the atmosphere.   Evaporation is the process in which water changes from liquid to gaseous form. Transpiration is the process by which plants give off water vapor during synthesis of plant tissue.

Evapo-transpiration, commonly termed **consumptive use,** refers to the total evaporation from all sources such as free water, ground, and plant-leaf surfaces.   On an annual basis, the consumptive use may vary from 15 in. per year for barren land to 35 in. per year for heavily forested areas.   Evapo-transpiration is important because, on a long-term basis, precipitation minus evapo-transpiration equals runoff.

Evaporation may occur from free-water, plant, or ground surfaces.   Of the three, free-water surface evaporation is usually the most important.   It must be considered in the design of a reservoir, especially if the reservoir is shallow, has a relatively large surface area, and is located in a semiarid or arid region.   Evaporation is a direct function of the wind and temperature and an inverse function of atmospheric pressure and amount of soluble solids in the water.

The rate of evaporation is dependent on the vapor-pressure gradient between the water surface and the air above it.   This relation is known as Dalton's law.   The **Meyer equation** [Eq. (21-125)] was developed from Dalton's law.   It is one of many evaporation formulas and is popular for making evaporation-rate calculations (V. T. Chow, "Handbook of Applied Hydrology," McGraw-Hill Book Company, New York).

$$E = C(e_w - e_a)\psi \qquad (21\text{-}125)$$

$$\psi = 1 + 0.1w \qquad (21\text{-}126)$$

where E = evaporation rate, in. per 30-day month
$\quad C$ = an empirical coefficient, equal to 15 for small, shallow pools and 11 for large, deep reservoirs
$\quad e_w$ = saturation vapor pressure, in. of mercury, corresponding to monthly mean air temperature observed at nearby stations for small bodies of shallow water or corresponding to water temperature instead of air temperature for large bodies of deep water
$\quad e_a$ = actual vapor pressure, in. of mercury, in the air based on monthly mean air temperature and relative humidity at nearby stations for small bodies of shallow water or based on information obtained about 30 ft above the water surface for large bodies of deep water
$\quad w$ = monthly mean wind velocity, mph at about 30 ft above ground
$\quad \psi$ = a wind factor

As an example of the evaporation that may occur from a large reservoir, the mean annual evaporation from Lake Mead is 6 ft.

Evaporation from free-water surfaces is usually measured with an evaporation pan. This pan is a standard size and is located on the ground near the body of water whose evaporation is to be determined. The depth of water in this pan is checked periodically and corrections made for factors other than evaporation that may have raised or lowered the water surface. A pan coefficient is then applied to the measured pan evaporation to get the reservoir evaporation.

The standard evaporation pan of the U.S. Weather Bureau, called a Class A Level Pan, is in widespread use. It is 4 ft in diameter and 10 in. deep. It is positioned 6 in. above the ground. Its pan coefficient is commonly taken as 0.70, although it may vary between 0.60 and 0.80, depending on the geographical region. Annual evaporation from the pan ranges from 25 in. in Maine and Washington to 120 in. along the Texas-Mexico and California-Arizona borders.

Evaporation rates from reservoirs may be reduced by spreading thin molecular films on the water surface. Hexadeconal, or cetyl alcohol, is one such film that has been effective on small reservoirs where there is little wind. On large reservoirs, wind tends to push the film to the shore. Since hexadeconal is removed by wind, birds, insects, aquatic life, and biologic attrition, it must be applied periodically for maximum effectiveness. Hexadeconal appears to have no adverse effects on either humans or wildlife.

Evaporation from ground surfaces is usually of minor importance, except in arid regions having high water tables and where it pertains to the determination of initial soil-moisture conditions in a runoff problem.

21-55. Runoff. This is the residual precipitation remaining after interception and evapotranspiration losses have been deducted. It appears in surface channels, natural or man-made, whose flow is of perennial or intermittent form. Classified by the path taken to a channel, runoff may be surface, subsurface, or groundwater flow.

Surface flow moves across the land as *overland flow* until it reaches a channel, where it continues as *channel* or *stream flow.* After joining stream flow, it combines with the other runoff components in the channel to form *total runoff.*

Subsurface flow, also known as *interflow, subsurface runoff, subsurface storm flow,* and *storm seepage,* infiltrates only the upper soil layers without joining the main groundwater body. Moving laterally, it may continue underground until it reaches a channel or returns to the surface and continues as overland flow. The time for subsurface flow to reach a channel depends on the geology of the area. Commonly, it is assumed that subsurface flow reaches a channel during or shortly after a storm. Subsurface flow may be the major portion of total runoff for moderate or light rains in arid regions, since surface flow under those conditions is reduced by unusually high evaporation and infiltration.

Groundwater flow, or **groundwater runoff,** is that flow supplied by deep percolation. It is the flow of the main groundwater body and requires long periods, perhaps several years, to reach a channel. Groundwater flow is responsible for the dry-weather flow of streams and remains practically constant during a storm. Groundwater flow is primarily the concern of water-supply engineers, while surface and subsurface flow are of interest to flood-control engineers.

In practice, **direct runoff** and **base flow** are the only two divisions of runoff used. The basis for this classification is travel time rather than path. Direct runoff leaves the basin during or shortly after a storm, whereas base flow from the storm may not leave the basin for months or even years.

Runoff is supplied by precipitation. The portion of precipitation that contributes entirely to direct runoff is called *effective precipitation,* or *effective rain* if the precipitation is rain. That portion of the precipitation that contributes entirely to surface runoff is called *excess precipitation,* or *excess rain.* Thus, effective rain includes subsurface flow, while excess rain is only surface flow.

The two major groups of factors that affect runoff are climatic characteristics and drainage-basin characteristics. A list of these in outline form follows. The number of factors is an indication of the complexity of accurately determining runoff.

1. Climatic characteristics
   a. Precipitation—form (rain, hail, snow, frost, dew), intensity, duration, time distribution, seasonal distribution, areal distribution, recurrence interval, antecedent precipitation, soil moisture, direction of storm movement
   b. Temperature—variation, snow storage, frozen ground during storms, extremes during precipitation
   c. Wind—velocity, direction, duration

    d. Humidity
    e. Atmospheric pressure
    f. Solar radiation
2. Drainage-basin characteristics
    a. Topographic—size, shape, slope, elevation, drainage net, general location, land use and cover, lakes and other bodies of water, artificial drainage, orientation, channels (size, shape of cross section, slope, roughness, length)
    b. Geologic—soil type, permeability, groundwater formations, stratification

21-56. Sources of Hydrologic Data. The importance of exhausting all possible sources of hydrologic data, both published and unpublished, as the first step in design of a hydraulic project cannot be overemphasized. The majority of hydrologic data is collected and published by governmental agencies, those of the Federal government being the largest and most important. A list of the various departments and bureaus in the Federal government that collect and publish hydrologic data is contained in a report, "Principal Federal Sources of Hydrologic Data," by the U.S. Inter-agency Committee on Water Resources, 1956. The first volume of the two-volume report lists the sources, indexed by the type of data, such as precipitation, runoff, and wind. The second volume contains data from numerous experimental stations, indexed by the experimental station that collected it. Another good reference is "Inventory of Unpublished Hydrologic Data," *Water Supply Paper* 837, U.S. Geological Survey, 1938. Unpublished data may be examined at the main office of the local agency involved.

The principal source of precipitation data is the U.S. Weather Bureau. Its extensive system of gages supplies complete precipitation data, as well as all other types of hydrologic data. These data are compiled and presented in monthly and yearly summaries in the Bureau's "Climatological Data." In addition to the monthly and yearly summaries, special-interest items, such as rainfall intensity for various durations and recurrence intervals, are published in Weather Bureau technical papers. The Weather Bureau also publishes records collected by the Forest Service, Tennessee Valley Authority, Bureau of Reclamation, Corps of Engineers, and Bureau of Plant Industry.

Other sources are *Water Bulletins* of the International Boundary Commission, the U.S. Agricultural Research Service, and various state and local agencies.

The principal source of runoff data is the *Water Supply Papers* of the U.S. Geological Survey. These papers contain records of daily flow, mean flow, yearly flow volume, extremes of flow, and statistical data pertaining to the entire record. Also included in the *Papers* are lists of reports covering unusually large floods and records of discharge collected by agencies other than the U.S. Geological Survey. The *Water Supply Papers* are published yearly in 14 parts. Each part is for an area whose boundaries coincide with natural-drainage features, as shown in Fig. 21-72.

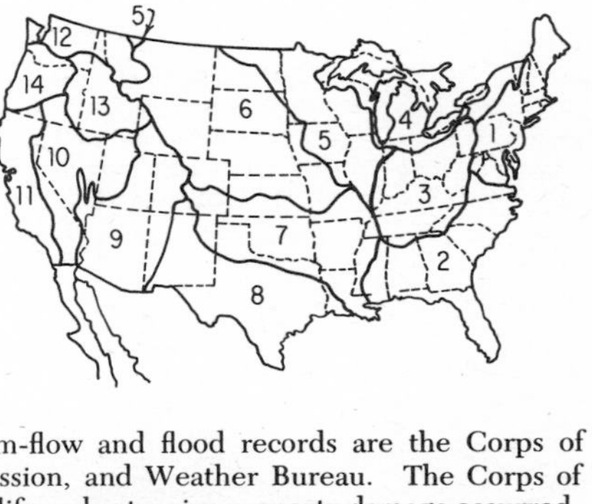

Fig. 21-72. Drainage subdivisions of the United States for stream-flow records published in *Water Supply Papers* of U.S. Geological Survey.

Other agencies that collect and publish stream-flow and flood records are the Corps of Engineers, TVA, International Boundary Commission, and Weather Bureau. The Corps of Engineers publishes data on floods in which loss of life and extensive property damage occurred. Less obvious sources of stream-flow data are water-right decrees by district courts, county records of water-right filings and State Engineer permits, and annual reports of various interstate-compact commissions.

21-57. Methods for Runoff Determinations. The method selected to determine runoff depends on its applicability to the area of concern, the quantity and type of data available, the detail required in the final answer, and the accuracy desired. Applicability depends on the characteristics of the particular area and the assumptions from which the method was developed.

Quantity and type of data available refer to the length, detail, and completeness of the hydrologic records, which may be either precipitation or stream flow. An example of the variation of detail in the final result may be found in the determination of flood runoff. Several methods yield only peak discharge, while others give the complete hydrograph. Accuracy is limited by cost and by assumptions made in the development of a method.

The methods that follow provide a convenient means for solving typical runoff problems encountered in water engineering. One method pertains to minor hydraulic structures, and a second, to major hydraulic structures. A minor structure is one of low cost and of relatively minor importance and presents small downstream damage potential. Typical examples are small highway and railroad culverts and low-capacity storm drains. Major hydraulic structures are characterized by their high cost, great importance, and large downstream damage potential. Typical examples of major hydraulic structures are large reservoirs, deep culverts under vital highways and railways, and high-capacity storm drains and flood-control channels.

Method for Determining Runoff for Minor Hydraulic Structures. The commonest means for determining runoff for minor hydraulic structures is the **rational formula**

$$Q = CIA \tag{21-127}$$

where Q = peak discharge, cfs
C = runoff coefficient = percentage of rain that appears as direct runoff
I = rainfall intensity, in. per hr
A = drainage area, acres

The assumptions inherent in the rational formula are:

1. The maximum rate of runoff for a particular rainfall intensity occurs if the duration of rainfall is equal to or greater than the time of concentration. The time of concentration is commonly defined as the time required for water to flow from the most distant point of a drainage basin to the point of flow measurement.

2. The maximum rate of runoff from a specific rainfall intensity whose duration is equal to or greater than the time of concentration is directly proportional to the rainfall intensity.

Table 21-16. Common Runoff Coefficients

Type of Drainage Area	Runoff Coefficient C
Business:	
Downtown areas	0.70–0.95
Neighborhood areas	0.50–0.70
Residential:	
Single-family areas	0.30–0.50
Multi-units, detached	0.40–0.60
Multi-units, attached	0.60–0.75
Suburban	0.25–0.40
Apartment dwelling areas	0.50–0.70
Industrial:	
Light areas	0.50–0.80
Heavy areas	0.60–0.90
Parks, cemeteries	0.10–0.25
Playgrounds	0.20–0.35
Railroad-yard areas	0.20–0.40
Unimproved areas	0.10–0.30
Streets:	
Asphaltic	0.70–0.95
Concrete	0.80–0.95
Brick	0.70–0.85
Drives and walks	0.75–0.85
Roofs	0.75–0.95
Lawns:	
Sandy soil, flat, 2%	0.05–0.10
Sandy soil, avg, 2–7%	0.10–0.15
Sandy soil, steep, 7%	0.15–0.20
Heavy soil, flat, 2%	0.13–0.17
Heavy soil, avg, 2–7%	0.18–0.22
Heavy soil, steep, 7%	0.25–0.35

3. The frequency of occurrence of the peak discharge is the same as that of the rainfall intensity from which it was calculated.

4. The peak discharge per unit area decreases as the drainage area increases, and the intensity of rainfall decreases as its duration increases.

5. The coefficient of runoff remains constant for all storms on a given watershed.

Since these assumptions apply reasonably well for urbanized areas with drainage facilities of fixed dimensions and hydraulic characteristics, the rational formula has gained widespread use in the design of drainage systems for these areas. Its simplicity and ease of application have resulted in its being used in rural areas where the assumptions are not so applicable.

The rational formula is criticized for expressing runoff as a fraction of rainfall rather than as rainfall minus losses, and for combining all the complex factors that affect runoff into a single coefficient. While these and similar criticisms are valid, use of a more complicated formula is not justified, because the time and money spent to obtain the necessary data would not be warranted for minor hydraulic structures.

Numerous refinements have been developed for the runoff coefficient. As an example, the Los Angeles County Flood Control District gives runoff coefficients as a function of the soil and area type and of the rainfall intensity for the time of concentration. Other similar refinements are possible if the resources are available. Careful selection of the runoff coefficient C will give values of peak runoff consistent with project significance. The values of C given in Table 21-16 for urban areas are commonly recommended design values (V. T. Chow, "Hydrologic Determination of Waterway Areas for the Design of Drainage Structures in Small Drainage Basins," *University of Illinois Engineering Experimental Station Bulletin* 426, 1962).

After selection of the design-storm frequency of occurrence, for example, a 50- or 100-year-frequency storm, the rainfall intensity I may be determined from any of a number of formulas or from a statistical analysis of rainfall data if enough are available. Chow lists 24 rainfall-intensity formulas of the form

$$I = \frac{KF^{n_1}}{(t + b)^n} \tag{21-128}$$

where I = rainfall intensity, in. per hr
K, b, n, and n_1 = respectively, coefficient, factor, and exponents depending on conditions that affect rainfall intensity
F = frequency of occurrence of rainfall, years
t = duration of storm, min
= time of concentration

Perhaps the most useful of these formulas is the **Steel formula:**

$$I = \frac{K}{t + b} \tag{21-129}$$

where K and b are dependent on the storm frequency and region of the United States (Fig. 21-73 and Table 21-17).

Table 21-17. Coefficients for Steel Formula

Frequency, years	Coefficients	Region						
		1	2	3	4	5	6	7
2	K	206	140	106	70	70	68	32
	b	30	21	17	13	16	14	11
4	K	247	190	131	97	81	75	48
	b	29	25	19	16	13	12	12
10	K	300	230	170	111	111	122	60
	b	36	29	23	16	17	23	13
25	K	327	260	230	170	130	155	67
	b	33	32	30	27	17	26	10
50	K	315	350	250	187	187	160	65
	b	28	38	27	24	25	21	8
100	K	367	375	290	220	240	210	77
	b	33	36	31	28	29	26	10

Fig. 21-73. Regions of the United States for use with the Steel formula.

Equation (21-129) gives the average maximum precipitation rates for durations up to 2 hr.

The time of concentration T_c at any point in a drainage system is the sum of the overland flow time; the flow time in streets, gutters, or ditches; and the flow time in conduits. Overland flow time may be determined from any number of nomographs that have been developed for the purpose. (See V. T. Chow, "Handbook of Applied Hydrology," McGraw-Hill Book Company, New York.) The flow time in gutters, streets, ditches, and conduits can be determined from a calculation of the average velocity using the Manning equation [Eq. (21-89)]. The time of concentration is usually expressed in minutes.

After determining the time of concentration, calculate the corresponding rainfall intensity from either Eq. (21-128) or Eq. (21-129), or any equivalent method. Then select the runoff coefficient from Table 21-16 and determine the peak discharge from Eq. (21-127).

Since the rational formula assumes a constant uniform rainfall for the time of concentration over the entire area, the area A must be selected so that this assumption applies with reasonable accuracy. Adhering to this assumption may necessitate subdividing the drainage area.

Method for Determining Runoff for Major Hydraulic Structures. The unit-hydrograph method, pioneered in 1932 by LeRoy K. Sherman, is a convenient, widely accepted procedure for determining runoff for major hydraulic structures. (Leroy K. Sherman, "Streamflow from Rainfall by Unit-graph Method," *Engineering News-Record*, vol. 108, pp. 501–505, January–June, 1932.) It permits calculation of the complete runoff hydrograph from any rainfall after the unit hydrograph has been established for the particular area of concern.

The **unit hydrograph** is defined as a runoff hydrograph resulting from a *unit storm*. A unit storm has practically constant rainfall intensity for its duration, termed a *unit period*, and a runoff volume of 1 in. (water with a depth of 1 in. over a unit area, usually 1 acre). Thus, a unit storm may have a 2-in.-per-hr effective intensity lasting ½ hr or a 0.2-in.-per-hr effective intensity lasting 5 hr. The significant part of the definition is not the volume but the constancy of intensity. Adjustments can be made within unit-hydrograph theory for situations where the runoff volume is different from 1 in., but corrections for highly variable rainfall rates cannot be made.

The unit hydrograph is similar in concept to determining a set of factors for a specific drainage basin. The set consists of one factor for each variable that affects runoff. The unit hydrograph is much quicker, easier, and more accurate than any such set of factors. The method is summarized by the formula

$$\text{Effective rain} \times \text{unit hydrograph} = \text{runoff} \tag{21-130}$$

The unit hydrograph, thus, is the link between rainfall and runoff. It may be thought of as an integral of the many complex factors that affect runoff. The unit hydrograph can be derived from rainfall and stream-flow data for a particular storm or from stream-flow data alone.

Assumptions made in the development of the unit-hydrograph theory are:

1. Rainfall intensity is constant for its duration or a specified period of time. This requires that a storm of short duration, termed a unit storm, be used for the derivation of the unit hydrograph.

2. The effective rainfall is uniformly distributed over the drainage basin. This specifies that the drainage area be small enough for the rainfall to be essentially constant over the entire area. If the watershed is very large, subdivision may be required. The unit-hydrograph theory is then applied to each subarea.

3. The base of the hydrograph of direct runoff is constant for any effective rainfall of unit duration. This needs no clarification except that the base of a hydrograph, that is, the time of storm runoff, is largely arbitrary, since it depends on the method of base-flow separation.

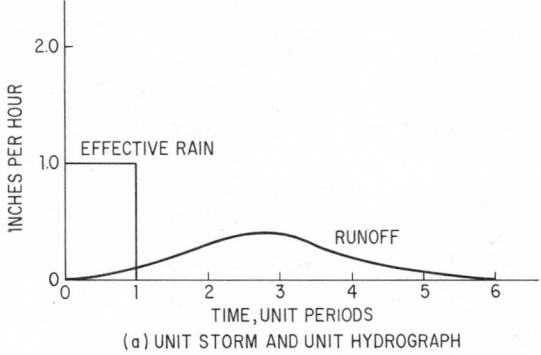

(a) UNIT STORM AND UNIT HYDROGRAPH

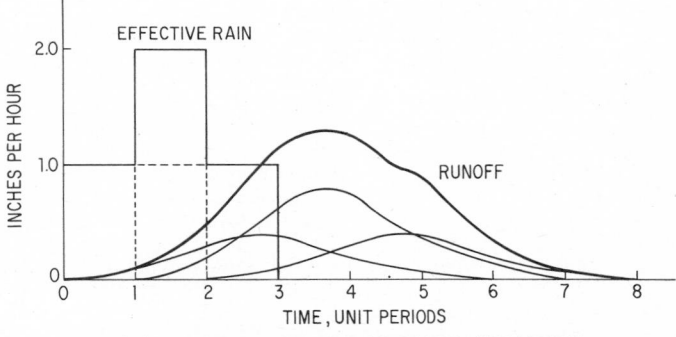

(b) COMPOSITE STORM AND COMPOSITE HYDROGRAPH

Fig. 21-74. Unit hydrograph (a) prepared for unit storm is used to develop composite hydrograph (b) for any storm.

4. "The ordinates of the direct runoff hydrographs of a common base time are directly proportional to the total amount of direct runoff represented by each hydrograph." (V. T. Chow, "Handbook of Applied Hydrology," McGraw-Hill Book Company, New York.) Illustrated in Fig. 21-74, this is basically the principle of superposition or proportionality. It enables calculation of the runoff for a storm of any intensity or duration from a unit storm, which is of fixed intensity and duration. A given storm may be resolved into a number of unit storms. Then, the runoff may be calculated by superimposing that number of unit hydrographs.

5. The hydrograph of direct runoff for a given period of rainfall reflects all the combined physical characteristics of the basin (commonly referred to as the *principle of time invariance*). This assumption implies that the characteristics of the drainage basin have not changed since the unit hydrograph was derived. Because this applies with varying degrees of accuracy to watersheds, the characteristics of the drainage basin must be fixed or specified. Daily and weekly variations in initial soil moisture are probably the greatest source of error in this method, since they are largely unknown. Man-made alterations and stream-flow conditions can be accounted for much more easily.

For ease of manipulation, the unit hydrograph is frequently expressed in histogram form as a **distribution graph** (Fig. 21-75). The distribution graph illustrates the percentages of total runoff that occur during successive unit periods. The ordinate for each unit period is the mean value of runoff for that period.

Since the unit hydrograph is derived for a unit storm of specific duration, it may be used only for storms divided into unit periods of that length. Usually, because of storm variations, the unit period must be different from that for which the unit hydrograph was derived. This requires the recalculation of the unit hydrograph for the new unit period. This is accomplished by offsetting two S hydrographs by a time equal to the duration of the desired unit

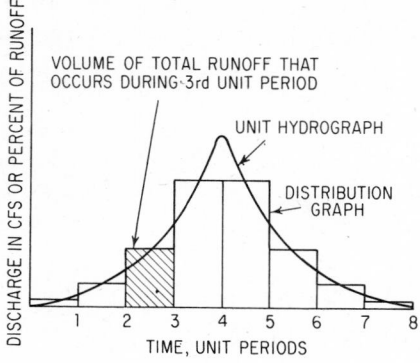

Fig. 21-75. Distribution graph represents unit hydrograph as a histogram.

period (Fig. 21-76).   An **S hydrograph** is a representation of the cumulative percentages of runoff that occur during a storm which has a continuous constant rainfall.   It is calculated by cumulatively plotting the distribution percentages that make up the distribution graph.   The distribution percentages for the new unit hydrograph are determined by taking the difference between mean ordinates for the two offset S hydrographs and dividing by the new unit period.

Transposition of a unit hydrograph from one basin to another similar basin may be made by correlating their respective shape and slope factors.   This method was developed by Franklin F. Snyder (*Transactions of the American Geophysical Union*, vol. 19, pt. I, pp. 447–454).   Also, since S hydrographs are a characteristic of a drainage basin, those from various basins may be compared to obtain an idea of the variations that might exist when transposing data from one basin to another.

In applying the unit-hydrograph method, a loss rate must be established to determine effective rain.   This loss, during heavy storms, is usually considered to be entirely infiltration. The infiltration capacity of a soil may be determined experimentally by lysimeter or infiltrometer tests.

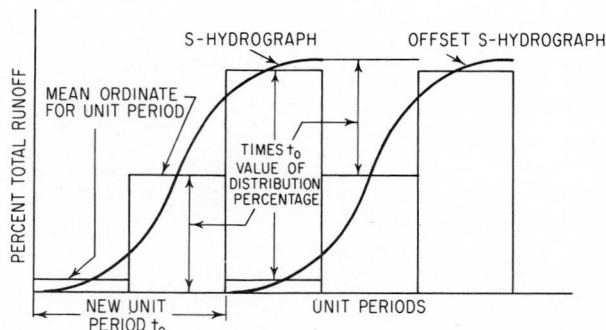

Fig. 21-76. Distribution percentages are determined from offset S hydrographs.

21-58. Groundwater.   Groundwater is subsurface water in porous strata within a zone of saturation.   It supplies about 20% of the United States water demand.   Where groundwater is to be used as a water-supply source, the extent of the groundwater basin and the rate at which continuing extractions may be made should be determined.

Aquifers are groundwater formations capable of furnishing an economical water supply. Those formations from which extractions cannot be made economically are termed **aquicludes.**

Permeability indicates the ease with which water moves through a soil and determines whether a groundwater formation is an aquifer or aquiclude.

The rate of movement of groundwater is given by **Darcy's law:**

$$Q = KIA \qquad (21\text{-}131)$$

where Q = flow rate, gpd
   K = coefficient of permeability
   I = hydraulic gradient
   A = cross-sectional area, perpendicular to the direction of flow

The standard coefficient of permeability used by the U.S. Geological Survey requires I to be in ft per ft and A in sq ft; the field coefficient of permeability, also used by the U.S. Geological Survey, requires I to be in feet per mile and A in foot-miles. When used with consistent units of I and A, the standard and field coefficients of permeability give Q in gpd. The standard coefficient equals the field coefficient multiplied by the ratio of the viscosity of the water in the stratum, for which the field coefficient was determined, to the viscosity of water at 60° F. Values of the standard coefficient of permeability vary from 0.0002 for tight clay to 90,000 for gravel, but the range for typical aquifers is only 10 to 5,000.

Transmissibility is another index for the rate of groundwater movement. It equals the product of the coefficient of permeability and the thickness of the aquifer. Transmissibility indicates for the aquifer as a whole what the coefficient of permeability indicates for the soil.

An aquifer whose water surface is subjected to atmospheric pressure and may rise and fall with changes in volume is a *free* or *unconfined aquifer*. An aquifer that contains water under hydrostatic pressure, because of impermeable layers above and below it, is a *confined* or *artesian aquifer*. If a well is drilled into an artesian aquifer, the water in this well will rise to a height corresponding to the hydrostatic pressure within the aquifer. Frequently, this hydrostatic pressure is sufficient to cause the water to jet beyond the ground surface into the atmosphere. An artesian aquifer is analogous to a large-capacity conduit with full flow in that extractions from it cause a decrease in pressure, rather than a change in volume. This is in contrast to a free aquifer, where extractions cause a decrease in the elevation of the groundwater table.

Groundwater Management. With increasing use being made of groundwater resources, effective groundwater management is an absolute necessity. Adequate management should include not only quantity but also quality. Quantity management consists of effective control over extractions and replenishment. Quality management, overlooked until very recently, consists of effective control over groundwater mineralization resulting from waste disposal, recycling, poor-quality replenishment waters, or other causes.

Several steps or investigations are necessary for the development of an effective management program. The first is a comprehensive geologic investigation of the groundwater basin to determine the characteristics of the aquifers. The second is a qualitative and quantitative hydrologic study of both surface water and groundwaters to determine historical surpluses and deficiencies, safe yield, and overdraft. (*Safe yield* is the magnitude of the annual extractions from an aquifer that can continue indefinitely without bringing some undesirable result. Deteriorating water quality, need for excessive pumping lifts, or infringement on the water rights of others are examples of undesirable results that could define safe yield. Regardless of how it is defined, safe yield applies only to a specific set of conditions based largely on judgment as to what is desirable. Extractions in excess of the safe yield are termed *overdrafts*.) In conjunction with the hydrologic study, present and future water demands should be determined. A detailed water quality study should be made of not only the groundwater within the basin but also all surface waters, wastewaters, and other waters that replenish the groundwater basin. Water quality and quantity problems should be identified.

Following the preceding preliminary work, alternative management plans should be formulated. These management plans should consider variations in the quantity of extractions; groundwater levels; quality, quantity, and location of artificial replenishment; source, quantity, and quality of water supply; and methods of wastewater disposal. All alternative plans must recognize all legal and jurisdictional constraints.

The final step is the operational-economic evaluation of the alternatives and the selection of a recommended groundwater management plan. Operations and economic studies are normally conducted by superimposing present and future conditions in each alternative plan on historical hydrologic conditions that occurred during a base period. (A *base period* is a period of time, usually a number of years, specifically chosen for detailed hydrologic analysis because conditions of water supply and climate during the period are equivalent to a mean of long-term conditions and because adequate data for such hydrologic analysis are available.)

Economic evaluation of alternative plans should consider cost of water supply facilities, cost of replenishment water, cost of wastewater disposal facilities, cost of pumping groundwater at the various operational levels considered, and indirect water-quality use costs, among others. (*Indirect water-quality use costs* are those indirect costs incurred by water distributors and consumers as a result of using water of different qualities. These costs include increased soap costs, water softening costs, and costs associated with the more rapid deterioration of plumbing and waterworks equipment—all of which increase as the hardness and salinity of the water increase.)

Operational studies should determine the most efficient manner of joint operation of surface and groundwater systems (conjunctive use).

Use of computers and the development of a mathematical model for the groundwater basin are almost essential because of the number of repetitive calculations involved.

Upon completion of the operational and economic studies, the most favorable management scheme should be selected as the recommended plan. This selection should be based not only on economic and operational considerations but also on social, institutional, legal, and environmental factors. The plan should be capable of being readily implemented, flexible enough to accommodate different growth rates, financially feasible, and generally acceptable to the water and wastewater agencies operating in the basin.

An operating agency should be designated or formed to implement the recommended plan. The agency should have adequate powers to control or cooperate in the control of surface-water supplies, groundwater recharge sites, surface-water delivery facilities, amount and location of groundwater extractions, and wastewater treatment and disposal facilities. The operating agency should develop a comprehensive monitoring network and a data collection and evaluation program to determine the effectiveness of the management plan and to implement any changes in the plan deemed necessary. This monitoring network may consist of selected wells where groundwater levels and chemical characteristics are measured and certain surface-water sampling locations where both quantitative and qualitative factors are measured. The program should also include quantitative evaluation of extractions, water used, wastewater disposed, and natural and artificial replenishment. Integration of the above data with the computer model of the groundwater basin is an efficient method of evaluating the groundwater management scheme.

("Ground Water Management," Manual and Report on Engineering Practice, No. 40, American Society of Civil Engineers, 1972; L. C. Fowler, "Groundwater Management for the Nation's Future—Ground-water Basin Operation," *Journal of the Hydraulics Division, American Society of Civil Engineers,* vol. 90, no. HY 4, p. 57, July, 1964; W. A. Hall and J. A. Dracup, "Water Resources Systems Engineering," McGraw-Hill Book Company, New York; L. Douglas James and R. R. Lee, "Economics of Water Resources Planning," McGraw-Hill Book Company, New York.)

WATER SUPPLY

21-59. Water Uses. A waterworks system is created or expanded to supply a sufficient volume of water at adequate pressure from the supply source to consumers for domestic, irrigation, industrial, fire-fighting, and sanitary purposes. A primary concern of the engineer is estimation of the quantity of potable water to be consumed by the community, since the engineer must design adequately sized components of the water-supply system. Water-supply facilities consist of collection, storage, transmission, pumping, distribution, and treatment works.

To assure continuous service to the consumer for fire-fighting and sanitary purposes in the event of an earthquake, fire, flood, or other unforeseen emergency, careful consideration must be given to the selection of standby equipment and alternative supplies of water. Maximum protection must be given to power sources and pumps that must be available to operate continuously during emergency conditions. A dependable supply with sufficient pressure for fighting fires considerably increases capital expenditures for system construction. The smaller the system, the larger will be the percent of the total cost chargeable to dependable fire flow.

The size of a proposed water-supply project is usually based on an average annual per capita consumption rate. Therefore, forecasts of population for the design period are of the greatest importance and must be made with care to insure that components for the project are of adequate size. Estimation of future population, however, is a very difficult task.

Several mathematical methods are available for use in predicting populations of cities. Some methods commonly used are arithmetical increase, percentage increase, decreasing percentage increase, graphical comparison with other cities, and the ratio method of comparing a community with a state or country of which the community is a part.   Great care and judgment must be exercised in population prediction, since many factors, such as industrial development, land speculation, geographical boundaries, and age of the city, may drastically alter mathematical estimates.

The total water supply of a city is usually distributed among the following four major classes of consumers: domestic, industrial, commercial, and public.

Domestic use consists of water furnished to houses, apartments, motels, and hotels for drinking, bathing, washing, sanitary, culinary, and lawn-sprinkling purposes.   Domestic use accounts for between 30 and 60% (50 to 60 gal per capita per day) of total water consumption in an average city.

Commercial water is used in stores and office buildings for sanitary, janitorial, and air-conditioning purposes.   Commercial use of water amounts to about 10 to 30% of total consumption.

Industrial uses of water are diverse but consist mainly of heat exchange, cooling, and cleaning. No direct relationship exists between the amount of industrial water used and the population of the community, but 20 to 50% of the total quantity of water used per capita per day is normally charged to industrial usage.   Usually the larger-sized cities have a high degree of industrialization and show a correspondingly greater percentage of total consumption as industrial water.

Public use of water for parks, public buildings, and streets contributes to the total amount of water consumed per capita. Fire demands are usually included in this class of water use. The total quantity of water used for fire fighting may not be large, but because of the high rate at which it is required, it may control the design of the facilities. About 5 to 10% of all water used is for public uses.

Waste and miscellaneous usage of water include that lost because of leakage in mains, meter malfunctions, reservoir evaporation, and unauthorized uses.   About 10 to 15% of total consumption may be charged to waste and miscellaneous uses.

21-60. Water Demand.   Many factors, such as the climate, size of the city, standard of living, degree of industrialization, type of service (metered or unmetered), lawn sprinkling, air conditioning, cost, pressure, and quality of the water, influence the demand rate for water.

Presence of industries usually increases the total per capita use of water but decreases the demand fluctuation.   A good estimate of the potential industrial water demand can be made by relating demand to the percent of land zoned for industrial use.

Small cities frequently have a low per capita demand for water, especially if portions of the city are unsewered.   Fluctuations in demand are greater in small cities, mainly because of the lack of large industries.   High standards of living increase water demand and fluctuations in rate of use.

Warm and dry climates have a higher rate of water consumption because of sprinkling and air conditioning.   Cold weather sometimes increases consumption, because water is allowed to run to prevent freezing of pipes.

Demand for water is related to water-service meters, cost, quality, and pressure.   Metering water reduces the quantity of water consumed by 10 to 25% because of the usual increase in total cost to consumers if they continue to use water at the unmetered rate.   High water pressures increase demand because of greater losses at leaking mains, valves, and faucets.   Normally, if the cost of water increases, the demand for it decreases.   Demand for water usually increases with an improvement in quality.

Demand rates vary with time of day, month, and year.   In Table 21-18, a comparison is made between water-demand rates for the city of Los Angeles in 1963 and a national average calculated

Table 21-18. Water-Demand Rates

	National avg		Los Angeles, Calif.	
	Gal per capita per day	% of avg annual rate	Gal per capita per day	% of avg annual rate
Avg day	160	100	175	100
Max day	265	165	280	160
Max hr	400	250	560	320

from data in *U.S. Public Health Service Report* 661, 1962. The higher consumption rates for Los Angeles are due primarily to the city's relatively warm, dry climate.

Demand-rate data, as presented in Table 21-18, are the average of a range of values, including some very high and very low rates due to variations in climatic conditions, degree of industrialization, time of day, etc. Examples of divergent average daily demand rates for various United States cities are: 230 gal per capita per day for Chicago, 210 gal per capita per day for Denver, 150 gal per capita per day for Baltimore, and 135 gal per capita per day for Kansas City, Mo. An example of a large deviation from the average value is the maximum hourly demand rate of 560 gal per capita per day at Los Angeles, compared with the national average of 400 gal per capita per day. The difference is due primarily to the great amount of lawn sprinkling in Los Angeles. Past water-demand records of both the city being considered and other cities of similar size, industrialization, climate, etc., should be considered and incorporated in demand-rate projections for water systems.

The total quantity of water used for fighting fires is normally quite small, but the demand rate is high. The fire demand as established by the American Insurance Association is

$$G = 1,020 \sqrt{P} \, (1 - 0.01 \sqrt{P}) \qquad (21\text{-}132)$$

where G = fire-demand rate, gpm

P = population, thousands

The required fire flows computed from this formula are given in Table 21-19. In calculating the total flow to be used in design, fire flow should be added to the average consumption for the maximum day.

Table 21-19. Required Fire Flow, Hydrant Spacing, and Fire Reserve Storage*

Population	Fire flow		Dura-tion, hr	Reserve storage, mg†	Avg area served per hydrant in high-value districts, sq ft	
	Gpm	Mgd†			Direct streams	Engine streams
1,000	1,000	1.4	4	0.3	100,000	120,000
2,000	1,500	2.2	6	0.6	90,000	
4,000	2,000	2.9	8	1.0	85,000	110,000
10,000	3,000	4.3	10	1.8	70,000	100,000
17,000	4,000	5.8	10	2.4	55,000	90,000
28,000	5,000	7.2	10	3.0	40,000	85,000
40,000	6,000	8.6	10	3.6	40,000	80,000
80,000	8,000	11.5	10	4.8	40,000	60,000
125,000	10,000	14.4	10	6.0	40,000	48,000
200,000	12,000	17.3	10	7.2	40,000	40,000

*American Insurance Association.

†Mgd = million gallons per day; mg = million gallons.

21-61. Water-Supply Sources.

The major sources of a water supply are surface water and groundwater. In the past, surface sources have included only the commonly occurring natural fresh waters, such as lakes, rivers, and streams, but with rapid population expansion and increased per capita water use associated with a higher standard of living, consideration must be given to desalination and waste-water reclamation as well.

In selection of a source of supply, the various factors to be considered are adequacy and reliability, quality, cost, legality, and politics. The criteria are not listed in any special order, since they are, to a large extent, interdependent. Cost, however, is probably the most important, because almost any source could be used if consumers were willing to pay a high enough price. In some local areas, as increasing demands exceed the capacity of existing sources, the increasing cost of each new supply focuses attention on reclamation of local supplies of waste water and desalination.

Adequacy of supply requires that the source be large enough to meet the entire water demand. Total dependence on a single source, however, is frequently undesirable, and in some cases, diversification is essential for reliability. The source must also be capable of meeting demands during power outages and natural or man-made disasters. The most desirable supplies from a reliability standpoint, in order, are (1) an inexhaustible supply, whether from surface or ground-water, which flows by gravity through the distribution system, (2) a gravity source supplemented by storage reservoirs, (3) an inexhaustible source that requires pumping, and (4) sources that require both storage and pumping.

Quality of the source determines both acceptability and cost; it varies considerably between regions. Preliminary estimates of quality can be made by examining the source, geology, and culture of the area.

Legality of supply is determined by doctrines and principles of water rights, such as appropriative, riparian, and ownership rights. Appropriation right gives the first right priority over later rights: "first in time means first in right." Riparian right permits the owner of land adjacent to a stream or lake to take water from that stream or lake for use on his or her land. Ownership right gives a landowner possession of everything below and above the land. Legality is especially important for groundwater supplies or where there is transfer of water from one watershed to another.

A political problem with water supply exists because political boundaries seldom conform to natural-drainage boundaries. This problem is especially acute in extensive water-importation plans, but it even exists in varying forms for waste-water reclamation and desalination projects.

Desalination processes are of two fundamental types: those that extract salt from the water, such as electrodialysis and ion exchange, and those that extract water from the salt, such as distillation, freezing, and reverse osmosis. The energy cost of the former processes is dependent on the salt concentration, and so they are used mainly for brackish water. The energy costs for the water-extraction processes are essentially independent of salinity; these processes are used for seawater conversion. Very large dual-purpose nuclear power and desalination plants, which take advantage of the economies realized by enormous facilities, have been proposed, but such plants are feasible only for those large urban areas located on coasts. Transmission and pumping costs make inland use uneconomical. Although desalination may have advantages as a local source, it is not, at present, a panacea that will irrigate the deserts.

Acceptance of waste-water reclamation as a water source for direct domestic use is hindered by public opinion and uncertainty regarding viruses. Much effort has been expended to solve these problems. But until such time as they are solved, waste-water reclamation will have only limited use for water supply.

(H. E. Babbitt, J. J. Doland, and J. L. Cleasby, "Water Supply Engineering," McGraw-Hill Book Company, New York.)

21-62. Quality Standards for Water. The U.S. Public Health Service has published a set of regulations, "Drinking Water Standards" (no. 956, 1962), for use in control of water quality for interstate carriers. These standards have been widely adopted voluntarily by both public and private utilities and have received the endorsement of the American Water Works Association as a minimum standard for all public water supplies in the United States. Similar standards have been developed by the World Health Organization to serve as standards for drinking-water quality at international ports ("International Standards for Drinking Water," World Health Organization, Geneva, Switzerland).

Following are extracts from the U.S. Public Health Service "Drinking Water Standards":

Source and Protection. The water supply should be obtained from the most desirable source feasible, and an effort should be made to prevent or control pollution of the source. If the source is not adequately protected against pollution by natural means, the supply shall be adequately protected by treatment.

Sanitary surveys shall be made of the water-supply system, from the source of supply to the connection of the customer's service piping, to locate and correct any health hazards that might exist. The frequency of these surveys shall depend upon the historical need.

Adequate capacity shall be provided to meet peak demands without development of low pressures and the possibility of backflow of polluted water from customer piping.

Bacteriological Quality. The major danger associated with drinking water is the possibility of its recent contamination by sewage containing human excrement. Such sewage may contain pathogenic bacteria capable of producing typhoid fever, cholera, or other enteric diseases. The organisms that have been most commonly employed as indicators of fecal pollution are *Escherichia coli* and the coliform group as a whole.

A standard sample for a bacteriological test consists of five portions of either 10- or 100-ml size. According to the U.S. Public Health Service not more than 10% of all the standard 10-ml portions examined per month shall show the presence of organisms of the coliform group. The presence of the coliform group in three or more 10-ml portions of a standard sample shall not be allowable if this occurs: in two consecutive samples; in more than one sample per month, when less than 20 are examined per month; or in more than 5% of the samples, when 20 or more are examined per month.

Not more than 60% of all the standard 100-ml portions examined per month shall show the presence of organisms of the coliform group. The presence of the coliform group in all five of the 100-ml portions of a standard sample shall not be allowable if this occurs: in two consecutive samples; in more than one sample per month, when fewer than five are examined per month; or in more than 20% of the samples, when five or more are examined per month.

When the membrane-filter technique is used, the arithmetic mean coliform density of all standard samples examined per month shall not exceed one per 100 ml. Coliform colonies per standard sample shall not exceed $^3/_{50}$, $^4/_{100}$, $^7/_{200}$, or $^{13}/_{500}$ ml in: two consecutive samples; more than one standard sample, when less than 20 are examined per month; or more than 5% of the standard samples, when 20 or more are examined per month.

When organisms of the coliform group occur in three or more of the 10-ml portions of a single standard sample, in all five of the 100-ml portions of a single standard sample, or exceed the given values for a standard sample with the membrane-filter test, then remedial measures shall be undertaken until daily samples from the same sampling point show at least two consecutive samples to be of satisfactory quality.

The minimum number of samples to be collected from the distribution system and examined each month should be in accordance with the population served. A minimum of two samples should be taken in any case, with 12 samples taken for 10,000 population, 100 for 100,000 population, 320 for 1,000,000 population, and 500 taken for 5,000,000 and over.

Physical Quality.   Samples should be collected at least once a week from representative points in the distribution system and examined for turbidity, color, threshold odor, and taste. The following limits should not be exceeded for aesthetic reasons:

> Turbidity . . . . 5 turbidity units
> Color . . . . 15 color units
> Threshold odor no. 3

("Standard Methods for the Examination of Water and Wastewater," American Public Health Association, American Water Works Association, Water Pollution Control Federation.)

Chemical Characteristics.   The chemical substances listed in Table 21-20 should not be present in a water supply in excess of the listed concentrations where more suitable supplies are, or can be made, available.

Table 21-20. Concentration Limit for Chemical Substances*

Substance	Concentration, mg per liter or ppm †
Alkyl benzene sulfonate (ABS)	0.5
Arsenic (As)	0.01
Chloride (Cl)	250
Copper (Cu)	1
Carbon chloroform extract (CCE)	0.2
Cyanide (CN)	0.01
Fluoride (F)	(See Table 21-21)
Iron (Fe)	0.3
Manganese (Mn)	0.05
Nitrate ‡ (NO_3)	45
Phenols	0.001
Sulfate (SO_4)	250
Total dissolved solids	500
Zinc (Zn)	5

*From "Drinking Water Standards," U.S. Public Health Service, no. 956, 1962.

†The concentration of these substances can be given either in milligrams per liter, mg per liter, or in parts per million, ppm, since the two notations give approximately equal answers for these small concentrations.

‡In areas in which the nitrate content of water is known to be in excess of the listed concentration, the public should be warned of the potential dangers of using the water for infant feeding, as it may give rise to infantile methoemoglobinaemia. This nitrate poisoning is serious and occasionally fatal in infants under one year of age.

Samples for chemical examination should be collected at least once every 3 months from supplies serving more than 50,000 inhabitants and at least twice a year from supplies serving up to 50,000 inhabitants.

Fluoride is considered an essential constituent of drinking water for prevention of tooth decay in children.   But the concentration should not average more than the appropriate upper limit in Table 21-21, since excess fluorides may give rise to dental fluorosis (spotting of the teeth) in children.   When fluoride is naturally present in drinking water, the concentration should not average more than the appropriate upper limit in Table 21-21.   Fluoride in average concentrations greater than twice the optimum values shall constitute grounds for rejection of the supply.

When fluoridation (supplementation of fluoride in drinking water) is practiced, the average fluoride concentration shall be kept within the upper and lower control limits given.   In addition to the above sampling schedule, fluoridated and defluoridated supplies shall be sampled with sufficient frequency to determine that the desired fluoride concentration is maintained.

Table 21-21. Allowable Fluoride Concentration*

Annual avg of max daily air temperatures†	Recommended control limits, fluoride concentrations, mg per liter or ppm		
	Lower	Optimum	Upper
50.0–53.7	0.9	1.2	1.7
53.8–58.3	0.8	1.1	1.5
58.4–63.8	0.8	1.0	1.3
63.9–70.6	0.7	0.9	1.2
70.7–79.2	0.7	0.8	1.0
79.3–90.5	0.6	0.7	0.8

*From "Drinking Water Standards," U.S. Public Health Service, no. 956, 1962.
†Based on temperature data obtained for a minimum of 5 years.

Samples should be collected and tested for the presence of toxic substances at least once every 3 months.   When subtolerance levels of these toxic substances are found to be present in a source of supply, samples should be collected and tested more frequently to determine if the concentration is approaching the specified limit.   The presence of these toxic substances in excess of the concentrations listed in Table 21-22 shall constitute grounds for rejection of the supply.

Table 21-22. Concentration Limits for Toxic Substances*

Substance	Concentration, mg per liter or ppm
Arsenic (As)	0.05
Barium (Ba)	1.0
Cadmium (Cd)	0.01
Chromium (hexavalent) (Cr^{6+})	0.05
Cyanide (CN)	0.2
Fluoride (F)	(See Table 21-21)
Lead (Pb)	0.05
Selenium (Se)	0.01
Silver (Ag)	0.05

*From "Drinking Water Standards," U.S. Public Health Service, no. 956, 1962.

Radioactivity.   The limiting values given for radioactivity are not to be taken as values that, if exceeded, would render the water unfit for drinking purposes.   Rather, these values indicate that if the levels observed are lower, then the water is safe for use without further consideration. If the levels exceed those given in Table 21-23, radiochemical analysis will be required to determine the nature of the radionuclides present and the safety of the water for use as a public supply.

The frequency of sampling and analysis for radioactivity shall be determined after consideration of the type and amount of radioactivity present.   Where concentrations of radium-226 or

Table 21-23. Limiting Values for Radioactivity, Picocuries (Curies × 10⁻¹²) per Liter*

Strontium-90	10
Radium-226	3
Gross beta concentration (where amounts of strontium-90 and alpha emitters are negligible	1,000

*From "Drinking Water Standards," U.S. Public Health Service, no. 956, 1962.

strontium-90 vary considerably, samples should be collected daily and may be composited over a period of not longer than 3 months before examination. Samples for determination of gross beta concentration should be taken daily and analyzed frequently.

21-63. Purposes of Water Treatment. Water is treated to remove disease-producing bacteria, unpleasant tastes and odors, particulate and colored matter, and hardness. Some of the commoner methods of treatment are plain sedimentation and storage, coagulation-sedimentation, slow and rapid sand filtration, disinfection, and softening (see also Art. 21-70).

21-64. Plain Sedimentation and Storage. Long-term storage of water reduces the amount of disease-producing bacteria and particulate matter. But economic conditions usually compel water purveyors to use more efficient methods of treatment.

Plain sedimentation is a process of removing particulate matter from water in a basin by reducing the flow-through velocity. Factors that affect the settling rate of particular matter suspended in water are size, shape, and specific gravity of the suspended particles; temperature and viscosity of the water; and size and shape of the settling basin.

The settling velocity v_s of spherically shaped particles in a viscous liquid can be found by use of Stokes' law if the Reynolds number $R = v\rho d/\mu$, calculated with $v = v_s$, is equal to or less than 1.

$$v_s = \frac{g(\rho_1 - \rho)d^2}{18\mu}$$ (21-133)

where v_s = settling velocity of particle, cm per sec
  g = acceleration due to gravity, cm per sec²
  μ = absolute viscosity of the fluid, dyne-sec per sq cm
  ρ_1 = density of particle, g per cu cm
  ρ = density of fluid, g per cu cm
  d = particle diameter, cm
If $R \geq 2,000$, Newton's law applies:

$$v_s = \frac{4g(\rho_1 - \rho)d}{3\rho C_D}$$ (21-134)

where C_D is the drag coefficient. Figure 21-77 shows a plot of C_D values vs. Reynolds numbers, to be used in Eq. (21-134).

In the region where $1.0 < R < 2,000$ there is a transition from Stokes' law to Newton's. The settling velocity in this region is somewhere between the values given by Newton's law and those given by Stokes' law; however, no exact expression has been developed to give the velocity.

Figure 21-78 shows the relationship of settling velocity to diameter of spherical particles with specific gravity S between 1.001 and 5.0.

The ideal settling basin (Fig. 21-79) is a sedimentation tank in which flow is horizontal, velocity is constant, and concentration of particles of each size is the same at all points of the vertical cross section at the inlet end. The basin has a volumetric capacity C, depth h_o, and width B. The surface loading rate or overflow velocity v_o, equal to the settling velocity of the smallest particle to be completely removed, can be determined by dividing the flow rate Q by the settling surface area A. For this ideal basin, the overflow velocity, therefore, is $v_o = Q/A = Q/BL_o$, where $Q = Bh_o V$, L_o is the length of settling zone, and V is the flow-through velocity. (Usually, v_o is expressed in gallons per day per square foot of surface area.) The detention time $t = h_o/v_o = L_o/V$ also equals the volumetric capacity C divided by the rate of flow Q.

Particles with a settling velocity $v_s \geq v_o$, and those that enter the settling zone between f and j with a settling velocity $v_s \geq (v_1 = h_1 V/L_o)$, though less than v_o, are removed in this basin. The particles with a settling velocity $v_s \leq v_1$ that enter the settling zone between f and e are not removed in this basin.

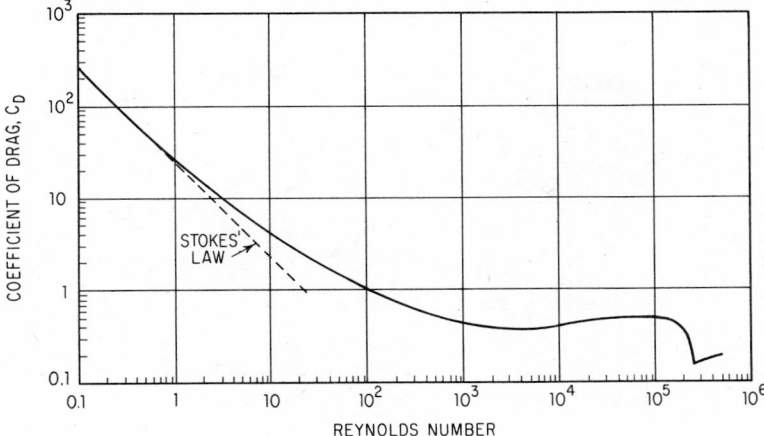

Fig. 21-77. Newton drag coefficients for spheres in fluids. (*Observed curves, after Camp, Transactions of the American Society of Civil Engineers, vol. 103, p. 897, 1946.*)

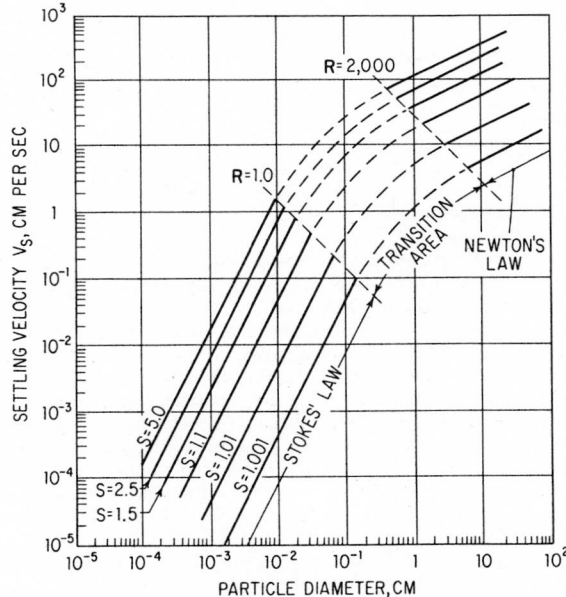

Fig. 21-78. Chart gives settling velocities of spherical particles with specific gravities S, at 10° C.

The efficiency of a sedimentation basin is the ratio of the flow-through period to the detention time. The flow-through period is the time required for a dye, salt, or other indicator to pass through the basin. Settling-basin efficiencies are reduced by many factors such as cross currents, short circuiting, and eddy currents. A well-designed tank should have an efficiency of 30 to 50%.

Some design criteria for sedimentation tanks are:

Period of detention—2 to 8 hr.

Length-to-width ratio of flow-through channel—3:1 to 5:1.

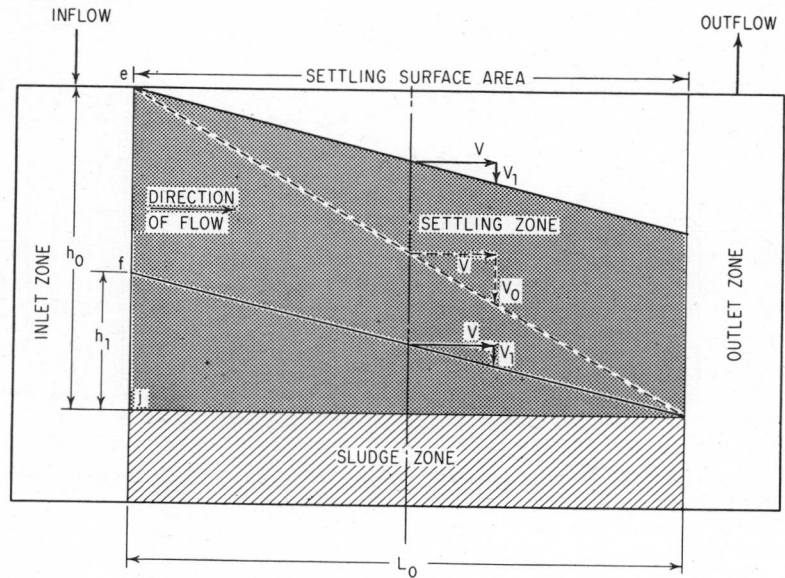

Fig. 21-79. Longitudinal section through an ideal settling basin.

Depth of basin—10 to 25 ft (15 ft average).
Width of flow-through channel—not over 40 ft (30 ft most common).
Diameter of circular tank—35 to 200 ft (most common, 100 ft).
Flow-through velocity—not to exceed 1.5 fpm (most common velocity, 1.0 fpm).
Surface loading or overflow velocity, gal per day per sq ft of surface area—between 500 and 2,000 for most settling basins.

(H. E. Babbitt, J. J. Doland, and J. L. Cleasby, "Water Supply Engineering," McGraw-Hill Book Company, New York; American Water Works Association, "Water Quality and Treatment," McGraw-Hill Book Company, New York; G. M. Fair, J. C. Geyer, and D. A. Okun, "Water and Wastewater Engineering," John Wiley & Sons, Inc., New York.)

21-65. Coagulation-Sedimentation.   (See also Art. 21-64.) To increase the settling rate and remove finely divided particles in suspension, coagulants are added to the water. Without coagulants, finely divided particles do not settle out, because of their high ratio of surface area to mass and the presence of negative charges on them. The velocity at which drag and gravitational forces are equal is very low, and the negative charges on the particles produce electrostatic forces of repulsion that tend to keep the particles separated and prevent agglomeration. When coagulating chemicals are mixed with water, however, they introduce highly charged positive nuclei that attract and neutralize the negatively charged suspended matter.

Iron and aluminum compounds are commonly used as coagulants because of their high positive ionic charge. The alkalinity of the water being treated must be high enough for an insoluble hydroxide or hydrate of these metals to form. These insoluble flocs of iron and aluminum, which combine with themselves and other suspended particles, precipitate out when a floc of sufficient size is formed.

The commoner coagulants are aluminum sulfate, commonly known as alum $[Al_2(SO_4)_3 \cdot 18H_2O]$; ferrous sulfate ($FeSO_4 \cdot 7H_2O$); ferric chloride ($FeCl_3$); and chlorinated copperas (a mixture of ferric chloride and ferrous sulfate). The type and amount of coagulant necessary to clarify a specified water depend on the qualities of water to be treated, such as pH, temperature, turbidity, color, and hardness. Jar tests are usually made in a laboratory to determine the optimum amount of coagulant.

The complete clarification process is usually divided into three stages: (1) rapid chemical mixing; (2) flocculation or slow stirring, to get the small floc to agglomerate; and (3) coagulation-sedimentation in low-flow-velocity settling basins. Rapid chemical mixing may be accomplished with many devices, such as mechanical stirrers, centrifugal pumps, and air jets. The time necessary for mixing ranges from a few seconds to 20 min. Flocculation or slow stirring increases floc size and speeds up settling. The speed of the agitators must be great enough, however, to cause contact between the small floc but not so great that the larger floc is broken up. Flocculator detention time should be in the 20- to 60-min range. The coagulation-sedimentation process takes place in a basin nearly identical to a plain sedimentation basin. The detention period for a sedimentation basin should be between 2 and 8 hr.

21-66. Sand Filtration. Passing water through a layer of sand removes much of the finely divided particulate matter and some of the larger bacteria. The filtering process has many components, such as physical straining, chemical and biological reactions, settling, and neutralization of electrostatic charges. Rapid and slow sand filters are the two types most commonly used.

The slow sand filter consists of an underdrained, watertight container containing a 2- to 4-ft layer of sand supported by a 6- to 12-in. layer of gravel. The effective size of the sand should be in the 0.25- to 0.35-mm range. (The *effective size* is the size of a sieve, in millimeters, that will pass 10%, by weight, of the sand. The *uniformity coefficient* is the ratio of the size of a sieve that will pass 60% of the sample to the effective size.) The uniformity coefficient of the sand should be less than 3. The sand is normally submerged under 4 or 5 ft of water. The water passes through the filter at a rate of 3 to 6 million gal per acre per day, depending on the turbidity. The slow filter is not commonly used in the United States because of the greater versatility and efficiency of rapid sand filters.

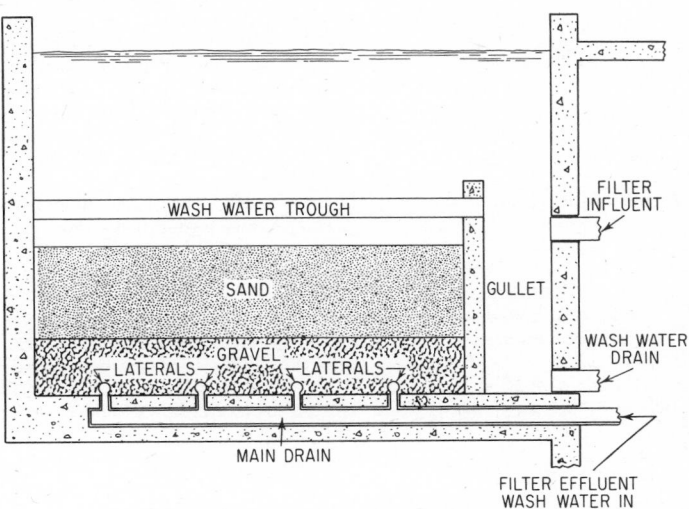

Fig. 21-80. Gravity-type rapid sand filter.

Rapid sand filtration is normally preceded by chemical treatment, such as flocculation-coagulation and disinfection; so the water can be passed through the sand at a higher rate. Usually, the effluent from a rapid filter needs further disinfection or chlorination, because the bacteria are not completely removed in this process. A diagram of a typical gravity-type rapid sand filter is shown in Fig. 21-80.

The normal order of flow through the varying components of the filter is from the clarifiers (settling tanks) to the top of the sand layer, through the sand and gravel layers, through the underdrain laterals to the main drain, and then through the controller to the clear well for storage. Wash (cleaning) water flow takes place in a reverse direction after the filter effluent line has been closed. The wash-water flow is through the main drain to the laterals, from the laterals upward through the gravel and sand to the wash-water troughs. The troughs carry the water to the gullet, which is drained to waste.

Some common design factors for rapid sand filters are:
Effective grain size—0.35 to 0.55 mm.
Uniformity coefficient—1.20 to 1.75.
Thickness of sand layer—24 to 30 in., depending on grain size.
Thickness of gravel layer—15 to 24 in.
Gravel size—from ⅛ to 1½ in.
Filtration rate—2 to 4 gpm per sq ft (125 to 250 million gal per acre per day).
Total depth of each basin—8 to 10 ft.
Maximum head loss allowed before washing sand—8 to 10 ft.
Sand expansion during washing—25 to 50%.
Wash-water rate—15 to 20 gpm per sq ft.
Distance from top edge of wash-water trough to top of unexpanded sand—24 to 30 in.
Length of filter runs between washings—12 to 72 hr.
Spacing between wash-water troughs—4 to 6 ft.
Ratio of length to width of each basin—1.25 to 1.35.

Rapid sand filters are operated until the particulate matter and unsettled floc cover the openings between the sand grains, creating a high head loss across the filter. This high head loss slows down the flow rate and may force some of the particulate matter through the sand and gravel layers. Filters are usually backwashed when the particulate-matter concentration increases in the filter effluent or when the head loss reaches 8 to 10 ft. Backwashing a filter consists of forcing filtered water through the filter from the drains upward to the wash-water troughs. The lightweight sediment is washed from the sand grains by the moving water and sometimes by other agitating devices, such as rakes, water sprays, and air jets. Filters must be washed thoroughly or difficulties with mud balls, bed cracking, or sand incrustation will be encountered.

Immediately after washing, filters pass water at a high rate, which produces an undertreated effluent. Either manual or automatic rate control must be used to prevent such an occurrence. Many treatment plants control the rate of filtration by using venturi-tube devices, which throttle the filter effluent line when there is high-velocity flow. As clogging begins to occur in the filter, the velocity of flow in the effluent line decreases, and the rate controller then opens to increase the velocity.

A negative head is produced on the filter when the head loss across the filter is greater than the depth of water on the sand. Negative heads can produce a condition known as air binding. It is caused by removal of dissolved gases from the water and subsequent formation of bubbles between sand grains. These bubbles appreciably decrease filter capacity.

The underdrains of a filter are commonly made of perforated pipe or porous plates. The underdrains should be arranged so that each area filters and distributes its proportionate share of water. The ratio of total area of perforations to the total filter-bed area is normally in the 0.002:1 to 0.005:1 range. The diameter of the perforations varies between ¼ and ¾ in.

Wash-water troughs should be evenly spaced, and water should not have to travel more than 3 ft horizontally to get to a wash-water gutter. The depth of water flow in a horizontal gutter may be calculated from

$$Q = 1.72by^{3/2} \qquad (21\text{-}135)$$

where Q = total flow received by trough, gpm
b = width of trough, in.
y = water depth at upstream end of trough, in.
The total depth of gutter can be found by adding 2 or 3 in. of freeboard to the calculated depth y.

Anthracite coal may be used in place of sand in gravity-type filters. The effective grain size of anthracite coal is greater than that of sand, thus permitting higher filtration rates and longer filter runs.

A *pressure filter* is composed of a gravity-filter medium enclosed in a watertight vessel. The filtering medium may be sand, diatomaceous earth, or anthracite coal. Pressure filters are primarily supplemental and are used for specialized industrial uses and for clarifying swimming-pool water.

Filter galleries are made up of horizontal, perforated or open-joint pipes, placed in shallow sand or gravel aquifers. Galleries typically are fed by diversion or pumping from streams into spreading basins with gravel or sand bottoms. Some, however, may be located in aquifers with high groundwater table. The filtered water may be pumped from the gallery or allowed to flow out one end by gravity.

(H. E. Babbitt, J. J. Doland, and J. L. Cleasby, "Water Supply Engineering," McGraw-Hill

Book Company, New York; American Water Works Association, "Water Quality and Treatment," McGraw-Hill Book Company, New York; G. M. Fair, J. C. Geyer, and D. A. Okun, "Water and Wastewater Engineering," John Wiley & Sons, Inc., New York.)

21-67. Water Softening.   Presence of the bicarbonates, carbonates, sulfates, and chlorides of calcium and magnesium in water causes hardness.   Three major classifications of hardness are: (1) carbonate (temporary) hardness caused by bicarbonates, (2) noncarbonate (permanent) hardness, and (3) total hardness.   Municipal treatment plants generally use either the lime-soda (precipitation) process or the base-exchange (zeolite) process to reduce the hardness of the water to below 100 mg per liter (about 100 ppm) of $CaCO_3$ equivalence.

In the lime-soda process, lime (CaO), hydrated lime [$Ca(OH)_2$], and soda ash (Na_2CO_3) are added to the water in sufficient quantities to reduce the hardness to an acceptable level.   The amounts of lime and soda ash required for softening to a residual hardness can be determined by use of chemical-equivalent weights, taking into account that commercial grades of lime and hydrated lime are about 90 and 68% CaO, respectively.   A residual hardness of 50 to 100 mg per liter as $CaCO_3$ remains in the treated water because of the very slight solubility of both $CaCO_3$ and $Mg(OH)_2$.   The hardness of water is normally expressed in grains per gallon (gpg) or milligrams per liter (mg per liter) of $CaCO_3$, where 1 gpg = 17.1 mg per liter.

Chemical equations for the commoner lime-soda softening processes are

$$CO_2 + CaO \rightarrow CaCO_3\downarrow \tag{21-136}$$

$$Ca(HCO_3)_2 + CaO \rightarrow 2CaCO_3\downarrow + H_2O \tag{21-137}$$

$$MgSO_4 + CaO + H_2O \rightarrow Mg(OH)_2\downarrow + CaSO_4(\text{soluble}) \tag{21-138}$$

$$CaSO_4(\text{soluble}) + Na_2CO_3 \rightarrow CaCO_3\downarrow + NaSO_4 \tag{21-139}$$

Since the carbonate and magnesium hydroxide particles settle out in sedimentation basins, facilities must be provided for particle removal and disposal.   Deposition of $CaCO_3$ and $Mg(OH)_2$ on sand grains, in clear wells, and in distribution pipes can be prevented by recarbonation with CO_2 before sand filter treatment.

Hardness in water can be reduced to zero by passing the water through a base-exchange or zeolite material.   These materials remove cations, such as calcium and magnesium, from water and replace them with soluble sodium and hydrogen cations.   Calcium can be removed from water as shown by the following reaction:

$$Ca^{++} + Na_2R \leftrightarrows CaR + 2Na^+ \tag{21-140}$$

where Ca^{++} is the calcium hardness ion removed, Na^+ is the sodium ion replacing the Ca^{++} in water, and R is the zeolite material.   The reaction can be reversed (from right to left) by increasing the Na^+ concentration to a high value, as generally is done in regeneration of the softening unit.

Sodium chloride (table salt) is commonly used to regenerate the unit.   Regeneration requires between 0.3 and 0.5 lb of salt per 1,000 grains of hardness removed.

Some hardness-removal capacities per cubic foot of base-exchange material are: natural zeolite—2,500 to 3,000 grains, synthetic zeolite—5,000 to 30,000 grains (1 lb = 7,000 grains).

(American Water Works Association; "Water Quality and Treatment," McGraw-Hill Book Company, New York.)

21-68. Disinfection with Chlorine.   Chlorine in either the liquid, gas, or hypochlorite form is the major chemical used for destroying bacteria in water supplies.   Other disinfectants are iodine, bromine, ozone, ultraviolet light, and lime,

The reaction of chlorine with water is

$$Cl_2 + H_2O \leftrightarrows H^+ + Cl^- + HOCl \tag{21-141}$$

The hypochlorous acid (HOCl) reacts with the organic matter in bacteria to form a chlorinated complex that destroys living cells.   The amount of chlorine (*chlorine dose*) added to the water depends on the amount of impurities to be removed and the desired residual of chlorine in the water.   A dose of 1 or 2 mg per liter of chlorine is usually sufficient to destroy all the bacteria and leave an adequate residual.   Chlorine residuals of 0.1 or 0.2 mg per liter are normally maintained in water-treatment-plant effluent streams as a factor of safety for the water as it travels to the consumer.

(American Water Works Association, "Water Quality and Treatment," McGraw-Hill Book Company, New York.)

21-69. Carbonate Stability.   Water may either corrode or place a protective carbonate film on the interior surfaces of pipes.   Which it does depends on the nature and amount of chemicals dissolved in the water (see W. F. Langlier, *Journal of American Water Works Association,* vol. 31, p. 1171, 1939, for an equation basing the carbonate stability on pH values).

An approximation of the stability of a water supply can be obtained by adding an excess of calcium carbonate powder to one-half of a water sample.   Stir or shake each half sample at 5-min intervals for about 1 hr.   Filter both solutions.   Then, either take the pH or determine the methyl orange alkalinity of each sample.   If the untreated water has a higher alkalinity or pH than the $CaCO_3$-treated water, the water is saturated with carbonate and may deposit protective films in pipes.   If the untreated water has a lower pH or alkalinity value than the treated water, the water is unsaturated with carbonate and may be corrosive.   If the pH or alkalinity is the same in both samples, the water is in equilibrium in regard to carbonates.

The greater the difference in either alkalinity or pH between the two samples, the greater the amount of either unsaturation or saturation with respect to carbonates.   If the untreated water has a much higher pH or alkalinity than the treated water, the water is highly saturated with carbonates.   It can cause a problem with heavy carbonate deposits in pipes and appurtenances of the purveyor and consumer.

(G. M. Fair, J. C. Geyer, and D. A. Okun, "Water and Wastewater Engineering," John Wiley & Sons, Inc., New York.)

21-70. Miscellaneous Treatment.   Many different methods of treatment are used to remove such undesirable elements as color, taste, odor, excessive fluorides, detergents, iron, and manganese.   Some special methods used are the application of activated carbon and boiling.

Activated carbon is commonly used for taste and odor removal.   The carbon can be applied as a powder to the water and later removed by a sand filter or the water can be passed through a bed of carbon.

Many specialized treatment problems exist that may be solved by methods given in water-treatment texts.

(American Water Works Association, "Water Quality and Treatment," McGraw-Hill Book Company, New York; G. M. Fair, J. C. Geyer, and D. A. Okun, "Water and Wastewater Engineering," John Wiley & Sons, Inc., New York.)

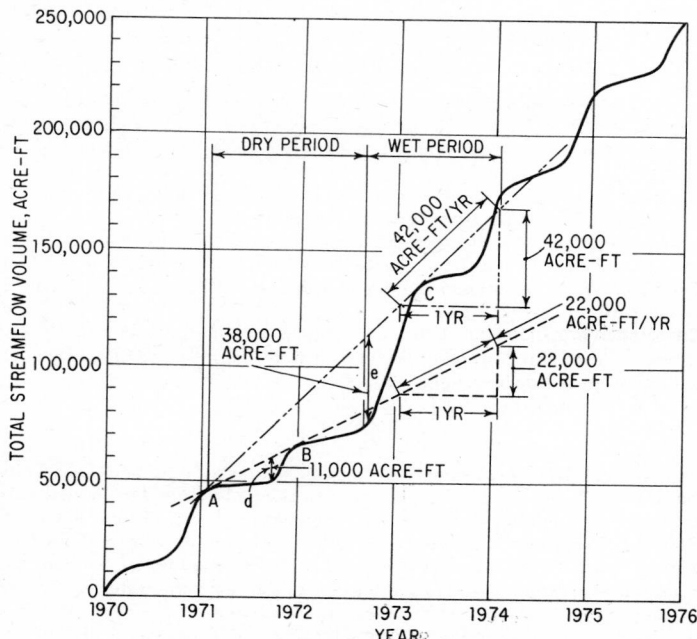

Fig. 21-81.   Mass diagram of stream flow.

21-71. Reservoirs. The basic purpose of impounding reservoirs is to hold runoff during periods of high runoff and release it during periods of low runoff. The specific functions of reservoirs are hydroelectric, flood control, irrigation, water supply, and recreation. Many large reservoirs are multipurpose. They pose additional problems, since the specific functions may dictate opposite design and operating criteria. Also, equitable cost allocation is more difficult. (V. T. Chow, "Handbook of Applied Hydrology," Mc-Graw-Hill Book Company, New York.)

Sizing of a reservoir for a project where the demand for water is much greater than the mean stream flow is an economic balance between benefits and costs. A preliminary study of available reservoir sites should be made to obtain the relative costs for various size reservoirs. From the mass diagram of stream flow, the dependable flow that can be obtained from various size reservoirs can be determined. An economic comparison should then be made of the benefits of various flows and the costs of various reservoirs. The reservoir size that will give the maximum benefit should be selected.

When the demand rate is known, as is the case for many water-supply projects, the required size of the reservoir can be determined directly from a mass diagram of stream flow.

The **mass diagram** (Fig. 21-81) is a graphical plot of total stream-flow volume against time. The slope of the curve is the rate of flow.

Selection of the critical period of years for a mass curve depends on the function of the reservoir. For a water-supply or hydroelectric development, minimum flows will be critical, whereas for flood-control reservoirs, maximum flows will govern.

Reservoir capacity for a certain demand can be obtained by drawing a line with a slope equal to the demand tangent to the mass curve at the beginning of a selected dry period, as shown by lines AB and AC in Fig. 21-81. The ordinates d and e represent the storage required to maintain demands AB and AC.

Once a reservoir site has been selected, **area-volume curves** (Fig. 21-82) are drawn to give the characteristics of the site. The plot of volume vs. water elevation is determined by planimetering the area of selected contours within the reservoir site and multiplying by the contour interval. Aerial mapping has made it possible to obtain accurate contour maps at only a fraction of the costs of older methods.

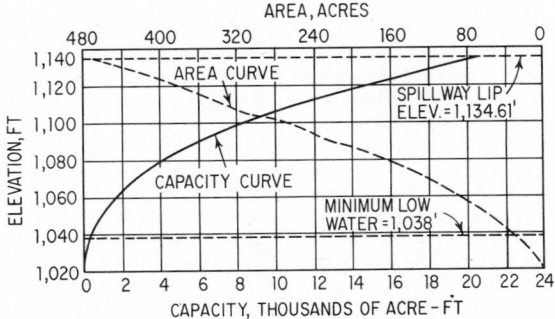

Fig. 21-82. Area-capacity curves for reservoir.

Another important consideration in the design of reservoirs is deposition of sediment (see Art. 21-49).

In selecting a site for a water-supply reservoir, give special attention to water quality. If possible, the watershed should be relatively uninhabited to reduce the amount of treatment required. (Water from practically all sources should be chlorinated in the distribution system to insure against pollution and contamination.) Shallow reservoirs usually give more problems with color, odor, and turbidity than deep reservoirs, particularly in warm climate or during warm seasons of the year. Runoff that is heavily laden with silt and debris should be diverted from the reservoir or treated before it is mixed with the water supply. Alum is mixed into reservoirs to reduce turbidity, and copper sulfate is used to kill vegetation.

In deep reservoirs, during the summer months the upper part of the reservoir will be warmed, while below a certain level the temperature may be many degrees cooler. The zone where the abrupt temperature change takes place, which may be only a few feet thick, is called the *thermo-*

cline. The waters above and below the thermocline circulate, but there is no circulation across this zone. The water in the lower level becomes low in dissolved oxygen and develops bad tastes and odors. When the temperature drops in the fall, the water at the upper level becomes heavier than the water at the lower level and the two levels become intermixed, causing tastes and odors to occur in the entire reservoir. To oxidize organic matter and prevent poor water quality in lower levels of reservoirs during summer months, chlorine or compressed air should be released at various points on the bottom of the reservoirs.

(H. E. Babbitt, J. J. Doland, and J. L. Cleasby, "Water Supply Engineering," McGraw-Hill Book Company, New York; R. K. Linsley and J. B. Franzini, "Water-Resources Engineering," McGraw-Hill Book Company, New York; G. M. Fair, J. C. Geyer, and D. A. Okun, "Water and Wastewater Engineering," John Wiley & Sons, Inc., New York.)

21-72. Water Distribution. A water-distribution system should reliably provide potable water in sufficient quantity and at adequate pressure for domestic and fire-protection purposes. To provide adequate domestic service, the pressure in the main at house service connections usually should not be below 45 psi. But if oversized plumbing is provided, 35 psi is adequate. In steep hillside areas, the system is usually divided into several different pressure zones, interconnected with pumps and pressure regulators. Since each additional zone causes increased expenses and decreased reliability, it is desirable to keep their number to a minimum. The American Water Works Association has recommended 60 to 75 psi as a desirable range for pressures; however, in areas of steep topography where local elevation differences may be over 1,000 ft, such a narrow range is not practical.

House plumbing is designed to withstand a maximum pressure of between 100 and 125 psi. When the pressure in distribution lines is above 125 psi, it is necessary to install pressure regulators at each house to prevent damage to appliances, such as water heaters and dishwashers.

Pressure requirements for fire fighting depend on the technique and equipment used. Four methods of supplying fire protection are:

1. Use of mobile pumpers which take water from a hydrant. This method is used in most large communities that have full-time, well-trained fire departments. The required pressure in the immediate area of the fire is 20 psi.

2. Maintenance of adequate pressure at all times in the distribution system to allow direct connection of fire hoses to hydrants. This technique is commonly used in small communities that do not have a full-time fire department and mobile pumpers. The pressure in the distribution system in the vicinity of a fire should be between 50 and 75 psi.

3. Use of stationary fire pumps located at various points in the distribution system, to boost the pressure during a fire and allow direct connection of hoses to hydrants. This method is not so reliable or so widely used as the first two.

4. Use of a separate high-pressure distribution system for fire protection only. There are only rare instances in high-value districts of large cities where this method is used, because the cost of a dual distribution system is usually prohibitive.

Distribution systems are usually laid out on a gridiron system with cross connections at various intervals. Dead-end pipes should be avoided. They cause water-quality problems.

Economic velocities are usually around 3 to 4 fps, although during fires they can be much higher. Two- and four-inch-diameter pipe can be used for short lengths in residential areas; however, the American Insurance Association requires 6-in. pipe for fire service in residential areas. Also, maximum length between cross connections is limited to 600 ft. In high-value districts, the AIA requires an 8-in. pipe, with cross connection at all intersecting streets. The AIA standards also require that gate valves be located so that no single case of pipe breakage, outside main arteries, requires shutting off from service an artery or more than 500 ft of pipe in high-valued districts, or more than 800 ft in any area. All small distribution lines branching from main arteries should be equipped with valves. ("Standard Schedule for Grading Cities and Towns of the United States with Reference to Their Fire Defenses and Physical Conditions," American Insurance Association.)

The cover required over distribution pipes depends on the climate, size of main, and traffic. In northern areas, frost penetration, which may be as deep as 7 ft, is usually the governing factor. In frost-free areas, a minimum of 24 in. is required by the AIA. If large mains are placed under heavy traffic, the stress produced by wheel loads should be investigated.

Maintenance of distribution systems involves keeping records, cleaning and lining of pipe, finding and repairing leaks, inspection of hydrants and valves, and many other functions necessary to eliminate problems in operation. Valves should be inspected annually and fire hydrants semiannually. Records of all inspections and repairs should be kept.

Unlined distribution pipes, after years of usage, lose much of their capacity because of corrosion and incrustations. Cleaning and lining with cement mortar restores the original capacity. Dead-end pipes should be flushed periodically to reduce the accumulation of rust and organic matter.

21-73. Distribution Reservoirs. (See also Art. 21-71.) The two main functions distribution reservoirs serve are to equalize supply and demand over periods of varying consumption and to supply water during equipment failure or for fire demand. Major sources of supply for some cities, such as New York, San Francisco, and Los Angeles, are hundreds of miles from the city. Because of the large cost of aqueducts, it is usually economical to size them for the mean annual flow and provide terminal storage for daily and seasonal fluctuations of demand. Terminal storage is also necessary because of the possibility of a failure along an aqueduct.

Since, for distribution systems, the maximum hourly demand may be several times the maximum daily demand, it is usually economical to have equalizing reservoirs at various points in the distribution system so that main supply lines, pumping plants, and treatment plants can be sized for maximum daily instead of maximum hourly demand. During hours of maximum demand, water flows from these reservoirs to the consumers. When the demand drops off, the flow refills the reservoir. A mass diagram (Art. 21-71) can be used to determine the required capacity of the reservoir.

Equalizing reservoirs are usually built at the opposite end of the system from the source of supply, so that during peak flows the maximum distance from the supply to the consumer is cut in half. It is necessary for an equalizing reservoir to have an elevation high enough to provide adequate pressure throughout the system served. For the correct hydraulic grade, it is necessary to build the reservoir above the area it serves. If the topography will not allow a surface reservoir, a standpipe or an elevated tank must be constructed. Standard elevated tanks are available in capacities up to 2 million gal.

21-74. Economic Sizing of Distribution Piping. In the design of any major project, the designer must choose the most economical of numerous alternatives. Most of these alternatives can be separated and studied individually. An example of two alternatives for a distribution system is one serving peak hourly demands totally by pumps and one doing it by pumps and equalizing reservoirs. The total costs of each plan should be compared by an annual or present-worth cost analysis.

A method of determining minimum cost that can readily be adapted to many problems is setting the first derivative of the total cost, taken with respect to the variable in question, equal to zero. In the sizing of pipes in a distribution system supplied by pumps, the total costs of the pipes, pumping plant, and energy may be expressed as an equation. To find the most economical diameter of pipe, the first derivative of the total cost, taken with respect to the pipe diameter, should be set equal to zero. The following equation for the most economical pipe diameter was derived in this manner:

$$D = 0.215 \left(\frac{fbQ_a^3 S}{aiH_a} \right)^{1/7} \qquad (21\text{-}142)$$

where D = pipe diameter, ft
f = Darcy-Weisbach friction factor
b = value of power, dollars per hp per year
Q_a = average discharge, cfs
S = allowable unit stress in steel, psi
a = in-place cost of steel, dollars per pound
i = yearly fixed charges for pipeline (expressed as a fraction of total capital cost)
H_a = average head on pipe, ft

21-75. Hydraulic Analysis of Distribution Piping. Adequate service requires a knowledge of the hydraulic grade at many points in a distribution system for various flows. Several methods, based on the following rules, have been developed for analysis of complex networks.

1. The head loss in a conduit varies as a power of the flow rate.
2. The algebraic sum of all flows into and out of any pipe junction equals zero.
3. The algebraic sum of all head losses between any two points is the same by any route, and the algebraic sum of all head losses around a loop equals zero.

A convenient device for simplifying complex networks of various size pipes is the *equivalent pipe*. For a series of different size pipes or several parallel pipes, one pipe of any desired diameter and one specific length or any desired length and one specific diameter can be substituted;

this will give the same head loss as the original for all flow rates if there are no take-outs or inputs between the two end points. In complex networks, the equivalent pipe is used mainly to simplify calculation.

Example 21-11. Determine the equivalent 8-in.-diameter pipe that will have the same loss of head as the sections of pipe from A to D in Fig. 21-83a.

First, transform pipes CD, AB, and BD into equivalent lengths of 8-in. pipe; then, transform the resulting sections into a single 8-in. pipe with the same head loss. The simplest way is to use a hydraulic slide rule or a nomograph (Fig. 21-84).

Assume any convenient flow through CD, say 500 gpm. Figure 21-84 indicates that loss of head in 1,000 ft of 6-in. pipe is 32 ft and in 1,000 ft of 8-in. pipe, 7.8 ft. Then, the equivalent length of 8-in. pipe for CD is $500 \times 32/7.8 = 2,050$ ft. Similarly, the equivalent pipe for AB should be 165 ft long, and for BD, 420 ft long. The network of 8-in. pipe is shown in Fig. 21-83b. It consists of pipe 1, $3,000 + 2,050 = 5,050$ ft long, connected in parallel to pipe 2, $165 + 420 = 585$ ft long.

To reduce the parallel pipes to an equivalent 8-in. pipe, assume a flow of 1,000 gpm through pipe 2. For this flow, the head loss in an 8-in. pipe per 1,000 ft is 29 ft. Hence the head loss in pipe 2 would be $29 \times 585/1,000 = 17$ ft. Since the pipes are connected in parallel, the head loss in pipe 1 also must be 17 ft, or 3.37 ft per 1,000 ft. The flow that will produce this head loss in an 8-in. pipe is 310 gpm (Fig. 21-84). The equivalent pipe, therefore, must carry $1,000 + 310 = 1,310$ gpm with a head loss of 17 ft. For a flow of 1,310 gpm, an 8-in. pipe would have a head loss of 48 ft in 1,000 ft, according to Fig. 21-84. For a loss of 17 ft, an 8-in. pipe would have to be $1,000 \times {}^{17}/_{48} = 350$ ft long. So the pipes between A and D in Fig. 21-83a are equivalent to a single 8-in. pipe 350 ft long.

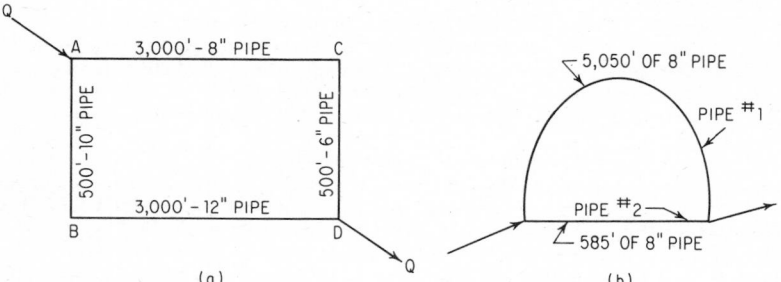

Fig. 21-83. Distribution loop (a) may be replaced by equivalent loop (b).

For complicated networks, the three most widely used methods of analysis are uncontrolled trial and error, analog computer, and Hardy Cross. The uncontrolled trial-and-error method consists of assuming flows in pipes and then checking the head losses to see if the assumed flows were correct. For relatively simple networks, an experienced designer can achieve good results by this method. The analog computer method is based on the fact that the flow of electric current in a circuit is very similar to the flow of water in pipes. The resistance of special elements in the electric circuit is analogous to the friction loss in pipes, the flow of current is analogous to the flow of water, and voltage is analogous to pressure. An electric circuit, representing a pipe network, may be set up and current and voltage measurements taken at various points to determine flows and pressures.

The **Hardy Cross method** is a controlled trial-and-error method. Flows are first assumed, then consecutive adjustments are computed to correct these assumed values. In most cases, sufficient accuracy can be obtained with three adjustments; however, there are rare cases where the computed adjustments do not approach zero. In these cases, an approximate method must be used.

Assumed flows in a loop are adjusted in accordance with the following equation:

$$\Delta Q = \frac{\Sigma KQ^n}{\Sigma n KQ^{n-1}} \tag{21-143}$$

where $KQ^n = h_f =$ loss of head due to friction. When the Hazen-Williams equation, used in

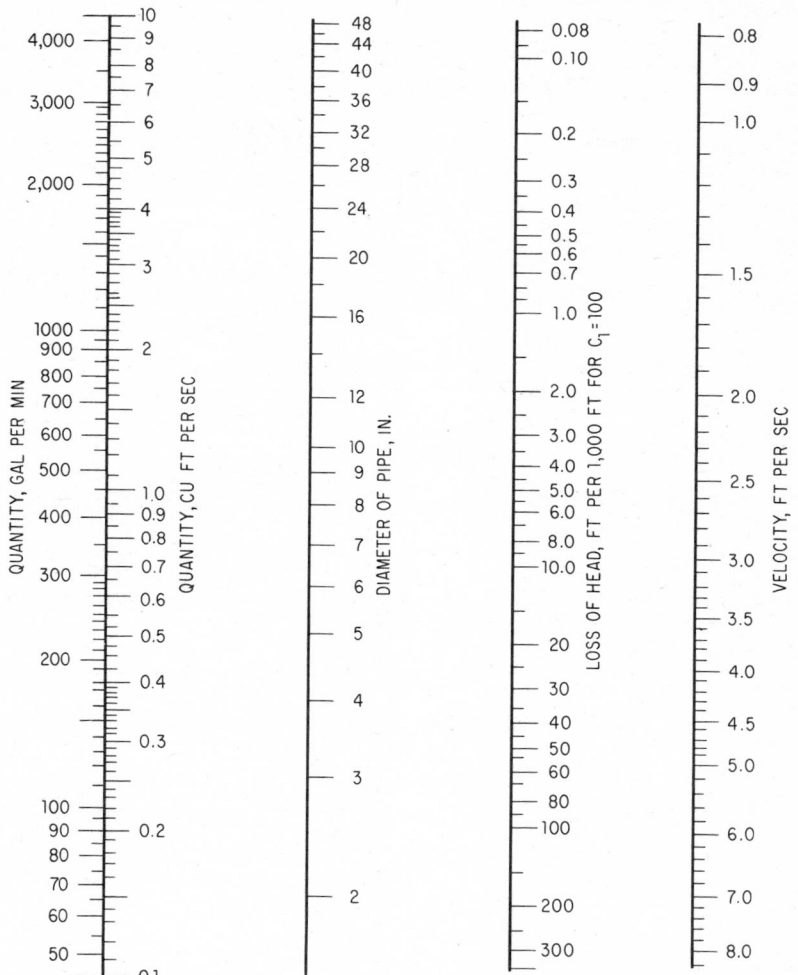

Fig. 21-84. Nomograph solves Hazen-Williams formula for pipe flow.

Example 21-11, is put in the form $h_f = KQ^n$, then $K = 4.727L/D^{4.87} C_1^{1.85}$ and $n = 1.85$. The expression ΣnKQ^{n-1} equals $\Sigma(nKQ^n/Q)$. In the Hazen-Williams formula $n = 1.85$ for all pipes and can therefore be taken outside the summation sign. Hence, the adjustment equation becomes

$$\Delta Q = \frac{\Sigma h_f}{n\Sigma(h_f/Q)} \qquad (21\text{-}144)$$

It is important that a consistent set of signs be used. The sign convention chosen for the following example makes clockwise flows and the losses from these flows positive, while counterclockwise flows and their losses are negative.

Example 21-12. Analyze the pipe network in Fig. 21-85 by the Hardy Cross method.

1. Transform the various sizes of pipes into equivalent pipes, all of the same diameter (in this case, 6 in.) with the same head-loss characteristics as the original pipe.

2. Number all the loops and pipes in the network for identification.

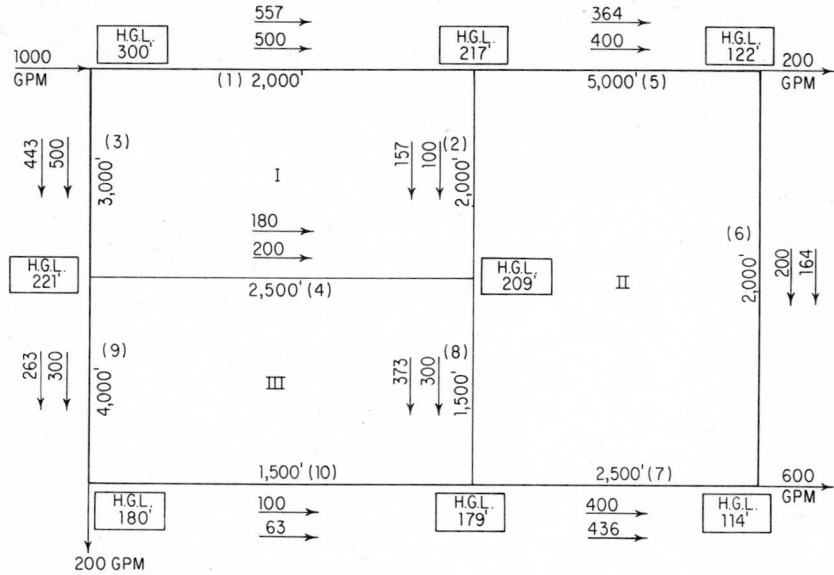

Fig. 21-85. Pipe network, in which all pipes are represented by equivalent lengths of 6-in.-diameter pipe with $C_1 = 100$, for use in Hardy Cross analysis. (See Example 21-12.)

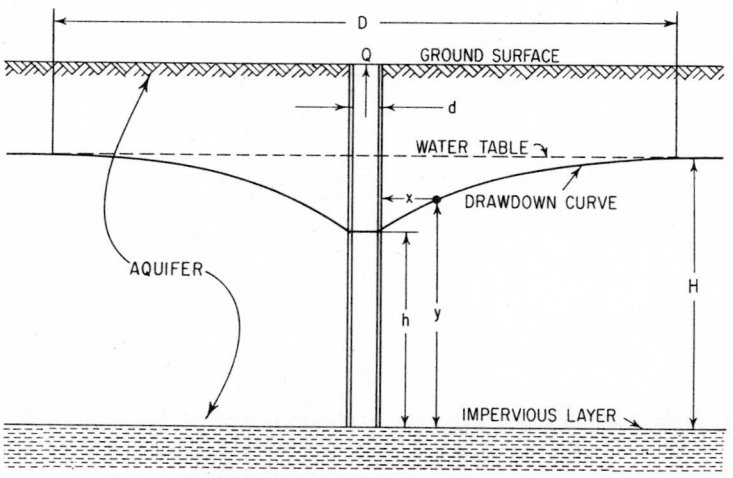

Fig. 21-86. Gravity well in a free aquifer.

3. Assume rates and direction of flows in each pipe. Experience is the best guide, but even an inexperienced designer can usually make a reasonable estimate. The assumed flows are shown in Fig. 21-85 as a labeled arrow adjacent to each pipe.

4. Fill in Table 21-24. Column 3 lists the equivalent length L of each pipe. In column 4, the assumed clockwise flows are positive and counterclockwise flows are negative. Note that the sum of the flows into each connection equals the sum of the flows out. The head losses in column 5 have the same sign as the corresponding Q values in column 4. The values in column 6 are summations of the positive and negative head losses around a loop. The values in column 7 are the head losses necessary to make the summation of head loss around a loop

Table 21-24. Hardy Cross Solution for Network in Fig. 21-85

Loop	Pipe	L, ft	Q, gpm	h_f, ft	Σh_f	Δh_f	h_f/Q	$n\Sigma(h_f/Q)$	ΔQ	$Q + \Delta Q$
(1)	(2)	(3)	(4)	(5)	(6)	(7)	(8)	(9)	(10)	(11)
I	1	2,000	+500	+65			+0.130			+557
	2	2,000	+100	+3.5	+68.5		−0.035			+157
	3	3,000	−500	−100		+46.5	+0.200		+57	−443
	4	2,500	−200	−15	−115.0		+0.075			−143
							0.440	× n = 0.815		
II	5	5,000	+400	+110			+0.275			+364
	6	2,000	+200	+12	+122		+0.060			+164
	2	2,000	−157	−8		−39	+0.051		−36	−193
	8	1,500	−300	−20			+0.067			−336
	7	2,500	−400	−55	−83		+0.137			×436
							0.590	× n = 1.09		
III	4	2,500	+143	+8			+0.056			+177
	8	1,500	+336	+24	+32		+0.072			+370
	9	4,000	−300	−50		+20	+0.167		+34	−266
	10	1,500	−100	−2	−52		+0.020			−66
							0.315	× n = 0.583		
I	1	2,000	+557	+82			0.147			+556
	2	2,000	+193	+11	+93		0.057			+192
	3	3,000	−443	−80		−1	0.181		−1	−444
	4	2,500	−177	−12	−92		0.068			−178
							0.453	× n = 0.837		
II	5	5,000	+364	+94			0.258			+366
	6	2,000	+164	+8	+102		0.049			+166
	2	2,000	−193	−11		+2	0.057		+2	−191
	8	1,500	−370	−28			0.076			−368
	7	2,500	−436	−65	−104		0.149			−434
							0.589	× n = 1.09		
III	4	2,500	+177	+12			+0.069			+180
	8	1,500	+370	+28	+40		+0.076			+373
	9	4,000	−266	−41		+2	+0.154		+3	−263
	10	1,500	−66	−1	−42		+0.015			−63
							0.314	× n = 0.580		

equal to zero. The values in column 8 are always positive. Column 9 is obtained with $n = 1.85$. The signs of the flows in column 10 are the same as the corresponding head-loss terms in column 7.

5. Loop I is adjusted first, using the assumed flow values of column 4. When loop II is adjusted, the assumed flow values are used, except for pipe 2. Since pipe 2 was adjusted in loop I, this more accurate value is used. The sign of the flow in pipe 2 in loop II is negative because the assumed flow in pipe 2 is counterclockwise for loop II. The most accurate value, which is usually the last adjusted value, is always used in column 4. After loop III is adjusted, the cycle may be repeated to obtain more accurate results.

One of the greatest assets of the Hardy Cross method is that it can be readily programmed for a digital computer.

21-76. Wells. A gravity well is a vertical hole penetrating an aquifer that has a free-water surface at atmospheric pressure (Fig. 21-86). A pressure or artesian well passes through an impervious stratum into a confined aquifer containing water at a pressure greater than atmospheric (Fig. 21-87). A flowing artesian well is an artesian well extending into a confined aquifer that is under sufficient pressure to cause water to flow above the casing head. A gallery or horizontal well is a horizontal or nearly horizontal tunnel, ditch, or pipe placed normal to groundwater flow in an aquifer.

When water is pumped from a well, the water level around the well draws down and forms a *cone of depression* (Fig. 21-86). The line of intersection between the cone of depression and the original water surface is called the *circle of influence*.

The steady flow rate Q can be found for a gravity well by using the **Dupuit formula:**

$$Q = \frac{1.36K(H^2 - h^2)}{\log (D/d)} \tag{21-145}$$

where Q = flow, gal per day
K = coefficient of permeability, gal per day per sq ft under a 1:1 hydraulic gradient
H = total depth of water from bottom of well to the free-water surface before pumping, ft
h = H minus the drawdown, ft
D = diameter of circle of influence, ft
d = diameter of well, ft

The steady flow, gpd, from an artesian well is given by

$$Q = \frac{2.73Kt(H - h)}{\log (D/d)} \tag{21-146}$$

where t is the thickness of confined aquifer, ft (Fig. 21-87).

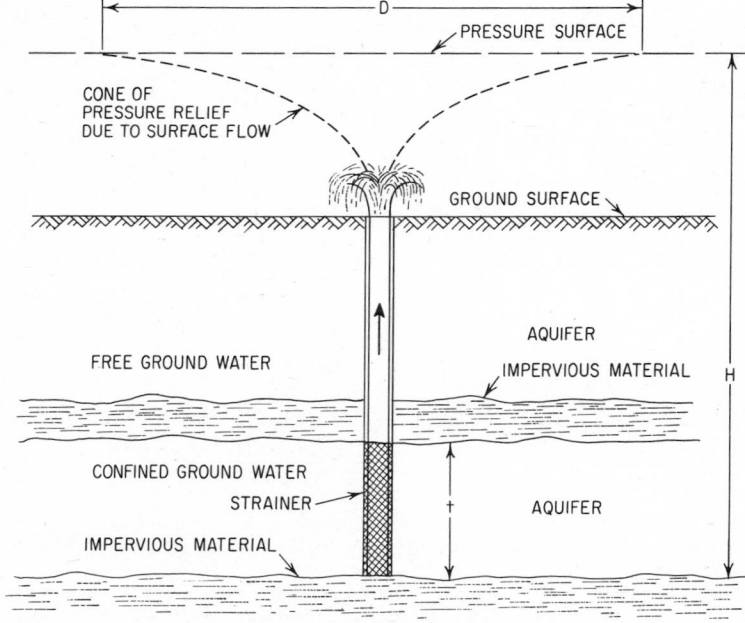

Fig. 21-87. Artesian well in a pressure aquifer.

A long time elapses between the beginning of pumping and establishment of a steady-flow condition (a circle of influence with constant diameter). Hence, correct values for drawdown and the circle of influence can be obtained only after long periods of continuous pumping.

Since nearly all soils are heterogeneous, pumping tests should be made in the field to determine the value of the coefficient of permeability K. A permeability analysis of a soil sample that is not representative of the soil throughout the aquifer would produce an unreliable value for K.

A nonequilibrium formula developed by Theis and a modified nonequilibrium formula produced by Jacob are used in analyzing well flow conditions where equilibrium has not been established. Both methods utilize a storage coefficient S and the coefficient of transmissibility

T to eliminate complications due to the time lag before reaching steady flow. (C. V. Theis, "The Significance of the Cone of Depression in Groundwater Bodies," *Economic Geology*, vol. 33, p. 889, December, 1938; C. E. Jacob, "Drawdown Test to Determine Effective Radius of Artesian Well," *Proceedings of the American Society of Civil Engineers*, vol. 72, no. 5, p. 629, 1940.)

Interference between two or more wells is caused by the overlapping of circles of influence. Drawdown for each interfering well is increased and the rate of water flow is decreased for each well in proportion to the degree of interference.   Interference between two or more closely spaced wells may increase to the extent that the system of wells produces one large cone of depression.

Wells may be classed by the method by which they are constructed and their depth.   Shallow wells (less than 100 ft deep) are usually dug, bored, or driven.   Deep wells (depth greater than 100 ft) are usually drilled by either the standard cable-tool, water-jet, hollow-core, or hydraulic rotary methods.

Essential well equipment consists of casing, screen, eductor or riser pipe, pump, and motor. The function of the casing is to keep the wall material and polluted water from entering the well and to prevent the leakage of good water from the well.

The screen is placed below the casing to contain the walls of the aquifer, to allow water to pass from the aquifer into the well, and to stop movement of the larger sand particles into the well. The pump, motor, and eductor pipe are utilized to move the water from the aquifer to the collecting lines at the ground surface.

(G. M. Fair, J. C. Geyer, and D. A. Okun, "Water and Wastewater Engineering," John Wiley & Sons, Inc., New York.)

21-77. Centrifugal Pumps.   The purpose of any pump is to transform mechanical or electrical energy into pressure energy.   The centrifugal pump, the commonest waterworks pump, accomplishes that in two steps.   The first transforms the mechanical or electrical energy into kinetic energy with a spinning element, or impeller.   The kinetic energy is then converted to pressure energy by diffuser vanes or a gradually diverging discharge tube, called a volute (Fig. 21-88).

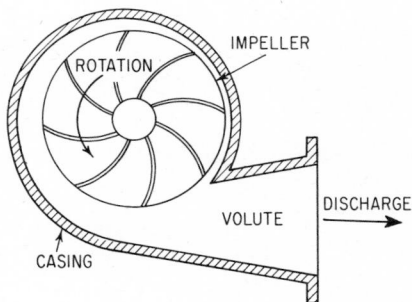

Fig. 21-88. Volute-type centrifugal pump.

Water enters at the center, or eye, of the impeller and is forced outward toward the casing by centrifugal force.   The discharge head of a centrifugal pump is a function of the impeller diameter and speed of rotation.

Design factors requiring consideration in the selection of a centrifugal pump are net positive suction head required, efficiency, horsepower, and the head-discharge relationship.

Net positive suction head $NPSH$ is the energy in the liquid at the center line of the pump.   To have practical meaning, it must be referred to as either the *required* or *available NPSH*.   Required NPSH is a characteristic of the pump and is given by the manufacturer.   Available NPSH is a characteristic of the system and is determined by the engineer.   It is the pressure in the liquid over and above its vapor pressure at the suction flange of the pump and is given, in feet, by

$$\text{Available NPSH} = 144 \frac{p_a - p_v}{w} - h_f + z \qquad (21\text{-}147)$$

where p_a = pressure, psia, on a free-water surface or at the center line of a closed conduit

p_v = vapor pressure, psia, of the water at its pumping temperature

h_f = friction loss in the suction line, ft of water

z = elevation difference, ft, between pump center line and water surface

w = unit weight of liquid, lb per cu ft

If the suction water surface is below the pump center line, z is negative. To prevent cavitation, it is necessary to have the available NPSH always greater than the required NPSH. For that reason, it is customary to analyze a required NPSH vs. discharge curve with the brake horse-power, head, and efficiency curves when selecting a pump.

The operating point of a centrifugal pump is determined by the intersection of the pump's head-capacity curve and the *system head curve*, as shown in Fig. 21-89. (Also included in Fig. 21-89 are the other curves used in pump selection.) A system head curve is a plot of the head losses in the system vs. pump discharge. This curve shows the head differential that must be supplied by the pump. In a typical water-system problem, there may be three or four pertinent system head curves corresponding to various consumption rates. The intersections of these curves with the head vs. Q curve define a range of operation rather than a single point.

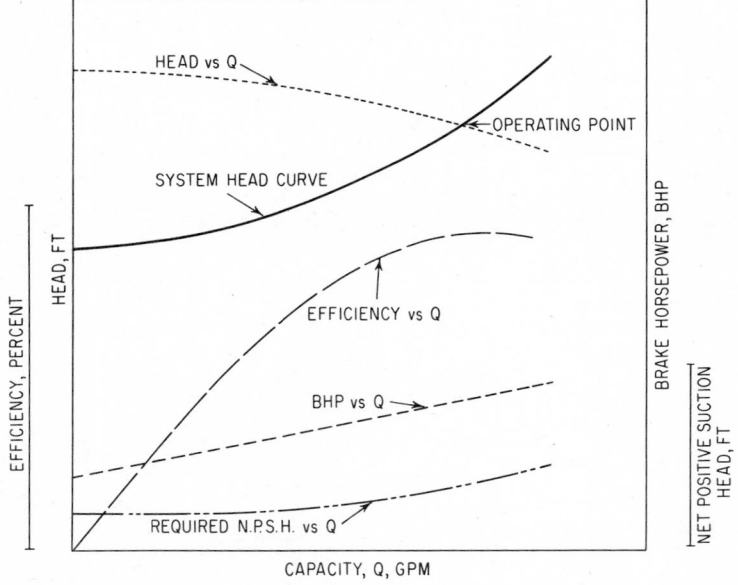

Fig. 21-89. Curves used in the selection of a centrifugal pump.

Selection of a centrifugal pump is largely a matter of matching one of the many pumps available to the system characteristics. In doing so, the point of maximum efficiency should be at or near the operating point. Centrifugal pumps are available in almost any capacity desired, with lifts of up to 700 ft per stage. Efficiencies may be as high as 93% for large pumps.

See also Art. 21-78.

(T. G. Hicks and T. Edwards, "Pump Application Engineering," McGraw-Hill Book Company, New York; I. J. Karassik, "Centrifugal Pumps," McGraw-Hill Book Company, New York.)

21-78. Well Pumps. These are classified as centrifugal, propeller, jet, helical, rotary, reciprocating, and air lift. Although centrifugal pumps (Art. 21-77) are the commonest for both shallow-well and deep-well pumps, circumstances may dictate one of the other types.

Centrifugal pumps are used in wells over 6 in. in diameter. They have capacities up to 4,000 or 5,000 gpm and heads up to 1,200 ft, depending on the number of stages. Efficiencies may be as high as 90% for the larger capacities; however, below 200 gpm, the maximum efficiency is 75 to 80%.

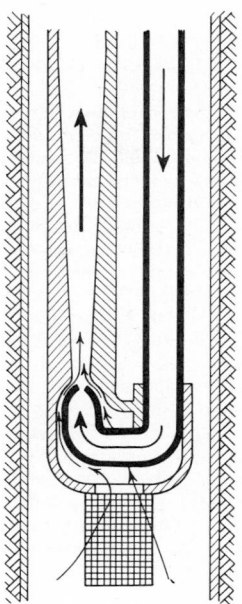

Fig. 21-90. Section through a jet pump (simplified).

Propeller pumps are an axial-flow type.   They are used in high-capacity low-head applications.

Jet pumps (Fig. 21-90) operate by discharging water through a nozzle and diverging conical tube, which are located at the well bottom.   The diverging conical tube creates lift by converting the high-velocity head to pressure head.   The suction connection is made between the nozzle and entrance to the diverging tube.   Jet pumps have low efficiencies.   They are used in small-capacity low-lift applications, especially where the water contains sand or other impurities.

Helical pumps are a positive-displacement type with a metal helical rotor rotating inside a rubber helical stator.   The screw action of the rotor forces water through the pump and up the discharge pipe.   Helical pumps are small-capacity high-lift pumps.   They may be used in wells over 4 in. in inside diameter.

Rotary pumps are also of the displacement type.   They have a fixed chamber in which gears, vanes, cams, or pistons rotate with very close tolerances.   These pumps have relatively constant partial-load efficiencies.   Full-load efficiencies range from 50 to 85%.   Because of the close tolerances, they can be used only for sediment-free water.

Reciprocating pumps, either hand- or motor-driven, utilize piston action to move water.   Their present-day use is primarily for small-capacity low-lift private applications.

Air-lift pumps generate lift by using air bubbles to decrease the specific weight of the column of water in the discharge pipe below that of the surrounding water in the well.   This creates a pressure differential that forces the water out of the well.   Air-lift pumps are the simplest and most foolproof of well pumps since they have no submerged moving parts.   They can be used in any well but have the disadvantage of efficiencies below 50%.

Specific speed N_s is a widely used criterion for pump selection.   It is the impeller speed corresponding to a discharge of 1 gpm at 1 ft of head for the most efficient design.

$$N_s = nQ^{1/2}H^{-3/4} \qquad (21\text{-}148)$$

where n = impeller speed, rpm
Q = discharge, gpm
H = head, ft

The favorable design range of N_s for radial-flow (centrifugal) pumps is from 1,500 to 4,100.   For N_s between 4,100 and 7,500, mixed-flow pumps having both radial and axial characteristics should be used, and for N_s above 7,500, axial-flow (propeller) pumps should be used.

Shallow-well pumps have their motors and impellers at ground level, so that the entire lift is suction.   Since excessive suction lifts cause cavitation, the lift is limited by atmospheric pressure and the velocity head at the impeller, which is a function of specific speed. At sea level, the maximum practical lift for a shallow-well pump is about 25 ft.

Deep-well pumps have their impellers close enough to the water surface to eliminate cavitation. The motor may be at ground level with a long shaft connecting it to the impellers, or it may be at the bottom of the well, below and directly adjacent to the impellers. The former type is called a *deep-well turbine pump* and the latter a *submersible pump*. Deep-well turbine pumps can be used only for straight wells. The pump shaft is supported at intervals of about 10 ft by rubber or metal bearings, which are water- or oil-lubricated, respectively. If sand is carried out with the water, an enclosed-shaft or submersible pump must be used to prevent bearing damage. Submersible pumps may be used in crooked wells. Other advantages include ease of increasing the well depth or lift and silent operation. One disadvantage is that the motors are difficult to reach for repairs.

(T. G. Hicks and T. Edwards, "Pump Application Engineering," McGraw-Hill Book Company, New York; I. J. Karassik, "Centrifugal Pumps," McGraw-Hill Book Company, New York; H. E. Babbitt, J. J. Doland, and J. L. Cleasby, "Water Supply Engineering," McGraw-Hill Book Company, New York.)

21-79. Pipe Materials. Cast iron, steel, concrete, and asbestos-cement are the commonest materials used in distribution pipes. Wood pipelines are still in existence, but wood is rarely used in new installations. Copper, lead, zinc, brass, bronze, and plastic are materials used in small pipes, valves, pumps, and other appurtenances. Common pipe-joint materials are lead, cement, sand, rubber, plastic, and sulfur compounds.

Cast iron is the commonest material for city water mains. Standard sizes range from 2 to 24 in. in diameter. Cast iron is resistant to corrosion and usually has good hydraulic characteristics. If it is cement-lined, the Hazen-Williams C value may be as high as 145. Iron tubercles may form in unlined cast-iron pipes and seriously affect flow capacity. Tuberculation can be prevented by lining with cement or tar materials. The relatively high cost of cast-iron pipe is only a slight disadvantage, because it is largely offset by the long average life of trouble-free service. Bell-and-spigot and flange are the commonest joints in cast-iron pipe.

Steel is commonly used for large pipelines and trunk mains, but rarely for distribution mains. Steel pipes with either longitudinal or spiral joints are formed at steel mills from flat sheets. The transverse joints between pipe sections are usually made by welding, riveting, bell-and-spigot with rubber gasket, sealed flanges, or Dresser-type couplings. Since steel is stronger than iron, thinner and lighter pipes can be used for the same pressures. Some disadvantages of thin steel pipe are inability to carry high external loads, possibility of collapse due to negative gage pressures, and high maintenance costs due to higher corrosion rates and thinner pipe walls. Steel pipes are usually corrosion-protected on both the outside and inside with coal tar or cement mortar. Under favorable conditions, the life of steel pipe is between 50 and 75 years.

Concrete pipe may be precast in sections and assembled on the job or cast in place. A machine that produces a monolithic, jointless concrete pipe without formwork has been developed for gravity-flow and low-pressure applications. Most of the precast-concrete pipe is reinforced or prestressed with steel. Concrete pipe may be made watertight by insertion of a thin steel cylinder in the pipe walls. High-strength wire is frequently wound around the thin steel cylinder for reinforcement. Concrete is placed inside and outside the steel cylinder to prevent corrosion and strengthen the pipe. Some advantages of concrete pipe are low maintenance cost, resistance to corrosion under normal conditions, low transportation costs for materials if water and aggregate are available locally, and ability to withstand external loads. Some disadvantages to be considered are leaching of free lime from the concrete, the tendency to leak under pressure due to the cracking and permeability of concrete, and corrosion in strong acids or alkalies.

Asbestos-cement pipe is made by compressing a mixture of asbestos fiber and portland cement to a dense mass which will withstand pressures of up to 200 psi. Asbestos-cement pipe is lightweight, corrosion-resistant, easily cut and tapped, easily joined together with rubber gaskets, and has a high water-carrying capacity (Hazen-Williams $C = 140$). A disadvantage of asbestos-cement pipe is that it is easily broken by unevenly distributed external loads.

(H. E. Babbitt, J. J. Doland, and J. L. Cleasby, "Water Supply Engineering," McGraw-Hill Book Company, New York.)

21-80. Corrosion. Many millions of dollars are expended every year to replace pipes, valves, hydrants, tanks, and meters that are destroyed by corrosion. Some of the causes of corrosion are the contact of two dissimilar metals with water or soil, stray electric currents, impurities and strains in metals, contact between acids and metals, bacteria in water, or soil-producing compounds that react with metals.

Electrochemical corrosion of a metal takes place when an electrolyte and two electrodes, an anode and a cathode, are present. (Water may serve as an electrolyte.) At the anode, the metal in contact with the electrolyte changes into a positively charged particle, which goes into solution or forms an oxide film. (The ease with which a metal changes to a metallic ion when it is in contact with water depends on its oxidation potential or solution pressure. Metals can be arranged in an electromotive series of decreasing oxidation potentials. Metals high in the electromotive series corrode more readily than metals located in a lower position.) For an iron pipe exposed to water, for example, the anode reaction is Fe (metal) → $Fe^{++} + 2e$. At the cathode, the metal having the excess electrons gives them up to a charged particle, such as hydrogen in solution: $2H^+ + 2e \rightarrow H_2$ (gas). If the hydrogen gas produced at the cathode is removed from the cathode by reaction with oxygen to produce water molecules or by water movement (depolarization), the corrosion process continues (Fig. 21-91).

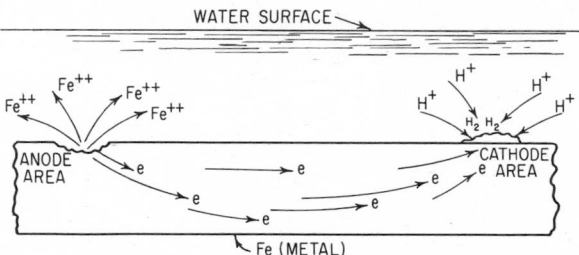

Fig. 21-91. Electrochemical corrosion of iron in low-pH water.

Indications of corrosion in an inaccessible iron or steel pipeline are discharges of rusty-colored water (due to the loosening of rust and scale) and metallic-tasting water. A marked decrease in capacity and pressure in a pipe section usually indicates tuberculation inside the line. Tuberculation is caused by the deposition and growth of insoluble iron compounds inside a pipe. Corrosion normally precedes deposition of scale because iron must be in solution to react with the basic substances and dissolved oxygen in the water to form scale. Iron-consuming bacteria in water can produce ferrous oxide directly if the iron concentration is about 2 ppm. A continuous supply of soluble iron in the presence of iron-consuming bacteria or dissolved oxygen and basic substances in the water increases the size of the tubercles. Tubercles may become so large and decrease the capacity in the pipe to such an extent that it has to be cleaned or replaced.

Some factors that influence the type and quantity of metallic corrosion are the presence of protective films, strains and cracks, undissolved impurities in the metal, moving water, ionic compounds in the water, high hydrogen-ion concentration, and alternate wetting and drying of the metal. Some metals form oxide films that act as protective layers for the metal. Aluminum, zinc, and chromium are examples of this type of metal. Strains, cracks, and undissolved impurities in a metal act as sites for corrosion. Agitation or movement of water increases the corrosion rate of a metal because the oxygen supply rate to the cathode and the removal rate of metal ions from the anode are increased. The presence of ionic compounds in the water speeds up corrosion, because the ions act as conductors of electricity, and the more ions, the faster electrons can move through the water. Alternate wetting and drying tends to break up the rust or oxide film. This facilitates penetration of the film by the oxygen and water and leads to increased corrosion. High hydrogen-ion concentrations increase corrosion rates because of the greater accessibility of the hydrogen ions to the cathode.

Corrosion may be prevented or retarded by proper selection of materials, use of protective coatings, and treatment of the water. In selecting materials, the engineer should take into account the characteristics of the water and soil conditions encountered. Protective coatings for metals may be metallic or nonmetallic. They may be applied on both the inside and outside surfaces of the pipe. Common nonmetallic coatings are cement and asphalt. Zinc and lead are examples of metallic coating materials used. Steel pipe dipped in zinc (galvanized) is commonly used for small service lines, although copper tubing is becoming increasingly popular for this purpose.

Also, to prevent corrosion, water may be treated with bases, such as soda ash, caustic soda, and lime, to reduce hydrogen-ion concentration and to induce precipitation of thin films of carbonates, hydroxides, oxides, etc., on the walls of the pipes. These thin films reduce the ability of water to corrode otherwise unprotected metal surfaces.

Electrochemical corrosion of external surfaces of pipelines and water tanks can be retarded by application of a direct current to the metal to be protected and to another metal that acts as a sacrificial anode (Fig. 21-92). The potential applied to or produced by the two metal surfaces must be large enough to make the protected metal act as a cathode. The sacrificial anode corrodes and must be replaced periodically. Zinc, magnesium, graphite, and aluminum alloys are commonly used for anode materials.

(American Water Works Association, "Water Quality and Treatment," McGraw-Hill Book Company, New York.)

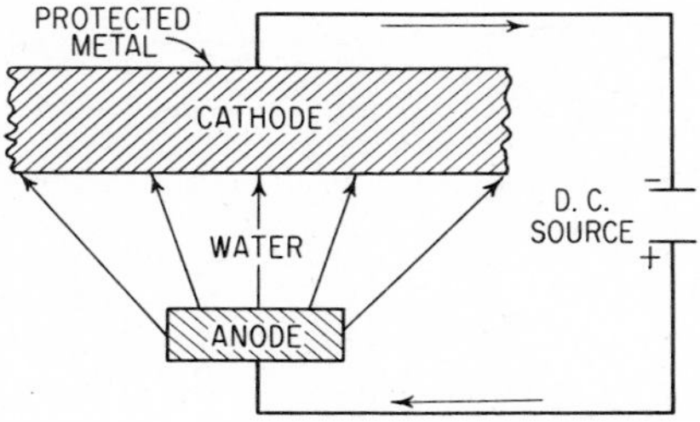

Fig. 21-92. Cathodic protection of a metal.

21-81. Valves. Water facilities use many different types of valves, which are generally classified according to the function they perform. The two major water-valve classifications are isolating and controlling.

Isolating valves are used for separating or shutting off sections of pipe, pumps, and control devices from the rest of the system for inspection and repair purposes. The major types of isolating valves are gate, plug, sluice gate, and butterfly.

A control valve is normally used for continuously controlling pressures and flow rates. Check, needle, globe, air-relief, pressure-regulating, pressure-relief, and altitude valves are usually considered as control valves.

Gate valves are the isolating valves used most often in distribution systems, primarily because of their low cost, availability, and low head loss when fully open. They have limited value as control or throttling devices, because of seat wear and the downstream deflection and chatter of the gate disk. Also, the open area and rate of flow through the valve are not proportional to the percentage opening of the valve when partly open. Corrosion, solids deposition, tubercle formation, large pressure differences, and thermal expansion produce difficulties in opening normally closed gate valves or in closing normally open gate valves. Periodic inspection and operation of valves that are infrequently operated will prevent many operational difficulties. Some of the larger gate valves have gear-reduction drives to permit manual operation. Very large valves are operated by hydraulic and electric power.

A *plug valve* may be used for both control and isolation purposes. It consists of a cylindrically shaped plug (with a rectangular slot or circular orifice) placed in a close-fitting cylindrical seat perpendicular to the direction of flow. Cone and spherical valves are special types of plug valves. Plug, cone, and spherical valves can all be fully closed or opened by a 90° rotation of the plug. The valves may or may not be lubricated (large iron valves usually are). Hydraulic or electric power is commonly utilized for operating the larger valves. Small plug valves are commonly used for isolation purposes on domestic and commercial service connections and are known as either service, curb, or corporation cocks. Usually, because the meter is not directly adjacent to the distribution pipe, three valves must be used, one at the service connection, one just upstream of the meter, and one between the meter and the customer's service line. Plug and cone valves are also used for throttling and remote-control shutoff purposes. Low head loss, in-service lubrication features, and easy, fast operation, even in the presence of unequal pressures across the valve, are the major advantages of plug-type valves. But these valves cost more than gate, globe, and butterfly valves.

Butterfly valves can be used for throttling and isolation purposes. The butterfly-valve mechanism consists of a relatively thin circular disk pivoted on a horizontal shaft. Hand or motor power, applied through a gear-reduction device, rotates the disk. Simplicity of construction and quick, easy operation are reasons these valves are replacing sluice gates and gate valves in many locations. Los Angeles, Calif., has replaced many sluice gates in reservoir towers with butterfly valves having seats of corrosion-resistant metal, rubber, or Neoprene. A disadvantage of butterfly valves is the higher cost relative to sluice gates or gate valves.

Sluice gates are mainly used on the sides of reservoir control towers and in open-channel structures where pressure on one side of the gate helps to seat it and prevent water leakage. Difficulties with leakage and corrosion of gate frames and stems are the main disadvantages of sluice gates. Low cost and ease of operation in open-channel flow conditions are the major advantages.

A *needle valve* is made of a streamlined plug or needle that fits into a small orifice with a carefully machined seat. Needle valves are used for accurate control of water flow because a large movement of the needle is necessary before any measurable change of flow rate takes place. Needle valves are not normally used for isolating purposes because of the high head losses produced by water flow through the small orifices. Large-sized needle valves are used for flow regulation under high heads, such as for free discharge from reservoirs. Interior-differential, tube, and hollow-jet are three of the commonest types of large needle valves.

Globe valves are commonly used in smaller sizes for domestic purposes. The valve mechanism consists of a screw-operated disk that is forced down on a circular seat. Because of high head losses, globe valves are rarely used for isolation purposes, but they are commonly used for pressure regulation in water-distribution systems. Many automatic control valves, such as pressure regulators and altitude, check, and relief valves, have globe-valve bodies with various types of control mechanisms.

Pressure-regulating valves are used to reduce pressures automatically. An air-relief and inlet valve serves the dual purpose of allowing air to either escape or enter a pipeline. Air that accumulates at high points in a pipe impedes water flow and should be allowed to escape through an air-relief valve placed at this location. Furthermore, draining water from low elevations in a pipeline may cause negative pressures at higher elevations and collapse a pipe. Air should be allowed to enter through air-relief and inlet valves at the high points to prevent this.

Pressure-relief valves are used to release excess pressure in an enclosure. Many times, these excess pressures are caused by sudden closure of a valve.

Altitude valves are used to control the water level of elevated reservoirs. A pressure-activated control closes the altitude valve when the tank is full and opens the valve to allow water to flow from the tank when pressure below the valve decreases.

Check valves are used in pipelines to allow for one-directional flow only. Check valves placed in centrifugal-pump suction lines are called foot valves. These valves hold water in the suction line and pump case so that the pump will not need manual priming when started. The commonest check valve is the swing type.

(H. E. Babbitt, J. J. Doland, and J. L. Cleasby, "Water Supply Engineering," McGraw-Hill Book Company, New York.)

21-82. Fire Hydrants. A fire hydrant normally consists of a cast-iron barrel and a gate or compression-type shutoff valve, which connects the barrel to the main. Two or more hose outlets are normally located in the barrel above the ground surface. Usually, an additional gate valve is required between the hydrant and the main to allow for shutoff and repair of the hydrant.

The number of 2½-in.-diameter hose outlets on a hydrant determines its class. For example, a hydrant with two hose outlets is called a two-way hydrant.

Fire-hydrant construction standards have been established by the American Water Works Association and the American Insurance Association. These standards relate the diameter of the barrel to the size of the main-valve opening. A barrel diameter of at least 4 in. is required for a two-way hydrant, 5 in. for a three-way hydrant, and 6 in. for a four-way hydrant. A minimum of two hose outlets is required on a fire hydrant. Where pumper service is necessary for adequate water pressure, a large pumper outlet must be furnished. This may take the place of one of the smaller 2½-in. hose outlets. The minimum allowable diameter for the pipe connection between the main and the hydrant is 6 in.

Fire hydrants usually are either dry or wet barrel, depending on the location of the main valve in the hydrant. The main valve in the dry-barrel type should be located below the frost line. When the valve is in a closed position, a drain should be open to prevent freezing of water in the barrel. The wet-barrel, or California type, hydrants have the main valve located near the hose outlets. Many fire hydrants have a safety joint above the ground surface to permit removal of the upper part of the barrel with a minimum loss of water.

Hose connections $3\frac{1}{16}$ in. in diameter with $7\frac{1}{2}$ threads per inch have been selected by the American Insurance Association as standard to allow for interchange of fire-fighting equipment between cities.

Friction losses should not exceed $2\frac{1}{2}$ psi in a hydrant and 5 psi between the main and outlet when flow is 600 gpm.

21-83. Metering Devices. Metering devices are classified as either velocity or displacement types. Velocity types measure the velocity of flow either directly by current-measuring devices or indirectly by venturi-principle devices and are usually calibrated to indicate the flow rate directly. The velocity-type metering devices are applied to measurement of flows in streams, rivers, and large pipes, such as trunk lines of distribution systems. Displacement-type metering devices indicate flow rate directly, by recording and integrating the rate at which their measuring chambers are filled and emptied. Weighing meters are also displacement-type metering devices, but they are used primarily in laboratories. Displacement types are used for the smaller flows in distribution systems, such as meters for individual customer connections.

Criteria for selection of a type of water meter include accuracy and range of measurement, amount of head loss through the meter, durability, simplicity and ease of repairs, and cost.

Velocity-type Metering Devices. Venturi meters, or modifications thereof, are the commonest velocity-type devices. These meters produce a regular and predictable fall in the hydraulic grade line that is related to flow rate. Three devices that operate on this principle are the venturi, nozzle, and orifice plate meters shown in Fig. 21-93.

Straightening vanes are installed upstream from these and other velocity-type meters if the pipe is of insufficient length to eliminate helical flow components caused by bends or other fittings.

The standard venturi meter (Fig. 21-93a) was developed to provide a device with minimum head loss. Since most of the loss is associated with the diffuser section, its angle is the major factor in determining the head loss.

Flow through a venturi meter is given by

$$Q = cKd_2^2 \sqrt{h_1 - h_2} \qquad (21\text{-}149)$$

$$K = \frac{4}{\pi} \sqrt{\frac{2g}{1 - (d_2/d_1)^2}} \qquad (21\text{-}150)$$

where Q = flow rate, cfs
c = empirical discharge coefficient dependent on throat velocity and diameter
d_1 = diameter of main section, ft
d_2 = diameter of throat, ft
h_1 = pressure in main section, ft of water
h_2 = pressure in throat section, ft of water
(For values of c and K for various throat diameters and velocities, see H. W. King and E. F. Brater, "Handbook of Hydraulics," McGraw-Hill Book Company, New York.)

As in venturi meters, flows through nozzle and orifice-plate meters are calculated from the pressure difference across the meters. Nozzle and orifice-plate meters are used where conservation of head is not the prime concern or where head dissipation is desired.

Current meters consist of either a propeller or a series of cups or vanes mounted on a shaft that is free to rotate under the action of the flowing water. The propeller type has its axis of rotation horizontal. It will not give accurate measurement unless the current velocity is parallel to the axis of rotation. The cup-type meter, called a Price meter, has a vertical axis of rotation and measures currents whose velocity is in any direction in a horizontal plane. However, vertical velocity components, which do not affect propeller meters, cause the Price meter to indicate greater-than-actual velocities. A clicking noise, made by the making and breaking of an electrical contact and picked up by a set of earphones, indicates the speed of rotation of the meter. The clicking noise occurs either once each revolution or once each five revolutions. Current meters are used almost exclusively for stream flow, although the propeller type is occasionally substituted for a venturi meter in pipe flow.

Displacement-type Meters. These may be piston, rotary, or nutating-disk types. The nutating disk is used, almost to the exclusion of the two other types, for metering domestic-service connections. Its widespread use stems from its simplicity of construction and long-term accuracy. The nutating-disk meter derives its name from the disk's nodding motion, which is similar to that of a top before it stops. The disk is kept in motion by successive volumes of water which enter above and below it. A hard rubber that softens at high temperature is usually used for the disk, and so a backflow-prevention device is required between a nutating-disk

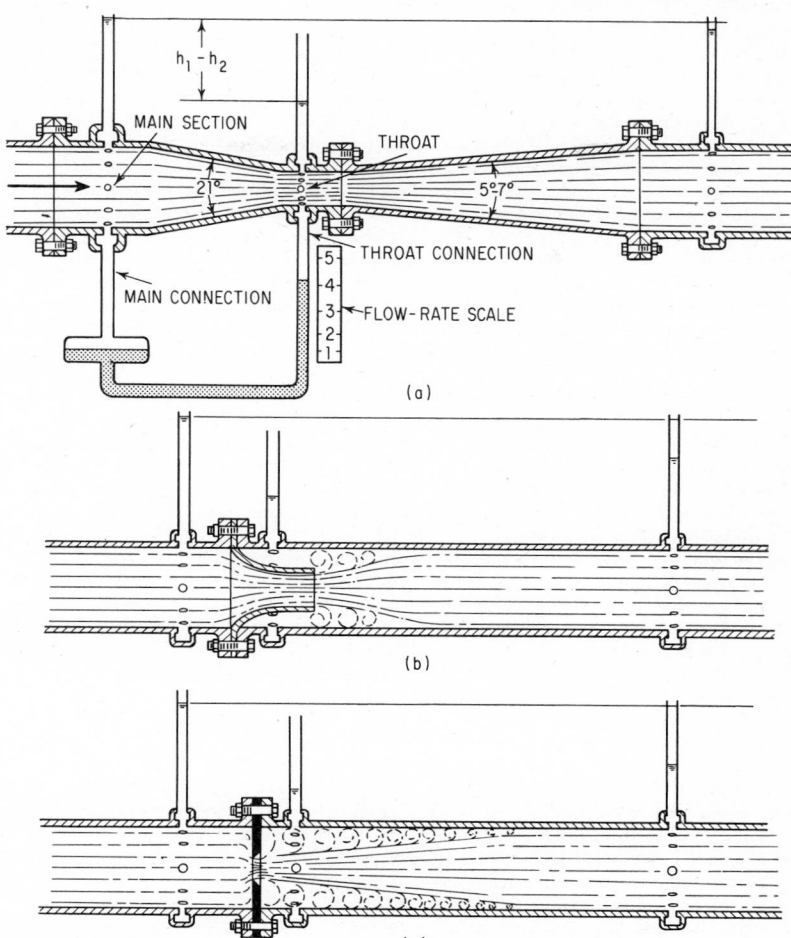

Fig. 21-93. Venturi-type metering devices: (*a*) Standard venturi meter; (*v*) nozzle meter; (*c*) orifice-plate meter.

meter and a water heater.   Error of nutating-disk meters is about 1.5% within the normal test-flow limits.

Compound meters contain separate measuring devices for both low and high flows.   They are usually a nutating-disk meter and a propeller-type current meter, respectively.   An automatic pressure-sensing device directs the flow through the appropriate meter.

(H. E. Babbitt, J. J. Doland, and J. L. Cleasby, "Water Supply Engineering," McGraw-Hill Book Company, New York.)

21-84. Water Rates and Financing.   The interests of the public and individual customers of water-supply systems can best be served by self-sustained, utility-type enterprises.   Rates charged to finance these systems should be based on sound engineering and economic principles and designed to avoid discrimination between classes of customers.   Gross revenue should cover operating and maintenance expenses, fixed charges on capital investment, and development of the system.   Billings for water should be based on metered use and such fixed charges as are required.   Rate structures are typically based on demand, load factors, fire use, peak rates of use, seasonal use, and similar items.   The system of accounting should conform to the legally established system of accounting prescribed for the utility, if any, or to some other recognized system.

Public ownership of urban water utilities is so prevalent as to be taken for granted (92 of the 100 largest water systems in the United States are publicly owned). Reasons for public ownership are the close relationship of quality of water to public health, the difficulty of separating the functional costs of supplying general customers from those of providing public fire protection, and the low return on capital investment. (From $9 to $12 is required to produce $1 in annual revenue for a water utility.) Whether ownership is private or public, the economic and fiscal requirements and the principles of rate making are similar.

Funds for current operations are derived from revenue for service provided to customers, charges to new customers for expansion or connection to the system, and proceeds from sales of bonds.

Revenue required for operations is that necessary to cover all costs of operation, including maintenance, taxes or payments in lieu of taxes, and interest on bonded debt; to cover depreciation; and to produce a net income that will provide for the orderly replacement and expansion of the system, and for repayment of borrowings. The provision for depreciation and proceeds from bonds may be used in conjunction with net income for expansion and replacement of the system.

Funds of publicly owned water utilities should be maintained in accounts separate from those of other governmental agencies and not diverted to uses unrelated to the public water supply. But reasonable payments in lieu of taxes or for services rendered can be made.

Financing operations under private ownership are not significantly different from financing of publicly owned utilities. The total revenue requirement not only must provide for all expenses of operations, including substantial income and property taxes, but it must also provide a fair return to the owners on their investment. The fair return, as determined by the regulatory body having jurisdiction, is generally a rate that will attract investor capital.

Whereas publicly owned utilities have a capital structure derived from bonded debt, net income reinvested in the business, and contributions from customers, a privately owned utility has a capital investment derived from sales of shares of stock, bonded indebtedness, and invested income. The particular weight of the several capital segments is subject to regulatory scrutiny. Many regulatory bodies have indicated that about one-half of new capital should be obtained from some form of debt issue and one-half from sale of equity shares.

A publicly owned utility may issue two types of bonds, depending on statutory restrictions: general obligation and revenue. General-obligation bonds have the advantage of greater security, since the credit of the community as a whole is pledged. They, therefore, usually carry a slightly lower interest rate, but since general electorate approval is required, the issuing process is complex and lengthy. Special arrangements must be made to provide repayment of principal, either from tax levies or from transfers from water revenues. Revenue bonds generally carry a somewhat higher interest rate, since only future income is pledged, but issuance is more easily accomplished when needed than is the case with general-obligation bonds. Expert bond counsel is recommended for either type of issue.

Investor-owned utilities similarly raise capital by issuing bonds; however, they usually carry a higher interest rate than bonds issued by publicly owned utilities. The main reason for this difference is that interest paid on bonds issued by publicly owned utilities is exempt from Federal and frequently from state income taxes, while interest on bonds issued by investor-owned utilities is not exempt.

Financing of extensions of a water system to provide facilities for new customers is a severe problem in areas of rapid growth. Publicly owned systems generally charge new customers some portion of the cost of these works, but the level of charge varies from nominal to full cost (usually 60 to 80% of the cost of minimum mains). The remainder of the cost is recovered from current net income or from proceeds of bonds sold. Financing for major new facilities is derived in part from sale of bonds.

Investor-owned utilities generally do not require significant customer contributions toward extension of the works, as do publicly owned utilities, since such contributions cannot be included in the investment on which a return may be earned. When customer contribution is required, there is ordinarily a formula for refund as other customers are added.

Investor-owned utilities are restrained by law from setting rates that result in undue discrimination. This term has been defined by the courts to mean that rates must be correlated with the cost of providing service. Complete elimination of discrimination would require setting each customer's rate separately; however, undue discrimination can be eliminated by establishing rates for the average customer in various classes. Although publicly owned utilities are not governed by the undue-discrimination law, they do have a moral and ethical obligation to set equitable rates.

Rates that are most commonly used today are flat rate, step rate, and block rate.

Flat rate is a monthly or quarterly charge that does not vary with the amount of water used. This type of charge tends to encourage waste. Although it has been commonly employed in small communities where water is not metered, flat rate is falling into disuse.

With *step rate*, customers are charged at one rate per 1,000 gal for all water used. The rate a customer pays decreases as the total quantity used increases. The major objection to this method is that a customer who uses a quantity slightly less than the point of rate change will pay more than the customer who uses a little more.

The *block rate schedule* consists of one rate for the first volume or block of water used per billing period and lesser rates for additional blocks. For example, a utility may charge 20 cents per 1,000 gal for the first 20,000 gal, 15 cents per 1,000 gal for the next 200,000 gal, and 10 cents for all usage above 220,000 gal. This type of pricing tends to discourage waste but does not restrict usage unnecessarily. Both the step and block rates can have a monthly service charge.

In fixing a system of rates, the supplier should consider the following: (1) cost of collection facilities, treatment chemicals, pumping energy, and, where applicable, buying water from a wholesale supplier; (2) cost of distribution and treatment facilities; and (3) cost, including metering and billing, of serving an individual customer. Cost component 1, called the commodity component, is directly dependent on total usage and therefore should be distributed equally to all water sold. Cost component 2, called the demand component, depends on the peak usage of a customer. If a customer's usage is zero during peak hour, it will not appreciably affect the cost or design of distribution facilities. Since peak-hour demands usually govern the design of a distribution system, this is a good criterion for allocating distribution costs. It is generally recognized that residential areas, where the majority of small users are, have very high ratios of peak demand to total usage and should, therefore, pay a major share of the demand component. Both the step and block rates attempt to allocate this cost to the small user by charging a higher rate for the first water sold to a customer and charging decreasing rates with increased usage. For most distribution systems, a large share of the demand component also should be allocated to fire service. The portion attributed to fire service is usually paid by taxes. Cost component 3, called the customer component, is usually distributed to the customer by a monthly service charge that depends only on the size of service. This charge is usually small.

HYDROELECTRIC POWER AND DAMS

21-85. Hydroelectric-Power Development.
Hydroelectric power is electrical power obtained from conversion of potential and kinetic energy of water. The potential energy of a volume of water is the product of its weight and the vertical distance it can fall:

$$P.E. = WZ \qquad (21\text{-}151)$$

where P.E. = potential energy

W = total weight of the water

Z = vertical distance the water can fall

Because the kinetic energy of the supply source is very small or zero in most *hydro power* (hydroelectric power) developments, the kinetic-energy term does not appear in power formulas.

Power is the rate at which energy is produced or utilized.

$$1 \text{ horsepower (hp)} = 550 \text{ ft-lb per sec}$$
$$1 \text{ kilowatt (kw)} = 738 \text{ ft-lb per sec}$$
$$1 \text{ hp} = 0.746 \text{ kw}$$
$$1 \text{ kw} = 1.341 \text{ hp}$$

Power obtained from water flow may be computed from

$$\text{hp} = \frac{\eta Qwh}{550} = \frac{\eta Qh}{8.8} \qquad (21\text{-}152a)$$

$$\text{kw} = \frac{\eta Qwh}{738} = \frac{\eta Qh}{11.8} \qquad (21\text{-}152b)$$

where kw = kilowatts

  hp = horsepower

  Q = flow rate, cfs

  w = unit weight of water = 62.4 lb per cu ft

  h = effective head = total elevation difference minus line losses due to friction and turbulence, ft

  η = efficiency of turbine and generator

Hydro plants can be classified on a basis of reservoir capacity and use as run-of-river hydro without storage, base-load plants, run-of-river plants with storage, and peak-load plants.

Run-of-River Hydro without Storage. This type of plant has no storage facilities. Power generation is totally dependent on the flow of the river. A development of this type is usually built for some other purpose, such as navigation, power production being only incidental.

The economics of a run-of-river hydro plant depend on the minimum flow of the river. If the minimum flow is very low, it will be necessary to invest money in steam-generation facilities to provide supplemental power during low-flow periods. Therefore, the value of the plant will be only the fuel saved that would otherwise be required for steam generation.

Base-Load Hydro Plants. This type is also a run-of-river hydro plant without storage, but it is located on a river that provides a minimum flow capable of serving the power demand without supplementary steam-generating facilities. The reliable plant capacity is set below the expected minimum flow in the river. This type of run-of-river hydro plant utilizes only a small proportion of the flow of a river. It must pass not only high seasonal flows but also the water it cannot utilize during hours of low power demand.

Cost of a base-load plant can be compared with the cost of the steam capacity that would be necessary to serve power demands if hydro generation were not developed.

Run-of-River Plants with Storage. A small amount of storage can greatly increase the reliable capacity of a hydro plant. The water not required for generation during hours of low power demand can be stored and used for generation during periods of peak demand.

Storage can be provided for a daily, weekly, or seasonal cycle. On a daily cycle, the required reservoir capacity is less than the river's daily flow volume. On a weekly cycle, the flow during the periods of low power demand on weekends is also stored to give additional capacity for peak periods during the week. On a seasonal cycle, the high flood flows are stored to be used during periods of low flow. The seasonal operation requires many times the storage necessary for weekly or daily operation and therefore may be uneconomical unless the reservoir is multipurpose. Then, part of its cost can be underwritten by flood-control or irrigation projects.

Peak-Load Plants. The power demand on an electrical system fluctuates from a daily high to a nightly low. Depending on the size of the utility and the type of customers served, peak demands may be several times the magnitude of the low demands encountered at night. These fluctuations in demand necessitate generation facilities whose full capacity is used only a few hours a day, during periods of peak power demand (Fig. 21-94).

Capacity factor is the percentage of the time the full capacity of a plant is used or the ratio of the average power the plant produces to the plant's capacity. It can be computed on a daily, weekly, or yearly basis.

Hydro plants that are used mainly to supply power for the periods of peak demand are generally called peak-load plants. The main classes of peak-load plants are pumped-storage plants and run-of-river plants with storage.

If sufficient generating capacity and reservoir storage are planned for a run-of-river hydro plant, only a relatively small supply of water is needed to produce a high generation capacity for a few hours duration. This enables a large utility to use steam generation at a high capacity factor where it is most efficient and to supply peak demands from hydro plants.

Pumped Storage. This provides a means of storing large quantities of energy, generated during periods when excess generating capacity is available, to be used at some future time. Water is pumped from a low reservoir to a higher one by energy from steam or base-load hydro when power demand is low. When needed, the water generates power by flowing through a turbine back into the low reservoir. Because of friction loss in the penstock and losses due to the imperfect efficiencies of pumps and turbines, only two-thirds of the energy required to pump the water is recovered.

The balance of energy between pumping and generating can be on a daily or weekly basis. But because the weekly cycle requires several times more reservoir storage than the daily cycle, the weekly cycle usually is not so economical.

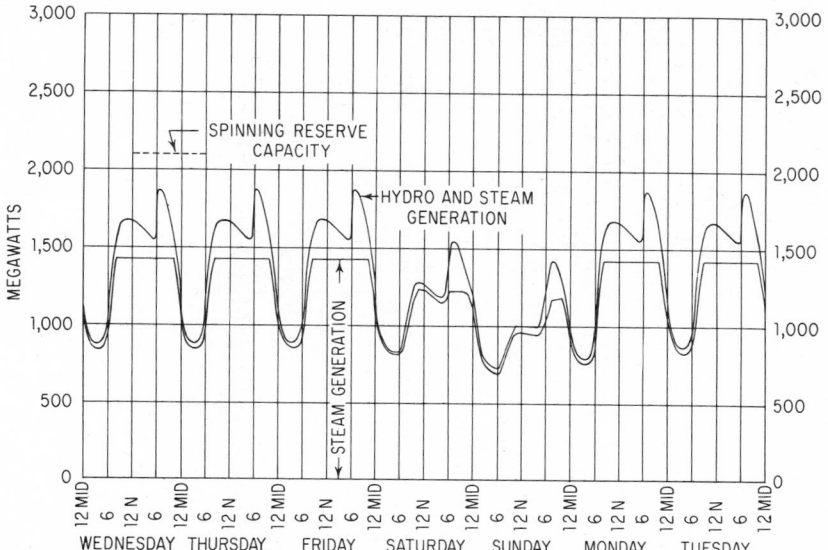

Fig. 21-94. Daily load curves for generating plants. (*City of Los Angeles, Department of Water and Power.*)

When pumped storage is operated at a high capacity factor to transfer large quantities of electrical energy from off-peak to peak, the energy loss may make it uneconomical. This undesirable energy-loss feature of pumped storage is overcome when it is used as reserve capacity.

Today's electrical systems require what is called *spinning reserve*. This is capacity above that necessary to serve the expected maximum load, ready instantly to generate power in case of failure of generating equipment or an unanticipated high power demand. Many utilities keep a spinning reserve capacity equal to the size of their largest single generating unit, or 15% of their maximum demand (Fig. 21-94).

(C, V. Davis and K. E. Sorensen, "Handbook of Applied Hydraulics," McGraw-Hill Book Company, New York.)

21-86. Dams. Dams are usually classified on the basis of the type of construction material or the method used to resist water pressure. The main classifications are gravity, arch, buttress, earth, and rock-fill.

Gravity dams are concrete or masonry dams that resist the forces acting on them entirely by their weight. Figure 21-95 shows the forces that act on a typical gravity dam. The largest force is usually the hydrostatic force of the water F_1. Its distribution is triangular, varying from zero at the top to full hydrostatic at the bottom. Force F_2 represents silt pressure, which results from deposition of silt at the base of the dam. This silt pressure can be calculated by Rankine's theory for earth pressure using the submerged weight of the silt.

Force F_3 represents ice pressure against the face of the dam. In cold climates, ice, which forms on the reservoir surface, expands when the temperature rises and exerts a force on the top of a dam. In the past, ice pressures as high as 50,000 psf have been used for the design of dams in the north; however, today it is realized these values are much too high. A method of calculating these forces, presented by Edwin Rose, gives values ranging from 2,000 to 10,000 psf, depending on the rate of temperature rise and restraining conditions at the edges of the reservoir. (E. Rose, "Thrust Exerted by Expanding Ice," *Proceedings of the American Society of Civil Enginers,* May, 1946.)

Practically all regions in the United States are subject to earthquakes of varying intensity. Earthquakes cause vertical and horizontal accelerations of the earth, which create forces on any object resting on it. The magnitude of these forces equals the mass of the object times the acceleration from the earthquake. These accelerations occur in every direction, and so the effect of the forces must be analyzed for all directions. Most dams in seismically active

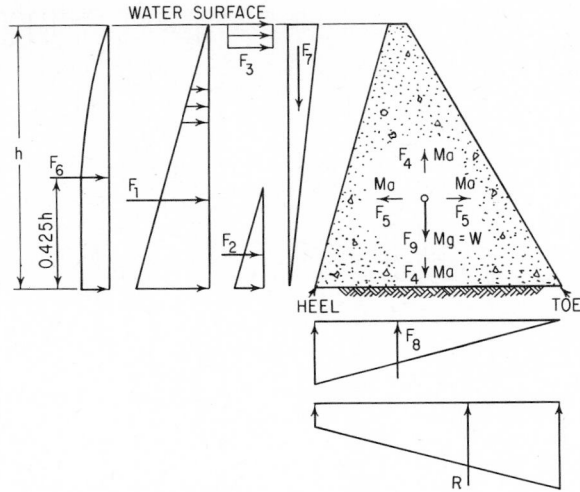

Fig. 21-95. Forces acting on a concrete gravity dam.

regions in the United States have been designed for an acceleration equal to 0.1g, where g is the acceleration due to gravity. The effect of accelerations on the dam is represented in Fig. 21-95 by forces F_4 and F_5. Force F_6 represents the inertial force of the water on the face of the dam. A close approximation of the force, given by Eq. (21-153), was developed by von Karman. ("Pressure on Dams during Earthquakes," discussion by von Karman, *Transactions of the American Society of Civil Engineers*, vol. 98, p. 434, 1933.)

$$F_6 = 0.555awh^2 \qquad (21\text{-}153)$$

where w = unit weight of water, lb per cu ft
$\quad a$ = acceleration due to the earthquake, ft per sec^2
$\quad h$ = depth of water behind dam, ft
The force F_6 acts at a point $0.425h$ above the base.

Force F_7 is due to the weight of water on an inclined face. Gravity dams usually have an inclined upstream face to facilitate construction.

Force F_8 represents an uplift force that acts on the undersurface of any section taken through the dam or under the base of the dam. This uplift is caused by the seepage of water through pores or imperfections in the foundations or through imperfectly bonded construction joints in the masonry. In the past, engineers assumed that, because of bearing contact, this pressure acted only on some percentage of the total area. Recent belief, however, is that uplift acts on 100% of the area of the base.

A process used to reduce uplift pressures calls for grouting along the heel and use of drains behind the grout. When the base is not drained, the uplift pressure is assumed to vary linearly from between full and one-half hydrostatic pressure at the heel to the full tailwater pressure at the toe.

Force F_9 represents the weight of the dam. It acts at the centroid of the cross-sectional area of the dam.

Summation of the vertical forces and of moments about any point yields the foundation pressure. The foundation pressure at the heel of the dam should be compressive. Hence, the resultant of all forces acting on the dam should fall within the middle third of the base of the dam.

The basic modes of failure possible for a gravity dam are by sliding along a horizontal plane, overturning by rotating about the toe, or failure of the foundation material. The first two depend mainly on the cross-sectional shape of the dam, whereas the third depends on both the cross-sectional shape and the foundation material.

Gravity dams can be built on earth foundations, but their height in these cases has been limited to around 65 ft. The main reason gravity dams are used is that they can pass large flood flows over their crest without damage. Their first cost and maintenance cost are usually greater than those of earth or rock-fill dams of comparable height and crest length.

Arch dams are concrete dams that carry the force of the water through arch action. The force is divided between elements: a series of horizontal arches that span between the abutments and a series of vertical cantilevers that are fixed at the foundation. The distribution of load between the arches and cantilevers is determined by the trial-load method. First, a division of the load is assumed and the deflections in the arches and cantilevers are computed. The deflection of an arch at any point must equal the deflection of the cantilever at the same point. If the deflections are not equal, a new division of the load is assumed and the deflections recalculated. This process is continued until equal deflections are obtained.

The external forces an arch dam must resist are basically the same as those on a gravity dam; however, their relative importance is much different. On arch dams, uplift is not so important, but ice loads and temperature stress are much more critical. Arch dams require much less concrete than gravity dams and usually have a much lower first cost. They are not suited to most sites, however, since they must be located in a relatively narrow canyon supported by good rock abutments.

Buttress dams consist of a watertight membrane supported by a series of buttresses at right angles to the axis of the dam. Although there are many types of buttress dams, those widely used are the *flat-slab* and the *multiple-arch*. These differ in that the water-supporting membrane for the flat-slab type is a continuous concrete slab spanning the buttresses, while in the multiple-arch, the membrane is a series of concrete arches. The multiple-arch requires less reinforcing steel and can span longer distances between buttresses, but its formwork is more expensive. The upstream face of a buttress dam is usually inclined at about 45°. The weight of the water on the face is necessary to increase the dam's resistance to sliding and overturning.

The forces acting on a buttress dam are exactly the same as those that act on a gravity dam. However, the vertical load of the water is much greater on a buttress dam, and uplift forces are smaller. The modes of failure are also the same, but the structural design is much more critical.

Although buttress dams usually require less than half the volume of concrete required by gravity dams, they are not necessarily less expensive, because of the large amount of formwork and reinforcing steel required. With the rapidly increasing cost of labor over the past several decades, the buttress dam has lost much of its earlier popularity.

Earth dams are designed to utilize materials available at the dam site. They can be constructed of almost any material with very primitive construction equipment. Successful earth dams have been built of gravel, sand, silt, rock flour, and clay. If a large quantity of pervious material, such as sand and gravel, is available and clayey materials must be imported, the dam would have a small impervious clay core, the material available locally making up the bulk of the dam. Concrete has been used for an impervious core, but it does not provide the flexibility of clay materials. If pervious material is not available, the dam can be constructed of clayey materials with underdrains of imported sand or gravel under the downstream toe to collect seepage and relieve pore pressures.

Slopes of an earth dam are rarely greater than 2 horizontal to 1 vertical and are usually about 3 to 1. The governing criterion is usually the stability of the slopes against slide-out failure.

Another factor that sometimes determines the steepness of the slopes is the amount of seepage that can be tolerated. If the dam is on a pervious foundation, it may be necessary to increase the base width to reduce seepage. The seepage may also be reduced by placing an impervious blanket on the upstream side of the dam to increase the seepage path or by using a cutoff wall in the foundation, such as sheetpiling or a clay-filled trench.

Earth dams can be built to almost any height and on foundations not strong enough for concrete dams. Improvements in earth-moving equipment have resulted in a decreased cost for earth dams, while rising labor costs have increased the cost for concrete dams.

Rock-fill dams usually consist of a dumped rock fill, a rubble cushion of laid-up stone on the upstream face, bonding into the dumped rock, and an upstream impervious facing, bearing on the rubble cushion, with a cutoff wall extending into the foundation. The dumped rock fill may consist of rocks varying in size from small fragments to boulders weighing as much as 25 tons. The fill is usually compacted by dropping the rock, sometimes from as high as 175 ft, onto the fill. Sluicing of the fill with high-pressure hoses is also used to wash fines from between contact points of the rock and reduce settlement. The rubble cushion consists of rocks individually placed to reduce the voids and provide support for the impervious facing. The facing is usually concrete, or wood over concrete, although steel has been used occasionally. The cutoff wall is usually concrete.

Rock-fill dams are generally designed empirically. Low rock-fill dams may have an upstream face as steep as ½ horizontal on 1 vertical. The downstream face is usually 1.3 on 1, the natural angle of repose of rock.   For dams over 200 ft high, both the upstream and downstream faces are usually on a slope of 1.3 on 1.

The major problem encountered in rock-fill dams is large settlements that occur after construction when the reservoir is first filled.   Vertical settlements and horizontal displacements in excess of 5% of the height of the dam have occurred; therefore, the impervious facing must be very flexible or damage will occur during settlement.   One solution to this problem has been to put a temporary facing on the dam and to replace it with a permanent facing after settlement has taken place.   Temporary facings are usually of wood.

Rock-fill dams are used extensively in remote locations where cement is expensive and the materials for an earth dam are not available.   Their cost compares favorably with that of concrete dams.   Leakage should be expected, but rock-fill dams are very stable and have been overtopped without suffering major damage.

(C. V. Davis and K. E. Sorensen, "Handbook of Applied Hydraulics," McGraw-Hill Book Company, New York.)

21-87. Hydraulic Turbines.   In the past, hydraulic power-generating machines meant a large number of different types of equipment.   Today, however, the turbine is the only type of importance in hydraulic power generation.   Its function is transformation of the kinetic and potential energy of water into useful work.

Turbines are classified as impulse turbines and reaction turbines.

Impulse turbines utilize the energy of water by first transforming it into kinetic energy.   This is achieved by free discharge of the water through a nozzle.   The nozzle is directed at buckets positioned along the perimeter of the water wheel.   The force of the water striking these buckets causes the wheel to rotate, providing power.

The only type of water wheel used today in impulse turbines was developed in 1880 by Pelton —the *Pelton wheel*.   The wheel is covered by a housing to prevent splashing and to guide the discharge after the water strikes the wheel.

In most impulse turbines, the water wheel rotates on a horizontal shaft and is acted on by the discharge from one or two nozzles.   But vertical shafts may be used with as many as six nozzles, to obtain a high efficiency for very low loads.   In such installations, efficiencies of 92% for full load and slightly below 90% for loads as low as one-quarter of full load have been obtained.

Impulse turbines are commonly used for heads greater than 1,000 ft.   (An impulse turbine at the Reisseck Power Plant in Austria operates under a net effective head of 5,800 ft.)   There is no lower limit of head for impulse turbines.   They have been used for heads as low as 50 ft; however, the reaction turbine is usually better suited to low heads at large flows.

Reaction Turbines.   There are two types of reaction turbines: the Francis turbine (Fig. 21-96) and the propeller-type turbine (Fig. 21-97).   In both, the flow from the headwater to the tailwater is in a closed conduit system.

The *Francis turbine* usually consists of four essential parts: scroll case, wicket gates, runner, and draft tube.

The *scroll case* transfers the water from the penstock (supply pipe) to the wicket gates and runner.   It distributes the water so that all points on the perimeter of the runner receive the same quantity of water.

The *wicket gates,* located just outside the perimeter of the runner, control the amount of water that enters the turbine.   When the power demand on the turbine changes, a governor actuates a mechanism that opens or closes the gates.

The *runner* is the part of the turbine that transforms the pressure and kinetic energy of the water into useful work.   As the water flows through the turbine, it changes direction.   This creates a force on the runner, causing it to rotate and turn the generator.

The *draft tube* is a conical tube with diverging sides.   It decelerates the flow discharged from the runner, so that the remaining kinetic energy may be regained by conversion into suction head.

Francis turbines have a maximum efficiency of about 94% when operated at or close to full load.   However, if the load drops below 50%, their efficiency decreases rapidly.   Francis turbines are commonly used for heads between 100 and 1,000 ft.   At heads above 1,000 ft, problems are encountered in controlling cavitation and in building a scroll case to take the high pressures.   At heads below 100 ft, the propeller-type turbine is usually more efficient.

Propeller Turbines.   There are two types of propeller turbines: the movable-blade type, called the Kaplan turbine, and the fixed-blade type.   The only difference between the two is that the pitch of the propeller blades is adjustable in a Kaplan turbine.

Fig. 21-96. Impeller of Francis turbine. (*Allis-Chalmers Mfg. Co.*)

The propeller turbine (Fig. 21-97) has the same basic parts as the Francis turbine: scroll case, wicket gates, runner, and draft tube. The basic difference between the Francis turbine and the propeller turbine is in the shape of the runner. The runner of a propeller-type turbine operates in the same manner as a fan or a ship's propeller. The water moving past the blades creates a force that causes the runner to rotate.

Propeller-type turbines are used for heads ranging from a few feet to about 100 ft. The Kaplan turbine has an efficiency of about 94% for full load and drops only to 92% for 40% load. The fixed-blade-type turbine also has an efficiency of about 94% for full load; however, its efficiency drops off rapidly below full load.

(C. V. Davis and K. E. Sorensen, "Handbook of Applied Hydraulics," McGraw-Hill Book Company, New York.)

21-88. Spillways. Any reservoir with an appreciable drainage area must have a spillway to discharge flood flows without damage to the dam and to keep the reservoir water surface below some predetermined level.

An **overflow spillway** allows water to pass over the crest of a section of the dam. This type of spillway is widely used for concrete dams because, if designed correctly, the dam will not be damaged by the water. To use an overflow spillway for earth or rock-fill dams, it is necessary to make the spillway a concrete gravity section. This may not be possible for high earth dams because the foundation may not be able to support a high concrete gravity section.

The discharge over an overflow spillway is given by the equation for discharge over a weir (Arts. 21-39 to 21-48). Since the discharge varies as the head to the $3/2$ power, overflow spillways keep the water level within close limits even when there is a large variation in flows.

It is desirable for an overflow spillway to have the form of the underside of the nappe of a sharp-crested weir. This type of spillway, called an *ogee spillway*, should be designed—as should all spillways—so that separation of the water from the face of the spillway will not occur. Thus, the danger of cavitation will be eliminated.

In a **chute spillway,** water flows over a crest into a steeply sloping, lined, open channel.   The flow is made supercritical to keep the size and length of the chute to a minimum.   Gradual vertical curves should be used in the chute to avoid separation of the flow from the channel bottom.

Chute spillways are commonly used for earth and rock-fill dams where the topography allows a chute to carry the water away from the toe to eliminate the danger of undermining.   The discharge over the crest is given by the equations for discharge over a weir or the entrance to an open channel.

In a **side-channel spillway,** the flow passes over a crest into a channel parallel to this crest. The crest is usually a concrete gravity section, although it can be concrete laid on the natural

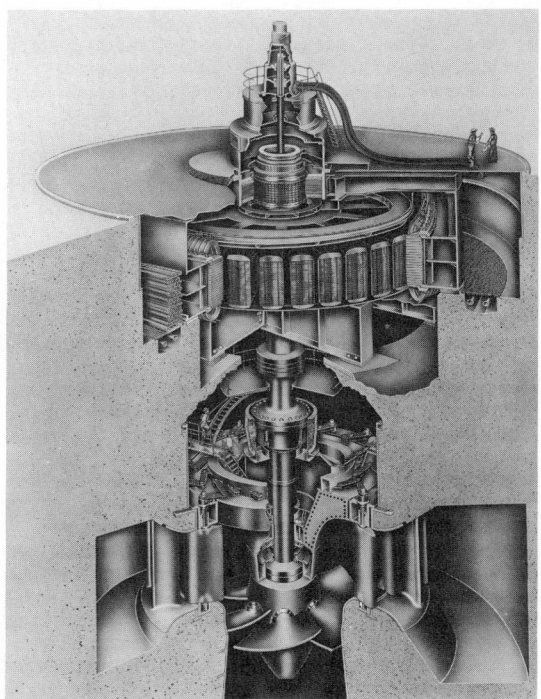

Fig. 21-97.   Propeller-type turbine.   (*Allis-Chalmers Mfg. Co.*)

embankment.   Side-channel spillways are often used in narrow canyons where it is not possible to obtain sufficient crest length for overflow or chute spillways.   The flow in the channel parallel to the crest is determined by applying the momentum principle in the direction of flow and assuming the energy of the water flowing over the crest is completely dissipated (U.S. Department of Interior, "Design of Small Dams," Government Printing Office, Washington, D.C.).

In a **shaft spillway,** sometimes termed a **morning-glory spillway,** the water flows over a circular weir into a vertical shaft.   The shaft terminates in a horizontal conduit that carries the water past the dam.   The weir can be sharp-crested, flared, or ogee in cross section.   (This type of spillway should not be constructed over or through earth dams.)   If the topography is not suitable for a chute or side-channel spillway, a shaft spillway may be the best alternative.

There are two conditions of discharge for a shaft spillway.   Both depend on the head on the weir.   When the head is relatively low, the discharge is governed by the flow over the weir, which is directly proportional to the $3/2$ power of the head on the weir. As the head increases, at some point the discharge will no longer be controlled by the amount of water that can flow over the weir but by the amount of water that can flow through the conduit. The discharge for this condition is directly proportional to the $1/2$ power of the elevation difference between the reservoir water level and the level of discharge of the spillway conduit. Once this second condition is reached, a large increase in head will cause only a small increase in flow. Since analytical analysis of discharge does not give good results on this type of spillway, model tests are usually employed.

A **siphon spillway** (Fig. 21-98) is a closed conduit for discharging water over or through a dam. The entrance to a siphon spillway is usually submerged below the normal water level so that it will not clog with debris or ice. The discharge end of the siphon is usually sealed by deflecting the flow across the barrel or by submerging it so that air cannot enter.

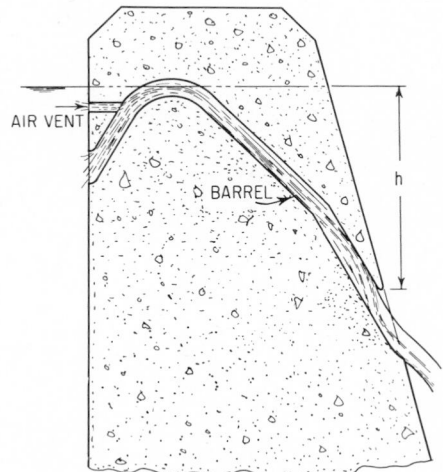

Fig. 21-98. Siphon spillway.

The air vent shown in Fig. 21-98 determines the reservoir level at which the siphon flow begins. When the reservoir water level rises above the vent, the siphon's intake is sealed. Water flowing over the crest of the siphon removes the air in the siphon and full flow begins. Because the flow depends on the siphoning action, siphon spillways hold the water level of a reservoir within close limits. But they are not good for handling large variations in flows, because their discharge is directly proportional to the square root of the head. They are relatively expensive because of the cost of forming the barrel.

(C. V. Davis and K. E. Sorensen, "Handbook of Applied Hydraulics," McGraw-Hill Book Company, New York.)

21-89. Intake Structures. The various functions an intake structure may serve include permitting withdrawal of water from various levels of a reservoir, controlling flow, excluding debris and ice from a conduit, and providing support for the conduit. The type of intake structure required depends on the functions and characteristics of the reservoir. The simplest type of intake is a block of concrete supporting the end of a conduit equipped with a bar screen to exclude foreign matter. In contrast, the intake towers at Hoover Dam, which serve 30-ft-diameter penstocks, are 395-ft-high concrete towers, with two 32-ft-diameter cylinder gates under a maximum head of over 300 ft.

Intake towers are commonly used where there is a large fluctuation in the water level of a reservoir or where it is necessary to control the quality of water used for a domestic supply. They are usually made of concrete and have ports at various levels to permit selection of water from different elevations. The ports are usually provided with gates or valves and some type of trash rack.

The main hydraulic consideration in the design of an intake is to keep losses to a minimum. To do this, the velocities through the trash racks should be kept less than 0.5 fps, and the standard rules for reducing hydraulic losses should be observed.

(C. V. Davis and K. E. Sorensen, "Handbook of Applied Hydraulics," McGraw-Hill Book Company, New York.)

21-90. Crest Gates. These include a number of different types of permanent and temporary devices that operate on the crest of spillways to increase the storage of a reservoir temporarily while control of spillway flows is retained. During periods of low flow when the full spillway capacity is not required, the additional head and storage gained with crest gates may be very valuable.

Flashboards and **stop logs** are the commonest types of crest gates used for small installations under low head. Flashboards are usually wood planks that span between vertical pipes that cantilever above the spillway crest. When the reservoir water surface reaches some predetermined level, the pipes will fail, allowing the full capacity of the spillway to be utilized.

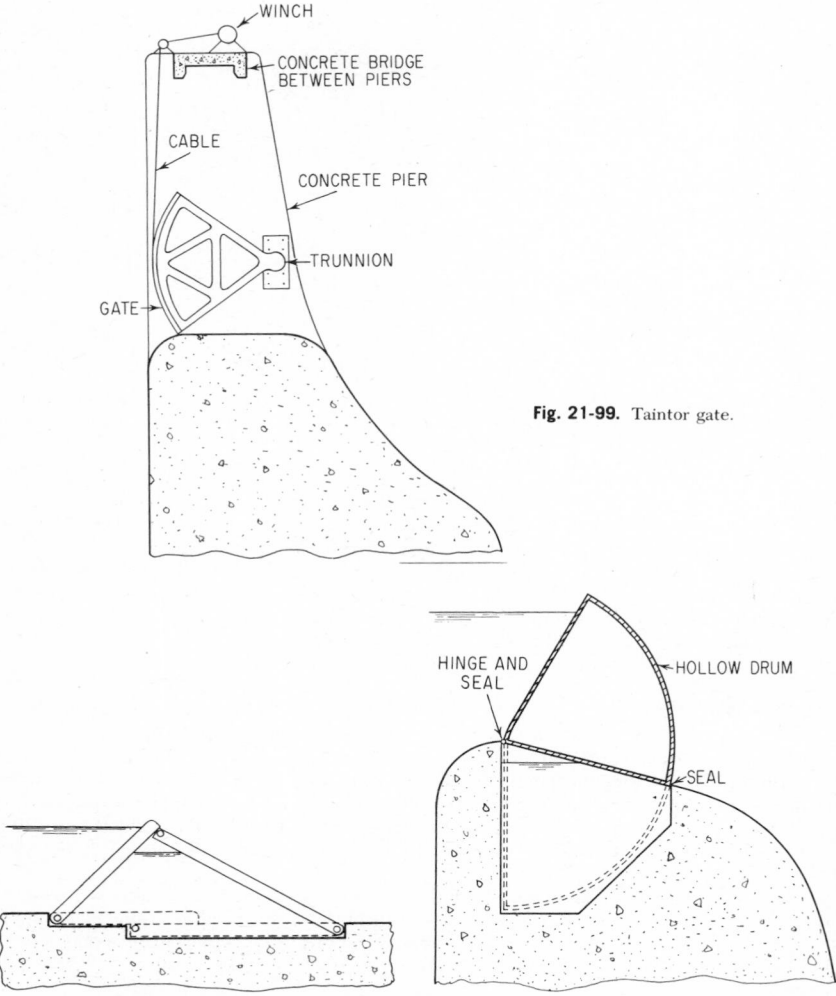

Fig. 21-99. Taintor gate.

Fig. 21-100. Bear-trap gate.

Fig. 21-101. Drum gate.

Stop logs are wood planks that span between slotted vertical piers that cantilever above the spillway crest.

On large stop-log installations, the hydrostatic force creates large frictional forces between the sliding element and the vertical guide, making removal difficult. These frictional forces make it necessary to use a type of gate that depends on rolling rather than sliding friction and operates freely under hydrostatic pressure.

Taintor gates and **sliding gates** mounted on low-friction roller bearings are the most widely used types of crest gates on major installations. In a taintor gate (Fig. 21-99) the friction is concentrated in the trunnion and does not affect the operation. Since flow passes under taintor and slide gates, there is a tendency for ice and trash to pile up against them, causing damage and hampering operation.

Bear-trap and **drum gates** allow the flow to pass over the top. The bear-trap gate consists of two leaves hinged as shown in Fig. 21-100. To raise a bear-trap gate, water is admitted to the space under the leaves to force the leaves up. The drum gate (Fig. 21-101) consists of a segment of a cylinder that is lowered into a recess in the crest when not in use. Because of the large recess required in the dam, drum gates are not suited to small dams.

(C. V. Davis and K. E. Sorensen, "Handbook of Applied Hydraulics," McGraw-Hill Book Company, New York; H. E. Babbitt, J. J. Doland, and J. L. Cleasby, "Water Supply Engineering," McGraw-Hill Book Company, New York.)

Section **22**

Environmental Engineering

WILLIAM T. INGRAM
Consulting Engineer, Whitestone, N.Y.

Environmental engineers are concerned with works developed to protect and promote public health and to improve the environment. Their practice includes surveys, reports, designs, reviews, management, operation, and investigations of such works. They also engage in research in engineering sciences and such related sciences as chemistry, physics, and microbiology to advance the objectives of protecting public health and controlling environment.

Environmental engineering deals with treatment and distribution of water supply; collection, treatment, and disposal of waste water; control of pollution in surface and underground waters; collection, treatment, and disposal of solid wastes; sanitary handling of milk and food; housing and institutional sanitation; rodent and insect control; control of atmospheric pollution; limitations on exposure to radiation; and other environmental factors affecting the health, comfort, safety, and well-being of people. This section, while covering primarily the aspects related to handling of liquid wastes, also deals briefly with other environment-related tasks, such as solid-waste handling and air pollution. (See also environmental discussions in Sec. 14 and subsequent sections.)

22-1. Types of Waste Water. Waste water is the liquid effluent of a community. This spent water is a combination of the liquid and water-carried wastes from residences, commercial buildings, industrial plants, and institutions, plus groundwater, surface water, or storm water.

Waste water may be grouped into four classes:

Class 1. Effluents that are nontoxic and not directly polluting but liable to disturb the physical nature of the receiving water; they can be improved by physical means. They include such effluents as cooling water from power plants.

Class 2. Effluents that are nontoxic but polluting because they have an organic content with high oxygen demand. They can be treated for removal of objectionable characteristics by biological methods. The main constituent of this class of effluent usually is domestic sewage. But the class also includes storm water and wastes from dairy product plants and other food factories.

Class 3. Effluents that contain poisonous materials and therefore are often toxic. They can be treated by chemical methods. When they occur, such effluents generally are included in industrial wastes, for example, those from metal finishing.

Class 4. Effluents that are polluting because of organic content with high oxygen demand and, in addition, are toxic. Their treatment requires a combination of chemical, physical, and biological processes. When such effluents occur, they generally are included in industrial wastes, for example, those from tanning.

Domestic sewage is collected from dwelling units, commercial buildings, and institutions of the community. It may include process wastes of industry, as well as groundwater infiltration and miscellaneous waste liquids. It is primarily spent water from building water supply, to which have been added the waste materials of bathroom, kitchen, and laundry.

Storm water is precipitation collected from property and streets and carrying with it the washings from surfaces.

Industrial wastes are primarily the specific liquid waste products collected from industrial processing, but may contain small quantities of domestic sewage. Such wastes vary with the process and contain some quantity of the material being processed or chemicals used for processing purposes. Industrial cooling water when mixed with process waste is also called industrial waste. (See also Art. 22-2.)

Combined wastes are the mixed discharge of domestic waste and storm water in a single pipeline. Industrial waste may or may not be found in a combined waste and can be carried apart from either in an industrial sewer.

22-2. Industrial Wastes. Industrial wastes, as distinguished from domestic wastes, are related directly to processing operations and usually are the liquid fraction of processing that has no further use in recovery of a product. These wastes may contain substances that, when discharged, cause some biological, chemical, or physical change in the receiving body of water.

Organic substances will exert a biochemical oxygen demand (BOD) of relatively high proportion compared with domestic waste. It is not unusual in food processing to have wastes with a BOD of 1,000 to 5,000 mg per liter, or in the processing of edible oils to have 10,000 to 25,000 mg per liter BOD.

The wastes may cause discoloration of a receiving stream, as in the release of dyes, or increase the temperature of the water, as in the case of a cooling tower or process-cooling-water discharges.

Chemicals in the waste may be toxic to aquatic life, to animals, or to human populations using the water, or may in some way affect water quality by imparting taste or odor. Phenols introduced into water in the parts-per-billion range can produce such marked taste that the water becomes unusable for many purposes. Some chemicals may stimulate aquatic growth, and algae populations in the receiving stream may be increased. These algae are detrimental to water quality, since they, too, produce taste, odor, color, and turbidity in the water.

Industrial wastes that contain large quantities of solids may produce objectionable and dangerous levels of sludge on the bottom of a stream or along the banks. These add to the chemical, biological, and physical degradation of the stream. Discharges containing oil may render bathing beaches useless, interfere with nesting water fowl, and present extra problems of removal in water-treatment processes.

Wastes containing acids or alkalies may attack pier structures and water craft and produce serious toxic effects on fish life.

Some wastes interfere with the normal processes of waste-water treatment and may, if mixed with municipal waste, render the whole treatment process inoperative. Pretreatment of industrial wastes is often required to protect the sewers and treatment plant maintained by a municipal agency. Treatment of industrial wastes to the degree required to protect a receiving body of water is a requirement in all states; it may range from neutralization and other simple primary treatments to complete treatment or, in some instances, even an advanced stage of treatment to remove trace chemicals (see also Art. 22-32).

(E. B. Besselievre and M. Schwartz, "Treatment of Industrial Wastes," McGraw-Hill Book Company, New York; N. L. Nemerow, "Liquid Wastes of Industry: Theories, Practices and Treatment," Van Nostrand Reinhold Company, New York; R. L. Culp and G. L. Culp, "Advanced Wastewater Treatment," Van Nostrand Reinhold Company, New York.)

22-3. Types of Sewers. A sewer is a conduit through which sewage, storm water, or other wastes flow. Sewerage is a system of sewers. Usually, it includes all the sewers between the ends of building-drainage systems and sewage-treatment plants or other points of disposal.

Sanitary sewers carry mostly domestic sewage. They may also receive some industrial wastes. But they are not designed for storm water or groundwater.

Storm sewers are designed specifically to convey storm water, street wash, and other surface water to disposal points.

Combined sewers are designed for both sewage and storm water. They cost less than separate sanitary and storm sewers, but disposal of the flow may create objectionable or hazardous conditions or may involve costly treatment. A large flow of water from a storm may make adequate sewage treatment impossible or increase its cost considerably.

Building sewers, or **house connections,** are pipes carrying sewage from the plumbing systems of buildings to a sewer or disposal point. In urban areas, the flow goes to a **common sewer,** which serves abutting property. This conduit may be a **lateral,** one that receives sewage only from house sewers. A **submain,** or **branch, sewer** takes the flow from two or more laterals. A **main,** or **trunk, sewer** handles the flow from two or more submains or a submain plus laterals. An **outfall sewer** extends from the end of a collection system or to a treatment plant disposal point.

An **intercepting sewer** receives dry-weather flow and specific, limited quantities of storm water from several combined sewers. A **storm-overflow sewer** carries storm-flow excess from a main or intercepting sewer to an independent outlet.

A **relief sewer** is one built to relieve an existing sewer with inadequate capacity.

Usually, sewage or storm-water flow does not completely fill the conduit. But all sewers may be filled at some time and must be capable of withstanding some hydraulic pressure. Some types are always under pressure: **Force mains** flow full under pressure from a pump. **Inverted siphons,** conduits that dip below the hydraulic grade line, also flow full and under pressure.

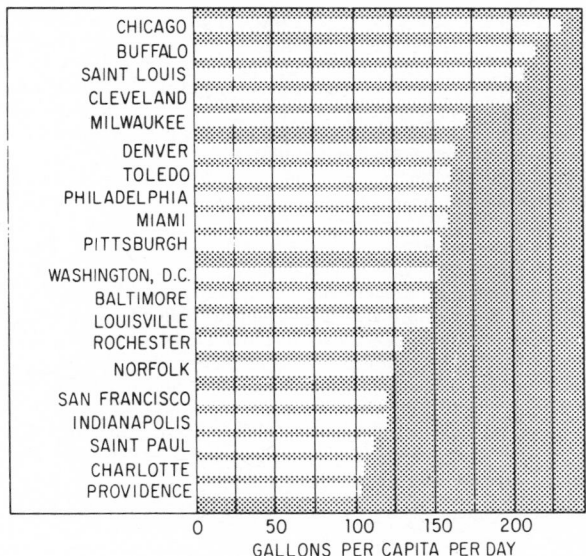

Fig. 22-1. Estimated water use in several major cities. (*Research Division, New York University, College of Engineering.*)

22-4. Estimating Quantity of Sewage. Before a sewer is designed, the community or area to be served should be studied for the purpose of estimating the type and quantity of flow to be handled. Design should be based on the flow estimated at some future time, 25 to 50 years ahead, or at completion of the development. Also, the engineer must have, in advance, policy decisions on whether separate or combined sewers will be built.

The quantity and flow patterns of domestic sewage are affected principally by population and population increase; population density and density change; water use, water demand, and water consumption; industrial requirements; commercial requirements; expansion of service geographically; groundwater geology of the area; and topography of the area.

Since sewage consists primarily of waste water, population and water consumption per capita are the more important factors. The quantity of sewage, however, generally is less than water consumption, since some portion of water used for fire fighting, lawn irrigation, street washing, industrial processing, and leakage does not reach the sewer. Some of these losses, however, may be offset by addition of water from private wells, groundwater infiltration, and illegal connections from roof drains. If the community to be served by the sewerage system already exists, the sewage flow may be estimated from the gallons per capita per day (gcd) of water being consumed. For a planned community, the estimate may be based on the gcd of water being consumed by an existing similar community. Old records may be used as a guide, but they are likely to give very unconservative figures for buildings with central air-conditioning systems that discharge cooling water into the sewers. On the average, communities with population under 1,000 use 60 gcd, while communities of 100,000 use about 140 gcd. [A 1957 study of large cities in the United States indicated a median consumption of 154 gcd and a median population of 658,000 (Fig. 22-1).] The demand for water has been increasing, and consequently wastewater discharges also are increasing. Though sewage flow may range from 70 to 130% of water consumption, designers often assume the average sewage flow equal to the average water consumption.

[H. E. Babbitt and E. R. Baumann, "Sewerage and Sewage Treatment," John Wiley & Sons, Inc., New York; H. S. Azad (ed.), "Industrial Wastewater Management Handbook," McGraw-Hill Book Company, New York; Metcalf and Eddy, Inc., "Wastewater Engineering," McGraw-Hill Book Company, New York; J. W. Clark, W. Viessman, Jr., and M. J. Hammer, "Water Supply and Pollution Control," Intext Educational Publishers, New York, N.Y.]

22-5. Infiltration into Sewers. Water may infiltrate sewers through poor joints, cracked pipes, walls of manholes, perforated manhole covers, and drains from flooded cellars. Sewers in wet ground with a high water table or close to stream beds will have more infiltration than sewers in other locations. Since infiltration increases the sewage load, it is undesirable. The sewer design should specify joints that will allow little or no infiltration, and the joints should be carefully made in the field.

Some specifications limit infiltration to 500 gal per day per in. diameter per mile. Often, enforcement agency specifications and requirements call for leakage tests. Some states limit the net leakage to 500 gal per day per in. diameter per mile for any section of the system.

22-6. Rate of Sewage Flow. Sewage flow varies with water use. But fluctuations tend to dampen out, since there is a time lag from the time of water use to the time the flow reaches the sewer mains. Hourly, daily, and seasonal fluctuations affect design of sewers, pumping stations, and treatment plants.

Daily and seasonal variations depend largely on community characteristics. In a residential district, greatest use of water is in the early morning; a pronounced peak usually occurs about 9 A.M., in the laterals. In commercial and industrial districts, where water is used all day long, a peak may occur during the day, but it is less pronounced. At the outfall, the peak flow probably will occur about noon. Wherever possible, measure flow in existing sewers and at treatment plants to determine actual variations in flow.

Weekend flows may be lower than weekday. Also, industrial operations of a seasonal nature influence the seasonal average. The seasonal average and annual average often are about equal in May and June. The seasonal average may rise to about 125% of the annual average in late summer and drop to about 90% at winter's end.

Peak flows may exceed 300% of average in laterals and 200% of average at the treatment plant. Several state health departments require that laterals and submains be designed for a minimum of 400 gcd, including normal infiltration (Art. 22-5); and main, trunk, and outfall sewers, for a minimum of 250 gcd, including normal infiltration, and any known substantial amounts of industrial waste.

("Design and Construction of Sanitary and Storm Sewers," Manual No. 37, American Society of Civil Engineers; Metcalf and Eddy, Inc., "Wastewater Engineering," McGraw-Hill Book Company, New York.)

22-7. Estimating Storm-Water Flow. An estimate of the quantity of storm water flowing into sewers during or following a period of rain is necessary for their design. Preparation of the estimate requires knowledge of intensity and duration of storms, distances the water will travel before reaching the sewers, permeability and slope of the drainage area, and shape and size of the drainage area. Estimating by the rational method (empirical) incorporates these general considerations in one equation:

$$Q = CIA \tag{22-1}$$

where Q = peak runoff, cfs
A = drainage area, acres

C = coefficient of runoff of the area

I = average rainfall rate, in. per hr, of rain producing runoff

A common value for C used for residential areas with considerable land in lawn, garden, and shrubbery is 0.30 to 0.40. In built-up areas, C may be taken as 0.70 to 0.90. (See also Art. 21-57.)

The **time of concentration** is the time required for the maximum runoff rate to develop at a point in a sewer. At an inlet to a sewer, time of concentration equals inlet time, the theoretical time required for a drop of water to flow to the inlet from the most distant point of the area served by the inlet. The time of concentration for a point in the first sewer entered equals the inlet time plus the time of flow in the sewer to that point. Where branches connect to a sewer, the longest time of concentration for all the branches is used in design.

Inlet time may range from 5 min for a steep slope on an impervious area to 30 min for a slightly sloped city street. Time of flow in a sewer (assumed to be flowing full) may be taken as the length of the sewer to the point of concentration divided by the velocity of flow. Flood crest and storage time while the sewer is filling usually are neglected. The effect of this approximation is calculation of a larger rate of flow, which provides a safety factor in design.

Critical duration of rainfall on a watershed is the time required to develop maximum runoff and therefore equals time of concentration. Observations indicate that rainfall rate I is a function of storm duration t, minutes. Therefore, I for design of storm sewers may be estimated from rate-duration curves or formulas by substituting time of concentration for t (Art. 21-57).

Rainfall-intensity values are selected on the basis of frequency as well as duration of storms that have occurred in the vicinity. Rainfalls that are exceeded only once in 10 years are called 10-year storms; once in 20 years, 20-year storms, etc. The designer has to decide for which frequency storm to design, and this involves a calculated risk combined with engineering judgment. For relatively inexpensive structures in residential areas, a 5-year storm may be used for design of storm sewers with reasonable safety. Where failure would endanger property, a 10-, 25-, or 50-year storm would be more conservative. A 50-year storm may be chosen if flooding would cause costly damage and disrupt essential activities. In such cases, cost-benefit studies may be made to guide selection of a suitable storm frequency.

Since storm-sewage flow is very large compared with dry-weather flow, combined sewers may be designed on the same basis as storm sewers. But cross sections and appurtenances of the sewers must be designed to handle the dry-weather flow efficiently.

(Metcalf & Eddy, Inc., "Wastewater Engineering," McGraw-Hill Book Company, New York; V. T. Chow, "Handbook of Applied Hydrology," McGraw-Hill Book Company, New York; "Design and Construction of Sanitary and Storm Sewers," Manual No. 37, American Society of Civil Engineers.)

22-8. Sewer Design. Before a sewer system can be designed, the quantities of sewage to be handled and the rates of flow must be estimated. This requires a comprehensive study of the community or area to be served (Arts. 22-4 to 22-7). Then, a preliminary layout of the sewerage can be made. Also, pipe sizes, slopes, and depths below grade can be tentatively selected. Preliminary drawings should include a plan of the proposed system and show, in elevation and plan, location of roads, streets, water courses, buildings, basements, underground utilities, and geology. In addition, construction costs should be estimated.

After the preliminary design has been accepted, a survey should locate, in plan and elevation, all existing structures that may affect the design. Preferably, borings should be taken to determine soil characteristics along sewer routes and at sites for structures in the system. Physical characteristics of the area, including contours, should be shown on a topographic map. Scale may be 1 in. to 200 ft, unless the number of details requires a larger scale. Contours at 5- or 10-ft intervals usually are satisfactory. Elevations of streets should be noted at intersections and abrupt changes in grade.

Sufficient cover should be placed over the tops of sewers to prevent damage from traffic loads. Also, the sewers should be below the frost level. Municipal and state regulations on cover should always be reviewed before a design for a specific location is undertaken.

The location of the sewers should be shown in elevation on profiles. Horizontal scale may be 1 in. in 40 ft or 1 in. in 100 ft, depending on the amount of detail. Vertical scale generally is ten times the horizontal.

The final design should include a general map of the whole area showing location of all sewers and structures and the drainage areas; detailed plans and profiles of sewers showing ground levels, sizes of pipe and slopes, and location of appurtenances; detailed plans of all appurtenances and structures; a complete report with necessary charts and tables to make clear the exact nature of the project; complete specifications; and a confidential estimate of costs for the owner or agency responsible for the project.

Extensive plans require tabulation of data beginning at the upper end of the system and proceeding downstream from manhole to manhole. The addition to flow from connecting sewers should be included.

For combined sewers, provision also must be made for handling dry-weather or sanitary flow at proper velocities in sewers that may carry large quantities of water after a storm. Design is complicated by the need for diversion of waters not flowing to a sewage-treatment plant. Diversion structures should be located at or near water courses into which storm water may be discharged. The effects of discharging polluted water, a combination of sanitary sewage and storm water, should be fully investigated.

Approval of a supervising government agency, such as the state health department or Federal environmental protection agency, usually must be obtained for the plans. Sewer designers should be familiar with requirements for sewers in the state in which work is to be done.

Design Flows.   Unless force mains are required, because sewage must be pumped, or inverted siphons are necessary, because of a drop in terrain, sewers usually are sized for open-channel flow. Maximum flow occurs when a conduit is not completely full. For example, for a circular pipe, maximum discharge takes place at about 0.9 of the total depth of the section. Sewers, however, should be designed to withstand some hydraulic pressure.

Sanitary sewers should be designed to carry peak design flow with a depth from one-half full for the smallest sewers to full. For storm sewers, common practice is to permit pipe to carry design flow at full depth.

Laterals may be designed for ultimate flow of the area to be served. Some designers size them to carry the expected flow while half full. This provides a safety factor. Submains may be designed for 10 to 40 years ahead. Trunk sewers may be planned for long periods, with provision made in design for parallel or separate routings of trunks of smaller size to be constructed as the need arises. The large sewers may be conservatively sized to carry the ultimate design flow while full. Appurtenances may have a different life, since replacement of mechanical equipment will be necessary. Usually, they are designed for 20 to 25 years ahead, and a timetable of additions during that period is then scheduled in an overall improvement plan.

In general, flow may be assumed uniform in straight sewers. Velocity changes, however, will occur at obstacles and changes in sewer cross section and should be considered in making hydraulic computations.

Velocity Formulas.   Velocity of flow, fps, in straight sewers without obstructions may be estimated with satisfactory accuracy from the Manning formula

$$V = \frac{1.5}{n} R^{2/3} S^{1/2} \qquad (22\text{-}2)$$

where n = coefficient dependent on roughness of conduit surface
R = hydraulic radius, ft = area, sq ft, of fluid divided by wetted perimeter, ft
S = energy loss, ft per ft of conduit length; approximately the slope of the conduit invert for uniform flow

(See also Art. 21-9.)   A common value for n is 0.013, suitable for well-laid brickwork, good concrete pipe, riveted steel pipe, and well-laid vitrified tile. Smaller values may be used for new, smooth pipe; but the roughness, and value of n, is likely to increase with age. For $n = 0.013$, the Manning formula becomes

$$V = 114 R^{2/3} S^{1/2} \qquad (22\text{-}3)$$

Table 22-1 lists slopes given by this formula for various velocities and hydraulic radii. The quantity, cfs, is given by

$$Q = AV \qquad (22\text{-}4)$$

where A = cross-sectional area of flow, sq ft

Minimum Velocity.   Velocity should be at least 2 fps in sanitary sewers to prevent settlement of solids. Slopes and cross sections of sewers should be chosen to give this velocity, or greater, for design flows. Greater velocities are desirable for storm and combined sewers because the flow may carry heavy sand and grit; 3 fps is desirable. Where sewers are sized for lower velocities than recommended minimums, provision for flushing and removal of obstructions should be made in the design.

Table 22-1. Slopes of Sewers, Ft per 1,000 Ft*

Velocity, fps	Hydraulic radius, ft							
	0.15	0.25	0.50	1.0	2.0	3.0	4.0	5.0
0.5	0.25	0.12	0.048					
1.0	0.96	0.48	0.19	0.077	0.03	0.02	0.01	
2.0	3.8	1.9	0.77	0.30	0.12	0.07	0.05	0.04
3.0	8.6	4.4	1.7	0.68	0.27	0.16	0.11	0.08
4.0	15.4	7.8	3.1	1.2	0.48	0.28	0.20	0.14
5.0	24.0	12.1	4.8	1.9	0.76	0.44	0.30	0.22
6.0	35.6	17.5	7.0	2.7	1.1	0.64	0.44	0.32
7.0	47.1	23.8	9.5	3.7	1.5	0.87	0.60	0.44
8.0	61.5	31.0	12.4	4.9	1.9	1.1	0.78	0.58

*From Manning formula [Eq. (22-3)] for $n = 0.013$. For other values of n, multiply slope given in table by $n/0.013$.

Slopes. Pipe slopes generally should exceed the minimum desirable for maintaining minimum velocity for design flow, since actual flows, especially before a development reaches its ultimate size, may be much smaller than design flow. Actual velocity then may be less than the cleaning velocity. For example, suppose a circular pipe is sized and sloped to handle design flow when flowing full at 3 fps. This velocity will also be maintained when the pipe is flowing half full to full. But if the depth of flow drops to one-third the diameter, the velocity will decrease to about 2.4 fps; and at a depth 0.2 of the diameter, velocity declines to about 1.8 fps.

Table 22-2 gives the hydraulic characteristics of circular pipe. It enables the quantity and velocity of flow to be computed for a circular pipe flowing partly full, when the respective values for the pipe flowing full are known. The quantity, cfs, for flow full may be estimated from

$$Q = \frac{0.463}{n} d^{8/3} S^{1/2} \tag{22-5}$$

and the velocity for flow full from

$$V = \frac{0.59}{n} d^{2/3} S^{1/2} \tag{22-6}$$

where d = inside diameter of pipe, ft

Table 22-2. Hydraulic Characteristics of a Circular Pipe

Depth of flow / Inside diameter	Partial area / Total area	Quantity, cfs, partly full / Quantity, cfs, flowing full	Velocity partly full / Velocity flowing full
0	0	0	0
0.05	0.019	0.005	0.25
0.10	0.052	0.021	0.40
0.15	0.094	0.049	0.52
0.20	0.143	0.088	0.62
0.25	0.196	0.137	0.70
0.30	0.252	0.195	0.77
0.35	0.312	0.262	0.84
0.40	0.374	0.336	0.92
0.45	0.437	0.416	0.95
0.50	0.500	0.500	1.00
0.60	0.627	0.671	1.07
0.70	0.748	0.837	1.12
0.80	0.858	0.977	1.14
0.90	0.950	1.062	1.12
0.95	0.982	1.073	1.09
1.00	1.000	1.000	1.00

Table 22-3 lists the quantities and slopes given by these formulas for various velocities and diameters. Information such as that in Tables 22-1 to 22-3 may be stored in computer memories for design use. Programs for application of the data are available commercially.

Table 22-3. Quantities, Velocities, and Slopes for Circular Sewers Flowing Full*

Dia, in.		Velocity, fps								
		0.5	1.0	2.0	3.0	4.0	5.0	6.0	7.0	8.0
8	Q†	0.17	0.35	0.70	1.1	1.4	1.8	2.1	2.4	2.8
	S‡	0.21	0.83	3.3	7.5	13.3	20.8	30.0	40.7	53.2
10	Q	0.27	0.55	1.1	1.6	2.2	2.7	3.3	3.8	4.4
	S	0.15	0.62	2.5	5.6	9.9	15.5	22.3	30.3	39.6
12	Q	0.39	0.79	1.6	2.4	3.1	3.9	4.7	5.5	6.3
	S	0.12	0.48	1.9	4.4	7.8	12.1	17.5	23.8	31.0
15	Q	0.61	1.2	2.5	3.7	4.9	6.1	7.4	8.6	9.8
	S	0.09	0.36	1.4	3.2	5.8	9.0	13.0	17.8	23.0
18	Q	0.88	1.8	3.5	5.3	7.1	8.8	10.6	12.4	14.2
	S	0.07	0.28	1.1	2.5	4.5	7.1	10.1	13.8	18.1
21	Q	1.2	2.4	4.8	7.2	9.6	12.0	14.4	17.8	19.2
	S	0.06	0.23	0.92	2.1	3.7	5.8	8.3	11.3	14.7
24	Q	1.6	3.1	6.3	9.4	12.6	15.7	18.8	22.0	25.2
	S	0.05	0.19	0.77	1.7	3.1	4.8	7.0	9.5	12.4
27	Q	2.0	4.0	8.0	11.9	15.9	19.9	23.9	27.9	31.9
	S	0.04	0.16	0.66	1.5	2.6	4.1	5.9	8.1	10.5
30	Q	2.5	4.9	9.8	14.7	19.6	24.5	29.4	34.4	39.3
	S	0.04	0.14	0.57	1.3	2.3	3.6	5.2	7.0	9.2
33	Q	3.0	5.9	11.9	17.8	23.8	29.7	35.7	41.7	47.6
	S	0.03	0.13	0.50	1.1	2.0	3.1	4.5	6.2	8.1
36	Q	3.5	7.1	14.1	21.2	28.3	35.4	32.4	49.5	56.6
	S	0.03	0.11	0.45	1.1	1.8	2.8	4.0	5.5	7.2
42	Q	4.8	9.6	19.2	28.9	38.4	48.1	57.7	67.3	76.9
	S		0.09	0.36	0.82	1.5	2.3	3.3	4.5	5.8
48	Q	6.3	12.6	25.2	37.7	50.3	62.8	75.4	88.0	101
	S		0.08	0.30	0.68	1.2	1.9	2.7	3.7	4.9
54	Q	8.0	15.9	31.8	47.7	63.6	79.5	95.4	111	127
	S		0.07	0.26	0.59	1.0	1.6	2.4	3.2	4.2
60	Q	9.8	19.6	39.2	58.8	78.5	98.1	118	137	157
	S		0.06	0.23	0.51	0.90	1.4	2.0	2.8	3.6
66	S	11.9	23.8	47.6	71.3	95.1	119	143	166	190
	S		0.05	0.20	0.45	0.80	1.2	1.8	2.4	3.2
72	Q	14.1	28.3	56.5	84.7	113	141	170	198	226
	S		0.04	0.17	0.40	0.71	1.1	1.6	2.2	2.8
78	Q	16.6	33.2	66.4	99.5	133	166	199	232	266
	S		0.04	0.16	0.36	0.64	0.99	1.4	2.0	2.5
84	Q	19.2	38.5	77.0	115	154	192	231	270	308
	S		0.04	0.14	0.33	0.58	0.91	1.3	1.8	2.3
90	Q	22.1	44.2	88.4	133	177	221	265	309	353
	S		0.03	0.13	0.30	0.53	0.83	1.2	1.6	2.1
96	Q	25.1	50.3	101	151	201	252	302	352	402
	S		0.03	0.12	0.27	0.48	0.76	1.1	1.5	1.9
108	Q	31.8	63.6	127	191	254	318	381	444	508
	S		0.03	0.10	0.23	0.41	0.64	0.93	1.3	1.7
120	Q	39.2	78.5	157	236	314	392	471	549	628
	S		0.02	0.09	0.20	0.36	0.56	0.81	1.1	1.5

*From Manning formula [Eqs. (22-5) and (22-6)] for $n = 0.013$. For other values of n, multiply slopes given in the table by $n/0.013$; multiply quantities and velocities by $0.013/n$.

†Q = quantity of flow, cfs.

‡S = slope, ft per 1,000 ft.

Minimum Pipe Size. In many cities, 8 in. is the minimum diameter of sewer permitted, and in large cities and metropolitan areas 10 in. may be the minimum. In any case, pipe smaller than 6 in. in diameter should not be used because of the possibility of stoppages.

Maximum Velocities. High velocities in sewers also should be avoided, because the solids carried in the flow may erode the conduit. A usual upper limit for sanitary sewers is 10 fps. For velocities in that range, though, lining at least the lower portion of the sewers with abrasion-resistant material, such as vitrified-tile blocks, is advisable. Maximum design velocities for storm sewers, however, may be much greater when such flows are likely to occur infrequently. Concrete channels have carried 40 fps without damage.

Energy Losses. The assumption of uniform, open-channel flow in sewer design implies that the hydraulic grade line, or water surface, will parallel the sewer invert. This may quite often be true. But where conditions exist that change the slope of the water surface, the carrying capacity of the sewer will change, regardless of the constancy of the invert slope. This should be taken into account in hydraulic computations for flow near intersections of large sewers, any structure combining the flow from two or more sources, interchange of velocity and pressure head, and submerged outlets at outfalls.

In curved sewer lines, allowance must be made for larger energy losses than in straight sewers. One way of doing this is to increase the value of n by, say, 0.003 to 0.005.

To account for the energy loss due to change in direction of sewers at manholes, the invert in the manhole may be dropped about 0.04 ft. If the sewer increases in size at the manhole, the crowns of the pipes may be kept aligned and the invert dropped accordingly, or the 0.8 depth points of the pipes may be set at the same elevation. The invert drop also may offset head losses due to size changes. Thus, it reduces the danger of the flow backing up and building up pressure.

Sewer Shapes. In selection of a sewer shape, designers sometimes favor one that permits higher velocities at both small and large flows. For example, an egg shape, with the small end down, offers a rapidly decreasing cross-sectional area for decreasing flows. Since, for a given quantity, velocity is inversely proportional to area, velocity in an egg shape does not fall off so rapidly with decreasing flow as in other shapes. But cost of constructing such curved sections may be higher than that for simpler shapes. Often, a compromise shape is chosen, one that has favorable hydraulic characteristics and relatively low cost.

For this reason, circular sewers generally are used, especially for prefabricated conduit. This shape provides the maximum cross-sectional area for the volume of material in the wall and has fair hydraulic properties (Table 22-2). But because of the roundness, there is added cost in bedding circular pipe compared with shapes with a flat bottom.

Figure 22-2 shows some typical shapes that have been used for large reinforced concrete sewers. The inverts usually are curved or incorporate a *cunette,* or small channel, to concentrate small flows to obtain desirable velocities.

Sewer Materials. Sewers should be made of materials resistant to corrosion and abrasion and with sufficient strength to resist hydraulic pressure, handling, and earth and traffic loads with economy. Materials meeting these requirements include salt-glazed vitrified clay, reinforced concrete, cast iron, galvanized iron, brick, asbestos-cement, coated steel, bituminized fiber, and plastics formulated for the purpose. Sewer pipe is covered by Federal standards and specifications of the American Society for Testing and Materials and American Pubic Works Association.

(G. M. Fair, J. C. Geyer, and D. A. Okun, "Water and Wastewater Engineering," John Wiley & Sons, Inc., New York; Metcalf & Eddy, Inc., "Wastewater Engineering," McGraw-Hill Book Company, New York; H. W. King and E. F. Brater, "Handbook of Hydraulics," McGraw-Hill Book Company, New York; "Design and Construction of Sanitary and Storm Sewers," Manual No. 37, American Society of Civil Engineers.)

22-9. Loads on Sewers. Sewers must be designed with adequate strength to withstand superimposed loads without crushing, collapsing, or through cracks. Usually, the loads are produced by earth pressure or loads transmitted through earth and may be assumed to be uniformly distributed.

Vertical earth loads on sewers may be estimated as indicated in Art. 7-14. Stresses in large sewers may be computed by elastic theory, and the sewers can be sized to resist these stresses. Standard culverts and sewer pipe generally may be selected with the aid of allowable-load tables prepared by the manufacturers.

22-10. Storm-Water Inlets. An inlet is an opening in a gutter or curb for passing storm-water runoff to a drain or sewer. In urban areas, inlets usually are positioned at street intersections to remove storm water before it reaches pedestrian crossings and so that water is never required to cross over the street crown to reach an inlet. If a block is more than 500 ft long, an inlet may be placed near the midpoint. Along rural highways, inlets generally are installed at low points.

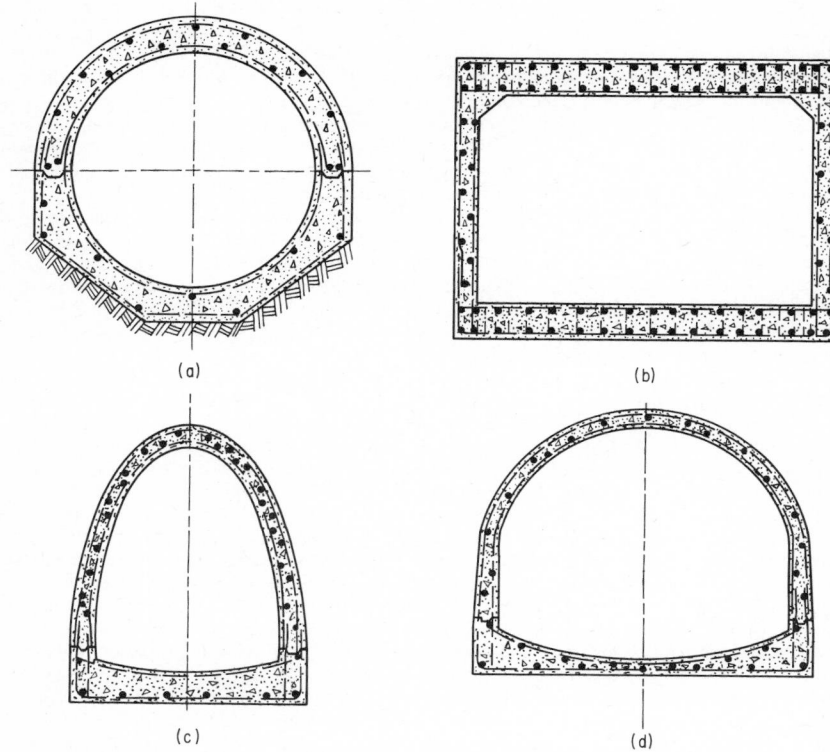

Fig. 22-2. Some shapes used for large reinforced concrete sewers: (*a*) Circular; (*b*) rectangular; (*c*) semielliptical; (*d*) horseshoe.

Spacings generally range from 300 ft for flat terrain and expressways to 600 ft. Often, however, the capacity of an inlet is increased by permitting some of the water to flow past to an inlet at a lower level. A common practice is to provide three inlets in each sag vertical curve, one at the low point and one on each side of it where the gutter elevation is about 0.2 ft higher. Several inlets also are necessary to reduce pondage where the drainage area would be too large for a single inlet in a valley.

Flow through an inlet is directed by a concrete or masonry enclosure to a pipe at the bottom (Fig. 22-3). The size of the enclosure generally is determined by the inlet length, which, in turn, is determined by the quantity of runoff to be drained, depth of water in the gutter at the inlet, and slope of the gutter. Runoff quantity can be estimated by use of the rational formula [Eq. (22-1)].

An inlet may be a curb opening, a gutter grating, or a combination of the two. Capacity of the curb-opening type when diverting 100% of gutter flow may be computed from

$$Q = 0.7L(a + y)^{3/2} \qquad (22\text{-}7)$$

where Q = quantity of runoff, cfs
  L = length of opening, ft
  a = depression in curb inlet, ft
  y = depth of flow at inlet, ft

In practice, the gutter may be depressed up to 5 in. below the normal gutter line along the length of the inlet. Slope of the gutter commonly is 1 in. in 12 in. The depression may extend up to 3 ft from the curb. Depth of flow in the gutter may be estimated from the Manning formula.

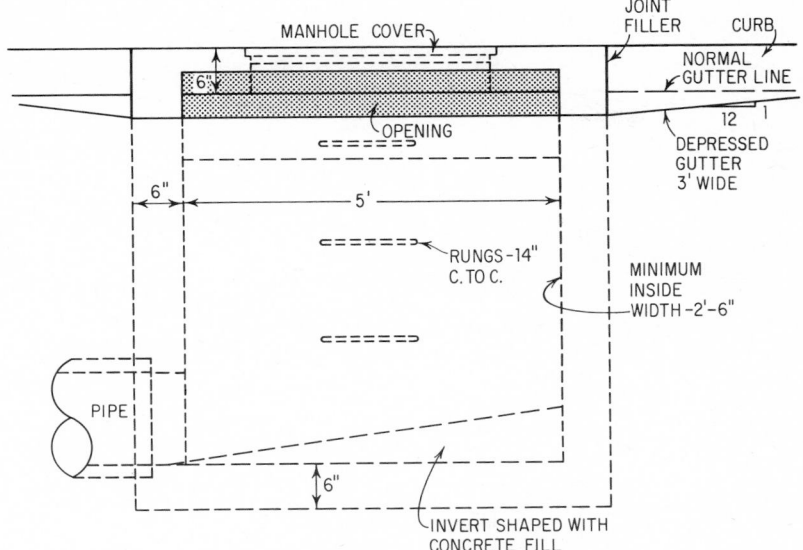

Fig. 22-3. Storm-water inlet with opening in curb.

Grate inlets should be placed with bars parallel to the flow. Length of opening should be at least 18 in., to allow the flow to fall clear of the downstream end of the slot. For depths of flow up to 0.4 ft, capacity of inlet may be calculated from the weir formula

$$Q = 3Py^{3/2} \tag{22-8}$$

where P = perimeter, ft, of grate opening over which water may flow, ignoring the bars. For depths of flow greater than 1.4 ft, capacity may be computed from the orifice formula

$$Q = 0.6A \sqrt{2gy} \tag{22-9}$$

where A = total area of clear opening, sq ft
$\quad g$ = acceleration due to gravity, 32 ft per sec^2
At depths between 0.4 and 1.4 ft, neither formula may be applicable because of turbulence. A rough estimate may be made by using the smaller of the values of Q obtained from Eqs. (22-8) and (22-9).

Combination inlets are desirable, especially at low points, because the curb opening provides relief from flooding if the grate becomes clogged. If the gutter grate is efficient, the combination inlet will have a capacity only slightly greater than a similar inlet with grate alone. Hence, only the grate capacity should be depended on in designing a combination inlet.

Catch basins (Fig. 22-4) are inlets with enclosures that permit debris to settle out before the water enters the sewer. With good sewer grades and careful construction, however, catch basins are unnecessary, because flow will be adequate to prevent debris from clogging the sewer. Also, since water trapped in catch basins may permit mosquitoes to hatch and may be a source of bad odors, simple inlets are preferable. Furthermore, catch basins are more expensive to maintain, because they must be cleaned frequently.

("Design and Construction of Sanitary and Storm Sewers," Manual No. 37, American Society of Civil Engineers; Metcalf & Eddy, Inc., "Wastewater Engineering," McGraw-Hill Book Company, New York; "The Design of Storm-Water Inlets," Storm Drainage Research Committee, Johns Hopkins University, Baltimore, Md., 1956.)

22-11. Manholes. A manhole is an enclosure of concrete or masonry for providing access to a sewer. The lower portion usually is cylindrical, with an inside diameter of at least 4 ft to allow adequate space for workers. The upper portion generally tapers to the opening to the street. About 2 ft in diameter, the opening is capped with a heavy cast-iron cover seated on a cast-iron

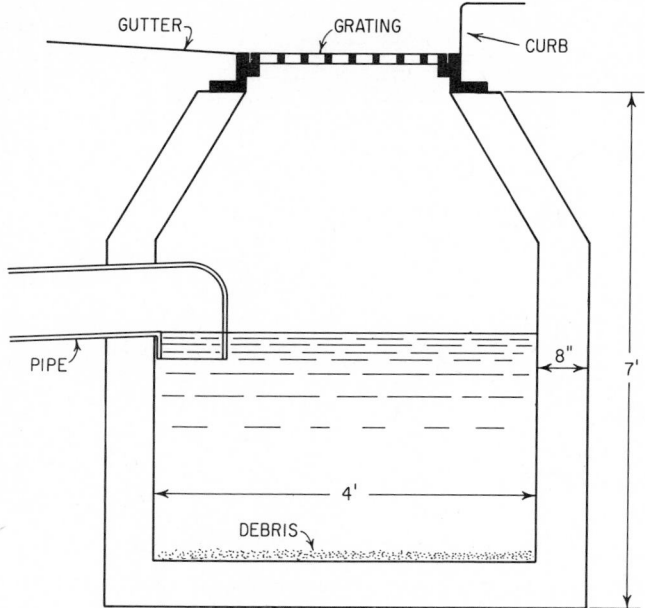

Fig. 22-4. Catch basin with grating inlet in gutter.

frame. Figure 22-5a shows a typical manhole for sewers up to about 60 in. in diameter, and Fig. 22-5b shows one type used for larger sewers.

Sewers are interrupted at manholes to permit inspection and cleaning. The flow passes through the manholes in channels at the bottom. Cast-iron rungs on the manhole walls enable workmen to climb down to the sewers.

For sewers up to about 60 in. in diameter, manholes are spaced 300 to 400 ft apart. They also are placed where sewers intersect or where there is a significant change in direction, grade, or pipe size. Since workers can walk through larger sewers, manholes for these may be spaced farther apart.

Drop manholes are used where one sewer joins another several feet below. The lower sewer enters the manhole at the bottom in the usual manner. The upper sewer, however, turns down sharply just outside the manhole and enters it at the bottom, where a channel feeds the flow to the main channel. To permit cleaning of the upper sewer from the manhole, the upper sewer also extends to the manhole at constant slope past the sharp drop through which the sewage flows. While the upper sewer could be brought down to the lower one more gradually, use of the drop manhole permits a more reasonable slope and thus saves considerable excavation. If the drop is less than 2 ft, however, a steeper slope for the upper sewer would usually be more economical.

Where a large quantity of sewage must be dropped a long distance, a wellhole may be used. The fall may be broken by staggered horizontal plates in the shaft or by a well or sump at the bottom from which the sewage overflows to a lower-level sewer. In a flight sewer, concrete steps break the fall.

Most street and highway departments and departments of public works have standard plans for manholes.

("Design and Construction of Sanitary and Storm Sewers," Manual No. 37, American Society of Civil Engineers.)

22-12. Sewer Outfalls.   Type of outfall depends on quantity of sewage to be discharged, degree of treatment of the sewage, and characteristics of the disposal source. The outlet should be located to avoid pollution of water supplies and creation of a nuisance. Submerged outlets away from shore are preferable to discharge along a shore or bank, which may create an unsightly appearance and odors. Currents should be strong enough to prevent buildup of sludge near the outlet. It should be protected against scour by its location or suitable construction. A flap valve

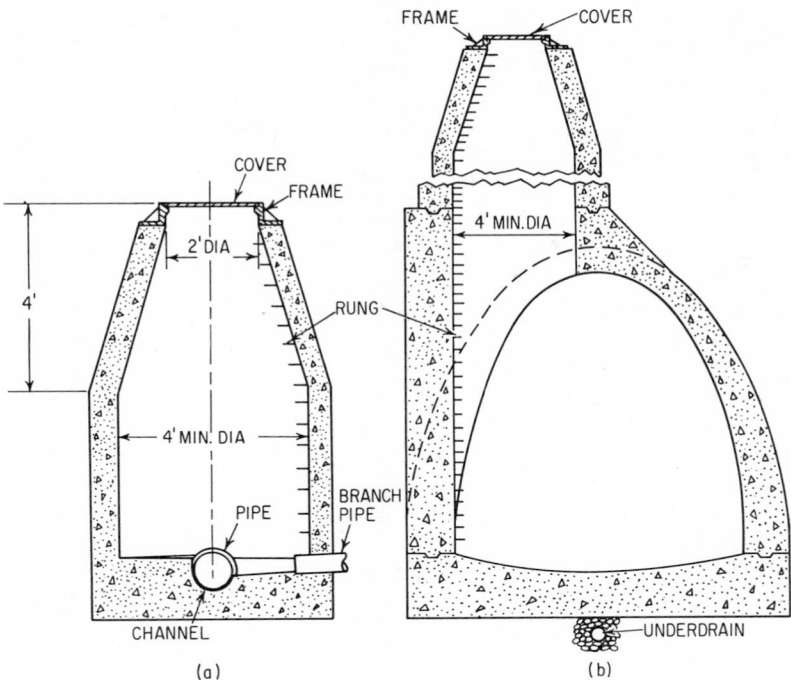

Fig. 22-5. Concrete manholes: (*a*) For sewer under 60 in. in diameter; (*b*) for large sewer.

or automatically closing gate is desirable at the outlet to prevent entrance of water into the sewer during highwater stages.

Outfalls in tidal waters require special investigations to insure suitable dispersion of the sewage and to avoid floating wastes at the water surface.   These outfalls often are constructed with a multiple discharge at the end, thus spreading the effluent over a large area and through a large volume of water.   Depth of water over the outfall must be sufficient to accomplish dispersion before currents can transport the concentrated effluent streams shoreward, over shellfish beds, or into shallow water.

The outfall may be laid on the bottom.   For protection against waves and scour, the pipe may be set in a trench or between two rows of piles and securely anchored.

An outlet discharging treated sewage into a small stream should be protected, by a concrete head wall and a concrete apron on the bank, against undercutting by the flow of the stream or sewage.   A similarly protected outlet may be used at a river bank for storm-water discharge from a combined sewer.   The dry-weather flow may be carried farther out into the river through a small pipe along the bottom.

22-13. Inverted Siphons (Sag Pipes).   These are sewers that dip below the hydraulic grade line.   They are used to avoid such obstructions as waterways, open-cut railways, subways, and extensive utility piping and structures.   After passing under an obstruction, the pipe is brought to grade to permit open-channel flow in the continuation, to keep down the amount of cut and thus the cost of installing the sewer.   The portion of the sewer below the hydraulic grade line flows full under pressure.   Hence, it must have tight joints and be made of a material suitable for this job, and it must be designed for the maximum expected pressure.

To prevent solids from being deposited and obstructing an inverted siphon, it should be sized and sloped to keep flow velocities as much above 3 fps as feasible.   Although experience has been good with a single pipe, 12 to 24 in. in diameter, carrying flows with such velocities, a pipe big enough to handle the maximum flow at an adequate velocity may carry small flows at undesirably low speeds.   In that case, two or more parallel pipes may be used instead of a single pipe.

An inlet chamber is constructed at the upstream end of the inverted siphon and an outlet chamber at the downstream end.   These chambers may be concrete enclosures, which may be entered through manholes extending to grade.   The inlet chamber for a multiple-pipe inverted siphon usually incorporates flow-regulating devices to control the flow to each pipe.   As a safety measure, the inlet chamber may also incorporate a bypass, or overflow, pipe to relieve the inlet should the inverted siphon be overloaded or obstructed.   In the outlet chamber, the inverts of the pipes merge into a single channel, which becomes the invert of the continuing sewer.   Provision should be made in the chambers for cleaning and repairing the pipes and for draining them for these purposes.   The designer should always investigate the hydraulic heads required in the inlet chamber to avoid surcharge on the upstream pipes.

Figure 22-6 shows a two-pipe inverted siphon, with stop planks for regulating flow in the inlet chamber.   When the smaller pipe is full, sewage overflows a stop plank, to enter the larger pipe.   Similarly, a three-pipe system may be used for a large combined sewer.   The smallest pipe may be assigned the minimum dry-weather flow; a larger pipe, the excess up to a specified percentage of the maximum flow; and the largest pipe, the remainder of the flow.   Built-in weirs may be used to regulate the flow to each pipe.

(E. W. Steel, "Water Supply and Sewerage," McGraw-Hill Book Company, New York.)

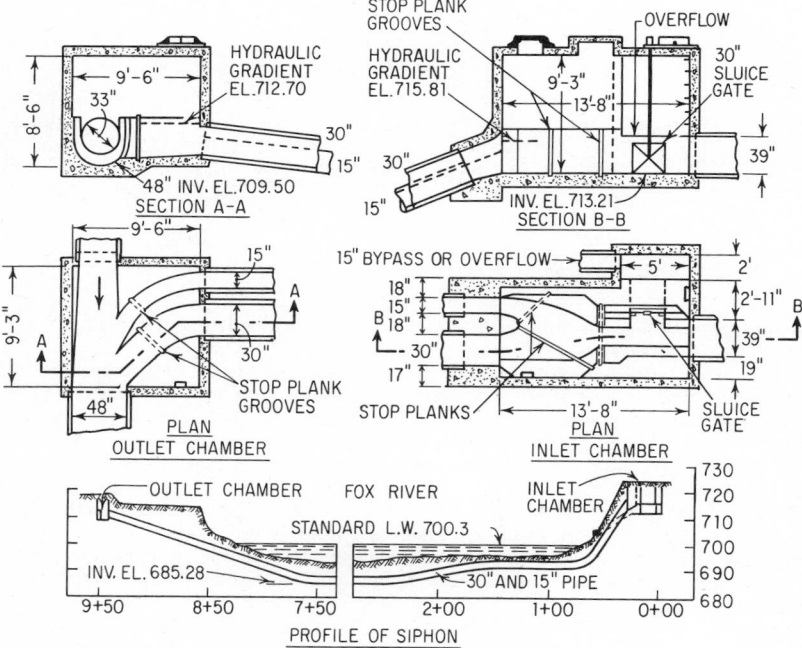

Fig. 22-6. Two-pipe inverted siphon at Appleton, Wis.   (*From E. W. Steel, "Water Supply and Sewerage," McGraw-Hill Book Company, New York.*)

22-14. Flow-regulating Devices in Sewers.   Sewerage systems often require some means for controlling flow, such as weirs, spillway siphons, and gates and valves.   The devices may be used to divert flow from one conduit to another or to distribute flow among several pipes.

A common application is control of flow in a combined sewer when the discharge goes through a treatment plant.   Treatment of maximum flow may not be economic, even if feasible.   So flow to the plant is limited, usually to twice the dry-weather flow.   For the purpose, a regulating device is installed in the sewer to permit the desired quantity to pass to the treatment plant.   The excess flow is diverted into other conduits or discharged untreated into a waterway.

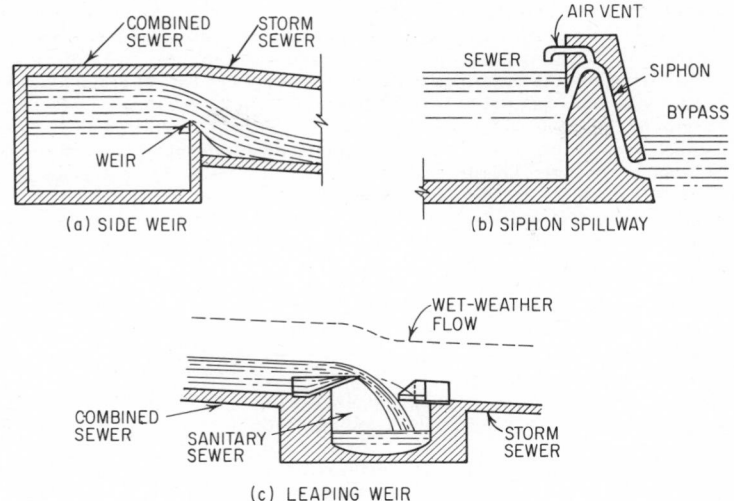

Fig. 22-7. Flow-regulating devices for sewers.

A simple device for such an application is a side weir, an overflow weir along the side wall of the combined sewer (Fig. 22-7a). Diversion, cfs, may be estimated from the Engels formula:

$$Q = 3.32 l^{0.83} h^{1.67} \tag{22-10}$$

where l = length of weir, ft

h = depth of flow over weir at downstream end, ft

Though simple to construct, side weirs may not control flow as closely as desired. Siphon spillways (Fig. 22-7b) are more effective, especially for large flows. The outlet may be placed considerably below the inlet (differences in elevation up to 33.9 ft at sea level under standard atmospheric conditions may be used). Siphons operate under higher heads than weirs and permit much larger flows. Control is better because siphons can be constructed to start or stop discharge at any desired depth of flow in the combined sewer.

Area, sq ft, of the siphon throat can be determined from

$$A = \frac{Q}{c\sqrt{2gh}} \tag{22-11}$$

where Q = discharge, cfs

c = coefficient of discharge, which varies from 0.6 to 0.8

g = acceleration due to gravity = 32.2 ft per sec^2

h = head, ft

For proper operation, the air vent should have an area of about $A/24$. The siphon inlet should be shaped to minimize entrance losses. The outlet should be completely submerged or sealed by the discharge.

A leaping weir, set in an invert of a combined sewer, permits low flows to drop through an opening into a sanitary sewer (Fig. 22-7c). Higher flows, having higher velocities, jump the opening and are discharged through a storm sewer. The opening may be made adjustable to correct for inaccuracies in computations based on theory.

Diversion of flow also may be accomplished with float-actuated gates and valves. For example, low flows may be permitted to reach an outfall through an opening controlled by a gate. When water reaches a predetermined level in a float chamber, a float valve closes the gate to divert the water to a bypass. The designer must provide access to diversion chambers for cleaning, since debris carried into the combined sewer will fill in the channel, clog the openings, and otherwise defeat the purpose of the flow-regulating device.

22-15. Sewer-Construction Methods. Sewers usually are placed in trenches. Occasionally, however, sewers may be constructed or installed in tunnels or laid at grade and covered with an embankment.

In trench construction, the sewer line is located with respect to an offset line, laid out with transit and tape sufficiently far away to avoid disturbance. The trench then is marked or staked out on the ground and excavated. For small sewers, both the vertical and the horizontal positions of the conduit in the trench may be determined with the aid of a string set at a convenient elevation across batter boards straddling the trench at 25- or 50-ft intervals. For large sewers, key points should be located with transit and tape.

Trench excavation may be done by hand or with powered equipment as described in Sec. 13. In rock, explosives should be avoided or used with great caution to avoid collapsing the trench or damaging nearby structures or utilities.

Experience generally will indicate whether the depth and type of soil require that the sides of the trench be sheeted and braced. If there is any doubt, sheeting should be used so as not to endanger workers. The sheeting methods described in Art. 7-12 are applicable to trench construction. Unless this is forbidden by specifications to prevent possible failures, the sheeting may be salvaged as backfilling proceeds.

Water may be drained, except in quicksand, by leading it to sumps and pumping it out. Wellpoints may be necessary to prevent quicksand from forming in a sandy trench bottom or to dry out the bottom.

The support for the sewer should be shaped to the conduit bottom, whether the support be the subgrade, a granular fill, or a concrete cradle. In rock, excavation should be carried to a depth of one-fourth the conduit diameter below the bottom of the conduit, but not less than 4 in. below. The space between the trench bottom and the conduit should be refilled with ¾-in. gravel or lean concrete (1:4½:9 mix), so that at least 120° of the pipe will be supported on it.

Pipelaying usually proceeds upgrade. Pipe is laid with bell ends upstream, to receive the spigots of subsequent sections. Grade usually is required to be within ½ in. of that specified.

Joints between lengths of pipe usually are calked with a gasket of plastic, twisted hemp, or oakum and a filling of plastic, bitumen, or portland-cement mortar (1:1 mix). Resilient joints are preferable to rigid types, which differential settlement may crack.

Feeder sewers come with Y or T stub branches for house sewer connections. If these connections are not made when the feeder is installed, a disk stopper is mortared into the bell of the stubs. Field notes should record the location of each branch so that it can be found when a connection has to be made in the future.

Backfilling should start as soon as possible. Earth should be placed and tamped evenly around the pipe to avoid disturbance of newly made joints and creation of high or unbalanced side pressures on the pipe. Material should be placed in layers not exceeding 6 in. in thickness and tamped lightly until about 2 ft of fill covers the top of the pipe. Walking on the fill above the pipe should be prohibited until at least 1 ft of fill has been placed over the pipe.

The upper portion of the backfill should be heavily tamped, to reduce future settlement, if the surface over the trench is to be paved. Backfill must be carefully placed throughout, and materials that may permit excessive settlement should not be used.

For trenches in fields, the backfill need not be tamped. After all the material previously excavated from the trench has been replaced, the resulting mound may be left to settle naturally.

Large sewers in trenches generally are constructed of reinforced concrete, cast in long, reusable forms. Often, the invert is concreted first. Then, the forms for the upper portion are supported on the hardened invert concrete.

For sewers in tunnels, the methods described in Sec. 20 may be used.

("Design and Construction of Sanitary and Storm Sewers," Manual No. 37, American Society of Civil Engineers.)

22-16. Pumping Stations for Sewage. Lift stations are used where it is necessary to pump sewage to a higher level. The installation may be underground or above grade, housed in a building. (For a discussion of sewage pumps, see Art. 22-17.)

Most installations have at least two pumps. One is available as a standby, ready to take over if the first should fail. Main pumping stations should have at least three pumps; with the largest pump out of service, the other two should be able to handle the design flow. Several pumps with different capacities permit flexibility of operation. The smallest pump should be able to handle minimum flow. The others can be brought on-stream in succession as flow increases.

At a small pumping station, sewage may flow into a manhole or a tank. A horizontal pump may be installed in a "dry" compartment alongside the manhole; or a vertical pump, on the roof of the tank (Fig. 22-8). At a large pumping station, sewage flows into a wet well. The pumps may be installed above or in an adjacent dry well.

Often, the pumps operate automatically when the liquid in the wet well reaches a selected level. (See, for example, Fig. 22-8.) The motors may be started and stopped by switches operated by a float rod, which rises and falls with the liquid level. Usually, two sources of electric power are provided to insure continuity of operation. If no attendants are present in an automatic station, provision should be made for an alarm to be sounded and recorded at a remote station when a pump fails or the liquid level rises above a selected elevation.

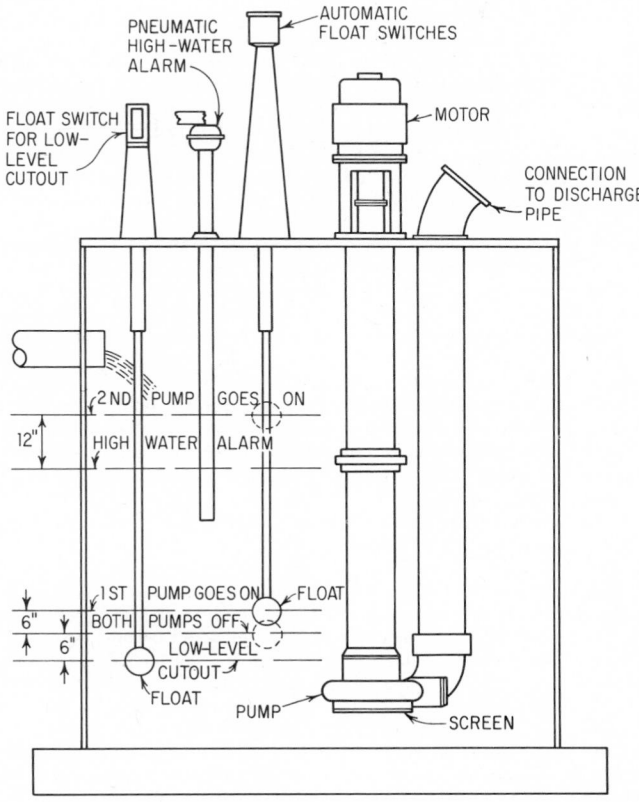

Fig. 22-8. Small automatic sewage pumping station.

Seepage in a dry well should be directed to a sump. It can be drained by one of the sewage pumps or by a special pump. This pump may also have a suction line to the wet well to drain it for cleaning and repair.

The wet well usually is small, to preclude septic action in the sewage; however, the well must be designed to handle maximum flow without flooding. The well should be vented to the outside, to prevent accumulation of odors. It may be divided into two interconnected compartments, which may be isolated for cleaning and repair by closure of a gate.

Pumps, though nonclogging, should be protected against debris in the sewage by a screen. For the purpose, a basket screen may be placed at the entering sewer, or a bar screen ahead of the wet well.

("Design and Construction of Sanitary and Storm Sewers," Manual No. 37, American Society of Civil Engineers.)

22-17. Sewage Pumps. Though sewage generally flows by gravity through conduit and treatment plants, pumping sometimes is required. Pumping may be the most economical means of conveying sewage past a hill, or the only way to get sewage from a cellar to a sewer at a higher level. Where desirable invert slopes would place a sewer far underground, making construction costs high, a more economical method is to raise the sewage in a pumping plant and then let it flow by gravity. Similarly, pumping may be necessary to give sufficient head for sewage to flow by gravity through a treatment plant.

"Nonclogging" centrifugal pumps are generally used. They are capable of passing solids with a maximum size of about 80% of the inside diameter of the pump suction and discharge pipes. These single-suction volute pumps may be bladeless or they may have two vanes. In some cases, grit chambers may be desirable ahead of pumps, to prevent accelerated wear in the pumps, and bar screens, perhaps mechanically cleaned, may be justified.

The pumps generally are driven by electric motors. Types preferred have a high efficiency over a wide range of operating conditions. But dependability is the most important characteristic. Also, slow-speed pumps are desirable for long life and less noise.

The shaft of the pump may be horizontal or vertical. Vertical pumps permit installation of motors above the pump pit where they are less likely to be damaged by floods.

Sewage ejectors operated by compressed air are an alternative to nonclogging centrifugal pumps. In buildings where compressed air is available, such ejectors may be used as sump pumps.

In a commonly used type of sewage ejector, the sewage flows into a storage chamber until it is full. During this stage, air is exhausted from the chamber as the liquid level rises. A float rod closes the air exhaust and opens a compressed-air inlet. The compressed air forces the sewage up the discharge pipe. When the storage chamber is emptied, the float valve shuts the compressed-air valve and opens the air exhaust. Check valves in inlet and discharge pipes prevent back flow.

(T. G. Hicks, "Pump Selection and Application," McGraw-Hill Book Company, New York.)

22-18. Characteristics of Domestic Sewage. Usually, sewage contains less than 0.1% of solid matter. Much of the flow looks like bath or laundry effluent, with garbage, paper, matches, rags, pieces of wood, and feces floating on top. Within a few hours at temperatures above 40° F, sewage becomes stale. Later, it may become septic, often with the odors of hydrogen sulfide, mercaptans, and other sulfur compounds predominating. The more putrescible compounds there are in sewage, the greater is its concentration or strength. In general, strength will vary with the amount of organic matter, water consumption per capita, and amount of industrial wastes.

Solids. Total solids in sewage comprise suspended and dissolved solids. About one-third of the total solids usually are in suspension. Suspended solids are those that can be filtered out on an asbestos mat. Usually more than half of these solids are organic material.

Suspended solids include settleable solids and colloids. Settleable solids precipitate out in sedimentation tanks in the usual detention periods. Colloids, mostly organic material, are smaller than 0.0001 mm in diameter and can remain in suspension indefinitely. They can pass through filter paper but are retained on a filtering membrane.

Dissolved solids are the residue from evaporation after removal of suspended solids.

Solids also may be classified as volatile or fixed. The loss of weight when dried solids are burned is attributed to the volatile solids, which are considered to be organic material. The residue comprises fixed solids, which are assumed to be inorganic.

Organic Content. The organic content of sewage may be classified as nitrogenous and nonnitrogenous. Principal nitrogenous compounds include proteins, urea, amines, and amino acids. Principal nonnitrogenous compounds comprise soaps, fats, and carbohydrates.

Analyses of Sewage. Tests are made on sewage to determine its strength, potential harmful effects in its disposal, and progress made in treating it. The most commonly made tests measure:

Suspended solids.
Biochemical oxygen demand (BOD).
Amounts of ammonia, which decrease as sewage is treated.
Nitrites and nitrates, which increase as sewage is treated.
Dissolved oxygen, which, for an effluent, indicates the efficiency of treatment.
Ether-soluble matter, or fats and greases, which can form a heavy scum.
pH value, which decreases, indicating greater acidity, as sewage becomes stale.
Chemical oxygen demand, which approximates the total oxidizable carbonaceous content.
Hydrogen sulfide, which indicates anaerobic decomposition.

Chlorine demand, the amount of chlorine added to sewage to produce a residual after a certain time, usually 15 min.

Bacteria and other microorganisms.

Coliform tests are usually required. Fecal coliform tests may be required when effluent is discharged into bathing or drinking waters or into tidal waters where shellfish are harvested.

Bacteria. These may be aerobic, requiring air for survival; anaerobic, thriving without air; or facultative, carrying on with or without air. (Some may be pathogenic, causers of intestinal diseases. If they are present, the effluent may have to be chlorinated or otherwise treated to eliminate such bacteria, depending on the method of disposal.) Bacteria are useful in stabilizing sewage, breaking it down into substances that do not decompose further.

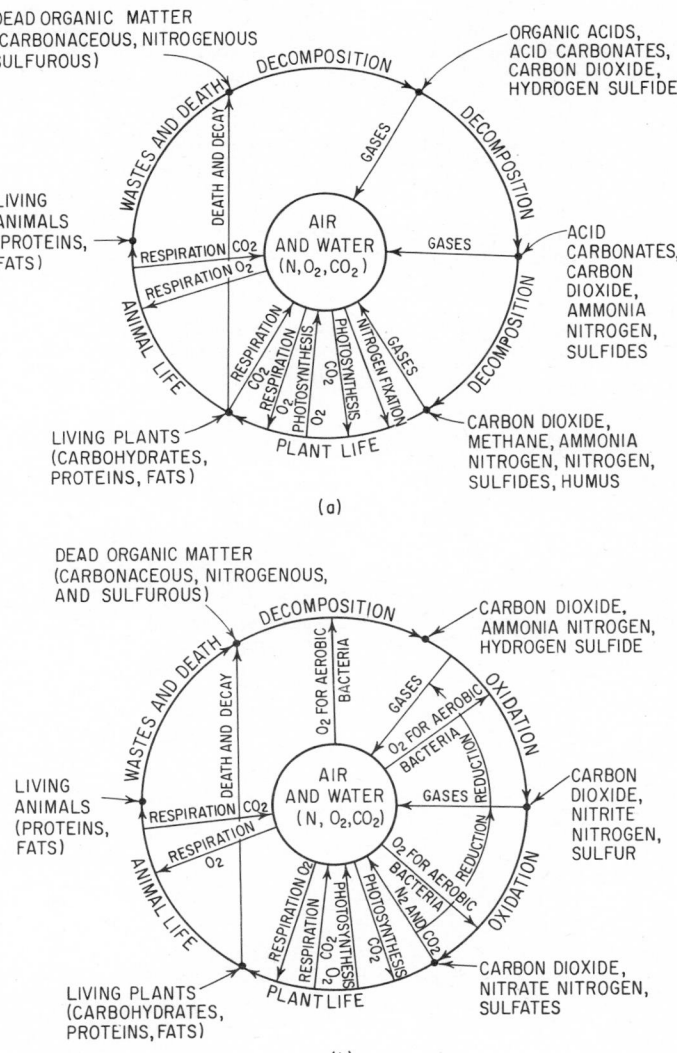

Fig. 22-9. Carbon, nitrogen, and sulfur cycles in (a) anaerobic decomposition; (b) aerobic decomposition. (From E. W. Steel, "Water Supply and Sewerage," McGraw-Hill Book Company, New York.)

Anaerobic bacteria are used in sludge digestion, the stabilization of organic material removed from sewage by sedimentation. Anaerobic stabilization takes longer than aerobic, is more sensitive to environmental conditions, and produces more disagreeable odors. Because the process is lengthy, it usually is not carried to complete stability, but to a stage where further decomposition proceeds slowly. Stabilization is part of a cycle in which the products of decomposition become food for plants, then in turn food for man and animals, and finally are reconverted into wastes (Fig. 22-9a).

Aerobic bacteria serve in self-purification of streams, trickling filters, and the activated-sludge method of treatment. In streams, oxygen may become available from several sources: absorption of air at the water surface; release by algae, which absorb carbon dioxide from decomposition; and production by decomposition of such compounds as nitrates and nitrites. In trickling filters, oxygen is supplied by allowing sewage to pass over filtering media while air circulates through the voids. In the activated-sludge process, oxygen is furnished by passing air through a mixture of sewage and previously activated sludge and by strongly agitating the mixture to dissolve air into the liquid. In aerobic stabilization also, decomposition occurs in steps and is part of a cycle (Fig. 22-9b). If the supply of oxygen is inadequate, however, anaerobic action will occur and disagreeable odors may be produced.

BOD. The amount of oxygen used during decomposition of organic material is the **biochemical oxygen demand (BOD)**. It is a measure of the amount of organic material present. BOD is determined by diluting a sewage sample with water with known dissolved-oxygen content and storing the mixture for 5 days at 20°C. The oxygen content at the end of the period is measured, and the difference is reported as the BOD.

At the end of a period of t days at 20°C,

$$\text{BOD} = O(1 - 10^{-K_1 t}) \tag{22-12}$$

where O = oxygen demand when $t = 0$ or at start of any oxidation period

K_1 = deoxygenation coefficient, usually about 0.1 for sewage, but may range from less than 0.05 to more than 0.2. For temperatures other than 20°C, multiply K_1 for 20°C by 1.047^{T-20}

T = temperature, °C

To obtain the initial oxygen demand at temperatures other than 20°C, multiply O for 20°C by $0.02T + 0.6$.

Table 22-4. Relative Stability of Treatment-Plant Effluent

Time at 20°C, or time required for decolorization of methylene blue, days	Proportion oxidized or relative stability, %	Time at 20°C, or time required for decolorization of methylene blue, days	Proportion oxidized or relative stability, %
0.5	11	8.0	84
1.0	21	9.0	87
1.5	30	10.0	90
2.0	37	11.0	92
2.5	44	12.0	94
3.0	50	13.0	95
4.0	60	14.0	96
5.0	68	16.0	97
6.0	75	18.0	98
7.0	80	20.0	99

Relative stability is a measure of the amount of oxygen needed for stabilization of a sewage-treatment-plant effluent. Table 22-4 shows how relative stability varies with storage time at 20°C. The table indicates that the aerobic process is nearly completed after 20 days. If the time required to exhaust oxygen from an effluent is known, the relative stability given by Table 22-4 also is taken as the percent of the initial oxygen demand O that has been satisfied.

The time can be determined by adding to a sample of an effluent a small amount of methylene blue, an aniline dye. On exhaustion of the oxygen in the sample, anaerobic bacteria become active. They release enzymes that remove the color from the dye. The time required at 20°C for this to take place may be used with Table 22-4 to determine the percent of organic material stabilized. For example, a sample that decolorizes in 5 days has a relative stability of 68%. Only 32% of the initial oxygen demand remains. Such an effluent may be stable enough to be discharged into a stream.

Since concentration and composition of sewage vary considerably throughout a day, care must be taken to obtain a representative sample for each type of test. Sampling and analyses should be made as directed in Standard Methods for the Examination of Water and Sewage, American Public Health Association, 1015 18th St., N.W., Washington, D.C. 20036; American Water Works Association, 6666 Quincy Ave., W., Denver, Col. 80235; Water Pollution Control Federation, 3900 Wisconsin Ave., Washington, D.C. 20016.

[Metcalf and Eddy, Inc., "Wastewater Engineering," McGraw-Hill Book Company, New York; H. E. Babbitt and E. R. Baumann, "Sewerage and Sewage Treatment," John Wiley & Sons, Inc., New York; H. S. Azad (ed.), "Industrial Wastewater Management Handbook," McGraw-Hill Book Company, New York; C. N. Sawyer, "Chemistry for Sanitary Engineers," McGraw-Hill Book Company, New York; R. E. McKinney, "Microbiology for Sanitary Engineers," McGraw-Hill Book Company, New York; G. M. Fair, J. C. Geyer, and D. A. Okun, "Water and Wastewater Engineering," John Wiley & Sons, Inc., New York.]

22-19. Sewage Treatment and Disposal. Because of the objectionable characteristics of raw sewage (Art. 22-18), disposal requires consideration of many factors, especially health hazards; odors, appearance, and other nuisance conditions; and economics. Rarely do conditions exist that permit low-cost disposal of raw sewage. Usually, some degree of treatment is necessary.

Several methods are used for disposal of sewage on land: oxidation ponds, or lagoons (Art. 22-30); irrigation; incineration (Art. 22-34); burial; composting; and dewatering and conversion into fertilizer.

Sewage irrigation is of importance because it permits reclamation of the water content, to replenish the groundwater. Surface, flood, or subsurface irrigation may be used. *Surface irrigation* discharges sewage on the ground. Part evaporates and part percolates into the ground, but a sizable amount remains on the surface and must be collected in surface drainage channels. For domestic sewage, the method is not efficient. A modification, *spray irrigation*, however, has been used successfully for some industrial wastes. *Flood irrigation* also discharges the sewage on the ground, but the sewage seeps down and is usually collected in underdrains. The soil acts as a filter and partly purifies the waste. But unless the sewage is treated before irrigation, odors and insects may be produced, the soil may become clogged by grease or soap, and surface and groundwater may become contaminated. Surface irrigation sometimes is used for watering and fertilizing crops. This application, however, may create potential health hazards unless treatment has stabilized and disinfected the effluent. Another form of irrigation, *subsurface irrigation*, often is used with cesspools (Art. 22-28) and septic tanks (Art. 22-27).

Self-purification. Sewage, with or without treatment, has been disposed of by dilution in a natural body of water. Partial or complete treatment then takes place in the water. Sometimes

Table 22-5. Solubility of Oxygen in Fresh Water at Sea Level

Temperature		Dissolved oxygen, ppm or mg per liter	Temperature		Dissolved oxygen, ppm or mg per liter
°C	°F		°C	°F	
1	33.8	14.23	16	60.8	9.95
2	35.6	13.84	17	62.6	9.74
3	37.4	13.48	18	64.4	9.54
4	39.2	13.13	19	66.2	9.35
5	41.0	12.80	20	68.0	9.17
6	42.8	12.48	21	69.8	8.99
7	44.6	12.17	22	71.6	8.83
8	46.4	11.87	23	73.4	8.68
9	48.2	11.59	24	75.2	8.53
10	50.0	11.33	25	77.0	8.38
11	51.8	11.08	26	78.8	8.22
12	53.6	10.83	27	80.6	8.07
13	55.4	10.60	28	82.4	7.92
14	57.2	10.37	29	84.2	7.77
15	59.0	10.15	30	86.0	7.63

self-purification occurs; more often, if the sewage has not had adequate treatment, the body of water becomes polluted. It may be unsafe for water supply and swimming. It may contaminate or kill fish and shellfish. And it may produce odors and have an unpleasant appearance. Therefore, treatment consistent with the self-purification characteristics of the body of water is desirable and is usually required by law. Secondary treatment now is required in most states. Requirements for tertiary treatment may be imposed to protect stream-water quality.

In polluted water, decomposition of organic matter utilizes oxygen from the water. If there is an adequate supply of oxygen, the biochemical oxygen demand (BOD) may be satisfied while enough dissolved oxygen remains to support fish life. If not, anaerobic decomposition will occur (Art. 22-18); the water becomes septic and malodorous and unable to support fish life.

Unpolluted water usually is saturated with oxygen. Table 22-5 shows the amount of oxygen that fresh water can hold in solution at various temperatures. The saturation quantity also depends on the concentration of dissolved substances. Salt water, for example, holds about 80% as much oxygen as fresh water.

Oxygen deficit D is the difference between saturation content and actual content, ppm or mg per liter. As oxygen is removed from the water, the loss is offset by absorption of atmospheric oxygen at the surface. The rate at which this reaeration occurs depends on deficit D, the amount of turbulence, and the ratio of volume of water to the surface area. At any time t, days,

$$D = \frac{K_1 O}{K_2 - K_1} (10^{-K_1 t} - 10^{-K_2 t}) + 10^{-K_2 t} D_o \qquad (22\text{-}13)$$

where K_1 = coefficient of deoxygenation [see Eq. (22-12)]

      K_2 = reaeration coefficient, which ranges from 0.05 to 0.5 at 20°C, depending on depth, velocity, and turbulence of the water. For temperatures other than 20°C, multiply K_2 by 1.047^{T-20}

      T = temperature, °C

      O = oxygen demand at $t = 0$, ppm or mg per liter

      D_O = oxygen deficit at point of pollution, or $t = 0$, ppm or mg per liter

(H. W. Streeter, "The Role of Atmospheric Reaeration of Sewage-polluted Streams," *Transactions, American Society of Civil Engineers*, vol. 89, p. 1355, 1926.)

In Fig. 22-10, the deoxygenation curve indicates the amount of dissolved oxygen remaining at any time as sewage with initial demand O stabilizes, if the supply of oxygen is not replenished. The reaeration curve shows the amount of new oxygen dissolved during the same period. The oxygen sag curve represents at any given time the dissolved oxygen present, the sum of the remaining oxygen after deoxygenation and the oxygen from reaeration. The oxygen deficit D, as given by Eq. (22-13), is the ordinate of the oxygen sag curve measured from the horizontal line representing oxygen content at saturation.

The lowest or critical point of the sag curve indicates the occurrence of minimum dissolved oxygen, or maximum deficit. The time at which this occurs may be calculated from

$$t_c = \frac{1}{K_1(f-1)} \log f \left[1 - \frac{D_o}{O} (f-1) \right] \qquad (22\text{-}14a)$$

where $f = K_2/K_1$ = self-purification coefficient

When $f = 1$,

$$t_c = \frac{0.434}{K_1} \left(1 - \frac{D_o}{O} \right) \qquad (22\text{-}14b)$$

The critical deficit is given by

$$D_c = \frac{O}{f} 10^{-K_1 t_c} \qquad (22\text{-}15)$$

The pollutional load O that a stream may absorb depends on the value of D_c, coefficients f and K_1, and the initial deficit D_o. The allowable value of D_c usually is established by law. The initial deficit is determined by existing pollution. The coefficients may be estimated from tests on the sewage and the body of water, or values may be assigned based on experience. Seasonal variations in temperature and water level or stream flow affect the amount of oxygen the water can hold and the amount of water available for dilution. Hence, the most critical conditions usually occur during summer, when rainfall is low and temperatures are high.

Self-purification is slower in lakes than in streams because of the low rate of dispersion of sewage. With turbulence usually not present, mixing of water and sewage in lakes depends

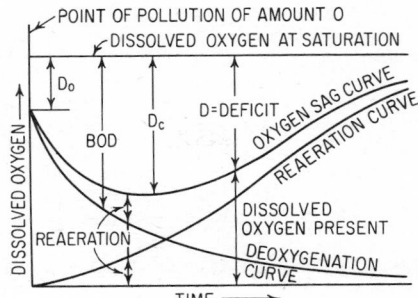

Fig. 22-10. Curves show variation in oxygen content of a stream below point of pollution.

mostly on currents and wind. Outfalls should be designed to take advantage of conditions encouraging dispersion, to prevent sludge buildup at the discharge.

In estuaries, tides complicate dispersion. They carry pollutants back and forth many times. Salinity, density, and currents may change with time. These factors may also affect dispersion in ocean waters. Special care is necessary in outfall design to promote mixing and to take advantage of currents.

Quality Standards.   Present legal standards for water quality for recreation and water supply are not uniform. A typical standard may limit coliforms to an average of 10 per milliliter; 5-day BOD to an average of 3 and a maximum of 6.5 mg per liter; and phenols to a maximum of 0.001 mg per liter. Dissolved oxygen may be required to be at least 5 and average 6.5 mg per liter, while pH must be between 6.5 and 8.5.

Stream Capacity.   A rough approximation of the capacity of a stream to absorb a pollutional load may be based on the **dilution factor,** the ratio of amounts of diluting water to sewage. The load may be estimated from the size of contributing population. For example, the 5-day BOD, lb per day per capita, may be assumed as 0.2 for domestic sewage, 0.3 for combined sewage, and 0.5 for combined sewage with large amounts of industrial waste.

Sometimes, the sewage concentration is expressed as **population equivalent,** the number of persons required to create the total oxygen demand of the sewage per day. For example, suppose domestic sewage has a BOD of 5,000 lb per day. The population equivalent then may be taken as $5,000/0.2 = 25,000$ persons.

As an example of the use of dilution factor, consider a residential community of 100,000 persons producing a sewage flow of 25 mgd to be disposed of in a river with no BOD and a dissolved-oxygen content of 10 ppm. Permissible oxygen content downstream is 6.5 ppm. What should the flow in the river be, and what is the dilution factor?

The total oxygen demand may be assumed to be $100,000 \times 0.2 = 20,000$ lb per day. Since a gallon of water weighs 8.33 lb, this for 25 mgd of sewage is equivalent to

$$\frac{20,000}{25 \times 8.33} = 96 \text{ ppm or mg per liter}$$

The required river flow Q, mgd, must supply this oxygen. Hence,

$$8.33(10 - 6.5)Q = 20,000 \qquad \text{and} \qquad Q = 686 \text{ mgd}$$

The dilution factor then is $686/25 = 27.4$.

The significance of this factor is questionable. A few early studies had indicated that nuisances would result with dilution factors for untreated sewage of less than 20 and might occur under unfavorable conditions with factors up to 40. These findings are no longer generally accepted. Use of Eq. (22-15) is preferred.

Types of Treatment.   The disposal problem generally makes some treatment of sewage necessary.

Sewage treatment is any process to which sewage is subjected to remove or alter its objectionable constituents and thus render it less offensive or dangerous. Treatment may be classified as preliminary, primary, secondary, or tertiary or complete, depending on the degree of processing.

Preliminary treatment may be the conditioning of industrial waste before discharge to remove or neutralize substances injurious to sewers and treatment processes, or it may be unit operations that prepare the waste for major treatment.

Primary treatment is the first and sometimes the only treatment of sewage. This process removes floating solids and suspended solids, both fine and coarse. If a plant provides only primary treatment, the effluent is considered only partly treated.

Secondary treatment applies biological methods to the effluent from primary treatment. Organic matter still present is stabilized by aerobic processes.

Tertiary or complete treatment removes a high percentage of suspended, colloidal, and organic matter. The sewage also may be disinfected.

Efficiency of treatment depends on quality of plant design and operation and on type and strength of sewage. Table 22-6 lists efficiencies for commonly used treatment methods in terms of percent reduction of suspended solids, bacteria, and BOD.

Table 22-6. Efficiencies of Sewage-Treatment Methods*

Type of treatment	% reduction		
	Suspended matter	BOD	Bacteria
Fine screens	5–20		10–20
Plain sedimentation	35–65	25–40	50–60
Chemical precipitation	75–90	60–85	70–90
Low-rate trickling filter, including presedimentation and final sedimentation	70–90+	75–90	90+
High-rate trickling filter, including presedimentation and final sedimentation	70–90	65–95	70–95
Conventional activated sludge, including presedimentation and final sedimentation	80–95	80–95	90–95+
High-rate activated sludge, including presedimentation and final sedimentation	70–90	70–95	80–95
Contact aeration, including presedimentation and final sedimentation	80–95	80–95	90–95+
Intermittent sand filtration, including presedimentation	90–95	85–95	95+
Chlorination:			
Settled sewage		†	90–95
Biologically treated sewage		†	98–99

*From E. W. Steel, "Water Supply and Sewerage," McGraw-Hill Book Company, New York.
†Reduction is dependent upon dosage.

[E. W. Steel, "Water Supply and Sewerage," McGraw-Hill Book Company, New York; G. M. Fair, J. C. Geyer, and D. A. Okun, "Water and Wastewater Engineering," John Wiley & Sons, Inc., New York; H. S. Azad (ed), "Industrial Wastewater Management Handbook," McGraw-Hill Book Company, New York; Metcalf and Eddy, Inc., "Wastewater Engineering," McGraw-Hill Book Company, New York; "Sewage Treatment Plant Design," Manual No. 36, American Society of Civil Engineers; "Sanitary Landfill," Manual No. 39, American Society of Civil Engineers.]

22-20. Sewage Pretreatment. The purpose of pretreatment is to remove from sewage coarse materials that may interfere with treatment, or do not respond to treatment, or may damage or clog pumps, pipes, valves, and nozzles. Various types of screening devices are used for the purpose. Generally, they are the first units in a treatment plant.

Racks are fixed screens composed of parallel bars, set vertically or sloped in the direction of flow, to catch debris. Coarse racks have spaces between the bars of 2 in. or more. They usually are used for large plants to protect sewage pumps. Medium racks, used more frequently, have bar spacings of ½ to 1½ in. They may be fixed or movable. Movable racks are three-sided cages. Sewage enters through the open side and leaves through the bars. One cage is periodically hoisted to the surface for manual cleaning, while sewage passes through a second cage. Fixed-bar racks may be manually or mechanically cleaned. The bars may be curved to the horizontal at the top to facilitate cleaning.

While a minimum velocity of about 2 fps is desirable in the approach channel to prevent sediment from clogging it, velocity through a rack should be lower, perhaps 0.5 to 1 fps, so that objects should not be forced through. This requires enlargement of the conduit in the vicinity of the rack. To allow for head loss through a rack, the conduit bottom may be lowered below the rack 3 to 6 in.

Fine screens, with uniform-size openings or slots ⅛ in. wide or less, have low efficiency for sewage treatment but are useful for removal of bulky and fibrous materials from industrial wastes. Generally, fine screens are movable and mechanically cleaned. Various types are used: rotating disk or drum, band, plate, or vibratory screens.

Screenings may be disposed of by burial, incineration, or digestion. Digestion of sludge will proceed normally when fine screenings are added in sludge-digestion tanks. In some treatment plants, screenings are passed through a grinder and returned to the flow, to settle out subsequently in a sedimentation tank. Screening and cutting are combined in such devices as comminutors, barminutors, and griductors. Their high-speed rotating edges cut through the sewage flow and chop and shred the solids, which then pass on to a sedimentation tank. Shearing-type units should be located after a grit chamber to prevent excessive wear of cutting edges.

Skimming tanks also may be placed ahead of sedimentation tanks. Skimmers remove oil and grease, which tend to form scum, clog fine screens, obstruct filters, and reduce the efficiency of activated sludge. Compressed air, applied through porous plates in the bottom of the tank, coagulates the grease and oil and causes them to rise to the surface. About 0.1 cu ft of air is required per gallon. Detention period ranges from 5 to 15 min. About 2 mg per liter of chlorine increases the efficiency of grease removal. After the effluent reaches the sedimentation tank, the coagulated material is removed with the scum or settled solids.

(Metcalf and Eddy, Inc., "Wastewater Engineering," McGraw-Hill Book Company, New York; E. B. Besselievre and M. Schwartz, "The Treatment of Industrial Wastes," McGraw-Hill Book Company, New York; "Chlorination of Sewage and Industrial Wastes," Manual of Practice 4, Federation of Sewage and Industrial Wastes Associations, Champaign, Ill.)

22-21. Sedimentation. At most sewage-treatment plants, sedimentation is a primary treatment. In activated-sludge plants, sedimentation is required after oxidation. It also is used after oxidation of sewage on trickling filters.

The major objective of sedimentation is removal of settleable solids. But often, some floating materials also are removed by clarifiers, skimming devices built into sedimentation tanks. These processes occur while sewage moves slowly through a settling basin.

Efficiency of a sedimentation tank depends on particle size, specific gravity, and settling velocity. It also depends on several other factors: concentration of suspended matter, temperature, surface area of the liquid, retention period, depth and shape of basin, baffling, total length of flow, wind, and biological effects. Density currents and short circuiting may negate theoretical detention computations. Improper baffling may have the effect of reducing the effective surface area of the liquid and creating dead or nonflow areas within the tank. In general, a settling tank of good design should have an efficiency in the upper range of that given in Table 22-6.

Settling velocity of a particle is a function of the specific gravity and diameter of the particle, and specific gravity and viscosity of fluid. Settling rates of particles larger than 200 microns are determined empirically. Sizes less than 200 microns settle in accordance with Stokes' law for drag of small settling spheres in a viscous fluid [Eq. (21-133)]. (See also Art. 21-64.)

Theoretically, if the forward motion of the water is less than the vertical settling rate of all the particles, they will settle some distance below the surface in a given time interval while in the tank. After that period, if the surface layer of water were removed, it would contain no solids.

Surface settling rate, or overflow rate, gal per sq ft of surface area per day, is a measure of the rate of flow through the basin when the rate of flow, cfs, equals the surface area, sq ft, times the settling velocity, fps, of the smallest particle to be removed. Hence, selection of an overflow or surface settling rate establishes a relationship between flow and area.

Detention period is the theoretical time water is detained in a basin. The average detention period is V/Q, where Q is the flow, mgd or cfs, and V, the basin volume. Since most of the settleable solids will settle out in 1 to 2 hr, long detention periods are not advantageous. In fact, they are undesirable because the sewage may become septic.

The flowing-through period is the time required for sewage to pass through the basin. This time may be estimated by adding sodium chloride to the influent and testing the effluent for increase of chloride. The flowing-through period should be at least 30% of the theoretical detention period. Dye may be used to follow the flow pattern.

Grit chambers (Fig. 22-11) are settling basins used to remove coarse inorganic solids. They may also trap heavier organic material, such as seeds. Grit chambers are necessary with combined sewers if the flow is to be treated. The wet-weather flow usually contains sand and grit, which must be removed to prevent damage to pumps and interference with sewage treatment.

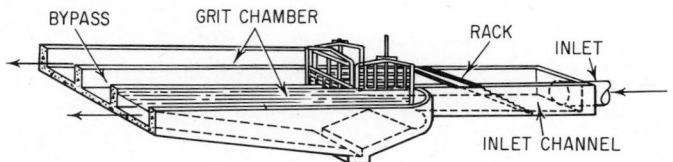

Fig. 22-11. Grit chamber.

Design of a grit chamber should insure settlement of all particles over 0.2 mm in size but should not remove organic solids. Flow should be fast enough to secure this result but without scouring solids already deposited. Scour will occur if the horizontal velocity, fps, of the sewage exceeds

$$v = 2.2 \sqrt{\frac{gd}{f} (s - 1)} \qquad (22\text{-}16)$$

where f = roughness coefficient (Darcy formula for flow in pipes) for the chamber (see Fig. 21-19, p. 21-20)

g = acceleration due to gravity, 32.2 ft per sec^2

d = particle diameter, ft

s = specific gravity of particle

Usually, grit chambers are designed for a flow of about 1 fps. Flow may be controlled by specially shaped gates or weirs to keep velocity constant. The material settling out may be removed manually or mechanically. Also, devices may be added to mechanically cleaned units to wash most of the organic material out of the grit.

A plain sedimentation tank is a settling basin where sedimentation is not aided by coagulants and the settled solids, or sludge, are not retained for digestion. Generally, sludge and scum are removed mechanically. Any of several methods may be used to remove light, suspended material.

Flocculent suspensions have little or no settling velocity. While they may occur in raw sewage, more frequently they are encountered when effluents from activated-sludge units undergo secondary settling. The suspensions may be removed by passing inflowing sewage upward through a blanket of the flocculent material (vertical-flow sedimentation tank). The objective is to produce a mechanical sweeping action in which small particles attach to larger particles, which then have sufficient weight to settle. Another removal method employs an inner chamber equipped with baffles that rotate and stir the liquid, to aid formation of larger, heavier floc. The same results also may be achieved by agitation with air. Some of the settled sludge is raised by air lift and mixed with the floc, to form a conglomerate with better settling characteristics. Variations utilizing the preceding principles have been introduced by several manufacturers, for example, the up-flow tube clarifier.

Design of a sedimentation tank should be based on the settling velocity of the smallest particle to be removed. Depth should be no larger than necessary for preventing scour and to accommodate cleaning mechanisms. Surface area of the liquid is more important than depth. So depth usually is held to 10 ft or less (at sidewalls). The surface-settling-rate requirement generally is 600 gal per sq ft per day for primary treatment alone and 800 to 1,000 for all other tanks. The detention period normally is 2 hr. These three design parameters must be adjusted, since each is dependent on the other for a given design flow (average daily flow for a plant).

Rectangular tanks are built in units with common walls. Width per unit ranges up to 25 ft. Minimum length should be at least 10 ft. The length-width ratio should not exceed 5:1. Final sizes may be determined by dimensions of available sludge-removal equipment.

Provision should be made for sludge removal on a regular schedule. If sludge is not removed, gasification occurs, and large blocks of sludge appear on the surface. These must then be broken up so that they will settle, or they must be removed by the scum-removal mechanism. In circular tanks, radial blades scrape the bottom to move the sludge to a central sludge hopper. In rectangular tanks, the hopper is located near the inlet end, since the heaviest sludge accumulation occurs in that region. Blades moving along the bottom against the flow of sewage push the sludge to the hopper. In some tanks, the same blades may be lifted to the surface, and, traveling with the sewage flow, move scum to the outlet end. There, the scum may be trapped by a baffle until taken out by a scum-removal device.

Many mechanical aids for use with sedimentation tanks are available commercially.   Manufacturers' literature should be carefully studied and specifications written to insure procurement of equipment exactly meeting design requirements.

Actual flowing-through time is influenced by inlet and outlet construction.   For circular tanks, inlets are submerged, at the center.   Sewage rises inside a baffle extending downward, to still the currents.   The outlet device nearly always is a circumferential weir adjusted to level after installation.   The weir may be sharp-edged and level or provided with V notches about 1 ft or less apart.   The notches permit more constant flow, since they are less affected by local differences in weir elevation and surface tension.   For rectangular tanks, inlets also may be submerged, but at one end.   More often, the sewage is brought to a trough that has a weir extending the width of the tank.   The flow then moves forward with less short circuiting.   At the outlet, to provide enough weir length, a launder is used.   This consists of a series of fingerlike shallow conduits set at water level and receiving flow from both sides.   Each finger is connected to a common discharge trough.   Normal weir loading should not exceed 10,000 gal per lin ft of weir per day in small plants, or 15,000 in units handling more than 1 mgd.

Chemical precipitation sometimes is used to improve the effluent from sedimentation.   The process is similar to that for water clarification.   The high cost of chemicals and the intermediate grade of treatment obtained with chemicals have kept the process from general use.   Chemical precipitation has, however, been found useful in specialized treatment.   Phosphorus removal, preparation of sludges for filtration or dewatering, and removal of trace metals are examples of such treatment.

Alum, ferric chloride, ferric sulfate, lime, sodium aluminate, ferrous chloride, ferrous sulfate, and polyelectrolytes are chemicals used to expedite precipitation.   The coagulation resulting is, in actuality, the result of a complex group of reactions involving the hydrolysis products of the added chemicals.   Effectiveness of the various chemicals depends on the conditions under which they are used and the types of wastes.

There has to be an optimum pH and an optimum dosage for efficient waste-water coagulation.   Consequently, dosages are often determined by trial (jar tests).   Measurement of zeta potential (an electrical potential related to particle stability and hence useful in controlling coagulation) and of phosphate content is also desirable.

Design requirements include rapid mixing, mixer-blade peripheral speeds of less than 5 ft per sec, control of slurry concentration, minimum sludge blanket levels, and controlled horizontal movement of clearer water by launder or weir spacing and by weir overflow rate control.   (See also Art. 21-65.)

(Metcalf and Eddy, Inc., "Wastewater Engineering," McGraw-Hill Book Company, New York; G. M. Fair, J. C. Geyer, and D. A. Okun, "Water and Wastewater Engineering," John Wiley & Sons, Inc., New York; "Sewage Treatment Plant Design," Manual No. 36, American Society of Civil Engineers, New York; "Water Treatment Plant Design Manual," American Water Works Association, Denver, Colo.)

22-22. Sewage Filtration.   Secondary treatments frequently employ oxidation to decompose and stabilize the putrescible matter remaining after primary treatments.   Filtration is one of these secondary treatments.   Others include the activated-sludge process, oxidation ponds, and irrigation.   These oxidation methods bring organic matter in sewage into immediate contact with microorganisms under aerobic conditions.   In filtration, the microorganisms coat the filtering media.   As the waste water flows through, adsorption occurs, and most of the organic materials are removed by contact with the coating.   The organisms decompose organic nitrogen compounds and destroy carbohydrates.   Efficiency of the method, as measured by reduction of biochemical oxygen demand (BOD), is high (Table 22-6).

Intermittent sand filters are sand beds, usually 2½ to 3 ft deep, with underdrains for collecting and carrying off the effluent.   Settled waste water, the effluent from a sedimentation tank, is applied to the sand surface in intermittent doses.   A rest period between doses allows time for air to assist in oxidation of the organic matter.   Application rates generally range from 20,000 gal per acre per day (gad) to 125,000 gad when the filters serve as a secondary treatment.   Rates may go as high as 0.5 million gal per acre per day (mgad) for tertiary treatments.

Sand for an intermittent filter should have a uniformity coefficient of 5 or less; 3.5 is preferred. (**Uniformity coefficient** is the ratio of the sieve size that will pass 60% of the material to the effective size of the sand.   **Effective size** is the size, mm, of the sieve that passes 10%, by weight, of the sand.)   The effective size of the sand should be between 0.2 and 0.5 mm.   A bed of gravel 6 to 12 in. thick usually underlies the sand.

A mat of solids forms on the filter surface and must be removed periodically. Generally, the mat can be scraped off when dry, but occasionally the top 6 in. or so of the filter material must be replaced.

In winter, there is danger that the sand surface will freeze. To keep the filter in operation, the bed may be ridged on 3-ft centers to support the ice while the sewage flows underneath it.

Trickling filters are beds of coarse aggregate over which settled sewage is sprayed. Filter media include gravel, crushed rock, ceramic shapes, slag, and plastics. Stone and crushed rock that do not fragment, flour, or soften on exposure to sewage are widely used. Generally, rock sizes are kept between 2 and 4 in. nominal diameter. Underdrains collect and carry off the effluent. Filters may be ventilated through the underdrain system or by other means, to supply air to the aerobic organisms.

Since suspended solids can clog filters, sedimentation of the sewage is desirable before it is fed to the filters. When, however, a waste, such as milk waste, contains a concentration of dissolved solids, it may be applied directly to a filter. In that case, preaeration is desirable, so that the waste contains some dissolved oxygen.

In time, oxidized matter breaks away from the filter media and is flushed from the filter with the effluent. Hence, the effluent is passed through a secondary settling basin, or clarifier. Design of these basins is similar to that of primary sedimentation tanks. Efficiency, or percent reduction of BOD, of a trickling filter generally is measured for both the filter and final sedimentation.

Trickling filters are classified as standard or low-rate, high-rate, and controlled.

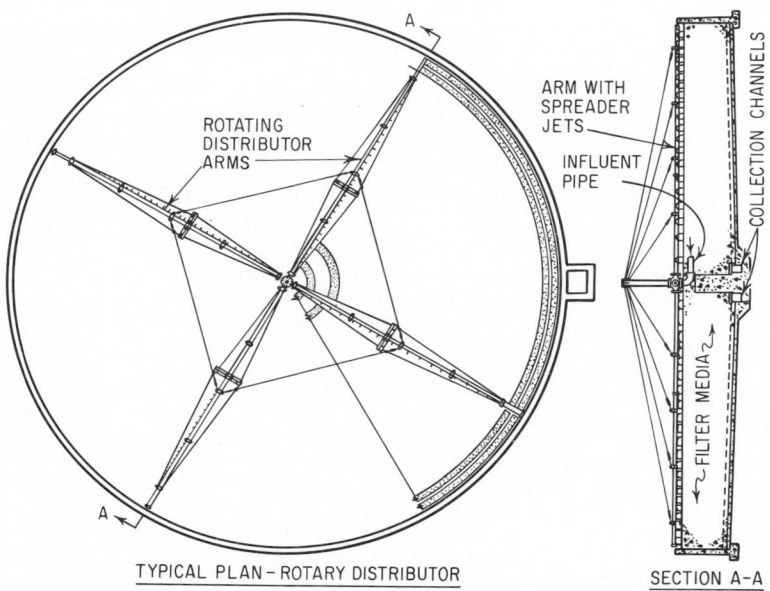

Fig. 22-12. Standard trickling filter with rotary distributor for applying sewage to surface of filter media. (*Pacific Flush Tank Company.*)

Standard filters were introduced in the United States early in the twentieth century. They consisted of an underdrained bed of stones, 6 to 8 ft deep. Settled sewage was distributed over the surface through fixed nozzles. Later, the fixed nozzles were superseded by a rotary distributor (Fig. 22-12). This type of distributor has two or four radial arms supported on a center pedestal. Jets of sewage from nozzles on the arms cause rotation. Thus, the filter surface is sprayed as the arms revolve. Dosing, as a result, is intermittent, though the interval between doses is short, often not more than 15 sec. A distributor may be kept rotating continuously by feeding the nozzles from a weir box or a dosing tank, with siphons or pumps. To accommodate rotary distribution, standard filters are built round in plan.

These low-rate filters are dosed at a rate of 1 to 4 mgad, substantially lower than that for high-rate filters.   Loading also may be expressed in terms of 5-day BOD, lb per acre-ft per day. Some state health departments limit the load on a standard filter to 400 to 600 lb per acre-ft per day.   The approximate load w to be applied to a filter, lb per day per acre-ft of filter volume, when the BOD of the sewage is known and a limit is specified for the BOD of the effluent, may be computed from

$$w = 13,840 \left(\frac{B}{A - B} \right)^2 \tag{22-17}$$

where A = 5-day BOD of the influent, mg per liter
$\qquad B$ = specified maximum BOD of effluent, mg per liter
High-rate filters receive a load three or more times greater than that usually applied to standard filters.   Usual rate is about 20 mgad, but rates from 9 to 44 mgad have been used.   Some state health departments limit the load to 2,000 to 5,000 lb of BOD per acre-ft per day.

Such high rates are feasible because the effluent is recirculated through the filter.   Recirculation reduces the load on the filter, seeds the media continuously with organisms, allows continuous dosage, offsets fluctuations in sewage flow, and reduces odors by freshening the influent. Several recirculation alternatives may be used.   For example, part of the filter effluent may be returned directly to the filter.   (Proponents of this method of recirculation claim direct return intensifies biological oxidation.)   Or part of the effluent of the filter or the final clarifier may be combined with the influent to the primary sedimentation tank.   Sometimes, dual recirculation is used: the filter effluent is returned to the primary sedimentation tank, while part of the final clarifier effluent is sent back through the filter.   In some cases, the sludge from the final clarifier is recirculated through the primary clarifier.

Two-stage filtration may be used when a better effluent is desired than can be obtained from a single filter.   For this purpose, two filters are connected in series.   Various recirculation methods may be used in this case also.

The recirculation ratio, or ratio of returned effluent to sewage influent, ranges from 1:1 to about 5:1.   At each passage, the amount of BOD removed decreases, because response to treatment decreases.   If the ratio of the decrease per passage to the BOD is given by k, then the number of effective passages of sewage through a filter may be computed from

$$F = \frac{1 + R}{(1 + kR)^2} \tag{22-18}$$

where R = recirculation ratio.   Under normal conditions, k may have a value of about 0.1.

The approximate load, lb of BOD per day per acre-ft of filter volume, to be applied to a single-stage high-rate filter or the first filter of a two-stage system, when the BOD of the sewage is known and a limit is specified for the BOD of the effluent, may be computed from

$$w = 13,840F \left(\frac{B}{A - B} \right)^2 \tag{22-19}$$

where A = 5-day BOD of the influent, mg per liter
$\qquad B$ = specified maximum BOD of effluent, mg per liter
The approximate load for a second-stage filter may be estimated from

$$w = 13,840 \left(\frac{B_1}{A_1} \right)^2 \left(\frac{B_2}{A_2 - B_2} \right)^2 F \tag{22-20}$$

where A_1 = 5-day BOD of influent of first-stage filter, mg per liter
$\qquad B_1$ = specified maximum BOD of effluent of first-stage filter, mg per liter
$\qquad A_2$ = 5-day BOD of influent of second-stage filter, mg per liter
$\qquad B_2$ = specified maximum BOD of effluent of final clarifier, mg per liter
$\qquad F$ = number of effective passages through the second-stage filter
Equations (22-17) to (22-20) are based on formulas recommended by a committee of the National Research Council ("Sewage Treatment at Military Institutions," *Sewage Works Journal*, vol. 18, no. 5, p. 794, September, 1946).

Sewage is sprayed over high-rate filters by rotary distributors or by a motor-driven disk that rains sewage continuously and uniformly over the surface.   Hence, the filters are built circular.

Controlled filters consist of sectionalized units combined into a deep filter.   The loading rate with no recirculation is ten to twelve times that for low-rate filters.

Essentials of this type of filter include sectional design, means for introduction and distribution of controlled quantities of waste water to top or upper sections of the filter, means for intro-

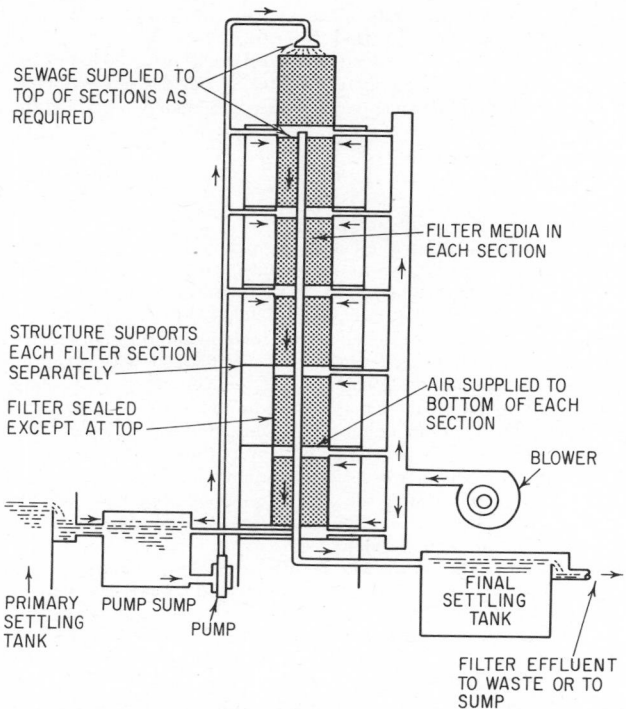

SEWAGE SUPPLIED TO TOP OF SECTIONS AS REQUIRED

FILTER MEDIA IN EACH SECTION

STRUCTURE SUPPORTS EACH FILTER SECTION SEPARATELY

FILTER SEALED EXCEPT AT TOP

AIR SUPPLIED TO BOTTOM OF EACH SECTION

BLOWER

PRIMARY SETTLING TANK

PUMP SUMP

PUMP

FINAL SETTLING TANK

FILTER EFFLUENT TO WASTE OR TO SUMP

Fig. 22-13. Controlled filtration applies sewage to tops of filter sections installed in sequence vertically. Each filter is sealed, except at the top, and has liquid inlets and outlets and air inlet.

duction of controlled quantities of air under each section of filter, temperature control between 15 and 30°C, and nonabsorbing filter media of sufficient uniformity to provide both media surface and void space (Fig. 22-13).

For domestic wastes having BOD values that do not limit the rate of absorption of oxygen, hydraulic loadings may be used as a primary design parameter according to the equation

$$n = C \left[\frac{V'(1 + R)}{Q} \right]^k \qquad (22\text{-}21)$$

where n = fraction of BOD remaining
C = constant
V' = total filter volume, thousands of cu ft
Q = daily flow, mgd
R = recirculation ratio
k = constant

Figure 22-14 can be used to select the constants C and k when $R = 0$. The V'/Q value may be read directly as the reciprocal of the filter hydraulic application rate L_H, million gal per 1,000 cu ft per day (mgtcfd), since

$$\frac{V'}{Q} = \frac{1}{L_H} \qquad (22\text{-}22)$$

When Eq. (22-22) is used in industrial-waste treatment, allowance must be made for organic loading and treatability of individual process wastes. Hence, it is advisable to develop pilot-plant information on filter application before final design.

Hydraulic surface loadings should always be greater than 70 mgad, to provide continuous washing or scouring of the filter. Unlike high-rate and low-rate filters, application of waste water must be continuous.

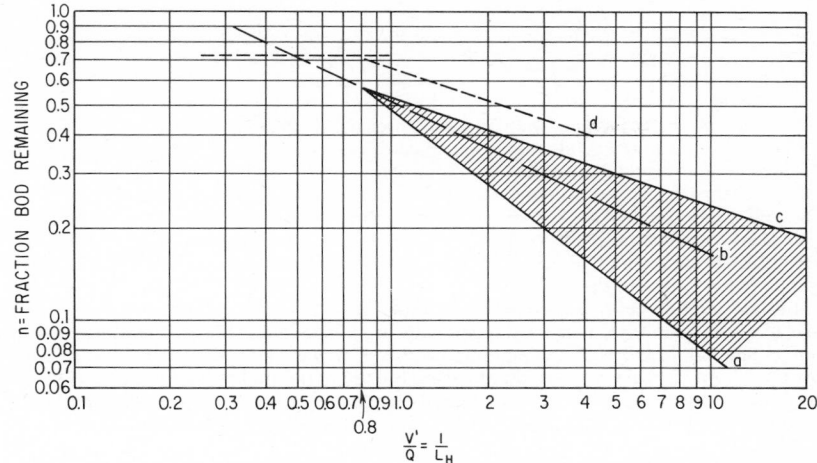

Fig. 22-14. Curves represent the equation $n = C(V'/Q)^k$ for controlled filtration with no recirculation, where n is the fraction of BOD remaining; V' is total filter volume, thousands of cu ft; Q is the daily flow, mgd; and C and k are constants.

Curve	C	k
a	0.48	− 0.795
b	0.51	− 0.482
c	0.52	− 0.343
d	0.65	− 0.343

(Metcalf and Eddy, Inc., "Wastewater Engineering," McGraw-Hill Book Company, New York; "Filtering Materials for Sewage Treatment Plants," Manual No. 13, American Society of Civil Engineers, "Sewage Treatment Plant Design," Manual No. 36, American Society of Civil Engineers; G. M. Fair, J. C. Geyer, and D. A. Okun, "Water and Wastewater Engineering," John Wiley & Sons, Inc., New York.)

22-23. Activated-Sludge Methods of Sewage Treatment.   Passing air bubbles through sewage coagulates colloids and grease, satisfies some of the biochemical oxygen demand (BOD), and reduces ammonia nitrogen a little.   Aeration also may prevent sewage from becoming septic in a following sedimentation tank.   But if sewage is mixed with previously aerated sludge and then aerated, as is done in activated-sludge methods of sewage treatment, the effectiveness of aeration is considerably improved.   Reduction of BOD and suspended solids in the conventional activated-sludge process, including presettling and final sedimentation, may range from 80 to 95% and of coliforms, from 90 to 95% (Table 22-6).   Furthermore, cost of constructing an activated-sludge plant may be competitive with other types of treatment plants producing comparable results.   Unit operating costs, however, are relatively high.

The activated-sludge method is a secondary biological treatment employing oxidation to decompose and stabilize the putrescible matter remaining after primary treatments.   Other oxidation methods include filtration, oxidation ponds, and irrigation.   These oxidation methods bring organic matter in sewage into immediate contact with microorganisms under aerobic conditions.

In a conventional activated-sludge plant, incoming sewage first passes through a primary sedimentation tank.   Activated sludge is added to the effluent from the tank, usually in the ratio of 1 part of sludge to 3 or 4 parts of settled sewage, by volume, and the mixture goes through an aeration tank.   In that tank, atmospheric air is mixed with the liquid by mechanical agitation, or compressed air is diffused in the fluid by various devices: filter plates, filter tubes, ejectors, and jets.   In either method, the sewage thus is brought into intimate contact with microorganisms contained in the sludge.   In the first 15 to 45 min, the sludge adsorbs suspended and colloidal solids.   As the organic matter is adsorbed, biological oxidation occurs.   The organisms in the sludge decompose organic nitrogen compounds and destroy carbohydrates. The process proceeds rapidly at first, then falls off gradually for 2 to 5 hr. After that, it continues at a nearly uniform rate for several hours. Generally, the aeration period ranges from 6 to 8 or more hours.

The aeration-tank effluent goes to a secondary sedimentation tank, where the fluid is detained, usually from 1½ to 2 hr, to settle out the sludge. The effluent from this tank is completely treated and, after chlorination, may be safely discharged.

About 25 to 35% of the sludge from the final sedimentation tank is returned for recirculation with incoming sewage. Sludge should not be detained in the tank. Frequent removal (at intervals of less than 1 hr) or continuous removal is necessary to avoid deaeration.

Overflow rates for final sedimentation normally range from about 800 gal per sq ft per day for small plants to 1,000 for plants of over 2-mgd capacity. Weir loadings preferably should not exceed 10,000 gal per lin ft per day. When tank volume required exceeds 2,500 cu ft, multiple sedimentation tanks are desirable.

Multiple aeration tanks are required when total tank volume exceeds 5,000 cu ft. Aeration tanks in which compressed air is used generally are long and narrow. To conserve space, the channel may be turned 180° several times, with a common wall between the flow in opposing directions. An air main is generally run along the top of the wall to feed diffusers (Fig. 22-15a and b) or porous plates (Fig. 22-15c) along its length. The air sets up a spiral motion in the liquid as it flows through the tanks. This agitation reduces air requirements.

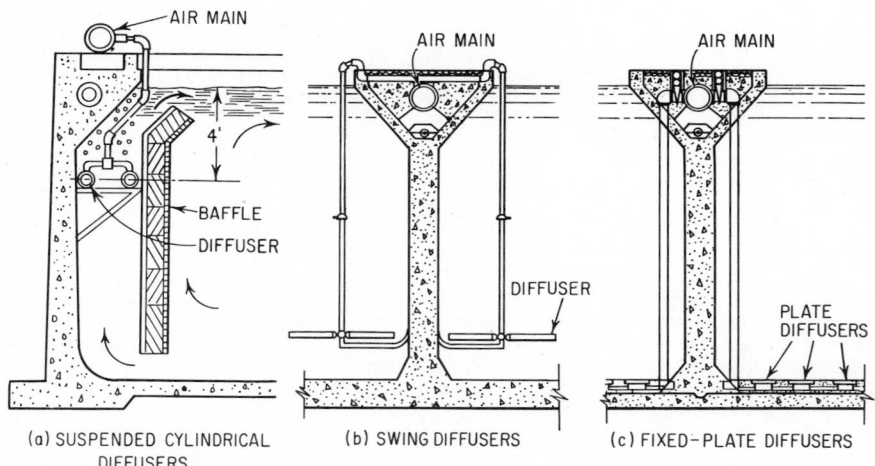

(a) SUSPENDED CYLINDRICAL DIFFUSERS   (b) SWING DIFFUSERS   (c) FIXED-PLATE DIFFUSERS

Fig. 22-15. Air main atop aeration-tank walls supplies air to diffusers in adjoining channels in which the mixture of activated sludge and sedimentation-tank effluent flows.

Width of channel ranges from 15 to 30 ft. Depth is about 15 ft.

Dissolved oxygen should be maintained at 2 ppm (mg per liter) or more. Air requirements normally range from 0.2 to 1.5 cu ft per gal of sewage treated. Most state authorities require a minimum of 1,000 cu ft of air per lb of applied BOD per day.

Mechanical aeration may be done in square, rectangular, or circular tanks, depending on the mechanism employed for agitation. In some plants, the fluid may be drawn up vertical tubes and discharged in thin sheets at the top, or the liquid may pass down draft tubes while air is bubbled through it. In both methods, agitation at the surface produced by the movement of the liquid increases aeration. Detention periods generally are longer, 8 hr or more, than for tanks with diffused air.

Several modifications of the activated-sludge method, seeking to improve performance or cut costs, are in use. These include modified, activated, tapered, and step aeration, and the Kraus, biosorption, and bioactivation processes.

Modified aeration decreases the aeration period to 3 hr or less and holds return sludge to a low proportion. Results are intermediate between primary sedimentation and full secondary treatment.

Activated aeration places aeration tanks in parallel. The activated sludge from one final sedimentation tank or group of such tanks is added to the influent of the aeration tanks. Other sludge is concentrated and removed. With much less air, results are better than with modified aeration.

Tapered aeration differs from conventional in that air diffusers are not uniformly spaced. Instead, more diffusers are placed near the inlet end of the aeration tanks than near the outlet. The theory is that oxygen demand is greater near the inlet, and so the efficiency of the treatment should be improved if more air is supplied there. However, results depend on degree of longitudinal mixing, rate of sludge return, and characteristics of recirculated matter, for example, air content of sludge or mixed liquor.

Step aeration adds sewage at four or more points in an aeration tank. Each increment reacts with sludge already present in the tank. Thus, air requirements are nearly uniform throughout the tank.

The **Kraus process** adds to the sewage an aerated mixture of activated sludge and material from sludge digester tanks. The **biosorption** process mixes sewage with sludge preaerated in a separate tank. The **bioactivation process** uses primary sedimentation, a trickling filter, and short secondary sedimentation, then adds activated sludge and passes the mixture through aeration and final sedimentation tanks.

Activated-sludge plants should be closely controlled for optimum performance. This requires frequent checking of the sludge content of the mixed liquor. Solids usually are limited to 1,500 to 2,500 ppm (mg per liter) in diffused-air plants and about 1,000 ppm when mechanical agitation is used. Settling characteristics of the sludge are indicated by the Mohlman index:

$$\text{Mohlman index} = \frac{\text{volume of sludge settled in 30 min, \%}}{\text{volume of suspended solids, \%}} \qquad (22\text{-}23)$$

A good settling sludge has an index below 100. An alternative measure is the sludge density index, 100 divided by the Mohlman index. Operating control may be maintained by holding a constant mixed-liquor suspended-solids (MLSS) or volatile-suspended-solids (MLVSS) concentration, by holding a constant ratio of food to microorganisms (F:M), or by holding a constant mean cell residence time (MCRT). The latter may be the simplest, because only suspended solids concentration in the aeration basin and in the waste activated sludge need be measured.

Sludge age is another important factor. It is the average time that a particle of suspended solids remains under aeration. Sludge age is measured by the ratio of dry weight of sludge in the aeration tank, lb, to the suspended-solids load, lb per day, of the incoming sewage. In a well-operated activated-sludge plant, sludge age is 3 to 5 days. But it may be only 0.3 days for a modified process that is well operated.

(Metcalf and Eddy, Inc., "Wastewater Engineering," McGraw-Hill Book Company, New York; "Sewage Treatment Plant Design," Manual No. 36, American Society of Civil Engineers; G. M. Fair, J. C. Geyer, and D. A. Okun, "Water and Wastewater Engineering," John Wiley & Sons, Inc., New York; L. G. Rich, "Unit Operations of Sanitary Engineering," John Wiley & Sons, Inc., New York.)

22-24. Contact Aeration. This is a secondary treatment similar to the activated-sludge method. Contact aeration also uses air diffusion to supply oxygen and to keep a suspension containing microorganisms thoroughly mixed with incoming sewage. In addition, active growths of microorganisms are maintained on plates of impervious material, such as asbestos cement, suspended in the mixing liquor of the aeration tank. Slime growth forms on the plates, and liquid passing by furnishes the organisms on the plates with nutrients. The organisms decompose organic nitrogen compounds and destroy carbohydrates.

Plates may be fixed or may rotate about a horizontal axis. As they rotate, biological growth adheres to them and is alternately immersed in waste liquid and exposed to the air. This alternation insures an aerobic condition for growth.

The aeration period in contact aerators may be 5 hr or more. Aeration usually is preceded by 1 hr of preaeration of the raw sewage and return sludge before primary settling. The load on the contact aerator is based on two factors: pounds per day of BOD per 1,000 sq ft of contact surfaces (6.0 or less) and pounds per day of BOD per 1,000 sq ft per hour of aeration (1.2 or less). About 1.5 cu ft of air per gal of flow is required. Overall plant efficiency may be about 90% BOD removal, with a higher percentage removal of suspended solids.

(E. W. Steel, "Water Supply and Sewerage," McGraw-Hill Book Company, New York.)

22-25. Sludge Treatment and Disposal. Sludge comprises the solids and accompanying liquids removed from sewage in screening and treating it. Solids are removed as screenings, grit, primary sludge, secondary sludge, and scum.   Often sludge treatment is necessary to make possible safe, economical disposal of these wastes.   The treatment to be selected depends on quantity and characteristics of the sludge, nature and cost of disposal, and cost of treatment.

Screenings are putrescible and offensive.   They may be disposed of by burning, burial, grinding and return to sewage, or grinding and transfer to a sludge digester.   The quantity of screenings is variable and dependent on sewage characteristics.   Coarse screenings may range from 0.3 to 5 cu ft per million gal (mg).   Fine screenings may range from 5 to 35 cu ft per mg.

Sand and other gritty materials also may be present in widely varying amounts.   Normally, the volume will be between 1 and 10 cu ft per mg.

Sludge varies in quantity and characteristics with the characteristics of the sewage and plant operations.   Usually, more than 90% is water containing suspended solids with a specific gravity of about 1.2.   Roughly, there may be about 0.20 lb of these solids per capita daily in sanitary sewage; 0.22 lb if a moderate amount of industrial wastes is present; 0.25 lb in effluents of combined sewers if considerable industrial wastes are present; and 0.32 to 0.36 lb if the sewage contains ground garbage also.

Primary sludge, derived from sedimentation tanks or the influent of digestion chambers of Imhoff tanks, is putrescible and odorous.   It is composed of gray, viscous identifiable solids and has a moisture content of 95% or more.   Primary treatment of 1 mg of sewage may produce about 2,500 gal of this sludge.

Trickling filter sludge is black or dark brown, granular or flocculent, and partly decomposed. It is not highly odorous when fresh.   Moisture content may be about 93%.   Passage of 1 mg of sewage through a trickling filter may produce about 500 gal of this sludge.

Activated sludge is dark to golden brown, granular or flocculent, and partly decomposed.   It has an earthy odor when fresh.   Moisture content may be about 98%.   Influent to an activated-sludge plant may yield about 13,500 gal of waste sludge per mg.

Chemical-precipitation sludge may have a solids content more than double that of sludge from primary sedimentation.   Normally, chemical precipitation from 1 mg of sewage will yield about 5,000 gal of sludge with moisture content of 95%.

Digested sludge, from septic, Imhoff, or separate digestion tanks, is very dark in color and has a homogeneous texture.   When wet, it has a tarry odor.   Roughly, treatment of 1 mg of sewage will produce 800 gal of digested sludge with a moisture content of about 90%.

Sludge digestion is the anaerobic decomposition of organic matter, which makes up about 70% of total solids, by weight, in sludge.   The process results in partial gasification, liquefaction, and mineralization of the solids.   It can be applied to treatment-process sludges other than chemical sludges and those containing substances toxic to sludge organisms, such as cyanides and chromium.   Advantages of sludge digestion include production of a stable, inoffensive sludge (if the process is continued long enough); 35 to 45% reduction of suspended solids; 55 to 75% reduction in dry weight of volatile matter; reduction in moisture content; and production of a sludge from which water may be more easily removed.   The digested sludge may be used as a soil conditioner and weak fertilizer under certain conditions.   Furthermore, gases produced during digestion may be used as fuel.

For sludge digestion, sludges are transferred to separate digestion tanks, unless Imhoff-type tanks or septic tanks are used.   While sludge decomposes in a digester, fresh sludge is added periodically.   Anaerobic bacteria attack the carbohydrates first, forming organic acids.   After this initial acid fermentation, acid digestion occurs.   Organisms living in the acid environment attack the organic acids and nitrogenous matter.   Then, a period of digestion, stabilization, and gasification takes place, in which the anaerobic bacteria feed on proteins and amino acids.   Volatile acids are reduced, and the pH rises.   In the final stage, methane fermentation occurs, with methane as the principal gaseous product.   Speed of digestion is indicated by the rate of gas formation.   Periodic removal of liquefied matter, excess liquor (or supernatant liquor), and digested solids makes room for fresh sludge.

All stages of the process proceed simultaneously in the tank.   Mingling of well-digested sludge with fresh sludge provides balance.   If the pH holds between 7.2 and 7.4, conditions for digestion will be most favorable.   Once achieved, balance may usually be maintained if addition of fresh solids is held to less than 4%, by weight, of the solids in the tank.

Speed of digestion depends on temperature (Fig. 22-16).   In conventional sludge digestion (mesophilic range), 100°F is the optimum temperature.   Between 110 and 140°F (thermophilic range), thermophilic, or heat-loving, bacteria become active and speed digestion even more, with an optimum temperature of 130°F.   Tanks usually are heated to hasten digestion.

Most states have established schedules of capacity requirements for digestion tanks, depending on type of sludge and whether or not the tanks are heated.   Typical requirements set a capacity, cu ft per capita, for heated tanks of 2 to 3 for primary sludges, 3 to 4 for mixtures of primary and standard-filter sludges, and 4 to 6 for activated sludge or mixtures of primary and high-rate filter sludges. Capacities of unheated tanks should be twice as great for each type of sludge.

Sludge-digestion tanks may be circular or rectangular in plan. They generally provide a means of manipulating the sludge. The system also may include preheater and heater equipment, recirculation pumps with sludge suction at several levels, supernatant-liquor drawoff at several levels, gas dome or collector, stirring mechanism, sludge rakes, and drawoff. The tank

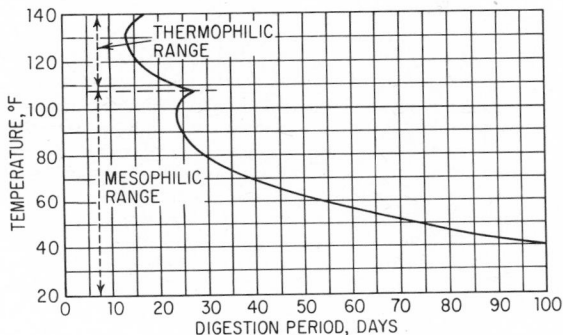

Fig. 22-16. Digestion period decreases with increasing temperature, reaching a minimum in the mesophilic range at about 100° F and in the thermophilic range at about 130° F.

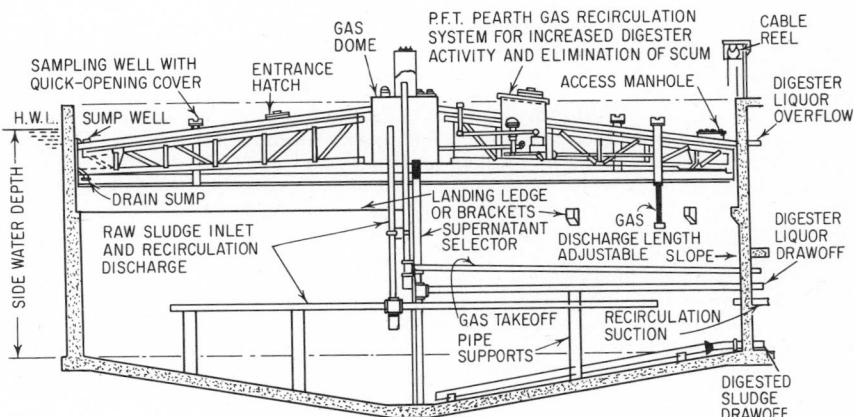

Fig. 22-17. Floating-cover sludge digester with gas recirculation. (*Pacific Flush Tank Company.*)

cover may be floating (Fig. 22-17) or fixed. With a fixed cover, when fresh sludge is added to a tank kept full, an equal volume of supernatant liquor must be removed. Addition of sludge creates currents, as a result of which the liquor being removed may carry off some of the sludge. A floating cover allows the liquor to be withdrawn before or after the fresh sludge enters the tank.

In multistage digestion, two or more digesters are placed in series. The sludge drawoff of each is fed to a subsequent one, and digested sludge is removed from the last. The system provides flexibility in manipulating and mixing sludges and in controlling supernatant liquor. Also, it may be possible to use a smaller tank than required for single-stage operation or, for a given-size tank, to retain solids longer. In two-stage digestion, good results may be obtained if less than 20%, by volume, of material transferred from the first to the second tank is the best-digested sludge and more than 80% is supernatant liquor with the lowest solid content.

Supernatant liquor, the liquid fraction in a digester, is high in solids and biochemical oxygen demand. It has an offensive odor. Withdrawn from a digester in small quantities at a level where the liquor contains relatively few solids, it is disposed of by insertion in the influent to a primary sedimentation tank.

Sludge-gas production under good operating conditions is about 12 cu ft per lb of volatiles destroyed. The gas is 60 to 70% methane, 20 to 30% carbon dioxide, plus minor amounts of other gases, including hydrogen sulfide. Fuel value of sludge gas usually ranges from 600 to 700 Btu per cu ft. The gas may be used at the treatment plant to operate auxiliary engines and provide heat for sludge-heating systems. Excess gas is burned.

Volume of sludge to be handled in a digester may be decreased substantially if the fresh sludge is first thickened. Sludge thickening is the concentration of primary- and secondary-treatment sludges in a thickening tank. In it, sludge settles to the bottom while being gently stirred and is pushed to a discharge pipe. The overflow, or decant liquor, much less objectionable than supernatant liquor, is discharged in the primary-sedimentation-tank influent.

Before disposal, digested sludge from relatively small treatment plants may be concentrated in drying beds. Area needed for this purpose is about 2 to 3 sq ft per capita (about three-fourths as much if the beds are covered). Beds consist of up to 12 in. of coarse sand over 12 in. of gravel. The natural earth bottom is sloped to underdrains, usually spaced about 30 ft apart. A bed may be from 20 to 30 ft wide and up to 125 ft long. It may be bounded or separated from an adjacent bed by a concrete wall extending about 15 in. above the sand surface.

The bed is dosed with sludge to a depth of 9 to 12 in. and allowed to drain and dry. A well-digested, granular sludge drains easily and reduces to a depth of 3 to 4 in. when dry (60 to 70% moisture content). Sludge removed from the bed has little or no odor. It may be used as a weak fertilizer or landfill.

Sludge processing may be required if the sludge is to be disposed of by other methods. One sludge-processing method is elutriation, or washing of sludge with plant effluent. This removes undesirable amino-ammonia nitrogen and reduces or eliminates the need for conditioning chemicals. After settlement, the washed sludge is drawn off for conditioning and filtration.

As an alternative, lime or ferric chloride may be used to prepare sludge for vacuum filtration.

Filtering reduces moisture content to 70 to 80%. Filter cakes are easier to handle than the digested sludge from digesters. (In some plants, raw sludge is conditioned and processed on various filters without digestion. Such sludge is offensive and is handled in the same manner as screenings.)

Filter rates range from 2.5 lb of dry solids per sq ft per hr for fresh or digested activated sludge to 8 for primary digested sludge. Usually, a filter is a hollow drum that rotates slowly about a horizontal axis in a basin of sludge. The filter is covered with wire, plastic, or cotton cloth or with flexible, metal, springlike coils. A vacuum in the compartmented interior of the drum holds sludge against the cover and separates water from the solids. As the drum rotates, the vacuum is released, and the concentrated sludge, forming a ¼-in.-thick cake, drops into a conveyor. The filtrate is returned to sewage influent or to elutriators.

Filter cake prepared from digested sludge may be used as fertilizer or landfill. Digested as well as undigested sludge, however, also may be disposed of by incineration. This requires auxiliary heat because the moisture content of the filter cake is high. Gas, including digester gas, oil, or coal, may be used as fuel.

Generally, incinerators used to burn sludge are multiple-hearth. Fed initially to the top hearth, sludge is pushed down to the next hearth by agitator arms as it dries. The heat drives off water and volatile gases, which are ignited by the high temperature. To avoid excessive odors, the temperature should be maintained at 1400° F or more. Ash residue is inert and may be used for fill or cover on sanitary landfill.

If digester gas is not available for fuel, cost of sludge incineration may be high. As an alternative, filter cake may be mixed with solid wastes and burned in a municipal incinerator, if it adjoins the treatment plant.

Also, sludge may be dried in a high-heat flash drier, instead of burned. The dried sludge can be used as fertilizer, after enrichment with chemicals, such as potash, which most digested sludges lack. Flash driers operate by mixing a portion of dried sludge with incoming wet sludge cake and introducing a high-velocity high-temperature gas stream. The dried material is separated from the gas in a cyclone separator and moved to storage. If a refuse incinerator is located at the sewage-treatment site, it can provide heat for sludge drying.

In an older method, sludge cake is dried in drum driers, long, rotating, horizontal cylinders. Large areas, much power, and considerable fuel are required.

(Metcalf and Eddy, Inc., "Wastewater Engineering," McGraw-Hill Book Company, New York; G. M. Fair, J. C. Geyer, and D. A. Okun, "Water and Wastewater Engineering," John Wiley & Sons, Inc., New York; "Sewage Treatment Plant Design," Manual No. 36, American Society of Civil Engineers; P. A. Vesilind, "Treatment and Disposal of Wastewater Sludges," Ann Arbor Publishers, Inc., Ann Arbor, Mich.)

22-26. Imhoff Tanks. Developed by Karl Imhoff in Germany for the Emscher sewage district, this type of tank has been widely used in the United States since 1907 for primary treatment of sewage. The tank permits both sedimentation and sludge digestion to take place. Sludge comprises the settled solids in sewage, and sludge digestion is the anaerobic decomposition of organic matter in sludge (Arts. 22-25).

Efficiency of Imhoff tanks is about the same as for plain sedimentation tanks. Imhoff effluents are suitable for treatment in trickling filters. Sludge digestion, however, may proceed much more slowly than in a separate digester. In an Imhoff tank, sludge digestion takes place without heat. Since rate of digestion decreases with drop in temperature (Fig. 22-16), lack of temperature control is a disadvantage, especially in regions where winters are cold.

Imhoff sludge has a tarlike odor and a black, granular appearance. It is dense. When withdrawn from a tank, it may have a moisture content of 90 to 95%. It dries easily, and when dry, it is comparatively odorless. It is an excellent humus but not a fertilizer.

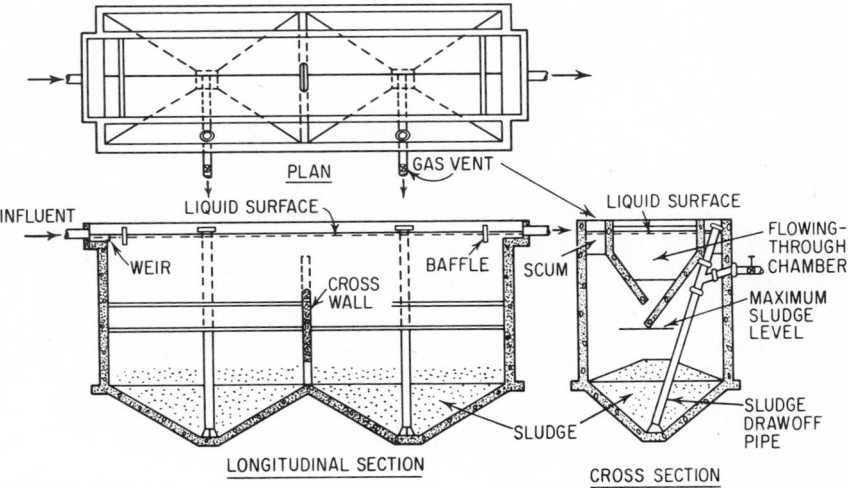

Fig. 22-18. Imhoff tank permits sedimentation of sewage in upper compartments, sludge digestion in lower.

Imhoff tanks are compartmented (Fig. 22-18). Sedimentation occurs in an upper, or flowing-through, chamber. Sludge settles into a lower chamber for digestion. To facilitate transfer of the settling solids, the flowing-through chamber has a smooth, sloping bottom (about 60° with the horizontal) with a slot at the lowest level. After particles pass through the slot, they are trapped in the lower chamber. Their path is obstructed either by overlapping walls at the slot, as shown in the cross section in Fig. 22-18, or by a triangular beam with an apex just below the slot.

As digestion proceeds in the lower chamber, scum is formed by rising sludge in which gas is trapped. The scum is directed to a scum chamber and gas vent alongside the upper chamber. As gases escape, sludge sinks back from the scum chamber to the lower chamber. (The gas vents occasionally may give off offensive odors.) The scum chamber should have a surface area 25 to 30% of the horizontal surface of the digestion chamber. Vents should be at least 24 in. wide. And top freeboard should be at least 2 ft to contain the scum. If foaming occurs at a gas vent, it can be knocked down with a water jet from a hose.

In the digestion chamber, sludge settles to the sloped bottom. After sufficient time has elapsed for anaerobic decomposition, the sludge is removed through drawoff pipes. Since the height of a tank usually is 30 to 40 ft, the sludge can be expelled under the hydraulic pressure of the liquid in the tank.   Ordinarily, sludge withdrawals are made twice a year.   With such a schedule, the digestion chamber may be designed for a capacity of 3 to 5 cu ft per capita of connected sewage load.   If, however, sludge removal is less frequent, or if industrial wastes with large quantities of solids are present in the sewage, the capacity should be greater.   Some chambers have been constructed with capacities up to 6.5 cu ft per capita.

Large tanks are provided with means for reversing flow in the upper chamber.   Since sedimentation generally is largest near an inlet, flow reversal permits a more even distribution of settled solids over the digestion chamber.

Detention period in the upper chamber usually is about 2½ hr.   Surface settling rate generally is 600 gal per sq ft per day.   The weir overflow rate normally does not exceed 10,000 gal per lin ft of weir per day.   Velocity of flow is held below 1 fps.

Length-width ratios of Imhoff tanks range from 3:1 to 5:1.   Depth to slot is about equal to the width.

Multiple units are preferable to a single large tank.   Sometimes, it also is expedient to set two flowing-through chambers above one digestion chamber.

22-27. Septic Tanks.   Like Imhoff tanks (Art. 22-26), septic tanks permit both sedimentation and sludge digestion.   But unlike Imhoff tanks, septic tanks do not provide separate compartments for these processes.   While undergoing anaerobic decomposition, the settled sludge is in immediate contact with sewage flowing through the tank.

Septic tanks have limited use in municipal treatment.   Their effluents are odorous, high in biochemical oxygen demand, and dangerous because of possible content of pathogenic organisms.   Septic tanks, however, are widely used for treatment of sewage from individual residences.   Such tanks also are used by isolated schools and institutions and for treatment of sanitary sewage at small industrial plants.

The tanks have a capacity of about 1 day's flow, plus storage capacity for sludge.   Design of residential tanks generally is based on 75 gal of sewage per person per day or 150 gal per bedroom per day.   If bedrooms are used as a criterion, allowance should be made for future conversion of some rooms into bedrooms.   If garbage grinders may be used, tank capacity should be increased (Table 22-7).   Most states set a minimum capacity of 500 gal for a single tank.   Some states require a second compartment of 300-gal capacity, separated from the first compartment by a vertical partition.   The partition has a horizontal slot, about 6 in. high, to permit passage of effluent from the first compartment.

Table 22-7. Minimum Capacities of Septic Tanks

Bedrooms	Persons	Liquid capacity, gal	
		Without garbage grinders*	With garbage grinders†
2 or fewer	4	500	750
3	6	600	900
4	8	750	1,000

*Add 150 gal for each bedroom over 4.
†Add 250 gal for each bedroom over 4.

A septic tank may be constructed of coated metal or reinforced concrete and should be watertight.   It should have a minimum liquid depth of 4 ft.   Length of a rectangular tank may be about twice the width.   Cast-in-place concrete tanks should be at least 6 in. thick, unless completely reinforced.   The top slab, at least, should be reinforced to support 150 psf.   The tank top should be between 12 and 24 in. below finished grade.   An opening at least 16 in. in diameter should be provided for a manhole.   The underside of the tank top should be at least 1 in. above the tops of partitions and baffles.   The invert of the inlet pipe should be at least 1 in., preferably 3 in., above the invert of the outlet.   When the length of a tank exceeds 9 ft, two compartments should be used.   Figure 22-19 shows a typical tank.

Residential septic tanks usually are buried in the ground and are forgotten until the system gives trouble because of clogging or overflow.   Actually, sludge should not be permitted to accumulate to a depth greater than that indicated in Table 22-8.

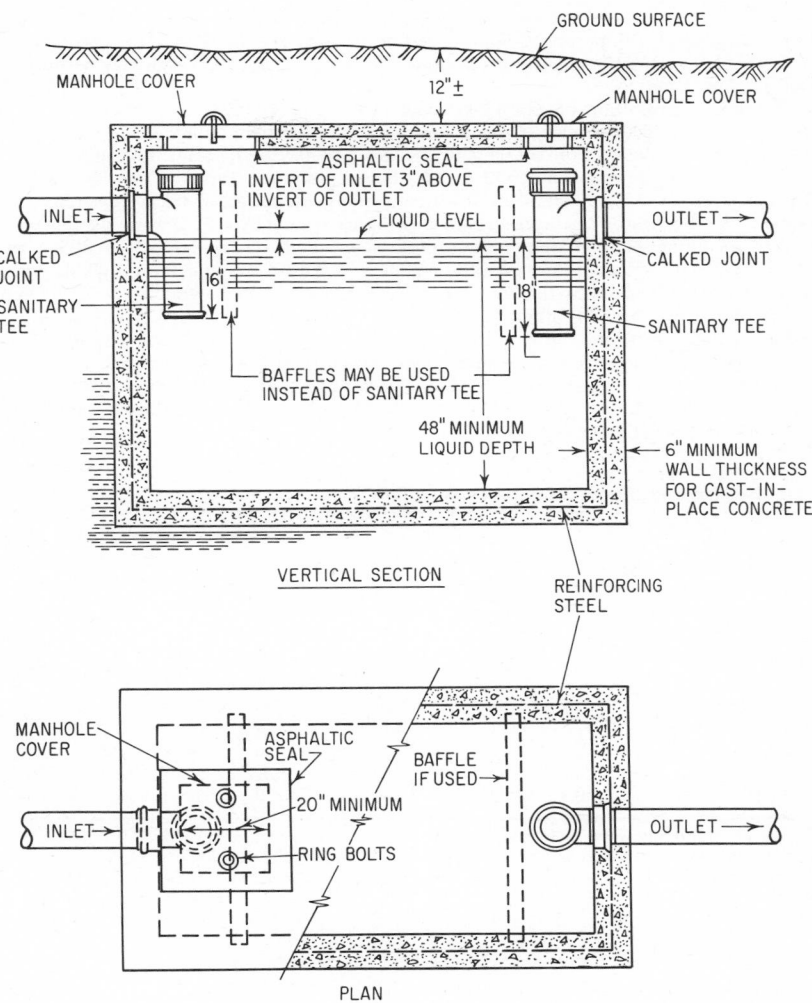

Fig. 22-19. Septic tank permits sedimentation and sludge digestion in the same compartment.

Table 22-8. Allowable Sludge Accumulation in Septic Tanks

Tank capacity, gal	Distance from bottom of outlet device to top of sludge, in., for liquid depth of	
	4 ft	5 ft
500	16	21
600	13	18
750	10	13
900	7	10
1,000	6	8

Commercial scavenger companies are available for sludge removal in most areas. Using a tank truck equipped with pumps, they remove the contents of a septic tank and cart them to a sewer manhole or a treatment plant for disposal. In rural areas, the sludge may be buried in an isolated site.

Municipal and institutional septic tanks are designed to hold 12 to 24 hr flow, plus stored sludge. For camps for 40 or more persons, septic tanks should have a liquid capacity of at least 25 gal per person served. For day schools, the capacity may be two-thirds as large.

For residential units, the main vent for the house plumbing normally provides adequate ventilation. For large septic tanks, however, separate vents for the tanks are desirable.

Septic-tank effluent may be disposed of in a leaching cesspool (Art. 22-28) or a tile field. The latter consists of lines of open-jointed tile or perforated pipe laid in trenches 18 to 30 in. deep. The lines receive the effluent from a distribution box, which distributes the liquid equally. From the box, the lines spread out, so that they are at least 6 ft apart. Lines should be of equal length, but none should be over 60 ft long.

Laid on a slight slope, not more than $1/16$ in. per ft, the tile or pipe is firmly set in a bed of crushed stone or washed gravel. The aggregate should extend 12 in. below and 2 in. above the conduit. The sewage, discharging from the openings, disperses over the entire trench bottom and seeps into the ground. The size of the tile field should be determined from the results of soil-percolation tests (Table 22-9).

Table 22-9. Required Sizes of Tile Fields for Septic-Tank Effluent

Soil-percolation rate, min*	Sewage application, gal per sq ft per day	Trench width, in.	Trench length, lin ft, for sewage loads, gpd, of			
			300	450	600	1,000
0–5	2.4	24	63	94	125	209
6–7	2.0	24	73	110	146	244
8–10	1.7	36	59	88	118	196
11–15	1.3	36	77	116	154	256
16–20	1.0	36	95	143	191	317
21–30	0.8	36	125	188	250	417
31–45†	0.6	36	167	250	334	555
46–60†	0.4	36	250	375	500	834

*Time for 1-in. drop in water level in soaked hole.

†If the percolation rate exceeds 60 min, the system is not suitable for a tile field. A rate over 30 min indicates borderline suitability for soil absorption; special care should be used in design and construction.

At least two soil-percolation tests should be made in the area of the tile field. To perform a test, dig a hole 8 in. in diameter or 12 in. square. It should extend 6 in. below the trench bottom or about 30 in. below the final ground surface. Place 2 in. of coarse sand or fine gravel in the bottom of the hole. Presoak the hole by filling it with water several hours before the test and again at the time of test and allowing the water to seep away. Remove any soil that falls into the hole. Pour clean water to a depth of 6 in. in the hole. Record the time, minutes, required for the water to drop 1 in. Repeat until the time for the water to drop from the 6- to 5-in. levels is about the same for two successive tests. Use the results of the last test as the stabilized rate. Alternative percolation-test methods have been developed for use where peculiar soil conditions exist.

Lots with less than 10 ft of soil above a rock formation usually are not suitable for construction of both sewage systems and well-water supplies, because of contamination hazards. Tile fields should not be constructed under driveways. The fields should be more than 100 ft away from any source of water supply, 20 ft from house foundation walls, and 10 ft from property lines. Trench bottoms should be at least 2 ft above groundwater, 5 ft above rock. Roof, footing, and basement drains should not be connected to septic tanks, or they will be overloaded with water not requiring treatment. Water from roof gutters and other storm water should be routed away from the tile field. This water would saturate the soil and interfere with proper operation of the field.

Where soil is impervious or nearly so, an underdrained tile field may be used. This, in reality, is a buried sand filter placed below the tile drainage system. The drainage tile is laid in trenches filled with gravel or other porous media. Underdrains at the bottom collect and convey the effluent to a central collection point. There, the waste may be either drained out by gravity, chlorinated and discharged to a body of water, or pumped to a discharge point.

("Sewage Disposal Systems for the Home," Part III, *Bulletin* 1, Department of Health, State of New York, Albany, N.Y.; "Studies on Household Sewage Disposal System," Parts 1, 2, and 3, Robert A. Taft Sanitary Engineering Center, U.S. Public Health Service; V. M. Ehlers and E. W. Steel, "Municipal and Rural Sanitation," McGraw-Hill Book Company, New York; V. T. Manas, "National Plumbing Code Handbook," McGraw-Hill Book Company, New York.)

22-28. Cesspools and Seepage Pits. A cesspool is a lined and covered hole in the ground into which sewage is discharged. It is used only when a sewerage system is not available. It may be watertight or leaching. A watertight cesspool retains sewage until the wastes are removed, by pumps or buckets. This type of cesspool is used only where no drainage into surrounding soil or rock is permitted. A leaching cesspool allows sewage to seep into the surrounding ground.

Seepage pits of similar construction may be used to supplement tile fields (Art. 22-27) or instead of such fields where conditions are favorable. The pits also may be used in series with cesspools or septic tanks, to drain overflow liquid into the surrounding soil. Results are similar to those obtained with septic tanks (Art. 22-27).

Use of a leaching cesspool for direct disposal should be restricted to a small family in a remote location where there is absorptive soil and no danger of groundwater pollution. Leaching cesspools and seepage pits should never be attempted in clay soils.

The bottom of a seepage pit should be at least 2 ft above groundwater and 5 ft above rock. Lots with less than 10 ft of soil above a rock formation generally are not suitable for construction of both seepage pits and well-water supplies, because of contamination hazards. Pits should be located more than 100 ft from a source of water supply, 20 ft from buildings, and 10 ft from property lines. Clear distance between two pits should be at least two times the diameter of the larger pit.

Size of seepage pit should be determined on the basis of 75 gal per person per day or 150 gal per bedroom per day. When bedrooms are used as a criterion, allowance should be made for future conversion of some rooms into bedrooms. The pit lining should be open-jointed or perforated to permit liquid to leak out. Wall area should be large enough to allow the soil to absorb the liquid without the pit overflowing. The required wall area, or effective absorption area, should be determined from soil-percolation tests (Table 22-10).

Table 22-10. Required Absorption Areas for Seepage Pits

Soil-percolation rate, min*	Sewage application, gal per sq ft per day	Required absorption area, sq ft, for sewage loads, gpd, of			
		300	450	600	1,000
0–5	3.2	94	141	188	313
6–10	2.3	130	196	261	435
11–15	1.8	167	250	334	555
16–20	1.5	200	300	400	666
21–30	1.1	273	409	545	911
31–45†	0.8	375	562	750	1,250
46–60†	0.5	600	900	1,200	2,000

*Time for 1-in. drop in water level in soaked hole.
†If the percolation rate exceeds 60 min, the system is not suitable for a seepage pit. A rate over 30 min indicates borderline suitability for soil absorption; special care should be used in design and construction.

Percolation tests for seepage pits are the same as for tile fields (Art. 22-27). The tests should be made, however, at half the depth and at the full estimated depth of the seepage pit. A larger excavation may be made for the upper portion of the hole, to facilitate execution of the test.

Table 22-11. Seepage-Pit Dimensions* for Required Absorption Area, Sq Ft

Depth, ft	Outside diameter, ft							
	5	6	7	8	9	10	11	12
3	47	57	66	75	85	94	104	113
4	63	75	88	101	113	126	138	151
5	79	94	110	126	141	157	173	188
6	94	113	132	151	169	189	207	226
7	110	132	154	176	197	220	242	263
8	126	151	176	201	225	252	276	302
9	141	170	198	226	254	283	310	339
10	157	189	220	251	282	314	346	377
11	173	207	242	276	310	346	380	415
12	188	226	263	302	339	377	415	453

*Outside diameter and effective depth.   Bottom area excluded from computations.

When the required absorption area has been obtained from Table 22-10, the outside diameter and effective depth of pit may be obtained from Table 22-11.   The lining generally is made of concrete block or precast-concrete sections.   Thickness should be at least 8 in.   With rectangular block, the bottom should not be more than 10 ft below grade; with interlocking block, not more than 15 ft.   For deeper pits, the lining should be structurally designed to resist saturated-earth pressures.   The top should have a watertight manhole and concrete cover.

Coarse gravel should be placed in the bottom of the pit to a depth of 6 in.   Backfill around the lining in the absorption area should be clean crushed stone or gravel, 1½ to 2 in. in diameter, to a thickness of at least 6 in.   A 2-in.-thick layer of straw should be placed on top of the gravel before soil is backfilled.

When a seepage pit is used at the end of a tile field, the pit wall should be at least 6 ft from the end of the trench.   The pipe connecting the end of the line with the pit should have tight joints.

(H. E. Babbitt, "Plumbing," McGraw-Hill Book Company, New York; V. T. Manas, "National Plumbing Code Handbook," McGraw-Hill Book Company, New York; "Sewage Disposal for the Home," Part III, *Bulletin* 1, Department of Health, State of New York, Albany, N.Y.)

22-29. Chemical Toilets.   Chemical toilets are sometimes substituted for a pit privy when there is danger of polluting groundwater.   Normally, a chemical toilet is used only when required by the health authority having jurisdiction.

The superstructure of the chemical toilet is ordinarily the same as that provided for a pit privy. In place of the pit, a watertight corrosionproof container is provided.   It contains caustic and has a capacity of about 100 gal per seat.   The receptacle is charged with at least 25 lb of caustic per seat.   The chemical is dissolved in 10 gal of water per seat.   The container normally is provided with an agitator that can be operated after each use.

Chemical toilets require cleaning and recharging every 6 months of use.   A manhole must be provided outside the building, to provide access to the container for cleaning purposes.

22-30. Oxidation Ponds.   These are artificial lagoons of sewage utilizing natural forces for purification.   Properly designed and maintained, they provide satisfactory treatment even for raw sewage.   Effluents are at least equal to those from sewage-treatment plants providing complete treatment.

Oxidation ponds are suitable for use where large areas of land are available at low cost.   Successful operation, however, usually requires relatively high temperatures and sunshine.   Ponds nevertheless are in use in northern states.   When the water surface freezes, effluents are poor, but the ice prevents odors.

Sewage treatment in oxidation ponds depends on aerobic decomposition of organic matter (Art. 22-18).   Bacterial decomposition of this matter releases carbon dioxide.   Algae develop, consume carbon dioxide, ammonia, and other waste products, and under proper climatic conditions release oxygen during daylight.   Oxygen also is dissolved from the atmosphere at the lagoon surface.   Hence, a large ratio of surface area to volume of liquid is desirable.   Aeration, however, may be used to increase the supply of oxygen, which decreases substantially at night and in cold weather where algae are depended on heavily for oxygen.   Aerated ponds are not so susceptible to climatic conditions as ordinary lagoons.

Pond depths normally range from 2.5 to 4 ft. With greater depths, septic conditions may develop at the bottom. Shallower ponds permit vegetation to emerge. This encourages mosquito breeding and obstructs movement of the water, which is desirable for solution of atmospheric oxygen.

Size of pond required may be conservatively estimated for southern areas at 0.003 acre per capita for raw sewage and 0.002 acre per capita for sewage with primary treatment. Another basis for design is a strength–surface-loading relationship in which 50 lb BOD per acre per day is considered satisfactory. About double these areas are required in northern regions.

For large installations, two or more ponds may be operated in series or in parallel. With series operation, aerobic conditions in the first pond can be improved by returning some of the effluent from the second pond.

An acceptable location for the inlet to a pond is its center. Effluent can be discharged at a convenient point along a bank.

(E. W. Steel, "Water Supply and Sewerage," McGraw-Hill Book Company, New York.)

22-31. Chlorination of Sewage. The major purpose of chlorinating treated sewage is to destroy pathogenic organisms. **Chlorine demand** of sewage or industrial waste is the difference between the amount of chlorine added and the residual after a short time. This interval usually is taken as 15 min, since this is the time required to kill nearly all the objectionable bacteria. Sufficient chlorine should be added to sewage to satisfy the demand and provide a residual of 2 ppm (mg per liter). The contact period should be at least 15 min at peak hourly flow or maximum pumping rate.

The following dosages, ppm or mg per liter, may be required for disinfection of treated sewage: primary sedimentation effluent, 20 or more; trickling-filter-plant effluent, 15; activated-sludge-plant effluent, 8; and sand-filter effluent, 6. Such disinfection is desirable and often mandatory where discharge of the effluent may pollute water supplies, shellfish beds, or beaches.

The 5-day biochemical oxygen demand (BOD) of sewage is reduced about 2 ppm for each ppm of chlorine added. A BOD reduction of 15 to 35% may be expected with residuals of 0.2 to 0.5 ppm after 10 min.

Chlorinators usually are used to feed chlorine to the sewage. Chlorine gas normally is dissolved in water, and the solution is pumped into the sewage in measured amounts, proportional to the flow. In small plants and some large plants, hypochlorinators may be used. These may feed sodium hypochlorite (laundry bleach) or calcium hypochlorite.

Chlorination should be done in a baffled contact tank, unless there will be sufficiently long contact time in a conduit or outfall before the chlorinated effluent is discharged. The accuracy of the chemical feeders should be checked daily by determining the weight of chlorine or hypochlorites used. In addition, the efficacy of dosages applied should be checked frequently by bacteriological tests.

Chlorine also may be useful in preventing odors at sewage-treatment plants. For this purpose, it may be added to primary influent to prevent odors in secondary treatment. Less chlorine will be required than if it is added to the secondary influent. Chlorination before primary sedimentation is not detrimental to sludge digestion.

Other uses of chlorine include neutralization of hydrogen sulfide, or prevention of its formation, where it may corrode concrete sewerage or structures; increasing the efficiency of air in grease removal in skimming tanks; control of ponding and filter-fly larvae on trickling filters; conditioning of sludge before dewatering; and treatment of industrial wastes.

(E. W. Steel, "Water Supply and Sewerage," McGraw-Hill Book Company, New York; "Chlorination of Sewage and Industrial Wastes," Manual of Practice 4, Federation of Sewage and Industrial Wastes Association, Champaign, Ill.; G. W. White, "Handbook of Chlorination," Van Nostrand Reinhold Company, New York.)

22-32. Industrial Waste Treatment. The treatment of industrial wastes (see Art. 22-2) is highly specialized. Selection of treatment processes must be engineered to the peculiar characteristics of a process waste. It is desirable, whenever possible, to reduce the volume of waste water requiring treatment or to separate wastes requiring intensive treatment from those requiring little or no treatment. Cooling water, for example, can be segregated from high-strength wastes, thereby reducing the size of the treatment plant.

Process wastes have a wide range of flow from hour to hour, depending on the operation. Hence, it may be necessary to provide equalization or holding tanks to produce a more uniform flow to be treated over a 24-hr period. This is more efficient than treatment units designed to handle maximum flows produced during an 8-hr shift. It is also possible with equalization tanks to mix wastes of different characteristics, such as acids and alkalies, and obtain a neutralized waste. Peak solids production and BOD may also be reduced or regulated.

Table 22-12. Types and Characteristics of Industrial Wastes*

Type of waste	Unit	Volume, gal per unit	BOD, lb per unit	Suspended solids, lb per unit	Population equivalent per unit
Canning					
Corn products	Ton	12,000	19.5	30.0	186
Beans	Case no. 2 cans	35	200.0	60.0	0.35
Peaches	Ton	2,610	29.2	13.0	280
Tomatoes	Ton	227	8.4	2.9	82
Milk products					
General dairy	1,000 lb raw milk	340	570	540	10
Fermentation					
Brewing	1 barrel beer	204	1.2	0.6	12
Laundry	100 lb dry wash	400±	1,250±	500±	20–25
Roofing					
Paperboard	Ton	36,075	18.2	144.0	125
General slaughterhouse	1 animal	360	7.7	3.2	74
Paper mill					
Paperboard	Ton pulp	14,000	121		84
Textile					
Cotton sizing	1,000 lb goods processed	60.0			2
Basic dyeing	1,000 lb goods processed	18,000			90
Rayon viscose	1,000 lb product	140	110	9.6	800
Wool dyeing and scouring	1,000 lb product	240,000	125		1,500
Vegetable oils					
Acidulating waste	1 ton oil	385	1	0.5	10

*From E. B. Besselievre and M. Schwartz, "The Treatment of Industrial Wastes," 2d ed., McGraw-Hill Book Company, New York.

Industrial wastes may be placed in general classifications, such as food processing, textile and apparel manufacture, chemical manufacture, and basic materials manufacture, including pulp and paper, iron and steel, metal plating, oil processing, glass, plastic, and rubber production and processing. Table 22-12 indicates some of the characteristics of waste typical of the several classifications. When preparing for treatment of any specific waste, an engineer should see that the waste is sampled over a sufficient time period to include major variations introduced by process operation.

Treatment of process wastes may require a series of methods selected to accomplish certain degrees of treatment that would ultimately produce an effluent acceptable for discharge to a receiving stream. These methods may include, in order:

Pretreatment to reduce temperature, neutralize the wastes, and remove fibers and other coarse solids by screening.

Primary treatment to remove settleable solids.

Secondary treatment by biological processes applied to biodegradable wastes.

Secondary treatment with chemicals for chemical conversion, precipitation and removal of solids, and oxidation or reduction of substances contained in the waste.

Preconditioning or secondary treatment by anaerobic digestion to produce a biochemical conversion of substances.

Ion exchange, dialysis, or evaporation to remove inorganic solids.

Chlorination for oxidation or disinfection purposes.

Various forms of irrigation, lagooning, or algal oxidation ponds.

It is frequently necessary to select theoretically best combinations of treatment for a process waste and to follow up the selection with pilot-plant operations to establish the parameters of design for the full-scale treatment plant. Employment of advanced waste treatment methods may be necessary for specified purposes, such as removal of trace metals, control of phosphorus- and nitrogen-bearing compounds, and reduction of excessive amounts of suspended solids. Several methods of treatment are described in Arts. 22-20 to 22-23, 22-25, and 22-31.

Radioactive wastes are subject to severe restrictions when the receiving body of water may be used for human consumption, recreational bathing, fish propagation for food, or plant irrigation. Federal and state regulations should be reviewed whenever radioactive wastes are to be disposed of. Permissible concentrations of radioactive material in water are usually specified in microcuries per milliliter of water. Procedures that have been used for the treatment of radioactive wastes include concentration and storage, and dilution and disposal. Burial after required decay may follow the first and discharge to sewers or streams may follow the latter. Low-activity material may be diluted, while high-activity material, requiring long storage periods, may be safely enclosed in containers and buried or stored in isolated caves or other underground facilities.

Concentration of radioactive wastes before storage may be accomplished by coprecipitation. The radioactive sludge concentrate is then removed, packaged, and buried.

Evaporation is widely used for concentration of low-activity wastes. Condensate may be released to a sewer. The sludge is transferred to polyethylene-lined drums for burial. Cation exchange with synthetic resins may be used on small liquid volumes having low solids concentration and low radioactivity levels.

(N. L. Nemerow, "Liquid Wastes of Industry: Theories, Practices and Treatment," Addison-Wesley Publishing Company, Inc., Reading, Mass.; E. B. Besselievre and M. Schwartz, "Treatment of Industrial Wastes," McGraw-Hill Book Company, New York; R. L. Culp and G. L. Culp, "Advanced Wastewater Treatment," Van Nostrand Reinhold Company, New York.)

22-33. Sanitary Landfills. Refuse collected from households, commercial establishments, and industrial plants must be disposed of at minimum cost and without creating health hazards or nuisances. One solution is a sanitary landfill. This requires daily compaction of refuse and daily placement of an earth cover 6 to 12 in. thick. The cover is increased to 2 ft when filling has been completed. The method is suitable where low-cost land is available within convenient hauling distance of the contributing population and good soil is available for the earth cover.

Refuse comprises all solid wastes except body wastes. It may consist of garbage, ashes, rubbish, street cleanings, dead animals, abandoned automobiles, and solid market and industrial wastes. Garbage consists of putrescible wastes resulting from processing, handling, preparation, cooking, and consumption of food. Rubbish consists of solid wastes other than ashes, body wastes, and garbage from domestic, commercial, and institutional sources.

About 14 acre-ft, including cover, per 10,000 population per year of operation will be required for sanitary landfill. Sufficient land should be available to insure area for a preplanned period of 5 to 10 years. The area needed can be derived from an estimate of volume required computed from

$$V = \frac{R}{D} \left(1 - \frac{P}{100}\right) + C_v \tag{22-24a}$$

where V = volume, cu yd per capita per year, of sanitary landfill

R = weight of refuse, lb per capita per year, to be handled at landfill

D = average density of refuse, lb per cu yd

P = percentage reduction of refuse volume from compaction

C_v = volume, cu yd, of cover material required (6- to 12-in.-thick intermediate layers, temporary sides, front slope, and top, and at least 24 in. on all finished surfaces)

C_v varies from 17% of the refuse volume for deep fills to 33% for shallow fills. It may be assumed at 25% in estimates. For this value, the required volume of landfill may be estimated from

$$V = 1.25 \frac{R}{D} \left(1 - \frac{P}{100}\right) \tag{22-24b}$$

Drainage of the site before, during, and after filling should be planned in advance. Provision should be made for windbreaks to keep dust, paper, and other light objects from being blown away from dumping areas and becoming a nuisance. Also, the final disposition of the site should be planned in advance.

Parks, recreational areas, and outdoor storage are suitable end uses for landfills. Choice of end use should be influenced by the uncertain settlement characteristics of such fills and the objectionable odors that may be released where excavations are made. While a covered fill may be odorless, excavation may be hazardous and expensive because of the presence of obnoxious toxic and inflammable gases produced by decomposing refuse. Light buildings may be constructed over old sanitary landfills if the surface is topped with gravel, crushed stone, or slag to permit gases to escape to the atmosphere. Even settlement of such buildings may be achieved with mat footings. Buildings may be erected on piles driven through the fill when suitable protection against gas entrapment is provided.

Soil used for cover should not have a high proportion of sand or clay, or operation of trucks will be hindered. Clay also is difficult to handle, and when dry, it cracks, providing openings for rodents, insects, and air. A sand-clay-loam mixture with about 50% sand has been found satisfactory.

Decomposition in a landfill is anaerobic and proceeds slowly. Even after 25 years, some of the organic matter may be unchanged. Preferably, 10 to 15 years of decomposition should be allowed before erecting buildings on a sanitary landfill.

Sanitary landfills may be carried out by the trench or area methods. Both require that refuse be compacted and covered daily. In the trench method, soil for covering a compacted windrow of refuse is obtained by digging a trench 15 to 25 ft wide, 100 to 400 ft long, and at least 3 ft deep, next to the windrow. This trench, in turn, is filled with refuse and covered with soil from an adjoining trench. The refuse should be placed in the trenches in layers 1 to 2 ft thick and 8 to 10 ft wide and compacted. Final height may be 6 to 10 ft. Only enough length of windrow should be built up in this manner daily so that full height is reached and sides and top are covered with soil at the end of a workday. In the area method, applicable to swamps, marshes, and below-grade terrain, compacted refuse is built up in layers 6 to 10 ft deep and covered daily with soil brought in from elsewhere.

Landfills must meet a number of restrictive requirements imposed by state regulatory agencies. The specific requirements in each state should be ascertained by the engineer. Usually, the engineer is required to submit a plan and report on the specific areas to be filled, schedule of filling, site preparation, sources and types of materials to be used as cover, and sub-base. The plan should also include details on application of cover material; composition of waste; final grades; handling of surface water and fill drainage, including the method of collection and treatment of leachate to prevent groundwater or surface-water pollution; erosion control; nuisance control; air-pollution prevention measures; method of record keeping; and, in general, any data required to insure that environmental impact (Arts. 22-36 to 22-40) will not be adverse or unacceptable to the enforcement agency.

(D. J. Hagerty, J. L. Pavoni, and J. E. Heer, Jr., "Solid Waste Management," Van Nostrand Reinhold Company, New York; American Public Works Association Committee on Refuse Disposal, "Municipal Refuse Disposal," Public Administration Service, 1313 E. 60th St., Chicago, Ill.)

22-34. Incineration of Refuse. Where land is costly or unavailable for sanitary landfill, municipalities may resort to incineration for refuse disposal. Refuse comprises all solid wastes except body wastes. The material is not homogeneous, and its characteristics vary considerably. Fuel value may range from 600 to 6,500 Btu per lb of refuse, as fired. Moisture content influences this value significantly.

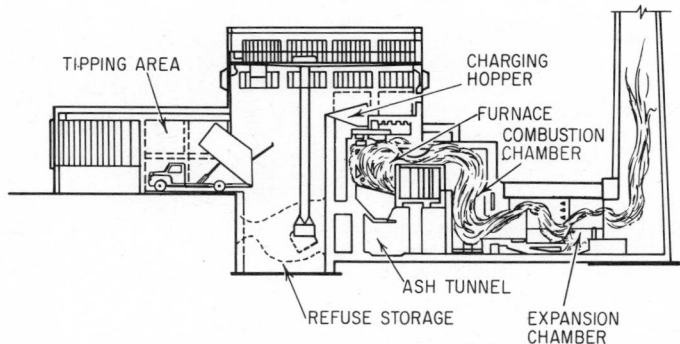

Fig. 22-20. Refuse incinerator.

In incineration, volatiles are driven off by destructive distillation. They ignite from the heat of a combustion chamber (Fig. 22-20). Gases produced pass through a series of oxidation changes in which time-temperature relationship is important. They must be heated above 1400° F to destroy odors. Combustion products ultimately discharge from a stack at 800° F or less, usually after passing through an expansion chamber, fly-ash collector, and wet scrubbers. Normally, only submicron- and the smaller micron-size particles should escape with the flue gases. Dust emission may be in the range of 2 to 3 lb per ton of refuse charged for a well-operated unit equipped with scrubbers.

Air needed ranges from 5 to 8 lb per lb of refuse burned. For nonhomogeneous wastes, up to 200% of theoretical requirements may be needed for combustion. The air provides oxygen for combustion of organic matter, helps dry wet refuse, mixes with organic gases, and cools the gases if too much dry material is being burned. Air should be passed over the refuse and through it from under the grates.

Incinerators generally are rated in accordance with the estimated weight of refuse they are capable of burning in 24 hr. Loading rates range up to slightly over 100 lb of refuse per hr per sq ft of grate area for incinerators with mechanical stoking. Small incinerators for apartment houses and institutions are loaded at much lower rates. Standards of the Incinerator Institute of America suggest loading rates for domestic refuse, lb per hr per sq ft, of 20 in 100-lb-per-hr burning units to 30 in 1,000-lb-per-hr burning units.

Several types of incinerators are available from manufacturers. Kiln shape may be round or rectangular. The kiln may be stationary or it may rotate about a horizontal axis. The hearth may be horizontal and fixed, with grates; traveling, with grates; multiple; step movement; or barrel-type rotary (Fig. 22-20). Some types have drying hearths. Feed may be continuous, stoker, gravity, or batch.

For rational design of incinerators, the engineer should know or estimate such characteristics of the refuse to be burned as weight, water content, percentage of combustible and inert material, and Btu content. Heat balance can be calculated from several estimates based on averages. Available heat from the refuse must be balanced against heat losses due to radiation, excess air, flue gas, and ash. Manufacturers of each type of incinerator recommend sizes for various conditions. Furnace volume may be approximated by allowing 20,000 Btu per cu ft, and grate area by allowing 300,000 Btu per sq ft. Secondary combustion chambers permit combustion to continue to completion. Volumes of such chambers range from 10 to 25 cu ft per ton of rated capacity. Expansion chambers and other air-cleaning devices remove fly ash and other particles carried out of the furnace by gases. Expansion chambers are desirable where a stack serves more than one furnace. Gas velocities in secondary chambers should not exceed 10 fps.

Stacks should be designed for gas velocities of about 25 fps with maximum air. As a rough approximation, 0.3 sq ft of stack area may be required per ton of rated capacity. Stack heights usually range from 100 to 180 ft. Height is desirable for creating natural draft and for dispersion of gases in the atmosphere.

(American Public Works Association, "Municipal Refuse Disposal," Public Administration Service, Chicago; R. C. Corey, "Principles and Practices of Incineration," John Wiley & Sons, Inc., New York.)

22-35. Air-Pollution Control.   Air pollution exists when one or more substances, such as dust, fumes, gas, mist, odor, smoke, or vapor, are present for a sufficient time in the atmosphere in quantities and with characteristics injurious to human, animal, or plant life or to property, or detrimental to comfortable enjoyment of life and property.

These pollutants derive from numerous sources. They may be roughly classified as natural, industrial, transportation, agricultural, commercial and domestic heat and power, municipal activities, and fallout.

Natural sources include water droplets or spray evaporation residues, windstorm dusts, meteoric dusts, surface detritus, and pollen from weeds.

Industrial sources include process waste discharges, ventilation products from local exhaust systems, and heat, power, and waste disposal by combustion processes.

Transportation sources include discharges from motor vehicles, rail-mounted vehicles, airplanes, and vessels.

Agricultural sources include applications of insecticides and pesticides and burning of vegetation.

Commercial and domestic heat and power sources include gas-, oil-, and coal-fired furnaces used to produce heat or power for dwellings, commercial establishments, and utilities.

Municipal activity sources include refuse disposal, liquid-waste disposal, road and street paving, and fuel-fired combustion operations.

Fallout comprises radioactive pollutants suspended in the air after a nuclear explosion. Since pollutants are contributed by many sources, air pollution is always present but in varying degrees. In effect, pollution from natural sources is a base line with which total pollution can be compared. The major problems of pollution are associated with community activity, rather than rural activity, because community air generally is more polluted.

Environment is made less desirable by pollutants. Hence, there is ample reason to conserve air as a resource, in many ways parallel to the need for conservation of water.

Air-pollution control requires knowledge of what constitutes an ideal atmosphere. This leads to establishment of criteria for clean air and standards setting limits on the permissible degree of pollution. Control also requires means for precise measurement of pollutants and practical methods for treating polluting sources to prevent undesirable emissions.

In addition to its adverse effects on health, air pollution also is objectionable because of its contribution to reduced visibility. In many parts of the world, burning of soft coal yields particles that combine with fog to produce smog, a mixture that at times reduces visibility to zero. Smog is created when microscopic water droplets condense about nucleating substances in the air to form aerosols. These are liquid or solid, submicron-size particles dispersed in a gaseous medium. In an atmosphere with an aerosol concentration of about 1 mg per cu m, visibility may be limited to 1,600 ft. There would be about 16,000 particles per milliliter restricting visibility by scattering light.

Coal is only one source of nucleating particles that are responsible for smog. Chemical conversion of reaction products in the air also produces nucleating substances that grow large enough to cause light scattering. Converted sulfur dioxide, too, becomes a nucleating substance as it oxidizes and hydrolizes to form sulfuric acid mist.

The most desirable means of controlling air pollution is to prevent contaminants from getting into the atmosphere. Complete elimination of air pollution, however, is not always practicable. But there are many means for reducing it. Sulfur dioxide release, for example, can be decreased by use of a fuel with low sulfur content. An industrial process with a gaseous effluent can be changed to eliminate the gaseous waste. Aerosols and particles can be removed from a gas stream by air-cleaning equipment.

Table 22-13. Approximate Sizes of Particles in Aerosols, Dusts, and Fumes

Type of Particle	Size Range, Microns
Tobacco smoke	0.01–0.2
Rosin smoke	0.01–1.1
Carbon black	0.01–0.3
Zinc oxide fumes	0.01–0.4
Magnesium oxide smoke	0.01–0.5
Metallurgical fumes	0.01–1.3
Viruses	0.01–0.05
Oil smoke	0.03–1.0
Pigments	0.09–8
Ammonium chloride fumes	0.1–1.4
Alkali fumes	0.1–1.6
Metallurgical dust	0.5–200
Sulfate mist	0.5–3
Spray-dried milk	1–10
Bacteria	1–12
Pulverized-coal fly ash	1–60
Fog	1–50
Sulfuric acid concentrator	1.1–11
Cement dust	5–200
Sulfide ore for flotation	8–300
Foundry dusts	8–1,000+
Stoker fly ash	10–900
Pulverized coal	10–500
Ground limestone	30–900
Mist	50–600

Sizes of substances to be eliminated (Table 22-13) are a major factor in selection of air-cleaning devices.   Coarse solids can be removed by screens.   Particles down to 10 microns in diameter can be settled out in settling chambers with expanding cross section for velocity reduction to under 10 fps.   Particles between 10 and 200 microns can be removed in cyclone separators, with an efficiency of 50 to 90%.   In this equipment, the gas to be cleaned is injected tangentially into a cylindrical chamber.   The gas spirals downward, then upward through the vortex at high velocity, and exits at the top.   Before the gas leaves, however, particles are centrifuged out, hit the side walls, and drop to the conical bottom of the chamber.

Particles 10 microns in diameter or smaller may be removed with filters made of cloth, metal, or glass fiber.   But air or gas velocities leaving such filters are low.   For dry fiber filters, efficiency may be only about 50%.   The efficiency of such filters, however, may be increased by application of a viscous coating, such as an oil with low volatility.   Filters made of cloth usually are tubular bags, which trap particles as air or gas passes through.   Many bags may be enclosed in a large chamber.   When loaded with dust, they are shaken, and the dust falls into a hopper.   Bag filters remove 99% of particles larger than 10 microns.   Filters packed with activated charcoal are used to adsorb gases.

Wet collectors or scrubbers (Figs. 22-21 and 22-22) remove particles 1 to 5 microns in size.   These devices also may remove water-soluble gases.   In a scrubber, the gas to be cleaned may pass through a countercurrent water flow.   The water may be sprayed or atomized.   The scrubber may have deflectors to improve mixing of the gas and water.   Chemicals may be added to the liquid to improve absorption.

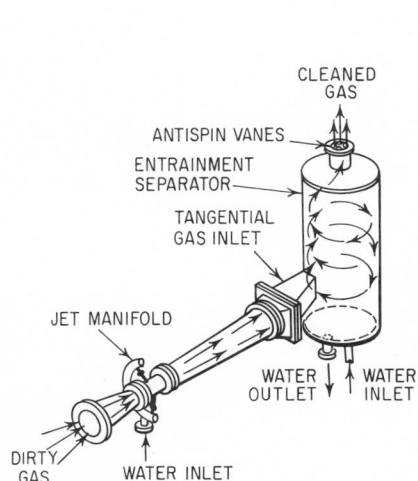

Fig. 22-21. Cyclonic spray scrubber. (*From W. L. Faith, "Air Pollution Control," John Wiley & Sons, Inc., New York.*)

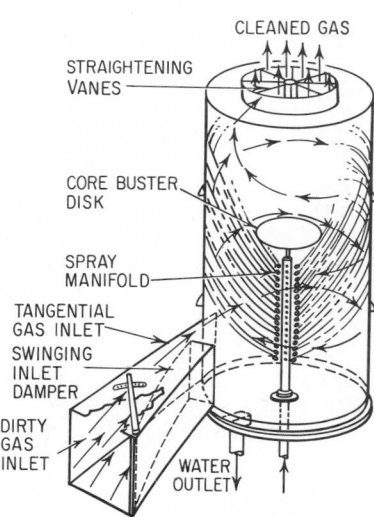

Fig. 22-22. Venturi scrubber. (*From W. L. Faith, "Air Pollution Control," John Wiley & Sons, Inc., New York.*)

Wet collectors often are used to clean air from kilns, roasters, and driers.   They also are used for processes producing fine dust, films, vapors, and mists in food, chemical, foundry, metal-working, and ceramic industries.   Scrubbers may be classified as dynamic precipitators, centrifugal collectors, orifice collectors, collectors with high-pressure nozzles, and packed towers.

In dynamic precipitators, dynamic or centrifugal forces, aided by water, clean the air.   In centrifugal collectors, centrifugal forces throw particles in the air against wetted collector surfaces.   After striking the surfaces, the particles fall to the bottom of the device and are removed.   Orifice collectors deliver large quantities of water to a collecting zone where dust is removed from the air by centrifugal force, impingement, or collision.   In collectors with high-pressure nozzles, air at 20,000 fpm or more and water under 250 psi or more jet through venturi tubes.

The water breaks into a fine mist, increasing the probability of contact with tiny particles.   The turbulence disperses the water, causes quick impact with dust in the air, and removes the particles.   In packed towers, dust particles are removed when air flows upward through the packing, which usually is in the form of ceramic saddles, while water flows downward.

Ionizable aerosols and particles down to 0.1 micron in size can be removed by electrostatic precipitators (Fig. 22-23), with an efficiency of 80 to 99%.   These devices ionize particles in a gas passing by high-voltage electrodes.   Oppositely charged plates trap the particles.   To rid the plates of the particles, the current to the plates is interrupted or the plates are rapped.

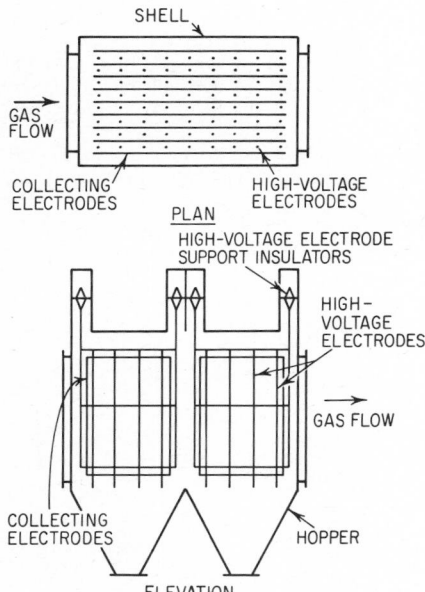

Fig. 22-23. Horizontal-flow precipitator.   (*From W. L. Faith, "Air Pollution Control," John Wiley & Sons, Inc., New York.*)

When pollutants cannot be completely eliminated at the source, air pollution may be reduced by keeping the concentration of the pollutants low by dispersing them.   Whether atmospheric dilution is a suitable solution depends on the meteorology of a region, local topography, and building configurations.   Basic meteorological conditions of the atmosphere that must be considered include wind speed, direction, and gustiness, and vertical temperature distribution.   Under some conditions, humidity also is important.   In general, diffusion theories predict that ground concentration of a gas or fine-particle effluent with very low subsidence velocity is inversely proportional to the mean wind speed.   Vertical temperature distribution determines the distance from a stack of given height at which maximum ground concentration occurs.   Raising the temperature of gas leaving a stack is equivalent to increasing stack height.

Gas does not normally come to the ground under inversion conditions.   It may accumulate aloft when the atmosphere is calm or nearly so and be brought down to the surface as the sun heats the ground in early morning.

Turbulence caused by buildings and topography usually is so complex that theoretical computation of the effect is impractical.   In some cases, however, model studies in wind tunnels have been used successfully to make predictions based on measurements of gas concentration and visible patterns of smoke.

The degree of air pollution at any time and place is determined by taking air samples and analyzing them.   Air-sampling methods may be classified as those sampling particles, inorganic metals and salts, inorganic gases, organic substances, and mixed miscellaneous substances.

Many automatic, recording, air-monitoring instruments are available.   They can be operated with few attendants and little manipulation.   It is generally necessary to calibrate automatic instruments against a standard wet chemical or physical measurement method.   Subsequent field calibration before, during, and after use may also be essential to maintain reliable test results.   Though there are many variations, particle-sampling devices generally use gravity or suction-type collection and pass the sample through thermal or electrostatic precipitators, impingers and impactors, cyclones, absorption and adsorption media, scrubbing apparatus, or filters of various materials, such as paper, glass, plastic, or cloth.

Several types of units with air pumps drawing air through paper tapes mounted on a spool are available.   The tape is moved automatically so that successive samples are taken for timed intervals on fresh paper.

In addition to standard wet chemical methods of measuring gases, there are many automatic or semiautomatic instruments designed to measure a spectrum (mass spectrometer) or one or more specific gases.   These include many different analytical principles, such as electrical conductivity; potentiometry; coulometry; flame ionization; thermal conductivity; heat of combustion; colorimetry; infrared, ultraviolet, and visible light photometry; gas chromatography; and electron capture.

Stack sampling requires special techniques and usually a train of sampling devices to measure particles and gases.

High-volume samplers are used at many sampling network stations in the United States. Electron microscopes may be used to examine aerosols and submicron particles.   Photoelectric meters are used to control alarm systems connected to stacks.   Combination instruments may be used for general sampling and location of emission sources.   Such devices measure wind direction and velocity and direct air samples into multiple sample units, each representing a wind-direction sector.

(P. L. Magill, F. R. Holden, and C. Ackley, "Air Pollution Handbook," McGraw-Hill Book Company, New York; M. B. Jacobs, "Chemical Analysis of Air Pollutants," John Wiley & Sons, Inc., New York; W. L. Faith and Arthur A. Atkisson, Jr., "Air Pollution," Wiley-Interscience, New York; Arthur C. Stern, ed., "Air Pollution," 3 vols., Academic Press, New York; Christian E. Junge, "Air Chemistry and Radioactivity," Academic Press, New York; D. L. Benchley, C. D. Turley, and R. F. Yarnac, "Industrial Source Sampling," Ann Arbor Science, Ann Arbor, Mich.; "Air Pollution Manual, Part II, Control Equipment," American Industrial Hygiene Association, 66 South Miller Road, Akron, Ohio 44313; J. A. Danielson, ed., "Air Pollution Engineering Manual," AP-40, Environmental Protection Agency, Office of Air Quality Planning and Standards, Research Triangle Park, N.C. 27711; "Air Sampling Instruments for Evaluation of Atmospheric Contaminants," American Conference of Governmental Industrial Hygienists, P.O. Box 1937, Cincinnati, Ohio 45201.)

22-36. Laws Concerned with Environmental Impact.   Consideration of the effect of sanitary works on and for the benefit of society has been part of the concept, planning, and design of sanitary works since the early history of sanitary engineering.

A definition of the term *sanitary engineer*, originally adopted by the Committee on Sanitary Engineering of the National Research Council in 1943, was revised by the Committee on Sanitary Engineering and Environment of the National Academy of Sciences-NRC in 1954.   The revised definition contains the following preamble:

"The professional occupational title *Sanitary Engineer* shall apply to a graduate of a full four-year, or longer, course leading to a Bachelor's or higher, degree at an educational institution of recognized standing with major study in engineering, who has fitted himself by suitable specialized training, study, and experience (a) to conceive, design, appraise, direct, and manage engineering works and projects developed, as a whole or in part, for the protection and promotion of the public health, particularly as it relates to the improvement of man's environment, and (b) to investigate and correct engineering works and other projects that are capable of injury to the public health by being or becoming faulty in conception, design, direction, or management."

The concept of that definition has been formulated into national policy and law.   The 1969 National Environmental Policy Act (NEPA) (P.L. 91-190 42 U.S.C. 4321 *et seq.*) was authorized in 1970 by Executive Order 11514 (35 FR 4247).   The President's order made preparation of environmental impact statements incumbent on Federal agencies, departments, and establishments in connection with proposals for legislative and other major Federal activities significantly affecting the quality of human environment.   A Council on Environmental Quality (CEQ) was appointed and prepared guidelines and specific policies to be implemented.   These appear in the Code of Federal Regulations Title 40, Chapter V, Part 1500, titled "Preparation of Environmental Impact Statements: Guidelines."

There followed in the next years a series of papers, brochures, etc., designed to interpret procedure for evaluating environmental impact.   One of the early descriptions of that procedure was published as Geological Circular No. 645 in 1971, "A Procedure for Evaluating Environmental Impact." In 1974, the U.S. Environmental Protection Agency developed a brief explanation of environmental impact statements that is succinct and to the point.

Since the requirement for filing environmental impact statements includes works that are financed in whole or in part by Federal or state agencies, there followed the immediate development of state regulations on environmental impact assessment. As of 1975, 28 states plus Puerto Rico had established requirements governing proposals for environmental impact statements, and many of the major counties, towns, and cities followed the practice.

For example, New York State's authority is contained in Part 615, Chapter VI, General Regulations, Title 6, Environmental Conservation. That state requires the filing of a written evaluation by a permit applicant, and the environmental impact assessment so filed must provide a description of the proposed project or development, and a detailed analysis of its environmental effects. Among other requirements of Part 615, permit areas requiring environmental-impact assessment statements include air-contamination-source construction, public water-supply approval, stream protection, municipal waste-disposal system construction, industrial waste-disposal system construction and, under specified conditions, future well installations on Long Island.

Since state or local guidelines may contain requirements in addition to those established by CEQ, engineers are advised to examine in detail the requirements in each state or local community for the filing of environmental impact statements.

22-37. When Impact Statements Are Required.   Actions for which agencies are required to prepare impact statements must be "major" and "environmentally significant." It is essential that a draft statement be prepared, as early as possible, by the project engineer or other appropriate authorized person for review and comment. The actions may include all or any of the following:

1. Agency recommendations on their own proposals for legislation.
2. Agency reports on legislation initiated elsewhere but concerning subject matter for which the agency has primary responsibility.
3. Projects and continuing activities that may be
    a. undertaken directly by an agency;
    b. supported in whole or in part through Federal contracts, grants, subsidies, loans or other forms of funding assistance; or
    c. part of a Federal lease, permit, license, certificate or other entitlement for use.
4. Decisions of policy, regulation, and proceduremaking.

Although it is possible that there can be exceptions, the following actions are generally considered major or environmentally significant:

1. Actions whose impact is significant and highly controversial on environmental grounds.
2. Actions that are precedents for much larger actions that may have considerable environmental impact.
3. Actions that are decisions in principle about major future courses of action.
4. Actions that are major because of the involvement of several Federal agencies, even though a particular agency's individual action is not major.
5. Actions whose impact includes environmentally beneficial as well as environmentally detrimental effects.

22-38. Contents of Environmental Impact Statements.   Environmental impact statements must assess in detail the potential environmental impact of a proposed action. The purpose of the statement is to disclose the environmental consequences of a proposed action. That disclosure is designed to alert the agency decision maker (local, state, or Federal, or any combination of these), the public, and, perhaps on major works, Congress and the President to environmental risks involved.

Environmental impact statements should present:

1. A detailed description of the proposed action, including information and technical data adequate to permit a careful assessment of environmental impact.
2. Discussion of the probable impact on the environment, including any impact on ecological systems and any direct or indirect consequences that may result from the action.
3. Any adverse environmental effects that cannot be avoided.
4. Alternatives to the proposed action that might avoid some or all of the adverse environmental effects, including analysis of costs and environmental impacts of these alternatives.
5. An assessment of the cumulative, long-term effects of the proposed action, including its relationship to short-term use of the environment versus the environment's long-term productivity.

6. Any irreversible or irretrievable commitment of resources that might result from the action or that would curtail beneficial use of the environment.

When the final statement is prepared it must also include any discussions, objections, or comments presented by Federal, state, and local agencies, private organizations, and individuals that addressed the subject during review of the draft statement.

22-39. Impact Statement Review.   In general, any Federal, state, or local agency that has jurisdiction by law or specific expertise with respect to any environmental impact involved must be consulted for comments.   The CEQ list of agencies to be consulted includes those having responsibilities for the following (state or local agencies may have additional agency review requirements):

Air quality and air pollution control
Weather modification
Environmental aspects of electric energy generation and transmission
Toxic materials
Pesticides
Herbicides
Transportation and handling of hazardous materials
Coastal areas: wetlands, estuaries, waterfowl refuges, and beaches
Historic and archeological sites
Flood plains and watersheds
Mineral land reclamation
Parks, forests, and outdoor recreational areas
Soil and plant life, sedimentation, erosion, and hydrologic conditions
Noise control and abatement
Chemical contamination of food products
Food additives and food sanitation
Microbiological contamination
Radiation and radiological health
Sanitation and waste systems
Shellfish sanitation
Transportation and air quality
Transportation and water quality
Congestion in urban areas, housing and building displacement
Environmental effects with special impact on low-income neighborhoods
Rodent control
Urban planning
Water quality and water pollution control
Marine pollution
River and canal regulation and stream channelization
Wildlife

In environmental engineering areas of activity, the principal government agency having responsibilities for reviewing impact statements is the Environmental Protection Agency, although many others (see NEPA Regulation 8, Appendix II) may also be involved.   As a matter of fact, any Federal agency having a jurisdiction that centers around air and water pollution, drinking water supplies, solid waste, pesticides, radiation and noise may be involved.   Hence, engineers should ascertain specifically any agencies in addition to EPA that may have review responsibilities.

Engineers should also determine to what extent the state agencies dealing with the above areas have jurisdiction.   In addition, engineers should check with the appropriate regional and municipal planning agencies.   (See also Sec. 14.)

22-40. How to Prepare an Impact Report.   There are several alternate formats of report that would contain all of the pertinent information required under the Federal guidelines.   One method that has had technical acceptance is the base matrix, in which a series of actions that are part of a proposed project are related to the characteristics and conditions of the environment that are affected.   Under each of the actions proposed, a ranking from 1 to 10 is placed to indicate impact magnitude, 10 being the highest order.   Correspondingly, under a diagonal in the box, a ranking from 1 to 10 can be inserted concerning the importance of a specific impact as related to an environmental condition.   Any suitable form of text that will discuss the significance of these two interrelated indices should be acceptable.   A sample matrix illustrating these points is shown in Fig. 22-24.

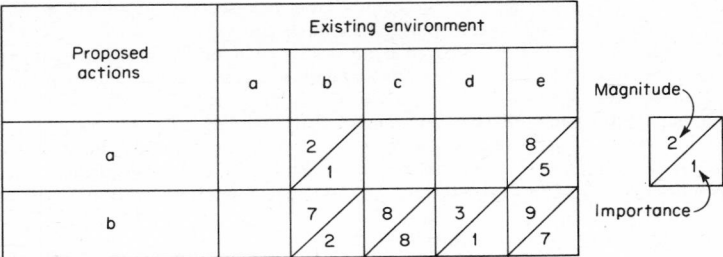

Fig. 22-24. Matrix used to demonstrate environmental impact of proposed actions.

One of the more complete diagrams for an information matrix was prepared by the United States Geological Survey in 1971. It appears as a separate attachment in Geological Survey Circular No. 645. The basis for the preparation of this matrix is as follows:

Left margin: Existing Characteristics and Conditions of the Environment

Top Margin: Proposed Actions that May Cause Environmental Impact

Existing Environment. Subdivision of *conditions* is listed as follows:

A. Physical and Chemical Characteristics
  1. Earth
     a. Mineral resources
     b. Construction material
     c. Soils
     d. Land form
     e. Force fields and background radiation
     f. Unique physical features
  2. Water
     a. Surface
     b. Ocean
     c. Underground
     d. Quality
     e. Temperature
     f. Recharge
     g. Snow, ice, and permafrost
  3. Atmosphere
     a. Quality (gases, particulates)
     b. Climate (micro, macro)
     c. Temperature
  4. Processes
     a. Floods
     b. Erosion
     c. Deposition (sedimentation, precipitation)
     d. Solution
     e. Sorption (ion exchange, complexing)
     f. Compaction and settling
     g. Stability (slides, slumps)
     h. Stress-strain (earthquake)
     i. Air movements
B. Biological Conditions
  1. Flora
     a. Trees
     b. Shrubs
     c. Grass
     d. Crops
     e. Microflora
     f. Aquatic plants
     g. Endangered species
     h. Barriers
     i. Corridors

2. Fauna
   a. Birds
   b. Land animals including reptiles
   c. Fish and shellfish
   d. Benthic organisms
   e. Insects
   f. Microfauna
   g. Endangered species
   h. Barriers
   i. Corridors
C. Cultural Factors
   1. Land Use
      a. Wilderness and open spaces
      b. Wetlands
      c. Forestry
      d. Grazing
      e. Agriculture
      f. Residential
      g. Commercial
      h. Industrial
      i. Mining and quarrying
   2. Recreation
      a. Hunting
      b. Fishing
      c. Boating
      d. Swimming
      e. Camping and hiking
      f. Picnicking
      g. Resorts
   3. Aesthetics and Human Interest
      a. Scenic views and vistas
      b. Wilderness qualities
      c. Open space qualities
      d. Landscape design
      e. Unique physical features
      f. Parks and reserves
      g. Monuments
      h. Rare and unique species or ecosystems
      i. Historical or archaeological sites and objects
      j. Presence of misfits
   4. Cultural Status
      a. Cultural patterns (life style)
      b. Health and safety
      c. Employment
      d. Population density
   5. Man-Made Facilities and Activities
      a. Structures
      b. Transportation network (movement, access)
      c. Utility networks
      d. Waste disposal
      e. Barriers
      f. Corridors
D. Ecological Relationships Such As:
      a. Salinization of water resources
      b. Eutrophication
      c. Disease-insect vectors
      d. Food chains
      e. Salinization of surficial material
      f. Brush encroachment
      g. Other
E. Others

Proposed Actions.   Subdivision of *proposed actions* is listed as follows:

A. Modification of Regime
   a. Exotic flora or fauna introduction
   b. Biological controls
   c. Modification of habitat
   d. Alteration of ground cover
   e. Alteration of ground water hydrology
   f. Alteration of drainage
   g. River control and flow modification
   h. Canalization
   i. Irrigation
   j. Weather modification
   k. Burning
   l. Surface or paving
   m. Noise and vibration

B. Land Transformation and Construction
   a. Urbanization
   b. Industrial sites and buildings
   c. Airports
   d. Highways and bridges
   e. Roads and trails
   f. Railroads
   g. Cables and lifts
   h. Transmission lines, pipelines, and corridors
   i. Barriers including fencing
   j. Channel dredging and straightening
   k. Channel revetments
   l. Canals
   m. Dams and impoundments
   n. Piers, seawalls, marinas, and sea terminals
   o. Offshore structures
   p. Recreational structures
   q. Blasting and drilling
   r. Cut and fill
   s. Tunnels and underground structures

C. Resource Extraction
   a. Blasting and drilling
   b. Surface excavation
   c. Subsurface excavation and retorting
   d. Well drilling and fluid removal
   e. Dredging
   f. Clear cutting and other lumbering
   g. Commercial fishing and hunting

D. Processing
   a. Farming
   b. Ranching and grazing
   c. Feed lots
   d. Dairying
   e. Energy generation
   f. Mineral processing
   g. Metallurgical industry
   h. Chemical industry
   i. Textile industry
   j. Automobile and aircraft
   k. Oil refining
   l. Food
   m. Lumbering
   n. Pulp and paper
   o. Product storage

E. Land Alteration
    a. Erosion control and terracing
    b. Mine sealing and waste control
    c. Strip mining rehabilitation
    d. Landscaping
    e. Harbor dredging
    f. Marsh fill and drainage

F. Resource Renewal
    a. Reforestation
    b. Wildlife stocking and management
    c. Ground water recharge
    d. Fertilization application
    e. Waste recycling

G. Changes in Traffic
    a. Railway
    b. Automobile
    c. Trucking
    d. Shipping
    e. Aircraft
    f. River and canal traffic
    g. Pleasure boating
    h. Trails
    i. Cables and lifts
    j. Communication
    k. Pipeline

H. Waste Emplacement and Treatment
    a. Ocean dumping
    b. Landfill
    c. Emplacement of tailings, spoil and overburden
    d. Underground storage
    e. Junk disposal
    f. Oil well flooding
    g. Deep well emplacement
    h. Cooling water discharge
    i. Municipal waste discharge including spray irrigation
    j. Liquid effluent discharge
    k. Stabilization and oxidation ponds
    l. Septic tanks, commercial and domestic
    m. Stack and exhaust emission
    n. Spent lubricants

I. Chemical Treatment
    a. Fertilization
    b. Chemical deicing of highways, etc.
    c. Chemical stabilization of soil
    d. Weed control
    e. Insect control (pesticides)

J. Accidents
    a. Explosions
    b. Spills and leaks
    c. Operational failure

K. Others

In dealing with any particular project, the engineer can select from this matrix on either margin those conditions and those actions that are applicable to the project. It is then possible for the engineer to prepare the assessment implied under Arts. 22-36 and 22-39 so that reviewing agencies can provide comments in an orderly fashion.   Within the format, it is important to present both present conditions and current trends, the alternate action proposed, and the impact either favorable or unfavorable that will result with and without the proposed action.   If unavoidable harm may result from the proposed action, the procedures for reducing the harmful effect, together with the ultimate benefits resulting even though some harm may be done, should be presented in full detail with objective substantiation of all statements.

It is very important that engineers present the multiphasic effects of the proposed project on air, water, and land characteristics, the biota, and on man-made structures, if any. It is also important to relate, in the discussion of impact, environmental interests related to recreation, education, science, history, and culture, as well as to overall community well-being. Health and safety considerations, both within the project and in any exterior community relationship, must be discussed.

As required by the Environmental Protection Agency, the report must assess:

1. The probable impact of the action.

2. The adverse environmental effect should the project be implemented.

3. The alternatives.

4. The relationship between the local short-term effect on environment, and maintenance of or increased benefit to the environment over the long term.

5. The commitments of resources that might be considered irreversible if the proposed action should take place.

22-41. Environmental Impact References.

Burchell, R. W., and D. Listokin, "The Environmental Impact Handbook," The Center for Urban Policy Research, Rutgers University, New Brunswick, N.J.; Camp, T. R., and R. L. Meserve, "Water and Its Impurities," Dowden, Hutchinson Ross, Inc., Stroudsburg, Pa.; Chanlett, E. T., "Environmental Protection," McGraw-Hill Book Company, New York, N.Y.; Gould, R. F., Ed., "Equilibrium Concepts in Natural Water Systems," Advances in Chemistry Series 67, American Chemical Society, Washington, D.C.; Hagerty, D. J., J. L. Pavoni, and J. E. Heer, Jr., "Solid Waste Management," Van Nostrand Reinhold Company, New York, N.Y.; Marquis, R. W., Ed., "Environmental Improvement (Air, Water and Soil)," U.S. Department of Agriculture, Washington, D.C.; National Academy of Sciences, "Principles for Evaluating Chemicals in the Environment," Washington, D.C.; National Research Council, "Waste Management and Control," Washington, D.C.; Public Administration Service, "Municipal Refuse Disposal," Chicago, Ill.; Seneca, J. J., and M.K. Taussig, "Environmental Economics," Prentice-Hall, Inc., Englewood Cliffs, N.J.; Sittig, M., "Environmental Sources and Emissions Handbook," Noyes Data Corp., Park Ridge, N.J.; Solnit, A., "The Job of the Planning Commissioner," University of California, Berkeley, Calif.; Vesilind, P. A., "Treatment and Disposal of Wastewater Sludges," Ann Arbor Science Publishers, Inc., Ann Arbor, Mich.

Harbor Engineering

ALONZO DeF. QUINN*

Consulting Engineer, Centerport, N.Y.

Harbor engineering involves planning and design of facilities for ships to discharge or receive cargo and passengers. These facilities include not only the harbor but its protection, if necessary, in the form of breakwaters; offshore moorings, marinas, and structures within the harbor for mooring the ships; port buildings for carrying on the commerce of the port; general cargo and bulk-cargo handling facilities; and many supplemental services.

23-1. Tides. The tide is the periodic rise and fall of ocean waters produced by the attraction of the moon and sun. Generally, the average interval between successive high tides is 12 hr 25 min, half the time between successive passages of the moon across a given meridian. The moon exerts a greater influence on the tides than the sun. This influence varies directly as the mass and inversely as the cube of the distance, and therefore the ratio is about 7:3.

The highest tides, which occur at intervals of half a lunar month, are called *spring tides*. They occur at or near the time when the moon is new or full, i.e., when the sun, moon, and earth fall in line, and the tide-generating forces of the moon and sun are additive. When the lines connecting the earth with the sun and the moon form a right angle, i.e., when the moon is in its quarters, then the actions of the moon and sun are subtractive, and the lowest tides of the month, the *neap tides*, occur.

Tidal waves are retarded by frictional forces as the earth revolves daily around its axis, and the tide tends to follow the direction of the moon. Thus, the highest tide for each location is not coincident with conjunction and opposition but occurs at some constant time after new and full moon. This interval, known as the *age* of the tide, may amount to as much as 2½ days.

Large differences in tidal range occur at different locations along the ocean coast. They arise because of secondary tidal waves set up by the primary tidal wave or mass of water moving around the earth. These movements are also influenced by the depth of shoaling water and configuration of the coast. The highest tides in the world occur in the Bay of Fundy, where a rise of 100 ft has been recorded. Inland and landlocked seas, such as the Mediterranean and the Baltic, have less than 1 ft of tide, and the Great Lakes are not noticeably influenced.

*With minor revisions for the second edition by Frederick S. Merritt, Consulting Engineer.

Tides that occur twice each lunar day are called *semidiurnal tides*. Since the lunar day, or time it takes the moon to make a complete revolution around the earth, is about 50 min longer than the solar day, the corresponding high tide on successive days is about 50 min later. In some places, such as Pensacola, Fla., only one high tide a day occurs. These tides are called *diurnal tides*. If one of the two daily high tides is incomplete, i.e., if it does not reach the height of the previous tide, as at San Francisco, then the tides are referred to as *mixed diurnal tides*.

Table 23-1. Mean and Spring Tidal Ranges for Some of the Major Ports of the World＊

	Mean range, ft	Spring range, ft
Anchorage, Alaska	26.7	29.6†
Antwerp, Belgium	15.7	17.8
Auckland, New Zealand	8.0	9.2
Baltimore, Md	1.1	1.3
Bilboa, Spain	9.0	11.8
Bombay, India	8.7	11.8
Boston, Mass	9.5	11.0
Buenos Aires, Argentina	2.2	2.4
Burntcoat Head, Nova Scotia (Bay of Fundy)	41.6	47.5
Canal Zone, Atlantic side	0.7	1.1†
Canal Zone, Pacific side	12.6	16.4
Capetown, Union of South Africa	3.8	5.2
Cherbourg, France	13.0	18.0
Dakar, Africa	3.3	4.4
Dover, England	14.5	18.6
Galveston, Tex	1.0	1.4†
Genoa, Italy	0.6	0.8
Gibraltar, Spain	2.3	3.1
Hamburg, Germany	7.6	8.1
Havana, Cuba	1.0	1.2
Hong Kong, China	3.1	5.3†
Honolulu, Hawaii	1.2	1.9†
Juneau, Alaska	14.0	16.6†
La Guaira, Venezuela		1.0†
Lisbon, Portugal	8.4	10.8
Liverpool, England	21.2	27.1
Manila, Philippines		3.3†
Marseilles, France	0.4	0.6
Melbourne, Australia	1.7	1.9
Murmansk, U.S.S.R.	7.9	9.9
New York, N.Y.	4.4	5.3
Osaka, Japan	2.5	3.3
Oslo, Norway	1.0	1.1
Quebec, Canada	13.7	15.5
Rangoon, Burma	13.4	17.0
Reikjavik, Iceland	9.2	12.5
Rio de Janeiro, Brazil	2.5	3.5
Rotterdam, Netherlands	5.0	5.4
San Diego, Calif	4.2	5.8†
San Francisco, Calif	4.0	5.7†
San Juan, Puerto Rico	1.1	1.3
Seattle, Wash	7.6	11.3†
Shanghai, China	6.7	8.9
Singapore, Malaya	5.6	7.4
Southampton, England	10.0	13.6
Sydney, Australia	3.6	4.5
Valparaiso, Chile	3.0	3.9
Vladivostok, U.S.S.R.	0.6	0.7
Yokohama, Japan	3.5	4.7
Zanzibar, Africa	8.8	12.4

＊Tide Tables. U.S. Coast and Geodetic Survey.
†Diurnal range.

There are other exceptional tidal phenomena.   For instance at Southampton, England, there are four daily high waters, occurring in pairs, separated by a short interval.   At Portsmouth, there are two sets of three tidal peaks per day.   *Tidal bores,* a regular occurrence at certain locations, are high-crested waves caused by the rush of flood tide up a river, as in the Amazon, or by the meeting of tides, as in the Bay of Fundy.

Tide tables have been published for most parts of the world.   Admiralty Tide Tables cover major ports in the United Kingdom and elsewhere, while the U.S. Coast and Geodetic Survey lists the tides for the major harbors in the United States and other parts of the world.   Table 23-1 gives the spring and mean tidal ranges for some major ports.

The rise of the tide is referred to some established datum of the charts, which varies in different parts of the world.   The British Admiralty charts use the level of mean low-water springs; in the United States, it is mean low water; in France and Spain, it is the lowest low water.

Mean high water is the average of the high water over a 19-year period, and *mean low water* is the average of the low water over a 19-year period.   *Higher high water* is the higher of the two high waters of any diurnal tidal day, and *lower low water* is the lower of the two low waters of any diurnal tidal day.   *Mean higher high water* is the average height of the higher high water over a 19-year period, and *mean lower low water* is the average height of the lower low waters over a 19-year period.   *Highest high water* and *lowest low water* are the highest and lowest, respectively, of the spring tides of record.   *Mean range* is the height of mean high water above mean low water.   The mean of this height is generally referred to as *mean sea* level.   *Diurnal range* is the difference in height between the mean higher high water and the mean lower low water.

WAVES

Water waves may be caused by certain artificial disturbances, such as moving vessels or explosions; or they may be caused by earthquakes, tides, or winds.   Wind produces the waves that have the most influence on the design of harbors and marine structures.

23-2. Waveform and Wave Generation.   Waves manifest themselves by curved undulations of the surface of the water.   They occur at periodic intervals, except for waves of translation and solitary waves or single waves of translation without any depression below still-water level.

Wave disturbance is felt to a considerable depth.   Therefore, depth of water has an effect on the character of the wave.   Deep-water waves are those that occur in water having a depth d greater than one-half the wave length L ($d > L/2$).   At such depths, the bottom does not have any significant influence on the motion of the water particles.   Shallow-water waves are those that occur in water having a depth less than one-half the wave length ($d < L/2$).   The influence of the bottom changes the form of orbital motion from circular to elliptical or near-elliptical.

Waves break when the forward velocity of the crest particles exceeds the velocity of propagation of the wave itself.   In deep water, this normally occurs when the wave height H (Fig. 23-1) exceeds one-seventh the wave length.   When the wave reaches shallow water where the depth

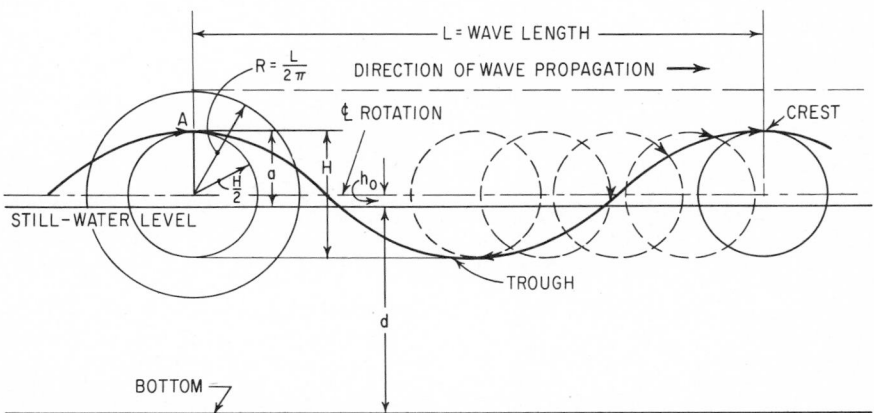

Fig. 23-1.   Particle at surface of wave in deep water rotates in a circle.

is equal to about $1.25H$, it will usually break. It may break in somewhat deeper water, however, depending on the strength of the wind and the condition of the bottom.

An unbroken wave is a wave of oscillation. Even after breaking in deep water, such a wave will usually re-form into an oscillatory wave of reduced height. It is only when it reaches shallow water and breaks without being able to re-form that it becomes a wave of translation, a familiar sight in the form of breakers along the shore. The only pure wave of translation is the solitary wave, a single crest of water above still-water level traveling without change of form at a constant speed, with a net displacement of water in the direction of wave travel. It is further characterized by its independence from wave length.

Figure 23-1 shows the oscillatory wave form and its characteristics. In deep water, each particle of water on the wave surface describes a circle. The radius of the circle is one-half the wave height about its normal center, midway between the crest and trough of the wave. The center of rotation is elevated above still-water level by the height h_o because the crest is at a greater distance above still water than the trough is below it (see also Art. 23-4). The difference depends on wave steepness. For a very steep wave, the proportion is about two-thirds above and one-third below still-water level.

At any instant, the arrows on the circular paths indicated by dotted circles in Fig. 23-1 show the relative position of the particles and their direction of motion in the formation of the wave. The heavy line connecting the arrows is the waveform at the surface of the water at that particular instant. The length between two consecutive crests is the wavelength L. And the height between the trough and the crest is the wave height or amplitude H. The waveform travels over the water surface, and the time for two consecutive crests to pass a point is the wave period T. The speed of the waveform v is called the wave velocity, or velocity of wave propagation. These various characteristics are related by the following equations:

$$v = \frac{L}{T} = \sqrt{\frac{gL}{2\pi}} = \frac{gT}{2\pi} \tag{23-1}$$

$$L = \frac{2\pi v^2}{g} = \frac{gT^2}{2\pi} \tag{23-2}$$

$$T = \sqrt{\frac{2\pi L}{g}} = \frac{2\pi v}{g} \tag{23-3}$$

where v = velocity of propagation of wave, fps
L = wavelength (distance between consecutive wave crests), ft
T = wave period (time for wave to travel L ft), sec
g = acceleration due to gravity, 32.2 ft per sec^2

If one characteristic is known, the others can be computed. Substitution of numerical values for π and g yields

$$v = 2.26 \sqrt{L} = 5.12T \tag{23-4}$$

$$L = 0.195v^2 = 5.12T^2 \tag{23-5}$$

$$T = 0.442 \sqrt{L} = 0.195v \tag{23-6}$$

Wave motion of water particles in shallow water ($d < L/2$), affected by the sea bottom, becomes nearly elliptical, the major axis being horizontal. The ratio of the major to the minor axis becomes greater with increased depth, until near the bottom the orbital motion is almost entirely horizontal. Unlike the circular orbital motion in deep water, in which the horizontal and vertical velocities are equal, the orbital velocity in shallow water is greater in a horizontal than in a vertical direction. The velocity of wave propagation v in shallow water is obtained by multiplying the velocity in deep water ($v = 2.26 \sqrt{L}$) by c, where $c = \sqrt{b/a}$, and b/a is the ratio of the vertical to the horizontal axis of the elliptical orbital motion, as given in Table 23-2. It should be noted that when the depth equals $L/2$, the ratio is close to unity, and the velocity is substantially that of a deep-water wave.

Whereas the motion of deep- and shallow-water waves maintains a certain degree of symmetry throughout the depth, breaking waves lose this characteristic. They are recognized by the rapid forward motion of the crest, while the lower part of the wave slowly moves in the opposite direction, there being a long, flat trough of relatively quiet water in back of the breaking crest. Unlike the wave travel, in form only, of a deep- or shallow-water wave, the breaking wave near

Table 23-2. Coefficients c for Velocity of Wave Propagation and μ for Orbital Velocity in Shallow Water*

d/L	c	μ	d/L	c	μ
0.05	0.552	1.814	0.30	0.977	1.023
0.10	0.746	1.340	0.35	0.988	1.013
0.15	0.858	1.165	0.40	0.994	1.007
0.20	0.922	1.085	0.45	0.997	1.004
0.25	0.958	1.044	0.50	0.998	1.002

*Multiply deep-water velocities by coefficients to obtain shallow-water velocities.

shore causes a net forward displacement of a mass of water, since the orbital velocity at the surface has exceeded the velocity of wave travel.

23-3. Forecasting Wave Height and Length.   The size of a wave for a particular location depends on the velocity, duration, and direction of the wind, the greatest distance over which the wind can act, and the depth of the water.

In determining the wave to be used in design of a structure at a particular location, only in exceptional cases will the designer be able to rely on a complete set of observations stretching over a sufficiently long period of time.   General observations have been made, and the order of maximum significant wave height is generally known for such bodies of water as the Atlantic Ocean (45 ft), the Pacific Ocean (60 ft), the Mediterranean Sea (20 ft), the Black Sea (30 ft), the Great Lakes (25 ft), the Gulf of Mexico (40 ft), and Lake Maracaibo (10 ft).   It is interesting to note that great expanses of water, such as oceans, do not necessarily produce waves of proportionally greater height than much smaller bodies, such as the Gulf of Mexico or the Great Lakes.   Ocean storms are generally more or less local, and the very long distance to land is not a controlling factor.   Their effects are sometimes felt long distances away in the form of *swells*, which are waves generated by storms occurring outside the area of observation.

Fetch is the horizontal extension, or width, of a storm's generating area for waves.   Thomas Stevenson, in 1864, established the first formulas for the relationship of fetch F, nautical miles, and wave height H, ft (1 nautical mile = 1.151 statute miles):

$$H = 1.5 \sqrt{F} \qquad \text{for long fetches } (F > 30 \text{ nautical miles}) \qquad (23\text{-}7)$$

and

$$H = 1.5 \sqrt{F} + 2.5 - \sqrt[4]{F} \qquad \text{for short fetches } (F < 30 \text{ nautical miles}) \qquad (23\text{-}8)$$

They were developed from observations on lakes but checked with waves on the coast of the North Sea.   They give the wave height only for maximum wind velocity in the observation area, since they do not include wind velocity as a variable.

D. A. Molitor, in "Wave Pressure on Sea-walls and Breakwaters" (*Proceedings of the American Society of Civil Engineers,* May, 1934), gave the following formulas for determining wave height for inland lakes.   They are based on the formulas of Thomas Stevenson but introduce wind velocity as a variable and use statute miles instead of nautical miles:

$$H = 0.15 \sqrt{UF} \qquad \text{for values of } F \text{ greater than 20 miles} \qquad (23\text{-}9)$$

$$H = 0.17 \sqrt{UF} + 2.5 - \sqrt[4]{F} \qquad \text{for values of } F \text{ less than 20 miles} \qquad (23\text{-}10)$$

where U = wind velocity, statute miles per hour
$\quad F$ = fetch, statute miles
$\quad H$ = wave height, ft

The ratio of wavelength L to height H depends on the wind velocity, duration of the storm, depth of water, and character of the bottom. L/H for inland lakes, in relatively shallow depth, is between 9 and 15, and for ocean waves, between 17 and 33.   The ratios are nearly inversely proportional to the wind intensity; the smaller ratios are for the stronger wind intensities.

In making observations on wave height for design purposes, keep in mind that in a train of waves the height of individual waves will vary greatly.   The average height of the highest one-third of the waves for a stated interval has been termed the *significant height*, and it has been found that the highest or maximum wave has a height of about 1.87 times the significant height.

The following tabulation from Engineer Manual, EM 1110-2-2904, "Design of Breakwaters and Jetties," Headquarters, Department of the Army, Office of the Chief of Engineers, Apr. 30, 1963, gives the ratios of commonly used wave-height parameters to significant height:

Significant height 1.0
Average height 0.6
Average height of highest 10% 1.3
Height of simple sine waves having same energy content as actual
    wave train 0.8
Height not exceeded more than 20% of the time 0.9
Height not exceeded more than 10% of the time 1.1
Height not exceeded more than 5% of the time 1.2
Height not exceeded more than 3% of the time 1.3
Height not exceeded more than 1% of the time 1.6
Average height of highest 1% 1.7

The following factors have an influence on generation and decay of waves, and data about these factors are needed to forecast the design wave.

Fetch.   Synoptic weather maps help in determining generating area and fetch for a given location.   At coastal points, fetch is limited by geographical barriers, such as islands and promontories, and it has to be determined for each direction.

Wind Velocity and Direction.   Waves are generated by transfer of energy from air moving over the water surface.   The transfer is effected in two ways: (1) The water surface reacts to small differences in pressure of the moving air, which creates the first variations in the water level. These are increased by the difference in pressure exerted by the moving wind on the back and on the front of the wave.   (2) Tangential stress occurs between air and water, which are in contact and moving at different speeds relative to each other.   Since both normal pressure and tangential stress are functions of wind velocity, it follows that wave characteristics also are functions of wind velocity.

The relationships among fetch, wind velocity, wave height, and wave period are expressed by the formulas

$$H = 0.0555UF^{0.5} \tag{23-11}$$

and

$$T = 5U^{0.5}F^{0.25} \tag{23-12}$$

where H = significant wave height in deep water, ft
    T = mean wave period, sec
    U = wind velocity, knots (nautical miles per hour)
    F = fetch, nautical miles (1 nautical mile = 1.151 statute miles)

Wind Duration.   Only after a wind has blown for some time will the wave generated by it attain the characteristics typical for the wind velocity.   The wave increases in height rapidly at first but grows at an ever-slower rate the longer the wind lasts.   However, it is normally assumed that the wind blows long enough for the maximum wave to develop for that particular wind velocity, unless observed wind data contradict this assumption.

Water Depth.   Waves generated in shallow waters are limited in height by two factors: bottom friction and breaking.   Bottom friction increases with growing waves, and eventually a steady state is reached where the energy transmitted by the wind is spent by bottom friction and no energy is left for wave growth.   The wave characteristics of the steady state are thus related to the water depth and, of course, to the wind velocity.   They also depend on the friction factor, which can be taken as $f = 0.01$ for most cases.   The wave height in shallow water has a further limitation.   Before the wave reaches the steady state, it may break.   This occurs when the wave reaches a height about 0.8 times the still-water depth.   Once a wave has started breaking, the additional energy transmitted to it by the wind is spent in its breaking crest, and no more growth can be achieved.

Another important consideration is the behavior of waves that are generated in deep water but move into shallow waters without any further wind acting on them.   In this case the bottom normally slopes upward, and there are two counteracting influences on the wave height.   One is the bottom friction, which tends to reduce the wave height, and the other is the shoaling effect, which tends to increase the wave height.   As waves move into shoaling waters, the wave energy becomes confined in ever-decreasing depths of water.   The amount of water on which the energy is acting becomes smaller, the energy per water particle increases, and the wave height increases.   On the other hand, the bottom friction increases as the water depth decreases.   The combined effect has been reported by C. L. Bretschneider and R. O. Reid ("Generation of Wind Waves over a Shallow Bottom," Beach Erosion Board, Corps of Engineers, U.S. Army, 1954).

Wave Decay.   As waves travel away from their generating area, they continuously lose energy.   Their height decreases, but their length, period, and velocity of progress increase. It is thus mainly air resistance that brings about the decay of waves; influence of molecular viscosity or internal friction is negligible.

Diffraction.   A typical example of wave diffraction occurs in a harbor protected by breakwaters except for the harbor entrance.   The gap in the breakwater, which serves as a shipping channel, admits a certain amount of waves into the otherwise still waters of the harbor.   These waves expand over the whole harbor area in wave fronts that have their center at the gap.   Since the wave fronts become ever more extensive as they travel away from the gap, their energy decreases.   Therefore, the farther away a structure is from the gap, the smaller are the waves acting on it.   The following equation by Stevenson may be used to approximate wave height within a harbor.   (Model tests will give a more accurate picture of wave conditions and are essential for studying various arrangements of breakwaters for important harbors.)

$$h_p = H \left[\sqrt{\frac{b}{B}} - 0.02 \sqrt[4]{D} \left(1 + \sqrt{\frac{b}{B}} \right) \right] \tag{23-13}$$

where h_p = height of reduced wave at any point p in harbor, ft
 H = height of wave at entrance, ft
 b = breadth of entrance, ft
 B = breadth of harbor at p, ft (= length of arc with radius D and center at middle of entrance)
 D = distance from entrance to p, ft

The equation does not apply to points less than 50 ft from the entrance.   The length of the arc B is measured between its points of intersection with the two diverging side walls of the harbor. In cases when one of the breakwaters meets the shore at a shorter distance, its line of direction must be extended landward to intersect the arc.

Refraction.   If a wave group travels in shallow water at an angle to the contour lines of the bottom, the waves in shallower water become shorter and their velocity decreases.   Since the portion of a wave in deeper water advances more rapidly than the portion in shallow water, the wave front turns and tends to become parallel to the bottom contours.   At a gently sloping beach, wave crests are always nearly parallel to the coastline, despite the angle they may form with it at sea.

Reflection.   Nonbreaking waves acting on a vertical wall, cliff, or steep beach do not lose their energy by the impact but are reflected.   They may form a standing wave or *clapotis*.   As a matter of principle, it is desirable to destroy the wave energy inside a harbor as soon as possible and not to allow it to be reflected back and forth.   There are many ways to minimize wave reflection in harbors.

Breakwaters must be aligned in such a way that waves reflected from them are not directed toward piers and other harbor installations.   Their interior slope can be designed to make the waves break, which spends the wave energy.   Special spending beaches can be planned at points of maximum wave action in a harbor.

(For more comprehensive coverage, including graphs, of the factors influencing generation and decay of waves, see Alonzo DeF. Quinn, "Design and Construction of Ports and Marine Structures," McGraw-Hill Book Company, New York.)

23-4. Wave Action on Vertical Walls.   Vertical walls may be mainly classified in two groups, breakwaters and seawalls.   **Breakwaters** are normally built in water deep enough to keep attacking waves from breaking.   **Seawalls,** on the other hand, are generally built at the top of a beach and are subjected to the action of breaking waves.   There is a considerable difference in wave pressure between the two types.   Breaking waves exert a much greater pressure.   The energy of the breaking wave is destroyed at the wall.   Therefore, such a wave transmits much greater energy to the wall than a nonbreaking wave, whose energy is mostly reflected by the wall.

Wave pressure against vertical walls consists of hydrostatic pressure, which varies as the wave rises and falls along the wall, and dynamic pressure exerted by the moving water particles. A number of theories and formulas have been developed over the years for determination of pressure of waves on vertical walls for the purely oscillatory wave condition.   The theory of Sainflou gives a close approximation to the actual wave force that acts against the vertical face of a wall or breakwater.

Robert Y. Hudson, hydraulic engineer, chief of Wave Action Section, U.S. Army Engineer Waterways Experiment Station, Vicksburg, Miss., compared the results of tests at the U.S. Waterways Experiment Station and overturning moments about the base of a vertical break-water calculated from the Sainflou theory (*Transactions of the American Society of Civil Engineers,* vol. 118, 1953).   For ratios of depth to wavelength d/L in the range of 0.1 to 0.2, often found in installations of breakwaters, the overturning moments obtained from the Sainflou theory agreed closely with the model test moments.   It is only when the value of d/L approaches 0.05 that a wide divergence exists.   For this ratio, the depth of water is so shallow that the wave approaches the breaking condition.   For example, if the depth of water d is 20 ft, the wave height H 20 ft, and the wave length L 400 ft, d/L will equal 0.05; but the wave will break when it approaches a depth equal to or a little greater than its height, and therefore the Sainflou theory should not be used.

A water particle at the surface of a deep-water wave oscillates about a point whose height above still-water level may be taken as

$$h_o = \frac{\pi H^2}{4L} \coth \frac{2\pi d}{L} \qquad (23\text{-}14)$$

Therefore, the crest height above still-water level $a = h_o + H/2$ (Fig. 23-1).   When this wave is reflected from a vertical wall, a clapotis or standing wave is created, and the height of the center

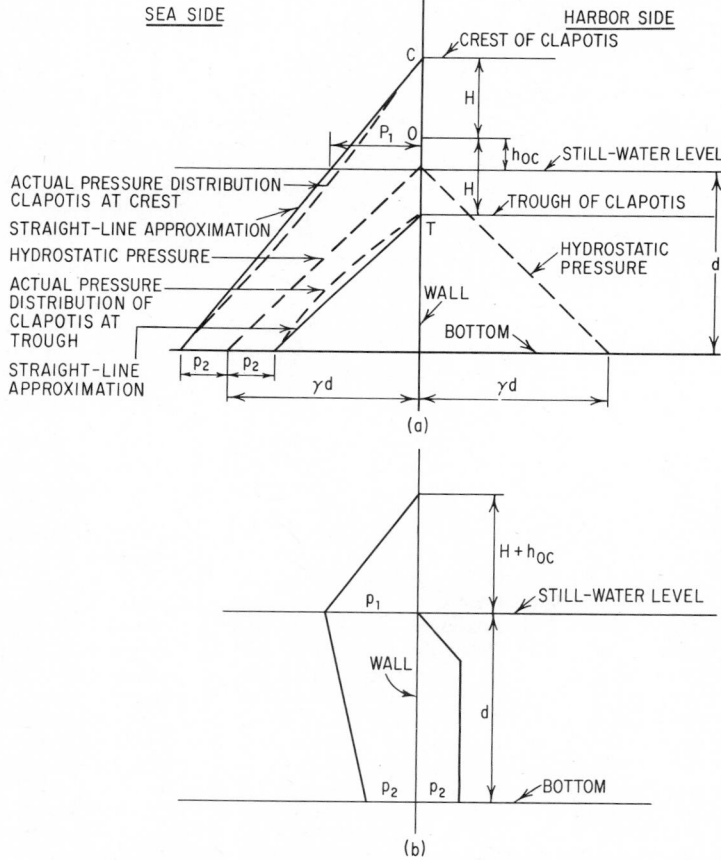

Fig. 23-2. Wave pressures on vertical walls, according to Sainflou.

of oscillation above still-water level is raised to

$$h_{oc} = \frac{\pi H^2}{L} \coth \frac{2\pi d}{L} \qquad (23\text{-}15)$$

or four times that for the unreflected wave. The wave height of the clapotis is $2H$, twice the wave height of the unreflected wave. The crest height above still-water level $a = H + 4h_0$, or slightly more than twice the crest height for the unreflected wave.

Sainflou computed a general formula for the pressure on a vertical wall, from which the pressure diagram shown in Fig. 23-2a can be constructed. The pressure, psf, at the bottom is

$$\gamma d \pm p_2 = \gamma d \pm \frac{\gamma H}{\cosh (2\pi d/L)} \qquad (23\text{-}16)$$

where γ = specific weight of water, lb per cu ft

p_2 = change in pressure, psf, from the hydrostatic pressure γd

The plus sign applies for the wave at crest position, the minus sign for the wave at trough position. Straight lines may be substituted for the actual pressure curves without affecting the results appreciably for calculation of pressure on a vertical wall. For the wave at crest position, the wave pressure at still-water level, by simple proportion, is

$$p_1 = (\gamma d + p_2) \frac{H + h_{oc}}{H + h_{oc} + d} \qquad (23\text{-}17)$$

Where still-water hydrostatic pressure acts on the opposite side of the wall, the resulting net pressure on the wall is as shown in Fig. 23-2b. The diagram on the left gives the pressure for the clapotis at crest position, and the diagram on the right gives the pressure for the clapotis at trough position.

Breaking Waves. Seawalls and the inshore end of breakwaters and jetties are usually subjected to breaking waves (Fig. 23-3a). The following method for determining these pressures

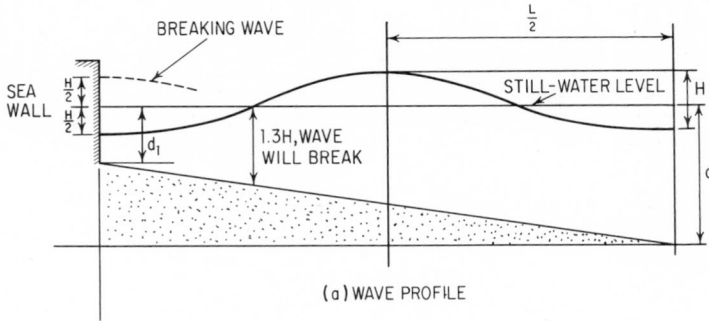

(a) WAVE PROFILE

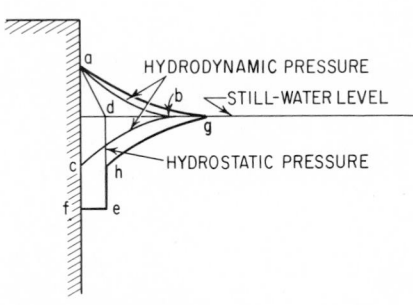

(b) PRESSURE DIAGRAMS

Fig. 23-3. Breaking waves exert hydrodynamic as well as hydrostatic pressures on vertical walls.

was developed by R. R. Miniken ("Winds, Waves and Maritime Structures," Charles Griffin & Company, Ltd., London).

According to Miniken, the total pressure on the wall from breaking waves is a combination of dynamic and hydrostatic pressures. The maximum dynamic pressure, psf, occurs at still-water level and is given by

$$p_1 = \frac{2\pi d_1}{Ld} \gamma Hg \frac{d + d_1}{2} \tag{23-18}$$

where d = deep-water depth, ft
d_1 = depth of water at structure, ft
γ = specific weight of water (64.4 lb per cu ft for salt water)
H = height of wave just breaking on structure, ft
L = wavelength, ft
g = acceleration of gravity, 32.2 ft per sec²

The pressure curves may be plotted, as shown in Fig. 23-3b, by considering that the peak dynamic pressure occurs at still-water level and diminishes rapidly to zero at the crest of the wave at a height of $H/2$ above peak pressure, and at a depth of $H/2$ below still-water level. The pressure curves are assumed to be parabolic. The total dynamic pressure on the wall is represented by the area within the curve $abca$ and is approximately $p_1 H/3$.

The total hydrostatic pressure of the water due to half the height of the wave above still-water level is represented by the area within the diagram $adef$ and is equal to

$$P = \frac{\gamma H}{2} \left(d_1 + \frac{H}{4} \right) \tag{23-19}$$

The total pressure is the sum of the pressures given by Eqs. (23-18) and (23-19) and is represented by the area within the diagram $aghef$.

23-5. Wave Action on Piles.   Waves acting on piles exert pressures that are the result of drag and inertial forces.

Drag Forces.   Flow with constant velocity (steady flow) of a fluid around an obstacle and the resultant forces (drag) exerted on the obstacle have been studied extensively. Normal and tangential stresses exist between the obstacle and the fluid. At low velocities, the fluid flows smoothly around the obstacle; the stresses are mainly tangential, or shear, stresses. The force is called "surface drag." At higher velocities, the smooth flow separates from the sides of the obstacle, creating areas of lower pressure downstream and along the sides of the obstacle where eddies and turbulence appear. The differences in pressure along the surface of the obstacle cause a different type of force, "form drag." The unit drag force, which comprises both the surface and form drag on a circular pile, can be computed from

$$f_D = C_D \frac{\gamma}{2g} Du^2 \tag{23-20}$$

where f_D = drag force, lb per lin ft of pile
C_D = drag coefficient
γ = unit weight of fluid (64.4 lb per cu ft for sea water)
D = diameter of pile, ft
g = acceleration due to gravity, 32.2 ft per sec²
u = velocity of fluid, fps

For steady flow, C_D is dependent on the shape, dimensions, and roughness of the pile and on the Reynolds number $\mathbf{R} = uD/v$, where v is the kinematic viscosity of the fluid. For water at about 70° F, $v = 10^{-5}$ ft² per sec and $\mathbf{R} = 10^5 uD$.

If the horizontal component u of the orbital velocity of a wave is substituted for the fluid velocity, the total drag force, lb, on the pile at any moment can be determined by integrating Eq. (23-20) over the whole height of the pile:

$$F_D = C_D \frac{\gamma}{2g} DH^2 K_D \tag{23-21}$$

$$K_D = \frac{1}{H^2} \int_{-d}^{\eta} \pm u^2 dz \tag{23-22}$$

where z = a vertical coordinate, ft
d = depth of water below still-water level, ft
η = distance of a surface particle above $(+\eta)$ or below $(-\eta)$ still-water level

The maximum value of the drag force in the direction of wave travel occurs when the crest passes the pile ($\eta = a$). The maximum value in the opposite direction occurs when the trough passes the pile ($\eta = a - H$).

Inertial Force.   In addition to the forces acting normally and tangentially on the surface of the pile, which are combined in the "drag force," the constantly accelerating or decelerating masses of water exert also a mass force, i.e., a kind of impact force, on the pile. This is termed inertial force. It can be computed numerically from

$$f_i = C_M \frac{\gamma\pi}{4g} D^2 \frac{du}{dt} \tag{23-23}$$

where f_i = inertial force, lb per lin ft of pile
  C_M = coefficient of mass
du/dt = horizontal fluid acceleration, ft per sec^2
The total inertial force, lb, on the pile at any moment is

$$F_i = C_M \frac{\gamma}{2g} D^2 H K_i \tag{23-24}$$

where
$$K_i = \frac{\pi}{2H} \int_{-d}^{\eta} \frac{du}{dt} \, dz \tag{23-25}$$

This force is zero at crest and trough positions. Maximum values occur between these two positions. Therefore, the total maximum force is less than the sum of the maximum values of F_D and F_i, since they occur at different times within the wave cycle. For high waves in shallow water, the drag force predominates, and the maximum force occurs at or near the crest of the waves. For low waves in deep water, the inertial force predominates, and the maximum force occurs at a phase angle of about 90° (one-quarter wavelength), or when the water surface at the pile is close to the still-water level.

Effect of Pile Shape.   Wave forces are smallest for piles of cylindrical cross section. For piles with flat or irregular surfaces, such as concrete and H piles, respectively, very little is known about the effect of the shape on drag and inertial forces.

(For more complete analyses and example of wave action on piles, including formulas and graphs for determining drag and inertial forces, see Alonzo DeF. Quinn, "Design and Construction of Ports and Marine Structures," McGraw-Hill Book Company, New York.)

HARBOR AND PORT PLANNING

23-6. Types of Harbors and Ports.   A harbor is a water area partly enclosed and so protected from storms as to provide safe and suitable accommodation for vessels seeking refuge, supplies, refueling, repairs, or the transfer of cargo.

Harbors may be classified as natural, seminatural, or artificial, and as harbors of refuge, military harbors, or commercial harbors. Commercial harbors may be either municipal or privately owned.

A **natural harbor** is an inlet or water area protected from storms and waves by the natural configuration of the land. Its entrance is so formed and located as to facilitate navigation while insuring comparative quiet within the harbor. Natural harbors are located in bays, tidal estuaries, and river mouths. Well-known natural harbors are New York, San Francisco, and Rio de Janeiro.

A **seminatural harbor** may be an inlet or a river sheltered on two sides by headlands requiring artificial protection only at the entrance. Next to a purely natural harbor, it forms the most desirable harbor site, other things being equal. Plymouth and Cherbourg take advantage of their natural location to become well-protected harbors by the addition of detached breakwaters at the entrances.

An **artificial harbor** is one protected from the effect of waves by breakwaters or one created by dredging. Buffalo, N.Y.; Matarani, Peru; Hamburg, Germany; and La Havre, France, are examples of artificial harbors.

A **harbor of refuge** may be used solely as a haven for ships in a storm, or it may be part of a commercial harbor. Sometimes an outer harbor serves as an anchorage, while a basin within the inner breakwater constitutes a commercial harbor. The essential features are good anchorage and safe and easy access from the sea during any condition of weather and state of tide. Well-known harbors of refuge are the one at Sandy Bay, near Cape Ann, Massachusetts, and that at the mouth of Delaware Bay. A fine example of a combined harbor of refuge and commercial harbor exists at Dover, England.

A **military harbor** or naval base has the purpose of accommodating naval vessels and serving as a supply depot. Guantanamo, Cuba; Hampton Roads, Va.; and Pearl Harbor, Hawaii, are some well-known naval bases.

A **commercial harbor** is one in which docks are provided with the necessary facilities for loading and discharging cargo. Drydocks are sometimes provided for ship repairs. Many commercial harbors are privately owned and operated by companies representing the steel, aluminum, copper, oil, coal, timber, fertilizer, sugar, fruit, chemical, and other industries. Municipal- or government-controlled harbors, often operated by port authorities, exist in many countries and are usually part of extensive port works, such as the harbors in New York, Los Angeles, and London.

A **port** is a harbor where marine terminal facilities are provided. These consist of piers or wharves at which ships berth while loading or unloading cargo, transit sheds and other storage areas where ships may discharge incoming cargo, and warehouses where goods may be stored for longer periods while awaiting distribution or sailing. The terminal must be served by railroad, highway, or inland-waterway connections. In this respect the area of influence of the port reaches out for a considerable distance beyond the harbor.

A **port of entry** is a designated location where foreign goods and foreign citizens may be cleared through a custom house.

A **free port** or **zone** is an isolated, enclosed, and policed area in or adjacent to a port of entry, without a resident population. Furnished with the necessary facilities for loading and unloading, for supplying fuel and ship's stores, for storing goods and reshipping them by land or water, a free port is an area within which goods may be landed, stored, mixed, blended, repacked, manufactured, and reshipped, without payment of duties and without the intervention of custom officials. The most important free port in Europe is Hamburg, which was originated about 1883 and has grown ever since.

A **marine terminal** is that part of a port or harbor that provides docking, cargo-handling, and storage facilities. When only passengers embark and disembark along with their baggage and miscellaneous small cargo, generally from ships devoted mainly to the carrying of passengers, it is called a *passenger terminal*. When the traffic is mainly cargo carried by freighters, although many of these ships may carry also a few passengers, the terminal is commonly referred to as a *freight* or *cargo terminal*. In many cases, it will be known as a *bulk cargo terminal*, where such products as petroleum, cement, and grain are stored and handled.

An **offshore mooring** is provided usually where it is not feasible or economical to construct a dock or provide a protected harbor. Such an anchorage consists of a number of anchorage units, each consisting of one or more anchors, chains, sinkers, and buoys to which the ship will attach its mooring lines. These anchorages are supplemented in most cases by the ship's bow anchors. Bulk cargo is usually transported to or from the ship by pipeline or trestle conveyor, while other cargo may be transferred by lighter.

An **anchorage area** is a place where ships may be held for quarantine inspection, to await docking space, sometimes while removing ballast in preparation for taking on cargo, or to await favorable weather conditions. Special anchorages are sometimes provided for ships carrying explosives or dangerous cargo and are usually so designated on harbor maps by name and depth of water.

A **turning basin** is a water area inside a harbor or an enlargement of a channel to permit the turning of a ship. When space is available, the area should have a radius of at least twice the length of the ship to permit either free turning or turning with the aid of tugs, if wind and water conditions require. When space is limited, the ship may be turned by warping around the end of a pier or turning dolphin, either with or without the use of its lines. In those cases, the turning basin will be much smaller and of a more triangular or rectangular shape.

23-7. Ship Characteristics. The length, beam, and draft of ships that will use a port have a direct bearing on the design of the approach channel, harbor, and marine-terminal facilities. The last is affected also by the type of vessel and its capacity or tonnage. These characteristics for representative ships of principal types are given in Tables 23-3 to 23-7, inclusive. The wind areas for selected ships are given in Table 23-8.

Displacement tonnage is the actual weight of the vessel, or the weight of water displaced when afloat, and may be either "loaded" or "light." *Displacement loaded* is the weight, in long tons, of the ship and its contents when fully loaded with cargo to the Plimsoll mark, or load line, painted on the hull of the ship (1 ton= 2,240 lb). The *Plimsoll mark*, used on British ships, and the load line, commonly used on American vessels, designate the depth under the maritime laws to which a ship may be loaded in different bodies of water during various seasons

Table 23-3. Characteristics of Tankers

Year built	Name or class	Length		Breadth	Depth	Draft loaded (summer)	Tonnage, long tons	
		Overall	Between perpendiculars				Deadweight	Displacement
1941	T-2 class tankers	523 ft 6 in.	503 ft 0 in.	68 ft 0 in.	39 ft 3 in.	30 ft 2 in.	16,350	21,880
1960	J. Paul Getty	844 ft 4 in.	808 ft 0 in.	110 ft 0 in.	61 ft 4 in.	46 ft 4 in.	73,900	97,000
1960	Universe Daphne	949 ft 9 in.	900 ft 0 in.	135 ft 0 in.	67 ft 6 in.	50 ft 11 in.	115,360	146,570
1961	Olympus	818 ft 11 in.	784 ft 1 in.	113 ft 6 in.	61 ft 6 in.	45 ft 10 in.	75,145	94,260
1961	Orion Hunter	860 ft 0 in.	820 ft 0 in.	104 ft 0 in.	60 ft 0 in.	43 ft 10 in.	67,208	86,800
1961	Naess Sovereign	874 ft 10 in.	833 ft 4 in.	122 ft 1 in.	64 ft 0 in.	48 ft 2 in.	90,200	113,900
1962	Esso Austria	849 ft 1 in.	809 ft 6 in.	116 ft 6 in.	60 ft 4 in.	45 ft 9 in.	78,566	99,676
1962	Manhattan	940 ft 5 in.	892 ft 0 in.	132 ft 0 in.	67 ft 6 in.	50 ft 1 in.	108,400	138,700
1962	Nissho Maru	954 ft 8 in.	905 ft 6 in.	141 ft 1 in.	72 ft 10 in.	54 ft 4 in.	130,217	160,673
1963	William M. Allen	824 ft 8 in.	782 ft 0 in.	116 ft 0 in.	56 ft 0 in.	43 ft 8 in.	69,480	90,333
1963	California Getty	835 ft 0 in.	794 ft 0 in.	122 ft 1 in.	65 ft 4 in.	48 ft 10 in.	90,324	110,145
1963	Esso Deutschland	855 ft 11 in.	820 ft 0 in.	125 ft 0 in.	62 ft 6 in.	47 ft 5 in.	90,187	111,610
1964	Esso Bayern	869 ft 6 in.	820 ft 0 in.	125 ft 0 in.	62 ft 6 in.	47 ft 11 in.	90,600	111,786
1965	Ionian Commander	775 ft 0 in.	738 ft 3 in.	104 ft 4 in.	53 ft 4 in.	40 ft 8 in.	60,032	73,841
1966	Ionic	835 ft 0 in.	784 ft 5 in.	124 ft 1 in.	57 ft 5 in.	43 ft 11 in.	84,227	100,587
1968	Kaimon Maru	984 ft 3 in.	935 ft 1 in.	158 ft 2 in.	78 ft 9 in.	59 ft 2 in.	175,891	205,096
1968	Universe Ireland	1,132 ft 10 in.	1,082 ft 8 in.	175 ft 0 in.	105 ft 0 in.	81 ft 5 in.	326,585	375,811
1969	Universe Kuwait	1,134 ft 10 in.	1,082 ft 3 in.	174 ft 10 in.	105 ft 0 in.	71 ft 11 in.	276,000	
	Planned	1,800 ft in.		283 ft in.		95 ft in.	1,000,000	

Table 23-4. Characteristics of Bulk Carriers (Ore, Coal, Etc.)

Year built	Name or class	Length		Breadth	Depth	Draft loaded (summer)	Tonnage, long tons	
		Overall	Between perpendiculars				Deadweight	Displacement
1960	Edward L. Ryerson	730 ft 0 in.	712 ft 0 in.	75 ft 0 in.	39 ft 0 in.	26 ft 6 in.*	26,055	34,135
1961	Timna	550 ft 2 in.	520 ft 0 in.	74 ft 0 in.	48 ft 4 in.	34 ft 7 in.	22,934	29,734
1961	Argonaftis	583 ft 6 in.	545 ft 0 in.	74 ft 8 in.	44 ft 0 in.	31 ft 6 in.	20,990	28,058
1961	Ore Venus	751 ft 0 in.	710 ft 0 in.	102 ft 0 in.	51 ft 6 in.	38 ft 0 in.	50,692	65,660
1962	Corsair	592 ft 2 in.	565 ft 0 in.	79 ft 0 in.	46 ft 8 in.	32 ft 9 in.	24,911	32,370
1962	Centauro	679 ft 0 in.	637 ft 0 in.	91 ft 10 in.	51 ft 6 in.	36 ft 1 in.	35,316	46,248
1962	Sonic	746 ft 1 in.	708 ft 8 in.	100 ft 5 in.	55 ft 7 in.	37 ft 11 in.	48,976	62,551
1963	Atlantic Eagle	625 ft 10 in.	589 ft 6 in.	75 ft 0 in.	46 ft 4 in.	33 ft 8 in.	23,670	31,947
1963	Archangel	628 ft 2 in.	589 ft 6 in.	75 ft 0 in.	46 ft 3 in.	33 ft 8 in.	23,960	31,994
1963	Aristeides	735 ft 1 in.	705 ft 0 in.	100 ft 8 in.	55 ft 6 in.	38 ft 0 in.	50,055	62,214
1963	Amalfi	753 ft 0 in.	700 ft 11 in.	98 ft 5 in.	55 ft 6 in.	39 ft 6 in.	46,730	60,122
1964	Dromon	643 ft 0 in.	600 ft 0 in.	76 ft 0 in.	45 ft 6 in.	33 ft 9 in.	27,480	35,883
1966	Cedros	995 ft 9 in.	939 ft 11 in.	142 ft 1 in.	81 ft 0 in.	62 ft 3 in.	170,418	200,242
1967	Alberto Lollighetti	709 ft 11 in.	656 ft 2 in.	93 ft 11 in.	57 ft 6 in.	38 ft 8 in.	44,477	55,399
1967	Leonidas D.	708 ft 9 in.	672 ft 7 in.	101 ft 9 in.	59 ft 1 in.	39 ft 8 in.	52,458	63,208
1968	Agamemnon	734 ft 9 in.	698 ft 10 in.	105 ft 10 in.	62 ft 6 in.	40 ft 2 in.	56,672	69,468
1968	Grischuna	742 ft 10 in.	710 ft 0 in.	102 ft 0 in.	57 ft 6 in.	42 ft 5 in.	60,639	72,120

*Fresh-water draft, Great Lakes.

Table 23-5. Characteristics of General Cargo Ships

Year built	Name or class	Length		Breadth	Depth	Draft loaded (summer)	Tonnage, long tons	
		Overall	Between perpendiculars				Deadweight	Displacement
1961	Export Agent	492 ft 6 in.	470 ft 0 in.	73 ft 0 in.	42 ft 2 in.	28 ft 2 in.	11,089	17,570
1961	Apollonia	505 ft 4 in.	475 ft 11 in.	66 ft 3 in.	41 ft 4 in.	30 ft 5 in.	14,974	20,274
1961	Philippine President Roxas	510 ft 2 in.	475 ft 9 in.	64 ft 0 in.	40 ft 4 in.	29 ft 7 in.	12,156	17,379
1961	Washington Mail	563 ft 8 in.	528 ft 6 in.	76 ft 0 in.	44 ft 6 in.	31 ft 7 in.	14,803	22,595
1962	Vasilios R.	526 ft 0 in.	492 ft 6 in.	67 ft 3 in.	42 ft 0 in.	31 ft 3 in.	15,450	20,728
1962	Pioneer Moon	560 ft 6 in.	530 ft 0 in.	75 ft 0 in.	42 ft 9 in.	31 ft 7 in.	13,583	21,053
1962	African Meteor	572 ft 6 in.	541 ft 0 in.	75 ft 0 in.	42 ft 10 in.	30 ft 10 in.	12,728	20,110
1963	Ashley Lykes	495 ft 0 in.	470 ft 0 in.	69 ft 0 in.	41 ft 6 in.	30 ft 1 in.	11,336	17,210
1963	C. E. Dant	565 ft 0 in.	528 ft 6 in.	76 ft 0 in.	44 ft 6 in.	31 ft 7 in.	14,376	22,629
1965	Gulf Merchant		470 ft 0 in.	69 ft 0 in.	41 ft 6 in.	30 ft 2 in.	11,368	17,210
1968	Alaskan Mail	605 ft 0 in.	582 ft 6 in.	82 ft 0 in.	46 ft 6 in.	35 ft 1 in.	22,208	31,995
1968	Genevieve Lykes	540 ft 0 in.	514 ft 11 in.	76 ft 0 in.	42 ft 8 in.	31 ft 8 in.	13,808	20,986
1968	Khian Wave	466 ft 9 in.	440 ft 0 in.	65 ft 0 in.	40 ft 6 in.	29 ft 9 in.	14,924	18,825

Table 23-6. Characteristics of Passenger Ships

Year built	Name	Length		Breadth	Depth	Draft loaded (summer)	Tonnage, long tons	
		Overall	Between perpendiculars				Deadweight	Displacement
1950	Independence	682 ft 6 in.	633 ft 0 in.	89 ft 0 in.	52 ft 11 in.	30 ft 2 in.	11,790	30,090
1952	United States	990 ft 0 in.	916 ft 10 in.	101 ft 7 in.		32 ft 0 in.		
1958	Argentina	617 ft 10 in.	570 ft 0 in.	84 ft 0 in.	45 ft 3 in.	27 ft 0 in.	10,170	22,700
1958	Santa Rosa	584 ft 0 in.	535 ft 2 in.	84 ft 0 in.	43 ft 1 in.	27 ft 2 in.	8,713	20,298
1960	Savannah (nuclear-powered)	595 ft 6 in.	545 ft 0 in.	78 ft 0 in.	41 ft 0 in.	29 ft 7 in.	10,190	21,800
1960	Leonardo da Vinci	767 ft 4 in.	677 ft 3 in.	91 ft 10 in.	50 ft 7 in.	31 ft 4 in.		32,787
1963	Galileo Galilei	700 ft 11 in.	619 ft 2 in.	93 ft 10 in.	47 ft 9 in.	28 ft 5 in.	9,331	26,894
1963	Guglielmo Marconi	700 ft 11 in.	619 ft 2 in.	93 ft 10 in.	47 ft 9 in.	28 ft 5 in.	9,490	26,900
1965	Oceanic	782 ft 4 in.	675 ft 11 in.	96 ft 6 in.	48 ft 3 in.	28 ft 4 in.	8,238	31,565
1965	Raffaello	904 ft 7 in.	800 ft 5 in.	101 ft 8 in.	51 ft 10 in.	30 ft 7 in.		41,328
1965	Michelangelo	904 ft 10 in.	800 ft 5 in.	101 ft 8 in.	51 ft 9 in.	30 ft 7 in.	8,886	41,328
1969	Queen Elizabeth II	963 ft 0 in.	887 ft 1 in.	105 ft 0 in.	56 ft 0 in.	32 ft 6 in.	15,724	48,886

of the year. *Displacement light* is the weight, in long tons, of the ship without cargo, fuel, and stores.

Deadweight tonnage is the carrying capacity of a ship in long tons and the difference between displacement light and displacement loaded to the Plimsoll mark or load line. It is the weight of cargo, fuel, and stores a ship carries when loaded to the load line, as distinguished from loaded to space capacity. This tonnage varies with latitude and season. It also depends on salinity of the water, because of the effect of temperature and salinity on the specific gravity and buoyancy of the water in which the vessel is operating. Unless otherwise indicated, deadweight tonnage is the mean of tropical, summer, and winter deadweight. Deadweight tonnage is indicated by weight, and gross tonnage by volume measurement; both indicate carrying capacity.

Ships are registered with gross or net tonnage expressed in units of 100 cu ft. **Gross tonnage** is the entire internal cubic capacity of a ship, and **net tonnage** is the gross tonnage less the space provided for the crew, machinery, engine room, and fuel.

Table 23-7. Characteristics of Large Ships of U.S. Navy

Year built	Type of ship, name or class	Length Overall	Length Water line	Breadth	Draft Standard	Draft Full load	Displacement, long tons Standard	Displacement, long tons Full load
	Battleships:							
1941	North Carolina Class	729 ft 0 in.	704 ft 0 in.	108 ft 0 in.	26 ft 8 in.	35 ft 0 in.	35,000	45,500
1942	South Dakota and							
	Indiana classes	680 ft 0 in.		108 ft 2 in.	26 ft 9 in.	37 ft 0 in.	35,000	44,500
1943–1944	Iowa Class	887 ft 3 in.	861 ft 3 in.	108 ft 0 in.		38 ft 0 in.	45,000	57,600
	Cruisers:							
1942–1945	Cleveland and Fargo classes	610 ft 0 in.	600 ft 0 in.	66 ft 0 in.	20 ft 0 in.	25 ft 0 in.	10,500	13,750
1942–1946	San Diego, Juneau, and Oakland classes	541 ft 0 in.		52 ft 10 in.	14 ft 9 in.	25 ft 0 in.	6,000	8,000
1943–1946	Baltimore and Oregon City classes	673 ft 6 in.		71 ft 0 in.		26 ft 0 in.	13,600	17,200
1944	Alaska Class	808 ft 6 in.		91 ft 0 in.		31 ft 6 in.	27,500	32,500
1948–1949	Des Moines Class	716 ft 6 in.		75 ft 4 in.		26 ft 0 in.	17,000	21,500
1948–1949	Worcester Class	679 ft 6 in.		70 ft 8 in.		26 ft 0 in.	14,700	18,500
1953	Northampton	676 ft 0 in.		71 ft 0 in.		29 ft 0 in.	13,000	17,200
	Aircraft carriers:							
1955	Forrestal	1,039 ft 0 in.		252 ft 0 in.		37 ft 0 in.	54,600	76,000
1956–1961	Saratoga and Kitty Hawk classes	1,046 ft 0 in.		252 ft 0 in.		37 ft 0 in.	56,000	78,700
1961	Enterprise (nuclear-powered)	1,100 ft 0 in.		252 ft 0 in.			74,700	85,350
1950–1959	Hancock and Oriskany classes	899 ft 0 in.		192 ft 0 in.		31 ft 0 in.	33,100	42,600
1952	Oriskany Class (axial)	899 ft 0 in.		152 ft 0 in.		31 ft 0 in.	33,100	40,800
1952	Antietam	899 ft 0 in.		154 ft 0 in.		31 ft 0 in.	30,000	38,000
1956–1960	Midway Class	974 ft 0 in.		210 ft 0 in.		36 ft 0 in.	51,000	62,000

Table 23-8. Wind Areas for Selected Ships*

Name	Type	Tonnage, long tons Deadweight	Tonnage, long tons Loaded displacement	Light draft wind area, sq ft
T-2 tankers	Tanker	16,350	21,880	19,000
Universe Ireland	Tanker	326,585	375,811	81,000
Midway Class	Aircraft carrier		55,000	66,000
North Carolina Class	Battleship		45,500	32,700
South Dakota and Indiana classes	Battleship		44,500	31,500
Iowa Class	Battleship		57,600	43,000
Cleveland and Fargo classes	Cruiser		13,750	26,000
San Diego, Juneau, and Oakland classes	Cruiser		8,000	20,000
Alaska Class	Cruiser		32,500	38,000
Des Moines Class	Cruiser		21,500	32,300
Worcester Class	Cruiser		18,500	28,000

*See also Art. 23-20.

Cargo or **freight tonnage,** a commercial expression, is the basis of the freight charge.   This tonnage may be measured by either weight or volume.   When 40 cu ft weighs 1 ton or less, the freight ton (2,240 lb) is 40 cu ft.   If, however, the cargo weighs more than 1 ton per 40 cu ft, the freight tonnage is the actual weight of the cargo.   Most ocean freight is accepted on a weight or volume basis at the shipping company's option.   Usually, whichever measurement yields the greatest revenue controls.   For instance, if the rate is $1.00 per cubic foot or $2.00 per 100 lb,

1 ton of freight by weight would cost $44.80 and 1 ton by measurement (40 cu ft) would cost $40. However, if the package measured 40 cu ft and weighed only ½ ton, the charge by measurement would still be $40, although by weight it would amount to only $22.40.

An ordinary seagoing vessel that can carry a nominal deadweight of 8,000 tons of cargo, fuel, and stores will have a displacement of about 11,500 tons, a gross of about 5,200 tons, and a net of about 3,200 tons.

The **draft** of a ship, expressed in relation to the displacement as loaded or light draft, is the depth of the keel of the ship below water level for the particular condition of loading.

Ballast is the weight added in the hold or ballast compartments of a ship to increase its draft after it has discharged its cargo and to improve its stability. It usually consists of water and is expressed in long tons. In an oceangoing tanker, salt-water ballast replaces a certain amount of petroleum when the ship is unloaded, whereas a dry-cargo or passenger vessel has separate compartments for ballast.

23-8. Harbor and Channel Lines.
The navigable waters of the United States are under the control of the Corps of Engineers, Department of the Army.

To define certain limits for channels and harbors, the following terms have become well established: A *bulkhead line* is the farthest line offshore to which a fill or solid structure may be constructed. Open-pier construction may extend outward from the bulkhead line to the *pierhead line*, beyond which no construction of any kind is allowed, except by special permit. This line is established to prevent piers from being constructed too far out into the water, since such construction might cause interference with navigation.

Pierhead lines may or may not coincide with *channel lines*, which define the limits of navigable channels that are dredged and maintained at established depths by the Federal government. These depths are usually referred to low water. Open water of navigable depth is called a *fairway*.

23-9. Harbor-Site Selection.
The decision to build a port, and its location, generally will be determined by factors having to do with need and economic justification, prospective volume of seaborne commerce, and availability of inland communications by both land and water. These considerations must precede the technical studies and planning of the port.

After the above studies have been made, the general location of the harbor, its principal use, and the type and tonnage of traffic to be handled may be established. The next step, which in some cases may be initiated during the above studies, is to make preliminary studies and layouts of the port in preparation for a complete site investigation. This investigation will gather all the information needed in making the final design of the port (see also Arts. 23-11 to 23-13).

Information for preliminary planning can usually be obtained from the following sources: U.S. Department of Commerce through the National Geodetic Survey, Navy Department through the Hydrographic Office, and Corps of Engineers, who have surveyed a great many U.S. navigable waters. Charts or information can be obtained from the Government Printing Office, Washington, D.C. 20402, or from the nearest U.S. District Engineer's Office. These charts are very valuable in the initial planning of the harbor because they give information on depth of water, general character of the bottom, and range of tides.

Meteorological data covering winds, temperature, and rainfall are published by the U.S. Weather Bureau, Washington, D.C. 20402. If there is no U.S. Weather Bureau near the site, this information may possibly be obtained from the nearest airport.

Tide and current tables are published by the National Geodetic Survey and can be obtained from the Government Printing Office.

If the port is to be located where none of the above information is available, it will be necessary to make a preliminary site reconnaissance. For the preliminary survey, aerial contour mapping may be a quick and convenient way of obtaining topography. Aerial photographs are useful, especially in examining the coast and adjacent shore for suitable locations of the port, if this has not been already fixed by other strategic considerations. Aerial photography will often show up shoals, reefs, mouths of rivers, and other important details along the shore. Soundings can be taken quite quickly with a fathometer. The depth and presence of rock, as well as the depth of overburden, can be determined by seismic equipment.

Unless the site is fixed by specific requirements of the port, several locations for the harbor will have to be studied. A comparison should determine the most protected location involving the least amount of dredging and with the most favorable bottom conditions. Also, the selected site should have a shore area suitable for development of the terminal facilities.

Depth of water, other things being equal, is a major factor in location of a port. A deep-water bay is, of course, ideal. But where a port must be located along an exposed coast, a study of hydrographic charts will generally indicate areas where the water is deep close to shore and other areas where the required harbor depth would not be reached for several thousand feet offshore. The latter might require a prohibitive amount of dredging.   However, at the deep-water location where the water a short distance offshore might be so deep that the construction cost of protective breakwaters would be prohibitive.   In cases where bulk materials are to be shipped, the solution may be an offshore anchorage with submarine pipeline if liquids are to be handled, or a lightly constructed trestle or ropeway if bulk materials are to be moved.

Bottom conditions are of utmost importance.   Underwater excavation of rock is very expensive and should be avoided, except in special cases where the rock may be needed for construction of the dock.   Also, a bottom consisting of a very deep bed of soft material, such as mud, silt, or clay, may be undesirable.   Although it can be removed easily with suction dredges, such poor foundation conditions would make construction of breakwaters and docks very expensive.

23-10. Harbor Layout.   The number and size of ships using a harbor determine its size to a large extent.   But existing site conditions also have an important influence.   Generally, unless

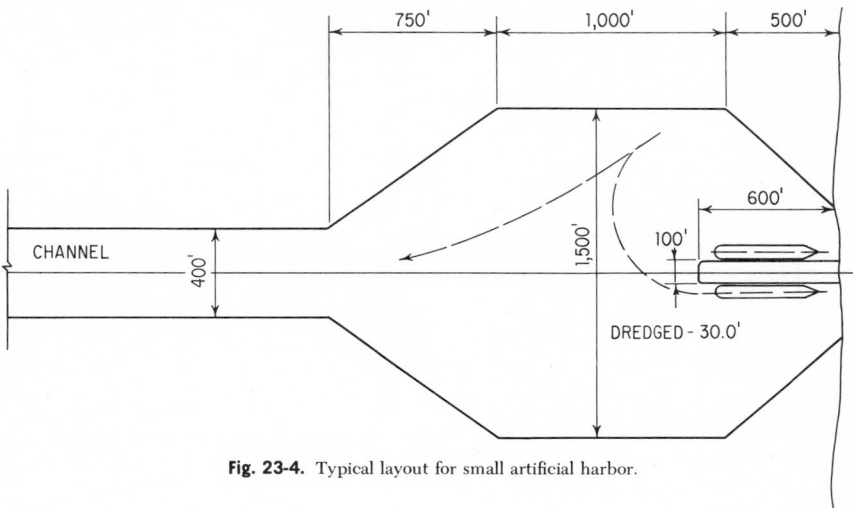

Fig. 23-4. Typical layout for small artificial harbor.

the harbor is a natural one, its size will be kept as small as feasible for safe and reasonably comfortable operations to take place.   Use of tugs to assist maneuvering of ships in docking may also influence the size of the harbor.

The usual minimum harbor area is the space required for docks plus a turning basin in front of them.   In some layouts, where a ship is turned by warping it around the end of the pier or turning dolphin, the harbor may be even smaller.   For instance, a minimum harbor with a single pier and turning basin and a long approach channel from the open sea (Fig. 23-4) can accommodate two 500-ft ships.   This artificial harbor may be formed by dredging a channel through shallow water, protected by offshore reefs and islands, and enlarging the inshore end to provide the minimum area of harbor that will meet the shipping requirements specified for the project. In leaving its berth, a ship must warp itself around the end of the pier so as not to have to back out through the long approach channel.

Another, less restricted harbor is a nearly square type of harbor, protected by two breakwater arms, with one opening.   The harbor has several docks and a turning basin with an area sufficient to inscribe a turning circle with a radius equal to twice the length of the largest ship.   This is the smallest radius a ship can comfortably turn on, under continuous headway, without the help of a tug.   Figure 23-5 shows such a harbor.

Breakwaters are required for protection of artificial and seminatural harbors.   Their location and extent depend on the direction of maximum waves, configuration of the shoreline, and minimum size of harbor required for the anticipated traffic in the port.   They may consist of two "arms" out from the shore, plus a single breakwater, more or less parallel to the shore, thereby providing two openings to the harbor; or the harbor may be protected with a single arm out from

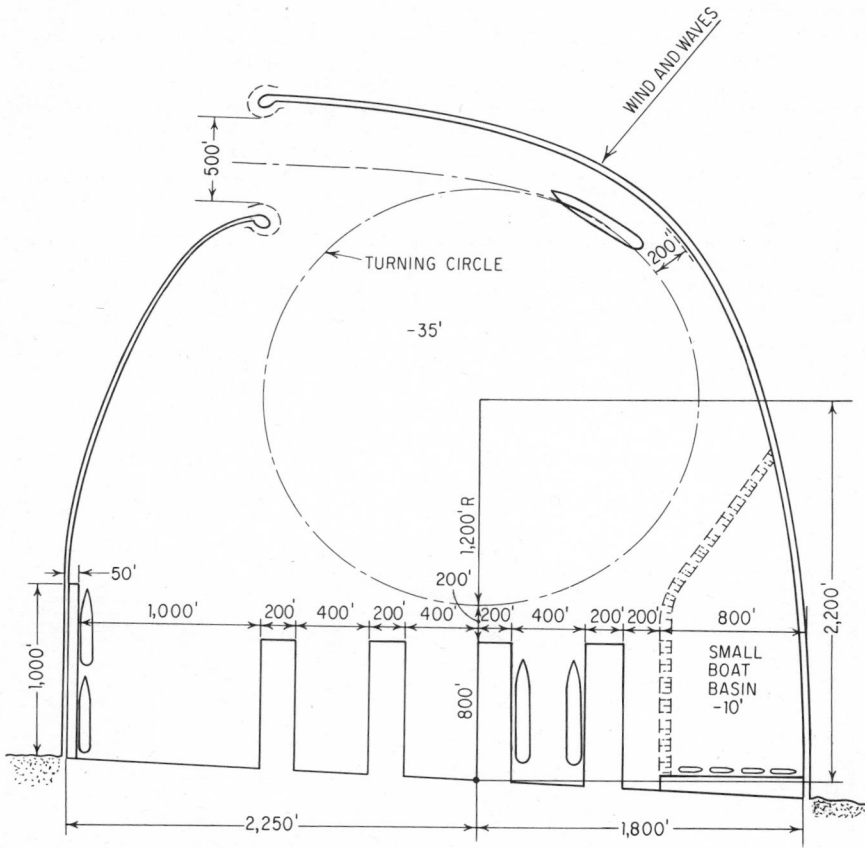

Fig. 23-5. Medium-size artificial harbor can have full-sized turning basin.

shore. Or the harbor may be protected by two arms converging near their outshore ends and overlapping to form a protected entrance to the harbor.

Selection of the most suitable arrangement of breakwaters depends principally on the direction of the maximum waves. The effectiveness of the chosen arrangement in quieting the harbor may be checked by model tests. For comfortable berthing, the wave height should not exceed 2 ft and winds should not exceed 10 to 15 mph. But wave heights up to 4 ft have been allowed where bulk cargo is being handled and where the wind direction is such as to hold a docked ship off the dock. In general, winds and current are more bothersome in docking a vessel when it is light than are relatively small harbor waves and may necessitate use of a tug.

Rarely will a location be found where the waves are from one direction only. Generally, it is better in a harbor having two openings for ships to enter from the direction of the minimum wind and waves and to leave toward the direction of the maximum wind and waves. On leaving the harbor, the ships usually have open water in which to maneuver, whereas on entering the harbor, they are immediately in a restricted area and must approach the docks at reduced speed and at a certain inclination to its face.

A single breakwater arm may be used where the waves predominate from one direction only. It also may serve where the configuration of the shoreline reduces the fetch in the opposite direction to such an extent that the wave-generating area is not sufficient to permit formation of bothersome waves within the harbor.

Harbor Entrances. To reduce wave height within a harbor, entrances should be no wider than necessary for safe navigation and for preventing dangerous currents when the tide is coming in and going out. The entrance width should be in proportion to the size of the harbor and the ships using it. In general, the following widths will be satisfactory: small harbors, 300 ft; medium harbors, 400 to 500 ft; and large harbors, 500 to 800 ft. When the entrance is between breakwaters with sloping faces, the width is measured at the required harbor or channel depth below low water. Thus, the entrances will be appreciably wider than the recommended widths at low-water level. In such cases, it is advisable to mark the full harbor depth of the entrance with buoys, one or more being placed on each side of the entrance channel.

The entrance should be on the lee side of the harbor, where possible. If the entrance must be located at the windward end of the harbor, breakwaters should overlap so that vessels may pass through the restricted entrance and be free to turn with the wind before being hit broadside by the waves. Also, the interior of the harbor will be protected from the waves.

When the entrance to a harbor is unobstructed, storm waves from the sea pass through the opening into the harbor. Unless they are reflected by a vertical surface, they will gradually decrease in height as they progress away from the entrance and as the harbor widens relative to the entrance width. Stevenson's equation [Eq. (23-13)] may be used to approximate wave height within the harbor, but model tests will give a more accurate picture of wave conditions and are essential for studying various arrangements of breakwaters for important harbors.

In tidal harbors where there are strong currents, the entrance width should be sufficient to prevent the velocity of the current through the opening at ebb tide from exceeding 4 fps; otherwise, it may affect navigation of ships and create scour at the base of adjacent breakwaters.

If waves pass through an entrance and strike a vertical face on the opposite side of the harbor, they reflect. The result is an increase in wave height within the harbor. This condition can be corrected by building wave-absorbing beaches, flat slopes of rock or granular material, in front of the vertical surface. However, when the vertical surface is a wharf or bulkhead used for berthing of ships, it is impossible to use a beach. Where conditions will permit use of a hollow structure, the outside vertical wall may be perforated or slotted and the energy of the waves absorbed in the chamber in back of the wall. Other means may be resorted to, such as installation of short wave-deflecting walls or wave traps along the approach channel to the dock.

Channel Depth. The harbor and approach channel for ideal operating conditions should be of sufficient depth to permit navigation at lowest low water when ships are fully loaded. This depth must include an allowance for the surge of the ship, which is about one-half the wave height, the out-of-trim or squat when in motion, and from 2- to 4-ft clearance under the keel, the larger figure being used when the bottom is of hard material such as rock. When there is a very soft mud bottom, a keel may at times touch bottom because of surge and squat without doing any damage to the ship, but it would be disastrous to have its fully loaded weight bump a hard rock bottom. Therefore, greater allowances must be made in computing the depth when the bottom is hard. Also the harbor and approach channel or approach sea lanes must be carefully swept to make sure that there are no obstructions, such as reefs or rocky pinnacles, boulders, or sunken ships, above the required depth for safe navigation. Since a good design is predicated on a maximum wave height in a harbor of not over 2 ft, allowing 1 ft for out-of-trim of ships, the maximum harbor depth below lowest low water then becomes the loaded draft plus 4 ft when the bottom is soft, or up to 6 ft when the bottom is rock.

The Panama Canal has a maintained dredged depth of 40 ft; the Delaware River, 40 ft; New York Harbor, 40 ft. The ports of Baltimore and Montreal have a maintained channel depth of 35 ft; that of Boston, 40 ft. In some harbors, some ships arrive or depart on the rising tide. But the advent of supertankers with deadweight tonnage over 300,000 tons and a draft exceeding 80 ft creates a need for deeper harbors. Otherwise, the cargo has to be transferred to smaller tankers or piped ashore through submarine lines from offshore anchorages.

Tides have a very important influence on harbor depth. Table 23-1 gives the tidal ranges in feet at principal ports throughout the world. It will be noted that the tidal range along the coasts of the United States seldom exceeds 10 ft, and therefore the harbors are dredged to provide the required depth for navigation at lowest low water. The condition is entirely different in the British Isles and on the western coast of Europe, where the port of Liverpool, England, has a spring tidal range of 27 ft; London, England, 20 ft; Calais, France, 20 ft; and others have even greater variations. In most cases, this fluctuation in sea level has resulted in the use of wet docks in all stages of the tide. These dock systems require entrance locks with massive gates, heavy swing or bascule bridges and the machinery for working them, pumping equipment, and other accessories. Since all this results in great cost, as well as continuing operational and maintenance expense, the question arises as to the limiting range for the natural tidal working of ports without recourse to enclosed docks. Generally, about 10 to 15 ft is considered to be the dividing point.

Docking Facilities. These may consist of a single pier or as many as a thousand piers. The number of berths depends on the anticipated number of ships that will use the port and the time it will take to discharge and take on cargo or passengers. This will vary for different kinds of cargo, but usually a vessel will not be in port more than 48 hr, and many bulk cargo ships are loaded in 24 hr or less.

Wharves and piers should be located in the most sheltered part of the harbor or along the lee side of the breakwaters. Where possible, piers should be so oriented as to have ships alongside headed as nearly into the wind and waves as possible. This is particularly important if the harbor is not well protected.

Onshore marine-terminal facilities may consist of one or more of the following, depending on the size of the port and the service it renders:

Transit sheds are located immediately in back of the apron on a pier or wharf. Their function is to store for a short period of time cargo awaiting loading or distribution after being unloaded from ships.

Warehouses may replace transit sheds at some marine terminals. But when used to supplement sheds, warehouses are usually located inland and not on the pier structure.

Bulk storage may be in open piles over conveyor tunnels, which may be covered with sheds when protection from the elements is required; in bins and silos or elevators (for grain storage); or in storage tanks (for liquids). These should be located as near the waterfront as possible, and sometimes directly alongside the wharf or pier, to enable direct loading into the hold of the ship.

A *terminal building* houses port-administration personnel and custom officials if a separate custom house is not provided. The terminal building should be located in a prominent and convenient location with respect to the docks.

Guard houses are located at strategic points in the port area, such as the entrance gates of highways and railways, entrances to piers or terminal areas, bonded storage, etc.

Stevedores' warehouses house cargo-handling gear, wash and locker rooms, and other facilities for stevedores.

Miscellaneous buildings and structures include a fire house and fire-fighting equipment, power plant, garages, repair shops, drydocks, marine railways, fishing piers, or yacht basins.

23-11. Hydrographic and Topographic Surveys. After preliminary layouts of a port have been completed and before the final design is started, it is necessary, in most instances, to obtain additional site information.

A *hydrographic survey*, if not already available, should be made to determine the elevations of the bottom of the body of water and should extend over an area somewhat larger than the proposed channel and harbor. In addition, the survey should locate the shoreline at low and high water and all structures or obstructions in the water and along the shore, such as sunken ships, reefs, or large rocks.

Determination of the relief of the bottom of the body of water is made by soundings or by the use of a fathometer designed for hydrographic surveys. The latter method is being used by the National Geodetic Survey and has superseded lead-line soundings to a large extent. A fathometer or depth-recording instrument is usually mounted in a motorboat, which is kept on course on established range lines, as the recording chart registers a natural profile of the bottom. The fathometer, when operated by experienced personnel and properly adjusted and calibrated daily, is superior to lead-line soundings in both accuracy and speed with which a survey can be made.

The depths of soundings are referred to water level at the time made and later corrected to the datum water level by means of tide gages or tide tables. Therefore, it is important to keep a record of the time and day the soundings are made.

Soundings should be made at about 25-ft intervals along lines from 50 to 100 ft apart, depending on the irregularity of the bottom. Closer spacing may be needed where greater detail is required to determine sharp changes in the profile of the bottom or to outline obstructions.

Soundings are plotted, usually relative to low-water datum, on a drawing (hydrographic map), which should show the datum, high- and low-water lines, contour lines of equal depth interpolated from the soundings, and principal land and water features. Contour depths may be in either feet, meters, or fathoms, although the last is not used generally for making harbor and marine-terminal studies and layouts. Since the sea bottom is usually less precipitous and the slopes more gentle and uniform than those on land, the scale of the hydrographic map may be somewhat smaller than would normally be used for plotting land topography. Unless the harbor area is very large, a scale of 1 in. = 200 ft or 1:2,000 in a proportional scale will be satisfactory.

It is desirable to have all the hydrography on one sheet, because this gives a better overall picture of the harbor. In general, the scale should be large enough so that not more than 10 contour lines, in 2-ft intervals, occur within 1 in.

If dredging of a harbor or channel is required, the material is usually measured in place to determine the quantity for payment. To determine this quantity, soundings on fixed sections are taken before and after dredging, and the changes in cross sections are determined by computation or planimeter. It is usually specified that payment will be made for material removed to a maximum of 2 ft below the required dredged bottom, but all material must be removed to at least the minimum depth specified.

A *topographic survey* of the marine-terminal area should be made, to obtain ground contours at 2- to 5-ft intervals. The larger figure is used where terrain is rough and in areas where there is to be little or no construction of importance. In building areas, elevations on 25-ft centers in two directions, with additional elevations at abrupt changes in ground, provide satisfactory information. Where there is dense ground cover, the cross-profile method is most suitable. The profiles may be made with level and tape or stadia, on about 100-ft centers, by clearing paths to permit an unobstructed line of sight. The ground between the 100-ft profiles should be examined, as far as possible, and any prominent irregularities in ground level estimated and noted, so that contours, which are interpolated from elevations along the profiles, can be estimated for the areas in between.

Topographic maps, in addition to showing the contours of the ground, should locate all borings and test pits, buildings, utilities, and any prominent landmarks. Contours generally are referred to high-water datum. The map scale should be such that the contour lines are not spaced closer than 30 to the inch. Where considerable detail is involved the scale should be 1 in. = 100 ft or 1 : 1,000 or less, but for small-scale maps 1 in. = 1,000 ft or 1 : 10,000 or more may be used.

23-12. Soil Investigations. For harbor and channel areas, borings or probings should be made at strategic points to obtain information on subsoil conditions at locations of breakwaters, piers, wharves, bulkheads, and other marine structures.

When dredging is involved, borings or probings should be taken on about 250- to 500-ft centers over the area to be dredged.

Borings that are made at the location of marine structures should be located along definite lines, such as the center line of a pier or breakwater. And the borings should be close enough to enable a reasonably accurate profile of the soil strata to be plotted. Usually, 100-ft centers will suffice for this purpose. If a structure is of considerable width, two or more lines of borings should be made so that transverse sections of the soil strata can be plotted.

Depth of the borings depends on the soil encountered and depth to bedrock. In most locations, a penetration of 150 ft below low-water level will encounter either rock or soil of suitable bearing value to support pile or caisson foundations. Generally, a penetration of 40 ft into firm material insures an adequate support for marine structures. For determining information on soil to be dredged, borings or probings need be carried only to a depth of 2 ft below dredged bottom. But if rock is encountered above this level, one or more of the borings should be drilled to a depth of 5 ft below dredged bottom and as much of the core recovered as possible. The core should be analyzed to determine the character of the rock, which influences the cost of its removal. For determining the elevations of the top of rock, jet probings may be used in place of borings. These jet probings are quicker and cheaper to perform.

Except for soil under breakwaters, the additional load imposed by open piers and other similar marine structures on the underlying soil is not large. Therefore, dry sample wash borings made with a 2½-in. casing usually provide adequate information. For obtaining dry samples, a split-barrel sampler is commonly used. However, in some locations where the soil is plastic, it is desirable to make undisturbed-soil sample borings and soil tests to determine the depths to which piles or cylinders should be driven. Likewise, where there is an appreciable load on the underlying plastic soil, such as under breakwaters or dock walls, or where sheetpile bulkheads or dock walls support the lateral pressures of fill and surcharge in back of them, it is desirable to make soil tests on undisturbed samples to determine the shearing strength and consolidation coefficient of the soil.

When undisturbed samples are to be taken in silts and clays, the casing should be preferably not less than 3½ or 4 in. in diameter, to permit use of a 3-in. thin-wall Shelby tube sampler.

To make borings and probings over water, it is necessary to have a small, flat-deck barge or pontoon on which to support the boring equipment. The pontoon is sometimes constructed of empty oil drums and wood framing (Fig. 23-6). Where the water is very rough, it is necessary

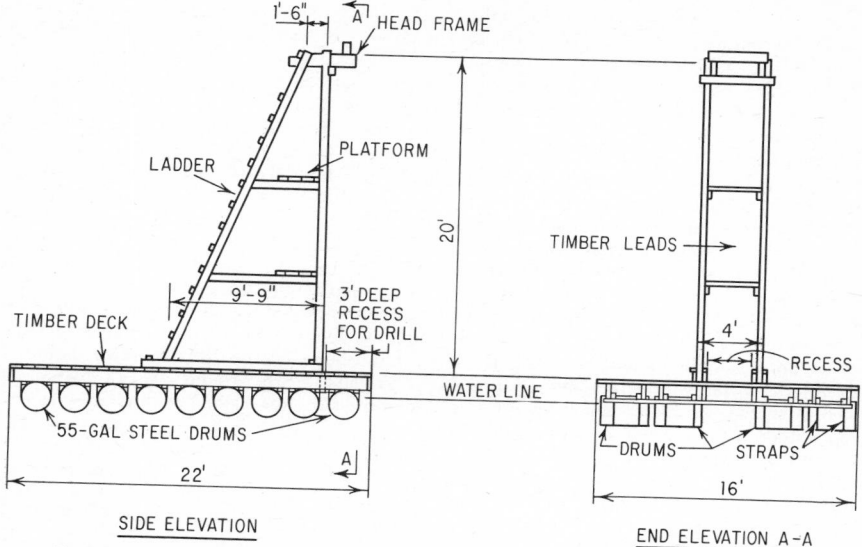

Fig. 23-6. Drill barge for making borings over water can be constructed in remote locations.

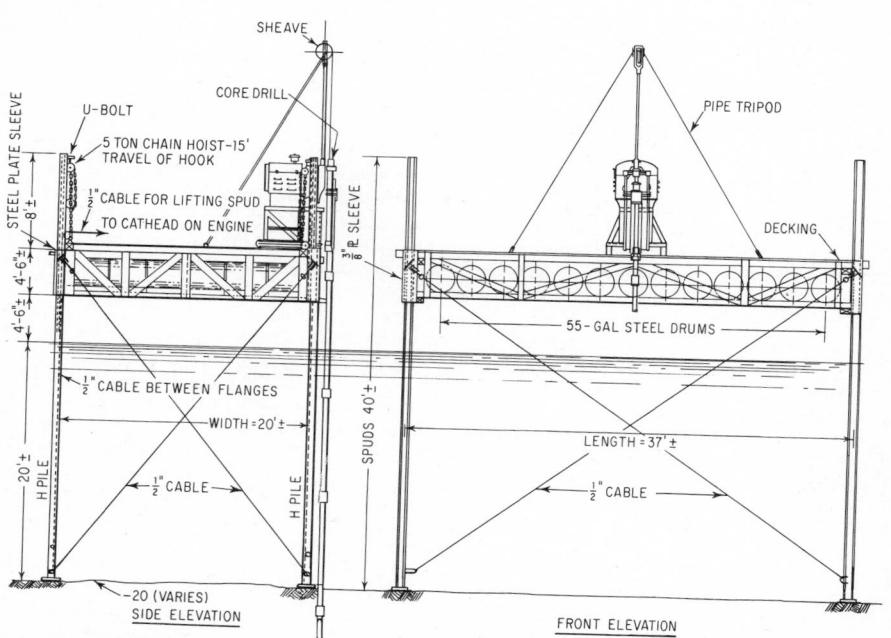

Fig. 23-7. Drill barge for making borings through rough water stands on spuds seated on the sea bottom.

to lower spuds onto the bottom and raise the barge above the reach of the waves.   A drill barge or pontoon equipped with spuds is shown in Fig. 23-7.

23-13. Current and Tidal Studies.   When investigating the site of a proposed port or harbor, the engineer should obtain information on the general direction and velocity of currents in the area.   One type of device used in making current observations is shown in Fig. 23-8.   This device or target consists of a surface float with a pole and flag, a submerged float that is moved by the current, and a counterweight consisting of a wire basket to which is added scrap metal in sufficient quantity that the surface float rides evenly on the water surface.

In making current observations, it is customary to lay out base lines on shore with a transit set up at each end of the lines.   The float is then dropped in the water beyond the area of the breakers and permitted to move in the direction of the current.   The transitmen sight on the flagpole at predetermined time intervals, and the course and speed of the float are determined by plotting the results of the observations.   Usually, a dozen or more of these tests are performed during the ebb and flow of the tide.   These tests, of course, should be performed during periods of relative calm.   If the observations are carefully made, the results will give the general direction and velocity of the currents in the area being investigated.

Tide Investigations.   If the National Geodetic Survey has a tide-gage station in the area, tide tables may be available.   Where this information is not available, it will be necessary to install a tide gage to determine the mean high- and mean low-water levels and to establish a datum for referencing the water level when making soundings.

The tide gage, in its simplest form, consists of a vertical post driven into the bank below lowest low-water level and graduated in feet or meters.   Where there is an extreme range of tide, more than one such indicator may be required.   They should be located across the bank at ascending levels, to cover the complete range of the tide.

In locations where there is a swell, the gage may consist of a rod with a float at its lower end and a pointer on its upper end.   The rod is enclosed in a pipe for protection from the waves.   Above

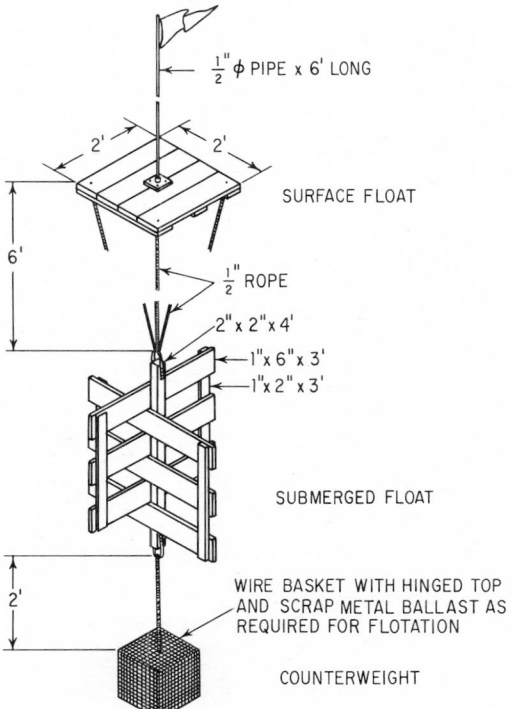

Fig. 23-8.   Device for making current observations.

the pipe, there is a graduated scale along which the pointer travels as the water level rises and falls. The bottom of the pipe should have a small hole to permit the water to seek its true level inside the pipe. Such a device for measuring the tide is usually mounted alongside a pier or wharf.

In important locations where it is desirable to have a continuous recording of the tide level over a long period of time, a self-recording apparatus should be installed. This may consist of a float hung in a stilling well with a small opening below the lowest wave trough. The float is connected to a self-recording instrument by a small cable or chain with a counterweight. The recording instrument may be set on a stand or table inside a building, or it may have a weather-proof housing, with a glass face, for outdoor mounting. The recording mechanism may be actuated by an electric or spring-wound clock, the recording pencil tracing a curve of the water level with respect to time on graph paper. All parts of the apparatus should be constructed of rustproof materials, for example, a hard-rubber float, stainless-steel cable, and an aluminum case to house the recording instrument.

23-14. Hydraulic Models. Probably no single factor has contributed so much to placing the design of harbors on a sound engineering basis as has the testing of hydraulic models. Hydraulic models of harbors are usually constructed to a linear scale of from $1:100$ to $1:150$, model to prototype, depending on the size of the harbor and the available wave-basin space. Models are designed and operated in accordance with Froude's model laws. The model-prototype relationships derived from the linear scale L_r of $1:150$ and a specific-weight scale γ_r of $1:1$ are as shown in Table 23-9.

Table 23-9. Model-Prototype Relationships

Characteristics	Dimensions	Model-prototype scales
Length	L	$L_r = 1:150$
Area	L^2	$A_r = L_r^2 = 1:22,500$
Volume	L^3	$\overline{V}_r = L_r^3 = 1:3,375,000$
Time	T	$T_r = L_r^{1/2} = 1:12.25$
Velocity	L/T	$V_r = L_r^{1/2} = 1:12.25$
Unit pressure	F/L^2	$P_r = L_r\gamma_r = 1:150$
Force	F	$F_r = L_r^3\gamma_r = 1:3,375,000$
Weight	F	$W_r = L_r^3\gamma_r = 1:3,375,000$

Models are constructed of concrete on a wave-basin floor, usually inside a large hangar or shed, which protects them from wind and rain, and are well lighted so that sharp photographs can be taken of wave patterns. Breakwaters are constructed of crushed stone, concrete rectangular blocks, or irregular shapes, such as tetrapods or tribars, to simulate the proposed construction. Waves are reproduced to scale by a movable, plunger-type wave machine which can rotate about a vertical axis so as to change the direction of the wave. Wave filters are usually placed in front of the machine. Their purpose is to diminish, as far as possible, the magnitude of reflected waves and serve the same function as absorbing beaches located in front of vertical walls. Wave heights are measured with electric wave-height gages set at key points in the harbor to measure the amplitude of the agitation.

(A. DeF. Quinn, "Design and Construction of Ports and Marine Structures," McGraw-Hill Book Company, New York.)

HARBOR PROTECTION—BREAKWATERS

A breakwater is a structure constructed to form an artificial harbor with a water area so protected from the effect of sea waves as to provide safe accommodation for shipping. There are two classes of breakwaters: One protects commercial harbors or their entrances, while the second shelters an anchorage or roadstead being used by vessels to escape the violence of storms or while awaiting orders and their turn to dock. Such an anchorage may be an outer harbor where there are no docks.

23-15. Types of Breakwaters. There are two main types of breakwaters, the mound type and the wall type. Falling under the first classification and identified more commonly by the materials out of which they are constructed are the following: natural rock, concrete block, a combination of rock and concrete block, and concrete tetrapods and tribars, or other irregular shapes.   These breakwaters may be supplemented in each case by concrete monoliths or sea-walls to break the force of the waves and to prevent splash and spray from passing over the top. In the second main classification of breakwaters there are such types as concrete-block gravity walls, concrete caissons, rock-filled sheetpile cells, rock-filled timber cribs, and concrete or steel sheetpile walls.

23-16. Mound Breakwaters.   The abundance of durable rock at economical costs has led to adoption of rock-mound breakwaters to a greater extent than any other type for the protection of harbors along the North American and South American coasts.   There are many variations in the classes of rock fill and the locations and proportions of these materials within a rock-mound breakwater.   But usually such a breakwater consists of three distinct parts: the armor (primary and secondary covers), the first underlayer, and the second underlayer or core.   A typical rock-mound breakwater cross section is shown in Fig. 23-9 for a no-overtopping requirement.

When natural rock is not available, or where it cannot be produced economically or in large enough size for armoring the breakwater, concrete blocks, or irregular concrete shapes, such as tetrapods, tribars, quadripods, or hexapods, may be used as armor.   A typical breakwater armored with concrete blocks laid pell-mell is shown in Fig. 23-10.   The weight of large artificial blocks of concrete is limited only by the equipment capable of handling them.   Except for this limitation, it is possible to design breakwaters for waves of any size; in recent ex-perience, these waves do not exceed 45 ft in height at any location where a breakwater would need to be constructed.   Concrete blocks of 50 to 60 tons are commonly used, and on rare occasions blocks weighing as much as 400 tons have been used.

Tetrapods are four-legged, truncated-cone-shaped, precast-concrete units (Fig. 23-11) developed by the NEYRPIC Hydraulic Laboratory, Grenoble, France, and licensed under the name Sotramer.   Units often used for large breakwaters weigh about 25 tons.

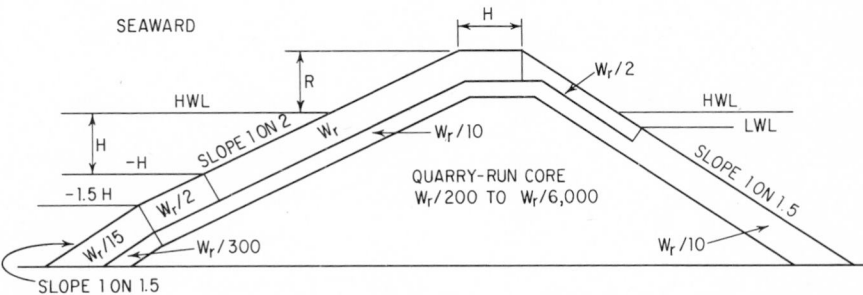

Fig. 23-9. A rubble-mound breakwater, in section, consists of a core of quarry-run rock protected by layers of armor rock, with weight of individual pieces as indicated [see Eq. (23-26)].   *HWL* = high water level; *LWL* = low water level; *R* = wave run-up.

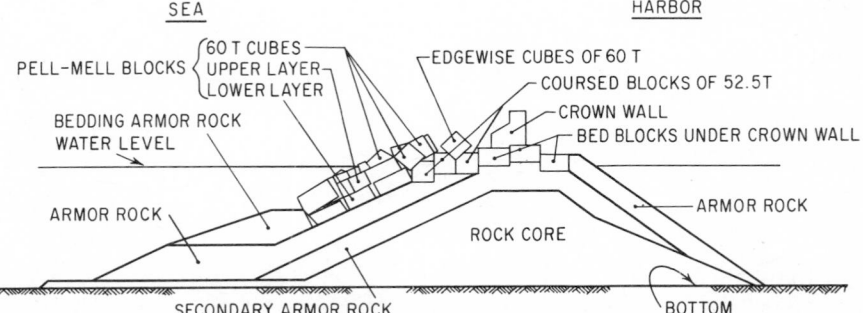

Fig. 23-10. Breakwater armored with concrete blocks laid pell-mell.

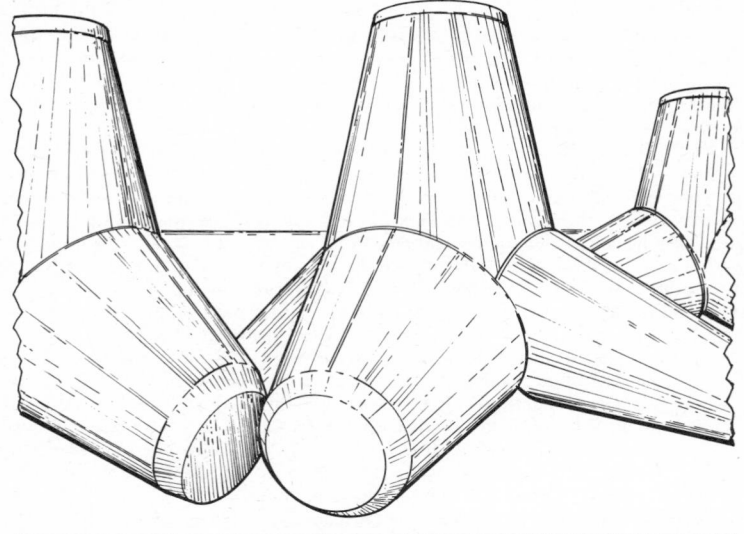

Fig. 23-11. Tetrapod armor unit. (*Courtesy of NEYRPIC, Grenoble, France.*)

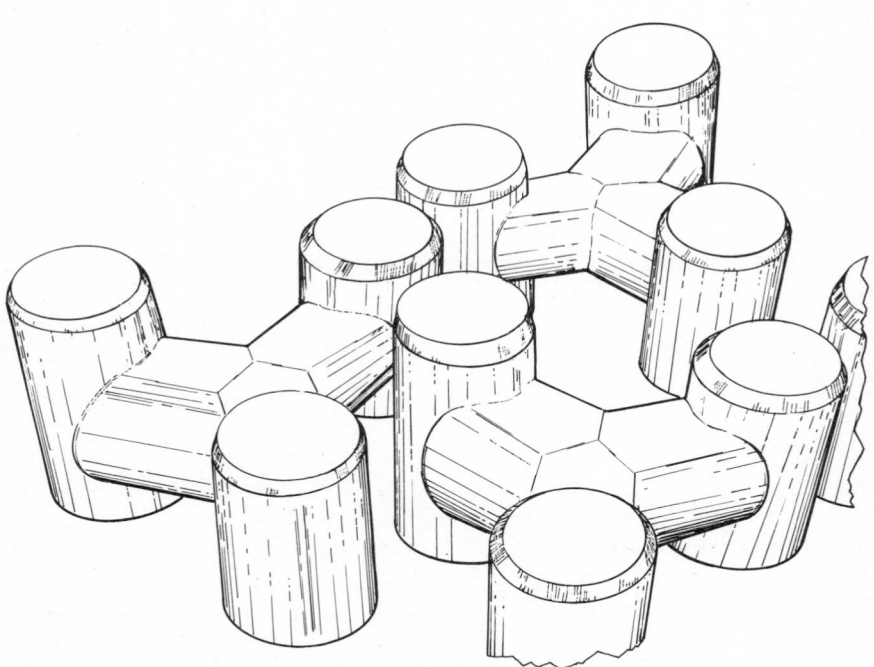

Fig. 23-12. Tribar armor unit. (*Courtesy of R. Q. Palmer, Consulting Engineer, Honolulu.*)

Tribars are special-shaped concrete armor units (Fig. 23-12) patented by Robert Q. Palmer, consulting engineer, Honolulu, Hawaii.

Quadripods are similar in shape to tetrapods except that the axes of the three legs forming the base are all in the same plane.

Hexapods are six-legged, truncated-cone-shaped units.

Dimensions and other properties of various armor units are given in Research Report 2-11, June, 1968, U.S. Army Engineer Waterways Experiment Station, Vicksburg, Miss.

Special-shaped concrete armor units, such as tetrapods and tribars, have the advantage over plain concrete blocks of permitting steeper slopes and units of lighter weight. This is due to their better "shape factor" and superior absorption of wave energy.

Unit Weights Required. Stability of a rock breakwater is dependent mainly on the weight and shape of the individual pieces of armor rock and the slope on which they are laid. But the units must be properly placed on the slope and interlocked with each other to form a stable and reasonably close-fitting envelope around the core. It can be proved analytically and by model tests that the weight required for the individual pieces of armor rock will vary with the degree of slope on which they are laid; i.e., steeper slopes require heavier rock, and flatter slopes, lighter rock.

Because of the complexity of wave action on a mound breakwater composed of irregular shapes of rock or concrete armor units, it is improbable that a single analytical and empirical formula can be developed to evaluate correctly all the variable conditions. But from tests begun in 1951 by the U.S. Army Engineer Waterways Experiment Station on the stability of rubble-mound breakwaters with quarry-stone armor units as cover-layer material, and from later, similar tests on irregular-shaped concrete units, Eq. (23-26) was developed by R. Y. Hudson, chief of the Wave Action Section. It has been generally accepted by engineers for the design of mound breakwaters using such armor units as quarrystones, tetrapods, quadripods, hexapods, tribars, modified cubes, and modified tetrahedrons.

$$W_r = \frac{\gamma_r H^3}{K_\Delta (S_r - 1)^3 \cot \alpha} \tag{23-26}$$

where W_r = weight of individual armor unit, lb
H = height of wave, ft
K_Δ = dimensionless, experimental coefficient that varies primarily with the shape of armor unit
α = angle of armor slope with horizontal
S_r = specific gravity of armor unit
γ_r = specific weight of armor, lb per cu ft

Table 23-10 gives suggested values of K_Δ for use in the above formula.

Table 23-10. Suggested Values of K_Δ for Use in Eq. (23-26)*

Armor unit	Method of placing	Layers, n	Coefficient K_Δ†			
			1	2	3	4
Smooth quarrystone	Pell-mell	2	2.6	2.5	2.4	2.0
Rough quarrystone	Pell-mell	2	3.5	3.0	2.9	2.0
Smooth quarrystone	Pell-mell	>3	3.2	3.0	2.9	
Rough quarrystone	Pell-mell	>3	4.3	4.0	3.8	
Modified cube	Pell-mell	2	7.5	7.0	5.0	
Tetrapod	Pell-mell	2	8.3	8.0	6.5	6.0
Quadripod	Pell-mell	2	8.3	8.0	6.5	6.0
Hexapod	Pell-mell	2	9.0	8.5	7.0	6.0
Tribar	Pell-mell	2	10.0	9.5	7.5	6.0
Tribar	Uniform	1	15.0	12.0	9.5	

*From *Miscellaneous Paper* 2-453, September, 1961, U.S. Army Engineer Waterways Experiment Station.

†1. For breakwater trunk in deep water of sufficient depth to prevent the breaking of waves, with no overtopping.

2. For breakwater trunk in shallow water and waves of such size that they break directly on the structure's slope.

3. For conical heads of breakwaters situated in water sufficiently deep to prevent the breaking of waves.

4. For conical heads of breakwaters subjected to forces from breaking waves.

The armor units for the primary cover layer, the weights of which are determined by Eq. (23-26), should be extended to a depth equal to the wave height H below high water level (HWL) when the breakwater is in deep water and to the bottom when it is in shallow water. In deep water, the slope for which the armor units are figured should extend to a depth of $1.5H$ below HWL. Below this depth, the slope can be reduced to 1 on 1½ (Fig. 23-9).

In deep water, the weight of armor units in the secondary cover layer between depths equal to H and $1.5H$ below HWL should be equal to one-half the weight of armor units in the primary cover layer. And below a depth of $1.5H$, the weight can be reduced to $W_r/15$.

The first underlayer, on which the armor units in the primary cover layer rest, should consist of at least two layers of rock weighing about $W_r/10$. Under the secondary cover layer, below a depth of $1.5H$, the weight may be reduced to $W_r/300$. The second underlayer, or core, should consist of rock weighing about $W_r/200$ to a depth of $1.5H$, and below this level the weight may be reduced to about $W_r/6,000$, or what is generally known as quarry-run material.

Miscellaneous Paper 2-453 of the U.S. Army Engineer Waterways Experiment Station, Vicksburg, Miss., gives the following formulas for thickness of cover layer and number of armor units for a given surface area:

$$r = nk_\Delta \left(\frac{W_r}{\gamma_r} \right)^{1/3}$$

(23-27)

$$N_r = Ank_\Delta \left(1 - \frac{P}{100} \right) \left(\frac{\gamma_r}{W_r} \right)^{2/3}$$

(23-28)

where r = thickness, ft, of n layers
W_r = weight of individual armor unit, lb
A = surface area, sq ft
P = average porosity of cover layer, percent (see Table 23-11)
k_Δ = experimental coefficient (see Table 23-11)
N_r = required number of individual armor units for a given surface area A
γ_r = specific weight of armor, lb per cu ft
n = layers of armor units

Table 23-11. Porosity of Cover-Layer Armor Units*

Armor unit	Method of placing	Layers, n	k_Δ	P, %
Quarrystone	Pell-mell	2	1.0	38
Quarrystone	Pell-mell	>3	1.0	40
Modified cube	Pell-mell	2	1.1	47
Tetrapod	Pell-mell	2	1.0	50
Quadripod	Pell-mell	2	1.0	50
Hexapod	Pell-mell	2	1.15	47
Tribar	Pell-mell	2	1.0	54
Tribar	Uniform	1	1.13	47

*From Miscellaneous Paper 2-453, September, 1961, U.S. Army Engineer Waterways Experiment Station.

Crest Elevation. Since the primary function of a breakwater is to provide the harbor with adequate protection from wave action, it is important to have the crest at an elevation that will prevent serious overtopping of the breakwater. Most waves will break on or just before they reach the armor slope and will run up the sloping surface.

Research Report 2-2, July, 1958, of the U.S. Army Engineer Waterways Experiment Station provides information, based on results of tests, on wave run-up on rubble-mound breakwaters. The report concludes that breakwater slope and wave steepness are the primary variables affecting wave run-up in water of depths corresponding to relatively large values of relative height H/d, where H = wave height, ft, and d = water depth, ft. The report also noted that wave run-up decreases with increasing values of wave steepness and flatness of slope. Wave run-up R, ft, is measured vertically above still-water level. The run-up factor R/H for smooth, impervious slopes obtained in model tests of Lake Okeechobee levees is about twice that obtained for the comparatively rough and porous slopes used in the stability tests of rubble-mound breakwaters.

For an average condition (where the ratio of wave height to length H/L equals 0.07) and for a slope of 1 on 2, the ratio R/H of the run-up on a rubble-mound breakwater with smooth, impervious slopes to the height of the wave is about 0.9. Therefore, if the crest is placed at a height equal to H above the highest tide level, the breakwater should be reasonably free of being overtopped.

23-17. Vertical-Wall Breakwaters. Vertical-wall breakwaters have been constructed of concrete blocks, concrete or steel caissons, rock-filled steel sheetpile cells, rock-filled timber cribs, and concrete or steel sheetpile walls.

Concrete blocks have been used more often in Europe than in the Western Hemisphere for construction of vertical-wall breakwaters. *Concrete caissons* have been used quite extensively for breakwaters in the Great Lakes, as well as for protection of harbors in Europe. Concrete caissons have the advantage of reducing considerably construction time on water. This is an important factor where the sea is rough and the working time of floating equipment is limited. A large amount of the work can be done on shore, and a period of relatively good weather and calm water can be selected for the actual installation. Breakwaters constructed of concrete blocks or caissons are usually founded on a base of rubble, unless the depth of water does not exceed 50 to 60 ft. In shallower depths, unless the bottom is extremely hard and resistant to scour, a gravity wall always should be placed on a foundation mat of rubble or other suitable material, of sufficient depth to distribute the load to a safe bearing pressure on the underlying soil. The mat should extend beyond the toe a sufficient distance to prevent scour and undermining of the breakwater. As a general rule, this distance should be not less than one-fourth the maximum wavelength, if scour is to be completely avoided.

Cellular sheetpile breakwaters have been used with considerable success on the Great Lakes. However, their use has never become widespread because they are difficult to construct in exposed locations. Where used, they are generally of the self-supporting type, i.e., each cell is stable by itself when filled with rock or other suitable material. The sheeting must extend to a sufficient depth below the harbor bottom to prevent undermining of the cell by erosion of the bottom. Minimum depth of penetration usually is not less than 10 ft, unless the bottom is rock or other very hard material. It is customary to place riprap against the toe of the sheeting to protect the bottom against erosion. The top of the sheeting should extend to twice the height of the maximum wave above high water. But the sheeting may terminate at or just above mean high water, with a seawall of cast-in-place concrete constructed to the required height. When the sheetpiling is extended to the full height, it may be capped with heavy rock, concrete blocks, or a cast-in-place concrete slab.

Steel and concrete sheetpile breakwaters are sometimes used where the bottom is of soft material that extends to a great depth. The sheetpiling is usually capped with concrete and supported by batter piles. This type of construction is suitable only when the height of wave does not exceed about 10 ft.

(A. DeF. Quinn, "Design and Construction of Ports and Marine Structures," McGraw-Hill Book Company, New York.)

MOORING STRUCTURES AND APPURTENANCES

23-18. Types of Marine Mooring Structures. A **dock,** in general, is a marine structure for mooring or tying up of vessels, loading and unloading cargo, or embarking and disembarking passengers. Often, piers, wharves, bulkheads, and, in Europe, jetties, quays, or quay walls are called docks. In Europe also, where there are large variations in tide level, a dock is commonly considered an artificial basin for vessels and is called a **wet dock.** When the basin is pumped out, it is termed a **dry dock.**

A **wharf** or **quay** is a dock that parallels the shore. It is generally contiguous with the shore but may not necessarily be so. On the other hand, a **bulkhead** or **quay wall,** while similar to a wharf and often referred to as such, is backed up by ground; the name is derived from the very nature of holding or supporting ground in back of it.

In many locations where industrial plants are to be built adjacent to water transportation, the ground will be low and marshy. It is therefore necessary to fill it in. The fill is often obtained by dredging the adjacent waterway, creating a navigable channel or harbor along the property. To retain the made ground, which will now be at a much higher elevation along the waterway, a bulkhead is usually installed. This, or a part of its length, may be used as a wharf for docking vessels if mooring appurtenances, paving, and facilities for handling and storing cargo are added. It is then termed a **bulkhead wharf.**

A **pier** or **jetty** is a dock that projects into the water.   Sometimes, it is referred to as a **mole.** When built in combination with a breakwater, it is termed a **breakwater pier.**   In contrast with a wharf, which can be used for docking on one side only, ships may use a pier on both sides.   But there are instances where only one side is used, owing to either the physical conditions of the site or the lack of need for additional berthing space.

A pier may be more or less parallel to the shore and connected to it by a mole or trestle, generally at right angles to the pier.   In this case, the pier is commonly referred to as a **T-head pier** or **L-shaped pier,** depending on whether the approach is at the center or at the end.   (See also Arts. 23-19 and 23-20.)

Dolphins are marine structures for mooring vessels.   They are commonly used in combination with piers and wharves to shorten the length of these structures.   Dolphins are a principal part of the fixed-mooring-berth type of installation used extensively in bulk-cargo loading and unloading installations.   Also, they are used for tying up ships and for transferring cargo between ships moored along both sides of the dolphins.   Dolphins are of two types: breasting and mooring.

Breasting dolphins, usually the larger of the two types, are designed to take the impact of a ship when docking and to hold the ship against a broadside wind.   Therefore, they are provided with fenders to absorb the impact of the ship and to protect the dolphin and ship from damage. Breasting dolphins usually have bollards or mooring posts to take the ship's lines, particularly

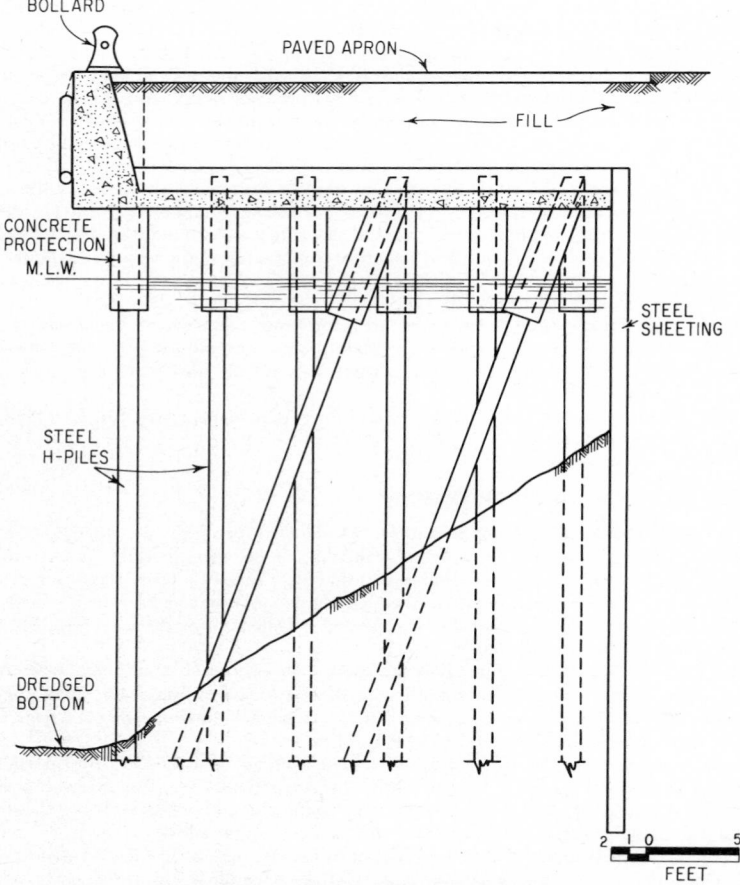

Fig. 23-13. Relieving-platform-type wharf.

springing lines for moving a ship along the dock or holding it against the current. These lines are not very effective in a direction normal to the dock, particularly when a ship is light.

To hold a ship against a broadside wind blowing in a direction away from the dock, additional dolphins must be provided off the bow and stern, some distance in back of the face of the dock. These are called *mooring dolphins*. They are not designed for the impact of the ship, since they are away from the face of the dock, where they will not be hit. If two mooring dolphins are to be used, they should be located about 45° off the bow and stern and permit mooring lines not less than 200 ft or more than 400 ft long. The largest ships may require two additional dolphins, off the bow and stern. These dolphins are usually located so that the mooring lines will be normal to the dock, which makes them most effective in holding the ship against an offshore wind. Mooring dolphins are provided with bollards or mooring posts and with capstans when heavy lines are to be handled. The maximum pull usually should not exceed 50 tons on a single line, or 100 tons on a single bollard if two lines are used. (See also Art. 23-22.)

A fixed mooring berth is a marine structure consisting of dolphins for tying up a ship and a platform for supporting the cargo-handling equipment. The platform is usually set back 5 to 10 ft from the face of the dolphins so that the ship will not come in contact with it. Therefore, the platform does not have to be designed to take the impact of the ship when docking.

Moorings for ships consist of ground tackle placed in fixed positions for attaching a ship's lines. Each unit of ground tackle consists of one or more anchors with chain, sinker, and buoy to which the ship's line is attached. These mooring units are usually located so as to take the bow and stern lines and, if the ship is large, one or more breasting lines. For some moorings, where the wind is in one direction, the ship may use its own bow anchor and the fixed tackle off the bow may be omitted. (See also Art. 23-24.)

23-19. Open and Closed Construction for Docks. Wharves, piers, bulkheads, and fixed mooring berths fall generally into two broad classifications: docks of open construction with their decks supported by piles or cylinders; and docks of closed or solid construction, such as sheetpile cells, bulkheads, cribs, caissons, and gravity (quay) walls.

Docks of open construction may be further subdivided into high-level decks and relieving-type platforms, in which the main structural slab is below the finished deck and the space between is filled to provide additional weight for stability (Fig. 23-13). High-level decks usually have a solid deck slab. But for oil piers the slab may be of skeleton construction and omitted at the pipeway (Fig. 23-14).

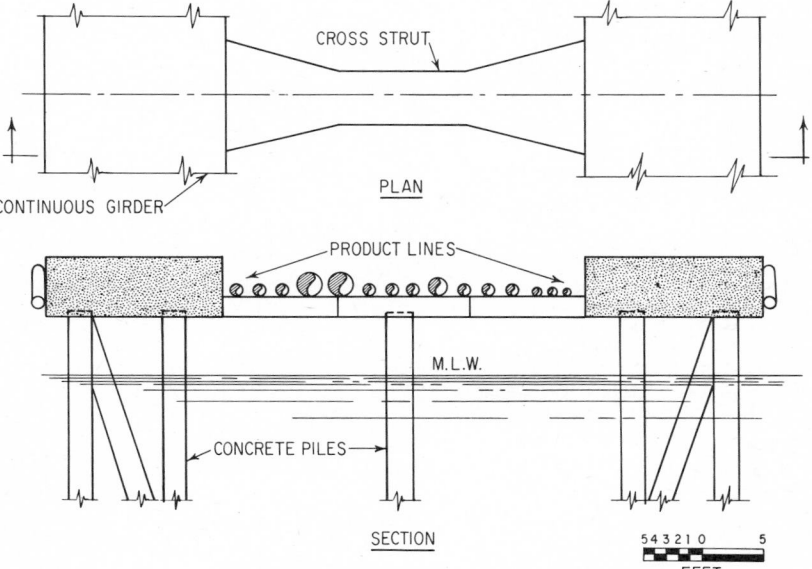

Fig. 23-14. Oil pier with skeleton framing.

Open-type Construction.   The deck may be of wood, usually creosoted, or reinforced concrete or a combination of concrete and steel or wood.   Precast- or prestressed-concrete slabs and beams also have proved to be an economical form of construction.

The deck may be supported on piles, which may be wood (usually creosoted), steel (H section or pipe), or reinforced concrete; or on large cylinders or caissons, which may be of steel or reinforced concrete.   Prestressed-concrete piles and cylinders have been used, particularly in deep water and where soft bottom conditions required very long foundation supports.   Prestressing simplifies the handling of long piles and reduces cracking, because the precompression is made sufficient to overcome the tensile bending stresses or to reduce them to an amount that will not cause the concrete to crack.

Solid-type Dock Construction.   Steel sheetpile cells are quite commonly used where depth of water does not exceed 50 ft and bottom conditions are suitable for support of gravity-type structures.   The cells are generally capped with a concrete slab and bulkhead wall above water level (Fig. 23-15).   Cells utilize flat-web steel piling, which acts in tension to retain the fill in-

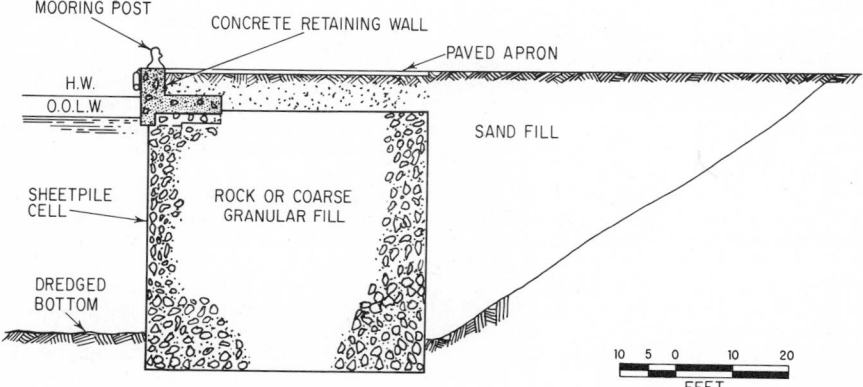

Fig. 23-15. Solid-type bulkhead wharf composed of rock-filled, steel sheetpile cells.

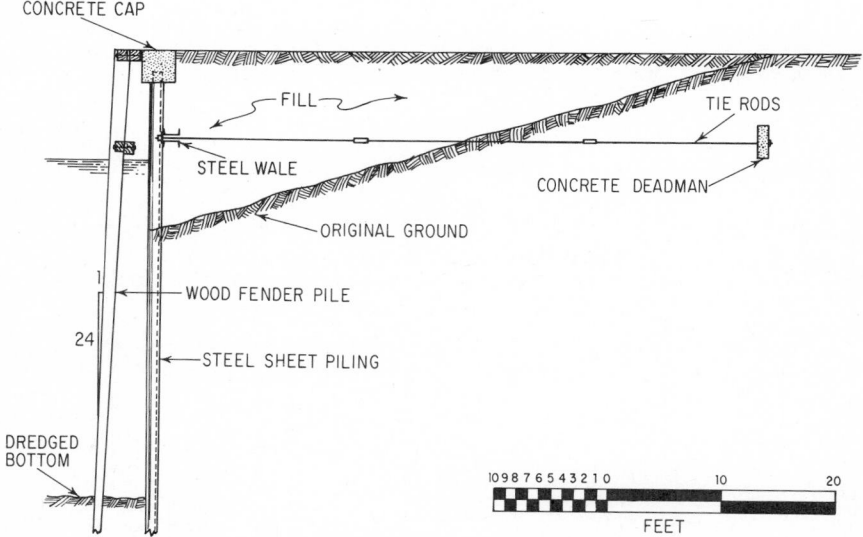

Fig. 23-16. Sheetpile bulkhead supported by tie rods and anchor wall.

side, thereby forming a gravity wall of sufficient weight and shearing strength to resist over-turning or sliding at the base.   Cells may be circular in shape, or they may have circular ends and straight walls.

Sheetpile bulkheads may be constructed of wood, steel, or concrete sheetpiles.   Sheetpiles may be supported by tie rods attached to an anchor wall or anchor piles located a safe distance in back of the face of the bulkhead (Fig. 23-16).   In shallow installations and where the bottom has good supporting value, the sheetpiles may be driven deep enough to act as a cantilever without additional support.

Concrete caissons are used quite extensively for construction of wharves or quay walls, especially in Europe.   Caissons may have open wells and cutting edges so that they may be sunk below the dredged bottom to obtain a firm support, or they may have a closed bottom (Fig. 23-17).   When they have a closed bottom, they are lowered onto a prepared foundation, usually consisting of a gravel or crushed-stone bed or leveling course.

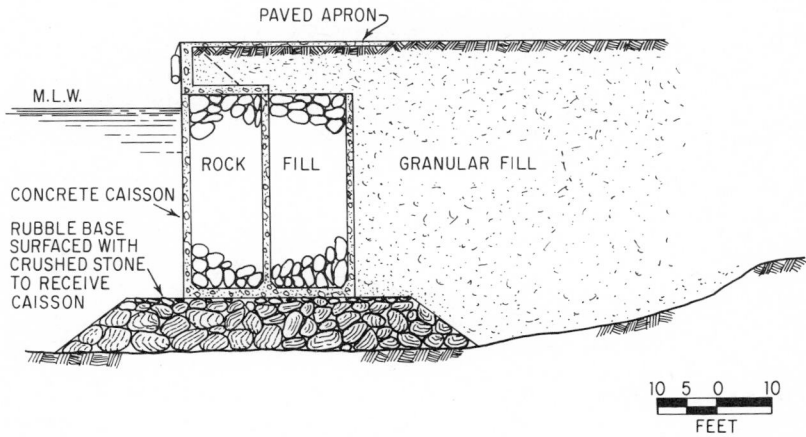

Fig. 23-17. Solid-type bulkhead wharf constructed of rock-filled concrete caissons with closed bottom.

Gravity-quay walls are usually constructed of heavy precast-concrete blocks.   (This is a type of construction seldom used in the United States.)   The individual blocks of concrete may weigh from 50 to 200 tons and are laid to give the wall a slight backward inclination.   The bottom course of blocks is usually laid on a rubble base, and a rock fill is placed in back of the wall to reduce the lateral earth pressure.   Above low-water level, the wall or parapet is usually constructed of cast-in-place concrete.

23-20. Design of General Cargo Terminals.   Figures 23-18 and 23-19 give the dimensions for two- and four-berth piers and slips, respectively.   The four-berth slip must be wide enough for maneuvering a ship with the aid of a tug in and out of the inside berth past ships moored at the outside berths.

Figure 23-20 shows dimensions for a wharf.   The area of each transit shed depends on the cargo storage capacity of the ship.   This capacity is based on measurement tons at 40 cu ft to the ton and an allowance of 50% for aisle space.   The cargo is assumed to be piled in the transit shed to a net average height of 13 ft 6 in.   For a typical dry-cargo ship, an area of 90,000 sq ft has been found to be about the minimum transit-shed space needed for one berth.   Loading platforms and truck areas are not included.   The width of apron *a* depends on the use of portal or semiportal cranes and the number of railroad tracks and truck lanes, if any.   Figure 23-21 gives the various widths of apron for different operating conditions.

A pier may be designed as a rigid structure in which lateral forces are taken by batter piles or by rigid-frame action.   Because of elastic deformation and bending, some movement may take place, but this is usually ignored in absorbing the impact of the ship.   Some installations are designed to be flexible, to absorb the docking impact.   Wood-pile clusters are an example of this type of flexibility.   They absorb the energy of impact through the large movement they are capable of undergoing without permanent distortion taking place.   Their use, however, is

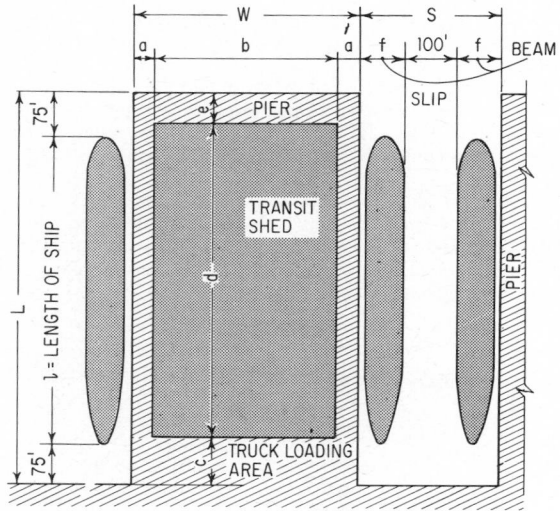

Fig. 23-18. Two-berth pier carries transit shed providing 90,000 sq ft of floor area per berth.   Length of shed $d = L - c - e$, and width $b = 90,000/d$.   For a, c, and e, see Fig. 23-21.

usually confined to docks for barges and small vessels.   Where large vessels are to be berthed against a flexible structure, it should be designed of structural-steel framing and steel piles to provide an adequate resisting force.   An example of this type of construction is shown in Fig. 23-22.

When the type of dock and its general construction features have been determined, it is necessary to establish the lateral and vertical loads for which the dock is to be designed.   These consist of the following:

Wind Forces.   Mooring lines, which pull the ship into or along the dock or hold it against the force of the wind or current, exert lateral forces on a dock.   The maximum wind force equals the exposed area, sq ft, of the broadside of the ship in a light condition, multiplied by the wind pressure, psf, to which a shape factor 1.3 is applied.   This is a combined factor that takes into consideration a reduction due to height and an increase for suction on the leeward side of the ship.   The wind force varies with the location but is usually assumed to be not less than 10 or more than 20 psf.   These pressures correspond to wind velocities of about 55 to 78 mph, respectively, based on the wind pressure $p = 0.00256v^2$, multiplied by the shape factor 1.3, where p is the pressure, psf, and v is the wind velocity, mph.   When ships are berthed on both sides of a pier, the total wind force acting on the pier due to the ships should equal that on one ship increased by 50% to allow for wind against the second ship.   A wind pressure higher than 20 psf against the side of a ship is not warranted in dock design because a ship would not remain alongside the dock, in a light condition, in a storm approaching hurricane intensity.   The ship would either put to sea or take on ballast, to reduce its exposed area to the wind.

Wind against a pier structure and a warehouse or transit shed on the pier may be more severe than wind against the ship, since the surface area may be larger and the wind intensity greater.   The wind pressure in this case should be figured for the maximum wind velocity in the area and the proper shape factor applied for the type of structure on the pier.   This factor may vary between 1.3 and 1.6.   The total wind pressure in a hurricane area where the wind velocity is figured at 125 mph may amount to as much as 64 psf.

Current Forces.   The force of the current, psf, equals $wv^2/2g$, where w is the weight, lb per cu ft, of water, v is the velocity of the current, fps, and g is 32.2 ft per sec^2.   For salt water this results in a pressure, psf, equal to v^2.   The velocity of current usually varies between 1 and 4 fps, which results in pressures of 1 to 16 psf, respectively.   Current pressure is applied to the area of a ship below the water line when the ship is fully loaded.   Since ships are generally berthed parallel to the current, this force is seldom a controlling factor.

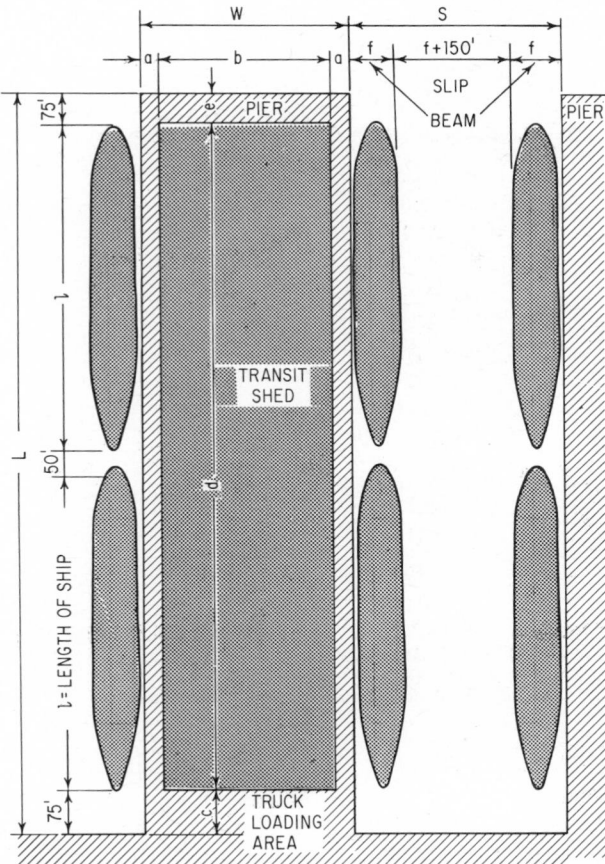

Fig. 23-19. Four-berth pier has transit shed with 90,000 sq ft of floor area per berth.   For *a*, *c*, and *e*, see Fig. 23-21.

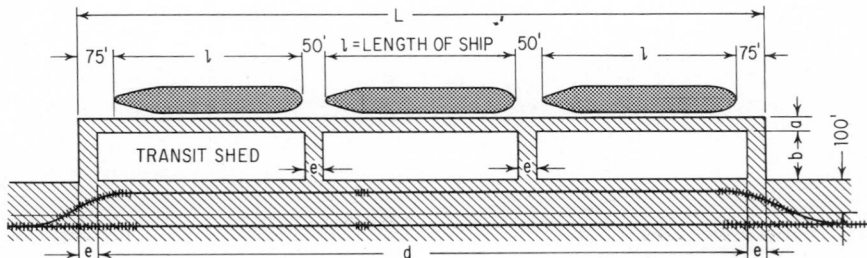

Fig. 23-20. Wharf, extending along the shore, provides ready access to railroad and trucking.   Transit sheds provide 90,000 sq ft of floor area per berth.   For *a*, *c*, and *e*, see Fig. 23-21.

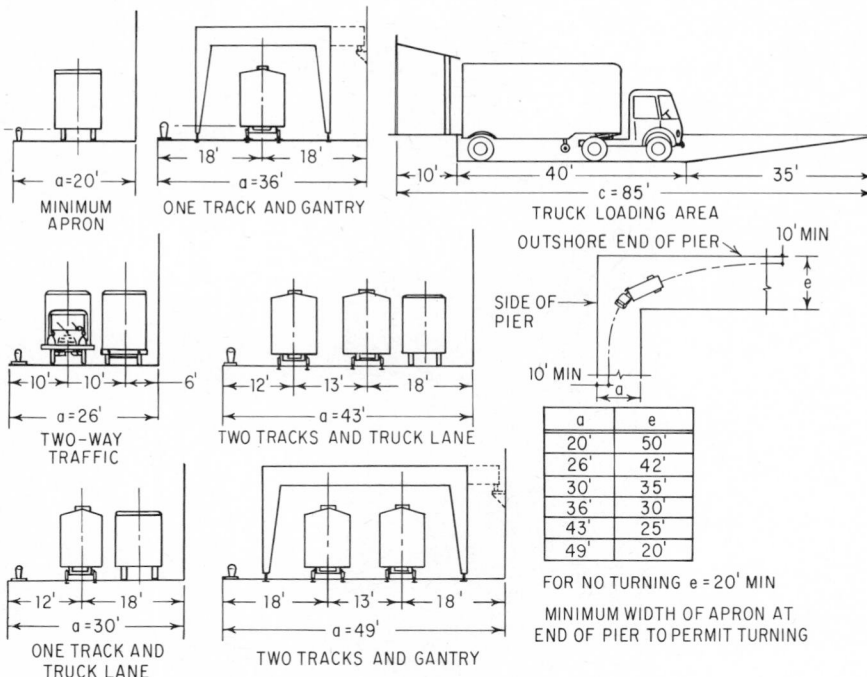

Fig. 23-21. Widths of dock apron for trucks and railroads.

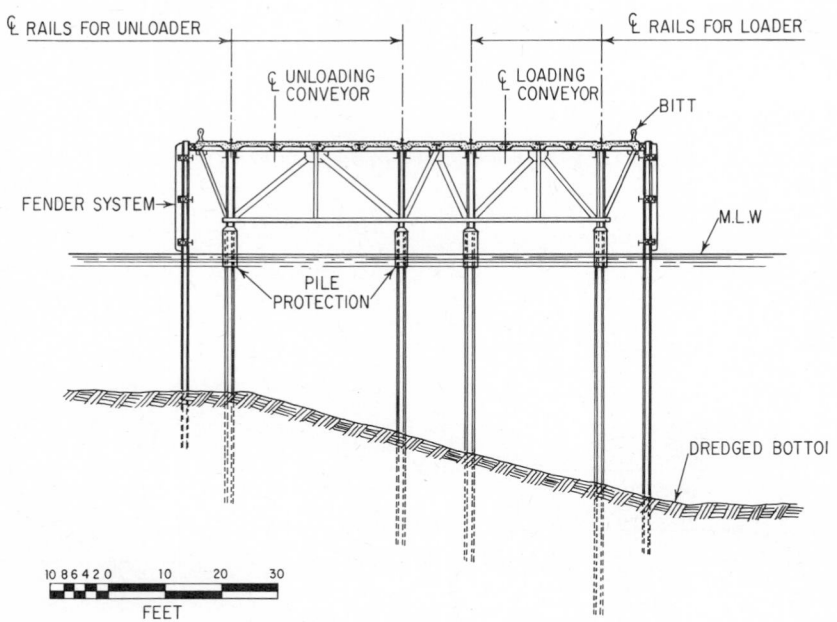

Fig. 23-22. Pier of rigid-frame construction.

Impact.   Docking impact is caused by a ship striking the dock when berthing. The assumption is usually made that the maximum impact to be considered is that produced by a ship fully loaded (displacement tonnage) striking the dock at an angle of 10° with the face of the dock, with a velocity normal to the dock of 0.25 to 0.5 fps (Fig. 23-23). A few installations have been designed for as much as 1.0 fps, but this may be excessive; it corresponds to a velocity of approach of about 3½ knots at an angle of 10° to the face of the dock, and such an impact could damage a ship.

Fender systems are designed to absorb the docking energy of impact. The resulting force to be resisted by a dock depends on the type and construction of the fender and the deflection of the dock if it is designed as a flexible structure. (See also Art. 23-21.)

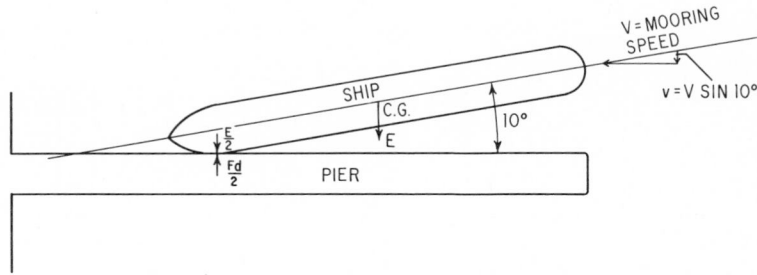

Fig. 23-23. Ship is assumed to strike pier at 10° angle for design against impact.

Earthquake Forces.   These have to be considered if a dock is in an area where seismic disturbances may occur. The horizontal force, applied at the dock's center of gravity, may vary between 0.025 and 0.15 of the acceleration of gravity g times the mass. The force also can be expressed as 0.025 to 0.15 of the weight, respectively. The weight to be used is the total dead load plus one-half the live load. Unless the dock is of massive or gravity-type construction, seismic effect on the design will usually be small, since the allowable stress, when combined with dead- and live-load stresses, may be increased by 33⅓%. If batter piles are used to take the lateral forces, they should be checked to see that they can carry the horizontal earthquake force without increasing the allowable loading by more than 33⅓%; otherwise, additional piles have to be added.

Gravity Loads.   These consist of the dead weight of the structure, or dead load, and the live load, which usually consists of a uniform load and wheel loads from trucks, railroad cars or locomotives, cargo-handling cranes, and equipment. The uniform live load may vary from 250 to 1,000 psf on the deck area. The smaller figure is used for oil docks and similar structures that handle bulk materials by conveyor or pipeline and where general cargo is of secondary importance. General-cargo piers usually are designed for heavier live loads, ranging from 600 to 800 psf. Piers handling heavy metals, such as copper ingots, may be designed for 1,000 psf or more. The uniform live load controls the design of the piles and pile caps, whereas the concentrated wheel loads, including impact, usually control the design of the deck slab and beams. A reduction of 33⅓% is sometimes made in the uniform live load in figuring the pile loads and in designing the pile caps or girders. This is based on the assumption that the entire deck area of adjoining bays will not be fully loaded at one time.

23-21. Dock Fenders.   The principal function of a dock fender is to prevent a ship or dock from being damaged during mooring. Under ideal conditions and under perfect control, a ship might approach a dock without striking a severe blow, but it is still essential to separate it from the dock with some form of rubbing strip of wood or rubber. Such a strip will prevent the ship's paint from being damaged because of the relative motion between dock and ship caused by wind and waves.

Types of Fenders.   In its simplest form, a fender may be a horizontal wood member or a number of vertical wood members or rubbing strips fastened to the deck or face of the dock. For vertical members, wood piles may be used or timbers terminating at the water level and hung from the deck (hung fender, Fig. 23-24). Care must be taken to see that the weight of the fender is supported on brackets from the face of the dock; otherwise, the long bolts that hold the timbers in place will bend and allow the fender to sag. The wood, in itself, can absorb

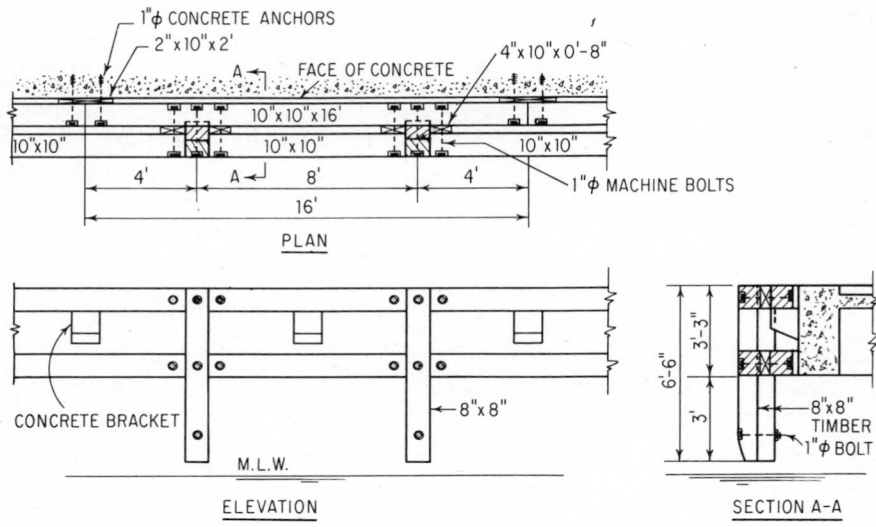

Fig. 23-24. Typical hung wood fender.

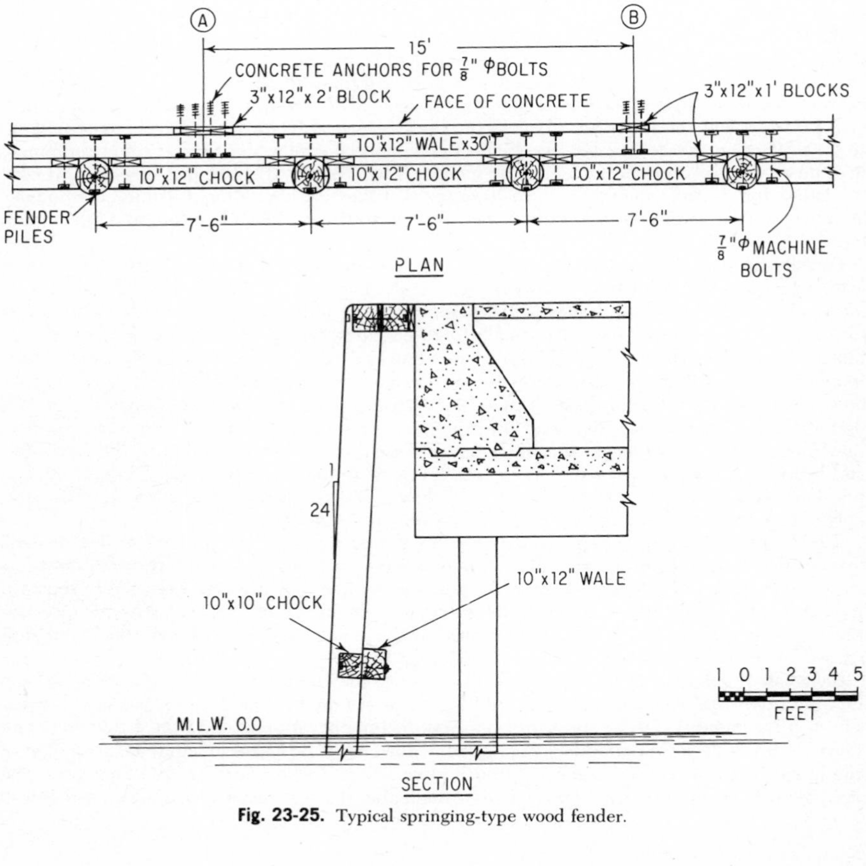

Fig. 23-25. Typical springing-type wood fender.

a certain amount of energy because it is compressed. If the wood members are built up to a substantial thickness, the force of impact will be considerably reduced.

Wood-fender piles, which are placed away from the dock on a slight batter, about 1 on 24, will absorb energy because of deflection when struck by a ship. However, for big ships, something additional is needed to absorb more energy, and various types of flexible-fender systems have been designed and have functioned with a considerable amount of success. Figure 23-25 shows a springing-type wood fender, in which energy absorption is obtained not only from the deflection of the wood piles but also from the deflection of a horizontal wood wale. This wale is blocked out from the deck so that the ship cannot strike at a point rigidly connected to the deck. The wood piles are placed at the quarter points of the span, and chocks are blocked out between the piles. The wood wale must be carefully proportioned to give the proper deflection without failing in shear or bending.

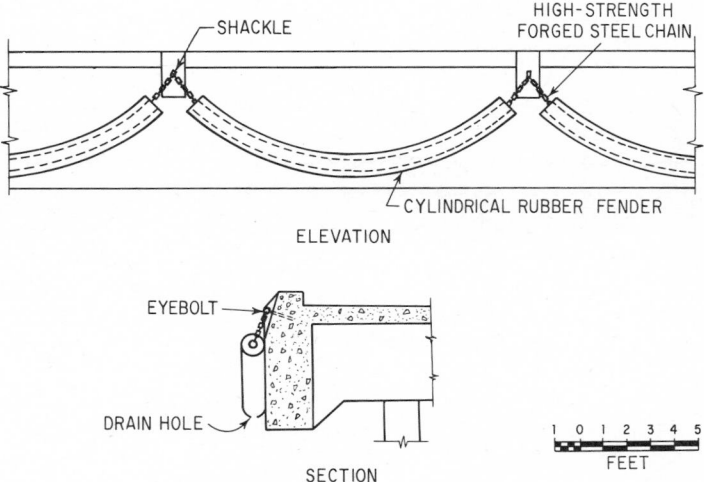

Fig. 23-26. Typical draped rubber fender.

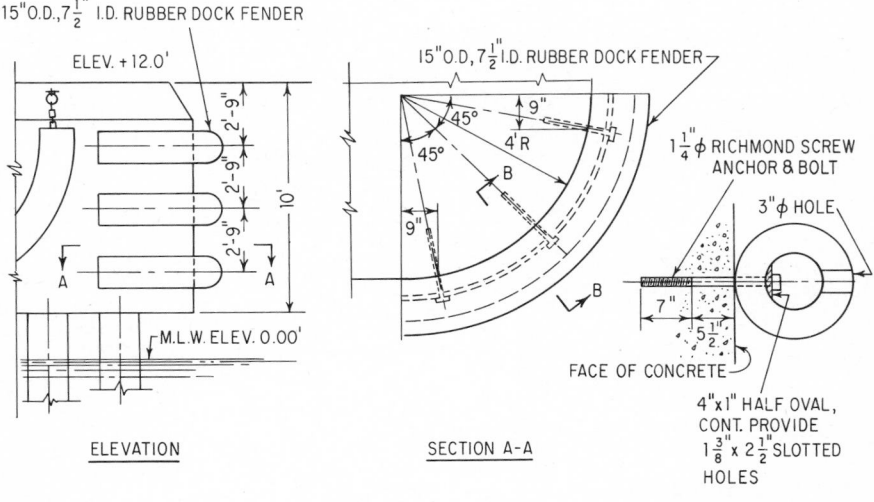

Fig. 23-27. Typical detail of rubber fender at curved corner.

In the design of fender systems primarily using wood, only stress-grade timber should be specified.   It is customary to creosote this wood to a retention of 12 to 16 lb per cu ft, or to refusal in the case of oak.   Boltholes should be drilled the same diameter as the bolts, and all holes on the face of the fender should be countersunk.   All cuts made in the field should be painted with creosote oil.   The remaining space around the heads of all countersunk bolts should be filled with mastic.   Fender piles should be creosoted to a retention of 14 to 20 lb per cu ft. All fender hardware should be galvanized.

Rubber is widely used for fender systems.   Rubber tires hung over the side of a dock are an example.

From the use of rubber tires have sprung hollow-cylindrical or rectangular fenders, rectangular rubber blocks, and the sandwich type known as the Raykin fender buffer.   The hollow-cylindrical type was originally used as a draped fender (Fig. 23-26).   This requires a solid fascia wall for a depth of at least 6 ft, because it is desirable to spread the load over at least a 3-ft height

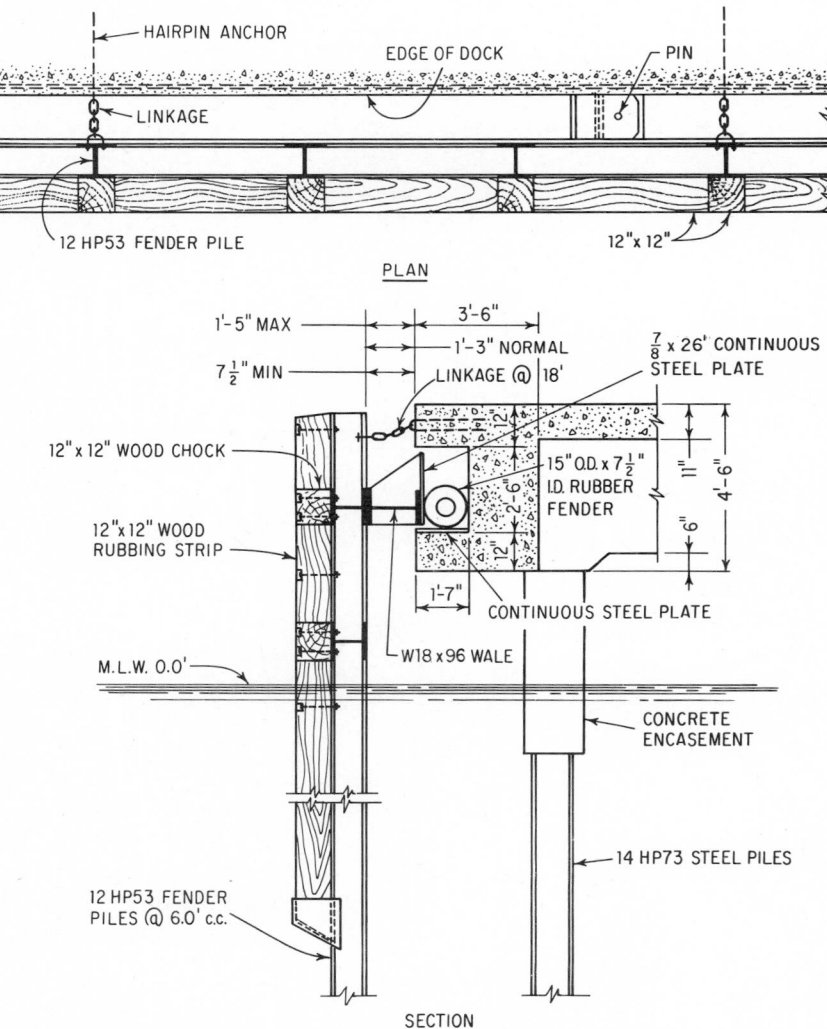

Fig. 23-28. Combination steel-beam and rubber fender system with steel fender piles.

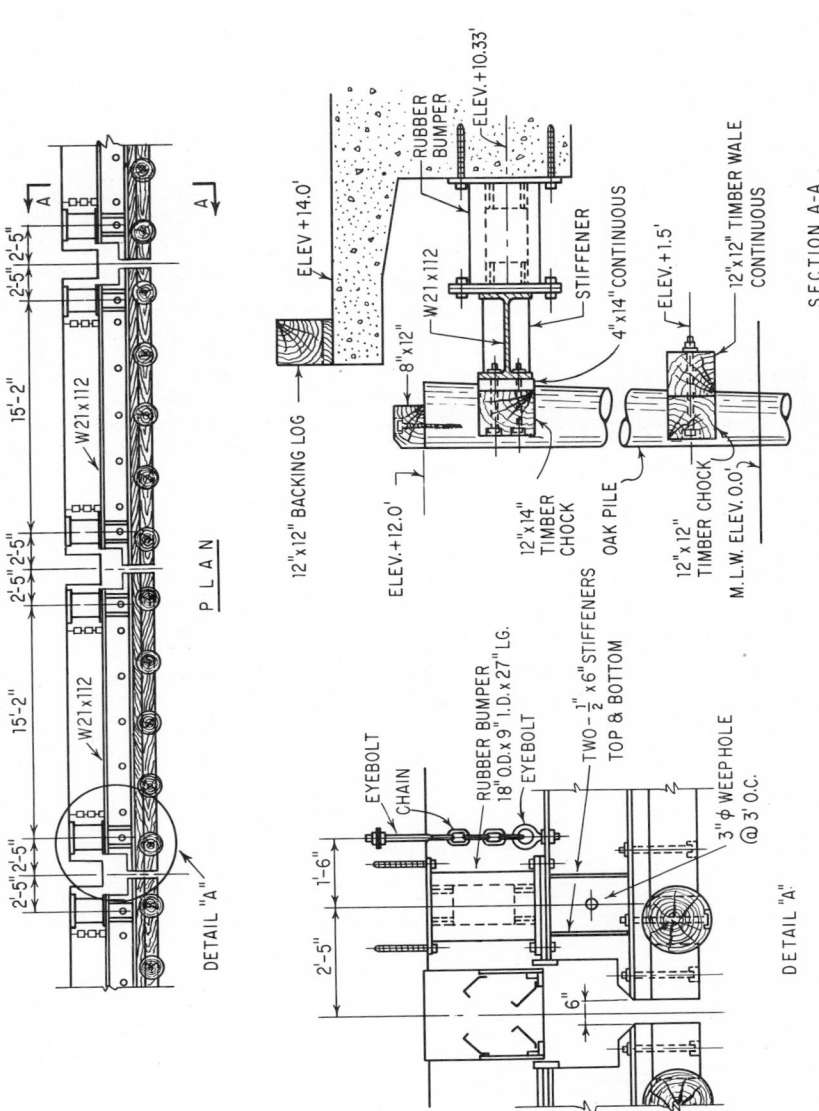

Fig. 23-29. Fender system constructed with cylindrical rubber blocks in end compression.

of the ship's plate.   In this respect, it is important to have the fenders precurved to the specified radius.   The draped fender is particularly adaptable to the solid type of dock construction,
such as sheetpile cells capped with a heavy concrete dock wall, quay walls, or bulkheads; or to
the open, relieving-platform type of construction with a deep fascia wall; or to the heavy-
concrete platform-type dolphins.   Draped rubber fenders are supported by wire rope or chain
attached to eyebolts set in the concrete dock wall.   The eyebolts are set in recesses so that
they do not protrude beyond the face of the wall.   The lowest point of each curved fender
should be provided with a drain hole.

Where it is not practical to drape the fender, as at a curved surface at the corner of a pier or
the rounded end of a dolphin, either a square or a cylindrical fender may be attached to the
concrete with bolts.   Curved plate washers or a continuous half-oval bar is placed under the
bolt heads (Fig. 23-27).   Holes are provided in the wall opposite the bolts to permit their
insertion.

Where it is not practical to use a deep dock-fascia beam or wall, a cylindrical rubber fender
may be placed in back of a horizontal steel beam to which are attached fender piles (Fig.
23-28).   If the steel beam is longer than about 30 ft, it should be articulated by inserting pin-
connected splices that will transmit shear but not moment.   Piles may be either wood or steel,
but steel piles should be provided with wood rubbing strips.

Cylindrical or rectangular rubber blocks may be used in compression behind steel wales or
piles to absorb a ship's impact (Fig. 23-29).

Raykin fender buffers consist of a series of connected sandwiches made of steel plates
cemented to layers of rubber (Fig. 23-30).

Lord flexible dock fenders of high-strength bonded rubber construction use a patented
"buckling column" principle to provide energy absorption.   Upon impact the units deflect, then
buckle in a given direction at a predetermined load.   The rubber member is bonded to flat
steel mounting plates which have holes to permit them to be bolted to the dock on one side
and a steel wale or fender pile on the other.

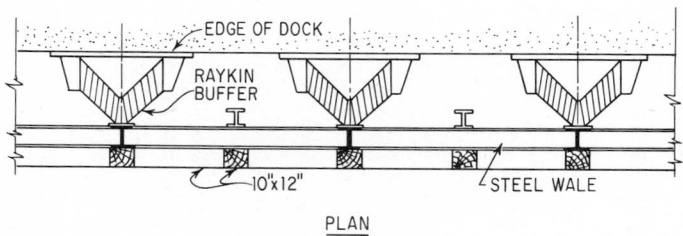

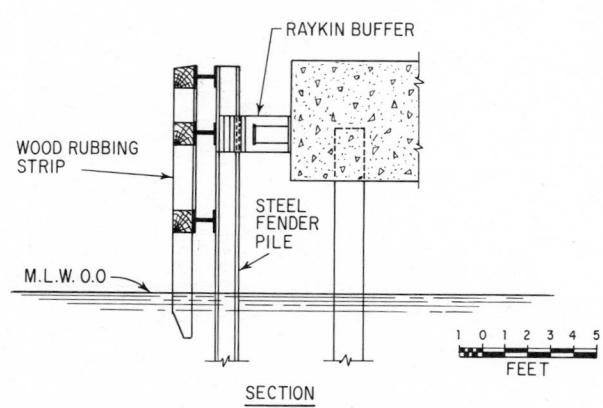

Fig. 23-30. Fender system with Raykin fender bumpers.

With the advent of supertankers, use of rubber fenders with low-reaction-force, high-energy-absorption characteristics has become necessary. Several types are available, generally combining elastomeric materials and steel.

For properties of rubber fenders, consult rubber manufacturers, such as Lord Manufacturing Co., General Tire and Rubber Co., Goodyear Tire and Rubber Co., Seibu Rubber Chemical Co., U.S. Rubber Co., and Yokohama Rubber Co. See also A. DeF. Quinn, "Design and Construction of Ports and Marine Structures," McGraw-Hill Book Company, New York.

Design of Fenders. The maximum impact caused by a ship striking a dock when berthing is based on certain assumptions as to the ship's angle and speed relative to the dock. It is customary to assume that the ship is fully loaded (displaced tonnage) and approaches at an angle of 10° to the face of the dock (Fig. 23-23). The bow of the ship will strike the fender, and only about one-half the tonnage will be effective in creating energy of impact to be absorbed by the fender and pier. The speed of approach must be assumed, and it is here that the greatest uncertainty exists, particularly since energy varies as the square of velocity. The speed of the ship must be converted into the component normal to the dock, and experience has indicated that this velocity will be between 0.15 and 1.0 fps. The latter figure corresponds to a velocity of approach of about 3½ knots at an angle of 10° to the face of the dock. In general, velocities of 0.5 to 1.0 fps normal to the dock are assumed for more exposed locations, where ships dock without the aid of tugs, and for ships of lighter tonnage. Velocities below 0.5 fps are applicable to heavier ships docking in protected locations or with the assistance of tugs.

The kinetic energy of impact $E = \frac{1}{2}Mv^2$. Substituting W/g for the mass M yields

$$E = \frac{Wv^2}{2g} \tag{23-29}$$

where E = energy, ft-tons (1 ton = 2,240 lb)

W = displaced weight of ship, long tons

v = velocity of ship normal to dock, fps

g = acceleration due to gravity, 32.2 ft per sec²

The energy to be absorbed by the fender system and dock is $\frac{1}{2}E$. The remaining half is assumed to be absorbed by the ship and water, because of the rotation of the center of mass of the ship around the point of contact of the bow with the fender. The resistance to impact increases from zero to a maximum. Hence, the work done by the pier is $\frac{1}{2}Fd = \frac{1}{2}E$, where F = force to be resisted, long tons, and d = distance through which the force moves, ft, and is the elastic compression of the fender or deflection of the fender and structure. The assumed d for timber is the thickness, ft, divided by 20. Fender systems are designed to absorb this energy, and the resulting force to be resisted by the dock depends on the type and construction of the fender and the deflection of the dock if it is designed as a flexible structure.

23-22. Dolphins. (See also Art. 23-18.) Dolphins are designed principally for horizontal loads of impact, wind, and current forces from a ship when docking or moored. These forces are determined in the same manner as for design of docks (Arts. 23-20 and 23-21).

Dolphins may be of the flexible or rigid type. Wood-pile clusters are examples of the former type. These are driven in clusters of 3, 7, 19, etc., piles, which are wrapped with galvanized cable (Fig. 23-31). The center pile of each cluster is usually permitted to extend about 3 ft above the other piles to provide a means of attaching a ship's mooring lines. A modification of this type of dolphin arranges the piles symmetrically and on a slight batter. They are bolted to wood cross members located just above low-water level, with wood framing at the top. Large steel cylinders and groups of steel-pipe piles have also been used to provide flexible dolphins.

In general, dolphins of the flexible type have been used for mooring small vessels, not exceeding 5,000 DWT (deadweight tonnage), or as an outer defense for protection of docks, or for breasting off somewhat larger vessels from loading platforms and structures not designed to take the impact of ships. Bottom soil conditions must be suitable for a flexible-type installation. If the soil is too soft, the dolphins or pile clusters will not rebound to their original positions after being struck by a vessel, and their energy-absorbing ability, which depends on deflection, will be gradually dissipated.

For larger cargo ships and tankers of the T-2 class (9,000 to 17,000 DWT) a wood-platform type of rigid dolphin, utilizing wood batter piles, may be used for mooring and breasting. Since the wood platform is relatively light in weight, its lateral stability depends to a large extent on the pullout value of the wood piles. In general, a lateral force of about 40 to 50 tons is about the most that a dolphin of this type can resist without becoming too large and unwieldy.

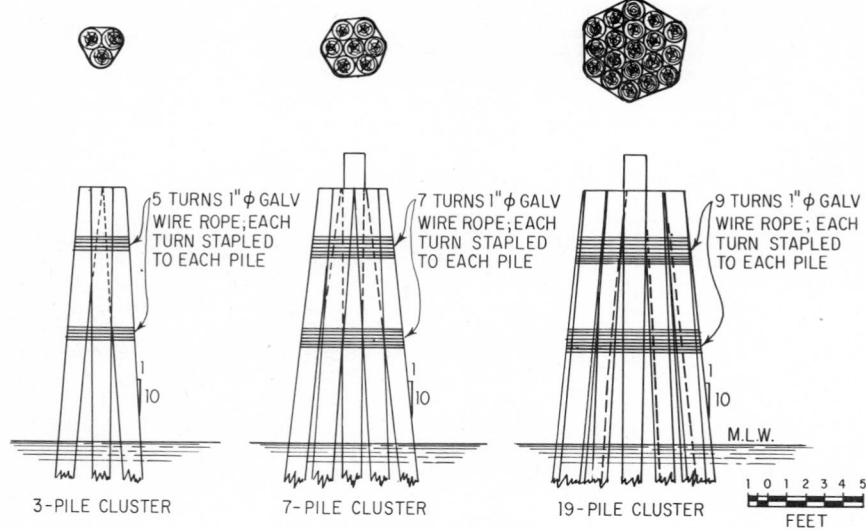

Fig. 23-31. Typical wood-pile dolphins.

If bottom soil conditions are suitable, sheetpile cells make excellent dolphins. They can be designed to withstand the forces from the largest ships, if provided with adequate fenders. Cells, because of their circular shape, are well suited for turning dolphins, for warping or turning a ship around at the end of a dock. Cellular dolphins are usually capped with a heavy concrete slab, to which the mooring post or bollard is anchored. When large ships are to be handled, a powered capstan should be provided to draw in the heavy wire-rope mooring lines.

For big ships, dolphins may be designed with heavy concrete platform slabs supported by vertical and batter piles, usually of steel or precast concrete. This type of dolphin with low-reaction-force, high-energy-absorption rubber fenders can take the docking and mooring forces from the largest supertankers. For this purpose, a large number of batter piles are required. The uplift from these piles, in turn, makes it necessary to have a considerable amount of deadweight, since the vertical piles will in general resist only a relatively small portion of the uplift. This deadweight is supplied by the concrete slab, which may be 5 to 6 ft thick. A sufficient number of vertical piles must be provided to support this deadweight. In addition, the vertical piles must not be spaced too far apart; otherwise, it will be difficult and expensive to provide forms for the concrete. When the depth of slab exceeds 4 ft 6 in., it is usually economical to cast the slab in two lifts, with horizontal shear keys at the construction joint. This greatly reduces the cost of the forms.

23-23. Moles, Trestles, and Catwalks. Many piers are located a considerable distance offshore. This is done to obtain water at the inshore end of the pier deep enough to accommodate the loaded draft of the largest vessel to use the pier, or where it is no longer economical to extend the required depth by dredging toward shore. In such instances, access from the mainland to the inshore end of the pier has to be provided by a mole or trestle, or a combination of both.

Moles. A mole is a fill, usually rock, extending out from shore. Side slopes of the fill are protected from erosion by riprap or armor rock. The upper surface of the fill is made wide enough and graded to accommodate the required facilities to serve the pier, such as roadway, sidewalk, railroad tracks, utilities, pipelines, and conveyor (Fig. 23-32).

In general, it is more economical to use a mole rather than a trestle out from shore to where the depth of water is about 10 ft. But suitable fill material must be readily available and the top of the mole should not be above the normal height of 12 to 15 ft above low water; that is, the maximum height of mole should not exceed about 25 ft. It must be kept in mind that 8 to 10 ft is about the minimum depth of water in which floating equipment can safely operate; therefore, if a trestle is to be constructed in shallower water, it will have to be done by the overhead method.

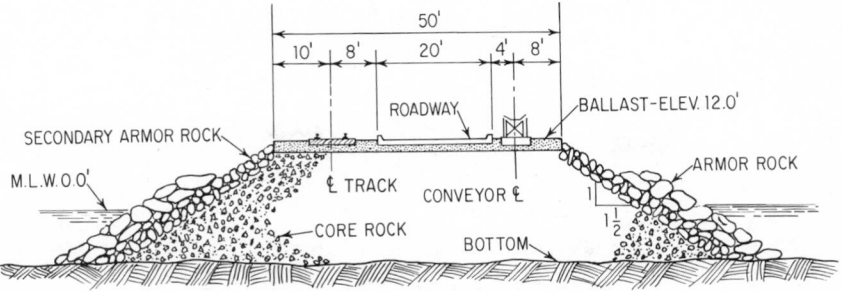

Fig. 23-32. Rock mole, in section, has rock core and armored slopes.

Moles should be constructed with stable slopes, protected with armor rock of a size depending on the degree of exposure. The interior fill or core, unless in a very protected area, should be of run-of-quarry rock so that it will not be washed away by waves and swells before it can be properly protected. In general, design requirements conform to those for breakwaters. Note that since the core material may have to support a roadway, tracks, etc., it must not become eroded or undergo future settlement.

Trestles. A trestle usually is of lighter construction than the pier, as it does not have to withstand a ship's docking and mooring forces. Vertical loads are the principal forces, although sufficient lateral stability has to be provided to resist current, wind, ice, and earthquake forces, and sway from equipment, where these exist. Generally, batter piles may be omitted where the height above firm bottom is 25 ft or less and the trestle is not subject to earthquake or unusual forces from equipment. In deeper water, batter piles are required, in all or in alternating bents. Vertical live loads may consist of one or more of the following: H-10 to H-20 trucks, railroad loads, mobile or truck cranes (deadweight only), pipelines filled with liquid, loaded conveyor, or 250 psf uniform live load.

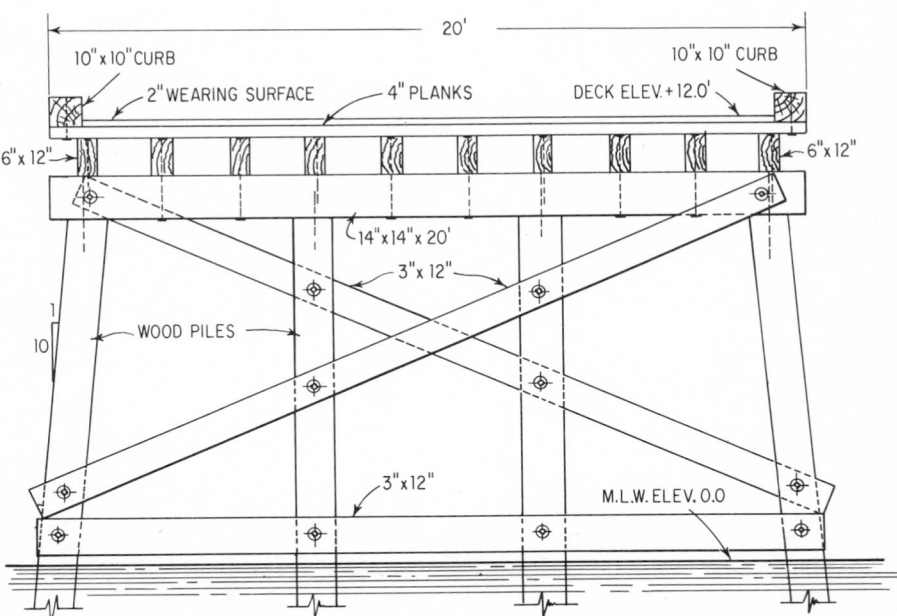

Fig. 23-33. Cross section of timber trestle on wood piles.

A typical cross section through a wood trestle is shown in Fig. 23-33. This type of construction is economical and particularly suitable for temporary or short-term use.

Catwalks. These are used to provide access to and between dolphins. These walkways provide a convenient means of running out ships' lines to their moorings; otherwise these must be handled by boat. A wood catwalk resembles the wood trestle in Fig. 23-33, and in its simplest form, it consists of a light wood walkway, with railing, supported on pairs of wood piles. A live load of 100 psf is usually adequate for design. Therefore, the piles are very lightly loaded. As a result, it is economical to use long spans of light steel or wood trusses where foundation conditions require long piles. These light trusses may have spans of about 80 to 100 ft. The length is limited by the lateral stiffness of the deck, since the walkway usually does not exceed 4 to 5 ft in width. Welded-pipe construction is ideal for these trusses. In locating the walkways, it is very important to place them far enough behind the dock or mooring that they will not be struck and damaged by a ship.

23-24. Offshore Moorings. A vessel may be moored offshore with its own anchors, or to a buoy or group of buoys, or by a combination of its own anchors and buoys. When moored with its anchor, it will swing about the anchor to be generally parallel with the wind or current. Such a mooring is not suitable for a sea loading or unloading cargo terminal if the wind and

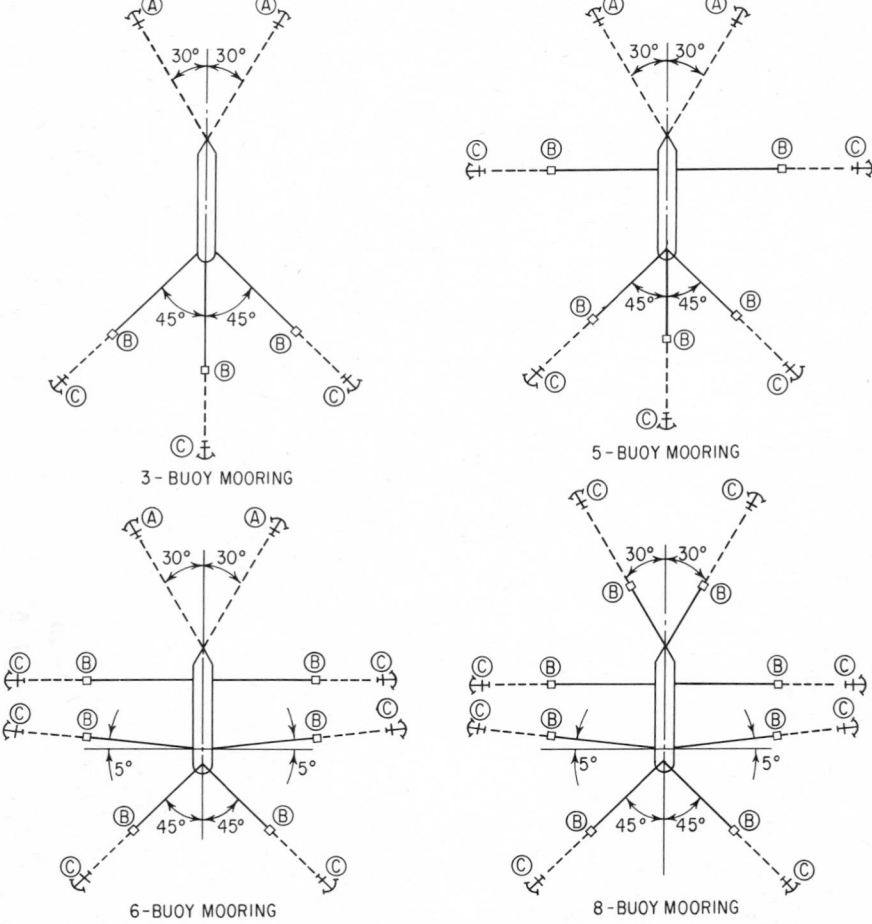

Fig. 23-34. Typical arrangements of offshore moorings. *A* represents ship's anchor; *B*, buoy; *C*, buoy anchor.

current vary in direction from time to time. However, single-buoy moorings which permit the loading hose to rotate as the ship swings around the loading or unloading point have been built for tankers.

There is a difference of opinion among sea-terminal operators as to the number of mooring buoys required at a sea berth. The result is that moorings at different terminals vary somewhat in number. They may range from two to eight buoys, the number being determined by the size of vessels to use the terminal; wind, current, and waves; condition of the sea bottom; and economic conditions.

Figure 23-34 shows different arrangements of moorings consisting of three to eight buoys, supplemented in most cases with either one or both of the ship's anchors. The commonest arrangement for vessels up to the T-2 tanker class is the three-buoy berth having two quarter lines and one stern line, supplemented by the ship's anchors placed at an angle of 30° to 45° off the bow. Where strong winds may occur broadside to the ship, two additional breasting lines may be added, making a five-buoy berth. For supertankers up to 100,000 DWT, six- to eight-buoy berths have been found necessary to assure a safe mooring. Where the prevailing wind direction may vary at different times of the year, it may be necessary to provide an eight-buoy berth with buoys for quarter lines at both ends of the berth, so as to tie up the vessel on either heading.

The components of a mooring-buoy unit or anchorage consist of the mooring buoy and sinkers, the anchor or anchors, and the connecting or ground chain between the anchors and mooring buoy. Figure 23-35 shows a typical mooring-buoy unit.

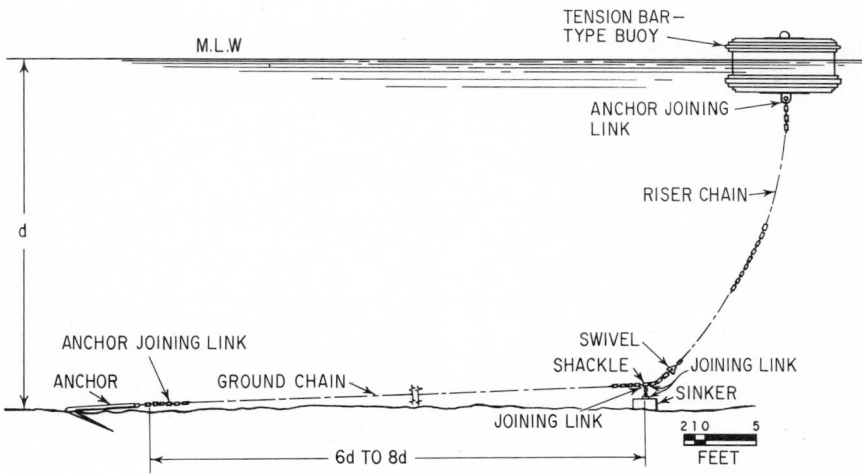

Fig. 23-35. Typical mooring-buoy unit or anchorage.

Mooring buoys are usually large cylindrical cans or drums which are provided with through stays. The mooring hook is secured at the top end and the anchor chain at the underwater end. The ship's line is attached to the mooring hook on the buoy, which is quite often a self-releasing type of hook. It enables the ship's mooring line to be unhooked from the buoy by someone on a launch or on the ship itself.

Mooring buoys are designed to have sufficient buoyancy to take the resultant downward component between the pull of the ship's mooring line and the ground chain. This force submerges the buoy to the general position shown in Fig. 23-36. To maintain the top of the buoy in a more or less horizontal position, the "Lamgar eccentric mooring buoy," patented by Lambert Garland Moorings, Ltd., may be used. Buoys of the drum type may be up to 18 ft in diameter and 9 ft in depth. The ratio of depth to diameter is usually made 1:2.

Mooring-buoy anchors are usually cast-steel Navy stockless or Danforth. The former has a holding power in firm sand of about $7W_a$ and the latter $65W_a^{0.82}$, if the load is applied at less than 3° above the horizontal, where W_a is the weight of the anchor in air. These values may

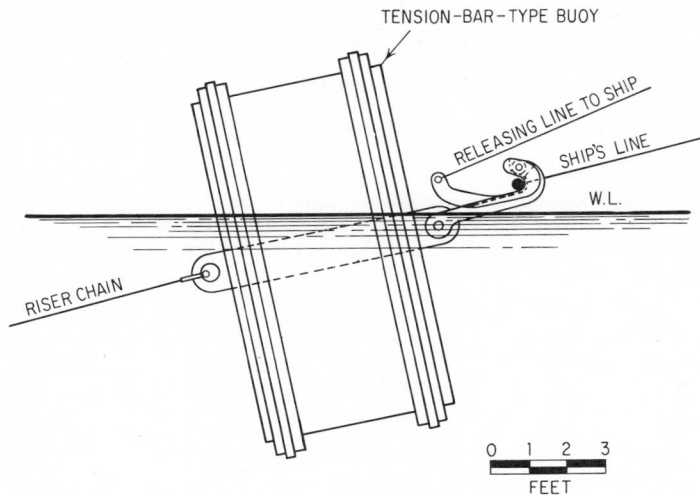

Fig. 23-36. Pull from ship's line partly submerges buoy.

be increased up to 50% where the bottom is a stiff plastic clay or reduced for soft mud to one-fourth those for firm sand.

Anchor chain most commonly used is cast-steel stud-link. The size generally does not exceed 2¾ in. for vessels up to the T-2 size (9,000 to 17,000 DWT) and 3⅜ in. for supertankers up to 100,000 DWT. The dip sections are sometimes made ⅛ in. larger in diameter to allow for wear.

Mooring wire rope is usually galvanized plow steel up to 1½ in. in diameter for the large supertankers. To obtain the maximum pull-out of the anchor, it must not be subjected to a pull more than 3° above the horizontal. With this as a basis, the length of chain can be obtained for a certain pull and depth of water. The horizontal distance from the buoy to the anchor usually varies between six and eight times the depth of the anchor below water level, the shorter distance being applicable if the concrete weight (sinker) for holding the buoy is attached to the ground chain.

A concrete weight is used to position the mooring buoy. It may be attached separately to the buoy by chain or to the ground chain at a distance from the buoy of about 1½ times the depth of water. The concrete weight, which may be from 2 to 10 tons, depending on the size of anchorage, reduces the length of chain required by adding to the deadweight of the chain.

23-25. Mooring Accessories. A large ship ties up to a dock with bow, stern, spring, and breast lines (Fig. 23-37). These lines are fastened to single or double bollards, which are

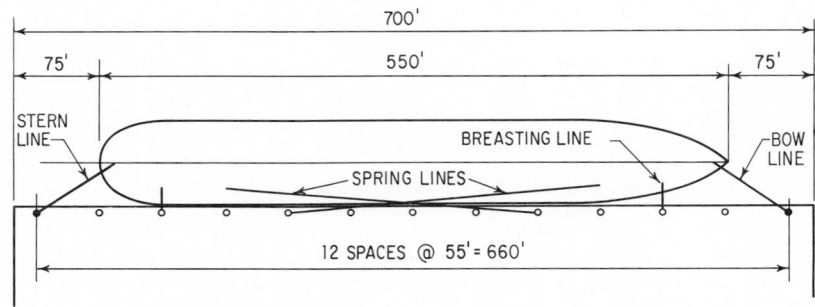

Fig. 23-37. Bow, stern, spring, and breasting lines tie ship to dock.

located along the face of the dock 50 to 80 ft apart (Fig. 23-38a and b). Larger fittings called corner mooring posts (Fig. 23-38c) are sometimes located at the outshore corners of a pier or at the ends of a wharf. They are used principally while bringing the ship into the dock or while it warps around the corner of the pier or turning dolphin.

Bollards or single bitts are usually designed to take line pulls of 35 tons, and corner mooring posts, of 50 tons. Special designs may be made for line pulls of up to 100 tons. The fittings are fastened on a concrete deck by galvanized bolts passing through pipe sleeves set in the concrete. Thus, the bolts can be removed at a later date if they are damaged. The base of the fitting is grouted in a recess formed in the deck, which permits the shear from the line pull to be transmitted directly in bearing to the concrete deck.

When large and long wire-rope lines are to be handled, particularly when attached to dolphins, the lines are pulled in to the mooring fittings with capstans (Fig. 23-38g), which may be either electrically or air operated. Where mooring lines are attached to buoys or dolphins that are reached only by a service boat, the steel hawsers may be provided with releasing hooks (Fig. 23-38h). These enable the ship's lines to be detached by tripping the hook with a small manila line from the ship.

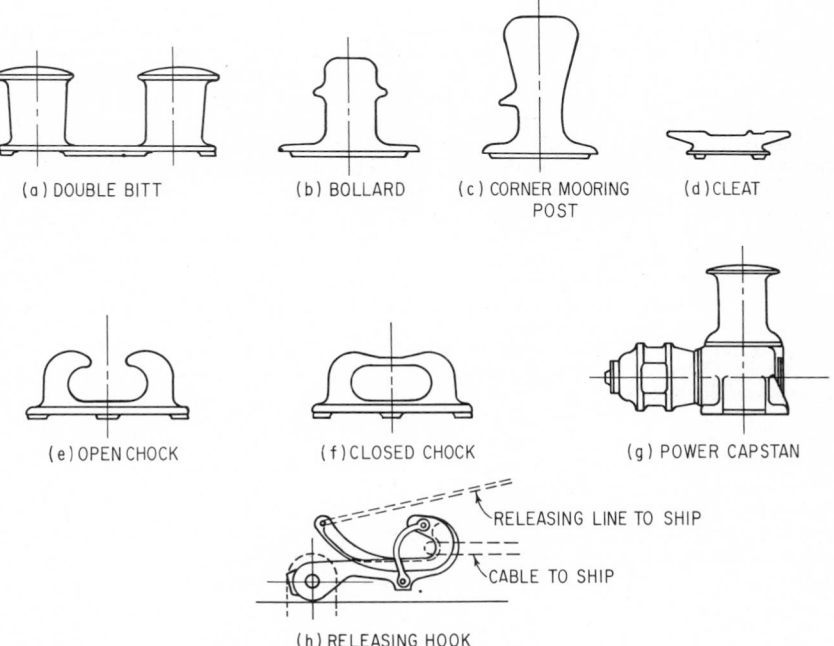

(a) DOUBLE BITT   (b) BOLLARD   (c) CORNER MOORING   (d) CLEAT
                                              POST

(e) OPEN CHOCK   (f) CLOSED CHOCK   (g) POWER CAPSTAN

RELEASING LINE TO SHIP

CABLE TO SHIP

(h) RELEASING HOOK

Fig. 23-38. Typical mooring accessories.

Small ships, tugs, and work boats are usually tied up to cleats (Fig. 23-38d), spaced about 30 to 40 ft apart along the face of the dock.

Open or closed chocks (Fig. 23-38e and f) are used for directing lines. Closed chocks are used when there is a change in the vertical as well as the horizontal direction of the line.

23-26. Shipping-Terminal Services. The following services are required at most shipping terminals: light and power, communication facilities, water supply and fire protection, and sewage disposal; also, steam, compressed air, and bunkering facilities may be needed.

Lighting. Shipping terminals are usually required to function at night as well as during daylight hours, and therefore, satisfactory illumination must be provided for night operation. For open working areas on a pier where ship loading or unloading occurs, a lighting intensity of at least 5 ft-c (foot-candles) should be maintained. This may be achieved with outdoor floodlights at strategic points. Warehouses or storage buildings in the terminal area should have incandescent or fluorescent lighting fixtures that will furnish at least 5 ft-c. Other port buildings or areas, such as administration, passenger waiting rooms, custom and security offices, and restaurant, require higher illumination levels, between 20 and 50 ft-c, depending on the type of work performed.

In addition to the open-working-area lights, which need be turned on only during night operating periods, security illumination must be provided for all roads and walkways every night that they are accessible. Here, an illumination level between 0.2 and 0.5 ft-c is generally supplied by street-lighting luminaires mounted on poles or standards.

Since a shipping terminal functions as part of a port system, the Coast Guard, or other local shipping authority, usually requires that specific signal lights for aiding navigation be located on the pier, or inshore, or mounted on buoys in the harbor. Where the location of these navigation lights has no readily available electric service, it is customary to furnish each navigation light with an independent low-discharge battery, which has a long life and provides assured continuity of operation.

Electric Power. In a shipping terminal, the mechanical equipment required for loading and unloading of vessels and for storage varies, depending on the kind of materials handled. It may include large ore bridges, traveling cranes, conveyors, pumps, hose-handling equipment, dump cars, or trucks. While some of these may be driven by diesel, gasoline, gas, or steam engines, electric motors usually are used because of better control. Where a large number of motors are required, one or more motor-control centers should be provided at convenient locations. Each contains the starters, protective devices, and relays required for a number of motors. Equipment operation is controlled from remote push-button stations or from limit switches located at the equipment so that automatic or sequential operation can be provided.

The electrical design of a shipping terminal must take into consideration the types of materials to be handled. Where explosive or combustible materials are involved, they may create atmospheric mixtures of hazardous gases, vapors, or dusts. Therefore, only specially designed electrical equipment may be used within areas designated as hazardous locations. The National Electrical Code requires adherence to special, rigid design standards for motors, starters, transformers, switches, lighting fixtures, receptacles, and other electrical equipment installed within designated hazardous locations. Generally described as "explosion-proof," this type of electrical equipment is required where the terminal handles gasoline, benzine, alcohol, or other flammable, volatile liquids.

While grounding the frames of electrical equipment and other non-current-carrying metal at any shipping terminal is a normal safety requirement, special grounding considerations are necessary where gasoline or other volatile flammable liquids are being pumped to or from a pier to which a vessel is tied. The danger arises from the fact that the vessel may have picked up an electric charge while sailing en route to its present location. This charge could be transmitted to the ship's hose, and when it is connected to the grounded coupling on the pier, a spark could ensue in the presence of a combustible atmospheric mixture, with disastrous results. To eliminate such possibility, a disconnecting ground switch should be provided on the pier. One terminal of the switch is permanently connected to the pier ground. The other terminal is connected to a flexible length of insulated cable. When a vessel is tied up, and before any hose connections are made, a clamp on the end of the insulated cable is connected to the steel hull of the ship, with the switch in the open position. The switch is then closed, thus grounding any static charge on the ship. The hose connections can then be safely made.

Communications. When a passenger or cargo vessel docks, ship-to-terminal communication is achieved by connecting a plug attached to a wire extension emanating from the ship's telephone system to one of the pier phone-outlet receptacles.

Potable Water Supply. Hose connections, usually 2½ in. in diameter, should be provided either amidship or fore and aft at the berthing space or spaces of the pier. On covered piers, the piping is usually confined within the shed, and the hose connections project through the walls. This arrangement is quite satisfactory if the ship's hose, when draped over the apron, does not interfere with pier operations. Should it interfere, hose connections along the outer edge of the apron are preferable. Connections of this type, as well as connections on open piers, are usually located below the deck in pits covered by removable grating or plates and designed for convenience of operation.

Supply piping on open piers may be supported either above or below the deck, depending on the design of the pier and its purpose. Where thermal expansion is a problem, proper provision must be made for expansion joints or loops. Expansion joints in the structure must also be considered when designing the piping. Piping with rigid joints requires nominal guiding, while piping with bell-and-spigot or mechanical friction joints should be guided at each joint and restrained at ends, elbows, and branches.

Ships' water supplies are usually metered, and a charge is made for water taken aboard. The meters should be easily accessible for reading by both terminal and ships' officers and therefore should be conveniently located on the pier. On covered piers, the meters are logically placed overhead indoors. On open piers, the meters should be located below deck in pits with proper covers so as not to interfere with pedestrian or vehicular traffic.

In cold climates, water lines should be protected against freezing by insulation and steam tracing or electric heating.

Fire Protection. Outdoor fire protection for terminal areas on land is no different from that for any other installation involving similar facilities. A water-distribution system, tailored to the layout of roads and buildings, should be looped, if possible, for maximum safety and assurance of continuity of service. Normally, the pipelines should be buried to a depth sufficient to prevent freezing in cold climates and to a nominal depth to prevent mechanical damage from traffic in any climate. Sectionalizing valves should be so located as to minimize isolated areas in the event part of the system must be shut off for damage or repairs. Conventional fire hydrants with 4½-in.-diameter pumper connections, as well as 2½-in.-diameter hose connections, are desirable. Where mobile pumpers are not used, the pumper connections may be eliminated.

Since the layout of a terminal may form an irregular pattern, it is difficult to define protection in terms of hydrant spacing. A good criterion is to locate hydrants so that any potential fire can be reached from two hydrants, each serving not more than 300 ft of hose. For warehouses, it is desirable to have four hydrants accessible, two on each side. Hydrants should not be located too close to buildings, since a fire therein may make such a hydrant unapproachable; 25 ft is the absolute minimum, but 50 ft or more is preferable.

Fire water requirements should be determined on the basis of the various hazards involved. However, 2,000 gpm for 4 hr is a fair minimum for developments involving large warehouses. This premise allows 1,000 gpm for sprinklers and 1,000 gpm for hose streams. The system should be designed to insure adequate residual pressure as follows: 10 psi at a pumper hydrant, 15 psi at the level of the highest sprinkler heads, and, where neither sprinklers nor pumpers are involved, sufficient pressure at the hose nozzles to throw a forceful stream on the tallest objective.

Where potable water is available in the vicinity, it is usually good policy to provide a combined system for both fire and drinking water. Since the demand for fire fighting is likely to exceed greatly the demand for other purposes, availability of adequate flows for fire is a primary consideration in system design. Where potable water is in limited supply, it may be necessary to provide separate systems and use seawater for fire fighting.

Fire protection on open piers usually can be provided by a single main with branches at strategic points along its length. Branch risers topped with a pair of 2½-in. hose valves are usually more practical than conventional hydrants. A lightweight hose cart or buggy, stored in a shed on the open pier, can readily be hand-drawn to the hose stations. Openings in the deck with removable covers permit use of fog nozzles in combating under-pier fires.

Foam is used for fighting petroleum fires. Various types of apparatus are available for this purpose. Foam can be produced from two separate dry chemicals, a single dry mixture, or concentrated solutions. Perhaps the simplest method of application is the eduction of solution from a pail into an eductor at the base of the foam nozzle, where water from the hose line mixes with the chemical to generate foam. Another system consists of a foam generator with a hopper for dry chemical. Here, the powder is drawn into the generator, where it mixes with the water, and foam is discharged through the hose line to the nozzle. Single or double hoppers are used for the mixed dry chemical or the separate chemicals, respectively. Where a foam truck is used, the concentrated solution is stored aboard in a fairly large tank. The liquid is metered to the generator through a proportioner, insuring proper mixture of chemical and water. Where water pressure is 125 psi, no pump is required. For lower pressures, a pump mounted on the truck boosts the pressure to the proper level.

Indoor fire protection naturally depends on the character of a building and its contents. A sprinkler system, hose stations, and chemical hand extinguishers may be used. Modern covered piers are frequently equipped with all three. In cold climates, water lines are steam traced, and sprinkler systems are of the dry-pipe type.

Sewage Disposal. Sanitary waste originating on a pier can sometimes be carried by gravity to a sewer on shore if the run is short and sufficient difference in elevation exists. In most cases, it is necessary to pump sewage back to shore.   This is accomplished by collecting the sewage in tanks or chambers under the deck and pumping off automatically with float-controlled sewage ejectors.

23-27. Corrosion Protection.   Protection of steel piling from corrosion can be accomplished by encasing the piles with concrete from the underside of the deck to 2 ft below low-water level. This is the area where the most severe corrosion is expected to take place.   Below low water, where corrosion is usually slow, the piles may be given two coats of bitumastic paint prior to driving.   Where more severe corrosion is expected, particularly in salt water, cathodic protection may be employed.

A schematic diagram of an external-impressed-current cathodic-protection installation, with two different methods of arranging the anodes, is shown in Fig. 23-39.   This is a typical application for protection of steel piles under a pier.   In a "distribution-anode" arrangement, the suspended, submerged graphite anodes are usually equally spaced under the full length of a pier.   In a "semiremote-anode-bed" arrangement, the underwater ground-bedded graphite anodes usually consist of one or more groups of anodes, with each anode in a supporting block of wood or concrete, resting on the bottom.   But the anode is held above the bottom to prevent its becoming covered with mud or silt.   The anode beds are generally located on both sides of the structure to be protected.

In the application of cathodic protection to submerged steel structures, such as piles and pipelines, the equipment most frequently selected to provide the required dc power consists primarily of a transformer and a rectifier mounted inside a protective housing.   This enclosure is usually weatherproof, although explosion-proof housings may be provided where required.

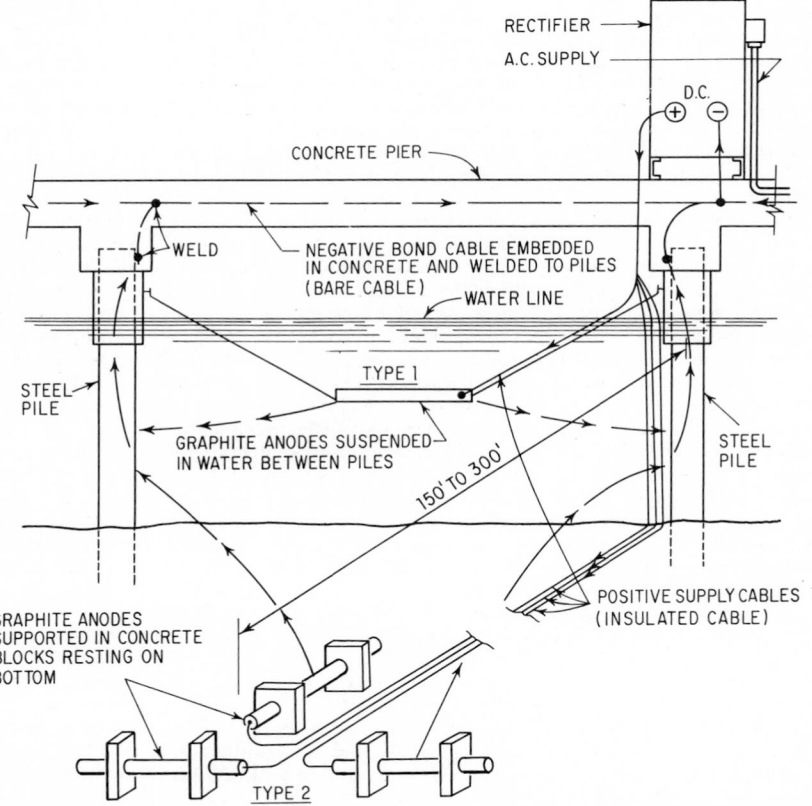

Fig. 23-39. Schematic diagram of cathodic-protection installation. Type 1 represents a distribution-anode arrangement; Type 2, a semiremote-anode-bed arrangement. Arrows show direction of current flow.

PORT BUILDINGS AND CARGO HANDLING

The buildings in a port may comprise one or more of the following: transit sheds, warehouses, cold-storage building, administration building, custom building, police station, guardhouses, stevedores' gear and change house, repair shop and garage, firehouse, and powerhouse. In addition, bulk-cargo shipping terminals may contain grain elevators, silos, storage tanks, and sheds for covering sugar, fertilizer, bauxite, etc.

Several of the facilities may be contained in one building; for instance, the transit shed may house the custom offices, the stevedores' gear room, and locker and washroom facilities. The transit shed is the logical location for these facilities, since it is the place where these services are required. Likewise, the transit shed may contain the administration and shipping companies' offices. On the other hand, the trend in large ports has been to place the general offices for port administration, shipping companies, customs, and port security within one centrally located administration building. It is here that one will find the port captain, chief customs inspector, chief of police, general manager of warehousing, personnel director, accounting department, paymaster, etc. Direct communication is maintained from here to all parts of the port: the offices in the transit sheds and warehouses, customs inspection counters and rooms, guardhouses, firehouse, etc.

23-28. Transit Sheds and Warehouses.   General cargo docks are provided with transit sheds for temporary storage of goods arriving by land and awaiting export and goods discharged from vessels and awaiting clearance through customs and distribution to points of destination by truck or railway.

Transit sheds should not be used as warehouses for long-term storage. The reasons: The space available alongside a berth is usually limited to that required for unloading and loading a single ship. The operation of a transit shed is entirely different from that of a warehouse, the former requiring a greater amount of aisle space for rapid handling of goods by mobile equipment. And economic considerations usually do not justify construction of warehouses on or alongside docks, since their structures are generally heavier than those required for transit sheds and soil conditions at these locations normally require expensive pile foundations.

A minimum area of 90,000 sq ft per berth is desirable for terminals at which an entire ship-load is to be handled. For smaller terminals where a ship may discharge and take on only a partial load, a proportionately smaller area may be used. The 90,000-sq ft requirement is based on storage area needed for discharging and loading a typical dry-cargo ship carrying 6,250 measurement tons of cargo. The total cargo to be handled per berth is therefore 12,500 measurement tons, which at 40 cu ft per ton will occupy a space of 500,000 cu ft. With the use of fork-lift trucks and with net storage height of cargo conservatively taken at 13 ft 6 in. (not including 18-in. thickness of pallets), a storage area equal to 37,000 sq ft, if packed solidly, would be required. But since some space is lost between stacks of pallets, this figure is increased about 25%, to 45,000 sq ft. However, the shed must have ample aisle space leading to the doors along both sides of the shed. It has been found that about 50% of the floor area must be allowed for this purpose if fork-lift trucks are to operate efficiently in sorting, stacking, and moving the cargo. Therefore, a gross area of 90,000 sq ft of transit shed is required per berth.

The length of the transit shed is generally governed by the length of berth. The shed should not be shorter than the out-to-out length between fore and aft hatches. Efficient moving of cargo from ship to shed and vice versa requires for each hatch at least one shed door, which should ideally be located opposite the hatch. Based on an average length of transit shed equal to 500 ft per berth, the required width will be 180 ft to obtain an area of 90,000 sq ft. Therefore, for a finger pier with berths on both sides a one-story transit shed must be 360 ft wide. Where space is not available for this width of shed, a two-story or multistory shed must be resorted to.

The clear height of transit shed should be not less than 16 ft and preferably 20 ft. If mobile cranes are to be used inside the shed, it will be desirable to have a clear height of 24 ft.

Floor live loads vary considerably, depending on the type of cargo and method of handling it. General cargo averages about 70 cu ft per long ton or about 32 lb per cu ft. In modern sheds and with fork-lift trucks, cargo may be piled on pallets to a net height of 13.5 ft, which would impose an average load of 432 psf on the floor slab. Over a floor slab panel, the unit weight might vary by as much as 50% from the average, but it will not be possible to stack the pallets close enough together to cover more than about 75% of the floor area. Hence, the unit load of 432 psf might be increased to 648 lb per sq ft but would probably not average more than 75% of this, or 486 psf, over the entire floor-slab panel. Therefore, a design uniform live load of 500 psf should be satisfactory for general-cargo transit-shed floors. This load may be either lighter or heavier for transit sheds or warehouses assigned to handle or store a specific commodity, such as cotton and wool (300 to 400 psf) and metal products (600 to 800 psf).

23-29. Types of Cargo. With respect to the type of handling equipment required, materials carried by ships can be classified as general cargo, bulk cargo, or containers.

General cargo includes items shipped as units, like automobiles and machinery, and materials in any kind of package, like bales, bags, barrels, or boxes. General cargo, as so defined, requires certain care in handling, to prevent damage, and in stowing, to trim the ship and prevent shifting. For handling of general cargo, see Art. 23-30.

Bulk cargo, on the other hand, can be defined as loose, unpackaged material that can be poured or pumped freely into a ship's holds. This category includes more or less free-flowing dry materials, like grain, ore, and coal, and also liquids, of which the most important are petroleum and petroleum products. Since bulk handling is faster and cheaper than unit or package handling, many products, formerly packaged, are now shipped in bulk. Examples of such materials are portland cement, sugar, orange juice, and wine. For material unsuited for bulk shipment, which was once packaged in a size that could be handled by one person, the trend is to larger packages suited for efficient handling by machine. For handling of bulk cargo, see Art. 23-31.

Containers are large enclosures containing cargo, usually transported over water on special ships, called containerships, and over land on truck trailers or by railroad, generally without unpacking of cargo at ports (see Art. 23-32).

23-30. General-Cargo Handling. The majority of general cargo is handled in lifts of 5 tons or less, generally considerably less.

Normal Lifts. Each hatch of a typical cargo vessel is equipped with a pair of cargo booms. For loading or unloading, one of these is stayed over the offshore edge of the hatch, the other overhanging the wharf. The cargo hook hangs from a link to which both hoisting lines are attached. By joint manipulation of two winches, the operator can drop the hook into either side of the hold and maneuver the load vertically and horizontally at high speeds.

Some multistory transit sheds are equipped with cargo beams supported on a framework running the length of the shed above the roof. If a line is reaved from one of the ship's winches through a block attached to a cargo beam instead of over the shore-side boom, goods may be transferred between the hold and an upper story of the shed. Use of ship's gear is the standard method for loading and unloading ships in the United States.

The fast-acting, revolving, level-luffing crane is the popular device for shiploading and unloading in European and many other world ports. The port of Hamburg, for example, has nearly 900 in use. The term *level-luffing* means that the crane is so rigged that the boom can be luffed (raised or lowered) without changing the height of the load. The fast action of the crane is partly due to the fact that the operator can hoist, swing, and luff all at once, and still keep the load under control. These cranes are mounted on portal or semiportal frames designed to clear railway or truck traffic on the apron. The frames move on tracks parallel to the wharf. The commonest crane sizes are in the capacity range of 3 to 5 tons. Special rubber-tired mobile cranes with gooseneck booms are also occasionally used for this purpose.

Heavy Lifts (50 Tons or More). There are three classes of equipment that are commonly used. Any one of these is necessarily much slower than the general-purpose equipment described. So ocean freight rates include a "heavy-lift charge" for items weighing more than a specified limit.

Most general-cargo ships are equipped, over at least one hold, with an additional boom, designed for lifting 50 tons or more. This boom can be swung over the ship's side by means of vang lines.

Fixed derricks on the wharf, to which the ship must be brought, or mobile cranes operating on the wharf deck are sometimes used. The latter equipment requires special design of the wharf to support the heavy concentrated loads imposed by the crawlers, wheels, or outriggers.

Some of the heaviest loads are handled by floating cranes or derricks. These normally operate on the offshore side of the ships, transferring cargo to barges, from which it must be rehandled.

Handling on Land. A limited amount of material can be hoisted directly between a ship and highway trucks or railroad cars on the wharf. However, the greater part of the contents of a general-cargo ship usually requires handling into and out of a transit shed for sorting and temporary storage.

The fork-lift truck is commonly used for cargo handling on the wharf. This versatile machine can pick up unit loads or palletized loads from the apron, run them into the shed, and there stack them up to 16 or 18 ft high. This stacking ability makes for efficient use of floor space.

Compact, agile, pneumatic-tired mobile cranes, some with hydraulically operated extensible booms, for operation in close quarters, are also used on wharves. They perform a function similar to that of the fork-lift trucks. Since the boom and sling occupy at least 3 or 4 ft above the load, they are not able to stack material up close to the underside of ceiling structures, as fork-lift trucks can. On the other hand, they are better able to handle long or awkwardly shaped objects.

Where distances between shipside and storage areas are too great for efficient use of fork-lift trucks, tractor-drawn trains of low-bed, small-wheeled trucks are used. The trucks are loaded or unloaded at shipside by the shiploading gear, and in the storage area by fork-lift trucks. These trains are also useful for transporting material that cannot be palletized or is otherwise unsuitable for fork-lift operation, for example, mixed cargo containing small packages consigned to many different addresses. In such a case, there is no avoiding a hand-sorting operation.

Conveyors are best suited for bulk-material handling. But certain types of conveyors have been found useful to some extent in general-cargo movements on the wharf. Goods in units small enough for a person to lift may be transported horizontally for short distances on portable roller or belt conveyors. Spiral chutes bring materials down from an upper to a lower floor. Portable, inclined-belt, or slat conveyors are used for stacking bags and other packages. Overhead chain or monorail conveyors provide horizontal transportation within sheds. Most of these functions, however, can be handled by fork-lift trucks. It is only on old wharves, unsuitable for fork-lift trucks, or for special, single-purpose operations, like banana handling, that the conveyor remains an important means of general-cargo handling.

23-31. Bulk-Cargo Handling. Most bulk materials are handled by conveyors or buckets and frequently by a combination of the two. Liquids may be pumped. Some lightweight, powdered, or fine granular materials, like cement and grain, can be transported pneumatically.

The types of conveyor most useful in the operation of a cargo terminal are belt conveyors, bucket elevators, and, less frequently, apron or pan conveyors, oscillating or vibrating conveyors, flight conveyors, and screw conveyors.

For rapid movement of a wide variety of powdered, granular, and lumpy materials, belt conveyors are the most versatile. They can carry large quantities for long distances, horizontally or up and down slopes of 15 to 20°. With appropriate auxiliary equipment, they can be loaded or discharged at their terminals or intermediate points. They are used to move material into and out of storage and into a ship's holds.

Materials can be stockpiled in open storage by a traveling stacker fed by a conveyor. The stacker has an inclined-boom conveyor, which sometimes is designed to rotate 360°, enabling the material to be stored in piles along both sides of the supply conveyor. As an alternative, materials may be elevated by inclined conveyor to a distributing conveyor above the storage pile. The distributing conveyor may be supported on a trestle or from the roof of the storage shed or silo. Materials may be reclaimed by means of a reclaiming conveyor in a tunnel underneath the storage pile. Or the materials may be loaded into hoppers that feed a conveyor above ground.

A conveyor from storage to shipside may be served by a stationary or traveling tower on the wharf. The tower may have a hinged or retractable boom conveyor supporting a chute at its end through which the materials drop into the hold of a ship.

Bucket elevators, usually of lower capacity than belts, convey material vertically or up steep inclines. They are used for operations like filling silos and, when mounted on a "marine leg," can be lowered into the ship's hold for unloading. Other types of conveyors are usually found in a port operation as auxiliary equipment, such as feeders, in a belt-conveyor system.

The clamshell bucket is the most used piece of equipment for high-speed unloading of bulk cargo. One type of bucket is designed for handling by ship's gear and another for handling by revolving cranes. But the greatest capacity is attained by a bucket working from a traveling trolley on the boom of an unloading tower on the wharf.

Towers may be stationary or traveling. The traveling type is a timesaver, because it can be moved from hatch to hatch faster than the ship can be moved to a new position in front of a fixed tower. For higher unloading speeds, two or more traveling towers can operate on one ship. Towers are generally equipped with hoppers into which the buckets may dump. The hoppers, in turn, feed materials into railroad cars or trucks or to a belt conveyor system for transfer to storage. Sometimes, the towers take the form of bridges extending back over inshore storage areas. In these cases, buckets can be used for reclaiming stored materials. Such buckets can be made with capacities up to about 25 tons of ore per bite.

Drag-scraper buckets are useful for storing and reclaiming bulk materials. The Hulett-type unloader is used extensively for unloading ore at Great Lakes ports.

For economical long-distance transportation of moderate quantities of bulk materials, aerial ropeways have some advantages. A ropeway is capable of delivering up to about 400 tons per hour over several miles of terrain impassable by other means. An appropriate application of a ropeway to marine purposes is loading of ships that are forced to anchor a long distance from shore because of shallow water.

Materials brought to a port by rail are frequently unloaded by car dumpers, which roll the cars over and pour out their contents. The material is usually received in a depressed hopper, from which it is transferred to storage by conveyors. One type of machine lifts the cars before dumping and delivers the material directly to a ship by means of an apron converging into a trimming chute.

Many ships carrying crushed stone and coal on the Great Lakes are loaded by conventional means but carry their own built-in unloading equipment. These ships have V-bottom hoppers in their holds. The hoppers have a series of bottom gates that feed materials to two longitudinal pan or belt conveyors or to drag scrapers operating in tunnels at the bottom of the ship. The materials are transferred at one end of the ship to bucket elevators. These, in turn, deliver the materials to a belt conveyor on a hinged boom capable of swinging out over either side of the ship. The boom conveyor can discharge directly to a storage pile, to a hopper feeding a conveyor system, or to barges.

Another form of self-unloading vessel is the cement barge, widely used for river traffic. In these barges, drag scrapers bring the cement to one end, from which it is discharged by cement pumps.

23-32. Container-handling Terminals. Containers used for transporting packaged cargo over water are generally similar to the "piggy-back" containers used successfully for many years for land transportation, and like them are limited in size by highway and railway clearance requirements. Ships' containers, however, may be designed for vertical stacking. They may be transported on conventional ships or on containerships designed specifically for standardized containers. Such special ships tend to be larger than general-cargo ships and may require berths up to 1,000 ft long, compared with about 600 ft for general-cargo ships.

Container handling may be classified as lift-on or roll-on, roll-off (ro-ro). In lift-on operation, containers are picked up, by, for example, a gantry crane, for transport to and from a ship. In port, they may be stored in marshaling yards on standard over-the-road chassis, on which they are moved to a pickup point; stored in marshaling yards on the ground and serviced by straddle or fork-lift trucks; or stored and retrieved automatically under computer control in multistory frames. In ro-ro operation, containers on wheels are rolled on and off ships by tractors and secured to the deck for shipment.

Standard containers are 8 ft square by 10, 20, 24, 30, 35, or 40 ft long. Maximum gross weight ranges from 10 to 30 long tons. For efficient handling of such containers, special equipment, such as gantry cranes, usually operating on the wharf side, is required.

Terminals for efficient handling of containers differ in some important respects from terminals for general or bulk cargo. Container terminals require much more open space, especially for truck or rail delivery of containers and for handling and storing them, and much less enclosed space. Some buildings may be required for offices and for whatever sorting and repacking may be required at the terminal, but generally containers may be stored outdoors. Because of the large land area required, wharf-type construction is preferable to piers or dolphin construction for ship berths. Also, wider wharfs are required.

A container terminal may require from 10 to 15 acres of land for 550- to 650-ft berths. For 750-ft berths, about 25 acres may be needed for a single berth, plus about 20 acres for each additional berth.

(A. DeF. Quinn, "Design and Construction of Ports and Marine Structures," McGraw-Hill Book Company, New York.)

NAVIGATION AIDS

Navigation aids are necessary in rivers, channels, and harbors, and along lake and ocean shores, to enable ships to travel safely and rapidly. The types of aids required vary with the kind of waterway they serve and their function. Navigation aids may be either floating or fixed structures equipped with the necessary type of beacon lighting, bells or other sound-warning devices, and radar reflectors. In United States waters, the type of aid, beacon lighting, sound-warning devices, numbering, and painting is governed by the U.S. Coast Guard.

23-33. Buoys.   The various types of floating buoys are spar, can, nun, spherical, lighted, sound-warning, etc.   Spar, can, and nun buoys are unlighted.   Typical buoys used as aids to navigation are shown in Fig. 23-40.   Their numbering, coloring, chart symbols, lighting, and other characteristics are indicated in Table 23-12.

23-34 Navigation Lights on Piers, Wharves, Dolphins, etc.   One marine beacon-navigation-light lantern is generally placed at each end of all piers, wharves, long mooring dolphins, etc., to outline their limits.   One light may be used on narrow dolphins or other narrow objects projecting into navigable waters.   These lights are generally fixed white lights and are fastened directly to the structures.   They are generally powered by shore electric current.

23-35. Fixed-Structure Beacon Lights on Breakwaters, Shore, etc.   Fixed-structure beacon lights are erected on the projecting ends of breakwaters, on salient points of land projecting into navigable waters, at harbor entrances, and on other points of special danger to shipping.   The

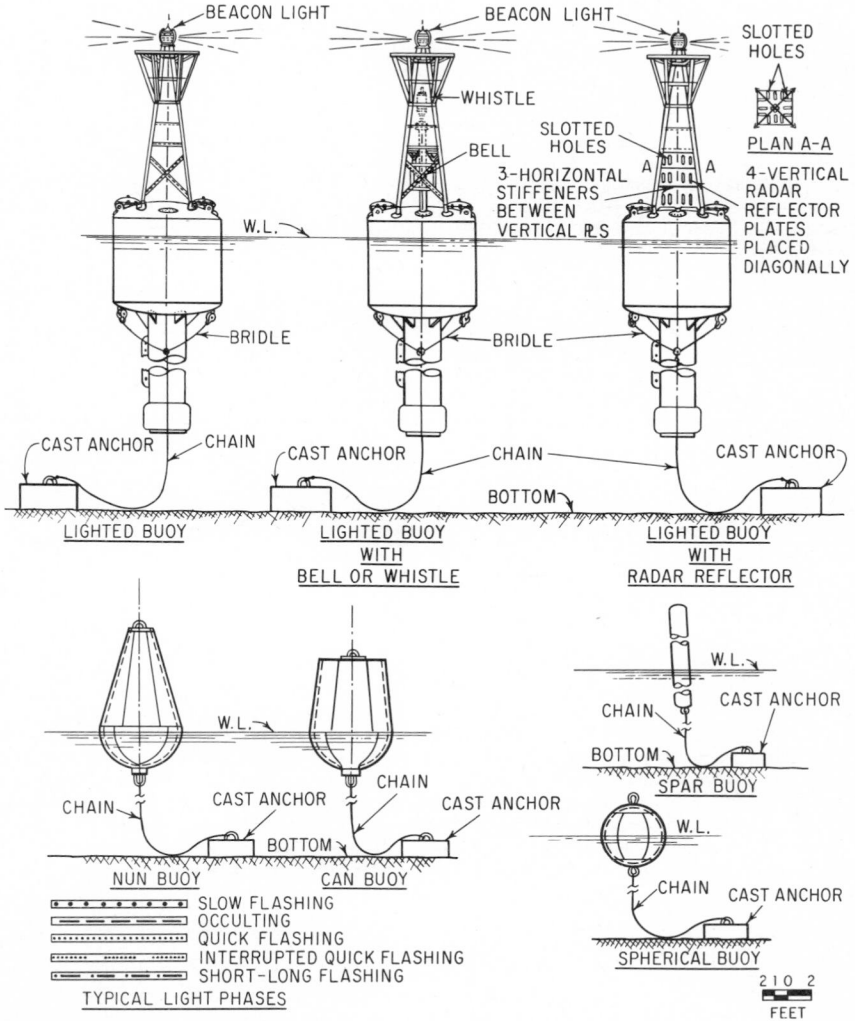

Fig. 23-40. Typical buoys used as navigation aids.   Characteristics of such buoys are given in Table 23-12.

Table 23-12. Characteristics of Buoys

TYPE OF BUOY AND LOCATION	BUOY OR LIGHTSHIP SYMBOL	NUMBERING	PAINT COLOR	LIGHT COLOR	USUAL LIGHT PHASES
LIGHTED - PORT		ODD	BLACK	WHITE OR GREEN	SLOW FLASHING, OCCULTING, QUICK FLASHING
LIGHTED-STARBOARD		EVEN	RED	WHITE OR RED	SLOW FLASHING, OCCULTING, QUICK FLASHING
LIGHTED BELL, GONG, OR WHISTLE - PORT	BELL, GONG, OR WHISTLE	ODD	BLACK	WHITE OR GREEN	SLOW FLASHING, OCCULTING, QUICK FLASHING
LIGHTED BELL, GONG, OR WHISTLE - STARBOARD	BELL, GONG, OR WHISTLE	EVEN	RED	WHITE OR RED	SLOW FLASHING, OCCULTING, QUICK FLASHING
LIGHTED-FAIRWAY- MIDCHANNEL	BW	OPTIONAL	BLACK & WHITE VERTICAL STRIPES	WHITE	SHORT- LONG FLASHING
LIGHTED-JUNCTION, ISOLATED DANGER, OR OBSTRUCTION	RB	OPTIONAL	RED & BLACK- HORIZONTAL BANDS	WHITE, RED, OR GREEN	INTERRUPTED QUICK FLASHING
BELL, GONG, OR WHISTLE-PORT	BELL, GONG, OR WHISTLE	ODD	BLACK	NONE	———
BELL, GONG, OR WHISTLE - STARBOARD	BELL, GONG, OR WHISTLE	EVEN	RED	NONE	———
CAN-PORT	C	ODD	BLACK	NONE	———
NUN - STARBOARD	N	EVEN	RED	NONE	———
SPHERICAL	SP	OPTIONAL	OPTIONAL	NONE	———
SPAR	S	OPTIONAL	OPTIONAL	NONE	———
CHECKERED		OPTIONAL	COLORS OPTIONAL- CHECKERED	NONE	———
FAIRWAY- MIDCHANNEL	BW	OPTIONAL	BLACK & WHITE VERTICAL STRIPES	NONE	———
JUNCTION, ISOLATED DANGER, OR OBSTRUCTION	RB	OPTIONAL	RED & BLACK- HORIZONTAL BANDS	NONE	———
QUARANTINE	Y	OPTIONAL	YELLOW	NONE	———
FISH TRAP	BW	OPTIONAL	WHITE & BLACK- HORIZONTAL BANDS	NONE	———
LIGHTSHIP		NAME	SUPERSTRUCTURE-WHITE MASTS & STACKS-BUFF	OPTIONAL	OPTIONAL

structures are metal-framed towers with marine beacon-light lanterns mounted on top. The lights must be high enough to be easily sighted by approaching vessels.

The lights may be fixed, occulting, or flashing, and of the color required. They may be powered by shore electric current, electric storage batteries, or acetylene gas. Radar reflector plates may be erected on the tower structure, where required. These towers are painted the colors required by their locations. They also may be galvanized and unpainted.

Lighthouses.   Lighthouses are tall tower structures with a marine beacon-light lantern on top. They are usually erected at points along the shore to guide shipping to a nearby port. They also are set on reefs, shoals, or other points of danger to shipping. They are usually constructed of masonry and are built to withstand heavy wave action and weather. The towers must be high enough, because of the earth's curvature, so that the beacon lights may be sighted by approaching vessels at a considerable distance offshore. Visibility of 20 miles is not unusual. Care should be taken that the lights are not so high that they are more frequently obscured by

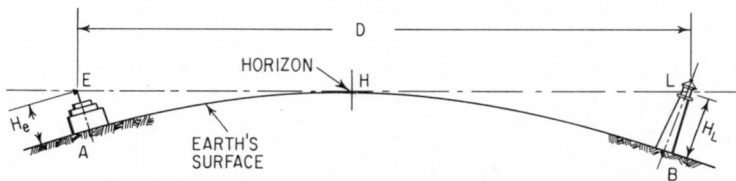

Fig. 23-41. Horizon determines range of visibility of lighthouse.

clouds, mist, etc., than those nearer to sea level would be. The required height of a lighthouse, H_L, ft, to be visible at a certain distance D, nautical miles, out to sea (Fig. 23-41) can be computed from $H_L = (\frac{7}{8}D - H_e^{1/2})^2$, where H_e is the height of the eye at sea, ft. The lights may be white or colored, and flashing or occulting, as required. They are powered by shore electric current, electric batteries, or acetylene gas. The flashing characteristics that distinguish many of the lighthouses are produced by revolving the entire lens by electric motor.

Large lighthouses are also equipped with fog signals produced by various types of sounding devices. The light- and fog-signal characteristics are distinct for each lighthouse and are given in Light Lists; many also are given on navigation charts.

23-36. Lightships. Where it is impracticable to build lighthouses, lightships serve the same purpose. Lightships vary in size from fully manned ones to small unmanned ones with automatic lights and fog signals, etc. Hulls of lightships in the United States are generally painted red, with the name of the station in white on both sides. Superstructures are white, masts and stacks buff. Hulls are built of steel. Propulsion is either steam or diesel. Auxiliary generators supply power for operation of the signals.

Lightships also show storm-warning signals, and observe all passing ships and any floating navigational aids in the vicinity. Lightships are usually held in position by a single anchor. Relief lightships, painted red, with the word "Relief" on the sides, replace regular lightships when they are being overhauled.

MARINAS

23-37. Characteristics of Marinas. Complete facilities for accommodating both local and transient pleasure boats are called marinas. They provide berthing space in calm water; electric power and fresh water while tied up at the berth; fueling and repair services; and on-shore power and fresh water while boats are berthed; fueling and repair services; and on-shore Many installations also provide equipment for removing and launching boats and space for storing them, either indoors or outdoors, during the winter. While most marinas cater to the local boating public, some provide for daily handling in and out of the water of small boats, trailer-hauled, and supply space for parking the trailers and cars.

A good marina also has an experienced, staffed, and well-equipped repair shop. During the winter months, motors are overhauled, batteries conditioned, and other maintenance services performed. Figure 23-42 shows a typical marina with complete berthing and on-shore facilities.

Marinas should be located in quiet water and protected from the wash of passing craft in the adjacent harbor or channel. Where these conditions do not prevail, it will be necessary to surround the berthing area with a protective bulkhead or breakwater.

In addition to mooring space, utility services must be provided. Each boat berth should be provided with a 110-volt single-phase 60-cycle power outlet. All electrical work should have explosion-proof fixtures, because of the danger of fire from gasoline vapor. It is usually worthwhile to consider using wrought iron for all conduit because of its resistance to corrosion. Lighting should be provided for water and land areas of the marina, and the lights should be shielded to prevent glare, which will interfere with night navigation. A wrought-iron fresh-water line should be provided, with taps and hoses to allow filling the boats' fresh-water tanks at their berths. Facilities should be provided also for removal of wastes and waste water from boats. Waste water should be treated in accordance with local and Federal health and environmental requirements.

Since fire is an ever-present danger in the vicinity of small craft with gasoline engines, great care must be taken to provide adequate fire protection. Each pier should have a 1½-in.-diameter fire line. Hose stations should not be farther apart than 150 ft. Each hose station should have 100 ft of hose, with dual-purpose fog nozzles and fog lances, and a large CO_2 fire extinguisher on a cart.

Fire is especially hazardous at the fueling pier. The fueling pier should preferably have a concrete deck. Heavy-timber fire walls should be carried down to low-water level, or if a floating pier is used, the fueling station should be on a separate float. The fueling pier and adjacent area must be conspicuously posted as a "NO SMOKING" area.

A fuel-handling facility usually consists of underground tanks on shore, with one or more tanks provided for diesel oil and for gasoline. The fuel lines on the pier should be of wrought iron. The metered fuel pumps should be provided with 25-ft-long hoses.

23-38. Pier Construction for Marinas. Timber and wood piling for walkways and berthing piers should be creosoted. Where there is a considerable range in tide level, the walkways

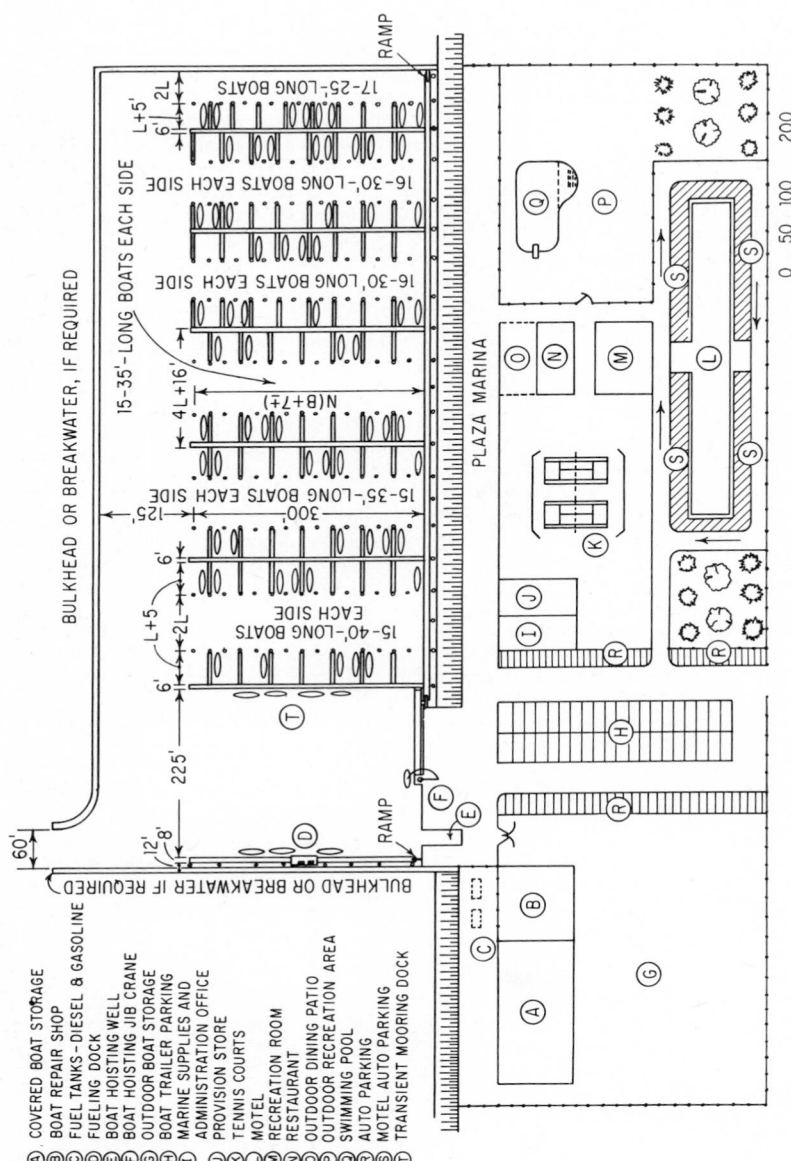

Fig. 23-42. Plan of typical marina. All mooring berths are floats fastened to piles not shown. Pier and berth lengths are given in terms of N, number of boats; B, beam of boat; and L, length of boat.

Ⓐ COVERED BOAT STORAGE
Ⓑ BOAT REPAIR SHOP
Ⓒ FUEL TANKS–DIESEL & GASOLINE
Ⓓ FUELING DOCK
Ⓔ BOAT HOISTING WELL
Ⓕ BOAT HOISTING JIB CRANE
Ⓖ OUTDOOR BOAT STORAGE
Ⓗ BOAT TRAILER PARKING
Ⓘ MARINE SUPPLIES AND
Ⓙ ADMINISTRATION OFFICE
Ⓚ PROVISION STORE
Ⓛ TENNIS COURTS
Ⓜ MOTEL
Ⓝ RECREATION ROOM
Ⓞ RESTAURANT
Ⓟ OUTDOOR DINING PATIO
Ⓠ OUTDOOR RECREATION AREA
Ⓡ SWIMMING POOL
Ⓡ AUTO PARKING
Ⓢ MOTEL AUTO PARKING
Ⓣ TRANSIENT MOORING DOCK

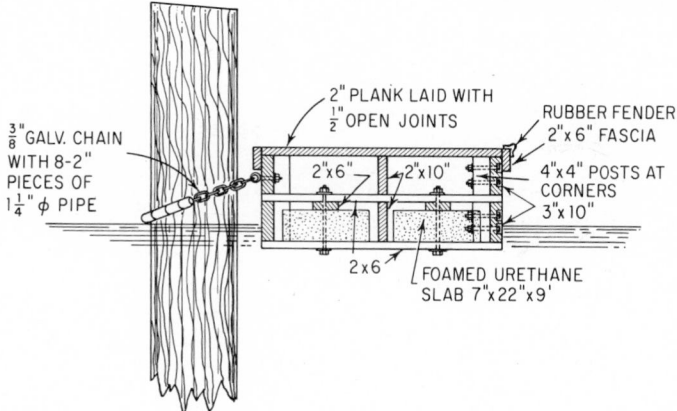

Fig. 23-43. Typical section through float. All bolts are ½ in. in diameter, galvanized. Painting requirements: two coats of antifouling paint on all submerged timber surfaces, two coats of marine primer on all exposed surfaces, two coats of nonskid enamel on deck, and two coats of marine enamel on all other exposed surfaces.

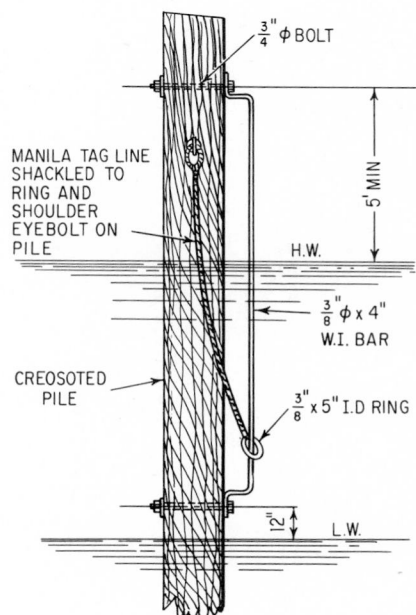

Fig. 23-44. Tie-up facility in tidal water. All hardware should be galvanized.

usually float and are attached to wood piles with chains so that they can rise and fall with the tide without moving out of position. Ramps provide access from shore to the floating walkways. The ramps are hinged at the top and are free to roll on the walkways as they rise and fall.

Floating mooring slips and walkways are made in sections, hinged at the joints and framed with timber (Fig. 23-43). Timber plank decks are laid with open joints. Buoyancy is provided by wrought-iron tanks or foamed-urethane slabs incorporated into the timber framing.

Mooring piles (Fig. 23-44) are usually provided with a ring with which to attach mooring lines. The ring can move up and down on a rod or cable attached to a pile as the tide rises and falls. All marine hardware should be hot-dipped galvanized.

Index

Index

3